Instructor's Manual
with Transparency Masters *for*

Microelectronic CIRCUITS

FIFTH EDITION

Adel S. Sedra
University of Waterloo

Kenneth C. Smith
University of Toronto

New York Oxford
OXFORD UNIVERSITY PRESS
2004

Oxford University Press

Oxford New York
Auckland Bangkok Buenos Aires Cape Town Chennai
Dar es Salaam Delhi Hong Kong Istanbul Karachi Kolkata
Kuala Lumpur Madrid Melbourne Mexico City Mumbai Nairobi
São Paulo Shanghai Taipei Tokyo Toronto

Published by Oxford University Press, Inc.
198 Madison Avenue, New York, New York 10016
www.oup.com

ISBN 0–19–517268–X

Cover Illustration: The chip shown is an inside view of a mass-produced surface-micromachined gyroscope system, integrated on a 3mm by 3mm die, and using a standard 3-μm 2-V BiCMOS process suited for the harsh automotive environment. This first single-chip gyroscopic sensor, in which micro-mechanical and electronic components are intimately entwined on the same chip, provides unprecedented performance through the use of a collection of precision-directed techniques, including emphasis on differential operation (both mechanically and electronically) bolstered by trimmable thin-film resistive components. This tiny, robust, low-power, angular-rate-to-voltage transducer, having a sensitivity of 12.5mV/°/s and resolution of 0.015°/s (or 50°/hour), has a myriad of applications—including automotive skid control and rollover detection, dead reckoning for GPS backup and robot motion control, and camera-field stabilization. The complete gyroscope package, weighing 1/3 gram with a volume of 1/6 cubic centimeter, uses 30mW from a 5-V supply. *Source:* John A. Geen, Steven J. Sherman, John F. Chang, Stephen R. Lewis; Single-chip surface micromachined integrated Gyroscope with 50°/h Allan deviation, IEEE Journal of Solid-State Circuits, vol. 37, pp. 1860 - 1866, December 2002. (Originally presented at ISSCC 2002.) Photographed by John Chang, provided by John Geen, both of Analog Devices, Micromachine Products Division, Cambridge, MA, USA.

Printing number: 9 8 7 6 5 4 3 2 1

Printed in the United States of America
on acid-free paper

Contents

Preface

This manual contains complete solutions for all 431 exercises and 1341 end-of-chapter problems included in the book *Microelectronic Circuits, Fifth Edition,* by Adel S. Sedra and Kenneth C. Smith. It also includes masters for overhead transparencies of 240 of the more complex diagrams.

We are grateful to Mandana Amiri, Karen Kozma, Shahriar Mirabbasi, Roberto Rosales, and John Wilson, who assisted in the preparation of this manual. We also acknowledge the contribution of Ralph Duncan and Brian Silveira to previous editions of this manual.

Communications concerning detected errors should be sent to the attention of the Engineering Editor, mail to Oxford University Press, 198 Madison Avenue, New York, New York, USA 10016, or e-mail to higher.education.us@oup.com. Needless to say, they would be greatly appreciated.

A website for the book is available at http://www.sedrasmith.org

Exercise Solutions

CHAPTER 1–EXERCISES

1.3 $f = \frac{1}{T} = \frac{1}{10^{-3}} = 1000$ Hz

$\omega = 2\pi f = 2\pi \times 10^3$ rad/s

1.4 (a) $T = \frac{1}{f} = \frac{1}{60}$ s $= 16.7$ ms

(b) $T = \frac{1}{f} = \frac{1}{10^{-3}} = 1000$ s

(c) $T = \frac{1}{f} = \frac{1}{10^6}$ s $= 1\ \mu$s

1.6 $P = \frac{1}{T}\int_0^T \frac{v^2}{R}\, dt$

$$= \frac{1}{T} \times \frac{V^2}{R} \times T = \frac{V^2}{R}$$

Alternatively,

$$P = P_1 + P_3 + P_5 + \cdots$$

$$= \left(\frac{4V}{\sqrt{2}\pi}\right)^2 \frac{1}{R} + \left(\frac{4V}{3\sqrt{2}\pi}\right)^2 \frac{1}{R} + \left(\frac{4V}{5\sqrt{2}\pi}\right)^2 \frac{1}{R} + \cdots$$

$$= \frac{V^2}{R} \times \frac{8}{\pi^2} \times \left(1 + \frac{1}{9} + \frac{1}{25} + \frac{1}{49} + \cdots\right)$$

It can be shown by direct calculation that the infinite series in the parentheses has a sum that approaches $\pi^2/8$; thus P becomes V^2/R as found from direct calculation.

Fraction of energy in fundamental $= 8/\pi^2 = 0.81$

Fraction of energy in first five harmonics

$$= \frac{8}{\pi^2}\left(1 + \frac{1}{9} + \frac{1}{25}\right) = 0.93$$

Fraction of energy in first nine harmonics

$$= \frac{8}{\pi^2}\left(1 + \frac{1}{9} + \frac{1}{25} + \frac{1}{49} + \frac{1}{81}\right) = 0.96$$

Note that 90% of the energy of the square wave is in the first three harmonics; that is, in the fundamental and the third harmonic.

1.7 (a) D can represent 15 distinct values between 0 and +15 V. Thus,

$v_A = 0$ V $\Rightarrow D = 0000$

$v_A = 1$ V $\Rightarrow D = 0001$

$v_A = 2$ V $\Rightarrow D = 0010$

$v_A = 15$ V $\Rightarrow D = 1111$

(b) (i) +1 V (ii) +2 V (iii) +4 V (iv) +8 V

(c) The closest discrete value represented by D is 5 V; thus $D = 0101$. The error is -0.2 V or $-0.2/5.2 \times 100 \simeq -4\%$.

1.8 Voltage gain = 20 log 100 = 40 dB

Current gain = 20 log 1000 = 60 dB

Power gain $= 10 \log A_p = 10 \log (A_v A_i)$

$= 10 \log 10^5 = 50$ dB

1.9 $P_{dc} = 15 \times 8 = 120$ mW

$$P_L = \frac{(6/\sqrt{2})^2}{1} = 18 \text{ mW}$$

$P_{dissipated} = 120 - 18 = 102$ mW

$$\eta = \frac{P_L}{P_{dc}} \times 100 = \frac{18}{120} \times 100 = 15\%$$

1.10 Refer to Example 1.4

$$v_O = 10 - 10^{-11} e^{40 v_I}$$

Let $v_I = V_I + v_i$ and $v_O = V_O + v_o = 5 + v_o$

Thus,

$$5 + v_o = 10 - 10^{-11} e^{40 V_I} e^{40 v_i}$$

$$= 10 - 5e^{40 v_i}$$

$$v_o = 5 - 5e^{40 v_i} \quad (1)$$

For $v_i = 1$ mV:

(a) Using the small-signal gain of -200 V/V,

$$v_o = -200 \times 1 \text{ mV} = -0.2 \text{ V}$$

(b) Using Eq. (1) above

$$v_o = 5 - 5e^{40 \times 0.001} = -0.204 \text{ V}$$

For $v_i = 5$ mV:

(a) Small-signal approx. $\Rightarrow v_o = -200 \times 5 = -1$ V.

(b) Transfer characteristic $\Rightarrow v_o = 5 - 5e^{40 \times 0.005} = -1.107$ V.

For $v_i = 10$ mV:

(a) Small-signal approx. $\Rightarrow v_o = -200 \times 10 = -2$ V.

(b) Transfer characteristic $\Rightarrow v_o = 5 - 5e^{40 \times 0.01} = -2.459$ V.

1.11 $v_o = 1 \times \frac{10}{10^6 + 10} \simeq 10^{-5}$ V $= 10\ \mu$V

$$P_L = v_o^2/R_L = \frac{(10 \times 10^{-6})^2}{10} = 10^{-11} \text{ W}$$

With the buffer amplifier:

$$v_o = 1 \times \frac{R_i}{R_i + R_s} \times A_{vo} \times \frac{R_L}{R_L + R_o}$$

$$= 1 \times \frac{1}{1+1} \times 1 \times \frac{10}{10+10} = 0.25 \text{ V}$$

$$P_L = \frac{v_o^2}{R_L} = \frac{0.25^2}{10} = 6.25 \text{ mW}$$

$$\text{Voltage gain} = \frac{v_o}{v_s} = \frac{0.25 \text{ V}}{1 \text{ V}} = 0.25 \text{ V/V}$$

$$= -12 \text{ dB}$$

$$\text{Power gain } (A_p) \equiv \frac{P_L}{P_i}$$

where $P_L = 6.25$ mW and $P_i = v_i i_i$,

$$v_i = 0.5 \text{ V} \quad \text{and} \quad i_i = \frac{1 \text{ V}}{1 \text{ M}\Omega + 1 \text{ M}\Omega} = 0.5 \ \mu\text{A}$$

Thus,

$$P_i = 0.5 \times 0.5 = 0.25 \ \mu\text{W}$$

and,

$$A_p = \frac{6.25 \times 10^{-3}}{0.25 \times 10^{-6}} = 25 \times 10^3$$

$$10 \log A_p = 44 \text{ dB}$$

1.12 Open-circuit (no load) output voltage $= A_{v_o} v_i$
Output voltage with load connected

$$= A_{v_o} v_i \frac{R_L}{R_L + R_o}$$

$$0.8 = \frac{1}{R_o + 1} \Rightarrow R_o = 0.25 \text{ k}\Omega = 250 \ \Omega$$

1.13 $A_{v_o} = 40 \text{ dB} = 100 \text{ V/V}$

$$P_L = \frac{v_o^2}{R_L} = \left(A_{v_o} v_i \frac{R_L}{R_L + R_o}\right)^2 \Big/ R_L$$

$$= v_i^2 \times \left(100 \times \frac{1}{1+1}\right)^2 \Big/ 1000 = 2.5 v_i^2$$

$$P_i = \frac{v_i^2}{R_i} = \frac{v_i^2}{10{,}000}$$

$$A_p \equiv \frac{P_L}{P_i} = \frac{2.5 v_i^2}{10^{-4} v_i^2} = 2.5 \times 10^4 \text{ W/W}$$

$$10 \log A_p = 44 \text{ dB}$$

1.17

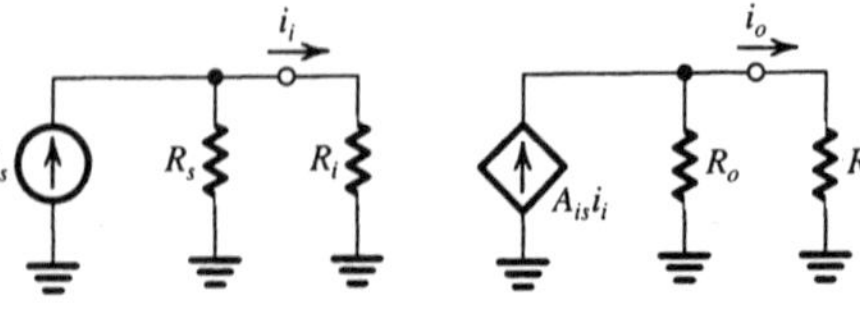

$$i_i = i_s \frac{R_s}{R_s + R_i}$$

$$i_o = A_{is} i_i \frac{R_o}{R_o + R_L} = A_{is} i_s \frac{R_s}{R_s + R_i} \frac{R_o}{R_o + R_L}$$

Thus,

$$\frac{i_o}{i_s} = A_{is} \frac{R_s}{R_s + R_i} \frac{R_o}{R_o + R_L}$$

1.18

$$v_i = v_s \frac{R_i}{R_i + R_s}$$

$$v_o = G_m v_i (R_o \,//\, R_L)$$

$$= G_m v_s \frac{R_i}{R_i + R_s} (R_o \,//\, R_L)$$

Thus,

$$\frac{v_o}{v_s} = G_m \frac{R_i}{R_i + R_s} (R_o \,//\, R_L)$$

1.20

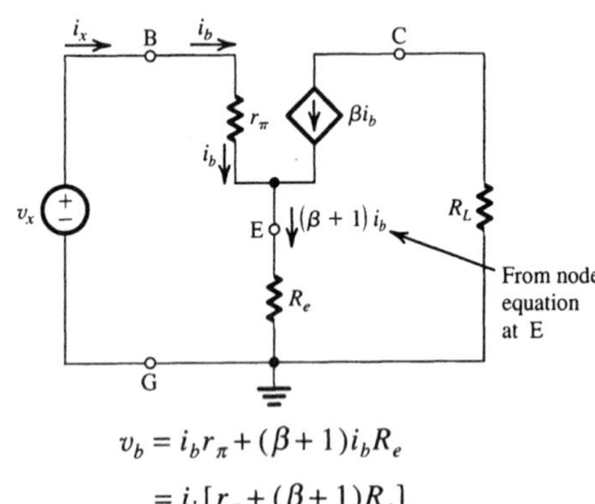

$$v_b = i_b r_\pi + (\beta + 1) i_b R_e$$

$$= i_b [r_\pi + (\beta + 1) R_e]$$

But $v_b = v_x$ and $i_b = i_x$, thus

$$R_{\text{in}} \equiv \frac{v_x}{i_x} = \frac{v_b}{i_b} = r_\pi + (\beta + 1) R_e$$

1.21

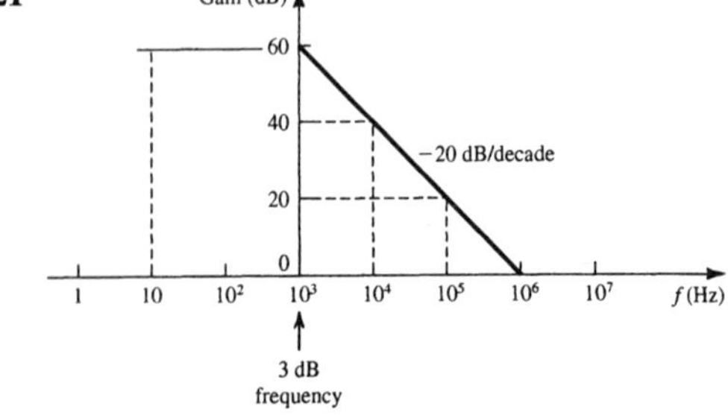

f	Gain
10 Hz	60 dB
10 kHz	40 dB
100 kHz	20 dB
1 MHz	0 dB

1.22

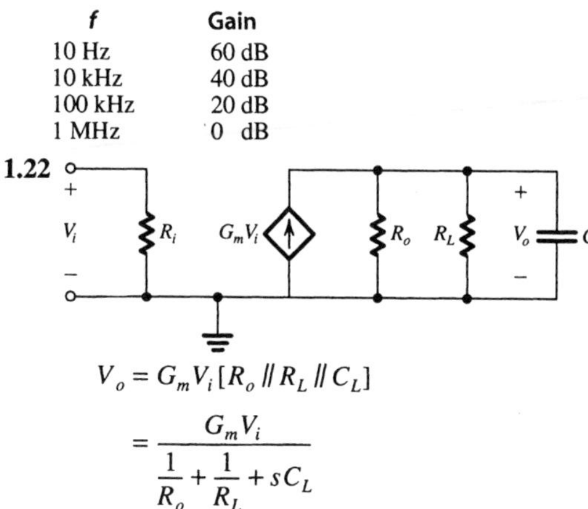

$$V_o = G_m V_i [R_o \,//\, R_L \,//\, C_L]$$

$$= \frac{G_m V_i}{\frac{1}{R_o} + \frac{1}{R_L} + sC_L}$$

Thus,

$$\frac{V_o}{V_i} = \frac{G_m}{\frac{1}{R_o} + \frac{1}{R_L}} \, \frac{1}{1 + \frac{sC_L}{\frac{1}{R_o} + \frac{1}{R_L}}}$$

which is of the STC LP type.

$$\text{DC gain} = \frac{G_m}{\frac{1}{R_o} + \frac{1}{R_L}} \geq 100$$

$$\frac{1}{R_o} + \frac{1}{R_L} \leq \frac{G_m}{100} = \frac{10}{100} = 0.1 \text{ mA/V}$$

$$\frac{1}{R_L} \leq 0.1 - \frac{1}{50} = 0.08 \text{ mA/V}$$

$$R_L \geq \frac{1}{0.08} \text{ k}\Omega = 12.5 \text{ k}\Omega$$

$$\omega_0 = \frac{1}{C_L}\left(\frac{1}{R_o} + \frac{1}{R_L}\right) \geq 2\pi \times 100 \text{ kHz}$$

$$C_L \leq \frac{\left(\frac{1}{50 \times 10^3} + \frac{1}{12.5 \times 10^3}\right)}{2\pi \times 10^5} = 159.2 \text{ pF}$$

1.23 Refer to Fig. E1.23

$$\frac{V_2}{V_s} = \frac{R_i}{R_s + \frac{1}{sC} + R_i} = \frac{R_i}{R_s + R_i} \, \frac{s}{s + \frac{1}{C(R_s + R_i)}}$$

which is a HP STC function.

$$f_{3\text{dB}} = \frac{1}{2\pi C(R_s + R_i)} \leq 100 \text{ Hz}$$

$$C \geq \frac{1}{2\pi(1+9)10^3 \times 100} = 0.16 \ \mu\text{F}$$

1.24 From Fig. 1.31(b): $V_{OH} = V_{DD} = 5$ V

From Fig. 1.31(c): $V_{OL} = V_{\text{offset}} + \frac{V_{DD} - V_{\text{offset}}}{R + R_{\text{on}}} R_{\text{on}}$

$$V_{OL} = 0.1 + \frac{5 - 0.1}{1 + 0.1} \times 0.1 = 0.55 \text{ V}$$

$$NM_H = V_{OH} - V_{IH} = 5 - 1.2 = 3.8 \text{ V}$$

$$NM_L = V_{IL} - V_{OL} = 0.8 - 0.55 = 0.25 \text{ V}$$

The inverter dissipates power in only one state: when the output is low. The power dissipated in this state is

$$P = V_{DD} \times \frac{V_{DD} - V_{\text{offset}}}{R + R_{\text{on}}} = 5 \times \frac{5 - 0.1}{1 + 0.1} = 22.3 \text{ mW}$$

Thus, the average power dissipation is

$$P_D = \tfrac{1}{2} P = 11.1 \text{ mW}$$

1.25 $P_D = fCV_{DD}^2$

$$= 50 \times 10^6 \times 2 \times 10^{-12} \times 5^2 = 2.5 \text{ mW}$$

Chapter 2 - Exercises

2.1

The minimum number of terminals required by a single op amp is five: two input terminals, one output terminal, one terminal for positive power supply and one terminal for negative power supply.

The minimum number of terminals required by a quad op amp is 14: each op amp requires two input terminals and one output terminal (accounting for 12 terminals for the four op amps). In addition, the four op amp can all share one terminal for positive power supply and one terminal for negative power supply.

2.2

We know that $v_3 = A(v_2 - v_1)$ and $A = 1000$. Therefore:

(a) $v_1 = v_2 - \frac{v_3}{A} = 0 - \frac{2}{10^3} = -0.002\,V = -2\,mV$

$\Rightarrow v_1 = 2\,mV$, $v_{Id} = v_2 - v_1 = 0 + 2\,mV \Rightarrow v_{Id} = +2\,mV$

$v_{Icm} = \frac{1}{2}(v_1 + v_2) = \frac{1}{2}(0 - 2\,mV) = -1\,mV$

(b) $v_1 = v_2 - \frac{v_3}{A} = 5 - \frac{-10}{1000} = 5.01\,V$

$v_{Id} = v_2 - v_1 = 5 - 5.01 = 0.01\,V = 10\,mV$

$v_{Icm} = \frac{1}{2}(v_1 + v_2) = \frac{1}{2}(5.01 + 5) = 5.005\,V \simeq 5\,V$

(c) $v_3 = A(v_2 - v_1) = 1000 \times (0.998 - 1.002) = -4\,V$

$v_{Id} = v_2 - v_1 = 0.998 - 1.002 = -4\,mV$

$v_{Icm} = \frac{1}{2}(v_1 + v_2) = \frac{1}{2}(1.002 + 0.998) = 1\,V$

(d) $v_2 = v_1 + \frac{v_3}{A} = -3.6 + \frac{-3.6}{1000} = -3.6036\,V$

$v_{Id} = v_2 - v_1 = -3.6036 - (-3.6) = -0.0036\,V = -3.6\,mV$

$v_{Icm} = \frac{1}{2}(v_1 + v_2) = -3.6018\,V \simeq -3.6\,V$

2.3

From Figure E2.3 we have: $v_3 = \mu v_d$ and

$v_d = (G_m v_2 - G_m v_1)R = G_m R(v_2 - v_1)$

Therefore:

$v_3 = \mu G_m R(v_2 - v_1)$

That is the open-loop gain of the op amp is $A = \mu G_m R$. For $G_m = 10\,mA/V$, $R = 10\,k\Omega$ and $\mu = 100$ we have:

$A = 100 \times 10 \times 10 = 10^4\,V/V$ or equivalently 80dB

2.4

The gain and input resistance of the inverting amplifier circuit shown in Figure 2.5 are $-\frac{R_2}{R_1}$ and R_1 respectively. Therefore, we have:

$R_1 = 100\,k\Omega$ and $-\frac{R_2}{R_1} = -10 \Rightarrow R_2 = 10R_1$

Thus:

$R_2 = 10 \times 100\,k\Omega = 1\,M\Omega$

2.5

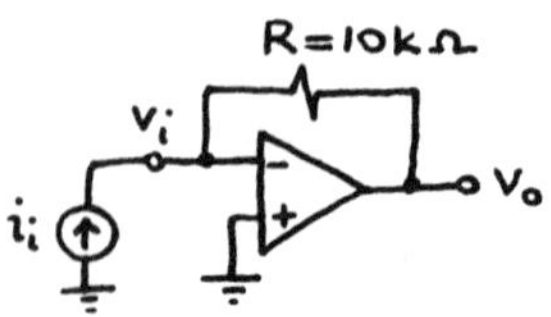

From Table 1.1 we have:

$R_m = \frac{v_o}{i_i}\Big|_{i_o=0}$, i.e. output is open circuit

The negative input terminal of the op amp, i.e., v_i is a virtual ground, thus $v_i = 0$

CONT.

$v_o = v_i - Ri_i = 0 - Ri_i = -Ri_i$

$R_m = \left.\frac{v_o}{i_i}\right|_{i_o=0} = \frac{-Ri_i}{i_i} = -R \Rightarrow R_m = -R = -10\,k\Omega$

$R_i = \frac{v_i}{i_i}$ and v_i is a virtual ground ($v_i = 0$),

thus $R_i = \frac{0}{i_i} = 0 \Rightarrow R_i = 0\,\Omega$

Since we are assuming that the op amp in this transresistance amplifier is ideal, the op amp has zero output resistance and therefore the output resistance of this transresistance amplifier is also zero. That is $R_o = 0\,\Omega$.

Connecting the signal source shown in Figure E2.5 to the input of this amplifier we have:

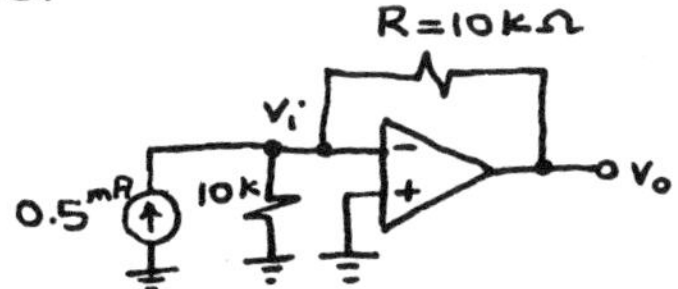

v_i is a virtual ground that is $v_i = 0$, thus the current flowing through the $10\,k\Omega$ resistor connected between v_i and ground is zero. Therefore

$v_o = v_i - R \times 0.5\,mA = 0 - 10\,k \times 0.5\,mA = -5\,V$

2.6

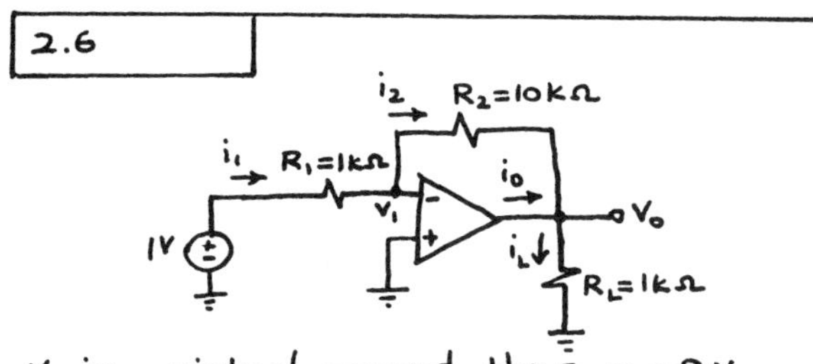

v_1 is a virtual ground, thus $v_1 = 0\,V$

$i_1 = \frac{1V - v_1}{R_1} = \frac{1-0}{1k\Omega} = 1\,mA$

Assuming an ideal op amp, the current flowing into the negative input terminal of the op amp is zero. Therefore, $i_2 = i_1 \Rightarrow$ $i_2 = 1\,mA$

$v_o = v_1 - i_2 R_2 = 0 - 1\,mA \times 10\,k\Omega = -10\,V$

$i_L = \frac{v_o}{R_L} = \frac{-10\,V}{1\,k\Omega} = -10\,mA$

$i_o = i_L - i_2 = -10\,mA - 1\,mA = -11\,mA$

voltage gain $= \frac{v_o}{1V} = \frac{-10\,V}{1\,V} = -10\,V/V$ or $20\,dB$

current gain $= \frac{i_L}{i_1} = \frac{-10\,mA}{1\,mA} = -10\,A/A$ or $20\,dB$

power gain $= \frac{P_L}{P_i} = \frac{-10V \times (-10\,mA)}{1V \times 1\,mA} = 100\,W/W$ or $20\,dB$

Note that power gain in dB is $10\log_{10}\left|\frac{P_L}{P_i}\right|$.

2.7

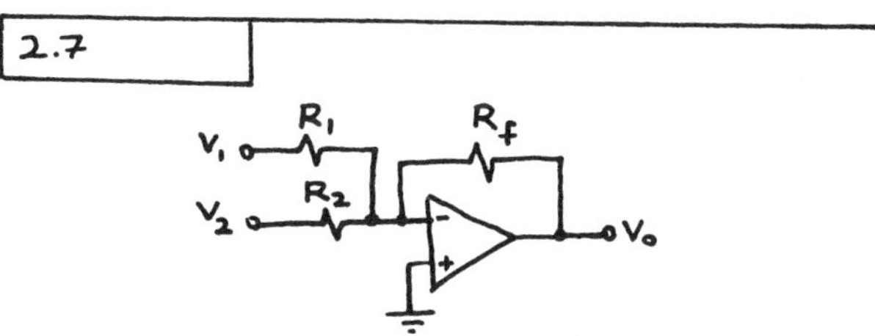

For the circuit shown above we have:

$v_o = -\left(\frac{R_f}{R_1}v_1 + \frac{R_f}{R_2}v_2\right)$

Since it is required that $v_o = -(v_1 + 5v_2)$, we want to have:

$\frac{R_f}{R_1} = 1$ and $\frac{R_f}{R_2} = 5$

It is also desired that for a maximum output voltage of 10V the current in the feedback resistor does not exceed 1mA.

Therefore $\frac{10V}{R_f} \leq 1\,mA \Rightarrow R_f \geq \frac{10V}{1\,mA} \Rightarrow$

$R_f \geq 10\,k\Omega$ Let us choose R_f to be $10\,k\Omega$, then $R_1 = R_f = 10\,k\Omega$ and $R_2 = \frac{R_f}{5} = 2\,k\Omega$

2.8

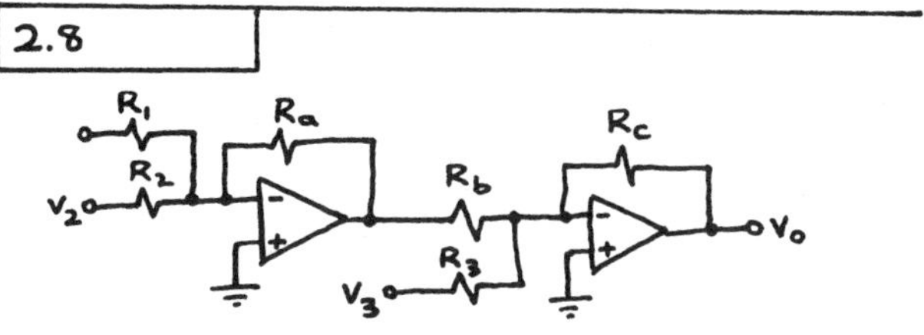

CONT.

$$v_o = \left(\frac{R_a}{R_1}\right)\left(\frac{R_c}{R_b}\right)v_1 + \left(\frac{R_a}{R_2}\right)\left(\frac{R_c}{R_b}\right)v_2 - \left(\frac{R_c}{R_3}\right)v_3$$

We want to design the circuit such that

$v_o = 2v_1 + v_2 - 4v_3$

Thus we need to have

$\left(\frac{R_a}{R_1}\right)\left(\frac{R_c}{R_b}\right) = 2$, $\left(\frac{R_a}{R_2}\right)\left(\frac{R_c}{R_b}\right) = 1$ and $\frac{R_c}{R_3} = 4$

From the above three equations, we have to find six unknown resistors, therefore, we can arbitrarily choose three of these resistors. Let us choose: $R_a = R_b = R_c = 10\,k\Omega$

Then we have

$R_3 = \frac{R_c}{4} = \frac{10}{4} = 2.5\,k\Omega$

$\left(\frac{R_a}{R_1}\right)\left(\frac{R_c}{R_b}\right) = 2 \Rightarrow \frac{10}{R_1} \times \frac{10}{10} = 2 \Rightarrow R_1 = 5\,k\Omega$

$\left(\frac{R_a}{R_2}\right)\left(\frac{R_c}{R_b}\right) = 1 \Rightarrow \frac{10}{R_2} \times \frac{10}{10} = 1 \Rightarrow R_2 = 10\,k\Omega$

2.9

Using the superposition principle, to find the contribution of v_1 to the output voltage v_o, we set $v_2 = 0$.

Then v_+ (the voltage at the positive input of the op amp is: $v_+ = \frac{3}{2+3} v_1 = 0.6 v_1$

Thus $v_o = \left(1 + \frac{9K}{1K}\right) v_+ = 10 \times 0.6 v_1 = 6 v_1$

To find the contribution of v_2 to the output voltage v_o we set $v_1 = 0$.

Then $v_+ = \frac{2}{2+3} v_2 = 0.4 v_2$

Hence $v_o = \left(1 + \frac{9K}{1K}\right) v_+ = 10 \times 0.4 v_2 = 4 v_2$

Combining the contributions of v_1 and v_2 to v_o we have $v_o = 6v_1 + 4v_2$

2.10

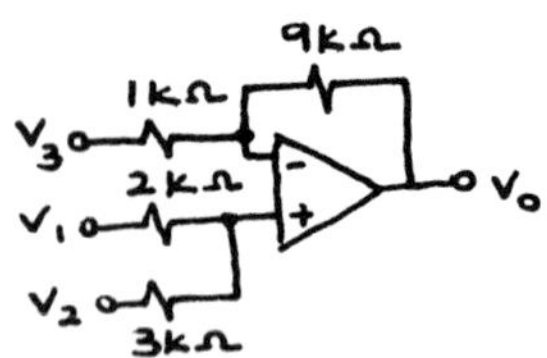

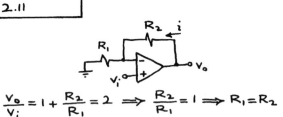

Using the superposition principle, to find the contribution of v_1 to v_o we set $v_2 = v_3 = 0$

Then we have (refer to the solution of exercise 2.9): $v_o = 6v_1$

To find the contribution of v_2 to v_o we set $v_1 = v_3 = 0$, then: $v_o = 4v_2$

To find the contribution of v_3 to v_o we set $v_1 = v_2 = 0$, then: $v_o = -\frac{9\,k\Omega}{1\,k\Omega} v_3 = -9v_3$

Combining the contributions of v_1, v_2, and v_3 to v_o we have: $v_o = 6v_1 + 4v_2 - 9v_3$

2.11

$\frac{v_o}{v_i} = 1 + \frac{R_2}{R_1} = 2 \Rightarrow \frac{R_2}{R_1} = 1 \Rightarrow R_1 = R_2$

If $v_o = 10\,V$ then it is desired that $i = 10\,\mu A$.

Thus, $i = \frac{10\,V}{R_1 + R_2} = 10\,\mu A \Rightarrow R_1 + R_2 = \frac{10\,V}{10\,\mu A}$

$R_1 + R_2 = 1\,M\Omega$ and $R_1 = R_2 \Rightarrow$

$R_1 = R_2 = 0.5\,M\Omega$

2.12

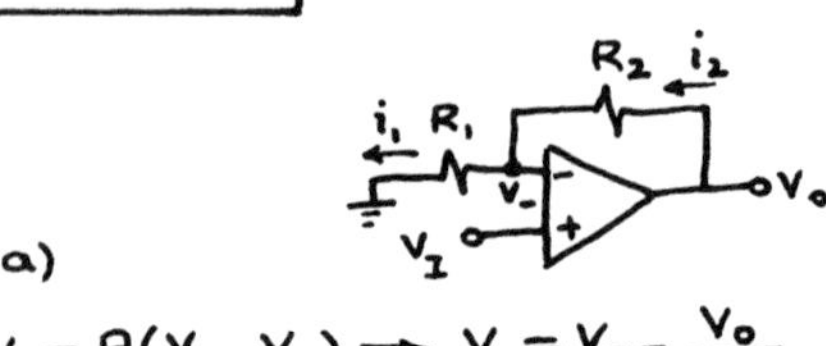

(a)

$v_o = A(v_I - v_-) \Rightarrow v_- = v_I - \frac{v_o}{A}$

CONT.

$$i_2=i_1 \Rightarrow \frac{v_o-v_-}{R_2}=\frac{v_-}{R_1} \Rightarrow \frac{v_o}{R_2}=\left(\frac{1}{R_2}+\frac{1}{R_1}\right)v_-$$

$$v_o=\left(1+\frac{R_2}{R_1}\right)v_-=\left(1+\frac{R_2}{R_1}\right)\left(v_I-\frac{v_o}{A}\right) \Rightarrow$$

$$v_o+\frac{1+R_2/R_1}{A}v_o=(1+R_2/R_1)v_I$$

$$\frac{v_o}{v_I}=\frac{1+R_2/R_1}{1+\frac{1+R_2/R_1}{A}} \Rightarrow G=\frac{1+R_2/R_1}{1+\frac{1+R_2/R_1}{A}}$$

(b) For $R_1=1k\Omega$ and $R_2=9k\Omega$, the ideal value for the closed-loop gain is $1+\frac{9}{1}$, that is 10. The actual closed-loop gain is $G=\frac{10}{1+\frac{10}{A}}$

If $A=10^3$ then $G=9.901$ and

$\epsilon=\frac{G-10}{10}\times 100=-0.99\% \simeq -1\%$

For $v_I=1V$, $v_o=G\times v_I=9.901V$ and

$v_o=A(v_+-v_-) \Rightarrow v_+-v_-=\frac{v_o}{A}=\frac{9.901}{1000}\simeq 9.9mV$

If $A=10^4$ then $G=9.99$ and $\epsilon=-0.1\%$

For $v_I=1V$, $v_o=G\times v_I=9.99V$, therefore,

$v_+-v_-=\frac{v_o}{A}=\frac{9.99}{10^4}=0.999\ mV\simeq 1\ mV$

If $A=10^5$ then $G=9.999$ and $\epsilon=-0.01\%$

For $v_I=1V$, $v_o=G\times v_I=9.999V$, thus,

$v_+-v_-=\frac{v_o}{A}=\frac{9.999}{10^5}=0.09999\ mV\simeq 0.1mV$

2.13

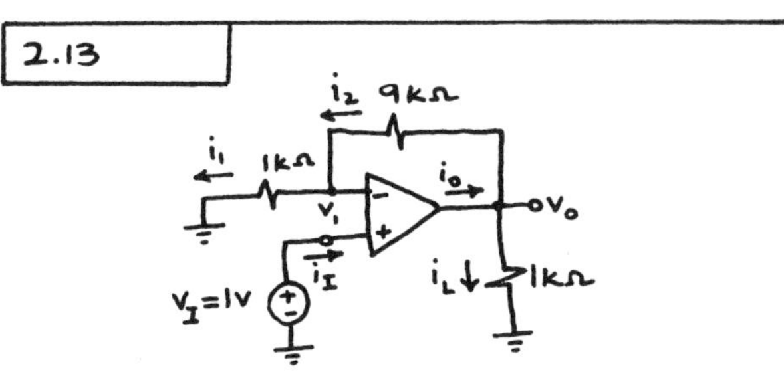

$i_I=0A$, $v_1=v_I=1V$, $i_1=\frac{v_1}{1k\Omega}=\frac{1V}{1k\Omega}=1mA$

$i_2=i_1=1mA$, $v_o=v_1+i_2\times 9k\Omega=1+1\times 9=10V$

$i_L=\frac{v_o}{1k\Omega}=\frac{10V}{1k\Omega}=10mA$, $i_o=i_L+i_2=11mA$

$\frac{v_o}{v_i}=\frac{10V}{1V}=10\frac{V}{V}$ or 20dB

$\frac{i_L}{i_I}=\frac{10mA}{0}=\infty$

$\frac{P_L}{P_I}=\frac{v_o\times i_L}{v_I\times I_i}=\frac{10\times 10}{1\times 0}=\infty$

2.14

(a) load voltage $=\frac{1k\Omega}{1k\Omega+1M\Omega}\times 1V\simeq 1mV$

(b) load voltage $=1V$

2.15

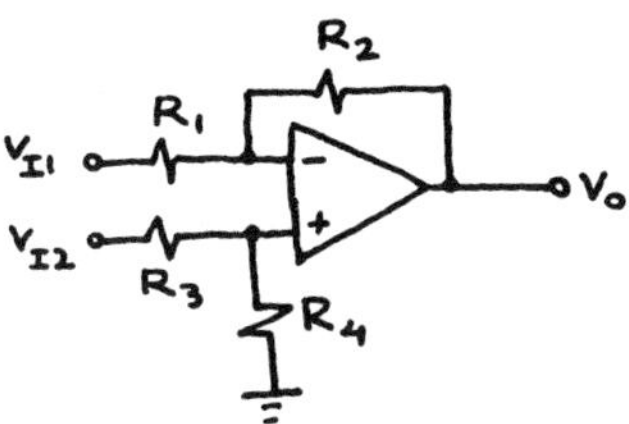

(a)

$R_1=R_3=2k\Omega$, $R_2=R_4=200k\Omega$

Since $R_4/R_3=R_2/R_1$ we have:

$A_d=\frac{v_o}{v_{I2}-v_{I1}}=\frac{R_2}{R_1}=\frac{200}{2}=100\ V/V$

(b) $R_{id}=2R_1=2\times 2k\Omega=4k\Omega$

Since we are assuming the op amp is ideal $R_o=0\Omega$

(c) $$A_{cm}=\frac{v_o}{v_{Icm}}=\left(\frac{R_4}{R_4+R_3}\right)\left(1-\frac{R_2}{R_1}\frac{R_3}{R_4}\right)$$

$$=\left(\frac{1}{1+\frac{R_3}{R_4}}\right)\left(1-\frac{R_2}{R_1}\frac{R_3}{R_4}\right)$$

$$=\frac{\frac{R_4}{R_3}-\frac{R_2}{R_1}}{\frac{R_4}{R_3}+1}$$

The worst case common-mode gain A_{cm} happens when $|A_{cm}|$ has its maximum value. If the resistors have 1% tolerance, we have $\frac{R_{4nom}(1-0.01)}{R_{3nom}(1+0.01)}\leqslant\frac{R_4}{R_3}\leqslant\frac{R_{4nom}(1+0.01)}{R_{3nom}(1-0.01)}$

CONT.

where R_{3nom} and R_{4nom} are nominal values for R_3 and R_4 respectively. We have: $R_{3nom} = 2k\Omega$ and $R_{4nom} = 200k\Omega$, thus,

$$\frac{200\times0.99}{2\times1.01} \leq \frac{R_4}{R_3} \leq \frac{200\times1.01}{2\times0.99}$$

$$98.02 \leq \frac{R_4}{R_3} \leq 102.02$$

Similarly, we can show that

$$98.02 \leq \frac{R_2}{R_1} \leq 102.02$$

Hence, $-102.02 \leq -\frac{R_2}{R_1} \leq -98.02$

Therefore,

$$-4 \leq \frac{R_4}{R_3} - \frac{R_2}{R_1} \leq 4 \Rightarrow \left|\frac{R_4}{R_3} - \frac{R_2}{R_1}\right| \leq 4$$

In the worst case

$$\frac{\left|\frac{R_4}{R_3} - \frac{R_2}{R_1}\right|}{1+\frac{R_4}{R_3}} \leq \frac{4}{1+98.02} \Rightarrow |A_{cm}| \leq 0.04$$

Note that the worst case A_{cm} happens when $\frac{R_4}{R_3} = 98.02$ and $\frac{R_2}{R_1} = 102.02$

The differential gain A_d of the amplifier is $A_d = \frac{R_2}{R_1}$, therefore, the corresponding value of CMRR for the worst case A_{cm} is:

$$CMRR = 20\log\frac{|A_d|}{|A_{cm}|} = 20\log\frac{102.02}{0.04} \Rightarrow$$

$$CMRR = 20\log(2550.5) \simeq 68dB$$

2.16

We choose $R_3 = R_1$ and $R_4 = R_2$. Then for the circuit to behave as a difference amplifier with a gain of 10 and an input resistance of $20k\Omega$ we require

$A_d = \frac{R_2}{R_1} = 10$ and $R_{id} = 2R_1 = 20k\Omega \Rightarrow$

$R_1 = 10k\Omega$ and $R_2 = A_dR_1 = 10\times10k\Omega = 100k\Omega$

Therefore, $R_1 = R_3 = 10k\Omega$ and $R_2 = R_4 = 100k\Omega$

2.17

For the instrumentation amplifier of Figure 2.20(b), we have:

$$v_{I1} = v_{Icm} - \frac{1}{2}v_{Id} = 5 - \frac{1}{2}\times0.01\sin\omega t \Rightarrow$$

$$v_{I1} = 5 - 0.005\sin\omega t$$

$$v_{I2} = v_{Icm} + \frac{1}{2}v_{Id} = 5 + 0.005\sin\omega t$$

$$v_-(\text{op amp } A_1) = v_{I1} = 5 - 0.005\sin\omega t$$

$$v_-(\text{op amp } A_2) = v_{I2} = 5 + 0.005\sin\omega t$$

The voltage at the output of op amp A_1 is: $v_{O1} = v_{I1} - R_2\times\frac{v_{Id}}{2R_1}$ where $v_{Id} = v_{I2} - v_{I1}$

$$v_{O1} = 5 - 0.005\sin\omega t - 500k\Omega\,\frac{0.01\sin\omega t}{1k\Omega}$$

$$v_{O1} = 5 - 5.005\sin\omega t$$

The voltage at the output of op amp A_2 is: $v_{O2} = v_{I2} + R_2\times\frac{v_{Id}}{2R_1} = 5 + 5.005\sin\omega t$

$v_+(\text{op amp } A_3) = v_{O2}\times\frac{R_4}{R_3+R_4}$ and

$R_3 = R_4 = 10k\Omega$, therefore

$$v_+(\text{op amp } A_3) = v_{O2}\times\frac{10}{10+10} = v_{O2}\times\frac{1}{2} \Rightarrow$$

$$v_+(\text{op amp } A_3) = 2.5 + 2.5025\sin\omega t$$

$$v_-(\text{op amp } A_3) = v_+(\text{op amp } A_3)$$
$$= 2.5 + 2.5025\sin\omega t$$

$$v_O = \frac{R_4}{R_3}\left(1+\frac{R_2}{R_1}\right)v_{Id}$$

$$= \frac{10k\Omega}{10k\Omega}\left(1+\frac{0.5M\Omega}{0.5k\Omega}\right)\times0.01\sin\omega t$$

$v_O = 10.01\sin\omega t$ Note: All node voltages are in volts.

2.18

From equation (2.28) we have:

$\omega_t = A_0\omega_b \Rightarrow f_t = A_0 f_b \Rightarrow f_b = \frac{f_t}{A_0}$, and we know $20\log A_0 = 106$ and $f_t = 3MHz$, therefore $f_b \simeq 15Hz$

By definition the open-loop gain (in dB) at f_b is: A_0(in dB) $- 3 = 106 - 3 = 103dB$

To find the open-loop gain at frequency f we can use equation (2.31) (especially when $f \gg f_b$ which is the case in this exercise) and write:

Open-loop gain at $f \simeq 20\log(\frac{f_t}{f})$

Therefore:

Open-loop gain at $300Hz = 20\log\frac{3MHz}{300} = 80dB$

Open-loop gain at $3KHz = 20\log\frac{3MHz}{3KHz} = 60dB$

Open-loop gain at $12KHz = 20\log\frac{3MHz}{12KHz} = 48dB$

Open-loop gain at $60KHz = 20\log\frac{3MHz}{60KHz} = 34dB$

2.19

Using equations (2.31) and (2.28) we have:

Open-loop gain (in dB) at $f = 20\log\frac{f_t}{f} = 20\log\frac{A_0 f_b}{f}$

We have: $40 = 20\log\frac{10^6 f_b}{10\times10^3} \Rightarrow f_b = 1Hz$

$f_t = A_0 f_b = 10^6 \times 1 = 1MHz$

gain-bandwidth product $= A_0 \times f_b = 1MHz$

gain at $1KHz = 20\log\frac{1MHz}{1KHz} = 60dB$

2.20

Since dc gain of the op amp is much larger than the dc gain of the designed non-inverting amplifier, we can use equation (2.35).

Therefore:

$f_{3dB} = \frac{f_t}{1 + \frac{R_2}{R_1}}$ and $1 + \frac{R_2}{R_1} = 100$ and $f_t = 2MHz$

Hence $f_{3dB} = \frac{2MHz}{100} = 20KHz$

2.21

For the input voltage step of magnitude V the output waveform will still be given by the exponential waveform of equation (2.40) if $\omega_t V \leq SR$

That is $V \leq \frac{SR}{\omega_t} \Rightarrow V \leq \frac{SR}{2\pi f_t}$

$V \leq 0.16v$, thus, the largest possible input voltage step is 0.16v.

From Appendix F we know that the 10% to 90% rise time of the output waveform of the form of equation (2.40) is $t_r \simeq 2.2\frac{1}{\omega_t}$

Thus, $t_r \simeq 0.35\mu s$

If an input step of amplitud 1.6v (10 times as large compared to the previous case) is applied, the the output is slew-rate limited and is linearly rising with a slope equal to the slew-rate, as shown in the following figure.

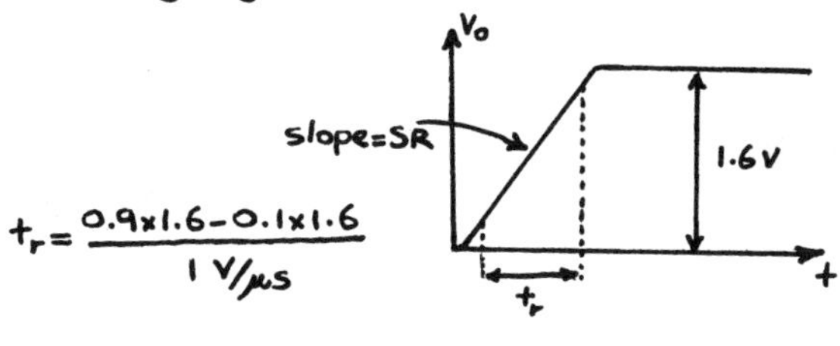

$t_r = \frac{0.9\times1.6 - 0.1\times1.6}{1\,V/\mu s}$

$\Rightarrow t_r = 1.28\mu s$

2.22

From equation (2.41) we have:

$f_M = \frac{SR}{2\pi V_{omax}} = 15.915\,KHz \simeq 15.9\,KHz$

Using equation (2.42), for an input sinusoid with frequency $f = 5f_M$, the maximum possible amplitude that can be accommodated at the output without incurring SR distortion is:

$V_o = V_{omax}\left(\frac{f_M}{5f_M}\right) = 10 \times \frac{1}{5} = 2V$ (peak)

2.23

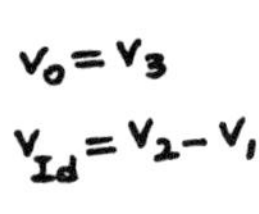

$V_o = V_3$

$V_{Id} = V_2 - V_1$

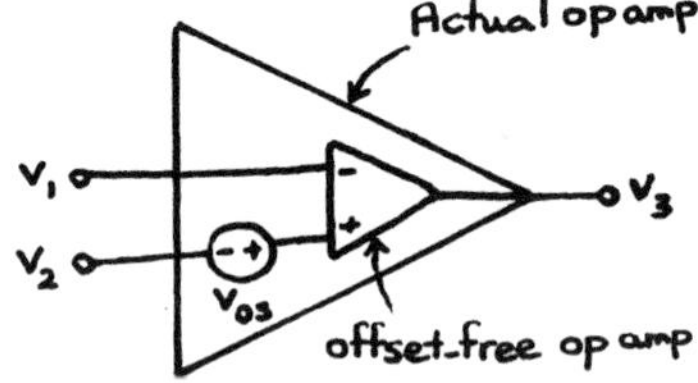

$V_{Id} = V_+ - V_{os} - V_-$

In order to have zero differential input for the offset-free op amp (i.e., $V_+ - V_- = 0$) we need $V_{Id} = V_+ - V_- - V_{os} = 0 - 5\text{mv} = -5\text{mv}$

Thus, the transfer characteristic v_o versus v_{Id} is:

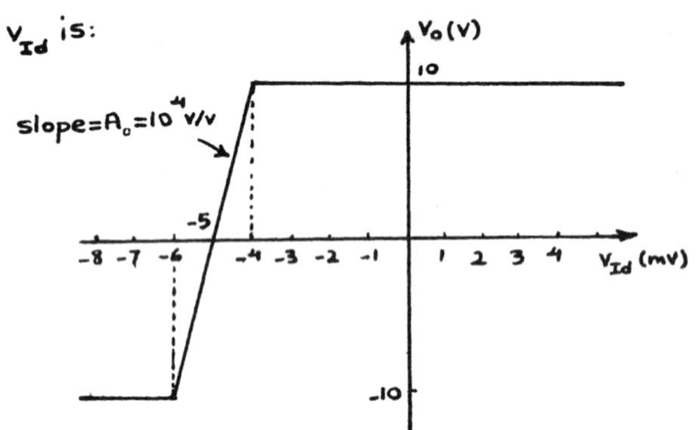

2.24

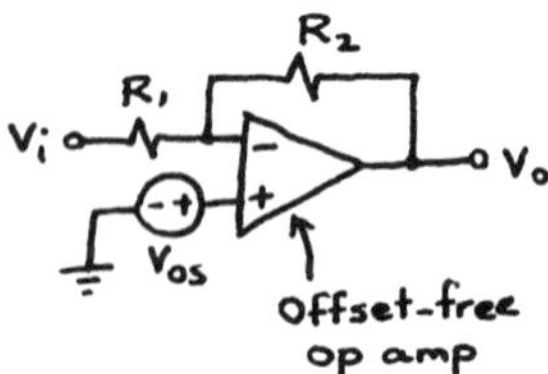

(a)

The output dc voltage due to the input offset is (from equation (2.43)):

$V_{o\,dc} = 3\,\text{mv}\,(1 + 1000) \simeq 3\,\text{V}$

Since the output saturation levels are ± 10v, the maximum amplitude of a sine-wave at the op amp output without clipping is $10 - 3 = 7$v which corresponds to the input sine-wave peak of $\frac{7\text{v}}{1000} = 7\text{mv}$

(b)

If the effect of the input offset is nulled at room temperature, then maximum input would be $\frac{10}{1000} = 10\text{mv}$

Temperature range of 0°C to 75°C corresponds to input offset voltage range of: $(0-25) \times 10\mu\text{V} = -250\mu\text{V}$ to $(75-25) \times 10\mu\text{V} = 500\mu\text{V}$

This input offset range corresponds to output dc levels of $-250\mu\text{V} \times (1+1000) \simeq -0.25\text{V}$ to $500\mu\text{V} \times (1+1000) \simeq 0.5\text{V}$

Thus the worst case output peak is $10 - 0.5 = 9.5\text{V}$ which corresponds to the input sine-wave peak of $\frac{9.5\text{V}}{1000} = 9.5\text{mV}$

2.25

C, R_1, R_2, V_{in}, V_o, V_{os}, offset-free op amp

(a) The equivalent circuit for determining dc output offset voltage is:

R_2, V_{os}, offset-free op amp

Thus, the dc output voltage due to the input offset is $V_{os} = 3\text{mv}$

Hence, the maximum amplitude of a sine wave at the op amp output without clipping is $10\text{V} - 3\text{mv} = 9.997\text{ V}$ which corresponds to the input sine-wave peak of

$\frac{9.997}{1000} \simeq 10\text{mV}$

Therefore, there is no need for offset trimming.

(b) The magnitude of the gain of this amplifier is

$$|A_d| = \frac{R_2}{\left|R_1 + \frac{1}{jC\omega}\right|} = \frac{R_2}{\sqrt{R_1^2 + \frac{1}{C^2\omega^2}}}$$

$$20\log|A_d| > 57 \Rightarrow |A_d| > 707.95$$

$$\frac{R_2}{\sqrt{R_1^2 + \frac{1}{C^2\omega^2}}} > 707.95 \Rightarrow \frac{1\text{M}\Omega}{\sqrt{1000^2 + \frac{1}{C^2\omega^2}}} > 707.95$$

$$\sqrt{1000^2 + \frac{1}{C^2\omega^2}} < 1412.5 \Rightarrow \frac{1}{C\omega} < 997.6$$

CONT.

$C > \frac{1}{997.6 \times \omega} \Rightarrow C > \frac{1}{997.6 \times 2\pi \times 100} \Rightarrow C > 1.59\mu F$

Hence, we can choose $C = 1.6\mu F$

2.26

From equation (2.44) we have:

$V_o = I_{B1} R_2 \simeq I_B R_2 = 100nA \times 1M\Omega = 0.1V$

From equation (2.46) the value of resistor R_3 (placed in series with positive input to minimize the output offset voltage) is:

$R_3 = R_1 \| R_2 = \frac{R_1 R_2}{R_1 + R_2} = \frac{10K\Omega \times 1M\Omega}{10K\Omega + 1M\Omega} = 9.9k\Omega$

$R_3 = 9.9k\Omega \simeq 10k\Omega$

With this value of R_3 the new value of the output dc voltage (using equation (2.47)) is:

$V_o = I_{os} R_2 = 10nA \times 10k\Omega \simeq 0.01V$

2.27

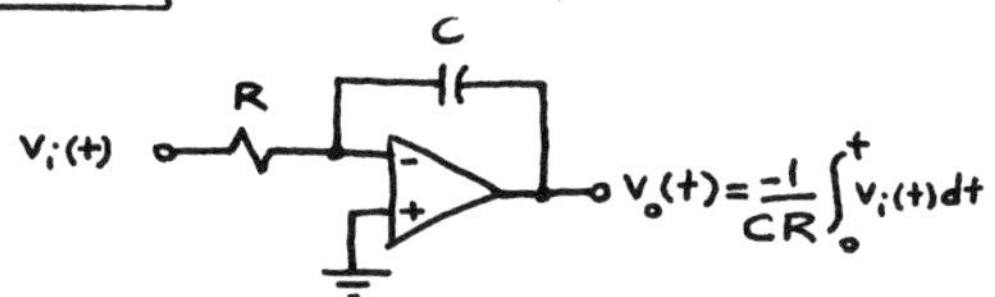

The waveforms for one period of the input and the output signals are shown below:

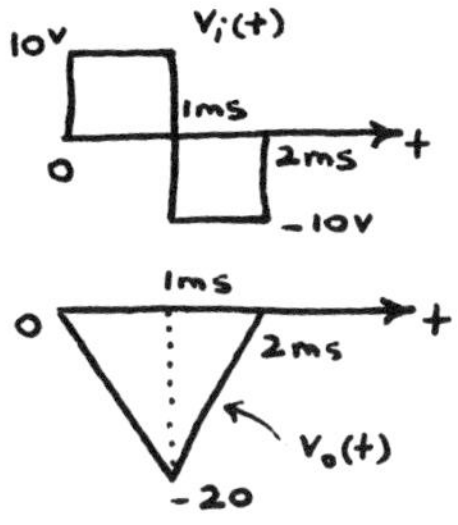

We have

$-20 = \frac{-1}{CR} \int_0^{1ms} 10\,dt$

$\Rightarrow -20 = \frac{-1}{CR} \times 10 \times 1ms$

$\Rightarrow CR = \frac{10}{20} \times 1ms = 0.5ms$

2.28

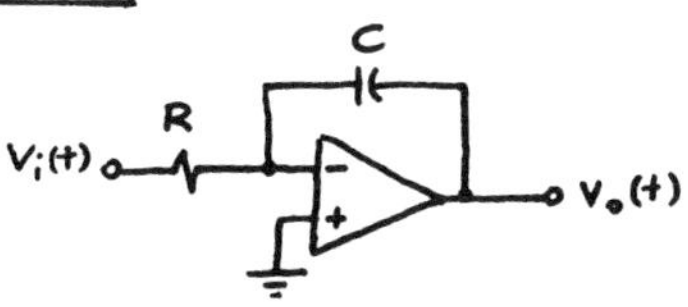

The input resistance of this inverting integrator is R, therefore, $R = 10k\Omega$

Since the desired integration time constant is $10^{-3}s$, we have: $CR = 10^{-3}s \Rightarrow$

$C = \frac{10^{-3}s}{10k\Omega} = 0.1\mu F$

From equation (2.50) the transfer function of this integrator is:

$\frac{V_o(j\omega)}{V_i(j\omega)} = -\frac{1}{j\omega CR}$

For $\omega = 10$ rad/s the integrator transfer function has magnitude

$\left|\frac{V_o}{V_i}\right| = \frac{1}{10 \times 10^{-3}} = 100\,V/V$ and phase $\phi = 90°$.

For $\omega = 1$ rad/s the integrator transfer function has magnitude

$\left|\frac{V_o}{V_i}\right| = \frac{1}{1 \times 10^{-3}} = 1000\,V/V$ and phase $\phi = 90°$.

Using equation (2.53) the frequency at which the integrator gain magnitude is unity is $\omega_{int} = \frac{1}{CR} = \frac{1}{10^{-3}} = 1000$ rad/s

2.29

Using equation (2.54) we have:

$V_o = V_{os} + \frac{V_{os}}{CR} t \Rightarrow 12 = 2mV + \frac{2mV}{1ms} t$

$\Rightarrow t = \frac{12V - 2mV}{2mV} \times 1ms \simeq 6s \Rightarrow t = 6s$

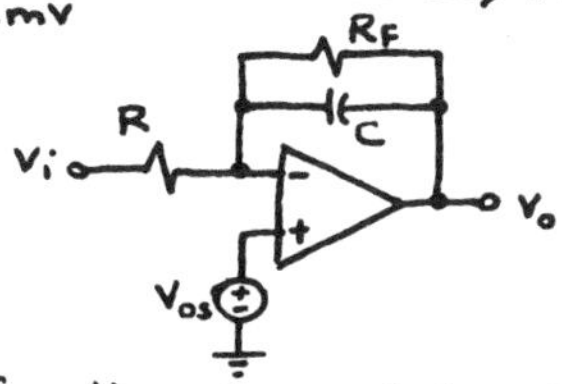

With the feedback resistor R_F, to have at least $\pm 10V$ of output signal swing available, we have to make sure that the output voltage due to V_{os} has a magnitude of at most 2V. From equation (2.43), we know that the output dc voltage due to V_{os}

CON'T.

is $V_o = V_{os}\left(1+\frac{R_F}{R}\right) \Rightarrow 2V = 2mV\left(1+\frac{R_F}{10k\Omega}\right)$

$1+\frac{R_F}{10k\Omega} = 1000 \Rightarrow R_F \simeq 10M\Omega$

The corner frequency of the resulting STC network is $\omega = \frac{1}{CR_F}$

We know $RC = 1ms$ and $R = 10k\Omega \Rightarrow C = 0.1\mu F$

Thus $\omega = \frac{1}{0.1\mu F \times 10M\Omega} = 1\ rad/s$

$f = \frac{\omega}{2\pi} = \frac{1}{2\pi} = 0.16Hz$

2.30

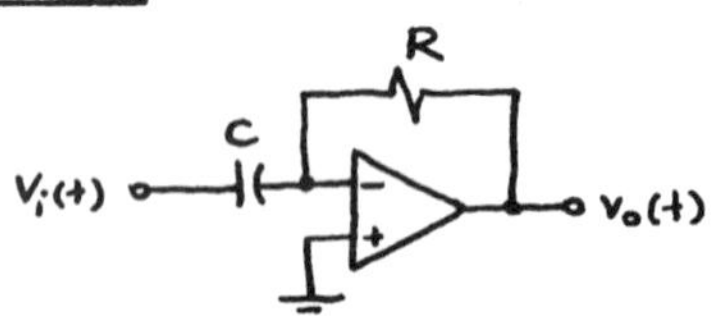

$C = 0.01\mu F$ is the input capacitance of this differentiator. We want $CR = 10^{-2}$ (the time constant of the differentiator), thus,
$R = \frac{10^{-2}}{0.01\mu F} = 1M\Omega$

From equation (2.57), we know that the transfer function of the differentiator is of the form $\frac{V_o(j\omega)}{V_i(j\omega)} = -j\omega CR$

Thus, for $\omega = 10\ rad/s$ the differentiator transfer function has magnitude
$\left|\frac{V_o}{V_i}\right| = 10 \times 10^{-2} = 0.1 V/V$ and phase $\phi = -90°$.

For $\omega = 10^3\ rad/s$ the differentiator transfer function has magnitude
$\left|\frac{V_o}{V_i}\right| = 10^3 \times 10^{-2} = 10\ V/V$ and phase $\phi = -90°$.

If we add a resistor in series with the capacitor to limit the high-frequency gain of the differentiator to 100, the circuit would be:

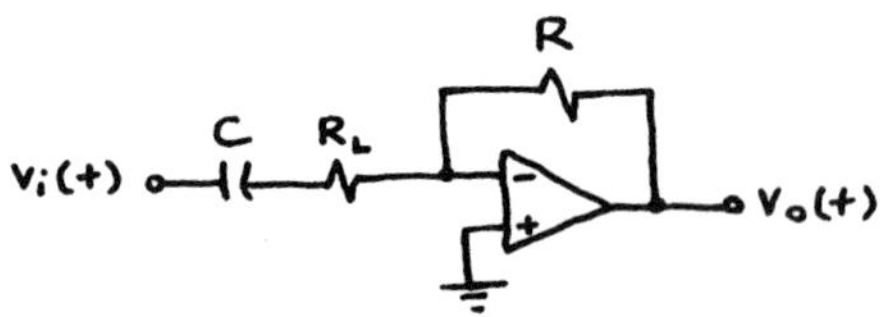

At high frequencies the capacitor C acts like a short circuit. Therefore, the high-frequency gain of this circuit is: $-\frac{R}{R_L}$. To limit the magnitude of this high-frequency gain to 100, we should have:

$\frac{R}{R_L} = 100 \Rightarrow R_L = \frac{R}{100} = \frac{1M\Omega}{100} = 10k\Omega$

CHAPTER 3 – EXERCISES

3.1

Refer to Fig 3.3 (a). For $v_I \geq 0$, the diode conducts and presents a zero voltage drop. Thus $v_O = v_I$. For $v_I < 0$, the diode is cut-off, zero current flows through R and $v_O = 0$. The result is the transfer characteristic in Fig. E3.1.

3.2

For v_I positive (i.e. in the positive half of the sinusoid), the diode conducts, resulting in $v_D = 0$. For v_I negative (during the negative half of the sinusoid) the diode is cut-off and

$$v_D = v_I - v_O = v_I - 0 = v_I.$$

Thus v_O is identical to the negative halves of v_I. The result is the waveform in Fig. E3.2.

3.3

$$\hat{i}_D = \frac{\hat{V}_I}{R} = \frac{10\text{V}}{1\text{k}\Omega} = \underline{\underline{10\text{ mA}}}$$

$$\text{dc component of } v_O = \frac{1}{\pi}\hat{V}_O$$

$$= \frac{1}{\pi}\hat{V}_I = \frac{10}{\pi}$$

$$= \underline{\underline{3.18\text{ V}}}$$

3.4

(a) 5V; 2.5kΩ; $\downarrow \frac{5-0}{2.5} = \underline{\underline{2\text{mA}}}$; $V = \underline{\underline{0\text{V}}}$

(b) 5V; 2.5kΩ; $\downarrow \underline{\underline{0\text{A}}}$; $V = \underline{\underline{5\text{V}}}$

(c) $I = \underline{\underline{0\text{A}}} \downarrow$; $V = -5\text{V}$; −5V

(d) $I = \frac{0+5}{2.5} \downarrow = \underline{\underline{2\text{mA}}}$; 2.5kΩ; −5V; $V = 0\text{V}$

(e) 3V, 2V, 1V; $V_O = \underline{\underline{3\text{V}}}$; 1kΩ; $I = \frac{3}{1} = \underline{\underline{3\text{mA}}}$

CONT.

(f)

5V

$1k\Omega$ ↓ $I = \frac{5-1}{1} = \underline{\underline{4mA}}$

+3V, +2V, +1V

+

$V = 1V$

−

3.5

$$V_{avg} = \frac{10}{\pi}$$

$$50 + R = \frac{\frac{10}{\pi} - 0}{1mA} = \frac{10}{\pi} k\Omega$$

$$\therefore R = \underline{\underline{3.133k\Omega}}$$

3.6

$$I = I_S e^{V/nV_T}$$

$$\therefore \frac{I_1}{I_2} = \frac{I_S}{I_S} e^{\frac{V_1 - V_2}{nV_T}}$$

$$\frac{10}{0.1} = e^{\Delta V/nV_T} = e^{\Delta V/1.5(0.025)}$$

$$\Delta V = \underline{\underline{172.5mV}}$$

3.7

$$i = I_S e^{V/nV_T}$$

$$10^{-3} = I_S e^{0.7/0.025} \Rightarrow I_S = 6.9 \times 10^{-16} A$$

$\therefore$ @

$$i = 0.1mA = 6.9 \times 10^{-16} e^{V/0.025}$$

$$\Rightarrow v = \underline{\underline{0.64V}}$$

$$i = 10mA = 6.9 \times 10^{-16} e^{V/0.025}$$

$$\Rightarrow v = \underline{\underline{0.76V}}$$

3.8

$$\Delta T = 125 - 25 = 100°C$$

$$I_S = 10^{-14} \times 1.15^{\Delta T}$$

$$= \underline{\underline{1.17 \times 10^{-8} A}}$$

3.9

At 20°C $\quad I = \frac{1V}{1M\Omega} = 1\mu A$

Since the reverse leakage current doubles for every 10°C increase, at 40° C

$$I = 4 \times 1\mu A = 4\mu A$$

$$\Rightarrow V = 4\mu A \times 1M\Omega = \underline{\underline{4.0V}}$$

@ 0°C $\quad I = \frac{1}{4}\mu A$

$$\Rightarrow V = \frac{1}{4} \times 1 = \underline{\underline{0.25V}}$$

3.10

(a) ITERATION

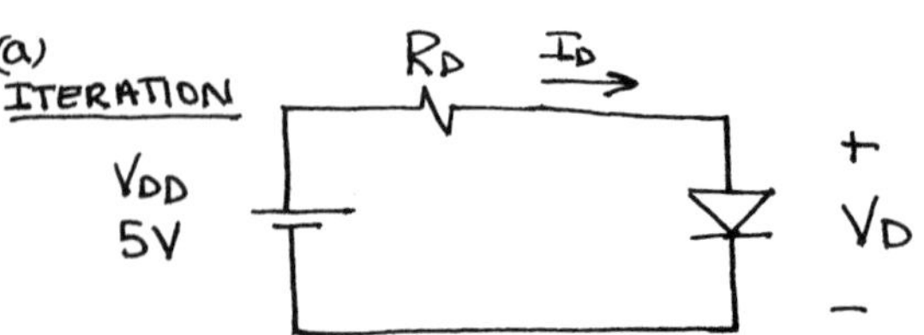

CONT.

(a) $i = I_S e^{V_D/nV_T}$

$$\frac{i_2}{i_1} = e^{\frac{V_{D2}-V_{D1}}{nV_T}}$$

$$V_{D2} = V_{D1} + nV_T \ln\left(\frac{i_2}{i_1}\right)$$

→ Given for this diode :-

$$V_{D2} = V_{D1} + 0.1 \log(i_2/i_1)$$

$$= 0.7 + 0.1 \log(i_2/1)$$

Using this we can iterate to find the diode voltage. Starting with a diode voltage of 0.7 V we have :

ITERATION #1

$$I_D = \frac{V_{DD}-0.7}{10} = \frac{5-0.7}{10} = 0.43\,mA$$

$$V_{D2} = 0.7 + 0.1 \log\frac{0.43}{1} = 0.663V$$

ITERATION #2

$$I_D = \frac{5-0.663}{10} = 0.434\,mA$$

$$V_{D2} = 0.663 + 0.1 \log\frac{0.434}{0.430} = 0.664V$$

ITERATION #3

$$I_D = \frac{5-0.664}{10} = 0.434\,mA$$

∴ Since the current has not changed we have reached convergence

$I_D = \underline{\underline{0.434mA}}$ $V_{D2} = \underline{\underline{0.664V}}$

(b) Piecewise Linear Model

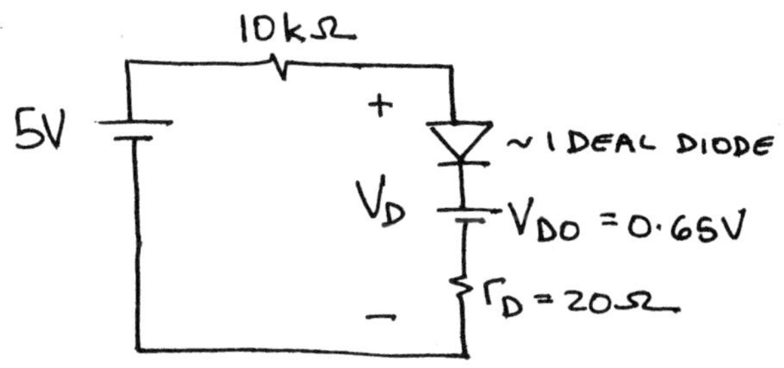

$$I_D = \frac{5-0.65}{10^4+20} = \underline{\underline{0.434mA}}$$

$$V_D = 0.65 + 0.434(10^{-3})20 = \underline{\underline{0.659V}}$$

(c) Constant Voltage Drop Model

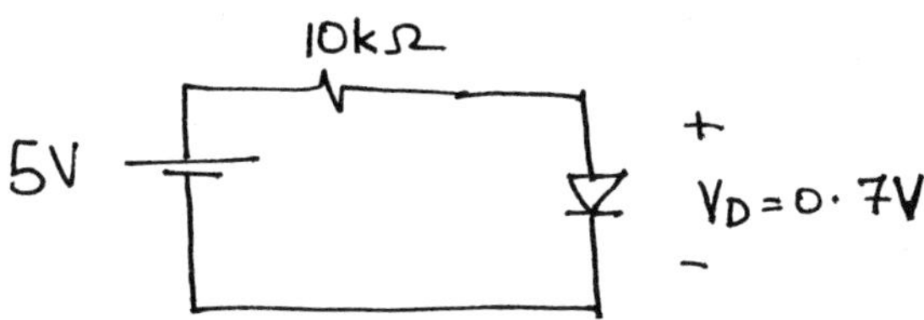

$V_D = \underline{\underline{0.7V}}$ — constant

$$I_D = \frac{5-0.7}{10} = \underline{\underline{0.43\,mA}}$$

3.11

The large diode is equivalent to 100 small diodes connected in parallel

IDEAL   0.65V $\frac{r_D}{100}$

V_{DO} does not change $V_{DO} = \underline{\underline{0.65V}}$

r_D reduces from 20Ω to $\underline{\underline{0.2\Omega}}$

3.12

10V, R, 2.4V, V_D

For an output voltage of 2.4V

$$V_D = \frac{2.4}{3} = 0.8\,V$$

∴ Using:

$$V_2 = V_1 + 0.1 \log \frac{I_2}{I_1}$$

$$0.8 = 0.7 + 0.1 \log \frac{I_2}{10^{-3}}$$

$$I_2 = 10\,mA$$

Thus: $R = \dfrac{10 - 2.4}{10} = \underline{\underline{760\,\Omega}}$

3.13

(a) 5V, 2.5kΩ

$\downarrow I = \dfrac{5 - 0.7}{2.5} = \underline{\underline{1.72\,mA}}$

$V = \underline{\underline{0.7\,V}}$

(b) 5V, 2.5kΩ

$I = \underline{\underline{0\,A}} \downarrow$

$V = \underline{\underline{5\,V}}$

(c) 2.5kΩ, −5V

$V = -\underline{\underline{5\,V}}$

$\downarrow I = \underline{\underline{0\,A}}$

(d) 2.5kΩ, −5V

$-0.7\,V = V$

$\downarrow I = \dfrac{-0.7 + 5}{2.5} = \underline{\underline{1.72\,mA}}$

(e)

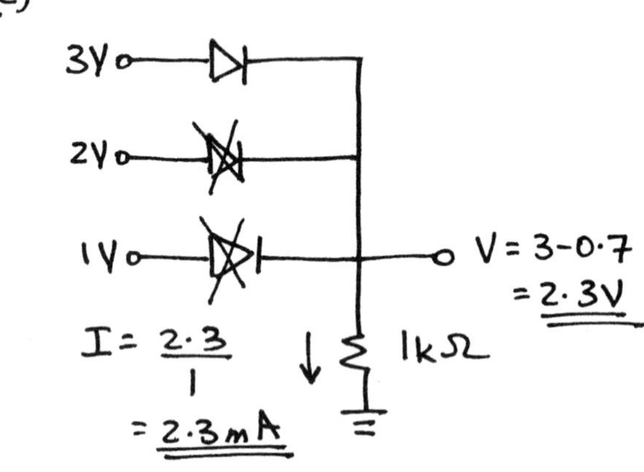

(f) 5V, 1kΩ, +3V, +2V, +1V

$I = \dfrac{5 - 1.7}{1} = \underline{\underline{3.3\,mA}}$

$V = 1 + 0.7 = \underline{\underline{1.7\,V}}$

3.14

$$r_d = nV_T / I_D = 0.025 / I_D$$

$I_D \Rightarrow$ 0.1mA, 1mA, 10mA

$r_d \Rightarrow$ $\underline{\underline{250\,\Omega}}$, $\underline{\underline{25\,\Omega}}$, $\underline{\underline{2.5\,\Omega}}$

3.15

(i) For the small signal model:

$$i_D = I_D + \frac{\Delta V}{r_d} \qquad n=2$$

$$= I_D + \frac{\Delta V\, I_D}{nV_T}$$

$$= 1 + \Delta V\left(\frac{1mA}{0.05}\right)$$

$$\Rightarrow \Delta i_D = \Delta V \times 20 \text{ mA}$$

(ii) For the exponential model

$$i_D = I_S\, e^{v/nV_T}$$

$$\frac{i_{D2}}{i_{D1}} = e^{\frac{v_2 - v_1}{V_T}} = e^{\Delta v/0.05}$$

$$\Rightarrow \Delta i_D = i_{D2} - i_{D1} = \left(e^{\Delta V/0.05} - 1\right) i_{D1} \text{ mA}$$

	ΔV(mV)	i_D(mA) small signal	i_D(mA) Exponential
(a)	−20	−0.4	−0.33
(b)	−10	−0.2	−0.18
(c)	−5	−0.10	−0.1
(d)	+5	+0.10	+0.11
(e)	+10	+0.2	+0.22
(f)	+20	+0.4	+0.49

3.16

15V, R, $V_O = 3V$, $I_L = 0$

For $\frac{\Delta V_O}{\Delta i_L} = \frac{0.04}{10^{-3}}$

$= 40\Omega$

$\therefore r_d = \frac{40}{4} = 10\Omega$

Since the small signal resistance for each diode, r_d, is 10Ω the dc bias current can be found

$$\frac{nV_T}{I_D} = r_d = 10 \qquad n=1$$

$$\Rightarrow I_D = \frac{0.025}{10} = 2.5\text{mA}$$

$$R = \frac{15-3}{2.5} = \underline{\underline{4.8\text{k}\Omega}}$$

At DC, the diode voltage for each diode is

$$V_D = \frac{3}{4} = 0.75V$$

$$I_D = I_S\, e^{V_D/nV_T}$$

using:

$$\frac{I_2}{I_1} = \frac{2.5}{1} = \frac{I_{S2}}{I_{S1}} e^{\frac{0.75-0.7}{0.025}}$$

where the relative junction area is:

$$\frac{I_{S2}}{I_{S1}} = \underline{\underline{0.34}}$$

3.17

For a zener diode

$$V_O = V_{Z0} + I_Z r_Z$$

$$10 = V_{Z0} + 0.01 \times 50$$

$$V_{Z0} = \underline{\underline{9.5V}}$$

For $I_Z = 5mA$

$$V_O = 9.5 + 0.005 \times 50 = \underline{\underline{9.75V}}$$

CONT.

3.18

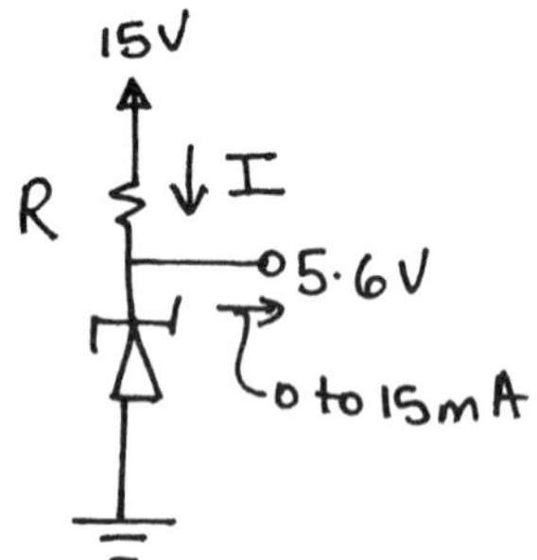

The minimum zener current should be $5 \times I_{ZK} = 5 \times 1 = 5mA$. Since the load current can be as large as 15mA, we should select R so that with $I_L = 15mA$, a zener current of 5mA is available. Thus the current should be 20mA leading to

$$R = \frac{15-5.6}{20mA} = \underline{\underline{470\Omega}}$$

Maximum power dissipated in the diode occurs when $I_L = 0$ is

$$P_{max} = 20 \times 10^{-3} \times 5.6 = \underline{\underline{112mV}}$$

3.19

15V

200Ω

$+$
V_Z
$-$

At no load $V_Z = \underline{\underline{5.1V}}$

For Line Regulation

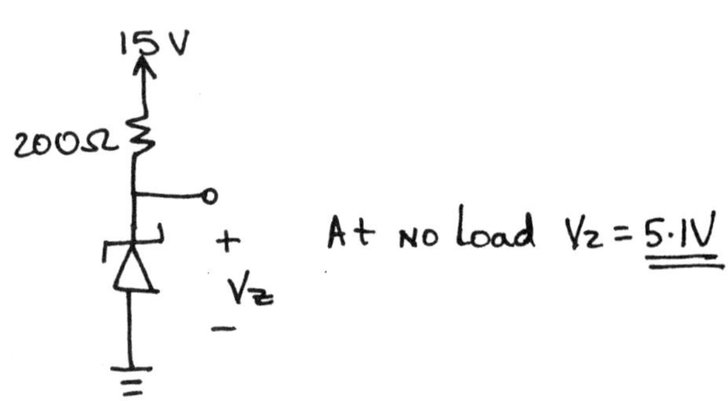

$$\text{Line Regulation} = \frac{v_o}{v_i} = \frac{7}{200+7} = \underline{\underline{33.8 \frac{mV}{V}}}$$

For Load Regulation:

200Ω

V_{Z0}

ΔI_L

r_Z

$$\frac{\Delta V_o}{\Delta I_L} = \frac{-\Delta I_L r_Z}{1mA}$$

$$= \underline{\underline{-7 \frac{mV}{mA}}}$$

3.20

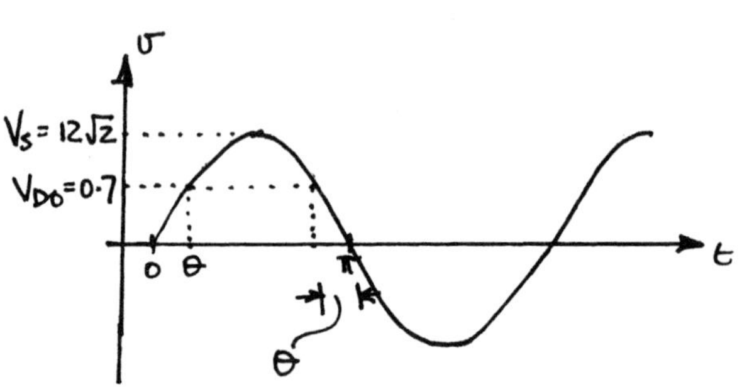

(a)

$v_S = V_S \sin \omega t$

at $\omega t = \theta \Rightarrow v_S = V_S \sin\theta = V_{DO}$

$$\theta = \sin^{-1}\left(\frac{V_{DO}}{V_S}\right) = \sin^{-1}\left(\frac{0.7}{12\sqrt{2}}\right)$$

$$= \underline{\underline{2.4^\circ}}$$

Conduction starts at $\theta = 2.4^\circ$ and terminates at $180^\circ - \theta$. Thus the total conduction angle is:

$$180 - 2\theta = \underline{\underline{175.2^\circ}}$$

(b) $V_{o,avg} = \frac{1}{2\pi}\int_{2\pi} V_s \sin\phi - V_{DO}\, d\phi$

$= \frac{1}{2\pi}\left[-V_s\cos\phi - V_{DO}\,\phi\right]_{\phi=\theta}^{\pi-\theta}$

$= \frac{1}{2\pi}\left[V_s\cos\theta - V_s\cos(\pi-\theta) - V_{DO}(\pi-2\theta)\right]$

but $\cos\theta \cong 1$ $\cos(\pi-\theta) \cong -1$

$\pi - 2\theta \cong \pi$

$\Rightarrow V_{o,avg} = \frac{2V_s}{2\pi} - \frac{\pi V_{DO}}{2\pi}$

$= \frac{V_s}{\pi} - \frac{V_{DO}}{2}$ Q.E.D.

$\therefore V_{o,avg} = \frac{12\sqrt{2}}{\pi} - \frac{0.7}{2} = 5.05V$

(c)

The peak diode current occurs at the peak diode voltage

$\hat{i}_D = \frac{V_s - V_{DO}}{r_D + R}$ neglecting r_D

$= \frac{12\sqrt{2} - 0.7}{100}$

$= 163$ mA

$PIV = +V_s = 12\sqrt{2} = 17V$

3.21

Refer to the full wave rectifier of fig. 3.26 (a)

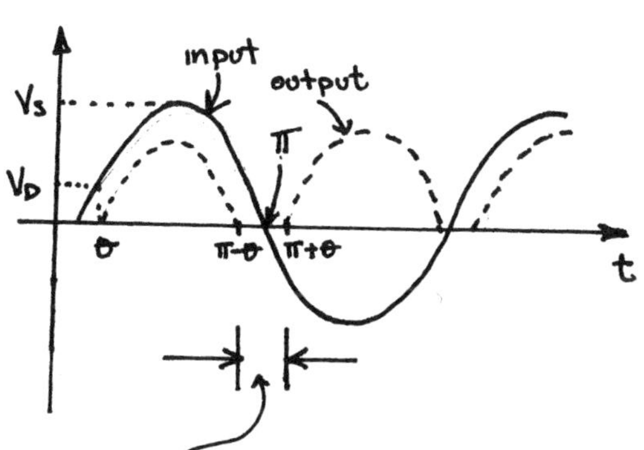

The output is zero between $(\pi+\theta)$ & $(\pi-\theta)$ = 2θ where θ is the angle at which the input signal reaches V_D $\Rightarrow$

$V_s \sin\theta = V_D$

$\theta = \sin^{-1}\left(\frac{V_D}{V_s}\right)$

$\therefore$ $2\theta = 2\sin^{-1}\left(\frac{V_D}{V_s}\right)$ Q.E.D.

(b) The average of the output signal is given by:

$V_{o,avg} = \frac{1}{2\pi}\int_{2\pi} V_s \sin\phi - V_D\, d\phi$

$= \left(\frac{1}{2\pi}\left[-V_s\cos\phi - \phi V_D\right]_{\phi=\theta}^{\pi-\theta}\right) \times 2$

$= 2\frac{V_s}{\pi} - V_D$ Q. E. D.

(c) Peak Current $= \left.\frac{V_s \sin\phi - V_D}{R}\right|_{\phi=\frac{\pi}{2}}$

$= \frac{V_s - V_D}{R} = \frac{12\sqrt{2} - 0.7}{100}$

$= 163mA$

The output is not zero for the angle $2(\pi - 2\theta)$

CONT.

Thus the fraction of the cycle during which $v_O > 0$ is

$$\frac{2(\pi - 2\theta)}{2\pi} \times 100$$

$$= \frac{2\pi - 4\sin^{-1}\left(\frac{0.7}{12\sqrt{2}}\right)}{2\pi} \times 100$$

$$= \underline{\underline{97.4\%}}$$

Average output voltage :

$$V_O = \frac{2V_s}{\pi} - V_D = \frac{2 \times 12\sqrt{2}}{\pi} - 0.7$$

$$= \underline{\underline{10.1\text{V}}}$$

Neglecting r_D, the peak diode current is given by

$$\hat{i}_D = \frac{V_s - V_{DO}}{R} = \frac{12\sqrt{2} - 0.7}{100} = \underline{\underline{163\text{mA}}}$$

$$PIV = V_s - V_D + V_s = 2 \times 12\sqrt{2} - 0.7 = \underline{\underline{33.2\text{V}}}$$

3.22

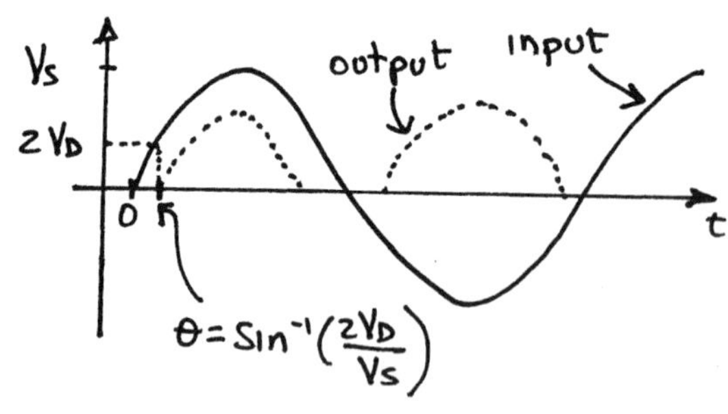

$$V_{O,avg} = \frac{1}{2\pi}\int V_s \sin\phi - 2V_D \, d\phi$$

$$= \frac{2}{2\pi}\left[-V_s\cos\phi - 2V_D\phi\right]_{\phi=\theta}^{\pi-\theta}$$

$$= \frac{1}{\pi}\left[2V_s - 2V_D(\pi - 2\theta)\right]$$

But $\cos\theta \approx 1$

$\cos(\pi - \theta) \approx -1$

$\pi - 2\theta \approx \pi$

$$\Rightarrow V_{O,avg} = \frac{2V_s}{\pi} - 2V_D$$

$$= \frac{2 \times 12\sqrt{2}}{\pi} - 1.4 = \underline{\underline{9.4\text{V}}}$$

(b) Peak diode current $= \frac{\text{Peak Voltage}}{R}$

$$= \frac{V_s - 2V_D}{R} = \frac{12\sqrt{2} - 1.4}{100}$$

$$= \underline{\underline{156\text{mA}}}$$

$$PIV = V_s - V_D = 12\sqrt{2} - 0.7 = \underline{\underline{16.3\text{V}}}$$

3.23

FULL WAVE Peak Rectifier :

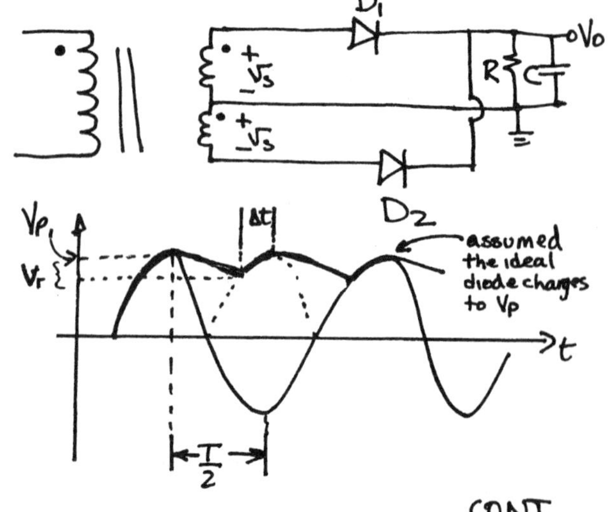

CONT.

The ripple voltage is the amount of discharge that occurs when the diodes are not conducting. The output voltage is given by:

$$v_O = V_p e^{-t/RC}$$

$$V_p - V_r = V_p e^{-\frac{T/2}{RC}} \leftarrow \text{discharge is only half the period.}$$

$$V_r = V_p\left(1 - e^{-\frac{T/2}{RC}}\right) \qquad e^{-\frac{T/2}{RC}} \approx 1 - \frac{T/2}{RC} \quad \text{for } CR \gg \frac{T}{2}$$

$$\cong V_p\left(1 - 1 + \frac{T/2}{RC}\right)$$

$$= \frac{V_p}{2fRC} \quad (a) \qquad \text{Q.E.D.}$$

To find the average current, note that the charge supplied during conduction is equivalent to the charge lost during discharge.

$$Q_{SUPPLIED} = Q_{LOST}$$

$$i_{Cav}\,\Delta t = CV_r \qquad \text{sub (a)}$$

$$(i_{D,av} - I_L)\,\Delta t = \cancel{C}\,\frac{V_p}{2fR\cancel{C}} = \frac{V_p}{2fR}$$

$$= \frac{V_p \pi}{\omega R}$$

$$i_{D,av} = \frac{V_p \pi}{\omega \Delta t\, R} + I_L$$

where $\omega\Delta t$ is the conduction angle. Note the conduction angle is the same expression as for the half wave rectifier and is given in EQ (3.30)

$$\omega\Delta t \cong \sqrt{\frac{2V_r}{V_p}} \quad (b)$$

Substituting for $\omega\Delta t$ we get:

$$\Rightarrow i_{D,av} = \frac{\pi V_p}{\sqrt{\frac{2V_r}{V_p}} \cdot R} + I_L$$

Since the output is approximately held at V_p, $\frac{V_p}{R} \approx I_L$. Thus:

$$\Rightarrow i_{D,av} \cong \pi I_L \sqrt{\frac{V_p}{2V_r}} + I_L$$

$$= I_L\left[1 + \pi\sqrt{\frac{V_p}{2V_r}}\right] \qquad \text{Q.E.D.}$$

If $t=0$ is at the peak, the maximum diode current occurs at the onset of conduction or at $t = -\omega\Delta t$. During conduction, the diode current is given by:

$$i_D = i_C + i_L$$

$$i_{D,max} = C\frac{dv_S}{dt} + i_L \qquad \text{assuming } i_L \text{ is const. } i_L \approx \frac{V_p}{R} = I_L$$

$$= C\frac{d}{dt}(V_p \cos\omega t) + I_L$$

$$= -C\sin\omega t \times \omega V_p + I_L$$

$$= -C\sin(-\omega\Delta t) \times \omega V_p + I_L$$

for a small conduction angle $\sin(-\omega\Delta t) \approx -\omega\Delta t$. Thus:

$$\Rightarrow i_{D,max} = C\omega\Delta t \times \omega V_p + I_L$$

sub (b) to get:

CONT.

$$i_{D,max} = C\sqrt{\frac{V_r 2}{V_p}}\, \omega V_p + I_L$$

sub $\omega = 2\pi f$ sub ⓐ for f

$$= 2\pi \frac{V_p}{2V_r RC}$$

$$\Rightarrow i_{D,max} = \cancel{C}\sqrt{\frac{2V_r}{V_p}}\, \frac{2\pi V_p^2}{2V_r R\cancel{C}} + I_L$$

$$= \pi \frac{V_p}{V_r} I_L \sqrt{\frac{2V_r}{V_p}} + I_L$$

$$= I_L\left[1 + \frac{\pi V_p}{V_r}\sqrt{\frac{2V_r}{V_p}}\right]$$

$$= I_L\left[1 + \pi\sqrt{\frac{2V_p}{V_r}}\right]$$

$$= I_L\left[1 + 2\pi\sqrt{\frac{V_p}{2V_r}}\right]$$ Q.E.D.

3.24

During discharge the output voltage may be expressed as:

$$v_o = (V_p - 2V_{DO})\, e^{-t/RC}$$

where

i) at the end of discharge
$v_o = V_p - V_r - 2V_{DO}$

ii) discharge occurs over half the period $\sim T/2$

iii) for $CR >> T/2$
$$e^{\frac{-T/2}{CR}} \approx 1 - \frac{T}{2CR}$$

Thus:

$$\Rightarrow V_p - V_r - 2V_{DO} = (V_p - 2V_{DO})\left(1 - \frac{T}{2RC}\right)$$

$$= V_p - \frac{V_p T}{2RC} - 2V_{DO} + \frac{2V_{DO} T}{2RC}$$

$$V_r = \frac{T}{2RC}(V_p - 2V_{DO})$$ ~ same as eq (3.28) but with V_p replaced with $V_p - 2V_{DO}$

$$\therefore V_r = 1 = \frac{1}{2 \times 60 \times C \times 100}(12\sqrt{2} - 1.6)$$

$$\Rightarrow C = \underline{1281\,\mu F}$$

The DC voltage at the output $\equiv$ average of the extremes of v_o

$$V_o = V_p - \tfrac{1}{2}V_r - 2V_{DO}$$

$$= 12\sqrt{2} - \tfrac{1}{2} - 1.6$$

$$= \underline{14.9\,V}$$

If you don't take the ripple voltage into account the DC output voltage is given by:

$$V_o = V_p - 2V_{DO} = 12\sqrt{2} - 1.6 = \underline{15.4\,V}$$

Assuming a constant load current

$$i_L = I_L = \frac{V_p - 2V_{DO}}{R} = \frac{12\sqrt{2} - 1.6}{100}$$

$$= \underline{0.15\,A}$$

Although not necessary the ripple voltage may be taken into account so that:

$$i_L = \frac{V_p - \tfrac{1}{2}V_r - 2V_{DO}}{R} = \frac{14.9}{100} = 0.149A$$

CONT.

Deriving the conduction angle (eq 3.30) but with the diode drop $2V_{DO}$ taken into account. At the end of conduction

$$(V_P - 2V_{DO})\cos\omega t = (V_P - 2V_{DO}) - V_r$$

For small angle $\cos(\omega\Delta t) \approx 1 - \frac{1}{2}(\omega\Delta t)^2$

$$\Rightarrow (V_P - 2V_{DO})\left(1 - \frac{1}{2}(\omega\Delta t)^2\right) = V_P - 2V_{DO} - V_r$$

$$\frac{V_P - 2V_{DO}}{2}(\omega\Delta t)^2 = V_r$$

$$\therefore \omega\Delta t = \sqrt{\frac{2V_r}{V_P - 2V_{DO}}}$$

same as eq (3.30) with V_P replaced with $V_P - 2V_{DO}$

$$= \sqrt{\frac{2}{12\sqrt{2} - 1.6}}$$

$$= \underline{\underline{0.36 \text{ rad}}} \text{ or } \underline{\underline{20.7^\circ}}$$

The average diode current can be found with Eq (3.34) with $V_P \longrightarrow V_P - 2V_{DO}$

$$i_{D,av} = I_L\left(1 + \pi\sqrt{\frac{(V_P - 2V_{DO})}{2V_r}}\right)$$

$$= I_L\left(1 + \pi\sqrt{\frac{(12\sqrt{2} - 1.6)}{2 \times 1}}\right)$$

The maximum diode current can be found with Eq (3.35) :

$$i_{D,max} = I_L\left(1 + 2\pi\sqrt{\frac{V_P - 2V_{DO}}{2V_r}}\right)$$

$$= I_L\left(1 + 2\pi\sqrt{\frac{12\sqrt{2} - 1.6}{2 \times 1}}\right)$$

The answers in the text used $I_L = 0.149A$ where the ripple voltage is taken into account. However since V_r is small the load current can be calculated without the ripple voltage being considered $I_L = 0.15A$. See calculations above.

with $I_L = 0.149A$

$$i_{D,av} = \underline{\underline{1.45A}}$$

$$i_{D,max} = \underline{\underline{2.74A}}$$

with $I_L = 0.15A$

$$i_{D,av} = 1.46\,A$$

$$i_{D,max} = 2.76A$$

$$PIV = V_S - V_{DO} = 12\sqrt{2} - 0.8 = \underline{\underline{16.2V}}$$

To allow for safety margins, we may specify the diode to be capable of a peak current of 3.5 to 4A and PIV of 20V

3.25

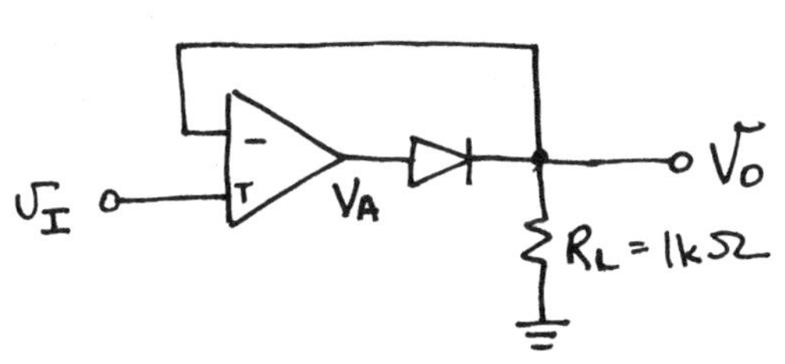

The diode is described by:

$$i_D = I_S e^{V_D/nV_T}$$

$$\frac{i_D}{1mA} = e^{\frac{V_D - 0.7}{nV_T}} \Rightarrow V_D - 0.7 = nV_T \ln\left(\frac{i_D}{1mA}\right)$$

$$= 0.1\log\left(\frac{i_D}{1mA}\right)$$

CONT.

• For $v_I = 10mV \rightarrow v_O = \underline{\underline{10mV}}$ as the output follows v_I.

$$i_R = \frac{10\times10^{-3}}{R} = \frac{10\times10^{-3}}{10^3} = 10\mu A$$

$\therefore i_D = 10\mu A$ so

$$v_D = 0.7 + 0.1\log\frac{10\times10^{-6}}{10^{-3}}$$

$$= 0.5V$$

$$V_A = 10mV + 0.5mV = \underline{\underline{0.51V}}$$

• For $v_I = 1V \rightarrow V_O = \underline{\underline{1V}}$

$$i_D = i_R = \frac{1}{10^3} = 10^{-3}A$$

$$V_D = 0.7 + 0.1\log\left(\frac{10^{-3}}{10^{-3}}\right) = 0.7V$$

$$V_A = 1 + 0.7 = \underline{\underline{1.7V}}$$

• For $v_I = -1V$ ~ the diode is cutoff

$\rightarrow V_O = \underline{\underline{0V}}$

$V_A = \underline{\underline{-12V}}$

3.26

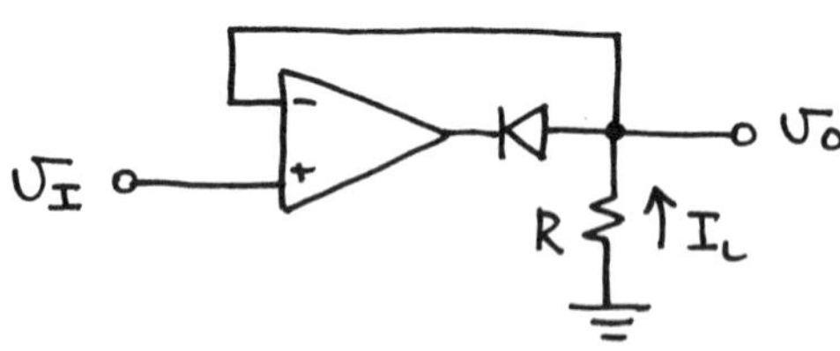

$v_I > 0$ ~ diode is cutoff

$v_O = \underline{\underline{0V}}$

$v_I < 0$ ~ diode conducts and opamp sinks load current.

$\underline{\underline{v_O = v_I}}$

3.27

◦ For $-5 \leq v_I \leq +5$ both diodes are cut off and $\underline{\underline{v_O = v_I}}$

◦ For $v_I \leq -5V$ ~ D_2 conducts

$$v_O = \frac{1}{2}(+v_I + 5) - 5$$

$$= \underline{\underline{-\frac{v_I}{2} - 2.5V}}$$

◦ For $v_I \geq 5V$ ~ D_1 conducts

$$v_O = \frac{1}{2}(v_I - 5) + 5$$

$$= \underline{\underline{\frac{v_I}{2} + 2.5V}}$$

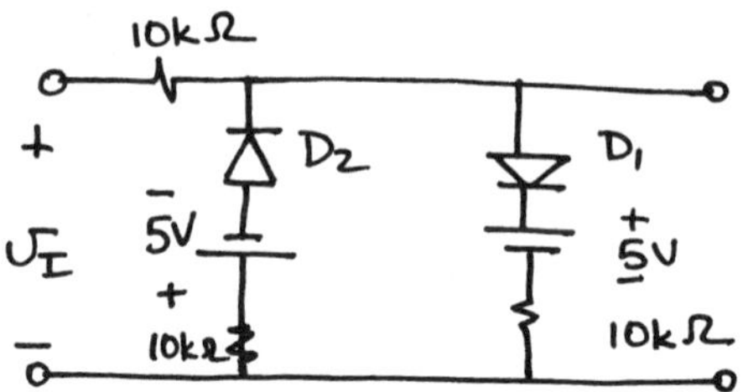

3.28

Reversing the diode results in the peak output voltage being clamped at 0V:

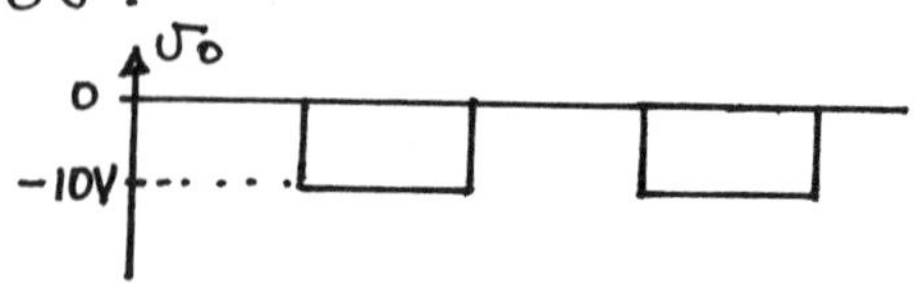

Here the dc component of $v_O = V_O = \underline{\underline{-5V}}$

3.29

Using eq (3.36) we have

$$n_i = \sqrt{BT^3 e^{-E_G/kT}} \quad \frac{\text{carriers}}{\text{cm}^3}$$

where $B = 5.4 \times 10^{31}$
$E_G = 1.12$ eV for silicon
$k = 8.62 \times 10^{-5} \frac{\text{eV}}{\text{K}}$
T = temperature in Kelvin

for $T = 250\text{ K} \sim n_i = 1.5 \times 10^{8} \frac{\text{carriers}}{\text{cm}^3}$

$T = 300\text{K} \sim n_i = 1.5 \times 10^{10} \frac{\text{carriers}}{\text{cm}^3}$

$T = 350\text{K} \sim n_i = 4.18 \times 10^{11} \frac{\text{carriers}}{\text{cm}^3}$

3.30

Given the dopant concentration is $N_D = 10^{17}$ the electron concentration is equivalent to

$N_D = 10^{17} \frac{\text{electrons}}{\text{cm}^3}$ at ALL temperatures

The concentration of holes can be found by:

$$p_{no} = \frac{n_i^2}{N_D} = \begin{cases} 0.225 \frac{\text{holes}}{\text{cm}^3} \text{ at } 250\text{K} \\ 2250 \frac{\text{holes}}{\text{cm}^3} \text{ at } 300\text{K} \\ 1.75 \times 10^6 \frac{\text{holes}}{\text{cm}^3} \text{ at } 350\text{K} \end{cases}$$

where n_i is obtained from exercise 3.30

3.31

Use eq (3.41)

(a) $$\rho = \frac{1}{q(p\mu_p + n\mu_n)}$$

$$= \frac{1}{1.6 \times 10^{-19}(1.5 \times 10^{10} \times 480 + 1.5 \times 10^{10} \times 1350)}$$

$$= 2.28 \times 10^5 \ \Omega \cdot \text{cm}$$

(b) $$\rho = \frac{1}{q(p\mu_{po} + n\mu_{no})}$$ ← thermal equilibrium

$$= \frac{1}{1.6 \times 10^{-19}\left(N_A \times 400 + \frac{n_i^2}{N_A} \times 1110\right)}$$

$$= \frac{1}{1.6 \times 10^{-19}\left(10^{16} \times 400 + \frac{(1.5 \times 10^{10})^2}{10^{16}} 1110\right)}$$

$$= 1.56 \ \Omega \cdot \text{cm}$$

3.32

From Eq (3.48)

$$V_o = V_T \ln\left(\frac{N_A N_D}{n_i^2}\right)$$

$$= 25 \ln\left(\frac{10^{17}\ 10^{16}}{(1.5 \times 10^{10})^2}\right)$$

$$= 728 \text{ mV}$$

CONT.

Width of depletion region is found from Eq. (3.50)

$$W_{dep} = \sqrt{\frac{2\epsilon_s}{q}\left(\frac{1}{N_A} + \frac{1}{N_D}\right)V_0}$$

$$= \sqrt{\frac{2\times 1.04\times 10^{-12}}{1.6\times 10^{-19}}\left(\frac{1}{10^{17}} + \frac{1}{10^{16}}\right)0.728}$$

$$= \underline{\underline{0.32\,\mu m}}$$

$\therefore W_{dep} = x_n + x_p$ where

$$\frac{x_n}{x_p} = \frac{N_A}{N_D} = \frac{10^{17}}{10^{16}} = 10$$

$$x_n = 10 x_p$$

$$11 x_p = 0.32\,\mu m \Rightarrow x_p = \underline{\underline{0.03\,\mu m}}$$

$$1.1 x_n = 0.32\,\mu m \Rightarrow x_n = \underline{\underline{0.29\,\mu m}}$$

3.33

(a) From Eq. 3.56

C_{jo} per cm^3 of junction area

$$= \sqrt{\frac{\epsilon_s q}{2} \cdot \frac{N_A N_D}{N_A + N_D} \cdot \frac{1}{V_0}}$$

$$= \sqrt{\frac{1.04\times 10^{-12}\times 1.6\times 10^{-19}}{2} \cdot \frac{10^{17}\,10^{16}}{10^{17}+10^{16}} \cdot \frac{1}{0.728}}$$

$$= 0.32\times 10^{-7}\ F/cm^2$$

$$= 0.32\times 10^{-7}\times 10^{-8}\ F/\mu m^2$$

$$= \underline{\underline{0.32\ fF/\mu m^2}}$$

(b) $C_{jo} = 0.32\times 10^{-15} \times 2500 = 0.8\,pF$

$$C_j = \frac{C_{jo}}{\left(1 + \frac{V_R}{V_0}\right)^m} = \frac{0.8}{\left(1 + \frac{2}{0.728}\right)^{1/2}}$$

$$= \underline{\underline{0.41\,pF}}$$

3.34

(a) Using Eq (3.64)

$$I_s = A q n_i^2 \left(\frac{D_P}{L_P N_D} + \frac{D_N}{L_N N_A}\right)$$

$$= 2500\times 10^{-8} \times 1.6\times 10^{-19} \times (1.5\times 10^{10})^2 \times \left(\frac{10}{5(10^{-4})\,10^{16}} + \frac{18}{10(10^{-4})\,10^{17}}\right)$$

$$= \underline{\underline{2\times 10^{-15}\ A}}$$

(b)

$$I \cong I_s e^{V/V_T}$$

$$V = V_T \ln\left(\frac{I}{I_s}\right) = 25 \ln\left(\frac{10^{-4}}{2\times 10^{-15}}\right)$$

$$= 615.9\,mV$$

$$\cong \underline{\underline{0.616\,V}}$$

(c) From the development leading to Eq (3.63) we find that

$$I_P = \frac{D_P}{L_P N_D} I \Big/ \left[\frac{D_P}{L_P N_D} + \frac{D_N}{L_N N_A}\right]$$

$$= \frac{2\times 10^{-12} \times 0.1\,mA}{2\times 10^{-12} + 1.8\times 10^{-13}} = \underline{\underline{91.7\,\mu A}}$$

$$I_n = I - I_p = 100\,\mu A - 91.7\,\mu A = \underline{\underline{8.3\,\mu A}}$$

CONT.

(d) $\tau_p = \frac{L_p^2}{D_p} = \frac{(5\times10^{-4})^2}{10} = \underline{\underline{25\text{ns}}}$

$\tau_n = \frac{L_n^2}{D_n} = \frac{(10\times10^{-4})^2}{18} = \underline{\underline{55.6\text{ns}}}$

(e)

$Q_p = \tau_p I_p = 25\times10^{-9} \times 91.7\times10^{-6}$

$= \underline{\underline{2.29\,pC}}$

$Q_n = \tau_n I_n = 55.6\times10^{-9} \times 8.3\times10^{-6}$

$= \underline{\underline{0.46\,pC}}$

$Q_{total} = Q_p + Q_n = 2.29 + 0.46$

$= 2.75\,pC$

$\tau_T = \frac{Q}{I} = \frac{2.75\times10^{-12}}{0.1\times10^{-3}} = \underline{\underline{27.5\text{ns}}}$

(f) $C_d = \left(\frac{\tau_T}{V_T}\right) I = \frac{27.5\times10^{-9}}{0.025} \times 0.1\times10^{-3}$

$= \underline{\underline{110\,pF}}$

Chapter 4 - Exercises

4.1

From Fig. 4.4, we see that the graph for which $v_{GS}=V_t+2^V$ has the point $V_{DS}=200\,mV$ and $i_D=0.4\,mA$. Thus the constant relating i_D and $(V_{GS}-V_t)V_{DS}$ is $\frac{0.4\,mA}{2^V\times 0.2V}=1\,mA/V^2$

$i_D=(V_{GS}-V_t)V_{DS}$ (mA)

$r_{DS}=\frac{V_{DS}}{i_D\,(mA)}=\frac{1}{V_{GS}-V_t}=\frac{1}{V_{ov}}$ (kΩ)

$V_{ov}=0.5V \Rightarrow r_{DS}=2K\Omega$

$V_{ov}=2^V \Rightarrow r_{DS}=0.5K\Omega \Rightarrow 0.5K\le r_{DS}\le 2K\Omega$

4.2

$C_{ox}=\frac{\varepsilon_{ox}}{t_{ox}}=\frac{3.45\times10^{-11}}{40\times10^{-9}}=0.86\times10^{-3}\,F/m^2=0.86\,fF/\mu m^2$

$K'_n=\mu_n C_{ox}=700\times10^{-4}\times0.86\times10^{-3}=60.2\,\mu A/V^2$

For $I_D=0.2mA$, $\frac{W}{L}=20$:

$V_{ov}^2=\frac{2I_D}{K'_n\frac{W}{L}}=\frac{2\times0.2^m}{60.2\times10^{-3}\times20}=0.33 \Rightarrow V_{ov}=0.58^V$

In saturation region: $V_{DS}\ge V_{ov}$

$V_{DSmin}=0.58V$

4.3

For the triode region: $i_D=K'_n\frac{W}{L}\left[(V_{GS}-V_t)V_{DS}-\frac{1}{2}V_{DS}^2\right]$

For small V_{DS}: $i_D=K'_n\frac{W}{L}(V_{GS}-V_t)V_{DS}$

or: $i_D=K'_n\frac{W}{L}V_{ov}V_{DS}$

$r_{DS}=\frac{V_{DS}}{i_D}=\frac{1}{K'_n\frac{W}{L}V_{ov}}$

IF $K'_n=100\,\mu A/V^2$ and $\frac{W}{L}=10$, $V_{ov}=0.5$, then

$r_{DS}=\frac{1}{100\times10^{-3}\times10\times0.5}=2K\Omega$

4.4

$V_{ov}=V_{GS}-V_t=1.5-0.7=0.8^V$

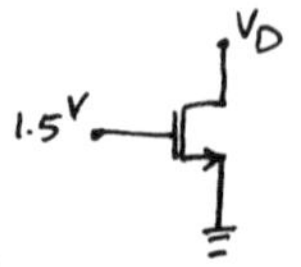

a) $V_D=0.5V$ then $V_{DS}<V_{ov}$

triode region

b) $V_D=0.9V$ then $V_D>V_{ov}$

saturation region

c) $V_D=3V$ then $V_D>V_{ov}$

saturation region.

4.5

a) Triode region: $i_D=K'_n\frac{W}{L}\left[V_{ov}\cdot V_{DS}-\frac{1}{2}V_{DS}^2\right]$

$i_D=100\times\frac{10}{1}\left[0.8\times0.5-\frac{1}{2}\times0.5^2\right]$

$i_D=275\,\mu A$

b) Saturation region: $i_D=\frac{1}{2}K'_n\frac{W}{L}V_{ov}^2$

$i_D=\frac{1}{2}\times100\times\frac{10}{1}\times0.8^2=320\,\mu A$

c) Saturation region: $i_D=\frac{1}{2}K'_n\frac{W}{L}V_{ov}^2$

$i_D=320\,\mu A$

4.6

For $V_{GS}=V_{DS}=1.2^V$, $V_{ov}=1.2-0.7=0.5^V$. Thus $V_{DS}>V_{ov}$ and the device operates in saturation region. ($i_D=100\,\mu A$)

For $V_{GS}=1.5^V$, $V_{ov}=1.5-0.7=0.8^V$, then $V_{DS}=3^V>V_{ov}$ and the device operates in saturation region.

$i_D=\frac{1}{2}K'_n\frac{W}{L}V_{ov}^2$

$\frac{i_{D1}}{i_{D2}}=\frac{V_{ov1}^2}{V_{ov2}^2} \Rightarrow \frac{100}{i_{D2}}=\frac{0.5^2}{0.8^2} \Rightarrow i_{D2}=256\,\mu A$

IF $V_{GS}=3.2^V$, then $V_{ov}=3.2-0.7=2.5^V$ and for small V_{DS} the device is in triode region.

($V_{DS}\le2.5^V$) $i_D\cong K'_n\frac{W}{L}V_{ov}\cdot V_{DS}$

$r_{DS}=\frac{V_{DS}}{i_D}=\frac{1}{K'_n\frac{W}{L}V_{ov}}$

Since we know for $i_D=100\,\mu A$, $V_{ov}=0.5^V$, then

$100\,\mu A=\frac{1}{2}K'_n\frac{W}{L}0.5^2 \Rightarrow K'_n\frac{W}{L}=800\,\mu A/V^2$

$r_{DS}=\frac{1}{800\times10^{-3}\times2.5}=0.5K\Omega=500\Omega$

4.7

$V_A=V'_AL=50\times0.8=40V$, $\lambda=\frac{1}{V_A}=0.025V^{-1}$

$V_{DS}=1^V>V_{ov}=0.5^V \Rightarrow$ Saturation: $I_D=\frac{1}{2}K'_n\frac{W}{L}V_{ov}^2(1+\lambda V_{DS})$

Cont.

$I_D = \frac{1}{2} \times 200 \times \frac{16}{0.8} \times 0.5^2 (1+0.025\times1) = 0.51\ mA$

$r_o = \frac{V_A}{I_D} = \frac{40}{0.51} = 78.4K\Omega \simeq 80K\Omega$

$r_o = \frac{\Delta V_{DS}}{\Delta I_D} \Rightarrow \Delta I_o = \frac{2^V}{80^K} = 0.025\ mA$

4.8

a) $V_{SG} \geq |V_t| \Rightarrow V_S - V_G \geq |V_t| \Rightarrow V_G \leq 5-1 \Rightarrow V_G \leq 4^V$

b) In triode region: $V_{DS} \geq V_{GS} - V_t$

$V_D - 5 \geq V_G - 5 - (-1) \Rightarrow V_D \geq V_G + 1$

c) In saturation region: $V_{DS} \leq V_{GS} - V_t \Rightarrow$

$V_D \leq V_G + 1$

d) $I_D = 75\mu A = \frac{1}{2} k'_n \frac{W}{L} V_{ov}^2 = \frac{1}{2} \times 60 \times 10 \times V_{ov}^2 \Rightarrow V_{ov} = 0.5^V$

$V_{ov} = V_{SG} - |V_t| \Rightarrow 0.5 = V_{SG} - 1 \Rightarrow 5 - V_G = 1.5 \Rightarrow$

$V_G = 3.5\,V$

To stay in saturation region: $V_{SD} \geq V_{ov}$ or

$V_D \leq 5 - 0.5 \Rightarrow V_D \leq 4.5V$

e) $\lambda = 0.02\,V^{-1} \Rightarrow r_o = \frac{1}{\lambda I_D} = \frac{1}{0.02 \times 75 \times 10^{-3}} = 667^{K\Omega} = 0.67^{M\Omega}$

f) For $V_D = 3V$, $V_{SD} = 2^V$ and $V_{SD} \geq V_{ov} = 0.5^V$, therefore the device operates in saturation:

$i_D = \frac{1}{2} k'_n \frac{W}{L} (V_{ov})^2 (1+\lambda V_{DS}) = 75\mu A \times (1+0.02\times2)$

$i_D = 78\mu A$

For $V_D = 0^V$, $V_{SD} = 5^V$ and $V_{SD} \geq V_{ov} = 0.5^V$, therefore the device operates in saturation region:

$i_D = 75(1+0.02\times5) = 82.5\mu A$

$r_o = \frac{\Delta V_{DS}}{\Delta i_D} = \frac{5-2}{82.5-78^{\mu}} = 0.67\,M\Omega$

This is the same as the value obtained for r_o in (e).

4.9

$V_t = V_{t_0} + \gamma\left[\sqrt{2\varphi_f + V_{SB}} - \sqrt{2\varphi_f}\right]$

$V_t = 0.8 + 0.4 \times \left[\sqrt{0.7+3} - \sqrt{0.7}\right] = 1.23V$

4.10

$I_D = \frac{1}{2} \mu_n C_{ox} \frac{W}{L} V_{ov}^2 \Rightarrow 0.3 = \frac{1}{2} \times \frac{60}{1000} \times \frac{120}{3} V_{ov}^2 \Rightarrow$

$V_{ov} = 0.5^V \quad \Rightarrow V_{GS} = V_{ov} + V_t = 0.5 + 1 = 1.5^V$

$V_S = -1.5^V \quad \Rightarrow R_S = \frac{V_S - V_{SS}}{I_D} = \frac{-1.5-(-2.5)}{0.3}$

$R_S = 3.33\,K\Omega$

$R_D = \frac{V_{DD} - V_D}{I_D} = \frac{2.5-0.4}{0.3} = 7K\Omega$

4.11

$I_D = 2 \times 80 = 160\mu A$, $V_D = 1^V \Rightarrow R = \frac{V_{DD}-V_D}{I_D} = \frac{3-1}{0.16} = 12.5^{K\Omega}$

$I_D = \frac{1}{2} k'_n \frac{W}{L} V_{ov}^2$, $V_{ov} = V_{GS} - V_t = 1 - 0.6 = 0.4V$ (unchanged)

$\frac{W}{L} = \frac{2I_D}{k'_n V_{ov}^2} = \frac{2\times0.16}{0.2\times0.4^2} = 10 \Rightarrow \frac{W}{L} = 10$

we can choose $\frac{W}{L} = \frac{8}{0.8}$ as an example.

4.12

$V_D = 1^V$, both transistors have the same V_{GS} and therefore same V_{ov}, thus Q_2 has the same I_D. $I_{D1} = I_{D2} = 80\mu A$

$V_D = V_{DD} - R_2 I_D = 3 - 20 \times 0.08 = 1.4V$

4.13

$R_D = 12.4 \times 2 = 24.8K\Omega$

$V_{GS} = 5^V$, Assume triode region:

$\left.\begin{array}{l} I_D = k'_n \frac{W}{L}\left[(V_{GS}-V_t)V_{DS} - \frac{1}{2}V_{DS}^2\right] \\ I_D = \frac{V_{DD} - V_{DS}}{R} \end{array}\right\} \Rightarrow$

$\frac{5 - V_{DS}}{24.8} = 1 \times \left((5-1)V_{DS} - \frac{V_{DS}^2}{2}\right) \Rightarrow V_{DS}^2 - 8.08V_{DS} + 0.4 = 0$

$\Rightarrow V_{DS} = 0.05\,V < V_{ov} \Rightarrow$ triode region.

$I_D = \frac{5-0.05}{24.8} = 0.2mA$

4.14

As indicated in Example 4.5, $V_D \geq V_G - V_t$ for the transistor to be in saturation region.

$V_{D\,min} = V_G - V_t = 5 - 1 = 4^V$

$I_D = 0.5mA \Rightarrow R_{D\,max} = \frac{V_{DD} - V_{D\,min}}{I_D} = \frac{10-4}{0.5} = 12K\Omega$

4.15

$I_D = 0.32\,mA = \frac{1}{2} k_n' \frac{W}{L} V_{ov}^2 = \frac{1}{2} \times 1 \times V_{ov}^2 \Rightarrow V_{ov} = 0.8^V$

$V_{GS} = 0.8 + 1 = 1.8V$

$V_G = V_S + V_{GS} = 1.6 + 1.8 = 3.4^V$

$R_{G2} = \frac{V_G}{I} = \frac{3.4}{1\mu} = \underline{3.4M\Omega}$, $R_{G1} = \frac{5-3.4}{1\mu} = \underline{1.6M\Omega}$

$R_S = \frac{V_S}{0.32} = 5K\Omega$

$V_{DS} \geq V_{ov} \Rightarrow V_D \geq V_{ov} + V_S \Rightarrow V_D \geq 0.8 + 1.6 = 2.4^V$

Assume $V_D = 3.4^V$, then $R_D = \frac{5-3.4}{0.32} = \underline{5K\Omega}$

4.16

$v_I = 0$: Since the circuit is perfectly symmetrical $V_O = 0$ and therefore $V_{GS} = 0$ which implies the transistors are turned off and $\underline{I_{DN} = I_{DP} = 0}$.

$V_I = 2.5^V$: If we assume that the NMOS is turned on, then v_O would be less than 2.5^V and this implies that PMOS is off ($V_{GSP} > 0$)

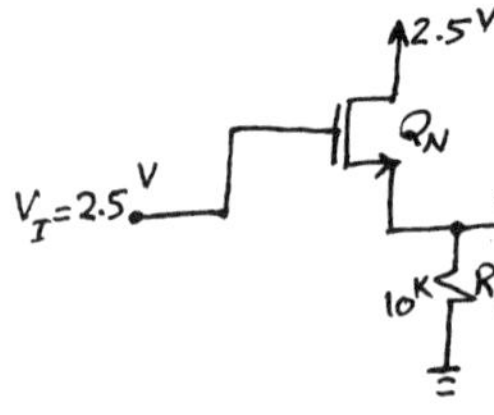

$I_{DN} = \frac{1}{2} k_n' \frac{W}{L} (V_{GS} - V_t)^2$

$I_{DN} = \frac{1}{2} \times 1 (2.5 - V_S - 1)^2$

$I_{DN} = 0.5 (1.5 - V_S)^2$

Also : $V_S = R_L I_{DN} = 10 I_{DN}$

$I_{DN} = 0.5 (1.5 - 10 I_{DN})^2$

$\Rightarrow 100 I_{DN}^2 - 32 I_{DN} + 2.25 = 0 \Rightarrow \underline{I_{DN} = 0.104\,mA}$

$I_{DP} = 0$, $V_O = 10 \times 0.104 = \underline{1.04V}$

$V_I = -2.5^V$: Again if we assume that Q_P is turned on, then $V_O > -2.5^V$ and $V_{GS1} < 0$ which implies the NMOS Q_N is turned off.

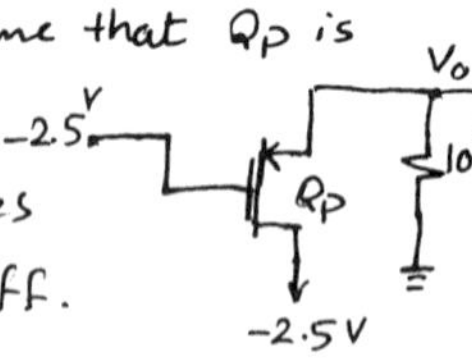

$I_{DN} = 0$

$I_{DP} = \frac{1}{2} k_n' \frac{W}{L} (V_{SG} - |V_{tp}|)^2 = \frac{1}{2} \times 1 \times (v_S + 2.5 - 1)^2$

$V_S = -10 I_{DP} \Rightarrow 2 I_{DP} = (-10 I_{DP} + 1.5)^2$

$\Rightarrow \underline{I_{DP} = 0.104\,mA}$ $\Rightarrow \underline{V_O = -10 \times 0.104 = -1.04^V}$

4.17

a) $V_{IQ} = V_{GSQ} = 1.816V$ Refer to Example 4.8

$V_{IB} = 2V$

$V_{OQ} = 4^V$

$V_{OB} = 1V$

b) By referring to Fig. 4.26c, negative peak of the output is $V_{OQ} - V_{OB} = 4 - 1 = 3V$ which corresponds to $V_{IB} - V_{IQ} = \underline{0.184^V}$ at the input.

c) The positive-output peak is $V_{DD} - V_{OQ} = 10 - 4 = 6^V$ which corresponds to $V_{IQ} - V_t = 1.816 - 1 = 0.816V$ at the input.

d) From (b) and (c) the peak voltage at the inut is 0.184V which translates to 3V at the output. This is a gain of $\frac{3}{0.184} \simeq 16.3\,V/V$. This gain is different than the value obtained in example 4.8 ($14.7 V/V$) due to the non-linear characteristic of the transfer function.

4.18

$v_O = V_{DD} - R_D i_D$ ①

$i_D = \frac{1}{2} k_n' \frac{W}{L} (v_I - V_t)^2 \Big\} \Rightarrow v_O = V_{DD} - \frac{1}{2} R_D k_n' \frac{W}{L} (v_I - V_t)^2$

Substitute $v_I = V_{IQ}$, $v_O = V_{OQ}$, $V_{OV} = V_{IQ} - V_t$

$\Rightarrow V_{OQ} = V_{DD} - \frac{1}{2} R_D k_n' \frac{W}{L} V_{OV}^2$

$\Rightarrow R_D k_n' \frac{W}{L} = \frac{2(V_{DD} - V_{OQ})}{(V_{OV})^2}$ ②

To derive an expression for A_v:

$A_v = \frac{\partial v_O}{\partial v_I}\Big|_{v_I = V_{IQ}} \overset{①}{=} -R_D k_n' \frac{W}{L} \underbrace{(V_{IQ} - V_T)}_{V_{OV}}$

Substitute from ②:

$A_v = -\frac{2(V_{DD} - V_{OQ})}{V_{OV}^2} \times V_{OV} = -\frac{2(V_{DD} - V_{OQ})}{V_{OV}} = \frac{-2V_{RD}}{V_{OV}}$

If we calculate A_v for values in Exercise 4.17:

$A_v = -\frac{2(10-4)}{0.816} = -14.7\,V/V$

Same as the result in Example 4.8.

4.19

$I_D = \frac{1}{2} k_n' \frac{W}{L}(V_{GS}-V_t)^2 \Rightarrow 0.5\,mA = \frac{1}{2}\times 1 (V_{GS}-1)^2$

$V_{GS} = 2V.$

IF $V_t = 1.5V$ then: $I_D = \frac{1}{2}\times 1\times(2-1.5)^2 = 0.125mA$

$\Rightarrow \frac{\Delta I_D}{I_o} = \frac{0.5-0.125}{0.5} = 0.75 = 75\%$

4.20

$R_D = \frac{V_{DD}-V_D}{I_D} = \frac{5-2}{0.5} = 6K\Omega \longrightarrow R_D = 6.2K\Omega$

$I_D = \frac{1}{2}k_n' \frac{W}{L} V_{OV}^2 \Rightarrow 0.5 = \frac{1}{2}\times 1 \times V_{OV}^2 \Rightarrow V_{OV} = 1^V$

$\Rightarrow V_{GS} = V_{OV} + V_t = 1+1 = 2V \Rightarrow V_S = -2V$

$R_S = \frac{V_S - V_{SS}}{I_D} = \frac{-2-(-5)}{0.5} = 6K\Omega \longrightarrow R_S = 6.2K\Omega$

IF we choose $R_D = R_S = 6.2K\Omega$ then I_D will slightly change:

$I_D = \frac{1}{2}\times 1\times (V_{GS}-1)^2$. Also $V_{GS} = -V_S = 5 - R_S I_D$

$2I_D = (4-6.2I_D)^2 \Rightarrow 38.44 I_D^2 - 51.6 I_D + 16 = 0$

$\Rightarrow I_D = 0.49mA$, $0.86mA$

$I_D = 0.86$ results in $V_S > 0$ or $V_S > V_G$ which is not acceptable, therefore $I_D = 0.49mA$

$V_S = -5 + 6.2\times 0.49 = -1.96V$

$V_D = 5 - 6.2\times 0.49 = +1.96V$

R_G should be selected in the range of $1M\Omega$ to $10M\Omega$ to have low current.

4.21

$I_D = 0.5mA = \frac{1}{2}k_n' \frac{W}{L} V_{OV}^2 \Rightarrow V_{OV}^2 = \frac{0.5\times 2}{1} = 1 \Rightarrow$

$V_{OV} = 1^V \Rightarrow V_{GS} = 1+1 = 2V = V_D \Rightarrow R_D = \frac{5-2}{0.5} = 6K\Omega$

$\Rightarrow R_D = 6.2K\Omega$ standard value. For this R_D we have to recalculate I_D:

$I_D = \frac{1}{2}\times 1\times (V_{GS}-1)^2 = \frac{1}{2}(V_{DD} - R_D I_D - 1)^2$

$(V_{GS} = V_D = V_{DD} - R_D I_D)$

$I_D = \frac{1}{2}(4-6.2I_D)^2 \Rightarrow I_D \cong 0.49mA$

$V_D = 5 - 6.2\times 0.49 = 1.96V$

4.22

Using Eq. 4.53 : $I = I_{REF}\frac{(W/L)_2}{(W/L)_1} \Rightarrow I_{REF} = 0.5\times\frac{1}{5}$

$\Rightarrow I_{REF} = 0.1mA$

$I_{REF} = 0.1 = \frac{1}{2}k_n'\left(\frac{W}{L}\right)_1 V_{OV}^2 \Rightarrow V_{OV}^2 = \frac{0.1\times 2}{0.8} = 0.25 \Rightarrow V_{OV} = 0.5^V$

$V_{GS} = V_{OV} + V_t = 1.5V \Rightarrow V_G = -5+1.5 = -3.5V$

$R = \frac{V_{DD}-V_G}{I_{REF}} = \frac{5-(-3.5)}{0.1} = 85K\Omega$

$V_{DS2} \geq V_{OV} \Rightarrow V_{DSmin} = V_{OV} = 0.5^V \Rightarrow V_{DMin} = -4.5^V$

4.23

a) $I_D = \frac{1}{2}k_n' \frac{W}{L}(V_{GS}-V_t)^2 = \frac{1}{2}\times 20^{\mu}\times 20\times(2-1)^2 = 200\mu A$

$V_D = V_{DD} - R_D I_D = 5 - 10\times 0.2 = 3V$

b) $g_m = \frac{2I_D}{V_{OV}} = \frac{2\times 0.2}{2-1} = 0.4\,mA/V$

c) $A_v = \frac{v_o}{v_{gs}} = -g_m R_D = -0.4\times 10 = -4V/V$

d) $v_{gs} = 0.2\sin\omega t \Rightarrow v_o = A_v v_{gs} = -4\times 0.2\sin\omega t$

$v_o = -0.8\sin\omega t$ $\quad v_{omax} = 3+0.8 = 3.8V$

$v_{omin} = V_D - 0.8 = 3 - 0.8 = 2.2V$

e) Eq. 4.57:

$i_D = \frac{1}{2}k_n'\frac{W}{L}(V_{GS}-V_t)^2 + k_n'\frac{W}{L}(V_{GS}-V_t)v_{gs} + \frac{1}{2}k_n'\frac{W}{L}v_{gs}^2$

For $v_{gs} = 0.2\sin\omega t$

$i_D = \frac{1}{2}\times 20\times 20(2-1)^2 + 20\times 20(2-1)0.2\sin\omega t + \frac{1}{2}\times 20\times 20 \times 0.2^2\times\sin^2\omega t$

$i_D^{(\mu A)} = 200 + 80\sin\omega t + 8\times\left(\frac{1}{2}-\frac{1}{2}\cos 2\omega t\right)$

$i_D^{(\mu A)} = 204 + 80\sin\omega t - 4\cos 2\omega t$

As we can see I_D is shifted from $200\mu A$ to $204\mu A$. Also there is a second harmonic term. $4\cos 2\omega t$ which has an amplitude of 4 compare to 80 which is the amplitude of the fundamental.

$\frac{4}{80} = 5\%$ = Second Harmonic Distortion

4.24

a) $g_m = \frac{2I_D}{V_{OV}}$ $\quad I_D = \frac{1}{2}\times k_n'\frac{W}{L}V_{OV}^2 = \frac{1}{2}\times 60\times 40\times(1.5-1)^2$

$I_D = 300\mu A = 0.3mA$, $V_{OV} = 0.5V$

$g_m = \frac{2\times 0.3}{0.5} = 1.2\,mA/V$, $r_o = \frac{V_A}{I_D} = \frac{15}{0.3} = 50K\Omega$

Cont.

$I_D = 0.5mA \Rightarrow g_m = \sqrt{2\mu_n C_{ox}\frac{W}{L}I_D} = \sqrt{2\times60\times40\times0.5\times10^{-3}}$

$g_m = 1.55\,mA/V$

$r_o = \frac{V_A}{I_D} = \frac{15}{0.5} = 30K\Omega$

4.25

$I_D = 0.1mA$, $g_m = 1mA/V$, $k_n' = 50\mu A/V^2$

$g_m = \frac{2I_D}{V_{ov}} \Rightarrow V_{ov} = \frac{2\times0.1}{1} = 0.2V$

$I_D = \frac{1}{2}k_n'\frac{W}{L}V_{ov}^2 \Rightarrow \frac{W}{L} = \frac{2I_D}{k_n' V_{ov}^2} = \frac{2\times0.1}{\frac{50}{1000}\times0.2^2} = 100$

4.26

$g_m = \mu_n C_{ox}\frac{W}{L}V_{ov}$ Same bias conditions, so same V_{ov} and also same L and g_m for both PMOS and NMOS.

$\mu_n C_{ox} W_n = \mu_p C_{ox} W_p \Rightarrow \frac{\mu_p}{\mu_n} = 0.4 = \frac{W_n}{W_p}$

$\Rightarrow \frac{W_p}{W_n} = 2.5$

4.27

$g_{mb} = \chi g_m = \frac{\gamma}{2\sqrt{2\phi_f+V_{SB}}}g_m \Rightarrow \chi = \frac{\gamma}{2\sqrt{2\phi_f+V_{SB}}} = \frac{0.5}{2\sqrt{0.6+4}}$

$\chi = 0.12$

4.28

$I_D = \frac{1}{2}k_p'\frac{W}{L}(V_{SG}-|V_t|)^2 = \frac{1}{2}\times60\times\frac{16}{0.8}\times(1.6-1)^2$

$I_D = 216\mu A$

$g_m = \frac{2I_D}{V_{ov}} = \frac{2\times216}{1.6-1} = 720\mu A/V = 0.72mA/V$

$\lambda = 0.04 \Rightarrow V_A' = \frac{1}{\lambda} = \frac{1}{0.04} = 25V/\mu m$

$r_o = \frac{V_A'\times L}{I_D} = \frac{25\times0.8}{0.216} = 92.6K\Omega$

4.29

$g_m r_o = \frac{2I_D}{V_{ov}}\times\frac{V_A}{I_D} = \frac{2V_A}{V_{ov}} = A_o$ $\quad V_A'\times L = V_A$

$L = 0.8\mu m \Rightarrow A_o = \frac{2\times12.5\times0.8}{0.2} = 100\,V/V$

4.30

$I = 0.5mA = \frac{1}{2}k_n'\frac{W}{L}V_{ov}^2 \Rightarrow V_{ov}^2 = \frac{2\times0.5}{1} = 1 \Rightarrow V_{ov} = 1V$

$V_{GS} = V_t + V_{ov} = 1+1.5 = 2.5V$

$V_D = V_{DD} - R_D I_D = 10 - 15\times0.5 = 2.5V$

$V_S = V_G - V_{GS} = -2.5V$

$g_m = \frac{2I_D}{V_{ov}} = \frac{2\times0.5}{1} = 1mA/V$

$r_o = \frac{V_A}{I_D} = \frac{75}{0.5} = 150K\Omega$

For saturation: $V_{DS} \geq V_{ov}$

$V_{DS} \geq 1 \Rightarrow V_{Dmin} = -2.5+1 = -1.5V$

So the drain can come down as low as $-1.5V$ from its bias point at $+2.5V$. This implies a negative signal swing of $2.5-(-1.5) = 4V$

4.31

a) $R_{sig} = 2\times100K = 200K\Omega$, Refer to Example 4.11

From Example 4.11, we have the following equivalent circuit:

R_{in} does not include R_{sig} and therefore it stays unchanged. $R_{in} = 400K\Omega$

$G_v = \frac{R_{in}}{R_{in}+R_{sig}}A_v = \frac{400}{400+200}\times8.75 = 5.83\,V/V$

$G_{vo} = \frac{R_i}{R_i+R_{sig}}A_{vo} = \frac{900K}{900+200}\times10 = 8.18\,V/V$

$G_v = G_{vo}\times\frac{R_L}{R_L+R_{out}} \Rightarrow 5.83 = \frac{10}{10+R_{out}}\times8.18 \Rightarrow$

$R_{out} = 4.03K\Omega$

Note that R_i and A_{vo} are the same as in Example 4.11, because they do not depend on R_{sig}.

b) $R_L = 20K\Omega$, $R_{sig} = 100K\Omega$

Since R_{sig} is the same, the values for R_o and R_{out} are the same as Example 4.11.

$R_{out} = 2.86K\Omega$

G_{vo} is voltage gain for $R_L = \infty$, thus it stays the same. $G_{vo} = 9V/V$, $G_v = G_{vo}\frac{R_L}{R_L+R_{out}} = 9\times\frac{20}{20+2.86}$

$G_v = 7.87V/V$.

$G_v = \frac{R_{in}}{R_{in}+R_{sig}}A_{vo}\frac{R_L}{R_L+R_o} \Rightarrow 7.87 = \frac{R_{in}}{R_{in}+100}\times10\times\frac{20}{20+1.43}$

Cont.

$R_{in} = 538\,k\Omega$

c) $R_L = 20\,k\Omega$, $R_{sig} = 200\,k\Omega$

R_{in} is the same as in (b): $R_{in} = 538\,k\Omega$

R_{out} is independent of R_L and therefore is the same as in (a): $R_{out} = 4.03\,k\Omega$

$$G_v = \frac{R_{in}}{R_{in}+R_{sig}} \times A_{vo} \times \frac{R_L}{R_L+R_o} = \frac{538}{538+200} \times 10 \times \frac{20}{20+4.03}$$

$G_v = 6.8\ V/V$

4.32

From Exercise 4.30: $g_m = 1\,mA/V$, $r_o = 150\,k\Omega$

$R_{in} = R_G = 4.7\,M\Omega$ (with or without r_o)

$A_{vo}\big|_{r_o=\infty} = -g_m R_L = -15 \times 1 = -15\ V/V$

$A_{vo}\big|_{r_o=150k} = -g_m (r_o \| R_L) = -13.6\ V/V$

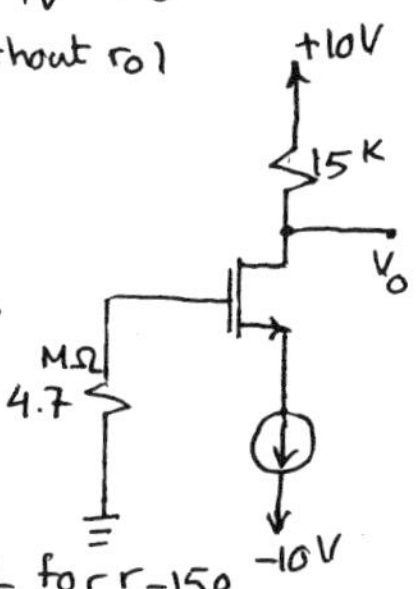

$R_{out} = R_L = 15\,k\Omega$ for $r_o = \infty$

$R_{out} = r_o \| R_L = 15^k \| 150^k = 13.6\,k\Omega$ for $r_o = 150\,k\Omega$

$$G_v = \frac{R_{in}}{R_{in}+R_{sig}} \times A_{vo} \times \frac{R_L}{R_L+R_o} = \frac{4.7^M}{4.7^M+0.1^M} \times -13.6 \times \frac{15}{15+13.6}$$

$G_v = -7\ V/V$

$G_v = \frac{v_o}{v_{sig}}$ therefore for $v_{sig} = 0.4$, $v_o = 0.4 \times 7 = 2.8^V$

The bias point is $V_{DS} = 5^V$ or $V_D = 2.5^V$, but as indicated in Exercise 4.30, maximum allowable output swing is 4^V, so $v_o = 2.8^V$ is acceptable.

The output is a 2.8V peak-to-peak sinusoid superimposed on $V_D = 2.5^V$.

4.33

$v_{sig} = 1.2^V$, $v_o = 2.8^V \Rightarrow G_v = \frac{v_o}{v_{sig}} = 2.33\ V/V$

In order to reduce the gain from 7 V/V to 2.33 V/V we can add R_s resistor in the source of the transistor:

From Eq. 4.90: $G_v = -\frac{R_G}{R_G+R_{sig}} \cdot \frac{g_m (R_L \| R_{out})}{1+g_m R_s}$

If we use $R_{out} = 13.6\,k\Omega$

From Exercise 4.32: $2.33 = \frac{4.7}{4.7+0.1} \cdot \frac{1 \times (15^k \| 13.6^k)}{1+1 \times R_s}$

$\Rightarrow R_s = 2\,k\Omega$

(Note that the effects of R_s in R_{out} is neglected.)

4.34

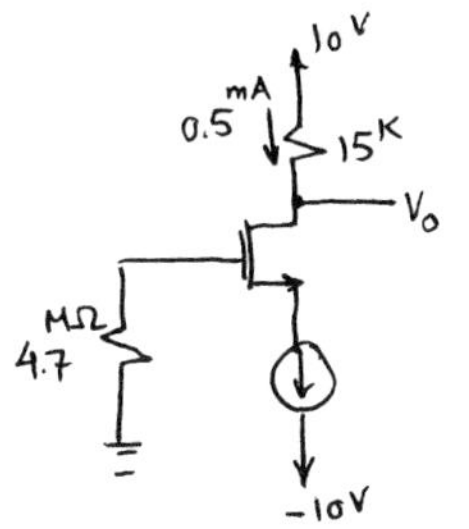

$g_m = 1\,mA/V$

$R_L = 15\,k\Omega$

$R_{in} = \frac{1}{g_m}$ (Eq. 4.91)

$R_{in} = \frac{1}{1m} = 1\,k\Omega$

$R_{out} = R_D = 15\,k\Omega$

$A_{vo} = g_m R_D = 1 \times 15 = +15\ V/V$

$A_v = g_m (R_D \| R_L) = 1 \times (15 \| 15) = +7.5\ V/V$

$G_v = \frac{R_{in}}{R_{in}+R_{sig}} A_v = \frac{1}{1+0.05} \times 7.5 = 7.14\ V/V$ for $R_{sig} = 50\,\Omega$

IF $R_{sig} = 1^k$: $G_v = \frac{1}{1+1} \times 7.5 = 3.75\ V/V$

IF $R_{sig} = 10^k$: $G_v = \frac{1}{1+10} \times 7.5 = 0.68\ V/V$

IF $R_{sig} = 100^k$: $G_v = \frac{1}{1+100} \times 7.5 = 0.07\ V/V$

4.35

a) $g_m = 1\,mA/V$, $r_o = 150\,k\Omega$, $R_{sig} = 1\,M\Omega$, $R_L = 15\,k\Omega$

$R_{in} = R_G = 4.7\,M\Omega$

$A_{vo} = \frac{r_o}{r_o + \frac{1}{g_m}}$ (Eq. 4.103) $\Rightarrow A_{vo} = \frac{150}{150+\frac{1}{1}} = 0.993\ V/V$

For $r_o \gg \frac{1}{g_m}$ or $r_o = \infty$, $A_{vo} \simeq 1\ V/V$

$A_v = \frac{R_L \| r_o}{R_L \| r_o + \frac{1}{g_m}}$ (Eq. 4.102)

For $r_o = \infty$: $A_v = \frac{R_L}{R_L + \frac{1}{g_m}} = 0.938\ V/V$

For $r_o = 150^K$: $A_v = \frac{15K \| 150K}{15K \| 150^K + 1} = 0.932\ V/V$

$R_o = \frac{1}{g_m} \| r_o$

For $r_o = \infty$ $R_o = 1\,k\Omega$

For $r_o = 150\,k\Omega$ $R_o = 0.993\,k\Omega$

Using Eq. 4.104: $G_v = \frac{R_G}{R_G+R_{sig}} \cdot \frac{R_L \| r_o}{R_L \| r_o + \frac{1}{g_m}}$

$$G_v = \frac{4.7}{4.7+1} \cdot \frac{15^K \| 150^K}{15^K \| 150^K + 1^K} = 0.768\ V/V$$

4.36

$C_{ox} = \frac{\varepsilon_{ox}}{t_{ox}} = \frac{3.45\times10^{-11}}{10\times10^{-9}} = 3.45\times10^{-3}\,F/m^2 = 3.45\,fF/\mu m^2$

$C_{ov} = WL_{ov}C_{ox} = 10\times0.05\times3.45 = 1.72fF$

$C_{gs} = \frac{2}{3}WLC_{ox} + C_{ov} = \frac{2}{3}\times10\times1\times3.45 + 1.72 = 24.72\,fF$

$C_{gd} = C_{ov} = 1.72fF$

$C_{sb} = \frac{C_{sbo}}{\sqrt{1+\frac{V_{SB}}{V_o}}} = \frac{10}{\sqrt{1+\frac{1}{0.6}}} = 6.1fF$

$C_{db} = \frac{C_{dbo}}{\sqrt{1+\frac{V_{DB}}{V_o}}} = \frac{10}{\sqrt{1+\frac{2+1}{0.6}}} = 4.1fF$

4.37

$f_T = \frac{g_m}{2\pi(C_{gs}+C_{gd})}$ Assume $\frac{W}{L}=12$

$g_m = \sqrt{2K'_n\frac{W}{L}I_D} = \sqrt{2\times160\times12\times100} = 619.7\mu A/V^2$

$g_m = 0.620\,mA/V$ Refer to Excercise(4.36)

$f_T = \frac{0.620^m}{2\pi(24.72+1.72)^f} = 3.7GHz$

4.38

$A_M = -\frac{R_G}{R_G+R_{sig}}g_m R'_L$ Refer to Example 4.12

$R'_L = 7.14K\Omega$, $g_m = 1mA/V$

$R_{sig} = 10K\Omega$

$A_M = \frac{-4.7M\Omega}{4.7+0.01M\Omega}\times1\times7.14 = -7.12V/V$

$f_H = \frac{1}{2\pi C_{in}(R_{sig}\|R_G)}$ $C_{in} = 4.26pF$

$f_H = \frac{1}{2\pi\times4.26(10^K\|4.7M)} = 3.7MHz$

4.39

$C_{gs} = 1pF$ $C_{eq} = (1+g_m R'_L)C_{gd} = (1+1\times7.14)C_{gd} = 8.14C_{gd}$

$f_T \geq 1MHz \Rightarrow \frac{1}{2\pi C_{in}(R_{sig}\|R_G)} \geq 1MHz$

$C_{in} = C_{gs} + C_{eq} = 1^p + 8.14C_{gd}^{(p)}$

$\frac{1}{2\pi(1+8.14C_{gd})^f(100^K\|4.7M)} \geq 1MHz \Rightarrow 1.63 \geq 1+8.14C_{gd}$

$C_{gd} < 0.077pF$ or $C_{gd} \leq 77fF$

4.40

$A_M = \frac{-R_G}{R_G+R_{sig}}\times g_m(R_L\|R_D) = -\frac{10}{10+0.1}\times2\times\frac{10^K}{2}$

$A_M = -9.9V/V$

$f_{P1} = \frac{1}{2\pi C_{C1}(R_G+R_{sig})} = \frac{1}{2\pi\times1^\mu\times(10+0.1^M)} = 0.016Hz$

$f_{P2} = \frac{1}{2\pi C_S/g_m} = \frac{1}{2\pi\times1^\mu/2^m} = 318Hz$

$f_{P3} = \frac{1}{2\pi C_{C2}(R_L+R_D)} = \frac{1}{2\pi\times1^\mu\times(10+10)} = 8Hz$

$f_L \simeq f_{P2} = 318\,Hz$

4.41

Eq. 4.149: $V_{IL} = \frac{1}{8}(3V_{DD}+2V_t) = \frac{1}{8}(3\times5+2\times1) = 2.1V$

Eq. 4.148: $V_{IH} = \frac{1}{8}(5V_{DD}-2V_t) = \frac{1}{8}(5\times5-2) = 2.9V$

$NMH = V_{OH}-V_{IH} = V_{DD}-V_{IH} = 5-2.9 = 2.1V$

$NML = V_{IL}-V_{OL} = V_{IL}-0 = 2.1V$

4.42

$i_o = i_{omax}$ at $v_o = 0.5V$

$i_{omax} = K'_n\frac{W}{L}((V_{GS}-V_t)V_{DS} - \frac{1}{2}V_{DS}^2)$

$i_{omax} = 20\times20((10-2)0.5 - \frac{1}{2}0.5^2) = 1.55mA$

4.43

For Q_N and Q_P matched: $K'_n\left(\frac{W}{L}\right)_n = K'_p\left(\frac{W}{L}\right)_p$

$80\times\frac{1.8}{1.2} = 27\frac{W_P}{1.2} \Rightarrow W_p = 5.4\mu m$

The output resistance in the low-output state is: $r_{DSN} = \frac{1}{K'_n\left(\frac{W}{L}\right)_n(V_{GS}-V_{tn})} = \frac{1}{80\times\frac{1.8}{1.2}\times(5-0.8)}$

$r_{DSN} = 2K\Omega$

4.44

AT $V_I = V_{th}$ BOTH Q_N AND Q_P ARE IN SATURATION, THUS:

$$\frac{1}{2} K'_n \left(\frac{W}{L}\right)_n (V_{th} - V_{tn})^2$$

$$= \frac{1}{2} K'_p \left(\frac{W}{L}\right)_p \left(V_{DD} - V_{th} - |V_{tp}|\right)^2$$

WE DEFINE

$$r = \frac{V_{th} - V_{tn}}{V_{DD} - V_{tn} - |V_{tp}|} = \sqrt{\frac{K'_p (W/L)_p}{K'_n (W/L)_n}}$$

THUS:

$$(V_{th} - V_{tn})^2 = r^2 (V_{DD} - V_{th} - |V_{tp}|)^2$$

$$V_{th} - V_{tn} = r(V_{DD} - V_{th} - |V_{tp}|)$$

$$V_{th}(1+r) = r(V_{DD} - |V_{tp}|) + V_{tn}$$

$$V_{th} = \frac{r(V_{DD} - |V_{tp}|) + V_{tn}}{r+1}$$

4.45

$$t_{PHL} = \frac{1.6C}{K'_n \left(\frac{W}{L}\right)_n V_{DD}} = \frac{1.6 \times 0.1 \times 10^{-12}}{20 \times 10^{-6} \times \frac{10}{5} \times 5} = 0.8\,ns$$

$$t_{PLH} = t_{PHL} = 0.8\,ns$$

$$t_P = \frac{1}{2}(t_{PHL} + t_{PLH}) = 0.8\,ns$$

4.46

$$t_P = \frac{1.6C}{K'_n \left(\frac{W}{L}\right)_n V_{DD}} = \frac{1.6 \times 15 \times 10^{-12}}{20 \times 10^{-6} \times 20 \times 10} = 6\,ns$$

4.47

PEAK CURRENT OCCURS AT $V_I = V_{th} = 5V$

$$i_{PEAK} = \frac{1}{2} K'_n \left(\frac{W}{L}\right)_n (V_{th} - V_{tn})^2$$

$$= \frac{1}{2} \times 20 \times 20 (5-2)^2$$

4.48

$$P_D = f C V_{DD}^2 = 2 \times 10^6 \times 15 \times 10^{-12} \times 100 = 3\,mW$$

$$I_{AV} = \frac{P_D}{V_{DD}} = \frac{3 \times 10^{-3}}{10} = 0.3\,mA$$

4.49

$$P_D|_{gate} = f C V_{DD}^2 = 10^8 \times 30 \times 10^{-15} \times 25 = 75\,\mu W$$

$$P_D|_{chip} = 0.3 \times 100\,000 \times 75 = 2.25\,W$$

4.50

$V_{GS} = +1V$, $V_t = -2V$

$$V_{GS} - V_t = 3V$$

TO OPERATE IN SATURATION REGION:

$$V_{DS\,min} = V_{GS} - V_t = 3V$$

$$i_D = \frac{1}{2} K'_n \frac{W}{L} (V_{GS} - V_t)^2 = \frac{1}{2} \times 2 \times 3^2 = 9\,mA$$

CHAPTER 5 EXERCISES

5.1

$\therefore i_C = I_S e^{v_{BE}/v_T}$

For $i_C = 0.1mA$

$$\frac{0.1mA}{1mA} = e^{\frac{v_{BE}-0.7}{0.025}}$$

$$v_{BE} = \underline{\underline{0.64V}}$$

For $i_C = 10mA$

$$\frac{10}{1} = e^{\frac{v_{BE}-0.7}{0.025}}$$

$$v_{BE} = \underline{\underline{0.76V}}$$

5.2

$$\therefore \alpha = \frac{\beta}{\beta+1}$$

α ranges from $\frac{50}{51}$ to $\frac{150}{151}$

OR $\underline{\underline{0.980 \text{ to } 0.993}}$

5.3

$$i_C = I_S e^{v_{BE}/v_T} = i_E - i_B$$

$$I_S e^{0.7/0.025} = (1.46 - 0.01446)10^{-3}$$

$$I_S = \underline{\underline{10^{-15}A}}$$

$$i_C = (1.46 - 0.01446)10^{-3} = \alpha\ 1.46 \times 10^{-3}$$

$$\alpha = \underline{\underline{0.99}}$$

$$i_C = \alpha i_E = \beta i_B$$

$$0.99 \times 1.46 = \beta \times 0.01446$$

$$\beta = 99.97 \cong \underline{\underline{100}}$$

5.4

$$\alpha = \frac{\beta}{\beta+1} \Rightarrow \beta = \frac{\alpha}{1-\alpha}$$

for $\alpha = 0.99$ $\beta = \frac{0.99}{1-0.99} = \underline{\underline{99}}$

$$i_B = \frac{i_C}{\beta} = \frac{10}{99} = \underline{\underline{0.1mA}}$$

for $\alpha = 0.98$ $\beta = \frac{0.98}{0.02} = \underline{\underline{49}}$

$$i_B = \frac{10}{49} = \underline{\underline{0.2mA}}$$

5.5

$$\alpha_F I_{SE} = \alpha_R I_{SC}$$

$$1 \times 10^{-15} = 0.01 I_{SC}$$

$$I_{SC} = \underline{\underline{10^{-13}A}}$$

$\frac{\alpha_F}{\alpha_R} = \frac{1}{0.01} = \underline{\underline{100}}$ times larger

$$\beta_R = \frac{\alpha_R}{1-\alpha_R} = \frac{0.01}{1-0.01} = \underline{\underline{0.01}}$$

5.6

$\alpha_F = 0.99$, $\alpha_R = 0.02$, $I_S = 10^{-15}A$

See calculations of Eq (5.31)(5.32)(5.33) respectively, below:

CONT.

$$i_E = \frac{I_s}{\alpha_F} e^{0.7/0.025} - I_s\left(1 - \frac{1}{\alpha_F}\right)$$

$$= 1.46 \times 10^{-3} - 1.1 \times 10^{-17}$$

$$= 1.461 \text{ mA}$$

$$i_C = I_s e^{0.7/0.025} + I_s\left(\frac{1}{\alpha_R} - 1\right)$$

$$= 1.446 \times 10^{-3} + 49 \times 10^{-15}$$

$$= 1.446 \text{ mA}$$

$$i_B = \frac{I_s}{\beta_F} e^{0.7/0.025} - I_s\left(\frac{1}{\beta_F} + \frac{1}{\beta_R}\right)$$

$$= \frac{I_s}{\beta_F} e^{0.7/0.025} - I_s\left(\frac{1-\alpha_F}{\alpha_F} + \frac{1-\alpha_R}{\alpha_R}\right)$$

$$= 1.4609 \times 10^{-5} - 49.01 \times 10^{-15}$$

$$\cong 0.015 \text{ mA}$$

5.7

(a) Using Eq (5.26) for i_C and Eq (5.27) for i_E we have

$$i_C = I_s\left(e^{v_{BE}/V_T} - 1\right) - \frac{I_s}{\alpha_R}\left(e^{v_{BC}/V_T} - 1\right)$$

$$\simeq I_s e^{v_{BE}/V_T} - \frac{I_s}{\alpha_R} e^{v_{BC}/V_T} \quad (1)$$

when all terms not involving exponents are ignored. Similarly:

$$i_E = \frac{I_s}{\alpha_F}\left(e^{v_{BE}/V_T} - 1\right) - I_s\left(e^{v_{BC}/V_T} - 1\right)$$

$$\cong \frac{I_s}{\alpha_F} e^{v_{BE}/V_T} - I_s e^{v_{BC}/V_T} \quad (2)$$

Manipulating this we get:

$$I_s e^{V_{BE}/V_T} = \alpha_F I_E + \alpha_F I_s e^{v_{BC}/V_T}$$

where $i_E = I_E$ denotes the fixed emitter current. Substitute this into (1) to get

$$i_C = \alpha_F I_E + I_s\left(\alpha_F e^{v_{BC}/V_T}\right) - \frac{I_s}{\alpha_R} e^{V_{BC}/V_T}$$

$$i_C = \alpha_F I_E + \left(I_s \alpha_F - \frac{I_s}{\alpha_R}\right) e^{v_{BC}/V_T}$$

$$= \alpha_F I_E + I_s\left(\alpha_F - \frac{1}{\alpha_R}\right) e^{v_{BC}/V_T}$$

Q.E.D.

(b) $I_s = 10^{-15}$ A $I_E = 1$ mA $\alpha_F \cong 1$
$\alpha_R = 0.01$

$$\therefore i_C = 10^{-3} + 10^{-15}(1 - 100)e^{v_{BC}/0.025}$$

v_{BC} (V)	i_C (mA)	
-1	1	← still in active mode
0.4	1	← ENTERING SATURATION ↓
0.5	0.95	
0.54	0.76	
0.57	0.20	

For $i_C = 0$

$$10^{-3} + 10^{-15}(1 - 100)e^{v_{BC}/0.025} = 0$$

$$v_{BC} = 0.576 \text{ V}$$

(c) i_B should equal i_E which is 1 mA. CHECKING with Eq (5.28) - complete Ebers-Moll model

$$i_B = \frac{I_s}{\beta_F}\left(e^{v_{BE}/V_T} - 1\right) + \frac{I_s}{\beta_R}\left(e^{v_{BC}/V_T} - 1\right)$$

(first term → 0)

Since $\alpha_F = 1$ $\beta_F = \infty$ - first term $\cong 0$

Using $\beta_R = \frac{\alpha_R}{1 - \alpha_R} = \frac{0.01}{1 - 0.01} = 0.01$

CONT.

Thus:

$$i_B = \frac{10^{-15}}{0.01}\left(e^{0.576/0.025} - 1\right)$$

$$= \underline{\underline{1\,mA}}$$

5.8

2mA, I_C, −10V

$\beta = 50$

$I_S = 10^{-14}\,A$

$$I_C = \alpha I_E = \frac{50}{51}\times 2 = \underline{\underline{1.96\,mA}}$$

$$I_C = I_S e^{v_{BE}/V_T} = 10^{-14} e^{v_{BE}/0.025} = 1.96\,mA$$

$$v_{BE} = \underline{\underline{0.650\,V}}$$

$$I_B = I_C/\beta = \frac{1.96}{50} = \underline{\underline{39.2\,\mu A}}$$

5.9

$$i_C = I_S e^{v_{BE}/V_T}$$

$$1.5 = 10^{-11} e^{v_{BE}/0.025}$$

$$v_{BE} = \underline{\underline{0.643\,V}}$$

5.10

10V, 5kΩ, 10kΩ, −10, $V_E = -0.7V$

$$i_C = \frac{50}{51}\times 0.93 = \underline{\underline{0.91118\,mA}}$$

$$I_E = \frac{10-0.7}{10} = \underline{\underline{0.93\,mA}}$$

$$i_B = 0.93 - 0.9118 = 18.3\,\mu A$$

$$V_C = 10 - 5(0.9118) = \underline{\underline{5.44\,V}}$$

5.11

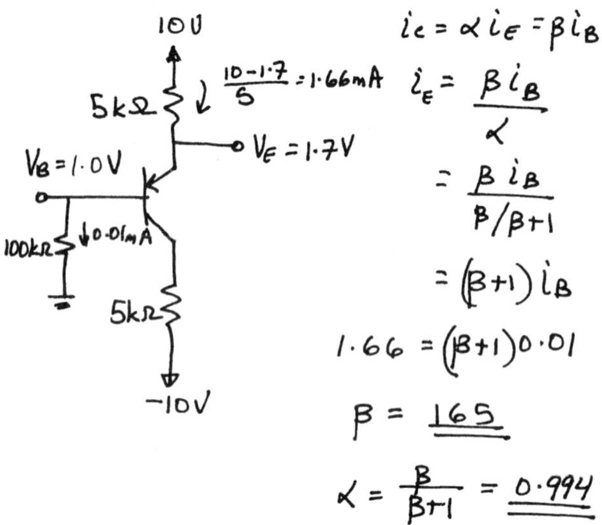

$$i_C = \alpha i_E = \beta i_B$$

$$i_E = \frac{\beta i_B}{\alpha} = \frac{\beta i_B}{\beta/\beta+1} = (\beta+1) i_B$$

$$1.66 = (\beta+1)0.01$$

$$\beta = \underline{\underline{165}}$$

$$\alpha = \frac{\beta}{\beta+1} = \underline{\underline{0.994}}$$

5.12

I_E = 2mA, 1kΩ, −5V

For a speific current v_{BE} changes by −2mV/°C

∴ For a 30° C increase in temperature

$$\Delta v_{BE} = -2\times 30 = \underline{\underline{-60\,mV}}$$

Since the current I_E doesnt change i_C doesnt change. $\Delta v_C = \underline{\underline{0\,V}}$

5.13

$I_E = 1\,mA$ $\quad v_{BE} = 0.7V$ with $v_{CB} = 0V$

$\alpha_F \cong 1 \quad \alpha_R = 0.1$

Use Eq (5.35) to find i_C

$$i_C = \alpha_F I_E - I_S\left(\frac{1}{\alpha_R} - \alpha_F\right)e^{v_{BC}/V_T}$$

CONT.

Use Eq (5.26) to find I_s.

$$i_E = \frac{I_s}{\alpha_F}\left(e^{v_{BE}/V_T} - 1\right) - I_s\underbrace{\left(e^{v_{BC}/V_T} - 1\right)}_{=0 \because V_{BC}=0}$$

$$10^{-3} = \frac{I_s}{\alpha_F}\left(e^{v_{BE}/0.025} - 1\right)$$

$$I_s = \frac{10^{-3} \times 1}{e^{0.7/0.025} - 1} = 6.91 \times 10^{-16} A.$$

For $I_E = 1mA$ $V_{BC} = 0$ & using Eq (5.35)

$$i_c = 1 \times 10^{-3} - 6.91 \times 10^{-16}\left(\frac{1}{0.1} - 1\right) \times 1$$
$$= 1mA.$$

(a) For $i_C = \frac{1}{2}mA$

$$i_c = 10^{-3} - 6.91 \times 10^{-16}(9)e^{v_{BC}/0.025} = \frac{1}{2} \times 10^{-3}$$

$v_{BC} = \underline{\underline{0.628V}}$

(b) For $i_C = 0$

$$i_c = 10^{-3} - 6.91 \times 10^{-16}(9)e^{v_{BC}/0.025} = 0$$

$v_{BC} = \underline{\underline{0.645V}}$

Repeating (a) & (b) for $\alpha_R = 0.01$

(a) For $i_C = \frac{1}{2}mA$

$$i_c = 10^{-3} - 6.91 \times 10^{-16}\left(\frac{1}{0.01} - 1\right)e^{v_{BC}/0.025} = \frac{1}{2} \times 10^{-3}$$

$v_{BC} = 0.568V$

(b) For $i_C = 0mA$

$$i_c = 10^{-3} - 6.91 \times 10^{-16}(99)e^{\frac{v_{BC}}{0.025}} = 0$$

$v_{BC} = \underline{0.585V}$

5.14

$r_o = V_A / I_C = 100/I_C$

$I_C = 0.1mA$ $r_o = \underline{\underline{1M\Omega}}$

$I_C = 1mA$ $r_o = \underline{\underline{100k\Omega}}$

$I_C = 10mA$ $r_o = \underline{\underline{10k\Omega}}$

5.15

$$r_o = \frac{V_A}{I_C} = \frac{100}{1mA} = 100k\Omega$$

$$\frac{\Delta v_C}{\Delta I_C} = r_o$$

$$\Delta I_C = \frac{\Delta v_C}{r_o} = \frac{11}{100k} = 0.11mA$$

$\therefore$ $i_C = 1 + 0.11 = \underline{\underline{1.11mA}}$

5.16

$\beta_F = 100$ $\alpha_R = 0.1$

$\beta_{FORCED} = 10$

$$\beta_R = \frac{\alpha_R}{1-\alpha_R} = 1/9$$

SATURATION

$I_B = 0.1mA$

I_{Csat}

$$v_{CE} = V_T \ln \frac{1}{\alpha_R}$$
$$= 0.025 \ln \frac{1}{0.1}$$
$$= \underline{\underline{58mV}}$$

v_{CEoff} v_{CEsat}

Using Fig (5.25) to find R_{CEsat}

$$R_{CEsat} = 1/\text{SLOPE} \times \beta_F I_B$$
$$= \frac{1}{10 \times 100 \times 0.1 \times 10^{-3}}$$
$$= \underline{\underline{10\Omega}}$$

CONT.

From Fig(5.25)

$$V_{CEoff} = V_T \ln\left(\frac{\beta_F}{\beta_R}\right) - 0.05$$

$$= 0.025 \ln\left(\frac{100}{1/9}\right) - 0.05$$

$$= \underline{\underline{120\,mV}}$$

Use the battery + resistance model in Fig (5.24c) to estimate V_{CEsat} ≡ the voltage at the edge of active mode.

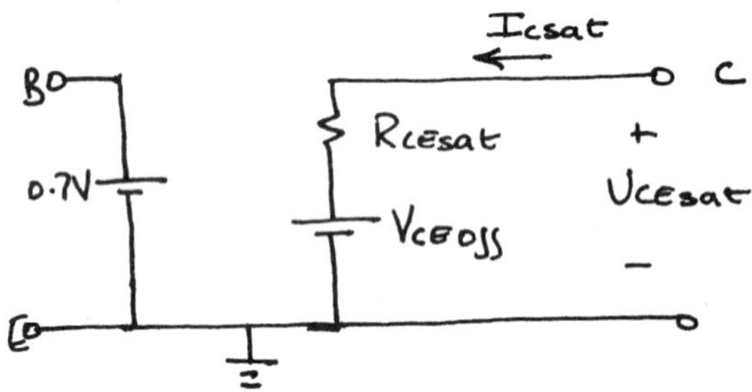

$$V_{CEsat} = V_{CEoff} + R_{CEsat} \times I_{Csat}$$

$$= 120\,mV + 10 \times \beta_{FORCED} I_B$$

$$= 0.12 + 10 \times 10 \times 0.1 \times 10^{-3}$$

$$= \underline{\underline{130\,mV}}$$

Using Eq (5.49)

$$V_{CEsat} = V_T \ln\left(\frac{1 + (\beta_{forced}+1)\frac{1}{\beta_R}}{1 - \beta_{forced}/\beta_F}\right)$$

$$= 0.025 \ln\left(\frac{1 + 11/\frac{1}{9}}{1 - 10/100}\right)$$

$$= \underline{\underline{118\,mV}}$$

Repeating for $\beta_{forced} = 20$

$R_{CEsat} = 10\Omega$ − NO CHANGE

$$V_{CEsat} = V_{CEoff} + R_{CEsat} I_{Csat}$$

$$= 0.120 + 10 \times \beta_{forced} I_B$$

$$= 0.120 + 10 \times 20 \times 0.1 \times 10^{-3}$$

$$= \underline{\underline{140\,mV}}$$

Using Eq (5.49)

$$V_{CEsat} = 0.025 \ln\left[\frac{1 + 21/\frac{1}{9}}{1 - 20/100}\right]$$

$$= \underline{\underline{137\,mV}}$$

5.17.

Transistor in saturation with constant base-current.

$$\text{SLOPE} = \frac{\Delta i_C}{\Delta v_{CE}} = \frac{3}{60} = \frac{1}{20}$$

$$R_{CEsat} = \frac{1}{\text{slope}} = \underline{\underline{20\Omega}}$$

$$V_{CEsat} = V_{CEoff} + I_{Csat} R_{CESAT}$$

$$0.170 = V_{CEoff} + 5 \times 10^{-3} \times 20$$

$$V_{CEoff} = \underline{\underline{70\,mV}}$$

5.18

10V

50μA

10 − 70 = $\underline{\underline{-60V}}$

5.19

Eq (5.56)

$A_v = -\frac{I_C R_C}{V_T}$

$= -\frac{10^{-3} R_C}{0.025} = -320$

$R_C = \underline{\underline{8\,k\Omega}}$

$V_C = 2 + \Delta v_o = 0.3$

$\Delta v_o = -1.7\,V$

If we assume linear operation

$\Delta v_{BE} = \frac{-1.7}{-320} = \underline{\underline{5.3\,mV}}$

5.20

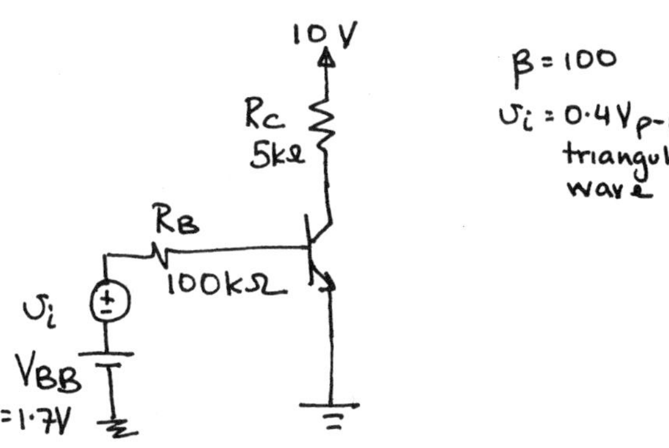

$\beta = 100$

$v_i = 0.4\,V_{p-p}$ triangular wave

From the slope of the load line through Q in Fig. 5.30(a)

$\frac{I_B}{V_{BB} - V_{BE}} = \frac{1}{R_B}$

$I_B = \frac{V_{BB} - V_{BE}}{I_B} = \frac{1.7 - 0.7}{100}$

$= \underline{\underline{10\,\mu A}}$

$\frac{\partial i_B}{\partial v_{BE}} = \frac{\partial}{\partial v_{BE}}\left(\frac{I_S}{\beta}\, e^{v_{BE}/V_T}\right)$

$= \left(\frac{I_S}{\beta}\, e^{v_{BE}/V_T}\right)\frac{1}{V_T}$

$= I_B / V_T$

$\therefore \text{slope}^{-1} = \frac{V_T}{I_B} = \frac{0.025}{10 \times 10^{-6}} = \underline{\underline{2.5\,k\Omega}}$

(c) Refer to Fig 5.30(a) from which we extract the sketch below:

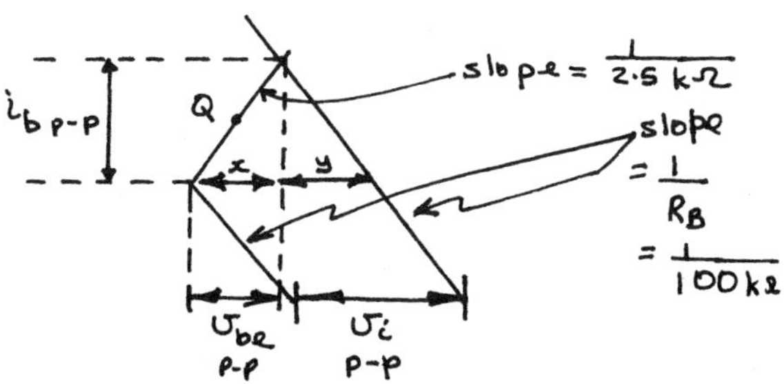

$v_{i\,p-p} = x + y$

$0.4 = i_b(2.5 \times 10^3) + i_b(100 \times 10^3)$

$i_b \cong \underline{\underline{4\,\mu A}}$

$v_{be\,p-p} = i_b \times 2.5 \times 10^3 = \underline{\underline{10\,mV}}$

(d) $I_C = \beta I_B = 100 \times 10\,\mu A = \underline{1\,mA}$

$V_{CE} = 10 - I_C R_C$

$= 10 - 1 \times 5$

$= \underline{\underline{5V}}$

(e) Peak to Peak value of

$i_c = \beta i_b = 100 \times 4\,\mu A = \underline{\underline{0.4\,mA}}$

$v_{ce} = i_c R_C = \underline{\underline{2\,V}}$

(f) Voltage gain $\equiv -\frac{v_{ce}}{v_i} = \frac{-2}{0.4} = \underline{\underline{-5\frac{V}{V}}}$

5.21

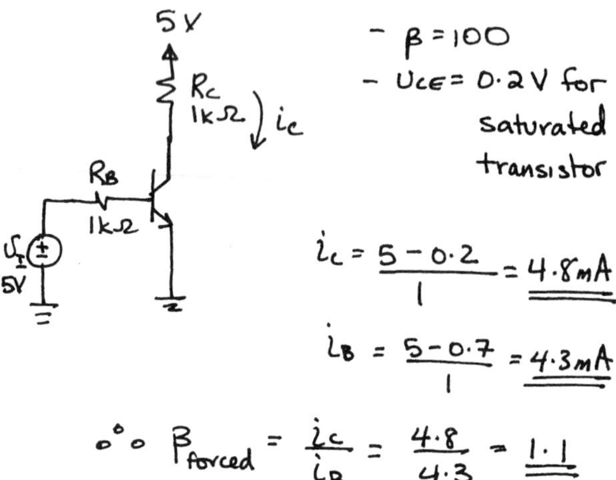

- $\beta = 100$
- $V_{CE} = 0.2\,V$ for saturated transistor

$$i_C = \frac{5-0.2}{1} = 4.8\,mA$$

$$i_B = \frac{5-0.7}{1} = 4.3\,mA$$

$$\therefore \beta_{forced} = \frac{i_C}{i_B} = \frac{4.8}{4.3} = 1.1$$

To bring the transistor to the edge of saturation but still in active mode you want $v_{CE} = v_C = 0.3\,V$

$$\Rightarrow i_C = \frac{5-0.3}{1} = 4.7\,mA$$

$$i_B = \frac{4.7}{100} = 0.047\,mA$$

$$\therefore R_B = \frac{5-0.7}{0.047} = 91.5\,k\Omega$$

5.22

10V, 4.7kΩ, V_B, 3.3kΩ

$\alpha \simeq 1$

To stay in active mode $v_{CB} \geq -0.4\,V$

$$i_E = \frac{V_B - 0.7}{3.3} = i_C$$

$$v_{CB} = v_C - v_B = 10 - i_C 4.7 - V_B$$

$$\Rightarrow -0.4 = 10 - \frac{4.7}{3.3}(V_B - 0.7) - V_B$$

$$\therefore V_B = 4.7\,V$$

5.23

10V, R_C, 0.5mA, 6V, 2V, R_E, 0.5mA

$\alpha \simeq 1$

$$i_E = \frac{6-0.7}{R_E} = 0.5$$

$$R_E = 11\,k\Omega$$

for $v_C = 8\,V$

$$R_C = \frac{10-8}{0.5} = 4\,k\Omega$$

5.24

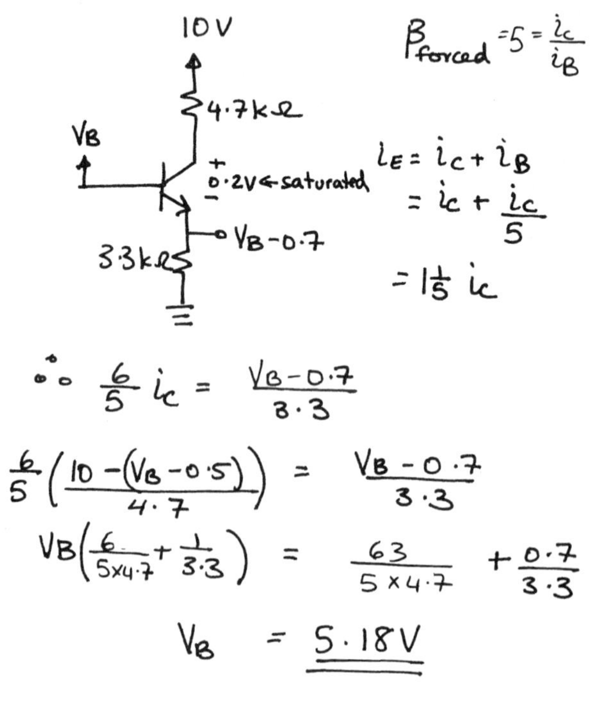

$$\beta_{forced} = 5 = \frac{i_C}{i_B}$$

$$i_E = i_C + i_B = i_C + \frac{i_C}{5} = 1\tfrac{1}{5}\, i_C$$

$$\therefore \frac{6}{5} i_C = \frac{V_B - 0.7}{3.3}$$

$$\frac{6}{5}\left(\frac{10-(V_B-0.5)}{4.7}\right) = \frac{V_B-0.7}{3.3}$$

$$V_B\left(\frac{6}{5\times4.7} + \frac{1}{3.3}\right) = \frac{63}{5\times4.7} + \frac{0.7}{3.3}$$

$$V_B = 5.18\,V$$

5.25

To stay in active mode $V \geq -0.4\,V$

$$\therefore V_C = 0 + 0.4 \Rightarrow R_C = \frac{0.4-(-10)}{4.6} = 2.26\,k\Omega$$

5.26

10V, R_E, ↓1mA, +4V−, −4V, R_C, ↓1mA, −10V

$\alpha \simeq 1$

assuming $V_{BE} = 0.7V$

$$R_E = \frac{10-0.7}{1} = \underline{\underline{9.3k\Omega}}$$

$$R_C = \frac{-4-(-10)}{1} = \underline{\underline{6k\Omega}}$$

5.27

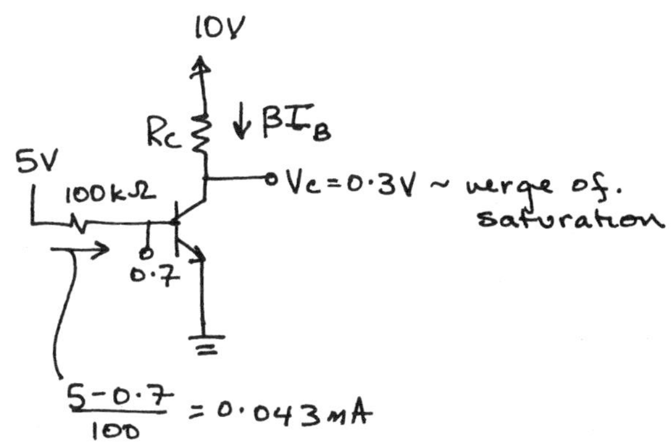

Design the transistor with the largest β & hence lowest Vc, and keep $V_C \geq 0.3V$ to keep transistor out of saturation.

$$R_C = \frac{10-0.3}{\beta I_B} = \frac{10-0.3}{150 \times 0.043}$$

$$= \underline{\underline{1.5k\Omega}}$$

Range of collector voltages is:

$0.3V$ to $10 - 50(0.043)R_C$ or

$\underline{\underline{0.3V \text{ to } 6.8V}}$

5.28

15V, R_C 5kΩ, 5V, 33.3kΩ, R_E 3kΩ, ①

NOTE $i_E = (\beta+1)\, i_B$

$\beta = 50$

Loop ①

$$5 - 33.3I_B - 0.7 - (\beta+1)I_B 3 = 0$$

$$I_B = \frac{5-0.7}{33.3 + 51(3)} = \underline{\underline{0.023\ mA}}$$

$$I_C = \beta I_B = 50 \times 0.023 = \underline{\underline{1.15mA}}$$

$$\% \text{ change} = \frac{1.15 - 1.28}{1.28} \times 100$$

$$= \underline{\underline{-10\%}}$$

5.29

Refer to Fig 5.41

$$\Sigma I = 0.103 + 1.252 + 2.78 = 4.135mA$$

$$P = 15 \times 4.136 = \underline{\underline{62mW}}$$

5.30

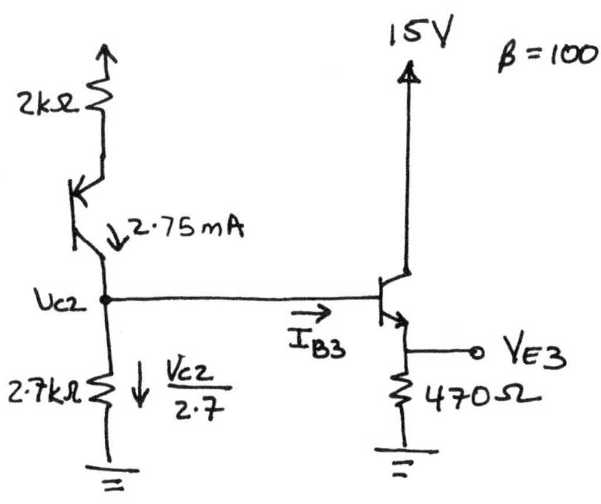

CONT.

$V_{E3} = V_{C2} - 0.7$

$I_{E3} = \frac{V_{C2} - 0.7}{0.470} = 101\, I_{B3}$

$= 101\left(2.75 - \frac{V_{C2}}{2.7}\right)$

$V_{C2}\left(\frac{1}{0.47} + \frac{101}{2.7}\right) = 101(2.75) + \frac{0.7}{0.47}$

$V_{C2} = \underline{\underline{7.06V}}$

$V_{E3} = V_{C2} - 0.7 = 7.06 - 0.7 = \underline{\underline{6.36V}}$

$I_{C3} = \alpha I_{E3} = \frac{100}{101}\left(\frac{6.36}{0.47}\right) = \underline{\underline{13.4mA}}$

5.31

Refer to Fig 5.42. $\beta_{min} = 30$

If $v_{IN} = 10V$, Q_1 is saturated as $v_{CB} \simeq -5V$

$\therefore$ Q_1 is on Q_2 is off

$\Rightarrow \underline{\underline{I_{C2} = 0A}}$

Σ voltage drops from input througe BE junction of Q_1

$10 - 0.7 - 10I_B - 1 \times I_E = 0$

$I_E = 9.3 - I_B \times 10$

$V_E = 5 - V_{CEsat} = 5 - 0.2 = \underline{\underline{4.8V}}$

$\&\ I_E = \frac{4.8}{1} = 9.3 - 10I_B$

$I_B = 0.45mA$

$I_C = I_E - I_B = 4.8 - 0.45$

$= \underline{\underline{4.35mA}}$

$V_B = 10 - 10I_B$

$= 10 - 10 \times 0.45$

$= \underline{\underline{5.5V}}$

5.32

Design 1: from Example 5.13 we know that

$R_B = 80K \| 40K = 26.6\,K\Omega$

$V_{BB} = 4V$, $V_E = 3.3V$

for nominal $I_E \simeq 1mA$

$\rightarrow R_E = 3K\Omega$

$$I_E = \frac{V_{BB} - V_{BE}}{R_E + R_B/(\beta+1)} \quad (1)$$

for $\beta = 50$ $I_E = 0.94\,mA$

$\beta = 150$ $I_E = 1.04\,mA$

$\Delta I_E = 1.04 - 0.94 = 0.1\,mA$

i.e 10% of 1mA

Design 2: from Example 5.13 we know that

$R_B = 8K \| 4K = 2.66\,K\Omega$

for nominal $I_E \simeq 1mA$

$\rightarrow R_E = 3.3K\Omega$.

Substituting in (1) for $\beta = 50$ and $\beta = 150$

$\beta = 50$ $I_E = 0.984\,mA$

$\beta = 150$ $I_E = 0.995\,mA$

$\Delta I_E = 0.011\,mA$

i.e. 1.1% of 1mA

5.33

+10V, R_C, 0V, R_B, R_E, −0.7V, 1mA, −5V

Common base: $R_B = 0\Omega$

$$R_E = \frac{5 - 0.7\,V}{1mA} = 4.3K\Omega$$

for high voltage gain $\rightarrow$ large V_{RC}

to allow 2V swing w/o falling into saturation

$V_{CE} = 2.4V$

$\Rightarrow V_C = -0.7 + 2.4 = 1.7V$

$$R_C = \frac{10 - 1.7}{0.99mA} = 8.4K\Omega$$

5.34

$V_{CE} = +2.4V$, $I_E = 1mA$

$V_{CC} = +10V$, $\beta = 100$

$$V_C = 2.4V = 0.7 + \frac{1m}{\beta+1} \cdot R_B$$

$\rightarrow R_B = 172\,K\Omega$

$$R_C = \frac{(10 - 2.4)V}{1mA} = 7.6\,K\Omega$$

5.35

+10V, 7.5K, $\beta = 100$, 100K, 1mA

$$I_B = \frac{1mA}{\beta+1} = 0.01m$$

$V_B = -I_B \cdot 100K$

$= -1V$

$V_C = 10 - I_C \cdot 7.5K$

$I_C = \alpha \cdot I_E$

$\alpha = 0.99$

$\Rightarrow V_C = +2.6V$

CONT.

For Fig. 5.47(b)
Applying Eqn. (5.77)

$$1\,mA = \frac{+10V + 10V - 0.7V}{R}$$

$\Rightarrow R = 19.3K\Omega$

5.36

$i_c = I_s e^{v_{BE}/V_T}$

$I_c = I_s e^{V_{BE}/V_T}$

$$g_m = \left.\frac{\partial i_c}{\partial v_{BE}}\right|_{i_c = I_c}$$

$$= \left.\frac{I_s}{V_T} \cdot e^{v_{BE}/V_T}\right|_{i_c = I_c}$$

$$= \left.\frac{i_c}{V_T}\right|_{i_c = I_c}$$

Thus, $g_m = \frac{I_c}{V_T}$ Q.E.D

5.37

$g_m = I_c / V_T$

$= 1/0.025 = 40mA/V$

$r_e = \frac{V_T}{I_E} \simeq \frac{V_T}{I_c} = \frac{25mV}{1.0mA} = 25\Omega$

$r_\pi = (\beta+1) r_e = \beta / g_m$

$= 100 / 40 \times 10^{-3} = 2.5K\Omega$

5.38

Voltage gain $\equiv \frac{v_c}{v_{be}} = -g_m \cdot R_c$

where $g_m = \frac{I_c}{V_T} = \frac{1mA}{25mV} = 40\frac{mA}{V}$

$\rightarrow \frac{v_c}{v_{be}} = -40 \times 10 = -400\, V/V$

$v_C(t) = V_C + v_c(t)$

$= (V_{CC} - R_C I_C) - 400\, v_{be}(t)$

$= (15 - 10 \times 1) - 400 \times 0.005 \sin \omega t$

$= 5 - 2 \sin \omega t$, Volts.

$i_B(t) = I_B + i_b(t)$

$= \frac{I_C}{\beta} + i_c(t)$

$= \frac{1000\mu A}{100} + \frac{g_m\, v_{be}(t)}{\beta}$

$= 10\mu A + \frac{40.10^{-3} \times 5.10^{-3} \sin \omega t}{100}$

$= 10 + 2 \sin \omega t$, μA

5.39

As in example 5.16:

$\alpha = 0.99$, $r_e = \frac{V_T}{I_E} = \frac{25m}{0.93m} = 27\Omega$

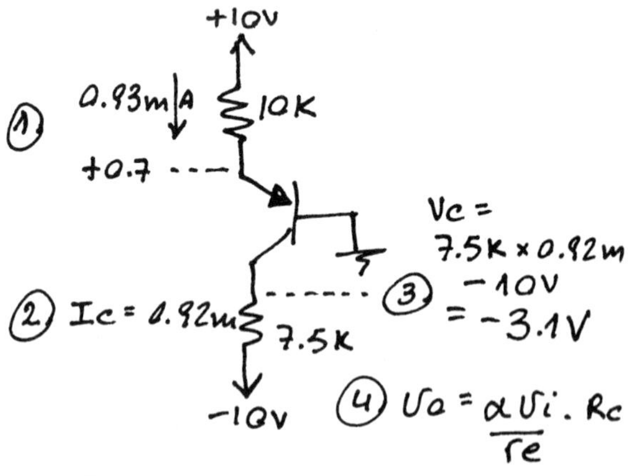

+10V

(1) 0.93mA

10K

+0.7

$V_C =$ $7.5K \times 0.92m$ $-10V$ $= -3.1V$ (3)

(2) $I_C = 0.92m$

7.5K

−10V

(4) $v_o = \alpha v_i \cdot \frac{R_c}{r_e}$

$A_v = \frac{v_o}{v_i} = \frac{0.99 \times 7.5K}{27} = 275\, V/V$

for $\hat{v}_i = 10mV \rightarrow \hat{v}_o = 2.75V$

$V_C + 2.75 = -0.35$
$V_C - 2.75 = -5.85$ } Active region.

5.40

(a) DC Analysis

+10V

8K $\downarrow I_C = \alpha I_E = 0.99mA$

① $1/101 \approx 0.01$

$V_C = 10 - 0.99 \times 8 \simeq 2V$ ④

② $V_B = -0.01 \times 10 = -0.1V$

10K

$V_E = -0.1 - 0.7 = -0.8V$ ③

1mA

(b) $g_m = \dfrac{I_C}{V_T} = \dfrac{0.99mA}{25mV} \simeq 40\dfrac{mA}{V}$

$r_\pi = \dfrac{\beta}{g_m} = \dfrac{100}{40mA/V} = 2.5K\Omega$

$r_o = \dfrac{V_A}{I_C} = \dfrac{100V}{0.99mA} = 100K\Omega$

(c)

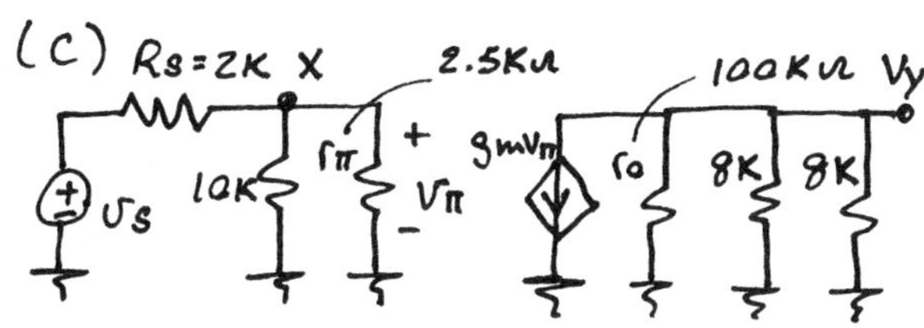

$\dfrac{v_\pi}{v_s} = \dfrac{(10 \| 2.5)}{(10 \| 2.5) + 2} = \dfrac{2}{2+2} = 0.5$

$V_y = -g_m v_\pi (r_o \| 8K \| 8K)$

$= -40 (100 \| 8 \| 8) v_\pi$

$= -153.8\, v_\pi$

$\dfrac{V_y}{V_s} = -77\ V/V$

Neglecting r_o:

$\dfrac{V_y}{V_s} = -0.5 \times 40 (8 \| 8) = -80\ V/V$

i.e. an error of $+3.9\%$

5.41

Refer to Fig. E5.41:

$I_C = \alpha I = 0.99 \times 1mA \sim 1mA$

$I_B = I_C/\beta = 0.01mA$

$V_C = 10 - 8K \times 1m = +2V$

$V_B = -0.01m \times 100K = -1V$

$V_E = -1 - 0.7 = -1.7V$

Allowable maximum swing are:

up: from V_C to V_{CC}

down: from V_C to V_C' such as $V_B - V_C' = 0.4$

$\beta = 100$: up, $V_{CC} - V_C = +8V$

down, $V_B - V_C' = 0.4$

$\rightarrow V_C' = -1.4$

swing $= -1.4 - 2 = -3.4V$

$\beta = 50$: $I_C \sim 1mA \rightarrow V_C = +2V$

$I_B = 0.0196m \rightarrow V_B = -1.96$

up, $+8V$

down, $(-1.96 - 0.4) - 2 = -4.4V$

$\beta = 200$: $I_C \sim 1mA \rightarrow V_C = +2V$

$I_B = 0.005mA \rightarrow V_B = -0.5$

up, $+8V$

down, $(-0.5 - 0.4) - 2 = -2.9V$

5.42

Refer to Example 5.17

The following parameters are independent of R_{sig} and R_L, and thus remain unchanged:

$R_i = 900K\Omega$ $\quad A_{is} = 572\ A/A$

$R_o = 1.43K\Omega$ $\quad G_m = 7mA/V$

$A_{vo} = 10\ V/V$

CONT.

(a) $R_{sig} = 200 K\Omega$ (doubled)
$R_L = 10 K\Omega$

Since R_{in} is independent of R_{sig} and R_L is unchanged then $R_{in} = \underline{400 K\Omega}$

$$G_v = \frac{R_{in}}{R_{in}+R_{sig}} \cdot A_{vo} \cdot \frac{R_L}{R_L+R_o} = \frac{400 \times 10 \times 10}{(400+200)(10+1.43)}$$

$G_v = \underline{5.83}$ V/V

$$G_{vo} = \frac{R_i}{R_i+R_{sig}} \cdot A_{vo} = \frac{900}{900+200} \times 10 = 8.18 \text{ V/V}$$

$$\rightarrow R_{out} = \frac{G_{vo}}{G_v} R_L - R_L = 10K\left(\frac{8.18}{5.83} - 1\right)$$

$R_{out} = \underline{4.03 K\Omega}$

(b) $R_{sig} = 100 K\Omega$, $R_L = 20 K\Omega$ (doubled)

Then R_{out} is the same as in Example 5.17; since R_{sig} is unchanged: $R_{out} = \underline{2.86} K\Omega$

$$G_{vo} = \frac{900}{900+100} \times 10 = 9 \text{ V/V}$$

$$\Rightarrow G_v = \frac{9 \times 20}{20 + 2.86} = \underline{7.87} \text{ V/V, also}$$

$$G_v = \frac{R_{in}}{R_{in}+R_{sig}} \cdot A_{vo} \cdot \frac{R_L}{R_L+R_o} \rightarrow R_{in} = 538 K\Omega$$

(c) $R_{in} = \underline{538} K\Omega$ as in (b)
$R_{out} = \underline{4.03} K\Omega$ (a) $\rightarrow G_{vo} = 8.18$ V/V
$\rightarrow G_v = \underline{6.8}$ V/V

5.43

From Exercise 5.41:
$I_C = 1 mA$, $I_B = 0.01 mA$
$\beta = 100$, $V_A = 100 V$
$R_C = 8K$, $R_B = 100 K\Omega$
$V_C = +2V$, $V_B = -1V$, $V_E = -1.7V$

① $R_{in} = R_B \| r_\pi$
$r_\pi = V_T / I_B = 2.5 K\Omega$

R_{in} (w/o R_B) $= r_\pi = \underline{2.5 K\Omega}$
R_{in} (w R_B) $= 2.5K \| 100K \simeq$
$= 2.439K \simeq 2.4K$

② $A_{vo} = -g_m (r_o \| R_C)$
$r_o = V_A / I_C = 100 K\Omega$ $\quad g_m = \frac{I_C}{V_T} = 40m$

A_{vo} (w/o r_o) $= -40m \times 8K = \underline{-320 \text{ V/V}}$

A_{vo} (w r_o) $= -40m \times (8K \| 100K)$
$= \underline{-296}$ V/V

③ $R_{out} = R_C \| r_o$
R_{out} (w/o r_o) $= \underline{8 K\Omega}$
R_{out} (w r_o) $= \underline{7.4 K\Omega}$

④ $A_{is} = -g_m \cdot R_{in}$
A_{is} (w/o R_B) $= -40m \times 2.5K$
$= \underline{100 \text{ A/A}}$
A_{is} (w R_B) $= -40m \times 2.439K$
$= -97.5 \simeq -98$ A/A

⑤ $A_v = -g_m (r_o \| R_C \| R_L)$
$A_v = -40 \frac{mA}{V} (100K \| 8K \| 5K)$

$= \underline{-119 \text{ V/V}}$

⑥ $$G_v = \frac{-\beta (R_C \| R_L \| r_o)}{r_\pi + R_{sig}}$$

$$= \frac{-100 \times (8K \| 5K \| 100K)}{2.5K + 5K}$$

$= \underline{-39 \text{ V/V}}$

⑦ $R_B \gg r_\pi \rightarrow v_i = v_\pi$

$$v_\pi = v_{sig} \left(\frac{r_\pi}{r_\pi + R_{sig}} \right)$$

for $v_\pi = 5mV$, $R_{sig} = 5K$ $\rightarrow v_{sig} = 15mV$

and: $\hat{v}_o = +g_m \cdot \hat{v}_\pi \times (r_o \| R_C \| R_L)$
$\simeq +40m \times 5m \times (R_C \| R_L)$
$= 0.6V$ peak //

5.44

$R_{sig} = 5K\Omega$, $R_L = 5K\Omega$
$R_{in} = 4 \times R_{sig} = 20K\Omega$

① $R_{in} = R_B \| R_{ib} = 100K \| R_{ib}$
for $R_{in} = 20K \longrightarrow R_{ib} = 25K\Omega$
$R_{ib} = (\beta+1)(r_\pi + R_e)$
$25K \approx 100 \times (25 + R_e)$
$\longrightarrow R_e = 225\Omega$

② $A_{vo} = -\dfrac{g_m . R_C}{1 + g_m R_e} = \dfrac{-40 \times 8}{1 + 40 \times 0.225}$
$A_{vo} = -32\ V/V$

③ $R_{out} = R_C = 8K\Omega$

④ $A_v \simeq -\dfrac{R_C \| R_L}{r_e + R_e} = -\dfrac{(8K \| 5K)}{25 + 225}$
$= -12.3 V/V$

⑤ $G_v = \dfrac{-\beta\,(R_C \| R_L)}{R_{sig} + (\beta+1)(r_e + R_e)}$
$= \dfrac{-100 \times (8K \| 5K)}{5K + 101 \times (25 + 225)} = -9.9 V/V$

⑥ $A_{is} = -\beta = -100$

⑦

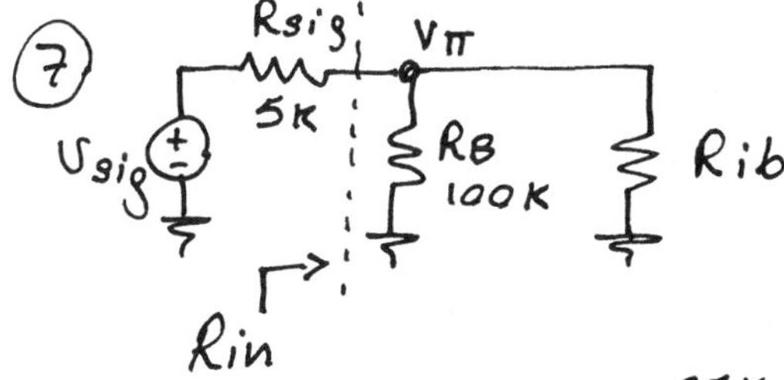

$R_{ib} = (\beta+1)(r_e + R_e)$ → 25K w Re ; → 2.5K w/o Re

$R_{in} = 100K \| R_{ib}$ → 20K w Re ; → 2.4K w/o Re

$v_\pi = v_{sig} \cdot \dfrac{R_{in}}{R_{si} R_{in}}$

$\Rightarrow \hat{v}_{sig} = \hat{v}_\pi \cdot \dfrac{(R_{in} + R_{sig})}{R_{in}}$

if $\hat{v}_\pi = 5m \Rightarrow \hat{v}_{sig} = \rightarrow 6.25mV$ (w Re) ; → 15mV (w/o Re)

⑧ $\hat{v}_o = \alpha\, \hat{i}_e \cdot (R_C \| R_L)$
$= \alpha \dfrac{\hat{v}_\pi}{r_e + R_e} \cdot (R_C \| R_L)$
$= 0.99 \times \left(\dfrac{5m}{25 + 225}\right) \cdot (8K \| 5K)$
$\hat{v}_o = 0.06$ peak

Notice that: $1 + g_m R_e$
$= 1 + 40m \times 225$
$= 10$
which is the factor by which the gain is reduced w. respect to Exercise 5.43.

5.45

$R_C = 8K\Omega$, $\beta = 100$, $I = 1mA$
$R_{sig} = 5K\Omega$, $R_L = 5K\Omega$

① $R_{in} = r_e = \dfrac{V_T}{I_E} = \dfrac{25mV}{1mA} = 25\Omega$

② $A_{vo} = g_m . R_C = \dfrac{1mA}{25mV} \times 8K$
$= +320\ V/V$

③ $R_o = R_C = 8K\Omega$

④ $A_v = g_m . (R_C \| R_L)$
$= 40m\,(8K \| 5K) = 123 \frac{V}{V}$

⑤ $\dfrac{v_i}{v_s} = \dfrac{r_e}{R_{sig} + r_e} = \dfrac{25}{5K + 25} = 0.005 \frac{V}{V}$

CONT.

⑥ $G_v = \alpha \dfrac{(R_C \| R_L)}{R_{sig} + r_e}$

$= 0.99 \times \dfrac{(8K \| 5K)}{5K + 25} = \underline{\underline{0.6 \, \frac{V}{V}}}$

⑦ If $G_v = 39 \rightarrow R_{sig} = ?$

$R_{sig} = \alpha \dfrac{(R_C \| R_L)}{G_v} - r_e$

$= \underline{\underline{54\Omega}}$

5.46

$R_{sig} = 50\Omega \Rightarrow R_{in} = r_e = 50\Omega$

$r_e = \dfrac{V_T}{I} \Rightarrow I = \dfrac{25mV}{50\Omega}$

$I = \underline{\underline{0.5mA}}$

$G_v = 100 \, V/V = 1 \times \dfrac{(R_C \| R_L)}{R_{sig} + r_e}$

$\Rightarrow \underbrace{R_C \| R_L}_{\text{Total } R \text{ in collector}} = 100 \times (50 + 50)$

$= \underline{\underline{10K\Omega}}$

5.47

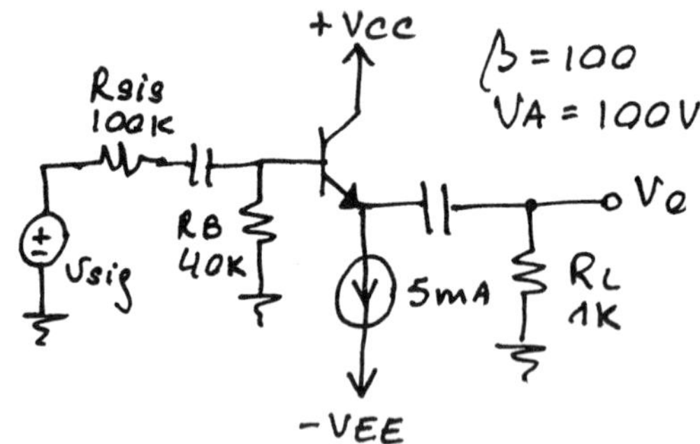

$I_C \simeq I_E \rightarrow I_C = 5mA$

$r_o = \dfrac{V_A}{I_C} = \dfrac{100V}{5mA} = 20K\Omega$

$r_e = \dfrac{V_T}{I} = \dfrac{25mV}{5mA} = 5\Omega$

① $R_{ib} = (\beta+1)\,[r_e + (r_o \| R_L)]$

$= 101 \cdot [5 + (20K \| 1K)]$

$= \underline{\underline{96.7K\Omega}}$

② $R_{in} = R_B \| R_{ib} = 40K \| 96.7K$

$= \underline{\underline{28.3K\Omega}}$

③ $G_v = \dfrac{R_B}{R_{sig} + R_B} \cdot \dfrac{(\beta+1)(r_o \| R_L)}{(R_{sig} \| R_B) + (\beta+1)(r_e + r_o \| R_L)}$

$= \dfrac{40}{(10+40)} \times \left[\dfrac{101 \times (20K \| 1K)}{(10K \| 40K) + 101\,(5 + 20K \| 1K)}\right]$

$\Rightarrow G_v = \underline{\underline{0.735 V/V}}$

④ $G_{vo} = \dfrac{R_B}{R_{sig} + R_B} \cdot \dfrac{r_o}{\dfrac{R_{sig} \| R_B}{\beta+1} + r_e + r_o}$

$= \dfrac{40}{10+40} \cdot \dfrac{20K}{\dfrac{(10K \| 40K)}{101} + 5 + 20K}$

$= \underline{\underline{0.8}} \, V/V$

⑤ $R_{out} = r_o \| \left(r_e + \dfrac{(R_{sig} \| R_B)}{\beta+1}\right)$

$= 20K \| \underbrace{\left(5 + \dfrac{(10K \| 40K)}{\beta+1}\right)}$

$\simeq 20K \| 84$

$= \underline{\underline{84\Omega}}$

⑥ $\hat{V}_o = I \cdot R_L = 5mA \times 1K\Omega = \underline{\underline{5V}}$

⑦ $\hat{V}_o = \hat{V}_\pi \times \dfrac{(R_L \| r_o)}{r_e}$

$\hat{V}_\pi = 10mV$

$\Rightarrow \hat{V}_o = \underline{\underline{1.9V}}$

⑧ $G_v = 0.8 \times \dfrac{101\,(20K \| R_L)}{8K + 101\,(5 + 20K \| R_L)}$

$R_L = 2K \rightarrow G_v = \underline{\underline{0.768 \, V/V}}$ CONT.

$R_L = 500 \rightarrow G_v = \underline{\underline{0.685}}$ V/V

5.48

$g_m = \frac{I_C}{V_T} = \frac{1mA}{25mV} = 40\frac{mA}{V}$

$C_{de} = \tau_F \cdot g_m = 20\times10^{-12}\times40\times10^{-3}$

$= \underline{\underline{0.8}}\ pF$

$C_{je} = 2C_{je0} = 2\times20 = \underline{\underline{40}}\ fF$

$C_\pi = C_{de} + C_{je} = \underline{\underline{0.84}}\ pF$

$C_\mu = \frac{C_{\mu 0}}{\left(1+\frac{V_{CB}}{V_{0c}}\right)^{m_{CBJ}}}$

$= \frac{20\,fF}{\left(1+\frac{2}{0.5}\right)^{0.33}} = \underline{\underline{12\,fF}}$

$f_T = \frac{g_m}{2\pi(C_\pi + C_\mu)}$

$= \frac{40\times10^{-3}}{2\pi(0.84+0.012)\times10^{-12}} = \underline{\underline{7.47}}\,GHz$

5.49

$|h_{fe}| \simeq \frac{f_T}{f} \rightarrow 10 = \frac{f_T}{50}$

$\Rightarrow f_T = \underline{500}\,MHz$

$f_T = \frac{g_m}{2\pi(C_\pi + C_\mu)}$

$C_\pi + C_\mu = \frac{40\times10^{-3}}{2\pi\times500\times10^6} = 12.7\,pF$

$C_\pi = 12.7 - C_\mu = 12.7 - 2 = \underline{\underline{10.7}}\,pF$

5.50

Diffusion component of C_π at I_C of 1mA

$= 10.7 - 2 = 8.7\,pF$

Since C_{de} is proportional to I_C, then:

$C_{de}(I_C = 0.1mA) = 0.87\,pF$

$C_\pi\ (I_C = 0.1mA) = 2.87\,pF$

$f_T(I_C = 0.1mA) = \frac{g_m}{2\pi(C_\pi + C_\mu)}$

$= \frac{4\times10^{-3}}{2\pi(2.87+2)\times10^{-12}}$

$= \underline{\underline{130.7}}\,MHz$

5.51

① $A_M = -39/2 = -19.5$ V/V

$A_M = \frac{-R_B}{R_B + R_{sig}} \cdot \frac{r_\pi \cdot g_m \cdot R_L'}{r_\pi + r_x + (R_B \| R_{sig})}$

$A_M = \frac{-100}{100+5} \cdot \frac{2.5\times40.10^{-3}\times R_L'}{(2.5+0.05+(100\|5))}$

$= -0.013\times R_L'$

$\Rightarrow R_L' = 1.5K\Omega = r_o \| R_C \| R_L$

$1.5K\Omega = (100\|8\|R_L)\,K\Omega$

$= 7.4K \| R_L$

$\longrightarrow R_L = \underline{\underline{1.9K\Omega}}$

② $f_H = \frac{1}{2\pi C_{in} \cdot R'_{sig}}$ $\quad R'_{sig} = 1.65K\Omega$

$C_{in} = C_\pi + C_\mu(1 + g_m R_L')$

$C_{in} = 7 + 1(1+40\times10^{-3}\times1.5\times10^3)$

$= 68\,pF$

$\Rightarrow f_H = \frac{1}{2\pi\,68p \cdot 1.65K} = \underline{\underline{1.42\,MHz}}$

5.52

$C_{C1} = C_E = C_{C2} = 1\mu F$

$g_m = 40\frac{mA}{V} \rightarrow I_C = 40m \times 25m = 1mA$

$r_\pi = 2.5K\Omega = \frac{\beta}{g_m} \Rightarrow \beta = 100$

$r_e = \frac{V_T}{I_E} = \frac{25mV}{1mA} = 25\Omega$

$$f_{P1} = \frac{1}{2\pi C_{C1}[R_B // r_\pi + R_{sig}]}$$

$$= \frac{1}{2\pi\, 1\mu\, [100K // 2.5K + 5K]}$$

$f_{P1} = \underline{\underline{21.4\,Hz}}$

$$f_{P2} = \frac{1}{2\pi . C_E\left[r_e + \frac{R_B // R_{sig}}{\beta + 1}\right]}$$

$$f_{P2} = \frac{1}{2\pi . 1\mu\left[25 + \frac{100K // 5K}{101}\right]}$$

$f_{P2} = \underline{\underline{2.2\,KHz}}$

$$f_{P3} = \frac{1}{2\pi . C_{C2} . (R_C + R_L)}$$

$$= \frac{1}{2\pi . 1\mu\, (8K + 5K)}$$

$f_{P3} = \underline{\underline{12.2\,Hz}}$

5.53

Refer to Fig. E5.53

$V_{OH} = V_{CC} - I R_C$

where $I = \frac{V_{CC} - V_{BE}}{R_C + \frac{R_B}{N}}$

thus,

$$V_{OH} = V_{CC} - R_C \cdot \frac{V_{CC} - V_{BE}}{R_C + \frac{R_B}{N}}$$

Q.E.D.

For $V_{CC} = +5V$, $R_C = 1K\Omega$
$N = 5$, $R_B = 10K\Omega$

$V_{OH} = 5 - 1 \times \frac{5 - 0.7}{1 + \frac{10}{5}} = \underline{\underline{3.6V}}$

Chapter 6 - Exercises

6.1

(a) The minimum value of I_D occurs when $V_{ov} = 0.2V$ and $\frac{W}{L} = 0.1$, that is

$$I_{Dmin} = \frac{1}{2}\mu_n C_{ox}\frac{W}{L}V_{ov}^2 \simeq 0.8\mu A$$

The maximum value of I_D occurs when $V_{ov} = 0.4V$ and $\frac{W}{L} = 100$, that is

$$I_{Dmax} = \frac{1}{2}\mu_n C_{ox}\frac{W}{L}V_{ov}^2 \simeq 3.1mA$$

(b) For a similar range of current in an npn transistor, we have

$$\frac{I_{Cmax}}{I_{Cmin}} = \frac{3.1mA}{0.8\mu A} = \frac{I_S e^{V_{BEmax}/V_T}}{I_S e^{V_{BEmin}/V_T}}$$

$$\Rightarrow e^{(V_{BEmax}-V_{BEmin})/V_T} = e^{\Delta V_{BE}/V_T} = \frac{3.1mA}{0.8\mu A}$$

$$\Delta V_{BE} = V_T \ln\left(\frac{3.1mA}{0.8\mu A}\right) \text{ and } V_T = 25mV$$

$$\Rightarrow \Delta V_{BE} = 207\ mV$$

6.2

For an NMOS fabricated in the $0.5\mu m$ process, with $\frac{W}{L} = 10$, we want to find the transconductance and the intrinsic gain obtained for the following drain currents: ($L = 0.5\mu m$)

$I_D = 10\mu A$, $g_m = \sqrt{2\mu_n C_{ox}\left(\frac{W}{L}\right)I_D}$, $\mu_n C_{ox} = 190\frac{\mu A}{V^2}$

$$g_m = \sqrt{2\times190\times10\times10} \simeq 0.2\frac{mA}{V}$$

$$r_o = \frac{V_A}{I_D} = \frac{V_A' L}{I_D} = \frac{20\times0.5}{10\mu A} = 1M\Omega$$

intrinsic gain $= g_m r_o = 0.2\frac{mA}{V}\times 1M\Omega = 200\frac{V}{V}$

For $I_D = 100\mu A$ we have:

$$g_m = \sqrt{2\mu_n C_{ox}\left(\frac{W}{L}\right)I_D} = \sqrt{2\times190\times10\times100}$$

$$g_m = 0.62\frac{mA}{V} \simeq 0.6\frac{mA}{V}$$

$$r_o = \frac{V_A' L}{I_D} = \frac{20\times0.5}{100\mu A} = 100\ k\Omega$$

$$g_m r_o = 0.62\frac{mA}{V}\times100k\Omega = 62\frac{V}{V}$$

For $I_D = 1mA$

$$g_m = \sqrt{2\mu_n C_{ox}\left(\frac{W}{L}\right)I_D} = \sqrt{2\times190\times10\times1} \simeq 2\frac{mA}{V}$$

$$r_o = \frac{V_A' L}{I_D} = \frac{20\times0.5}{1mA} = 10k\Omega$$

$$g_m r_o = 2\frac{mA}{V}\times10k\Omega = 20\frac{V}{V}$$

6.3

For an NMOS fabricated in the $0.5\mu m$ CMOS technology specified in Table 6.1 with $L = 0.5\mu m$, $W = 5\mu m$, and $V_{ov} = 0.3V$ we have

$$I_D = \frac{1}{2}\mu_n C_{ox}\frac{W}{L}V_{ov}^2 = \frac{1}{2}\,190\frac{\mu A}{V^2}\times\frac{5}{0.5}\times0.3^2$$

$$I_D = 85.5\mu A$$

$$g_m = \frac{2I_D}{V_{ov}} = \frac{2\times85.5\mu A}{0.3V} = 0.57\frac{mA}{V}$$

$$r_o = \frac{V_A' L}{I_D} = \frac{20\times0.5}{85.5\mu A} \simeq 117\ k\Omega$$

$$A_o = g_m r_o = 66.7\ V/V$$

$$C_{gs} = \frac{2}{3}WLC_{ox} + C_{ov} = \frac{2}{3}\times5\times0.5\times3.8 + 0.4\times5$$

$$C_{gs} = 8.3fF,\quad C_{gd} = C_{ov}\times W = 0.4\times5 = 2fF$$

$$f_T = \frac{g_m}{2\pi(C_{gs}+C_{gd})} = \frac{0.57\frac{mA}{V}}{2\pi(8.3+2)}$$

$$f_T = 8.8\ GHz$$

6.4

In the current source of Example 6.4 we have $I_o = 100\mu A$ and we want to reduce the change in output current, ΔI_o, corresponding to a 1V change in output voltage, ΔV_o, to 1% of I_o. That is

$$\Delta I_o = \frac{\Delta V_o}{r_{o2}} = 0.01 I_o \Rightarrow \frac{1V}{r_{o2}} = 0.01 \times 100\mu A$$

$$r_{o2} = \frac{1V}{1\mu A} = 1M\Omega$$

$$r_{o2} = \frac{V_A' \times L}{I_o} \Rightarrow 1M\Omega = \frac{20 \times L}{100\mu A} \Rightarrow$$

$$L = \frac{100V}{20V/\mu m} = 5\mu m$$

To keep V_{ov} of the matched transistors the same as that of Example 6.4, $\frac{W}{L}$ of the transistors should remain the same. Therefore

$$\frac{W}{5\mu m} = \frac{10\mu m}{1\mu m} \Rightarrow W = 50\mu m$$

So the dimensions of the matched transistors Q_1 and Q_2 should be changed to: $W = 50\mu m$ and $L = 5\mu m$

6.5

For the circuit of Figure 6.7 we have:

$$I_2 = I_{REF}\frac{(W/L)_2}{(W/L)_1}, \quad I_3 = I_{REF}\frac{(W/L)_3}{(W/L)_1} \text{ and}$$

$$I_5 = I_4\frac{(W/L)_5}{(W/L)_4}$$

Since all channel lengths are equal ($L_1 = L_2 = \dots = L_5 = 1\mu m$) and $I_{REF} = 10\mu A$, $I_2 = 60\mu A$, $I_3 = 20\mu A$, $I_4 = I_3 = 20\mu A$ and $I_5 = 80\mu A$, we have:

$$I_2 = I_{REF}\frac{W_2}{W_1} \Rightarrow \frac{W_2}{W_1} = \frac{I_2}{I_{REF}} = \frac{60}{10} = 6$$

$$I_3 = I_{REF}\frac{W_3}{W_1} \Rightarrow \frac{W_3}{W_1} = \frac{I_3}{I_{REF}} = \frac{20}{10} = 2$$

$$I_5 = I_4\frac{W_5}{W_4} \Rightarrow \frac{W_5}{W_4} = \frac{I_5}{I_4} = \frac{80}{20} = 4$$

In order to allow the voltage at the drain of Q_2 to go down to within 0.2V of the negative supply voltage we need $V_{OV2} = 0.2V$

$$I_2 = \frac{1}{2}\mu_n C_{ox}\left(\frac{W}{L}\right)_2 V_{OV2}^2 = \frac{1}{2}k_n'\left(\frac{W}{L}\right)_2 V_{OV2}^2$$

$$60\mu A = \frac{1}{2}\, 200\frac{\mu A}{V^2}\left(\frac{W}{L}\right)_2 (0.2)^2 \Rightarrow$$

$$\left(\frac{W}{L}\right)_2 = \frac{120}{200 \times (0.2)^2} = 15 \Rightarrow W_2 = 15 \times L_2$$

$$W_2 = 15\mu m, \quad \frac{W_2}{W_1} = 6 \Rightarrow W_1 = \frac{W_2}{6} = 2.5\mu m$$

$$\frac{W_3}{W_1} = 2 \Rightarrow W_3 = 2 \times W_1 = 5\mu m$$

In order to allow the voltage at the drain of Q_5 to go up to within 0.2V of positive supply we need $V_{OV5} = 0.2V$

$$I_5 = \frac{1}{2}k_p'\left(\frac{W}{L}\right)_5 V_{OV5}^2 \Rightarrow$$

$$80\mu A = \frac{1}{2}\, 80\frac{\mu A}{V^2}\left(\frac{W}{L}\right)_5 (0.2)^2 \Rightarrow$$

$$\left(\frac{W}{L}\right)_5 = \frac{2 \times 80}{80 \times (0.2)^2} = 50 \Rightarrow W_5 = 50L_5$$

$$W_5 = 50\mu m$$

$$\frac{W_5}{W_4} = 4 \Rightarrow W_4 = \frac{50\mu m}{4} = 12.5\mu m$$

Thus: $W_1 = 2.5\mu m$, $W_2 = 15\mu m$, $W_3 = 5\mu m$, $W_4 = 12.5\mu m$ and $W_5 = 50\mu m$

6.6

From equation (6.24) we have:

$$I_o = I_{REF}\left(\frac{m}{1 + \frac{m+1}{\beta}}\right)\left(1 + \frac{V_o - V_{BE}}{V_{A2}}\right)$$

CONT.

$$I_o = 1\,mA \left(\frac{1}{1+\frac{1+1}{100}}\right)\left(1+\frac{5-0.7}{100}\right) = 1.02\,mA$$

$I_o = 1.02\,mA$

$$R_o = r_{o2} = \frac{V_A}{I_o} = \frac{100V}{1.02mA} = 98k\Omega \simeq 100k\Omega$$

6.7

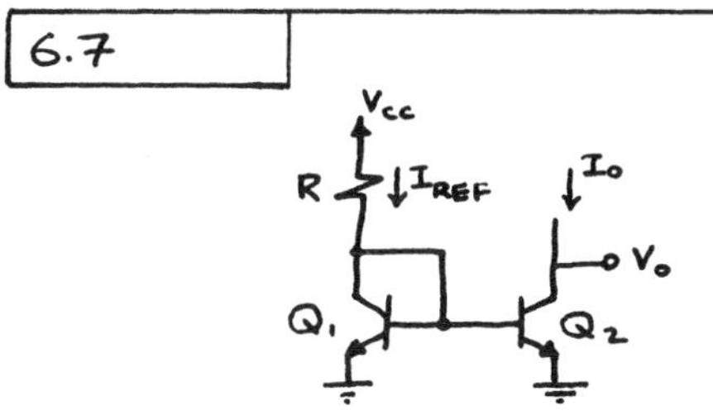

From equation (6.26) we have:

$$I_o = \frac{I_{REF}}{1+(2/\beta)}\left(1+\frac{V_o - V_{BE}}{V_A}\right) \Rightarrow$$

$$0.5mA = \frac{I_{REF}}{1+2/100}\left(1+\frac{2-0.7}{50}\right) \Rightarrow$$

$$I_{REF} = 0.5\,mA\,\frac{1.02}{1.026} = 0.497\,mA$$

$$I_{REF} = \frac{V_{CC}-V_{BE}}{R} \Rightarrow R = \frac{V_{CC}-V_{BE}}{I_{REF}}$$

$$R = \frac{5-0.7}{0.497\,mA} = \frac{4.3}{0.497} = 8.65\,k\Omega$$

$$V_{omin} = V_{CESAT} = 0.3\,V$$

For $V_o = 5v$, from equation (6.26) we have:

$$I_o = \frac{I_{REF}}{1+(2/\beta)}\left(1+\frac{V_o-V_{BE}}{V_A}\right)$$

$$I_o = \frac{0.497}{1+2/100}\left(1+\frac{5-0.7}{50}\right) = 0.53\,mA$$

6.8

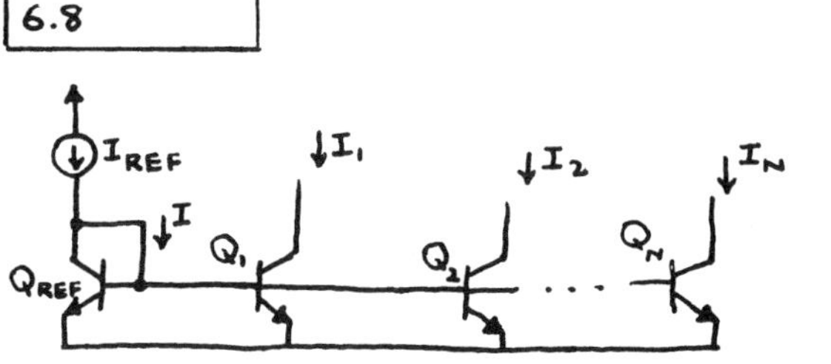

Ignoring the effect of finite output resistances, we have

$$I_1 = I_2 = \ldots = I_N = I_{CQ_{REF}}$$

$$I_{CQ_{REF}} + I = I_{REF} \quad (*)$$

$$I = I_{BQ_{REF}} + I_{B1} + \cdots + I_{BN}$$

$$I = \frac{I_{CQREF}}{\beta} + \frac{I_{C1}}{\beta} + \cdots + \frac{I_{CN}}{\beta}$$

$$I = I_{CQREF}\underbrace{\left(\frac{1}{\beta}+\frac{1}{\beta}+\cdots+\frac{1}{\beta}\right)}_{N+1\ \text{times}}$$

$$I = I_{CQREF}\frac{N+1}{\beta}$$

From (*) we have:

$$I_{CQREF} + I = I_{REF} \Rightarrow$$

$$I_{REF} = I_{CQREF} + I_{CQREF}\frac{N+1}{\beta}$$

$$\Rightarrow I_{CQREF} = \frac{I_{REF}}{1+\frac{N+1}{\beta}}$$

Thus:

$$I_1 = I_2 = \ldots = I_N = \frac{I_{REF}}{1+\frac{N+1}{\beta}}$$

For an error not exceeding 10% we need:

$$\frac{I_{REF}}{1+\frac{N+1}{\beta}} \geqslant I_{REF}(1-0.1)$$

$$\frac{I_{REF}}{1+\frac{N+1}{\beta}} \geqslant 0.9\,I_{REF} \Rightarrow \frac{1}{1+\frac{N+1}{\beta}} \geqslant 0.9$$

$$\Rightarrow 1+\frac{N+1}{\beta} \leqslant \frac{1}{0.9} \Rightarrow 1+\frac{N+1}{\beta} \leqslant 1.11$$

$$\frac{N+1}{\beta} \leqslant 0.11 \Rightarrow N+1 \leqslant 0.11\beta \Rightarrow$$

$$N+1 \leqslant 11 \Rightarrow N \leqslant 10$$

The maximum number of outputs for

CONT.

an error not exceeding 10% is $N=10$. If the error is required to be less than 10% then we need $N<10$. In this case the maximum number of outputs for an error of less than 10% is $N=9$.

6.9

Using equations (6.29) and (6.31), we can write the general form of the transfer function of a direct-coupled amplifier as:

$A(s) = \dfrac{A_{dc}}{1+\frac{s}{2\pi f_{3dB}}}$ where A_{dc} is the dc gain of the amplifier and f_{3dB} is the upper 3dB frequency of the amplifier. In this case we have $A_{dc}=1000$ and $f_{3dB}=100\text{ KHz}=100\times10^3\text{ Hz}=10^5\text{ Hz}$

Therefore, $A(s)=\dfrac{1000}{1+\frac{s}{2\pi\times10^5}}$

The gain-bandwidth product in Hertz is: $1000\times10^5=10^8\text{ Hz}$

6.10

For this amplifier we have:

$$H(s)=\frac{A_M}{(1+\frac{s}{\omega_{p1}})(1+\frac{s}{\omega_{p2}})}$$

By definition at $\omega=\omega_H$ we have

$$|H(j\omega_H)|^2=\frac{A_M^2}{2}\Rightarrow$$

$$\frac{A_M^2}{(1+(\frac{\omega_H}{\omega_{p1}})^2)(1+(\frac{\omega_H}{\omega_{p2}})^2)}=\frac{A_M^2}{2}\Rightarrow$$

$$\left[1+\left(\frac{\omega_H}{\omega_{p1}}\right)^2\right]\left[1+\left(\frac{\omega_H}{\omega_{p2}}\right)^2\right]=2$$

If $\omega_{p2}=K\omega_{p1}$ and $\omega_H=0.9\,\omega_{p1}$, then

$$\left[1+\left(\frac{0.9\omega_{p1}}{\omega_{p1}}\right)^2\right]\left[1+\left(\frac{0.9\omega_{p1}}{K\omega_{p1}}\right)^2\right]=2$$

$$(1+0.9^2)(1+(\frac{0.9}{K})^2)=2$$

$$1+(\frac{0.9}{K})^2=1.1\Rightarrow(\frac{0.9}{K})^2=0.1\Rightarrow K=2.78$$

If $\omega_H=0.99\,\omega_{p1}$, then:

$$\left[1+\left(\frac{0.99\omega_{p1}}{\omega_{p1}}\right)^2\right]\left[1+\left(\frac{0.99\omega_{p1}}{K\omega_{p1}}\right)^2\right]=2$$

$$(1+0.99^2)(1+(\frac{0.99}{K})^2)=2\Rightarrow$$

$$1+(\frac{0.99}{K})^2=1.01\Rightarrow(\frac{0.99}{K})^2=0.01\Rightarrow$$

$K=9.88$

6.11

From Exercise (6.10) we have:

$$\left[1+\left(\frac{\omega_H}{\omega_{p1}}\right)^2\right]\left[1+\left(\frac{\omega_H}{\omega_{p2}}\right)^2\right]=2 \text{ and } \omega_{p2}=K\omega_{p1}$$

$$K=1\Rightarrow\left[1+\left(\frac{\omega_H}{\omega_{p1}}\right)^2\right]\left[1+\left(\frac{\omega_H}{\omega_{p1}}\right)^2\right]=2$$

$$1+(\frac{\omega_H}{\omega_{p1}})^2=\sqrt{2}\Rightarrow(\frac{\omega_H}{\omega_{p1}})^2=\sqrt{2}-1$$

$\omega_H=\sqrt{\sqrt{2}-1}\;\omega_{p1}=0.64\,\omega_{p1}$ (exact value)

Using Equation 6.36 (note that in this case the zeros are at $s=\infty$) we have:

$$\omega_H=1\Big/\sqrt{\left(\frac{1}{\omega_{p1}^2}+\frac{1}{\omega_{p2}^2}\right)}=1\Big/\sqrt{\frac{1}{\omega_{p1}^2}+\frac{1}{K^2\omega_{p1}^2}}$$

$$\omega_H=\omega_{p1}\Big/\sqrt{1+\frac{1}{K^2}}$$

For $K=1\Rightarrow\omega_H=\frac{1}{\sqrt{2}}\omega_{p1}=0.71\,\omega_{p1}$

For the case of $K=2$, the exact value of ω_H can be found from the following

CONT.

equation:

$$\left[1+\left(\frac{\omega_H}{\omega_{P1}}\right)^2\right]\left[1+\left(\frac{\omega_H}{K\omega_{P1}}\right)^2\right]=2$$

Assuming $\frac{\omega_H}{\omega_{P1}}=X$ we have

$$(1+X^2)\left(1+\frac{X^2}{k^2}\right)=2 \Rightarrow$$

$$\frac{1}{k^2}X^4+\left(1+\frac{1}{k^2}\right)X^2+1=2 \Rightarrow$$

$$X^4+(k^2+1)X^2-k^2=0 \Rightarrow$$

$$X^2=\frac{-(k^2+1)+\sqrt{(k^2+1)^2+4k^2}}{2}$$

$$\frac{\omega_H}{\omega_{P1}}=\sqrt{\frac{-(k^2+1)+\sqrt{(k^2+1)^2+4k^2}}{2}} \quad (*)$$

For $k=2 \Rightarrow \frac{\omega_H}{\omega_{P1}}=0.84 \Rightarrow \omega_H=0.84\omega_{P1}$

In this case, the approximate value of ω_H is:

$$\omega_H=\omega_{P1}\Big/\sqrt{1+\frac{1}{k^2}}=0.89\omega_{P1}$$

For $k=4$, using equation(*), the exact value of ω_H is:

$\omega_H=0.95\omega_{P1}$

In this case, the approximate value of ω_H is:

$$\omega_H=\omega_{P1}\Big/\sqrt{1+\frac{1}{k^2}}=0.97\omega_{P1}$$

6.12

For the amplifier in Example 6.6 we have $A_M=-10.8$ V/V and $f_H \simeq 128.3$ kHz, therefore, the gain-bandwidth product is: $10.8\times128.3=1.3856\text{ MHz}\simeq 1.39\text{ MHz}$

Now we want to find the value of R'_L that will result in $f_H=180$ kHz. Referring to the solution of Example 6.6, we have:

$$\tau_{gs}+\tau_{gd}\simeq\frac{1}{\omega_H}=\frac{1}{2\pi f_H}$$

$$\tau_{gs}+\tau_{gd}=\frac{1}{2\pi\times180\text{kHz}}=884.2\text{ nsec}$$

$\tau_{gs}=80.8\text{ nsec} \Rightarrow \tau_{gd}=884.2-80.8$

$\tau_{gd}=803.4$ nsec

$\tau_{gd}=R_{gd}C_{gd}=(R'+R'_L+g_mR'_LR')C_{gd}$

$R'=R_{in}\|R_{sig}=80.8\text{ k}\Omega$, $g_m=4\frac{\text{mA}}{\text{V}}$, $C_{gd}=1\text{pF}$

Thus

$$803.4\text{ nsec}=(80.8\text{k}\Omega+R'_L+323.2R'_L)\times1\text{pF}$$

$$\Rightarrow 324.2R'_L=\frac{803.4\text{nsec}}{1\text{pF}}-80.8\text{k}\Omega$$

$$\Rightarrow R'_L=\frac{722.6\text{k}\Omega}{324.2}=2.23\text{k}\Omega$$

$\Rightarrow R'_L=2.23\text{k}\Omega$

For this value of R'_L we have

$$A_M=-\frac{R_{in}}{R_{in}+R_{sig}}(g_mR'_L)$$

$$A_M=-\frac{420}{420+100}\times4\times2.23=-7.2\text{ V/V}$$

Therefore, the gain-bandwidth product is: $7.2\times180\text{ kHz}=1.296\text{ MHz}\simeq1.3\text{ MHz}$

6.13

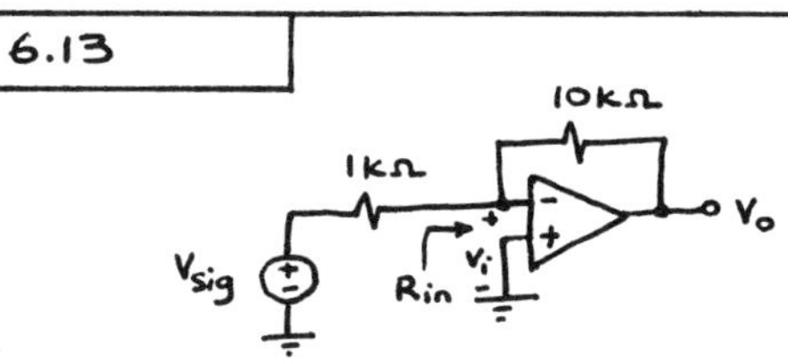

Using Miller's theorem we have

$R_{in}=\frac{10\text{k}\Omega}{A+1}$, $V_i=\frac{R_{in}}{R_{in}+1\text{k}\Omega}V_{sig}$ and $V_o=-AV_i$

Asumming $V_{sig}=1$V we have:

A (V/V)	R_{in} (Ω)	V_i (mV)	V_o (V)	V_o/V_{sig} (V/V)
10	909	476	-4.76	-4.76
100	99	90	-9	-9
1000	9.99	9.9	-9.9	-9.9
10000	1	0.999	-9.99	-9.99

6.14

$$g_m = \frac{2I_D}{V_{ov}} = \frac{2 \times 100\mu A}{0.25V} = 0.8\ \frac{mA}{V}$$

$$r_o = \frac{V_A' L}{I_D} = \frac{20 \times 0.4}{100\mu A} = 80\ k\Omega$$

$$A_o = g_m r_o = 0.8\ \frac{mA}{V} \times 80k\Omega = 64\ \frac{V}{V}$$

$$I_D = \frac{1}{2} k_n' \frac{W}{L} V_{ov}^2 \Rightarrow 100\mu A = \frac{200 \frac{\mu A}{V^2} \frac{W}{0.4\mu m} (0.25)^2}{2}$$

$W = 6.4\mu m$

For $L = 0.8\mu m$ we have:

$$g_m = \frac{2I_D}{V_{ov}} = \frac{2 \times 100\mu A}{0.25V} = 0.8\ \frac{mA}{V}$$

$$r_o = \frac{V_A' L}{I_D} = \frac{20 \times 0.8}{100\mu A} = 160k\Omega$$

$$A_o = g_m r_o = 0.8\ \frac{mA}{V} \times 160k\Omega = 128\ \frac{V}{V}$$

$$I_D = \frac{1}{2} k_n' \frac{W}{L} V_{ov}^2 \Rightarrow W = \frac{2 I_D L}{k_n' V_{ov}^2}$$

$$W = \frac{2 \times 100\mu A \times 0.8\mu m}{200 \frac{\mu A}{V^2} \times (0.25)^2} = 12.8\mu m$$

6.15

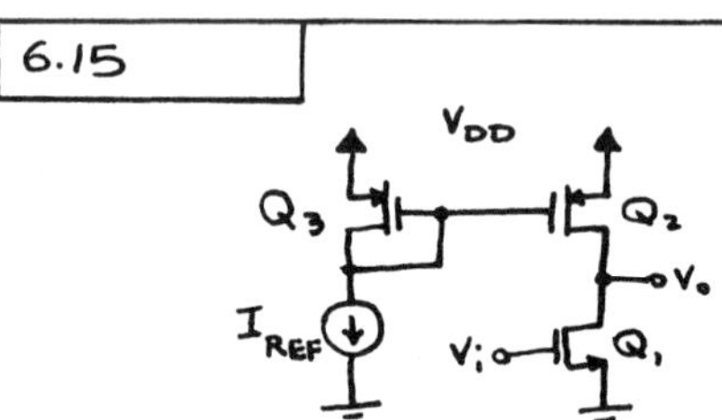

Since all transistors, in particular Q_3 and Q_2 have the same $\frac{W}{L} = \frac{7.2\mu m}{0.36\mu m}$ we have

$$I_{REF} = I_{D3} = I_{D2} = I_{D1} \Rightarrow I_{D1} = I_{REF} = 100\mu A$$

$$g_{m1} = \sqrt{2\mu_n C_{ox} \left(\frac{W}{L}\right)_1 I_{D1}} = \sqrt{2 k_n' \left(\frac{W}{L}\right)_1 I_{D1}}$$

$$g_{m1} = \sqrt{2 \times 387\ \mu A/V^2 \times \frac{7.2}{0.36} \times 100\mu A} \simeq 1.25\ \frac{mA}{V}$$

$$r_{o1} = \frac{V_{An}' \times L_1}{I_{D1}} = \frac{5 \times 0.36}{0.1mA} = 18k\Omega$$

$$r_{o2} = \frac{|V_{Ap}| L_2}{I_{D2}} = \frac{6 \times 0.36}{0.1mA} = 21.6k\Omega$$

$$A_v = -g_{m1}(r_{o1} \| r_{o2}) = -1.25 \times (18k\Omega \| 21.6\ k\Omega)$$

$$A_v = -1.25 \times 9.82 = -12.3\ V/V$$

6.16

$$I = 0.1mA,\quad R_i = r_\pi = \frac{\beta}{g_m} = \frac{\beta V_T}{I}$$

$$R_i = \frac{100 \times 25mV}{0.1mA} = 25k\Omega$$

$$r_{o1} = \frac{V_A}{I} = \frac{50V}{0.1mA} = 500k\Omega = 0.5M\Omega$$

$$r_{o2} = \frac{V_A}{I} = \frac{50V}{0.1mA} = 500k\Omega = 0.5M\Omega$$

$$g_m = \frac{I}{V_T} = \frac{0.1mA}{25mV} = 4\ \frac{mA}{V}$$

$$A_o = g_m r_{o1} = 4\ \frac{mA}{V} \times 500k\Omega = 2000\ \frac{V}{V}$$

$$A_v = -g_m(r_{o1} \| r_{o2}) = -4\ (500 \| 500)$$

$$A_v = -1000\ \frac{V}{V}$$

6.17

Referring to the solution of Example 6.9 the value of f_H determined by the exact analysis is:

$f_H = f_{P1} = 145.3\ MHz$

Also, $A_M = -12.3\ V/V$, therefore, the gain-bandwidth product (GBW) is:

$12.3 \times 145.3 \simeq 1.79\ GHz$

Since GBW is less than $f_{P2} = 2.45\ GHz$ and $f_Z = 40\ GHz$, therefore it is a good approximation for the unity gain frequency.

6.18

Referring to the solution of Example 6.9, if a load resistor is connected at the output, halving the value of R'_L, then we have $R'_L = \frac{r_{o1} \| r_{o2}}{2}$ and therefore

$$|A_M| = g_m \frac{r_{o1} \| r_{o2}}{2} = 1.25 \times \frac{9.82}{2} \simeq 6.15 \frac{V}{V}$$

Using equation 6.66 and assuming $f_H \simeq f_{P1}$ we have

$$f_H \simeq \frac{1}{2\pi\{[C_{gs} + C_{gd}(1 + g_m R'_L)]R_{sig} + (C_L + C_{gd})R'_L\}}$$

$$f_H \simeq \frac{1}{2\pi\{[20fF + 5fF(1 + 6.15)]10k\Omega + (25+5)\frac{9.82}{2}\}}$$

$$f_H \simeq 226\text{MHz}$$

$$f_t \simeq |A_M| f_H = 6.15 \times 226\text{MHz} = 1.39\text{GHz}$$

6.19

Referring to the solution of Example 6.9, if the amplifying transistor is operated at double the value of V_{ov} by using $I_D = 400\mu A$, we have:

$$V_{ov} = 2 \times 0.16 = 0.32\text{V}$$

$$g_m = \frac{I_D}{V_{ov}/2} = \frac{400\mu A}{0.32/2} = 2.5\text{mA/V}$$

$$r_{o1} = \frac{V'_{An} \times L_1}{I_D} = \frac{5 \times 0.36}{0.4\text{mA}} = 4.5\text{k}\Omega$$

$$r_{o2} = \frac{V'_{Ap} \times L_2}{I_D} = \frac{6 \times 0.36}{0.4\text{mA}} = 5.4\text{k}\Omega$$

$$R'_L = r_{o1} \| r_{o2} = 2.46\text{k}\Omega$$

$$|A_M| = g_m R'_L = 2.5 \times 2.46 = 6.15 \frac{V}{V}$$

Using equation 6.66 and assuming $f_H \simeq f_{P1}$ we have

$$f_H \simeq \frac{1}{2\pi\{[C_{gs} + C_{gd}(1 + g_m R'_L)]R_{sig} + (C_L + C_{gd})R'_L\}}$$

$$f_H \simeq \frac{1}{2\pi\{[20fF + 5fF(1 + 6.15)]10k\Omega + (25+5) \times 2.46\}}$$

$$f_H = 252\text{MHz} \Rightarrow f_{P1} \simeq f_H = 252\text{MHz}$$

$$f_t \simeq |A_M| f_H = 6.15 \times 252\text{MHz} = 1.55\text{GHz}$$

6.20

$$r_{o\,npn} = \frac{V_{An}}{I} = \frac{130V}{1mA} = 130\text{k}\Omega$$

$$r_{o\,pnp} = \frac{|V_{Ap}|}{I} = \frac{50V}{1mA} = 50\text{k}\Omega$$

$$R'_L = r_{o\,npn} \| r_{o\,pnp} = 130\text{k}\Omega \| 50\text{k}\Omega$$

$$R'_L = 36\text{k}\Omega$$

$$g_m = \frac{I}{V_T} = \frac{1mA}{25mV} = 40\frac{mA}{V}$$

$$r_\pi = \frac{\beta}{g_m} = \frac{200}{40\frac{mA}{V}} = 5\text{k}\Omega$$

(a)

From equation (6.70) we have:

$$A_M = -\frac{r_\pi}{R_{sig} + r_x + r_\pi}(g_m R'_L)$$

$$= -\frac{5}{36 + 0.2 + 5}(40 \times 36k\Omega) \simeq -175\frac{V}{V}$$

$$A_M = -175\frac{V}{V}$$

(b) Using Miller's theorem we have:

$$C_{in} = C_\pi + C_\mu(1 + g_m R'_L)$$

$$= 16pF + 0.3pF(1 + 40 \times 36) = 448^{pF}$$

$$C_{in} = 448\text{ pF}$$

$$f_H \simeq \frac{1}{2\pi C_{in} R'_{sig}} = \frac{1}{2\pi C_{in}[r_\pi \| (R_{sig} + r_x)]}$$

$$f_H = \frac{1}{2\pi \times 448pF \underbrace{[5 \| (36 + 0.2)]}_{\simeq 4.3k\Omega}} \simeq 82.6\text{KHz}$$

(c) Using the method of open-circuit time constants, from equation (6.73) we have:

$$\tau_H = C_\pi R'_{sig} + C_\mu[(1 + g_m R'_L)R'_{sig} + R'_L] + C_L R'_L$$

CONT.

We have $R'_{sig} = r_\pi \| (R_{sig} + r_x) \simeq 4.3\ k\Omega$

$R'_L = r_{onpn} \| r_{opnp} = 36\ k\Omega$

Thus:

$$\tau_H = 16 \times 4.3 + 0.3[(1 + 40 \times 36)\,4.3 + 36] + 5 \times 36$$

$$\tau_H = 2.12\text{nsec}$$

$$f_H = \frac{1}{2\pi\tau_H} = 75.1\ kHz$$

(d) Using equations (6.75), (6.76), and (6.77) we have:

$$f_Z = \frac{1}{2\pi}\frac{g_m}{C_\mu} = \frac{1}{2\pi}\frac{40\frac{mA}{V}}{0.3pF} = 21.2\ GHz$$

$$f_{P1} \simeq \frac{1}{2\pi}\frac{1}{[C_\pi + C_\mu(1 + g_m R'_L)]R'_{sig} + (C_L + C_\mu)R'_L}$$

$$\Rightarrow f_{P1} = 75.1\ kHz$$

$$f_{P2} \simeq \frac{1}{2\pi}\frac{[C_\pi + C_\mu(1 + g_m R'_L)]R'_{sig} + (C_L + C_\mu)R'_L}{[C_\pi(C_L + C_\mu) + C_L C_\mu]R'_{sig}R'_L}$$

$$f_{P2} = 25.2\ MHz$$

Since $f_{P1} \ll f_Z$ and $f_{P1} \ll f_{P2}$, thus $f_H \simeq f_{P1} = 75.1\ kHz$

(e) $f_+ = |A_M|\, f_H = 175 \times 75.1\ kHz = 13.1\ MHz$

$f_+ = 13.1\ MHz$

6.21

Referring to the solution of Example 6.10 we have $f_+ = |A_M|\, f_H$. Since $|A_M|$ remains the same as that of the example, to place f_+ at 2 GHz we need

$$2\ GHz = f_+ = |A_M|\, f_H = \frac{|A_M|}{2\pi(C_L + C_{gd})R'_L}$$

$$\Rightarrow C_L = \frac{|A_M|}{2\pi R'_L f_+} - C_{gd} = \frac{12.3}{2\pi \times 9.82\ k\Omega \times 2\ GHz} - 5fF$$

$$\Rightarrow C_L \simeq 94.4\ fF$$

6.22

For a CS amplifier fed with $R_{sig} = 0$ we know (equation 6.80) that:

$$f_+ = \frac{g_m}{2\pi(C_L + C_{gd})}$$

and $f_Z = \dfrac{g_m}{2\pi C_{gd}}$

Therefore,

$$\frac{f_Z}{f_+} = \frac{g_m/(2\pi C_{gd})}{g_m/[2\pi(C_L + C_{gd})]} = \frac{C_L + C_{gd}}{C_{gd}}$$

$$\frac{f_Z}{f_+} = \frac{C_L}{C_{gd}} + 1 \Rightarrow \frac{f_Z}{f_+} = 1 + \frac{C_L}{C_{gd}}$$

6.23

With the capacitance $C_L = 5fF$ at the output, from equation 6.109 we know:

$$\tau_H = \frac{1}{2\pi f_H} = C_{gs}R_{gs} + (C_{gd} + C_L)R_{gd}$$

$$\tau_H = \underbrace{C_{gs}R_{gs} + C_{gd}R_{gd}}_{\tau_H \text{ in Example 6.11}} + C_L R_{gd}$$

Thus $\tau_H = 435ps + 5fF \times 75k\Omega$

$\tau_H = 435ps + 375ps = 810ps$

$$f_H = \frac{1}{2\pi\tau_H} \simeq 196\ MHz$$

6.24

Referring to the solution of Example 6.11 we have:

$$g_m + g_{mb} = 1.25 + 0.2 \times 1.25 = 1.5\ \frac{mA}{V}$$

$$A_{vo} = 1 + (g_m + g_{mb})r_o = 1 + 1.5 \times 18 = 28\ \frac{V}{V}$$

$$R_{in} = \frac{r_o + R_L}{A_{vo}} = \frac{18 + 10}{28} = 1\ k\Omega$$

$$R_{out} = r_o + A_{vo}R_s = 18 + 28 \times 1 = 46\ k\Omega$$

$$G_v = G_{vo}\frac{R_L}{R_L + R_{out}} = A_{vo}\frac{R_L}{R_L + R_{out}} = 28\,\frac{10}{10 + 46}$$

CONT.

$$G_v = 28 \times \frac{10}{56} = 5\frac{V}{V}$$

$$G_{is} = \frac{A_{vo} R_s}{R_{out}} = \frac{28 \times 1}{46} = 0.61 \text{ A/A}$$

$$G_i = G_{is} \frac{R_{out}}{R_{out} + R_L} = 0.61 \frac{46}{46 + 10} = 0.5 \text{ A/A}$$

$R_{gs} = R_s \| R_{in} = 1k\Omega \| 1k\Omega = 0.5k\Omega$

$R_{gd} = R_L \| R_{out} = 10k\Omega \| 46k\Omega = 8.2k\Omega$

$\tau_H = C_{gs} R_{gs} + C_{gd} R_{gd} = 20 \times 0.5 + 5 \times 8.2$

$\tau_H = 51 \text{ ps}$

$$f_H \simeq \frac{1}{2\pi \tau_H} = \frac{1}{2\pi \times 51\text{ps}} = 3.1 \text{ GHz}$$

6.25

We have: $r_o = \frac{V_A}{I} = \frac{100V}{1mA} = 100k\Omega$

$$g_m = \frac{I}{V_T} = \frac{1mA}{25mV} = 40 \frac{mA}{V}$$

$$r_\pi = \frac{\beta}{g_m} = \frac{100}{40} = 2.5k\Omega$$

$$r_e = \frac{r_\pi}{\beta + 1} = 24.8\Omega$$

From equation (6.111) we have

$$R_{in} = \frac{r_o + R_L}{1 + \frac{r_o}{r_e} + \frac{R_L}{(\beta+1) r_e}} = \frac{100k\Omega + 1000k\Omega}{1 + \frac{100k\Omega}{24.8\Omega} + \frac{1000k\Omega}{2.5k\Omega}}$$

$R_{in} \simeq 250\Omega$

$A_{vo} = 1 + g_m r_o = 1 + 40 \times 100 = 4001 \text{ V/V}$

$R_o = r_o = 100k\Omega$

$$A_v = A_{vo} \frac{R_L}{R_L + R_o} = 4001 \times \frac{1000}{1000 + 100} = 3637 \frac{V}{V}$$

$R_{out} = r_o + (1 + g_m r_o) R'_e$ and $R'_e = R_e \| r_\pi$

$R_{out} = 100 + (1 + 40 \times 100)(1 \| 2.5) \simeq 2.94 M\Omega$

$$G_{vo} = \frac{r_\pi}{r_\pi + R_e} A_{vo} = \frac{2.5}{2.5 + 1} \times 4001 \simeq 2858 \frac{V}{V}$$

$$G_v = G_{vo} \frac{R_L}{R_L + R_{out}} = 2858 \times \frac{1000}{1000 + 2940} \simeq 727 \frac{V}{V}$$

$$G_v = \frac{V_o}{V_{sig}} \Rightarrow V_o = G_v V_{sig}$$

Thus for a 5mV peak sine wave at the V_{sig}, the output is a sine wave with $727 \times 5mV \simeq 3.64$ V peak.

6.26

Referring to Figure 6.36(a) the minimum value of V_{BIAS} required for a cascode amplifier is:

$V_{BIAS} = V_{ov1} + V_{GS2}$

Since $V_{GS1} = V_{GS2} = V_{GS}$ and $V_{ov1} = V_{ov2} = V_{ov}$ we can write:

$V_{BIAS} = V_{ov} + V_{GS} = V_{ov} + V_{ov} + V_{tn} = 2V_{ov} + V_{tn}$

$$I_D = \frac{1}{2} \mu_n C_{ox} \frac{W}{L} V_{ov}^2 \Rightarrow V_{ov} = \sqrt{\frac{2I_D}{\mu_n C_{ox} \frac{W}{L}}}$$

$V_{ov} = 0.26V$

$V_{BIAS} = 2V_{ov} + V_{tn} = 2 \times 0.26 + 0.6 = 1.12V$

6.27

$A_{o1} = g_{m1} r_{o1}$, Assuming $I_D = 100\mu A$ we have:

$$g_{m1} = \sqrt{2\mu_n C_{ox} \frac{W}{L} I_D} = 0.62 \frac{mA}{V}$$

$$r_{o1} = \frac{V_A}{I_D} = \frac{V'_A L}{I_D} = \frac{20 \times 0.5}{0.1mA} = 100k\Omega$$

$$A_{o1} = g_{m1} r_{o1} = 0.62 \times 100 = 62 \frac{V}{V}$$

$$A_{vo2} = 1 + (g_{m2} + g_{mb2}) r_{o2}$$

$g_{m2} = g_{m1} = 0.62 \frac{mA}{V}$, $g_{mb2} = \chi g_{m2} = 0.124 \frac{mA}{V}$

$r_{o2} = r_{o1} = 100k\Omega$

$A_{vo2} \simeq 75 \text{ V/V}$

$A_{vo} = -A_{o1} A_{vo2} = -62 \times 75 = -4650 \frac{V}{V}$

CONT.

$R_{out1} = r_{o1} = 100\,k\Omega$

$$R_{in2} = \frac{1}{g_{m2}+g_{mb2}} + \frac{R_L}{A_{vo2}}$$

$R_L = R_{out}$ and $R_{out} = r_{o2} + A_{vo2} r_{o1} \Rightarrow$

$R_{out} = 100 + 75 \times 100 = 7.6\,M\Omega$

$R_L = R_{out} = 7.6\,M\Omega$

$$R_{in2} = \frac{1}{g_{m2}+g_{mb2}} + \frac{R_L}{A_{vo2}} = \frac{1}{0.62+0.2\times0.62} + \frac{7.6M}{75}$$

$R_{in2} \simeq 103\,k\Omega$

$$R_{d1} = r_{o1} \| \left[\frac{1}{g_{m2}+g_{mb2}} + \frac{R_L}{A_{vo2}}\right] = r_{o1} \| R_{in2}$$

$R_{d1} = 100 \| 103 = 50.7\,k\Omega$

$$A_v = A_{vo}\frac{R_L}{R_L+R_{out}} \text{ and } R_L = R_{out} \Rightarrow$$

$$A_v = A_{vo}\frac{R_L}{R_L+R_L} = \frac{A_{vo}}{2} = \frac{-4650}{2} = -2325\frac{V}{V}$$

Using equation 6.139 we have:

$$f_t \simeq \frac{1}{2\pi}\frac{g_m}{C_L+C_{gd}+C_{db}} = \frac{1}{2\pi}\frac{0.62\frac{mA}{V}}{5fF+2fF+3fF}$$

$\Rightarrow f_t = 9.9\,GHz$

$$f_t \simeq f_H \times |A_v| \Rightarrow f_H \simeq \frac{f_t}{|A_v|} = \frac{9.9}{2325} = 4.3\,MHz$$

6.28

Referring to the solution of Exercise 6.20 we have $g_m = 40\frac{mA}{V}$ and $r_\pi = 5\,k\Omega$

Note that for the cascode amplifier considered in this exercise:

$r_{\pi1} = r_{\pi2} = r_\pi = 5\,k\Omega$ and $g_{m1} = g_{m2} = g_m = 40\frac{mA}{V}$

$R_{in} = r_{\pi1} + r_x = 5\,k\Omega + 0.2\,k\Omega = 5.2\,k\Omega$

$A_o = g_m r_o = 40 \times 130 = 5200\frac{V}{V}$

$R_{out1} = r_{o1} = r_o = 130\,k\Omega$

$$R_{in2} \simeq r_{e2}\frac{r_{o2}+R_L}{r_{o2}+\frac{R_L}{\beta+1}} = \frac{5k}{200+1} \times \frac{130+50}{130+\frac{50}{201}}$$

$R_{in2} \simeq 35\,\Omega$

$R_{out} \simeq \beta_2 r_{o2} = 200 \times 130\,k\Omega = 26\,M\Omega$

$$A_M \simeq -\frac{r_\pi}{r_\pi + r_x + R_{sig}} g_m(\beta r_o \| R_L)$$

$A_M \simeq -242\frac{V}{V}$

To calculate f_H we use the method of open-circuit time constants. From Figure 6.42 we have:

$R'_{sig} = r_{\pi1} \| (r_{x1} + R_{sig}) = 5k \| (0.2+36)$

$R'_{sig} = 4.4\,k\Omega$

$R_{\pi1} = R'_{sig} = 4.4\,k\Omega$

$$R_{o1} = r_{o1} \| \underbrace{\left[r_{e2}\left(\frac{r_{o2}+R_L}{r_{o2}+\frac{R_L}{\beta_2+1}}\right)\right]}_{R_{in2}}$$

$R_{o1} = 130k \| 35\,\Omega \simeq 35\,\Omega$

$R_{\mu1} = R'_{sig}(1 + g_{m1}R_{o1}) + R_{o1}$

$R_{\mu1} = 10.6\,k\Omega$

$$\tau_H = C_{\pi1}R_{\pi1} + C_{\mu1}R_{\mu1} + (C_{cs1}+C_{\pi2})R_{o1} + (C_L + C_{cs2} + C_{\mu2})(R_L \| R_{out})$$

$$\tau_H = 16pF \times 4.4k\Omega + 0.3pF \times 10.6k\Omega + (0+16pF)\times 35\Omega + (5pF + 0 + 0.3pF)(50k \| 26M\Omega)$$

$\tau_H = 339\,ns$

$$f_H = \frac{1}{2\pi\tau_H} = \frac{1}{2\pi \times 339ns} \simeq 469\,kHz$$

$f_t \simeq |A_M| f_H = 242 \times 469\,kHz \simeq 113.5\,MHz$

Compared to the CE amplifier in Exercise 6.20, $|A_M|$ has increased from $175\frac{V}{V}$ to $242\,V/V$, f_H has increased from 75.1 kHz to 469 kHz and f_t has increased from 13.1 MHz to 113.5 MHz. To increase f_H to 1 MHz we need $\tau_H = \frac{1}{2\pi f_H} = 159$ ns, thus,

$$16pF \times 4.4k\Omega + 0.3pF \times 10.6k\Omega + 16pF \times 35\Omega + (C_L + 0.3pF)(50k \| 26M\Omega) = 159ns$$

$\Rightarrow C_L = 1.4\,pF$

6.29

(a) $I_{D1} = I$ and $I_{D2} = I$. Since $V_{ov1} = V_{o2} = 0.2^{V}$ we have:

$$\frac{I_{D2}}{I_{D1}} = \frac{\frac{1}{2}\mu_p C_{ox} (\frac{W}{L})_2 V_{ov2}^2}{\frac{1}{2}\mu_n C_{ox} (\frac{W}{L})_1 V_{ov1}^2} = \frac{I}{I}$$

$$\Rightarrow \frac{k'_p (\frac{W}{L})_2}{k'_n (\frac{W}{L})_1} = 1 \Rightarrow (\frac{W}{L})_2 = \frac{k'_n}{k'_p} (\frac{W}{L})_1$$

$$\Rightarrow (\frac{W}{L})_2 = 4 (\frac{W}{L})_1$$

(b) Since current source I_1 is a simple single transistor circuit studied in Section 6.2, for proper operation it requires $V_{ov} = 0.2V$ across it.
If a 0.1 V peak-to-peak signal swing is to be allowed at the drain of Q_1, then the highest dc bias voltage that can be used at that point is:

$$V_{DD} - V_{ov} - \frac{1}{2} V_{peak\text{-}to\text{-}peak} = 1.8 - 0.2 - \frac{1}{2} \times 0.1 = 1.55V$$

(c) $V_{SG2} = V_{ov} + |V_{tp}| = 0.2 + 0.5 = 0.7V$
Therefore, the largest value to which V_{BIAS} can be set is $1.55 - 0.7 = 0.85V$

(d) Since current source I_2 is a telescopic cascode current source the minimum dc voltage required across it for its proper operation is $V_{ov} + V_{ov}$ that is $2V_{ov} = 2 \times 0.2 = 0.4V$

(e) From the results of parts (c) and (d) the allowable range of signal swing at the output is from 0.4V to $1.55 - V_{ov}$ that is 1.35V

6.30

For the circuit in Figure 6.46(a) we have: $G_m \simeq g_{m1} = \sqrt{2\mu_n C_{ox} (\frac{W}{L})_1 I_{D1}}$

$$\Rightarrow G_m \simeq \sqrt{2 \times 200 \frac{\mu A}{V^2} \times 10 \times 100 \mu A}$$

$$G_m \simeq 0.63 \frac{mA}{V}$$

$$R_{out} \simeq \beta_2 r_{o2} = 100 \times \frac{V_A}{I} = 100 \times \frac{50}{0.1} = 50 M\Omega$$

$$A_{vo} = -G_m R_{out} \simeq -3.2 \times 10^4 \frac{V}{V}$$

For the circuit in Figure 6.46(b) we have:

$$G_m \simeq g_{m1} = \frac{I}{V_T} = \frac{0.1mA}{25mV} = 4 \frac{mA}{V}$$

$$R_{out} \simeq A_{03} \times \beta_2 r_{o2} = g_{m3} r_{o3} \beta_2 r_{o2}$$

$$R_{out} \simeq 1575 M\Omega$$

$$A_{vo} = -G_m R_{out} = -6.3 \times 10^6 \frac{V}{V}$$

6.31

(a) $A_M = -g_m R'_L = -g_m (R_L \| r_o)$

$$A_M = -2 \frac{mA}{V} (20k\Omega \| 20k\Omega) = -2 \frac{mA}{V} \times 10k\Omega$$

$$A_M = -20 \frac{V}{V}$$

To calculate τ_H using the method of open circuit time constants we can write (equation 6.57):

$$\tau_H = C_{gs} R_{sig} + C_{gd} [R_{sig}(1 + g_m R'_L) + R'_L] + C_L R'_L$$

$$\tau_H = 20fF \times 20k\Omega + 5[20(1+20) + 10] + 5 \times 10$$

$$\tau_H = 2.6 ns \Rightarrow f_H = \frac{1}{2\pi \tau_H} = 61.2 MHz$$

gain-bandwidth product $= 20 \times 61.2 = 1.22GHz$

(b) With $R_S = \frac{2}{g_m + g_{mb}}$ connected in the source terminal we have:

CONT.

$R_{out} \simeq r_o[1+(g_m+g_{mb})R_S] = r_o(1+2) = 3r_o$

$R_{out} \simeq 3 \times 20k\Omega = 60k\Omega$

From equation (6.145) we know:

$$A_M = -A_{vo}\frac{R_L}{R_L+R_{out}} = -g_m r_o \frac{R_L}{R_L+R_{out}}$$

$$A_M = -2\times 20\,\frac{20}{20+60} = -10\,\frac{V}{V}$$

$$G_m = \frac{g_m}{1+(g_m+g_{mb})R_S} = \frac{g_m}{1+2} = \frac{g_m}{3} = \frac{2}{3}\,\frac{mA}{V}$$

Using equations 6.148 to 6.152 we have:

$R_L' = R_L \| R_{out} = 20k \| 60k = 15k\Omega$

$R_{gd} = R_{sig}(1+G_m R_L') + R_L' = 235k\Omega$

$$R_{gs} \simeq \frac{R_{sig}+R_S}{1+(g_m+g_{mb})R_S\frac{r_o}{r_o+R_L}} = \frac{R_{sig}+R_S}{1+2\frac{20}{20+20}}$$

$$R_{gs} \simeq \frac{R_{sig}+R_S}{2}$$

Assuming X is small, that is $g_{mb} \ll g_m$

then $R_S = \frac{2}{g_m+g_{mb}} \simeq \frac{2}{g_m} = 1k\Omega$

$$R_{gs} \simeq \frac{20+1}{2} = 10.5k\Omega$$

$$\tau_H = C_{gs}R_{gs} + C_{gd}R_{gd} + C_L R_L'$$

$\tau_H = 20fF \times 10.5 + 5fF \times 235 + 5\times 15$

$\tau_H = 1.46ns \Rightarrow f_H = \frac{1}{2\pi\tau_H} = 109\,MHz$

gain-bandwidth product $= 10\times 109MHz \simeq 1.1GHz$

6.32

$r_o = \frac{V_A}{I} = \frac{100V}{1mA} = 100k\Omega$, $g_m = \frac{I}{V_T} = 40\frac{mA}{V}$

$r_\pi = \frac{\beta}{g_m} = 2.5k\Omega$, $r_e = \frac{r_\pi}{\beta+1} = 24.75\Omega$

Since $R_L = 2r_o$ (thus R_L is on the order of r_o) and $R_e \ll r_o$, using equation 6.158 we have:

$$R_{in} \simeq (\beta+1)r_e + (\beta+1)R_e\frac{1}{1+\frac{R_L}{r_o}}$$

$$R_{in} \simeq 101\times 24.75\Omega + 101\times 75\Omega \times \frac{1}{1+\frac{2r_o}{r_o}}$$

$R_{in} \simeq 5k\Omega$

From equation (6.160) we have:

$R_o \simeq r_o(1+g_m R_e')$ where $R_e' = R_e \| r_\pi \simeq R_e$

$R_o \simeq 100(1+40\times 0.075) = 400k\Omega$

$A_{vo} \simeq -g_m r_o = -40\times 100 = -4000\frac{V}{V}$

$$G_m = \frac{g_m}{1+g_m R_e} = \frac{40}{1+40\times 0.075} = \frac{40}{4} = 10\frac{mA}{V}$$

$$\frac{V_o}{V_{sig}} = \frac{R_{in}}{R_{in}+R_{sig}}A_{vo}\frac{R_L}{R_L+R_{out}}, \text{ assuming } R_{out} \simeq R_o$$

$$\frac{V_o}{V_{sig}} = \frac{5k\Omega}{5k\Omega+5k\Omega}\times\left(-4000\frac{V}{V}\right)\frac{2\times 100}{2\times 100+400}$$

$$\frac{V_o}{V_{sig}} \simeq -667\frac{V}{V}$$

6.33

For the source follower to provide a dc level shift of 0.9V we need $V_{GS} = 0.9V$

Therefore, $V_{ov} = V_{GS} - V_{tn} = 0.9 - 0.6 = 0.3V$

$$I = \frac{1}{2}k_n'\left(\frac{W}{L}\right)V_{ov}^2 = \frac{1}{2}\times 200\frac{\mu A}{V^2}\left(\frac{20}{0.5}\right)(0.3)^2$$

$I = 360\mu A$

$$g_m = \frac{2I}{V_{ov}} = \frac{2\times 360\mu A}{0.3V} = 2.4\frac{mA}{V}$$

$$g_{mb} = X g_m = 0.2\times 2.4 = 0.48\frac{mA}{V}$$

$$r_o = \frac{V_A' L}{I} = \frac{20\times 0.5}{360\mu A} = 27.8\,k\Omega$$

$$A_{vo} = \frac{g_m r_o}{1+(g_m+g_{mb})r_o} = 0.82\frac{V}{V}$$

$$R_o = \frac{1}{g_m+g_{mb}} \| r_o = 343\Omega$$

$$A_v = A_{vo}\frac{R_L}{R_L+R_{out}} \simeq 0.79\,V/V$$

6.34

$g_{mb} = \chi g_m = 0.25 \frac{mA}{V}$

$R_L' = R_L \| r_o \| \frac{1}{g_{mb}} = 10k \| 20k \| \frac{1}{0.25}$

$R_L' = 2.5 k\Omega$

From equation (6.166) we have

$A_v = \frac{g_m R_L'}{1 + g_m R_L'} = \frac{1.25 \times 2.5}{1 + 1.25 \times 2.5} = 0.76 \frac{V}{V}$

$f_T = \frac{1}{2\pi} \frac{g_m}{C_{gs} + C_{gd}} = \frac{1}{2\pi} \frac{1.25 \frac{mA}{V}}{20fF + 5fF} \simeq 8 GHz$

$f_Z = \frac{1}{2\pi} \frac{g_m}{C_{gs}} = \frac{1}{2\pi} \frac{1.25}{20fF} \simeq 10 GHz$

$R_{gd} = R_{sig} = 20 k\Omega$

$R_{gs} = \frac{R_{sig} + R_L'}{1 + g_m R_L'} = \frac{20 + 2.5}{1 + 1.25 \times 2.5} = 5.45 k\Omega$

$R_{C_L} = R_L \| R_o = R_L \| \frac{1}{g_m + g_{mb}} \| r_o \simeq 0.61 k\Omega$

$\tau_{gd} = R_{gd} C_{gd} = 100 ps$

$\tau_{gs} = R_{gs} C_{gs} = 109 ps$

$\tau_{C_L} = R_{C_L} C_L \simeq 9 ps$

$\tau_H = \tau_{gd} + \tau_{gs} + \tau_{C_L} = 218 ps$

Percentage contribution of time constants associated with C_{gd}, C_{gs}, and C_L to τ_H are $\frac{100}{218} \simeq 46\%$, $\frac{109}{218} = 50\%$ and $\frac{9}{218} \simeq 4\%$

$f_H = \frac{1}{2\pi\tau_H} = \frac{1}{2\pi \times 218 ps} \simeq 730 MHz$

6.35

We have $r_\pi = \frac{\beta V_T}{I} = \frac{100 \times 25mV}{1mA} = 2.5k\Omega$

$g_m = \frac{I}{V_T} = \frac{1mA}{25mV} = 40 mA/V$

Using equation (5.144) noting that in this case $R_B \to \infty$, we have:

$G_v = \frac{(\beta+1)(r_o \| R_L)}{R_{sig}' + (\beta+1)[r_e + (r_o \| R_L)]}$ and $r_e = \frac{r_\pi}{\beta+1}$

$R_{sig}' = R_{sig} + r_x$ Assume $r_x = 100\Omega$

$\Rightarrow G_v \simeq 0.965 \frac{V}{V}$

From equation (6.178) we have:

$f_Z = \frac{1}{2\pi C_\pi r_e} = \frac{1}{2\pi C_\pi \frac{r_\pi}{\beta+1}} = \frac{\beta+1}{2\pi C_\pi r_\pi}$

We know $f_T = \frac{g_m}{2\pi(C_\pi + C_\mu)} \Rightarrow$

$C_\pi = \frac{g_m}{2\pi f_T} - C_\mu = \frac{40 \frac{mA}{V}}{2\pi \times 400MHz} - 2pF$

$C_\pi = 14 pF$

Thus $f_Z = \frac{\beta+1}{2\pi C_\pi r_\pi} = 459 MHz$

From equation (6.179) we have:

$R_\mu = R_{sig}' \| [r_\pi + (\beta+1) R_L']$

$R_{sig}' = R_{sig} + r_x = 1k\Omega + 0.1k\Omega = 1.1k\Omega$

$R_L' = R_L \| r_o = 1.1k\Omega \| 100k\Omega = 0.99 k\Omega$

$R_\mu = 1.1 k\Omega \| [2.5 + (100+1) 0.99]$

$R_\mu = 1.09 k\Omega$

From equation (6.180) we have:

$R_\pi = \frac{R_{sig}' + R_L'}{1 + \frac{R_{sig}'}{r_\pi} + \frac{R_L'}{r_e}} = \frac{1.1 + 0.99}{1 + \frac{1.1}{2.5} + \frac{0.99}{\frac{2.5}{101}}}$

$R_\pi \simeq 51\Omega$

From equation (6.181) we have:

$f_H = \frac{1}{2\pi(C_\mu R_\mu + C_\pi R_\pi)}$

$f_H = 55 MHz$

6.36

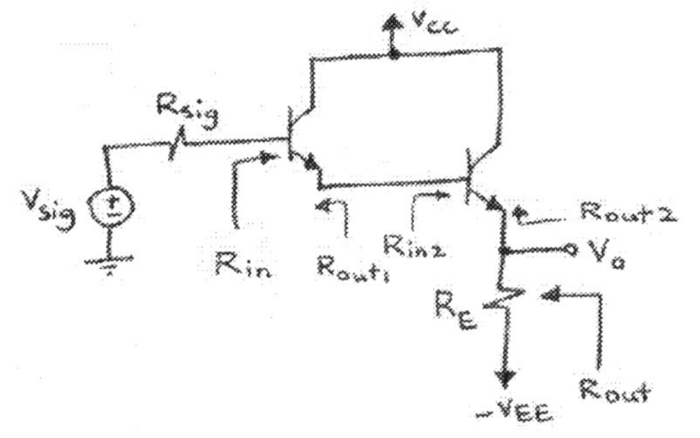

CONT

We have:

$$R_{in2} = r_{\pi 2} + (\beta_2+1)R_E = (\beta_2+1)(r_{e2}+R_E)$$

$$R_{in} = r_{\pi 1} + (\beta_1+1)R_{in2}$$
$$= (\beta_1+1)r_{e1} + (\beta_1+1)(\beta_2+1)(r_{e2}+R_E)$$

$$R_{in} = (\beta_1+1)\left[r_{e1} + (\beta_2+1)(r_{e2}+R_E)\right]$$

$$R_{out1} = \frac{r_{\pi 1}+R_{sig}}{\beta_1+1} = \frac{r_{\pi 1}}{\beta_1+1} + \frac{R_{sig}}{\beta_1+1} = r_{e1} + \frac{R_{sig}}{\beta_1+1}$$

$$R_{out2} = \frac{r_{\pi 2}+R_{out1}}{\beta_2+1} = r_{e2} + \frac{R_{out1}}{\beta_2+1}$$

$$R_{out2} = r_{e2} + \frac{r_{e1}+R_{sig}/(\beta_1+1)}{\beta_2+1}$$

$$R_{out} = R_E \| R_{out2}$$

$$R_{out} = R_E \| \left[r_{e2} + \frac{r_{e1}+R_{sig}/(\beta_1+1)}{\beta_2+1}\right]$$

$$\frac{V_o}{V_{sig}} = \frac{R_E}{R_E+R_{out2}} = \frac{R_E}{R_E+r_{e2}+\left[r_{e1}+\frac{R_{sig}}{\beta_1+1}\right]/(\beta_2+1)}$$

$$I_{E2} = 5\,mA \Rightarrow I_{E1} = I_{B2} = \frac{I_{E2}}{\beta+1} = 49.5\,\mu A$$

$$r_{e2} = \frac{V_T}{I_{E2}} = \frac{0.025\,V}{5\,mA} = 5\,\Omega$$

$$r_{e1} = \frac{V_T}{I_{E1}} = \frac{0.025\,V}{49.5\,\mu A} = 505\,\Omega$$

$R_{in} = 10.3\,M\Omega$, $R_{out} \simeq 20\,\Omega$ and

$$\frac{V_o}{V_{sig}} = 0.98\,\frac{V}{V}$$

6.37

From equation (6.182) we have:

$$R_{in} = (\beta_1+1)(r_{e1}+r_{e2})$$

Since in this case $r_{e1}=r_{e2}=r_e$ and $\beta_1=\beta_2=\beta$ we have

$$R_{in} = r_\pi + r_\pi = 2r_\pi = \frac{2\beta V_T}{I_c} \simeq 10\,k\Omega$$

From equation (6.185) we have:

$$\frac{V_o}{V_{sig}} = \frac{1}{2}\left(\frac{R_{in}}{R_{in}+R_{sig}}\right) g_m R_L$$

$$\frac{V_o}{V_{sig}} = \frac{1}{2}\left(\frac{10}{10+10}\right)\frac{V_T}{I_c} R_L = 50\,\frac{V}{V}$$

From equation (6.186) we have

$$f_{p1} = \frac{1}{2\pi\left(\frac{C_\pi}{2}+C_\mu\right)(R_{sig}\|2r_\pi)}$$

$$f_{p1} = \frac{1}{2\pi\left(\frac{6pF}{2}+2pF\right)(10k\|10k)} \simeq 6.4\,MHz$$

From equation (6.187) we have

$$f_{p2} = \frac{1}{2\pi C_\mu R_L} = \frac{1}{2\pi \times 2pF \times 10k} \simeq 8\,MHz$$

Therefore, the transfer function of this CC-CB amplifier is:

$$A(s) = \frac{A_M}{\left(1+\frac{s}{2\pi f_{p1}}\right)\left(1+\frac{s}{2\pi f_{p2}}\right)}$$

$$|A(s)|_{s=j\omega_H} = \frac{|A_M|}{\sqrt{2}} \quad \text{or} \quad |A(s)|^2_{s=j\omega_H} = \frac{A_M^2}{2}$$

Thus:

$$\frac{A_M^2}{\left(1+\frac{(2\pi f_H)^2}{(2\pi f_{p1})^2}\right)\left(1+\frac{(2\pi f_H)^2}{(2\pi f_{p2})^2}\right)} = \frac{A_M^2}{2}$$

$$\left(1+\frac{f_H^2}{f_{p1}^2}\right)\left(1+\frac{f_H^2}{f_{p2}^2}\right) = 2$$

Solving this equation for f_H we have: $f_H \simeq 4.6\,MHz$

Using the approximate formula in equation (6.188), we have:

$$f_H \simeq \frac{1}{\sqrt{\frac{1}{f_{p1}^2}+\frac{1}{f_{p2}^2}}} \simeq 5\,MHz$$

6.38

$I_D = I_{REF} = 100\mu A$

$I_D = \frac{1}{2}\mu_n C_{ox}(\frac{W}{L})V_{ov}^2 \Rightarrow V_{ov} = \sqrt{\frac{2I_D}{\mu_n C_{ox}(\frac{W}{L})}}$

$V_{ov} \simeq 0.23V$

The minimum dc voltage required at the output is $V_t + 2V_{ov} = 0.96\ v$

$r_{o2} = r_{o3} = \frac{V_A' \times L}{I_D} = \frac{5 \times 0.36}{0.1mA} = 18k\Omega$

$g_{m3} = \frac{I_D}{V_{ov}/2} = \frac{2 \times 0.1}{0.23} = 0.87\ \frac{mA}{V}$

From equation 6.190 we have

$R_o \simeq g_{m3} r_{o3} r_{o2} = 0.87 \times 18 \times 18 \simeq 282k\Omega$

6.39

For the Wilson mirror from equation (6.193) we have:

$\frac{I_o}{I_{REF}} = \frac{1}{1 + \frac{2}{\beta(\beta+2)}} = 0.9998$

Thus $\frac{|I_o - I_{REF}|}{I_{REF}} \times 100 = 0.02\%$

whereas for the simple mirror from equation (6.21) we have:

$\frac{I_o}{I_{REF}} = \frac{1}{1 + \frac{2}{\beta}} = 0.98$

Hence $\frac{|I_o - I_{REF}|}{I_{REF}} \times 100 = 2\%$

For the Wilson current mirror we have $R_o = \frac{\beta r_o}{2} = \frac{100 \times 100}{2} = 5M\Omega$ and for the simple mirror $R_o = r_o = 100k\Omega$

6.40

For the two current sources designed in Example 6.14, we have:

$g_m = \frac{I_C}{V_T} = \frac{10\mu A}{25mV} = 0.4\ \frac{mA}{V}$ and

$r_o = \frac{V_A}{I_C} = \frac{100v}{10\mu A} = 10M\Omega,\ r_\pi = \frac{\beta}{g_m} = 250k\Omega$

For the current source in Figure 6.63a we have

$R_o = r_{o2} = r_o = 10M\Omega$

For the current source in Figure 6.63b from equation (6.200) we have:

$R_o \simeq [1 + g_m(R_E \| r_\pi)]r_o$

In Example 6.14, $R_E = R_3 = 11.5k\Omega$, therefore,

$R_o \simeq [1 + 0.4\frac{mA}{V}(11.5k\Omega \| 250k\Omega)]10M\Omega$

$\Rightarrow R_o = 54M\Omega$

Chapter 7 - Exercises

7.1

(a) $V_{OV} = \sqrt{I / k_n'(W/L)}$

$= \sqrt{0.4/4} = \sqrt{0.1} = \underline{0.316\,V}$

$V_{GS} = V_{OV} + V_t$

$= 0.316 + 0.5 \cong \underline{0.82\,V}$

(b) Refer to Fig 7.3 (a)

$V_{S1} = V_{S2} = 0 - V_{GS} = \underline{-0.82\,V}$

$I_{D1} = I_{D2} = I/2 = \underline{0.2\,mA}$

$V_{D1} = V_{D2} = V_{DD} - \frac{I}{2} \times R_D$

$= 1.5 - 0.2 \times 2.5 = \underline{1\,V}$

(c) Refer to Fig 7.3 (b)

$V_{S1} = V_{S2} = 1 - 0.82 = \underline{0.18\,V}$

$V_{D1} = V_{D2} = 1.5 - \frac{I}{2} \times 2.5K$

$= 1.5 - 0.2 \times 2.5 = \underline{1\,V}$

$I_{D1} = I_{D2} = I/2 = \underline{0.2\,mA}$

(d) Refer to Fig. 7.3 (c)

$V_{S1} = V_{S2} = -0.2 - 0.82 = \underline{-1.02\,V}$

$I_{D1} = I_{D2} = I/2 = \underline{0.2\,mA}$

$V_{D1} = V_{D2} = 1.5 - 0.2 \times 2.5 = \underline{1\,V}$

(e) $V_{CM\,max} = V_t + V_{DD} - \frac{I}{2} R_D$

$\rightarrow 0.5 + 1.5 - 0.2 \times 2.5 = \underline{+1.5\,V}$

(f) $V_{CM\,min} = -V_{SS} + V_{CS} + V_t + V_{OV}$

$= -1.5 + 0.4 + 0.5 + 0.316$

$= \underline{-0.28\,V}$

7.2

(a) The value of v_{id} that causes Q_1 to conduct the entire current is $\sqrt{2}\,V_{OV}$

$\rightarrow \sqrt{2} \times 0.316 = \underline{0.45\,V}$

then, $V_{D1} = V_{DD} - I \times R_D$

$= 1.5 - 0.4 \times 2.5 = \underline{0.5\,V}$

$V_{D2} = V_{DD} = \underline{+1.5\,V}$

(b) For Q_2 to conduct the entire current:

$v_{id} = -\sqrt{2}\,V_{OV} = \underline{-0.45\,V}$

then,

$V_{D1} = V_{DD} = \underline{+1.5\,V}$

$V_{D2} = 1.5 - 0.4 \times 2.5 = \underline{0.5\,V}$

(c) Thus the differential output range is:

$V_{D2} - V_{D1}$: from $1.5 - 0.5 = \underline{+1\,V}$ to $0.5 - 1.5 = \underline{-1\,V}$

7.3

Refer to answer table for Exercise 7.3. where values were obtained in the following way:

$V_{OV} = \sqrt{I / k\,W/L} \rightarrow \frac{W}{L} = \frac{I}{k V_{OV}^2}$

$g_m = \frac{I}{V_{OV}}$

CONT.

$$\left(\frac{v_{id}/2}{V_{ov}}\right)^2 = 0.1 \rightarrow v_{id} = 2V_{ov}\sqrt{0.1}$$

7.4

$$V_{ov} = \sqrt{I/k_n'(W/L)}$$
$$= \sqrt{0.8m/(0.2m \times 100)}$$
$$= \underline{0.2V}$$

$$g_m = \frac{2I_D}{V_{ov}} = \frac{2(I/2)}{V_{ov}} = \frac{0.8mA}{0.2V}$$
$$g_m = \underline{4\,mA/V}$$

$$r_o = \frac{V_A}{I_D} = \frac{20V}{(0.8m/2)A} = \underline{50K\Omega}$$

$$A_d = \frac{v_{o2}-v_{o1}}{v_{id}} = g_m(R_D \parallel r_o)$$
$$= 4\frac{mA}{V}(5K \parallel 50K)$$
$$= \underline{18.2\ V/V}$$

7.5

From Exercise 7.4 we have:
$V_{ov} = 0.2V$
$g_m = 4\frac{mA}{V}$

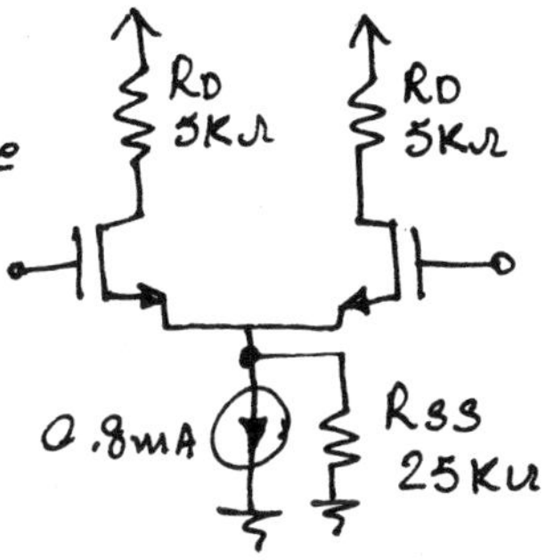

Since no value of V_A is specified r_o is neglected.

(a) Single-ended output:

From Eqn. (7.42)

$$|A_d| = \frac{1}{2}g_m \times R_D = \frac{4\times5}{2} = \underline{10\,V/V}$$

From Eqn. (7.41)

$$|A_{cm}| = \frac{R_D}{2R_{SS}} = \frac{5}{2\times25} = \underline{0.1\ V/V}$$

$$CMRR = \left|\frac{A_d}{A_{cm}}\right| = \frac{10}{0.1} = 100\,V/V$$

$$\rightarrow 20\log CMRR = \underline{40dB}$$

(b) Differential output:

From Eqn. (7.45)

$$|A_d| = g_m R_D = 4\times5 = \underline{20\ V/V}$$

From Eqn. (7.44)

$$|A_{cm}| = \underline{0}\ V/V$$

$$\rightarrow CMRR = 20/0 = \underline{\infty}$$

(c) Differential output and 1% mismatch in R_D

$$\Delta R = 0.01 \times R_D = 0.01 \times 5K$$
$$= 50\Omega$$

From Eqn. (7.52)

$$|A_d| \simeq g_m \cdot R_D = \underline{20\,V/V}$$

From Eqn. (7.50)

$$|A_{cm}| = \frac{\Delta R_D}{2R_{SS}} = \frac{50}{2\times25K} = \underline{0.001\frac{V}{V}}$$

$$CMRR = \frac{20}{0.001} = 20{,}000$$

$$\rightarrow 20\log CMRR = \underline{86dB}$$

7.6

From Exercise 7.5:
$g_m = 4mA/V$; $R_{SS} = 25K\Omega$
Then using Eqn. (7.66)

$$CMRR = \frac{2g_m \cdot R_{SS}}{\left(\frac{\Delta g_m}{g_m}\right)}$$

CONT.

$\rightarrow CMRR = \dfrac{2 \times 4 \times 25}{0.01} = 20{,}000$

$20 \log CMRR = \underline{\underline{86 dB}}$

7.7

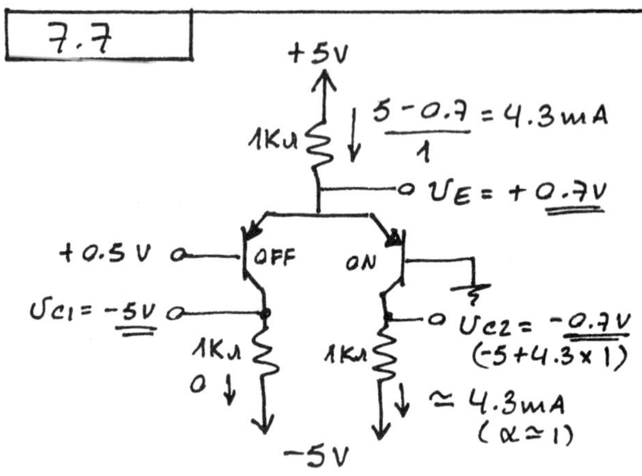

7.8

Substituting $i_{E1} + i_{E2} = I$ in Eqn. (7.69) yields

$$i_{E1} = \frac{I}{1 + e^{(v_{B2} - v_{B1})/V_T}}$$

$$0.99 I = \frac{I}{1 + e^{(v_{B2} - v_{B1})/V_T}}$$

$$v_{B1} - v_{B2} = -V_T \ln\left(\frac{1}{0.99} - 1\right)$$

$$= -25 \ln(1/99)$$

$$= 25 \ln(99) = \underline{\underline{115 mV}}$$

7.9

(a) The DC current in each transistor is 0.5mA. Thus V_{BE} for each will be

$$V_{BE} = 0.7 + 0.025 \ln\left(\frac{0.5}{1}\right)$$

$$= 0.683V$$

$\rightarrow V_E = 5 - 0.683 = +\underline{4.317V}$

(b) $g_m = \dfrac{I_C}{V_T} = \dfrac{0.5}{0.025} = \underline{\underline{20 \frac{mA}{V}}}$

(c) $i_{C1} = 0.5 + g_{m1} \cdot \Delta v_{BE1}$

$= 0.5 + 20 \times 0.005 \sin(2\pi \times 1000t)$

$= \underline{\underline{0.5 + 0.1 \sin(2\pi \times 1000t)}}$, mA

$i_{C2} = \underline{\underline{0.5 - 0.1 \sin(2\pi \times 1000t)}}$, mA

(d) $v_{C1} = (V_{CC} - I_C R_C) - \ldots$
$0.1 \times R_C \sin(2\pi \times 1000t)$

$= (15 - 0.5 \times 10) - 0.1 \times 10 \sin(2\pi 1000t)$

$= \underline{\underline{10 - 1 \sin(2\pi \times 1000t)}}$, V

$v_{C2} = \underline{\underline{10 + 1 \sin(2\pi \times 1000t)}}$, V

(e) $v_{C2} - v_{C1} = \underline{\underline{2 \cdot \sin(2\pi \times 1000t)}}$, V

(f) Voltage gain $\equiv \dfrac{v_{C2} - v_{C1}}{v_{B1} - v_{B2}}$

$= \dfrac{2 V \text{ peak}}{0.1 V \text{ peak}} = \underline{\underline{200}} V/V$

7.10

From Exercise 7.4:

$V_{OV} = 0.2 V$

Using Eqn. (7.112) we obtain V_{OS} due to $\Delta R_D / R_D$ as:

$$V_{OS} = \left(\frac{V_{OV}}{2}\right) \cdot \left(\frac{\Delta R_D}{R_D}\right)$$

$$= \frac{0.2}{2} \times 0.02 = 0.002 V$$

i.e $2mV$

To obtain V_{OS} due to $\dfrac{\Delta W/L}{W/L}$ use Eqn. (7.117)

$$V_{OS} = \left(\frac{V_{OV}}{2}\right)\left(\frac{\Delta W/L}{W/L}\right)$$

CONT.

$\rightarrow V_{os} = \left(\frac{0.2}{2}\right) \times 0.02 = 0.002$

$\rightarrow 2mV$

The offset voltage arising from ΔV_t is obtained from Eqn. (7.120)

$$V_{os} = \Delta V_t = 2mV$$

Finally, from Eqn. (7.121) the total input offset is:

$$V_{os} = \left[\left(\frac{V_{ov}}{2}\frac{\Delta R_D}{R_D}\right)^2 + \left(\frac{V_{ov}}{2}\frac{\Delta W/L}{W/L}\right)^2 + \ldots + (\Delta V_t)^2\right]^{1/2}$$

$$= \sqrt{(2\times10^{-3})^2 + (2\times10^{-3})^2 + (2\times10^{-3})^2}$$

$$= \sqrt{3\times(2\times10^{-3})^2}$$

$$= 3.46mV$$

7.11

From Eqn. (7.131)

$$V_{os} = V_T\sqrt{\left(\frac{\Delta R_C}{R_C}\right)^2 + \left(\frac{\Delta I_S}{I_S}\right)^2}$$

$$= 25\sqrt{(0.02)^2 + (0.1)^2}$$

$$= 2.5mV$$

$$I_B = \frac{100}{2(\beta+1)} = \frac{100}{2\times101} \cong 0.5\mu A$$

$$I_{os} = I_B\left(\frac{\Delta\beta}{\beta}\right)$$

$$= 0.5\times0.1\mu A = 50nA$$

7.12

$$(W/L)_n \times \mu_n C_{ox} = 0.2m \times 100 = 20\frac{mA}{V}$$

$$(W/L)_p \times \mu_p C_{ox} = 0.1m \times 200 = 20\frac{mA}{V}$$

Since all transistors have the same drain current ($I/2$) and the same product $W/L \times \mu C_{ox}$, then all transconductances g_m are identical.

$$|V_{ov}| = \sqrt{\frac{I_D}{20mA/V}} = \sqrt{\frac{0.8mA}{20mA/V}} = 0.2V$$

thus,

$$g_m = \frac{I_D}{V_{ov}} = \frac{(0.8mA/2)}{0.2V} = 4\frac{mA}{V}$$

From Eqn. (7.142)

$$G_m = g_m = 4mA/V$$

$$R_o = r_{o2} \parallel r_{o4}$$

$$r_{o2} = \frac{V_{An}}{I_{D2}} = \frac{20}{(0.8m/2)} = 50K\Omega$$

$$r_{o4} = \frac{V_{Ap}}{I_{D4}} = \frac{20}{(0.8m/2)} = 50K\Omega$$

thus,

$$R_o = 50\parallel50 = 25K\Omega$$

From Eqn. (7.146)

$$A_d = G_m R_o = 4\frac{mA}{V} \times 25K\Omega = 100\frac{V}{V}$$

From Eqn. (7.153)

$$A_{cm} \simeq \frac{1}{2g_{m3}R_{SS}} = \frac{1}{2\times4\times25} = 0.005 V/V$$

$$CMRR = \frac{|A_d|}{|A_{cm}|} = \frac{100}{0.005} = 20{,}000$$

$\longrightarrow 86dB$

7.13

From Eqn. (7.161): $G_m = g_m$

$$g_m \simeq \frac{I/2}{V_T} = \frac{(0.8mA/2)}{25mV} = 16\frac{mA}{V}$$

From Eqn. (7.164):

$$R_o = r_{o2} \parallel r_{o4}$$
$$= \frac{V_A}{I_{C2}} \parallel \frac{V_A}{I_{C4}} \simeq \frac{1}{2}\frac{V_A}{I/2}$$
$$= \frac{100\,V}{0.8mA} = 125\,K\Omega$$

$$A_d = G_m \times R_o = 16 \times 125 = 2000\frac{V}{V}$$

From Eqn. (7.167)

$$R_{id} = 2 \times r_\pi$$
$$\simeq 2 \times \frac{V_T \beta_n}{(I/2)} = \frac{2 \times 25m \times 160}{(0.8m/2)}$$
$$= 20K\Omega$$

For a simple current mirror the output resistance (thus R_{EE}) is r_o.

$$\rightarrow R_{EE} = \frac{V_A}{I} = \frac{100V}{0.8mA} = 125K\Omega$$

From Eqn. (7.172):

$$A_{cm} = \frac{-r_{o4}}{\beta_3 R_{EE}}$$

If: $\beta_p = 100 \rightarrow \beta_3 = 100$

$$\rightarrow A_{cm} = \frac{-2 \times 125K}{100 \times 125K} = -0.02\frac{V}{V}$$

and, $CMRR = \left|\frac{2000}{0.02}\right| = 100,000$

i.e $100dB$

7.14

$$G_m = g_{m_{1,2}} = \frac{I/2}{V_T} = \frac{1m/2}{25m} = 20\frac{mA}{V}$$

$$r_{o4} = r_{o5} = \frac{V_A}{I/2} = \frac{100V}{0.5mA} = 200K\Omega$$

$$\rightarrow R_{o4} = \beta_4 r_{o4} = 50 \times 200K = 10M\Omega$$
$$R_{o5} = \beta_5 \frac{r_{o5}}{2} = 100 \times \frac{200K}{2} = 10M\Omega$$

From Eqn. (7.179)

$$R_o = [\beta_4 r_{o4} \parallel \beta_5 \frac{r_{o5}}{2}]$$
$$= (10 \parallel 10)M\Omega = 5M\Omega$$

$$A_d = g_m \times R_o = 20 \times 5000 = 10^5\,V/V$$

i.e. $100dB$

7.15

(a) $V_{ov} = \sqrt{\frac{I}{(W/L)k'_n}} = \sqrt{\frac{0.8mA}{100 \times 0.2\frac{mA}{V}}}$

$= 0.2V$

$$g_m = \frac{I}{V_{ov}} = \frac{0.8mA}{0.2V} = 4mA/V$$

(b) $A_d = g_m(R_D \parallel r_o)$

where $r_o = \frac{V_A}{(I/2)} = \frac{20V}{0.4mA} = 50K\Omega$

$$\Rightarrow A_d = 4m(5K \parallel 50K)$$
$$= 18.2\,V/V$$

(c) For a CS amplifier, when R_{sig} is low:

$$f_H = \frac{1}{2\pi(C_L + C_{gd})R_L'} \quad (6.79)$$

CONT.

where: $R_L' = R_D \| r_o$
$= 5K \| 50K$
$= 4.545K\Omega$.

and,
$C_L = 100fF + C_{db}$
(Since for a grounded source C_{db} is in parallel with the load R and C.

thus,
$$f_H = \frac{1}{2\pi(100f + 10f) \times 4.545K} = \underline{\underline{291.8\,MHz}}$$

(d) Using the open-circuit time-constants method for $R_S = 10K\Omega$:

$$f_H = 1/(2\pi \tau_H) \quad \text{(Eqn. 6.58)}$$

where: (from Eqn. 6.57)
$$\tau_H = C_{gs}.R_S + C_{gd}[R_S(1 + g_m R_L') + .. + R_L'] + C_L R_L'$$

thus,
$$\tau_H = 50f \times 10K .. + 10f[10K(1 + 4m \times 4.545K) + 4.545K] + (100f + 10f).4.545K.$$

$$\tau_H = 0.5ns + 1.96ns + 0.5ns = 2.96ns$$

$$\Rightarrow f_H = 1/2\pi \times 2.96ns = \underline{\underline{53.7\,MHz}}$$

7.16

From Eqn. (7.184):
$$f_Z = \frac{1}{2\pi.C_{SS}.R_{SS}} = \frac{1}{2\pi.(0.4p)25K} = \underline{\underline{15.9\,MHz}}$$

7.17

For a loaded bipolar differential amplifier:
$$A_d = \frac{1}{2} g_m.r_o$$
where,
$$g_m = \frac{I/2}{V_T} = \frac{0.5mA}{25mV} = 20mA/V$$
$$r_o = \frac{V_A}{I/2} = \frac{100V}{0.5mA} = 200K\Omega$$
$$\Rightarrow A_d = \frac{1}{2} \times 20\frac{mA}{V} \times 200K\Omega = \underline{\underline{2000\,V/V}}$$

The dominant pole is set by the output load capacitance

$$f_Z = \frac{1}{2\pi.C_L(r_{o2} \| r_{o4})} = \frac{1}{2\pi \times 2pF \times (200K \| 200K)\Omega} = 0.796MHz \simeq \underline{\underline{0.8MHz}}$$

7.18

Refer to Fig.(7.40)

(a) Using Eqn. (7.199)

CONT.

$$I_6 = \frac{(W/L)_6}{(W/L)_4} (I/2)$$

$$\Rightarrow 100 = \frac{(W/L)_6}{100} \times 50$$

thus, $(W/L)_6 = \underline{\underline{200}}$

Using. Equ. (7.200)

$$I_7 = \frac{(W/L)_7}{(W/L)_5} (I)$$

$$\Rightarrow 100 = \frac{(W/L)_7}{200} \times 100$$

thus, $(W/L)_7 = \underline{\underline{200}}$

(b) For Q_1,

$$I = \frac{1}{2} \mu_p C_{ox} \left(\frac{W}{L}\right)_1 V_{ov_1}^2$$

$$\Rightarrow V_{ov_1} = \sqrt{\frac{50}{\frac{1}{2} \times 30 \times 200}} = \underline{\underline{0.129V}}$$

Similarly for Q_2, $V_{ov_2} = \underline{\underline{0.129V}}$

For Q_6,

$$100 = \frac{1}{2} \times 90 \times 200 \, V_{ov_6}^2$$

$$\Rightarrow V_{ov_6} = \underline{\underline{0.105V}}$$

(c) $g_m = \dfrac{2 I_D}{V_{ov}}$

	I_D	V_{ov}	g_m
Q_1	$50\mu A$	$0.129V$	$\underline{\underline{0.775}}$ mA/V
Q_2	$50\mu A$	$0.129V$	$\underline{\underline{0.775}}$ mA/V
Q_6	$100\mu A$	$0.105V$	$\underline{\underline{1.90}}$ mA/V

(d) $r_{o2} = 10/0.05 = \underline{\underline{200k\Omega}}$
$r_{o4} = 10/0.05 = \underline{\underline{200k\Omega}}$
$r_{o6} = 10/0.1 = \underline{\underline{100k\Omega}}$
$r_{o7} = 10/0.1 = \underline{\underline{100k\Omega}}$

(e) Equ. (7.197):

$$A_1 = -g_{m1} (r_{o2} \| r_{o4})$$
$$= -0.775 (200 \| 200) = \underline{\underline{-77.5 V/V}}$$

Equ. (7.198):

$$A_2 = -g_{m6} (r_{o6} \| r_{o7})$$
$$= -1.90 \times (100 \| 100)$$
$$= \underline{\underline{-95}} V/V$$

Overall voltage gain is:

$A_1 \times A_2 = 77.5 \times 95 = \underline{\underline{7363}}$ V/V

7.19

From Equ. (7.211) $\omega_t = \dfrac{G_{m1}}{C_C}$

where, from Table 7.1,

$G_{m1} = g_{m1} = g_{m2} = 0.3$ mA/V

thus, for $f_T = 10MHz$

$$C_C = \frac{0.3mA/V}{2\pi \times 10 \times 10^6} = \underline{\underline{4.8pF}}$$

From Equ. (7.206)

$$f_Z = \frac{G_{m2}}{2\pi . C_C}$$

$G_{m2} = g_{m6} = 0.6mA/V$

$$\rightarrow f_Z = \frac{0.6mA/V}{2\pi \times 4.8pF} = \underline{\underline{20MHz}}$$

From Equ. (7.210)

$$f_{P2} = \frac{G_{m2}}{2\pi \times C_2} = \frac{0.6mA/V}{2\pi \times 2pF}$$
$$= \underline{\underline{48MHz}}$$

7.20

Using the Equ. following (7.215)

CONT.

$$R_B = \frac{2}{\sqrt{2\mu_n C_{ox}\left(\frac{W}{L}\right)_{12} \cdot I_B}} \cdot \left(\sqrt{\frac{(W/L)_{12}}{(W/L)_{13}}} - 1\right)$$

$$= \frac{2}{\sqrt{2 \times 90 \times 80 \times 10}} \left(\sqrt{\frac{80}{20}} - 1\right)$$

$$= \underline{\underline{5.27 k\Omega}}$$

$$g_{m12} = \sqrt{2\mu_n C_{ox} (W/L)_{12} I_B}$$

$$= \sqrt{2 \times 90 \times 80 \times 10}$$

$$= 379 \mu A/V$$

$$= \underline{\underline{0.379}} mA/V$$

7.21

$I_D = 90 \mu A$
$\mu_n C_{ox} = 160 \mu A/V^2$
$\mu_p C_{ox} = 40 \mu A/V^2$

For Q_8 and Q_9: $W/L = 40/0.8$ (as given in Example 7.3)

$$|V_{ov}| = \sqrt{\frac{2I_D}{\mu_p C_{ox} (W/L)}}$$

$$\rightarrow |V_{ov}|_{8,9} = \sqrt{\frac{2 \times 90\mu}{40\mu \times \frac{40}{0.8}}} = 0.3V$$

then,

$$g_{m8,9} = \frac{2I_D}{|V_{ov}|} = \frac{2 \times 90 \mu A}{0.3V}$$

$$= 0.6 mA/V$$

Since g_m of Q_{10}, Q_{11}, and Q_{13} are identical to g_m of Q_8 and Q_9 then:

$$V_{ov13} = 0.3V$$

Thus, for Q_{13}

$$(0.3)^2 = \frac{2 \times 90\mu}{160\mu (W/L)_{13}}$$

$$\rightarrow (W/L)_{13} = 12.5$$

i.e. (10/0.8)

Since Q_{12} is 4 times as wide as Q_{13}, then

$$\left(\frac{W}{L}\right)_{12} = \frac{4 \times 10}{0.8} = \frac{40}{0.8}$$

$$R_B = \frac{2}{\sqrt{2\mu_n C_{ox} \left(\frac{W}{L}\right)_{12} I_B}} \cdot \left(\sqrt{\frac{(W/L)_{12}}{(W/L)_{13}}} - 1\right)$$

$$= \frac{2}{\sqrt{2 \times 160\mu \times \frac{40}{0.8} \times 90\mu}} \cdot \left(\sqrt{\frac{40/0.8}{12.5}} - 1\right)$$

$$\rightarrow R_B = \underline{\underline{1.67}} k\Omega$$

The voltage drop on R_B is:
$1.67 k\Omega \times 90 \mu A = \underline{\underline{150}} mV$

$$V_{ov12} = \sqrt{\frac{2 \times 90\mu}{160\mu \times \frac{40}{0.8}}} = 0.15V$$

$$V_{ov12} = V_{GS12} - V_{tn}$$

$$V_{GS12} = 0.15 + 0.7 = 0.85V$$

thus, $V_{G12,13} = V_{GS12} + I_B R_B - V_{SS}$

$$= 0.85 + 0.15 - 2.5$$

$$= \underline{\underline{-1.5}} V$$

$$V_{ov11} = |V_{ov8}| = 0.3V$$

$$\rightarrow V_{GS11} = 0.3 + 0.7 = 1V$$

$$V_{G11} = -1.5 + 1 = \underline{\underline{-0.5}} V$$

Finally,

$$V_{G8} = V_{DD} - V_{SG8} = +2.5 + (-0.3 - 0.8)$$

$$= +\underline{\underline{1.4}} V$$

7.22

$R_{id} = 20.2 k\Omega$
$A_{vo} = 8513 V/V$

CONT.

$R_o = 152\,\Omega$

with $R_S = 10\,K\Omega$ and $R_L = 1\,K\Omega$

$$A_v = \frac{20.2}{20.2+10} \times 8513 \times \frac{1}{(1+0.152)}$$

$$= \underline{\underline{4943}}\ V/V$$

7.23

$$\frac{i_{e8}}{i_{b8}} = \beta_8 + 1 = \underline{\underline{101}}$$

$$\frac{i_{b8}}{i_{c7}} = \frac{R_5}{R_5 + R_{i4}} = \frac{15.7}{15.7+303.5} = \underline{\underline{0.0492}}$$

$$\frac{i_{c7}}{i_{b7}} = \beta_7 = \underline{\underline{100}}$$

$$\frac{i_{b7}}{i_{c5}} = \frac{R_3}{R_3 + R_{i3}} = \frac{3}{3+234.8} = \underline{\underline{0.0126}}$$

$$\frac{i_{c5}}{i_{b5}} = \beta_5 = \underline{\underline{100}}$$

$$\frac{i_{b5}}{i_{c2}} = \frac{R_1+R_2}{R_1+R_2+R_{i2}} = \frac{40}{40+5.05} = \underline{\underline{0.8879}}$$

$$\frac{i_{c2}}{i_i} = \beta_2 = \underline{\underline{100}}$$

Thus the overall current gain is:

$$\frac{i_{e8}}{i_i} = 101 \times 0.0492 \times 100 \times 0.0126 \times 100 \ldots \times 0.8879 \times 100$$

$$= \underline{\underline{55593}}\ A/A$$

and the overall voltage gain is

$$\frac{v_o}{v_{id}} = \frac{R_6}{R_{i1}} \cdot \frac{i_{e8}}{i_i}$$

$$= \frac{3}{20.2} \times 55593 = \underline{\underline{8256}}\ V/V$$

7.24

To obtain R_{eq}:

$$R_{eq} = R_2 \,\|\, r_{o2} \,\|\, r_{\pi 5}$$

From Fig. 7.43: $R_2 = 20\,K\Omega$

$$r_{o2} = \frac{V_A}{I_{C2}} \simeq \frac{100V}{0.25\,mA} = 400\,K\Omega$$

$$r_{\pi 5} = (\beta+1)\frac{V_T}{I_5} = 101 \times \frac{25\,mV}{1\,mA}$$

$$= 2525\,\Omega$$

thus,

$$R_{eq} = 20K \,\|\, 400K \,\|\, 2525$$

$$= \underline{\underline{2.2\,K\Omega}}$$

To obtain C_{eq}:

$$C_{eq} = C_{\mu 2} + C_{\pi 5} + C_{\mu 5}(1 + g_{m_5} R_{L5})$$

$C_{\mu 2} = C_{\mu 5} = 2\,pF$

$R_{L5} \simeq R_3 = 3\,K\Omega$ (from Fig 7.43)

$$g_{m5} = \frac{I_5}{V_T} = \frac{1\,mA}{25\,mV} = 40\,\frac{mA}{V}$$

$$C_{\pi 5} + C_{\mu 5} = \frac{g_{m5}}{2\pi f_T}$$

$$\Rightarrow C_{\pi 5} = \frac{40m}{2\pi \times 400M} - 2p = 14\,pF$$

thus,

$$C_{eq} = 2pF + 14pF + 2pF(1 + 40 \times 3)$$

$$= \underline{\underline{258\,pF}}$$

Finally,

$$f_P = \frac{1}{2\pi \cdot R_{eq} \cdot C_{eq}}$$

$$= \frac{1}{2\pi \times 2.2K \times 258p}$$

$$= \underline{\underline{280\,KHz}}$$

Chapter 8 - Exercises

Ex 8.1

(a) $\beta = R_1/(R_1+R_2)$

(b) $A_f = \dfrac{A}{1+A\beta} = \dfrac{10^4}{1+10^4\beta} = 10$

$\Rightarrow \beta = 0.0999$

$\dfrac{1}{\beta} = 1 + \dfrac{R_2}{R_1} \Rightarrow \dfrac{R_2}{R_1} = \dfrac{1}{\beta} - 1 = 9.01$

(c) Amount of feedback = $(1+A\beta)$

$\Rightarrow 20\log(1+A\beta) = 60dB$

(d) $V_s = 1$: $V_o = A_f \times V_s = 10v$

$V_f = \beta V_o = 0.0999 \times 10 = 0.999v$

$V_i = V_s - V_f = 10 - 0.999 = 0.001v$

(e) $A_f = \dfrac{0.8 \times 10^4}{1+(0.8\times10^4)(0.0999)} = 9.9975$

Thus A_f decreased by about 0.02%

Ex 8.2

$\beta = R_1/(R_1+R_2) = 1/(1+9) = 0.1$

$A_f = \dfrac{A_0}{1+A_0\beta} = \dfrac{10^4}{1+10^4(0.1)} = 9.999$

$f_{Hf} = f_H(1+A_0\beta) = 100(1+10^4/10) = 100.1kHz$

Ex 8.3

$V_o = V_s \dfrac{A_1A_2}{1+A_1A_2\beta} + V_n \dfrac{A_1}{1+A_1A_2\beta} = V_{sf} + V_{nf}$

$= \dfrac{(1\times100)\times V_s}{1+(100\times1)\times1} + \dfrac{1 \times V_n}{1+(100\times1)\times1}$

$= 0.99 + 0.0099$

Thus $V_{sf} \approx 1v$ and $V_{nf} \approx 0.01v$

New S/N ratio $\approx 100/1$

an improvement of $20\log(100/1) = 40dB$

Ex 8.4

From Example 8.1 $A_0 \approx 6000$, $\beta = 10^{-3}$

$(1+A\beta) = (1+(6\times10^3)\times10^{-3}) = 7$

$\therefore f_{Hf} = f_H(1+A\beta) = 1\times7 = 7kHz$

Ex 8.5

$I_{E1} = I_{E2} = 0.5mA$

$V_{C2} = 10.7 - 0.5\times20 = +0.7v$

$V_O = 0.7 - V_{BE3} = 0$

$I_{E3} = 5mA$

$r_{e1} = r_{e2} = V_A/I = 50\Omega$, $r_{e3} = 5\Omega$

A. Circuit

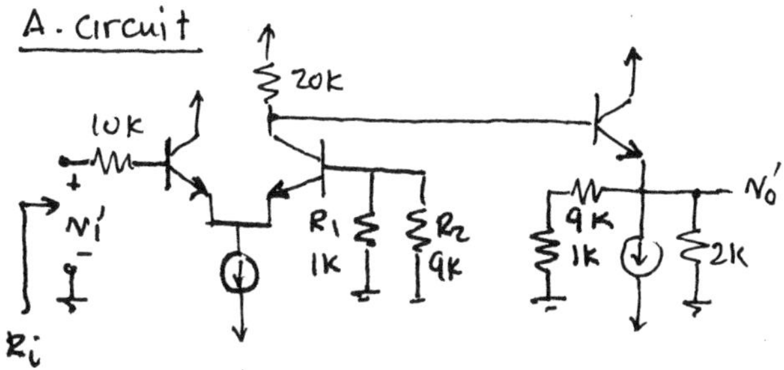

$A = \dfrac{V_o'}{V_i'} = \dfrac{[20\|(\beta_2+1)(r_{e3}+(2\|10)]}{r_{e1}+r_{e2}+\frac{10}{\beta_1+1}+\frac{(1)\|9)}{\beta_2+1}} \times \dfrac{(2\|10)}{r_{e3}+(2\|10)}$

$= 85.7 V/V$

$R_i = R_s + (\beta+1)(r_{e1}+r_{e2}) + R_E\|R_4$

$= 10 + 101(50+50) + (1\|9) = 21k\Omega$

$R_O = 2\|10\|[r_{e3} + \frac{20}{\beta_2+1}] = 181\Omega$

B - Circuit

$\beta = V_f'/V_o'$

$= \dfrac{1}{9+1} = 0.1 V/V$

$A_f = \dfrac{V_o}{V_s} = \dfrac{A}{1+A\beta} = \dfrac{85.7}{1+85.7\times0.1} = 8.96 V/V$

$R_{if} = R_i(1+A\beta) = 21\times9.57 = 201k\Omega$

$R_{IN} = R_{if} - R_s = 201 - 10 = 191k\Omega$

$R_{of} = (R_{out}\|R_L) = \dfrac{R_o}{1+A\beta} = \dfrac{181}{9.57} = 18.8\Omega$

$\Rightarrow R_{out} = 19.1\Omega$

Ex 8.6

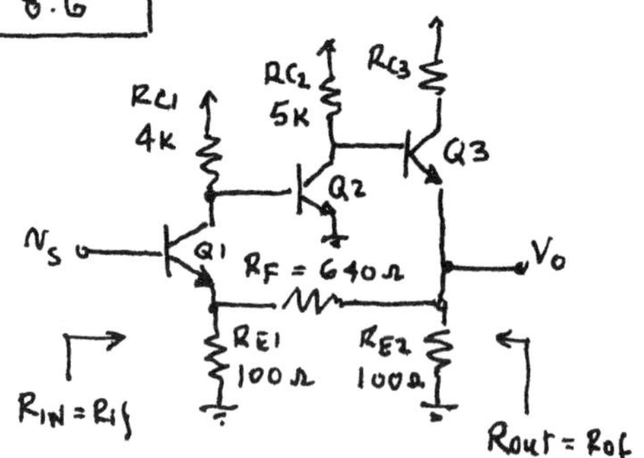

$I_{C1} = 0.6\,mA$, $I_{C2} = 1\,mA$, $I_{C3} = 4\,mA$

$r_{e1} = 41.7\,\Omega$, $r_{\pi 2} = 2.5K$, $r_{e3} = 6.25\,\Omega$

Series-Shunt Feedback

A-circuit

V_s', R_{C1}, I_{C1}, I_{B2}, R_{C2}, I_{B3}, R_{C3}, I_{E3}, V_o', R_F, R_{E1}, R_{E2}, R_i, R_o

$$I_{C1} = \frac{(\beta/\beta+1)V_s'}{r_{e1} + [R_{E1} \| (R_F + R_{E2})]}$$

$$= \frac{0.99\,V_s'}{0.0417 + [0.1 \| (0.64 + 0.1)]} = 7.627\,V_s'$$

$$I_{b2} = \frac{I_{C1} R_{C1}}{R_{C1} + r_{\pi 2}} = I_{C1}\frac{4}{4+2.5} = 0.7826\,I_{C1}$$

$$I_{C2} = \beta_2 I_{b2} = 100\,I_{b2}$$

$$I_{b3} = \frac{I_{C2} R_{C2}}{R_{C2} + (\beta_3+1)[r_{e3} + (R_{E2} \| (R_F + R_{E1}))]}$$

$$= \frac{5}{5 + 101[0.00625 + (0.1 \| (0.64 + 0.1))]}$$

$$= 0.344\,I_{C2}$$

$$I_{e3} = (\beta_3 + 1) I_{b3} = 101\,I_{b3}$$

$$V_o' = I_{E3}[R_{E2} \| (R_F + R_{E1})]$$

$$= 0.0881\,I_{e3}$$

Hence, combining above $V_o' = 1827.5\,V_s'$

Thus $A \equiv \dfrac{V_o'}{V_s'} = 1827.5$ V/V

B-CIRCUIT

V_f', R_F, R_{E1}, V_o'

$$\beta \equiv \frac{V_f'}{V_o'} = \frac{R_{E1}}{R_{E1} + R_F}$$

$$\Rightarrow \beta = (100/740) = 0.135 \text{ V/V}$$

Then $A\beta = 247$

(within calc error of 246.3 in Example 8.2)

$$A_f \equiv \frac{V_o}{V_s} = \frac{A}{1 + A\beta} = \frac{1827.5}{1 + 247} = 7.4 \text{ V/V}$$

$$R_o = R_{E2} \| (R_F + R_{E1}) \| [r_{e3} + R_{C2}/(\beta_3+1)]$$

$$= 100 \| (640 + 100) \| [6.25 + 5000/101]$$

$$= 34.1\,\Omega$$

$$R_{OUT} = R_{of} = \frac{R_o}{1 + A\beta} = \frac{34.1}{1 + 247} = 0.14\,\Omega$$

Ex 8.7

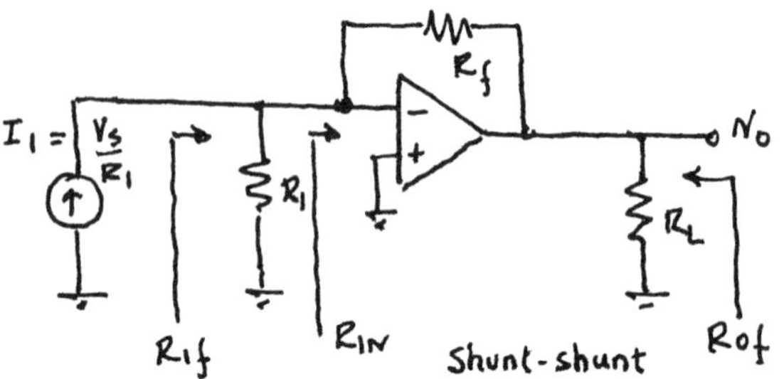

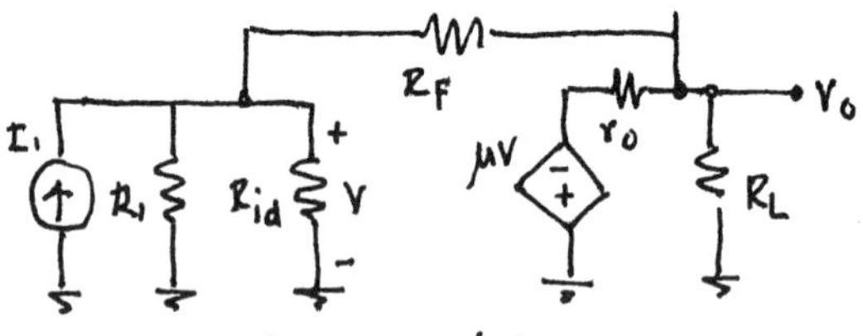

Equivalent circuit

A. circuit

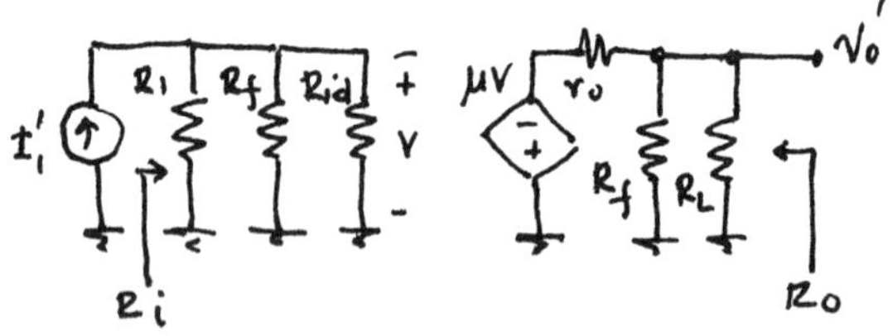

$$A \equiv \frac{V_o'}{I_i'} = -\mu V \frac{(R_L \| R_f)}{(R_L \| R_f) + r_o} \times \frac{1}{I_i'}$$

$$= -\mu (R_s \| R_f \| R_{id}) \times \frac{(R_L \| R_f)}{(R_L \| R_f) + r_o}$$

$$= -10^4 (1K \| 100 \| 1M) \times \frac{(2K \| 2M)}{(2K \| 2M) + 1K}$$

$$= -6589\,K\Omega$$

$$R_i = R_s \| R_{id} \| R_f = 989.1\,\Omega$$

$R_o = R_L \| R_f \| r_o = 2 \| 1000 \| 1 = 666.2\,\Omega$

β-circuit

$\beta \equiv \frac{I_f'}{V_o'} = -\frac{1}{R_F}$

$= -10^{-6}\,\mho$

$[1 + A\beta] = 1 + 6589 \times 10^3 \times 10^{-6} = 7.589$

$A_f \equiv \frac{V_o}{I_s} = \frac{A}{1+A\beta} = \frac{-6589}{7.589} \approx -870\ \text{k}\Omega$

$\frac{V_o}{V_s} = \frac{V_o}{I_s R_s} = \frac{-870}{1} = -870\ \text{V/V}$

$R_{if} = R_i/(1+A\beta) = 989.1/7.589 = 130.3\,\Omega$

$\frac{1}{R_{if}} = \frac{1}{R_{IN}} + \frac{1}{R_S} \Rightarrow R_{IN} \approx 150\,\Omega$

$R_{of} = (R_{OUT} \| R_L) = R_o/(1+A\beta) = 82.8\,\Omega$

$\Rightarrow R_{OUT} = 92\,\Omega$

Ex 8.8

Use circuit of Fig 8.25(b)

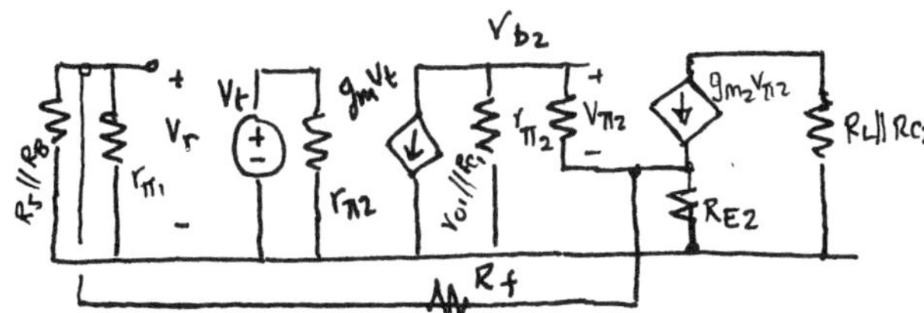

$A\beta \equiv -\frac{V_r}{V_t} = \left(-\frac{V_{b2}}{V_t}\right)\left(\frac{V_{e2}}{V_{b2}}\right)\left(\frac{V_r}{V_{e2}}\right)$

$= g_{m1}\left[r_{o1} \| R_{C1} \| [r_{\pi 2} + (\beta+1)(R_{E2} \| R_F + (R_S \| R_B \| r_{\pi 1}))]\right]$

$\times \frac{R_{E2} \| [R_f + (R_B \| R_S \| r_{\pi 1})]}{R_{E2} \| [R_f + (R_B \| R_S \| r_{\pi 1})] + r_{e2}}$

$\times \frac{(R_S \| R_B \| r_{\pi 1})}{(R_S \| R_B \| r_{\pi 1}) + R_f}$ QED

Now subst values from Example 8.4

$A\beta = 40[100 \| 10 \| [2.5 + 101(1.3 \| (10 + (10 \| 100 \| 15 \| 2.5)..)]$

$\times \frac{1.3 \| (10 + (100 \| 15 \| 10 \| 2.5))}{1.3 \| (10 + (100 \| 15 \| 10 \| 2.5) + 0.25}$

$\times \frac{(10 \| 100 \| 15 \| 2.5)}{(10 \| 100 \| 15 \| 2.5) + 10}$

$= 338.2 \times 0.986 \times 0.148$

≈ 49.3

(cf 52.1)

Ex 8.9

In Ex 8.7 above $|1 + A\beta| = 7.589$

$\Rightarrow A\beta = 7.589 - 1 = 6.589$

Ex 8.10

$$A(j\omega) = \left(\frac{10}{1 + j\omega/10^4}\right)^3$$

Thus, $\phi = -3\tan^{-1}(\omega/10^4)$

At ω_{180}, $\phi = 180° \Rightarrow \tan^{-1}(\omega_{180}/10^4) = 60°$

$(\omega_{180}/10^4) = \sqrt{3} \Rightarrow \omega_{180} = \sqrt{3} \times 10^4$ rad/s

Amplifier stable if $|A\beta| < 1$ at ω_{180}

When $|A\beta| = 1$: $\beta_{cr} = \frac{1}{|A(j\omega_{180})|}$

$\therefore \beta_{cr} = \frac{1}{1000/((1+(\sqrt{3})^2))^{3/2}} = 0.008$

Ex 8.11

Pole is shifted by factor $(1 + A_0\beta)$

$= 1 + 10^5 \times 0.01 = 1001$

$f_{pf} = f_p(1 + A_0\beta) = 100 \times 1001 = 100.1$ kHz

For closed loop gain $= 1$, $\beta = 1$

$f'_{pf} = f''_p(1 + A_0\beta) = 10^5(1001) = 10^7$ Hz.

Ex 8.12

From Eqn 8.63 poles will coincide when

$(\omega_{p1} + \omega_{p2})^2 - 4(1 + A_0\beta)\omega_{p1}\omega_{p2} = 0$

Using $A_0 = 100$, $\omega_{p1} = 10^4$, $\omega_{p2} = 10^6$ rad/s

$(10^4 + 10^6)^2 - 4(1 + 100\beta) \times 10^{10} = 0$

$1 + 100\beta = (1.01)^2 \times 100/4$

$\Rightarrow \beta = 0.245$

Corresponding $Q = 0.5$

For maximally flat response $Q = 0.707$

and $\frac{1}{\sqrt{2}} = \frac{\sqrt{(1+100\beta) \times 10^{10}}}{10^4 + 10^6} \Rightarrow \beta = 0.5$

corresponding gain is

$A = \frac{A_0}{1 + A_0\beta} = \frac{100}{1 + 100 \times 0.5} = 1.96\ \text{V/V}$

Ex 8.13

Closed loop poles are found using

$$1 + A(s)\beta = 0$$

$$1 + \frac{10^3}{(1+s/10^4)^3}\beta = 0$$

$$(1+s/4)^3 + 10^3\beta = 0$$

$$\frac{s^3}{10^{12}} + \frac{3s^2}{10^8} + \frac{3s^1}{10^4} + (1+100\beta) = 0$$

$$\equiv s_n^3 + 3s_n + 3s_n + (1+100\beta) = 0 \quad \text{for } s_n = \frac{s}{10^4}$$

Roots of this cubic equation are:

$$-(1+10\beta^{1/3}),\ -1+5\beta^{1/3} \pm j\,5\sqrt{3}\,\beta^{1/3}$$

Amplifier becomes unstable when complex poles lie on $j\omega$ axis, i.e. when $\beta = \beta_{cr}$

$$10\,\beta_{cr}^{1/3} = \frac{1}{\cos 60^\circ} = 2 \Rightarrow \beta_{cr} = 0.008$$

Ex 8.14

$$A = \frac{A_0}{1+j\frac{\omega}{\omega_p}} = \frac{A_0}{1+jf/f_p} = \frac{10^5}{1+jf/10}$$

$$\beta = 0.01$$

$$|A\beta| = \frac{10^5 \times 0.01}{\sqrt{1+f^2/100}} = 1$$

thus $1+f^2/100 = 10^6 \Rightarrow f \approx 10^4\ \text{Hz}$

At $f = 10^4\ \text{Hz}$

$$\phi = -\tan^{-1}(10^4/10) \approx -90^\circ$$

making phase margin $180 - 90 = 90^\circ$

Ex 8.15

From Eqn. 8.76 $|Af(j\omega_2)| = \dfrac{1/\beta}{|1+e^{-j\theta}|}$

and $\frac{1}{\beta} \approx$ low frequency gain

$\theta = 180^\circ -$ Phase Margin

For PM $= 30^\circ$, $\theta = 150^\circ$

$$|A_f(j\omega_1)|/(1/\beta) = 1.93$$

For PM $= 60^\circ$, $\theta = 120^\circ$

$$|A_f(j\omega_2)|/(1/\beta) = 1.0$$

For PM $= 90^\circ$, $\theta = 90^\circ$

$$|A_f(j\omega_{90})|/(1/\beta) = 0.707$$

Ex 8.16

$$\beta = \frac{1/sC}{R + 1/sC} = \frac{1}{1+sCR}$$

$$\left|\frac{1}{\beta}\right| = \sqrt{1+(\omega CR)^2}$$

$$\therefore \frac{1}{2\pi CR} \leq 1\ \text{Hz}$$

$$CR \leq \frac{1}{2\pi}$$

thus $CR \geq 0.159\,s$.

C, R; 20log|A|; 100dB; +20dB/dec; 20log(1/|β|); 0dB; f; 1Hz; 10^5 Hz

Rate of closure = 20dB/dec

Ex 8.17

Must place new dominant pole at $f_D = \dfrac{f_P}{A}$

$$= \frac{10^6}{10^4}$$

$$\therefore f_D = 100\ \text{Hz}$$

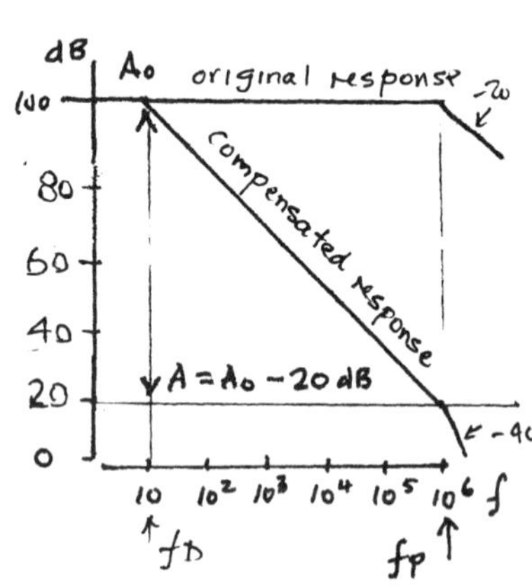

Ex 8.18

The pole must be moved f_{P1} to f_D

where $f_D = \dfrac{\text{Frequency of 2nd pole}}{A_0 \div A_F}$

$$= \frac{10 \times 10^6}{10^4} \leftarrow (100\text{dB} - 20\text{dB})$$

$$= 10^3\ \text{Hz}$$

The capacitance at the controlling node must be increased by same factor as f is lowered.

$$\therefore C_{new} = C_{old} \times 1000$$

CHAPTER 9 – EXERCISES

Ex. 9.1

$V_{ICM}(max) \le V_{DD} - |V_{OV5}| - |V_{tp}| - |V_{OV1}|$
$\le +1.65 - 0.3 - 0.5 - 0.3$
$\le +0.55V$

$V_{ICM}(min) \ge -V_{SS} + V_{OV3} + V_{tn} - |V_{tp}|$
$\ge -1.65 + 0.3 + 0.5 - 0.5$
$\ge -1.35V$

$V_O(max) \le V_{DD} - |V_{OV7}|$
$\le +1.65 - 0.5$
$\le +1.15V$

$V_O(min) \ge -V_{SS} + V_{OV6}$
$\ge -1.65 + 0.5$
$\ge +1.15V$

Ex. 9.2

$|V_A| = 20V,\ I_6 = 0.5mA,\ V_{OV1} = 0.2V,\ V_{OV6} = 0.5V$

$I = K(V_{OV})^2$

For Q6 : $0.5 = K(0.5)^2 \Rightarrow K = 2mA/V^2$

For Q2 : $I_2 = 2(0.2)^2 \Rightarrow I_2 = 0.8mA$

$g_m = \frac{I}{V_{OV}} \Rightarrow g_{m2} = \frac{0.8}{0.2} = 4mA/V$

$\Rightarrow g_{m6} = \frac{0.5}{0.5} = 1mA/V$

$r_o = \frac{V_A}{I} \Rightarrow r_{o2} = \frac{20}{0.8} = 25K\Omega$

$\Rightarrow r_{o6} = r_{o7} = \frac{20}{0.5} = 40K\Omega$

$A_1 = -g_{m2} r_{o2} = -4 \times 25 = -100V/V$
$A_2 = -g_{m6} r_{o6} = -1 \times 40 = -40V/V$
$A = A_1 A_2 = (-100)(-40) = +4000V/V$

$R_o = (r_{o6} \| r_{o7}) = \frac{40K}{2} = 20K\Omega$

Ex. 9.3

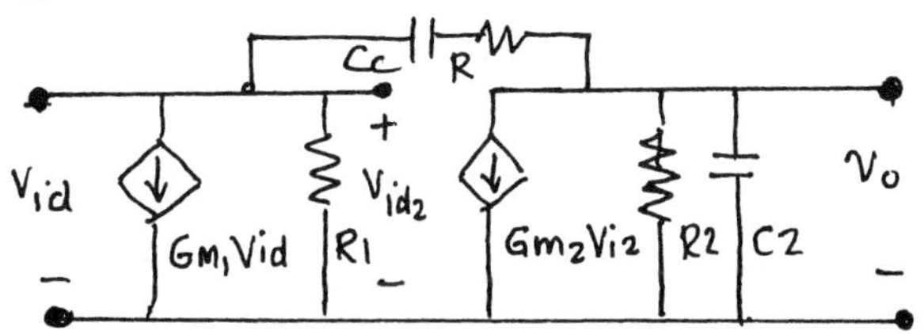

$G_{m1} = 1mA/V \quad G_{m2} = 2mA/V \quad C_2 = 1p$
$r_{o2} = r_{o4} = 100K, \quad r_{o6} = r_{o7} = 40K$

(a) $f_t = \frac{G_{m1}}{2\pi C_c} = \frac{1mA/V}{2\pi C_c} = 100MHz$

$\Rightarrow C_c = 1.6pF$

$A_1 = -G_{m1} R_1 = (1\times10^{-3})(r_{o2} \| r_{o1})$
$= (1\times10^{-3})(100K/2) = -50\ V/V$
$A_2 = -G_{m2} R_2 = (2\times10^{-3})(r_{o6} \| r_{o7})$
$= (2\times10^{-3})(40K/2) = -40V/V$
$A = A_1 \cdot A_2 = (-50)(-40) = +2000$
$f_p = f_t/A = 100\times10^6/2\times10^3 \equiv 50KHz$

(b) to move zero to $s = \infty$

$R = \frac{1}{G_{m2}} = \frac{1}{2\times10^{-3}} = 500\Omega$

$f_{p2} \approx \frac{G_{m2}}{2\pi C_2} = \frac{0.2\times10^{-3}}{2\pi\ 10^{-9}} = 318\times10^6 Hz$

$\theta = \tan^{-1}\frac{f_t}{f_p} = \tan^{-1}\frac{100\times10^6}{318\times10^6} = 17.4^\circ$

$PM = 90 - \theta = 72.6^\circ$

Ex. 9.4

Find SR for $f_t = 100MHz$
$V_{OV1} = 0.2V$

$SR = 2\pi f_t V_{OV} = 2\pi \times 100\times10^6 \times 0.2$
$= 125.67 \equiv 126V/\mu s$

$SR = \frac{I}{C_c} \Rightarrow I = SR \times C_c$

$\therefore I = 126\times10^6 \times 1.6\times10^{-12}$
$= 200\mu A$

Ex. 9.5

$V_{ICM}(max) \leq V_{DD} - V_{OV9} + V_{tn}$
$\leq +1.65 - 0.3 + 0.5 = +1.85V$

$V_{ICM}(min) \geq -V_{SS} + V_{OV11} + V_{OV1} + V_{tn}$
$\geq -1.65 + 0.3 + 0.3 + 0.5 = -0.55V$

$V_O(max) \leq V_{DD} - |V_{OV9}| - |V_{OV}|$
$\leq +1.65 - 0.3 - 0.3 = +1.05V$

$V_O(min) \geq -V_{SS} + V_{OV11} + V_{OV1} + V_{tn}$
$\geq -1.66 + 0.3 + 0.3 + 0.5 = -0.55V$

Ex. 9.6

$|V_A| = 20V$, $V_{OV} = 0.2V$, $I = 100\mu A$

$G_m = \frac{2I}{V_{OV}} = \frac{2 \times 100 \times 10^{-6}}{0.2} = 1.0\ mA/V$

$r_o = \frac{V_A}{I} = \frac{20 \times 10^6}{100} \Rightarrow 200k\Omega$

$R_o = [g_{m4} r_{o4}(r_{o2} \| r_{o10})] \| [g_{m6} r_{o6} r_{o8}]$
$= g_m r_o^2 [\frac{1}{2}(1 \| 2)]$
$= 1.0 \times 200^2 \times 1/3 \times 10^6 \equiv 13.33M\Omega$

$A = G_m R_o = 1.0 \times 10^{-3} \times 13.33 \times 10^6$
$= 13.33 \times 10^3\ V/V$

Ex. 9.7

Given: all $V_{OV} = 0.3V$, $|V_t| = 0.7V$
$V_{DD} = V_{SS} = 2.5V$

(a) $V_{IC}(max)$ for NMOS

$V_{ICM}(max)_N \leq V_{DD} - V_{OV} + V_T$
$\leq +2.5 - 0.3 + 0.7 = +2.9V$

$V_{ICM}(min)_N \geq -V_{SS} + V_{OV} + V_{OV} + V_T$
$\geq -2.5 + 0.3 + 0.3 + 0.7 = -1.2V$

$\therefore\ -1.2V \leq (V_{ICM})_N \leq +2.9V$

(b) By sym.

$-2.9V \leq (V_{ICM})_P \leq +1.2V$

(c) $-1.2V \leq (V_{ICM})_{BOTH} \leq +1.2V$

(d) $-2.9V \leq (V_{ICM})_{overall} \leq +2.9V$

Ex. 9.8

$I_1 = \frac{1}{2} k (W/L)(V_{GS1} - V_T)^2$
$I_2 = \frac{1}{2} k (W4/L)(V_{GS2} - V_T)^2$

For $I_1 = I_2$:

$(V_{GS1} - V_T)^2 = 4(V_{GS2} - V_T)^2$

i.e. $V_{GS1} - V_T = 2(V_{GS1} - V_T)$

or $V_{GS1} = 2V_{GS2} + V_T$

QED

Ex. 9.9

npn: $I_S = 10^{-14}A$, $\beta = 200$, $V_A = 125V$
pnp: $I_S = 10^{-14}A$, $\beta = 50$, $V_A = 50V$

$I = I_S e^{V_{BE}/V_T}$
$\Rightarrow V_{BE} = V_T \ln \frac{I}{I_S}$

$V_{BE} = 25mV \ln \frac{10^{-3}}{10^{-14}} = 633mV$

$g_m = \beta / r_\pi = 200/5K = 40\ mA/V$
$r_\pi \approx \beta r_e \approx 200 \times 25 \Rightarrow 5K\Omega$
$r_e = \frac{V_T}{I} = \frac{25mV}{1mA} = 25\Omega$
$r_o = \frac{V_A}{I_C} = \frac{125V}{1mA} = 125K$

Ex. 9.10

$I = I_S e^{V_B/V_T} \Rightarrow V_{BE} = V_T \ln(I/I_S)$

and $I_3 = I_4$, $I_1 = I_2$

From cct: $V_{BE1} + V_{BE2} = V_{BE3} + V_{BE4}$

$V_T \ln\left[\frac{I_1}{I_{S1}}\right] + V_T \ln\left[\frac{I_2}{I_{S2}}\right] = V_T \ln\left[\frac{I_3 I_4}{I_{S3} I_{S4}}\right]$

$\therefore \ln\left[\frac{I_1^2}{I_{S1} \cdot I_{S2}}\right] = \ln\left[\frac{I_3^2}{I_{S3} I_{S4}}\right]$

$\therefore I_3 = I_1 \left[\frac{I_{S3} I_{S4}}{I_{S1} I_{S2}}\right]$

QED

Ex. 9.11

$V_{BE} = 0.7V$ for $I_C = 1mA$ for Q_{11}
$I_C = 10\mu A$ for Q_{10}

$V_{BE10} = 0.7 + V_T \ln\left[\frac{10\mu A}{1mA}\right] = 0.585V$

Voltage across $R_4 = V_{BE10} - V_{BE11}$
$= 0.7 - 0.585 = 0.115V$

$\Rightarrow I_{R4} = \frac{V_{R4}}{R_4} \Rightarrow R_4 = \frac{0.115V}{10\mu A} \Rightarrow 11.5K\Omega$

Ex. 9.12

$V_{ICM}(max) = V_{DD} - V_{BE8} - V_1 sat + V_{BE}$
$= +15 - 0.6 - 0.3 + 0.6$
$= +14.7V$

$V_{ICM}(min) = -V_{SS} + V_{BE5} + V_{BE7} + V_3 sat + V_{BE1}$
$= -15 + 0.6 + 0.6 + 0.3 + 0.6$
$= -12.9V$

(neglecting $R_1 + R_2$ drops)

Ex. 9.13

Assume Q18, Q19 have normal area
Q_{14}, Q_{20} have area = 3 x normal

$I_{14} = 0.25 I_{REF}\sqrt{\frac{I_{S14} \cdot I_{S20}}{I_{SA} \cdot I_{SB}}}$

$= 0.25(730)\sqrt{\frac{3I}{I} \cdot \frac{3I}{I}}$

$= 0.25(730)\,3 = 0.548mA$

Assumes $I_{REF} = 730\mu A$.

Ex. 9.14

Assume $I_{C7} \approx I_{C5} \approx I_{C6} = 9.5\mu A$

(a) $V_{B6} = I_E(R_2 + r_{e6}) = i_e(1 + 2.63)$
$= 3.63 I_3$

(b) $I_{E7} = \frac{V_{B6}}{R_3} + I_{B5} + I_{B6}$
$= \frac{V_{B6}}{R_3} + \frac{2I_E}{\beta+1} = \frac{3.63 I_E}{50} + \frac{2I_E}{201}$
$= 0.08 I_E$

(c) $I_{B7} = \frac{I_{E7}}{\beta+1} \approx \frac{0.08 I_E}{201} \approx 0.0004 I_E$

(d) $V_{B7} = V_{B6} + I_{E7} r_{e7}$
$= 3.63 I_E + 0.08 I_E \times \frac{25mV}{9.5\mu A}$
$= 3.84K\Omega \times I_E$

(e) $R_{IN} = \frac{(\beta+1)}{\beta}\frac{V_{B7}}{I_E} \approx 3.84K\Omega$

Ex. 9.15

Isolating the common-mode half circuit:

V_{ICM}, Q_1, i_e, Q_3, $2R_o$

$i = \frac{\beta}{\beta+1} i_e \approx i_e$

$= \frac{V_{ICM}}{r_{e1} + r_{e3} + \frac{2R_o}{1+\beta_P}}$

QED

Ex. 9.16

Let $R_1 = R$, $R_2 = R + \Delta R$
Assume $\beta >> 1$ and $r_{e5} = r_{e6}$

$V_{B5} = V_{B6} = i(r_{e5} + R_1)$
$= i(r_{e6} + R_2)$

$\therefore i_{C6} \Rightarrow \frac{i(r_{e5} + R)}{(r_{e5} + R + \Delta R)}$

$i_o = i_{C6} - i = i$

$$\therefore i_6 = i\left[\frac{r_{e5}+R}{r_{e5}+R+\Delta R}\right] - i$$

$$= i\left[\frac{(r_{e5}+R)-(r_{e5}+R+\Delta R)}{r_{e5}+R+\Delta R}\right]$$

$$= i\,\frac{\Delta R}{r_{e5}+R+\Delta R} \qquad \text{QED}$$

Ex. 9.17

Assume results of Ex 9.15 & 9.16

and $\Delta R << (R+r_e)$, $\beta >> 1$

$R_o/(\beta_p+1) >> (r_{e1}+r_{e3})$

New $G_{mcm} \equiv \frac{|i_o|}{v_{Icm}}$

$$\Rightarrow \left|\frac{-\Delta R}{R+r_{e5}+\Delta R}\cdot\frac{1}{r_{e1}+r_{e2}+\frac{2R_o}{\beta_p+1}}\right|$$

$$\Rightarrow \left|-\frac{\Delta R}{R+r_{e5}}\cdot\frac{\beta_p}{2R_o}\right| \text{ (approx)} \qquad \text{QED}$$

Ex. 9.18

Neglecting r_μ and using $I_{C9} = 19\,\mu A$ (Table)

$$r_{o9} = \frac{V_A}{I_{C9}} = \frac{50}{19\,\mu A} = 2.63\,M\Omega$$

(No R_{E9}) $\therefore R_{o9} = 2.63\,M\Omega$

For Q_{10}: $R_{o_{10}} = r_o[1+g_m(R_E\|r_\pi)]$

$$= 2.63[1+0.76(5\|2.63M)]$$

$$= 31.1\,M\Omega$$

Then $R_o = r_{o9}\|R_{o10}$

$$= 2.63\|31.1 = 2.43\,M\Omega$$

Ex. 9.19

For $\beta_p = 50$, $\Delta R/R = 0.02$, $R = 1K$

$$G_{mcm} \approx \frac{\Delta R}{R+r_{e5}}\cdot\frac{\beta_p}{2R_o}$$

$$= \frac{0.02}{1+2.63K/1K}\cdot\frac{50}{2\times 2.43} = 0.057\,\mu A/V$$

Ex. 9.20

$G_{mcm} = 0.057\,\mu A/V$, $G_{m1} = \frac{1}{5.26}\,mA/V$

Thus $\text{CMRR} = 20\log\left[\frac{1/5.26\times 10^{-3}}{0.057\times 10^{-6}}\right]$

$$= 70.5\,dB$$

Ex. 9.21

Assume loop gain $\approx \beta_p$

$$\text{CMRR} \rightarrow 20\log\left[\frac{G_{m1}}{G_{mcm}}\times\beta_p\right]$$

$$= 20\log\left[\frac{G_{m1}}{G_{mcm}}\right] + 20\log\beta_p$$

$$= 70.5 + 20\log 50 = 104.6\,dB$$

Ex. 9.22

$\beta_{16} = \beta_{17} = 200$

$$r_{e16} = \frac{25\,mV}{16.2\,\mu A} = 1.54\,K\Omega$$

$$r_{e17} = \frac{25\,mV}{0.55\,mA} = 45.5\,\Omega$$

$R_8 = 100\,\Omega$, $R_9 = 50\,K\Omega$

Subst. into Eq. (9.77)

$$R_{i2} = 201[1.54 + 50\|(201\times 0.0455)]$$

$$\approx 4\,M\Omega$$

Ex. 9.23

$$i_{c17} = \frac{\beta}{\beta+1}\cdot\frac{v_{b17}}{r_{e17}+R_8} \approx \frac{v_{b17}}{45.5+100} = \frac{v_{b17}}{145.5}$$

$$v_{b17} = v_{i2}\,\frac{(R_9\|R_{i17})}{(R_9\|R_{i17})+r_{e16}}$$

needs $R_{i17} = (\beta+1)(r_{e17}+R_8)$

$$= 201(45.5+100) = 29.2\,K\Omega$$

$$\therefore v_{b17} \approx v_{i2}\times 0.92$$

Hence $G_{m2} = \frac{i_{c19}}{v_{12}} \approx \frac{v_{b17}}{145.5} \times \frac{0.92}{v_{b17}}$

$= 6.3 \text{ mA/V}$

$(\approx 6.5 \text{mA/V})$

Ex.9.24

$R_{o2} = R_{o13B} \| R_{o17}$

where $R_{o13B} = r_{o13B} = \frac{50V}{0.55mA} = 90.9K\Omega$

$R_{o17} = r_{o17}(1+g_{m17}(R_8 \| r_{\pi17}))$

$r_{o17} = \frac{125V}{0.55\,mA} = 227.3K\Omega$

$g_{m17} = \frac{0.55mA}{0.025mV} = 22mA/V$

$r_{\pi17} = \frac{\beta}{g_m} = \frac{200}{22} = 9.09K\Omega$

$R_8 = 100\Omega$

Thus $R_{o17} \Rightarrow 722K\Omega$

Hence $R_{o2} = 90.9 \| 722K \approx 81K\Omega$

Ex.9.25

$\frac{I_t}{V_t} = \frac{1}{18.8} + 6.6 \times 0.917 = 0.05 + 6.05$

$= 6.11$

Open-circuit voltage gain

$(A_2)_{o/c} = -G_{m2} R_{o2}$

$= -6.5 \times 8.1 = -526.5 \text{ V/V}$

Ex.9.26

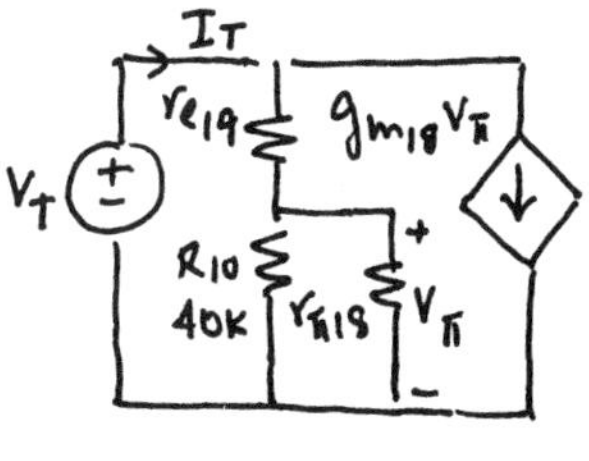

$r_{e19} = \frac{25mV}{16\mu A}$

$= 1.56K$

$r_{\pi18} = \frac{200}{40 \times 0.165}$

$= 30.3K\Omega$

$I_T = \frac{V_T}{r_{e19} + (R_{10} \| r_{\pi18})} + g_{m18} \frac{V_T (R_{10} \| r_{\pi18})}{r_{e19} + (R_{10} \| r_{\pi18})}$

$\therefore I_T = V_T \left[\frac{1}{18.8K} + \frac{6.6 \times 0.917}{18.8K} \right]$

$= V_T [0.05 + 6.05]$

$\Rightarrow R_T = \frac{V_T}{I_T} = 163\Omega$

Ex.9.27

$R_o \approx r_{e14} + \frac{\left[\frac{R_{o2}}{\beta_{23}} + r_{e23} + R\right]}{\beta_{14}+1}$

$\approx \frac{0.025mV}{5mA} + \frac{\left[\frac{81K}{51} + \frac{0.025}{0.18 \times 10^{-3}} + 163\right]}{201}$

$\approx 5 + 9.4 \approx 14.4\Omega$

Ex.9.28

$SR = 0.63 V/\mu s$

For $v = 10 \sin \omega t$

$\frac{dv}{dt} = 10\omega \cos t \Rightarrow \left|\frac{dv}{dt}\right|_{max} = 10\omega \text{ V/s}$

$SR = 10\omega = 10 \times 2\pi f_{max}$

$\Rightarrow f_{max} = \frac{SR}{20\pi} = \frac{0.63}{20\pi} \approx 10KHz.$

Ex.9.29

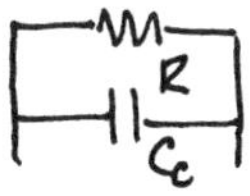

$A_0 = 2.43147 \times 10^5$

$G_{m1} = \frac{1}{5.26} \times 10^{-3}$

$\therefore A_0 = G_{m1} R$

$\therefore R = A_0 / G_{m1} \Rightarrow 1279 \text{ MHz}$

Ex.9.30

$SR = \frac{2I}{C_c}$, $\omega_t = \frac{G_{m1}}{C_c}$

$\Rightarrow SR = \frac{2I}{G_{m1}} \cdot \omega_t$

With R_E inserted in emitters of Q_3, Q_4

$$G_{M1} = 2 \times \frac{1}{4r_e + 2R_E} = \frac{1}{2r_e + R_E}$$

$$= \frac{1}{\frac{2 \times 0.025mV}{I} + R_E} = \frac{I}{2V_T + IR_E}$$

for $I = 9.5 \times 10^{-6}A$:

$$R_E = \frac{0.050}{9.5 \times 10^{-6}} = 5.26k\Omega$$

now $SR = \frac{2I\omega t}{I} \times (2V_T + IR_E)$

$$= 4[V_T + IR_E/2]\omega t \quad \text{QED}$$

new C_c: $\frac{G^*_{M1}}{C_c} = \frac{G_{M1}}{2C_c}$

∴ C_c must be reduced × factor of 2

$$C_{c\,new} = \frac{C_{c\,old}}{2} = \frac{30}{2} = 15pF$$

Gain $A \propto G_{M1}$ ∴ A also halved

$$A_{new} = A_{old} - 6dB = 101.7dB$$

$f_p = f_t/A$ a A has been halved

$$f_{p\,new} = 2 \times f_{p\,old} = 8.2Hz$$

Ex. 9.31

From text: 8 bit has 255 steps and defines resolution as $\frac{V_{max}}{2^N - 1} = V_{step}$.

Thus Resolution $= V_{step} \Rightarrow \frac{10}{255} \doteq 0.0392V$

However: 8 bit converter has 256 ranges which must be $\frac{V_{max}}{2^N}$ apart. → $V_{step} = \frac{10}{256}$

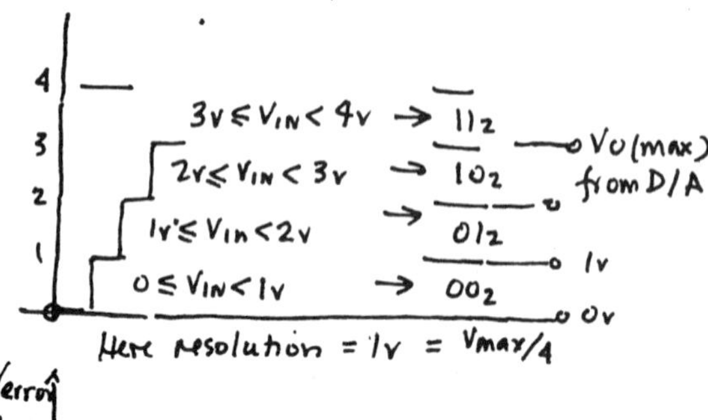

$0 \leq$ error < 1 LSB in each range

Using $V_{step} = \frac{V_{max}}{2^N - 1} = \frac{10}{255}$

Then $V_{IN} = 6v \equiv \frac{6 \times 255}{10} = 153$ steps

$153_{10} \equiv 128 + 16 + 8 + 1$

∴ Binary output $= 10011001_2$

And $V_{IN} = 6.2v \equiv \frac{6.2 \times 255}{10} = 158.1$ steps

$158_{10} = 128 + 16 + 8 + 4 + 2$

∴ Binary → 10011110_2

Now 158_{10} steps → $6.196v$

Absolute error $= 6.200 - 6.196v$

$= {}^{-}0.004v$ (too low)

% of input $= \frac{{}^{-}0.004}{6.2} \times 100 \approx {}^{-}0.064\%$

However using $V_{step} = 0.0392v$

Make $6.2v \Rightarrow 158 \times 0.0392v$

$= 6.1936v$

Absolute error $= 6.200 - 6.1936$

$= -0.0064v$

instead of $-0.004v$ above.

Error caused by approximation

$\frac{10}{255} \doteq 0.0392156\ldots$

Correct answers

Resolution $= V_{max}/2^N = 10/256 \approx 0.0390\ldots$

but safer to use ratio $= 10/256$

$6.2v \Rightarrow 158.71 \times 10/256 = 6.1719v$

Absolute error $= -0.0281V$

% of input $= -0.45 \approx -\frac{1}{2}\%$

% of F.S $= -0.28 \approx -\frac{1}{4}\%$

Consistent with <1 part in 256

[apparently larger error is due to V_{IN} falling different places within range 158.1 v. 158.71 steps but close enough to give ≈ 158. Also 0.71 > ½ LSB]

One must exercise care in assessing errors.

± ½ LSB can be obtain by converting (V_{IN} + ½LSB)

Ex. 9.32

Consider a simple 3 bit DAC

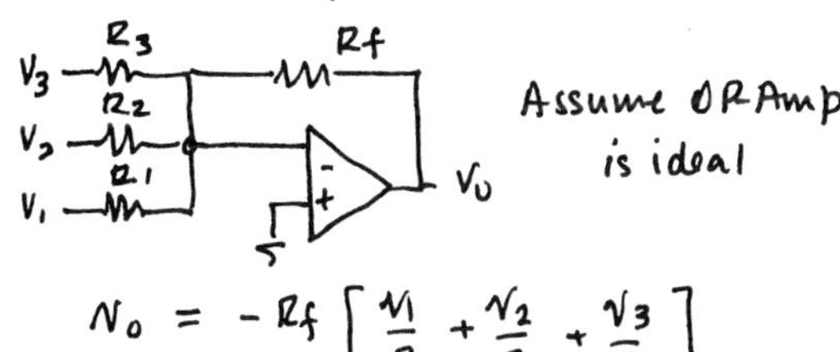

Assume OPAmp is ideal

$$V_o = -R_f\left[\frac{V_1}{R_1} + \frac{V_2}{R_2} + \frac{V_3}{R_3}\right]$$

if $V_1 = V_2 = V_3 = V$ and $R_3 = 2R_2 = 4R_1$

then $V_O = -R_f.V\left[\frac{1}{R} + \frac{1}{2R} + \frac{1}{4R}\right]$

$$V_{o\,max} = -\frac{R_f.V}{R}\left[\frac{4R + 2R + R}{4R}\right]$$

$$= -\frac{R_f.V\,2}{R}\left[\frac{1}{2} + \frac{1}{4} + \frac{1}{8}\right]$$

Then R_1 switches half, R_2 quarter...

and $V_O = -V_{MAX}\left[\frac{1}{2} + \frac{1}{4} + \frac{1}{8}\right]$

Ratio of resistors $\equiv \frac{8}{2} = 4 = 2^{(N-1)}$

∴ For 12 bit DAC

Ratio of R's $= 2^{(12-1)} = 2^{11} = 1024$

Ex. 9.33

From Ex 9.2 MSDigit switches $\frac{1}{2}V_{max}$ (or $\frac{1}{2}I_{max}$) ≡ $\frac{1}{2}$ number of steps

For 12 bit DAC there are $2^{12} = 2048$ steps and MSD has weight ≅ 1024 steps while LSD ≡ 1 step.

Here $0.5\mu A \equiv \frac{1}{4}$ LSB ∴ $2\mu A \equiv 1$ LSB

and thus MSD ≡ 2 x 1024 = 2048 μA

⇒ 2.048 mA

Ex 9.34

When only C_5 is connected to V_{REF}.
then voltage divider provides

$$V = V_{REF} \times \frac{C/16}{2C} = 4 \times \frac{1}{32} = \frac{1}{8}V$$

For 5-bit A/D converter 1LSD corresponds to $V_{max}/2^n$. Since $S_5 \equiv$ LSD, then

$$V_{max} = 2^n \times 1/8 = 32 \times 1/8 = 4V$$

[V_{max} applies to V_{IN}. Max $V_o = (4 - \frac{1}{8})V$]

Ex. 9.35

The max quantization error for a A/D converter depends upon how it is set up.

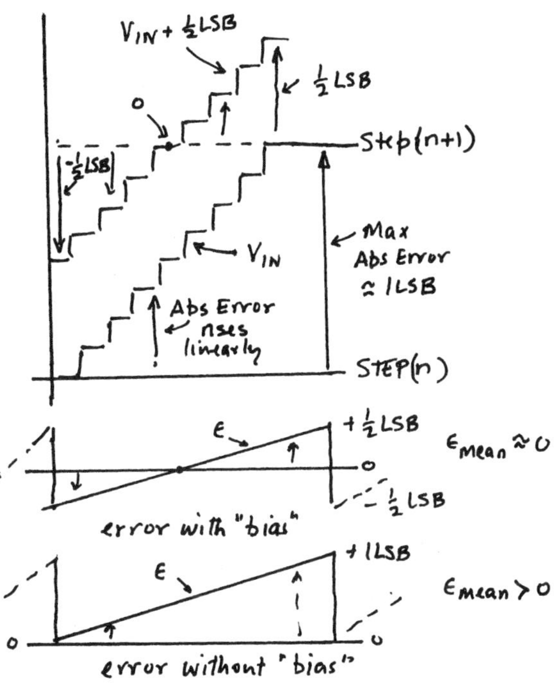

A "good" A/D will give ± ½ LSB

For a full-scale input voltage $= V_1(max)$
max. converted output voltage
$= V_1(max) - 1LSB$

An error of ½LSB $\equiv \frac{V_1(max)}{2 \times 2^N} = \frac{V_{FS}}{2^{N+1}}$

CHAPTER 10 - EXERCISES

10.1

PROPAGATION DELAY = 437 ps

10.2

PERCENTAGE REDUCTION = 50%.
$C = 4.225\,fF$
$t_{PHL} = 15.8\,ps$
$t_{PLH} = 20.5\,ps$
$t_P = 18.1\,ps$

10.3

POWER DISSIPATION = $19.5\,\mu W$

10.4

(a) NMOS DEVICES: $W/L = 0.75/0.5$
PMOS DEVICES: $12/0.5$
(b) NMOS DEVICES: $W/L = 3/0.5$
PMOS DEVICES: $3/0.5$
NOR AREA/NAND AREA = 2.125

10.5

REFER TO FIG. 10.19.

(a) THE MINIMUM CURRENT AVAILABLE TO CHARGE A LOAD CAPACITANCE IS THAT PROVIDED BY A SINGLE PMOS DEVICE. THE MAXIMUM CURRENT AVAILABLE TO CHARGE A LOAD CAPACITANCE IS THAT PROVIDED BY FOUR PMOS TRANSISTORS. THUS, THE RATIO IS $\underline{4}$.

(b) THERE IS ONLY ONE POSSIBLE CONFIGURATION (OR PATH) FOR CAPACITOR DISCHARGE. THUS THE MINIMUM AND MAXIMUM CURRENTS ARE THE SAME ⇒ RATIO IS $\underline{\underline{1}}$.

10.6

$(W/L)_n = 1.5 \quad (W/L)_p = 0.32$

$t_{PLH} = 0.5\,ns$
$t_{PHL} = 0.03\,ns$
THE NOISE MARGINS WILL <u>NOT</u> CHANGE.

10.7

$V_{OL} = 0.28\,V$
$NM_L = 0.59\,V$
$NM_H = 0.89\,V$
$(W/L)_p = 1.44$
$I_{stat} = 95.3\,\mu A$
$P_D = 0.24\,mW$
$t_{PLH} = 0.11\,ns$
$t_{PHL} = 0.03\,ns$
$t_P = 0.07\,ns$

10.8

Refer to Fig. 10.29(a).

$$i_{DN}(0) = \frac{1}{2} k_n (V_{DD} - V_{t0})^2 = \frac{1}{2} \times 50 \times \frac{4}{2}(5-1)^2 = \underline{\underline{800\ \mu A}}$$

$$i_{DP}(0) = \frac{1}{2} k_p (V_{DD} - V_{t0})^2 = \frac{1}{2} \times 20 \times \frac{4}{2}(5-1)^2 = \underline{\underline{320\ \mu A}}$$

Thus, $i_C(0) = 800 + 320 = 1120\ \mu A$

To obtain $i_{DN}(t_{PLH})$ we note that the situation is identical to that in Example 13.4 and thus we can use the results found in part (c) (page 1084),

$$i_{DN}(t_{PLH}) = \underline{50\ \mu A}$$

$$i_{DP}(t_{PLH}) = k_p\left[(V_{DD}-V_{t0})\frac{V_{DD}}{2} - \frac{1}{2}\left(\frac{V_{DD}}{2}\right)^2\right] = 20 \times \frac{4}{2}\left[(5-1)\,2.5 - \frac{1}{2}(2.5)^2\right] = \underline{\underline{275\ \mu A}}$$

Thus, $i_C(t_{PLH}) = 50 + 275 = 325\ \mu A$

$$i_C|_{av} = \frac{1}{2}(1120 + 325) = 722.5\ \mu A$$

$$t_{PLH} = \frac{C(V_{DD}/2)}{i_C|_{av}} = \frac{70\times10^{-15}\times 2.5}{722.5\times10^{-6}} = \underline{\underline{0.24\ ns}}$$

CONT.

10.8 CONT.

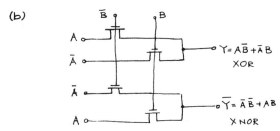

(b) Refer to Fig. 10.29(b).

$$i_{DN}(0) = \frac{1}{2} \times 50 \times \frac{4}{2}(5-1)^2 = \underline{\underline{800\ \mu A}}$$

$$i_{DP}(0) = \frac{1}{2} \times 20 \times \frac{4}{2}(5-1)^2 = \underline{\underline{320\ \mu A}}$$

Thus, $i_C(0) = 800 + 320 = 1120\ \mu A$

$$i_{DN}(t_{PHL}) = k_n\left[(V_{DD}-V_{t0})\frac{V_{DD}}{2} - \frac{1}{2}\left(\frac{V_{DD}}{2}\right)^2\right]$$

$$= 50 \times \frac{4}{2}\left[(5-1)\frac{5}{2} - \frac{1}{2}\left(\frac{5}{2}\right)^2\right] = \underline{\underline{688\ \mu A}}$$

To find $i_{DP}(t_{PHL})$ we first determine V_{tp} with $v_O = \frac{V_{DD}}{2}$ which corresponds to $|V_{SB}| = \frac{V_{DD}}{2}$

$$|V_{tp}| = V_{t0} + \gamma\left[\sqrt{\frac{V_{DD}}{2} + 2\phi_f} - \sqrt{2\phi_f}\right]$$

$$= 1 + 0.5\left[\sqrt{2.5+0.6} - \sqrt{0.6}\right] = 1.49\ V$$

Thus, $i_{DP}(t_{PHL}) = \frac{1}{2} \times 20 \times \frac{4}{2}(2.5 - 1.49)^2 = \underline{\underline{20\ \mu A}}$

$$i_C(t_{PHL}) = 688 + 20 = 708\ \mu A$$

$$i_C|_{av} = \frac{1120 + 708}{2} = 914\ \mu A$$

$$t_{PHL} = \frac{70 \times 10^{-15} \times 2.5}{914 \times 10^{-6}} = \underline{\underline{0.19\ ns}}$$

Q_P will turn off at $v_O = |V_{tp}|$ where

$$|V_{tp}| = V_{t0} + \gamma\left[\sqrt{V_{DD} - |V_{tp}| + 2\phi_f} - \sqrt{2\phi_f}\right]$$

$$= 1 + 0.5\left[\sqrt{5.6 - |V_{tp}|} - \sqrt{0.6}\right]$$

which results in a quadratic equation in $|V_{tp}|$ whose solution is

$$|V_{tp}| = \underline{\underline{1.6\ V}}$$

(c)
$$t_P = \frac{1}{2}(t_{PLH} + t_{PHL}) = \frac{1}{2}(0.24 + 0.19) = \underline{\underline{0.22\ ns}}$$

10.9

(a)

B̄, B, A, B inputs

Y = A + B
OR

Ā, B̄ inputs

$\bar{Y} = \bar{A}\,\bar{B}$
NOR

(b)

B̄, B, A, Ā inputs

$Y = A\bar{B} + \bar{A}B$
XOR

Ā, A inputs

$\bar{Y} = \bar{A}\bar{B} + AB$
XNOR

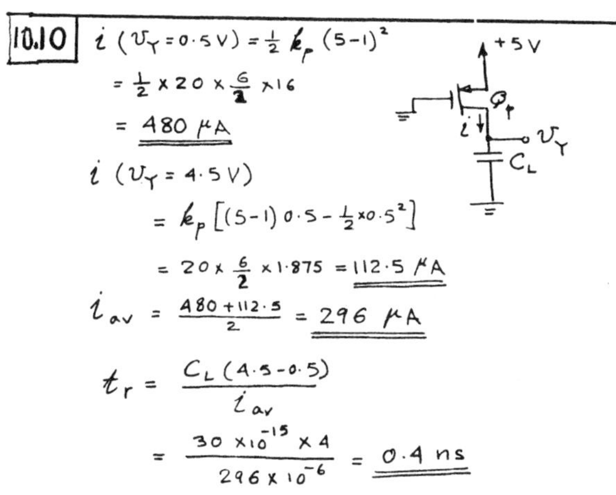

10.10

$$i(v_Y = 0.5\,V) = \frac{1}{2}k_p(5-1)^2 = \frac{1}{2} \times 20 \times \frac{6}{2} \times 16 = \underline{\underline{480\ \mu A}}$$

$$i(v_Y = 4.5\,V) = k_p\left[(5-1)0.5 - \frac{1}{2} \times 0.5^2\right] = 20 \times \frac{6}{2} \times 1.875 = \underline{\underline{112.5\ \mu A}}$$

$$i_{av} = \frac{480 + 112.5}{2} = \underline{\underline{296\ \mu A}}$$

$$t_r = \frac{C_L(4.5 - 0.5)}{i_{av}} = \frac{30 \times 10^{-15} \times 4}{296 \times 10^{-6}} = \underline{\underline{0.4\ ns}}$$

10.11

A, B, C, D, ϕ_e transistors each 4/2, output v_Y, C_L

+5V, i, v_Y, C_L

$$\left(\frac{W}{L}\right)_{eq} = \frac{1}{5}\,\frac{4}{2} = \underline{\underline{0.4}}$$

$$i(v_Y = 5\,V) = \frac{1}{2}k_n(5-1)^2 = \frac{1}{2} \times 50 \times 0.4 \times 16 = \underline{\underline{160\ \mu A}}$$

$$i(v_Y = 2.5\,V) = k_n\left[(5-1)2.5 - \frac{1}{2} \times 2.5^2\right] = 50 \times 0.4\,(10 - 3.125) = \underline{\underline{137.5\ \mu A}}$$

$$i_{av} = \frac{160 + 137.5}{2} = 149\ \mu A$$

$$t_{PHL} = \frac{C_L(V_{DD}/2)}{i_{av}} = \frac{30 \times 10^{-15} \times 2.5}{149 \times 10^{-6}} = \underline{\underline{0.5\ ns}}$$

10.12 Refer to Fig. E13.12.

(a) $\left(\frac{W}{L}\right)_{eq1} = \frac{1}{2}\left(\frac{W}{L}\right) = \frac{1}{2} \times \frac{4}{2} = \underline{\underline{1}}$

$\left(\frac{W}{L}\right)_{eq2} = \frac{1}{2}\left(\frac{W}{L}\right) = \frac{1}{2} \times \frac{4}{2} = \underline{\underline{1}}$

(b) $i_{D1}(v_{Y1} = V_{DD}) = \frac{1}{2} k_n'\left(\frac{W}{L}\right)_{eq1}(V_{DD} - V_t)^2$

$= \frac{1}{2} \times 50 \times 1 \, (5-1)^2$

$= \underline{\underline{400\ \mu A}}$

$i_{D1}(v_{Y1} = V_t) = k_n'\left(\frac{W}{L}\right)_{eq1}\left[(V_{DD} - V_t)V_t - \frac{1}{2}V_t^2\right]$

$= 50 \times 1 \left[(5-1)1 - \frac{1}{2} \times 1\right]$

$= \underline{\underline{175\ \mu A}}$

$i_{D1}|_{av.} = \frac{400 + 175}{2} = \underline{\underline{288\ \mu A}}$

(c) $i_{D1}|_{av}\, \Delta t = C_{L1}\, \Delta v_{Y1}$

$\Delta t = \frac{C_{L1}(V_{DD} - V_t)}{i_{D1}|_{av}} = \frac{40 \times 10^{-15} \times 4}{288 \times 10^{-6}}$

$= \underline{\underline{0.56 \text{ ns}}}$

(d) Following the hint we assume that Q_{eq} remains saturated during Δt.

$i_{D2}|_{av} = i_{D2}(v_{Y1} = 3\text{ V}) = \frac{1}{2} k_n'\left(\frac{W}{L}\right)_{eq2}(3-1)^2$

$i_{D2}|_{av} = \frac{1}{2} \times 50 \times 1 \, (3-1)^2$

$= \underline{\underline{100\ \mu A}}$

(e) $\Delta v_{Y2} = -\frac{i_{D2}|_{av.}\, \Delta t}{C_{L2}}$

$= -\frac{100 \times 10^{-6} \times 0.56 \times 10^{-9}}{40 \times 10^{-15}}$

$= \underline{\underline{-1.4\text{ V}}}$

Thus, v_{Y2} decreases to $\underline{\underline{3.6\text{ V}}}$.

CHAPTER 11 – EXERCISES

11.1 Refer to Fig. 11.4. Replace Q_5 and Q_6 with an equivalent transistor having a W/L equal to half that of Q_5 (or Q_6). With the input at $V_{DD}/2$ and $v_{\bar{\phi}} = V_{DD}/2$, the equivalent transistor will be in saturation, thus

$$\frac{1}{2} k_n' \times \frac{1}{2}\left(\frac{W}{L}\right)_5 \left(\frac{V_{DD}}{2} - V_t\right)^2$$

$$= k_p'\left(\frac{W}{L}\right)_2 \left[(V_{DD}-V_t)\frac{V_{DD}}{2} - \frac{1}{2}\left(\frac{V_{DD}}{2}\right)^2\right]$$

$$\frac{1}{2} \times 50 \times \frac{1}{2} \times \left(\frac{W}{L}\right)_5 (2.5-1)^2 = 20 \times \frac{10}{2}\left[4 \times 2.5 - \frac{1}{2} \times 2.5^2\right]$$

$$\Rightarrow \left(\frac{W}{L}\right)_5 = \underline{\underline{24.4}}$$

11.2 (a) At $t=0$, $v_{\bar{\phi}} = V_{DD}$

$i_2 = 0$

$$i_1 = \frac{1}{2} k_n' \left(\frac{W}{L}\right)_{eq} (V_{DD} - V_t)^2$$

$$= \frac{1}{2} \times 50 \times 4 \, (5-1)^2$$

$$= 1.6 \text{ mA}$$

Thus, $i_C(t=0) = 1.6$ mA

At $t = t_{PHL}$, $v_{\bar{\phi}} = \frac{V_{DD}}{2}$

$$i_2 = k_p'\left(\frac{W}{L}\right)_2 \left[(V_{DD}-V_t)\frac{V_{DD}}{2} - \frac{1}{2}\left(\frac{V_{DD}}{2}\right)^2\right]$$

$$= 20 \times \frac{10}{2}\left[(5-1)\, 2.5 - \frac{1}{2}(2.5)^2\right]$$

$$= 687.5$$

$$i = k_n'\left(\frac{W}{L}\right)_{eq} \left[(V_{DD}-V_t)\frac{V_{DD}}{2} - \frac{1}{2}\left(\frac{V_{DD}}{2}\right)^2\right]$$

$$= 50 \times 4 \left[(5-1)\, 2.5 - \frac{1}{2}(2.5)^2\right]$$

$$= 1375 \ \mu\text{A}$$

Thus, $i_C(t_{PHL}) = 1375 - 687.5 = 687.5 \ \mu$A

$$i_{C|av} = \frac{1.6 + 0.6875}{2} = 1.14 \text{ mA}$$

$$t_{PHL} = \frac{C\,(V_{DD}/2)}{i_{C|av}} = \frac{50 \times 10^{-15} \times 2.5}{1.14 \times 10^{-3}}$$

$$= \underline{\underline{0.11 \text{ ns}}}$$

(b) An estimate of the propagation delay through the inverter whose output is v_ϕ can be found using Eq. (13.19),

$$t_{PLH} = \frac{1.7\, C}{k_p'\left(\frac{W}{L}\right)_p V_{DD}}$$

$$= \frac{1.7 \times 50 \times 10^{-15}}{20 \times 10^{-6} \times 5 \times 5} = \underline{\underline{0.17 \text{ ns}}}$$

(c) The minimum width of the set pulse can be found by adding the delay times found in (a) and (b),

$$T_{min} = 0.11 + 0.17 = \underline{\underline{0.28 \text{ ns}}}$$

11.3

$$T = C(R + R_{on}) \ln\left[\frac{R}{R+R_{on}} \frac{V_{DD}}{V_{DD}-V_{th}}\right]$$

For $V_{th} = V_{DD}/2$ and $R_{on} \ll R$,

$$T \simeq CR \ln 2$$

$$= \underline{\underline{0.69\, CR}}$$

11.4 We choose values according to

$$T = 0.69\, CR = 10 \ \mu\text{s}$$

but with an eye on making R_{on} negligible. This can be achieved by selecting $R \gg R_{on}$

We do not, however, want to choose R too large for C becomes small and the operation of the one-shot becomes affected by the inevitable stray capacitances present in the circuit. Selecting $\underline{C = 1 \text{ nF}}$ yields

$$R = \frac{10 \times 10^{-6}}{0.69 \times 10^{-9}} = \underline{\underline{14.5 \text{ k}\Omega}}$$

To determine the effect of an R_{on} as high as 1 kΩ we use the formula

$$T = C(R + R_{on}) \ln\left[\frac{R}{R+R_{on}} \times 2\right]$$

$$= 10^{-9}\,(14.5+1)\,10^3 \ln\left(\frac{2 \times 14.5}{15.5}\right)$$

$$= 9.7 \ \mu\text{s}$$

Thus the error introduced by neglecting R_{on} in the design is $\frac{10-9.7}{10} \times 100 = \underline{\underline{3\%}}$.

11.5 Refer to the waveforms in Fig. 11.15(b)

During the capacitor charging,

$$v_{Y1} = v(\infty) - [v(\infty) - v(0)]\, e^{-t/\tau}$$

$$= V_{DD} - (V_{DD} - 0)\, e^{-t/\tau}$$

$$= V_{DD}\,(1 - e^{-t/\tau}) \quad \text{where } \tau = CR$$

The charging process terminates when v_{Y1} reaches V_{th}. Let this be $t = T_1$, thus

$$V_{th} = V_{DD}\,(1 - e^{-T_1/CR})$$

$$\Rightarrow T_1 = CR \ln\left(\frac{V_{DD}}{V_{DD} - V_{th}}\right) \qquad (1)$$

Now, for the capacitor discharge we can write

$$v_{Y1} = V_{DD}\, e^{-t/CR}$$

where we assumed $t = 0$ be the instant at which the discharge begins. The discharge process terminates when v_{Y1} reaches V_{th}. Let this be $t = T_2$, thus

$$V_{th} = V_{DD}\, e^{-T_2/CR}$$

$$\Rightarrow T_2 = CR \ln\left(\frac{V_{DD}}{V_{th}}\right) \qquad (2)$$

The period of oscillation T can be obtained as the sum of T_1 and T_2,

$$T = T_1 + T_2 = CR \ln\left(\frac{V_{DD}}{V_{DD} - V_{th}} \cdot \frac{V_{DD}}{V_{th}}\right)$$

Q.E.D.

11.6

$$f = \frac{1}{2 \times 5 t_p}$$

$$= \frac{1}{2 \times 5 \times 10^{-9}} = \underline{100\ \text{MHz}}$$

11.7

1024 rows ⇒ 10 bits for the row address

128 columns ⇒ 7 bits for the column address

32 blocks ⇒ 5 bits for the block address

11.8

$$v_W = V_{DD}\,(1 - e^{-t/CR})$$

$$\frac{V_{DD}}{2} = V_{DD}\,(1 - e^{t_d/CR})$$

$$t_d = CR \ln 2$$

$$= 2 \times 10^{-12} \times 5 \times 10^{3} \times 0.69$$

$$= \underline{6.9\ \text{ns}}$$

11.9 Refer to the circuit in Fig. 11.20(b).

(a) At $t = 0$,

$I_4 = \underline{0}$ (the voltage between its drain and source is zero)

$$I_6 = \tfrac{1}{2} k_n' \left(\tfrac{W}{L}\right)_6 (V_{DD} - 0 - V_{t0})^2$$

$$I_6 = \tfrac{1}{2} \times 50 \times \tfrac{10}{2}\,(5-1)^2$$

$$= \underline{2\ \text{mA}}$$

Thus, $I_{C\phi} = I_6 - I_4 = \underline{2\ \text{mA}}$

(b) At $t = \Delta t$,

$$I_4 = \tfrac{1}{2} k_p' \left(\tfrac{W}{L}\right)_4 \left(V_{DD} - \tfrac{V_{DD}}{2} - |V_{t0}|\right)^2$$

$$= \tfrac{1}{2} \times 20 \times \tfrac{10}{2}\,(5 - 2.5 - 1)^2$$

$$= \underline{0.11\ \text{mA}}$$

$$I_6 = k_n' \left(\tfrac{W}{L}\right)_6 \left[(V_{GS} - V_{t0})\frac{V_{DD}}{2} - \frac{1}{2}\left(\frac{V_{DD}}{2}\right)^2\right]$$

$$= 50 \times \tfrac{10}{2}\left[4 \times 2.5 - \tfrac{1}{2} \times 2.5^2\right]$$

$$= \underline{1.72\ \text{mA}}$$

Thus, $I_{C\phi} = 1.72 - 0.11 = \underline{1.61\ \text{mA}}$

(c) $I_{C\phi}|_{av} = \tfrac{1}{2}\,(2 + 1.61) \simeq 1.8\ \text{mA}$

(d)

$$\Delta t \simeq \frac{C\,\Delta V}{I_{C\phi}|_{av}} = \frac{C V_{DD}/2}{I_{C\phi}|_{av}}$$

$$= \frac{50 \times 10^{-15} \times 2.5}{1.8 \times 10^{-3}}$$

$$= \underline{69.4\ \text{ps}}$$

11.10

We use Eqs. (11.8) and (11.9)

$$\Delta V(1) \simeq \frac{C_S}{C_B}\left(\frac{V_{DD}}{2} - V_t\right)$$

$$= \frac{30 \times 10^{-15}}{1 \times 10^{-12}}\left(\frac{5}{2} - 1.5\right)$$

$$= \underline{30\ \text{mV}}$$

$$\Delta V(0) \simeq -\frac{C_S}{C_B}\frac{V_{DD}}{2}$$

$$= -\frac{30 \times 10^{-15}}{1 \times 10^{-12}}\,\frac{5}{2}$$

$$= \underline{-75\ \text{mV}}$$

11.11 Area of Storage array

$= 64 \times 1024 \times 1024 \times 2$

$= 134217728\ \mu m^2$

$= 134.2\ mm^2$

$= \underline{11.6\ mm \times 11.6\ mm}$

Total chip area $= 1.3 \times 134.2 = 174.46\ mm^2$

$= \underline{13.2\ mm \times 13.2\ mm}$

11.12 Refer to Example 11.3

Since Δt is proportional to $\tau = C/G_m$, we can reduce Δt from 6.65 ns to 4 ns by decreasing τ by the same factor. This in turn can be achieved by increasing G_m by the same factor $(6.65 \div 4 = 1.66)$. Sine $G_m = g_{mn} + g_{mb}$, both g_{mn} and g_{mb} have to be increased by the factor 1.66 (this is because we are asked to maintain the matched design of the inverter). The increase in g_m values can be achieved through a corresponding increase in the (W/L) ratios, thus

$$\left(\frac{W}{L}\right)_n = \frac{12}{4} \times 1.66 = \underline{5}$$

$$\left(\frac{W}{L}\right)_p = \frac{30}{4} \times 1.66 = \underline{12.5}$$

11.13 Refer to Example 11.3

With $\Delta V = 0.05$ V,

$$4.5 = 2.5 + 0.05\, e^{\Delta t/2.22}$$

$$\Delta t = 2.22\, \ln\left(\frac{2}{0.05}\right)$$

$$= \underline{8.19 \text{ ns}}$$

i.e. it increases by about 23%.

11.14 Refer to Fig. 11.26 Our decoder is an extension of that shown: We have M bits in the address (as opposed to 3) and correspondingly there will be 2^M word lines. Now, each of the 2^M word lines is connected to M NMOS devices and to one PMOS transistor. Thus the total number of devices required is

$$M\, 2^M \text{ (NMOS)} + 2^M \text{(PMOS)}$$

$$= 2^M (M+1)$$

11.15 Refer to Fig. 11.28: Our tree decoder will have 2^N bit lines. Thus it will have N levels: At the first levels there will be 2 transistors, at the second 2^2, ..., at the Nth level there will be 2^N transistors. Thus the total number of transistors can be found as

$$\text{Number} = 2 + 2^2 + 2^3 + \cdots + 2^N$$

$$= 2\underbrace{\left(1 + 2 + 2^2 + \cdots + 2^{N-1}\right)}$$

Geometric Series $r = 2$

$$\text{Sum} = \frac{r^N - 1}{r-1} = \frac{2^N - 1}{2-1} = 2^N - 1$$

Thus,

$$\text{Number} = \underline{2(2^N - 1)}$$

11.16 (a) $I_{av} = k_p'\left(\frac{W}{L}\right)_p\left[(5-1)2.5-\frac{1}{2}2.5^2\right]$

$= 20 \times \frac{24}{2}\left[10-3.125\right]$

$= 1.65$ mA

Thus, $t_{charging} = \frac{2 \times 10^{-12} \times 5}{1.65 \times 10^{-3}} = \underline{\underline{6.1 \text{ ns}}}$

(b)

$t_r \simeq 2.2\tau$

$= 2.2 \times 3 \times 10^{-12} \times 3 \times 10^{3}$

$= \underline{\underline{19.8 \text{ ns}}}$

(c) In one time-constant the voltage reached is $V_{DD}(1-e^{-1}) = 0.632\,V_{DD}$

$= 3.16$ V

$I_D = \frac{1}{2}k_n'\left(\frac{W}{L}\right)_n(3.16-1)^2$

$= \frac{1}{2} \times 50 \times \frac{6}{2} \times 2.16^2$

$= 0.35$ mA

$\Delta t = \frac{C_B \Delta V}{I_D}$

$= \frac{2 \times 10^{-12} \times 0.5}{0.35 \times 10^{-3}} = \underline{\underline{2.9 \text{ ns}}}$

11.17

$V_{OH} = 0$

$V_{OL} = -0.88$ V

SHOULD BE SHIFTED BY −0.88V

$V_{OH} = -0.88$ V AFTER SHIFTING

$V_{OL} = -1.76$ V AFTER SHIFTING

11.18 Refer to Fig. E11.18. Neglecting the base current of Q_1, the current through R_1, D_1, D_2 and R_2 is

$I = \frac{5.2 - V_{D1} - V_{D2}}{R_1 + R_2}$

$= \frac{5.2 - 0.75 - 0.75}{0.907 + 4.98} = 0.6285$ mA

Thus,

$V_B = -IR_1 = -0.57$ V

$V_R = V_B - V_{BE1} = -0.57 - 0.75 = \underline{\underline{-1.32 \text{ V}}}$

11.19 Refer to Fig. 11.34

$I_E = \frac{V_R - V_{BE}|_{Q_R} - (-V_{EE})}{R_E}$

$= \frac{-1.32 - 0.75 + 5.2}{0.779} \simeq \underline{\underline{4 \text{ mA}}}$

$V_C|_{Q_R} = -\alpha \times 4 \times R_{C2} \simeq -4 \times 0.245 \simeq \underline{\underline{-1 \text{ V}}}$

$V_C|_{Q_A, Q_B} = \underline{\underline{0 \text{ V}}}$ (because the current through R_{C1} is zero)

11.20

REFER TO FIG. 11.36.

For $v_I = V_{IL}$, $I_{Q_R} = 99\, I_{Q_A}$,

$$I_E = \frac{-1.32 - V_{BE}|_{Q_R} + 5.2}{0.779}$$

Assume $V_{BE}|_{Q_R} = 0.75$ V, $I_E = 4.018$ mA

$$I_{Q_R} \simeq 0.99 \times 4.018 = 3.98 \text{ mA}$$

Thus a better estimate of $V_{BE}|_{Q_R}$ is

$$V_{BE}|_{Q_R} = 0.75 + 0.025 \ln\left(\frac{3.98}{1}\right)$$

$$= 0.785 \text{ V}$$

and correspondingly,

$$I_E = \frac{-1.32 - 0.785 + 5.2}{0.779} = \underline{\underline{3.97 \text{ mA}}}$$

For $v_I = -1.32$ V, $I_{Q_R} = I_{Q_A} = I_E/2$,

$$I_E = \frac{-1.32 - 0.75 + 5.2}{0.779} = 4.018 \text{ mA}$$

Thus a better estimate for $V_{BE}|_{Q_R}$ is

$$V_{BE}|_{Q_R} = 0.75 + 0.025 \ln\left(\frac{2.009}{1}\right)$$

$$= 0.767 \text{ V}$$

and correspondingly,

$$I_E = \underline{\underline{4.00 \text{ mA}}}$$

For $v_I = V_{IH} = -1.205$ V,

$$I_{Q_A} = 99\, I_{Q_R},$$

$$I_E = \frac{-1.205 - 0.75 + 5.2}{0.779} = 4.166 \text{ mA}$$

Thus a better estimate for $V_{BE}|_{Q_A}$ is

$$V_{BE}|_{Q_A} = 0.75 + 0.025 \ln\left(\frac{0.99 \times 4.166}{1}\right)$$

$$= 0.788 \text{ V},$$

and correspondingly

$$I_E = \frac{-1.205 - 0.788 + 5.2}{0.799} = \underline{\underline{4.12 \text{ mA}}}$$

At $v_I = V_R$, $I_{Q_R} = \frac{1}{2} I_E = 2$ mA.

Thus,

$$V_C|_{Q_R} \simeq -2 \times 0.245 = -0.49 \text{ V}$$

$$v_{OR} = -0.49 - 0.75 = -1.24 \text{ V}$$

$$I_E|_{Q_2} = \frac{-1.24 + 2}{0.05} = 15.2 \text{ mA}$$

A better estimate for $V_{BE}|_{Q_2}$ is

$$V_{BE}|_{Q_2} = 0.75 + 0.025 \ln\left(\frac{15.2}{1}\right)$$

$$= 0.818 \text{ V}$$

Thus a better estimate for v_{OR} is

$$v_{OR} = -0.49 - 0.818 = \underline{\underline{-1.31 \text{ V}}}$$

11.21

REFER TO FIG. 11.40. for $v_I = V_{IH} = -1.205$ the value of I_E was found in Exercise 14.26 to be 4.12 mA. Thus $V_C|_{Q_A} \simeq -0.22 \times 4.12$

$$= -0.906 \text{ V}$$

$$v_{NOR} \simeq -0.906 - 0.75 = -1.656 \text{ V}$$

$$I|_{Q_3} = \frac{-1.656 + 2}{0.05} = 6.88 \text{ mA}$$

A better estimate for $V_{BE}|_{Q_3}$ is

$$V_{BE}|_{Q_3} = 0.75 + 0.025 \ln\left(\frac{6.88}{1}\right)$$

$$= 0.798 \text{ V}$$

and correspondingly

$$v_{NOR} = -0.906 - 0.798 = \underline{\underline{-1.704 \text{ V}}}$$

(b) for $v_I = V_{OH} = -0.88$ V,

$$I_E \simeq \frac{-0.88 - 0.75 + 5.2}{0.779} = 4.58 \text{ mA}$$

A better estimate for $V_{BE}|_{Q_A}$ is

$$V_{BE}|_{Q_A} = 0.75 + 0.025 \ln\left(\frac{4.58}{1}\right) = 0.788 \text{ V}$$

Thus,

$$I_E = \frac{-0.88 - 0.788 + 5.2}{0.779} = 4.53 \text{ mA}$$

$$V_C|_{Q_A} = -0.22 \times 4.53 = -1 \text{ V}$$

$$v_{NOR} = -1 - 0.75 = -1.75 \text{ V}$$

$$I|_{Q_R} = \frac{-1.75 + 2}{0.05} = 5 \text{ mA}$$

$$V_{BE}|_{Q_3} = 0.75 + 0.025 \ln\left(\frac{5}{1}\right)$$

$$= 0.79 \text{ V}$$

$$v_{NOR} = -1 - 0.79 = \underline{\underline{-1.79 \text{ V}}}$$

CONT.

11.21 CONT.

(c) The input resistance into the base of Q_3 is $(\beta+1)\,[r_{e3} + R_T]$

$$= 101\left[\frac{25}{5} + 50\right] = 5.55\ k\Omega$$

$$\frac{v_c|_{Q_A}}{v_i} = -\frac{(5.55\ k\Omega \,//\, 0.22\ k\Omega)}{r_e|_{Q_A} + R_E}$$

$$= \frac{-5.55 // 0.22}{\left(\frac{25}{4.53} + 779\right)\times 10^{-3}} = -0.269$$

$$\frac{v_{noR}}{v_c|_{Q_A}} = \frac{50\Omega}{50\Omega + 5\Omega} = 0.909$$

Thus,

$$\frac{v_{noR}}{v_c|_{Q_A}} = -0.269 \times 0.909 = \underline{\underline{-0.24\ V/V}}$$

d) See figure →

Assume $V_{BE} \simeq 0.79$ V (because the current will be 4 to 5 mA).

At the verge of saturation,

$I_C = \alpha I_E = 0.99\, I_E$

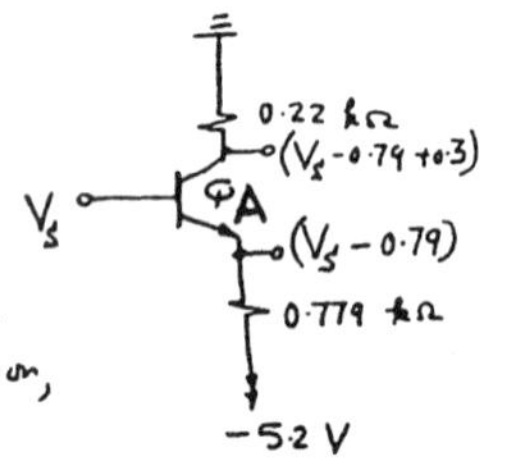

Circuit labels: 0.22 kΩ; $(V_S - 0.79 + 0.3)$; V_S; Q_A; $(V_S - 0.79)$; 0.779 kΩ; −5.2 V

Thus,

$$\frac{0 - V_S + 0.79 - 0.3}{0.22} = 0.99\ \frac{V_S - 0.79 + 5.2}{0.779}$$

$$\Rightarrow V_S = \underline{\underline{-0.58\ V}}$$

11.22

Refer to Fig. 11.34. For the reference circuit, the current through R_1, D_1, D_2 and R_2 is

$$\frac{5.2 - 2\times 0.75}{4.98 + 0.907} = 0.629\ mA$$

$V_B|_{Q_1} = -0.57$ V $\quad V_R = -0.57 - 0.75 = -1.32$ V

$$I_E|_{Q_1} = \frac{-1.32 + 5.2}{6.1} = 0.636\ mA$$

Thus the reference circuit draws a current of $(0.629 + 0.636) = 1.265$ mA from the 5.2-V supply. It follows that the power dissipated in the reference circuit is $1.265 \times 5.2 = 6.6$ mW. Since the reference circuit supplies four gates, the dissipation attributed to a gate is

$$\frac{6.6}{4} = 1.65\ mW$$

In addition, the gate draws a current $I_E \simeq 4$ mA from the 5.2-V supply. Thus the total power dissipation/gate is

$$P_D = 4\times 5.2 + 1.65 = \underline{22.4\ mW}$$

11.23

$$I_1 = \tfrac{1}{2}\mu_p C_{ox}\left(\tfrac{W}{L}\right)_1 (5 - 2.5 - 0.6)^2$$

$$I_2 = \tfrac{1}{2}\mu_n C_{ox}\left(\tfrac{W}{L}\right)_2 (2.5 - 0.7 - 0.6)^2$$

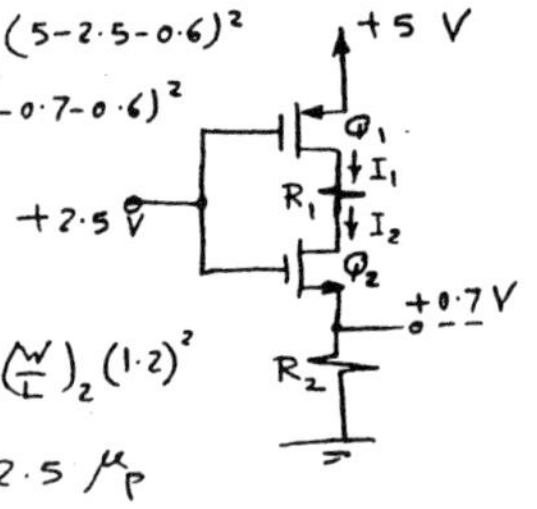

Equating I_1 & I_2 results in

$$\mu_p\left(\tfrac{W}{L}\right)_1 \times 1.9^2 = \mu_n\left(\tfrac{W}{L}\right)_2 (1.2)^2$$

Substituting $\mu_n = 2.5\,\mu_p$ and recalling that $L_1 = L_2$,

$$W_1 \times 1.9^2 = 2.5\, W_2 \times 1.44$$

$$\Rightarrow \frac{W_2}{W_1} = \underline{\underline{1}}$$

Circuit labels: +5 V; Q_1; I_1; R_1; I_2; +2.5 V; Q_2; +0.7 V; R_2

CHAPTER 12- EXERCISES

12.1

$A = -20 \log |T| \quad [dB]$

$\lvert T \rvert$ =	1	0.99	0.9	0.8	0.7	0.5	0.1	0
A ≃	0	0.1	1	2	3	6	20	∞

12.2

$A_{max} = 20 \log 1.05 - 20 \log 0.95 = \underline{\underline{0.9 dB}}$

$A_{min} = 20 \log (1/0.01) = \underline{\underline{40 \text{ dB}}}$

12.3

$$T(s) = k \frac{(s+j2)(s-j2)}{(s+1/2+j\sqrt{3}/2)(s+1/2-j\sqrt{3}/2)}$$

$$= k \frac{(s^2+4)}{s^2+s+1/4+3/4}$$

$$= k \frac{(s^2+4)}{s^2+s+1}$$

$$T(0) = k \frac{4}{1} = 1$$

$$k = 1/4$$

$$\therefore T(s) = \frac{1}{4} \frac{s^2+4}{s^2+s+1}$$

12.4

$$T(s) = k \frac{s(s^2+4)}{(s+0.1+j8)(s+0.1-j8)(s+0.1+j1.2)(s+0.1-j1.2)}$$

$$= k \frac{s(s^2+4)}{(s^2+0.2s+0.65)(s^2+0.2s+1.45)}$$

12.5

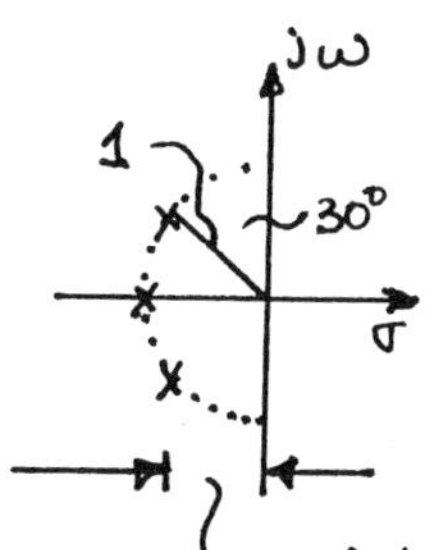

As shown, the pair of complex poles has $\omega_0 = 1$ and $Q = 1$

$$\omega_0/2Q = 1 \cos 60 = 1/2$$

$$\frac{1}{2Q} = \frac{1}{2}$$

$$Q = 1$$

$$\therefore T(s) = k \frac{1}{(s+1)(s^2+s+1)}$$

Since $T(0) = 1$, $k = 1$

Thus: $$T(s) = \underline{\underline{\frac{1}{(s+1)(s^2+s+1)}}}$$

$$T(j\omega) = \frac{1}{\sqrt{1+\omega^2}\sqrt{(1-\omega^2)^2+\omega^2}}$$

$$= \frac{1}{\sqrt{(1-\omega^4)(1-\omega^2)+\omega^2(1+\omega^2)}}$$

$$= \frac{1}{\sqrt{1-\omega^4-\omega^2+\omega^6+\omega^2+\omega^4}}$$

$$= \frac{1}{\sqrt{1+\omega^6}} \quad \text{Q.E.D.}$$

Thus: $$\frac{1}{\sqrt{2}} = \frac{1}{(1+\omega_{3dB}^6)^{1/2}} \Rightarrow \omega_{3dB} = \underline{\underline{1 \text{ rad/s}}}$$

$$A(3) = -20 \log \frac{1}{\sqrt{1+3^6}} = \underline{\underline{28.6 dB}}$$

12.6

$$\epsilon = \sqrt{10^{A_{max}/10} - 1} = \sqrt{10^{1/10} - 1} = 0.5088$$

$$|T(j\omega)| = \frac{1}{\sqrt{1 + \epsilon^2 \left(\frac{\omega}{\omega_p}\right)^{2N}}}$$

$$A(\omega_s) = -20\log|T(j\omega_s)|$$

$$= 10\log\left[1 + \epsilon^2\left(\frac{\omega_s}{\omega_p}\right)^{2N}\right]$$

Thus, $10\log\left[1 + 0.5088^2 \times 1.5^{2N}\right] \geq 30$

$N = 10$ LHS $= 29.35$ dB

$N = 11$ LHS $= 32.87$ dB

$\therefore$ Use $N = 11$ and obtain

$A_{min} = \underline{\underline{32.87 \text{ dB}}}$

For A_{min} to be exactly 30dB

$$10\log\left[1 + \epsilon^2 \times 1.5^{22}\right] = 30$$

$$\epsilon = 0.3654 \Rightarrow A_{max} = 20\log\sqrt{1 + 0.3654^2}$$

$$= \underline{\underline{0.54 \text{ dB}}}$$

12.7

The real pole is at $s = \underline{\underline{-1}}$

$\omega_p\left(\frac{1}{\epsilon}\right)^{1/3} = 1$

$j\omega$, σ, $60°$, $30°$

The complex conjugate poles are at

$s = -\cos 60° \pm j\sin 60°$

$= \underline{\underline{-0.5 \pm j\sqrt{3}/2}}$

$$T(s) = \frac{1}{(s+1)(s+0.5+j\sqrt{3}/2)(s+0.5-j\sqrt{3}/2)}$$

$$= \frac{1}{(s+1)(s^2+s+1)} \quad \text{(for DC gain} = 1\text{)}$$

12.8

$N = 5.$

ω_p

$$|T(j\omega)| = \frac{1}{\sqrt{1 + \epsilon^2\cos^2\left[N\cos^{-1}\left(\frac{\omega}{\omega_p}\right)\right]}}$$

for $\omega \leq \omega_p$.

Peaks are obtained when

$$\cos^2\left[N\cos^{-1}\left(\frac{\omega}{\omega_p}\right)\right] = 0$$

$$\cos^2\left[5\cos^{-1}(\hat{\omega}/\omega_p)\right] = 0$$

$$5\cos^{-1}(\hat{\omega}/\omega_p) = (2k+1)\frac{\pi}{2}, \quad k = 0, 1, 2$$

$$\therefore \hat{\omega} = \omega_p\cos\left[\frac{(2k+1)\pi}{10}\right], \quad k = 0, 1, 2$$

$$\hat{\omega}_1 = \omega_p\cos(\pi/10) = \underline{\underline{0.95\,\omega_p}}$$

$$\hat{\omega}_2 = \omega_p\cos\left(\tfrac{3}{10}\pi\right) = \underline{\underline{0.59\,\omega_p}}$$

$$\hat{\omega}_3 = \omega_p\cos\left(\tfrac{5}{10}\pi\right) = 0$$

Valleys are obtained when

$$\cos^2\left[N\cos^{-1}\left(\frac{\omega}{\omega_p}\right)\right] = 1$$

$$5\cos^{-1}(\omega/\omega_p) = k\pi, \quad k = 0, 1, 2$$

$$\therefore \check{\omega} = \omega_p\cos\left(\frac{k\pi}{5}\right), \quad k = 0, 1, 2$$

CONT.

$\check{\omega}_1 = \omega_p \cos 0 = \underline{\underline{\omega_p}}$

$\check{\omega}_2 = \omega_p \cos \frac{\pi}{5} = \underline{\underline{0.81\,\omega_p}}$

$\check{\omega}_3 = \omega_p \cos 2\pi/5 = \underline{\underline{0.31\,\omega_p}}$

12.9

$\epsilon = \sqrt{10^{A_{max}/10} - 1} = \sqrt{10^{0.5/10} - 1} = 0.3493$

$A(\omega_s) = 10\log\left[1 + \epsilon^2 \cosh^2\left(N \cosh^{-1} \frac{\omega_s}{\omega_p}\right)\right]$

$= 10\log\left[1 + 0.3493^2 \cosh^2(7\cosh^{-1} 2)\right]$

$= \underline{\underline{64.9\,dB}}$

For $A_{max} = 1$ dB, $\epsilon = \sqrt{10^{0.1} - 1} = 0.5088$

$A(\omega_s) = 10\log\left[1 + 0.5088^2 \cosh^2(7\cosh^{-1} 2)\right]$

$= \underline{\underline{68.2\,dB}}$

This is an increase of $\underline{\underline{3.3\,dB}}$

12.10

$\epsilon = \sqrt{10^{1/10} - 1} = 0.5088$

(a) For the Chebyshev Filter:

$A(\omega_s) = 10\log\left[1 + 0.5088^2 \cosh^2(N\cosh^{-1} 1.5)\right] \geq 50\,dB$

$N = 7.4$ ∴ Choose $N = \underline{\underline{8}}$

Excess Attenuation =

$10\log\left[1 + 0.5088^2 \cosh^2(8\cosh^{-1} 1.5)\right] - 50$

$= 55 - 50 = \underline{\underline{5\,dB}}$

(b) For a Butterworth Filter

$\epsilon = 0.5088$

$A(\omega_s) = 10\log\left[1 + \epsilon^2 \left(\frac{\omega_s}{\omega_p}\right)^{2N}\right]$

$= 10\log\left[1 + 0.5088^2 (1.5)^{2N}\right] \geq 50$

$N = 15.9$ ∴ Choose $N = \underline{\underline{16}}$

Excess attenuation =

$10\log\left[1 + 0.5088^2 (1.5)^{32}\right] - 50 = \underline{\underline{0.5\,dB}}$

12.11

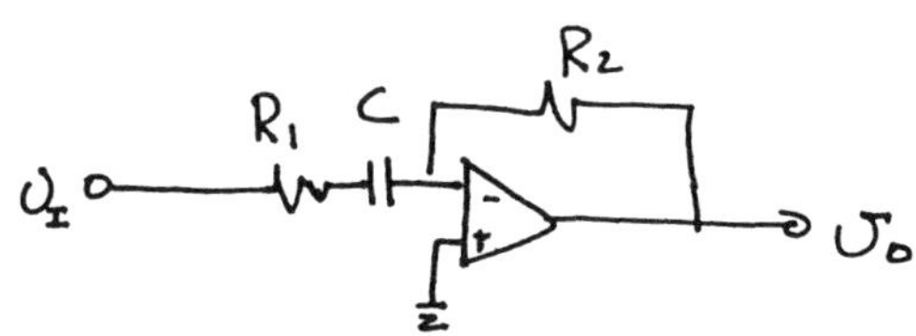

$10^4 = \frac{1}{CR_1}$

$R_1 = 10\,k\Omega$

$C = \underline{\underline{0.01\,\mu F}}$

H.f. Gain $= -\frac{R_2}{R_1} = -10$

$R_2 = \underline{\underline{100\,k\Omega}}$

12.12

Refer to Fig. 12.14

$\omega_0 = \frac{1}{CR} = 10^3$ rad/s

For R arbitrarily selected to be $10\,k\Omega$ $C = \frac{1}{10^3 \times 10^4} = \underline{\underline{0.1\,\mu F}}$

CONT.

The two resistors labelled R_1 can also be selected to be

$\underline{10k\Omega}$ each.

12.13

$$T(s) = \frac{\omega_0^2}{s^2 + s\sqrt{2}\omega_0 + \omega_0^2} \quad \text{(for dc gain = 1)}$$

$$|T(j\omega)| = \frac{\omega_0^2}{\sqrt{(\omega_0^2-\omega^2)^2 + 2\omega_0^2\omega^2}}$$

$$= \frac{\omega_0^2}{\sqrt{\omega_0^4 + \omega^4}}$$

$$= \frac{\omega_0^2}{\sqrt{1 + \left(\frac{\omega}{\omega_0}\right)^4}}$$

At $\omega = \omega_0$, $|T| = 1/\sqrt{2}$ which is 3dB below the value at dc (unity) Q.E.D.

12.14

$\omega_0 = 10^5$ rad/s

$\omega_0/Q = 10^3$ rad/s

Thus $T(s) = \dfrac{10^4 s}{s^2 + 10^3 s + 10^{10}}$ (selected to yield a centre-frequency gain of 10.)

12.15

(a)

A [dB] {0

BWa

ω_1 $\omega_n = \omega_0$ ω_2

$$T(s) = \frac{s^2 + \omega_0^2}{s^2 + s\,\omega_0/Q + \omega_0^2}$$

$$|T(j\omega)| = \frac{\omega_0^2 - \omega^2}{\sqrt{(\omega_0^2-\omega^2)^2 + \frac{\omega^2\omega_0^2}{Q^2}}}$$

$$= \frac{1}{\sqrt{1 + \frac{\omega_0^2\omega^2}{(\omega_0^2-\omega^2)^2 Q^2}}}$$

For any two frequencies ω_1 and ω_2 at which $|T|$ is the same

$$\frac{\omega_1^2 \omega_0^2}{(\omega_0^2-\omega_1^2)^2} = \frac{\omega_2^2\omega_0^2}{(\omega_0^2-\omega_2^2)^2}$$

$$\omega_1(\omega_0^2-\omega_2^2) = \omega_2(\omega_0^2-\omega_1^2)$$

$$\Rightarrow \omega_1\omega_2 = \omega_0^2 \quad (1)$$

Now to obtain attenuation $\geq A$ dB at ω_1 and ω_2 where $\omega_2 - \omega_1 = BWa$

$$10\log\left[1 + \frac{\omega_0^2\omega_1^2}{(\omega_0^2-\omega_1^2)^2 Q^2}\right] \geq A$$

$$\frac{\omega_1\omega_0}{\omega_0^2-\omega_1^2}\frac{1}{Q} \geq \sqrt{10^{A/10}-1} \quad \text{SUB (1)}$$

$$\frac{\omega_1\omega_0}{\omega_1\omega_2-\omega_1^2}\frac{1}{Q} \geq \sqrt{10^{A/10}-1}$$

$$\frac{\omega_0}{\omega_2-\omega_1}\frac{1}{Q} \geq \sqrt{10^{A/10}-1}$$

$$\frac{\omega_0}{BWa}\frac{1}{Q} \geq \sqrt{10^{A/10}-1}$$

$$\Rightarrow Q \leq \frac{\omega_0}{BWa\sqrt{10^{A/10}-1}} \quad \text{Q.E.D.}$$

(b) For $A = 3$dB

$$Q = \frac{\omega_0}{BW_3\sqrt{10^{0.3}-1}} = \frac{\omega_0}{BW_3}$$

OR $BW_3 = \omega_0/Q$ Q.E.D.

12.16

From Fig 12.16(e)

$$\omega_{max} = \omega_o \sqrt{\frac{(\omega_n/\omega_o)^2\left(1-\frac{1}{2Q^2}\right)-1}{(\omega_n/\omega_o)^2+\frac{1}{2Q^2}-1}}$$

For $\omega_o = 1\text{rad/s}$, $\omega_n = 1.2$ rad/s $Q=10$

& dc gain $= |a_2|\left(\frac{\omega_n^2}{\omega_o^2}\right) = 1$

$$|a_2| = \omega_o^2/\omega_n^2 = \frac{1}{1.44}$$

$$\omega_{max} = 1\sqrt{\frac{1.44\left(1-\frac{1}{200}\right)-1}{1.44+\frac{1}{200}-1}}$$

$$= \underline{\underline{0.986 \text{ rad/s}}}$$

$$|T(j\omega_{max})| = \frac{a_2|(\omega_n^2-\omega_{max}^2)|}{\sqrt{(\omega_o^2-\omega_{max}^2)^2+\left(\frac{\omega_o\omega_{max}}{Q}\right)^2}}$$

$$= \underline{\underline{3.17}}$$

$$|T(j\infty)| = a_2 = \frac{1}{1.44} = \underline{\underline{0.69}}$$

12.17

maximally flat $\Rightarrow Q = 1/\sqrt{2}$

$\omega_o = 2\pi \times 100 \times 10^3$

Arbitrarily selecting $R = \underline{\underline{1k\Omega}}$

$$Q = \omega_o CR \Rightarrow C = \frac{1}{\sqrt{2}\times 2\pi 10^5 \times 10^3}$$

$$= \underline{\underline{1125\, pF}}$$

Also $Q = \frac{R}{\omega_o L}$

$$\therefore L = \frac{R}{\omega_o Q} = \frac{10^3}{2\pi 10^5 \times 1/\sqrt{2}} = \underline{\underline{2.25\text{mH}}}$$

12.18

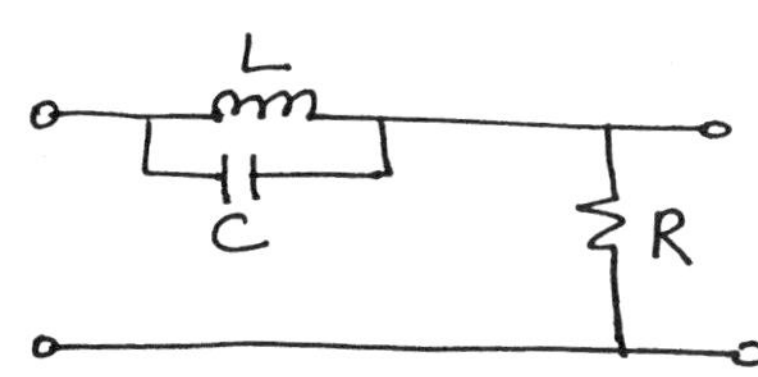

From Exercise 12.16 above

3dB bandwidth $= \omega_o/Q$

$$2\pi 10 = 2\pi 60/Q \Rightarrow Q = 6$$

$$Q = \omega_o CR$$

$$6 = 2\pi 60 \times C \times 10^4 \Rightarrow C = \underline{\underline{1.6\,\mu F}}$$

$$Q = \frac{R}{\omega_o L}$$

$$L = \frac{R}{\omega_o Q} = \frac{10^4}{2\pi 60 \times 6} = \underline{\underline{4.42\, H}}$$

12.19

$f_o = 10\text{kHz}$ $\quad \Delta f_{3dB} = 500\text{Hz}$

$$Q = \frac{f}{\Delta f_{3dB}} = \frac{10^4}{500} = 20$$

Using the data at the top of Table 12.1:

$$C_A = C_6 = \underline{\underline{1.2\, nF}}$$

$$R_1 = R_2 = R_3 = R_5 = \frac{1}{\omega_o C} = \frac{1}{2\pi 10^4 \times 1.2\times10^{-9}}$$

$$= \underline{\underline{13.26\, k\Omega}}$$

CONT.

$R_6 = Q/\omega_0 C = \dfrac{20}{2\pi 10^4 \times 1.2\times 10^{-9}} = \underline{\underline{265\,k\Omega}}$

Now using the data in Table 12.1 for the bandpass case

K = centre-frequency gain $= 10$

$1 + r_2/r_1 = 10$

Selecting $r_1 = \underline{10\,k\Omega}$ then $r_2 = \underline{90\,k\Omega}$

12.20

Eq (12.25) ~ $\omega_p = 2\pi 10^4$ ~

$$T(s) = \frac{\omega_p^5}{8.1408(s + 0.2895\omega_p)} \times \frac{1}{(s^2 + s0.4684\omega_p + 0.4293\omega_p^2)} \times \frac{1}{(s^2 + s0.1789\omega_p + 0.9883\omega_p^2)}$$

The circuit consists of 3 sections in cascade:

(a) First Order Section

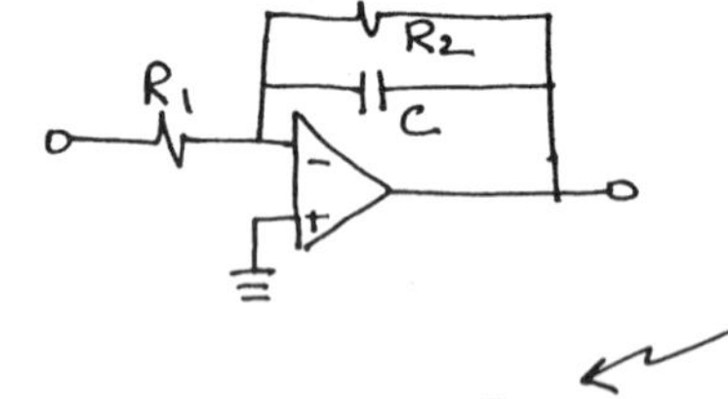

$$T(s) = \frac{-0.2895\omega_p}{s + 0.2895\omega_p}$$

the numerator coefficient was set so that the dc gain = 1

Let $R_1 = \underline{\underline{10\,k\Omega}}$

dc gain $= R_2/R_1 = 1 = R_2 = \underline{\underline{10\,k\Omega}}$

as $j\omega \to \infty$

$$|T(j\omega)| \to \frac{0.2895\omega_p}{\omega} = \frac{1}{\omega C R_1}$$

$$C = \frac{1}{0.2895 \times 2\pi 10^4 \times 10^4} = \underline{\underline{5.5\,nF}}$$

(b) Second-Order section with transfer-function:

$$T(s) = \frac{0.4295\omega_p^2}{s^2 + 0.4684\omega_p + 0.4293\omega_p^2}$$

where the numerator coefficient was selected to yield a dc gain of unity.

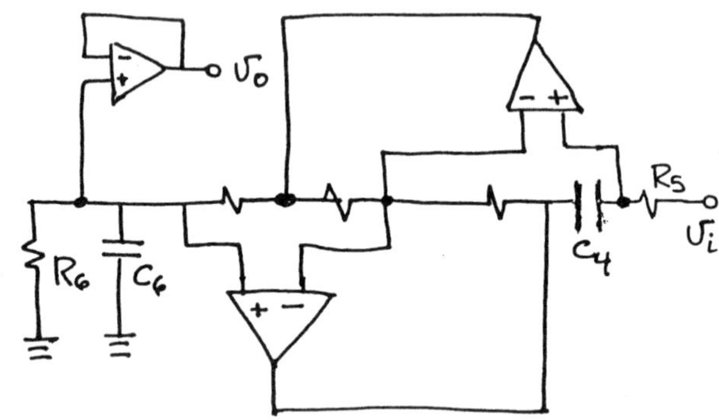

Select $R_1 = R_2 = R_3 = R_5 = \underline{\underline{10\,k\Omega}}$

$$\Rightarrow C = \frac{1}{\sqrt{0.4293} \times 2\pi 10^4 \times 10^4} = 2.43\,nF$$

$C_4 = C_6 = C = \underline{\underline{2.43\,nF}}$

$$Q = \frac{\sqrt{0.4293}\,\omega_p}{0.4684\,\omega_p} = 1.4 \Rightarrow R_6 = \frac{Q}{\omega_0 C} = \underline{\underline{14\,k\Omega}}$$

(c) Second-Order Section with Transfer-function:

$$T(s) = \frac{0.9883\,\omega_p^2}{s^2 + s0.1789\omega_p + 0.9883\omega_p^2}$$

CONT.

The circuit is similar to that in (b) above but with

$R_1 = R_2 = R_3 = R_5 = \underline{10\,k\Omega}$

$C_4 = C_6 = \frac{1}{\omega_0 \times 10^4} = \frac{1}{\sqrt{0.9883} \times 2\pi 10^4 \times 10^4}$

$= \underline{1.6\,nF}$

$Q = \frac{\sqrt{0.9883}}{0.1789} = 5.56$

Thus $R_6 = Q/\omega_0 C = \underline{55.6\,k\Omega}$

Placing the three sections in cascade, i.e. connecting the output of the first-order section to the input of the second-order section in (b) and the output of of section (b) to the input of (c) results in the overall transfer function in eq.(12.25).

12.21

Refer to the KHN circuit in Fig.12.24(a)
Choosing C = 1nF

$R = \frac{1}{\omega_0 C} = \frac{1}{2\pi\, 10^4 \times 10^{-9}} = \underline{15.9\,k\Omega}$

Using Eq(12.62) and selecting $R_1 = \underline{10\,k\Omega}$

$R_f = R_1 = \underline{10\,k\Omega}$

Using Eq(12.63) and setting $R_2 = \underline{10\,k\Omega}$

$R_3 = R_2(2Q-1) = 10(2\times 2-1) = \underline{30\,k\Omega}$

High frequency gain $= K = 2 - 1/Q = \underline{1.5\,V/V}$

The transfer function to the output of the first integrator is

$$\frac{V_{bp}}{V_i} = -\frac{1}{sCR} = \frac{V_{hp}}{V_i} = \frac{sK/CR}{s^2 + s\frac{\omega_0}{Q} + \omega_0^2}$$

Thus the centre-frequency gain

$= \frac{K}{CR}\frac{Q}{\omega_0} = KQ = 1.5 \times 2 = \underline{3\,V/V}$

12.22

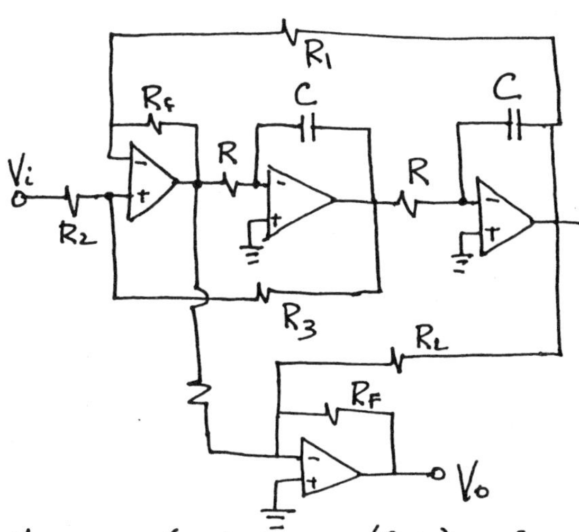

$$\frac{V_o}{V_i} = -K\frac{(R_F/R_H)s^2 + (R_F/R_L)\omega_0^2}{s^2 + s\,\omega_0/Q + \omega_0^2}$$

given $C = \underline{1nF}$ $R_L = \underline{10\,k\Omega}$

$R = \frac{1}{\omega_0 C} = \frac{1}{2\pi 5\times 10^3 \times 10^{-9}} = \underline{31.83\,k\Omega}$

$R_1 = 10\,k\Omega \Rightarrow R_F = \underline{10\,k\Omega}$

$R_2 = 10\,k\Omega \Rightarrow R_3 = R_2(2Q-1)$

$= 10(10-1) = \underline{90\,k\Omega}$

$\frac{R_H}{R_L}\omega_0^2 = \omega_n^2 \Rightarrow R_H = 10\left(\frac{8}{5}\right)^2$

$= \underline{25.6\,k\Omega}$

DC gain $= K\frac{R_F}{R_L} = \left(2 - \frac{1}{Q}\right)\frac{R_F}{R_L} = 3$

$R_F = \frac{3 \times 10}{2 - 1/5} = \underline{16.7\,k\Omega}$

12.23

Refer to Fig 12.25 (b)

$CR = 1/\omega_0 \Rightarrow C = \dfrac{1}{2\pi\, 10^4 \times 10^4} = \underline{\underline{1.59\,nF}}$

$R_d = QR = 20 \times 10 = \underline{\underline{200\,k\Omega}}$

Centre frequency gain = $KQ = 1$

$\therefore K = 1/Q = 1/20$

$R_g = R/K = 20R = \underline{\underline{200k\Omega}}$

12.24

Refer to Fig 12.26 and Table 12.2

$C = 10\,nF$

$R = \dfrac{1}{\omega_0 C} = \dfrac{1}{10^4 \times 10 \times 10^{-9}} = \underline{10\,k\Omega}$

$QR = 5 \times 10 = \underline{\underline{50k\Omega}}$

$C_1 = C \times \text{flat gain} = 10 \times 1 = \underline{\underline{10\,nF}}$

$R_1 = \underline{\underline{\infty}}$

$R_2 = R/\text{gain} = R/1 = \underline{10\,k\Omega}$

$r = \underline{\underline{10\,k\Omega}}$

$R_3 = \dfrac{Qr}{\text{gain}} = \dfrac{5 \times 10}{1} = \underline{\underline{50\,k\Omega}}$

12.25

From Eq (12.76)

$CR = \dfrac{2Q}{\omega_0} = \dfrac{2 \times 1}{10^4} = 2 \times 10^4\, s$

For $C = C_1 = C_2 = 1nF$

$R = \dfrac{2 \times 10^{-4}}{10^{-9}} = \underline{\underline{200\,k\Omega}}$

Thus $R_3 = \underline{\underline{200k\Omega}}$

From Eq. (12.75)

$m = 4Q^2 = 4$

Thus, $R_4 = \dfrac{R}{m} = \dfrac{200}{4} = \underline{\underline{50k\Omega}}$

12.26

The transfer function of the feedback network is given in Fig. (12.28a). The poles are the roots of the denominator polynomial,

$$s^2 + s\left(\frac{1}{C_1 R_3} + \frac{1}{C_2 R_3} + \frac{1}{C_1 R_4}\right) + \frac{1}{C_1 C_2 R_3 R_4} = 0$$

For $C_1 = C_2 = 10^{-9}\,F$, $R_3 = 2 \times 10^5\,\Omega$ & $R_4 = 5 \times 10^4\,\Omega$

$$s^2 + s\left(\frac{2}{10^{-9} \times 2 \times 10^5} + \frac{1}{10^{-9} \times 5 \times 10^4}\right) + \frac{1}{10^{-18}\, 10^{10}} = 0$$

$$s^2 + s\,(3 \times 10^4) + 10^8 = 0$$

$$s = \frac{-3 \times 10^4 \pm \sqrt{9 \times 10^8 - 4 \times 10^8}}{2}$$

$$= \underline{\underline{-0.382 \times 10^4 \text{ and } -2.618 \times 10^4 \; \text{rad/s}}}$$

12.27

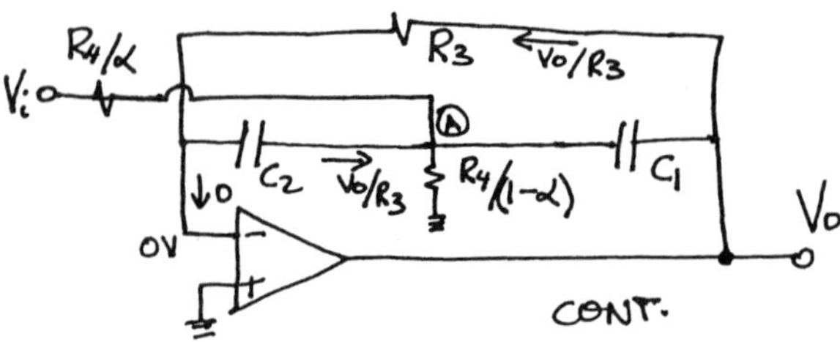

CONT.

$$V_A = 0 - \frac{V_o}{sC_2R_3}$$

ΣI at Ⓐ

$$\frac{V_o}{R_3} + sC_1(V_o - V_A) + \frac{-V_A}{R_4/(1-\alpha)} + \frac{V_i - V_A}{R_4/\alpha} = 0$$

$$\frac{V_o}{R_3} + sC_1V_o + \frac{sC_1V_o}{sC_2R_3} + \frac{(1-\alpha)V_o}{sC_2R_3R_4} + \frac{\alpha V_i}{R_4} + \frac{\alpha V_o}{sC_2R_3R_4} = 0$$

$$\frac{V_o}{V_i} = \frac{-\alpha/R_4}{sC_1 + 1/R_3 + \frac{C_1}{C_2R_3} + \frac{1}{sC_2R_3R_4}}$$

$$= \frac{-s\,\alpha/(C_1R_4)}{s^2 + s\left(\frac{1}{C_1R_3} + \frac{1}{C_2R_3}\right) + \frac{1}{C_1C_2R_3R_4}}$$

This is a bandpass function whose poles are identical to the zeros of $t(s)$ in Fig. (12.28a).

For $C_1 = C_2 = 10^{-9}$F, $R_3 = 2\times10^5\,\Omega$ & $R_4 = 5\times10^4\,\Omega$

$$\frac{V_o}{V_i} = \frac{-s\times2\times10^4\times\alpha}{s^2 + s\times10^4 + 10^8}$$

For unity centre-frequency gain

$2\times10^4\times\alpha = 10^4 \Rightarrow \alpha = 0.5$

Thus $\frac{R_4}{\alpha} = \underline{\underline{100k\Omega}}$; $\frac{R_4}{1-\alpha} = \underline{\underline{100k\Omega}}$

12.28

$$V_A = V_o + sC_3V_oR_2$$
$$= V_o(1 + sC_3R_2)$$

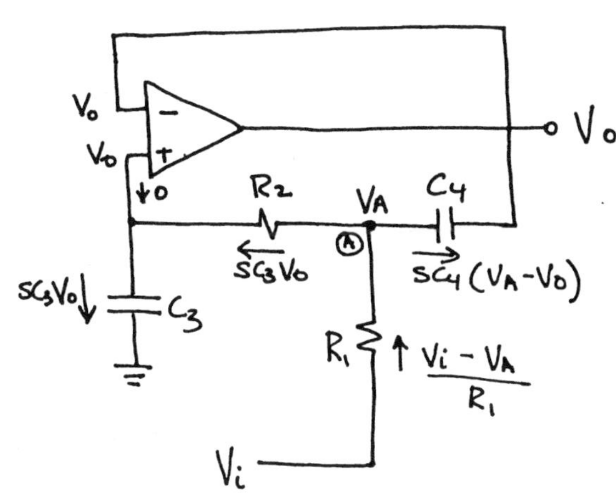

ΣI at Ⓐ

$$sC_3V_o - \frac{V_i}{R_1} + \frac{V_o}{R_1} + \frac{V_o}{R_1}sC_3R_2 + V_osC_4(sC_3R_2)$$

$$\frac{V_i}{R_1} = V_o\left[s^2C_3C_4R_2 + \frac{sC_3R_2}{R_1} + sC_3 + \frac{1}{R_1}\right]$$

$$\frac{V_o}{V_i} = \frac{1/C_3C_4R_1R_2}{s^2 + s\frac{1}{C_4R_2}\left(1 + \frac{R_2}{R_1}\right) + \frac{1}{C_3C_4R_1R_2}}$$

$$\omega_0 = \frac{1}{\sqrt{C_3C_4R_1R_2}} \quad \text{as in Eq.(12.77)}$$

$$Q = \frac{1}{\sqrt{C_3C_4R_1R_2}}\;\frac{C_4}{\left(\frac{1}{R_1}+\frac{1}{R_2}\right)} \quad \text{as in Eq (12.78)}$$

D.C. gain = 1 Q.E.D.

12.29

From Fig 12.34 (c)

$R_1 = R_2 = R = \underline{10k\Omega}$

$CR = 2Q/\omega_0$

$$C = \frac{2Q}{\omega_0 R} = \frac{2/\sqrt{2}}{2\pi(4\times10^3)\,10^4} = \underline{\underline{5.63nF}}$$

$m = 4Q^2 = 4\left(\frac{1}{2}\right) = 2$

$C_3 = \underline{\underline{2.81nF}}$

12.30

Refer to the results in Example 12.3

(a) $\Delta R_3/R_3 = +2\%$

$S^{\omega_0}_{R_3} = -1/2 \Rightarrow \frac{\Delta\omega_0}{\omega_0} = -\frac{1}{2}\times 2 = \underline{\underline{-1\%}}$

$S^{Q}_{R_3} = \frac{1}{2} \Rightarrow \Delta Q/Q = \frac{1}{2}\times 2 = \underline{\underline{1\%}}$

(b) $\Delta R_4/R_4 = 2\%$

$S^{\omega_0}_{R_4} = -1/2 \Rightarrow \frac{\Delta\omega_0}{\omega_0} = \underline{\underline{-1\%}}$

$S^{Q}_{R_4} = -1/2 \Rightarrow \Delta Q/Q = -\frac{1}{2}\times 2 = \underline{\underline{-1\%}}$

(c) Combining the results in (a) & (b)

$\Delta\omega_0/\omega_0 = -1-1 = \underline{\underline{-2\%}}$

$\frac{\Delta Q}{Q} = 1-1 = \underline{\underline{0\%}}$

(d) Using the results in (c) for both resistors being 2% high we have:

$$\frac{\Delta\omega_0}{\omega_0} = S^{\omega_0}_{C_1}\frac{\Delta C_1}{C_1} + S^{\omega_0}_{C_2}\frac{\Delta C_2}{C_2} - 2$$

$$= \frac{-1}{2}(-2) + \frac{-1}{2}(-2) - 2$$

$$= 2-2 = \underline{\underline{0\%}}$$

$$\frac{\Delta Q}{Q} = S^{Q}_{C_1}\frac{\Delta C_1}{C} + S^{Q}_{C_2}\frac{\Delta C_2}{C_2} + 0$$

$$= 0(-2) + (0)(-2) + 0$$

$$= \underline{\underline{0\%}}$$

12.31

From Eq (12.96) & (12.97)

$$C_3 = C_4 = \omega_0 T_c C$$

$$= 2\pi 10^4 \times \frac{1}{200\times 10^3} \times 20$$

$$= \underline{\underline{6.283\ pF}}$$

From Eq. (12.99)

$$C_5 = \frac{C_4}{Q} = \frac{6.283}{20} = \underline{\underline{0.314\ pF}}$$

From Eq (12.100)

$$\text{Centre-frequency gain} = \frac{C_6}{C_5} = 1$$

$$C_6 = C_5 = \underline{\underline{0.314\ pF}}$$

12.32

$$R_p = \omega_0 L Q_0 = 2\pi 10^6 \times 3.2\times 10^{-6} \times 150 = \underline{\underline{3k\Omega}}$$

$$R = R_L \,//\, r_o \,//\, R_p = 2k\Omega \Rightarrow R_L = \underline{\underline{15k\Omega}}$$

12.33

$$Q = (R_1 \,//\, R_{in})/\omega_0 L$$

$$= \frac{10^3 // 10^3}{2\pi \times 455\times 10^3 \times 5\times 10^{-6}} = 35$$

$$BW = f_0/Q = 455/35 = \underline{\underline{13\ kHz}}$$

$$C_1 + C_{in} = \frac{1}{\omega_0^2 L}$$

$$= \frac{1}{(2\pi\times 455\times 10^3)^2 \times 5\times 10^{6}}$$

$$= \underline{\underline{24.47\ nF}}$$

$$C_1 = 24.47 - 0.2 = \underline{\underline{24.27\ nF}}$$

12.34

To just meet specifications

$$Q = \frac{f_0}{BW} = \frac{455}{10} = 45.5$$

$$\therefore \frac{R_1 \| n^2 R_{in}}{\omega_0 L} = 45.5$$

$$R_1 \| n^2 R_{in} = 45.5 \times 455 \times 10^3 \times 5 \times 10^{-6}$$
$$= 650\,\Omega$$
$$n^2 R_{in} = 1.86\ k\Omega$$
$$n = \sqrt{\frac{1.86}{1}} = \underline{\underline{1.36}}$$

$$C_1 + \frac{C_{in}}{n^2} = \frac{1}{\omega_0^2 L} = 24.47$$
$$C_1 = \underline{\underline{24.36\,nF}}$$

At resonance, the voltage developed across R_1 is $I\left(R_1 \| n^2 R_{in}\right)$. Thus, $V_{be} = IR/n$ & $I_c = g_m V_{be} = g_m IR/n$, hence

$$\frac{I_c}{I} = g_m R/n = \frac{40 \times 0.65}{1.36} = \underline{\underline{19.1\ \frac{A}{A}}}$$

12.35

$$200 = f_0/Q\,\sqrt{2^{1/2}-1} \qquad Eq\ (12.110)$$
$$\frac{f_0}{Q} = \underline{\underline{310.8\ kHz}}$$

$$C = \frac{1}{\omega_0^2 L} = \frac{1}{(2\pi\, 10.7\times10^6)^2 \times 3\times10^{-6}}$$
$$= \underline{\underline{73.7\,pF}}$$

$$\frac{\omega_0}{Q} = \frac{1}{CR}$$

$$R = \frac{1}{73.7\times10^{-12} \times 2\pi \times 310.8\times10^3} = \underline{\underline{6.95\ k\Omega}}$$

12.36

$$f_{01} = f_0 + \frac{2\pi B}{2\sqrt{2}} \qquad Eq\ (12.115)$$
$$= 10.7\,MHz + \frac{200\,kHz}{2\sqrt{2}} = \underline{\underline{10.77\,MHz}}$$

$$B_1 = B/\sqrt{2} = 200/\sqrt{2} = \underline{\underline{141.4\ kHz}}$$

$$f_{02} = f_0 - \frac{2\pi B}{2\sqrt{2}} \qquad Eq.\,(12.116)$$
$$= 10.7\ MHz - 200/2\sqrt{2} = \underline{\underline{10.63\,MHz}}$$

$$B_2 = \frac{200}{\sqrt{2}} = \underline{\underline{141.4\ kHz}}$$

For Stage 1

$$C = \frac{1}{\omega_{01}^2 L} = \frac{1}{(2\pi\, 10.77\times10^6)^2 \times 3\times10^{-6}} = \underline{\underline{72.8\,pF}}$$

$$R = \frac{1}{CB_1} = \frac{1}{72.8\times10^{-12} \times 141.4\times2\pi\,10^3} = \underline{\underline{15.5\,k\Omega}}$$

For Stage 2

$$C = \frac{1}{\omega_{02}^2 L} = \frac{1}{(2\pi\, 10.63\times10^6)^2\, 3\times10^{-6}} = \underline{\underline{74.7\,pF}}$$

$$R = \frac{1}{CB_2} = \frac{1}{74.7\times10^{-12} \times 141.4\times2\pi\,10^3} = \underline{\underline{15.1\ k\Omega}}$$

12.37

Gain of stagger-tuned amplifier at f_0 is proportional to

$$\frac{1}{\sqrt{2}} R_{stage1} \times \frac{1}{\sqrt{2}} R_{stage2}$$
$$= \frac{1}{2} \times 15.5 \times 15.1 = 117$$

Gain of synchronous-tuned amplifier at $f_0 \propto R_{stage1} \times R_{stage2}$
$$= 6.95 \times 6.95$$
$$= 48.3$$

$$\therefore \text{Ratio} = \frac{117}{48.3} = \underline{\underline{2.42}}$$

CHAPTER 13 - EXERCISES

13.1

Pole frequency $f_0 = 1\text{kHz}$

Centre frequency gain $= \dfrac{1}{\text{AMPLIFIER GAIN}}$

$= \underline{\tfrac{1}{2}\text{V/V}}$

13.2

$L_+ = V R_4/R_5 + V_D(1 + R_4/R_5)$

$= 15(3/9) + 0.7(1 + 3/9)$

$= 5 + 0.93 = \underline{+5.93\text{V}}$

$L_- = -V\dfrac{R_3}{R_2} - V_D(1 + R_3/R_2)$

$= -15 \times \dfrac{3}{9} - 0.7(1 + 3/9)$

$= \underline{-5.93\text{ V/V}}$

Limiter gain $= \dfrac{-R_f}{R_1} = \dfrac{-60}{30}$

$= \underline{-2\text{V/V}}$

Thus limiting occurs at $\dfrac{\pm 5.93}{2}$

$= \underline{\pm 2.97\text{V}}$

Slope in the limiting regions

$= -\dfrac{R_f \| R_4}{R_1} = -\dfrac{60\|3}{30} = \underline{-0.095\dfrac{\text{V}}{\text{V}}}$

13.3

(a) $L(s) = \left(1 + \dfrac{R_2}{R_1}\right)\dfrac{Z_p}{Z_p + Z_s}$

$= \left(1 + \dfrac{R_2}{R_1}\right)\left(\dfrac{1}{1 + Z_s Y_p}\right)$

$= \left(1 + \dfrac{20.3}{10}\right)\left(\dfrac{1}{1 + (R + \frac{1}{sC})(\frac{1}{R} + sC)}\right)$

$= \dfrac{3.03}{3 + sCR + \dfrac{1}{sCR}}$

where $R = 10\text{k}\Omega$ and $C = 16\text{nF}$

Thus

$$L(s) = \frac{3.03}{3 + s\,16 \times 10^{-5} + \dfrac{1}{s \times 16 \times 10^{-5}}}$$

The closed loop poles are found by setting $L(s) = 1$, that is, they are the values of s, satisfying

$$3 + s \times 16 \times 10^{-5} + \frac{1}{s \times 16 \times 10^{-5}} = 3.03$$

$\Rightarrow s = \dfrac{10^5}{16}(0.015 \pm j)$

(b) The frequency of oscillation is $(10^5/16)$ rad/s or $\underline{1\text{ kHz}}$

(c) Refer to fig. 13.5. At the positive peak $\hat{V}_o$, the voltage at node b will be one diode drop (0.7V) above the voltage V_1 which is about 1/3 of $\hat{V}_0$; thus $V_b = 0.7 + \hat{V}_0/3$. Now if we neglect the current through D_2 in comparison with the currents through R_5 and R_6 we find that

$$\frac{\hat{V}_o - V_b}{R_5} \cong \frac{V_b - (-15)}{R_6}$$

CONT

Thus,

$$\frac{\hat{V}_o - V_b}{1} = \frac{V_b + 15}{3}$$

$$\hat{V}_o = \frac{4}{3}V_b + 5$$

$$\hat{V}_o = \frac{4}{3}\left(0.7 + \frac{\hat{V}_o}{3}\right) + 5$$

$$\Rightarrow \hat{V}_o = 10.68\text{V}$$

From symmetry, we see that the negative peak is equal to the positive peak. Thus the output peak-to-peak voltage is 21.36V

13.4

a) for oscillations to start, $R_2/R_1 = 2$ thus the potentiometer should be set so that its resistance to ground is 20kΩ

(b) $f_0 = \frac{1}{2\pi RC} = \frac{1}{2\pi\ 10\times10^3 \times 16\times10^{-9}}$

$= 1\text{kHz}$

13.5

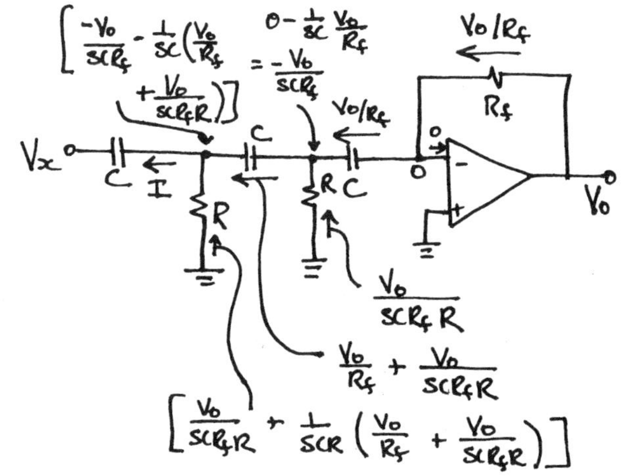

Working from the output back to the input and continuing the equations we get:

$$I = \frac{V_o}{R_f} + \frac{V_o}{sCR_fR} + \frac{V_o}{sCR_fR} + \frac{1}{sCR}\left(\frac{V_o}{R_f} + \frac{V_o}{sCR_fR}\right)$$

$$V_x = \frac{-V_o}{sCR_f} - \frac{1}{sC}\left(\frac{V_o}{R_f} + \frac{V_o}{sCR_fR}\right) - \frac{I}{sC}$$

$$V_x = -\frac{V_o}{sCR_f}\left(2 + \frac{1}{sCR}\right)$$

$$\frac{-V_o}{sCR_f}\left[1 + \frac{1}{sCR} + \frac{1}{sCR} + \frac{1}{sCR}\left(1 + \frac{1}{sCR}\right)\right]$$

$$= \frac{-V_o}{sCR_f}\left(3 + \frac{4}{sCR} + \frac{1}{s^2C^2R^2}\right)$$

Thus:

$$\frac{V_o}{V_x} = \frac{-sCR_f}{3 + \frac{4}{sCR} + \frac{1}{s^2C^2R^2}}$$

$$\frac{V_o}{V_x}(j\omega) = \frac{-j\omega CR_f}{4 + j\left(3\omega CR - \frac{1}{\omega CR}\right)}$$

13.6

The circuit will oscillate at the value of ω that makes $\frac{V_o}{V_x}(j\omega)$ a real number.

It follows that ω_0 is obtained from

$$3\omega_0 CR = \frac{1}{\omega_0 CR} \Rightarrow \omega_0 = \frac{1}{\sqrt{3}CR}$$

Thus, $f_0 = \frac{1}{2\pi\sqrt{3}\times16\times10^{-9}\times10\times10^3}$

$= 574.3\text{Hz}$

For oscillations to begin, the magnitude of $\frac{V_e}{V_x}(j\omega)$ should equal to (or greater than) unity, that is

$$\frac{\omega_0^2 C^2 R R_f}{4} \geq 1$$

Thus the minimum value of R_f is

$$R_f = \frac{4}{\omega_0^2 C^2 R} = \frac{4R}{\omega_0^2 C^2 R^2} = \frac{4R}{1/3}$$

$$= 12R \quad \text{or} \quad \underline{\underline{120\,k\Omega}}$$

13.7

$$\omega_0 = \frac{1}{CR} \Rightarrow CR = \frac{1}{2\pi 10^3}$$

For $C = 16\,nF$ $R = \underline{10k\Omega}$

∴ the output is twice as large as the voltage across the resonator, the peak-to-peak amplitude is

$$\frac{4V}{\pi} = \frac{4(2\times 1.4)}{\pi} = \underline{\underline{3.6V}}$$

13.8

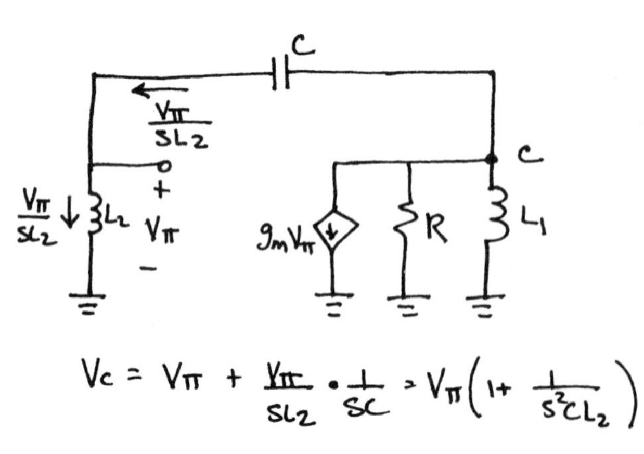

$$V_c = V_\pi + \frac{V_\pi}{sL_2} \cdot \frac{1}{sC} = V_\pi\left(1 + \frac{1}{s^2 C L_2}\right)$$

Node equation at collector:

$$\frac{V_\pi}{sL_2} + g_m V_\pi + \frac{V_c}{R} + \frac{V_c}{sL_1} = 0$$

$$\frac{V_\pi}{sL_2} + g_m V_\pi + \frac{V_\pi}{R}\left(1 + \frac{1}{s^2 C L_2}\right) + \frac{V_\pi}{sL_1}\left(1 + \frac{1}{s^2 C L_2}\right) = 0$$

Since $V_\pi \neq 0$, (oscillations have started) it can be eliminated resulting in

$$s^3 L_1 L_2 C\left(g_m + \frac{1}{R}\right) + s^2\left(L_1 C + L_2 C\right) + s\frac{L_1}{R} + 1 = 0$$

Substituting $s = j\omega$

$$\left[1 - \omega^2 C(L_1 + L_2)\right] + j\omega\left[\frac{L_1}{R} - \left(g_m + \frac{1}{R}\right)\times \omega^2 L_1 L_2 C\right] = 0$$

$$R_E = 0 \Rightarrow \omega_0 = \frac{1}{\sqrt{C(L_1 + L_2)}} \quad \text{Q.E.D.}$$

$$I_m = 0 \Rightarrow g_m R + 1 = \frac{1}{\omega_0^2 L_2 C} = \frac{L_1 + L_2}{L_2}$$

$$\Rightarrow g_m R = L_1/L_2$$

For oscillations to start

$$g_m R > L_1/L_2 \quad \text{Q.E.D.}$$

13.9

$$R = \frac{Q}{\omega_0 C_1} \parallel R_L \parallel r_o$$

$$= \frac{100}{10^6 \times 10^{-8}} \parallel 2\times 10^3 \parallel 100\times 10^3$$

$$= 10 \parallel 2 \parallel 100 = 1.64\,k\Omega$$

CONT.

$$\frac{C_2}{C_1} = g_m R = 40 \times 1.64 = 65.6$$

$$C_2 = 65.6 \times 0.01 = \underline{\underline{0.66\,\mu F}}$$

$$L = \frac{1}{\omega_0^2 \frac{C_1 C_2}{C_1 + C_2}}$$

$$= \frac{1}{10^{12} \times \frac{0.01 \times 0.66 \times 10^{-6}}{0.01 + 0.66}} \cong \underline{\underline{100\,\mu H}}$$

13.10

from Eq(13.24)

$$f_s = \frac{1}{2\pi\sqrt{LC_s}} = \frac{1}{2\pi\sqrt{0.52 \times 0.012 \times 10^{-12}}}$$

$$= \underline{\underline{2.015\text{ MHz}}}$$

From Eq (13.25)

$$f_p = \frac{1}{2\pi\sqrt{L\frac{C_s C_p}{C_s + C_p}}}$$

$$= \frac{1}{2\pi\sqrt{0.52 \times \frac{0.012 \times 4 \times 10^{-12}}{0.012 + 4}}}$$

$$= \underline{\underline{2.018\text{ MHz}}}$$

$$Q = \frac{\omega_0 L}{r} \cong \frac{\omega_s L}{r}$$

$$= \frac{2\pi \times 2.015 \times 10^6 \times 0.52}{120}$$

$$\cong \underline{\underline{55,000}}$$

13.11

$$V_{TH} = V_{TL} = \beta |L \pm|$$

$$5 = \frac{R_1}{R_1 + R_2} \times 13$$

$$\frac{R_2}{R_1} = 1.6$$

$$R_2 = \underline{\underline{16\,k\Omega}}$$

13.12

$$V_{TH} - V_{TL} = \frac{R_1}{R_2}|L|$$

$$5 = \frac{R_1}{R_2} \times 10$$

$$R_2 = 2R_1$$

Possible choice $R_1 = \underline{\underline{10\,k\Omega}}$ $R_2 = \underline{\underline{20\,k\Omega}}$

13.13

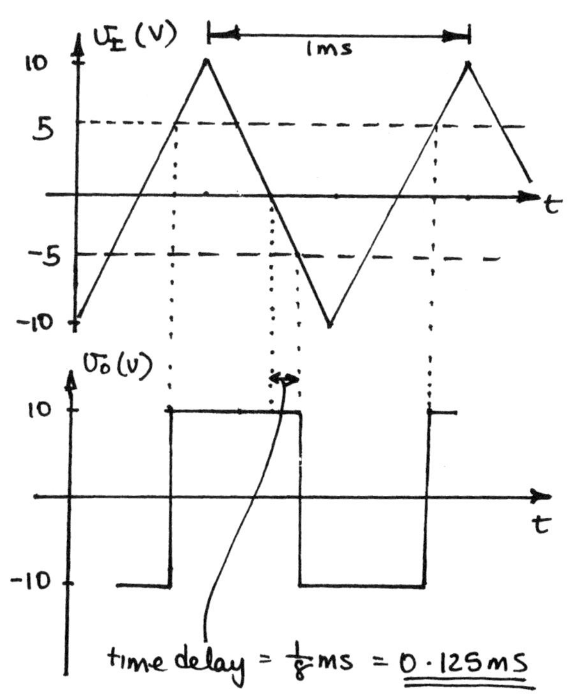

time delay $= \frac{1}{8}$ms $= \underline{\underline{0.125\text{ms}}}$

13.14

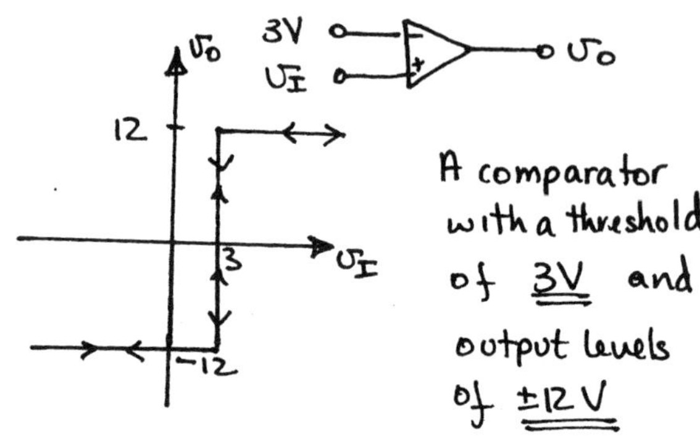

A comparator with a threshold of <u>3V</u> and output levels of <u>±12V</u>

13.15

$$|V_T| = \frac{100}{2} = 50\text{mV}$$

$$50 \times 10^{-3} = 10\,\frac{R_1}{R_2}$$

$$\frac{R_2}{R_1} = \frac{10}{0.05}$$

$$R_2 = 200 R_1$$

for $R_1 = 1\text{k}\Omega$ $\quad R_2 = \underline{200\text{k}\Omega}$

13.16

$$\beta = \frac{R_1}{R_1+R_2} = \frac{100}{100+1000} = 0.091\,\frac{V}{V}$$

$$T = 2\tau \ln \frac{1+\beta}{1-\beta}$$

$$= 2 \times 0.01 \times 10^{-6} \times 10^{6} \times \ln\left(\frac{1.091}{1-0.091}\right)$$

$$= 0.00365\,\text{s}$$

$$f_0 = \frac{1}{T} = \underline{274\,\text{Hz}}$$

13.17

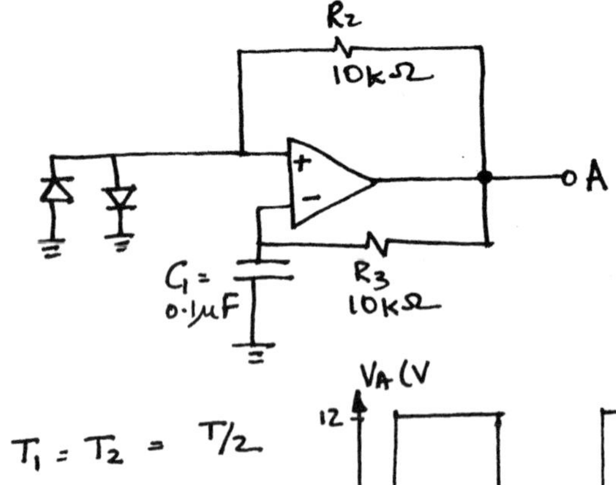

$T_1 = T_2 = T/2$

During T_1

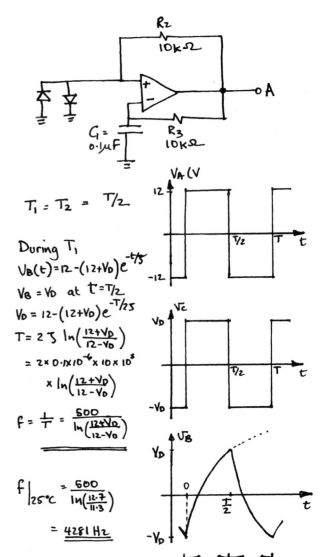

$V_B(t) = 12 - (12+V_D)e^{-t/\tau}$

$V_B = V_D$ at $t = T/2$

$V_D = 12 - (12+V_D)e^{-T/2\tau}$

$T = 2\tau \ln\left(\frac{12+V_D}{12-V_D}\right)$

$= 2 \times 0.1 \times 10^{-6} \times 10 \times 10^{3} \times \ln\left(\frac{12+V_D}{12-V_D}\right)$

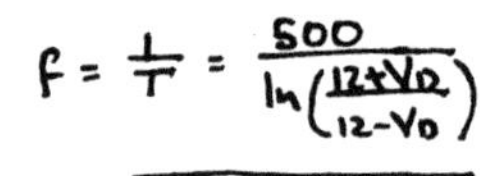

$$f = \frac{1}{T} = \frac{500}{\ln\left(\frac{12+V_D}{12-V_D}\right)}$$

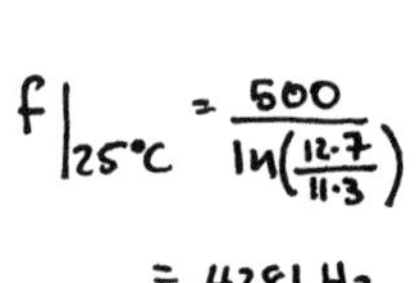

$$f\Big|_{25°C} = \frac{500}{\ln\left(\frac{12.7}{11.3}\right)} = \underline{4281\,\text{Hz}}$$

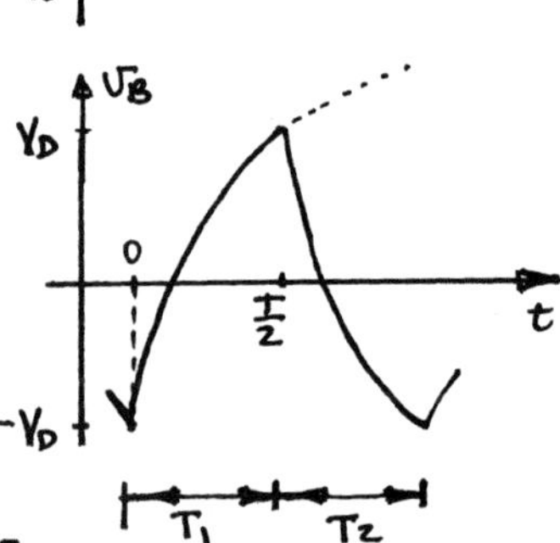

At 0°C, $V_D = 0.7 + 0.05 = 0.75$ V

$$f\Big|_{0°C} = \frac{500}{\ln\left(\frac{12.75}{11.25}\right)} = \underline{3,995\,\text{Hz}}$$

At 50°C, $V_D = 0.7 - 0.05 = 0.65$V

$$f\Big|_{50°C} = \frac{500}{\ln\left(\frac{12.65}{11.35}\right)} = \underline{4,611\,\text{Hz}}$$

At 100°C, $V_D = 0.7 - 0.15 = 0.55$ V

$$f\Big|_{100°C} = \frac{500}{\ln\left(\frac{12.55}{11.45}\right)} = \underline{5,451\,\text{Hz}}.$$

13.18

To obtain a triangular waveform with 10-V peak-to-peak amplitude we should have

$$V_{TH} = -V_{TL} = 5V$$

But $V_{TL} = -L_+ \frac{R_1}{R_2}$

Thus $-5 = -10 \times 10/R_2$

$R_2 = \underline{\underline{20k\Omega}}$

For 1kHz frequency, T = 1ms.

Thus,

$$T/2 = 0.5\times10^{-3} = CR \frac{V_{TH} - V_{TL}}{L_+}$$

$$= 0.01 \times 10^{-6} \times R \times 10/10$$

$R = \underline{\underline{50 k\Omega}}$

13.19

Using Eq (13.37)

$$100\times10^{-6} = 0.1\times10^{-6} \times R_3 \ln\left(\frac{12.7}{10.8}\right)$$

$R_3 = \underline{\underline{617 1\Omega}}$

13.20

$T = 1.1CR \Rightarrow R = T/1.1C = \underline{\underline{9.1k\Omega}}$

13.21

$$T = 0.69C(R_A + 2R_B)$$

$$\frac{1}{100\times10^3} = 0.69\times10^3 \times 10^{-12}(R_A + 2R_B)$$

$$\Rightarrow R_A + 2R_B = \frac{1}{0.69\times10^{-4}} = 14.49k\Omega \quad (1)$$

Using Eq (13.45)

$$0.75 = \frac{R_A + R_B}{R_A + 2R_B}$$

$R_A + R_B = 0.75 \times 14.44 = 10.88 k\Omega$ (2)

(1) − (2) ⇒ $R_B = \underline{\underline{3.61k\Omega}}$

Now, substituting into (2)

$R_A = \underline{\underline{7.27 k\Omega}}$

Use $7.2k\Omega$ and $3.6k\Omega$, standard 5% resistors.

13.22

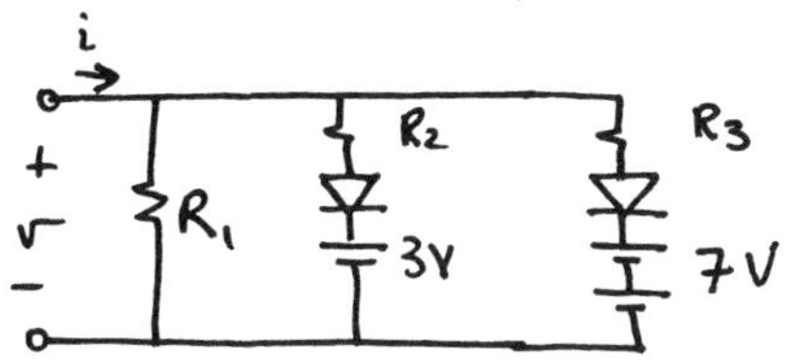

$$i = 0.1v^2$$

At $v = 2V$, $i = 0.4mA$

Thus $R_1 = \frac{2}{0.4} = \underline{\underline{5k\Omega}}$

For $3V \leq V \leq 7V$

$$i = \frac{v}{R_1} + \frac{v-3}{R_2}$$

To obtain a perfect match at $v = 4V$ (i.e. to obtain $i = 1.6mA$)

$$1.6 = \frac{4}{5} + \frac{4-3}{R_2}$$

$R_2 = 1.25k\Omega$

for $v \geq 7V$

$$i = \frac{v}{R_1} + \frac{v-3}{R_2} + \frac{v-7}{R_3}$$

CONT.

To obtain a perfect match at $v = 8\,V$ we must have to select R_3 so that $i = 6.4\ mA$,

$$6.4 = \frac{8}{5} + \frac{8-3}{1.25} + \frac{8-7}{R_3}$$

$$\Rightarrow R_3 = \underline{\underline{1.25\ k\Omega}}$$

At $v = 3V$, the circuit provides $i = \frac{3}{5} = 0.6\ mA$ while ideally $i = 0.1 \times 9 = 0.9\ mA$. Thus the error is $\underline{\underline{-0.3\ mA}}$

* At $V = 5V$, the circuit provides $i = \frac{5}{5} + \frac{5-3}{1.25} = 2.6\ mA$, while ideally $i = 0.1 \times 25 = 2.5\ mA$. Thus the error is $+0.1\ mA$.

* At $v = 7V$, the circuit provides $i = \frac{7}{5} + \frac{7-3}{1.25} = 4.6\ mA$, while ideally $i = 0.1 \times 49 = 4.9\ mA$. Thus the error is $\underline{\underline{-0.3\ mA}}$

* At $v = 10V$, the circuit provides, $i = \frac{10}{5} + \frac{10-3}{1.25} + \frac{10-7}{1.25} = 10\ mA$, while ideally $i = 10\ mA$. Thus the error is $\underline{\underline{0\ A}}$

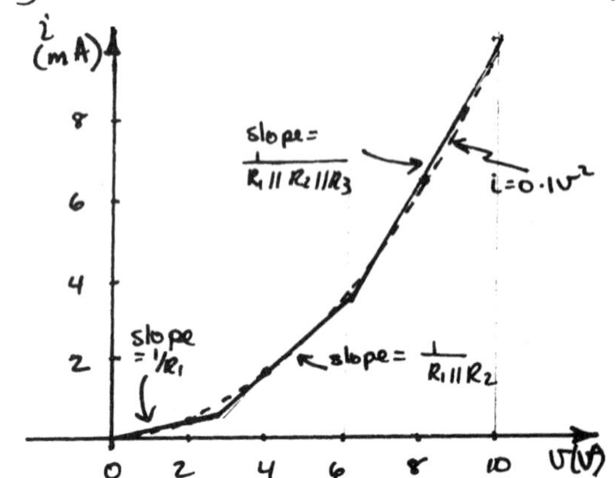

13.23

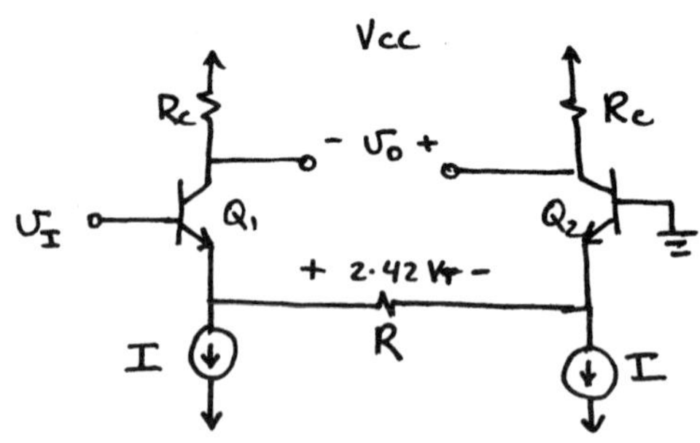

$$I_{E1} = I + (2.42 V_T)/R$$

$$= I\left[1 + \frac{2.42 V_T}{IR}\right]$$

$$= I\left[1 + \frac{2.42 V_T}{2.5 V_T}\right]$$

$$= I\left(1 + \frac{2.42}{2.5}\right)$$

$$I_{C1} \cong I\left(1 + \frac{2.42}{2.5}\right)$$

$$I_{C2} \cong I\left(1 - \frac{2.42}{2.5}\right)$$

$$v_O = (V_{CC} - I_{C2} R_C) - (V_{CC} - I_{C1} R_C)$$

$$= (I_{C1} - I_{C2}) R_C$$

$$= I R_C \times 2 \times \frac{2.42}{2.5}$$

$$= 0.25 \times 10 \times 2 \times \frac{2.42}{2.5} = \underline{\underline{4.84 V}}$$

13.24

∵ the opamp is ideal $v_O = v_I$ for $v_I > 0$.

CONT.

$v_I = 10\,mV \quad v_O = \underline{\underline{10\,mV}}$

$i_D = \frac{10\,mV}{R} = 10\,\mu A$

Given ⇒ $i_D = 1\,mA \quad 0.1\,mA \quad 10\,\mu A$

$v_D = 0.7V \quad 0.6V \quad 0.5V$

Thus, $V_D = 0.5\,V$ so

$V_A = v_O + V_D = \underline{\underline{0.51V}}$

$v_I = 1V \Rightarrow v_O = 1V$

$i_D = 1mA, \quad v_D = 0.7V \quad v_A = \underline{\underline{1.7V}}$

$v_I = -1V$ ~ The negative feedback loop is not operative.

$v_O = \underline{\underline{0V}} \quad v_A = \underline{\underline{-12\,V}}$

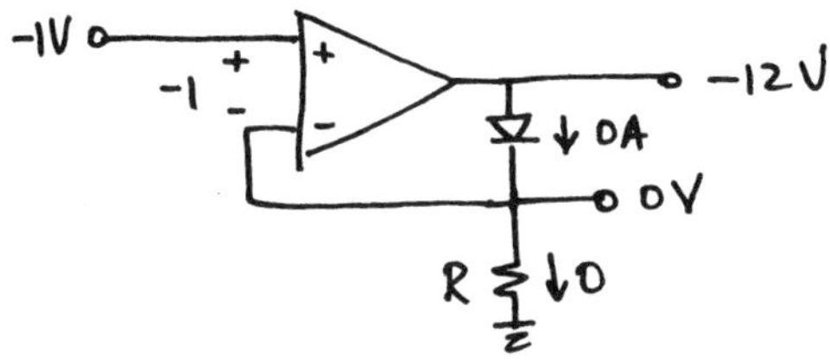

13.25

For the diode to conduct and close the negative feedback loop, v_O must be negative, in which case, the negative feedback causes a virtual short circuit to appear between the input terminals of the op amp and thus $v_O = v_I$.

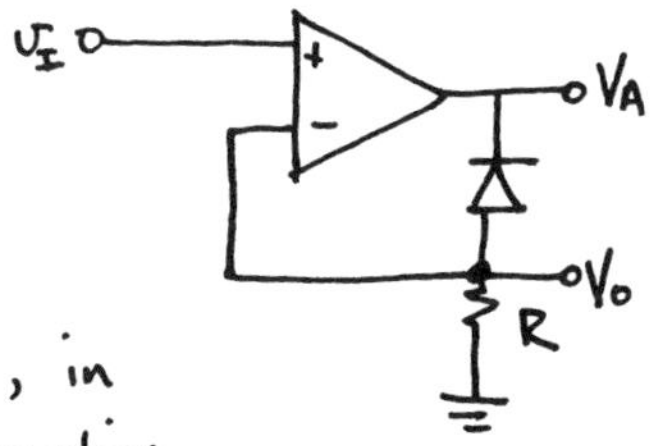

For positive v_I, the op amp saturates in the positive saturation level. The diode will be cut off and $v_O = 0$.

In summary

$v_O = 0$ for $v_I \geq 0$

$\underline{\underline{v_O = v_I \text{ for } v_I \leq 0}}$

13.26

Refer to Fig(13.34)

For $v_I = +1V$:

D_2 will conduct and close the negative feedback loop around the op amp. $v_- = 0$, the current through R_1 and D_2 will be 1mA. Thus the voltage at the op amp output, $v_A = \underline{\underline{-0.7V}}$ which will set D_1 off and no current will flow through R_2. Thus $v_O = \underline{\underline{0V}}$

For $v_I = -10\,mV$

D_1 will conduct through R_2 & R_1 to v_I. The negative feedback loop of the op amp will thus be closed and a virtual ground will appear at the inverting input terminal. D_2 will be cutoff. The current through R_1, R_2 and D_1 will be $\frac{10\,mV}{1\,k\Omega} = 10\,\mu A$. Thus the diode, D_1, voltage will be 0.5V.

$v_O = 0 + 10\mu A \times 10\,k\Omega = \underline{\underline{+0.1V}}$

$v_A = v_{D1} + v_O = 0.5 + 0.1 = \underline{\underline{0.6V}}$

CONT.

For $v_I = -1V$

This is similar to the case when $v_I = -10mV$. The current through R_1, R_2, D_1 will be

$$I = \frac{1}{1k\Omega} = 1mA$$

$\therefore$ $v_{D1} = 0.7V$

$$v_O = 0 + 1mA \times 10k\Omega = \underline{10V}$$

$$v_A = 10 + v_{D1} = \underline{\underline{10.7V}}$$

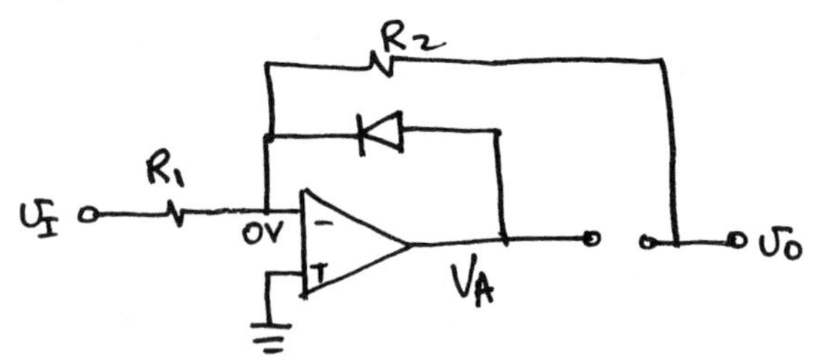

13.27

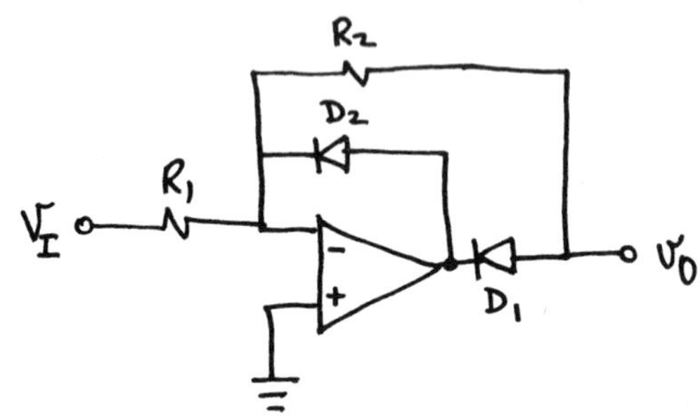

$v_I > 0$

Current flows from v_I through R_1, R_2, D_1 into the output terminal of the opamp. v_O goes negative and is thus off. The following circuit results:

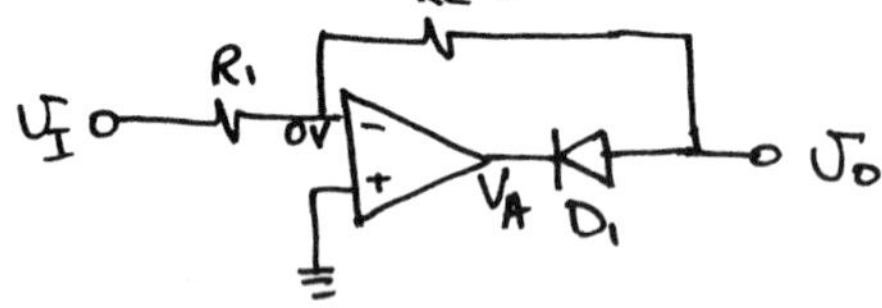

$$\frac{v_O}{v_I} = -\frac{R_2}{R_1}$$

$v_I < 0$ ~ D_2 on

~ v_O goes +ve & turns D_1 off

~ no current flows through R_2 = $v_O = \underline{0V}$

~ $V_A = 0.7V$

13.28

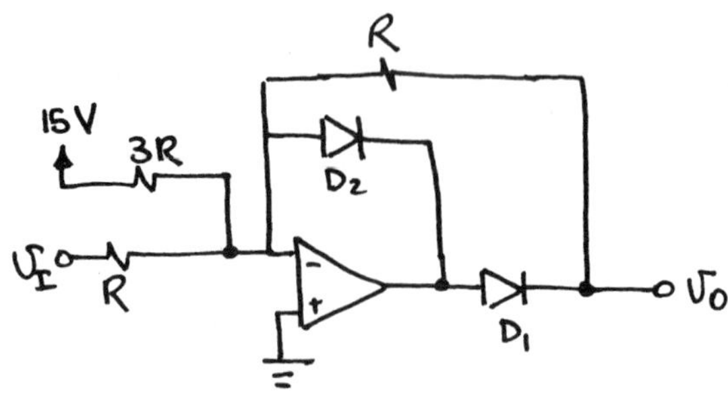

$v_I > 0$ – Equivalent Circuit

~ D_2 on, D_1 off

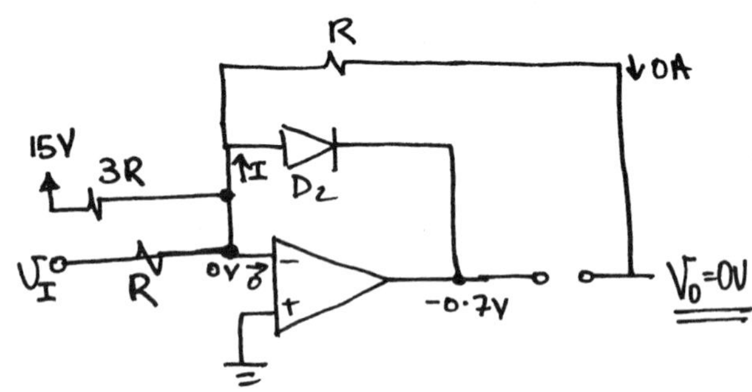

$$I = \frac{15}{3R} + \frac{v_I}{R}$$

As v_I goes negative, the above circuit holds so that $v_O = 0$. This occurs as the 15V supply sources the current I even for small negative v_I. This situation remains the case until $I = 0$

$$\therefore \frac{15}{3R} + \frac{v_I}{R} = 0$$

$$v_I = \underline{\underline{-5V}}$$

CONT.

$v_I < -5V$ – D_2 off D_1–on.

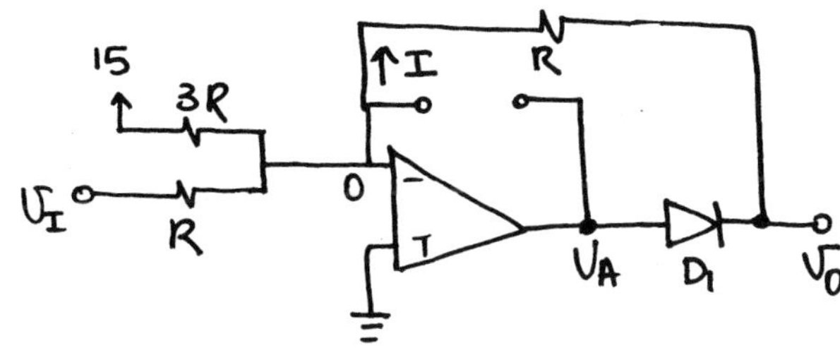

$$v_O = 0 - IR$$
$$= 0 - \left(\frac{15}{3R} + \frac{v_I}{R}\right) R$$
$$= -v_I - 5$$

note $v_A = v_O + 0.7 = -v_I - 4.3 > 0$

$\therefore$ $v_{D2} = 0 - v_A < 0$ – D_2 off!

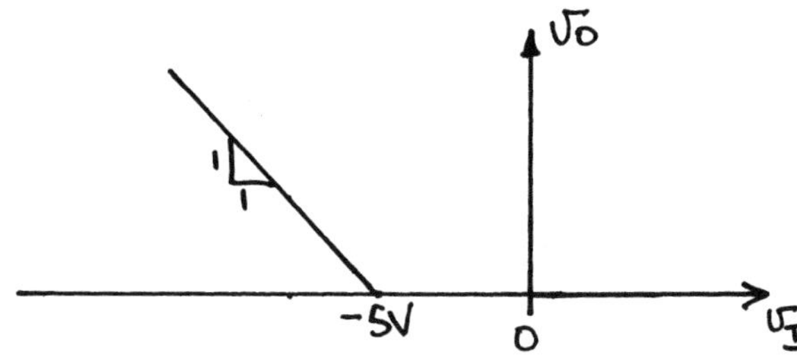

13.29

a) $v_I = 0.1V$

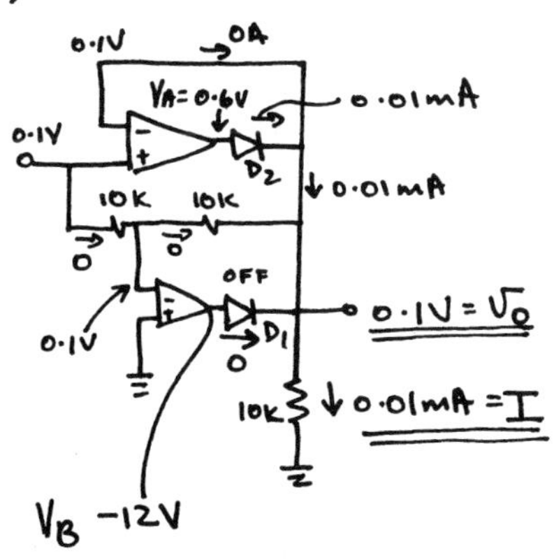

NB. For all circuits, currents are given in mA, resistance in kΩ & voltages in V.

b) $v_I = 1V$ ~ similar to the circuit in (a) but with all of the ungrounded opamp input terminals at $v_I = 1V$

$v_O = \underline{1V}$

$I = 1/10k\Omega = \underline{0.1mA}$

$$v_A = 1 + v_{D2}$$
$$= 1 + 0.7 + 0.1 \log\left(\frac{0.1}{1}\right)$$
$$= \underline{1.6V}$$

(c) $v_I = 10V$ ~ similar to (a) & (b) ~ all input terminals (not grounded) of opamps is equal to 10V.

$v_O = \underline{10V}$

$I = \frac{10}{10} = \underline{1mA}$ ~ diode voltages = 0.7V

$V_A = v_O + v_{D2} = 10 + 0.7 = \underline{10.7V}$

d) $v_I = -0.1V$

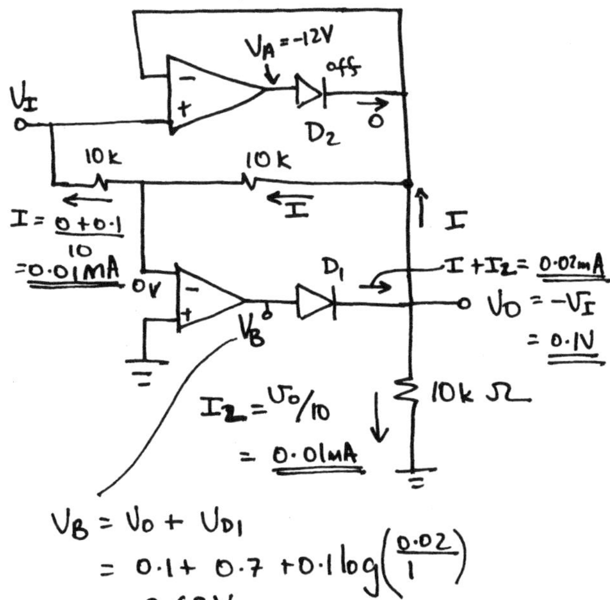

$$V_B = V_O + V_{D1}$$
$$= 0.1 + 0.7 + 0.1 \log\left(\frac{0.02}{1}\right)$$
$$= \underline{0.63V}$$

(e) $v_I = -1V$ - use circuit in (d)

$I = \underline{0.1mA}$

$V_O = -v_I = \underline{1V}$

$I_2 = \underline{0.1mA}$

$I_{D1} = I + I_2 = \underline{0.2mA}$

$V_B = V_O + V_{D1}$

$= 1 + 0.7 + 0.1\log\left(\frac{0.2}{1}\right)$

$= \underline{\underline{1.63V}}$

(f) $v_I = -10V$ ~ use circuit (d)

$I = \underline{1.0mA}$

$v_O = -v_I = \underline{10V}$

$I_2 = \underline{1.0\ mA}$

$I_{D1} = \underline{2mA}$

$V_B = V_O + V_{D1} = 10 + 0.7 + 0.1\log\left(\frac{2}{1}\right)$

$= \underline{\underline{10.73V}}$

13.30

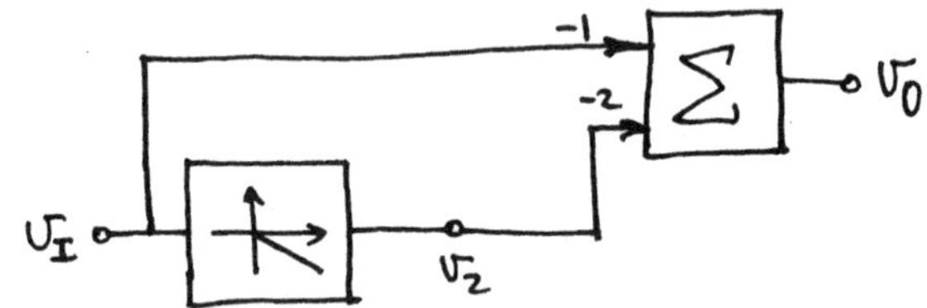

For $v_I \geq 0$, i.e. $v_I = |v_I|$,

$v_2 = -|v_I|$ and

$v_O = -|v_I| - 2 \times -|v_I| = +|v_I|$

For $v_I \leq 0$, i.e. $v_I = -|v_I|$

$v_2 = 0$, $v_O = -1 \times -|v_I|$

$= +|v_I|$

Thus, the block diagram implements the absolute value operation.

Using the circuit of Fig (13.34a), with the diodes reversed, to implement the half-wave rectifier, and a weighted summer results in the circuit shown below.

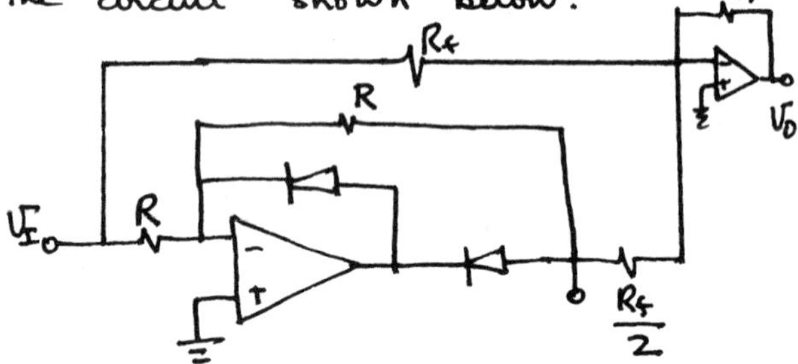

Use $R = R_f = 10k\Omega$

13.31

v_A is a sinusoid of 5-V rms (peak voltage of $5\sqrt{2}$) The average current through the meter will be $\frac{2}{\pi} \times \frac{5\sqrt{2}}{R}$. To obtain full-scale reading, this current must be equal to 1mA. Thus $\frac{2}{\pi} \times \frac{5\sqrt{2}}{R} = 1mA$, which leads to $R = \underline{4.5k\Omega}$

v_C will be maximum when v_A is at its positive peak, i.e. $v_A = 5\sqrt{2}$ V. At this value of v_A, we obtain

$v_C = V_{D1} + V_M + V_{D3} + V_R$ where

$V_{D1} = V_{D3} \simeq 0.7V$ and

$V_M = \frac{5\sqrt{2}}{4.5} \times 0.05 = 0.08V$

Thus $v_C|_{max} = 0.7 + 0.8 + 0.7 + 5\sqrt{2} = \underline{\underline{8.55V}}$

Similarly we can calculate:

$v_C|_{min} = \underline{\underline{-8.55V}}$

Chapter 14 — Exercises

1 $I = \frac{|-V_{CC} + V_{CEsat}|}{R_L}$

$= \frac{|-15+0.2|}{1} = 14.8\ mA$

$R = \frac{V_{CC} - V_D}{I}$

$= \frac{15-0.7}{14.8} = \underline{0.97\ k\Omega}$

Output voltage swing $= \underline{-14.8\ V\ to\ +14.8\ V}$

Min. emitter current $= \underline{0\ mA}$

Max. emitter current $= 2I$

$= 2 \times 14.8 = \underline{29.6\ mA}$

2 At $v_O = -10\ V$, the load current is $-10\ mA$ and the emitter current of Q_1 is $14.8 - 10 = 4.8\ mA$. Thus

$v_{BE1} = 0.6 + 0.025\ ln\left(\frac{4.8}{1}\right)$

$= 0.64\ V$

Thus, $v_I = -10 + 0.64 = \underline{-9.36\ V}$

At $v_O = 0\ V$, $i_L = 0$ and $i_{E1} = 14.8\ mA$

Thus, $v_{BE1} = 0.6 + 0.025\ ln\ \frac{14.8}{1}$

$= 0.67\ V$

$v_I = \underline{+0.67\ V}$

At $v_O = +10\ V$, $i_L = 10\ mA$ and $i_{E1} = 24.8\ mA$

Thus, $v_{BE1} = 0.6 + 0.025\ ln(24.8)$

$= 0.68\ V$

$v_I = \underline{10.68\ V}$

To calculate the incremental voltage gain we use

$\frac{v_o}{v_i} = \frac{R_L}{R_L + r_{e1}}$

At $v_O = -10\ V$, $i_{E1} = 4.8\ mA$ and $r_{e1} = \frac{25}{4.8} = 5.2\ \Omega$

Thus, $\frac{v_o}{v_i} = \frac{1}{1+0.0052} = \underline{0.995\ V/V}$

Similarly, at $v_O = 0\ V$, $r_{e1} = \frac{25}{14.8} = 1.7\ \Omega$

and, $\frac{v_o}{v_i} = \frac{1}{1+0.0017} = \underline{0.998\ V/V}$

At $v_O = +10\ V$, $i_{E1} = 24.8\ mA$ and $r_{e1} = 1\ \Omega$

Thus, $\frac{v_o}{v_i} = \frac{1}{1+0.001} = \underline{0.999\ V/V}$

3 For $v_O = 0\ V$,

$P_{D1} = V_{CC}\ I = 10 \times 0.1 = \underline{1\ W}$

$P_{D2} = V_{CC}\ I = 10 \times 0.1 = \underline{1\ W}$

For a 10-V amplitude output sinusoid, the waveform shown in Fig. 9.4 apply and the average power dissipation in Q_1 is seen to be $\frac{1}{2} V_{CC}\ I = \frac{1}{2} \times 10 \times 0.1 = \underline{0.5\ W}$. Transistor Q_2 carries a constant current I and has an average v_{CE} of 15 V, thus the average power dissipated in Q_2 remains unchanged at $\underline{1\ W}$. The load power is

$P_L = \frac{(\hat{V}_o/\sqrt{2})^2}{R_L} = \frac{1}{2}\ \frac{100}{100} = \underline{0.5\ W}$

4 $P_L = \frac{(8/\sqrt{2})^2}{100} = \underline{0.32\ W}$

$P_+ = 10 \times 0.1 = 1\ W$

$P_- = 10 \times 0.1 = 1\ W$

$P_{supplies} = \underline{2\ W}$

$\eta = \frac{P_L}{P_S} \times 100$

$= \frac{0.32}{2} \times 100 = \underline{16\ \%}$

5 (a) $P_L = \frac{1}{2}\frac{\hat{V}_o^2}{R_L}$

$= \frac{1}{2}\ \frac{(4.5)^2}{4} = \underline{2.53\ W}$

(b) $P_+ = P_- = V_{CC} \times \frac{1}{\pi}\ \frac{\hat{V}_o}{R_L}$

$= 6 \times \frac{1}{\pi} \times \frac{4.5}{4} = \underline{2.15\ W}$

(c) $\eta = \frac{P_L}{P_S} = \frac{2.53}{2 \times 2.15} \times 100$

$= \underline{59\ \%}$

(d) Peak input currents $= \frac{1}{\beta+1}\ \frac{\hat{V}_o}{R_L}$

$= \frac{1}{51} \times \frac{4.5}{4}$

$= \underline{22.1\ mA}$

(e) Using Eq. (9.22)

$P_{DNmax} = P_{DPmax} = \frac{V_{CC}^2}{\pi^2 R_L}$

$= \frac{6^2}{\pi^2 \times 4} = \underline{0.91\ W}$

14.6 Under quiescent conditions

$i_N = i_P = I_Q$

and $v_O = v_I = 0$

$I_Q = I_S \, e^{|v_{BE}|/V_T}$

$= I_S \, e^{V_{BB}/2V_T}$

$V_{BB} = 2 V_T \ln\left(\frac{I_Q}{I_S}\right)$

$= 2 \times 0.025 \ln\left(\frac{2\times10^{-3}}{10^{-13}}\right)$

$= \underline{\underline{1.186 \text{ V}}}$

We shall illustrate the construction of the table by calculating in detail one of its rows. For $v_O = +1$ V :

$i_L = \frac{1}{0.1} = 10$ mA

Substituting in Eq.(9.27) gives

$i_N^2 - 10\, i_N - 4 = 0$

$\Rightarrow i_N = \underline{\underline{10.39 \text{ mA}}}$

$i_P = i_N - i_L = 10.39 - 10 = \underline{\underline{0.39 \text{ mA}}}$

$v_{BEN} = V_T \ln\left(\frac{i_N}{I_S}\right)$

$= 0.025 \times \ln\left(\frac{10.39\times10^{-3}}{10^{-13}}\right)$

$= \underline{\underline{0.634 \text{ V}}}$

$v_{EBP} = V_T \ln\left(\frac{i_P}{I_S}\right)$

$= 0.025 \ln\left(\frac{0.39\times10^{-3}}{10^{-13}}\right) = \underline{\underline{0.552 \text{V}}}$

Note that, $v_{BEN} + v_{BEP} = V_{BB}$ (as should be expected)

$v_I = v_O + v_{BEN} - V_{BB}/2$

$= 1 + 0.634 - \frac{1.186}{2}$

$= \underline{\underline{1.041 \text{ V}}}$

$\frac{v_O}{v_I} = \frac{1}{1.041} = \underline{\underline{0.96}}$

From Eq. (9.31),

$R_{out} = \frac{V_T}{i_P + i_N}$

$= \frac{25 \text{ mV}}{(0.39 + 10.39) \text{ mA}} = \underline{\underline{2.32 \,\Omega}}$

$\frac{v_o}{v_i} = \frac{R_L}{R_L + R_{out}} = \frac{100}{100 + 2.32}$

$= \underline{\underline{0.98 \text{ V/V}}}$

14.7 Refer to Fig.(9.14). For $v_O = +10$ V,

$i_L = \frac{10}{100} = 0.1$ A. As a first approximation $i_N \cong 0.1$ A and $i_P = 0$. Thus $i_{BN} \simeq \frac{100}{51} \simeq 2$ mA

Since $I_{bias} = 3$ mA it follows that the current through the diodes will decrease to 1 mA. Thus

$$V_{BB} = 2 V_T \ln\left(\frac{10^{-3}}{\frac{1}{3}\times10^{-13}}\right) \quad (1)$$

But, $V_{BB} = v_{BEN} + v_{EBP}$

$= V_T \ln\left(\frac{i_N}{I_S}\right) + V_T \ln\frac{(i_N - i_L)}{I_S}$

$$= V_T \ln\left[\frac{i_N(i_N - 0.1)}{10^{-26}}\right] \quad (2)$$

Equating the RHS of (1) and (2) gives

$\left(\frac{10^{-3}}{\frac{1}{3}\times10^{-13}}\right)^2 = \frac{i_N(i_N - 0.1)}{10^{-26}}$

$i_N(i_N - 0.1) = 9\times10^{-6}$

If i_N is in mA then

$i_N(i_N - 100) = 9$

$i_N^2 - 100\, i_N - 9 = 0$

$\Rightarrow i_N = \underline{\underline{100.1 \text{ mA}}}$ $\quad i_P = \underline{\underline{0.1 \text{ mA}}}$

For $v_O = -10$ V, $i_L = -100$ mA

As a first approximation, $i_P \cong 100$ mA and $i_N \simeq 0$. Thus $i_{BN} \simeq 0$ and the current through the diodes will be equal to I_{bias},

$$V_{BB} = 2 V_T \ln\left(\frac{3\times10^{-3}}{\frac{1}{3}\times10^{-13}}\right) \quad (3)$$

But, $V_{BB} = V_T \ln\left(\frac{i_N}{10^{-13}}\right) + V_T \ln\left(\frac{i_P}{10^{-13}}\right)$

$= V_T \ln\left(\frac{i_P - 0.1}{10^{-13}}\right) + V_T \ln\left(\frac{i_P}{10^{-13}}\right)$

$$= V_T \ln\left[\frac{i_P(i_P - 0.1)}{10^{-26}}\right] \quad (4)$$

Equating the RHS of (3) and (4) gives

$\left(\frac{3\times10^{-3}}{\frac{1}{3}\times10^{-13}}\right)^2 = \frac{i_P(i_P - 0.1)}{10^{-26}}$

$i_P(i_P - 0.1) = 81\times10^{-6}$

If i_P is in mA,

$i_P(i_P - 100) = 81$

$i_P^2 - 100\, i_P - 81 = 0$

$\Rightarrow i_P = \underline{\underline{100.8 \text{ mA}}}$

$i_N = \underline{\underline{0.8 \text{ mA}}}$

14.8 $\Delta I_C = g_m \times 2\ mV/°C \times 5°C$, mA

where g_m is in mA/mV,

$$g_m = \frac{10\ mA}{25\ mV} = 0.4\ mA/mV$$

Thus, $\Delta I_C = 0.4 \times 2 \times 5 = \underline{\underline{4\ mA}}$

14.9 Refer to Fig. 9.15.

(a) To obtain a terminal voltage of 1.2V, and since β_1 is very large, it follows that $V_{R_1} = V_{R_2} = 0.6\ V$. Thus $I_{C_1} = 1\ mA$

$$I_R = \frac{1.2\ V}{R_1 + R_2} = \frac{1.2}{2.4} = 0.5\ mA$$

Thus,

$$I = I_{C1} + I_R = \underline{1.5\ mA}$$

(b) For $\Delta V_{BB} = +50\ mV$:

$V_{BB} = 1.25\ V$ $\quad I_R = \frac{1.25}{2.4} = 0.52\ mA$

$V_{BE} = \frac{1.25}{2} = 0.625\ V$

$I_{C1} = 1 \times e^{\Delta V_{BE}/V_T} = e^{0.025/0.025}$

$= 2.72\ mA$

$I = 2.72 + 0.52 = \underline{\underline{3.24\ mA}}$

For $\Delta V_{BB} = +100\ mV$:

$V_{BB} = 1.3\ V$ $\quad I_R = \frac{1.3}{2.4} = 0.54\ mA$

$V_{BE} = \frac{1.3}{2} = 0.65\ V$

$I_{C1} = 1 \times e^{\Delta V_{BE}/V_T} = 1 \times e^{0.05/0.025}$

$= 7.39\ mA$

$I = 7.39 + 0.54 = \underline{\underline{7.93\ mA}}$

For $\Delta V_{BB} = +200\ mV$:

$V_{BB} = 1.4\ V$ $\quad I_R = \frac{1.4}{2.4} = 0.58\ mA$

$V_{BE} = 0.7\ V$

$I_{C1} = 1 \times e^{0.1/0.025} = 54.60\ mA$

$I = 54.60 + 0.58 = \underline{55.18\ mA}$

For $\Delta V_{BB} = -50\ mV$:

$V_{BB} = 1.15\ V$ $\quad I_R = \frac{1.15}{2.4} = 0.48\ mA$

$V_{BE} = \frac{1.15}{2}$

$= 0.575\ V$

$I_{C1} = 1 \times e^{-0.025/0.025} = 0.37\ mA$

$I = 0.48 + 0.37 = \underline{\underline{0.85\ mA}}$

For $\Delta V_{BB} = -100\ mV$:

$V_{BB} = 1.1\ V$ $\quad I_R = \frac{1.1}{2.4} = 0.46\ mA$

$V_{BE} = 0.55\ V$

$I_{C1} = 1 \times e^{-0.05/0.025} = 0.13\ mA$

$I = 0.46 + 0.13 = \underline{\underline{0.59\ mA}}$

For $\Delta V_{BB} = -200\ mV$:

$V_{BB} = 1.0\ V$ $\quad I_R = \frac{1}{2.4} = 0.417\ mA$

$V_{BE} = 0.5\ V$

$I_{C1} = 1 \times e^{-0.1/0.025} = 0.018\ mA$

$I = \underline{0.43\ mA}$

14.10

$T_J - T_A = \theta_{JA} P_D$

$200 - 25 = \theta_{JA} \times 50$

$\theta_{JA} = \frac{175}{50} = 3.5\ °C/W$

But, $\theta_{JA} = \theta_{JC} + \theta_{CS} + \theta_{SA}$

$3.5 = 1.4 + 0.6 + \theta_{SA}$

$\Rightarrow \theta_{SA} = \underline{1.5°C/W}$

$T_J - T_C = \theta_{JC} \times P_D$

$T_C = T_J - \theta_{JC} P_D$

$= 200 - 1.4 \times 50$

$= \underline{\underline{130°C}}$

14.11

(a) From symmetry we see that all transistors will conduct equal currents and have equal V_{BE}'s. Thus $v_O = 0$ V.

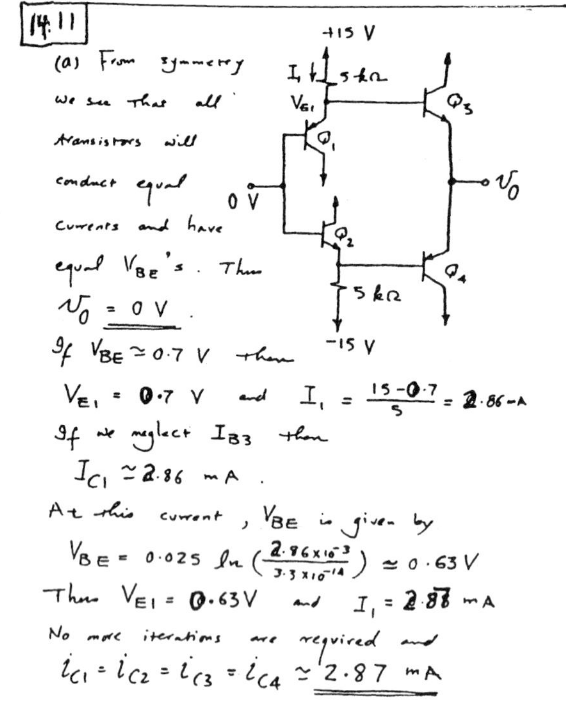

If $V_{BE} \simeq 0.7$ V then

$V_{E1} = 0.7$ V and $I_1 = \frac{15 - 0.7}{5} = 2.86$ mA

If we neglect I_{B3} then

$I_{C1} \simeq 2.86$ mA.

At this current, V_{BE} is given by

$$V_{BE} = 0.025 \ln\left(\frac{2.86 \times 10^{-3}}{3.3 \times 10^{-14}}\right) \simeq 0.63 \text{ V}$$

Thus $V_{E1} = 0.63$ V and $I_1 = 2.88$ mA

No more iterations are required and

$$i_{C1} = i_{C2} = i_{C3} = i_{C4} \simeq 2.87 \text{ mA}$$

(b) For $v_I = +10$ V:

To start the iterations let $V_{BE1} \simeq 0.7$ V

Thus,

$V_{E1} = 10.7$ V

and,

$I_1 = \frac{15 - 10.7}{5} = 0.86$ mA

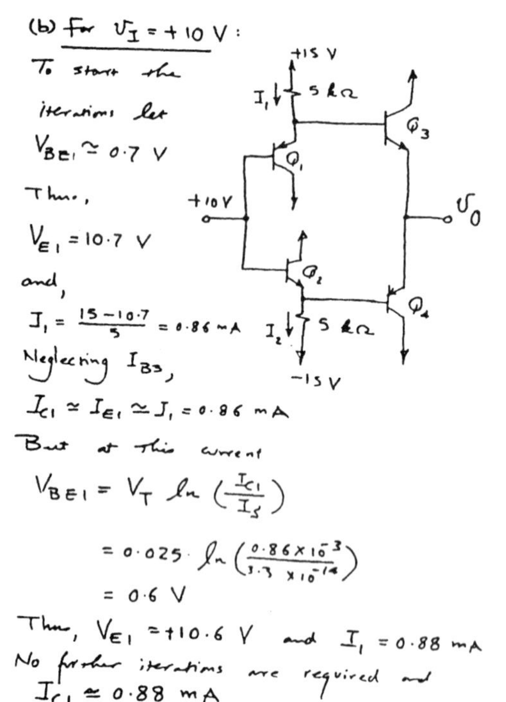

Neglecting I_{B3},

$I_{C1} \simeq I_{E1} \simeq I_1 = 0.86$ mA

But at this current

$$V_{BE1} = V_T \ln\left(\frac{I_{C1}}{I_S}\right)$$

$$= 0.025 \cdot \ln\left(\frac{0.86 \times 10^{-3}}{3.3 \times 10^{-14}}\right)$$

$$= 0.6 \text{ V}$$

Thus, $V_{E1} = +10.6$ V and $I_1 = 0.88$ mA

No further iterations are required and

$I_{C1} \simeq 0.88$ mA.

To find I_{C2} we use an identical procedure:

$V_{BE2} \simeq 0.7$ V

$V_{E2} = 10 - 0.7 = +9.3$ V

$I_2 = \frac{9.3 - (-15)}{5} = 4.86$ mA

$V_{BE2} = 0.025 \ln\left(\frac{4.86 \times 10^{-3}}{3.3 \times 10^{-14}}\right)$

$= 0.643$ V

$V_{E2} = 10 - .643 = +9.357$

$I_2 = 4.87$ mA

$I_{C2} \simeq 4.87$ mA

Finally,

$$I_{C3} = I_{C4} = 3.3 \times 10^{-14} e^{V_{BE}/V_T}$$

where

$$V_{BE} = \frac{V_{E1} - V_{E2}}{2} = 0.62 \text{ V}$$

Thus, $I_{C3} = I_{C4} = 1.95$ mA

The symmetry of the circuit enables us to find the values for $v_I = -10$ V as follows:

$I_{C1} = 4.87$ mA $\quad I_{C2} = 0.88$ mA

$I_{C3} = I_{C4} = 1.95$ mA

For $v_I = +10$ V, $v_O = V_{E1} - V_{BE3}$

$= 10.6 - 0.62 = +9.98$ V

For $v_I = -10$ V, $v_O = V_{E1} - V_{BE3}$

$= -9.357 - 0.62 = -9.98$ V

(c) For $v_I = +10$ V,

$v_O \simeq 10$ V

$I_L \simeq 100$ mA

$I_{C3} \simeq 100$ mA

$I_{B3} = \frac{100}{201}$

$\simeq 0.5$ mA

Assuming that V_{BE1} has not changed much from 0.6 V, the

14.11 CONT.

$V_{E1} \simeq 10.6\ \text{V}$

$I_1 = \frac{15-10.6}{5} = 0.88\ \text{mA}$

$I_{E1} = I_1 - I_{B3} = 0.88 - 0.5 = 0.38\ \text{mA}$

$I_{C1} \simeq 0.38\ \text{mA}$

$V_{BE1} = 0.025\ \ln\left(\frac{0.38 \times 10^{-3}}{3.3 \times 10^{-14}}\right)$

$= 0.58\ \text{V}$

$V_{E1} = 10.88\ \text{V}$

$I_1 = \frac{15 - 10.58}{5} = 0.88\ \text{mA}$

Thus, $I_{C1} \simeq \underline{\underline{0.38\ \text{mA}}}$

Now for Q_2 we have:

$V_{BE2} \simeq 0.643\text{V}$

$V_{E2} = 10 - 0.643 = 9.357$

$I_2 = 4.87\ \text{mA}$

$I_{B4} \simeq 0$

$I_{C2} \simeq \underline{\underline{4.87\ \text{mA}}}$ (as in (b))

Assuming that $I_{C3} \simeq \underline{\underline{100\ \text{mA}}}$,

$V_{BE3} = 0.025\ \ln\left(\frac{100 \times 10^{-3}}{3.3 \times 10^{-14}}\right)$

$= 0.72\ \text{V}$

Thus, $v_O = V_{E1} - V_{BE3}$

$= 10.58 - 0.72 = \underline{\underline{+9.86\ \text{V}}}$

$V_{BE4} = v_O - V_{E2}$

$= 9.86 - 9.36 = 0.5\ \text{V}$

Thus, $I_{C4} = 3.3 \times 10^{-14}\ e^{0.5/0.025}$

$\simeq \underline{\underline{0.02\ \text{mA}}}$

From symmetry we find the values for the case $v_I = -10\ \text{V}$ as,

$I_{C1} = \underline{\underline{4.87\ \text{mA}}}$ $\quad I_{C2} = \underline{\underline{0.38\ \text{mA}}}$

$I_{C3} = \underline{\underline{0.02\ \text{mA}}}$ $\quad I_{C4} = \underline{\underline{100\ \text{mA}}}$

$v_O = \underline{\underline{-9.86\ \text{V}}}$.

14.12 (a)

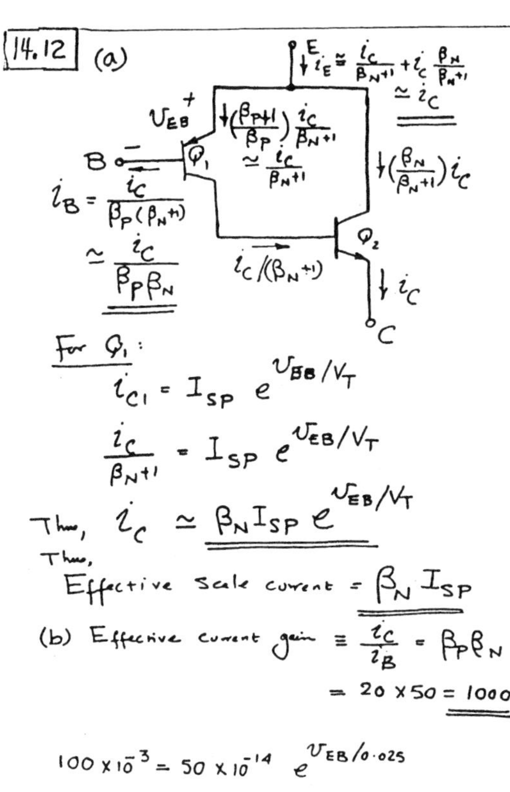

For Q_1:

$i_{C1} = I_{SP}\ e^{v_{EB}/V_T}$

$\frac{i_C}{\beta_N+1} = I_{SP}\ e^{v_{EB}/V_T}$

Thus, $i_C \simeq \underline{\underline{\beta_N I_{SP}\ e^{v_{EB}/V_T}}}$

Thus,

Effective scale current $= \underline{\underline{\beta_N I_{SP}}}$

(b) Effective current gain $\equiv \frac{i_C}{i_B} = \beta_P \beta_N$

$= 20 \times 50 = \underline{\underline{1000}}$

$100 \times 10^{-3} = 50 \times 10^{-14}\ e^{v_{EB}/0.025}$

$v_{EB} = 0.025\ \ln\ (2 \times 10^{11})$

$= \underline{\underline{0.651\ \text{V}}}$

14.13 Refer to Fig. 9.28.

We require Q_5 to conduct a collector current of 2 mA when its $v_{BE} = 150\ \text{mA} \times R_{E1}$:

$v_{BE5} = V_T\ \ln\left(\frac{2 \times 10^{-3}}{10^{-14}}\right)$

$= 0.651\ \text{V}$

Thus, $R_{E1} = \frac{0.651}{0.150} = \underline{\underline{4.3\ \Omega}}$

For a peak output current of 100 mA, a $\underline{\underline{430\ \text{mV}}}$ voltage drop develops across R_{E1}. Thus Q_5 will conduct a collector current of

$i_{C5} = 10^{-14}\ e^{0.43/0.025} = 2.95 \times 10^{-7}\ \text{A}$

$\simeq \underline{\underline{0.3\ \mu\text{A}}}$

14.14 Total current out of node B $= \frac{2v_i}{R_3} + \frac{v_o}{R_2}$

Thus

$$\left(\frac{2v_i}{R_3} + \frac{v_o}{R_2}\right)R = -\frac{v_o}{A}$$

$$\Rightarrow \quad v_o\left(\frac{1}{A} + \frac{R}{R_2}\right) = \frac{-2R}{R_3} v_i$$

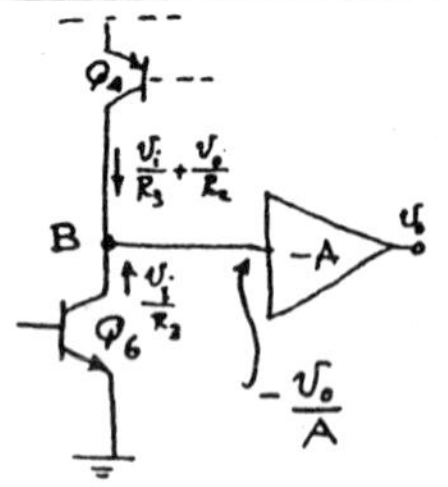

$$\frac{v_o}{v_i} = \frac{-\frac{2R}{R_3}}{\frac{1}{A} + \frac{R}{R_2}}$$

$$= \frac{-2R_2/R_3}{1 + (R_2/AR)} \qquad \text{Q.E.D}$$

For $AR \gg R_2$

$$\frac{v_o}{v_i} \simeq -\frac{2R_2}{R_3}$$

14.15

$$P_{Dmax} = \frac{T_{Jmax} - T_A}{\theta_{JA}}$$

$$= \frac{150 - 50}{35} = 2.9\text{ W}$$

14.16 From Fig. 9.32 we see that for $P_{dissipation}$ to be less than 2.9 W, a maximum supply voltage of 20 V is called for. The 20-V-supply curve intersects the 3% distortion line at a point for which the output power is 4.2 W. Since

$$P_L = \frac{(\hat{V}_o/\sqrt{2})^2}{R_L}$$

thus $\hat{V}_o = \sqrt{4.2 \times 2 \times 8} = 8.2\text{ V}$

or 16.4 V peak-to-peak

14.17 Voltage gain $= 2K$

where $K = \frac{R_4}{R_3} = 1 + \frac{R_2}{R_1} = 1.5$

Thus, $A_v = 3\text{ V/V}$

Input resistance $= R_3 = 10\text{ k}\Omega$

Peak-to-peak $v_O = 3 \times 20 = 60\text{ V}$

Peak load current $= \frac{30\text{ V}}{8\ \Omega} = 3.75\text{ A}$

$$P_L = \frac{(30/\sqrt{2})^2}{8} = 56.25\text{ W}$$

14.18 We wish to make

$$\frac{\partial V_{GG}}{\partial T} = -3 - 3 = -6\text{ mV/°C}$$

but from Eq. (9.49)

$$\frac{\partial V_{GG}}{\partial T} = \left(1 + \frac{R_3}{R_4}\right)\frac{\partial V_{BE6}}{\partial T}$$

Thus

$$-6 = \left(1 + \frac{R_3}{R_4}\right) \times -2$$

$$\Rightarrow \frac{R_3}{R_4} = 2$$

14.19 $I_{DN} = I_{DP} = \frac{1}{2}\mu C_{ox}\frac{W}{L}(|V_{GS}| - V_t)^2$

$$0.1 = 1 \times (|V_{GS}| - 3)^2$$

$$\Rightarrow V_{GS} = 3.32\text{ V}$$

$$V_{GG} = 2 \times 3.32 = 6.64\text{ V}$$

$$R = \frac{V_{GG}}{20\text{ mA}} = \frac{6.64}{20} = 332\ \Omega$$

From Eq. (9.48)

$$V_{GG} = \left(1 + \frac{R_3}{R_4}\right)V_{BE6} + \left(1 + \frac{R_1}{R_2}\right)V_{BE5} - 4V_{BE}$$

$$6.64 = 3 \times 0.7 + \left(1 + \frac{R_1}{R_2}\right) \times 0.7 - 4 \times 0.7$$

$$\Rightarrow \frac{R_1}{R_2} = 9.5.$$

Problem Solutions

CHAPTER 1—PROBLEM SOLUTIONS

1.1 (a) $I = \frac{V}{R} = \frac{10 \text{ V}}{1 \text{ k}\Omega} = 10 \text{ mA}$

(b) $R = \frac{V}{I} = \frac{10 \text{ V}}{1 \text{ mA}} = 10 \text{ k}\Omega$

(c) $V = IR = 10 \text{ mA} \times 10 \text{ k}\Omega = 100 \text{ V}$

(d) $I = \frac{V}{R} = \frac{10 \text{ V}}{100 \ \Omega} = 0.1 \text{ A}$

Note: Volts, milliamps, and kilo-ohms constitute a consistent set of units.

1.2 (a) $P = I^2R = (30 \times 10^{-3})^2 \times 1 \times 10^3 = 0.9 \text{ W}$
Thus, R should have a 1-W rating.

(b) $P = I^2R = (40 \times 10^{-3})^2 \times 1 \times 10^3 = 1.6 \text{ W}$
Thus, the resistor should have a 2-W rating.

(c) $P = I^2R = (3 \times 10^{-3})^2 \times 10 \times 10^3 = 0.09 \text{ W}$
Thus, the resistor should have a $\frac{1}{8}$-W rating.

(d) $P = I^2R = (4 \times 10^{-3})^2 \times 10 \times 10^3 = 0.16 \text{ W}$
Thus, the resistor should have a $\frac{1}{4}$-W rating.

(e) $P = V^2/R = 20^2/(1 \times 10^3) = 0.4 \text{ W}$
Thus, the resistor should have a $\frac{1}{2}$-W rating.

(f) $P = V^2/R = 11^2/(1 \times 10^3) = 0.121 \text{ W}$

Thus, a rating of $\frac{1}{8}$ W should theoretically suffice though $\frac{1}{4}$ W would be prudent to allow for consistent tolerances and measurement errors.

1.3 (a) $V = IR = 10 \text{ mA} \times 1 \text{ k}\Omega = 10 \text{ V}$
$P = I^2R = (10 \text{ mA})^2 \times 1 \text{ k}\Omega = 100 \text{ mW}$

(b) $R = V/I = 10 \text{ V}/1 \text{ mA} = 10 \text{ k}\Omega$
$P = VI = 10 \text{ V} \times 1 \text{ mA} = 10 \text{ mW}$

(c) $I = P/V = 1 \text{ W}/10 \text{ V} = 0.1 \text{ A}$
$R = V/I = 10 \text{ V}/0.1 \text{ A} = 100 \ \Omega$

(d) $V = P/I = 0.1 \text{ W}/10 \text{ mA} = 100 \text{ mW}/10 \text{ mA} = 10 \text{ V}$
$R = V/I = 10 \text{ V}/10 \text{ mA} = 1 \text{ k}\Omega$

(e) $P = I^2R \Rightarrow I = \sqrt{P/R}$
$I = \sqrt{1000 \text{ mW}/1 \text{ k}\Omega} = 31.6 \text{ mA}$
$V = IR = 31.6 \text{ mA} \times 1 \text{ k}\Omega = 31.6 \text{ V}$

Note: V, mA, kΩ, and mW constitute a consistent set of units.

1.4

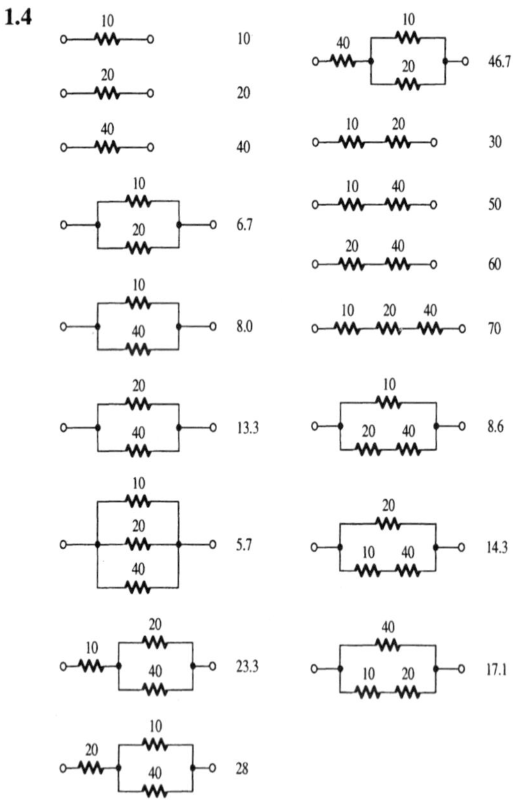

Thus, there are 17 possible resistance values.

1.5 Shunting the 10 kΩ by a resistor of value R result in the combination having a resistance R_{eq},

$$R_{eq} = \frac{10R}{R + 10}$$

Thus, for a 1% reduction,

$$\frac{R}{R + 10} = 0.99 \Rightarrow R = 990 \text{ k}\Omega$$

For a 5% reduction,

$$\frac{R}{R + 10} = 0.95 \Rightarrow R = 190 \text{ k}\Omega$$

For a 10% reduction,

$$\frac{R}{R + 10} = 0.90 \Rightarrow R = 90 \text{ k}\Omega$$

For a 50% reduction,

$$\frac{R}{R + 10} = 0.50 \Rightarrow R = 10 \text{ k}\Omega$$

Shunting the 10 kΩ by

(a) 1 MΩ result in

$$R_{eq} = \frac{10 \times 1000}{1000 + 10} = \frac{10}{1.01} = 9.9 \text{ k}\Omega, \text{ a 1\% reduction;}$$

(b) 100 kΩ results in

$$R_{eq} = \frac{10 \times 100}{100 + 10} = \frac{10}{1.1} = 9.09 \text{ k}\Omega, \text{ a 9.1\% reduction;}$$

(c) 10 kΩ results in

$$R_{eq} = \frac{10}{10 + 10} = 5 \text{ k}\Omega, \text{ a 50\% reduction.}$$

1.6 $V_O = V_{DD}\dfrac{R_2}{R_1 + R_2}$

To find R_O, we short circuit V_{DD} and look back into node X,

$$R_O = R_2 \,//\, R_1 = \frac{R_1 R_2}{R_1 + R_2}$$

1.7

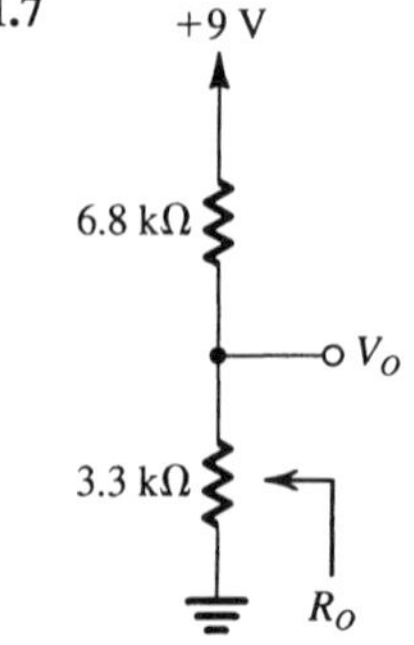

$$V_O = 9\frac{3.3}{3.3 + 6.8}$$

$$= 2.94\text{ V}$$

$$R_O = 2.22\text{ k}\Omega$$

For ±5% resistor tolerance the extreme values of V_O are

$$V_{O\text{-low}} = 9\frac{3.3(1-0.05)}{3.3(1-0.05) + 6.8(1+0.05)}$$

$$= 2.75\text{ V}$$

$$V_{O\text{-high}} = 9\frac{3.3(1+0.05)}{3(1+0.05) + 6.8(1-0.05)}$$

$$= 3.14\text{ V}$$

The extreme values of R_O are

$$R_{O\text{-low}} = \frac{3.3(1-0.05) \times 6.8(1-0.05)}{3.3(1-0.05) + 6.8(1-0.05)}$$

$$= 2.22(1-0.05) = 2.11\text{ k}\Omega$$

$$R_{O\text{-high}} = 2.22(1+0.05) = 2.33\text{ k}\Omega$$

1.8

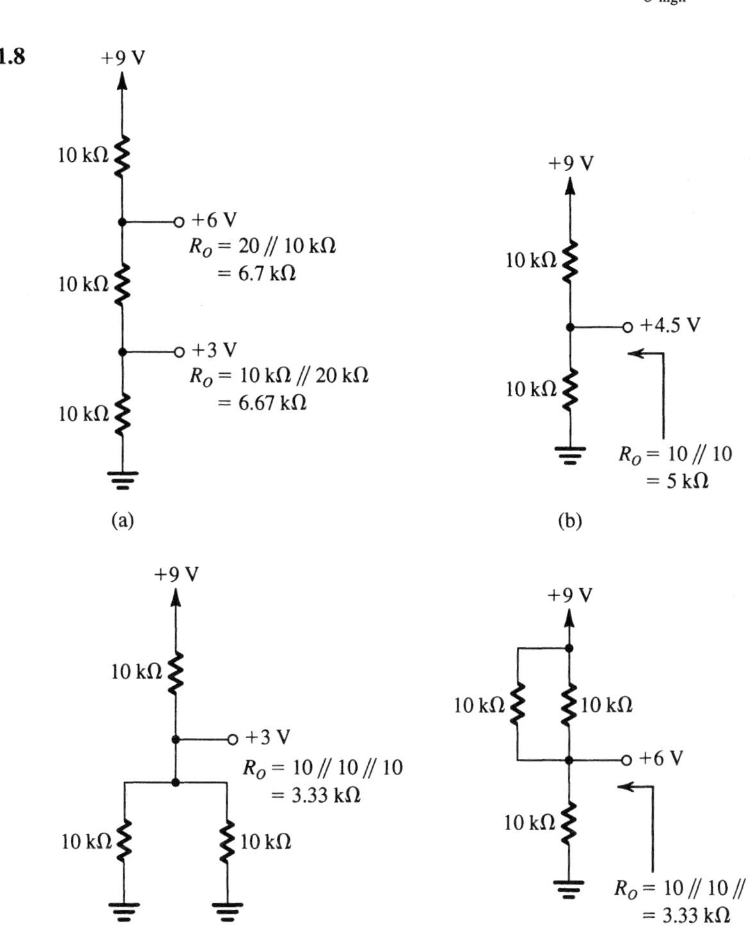

Voltages generated:

+3 V (two ways: (a) and (c) with (c) having lower output resistance)
+4.5 V (b)
+6 V (two ways: (a) and (d) with (d) having a lower output resistance)

1.9

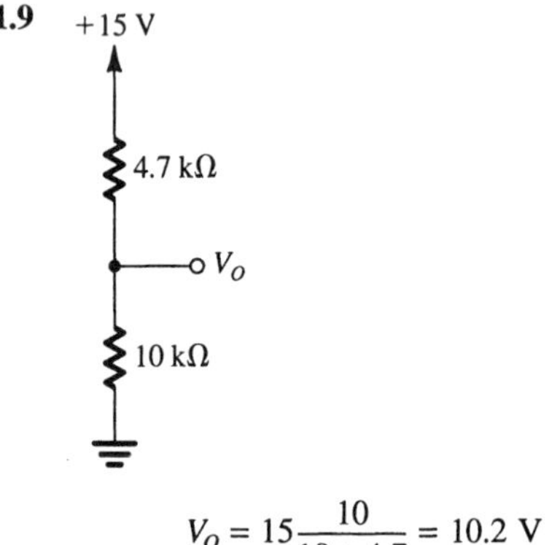

$$V_O = 15\frac{10}{10+4.7} = 10.2 \text{ V}$$

To reduce V_O to 10.00 V we shunt the 10-kΩ resistor by a resistor R whose value is such that $10 \,//\, R = 2 \times 4.7$.

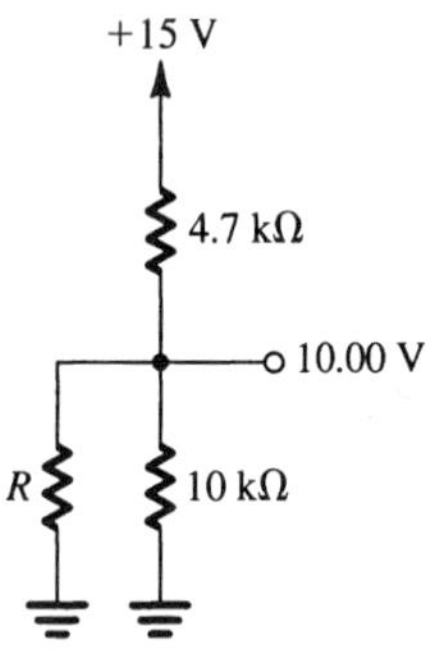

Thus

$$\frac{1}{10} + \frac{1}{R} = \frac{1}{9.4}$$

$$\Rightarrow R = 156.7 \simeq 157 \text{ k}\Omega$$

Now,

$$R_O = 10 \text{ k}\Omega \,//\, R \,//\, 4.7 \text{ k}\Omega$$

$$= 9.4 \,//\, 4.7 = \frac{9.4}{3} = 3.133 \text{ k}\Omega$$

To make $R_O = 3.33$ we add a series resistance of approximately 200 Ω, as shown.

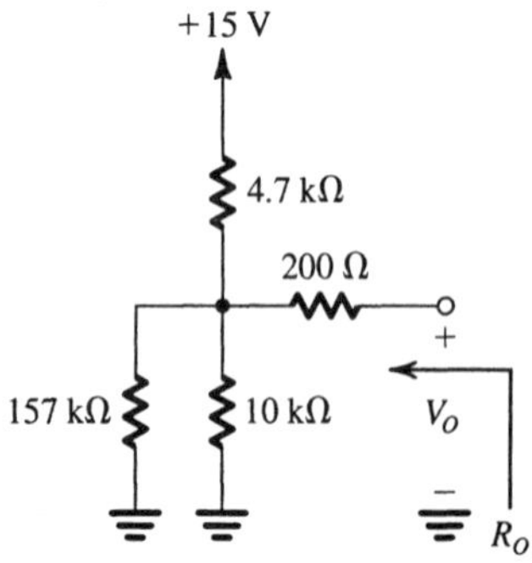

To obtain V_O = 10.00 V and R_O = 3 kΩ we have to shunt both the 4.7-kΩ and the 10-kΩ resistors as shown. To yield an output voltage V_O = 10.00 V we must have

$$\underbrace{(R_2 \,//\, 10)}_{R_2'} = \underbrace{2(R_1 \,//\, 4.7)}_{R_1'}$$

$$R_2' = 2R_1' \qquad (1)$$

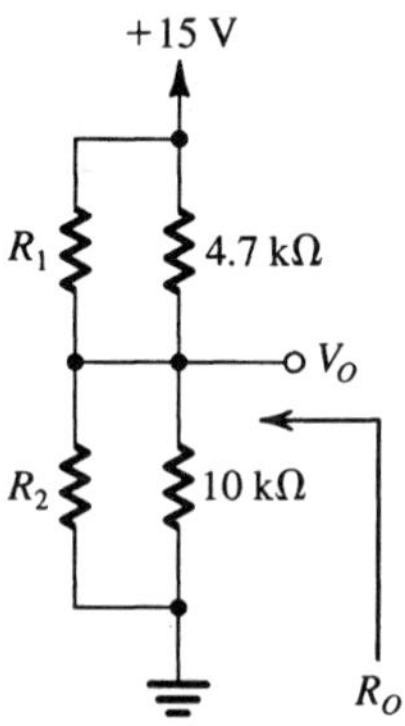

For R_O = 3 kΩ we must have

$$R_1' \,//\, R_2'' = 3 \qquad (2)$$

Solving (1) and (2) yields

$$R_1' = 4.5 \text{ k}\Omega$$

$$R_2' = 9.0 \text{ k}\Omega$$

which can be used to find R_1 and R_2 respectively,

$$R_1 = 157 \text{ k}\Omega$$

$$R_2 = 90 \text{ k}\Omega$$

1.10

$$V = I(R_1 \,//\, R_2)$$

$$= I\frac{R_1 R_2}{R_1 + R_2}$$

$$I_1 = \frac{V}{R_1} = I\frac{R_2}{R_1 + R_2}$$

$$I_2 = \frac{V}{R_2} = I\frac{R_1}{R_1 + R_2}$$

1.11 Connect a resistor R in parallel with R_L. To make $I_L = 0.2I$ (and thus the current through R, $0.8I$), R should be such

$$0.2I \times 1\text{ k}\Omega = 0.8IR$$

$$\Rightarrow R = 250\ \Omega$$

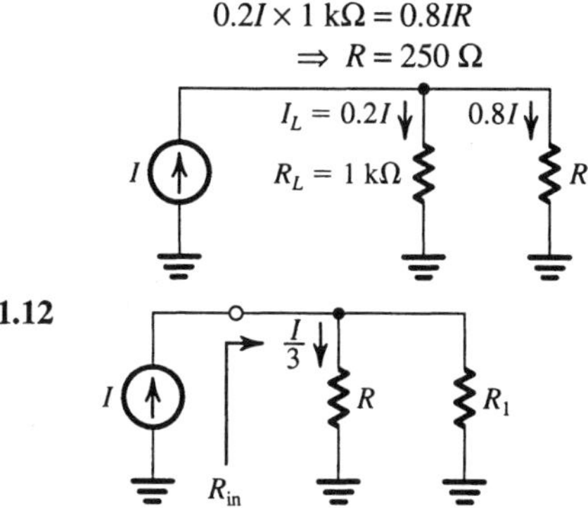

1.12

To make the current through R equal to $I/3$ we shunt R by a resistance R_1 of value such that the current through it will be $2I/3$; thus

$$\frac{I}{3}R = \frac{2I}{3}R_1 \Rightarrow R_1 = \frac{R}{2}$$

The input resistance of the divider, R_{in}, is

$$R_{in} = R \,//\, R_1 = R \,//\, \frac{R}{2} = \frac{1}{3}R$$

Now if R_1 is 10% too high, i.e.,

$$R_1 = 1.1\frac{R}{2}$$

the problem can be solved in two ways:

(a) Connect a resistor R_2 across R_1 of value such that $R_1 \,//\, R_2 = R/2$, thus

$$\frac{R_2(1.1R/2)}{R_2 + (1.1R/2)} = \frac{R}{2}$$

$$1.1R_2 = R_2 + \frac{1.1R}{2}$$

$$\Rightarrow R_2 = \frac{11R}{2} = 5.5R$$

$$R_{in} = R \,//\, \frac{1.1R}{2} \,//\, \frac{11R}{2}$$

$$= R \,//\, \frac{R}{2} = \frac{R}{3}$$

(b) Connect a resistor in series with the load resistor R so as to raise the resistance of the load branch by 10%, thereby restoring the current division ratio to its desired value. The added series resistance must be 10% of R i.e., $0.1R$.

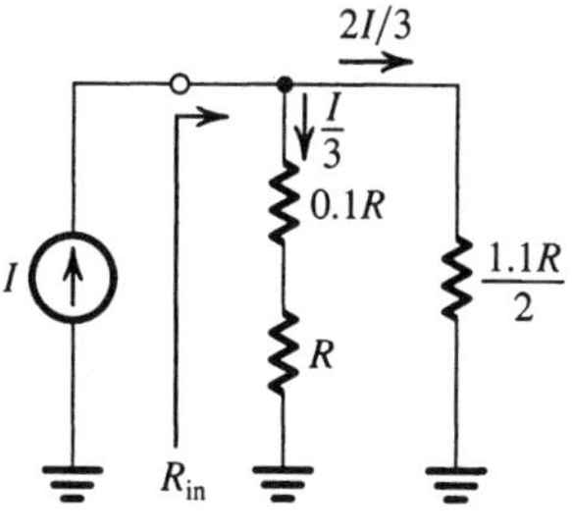

$$R_{in} = 1.1R \,//\, \frac{1.1R}{2}$$

$$= \frac{1.1R}{3}$$

i.e., 10% higher than in case (a).

1.13 If $R_L = 10$ kΩ then a voltage of 0 to 10 V may develop across the source. To limit the voltage to the specified maximum of 1 V, we have to shunt R_L with a resistor R whose value is such that the parallel combination of R_L and R is ≤1 kΩ. Thus,

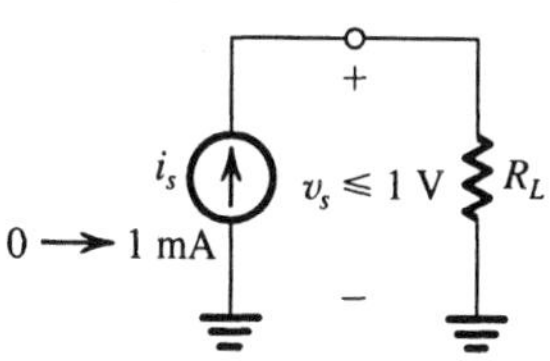

$$\frac{RR_L}{R + R_L} \leq 1$$

$$R \leq 1.111\text{ k}\Omega$$

$$\Rightarrow R \simeq 1.1\text{ k}\Omega$$

The resulting circuit, utilizing only one additional resistor of value 1.1 kΩ creates a current divider across the source.

1.14

1.15

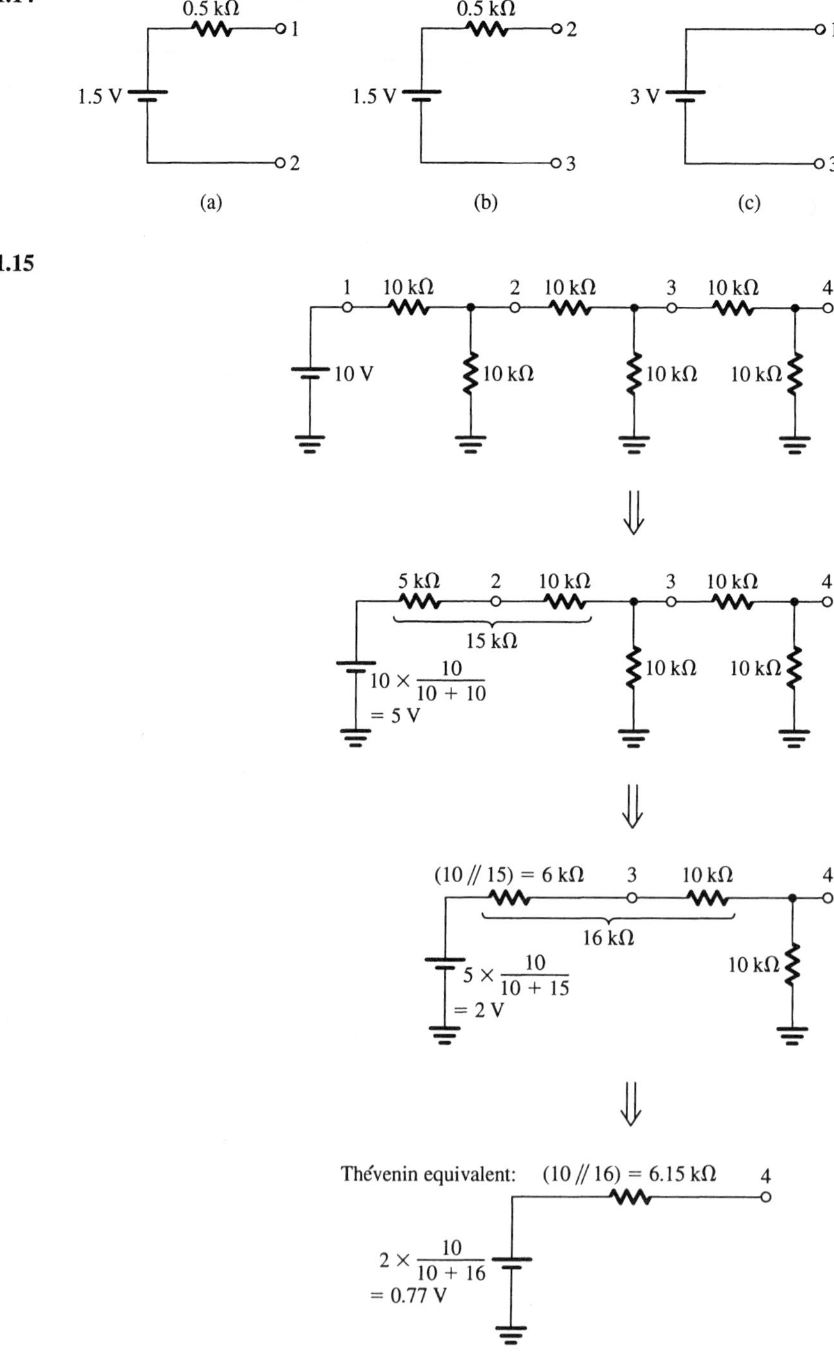

0.5 kΩ
1
1.5 V
2
(a)
0.5 kΩ
2
1.5 V
3
(b)
1
3 V
3
(c)
1
10 kΩ
2
10 kΩ
3
10 kΩ
4
10 V
10 kΩ
10 kΩ
10 kΩ
5 kΩ
2
10 kΩ
3
10 kΩ
4
15 kΩ
10 × 10/(10 + 10)
= 5 V
10 kΩ
10 kΩ
(10 // 15) = 6 kΩ
3
10 kΩ
4
16 kΩ
5 × 10/(10 + 15)
= 2 V
10 kΩ
Thévenin equivalent:
(10 // 16) = 6.15 kΩ
4
2 × 10/(10 + 16)
= 0.77 V

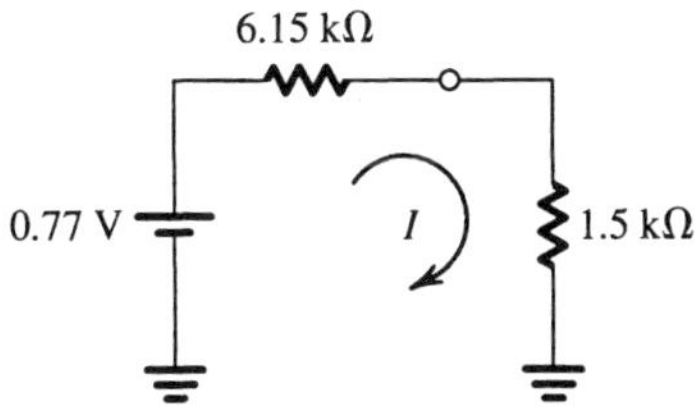

Now, when a resistance of 1.5 kΩ is connected between 4 and ground,

$$I = \frac{0.77}{6.15 + 1.5}$$

$$= 0.1 \text{ mA}$$

1.16 (a) Node equation at the common node yields

$$I_3 = I_1 + I_2$$

Using the fact that the sum of the voltage drops across R_1 and R_3 equals 15 V, we write

$$15 = I_1R_1 + I_3R_3$$

$$= 10I_1 + (I_1 + I_2) \times 2$$

$$= 12I_1 + 2I_2$$

+15 V
+10 V
R_1 10 kΩ
R_2 5 kΩ
I_2 V I_1
I_3
R_3 2 kΩ

That is,

$$12I_1 + 2I_2 = 15 \quad (1)$$

Similarly, the voltage drops across R_2 and R_3 add up to 10 V, thus

$$10 = I_2R_2 + I_3R_3$$

$$= 5I_2 + (I_1 + I_2) \times 2$$

which yields

$$2I_1 + 7I_2 = 10 \quad (2)$$

Equations (1) and (2) can be solved together by multiplying (2) by 6,

$$12I_1 + 42I_2 = 60 \quad (3)$$

Now, subtracting (1) from (3) yields

$$40I_2 = 45$$

$$\Rightarrow I_2 = 1.125 \text{ mA}$$

Substituting in (2) gives

$$2I_1 = 10 - 7 \times 1.125 \text{ mA}$$

$$\Rightarrow I_1 = 1.0625 \text{ mA}$$

$$I_3 = I_1 + I_2$$

$$= 1.0625 + 1.1250$$

$$= 1.1875 \text{ mA}$$

$$V = I_3R_3$$

$$= 1.1875 \times 2 = 2.3750 \text{ V}$$

To summarize:

$$I_1 \simeq 1.06 \text{ mA} \qquad I_2 \simeq 1.13 \text{ mA}$$

$$I_3 \simeq 1.19 \text{ mA} \qquad V \simeq 2.38 \text{ V}$$

(b) A node equation at the common node can be written in terms of V as

$$\frac{15 - V}{R_1} + \frac{10 - V}{R_2} = \frac{V}{R_3}$$

Thus,

$$\frac{15 - V}{10} + \frac{10 - V}{5} = \frac{V}{2}$$

$$\Rightarrow 0.8V = 3.5$$

$$\Rightarrow V = 2.375 \text{ V}$$

Now, I_1, I_2, and I_3 can be easily found as

$$I_1 = \frac{15 - V}{10} = \frac{15 - 2.375}{10} = 1.0625 \text{ mA} \simeq 1.06 \text{ mA}$$

$$I_2 = \frac{10 - V}{5} = \frac{10 - 2.375}{5} = 1.125 \text{ mA} \simeq 1.13 \text{ mA}$$

$$I_3 = \frac{V}{R_3} = \frac{2.375}{2} = 1.1875 \text{ mA} \simeq 1.19 \text{ mA}$$

Method (b) is much preferred; faster, more insightful and less prone to errors. In general, one attempts to identify the least possible number of variables and write the corresponding minimum number of equations.

1.17 See diagram

(1 // 1.2) 1 I_5 2 (9.1 // 11)
0.545 kΩ R_5 3 kΩ 4.980 kΩ
$9 \times \frac{1.2}{1.2 + 1}$ = 4.909 V
$9 \times \frac{11}{11 + 9.1}$ = 4.925 V

$$I_5 = \frac{4.925 - 4.909}{4.98 + 3 + 0.545} = 1.88 \ \mu\text{A}$$

$$V_5 = 1.88 \ \mu\text{A} \times 3 \text{ k}\Omega = 5.64 \text{ mV}$$

1.18 From the symmetry of the circuit, there will be no current in R_5. (Otherwise the symmetry would be violated.) Thus each branch will carry a current $V_x/2\text{ k}\Omega$ and I_x will be the sum of the two current,

$$I_x = \frac{2V_x}{2\text{ k}\Omega} = \frac{V_x}{1\text{ k}\Omega}$$

$I_x = 2V_x/2\text{ k}\Omega$

$V_x/2\text{ k}\Omega$ $\quad$ $V_x/2\text{ k}\Omega$ $\quad$ V_x

R_1 1 kΩ $\quad$ R_3 1 kΩ

$V_x/2$ $\quad$ R_5 1 kΩ $\quad$ 0 $\quad$ $V_x/2$

R_2 1 kΩ $\quad$ R_4 1 kΩ

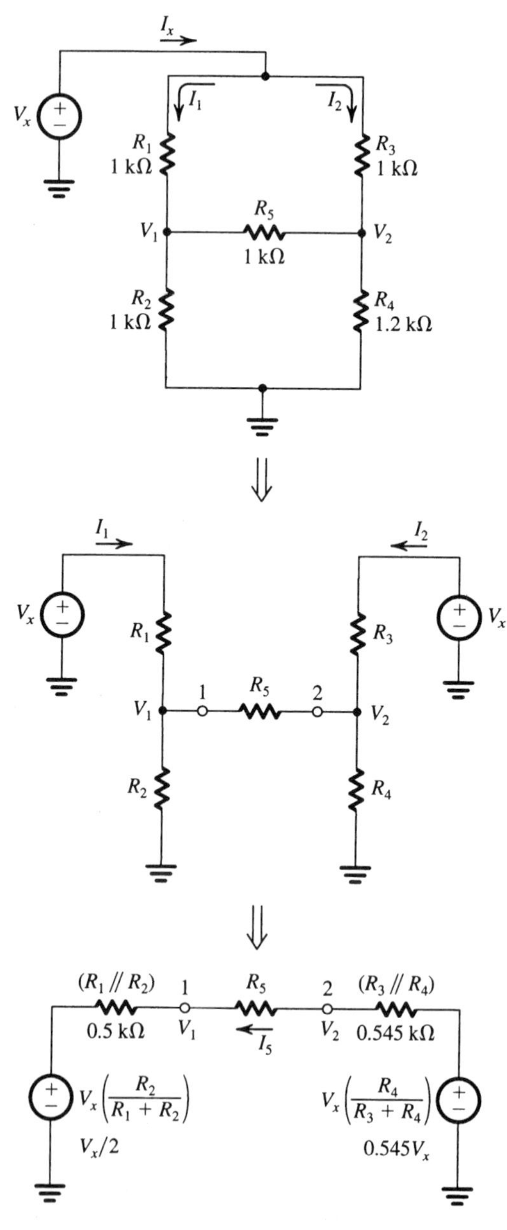

Thus,

$$R_{eq} \equiv \frac{V_x}{I_x} = 1\text{ k}\Omega$$

Now, if R_4 is raised to 1.2 kΩ the symmetry will be broken. To find I_x we use Thévenin's theorem as follows:

$$I_5 = \frac{0.545V_x - 0.5V_x}{0.5 + 1 + 0.545} = 0.022V_x$$

$$V_1 = \frac{V_x}{2} + 0.022V_x \times 0.5$$

$$= 0.5V_x \times 1.022 = 0.511V_x$$

$$V_2 = V_1 + I_5R_5 = 0.533V_x$$

$$I_1 = \frac{V_x - V_1}{1\text{ k}\Omega} = 0.489V_x$$

$$I_2 = \frac{V_x - V_2}{1\text{ k}\Omega} = 0.467V_x$$

$$I_x = I_1 + I_2 = 0.956V_x$$

$$\Rightarrow R_{eq} \equiv \frac{V_x}{I_x} = 1.05\text{ k}\Omega$$

1.19 (a) $T = 10^{-4}$ ms $= 10^{-7}$ s

$$f = \frac{1}{T} = 10^7 \text{ Hz}$$
$$\omega = 2\pi f = 6.28 \times 10^7 \text{ Hz}$$

(b) $f = 1$ GHz $= 10^9$ Hz

$$T = \frac{1}{f} = 10^{-9} \text{ s}$$
$$\omega = 2\pi f = 6.28 \times 10^9 \text{ rad/s}$$

(c) $\omega = 6.28 \times 10^2$ rad/s

$$f = \frac{\omega}{2\pi} = 10^2 \text{ Hz}$$
$$T = \frac{1}{f} = 10^{-2} \text{ s}$$

(d) $T = 10$ s

$$f = \frac{1}{T} = 10^{-1} \text{ Hz}$$
$$\omega = 2\pi f = 6.28 \times 10^{-1} \text{ rad/s}$$

(e) $f = 60$ Hz

$$T = \frac{1}{f} = 1.67 \times 10^{-2} \text{ s}$$
$$\omega = 2\pi f = 3.77 \times 10^2 \text{ rad/s}$$

(f) $\omega = 1$ krad/s $= 10^3$ rad/s

$$f = \frac{\omega}{2\pi} = 1.59 \times 10^2 \text{ Hz}$$
$$T = \frac{1}{f} = 6.28 \times 10^{-3} \text{ s}$$

(g) $f = 1900$ MHz $= 1.9 \times 10^9$ Hz

$$T = \frac{1}{f} = 0.526 \times 10^{-9} \text{ s}$$
$$\omega = 2\pi f = 1.194 \times 10^9 \text{ rad/s}$$

1.20 (a) $Z = 1$ kΩ at all frequencies

(b) $Z = 1/j\omega C = -j\dfrac{1}{2\pi f \times 10 \times 10^{-9}}$

At $f = 60$ Hz, $Z = -j265$ kΩ
At $f = 100$ kHz, $Z = -j159$ Ω
At $f = 1$ GHz, $Z = -j0.016$ Ω

(c) $Z = 1/j\omega C = -j\dfrac{1}{2\pi f \times 2 \times 10^{-12}}$

At $f = 60$ Hz, $Z = -j1.33$ GΩ
At $f = 100$ kHz, $Z = -j0.8$ MΩ
At $f = 1$ GHz, $Z = -j79.6$ Ω

(d) $Z = j\omega L = j2\pi fL = j2\pi f \times 10 \times 10^{-3}$

At $f = 60$ Hz, $Z = j3.77$ Ω
At $f = 100$ kHz, $Z = j6.28$ kΩ
At $f = 1$ GHz, $Z = j62.8$ MΩ

1.21 (a) $Z = R + \dfrac{1}{j\omega C}$

$$= 10^3 + \frac{1}{j2\pi \times 10 \times 10^3 \times 10 \times 10^{-9}}$$
$$= (1 - j1.59) \text{ k}\Omega$$

(b) $Y = \dfrac{1}{R} + j\omega C$

$$= \frac{1}{10^3} + j2\pi \times 10 \times 10^3 \times 0.01 \times 10^{-6}$$
$$= 10^{-3}(1 + j0.628) \text{ ℧}$$
$$Z = \frac{1}{Y} = \frac{1000}{1 + j0.628}$$
$$= \frac{1000(1 - j0.628)}{1 + 0.628^2}$$
$$= (717.2 - j450.4)\ \Omega$$

(c) $Y = \dfrac{1}{R} + j\omega C$

$$= \frac{1}{100 \times 10^3} + j2\pi \times 10 \times 10^3 \times 100 \times 10^{-12}$$
$$= 10^{-5}(1 + j0.628)$$
$$Z = \frac{10^5}{1 + j0.628}$$
$$= (71.72 - j45.04) \text{ k}\Omega$$

(d) $Z = R + j\omega L$

$$= 100 + j2\pi \times 10 \times 10^3 \times 10 \times 10^{-3}$$
$$= 100 + j6.28 \times 100$$
$$= (100 + j628)\ \Omega$$

1.22

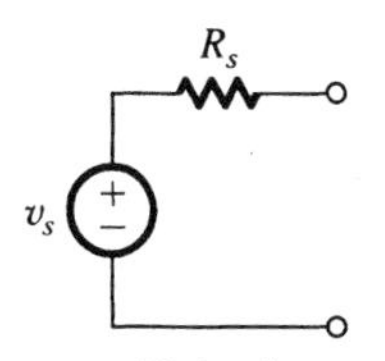

Thévenin Equivalent

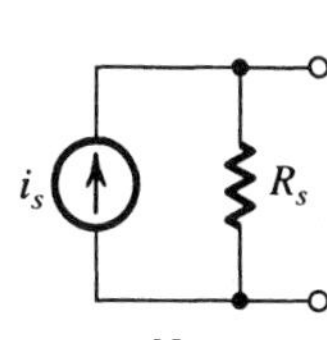

Norton Equivalent

$$v_{oc} = v_s$$
$$i_{sc} = i_s$$
$$v_s = i_s R_s$$

Thus,

$$R_s = \frac{v_{oc}}{i_{sc}}$$

(a) $v_s = v_{oc} = 10\text{ V}$

$i_s = i_{sc} = 100\ \mu\text{A}$

$$R_s = \frac{v_{oc}}{i_{sc}} = \frac{10\text{ V}}{100\ \mu\text{A}} = 0.1\text{ M}\Omega = 100\text{ k}\Omega$$

(b) $v_s = v_{oc} = 0.1\text{ V}$

$i_s = i_{sc} = 10\ \mu\text{A}$

$$R_s = \frac{v_{oc}}{i_{sc}} = \frac{0.1\text{V}}{10\ \mu\text{A}} = 0.01\text{ M}\Omega = 10\text{ k}\Omega$$

1.23

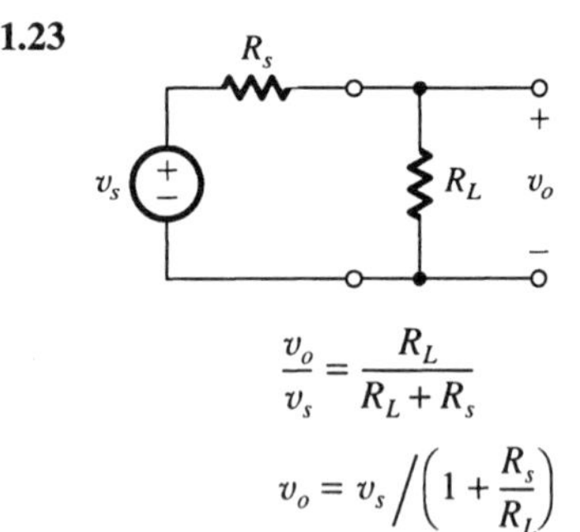

$$\frac{v_o}{v_s} = \frac{R_L}{R_L + R_s}$$

$$v_o = v_s \Big/ \left(1 + \frac{R_s}{R_L}\right)$$

Thus,

$$\frac{v_s}{1 + \dfrac{R_s}{100}} = 30 \qquad (1)$$

and

$$\frac{v_s}{1 + \dfrac{R_s}{10}} = 10 \qquad (2)$$

Dividing (1) by (2) gives

$$\frac{1 + (R_s/10)}{1 + (R_s/100)} = 3$$
$$\Rightarrow R_s = 28.6\text{ k}\Omega$$

Substituting in (2) gives

$$v_s = 38.6\text{ mV}$$

The Norton current i_s can be found as

$$i_s = \frac{v_s}{R_s} = \frac{38.6\text{ mV}}{28.6\text{ k}\Omega} = 1.35\ \mu\text{A}$$

1.24 The observed output voltage is 1 mV/°C which is one half the voltage specified by the sensor, presumably under open-circuit conditions that is without a load connected. It follows that that sensor internal resistance must be equal to R_L, i.e., 10 kΩ.

1.25

$$v_o = v_s - i_o R_s$$

$$v_o = (i_s - i_o)R_s$$
$$= i_s R_s - i_o R_s$$
$$v_o = v_s - i_o R_s$$

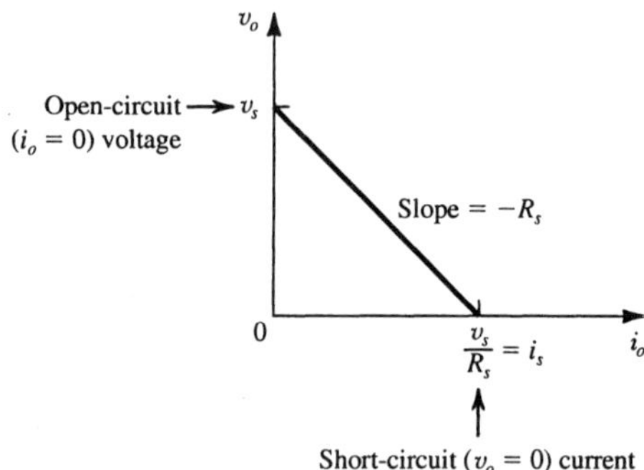

1.26

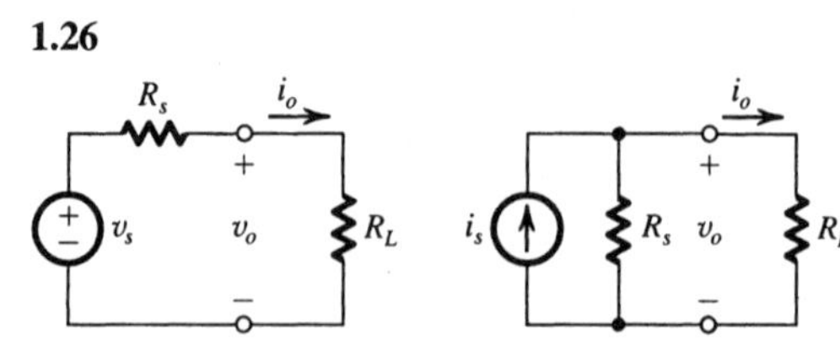

R_L represents the input resistance of the processor

For $v_o = 0.9\,v_s$,

$$0.9 = \frac{R_L}{R_L + R_s} \quad \Rightarrow \quad R_L = 9R_s$$

For $i_o = 0.9\,i_s$,

$$0.9 = \frac{R_s}{R_s + R_L} \quad \Rightarrow \quad R_L = R_s/9$$

1.27

Case	ω (rad/s)	f (Hz)	T (s)
a	6.28×10^9	1×10^9	1×10^{-9}
b	1×10^9	1.59×10^8	6.28×10^{-9}
c	6.28×10^{10}	1×10^{10}	1×10^{-10}
d	3.77×10^2	60	1.67×10^{-2}
e	6.28×10^3	1×10^3	1×10^{-3}
f	6.28×10^6	1×10^6	1×10^{-6}

1.28 (a) $V_{peak} = 117 \times \sqrt{2} = 165$ V

(b) $V_{rms} = 33.9/\sqrt{2} = 24$ V

(c) $V_{peak} = 220 \times \sqrt{2} = 311$ V

(d) $V_{peak} = 220 \times \sqrt{2} = 311$ kV

1.29 (a) $v = 10 \sin(2\pi \times 10^4 t)$, V

(b) $v = 120\sqrt{2} \sin(2\pi \times 60)$, V

(c) $v = 0.1 \sin(1000t)$, V

(d) $v = 0.1 \sin(2\pi \times 10^{+3} t)$, V

1.30 Comparing the given waveform to that described by Eq. 1.2 we observe that the given waveform has an amplitude of 0.5 V (1 V peak-to-peak) and its level is shifted up by 0.5 V (the first term in the equation). Thus the waveform look as follows

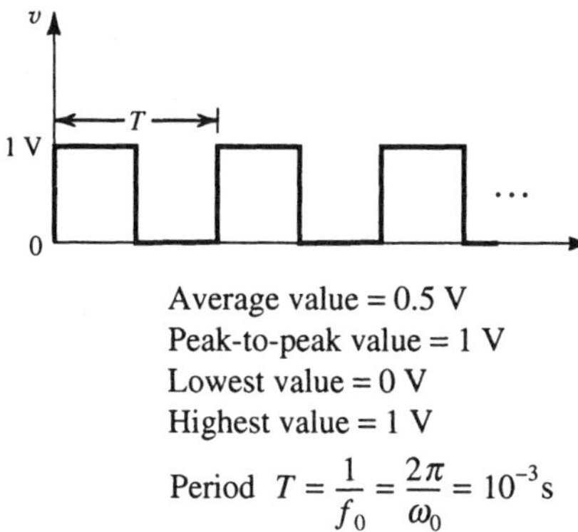

Average value = 0.5 V
Peak-to-peak value = 1 V
Lowest value = 0 V
Highest value = 1 V

$$\text{Period } T = \frac{1}{f_0} = \frac{2\pi}{\omega_0} = 10^{-3}\text{s}$$

1.31 The two harmonics have the ratio $126/98 = 9/7$. Thus, these are the 7th and 9th harmonics. From Eq. 1.2 we note that the amplitudes of these two harmonics will have the ratio 7 to 9, which is confirmed by the measurement reported. Thus the fundamental will have a frequency of 98/7 or 14 kHz and peak amplitude of $63 \times 7 = 441$ mV. The rms value of the fundamental will be $441/\sqrt{2} = 312$ mV. To find the peak-to-peak amplitude of the square wave we note that $4V/\pi = 441$ mV. Thus,

$$\text{Peak-to-peak amplitude} = 2V = 441 \times \frac{\pi}{2} = 693 \text{ mV}$$

$$\text{Period } T = \frac{1}{f} = \frac{1}{14 \times 10^3} = 71.4\ \mu\text{s}$$

1.32 To be barely audible by a relatively young listener, the 5th harmonic must be limited to 20 kHz; thus the fundamental will be 4 kHz. At the low end, hearing extends down to about 20 Hz. For the fifth and higher to be audible the fifth must be no lower than 20 Hz. Correspondingly, the fundamental will be at 4 Hz.

1.33 If the amplitude of the square wave is V_{sq} then the power delivered by the square wave to a resistance R will be V_{sq}^2/R. If this power is to equal that delivered by a sine wave of peak amplitude $\hat{V}$ then

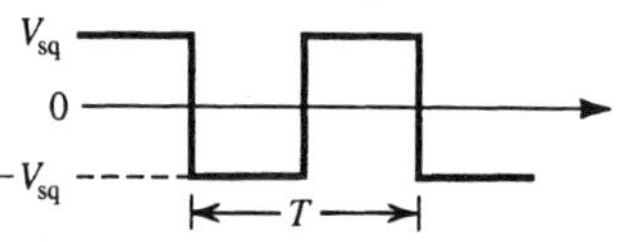

$$\frac{V_{sq}^2}{R} = \frac{(\hat{V}/\sqrt{2})^2}{R}$$

Thus, $V_{sq} = \hat{V}/\sqrt{2}$. This result is independent of frequency.

1.34

Decimal	Binary
0	0
5	101
8	1000
25	11001
57	111001

1.35

b_3 b_2 b_1 b_0	Value Represented
0 0 0 0	+0
0 0 0 1	+1
0 0 1 0	+2
0 0 1 1	+3
0 1 0 0	+4
0 1 0 1	+5
0 1 1 0	+6
0 1 1 1	+7
1 0 0 0	−0
1 0 0 1	−1
1 0 1 0	−2
1 0 1 1	−3
1 1 0 0	−4
1 1 0 1	−5
1 1 1 0	−6
1 1 1 1	−7

Note that there are two possible representation of zero: 0000 and 1000. For a 0.5-V step size, analog signals in the range ±3.5 V can be represented

Input	Steps	Code
+2.5 V	+5	0101
−3.0 V	−6	1110
+2.7	+5	0101
−2.8	−6	1110

1.36 (a) For N bits there will be 2^N possible levels, from 0 to V_{FS}. Thus there will be $(2^N - 1)$ discrete steps from 0 to V_{FS} with the step size given by

$$\text{Step size} = \frac{V_{FS}}{2^N - 1}$$

This is the analog change corresponding to a change in the LSB. It is the value of the resolution of the ADC.

(b) The maximum error in conversion occurs when the analog signal value is at the middle of a step. Thus the maximum error is

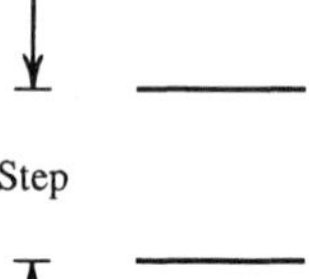

$$\frac{1}{2} \times \text{step size} = \frac{1}{2}\frac{V_{FS}}{2^N - 1}$$

This is known as the quantization error.

(c)
$$\frac{10\text{ V}}{2^N - 1} \leq 5\text{ mV}$$

$$2^N - 1 \geq 2000$$

$$2^N \geq 2001 \Rightarrow N = 11$$

For $N = 11$,

$$\text{Resolution} = \frac{10}{2^{11} - 1} = 4.9\text{ mV}$$

$$\text{Quantization error} = \frac{4.9}{2} = 2.4\text{ mV}$$

1.37 When $b_i = 1$, the ith switch is in position 1 and a current $(V_{\text{ref}}/2^iR)$ flows to the output. Thus i_O will be the sum of all the currents corresponding to "1" bits, i.e.,

$$i_O = \frac{V_{\text{ref}}}{R}\left(\frac{b_1}{2^1} + \frac{b_2}{2^2} + \cdots + \frac{b_N}{2^N}\right)$$

(b) b_N is the LSB
b_1 is the MSB

(c) $$i_{O\max} = \frac{10\text{ V}}{5\text{ k}\Omega}\left(\frac{1}{2^1} + \frac{1}{2^2} + \frac{1}{2^3} + \frac{1}{2^4} + \frac{1}{2^5} + \frac{1}{2^6}\right)$$
$$= 1.96875\text{ mA}$$

Corresponding to the LSB changing from 0 to 1 the output changes by $10/5 \times 1/2^6 = 0.03125$ mA.

1.38 There will be 44,100 samples per second with each sample represented by 16 bits. Thus the throughput or speed will be $44{,}100 \times 16 = 7.056 \times 10^5$ bits per second.

1.39 (a) $$A_v = \frac{v_o}{v_i} = \frac{10\text{ V}}{100\text{ mV}} = 100\text{ V/V}$$

or, $20 \log 100 = 40$ dB

$$A_i = \frac{i_o}{i_i} = \frac{v_o/R_L}{i_i} = \frac{10\text{ V}/100\ \Omega}{100\ \mu\text{A}} = \frac{0.1\text{ A}}{100\ \mu\text{A}}$$
$$= 1000\text{ A/A}$$

or, $20 \log 1000 = 60$ dB

$$A_p = \frac{v_o i_o}{v_i i_i} = \frac{v_o}{v_i} \times \frac{i_o}{i_i} = 100 \times 1000$$
$$= 10^5\text{ W/W}$$

or $10 \log 10^5 = 50$ dB

(b) $$A_v = \frac{v_o}{v_i} = \frac{2\text{ V}}{10\ \mu\text{V}} = 2 \times 10^5\text{ V/V}$$

or, $20 \log 2 \times 10^5 = 106$ dB

$$A_i = \frac{i_o}{i_i} = \frac{v_o/R_L}{i_i} = \frac{2\text{ V}/10\text{ k}\Omega}{100\text{ nA}}$$
$$= \frac{0.2\text{ mA}}{100\text{ nA}} = \frac{0.2 \times 10^{-3}}{100 \times 10^{-9}} = 2000\text{ A/A}$$

or $20 \log A_i = 66$ dB

$$A_p = \frac{v_o i_o}{v_i i_i} = \frac{v_o}{v_i} \times \frac{i_o}{i_i}$$
$$= 2 \times 10^5 \times 2000$$
$$= 4 \times 10^8\text{ W/W}$$

or $10 \log A_p = 86$ dB

(c) $$A_v = \frac{v_o}{v_i} = \frac{10\text{ V}}{1\text{ V}} = 10\text{ V/V}$$

or, $20 \log 10 = 20$ dB

$$A_i = \frac{i_o}{i_i} = \frac{v_o/R_L}{i_i} = \frac{10\text{ V}/10\ \Omega}{1\text{ mA}}$$
$$= \frac{1\text{ A}}{1\text{ mA}} = 1000\text{ A/A}$$

or, 20 log 1000 = 60 dB

$$A_p = \frac{v_o i_o}{v_i i_i} = \frac{v_o}{v_i} \times \frac{i_o}{i_i}$$

$$= 10 \times 1000 = 10^4 \text{ W/W}$$

or $10 \log_{10} A_p = 40$ dB

1.40

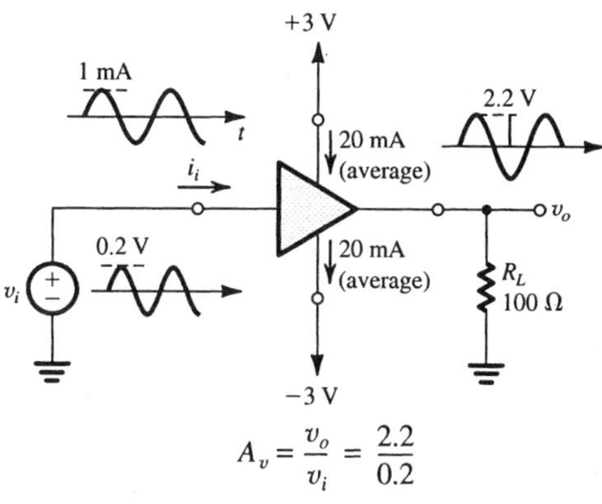

$$A_v = \frac{v_o}{v_i} = \frac{2.2}{0.2}$$

$$= 11 \text{ V/V}$$

or 20 log 11 = 20.8 dB

$$A_i = \frac{i_o}{i_i} = \frac{2.2 \text{ V}/100\ \Omega}{1 \text{ mA}}$$

$$= \frac{22 \text{ mA}}{1 \text{ mA}} = 22 \text{ A/A}$$

or, $20 \log A_i = 26.8$ dB

$$A_p = \frac{p_o}{p_i} = \frac{(2.2/\sqrt{2})^2/100}{\frac{0.2}{\sqrt{2}} \times \frac{10^{-3}}{\sqrt{2}}}$$

$$= 242 \text{ W/W}$$

or, $10 \log A_p = 23.8$ dB

Supply power = 2 × 3 V × 20 mA = 120 mW

$$\text{Output power} = \frac{v_{orms}^2}{R_L} = \frac{(2.2/\sqrt{2})^2}{100\ \Omega} = 24.2 \text{ mW}$$

$$\text{Input power} = \frac{24.2}{242} = 0.1 \text{ mW} \quad \text{(negligible)}$$

Amplifier dissipation ≃ Supply power − Output power

= 120 − 24.2 = 95.8 mW

$$\text{Amplifier efficiency} = \frac{\text{Output power}}{\text{Supply power}} \times 100$$

$$= \frac{24.2}{120} \times 100 = 20.2\%$$

1.41 For $V_{DD} = 5$ V:
The largest undistorted sine-wave output is of 3.8-V peak amplitude or $3.8/\sqrt{2} = 2.7\,V_{\text{rms}}$. Input needed is 5.4 mV $_{\text{rms}}$.

Supplies are V_{DD} and $-V_{DD}$

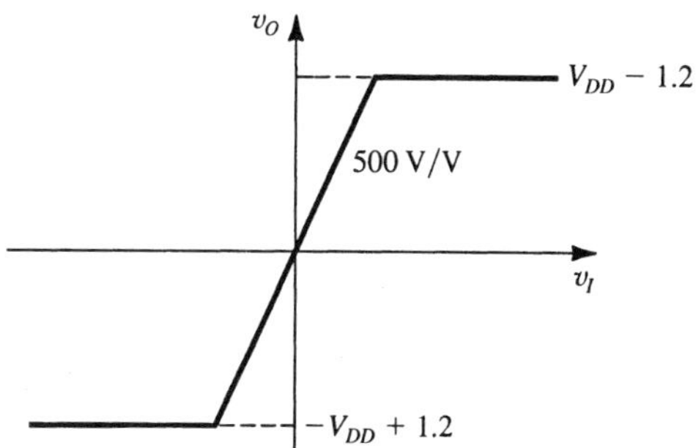

For $V_{DD} = 10$ V, the largest undistorted sine-wave output is of 8.8-V peak amplitude or 6.2 V_{rms}. Input needed is 12.4 mV_{rms}.

For $V_{DD} = 15$ V, the largest undistorted sine-wave output is of 13.8-V peak amplitude or 9.8 V_{rms}. The input needed is 9.8 V/500 = 19.6 mV_{rms}.

1.42 (a) For an output whose extremes are just at the edge of clipping, i.e., an output of 9-V_{peak}, the input must have 9 V/1000 = 9 mV_{peak}.

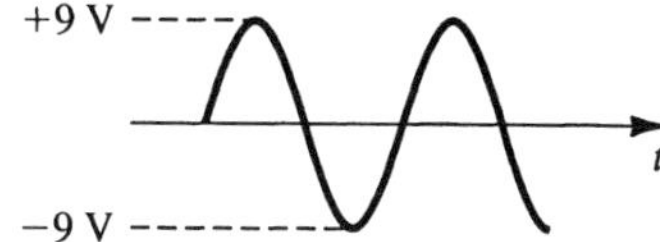

(b) For an output that is clipping 90% of the time, $\theta = 0.1 \times 90° = 9°$ and $V_p \sin 9° = 9 \text{ V} \Rightarrow V_p = 57.5$ V which of course does not occur as the output saturates at ±9 V. To produce this result, the input peak must be 57.5/1000 = 57.5 mV.

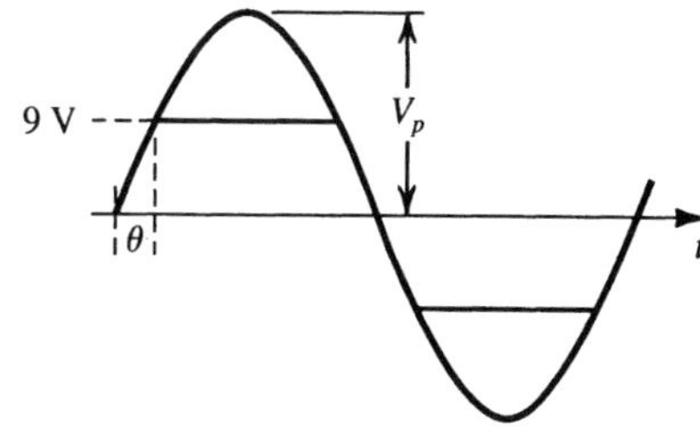

(c) For an output that is clipping 99% of the time, $\theta = 0.01 \times 90° = 0.9°$

$$V_p \sin 0.9° = 9 \text{ V}$$

$$\Rightarrow V_p = 573 \text{ V}$$

and the input must be 573 V/1000 or 0.573 V_{peak}.

1.43 When the amplifier is biased at 4 V (i.e., at point Q_1), the maximum possible amplitude of a sine-wave output without clipping is $(4 - 1.5) = 2.5$ V_{peak}.

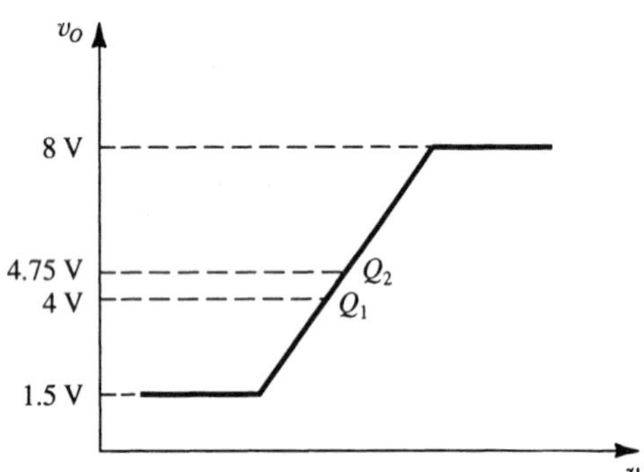

To obtain the largest undistorted sine wave output possible with this amplifier, it must be biased halfway between the saturation levels, i.e., at $V_O = (8 + 1.5)/2 = 4.75$ V (point Q_2) and the resulting output will have a peak value of $8 - 4.75 = 4.75 - 1.50 = 3.25$ V.

1.44 $v_O = 10 - 5(v_I - 2)^2, \quad 2 \le v_I \le v_O + 2, \quad v_O \ge 0$

(a) For $v_I \le 2$ V, $v_O = 10$ V.

The upper limit on v_I is found by substituting $v_I = v_O + 2$, that is, $v_O = v_I - 2$ in the transfer characteristic. The result is $v_O = 10 - 5v_O^2$, whose solution is $v_O = 1.317$ V and the corresponding $v_I = 3.317$ V. To obtain a sketch of v_O versus v_I, we evaluate v_O for values of v_I in the range 2 V to 3.317 V. The result is the following sketch:

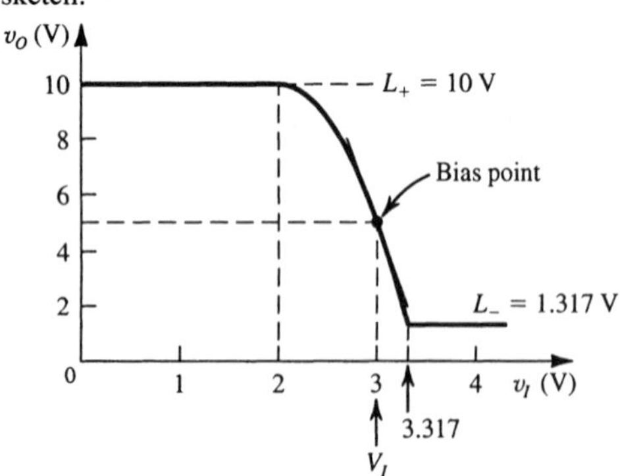

(b) To obtain $V_O = 5$ V we bias at $V_I = 3$ V.

(c) Small-signal gain at bias point $= \left.\dfrac{\partial v_O}{\partial v_I}\right|_{V_I = 3\text{V}}$

$$= -5 \times 2(V_I - 2) = -10 \text{ V/V}$$

(d) $v_I = 3 + V_i \cos \omega t$

$$v_O = 10 - 5(3 + V_i \cos \omega t - 2)^2$$

$$= 10 - 5(1 + 2V_i \cos \omega t + V_i^2 \cos^2 \omega t)$$

$$= 5 - 10V_i \cos \omega t - 5V_i^2\left(\frac{1}{2} + \frac{1}{2}\cos 2\omega t\right)$$

$$= \underbrace{(5 - 2.5V_i^2)}_{\text{dc}} - \underbrace{10V_i \cos \omega t}_{\text{Fundamental}} - \underbrace{2.5V_i^2 \cos 2\omega t}_{\text{2nd harmonic}}$$

For 1% second-harmonic distortion: $2.5V_i^2/10V_i = 0.01$

Thus,

$$V_i = \frac{10 \times 0.01}{2.5} \text{ V} = 40 \text{ mV}$$

1.45

$$\frac{v_o}{v_s} = \frac{R_i}{R_i + R_s} \times A_{v_o} \times \frac{R_L}{R_L + R_o}$$

(a) $\dfrac{v_o}{v_s} = \dfrac{10R_s}{10R_s + R_s} \times A_{v_o} \times \dfrac{10R_o}{10R_o + R_o}$

$$= \frac{10}{11} \times 10 \times \frac{10}{11} = 8.26 \text{ V/V}$$

or, 20 log 8.26 = 18.3 dB

(b) $\dfrac{v_o}{v_s} = \dfrac{R_s}{R_s + R_s} \times A_{v_o} \times \dfrac{R_o}{R_o + R_o}$

$$= 0.5 \times 10 \times 0.5 = 2.5 \text{ V/V}$$

or, 20 log 2.5 = 8 dB

(c) $\dfrac{v_o}{v_s} = \dfrac{R_s/10}{(R_s/10) + R_s} \times A_{v_o} \times \dfrac{R_o/10}{(R_o/10) + R_o}$

$$= \frac{1}{11} \times 10 \times \frac{1}{11} = 0.083 \text{ V/V}$$

or 20 log 0.083 = −21.6 dB

1.46 $20 \log A_{v_o} = 40$ dB $\Rightarrow A_{v_o} = 100$ V/V

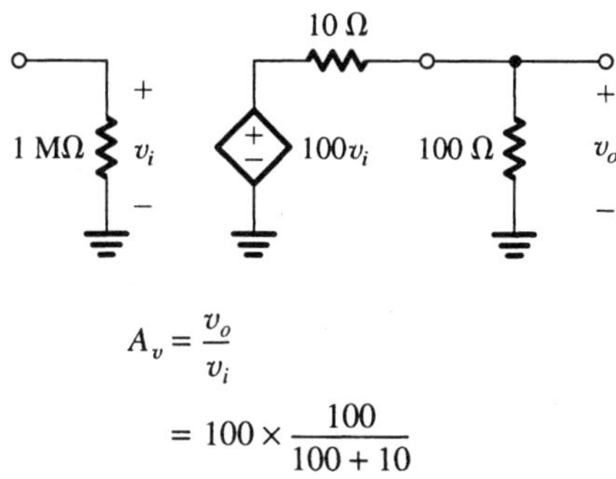

$$A_v = \frac{v_o}{v_i}$$

$$= 100 \times \frac{100}{100 + 10}$$

$$= 90.9 \text{ V/V}$$

or, 20 log 90.9 = 39.1 dB

$$A_p = \frac{v_o^2/100\ \Omega}{v_i^2/1\ \text{M}\Omega} = A_v^2 \times 10^4 = 8.3 \times 10^7 \text{ W/W}$$

or 10 log $(8.3 \times 10^7) = 79.1$ dB.

For a peak output sine-wave current of 100 Ω, the peak output voltage will be 100 mA × 100 Ω = 10 V. Correspondingly v_i will be a sine wave with a peak value of 10 V/A_v = 10/90.9 or an rms value of $10/(90.9 \times \sqrt{2}) = 0.08$ V.

Corresponding output power $= (10/\sqrt{2})^2/100\ \Omega$

$$= 0.5 \text{ W}$$

1.47

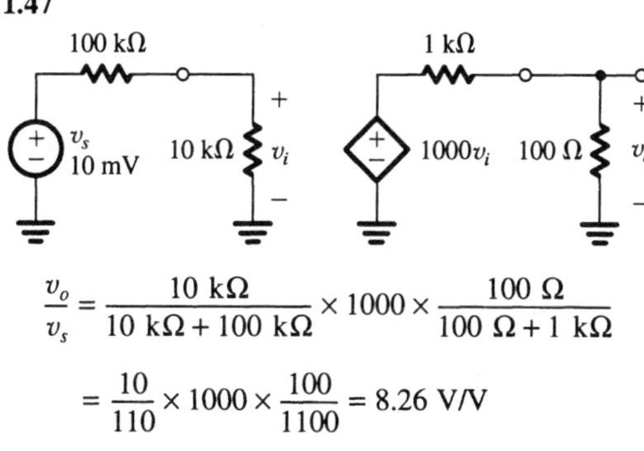

$$\frac{v_o}{v_s} = \frac{10\ \text{k}\Omega}{10\ \text{k}\Omega + 100\ \text{k}\Omega} \times 1000 \times \frac{100\ \Omega}{100\ \Omega + 1\ \text{k}\Omega}$$

$$= \frac{10}{110} \times 1000 \times \frac{100}{1100} = 8.26 \text{ V/V}$$

The signal loses about 90% of its strength when connected to the amplifier input (because $R_i = R_s/10$). Also, the output signal of the amplifier loses approximately 90% of its strength when the load is connected (because $R_L = R_o/10$). Not a good design! Nevertheless, if the source were connected directly to the load,

$$\frac{v_o}{v_s} = \frac{R_L}{R_L + R_s}$$

$$= \frac{100\ \Omega}{100\ \Omega + 100\ \text{k}\Omega}$$

$$\simeq 0.001 \text{ V/V}$$

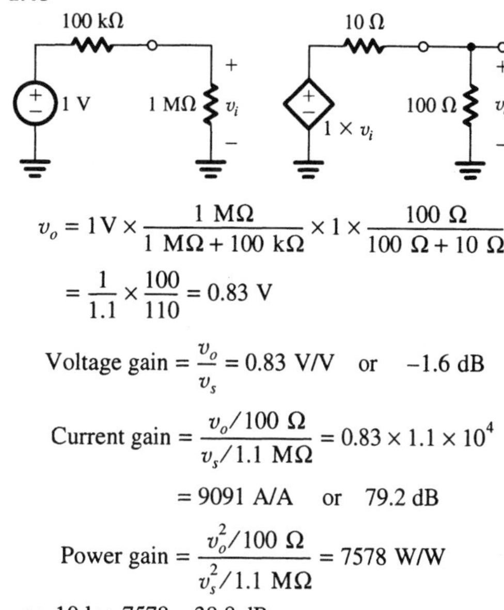

which is clearly a much worse situation. Indeed inserting the amplifier increases the gain by a factor 8.3/0.001 = 8300.

1.48

100 kΩ, 1 V, 1 MΩ, v_i, 10 Ω, 1 × v_i, 100 Ω, v_o

$$v_o = 1\text{V} \times \frac{1\ \text{M}\Omega}{1\ \text{M}\Omega + 100\ \text{k}\Omega} \times 1 \times \frac{100\ \Omega}{100\ \Omega + 10\ \Omega}$$

$$= \frac{1}{1.1} \times \frac{100}{110} = 0.83 \text{ V}$$

Voltage gain $= \frac{v_o}{v_s} = 0.83$ V/V or -1.6 dB

Current gain $= \frac{v_o/100\ \Omega}{v_s/1.1\ \text{M}\Omega} = 0.83 \times 1.1 \times 10^4$

$= 9091$ A/A or 79.2 dB

Power gain $= \frac{v_o^2/100\ \Omega}{v_s^2/1.1\ \text{M}\Omega} = 7578$ W/W

or 10 log 7578 = 38.8 dB

(This takes into acct. the power dissipated in the internal resistance of the source.)

1.50 Case (a) S-A-B-L

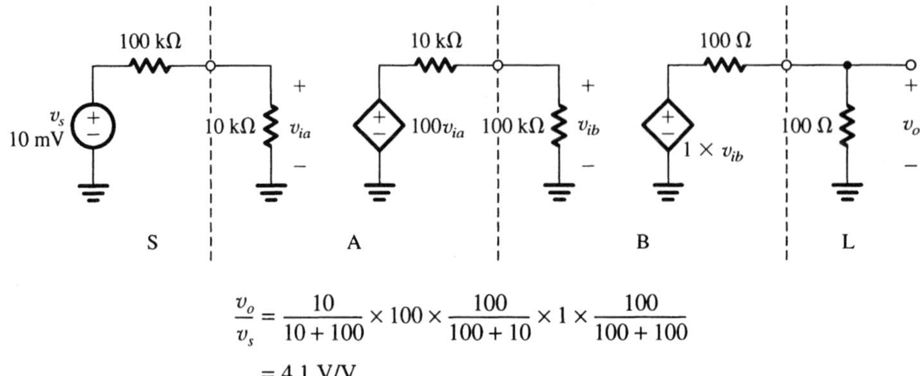

$$\frac{v_o}{v_s} = \frac{10}{10+100} \times 100 \times \frac{100}{100+10} \times 1 \times \frac{100}{100+100}$$

$$= 4.1 \text{ V/V}$$

$$\text{or} \quad 20 \log 4.1 = 12.3 \text{ dB}$$

Case (b) S-B-A-L

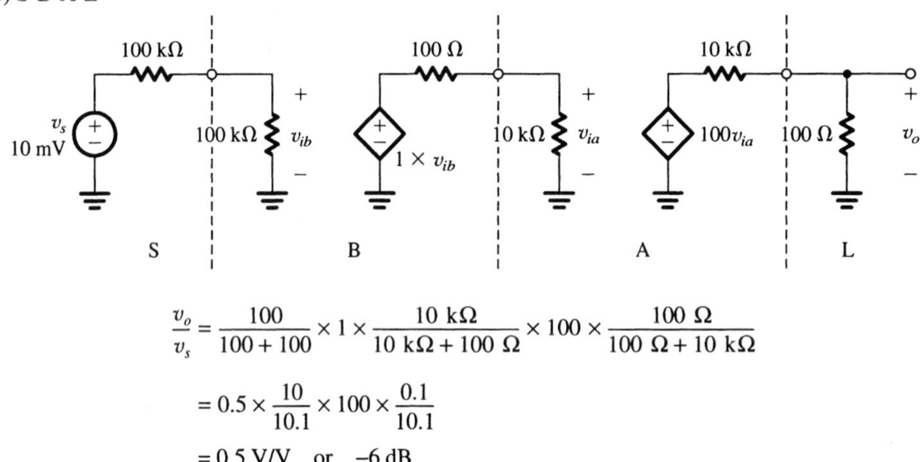

$$\frac{v_o}{v_s} = \frac{100}{100+100} \times 1 \times \frac{10 \text{ k}\Omega}{10 \text{ k}\Omega + 100 \text{ }\Omega} \times 100 \times \frac{100 \text{ }\Omega}{100 \text{ }\Omega + 10 \text{ k}\Omega}$$

$$= 0.5 \times \frac{10}{10.1} \times 100 \times \frac{0.1}{10.1}$$

$$= 0.5 \text{ V/V} \quad \text{or} \quad -6 \text{ dB}$$

Thus, obviously case (a) i.e., SABL is preferred.

1.51

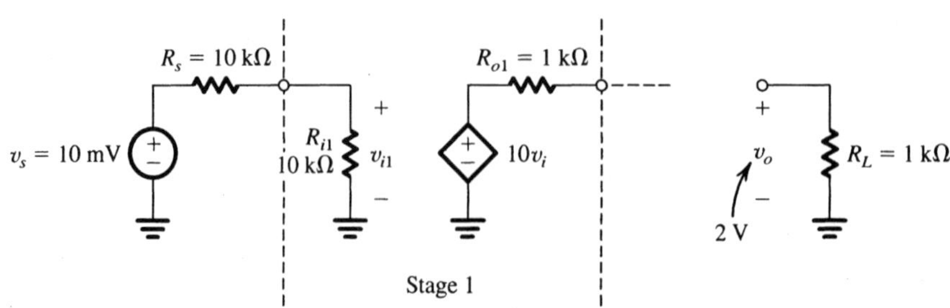

Required overall voltage gain = 2 V/10 mV = 200 V/V. Each stage is capable of providing a *maximum* voltage gain of 10 (the open-circuit gain value). For n stages in cascade the maximum (unattainable) voltage gain in 10^n. We thus see that we need at least 3 stages. For 3 stages, the overall voltage gain obtained is

$$\frac{v_o}{v_s} = \frac{10}{10+10} \times 10 \times \frac{10}{1+10} \times 10 \times \frac{10}{1+10} \times 10 \times \frac{1}{1+1}$$

$$= 206.6 \text{ V/V}$$

Thus, three stages suffice and provide a gain slightly larger than required. The output voltage actually obtained is 10 mV × 206.6 = 2.07 V.

1.53

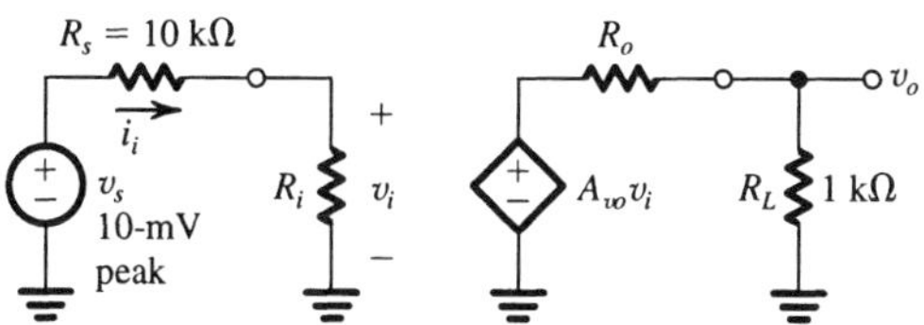

(a) Required voltage gain $\equiv \frac{v_o}{v_s}$

$$= \frac{3 \text{ V}}{0.01 \text{ V}} = 300 \text{ V/V}$$

(b) The smallest R_i allowed is obtained from

$$0.1\ \mu\text{A} = \frac{10 \text{ mV}}{R_s + R_i} \Rightarrow R_s + R_i = 100 \text{ k}\Omega$$

Thus $R_i = 90$ kΩ.

For $R_i = 90$ kΩ, $i_i = 0.1\ \mu$A peak, and

$$\text{Overall current gain} = \frac{v_o/R_L}{i_i}$$

$$= \frac{3 \text{ mA}}{0.1\ \mu\text{A}} = 3 \times 10^4 \text{A/A}$$

$$\text{Overall power gain} \equiv \frac{v_{o\text{rms}}^2/R_L}{v_{s(\text{rms})} \times i_{i(\text{rms})}}$$

$$= \frac{\left(\frac{3}{\sqrt{2}}\right)^2 / 1000}{\left(\frac{10 \times 10^{-3}}{\sqrt{2}}\right) \times \left(\frac{0.1 \times 10^{-6}}{\sqrt{2}}\right)}$$

$$= 9 \times 10^6 \text{ W/W}$$

(This takes into acct. the power dissipated in the internal resistance of the source.)

(c) If $(A_{v_o} v_i)$ has its peak value limited to 5 V, the largest value of R_o is found from

$$5 \times \frac{R_L}{R_L + R_o} = 3 \Rightarrow R_o = \frac{2}{3}R_L = 667\ \Omega$$

(If R_o were greater than this value, the output voltage across R_L would be less than 3 V.)

(d) For $R_i = 90$ kΩ and $R_o = 667$ Ω, the required value of A_{v_o} can be found from

$$300 \text{ V/V} = \frac{90}{90+10} \times A_{v_o} \times \frac{1}{1+0.667}$$

$$\Rightarrow A_{v_o} = 555.7 \text{ V/V}$$

(e) $R_i = 100$ kΩ $(1 \times 10^5\ \Omega)$

$R_o = 100$ Ω $(1 \times 10^2\ \Omega)$

$$300 = \frac{100}{100+10} \times A_{v_o} \times \frac{1000}{1000+100}$$

$$\Rightarrow A_{v_o} = 363 \text{ V/V}$$

1.54

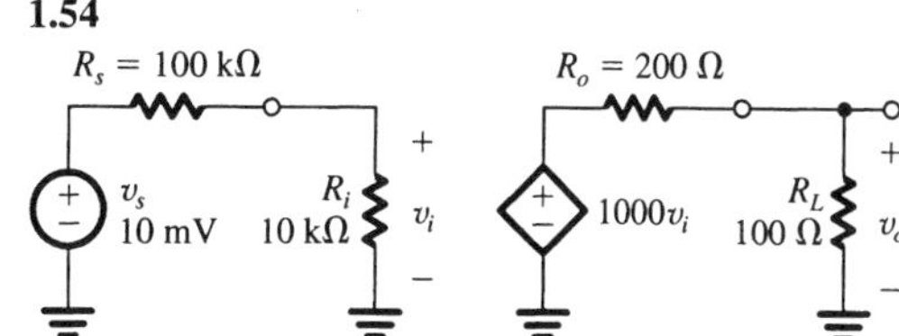

(a) $v_o = 10 \text{ mV} \times \frac{10}{10+100} \times 1000 \times \frac{100}{100+200}$

$= 303$ mV

(b) $\frac{v_o}{v_s} = \frac{303 \text{ mV}}{10 \text{ mV}} = 30.3$ V/V

(c) $\frac{v_o}{v_i} = 1000 \times \frac{100}{100+200} = 333.3$ V/V

(d)

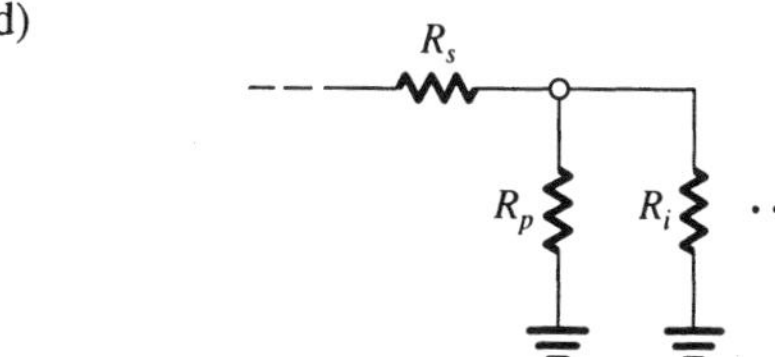

Connect a resistance R_p in parallel with the input and select its value from

$$\frac{(R_p \,//\, R_i)}{(R_p \,//\, R_i) + R_s} = \frac{1}{2}\frac{R_i}{R_i + R_s}$$

$$\Rightarrow 1 + \frac{R_s}{R_p \,//\, R_i} = 22 \Rightarrow R_p \,//\, R_i = \frac{R_s}{21} = \frac{100}{21}$$

$$\Rightarrow \frac{1}{R_p} + \frac{1}{R_i} = \frac{21}{100}$$

$$R_p = \frac{1}{0.21 - 0.1} = 9.1 \text{ k}\Omega$$

1.55

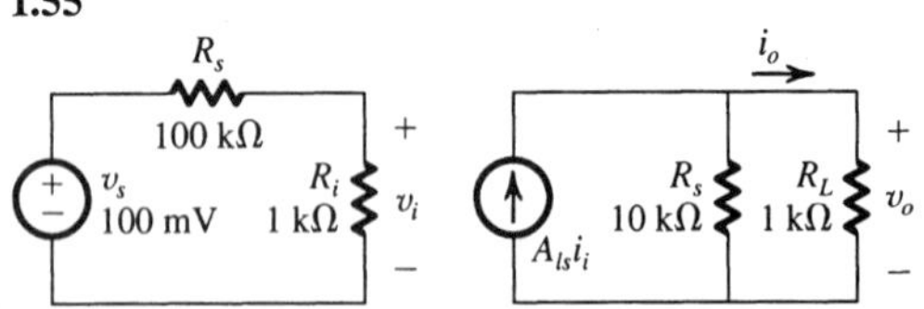

(a) Current gain $= \dfrac{i_o}{i_i}$

$$= A_{is}\frac{R_o}{R_o + R_L}$$

$$= 100\frac{10}{11}$$

$$= 90.9 \frac{\text{A}}{\text{A}} = 39.2 \text{ dB}$$

(b) Voltage gain $= \dfrac{v_o}{v_s}$

$$= \frac{i_o}{i_i}\frac{R_s}{R_s + R_i}$$

$$= 90.9 \times \frac{1}{101}$$

$$= 0.9 \text{ V/V} = -0.9 \text{ dB}$$

(c) Power gain $= A_p = \dfrac{v_o i_o}{v_s i_i}$

$$= 0.9 \times 90.9$$

$$= 81.8 \text{ W/W} = 19.1 \text{ dB}$$

1.56

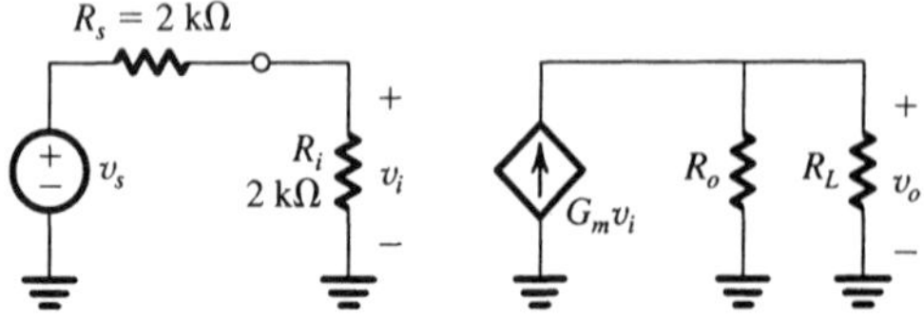

$G_m = 40$ mA/V

$R_o = 20$ kΩ

$R_L = 1$ kΩ

$$v_i = v_s\frac{R_i}{R_s + R_i}$$

$$= v_s\frac{2}{2+2} = \frac{v_s}{2}$$

$$v_o = G_m v_i (R_L \,//\, R_o)$$

$$= 40\frac{20 \times 1}{20 + 1} v_i$$

$$= 40\frac{20}{21}\frac{v_s}{2}$$

Overall voltage gain $\equiv \dfrac{v_o}{v_s} = 19.05$ V/V

1.57 A voltage amplifier is required.

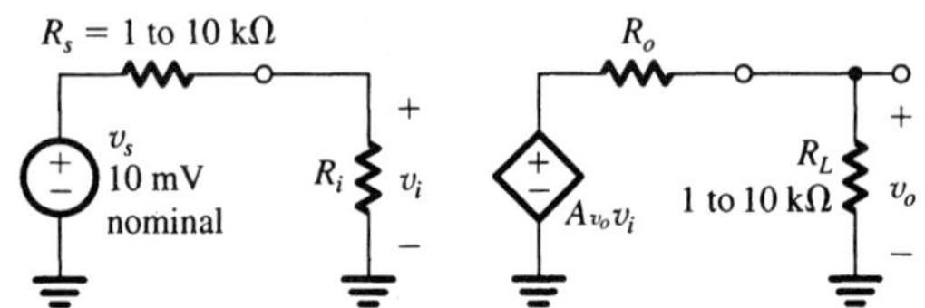

To limit the change in v_o to 10% as R_s varies from 1 to 10 kΩ we select R_i sufficiently large;

$$R_i \geq 10R_{s\max}$$

Thus $R_i = 100$ kΩ.

To limit the change in v_o corresponding to R_L varying in the range 1 to 10 kΩ, to 10%, we select R_o sufficiently small;

$$R_o \leq R_{L\min}/10$$

Thus,

$$R_o = 100 \ \Omega$$

$$v_{o\min} = v_s\frac{R_i}{R_i + R_{s\max}}A_{v_o}\frac{R_{L\min}}{R_{L\min} + R_o}$$

$$1 = 0.01\frac{100}{100 + 10}A_{v_o}\frac{1000}{1000 + 100}$$

$$\Rightarrow A_{v_o} = 121 \text{ V/V}$$

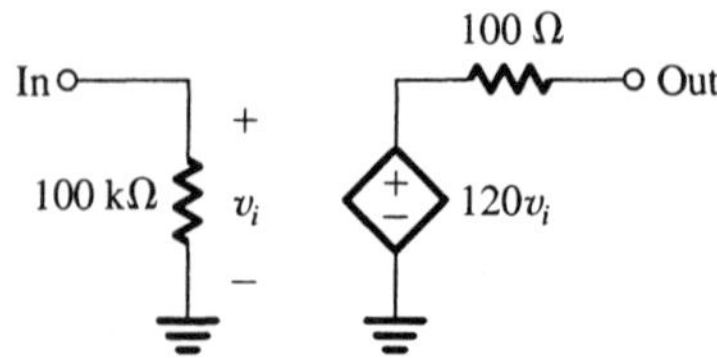

1.58 Current amplifier.

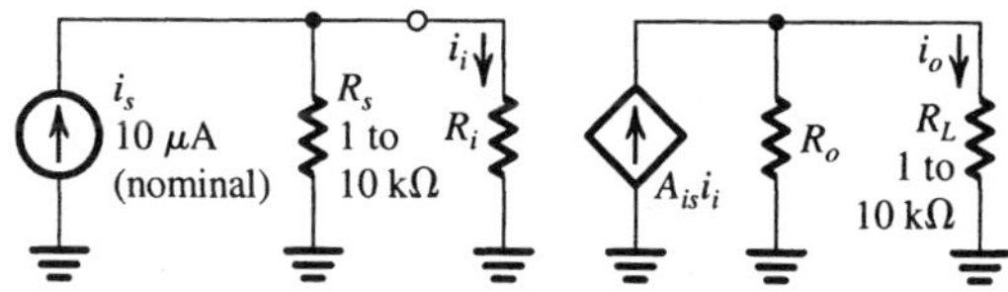

To limit the change in i_o resulting from R_s varying over the range 1 to 10 kΩ, to 10% we select R_i sufficiently low so that,

$$R_i \le R_{s\min}/10$$

Thus, $R_i = 100\ \Omega$

To limit the change in i_o as R_L changes from 1 to 10 kΩ, to 10% we select R_o sufficiently large;

$$R_o \ge 10 R_{L\max}$$

Thus, $R_o = 100\ \text{k}\Omega$

Now for $i_s = 10\ \mu\text{A}$,

$$i_{o\min} = 10^{-5}\frac{R_{s\min}}{R_{s\min}+R_i}A_{is}\frac{R_o}{R_o+R_{L\max}}$$

Thus,

$$10^{-3} = 10^{-5}\frac{1000}{1000+100}A_{is}\frac{100}{100+10}$$

$$\Rightarrow A_{is} = 121\ \text{A/A}$$

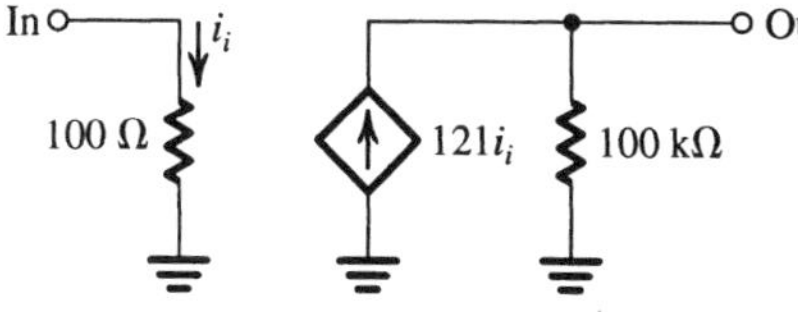

1.59 Transconductance amplifier.

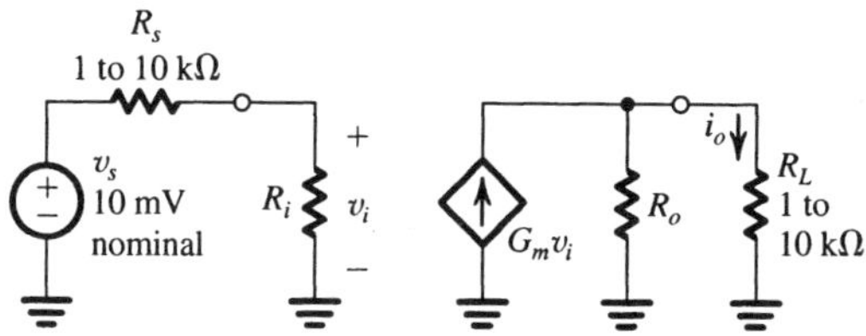

For R_s varying in the range 1 to 10 kΩ, and Δi_o limited to 10% we have to select R_i sufficiently large;

$$R_i \ge 10R_{s\max}$$

$$R_i = 100\ \text{k}\Omega$$

For R_L varying in the range 1 to 10 kΩ, the change in i_o can be kept to 10% if R_o is selected sufficiently large;

$$R_o \ge R_{L\max}$$

Thus $R_o = 100\ \text{k}\Omega$

For $v_s = 10$ mV,

$$i_{o\min} = 10^{-2}\frac{R_i}{R_i+R_{s\max}}G_m\frac{R_o}{R_o+R_{L\max}}$$

$$10^{-3} = 10^{-2}\frac{100}{100+10}G_m\frac{100}{100+10}$$

$$G_m = 1.21\times 10^{-1}\ \text{A/V}$$

$$= 121\ \text{mA/V}$$

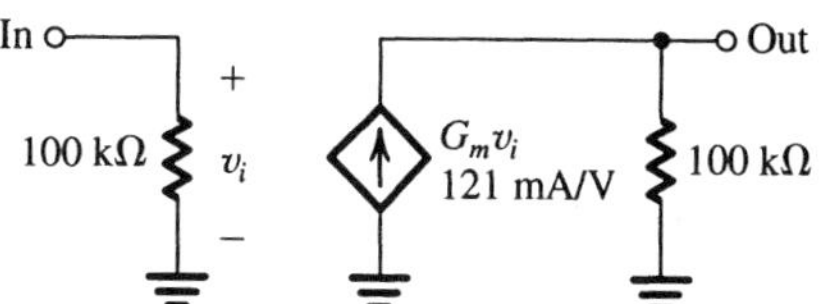

1.60 Transresistance amplifier

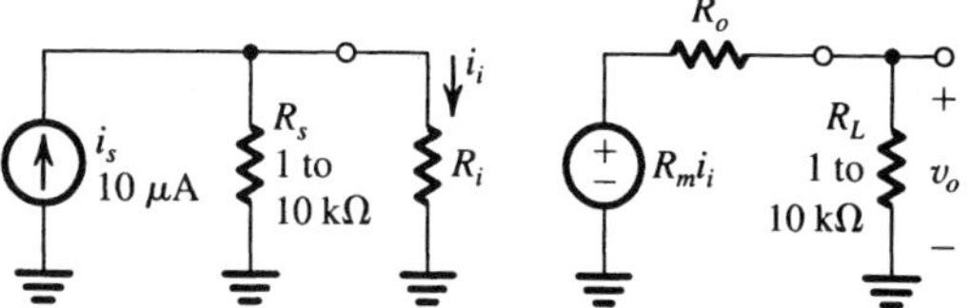

To limit Δv_o to 10% corresponding to R_s varying in the range 1 to 10 kΩ, we select R_i sufficiently low;

$$R_i \le \frac{R_{s\min}}{10}$$

Thus, $R_i = 100\ \Omega$

To limit Δv_o to 10% while R_L varies over the range 1 to 10 kΩ, we select R_o sufficiently low;

$$R_o \le R_{L\min}/10$$

Thus, $R_o = 100\ \Omega$

Now, for $i_s = 10\ \mu\text{A}$,

$$v_{o\min} = 10^{-5}\frac{R_{s\min}}{R_{s\min}+R_i}R_m\frac{R_{L\min}}{R_{L\min}+R_o}$$

$$1 = 10^{-5}\frac{1000}{1000+100}R_m\frac{1000}{1000+100}$$

$$\Rightarrow R_m = 1.21\times 10^5$$

$$= 121\ \text{k}\Omega$$

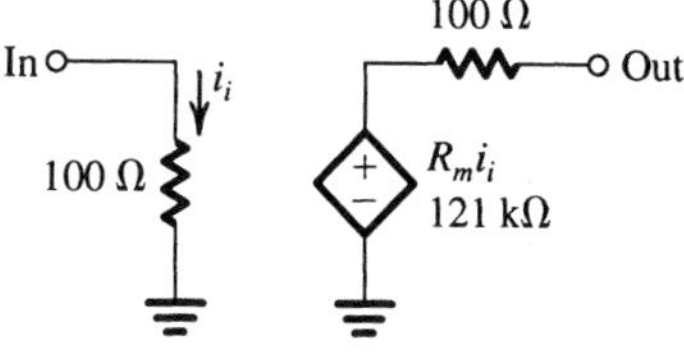

1.64

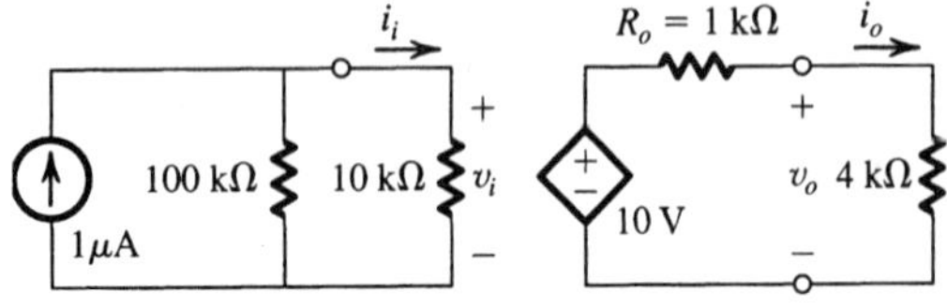

$$R_o = \frac{\text{Open-circuit output voltage}}{\text{Short-circuit output current}} = \frac{10\ \text{V}}{10\ \text{mA}} = 1\ \text{k}\Omega$$

$$v_o = 10 \times \frac{4}{1+4} = 8\ \text{V}$$

$$A_v = \frac{v_o}{v_i} = \frac{8}{1 \times 10^{-3} \times (100 \,//\, 10) \times 10^3}$$

$$= 888\ \text{V/V} \quad \text{or} \quad 58.9\ \text{dB}$$

$$A_i = \frac{i_o}{i_i} = \frac{v_o/R_L}{10^{-3} \times \frac{100}{100+10}} = \frac{8/(4 \times 10^3)}{10^{-3} \times \frac{100}{110}}$$

$$= 2200\ \text{A/A} \quad \text{or} \quad 66.8\ \text{dB}$$

$$A_i = \frac{v_o^2/R_L}{i_i^2 R_i} = \frac{8^2/(4 \times 10^3)}{\left(10^{-3} \times \frac{100}{100+10}\right)^2 10 \times 10^3}$$

$$= 19.36 \times 10^5\ \text{W/W} \quad \text{or} \quad 62.9\ \text{dB}$$

$$\text{Overall current gain} \equiv \frac{i_o}{1\,\mu\text{A}}$$

$$= \frac{v_o/R_L}{1\,\mu\text{A}} = \frac{8/(4 \times 10^3)}{10^{-3}}$$

$$= 2000\ \text{A/A} \quad \text{or} \quad 66\ \text{dB}$$

1.65 Using the voltage divider rule

R_s, V_s, R_i, C_i, V_i

$$Z_i = R_i \,//\, \frac{1}{sC_i}$$

$$Y_i = \frac{1}{R_i} + sC_i$$

$$\frac{V_i}{V_s} = \frac{Z_i}{Z_i + R_s}$$

$$= \frac{1}{1 + R_s Y_i}$$

$$= \frac{1}{1 + R_s\left(\frac{1}{R_i} + sC_i\right)}$$

$$= \frac{1}{1 + \frac{R_s}{R_i} + sC_i R_s} = \frac{1\Big/\left(1 + \frac{R_s}{R_i}\right)}{1 + sC_i \frac{R_s}{1 + \frac{R_s}{R_i}}}$$

$$= \frac{1}{1 + \frac{R_s}{R_i}} \; \frac{1}{1 + sC_i\left(\frac{R_s R_i}{R_s + R_i}\right)}$$

$$= \frac{R_i}{R_i + R_s} \frac{1}{1 + sC_i(R_i \,//\, R_s)}$$

This transfer function is of the STC low-pass type with a dc gain $K = R_i/(R_i + R_s)$ and a 3-dB frequency $\omega_0 = 1/C_i(R_i \,//\, R_s)$.

For $R_s = 20\ \text{k}\Omega$, $R_i = 80\ \text{k}\Omega$, and $C_i = 5$ pF,

$$\omega_0 = \frac{1}{5 \times 10^{-12} \times \frac{20 \times 80}{20 + 80} \times 10^3} = 1.25 \times 10^7\ \text{rad/s}$$

$$f_0 = \frac{\omega_0}{2\pi} = \frac{1.25 \times 10^7}{2\pi} \simeq 2\ \text{MHz}$$

1.66 Using the voltage-divider rule,

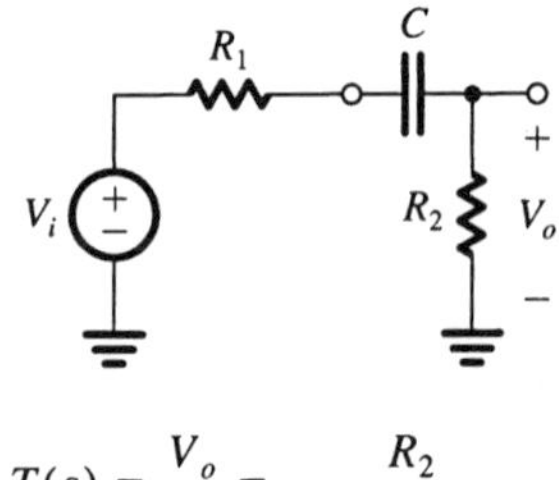

$$T(s) = \frac{V_o}{V_i} = \frac{R_2}{R_2 + R_1 + \frac{1}{sC}}$$

$$= \frac{R_2}{R_1 + R_2} \; \frac{s}{s + \frac{1}{C(R_1 + R_2)}}$$

which from Table 1.2 is of the high-pass type with

$$K = \frac{R_2}{R_1 + R_2}$$

and

$$\omega_0 = \frac{1}{C(R_1 + R_2)}$$

As a further verification that this is a high-pass network and $T(s)$ is a high-pass transfer function we observe as at $s = 0$, $T(s) = 0$; and that as $s \to \infty$, $T(s) = R_2/(R_1 + R_2)$. Also, from the circuit observe as at $s \to \infty$, $(1/sC) \to 0$ and $V_o/V_i = R_2/(R_1 + R_2)$. Now, for $R_1 = 10\ \text{k}\Omega$, $R_2 = 40\ \text{k}\Omega$, and $C = 0.1\ \mu\text{F}$,

$$f_0 = \frac{\omega_0}{2\pi} = \frac{1}{2\pi \times 0.1 \times 10^{-6}(10+40) \times 10^3}$$

$$= 31.8\ \text{Hz}$$

$$|T(j\omega_0)| = \frac{K}{\sqrt{2}} = \frac{40}{10+40}\frac{1}{\sqrt{2}} = 0.57\ \text{V/V}$$

1.67 Using the voltage divider rule,

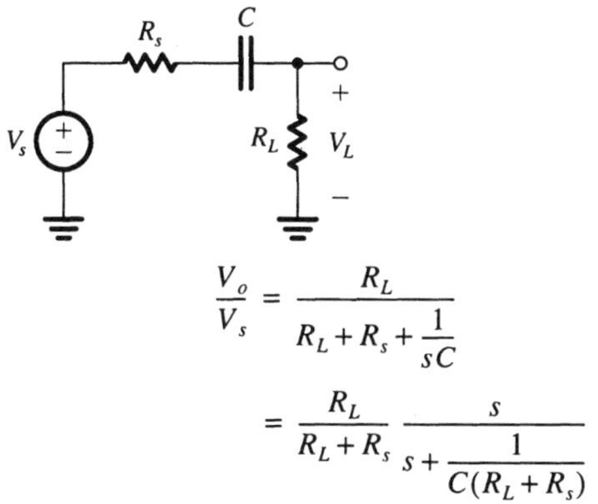

$$\frac{V_o}{V_s} = \frac{R_L}{R_L + R_s + \frac{1}{sC}}$$

$$= \frac{R_L}{R_L + R_s} \frac{s}{s + \frac{1}{C(R_L + R_s)}}$$

which is of the high-pass STC type (see Table 1.2) with

$$K = \frac{R_L}{R_L + R_s} \qquad \omega_0 = \frac{1}{C(R_L + R_s)}$$

For $f_0 \leq 10$ Hz

$$\frac{1}{2\pi C(R_L + R_s)} \leq 10$$

$$\Rightarrow \quad C \geq \frac{1}{2\pi \times 10(20+5) \times 10^3}$$

Thus, the smallest value of C that will do the job is $C = 0.64\ \mu\text{F}$.

1.68 The given measured data indicate that this amplifier has a low-pass STC frequency response with a low-frequency gain of 40 dB, and a 3-dB frequency of 10^4 Hz. From our knowledge of the Bode plots for low-pass STC networks (Figure . . .) we can complete the Table entries and sketch the amplifier frequency response.

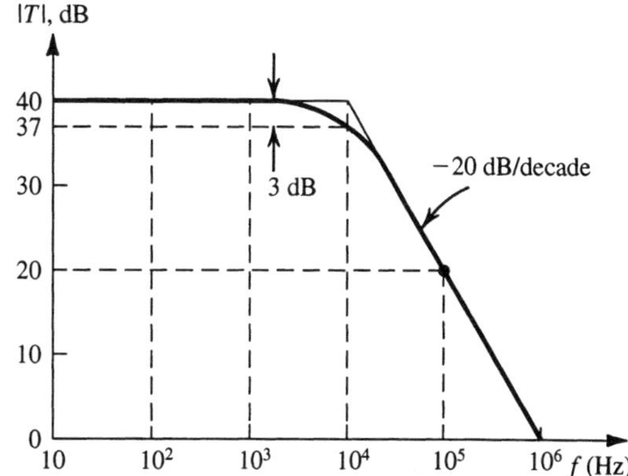

f (Hz)	$\|T\|$ (dB)	$\angle T$ (degrees)
1000	40	−5.7°
10^4	37	−45°
10^5	20	−84.3°
10^6	0	−90°

1.69 From our knowledge of the Bode plots of STC low-pass and high-pass networks we see that this amplifier has a mid-band gain of 40 dB, a low-frequency response of the high-pass STC type with $f_{3dB} = 10^2$ Hz, and a high-frequency response of the low-pass STC type with $f_{3dB} = 10^6$ Hz. We thus can sketch the amplifier frequency response and complete the table entries as follows

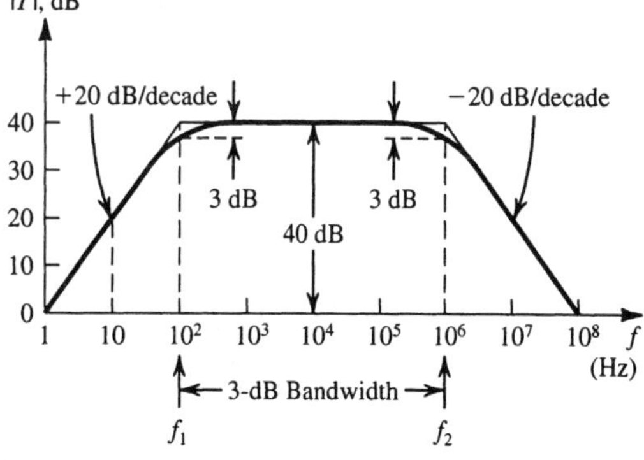

f (Hz)	1	10	10^2	10^3	10^4	10^5	10^6	10^7	10^8
$\|T\|$ (dB)	0	20	37	40	40	40	37	20	0

1.72 Since the overall transfer function is that of three identical STC LP circuits in cascade (but with no loading effects since the buffer amplifiers have input and zero output resistances) the overall gain will drop by 3 dB below the value at dc at the frequency for which the gain of each STC circuit is 1 dB down. This frequency is found as follows: The transfer function of each STC circuit is

$$T(s) = \frac{1}{1 + \frac{s}{\omega_0}}$$

where

$$\omega_0 = 1/CR$$

Thus,

$$|T(j\omega)| = \frac{1}{\sqrt{1 + \left(\frac{\omega}{\omega_0}\right)^2}}$$

$$20 \log \frac{1}{\sqrt{1 + \left(\frac{\omega_{1\text{-dB}}}{\omega_0}\right)^2}} = -1$$

$$\Rightarrow 1 + \left(\frac{\omega_{1\text{-dB}}}{\omega_0}\right)^2 = 10^{0.1}$$

$$\omega_{1\text{-dB}} = 0.51\,\omega_0$$

$$\omega_{1\text{-dB}} = 0.51/CR$$

1.73 $R_s = 100\ \text{k}\Omega$, since the 3-dB frequency is reduced by a very high factor (from 6 MHz to 120 kHz) C_2 must be much larger than C_1. Thus, neglecting C_1 we find C_2 from

$$120\ \text{kHz} \simeq \frac{1}{2\pi C_2 R_s}$$

$$= \frac{1}{2\pi C_2 \times 10^5}$$

$$\Rightarrow C_2 = 13.3\ \text{pF}$$

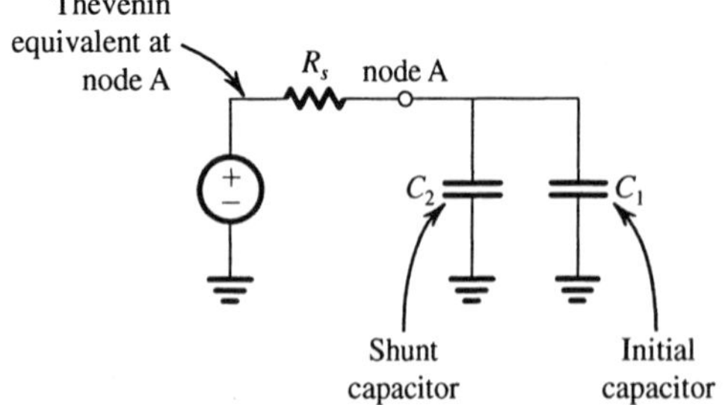

If the original 3-dB frequency (6 MHz) is attributable to C_1 then

$$6\ \text{MHz} = \frac{1}{2\pi C_1 R_s}$$

$$\Rightarrow C_1 = \frac{1}{2\pi \times 6 \times 10^6 \times 10^5}$$

$$= 0.26\ \text{pF}$$

1.74

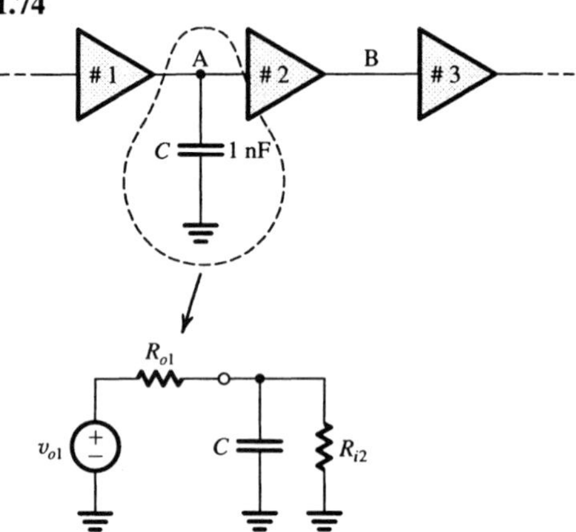

Since when C is connected the 3-dB frequency is reduced by a large factor, the value of C must be much larger than whatever parasitic capacitance originally existed at node A (i.e., between A and ground). Furthermore, it must be that C is now the dominant determinant of the amplifier 3-dB frequency (i.e., it is dominating over whatever may be happening at node B or anywhere else in the amplifier). Thus, we can write

$$150\ \text{kHz} = \frac{1}{2\pi C(R_{o1} \,//\, R_{i2})}$$

$$\Rightarrow (R_{o1} \,//\, R_{i2}) = \frac{1}{2\pi \times 150 \times 10^3 \times 1 \times 10^{-9}}$$

$$= 1.06\ \text{k}\Omega$$

Now $R_{i2} = 100\ \text{k}\Omega$,

Thus $R_{o1} = 1.07\ \text{k}\Omega$
Similarly, for node B,

$$15\ \text{kHz} = \frac{1}{2\pi C(R_{o2} \,//\, R_{i3})}$$

$$\Rightarrow R_{o2} \,//\, R_{i3} = \frac{1}{2\pi \times 15 \times 10^3 \times 1 \times 10^{-9}}$$

$$= 10.6\ \text{k}\Omega$$

$$R_{o2} = 11.9\ \text{k}\Omega$$

She should connect a capacitor of value C_p to node B where C_p can be found from,

$$10 \text{ kHz} = \frac{1}{2\pi C_p(R_{o2} \parallel R_{i3})}$$

$$\Rightarrow C_p = \frac{1}{2\pi \times 10 \times 10^3 \times 10.6 \times 10^3}$$

$$= 1.5 \text{ nF}$$

Note that if she chooses to use node A she would need to connect a capacitor 10 time larger!

1.75

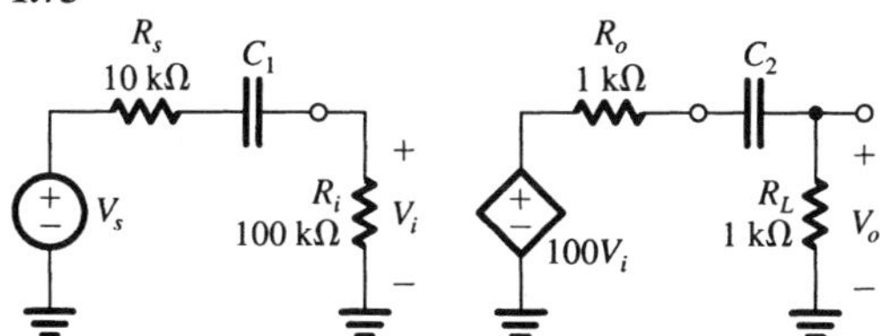

For the input circuit, the corner frequency f_{01} is found from

$$f_{01} = \frac{1}{2\pi C_1(R_s + R_i)}$$

For $f_{01} \leq 100$ Hz,

$$\frac{1}{2\pi C_1(10 + 100) \times 10^3} \leq 100$$

$$\Rightarrow C_1 \geq \frac{1}{2\pi \times 110 \times 10^3 \times 10^2} = 4.4 \times 10^{-8}$$

Thus we select $C_1 = 1 \times 10^{-7}$ F $= 0.1$ μF. The actual corner frequency resulting from C_1 will be

$$f_{01} = \frac{1}{2\pi \times 10^{-7} \times 110 \times 10^3} = 14.5 \text{ Hz}$$

For the output circuit,

$$f_{02} = \frac{1}{2\pi C_2(R_o + R_L)}$$

For $f_{02} \leq 100$ Hz,

$$\frac{1}{2\pi C_2(1 + 1) \times 10^3} \leq 100$$

$$\Rightarrow C_2 \geq \frac{1}{2\pi \times 2 \times 10^3 \times 10^2} = 0.8 \times 10^{-6}$$

Select $C_2 = 1 \times 10^{-6} = 1$ μF

This will place the corner frequency at

$$f_{02} = \frac{1}{2\pi \times 10^{-6} \times 2 \times 10^3} = 80 \text{ Hz}$$

$$T(s) = 100 \frac{s}{\left(1 + \frac{s}{2\pi f_{01}}\right)} \frac{s}{\left(1 + \frac{s}{2\pi f_{02}}\right)}$$

1.76 The LP factor $1/(1 + jf/10^4)$ results in a Bode plot like that in Fig. 1.23(a) with the 3 dB frequency $f_0 = 10^4$ Hz. The high-pass factor $1/(1 + 10^4/jf)$ results in a Bode plot like that in Fig. 1.24(a) with the 3 dB frequency $f_0 = 10^4$ Hz.

The Bode plot for the overall transfer function can be obtained by summing the dB values of the two individual plots and then raising the resulting plot vertically by 40 dB (corresponding to the factor 100 in the numerator). The result is as follows:

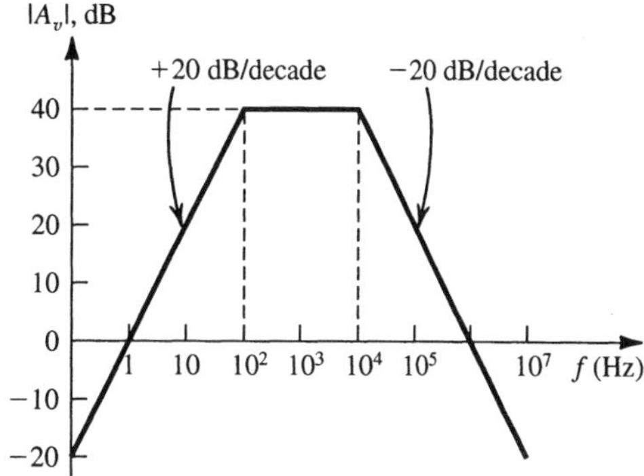

$f =$	10	10^2	10^3	10^4	10^5	10^6	10^7	(Hz)
$\lvert A_v \rvert \simeq$	20	40	40	40	20	0	−20	(dB)
		37		37				

Better approximation
(3-dB frequencies)

Bandwidth $= 10^4 - 10^2 = 9900$ Hz

1.77

$$T_i(s) = \frac{V_i(s)}{V_s(s)} = \frac{1/sC_1}{1/sC_1 + R_1} = \frac{1}{sC_1R_1 + 1} \quad \text{LP}$$

$$3 \text{ dB frequency} = \frac{1}{2\pi C_1 R_1} = \frac{1}{2\pi 10^{-11} 10^6} = 15.9 \text{ Hz}$$

For $T_o(s)$, the following equivalent circuit can be used:

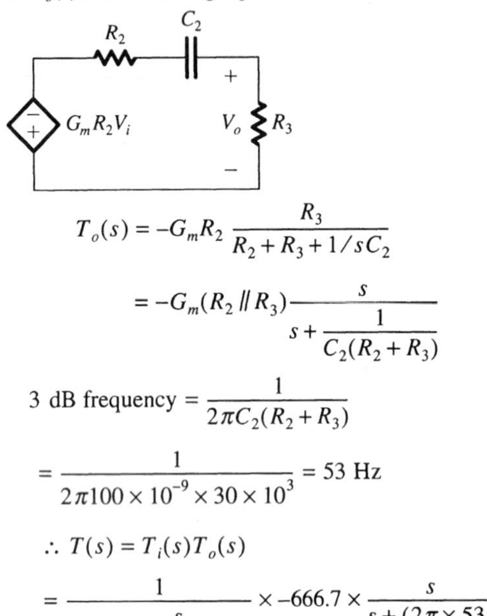

$$T_o(s) = -G_m R_2 \frac{R_3}{R_2 + R_3 + 1/sC_2}$$

$$= -G_m (R_2 \,//\, R_3) \frac{s}{s + \dfrac{1}{C_2(R_2 + R_3)}}$$

$$3 \text{ dB frequency} = \frac{1}{2\pi C_2(R_2 + R_3)}$$

$$= \frac{1}{2\pi 100 \times 10^{-9} \times 30 \times 10^3} = 53 \text{ Hz}$$

$$\therefore \; T(s) = T_i(s) T_o(s)$$

$$= \frac{1}{1 + \dfrac{s}{2\pi \times 15.9 \times 10^3}} \times -666.7 \times \frac{s}{s + (2\pi \times 53)}$$

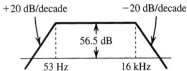

+20 dB/decade
−20 dB/decade
56.5 dB
53 Hz
16 kHz

Bandwidth = 16 kHz − 53 Hz ≃ 16 kHz

1.78

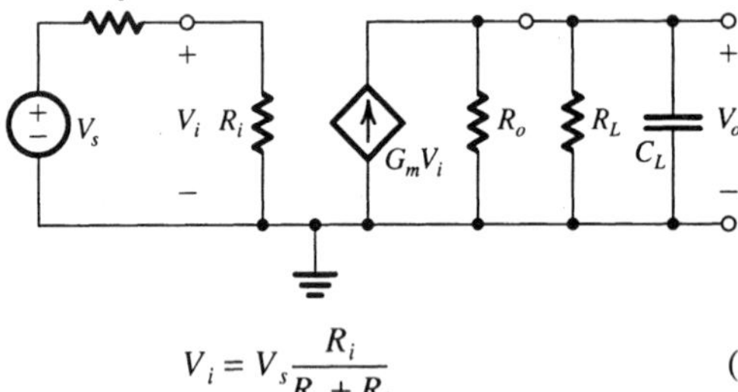

$$V_i = V_s \frac{R_i}{R_s + R_i} \qquad (1)$$

To satisfy constraint (1), namely

$$V_i \ge \left(1 - \frac{x}{100}\right) V_s$$

We substitute in Eq.(1) to obtain

$$\frac{R_i}{R_s + R_i} \ge 1 - \frac{x}{100}$$

Thus

$$\frac{R_s + R_i}{R_i} \le \frac{1}{1 - \dfrac{x}{100}}$$

$$\frac{R_s}{R_i} \le \frac{1}{1 - \dfrac{x}{100}} - 1 = \frac{\dfrac{x}{100}}{1 - \dfrac{x}{100}}$$

which can be expressed as

$$\frac{R_i}{R_s} \ge \frac{1 - \dfrac{x}{100}}{\dfrac{x}{100}}$$

resulting in

$$R_i \ge R_s\left(\frac{100}{x} - 1\right) \qquad (1)$$

The 3-dB frequency is determined by the parallel RC circuit at the output,

$$f_0 = \frac{1}{2\pi}\omega_0 = \frac{1}{2\pi} \frac{1}{C_L(R_L \,//\, R_o)}$$

Thus,

$$f_0 = \frac{1}{2\pi C_L}\left(\frac{1}{R_L} + \frac{1}{R_o}\right)$$

To obtain a value for f_0 greater than a specified value $f_{3\text{dB}}$ we select R_o so that

$$\frac{1}{2\pi C_L}\left(\frac{1}{R_L} + \frac{1}{R_o}\right) \ge f_{3\text{dB}}$$

$$\frac{1}{R_L} + \frac{1}{R_o} \ge 2\pi C_L f_{3\text{dB}}$$

$$\frac{1}{R_o} \ge 2\pi C_L f_{3\text{dB}} - \frac{1}{R_L}$$

$$R_o \le \frac{1}{2\pi f_{3\text{dB}} C_L - \dfrac{1}{R_L}} \qquad (2)$$

To satisfy constraint (3), we first determine the dc gain as

$$\text{dc gain } = \frac{R_i}{R_s + R_i} G_m (R_o \,//\, R_L)$$

For the dc gain to be greater than a specified value A_0,

$$\frac{R_i}{R_s + R_i} G_m (R_o \,//\, R_L) \ge A_0$$

The first factor on the LHS is (from constraint (1))

greater or equal to $(1 - x/100)$. Thus

$$G_m \geq \frac{A_0}{\left(1 - \frac{x}{100}\right)(R_o \,//\, R_L)} \quad (3)$$

Substituting $R_s = 10$ kΩ and $x = 20\%$ in (1) results in

$$R_i \geq 10\left(\frac{100}{20} - 1\right) = 40 \text{ k}\Omega$$

Substituting $f_{3dB} = 3$ MHz, $C_L = 10$ pF and $R_L = 10$ kΩ in Eq. (2) results in

$$R_o \leq \frac{1}{2\pi \times 3 \times 10^6 \times 10 \times 10^{-12} - \frac{1}{10^4}} = 11.3 \text{ k}\Omega$$

Substituting $A_0 = 80$, $x = 20\%$, $R_L = 10$ kΩ, and $R_o = 11.3$ kΩ, eq. (3) results in

$$G_m \geq \frac{80}{\left(1 - \frac{20}{100}\right)(10 \,//\, 11.3) \times 10^3} = 18.85 \text{ mA/V}$$

If the more practical value of $R_o = 10$ kΩ is used then

$$G_m \geq \frac{80}{\left(1 - \frac{20}{100}\right)(10 \,//\, 10) \times 10^3} = 20 \text{ mA/V}$$

1.79 Using the voltage-divider rule we obtain

$$\frac{V_o}{V_i} = \frac{Z_2}{Z_1 + Z_2}$$

where

$$Z_1 = R_1 \,//\, \frac{1}{sC_1} \quad \text{and} \quad Z_2 = R_2 \,//\, \frac{1}{sC_2}.$$

It is obviously more convenient to work in terms of admittances. Therefore we express V_o/V_i in the alternate form

$$\frac{V_o}{V_i} = \frac{Y_1}{Y_1 + Y_2}$$

and substitute $Y_1 = (1/R_1) + sC_1$ and $Y_2 = (1/R_2) + sC_2$ to obtain

$$\frac{V_o}{V_i} = \frac{\frac{1}{R_1} + sC_1}{\frac{1}{R_1} + \frac{1}{R_2} + s(C_1 + C_2)}$$

$$= \frac{C_1}{C_1 + C_2} \frac{s + \frac{1}{C_1 R_1}}{s + \frac{1}{(C_1 + C_2)}\left(\frac{1}{R_1} + \frac{1}{R_2}\right)}$$

This transfer function will be independent of frequency (s) if the second factor reduces to unity. This in turn will happen if

$$\frac{1}{C_1 R_1} = \frac{1}{C_1 + C_2}\left(\frac{1}{R_1} + \frac{1}{R_2}\right)$$

which can be simplified as follows

$$\frac{C_1 + C_2}{C_2} = R_1\left(\frac{1}{R_1} + \frac{1}{R_2}\right) \quad (1)$$

$$1 + \frac{C_2}{C_1} = 1 + \frac{R_1}{R_2}$$

or

$$C_1 R_1 = C_2 R_2$$

When this condition applies, the attenuator is said to be compensated, and its transfer function is given by

$$\frac{V_o}{V_i} = \frac{C_1}{C_1 + C_2}$$

which, using Eq. (1) above can be expressed in the alternate form

$$\frac{V_o}{V_i} = \frac{1}{1 + \frac{R_1}{R_2}} = \frac{R_2}{R_1 + R_2}$$

Thus when the attenuator is compensated ($C_1 R_1 = C_2 R_2$) its transmission can be determined either by its two resistors R_1, R_2 or by its two capacitors. C_1, C_2, and the transmission is *not* a function of frequency.

1.80 The HP STC circuit whose response determines the frequency response of the amplifier in the low-frequency range has a phase angle of 11.4° at $f = 100$ Hz. Using the equation for $\angle T(j\omega)$ from Table 1.2 we obtain

$$\tan^{-1}\frac{f_0}{100} = 11.4° \Rightarrow f_0 = 20.16 \text{ Hz}.$$

The LP STC circuit whose response determines the amplifier response at the high-frequency end has a phase angle of −11.4° at $f = 1$ kHz. Using the relationship for $\angle T(j\omega)$ given in Table 1.2 we obtain for the LP STC circuit.

$$-\tan^{-1}\frac{10^3}{f_0} = -11.4° \Rightarrow f_0 = 4959.4 \text{ Hz}$$

At f = 100 Hz the drop in gain is due to the HP STC network, and thus its value is

$$20 \log \frac{1}{\sqrt{1+\left(\frac{20.16}{100}\right)^2}} = -0.17 \text{ dB}$$

Similarly, at f = 1 kHz the drop in gain is caused by the LP STC network. The drop in gain is

$$20 \log \frac{1}{\sqrt{1+\left(\frac{1000}{4959.4}\right)^2}} = -0.17 \text{ dB}$$

The gain drops by 3 dB at the corner frequencies of the two STC networks, that is, at f = 20.16 Hz and f = 4959.4 Hz.

1.81 $NM_H = V_{OH} - V_{IH} = 3.3 - 1.7 = 1.6$ V

$NM_L = V_{IL} - V_{OL} = 1.3 - 0 = 1.3$ V

1.82

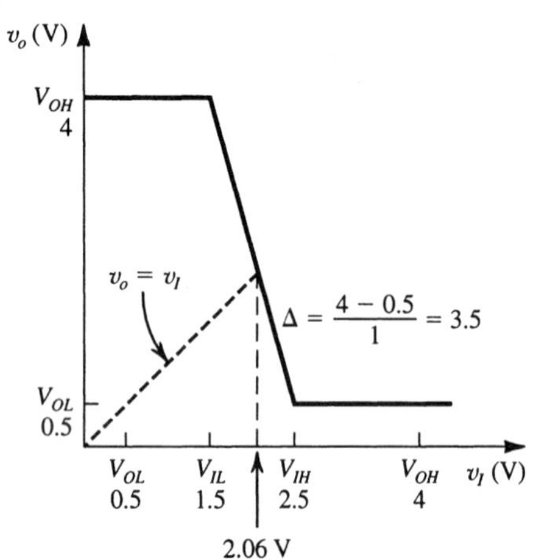

(a) $NM_H = V_{OH} - V_{IH} = 4 - 2.5 = 1.5$ V

$NM_L = V_{IL} - V_{OL} = 1.5 - 0.5 = 1$ V

(b) In the transition region

$$V_O = 4 - 3.5(V_I - 1.5)$$

$$= 9.25 - 3.5V_I$$

If

$$V_O = V_I \Rightarrow 4.5V_O = 9.25$$

$$V_O = V_I = 2.06 \text{ V}$$

(c) Slope = −3.5 V/V

1.83

$NM_H = V_{OH} - V_{IH} = 0.8\ V_{DD} - 0.6\ V_{DD} = 0.2V_{DD}$

$NM_L = V_{IL} - V_{OL} = (0.4 - 0.1)V_{DD} = 0.3V_{DD}$

width of transition region $= V_{IH} - V_{IL} = 0.2V_{DD}$

for a minimum NM of 1 V $\Rightarrow 0.2V_{DD} = 1$

$V_{DD} = 5$ V

1.84

(a) Worse case $NM_H = V_{OH,\,\min} - V_{IH} = 2.4 - 2 = 0.4$ V
Worse case $NM_L = V_{IL,\,\max} - V_{OL} = 0.8 - 0.4 = 0.4$ V

(b) $P_{D_{av}} = \frac{1}{20}[5 \times 3 + 5 \times 1] = 10$ mW

(c) Dynamic power dissipation $= f C V_{DD}^2 =$

$10^6 \times 45 \times 10^{-12} \times 25 = 1.13$ mW

(d) $t_P(\text{typical}) = \frac{1}{2}(t_{PHL} + t_{PLH}) = \frac{1}{2}(7 + 11) = 9$ ns

$t_P(\text{maximum}) = \frac{1}{2}(15 + 22) = 18.5$ ns

1.85

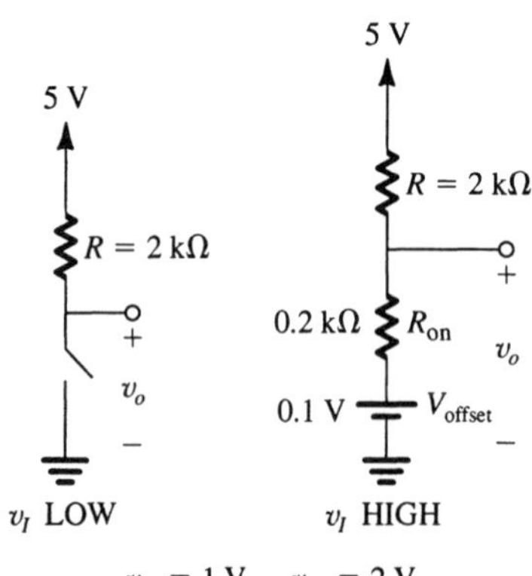

(a) $V_{OL} = \frac{5 - 0.1}{2.2} = 0.2 + 0.1 = 0.545$ V

$V_{OH} = 5$ V

$NM_H = V_{OH} - V_{IH} = 3$ V

$NM_L = V_{IL} - V_{OL} = 0.455$ V

(b) $V_{OH} = 5 - N\,(0.2 \times 10^{-3})R = 5 - 0.4N$

$NM_H = 5 - 0.4N - 2 = 3 - 0.4N = 0.455 \quad \therefore\ N = 6$

(c) (i) $P_{D_{v_o,\text{LOW}}} = (5-0.1)^2/2.2\text{ k}\Omega = 10.9\text{ mW}$

(ii) $P_{D_{v_o,\text{HIGH}}} = 5\times(0.2\times 6) = 6\text{ mW}$

1.86 (a) $V_{OL}=0 \quad V_{OH}=5 \quad NM_L = V_{IL}-V_{OL} = 2.5-0 = 2.5\text{ V}$

$NM_H = V_{OH}-V_{IH} = 5-2.5 = 2.5\text{ V}$

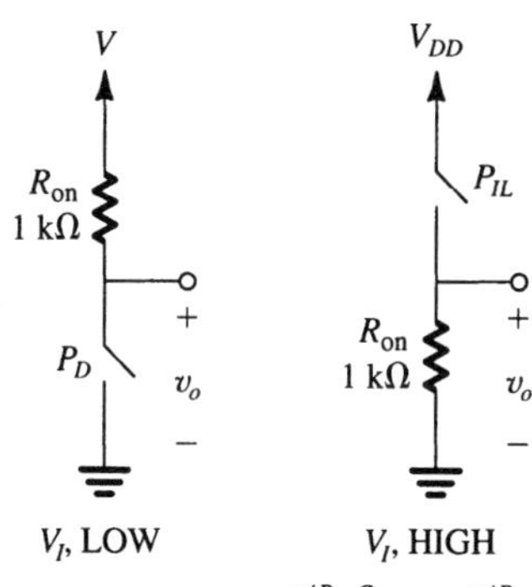

(b) $V_O(t) = 0-(0-5)e^{-t/R_{on}C} = 5e^{-t/R_{on}C}$

For $t_{PHL} \Rightarrow V_O(t) = 5e^{-t_{PHL}/R_{on}C} = \frac{1}{2}(5) = 2.5$

$t_{PHL} = -(10^3)(10^{-n})\ln\frac{2.5}{5} = 0.69\text{ ns}$

For t_{THL} $V_O(t) = 5e^{-t_H/R_{on}C} = 4.5\text{ V}$

$t_1 = 0.01\text{ ns} \quad V_O(t) = 5e^{-t_L/R_{on}C} = 0.5\text{ V}$

$t_2 = 2.3\text{ ns}$

$\therefore \quad t_{THL} = t_2 - t_1 = 2.2\text{ ns}$

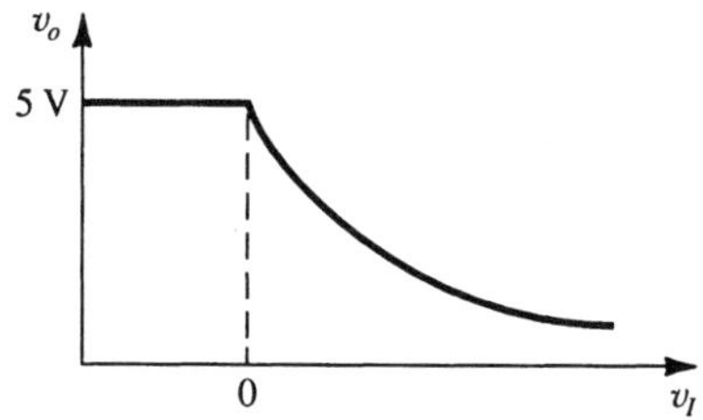

(c) $V_O(t) = 5-(5-0)e^{-t/R_{on}C} = 5-5e^{-t/R_{on}C}$

$V_O = 5-5e^{-t_{PLH}/R_{on}C} = 2.5\text{ ns}$

$t_{PLH} = 0.69\text{ ns}$

For t_{TLH},

$V_O(t) = 5-5e^{-t_1/R_{on}C} = 0.5 \Rightarrow t_1 = 0.10\text{ ns.}$

$V_O(t) = 5-5e^{-t_2/R_{on}C} = 4.5 \Rightarrow t_2 = 2.3\text{ ns.}$

$t_{TLH} = 2.3-0.1 = 2.2\text{ ns}$

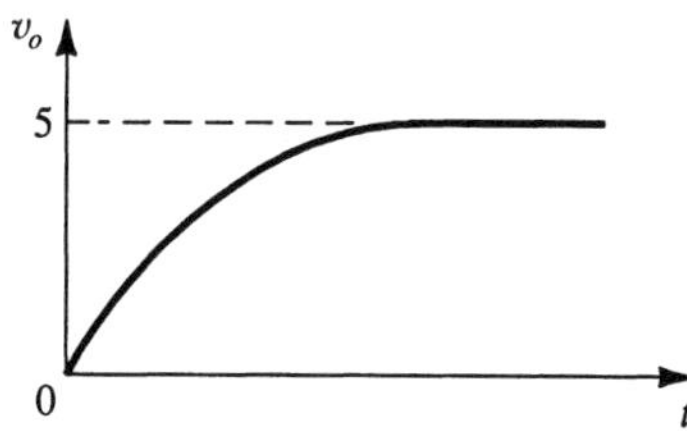

1.87

$V_{OH} = 5\text{ V}$

$V_{OL} = 5-2\times 1 = 3\text{ V}$

1.88 $P_{\text{dynamic}} = fCV_{DD}^2 = 100\times 10^6\times 10\times 10^{-12}\times 25 = 25\text{ mW}$

$P = V_{DD}I_{\text{avg}} = 5I_{\text{avg}} = 25\text{ mW}$

$I_{\text{avg}} = 5\text{ mA}$

1.89

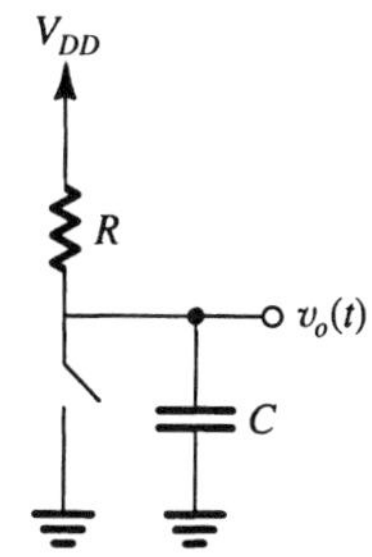

$v_O(t)$ begins at V_{OL} and rises toward V_{OH} (in this case $V_{OH} = V_{DD}$) according to

$v_O(t) = v_\infty - (v_\infty - v_{ot})e^{-t/CR}$

$= V_{OH} - (V_{OH} - V_{OL})e^{-t/CR}$

$= V_{OH} - (V_{OH} - V_{OL})e^{-t/\tau_1}, \ \tau_1 = CR$ Q.E.D.

$v_O(t)$ reaches $\frac{1}{2}(V_{OH}+V_{OL})$ at $t = t_{PLH}$,

$\frac{1}{2}(V_{OH}+V_{OL}) = V_{OH}-(V_{OH}-V_{OL})e^{-t_{PLH}/\tau_1}$

$\Rightarrow t_{PLH} = \tau_1 \ln 2 = 0.69CR$ Q.E.D.

(b)

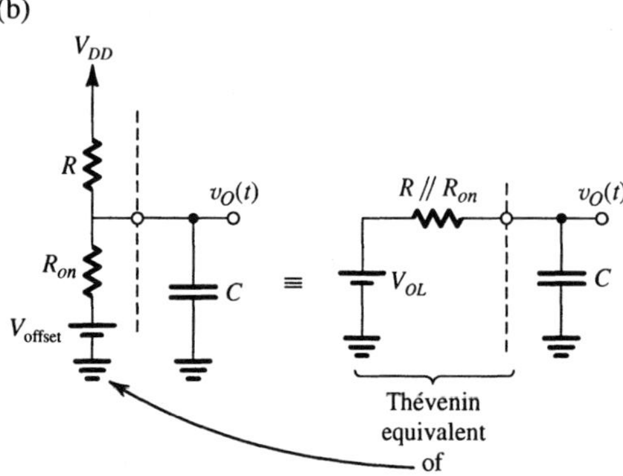

$$V_{OL} = V_{DD}\frac{R_{on}}{R+R_{on}} + V_{offset}\frac{R}{R+R_{on}}$$

$$v_O(t) = v_\infty - (v_\infty - v_{ot})e^{-t/\tau_2},\ \tau_2 = C(R // R_{on})$$

$$= V_{OL} - (V_{OL} - V_{OH})e^{-t/\tau_2}$$

$$= V_{OL} + (V_{OH} - V_{OL})e^{-t/\tau_2} \qquad \text{Q.E.D.}$$

$$v_O(t_{PHL}) = \frac{1}{2}(V_{OL} + V_{OH})$$

Thus,

$$t_{PHL} = \tau_2 \ln 2 = 0.69C\,(R // R_{on})$$

$$\simeq 0.69\,CR_{on}, \qquad R_{on} \ll R \qquad \text{Q.E.D.}$$

(c) $\tau_P = \frac{1}{2}(t_{PLH} + t_{PLH}) = \frac{1}{2} \times 0.69(CR + CR_{on})$

$$\simeq 0.35CR, \quad \text{for} \quad R_{on} \ll R \qquad \text{Q.E.D.}$$

(d) Static power is dissipated only when the output is low, in which case

$$P = \frac{V_{DD}^2}{R+R_{on}} \simeq \frac{V_{DD}^2}{R}, \quad \text{for } R_{on} \ll R,$$

and assuming that $V_{offset} \ll V_{DD}$. Thus if the inverter spends only half the time in this low-output state,

$$P = \frac{1}{2}\frac{V_{DD}^2}{R}$$

(e) Large R results in low P but high τ_P and vice-versa. For $V_{DD} = 5$ V and $C = 10$ pF,

- $\tau_P \le 10$ ns $\Rightarrow$ $0.35CR \le 10$ ns
  $\Rightarrow R \le 2875\ \Omega$
- $P \le 10$ mW $\Rightarrow$ $\frac{1}{2} \times \frac{25}{R} \le 10 \times 10^{-3}$
  $\Rightarrow R \ge 1250\ \Omega$

Thus,

$$1250 \le R \le 2875\ \Omega$$

Selecting $R = 2$ kΩ,

$$\tau_P = 0.35 \times 10 \times 10^{-12} \times 2 \times 10^3 = 7 \text{ ns}$$

$$P = \frac{1}{2} \times \frac{25}{2} = 6.25 \text{ mW}$$

Unnumbered 1.48

$$\frac{v_{i1}}{v_i} = 0.5 \qquad \frac{v_{i2}}{v_{i1}} = 100\,\frac{1000}{1001} \simeq 100$$

$$\frac{v_{i3}}{v_{i2}} = 10 \times \frac{10}{11} = 9.09$$

$$\frac{v_L}{v_{i3}} = \frac{10}{11} = 0.909$$

$$A_v = 826.3$$

$$\frac{v_o}{v_i} = 413.1$$

$$0.5 \times 100 \times 10 \times \frac{10}{11} = 450$$

$$P_L = 0.5 \text{ W} = v_o^2/R_L$$

$$v_o = \sqrt{0.5 \times 100} = 7.07 \text{ V}$$

Unnumbered 1.50

Required overall voltage gain $= 7.07/0.03 = 235.7 \simeq 235$ V/V.

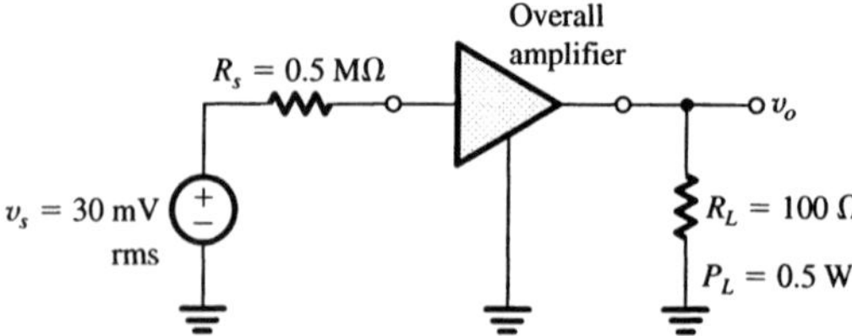

In order to avoid the loss of more than 2/3 of the signal strength in coupling the source to the first stage of the amplifier, we have use the type (1), high-input resistance amplifier as the input stage. Both type 2 and type 3 do not satisfy this requirement. The type (1) amplifier has an open-circuit voltage gain of 10 and thus connect by itself satisfy the overall gain requirement.

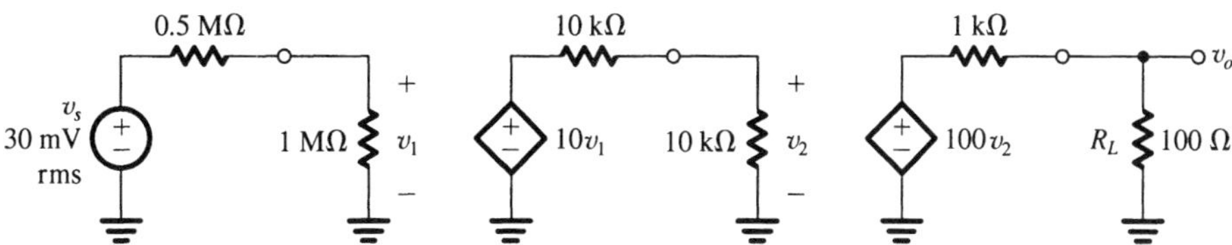

We next consider cascading the type (1) input stage with the type (2), high-gain stages. The result would be

Overall voltage gain

$$= \frac{1}{1+0.5} \times 10 \times \frac{10}{10+10} \times 100 \times \frac{100}{100+1000}$$

$$= 30.3 \text{ V/V}$$

Thus, this cascade amplifier does not meet specs. The reason is obvious from the gain calculation: the output resistance, 1 kΩ, is too high for feeding a load of 100 Ω and indeed results in a loss of gain by a factor of 11!

That's where the type (3) amplifier stage can be beneficial. While its open-circuit gain is only 1, its output resistance is 20 Ω, five times lower than the 100 Ω load. The overall amplifier would then look as follows:

$$\text{Overall voltage gain} \equiv \frac{v_o}{v_s}$$

$$= \frac{1}{1+0.5} \times 10 \times \frac{10}{10+10} \times 100 \times \frac{10}{10+1} \times 1 \times \frac{100}{100+20}$$

$$= 252.5 \text{ V/V}$$

which meets the specified gain of 235 V/V.

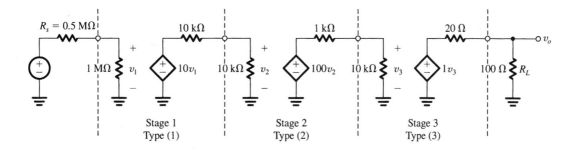

R_s = 0.5 MΩ, 1 MΩ, v_1, 10 kΩ, $10v_1$, 10 kΩ, v_2, 1 kΩ, $100v_2$, 10 kΩ, v_3, 20 Ω, $1v_3$, 100 Ω, R_L, v_o

Stage 1 Type (1) | Stage 2 Type (2) | Stage 3 Type (3)

Chapter 2 – Problems

2.1

The minimum number of pins required by dual-op-amp is 8. Each op-amp has 2 input terminals (4 pins) and one output terminal (2 pins). Another 2 pins are required for power.
Similarly, the minimum number of pins required by quad-op-amp is 14:
$4\times 2 + 4\times 1 + 2 = 14$

2.2

Refer to Fig. P2.2. $v_+ = v_I \dfrac{1K\Omega}{1M\Omega + 1K\Omega} = \dfrac{4}{1001}\,V$

$v_o = A v_+ \Rightarrow A = \dfrac{4}{4/1001} = 1001$

2.3

The voltage at the positive input has to be $-3.000V$.

$v_+ = -3.020\,V$, $A = \dfrac{v_o}{(v_+ - v_-)} = \dfrac{-2}{-3.020-(-3)} = 100$

2.4

#	v_1	v_2	$v_d = v_2 - v_1$	v_o	v_o/v_d
1	0.00	0.00	0.00	0.00	–
2	1.00	1.00	0.00	0.00	–
3	(a)	1.00	(b)	1.00	
4	1.00	1.10	0.10	10.1	101
5	2.01	2.00	-0.01	-0.99	99
6	1.99	2.00	0.01	1.00	100
7	5.10	(c)	(d)	-5.10	

experiments 4, 5, 6 show that the gain is approximately 100 V/V. The missing entry for experiment #3 can be predicted as follows:

(b) $v_d = \dfrac{v_o}{A} = \dfrac{1.00}{100} = 0.01\,V.$

(a) $v_1 = v_2 - v_d = 1.00 - 0.01 = 0.99\,V$

The missing entries for experiment #7:

(d) $v_d = \dfrac{-5.10}{100} = -0.051\,V$

(c) $v_2 = v_1 + v_d = 5.10 - 0.051 = 5.049\,V$

All the results seem to be reasonable.

2.5

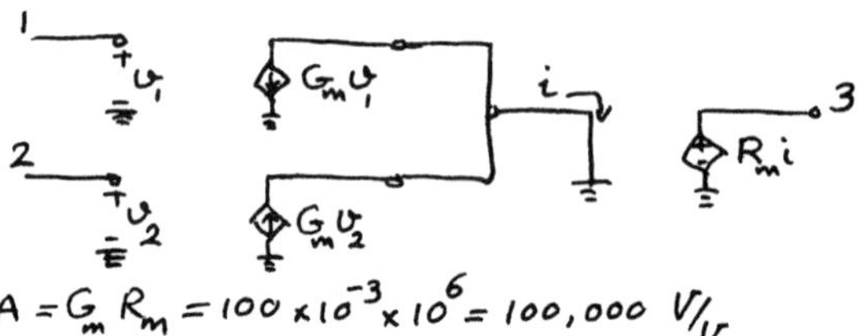

$A = G_m R_m = 100\times 10^{-3}\times 10^{6} = 100{,}000\ V/V$

2.6

$v_{CM} = 1V \sin(2\pi 60)t \qquad = \frac{1}{2}(v_1 + v_2)$

$v_d = 0.01 \sin(2\pi 1000)t \quad = v_2 - v_1$

$v_1 = v_{CM} - v_d/2 = \sin(120\pi)t - 0.005 \sin 2000\pi t$

$v_2 = v_{CM} + v_d/2 = \sin 120\pi t + 0.005 \sin 2000\pi t$

2.7

$v_d = R(G_{m2} v_2 - G_{m1} v_1)$ Refer to Fig. 2.4.

$v_o = v_3 = \mu v_d = \mu R(G_{m2} v_2 - G_{m1} v_1)$

$v_o = \mu R(G_m v_2 + \frac{1}{2}\Delta G_m v_2 - G_m v_1 + \frac{1}{2}\Delta G_m v_1)$

$v_o = \mu R G_m \underbrace{(v_2 - v_1)}_{v_{Id}} + \frac{1}{2}\mu R \Delta G_m \underbrace{(v_1 + v_2)}_{2v_{ICM}}$

we have $v_o = A_d v_{Id} + A_{cm} v_{ICM}$

$\Rightarrow A_d = \mu R G_m, \quad A_{CM} = \mu R \Delta G_m$

$CMRR = 20 \log \left|\dfrac{A_d}{A_{CM}}\right| = 20 \log \dfrac{G_m}{\Delta G_m}$

cont.

$20\log_{10} A_d = 80\,dB \Rightarrow A_d = 10^4$

$\frac{A_{CM}}{A_d} = \frac{\Delta G_m}{G_m} \Rightarrow A_{cm} = 10^4 \times \frac{0.1}{100} = 10$

$CMRR = 20 \log \frac{G_m}{\Delta G_m} = 20 \log \frac{1}{0.1/100} = 60$

2.8

circuit	v_o/v_i (V/V)	R_{in} (kΩ)	
a	$\frac{-100}{10} = -10$	10	
b	−10	10	
c	−10	10	virtual ground
d	−10	10	no current in 10kΩ

2.9

closed loop gain = −1 V/V. for $v_i = 5V \Rightarrow v_o = -5\,V$

Gain would be in the range of $\frac{-0.95}{1.05}$ to $\frac{-1.05}{0.95}$: $-0.9 < G < -1.1$

for $v_i = 5 \Rightarrow -4.5 < v_o < -5.5V$

2.10

There are four possibilities:

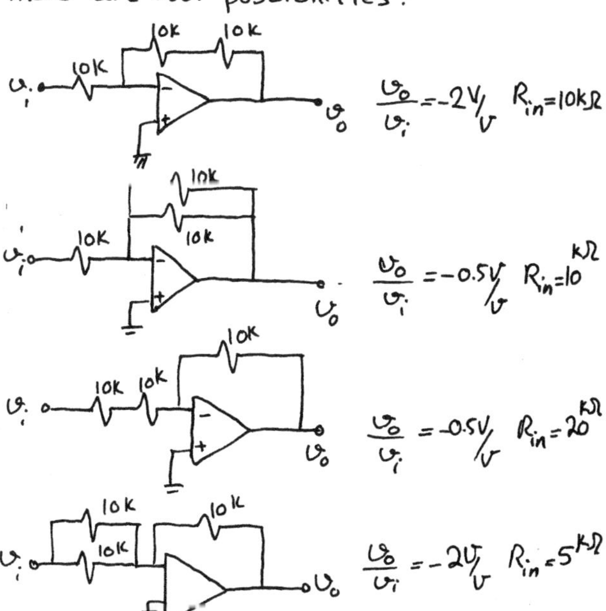

$\frac{v_o}{v_i} = -2\,V/V \quad R_{in} = 10k\Omega$

$\frac{v_o}{v_i} = -0.5\,V/V \quad R_{in} = 10\,k\Omega$

$\frac{v_o}{v_i} = -0.5\,V/V \quad R_{in} = 20\,k\Omega$

$\frac{v_o}{v_i} = -2\,V/V \quad R_{in} = 5\,k\Omega$

2.11

a. $G = -1\,V/V$

b. $G = -10\,V/V$

c. $G = -0.1\,V/V$

d. $G = -100\,V/V$

e. $G = -10\,V/V$

2.12

a. $G = -1\,V/V = -\frac{R_2}{R_1} \Rightarrow R_1 = R_2 = 10k\Omega$

b. $G = -2\,V/V = -\frac{R_2}{R_1} \Rightarrow R_1 = 10k\Omega,\ R_2 = 20k\Omega$

c. $G = -0.5\,V/V = -\frac{R_2}{R_1} \Rightarrow R_1 = 20k\Omega,\ R_2 = 10k\Omega$

d. $G = -100\,V/V = -\frac{R_2}{R_1} \Rightarrow R_1 = 10k\Omega,\ R_2 = 1M\Omega$

2.13

$\frac{v_o}{v_i} = -5 = -\frac{R_2}{R_1} \Rightarrow R_2 = 5R_1$

$R_1 + R_2 = 120k\Omega \Rightarrow 5R_1 + R_1 = 120k\Omega \Rightarrow$

$R_1 = 20k\Omega \Rightarrow R_2 = 100k\Omega$

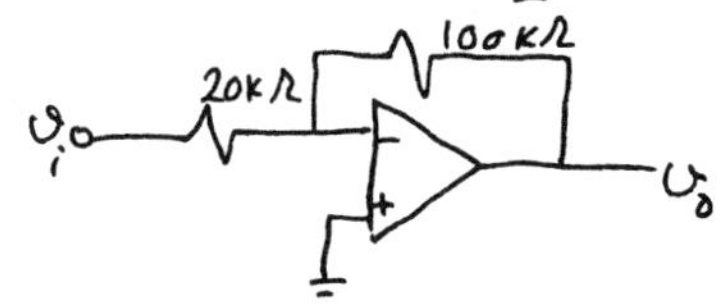

2.14

$20\log|G| = 26dB \Rightarrow G = -19.95\,V/V = \frac{v_o}{v_i} = -\frac{R_2}{R_1}$

$\Rightarrow R_2 = 19.95 R_1 \leq 10M\Omega$

For largest possible input resistance, select

$R_2 = 10M\Omega \Rightarrow R_1 \simeq 500k\Omega$

$R_{in} = 500k\Omega$

2.15

$G = \frac{v_o}{v_i} = \frac{-R_2}{R_1} = \frac{-100}{10} = -10$

$v_{low} = -10V,\ v_{high} = 0,\ v_{avg} = -5V$

0 ... −10V

2.16

$$\frac{v_o}{v_i} = -\frac{R_2}{R_1} \Rightarrow v_o = -1 \times -\frac{10k\Omega}{1k\Omega} = 10V$$

$i_2 = \frac{v_o}{2k\Omega} = 5mA$

$i_1 = i_3 = \frac{v_o}{10k\Omega} = 1mA$

$i_4 = i_2 - i_3 = 4mA$ This additional current comes from the output of the op amp.

2.17

$$|Gain| = \frac{R_2}{R_1} = \frac{R_2(1+x/100)}{R_1(1+x/100)} \sim \frac{R_2}{R_1}\left(1 \pm \frac{2x}{100}\right)$$

for small x

$\Rightarrow 2x\%$ is the tolerance on the closed loop gain (G).

$G = -100\,V/V$, $x = 5 \Rightarrow -110 < G < -90$

or more precisely: $-100 \times \frac{105}{95} < G < -100\,\frac{95}{105}$

$-110.5 < G < -90.5$

2.18

$$G = \frac{v_o}{v_i} = -\frac{R_2}{R_1} \Rightarrow \frac{R_2}{R_1} = \frac{5}{15}$$

$v_1 = 0V$, $v_2 = v_o = 5V$

For $\pm 1\%$ on R_1, R_2: $R_1 = 15 \pm 0.15\,k\Omega$

$R_2 = 5 \pm 0.05\,k\Omega$

$$v_o = v_i \frac{-R_2}{R_1} = 15\frac{R_2}{R_1} \Rightarrow 15 \times \frac{4.95}{15.15} \leq v_o \leq 15 \times \frac{5.05}{14.85}$$

$\Rightarrow 4.9V \leq v_o \leq 5.1V$

For $v_i = -15 \pm 0.15\,V$ $\quad 14.85 \times \frac{4.95}{15.15} \leq v_o \leq 15.15 \times \frac{5.05}{14.85}$

$\Rightarrow 4.85V \leq v_o \leq 5.15V$

2.19

$$v_1 = \frac{-v_o}{A} = -\frac{v_o}{200}$$

$\frac{v_o}{v_i} = -50\,V/V$

$$\frac{v_i - \left(\frac{-v_o}{A}\right)}{R_1} = \frac{\left(\frac{-v_o}{A} - v_o\right)}{100k\Omega} \Rightarrow R_1 = 100K \times \frac{\frac{v_o}{200} - \frac{v_o}{50}}{\frac{-v_o}{200} - v_o}$$

$\Rightarrow R_1 = 100K \times \frac{3}{201} = 1.49k\Omega$

Shunt Resistor R_a: $R_a \,//\, 2k\Omega = 1.49K$

$$\frac{R_a \times 2}{R_a + 2} = 1.49 \Rightarrow R_a = 5.84k\Omega$$

2.20

(a)

$$\frac{v_o}{v_i} = -\frac{R_2}{R_1} \Rightarrow -100\,V/V = \frac{-R_2}{1k\Omega} \Rightarrow R_2 = 100k\Omega$$

(b) $A = 1000\,V/V$

$v_1 = \frac{-v_o}{A}$

$$\frac{v_i - v_1}{R_1} = \frac{v_1 - v_o}{R_2}$$

$$\frac{v_o}{v_i} = \frac{-R_2/R_1}{1 + (1 + \frac{R_2}{R_1})/A} = \frac{-100}{1 + \frac{101}{1000}} = -90.8\,V/V$$

$\Rightarrow \frac{v_o}{v_i} = -90.8\,V/V$

(c) Assume $R_1' = R_x \,//\, R_1$ when $R_1 = 1k\Omega$

$\frac{v_o}{v_i} = -100\,V/V$

$$\frac{v_i - v_1}{R_1'} = \frac{v_1 - v_o}{R_2} \Rightarrow R_1' = R_2 \times \left(\frac{-v_o}{100} - \frac{-v_o}{1000}\right) \Big/ \left(\frac{-v_o}{1000} - v_o\right)$$

$$R_1' = \frac{1 - 0.1}{1.001} = 0.899\,k\Omega = \frac{R_1 R_x}{R_1 + R_x} = \frac{R_x}{1 + R_x}$$

$\Rightarrow R_x = 8.9\,k\Omega \simeq 8.87\,k\Omega \pm 1\%$

2.21

Voltage of the inverting input terminal

Cont.

will vary from $\frac{-10V}{1000}$ to $\frac{+10V}{1000}$. Thus the virtual ground will depart from the ideal voltage of zero by a maximum of $\pm 10\,mV$.

2.22

a) For $A = \infty$: $v_i = 0$

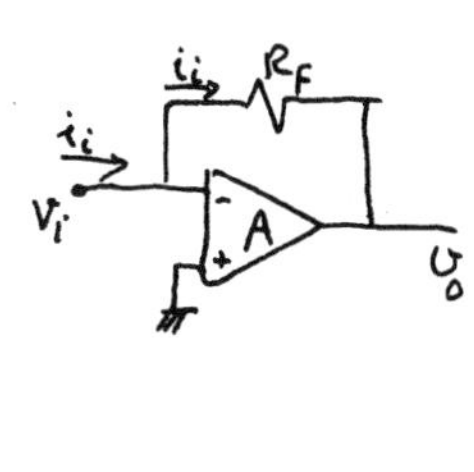

$v_o = -i_i R_F$

$R_m = \frac{v_o}{i_i} = -R_F$

$R_{in} = \frac{v_i}{i_i} = 0$

b) For A = finite : $v_i = -\frac{v_o}{A}$, $v_o = v_i - i_i R_F$

$\Rightarrow v_o = -\frac{v_o}{A} - i_i R_F \Rightarrow R_m = \frac{v_o}{i_i} = -\frac{R_F}{1+\frac{1}{A}}$

$R_i = \frac{v_i}{i_i} = \frac{R_F}{1+A}$

2.23

$v_o = -Av_- = v_- - i_i R_2$

$i_i R_2 = (1+A) v_-$

$v_- = \frac{i_i R_2}{1+A}$

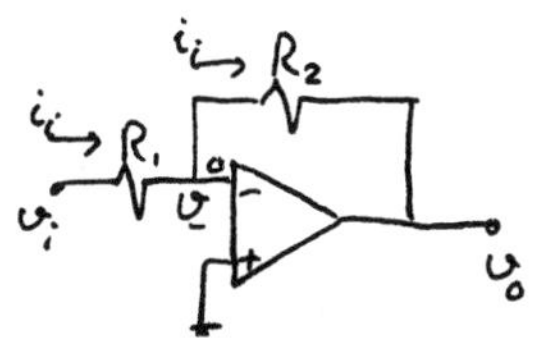

Now: $v_i = i_i R_1 + v_- = i_i R_1 + i_i \frac{R_2}{1+A}$

$R_{in} = \frac{v_i}{i_i} = R_1 + \frac{R_2}{1+A}$

2.24

$G = \frac{-R_2/R_1}{1 + \frac{1+R_2/R_1}{A}}$ $\quad$ Gain Error $\epsilon = \left(1+\frac{R_2}{R_1}\right)/A \times 100$

ϵ	0.1%	1%	10%
A	$1000(1+\frac{R_2}{R_1})$	$100(1+\frac{R_2}{R_1})$	$10(1+\frac{R_2}{R_1})$

Gain with R_{1a} :

$G \simeq \frac{\frac{R_2}{R_1}\left(1+\frac{R_1}{R_{1a}}\right)}{1 + \frac{1+R_2/R_1}{A}}$

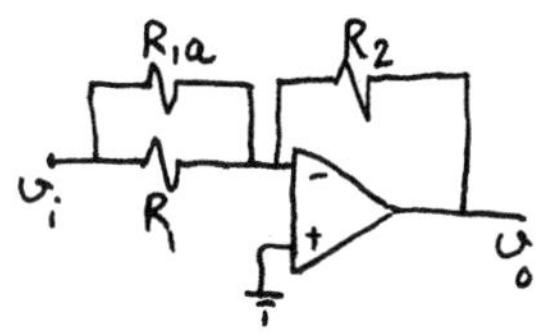

where we have neglected the effect of R_{1a} on the error on the denominator. To restore the gain to its nominal value of R_2/R_1 we use:

$\frac{R_1}{R_{1a}} = \frac{1+R_2/R_1}{A} = \frac{\epsilon}{100} \Rightarrow R_{1a} = \frac{100 R_1}{\epsilon}$

ϵ	0.1%	1%	10%
R_{1a}	$1000R_1$	$100R_1$	$10R_1$

2.25

$R_1' = R_1 \| R_c$ $\qquad$ $G' = \frac{-R_2/R_1'}{1+\frac{1+R_2/R_1'}{A}}$

$G = -\frac{R_2}{R_1}$

In order for $G' = G$: $G = \frac{-R_2/R_1'}{1+\frac{1+R_2/R_1'}{A}} = \frac{-R_2}{R_1}$

$R_1' = \frac{R_1 R_c}{R_1 + R_c}$

$\Rightarrow \frac{R_1+R_c}{R_1 R_c} = \frac{1}{R_1}\left(1+\frac{1+R_2\frac{(R_1+R_c)}{R_1 R_c}}{A}\right)$

$(R_1+R_c)A = AR_c + R_c + \frac{R_2}{R_1}(R_1+R_c)$

$R_1 A = R_c + GR_1 + GR_c$

$\frac{R_c}{R_1} = \frac{A-G}{1+G}$

2.26

$G = \frac{-R_2/R_1}{1+\frac{1+R_2/R_1}{A}}$ $\qquad$ $G_{nominal} = \frac{-R_2}{R_1}$

$\epsilon = \left|\frac{G - G_{nominal}}{G_{nominal}}\right| = \left|\frac{G}{G_{nominal}} - 1\right|$

$\epsilon = \left|\frac{1}{1+\frac{1+R_2/R_1}{A}} - 1\right| = \left|\frac{-\frac{1+R_2/R_1}{A}}{1+\frac{1+R_2/R_1}{A}}\right| = \frac{1}{\frac{A}{1+\frac{R_2}{R_1}} + 1}$

which can be rearranged to yield:

$\frac{A}{1+\frac{R_2}{R_1}} + 1 = \frac{1}{\epsilon} \Rightarrow A = \left(1+\frac{R_2}{R_1}\right)\left(\frac{1}{\epsilon} - 1\right)$

or $A = (1 - G_{nominal})\left(\frac{1}{\epsilon} - 1\right)$

For $G_{nominal} = -100\ V/V$ and $\epsilon = 10\% = 0.1$

$A = (1+100)\left(\frac{1}{0.1} - 1\right) = 909\ V/V$

This is the minimum required value for A.

2.27

$|G| = \frac{R_2/R_1}{1+\frac{1+R_2/R_1}{A}}$ $\qquad A \rightarrow A(1-\frac{x}{100})$ $\qquad G \rightarrow G'$

$|G'| = \frac{R_2/R_1}{1+\frac{1+R_2/R_1}{A(1-\frac{x}{100})}}$

For $|G'| = |G|\,(1-\frac{x}{100K})$

$$\frac{R_2/R_1}{1+\frac{1+R_2/R_1}{A(1-x/100)}} = \frac{R_2/R_1}{1+\frac{1+R_2/R_1}{A}}\left(1-\frac{x}{100K}\right)$$

$$1+\frac{1+R_2/R_1}{A(1-\frac{x}{100})} = \left(1+\frac{1+R_2/R_1}{A}\right)\Big/\left(1-\frac{x}{100K}\right)$$

$$1-\frac{x}{100K} + \frac{1+R_2/R_1}{A}\;\frac{1-x/100K}{1-x/100} = 1+\frac{1+R_2/R_1}{A}$$

$$\frac{1+R_2/R_1}{A}\;\frac{1-x/100K-1+x/100}{1-x/100} = \frac{x}{100K}$$

$$A = \frac{-1+K}{1-\frac{x}{100}}\left(1+R_2/R_1\right) = \left(\frac{K-1}{1-\frac{x}{100}}\right)\left(1+\frac{R_2}{R_1}\right)$$

For $\frac{R_2}{R_1}=100$ $\quad x=50 \quad K=100$: $A = \frac{99}{0.5}\times 101 = 19998$

$A \simeq 2\times 10^4\ V/V$

Thus for $A = 2\times10^4\ V/V$, a reduction of 50% results in only 0.5% reduction of the closed loop gain whose nominal value is $\frac{R_2}{R_1}$ (100).

2.28

From the results of example 2.2, the gain of the circuit in fig. 2.8 is given by:

$$\frac{v_o}{v_i} = -\frac{R_2}{R_1}\left(1+\frac{R_4}{R_2}+\frac{R_4}{R_3}\right)$$

For $R_1 = R_2 = R_4 = 1M\Omega \Rightarrow \frac{v_o}{v_i} = -(1+1+\frac{1}{R_3})$

a) $\frac{v_o}{v_i} = -10\ V/V \Rightarrow 10 = 2+\frac{1}{R_3} \Rightarrow R_3 = \frac{1}{8}M\Omega = 125\ k\Omega$

b) $\frac{v_o}{v_i} = -100\ V/V \Rightarrow 100 = 2+\frac{1}{R_3} \Rightarrow R_3 = \frac{1}{98}M\Omega = 10.2\ k\Omega$

c) $\frac{v_o}{v_i} = -2\ V/V \Rightarrow 2 = 2+\frac{1}{R_3} \Rightarrow R_3 = \infty$: eliminate R_3.

2.29

$R_2/R_1 = 1000$, $R_2 = 100k\Omega \Rightarrow R_1 = 100\Omega$

a) $R_{in} = R_1 = 100\Omega$

b) $\frac{v_o}{v_i} = \frac{-R_2}{R_1}\left(1+\frac{R_4}{R_2}+\frac{R_4}{R_3}\right) = -1000$

IF $R_2 = R_1 = R_4 = 100K \Rightarrow R_3 = \frac{100K}{1000-2} \simeq 100\ \Omega$

$R_{in} = R_1 = 100k\Omega$

2.30

$v_x = 0 - i_1 R_2$, $\quad i_1 = \frac{v_I}{R_1} \Rightarrow v_x = -v_I\frac{R_2}{R_1}$

$\frac{v_x}{v_I} = -\frac{R_2}{R_1}$

$$v_x = v_o\,\frac{R_2\|R_3}{R_2\|R_3+R_4} = v_o\,\frac{R_2R_3}{R_2R_3+R_4R_2+R_4R_3}$$

$$\frac{v_o}{v_x} = \frac{R_2R_3+R_2R_4+R_3R_4}{R_2R_3} = 1+\frac{R_4}{R_3}+\frac{R_4}{R_2}$$

$$\frac{v_o/v_x}{v_I/v_x} = \frac{v_o}{v_i} = \frac{(1+R_4/R_3+R_4/R_2)}{-R_1/R_2} \Rightarrow$$

$$\frac{v_o}{v_i} = -\frac{R_2}{R_1}\left(1+\frac{R_4}{R_3}+\frac{R_4}{R_2}\right)$$

2.31

a) $R_1 = R$

$R_2 = R\|R + \frac{R}{2} = \frac{R}{2}+\frac{R}{2} = R$

$R_3 = R_2\|R + \frac{R}{2} = R\|R+\frac{R}{2} = R$

$R_4 = R_3\|R + \frac{R}{2} = R\|R+\frac{R}{2} = R$

b) $v_1 = RI = RI_1 \Rightarrow I_1 = I$

$I_{12} = I+I = 2I \Rightarrow v_1 + 2I\times\frac{R}{2} = RI_2$

$RI + RI = RI_2 \Rightarrow I_2 = 2I$

$I_3 = I_2 + I_{12} = 4I \Rightarrow v_2 + 4I\times\frac{R}{2} = RI_3$

$R\times 2I + 4I\times\frac{R}{2} = RI_3 \Rightarrow I_3 = 4I$, $I_4 = -(4I+4I)$

$I_4 = 8I$

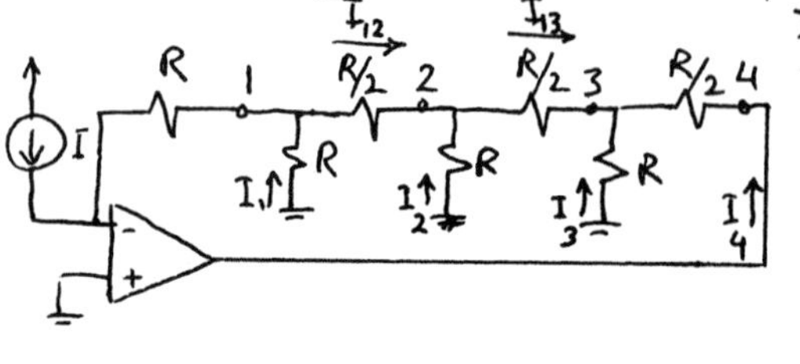

Cont.

c) $v_1 = I_1 R = IR$
$v_2 = I_2 R = 2IR$
$v_3 = -I_3 R = -4IR$
$v_4 = -I_3 R + I_4 \frac{R}{2} = -4IR - 8I\frac{R}{2} = -8IR$

2.32

a) $I_1 = \frac{1V}{10k\Omega} = 0.1mA$
$I_2 = I_1 = 0.1\ mA$, $I_2 \times 10k\Omega = I_3 \times 100\Omega \Rightarrow I_3 = 10^{mA}$
$v_x = 10mA \times 100\Omega = 1V$

b) $v_x = R_L I_L + v_o$, $I_L = I_2 + I_3 = 10.1mA$
$1V = R_L \times 10.1mA + v_o$
$R_L = \frac{1 - v_o}{10.1} \Rightarrow R_{Lmax} = \frac{1 - v_{omin}}{10.1} = \frac{14}{10.1}$
$R_{Lmax} =$

c) $100\Omega \leq R_L \leq 1k\Omega$
I_L stays fixed at 10.1mA
$v_o = v_x - R_L I_L = 1 - R_L \times 10.1 \Rightarrow -9.1^V \leq v_o \leq -0.01^V$

2.33

a) $\frac{i_L}{i_i} = 20 \Rightarrow i_L = 20 i_I$

$-10k\Omega \times i_I = R(i_I - i_L)$
$R = \frac{10k\Omega \times i_I}{20 i_I - i_I} = 0.53k\Omega$

b) $R_L = 1k\Omega$ $-12^V \leq v_o \leq 12^V$
$v_o = R_L i_L + 10k\Omega \times i_I = i_I (1k\Omega \times \frac{i_L}{i_I} + 10^{k\Omega})$
$v_o = i_I (1 \times 20 + 10) = 30 i_I$
$i_I = \frac{v_o}{30} \Rightarrow \frac{-12}{30} \leq i_I \leq \frac{12}{30} \Rightarrow -0.4_{mA} \leq i_I \leq 0.4_{mA}$

c) $R_I = \frac{v_I}{i_I} = \frac{0}{i_I} = 0$
$v_- = 0 \Rightarrow i = 0$
$\Rightarrow i_I = 1mA$
From part a: $i_L = 20 \times i_I = 20\ mA$

2.34

$R_2 \gg R_3$, if we ignore the current accoss R_2 : $v_A = \frac{v_o R_3}{R_3 + R_4}$

$\frac{v_I}{R_1} = \frac{0 - v_A}{R_2} \Rightarrow v_A = -\frac{R_2 v_I}{R_1}$

$v_o \frac{R_3}{R_3 + R_4} = -\frac{R_2}{R_1} \times v_I \Rightarrow \frac{v_o}{v_I} = \frac{-R_2}{R_1}(1 + \frac{R_4}{R_3})$

Now if we recalculate v_A considering that there is a voltage divider between R_4 and $R_3 \| R_2$:

$v_A = v_o \frac{R_3 \| R_2}{R_4 + R_3 \| R_2} = v_o \frac{R_3 R_2}{R_4(R_3 + R_2) + R_2 R_3}$

$v_A = v_o \frac{R_2 R_3}{R_3 R_4 + R_2 R_4 + R_2 R_3}$

$v_A = v_o \frac{1}{R_4/R_2 + \frac{R_4}{R_3} + 1}$

$v_A = -\frac{R_2}{R_1} . v_I \Rightarrow \frac{v_o}{v_I} = \frac{-R_2}{R_1}(\frac{R_4}{R_2} + \frac{R_4}{R_3} + 1)$

Same as example 2.2.

2.35

$R_I = 100k\Omega$ $-10 \leq \frac{v_o}{v_i} \leq -1\ V/V$

$R_I = R_1 = 100k\Omega$

$\frac{v_o}{v_i} = \frac{-R_2}{R_1}(\frac{R_4}{R_3} + \frac{R_4}{R_2} + 1)$

$R_4 = 0 \Rightarrow \frac{v_o}{v_i} = \frac{-R_2}{R_1} = -1 \Rightarrow R_2 = 100k\Omega$

$R_4 = 10^{k\Omega} \Rightarrow \frac{v_o}{v_i} = -10 = -1 \times (\frac{10^{k\Omega}}{R_3} + \frac{10^{k\Omega}}{100^{k\Omega}} + 1)$

$+10 = (\frac{10}{R_3} + 1.1) \Rightarrow R_3 = 1.12k\Omega$

Potentiometer in the middle: $\frac{v_o}{v_i} = -1(\frac{5}{5 + R_3} + \frac{5}{100} + 1)$
$\frac{v_o}{v_i} = -1.87 V/V$

2.36

According to eq. 2.7:

$v_o = -\left(\frac{R_F}{R_1}v_1 + \frac{R_F}{R_2}v_2 + \ldots \frac{R_F}{R_n}v_n\right)$

$v_o = -\left(\frac{50k}{100k}v_1 + \frac{50k}{100k}v_1 + \frac{50k}{100k}v_2\right)$

$v_o = -\left(v_1 + \frac{v_2}{2}\right) \qquad v_1 = 3, v_2 = -3 \Rightarrow v_o = -1.5v$

2.37

we choose the weighted summer configuration

$v_o = -\left[4v_1 + \frac{v_2}{3}\right]$

$i_1 = \frac{v_1}{R_1} \qquad i_2 = \frac{v_2}{R_2}$

$i_1, i_2 \leq 0.1\,mA$ for $v_1, v_2 = 1V$

$R_1, R_2 \geq 10k\Omega$

$\frac{R_F}{R_1} = 4$, if $R_1 = 10k\Omega \Rightarrow R_F = 40k\Omega$

$\frac{R_F}{R_2} = \frac{1}{3} \Rightarrow R_2 = 120k\Omega$

2.38

$v_o = -(2v_1 + 4v_2 + 8v_3)$

$R_1, R_2, R_3 \geq 10k\Omega$

$\frac{R_F}{R_1} = 2, \quad \frac{R_F}{R_2} = 4 \quad \frac{R_F}{R_3} = 8$

$R_3 = 10k\Omega \Rightarrow R_F = 80k\Omega$

$R_2 = 20k\Omega$

$R_1 = 40k\Omega$

2.39

a) $v_o = -(v_1 + 2v_2 + 3v_3)$

$\frac{R_F}{R_1} = 1 \Rightarrow R_1 = 10k\Omega, \quad \frac{R_F}{R_2} = 2 \Rightarrow R_2 = 5k\Omega$

$\frac{R_F}{R_3} = 3 \Rightarrow R_3 = \frac{10}{3}k\Omega$

$R_{I1} = 10k\Omega$

$R_{I2} = 5k\Omega$

$R_{I3} = 3.3K$

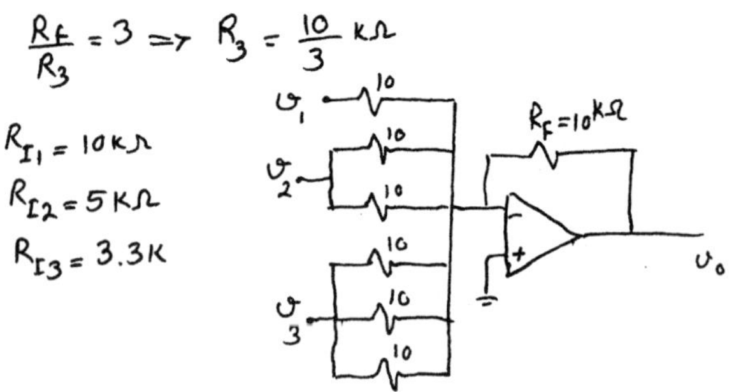

b) $v_o = -(v_1 + v_2 + 2v_3 + 2v_4)$

$\frac{R_F}{R_1} = 1 \Rightarrow R_1 = 10k\Omega$

$\frac{R_F}{R_2} = 1 \Rightarrow R_2 = 10k\Omega$

$\frac{R_F}{R_3} = 2 \Rightarrow R_3 = \frac{10}{2}k\Omega$

$\frac{R_F}{R_4} = 2 \Rightarrow R_4 = \frac{10}{2}k\Omega$

$R_{I1} = R_{I2} = 10k\Omega$

$R_{I3} = R_{I4} = 5k\Omega$

c) $v_o = -(v_1 + 5v_2)$

$R_1 = 10k$

$R_2 = \frac{10k}{5}$

$R_{I1} = 10k\Omega$

$R_{I2} = 2k\Omega$

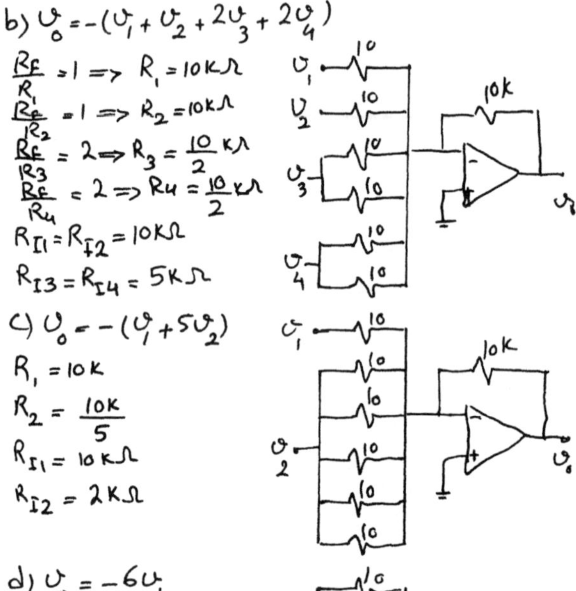

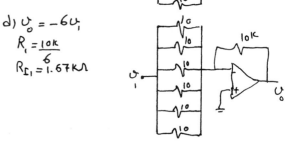

d) $v_o = -6v_1$

$R_1 = \frac{10k}{6}$

$R_{I1} = 1.67k\Omega$

Suggested configurations:

$v_o = -(2v_1 + 2v_2 + 2v_3)$

$v_o = -(3v_1 + 3v_2)$

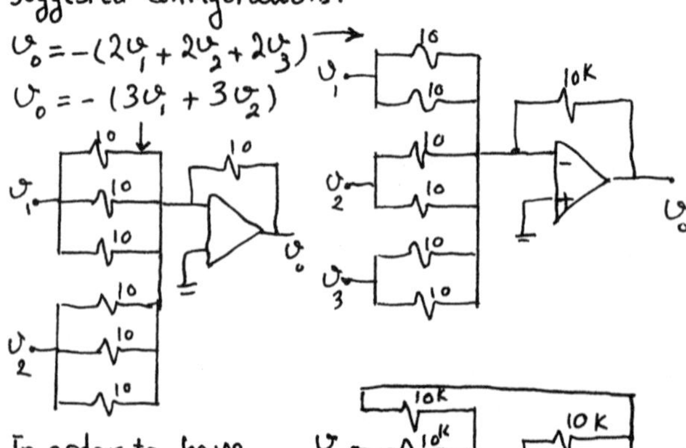

In order to have coeficient = 0.5, connect one of the input resistors to v_o. $\frac{v_o}{v_1} = 0.5$

2.40

The output signal should be:

$v_o = -5\sin\omega t - 5$

if we assume: $v_1 = 5\sin\omega t$, $v_2 = 2V$ $\Big\}\ v_o = -(v_1 + 2.5v_2)$

In a weighted summer configuration:

$\frac{R_f}{R_1} = +1 \qquad \frac{R_f}{R_2} = 2.5$

$R_2 = 10K\Omega \Rightarrow R_f = 25K = R_1$

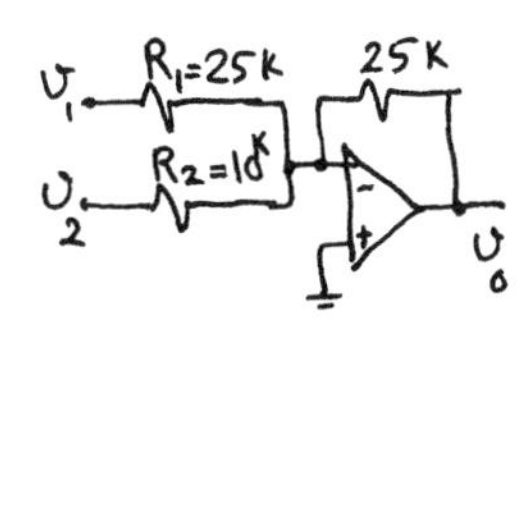

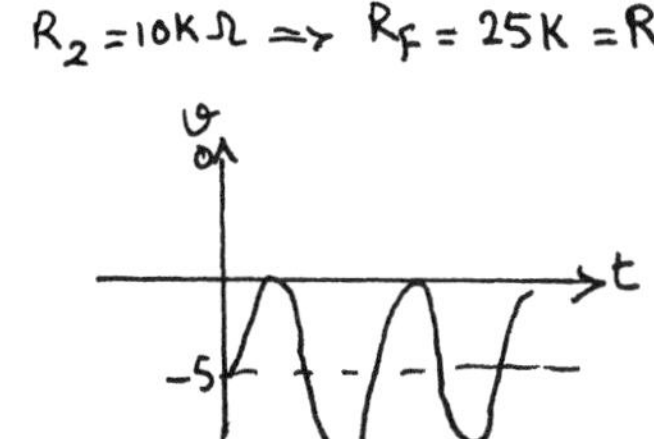

2.41

$v_o = v_1 + 2v_2 - 3v_3 - 4v_4$: Consider Fig. 2.11.

According to eq. 2.8 for a weighted summer circuit:

$v_o = v_1 \frac{R_a}{R_1}\frac{R_c}{R_b} + v_2 \frac{R_a}{R_2}\frac{R_c}{R_b} - v_3 \frac{R_c}{R_3} - v_4 \frac{R_c}{R_4}$

$\frac{R_a}{R_1}\frac{R_c}{R_b} = 1, \quad \frac{R_a}{R_2}\frac{R_c}{R_b} = 1, \quad \frac{R_c}{R_3} = 3, \quad \frac{R_c}{R_4} = 4$

assume:

$R_4 = 10K\Omega \Rightarrow R_c = 40K\Omega \Rightarrow R_3 = \frac{40}{3} = 13.3K\Omega$

$\frac{R_a}{R_1} \times \frac{40}{R_b} = 1 \qquad \frac{R_a}{R_2} \times \frac{40}{R_b} = 1$

$R_b = 40K\Omega, \quad R_1 = R_2 = R_a = 10K\Omega$

2.42

$v_1 = 3\sin(2\pi \times 60t) + 0.01\sin(2\pi \times 1000t)$

$v_2 = 3\sin(2\pi \times 60t) - 0.01\sin(2\pi \times 1000t)$

we want to have: $v_o = 10v_1 - 10v_2$

we use the circuit in Fig. 2.11.

According to Eq. 2.8:

$v_o = v_1 \frac{R_a}{R_1}\frac{R_c}{R_b} - v_3 \frac{R_c}{R_3}$

$\frac{R_a}{R_1}\frac{R_c}{R_b} = 10, \quad \frac{R_c}{R_3} = 10$, if $R_3 = 10K\Omega \Rightarrow R_c = 100K\Omega$

$\Rightarrow \frac{R_a}{R_1} \times \frac{100K\Omega}{R_b} = 10 \Rightarrow R_a = R_1 = R_b = 10k\Omega$

$R_c = 100k\Omega$

$v_o = 10v_1 - 10v_2 = 10 \times 0.02\sin 2\pi \times 1000t$

$v_o = 0.2\sin(2\pi \times 1000t) \qquad -0.2V \le v_o \le 0.2V$

2.43

This is a weighted summer circuit:

$v_o = -\left(\frac{R_f}{R_0}v_0 + \frac{R_f}{R_1}v_1 + \frac{R_f}{R_2}v_2 + \frac{R_f}{R_3}v_3\right)$

we may write: $v_0 = 5^V \times a_0$, $v_2 = 5^V \times a_2$, $v_1 = 5^V \times a_1$, $v_3 = 5^V \times a_3$

$v_o = -R_f\left(\frac{5a_0}{80K} + \frac{5}{40K}a_1 + \frac{5}{20K}a_2 + \frac{5}{10K}a_3\right)$

$v_o = -R_f\left(\frac{a_0}{16} + \frac{a_1}{8} + \frac{a_2}{4} + \frac{a_3}{2}\right)$

$v_o = -\frac{R_f}{16}(2^0 a_0 + 2^1 a_1 + 2^2 a_2 + 2^3 a_3)$

$-12^V \le v_o \le 0 \Rightarrow \frac{R_f}{16}(2^0 \times 1 + 2 \times 1 + 2^2 \times 1 + 2^3 \times 1) =$

$= \frac{15 R_f}{16} = 12$ when $a_0 = a_1 = a_2 = a_3 = 1$ we have the peak value at v_o.

$\Rightarrow R_f = 12.8k\Omega$

2.44

a) $\frac{v_o}{v_i} = 1 = 1 + \frac{R_2}{R_1} \Rightarrow R_2 = 0, R_1 = 10k\Omega$

b) $\frac{v_o}{v_i} = 2 = 1 + \frac{R_2}{R_1} \Rightarrow R_1 = R_2 = 10k\Omega$

Cont.

c) $\frac{v_o}{v_i} = 101\ V/V = 1 + \frac{R_2}{R_1} \Rightarrow$ if $R_1 = 10\,k\Omega \Rightarrow R_2 = 1\,M\Omega$

d) $\frac{v_o}{v_i} = 100\ V/V = 1 + \frac{R_2}{R_1} \Rightarrow$ if $R_1 = 10\,k\Omega \Rightarrow R_2 = 990\,k\Omega$

2.45

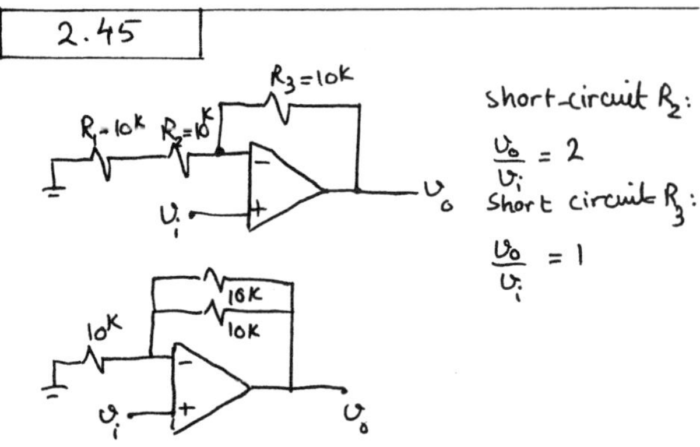

short-circuit R_2:

$\frac{v_o}{v_i} = 2$

Short circuit R_3:

$\frac{v_o}{v_i} = 1$

2.46

$v_+ = v_- = V = R \times i$, $i = 100\,\mu A$ when $V = 10\,V$

$\Rightarrow R = \frac{10}{0.1\,mA} = 100\,k\Omega$

As indicated, i only depends on R and V and the meter resistance does not affect i.

2.47

Refer to the circuit in P2.47:

a) using superposition, we first set $v_{P_1} = v_{P_2} = \dots = 0$

The output voltage that results in response to $v_{N1}, v_{N2}, \dots v_{Nn}$ is:

$$v_{ON} = -\left[\frac{R_F}{R_{N1}}v_{N1} + \frac{R_F}{R_{N2}}v_{N2} + \dots + \frac{R_F}{R_{Nn}}v_{Nn}\right]$$

Then we set $v_{N1} = v_{N2} = \dots = 0$, then:

$R_N = R_{N1} // R_{N2} || R_{N3} || \dots || R_{Nn}$

The circuit simplifies to:

$v_{OP} = \left(1 + \frac{R_F}{R_N}\right) \times$

$$\left(v_{P_1}\frac{1/R_{P1}}{\frac{1}{R_{P1}}+\frac{1}{R_{P2}}+\dots+\frac{1}{R_{Pn}}} + v_{P_2}\frac{1/R_{P2}}{\frac{1}{R_{P_1}}+\dots+\frac{1}{R_{Pn}}} + \dots v_{Pn}\frac{1/R_{Pn}}{1/R_{P_1}+\dots+\frac{1}{R_{Pn}}}\right)$$

$$v_{OP} = \left(1 + \frac{R_F}{R_N}\right)\left(v_{P_1}\frac{R_P}{R_{P_1}} + v_{P_2}\frac{R_P}{R_{P_2}} + \dots + \frac{R_P}{R_{P_n}}v_{P_n}\right)$$

where:

$$R_P = R_{P_1} || R_{P2} || \dots || R_{Pn}$$

when all inputs are present:

$$v_O = v_{ON} + v_{OP} = -\left(\frac{R_F}{R_{N_1}}v_{N_1} + \frac{R_F}{R_{N2}}v_{N2} + \dots\right) + \left(1 + \frac{R_F}{R_N}\right)\left(\frac{R_P}{R_{P_1}}v_{N_1} + \frac{R_P}{R_{P_2}}v_{N2} + \dots\right)$$

b) $v_O = -2v_{N1} + v_{P_1} + 2v_{P_2}$

$\frac{R_F}{R_{N1}} = 2$ $\quad R_{N1} = 10\,K\Omega \Rightarrow R_F = 20\,K\Omega$

$\left(1 + \frac{R_F}{R_N}\right)\left(\frac{R_P}{R_{P_1}}\right) = 1 \Rightarrow 3\frac{R_P}{R_{P_1}} = 1 \Rightarrow R_{P2} = \frac{R_{P1}}{2}$

$\left(1 + \frac{R_F}{R_N}\right)\left(\frac{R_P}{R_{P_2}}\right) = 2 \Rightarrow 3\frac{R_P}{R_{P_2}} = 2 \Rightarrow R_{P2} = \frac{R_{P1}}{2}$

where $R_P = \frac{R_{P_1} \times R_{P2}}{R_{P_1} + R_{P2}}$ (ignoring $R_{P\phi}$)

Note that if the results from the last 2 constraints differ, we would use an additional resistor connected from the positive input to ground. ($R_{P\phi}$)

2.48

$v_O = v_{I1} + 3v_{I2} - 2(v_{I3} + 3v_{I4})$

Refer to P2.47.

$\frac{R_F}{R_{N3}} = 2$ if $R_{N3} = 10\,k\Omega \Rightarrow R_F = 20\,k\Omega$

$\frac{R_F}{R_{N4}} = 6 \Rightarrow R_{N4} = \frac{20}{6} = 3.3\,k\Omega$

$R_N = R_{N3} // R_{N4} = 10K // 3.3K = 2.48\,k\Omega$

$\left(1 + \frac{R_F}{R_N}\right)\frac{R_P}{R_{P_1}} = 1 \Rightarrow \left(1 + \frac{20}{2.48}\right)\frac{R_P}{R_{P_1}} = 1 \Rightarrow 9.06R_P = R_{P_1}$

$R_P = R_{P_1} || R_{P_2} || R_{P\phi} \Rightarrow R_P = \frac{1}{\frac{1}{R_{P1}} + \frac{1}{R_{P2}} + \frac{1}{R_{P\phi}}}$

$\left(1 + \frac{R_F}{R_N}\right)\frac{R_P}{R_{P2}} = 3 \Rightarrow 9.06\frac{R_P}{R_{P2}} = 3 \Rightarrow R_{P_2} \simeq 3R_P$

$R_{P1} // R_{P2} = \frac{9 \times 3R_P}{9+3} = 2.25R_P$, $R_P = 2.25R_P // R_{P\phi}$

Cont.

$2.25 R_p + R_{P_0} = 2.25 R_{P_0} \Rightarrow R_{P_0} = 1.8 R_p$

if $R_p = 10 K\Omega \Rightarrow R_{P_0} = 18 K\Omega$

$R_{P_1} = 9 \times 10K = 90 K\Omega$

$R_{P_2} = 3 \times 10K = 30 K\Omega$

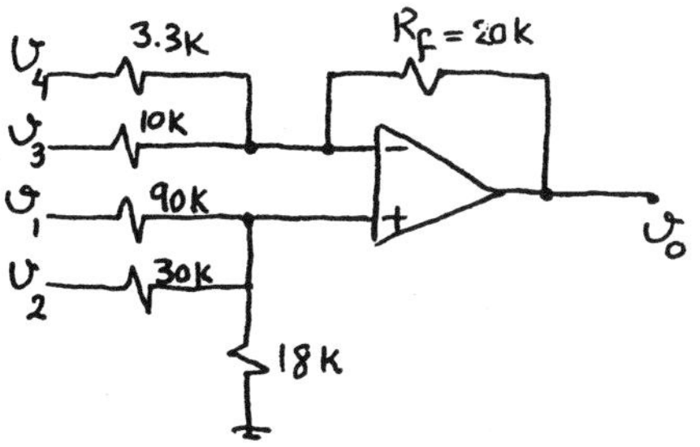

2.49

$$v_+ = v_I \frac{R_4}{R_3 + R_4} = v_-$$

$$\frac{v_-}{R_1} = \frac{v_0 - v_-}{R_2} \Rightarrow v_0 = v_-\left(1 + \frac{R_2}{R_1}\right)$$

from the two above equations :

$$\frac{v_O}{v_I} = \left(1 + \frac{R_2}{R_1}\right)\left(\frac{R_4}{R_3 + R_4}\right) = \frac{1 + R_2/R_1}{1 + R_3/R_4}$$

2.50

Refer to Fig. 2.50. Setting $v_2 = 0$, we obtain the output component due to v_1 as:

$v_{O1} = -20v$

Setting $v_1 = 0$, we obtain the output component due to v_2 as:

$$v_{O2} = v_2\left(1 + \frac{20R}{R}\right)\left(\frac{20R}{20R + R}\right) = 20 v_2$$

The total output voltage is :

$v_O = v_{O1} + v_{O2} = 20(v_2 - v_1)$

For $v_1 = 10 \sin 2\pi \times 60t - 0.1 \sin(2\pi \times 1000t)$

$v_2 = 10 \sin 2\pi \times 60t + 0.1 \sin(2\pi \times 1000t)$

$v_O = 4 \sin(2\pi \times 1000t)$

2.51

$$\frac{v_O}{v_I} = 1 + \frac{R_2}{R_1} = 1 + \frac{(1-x)}{x} = 1 + \frac{1}{x} - 1 = \frac{1}{x}$$

$0 \leq x \leq 1 \Rightarrow 1 \leq \frac{v_O}{v_I} \leq \infty$

if we add a resistor on the ground path :

$$\frac{v_O}{v_I} = 1 + \frac{(1-x) \times 10K}{x \times 10K + R}$$

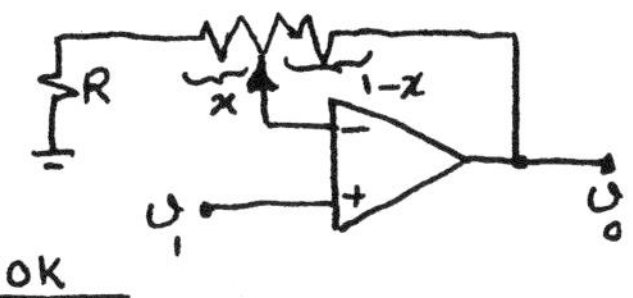

$\text{Gain}_{max} = 21$ when

$x = 0 \Rightarrow 21 = 1 + \frac{10K}{R}$

$\Rightarrow R = \frac{10K}{20} = 0.5 K\Omega$

2.52

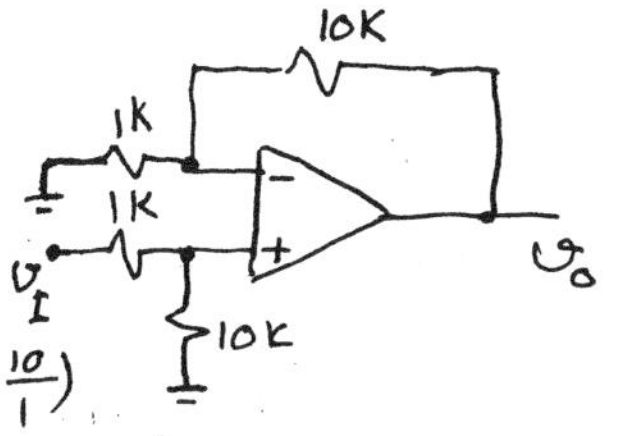

$$v_O = v_I \frac{10}{1 + 10}\left(1 + \frac{10}{1}\right)$$

$v_O = 10 v_I$

2.53

a) Source is connected directly.

$v_O = 10 \times \frac{1}{101} = 0.099V$

$i_L = \frac{v_O}{1K} = \frac{0.099}{1} = 0.099 mA$

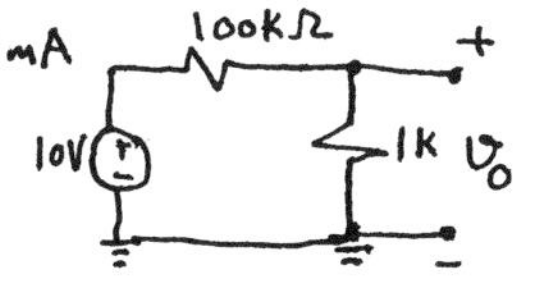

current supplied by the source is 0.099 mA.

b) inserting a buffer

$v_O = 10V$

$i_L = \frac{10V}{1K} = 10 mA$

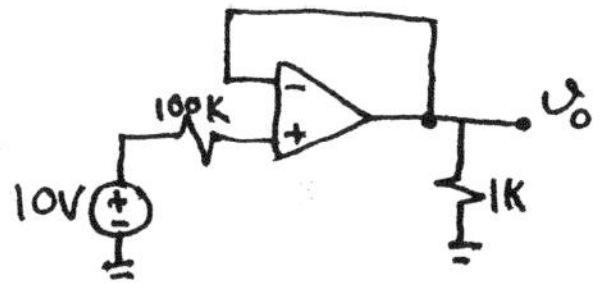

current supplied by the source is 0.

The load current i_L comes from the power supply of the op-amp.

2.54

$v_O = v_I - \frac{v_O}{A}$

$\frac{v_O}{v_I} = \frac{1}{1+\frac{1}{A}}$

error of Gain magnitude

$\frac{\left|\frac{v_O}{v_I} - 1\right|}{1} = -\frac{1}{A+1}$

A (V/V)	1000	100	10
$\frac{v_O}{v_I}$ (V/V)	0.999	0.990	0.909
Gain Error	−0.1%	−1%	−9.1%

2.55

for an inverting amplifier:

$R_i = R_1$, $G = -\frac{R_2}{R_1}$

for a non-inverting amplifier:

$R_i = \infty$ $G = 1 + \frac{R_2}{R_1}$

Case	Gain V/V	Rin	R_1	R_2
a	−10	10K	10K	100K
b	−1	100K	100K	100K
c	−2	50K	50K	100K
d	+1	∞	10K	10K
e	+2	∞	10K	10K
f	+11	∞	10K	100K
g	−0.5	10K	10K	5K

2.56

$A = 50\,V/V$ $1 + \frac{R_2}{R_1} = 10\,V/V$

if $R_1 = 10K\Omega \Rightarrow R_2 = 90K$

According to Eq. 2.11: $G = \frac{v_O}{v_I} = \frac{1+\frac{R_2}{R_1}}{1+\frac{1+R_2/R_1}{A}}$

$G = \frac{1+90/10}{1+\frac{1+90/10}{50}} = \frac{10}{1.2} = 8.33\,V/V$

In order to compensate the gain drop, we can shunt a resistor with R_1.

Compensated:

R_{sh}: $10 = \frac{1 + \left(\frac{90}{10} + \frac{90}{R_{sh}}\right)}{1 + \frac{1+\frac{90}{10}+\frac{90}{R_{sh}}}{50}} \Rightarrow$

$10 \times (510 R_{sh} + 90R_{sh} + 900) = 50 \times (10R_{sh} + 90R_{sh} + 900)$

$100R_{sh} = 3600 \Rightarrow R_{sh} = 36K\Omega$

if A = 100 then:

$G_{uncompensated} = \frac{1+\frac{90}{10}}{1+\frac{1+90/10}{100}} = \frac{10}{1.1} = 9.09\,V/V$

$G_{compensated} = \frac{1+\frac{90}{10}+\frac{90}{36}}{1+\frac{1+\frac{90}{10}+\frac{90}{36}}{100}} = \frac{12.5}{1.125} = 11.1\,V/V$

2.57

$G = \frac{G_0}{1+\frac{G_0}{A}}$, $\frac{G_0 - G}{G_0} \times 100 = \frac{G_0/A \times 100}{1+\frac{G_0}{A}} \leq x$

or $\frac{1+\frac{G_0}{A}}{G_0/A} \geq \frac{100}{x} \Rightarrow \frac{A}{G_0} \geq \underbrace{\left(\frac{100}{x} - 1\right)}_{F}$

$\Rightarrow A \geq G_0 F$ where $F = \frac{100}{x} - 1 \simeq \frac{100}{x}$

x	0.01	0.1	1	10
F	10^4	10^3	10^2	10

Thus for:

x = 0.01:	G_0 (V/V)	1	10	10^2	10^3	10^4
	A (V/V)	10^4	10^5	10^6	10^7	10^8

(10^7, 10^8: too high to be practical)

x = 0.1:	G_0 (V/V)	1	10	10^2	10^3	10^4
	A (V/V)	10^3	10^4	10^5	10^6	10^7
x = 1:	G_0 (V/V)	1	10	10^2	10^3	10^4
	A (V/V)	10^2	10^3	10^4	10^5	10^6
x = 10:	G_0 (V/V)	1	10	10^2	10^3	10^4
	A (V/V)	10	10^2	10^3	10^4	10^5

2.58

for non_inverting amplifier, Eq. 2.11:

$G = \frac{G_o}{1+\frac{G_o}{A}}$, $\varepsilon = \frac{G_o - G}{G_o} \times 100$

for inverting amplifier, Eq. 2.5:

$G = \frac{G_o}{1+\frac{1-G_o}{A}}$, $\varepsilon = \frac{G_o - G}{G_o} \times 100$

Case	G_o (V/V)	A (V/V)	G (V/V)	ε %
a	−1	10	−0.83	16
b	1	10	0.91	9
c	−1	100	−0.98	2
d	10	10	5	50
e	−10	100	−9	10
f	−10	1000	−9.89	1.1
g	+1	2	0.67	33

2.59

Refer to fig. P2.59, when potentiometer is set to the bottom:

$v_o = v_+ = -15 + \frac{30 \times 20}{20+100+20} = -10.714\,V$

when set to the top: $v_o = -15 + \frac{30 \times 120}{20+100+20} = 10.714V$

$\Rightarrow -10.714 \leq v_o \leq +10.714$

pot has 20 turn, each turn: $\Delta v_o = \frac{2 \times 10.714}{20} = 1.07V$

2.60

Refer to Fig. 2.16. Notice that similar to eq. 2.15 we have: $\frac{R_4}{R_3} = \frac{R_2}{R_1} = \frac{100}{10}$. therefore according to 2.16:

$v_o = \frac{R_2}{R_1} v_{Id} \Rightarrow A = \frac{R_2}{R_1} = 10\ V/V$

According to 2.20: $R_{id} = 2R_1 = 20k\Omega$

If $\frac{R_2}{R_1}$, $\frac{R_4}{R_3}$ were different by 1%:

$\frac{R_2}{R_1} = 0.99 \frac{R_4}{R_3}$

Refer to eq. 2.19: $A_{cm} = \frac{R_4}{R_4+R_3}\left(1 - \frac{R_2}{R_1} \cdot \frac{R_3}{R_4}\right)$

$A_{cm} = \frac{100}{100+10}(1-0.99) = 0.009$

$CMRR = 20 \log \frac{|A_d|}{|A_{cm}|}$, so let's calculate A_d

$A_d = \frac{v_o}{v_{id}}$ if we apply superposition:

$v_{O1} = -\frac{R_2}{R_1} v_{I1}$ $\quad v_{O2} = v_{I2} \frac{R_4}{R_3+R_4}\left(1+\frac{R_2}{R_1}\right)$

$v_o = v_{O2} + v_{O1} = v_{I2} \frac{R_4/R_3}{1+\frac{R_4}{R_3}}\left(1+\frac{R_2}{R_1}\right) - \frac{R_2}{R_1} v_{I1}$

Replace $\frac{R_2}{R_1} = 0.99 \frac{R_4}{R_3} \Rightarrow \frac{R_2}{R_1} = 0.99 \times \frac{100}{10} = 9.9$

$v_o = v_{I2} \frac{10}{1+10}(1+9.9) - v_{I1} 9.9 = 9.9(v_{I2} - v_{I1})$

$\frac{v_o}{v_{Id}} = 9.9 = A_d \Rightarrow CMRR = 20 \log \frac{9.9}{0.009} = 60.8$

CMRR = 60.8

2.61

If we assume $R_3 = R_1$, $R_4 = R_2$, then eq. 2.20: $R_{id} = 2R_1 \Rightarrow R_1 = \frac{20}{2} = 10k\Omega$ (Refer to Fig. 2.16)

a) $A_d = \frac{R_2}{R_1} = 1\ V/V \Rightarrow R_2 = 10k\Omega$

$R_1 = R_2 = R_3 = R_4 = 10k\Omega$

b) $A_d = \frac{R_2}{R_1} = 2\ V/V \Rightarrow R_2 = 20k\Omega = R_4$

$R_1 = R_3 = 10k\Omega$

c) $A_d = \frac{R_2}{R_1} = 100\ V/V \Rightarrow R_2 = 1M\Omega = R_4$

$R_1 = R_3 = 10k\Omega$

d) $A_d = \frac{R_2}{R_1} = 0.5\ V/V \Rightarrow R_2 = 5k\Omega = R_4$

$R_1 = R_3 = 10k\Omega$

2.62

Refer to Fig P2.62:

Cont.

Considering that $v_- = v_+$:

$$v_1 + \frac{v_o - v_1}{2} = \frac{v_2}{2} \Rightarrow v_o = v_2 - v_1$$

v_1 only: $R_I = \frac{v_1}{I} = R$

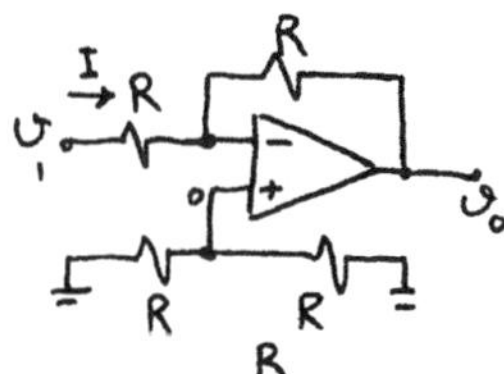

v_2 only: $R_I = \frac{v_2}{I} = 2R$

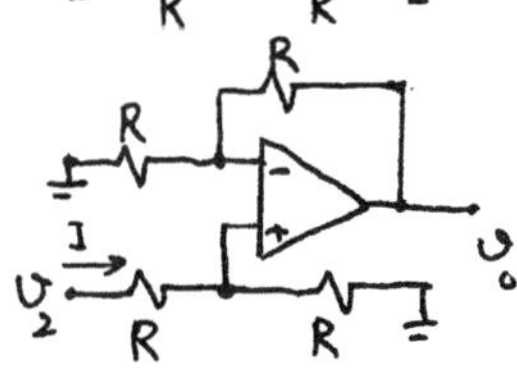

v_s between 2 terminals:

$R_I = \frac{v}{I} = 2R$

$v_+ = v_- = 0$

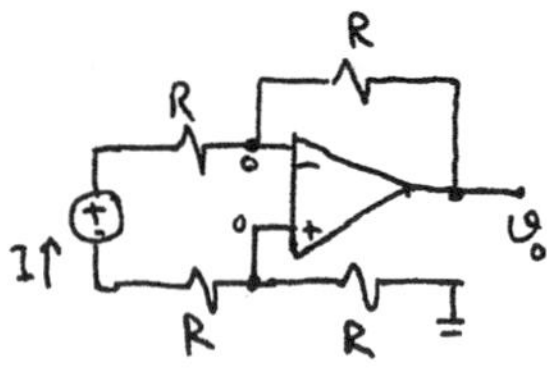

v_s connected to both v_1 & v_2:

$R_I = \frac{v}{I} = R$

$v_+ = v_- = \frac{v_I}{2}$

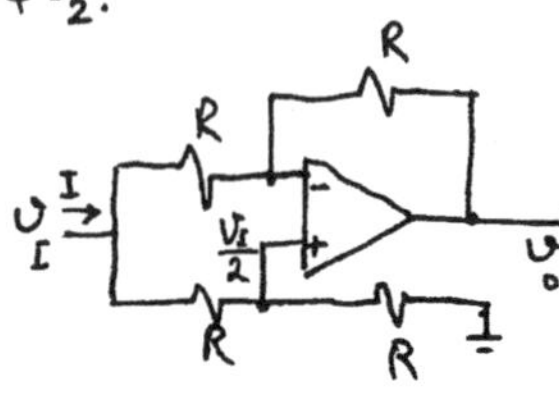

2.63

$v_+ = v_{CM}\frac{R_4}{R_3 + R_4}$

$v_+ = v_-$

$i_2 = \frac{v_{CM}}{R_3 + R_4}$

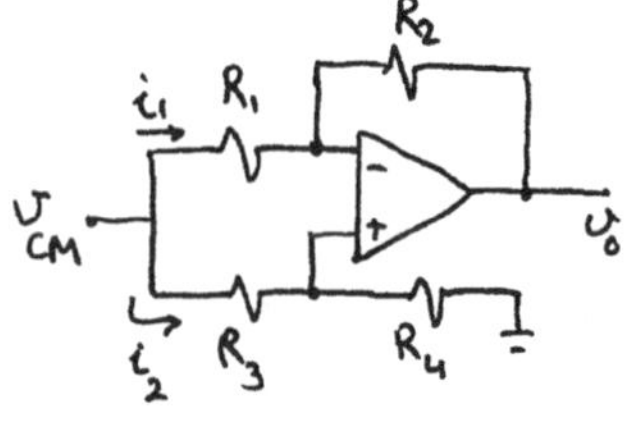

$$i_1 = \frac{v_{CM}}{R_1} - \frac{v_{CM} R_4}{R_3 + R_4} \times \frac{1}{R_1} = \frac{v_{CM}}{R_1}\frac{R_3}{R_3 + R_4}$$

$$i = i_1 + i_2 = \frac{v_{CM}}{R_1}\frac{R_3}{R_3 + R_4} + \frac{v_{CM}}{R_3 + R_4}$$

$$\frac{1}{R_I} = \frac{i}{v_{CM}} = \frac{1}{R_1}\frac{1}{\frac{R_4}{R_3} + 1} + \frac{1}{R_3 + R_4}$$

if we replace $\frac{R_4}{R_3}$ with $\frac{R_2}{R_1}$: $\left(\frac{R_4}{R_3} = \frac{R_2}{R_1}\right)$

$$\frac{1}{R_I} = \frac{1}{R_1 + R_2} + \frac{1}{R_3 + R_4} \Rightarrow R_I = (R_1 + R_2) \| (R_3 + R_4)$$

2.64

In order to have an ideal differential amp:

$$\frac{R_5 + R_1}{R_2} = \frac{R_5 + R_3}{R_4}$$

$$\frac{R_5/R_1 + 1}{R_2/R_1} = \frac{R_5/R_3 + 1}{R_4/R_3}$$

Since $\frac{R_2}{R_1} = \frac{R_4}{R_3}$:

$$\frac{R_5}{R_1} + 1 = \frac{R_5}{R_3} + 1 \Rightarrow R_1 = R_3 \Rightarrow R_2 = R_4$$

2.65

Refer to eq. 2.19 and Fig. P2.62:

$$A_{cm} = \frac{v_o}{v_{ICM}} = \frac{R_4}{R_3 + R_4}\left(1 - \frac{R_2}{R_1}\frac{R_3}{R_4}\right)$$

The worst case is when A_{cm} has its maximum value.

$$A_{cm} = \frac{1}{\frac{R_3}{R_4} + 1}\left(1 - \frac{R_2}{R_1}\frac{R_3}{R_4}\right)$$

Max $A_{cm} \Rightarrow \frac{R_3}{R_4}$ has to be at its minimum value and also $\frac{R_2}{R_1}$ has to be minimum.

$$\frac{100-x}{100+x} \leq \frac{R_3}{R_4} \leq \frac{100+x}{100-x} \qquad \frac{100-x}{100+x} \leq \frac{R_2}{R_1} \leq \frac{100+x}{100-x}$$

so if $\frac{R_3}{R_4} = \frac{100-x}{100+x}$ & $\frac{R_2}{R_1} = \frac{100-x}{100+x}$

$$A_{cm\,Max} = \frac{1}{\frac{100-x}{100+x} + 1}\left(1 - \frac{100-x}{100+x}\,\frac{100-x}{100+x}\right)$$

$$A_{cm\,Max} = \frac{1}{200}\,\frac{(100+x)^2 - (100-x)^2}{100+x} = \frac{2x}{100+x} \simeq \frac{x}{50}$$

x	0.1	1	5
$A_{cm\,Max}$	0.002	0.02	0.1

$CMRR = 20\log\left|\frac{A_d}{A_{cm}}\right|$. Now we have to calculate A_d based on values we chose for $R_1 - R_4$ that gave us $A_{cm\,Max}$.

$R_2 = R_3 = 100 - x \qquad R_1 = R_4 = 100 + x$

$v_o = v_{o1} + v_{o2}$ by applying superposition

$$v_o = -\frac{R_2}{R_1}v_1 + v_2\frac{R_4}{R_3 + R_4}\left(1 + \frac{R_2}{R_1}\right)$$

$$v_o = -\frac{100-x}{100+x}v_1 + v_2\frac{100+x}{200}\left(1 + \frac{100-x}{100+x}\right)$$

$$v_o = -\frac{100-x}{100+x}v_1 + v_2$$

if we consider $\frac{100-x}{100+x} \simeq 1 \Rightarrow \frac{v_o}{v_{id}} \simeq 1$

Cont.

$$CMRR = 20\log\frac{A_d}{A_{cm}} = 20\log\frac{1}{x/50} = 20\log\frac{50}{x}$$

x	0.1	1	5
CMRR	54db	34db	20db

2.66

Refer to Fig. 2.16 and eq. 2.19:

$$A_{cm} = \frac{R_4}{R_3+R_4}\left(1 - \frac{R_2}{R_1}\frac{R_3}{R_4}\right)$$

In order to calculate A_d, we use superposition principle:

$$v_o = v_{o1} + v_{o2} = -\frac{R_2}{R_1}v_1 + v_2\frac{R_4}{R_3+R_4}\left(1+\frac{R_2}{R_1}\right)$$

then replace $v_1 = v_{cm} - \frac{v_d}{2}$

$v_2 = v_{cm} + \frac{v_d}{2}$

$$v_o = \frac{-R_2}{R_1}v_{cm} + \frac{R_2}{R_1}v_d/2 + v_{cm}\frac{1+R_2/R_1}{1+\frac{R_3}{R_4}} + \frac{v_d}{2}\frac{1+R_2/R_1}{1+\frac{R_3}{R_4}}$$

$$v_o = \underbrace{\frac{R_2}{2R_1}\left[1+\frac{R_1/R_2+1}{R_3/R_4+1}\right]}_{A_d}v_d + \frac{R_2}{R_1}\left[-1+\frac{R_1/R_2+1}{R_3/R_4+1}\right]v_{cm}$$

$$CMRR = 20\log\left|\frac{A_d}{A_{cm}}\right| = 20\log\frac{\frac{R_2}{2R_1}\left[1+\frac{R_1/R_2+1}{R_3/R_4+1}\right]}{\frac{1}{\frac{R_3}{R_4}+1}\left(1-\frac{R_2}{R_1}\frac{R_3}{R_4}\right)}$$

$$CMRR = 20\log\left|\frac{\frac{1}{2}\frac{R_2}{R_1}\left[2+\frac{R_1}{R_2}+R_3/R_4\right]}{1-\frac{R_2}{R_1}\cdot\frac{R_3}{R_4}}\right|$$

$$CMRR = 20\log\left|\frac{1+\frac{1}{2}\frac{R_1}{R_2}+\frac{1}{2}\frac{R_3}{R_4}}{\frac{R_1}{R_2}-\frac{R_3}{R_4}}\right|$$

for worst case, or minimum CMRR we have to maximize the denominator, which means:

$R_1 = R_{1n}(1+\varepsilon)$ $\quad R_3 = R_{3n}(1-\varepsilon)$

$R_2 = R_{2n}(1-\varepsilon)$ $\quad R_4 = R_{4n}(1+\varepsilon)$

also: $\frac{R_{2n}}{R_{1n}} = \frac{R_{4n}}{R_{3n}} = K$

$$CMRR = 20\log\left|K\frac{1+\frac{1}{2K}\frac{1+\varepsilon}{1-\varepsilon}+\frac{1}{2K}\frac{1-\varepsilon}{1+\varepsilon}}{\frac{1+\varepsilon}{1-\varepsilon}-\frac{1-\varepsilon}{1+\varepsilon}}\right|$$

$$CMRR = 20\log\left|\frac{K(1-\varepsilon^2)+(1+\varepsilon^2)}{4\varepsilon}\right| \simeq 20\log\left|\frac{K+1}{4\varepsilon}\right|$$

for $\varepsilon^2 \ll 1$.

if $K = A_{d\,\text{Ideal}} = 100$, $\varepsilon = 0.01$

$$CMRR = 20\log\frac{101}{0.04} = 68\text{db}$$

2.67

$A_d = 100$

we assume $\frac{R_2}{R_1} = \frac{R_4}{R_3}$ then $A_d = \frac{R_2}{R_1} = K$

$K \simeq 100$

$R_{id} = 2R_1 = 20\text{K}\Omega \Rightarrow R_1 = 10\text{K}\Omega$

$CMRR = 80\text{db} = 20\log\frac{A_d}{A_{cm}} \Rightarrow \frac{A_d}{A_{cm}} = 10^4$

$\Rightarrow A_{cm} = 0.01$

$A_d = 100 \simeq \frac{R_2}{R_1} \Rightarrow R_2 = 1\text{M}\Omega$

Refer to P2.66: $\quad CMRR = 20\log\frac{K+1}{4\varepsilon}$

$CMRR = 10^4$

$\Rightarrow \varepsilon = 10^{-2}\times 0.25$

we assumed earlier $\frac{R_2}{R_1} = \frac{R_4}{R_3}$ then

$\frac{R_4}{R_3} \simeq 100 \Rightarrow$ if $R_3 = 10\text{K} \pm \varepsilon$

$\Rightarrow R_4 = 1\text{M}\Omega \pm \varepsilon$ $\quad \varepsilon = 0.25\%$

$R_2 = 1\text{M}\Omega \pm \varepsilon$

$R_1 = 10\text{K} \pm \varepsilon$

2.68

Referto Fig. P2.68 and Eq. 2.19:

$$A_{cm} = \frac{R_4}{R_3+R_4}\left(1-\frac{R_2}{R_1}\frac{R_3}{R_4}\right) = \frac{100}{100+100}\left(1-\frac{100.100}{100.100}\right)$$

$A_{cm} = 0$

Refer to 2.17: $\frac{R_2}{R_1} = \frac{R_4}{R_3}$

$\Rightarrow A_d = \frac{R_2}{R_1} = 1$

b) Since $A_{cm} = 0$, then if we apply V_{cm} to V_{I1} and V_{I2}, $v_o = 0$.

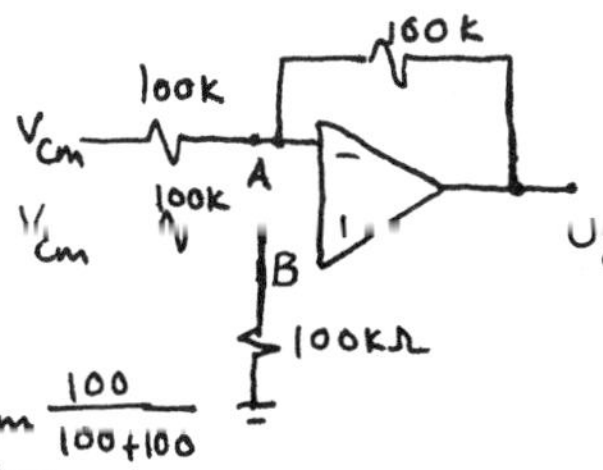

Therefore, $V_A = v_{cm}\frac{100}{100+100}$

$V_A = \frac{v_{cm}}{2}$

Similarly, $v_B = \frac{v_{cm}}{2}$

We know $V_A = V_B$ and $-2.5 \leq v_A \leq 2.5$

$\Rightarrow -5 \leq v_{cm} \leq 5$

c) we apply the superposition principle to calculate A_d.

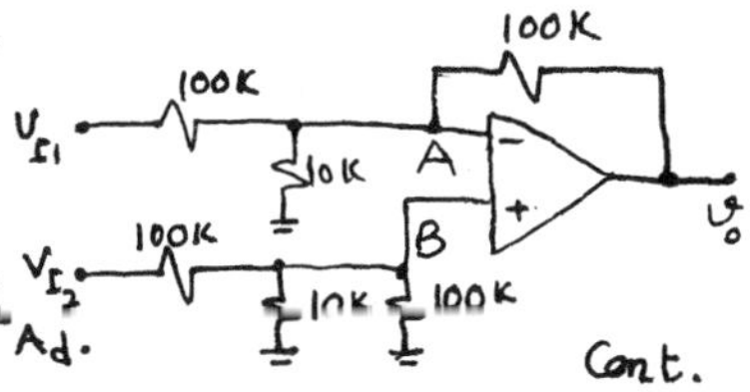

Cont.

v_{O1} is the output voltage when $v_{I2}=0$

v_{O2} is the output voltage when $v_{I1}=0$

$v_O = v_{O1} + v_{O2}$

$v_{O1} = \frac{-R_2}{R_1} v_{I1} = -v_{I1}$

$$v_{O2} = v_{I2} \frac{100k\Omega \| 10k}{100k \| 10k + 100} \left(1 + \frac{100k}{100k \| 10k}\right)$$

$v_{O2} = v_{I2} \times 1$

$\Rightarrow v_O = v_{O1} + v_{O2} = -v_{I1} + v_{I2} \Rightarrow A_d = 1$

Now we calculate A_{cm}:

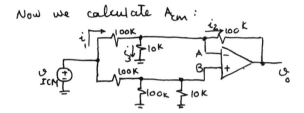

$$v_B = v_{ICM} \frac{100k \| 10k}{100k \| 10k + 100k}, \quad v_A = v_B$$

$$i_1 = \frac{v_{ICM} - v_A}{100k}$$

$v_O = v_A - 100k \times i_2$ and $i_2 = i_1 - i_3 = i_1 - \frac{v_A}{10k}$

$v_O = v_A - 100k\, i_1 + 10 \times v_A$

$v_O = v_A - v_{ICM} + v_A + 10 \times v_A$

$$v_A = v_B \Rightarrow v_O = v_{ICM}\left(-1 + 12 \frac{100k \| 10k}{100k \| 10k + 100k}\right)$$

$$\frac{v_O}{v_{ICM}} = A_{cm} = 0$$

Now we calculate v_{ICM} range:

$$-2.5 \leq v_B \leq 2.5 \Rightarrow -2.5 \leq v_{ICM} \times \frac{100k \| 10k}{100k \| 10k + 100k} \leq 2.5$$

$-30^V \leq v_{ICM} \leq 30^V$

2.69

Refer to Fig. P2.69; we use superposition:

$v_O = v_{O1} + v_{O2}$

calculate v_{O1}: $v_+ = \frac{\beta v_{O1}}{2} = v_-$

$$\frac{v_1 - \frac{\beta v_{O1}}{2}}{R} = \frac{\frac{\beta v_{O1}}{2} - v_O}{R} \Rightarrow v_{O1} = \frac{v_1}{\beta - 1}$$

Calculate v_{O2}:

$$v_- = \frac{v_{O2}}{2} = v_+ \Rightarrow v_2 - \frac{v_{O2}}{2} = \frac{v_{O2}}{2} - \beta v_{O2}$$

$$\Rightarrow v_{O2} = \frac{v_2}{1-\beta}$$

$$v_O = v_{O1} + v_{O2} = \frac{v_1}{\beta - 1} + \frac{v_2}{1-\beta} = \frac{1}{1-\beta}(v_2 - v_1)$$

$$A_d = \frac{v_O}{v_2 - v_1} = \frac{1}{1-\beta}$$

$$A_d = 10\,V/V \Rightarrow \beta = 0.9 = \frac{R_5}{R_5 + R_6}$$

$R_{id} = 2R = 2M\Omega \Rightarrow R = 1M\Omega$

$R_5 + R_6 \leq \frac{R}{100} \Rightarrow R_5 + R_6 \leq 10k\Omega$

$R_5 = 6.8k\Omega \quad R_6 = 680\Omega \Rightarrow \beta = \frac{6.8}{6.8 + 0.68} \cong 0.9$

2.70

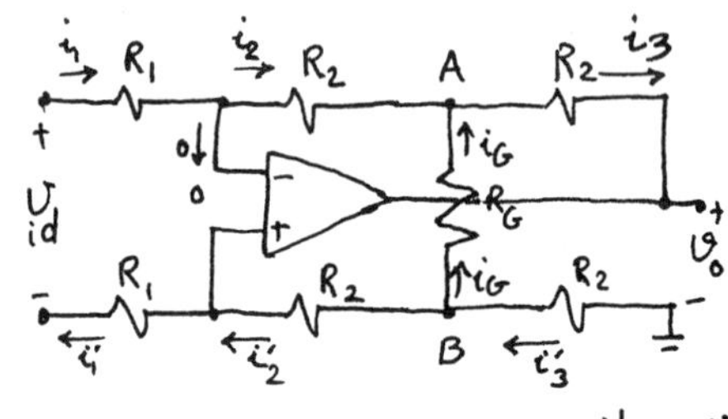

$v_+ = v_-$ So we can consider v_+, v_- a virtual short: $i_1 = v_{id}/2R_1 \Rightarrow i_2 = \frac{v_{id}}{2R_1}$

$i'_1 = i'_2 = \frac{v_{id}}{2R_1}$

then: $i_2 R_2 + v_{AB} + i'_2 R_2 = 0 \Rightarrow v_{AB} = \frac{-v_{id}}{R_1} R_2$

$$i_G = \frac{v_{id}}{R_G} \times \frac{R_2}{R_1}$$

$$i_3 = i_2 + i_G = \frac{v_{id}}{2R_1} + \frac{v_{id}}{R_G}\frac{R_2}{R_1}$$

$i'_3 = i_G + i'_2 = i_3$

$\Rightarrow v_O = -[i'_3 R_2 + v_{BA} + i_3 R_2]$

$v_O = -[2 i_3 R_2 + v_{BA}]$

$$v_O = -\left[\frac{2 v_{id}}{2R_1} R_2 + 2 v_{id} \frac{R_2}{R_1}\frac{R_2}{R_G} + \frac{v_{id}}{R_1} R_2\right]$$

$$\frac{v_O}{v_{id}} = A_d = -2\frac{R_2}{R_1}\left[1 + \frac{R_2}{R_G}\right]$$

2.71

a)

Refer to Eq. 2.17: $A_d = \frac{R_2}{R_1} = 1$. Connect C and O together.

Cont.

a)

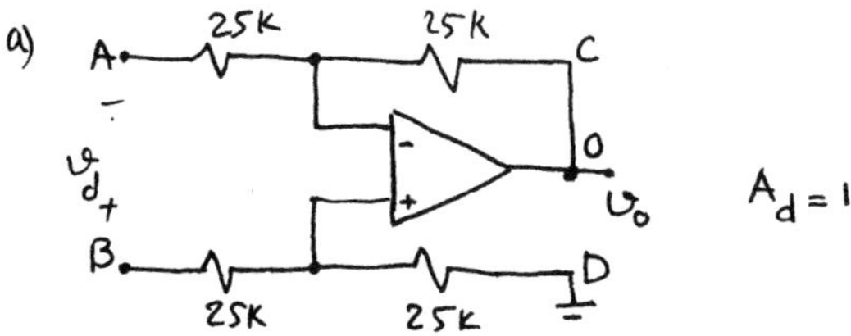

$A_d = 1$

b) $\frac{v_o}{v_i} = -1\ V/V$

i)

ii) $\frac{v_o}{v_i} = +1\ V/V$

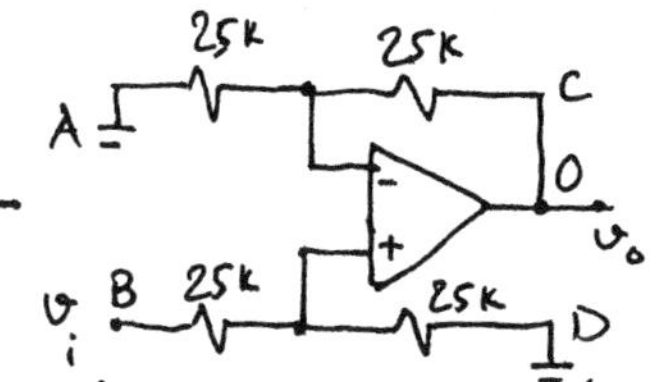

The circuit on the left ideally has infinite input resistance.

iii) $\frac{v_o}{v_i} = +2\ V/V$

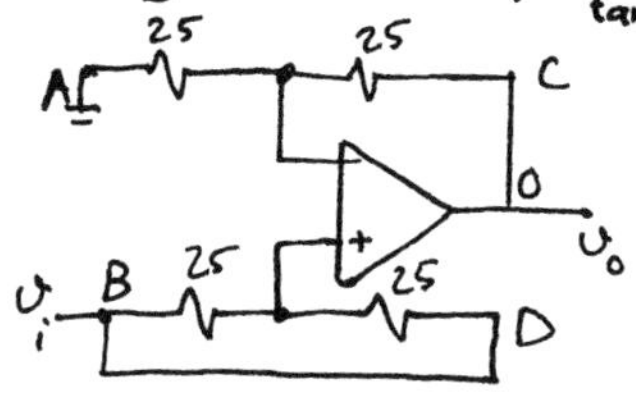

iv) $\frac{v_o}{v_i} = +\frac{1}{2}\ V/V$

$v_+ = \frac{v_i}{2} = v_o$

$\Rightarrow \frac{v_o}{v_i} = \frac{1}{2}$

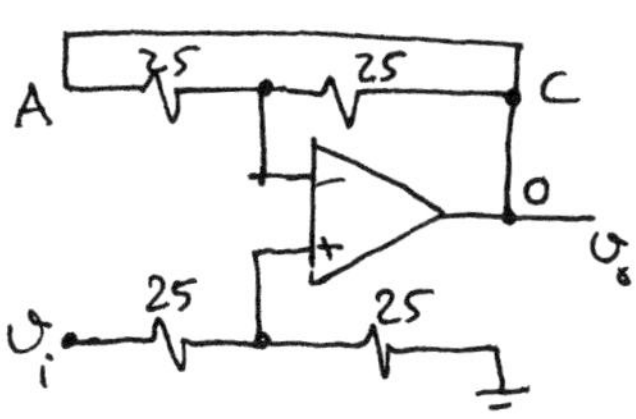

2.72

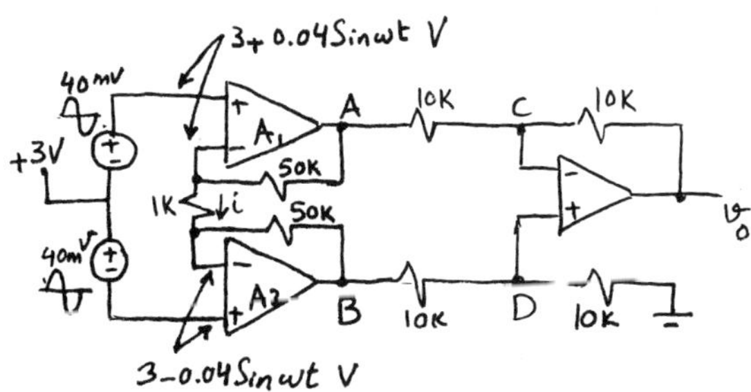

$i = \frac{3+0.04\sin\omega t-(3-0.04\sin\omega t)}{1K} = 0.08\sin\omega t,\ mA$

$v_A = 3+0.04\sin\omega t + 50K\times i = 3+4.04\sin\omega t,\ V$

$v_B = 3-0.04\sin\omega t - 50K\times i = 3-4.04\sin\omega t,\ V$

$v_C = v_D = \frac{1}{2}v_B = 1.5-2.02\sin\omega t,\ V$

$v_o = v_B - v_A = -8.08\sin\omega t,\ V$

2.73

Refer to Fig. 2.20.a.

The gain of the first stage is: $(1+\frac{R_2}{R_1}) = 101$

If the opamps of the first stage saturate at $\pm 14\ V$: $-14 \le v_1 \le +14V \Rightarrow -14 \le 101\, v_{icm} \le +14$

$\Rightarrow -0.14 \le v_{ICM} \le 0.14$

As explained in the text, the disadvantage of circuit in Fig. 2.20.a is that v_{ICM} is amplified by a gain equal to $v_{id}, (1+\frac{R_2}{R_1})$ in the first stage and therefore a very small v_{icm} range is acceptable to avoid saturation.

b) In Fig. 2.20b, when v_{ICM} is applied, v_- for both A_1 & A_2 is the same and therefore no current flows through $2R_1$. This means voltage at the output of A_1 and A_2 is the same as V_{ICM}.

$-14 \le v_o \le 14 \Rightarrow -14 \le v_{ICM} \le 14$

This circuit allows for bigger range of v_{icm}.

2.74

$v_{i1} = v_{cm} - v_d/2$

$v_{i2} = v_{cm} + v_d/2$

Refer to Fig. 2.20.a.

output of the first stage: $(1+\frac{R_2}{R_1})(v_{cm} - v_d/2)$

$v_{o1} = (1+\frac{R_2}{R_1})(v_{cm} - v_d/2)$

$v_{o2} = (1+\frac{R_2}{R_1})(v_{cm} + v_d/2)$

$v_{o2} - v_{o1} = (1+\frac{R_2}{R_1})\, v_d \Rightarrow A_{d(1)} = 1+\frac{R_2}{R_1}$

$\frac{v_{o2}+v_{o1}}{2} = (1+\frac{R_2}{R_1})\, v_{cm} \Rightarrow A_{cm(1)} = 1+\frac{R_2}{R_1}$

$CMRR = 20\log\left|\frac{A_d}{A_{cm}}\right| = 0$ (First stage)

Now Consider Fig. 2.20.b

$v_{o1} = v_{i1} + R_2 \times \frac{(v_{i1}-v_{i2})}{2R_1}$

$v_{o1} = v_{cm} - v_d/2 + \frac{R_2}{2R_1}(-v_d)$

Cont.

$$v_{o1} = v_{cm} - \frac{v_d}{2}\left(1+\frac{R_2}{R_1}\right)$$

$$v_{o2} = v_{i2} - R_2 \times \frac{v_{i1}-v_{i2}}{2R_1} = v_{cm} + \frac{v_d}{2} + R_2 \cdot \frac{v_d}{2R_1}$$

$$v_{o2} = v_{cm} + \frac{v_d}{2}\left(1+\frac{R_2}{R_1}\right)$$

$$v_{o2} - v_{o1} = v_d\left(1+\frac{R_2}{R_1}\right) \Rightarrow A_{d(1)} = 1+\frac{R_2}{R_1}$$

$$\frac{v_{o2}+v_{o1}}{2} = v_{cm} \Rightarrow A_{cm(1)} = 1$$

$$CMRR = 20\log\left|\frac{A_d}{A_{cm}}\right| = 20\log\left(1+\frac{R_2}{R_1}\right)$$

In 2.20.b, the common mode voltage is not amplified and it is only propagated to the outputs of the first stage.

2.75

Refer to eq. 2.22:

$$A_d = \frac{R_4}{R_3}\left(1+\frac{R_2}{R_1}\right) = \frac{100^K}{100K}\left(1+\frac{100K}{5^K}\right) = 21\ V/V$$

$A_{cm} = 0$

$$CMRR = 20\log\left|\frac{A_d}{A_{cm}}\right| = \infty$$

If all resistors are $\pm 1\%$:

$A_d \simeq 21$

In order to calculate A_{cm}, apply V_{cm} to both inputs and note that V_{cm} will appear at both output terminals of the first stage. Now we can evaluate v_o by analyzing the second stage as was done in problem 2.65. In P2.65 we showed that if each 100K resistor has $\pm x\%$ tolerance, A_{cm} of the differential amplifier is: $A_{cm} = \frac{v_o}{v_{cm}} = \frac{x}{50}$. Therefore the overall A_{cm} is also $\frac{x}{50}$.

$x = 1 \Rightarrow A_{cm} = \frac{1}{50} = 0.02$

$$CMRR = 20\log\frac{21}{0.02} = 60\,db$$

If $2R_1 = 1K\Omega$: $A_d = \frac{R_4}{R_3}\left(1+\frac{R_2}{R_1}\right) = 201\ V/V$

$A_{cm} = 0.02$ unchanged

$$CMRR = 20\log\frac{201}{0.02} = 80\,db$$

Conclusion: Large CMRR can be achieved by having relatively Large A_d in the first stage.

2.76

$A_{d(2)}$ of the second stage is $\frac{R_4}{R_3} = 0.5$

$R_4 = 100K\Omega$, $R_3 = 200K\Omega$

we use a series configuration of R_{1F} (Fixed) and R_1 (pot): $R_1 = 100K$ pot

Minimum gain $= 0.5\left(1+\frac{R_2}{R_1}\right) = 0.5\left(1+\frac{R_2}{\frac{100^K+R_1}{2}}\right)$

$1 \le A_d \le 100 \Rightarrow 1 = 0.5\left(1+\frac{2R_2}{R_{1F}+100K}\right)$

$\Rightarrow R_{1F} + 100 = 2R_2$ ①

Maximum gain $= 100 = 0.5\left(1+\frac{R_2}{R_{1F}/2}\right) \Rightarrow$

$2R_2 = 199R_{1F}$ ②

①, ② $\Rightarrow R_{1F} = 0.505\,K\Omega \simeq 0.5K\Omega$

$R_2 = 50.25\,K\Omega \simeq 50K\Omega$

2.77

a) $\frac{v_B}{v_A} = 1+\frac{20}{10} = 3\ V/V$, $\frac{v_C}{v_A} = -\frac{30}{10} = -3\ V/V$

b) $v_O = v_B - v_C = 6\,v_A \Rightarrow \frac{v_O}{v_A} = 6\ V/V$

$v_A(V)$

0.5

-0.5

t

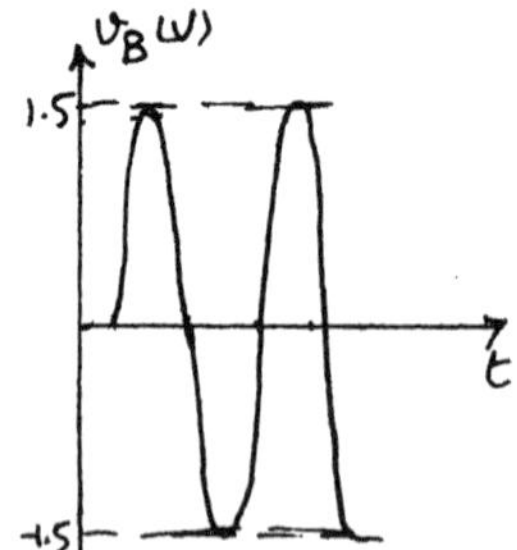

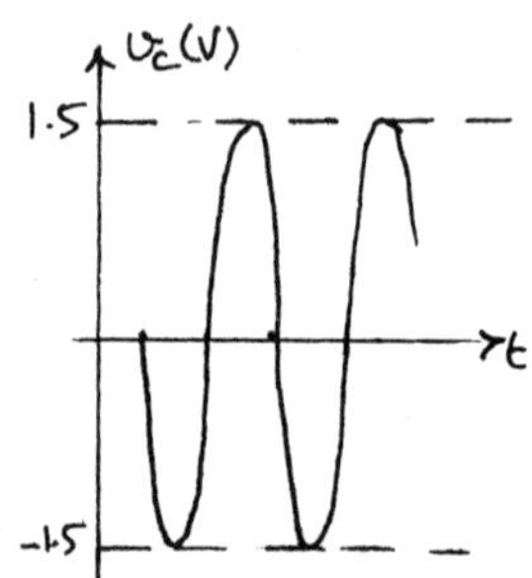

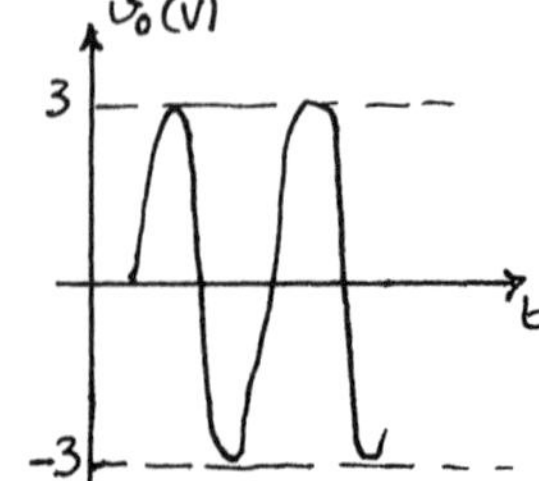

c) v_B and v_C can be $\pm 14^V$ or $28V$ p-p.

$-28 \le v_o \le 28$ or 56^V p-p.

$v_{o\,rms} = 19.8\,V = \frac{28}{\sqrt{2}}$

2.78

Refer to Fig. P.2.78.a:
Since the inputs of the op-amp donot draw any current, v_I appears across R:

$i_o = \frac{v_I}{R}$

Fig. P2.78.B

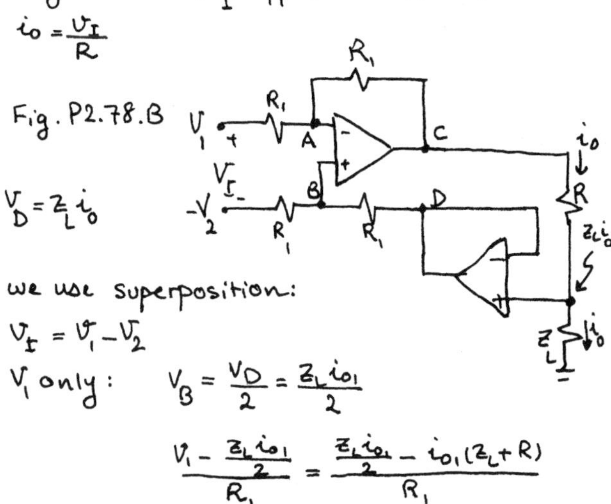

$V_D = Z_L i_o$

we use superposition:

$v_I = v_1 - v_2$

v_1 only: $V_B = \frac{V_D}{2} = \frac{Z_L i_{o1}}{2}$

$$\frac{V_1 - \frac{Z_L i_{o1}}{2}}{R_1} = \frac{\frac{Z_L i_{o1}}{2} - i_{o1}(Z_L + R)}{R_1}$$

$\Rightarrow V_1 = i_{o1} R \Rightarrow i_{o1} = \frac{V_1}{R}$

Now if only $(-v_2)$ is applied:

$V_B = \frac{-V_2 + Z_L i_{o2}}{2}$, $V_A = \frac{i_{o2} \times (R + Z_L)}{2}$

$v_A = v_B \Rightarrow -v_2 + Z_L i_{o2} = i_{o2} R + i_{o2} Z_L$

$-v_2 = i_{o2} R \Rightarrow i_{o2} = \frac{-V_2}{R}$

The total current due to both sources is:

$i_o = i_{o1} + i_{o2} = \frac{v_1}{R} - \frac{v_2}{R} = \frac{v_I}{R}$

The circuit in Figure P2.78(a) has ideally infinite input resistance, and it requires that both terminals of Z_L be availabe, while the other circuit has finite input resistance with one side of Z_L grounded.

2.79

A_o	f_b (HZ)	f_t (HZ)
10^5	10^2	10^7
10^6	1	10^6
10^5	10^3	10^8
10^7	10^{-1}	10^6
2×10^5	10	2×10^6

eq. 2.28:

$\omega_t = A_o \omega_b$

$\Rightarrow f_t = A_o f_b$

2.80

Eq. 2.25: $A = \frac{A_o}{1 + j\,\omega/\omega_b} \Rightarrow |A| = \frac{|A_o|}{\sqrt{1 + (\frac{f}{f_b})^2}}$

$A_o = 86\,db$, $A = 40\,db$ @ $f = 100\,KHZ$

$20 \log \sqrt{1 + (\frac{f}{f_b})^2} = 20 \log \frac{|A_o|}{|A|} = 20 \log A_o - 20 \log A$

$= 86 - 40 = 46\,db$

$1 + (\frac{100\,KHZ}{f_b})^2 = (199.5)^2 \Rightarrow f_b = 0.501\,KHZ$

$f_b = 501\,HZ$

$f_t = A_o f_b = \underbrace{1.995 \times 10^4}_{86db} \times 501 = 9.998\,MHZ \simeq 10\,MHZ$

2.81

$A_o = 8.3 \times 10^3\,V/V$

Eq. 2.25: $A = \frac{A_o}{1 + j\frac{f}{f_b}}$

$f_t = A_o f_b$

$5.1 \times 10^3 = \frac{8.3 \times 10^3}{\sqrt{1 + (\frac{100KH}{f_b})^2}} \Rightarrow 1 + (\frac{100\,KHZ}{f_b})^2 = 2.65$

$f_b = 60.7\,KHZ$

$f_t = A_o f_b = 8.3 \times 10^3 \times 60.7^K = 503\,MHZ$

2.82

we have:

$A_{o(db)} = 20db + A_{(db)}$ $20db = 20 \log 10 \Rightarrow A_o = 10 \times A$

a) $A_o = 10 \times 3 \times 10^5 = 3 \times 10^6\,HZ\ V/V$

$A = \frac{A_o}{1 + jf/f_b} \Rightarrow |1 + j f/f_b| = \frac{A_o}{A} = 10 \Rightarrow \frac{6 \times 10^2}{f_b} = \sqrt{99}$

$\Rightarrow f_b = 60.3\,HZ$

$f_t = A_o f_b = 3 \times 10^6 \times 60.3 = 180.9\,MHZ$

b) $A = 50 \times 10^5\,V/V \Rightarrow A_o = 10 \times 50 \times 10^5 = 50 \times 10^6\,V/V$

$|1 + \frac{jf}{f_b}| = \frac{A_o}{A} = 10 \Rightarrow \frac{10^{HZ}}{f_b} = \sqrt{99} \Rightarrow f_b = 1\,HZ$

$f_t = A_o f_b = 50\,MHZ$

c) $A = 1500\,V/V \Rightarrow A_o = 15000\,V/V$

Cont.

$$\left|1+\frac{jf}{f_b}\right| = 10 \Rightarrow \frac{0.1\times10^6}{f_b} = \sqrt{99} \Rightarrow f_b = 10KHz$$
$f_t = 15000\times10^K = 150MHz$

d) $A_o = 10\times100 = 1000\ V/V_q$
$$\left|1+\frac{jf}{f_b}\right| = 10 \Rightarrow \frac{0.1\times10^6}{f_b} = \sqrt{99} \Rightarrow f_b = 10\ MHz$$
$f_t = 1000\times10^{MHz} = 10GHz$

e) $A_o = 25\ V/mV \times 10 = 25\times10^4\ V/V$
$$\left|1+\frac{jf}{f_b}\right| = 10 \Rightarrow \frac{25KHz}{f_b} = \sqrt{99} \Rightarrow f_b = 2.51^{KHz}$$
$f_t = A_o f_b = 25\times10^4 \times 2.51\times10^3 = 627.5MHz$

2.83

$G_{nom} = -\frac{R_2}{R_1} = -20 \qquad A_o = 10^4\ V/V \qquad f_t = 10^6 Hz$

Eq. 2.35 : $\omega_{3db} = \frac{\omega_t}{1+R_2/R_1} = \frac{2\pi\times10^6}{1+20} = 2\pi\times47.6\ KHz$

$f_{3db} = 47.6KHz$

Eq. 2.34 : $\frac{V_o}{V_i} \simeq \frac{-R_2/R_1}{1+\frac{s}{\omega_t/(1+\frac{R_2}{R_1})}} = \frac{-20}{1+\frac{2\pi s}{2\pi\times10^6}}$

$f = 0.1 f_{3db} \Rightarrow \left|\frac{V_o}{V_i}\right| = \frac{-20}{\sqrt{1+(0.1)^2}} = -19.9\ V/V$

$f = 10 F_{3db} \Rightarrow \left|\frac{V_o}{V_i}\right| = \frac{-20}{\sqrt{1+100}} = 1.99\ V/V$

2.84

$1+\frac{R_2}{R_1} = 100\ V/V \quad , f_t = 20MHz$

$f_{3db} = \frac{f_t}{1+\frac{R_2}{R_1}} = 200\ KHz$

$G(j\omega) = \frac{100}{1+j\,f/f_{3db}} \Rightarrow \varphi = -\tan^{-1}\frac{f}{f_{3db}} =$

$\varphi = -6^\circ \Rightarrow f = f_{3db}\times\tan 6^\circ = 21KHz$

$\varphi = -84^\circ \Rightarrow f = f_{3db}\times\tan 84^\circ = 1.9MHz$

2.85

a) $\frac{-R_2}{R_1} = -100\ V/V \quad , f_{3db} = 100KHz$

Eq. 2.35 : $\omega_t = \omega_{3db}\left(1+\frac{R_2}{R_1}\right) \Rightarrow f_t = 100^K\times101 = 10.1\ MHz$

b) $1+\frac{R_2}{R_1} = 100\ V/V \qquad f_{3db} = 100KHz$

$f_t = f_{3db}\left(1+\frac{R_2}{R_1}\right) = 10\ MHz$

c) $1+\frac{R_2}{R_1} = 2\ V/V \qquad f_{3db} = 10MHz$

$f_t = 10MHz\times2 = 20MHz$

d) $-\frac{R_2}{R_1} = -2\ V/V \qquad f_{3db} = 10MHz$

$f_t = 10MHz(1+2) = 30MHz$

e) $-\frac{R_2}{R_1} = -1000\ V/V \qquad f_{3db} = 20KHz$

$f_t = 20K(1+1000) = 20.02MHz$

f) $1+\frac{R_2}{R_1} = 1\ V/V \qquad f_{3db} = 1MHz$

$f_t = 1M\times1 = 1MHz$

g) $-\frac{R_2}{R_1} = -1 \qquad f_{3db} = 1MHz$

$f_t = 1M(1+1) = 2MHz$

2.86

$1+\frac{R_2}{R_1} = 100\ V/V \qquad f_{3db} = 8KHz$

$f_t = 8\times100 = 800KHz$

for $f_{3db} = 20KHz$: $G_o = \frac{800}{20} = 40\ V/V$

2.87

$f_{3db} = f_t = 1MHz$

$|G| = \frac{1}{\sqrt{1+\left(\frac{f}{f_{3db}}\right)^2}} = \frac{1}{\sqrt{1+f^2}} \qquad f\ \text{in MHz}$

$|G| = 0.99 \Rightarrow f = 0.142MHz$

The follower behaves like a low_pass STC circuit with a time constant $\tau = \frac{1}{\omega_{3db}}$

Thus: $\tau = \frac{1}{2\pi\times10^6} = \frac{1}{2\pi}\ \mu s$

$t_r = 2.2\tau = 0.35\mu s$ (Refer to Appendix F)

2.88

$1+\frac{R_2}{R_1}=10\ V/V \qquad R_1=1K\Omega \quad R_2=9K\Omega$

If we consider 5τ the time that it takes for the output voltage to reach 99% of its final value, then: $5\tau=100ns \Rightarrow \tau=20ns$

$\tau=\frac{1}{\omega_{3db}} \Rightarrow \omega_{3db}=50\times10^6 \Rightarrow f_{3db}=7.96MHZ$

$f_t=(1+\frac{R_2}{R_1})f_{3db}=10\times7.96=79.6\ MHZ$

2.89

a) Assume two identical stages, each with a gain function: $G=\frac{G_0}{1+j\frac{\omega}{\omega_1}}=\frac{G_0}{1+jf/f_1}$

$G=\frac{G_0}{\sqrt{1+(\frac{f}{f_1})^2}}$

Overall gain of the cascade is $\frac{G_0^2}{1+(\frac{f}{f_1})^2}$

The gain will drop by 3db when:

$1+(\frac{f_{3db}}{f_1})^2=\sqrt{2}$, Note $3db=20\log\sqrt{2}$

$f_{3db}=f_1\sqrt{\sqrt{2}-1}$

b) $40db=20\log G_0 \Rightarrow G_0=100=1+\frac{R_2}{R_1}$

$f_{3db}=\frac{f_t}{1+\frac{R_2}{R_1}}=\frac{1\ MHZ}{100}=10KHZ$

c) Each stage should have 20db gain or $1+\frac{R_2}{R_1}=10$ and therefore a 3db frequency of: $f_1=\frac{10^6}{10}=10^5\ HZ$.

The overall $f_{3db}=10^5\sqrt{\sqrt{2}-1}=64.35KHZ$

which is 6 times greater than the bandwidth achieved using single op-amp. (case b above)

2.90

$f_t=100\times5=500MHZ$. if single op-amp is used.

with op-amp that has only $f_t=40MHZ$, the possible closed loop gain at 5MHZ is:

$|A|=\frac{40}{5}=8\ V/V$.

To obtain an overall gain of 100, three such amplifiers cascaded, would be required. Now, if each of the 3 stages, has a low-frequency (d) closed loop gain K, then its 3db frequency will be $\frac{40}{K}$ MHZ. Thus for each stage the closed loop gain is: $|G|=\frac{K}{\sqrt{1+(\frac{f}{\frac{40}{K}})^2}}$

which at f=5MHZ becomes:

$|G_{5MHZ}|=\frac{K}{\sqrt{1+(\frac{K}{8})^2}}$.

The overall gain of 100: $100=\left[\frac{K}{\sqrt{1+(\frac{K}{8})^2}}\right]^3$

$K=5.7$

Thus for each cascade stage: $f_{3db}=\frac{40}{5.7}$

$f_{3db}=7MHZ$

The 3-db frequency of the overall amplifier, f_1, can be calculated as:

$\left[\frac{5.7}{\sqrt{1+(\frac{f_1}{7})^2}}\right]^3=\frac{(5.7)^3}{\sqrt{2}} \Rightarrow f_1=3.6MHZ$

2.91

a) $\frac{R_2}{R_1}=K \qquad f_{3db}=\frac{f_t}{1+\frac{R_2}{R_1}}=\frac{f_t}{1+K}$

$GBP=Gain\times f_{3db}$

$GBP=K\frac{f_t}{1+K}$

b) $1+\frac{R_2}{R_1}=K \qquad f_{3db}=\frac{f_t}{K}$

$GBP=K\frac{f_t}{K}=f_t$

The non-inverting amplifier realizes a higher GBP and it's independent of K.

2.92

To find f_{3db} we use superposition:

set V2 = 0

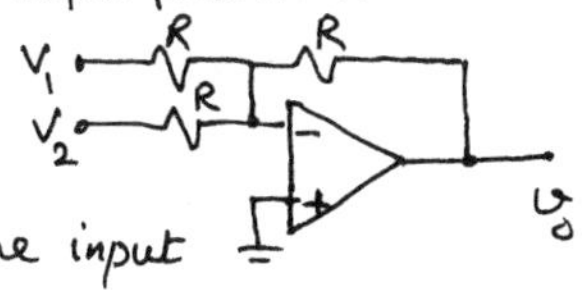

Now using Thevenin's Theorem to simplify the input circuit results in:

Cont.

$$\frac{V_o}{V_1/2} = \frac{-R/R/2}{1+s\,\frac{1+R/R/2}{\omega_t}}$$

which gives:

$$\frac{V_o}{V_1} = \frac{-1}{1+S/(\omega_t/3)}$$

Thevenin's equivalent

$f_{3db} = \frac{f_t}{3}$. Similar results can be obtained for $\frac{V_o}{V_2}$.

2.93

The peak value of the largest possible sine wave that can be applied at the input without output clipping is: $\frac{\pm 12V}{100} = 0.12V = 120mV$

rms value $= \frac{120}{\sqrt{2}} = 85mV$

2.94

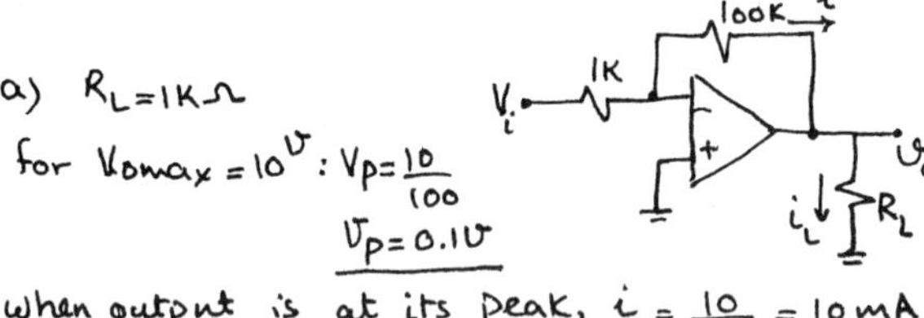

a) $R_L = 1K\Omega$

for $V_{omax} = 10^V$: $V_p = \frac{10}{100}$

$V_p = 0.1V$

when output is at its peak, $i_L = \frac{10}{1K} = 10mA$

$i = \frac{10}{100K} = 0.1mA$. therefore $i_o = 10 + 0.1 = 10.1$ mA is well under $i_{omax} = 20mA$.

b) $R_L = 100\Omega$

If output is at its peak: $i_L = \frac{10^V}{0.1} = 100mA$ which exceeds $i_{omax} = 20mA$. Therefore Vo cannot go as high as 10^V. instead:

$20mA = \frac{V_o}{100\Omega} + \frac{V_o}{100K} \Rightarrow V_o = \frac{20}{10.01} = 2^V$

$V_p = \frac{2}{100} = 0.02V = 20mV$

c) $R_L = ?$ $\quad i_{omax} = 20mA = \frac{10^V}{R_{min}} + \frac{10^V}{100^K}$

$20 - 0.1 = \frac{10}{R_{Lmin}} \Rightarrow R_{Lmin} = 502\Omega$

2.95

The output is triangular with the slew rate of $20V/\mu s$. In order to reach $3V$, it takes $\frac{3}{20}\mu s = 0.15\mu s = 150ns$.

Therefore the minimum pulse width is 150ns.

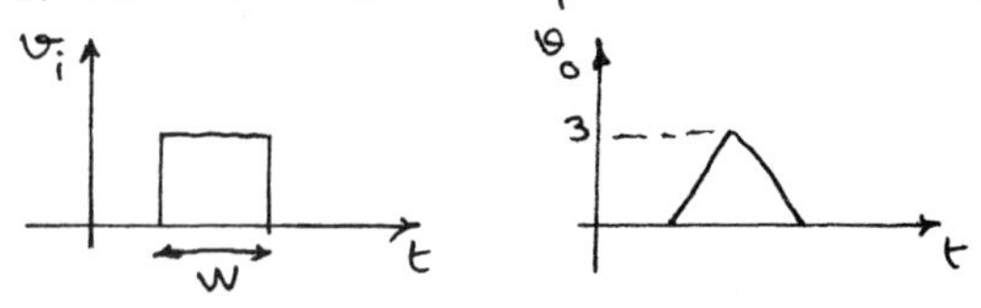

2.96

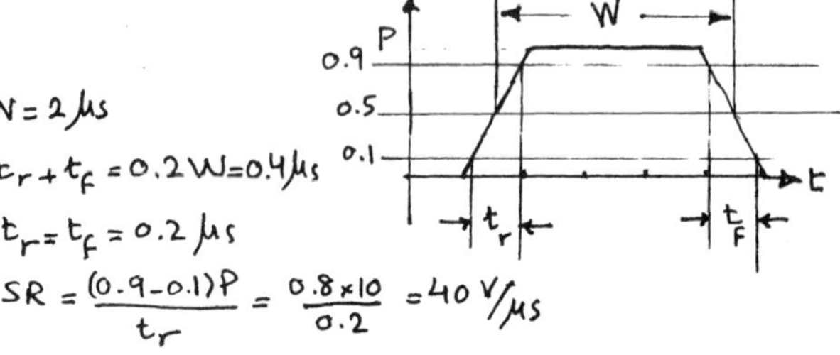

$W = 2\mu s$

$t_r + t_f = 0.2W = 0.4\mu s$

$t_r = t_f = 0.2\mu s$

$SR = \frac{(0.9-0.1)P}{t_r} = \frac{0.8 \times 10}{0.2} = 40 V/\mu s$

2.97

Slope of the triangle wave $= \frac{20V}{T/2} = SR$

Thus $\frac{20}{T} \times 2 = 10V/\mu s$

$\Rightarrow T = 4\mu s$ or $f = \frac{1}{T} = 250KHZ$

for a Sine wave $v_o = \hat{v}_o \sin(2\pi \times 250 \times 10^3 t)$

$\left.\frac{dv_o}{dt}\right|_{max} = 2\pi \times 250 \times 10^3 \hat{v}_o = SR$

$\Rightarrow \hat{v}_o = \frac{10 \times 10^6}{2\pi \times 10^3 \times 250} = 6.37V$

2.98

$v_o = 10 \sin\omega t \Rightarrow \frac{dv_o}{dt} = 10\omega \cos\omega t \Rightarrow \left.\frac{dv_o}{dt}\right|_{max} = 10\omega$

The highest frequency at which this output is possible is that for which:

$\left.\frac{dv_o}{dt}\right|_{max} = SR \Rightarrow 10\omega_{max} = 60 \times 10^{+6} \Rightarrow \omega_{max} = 6 \times 10^5$

$\Rightarrow f_{max} = 95.5 KHZ$.

2.99

a) $v_i = 0.5$, $V_o = 10 \times 0.5 = 5V$

Cont.

Output distortion will be due to slew Rate limitation and will occur at the frequency for which $\left.\frac{dv_o}{dt}\right|_{max} = SR$

$$\omega_{max} \times 5 = \frac{1}{10^{-6}} = 2\times10^5 \text{ rad/s} \Rightarrow f_{max} = 31.8\text{KHz}$$

b) The output will distort at the value of V_i that results in $\left.\frac{dv_o}{dt}\right|_{max} = SR$.

$v_o = 10\,v_i \sin 2\pi\times20\times10^3$

$\left.\frac{dv_o}{dt}\right|_{max} = 10\,v_i \times 2\pi \times 20\times10^3$

Thus $v_i = \frac{1/10^{-6}}{10\times2\pi\times20\times10^3} = 0.795\text{ V}$

c) $v_i = 50\text{mV}$ $V_o = 500\text{mV} = 0.5\text{V}$

Slew rate begins at the frequency for which $\omega\times0.5 = SR$

which gives $\omega = \frac{1/10^{-6}}{0.5} = 2\times10^6 \text{ rad/s}$ or $f = 318.3$ KHz

However the small signal 3db frequency is

$$f_{3db} = \frac{f_t}{1+\frac{R_2}{R_1}} = \frac{2\times10^6}{10} = 200\text{KHz}$$

Thus the useful frequency range is limited at 200KHz.

d) for $f = 5\text{KHz}$, the slew Rate limitation occurs at the value of v_i given by

$$\omega\times10\,V_i = SR \Rightarrow V_i = \frac{1/10^{-6}}{2\pi\times5\times10^3\times10} = 3.18\text{V}$$

Such an input voltage, however would ideally result in an output of 31.8V which exceeds V_{omax}. Thus $V_{imax} = \frac{V_{omax}}{10} = 1\text{V}$ peak.

2.100

$$v_o = v_{os}\left(1+\frac{R_2}{R_1}\right) \Rightarrow -0.3 = V_{os}\left(1+\frac{100}{1}\right) \approx 3\text{mV}$$

2.101

$v_{os} = \pm2\text{mV}$

$v_o = 0.01\sin\omega t \times 200 + v_{os}\times200 = 2\sin\omega t \pm 0.4\text{ V}$

2.102

Output DC offset, $v_{os} = 3\text{mV}\times1000 = 3\text{V}$

Therefore the maximum amplitude of an input sinusoid is the one that results in an output peak amplitude of $13-3 = 10\text{V} \Rightarrow v_i = \frac{10}{1000} = 10\text{ mV}$

If the amplifier is capacity coupled, then:

$v_{imax} = \frac{13}{1000} = 13\text{mV}$

2.103

$v_{os} = \frac{1.4}{100} = 1.4\text{ mV}$

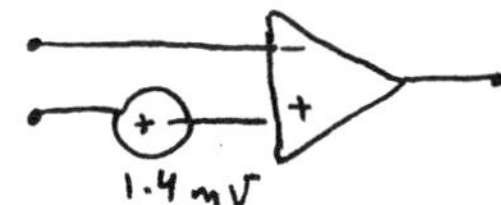

2.104

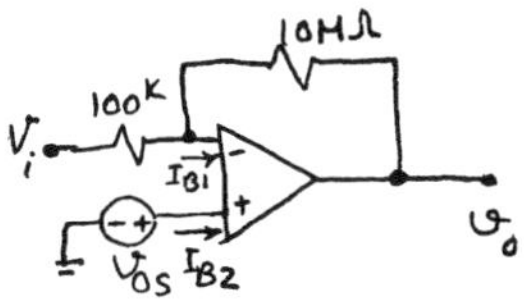

a) $I_B = (I_{B1} + I_{B2})/2$

open input:

$v_o = V_+ + R_2 I_{B_1} = v_{os} + R_2 I_{B1}$

$9.31 = v_{os} + 10000\,I_{B1}$ ①

input connected to ground:

$v_o = V_+ + R_2\left(I_{B1} + \frac{v_{os}}{R_1}\right) = v_{os}\left(1+\frac{R_2}{R_1}\right) + R_2 I_{B1}$

$9.09 = v_{os}\times101 + 10000\,I_{B1}$ ②

①, ② $\Rightarrow 100\,v_{os} = -0.22 \Rightarrow v_{os} = -2.2\text{mV}$

$\Rightarrow I_{B_1} = 930\text{nA}$

$I_B \approx I_{B_1} = 930\text{nA}$

b) $v_{os} = -2.2\text{ mV}$

c) In this case, Since R is too large, we may ignore v_{os} compare to the voltage drop across R.

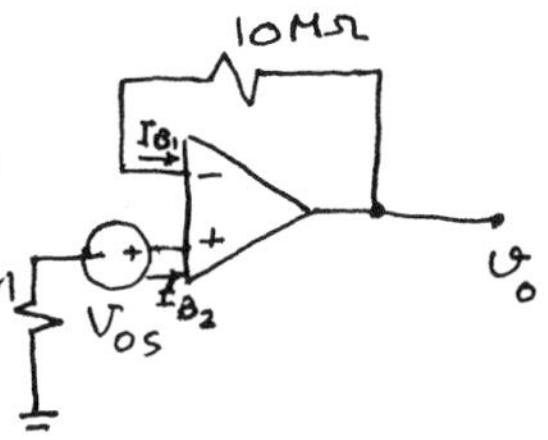

$v_{os} \ll R I_B$, Also Eq. 2.46 holds : $R_3 = R_1 \| R_2$

therefore from Eq. 2.47: $v_o = I_{os}\times R_2 \Rightarrow I_{os} = \frac{-0.8}{10^M}$

$I_{os} = -80\text{ nA}$

2.105

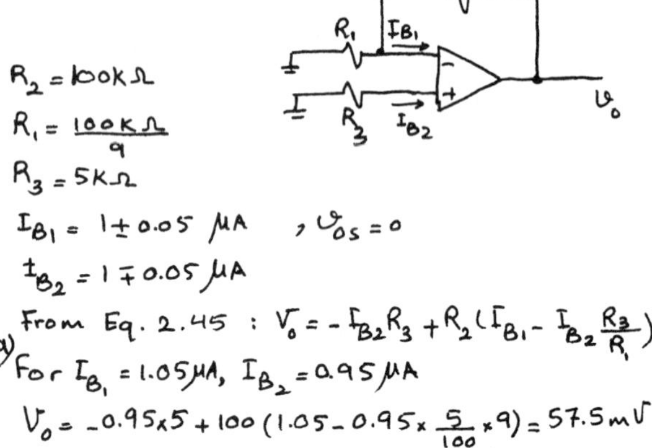

$R_2 = 100k\Omega$

$R_1 = \frac{100k\Omega}{9}$

$R_3 = 5k\Omega$

$I_{B1} = 1 \pm 0.05\ \mu A$, $v_{os} = 0$

$I_{B2} = 1 \mp 0.05\ \mu A$

From Eq. 2.45 : $V_o = -I_{B2}R_3 + R_2(I_{B1} - I_{B2}\frac{R_3}{R_1})$

a) For $I_{B1} = 1.05\mu A$, $I_{B2} = 0.95\mu A$

$V_o = -0.95 \times 5 + 100(1.05 - 0.95 \times \frac{5}{100} \times 9) = 57.5mV$

b) For $I_{B1} = 0.95\mu A$, $I_{B2} = 1.05\mu A$

$V_o = -1.05 \times 5 + 100(0.95 - 1.05 \times \frac{5}{100} \times 9) = 42.5mV$

$\Rightarrow 42.5^{mV} \leq V_o \leq 57.5^{mV}$

From the discussion in the text we know that to minimize the dc output voltage resulting from the input bias current, we should make the total DC resistance in the inputs of the op-amp equal. Currently, the negative input sees a resistance of $R_1 \| R_2 = \frac{100}{9} \| 100 = 10k\Omega$ while the positive input terminal sees $5k\Omega$ source resistance. Therefore we should add $5k\Omega$ series resistor to the positive input terminal to make the effective Resistance $5k\Omega + 5k\Omega = 10^{k\Omega}$

The resulting V_o can be found as follows:

$V_o = -I_{B2} \times 10 + 100(I_{B1} - I_{B2}\frac{10}{100/9}) = (I_{B1} - I_{B2}) \times 100$

$V_o = I_{os} \times 100 = \pm 0.1 \times 100 = \pm 10mV$

$V_o = \pm 10mV$

If the signal source resistance is $15k\Omega$, then the resistances can be equalized by adding a $5k\Omega$ resistor in series with the negative input load of the op-amp.

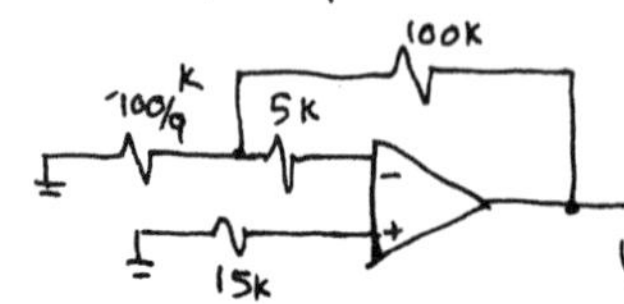

2.106

$R_2 = R_3 = 100k\Omega$

$1 + \frac{R_2}{R_1} = 200$

$R_1 = \frac{100k}{199} = 502\Omega \simeq 500\Omega$

$\frac{1}{R_1C_1} = 2\pi \times 100 \Rightarrow C_1 = \frac{1}{500 \times 2\pi \times 100} = 3.18\ \mu F$

$\frac{1}{R_3C_2} = 2\pi \times 10 \Rightarrow C_2 = \frac{1}{100^k \times 2\pi \times 10} = 0.16\ \mu F$

2.107

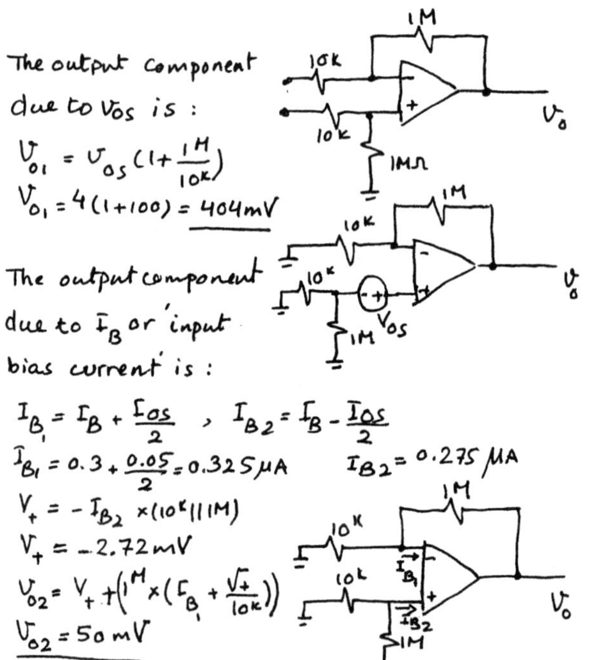

The output component due to Vos is :

$V_{o1} = V_{os}(1 + \frac{1M}{10K})$

$V_{o1} = 4(1 + 100) = 404mV$

The output component due to I_B or input bias current is :

$I_{B1} = I_B + \frac{I_{os}}{2}$, $I_{B2} = I_B - \frac{I_{os}}{2}$

$I_{B1} = 0.3 + \frac{0.05}{2} = 0.325\mu A$ $I_{B2} = 0.275\ \mu A$

$V_+ = -I_{B2} \times (10^k \| 1M)$

$V_+ = -2.72mV$

$V_{o2} = V_+ + (1^M \times (I_{B1} + \frac{V_+}{10k}))$

$V_{o2} = 50mV$

The worst case (largest) DC offset voltage at the output is $404 + 50 = 454mV$

2.108

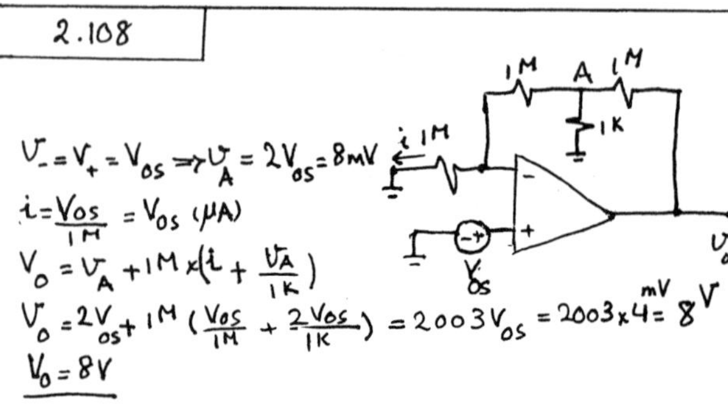

$V_- = V_+ = V_{os} \Rightarrow V_A = 2V_{os} = 8mV$

$i = \frac{V_{os}}{1M} = V_{os}\ (\mu A)$

$V_o = V_A + 1M \times (i + \frac{V_A}{1K})$

$V_o = 2V_{os} + 1M(\frac{V_{os}}{1M} + \frac{2V_{os}}{1K}) = 2003V_{os} = 2003 \times 4 = 8^{mV}\ V$

$V_o = 8V$

Cont.

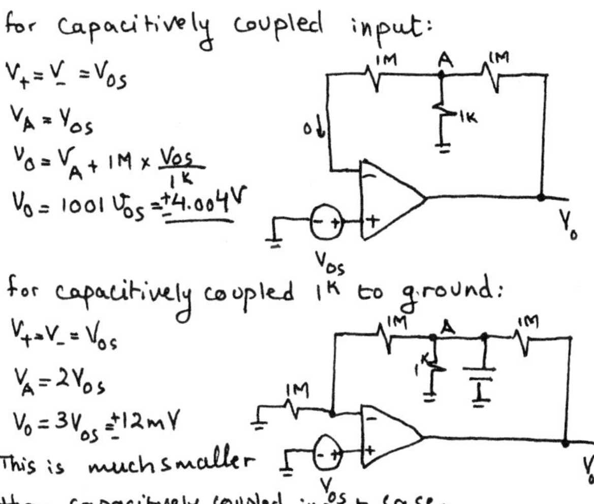

for capacitively coupled input:

$V_+ = V_- = V_{OS}$

$V_A = V_{OS}$

$V_O = V_A + 1M \times \frac{V_{OS}}{1K}$

$V_O = 1001\, V_{OS} = \pm 4.004V$

for capacitively coupled 1K to ground:

$V_+ = V_- = V_{OS}$

$V_A = 2V_{OS}$

$V_O = 3V_{OS} = \pm 12mV$

This is much smaller than capacitvely coupled input case.

2.109

At $0^\circ C$, we expect $\pm 10 \times 25 \times 1000^{\mu} = \pm 250mV$

At $75^\circ C$, we expect $\pm 10 \times 50 \times 1000^{\mu} = \pm 500mV$

We expect these quantities to have opposite polarities.

2.110

$100 = 1 + \frac{R_2}{R_1} \Rightarrow R_1 = 10.1^{K\Omega}$

a) $V_O = 100 \times 10^{-9} \times 1 \times 10^6 = 0.1^V$

b) Largest output offset is:

$V_O = 1^{mV} \times 100 + 0.1^V = 200mV = 0.2^V$

c) For bias current compensation we connect a resistor R_3 in series with the positive input terminal of the op-amp, with: $R_3 = R_1 \| R_2$

$I_{OS} = \frac{100}{10} = 10nA$ $\qquad R_3 = 10.1^K \| 1^M = 10K\Omega$

The offset current alone results in an output offset voltage of $I_{OS} \times R_2 = 10 \times 10^{-9} \times 1 \times 10^6 = 10^{mV}$

d) $V_O = 100mV + 10mV = 110mV$

2.111

$R_3 = R_1 \| R_2 = 9.9^{K\Omega}$

(Refer to 2.46)

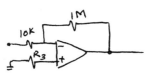

$V_O = I_{OS} R_2$ Eq. 2.47

$V_O = 0.21 = I_{OS} \times 1^M \Rightarrow I_{OS} = 0.21\,\mu A$

If $V_{OS} = 1mV$

$V_+ = -I_{B2} R_3 \mp V_{OS}$

$$I_{B_1} = \frac{R_3 I_{B2} \pm V_{OS}}{R_1} + \frac{0.21 + R_3 I_{B2} \pm V_{OS}}{R_2}$$

$$I_{B_1} = R_3 I_{B2}\left(\frac{1}{R_1} + \frac{1}{R_2}\right) \pm V_{OS}\left(\frac{1}{R_1} + \frac{1}{R_2}\right)$$

$$\frac{1}{R_3} = \frac{1}{R_1} + \frac{1}{R_2} \Rightarrow I_{B_1} - I_{B2} = \pm V_{OS}\left(\frac{1}{R_1} + \frac{1}{R_2}\right)$$

$$\Rightarrow I_{OS} = \frac{\pm 1^{mV}}{9.9K} = \pm 0.1\,\mu A$$

If we apply the same current as I_{OS} to the other end of R_3, then it will cancel out the offset current effect on the output. $\pm 0.1\mu A$

Now if we use $\pm 15V$ supplies:

2.112

$$\frac{V_O}{V_i} = -\frac{1}{SCR} = \frac{-1}{j\omega CR} = \frac{1}{-j\omega \times 10 \times 10^{-9} \times 100 \times 10^3}$$

$$\frac{V_O}{V_i} = -\frac{10^3}{j\omega}$$

a) $\frac{V_O}{V_i} = 1 \Rightarrow \omega = 1^{KRad/s} \Rightarrow f = 159HZ$

b) $\frac{1}{j}$ indicates 90° lag, but since its $\frac{-1}{j}$, it results in output leading the input by 90°

c) $\frac{V_O}{V_i} = \frac{-10^3}{j\omega}$ if frequency if lowered by a factor of 10, then the output would increase by a factor of 10.

Cont.

d) The phase doesnot change and the output still leads the input by 90°

2.113

$R_{in} = R = 100K\Omega$

$CR = 1s \Rightarrow C = \frac{1}{100\times10^3} = 10\mu F$

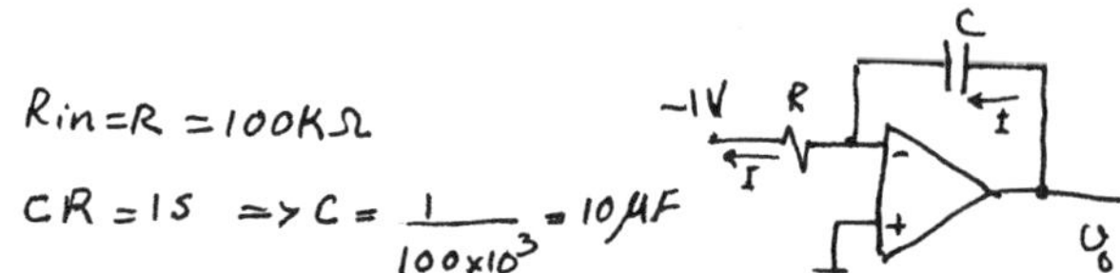

with a $-1V$ dc input applied, the capacitor charges with a constant current:

$I = \frac{1V}{R} = 0.01\ mA$ and its voltage rises linearly:

$v_o(t) = -10 + \frac{1}{C}\int_0^t I dt = -10 + \frac{I}{C}t = -10 + \frac{t}{RC}$

the voltage reaches $0V$ at $t = 10RC = 10s$ and it reaches $10V$ at $t = 20s$

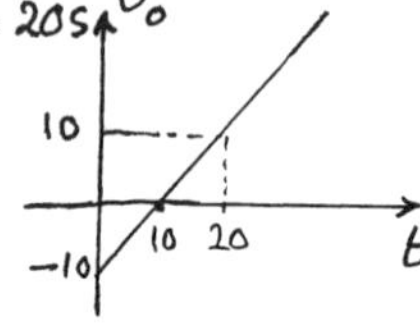

2.114

$|T| = \frac{1}{\omega RC}$ If $|T| = 100V/_V$ for $f = 1^{KHZ}$, then for $|T| = 1V/_V$, f has to be $1^K \times 100 = 100KHZ$.

Also $RC = \frac{1}{\omega \cdot T} = \frac{1}{2\pi\times1^K\times100} = 1.59\mu s$

2.115

$R_{in} = R$, Thus $R = 100k\Omega$.

$|T| = \frac{1}{\omega RC} = 1$ at $\omega = \frac{1}{RC}$.

$\omega = 1000 = \frac{1}{RC} \Rightarrow C = \frac{1}{1000\times100k} = 10^{nF}$

with a 2V-2ms pulse at the input, the output falls linearly until t=2ms at which $v_o = V_1$, $v_o = \frac{-I}{C}t = \frac{-2}{RC}t = -2t$ Volts where t in ms

Thus $V_1 = -4V$

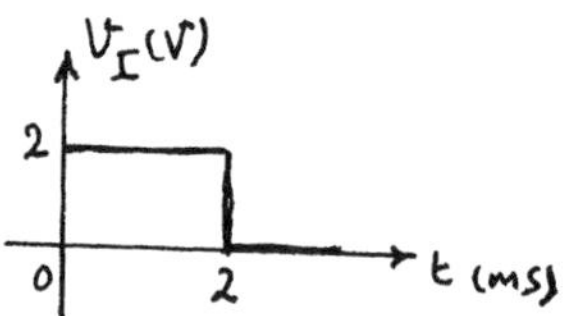

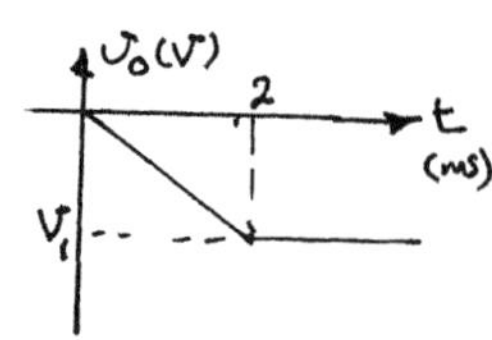

with $v_I = 2\sin 1000t$ applied at the input,

$v_o(t) = 2\times\frac{1}{1000\times10^{-3}}\sin(1000t + 90^\circ)$

$v_o(t) = 2\sin(1000t + 90^\circ)$

2.116

$R_{in} = R = 20k\Omega$

$|T| = \frac{1}{\omega RC} = 1$ at $\omega = 2\pi\times10^{KHZ} \Rightarrow C = \frac{1}{2\pi\times10^K\times20^K}$

$C = 0.796\ nF$

Refer to discussion in page 110:

$\frac{V_o}{V_i} = \frac{R_F/R}{1+sCR_F}$ and the finite dc gain is $\frac{-R_F}{R}$. Therefore for 40db gain or equivalently $100V/_V$ we have: $\frac{-R_F}{R} = -100\ V/_V$

$\Rightarrow R_F = 100\times20K = 2M\Omega$

The corner frequency $\frac{1}{CR_F}$ is: $\frac{1}{0.796^n\times2^M} = 628^{HZ}$

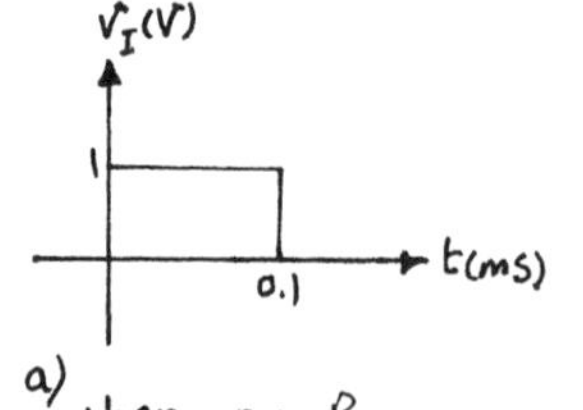

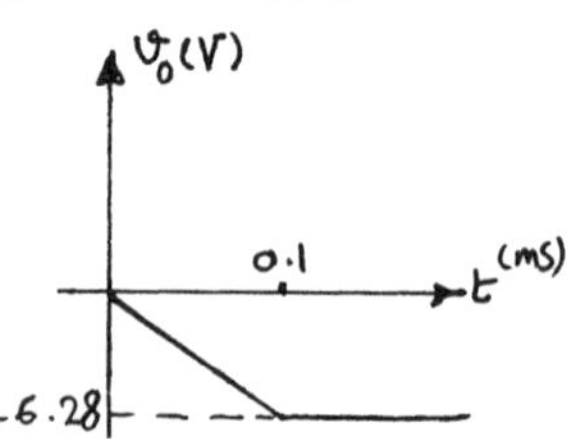

a) when no R_F

$v_o(t) = \frac{-1}{RC}\int_0^t 1.dt = -62.8t \quad 0 \leq t \leq 0.1ms$

$v_o(0.1) = -6.28V$

b) with R_F: $v_o(t) = V_o(\infty)(1 - e^{-t/CR_F})$

(Refer to pg. 112)

$v_o(\infty) = -I\times R_F = -\frac{1V}{20k}\times2^M = -100V$

$v_o(t) = -100(1 - e^{-t/1.6})$

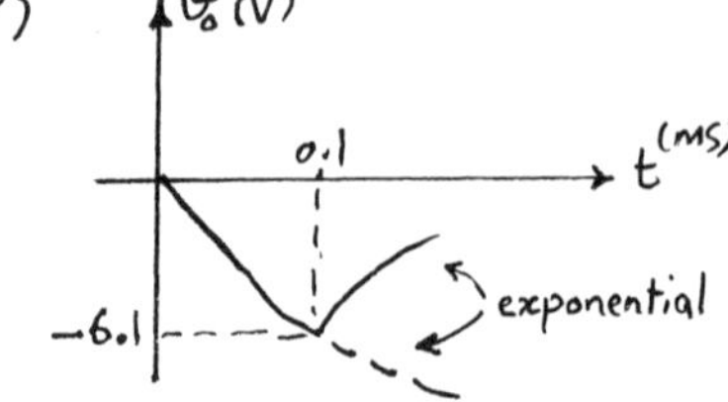

2.117

For $0 \le t \le 0.5$ ms: $v_O(t) = v_O(0) - \frac{1}{RC}\int_0^t v_I\,dt$

$v_O(t) = 0 - \frac{t}{RC} = -\frac{t}{1\,ms}$

$v_O(0.5) = -0.5V$

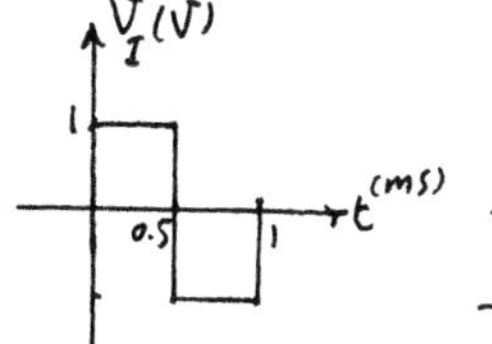

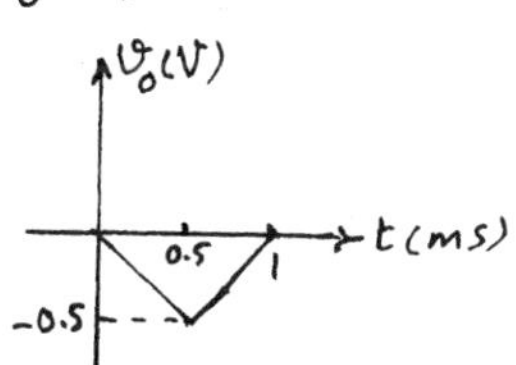

For $0.5 \le t \le 1$ ms: $v_O(t) = v_O(0.5) - \frac{1}{RC}\int_{0.5}^{t} -1\,dt$

$v_O(t) = -0.5 + \frac{1}{RC}(t - 0.5)$

$v_O(1^{ms}) = -0.5 + \frac{0.5}{1} = 0V$

Another way of thinking about this circuit is as follows:

For $0 \le t \le 0.5$ms a current $I = \frac{1V}{R}$ flows through R and C in the direction indicated on the diagram. At time t we write:

$I.t = -C\,v_O(t) \Rightarrow v_O(t) = -\frac{I}{C}t = \frac{-1}{RC}t$

which indicates that the output voltage is linearly decreased, reaching −0.5V at t=0.5ms.

Then for $0.5 \le t \le 1$ms, the current flows in the opposite direction and v_O rises linearly reaching 0V at t=1ms.

For $V_I = \pm 2V$: we obtain the following waveform: (assuming time constant is the same)

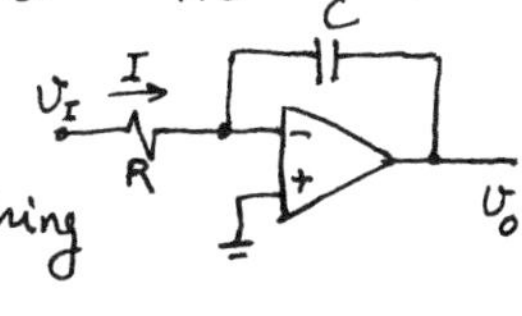

If RC is also doubled, then the waveform becomes the same as the first case where $V_I = \pm 1V$ and $RC = 1$ms.

2.118

Each pulse lowers the output voltage by:

$$\Delta v_O = \frac{1}{RC}\int_0^{10ms} 1.\,dt = \frac{10\mu s}{RC} = \frac{10\mu s}{1ms} = 10mV$$

Therefore a total of 100 pulses are required to cause a change of 1V in $v_O(t)$.

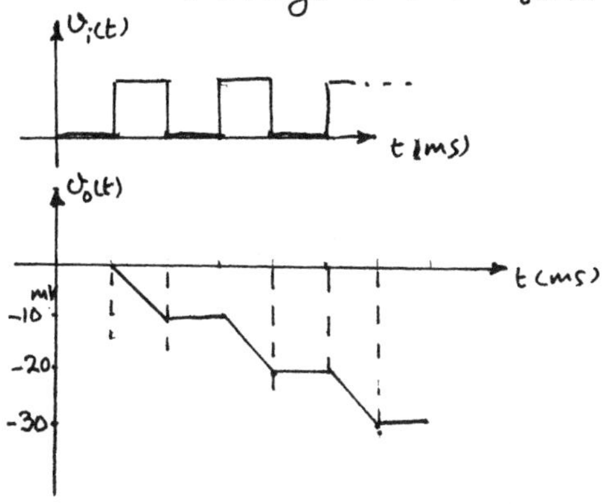

2.119

Refer to Fig. P2.119.

$$\frac{V_o}{V_i} = -\frac{Z_2}{Z_1} = -\frac{Y_1}{Y_2} = -\frac{1/R_1}{1/R_2 + SC} = -\frac{R_2/R_1}{1 + SCR_2}$$

which is an STC LP circuit with a dc gain of $-\frac{R_2}{R_1}$ and a 3-db frequency $\omega_o = \frac{1}{CR_2}$.

The input resistance equal to R_1. So for:

$R_i = 1K \Rightarrow R_1 = 1K\Omega$ and for dc-gain of 20db or 10: $\frac{R_2}{R_1} = 10 \Rightarrow R_2 = 10\ K\Omega$

for 3db frequency of 4KHz: $\omega_o = 2\pi \times 4 \times 10^3 = \frac{1}{CR_2}$

$\Rightarrow C = 4nF$

the unity gain frequency is (odb) is 40 KHZ

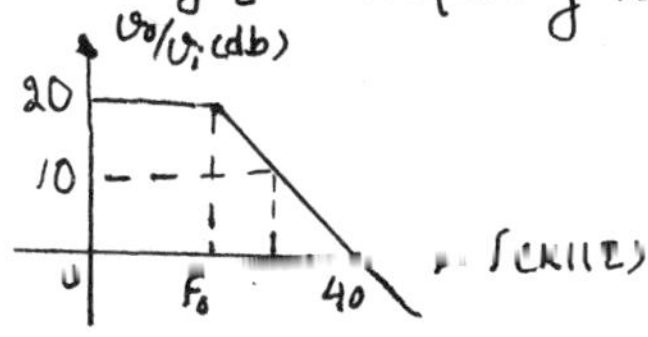

2.120

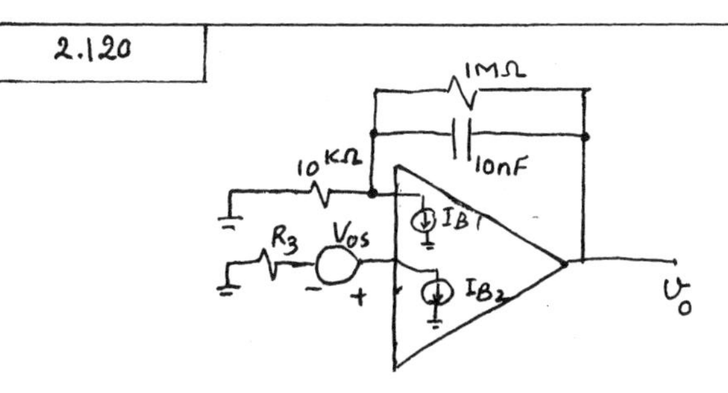

Cont.

a) To compensate for the effect of dc bias current I_B, we can consider the following model

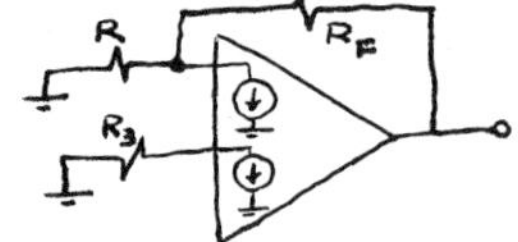

Similar to the discussion leading to equation(2.46) we have: $R_3 = R || R_F = 10k\Omega || 1M\Omega \Rightarrow R_3 = 9.9k\Omega$

(b) As discussed in Section 2.8.2 the dc output voltage of the integrator when the input is grounded is: $V_o = V_{os}(1 + \frac{R_F}{R}) + I_{os}R_F$

$V_o = 3mV(1 + \frac{1M\Omega}{10k\Omega}) + 10nA \times 1M\Omega = 0.303V + 0.01V$

$V_o = 0.313V$

2.121

$\frac{v_o}{v_i}(s) = -SRC = -S \times 0.01 \times 10^{-6} \times 10 \times 10^3 = -10^{-4}S$

$\frac{v_o}{v_i}(j\omega) = -j\omega \times 10^{-4} \Rightarrow \left|\frac{v_o}{v_i}\right| = -\omega \times 10^{-4} \Rightarrow$

$\left|\frac{v_o}{v_i}\right| = 1$ when $\omega = 10^4$ Rad/s or $f = 1.59KHZ$

for an input 10 times this frequency, the output will be 10 times as large as the input: 10V peak-to-peak. The $(-j)$ indicates that the output Lags the input by 90°. Thus $v_o(t) = -5\sin(10^5t + 90^\circ)$ volts

2.122

$v_o = -CR\frac{dv_i}{dt}$

therefore:

For $0 \leq t \leq 0.5$:

$v_o = -1ms \times \frac{1V}{0.5ms} = -2V$

and $v_o = 0$ otherwise

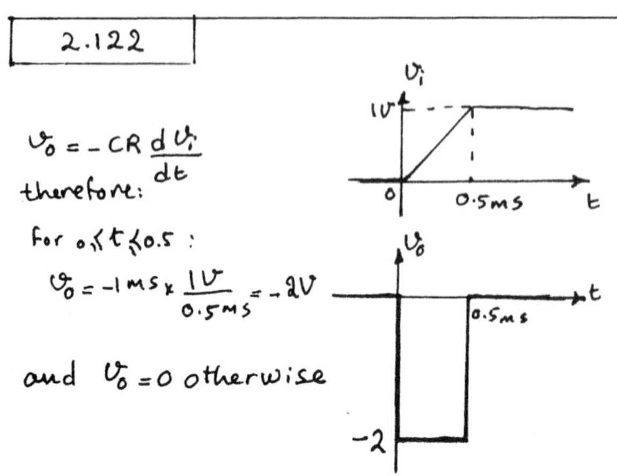

2.123

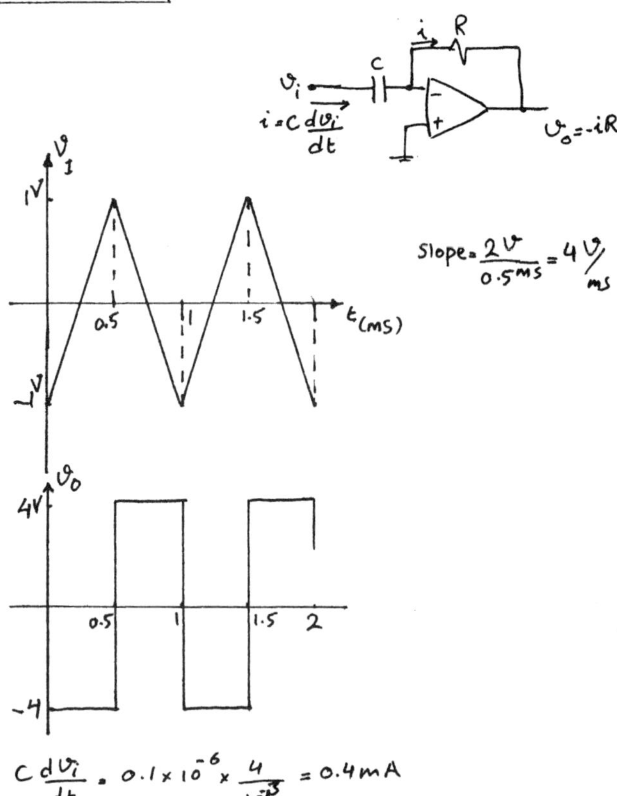

$C\frac{dv_i}{dt} = 0.1 \times 10^{-6} \times \frac{4}{10^{-3}} = 0.4mA$

Thus the peak value of the output square wave is $0.4mA \times 10k\Omega = 4V$. The frequency of the output is the same as the input (1KHz).

The average value of the output is 0.

To increase the value of the output to 10V, R has to be increased to $\frac{10}{4} = 2.5$, i.e $25k\Omega$.

When a 1-KHz, 1V peak input sine wave is applied

$v_i = \sin(2\pi \times 1000t)$

a sinusoidal signal appears at the output. It can be determined by one of the following methods:

a) $v_o(t) = -RC\frac{dv_i}{dt} = -0.1 \times 10^{-6} \times 10 \times 10^3 \frac{dv_i}{dt} = -10^{-3}\frac{dv_i}{dt}$

$v_o(t) = -10^{-3} \times 2\pi \times 1000 \times \cos(2\pi \times 1000t)$

$v_o(t) = -2\pi\cos(2\pi \times 1000t)$

Thus the peak amplitude is 6.28V and the negative peaks occur at $t = 0, \frac{2\pi}{2\pi \times 1000}, \ldots$

Cont.

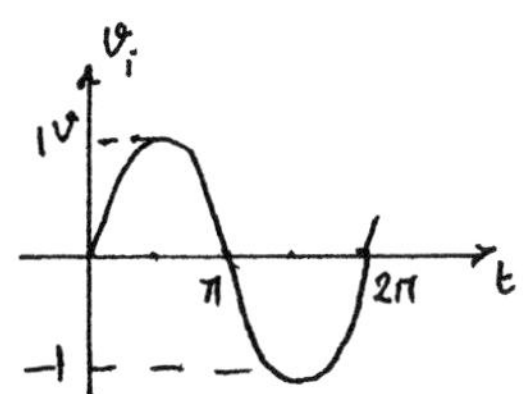

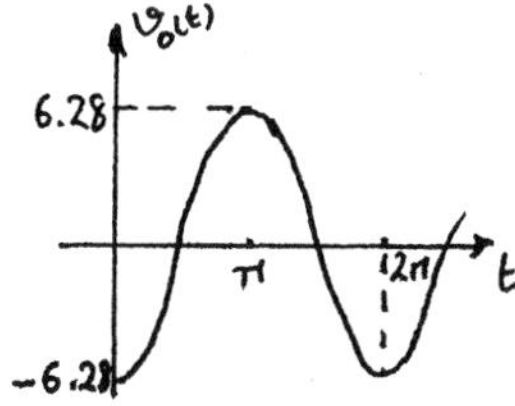

b) $\frac{v_o}{v_i} = -SRC \Rightarrow \frac{v_o}{v_i}(j\omega) = -j\omega RC \Rightarrow v_o(t) = -j\omega RC\, v_i(t)$

the output is inverted and has 90° phase shift, due to (-j) factor.

$v_o(t) = -(\omega RC) \times 1 \sin(2\pi \times 1000t + 90°)$

$v_o(t) = -6.28 \sin(2\pi \times 1000t + 90°)$

$v_o(t) = -6.28 \cos(2\pi \times 1000t)$

Same as before.

c) The peaks of the output waveform are equal to RC×(maximum slope of input wave). Since the maximum slope occurs at the zero crossings, its value is $2\pi \times 1000$. Thus the peak output $= 2\pi \times 1000 \times RC = 6.28V$

The negative peak occurs at $\omega t = 0, 2\pi, \ldots$

2.124

$RC = 10^{-3}s$ when $C = 10nF \Rightarrow R = 100K\Omega$

$\frac{v_o}{v_i} = -SRC \quad \frac{v_o}{v_i}(j\omega) = -j\omega RC \quad \varphi = -90°$ always

$\left|\frac{v_o}{v_i}\right| = 1 \Rightarrow \omega_{unity} = \frac{1}{RC} = 1\,\text{Krad/s}$ Gain is 10 times the unity gain, when the frequency is 10 times the unity gain frequency. Similarly for $\omega = \frac{1}{10}$ Krad/s, gain is $0.1 V/V$. (for $\omega = 10$ krad/s, gain $= 10 V/V$)

for high frequencies C is short-circuited,

$\frac{V_o}{V_i} = \frac{-R}{R_1} = -100 \Rightarrow R_1 = 1K\Omega$

$\frac{v_o}{v_i} = \frac{-RCS}{R_1CS+1} = \frac{-10^{-3}S}{10^{-5}S+1} \Rightarrow \omega_{3db} = 100\,\text{Krad/s}$ or $f_{3db} = 15.9$ KHz

for unity gain: $|10^{-3}S| = |10^{-5}S + 1| \Rightarrow \omega_H = 1.01$ Krad/s

if $\omega = 10.1$ Krad/s : $\left|\frac{v_o}{v_i}\right| = \frac{10.1}{1.01} = 10, \quad \varphi = -95.77°$

2.125

Refer to Fig. P2.125:

$$\frac{v_o}{v_i} = -\frac{Z_2}{Z_1} = \frac{-R_2}{R_1 + \frac{1}{SC}} = \frac{-\left(\frac{R_2}{R_1}\right)S}{S + \frac{1}{R_1C}}$$

which is the transfer function of an STC HP filter with a high frequency gain $K = -\frac{R_2}{R_1}$ and a 3-db frequency $\omega_0 = \frac{1}{R_1C}$

The high-frequency input impedance approaches R_1. (as $\frac{1}{j\omega C}$ becomes neglibibly small) So we can select $R_1 = 10K\Omega$

To Obtain a high-frequency gain of 40db (i.e. 100) : $\frac{R_2}{R_1} = 100 \Rightarrow R_2 = 1M\Omega$.

For a 3-db frequency of 1000 Hz:

$\frac{1}{R_1C} = 2\pi \times 1000 \Rightarrow C = 15.9nF$

from the Bode-diagram below, we see that $\left|\frac{v_o}{v_i}\right|$ reduces to unity at $f = 0.01 f_0 = 10$ Hz

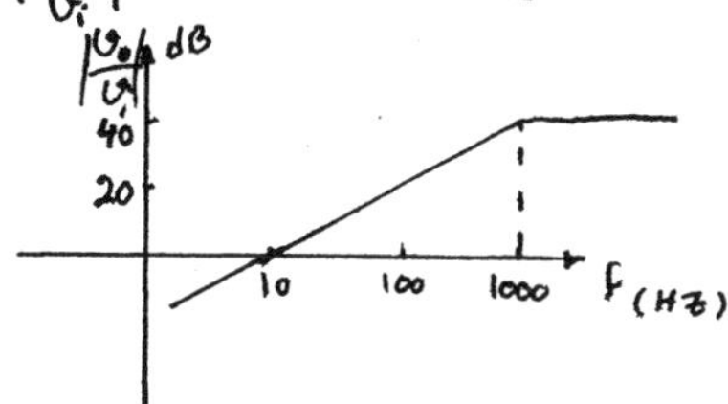

2.126

Refer to the circuit in Fig. P2.126:

$$\frac{v_o}{v_i} = -\frac{Z_2}{Z_1} = -\frac{1}{Z_1 Y_2} = -\frac{1}{\left(R_1 + \frac{1}{SC_1}\right)\left(\frac{1}{R_2} + SC_2\right)}$$

$$\frac{v_o}{v_i} = -\frac{R_2/R_1}{\left(1 + \frac{1}{R_1C_1S}\right)(1 + SR_2C_2)}$$

$$\frac{v_o}{v_i}(j\omega) = \frac{-R_2/R_1}{\left(1 + \frac{1}{j\omega R_1C_1}\right)(1 + j\omega R_2C_2)} = \frac{-R_2/R_1}{\left(1 + \frac{\omega_1}{j\omega}\right)\left(1 + j\frac{\omega}{\omega_2}\right)}$$

where $\omega_1 = \frac{1}{R_1C_1}$, $\omega_2 = \frac{1}{R_2C_2}$

a) for $\omega \ll \omega_1 \ll \omega_2$

$$\frac{v_o}{v_i}(j\omega) \simeq \frac{-R_2/R_1}{\left(1 + \frac{\omega_1}{j\omega}\right)} \simeq \frac{-R_2/R_1}{\omega_1/j\omega} = -j\frac{R_2}{R_1}\frac{\omega}{\omega_1}$$

Cont.

b) for $\omega_1 \ll \omega \ll \omega_2$

$$\frac{V_o}{V_i}(j\omega) \simeq \frac{-R_2}{R_1}$$

c) for $\omega \gg \omega_2$ and $\omega_2 \gg \omega_1$:

$$\frac{V_o}{V_i}(j\omega) \simeq \frac{-R_2/R_1}{1 + j\omega/\omega_2} \simeq \frac{-R_2/R_1}{j\,\omega/\omega_2} = j\left(\frac{R_2}{R_1}\right)\left(\frac{\omega_2}{\omega}\right)$$

from the results of a), b) and c) we can draw the Bode-plot:

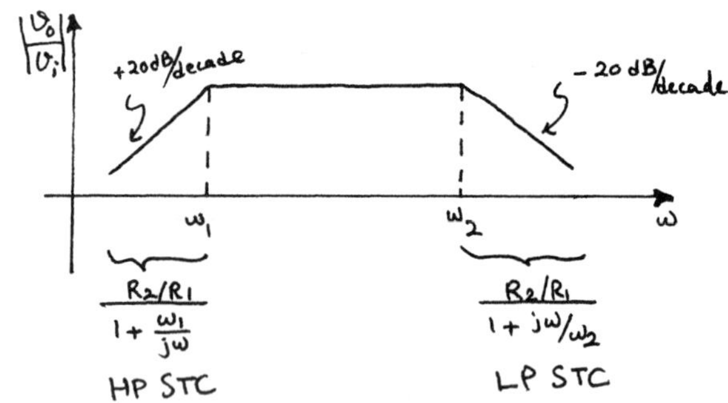

$$\frac{R_2/R_1}{1 + \frac{\omega_1}{j\omega}} \qquad\qquad \frac{R_2/R_1}{1 + j\omega/\omega_2}$$

HP STC $\qquad\qquad$ LP STC

Design: $\frac{R_2}{R_1} = 1000$ (60dB gain in the mid-frequency range)

Rin for $\omega \gg \omega_1$ $= R_1 = 1K\Omega \Rightarrow R_2 = 1M\Omega$

$f_1 = 100HZ \Rightarrow \omega_1 = 2\pi \times 100 = \frac{1}{R_1 C_1} \Rightarrow C_1 = 1.59\mu F$

$f_2 = 10KHZ \Rightarrow \omega_2 = 2\pi \times 10 \times 10^3 = \frac{1}{R_2 C_2} \Rightarrow C_2 = 15.9\,pF$

CHAPTER 3 - PROBLEMS

3.1

1Ω

1.5V

I

$+ \; V_D \; -$

The diode can be reverse-biased and thus no current would flow, or forward-biased where current would flow.

(a) Reverse biased $I = 0A \quad V_D = 1.5V$

(b) Forward biased $I = 1.5A \quad V_D = 0V$

3.2

(a) Diode is conducting and thus has a 0V drop across it. Consequently

$V = \underline{\underline{-3V}}$

$$I = \frac{3-(-3)}{10k\Omega} = \underline{\underline{0.6mA}}$$

(b) Diode is cut off.

$V = \underline{\underline{3V}} \qquad I = \underline{\underline{0A}}$

(c) Diode is conducting

$V = \underline{\underline{3V}}$

$$I = \frac{3-(-3)}{10k\Omega} = \underline{\underline{0.6mA}}$$

(d) Diode is cut off.

$V = \underline{\underline{-3V}} \qquad I = \underline{\underline{0A}}$

3.3

(a)

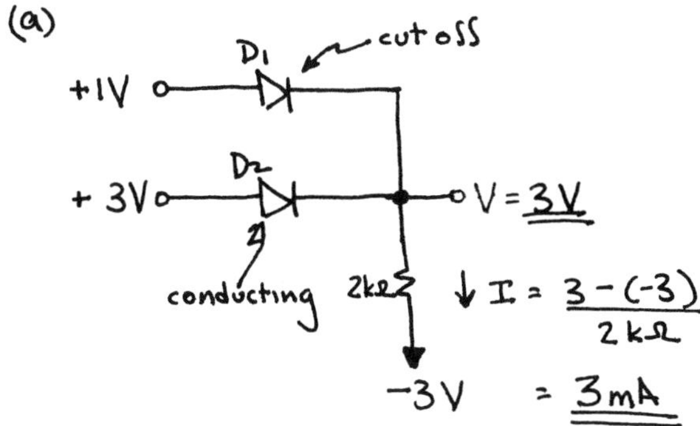

$$I = \frac{3-(-3)}{2k\Omega} = \underline{\underline{3mA}}$$

(b)

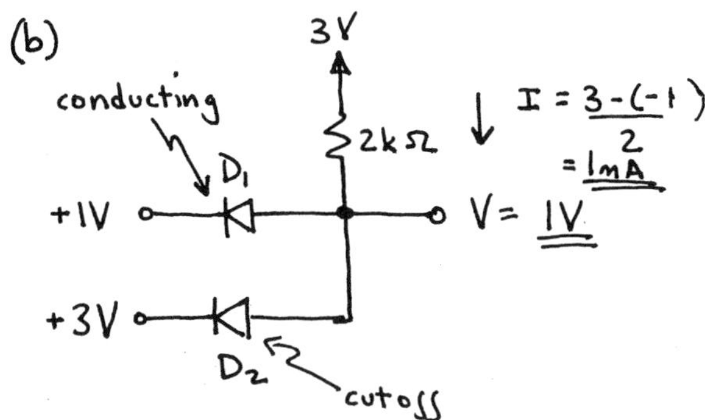

$$I = \frac{3-(-1)}{2} = \underline{\underline{1mA}}$$

3.4

(a)

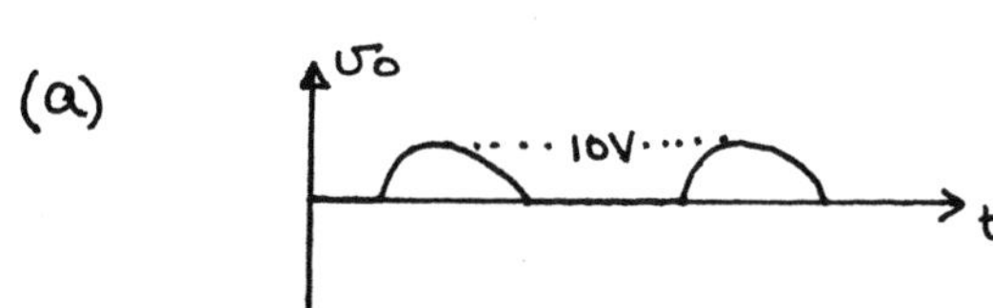

$V_{p+} = \underline{\underline{10V}} \qquad V_{p-} = \underline{\underline{0V}}$

$f = 1kHz.$

(b)

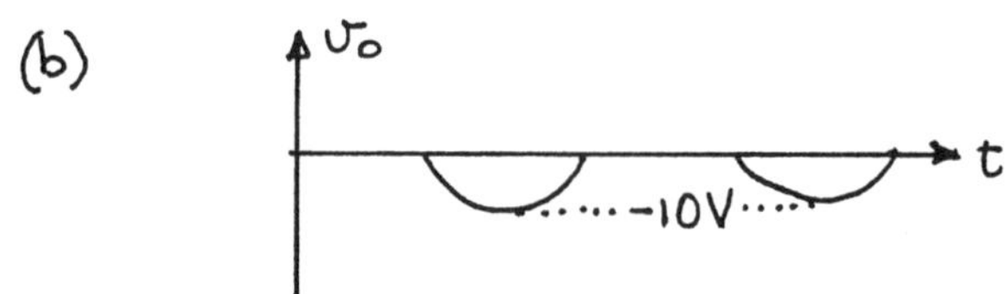

$V_{p+} = \underline{\underline{0\,V}}$ $\quad V_{p-} = \underline{\underline{-10V}}$

$f = 1\,kHz$

(c)

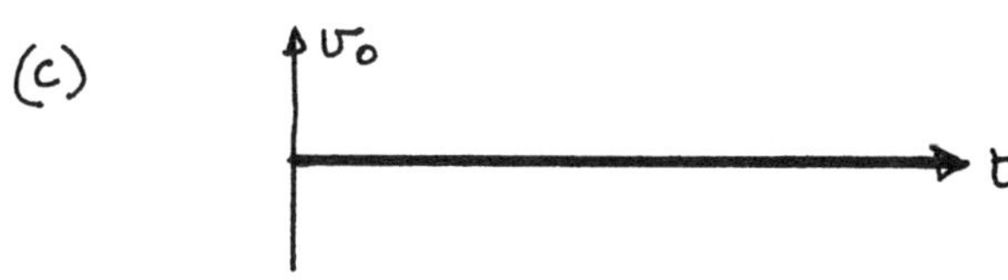

$v_O = \underline{\underline{0V}}$

Neither D_1 nor D_2 conducts so there is no output.

(d)

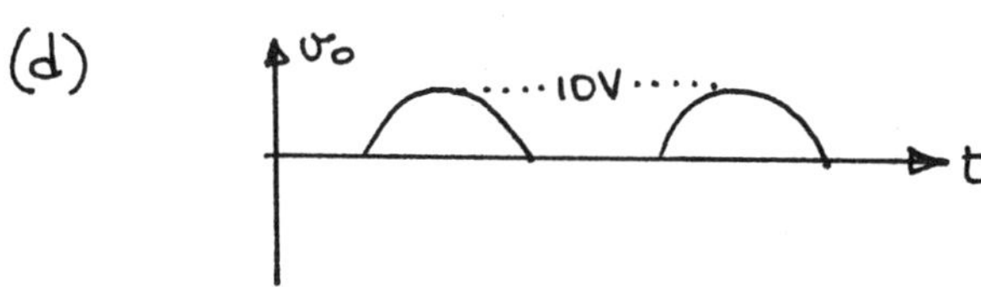

$V_{p+} = \underline{\underline{10V}}$ $\quad V_{p-} = \underline{\underline{0\,V}}$ $\quad f = 1\,kHz$

Both D_1 and D_2 conduct when $v_I > 0$

(e)

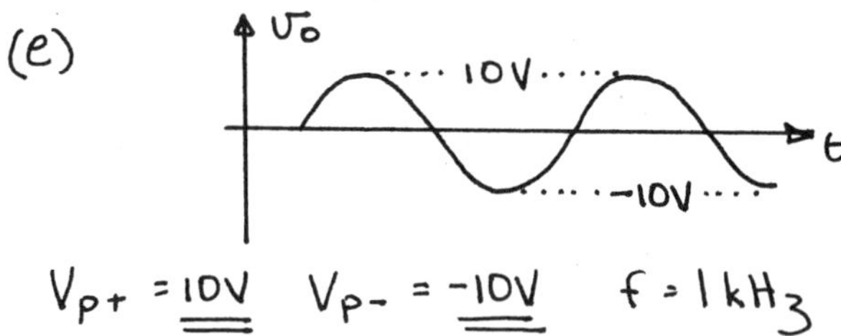

$V_{p+} = \underline{\underline{10V}}$ $\quad V_{p-} = \underline{\underline{-10V}}$ $\quad f = 1\,kHz$

D_1 conducts when $v_I > 0$ and D_2 conducts when $v_I < 0$. Thus the output follows the input.

(f)

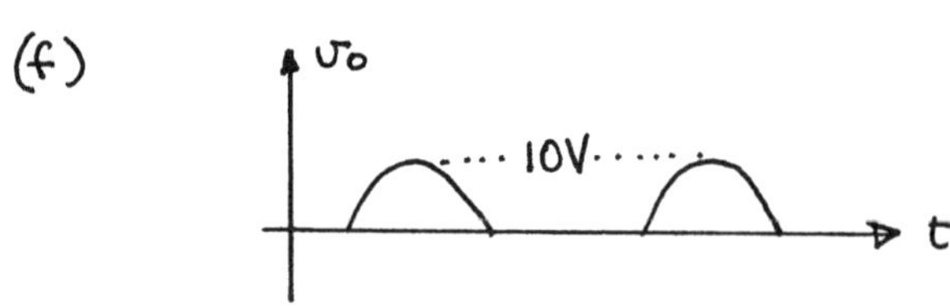

$V_{p+} = \underline{\underline{10V}}$ $\quad V_{p-} = \underline{\underline{0V}}$ $\quad f = 1\,kHz$

- D_1 is cutoff when $v_I < 0$

(g)

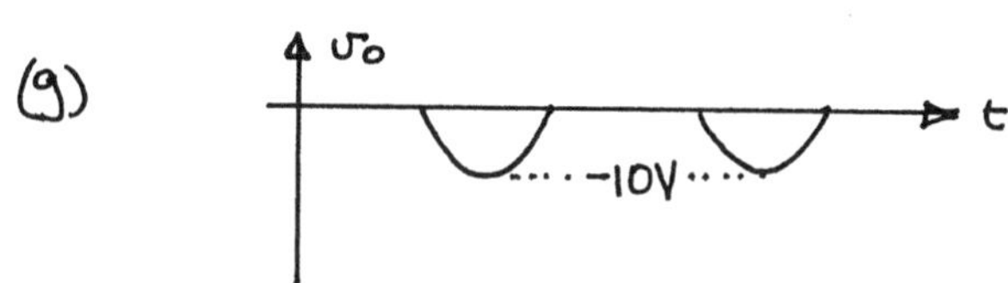

$V_{p+} = \underline{\underline{0V}}$ $\quad V_{p-} = \underline{\underline{-10V}}$ $\quad f = 1\,kHz$

D_1 shorts to ground when $v_I > 0$ and is cut off when $v_I < 0$ whereby the output follows v_I.

(h) $v_O = 0V$, t

$v_O = \underline{\underline{0V}}$ ~ The output is always shorted to ground as D_1 conducts when $v_I > 0$ and D_2 conducts when $v_I < 0$.

(i)

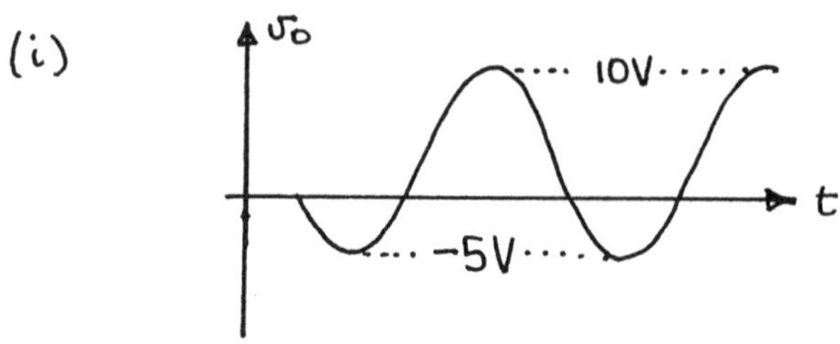

$V_{p+} = \underline{\underline{10V}}$ $\quad V_{p-} = \underline{\underline{-5V}}$ $\quad f = 1\,kHz$

- When $v_I > 0$, D_1 is cutoff and v_O follows v_I.

- When $v_I < 0$, D_1 is conducting and the circuit becomes a voltage divider where the negative peak is

$$\frac{1k\Omega}{1k\Omega + 1k\Omega} \cdot -10V = -5V$$

(j)

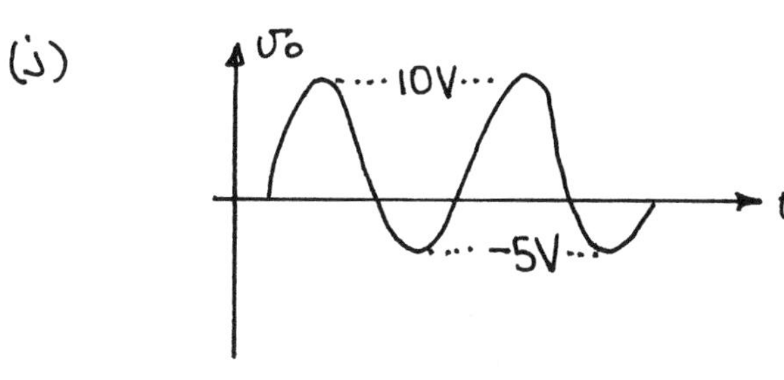

$V_{p+} = \underline{\underline{10V}}$ $V_{p-} = \underline{\underline{-5V}}$ $f = 1kHz$

- When $v_I > 0$, the output follows the input as D_1 is conducting.
- When $v_I < 0$, D_1 is cut off and the circuit becomes a voltage divider.

(k)

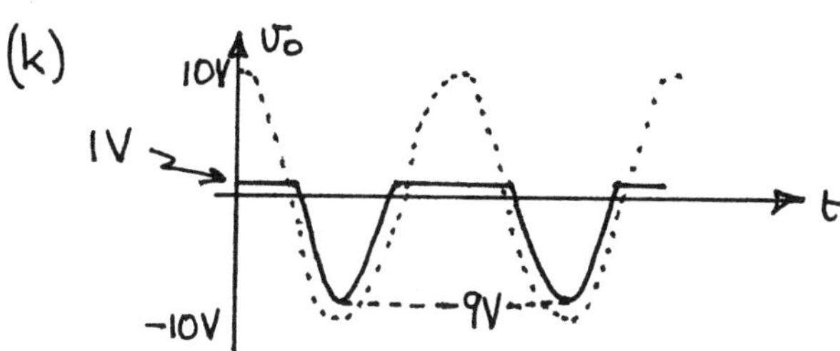

$V_{p+} = \underline{\underline{1V}}$ $V_{p-} = \underline{\underline{-9V}}$ $f = 1kHz$

- When $v_I > 0$, D_1 is cutoff and D_2 is conducting. The output becomes 1V.
- When $v_I < 0$, D_1 is conducting and D_2 is cutoff. The output becomes :-

$$v_O = v_I + 1V.$$

CHAPTER 3 PROBLEMS

3.5

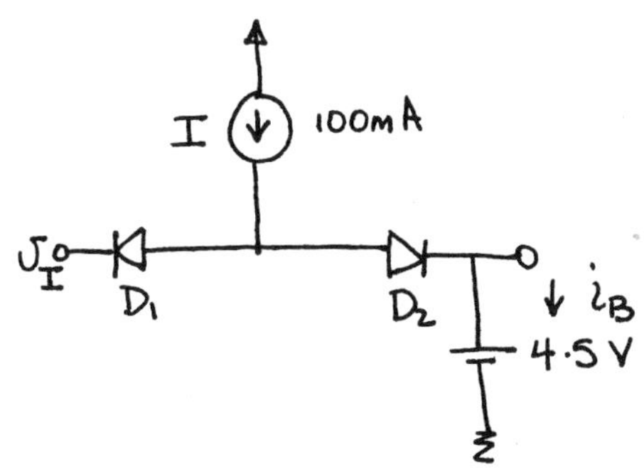

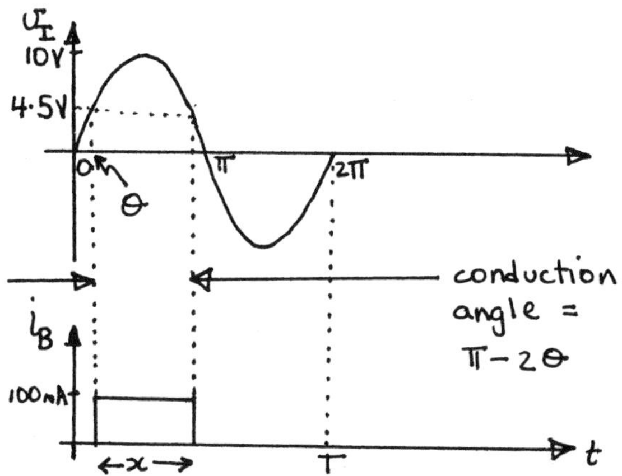

- When $v_I < 4.5V$ D_1 conducts and D_2 is cutoff so $i_B = 0A$. For $v_I > 4.5V$ D_2 conducts and D_1 is cut off thus disconnecting the input v_I. All of the current then flows through the battery.

$$10 \sin\theta = 4.5\ V$$
$$\theta = \sin^{-1}(4.5/10)$$

conduction angle $= \pi - 2\theta$

Fraction of cycle that $i_B = \underline{\underline{100mA}}$ is given by :-

$$x = \frac{\pi - 2\theta}{2\pi} = 0.35$$

$$i_{B\,avg} = \frac{1}{T}\int_T i_B \, dt$$

$$= \frac{1}{T}\left[100 \cdot 0.35T\right]$$

$$= \underline{\underline{35\,mA}}$$

If v_I is reduced by 10% the peak value of i_B remains the same

$$i_{B\,peak} = \underline{100\,mA}$$

but the fraction of the cycle for conduction changes

$$x = \frac{\pi - 2\theta}{2\pi} = \frac{\pi - 2\sin^{-1}(4.5/9)}{2\pi}$$

$$= \frac{1}{3}$$

Thus:

$$i_{B\,avg} = \frac{1}{T}\left[100 \cdot \frac{T}{3}\right]$$

$$= \underline{\underline{33.3\,mA}}$$

3.6

A	B	x	y
0	0	0	0
0	1	0	1
1	0	0	1
1	1	1	1

$x = \underline{\underline{AB}}$ $y = \underline{\underline{A + B}}$

CHAPTER 3 PROBLEMS

- x and y are the same for $A = B$
- x and y are opposite if $A \neq B$

3.7

$$\frac{5 - 0}{R} \leq 0.1\,mA$$

$$R \geq 5/0.1 = \underline{\underline{50\,k\Omega}}$$

3.8

The maximum input current occurs when one input is low and the other two are high.

$$\frac{5 - 0}{R} \leq 0.1\,mA$$

$$R \geq \underline{\underline{50\,k\Omega}}$$

3.9

(a)

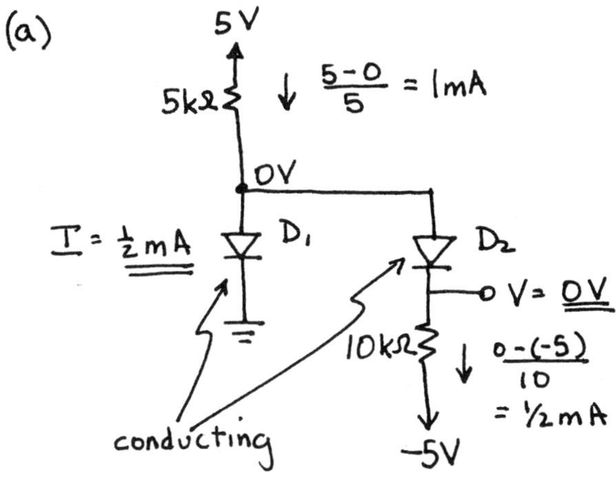

(b)

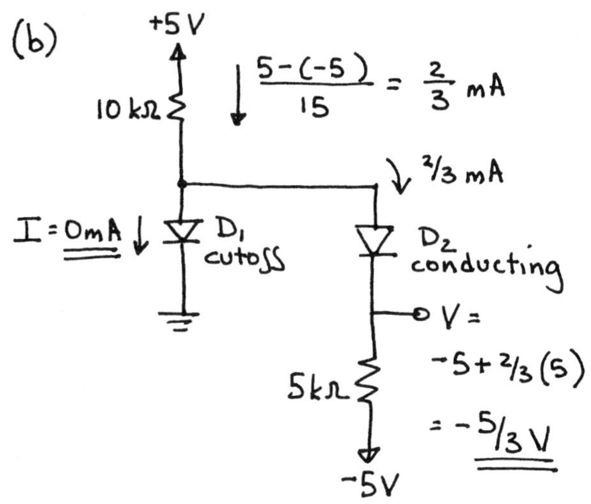

3.10

(a)

$(10 \| 20)\text{k}\Omega$

$9\left(\frac{20}{30}\right) = 6\text{V}$ I, 20kΩ, V

$$I = \frac{6}{(10\|20) + 20} = \underline{\underline{0.225\text{mA}}}$$

$$V = \frac{20}{(10\|20) + 20} \times 6 = \underline{\underline{4.5\text{V}}}$$

(b)

$V = 5/2 - 9/2 = \underline{\underline{-2\text{V}}}$

− V + cutoff

$10\|10 = 5\text{k}\Omega$ $10\|10 = 5\text{k}\Omega$

$\frac{10 \times 9}{10+10} = 4.5\text{V}$ $I = \underline{\underline{0\text{A}}}$ $\frac{5\,(10)}{10+10} = 2.5\text{V}$

3.11

$$R \geq \frac{120\sqrt{2}}{50} \geq \underline{\underline{3.4\text{k}\Omega}}$$

The largest reverse voltage appearing across the diode is equal to the peak input voltage

$$120\sqrt{2} = \underline{\underline{169.7\text{V}}}$$

3.12

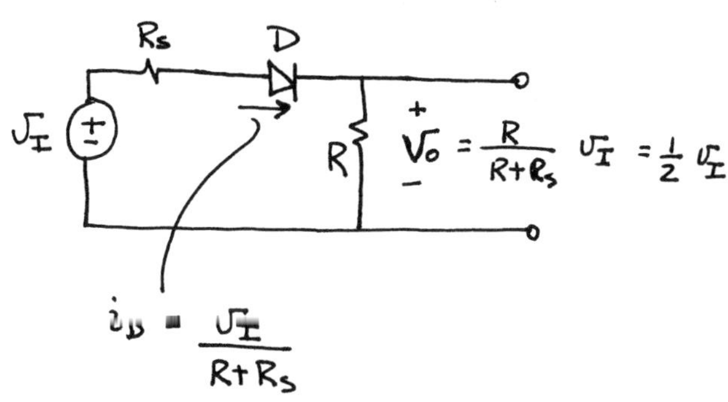

D starts to conduct when $v_I > 0$

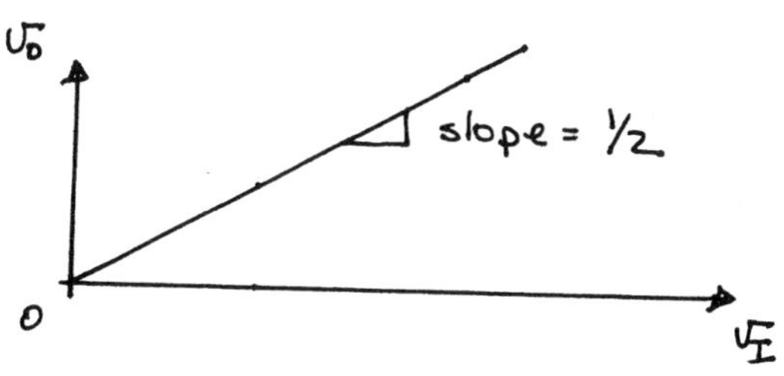

3.13

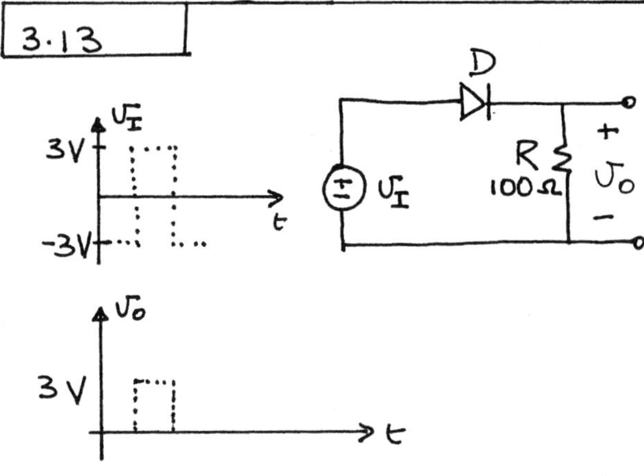

$v_{o,\,peak} = 3\text{V}$

$$v_{o,\,avg} = \frac{1}{T}\int v_o \, dt$$

$$= \frac{1}{T}\left[3\,\frac{T}{2}\right] = \underline{\underline{3/2\ \text{V}}}$$

$$i_{D,\,peak} = 3/100 = \underline{\underline{30\ \text{mA}}}$$

$$i_{D,\,avg} = \frac{3/2}{100} = \underline{\underline{15\ \text{mA}}}$$

The maximum reverse diode voltage is $\underline{\underline{3\text{ V}}}$

3.14

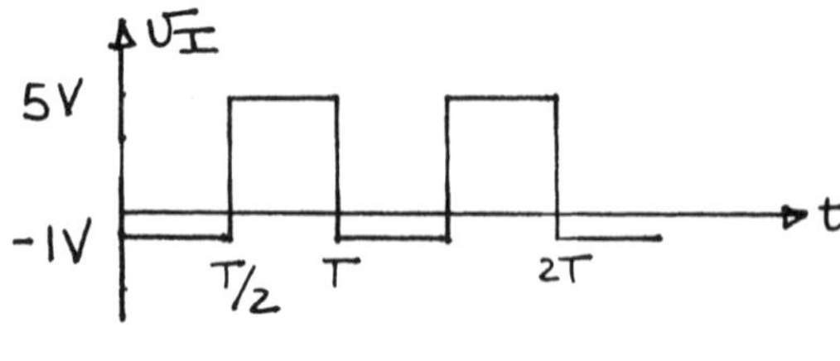

$v_{o,\,peak} = \underline{\underline{5\text{V}}}$

$v_{o,\,avg} = \underline{\underline{2.5\text{V}}}$

$$i_{D,\,peak} = \frac{v_{o,\,peak}}{100} = \underline{\underline{50\text{mA}}}$$

$$i_{D,\,avg} = i_{D,\,peak}/2 = \underline{\underline{25\text{mA}}}$$

maximum reverse voltage = $\underline{\underline{1\text{V}}}$

3.15

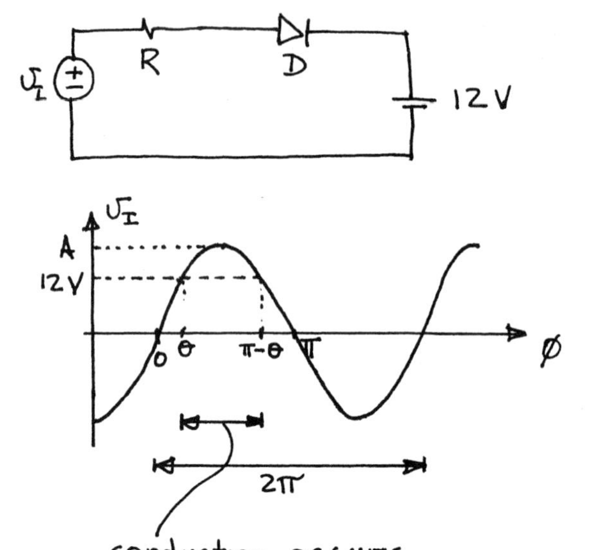

$v_I = A \sin\theta = 12$ ~ conduction across D occurs

For a conduction angle $(\pi - 2\theta)$ that is 20% of a cycle

$$\frac{\pi - 2\theta}{2\pi} = \frac{1}{5}$$

$$\theta = 0.3\pi$$

$$A = 12/\sin\theta = 14.83\text{V}$$

∴ Peak-to-peak sine wave voltage

$$= 2A = \underline{\underline{29.67\text{V}}}$$

Given the average diode current to be

$$\frac{1}{2\pi}\int_0^{2\pi} \frac{A\sin\phi - 12}{R}\, d\phi = 100\text{mA}$$

$$\frac{1}{2\pi}\left[\frac{-14.83\cos\phi - 12\phi}{R}\right]_{\phi=0.3\pi}^{\phi=0.7\pi} = 0.1$$

$R = \underline{\underline{3.75\,\Omega}}$

Peak diode current $= \dfrac{A-12}{R} = \underline{\underline{0.75\text{A}}}$

Peak reverse voltage $= A+12 = \underline{\underline{26.83\text{V}}}$

For resistors specified to only one significant digit and peak-to-peak voltage to the nearest volt then choose $A = 15$ so the peak-to-peak sine wave voltage $= \underline{\underline{30\text{V}}}$ and $R = \underline{\underline{3\,\Omega}}$

Conduction starts at $v_I = A\sin\theta = 12$

$15\sin\theta = 12$

$\theta = 0.93\,\text{rad}$

Conduction stops at $\pi - \theta$

$\therefore$ Fraction of cycle that current flows is $\dfrac{\pi - 2\theta}{2\pi} \times 100 = 20.5$

$\simeq \underline{\underline{20\%}}$

Average diode current $=$

$$\frac{1}{2\pi}\left[\frac{-15\cos\phi - 12\phi}{3}\right]_{\phi=0.93}^{2.21} = \underline{\underline{136\,\text{mA}}}$$

Peak diode current

$= \dfrac{15-12}{3} = \underline{\underline{1\,\text{A}}}$

Peak reverse voltage $=$

$A + 12 = \underline{\underline{27\,\text{V}}}$

3.16

V	RED	GREEN	
3V	ON	OFF	- D_1 conducts
0	OFF	OFF	- No current flows
-3V	OFF	ON	- D_2 conducts

3.17

$V_T = kT/q$ where $k = 1.38\times10^{-23}\,\text{J/K}$

$T = 273 + x\,°\text{C}$

$q = 1.60\times10^{-19}\,\text{C}$

x [°C]	V_T [mV]
−40	20
0	23.5
40	27
150	36.5

for $V_T = 25\,\text{mV}$

$T = \underline{\underline{16.8\,°\text{C}}}$

3.18

$i = I_S e^{v/2\times0.025}$

$\therefore\ 1000\,I_S = I_S e^{v/0.05}$

$v = \underline{\underline{0.345\,\text{V}}}$

at $v = 0.7\,\text{V}$

$i = I_S e^{0.7/0.05} = \underline{\underline{1.2\times10^{6}\,I_S}}$

3.19

$$i_1 = I_s e^{0.7/V_T} = 10^{-3}$$

$$i_2 = I_s e^{0.5/V_T}$$

$$\frac{i_2}{i_1} = \frac{i_2}{10^{-3}} = e^{\frac{0.5-0.7}{0.025}}$$

$$i_2 = \underline{\underline{0.335\ \mu A}}$$

3.20

$$i = I_s e^{v/nV_T} = I_s e^{0.7/0.025} = 5(10^{-3})$$

$$I_s = 5(10^{-3})\, e^{-0.7/0.025} = \underline{\underline{3.46 \times 10^{-15} A}}$$

v	i
0.71 V	7.46 mA
0.8 V	273.21 mA
0.69 V	3.35 mA
0.6 V	91.65 μA

Let $i_1 = I_s e^{v_1/0.025}$

$$i_2 = 10 i_1 = I_s e^{v_2/0.025}$$

$$\frac{i_2}{i_1} = 10 = e^{\frac{v_2 - v_1}{0.025}}$$

$$\therefore \Delta v = v_2 - v_1 = \underline{\underline{57.56\,mV}}$$

3.21

To calculate I_s use

$$I_s = I e^{-V/nV_T} = I e^{-V/n \times 0.025}$$

To calculate the voltage at 1% of the measured current use

$i_2 = 0.01\, i_1$ so,

$$\frac{i_2}{i_1} = 0.01 = e^{\frac{V_2 - V_1}{nV_T}}$$

$$V_2 = V_1 + nV_T \ln 0.01$$

$$= V + n(0.025)\ln(0.01)$$

V [V]	I	I_s n=1 [A]	I_s n=2 [A]	V n=1 [V]	V n=2 [V]
0.7	1 A	6.91×10^{-13}	8.32×10^{-7}	0.585	0.470
0.650	1 mA	5.11×10^{-15}	2.26×10^{-9}	0.535	0.420
0.650	10 μA	5.11×10^{-17}	2.26×10^{-11}	0.535	0.420
0.7	10 mA	6.91×10^{-15}	8.32×10^{-9}	0.584	0.470

3.22

Let $I_1 = I_s e^{V_1/nV_T}$ and

$$I_2 = I_s e^{V_2/nV_T} = I_1/10$$

Calculate n by :-

$$\frac{I_2}{I_1} = e^{\frac{V_2 - V_1}{nV_T}}$$

$$n = \frac{1}{V_T}\left[\frac{V_2 - V_1}{\ln I_2/I_1}\right] = \frac{1}{0.025}\left[\frac{V_2 - V_1}{\ln 0.1}\right]$$

Calculate I_s by :-

$$I_s = I_1 e^{-V_1/nV_T}$$

Calculate the diode voltage at $10I_1$ by :- $V_3 = nV_T \ln \frac{10 I_1}{I_s}$

I	V_1 [V]	V_2 [V]	n	I_s [A]	V_3 [V]
10mA	0.7	0.6	1.737	10^{-9}	0.8
1mA	0.7	0.6	1.737	10^{-10}	0.8
10A	0.8	0.7	1.737	10^{-7}	0.9
1 mA	0.7	0.58	2.085	1.47×10^{-9}	0.82
10μA	0.7	0.64	1.042	2.15×10^{-17}	0.7

3.23

∴ The voltage across each diode is $V_0/3$

$$I = I_s e^{\frac{V_0/3}{nV_T}} = 10^{-14} e^{\frac{2/3}{0.025}}$$

$$= \underline{\underline{3.81\,mA}}$$

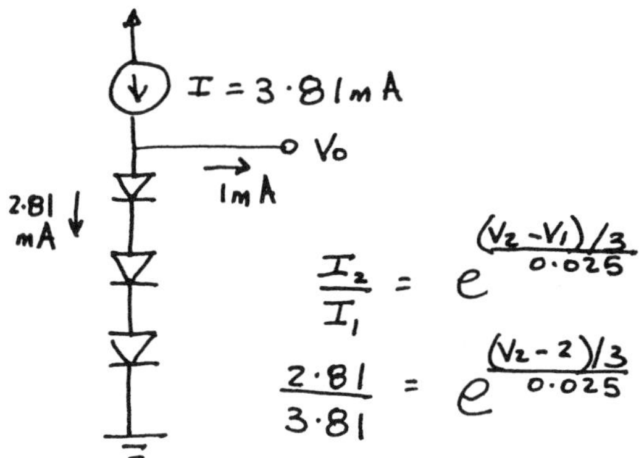

$$\frac{I_2}{I_1} = e^{\frac{(V_2-V_1)/3}{0.025}}$$

$$\frac{2.81}{3.81} = e^{\frac{(V_2-2)/3}{0.025}}$$

$$\Delta V = V_2 - 2 = \underline{\underline{-22.8\,mV}}$$

3.24

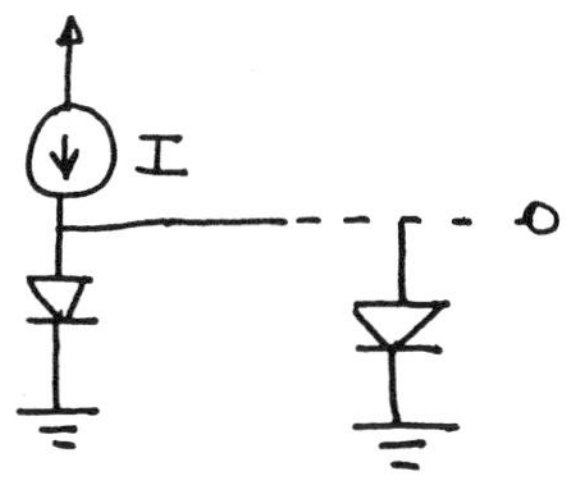

With one diode the current through it is

$$I = I_s e^{V_1/nV_T}$$

With two diodes in parallel, the current splits between each diode so that the diodes each has half the current

$$\frac{I}{2} = I_s e^{V_2/nV_T}$$

$$\therefore \frac{I/2}{I} = e^{\frac{V_2-V_1}{nV_T}}$$

The change in voltage is

$$\Delta V = V_2 - V_1 = nV_T \ln\left(\tfrac{1}{2}\right) = \underline{\underline{-17.3\,mV}}$$

3.25

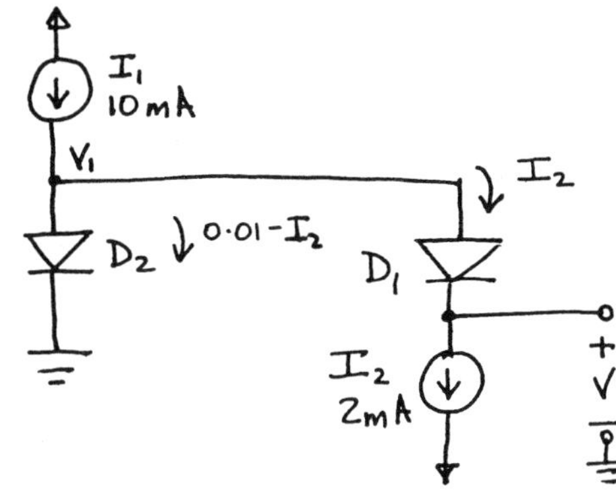

The current through D_1 is

$$10 I_s e^{\frac{V_1-V}{nV_T}} = I_2 \quad (A)$$

The current through D_2 is

$$I_s e^{\frac{V_1}{nV_T}} = 0.01 - I_2$$

$$I_s = (0.01 - I_2)\, e^{\frac{-V_1}{nV_T}} \quad (B)$$

(B) → (A)

$$10\,(0.01 - I_2)\, e^{\frac{-V}{nV_T}} = I_2$$

$$V = -V_T \ln\left(\frac{I_2}{10(0.01 - I_2)}\right)$$

$$= 0.025 \ln\left(\frac{2}{10(8)}\right) = \underline{\underline{92.2\text{ mV}}}$$

For $V = 50\text{ mV}$

$$-V_T \ln\left(\frac{I_2}{10(10 - I_2)}\right) = 50\times10^{-3}$$

$$I_2 = 10(10 - I_2)e^{-2}$$

$$I_2(1 + 10e^{-2}) = 100e^{-2}$$

$$I_2 = \underline{\underline{5.75\text{ mA}}}$$

3.26

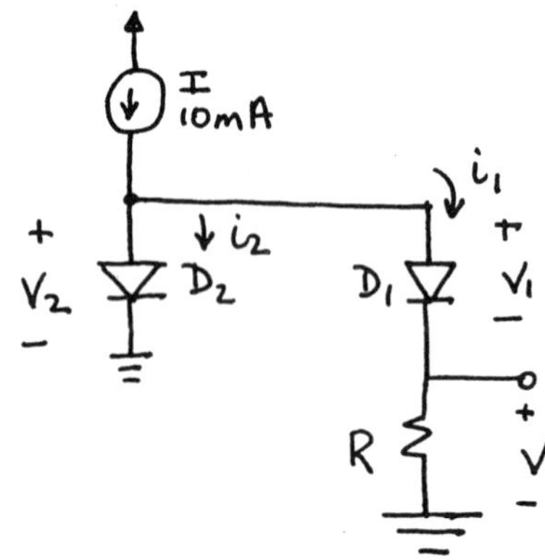

Given for each diode

$i = I_S e^{V/nV_T} \Rightarrow 10\times10^{-3} = I_S e^{0.7/n\times0.025}$ ①

$100\times10^{-3} = I_S e^{0.8/n\times0.025}$ ②

②/① $\quad 10 = e^{0.1/n(0.025)}$

$n = 1.737$

$$V = V_2 - V_1 = nV_T \ln(i_2/i_1) = 80\text{ mV}$$

$$1.737(25\times10^{-3}) \ln\left(\frac{0.01 - i_1}{i_1}\right) = 80$$

$i_1 = 1.4\text{ mA}$

$R = 80/i_1 = 80/1.4 = \underline{\underline{57.1\,\Omega}}$

3.27

At a constant temperature, the diode voltage drop changes with current according to

$$\Delta V = V_T \ln\left(\frac{I_2}{I_1}\right)$$

where

$$V_T = \frac{kT}{q} = \frac{1.38\times10^{-23}(273 + \text{Temp. }°C)}{1.6\times10^{-19}}$$

Thus:

Temp (°C)	0	50	75	100	−50
V_T (mV)	23.5	27.9	30	32.2	19.2

At a constant current, the diode voltage drop changes with temperature according to

$$\Delta V = -2\text{ (mV)} \times \text{Temperature change (°C)}$$

Thus:

(a) 620 mV at 10 μA and 0°C
728 mV at 1 mA and 0°C
678 mV at 1 mA and 25°C

3.28

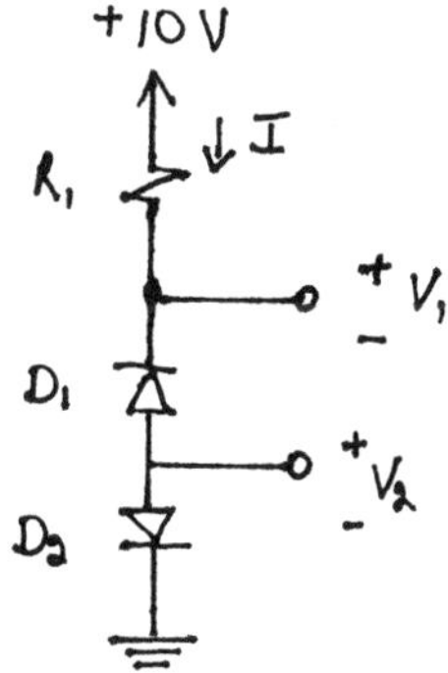

At 20°C:

$V_{R1} = V_2 = 520 \text{ mV}$

$R_1 = 520 \text{ k}\Omega$

$I = \dfrac{520 \text{ mV}}{520 \text{ k}\Omega} = 1 \mu A$

At 40°C, $I = 4\mu A$

$V_2 = 480 + 2.3 \times 1 \times 25 \log 4$
$= 514.6 \text{ mV}$

$V_{R1} = 4\mu A \times 520 \text{ k}\Omega = \underline{\underline{2.08 V}}$

At 0°C, $I = \frac{1}{4}\mu A$

$V_2 = 560 - 2.3 \times 1 \times 25 \log 4$
$= \underline{\underline{525.4 \text{ mV}}}$

$V_{R1} = \frac{1}{4} \times 520 = \underline{\underline{0.13 V}}$

3.29

The voltage drop = 700 − 580 = 120 mV
Since the diode voltage decreases by approximately 2mV for every 1°C increase in temperature, the junction temperature must have increased by

$\frac{120}{2} = \underline{\underline{60 °C}}$

Power being dissipated =
$580 \times 10^{-3} \times 15 = \underline{\underline{8.7 W}}$

Thermal Resistance = temperature rise/watt
$= 60/8.7 = \underline{\underline{6.9 °C/W}}$

3.30

$i = I_S e^{v/nV_T}$

$10 = I_S e^{0.8/2(0.025)}$

$I_S = 1.12 \times 10^{-6} A$

For current varying between $i_1 = 0.5 mA$ to $i_2 = 1.5 mA$, the voltage varies from

$v_1 = 2(0.025) \ln\left(\frac{0.5 \times 10^{-3}}{1.12 \times 10^{-6}}\right) = \underline{\underline{0.305 V}}$

to:

$v_2 = 2(0.025) \ln\left(\frac{1.5 \times 10^{-3}}{1.12 \times 10^{-6}}\right) = \underline{\underline{0.360 V}}$

∴ the voltage decreases by approximately 2mV for every 1°C increase in temperature, the voltage may vary by ∓50mV for the ±25°C temperature variation.

3.31

$i = I_S e^{v/nV_T}$

$\frac{I_{S2}}{I_{S1}} = \frac{1}{0.1 \times 10^{-3}} = 10^4$

For identical currents

$I_{S1} e^{v_1/nV_T} = I_{S2} e^{v_2/nV_T}$

$e^{\frac{v_1 - v_2}{nV_T}} = 10^4$

$v_1 - v_2 = nV_T \ln 10^4$

$= 25 \times 10^{-3} \ln 10^4$

$= +0.23\text{V}$

I.E. THE VOLTAGE DIFFERENCE BETWEEN THE TWO DIODES IS +0.23 V INDEPENDENT OF THE CURRENT. HOWEVER, SINCE THE TWO CURRENTS CAN VARY BY A FACTOR OF 3 (0.5 mA TO 1.5 mA) THE DIFFERENCE VOLTAGE WILL BE:

$0.23\text{ V} \pm nV_T \ln 3 = 0.23\text{ V} \pm 2.75\text{ mV}$

SINCE TEMPERATURE CHANGE AFFECTS BOTH DIODES SIMILARLY THE DIFFERENCE VOLTAGE REMAINS CONSTANT.

3.32

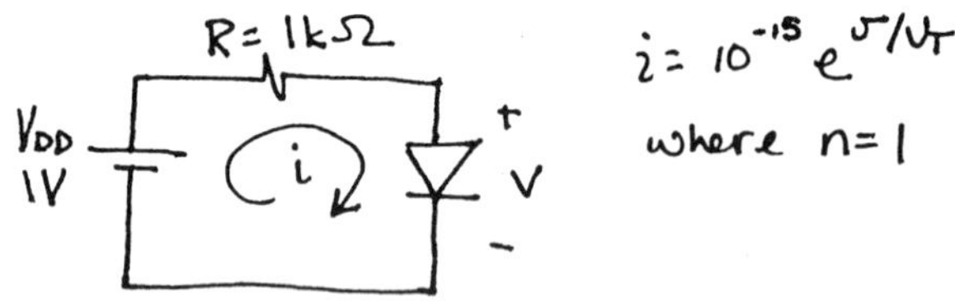

$i = 10^{-15} e^{v/V_T}$

where $n = 1$

$v = 0.7\text{V} \quad i = 1.45\text{ mA}$

$v = 0.6\text{V} \quad i = 0.026\text{ mA}$

A sketch of the graphical construction to determine the operating point is shown below.

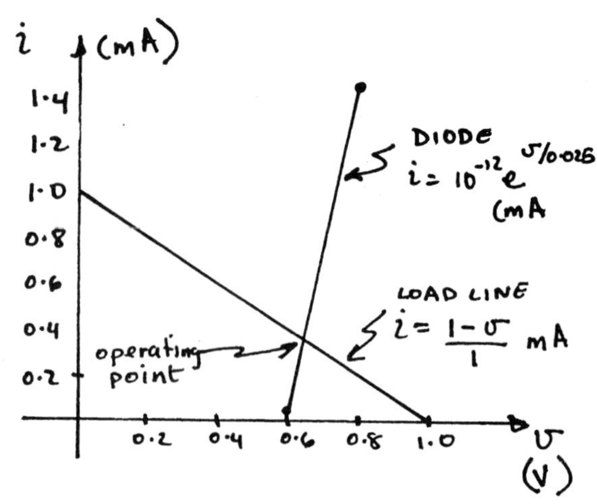

From the above sketch we see that the operating point must lie between $v = 0.6$ and 0.7 V and $i \approx 0.3$ to 0.4 mA. To find the point more accurately an enlarged graph is plotted.

For $i = 0.3\text{mA} = 10^{-12} e^{v/0.025}$

$\Rightarrow v = 660.7\text{ mV}$

For $i = 0.4\text{mA} = 10^{-12} e^{v/0.025}$

$\Rightarrow v = 667.9\text{ mV}$

For the load line:

$v = 660\text{mV} \Rightarrow i = 0.34\text{mA}$

$v = 670\text{mV} \Rightarrow i = 0.33\text{mA}$

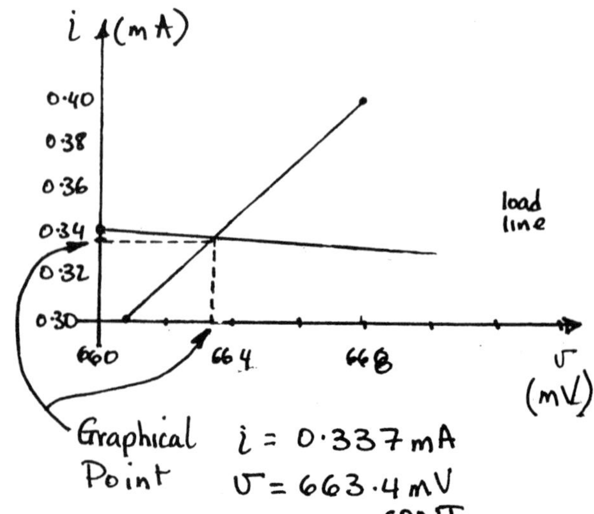

Graphical Point $i = 0.337\text{mA}$

$v = 663.4\text{mV}$

CONT.

Comparing the graphical results to the exponential model gives:

At $i = 0.337\,mA = 10^{-12} e^{v/0.025}$

$\Rightarrow v = 663.6\,mV$

which is only $(663.6 - 663.4) = \underline{\underline{0.2\,mV}}$ greater than the value found graphically!

3.33

Iterative Analysis:

1kΩ

+ 1V −

i

+ v −

$I_S = 10^{-12}\,mA$

$n = 1$

#1 $v = 0.7V \quad i = \frac{1-0.7}{1} = 0.3\,mA$

#2 $v = 0.25\ln\left(\frac{0.3}{10^{-12}}\right) = 0.6607V$

$i = \frac{1-0.6607}{1} = 0.3393\,mA$

#3 $v = 0.25\ln\left(\frac{0.3393}{10^{-12}}\right) = \underline{\underline{0.6638V}}$

$i = \frac{1-0.6638}{1} = \underline{\underline{0.3362}}$

$\therefore$ i did not change by much stop here.

3.34

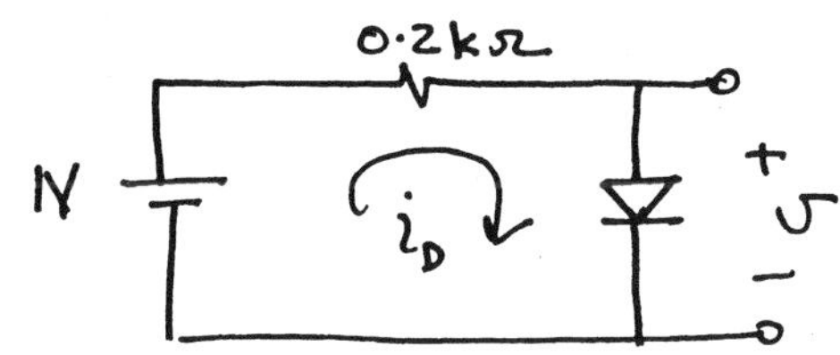

(a) $i_D = \frac{1-0.7}{0.2} = \underline{\underline{1.5mA}}$

(b) Iterative Analysis given $v_D = 0.7V$ at $i_D = 1mA$

#1 $v = 0.7V \quad i_D = \frac{1-0.7}{0.2} = 1.5mA$

#2 $\therefore i = I_S e^{v/nV_T} \qquad n = 2$

$\frac{i_2}{i_1} = e^{\frac{v_2 - v_1}{0.05}}$

thus $v_2 = v_1 + 0.05\ln i_2/i_1$

$\therefore$ for $i = 1.5\,mA$

$v = 0.7 + 0.05\ln\frac{1.5}{1} = 0.720V$ & $i_D = \frac{1-0.720}{0.2} = 1.4mA$

#3

$v = 0.720 + 0.05\ln\left(\frac{1.4}{1.5}\right) = 0.716V$ & $i_D = \frac{1-0.716}{0.2} = 1.42mA$

#4

$v = 0.716 + 0.05\ln\left(\frac{1.42}{1.4}\right) = \underline{\underline{0.716V}}$ & $i_D = \underline{\underline{1.42mA}}$

3.35

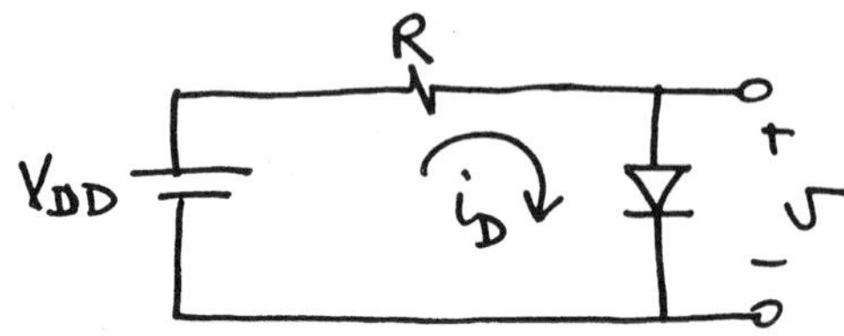

Derivation of iterative equation

$$i_D = I_s e^{v/V_T}$$

$$\frac{i_{D2}}{i_{D1}} = e^{\frac{v_2 - v_1}{n V_T}}$$

$$v_2 = v_1 + n V_T \ln\left(\frac{i_{D2}}{i_{D1}}\right)$$

$$= v_1 + \Delta V \log\left(\frac{i_{D2}}{i_{D1}}\right)$$

(a) $V = 0.7\text{ V}$ $\quad i_D = \frac{10 - 0.7}{9.3} = 1\text{ mA}$

(b) $v = 0.7\text{V}$ $\quad i_D = \frac{3 - 0.7}{2.3} = 1\text{mA}$

~ for both these cases the diode is rated at 1mA for 0.7V so stop.

(c) $V_{DD} = 2\text{V}$ $\quad R = 2\text{k}\Omega$

#1 $v = 0.7\text{V}$ $\quad i_D = \frac{2 - 0.7}{2} = 0.65\text{mA}$

#2 $v = 0.7 + 0.1\log\left(\frac{0.650}{10}\right) = 0.581\text{V}$

$i_D = \frac{2 - 0.581}{2} = 0.709\text{ mA}$

#3 $v = 0.581 + 0.1\log\left(\frac{0.709}{0.650}\right)$

$= \underline{0.584\text{V}}$

$i_D = \frac{2 - 0.584}{2} = \underline{\underline{0.708\text{mA}}}$

(d) $V_{DD} = 2\text{V}$ $\quad R = 2\text{k}\Omega$

#1 $V = 0.7\text{V}$ $\quad i_D = \frac{2 - 0.7}{2}$

$= 0.650\text{ mA}$

#2.

$V = 0.7 + 0.1\log\left(\frac{0.650}{1}\right)$ $\quad i_D = \frac{2 - 0.681}{2}$

$= 0.681\text{V}$ $\quad = 0.659\text{ mA}$

#3

$V = 0.681 + 0.1\log\left(\frac{0.659}{0.650}\right)$ $\quad i_D = \frac{2 - 0.682}{2}$

$= \underline{\underline{0.682\text{V}}}$ $\quad = \underline{\underline{0.659\text{mA}}}$

(e) $V_{DD} = 1\text{V}$ $\quad R = 0.3\text{k}\Omega$

#1 $v = 0.7\text{V}$ $\quad i_D = \frac{1 - 0.7}{0.3} = 1\text{mA}$

#2

$v = 0.7 + 0.1\log\frac{1}{10}$ $\quad i_D = \frac{1 - 0.6}{0.3} = 1.333\text{mA}$

$= 0.6\text{V}$

#3

$v = 0.6 + 0.1\log\frac{1.333}{1}$ $\quad i_D = \frac{1 - 0.612}{0.3} = 1.293\text{ mA}$

$= 0.612\text{V}$

#4

$v = 0.612 + 0.1\log\frac{1.293}{1.333}$ $\quad i_D = \frac{1 - 0.611}{0.3} = 1.297\text{mA}$

$= 0.611\text{V}$

#5

$v = 0.611 + 0.1\log\frac{1.297}{1.293}$ $\quad i_D = \underline{\underline{1.297\text{mA}}}$

$= \underline{\underline{0.611\text{ V}}}$

(f) $V_{DD} = 1\text{V}$ $\quad R = 0.3\text{k}\Omega$

#1 $v = 0.7\text{ V}$ $\quad i_D = \frac{1 - 0.7}{0.3} = 1\text{mA}$

CONT.

#2

$v = 0.7 + 0.06 \log \frac{1}{10}$ $\quad i_D = \frac{1-0.640}{0.3}$

$= 0.640\text{ V}$ $\quad = 1.2\text{ mA}$

#3

$v = \quad + 0.06 \log \frac{1.2}{1}$ $\quad i_D = \frac{1-0.645}{0.3}$

$= 0.645\text{ V}$ $\quad = 1.183\text{ mA}$

#4

$v = \quad + 0.06 \log \frac{1.183}{1.2}$

$= \underline{\underline{0.645\text{ V}}}$ $\quad i_D = \underline{\underline{1.183\text{ mA}}}$

(g) $V_{DD} = 1\text{ V}$ $\quad R = 0.3\text{ k}\Omega$

#1 $V = 0.7$ $\quad i_D = \frac{1-0.7}{0.3} = 1\text{ mA}$

#2

$v = 0.7 + 0.12 \log \frac{1}{10}$ $\quad i_D = \frac{1-0.580}{0.3}$

$= 0.580\text{ V}$ $\quad = 1.381\text{ mA}$

#3

$v = 0.680 + 0.12 \log \frac{1.381}{1}$ $\quad i_D = \frac{1-0.697}{0.3}$

$= 0.697\text{ V}$ $\quad = 1.343\text{ mA}$

#4

$v = 0.697 + 0.12 \log \frac{1.343}{1.381}$ $\quad i_D = \frac{1-0.596}{0.3}$

$= 0.596\text{ V}$ $\quad = 1.347\text{ mA}$

#5

$v = 0.596 + 0.12 \log \frac{1.347}{1.343}$

$= \underline{\underline{0.596\text{ V}}}$ $\quad i_D = \underline{\underline{1.347\text{ mA}}}$

(h) $V_{DD} = 0.5\text{ V}$ $\quad R = 30\text{ k}\Omega$

#1 let $v = 0.4\text{ V}$ $\quad i_D = \frac{0.5-0.4}{30}$

$= 3.333\ \mu\text{A}$

#2

$v = 0.7 + 0.1 \log \frac{3.333 \times 10^{-3}}{10}$ $\quad i_D = \frac{0.5-0.352}{30}$

$= 0.352\text{ V}$ $\quad = 4.933\ \mu\text{A}$

#3

$v = 0.352 + 0.1 \log \frac{4.933}{3.333}$ $\quad i_D = \frac{0.5-0.369}{30}$

$= 0.369\text{ V}$ $\quad = 4.367\ \mu\text{A}$

#4

$v = 0.369 + 0.1 \log \frac{4.367}{4.933}$ $\quad i_D = \frac{0.5-0.364}{30}$

$= 0.364\text{ V}$ $\quad = 4.533\ \mu\text{A}$

#5

$v = 0.364 + 0.1 \log \left(\frac{4.533}{4.367}\right)$ $\quad i_D = \frac{0.5-0.366}{30}$

$= 0.366\text{ V}$ $\quad = 4.467\ \mu\text{A}$

#6

$v = 0.366 + 0.1 \log \frac{4.467}{4.533}$ $\quad i_D = \frac{0.5-0.365}{30}$

$= 0.365\text{ V}$ $\quad = 4.5\ \mu\text{A}$

#7

$v = 0.365 + 0.1 \log \frac{4.5}{4.467}$

$= \underline{\underline{0.365\text{ V}}}$ $\quad i_D = \underline{\underline{4.5\ \mu\text{A}}}$

3.36

10V

R

3V

i_D

V_D

$V_D = \frac{3}{4} = 0.75\text{ V}$

$i_D = I_S e^{V_D/nV_T}$

$\frac{i_{D2}}{i_{D1}} = e^{\frac{V_{D2}-V_{D1}}{nV_T}}$

CONT.

$$\therefore\ i_D = i_{D2} = i_{D1}\, e^{\frac{V_{D2}-V_{D1}}{nV_T}}$$

$$= 1 \times e^{\frac{0.75-0.7}{1\times 0.025}}$$

$$= 7.389\ mA$$

$$\therefore\ R = \frac{10-3}{I_D} = \frac{10-3}{7.389} = \underline{\underline{0.947k\Omega}}$$

3.37

Piecewise linear model:

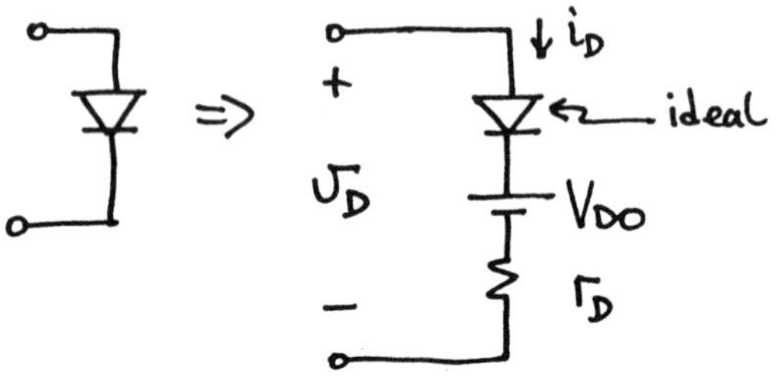

Given $n=2$, $V_D=0.7V$ $i_D=1mA$

The current throug the diode is given by:

$$i_D = \frac{V_D - V_{DO}}{r_D}$$ ~ need to find the parameters V_{DO} and r_D

Using the exponential model to find the diode voltage at 10mA

$$\frac{i_{D2}}{i_{D1}} = e^{\frac{V_{D2}-V_{D1}}{nV_T}}$$

$$V_{D2} = nV_T \ln\left(\frac{i_{D2}}{i_{D1}}\right) = 0.05\ln\left(\frac{10}{1}\right)$$

$$= 0.815V$$

Finding V_{DO} & r_D using the given facts:

$$i_D = \frac{0.7 - V_{DO}}{r_D} = 1mA \quad ①$$

$$i_D = \frac{0.815 - V_{DO}}{r_D} = 10mA \quad ②$$

$$②/① \Rightarrow \frac{0.815 - V_{DO}}{0.7 - V_{DO}} = 10$$

$$V_{DO} = \underline{0.687V}$$

$$r_D = \frac{0.7 - V_{DO}}{1} = \underline{\underline{12.8\ \Omega}}$$

Using the piecewise linear model

$$i_D = \frac{V_D - 0.687}{12.8} \Rightarrow V_D = 0.687 + 12.8\, i_D$$

Using the exponential model

$$\frac{i_{D2}}{i_{D1}} = e^{\frac{V_D - 0.7}{0.05}} \Rightarrow V_D = 0.7 + 0.05\ln\left(\frac{i_{D2}}{1}\right)$$

i_D (mA)	PIECEWISE LINEAR V_D (V)	EXPONENTIAL V_D (V)	Error (mV)
0.5	0.693	0.655	28
5	0.751	0.780	−29.5
14	0.866	0.832	34

3.38

Looking at the copy of FIG 3.12 below, we see at

$$i_D = 1mA \rightarrow V_D = 0.7V$$

$$i_D = 10mA \rightarrow V_D = 0.8V$$

$$\therefore\ \text{slope} = \frac{1}{r_D} = \frac{10-1}{0.8-0.7} = 90\frac{mA}{V}$$

$$\therefore\ r_D = \frac{1}{90\times10^{-3}} = \underline{\underline{11.1\Omega}}$$

To find v_{DO}:

$$i_D = \frac{v_D - V_{DO}}{r_D}$$

$$10^{-3} = \frac{0.7 - V_{DO}}{11.1} \Rightarrow V_{DO} = 0.689\text{V}$$

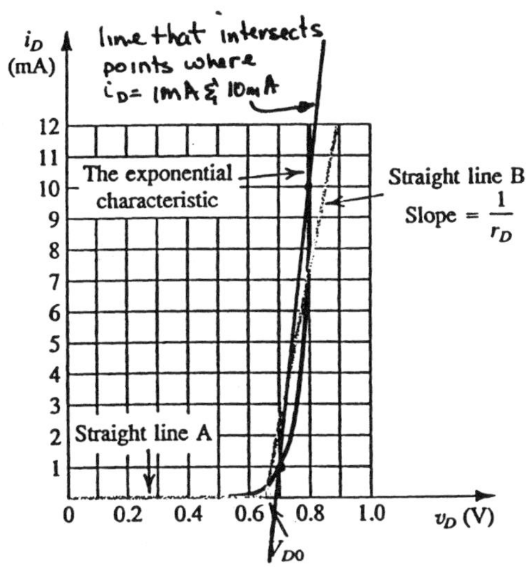

3.39

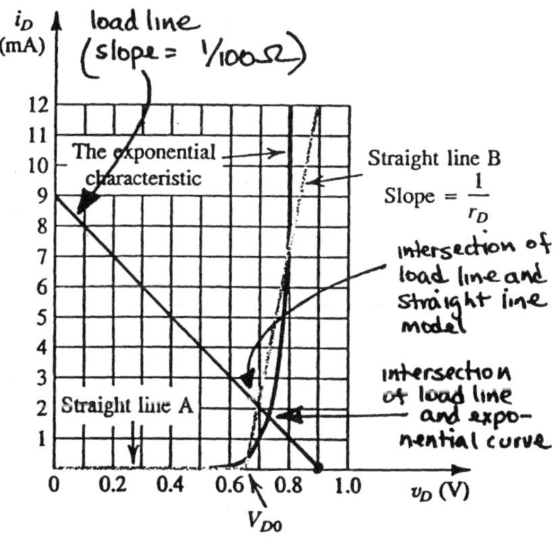

(a) The load line intersects the exponential model at:

$v_D = \underline{0.73\text{V}}$ $\quad i_D = \underline{1.7\text{mA}}$

(b) The load line intersects the straight-line model at

$v_D = \underline{0.7\text{V}}$ $\quad i_D = \underline{2\text{mA}}$

3.40

Calculating the parameters r_D & V_{DO} for the battery plus resistor model

$$i_D = I_S e^{v_D/nV_T} \qquad n = 1$$

For $i_D = 0.1\, I_D$

$$v_{D2} = 0.7 + 0.025 \ln(0.1) = \underline{0.642\text{V}}$$

For $i_D = 10 I_D$

$$v_{D3} = 0.7 + 0.025 \ln(10) = \underline{0.758\text{V}}$$

Note that since the specifications for all of the diodes are given for 0.7V, the end voltages are the same as the voltage change for relative currents are independent to I_D & I_S.

$\therefore$ $v_{D2} = V_{DO} + i_{D2} r_D$ ①

$\quad v_{D3} = V_{DO} + i_{D3} r_D$ ②

② → ①

$$v_{D3} - v_{D2} = (i_{D3} - i_{D2}) r_D$$

CONT.

$$0.758 - 0.642 = (10 I_D - 0.1 I_D)\, r_D$$

$$0.166 = 9.9\, I_D r_D$$

$$r_D = \frac{0.0117}{I_D} \quad (3)$$

for (a) $I_D = 1\text{mA}$ $\quad r_D = \frac{0.0117}{1} = \underline{11.7\Omega}$

(b) $I_D = 1\text{A}$ $\quad r_D = \frac{0.0117}{1} = \underline{0.0117\Omega}$

(c) $I_D = 10\mu\text{A}$ $\quad r_D = \frac{0.0117}{10\mu A} = \underline{1.17k\Omega}$

(3) → (1)

$$0.642 = V_{DO} + 0.1 I_D \times \frac{0.0117}{I_D}$$

$$= V_{DO} + 0.00117$$

$V_{DO} = \underline{0.641V}$ ← same for all diodes

3.41

Since for a current of 10 mA, the diode voltage is $\underline{0.8V}$, this would be a suitable choice for the constant-voltage-drop model.

3.42

Constant Voltage drop Model:

Using $v_D = 0.7V \Rightarrow i_{D1} = \frac{V - 0.7}{R}$

Using $v_D = 0.6V \Rightarrow i_{D2} = \frac{V - 0.6}{R}$

For the difference in currents to vary by only 1% ⇒

$$i_{D2} = 1.01\, i_{D1}$$

$$V - 0.6 = 1.01\,(V - 0.7)$$

$$V = \underline{10V}$$

For $V = 2V$ & $R = 1k\Omega$

At $V_D = 0.7V$ $\quad i_{D1} = \frac{2-0.7}{1} = 1.3mA$

$V_D = 0.6V$ $\quad i_{D2} = \frac{2-0.6}{1} = 1.4mA$

$$\frac{i_{D2}}{i_{D1}} = \frac{1.4}{1.3} = 1.08$$

Thus the percentage difference is $\underline{8\%}$

3.43

Since $2V_D = 1.4V$ is close to the required 1.25V, use N parallel pairs of diodes to split the current evenly.

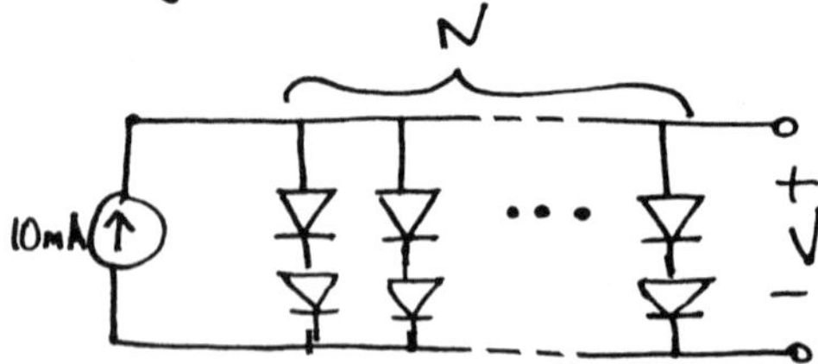

$$\therefore V = 2\left[0.7 + 0.1 \log \frac{10/N}{20}\right] = 1.25V$$

$N = 2.8$ ⇒ Use $\underline{3 \text{ sets of diodes}}$

$$V = 2\left(0.7 + 0.1 \log \frac{10/3}{20}\right) = \underline{1.244V}$$

3.44

Piecewise linear model in a half-wave rectifier.

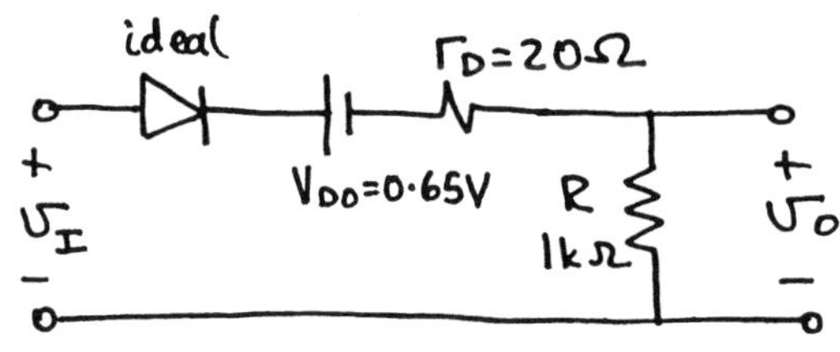

$$v_O = \frac{v_I - V_{DO}}{r_D + R} R, \quad \text{for } v_I \geq V_{DO}$$

Thus, $v_O = 0.98(v_I - 0.65)$ for $v_I \geq 0.65$

$v_O = 0$ for $v_I < 0.65V$

Sketch of transfer characteristic :-

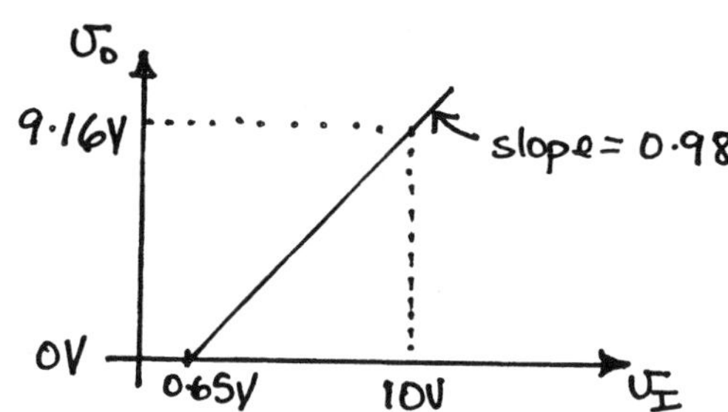

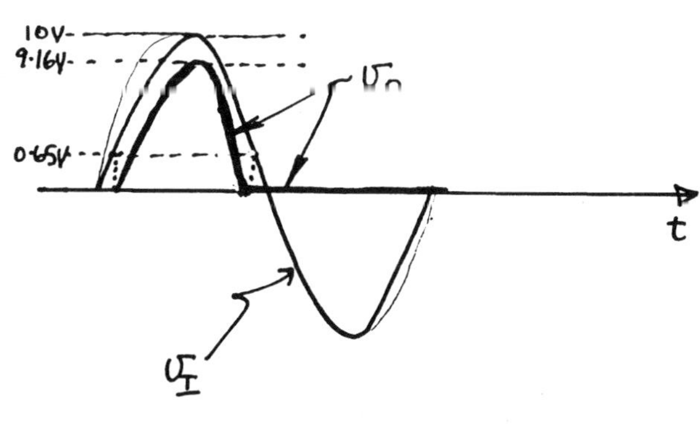

3.45

Refer to example 3-2 ~ CONSTANT VOLTAGE DROP MODEL

(a)

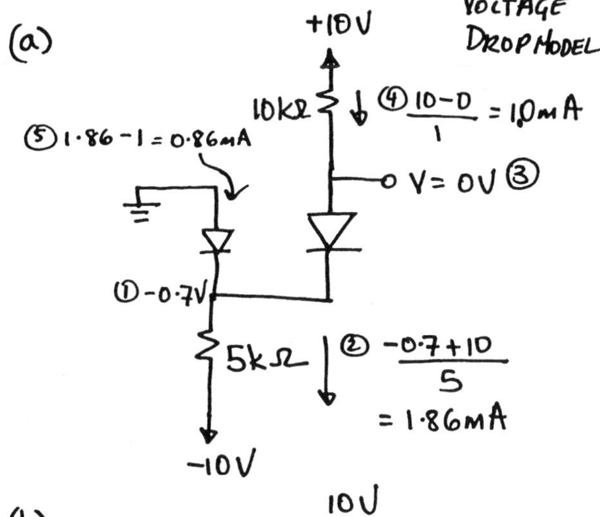

(b)

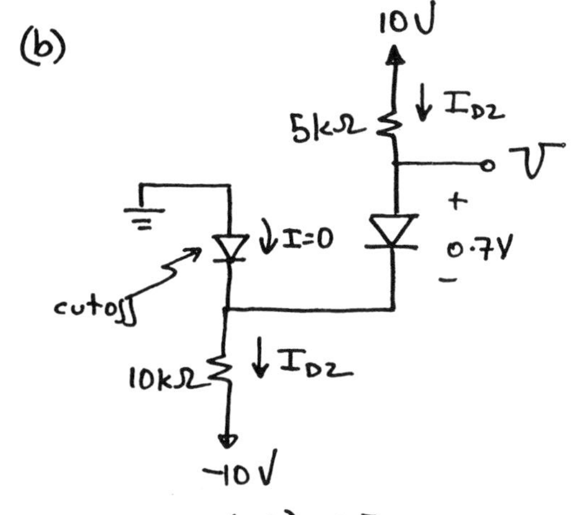

$$I_{D2} = \frac{10 - (-10) - 0.7}{15} = 1.29\text{mA}$$

$$V = -10 + 1.29(10) + 0.7 = \underline{3.6V}$$

3.46

(a)

3V

10kΩ

$V = -3 + 0.7 = \underline{-2.3V}$

$$I = \frac{3 + 2.3}{10}$$

$= \underline{0.53\text{mA}}$

3V

CONT.

(b)

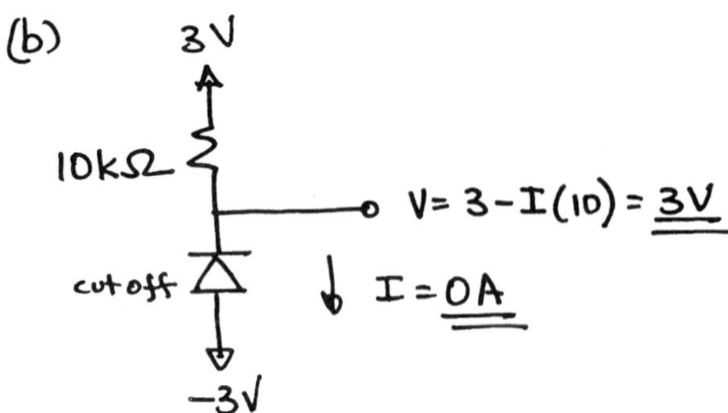

(c)

3V

↓ $I = \frac{2.3+3}{10} = 0.53\text{mA}$

$3 - 0.7 = 2.3\text{V}$

10kΩ

−3V

(d)

3V

cutoff

↓ $I = 0\text{A}$

$V = -3\text{V}$

−3V

3.47

(a)

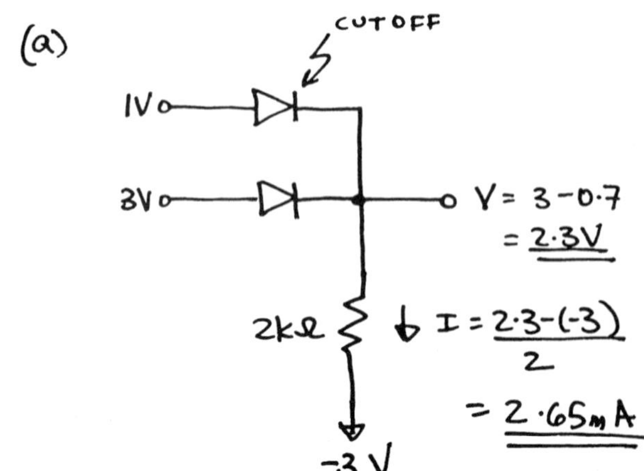

(b)

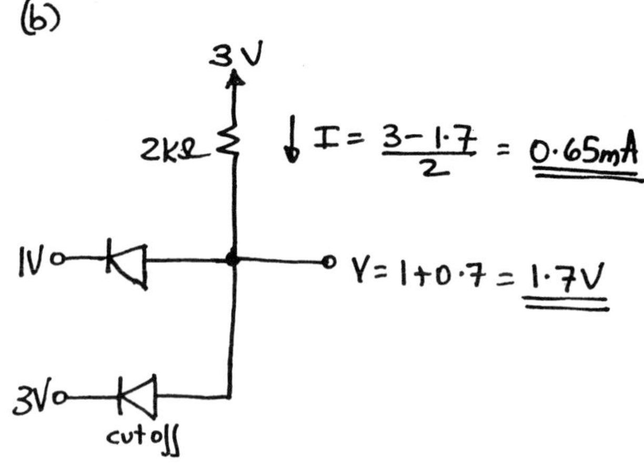

3.48

(a)

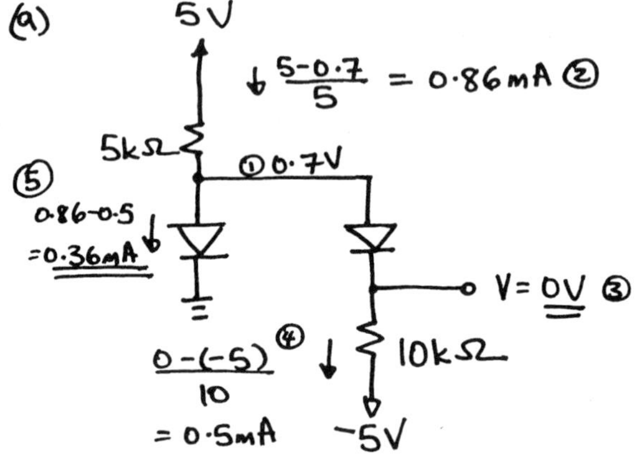

(b)

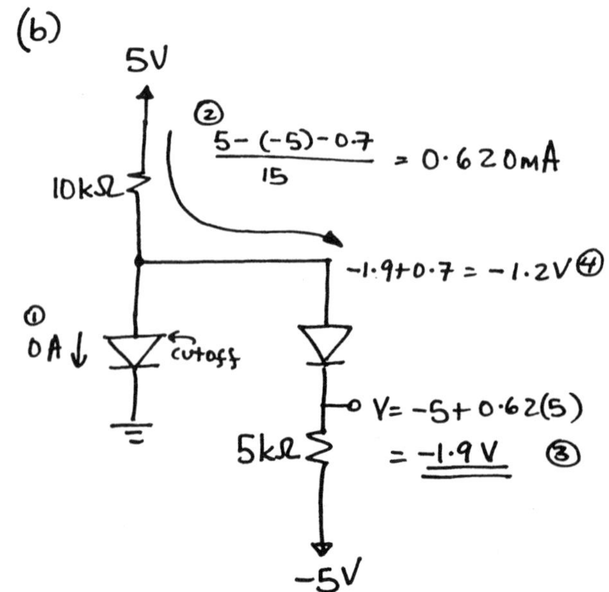

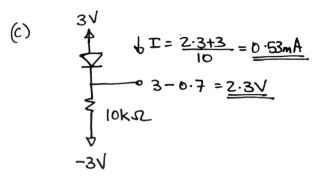

3.49

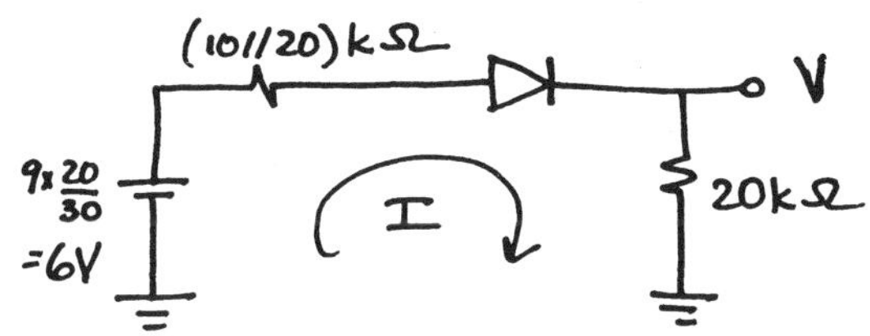

$$I = \frac{6-0.7}{(101/20)+20} = \underline{\underline{0.199\,mA}}$$

$$V = 20I = \underline{\underline{3.98V}}$$

(b)

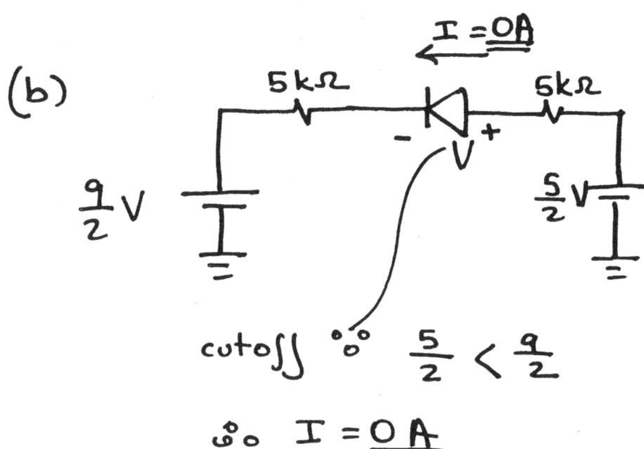

cutoff ∴ $\frac{5}{2} < \frac{9}{2}$

∴ $I = \underline{\underline{0A}}$

$$V = 5/2 - 9/2 = \underline{\underline{-2V}}$$

3.50

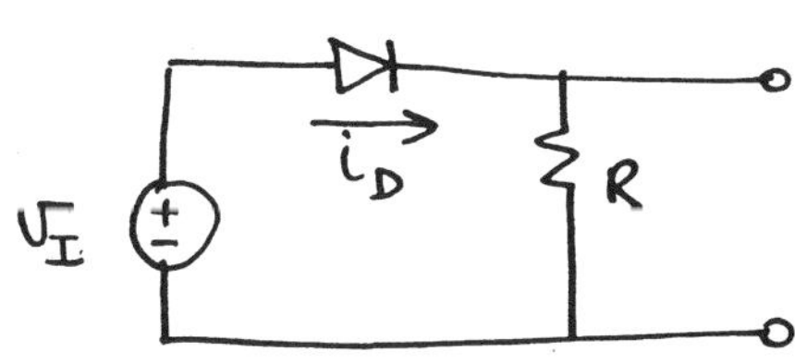

$$i_{D,peak} = \frac{v_{I,peak} - 0.7}{R} \leq 50$$

$$R \geq \frac{120\sqrt{2} - 0.7}{50} = \underline{\underline{3.38k\Omega}}$$

Reverse voltage $= 120\sqrt{2} = 169.7V$.
The design is essentially the same since the supply voltage $>> 0.7V$

3.51

Battery plus resistance model

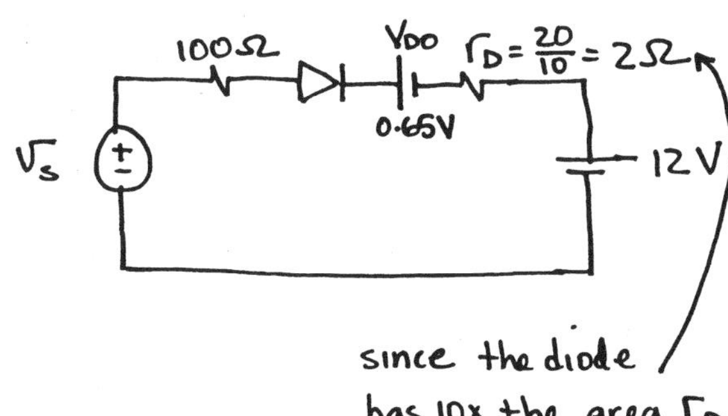

since the diode has 10x the area r_D is 1/10 as big.

$$i_D = I_s e^{v/nV_T} = \frac{V_s - V_{DO} - 12}{100 + 2}$$

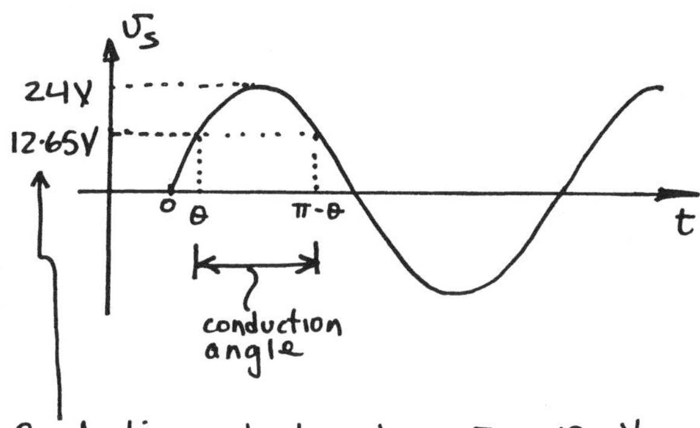

Conduction starts when $v_s > 12 + V_{DO}$
$v_s > 12.65V$

∴ $24 \sin\theta = 12.65°$
$\theta = 0.555$ rad

Conduction angle $= \pi - 2\theta$
$= 2.031$ rad.

Fraction of cycle for conduction $= \frac{2.031}{2\pi} - \underline{0.323}$

CONT.

$$i_{D,peak} = \frac{24 - 12.65}{100 + 2} = \underline{0.111\ A}$$

Maximum reverse voltage occurs across the diode when v_S is at its negative peak and is equal to:

$$24 + 12 = \underline{36V}$$

3.52

Using the exponential model

$$i_D = I_S e^{\Delta V / nV_T}$$

FOR A +10mV CHANGE

$$\frac{i_{D2}}{i_{D1}} = e^{\Delta V/nV_T} = e^{0.01/n(0.025)}$$

$$= \begin{cases} 1.492 & \sim n=1 \\ 1.221 & \sim n=2 \end{cases}$$

$$\% \text{ CHANGE} = \frac{i_{D2} - i_{D1}}{i_{D1}} \times 100$$

$$= \begin{cases} (1.492 - 1) \times 100 = \underline{+49.2\%} & n=1 \\ (1.221 - 1) \times 100 = \underline{22.1\%} & n=2 \end{cases}$$

FOR A −10mV CHANGE

$$\frac{i_{D2}}{i_{D1}} = 10^{-0.01/n(0.025)} = \begin{cases} 0.670 & n=1 \\ 0.819 & n=2 \end{cases}$$

$$\% \text{CHANGE} = \begin{cases} (0.670 - 1)100 = \underline{-33\%} & n=1 \\ (0.819 - 1)100 = \underline{-18\%} & n=2 \end{cases}$$

For a current change limited to ±10%

$$\frac{i_{D2}}{i_{D1}} = e^{\Delta V / n \times 0.025} = 0.9 \text{ to } 1.1$$

$$\Delta V = \begin{cases} \underline{-2.634mV \text{ to } 2.383mV} & n=1 \\ \underline{-5.268\,mV \text{ to } 4.766\,mV} & n=2 \end{cases}$$

3.53

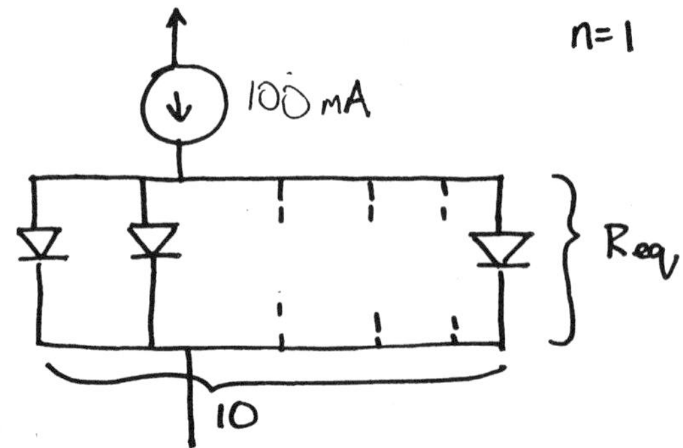

Each diode has the current

$$i_D = \frac{0.1}{10} = 0.01\,A$$

Each diode has a small-signal resistance

$$r_d = \frac{nV_T}{I_D} = \frac{0.025}{0.01} = \underline{2.5\Omega}$$

$$R_{eq} = r_d/10 = \underline{0.25\,\Omega}$$

For one diode conducting 0.1 A

$$r_d = nV_T/0.1 = \frac{0.025}{0.1} = \underline{0.25\Omega}$$

This is the same as R_{eq}. We can think of the parallel connection as equivalent to a single diode having 10 times the junction area of each diode. This large diode is fed with 10x the current (or 0.1A) and this exhibits the same small-signal resistance as 10 parallel smaller diodes.

Now consider the series resistance of 0.2Ω to connect a diode. For the parallel combination above:

$$R_{eq} = \frac{1}{10}(0.2 + 2.5) = 0.27\,\Omega$$

To have an equivalent resistance, the single diode conducting all of the 0.1A would need a series resistance 10× as small or 0.02Ω. Specifically:

$$r_{out} = r_s + r_d = 0.27$$

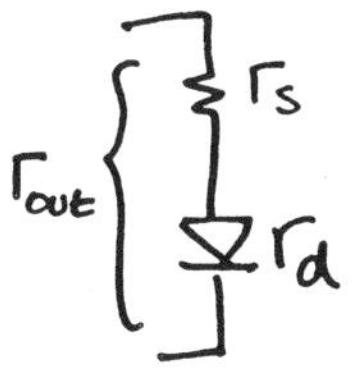

$$= r_s + \frac{nV_T}{I_D} = 0.27$$

$$= r_s + 0.25 = 0.27$$

$$r_s = 0.27 - 0.25 = \underline{\underline{0.02\,\Omega}}$$

3.54

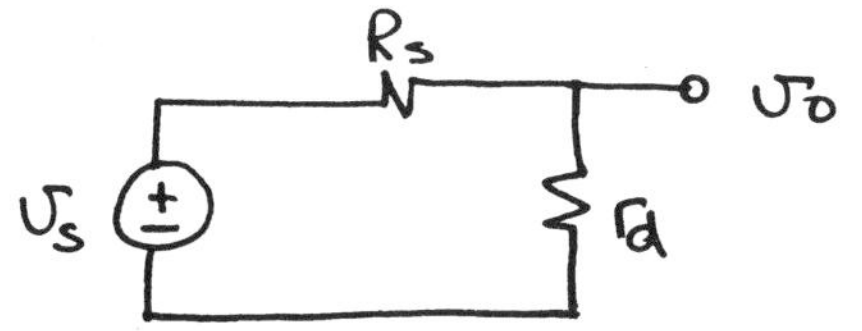

SMALL SIGNAL EQUIVALENT CIRCUIT

To find the small-signal response, v_o, open the dc current source I, and short the capacitors C_1 and C_2. Also replace the diode with its small signal resistance:

$$r_d = \frac{nV_T}{I} \qquad n=2$$

Now:

$$v_o = v_s \frac{r_d}{r_d + R_s}$$

$$= v_s \frac{nV_T/I}{\frac{nV_T}{I} + R_s} = v_s \frac{nV_T}{nV_T + IR_s}$$

Q.E.D.

$$v_o = 10\text{mV}\,\frac{0.05}{0.05 + 10^3 I}$$

$$= \begin{cases} 0.476\text{ mV} & \sim I = 1\text{mA} \\ 3.333\text{ mV} & \sim I = 0.1\text{mA} \\ 9.804\text{ mV} & \sim I = 1\mu\text{A} \end{cases}$$

For $v_o = \frac{1}{2} v_s = v_s \times \frac{0.05}{0.05 + 10^3 I}$

$$\underline{\underline{I = 50\,\mu\text{A}}}$$

3.55

$R_s = 10\text{k}\Omega$, $n = 1$, a 1mA diode

$$v_o/v_I = \frac{0.025}{0.025 + R_s I}$$

$$= \frac{0.025}{0.025 + 10^4 I} \quad (1)$$

For the current change limited to ±10% of I & using the exponential model we get

$$\frac{i_{D2}}{i_{D1}} = e^{\Delta V/nV_T} = 0.9 \text{ to } 1.1$$

CONT.

$\Delta v = -2.63\text{ mV to } 2.38\text{ mV}$

This is the amount the output will vary for a 10% change in diode current. Divide this by the specific gains given in the problem to find the limit on the input signal.

$$\Delta v_S = \frac{\Delta v_O}{v_O/v_I}$$

$$= \frac{-2.63\text{ mV}}{\Delta v_O/v_I} \text{ to } \frac{2.38\text{ mV}}{v_O/v_I} \quad (2)$$

v_O/v_I	I use (1) (mA)	v_S (using (2)) (mV)
0.5	0.0025	5.26 to 4.76
0.1	0.0225	26.3 to 23.8
0.01	0.25	263 to 238
0.001	2.5	2630 to 2380

3.56

$n=1$

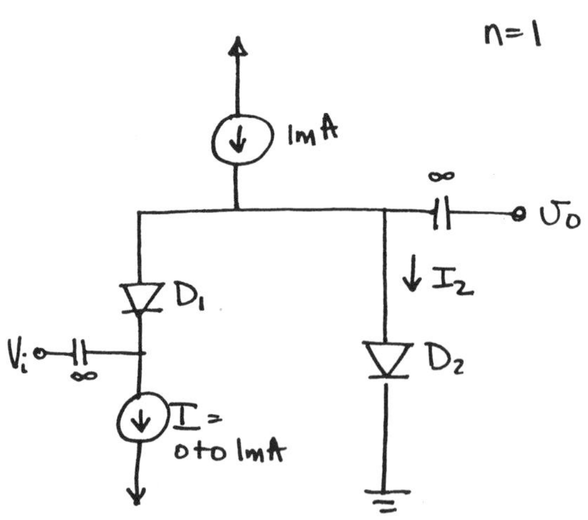

small signal model when D_1 & D_2 are conducting

v_i — r_{d1} — v_O, r_{d2} to ground

(a) $I = 0\,\mu A$

D_1 - cutoff $\Rightarrow \dfrac{v_O}{v_I} = 0\text{ V/V}$

$I_Z = 1\text{mA}$

(b) $I = 1\,\mu A \quad I_Z = 999\,\mu A$

$$r_{d1} = \frac{nV_T}{I} = \frac{0.025}{I} = 25\,\Omega \qquad r_{d2} = \frac{0.025}{999\times10^{-6}} = 25.025\,\Omega$$

$$\frac{v_O}{v_I} = \frac{r_{d2}}{r_{d1}+r_{d2}} = 0.001\,\frac{V}{V}$$

(c) $I = 10\,\mu A \quad I_Z = 990\,\mu A$

$$r_{d1} = \frac{0.025}{10\times10^{-6}} = 2.5\text{k}\Omega \qquad r_{d2} = \frac{0.025}{990\times10^{-6}} = 25.25\,\Omega$$

$$v_O/v_I = 0.01\text{ V/V}$$

(d) $I = 100\,\mu A \quad I_Z = 900\,\mu A$

$$r_{d1} = \frac{0.025}{100\times10^{-6}} = 250\,\Omega \qquad r_{d2} = \frac{0.025}{990\times10^{-6}} = 27.78\,\Omega$$

$$\frac{v_O}{v_I} = 0.1\text{ V/V}$$

(e) $I = 500\,\mu A \quad I_Z = 500\,\mu A$

$$r_{d1} = r_{d2} = \frac{0.025}{500\times10^{-6}} = 50\,\Omega$$

$$\frac{v_O}{v_I} = \frac{1}{2}\text{ V/V}$$

(f) $I = 600\,\mu A \quad I_Z = 400\,\mu A$

$$r_{d1} = \frac{0.025}{600\times10^{-6}} = 41.67\,\Omega \qquad r_{d2} = \frac{0.025}{400\times10^{-6}} = 62.5\,\Omega$$

3.56 CONT.

WHEN THE BIAS CURRENT IN EACH DIODE IS $\geq 10\,\mu A$, THE DIODE RESISTANCE WILL BE $\leq 2.5\,k\Omega$. TO LIMIT THE CURRENT SIGNAL TO A MAXIMUM OF 10% OF BIAS, THE CURRENT SIGNAL MUST BE $\leq 1\,\mu A$. THUS, THE SIGNAL VOLTAGE ACROSS THE "STARVED" DIODE WILL BE 2.5 mV WHICH IS APPROXIMATELY THE VALUE TO WHICH THE INPUT SIGNAL SWING SHOULD BE LIMITED.

3.57

(a) $$\frac{v_o}{v_i} = \frac{R}{R + (2r_d \,//\, 2r_d)}$$

$$= \frac{R}{R + r_d}$$

WHERE $r_d = \dfrac{V_T}{I/2} = \dfrac{2V_T}{I}$

$$= \frac{0.05\,V}{I}$$

r_d r_d r_d r_d v_i v_o R 10 kΩ

I (mA)	v_o/v_i (V/V)
0	0
10^{-3}	0.167
0.01	0.667
0.1	0.952
1.0	0.995
10	0.9995

(b) IF THE SIGNAL CURRENT IS TO BE LIMITED TO $\pm 10 I$, THE CHANGE IN DIODE VOLTAGE Δv_D CAN BE FOUND FROM

$$\frac{i_D}{I} = e^{\Delta v_D / nV_T} = 0.9 \text{ TO } 1.1$$

THUS, FOR $n = 1$

$\Delta v_D = -2.63$ mV TO $+2.38$ mV

OR APPROXIMATELY ± 2.5 mV

CONT.

3.57 CONT.

(b CONT.) FOR THE DIODE CURRENT TO REMAIN WITHIN $\pm 10\%$ OF THEIR dc BIAS CURRENTS, THE SIGNAL VOLTAGE ACROSS EACH DIODE MUST BE LIMITED TO 2.5 mV. NOW, IF $v_{i\,PEAK} = 10$ mV WE CAN OBTAIN THE FOLLOWING SITUATION

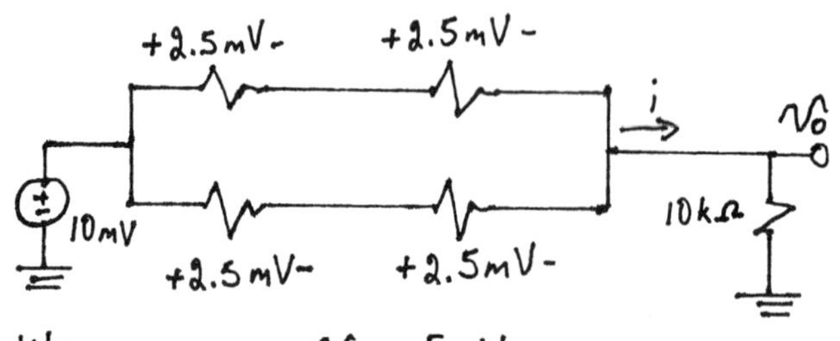

WE SEE THAT $V_o = 5$ mV AND

$$i = \frac{5\,mV}{10\,k\Omega} = 0.5\,\mu A.$$

THUS, EACH DIODE IS CARRYING A CURRENT SIGNAL OF 0.25 mA. FOR THIS TO BE AT MOST 10% OF THE dc CURRENT, THE dc CURRENT IN EACH DIODE MUST BE AT LEAST 2.5 μA. IT FOLLOWS THAT THE MINIMUM VALUE OF I MUST BE **5 μA.**

(c) FOR $I = 1$ mA, $I_D = 0.5$ mA, AND FOR MAXIMUM SIGNAL OF 10%, $I_D = 0.05$ mA. THUS $i_D = 2i_d = 0.1$ mA AND THE CORRESPONDING MAXIMUM v_o IS $0.1\,mA \times 10\,k\Omega = 1\,V$.

THE CORRESPONDING PEAK INPUT CAN BE FOUND BY DIVIDING V_o BY THE TRANSMISSION FACTOR OF 0.995, THUS

$$v_{i\,MAX} = \frac{1\,V}{0.995\,V} = \underline{1.005\,V}$$

SEE FIGURE.
EACH DIODE HAS $r_d = 50\,\Omega$

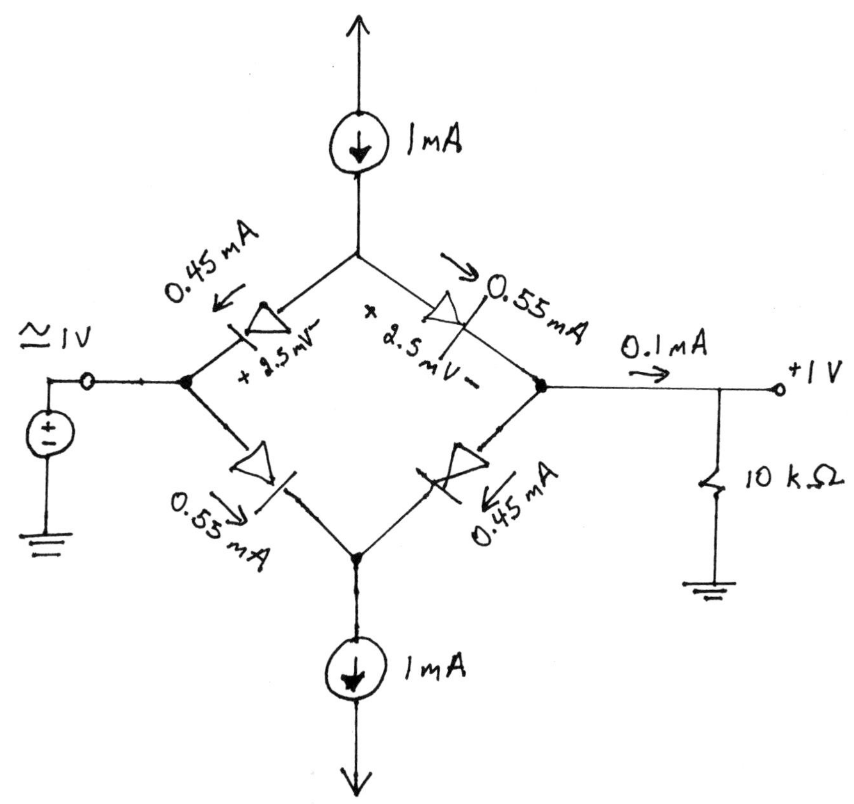

3.58

Opening the current source we get the following small-signal circuit:
$(n=1)$

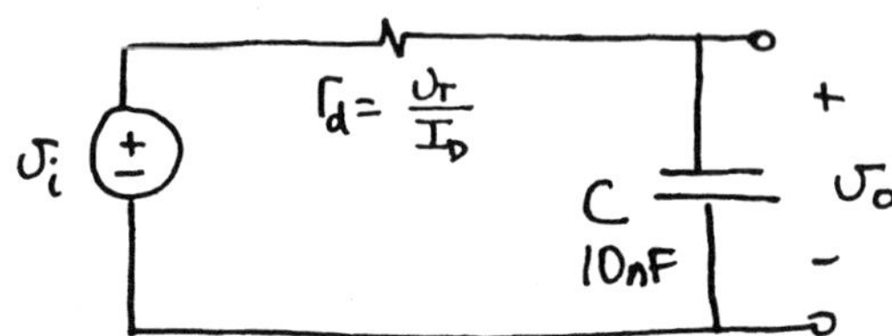

$$\frac{v_o}{v_i} = \frac{1/sC}{1/sC + r_d} = \frac{1}{1 + sCr_d}$$

$$\text{Phase Shift} = -\tan^{-1}\left(\frac{\omega C r_d}{1}\right)$$

$$= -\tan^{-1}\left(2\pi\, 10^{5} \times 10 \times 10^{-9} \times 0.025/I\right)$$

For a phase shift of -45° we have

$$2\pi 10^{5} \times 10(10^{-9}) \times \frac{0.025}{I} = 1$$

$$I = \underline{\underline{157\,\mu A}}$$

Range of phase shift for $I = 15.7\,\mu A$ to $1570\,\mu A$ is:

$$\underline{\underline{-84.3^\circ \text{ to } -5.71^\circ}}$$

3.59

V^+

R

V_o

= small signal model ⇒ V_i^+, R, r_d, V_o

CONT.

(a) $\frac{\Delta V_o}{\Delta V^+} = \frac{r_d}{r_d + R} = \frac{nV_T/I}{nV_T/I + R}$

$= \frac{nV_T}{nV_T + IR}$ where at No load $I = \frac{V^+ - 0.7}{R}$

$= \frac{nV_T}{nV_T + V^+ - 0.7}$ Q.E.D.

(b) For m diodes in series use

$I = \frac{V^+ - m \times 0.7}{R}$

Thus:

$\frac{\Delta V_o}{\Delta V^+} = \frac{m r_d}{m r_d + R} = \frac{m(nV_T)}{m(nV_T) + IR}$

$= \frac{m(nV_T)}{m(nV_T) + V^+ - 0.7m}$

(c) Line Regulation for $V^+ = 10V$, $n=2$

i) $m=1$ $\frac{\Delta V_o}{\Delta V^+} = 5.35\,mV/V$

ii) $m=3$ $\frac{\Delta V_o}{\Delta V^+} = 18.63\,mV/V$

3.60

$V+$, R, I_L, $+$, V_o, $-$ small signal model ⇒ V_o, R, r_d, I_L

$\Delta V_o = -I_L (R \| r_d)$

$\frac{\Delta V_o}{I_L} = -(R \| r_d)$ Q.E.D.

(b) Given at DC $I_D = \frac{V^+ - 0.7}{R}$

Also $r_d = \frac{nV_T}{I_D}$

We have:

$\frac{\Delta V_o}{I_L} = -\frac{1}{\frac{1}{R} + \frac{1}{r_d}}$

$= -\frac{1}{\frac{I_D}{V^+ - 0.7} + \frac{I_D}{nV_T}}$

$= -\frac{nV_T}{I_D} \frac{1}{1 + \frac{nV_T}{V^+ - 0.7}}$

$= \frac{-nV_T}{I_D} \frac{V^+ - 0.7}{V^+ - 0.7 + nV_T}$ Q.E.D.

for $\frac{\Delta V_o}{I_L} \leq 5 \frac{mV}{mA}$

$\frac{-2 \times 0.025}{I_D} \times \frac{10 - 0.7}{10 - 0.7 + 0.05} \leq \frac{5 \times 10^{-3}}{10^{-3}}$

$I_D \geq 9.947\,mA \Rightarrow I_D = 10\,mA$

$R = \frac{V^+ - 0.7}{I_D} = \frac{10 - 0.7}{10} = 930\,\Omega$

Thus the diode should be a 10mA diode.

(c) For m diodes

$I_D = \frac{V^+ - 0.7m}{R}$ & $r_d = \frac{m(nV_T)}{I_D}$

CONT.

$$\frac{\Delta V_O}{I_L} = \frac{-1}{\frac{1}{R} + \frac{1}{r_d}}$$

$$= \frac{-1}{\frac{I_D}{V^+ - 0.7m} + \frac{I_D}{mnV_T}}$$

$$= -\frac{mnV_T}{I_D} \frac{1}{\frac{mnV_T}{V^+ - 0.7m} + 1}$$

$$= -\frac{mnV_T}{I_D} \frac{V^+ - 0.7m}{V^+ - 0.7m + mnV_T}$$

3.61

3.62

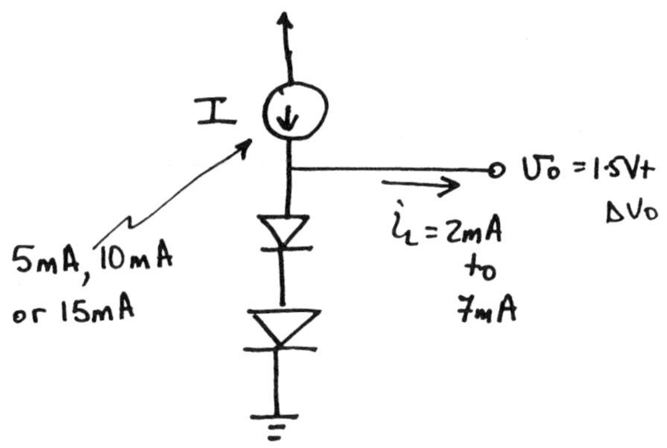

For a load current of 2 to 7mA, I must be greater than 7 mA. Thus the 5 mA source would not do.

We are left to choose between the 10 and 15 mA sources. The 15 mA source provides lower load regulation because the diodes will have more current flowing through them at all times. This is shown below:

Load Regulation if $I = 10mA$

use $\dfrac{i_{D2}}{i_{D1}} = e^{\Delta V/2 \times nV_T}$ ← 2 diodes

$\therefore\ e^{\Delta V/0.05 \times 2} = 3/10$ to $8/10$

$\Delta V_o = -120mV$ to $-22.3mV$

$\therefore$ The peak to peak ripple is

$-120 - (-22.3) \approx -100mV$

$$\text{Load Regulation} = \frac{\Delta V_o}{I_L} = \frac{-100}{5}$$

$$= -20\ \frac{mV}{mA}$$

Load Regulation for $I = 15mA$. Here the current through the diodes change from 8 to 13 mA corresponding to

$$\Delta V_o = 0.1 \ln(8/13)$$

$$= -49mV$$

$$\text{Load Regulation} = \frac{-49}{5} \approx -10\frac{mV}{mA}.$$

The obvious disadvantage of using the 15 mA supply is the requirement of higher current and higher power dissipation.

Alternate solution of Line Regulation using the small signal model

open current source

$I_L = 2$ to $7mA$ or $4.5 \pm 2.1mA$

V_o

$2r_d$

$$\text{Load Regulation} = \frac{\Delta V_o}{I_L} = -2r_d = \frac{-2nV_T}{I_D}$$

Where the bias current $I_D = 10 - 4.5$ for the 10 mA source.

$$\Rightarrow \frac{\Delta V_o}{I_L} = \frac{-2 \times 2 \times 0.025}{10 - 4.5} = -18.2\ \frac{mV}{mA}$$

For 15 mA source $I_D = 15 - 4.5$

$$\frac{\Delta V_o}{\Delta I_L} = \frac{-0.1}{15-4.5} = -9.5\frac{mV}{mA}$$

CONT.

Sketch of output:-

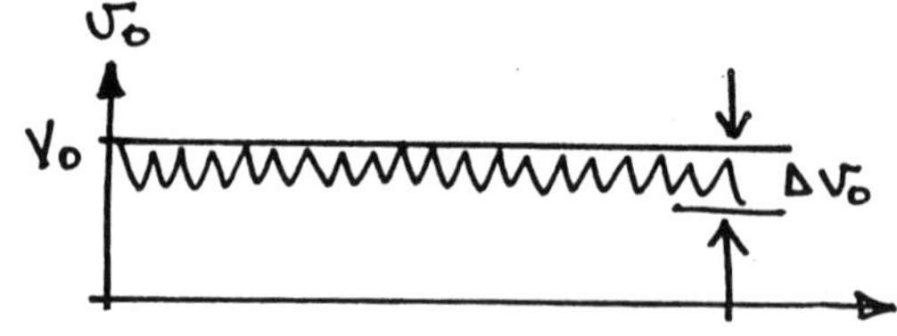

3.63

3.64

(a) $V_Z = V_{Z0} + r_Z I_{ZT}$

$10 = 9.6 + r_Z \times 50\times10^{-3}$

$r_Z = \underline{\underline{8\,\Omega}}$

Power rating:

$V_Z = V_{Z0} + r_Z \times 2I_{ZT}$

$= 9.6 + 8\times100\times10^{-3}$

$= 10.4\,V$

$P = 10.4\times100\times10^{-3} = \underline{\underline{1.04\,W}}$

(b) $V_Z = V_{Z0} + r_Z I_{ZT}$

$9.1 = V_{Z0} + 30\times10\times10^{-3}$

$V_{Z0} = \underline{\underline{8.8\,V}}$

$V_Z = 8.8 + 30\times20\times10^{-3} = 9.4\,V$

$P = 9.4\times20\times10^{-3} = \underline{\underline{188\,mW}}$

(c) $6.8 = 6.6 + 2\times I_{ZT}$

$I_{ZT} = \underline{\underline{100\,mA}}$

$V_Z = 6.6 + 2\times200\times10^{-3} = 7\,V$

$P = 7\times200\times10^{-3} = \underline{\underline{1.4\,W}}$

(d) $18 = 17.2 + r_Z\times5\times10^{-3}$

$r_Z = \underline{\underline{160\,\Omega}}$

$V_Z = 17.2 + 160\times10\times10^{-3} = 18.8\,V$

$P = 18.8\times10\times10^{-3} = \underline{\underline{188\,mW}}$

(e) $7.6 = V_{Z0} + 1.5\times200\times10^{-3}$

$V_{Z0} = \underline{\underline{7.2\,V}}$

$V_Z = 7.2 + 1.5\times400\times10^{-3} = 7.8\,V$

$P = 7.8\times400\times10^{-3} = \underline{\underline{3.12\,W}}$

3.65

(a) Three 6.8 V zeners provide $3\times6.8 = 20.4\,V$ with $3\times10 = 30\,\Omega$ resistance. Neglecting R, we have

Load Regulation $= \underline{\underline{-30\,mV/mA}}$.

(b) For 5.1 V zeners we use 4 diodes to provide 20.4 V with $4\times30 = 120\,\Omega$ resistance.

Load regulation $= \underline{\underline{-120\,mV/mA}}$

3.66

Small signal model for line regulation:

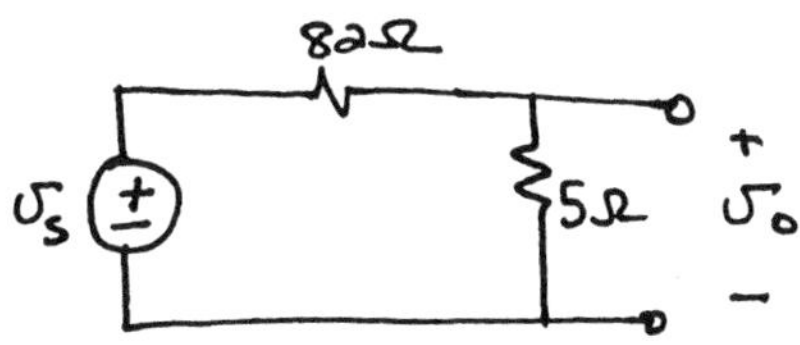

$$\frac{\Delta v_o}{\Delta v_s} = \frac{5}{5+82}$$

$$\Delta v_o = \frac{5}{87}\times\Delta v_s$$

$$= \frac{5}{87}\times1.3$$

$$= \underline{\underline{74.7\,mV}}$$

3.67

$$V_Z = V_{ZO} + r_Z I_{ZT}$$

$$9.1 = V_{ZO} + 5 \times 28 \times 10^{-3}$$

$$V_{ZO} = 8.96\text{V}$$

$$V_Z = V_{ZO} + 5I_Z = 8.96 \times 5I_Z$$

FOR $I_Z = 10\text{mA}$ $\quad V_Z = \underline{\underline{9.01\text{V}}}$

FOR $I_Z = 100\text{mA}$ $\quad V_Z = \underline{\underline{9.46\text{V}}}$

3.68

$r_z = 30\,\Omega$

$I_{ZK} = 0.5\text{ mA}$

$V_Z = 7.5\text{ V}$

$I_Z = 12\text{ mA}$

$$7.5 = V_{Z0} + 12 \times 30 \times 10^{-3}$$

$$\Rightarrow V_{Z0} = 7.14\text{ V}$$

$$I_Z = \frac{7.5}{1.2} = 6.25\text{ mA}$$

SELECT $I = 10\text{ mA}$
SO THAT $I_Z = 3.7\text{ mA}$

WHICH IS $> I_{ZK}$

$$R = \frac{10 - 7.5}{10} = \underline{\underline{250\,\Omega}}$$

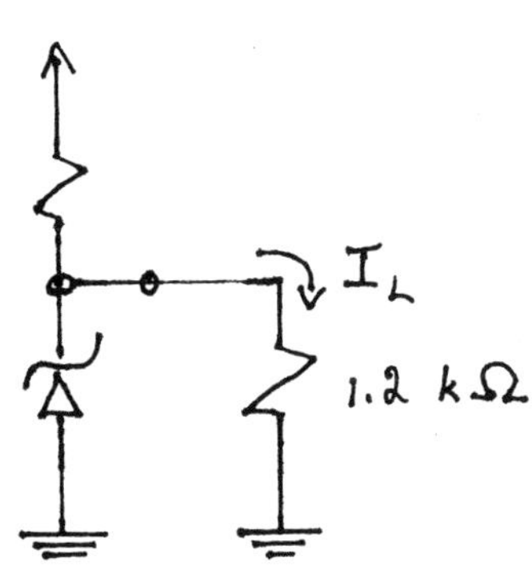

FOR $\Delta V^+ = \pm 1\text{V}$

$$\Delta V_o = \pm 1 \times \frac{1.2 /\!/ 0.03}{0.250 + (1.2 /\!/ 0.03)}$$

$$= \pm 0.1\text{ V}$$

THUS $V_o = +7.4\text{ V}$ TO $+7.6\text{ V}$
WITH $V^+ = 11\text{ V}$ AND $I_L = 0$

$$V_o = V_{Z0} + \frac{11 - V_o}{0.25} \times 0.03$$

$$\Rightarrow V_o = \underline{\underline{7.55\text{ V}}}$$

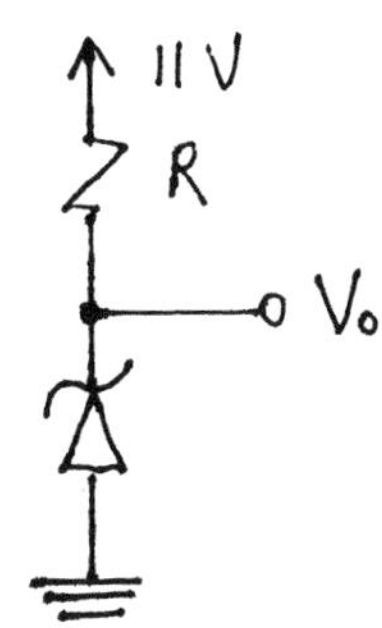

9V

$\frac{9 - 7.155}{0.25} = 7.38\text{ mA}$

250 Ω

$7.14 + 0.03 \times 0.5 = 7.155\text{V}$

0.5 mA

$R_{L_{MIN}}$

$$R_{L_{MIN}} = \frac{7.155}{7.38 - 0.5} = \underline{\underline{1.04\text{ k}\Omega}}$$

3.69

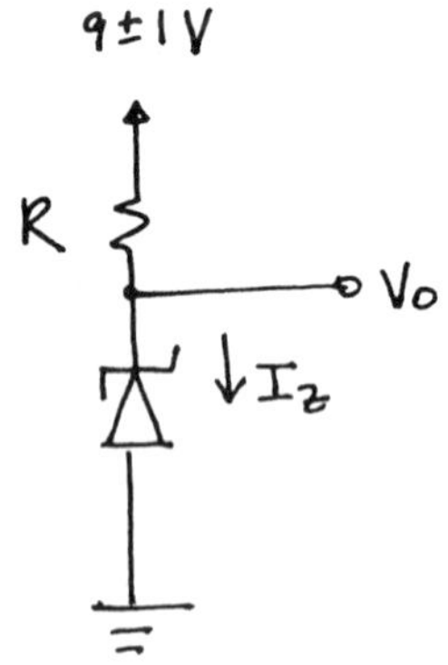

Given Parameters

$V_Z = 6.8V$, $r_z = 5\Omega$,

$I_Z = 20mA$

By knee

$I_{ZK} = 0.25mA$

$r_z = 750\Omega$

First Design – 9V supply can easily supply current.

Let $I_Z = 20mA$ ~ well above knee

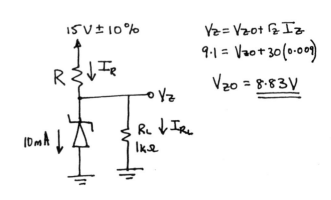

$$\therefore R = \frac{9-0.68}{20} = \underline{\underline{110\Omega}}$$

$$\text{Line Regulation} = \frac{\Delta v_O}{\Delta v_S} = \frac{r_z}{r_z + R}$$

$$= \frac{5}{5+110}$$

$$= \underline{\underline{43.5 \ \frac{mV}{V}}}$$

Second Design ~ limited current from 9V supply

$I_Z = 0.25mA$

$V_Z = V_{ZK} \simeq V_{Z0}$ – calculate V_{Z0} from

$V_Z = V_{Z0} + r_z I_{ZT}$

$6.8 = V_{Z0} + 5 \times 0.02$

$V_{Z0} = 6.7V$

$$\therefore R = \frac{9-6.7}{0.25} = \underline{\underline{9.2k\Omega}}$$

$$\text{Line Regulation} = \frac{\Delta v_O}{\Delta v_S} = \frac{750}{750+9200}$$

$$= \underline{\underline{75.4 \ \frac{mV}{V}}}$$

3.70

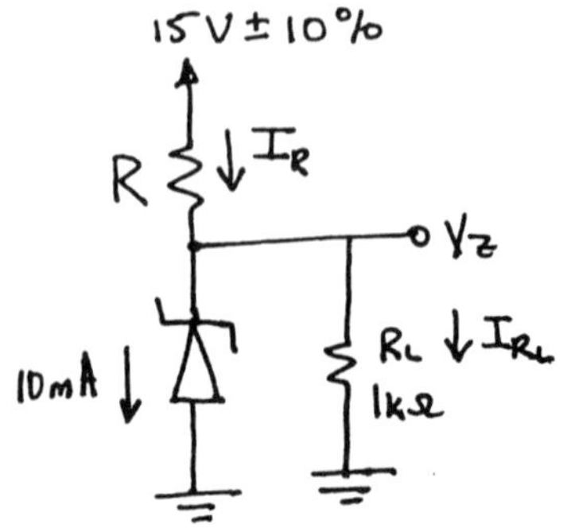

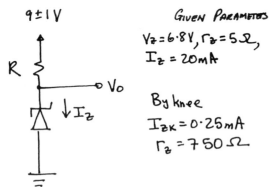

$V_Z = V_{Z0} + r_z I_Z$

$9.1 = V_{Z0} + 30(0.009)$

$V_{Z0} = \underline{\underline{8.83V}}$

CONT.

$V_Z = 8.83 + 30(0.01) = 9.13\,V$

$I_{R_L} = 9.13/1k\Omega = 9.13\,mA$

$I_R = 10 + 9.13 = \underline{\underline{19.13\,mA}}$

$\therefore\ R = \frac{15 - 9.13}{19.13} = 306.8\,\Omega$

$\cong \underline{\underline{300\,\Omega}}$

$V_Z = 8.83 + 30\left(\frac{15 - V_Z}{300} - \frac{V_Z}{1000}\right)$

$= 10.33 - V_Z/10 - 3/100\,V_Z$

$V_Z = 9.14\,V$

$V_Z = 8.83 + 30\left(\frac{15 \pm 1.5 - V_Z}{300} - \frac{V_Z}{1000}\right)$

$= \frac{1}{1.13}\left[8.83 + 1.5 \pm 0.15\right] = 9.14 \pm 0.13\,V$

$\therefore$ $\underline{\underline{\pm 0.13\,V}}$ variation in output voltage

Halfing the load current $\equiv R_L$ doubling

$V_Z = 8.83 + 30\left(\frac{15 - V_Z}{300} - \frac{V_Z}{2000}\right)$

$= \frac{10.33}{1.115} = 9.26\,V$

$\therefore\ 9.26 - 9.14 = 0.12\,V$ increase in output voltage.

At the edge of the breakdown region

$V_Z \cong V_{Z0} = 8.83\,V \qquad I_{ZK} = 0.3\,mA$

$R_L = \frac{8.83}{\frac{13.5 - 8.83}{300} - 0.0003}$

$= \underline{\underline{578\,\Omega}}$

13.5V

300Ω

8.83V

0.3mA

Lowest output voltage $= 8.83\,V$

Line Regulation $= \frac{r_Z}{R + r_Z} = \frac{30}{300 + 30}$

$= 90\ \frac{mV}{V}$

Load Regulation $= -(r_Z \parallel R) = \underline{\underline{-29.1\ \frac{mV}{mA}}}$

3.71

(a) $V_{ZT} = V_{Z0} + r_Z I_{ZT}$

$10 = V_{Z0} + 7(0.025)$

$\Rightarrow V_{Z0} = \underline{\underline{9.825\,V}}$

(b) The minimum zener current of 5mA occurs when $I_L = 20\,mA$ and V_S is at its minimum of $20(1 - 0.25) = 15\,V$. See the circuit below:

15V

R

20mA

5mA

R_L

$R \leq \frac{15 - (V_{Z0} + r_Z I_Z)}{20 + 5}$

$\leq \frac{15 - (9.825 + 7(0.005))}{25}$

$\leq 205.6\,\Omega.$

$\therefore$ Use $R = \underline{\underline{205\,\Omega}}$

(c) Line Regulation $= \frac{7}{205 + 7} = \underline{\underline{33\ \frac{mV}{V}}}$

$\pm 25\%$ change in $V_S \equiv \pm 5\,V$

V_O changes by $\pm 5 \times 33 = \underline{\underline{\pm 165\,mV}}$

corresponding to $\frac{\pm 165}{10} \times 100 = \underline{\underline{\pm 1.65\%}}$

CONT.

(d) Load Regulation $= -(r_z \| R)$

$= -(7 \| 205) = -6.77\,\Omega$

or -6.77 V/A

$\Delta V_O = -6.77 \times 20\,\text{mA} = -135.4\,\text{mV}$

corresponding to $\frac{-0.1354}{10} \times 100 = -1.35\%$

(e) The maximum zener current occurs at no load $\equiv I_L = 0$ and the supply at $20 + 1/4(20) = 25\,\text{V}$.

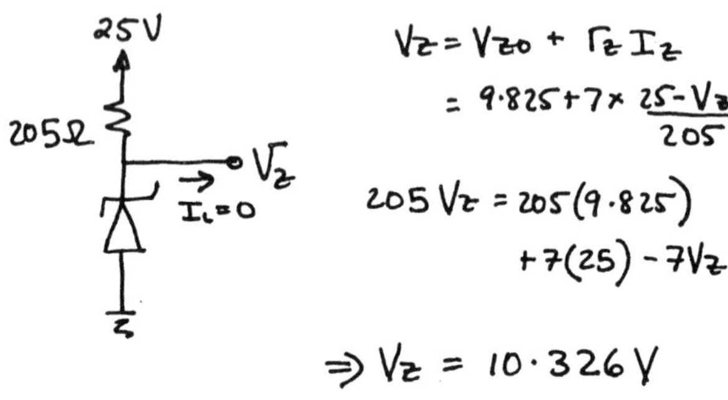

$V_Z = V_{Z0} + r_Z I_Z$

$= 9.825 + 7 \times \frac{25 - V_Z}{205}$

$205 V_Z = 205(9.825) + 7(25) - 7V_Z$

$\Rightarrow V_Z = 10.326\,\text{V}$

$P_Z = 10.326 \times \left(\frac{25 - 10.326}{205}\right)$

$= 739.4\,\text{mW}$

Alternate circuit to calculate V_Z

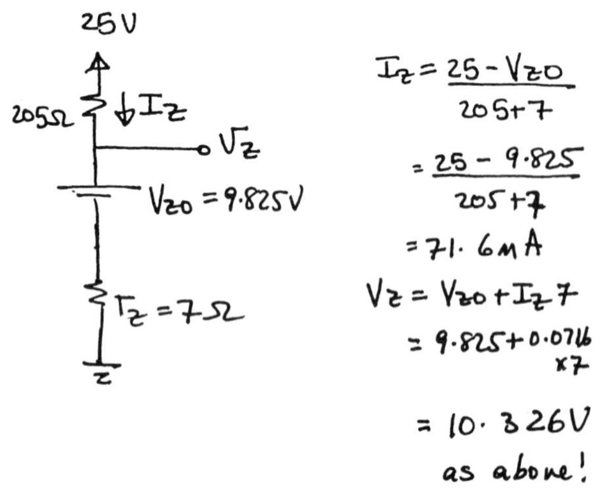

$I_Z = \frac{25 - V_{Z0}}{205 + 7}$

$= \frac{25 - 9.825}{205 + 7}$

$= 71.6\,\text{mA}$

$V_Z = V_{Z0} + I_Z 7$

$= 9.825 + 0.0716 \times 7$

$= 10.326\,\text{V}$

as above!

3.72

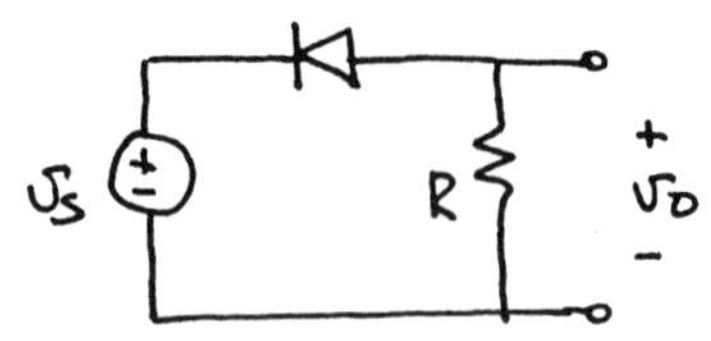

Using the constant voltage drop model:

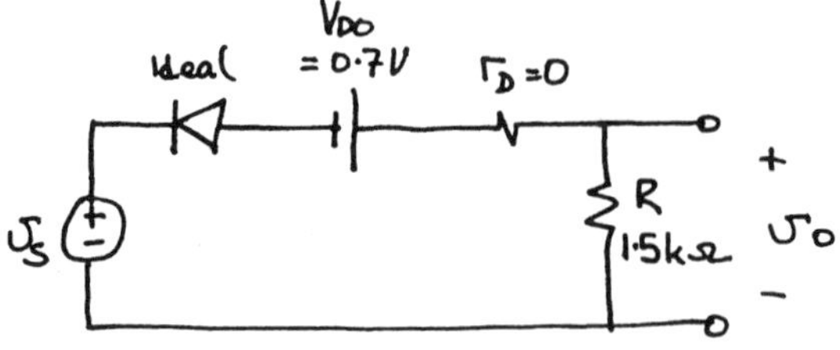

(a) $v_O = v_S + 0.7\,\text{V}$, for $v_S \le -0.7\,\text{V}$

$v_O = 0$, for $v_S \ge -0.7\,\text{V}$

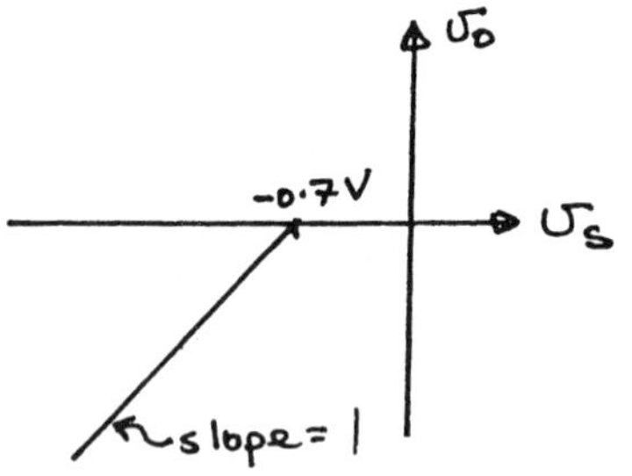

(b)

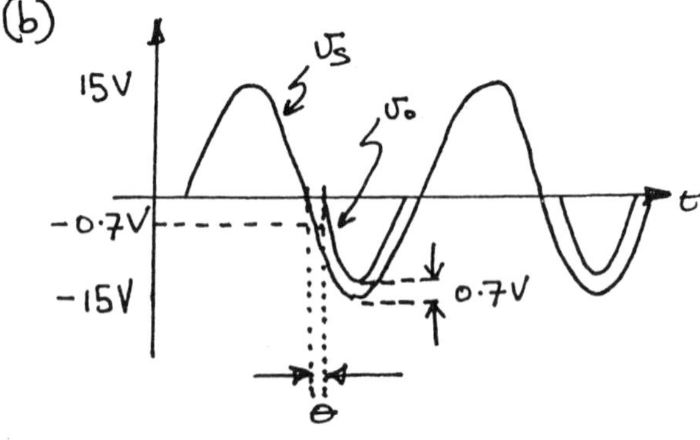

(c) The diode conducts at an angle $\theta = \sin^{-1}\frac{0.7}{15} = 2.67°$ & stops at $\pi - \theta = 177.33°$

CONT.

Thus the conduction angle is $\pi - 2\theta$

$= 174.66°$ or 3.05 rad.

$$v_{O,avg} = \frac{-1}{2\pi}\int_{\theta}^{\pi-\theta} 15\sin\phi - 0.7\, d\phi$$

$$= \frac{-1}{2\pi}\Big[-15\cos\phi - 0.7\phi\Big]_{\theta}^{\pi-\theta}$$

$$= \frac{-1}{2\pi}\Big[15\times 2\cos\theta - 0.7(\pi - 2\theta)\Big]$$

$$= \underline{\underline{-4.43\text{V}}}$$

(d) Peak current in diode is:

$$\frac{15-0.7}{1.5\times10^{3}} = \underline{\underline{9.5\text{mA}}}$$

(e) PIV occurs when v_S is at its +ve peak and $v_O = 0$.

$$\text{PIV} = \underline{\underline{15\text{V}}}$$

3.73

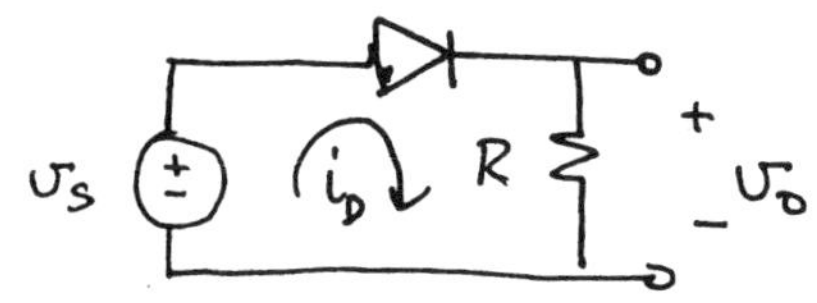

$$i_D = I_S e^{v_D/nV_T}$$

$$\frac{i_D}{i_D(1\text{mA})} = e^{\frac{v_D - v_D(\text{at } 1\text{mA})}{nV_T}}$$

$$v_D = v_D(\text{at } 1\text{mA}) + nV_T \ln\left(i_D/10^{-3}\right)$$

$$= \quad '' \quad + nV_T \ln\left(\frac{v_O/R}{10^{-3}}\right)$$

$$= \quad '' \quad + nV_T \ln\left(v_O/R\right) \quad R \text{ in k}\Omega$$

$$v_O = v_S - v_D$$

$$= v_S - v_D(\text{at } 1\text{mA}) - nV_T \ln\left(\frac{v_O}{R}\right)$$

QED.

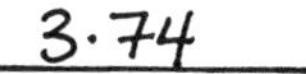

3.74

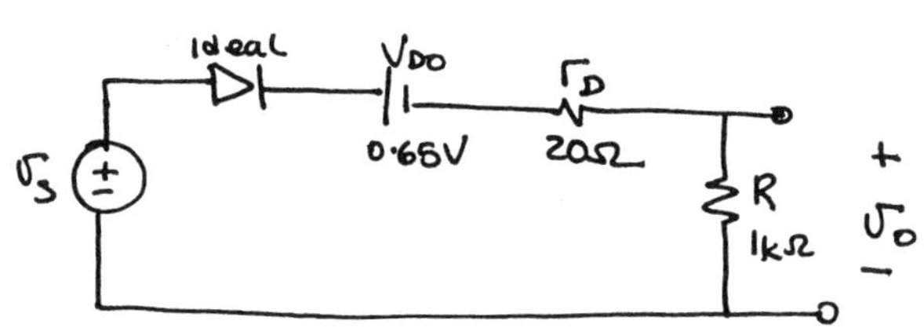

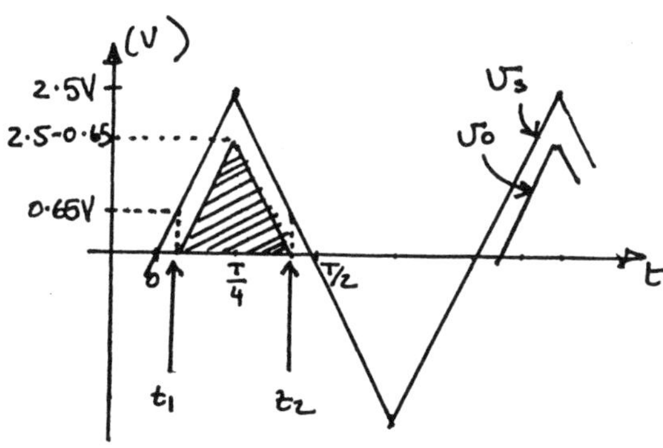

Find t_1 & t_2 by:

$$\frac{2.5}{T/4} = \frac{0.65}{t_1} \Rightarrow t_1 = 0.065T$$

$$t_2 = T/2 - 0.065T = 0.435T$$

$$v_O\big|_{avg} = \frac{1}{T}\int_T \frac{R}{R+r_D}(v_S - 0.65)\, dt$$

$$= \frac{1}{T}\frac{R}{R+r_D}\left(\text{AREA OF SHADED } \triangle\right)$$

$$= \frac{1}{T}\frac{R}{R+r_D}(2.5-0.65)\left(\frac{T}{4} - 0.065T\right)$$

$$= \frac{1000}{1020}(0.342) = \underline{\underline{0.335\text{V}}}$$

3.75

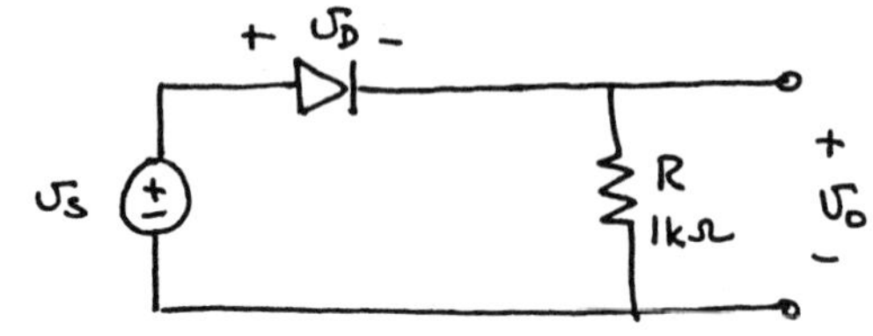

$$i_D = I_S e^{v_D/nV_T}$$

$$\frac{i_D}{1\text{mA}} = e^{\frac{v_D - 0.7}{nV_T}}$$

$$v_D - 0.7 = nV_T \ln\left(i_D/10^{-3}\right) = 0.1 \log\left(\frac{i_D}{10^{-3}}\right)$$

$$v_D = 0.7 + 0.1 \log\left(\frac{v_O}{R}\right) \quad R \text{ in k}\Omega$$

$$= 0.7 + 0.1 \log\left(\frac{v_O}{1}\right)$$

v_O (V)	v_D (V)	$v_S = v_D + v_O$ (V)
0.10	0.6	0.7
0.5	0.67	1.17
1	0.7	1.7
2	0.73	2.73
5	0.77	5.77
10	0.8	10.8

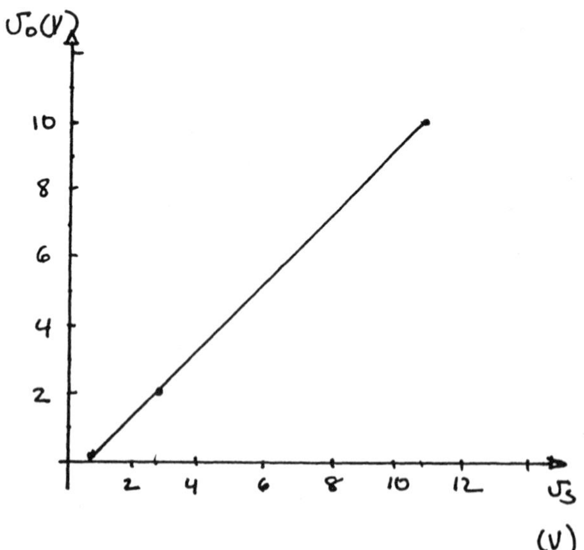

3.76

$$\hat{V}_O = 12\sqrt{2} - 0.7 = \underline{\underline{16.27\text{V}}}$$

Conduction begins at

$$v_S = 12\sqrt{2} \sin\theta = 0.7$$

$$\theta = \sin^{-1}\left(\frac{0.7}{12\sqrt{2}}\right)$$

$$= 0.0412 \text{ rad}$$

Conduction ends at $\pi - \theta$

∴ Conduction angle $= \pi - 2\theta = \underline{\underline{3.06 \text{ rad}}}$

The diode conducts for

$$\frac{3.06}{2\pi} \times 100 = \underline{\underline{48.7\%}} \text{ of the cycle}$$

$$V_{O,avg} = \frac{1}{2\pi} \int_{\theta}^{\pi-\theta} 12\sqrt{2}\sin\phi - 0.7 \, d\phi$$

$$= \underline{\underline{5.06\text{V}}}$$

$$i_{D,avg} = \frac{v_{O,avg}}{R} = \underline{\underline{5.06\text{mA}}}$$

3.77

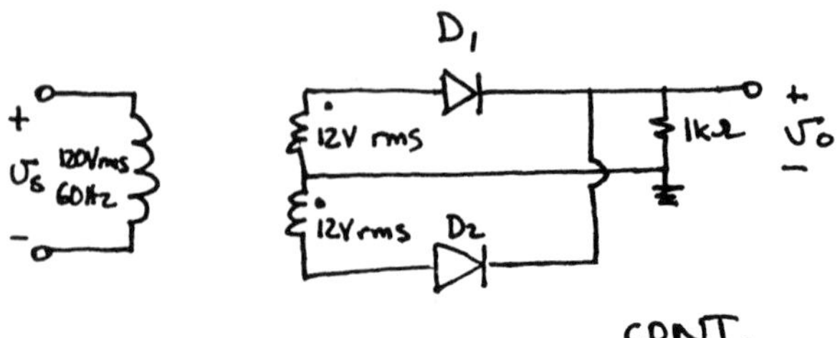

CONT.

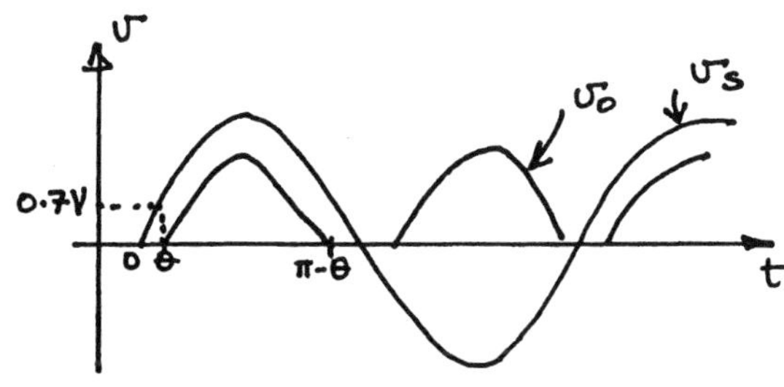

$\hat{v}_O = 12\sqrt{2} - V_{DO} = \underline{\underline{16.27V}}$

Conduction starts at $\theta = \sin^{-1}\frac{0.7}{12\sqrt{2}}$
$= 0.0412$ rad

and ends at $\pi - \theta$. Conduction angle $= \pi - 2\theta = 3.06$ rad in each half cycle. Thus the fraction of a cycle for which one of the two diodes conduct $= \frac{2(3.06)}{2\pi} \times 100$

$= \underline{\underline{97.4\%}}$

Note that during 97.4% of the cycle there will be conduction. However each of the two diodes conducts for only half the time, i.e. for 48.7% of the cycle.

$$v_{O,avg} = \frac{1}{\pi}\int_{\theta}^{\pi-\theta} 12\sqrt{2}\sin\phi - 0.7 \; d\phi$$

$= \underline{\underline{10.12V}}$

$$i_{D,avg} = \frac{10.12}{1k\Omega} = \underline{\underline{10.12mA}}$$

3.78

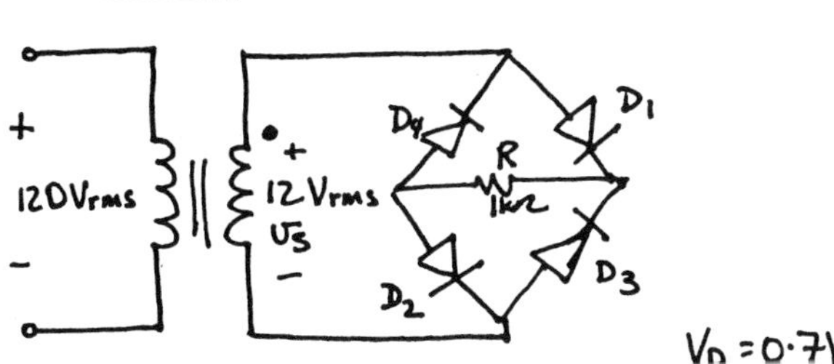

$V_D = 0.7V$

Peak voltage across $R = 12\sqrt{2} - 2V_D$
$= 12\sqrt{2} - 1.4$
$= 15.57V$

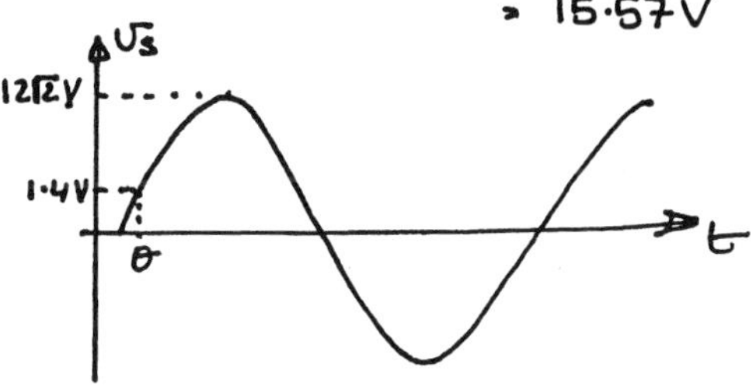

$\theta = \sin^{-1}\frac{1.4}{12\sqrt{2}} = 0.0826$ rad

Fraction of cycle that D_1 & D_2 conduct is $\frac{\pi - 2\theta}{2\pi} \times 100 = \underline{\underline{47.4\%}}$

Note D_3 & D_4 conduct in the other half cycle so that there is $2(47.4) =$ 94.8% conduction interval.

$$v_{O,avg} = \frac{2}{2\pi}\int_{\theta}^{\pi-\theta} 12\sqrt{2}\sin\phi - 2V_D \; d\phi$$

$$= \frac{1}{\pi}\left[-12\sqrt{2}\cos\phi - 1.4\,\phi\right]_{\theta}^{\pi-\theta}$$

$$= \frac{2(12\sqrt{2}\cos\theta)}{\pi} - \frac{1.4(\pi-2\theta)}{\pi}$$

$= \underline{\underline{9.44V}}$

$$i_{R,avg} = \frac{v_{O,avg}}{R} = \frac{9.44}{1} = 9.44\,mA$$

3.79

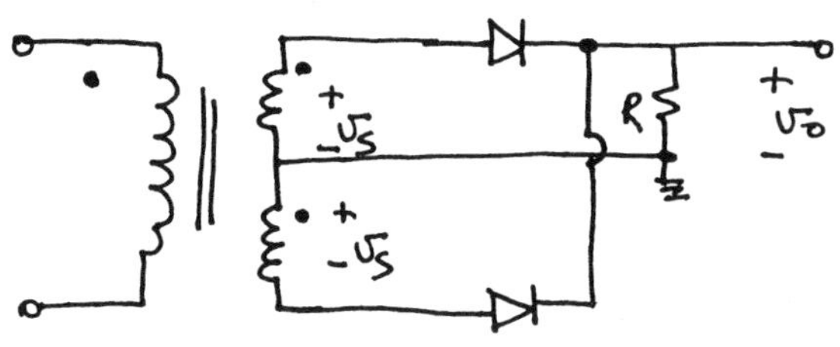

CONT.

For $V_{DO} \ll V_s$,

$V_{O,avg} \simeq \frac{2}{\pi} V_s - V_{DO}$

(a) For $V_{O,avg} = 10V$

$10 = \frac{2}{\pi} V_s - 0.7$

$\Rightarrow V_s = 16.81V$

Line peak $= 120\sqrt{2}$

Thus,

turns ratio to each half of the secondary $= \frac{120\sqrt{2}}{16.81} = 10.1:1$

OR $5.05:1$ centre tapped

(b) For $V_{O,avg} = 100V$

$V_s = \frac{\pi}{2}(100.7) = 158.2V$

Turns Ratio $= \frac{120\sqrt{2}}{158.2} = 1.07:1$ to each half

OR $0.535:1$ centre tapped

3.80

Refer to Fig 3.27.

For $2V_{DO} \ll V_s$,

$v_{O,avg} = \frac{2}{\pi} V_s - 2V_{DO} = \frac{2}{\pi} V_s - 1.4$

(a) For $v_{O,avg} = 10V$

$V_s = \frac{\pi}{2} \times 11.4 = 17.91V$

Turns Ratio $= \frac{120\sqrt{2}}{17.91} = 9.477$ to 1

(b) For $v_{O,avg} = 100V$

$V_s = \frac{\pi}{2} \times 101.4 = 159.3V$

Turns Ratio $= \frac{120\sqrt{2}}{159.3} = 1.065$ to 1

3.81

$120\sqrt{2} \pm 10\%$: $24\sqrt{2} \pm 10\%$

$\Rightarrow$ turns Ratio = 5 : 1

$v_s = \frac{24\sqrt{2}}{2} \pm 10\%$

$PIV = 2V_s|_{max}\ V_{DO}$

$= 2 \times \frac{24\sqrt{2}}{2} \times 1.1 - 0.7$

$= 36.6V$

Using a factor of 1.5 for safety we select a diode having a PIV rating of 55V

3.82

The circuit is a full wave rectifier with centre tapped secondary winding. The circuit can be analyzed by looking at v_O^+ and v_O^- separately.

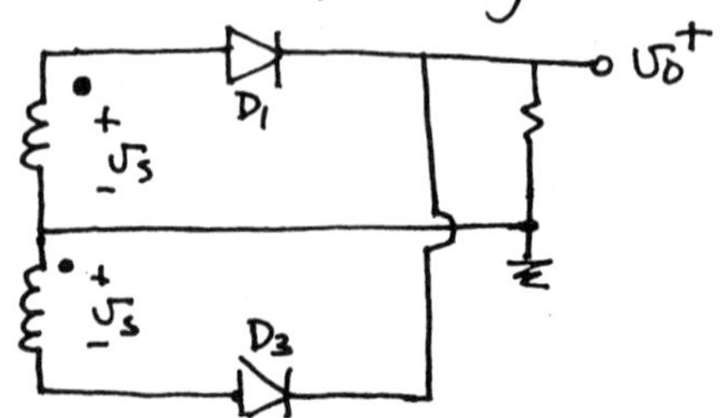

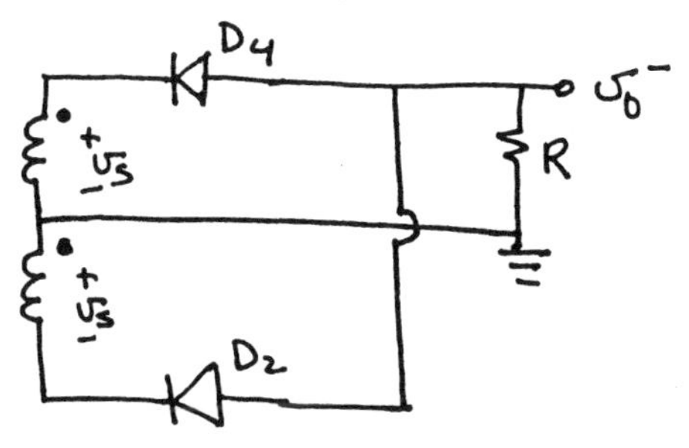

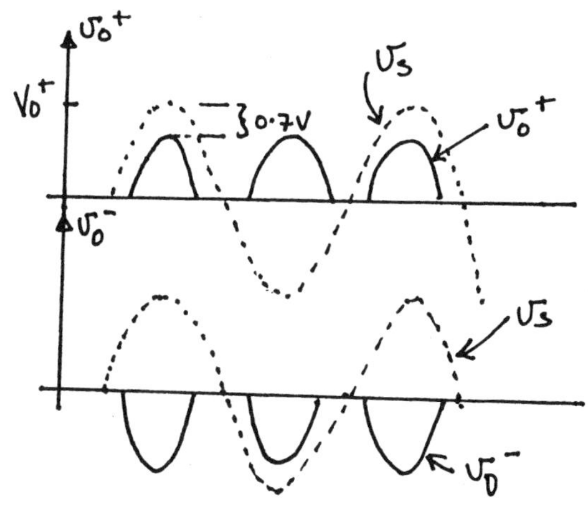

$$v_{O,avg} = \frac{1}{2\pi}\int V_S \sin\phi - 0.7\, d\phi = 15$$

$$= \frac{2V_S}{\pi} - 0.7 \approx 15 \qquad \text{assumed } V_S >> 0.7V$$

$$V_S = \frac{15+0.7}{2}\pi = 24.66V$$

Thus voltage across secondary winding $= 2V_S = \underline{\underline{49.32V}}$

Looking at D_4

$$PIV = V_S - V_O^-$$

$$= V_S + (V_S - 0.7)$$

$$= 2V_S - 0.7$$

$$= \underline{\underline{48.6V}}$$

If choosing a diode, allow a safety margin of $1.5 PIV = \underline{\underline{73V}}$

3.83

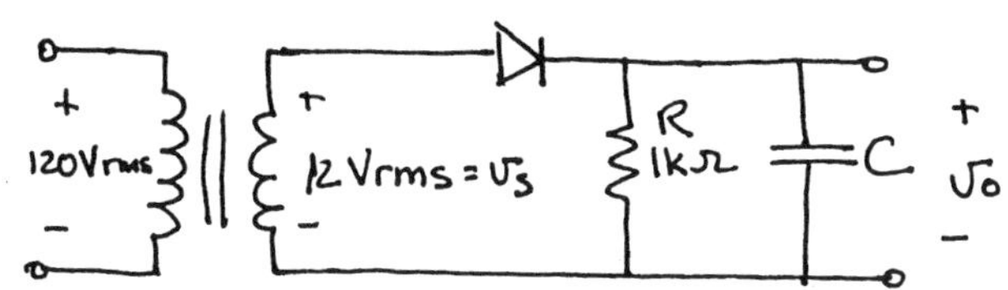

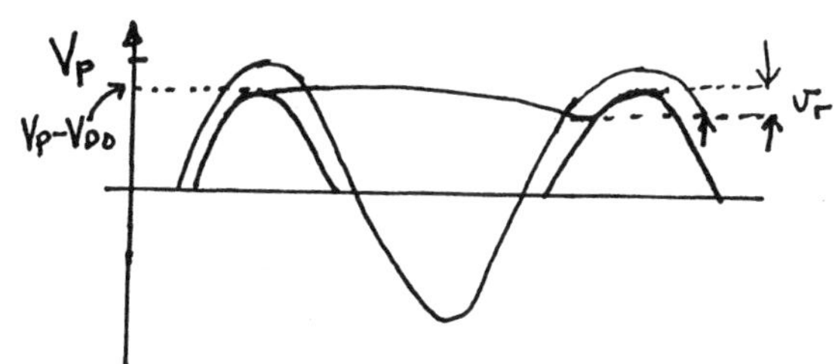

(i) $V_r \cong (V_P - V_{DO})\frac{T}{CR}$ Eq. (3.28)

$$0.1(V_P - V_{DO}) = (V_P - V_{DO})\frac{T}{CR}$$

$$C = \frac{1}{0.1 \times 60 \times 10^3} = \underline{\underline{166.7\mu F}}$$

(ii) For $V_r = 0.01(V_P - V_{DO}) = (V_P - V_{DO})\frac{T}{CR}$

$$C = \underline{\underline{1667\ \mu F}}$$

(a)

(i) $v_{O,avg} = V_P - V_{DO} - \frac{1}{2}V_r$

$$= 12\sqrt{2} - 0.7 - \frac{1}{2}(12\sqrt{2} - 0.7)0.1$$

$$= (12\sqrt{2} - 0.7)(1 - \frac{0.1}{2})$$

$$= \underline{\underline{15.5V}}$$

ii) $v_{O,avg} = (12\sqrt{2} - 0.7)(1 - \frac{0.01}{2})$

$$= \underline{\underline{16.19V}}$$

CONT.

(b)

i) Using eq (3.30) we have the conduction angle =

$$\omega \Delta t \cong \sqrt{2V_r/(V_p - V_{DO})}$$

$$= \sqrt{\frac{2 \times 0.1 (V_p - 0.7)}{(V_p - 0.7)}}$$

$$= \sqrt{0.2}$$

$$= 0.447 \text{ rad}$$

∴ Fraction of cycle for conduction $= \frac{0.447}{2\pi} \times 100$

$$= 7.1\%$$

ii) $\omega \Delta t \simeq \sqrt{\frac{2 \times 0.01 (V_p - 0.7)}{V_p - 0.7}} = 0.141 \text{ rad}$

Fraction of cycle $= \frac{0.141}{2\pi} \times 100 = 2.25\%$

(c)(i) Use Eq (3.31)

$$i_{D,avg} = I_L \left(1 + \pi \sqrt{\frac{2(V_p - V_{DO})}{V_r}}\right)$$

$$= \frac{V_{o,avg}}{R} \left(1 + \pi \sqrt{\frac{2(V_p - V_{DO})}{0.1(V_p - V_{DO})}}\right)$$

$$= \frac{15.5}{10^3} \left(1 + \pi \sqrt{\frac{2}{0.1}}\right)$$

$$= 233 \text{ mA}$$

ii) $i_{D,avg} = \frac{16.19}{10^3} \left(1 + \pi \sqrt{200}\right)$

$$= 735 \text{ mA}$$

NB Text uses $I_L \cong V_P/R = \frac{V_P - V_{DO}}{R}$

but here we used $i_{L,avg} = \frac{V_P - V_{DO} - \frac{1}{2}V_r}{R}$

which is more accurate.

(d) i) $i_{D,peak} = I_L \left(1 + 2\pi \sqrt{\frac{2(V_p - V_{DO})}{V_r}}\right)$

$$= \frac{15.42}{10^3} \left(1 + 2\pi \sqrt{2/0.1}\right)$$

$$= 449 \text{ mA}$$

ii) $i_{D,peak} = \frac{16.19}{10^3} \left(1 + 2\pi \sqrt{\frac{2}{0.01}}\right)$

$$= 1455 \text{ mA}$$

3.84

i) $V_r = 0.1(V_p - V_{DO}) = \frac{(V_p - V_{DO})}{2fCR}$

the factor of 2 accounts for discharge occuring only half of the period $T/2 = \frac{1}{2f}$

$$C = \frac{1}{(2fR)0.1} = \frac{1}{2(60)10^3 \times 0.1} = 83.3\,\mu F$$

ii) $C = \frac{1}{2(60)10^3 \, 0.01} = 833\,\mu F$

(a) i) $V_o = V_p - V_{DO} - \frac{1}{2}V_r$

$$= (V_p - V_{DO})\left(1 - \frac{0.1}{2}\right)$$

$$= (16.27)\left(1 - \frac{0.1}{2}\right)$$

$$= 15.5 \text{ V}$$

ii) $V_o = (16.27)\left(1 - \frac{0.01}{2}\right) = 16.19 \text{ V}$

(b)

(i) Fraction of cycle $= \frac{2\omega \Delta t}{2\pi} \times 100$

$$= \frac{\sqrt{2V_r/(V_p - V_{DO})}}{\pi} \times 100$$

$$= \frac{1}{\pi}\sqrt{2(0.1)} \times 100 = 14.2\%$$

CONT.

ii) Fraction of Cycle $= \frac{2\sqrt{2(0.01)}}{2\pi} \times 100$

$= \underline{\underline{4.5\%}}$

(c) Use eq (3.34)

i) $i_{D,avg} = I_L\left(1+\pi\sqrt{\frac{V_p - V_{DO}}{2V_r}}\right)$

$= \frac{15.5}{1}\left(1+\pi\sqrt{\frac{1}{2(0.1)}}\right) = \underline{\underline{124.4\,mA}}$

ii) $i_{D,avg} = \frac{16.19}{1}\left(1+\pi\frac{1}{\sqrt{2(0.01)}}\right) = \underline{\underline{376\,mA}}$

(d) Use eq (3.35)

(i) $\hat{i}_D = I_L\left(1+2\pi\frac{1}{\sqrt{2(0.1)}}\right) = \underline{\underline{233\,mA}}$

ii) $\hat{i}_D = I_L\left(1+2\pi\frac{1}{\sqrt{0.02}}\right) = \underline{\underline{735\,mA}}$

3.85

i) $V_r = 0.1(V_P - V_{DO}\times 2) = \frac{V_P - 2V_{DO}}{2fCR}$

discharge occurs only over $\frac{1}{2}T = \frac{1}{2f}$

$C = \frac{(V_P - 2V_{DO})}{(V_P - 2V_{DO})}\frac{1}{2(0.1)fR} = \underline{\underline{83.3\,\mu F}}$

ii) $C = \frac{1}{2(0.01)fR} = \underline{\underline{833\,\mu F}}$

(b) i) Fraction of cycle $= \frac{2\omega\Delta t}{2\pi} \times 100$

$= \frac{\sqrt{2(0.1)}}{\pi} \times 100 = \underline{\underline{14.2\%}}$

ii) Fraction of cycle $= \frac{\sqrt{2(0.01)}}{\pi} \times 100 = \underline{\underline{4.5\%}}$

(c) i)

$i_{D,avg} = \frac{14.79}{1}\left(1+\pi\sqrt{\frac{1}{0.2}}\right) = \underline{\underline{119\,mA}}$

ii) $i_{D,avg} = \frac{15.49}{1}\left(1+\pi/\sqrt{0.02}\right) = \underline{\underline{356\,mA}}$

(d)

(i) $\hat{i}_D = \frac{14.79}{1}\left(1+2\pi\sqrt{\frac{1}{0.2}}\right) = \underline{\underline{223\,mA}}$

(ii) $\hat{i}_D = \frac{15.49}{1}\left(1+2\pi\sqrt{\frac{1}{0.02}}\right) = \underline{\underline{704\,mA}}$

3.86

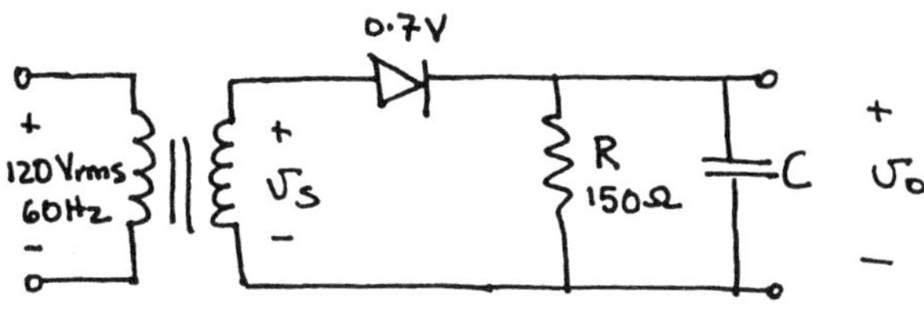

$v_{O,peak} = V_p - V_{DO} = 16$

$V_p = 16.7\ V$

$V_{rms} = \frac{16.7}{\sqrt{2}} = 11.8$

(b) $V_r = (V_P - V_{DO})\frac{T}{CR}$ Eq (3.28)

$2 = \frac{16}{60\times C\times 150}$

$C = 889\,\mu F$

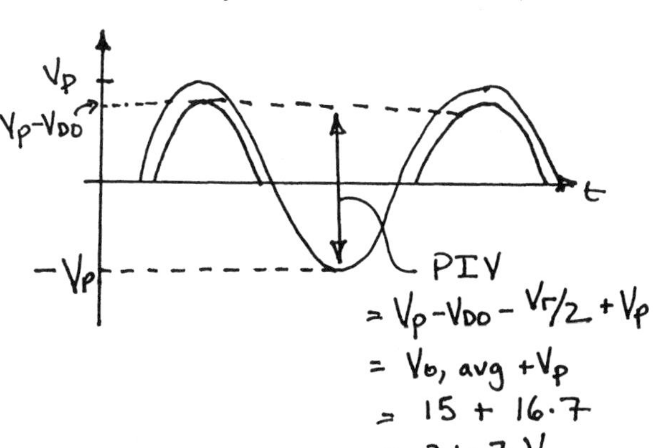

$= V_P - V_{DO} - V_r/2 + V_P$

$= V_{O,avg} + V_P$

$= 15 + 16.7$

$= 31.7\ V$

For a 50% safety margin PIV $= 1.5\times 31.7$

$= \underline{\underline{47.6\,V}}$

CONT.

(d)

$$i_{D,avg} = I_L\left(1+\pi\sqrt{\frac{2(V_p-V_{D0})}{V_r}}\right)$$

using $I_L = \frac{v_{O,avg}}{R} = \frac{15}{R}$ we have

$$i_{D,avg} = \frac{15}{150}\left(1+\pi\sqrt{\frac{2(16)}{2}}\right)$$

$= \underline{\underline{1.36A}}$

(e) $i_{D,peak} = I_L\left(1+2\pi\sqrt{\frac{2(V_p-V_{D0})}{V_r}}\right)$

$$= \frac{15}{150}\left(1+2\pi\sqrt{\frac{2(16)}{2}}\right)$$

$= \underline{\underline{2.61\ A}}$

3.87

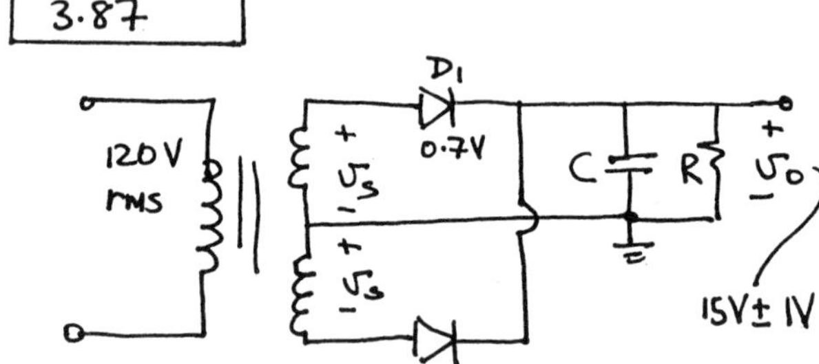

(a) $\hat{v}_O = 16V$

$\therefore \hat{v}_S = 16 + V_{D0} = 16.7V$

RMS Voltage across secondary

$= \frac{2\times16.7}{\sqrt{2}}$

$= \underline{\underline{23.6V}}$

(b) Using Eq. (3.28)

$$V_r = \frac{V_p}{2fCR} = \frac{16}{2\times60\times C\times150} = 2$$

$C = \underline{\underline{444.4\ \mu F}}$

(c)

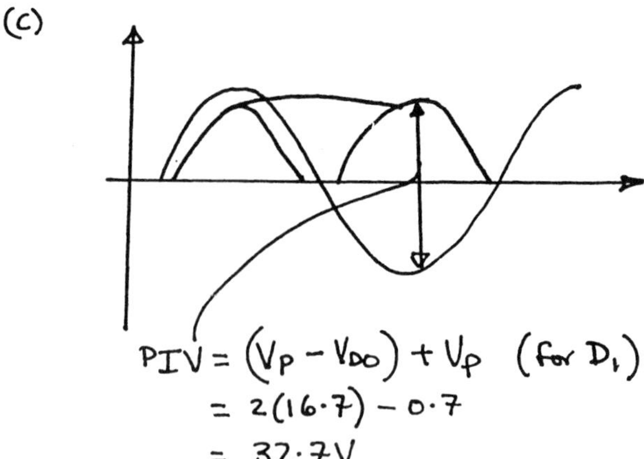

PIV $= (V_p - V_{D0}) + V_p$ (for D_1)

$= 2(16.7) - 0.7$

$= \underline{\underline{32.7V}}$

$\therefore$ Using a 50% margin

PIV $= 1.5(32.7) = \underline{\underline{49V}}$

(d) Using Eq (3.34)

$$i_{D,avg} = I_L\left(1+\pi\sqrt{V_p/2V_r}\right)$$

$$= \frac{15}{150}\left(1+\pi\sqrt{\frac{16}{2\times2}}\right)$$

$= \underline{\underline{0.73\ A}}$

(e) Using Eq. (3.35)

$$i_{D,max} = I_L\left(1+2\pi\sqrt{V_p/2V_r}\right)$$

$$= \frac{15}{150}\left(1+2\pi\sqrt{\frac{16}{2\times2}}\right)$$

$= \underline{\underline{1.36A}}$

3.88

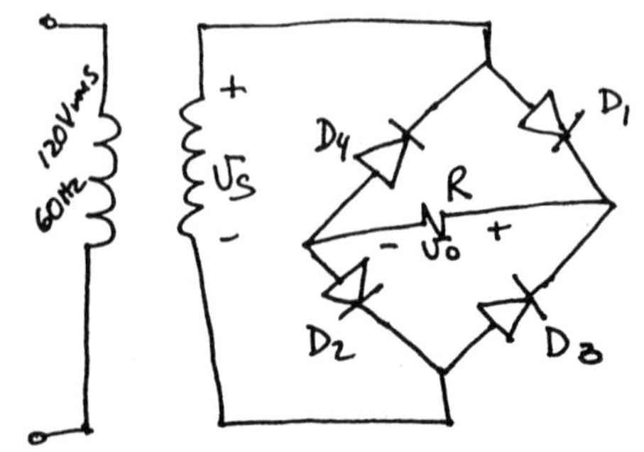

$v_O = 15 \pm 1V$, $R = 150\Omega$

CONT.

$\hat{v}_o = 16V$

$\hat{v}_s = 16 + 2V_{DO} = 17.4V$

RMS Secondary Voltage $= \frac{17.4}{\sqrt{2}} = 12.3V$

(b) $V_r = \frac{V_p}{2fCR}$

$2 = \frac{16}{2 \times 60 \times C \times 150}$

$C = 444.4\,\mu F$

note: we got the same value for C as the full wave rectifier as discharge is over the same amt of time T/2 and the peak is the same - 16V.

c)

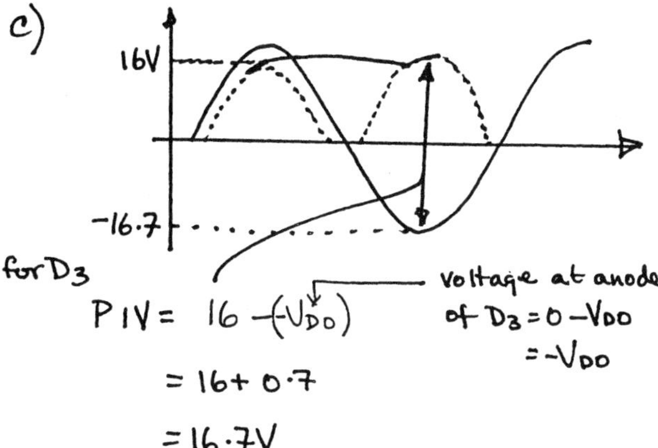

for D_3

$PIV = 16 - (-V_{DO})$ ← Voltage at anode of $D_3 = 0 - V_{DO} = -V_{DO}$

$= 16 + 0.7$

$= 16.7V$

Allowing a 50% margin $= 16.7 \times 1.5 = 25V$

(d) Using 3.34

$i_{D,avg} = I_L\left(1 + \pi\sqrt{V_p/2V_r}\right)$

$= \frac{15}{150}\left(1 + \pi\sqrt{16/2\times2}\right)$

$= 0.73A$

(e) Using (3.35)

$i_{D,max} = I_L\left(1 + 2\pi\sqrt{V_p/2V_r}\right)$

$= \frac{15}{150}\left(1 + 2\pi\sqrt{16/2\times2}\right)$

$= 1.36A$

3.89

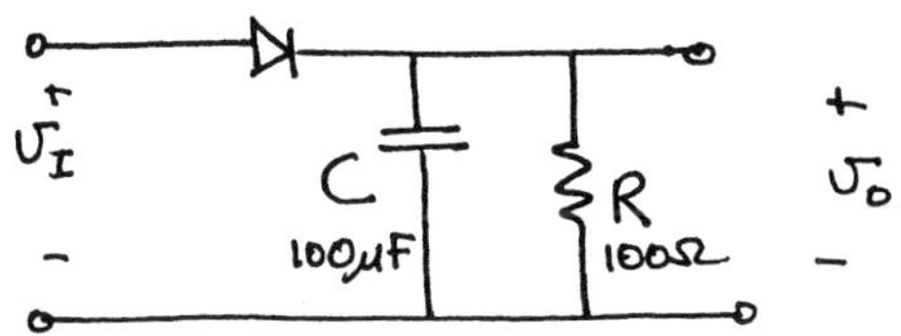

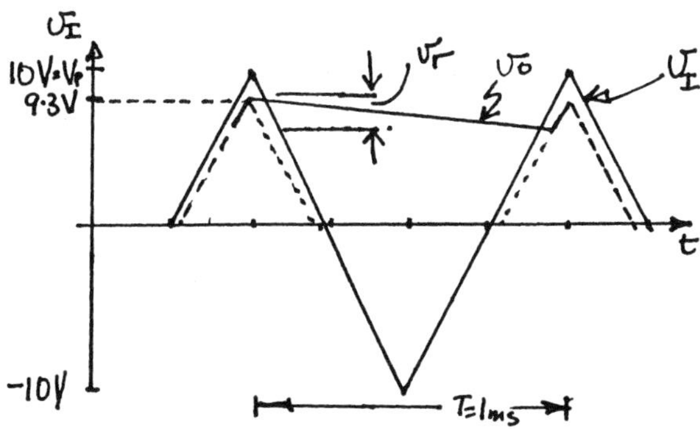

During the diode's off interval, the capacitor discharges through the resistor R according to:

$v_o = 9.3\, e^{-t/RC} \cong 9.3(1 - T/CR)$

$\therefore V_r = 9.3 - 9.3(1 - T/CR)$

$= 9.3T/CR$

$= \frac{9.3}{fCR}$ NB this is Eq (3.38)

$= 0.93V$

$v_{o,avg} = V_o - V_{DO} - \frac{1}{2}V_r$

$= 9.3 - \frac{1}{2}0.93$

$= 8.84V$

(b) 10V, 9.3, $9.3 - V_r = 8.4$, 0 } $V_r = 0.93V$

T/4

$\Delta t \Rightarrow \frac{10}{T/4} = \frac{0.93}{\Delta t}$

$\Delta t = 0.02325T$

$= 0.02325\ ms$

CONT.

(c) $\therefore$ Charge gained during conduction = Charge lost during discharge

$$i_{C,avg}\,\Delta t = C\,V_r$$

$$i_{C,avg} = \frac{C V_r}{\Delta t} = \frac{100\times10^{-6}\times 0.93}{0.02325\times10^{-3}}$$

$$= 4.0\text{ A}$$

$$i_{D,avg} \cong i_{C,avg} + i_{L,avg} \qquad \leftarrow \frac{V_{O,avg}}{R}$$

$$\cong 4.0 + \frac{8.84}{100} = \underline{\underline{4.09\text{A}}}$$

(d) $i_{C,max} = C\,\frac{\delta V_O}{\partial t}\Big|_{\text{at onset of conduction}}$

$$= C\,\frac{\delta V_I}{\partial t}$$

$$= 100\times10^{-6}\times 40\times10^{3}$$

$$= \underline{\underline{4\text{A}}}$$

$$i_{D,max} = i_{C,max} + i_{L,max}$$

$$= 4 + V_{O,max}/100$$

$$= 4 + 9.3/100$$

$$= 4.09\text{ A}.$$

Note that in this case $i_{D,avg} = i_{D,max}$ owing to the linear input (i_C is constant and i_L is approximately constant).

3.90

Refer to Fig P3.82 and let capacitor C be connected across each of the load resistors R. The two supplies, V_O^+ and V_O^- are identical. Each is a full-wave rectifier similar to that based on the centre-tapped-transformer circuit. For each supply, the dc output is 15 V and the ripple is 1V peak-to-peak. Thus $V_O = 15 \pm \frac{1}{2}$ V. It follows that the peak value of V_S must be $15.5 + 0.7 = 16.2$V.

$\therefore$ Voltage across secondary $= 2(16.2) = 32.4$ V

RMS across secondary $= \frac{32.4}{\sqrt{2}} = \underline{\underline{22.9\text{ V}_{rms}}}$

Turns Ratio $= \frac{120}{22.9} = 5.24 : 1$

Use Eq. (3.35) to find

$$i_{D,max} = I_L\left(1 + 2\pi\sqrt{V_P/2V_r}\right)$$

$$= 0.2\left(1 + 2\pi\sqrt{15.5/2}\right)$$

$$= \underline{\underline{3.70\text{A}}}$$

$$V_r = \frac{V_P}{2fCR} = 1 \qquad \text{Eq (3.28)}$$

DISCHARGE OCCURS OVER $T/2 = \frac{1}{2f}$

$$\Rightarrow C = \frac{15.5}{2\times60\times75}$$

where $200\text{mA} = \frac{15}{R}$, $R = \frac{15}{0.2} = 75\Omega$

$$= \underline{\underline{1722\,\mu\text{F}}}$$

Consider D_3 when looking at PIV

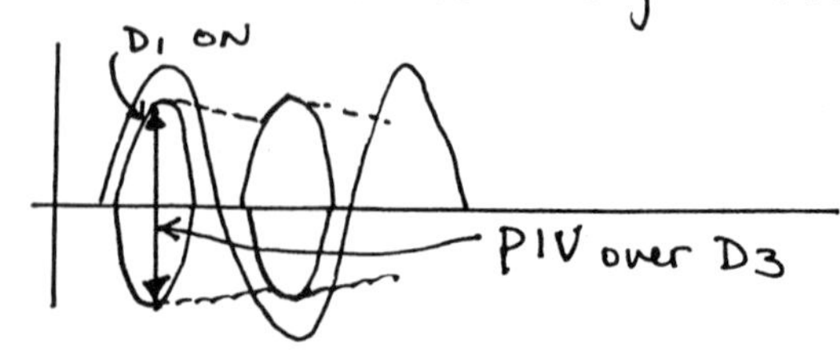

CONT.

$$PIV = \hat{v}_O + \hat{v}_S$$
$$= 15.5 + 16.2 = 31.7V.$$

Allowing for 50% safety margin

$$PIV = 1.5 \times 31.7 = \underline{\underline{47.6V}}$$

Use Eq (3.34) to find

$$i_{D,avg} = I_L \left(1 + \pi\sqrt{V_P/2V_r}\right)$$
$$= 0.2\left(1 + \pi\sqrt{15.5/2}\right)$$
$$= \underline{\underline{1.95A}}$$

3.91

$v_O = v_I\,(1 + R/R)$
$= 2v_I$ when the diode is conducting.

(a) $v_I = +\underline{\underline{1V}}$ $v_O = \underline{\underline{2V}}$ $v_A = \underline{\underline{1.7V}}$ $v_- = v_I = \underline{\underline{1V}}$

b) $v_I = \underline{\underline{2V}}$ $v_O = \underline{\underline{4V}}$ $v_A = \underline{\underline{4.7V}}$ $v_- = \underline{\underline{2V}}$

c) $v_I = -1V$ $v_A = \underline{\underline{-12V}}$ ~ diode is cut off
$v_O = \underline{\underline{0V}}$
$v_- = \underline{\underline{0V}}$

d) $v_I = -2V$ $v_A = \underline{\underline{-12V}}$ $v_O = \underline{\underline{0V}}$ $v_- = \underline{\underline{0V}}$

$v_{O,avg} = \underline{\underline{5V}}$

3.92

$v_I > 0$ D_1 conducts D_2 cutoff

$v_I < 0$ D_1 cut off
D_2 conducts ~ $\frac{v_O}{v_I} = -1$

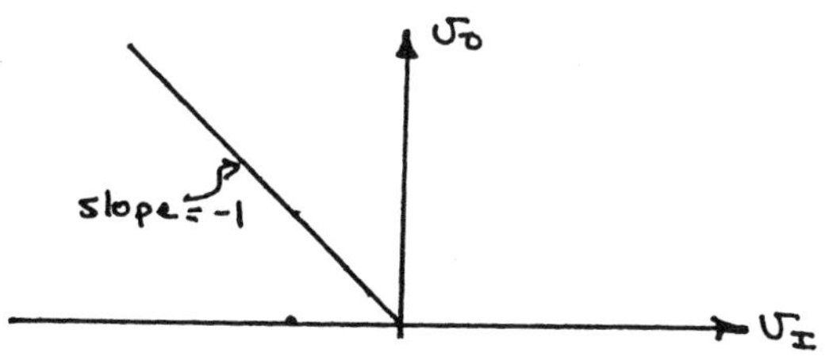

a) $v_I = 1V$ $v_O = \underline{\underline{0V}}$
$v_A = \underline{\underline{-0.7V}}$ - keeps D_2 off so no current flows through R
$\Rightarrow v_- = \underline{\underline{0V}}$ ~ virtual gnd as fdbk is closed through D_1

(b) $v_I = 2V$
$v_O = \underline{\underline{0V}}$
$v_A = \underline{\underline{-0.7V}}$
$v_- = \underline{\underline{0V}}$

(c) $v_I = -1V$
$v_O = \underline{\underline{1V}}$
$v_A = \underline{\underline{1.7V}}$
$v_- = 0V$ ~ virtual gnd as negative feedback is closed through R.

(d) $v_I = -2V \Rightarrow v_O = \underline{\underline{2V}}$
$v_A = \underline{\underline{2.7V}}$
$v_- = \underline{\underline{0V}}$

3.93

(a)

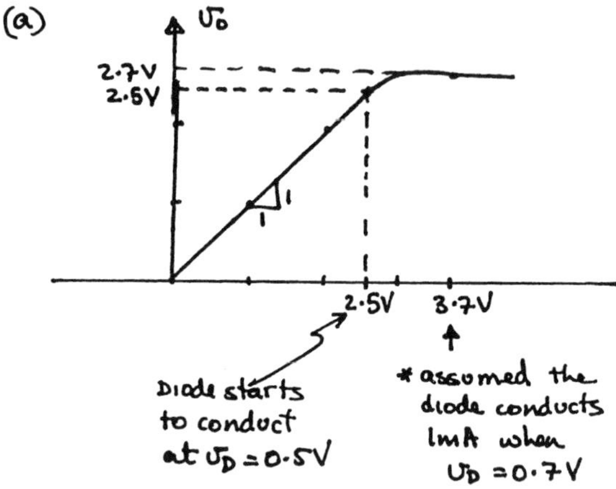

(b)

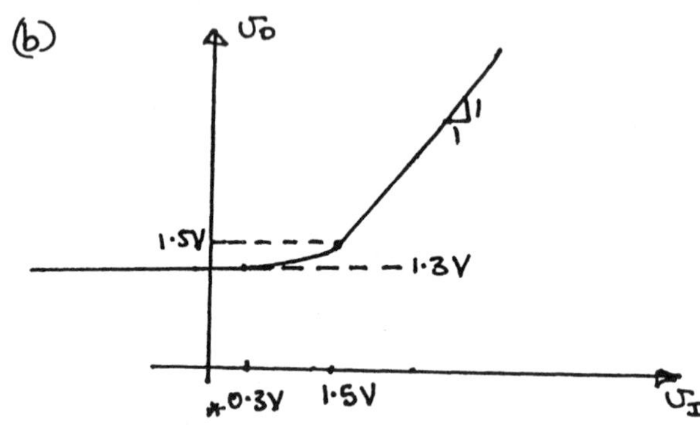

(c)

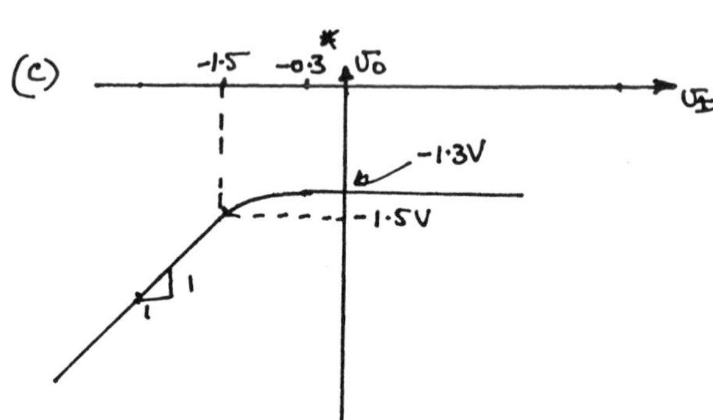

(d)

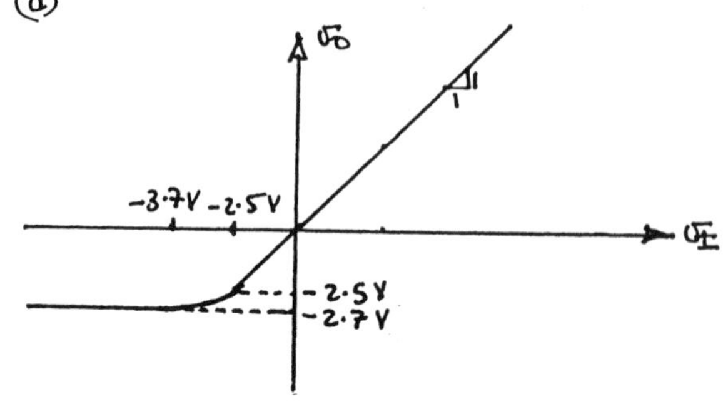

3.94

(a) v_I < 2.65V ~ the diode is off and the circuit reduces to:

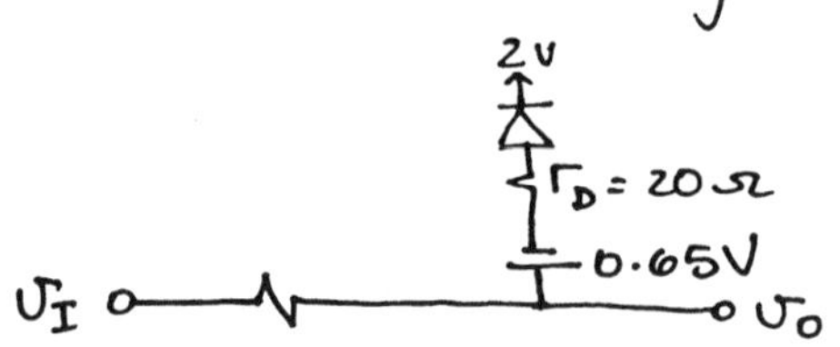

With the diode conducting we have:

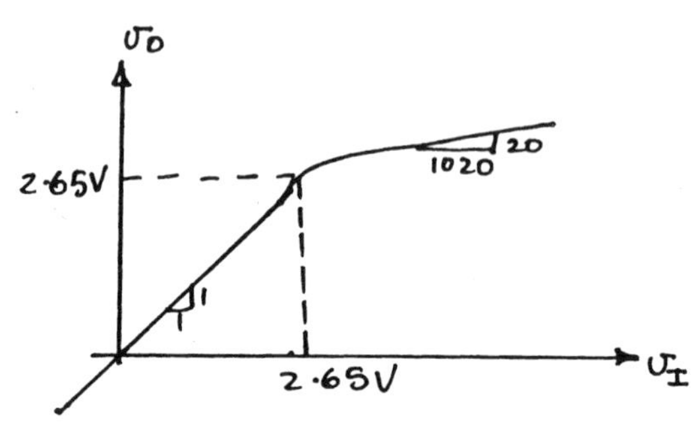

Similarily for (b), (c,) (d)

(b)

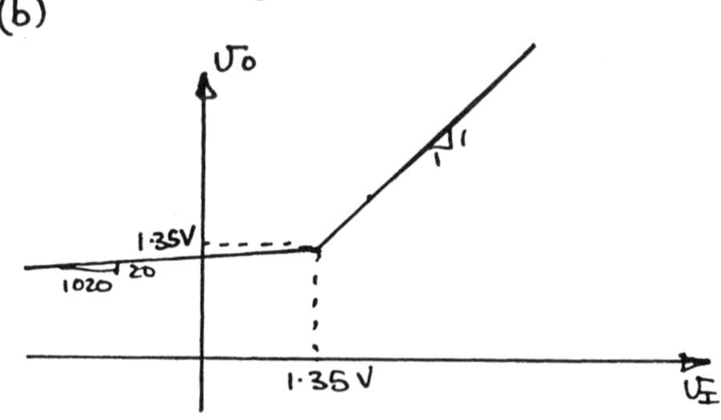

(c)

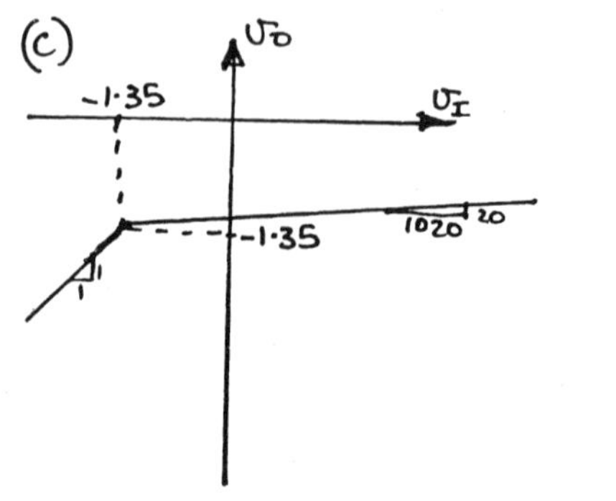

(d)

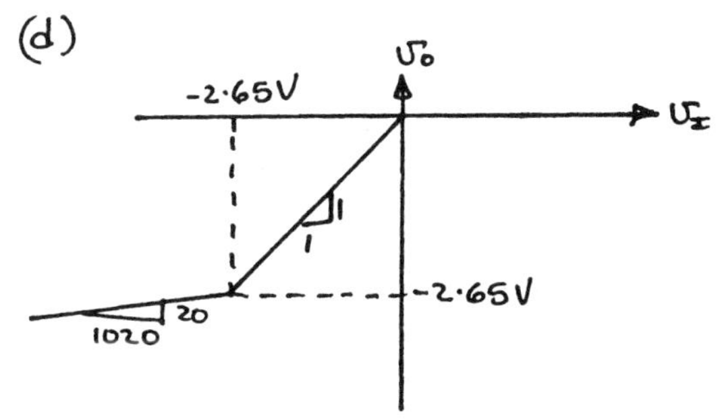

3.95

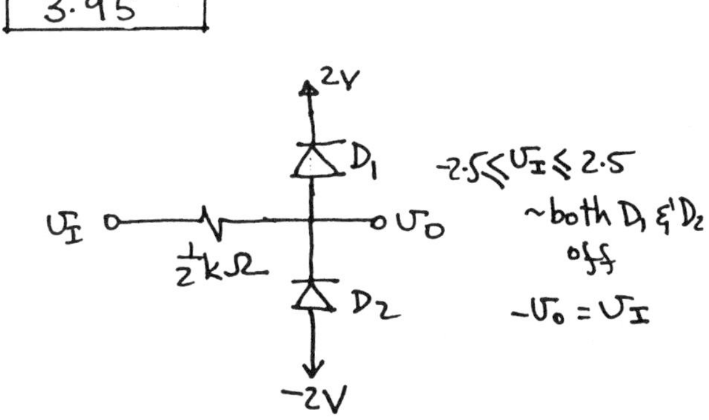

$-2.5 \leq v_I \leq 2.5$

~ both D_1 & D_2 off

$v_O = v_I$

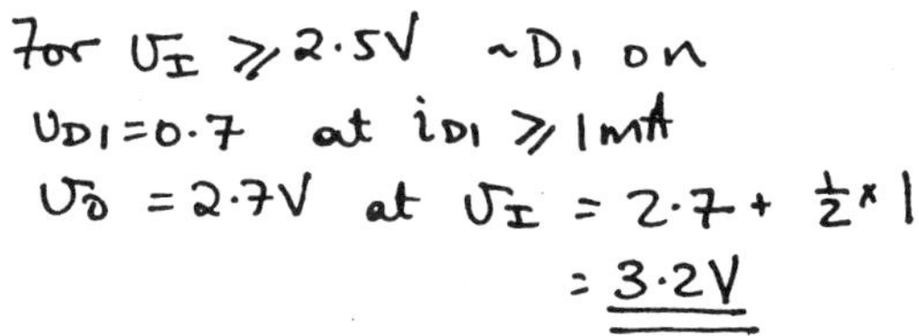

For $v_I \geq 2.5V$ ~ D_1 on

$v_{D1} = 0.7$ at $i_{D1} \geq 1mA$

$v_O = 2.7V$ at $v_I = 2.7 + \frac{1}{2} \times 1$

$= \underline{3.2V}$

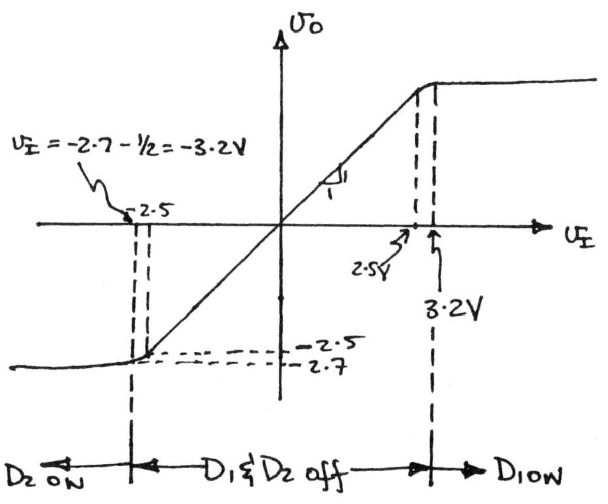

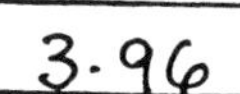

3.96

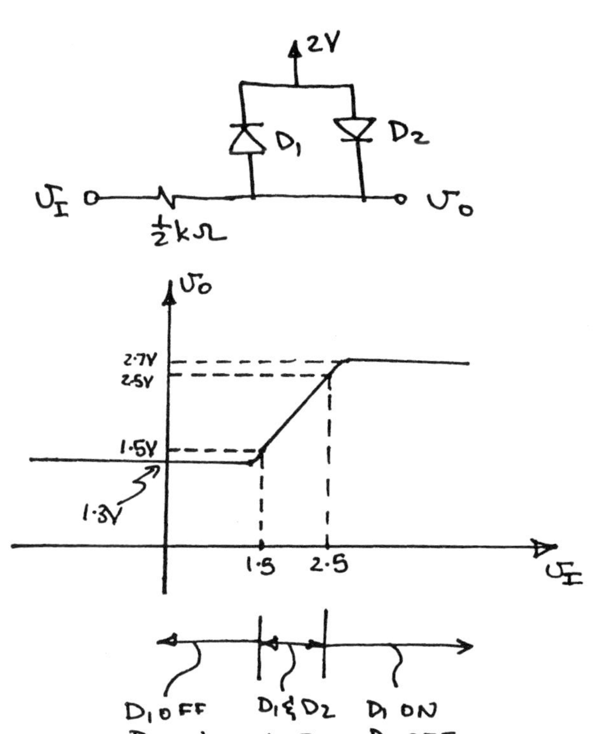

3.97

For each diode

For the zener diode

$8.2 = V_{Z0} + 10 \times 10^{-3} \times 20$

$V_{Z0} = 8.0V$

The limiter thresholds are

$\pm(2 \times 0.65 + 8.0) = \pm 9.3V$

For $v_I > 9.3$ (as well as for $v_I < -9.3$)

$$\frac{\partial v_O}{\partial v_I} = \frac{r_{D1} + r_Z + r_{D2}}{1k\Omega + r_{D1} + r_Z + r_{D2}} = \frac{3(20)}{1k\Omega + 3(20)} = 0.057 \frac{V}{V}$$

CONT.

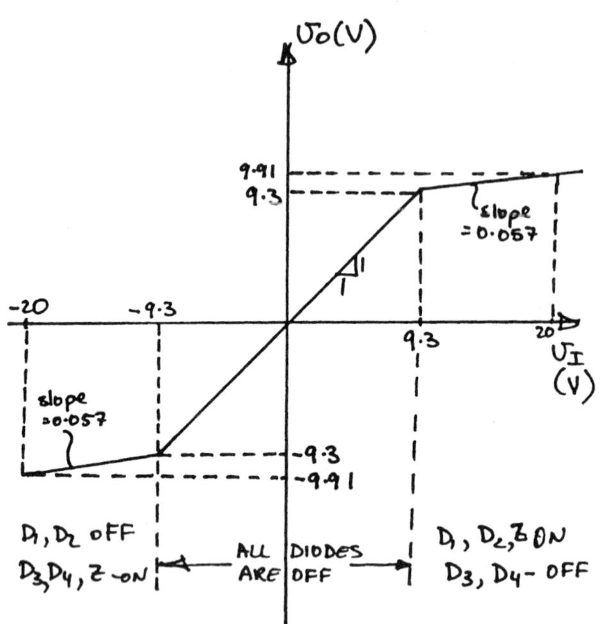

3.98

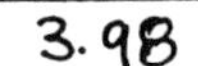

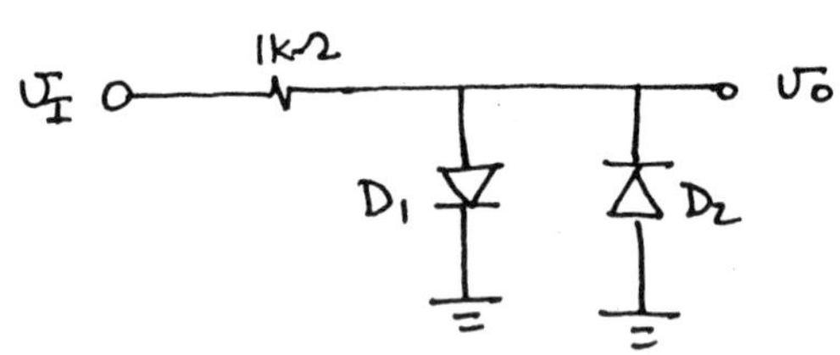

for D_1

Given $\dfrac{i_D}{1\,mA} = e^{\frac{v_O - 0.7}{nV_T}}$

$$(v_O - 0.7) = nV_T \ln(i_D/1\,mA)$$

$$= 0.1 \log\left(\frac{i_D}{10^{-3}}\right) \leftarrow \therefore \text{can find } v_O \text{ from } i_D$$

$$\therefore\ i_D = 10^{-3} \times 10^{\frac{v_O - 0.7}{0.1}}$$

$$= 10^{-3} \times 10^{10(v_O - 0.7)}$$

$$\& \quad v_I = v_O + i_D \times 10^3$$

$$= v_O + 10^{10(v_O - 0.7)}$$

for D_2: $\quad v_I = v_O - 10^{-10(v_O - 0.7)}$

v_O (V)	v_I (V)	
0.5	0.510	D_1 ON
0.6	0.7	
0.7	1.7	
0.8	10.7	
0	0	
-0.5	-0.51	D_2 ON
-0.6	-0.7	
-0.7	-1.7	
-0.8	-10.7	

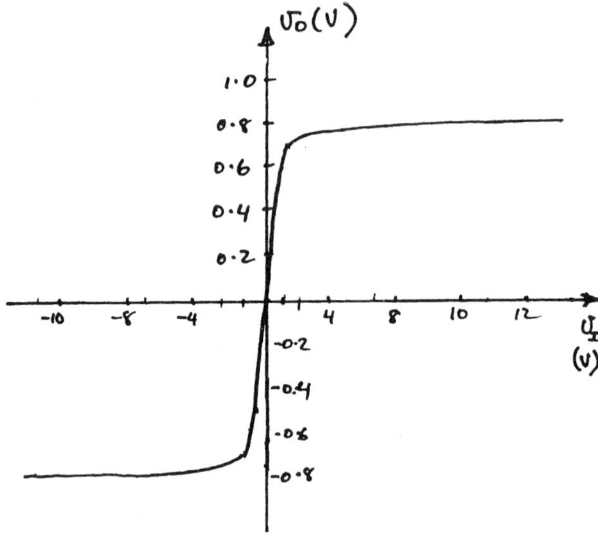

The limiter is fairly hard with a gain $K \cong 1$

$L_+ \cong \underline{\underline{0.8\,V}}$, $\quad L_- \cong \underline{\underline{-0.8\,V}}$

3.99

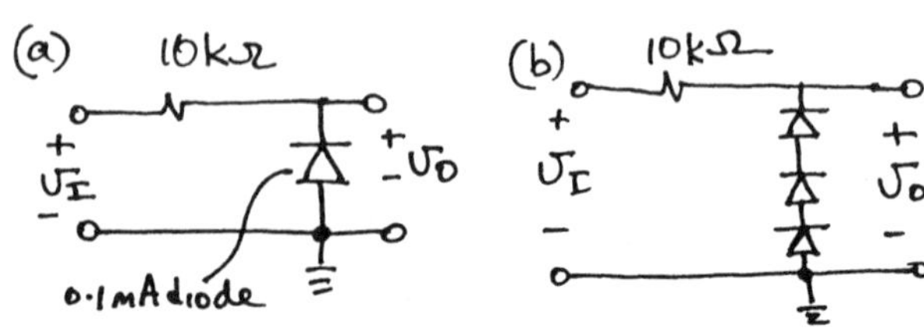

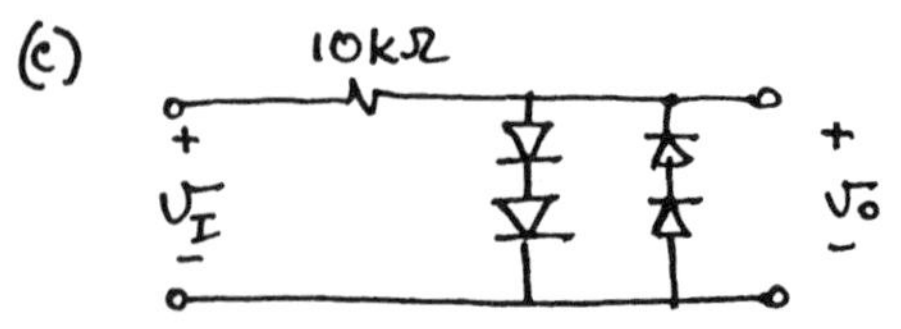

3.100

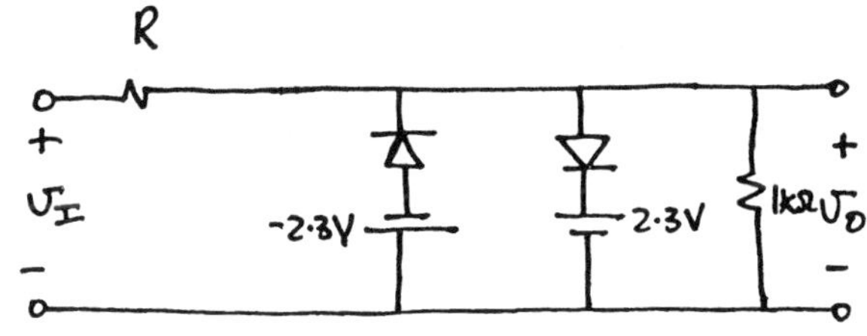

In the limiting region

$$\frac{v_O}{v_I} = \frac{1000}{1000+R} \geq 0.95$$

$R \leq \underline{\underline{52.6\ \Omega}}$

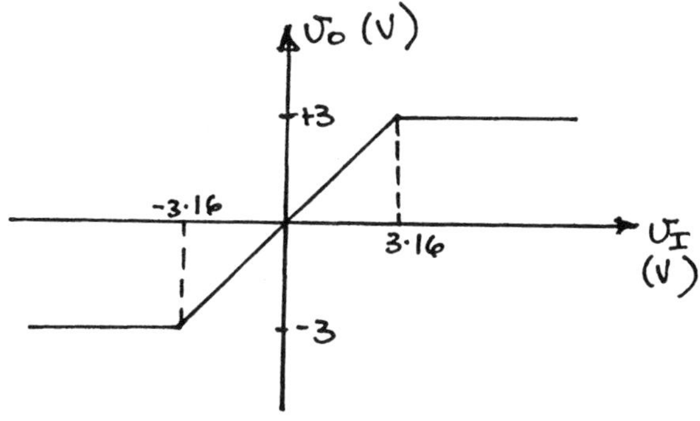

3.101

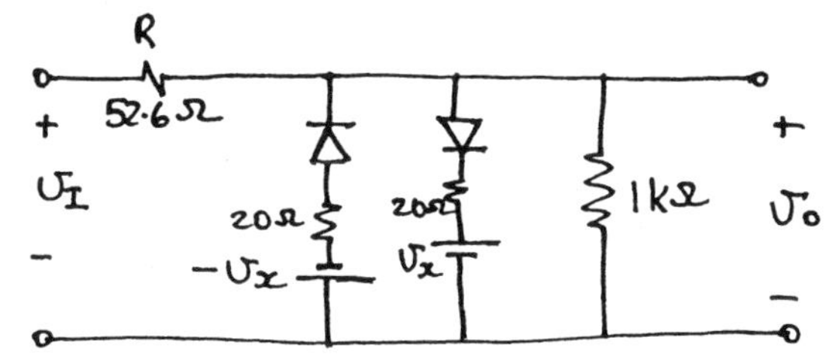

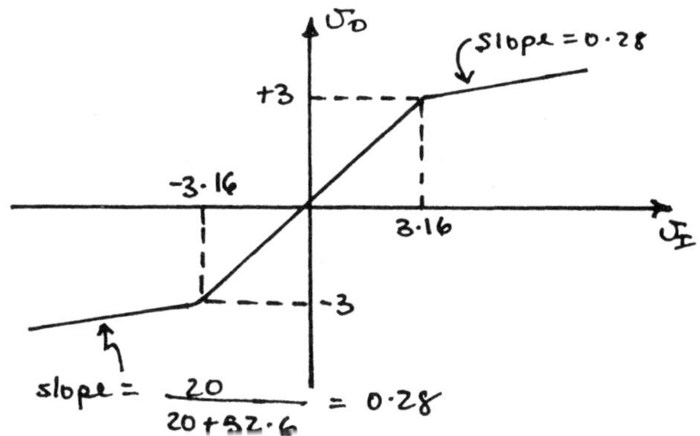

$$\text{slope} = \frac{20}{20+52.6} = 0.28$$

$$\frac{0.2}{10\times10^{-3}} = 20\ \Omega$$

At the verge of limiting in the positive direction we have :-

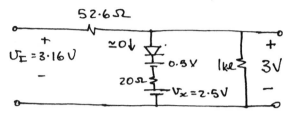

For $v_I = 10V$

$v_O \simeq 3 + 0.28(10 - 3.16)$

$= \underline{\underline{4.9V}}$

For $v_I = -10V$

$v_O = \underline{\underline{-4.9V}}$

3.102

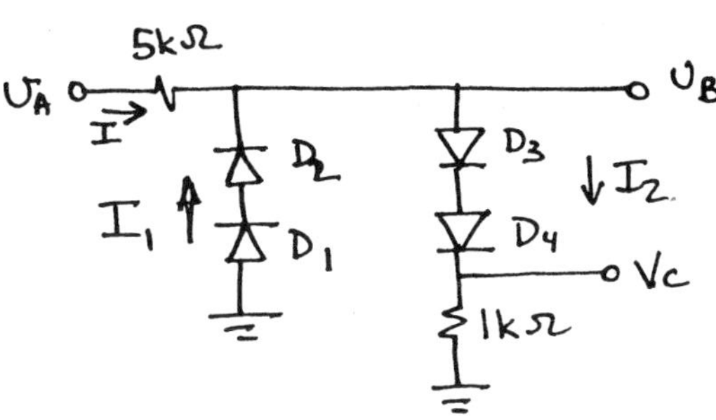

$v_D = 0.7 + 0.1 \log(i_D/0.1)$

(a) For $v_A > 0$ $\quad D_1, D_2$ off $\Rightarrow I_1 = 0$

$I = I_2 = v_C/1k\Omega \quad v_A = V_B + I_2 5$

(b) For $v_I < 0$ $\quad D_3, D_4$ off $\Rightarrow V_C = 0$

$I = -I_1 \quad v_B = -(v_{D1} + v_{D2})$

$v_A = -(V_B + 5I_1)$

CONT.

(a) List of points for $v_A > 0$

v_C (V)	I_2 (mA)	v_{D3}, v_{D4} (V)	$v_B = v_C + V_{D3} + V_{D4}$ (V)	v_A (V)
0.0001	0.0001	0.4	0.8	0.8
0.001	0.001	0.5	1.00	1.01
0.01	0.1	0.6	1.21	1.26
0.1	0.1	0.7	1.50	1.90
0.2	0.2	0.73	1.66	2.66
0.3	0.3	0.75	1.80	3.30
0.4	0.4	0.76	1.92	3.92
0.5	0.5	0.77	2.04	4.54
0.6	0.6	0.78	2.16	5.16

(b) List of Points for $v_A < 0$

I_1 (mA)	v_{D1}, v_{D2} (V)	v_B (V)	v_A (V)
0.0001	0.4	−0.80	−0.80
0.001	0.5	−1.00	−1.01
0.01	0.6	−1.20	−1.25
0.10	0.7	−1.40	−1.90
0.20	0.73	−1.46	−2.46
0.30	0.75	−1.50	−3.00
0.40	0.76	−1.52	−3.52
0.50	0.77	−1.54	−4.04
0.60	0.78	−1.56	−4.56
0.70	0.785	−1.57	−5.07

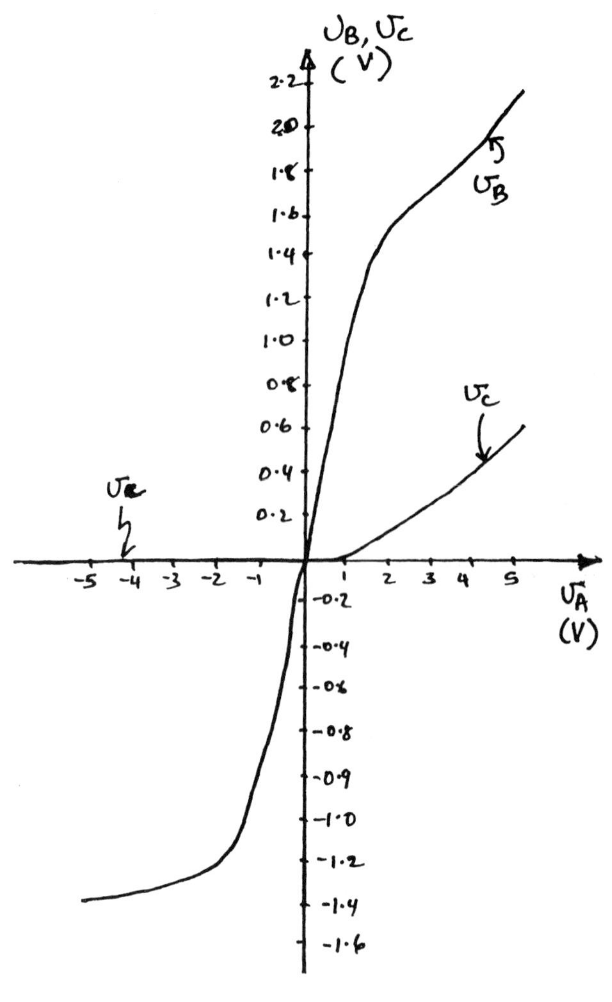

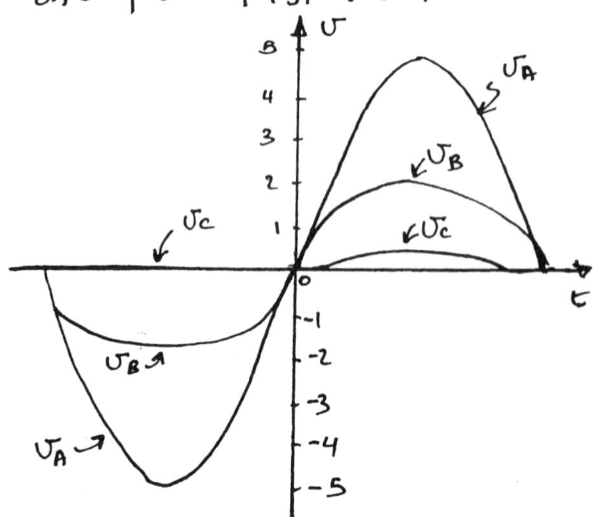

3.103

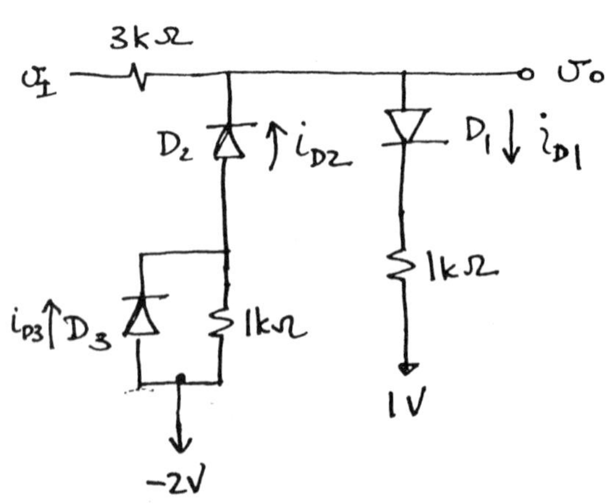

At currents $i_{D1} > 1\text{mA}$, $v_{D1} \simeq 0.7\text{V}$

Let $v_{D1} = 0.71\text{V}$ $\quad v_I > 5.7\text{V}$

CONT.

$$\upsilon_O = 1.71 + i_{D1} \times 1k\Omega$$

$$= 1.71 + \left(\frac{\upsilon_I - 1.71}{4}\right) \times 1$$

$$= \frac{\upsilon_I}{4} + 1.2825 \qquad \text{NB slope} = 1/4$$

∴ for $\upsilon_I > 5V$ slope $\upsilon_O/\upsilon_I \approx 1/4$

υ_I(V)	υ_O (V)
5.8	2.7325
6.0	2.7825
7.0	3.0325
8.0	3.2825
9.0	3.5325
10.0	3.7825

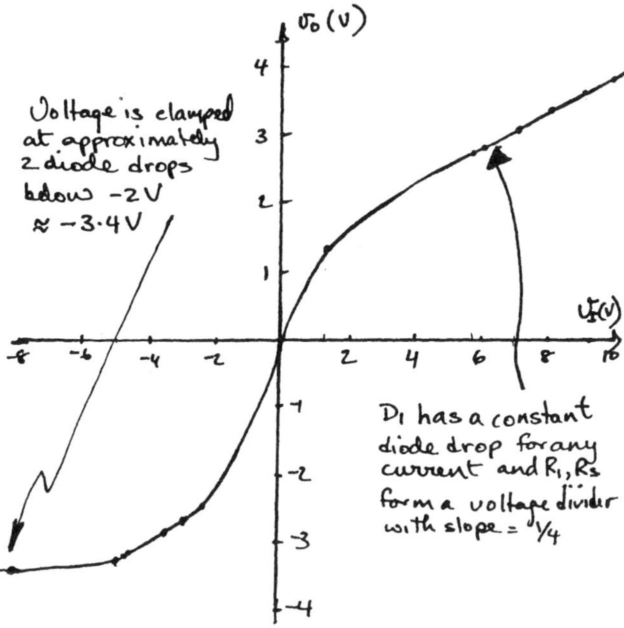

υ_O (V)
υ_I(V)
Voltage is clamped at approximately 2 diode drops below −2V ≈ −3.4V
D1 has a constant diode drop for any current and R1, R5 form a voltage divider with slope = 1/4

where points for $-8 \le \upsilon_I \le 6$ V are calculated as shown below:

$i_D = 1mA$ at $\upsilon_D = 0.7V$ $\quad n=1$

$$i_D = I_S e^{0.7/0.025} = 10^{-3}$$

$$I_S = 6.914 \times 10^{-16} A.$$

∴ For Diodes use $i_D = 6.914 \times 10^{-16} e^{\upsilon_D/0.025}$

D_1 conducting $\quad i_{D2} = 0$

i_{D1} (A)	υ_{D1} (V)	υ_O (V)	$\upsilon_I = (4k)i_{D1} + \upsilon_{D1} + 1$ (V)
10^{-10}	0.297	1.297	1.297 ← even at small i_{D1} $\upsilon_O > 1V$, $\upsilon_O \approx \upsilon_I$ since $i_{D1} \approx 0$
10^{-6}	0.527	1.528	1.5313
10^{-5}	0.584	1.595	1.625
10^{-4}	0.64	1.742	2.042
10^{-3}	0.70	2.7	5.7
0.2×10^{-2}	0.74	6.74	12.74
10^{-2}	0.758	11.75	41.75

For the D_2, D_3 arm conducting use the following equations:
Note $\upsilon_I < -2.5V$

Starting with a value for V_A ① we have

$$V_{D3} = V_A + 2$$

$$i_{D3} = I_S e^{\upsilon_{D3}/0.025} \qquad ②$$

$$i_{D2} = i_{D3} + \frac{V_A + 2}{1} \qquad ③$$

$$\upsilon_{D2} = 0.025 \ln\left(\frac{i_{D2}}{6.914\times10^{-16}}\right) \qquad ④$$

$$\upsilon_O = \upsilon_A - \upsilon_{D2} \qquad ⑤$$

$$\upsilon_I = \upsilon_O - i_{D2} \times 3k\Omega \qquad ⑥$$

① υ_A (V)	② i_{D3} (A)	③ i_{D2} (A)	④ υ_{D2} (V)	⑤ υ_O (V)	⑥ υ_I (V)
−2.001	7×10^{-16}	10^{-6}	0.527	−2.528	−2.531 Ⓐ
−2.01	10^{-15}	10^{-5}	0.585	−2.595	−2.625
−2.10	3.8×10^{-14}	10^{-4}	0.642	−2.724	−3.024
−2.20	2×10^{-12}	0.2×10^{-3}	0.659	−2.859	−3.459
−2.5	33μA	0.5×10^{-3}	0.682	−3.128	−4.628 Ⓑ
−2.6	18μA	0.6×10^{-3}	0.687	−3.287	−5.087
−2.7	1mA	1.7×10^{-3}	0.713	−3.413	−8.516 Ⓒ
−2.71	1.5mA	2.2mA	0.720	−3.43	−10

Ⓐ for small i_{D2}, D_3 is off and D_2 is on
∴ i_2 flows through 1kΩ resistor

CONT.

(B) 0.5 V drop across D_3 causes D_3 to start to conduct

(C) $U_A = -2.7V$

The 0.7 voltage across D_3 clamps the voltage across R_3 so that D_3 control the current i_{D2}

3.104

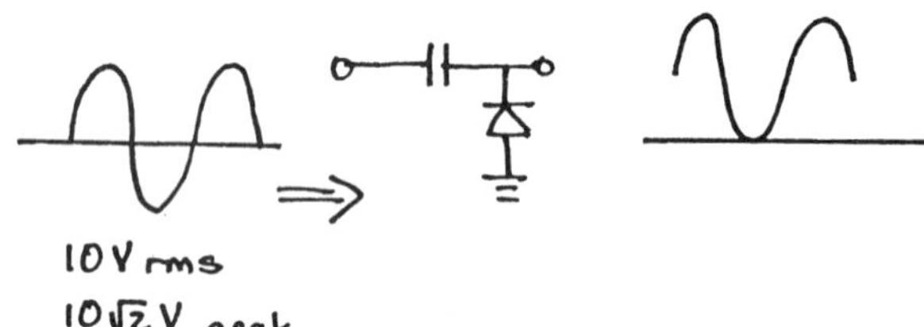

10V rms

$10\sqrt{2}$ V, peak

Average (dc) value of output $= 10\sqrt{2}$

$= \underline{\underline{14.14V}}$

3.105

(a)

(b)

(c)

(d)

(e) +10V, −10V

(f) Here there are two different time constants involved. To calculate the output levels we shall exagerate the discharge and charge waveforms.

During T_1 $v_O = V_1 e^{-t/RC}$

At $t = T_1 = T$ $v_O = V_1'$

$= V_1 e^{-T/RC}$

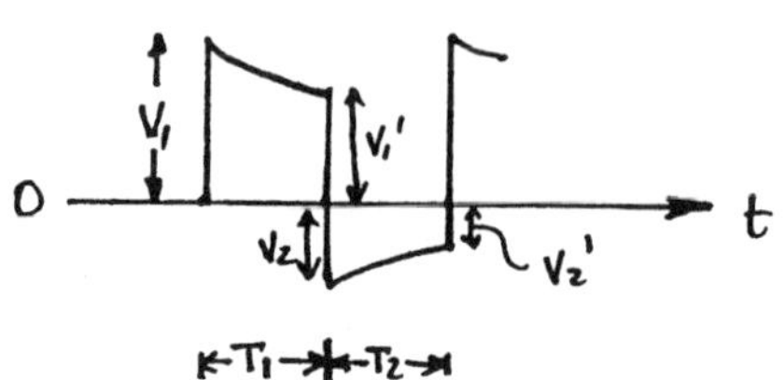

where for $T \ll CR$

$V_1' \cong V_1(1 - T/CR) = V_1(1-\alpha)$ where $\alpha \ll 1$

During the period T_2

$|v_O| = |V_2| e^{-t/CR/2}$

at the end of T_2 $t = T$, $v_O = V_2'$

where

$V_2' = |V_2| e^{-T/CR/2}$

$\cong |V_2|\left(1 - \frac{T}{CR/2}\right) = |V_2|(1-2\alpha)$

Now $V_1' + |V_2| = 20 \Rightarrow V_1 + |V_2| - \alpha V_1 = 20$ ①

and $|V_2'| + V_1 = 20 \Rightarrow V_1 + |V_2| - 2\alpha|V_2| = 20$ ②

from ① & ② we find that

$V_1 = 2V_2$

Then using ① and neglecting αV_1 yields

$3|V_2| = 20 \Rightarrow |V_2| = \underline{\underline{6.67V}}$

$V_1 = \underline{\underline{13.33V}}$

The result is:

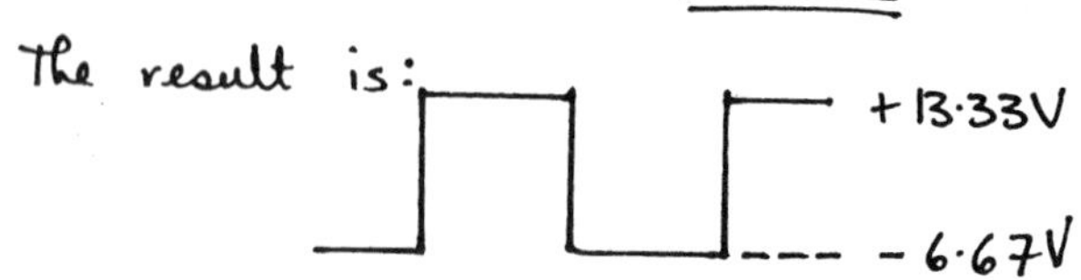

CONT.

(g) +18 V

−2V

(h) Using a method similar to that employed for case (f) above we obtain.

3.106

$$n_i^2 = BT^3 e^{-E_G/kT}$$

$B = 5.4 \times 10^{31}$

$E_G = 1.12\,eV$ for silicon

$k = 8.62 \times 10^{-5}\ \frac{eV}{K}$

Fraction of ionized atoms $= n_i / 5 \times 10^{22}$; $T = 273 + x°C$

T (K)	n_i (carriers/cm³)	Fraction of ionized atoms.
203	2.68×10^5	5.37×10^{-18}
273	1.53×10^9	3.07×10^{-14}
293	8.65×10^9	1.73×10^{-13}
373	1.44×10^{12}	2.89×10^{-11}
398	4.75×10^{12}	9.51×10^{-11}

3.107

(a) intrinsic silicon

$$\rho = \frac{1}{q(p\mu_p + n\mu_n)}$$

$$= \frac{1}{1.6 \times 10^{-19}(1.5 \times 10^{10} \times 480 + 1.5 \times 10^{10} + 1350)}$$

$$= 2.28 \times 10^5\ \Omega \cdot cm$$

$$R = \rho\, L/A = 2.28 \times 10^5 \frac{10 \times 10^{-4}}{\underbrace{3.4 \times 10^{-4}}_{10^4/3}} = \underline{\underline{7.6 \times 10^9\ \Omega}}$$

(b)

$$\rho = \frac{1}{q(p_{no}\mu_p + n_{no}\mu_n)}$$

$$= \frac{1}{(1.6 \times 10^{-19})\left(\frac{(1.5 \times 10^{10})^2}{10^{16}} \times \underbrace{\frac{1200}{2.5}}_{\mu_p} + 10^{16} \times 1200\right)}$$

$$= 0.521\ \Omega \cdot cm$$

$$R = \rho \times 10^4/3 = \underline{\underline{1.74\ k\Omega}}$$

(c) n doped $N_D = n_{no} = 10^{18}$

$n_{po} = n_i^2 / n_{no}$

$$\rho = \frac{1}{1.6 \times 10^{-19}\left(\frac{(1.5 \times 10^{10})^2}{10^{18}} \times \frac{1200}{2.5} + 10^{18} \times 1200\right)}$$

$$= 5.21 \times 10^{-3}\ \Omega \cdot cm$$

$$R = \rho \frac{10^4}{3} = \underline{\underline{17.4\ \Omega}}$$

(d) P doped $N_A = P_{PO} = 10^{10}/cm^3$

$$\rho = \frac{1}{q(P_{PO}\mu_p + n_{PO}\mu_n)}$$

$$= \frac{1}{1.6 \times 10^{-19}\left(10^{10} \times \frac{1200}{2.5} + \frac{(1.5 \times 10^{10})^2}{10^{10}} \times 1200\right)}$$

$$= 2.05 \times 10^4\ \Omega \cdot cm$$

$$R = \rho \frac{10^4}{3} = \underline{\underline{68.35\ k\Omega}}$$

(e) $R = 2.8 \times 10^4/3 = \underline{\underline{0.933\ \Omega}}$

3.108

$$p_{no} = \frac{n_i^2}{N_D} = \frac{(1.5 \times 10^{10})^2}{10^{16}} = 2.25 \times 10^4\ /cm^3$$

CONT.

$$\frac{dP}{dx} = \frac{1000\,P_{no} - P_{no}}{W}$$

$$\cong \frac{2.26\times10^{-7}}{5\times10^{-4}}$$

$$J_P = -qD_P\frac{dP}{dx}$$

$$= -1.6\times10^{-19}\times12\times\frac{-2.25\times10^{-1}}{5\times10^{-4}}$$

$$= \underline{\underline{8.64\times10^{-8}\ A/cm^2}}$$

3.109

$$v_{drift,p} = \mu_p E$$

$$= 480\,\frac{cm^2}{V\cdot s}\times\frac{1}{10^{-3}}\,\frac{V}{cm}$$

$$= \underline{\underline{4.8\times10^{5}\ cm/s}}$$

$$v_{drift,n} = \mu_n E$$

$$= 1350\,\frac{cm^2}{V\cdot s}\times\frac{1}{10^{-3}}\,\frac{V}{cm}$$

$$= \underline{\underline{1.35\times10^{6}\ cm/s}}$$

3.110

$$J_{drift} = q(n\mu_n + p\mu_p)E \quad A/cm^2$$

$$I_{drift} = q(n\mu_n + p\mu_p)E\cdot A$$

$$= 1.6\times10^{-19}(10^{5}\cdot1350 + 10^{15}\cdot480)\times\frac{1}{10^{-3}}\times(5\times4\times10^{-8})$$

$$= \underline{\underline{15.36\ \mu A}}$$

3.111

$$E = \frac{1V}{10\times10^{-4}} = \frac{1}{10^{-3}}\,\frac{V}{cm}$$

$n_{no} = N_D \qquad P_{no} = \frac{n_i^2}{N_D}$ assume $P_{no} \ll n_{no}$

$$\therefore J_{drift} = q(\mu_n n + \mu_p p)E = \frac{10^{-3}}{10^{-8}}\,\frac{A}{cm^2}$$

$$\cong 1.6\times10^{-19}(1350\,N_D)\frac{1}{10^{-3}} = 10^{5}$$

$$N_D = 4.6\times10^{17}\ /cm^3$$

3.112

$P_{po} = N_A = 10^{16}\ /cm^3$ at all temperatures

At $25^\circ C \cong 300K$

$$n_i^2 = BT^3 e^{-E_G/kT}$$

$$= 5.4\times10^{31}(300)^3 e^{-1.12/8.62\times10^{-5}(300)}$$

$$= 2.26\times10^{20}$$

$$n_{po} = \frac{n_i^2}{N_A} = \underline{\underline{2.26\times10^{4}\ /cm^3}}$$

At $125^\circ C \cong 400K$

$$n_i^2 = 5.4\times10^{31}(400)^3 e^{1.12/8.62\times10^{-5}(400)}$$

$$= 2.7\times10^{25}$$

$$n_{po} = \frac{n_i^2}{N_A} = \underline{\underline{2.7\times10^{9}\ /cm^3}}$$

3.113

DOPING CONCENTRATION	μ_n $cm^2/V\cdot s$	μ_p $cm^2/V\cdot s$	D_n cm^2/s	D_p cm^2/s
INTRINSIC	1350	480	34	12
10^{16}	1100	400	28	10
10^{17}	700	260	18	6
10^{18}	360	150	9	4

where $D_n = V_T \mu_n = \underline{\underline{0.025\,\mu_n}}$

$D_p = V_T \mu_p = \underline{\underline{0.025\,\mu_p}}$

3.114

$V_0 = V_T \ln\left(\frac{N_A N_D}{n_i^2}\right) = 0.025 \ln\left(\frac{10^{16}\, 10^{16}}{10^{10}}\right)$

$= 1.27\,V$

$W_{dep} = x_n + x_p = \sqrt{\frac{2\epsilon_s}{q}\left(\frac{1}{N_A} + \frac{1}{N_D}\right) V_0}$

$= \sqrt{\frac{2(11.7)\ 8.85 \times 10^{-14}}{1.6 \times 10^{-19}}\left(\frac{1}{10^{16}} + \frac{1}{10^{16}}\right) 1.27}$

$= 57 \times 10^{-6}$ cm

$= \underline{\underline{0.57\,\mu m}}$

$\frac{x_n}{x_p} = \frac{N_A}{N_D} \Rightarrow x_n = x_p = \underline{\underline{0.28\,\mu m}}$

$q_J = q_N = q_P = q \frac{N_A N_D}{N_A + N_D} A\, W_{dep}$

$= \underline{\underline{45.6 \times 10^{-15}\ C}}$

$C_j = \frac{\epsilon_s A}{W_{dep}} = \frac{11.7(8.85 \times 10^{-14}) \times 100 \times 10^{-8}}{0.57 \times 10^{-4}}$

$= \underline{\underline{18.2\,fF}}$

3.115

$V_0 = V_T \ln\left(\frac{N_A N_D}{n_i^2}\right) = 0.025 \ln\left(\frac{10^{16}\, 10^{15}}{10^{20}}\right)$

$= 0.633\,V$

$W_{dep} = x_n + x_p = \sqrt{\frac{2\epsilon_s}{q}\left(\frac{1}{N_A} + \frac{1}{N_D}\right)(V_0 + V_R)}$

$W_{dep} = \sqrt{\frac{2(11.7)(8.85 \times 10^{-14})}{1.6 \times 10^{-19}}\left(\frac{1}{10^{16}} + \frac{1}{10^{15}}\right)(0.633 + 5)}$

$= 2.83 \times 10^{-4}$ cm $= \underline{\underline{2.83\,\mu m}}$

$x_n = \frac{N_A}{N_D} x_p = 10\, x_p$

$\therefore\ x_p = \frac{2.83}{11} = \underline{\underline{0.26\,\mu m}}$

$x_n = \underline{\underline{2.57\,\mu m}}$

$q_J = q_N = q_P = q \frac{N_A N_D}{N_A + N_D} A\, W_{dep}$

$= 1.6 \times 10^{-19}\left(\frac{10^{16}\, 10^{15}}{10^{16} + 10^{15}}\right) 400 \times 10^{-8} \times 2.83 \times 10^{-4}$

$= \underline{\underline{1.65 \times 10^{-13}\ C}}$

$C_j = \frac{\epsilon_s A}{W_{dep}} = \frac{11.7(8.84 \times 10^{-14})\ 400 \times 10^{-8}}{2.83 \times 10^{-4}}$

$= 14.6 \times 10^{-15} = \underline{\underline{14.6\,fF}}$

3.116

$q_J = q N_x A = 1.6 \times 10^{-19} \times 10^{16} \times 0.1 \times 10^{-4} \times 100 \times 10^{-8}$

$= \underline{\underline{16\ fC}}$

3.117

$q_J = q \frac{N_A N_D}{N_A + N_D} A\, W_{dep}$

$= q \frac{N_A N_D}{N_A + N_D} A \sqrt{\frac{2\epsilon_s}{q}\left(\frac{1}{N_A} + \frac{1}{N_D}\right)(V_0 + V_R)}$

CONT.

$$\frac{dQ_S}{dV_R} = q\frac{N_A N_D}{N_A+N_D} A \times \frac{1}{q\sqrt{\frac{2\epsilon_s}{q}\left(\frac{1}{N_A}+\frac{1}{N_D}\right)(V_0+V_R)}} \times \frac{q\epsilon_s}{q}\left(\frac{1}{N_A}+\frac{1}{N_D}\right)$$

$$= \frac{\epsilon_s A}{W_{dep}}$$

3.118

$$C_j = \frac{C_{j0}}{(1+V_R/V_0)^m}$$

$$V_R = 1V \quad C_j = \frac{0.6\,pF}{(1+1/0.75)^{1/3}} = \underline{\underline{0.45pF}}$$

$$V_R = 10V \quad C_j = \frac{0.6\,pF}{(1+10/0.75)^{1/3}} = \underline{\underline{0.25pF}}$$

3.119

$$V_Z = 10$$

$$I_D = \frac{1}{2}\left(\frac{0.25}{10}\right) = \underline{\underline{12.5mA}}$$

Since breakdown occurs only half the time, the average breakdown current can be twice the continuous value i.e. it can be 25 mA if the dissipation is limited to half the rated value of 50 mA is the dissipation is allowed to rise to the rated value.

3.120

$$I_p = Aqn_i^2\left(\frac{D_p}{L_p N_D}\right)\left(e^{V/V_T}-1\right)$$

$$I_n = Aqn_i^2\left(\frac{D_n}{L_n N_A}\right)\left(e^{V/V_T}-1\right)$$

$$\therefore \frac{I_p}{I_n} = \frac{D_p L_n N_A}{D_n L_p N_D} = \frac{10\times10\times10^{18}}{20\times5\times10^{16}} = \underline{\underline{100}}$$

$$I = I_p + I_n = 100 I_n + I_n = 1mA$$

$$I_n = 1/101 = \underline{\underline{9.9\,\mu A}}$$

$$I_p = 100(9.9) = \underline{\underline{990\,\mu A}}$$

3.121

$$I_p = Aqn_i^2\frac{D_p}{L_p N_D}\left(e^{V/V_T}-1\right)$$

$$I_n = Aqn_i^2\frac{D_n}{L_n N_A}\left(e^{V/V_T}-1\right)$$

For p^+-n $\quad N_A >> N_D$

$$\therefore I \cong I_p = Aqn_i^2\frac{D_p}{L_p N_D}\left(e^{V/V_T}-1\right)$$

$$I_S \cong Aqn_i^2\frac{D_p}{L_p N_D} = Aqn_i^2\left(\frac{D_p}{\sqrt{D_p\tau_p}\,N_D}\right)$$

$$= 10^4(10^{-8})\,1.6\times10^{-19}\frac{(1.5\times10^{10})^2\quad 10}{\sqrt{10\times0.1\times10^{-6}}\times5\times10^{16}}$$

$$= \underline{\underline{0.72\times10^{-15}\,A}}$$

CONT.

$$I = I_S\left(e^{V/V_T} - 1\right) = 0.2\times10^{-3}$$

$$0.72\times10^{-15}\left(e^{V/0.025} - 1\right) = 0.2\times10^{-3}$$

$$V = \underline{0.684\,V}$$

Excess minority charge

$$= Q_p + Q_n$$
$$= \tau_p I_p + \tau_n I_n \cong \tau_p I_p$$
$$= (0.1\times10^{-6})(0.2\times10^{-3}) = \underline{\underline{2\times10^{-11}\,C}}$$

$$C_d = \frac{\tau_T I}{V_T} \simeq \frac{\tau_p I}{V_T}$$

$$= 0.1\times10^{-6}\times\frac{0.2\times10^{-3}}{0.025} = \underline{\underline{800\,pF}}$$

3.122

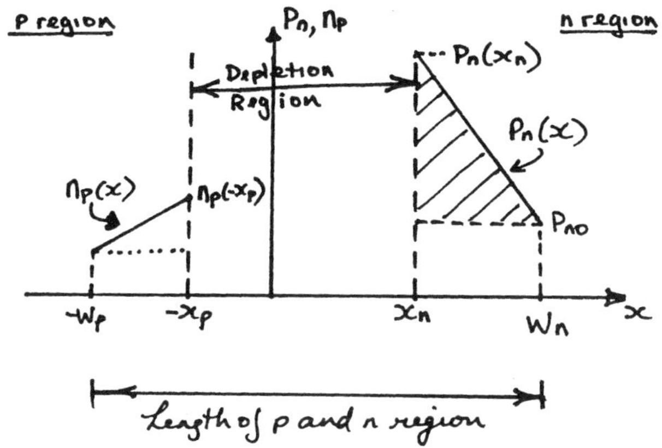

(a) First consider I_p:

$$p_n(x_n) = p_{n0}\,e^{V/V_T},$$
$$I_p = A J_p, \quad \text{and}$$
$$p_{n0} = \frac{n_i^2}{N_D}$$

Thus, $I_p = Aqn_i^2 \dfrac{D_p}{(W_n - x_n)N_D}\left(e^{V/V_T} - 1\right)$

& $I_n = Aqn_i^2 \dfrac{D_n}{(W_p - x_p)N_A}\left(e^{V/V_T} - 1\right)$

Thus: $I = I_p + I_n$

$$= Aqn_i^2\left[\frac{D_p}{(W_n - x_n)N_D} + \frac{D_N}{(W_p - x_p)N_A}\right]\times\left(e^{V/V_T} - 1\right)$$

Q.E.D

The excess charge, Q_p, can be found by multiplying the area of the shaded triangle of the $p_n(x)$ distribution graph by Aq.

$$Q_p = Aq\times\frac{1}{2}\left[p_n(x_n) - p_{n0}\right]\left[W_n - x_n\right]$$

$$= \frac{1}{2}Aq\,p_{n0}\left(e^{V/V_T} - 1\right)(W_n - x_n)$$

$$= \frac{1}{2}Aq\frac{n_i^2}{N_D}(W_n - x_n)\left(e^{V/V_T} - 1\right)$$

$$= \frac{1}{2}\frac{(W_n - x_n)^2}{D_p}I_p$$

$$\simeq \frac{1}{2}\frac{W_n^2}{D_p}I_p \quad \text{for } W_n \gg x_n$$

Q.E.D.//

(c) $C_d = \dfrac{dQ}{dV} = \tau_T\dfrac{dI}{dV}$ But $I = I_S\left(e^{V/V_T} - 1\right)$

$$\frac{dI}{dV} = \frac{I_S e^{V/V_T}}{V_T} \simeq I/V_T$$

$$\therefore C_d \simeq \frac{\tau_T I}{V_T}$$

(d) $C_d = \dfrac{1}{2}\dfrac{W_n^2}{10}\times\dfrac{10^{-3}}{0.025} = 8\times10^{-12}$

$$W_n = \underline{\underline{63.2\,\mu m}}$$

Chapter 4 - Problems

4.1

The capacitance per unit area is: $C_{ox} = \frac{\varepsilon_{ox}}{t_{ox}}$

$\varepsilon_{ox} = 3.45 \times 10^{-11}$ F/m

$t_{ox} = 5nm \Rightarrow C_{ox} = \frac{3.45 \times 10^{-11}}{5 \times 10^{-9}} = 6.9\ fF/\mu m^2$

$t_{ox} = 40nm \Rightarrow C_{ox} = 0.86\ fF/\mu m^2$

For 1PF capacitance, we require an area A:

$A = \frac{10^{-12}}{6.9 \times 10^{-15}} = 145\ \mu m^2$ for $t_{ox} = 5nm$

$A = \frac{10^{-12}}{0.86 \times 10^{-15}} = 1163\ \mu m^2$ for $t_{ox} = 20nm$

For a square plate capacitor of 10PF:

$A = 10 \times 145 = 1450\ \mu m^2$ or $38 \times 38\ \mu m^2$ square for $t_{ox} = 5^{nm}$

$A = 10 \times 1163 = 11630\ \mu m^2$ or $108 \times 108\ \mu m^2$ square for $t_{ox} = 20^{nm}$

4.2

Drain current is directly proportional to the width of the chanel. Therefore if width is 10 times greater, then i_D would be 10 times greater as well.

K = Constant of proportionality = $\frac{1}{0.5 \times 0.2} = 10\ mA/V^2$

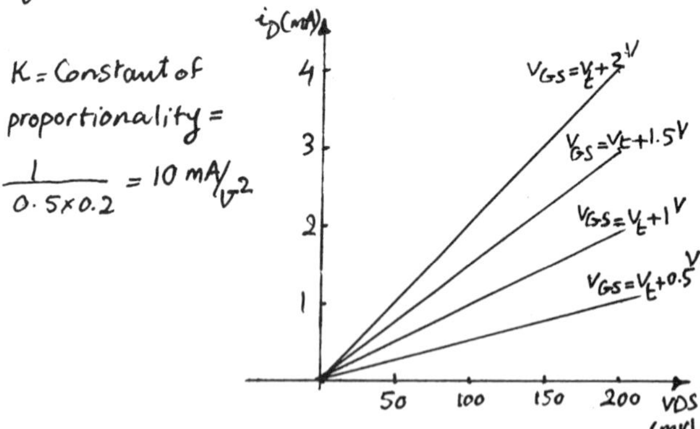

$r_{DS} = \frac{i_D}{V_{DS}} = \frac{1}{0.2} = 5^{K\Omega}$ for $V_{ov} = 0.5^V$

$r_{DS} = 10\ K\Omega$ for $V_{ov} = 1^V$

$r_{DS} = 15^{K\Omega}$ for $V_{ov} = 1.5^V$

$r_{DS} = 20K\Omega$ for $V_{ov} = 2^V$

$5^{K\Omega} \leq r_{DS} \leq 20K\Omega$ for $0.5^V \leq V_{ov} \leq 2^V$

4.3

eq. 4.6a : $i_D = \frac{1}{2} k_n' \frac{W}{L} (V_{GS} - V_t)^2 \quad k_n' = \mu_n C_{ox}$

for equal drain currents:

$\mu_n C_{ox} \frac{W_n}{L} = \mu_p C_{ox} \frac{W_p}{L} \Rightarrow \frac{W_p}{W_n} = \frac{\mu_n}{\mu_p} = \frac{1}{0.4} = 2.5$

4.4

for small V_{DS}: $i_D \simeq k_n' \frac{W}{L} (V_{GS} - V_t) V_{DS}$

$r_{DS} = \frac{V_{DS}}{i_D} = \frac{1}{k_n' \frac{W}{L}(V_{GS} - V_t)} = \frac{1}{50 \times 10^{-6} \times 20 \times (5 - 0.8)}$

$r_{DS} = 238\Omega \qquad v_{DS} = r_{DS} \times i_D = 238mV$

for the same performance of a p-channel device: $\frac{W_p}{W_n} = \frac{\mu_n}{\mu_p} = 2.5 \Rightarrow \frac{W_p}{L} = \frac{W_n}{L} \times 2.5 = 20 \times 2.5$

$\Rightarrow \frac{W_p}{L} = 50$

4.5

Eq. 4.5a : $i_D = k_n' \frac{W}{L}\left[(V_{GS} - V_t)V_{DS} - \frac{1}{2} V_{DS}^2\right]$ triode region

for small V_{DS}:

$i_D \simeq k_n' \frac{W}{L} (V_{GS} - V_t) V_{DS}$

$r_{DS} = \frac{V_{DS}}{i_D} = \frac{1}{k_n' \frac{W}{L} (V_{GS} - V_t)}$

for $r_{DS} = 1K\Omega$:

$k' = 100\ \mu A/V^2$: $\quad 1000 = \frac{1}{100 \times 10^{-6} \frac{W}{1} \times (5 - 0.8)}$

$\Rightarrow W = 2.4\mu m$

4.6

a) $C_{ox} = \frac{\varepsilon_{ox}}{t_{ox}} = \frac{3.45 \times 10^{-11}}{15 \times 10^{-9}} = 2.3\ fF/\mu m^2$

$k_n' = \mu_n C_{ox} = 550 \times 10^{-4} \times 2.3 \times 10^{-3} = 126.5\ \mu A/V^2$

b) $i_D = \frac{1}{2} k_n' \frac{W}{L} (V_{GS} - V_t)^2 \Rightarrow 100 = \frac{1}{2} \times 126.5 \times \frac{16}{0.8} (V_{GS} - 0.7)^2$

$V_{GS} - 0.7 = 0.28 \Rightarrow V_{OV} = 0.28V$

$V_{GS} = 0.98V$

$V_{DS\,min} = V_{GS} - V_t = 0.28V$

c) for small V_{DS}: (triode region) $i_D \simeq k_n' \frac{W}{L} V_{ov} \cdot V_{DS}$

Cont.

$$r_{DS} = \frac{v_{DS}}{i_D} = \frac{1}{k'_n \frac{W}{L} V_{OV}} = \frac{1}{126.5\times10^{-6}\times\frac{16}{0.8}V_{OV}} = 1000$$

$\Rightarrow V_{OV} = 0.4V$

$V_{GS} = V_{OV} + V_t = 0.4 + 0.7 = 1.1V$

4.7

$$k'_n = \mu_n C_{ox} = \mu_n \frac{\varepsilon_{ox}}{t_{ox}} = 650\times10^{-4}\times\frac{3.45\times10^{-11}}{20\times10^{-9}} = 112.1\,\mu A/V^2$$

a) triode region: $v_{DS} < V_{GS} - V_t$

$$i_D = k'_n \frac{W}{L}\left[(V_{GS}-V_t)V_{DS} - \frac{1}{2}V_{DS}^2\right]$$

$$i_D = 112.1\times10^{-6}\times10\times\left[(5-0.8)\times1 - \frac{1}{2}\times1^2\right] = 4.15\,mA$$

b) edge of saturation region: $V_{DS} = V_{GS} - V_t$

$$i_D = \frac{1}{2}k'_n \frac{W}{L}(V_{GS}-V_t)^2 = \frac{1}{2}\times112.1\times10^{-6}\times10\times(1.2)^2 = 0.8\,mA$$

c) triode region: $V_{DS} < V_{GS} - V_t$

$$i_D = 112.1\times10^{-6}\times10\left[(5-0.8)\times0.2 - \frac{1}{2}\times0.2^2\right] = 0.92\,mA$$

d) Saturation region: $v_{DS} > V_{GS} - V_t$

$$i_D = \frac{1}{2}\times112.1\times10^{-6}\times10\times(5-0.8)^2 = 9.9\,mA$$

4.8

Refer to Fig. 4.11b, $i_D \propto \frac{W}{L}$ so if W is halved, then i_D would also be halved. The vertical axis, i_D, has to be divided by 2:

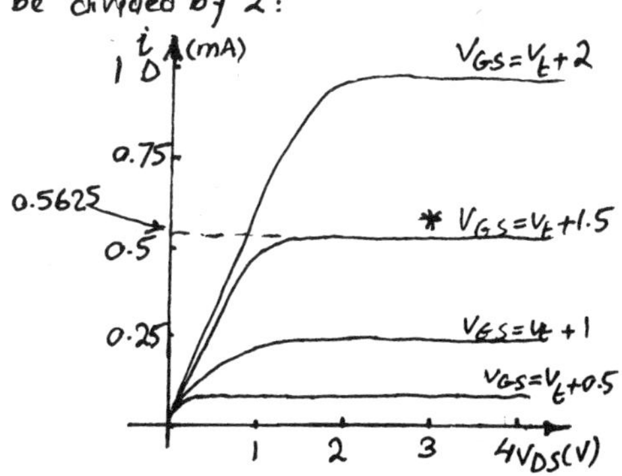

For $V_{OV} = 1.5^V$, by looking at the corresponding curve, $V_{GS} = V_t + 1.5$, we observe that:

$i_D = 0.5625$ mA

4.9

4.10

$i_D = \frac{1}{2}k'_n \frac{W}{L}V_{OV}^2$

assuming $k'_n \frac{W}{L} = 1$:

$i_D = \frac{1}{2}V_{OV}^2$

for a device with half the width, i_D is divided by 2.

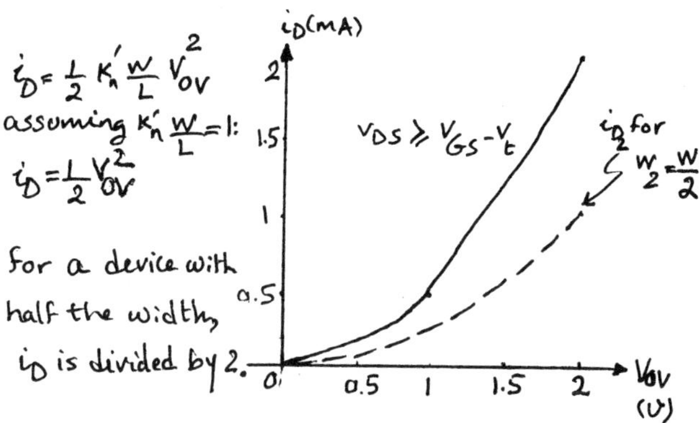

This graph is not dependent on V_t, while Fig. 4.12 is dependent on V_t.

4.11

Eq. 4.13: $r_{DS} = \left[k'_n \frac{W}{L}(V_{GS}-V_t)\right]^{-1}$ therefore:

$$\frac{r_{DS1}}{r_{DS2}} = \frac{V_{GS2}-V_t}{V_{GS1}-V_t} \Rightarrow \frac{1000}{200} = \frac{V_{GS2}-1}{1.5-1} \Rightarrow V_{GS2} = 3.5^V$$

Now for a device with twice the width:

Cont.

$$\frac{r_{DS1}}{r_{DS2}} = \frac{W_2(V_{GS2}-V_t)}{W_1(V_{GS2}-V_t)}$$

for $V_{GS}=1.5V$ $\frac{r_{DS1}}{r_{DS2}} = 2 \Rightarrow r_{DS2} = \frac{1000}{2} = 500\Omega$

for $V_{GS}=3.5V$ $r_{DS2} = \frac{200}{2} = 100\Omega$

4.12

$i_D = \frac{1}{2}k'_n\frac{W}{L}(V_{GS}-V_t)^2 \Rightarrow 0.2\times10^{-3} = \frac{1}{2}\times0.1\times10^{-3}(V_{GS}-1)^2$

$V_{GS}-1=2 \Rightarrow V_{GS}=3V$

$V_{DSmin} = V_{GS}-V_t = 3-1 = 2V$

for $i_D = 0.8mA$: $0.8 = \frac{1}{2}\times0.1(V_{GS}-1)^2$

$V_{GS}-1 = 4 \Rightarrow V_{GS} = 5V$

$V_{DSmin} = V_{GS}-V_t = 5-1 = 4V$

4.13

$V_{GS}=V_{DS}$ indicates operation in saturation mode: $i_D = \frac{1}{2}k'_n\frac{W}{L}(V_{GS}-V_t)^2$

$$\left.\begin{aligned}4 &= \frac{1}{2}k'_n\frac{W}{L}(5-V_t)^2\\ 1 &= \frac{1}{2}k'_n\frac{W}{L}(3-V_t)^2\end{aligned}\right\} \Rightarrow 4 = \frac{(5-V_t)^2}{(3-V_t)^2} \Rightarrow$$

$(5-V_t) = 2(3-V_t) \Rightarrow V_t = 1V$, $k'_n\frac{W}{L} = 0.5\,mA/V^2$

4.14

$i_D = \frac{1}{2}k'_n\frac{W}{L}(V_{GS}-V_t)^2 \Rightarrow 0.8 = \frac{1}{2}\times50\times10^{-3}\frac{W}{L}(5-1)^2$

$\frac{W}{L} = 2 \Rightarrow W = 2\times2 = 4\mu m$

4.15

In triode region: $i_D = k'_n\frac{W}{L}\left[(V_{GS}-V_t)V_{DS}-\frac{V_{DS}^2}{2}\right]$

$\frac{i_{D1}}{i_{D2}} = \frac{60\mu A}{160\mu A} = \frac{(2-V_t)0.1-0.01/2}{(4-V_t)0.1-0.01/2} \Rightarrow V_t = 0.75V$

if $k'_n = 50\mu A/V^2$: $60 = 50\frac{W}{L}\left[(2-0.75)0.1-\frac{0.01}{2}\right]$

$\frac{W}{L} = 10$

If $V_{GS}=3V$, $V_{DS}=0.15V$ then $i_D = 50\times10\times\left[2.25\times0.15-\frac{0.15^2}{2}\right]$

$i_D = 163.125\,\mu A$

IF $V_{GS}=3V$, the channel reach pinchoff at $V_{DS}=V_{GS}-V_t = 3-0.75 = 2.25V$ for which:

$i_D = \frac{1}{2}\times k'_n\frac{W}{L}(V_{GS}-V_t)^2 = \frac{1}{2}\times50\times10\times2.25^2 = 1.3mA$

$i_D = 1.3mA$

4.16

For the channel to remain continous:

$V_{DS} \leq V_{GS}-V_t \Rightarrow V_{DSmax} = 1.5-0.8 = 0.7V$

4.17

Eq. 4.15: $r_{DS} = \left[k'_n\frac{W}{L}V_{ov}\right]^{-1} = \frac{1}{50\times\frac{100}{5}(V_{GS}-1)}\,M\Omega$

$r_{DS} = \frac{1}{V_{GS}-1}\,K\Omega$

$V_{GS} = 1.1V \Rightarrow r_{DS} = 10K\Omega$

$V_{GS} = 11V \Rightarrow r_{DS} = 100\Omega$ $\Rightarrow 100\Omega \leq r_{DS} \leq 10K\Omega$

a) $r_{DS} \propto \frac{1}{W}$ so if W is halved, r_{DS} is doubled:

$200\Omega \leq r_{DS} \leq 20K\Omega$

b) $r_{DS} \propto L$ so if L is halved, r_{DS} is also halved: $50\Omega \leq r_{DS} \leq 5K\Omega$

c) $r_{DS} \propto \frac{L}{W}$ so if both W and L are halved, $\frac{W}{L}$ stays unchanged and so does r_{DS}.

$100\Omega \leq r_{DS} \leq 10K\Omega$

4.18

According to eq. 4.17: $V_{GD} \leq V_t$ for saturation region. Since V_{GD} is zero in Fig. P4.18, then the device is always in saturation region:

$i_D = i = \frac{1}{2}k'_n\frac{W}{L}(V_{GS}-V_t)^2$

if we replace V_{GS} by v and $k' = \frac{1}{2}k'_n$

$i = k'\frac{W}{L}(v-|V_t|)^2$

$r = \left(\frac{\partial i}{\partial v}\right)^{-1} = \frac{1}{2k'\frac{W}{L}(v-|V_t|)} = \frac{1}{k'_n\frac{W}{L}V_{ov}}$ where $v = |V_t| + V_{ov}$

4.19

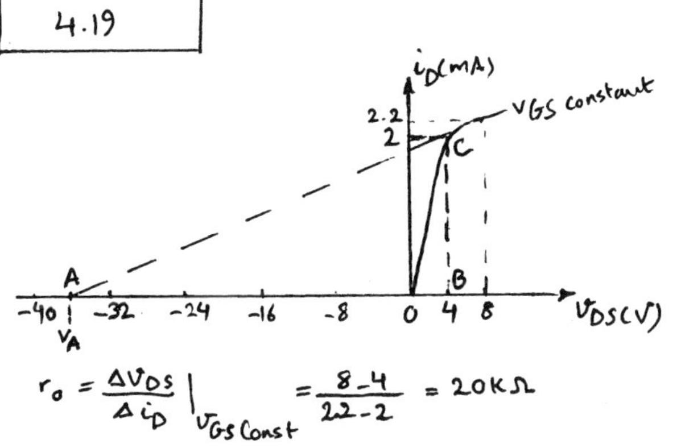

$r_o = \frac{\Delta V_{DS}}{\Delta i_D}\Big|_{V_{GS}\,Const} = \frac{8-4}{2.2-2} = 20K\Omega$

To calculate V_A, consider the ABC triangle:

$V_A + 4 = 2mA \times r_o = 2 \times 20 = 40V \Rightarrow V_A = 36V$

$\lambda = \frac{1}{V_A} = 0.028\ V^{-1}$

4.20

Eq. 4.26 ; $r_o = \frac{V_A}{i_D} = \frac{50}{i_D}$ $\quad 0.1^{mA} \leq i_D \leq 1^{mA} \Rightarrow 50^{K\Omega} \leq r_o \leq 500^{K\Omega}$

$r_o = \frac{\Delta V_{DS}}{\Delta i_D} \Rightarrow \Delta i_D = \frac{\Delta V_{DS}}{r_o} = \frac{1}{r_o}$

$i_D = 0.1\,mA$	$\Delta i_D = 2\mu A$	$\frac{\Delta i_D}{i_D} = 2\%$
$i_D = 1\,mA$	$\Delta i_D = 20\mu A$	$\frac{\Delta i_D}{i_D} = 2\%$

4.21

$V_A = V_A' L$ where V_A' is completely process dependent. Also, $r_o = \frac{V_A}{i_D}$. therefore to achieve desired r_o (which is 4 times larger) we should increase L. $L = 4 \times 2 = 8\mu m$

In order to keep I_D unchanged, $\frac{W}{L}$ ratio has to stay unchanged. Therefore:

$W = 4 \times 10 = 40\mu m$. (So $\frac{W}{L}$ is kept at 5)

$V_A = r_o i_D = 0.5M \times 100\mu = 50V$ (For standard)

$V_A = 4 \times 0.5M \times 100 = 200V$ (For new device)

4.22

$\lambda = 0.02\ V^{-1} \Rightarrow V_A = 50V$ for $L = 1\mu m$

$V_A = V_A' L \Rightarrow V_A' = 50V$

for $L = 3\mu m$: $V_A = 50 \times 3 = 150V$

$r_o = \frac{V_A}{I_D} = \frac{150}{0.08} = 1875K\Omega$

$r_o = \frac{\Delta V_{DS}}{\Delta i_D} \Rightarrow \Delta i_D = \frac{\Delta V_{DS}}{r_o} = \frac{5-1}{1875} = 2.13\mu A$

for V_{DS} raised from 1^V to 5^V, i_D increases from $80\mu A$ to $82.13\mu A$.

$\frac{\Delta i_D}{i_D} = 2.7\%$ Change in i_D.

In order to reduce $\frac{\Delta i_D}{i_D}$ by a factor of 2, Δi_D has to be halved, or equivalently r_o has to be doubled. In order to double r_o, V_A has to be doubled and this can be done by doubling the length. $L = 2 \times 3 = 6\mu m$

4.23

$V_A = V_A' L = 20 \times 1.6 = 32V$, $\lambda = \frac{1}{V_A} = 0.031\ V^{-1}$

$V_{DS} = 2^V > (V_{GS} - V_t) = 0.5V \Rightarrow$ device in saturation region

$i_D = \frac{1}{2} k_n' \frac{W}{L} V_{OV}^2 = \frac{1}{2} \times 130 \times \frac{16}{1.6}\ 0.5^2 = 162.5\mu A$

$r_o = \frac{V_A}{i_D} = \frac{32}{162.5\mu} = 197K\Omega$

$r_o = \frac{\Delta V_{DS}}{\Delta i_D} \Rightarrow 197K\Omega = \frac{1}{\Delta i_D} \Rightarrow \Delta i_D = 5.1\mu A$

4.24

MOS	1	2	3	4
$\lambda(V^{-1})$	0.02	0.01	0.1	0.005
V_A (V)	50	100	10	200
I_D (mA)	5	3.33	0.1	0.2
r_o (KΩ)	10	30	100	1000

$r_o = \frac{V_A}{I_D}$, $\lambda = \frac{1}{V_A}$

4.25

$V_A = \frac{1}{\lambda} = 100V$ $\quad V_A = V_A' L$ if L is doubled, so is V_A.

$V_A = 2 \times 100 = 200V$, $\lambda = \frac{1}{200} = 0.005\ V^{-1}$

a) if V_{GS} is fixed, then i_D is halved, as a result of doubling L. $I_D \propto \frac{1}{L} \Rightarrow I_D = \frac{1}{2} = 0.5mA$

Cont.

$r_o = \frac{V_A}{I_D} = \frac{200}{0.5} = 400\,k\Omega$

b) if I_D is fixed : $r_o = \frac{V_A}{I_D} = \frac{200}{1} = 200\,k\Omega$

4.26

+5V

$V_{GS} = -5V$

To operate in saturation:

$V_{DS} \leq V_{GS} - V_t$ or $V_{DS} \leq -5-(-1.5)$ or $V_{DS} \leq -3.5$

a) $V_D = 4^V \Rightarrow V_{DS} = -1V > -3.5^V \Rightarrow$ triode region

$i_D = k_p' \frac{W}{L}\left[(V_{GS}-V_t)V_{DS} - \frac{1}{2}V_{DS}^2\right]$ eq. 4.29

$i_D = 80\left[(-5-(-1.5))\times(-1) - \frac{1}{2}\times 1\right] = 0.24\,mA$

b) $V_D = 1.5^V \Rightarrow V_{DS} = -3.5^V = V_{GS} - V_t \Rightarrow$ edge of saturation

$i_D = \frac{1}{2}k_p' \frac{W}{L}(V_{GS}-V_t)^2(1+\lambda V_{DS})$ eq. 4.32

$i_D = \frac{80}{2}(3.5)^2(1+0.02\times 3.5) = 0.52\,mA$

c) $V_D = 0 \Rightarrow V_{DS} = -5^V < -3.5 \Rightarrow$ saturation

$i_D = \frac{1}{2}\times 80\times(3.5)^2(1+0.02\times 5) = 0.54\,mA.$

d) $V_D = -5V \Rightarrow V_{DS} = -10^V < -3.5 \Rightarrow$ saturation

$i_D = \frac{1}{2}\times 80(3.5)^2(1+0.02\times 10) = 0.59\,mA$

4.27

$V_{GS} = -3V$ $V_{SG} = 3V$ $V_t = -1V$

$V_{DS} = -4V$ $V_{SD} = 4V$ $V_A = -50^V$ $\lambda = -0.02\,V^{-1}$

$i_D = \frac{1}{2}k_p' \frac{W}{L}(V_{GS}-V_t)^2(1+\lambda V_{DS})$

$3 = \frac{1}{2}k_p' \frac{W}{L}(-3+1)^2(1+0.02\times 4) = 2.16\,k_p' \frac{W}{L}$

$k_p' \frac{W}{L} = 1.39\,mA/V^2$

4.28

Eq. 4.34 : $\gamma = \frac{\sqrt{2qN_A\varepsilon_s}}{C_{ox}} \Rightarrow N_A = \frac{\gamma^2 C_{ox}^2}{2q\varepsilon_s}$

$t_{ox} = 20nm$ $C_{ox} = \frac{\varepsilon_{ox}}{t_{ox}} = \frac{3.45\times10^{-11}}{20\times10^{-9}} = 1.73\,fF/\mu m^2$

$N_A = \frac{0.5^2\times 1.73^2\times 10^{-6}}{2\times1.6\times10^{-19}\times11.7\times8.854\times10^{-14}\times10^2} = 2.3\times10^{22}/m^3$

$N_A = 2.3\times10^{16}\,/cm^3$

if $t_{ox} = 100nm$ $\frac{\gamma_1}{\gamma_2} = \frac{C_{ox2}}{C_{ox1}} = \frac{t_{ox1}}{t_{ox2}} \Rightarrow \frac{0.5}{\gamma_2} = \frac{20}{100}$

$\Rightarrow \gamma_2 = 2.5V^{\frac{1}{2}}$

If γ is kept at $0.5V^{\frac{1}{2}}$, then when t_{ox} is changed:

$\frac{\sqrt{N_{A1}}}{C_{ox1}} = \frac{\sqrt{N_{A2}}}{C_{ox2}} \Rightarrow \frac{\sqrt{N_{A1}}}{\sqrt{N_{A2}}} = \frac{C_{ox1}}{C_{ox2}} = \frac{t_{ox2}}{t_{ox1}} = \frac{100}{20} = 5$

$\sqrt{N_{A2}} = \frac{\sqrt{N_{A1}}}{5} \Rightarrow N_{A2} = \frac{N_{A1}}{25} = 9.2\times10^{14}/cm^3$

4.29

Eq. 4.33 $V_t = V_{t_0} + \gamma\left[\sqrt{2\varphi_F + V_{SB}} - \sqrt{2\varphi_F}\right]$

$V_{SB} = 0^V \Rightarrow V_t = V_{t_0} = 1V$

$V_{SB} = 4^V \Rightarrow V_t = 1 + 0.5\left[\sqrt{0.6+4} - \sqrt{0.6}\right] = 1.69^V$

$1^V \leq V_t \leq 1.69^V$

If t_{ox} is increased by a factor of 4:

from Eq. 4.34 $\gamma \propto \frac{1}{C_{ox}}$ or $\gamma \propto t_{ox}$ or $\frac{\gamma_1}{\gamma_2} = \frac{t_{ox1}}{t_{ox2}}$

$\Rightarrow \gamma_2 = 4\times0.5 = 2V^{1/2}$

$V_t = 1 + 2(\sqrt{0.6+4} - \sqrt{0.6}) = 3.74V$ for $V_{SB} = 2^V$

$V_t = 1^V$ For $V_{SB} = 0$

Therefore for $\gamma = 2$: $1 \leq V_t \leq 3.74^V$

4.30

$|V_{SB}| = 3^V$ Eq. 4.33 $V_t = V_{t_0} + \gamma\left[\sqrt{2\varphi_F + V_{SB}} - \sqrt{2\varphi_F}\right]$

$V_t = -1 - 0.5\left[\sqrt{0.6+3} - \sqrt{0.6}\right] = -1.56V$

4.31

a) $i_D = \frac{1}{2}k_n' \frac{W}{L}(V_{GS}-V_t)^2$

$\frac{\partial i_D}{\partial T} = \frac{1}{2}\frac{W}{L}\frac{\partial k_n'}{\partial T}(V_{GS}-V_t)^2 - k_n' \frac{W}{L}(V_{GS}-V_t)\frac{\partial V_t}{\partial T}$

$\frac{\partial i_D/i_D}{\partial T} = \frac{\partial k_n'/k_n'}{\partial T} - 2\frac{\partial V_t}{\partial T}\frac{1}{V_{GS}-V_t}$

Cont.

b) $\frac{\partial V_t}{\partial T} = -2\,mV/^\circ C$ For $\frac{\partial i_D/i_D}{\partial T} = -0.2\% = -0.002/^\circ C$

$$-0.002 = \frac{\partial k_n'/k_n'}{\partial T} - 2\times(-2^{mV})\frac{1}{5-1} \Rightarrow$$

$$\frac{\partial k_n'/k_n'}{\partial T} = -0.003\,/^\circ C = -0.3\%/^\circ C$$

4.32

Case	Transistor	V_S	V_G	V_D	I_D	type	mode	$\mu C_{ox}\frac{W}{L}$	V_t
a	1	0	2	5	100	N	Sat.	200	1
	1	0	3	5	400	N	Sat.	200	1
b	2	5	3	-45	50	P	Sat.	400	-1.5
	2	5	2	-05	450	P	Sat.	400	-1.5
c	3	5	3	4	200	P	Sat.	400	-1
	3	5	2	0	800	P	Sat.	400	-1
d	4	-2	0	0	72	N	Sat.	100	0.8
	4	-4	0	-3	270	N	Triode	100	0.8

Case a) transistor 1: $V_{GS} = 2^V$ $V_{DS} = 5^V$ $I_D = 100\mu A$

This must be an NMOS operating in Saturation.

$$I_D = 100 = \frac{1}{2}\mu_n C_{ox}\frac{W}{L}(2-V_t)^2$$

when $V_{GS}=3V$: $400 = \frac{1}{2}\mu_n C_{ox}\frac{W}{L}(3-V_t)^2 \Rightarrow 4 = \frac{(3-V_t)^2}{(2-V_t)^2}$

$$\Rightarrow 2(2-V_t) = 3-V_t \Rightarrow V_t = 1^V,\ \mu_n C_{ox}\frac{W}{L} = 200\frac{\mu A}{V^2}$$

Case b) transistor 2: $V_{GS} = 3-5 = -2^V$ $V_{DS} = -9.5V$

Therefore $V_{DS} < V_{GS} - V_t$ regardless of value of V_t, and the device operates in saturation. (PMOS)

$$I_D = 50 = \frac{1}{2}\mu_p C_{ox}\frac{W}{L}(-2-V_t)^2$$
$$450 = \frac{1}{2}\mu_p C_{ox}\frac{W}{L}(-3-V_t)^2 \Rightarrow 9 = \frac{(3+V_t)^2}{(2+V_t)^2}$$

$$\Rightarrow 3(2+V_t) = 3+V_t \Rightarrow V_t = -1.5V,\ \mu_p C_{ox}\frac{W}{L} = 400\frac{\mu A}{V^2}$$

case c) transistor 3: $V_{GS} = -2^V$ $V_{DS} = -1^V \Rightarrow$ PMOS

This device can be either in saturation or triode region. First, we assume saturation region:

$$I_D = 200 = \frac{1}{2}\mu_p C_{ox}\frac{W}{L}(-2-V_t)^2$$
$$800 = \frac{1}{2}\mu_p C_{ox}\frac{W}{L}(-3-V_t)^2 \Rightarrow 4 = \frac{(3+V_t)^2}{(2+V_t)^2}$$

$$\Rightarrow 2(2+V_t) = 3+V_t \Rightarrow V_t = -1V,\ \mu_p C_{ox}\frac{W}{L} = 400\frac{\mu A}{V^2}$$

So our assumption was right:

$$(V_{DS} = -1^V) \leq (V_{GS} - V_t) = -2+1 = -1^V$$

edge of saturation

Case d) transistor 4: $V_{GS} = 2^V$ $V_{DS} = 2^V \Rightarrow$ NMOS

So $V_{DS} > V_{GS} - V_t \Rightarrow$ Saturation region

① $I_D = 72 = \frac{1}{2}\mu_n C_{ox}\frac{W}{L}(2-V_t)^2$

But for $V_{GS} = 4^V, V_{DS} = 1^V$ and considering that $V_t < 2V$, then the device is in triode region:

$$270 = \mu_n C_{ox}\frac{W}{L}\left[(4-V_t)\times 1 - \frac{1}{2}\times 1\right]$$

② $270 = \mu_n C_{ox}\frac{W}{L}(3.5-V_t)$

①,② $\Rightarrow V_t = 0.8V$ $\mu_n C_{ox}\frac{W}{L} = 100\,\mu A/V^2$

4.33

a) $I_D = \frac{1}{2}k_n'\frac{W}{L}(V_{GS}-V_t)^2 \Rightarrow 2 = \frac{1}{2}k_n'\frac{W}{L}(3-1)^2$

$\Rightarrow k_n'\frac{W}{L} = 1\,mA/V^2$

$V_1 = V_{DS} = 3V$

b) $V_2 = V_S = V_D - V_{DS} = 1-3 = -2^V$

c) $V_3 = V_S = V_D - V_{DS} = 0-(-3) = 3^V$

d) $V_4 = V_D = V_S + V_{DS} = 5+(-3) = 2^V$

In order to calculate R_{max} that can be inserted in series with the drain, V_{DS} has to be equal to $V_{GS} - V_t$, so that the device is operating on the edge of saturation: $|V_{DS}| = 3-1 = 2^V$. Note that since i_D is the same, V_{GS} stays the same.

a) $R_{D\,max} = \frac{3-2}{2^{mA}} = 0.5K\Omega$

b) $V_2 = -2^V \Rightarrow V_D = -2+2 = 0 \Rightarrow R_{D\,max} = \frac{1}{2} = 0.5^{K\Omega}$

Note that V_2 is Fixed through $V_{GS} = 3V$.

c) $V_{GS} = -3V \Rightarrow V_S = V_3 = 3^V$. Now for V_{DS} to be $-2V$, V_D has to be 1^V.

$$R_{D\,Max} = \frac{1^V}{2^{mA}} = 0.5k\Omega$$

d) $V_{GS} = -3V \Rightarrow V_G = V_4 = 2^V$. Adding the resistor between V_4 and drain means that V_D has to be $5-2 = 3^V$ and this leaves 1^V voltage drop on the resistor: $R_{D\,max} = \frac{1}{2} = 0.5k\Omega$

In order to calculate the Largest Resistor added to the gates, note that since the gate doesn't draw any current, the value of the resistor is immaterial.

Cont.

Now we calculate R_{Smax}, assuming that the voltage drop across the current source is at least 2^V:

a) $V_1 = 8^V$ then $V_{GS} = 3^V \Rightarrow V_S = 8 - 3 = 5^V$

$R_{SMax} = \frac{5}{2} = 2.5 K\Omega$

b) $V_2 = -9 + 2 = -7^V$, $V_S = 1 - |V_{GS}| = -2^V$

$R_{Smax} = \frac{-2-(-7)}{2} = 2.5 K\Omega$

c) $V_3 = 10 - 2 = 8^V$, $V_S = 0 + |V_{GS}| = 3^V$

$R_{Smax} = \frac{8-3}{2} = 2.5 K\Omega$

d) $V_4 = -5 + 2 = -3^V$; $V_S = -3 + |V_{GS}| = 0^V$

$R_{Smax} = \frac{5}{2} = 2.5 K\Omega$

4.34

$I_D = \frac{V_{DD} - V_D}{R_D} \Rightarrow \frac{5-0}{R_D} = 1^{mA} \Rightarrow R_D = 5 K\Omega$

$V_D = V_G \Rightarrow$ saturation

therefore: $i_D = \frac{1}{2} k_n' \frac{W}{L} (V_{GS} - V_t)^2$

$1 = \frac{1}{2} \times 60 \times 10^{-3} \times \frac{100}{3} (V_{GS} - 1)^2$

$\Rightarrow V_{GS} = 2V \Rightarrow V_S = -2^V$

$R_S = \frac{-2-(-5)}{1} = 3 K\Omega$

4.35

a) $i_D = \frac{1}{2} \mu_n C_{ox} \frac{W}{L} (V_{GS} - V_t)^2$

$i_{D_1} = 0.2^m = \frac{1}{2} \times 200 \times 10^{-6} \times \frac{8}{0.8} (V_{GS_1} - 0.6)^2$

$V_{GS_1} - 0.6 = \sqrt{0.2}$

$V_{GS_1} = 1.05 V$

$R = \frac{3 - 1.05}{0.2} = 9.75 K\Omega$

$V_{DD} = 3V$, 20^K R_2, R, Q_2, Q_1

b) For Q_2 to conduct 0.5mA:

$R_2 = \frac{3-1}{0.5} = 4 K\Omega$

Since the gates are connected together, both trasistors have the same V_{GS} and hence same i_D. In order to conduct 0.5mA or multiply i_D by 5: $\frac{W_2}{W_1} = 5 \Rightarrow W_2 = 5 \times 8$

$W_2 = 40 \mu m$

4.36

Refer to Fig. P4.36.

$R = \frac{3.5}{0.115} = 3.04 K\Omega$

$0.115_{mA} = \frac{1}{2} \times 60 \times 10^{-3} \times \frac{W}{0.8} (-1.5 - (-0.7))^2 \Rightarrow W = 4.8 \mu m$

4.37

Refer to Fig. P4.37

$V_{GS1} = 1.5V$, $120 \mu A = \frac{1}{2} \times 120 \times \frac{W_1}{1} (1.5 - 1)^2$

$\Rightarrow W_1 = 8 \mu m$

$V_{GS2} = 3.5 - 1.5 = 2^V$, $120 = \frac{1}{2} \times 120 \frac{W_2}{1} (2-1)^2$

$\Rightarrow W_2 = 2 \mu m$

$R = \frac{5 - 3.5}{0.120} = 12.5 K\Omega$

4.38

Refer to Fig. P4.38.

$V_{GS1} = 1.5^V$ $\quad 120 \mu A = \frac{1}{2} \times 120 \times \frac{W_1}{1} (1.5 - 1)^2 \Rightarrow W_1 = 8 \mu m$

$V_{GS2} = 2^V$ $\quad 120 \mu A = \frac{1}{2} \times 120 \frac{W_2}{1} (2-1)^2 \Rightarrow W_2 = 2 \mu m$

$V_{GS3} = 1.5^V$ $\quad W_3 = 8 \mu m$

4.39

Refer to Fig. 4.23a. $V_{GS} = 5 - 6I_D \Rightarrow$

$I_D = \frac{1}{2} \times 2 \times (5 - 6I_D - 2)^2 \Rightarrow 36I_D^2 - 37I_D + 9 = 0$

$\Rightarrow I_D = 0.395^{mA}$ or 0.633 mA

$I_D = 0.633$mA is not acceptable, because it results in $V_{GS} = 1.2^V$ which is lower than $V_t = 2^V$.

Therefore $I_D = 0.395$ mA $\cong 0.4$mA

$V_D = 10 - 6I_D = 7.6^V$

These results show that although V_t and $k_n' \frac{W}{L}$ are doubled, I_D is only decreased by 20% and V_D is only increased by 8.5%. Therefore the conclusion is

Cont.

that the circuit is tolerant to changes in device parameters.

4.40

$V_G = 6^V \Rightarrow R_{G_1} = 4^{M\Omega},\ R_{G_2} = 6^{M\Omega}$

+10V, R_{G_1}, R_S, $+6^V$, $+5^V$, R_{G_2}, R_D

$I_D = 0.5mA \Rightarrow R_D = \frac{5}{0.5} = 10K\Omega$

$I_D = 0.5 = \frac{1}{2} \times 1 \times (V_{GS} + 1.5)^2$

$V_{GS} = -2.5V$ or $-0.5V$

($V_{GS} = -0.5V$ is rejected, because it is less than V_t)

$V_{GS} = -2.5V \Rightarrow V_S = 8.5^V \Rightarrow R_S = \frac{10-8.5}{0.5} = 3K\Omega$

4.41

Refer to Fig. P4.41.

$V_I = V_{GS} = 5V$, $V_O = V_{DS} = 0.05V$

$r_{DS} = 50\Omega = \frac{V_{DS}}{I_D} \Rightarrow I_D = \frac{0.05}{50} = 0.001^A = 1mA$

$R = \frac{V_{DD} - V_O}{I_D} = \frac{5 - 0.05}{1} = 4.95K\Omega$

$V_{DS} < V_{GS} - V_t \Rightarrow$ triode region

$I_D = K_n' \frac{W}{L}\left[(V_{GS} - V_t)V_{DS} - \frac{V_{DS}^2}{2}\right]$

$1 = 100 \times 10^{-3} \frac{W}{L}\left[(5-1) \times 0.05 - \frac{0.05^2}{2}\right] \Rightarrow \frac{W}{L} = 50$

4.42

Refer to Fig. P4.42.

In circuit a: $V_2 = 10 - 4 \times 2 = 2^V$

assume saturation: $I_D = 2 = \frac{1}{2} \times 1 \times (V_{GS} - 2)^2 \Rightarrow V_{GS} = 4^V$

$\Rightarrow V_1 = -4^V$, $V_{DS} = 6V > V_{GS} - V_t$ So our assumption was correct.

In circuit b: $I_D = 1 = \frac{1}{2} \times 1 \times (V_{GS} - 2)^2 \Rightarrow V_{GS} = 3.41^V$

$V_3 = 3.41^V$

In Circuit c: $I_D = 2mA \Rightarrow V_{GS} = -4^V \Rightarrow V_S = 4^V = V_4$

$V_5 = -10 \times 2.5 \times 2 = -5^V$

In Circuit d: $I_D = 2mA \Rightarrow V_{GS} = -4^V \Rightarrow V_6 = 6^V$

$\Rightarrow V_7 = V_6 - 4 = 2^V$

If we replace the current source with a resistor in each of those circuits:

in circuit a:

$R = \frac{-4-(-10)}{2} \simeq 3.01^{K\Omega}$ (by looking at the table for 1% resistors)

Now recalculate I_D: $I_D = \frac{1}{2} \times 1 \times (V_{GS} - V_t)^2$

$V_{GS} - V_t = 0 - (-10 + 3.01 I_D) - 2 = 8 - 3.01 I_D$ $\Big\} \Rightarrow$

$2I_D = (8 - 3.01 I_D)^2 \Rightarrow I_D = 1.99mA \Rightarrow V_2 = 2.04^V$

$V_1 = -4.01^V$

in circuit b:

$R = \frac{10 - 3.41}{1} = 6.59^K \simeq 6.65^{K\Omega}$

Then $V_{GS} = 10 - 6.65I$

$I = \frac{1}{2} \times 1(10 - 6.65I - 2)^2 \Rightarrow I = 0.99mA$

$V_3 = 10 - 6.65 \times 0.99 = 3.41^V$

in circuit c:

$R = \frac{10-4}{2} \simeq 3.01K\Omega$, $V_{GS} = -(10 - 3.01I)$

$I = \frac{1}{2} \times 1 \times (-10 + 3.01I + 2)^2 \Rightarrow I_D = 1.99mA$

$V_4 = 10 - 3.01 \times 1.99 = 4.01V$

$V_5 = -10 + 2.5k \times 1.99 = -5.03V$

in circuit d:

$R = \frac{2}{2} = 1K$ so V_7 is still 2^V.

4.43

a) $V_{GS} = -V_1$. $10\mu A = \frac{1}{2} \times 0.4 \times 10^3 (V_{GS} - 1)^2 \Rightarrow$

$V_{GS} = 1.22^V \Rightarrow V_1 = -1.22V$

b) $100\mu A = \frac{1}{2} \times 0.4 \times 10^3 \times (V_{GS} - 1)^2 \Rightarrow V_{GS} = 1.71^V,\ V_2 = -1.71^V$

c) $1 = \frac{1}{2} \times 0.4 \times (V_{GS} - 1)^2 \Rightarrow V_{GS} = 3.23^V \Rightarrow V_3 = -3.23V$

d) $10 = \frac{1}{2} \times 0.4 \times 10^3 (V_{GS} - 1)^2 \Rightarrow V_{GS} = 1.22^V \Rightarrow V_4 = 1.22V$

e) $1 = \frac{1}{2} \times 0.4 (V_{GS} - V_t)^2 \Rightarrow V_{GS} = 3.24^V \Rightarrow V_5 = 3.24V$

f) $I = \frac{1}{2} \times 0.4 \times (5 - 100I - 1)^2 \Rightarrow I = 0.045mA, 0.036mA$

$V_6 = 5 - 100 \times 0.036 = 1.4^V$

g) $I = \frac{1}{2} \times 0.4 \times (5 - 1 \times I - 1)^2 \Rightarrow I = 1.38\,mA$

$V_7 = 5 - 1.38 \times 1 = 3.62^V$

h) $I = \frac{1}{2} \times 0.4 \times (5 - 100 - I)^2 \Rightarrow I = ~~0.045^{mA}~~, 0.036^{mA}$

$V_8 = -5 + 100 \times 0.036 = -1.4^V$

Note that $I = 0.045^{mA}$ in circuits h and f is not acceptable, because it results in $V_{GS} < V_t$ that is not physically possible.

4.44

Refer to Fig. P4.44.

a) $V_{GS2} = -V_2$, $I = \frac{V_2 - (-5)}{1K} = \frac{1}{2} \times 2 \times (-V_2 - 1)^2$

$\Rightarrow V_2 + 5 = V_2^2 + 2V_2 + 1 \Rightarrow V_2^2 + V_2 - 4 = 0 \Rightarrow$ ~~$V_2 = 1.55^V$~~, $V_2 = -2.56^V$

$V_2 = 1.55^V$ is not acceptable because it results in $V_{GS} < 0$ that is not possible for an NMOS.

Therefore $V_2 = -2.56V$

$i_{D1} = i_{D2} \Rightarrow \frac{V_2 - (-5)}{1K} = \frac{1}{2} \times 2 (5 - V_1 - 1)^2 \Rightarrow$

$2.44 = (4 - V_1)^2 \Rightarrow 4 - V_1 = \pm 1.56^V \Rightarrow V_1 = 2.44^V$

$V_1 = 5.56^V$ X

The second answer results in $V_{GS} = 5 - 5.56 < 0$ which is not acceptable. Therefore $V_1 = 2.44^V$

b) $\frac{10 - V_3}{1K} = \frac{V_5}{1K} = i_D \Rightarrow 10 - V_3 = V_5$ ①

$i_{D1} = \frac{V_5}{1K} = \frac{1}{2} \times 2 \times (V_3 - V_4 - 1)^2 \Rightarrow V_5 = (V_3 - V_4 - 1)^2$ ②

$i_{D2} = V_5/1K = \frac{1}{2} \times 2 \times (V_4 - V_5 - 1)^2 \Rightarrow V_5 = (V_4 - V_5 - 1)^2$ ③

②,③ $\Rightarrow V_3 - V_4 - 1 = V_4 - V_5 - 1 \Rightarrow V_5 = 2V_4 - V_3$ ④

①,④ $\Rightarrow 2V_4 - V_3 = 10 - V_3 \Rightarrow V_4 = 5^V$

③ $\Rightarrow V_5 = (4 - V_5)^2 \Rightarrow V_5^2 - 9V_5 + 16 = 0 \Rightarrow V_5 = 6.55^V$ X, $V_5 = 2.45^V$

$V_5 = 6.55$ results in $i_D = 6.55mA$, $V_3 = 4.45^V$ and this is not physicly possible. So $V_5 = 2.45^V$

$V_3 = 10 - 2.45 = 7.55V$

4.45

The PMOS transistor operates in saturation region if $V_{SD} \geq V_{SG} - |V_t|$ or $V_{SD} \geq V_{SG} - 1$

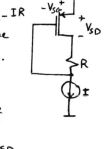

Also, $V_{SD} + IR = V_{SG} \Rightarrow V_{SD} = V_{SG} - IR$

$\Rightarrow IR \leq |V_t|$ for PMOS to be in saturation.

a) $R = 0 \Rightarrow IR = 0 < |V_t|$

saturation:

$I = 100 = \frac{1}{2} \times 8 \times 25 \times (V_{SG} - |V_t|)^2$

$V_{SG} - 1 = \pm 1 \Rightarrow V_{SG} = 2^V = V_{SD}$

b) $R = 10K\Omega \Rightarrow IR = 10 \times 0.1 = 1^V \Rightarrow$ saturation

$V_{SG} = 2^V \Rightarrow V_{SD} = 2 - 1 = 1^V$

c) $R = 30K\Omega \Rightarrow IR = 30 \times 0.1 = 3^V \Rightarrow$ triode region

$100 = 8 \times 25 \left[(V_{SG} - |V_t|) V_{SD} - \frac{1}{2} V_{SD}^2 \right]$

$0.5 = \left[(V_{SG} - 1)(V_{SG} - 3) - \frac{1}{2} (V_{SG} - 3)^2 \right]$

$0.5 = 0.5 V_{SG}^2 - V_{SG} - 1.5 \Rightarrow V_{SG}^2 - 2V_{SG} - 4 = 0$

$V_{SG} = 3.24V$, -1.2^V X

$V_{SD} = 3.24 - 3 = 0.24V$

d) $R = 100K\Omega \Rightarrow IR = 100 \times 0.1 = 10^V \Rightarrow$ triode region

$100 = 8 \times 25 \left[(V_{SG} - 1)(V_{SG} - 10) - \frac{1}{2} (V_{SG} - 10)^2 \right]$

$0.5 = 0.5 V_{SG}^2 - V_{SG} - 40 \Rightarrow V_{SG}^2 - 2V_{SG} - 81 = 0$

$V_{SG} = 10.1^V \Rightarrow V_{SD} = 0.1^V$

e) $V_{SD} = V_{SG}$ when $R = 0$

f) $V_{SD} = V_{SG}/2$ For the device in saturation, $V_{SG} = 2^V$ (for $I = 100\mu A$), therefore $V_{SD} = \frac{V_{SG}}{2} = 1^V$ implies : $IR = 1^V \Rightarrow R = 10K\Omega$

g) $V_{SD} = \frac{V_{SG}}{10}$ In saturation: $V_{SG} = 2^V$, $V_{SD} = 0.2^V$

$\Rightarrow V_{SD} < V_{SG} - |V_t| \Rightarrow$ therefore the device can't be in saturation and it is in triode region : $100 = \frac{1}{2} \times 8 \times 25 \left[(V_{SG} - 1)\left(\frac{V_{SG}}{10}\right) - \left(\frac{V_{SG}}{10}\right)^2 \frac{1}{2} \right]$

$0.5 = \frac{V_{SG}^2}{10} - \frac{V_{SG}^2}{200} - 0.1 V_{SG} \Rightarrow 19 V_{SG}^2 - 20 V_{GS} - 100 = 0$

$\Rightarrow V_{SG} = 2.88V$, -1.82^V X

$V_{SD} = \frac{2.88}{10} = 0.29^V \Rightarrow IR = 2.88 - 0.29 = 2.59^V$

$\Rightarrow R = 25.9K\Omega$

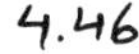

4.46

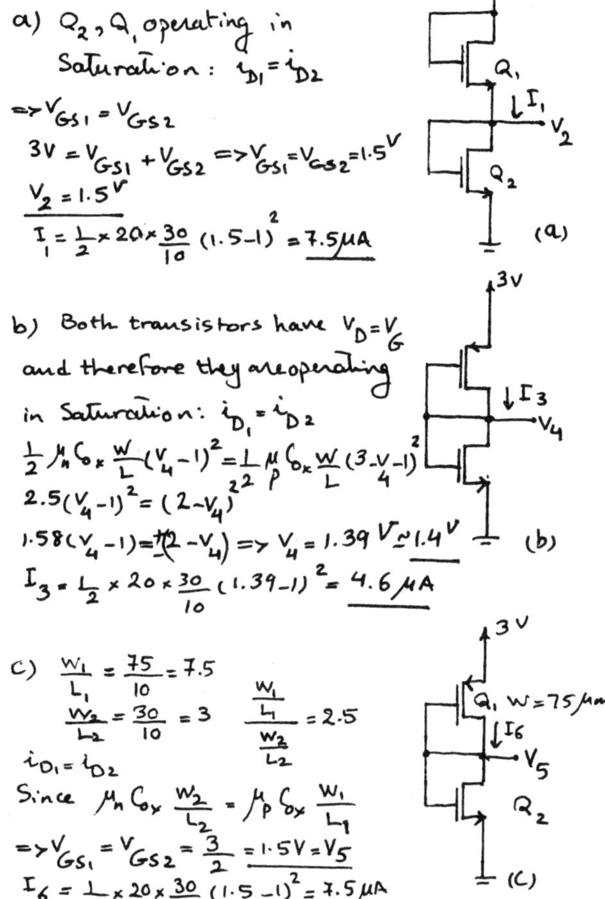

a) Q_2, Q_1 operating in Saturation: $i_{D1} = i_{D2}$

$\Rightarrow V_{GS1} = V_{GS2}$

$3V = V_{GS1} + V_{GS2} \Rightarrow V_{GS1} = V_{GS2} = 1.5^V$

$V_2 = 1.5^V$

$I_1 = \frac{1}{2} \times 20 \times \frac{30}{10}(1.5-1)^2 = \underline{7.5 \mu A}$

b) Both transistors have $V_D = V_G$ and therefore they are operating in Saturation: $i_{D_1} = i_{D2}$

$\frac{1}{2}\mu_n C_{ox}\frac{W}{L}(V_4-1)^2 = \frac{1}{2}\mu_p C_{ox}\frac{W}{L}(3-V_4-1)^2$

$2.5(V_4-1)^2 = (2-V_4)^2$

$1.58(V_4-1) = \pm(2-V_4) \Rightarrow V_4 = 1.39\ V \simeq 1.4^V$

$I_3 = \frac{1}{2} \times 20 \times \frac{30}{10}(1.39-1)^2 = \underline{4.6 \mu A}$

c) $\frac{W_1}{L_1} = \frac{75}{10} = 7.5$ $\quad \frac{W_2}{L_2} = \frac{30}{10} = 3$ $\quad \frac{W_1/L_1}{W_2/L_2} = 2.5$

$i_{D1} = i_{D2}$

Since $\mu_n C_{ox}\frac{W_2}{L_2} = \mu_p C_{ox}\frac{W_1}{L_1}$

$\Rightarrow V_{GS_1} = V_{GS2} = \frac{3}{2} = 1.5V = V_5$

$I_6 = \frac{1}{2} \times 20 \times \frac{30}{10}(1.5-1)^2 = \underline{7.5 \mu A}$

4.47

Refer to Fig. P4.47.

Since $V_{G_1} = V_{D1}$ then Q_1 is in saturation. We assume that Q_2 is also in saturation, then because $i_{D_1} = i_{D2}$, V_{GS1} would be equal to V_{GS2}.

$V_{GS_1} = V_{GS2} = \frac{5}{2} = 2.5V$

$I_1 = \frac{1}{2} \times 50 \times \frac{10}{1}(2.5-1)^2 = 562.5 \mu A$

$V_{GS3} = V_{GS1} = 2.5V$. Since Q_3 and Q_4 have the same drain current, then $V_{GS3} = V_{GS4} = 2.5^V$.

This is based on the assumption that Q_3 & Q_4 are saturated:

$V_{GS3} = V_{GS1} \Rightarrow I_2 = I_{GS3} = I_{GS1} = \underline{562.5 \mu A}$

$V_2 = 5 - 2.5 = \underline{2.5V}$

Now if Q_3 and Q_4 have $W = 100 \mu m$ then:

$I_2 = \frac{1}{2} \times 50 \times \frac{100}{1}(2.5-1)^2 = 5.625\ mA$ or

$\frac{I_{Q3}}{I_{Q_1}} = \frac{W_3}{W_1} = \frac{100}{10} \Rightarrow I_{Q_3} = 10 \times 562.5 \mu A = 5.625 mA$

4.48

a) $\left(\frac{W}{L}\right)_1 = \left(\frac{W}{L}\right)_2 = 20$

$V_{GS1} = V_{GS2} = -V_3$

Assuming that the transistors are in saturation, since K_n' and $\frac{W}{L}$ are the same for both Q_2 and Q_1 we can write: $i_{D1} = i_{D2} = \frac{200}{2} = 100 \mu A$

$100 \mu A = \frac{1}{2} \times 100 \times 20 \times (V_{GS}-1)^2 \Rightarrow 0.1 = (V_{GS}-1)^2 \Rightarrow$

$V_{GS} = 1.32^V = -V_3 \Rightarrow V_3 = -1.32V$

$V_1 = 5 - 40 \times 100^{\mu} = \underline{1\ V}$ $\qquad \underline{V_2 = 1V}$

b) $\left(\frac{W}{L}\right)_1 = \left(\frac{W}{L}\right)_2 \times 1.5 = 20$ $\qquad$ assume Q_1, Q_2 are saturated:

$V_{GS_1} = V_{GS2} = -V_3 \Rightarrow \frac{i_{D1}}{i_{D2}} = \frac{(W/L)_1}{(W/L)_2} = 1.5$

$\left.\begin{array}{l} i_{D1} = 1.5 i_{D2} \\ i_{D1} + i_{D2} = 0.2mA \end{array}\right\} \Rightarrow i_{D1} = 120 \mu A,\ i_{D2} = 80 \mu A$

$i_{D1} = 120 = \frac{1}{2} \times 100 \times 20 \times (V_{GS_1}-1)^2 \Rightarrow V_{GS_1} = 1.35^V$

$\Rightarrow V_3 = -1.35^V$

$V_1 = 5 - 40 \times 0.12 = \underline{0.2^V}$ $\qquad$, $V_2 = 5 - 40 \times 0.8 = \underline{1.8^V}$

$V_{DS1} = 0.2 + 1.35 = 1.55^V \quad > V_{GS} - 1$

$V_{DS2} = 1.8 + 1.35 = 3.15V \quad > V_{GS} - 1$

This Confirms that both transistors are indeed saturated and our assumption was correct.

4.49

a)

Point A : $V_{IA} = V_t = 1^V$, $V_{OA} = V_{DD} = 5^V$

For $V_i < V_t$, the transistor is not on, $V_{GS} < V_t$.

Point A is when $V_{GS} = V_t$ and the transistor turns on. As V_i increases, the i_D increases and v_o decreases. v_o decreases to the point that it is below V_I by V_t volts. At this point, B, the MOSFET enters the triode region: $V_{OB} = V_{IB} - V_t$ or $V_{DS} = V_{GS} - V_t$. So at point B: $I = \frac{V_{DD} - V_{OB}}{R}$

$I = \frac{V_{DD} - (V_{GS} - V_t)}{R} = \frac{1}{2} \times K_n' \frac{W}{L}(V_{GS} - V_t)^2$

$\frac{5 - V_{GS} + 1}{24} = \frac{1}{2} \times 1 \times (V_{GS} - 1)^2 \Rightarrow 12V_{GS}^2 - 23V_{GS} + 6 = 0$

Cont.

$V_{GS} = 1.61V \Rightarrow V_I = 1.61^V \quad v_O = 1.61 - 1 = 0.61^V$

Point B: $V_{OB} = 0.61^V \quad V_{IB} = 1.61V$

b) $I_Q = \frac{1}{2} \times 1 \times 0.5^2 = 0.125mA$

$V_{OQ} = 5 - 24 \times 0.125 = 2^V$

$V_{IQ} = V_{GS} = V_{OV} + V_t = 0.5 + 1 = 1.5^V$

Now to calculate the incremental gain

A_v at this bias point, From equation 4.41, we have: $A_v = -2\, V_{RD}/V_{OV} = \frac{-2(V_{DD}-V_{OQ})}{V_{OV}}$

$A_v = \frac{-2(5-2)}{0.5} = -12\,V/V$

c) $V_{IQ} = 1.5^V$, $V_t = 1^V$, $V_{IB} = 1.61^V$. Thus the largest amplitude of a sine wave that can be applied to the input while the transistor remain in saturation is: $1.61 - 1.5 = 0.11V$

The amplitude of the output voltage signal that results is approximately equal to $V_{OQ} - V_{OB} = 2 - 0.61 = 1.39^V$. The gain implied by this amplitudes is:

$gain = \frac{-1.39}{0.11} = 12.64\,V/V$

This gain is 5.3% different from the incremental gain calculated in part(b). This difference is due to the fact that the segment of the voltage transfer curve considered here is not perfectly linear.

4.50

$v_{DS} = 1^V = V_{OQ} \Rightarrow I_D = \frac{V_{DD} - V_{OQ}}{R} = \frac{10-1}{18} = 0.5mA$

$I_D = 0.5 = \frac{1}{2} \times 1 \times V_{OV}^2 \Rightarrow V_{OV} = \sqrt{2I_D} = 1V$

$V_{OV} = V_{GS} - V_t \Rightarrow V_{GS} = 2^V$

$A_v = \frac{-2\,V_{RD}}{V_{OV}} = -\frac{2 \times 18 \times 0.5}{1} = -18\,V/V$

$v_o^+ = V_{DD} - V_{OQ} = 10 - 1 = 9^V$

$v_o^- = V_{OQ} - V_{OB}$

In order to find V_{OB} where $V_{DS} = V_{GS} - V_t$ or $V_{OB} = V_{IB} - V_t$: $v_O = v_{DD} - R_D I_D$

$v_{OB} = 10 - 18 \times \frac{1}{2} \times 1\, V_{OB}^2 \Rightarrow 9V_{OB}^2 + V_{OB} - 10 = 0 \Rightarrow V_{OB} = 1^V$

$v_o^- = V_{OQ} - V_{OB} = 1 - 1 = 0$

Using the same formulas we can calculate the rest of the values for $V_{DS} = 2, 3, \ldots, 10^V$

V_{DS} (V)	I_D (mA)	V_{OV} (V)	V_{GS} (V)	A_v (V/V)	v_o^+ (V)	v_o^- (V)
1	0.5	1	2	18	9	0
2	0.44	0.94	1.94	17.02	8	1
3	0.39	0.88	1.88	15.9	7	2
4	0.33	0.81	1.81	14.8	6	3
5	0.28	0.75	1.75	13.3	5	4
6	0.22	0.66	1.66	12.1	4	5
7	0.17	0.58	1.58	10.3	3	6
8	0.11	0.47	1.47	8.5	2	7
9	0.06	0.35	1.35	5.7	1	8
10	0	0	$\leq 1^V$	0	0	0

4.51

$R_D = 20k\Omega$, $V_{RD} = 2^V \Rightarrow I_D = 0.1mA$

$A_v = -\frac{2\,V_{RD}}{V_{OV}} \Rightarrow -10 = -\frac{2 \times 2}{V_{OV}} \Rightarrow V_{OV} = 0.4^V$

$V_{GS} = 1.2V \Rightarrow V_t = 1.2 - 0.4 = 0.8V$

$I_D = \frac{1}{2} k_n' \frac{W}{L} V_{OV}^2 \Rightarrow 0.1 = \frac{1}{2} \times 50 \times 10^{-3} \frac{W}{L}\, 0.4^2$

$\Rightarrow \frac{W}{L} = 25$

4.52

Eq. 4.41 : $A_v = -\frac{2(V_{DD} - V_{OQ})}{V_{OV}}$

a) $A_{v\,max} = -\frac{2(V_{DD} - 0)}{V_{OV\,min}} = -\frac{2 \times 5}{0.2} = 50\,V/V$

b)

c)

4.53

From Figure 4.26(c) we have: $\hat{v}_o = V_{DS} - V_{OB}$

Assuming linear operation around the bias point:

$$A_v = -\frac{\hat{v}_o}{V_{IB} - V_{IQ}} = -\frac{\hat{v}_o}{V_{OB} + V_t - V_{IQ}} = -\frac{\hat{v}_o}{V_{OB} + V_t - (V_{ov} + V_t)}$$

$$A_v = -\frac{\hat{v}_o}{V_{OB} - V_{ov}} = -\frac{\hat{v}_o}{V_{OB} - V_{DS} + V_{DS} - V_{ov}} = -\frac{\hat{v}_o}{-\hat{v}_o + V_{DS} - V_{ov}}$$

$$-A_v\hat{v}_o + A_v(V_{DS} - V_{ov}) = -\hat{v}_o \Rightarrow \hat{v}_o(1 - A_v) = -A_v(V_{DS} - V_{ov})$$

$$\hat{v}_o = \frac{A_v(V_{DS} - V_{ov})}{A_v - 1} \Rightarrow \hat{v}_o = \frac{V_{DS} - V_{ov}}{1 - \frac{1}{A_v}}$$

V_{DS} (V)	A_v (V/V)	$\hat{v}_o$ (V)	$\hat{v}_i$ (V)
1	-16	0.47	0.029
1.5	-14	0.93	0.066
2	-12	1.38	0.12
2.5	-10	1.81	0.18

$I_D = \frac{1}{2} \times 1 \times (0.5)^2 = 0.125\,mA$

$R_D = \frac{V_{DD} - V_{DS}}{I_D} = \frac{5-1}{0.125} = 32\,k\Omega$

4.54

$$i_{D1} = i_{D2} \Rightarrow \frac{1}{2}k_n'\left(\frac{W}{L}\right)_1 (V_{GS1} - V_t)^2 = \frac{1}{2}k_n'\left(\frac{W}{L}\right)_2 (V_{GS2} - V_t)^2$$

Note that $V_{GS1} = V_I$

$V_{GS2} = V_{DD} - V_O$

$$\Rightarrow \left(\frac{W}{L}\right)_1 (V_I - V_t)^2 = \left(\frac{W}{L}\right)_2 (V_{DD} - V_O - V_t)^2$$

$$\sqrt{\left(\frac{W}{L}\right)_1 / \left(\frac{W}{L}\right)_2}\,(v_I - V_t) = V_{DD} - v_O - V_t$$

$$v_O = V_{DD} - V_t + \sqrt{\frac{(W/L)_1}{(W/L)_2}}\,V_t - \sqrt{\frac{(W/L)_1}{(W/L)_2}}\,v_I$$

IF $\left(\frac{W}{L}\right)_1 = \frac{50}{0.5} = 100$ $\quad \left(\frac{W}{L}\right)_2 = \frac{5}{0.5} = 10$

then:

$$A_v = \frac{\partial v_O}{\partial v_I} = -\sqrt{\frac{100}{10}} = -3.16\,V/V$$

4.55

$$I_D = 2\,mA = \frac{1}{2} \times 80 \times 10^{-3} \times \frac{240}{6} \times (V_{GS} - 1.2)^2 \Rightarrow$$

$V_{GS} = 2.32\,V$ $\qquad$ Refer to Fig. 4.30C

$R_D I_D = \frac{15}{3} = 5^V \Rightarrow R_D = \frac{5}{2^{mA}} = 2.5^{K\Omega}$

$R_S I_D = 5^V \Rightarrow R_S = \frac{5}{2} = 2.5\,K\Omega$

$V_G = 5 + V_{GS} = 7.32V$

$\frac{15}{R_{G1} + R_{G2}} \times R_{G2} = 7.32$ $\qquad R_{G1} = 22M\Omega \Rightarrow R_{G2} = 20.97\,M\Omega$

$V_{DS} = 5^V$

at the edge of saturation $V_{DS} = V_{GS} - V_t$ or $V_{DS} = 2.32 - 1.2 = 1.12^V$. So V_{DS} is $5 - 1.12 = 3.88^V$ away from the edge of saturation.

4.56

Refer to Fig. 4.30e

$I_D = 2mA = \frac{1}{2}k_n'\frac{W}{L}V_{ov}^2 \Rightarrow 2 = \frac{1}{2} \times 50 \times 10^{-3} \times \frac{200}{4} V_{ov}^2$

$V_{ov} = 1.26\,V$

$V_{DS} = V_{ov}$ edge of triode

Midway of cutoff ($V_{DS} = V_{DD}$) and beginning of triode operation ($V_{DS} = V_{ov}$) is when $V_{DS} = \frac{30 + 1.26}{2}$

$V_{DS} = 15.63V$

$V_{GS} = 2.32\,V \rightarrow V_S = -2.32V \Rightarrow R_S = \frac{-2.32 + 15}{2}$

$R_S = 6.34\,k\Omega$

$V_D = V_S + V_{DS} = -2.32 + 15.63 = 13.31\,V \Rightarrow R_D = \frac{15 - 13.31}{2}$

$R_D = 0.85\,k\Omega$

4.57

$V_G = 12 \times \frac{2.2}{2.2 + 5.6} = 3.4^V$

$k_n'\frac{W}{L} = 220$ to $380\,\mu A/V^2$

$V_t = 1.3$ to $2.4V$

$I_D = \frac{1}{2} \times k_n'\frac{W}{L}(3.4 - V_t)^2$

$I_{Dmin} = \frac{1}{2} \times 220(3.4 - 2.4)^2 = 110\,\mu A$

$I_{Dmax} = \frac{1}{2} \times 380(3.4 - 1.3)^2 = 838\,\mu A$

to limit I_{Dmax} to $150\mu A$:

$150 = \frac{1}{2} \times 380(3.4 - 0.15R_S - 1.3)^2$

$R_S = 8.1\,K\Omega$

Select $R_S = 8.2\,K\Omega$

$I_{Dmax} = \frac{1}{2} \times 380 \times (3.4 - I_{Dmax} \times 8.2 - 1.3)^2$

$I_{Dmax} = 0.15\,mA$ or ~~0.4~~ mA

The second answer results in negative V_{GS}

12V, 5.6 MΩ, 2.2 MΩ, 3.4V, R_S

Cont.

and therefore it is not acceptable.

$I_{Dmin} = \frac{1}{2} \times 0.22 \times (3.4 - 8.2 I_{Dmin} - 2.4)^2$

$I_{Dmin} = 0.04 mA$

4.58

$V_t = 2^V$, $K'_n \frac{W}{L} = 2 mA/V^2$

$I_D = \frac{1}{2} \times 2 \times (4 - I_D \times 1 - 2)^2$

$I_D = 4 + I_D^2 - 4I_D \Rightarrow I_D = 1^{mA}, \cancel{4^{mA}}$

$I_D = 4^{mA}$ results in $V_{GS} = 0$ which is not acceptable, therefore $I_D = 1 mA$.

For $K' \frac{W}{L}$ 50% larger, i.e. $K' \frac{W}{L} = 3 mA/V^2$

$I_D = \frac{1}{2} \times 3 (4 - I_D - 2)^2 \Rightarrow I_D = 1.13 mA$

I_D increases by 13%.

4.59

Refer to fig. 4.30.c

$V_{GS} = 5 - 2 = 3V$, $I_D = \frac{V_S}{R_S} = \frac{2}{1} = 2 mA$

$I_D = 2 = \frac{1}{2} \times 2 \times (3 - V_t)^2 \Rightarrow \pm 1.41 = 3 - V_t \Rightarrow V_t = 1.59^V$

For a device with $V_t = 1.59 - 0.5 = 1.09^V$:

$I_D = \frac{1}{2} \times 2 \times (5 - I_D \times 1 - 1.09)^2 \Rightarrow I_D = 2.37^{mA}$

$V_S = 2.37V$

4.60

To maximize gain, we design for the lowest possible V_D consistant with allowing a 2V p-p signal swing. $V_{Dmin} = V_D - 1$

$V_{Dmin} = V_G - V_t = 0 - 2$

$V_D - 1 = -2 \Rightarrow V_D = -1V \Rightarrow R_D = \frac{10 - (-1)}{1 mA} = 11 K\Omega$

$I_D = \frac{1}{2} \times 2 [0 - (-10 + 1 \times R_S) - 2]^2 = 1 \Rightarrow 1 = (8 - R_S)^2$

$R_S = 7 K\Omega$

4.61

For P-channel MOSFET to be 1V from the edge of saturation: $V_{DS} = V_{GS} - V_t - 1$. Since $V_D = 3V$ and $I_D = 1mA$: $R_D = \frac{3}{1} = 3K\Omega$, $R_1 + R_2 = \frac{10V}{10\mu A} = 1M\Omega$

a) $|V_t| = 1V$, $K'_p \frac{W}{L} = 0.5 mA/V^2$

$I_D = 0.5 \times \frac{1}{2} \times (V_{GS} + 1)^2 = 1 \Rightarrow V_{GS} = -3V$

$V_{DS} = -3 + 1 - 1 = -3V$

$V_S = 6V$, $V_G = 3V$ $R_S = \frac{10 - 6}{1} = 4K\Omega$

$\frac{R_2}{R_1 + R_2} = \frac{3}{10} \Rightarrow R_1 = 0.7 M\Omega$, $R_2 = 0.3 M\Omega$

b) $|V_t| = 2V$, $K'_p \frac{W}{L} = 1.25 mA/V^2$

$1 = \frac{1}{2} \times 1.25 (V_{GS} + 2)^2 \Rightarrow V_{GS} = -3.26^V$ or -0.74^V

the second answer is not acceptable $|V_{GS}| < |V_t|$

$V_{GS} = -3.26V$

$V_{DS} = -3.26 + 2 - 1 = -2.26V$

$V_S = 3 + 2.26 = 5.26 V$

$V_G = 2V$

$R_S = \frac{10 - 5.26}{1} = 4.74 K\Omega$, $R_D = 3K\Omega$

$\frac{R_2}{R_1 + R_2} = \frac{2}{10} \Rightarrow R_2 = 0.2 M\Omega$ $R_1 = 0.8 M\Omega$

4.62

$K = \frac{1}{2} K' \frac{W}{L}$

a) $I_D = \frac{1}{2} K'_n \frac{W}{L} (V_{GS} - V_t)^2$

$I_D = K (0 + V_{SS} - R_S I_D - V_t)^2$

$\frac{\partial I_D}{\partial K} = (V_{SS} - R_S I_D - V_t)^2 + 2K (V_{SS} - R_S I_D - V_t)(-R_S) \frac{\partial I_D}{\partial K}$

$\frac{\partial I_D}{\partial K} = \frac{I_D}{K} - 2R_S \sqrt{\frac{I_D}{K}} K \frac{\partial I_D}{\partial K}$

$\frac{\partial I_D}{\partial K} (1 + 2\sqrt{K I_D} R_S) = \frac{I_D}{K} \Rightarrow S_K^{I_D} = \frac{\partial I_D}{\partial K} \frac{K}{I_D} = \frac{1}{1 + 2\sqrt{K I_D} R_S}$

b) $K = 100 \mu A/V^2$, $\frac{\Delta K}{K} = \pm 10\%$, $V_t = 1V$, $I_D = 100 \mu A$

$\frac{\Delta I_D}{I_D} = \pm 1\%$

$S_K^{I_D} \simeq \frac{\partial I_D / I_D}{\partial K / K} = \frac{1}{10} = 0.1 = \frac{1}{1 + 2\sqrt{100 \times 10^{-3} \times 100 \times 10^{-3}} R_S}$

$\Rightarrow R_S = 45 K\Omega$

Cont.

$100 = 100\,(V_{GS}-1)^2 \Rightarrow V_{GS} = 2V$

$V_{GS} = V_{SS} - I_D R_S \Rightarrow 2 = V_{SS} - 0.1\times 45 \Rightarrow V_{SS} = 6.5^V$

c) For $V_{SS} = 5V$:

$R_S = \frac{-V_{GS}+V_{SS}}{I_D} = \frac{-2+5}{0.1} = 30k\Omega$

$S_k^{I_D} = \frac{1}{1+2\sqrt{0.1\times 0.1}\times 30} = \frac{1}{1+2\times 0.1\times 30} = 0.14$

Therefore for $\frac{\Delta k}{k} = \pm 10\%$, $\frac{\Delta I_D}{I_D} = \pm 1.4\%$

4.63

Both cases are in saturation region, because $V_{DG} \geq V_t$.

$V_D = 10 - 5\times 1 = 5^V$

a) $1 = \frac{1}{2}\times 0.5\times (V_{GS}-1)^2 \Rightarrow V_{GS} = 3V$, $V_S = -3V$

$V_{DS} = 8V$

b) $1 = \frac{1}{2}\times 1.25\times (V_{GS}-2)^2 \Rightarrow V_{GS} = 3.3V$, $V_S = -3.3^V$

$V_{DS} = 8.3V$

4.64

$V_D = V_G = V_{GS}$ Refer to Fig. 4.32

to operate in saturation:

$V_{DS} \geq V_{GS} - V_t \Rightarrow V_{DG} \geq -V_t$

$V_{DG} = 0$

10V, $R_D = 10^K$, $R_G = 10M$

a) $\frac{10 - V_D}{10} = \frac{1}{2}\times 0.5\times (V_D - 1)^2 \Rightarrow V_D = 2.7^V$

$V_G = 2.7^V$

b) $\frac{10 - V_D}{10} = \frac{1}{2}\times 1.25\,(V_D - 2)^2 \Rightarrow V_D = 3.05V$

$V_G = 3.05V$

4.65

9V, R_D, I_D, R_G

For $I_D = 0.2mA$:

$0.2 = \frac{1}{2}\times 0.4\times (V_{GS}-1)^2$

$V_{GS} = 2V$, $V_D = V_G = V_{GS} = 2V$

$R_D = \frac{9-2}{0.2} = 35K\Omega$

Select $R_D = 36\,K\Omega \Rightarrow \frac{9-V_D}{R_D} = \frac{1}{2}\times 0.4\,(V_D - 1)^2$

$\frac{9-V_D}{36} = 0.2\,(V_D - 1)^2 \Rightarrow V_D = 2V$, $I_D = 0.21\,mA$.

4.66

$I_D = 2 = \frac{1}{2}\times 3.2\times (V_{GS}-1.2)^2$

$V_{GS} - 1.2 = 1.12 \Rightarrow V_{GS} = 2.32^V$

$V_G = 2.32^V$

$V_{DSmin} = V_{GS} - V_t = 1.12^V$

$V_{DS} = V_{DSmin} + 2 = 3.12^V$

$R_{G_2} = 22M\Omega \Rightarrow I = \frac{2.32}{22} = 0.11\mu A$

$R_{G_1} = \frac{3.12 - 2.32}{0.11} = 7.58M\Omega$

$R_D = \frac{6 - 3.12}{2 + 0.11\times 10^{-3}} = 1.44K\Omega$

$V_{DD} = 6^V$, R_D, R_{G1}, R_{G2}, I, $I_D = 2^{mA}$

4.67

Eq. 4.57 implies:

$i_D = I_D + K_n' \frac{W}{L}(V_{GS} - V_t)\,v_{gs}\sin\omega t + \frac{1}{2}K_n'\frac{W}{L}v_{gs}^2 \sin^2\omega t$

$i_D = I_D + K_n' \frac{W}{L}(V_{GS} - V_t)\,V_{gs}\sin\omega t + \frac{1}{2}K_n'\frac{W}{L}V_{gs}^2\left(\frac{1-\cos 2\omega t}{2}\right)$

$i_D = I_D + K_n' \frac{W}{L}(V_{GS} - V_t)\,V_{gs}\sin\omega t + \frac{1}{4}K_n'\frac{W}{L}V_{gs}^2 - \frac{V_{gs}^2}{4}K_n'\frac{W}{L}\cos 2\omega t$

Second Harmonic Distortion $= \frac{1/4\, K_n'\, W/L\, V_{gs}^2}{K_n'\, W/L\,(V_{GS}-V_t)\,V_{gs}} \times 100$

$= \frac{1}{4}\,\frac{V_{gs}}{V_{GS}-V_t}\times 100$

Second Harmonic Distortion $= \frac{1}{4}\,\frac{V_{gs}}{V_{ov}}\times 100$

For $V_{gs} = 10mV$ $\frac{1}{4}\times\frac{10\times 10^{-3}}{V_{ov}}\times 100 \leq 1$

$\Rightarrow V_{ov} \geq 0.25V \Rightarrow V_{ov\,min} = 0.25V$

4.68

$I_D = \frac{1}{2}K_n'\frac{W}{L}V_{ov}^2 \Rightarrow I_D = \frac{1}{2}\times 2\times 1^2 = 1mA$

$i_D = \frac{1}{2}\times 2\times (1+0.1)^2 = 1.21mA$ $(v_{gs} = 0.1V)$

$i_D = 1.21 - 1 = 0.21mA$

If $v_{gs} = -0.1^V \Rightarrow i_D = \frac{1}{2}\times 2(1-0.1)^2 = 0.81mA$

$i_D = 0.81 - 1 = -0.19\,mA$

For positive increment : $g_m = \frac{\Delta i_D}{\Delta v_{gs}} = \frac{0.21}{0.1} = 2.1\,mA/V$

For negative increment : $g_m = \frac{0.19}{0.1} = 1.9mA/V$

An estimate of $g_m = \frac{2.1+1.9}{2} = 2\,mA/V$

Eq. 4.62 $g_m = K_n'\frac{W}{L}V_{ov} = 2\times 1 = 2\,mA/V$. Same as estimate!

4.69

a) $I_D = \frac{1}{2} \times 1 \times (4-2)^2 = 2\,mA$

$V_D = V_{DD} - R_D I_D = 10 - 2 \times 3.6$

$V_D = 2.8\,V$

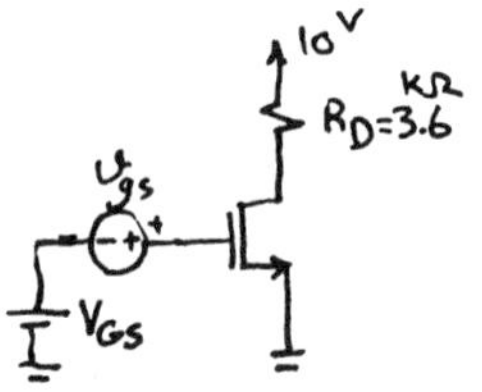

b) $g_m = k_n' \frac{W}{L} V_{ov} = 1 \times (4-2) = 2\,mA/V$

c) $A_v = \frac{v_d}{v_{gs}} = -g_m R_D = -2 \times 3.6 = -7.2\,V/V$

d) $r_o \simeq \frac{1}{\lambda I_D} = \frac{1}{0.01 \times 2} = 50\,k\Omega$

$A_v = \frac{v_d}{v_{gs}} = -g_m (R_D \| r_o) = -2(3.6 \| 50) = -6.7\,V/V$

4.70

$g_m R_D = 5 \Rightarrow g_m = \frac{5}{50} = 0.1\,mA/V$

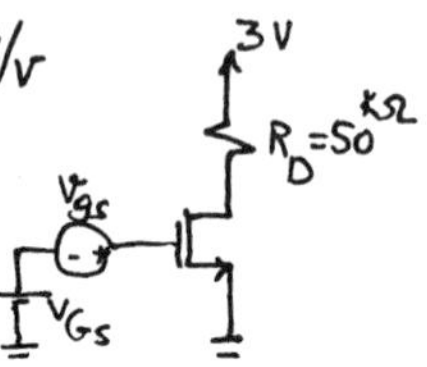

For 0.5V output signal and a gain of 5 V/V, $v_{gs} = \frac{0.5}{5} = 0.1\,V$

So we can write $V_{DS} - 0.5 \geq V_{GS} + 0.1 - V_t$

or $V_{DS} \geq V_{GS} + 0.6 - 0.8 \Rightarrow V_{DS} \geq V_{GS} - 0.2$

Also, from the other side: $V_{DS} + 0.5 \leq V_{DD}$

or $V_{DS} \leq 3 - 0.5 \Rightarrow V_{DS} \leq 2.5\,V$

We design the circuit for lowest possible VDS that guarantees the device operation in saturation: $V_{DS} = V_{GS} - 0.2$

$V_{DS} = V_{DD} - R_D I_D \Rightarrow V_{GS} - 0.2 = 3 - 50 \times I_D$

$\Rightarrow I_D = \frac{3.2 - V_{GS}}{50}$

Also, from eq. 4.71: $g_m = \frac{2 I_D}{V_{GS} - V_t} = 0.1$

$0.1 = \frac{2}{V_{GS} - 0.8} \times \frac{3.2 - V_{GS}}{50}$

$\Rightarrow V_{GS} = 1.49\,V$, $I_D = 0.034\,mA$

$V_{DS} = 1.49 - 0.2 = 1.29\,V$ $\quad V_{OV} = 1.49 - 0.8 = 0.69\,V$

$\frac{W}{L} = \frac{I_D}{\frac{1}{2} k_n' V_{OV}^2} = \frac{0.034 \times 10^3}{\frac{1}{2} \times 100 \times 0.69^2} = 1.43$

$\frac{W}{L} = 1.43$

4.71

$A_v = -g_m R_D$,
$g_m = \frac{2 I_D}{V_{ov}}$ eq. 4.71 $\Big\} \Rightarrow A_v = -\frac{2 R_D I_D}{V_{ov}} = -\frac{2(V_{DD} - V_D)}{V_{ov}}$ (1)

Minimum v_{DS} for edge of saturation:

$v_{DS} \geq v_{GS} - V_t$ or $v_{DSmin} = v_{GSmax} - V_t$

$V_{DS} - |A_v| \hat{v}_i = V_{GS} + \hat{v}_i - V_t$

IF we replace A_v with (1):

$V_D - \frac{2(V_{DD} - V_D)}{V_{ov}} \hat{v}_i = V_{ov} + \hat{v}_i$

$\Rightarrow V_D \left(1 + \frac{2 \hat{v}_i}{V_{ov}}\right) = V_{ov} + \hat{v}_i + \frac{2 V_{DD} \hat{v}_i}{V_{ov}}$

$$V_D = \frac{V_{ov} + \hat{v}_i + 2 V_{DD} (\hat{v}_i / V_{ov})}{1 + 2(\hat{v}_i / V_{ov})}$$

$V_{DD} = 3\,V$, $\hat{v}_i = 20\,mV$ $\quad m = 10 = \frac{V_{ov}}{\hat{v}_i} \Rightarrow V_{ov} = 0.2\,V$

$V_D = \frac{0.2 + 0.02 + 2 \times 3 \times 10^{-1}}{1 + 2 \times 0.1} = 0.68\,V$

$A_v = -2 \frac{(3 - 0.68)}{0.2} = -23.2\,V/V$

IF $I_D = 100\,\mu A = 0.1\,mA$:

$A_v = -\frac{2 R_D I_D}{V_{ov}} \Rightarrow 23.2 = \frac{2 \times R_D \times 0.1}{0.2} \Rightarrow$

$R_D = 23.2\,k\Omega$

$I_D = \frac{1}{2} k_n' \frac{W}{L} V_{ov}^2 \Rightarrow 0.1 = \frac{1}{2} \times 100 \times 10^{-3} \frac{W}{L} 0.2^2$

$\Rightarrow \frac{W}{L} = 50$

4.72

$k_n' = \mu_n C_{ox} = 500 \times 10^8 \times 0.4 \times 10^{-15} = 20\,\mu A/V^2$, $k_p' = 10\,\mu A/V^2$

Case	Type	I_D (mA)	$\|V_{GS}\|$ (V)	$\|V_t\|$ (V)	V_{OV} (V)	W (μm)	L (μm)	$\frac{W}{L}$	$k'\frac{W}{L}$ (mA/V²)	g_m (mA/V)
a	N	1	3	2	1	100	1	100	2	2
b	N	1	1.2	0.7	0.5	50	$\frac{1}{8}$	400	8	4
c	N	10	?	?	2	250	1	250	5	10
d	N	0.5	?	?	0.5	?	?	200	4	2
e	N	0.1	?	?	1.41	10	2	5	0.1	0.14
f	N	0.1	1.8	0.8	1	40	4	10	0.2	0.2
g	P	1	?	?	2	?	?	25	0.25	1
h	P	1	3	1	2	?	?	50	0.5	1
i	P	10	?	?	1	4000	2	2000	20	20

Cont.

Case	Type	I_D	V_{GS}	V_t	V_{OV}	W	L	$\frac{W}{L}$	$\frac{k'W}{L}$	g_m
j	P	10	?	?	4	?	?	125	1.25	5
K	P	0.05	?	?	1	30	3	10	0.1	0.1
L	P	0.1	?	?	5	?	?	0.8	$\frac{8}{1000}$	0.04

4.73

Eq. 4.70 : $g_m = \sqrt{2k'_n \frac{W}{L} I_D} \Rightarrow \frac{W}{L} = \frac{g_m^2}{2k'_n I_D}$

$\frac{W}{1} = \frac{1}{2 \times 50 \times 10^{-3} \times 0.5} \Rightarrow W = 20\ \mu m$

$g_m = \frac{2I_D}{V_{ov}} \Rightarrow V_{ov} = \frac{2 \times 0.5}{1} = 1 \Rightarrow V_{GS} = 1 + V_t = 1.7^V$

4.74

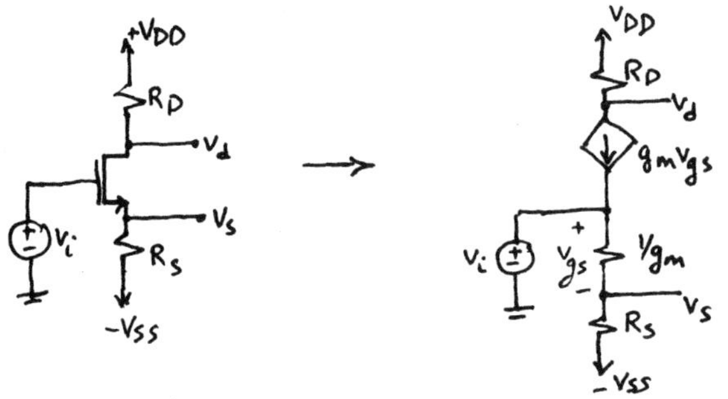

$\frac{v_s}{v_i} = \frac{R_s}{R_s + \frac{1}{g_m}} = \frac{R_s g_m}{R_s g_m + 1}$

$\frac{v_d}{v_i} = \frac{-g_m v_{gs} R_D}{v_i} = -g_m R_D \frac{1/g_m}{1/g_m + R_s} = \frac{-g_m R_D}{1 + g_m R_s}$

4.75

$r_o \simeq \frac{V_A}{I_D} = \frac{50}{0.5} = 100\,K\Omega$

$g_m = \frac{2I_D}{V_{GS} - V_t}$, $V_{GS} = V_{DS} = 2^V$

$g_m = \frac{2 \times 0.5}{2 - 0.9} = 0.91\,mA/V$

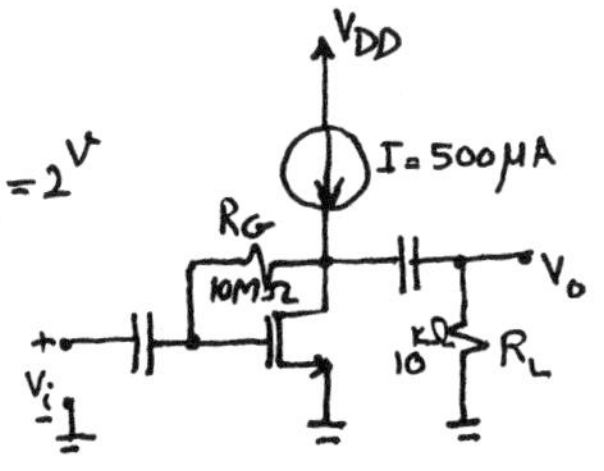

$\frac{v_o}{v_i} = -g_m (r_o \| R_L) = -0.91(100^k \| 10^k) = -8.3\,V/V$

For I=1mA or twice the current:

$\frac{I_{D1}}{I_{D2}} = \frac{(V_{GS_1} - V_t)^2}{(V_{GS_2} - V_t)^2} \Rightarrow V_{GS2} = V_t + \sqrt{2}\,(V_{GS_1} - V_t)$

$V_{GS2} = 0.9 + \sqrt{2}\,(2 - 0.9) = 2.5^V$

$\frac{g_{m1}}{g_{m2}} = \sqrt{\frac{I_{D1}}{I_{D2}}} \Rightarrow g_{m2} = \sqrt{2}\,g_{m1} = 1.3\,mA/V$

$\frac{r_{o1}}{r_{o2}} = \frac{I_{D2}}{I_{D1}} \Rightarrow r_{o2} = \frac{100}{2} = 50K\Omega$

$A_v = -1.3 \times (50K \| 10K) = -10.8\,V/V$

4.76

NMOS: $g_m = \sqrt{2 k'_n \frac{W}{L} I_D} = \sqrt{2 \times 90 \times 10^{-3} \times \frac{20}{2} \times 0.1} = 0.42\,\frac{mA}{V}$

$r_o = \frac{|V_A|}{I_D} = \frac{8 \times 2}{0.1} = 160K\Omega$

$\chi = \frac{\gamma}{2\sqrt{2\phi_f + |V_{SB}|}} = \frac{0.5}{2\sqrt{2 \times 0.34 + 1}} = 0.2$

$g_{mb} = \chi g_m = 0.2 \times 0.42 = 0.084\,mA/V$

$g_m = \frac{2I_D}{V_{ov}} \Rightarrow V_{ov} = \frac{2 \times 0.1}{0.42} = 0.48V$

PMOS: $g_m = \sqrt{2 \times 30 \times 10^{-3} \times \frac{20}{2} \times 0.1} = 0.24\,mA/V$

$r_o = \frac{|V_A|}{I_D} = \frac{12 \times 2}{0.1} = 240K\Omega$

$\chi = 0.2 \Rightarrow g_{mb} = 0.2 \times 0.24 = 0.048\,mA/V$

$V_{ov} = \frac{2 \times 0.1}{0.24} = 0.83V$

4.77

Refer to Fig. P4.77.

a) $V_G = 15 \times \frac{5}{10+5} = 5^V$ $V_s = 3 \times I_D \Rightarrow V_{GS} = 5 - 3I_D$

$I_D = \frac{1}{2} \times k' \frac{W}{L} (V_{GS} - V_t)^2 \Rightarrow I_D = \frac{1}{2} \times 2 \times (5 - 3I_D - 1)^2$

$\Rightarrow 16 - 25I + 9I^2 = 0 \Rightarrow I = 1mA$

$V_{GS} = 2V$ $V_D = 15 - 7.5 = 7.5^V$

b) $g_m = \frac{2I_D}{V_{ov}} = \frac{2 \times 1}{2 - 1} = 2\,mA/V$ $r_o = \frac{V_A}{I_D} = 100K\Omega$

c) $R_{in} = 5^M \| 10^M = 3.33M\Omega$

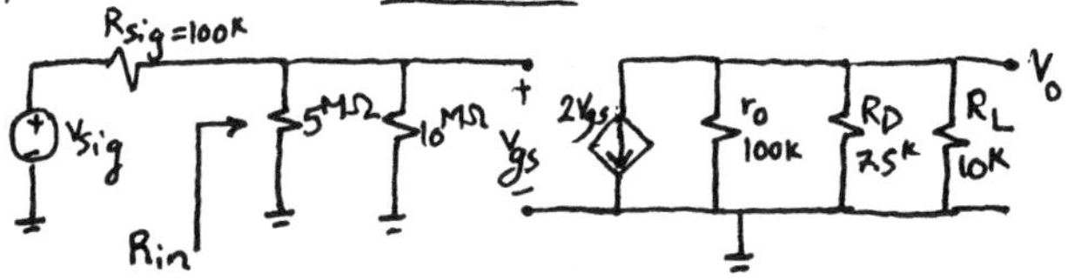

d) $\frac{V_{gs}}{V_{sig}} = \frac{R_{in}}{R_{sig} + R_{in}} = \frac{3.33}{0.1 + 3.33} = 0.97\,V/V$

$\frac{V_o}{V_{gs}} = 2 \times (r_o \| R_D \| R_L) = 2 \times 4.1^K = 8.2\,V/V$ $\frac{V_o}{V_{sig}} = 7.95\,V/V$

4.78

$i_D = \frac{1}{2} k'_n \frac{W}{L} v_{OV}^2$

The equation for the tangent line passing from point A (V_{OV}, I_D) can be written as:

$$(i_D - I_D) = \left.\frac{\partial i_D}{\partial v_{OV}}\right|_{\substack{i=I_D \\ v=V_{OV}}} (v_{OV} - V_{OV})$$

$$\left.\frac{\partial i_D}{\partial v_{OV}}\right|_{i_D=I_D} = k'_n \frac{W}{L} V_{OV}$$

$$\Rightarrow i_D - I_D = k'_n \frac{W}{L} V_{OV} (v_{OV} - V_{OV})$$

$$i_D - \frac{1}{2} k'_n \frac{W}{L} V_{OV}^2 = k'_n \frac{W}{L} V_{OV} (v_{OV} - V_{OV})$$

The tangent intersects the v_{OV} axis at $i_D = 0$

$$0 - \frac{1}{2} k'_n \frac{W}{L} V_{OV}^2 = k'_n \frac{W}{L} V_{OV} (v_{OV} - V_{OV})$$

$$-\frac{V_{OV}}{2} = v_{OV} - V_{OV} \Rightarrow v_{OV} = \frac{V_{OV}}{2}$$

The slope of the tangent is g_m and it is equal to: $g_m = \frac{\Delta i_D}{\Delta v_{OV}} = \frac{I_D}{V_{OV} - \frac{V_{OV}}{2}} = \frac{I_D}{\frac{V_{OV}}{2}} = \frac{2I_D}{V_{OV}}$

4.79

For this common-source amplifier we have:

$g_m = 2\frac{mA}{V}$, $r_o = 50k\Omega$, $R_D = 10k\Omega$

$R_G = 10M\Omega$, $R_{sig} = 0.5M\Omega$ and $R_L = 20k\Omega$

From equation 4.82 we have:

$$G_v = -\frac{R_G}{R_G + R_{sig}} g_m (r_o \| R_D \| R_L) = -\frac{10 \times 2}{10 + 0.5} (50^k \| 10^k \| 20^k)$$

$G_v = -11.2\,V/V$

4.80

a) $A_{vo} = -2\frac{(V_{DD} - V_D)}{V_{OV}} = -2\frac{(10 - 2.5)}{1} = -15\,V/V$

b) if V_{OV} is halved ($V_{OV} = 0.5$) then I_D is divided by 4, i.e. $I_D = \frac{0.5}{4} = 0.125\,mA$

Since V_D is kept unchanged at 2.5^V then:

$R_D = \frac{10 - 2.5}{0.125} = 60k\Omega$, $g_m = \frac{2I_D}{V_{OV}} = 0.5\frac{mA}{V}$

$r_o = \frac{V_A}{I_D} \Rightarrow r_o = 4 \times r_{o1} = 4 \times \frac{75}{0.5} = 600k\Omega$

$A_{vo} = -15 \times 2 = -30\,V/V$ (without r_o)

c) If we take r_o into account:

$A_{vo} = -g_m (r_o \| R_D) = -0.5(600^K \| 60^K) = -27.3\,V/V$

$R_{out} = R_D \| r_o = 600^K \| 60^K = 54.5k\Omega$

d) $R_{in} = R_G = 4.7M\Omega$

$R_o = R_{out} = 54.5k\Omega$

$G_v = \frac{R_{in}}{R_{in} + R_{sig}} A_{vo} \frac{R_L}{R_L + R_o} = \frac{4.7}{4.7 + 0.1} \times 27.3 \times \frac{15}{15 + 54.5}$

$G_v = 5.77\,V/V$

e) As we can see by reducing V_{OV} to half of its value or equivalently multiplying drain current by 4, A_{vo} is almost doubled, while R_{out} is multiplied by 4.

As a result G_v which is proportional to both A_{vo} and $\frac{1}{R_{out}}$ is only slightly reduced.

(G_v was $-7\,V/V$ before and it is $5.8\,V/V$ now)

4.81

For an NMOS common-gate amplifier with $g_m = 5\,mA/V$, $R_D = 5k\Omega$, $R_L = 2k\Omega$, $R_{sig} = 200\Omega$ we have:

$R_{in} = \frac{1}{g_m} = \frac{1}{5} = 0.2k\Omega = 200\Omega$

From Eq. 4.96b we know that the overall voltage gain of this amplifier is:

$$G_v = \frac{g_m (R_D \| R_L)}{1 + g_m R_{sig}} = \frac{5(5K \| 2K)}{1 + 5 \times 0.2} = 3.57\,V/V$$

If we increase the bias current by a factor of 4, while maintaining the other parameter constant (assuming linear operation), we have:

$g_m = \sqrt{2 k'_n \frac{W}{L} I_D} \Rightarrow \frac{g_{m2}}{g_{m1}} = \sqrt{\frac{I_{D2}}{I_{D1}}} = \sqrt{4} = 2$

$g_m = 2 \times 5 = 10\,mA/V$

$R_{in} = \frac{1}{g_m} = 0.1\,k\Omega = 100\Omega$

$$G_v = g_m \frac{R_D \| R_L}{1 + g_m R_{sig}} = 10 \frac{(5^K \| 2^K)}{1 + 10 \times 0.2} = 4.76\,V/V$$

4.82

Refer to Eq. 4.90 : $G_v = \frac{R_G}{R_G + R_{sig}} \frac{g_m (R_D \| R_L)}{1 + g_m R_s}$

$G_{v1} = -16\ V/V$ reduced by factor of 4:

$\frac{G_{v2}}{G_{v1}} = \frac{1}{1 + g_m R_s} \Rightarrow \frac{1}{4} = \frac{1}{1 + 2R_s} \Rightarrow R_s = 1.5K\Omega$

4.83

Eq. 4.90 : $G_v = \frac{R_G}{R_G + R_{sig}} \frac{g_m (R_D \| R_L)}{1 + g_m R_s}$

$G_v = -10\ V/V$ for $R_s = 1K\Omega$ and $G_v = -20\ V/V$ for $R_s = 0$

$\frac{G_{v1}}{G_{v2}} = \frac{1 + g_m R_{s2}}{1 + g_m R_{s1}} \Rightarrow \frac{10}{20} = \frac{1 + g_m \times 0}{1 + g_m \times 1} \Rightarrow 1 + g_m = 2$

$g_m = 1\ mA/V$

In order to have $G_v = -8\ V/V$:

$\frac{G_{v1}}{G_{v3}} = \frac{20}{8} = \frac{1 + g_m R_{s3}}{1 + 0} \Rightarrow 1 + R_{s3} = 2.5 \Rightarrow R_{s3} = 3.5^{K\Omega}$

4.84

$A_{vo} = 0.98\ V/V$ and $A_v = \frac{0.98}{2} = 0.49\ V/V$ for $R_L = 500^{\Omega}$

Eq. 4.102a: $A_v \simeq \frac{R_L}{R_L + \frac{1}{g_m}} \Rightarrow 0.49 = \frac{0.5}{0.5 + \frac{1}{g_m}} \Rightarrow$

$g_m = 1.92\ mA/V$

Eq. 4.103 : $A_{vo} = \frac{r_o}{r_o + \frac{1}{g_m}} \Rightarrow 0.98 = \frac{r_o}{r_o + 0.52}$

$\Rightarrow r_o = 25.5\ K\Omega$

4.85

we have $g_m = 5\ mA/V$ and $r_o = 20\ K\Omega$. From Eq. 4.103 we know: $A_{vo} = \frac{r_o}{r_o + \frac{1}{g_m}} = \frac{20}{20 + \frac{1}{5}} = 0.99\ V/V$

$A_{vo} = 0.99\ V/V$

From Eq. 4.105: $R_{out} = \frac{1}{g_m} \| r_o = \frac{1}{5}^K \| 20^K = 198\Omega$

$R_{out} \simeq 200\Omega$

with $R_L = 1K\Omega$ we have:

$A_v = \frac{R_L \| r_o}{(R_L \| r_o) + \frac{1}{g_m}} = \frac{1^K \| 20^K}{(1^K \| 20^K) + \frac{1}{5}}$

$\Rightarrow A_v = 0.83\ V/V$

4.86

$R_{i2} = \frac{1}{g_{m2}} = 50\Omega \Rightarrow g_{m2} = \frac{1}{0.05} = 20\ mA/V$

if Q_1 is biased the same as Q_2, then $g_{m1} = g_{m2}$

$i_{D1} = g_{m1} v_i = 20 \times 5^{mV} = 100\ \mu A = 0.1\ mA$

$v_{D1} = i_{D1} \times 50^{\Omega} = 0.5^V$

In order to have $v_{d2} = v_o = 1^V$:

$v_o = i_D R_D \Rightarrow 1 = 0.1 \times R_D \Rightarrow R_D = 10K\Omega$

4.87

a) $I_D = 0.1 = \frac{1}{2} \times 0.8 \times V_{ov}^2 \Rightarrow V_{ov} = 0.5^V$

$\Rightarrow V_{GS} = 0.5 + 1 = 1.5^V$

$V_G = 0 \Rightarrow V_S = -1.5V$

$R_S = \frac{-1.5 - (-5)}{0.1} = 35K\Omega$

$V_{DS} = 5 - R_D \times 0.1$

Largest possible R_D is achieved for $V_{DS\,min}$

$V_{DS} \geq V_{GS} - V_t \Rightarrow V_{DS\,min} = V_{ov} \Rightarrow V_{DS} - 1 = V_{ov}$

$\Rightarrow V_{DS} = 1 + 0.5 = 1.5V \Rightarrow R_D = \frac{5 - 1.5}{0.1} = 35K\Omega$

$R_G = 10M\Omega$.

5, R_D, ∞, Y, ∞, X, ∞, Z, R_G, R_S, -5

b) $g_m = \frac{2 I_D}{V_{ov}} = \frac{2 \times 0.1}{0.5} = 0.4\ mA/V$

$r_o = \frac{V_A}{I_D} = \frac{40}{0.1} = 400K\Omega$

c) IF Z is grounded then the circuit becomes a common-source configuration. The voltage gain according to Eq. 4.82:

$G_v = -\frac{R_G}{R_G + R_{sig}} g_m (r_o \| R_D \| R_L)$

$G_v = \frac{10M}{10^M + 1^M} \times 0.4 \times (400^K \| 35^K \| 40^K) = 6.5\ V/V$

$G_v = 6.5\ V/V$

d) IF Y is grounded, then the circuit becomes a source follower configuration.

Eq. 4.103 : $A_{vo} = \frac{r_o}{r_o + \frac{1}{g_m}} = \frac{400}{400 + \frac{1}{0.4}} = 0.99\ V/V$

$R_{out} = \frac{1}{g_m} \| r_o = \frac{1}{0.4} \| 400$

$R_{out} = 2.48K\Omega$

Cont.

e) If x is grounded, the circuit becomes a common-gate configuration.

$R_{in} = \frac{1}{g_m} \| R_S = 35K \| \frac{1}{0.4} = 2.33K\Omega$

Eq. 4.98: $i_i = i_{sig} \frac{R_{sig}}{R_{sig}+R_{in}}$ ⇒

$$i_i = 10\mu A \frac{100^K}{100^K+2.33^K} = 9.77\mu A$$

$v_y = R_D \times i_i = 35 \times 9.77\mu A = 0.34V$

4.88

a) Fig. P4.88a is a source follower:

Eq. 4.103: $A_{v_o} = \frac{r_o}{r_o+\frac{1}{g_m}}$, $r_o \gg \frac{1}{g_m} \Rightarrow A_{v_o} \simeq 1 V/V$

$R_{out} = \frac{1}{g_m} = \frac{1}{5} = 0.2K\Omega$

b) Fig. P4.88b is a common-gate configuration:

$R_{in} = \frac{1}{g_m} = \frac{1}{5} = 0.2K\Omega$

Eq. 4.94: $A_v = g_m (R_D \| R_L) = 5(5^K \| 2^K) = 7.1 V/V$

c) If we connect both stages together, then:

for the first stage: $A_v = A_{v_o} \frac{R_L}{R_L+R_{out}}$

where R_L is in fact R_{in} of the second stage

Therefore: $A_{v_1} = 1 \times \frac{0.2K}{0.2+0.2} = 0.5 V/V$

For the second stage: $A_{v_2} = 7.1 V/V$

Overall gain $A_v = A_{v_1} A_{v_2} = 7.1 \times 0.5 = 3.55 V/V$

4.89

4.90

$C_{ox} = \frac{\varepsilon_{ox}}{t_{ox}} = \frac{3.45\times10^{-11}}{8\times10^{-9}} = 4.3\times10^{-3}\,F/m^2 = 4.3\ fF/\mu m^2$

$k'_n = \mu_n C_{ox} = 450\times10^{-4}\times4.3\times10^{-3} = 193.5\,\mu A/V^2$

$I_D = 100\,\mu A = \frac{1}{2}\times193.5\times\frac{20}{1}V_{ov}^2 \Rightarrow V_{ov} = 0.23\,V$

$V_{DS} = 1.5^V > V_{ov} \Rightarrow$ Saturation

$g_m = \frac{2I_D}{V_{ov}} = 880\,\mu A/V$, $r_o = \frac{1}{\lambda I_D} = \frac{1}{0.05\times0.1} = 200^{k\Omega}$

$\chi = \frac{\gamma}{2\sqrt{2\phi_f+V_{SB}}} = \frac{0.5}{2\sqrt{0.65+1}} = 0.19$

$g_{mb} = \chi g_m = 167.2\,\mu A/V$

$C_{ov} = W L_{ov} C_{ox} = 20\times0.05\times4.3 = 4.3\,fF$

$C_{gs} = \frac{2}{3}WLC_{ox} + C_{ov} = \frac{2}{3}\times20\times1\times4.3 + 4.3 = 61.6\,fF$

$C_{gd} = C_{ov} = 4.3\,fF$

$C_{sb} = \frac{C_{sb0}}{\sqrt{1+\frac{V_{SB}}{V_0}}} = \frac{15}{\sqrt{1+\frac{1}{0.7}}} = 9.6\,fF$

$C_{db} = \frac{C_{db0}}{\sqrt{1+\frac{V_{DB}}{V_0}}} = \frac{15}{\sqrt{1+\frac{(1+1.5)}{0.7}}} = 7\,fF$

4.91

$g_m = \frac{2I_D}{V_{ov}} = \frac{2\times0.1}{0.25} = 0.8\,mA/V$

$f_T = \frac{g_m}{2\pi(C_{gs}+C_{gd})} = \frac{0.8\times10^{-3}}{2\pi(20+5)\times10^{-15}} = 5.1\,GHz$

4.92

$f_T = \frac{g_m}{2\pi(C_{gs}+C_{gd})}$, $g_m = \sqrt{2\mu_n C_{ox}\frac{W}{L}I_D}$

Also $C_{gs} \simeq \frac{2}{3}WLC_{ox}$, if $C_{gs} \gg C_{gd}$ then we can ignore C_{gd}. If we replace for g_m and C_{gs} in the f_T formula, we have:

$f_T = \frac{\sqrt{2\mu_n C_{ox} W/L\, I_D}}{2\pi\times\frac{2}{3}WLC_{ox}} = \frac{1.5}{\pi L}\sqrt{\frac{\mu_n I_D}{2C_{ox}WL}}$

Therefore we can see that the higher the current I_D, then the higher is the f_T. Also the frequency is reverse proportional to the size of the device, i.e. higher frequencies are achievable for smaller devices.

4.93

$f_T = \frac{g_m}{2\pi(C_{gs}+C_{gd})}$ ①

For $C_{gs} \gg C_{gd}$ and the overlap capacitance of C_{gs} negligibly small: $C_{gs} \simeq \frac{2}{3}WLC_{ox}$

Also $g_m = \frac{2I_D}{V_{ov}} = k'_n\frac{W}{L}V_{ov}$

If we substitude g_m and C_{gs} in ① from the above formulas: $f_T = k'_n\frac{W}{L}V_{ov}\frac{1}{2\pi\times\frac{2}{3}WLC_{ox}}$

$\Rightarrow f_T = \frac{3\mu_n V_{ov}}{4\pi L^2}$

Therefore, for a given device f_T is proportional to V_{ov}. $f_T \propto V_{ov}$

For $L = 1\,\mu m$, $V_{ov} = 0.25$:

$f_T = \frac{3\times450\times10^{-4}\times0.25}{4\times\pi\times1\times10^{-12}} = 2.7\,GHz$

For $V_{ov} = 0.5V$: $\frac{f_{T1}}{f_{T2}} = \frac{V_{ov1}}{V_{ov2}} \Rightarrow f_{T_2} = 2.7\times\frac{0.5}{0.25}$

$f_T = 5.4\,GHz$

4.94

$A_M = -27\,V/V$, $C_{gs} = 0.3\,pF$, $C_{gd} = 0.1\,pF$

Common-Source configuration.

Eq. 4.127: $C_{in} = C_{gs} + C_{gd}(1+g_m R'_L)$.

Also $A_M = -g_m R'_L$, therefore:

$C_{in} = 0.3 + 0.1\times(1+27) = 3.1\,pF$

Now to find the range of R_{sig} that results in 3-db frequencies over 10MHZ, we use eq. 4.132: $f_H = \frac{1}{2\pi C_{in}R'_{sig}}$

If we neglect R_G effect then $R'_{sig} \simeq R_{sig}$.

$f_H \geq 10^{MHZ} \Rightarrow \frac{1}{2\pi\times3.1\times10^{-12}\times R_{sig}} \geq 10^{MHZ}$

$\Rightarrow R_{sig} \leq 5.1^{k\Omega}$

4.95

$R_{sig} = 100^{k\Omega}$, $R_{in} = 100k\Omega$, $C_{gs} = 1pF$, $C_{gd} = 0.2pF$

Cont.

$$A_M = -\frac{R_G}{R_G + R_{sig}} g_m (r_o \| R_D \| R_L) \quad \text{(Eq. 4.119)}$$

Also $R_{in} = 100K\Omega = R_G$

$$A_M = \frac{-100}{100+100} 3(50^K \| 8K \| 10^K) = -6.1 V/V$$

$$f_H = \frac{1}{2\pi C_{in} R'_{sig}} \quad \text{(Eq. 4.132)}$$

$R'_{sig} = R_{sig} \| R_G = 100 \| 100 = 50K\Omega$

$C_{in} = C_{gs} + C_{gd}(1 + g_m R'_L)$

$R'_L = r_o \| R_D \| R_L = 4.1K\Omega$

$C_{in} = 1 + 0.2(1 + 3\times 4.1) = 3.66pF$

Now we can calculate f_H:

$$f_H = \frac{1}{2\pi \times 3.66\times 10^{-12} \times 50\times 10^3} = 870^{KHZ}$$

In order to double f_H, we have to either decrease C_{in} (by reducing R_{out}) or reduce R'_{sig} by reducing R_{in}.

If we reduce $R_{out} = R_D \| r_o$:

$$\frac{f_{H2}}{f_{H1}} = \frac{C_{in1}}{C_{in2}} \Rightarrow 2 = \frac{3.66\,PF}{1 + 0.2(1 + 3\times R'_L)}$$

$\Rightarrow R'_L = 1.27K\Omega \qquad R'_L = R_{out} \| R_L = R_{out} \| 10^k$

$\Rightarrow R_{out} = 1.45K\Omega$

Therefore in order to double f_H to $870\times 2 = 1.74^{MHZ}$, we have to reduce $R_{out} = r_o \| R_D$ to $1.45^{K\Omega}$ or equivalently reducing R_D to $1.5K\Omega$. The new midband gain would be:

$$\frac{A_{M2}}{A_{M1}} = \frac{R'_{L2}}{R'_{L1}} \Rightarrow A_{M2} = -6.1\times\frac{1.27}{4.1} = -1.9 V/V$$

Gain is almost reduced by a factor of 3.

If we reduce $R_{in} = R_G$:

$$\frac{f_{H2}}{f_{H1}} = \frac{R'_{sig1}}{R'_{sig2}} \Rightarrow 2 = \frac{50^K}{R'_{sig2}} \Rightarrow R'_{sig2} = 25^{K\Omega}$$

$\Rightarrow 25K\Omega = 100K \| R_G \Rightarrow R_G = 33^{K\Omega} = R_{in}$

Therefore in order to double f_H, R_{in} is reduced by a factor of 3, from $100K\Omega$ to $33K\Omega$.

The new midband gain would be:

$$\frac{A_{M2}}{A_{M1}} = \frac{R_{G2}}{R_{G1}} \cdot \frac{R_{G1} + R_{sig}}{R_{G2} + R_{sig}} \Rightarrow A_{M2} = -6.1\times\frac{1}{3}\times\frac{100+100}{33+100}$$

$A_{M2} = -3.06\,V/V$

Gain is almost reduced by a factor of 2.

4.96

$R_{in} = 2M\Omega$, $g_m = 4mA/V$, $r_o = 100K\Omega$, $R_D = 10K\Omega$
$C_{gs} = 2PF$, $C_{gd} = 0.5pF$, $R_{sig} = 500K\Omega$, $R_L = 10K\Omega$

a) using Eq. 4.119 and noting that $R_G = R_{in}$, we have:

$$A_M = -\frac{R_G}{R_G + R_{sig}} g_m (r_o \| R_D \| R_L) = -\frac{2\times 4}{2+0.5}(100^K \| 10^K \| 10^K)$$

$A_M = -15.2\,V/V$

b) Eq. 4.132: $f_H = \dfrac{1}{2\pi C_{in} R'_{sig}}$ where

$C_{in} = C_{gs} + C_{gd}(1 + g_m(r_o \| R_D \| R_L))$, $R'_{sig} = R_{sig} \| R_G$

$C_{in} = 2 + 0.5(1 + 4\times(100^K \| 10^K \| 10^K)) = 12.02\,pF$

$R'_{sig} = 0.5^M \| 2^{M\Omega} = 0.4M\Omega = 400K\Omega$

$$f_H = \frac{1}{2\pi \times 12.02\times 10^{-12} \times 400\times 10^3} = 33.1^{KHZ}$$

4.97

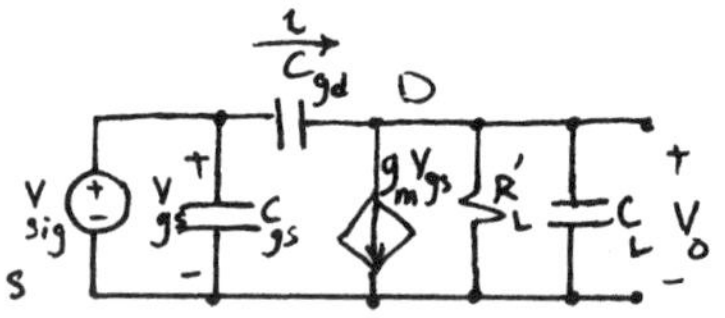

If we write KCL at node D:

$$i = g_m V_{gs} + \frac{V_o}{R'_L} + V_o C_L s$$

then: $V_{sig} = i \times \dfrac{1}{C_{gd}s} + v_o$

$$V_{sig} = \left(g_m V_{gs} + \frac{V_o}{R'_L} + V_o C_L s\right)\frac{1}{C_{gd}s} + v_o, \quad V_{sig} = V_{gs}$$

$$v_{sig}\left(1 - \frac{g_m}{C_{gd}s}\right) = v_o\left(1 + \frac{1}{R'_L C_{gd} s} + \frac{C_L s}{C_{gd} s}\right)$$

$$\frac{v_o}{v_{sig}} = -g_m R'_L \frac{(1 - s(C_{gd}/g_m))}{R'_L C_{gd} s + 1 + R'_L C_L s}$$

$$\frac{v_o}{v_{sig}} = -g_m R'_L \frac{1 - s\, C_{gd}/g_m}{1 + s(C_L + C_{gd})R'_L}$$

$$\text{If } (g_m/C_{gd}) \gg \omega \Rightarrow \frac{v_o}{v_{sig}} = \frac{-g_m R'_L}{1 + s(C_L + C_{gd})R'_L}$$

For $C_{gd} = 0.5\,pF$, $C_L = 2pF$, $g_m = 4mA/V$, $R'_L = 5K\Omega$

$$\frac{V_o}{V_{sig}} = \frac{A_M}{1 + s/\omega_H} \Rightarrow \begin{cases} A_M = -g_m R'_L = -4\times 5 = -20 V/V \\ f_H = \dfrac{1}{2\pi(C_L + C_{gd})R'_L} \end{cases}$$

$$\Rightarrow f_H = \frac{10^{12}}{2\pi \times (2+0.5)\times 5\times 10^3}$$

$f_H = 12.7MHZ$

$g_m/C_{gd} = \dfrac{4}{0.5} = 8^{GRad/s} \gg \omega_H$

4.98

Eq. 4.134 $\omega_{P1} = \frac{1}{C_{C1}(R_G + R_{sig})}$

For $f_P = 10\,Hz$:

$$2\pi \times 10 = \frac{1}{C_{C1}(1+1)\times 10^6} \Rightarrow C_{C1} = 7.96\,nF$$

To ensure that f_P is not exceeding 10Hz, C_{C1} has to be greater than 7.96nF or: $C_{C1} = 8\,nF$. If $C_{C1} = 8\,nF \Rightarrow f_P = 9.95\,Hz$

To lower f_P, R_G has to be increased. For the lowest possible f_P, the largest available R_G has to be used, which in this case is $10 \times 1 = 10\,M\Omega$

To calculate the new f_P for $R_G = 10^{M\Omega}$:

$$\frac{f_{P2}}{f_{P1}} = \frac{R_{G1} + R_{sig}}{R_{G2} + R_{sig}} \Rightarrow f_{P2} = 9.95\,\frac{1+1}{10+1} = 1.81^{HZ}$$

f_P is reduced by a factor of 5.5.

4.99

$I_D = 1\,mA$, $g_m = 1\,mA/V$

using eq. 4.89 we have:

$$A_M = \frac{-g_m R_D}{1 + g_m R_S} = -\frac{1\times 10}{1+1\times 6}$$

$A_M = 1.43\,V/V$

$$f_L = \frac{1}{2\pi(\frac{1}{g_m} \| R_S)C_S} = 10\,Hz$$

$$C_S = \frac{1}{2\pi \times 10(1^K \| 6^K)} = 18.57\,\mu F$$

V_{DD}, $R_D = 10\,k\Omega$, V_o, C_S, V_i, $R_S = 6\,k\Omega$

4.100

$$f_{C_{C2}} = \frac{1}{2\pi C_{C2}(R_L + R_D \| r_o)} \leq 10^{HZ}$$

$$\Rightarrow C_{C2} \geq \frac{1}{10 \times 2\pi \times (10^K + 15^K \| 150^K)} \Rightarrow C_{C2} \geq 0.67\,\mu F$$

$$\Rightarrow C_{C2} = 0.7\,\mu F \Rightarrow f_{C_{C2}} = 9.62\,Hz$$

If I_D is doubled with both r_o and R_D halved:

$$f_{C_{C2}} = \frac{1}{2\pi \times 0.7^{\mu}(10^K + \frac{15^K}{2} \| \frac{150^K}{2})} = 13.5^{HZ}$$

For higher-power designs, where I_D is increased and consequently r_o and R_D are reduced. For smallest r_o and R_D where $r_o \| R_D \ll R_L$, R_L becomes dominant in determining the corner frequency:

$$f_{C_{C2}max} = \frac{1}{2\pi C_{C2}(R_L)} = 22.75\,Hz$$

4.101

Refer to Fig. P4.101: $g_m = 1\,mA/V$

$$A_M = -\frac{R_G}{R_G + R_{sig}}\, g_m (R_D \| R_L) \quad \text{where } R_G = 10^M \| 47^M$$

$R_G = 8.25^{M\Omega}$

$$A_M = -\frac{8.25}{8.25+0.1} \times 1 \times (4.7^K \| 10^K) = -3.16\,V/V$$

$$f_{P1} = \frac{1}{2\pi C_{C1}(R_G + R_{sig})} \quad \text{(Eq. 4.134)}$$

$$f_{P1} = \frac{1}{2\pi \times 0.01\times 10^{-6} \times (8.25+0.1)\times 10^6} = 1.9\,Hz$$

$$f_{P2} = \frac{1}{2\pi C_S (R_S \| \frac{1}{g_m})} = \frac{1}{2\pi \times 10 \times 10^{-6} \times (2^K \| \frac{1}{1}^K)} = 23.9^{HZ}$$

$$f_{P3} = \frac{1}{2\pi C_{C2}(R_D + R_L)} = \frac{1}{2\pi \times 0.1\times 10^{-6} \times (4.7+10)\times 10^3} = 108.3\,Hz$$

$f_L \simeq 108.3\,Hz$

4.102

If $g_m = 1\,mA/V$ and $r_o = 100\,k\Omega$:

$$A_M = -\frac{R_G}{R_G + R_{sig}}\, g_m (r_o \| R_D \| R_L) \quad \text{where } R_G = 10^M \| 47^M$$

$R_G = 8.25^{M\Omega}$

$$A_M = -\frac{8.25}{8.25+0.1}\, 1\,(100^K \| 4.7^K \| 10^K) = -3.06\,V/V$$

$$f_H = \frac{1}{2\pi C_{in} R'_{sig}} \quad \text{where } R'_{sig} = R_{sig} \| R_G = 0.1^M \| 8.25^{M\Omega}$$

$R'_{sig} = 0.1\,M\Omega$

$$C_{in} = C_{gs} + C_{gd}(1 + g_m (r_o \| R_D \| R_L))$$

$$C_{in} = 1 + 0.2(1 + 1(100^K \| 4.7^K \| 10^K)) = 1.82\,pF$$

$$f_H = \frac{1}{2\pi \times 1.82\times 10^{-12} \times 0.1\times 10^6} = 875\,kHz$$

4.103

$A_M = -\frac{R_G}{R_G + R_{sig}} g_m (r_o \| R_D \| R_L) = -\frac{2\times3}{2+0.5}(20^K \| 10^K)$
$(r_o = \infty)$

$A_M = -16 V/V$

Using equations 4.134, 4.136 and 4.138 we have:

$f_{P1} = \frac{1}{2\pi C_{C1}(R_G + R_{sig})} \Rightarrow 3^{Hz} = \frac{1}{2\pi \times C_{C1}(2+0.5)\times10^6}$

$\Rightarrow C_{C1} = 21.2\,nF$

$f_{P2} = \frac{g_m}{2\pi C_S} \Rightarrow 50^{Hz} = \frac{3\times10^{-3}}{2\pi \times C_S} \Rightarrow C_S = 9.6\mu F$

$f_{P3} = \frac{1}{2\pi C_{C2}(R_D + R_L)} \Rightarrow 10^{Hz} = \frac{1}{2\pi C_{C2}(10^K + 20^K)}$

$\Rightarrow C_{C2} = 0.5\mu F$

$f_L = 50^{Hz}$

4.104

Refer to Fig. P4.104.

$f_{P_1} = \frac{1}{2\pi C_{C_1} R_{in}} = 1^{Hz}$ (a decade lower than 10^{Hz})

$f_{P2} = \frac{1}{2\pi C_{C2}(R_L + R_D \| R_O)} = 10^{Hz}$

$f_{P1} = 1^{Hz} = \frac{1}{2\pi C_{C1} \times 2.33\times10^6} \Rightarrow C_{C1} = 68.3\,nF$

$f_{P2} = 10^{Hz} = \frac{1}{2\pi C_{C2}(10^K + 10^K \| 47^K)} \Rightarrow C_{C2} = 873\,nF$

4.105

Note that $k_n' \frac{W_n}{L} = k_p' \frac{W_p}{L}$ (matched)

a) when V_o is low: (eq. 4.140)

$r_{DSN} = \frac{1}{k_n'(\frac{W}{L})_n (V_{DD} - V_{tn})} = \frac{1}{120\times10^{-6}\times\frac{1.2}{0.8}\times(3-0.7)}$

$r_{DSN} = 2.4\,K\Omega$

when V_o is high (V_{OH}): (eq. 4.141)

$r_{DSP} = \frac{1}{k_p'(\frac{W}{L})_p (V_{DD} - |V_{tp}|)} = \frac{1}{60\times10^{-6}\times\frac{2.4}{0.8}\times(3-0.7)}$

$r_{DSP} = 2.4\,K\Omega$

b) $I_{max} = k_n'(\frac{W}{L})_n (V_{DD} - V_{tn}) \times 0.1$

$I_{max} = 120\times10^{-6} \times \frac{1.2}{0.8} \times (3-0.7)\times 0.1 = 41.4\mu A$

c) Eq. 4.148: $V_{IH} = \frac{1}{8}(5V_{DD} - 2V_t)$

$V_{IH} = \frac{1}{8}(5\times3 - 2\times0.7) = 1.7V$

Eq. 4.149: $V_{IL} = \frac{1}{8}(3V_{DD} + 2V_t) = 1.3^V$

$N_{MH} = V_{OH} - V_{IH} = 3 - 1.7 = 1.3V$

$N_{ML} = V_{IL} - V_{OL} = 1.3^V - 0 = 1.3^V$

4.106

From the formula given in Excercise 4.44 we have: $V_{th} = \frac{r(V_{DD} - |V_{tp}|) + V_{tn}}{1+r}$

where $r = \sqrt{\frac{k_p'(\frac{W}{L})_p}{k_n'(\frac{W}{L})_n}}$

We have $V_{tn} = |V_{tp}| = 0.7V$, $V_{DD} = 3^V$, $k_n' = 120\mu A/V^2$

$k_p' = 60\mu A/V^2$. Thus: for $(\frac{W}{L})_p = (\frac{W}{L})_n$ we have:

$r = \sqrt{\frac{60}{120}} = 0.71$

$V_{th} = \frac{0.71(3-0.7)+0.7}{1+0.71} = 1.36^V$

For $(\frac{W}{L})_p = 2(\frac{W}{L})_n$ (the matched case)

we have: $r = \sqrt{\frac{60}{120}\times2} = 1$

$V_{th} = \frac{1\times(3-0.7)+0.7}{1+1} = 1.5^V$

For $(\frac{W}{L})_p = 4(\frac{W}{L})_n$ we have: $r = \sqrt{\frac{60\times4}{120}} = 1.41$

$V_{th} = \frac{1.41\times(3-0.7)+0.7}{1+1.41} = 1.64^V$

The results are summarized in the following table:

$(\frac{W}{L})_p = (\frac{W}{L})_n$	$V_{th} = 1.36^V$
$(\frac{W}{L})_p = 2(\frac{W}{L})_n$	$V_{th} = 1.5^V$
$(\frac{W}{L})_p = 4(\frac{W}{L})_n$	$V_{th} = 1.64^V$

4.107

equal sizes NMOS and PMOS, but $k'_n = 2k'_p$

$V_t = 0.7^V$

for V_{IH} : Q_N in triode and Q_P in saturation

$$k'_n\left(\frac{W}{L}\right)_n\left[(v_I - V_t)v_O - \frac{1}{2}v_O^2\right] = \frac{1}{2}k'_p\left(\frac{W}{L}\right)_p(V_{DD} - v_I - V_t)^2$$

$$4(v_I - V_t)v_O - 2v_O^2 = (V_{DD} - V_I - V_t)^2 \quad ①$$

Differentiating both sides relative to v_I results in:

$$\frac{\partial}{\partial v_i} : 4(v_I - V_t)\frac{\partial v_O}{\partial v_i} + 4v_O - 4v_O\frac{\partial v_O}{\partial v_i} = 2(V_{DD} - V_I - V_t)(-1)$$

substitude the values together with $V_I = V_{IH}$, $\frac{\partial v_O}{\partial v_i} = -1$:

$$4(V_{IH} - 0.7)(-1) + 4V_O + 4V_O = 2(V_{IH} - 3 + 0.7)$$

$$V_{IH} = \frac{8v_O + 7.4}{6} = 1.33v_O + 1.23 \quad ②$$

From ① : $4(V_{IH} - 0.7)v_O - 2v_O^2 = (3 - V_{IH} - 0.7)^2$

③ $4(V_{IH} - 0.7)v_O - 2v_O^2 = (2.3 - V_{IH})^2$

Solving ② and ③ : $1.55v_O^2 + 4.97v_O - 1.14 = 0$

$\Rightarrow v_O = 0.22V$

$V_{IH} = 1.52V$

For V_{IL} : Q_N is in saturation and Q_p in triode.

$$\frac{1}{2}k'_n\left(\frac{W}{L}\right)_n(v_I - V_t)^2 = k'_p\left(\frac{W}{L}\right)_p\left[(V_{DD} - V_I - V_t)(V_{DD} - v_O) - \frac{1}{2}(V_{DD} - V_O)^2\right]$$

$$(V_I - 0.7)^2 = (3 - V_I - 0.7)(3 - v_O) - \frac{1}{2}(3 - v_O)^2 \quad ①$$

$$(V_I - 0.7)^2 = (2.3 - V_I)(3 - V_O) - \frac{1}{2}(3 - v_O)^2$$

$$\frac{\partial}{\partial v_I} \Rightarrow 2(v_I - 0.7) = (2.3 - v_I)\left(-\frac{\partial v_O}{\partial v_I}\right) - (3 - V_O) + (3 - v_O)\frac{\partial v_O}{\partial v_I}$$

$v_I = V_{IL}$ and $\frac{\partial v_O}{\partial v_I} = -1$

$$2V_{IL} - 1.4 = 2.3 - V_{IL} - 3 + v_O - 3 + v_O$$

$$V_{IL} = \frac{2}{3}v_O - 1.15$$

From ① : $(V_{IL} - 0.7)^2 = (2.3 - V_{IL})(3 - v_O) - \frac{1}{2}(3 - v_O)^2$

$$(0.66v_O - 1.85)^2 = (3.45 - 0.66v_O)(3 - v_O) - \frac{1}{2}(3 - v_O)^2$$

$v_O = 2.96V$

$V_{IL} = 0.81V$

$NMH - 3 - 1.52 = 1.48^V$, $NM_L = 0.81 - 0 = 0.81^V$

4.108

Matched MOSFETs with $V_t = 1^V$, $V_{DD} = 10^V$.

$V_{IL} = \frac{1}{8}(3V_{DD} + 2V_t)$ Eq. 4.149

$V_{IL} = \frac{1}{8}(3\times10 + 2\times1) = 4^V$

$V_{IH} = \frac{1}{8}(5V_{DD} - 2V_t)$ Eq. 4.148

$V_{IH} = \frac{1}{8}(5\times10 - 2\times1) = 6^V$

$NMH = V_{OH} - V_{IH} = 10 - 6 = 4^V$

$NM_L = V_{IL} - V_{OL} = 4 - 0 = 4^V$

For $V_{DD} = 15^V$

$V_{IL} = \frac{1}{8}(3\times15 + 2\times1) = 5.875V$

$V_{IH} = \frac{1}{8}(5\times15 - 2\times1) = 9.125V$

$NM_H = 15 - 9.125 = 5.875V$

$NM_L = 5.875 - 0 = 5.875V$

4.109

$V_t = 0.5V$

$$I_{max} = k'_n\left(\frac{W}{L}\right)_n\left[(V_{DD} - V_{tn})\times0.5 - \frac{1}{2}\times0.5^2\right]$$

$$I_{max} = 20\times20\left[(10 - 0.5)\times0.5 - 0.125\right]$$

$I_{max} = 185\mu A$

$V_t = 1.5V \Rightarrow I_{max} = 165\mu A$

$V_t = 2^V \Rightarrow I_{max} = 155\mu A$

4.110

$$i_{Dn} = k'_n\left(\frac{W}{L}\right)_n\left[(V_{DD} - V_t)v_O - \frac{1}{2}v_O^2\right]$$

$$i_{Dnmax} = k'_n\left(\frac{W}{L}\right)_n\left[(V_{DD} - 0.2V_{DD})\times0.1V_{DD} - \frac{1}{2}\times0.01V_{DD}^2\right]$$

$$i_{Dnmax} = k'_n\left(\frac{W}{L}\right)_n \times 0.075V_{DD}^2.$$

IF $V_{DD} = 3V$, $k'_n = 120\mu A/V^2$, $L_n = 0.8\mu m$, $i_{Dmax} = 1mA$

$$i_{Dnmax} = 1 = 120\times10^{-3}\times\frac{W_n}{0.8}\times0.075\times9$$

$W_n = 9.9 \simeq 10\mu m$

4.111

we use the I_{peak} equation given in Table 4.6:

$I_{peak} = \frac{1}{2} k'_n \left(\frac{W}{L}\right)_n \left(\frac{V_{DD}}{2} - V_{tn}\right)^2$

$I_{peak} = \frac{1}{2} \times 120 \times \frac{1.2}{0.8} \left(\frac{3}{2} - 0.7\right)^2$

$I_{peak} = 57.6 \mu A$

4.112

Eq. 4.156 : $t_{PHL} = \frac{2C}{k'_n\left(\frac{W}{L}\right)_n (V_{DD}-V_t)}\left[\frac{V_t}{V_{DD}-V_t} + \frac{1}{2}\ln\left(\frac{3V_{DD}-4V_t}{V_{DD}}\right)\right]$

$\Rightarrow t_{PHL} = \frac{2 \times 0.05 \times 10^{-12}}{120 \times 10^{-6} \times \frac{1.2}{0.8} \times (3-0.7)}\left[\frac{0.7}{3-0.7} + \frac{1}{2}\ln\left(\frac{3\times 3 - 4\times 0.7}{3}\right)\right]$

$t_{PHL} = 0.16 ns$

If we use the estimation formula of Eq. 4.157:

$t_{PHL} = \frac{1.6C}{k'_n\left(\frac{W}{L}\right)_n V_{DD}} = \frac{1.6 \times 10^{-12} \times 0.05}{120 \times 10^{-6} \times \frac{1.2}{0.8} \times 3} = 0.15 ns$

The value obtained using the estimation formula is 6% lower. Eq. 4.157 is for $V_t \simeq 0.2 V_{DD}$, but in our case $V_t = 0.7 \simeq 0.23 V_{DD}$.

4.113

We use the estimation formulas for t_{PHL} and t_{PLH} provided in Table 4.6:

$t_{PLH} = \frac{1.6C}{k'_p\left(\frac{W}{L}\right)_p V_{DD}} = \frac{1.6 \times 0.05 \times 10^{-12}}{60 \times 10^{-6} \times \left(\frac{W}{L}\right)_p \times 3} = \frac{444}{\left(\frac{W}{L}\right)_p} \times 10^{-12} \leq 60^{ps}$

$\Rightarrow \frac{444}{\left(\frac{W}{L}\right)_p} \leq 60 \Rightarrow \left(\frac{W}{L}\right)_p \geq 7.41 \Rightarrow W_p \geq 5.93 \mu m$

$\left(\frac{W}{L}\right)_p = 2\left(\frac{W}{L}\right)_n \Rightarrow W_p = 2W_n \Rightarrow W_n \geq 2.97 \mu m$

We choose $W_p = 6 \mu m$ and $W_n = 3 \mu m$

4.114

At $V_I = V_O = \frac{V_{DD}}{2}$ both Q_N and Q_P are operating in saturation with I_D given by:

$I_D = \frac{1}{2} k'_n \left(\frac{W}{L}\right)_n \left(\frac{V_{DD}}{2} - V_t\right)^2$

Thus each device has g_m given by $g_m = \frac{2I_D}{V_{GS}-V_t}$ or $g_m = \frac{2I_D}{V_{DD}/2 - V_t}$ and r_o given by $r_o = \frac{|V_A|}{I_D}$ where we have assumed matched devices: $g_{mn} = g_{mp} = g_m$, $r_{op} = r_{on} = r_o$.

From the small signal equivalent circuit:

$A_v = \frac{v_o}{v_i}$

$A_v = -(g_{mn} + g_{mp})(r_{on} \| r_{op})$

$A_v = -2g_m \frac{r_o}{2} = -g_m r_o$

Therefore:

$A_v = -\frac{2|V_A|}{\frac{V_{DD}}{2} - V_t}$

b) $V_I = V_O = \frac{V_{DD}}{2} = 1.5^V$

$A_v = -\frac{2 \times 50}{\frac{3}{2} - 0.7} = -\frac{100}{0.8} = -125 V/V$

$R_{in} = \frac{v_i}{i_i} = \frac{v_i}{\frac{v_i - v_o}{R_G}} = \frac{R_G}{1 - A_v}$

$R_{in} = \frac{10 \times 10^3}{1 + 125} = 79.4 k\Omega$

$R_{in} \simeq 80 k\Omega$

4.115

we have $k'_n \frac{W}{L} = 2 mA/V^2$, $V_t = -3V$

$V_{GS} = 0 > V_t \Rightarrow$ device is on

$V_{GS} - V_t = 3^V$

a) $V_D = 0.1V$ $V_{DS} = 0.1 < V_{GS} - V_t \Rightarrow$ triode region

$i_D = k'_n \frac{W}{L}\left[(V_{GS}-V_t)V_{DS} - \frac{1}{2}V_{DS}^2\right]$

$i_D = 2 \times \left[3 \times 0.1 - \frac{1}{2} 0.1^2\right] = 0.59 mA$

b) $V_D = 1^V \Rightarrow V_{DS} = 1^V < V_{GS} - V_t \Rightarrow$ triode region

$i_D = 2 \times \left[3 \times 1 - \frac{1}{2} \times 1\right] = 5 mA$

c) $V_D = 3^V \Rightarrow V_{DS} = 3^V = V_{GS} - V_t \Rightarrow$ edge of saturation

$i_D = \frac{1}{2} k'_n \frac{W}{L} (V_{GS} - V_t)^2 = \frac{1}{2} \times 2 \times 3^2 = 9 mA$

d) $V_D = 5^V$, $V_{DS} = 5^V > V_{GS} - V_t \Rightarrow$ saturation

$i_D = \frac{1}{2} \times 2 \times 3^2 = 9 mA$

4.116

$V_{GS}=0 \quad V_t=-2V \quad V_{GS}-V_t=2V$

* $V_{DS}=1V$: $V_{DS}<V_{GS}-V_t \Rightarrow$ triode

$i_D=K_n'\frac{W}{L}\left[(V_{GS}-V_t)V_{DS}-\frac{1}{2}V_{DS}^2\right]=200\left[2\times1-\frac{1}{2}\times1\right]$

$i_D=300\mu A$

If W is doubled, with L the same: $\frac{i_{D2}}{i_{D1}}=\frac{W_2}{W_1}=2$

$\Rightarrow i_{D2}=2\times300=600\mu A$

If W is doubled and L is also doubled, then $\frac{W}{L}$ remains the same and therefore $i_D=300\mu A$

* If $V_{DS}=2V$, then $V_{DS}=V_{GS}-V_t \Rightarrow$ edge of saturation

$i_D=\frac{1}{2}K_n'\frac{W}{L}(V_{GS}-V_t)^2(1+\lambda V_{DS})=\frac{1}{2}\times200\times2^2\times(1+0.02\times2)$

$i_D=416\mu A$

If W is doubled and L is the same, then:

$i_D=2\times416=832\mu A$

If W is doubled and L is also doubled, then $\frac{W}{L}$ stays the same but $V_A=V_A'L$ is doubled or equivalently λ is halved, thus:

$i_D=\frac{1}{2}\times200\times2^2\times(1+0.01\times2)=408\mu A$

* If $V_{DS}=3V$, then $V_{DS}>V_{GS}-V_t \Rightarrow$ saturation

$i_D=\frac{1}{2}\times200\times2^2\times(1+0.02\times3)=424\mu A$

if W is doubled and L is the same: $i_D=2\times424 \Rightarrow$

$i_D=848\mu A$

if W is doubled and L is also doubled, then $\frac{W}{L}$ stays unchanged, V_A is doubled, λ is halved:

$\frac{i_{D2}}{i_{D1}}=\frac{(1+\lambda_2V_{DS})}{(1+\lambda_1V_{DS})} \Rightarrow i_{D2}=424\times\frac{(1+0.01\times3)}{(1+0.02\times3)}=412\mu A$

* if $V_{DS}=10V$, then $V_{DS}>V_{GS}-V_t \Rightarrow$ saturation

$i_D=\frac{1}{2}\times200\times2^2\times(1+0.02\times10)=480\mu A$

if W is doubled and L is the same: $i_D=960\mu A$

if W is doubled and L is doubled: $i_D=480\,\frac{(1+0.01\times10)}{1+0.02\times10}$

$i_D=440\mu A$

4.117

$V_{GS}=V_{DS}=v$. for $v\geq V_t$ the device will be conducting and since $V_{DS}=v<v-V_t$, the device will be operating in the triode region and thus

$i=K_n'\frac{W}{L}\left[(v-V_t)v-\frac{1}{2}v^2\right]$

$i=\frac{1}{2}K_n'\frac{W}{L}(v^2-2V_tv) \quad v\geq V_t$

For $v\leq V_t$, the source and drain exchange roles and the MOSFET operates with $v_{GS}=0$ and $v_{DS}=-v\geq -V_t$ which implies saturation region operation with $i_D=\frac{1}{2}K_n'\frac{W}{L}V_t^2$ (for $v\leq V_t$)

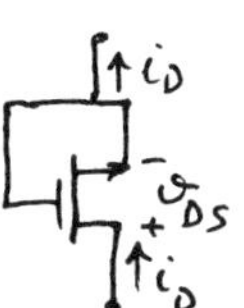

For $V_t=-2V$ and $K_n'\frac{W}{L}=2\,mA/V^2$

$i=v^2+4v \quad$ for $v\geq-2V$

$i=-4\,mA \quad$ for $v\leq-2V$

4.118

$K_n'\frac{W}{L}=4\,mA/V^2$, $V_t=-2V$

Since $V_{DG}\leq|V_t|$, the MOSFET will be operating in the triode region:

$i_D=K_n'\frac{W}{L}\left[(V_{GS}-V_t)V_{DS}-\frac{1}{2}V_{DS}^2\right]$

$2=4\left[(-V_S+2)(1-V_S)-\frac{1}{2}(1-V_S)^2\right]$

$2=+4V_S^2+8-12V_S-2V_S^2-2+4V_S$

$2V_S^2-8V_S+4=0 \Rightarrow V_S^2-4V_S+2=0$

$V_S=3.4V$ or $0.59V$

The first answer is not physically meaningful, for it results in $V_{GS}=-3.4<V_t$ and that implies cutoff. Therefore: $V_S=0.59V$

4.119

$i_D=\frac{1}{2}K_n'\frac{W}{L}\left[V_{GS}-V_t\right]^2$, $\lambda=0$

① $1=\frac{K_n'}{2}\frac{W}{L}(-1-V_t)^2$

$9=\frac{K_n'}{2}\frac{W}{L}(1-V_t)^2$

$\Rightarrow 9=\frac{(1-V_t)^2}{(1+V_t)^2}\Rightarrow \pm3=\frac{1-V_t}{1+V_t}$

$\Rightarrow V_t=-0.5V, -2V$

Cont.

$V_t = -0.5V$ is not acceptable, for it results in $V_{GS} = -1V < V_t$. Therefore $V_t = -2V$.

If we replace this in ①:

$1 = \frac{1}{2}k_n'\frac{W}{L}(-1+2)^2 \Rightarrow k_n'\frac{W}{L} = 2\,mA/V^2$

$I_{DSS} = \frac{1}{2}k_n'\frac{W}{L}V_t^2 = \frac{1}{2}\times 2\times 2^2 = 4\,mA$

4.120

$I_{DSS} = \frac{1}{2}k_n'\frac{W}{L}V_t^2 \Rightarrow 4 = \frac{1}{2}k_n'\frac{W}{L}\times 2^2 \Rightarrow k_n'\frac{W}{L} = 2\,mA/V^2$

Refer to Fig. P4.120

Consider Q_1: $V_{GS1} = -I_D R_1$

$I_{D1} = \frac{1}{2}k_n'\frac{W}{L}(V_{GS1}-V_t)^2 \Rightarrow 1 = \frac{1}{2}\times 2\times(V_{GS}+2)^2 \Rightarrow$

$V_{GS} = -1V$

$R_1 = \frac{1V}{1mA} = 1k\Omega$, $R_1 = R_2 = 1k\Omega$

$V_E = 6V \Rightarrow R_3 = \frac{10-6}{1mA} = 4k\Omega$

$V_C = V_A - V_{GS2} - I_{D2}R_2 = 0 - V_{GS1} - I_{D2}R_2 = 0-(-1)-1\times 1$

$V_C = 0V$ (for $V_A = 0$)

Now if $V_A = \pm 1V$: $v_C = V_A - V_{GS1} - R_2 I_2 = V_A$

$\Rightarrow V_C = \pm 1V$

Since $v_C = v_A$, the source follower has zero offset. Q_2 enters the triode region when:

$V_E = V_A + |V_t| \Rightarrow V_A = 6-2 = 4V$

Q_1 enters the triode region when $v_C = -10+2$

$v_C = -8V$ which corresponds to $V_A = -8V$

4.121

a) $v_{sig} = v_i$

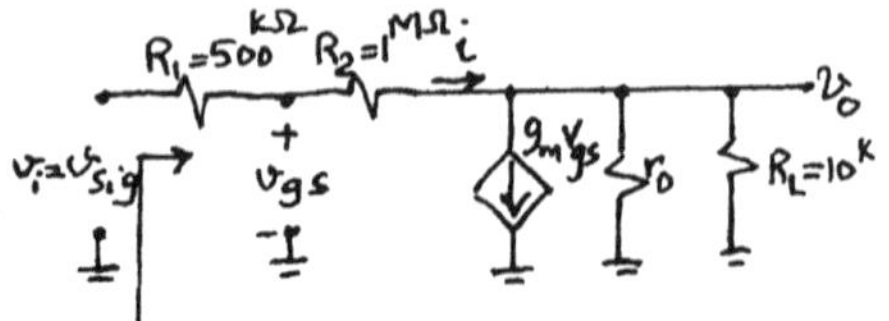

If we write KCL at the output node:

$$\frac{v_i - v_o}{R_1+R_2} = g_m v_{gs} + \frac{v_o}{r_o} + \frac{v_o}{R_L} \quad ①$$

Also note that $v_{gs} = v_i + R_1\frac{v_o - v_i}{R_1+R_2}$

$v_{gs} = \frac{R_2}{R_1+R_2}v_i + \frac{R_1}{R_1+R_2}v_o$

By substituting for v_{gs} in ①:

$$\frac{v_i - v_o}{R_1+R_2} = g_m\frac{R_2}{R_1+R_2}v_i + \frac{g_m R_1}{R_1+R_2}v_o + \frac{v_o}{R_L} + \frac{v_o}{r_o}$$

$$v_i\left(\frac{1-g_mR_2}{R_1+R_2}\right) = v_o\left(\frac{1+g_mR_1}{R_1+R_2} + \frac{1}{r_o} + \frac{1}{R_L}\right)$$

$$\frac{v_o}{v_i} = \frac{1-g_mR_2}{1+g_mR_1+\frac{(R_1+R_2)}{r_o\|R_L}}$$

$g_m = 1\,mA/V$, $r_o = 100k\Omega$, $R_1 = 0.5M\Omega$, $R_2 = 1M\Omega$

$R_L = 10k\Omega$

$$\frac{v_o}{v_i} = \frac{1-1\times 1000}{1+1\times 500+\frac{1500}{10k\|100k}} = -1.5\,V/V$$

Now we calculate $R_{in} = \frac{v_i}{i_i}$:

$v_i = (R_1+R_2)i_i + (i_i - g_m v_{gs})(r_o\|R_L)$

$v_{gs} = v_i - R_1 i_i$

$v_i = (R_1+R_2)i_i + (i_i - g_m v_i + g_m R_1 i_i)(r_o\|R_L)$

$$\frac{v_i}{i_i} = R_{in} = \frac{R_1+R_2+(1+g_mR_1)(r_o\|R_L)}{1+g_m(r_o\|R_L)}$$

$$R_{in} = \frac{500+1000+(1+1\times 500)\times(10k\|100k)}{1+1(10k\|100k)}$$

$R_{in} = 600k\Omega$

b) If we write a KCL for node G:

$$① \quad \frac{v_{gs}+v_i}{R_1} + g_m v_{gs} + \frac{v_o - v_i}{r_o} + \frac{v_o}{R_L} = 0$$

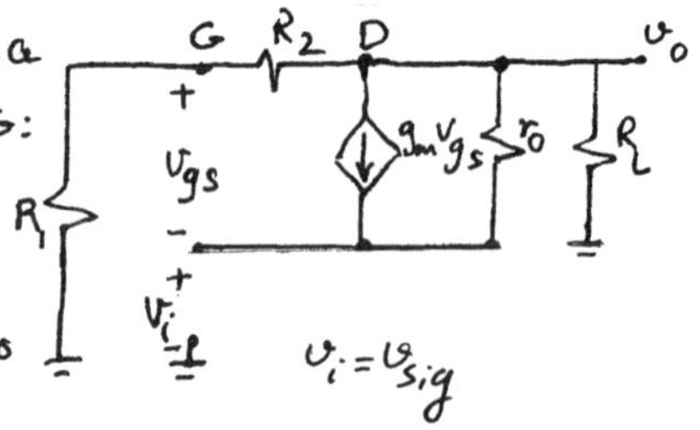

$v_i = v_{sig}$

Also: $v_{gs} = v_o\frac{R_1}{R_1+R_2} - v_i$

substitute for v_{gs} in ①:

$$\frac{v_o}{R_1+R_2} - \frac{v_i}{R_1} + \frac{v_i}{R_1} + g_m\frac{R_1}{R_1+R_2}v_o - g_m v_i + \frac{v_o}{r_o} - \frac{v_i}{r_o} + \frac{v_o}{R_L} = 0$$

$$v_o\left(\frac{1+g_mR_1}{R_1+R_2} + \frac{1}{r_o} + \frac{1}{R_L}\right) = v_i\left(+g_m + \frac{1}{r_o}\right)$$

$$\frac{v_o}{v_i} = \frac{g_m + \frac{1}{r_o}}{\frac{g_mR_1+1}{R_1+R_2} + \frac{1}{r_o} + \frac{1}{R_L}}$$

$$\frac{v_o}{v_i} = \frac{1+\frac{1}{100}}{\frac{1\times 500+1}{500+1000} + \frac{1}{100} + \frac{1}{10}} = 2.27\,V/V$$

Now we find $R_i = \frac{v_i}{i_i}$:

Cont.

$$i_i = \frac{v_o}{R_1+R_2} + \frac{v_o}{R_L}$$ (Note that (R_1+R_2) is parallel to R_L)

$$i_i = v_o\left(\frac{1}{R_1+R_2} + \frac{1}{R_L}\right)$$

$$\frac{i_i}{v_i} = \frac{v_o}{v_i}\left(\frac{1}{R_1+R_2} + \frac{1}{R_L}\right)$$

$$R_i = \frac{v_i}{i_i} = \frac{1}{\frac{v_o}{v_i}\left(\frac{1}{R_1+R_2} + \frac{1}{R_L}\right)}$$

$$R_i = \left(\frac{g_m R_1 + 1}{R_1+R_2} + \frac{1}{r_o} + \frac{1}{R_L}\right) \Big/ \left(g_m + \frac{1}{r_o}\right)\left(\frac{1}{R_1+R_2} + \frac{1}{R_L}\right)$$

$$R_i = \left(\frac{1\times 500+1}{500+1000} + \frac{1}{100} + \frac{1}{10}\right) \Big/ \left(1 + \frac{1}{100}\right)\left(\frac{1}{500+1000} + \frac{1}{10}\right)$$

$$R_i = 4.37\,k\Omega$$

4.122

For DC-bias analysis purposes, both circuits are basically the same.

$$g_m = \frac{2I_D}{V_{OV}} \Rightarrow V_{OV} = \frac{2\times 0.1}{1} = 0.2^V$$

$$I_D = \frac{1}{2}k_n'\frac{W}{L}V_{OV}^2 \Rightarrow k_n'\frac{W}{L} = \frac{2\times 0.1}{0.2^2} = 5\,mA/V^2$$

$$r_o = \frac{V_A}{I_D} \Rightarrow V_A = 100\times 0.1 = 10^V$$

In order to find the required value for V_{DD}:

$$V_{OV} = V_{GS} - V_t \Rightarrow V_{GS} = 0.2 + 0.6 = 0.8V$$

$$V_{GS} = R_1 I_{R_1} \Rightarrow I_{R_1} = \frac{0.8}{500k} = 1.6\,\mu A$$

$$V_{DS} = (R_1 + R_2)I_{R_1} = (1+0.5)\times 1.6 = 2.4V$$

$$V_{DD} = V_{DS} + (I_{R_1} + I_D)R_L = 2.4 + (0.0016 + 0.1)\times 10^k$$

$$V_{DD} = 3.416V$$

4.123

$V_{R_1} = V_{GS} = 0.8V = R_1 I_{R_1}$, $I_{R_1} = 0.01\times I_{D2} = 0.01mA$

$\Rightarrow 0.8 = R_1 \times 0.01 \Rightarrow R_1 = 80k\Omega$

$(R_1+R_2)\times 0.01 = 2^V \Rightarrow R_2 = 120k\Omega$

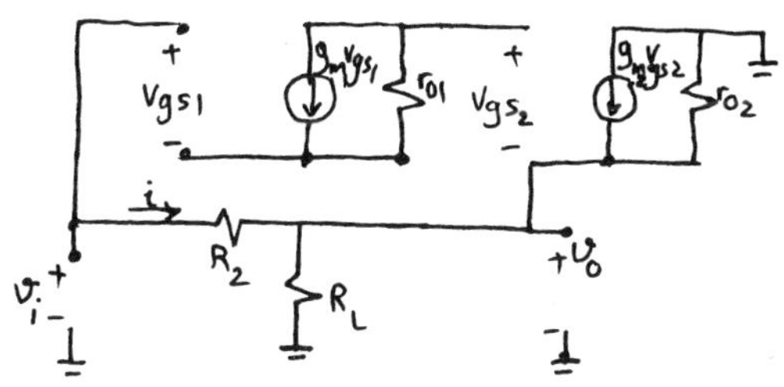

$$v_{gs1} = v_i$$

$$v_i = R_2 i + v_o \quad (1)$$

$$v_{gs2} = -g_{m1}v_{gs1}r_{o1} - v_o$$

$$i = \frac{v_o}{R_L} - g_{m2}v_{gs2} + \frac{v_o}{r_{o2}}$$

Substitute i in (1):

$$v_i = R_2/R_L\, v_o + g_{m2}R_2 g_{m1} r_{o1} v_i + g_{m2}R_2 v_o + \frac{v_o}{r_{o2}}R_2$$

$$\frac{v_o}{v_i} = \frac{1 - g_{m2}R_2 g_{m1} r_{o1}}{\frac{R_2}{R_L} + g_{m2}R_2 + \frac{R_2}{r_{o2}}}$$

We calculate the parameters:

$$g_{m1} = \frac{2I_D}{V_{OV}} = \frac{2\times 0.01}{0.8-0.6} = 0.1\,mA/V$$

$$g_{m2} = \frac{2\times 1}{0.2} = 10\,mA/V$$

$$r_{o1} = \frac{V_A}{I_D} = \frac{20}{0.01} = 2000^{k\Omega} = 2M\Omega$$

$$r_{o2} = \frac{20}{1} = 20k\Omega$$

$$\frac{v_o}{v_i} = \frac{1 - 10\times 120\times 0.1\times 2000}{\frac{120}{80} + 10\times 120 + \frac{120}{20}} = -198.8\,V/V$$

We can write: $v_i = v_o + R_2 i_i \Rightarrow 1 = \frac{v_o}{v_i} + R_2\frac{i_i}{v_i}$

$$\Rightarrow R_{in} = \frac{v_i}{i_i} = \frac{R_2}{1 - \frac{v_o}{v_i}}$$

$$R_{in} = \frac{120^k}{1+198.8} = 0.6k\Omega$$

$$\frac{v_o}{v_{sig}} = \frac{R_{in}}{R_1 + R_{in}}\frac{v_o}{v_i} = \frac{0.6^k}{80+0.6}\times(-198.8) = -1.48\,V/V$$

This two stage amplifier consisting of a CS stage followed by a source follower stage with a relatively high gain $\left(\frac{v_o}{v_i}\right)$ acts similar to an op-amp in an inverting configuration. Also note that $\frac{v_o}{v_{sig}} \simeq \frac{-R_2}{R_1} = -1.5\,V/V$

Cont.

In order to increase the gain magnitude to +5 V/V, we need to reduce the effective source resistance R_1. To do so, the following configuration can be used:

To calculate R_S: $R_1' = R_1 \,||\, R_S$ (midband)

R_1, R_S, C_S

$$\frac{v_o}{v_{sig}} = \frac{R_{in}}{R_1' + R_{in}} \frac{v_o}{v_i}$$

$$-5 = \frac{0.6}{R_1' + 0.6} \times (-198.8) \Rightarrow R_1' = 23.3\,k\Omega$$

$$R_1' = 23.3^k = 80^k \,||\, R_S \Rightarrow R_S \simeq 33\,k\Omega$$

4.124

$$I_1 = 10\,\mu A = I_{D1} = \frac{1}{2} \times 200 \times \left(\frac{W}{L}\right)_1 (0.8-0.6)^2$$

$$\Rightarrow \left(\frac{W}{L}\right)_1 = 2.5$$

$$I_{D2} = 1\,mA = \frac{1}{2} \times \frac{200}{1000} \times \left(\frac{W}{L}\right)_2 \times (0.8-0.6)^2$$

$$\Rightarrow \left(\frac{W}{L}\right)_2 = 250$$

CHAPTER 5 PROBLEMS

5.1

Case	Mode
1	active
2	saturation
3	active
4	saturation
5	inverted active mode
6	active
7	cut-off
8	cut-off

5.2

Using eq. (5.4)

$$I_S = \frac{A_E q D_n n_i^2}{N_A W}$$

$$= \frac{10 \times 10 \times 10^{-8} \times 1.6 \times 10^{-19} \times 21.3 \times 1.5^2 \times 10^{20}}{10^{17} \times 1 \times 10^{-4}}$$

$$= \underline{\underline{7.7 \times 10^{-17} \text{ A}}}$$

$$\beta = \frac{1}{\frac{D_p N_A W}{D_n N_D L_p} + \frac{1}{2}\frac{W^2}{D_n \tau_b}}$$

$$= \frac{1}{\frac{D_p N_A W}{D_n N_D L_p} + \frac{1}{2}\frac{W^2 D_n}{D_n L_n^2}} \qquad \text{Eq (5.12)}$$

$$= \frac{1}{\frac{1.7 \times 10^{17} W}{21.3 \times 10^{19} \times 0.6 \times 10^{-14}} + \frac{1}{2}\frac{W^2}{19^2 \times 10^{-8}}}$$

(a) $W = 10^{-4}$ cm $I_S = \underline{\underline{7.7 \times 10^{-17} \text{A}}}$ $\beta = \underline{\underline{368}}$

(b) $W = 2 \times 10^{-4}$ cm $I_S = \underline{\underline{3.8 \times 10^{-17} \text{A}}}$ $\beta = \underline{\underline{122}}$

(c) $W = 5 \times 10^{-4}$ cm $I_S = \underline{\underline{1.5 \times 10^{-17} \text{A}}}$ $\beta = \underline{\underline{24.2}}$

5.3

$$i_C = I_S e^{v_{BE}/V_T}$$

For Device #1

$$0.2 \times 10^{-3} = I_{S1} e^{0.72/0.025}$$

$$I_{S1} = \underline{\underline{6.214 \times 10^{-17} \text{ A}}}$$

For Device #2

$$12 \times 10^{-3} = I_{S2} e^{0.72/0.025}$$

$$I_{S2} = \underline{\underline{3.728 \times 10^{-15} \text{ A}}}$$

Since $I_S \propto A$, the relative junction areas is:

$$\frac{A_2}{A_1} = \frac{I_{S2}}{I_{S1}} = \frac{i_{C2}}{i_{C1}} = \frac{12}{0.2} = \underline{\underline{60}}$$

5.4

$$i_C = \beta i_B$$

$$400 = \beta \times 7.5$$

$$\beta = \frac{400}{7.5} = \underline{\underline{53.3}}$$

$$\alpha = \frac{\beta}{\beta + 1} = \frac{53.3}{54.3} = \underline{\underline{0.982}}$$

5.5

Use $\beta = \alpha/(1-\alpha)$

α	β
0.5	1
0.8	4
0.9	9
0.95	19

α	β
0.99	99
0.995	199
0.999	999

5.6

Use $\alpha = \frac{\beta}{\beta+1}$

β	α	β	α
1	0.5	100	0.9907
2	0.6667	200	0.9950
10	0.9091	1000	0.9990
20	0.9524	2000	0.9995

5.7

(a) $I_C = \beta I_B$

$10^{-3} = \beta \times 50 \times 10^{-6}$

$\beta = \underline{20}$

$\alpha = \frac{\beta}{\beta+1} = \frac{20}{21} = \underline{0.9524}$

$I_C = \alpha I_E$

$I_E = \frac{I_C}{\alpha} = \frac{10^{-3}}{0.9524} = \underline{1.05\,mA}$

$I_C = I_S e^{V_{BE}/V_T}$

$I_S = I_C e^{-V_{BE}/V_T} = 10^{-3} e^{-0.69/0.025} = \underline{1.03 \times 10^{-15}\,A}$

(b) $I_C = \alpha I_E$

$10^{-3} = \alpha \times 1.07 \times 10^{-3} \Rightarrow \alpha = \underline{0.9346}$

$\beta = \frac{\alpha}{1-\alpha} = \underline{14.286}$

$I_B = I_C/\beta = \frac{10^{-3}}{14.286} = \underline{69.998\,\mu A}$

$I_C = I_S e^{v_{BE}/V_T}$

$I_S = 10^{-3} e^{-0.69/0.025} = \underline{1.03 \times 10^{-15}}$

(c) $I_C = \alpha I_E = \beta I_B$

$\frac{\beta}{\beta+1} I_E = \beta I_B$

$I_E = (\beta+1) I_B$

$0.137 \times 10^{-3} = (\beta+1) 7 \times 10^{-6} \Rightarrow \beta = \underline{18.571}$

$\alpha = \frac{\beta}{\beta+1} = \underline{0.9489}$

$I_C = \beta I_B = 18.571 \times 7 \times 10^{-6} = \underline{0.130\,mA}$

$I_C = I_S e^{V_{EB}/V_T}$

$I_S = 0.130 \times 10^{-3} e^{-0.58/0.025} = \underline{10.922 \times 10^{-15}}$

(d) $I_E = I_C + I_B = 10.1 + 0.120 = \underline{10.22\,mA}$

$I_C = \alpha I_E$

$\alpha = I_C / I_E = \frac{10.1}{10.22} = \underline{0.9883}$

$\beta = I_C / I_B = 10.1 / 0.120 = \underline{84.167}$

$I_C = 10.1 \times 10^{-3} = I_S e^{0.78/0.025}$

$I_S = \underline{284.66 \times 10^{-18}\,A}$

(e) $I_E = I_C + I_B$

$I_C = I_E - I_B = 75 - 1.05 = \underline{73.95\,mA}$

$\alpha = I_C / I_E = 73.95/75 = \underline{0.986}$

$\beta = I_C / I_B = 73.95/1.05 = \underline{70.429}$

$I_C = 73.95 \times 10^{-3} = I_S e^{0.82/0.025}$

$I_S = \underline{4.208 \times 10^{-16}\,A}$

5.8

$i_C = I_S e^{v_{BE}}$

$10 \times 10^{-3} = I_S e^{0.76/0.025} \Rightarrow I_S = 6.273 \times 10^{-16}\,A$

For $v_{BE} = 0.7V \Rightarrow i_C = 6.273 \times 10^{-16} e^{0.7/0.025} = \underline{0.907\,mA}$

For $i_C = 10\,\mu A \Rightarrow 10 \times 10^{-6} = 6.273 \times 10^{-16} e^{v_{BE}/0.025}$

$\therefore\ v_{BE} = \underline{0.587V}$

Alternate way - without calculating I_S

For $v_{BE} = 0.7V$

$$\frac{i_C}{10mA} = e^{\frac{0.7-0.76}{0.025}}$$

$\therefore i_C = \underline{\underline{0.907mA}}$

For $i_C = 10\mu A$

$$\frac{10 \times 10^{-6}}{10 \times 10^{-3}} = e^{\frac{v_{BE}-0.76}{0.025}}$$

$v_{BE} = \underline{\underline{0.587V}}$

5.9

$\beta = \frac{\alpha}{1-\alpha}$, for $\alpha \rightarrow \alpha + \Delta\alpha$

$\beta \rightarrow \beta + \Delta\beta$

$$\beta + \Delta\beta = \frac{\alpha + \Delta\alpha}{1-\alpha-\Delta\alpha}$$ solve for $\Delta\beta$

$$\Delta\beta = \frac{\alpha+\Delta\alpha}{1-\alpha-\Delta\alpha} - \frac{\alpha}{1-\alpha}$$

$$= \frac{\alpha}{1-\alpha} \cdot \frac{1+\Delta\alpha/\alpha}{1-\frac{\Delta\alpha}{1-\alpha}} - \frac{\alpha}{1-\alpha}$$

$$= \beta \times \frac{\Delta\alpha/\alpha + \Delta\alpha/1-\alpha}{1-\frac{\Delta\alpha}{1-\alpha}}$$

Thus,

$$\frac{\Delta\beta}{\beta} = \frac{\Delta\alpha}{\alpha} \cdot \frac{1+\frac{\alpha}{1-\alpha}}{1-\frac{\Delta\alpha}{1-\alpha}}$$

$$= \frac{\Delta\alpha}{\alpha} \frac{1-\alpha+\alpha}{1-\alpha-\Delta\alpha}$$

$$= \frac{\Delta\alpha}{\alpha} \frac{1}{1-\alpha-\Delta\alpha}$$

for small $\Delta\alpha$, $\frac{1}{1-\alpha-\Delta\alpha} \simeq \beta$

$$\therefore \frac{\Delta\beta}{\beta} \simeq \left(\frac{\Delta\alpha}{\alpha}\right)\beta$$

Q.E.D.

5.10

9V

50μA

$\beta = 60$ to 300

$I_C = \beta I_B$ ranges from

$= 60 \times 50\mu A$ to $300 \times 50\mu A$

$= \underline{\underline{3mA \text{ to } 15mA}}$

$I_E = I_C + I_B$ ranges from

$= \underline{\underline{3.05mA \text{ to } 15.05mA}}$

Max Power $= 9 \times I_{C max} = 9 \times 15$

$= \underline{135mW}$

5.11

$i_C = I_S e^{v_{BE}/V_T}$

$10 \times 10^{-3} = I_S e^{0.7/0.025}$

$\Rightarrow I_S = 6.91 \times 10^{-15}$ A

$\alpha = \frac{i_C}{i_C + i_B} = \frac{10}{10+0.1} = 0.990$

$\beta = \frac{i_C}{i_B} = \frac{10}{0.1} = 100$

$\frac{I_S}{\alpha} = 6.98 \times 10^{-15}$ A

$\frac{I_S}{\beta} = 6.91 \times 10^{-17}$ A

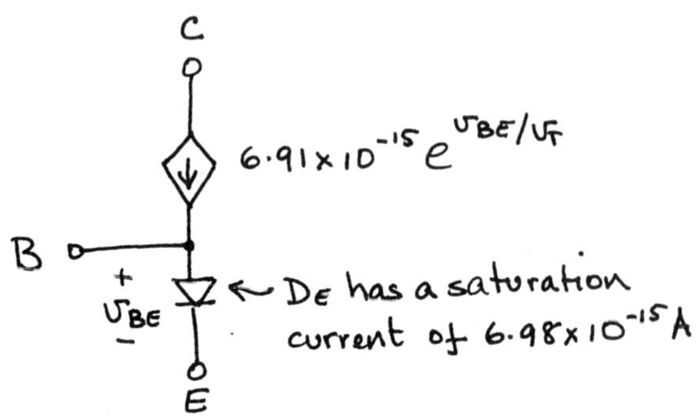

B
C
$100\,i_B$
E

D_B has a saturation current of 6.91×10^{-17}A

5.12

10V
2kΩ
$\alpha_F i_E$
i_B
i_E
+
V_{BE}
−
D_E
$I_{SE} = \frac{I_s}{\alpha} = \frac{10^{-15}}{0.99}$
3 mA

$\beta = 100,\ I_s = 10^{-15}$

$\Rightarrow \alpha = \frac{100}{101} = 0.99$

$$i_E = I_{SE}\, e^{v_{BE}/V_T}$$

$$3\times10^{-3} = \frac{10^{-15}}{0.99}\, e^{v_{BE}/0.025}$$

$$v_{BE} = 0.718\text{V}$$

$$V_E = 0 - 0.718 = \underline{-0.718\text{V}}$$

$$i_C = \alpha\, i_E = 0.99\times3\text{mA}$$

$$V_C = 10 - 2i_C = 10 - 2\times0.99\times3$$

$$= \underline{4.06\text{ V}}$$

5.13

$\beta_F = 100 \quad \alpha_R = 0.01 \quad I_s = 10^{-15}$A

$\Rightarrow \alpha_F = 0.99 \quad \beta_R = 0.111$

(a) Forward Active Mode Operation

$I_B = 10\mu A \quad V_{CB} = 1V \quad V_{BE}, I_C, I_E = ?$

Using the Ebers Moll model we have from eq. (5.28)

$$i_B = \frac{I_s}{\beta_F}\left(e^{v_{BE}/V_T} - 1\right) + \frac{I_s}{\beta_R}\left(e^{v_{BC}/V_T} - 1\right)$$

NB. NEGATIVE IN FORWARD ACTIVE MODE

$$10\times10^{-6} = \frac{10^{-15}}{100}\left(e^{v_{BE}/0.025} - 1\right) + \frac{10^{-15}}{0.111}\left(e^{\frac{-1}{0.025}} - 1\right)$$

($9.11\times10^{-16} \approx 0$)

$$= \underline{0.691\text{V}}$$

$$I_C = I_s\, e^{v_{BE}/V_T} = 10^{-15}\, e^{\frac{0.691}{0.025}}$$

$$= \underline{1\text{mA}}$$ ← NB $I_C = \beta I_B = 100 I_B$

$$I_E = I_C + I_B = \underline{1.01\text{mA}}$$

(b) Reverse Active Mode Operation

0.691V
−
+
i_C
+
−1V
−

Eq (5.28)

this term dominates in reverse active mode

$$i_B = \frac{I_s}{\beta_F}\left(e^{v_{BE}/v_T}-1\right) + \frac{I_s}{\beta_R}\left(e^{v_{BC}/v_T}-1\right)$$

$$= \underbrace{\frac{10^{-15}}{100}\left(e^{\frac{-1}{0.025}}-1\right)}_{\approx 0} + \frac{10^{-15}}{0.111}\left(e^{\frac{0.691}{0.025}}-1\right)$$

$$= -10^{-17} \quad + \quad 0.00909$$

$$= \underline{\underline{9.08\,mA}}$$

Eq (5.27)

$$i_C = I_s\left(e^{v_{BE}/v_T}-1\right) - \frac{I_s}{\alpha_R}\left(e^{v_{BC}/v_T}-1\right)$$

$$= \underbrace{10^{-15}\left(e^{-1/0.025}-1\right)}_{\approx 0} - \frac{10^{-15}}{0.1}\left(e^{\frac{0.691}{0.025}}-1\right)$$

$$= -10^{-15} \quad - 0.01009$$

$$= \underline{\underline{-10.09\,mA}}$$

$$i_E = i_C + i_B = \underline{-1.01\,mA}$$

Check with Eq (5.26) - Ebers Moll

$$i_E = \underbrace{\frac{I_s}{\alpha_F}\left(e^{-1/0.025}-1\right)}_{\approx 0} - I_s\left(e^{\frac{0.691}{0.025}}-1\right)$$

$$= \underline{\underline{-1.01\,mA}}$$

5.14

$\alpha_F \cong 1$

$\therefore \beta_F = \infty$

i_C

$i_B = 1mA$ i_E

$$\alpha_F I_{SE} = \alpha_R I_{SC} = I_s$$

Since $I_s \propto$ Area

$$\frac{I_{SC}}{I_{SE}} = \frac{\alpha_F}{\alpha_R} = 10$$

$$\alpha_R = \frac{\alpha_F}{10} = \frac{1}{10}, \quad \beta_R = \frac{\alpha_R}{1-\alpha_R} = \frac{1}{9}$$

Using Eq (5.26)

$$i_E = \frac{I_s}{\alpha_F}\left(e^{v_{BE}/v_T}-1\right) - I_s\left(e^{v_{BC}/v_T}-1\right)$$

$$= \underline{\underline{0A}} \quad - \text{Since } V_{BE} = V_{BC}$$

Using Eq (5.28)

$$i_B = \frac{I_s}{\beta_F}\left(e^{v_{BE}/v_T}-1\right) + \frac{I_s}{\beta_R}\left(e^{v_{BC}/v_T}-1\right)$$

$$= I_s\left(\underbrace{\frac{1}{\beta_F}}_{\approx 0} + \frac{1}{\beta_R}\right)\left(e^{v_{BE}/v_T}-1\right)$$

$$\therefore I_s\left(e^{v_{BE}/v_T}-1\right) = \beta_R i_B = \frac{10^{-3}}{9}$$

Using Eq (5.27)

$$i_C = I_s\left(e^{v_{BE}/v_T}-1\right) - \frac{I_s}{1/10}\left(e^{v_{BC}/v_T}-1\right)$$

$$= I_s(1-10)\left(e^{v_{BE}/v_T}-1\right)$$

$$= -9\times 10^{-3}/9 = \underline{\underline{-1\,\mu A}}$$

5.15

Using the complete Ebers Moll equations we have:

from Eq (5.26)

$$i_E = \frac{I_S}{\alpha_F}\left(e^{V_{BE}/V_T}-1\right) - I_S\left(e^{V_{BC}/V_T}-1\right)$$

$$\Rightarrow I_S\left(e^{V_{BE}/V_T}-1\right) = \alpha_F\left(i_E + I_S\left(e^{\frac{V_{BC}}{V_T}}-1\right)\right)$$

substitute this expression in Eq (5.27):

$$i_C = I_S\left(e^{\frac{V_{BE}}{V_T}}-1\right) - \frac{I_S}{\alpha_R}\left(e^{V_{BC}/V_T}-1\right) \quad (A)$$

$$= \alpha_F\left(i_E + I_S\left(e^{\frac{V_{BC}}{V_T}}-1\right)\right) - \frac{I_S}{\alpha_R}\left(e^{\frac{V_{BC}}{V_T}}-1\right)$$

$$= \alpha_F i_E + I_S\left(\alpha_F e^{V_{BC}/V_T} - \alpha_F - \frac{e^{V_{BC}/V_T}}{\alpha_R} + \frac{1}{\alpha_R}\right)$$

SEE FIG 5.9

$i_E = I_E$ since the npn transistor is fed with a constant current

$$= \alpha_F I_E - I_S\left(\frac{1}{\alpha_R} - \alpha_F\right)e^{\frac{V_{BC}}{V_T}} + I_S\left(\frac{1}{\alpha_R} - \alpha_F\right)$$

$\alpha_F \simeq 1$

- neglect as $I_S << 1$
- this is the same as neglecting terms not involving exponents

Thus,

$$i_C = I_E - I_S\left(\frac{1}{\alpha_R} - \alpha_F\right)e^{V_{BC}/V_T} \quad (B)$$

Q.E.D.

(b) $I_S = 10^{-15}$ $\alpha_F \cong 1$ $\alpha_R = 0.1$

$\beta_F = \infty$

from (B)

$$i_C = I_E - 9\times10^{-15}e^{V_{BC}/0.025} \quad (C)$$

This expression is used to create the following tables and hence the curves of i_C vs V_{BC}.

V_{BC} (V)	$i_E = 0.1$mA i_C (mA)	$i_E = 0.5$mA i_C (mA)	$i_E = 1$mA i_C (mA)
0.7	-12.92	-12.52	-12.02
0.6	-0.14	0.26	0.762
0.5	+0.1	0.5	1
0.45	+0.1	0.5	1
0.4	0.1	0.5	1
0.35	0.1	0.5	1
0.3	0.1	0.5	1
0	0.1	0.5	1
-1	0.1	0.5	1
-2	0.1	0.5	1

V_{BC} is getting smaller so less current is subtracted from I_E. $\therefore i_C \to I_E$

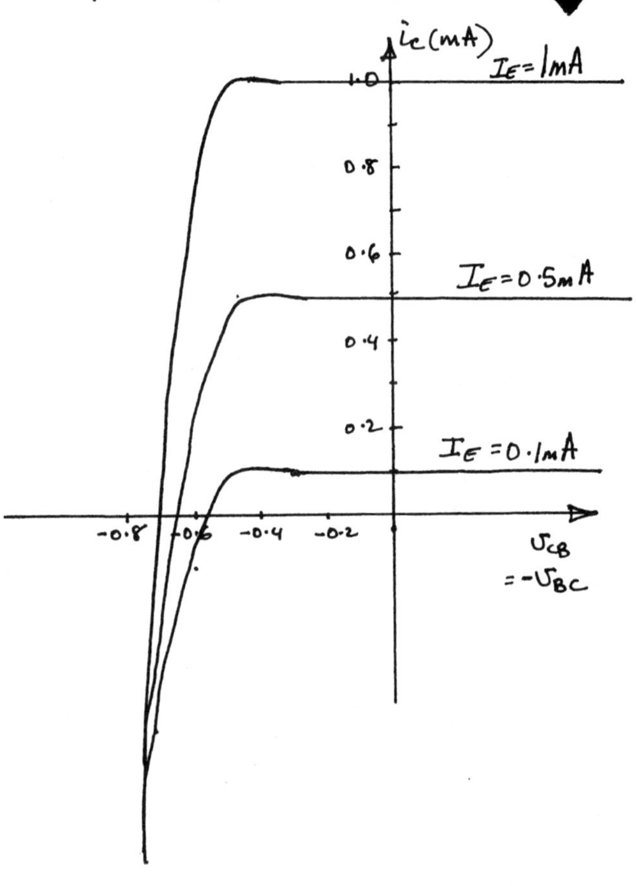

For a particular I_E and i_c, v_{BC} can be calculated using (C)

$$i_c = I_E - 9\times10^{-15} e^{v_{BC}/0.025} \quad (1)$$

Having calculated v_{BC}, v_{BE} can be calculated using (A)

$$i_c = 10^{-15}\left(e^{\frac{v_{BE}}{V_T}} - 1\right) - 10^{-14}\left(e^{\frac{v_{BC}}{V_T}} - 1\right) \quad (2)$$

Now v_{CE} can be calculated with

$$v_{CE} = -v_{BC} + v_{BE} \quad (3)$$

Thus,

for $I_E = 0.1\,mA$

i_c (mA)	v_{BC}(1) (V)	v_{BE}(2) (V)	v_{CE}(3) (V)
$0.5I_E$	0.561	0.644	0.083
0	0.578	0.635	0.057

For $I_E = 0.5\,mA$

i_c (mA)	v_{BC} (1) (V)	v_{BE}(2) (V)	v_{CE} (3) (V)
$0.5I_E$	0.601	0.674	0.073
0	0.618	0.675	0.057

for $I_E = 1\,mA$

i_c (mA)	v_{BC}(1) (V)	v_{BE}(2) (V)	v_{CE}(3) (V)
$0.5I_E = \frac{1}{2}$	0.618	0.691	0.073
0	0.638	0.696	0.058

5.16

$\beta = 40$

$\alpha_F = \frac{40}{41}$

$I_s = 10^{-13}$

$$i_E = \frac{I_s}{\alpha} e^{V_{EB}/V_T} = I_s e^{V_{EB}/V_T} + 0.02\times10^{-3}$$

$$I_s e^{V_{EB}/V_T}\left(\frac{1}{\alpha} - 1\right) = 0.02\times10^{-3}$$

$$10^{-13} e^{V_{EB}/0.025}\left(\frac{41}{40} - 1\right) = 0.02\times10^{-3}$$

$$V_{EB} = 0.570V \Rightarrow V_B = \underline{\underline{-0.570V}}$$

$$i_E = \frac{I_s}{\alpha} e^{V_{EB}/V_T} = \frac{10^{-13}}{40} e^{\frac{0.57}{0.025}}$$

$$= \underline{\underline{0.82\,mA}}$$

$$i_c = \alpha i_E \Rightarrow V_c = -10 + \alpha i_E \times 10$$

$$= -10 + \frac{40}{41} \times 0.82 \times 10$$

$$= \underline{-2V}$$

5.17

$\therefore\ i_c = I_s e^{V_{EB}/V_T}$

Use $\frac{i_c}{1A} = e^{\frac{V_{EB} - 0.8}{0.025}}$

to calculate v_{EB} for a particular i_c

CONT.

for $i_C = 10\,mA$ $\quad v_{EB} = 0.685V$

$i_C = 5\,A$ $\quad v_{EB} = 0.840V$

5.18

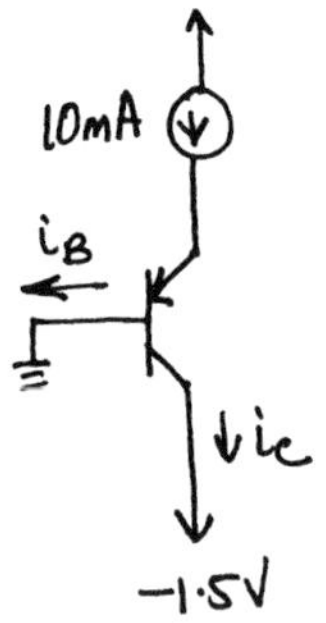

$\beta = 10$

$i_C = \alpha i_E = \frac{10}{11} \times 10 = 9.09\,mA$

$i_B = i_E - i_C = 0.91\,mA$

$i_C = I_S e^{V_{EB}/V_T}$

$9.09 \times 10^{-3} = 10^{-16} e^{V_{EB}/0.025}$

$V_E = V_{EB} = 0.803V$

For $\beta = 1000$

$i_C = \frac{\beta}{\beta+1} i_E = \frac{1000}{1001} \times 10 = 9.99\,mA$

5.19

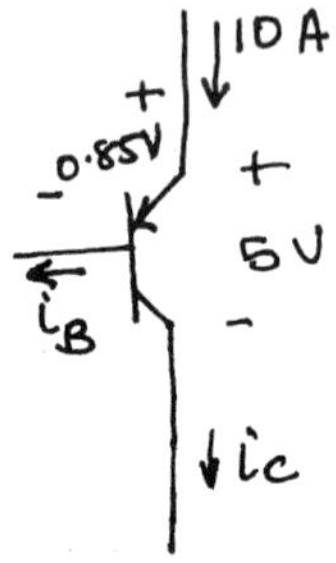

for $\beta = 15$

$i_E = (\beta+1)\, i_B$

$10 = (\beta+1)\, i_B$

$i_B = \frac{10}{16} = 0.625A$

Calculating I_{S1}

$i_C = \frac{\beta}{\beta+1} i_E = I_{S1} e^{v_{EB}/V_T}$

$\frac{15}{16} \times 10 = I_{S1} e^{0.85/0.025}$

$I_{S1} = 1.608 \times 10^{-14} A$

Compare this to

$I_{S2} = i_C e^{-v_{EB}/V_T}$

$= 10^{-3} e^{-0.7/0.025}$

$= 6.914 \times 10^{-16}$

$\therefore$ $I_S \propto$ Area

$\frac{\text{Area 1}}{\text{Area 2}} = \frac{I_{S1}}{I_{S2}} = \frac{1.608 \times 10^{-14}}{6.914 \times 10^{-16}}$

$= 23.3$ times larger

5.20

(a) $I_1 = \frac{10.7 - 0.7}{10} = 1mA$

(b)

$V_2 = 12 - 5.6 \times 2.5 = -2V$

$\frac{-4+10}{2.4} = 2.5\,mA$

(c)

$I_3 = 1mA$ $\quad \therefore \alpha \approx 1$

$V_4 = 1 - 0 = 1V$

$\frac{0-(-10)}{10} = 1mA$

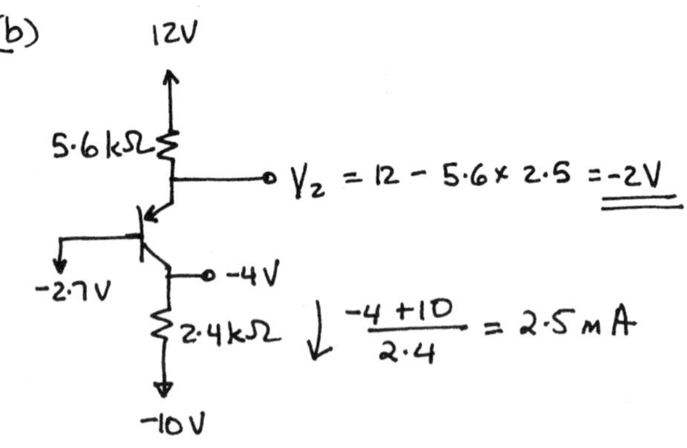

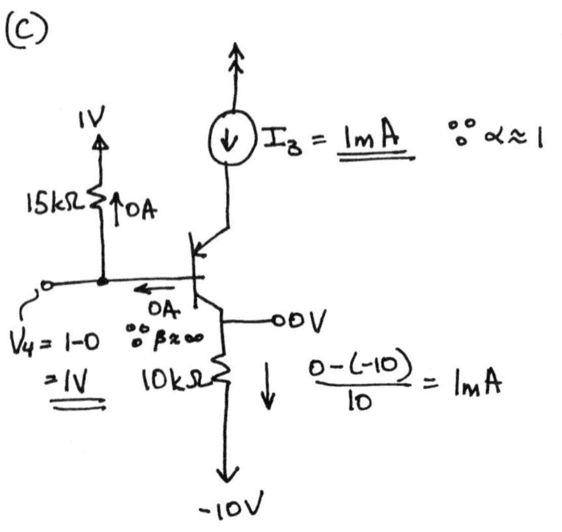

CONT

(d)

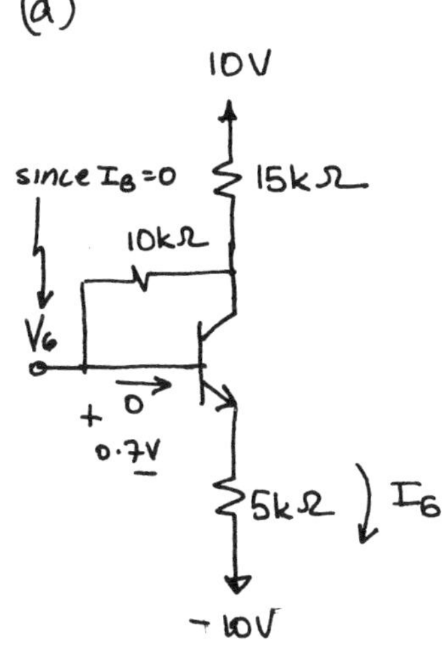

$$I_E = I_C$$

$$\frac{V_6 - 0.7 + 10}{5} = \frac{10 - V_6}{15}$$

$$15V_6 + 139.5 = 50 - 5V_6$$

$$V_6 = \underline{\underline{-4.475V}}$$

$$I_5 = \frac{V_6 - 0.7 + 10}{5}$$

$$= \frac{-4.475 - 0.7 + 10}{5} = \underline{\underline{0.965mA}}$$

5.21

(a)

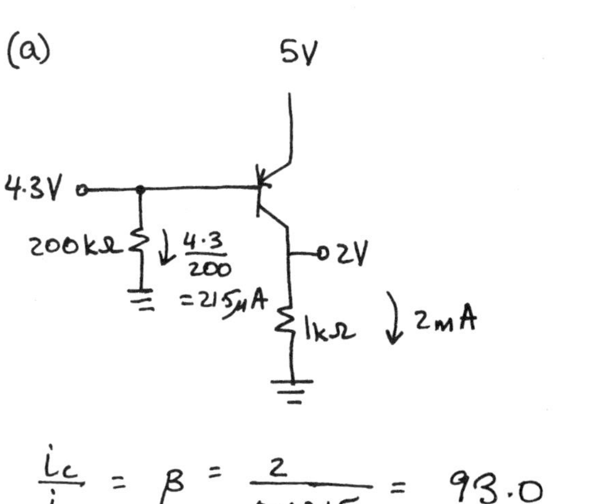

$$\frac{i_C}{i_B} = \beta = \frac{2}{0.0215} = \underline{\underline{93.0}}$$

(b)

5V

$\frac{4.3 - 2.3}{20} = 0.1mA$

+4.3V

i_C

2.3V

20kΩ

230Ω $\downarrow \frac{2.3}{0.23} = \underline{10mA}$

$\therefore 10mA - 0.1mA = i_C$

$9.9mA = i_C$

$$\frac{i_C}{i_B} = \frac{9.9}{0.1} = \beta$$

$$\beta = \underline{\underline{99}}$$

(c)

10V

1kΩ $\downarrow \frac{10-7}{1} = 3mA$

i_B

7V

6.3V

$i_E = 3mA$

100kΩ

i_C

V_C

i_B

1kΩ $\downarrow i_C + i_B = i_E = 3mA$

$$V_C = 3 \times 1 = 3V$$

$$i_B = \frac{6.3 - 3}{100} = \frac{3.3}{100}$$

$$\beta + 1 = \frac{i_E}{i_B} = \frac{3 \times 100}{3.3}$$

$$\beta = \frac{300}{3.3} - 1 = \underline{\underline{89.9}}$$

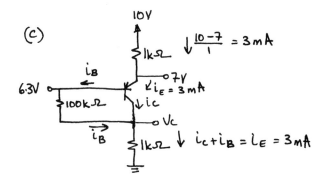

5.22

15V

$\alpha = 1$

R_C 5.1kΩ

$V_C = 15 - 5.1 \times i_E = 15 - 5.1 \times 2.1$

$= \underline{\underline{4.3V}}$

−0.7V

R_E 6.8kΩ

$\downarrow i_E = i_C = \frac{-0.7 + 15}{6.8} = \underline{\underline{2.1mA}}$

−15

5.23

$I_C = 5mA,\ V_C = 3V$

15V, R_C, 3V, R_E, −15V

$\downarrow\ R_C = \frac{15-3}{5} = \underline{\underline{2.4k\Omega}}$

$\sim$ assume $V_E = -0.7V$

$R_E = \frac{-0.7-(-15)}{5/\left(\frac{100}{101}\right)} = \underline{\underline{2.83\ k\Omega}}$

5.24

(a) $V_B = \underline{0V}$

$V_E = V_B - 0.7 = \underline{-0.7V}$

$I_E = \frac{-0.7+3}{2.2} = \underline{\underline{1.05mA}}$

$I_C = \frac{30}{31} I_E = \underline{\underline{1.02mA}}$

$V_C = 3 - 1.02 \times 2.2 = \underline{\underline{0.756V}}$

$I_B = \frac{I_C}{\beta} = \frac{1.02}{30} = \underline{\underline{0.034mA}}$

(b) $V_B = \underline{0V}$

$V_E = V_B + 0.7 = \underline{0.7V}$

$I_E = \frac{3-V_E}{1} = \frac{3-0.7}{1} = \underline{\underline{2.3mA}}$

$I_C = \alpha I_E = \frac{30}{31} \times 2.3 = \underline{\underline{2.23mA}}$

$V_C = -3 + 1 \times I_C = -3 + 2.23$

$= \underline{\underline{-0.77V}}$

$I_B = \frac{I_C}{\beta} = \frac{2.23}{30} = \underline{\underline{0.0743mA}}$

(c) $V_B = \underline{3V}$

$V_E = V_B + 0.7 = \underline{\underline{3.7V}}$

$I_E = \frac{9-V_E}{1.1} = \frac{9-3.7}{1.1} = \underline{\underline{4.82mA}}$

$I_C = \alpha I_E = \frac{30}{31} \times 4.82 = \underline{\underline{4.66mA}}$

$V_C = I_C \times 0.56 = \underline{\underline{2.62V}}$

$I_B = \frac{I_C}{\beta} = \frac{4.66}{30} = \underline{\underline{0.155mA}}$

(d) $V_B = \underline{3V}$

$V_E = 3 - 0.7 = \underline{2.3V}$

$I_E = V_E/0.47 = 2.3/0.47 = \underline{\underline{4.89mA}}$

$I_C = \alpha I_E = \frac{30}{31} \times 4.89 = 4.73mA$

$V_C = 9 - 1 \times I_C = 9 - 4.73 = \underline{\underline{4.22V}}$

$I_B = I_C/\beta = \frac{4.73}{30} = \underline{\underline{0.158mA}}$

5.25

Instead of V_{BE} being independent of the current level $V_{BE} = 0.7V$ only at $I_C = 1mA$. If I_C is different V_{BE} will change.

(a) $V_E = -V_{BE}$

$I_C = \alpha I_E = \frac{30}{31}\left(\frac{-V_{BE}+3}{2.2}\right)$

$= \frac{-30V_{BE}+90}{31(2.2)}$

$\therefore \frac{I_C}{1mA} = e^{\frac{V_{BE}-0.7}{0.025}}$ we have

$$\frac{-30V_{BE}+90}{31(2.2)} \Big/ 1 = e^{\frac{V_{BE}-0.7}{0.025}}$$

CONT.

$$V_{BE} = 0.025 \ln\left(\frac{90-30V_{BE}}{31\times 2.2}\right) + 0.7$$

Iteration #1

$V_{BE} = 0.7 \rightarrow V_{BE} = 0.70029$
$\simeq 0.7V$

$\therefore$ All the values calculated in 5.24(a) remain the same. This makes sense as I_c in 5.24 (a) was 1.02mA which has a rated V_{BE} voltage of 0.7V. Specifically we have:

$V_B = \underline{0V}$ $\qquad I_B = \underline{0.034mA}$
$V_E = -\underline{0.7V}$ $\qquad I_E = \underline{1.05mA}$
$V_c = \underline{0.756}V$ $\qquad I_c = \underline{1.02mA}$.

(b) Continuing to use the iterative process on relative currents

$$\frac{I_c}{1mA} = e^{\frac{V_{BE}-0.7}{0.025}} \quad \text{we have:}$$

$$\frac{I_c}{1} = \frac{\alpha I_E}{1} = e^{\frac{V_{EB}-0.7}{0.025}}$$

$$\frac{30}{31}\left(\frac{3-V_{EB}}{1}\right) = e^{\frac{V_{EB}-0.7}{0.025}}$$

$$V_{EB} = 0.7 + 0.025 \ln\left(\frac{90-30V_{EB}}{31}\right)$$

By Iteration

$V_{EB} = 0.7V \rightarrow 0.72V \rightarrow 0.72V$

$V_E = \underline{0.72V}$ $\qquad I_E = \frac{3-0.72}{1} = \underline{2.28mA}$

$I_c = \frac{30}{31} I_E = \underline{2.21mA}$

$V_c = -3 + 2.21 = \underline{-0.79V}$

$V_B = 0$ $\qquad I_B = I_c/\beta = \frac{2.21}{30} = \underline{0.074mA}$

(c) $I_c = \alpha I_E = e^{\frac{V_{EB}-0.7}{0.025}}$

$$\frac{30}{31}\left(\frac{9-(3+V_{EB})}{1}\right) = e^{\frac{V_{EB}-0.7}{0.025}}$$

$$V_{EB} = 0.7 + 0.025 \ln\left[\frac{180-30V_{EB}}{31}\right]$$

By Iteration

$V_{EB} = 0.7 \rightarrow 0.741 \rightarrow 0.741$

$\therefore V_E = \underline{3.741V}$ $\qquad I_E = \frac{9-3.741}{1.1} = \underline{4.78mA}$

$I_c = \frac{30}{31} I_E = \underline{4.63mA}$ $\qquad V_c = 0.56 I_c = \underline{2.59V}$

$I_B = \frac{I_c}{30} = \underline{0.154mA}$ $\qquad V_B = \underline{3V}$

(d) $I_c = \alpha I_E = e^{\frac{V_{BE}-0.7}{0.025}}$

$$\frac{30}{31}\left(\frac{3-V_{BE}}{0.47}\right) = e^{\frac{V_{BE}-0.7}{0.025}}$$

$$V_{EB} = 0.7 + \ln\left[\frac{90-30V_{BE}}{31\times 0.47}\right]$$

By Iteration:

$V_{BE} = 0.7 \rightarrow 0.739 \rightarrow 0.738 \rightarrow 0.738$

$\therefore V_E = 3 - 0.738 = \underline{2.26V}$

$I_E = V_E/0.47 = \underline{4.81mA}$

CONT.

$I_C = \frac{30}{31} I_E = \underline{4.65\,mA}$ $\quad V_C = 9 - I_C \times 1 = \underline{4.35V}$

$I_B = \frac{I_C}{30} = \underline{0.155\,mA}$ $\quad V_B = \underline{3V}$

$I_E = 0.68\,mA$

$I_B = 0 \quad V_B = \underline{0}$

$V_C = 9 - 10 I_E = 9 - 6.8 = \underline{2.2V}$

5.26

FIRST WITH FINITE β

9V

10kΩ

$I_B = \frac{1.5}{10} = 0.15\,mA$

V_C

$V_B = -1.5V$

10kΩ

$V_E = -1.5 - 0.7 = \underline{-2.2V}$

10kΩ

$\downarrow I_E = \frac{-2.2+9}{10} = 0.68\,mA$

−9V

$$\frac{I_E}{I_B} = \beta + 1$$

$$\frac{0.68}{0.15} = \beta + 1 \Rightarrow \beta = \underline{3.53}$$

$$\therefore \alpha = \frac{\beta}{\beta+1} = \underline{0.779}$$

$$V_C = 9 - 10(\alpha I_E)$$
$$= 9 - 10(0.779 \times 0.68)$$
$$= \underline{3.7V}$$

SECOND WITH INFINITE β

$\beta = \infty \quad V_{BE} = 0.7V \Rightarrow V_E = \underline{-2.2V}$

$\alpha = 1$

5.27

I_{CBO} doubles for every 10°C rise in temperature.

Thus if $I_{CBO} = 20\,nA$ at 25°C

At 85°C $\quad I_{CBO} = 2^{\frac{85-25}{10}} \times 20 = \underline{1280\,nA}$

5.28

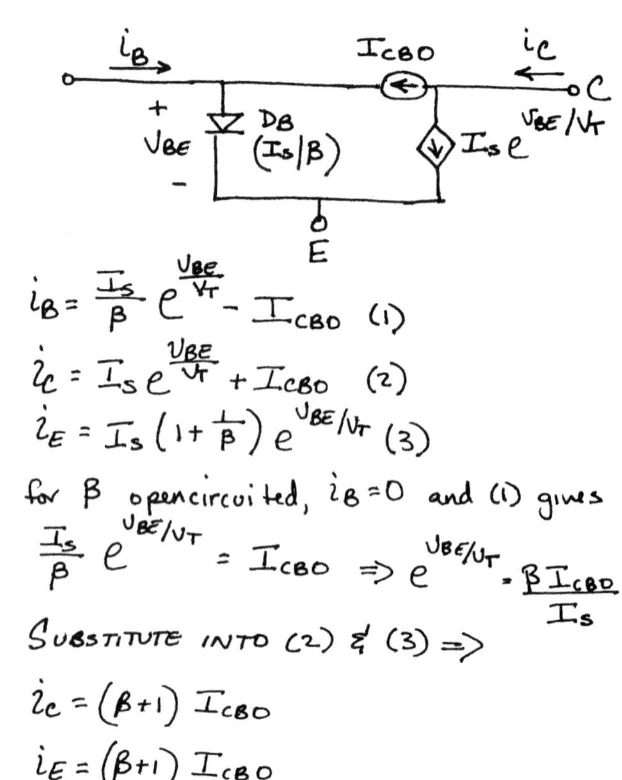

$$i_B = \frac{I_S}{\beta} e^{\frac{V_{BE}}{V_T}} - I_{CBO} \quad (1)$$

$$i_C = I_S e^{\frac{V_{BE}}{V_T}} + I_{CBO} \quad (2)$$

$$i_E = I_S\left(1 + \frac{1}{\beta}\right) e^{V_{BE}/V_T} \quad (3)$$

for B opencircuited, $i_B = 0$ and (1) gives

$$\frac{I_S}{\beta} e^{V_{BE}/V_T} = I_{CBO} \Rightarrow e^{V_{BE}/V_T} = \frac{\beta I_{CBO}}{I_S}$$

SUBSTITUTE INTO (2) & (3) ⇒

$$i_C = (\beta+1) I_{CBO}$$

$$i_E = (\beta+1) I_{CBO}$$

5.29

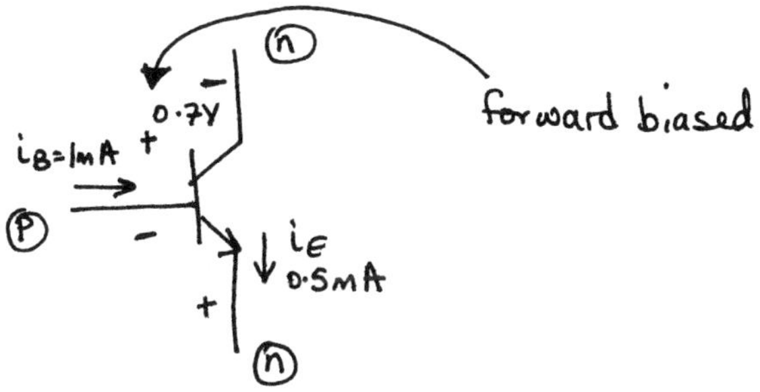

Reverse Active Mode – see Table 5.1

Use Ebers Moll Model:

Eq (5.26)

$$i_E = \underbrace{\frac{I_S}{\alpha_F}\left(e^{V_{BE}/0.7} - 1\right)}_{\text{NEGLECT}} - I_S\left(e^{\frac{V_{BC}}{V_T}} - 1\right)$$

I_S is small; NEGATIVE AND THUS VERY SMALL

Eq (5.28)

$$i_B = \underbrace{\frac{I_S}{\beta_F}\left(e^{\frac{V_{BE}}{V_T}} - 1\right)}_{\text{NEGLECT THIS TERM}} + \frac{I_S}{\beta_R}\left(e^{\frac{V_{BC}}{V_T}} - 1\right)$$

∴ Neglecting the first terms in the expressions for i_E and i_B we have

$$i_E = -I_S\left(e^{\frac{V_{BC}}{V_T}} - 1\right) = 0.5 \times 10^{-3}$$

$$i_B = \frac{I_S}{\beta_R}\left(e^{\frac{V_{BC}}{V_T}} - 1\right) = 10^{-3}$$

$$\left|\frac{i_E}{i_B}\right| = \underline{\underline{\beta_R = 1/2}} \qquad \underline{\underline{\alpha_R = \frac{\beta_R}{\beta_R + 1} = \frac{1}{3}}}$$

5.30

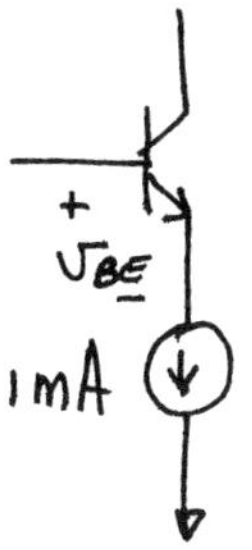

V_{BE} changes by −2 mV/°C

GIVEN $V_{BE} = 0.69$V at 25°C

Thus,

at 0°C $V_{BE} = 0.69 - 2\times10^{-3}(-25)$

$= \underline{\underline{0.74\text{V}}}$

At 100°C $V_{BE} = 0.69 - 2\times10^{-3}(75)$

$= \underline{\underline{0.54\text{V}}}$

5.31

GIVEN $i_E = 0.5$mA, $V_{EB} = 0.692$V } AT 20°C

(a) The junction temperature rises to 50°C

$V_{EB} = 0.692 - 2\times10^{-3}(50-20)$

$= \underline{\underline{0.632\text{V}}}$

(b) The Base-Emitter Voltage is fixed $V_{EB} = 0.7$V at ALL TEMPERATURES

At 20°C ~ $i_E = 0.5$mA at $V_{EB} = 0.692$V

Thus for $V_{EB} = 0.7$V we have

$$\frac{i_E}{0.5\times10^{-3}} = e^{\frac{0.7-0.692}{0.025}}$$

$i_E = \underline{\underline{0.689\text{mA}}}$

CONT.

Now if $T = 50°C$ & $V_{EB} = 0.7V$

from (a) we see that at 50°C,

$I_E = 0.5mA, \quad V_{EB} = 0.632V$

Therefore for $V_{EB} = 0.7V$

$$\frac{i_E}{0.5\times10^{-3}} = e^{\frac{0.7-0.632}{0.025}}$$

$$i_E = \underline{7.59mA}$$

5.32

↓ 10mA

+ 0.7V −

$$\frac{i_C}{10mA} = e^{\frac{v_{BE}-0.7}{0.025}}$$

$$= e^{\frac{0.5-0.7}{0.025}}$$

$$i_C = \underline{3.35\,\mu A}$$

i_E

v_{BE}

0.5 0.7

Notice the current drops significantly at $v_{BE} = 0.5V$ See Fig. 5.16

5.33

v_{BE} changes by −2mV/°C FOR A PARTICULAR CURRENT. Given that at 25°C $v_{BE} = 0.7V$ and $i_C = 10mA$

Thus

@ −25°C $v_{BE} = 0.7 - 2\times10^{-3}(-50)$

$= \underline{0.8V}$ and $i_C = 10mA$

@ 125°C $v_{BE} = 0.7 - 2\times10^{-3}(100)$

$= \underline{0.5V}$ and $i_C = 10mA$

5.34

Using the complete Ebers Moll Model in eqs. (5.26) and (5.27) and neglecting terms not containing exponentials the i_C dependence on v_{BC} with fixed $i_E = I_E$ can be derived :-

$$i_C = I_S\left(e^{\frac{V_{BE}}{V_T}}-1\right) - \frac{I_S}{\alpha_R}\left(e^{\frac{V_{BC}}{V_T}}-1\right)$$

$$\simeq I_S e^{V_{BE}/V_T} - \frac{I_S}{\alpha_R} e^{V_{BC}/V_T} \quad (1)$$

$$i_E = \frac{I_S}{\alpha_F}\left(e^{V_{BE}/V_T}-1\right) - I_S\left(e^{V_{BC}/V_T}-1\right)$$

$$\simeq \frac{I_S}{\alpha_F} e^{V_{BE}/V_T} - I_S e^{v_{BC}/V_T}$$

Noting the emitter current as being fixed at I_E we get

$$I_S e^{v_{BE}/V_T} = \alpha_F I_E + \alpha_F I_S e^{v_{BC}/v_T}$$

Substituting this into (1) we get :

$$i_C = \alpha_F I_E + \alpha_F I_S e^{v_{BC}/v_T} - \frac{I_S}{\alpha_R} e^{v_{BC}/v_T}$$

$$= \underline{\alpha_F I_E + I_S\left(\alpha_F - 1/\alpha_R\right) e^{v_{BC}/v_T}}$$

$\equiv$ Eq.(5.35)

Q.E.D.

5.35

See details in problem 5.15(b).

5.36

For the saturated transistor in Fig P.5.36

$V_{BC} > 0 \Rightarrow e^{V_{BC}/V_T} >> 1$

$V_{BE} > 0 \Rightarrow e^{V_{BE}/V_T} >> 1$

Therefore only exponential terms in the complete Ebers Moll model are kept. Specifically in Eq(5.27) we have

$$i_C = I_S e^{V_{BE}/V_T} - \frac{I_S}{\alpha_R} e^{V_{BC}/V_T}$$

and in Eq (5.26) we have

$$i_E = \frac{I_S}{\alpha_F} e^{V_{BE}/V_T} - I_S e^{V_{BC}/V_T}$$

($\alpha_F \approx 1$)

Noting that $V_{BE} = V_{BC} + V_{CE,sat}$ we get:

$$\frac{i_C}{i_E} = \frac{I_S e^{\frac{V_{BC}+V_{CEsat}}{V_T}} - \frac{I_S}{\alpha_R} e^{\frac{V_{BC}}{V_T}}}{I_S e^{\frac{V_{BC}+V_{CE,sat}}{V_T}} - I_S e^{\frac{V_{BC}}{V_T}}}$$

$$= \frac{e^{V_{CEsat}/V_T} - \frac{1}{\alpha_R}}{e^{\frac{V_{CE,sat}}{V_T}} - 1}$$

$$\left(e^{\frac{V_{CE,sat}}{V_T}} - 1\right) \frac{I_{Csat}}{I_E} = e^{\frac{V_{CE,sat}}{V_T}} - \frac{1}{\alpha_R}$$

$$e^{\frac{V_{CE,sat}}{V_T}} \left[1 - \frac{I_{Csat}}{I_E}\right] = \frac{1}{\alpha_R} - \frac{I_{Csat}}{I_E}$$

$$V_{CE,sat} = V_T \ln \left[\frac{1/\alpha_R - I_{Csat}/I_E}{1 - \frac{I_{Csat}}{I_E}}\right]$$ Q.E.D.

Given $\alpha_R = 0.1$ then

$$V_{CE,sat} = 0.025 \ln \left[\frac{10 - \frac{I_{Csat}}{I_E}}{1 - I_{Csat}/I_E}\right]$$

I_{Csat}/I_E	$V_{CE,sat}$ (V)
0.9	0.113
0.5	0.0736
0.1	0.06
0	0.058

5.37

Eq(5.36):

$$i_C = I_S e^{\frac{V_{BE}}{V_T}} \left(1 + \frac{V_{CE}}{V_A}\right) \leftarrow I_S = 10^{-15} A, \; V_A = 100V$$

$$= 10^{-15} e^{V_{BE}/0.025} \left(1 + \frac{V_{CE}}{100}\right)$$

V_{BE} =	0.65V	0.70V	0.72V	0.73V	0.74V
V_{CE} (V)	i_C (mA)	i_C (mA)	i_C (mA)	i_C (mA)	i_C (mA)
0.2	0.1961	1.4491	3.2251	4.8113	7.1777
0.8	0.1973	1.4578	3.2445	4.8402	7.2207
1.0	0.1977	1.4607	3.2509	4.8498	7.2350
4.0	0.2036	1.5041	3.3475	4.9938	7.4499
10	0.2153	1.5909	3.5406	5.2819	7.8797
12	0.2192	1.6198	3.6409	5.3780	8.0230
13	0.2936	1.6632	3.7015	5.5220	8.2379

To find the intercept of the straight-line characteristics on the i_C axis, we substitute $V_{CE} = 0$ and evaluate $i_C = 10^{-15} e^{V_{BE}/0.025}$ A for the given value of V_{BE}. The slope of each straight line is equal to this value divided by 100V (V_A). Thus we obtain

CONT.

v_{BE} (V)	0.65	0.70	0.72	0.73	0.74
Intercept (mA)	0.2	1.45	3.22	4.80	7.16
Slope (mA/V)	0.002	0.015	0.032	0.048	0.072

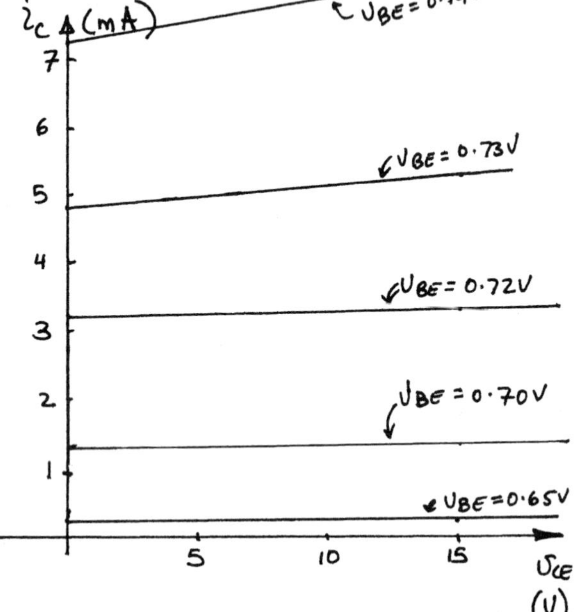

5.38

$$r_o = \frac{1}{3\times10^{-5}} = \underline{33.3k\Omega}$$

$$V_A = r_o I_c = 33.3\times10^{3} \times 3\times10^{-3}$$

$$= \underline{100V}$$

$$r_o = \frac{V_A}{I_c} = \frac{100}{30} = \underline{3.3k\Omega}$$

5.39

$$r_o = V_A/I_C = 200/I_C$$

@ $I_C = 1mA$ $\quad r_o = \underline{200k\Omega}$

@ $I_C = 100\mu A$ $\quad r_o = \frac{200}{0.1} = \underline{2.0M\Omega}$

5.40

$v_{BE} = 0.72V$ - $i_C = 1.8mA$ $\quad v_{CE} = 2V$

$i_C = 2.4mA$ $\quad v_{CE} = 14V$

$$r_o = \frac{\Delta v_{CE}}{\Delta i_C} = \frac{14-2}{2.4-1.8} = 20k\Omega$$

Near saturation $v_{CE} = 0.3V$

$$\therefore \quad \frac{\Delta v_{CE}}{\Delta i_C} = \frac{0.3-2}{i_C - 1.8} = 20$$

$$i_C = \underline{1.72mA}$$

Calculating v_{CE} for $i_C = 2.0mA$

$$\frac{\Delta v_{CE}}{\Delta i_C} = r_o$$

$$\frac{v_{CE}-2}{2-1.8} = 20 \Rightarrow v_{CE} = \underline{6V}$$

Take the ratio of currents to find the Early Voltage (with Eq 5.36)

$$\frac{2.4}{1.8} = \underbrace{e^{\frac{V_{BE}-V_{BE}}{V_T}}}_{=1}\left(\frac{1+14/V_A}{1+2/V_A}\right)$$

$$2.4 + \frac{4.8}{V_A} = 1.8 + \frac{25.2}{V_A}$$

$$V_A = \underline{34V}$$

$$r_o = \frac{V_A}{I_c'}$$

where I_c' is the current near saturation ⟺ active boundary. As calculated above $I_c' = 1.72mA$

$$= \frac{34}{1.72}$$

$$= \underline{19.8k\Omega}$$ ← ≈ to the above calculation of 20kΩ.

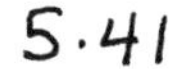

5.41

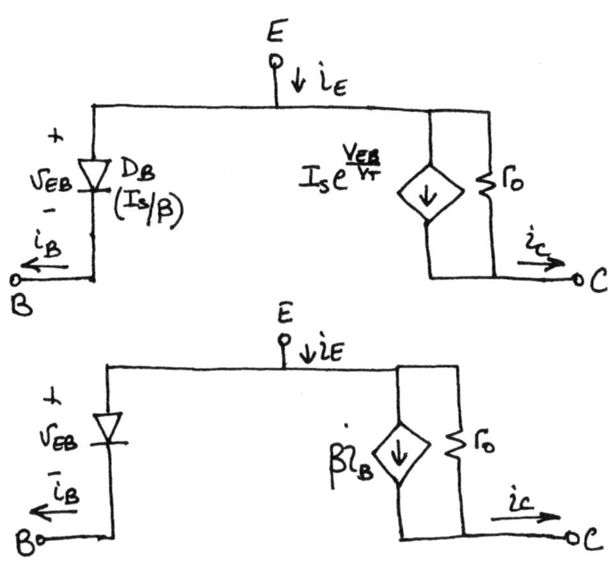

5.42

Large signal or DC β:

$$h_{FE} = \frac{i_C}{i_B} = \frac{1.2\text{mA}}{8\mu\text{A}} = \underline{\underline{150}}$$

Small signal $\quad h_{fe} = \dfrac{0.1\text{mA}}{0.8\mu\text{A}} = \underline{\underline{125}}$

$$r_o = \frac{V_A}{I_C} = \frac{100\text{V}}{1.2\text{mA}} = 83.3\text{k}\Omega$$

$$\Delta i_C = h_{fe}\,\Delta i_B + \frac{\Delta v_{CE}}{r_o}$$

$$= 125 \times 2\mu\text{A} + 2/83.3\text{k}\Omega = 0.274\text{mA}$$

$\therefore\ i_C = 1.2\text{mA} + \Delta i_C = \underline{\underline{1.474\text{mA}}}$

5.43

		$-55°C$	$25°C$	$125°C$
$I_C = 100\mu A$	$\beta \Rightarrow$	100	187	300
$I_C = 10mA$	$\beta \Rightarrow$	78	178	322

For $I_C = 100\mu A$

Temp coef below $25°C = \dfrac{187-100}{25+55} = \underline{\underline{1.09/°C}}$

Temp coef above $25°C = \dfrac{300-187}{125-25} = \underline{\underline{1.13/°C}}$

For $I_C = 10mA$

Temp coef below $25°C = \dfrac{178-78}{25+55} = \underline{\underline{1.25/°C}}$

Temp. coef above $25°C = \dfrac{322-178}{125-25} = \underline{\underline{1.44/°C}}$

5.44

i, +, $v = v_{BE}$, −

$v_{CB} = 0$

$\therefore$ active mode operation

Use Eq. (5.27) & (5.28) but neglect $(e^{v_{BC}/V_T} - 1) = 0$ terms as $v_{BC} = 0$.

$$i = i_C + i_B$$

$$= I_S\left(e^{v_{BE}/V_T} - 1\right) + \frac{I_S}{\beta_F}\left(e^{v_{BC}/V_T} - 1\right)$$

$$= I_S\underbrace{\left(e^{v_{BE}/V_T} - 1\right)}_{\approx e^{v/V_T}}\left(1 + 1/\beta_F\right) \quad \frac{\beta_F + 1}{\beta_F} = \frac{1}{\alpha_F} \approx 1$$

$$= \underline{\underline{I_S\, e^{v/V_T}}} \qquad \text{Q.E.D.}$$

5.45

I_B

$$v_{CEsat} = v_{CEoff} + I_{Csat} R_{Csat}$$

$I_{Csat} = 0$ $\therefore$ collector is open

$$= V_T \ln\left(\frac{1}{\alpha_R}\right)$$

$$= 0.025 \ln\frac{1}{0.2} = \underline{\underline{40.2\text{mV}}}$$

5.46

$\beta_{forced} = 20$

$\beta_F = 50, \ \alpha_F = 0.98$

$\beta_R = 0.2, \ \alpha_R = 0.167.$

Eq (5.48)

$$R_{CEsat} = \frac{1}{10\beta_F I_B} = \frac{1}{10 \times 50 \times 0.1} = \underline{\underline{20\,\Omega}}$$

Eq (5.49)

$$V_{CE,sat} = V_T \ln\left[\frac{1 + \frac{\beta_{forced}+1}{\beta_R}}{1 - \beta_{forced}/\beta_F}\right]$$

$$= 0.025 \ln\left[\frac{1 + 21/0.2}{1 - 20/50}\right] = \underline{\underline{0.187\,V}}$$

5.47

Using Eq (5.47) which gives the collector current for a transistor in saturation and letting $a = \frac{i_C}{\beta_F I_B}$

we have:

$$i_C = \beta_F I_B \left(\frac{e^{\frac{V_{CE}}{V_T}} - 1/\alpha_R}{e^{V_{CE}/V_T} + \frac{\beta_F}{\beta_R}}\right)$$

solve for V_{CE} & take derivative

$$\left.\frac{\partial V_{CE}}{\partial i_C}\right|_{i_B = I_B,\ i_C = I_{Csat}} = R_{CEsat}$$

$$\therefore \frac{i_C}{\beta_F I_B}\left(e^{V_{CE}/V_T} + \frac{\beta_F}{\beta_R}\right) = e^{\frac{V_{CE}}{V_T}} - \frac{1}{\alpha_R}$$

$$e^{V_{CE}/V_T}\left(\frac{i_C}{\beta_F I_B} - 1\right) = -\frac{\beta_F}{\beta_R}\frac{i_C}{\beta_F I_B} - \frac{1}{\alpha_R}$$

$$e^{\frac{V_{CE}}{V_T}}(a-1) = -\frac{\beta_F}{\beta_R}a - \frac{1}{\alpha_R}$$

$$V_{CE} = V_T \ln\left[\frac{a\frac{\beta_F}{\beta_R} + \frac{1}{\alpha_R}}{1 - a}\right]$$

Now:

$$R_{CEsat} = \left.\frac{\partial V_{CE}}{\partial i_C}\right|_{i_C = I_{Csat}}$$

$$= \frac{\partial V_{CE}}{\partial a}\frac{\partial a}{\partial i_C} = \frac{\partial V_{CE}}{\partial a} \times \frac{1}{\beta_F I_B}$$

$$= V_T\left[\frac{1-a}{a\frac{\beta_F}{\beta_R} + \frac{1}{\alpha_R}} \times \left(\frac{\beta_F/\beta_R}{1-a} + \frac{a\beta_F/\beta_R + \frac{1}{\alpha_R}}{(1-a)^2}\right)\right]\frac{1}{\beta_F I_B}$$

$$= \frac{V_T}{\beta_F I_B}\left[\underbrace{\frac{\beta_F/\beta_R}{a\beta_F/\beta_R + \frac{1}{\alpha_R}}}_{\approx \frac{1}{a}} + \frac{1}{1-a}\right]$$

$\because \frac{\beta_F}{\beta_R}a >> \frac{1}{\alpha_R}$

$$\cong \frac{V_T}{\beta_F I_B}\left[\frac{1 - a + a}{a(1-a)}\right]$$

But at $i_C = I_{Csat}$

$$a = \frac{I_{Csat}}{\beta_F I_B} = x$$

$$= \underline{\underline{\frac{V_T}{\beta_F I_B} \times \frac{1}{x(1-x)}}}$$

Q.E.D.

Using this expression to find R_{CEsat} at $\beta_{forced} = \beta_F/2 \Rightarrow x = \frac{\beta_{forced}}{\beta_F} = \frac{1}{2}$

$$R_{CEsat} = \frac{V_T}{\beta_F I_B} \times \frac{1}{\frac{1}{2}(1 - \frac{1}{2})}$$

$$= \frac{4 \times 0.025}{\beta_F I_B}$$

$$= \frac{1}{10\beta_F I_B}$$

← same as eq (5.48)

5.48

$\beta_F = 70 \rightarrow \alpha_F = 70/71$

$\beta_R = 0.7 \rightarrow \alpha_R = \frac{0.7}{1.7}$

$I_B = 2\,mA$

For $i_C = 3\,mA$

$\beta_{forced} = \frac{3}{2} = 1.5$

Note the small β_{forced} means we are deeper in saturation. Hence Eq. (5.48) which approximates the slope at $i_C = \frac{\beta_F I_B}{2}$ is not useful here!

Thus we have to use the expression in Problem 5.47

$$R_{CEsat} = \frac{V_T}{\beta_F I_B}\left[\frac{1}{\frac{\beta_{forced}}{\beta_F}\left(1 - \frac{\beta_{forced}}{\beta_F}\right)}\right]$$

$$= \frac{0.025}{70 \times 2 \times 10^{-3}}\left[\frac{1}{\frac{1.5}{70}\left(1 - \frac{1.5}{70}\right)}\right]$$

$$= \underline{\underline{8.516\,\Omega}}$$

Using Eq (5.49) to calculate V_{CEsat}

$$V_{CEsat} = V_T \ln\left[\frac{1 + \frac{\beta_{forced}+1}{\beta_R}}{1 - \beta_{forced}/\beta_F}\right]$$

$$= 0.025 \ln\left[\frac{1 + \frac{2.5}{0.7}}{1 - 1.5/70}\right]$$

$$= 38.5\,mV$$

$\therefore$ $V_{CEsat} = V_{CEoff} + R_{CEsat} I_{Csat}$

$38.5 = V_{CEoff} + 8.516 \times 3 \times 10^{-3}$

$V_{CEoff} = \underline{\underline{12.95\,mV}}$

Now for $i_C = 0.3\,mA$

$$\beta_{forced} = \frac{0.3}{2} = 0.15$$

$$R_{CE,sat} = \frac{V_T}{\beta_F I_B}\left[\frac{1}{\frac{\beta_{forced}}{\beta_F}\left(1 - \frac{\beta_{forced}}{\beta_F}\right)}\right]$$

$$= \frac{0.025}{70 \times 2 \times 10^{-3}}\left[\frac{1}{\frac{0.15}{70}\left(1 - \frac{0.15}{70}\right)}\right]$$

$$= 83.51\,\Omega$$

$$V_{CEsat} = 0.025 \ln\left[\frac{1 + \frac{\beta_{forced}+1}{\beta_R}}{1 - \beta_{forced}/\beta_F}\right]$$

$$= 0.025 \ln\left[\frac{1 + \frac{1.15}{0.7}}{1 - 0.15/70}\right]$$

$$= 24.35\,mV$$

Thus $V_{CE,off}$ can be calculated from

$$V_{CE,sat} = V_{CEoff} + R_{CEsat} I_{Csat}$$

$$V_{CEoff} = V_{CEsat} - R_{CEsat} I_{Csat}$$

$$= 83.51 \times 0.3 \times 10^{-3}$$

$$= \underline{\underline{-0.7\,mV}}$$

5.49

$\beta_F = 150 \rightarrow \alpha_F = \frac{150}{151}$

$\alpha_F I_{SE} = \alpha_R I_{SC} = I_S$

$\frac{A_{CBJ}}{A_{EBJ}} \propto \frac{I_{SC}}{I_{SE}} = \frac{\alpha_F}{\alpha_R} = 10$

$\therefore \alpha_R = \frac{\alpha_F}{10} = 0.0993$

$\beta_R = \frac{\alpha_R}{1-\alpha_R} = 0.110$

Calculate V_{CEsat} with

$$V_{CEsat} = V_T \ln\left[\frac{1 + \frac{\beta_{forced}+1}{\beta_R}}{1 - \beta_{forced}/\beta_F}\right]$$

$$= 0.025 \ln\left[\frac{1 + \frac{\beta_{forced}+1}{0.110}}{1 - \beta_{forced}/\beta_F}\right]$$

$\frac{\beta_{forced}}{\beta_F}$	β_{forced}	V_{CEsat} (V)
0.99	148.5	0.296
0.95	142.5	0.254
0.9	135	0.236
0.5	75	0.180
0.1	15	0.127
0.01	1.5	0.079
0	0	0.058

As β_{forced} is decreased, V_{CEsat} decreases and the BJT is drawn further into saturation.

5.50

$V_{BE} = 0.72\text{V}$ at $i_C = 0.6$ mA

(a)

$\beta_F = 150 \rightarrow \alpha_F = \frac{150}{151} = \underline{\underline{0.993}}$

$\frac{A_{CBJ}}{A_{EBJ}} \propto \frac{I_{SC}}{I_{SE}} = \frac{\alpha_F}{\alpha_R} = 20$

$\Rightarrow \alpha_R = \frac{\alpha_F}{20} = \frac{0.993}{20} = \underline{\underline{0.0497}}$

$\therefore \beta_R = \underline{\underline{0.0523}}$

(b) $i_C = 5$mA ~ assuming active mode

Using the Ebers Moll model in active mode - Eq (5.32)

$$i_C = I_S e^{V_{BE}/V_T} + I_S\left(\frac{1}{\alpha_R} - 1\right)$$

$$0.6 \times 10^{-3} = I_S\left(e^{0.72/0.025} + \frac{1}{0.0497} - 1\right)$$

$$I_S = 1.86 \times 10^{-16}$$

Therefore at $i_C = 5$mA we have

$$5 \times 10^{-3} = 1.86 \times 10^{-16}\left(e^{V_{BE}/0.025} + \frac{1}{0.0497} - 1\right)$$

$V_{BE} = \underline{\underline{0.773\text{V}}}$

Use Eq (5.33) to calculate i_B

$$i_B = \frac{I_S}{\beta_F} e^{V_{BE}/V_T} - I_S\left(\frac{1}{\beta_F} + \frac{1}{\beta_R}\right)$$

$$= \frac{1.86 \times 10^{-16}}{150} e^{\frac{0.773}{0.025}} - 1.86 \times 10^{-16}\left(\frac{1}{150} + \frac{1}{0.0523}\right)$$

$= \underline{\underline{33.2\mu A}}$

CONT.

This value for i_B makes sense as $\frac{i_C}{\beta_F} = \frac{5}{150} = 33.3\mu A$.

It is a little less due to the reverse leakage currents.

(c) The base current is doubled

$$\left.\begin{array}{l} i_B = 66.4\mu A \\ i_C = 5mA \end{array}\right\} \beta_{forced} = \frac{i_C}{i_B} = \frac{I_{Csat}}{i_B}$$

$$= \frac{5}{0.0664}$$

$$= \underline{\underline{75.3}}$$

Note β_{forced} is not small so we are not deep in saturation. Hence we can use Eq (5.48) to calculate R_{CEsat}:

$$R_{CEsat} = \frac{1}{10\beta_F I_B} = \frac{1}{10 \times 150 \times 66.4\mu A}$$

$$= \underline{\underline{10.04\Omega}}$$

checking this result with the expression from PROB. 5.47 we get the same result:

$$R_{CEsat} = \frac{V_T}{\beta_F I_B}\left(\frac{1}{\frac{\beta_{forced}}{\beta_F}\left(1 - \frac{\beta_{forced}}{\beta_F}\right)}\right)$$

$$= \frac{0.025}{150 \times 66.4 \times 10^{-6}}\left(\frac{1}{\frac{75.3}{150}\left(1 - \frac{75.3}{150}\right)}\right)$$

$$= 10.04\Omega$$

From Eq (5.49)

$$|V_{CEsat}| = V_T \ln\left[\frac{1 + \frac{\beta_{forced}+1}{\beta_R}}{1 - \beta_{forced}/\beta_F}\right]$$

$$|V_{CEsat}| = 0.025\ln\left[\frac{1 + 76.3/0.0523}{1 - 75.3/150}\right]$$

$$= \underline{\underline{0.1996V}}$$

Use the complete Ebers Moll equations to calculate v_{BE} and v_{BC}.

Note that
$$v_{CE} = v_{CB} + v_{BE}$$
$$= -v_{BC} + v_{BE}$$

$$\therefore v_{BE} = v_{CEsat} + v_{BC}$$

Eq (5.27):

$$i_C = I_S\left(e^{\frac{v_{BE}}{V_T}} - 1\right) - \frac{I_S}{\alpha_R}\left(e^{\frac{v_{BC}}{V_T}} - 1\right)$$

$$\frac{i_C}{I_S} = e^{\frac{V_{CEsat} + v_{BC}}{V_T}} - 1 - \frac{e^{v_{BC}/V_T}}{\alpha_R} + \frac{1}{\alpha_R}$$

$$\frac{5 \times 10^{-3}}{1.86 \times 10^{-16}} = e^{\frac{0.1996}{0.025}} e^{\frac{v_{BC}}{V_T}} - \frac{e^{v_{BC}/V_T}}{\alpha_R} - 1 + \frac{1}{\alpha_R}$$

$$v_{BC} = 0.025\ln\left[\frac{\frac{5 \times 10^{-3}}{1.86 \times 10^{-16}} + 1 - \frac{1}{0.0497}}{e^{\frac{0.1996}{0.025}} - \frac{1}{0.0497}}\right]$$

$$= \underline{\underline{0.5736}}$$

Now

$$v_{BE} = v_{CEsat} + v_{BC}$$
$$= 0.1996 + 0.5736$$
$$= \underline{\underline{0.7732V}}$$

5.51

Eq (5.49)

$$V_{CEsat} = V_T \ln\left[\frac{1 + \dfrac{\beta_{forced}+1}{\beta_R}}{1 - \beta_{forced}/\beta_F}\right]$$

For E grounded and C open: $\beta_{forced} = 0$

$$\therefore\ 60 = 25 \ln\left[\frac{1 + 1/\beta_R}{1-0}\right]$$

$$\ln\left(1+\frac{1}{\beta_R}\right) = \frac{60}{26} = 2.4$$

$$\beta_R = \underline{\underline{0.01}}$$

For C grounded and E open

$\beta_{forced} = 0 \quad \beta_R^* = \beta_F \quad \beta_F^* = \beta_R$

$$\therefore\ 1 = 25 \ln\left[\frac{1 + \dfrac{0+1}{\beta_F}}{1-0}\right]$$

$$\ln\left(1+\frac{1}{\beta_F}\right) = \frac{1}{26}$$

$$\beta_F = \underline{\underline{24.5}}$$

Check:

$$-(-1) = 25 \ln\left[\frac{1+\dfrac{1}{24.5}}{1-0}\right] = 1 mV.$$

5.52

Information given (where $\beta_{forced} = I_C/I_B$)

I_B (mA)	I_C (mA)	β_{forced}	V_{CEsat} (mV)
0.5	10	20	140
0.5	20	40	170

Given information displayed pictorally

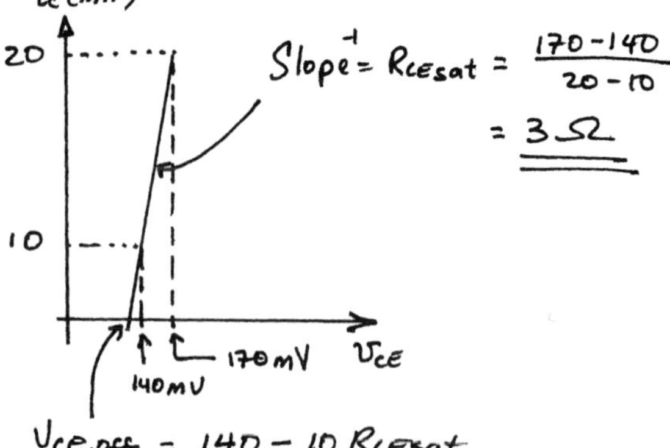

$V_{CE,off} = 140 - 10 R_{CEsat}$

$= 140 - 10 \times 3$

$= \underline{\underline{110 \text{ mV}}}$

Inserting the given data into Eq (5.49) gives the following two equations

$$140 = 25 \ln\left[\frac{1+\dfrac{21}{\beta_R}}{1 - 20/\beta_F}\right] \quad (1)$$

$$170 = 25 \ln\left[\frac{1 + 41/\beta_R}{1 - 40/\beta_F}\right] \quad (2)$$

which can be solved together to yield: $\beta_F = \underline{\underline{68.2}}$ $\quad \beta_R = \underline{\underline{0.111}}$

5.53

50V, 10kΩ, 20kΩ

$BV_{CBO} = 30V$

$$V_B = \frac{50-30}{30} \times 20 = \underline{\underline{13.3V}}$$

$$V_C = 13.3 + 30 = \underline{\underline{43.3V}}$$

$$V_E = 13.3 - 0.7 = \underline{\underline{12.6V}}$$

5.54

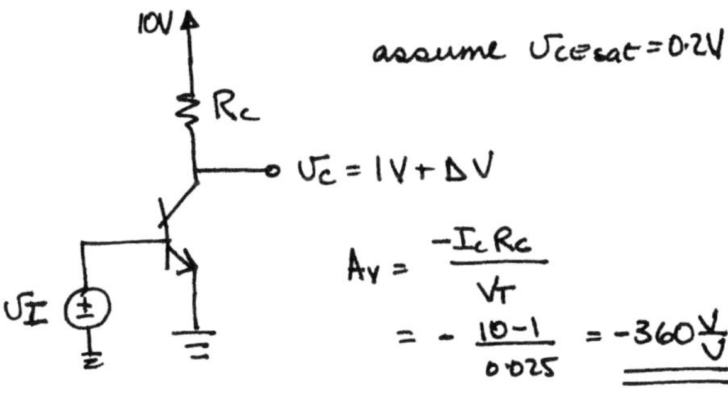

assume $v_{CEsat} = 0.2V$

$v_C = 1V + \Delta V$

$$A_V = \frac{-I_C R_C}{V_T} = -\frac{10-1}{0.025} = -360\,\frac{V}{V}$$

At the verge of saturation $v_{CEsat} = 0.3V$

$\therefore v_C = 1 + \Delta V = 0.3$

$$\Delta v = -0.7V$$

$\therefore v_O = 0.3V \quad i_C = \frac{10-0.3}{R_C}$

$$\frac{i_{C2}}{i_{C1}} = \frac{9.7/R_C}{(10-1)/R_C} = e^{\Delta v/V_T}$$

$\therefore$ Maximum Input signal

$$\Delta v = 0.025 \ln \frac{9.7}{9} = 1.87mV$$

If we assume linear operation right to saturation we can use the gain A_V to calculate the maximum input swing. Thus for an output swing $\Delta v_O = 0.8$ we have

$$\Delta v_i = \frac{-\Delta v_O}{A_V} = \frac{-0.7}{-360} = 1.94 mV$$

5.55

10V

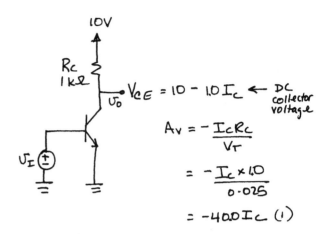

R_C 1kΩ

v_O $V_{CE} = 10 - 1.0 I_C$ ← DC collector voltage

v_I

$$A_V = -\frac{I_C R_C}{V_T} = -\frac{I_C \times 1.0}{0.025} = -40.0 I_C \quad (1)$$

- Assuming the output voltage $v_O = 0.3V$ is the lowest v_{CE} to stay out of saturation.

$$\therefore v_O = 0.3 = 10 - i_C R_C = 10 - I_C R_C + \Delta v_O$$

$$\Delta v_O = -10 + 0.3 + I_C \times 1 \quad (2)$$

- Max output voltage before the transistor is cut off

$$V_{CE} + \Delta v_O = V_{CC}$$

$$\Delta v_O = V_{CC} - V_{CE} = 10 - 10 + 1.0 I_C = 1.0 I_C \quad (3)$$

Use (1) to calculate the gain and (2), (3) to calculate the output limits in order to stay in active mode for a particular bias current I_C.

I_C (mA)	A_V (V/V)	ΔV_O (V)
1	−40	−8. to 1
2	−80	−7. to 2
5	−200	−4.7 to 5
8	−320	−1.7 to 8
9	−360	−0.7 to 9

5.56

Since we are assuming linear operation we don't have to go to $i_c = I_s e^{V_{BE}/V_T}$ equation.

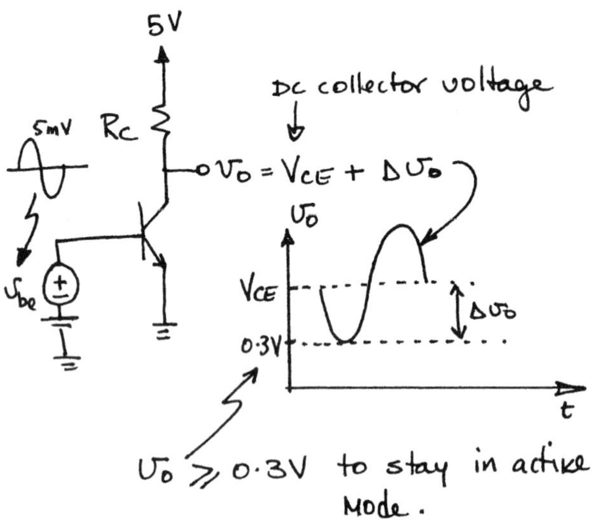

$v_O \geq 0.3V$ to stay in active Mode.

$$A_v = -\frac{I_C R_C}{V_T} = -\frac{V_{CC} - V_{CE}}{V_T}$$

On the verge of saturation

$$V_{CE} - \Delta v_O = 0.3V$$

for linear opearation $\Delta v_O = A_v v_{be}$

$$V_{CE} - |A_v v_{be}| = 0.3$$

$$(5 - I_C R_C) - A_v \times 5 \times 10^{-3} = 0.3$$

$$5 - |A_v V_T| - |A_v \times 5 \times 10^{-3}| = 0.3$$

$$|A_v (0.025 + 0.005)| = 5 - 0.3$$

$$|A_v| = 156.67$$ Note A_v is negative.

$$\therefore A_v = -156.67 \text{ V/V}$$

Now we can find the dc collector voltage. Refer to sketch of the output voltage, we see that

$$|\Delta v_O| = |(A_v \times 0.005)|$$

$$\therefore V_{CE} = 0.3 + |A_v| 0.005$$

$$= \underline{\underline{1.08V}}$$

5.57

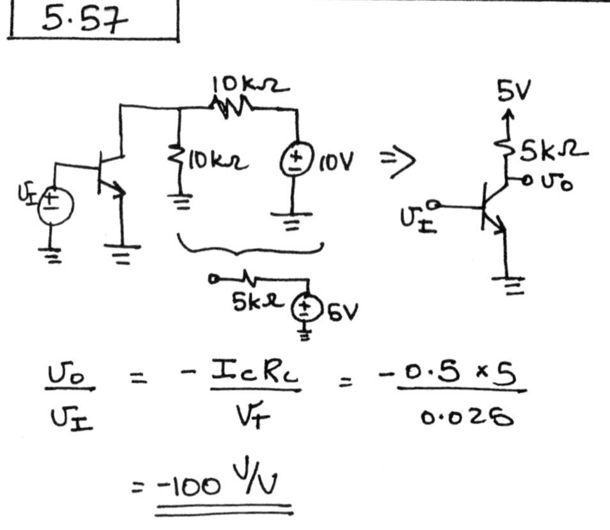

$$\frac{v_O}{v_I} = -\frac{I_C R_C}{V_T} = \frac{-0.5 \times 5}{0.025}$$

$$= \underline{\underline{-100 \text{ V/V}}}$$

5.58

(a)

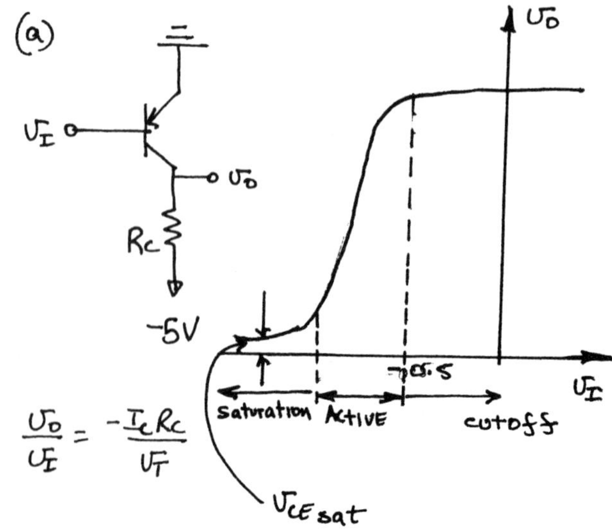

$$\frac{v_O}{v_I} = -\frac{I_C R_C}{V_T}$$

CONT.

(b)

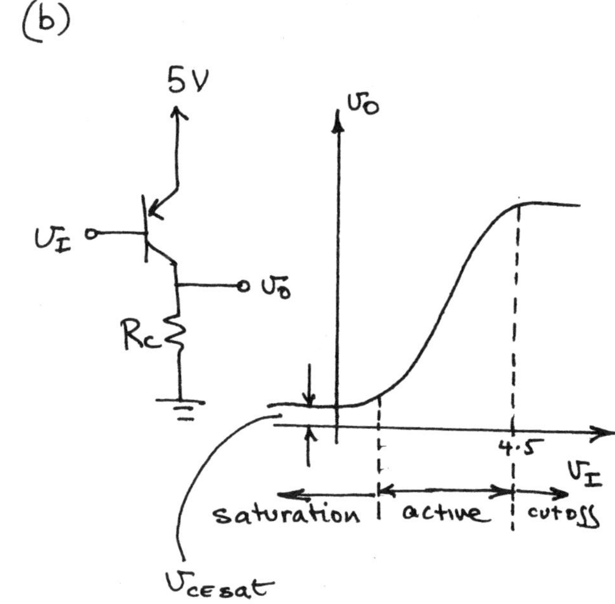

5.59

Including the Early effect we note that:

$$i_C = I_S e^{v_{BE}/V_T}\left(1 + v_{CE}/V_A\right)$$

Also, note $I_C = I_S e^{v_{BE}/V_T}$ Eq (5.38b) is the value of the collector current with the Early voltage neglected.

Starting with the voltage at the collector we have:

$$v_O = V_{CC} - i_C R_C$$

$$= V_{CC} - R_C I_S e^{v_{BE}/V_T}\left(1 + \frac{v_{CE}}{V_A}\right)$$

Take derivative to get gain A_v

$$A_v = \frac{\partial v_O}{\partial v_I}$$

$$= -R_C I_S\left[\frac{e^{v_{BE}/V_T}}{V_T}\left(1 + \frac{v_{CE}}{V_A}\right) + \frac{e^{v_{BE}/V_T}}{V_A}\frac{\partial v_{CE}}{\partial v_I}\right]$$

$$A_v = -\frac{R_C I_S}{V_T} e^{v_{BE}/V_T}\left[1 + v_{CE}/V_A + \frac{V_T}{V_A}\frac{\partial v_{CE}}{\partial v_I}\right]$$

$$= \frac{-R_C I_C}{V_T}\left[1 + \frac{v_{CE}}{V_A} + \frac{V_T}{V_A} A_v\right]$$

$$-A_v\left[\frac{1}{\frac{R_C I_C}{V_T}} + \frac{V_T}{V_A}\right] = 1 + \frac{v_{CE}}{V_A} = \frac{V_A + v_{CE}}{V_A}$$

$$-A_v\left[\frac{V_A + R_C I_C}{\frac{R_C I_C V_A}{V_T}}\right] = \frac{V_A + V_{CE}}{V_A}$$

$$-A_v \Big/ \frac{R_C I_C}{V_T} = \frac{V_A}{V_A + R_C I_C} \times \frac{V_A + v_{CE}}{V_A}$$

$$= \frac{V_A + V_{CE}}{V_A + R_C I_C} \quad \div \text{ top + bottom by } V_A + V_{CE}$$

$$= \frac{1}{\frac{V_A}{V_A + v_{CE}} + \frac{R_C I_C}{V_A + V_{CE}}}$$

This term is ≅ 1 ∵ $V_A >> v_{CE}$

$$\therefore A_v \cong \left[\frac{-R_C I_C / V_T}{\left(1 + \frac{R_C I_C}{V_A + v_{CE}}\right)}\right]$$

Q.E.D.

For $V_{CC} = 5V$ $V_{CE} = 2.5V$ $V_A = 100V$

Ignoring the Early Voltage:

$$A_v = \frac{-I_C R_C}{V_T} = \frac{V_{CC} - V_{CE}}{V_T} = \frac{5 - 2.5}{0.025} = 100\,\frac{V}{V}$$

With the Early voltage

$$A_v \cong \frac{-I_C R_C / V_T}{1 + \frac{R_C I_C}{V_A + v_{CE}}}$$

CONT.

But $v_{CE} = 0.5V$ & $\frac{I_C R_C}{V_T} = 100$ as shown above.

$$\therefore A_v = \frac{-100}{1 + \frac{2.5}{100+2.5}}$$

$$= -97.7 \text{ V/V}$$

5.60

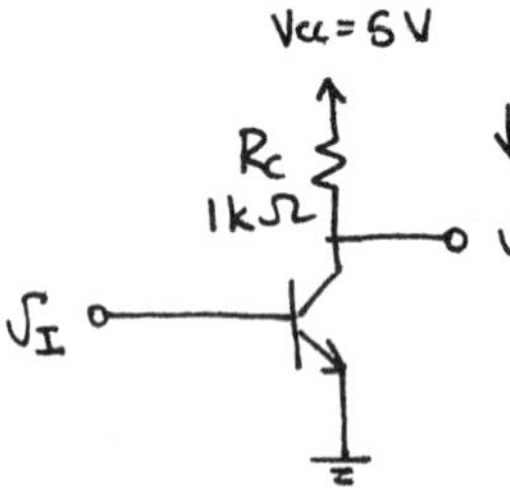

$V_{CC} = 5V$, R_C $1k\Omega$, v_I, $v_O = 2 \pm \Delta v_O$

$$I_C = \frac{5-2}{1} = 3mA$$

Small signal voltage gain

$$A_v = \frac{-I_C R_C}{V_T} = \frac{-3\times1}{0.025} = -120 \text{ V/V}$$

If we assume linear operation around the bias point we can use A_v.

$\Delta v_O = -120 \times 5 \times 10^{-3} = -0.6V$ for $\Delta v_{BE} = 5mV$

$\Delta v_O = 120 \times 5 \times 10^{-3} = 0.6V$ for $\Delta V_{BE} = -5mV$

Using the transistor exponential characteristics as in Example 5.2: for $\Delta v_{BE} = 5mV$:

$$\frac{i_C}{3mA} = e^{\Delta v_{BE}/V_T} = e^{5/25}$$

$$i_C = 3.66mA$$

$$v_O = 5 - 3.66 = 1.34\ V$$

$$\Delta v_O = 1.34 - 2 = -0.66V$$

For $\Delta v_{BE} = -5mV$

$$i_C = 3 \times e^{-5/25} = 2.46mA$$

$$v_O = 5 - 2.46\times1 = 2.54V$$

$$\Delta v_O = 2.54 - 2 = 0.54V$$

Δv_{BE} (mV)	Δv_O(V) Exponential	Δv_O(V) Linear
+0.5	−0.66	−0.6
−0.5	+0.54	+0.6

5.61

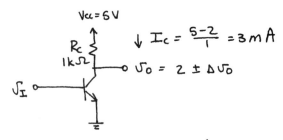

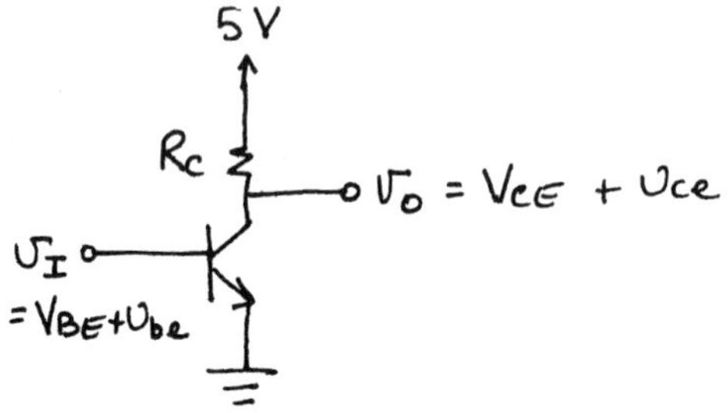

5V, R_C, $v_O = V_{CE} + v_{ce}$, $v_I = V_{BE} + v_{be}$

(a) For maximum gain you would bias at the largest current since $A_v = -I_C R_C / V_T$. This also means you would bias at the edge of saturation

$$A_v = -\frac{V_{CC} - V_{CEsat}}{V_T} = -\frac{5-0.3}{0.025} = -188 \text{ V/V}$$

However any signal swing at the output would automatically drive it into saturation.

(b) for $A_v = -100$ V/V

$$A_v = \frac{V_{CC} - V_{CE}}{V_T} = \frac{5 - V_{CE}}{V_T} = 100$$

$$V_{CE} = 2.5V$$

CONT.

(c) For a dc collector current of 0.5mA

$$R_c = \frac{5-2.5}{0.5} = \underline{\underline{5\,k\Omega}}$$

(d) $I_s = 10^{-15}A \Rightarrow$

$$I_c = I_s e^{V_{BE}/V_T}$$

$$0.5\times10^{-3} = 10^{-15} e^{V_{BE}/0.025}$$

$$V_{BE} = \underline{\underline{0.673V}}$$

(e) If we assume linear operation we can use A_v to find the output change for $v_{be} = 5mV$

$$v_{ce} = A_v v_{be} = -100 \times 0.005$$

$$= -0.5\,V \sim \text{peak sine wave.}$$

$\therefore$ the output is a $\underline{\underline{0.5V_p \text{ sine wave}}}$

(f) for $v_{ce} = 0.5$

$$i_c = \frac{0.5}{5} = \underline{0.1mA \text{ peak}}$$

This current is superimposed on I_c.

(g) $I_B = I_c/\beta = \frac{0.5}{100} = \underline{\underline{0.005mA}}$

$$i_b = \frac{i_c}{\beta} = \frac{0.1}{100} = \underline{\underline{0.001\,mA_p}}$$

(h) $r_{in} = \frac{v_{be}}{i_b} = \frac{0.005}{0.001\times10^{-3}}$

$$= \underline{\underline{5k\Omega}}$$

(i) See sketches that follow:

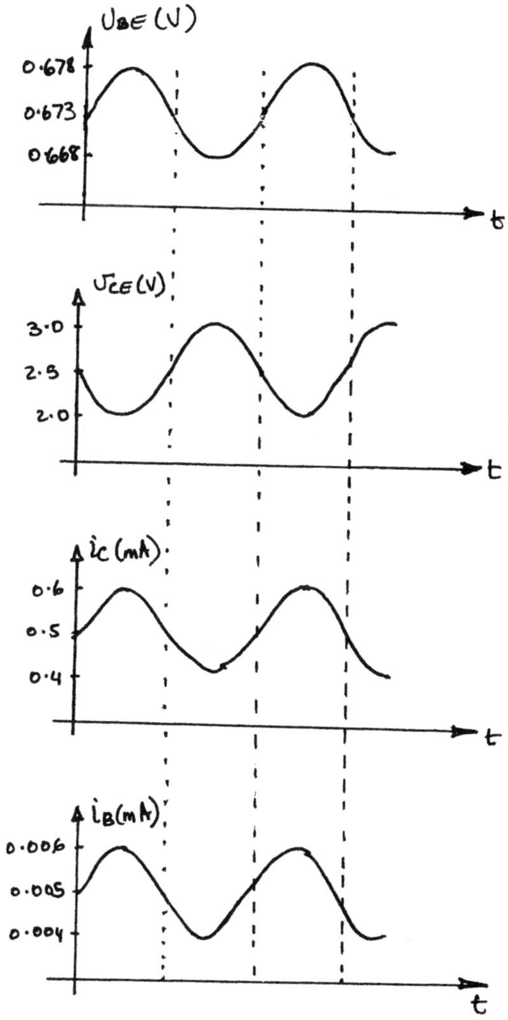

5.62

Eq (5.56)

$$A_v = \frac{v_{ce}}{v_{be}} = -\frac{I_c R_c}{V_T}$$

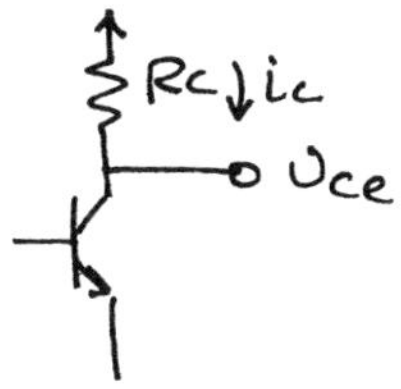

But $v_{ce} = -i_c R_c$

$$\therefore -\frac{i_c R_c}{v_{be}} = \frac{-I_c R_c}{V_T}$$

Now $g_m = \frac{\text{output current}}{\text{input voltage}} = \frac{i_c}{v_{be}}$

$$\therefore g_m R_c = \frac{I_c R_c}{V_T}$$

$$\underline{\underline{g_m = I_c/V_T}}$$

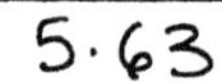
5.63

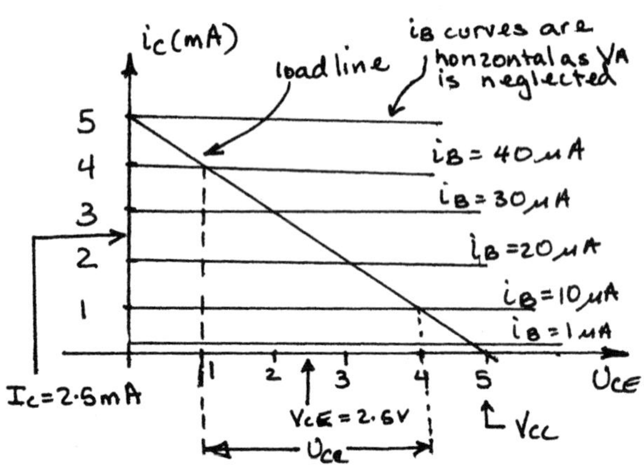

For i_B varying from 10 μA to 40 μA, i_C varies from 1 mA to 4 mA ($\beta = 100$), and $v_C = V_{CC} - R_C i_C$ varies from 4 to 1 V. Thus the peak-to-peak collector voltage swing is

$$v_{ce} = \underline{\underline{3V\ p\text{-}p}}$$

For $V_{CE} = \frac{1}{2} V_{CC} = 2.5V$

$$I_C = \underline{\underline{2.5\text{mA}}}$$

$$\& \; I_B = \frac{2.5}{100} = \underline{\underline{25\ \mu A}}$$

$$V_{BB} = V_{BE} + I_B R_B$$
$$= 0.7 + 0.025 \times 100$$
$$= \underline{3.2V}$$

5.64

See the graphical construction that follows. For this circuit:

$V_{CC} = 10V$ $\qquad \beta = 100$

$R_C = 1k\Omega$ $\qquad V_A = 100V$

$I_B = 50\mu A$ - dc bias $\qquad i_C = \beta i_B$ at $V_{CE} = 0$

$\therefore I_C = 50 \times 100$

$= 5mA$ - dc bias

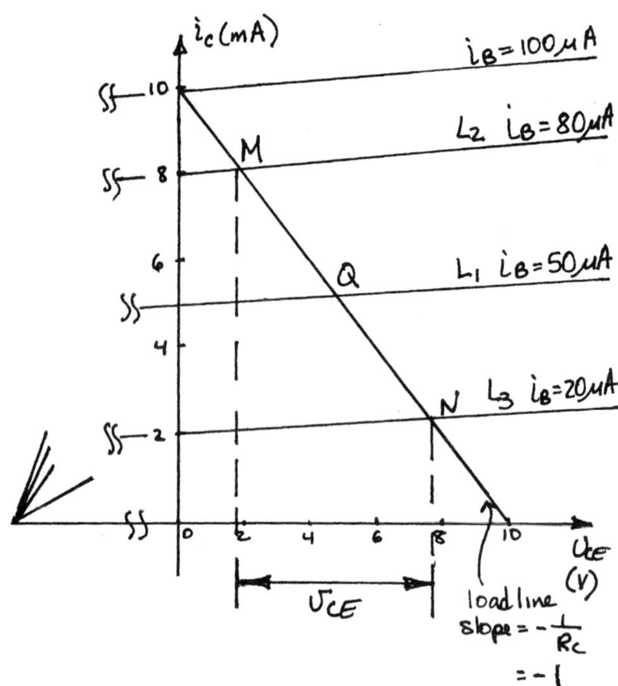

- Given the base bias current of 50 μA the dc or bias point of the collector current I_C & voltage V_{CE} can be found from the intersection of the Load Line & the transistor line L_1 of $i_B = 50\mu A$. Specifically:

$$\text{Eq of } L_1 \Rightarrow i_C = I_C(1 + v_{CE}/V_A)$$
$$= 5(1 + v_{CE}/100)$$
$$= 5 + 0.05 v_{CE}$$

$$\text{Eq of load line} \Rightarrow i_C = \frac{V_{CC} - v_{CE}}{R_C}$$
$$= 10 - v_{CE}$$

$$\therefore \; 10 - v_{CE} = 5 - 0.05 v_{CE}$$
$$V_{CE} = v_{CE} = \underline{\underline{4.76\ V}}$$

$$I_C = i_C = 10 - v_{CE} = \underline{\underline{5.24\ mA}}$$

Now for a signal of 30 μA peak superimposed on $I_B = 50\mu A$, the operating point moves along the

CONT.

load line between points N and M. To obtain the coordinates of point M, we solve the load line and line L_2 to find the intersection M and the load line and line L_3 to find N:

FOR POINT M:

$i_c = 8 + 8/100\, v_{CE}$ & $i_c = 10 - v_{CE}$

$\therefore\ i_c|_M = 8.15\text{mA} \quad V_{CE}|_M = 1.85\text{V}$

FOR POINT N:

$i_c = 2 + 0.02\, v_{CE}$ & $i_c = 10 - v_{CE}$

$v_{CE}|_N = 7.84\text{V} \qquad i_c|_N = 2.16\text{mA}$

Thus the collector current varies as follows

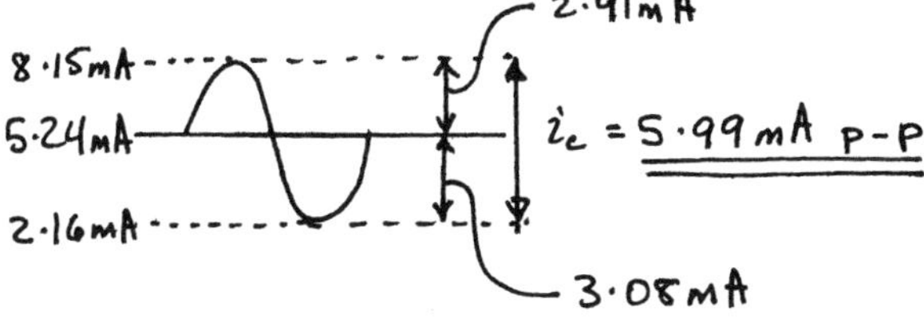

And the collector voltage varies as:

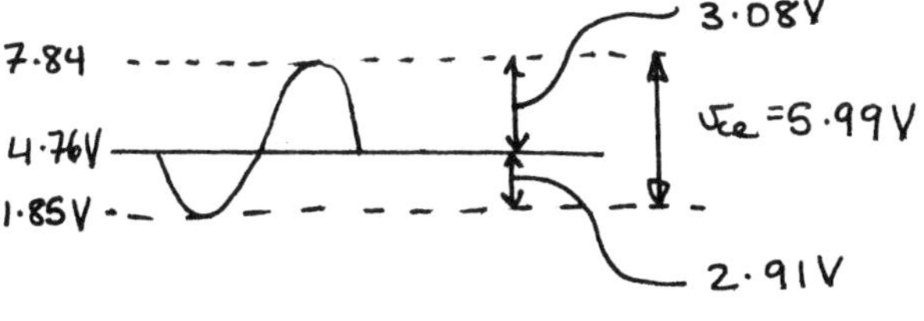

5.65

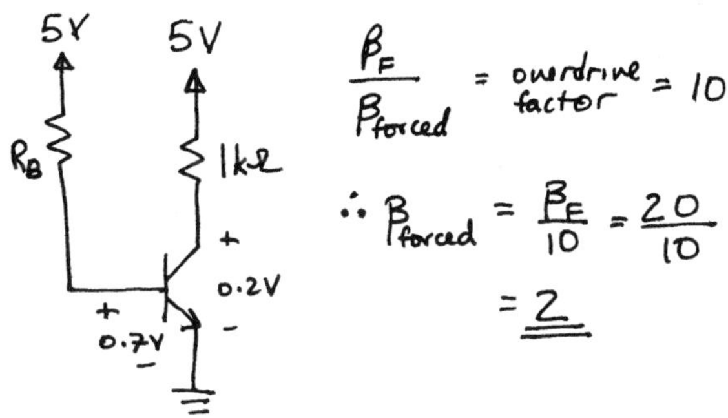

$\frac{\beta_F}{\beta_{forced}}$ = overdrive factor = 10

$\therefore \beta_{forced} = \frac{\beta_F}{10} = \frac{20}{10} = 2$

for $V_{CEsat} = 0.2\text{V}$

$5 - I_{csat} \times 1 = 0.2$

$I_{csat} = 5 - 0.2 = 4.8\text{mA}$

$$I_B = \frac{I_{csat}}{\beta_{forced}} = \frac{4.8}{2} = 2.4\text{mA}$$

$$\therefore R_B = \frac{5 - 0.7}{2.4} = \frac{4.3}{2.4} = 1.8\text{k}\Omega$$

5.66

want $\beta_{forced} = 10$

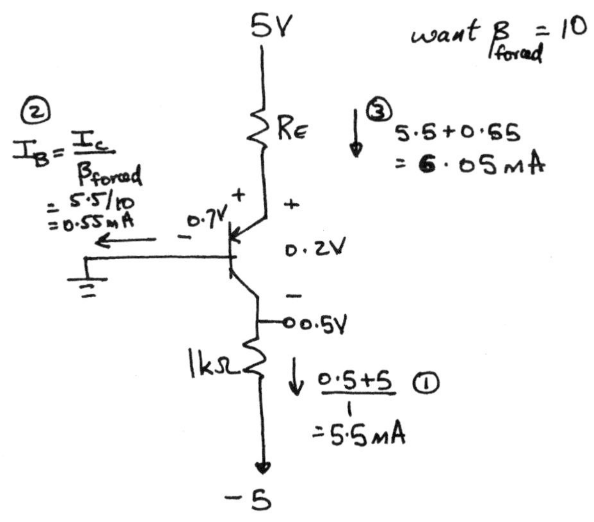

$$R_E = \frac{5 - 0.7}{6.05} = 710\,\Omega$$

5.67

(a)

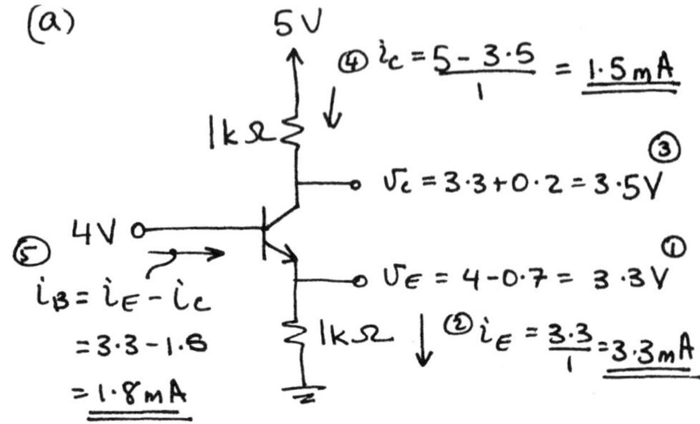

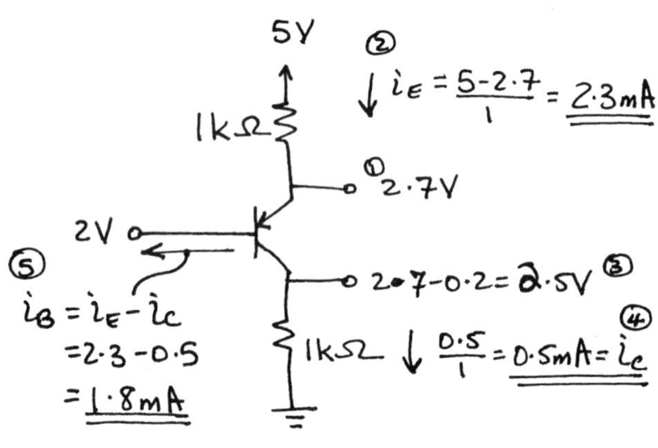

5.68

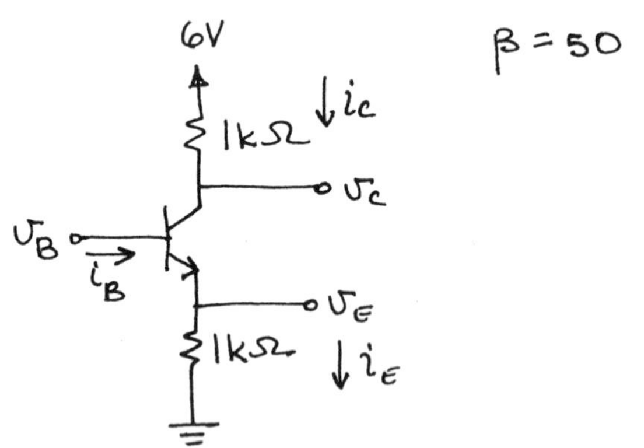

$\beta = 50$

Since conduction does not start until $V_{BE} = 0.5V$, i_C & i_E are essentially 0 for $V_B \leq 0.5V$

For:

$V_B = 1V$

$V_E = 1 - 0.3 = 0.7V$

$i_E = \frac{0.3}{1} = 0.3mA$

$i_C = \frac{\beta}{\beta+1} i_E$

$= \frac{50}{51} \times 0.3$

$= 0.2941mA$

$V_C = 6 - 0.29 \times 1$

$= 5.71V$

$i_B = i_E - i_C \; (\equiv \frac{i_C}{\beta})$

$= 0.0059mA$

$V_B = 3V$

$V_E = 3 - 0.7 = 2.3V$

$i_E = 2.3mA$

$i_C = \frac{\beta}{\beta+1} i_E$

$= \frac{50}{51} \times 2.3$

$= 2.2549$

$V_C = 6 - 2.25$

$= 3.75V$

$i_B = 2.3 - 2.25$

$= 0.0451mA$

- let saturation begins at $V_B = x$
- at saturation starting $V_{BC} = 0.4V$

$V_C = x - 0.4 \Rightarrow i_C = \frac{6 - x + 0.4}{1}$

$V_E = x - 0.7 \Rightarrow i_E = \frac{x - 0.7}{1}$

Since we are still in active mode $i_C = \frac{50}{51} i_E$

$(6.4 - x) = \frac{50}{51}(x - 0.7)$

$101x = 361.4$

$\therefore x = V_B = 3.58V$

$i_C = 6 - x + 0.4 = 2.82mA$

$i_B = \frac{i_C}{\beta} = \frac{2.82}{50} = 0.056mA$

For $V_B = 4V$ and $6V$, the transistor is in saturation. Assume $V_{BE} = 0.7V$ & $V_{CEsat} = 0.7 - 0.6 = 0.1V$

CONT.

For:

$V_B = 4V$

$V_E = 4 - 0.7 = 3.3V$

$i_E = \frac{3.3}{1} = \underline{\underline{3.3\,mA}}$

$V_C = 3.3 + 0.1 = \underline{\underline{3.4V}}$

$i_C = \frac{6 - 3.4}{1} = \underline{\underline{2.6mA}}$

$i_B = i_E - i_C$
$= 3.3 - 2.6$
$= \underline{\underline{0.7mA}}$

$V_B = 6V$

$V_E = 6 - 0.7 = \underline{\underline{5.3V}}$

$i_E = \frac{5.3}{1} = \underline{\underline{5.3mA}}$

$V_C = 5.3 + 0.1 = \underline{\underline{5.4V}}$

$i_C = \frac{6 - 5.4}{1}$
$= \underline{\underline{0.6mA}}$

$i_B = 5.3 - 0.6$
$= \underline{\underline{4.7mA}}$

CIRCUIT IN SATURATION:

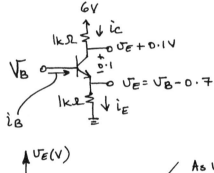

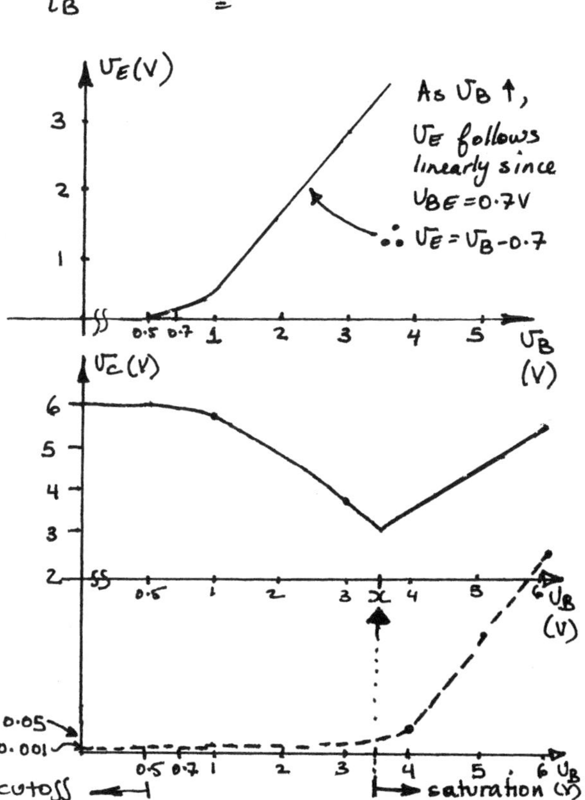

Notice for $v_B \leq 0.5V$ the transistor is essentially cutoff. Then at saturation, as v_B increases, both v_E & v_C increases linearly. This occurs as $v_{BE} = 0.7V$ & $V_{CE} = 0.1V$ in saturation.

5.69

(a) $V_B = 2V$

$V_E = 2 - 0.7 = \underline{\underline{1.3V}}$

$I_E = \frac{V_E}{1} = 1.3mA$

$I_C \cong 1.3mA$

$V_C = 5 - 1.3 = \underline{\underline{3.7V}}$

(b) $V_B = 1V$

$V_E = 1 - 0.7 = \underline{\underline{0.3V}}$

$I_E \cong I_C = 0.3mA$

$V_C = 5 - 0.3$
$= \underline{\underline{4.7V}}$

(c) $V_B = 0V$ - cutoff

$V_E = \underline{0V}$

$I_E = 0A$

$V_C = \underline{5V}$

5.70

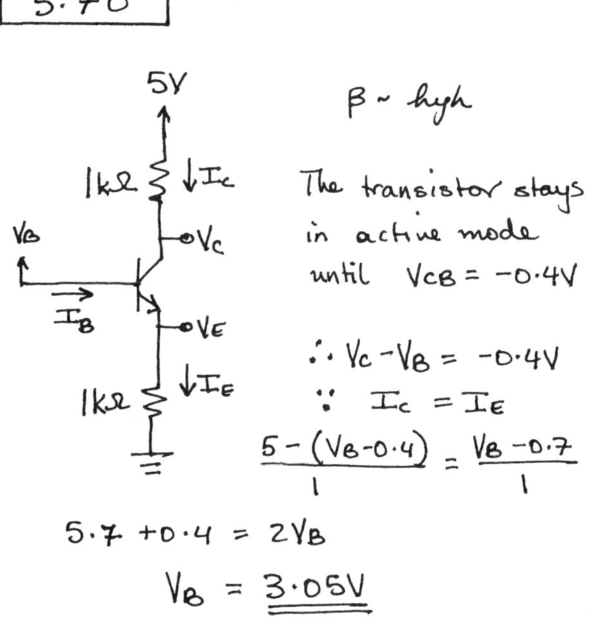

$\beta \sim$ high

The transistor stays in active mode until $V_{CB} = -0.4V$

∴ $V_C - V_B = -0.4V$

∵ $I_C = I_E$

$$\frac{5 - (V_B - 0.4)}{1} = \frac{V_B - 0.7}{1}$$

$5.7 + 0.4 = 2V_B$

$V_B = \underline{\underline{3.05V}}$

CONT.

For operation with $\beta_{forced} = 1 \Rightarrow$

$$\beta_{forced} = \frac{I_{Csat}}{I_B} = 1 \quad \left\{\begin{matrix}\text{assume} \\ V_{CE} = 0.2V \\ V_{CB} = -0.5V\end{matrix}\right\}$$

$\therefore I_{Csat} = I_B$

$I_E = I_{Csat} + I_B = 2I_{Csat}$

$$\frac{V_B - 0.7}{1} = 2 \times \frac{5 - (V_B - 0.5)}{1}$$

$$V_B - 0.7 = 2(5.5 - V_B)$$

$$V_B = \underline{\underline{3.9V}}$$

5.71

V_B: $-1, 0, 1$; $6V$; $2k\Omega$; $2k\Omega$; $-3V$; $\beta = \infty$; $I_C \cong I_E$

$V_C = 6 - 2I_E = \underline{\underline{4.7V}}, \underline{\underline{3.7V}}, \underline{\underline{2.7V}}$

$V_E = V_B - 0.7 = \underline{\underline{-1.7}}, \underline{\underline{-0.7}}, \underline{\underline{0.3}}$

$I_E = \frac{V_E + 3}{2} = 0.65, 1.15, 1.65 mA$

- Want V_B when $I_E = \frac{1}{10} \times 1.15mA = 0.115mA$

$V_E = -3 + 0.115 \times 2 = -2.77V$

$V_B = V_E + 0.7 = \underline{\underline{-2.07V}}$

- Want V_B at the edge of conduction
At the edge of conduction assume $V_{BE} = 0.5V$

$\therefore V_B - 0.5 - 2I_E + 3 = 0 \leftarrow I_E = 0$ at edge of conduction

$V_B = \underline{\underline{-2.5V}}$

$V_E = V_B - 0.5 = \underline{\underline{-3V}}$

$I_C \cong 0A \quad \therefore V_C = \underline{\underline{6V}}$

At saturation assume $V_{CE} = 0.2V$, $V_{CB} = -0.5V$

$$\therefore I_E = \frac{V_B - 0.7 + 3}{2} \cong I_C = \frac{6 - (V_B - 0.5)}{2}$$

$\therefore V_B + 2.3 = 6.5 - V_B$

$V_B = \underline{\underline{2.1V}}$

$V_E = 2.1 - 0.7 = \underline{\underline{1.4V}} \quad V_C = V_B - 0.5 = \underline{\underline{1.6V}}$

- Want V_B at $\beta_{forced} = 2$, $V_{CE} = 0.2V$, $V_{CB} = -0.5V$

$$\beta_{forced} = \frac{I_{Csat}}{I_B} = 2$$

$$I_E = I_B + I_{Csat} = \frac{I_{Csat}}{2} + I_{Csat}$$

$$= 3/2\, I_{Csat}$$

$V_E = V_B - 0.7 = -3 + 2I_E$
$= -3 + 3I_{Csat}$

$$I_{Csat} = \frac{2.3 + V_B}{3}$$

$$I_C = \frac{V_{CC} - (V_B - 0.5)}{2} = I_{Csat}$$

$$6.5 - V_B = \frac{2}{3}(2.3) + \frac{2}{3}V_B$$

$$V_B = \frac{6.5 - \frac{2(2.3)}{3}}{1\,2/3} = \underline{\underline{2.98V}}$$

5.72

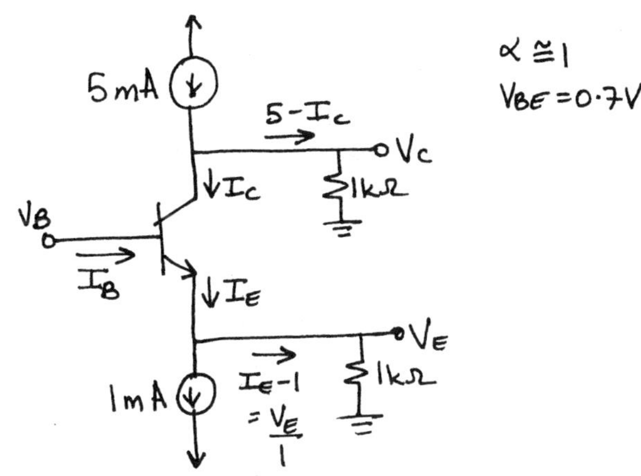

$\alpha \cong 1$

$V_{BE} = 0.7V$

Under normal conduction & $V_B = 0V$

$V_E = 0 - 0.7 = \underline{-0.7V}$

$I_E = 1 - \frac{0.7}{1} = 0.3mA \cong I_C$

$V_C = (5 - 0.3)\,1 = \underline{\underline{4.7V}}$

At cutoff ~ all of the current in the 1mA source is supplied through R_E
~ $V_{BE} = 0.5V$

$\therefore V_E = -1 \times 1 = -1V$ & $I_E = I_C = 0$

$\therefore V_B = V_E + 0.5 = -1 + 0.5$
$= \underline{\underline{-0.5V}}$

- For Saturation $V_{CE} = 0.2V$, $V_{CB} = -0.5V$

$V_C = V_E + 0.2$

$I_C = 5 - V_C/1 = 5 - \left(\frac{V_E - 0.2}{1}\right)$

$= 5.2 - V_E$

$I_E = \frac{V_E}{1} + 1$

At the edge of saturation $I_E \cong I_C$ still!

$\therefore 5.2 - V_E = V_E + 1$

$V_E = 4.2/2 = \underline{\underline{2.1\ V}}$

$V_B = V_E + 0.7 = \underline{\underline{2.8V}}$

$V_C = V_E + 0.2 = \underline{\underline{2.3V}}$

5.73

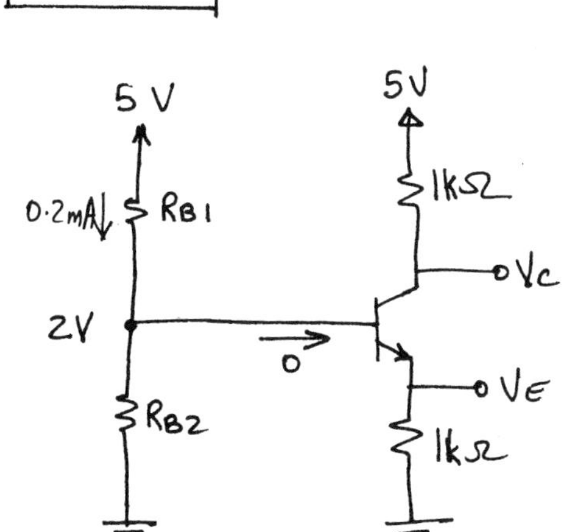

For $\beta = \infty$

$\frac{5}{R_{B1} + R_{B2}} = 0.2$ & $\frac{R_{B2}}{R_{B1} + R_{B2}} 5 = 2$

$R_{B1} + R_{B2} = 25k\Omega$ → sub

$\Rightarrow \frac{R_{B2}}{25} \times 5 = 2$

$R_{B2} = \underline{\underline{10k\Omega}}$ $\quad R_{B1} = \underline{\underline{15k\Omega}}$

Now for $\beta = 100$, use Thevenin's to obtain:

5V
2V
$R_{B1} \| R_{B2}$
$10 \| 15$
$= 6k\Omega$
1kΩ
I_C
Vc
1kΩ
(1)

Loop (1) $\quad 2 - 6\left(\frac{I_E}{\beta+1}\right) - 0.7 - I_E(1) = 0$, with $\beta + 1 = 101$

$I_E = 1.29mA$

CONT.

$I_C = \frac{100}{101} I_E = \frac{100}{101} \times 1.29 = \underline{\underline{1.28\text{mA}}}$

$V_C = 5 - 1.28(1) = \underline{\underline{3.72\text{V}}}$

5.74

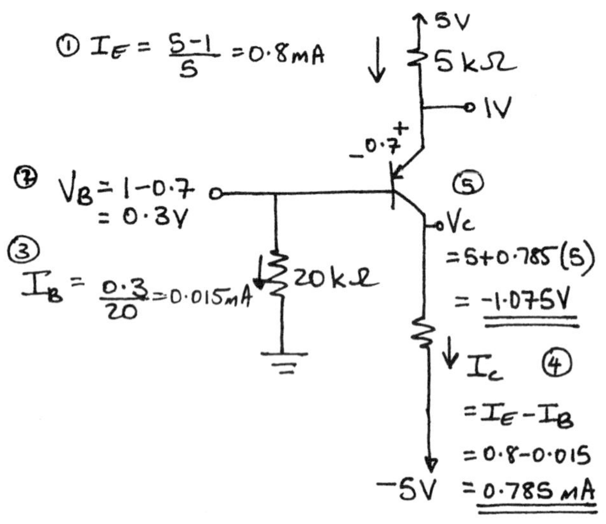

① $I_E = \frac{5-1}{5} = 0.8\text{mA}$
5V
5kΩ
1V
0.7
② $V_B = 1 - 0.7 = 0.3\text{V}$
⑤ $V_C = 5 + 0.785(5) = -1.075\text{V}$
③ $I_B = \frac{0.3}{20} = 0.015\text{mA}$
20kΩ
I_C ④ $= I_E - I_B = 0.8 - 0.015 = 0.785\text{ mA}$
−5V

⑥ $\beta = \frac{I_C}{I_B} = \frac{0.785}{0.015} = \underline{\underline{52.3}}$

⑦ $\alpha = \frac{I_C}{I_E} = \frac{0.785}{0.8} = \underline{\underline{0.98}}$

5.75

$\alpha \cong 1$

9V, $I_E = 2\text{mA}$, R_E, V_E, 0.7V, V_B, 4.5V, V_C, R_C, $\cong 2\text{mA}$, −9V

$9 - 2R_E - 0.7 - 4.5 - 2R_C + 9 = 0$

$6.4 - R_E - R_C = 0$

LET $V_B = 0\text{V}$

$R_E = \frac{9 - 0.7}{2} = \underline{\underline{4.15\text{k}\Omega}}$

$R_C = \frac{-4.5 + 9}{2} = \underline{\underline{2.25\text{k}\Omega}}$

Using 5% resistor values

$R_E = 3.9\text{k}\Omega \qquad R_C = 2.2\text{k}\Omega$

$I_E = \frac{9 - 0.7}{3.9} = \underline{\underline{2.12\text{mA}}}$

$V_C = -9 + 2.12 \times 2.2 = -4.3\text{V}$

$\therefore V_{BC} = \underline{\underline{4.3\text{V}}}$

5.76

$\beta = 30$

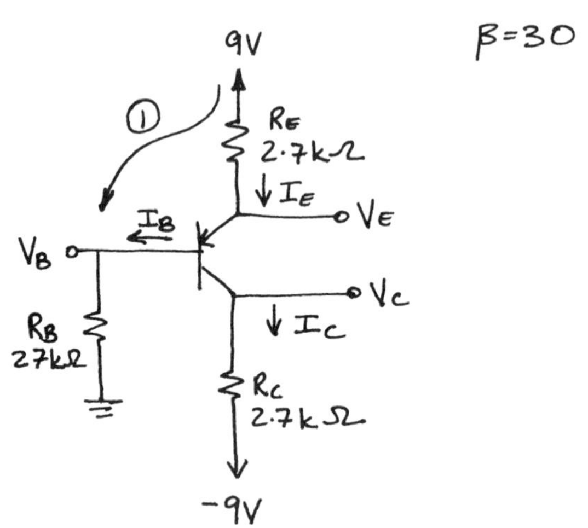

Loop ① $9 - 2.7 I_E - 0.7 - \frac{I_E}{31} R_B = 0$

$I_E = 2.3243\text{ mA}$

$V_B = R_B \times I_E/31 = \underline{\underline{2.02\text{V}}}$

$V_E = 9 - 2.7 I_E = \underline{\underline{2.72\text{V}}}$

$V_C = -9 + \frac{30}{31} I_E (2.7) = \underline{\underline{-2.93\text{V}}}$

FOR $R_B = 270\text{k}\Omega$

Loop ① $9 - 2.7 I_E - 0.7 - \frac{R_B}{31} I_E = 0$

$I_E = 0.7274\text{ mA}$

CONT.

$V_B = R_B \times \frac{I_E}{31} = \underline{\underline{6.34V}}$

$V_E = 9 - 2.7 I_E = \underline{\underline{7.04V}}$

$V_C = \frac{30}{31} I_E (2.7) - 9 = \underline{\underline{-7.10V}}$

To return the voltages to the ones first calculated we have

Loop ① $\sim I_E = 2.3243\ mA$

$$9 - 2.7 I_E - 0.7 - \frac{270}{\beta+1} I_E = 0$$

$\beta = \underline{\underline{309}}$

5.77

Using the values from the first part of P5.76 and for the edge of saturation $V_{BC} > -0.4V$

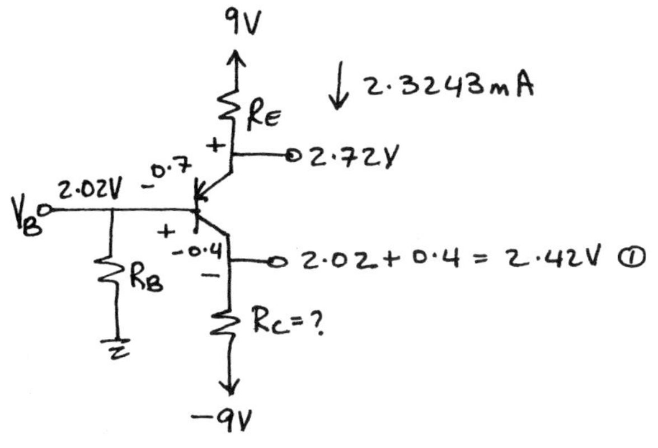

CIRCUIT AT THE EDGE OF SATURATION

$I_C = \frac{30}{31} I_E = \frac{30}{31} \times 2.3243$

$R_C = \frac{2.42 + 9}{30/31 \times 2.3243} = \underline{\underline{5.08k\Omega}}$

5.78

$\beta = 100$

(a) $\underline{R_B = 100k\Omega}$ — ∵ R_B is large assume active mode.

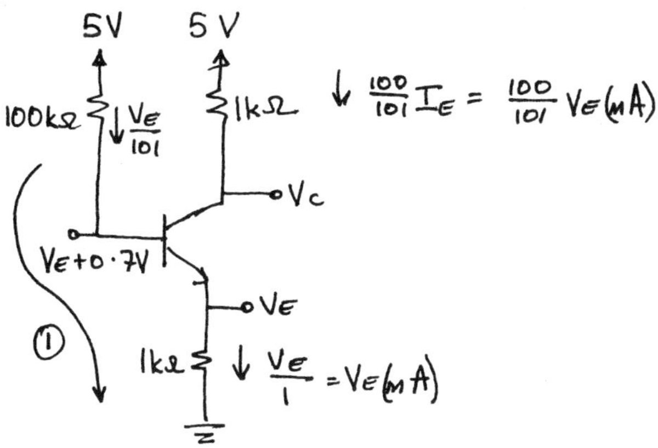

Loop ①

$$5 - \frac{V_E}{101} \times 100 - 0.7 - V_E \times 1 = 0$$

$V_E = \underline{\underline{2.16V}}$

$V_B = V_E + 0.7 = \underline{\underline{2.86V}}$

$V_C = 5 - 1 \times \frac{100}{101} V_E = \underline{\underline{2.86V}}$

Thus the BJT is in active mode as assumed.

(b) $\underline{R_B = 10k\Omega}$ - assume saturation

5V

$I_B = \frac{5 - (V_E + 0.7)}{R_B}$

$I_C = \frac{5 - (V_E + 0.2)}{1}$

$I_E = \frac{V_E}{1} = I_B + I_C$

RB, 1kΩ, I_C, I_B, + 0.2V −, + 0.7 −, VE, 1kΩ, I_E

$\therefore V_E = \frac{4.3 - V_E}{10} + 4.8 - V_E$

$10V_E + V_E + 10V_E = 4.3 + 48$

CONT

$V_E = \underline{2.49\text{V}}$

$V_C = 2.49 + 0.2 = \underline{\underline{2.69\text{V}}}$

$V_B = V_E + 0.7 = \underline{\underline{3.19\text{V}}}$

Check: $I_C = \dfrac{5 - 2.69}{1} = 2.31\text{mA}$

$I_B = \dfrac{5 - 3.19}{10} = 0.181\text{mA}$

$\dfrac{I_C}{I_B} = \dfrac{2.31}{0.181} = 12.76 < 100$

Hence we are in saturation as assumed!

(C) $\underline{R_B = 1\text{k}\Omega}$ - expect saturation
- use circuit in (b)

$I_B = \dfrac{5 - (V_E + 0.7)}{R_B} = \dfrac{4.3 - V_E}{1}$

$I_C = \dfrac{5 - (V_E + 0.2)}{1} = \dfrac{4.8 - V_E}{1}$

$I_E = I_B + I_C = V_E$

$4.3 - V_E + 4.8 - V_E = V_E$

$V_E = \underline{\underline{3\text{V}}}$

$V_B = \underline{\underline{3.7\text{V}}}$

$V_C = \underline{\underline{3.2\text{V}}}$

Check $I_B = 4.3 - 3 = 1.3\text{mA}$

$I_C = 4.8 - 3 = 1.8\text{mA}$

$\dfrac{I_C}{I_B} = \dfrac{1.8}{1.3} = 1.4 < 100$ ∴ SATURATION AS ASSUMED

5.79

(a)

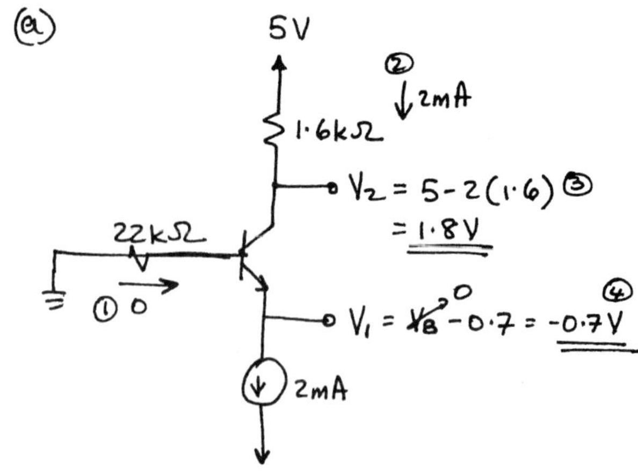

(b)

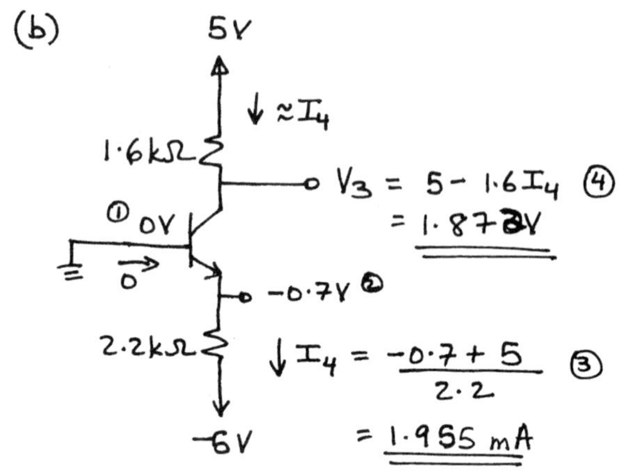

see below for part (c)

(d)

5V

3.3kΩ $\downarrow \dfrac{5 - 1.9}{3.3} = 0.939\text{mA}$ ③

$V_8 = 1.2 + 0.7 = \underline{\underline{1.9\text{V}}}$ ②

1.2V

51kΩ

0

① 1.2V

$V_9 = -5 + 0.939(5.1)$
$= \underline{\underline{-0.209\text{V}}}$ ⑤

5.1kΩ $\downarrow \approx 0.939\text{mA}$ ④

−5V

CONT.

(c)

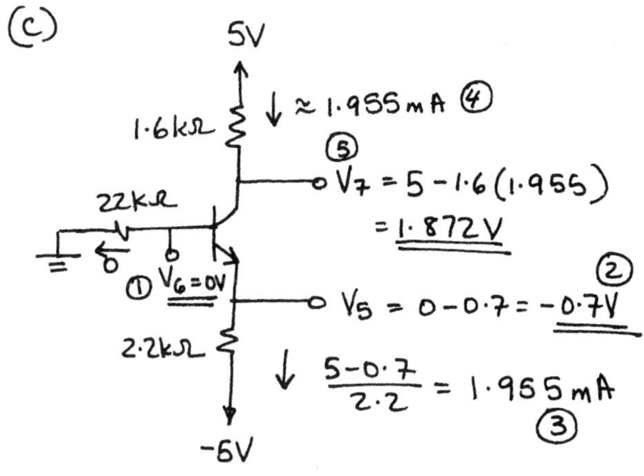

(e)

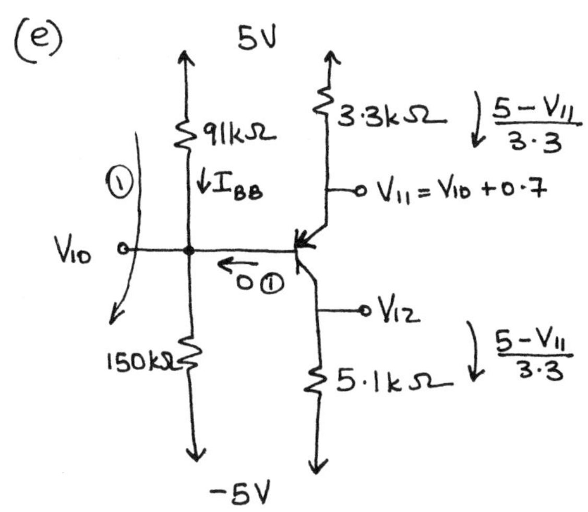

Loop ①

$$5 - 91 I_{BB} - 150 I_{BB} + 5 = 0$$

$$I_{BB} = \frac{10}{91+150}$$

$$V_{10} = -5 + 150 I_{BB}$$

$$= -5 + \frac{150}{91+150} \times 10$$

$$= \underline{\underline{1.224V}}$$

$$V_{11} = V_{10} + 0.7 = \underline{\underline{1.924V}}$$

$$\therefore I_C \approx I_E = \frac{5 - V_{11}}{3.3}$$

$$V_{12} = -5 + \left(\frac{5-V_{11}}{3.3}\right) 5.1 = \underline{\underline{-0.246V}}$$

5.80

(a)

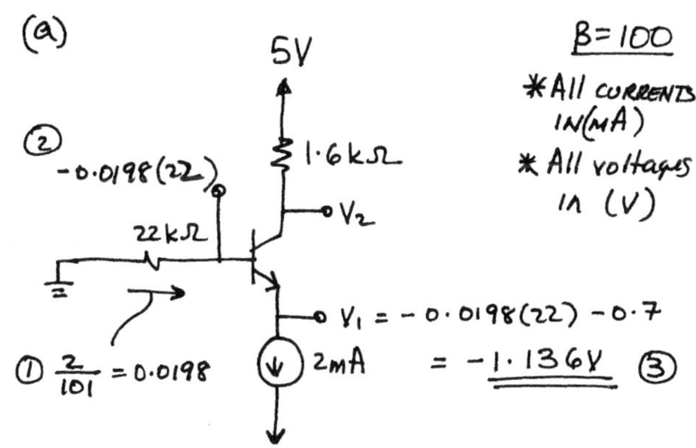

④ $V_2 = 5 - 2\left(\frac{100}{101}\right)1.6 = \underline{\underline{1.832V}}$

(b)

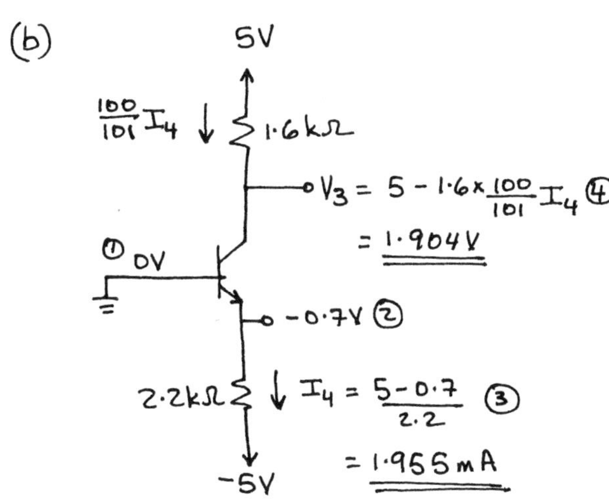

(c)

5V

1.6kΩ ④

$V_7 = 5 - \frac{100}{101} I_E \times 1.6$

$= \underline{\underline{2.183V}}$

22kΩ

V_6

① 2.2kΩ

$V_5 = -5 + 2.2 I_E$ ②

$= \underline{\underline{-1.087\ V}}$

−5V

Loop ① $0 - \frac{I_E}{101} 22 - 0.7 - 2.2 I_E + 5 = 0$

$I_E = 1.778\,mA$

③ $V_6 = V_5 + 0.7 = \underline{\underline{-0.387\ V}}$

CONT

(d)

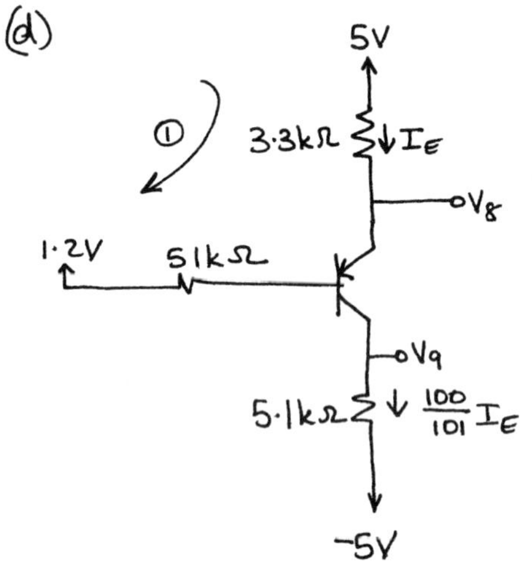

Loop ①

$$5 - 3.3I_E - 0.7 - \frac{I_E}{101}51 - 1.2 = 0$$

$$I_E = 0.8147\ mA$$

$$V_8 = 5 - 3.3\,I_E = \underline{\underline{2.3114\ V}}$$

$$V_9 = -5 + 5.1 \times \frac{100}{101} I_E = \underline{\underline{-0.8862V}}$$

(e) Use Thévenin's theorem to simplify the bias network:

$$V_{BB} = -5 + \frac{150}{150+91} \times 10 = 1.224\ V$$

$$R_{BB} = 150 \| 91 = 56.64\,k\Omega$$

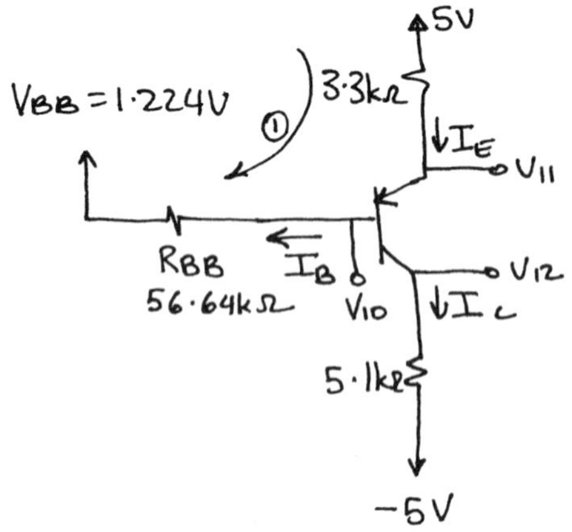

Loop ①

$$5 - 3.3I_E - 0.7 - \frac{I_E}{101} R_{BB} - 1.224 = 0$$

$$I_E = 0.7967\ mA$$

$$V_{11} = 5 - 3.3I_E = \underline{\underline{2.371V}}$$

$$V_{12} = \frac{100}{101} I_E \times 5.1 - 5 = \underline{\underline{-0.977V}}$$

$$V_{10} = V_{11} - 0.7 = \underline{\underline{1.67V}}$$

5.81

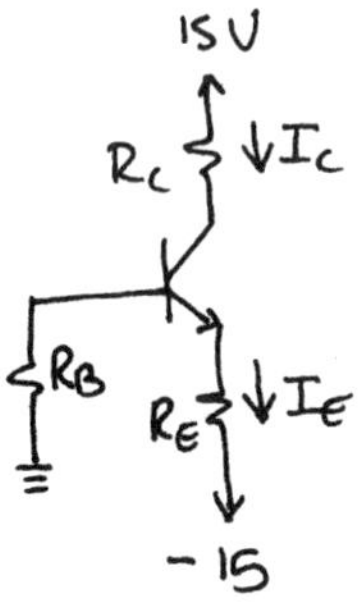

Nominal $\beta = 100$.

Thus,

nominal $\alpha = \frac{100}{101} = 0.99$

nominal $I_E = 1mA$

nominal $I_C = 0.99mA$

nominal $V_C = 5V$

Thus, $R_C = \frac{15-5}{0.99} = 10.1k\Omega \overset{use}{\Longrightarrow} \underline{\underline{10k\Omega}}$

$$I_E = 1 = \frac{15 - 0.7}{R_E + \frac{R_B}{\beta+1}}$$

$$= \frac{14.3}{R_E + \frac{R_B}{101}}$$

$$\Rightarrow R_E + \frac{R_B}{101} = 14.3 \qquad (1)$$

As β varies from 50 to 150, need to limit the variation of I_E to $\pm 10\%$ of 1mA. One can reason that the maximum variation in I_E occurs for $\beta = 50$ (as opposed to $\beta = 150$). To see this more that when β decreases from 100 to 50 the base current doubles while a change in β from

CONT.

100 to 150 causes the base current to decrease to 2/3 its nominal value. Thus our decision will be based on imposing the 10% limit for $\beta = 50$.

$$0.9 = \frac{14.3}{R_E + \frac{R_B}{\beta+1}} = \frac{14.3}{R_E + \frac{R_B}{51}}$$

$$R_E + \frac{R_B}{51} = 15.89 \quad (2)$$

$$(2) - (1) \Rightarrow R_B\left(\frac{1}{51} - \frac{1}{101}\right) = 1.59$$

$$\Rightarrow R_B = 163.8\,k\Omega \overset{\text{USE}}{\Rightarrow} \underline{\underline{164\,k\Omega}}$$

Sub into (1) gives

$$R_E = 12.7\,k\Omega \overset{\text{USE}}{\Rightarrow} \underline{\underline{13\,k\Omega}}$$

To find the expected range of I_C & V_C corresponding to β variation from 50 to 150 we use

$$I_C = \alpha \frac{14.3}{R_E + \frac{R_B}{\beta+1}}$$

for $\beta = 50$: $I_C = \frac{50}{51} \cdot \frac{14.3}{13 + \frac{164}{51}} = \underline{\underline{0.864\,mA}}$

$$V_C = 15 - 0.864 \times 10 = \underline{\underline{6.36\,V}}$$

for $\beta = 150$: $I_C = \frac{150}{151} \times \frac{14.3}{13 + \frac{164}{151}}$

$$= \underline{\underline{1.008\,mA}}$$

$$V_C = 15 - 1.008 \times 10 = \underline{\underline{4.92\,V}}$$

5.82

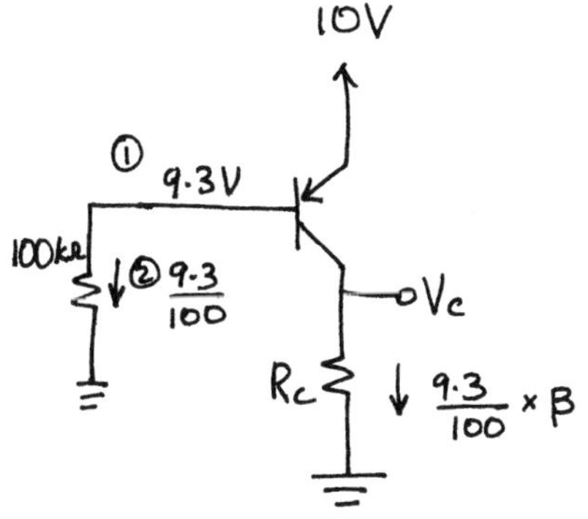

For $V_C = 5\,V = \frac{9.3}{100} \times \beta \times R_C \qquad \beta = 50$

$$R_C = \frac{500}{9.3 \times 50} = \underline{\underline{1.08\,k\Omega}}$$

For $\beta = 100$

$$V_C = \frac{9.3}{100} \times \beta \times R_C = \frac{9.3}{100} \times 100 \times 1.08$$

$$= \underline{\underline{10.04\,V}} \leftarrow V_{BC} = 9.3 - 10.04 = -0.74$$

Since $V_{BC} < -0.4\,V$ the transistor $\underline{\underline{\text{saturates}}}$!

5.83

For $\beta = \infty$ and R open:

$i_{B1} = i_{B2} = 0$, $\alpha_1 = \alpha_2 = 1$

+9V, 80K, I_{D1}, D1, 40K

$I_{D1} = \frac{9 - 0.7}{80K + 40K} = 0.07mA$

$V_{B1} = 0.7 + 0.07m \times 40K = 3.5V$

$V_{E1} = V_{B1} - 0.7V = 3.5 - 0.7 = 2.8V$

$I_{E1} = \frac{2.8V}{2K} = 1.4mA$

$I_{E1} = I_{C1}$ since $\alpha = 1$

$V_{C1} = 9V - 2K \times 1.4m - 0.7 = 5.5V$

$V_{CB} = 2V \rightarrow$ Transistor is in active mode.

$V_{B2} = V_{C1} = 5.5V$

$V_{E2} = 5.5 + 0.7 = 6.2V$

$I_{E2} = \frac{9 - 6.2}{100} = 28mA$

$I_{E2} = I_{C2}$ since $\alpha = 1$

$\rightarrow V_{C2} = 28mA \times 100\Omega = 2.8V$

For $\beta = \infty$ and R connected:

Still: $V_{B1} = 3.5V$, $V_{E1} = 2.8V$

2K, + Same potential, 100, + 0.7 −, +3.5V, +2.8V, 2K, 2K, 100

Since the voltage across the top two resistors is equal:

$I_{C1} \times 2K = I_{E2} \times 100$

$I_{E2} = 20 I_{C1}$

also: $I_{C1} = I_{E1}$, $I_{C2} = I_{E2}$

+2.8V, I_{C1}, $20 I_{C1}$, V_{C2}, 2K, $1.4mA = \frac{2.8V}{2.4K}$, 2.4K, 100, $\frac{V_{C2}}{100}$

$1.4m - I_{C1} = 20 I_{C1} - \frac{V_{C2}}{100}$

$\rightarrow V_{C2} = 100 \times (21 . I_{C1} - 1.4m)$ ①

also:

$\frac{V_{C2} - 2.8V}{2K\Omega} = 1.4mA - I_{C1}$

$\rightarrow V_{C2} = 5.6 - I_{C1} \times 2K$ ②

Solving for I_{C1} from ① & ②

$I_{C1} = 1.4mA$

Substituting in either ① v ②

$V_{C2} = 2.8V$

and: $V_{E2} = 9 - 100 \times 1.4m = 8.86V$

$V_{B2} = V_{C1} = 8.86 - 0.7 = 8.16V$

For $\beta = 100$ and R open:

In the previous two cases $I_{D1} = 0.07mA$, $I_{E1} = 1.4mA$

if $\beta = 100 \rightarrow I_{B1} \simeq 0.014mA$ which is a significant amount compared to $0.07m$

$\rightarrow$ must be taken into account

The bottom two resistors have equal voltage drops thus,

CONT.

$2K \times I_{E1} = 40K \times I_{D1}$

$\longrightarrow I_{D1} = 0.05 \times I_{E1}$ (3)

also: $\underbrace{I_{D1} + I_{B1}}_{\text{FROM (3)}} = \dfrac{9 - V_{B1}}{80K}$

$I_{E1}\left(\dfrac{1}{\beta+1} + 0.05\right)$

for $\beta = 100$:

$0.06 \times I_{E1} = \dfrac{9 - V_{B1}}{80K}$

$\rightarrow V_{B1} = 9 - 4800 \times I_{E1}$ (4)

also: $V_{B1} = 0.7 + I_{E1} \times 2K$ (5)

From (4) ∧ (5) $\begin{cases} V_{B1} = 3.14V \\ I_{E1} = 1.22mA \end{cases}$

$V_{E1} = 1.22m \times 2K \rightarrow V_{E1} = \underline{\underline{2.44V}}$

$I_{C1} = \alpha I_E = 0.99 \times 1.22m$
$= 1.2mA$

Again: voltage drop on top two resistors is equal

$2K \cdot I_{D2} = 100 \cdot I_{E2}$

$I_{D2} = 0.05\ I_{E2}$

but $I_{D2} = 1.2m - \dfrac{I_{E2}}{\beta+1}$

$\Rightarrow 1.2mA = \underbrace{\left(0.05 + \dfrac{1}{\beta+1}\right)}_{0.06} I_{E2}$

$I_{E2} = 20mA$

$V_{E2} = 9 - 100 \times 20m = \underline{\underline{7V}}$

$V_{B2} = V_{C1} = 7 - 0.7 = \underline{\underline{6.3V}}$

$I_{C1} = \alpha I_{E1} = 19.8mA$

$V_{C2} = 100 \times 19.8mA = \underline{\underline{1.98V}}$

__For $\beta = 100$ and R connected:__

To simplify the solution: assume i on R is $<< I_{E1} \rightarrow V_{E1} = 2.44V$

From top of circuit:

$I_{E2} = I_{C1}/0.06$

$I_{C2} = \dfrac{\alpha^2}{0.06} \cdot I_{E1}$

$I_{C2} = 16.3 \times I_{E1}$

To obtain I_{E1}:

$1.22m - I_{E1} = 16.3 I_{E1} - \dfrac{V_{C2}}{100}$

$V_{C2} = 100\,(17.2 I_{E1} - 1.22m)$ (6)

also:

$\dfrac{V_{C2} - 2.44}{2K} = 1.22m - I_{E1}$

$\rightarrow V_{C2} = 2.44 - 2000 I_{E1} + 2.44$
$= 4.88 - 2 \times 10^3 \cdot I_{E1}$ (7)

From (7) ∧ (8): $I_{E1} = 1.34mA$
$V_{C2} = \underline{\underline{2.18V}}$

thus,

$I_{C1} = 1.33mA$

$I_{E2} = 22.1mA$

$V_{E2} = 9 - 22.1m \times 100 = \underline{\underline{6.79V}}$

$V_{B2} = 6.79 - 0.7 = \underline{\underline{6.09V}}$

To confirm initial assumption on i of R:

$\dfrac{2.44 - 2.18}{2K} = 0.13mA$

which is 10 times smaller than I_{E1}

5.84

(a) $\beta = \infty$

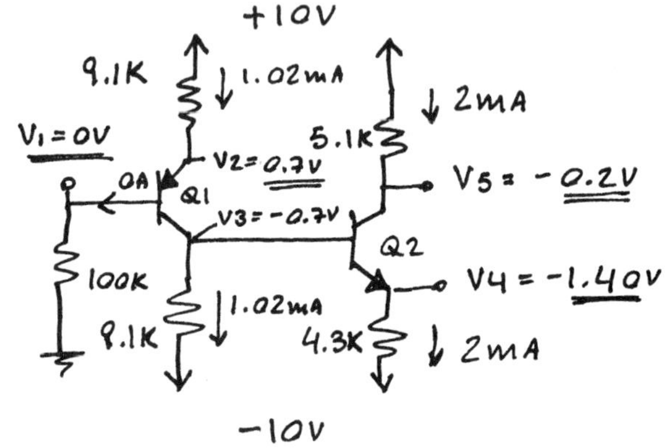

$V_1 = 0V$, $V_2 = 0.7V$, $V_3 = -0.7V$, $V_5 = -0.2V$, $V_4 = -1.40V$

(b) $\beta = 100$

$V_1 = 0.91V$, $V_2 = 1.61V$, $V_3 = -1.87V$, $V_5 = 1.27V$, $V_4 = -2.57V$

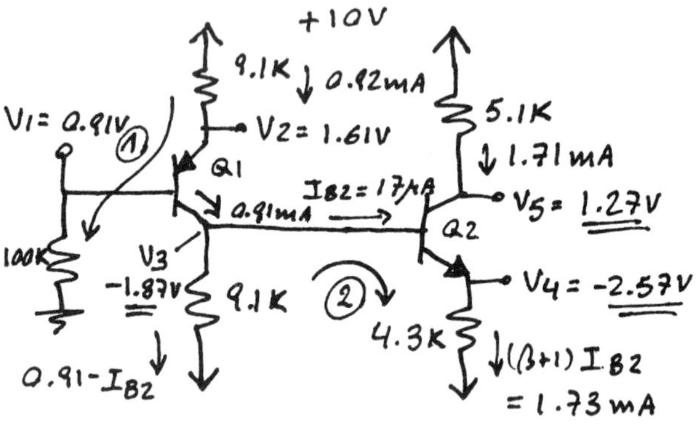

① $10 - 9.1(\beta+1)I_{B1} - 0.7 - 100 I_{B1} = 0$

$\rightarrow I_{B1} = 0.009\ mA$

② $(0.91 - I_{B2}) \times 9.1 = 0.7 + (\beta+1) I_{B2} \times 4.3$

$\rightarrow I_{B2} = 0.017\ mA$

$\rightarrow I_{E2} = 1.73\ mA$

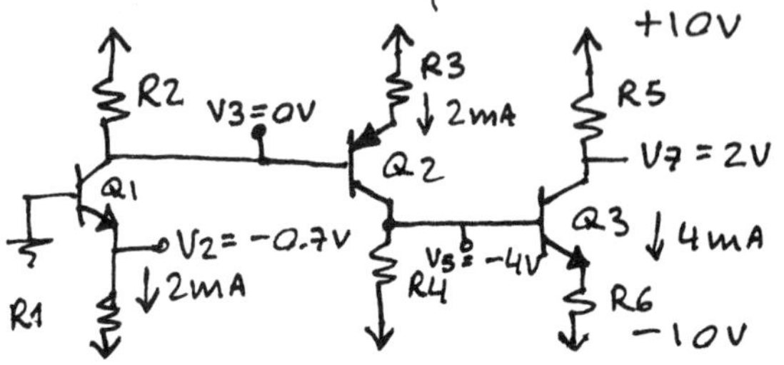

$R_1 = \frac{9.3}{2} = 4.7K\Omega$

$R_2 = \frac{10}{2} = 5 \rightarrow 5.1K\Omega$

$R_3 = \frac{9.3}{2} = 4.7K\Omega$

$R_4 = \frac{6}{2} = 3K\Omega$

$R_5 = \frac{8}{4} = 2K\Omega$

$R_6 = \frac{10-4.7}{4} = 1.3K\Omega$

$\beta = 100$

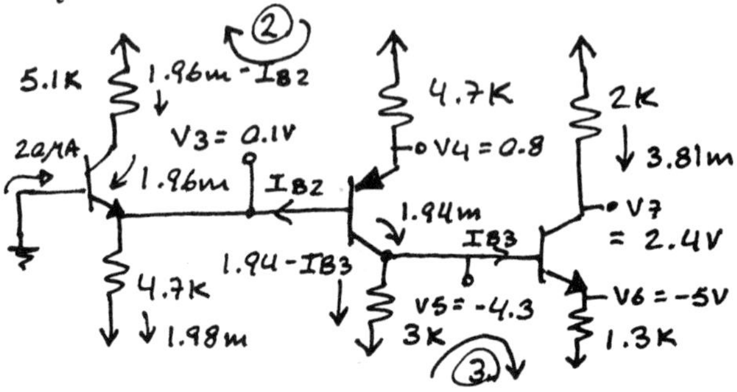

② $(1.96 - I_{B2}) \times 5.1 = (\beta+1) I_{B2} \times 4.7 + 0.7$

$I_{B2} = 0.0194\ mA$

$I_{E2} = 1.96\ mA$

$V_3 = 0.1V \qquad V_4 = 0.8V$

③ $(1.94 - I_{B3}) \times 3 = 0.7 + 1.3 \times (\beta+1) \cdot I_{B3}$

$I_{B3} = 0.038\ mA$

$I_{E3} = 3.85\ mA$

$V_5 = -4.3V \qquad V_6 = -5V$

$V_7 = 2.4V$

5.85

$\beta = \infty$

$V_3 = 0V$, $V_2 = -0.7V$, $V_7 = 2V$, $V_5 = -4V$

5.86

$v_I = 0\text{V}$

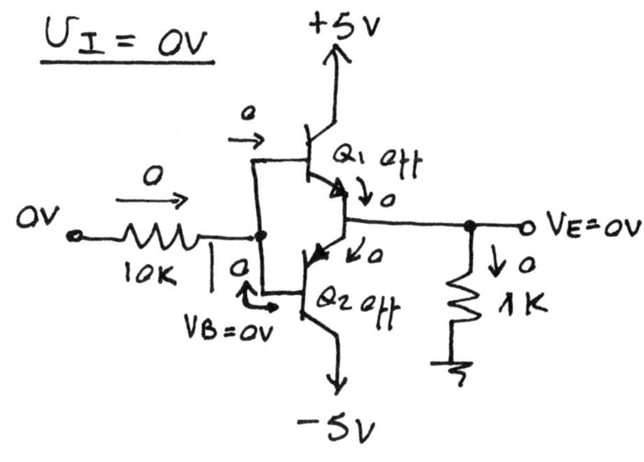

$v_I = +3\text{V}$

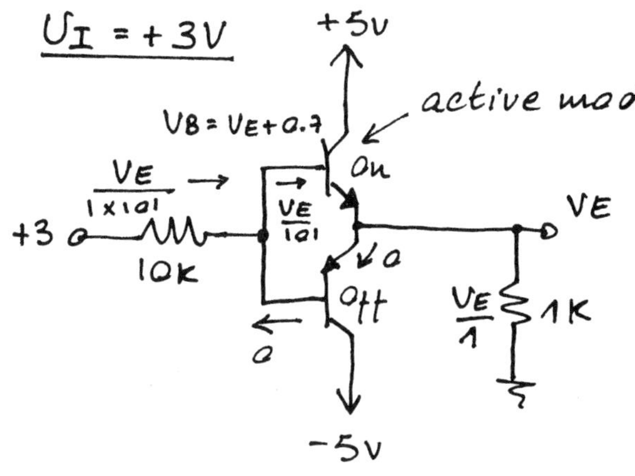

$$3 = \frac{V_E}{101} \times 10 + 0.7 + V_E$$

$$\Rightarrow V_E = \underline{\underline{2.09\text{V}}}$$

$$V_B = \underline{\underline{2.79\text{V}}}$$

$v_I = -5\text{V}$

+5V

0 → off

−5, 10K

I_E ← on, active mode

$\frac{I_E}{101}$

VE, 1K

−5V

$$I_E = \frac{5 - 0.7}{1 + 10/101} = 3.91\text{mA}$$

$$V_E = -3.91\text{V}$$

$$V_B = \underline{\underline{-4.61\text{V}}}$$

$v_I = -10\text{V}$

+5V

$\frac{-5.5-(-10)}{10} = 0.45\text{mA}$

−10V, 10K, off

$V_E = -4.8\text{V}$

on, + 0.2V, 1K, 4.8mA

$V_B = -5.5\text{V}$, saturated

$$\frac{I_C}{I_B} = \frac{4.35}{0.45} = 9.7 < 100$$

thus, Q2 is saturated as assumed

$$V_E = \underline{\underline{-4.8\text{V}}} \quad V_B = \underline{\underline{-5.5\text{V}}}$$

5.87

(a)

$\frac{10 - V_C}{20} = (0.5 - 0.05V_C)\text{mA}$, 20K

10K, $\frac{10 - V_C}{10} = (1 - 0.1V_C)\text{mA}$

V_C, 1K, $\left(\frac{V_C}{1}\right)\text{mA}$

$$(0.5 - 0.005V_C) + (1 - 0.1V_C) = V_C$$

$$V_C = \underline{\underline{1.3\text{V}}}$$

$$I_C = \frac{10 - 1.3}{10} = 0.87\text{mA}$$

$$I_B = \frac{10 - 1.3}{20} = 0.435\text{mA}$$

$$\text{thus } \beta_{forced} = \frac{0.87}{0.435} = \underline{\underline{2}}$$

CONT.

(b)

$$10 - V_C = (0.1V_C + 1) + (V_C)$$

$$\Rightarrow V_C = +4.29\,V$$

$I_C = 4.29\,mA$

$$I_B = \frac{4.29 + 10}{10} = 1.43\,mA$$

$$\beta_{forced} = \frac{4.29}{1.43} = 3$$

(c)

Node equation:

$$\frac{10-V}{10} + \frac{10-V}{30} = 0.1V + 0.1V + 0.1V$$

$$30 - 3V + 10 - V = 9V$$

$$40 = 13V$$

$$\Rightarrow V = 3.08\,V$$

thus, $V_{C3} \simeq V_{C4} \simeq 3.08\,V$

$I_{B3} = 0.1V = 0.308\,mA$

$$I_{E3} = \frac{10 - 3.08}{10} \simeq 0.692\,mA$$

$I_{C3} = 0.692 - 0.308 = 0.384\,mA$

$$\beta_{3\,forced} = \frac{0.384}{0.308} = 1.25$$

$$I_{C4} = \frac{10 - 3.08}{30} = 0.231\,mA$$

$I_{E4} = 0.1V = 0.308\,mA$

$I_{B4} = 0.308 - 0.231 = 0.077\,mA$

$$\beta_{4\,forced} = \frac{0.231}{0.077} = 3$$

5.88

(1)

$$\frac{5 \times R_{B2}}{R_{B1} + R_{B2}} = 0.690$$

$\rightarrow 5R_{B2} = 0.69R_{B1} + 0.69R_{B2}$

$\rightarrow 4.31R_{B2} = 0.69R_{B1}$

$$\Rightarrow \frac{R_{B1}}{R_{B2}} = 6.24$$

(2) Since $V_{BE} = \dfrac{5R_{B2}}{R_{B1} + R_{B2}}$

If both R_{B2} & R_{B1} are at 0.99 or 1.01 of their nominal value $\rightarrow$ V_{BE} will not be affected.
We must consider the cases when one resistor is at 0.99 and the other at 1.01 of their nominal value.

If: $R_{B2}' = 1.01\,R_{B2}$
$R_{B1}' = 0.99\,R_{B1}$

$\Rightarrow V_{BE} = 0.702\,V$

If: $R_{B2}' = 0.99\,R_{B2}$
$R_{B1}' = 1.01\,R_{B1}$

$\Rightarrow V_{BE} = 0.678\,V$

thus V_{BE} ranges from 0.678V to 0.702V

CONT.

For I_C: $I_C = I_S e^{V_{be}/V_T}$

for $V_{be} = 0.690 \rightarrow I_C = 1mA$

$\Rightarrow I_S = 1.032 \times 10^{-15}$

for $V_{BE} = 0.678 \rightarrow I_C = 0.618m$

$V_{BE} = 0.702 \rightarrow I_C = 1.62m$

I_C ranges from 0.618 mA to 1.62 mA.

(3) If $R_C = 3K\Omega$

$V_{CE} = 5 - 3K \times 0.62m = 3.14V$

$V_{CE} = 5 - 3K \times 1.62m = 0.14V$

This circuit is too sensitive to parameter variations as shown here for a 1% resistor tolerance.

5.89

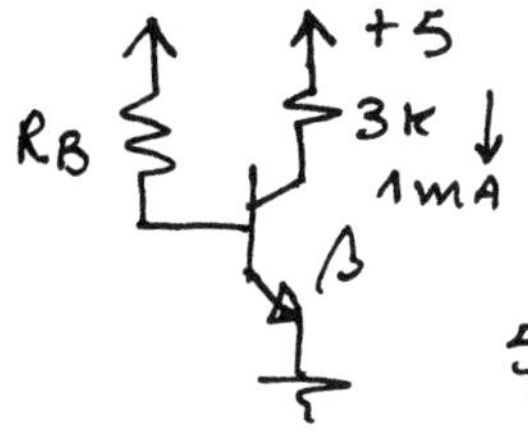

+5, R_B, 3K, 1mA, β

$R_B = ?$ if $\beta = 100$

$I_B \times \beta = I_C$

$\frac{5-0.7}{R_B} = \frac{1m}{100}$

$\rightarrow R_B = 430K\Omega$

$V_{CE} = 5V - 3K \cdot 1mA = 2V$

If $\beta = 50$: $I_C = \frac{5-0.7}{430K} \times 50$

$I_C = 0.50mA$

$\Rightarrow V_{CE} = 5 - 3K \times 0.5m = +3.5V$

If $\beta = 150$: $I_C = 1.5mA$

$V_{CE} = 0.5V$

This design is too sensitive to variations of β.

5.90

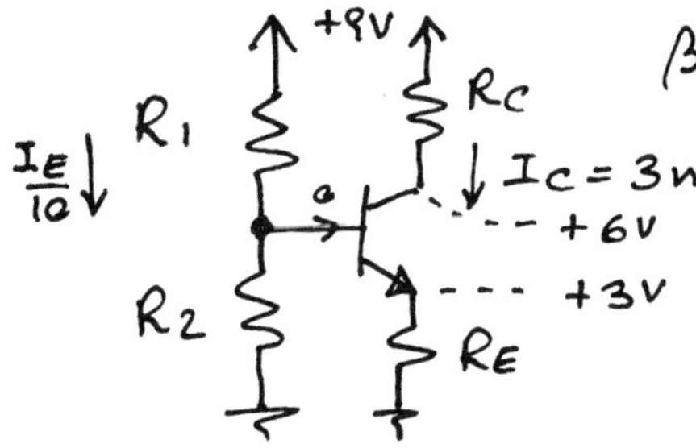

+9V, R_1, R_C, $\frac{I_E}{10}$, $I_C = 3mA$, +6V, +3V, R_2, R_E

$\beta = \infty$

$R_C = \frac{3V}{3mA} = 1K\Omega$

$R_E = \frac{3V}{3mA} = 1K\Omega$

$V_b = 0.7 + 3 = 3.7V$

$R_1 = \frac{9 - 3.7}{I_E/10} = 17.7K\Omega$

$9V = (R_1 + R_2)\frac{I_E}{10} \rightarrow R_2 = 12.3K$

Choose suitable 5% resistors

$R_1 = 17.7K \rightarrow 18K\Omega$

$R_2 = 12.3K \rightarrow 13K\Omega$

$R_1 = R_2 = 1K$

$V_{BB} = \frac{9 \times 13}{18+13} = 3.77V$

For these values of R and $\beta = 90$: $R_B = 18 \parallel 13 = 7.55K\Omega$

$I_E = \frac{3.77 - 0.7}{1K + \frac{7.55K}{91}} = 2.83mA$

$\alpha = 0.989 \Rightarrow I_C = 2.80mA$

If R_E is reduced by $\sim \frac{7.55K}{91}$

$\rightarrow R_E = 910\Omega$

$\Rightarrow I_E = 3.09mA$

$I_C = 3.05mA$.

5.91

For $\beta = \infty$ $\quad I_B = 0$, $I_E = 3mA$

$$R_1 = \frac{9 - 3.7}{I_E/2} = 3.5\,K\Omega$$

$$9V = (R_1 + R_2)\frac{I_E}{2} \Rightarrow R_2 = 2.5\,K\Omega$$

Suitable 5% resistors:

$R_1 = 3.5K \rightarrow 3.3\,K\Omega$

$R_2 = 2.5K \rightarrow 2.4\,K\Omega$

$R_E = R_C = 1\,K\Omega$

$$V_{BB} = \frac{9 \times 2.4}{3.3 + 2.4} = 3.79\,V$$

For $\beta = 90$:

$$R_B = (3.3 \| 2.4)K = 1.39\,K\Omega$$

$$I_E = \frac{3.79 - 0.7}{1K + \frac{1.39K}{91}} = 3.04\,mA$$

No need to adjust R_E. As the current from the voltage divider increases the effect of I_B is reduced.

5.92

$$I_E = \frac{V_{BB} - V_{BE}}{R_E + \frac{R_B}{\beta + 1}}$$

(a) For $\beta = 100$, varying between 50 and 150 the maximum deviation in I_E (from the nominal value obtained for $\beta = 100$) occurs at the low end of β values ($\beta = 50$). Thus, to keep I_E within $\pm 5\%$ of nominal we must impose the constraint

$$I_E(\beta = 50) \geq 0.95\, I_E(\beta = 100)$$

or,
$$\frac{V_{BB} - V_{BE}}{R_E + \frac{R_B}{51}} \geq 0.95\, \frac{V_{BB} - V_{BE}}{R_E + \frac{R_B}{101}}$$

or,
$$R_E + \frac{R_B}{101} \geq 0.95\left(R_E + \frac{R_B}{51}\right)$$

$$0.05\,R_E \geq R_B\left(\frac{0.95}{51} - \frac{1}{101}\right)$$

$$\Rightarrow \frac{R_B}{R_E} \leq 5.73$$

Thus, the largest ratio of R_B/R_E is 5.73

(b) $I_E \cdot R_E = V_{CC}/3$

$$\rightarrow \frac{V_{BB} - V_{BE}}{R_E + \frac{R_B}{\beta + 1}} \cdot R_E = \frac{V_{CC}}{3}$$

$$\frac{V_{BB} - 0.7}{1 + \frac{R_B}{R_E} \cdot \frac{1}{\beta + 1}} = \frac{V_{CC}}{3}$$

$$V_{BB} = \frac{1}{3} V_{CC}\left(1 + \frac{5.73}{101}\right) + 0.7$$

$$\Rightarrow \underline{\underline{V_{BB} = 0.35\, V_{CC} + 0.7}}$$

(c) $V_{CC} = 10V$

$$V_{BB} = 0.35 \times 10 + 0.7 = 4.2V$$

$$\rightarrow \frac{R_2}{R_1 + R_2} \times 10 = 4.2$$

$$\frac{R_2}{R_1 + R_2} = 0.42 \quad (1)$$

$$I_E \cdot R_E = \frac{1}{3} V_{CC}$$

CONT.

$2 \times R_E = \frac{1}{3} \times 10$

$\Rightarrow R_E = \underline{\underline{1.67\ k\Omega}}$

$R_B = 5.73 \times 1.67 = 9.55\ k\Omega$

$\frac{R_1 \cdot R_2}{R_1 + R_2} = 9.55$

Substituting from ① gives

$R_1 = \frac{9.55}{0.42} = \underline{\underline{22.7\ k\Omega}}$

5.93

(a)

+3V, R_C, 3mA, 0, 0V, -0.7, R_B, R_E, -3V

$\beta = \infty,\ I_B = 0$

$I_E = I_C$

$3V = R_C \times 3mA$

$\rightarrow R_C = \underline{\underline{1k\Omega}}$

$(+3 - 0.7) = R_E \times 3mA$

$\rightarrow R_E = \underline{\underline{767\Omega}}$

(b) $\beta = 90 \quad \frac{V_{RE}}{10} = V_{RB}$

$I_B \cdot R_B = \frac{I_E \cdot R_E}{10}$

$\rightarrow \frac{I_E}{(\beta+1)} \cdot R_B = \frac{I_E \cdot R_E}{10}$

$\rightarrow R_B = \frac{(\beta+1)}{10} R_E$ ①

also, $0 = V_{RB} + 0.7 + V_{RE} - 3$

$2.3 = \frac{V_{RE}}{10} + V_{RE}$

$\rightarrow V_{RE} = \frac{2.3}{1.1} = 2.09V$

$2.09 = I_E \times R_E$ ②

but: $I_E = \frac{I_C}{\alpha} = \frac{3mA}{0.989} = 3.033m$

Substituting in ②:

$R_E = 689\Omega$

from ①:

$R_B = \underline{\underline{6269\Omega}}$

(c) Standard 5% values:

$R_C = 1k\Omega$

$R_E = 689\Omega \rightarrow 680\Omega$

$R_B = 6269\Omega \rightarrow 6.2k\Omega$

(d) $\beta = \infty$: $I_B = 0$

$I_C = I_E$

$\underline{V_B = 0}$

$\underline{V_E = -0.7}$

$I_E = \frac{3 - 0.7}{R_E} = \frac{3 - 0.7}{680} = \underline{\underline{3.38mA}}$

$V_C = 3 - 3.38m \times 1K = \underline{\underline{-0.38V}}$

For $\beta = 90$:

$I_E = \frac{2.3}{680 + \frac{6.2K}{91}} = 3.07mA$

$I_C = \alpha I_E = \underline{\underline{3.04mA}}$

$V_B = \frac{R_B \cdot I_E}{\beta+1} = -0.209$

$V_E = -0.209 - 0.7 = \underline{\underline{0.909\ V}}$

$V_C = 3 - I_C \cdot R_C$

$= 3 - 3.04 \times 1 = \underline{\underline{-0.04V}}$

5.94

$V_E = -0.7V$

To obtain $I_E = 1mA$

$R_E = \frac{-0.7 - (-5)}{1}$

$= 4.3k\Omega$

+5V, R_C, ∞, R_E, -5V

CONT.

To maximize gain while allowing ±1V signal at collector, design for a dc collector voltage of +1V. Thus,

$$R_C = \frac{5-1}{I_C} \simeq \frac{4}{1} = \underline{\underline{4\,K\Omega}} \quad (\text{for } \alpha = 1)$$

For 100°C rise in temperature, V_{BE} decreases by $2 \times 100 = 200mV$ and thus I_E increases by $\frac{0.2V}{R_E}$

$$= \frac{0.2V}{4.3K\Omega} = 0.047\,mA$$

i.e. an increase of $\underline{\underline{4.7\%}}$

The change in β from 50 to 150 causes α to change from 0.980 to 0.993 which implies an increase in collector current of $\underline{1.3\%}$

Thus the overall increase in I_C is $\underline{\underline{6\%}}$

5.95

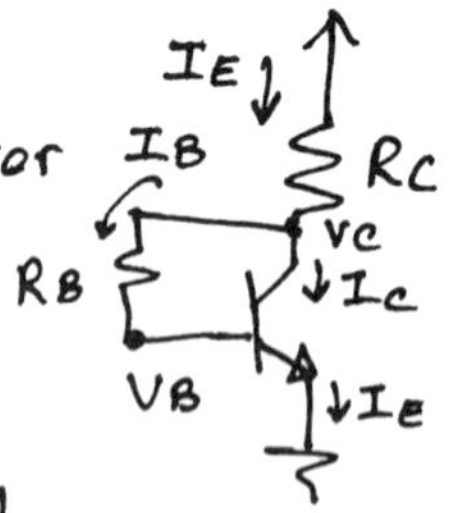

To allow a collector voltage swing of ±1V, we design for:

$$V_C = V_B + 1$$
$$= 0.7 + 1 = 1.7V$$

$I_E = 0.5mA$

$$\rightarrow R_C = \frac{5-1.7}{0.5} = \underline{\underline{6.6\,K\Omega}}$$

For $\beta = 100$:

$$I_B = \frac{I_E}{\beta+1} = \frac{0.5}{101} \simeq 5\mu A$$

$$I_B \cdot R_B = 1V$$

$$R_B = \frac{1V}{5\mu A} = \frac{1}{5}\,M\Omega = \underline{\underline{200K\Omega}}$$

Now, if the BJT used has $\beta = 50$, the emitter current resulting can be found from Eq (5.94)

$$I_E = \frac{V_{CC} - V_{BE}}{R_C + \frac{R_B}{\beta+1}}$$

$$= \frac{5-0.7}{6.6 + \frac{200}{51}} = \underline{\underline{0.41mA}}$$

and $I_B = \frac{0.41}{51} \simeq 8\mu A$

Thus the collector will be higher than the base by $8 \times 0.2 = 1.6V$, allowing for a $\underline{\underline{\pm 1.6V}}$ signal swing at the collector.

For $\beta = 150$:

$$I_E = \frac{5-0.7}{6.6 + \frac{200}{151}} = 0.54mA$$

$$I_B = \frac{0.54}{151} = 36\mu A$$

Thus the collector voltage will be higher than that of the base by $3.6 \times 0.2 = 0.72V$ allowing for only $\underline{\underline{\pm 0.72V}}$ signal swing.

5.96

(a)

$$I_E = \frac{1.5}{R_C} = \frac{I_C}{\alpha} \Rightarrow R_C = \frac{1.5\,\alpha}{3\,mA}$$

+3V, I_E, R_C, $V_C = 1.5V$, R_B, $I_C = 3mA$, $\beta = 90$

$R_C = 495\,\Omega$

$1.5 = R_B I_B + 0.7$

$$R_B = \frac{1.5 - 0.7}{I_C/\beta} = 24K\Omega$$

(b) Standard 5% values

$R_B = 24K\Omega$

$R_C = 495\Omega \rightarrow 510\Omega$

then, $$I_E = \frac{3 - 0.7}{510 + \frac{24K}{91}} = 2.97mA$$

$I_C = I_E \cdot \alpha = 2.93mA$

$V_C = 3 - 2.97m \times 510 = 1.48V$

(c) $\beta = \infty \quad I_B = 0,\ V_B = 0.7V$

$$I_C = I_E = \frac{3 - 0.7}{510} = 4.5mA$$

$V_C = 3 - 4.5m \times 510 = 0.7V$

(d)

+3V, R_C, $I_C + 2I_B$, $V_C = 1.5V$, $2I_B$, $I_C = 3mA$, R_{B1}, I_B, $\beta = 90$, R_{B2}, I_B

$$I_B = \frac{I_C}{\beta} = \frac{3mA}{90} = 0.033mA$$

$1.5V = 2I_B \times R_{B1} + 0.7$

$\rightarrow R_{B1} = 12.1K\Omega$

$0.7 = I_B \times R_{B2}$

$\rightarrow R_{B2} = 21.2K\Omega$

on R_C:

$I_C + 2I_B = 3.066mA$

$$R_C = \frac{1.5V}{3.066mA} = 489\Omega$$

Standard 5% values:

$R_{B1} = 12.1K \rightarrow 12K\Omega$

$R_{B2} = 21.2K \rightarrow 20K$

$R_C = 489.2 \rightarrow 480\Omega$

Re-evaluate if $\beta = \infty$:

$I_B = 0$

+3V, 480Ω, 12K, 0, 20K

① $\frac{0.7}{20K} = 35\mu A$ (0.7)

② $12K \times 35\mu A = 0.42V$

③ $0.42 + 0.7 = 1.12V$

④ $\frac{3V - 1.12}{480} = 4mA$

5.97

$I_B = I_C/\beta = 3mA/90 = 0.033m$

$V_C = R_B \cdot I_B + 0.7$

$V_C = 1.5V \rightarrow R_B = 24.2K\Omega$

$$I_E = \frac{I_C}{\alpha} = 3.03mA$$

$I = I_C - I_B \equiv I_E$

$I = 3.03mA$

5.98

$$= \; \frac{I_E}{\beta+1}, \quad R_1 \| R_2, \quad I_o = \alpha I_E, \quad I_E, \quad R_E, \quad V_{CC}\frac{R_2}{R_1+R_2}$$

$$V_{CC}\frac{R_2}{R_1+R_2} = \frac{I_E}{\beta+1}(R_1\|R_2) + V_{BE} + I_E R_E$$

$$\Rightarrow I_E = \frac{V_{CC}\frac{R_2}{R_1+R_2} - V_{BE}}{R_E + \frac{R_1\|R_2}{\beta+1}}$$

Thus,

$$I_o = \alpha I_E = \alpha \cdot \frac{\left(V_{CC}\frac{R_2}{R_1+R_2} - V_{BE}\right)}{R_E + \frac{R_1\|R_2}{\beta+1}}$$

Q.E.D.

5.99

For $\beta = \infty$:

$$R_{E2} = \frac{2}{0.1} = 20\,K\Omega$$

$$R_C = \frac{5-0.8}{0.1} = 42\,K\Omega$$

Select $R_C = 43K$

Also $V_{R1} = 2.3V$

$V_{R2} = 2.7V$

$$\rightarrow \frac{R_1}{R_2} = \frac{2.3}{2.7}$$

$$\Rightarrow R_1 = \frac{23}{27} \cdot R_2$$

+5V, R_C, 0V, Q_1, R_B, +1.5V −, −0.7V, R_1, −2.3V, Q_2, −3V, R_2, + 2V −, R_{E2}, 0.1mA, −5V, −5V

For $\beta = 50$

$5R_1/(R_1+R_2)$, $I_{E2}/(\beta+1)$, $R_1\|R_2$, Q2, 20K, I_{E2}, −5V

+5V, 43KΩ, Q1, R_1, Q2, I_{E2}, R_2, 20KΩ, −5V, −5V

$$I_{E2} = \frac{-5 \cdot \frac{R_1}{R_1+R_2} - 0.7 + 5}{20 + \frac{R_1\|R_2}{\beta+1}}$$

$$= \frac{5 \cdot \frac{R_1}{R_1+R_2} - 0.7}{20 + \frac{1}{51}\cdot\frac{R_1 R_2}{R_1+R_2}}$$

Substitute $R_1 = \frac{23}{27}R_2$

$$I_{E2} = \frac{\frac{5}{1+23/27} - 0.7}{20 + \frac{1}{51}\cdot\frac{23/27\, R_2}{1+23/27}}$$

$$= \frac{2}{20\left[1 + \frac{1}{20\times 51} \times \frac{23R_2}{50}\right]}$$

$$V_{RE2} = I_{E2} \times 20 = \frac{2}{1 + \frac{0.46}{51\times 20}\cdot R_2}$$

For a reduction less than 5%

$$\frac{2}{1 + \frac{0.46 R_2}{51\times 20}} \geq 0.95\times 2$$

$$R_2 \leq \left(\frac{1}{0.95} - 1\right)\frac{51\times 20}{0.46} = 116.7\,K\Omega$$

CONT.

Select $R_2 = \underline{\underline{120 k\Omega}}$

$$R_1 = \frac{23}{27} R_2 = 102.2 k\Omega$$

Select $R_1 = \underline{\underline{100 k\Omega}}$

For these values,

$$\beta = \infty : I_{E2} = \frac{-5 \times \frac{100}{220} - 0.7 + 5}{20}$$

$$= \frac{2.027}{20} = 0.101 mA$$

$$V_{RE2} = 2.027 V$$

$$\beta = 50 : I_{E2} = \frac{2.027}{20 + \frac{120 \times 100}{220 \times 51}}$$

$$I_{E2} = 0.096 mA$$

$$I_{C2} = 0.98 \times 0.096 = 0.094 mA$$

$$I_{C1} = 0.98 \times 0.094 = 0.092 mA$$

To determine R_B:

$$V_{B1} = -I_{B1} . R_B = -\frac{I_{C1}}{\beta} . R_B$$

$$= -\frac{0.094}{50} . R_B$$

$$V_{E1} = V_{B1} - 0.7 = -\frac{0.094}{50} . R_B - 0.7$$

$$V_{C1} = 5 - R_C I_{C1} = 5 - 43 \times 0.092$$

$$= 1.044 V$$

$$V_{CE1} = 1.044 + \frac{0.094}{50} . R_B + 0.7$$

$$= 2.5 V$$

$\Rightarrow R_B = 402 k\Omega$

Select $R_B = \underline{\underline{390 k\Omega}}$

Now,

For $\beta = 50$: $I_{C1} = \underline{\underline{0.092 mA}}$

$$V_{B1} = -\frac{0.092}{50} \times 390 = -0.717 V$$

$$V_{E1} = -1.417 V$$

$$V_{CE1} = 1.044 + 1.417 = 2.46 \simeq \underline{\underline{2.5 V}}$$

For $\beta = 100$:

$$I_{E2} = \frac{2.027}{\frac{R_1 R_2}{R_1 + R_2} . \frac{1}{101} + 20} = 0.099 mA$$

$$I_{C1} = 0.099 \times 0.99 \times 0.99 = \underline{\underline{0.097 mA}}$$

$$V_{C1} = 0.829 V$$

$$V_{B1} = -0.378 V$$

$$V_{E1} = -1.078 V$$

$$V_{CE1} = \underline{\underline{1.91 V}}$$

For $\beta = 200$:

$$I_{E2} = \frac{2.027}{\frac{R_1 R_2}{R_1 + R_2} \cdot \frac{1}{201} + 20} = 0.1 mA$$

$$I_{C1} = 0.1 \times 0.995 \times 0.995 = \underline{\underline{0.099 mA}}$$

$$V_{C1} = 0.743 V$$

$$V_{B1} = -0.193 V$$

$$V_{E1} = -0.893 V$$

$$V_{CE1} = 1.636 \simeq \underline{\underline{1.64 V}}$$

5.100

Refer to Fig P5.100

Assuming $\beta = \infty \rightarrow I_{B3} = 0$

$$V_{B3} = \frac{V_{CC} - V_{BE1} - V_{BE2}}{R_1 + R_2} \times R_2 \ldots$$

$$+ V_{BE1} + V_{BE2}$$

$$V_{E3} = V_{B3} - V_{BE3}$$

$$I_O = \frac{\alpha V_{E3}}{R_E}$$

$$I_O = \frac{\alpha}{R_E} \left[\frac{(V_{CC} - V_{BE1} - V_{BE2}) R_2}{R_1 + R_2} + V_{BE1} + V_{BE2} - V_{BE3} \right]$$

CONT.

Now, if the circuit is designed so that all three transistors conduct equal currents, then $V_{BE1} = V_{BE2} = V_{BE3} = V_{BE}$

and I_O becomes:

$$I_O = \frac{\alpha}{R_E}\left[(V_{CC} - 2V_{BE})\frac{R_2}{R_1+R_2} + V_{BE}\right]$$

To eliminate the V_{BE} terms in this equation we select $\underline{R_1 = R_2}$, resulting in

$$I_O = \frac{\alpha V_{CC}}{2R_E} \quad \text{Q.E.D.}$$

Now, to make the current through Q_1 and Q_2 equal to that through Q_3 (which is I_O/α)

$$\frac{V_{CC} - 2V_{BE}}{2R_1} = \frac{V_{CC}}{2R_E}$$

$$\Rightarrow R_1 = R_2 = R_E\left(1 - \frac{2V_{BE}}{V_{CC}}\right)$$

For $V_{CC} = 10\text{V}$, $\alpha = 1$
$I_O = 0.5\text{mA}$

$$0.5 = \frac{10}{2R_E} \Rightarrow R_E = \underline{10\text{K}\Omega}$$

$$R_1 = R_2 = 10\left(1 - \frac{2\times 0.7}{10}\right)$$

$$= \underline{8.6\text{K}\Omega}$$

$$V_{C3}|_{min} = I_O . R_E + V_{BE}$$
$$= 0.5\times 10 + 0.7$$
$$= \underline{5.7\text{V}}$$

5.101

Refer to Fig. P5.101

$$I_O = 2\text{mA} = \alpha \times \frac{5-0.7}{R} \simeq \frac{4.3}{R}$$

$$\Rightarrow R = \underline{2.15\text{K}\Omega}$$

$V_C\,min = \underline{0\text{V}}$ (In actual practice, $V_C\,min \simeq 0.4\text{V}$)

5.102

(a) Using the exponential characteristic:

$$i_C = I_C e^{v_{be}/V_T}$$

thus,

$$i_c = I_C e^{v_{be}/V_T} - I_C$$

giving

$$\frac{i_c}{I_C} = e^{v_{be}/V_T} - 1$$

(b) Using small-signal approximation:

$$i_c = g_m v_{be} = \frac{I_C}{V_T} . v_{be}$$

Thus,

$$\frac{i_c}{I_C} = \frac{v_{be}}{V_T}$$

See table below

For signals of ±5mV, the error introduced by the small-signal approximation is 10%.
The error increases to above 20% for signals of ±10mV.

CONT.

v_{be} (mV)	i_c/I_C Expon.	i_c/I_C small-signal.	% Error
+1	+0.041	+0.040	−2
−1	−0.039	−0.040	+2
+2	+0.083	+0.080	−4
−2	−0.077	−0.080	+4
+5	+0.221	+0.200	−9.5
−5	−0.181	−0.200	+10.3
+8	+0.377	+0.320	−15.2
−8	−0.274	−0.320	+16.8
+10	+0.492	+0.400	−18.7
−10	−0.330	−0.400	+21.3
+12	+0.616	+0.480	−22.1
−12	−0.381	−0.480	+25.9

5.103

+15V, 10K, I_C 1mA, V_C, +0.700

$V_C = 15 - 1 \times 10$
$= 5V$

$v_{BE} = 0.7 + v_{be}$
where $v_{be} = 0.005V$

$i_C \cong I_C \left(1 + \dfrac{v_{be}}{V_T}\right)$ Eq. (5.83)

$i_C \simeq I_C + i_c$ where:

$i_c = \dfrac{1m \times 0.005}{25m} = 0.2m$

$i_C = 1mA + 0.2mA$

$V_C = V_{CC} - i_C R_C$ Eq. (5.101)

$\Rightarrow V_C - \underbrace{i_c R_C}_{0.2m \times 10K}$

$V_C = 5V - 2V$

$\text{gain} = \dfrac{-2V}{0.005V} = -400\ V/V$

while $-g_m . R_C = -\dfrac{1m}{25m} . 10K = -400\ V/V$

5.104

$g_m = \dfrac{I_C}{V_T} = \dfrac{1.2mA}{25mV} = 48 \dfrac{mA}{V}$

$r_\pi = \dfrac{\beta}{g_m} = \dfrac{120}{48 \times 10^{-3}} = 2.5K\Omega$

$r_e = \dfrac{r_\pi}{\beta + 1} = \dfrac{2500}{121} = 20.6\Omega$

For a bias current of 120μA i.e. 10 times lower:

$g_m = \dfrac{48}{10} = 4.8\ mA/V$

$r_\pi = 10 \times 2.5 = 25K\Omega$

$r_e = 10 \times 20.6 = 206\Omega$

5.105

$I_C = 2mA \Rightarrow g_m = \dfrac{2mA}{25mV}$

$g_m = 80mA/V$

$r_e = \dfrac{V_T}{I_E}$, $I_E = I_C \dfrac{(\beta + 1)}{\beta}$

$I_E = 2mA \times \dfrac{51}{50} = 2.04mA$

$r_e = \dfrac{25m}{2.04m} = 12.25\Omega$

$r_\pi = \dfrac{\beta}{g_m} = \dfrac{50}{80 \times 10^{-3}} = 625\Omega$

gain: $-g_m \times R_C$

For $R_C = 5K\Omega$ and $\hat{v}_{be} = 5mV$

$\hat{v}_o = -80m \times 5K \times 5mV$
$= -2V$

5.106

$g_m = 50\frac{mA}{V} = \frac{I_C}{V_T}$

$\Rightarrow I_C = g_m \times V_T = 50m \times 25m$
$= 1.25 mA$

$r_\pi = 2K = \frac{\beta}{g_m} \Rightarrow \beta = 2K \times 50m$

$\beta = \underline{\underline{100}} \longrightarrow \alpha = \frac{100}{101} = 0.99$

$I_E = \frac{I_C}{\alpha} = \frac{1.25mA}{0.99} = \underline{\underline{1.26mA}}$

5.107

g_m varies from: $1.2 \times 60 = 72\frac{mA}{V}$
to $0.8 \times 60 = 48\frac{mA}{V}$

β varies from 50 to 200

$R_{in}|_{base} = r_\pi = \beta / g_m$

Largest value: $r_\pi = \frac{\beta_{max}}{g_{m\,min}} = \frac{200}{48m}$
$= \underline{4.2K\Omega}$

Smallest value: $r_\pi = \frac{\beta_{min}}{g_{m\,max}} = \frac{50}{72m}$
$= \underline{\underline{694\Omega}}$

5.108

Refer to Fig. 5.48.

$V_C = 2V \Rightarrow I_C = \frac{V_{CC} - V_C}{R_C}$

$I_C = \frac{5-2}{3K} = 1mA$

$g_m = \frac{I_C}{V_T} = \frac{1m}{25m} = 40\frac{mA}{V}$

$i_C(t) = I_C + g_m v_{be}(t)$
$= 1m + 40.10^{-3} \times 0.005 \sin\omega t$
$= \underline{\underline{1 + 0.2 \sin\omega t, \; mA}}$

$v_C(t) = 5 - R_C i_C(t)$
$= \underline{\underline{2 - 0.6 \sin\omega t, \; V}}$

$i_B(t) = i_C(t) / \beta$
$= \frac{1 + 0.2 \cdot \sin\omega t}{100}, mA$
$= \underline{\underline{10 + 2 \sin\omega t, \; \mu A}}$

Voltage gain $= \frac{-0.6}{0.005} = \underline{\underline{-120}}\; V/V$

5.109

$i_C = I_C + g_m \hat{v}_{be} \sin\omega t$

$v_C = V_{CC} - I_C R_C - g_m \hat{v}_{be} R_C \sin\omega t$

To maintain BJT in active region, $v_C \geq v_{BE}$, thus
$V_{CC} - I_C R_C - g_m R_C \hat{v}_{be} \geq V_{BE} + \hat{v}_{be}$

To obtain the largest possible output signal we design such that this constraint is satisfied with the equality sign, that is:

$V_{CC} - R_C I_C - g_m R_C \hat{v}_{be} = V_{BE} + \hat{v}_{be}$

substituting $g_m = \frac{I_C}{V_T}$, gives.

$V_{CC} - R_C I_C - R_C I_C \frac{\hat{v}_{be}}{V_T} = V_{BE} + \hat{v}_{be}$

$\Rightarrow R_C I_C \left(1 + \frac{\hat{v}_{be}}{V_T}\right) = V_{CC} - V_{BE} - \hat{v}_{be}$

CONT.

$$R_C I_C = \frac{(V_{CC} - V_{BE} - \hat{V}_{be})}{\left(1 + \frac{\hat{V}_{be}}{V_T}\right)} \quad \text{Q.E.D.}$$

$$\text{Voltage gain} = -g_m \cdot R_C$$
$$= -\frac{I_C}{V_T} \cdot R_C$$
$$= -\frac{V_{CC} - V_{BE} - \hat{V}_{be}}{V_T + \hat{V}_{be}}$$

For $V_{CC} = 5V$, $V_{BE} = 0.7V$ and $\hat{V}_{be} = 5mV$

$$R_C I_C = \frac{5 - 0.7 - 0.005}{1 + \frac{0.005}{0.025}} = 3.6V$$

Thus,

$$V_C = 5 - 3.6 = +1.4V$$

Amplitude of output signal is

$$= 1.4 - (V_{BE} + \hat{V}_{be})$$
$$= 1.4 - 0.7 - 0.005$$
$$= 0.695\ V$$

$$\text{Voltage gain} = -\frac{0.695}{0.005} = -139\ V/V$$

Check

$$\text{Voltage gain} = -\frac{(5 - 0.7 - 0.005)}{0.025 + 0.005}$$
$$= -143\ V/V$$

The difference is caused by decimal rounding-up of $R_C I_C$.

Otherwise:

$$\text{Voltage gain} = -\frac{0.716}{0.005}$$
$$= -143\ V/V$$

5.110

	a	b	c	d	e	f	g
α	1.000	0.990	0.98	1	0.990	0.90	0.941
β	∞	100	50	∞	100	9	16
I_C (mA)	1.00	0.99	1.00	1.00	0.248	4.5	17.5
I_E (mA)	1.00	1.00	1.02	1.00	0.25	5	18.6
I_B (mA)	0	0.010	0.020	0	0.002	0.5	1.10
g_m (mA/V)	40	39.6	40	40	0.01	180	700
r_e (Ω)	25	25	24.5	25	100	5	1.34
r_π (Ω)	∞	2.5K	1.25K	∞	10.1K	50	22.7

5.111

$$I_C = 1mA, \quad \beta = 120 \quad \alpha = 0.992$$

$$g_m = \frac{I_C}{V_T} = \frac{1}{25} = 40\frac{mA}{V}$$

$$r_\pi = \frac{\beta}{g_m} = \frac{120}{40 \times 10^{-3}} = 3K\Omega$$

$$r_e = \frac{V_T}{I_E} = \frac{\alpha}{g_m} = \frac{0.992}{40 \times 10^{-3}} = 24.8\Omega$$

The four equivalent circuit models are:

B, C, E, r_π 3KΩ, v_{be}, $g_m v_{be}$, $0.04 v_{be}$

B, C, E, i_b, r_π 3KΩ, βi_b, $120 i_b$

C, B, E, $g_m v_{be}$, $0.04 v_{be}$, v_{be}, r_e 24.8Ω

C, B, E, αi_e, $0.992 \cdot i_e$, i_e, r_e 24.8Ω

5.112

Refer to Fig P5.112

β very high $\rightarrow \alpha = 1$

$I_C = I_E = 0.5\text{mA}$

$V_C = 5 - 7.5 \times 0.5 = +\underline{\underline{1.25V}}$

$$g_m = \frac{I_C}{V_T} = \frac{0.5m}{25m} = \underline{\underline{20\frac{mA}{V}}}$$

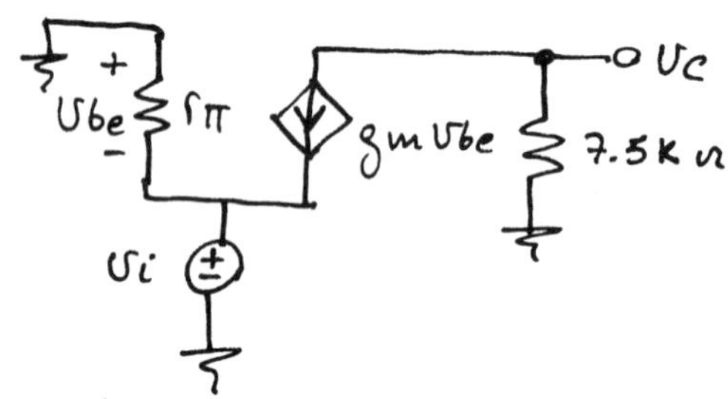

Observe that $v_{be} = -v_i$
the output voltage v_C is found from:

$v_C = -g_m v_{be} \times 7.5K$

Thus the voltage gain is

$$\frac{v_C}{v_i} = g_m \times 7.5K$$

$$= 20 \times 7.5 = \underline{\underline{150}}\text{V/V}$$

5.113

$$\frac{v_C}{v_{be}} = -g_m R_C \Rightarrow v_{be} = \frac{1}{50 \times 2}$$

$$= \underline{10\text{mV}} \text{ pk-to-pk}$$

$$i_b = \frac{v_{be}}{r_\pi} = \frac{10 \times 10^{-3}}{\beta/g_m} = \frac{0.01}{100/0.05}$$

$i_b = \underline{\underline{0.005}}\text{mA pk-to-pk}$

5.114

R_S
10K
$R_L = 10K$

$g_m = 40\frac{mA}{V}$, $r_\pi = 2.5K$

$r_\pi = \beta/g_m \Rightarrow \beta = 2.5K \times 40m$
$= \underline{\underline{100}}$

$R_S = 10K$ B C
v_S + v_{be} − r_π 2.5K $g_m v_{be}$ R_L 10k
E

$$v_{be} = \frac{v_S \cdot r_\pi}{R_S + r_\pi} = 0.2 \times v_S$$

$v_C = -g_m v_{be} \cdot R_L$
$= -g_m (0.2 v_S) \cdot R_L$

$v_C = -80\, v_S$

$\Rightarrow$ gain $\frac{v_C}{v_S} = \underline{\underline{-80}}$

To double the gain:

If I_C is fixed $\rightarrow g_m$ does not change.

$\rightarrow \frac{v_{be}}{v_S} = 0.4 \quad (2 \times 0.2)$

$\rightarrow \frac{r_\pi}{R_S + r_\pi} = 0.4 \quad \wedge R_S = 10K$

$\Rightarrow r_\pi = 6.6K\Omega$

Since $r_\pi = \frac{\beta}{g_m} \Rightarrow \beta = 6.6K \times 40m$

$\beta = \underline{\underline{264}}$

5.115

Refer to Fig. P5.115

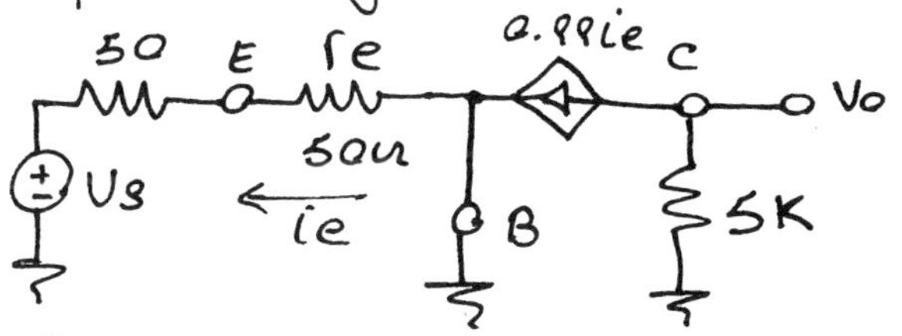

50 E re 0.99ie C Vo

50Ω ie B 5K Vs

$$r_e = \frac{V_T}{I_E} = \frac{25mV}{0.5mA} = 50\Omega$$

$$R_{in} = R_s + r_e = \underline{\underline{100\Omega}}$$

$$V_o = -0.99\, i_e \times 5K$$

but: $i_e = \frac{-V_s}{R_{in}} = \frac{-V_s}{100}$

$$\Rightarrow V_o = +\frac{0.99 \times 5K \cdot V_s}{100}$$

$$\frac{v_o}{v_s} = \underline{\underline{49.5}}\ V/V$$

5.116

Refer to Fig P5.116

$\beta = 200 \rightarrow \alpha = 0.995$

$I_C = \alpha I_E = 0.995 \times 10m$

$= 9.95mA$

$V_C = 9.95m \times 100 = \underline{\underline{0.995V}}$

$I_B \cong \frac{10m}{200} = 0.05mA$

$V_B = 1.5 - 10K \times 0.05m$

$= 1V$

$\Rightarrow V_{BC} = +0.005$

$\rightarrow$ Active region.

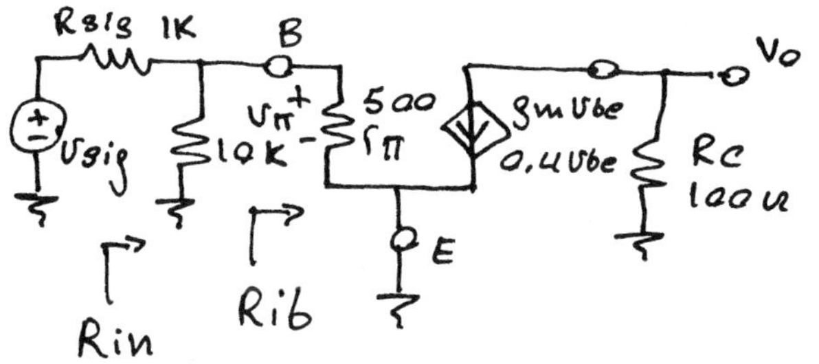

Rsig 1K B Vo

Vsig 10K v_π 500 r_π $g_m v_{be}$ 0.4v_{be} Rc 100Ω

E Rin Rib

$$g_m = \frac{I_C}{V_T} = \frac{9.95m}{25m} = 0.4\ A/V$$

$$r_\pi = \frac{\beta}{g_m} = \frac{200}{0.4} = 500\ \Omega$$

$R_{ib} = r_\pi = \underline{\underline{500\Omega}}$

$R_{in} = 10K \parallel r_\pi = \underline{\underline{476\Omega}}$

$$v_{be} = v_{sig} \times \frac{R_{in}}{R_{sig} + R_{in}} = v_{sig} \times 0.32$$

also:

$v_o = -g_m v_{be} \cdot R_C$

$= -g_m R_C \times 0.32\, v_{sig}$

$= -0.4 \times 100 \times 0.32\, v_{sig}$

$= -12.8\, v_{sig}$

$$\Rightarrow \text{gain } \frac{v_o}{v_s} = -12.8 \simeq -13\ V/V$$

If $v_o = \pm 0.4V$:

$$\hat{v}_s = \frac{\hat{v}_o}{13} = 30mV$$

$\hat{v}_{be} = 0.32 \times 30m = 9.8mV$

5.117

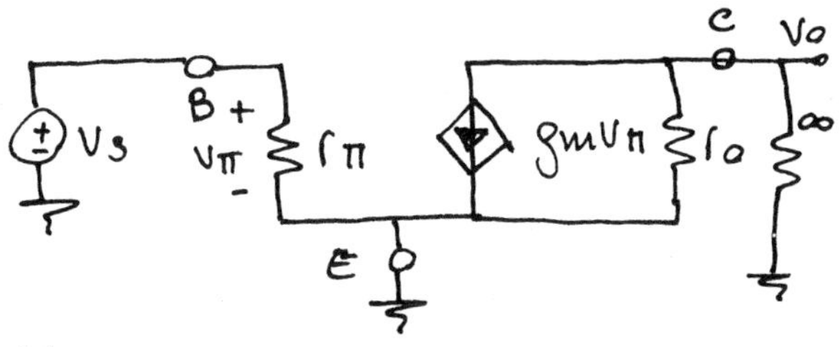

C Vo

Vs B + V_π − r_π $g_m V_\pi$ r_o ∞

E

$$V_s = V_\pi \Rightarrow \frac{V_o}{V_s} = -g_m \times r_o$$

CONT.

but: $r_o = \frac{V_A}{I_C} = \frac{V_A}{V_T \cdot g_m}$

$\Rightarrow \frac{V_o}{V_s} = \frac{-V_A}{V_T}$

if $V_A = 25V \Rightarrow \frac{V_o}{V_s} = -1000 \frac{V}{V}$

if $V_A = 250V \Rightarrow \frac{V_o}{V_s} = -10{,}000 \frac{V}{V}$

5.118

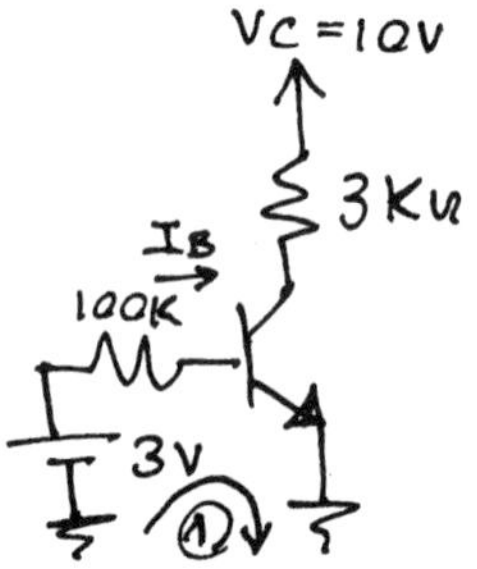

$V_C = 10V$, $3K\Omega$, I_B, $100K$, $3V$, ①

DC Analysis:

① $I_B = \frac{3-0.7}{100}$

$I_B = 0.023\,mA$

Saturation begins to occur when $V_C \leq 0.7V$

$\therefore I_C \geq \frac{10-0.7}{3} = 3.1\,mA$

$I_C = \beta I_B \rightarrow \beta \geq \frac{3.1}{0.023} = 135$

$\beta = 25$:

$r_e = \frac{V_T}{I_E} = \frac{V_T}{(\beta+1)I_B} = \frac{25\times10^{-3}}{26\times0.023\times10^{-3}}$

$r_e = 41.8\Omega$

$g_m = \frac{\alpha}{r_e} = \frac{25/26}{41.8} = 23\frac{mA}{V}$

5.119

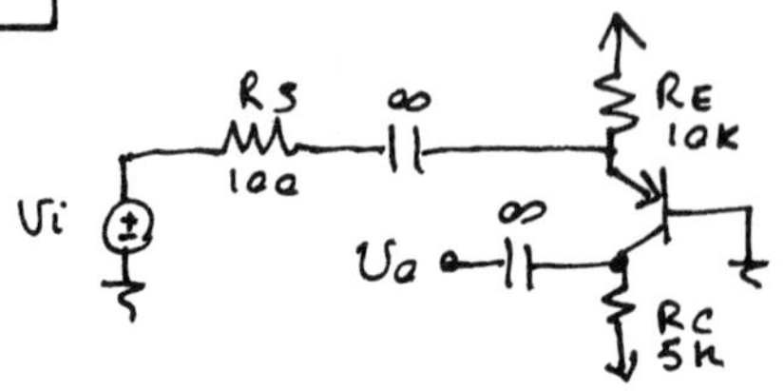

R_S 100Ω, ∞, R_E $10k$, V_i, V_o, ∞, R_C $5k$

$\frac{V_o}{V_i} = \alpha \frac{r_e \| R_E}{R_E \| r_e + R_S} \cdot \frac{1}{r_e} \cdot R_C$

$= 0.99 \frac{(27\|10k)}{(27\|10k)+100} \times \frac{1}{27} \times 5\times10^3$

$= 38.9 \simeq 39\,V/V$

The amplifier clips when the peak output signal exceeds the difference between V_C (−5.4V) and the negative supply (−10V); see Fig. 5.56.

That is when the peak output signal is 4.6V

This corresponds to a peak input signal of

$\hat{V}_i = \frac{4.6}{39} = 118\,mV$

5.120

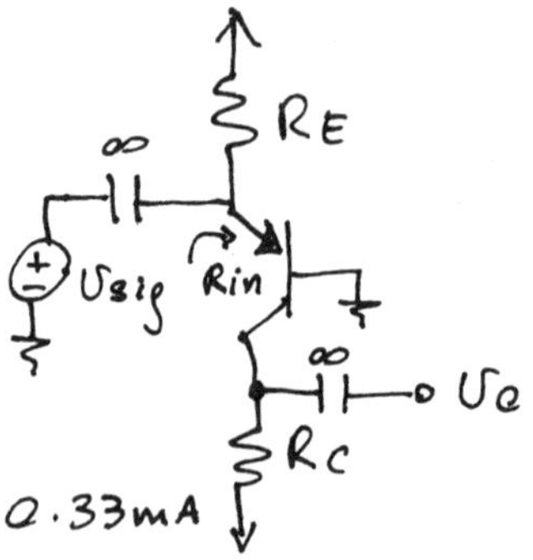

R_E, ∞, V_{sig}, R_{in}, ∞, V_o, R_C

$R_{in} = r_e \| R_E$
$\simeq r_e$
$= 75\Omega$

$I_E = \frac{25mV}{75\Omega} = 0.33\,mA$

$R_E = \frac{10-0.7}{0.33} = 28\,K\Omega$

$n = 2.8$

$R_C = 14\,K\Omega$

$\frac{V_o}{V_i} = \frac{\alpha R_C}{r_e} = \frac{14}{0.075} = 187\,V/V$

5.121

(a) Using the voltage divider rule:

$$\frac{v_e}{v_b} = \frac{R_e}{R_e + r_e} \quad \text{Q.E.D.}$$

(b) Node equation at B:

$$i_b = \frac{v_{be}}{r_e} - g_m v_{be}$$

$$= \frac{v_{be}}{r_e}(1 - g_m r_e)$$

$$= \frac{v_{be}}{r_e}\left(1 - g_m \frac{\alpha}{g_m}\right)$$

$$= \frac{v_{be}}{r_e}(1 - \alpha)$$

$$= \frac{v_{be}}{r_e}\left(1 - \frac{\beta}{\beta+1}\right)$$

$$= \frac{v_{be}}{r_e} \cdot \frac{1}{(\beta+1)}$$

But, from voltage-divider rule

$$v_{be} = v_b \cdot \frac{r_e}{r_e + R_e}$$

$$\Rightarrow i_b = \frac{1}{(\beta+1) r_e} \cdot \frac{v_b \cdot r_e}{r_e + R_e}$$

from which we find

$$R_{in} = \frac{v_b}{i_b} = (\beta+1)(R_e + r_e)$$

Q.E.D.

For $R_e = 1\,k\Omega$, $\beta = 100$ and $I_E = 1\,mA$

$$r_e = \frac{V_T}{I_E} = \frac{25\,mV}{1\,mA} = 25\,\Omega$$

Thus,

$$\frac{v_e}{v_b} = \frac{1000}{1000 + 25} = 0.976\ V/V$$

$$R_{in} = (100+1)(1000+25)\,\Omega$$

$$= 101 \times 1.025\ k\Omega$$

$$= 103.5\ k\Omega$$

5.122

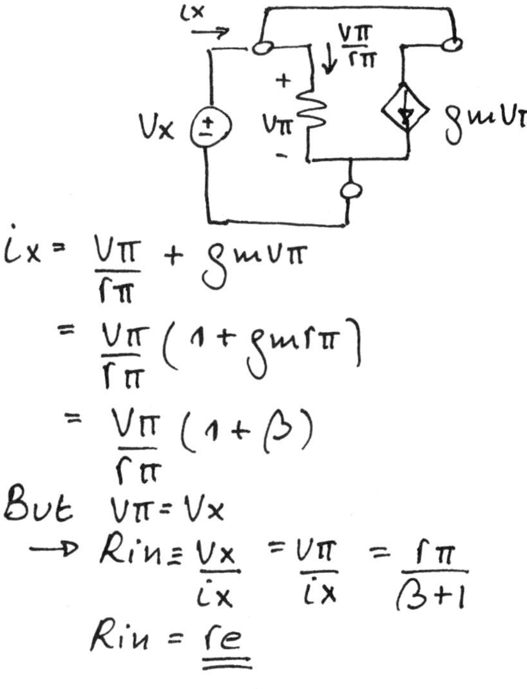

$$i_x = \frac{v_\pi}{r_\pi} + g_m v_\pi$$

$$= \frac{v_\pi}{r_\pi}(1 + g_m r_\pi)$$

$$= \frac{v_\pi}{r_\pi}(1 + \beta)$$

But $v_\pi = v_x$

$$\rightarrow R_{in} = \frac{v_x}{i_x} = \frac{v_\pi}{i_x} = \frac{r_\pi}{\beta+1}$$

$$R_{in} = r_e$$

5.123

$R_{in} = R_E \parallel r_e \qquad r_e \simeq 100\,\Omega$

thus, $\frac{V_T}{I_E} = 100 \rightarrow I_E = 0.25\,mA$

$$V_E = 0.7\,V$$

$$R_E = \frac{10 - 0.7}{1m} = 9.3\,k\Omega$$

CONT.

Selection of a value for Rc:

The voltage gain is directly proportional to Rc,

$$\frac{v_o}{v_s} = \frac{v_e}{v_s} \cdot \frac{v_o}{v_e}$$

$$= \frac{R_{in}}{R_s + R_{in}} \cdot \frac{\alpha R_C}{r_e}$$

$$\simeq \frac{100}{100+100} \cdot \frac{R_C}{0.1}$$

$= 5R_C$, R_C in $k\Omega$.

For an emitter-base signal as large as 10mV, the signal at the collector will be $g_m R_C \times 0.010$ volts. Thus the maximum collector voltage in the positive direction will be:

$$v_C|_{max} = V_C + 0.01\, g_m \cdot R_C$$

$$= -10 + I_C R_C + 0.01 \times \frac{1}{0.1} \times R_C$$

$$= -10 + 0.25 R_C + 0.1 R_C$$

$$= -10 + 0.35 R_C$$

To prevent saturation, $v_C|_{max} \leq V_B$ which is 0V.

Thus to obtain maximum gain while allowing an emitter-base signal as large as 10mV and at the same time keeping the transistor in the active mode we select Rc from:

$$-10 + 0.35 R_C = 0$$

$$\Rightarrow R_C = \underline{\underline{28.6\ k\Omega}}$$

Voltage gain $= \frac{v_o}{v_s} = 5R_C = 143\ V/V$

5.124

Refer to Fig P5.124

For large β, the DC base current will be ~0

Thus the DC voltage at the base can be found directly using the voltage-divider rule

$$V_B = 15 \cdot \frac{100}{100+100} = 7.5V$$

If: $V_{BE} = 0.7$

$$V_E = 7.5 - 0.7 = 6.8V$$

$$\rightarrow I_E = \frac{6.8V}{6.8k\Omega} = \underline{\underline{1mA}}$$

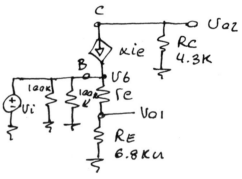

C, v_{o2}, αi_e, Rc 4.3K, B, v_b, 100K, 100K, r_e, v_i, v_{o1}, RE 6.8KΩ

$v_b = v_i$

$$\rightarrow \frac{v_{o1}}{v_i} = \frac{R_E}{R_E + r_e} \quad \text{Q.E.D.}$$

Also,

$$i_e = \frac{v_b}{r_e + R_E} = \frac{v_i}{r_e + R_E}$$

and,

$$v_{o2} = -\alpha i_e \cdot R_C$$

$$= \frac{-\alpha R_C v_i}{r_e + R_E}$$

Thus,

$$\frac{v_{o2}}{v_i} = \frac{-\alpha R_C}{R_E + r_e} \quad \text{Q.E.D.}$$

CONT.

Substituting $r_e = \frac{V_T}{I_E} = 25\Omega$

and $R_E = 6.8K\Omega$, $R_C = 4.3K\Omega$ and $\alpha \simeq 1$ gives

$$\frac{v_{o1}}{v_i} = \frac{6.8}{0.025+6.8} = \underline{\underline{0.996\ V/V}}$$

$$\frac{v_{o2}}{v_i} = -\frac{4.3}{6.8+0.025} = \underline{\underline{0.63\ V/V}}$$

If the node labeled v_{o2} is connected to ground:
$R_E = 0$

$$\frac{v_{o2}}{v_i} = -\frac{\alpha R_C}{r_e}$$

5.125

If: $R_i = 10K\Omega$, $A_{vo} = 100\ V/V$
$R_o = 100\Omega$, $R_{sig} = 2K\Omega$ and
$R_{in} = 8K\Omega$ when $R_L = 1K\Omega$
then:

$$G_m = \frac{A_{vo}}{R_o} = \frac{100}{100} = \underline{\underline{1 A/V}}$$

$$A_v = A_{vo} \cdot \frac{R_L}{R_L+R_o} = \frac{100 \times 1K}{1K+100} = \underline{\underline{91 V/V}}$$

$$G_{vo} = \frac{R_i}{R_i+R_{sig}} \cdot A_{vo} = \frac{10K}{10K+2K} \times 100 = \underline{\underline{83.3\ V/V}}$$

$$G_v = \frac{R_{in}}{R_{in}+R_{sig}} \times A_{vo} \times \frac{R_L}{R_L+R_o}$$

$$= \frac{8K}{8K+2K} \times 100 \times \frac{1K}{1K+100} = \underline{\underline{72.7\ \frac{V}{V}}}$$

$$R_{out} = \frac{G_{vo}}{G_v} \cdot R_L - R_L = \frac{8.3}{72.7} \times 1K - 1K$$

$$= \underline{\underline{146\Omega}}$$

$$A_i = \frac{v_o/R_L}{v_i/R_{in}} = A_v \cdot \frac{R_{in}}{R_L} = 91 \times \frac{8K}{1K}$$

$$= 728\ A/A \ //$$

5.126

Refer to Fig. P5.126
$R_i = 900K\Omega$
$A_{vo} = 10\ V/V$
$R_o = 1.43K\Omega$
$R_{in} = 400K\Omega$ when $R_L = 10K\Omega$

(a) $i_i \equiv v_i / R_{in}$

$$\Rightarrow \frac{v_i}{R_{in}} = \frac{v_i}{R_i} - f \cdot i_o$$

where $i_o = \frac{A_{vo} v_i}{R_o + R_L}$

$$\rightarrow \frac{v_i}{R_{in}} = \frac{v_i}{R_i} - f \cdot \frac{A_{vo} v_i}{R_o+R_L}$$

Solving for f:

$$f = \left(\frac{1}{R_i} - \frac{1}{R_{in}}\right) \cdot \frac{R_o+R_L}{A_{vo}}$$

Thus,

$$f = \left(\frac{1}{900K} - \frac{1}{400K}\right) \cdot \frac{(1.43K+10K)}{10}$$

$$f = \underline{\underline{-1.6 \times 10^{-3}}}$$

(b)

$$R_{out} = \frac{v_x}{i_x}\bigg|_{v_{sig}=0}$$

If $v_{sig} = 0$: on the left hand side

$$v_i = f \cdot i_o\ (R_{sig} \| R_i)$$

$$v_i = (-1.6 \times 10^{-3} \times (100K \| 900K)\, i_o$$

$$= -144 \cdot i_o$$

On the right hand side:

$$V_x = R_o i_x + A_{vo} \cdot v_i$$

$$= R_o \cdot i_x + A_{vo}(-144\, i_o)$$

$$= R_o\, i_x + A_{vo}(+144)\, i_x$$

CONT.

$\rightarrow V_x = i_x\,(R_o + A_{vo} \times 144)$

$R_{out} = \frac{V_x}{i_x} = 1.43K + 10 \times 144$

$R_{out} = \underline{\underline{2.87\,K\Omega}}$

Which is the same value as in Example 5.17

5.127

$$G_v = \frac{R_{in}}{R_{in}+R_{sig}} \cdot A_{vo} \cdot \frac{R_L}{R_L+R_o} = G_{vo} \cdot \frac{R_L}{R_L+R_{out}}$$

but $G_{vo} = \frac{R_i}{R_i+R_{sig}} \cdot A_{vo}$

Substituting G_{vo}:

$$\frac{R_{in}}{R_{in}+R_{sig}} \cdot A_{vo} \cdot \frac{R_L}{R_L+R_o} = \frac{R_i}{R_i+R_{sig}} \cdot A_{vo} \cdot \frac{R_L}{R_L+R_{out}}$$

$$\rightarrow \frac{R_{in}}{R_i} \frac{(R_{sig}+R_i)}{(R_{sig}+R_{in})} = \frac{R_L+R_o}{R_L+R_{out}}$$

Q.E.D

(a) for $R_L = \infty$

$$\frac{R_{in}}{R_i} \frac{(R_{sig}+R_i)}{(R_{sig}+R_{in})} = 1$$

$$R_{in}R_{sig} + R_{in}R_i = R_iR_{sig} + R_iR_{in}$$

$$R_{in} \cdot R_{sig} = R_i \cdot R_{sig}$$

$$\rightarrow R_{in} = R_i$$

Q.E.D

(b) for $R_{sig} = 0$

$$\frac{R_{in}}{R_i} \cdot \frac{R_i}{R_{in}} = \frac{R_L+R_o}{R_L+R_{out}}$$

$$\rightarrow R_{out} = R_o$$

Q.E.D.

(c) For $R_{sig} = \infty$

$$\frac{R_{in}}{R_i} = \frac{R_L+R_o}{R_L+R_{out}}$$

$$R_{out} = (R_L+R_o)\frac{R_i}{R_{in}} - R_L \quad //$$

In Example 5.17:

$$R_{out} = (10K + 1.43K)\frac{900}{400} - 10K$$

$$= \underline{\underline{15.7\,K\Omega}}$$

5.128

Refer to Fig. 5.60(a)

$I_C = 0.2\,mA \Rightarrow I_E = \frac{I_C}{\alpha}$

$I_E = \frac{0.2}{\beta} \cdot (\beta+1) = 0.202\,mA$

$r_e = \frac{V_T}{I_E} = \frac{0.025}{0.202m} = 123.76\,\Omega$

$R_i = (\beta+1)r_e = \underline{\underline{12{,}5K\,\Omega}}$

$$\frac{v_o}{v_s} = -g_m(R_c \| r_o) \cdot \frac{R_i}{R_i+R_s}$$

$$\simeq -\frac{\alpha}{r_e} \cdot R_c \cdot \frac{R_i}{R_i+R_s}$$

$$= -\frac{100}{101} \times \frac{24K}{123.76} \times \frac{12.5K}{12.5K+10K}$$

$\frac{v_o}{v_s} = \underline{\underline{-106.7\ V/V}}$

$R_o = R_c \| r_o \simeq R_c = \underline{24\,K\Omega}$

$$\frac{v_o}{v_s} = \left.\frac{v_o}{v_s}\right|_{no\ load} \times \frac{10}{10+24}$$

$$= -106.7 \times \frac{10}{34} = \underline{\underline{-31.4\ V/V}}$$

5.129

$I_C = 0.2\text{mA} \Rightarrow r_e = 123.76\Omega$

$R_i = (\beta+1)(r_e + R_E)$

$= (101)(123.76 + 125) = \underline{\underline{25.1\text{K}\Omega}}$

$$\frac{v_o}{v_s} = \frac{-\alpha R_C}{r_e + R_E} \times \frac{R_i}{R_i + R_s}$$

$$= \frac{-100}{101} \times \frac{24\text{K}}{(123.76 + 125)} \times \frac{25.1\text{K}}{25.1\text{K} + 10\text{K}}$$

$$= \underline{\underline{-68.30}} \text{ V/V}$$

$R_o = R_C = \underline{\underline{24\text{K}\Omega}}$

With 10KΩ load

$$\frac{v_o}{v_s} = -68.30 \times \frac{10}{10+24} = \underline{\underline{-20 \frac{V}{V}}}$$

Without Re $\quad V_\pi \leq 5\text{mV}$

$$V_\pi = \frac{(\beta+1) r_e}{(\beta+1) r_e + R_s} \cdot V_s$$

$$= \frac{(101)\cdot(123.76)}{(101)\cdot(123.76) + (10\text{K})} \cdot V_s \leq 5\text{mV}$$

$\Rightarrow V_s \leq \underline{\underline{9\text{mV}}}$

With Re

$$V_\pi = \frac{(\beta+1) r_e \cdot v_s}{(\beta+1)(r_e + R_E) + R_s} \leq 5\text{mV}$$

$$V_s \leq 5\text{m}\left[\frac{(101)(123.76 + 125) + 10\text{K}}{101 \times 123.76}\right]$$

$V_s \leq \underline{\underline{14\text{mV}}}$

5.130

Refer to Fig. P5.130

$$I_E = \frac{V_{BB} - V_{BE}}{R_E + R_B/(\beta+1)}$$

where, $V_{BB} = V_{CC} \cdot \frac{R_2}{R_1 + R_2}$

$$= 9 \cdot \frac{15}{27+15} = 3.21\text{V}$$

$R_B = R_1 \| R_2 = 15 \| 27 = 9.64\text{ K}\Omega$

Thus, $I_E = \dfrac{3.21 - 0.7}{1.2 + \dfrac{9.64}{101}} = \underline{\underline{1.94\text{mA}}}$

Rs, Vs, RB, rπ, +, vπ, -, gmvπ, ro, RC, RL, Vo

$$g_m = \frac{I_C}{V_T} = \frac{0.99 \times 1.94}{0.025} = 76.8 \frac{\text{mA}}{\text{V}}$$

$$r_\pi = \frac{\beta}{g_m} = \frac{100}{76.8} = 1.3\text{K}\Omega$$

$$r_o = \frac{V_A}{I_C} = \frac{100}{0.99 \times 1.94} = 52.1\text{ K}\Omega$$

$R_i = R_B \| r_\pi = 9.64 \| 1.3 = \underline{\underline{1.15\text{K}\Omega}}$

$G_m = -g_m = \underline{\underline{-76.8 \frac{\text{mA}}{\text{V}}}}$

$R_o = R_C \| r_o = 2.2 \| 52.1 = \underline{\underline{2.11\text{ K}\Omega}}$

Rs, ii, Vs, Ri, +, vi, -, Gmvi, Ro, io, RL, Vo

Equivalent circuit

$$A_v \equiv \frac{v_o}{v_s} = \frac{v_i}{v_s} \cdot \frac{v_o}{v_i}$$

$$= \frac{R_i}{R_s + R_i} \cdot \frac{G_m (R_o \| R_L) v_i}{v_i}$$

$$= \frac{-1.15}{10 + 1.15} \times 76.8 \times (2.11 \| 2)$$

$$= \underline{\underline{-8.13}} \text{ V/V}$$

$$A_i = \frac{i_o}{i_i} = \frac{v_o / R_L}{v_s / (R_s + R_i)}$$

CONT.

$$\rightarrow A_i = \frac{v_o}{v_s} \cdot \frac{R_s + R_i}{R_L}$$

$$= -8.13 \times \frac{(10 + 1.15)}{2}$$

$$= \underline{\underline{-45.3}} \text{ A/A}$$

5.131

Refer to Fig. P5.130.

$V_{CC} = 9V \quad V_{BB} = \frac{1}{3} V_{CC} = 3V$

Neglecting the base current,

$$R_1 + R_2 = \frac{9}{0.2} = 45 \text{ k}\Omega$$

$$\frac{R_2}{R_1 + R_2} = \frac{1}{3}$$

$$\Rightarrow R_2 = \underline{15 \text{ k}\Omega}, \quad R_1 = \underline{\underline{30 \text{ k}\Omega}}$$

$$R_B = R_1 \parallel R_2 = \frac{30 \times 15}{45} = 10 \text{ k}\Omega$$

$$I_E = \frac{V_{BB} - V_{BE}}{R_E + \frac{R_B}{\beta + 1}}$$

$$2 = \frac{3 - 0.7}{R_E + 10/101} \Rightarrow R_E = \underline{\underline{1.05 \text{ k}\Omega}}$$

Use $R_E = \underline{\underline{1 \text{ k}\Omega}}$

The resulting I_E will be

$$I_E = \frac{3 - 0.7}{1 + 10/101} = 2.09 \text{ mA}$$

$$I_C = \alpha I_E = 0.99 \times 2.09 = 2.07 \text{ mA}$$

$$g_m = \frac{I_C}{V_T} = \frac{2.07}{0.025} = 82.9 \frac{\text{mA}}{\text{V}}$$

$$r_\pi = \frac{\beta}{g_m} = \frac{100}{82.9} = 1.21 \text{ k}\Omega$$

$$r_o = \frac{V_A}{I_C} = \frac{100}{2.07} = 48.3 \text{ k}\Omega.$$

$$\frac{v_o}{v_s} = \frac{v_\pi}{v_s} \cdot \frac{v_o}{v_\pi} = \frac{R_i}{R_s + R_i} \cdot \frac{-g_m v_\pi R}{v_\pi}$$

$$= \frac{-1.08}{10 + 1.08} \times 82.9 \times R$$

To obtain $\frac{v_o}{v_s} = -8 \frac{\text{V}}{\text{V}}$ we use:

$$R = \frac{8 \times 11.08}{1.08 \times 82.9} = 0.99 \text{ k}\Omega$$

Now $R = r_o \parallel R_C \parallel R_L$

$$0.99 = 48.3 \parallel R_C \parallel 2$$

$$\Rightarrow R_C = 2.04 \text{ k}\Omega$$

use $R_C = \underline{\underline{2 \text{ k}\Omega}}$

Check: $V_C = 9 - 2.07 \times 2 = 4.86V$ while $V_B \simeq 3V$. Thus in active mode as assumed.

5.132

Refer to Fig. P5.130

$$V_{BB} = 9 \cdot \frac{47}{82 + 47} = 3.28 \text{ V}$$

$$R_B = 47 \parallel 82 = 29.88 \text{ k}\Omega$$

$$I_E = \frac{3.28 - 0.7}{3.6 + \frac{29.88}{101}} = 0.66 \text{ mA}$$

$$I_C = 0.99 \times 0.66 = 0.65 \text{ mA}$$

$$g_m = \frac{0.65}{0.025} = 26 \frac{\text{mA}}{\text{V}}$$

$$r_\pi = \frac{100}{26} = 3.85 \text{ k}\Omega$$

$$r_o = \frac{100}{0.66} = 151.5 \text{ k}\Omega$$

CONT.

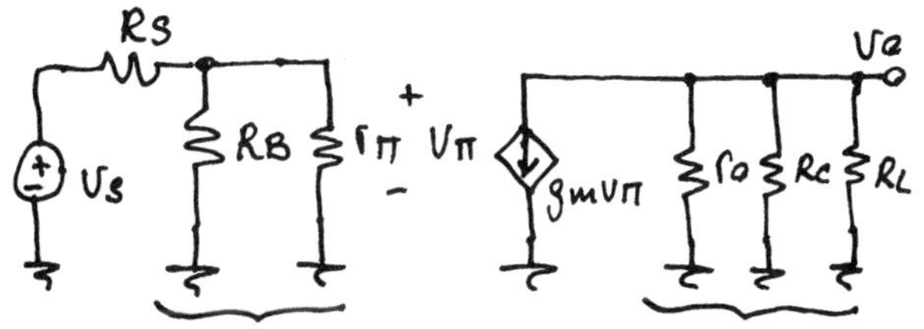

$R_i = 29.88 \parallel 3.85 = 3.41K\Omega$ $151.5 \parallel 6.8 \parallel 2 = 1.53K\Omega$

$$A_v = \frac{V_o}{V_s} = \frac{3.41}{10+3.41} \times -26 \times 1.53$$

$= \underline{-10.1\ V/V}$ Which is about 25% higher than in the original design. The improvement is not as large as might have been expected because although R_i increases, g_m decreases by about the same factor. Indeed most of the improvement is due to the increase in R_c and hence in the effective load resistance.

5.133

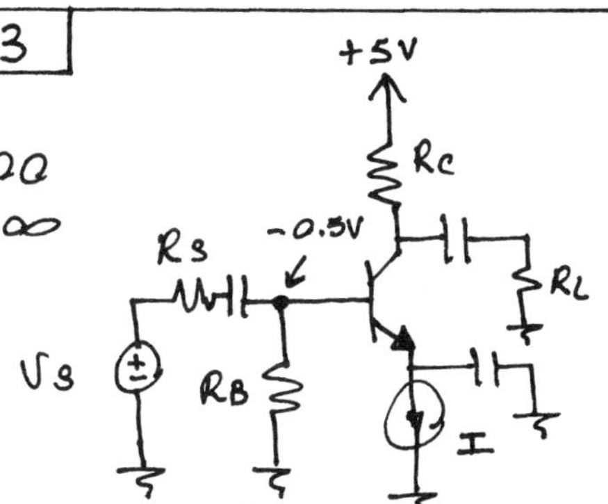

$\beta = 100$
$r_o = \infty$

$R_{in} = 5K\Omega$, $R_{in} = R_B \parallel r_\pi$

$$\Rightarrow 5K = \frac{R_B \cdot r_\pi}{R_B + r_\pi}$$

$5K r_\pi = R_B(r_\pi - 5K)$

but: $r_\pi = \frac{V_T}{I_B}$ and $R_B \cdot I_B = 0.5$

$$\rightarrow 5K \cdot \frac{V_T}{I_B} = \frac{0.5}{I_B}(r_\pi - 5K)$$

thus, $r_\pi = 5250\Omega$
then $R_B = 105K$
choose $R_B = \underline{100K\Omega}$

and $I_B = 4.76\mu A$
$I_E = (\beta+1) I_B = 101 \times 4.76\mu A$
$I_E = 0.48mA$
$I = I_E \rightarrow I \simeq \underline{0.5mA}$

To avoid saturation:
$V_C - V_B \geq -0.5$
$V_C = 5V - R_C[I_C + g_m v_{be}]$

$I_C = I \cdot \alpha = 0.5m \times 100/101 = 0.49mA$

$$g_m = \frac{V_T}{I_C} = \frac{25m}{0.49m} \simeq 50\frac{mA}{V}$$

$v_{be} = 0.005V$

$\rightarrow V_C = 5 - R_C[0.49m + 50m \times 5m] = 5 - 0.74 \times 10^{-3} \times R_C$

Then:
$V_C - V_B = (5 - 0.74m R_C) - (-0.5 + v_{be}) = 5.495 - 0.74m R_C \geq -0.5$

$R_C \leq \underline{8.1K\Omega}$.

Base-to-Collector open circuit gain:

$$\frac{v_c}{v_b} = -g_m R_C = -50m \times 8.1K = \underline{-405V/V}$$

For $R_s = 10K$, $R_L = 10K$

$$\frac{v_o}{v_b} = -g_m(R_C \parallel R_L) = -50m \times 4.47K = -223V/V$$

$$\frac{v_c}{v_s} = \frac{v_b}{v_s} \cdot \frac{v_o}{v_b} = \frac{5}{5+10} \times -223 = \underline{-74.3V/V}$$

5.134

Refer to Fig P5.134

$I_E = 0.5\,mA$

(a) $I_E = \dfrac{15 - 0.7}{R_E + \dfrac{R_s}{\beta+1}}$

$0.5 = \dfrac{14.3}{R_E + \dfrac{2.5}{100}}$

$\Rightarrow R_E = \underline{\underline{28.57\,K\Omega}}.$

(b) $V_C = 15 - R_C . I_C$

$5 = 15 - R_C \times 0.99 \times 0.5m$

$\Rightarrow R_C = 20.2\,K\Omega$

$\simeq \underline{\underline{20\,K\Omega}}$

(c)

$R_L = 10\,K\Omega\ ,\ R_s = 2.5K$

$r_o = 200\,K\Omega$

$g_m = \dfrac{I_C}{V_T} \simeq \dfrac{0.5m}{25m} = 20\,\dfrac{mA}{V}$

$r_\pi = \dfrac{\beta}{g_m} = \dfrac{100}{20} = 5\,K\Omega$

$A_v = \dfrac{v_o}{v_s} = \dfrac{v_\pi}{v_s} \times \dfrac{v_o}{v_\pi}$

$= \dfrac{r_\pi}{r_\pi + R_s} \times -g_m\,(r_o /\!/ R_C /\!/ R_L)$

$= -\dfrac{5}{5+2.5} \times 20\,(200 /\!/ 20 /\!/ 10)$

$= \underline{\underline{-86}}\ V/V$

5.135

Refer to Fig. P5.135

(a) For each transistor

$V_{BB} = 15 \times \dfrac{47}{100+47} = 4.8\,V$

$R_B = R_1 /\!/ R_2 = 100 /\!/ 47 = 32\,K\Omega$

$I_E = \dfrac{4.8 - 0.7}{3.9 + \dfrac{32}{101}} = 0.97\,mA$

$I_C = 0.99 \times 0.97 = \underline{\underline{0.96\,mA}}$

$V_C = V_{CC} - I_C \times R_C$

$= 15 - 0.96 \times 6.8 = \underline{\underline{8.5\,V}}$

(b)

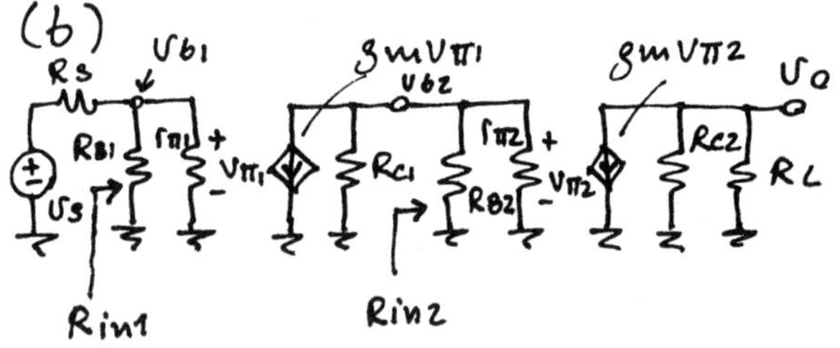

$R_{B1} = R_{B2} = R_B = 32\,K\Omega$

$g_{m1} = g_{m2} = \dfrac{0.96}{0.025} = 38.4\,\dfrac{mA}{V}$

$r_{\pi 1} = r_{\pi 2} = \dfrac{100}{38.4} = 2.6\,K\Omega$

$R_{C1} = R_{C2} = 6.8\,K\Omega$

$r_{o1} = r_{o2} = \infty$

(c) $R_{in1} = R_{B1} /\!/ r_{\pi 1}$

$= 32 /\!/ 2.6 = \underline{\underline{2.4\,K\Omega}}$

$\dfrac{v_{b1}}{v_s} = \dfrac{R_{in1}}{R_s + R_{in1}}$

$= \dfrac{2.4}{5 + 2.4} = \underline{\underline{0.32}}\ V/V$

(d) $R_{in2} = R_{B2} /\!/ r_{\pi 2}$

$= 32 /\!/ 2.6 = \underline{\underline{2.4\,K\Omega}}$

CONT.

$v_{b2} = -g_{m1} v_{\pi 1} (R_{C1} \| R_{in2})$
$= -38.4 \, v_{b1} (6.8 \| 2.4)$

$\frac{v_{b2}}{v_{b1}} = \underline{\underline{-68.1}} \text{ V/V}$

(e) $v_o = -g_{m2} v_{\pi 2} (R_{C2} \| R_L)$
$= -38.4 \, v_{b2} (6.8 \| 2)$

$\frac{v_o}{v_{b2}} = \underline{\underline{-59.3}} \text{ V/V}$

(f) $\frac{v_o}{v_s} = \frac{v_{b1}}{v_s} \times \frac{v_{b2}}{v_{b1}} \times \frac{v_o}{v_{b2}}$
$= 0.32 \times -68.1 \times -59.3$
$= \underline{\underline{1292}} \text{ V/V}$

5.136

Refer to the circuit in Fig. P5.136

$R_{in} = (\beta+1)(r_e + 250)$

$\beta = 100 \quad r_e = \frac{V_T}{I_E} = \frac{0.025}{0.1} = 250\,\Omega$

$R_{in} = 101 \times (250 + 250)$
$= \underline{\underline{50.5\,K\Omega}}$

$\frac{v_b}{v_s} = \frac{R_{in}}{R_s + R_{in}} = \frac{50.5}{20 + 50.5}$
$= 0.72 \text{ V/V}$

$\frac{v_o}{v_b} = \frac{-\alpha (20 \| 20)}{(r_e + R_E)}$
$= \frac{-0.99 \times 10}{0.250 + 0.250} = \underline{\underline{-19.8}} \text{ V/V}$

Thus, $\frac{v_o}{v_s} = 0.72 \times -19.8 = \underline{\underline{-14.2}} \text{ V/V}$

For $v_{be} = 5\text{mV}$, $v_e = 5\text{mV}$ also (since $R_e = r_e = 250\,\Omega$)

Thus,
$v_b = 5 + 5 = 10\text{mV}$
$v_s = \frac{10\text{m}}{0.72} = \underline{\underline{13.88\text{mV}}}$
$v_o = 13.88 \times 14.2 = \underline{\underline{197.2\text{mV}}}$

5.137

(a) $I_C = 0.99 \times 0.5 \text{ mA}$
$= 0.495\text{mA}$

$V_C = I_E R_E + V_{BE} \ldots + I_B R_B$
$= 0.5 \times 0.175 + 0.7 + 0.005 \times 300$
$= \underline{\underline{2.28\text{V}}}$

0.5mA
V_C
300K
0.5mA
250

(b) $i_e = \frac{v_i}{r_e + R_E}$

$r_e = \frac{V_T}{I_E} = 50\,\Omega$

C
V_o
30K
αi_e
300 KΩ
v_i
r_e
E
i_e
250Ω

$\rightarrow i_e = \frac{v_i}{50 + 250}$

$i_e = \frac{v_i}{300}$

Node equation at c:

$\frac{v_o - v_i}{300K} + \alpha i_e + \frac{v_o}{30K} = 0$

$\frac{v_o - v_i}{300K} + \frac{\alpha v_i}{(250+50)} + \frac{v_o}{30K} = 0$

$\Rightarrow \frac{v_o}{v_i} = \underline{\underline{-90 \text{ V/V}}}$

5.138

(a) Without R_e,

$A_v \simeq \frac{-\beta (R_C \| R_L \| r_o)}{r_\pi + R_s}$ CONT.

Since no value is specified for r_o (or V_A) we shall neglect its effect, thus

$$A_v = -\beta \frac{(R_C \| R_L)}{r_\pi + R_s}$$

Substituting $r_\pi = \beta / g_m$ yields

$$A_v = \frac{-(R_C \| R_L)}{\frac{\beta}{\beta+1} \cdot \frac{1}{g_m} + \frac{R_s}{\beta}}$$

$$= \frac{-(R_C \| R_L)}{r_e + \frac{R_s}{\beta}}$$

Maximum gain is obtained at the high β,

$$A_{v\,max} = -\frac{(R_C \| R_L)}{0.050 + \frac{10}{150}}$$

$r_e \rightarrow$ (0.050)

$$= -8.57\,(R_C \| R_L)$$

The minimum gain is obtained for β at its lowest value

$$A_{v\,min} = -\frac{(R_C \| R_L)}{0.050 + \frac{10}{50}}$$

$$= -4\,(R_C \| R_L)$$

Thus, $\frac{A_{v\,max}}{A_{v\,min}} = \frac{8.57}{4} = \underline{\underline{2.14}}$

(b) With R_e

$$A_v \simeq -\beta \frac{(R_C \| R_L)}{r_\pi (1 + g_m R_e) + R_s}$$

$$= -\frac{(R_C \| R_L)}{\frac{1}{g_m}(1 + g_m R_e) + \frac{R_s}{\beta}}$$

$$A_v = \frac{-(R_C \| R_L)}{\frac{1}{g_m} + R_e + \frac{R_s}{\beta}}$$

$$\frac{A_{v\,max}}{A_{v\,min}} = \frac{\frac{1}{g_m} + R_e + \frac{R_s}{\beta_{min}}}{\frac{1}{g_m} + R_e + \frac{R_s}{\beta_{max}}}$$

Thus,

$$1.2 = \frac{50 + R_e + \frac{10,000}{50}}{50 + R_e + \frac{10,000}{150}}$$

$$\Rightarrow R_e = \underline{\underline{550\,\Omega}}$$

(c) For $\beta = 100$:

(i) Without R_e,

$$A_v = \frac{-(R_C \| R_L)}{r_e + R_s/\beta}$$

$$= \frac{-(R_C \| R_L)}{0.050 + 10/100}$$

$$= -6.66\,(R_C \| R_L)$$

(ii) With $R_e = 550\,\Omega$

$$A_v \simeq \frac{-(R_C \| R_L)}{0.050 + 0.550 + \frac{10}{100}}$$

$$= -1.43\,(R_C \| R_L)$$

Thus including R_e reduces the gain by a factor of $\frac{6.66}{1.43} = \underline{\underline{4.6}}$

5.139

$R_i = r_e$

$r_e = \frac{25m}{I} = 100$

$I = \frac{25m}{100} = 0.25m$

$\frac{V_o}{V_s} = -\alpha \frac{(R_C \| R_L)}{R_s + r_e}$

assuming $\alpha = 1$

$\frac{V_o}{V_s} = \frac{10 \| 10}{(0.100)\,2} = -25 \frac{V}{V}$

5.140

$\alpha \simeq 1$

$R_i = \frac{V_T}{I} = 50\Omega \Rightarrow I = 0.5mA$

$V_C = 5 - 0.5 . R_C$

$V_{C\,min} = V_C - 0.01\, g_m\,(R_C \| 1K)$

To prevent saturation $V_{C\,min} = 0$

$\rightarrow 0 = V_C - 0.01 \times 20\,(R_C \| 1)$

$= 5 - 0.5 \frac{R_C}{R_C + 1}$

$5R_C + 5 - 0.5R_C^2 - 0.5R_C - 0.2R_C = 0$

$0.5R_C^2 - 4.3R_C + 5 = 0$

$R_C = \frac{4.3 + \sqrt{4.3^2 + 10}}{1}$

$= 9.64 K\Omega$

Select $R_C = 9.1K\Omega$

$V_C = 0.45V$

$\frac{V_o}{V_s} = \frac{R_i}{R_s + R_i}\, g_m\,(R_C \| 1)$

$= \frac{50}{50+50} \times 20 \times (9.1 \| 1)$

$= 9\ V/V$

For $V_{be\,max} = 10mV$

$V_{s\,max} = 20mV$

$V_{c\,max} = 180mV$

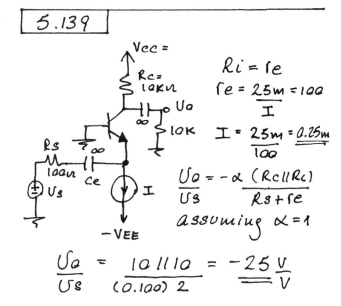

Thus the collector voltage swings from

$(0.45 - 0.18)V$ to $(0.45 + 0.18)V$

i.e from 0.27V to 0.63V

5.141

Refer to Fig P5.141

$R_i = r_e = \frac{V_T}{I_E} = \frac{V_T}{0.5} = 50\Omega$

To find the voltage gain V_o/V_s first note that

$\frac{V_e}{V_s} = \frac{R_i}{R_s + R_i} = \frac{50}{50+50} = 0.5$

Then,

$\frac{V_c}{V_e} = \alpha \times \frac{(\text{Total resistance at c})}{r_e}$

$\simeq \frac{1 \times (100K\Omega \| 1K\Omega)}{50\Omega}$

$= 19.8\ V/V$

Thus, $\frac{V_o}{V_s} = 19.8 \times 0.5 = 9.9\ V/V$

5.142

$V_A = 100 \Rightarrow r_o = \frac{V_A}{I_o} = \frac{100}{\frac{\beta}{\beta+1} \times 1}$

$r_o = 101 K\Omega$

(a) $R_{in} = R_B \| [(\beta+1)(r_e + (R_L \| r_o))]$

$R_{in} = 100K \| [(101)(25 + (1K \| 101K))]$

$R_{in} = \underline{50.6 K\Omega}$

$\frac{v_b}{v_s} = \frac{R_{in}}{R_{in}+R_s} = \underline{0.717}\ V/V$

$\frac{v_o}{v_s} = \frac{v_b}{v_s} \cdot \frac{(R_L \| r_o)}{(R_L \| r_o) + r_e} = \underline{0.975}\ V/V$

(b) The peak value of v_o occurs when the current flowing from ground through R_L equals the bias current I on a negative signal swing

$\therefore \frac{\hat{v}_o}{R_L} = I = 1 mA$

$\hat{v}_o = 1V \Rightarrow \hat{v}_s = \frac{1}{0.975} = \underline{1.025 V}$

$v_\pi = v_s' \times \frac{(\beta+1) r_e}{(\beta+1) r_e + (\beta+1)(R_L \| r_o)}$: where v_s' is

$v_s' = v_s \times \frac{R_{in}}{R_{in}+R_s} = v_s \times 0.717$

$v_\pi = \frac{(1.025 \times 0.717) \times 101 \times 25}{101[25 + 990]}$

$= \underline{18.10 mV}$

(c) $\hat{v}_\pi = 10 mV \Rightarrow$

$v_s = \frac{10m \times 101 [25 + 990]}{0.717 \times 101 \times 25}$

$= \underline{0.566 V}$

$v_o = \frac{v_o}{v_s} \times 0.566 = \underline{0.552}\ V$

(d) $R_o = \left(r_e + \frac{(R_s \| R_B)}{\beta+1}\right) \| r_o$

$= \left(25 + \frac{(20K \| 100K)}{101}\right) \| 101K$

$= \underline{190 \Omega}$

Open circuit voltage gain

$A_v|_{R_L=\infty} = \frac{R_B}{R_s+R_B} \cdot \frac{r_o}{r_o + r_e + \frac{R_s \| R_B}{\beta+1}}$

$= \frac{100}{20+100} \cdot \frac{101K}{101K + 25 + \frac{(100 \| 20)K}{101}}$

$= \underline{0.832}\ V/V$

$A_v = A_v|_{R_L=\infty} \times \frac{R_L}{R_L+R_o}$

$= 0.832 \times \frac{500}{500+190}$

$= \underline{0.603}\ V/V$

5.143

Refer to Fig. P5.143

(a) $I_E = \frac{9 - 0.7}{1 + 100/(\beta+1)}$ CONT.

For $\beta = 40$, $I_E = \dfrac{8.3}{1+\frac{100}{41}} = 2.41\,mA$

$V_E = 1 \times 2.41 = 2.41\,V$

$V_B = 2.41 + 0.7 = 3.11\,V$

For $\beta = 200$, $I_E = \dfrac{8.3}{1+\frac{100}{201}} = 5.54\,mA$

$V_E = +5.54\,V$

$V_B = +6.24\,V$

(b) $R_i = 100\,K\Omega \,//\, (\beta+1)[r_e + (1//1)]$
$= 100 \,//\, (\beta+1)[r_e + 0.5]$

For $\beta = 40$, $I_E = 2.41\,mA$
$\rightarrow r_e = 10.37\,\Omega$

thus $R_i = 100 \,//\, 41 \times (0.01037 + 0.5)$
$= 100 \,//\, 21$
$= 17.3\,\Omega$

For $\beta = 200$, $I_E = 5.54\,mA$
$\rightarrow r_e = 4.51\,\Omega$

thus $R_i = 100 \,//\, 201(0.0045 + 0.5)$
$= 100 \,//\, 101.4$
$= 50.3\,K\Omega$

(c) $\dfrac{v_o}{v_s} = \dfrac{v_b}{v_s} \cdot \dfrac{v_o}{v_b}$
$= \dfrac{R_i}{R_s + R_i} \cdot \dfrac{(1//1)}{(1//1) + r_e}$

For $\beta = 40$,
$\dfrac{v_o}{v_s} = \dfrac{17.3}{10 + 17.3} \times \dfrac{0.5}{0.5 + 0.01037}$
$= 0.621\,V/V$

For $\beta = 200$,
$\dfrac{v_o}{v_s} = \dfrac{50.3}{10 + 50.3} \cdot \dfrac{0.5}{0.5 + 0.0045}$
$= 0.827\,V/V$

5.144

Refer to Fig. P5.144

$I_E = \dfrac{5 - 0.7}{3.3 + \frac{100}{101}} = 1.00\,mA$

$r_e = \dfrac{25}{1.00} = 25\,\Omega$

$R_i = (\beta+1)[r_e + (3.3 // 1)]$
$= 80.0\,K\Omega$

$\dfrac{v_o}{v_s} = \dfrac{v_b}{v_s} \cdot \dfrac{v_o}{v_b} = \dfrac{R_i}{R_s + R_i} \; \dfrac{(3.3 // 1)}{r_e + (3.3 // 1)}$

Thus,
$\dfrac{v_o}{v_s} = \dfrac{80}{100 + 80} \times \dfrac{(3.3 // 1)}{0.025 + (3.3 // 1)}$
$= 0.430\,V/V$

$\dfrac{i_o}{i_i} = \dfrac{v_o / R_L}{v_s / (R_s + R_i)}$
$= \dfrac{v_o}{v_s} \cdot \dfrac{(R_s + R_i)}{R_L}$
$= 0.43 \times \dfrac{(100 + 80)}{1}$
$= 77.4\,A/A$

$R_{out} = 3.3 \,//\, \left[r_e + \dfrac{100}{\beta+1}\right]$
$= 3.3 \,//\, \left[0.025 + \dfrac{100}{101}\right]$
$= 0.776\,K\Omega$

5.145

Refer to Fig 5.63(a)

$R_o = r_o \,//\, [r_e + 10K/\beta+1]$

CONT.

$$R_o = \frac{V_A}{I_C} \,//\, \left[\frac{V_T}{I_E} + \frac{10K\Omega}{\beta+1}\right]$$

$$R_o = \frac{125}{0.99\times 2.5} \,//\, \left[\frac{0.025}{2.5} + \frac{10}{100+1}\right]$$

$$= 50.5 \,//\, [0.010 + 0.099]$$

$$= 0.109 K\Omega = \underline{\underline{109\Omega}}$$

With no load:

$$R_i = (\beta+1)[r_e + r_o]$$

For $\beta = 100$, $r_e = 10\Omega$, $r_o = 50.5 K\Omega$

$$R_i = 101 \times (0.010 + 50.5)$$

$$= \underline{\underline{5.1 M\Omega}}$$

$$\frac{v_o}{v_s} = \frac{v_b}{v_s} \cdot \frac{v_o}{v_b} = \frac{R_i}{R_s + R_i} \times \frac{r_o}{r_o + r_e}$$

$$= \frac{5100}{10+5100} \cdot \frac{50.5}{50.5+0.01}$$

$$= \underline{\underline{0.998 V/V}}$$

With $R_L = 1K\Omega$:

$$\frac{v_o}{v_s} = 0.998 \times \frac{R_L}{R_L + R_o}$$

$$= 0.998 \times \frac{1}{1+0.109} = \underline{\underline{0.890}} V/V$$

Largest negative output signal occurs when the BJT cuts off, thus:

$$v_{omin} = -2.5mA \times 1K\Omega$$

$$= \underline{\underline{-2.5V}}$$

Largest positive output signal occurs when $v_B = +3.4V$. The corresponding value of V_E is $\simeq 3.4 - 0.7 = +2.7V$ (where we have neglected the signal component of v_{BE})

Now, the DC level at the emitter with zero input signal is

$$V_E = -10K\Omega \times I_B - 0.7$$

$$= -10 \times \frac{2.7}{101} - 0.7$$

$$\simeq -0.97 V$$

Thus the signal component is:

$$v_{e\,max} = V_E - V_E$$

$$= 2.5 - (-0.97)$$

$$= \underline{\underline{3.47V}}$$

5.146

R_{sig}, v_{sig}, R_B, I, ∞, v_o, R_L

$$G_v\Big|_{R_L=\infty} = \frac{R_B}{R_{sig}+R_B} \cdot \frac{r_o}{(R_{sig}//R_B + r_e + r_o)} = 0.99$$

$\underbrace{}$ $\simeq 1$ for very large r_o

$$\Rightarrow \frac{R_B}{R_{sig}+R_B} = 0.99$$

$$\rightarrow \frac{R_B}{10K + R_B} = 0.99 \Rightarrow R_B = 990K\Omega$$

$$R_{out} = r_o \,||\, \left(r_e + \frac{R_s // R_B}{\beta+1}\right)$$

for r_o large: $R_{out} \simeq r_e + \dfrac{R_s//R_B}{\beta+1}$

$$r_e + \frac{(10//990)K}{\beta+1} = 200\Omega$$

$$r_e + \frac{(20//990)K}{\beta+1} = 300\Omega$$

CONT.

Then,

$$r_e + \frac{9.9K}{\beta+1} = 200 \quad (1)$$

$$r_e + \frac{19.6K}{\beta+1} = 300 \quad (2)$$

Solving Eqs. (1) and (2)

$$\beta+1 = 97 \rightarrow \beta = 96$$

and $r_e = 98\,\Omega$

If: $R_{sig} = 30K\Omega$ and $R_L = 1K\Omega$

$$G_v = \frac{R_B}{R_{sig}+R_B} \cdot \frac{(r_o \| R_L)}{\frac{R_{sig}\|R_B}{\beta+1} + r_e + (r_o\|R_L)}$$

If r_o is large:

$$G_v = \frac{R_B}{R_{sig}+R_B} \cdot \frac{R_L}{\frac{R_{sig}\|R_B}{\beta+1} + r_e + R_L}$$

$$= \frac{990}{30+990} \cdot \frac{1}{\frac{(30\|990)}{97} + 0.098 + 1}$$

$$= \underline{\underline{0.7}}\ V/V$$

5.147

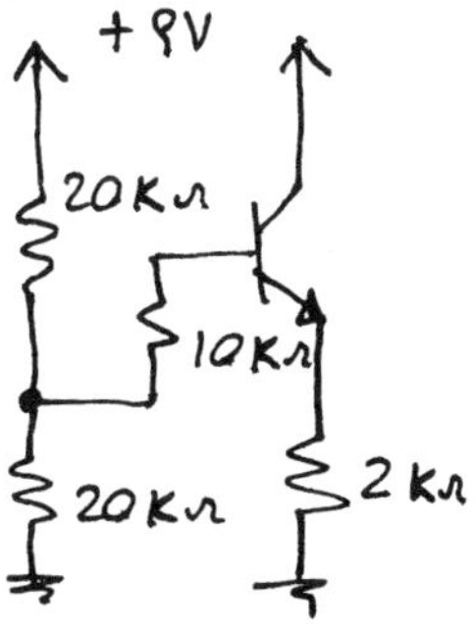

(a) DC Analysis

$$I_E = \frac{4.5-0.7}{2 + \frac{10+10}{101}}$$

$$= \underline{\underline{1.73\,mA}}$$

$$I_C = 0.99 \times 1.73 = \underline{\underline{1.71\,mA}}$$

$$g_m = \frac{I_C}{V_T} = 68.5\,mA/V$$

$$r_e = \frac{V_T}{I_E} = \underline{\underline{14.5\,\Omega}}$$

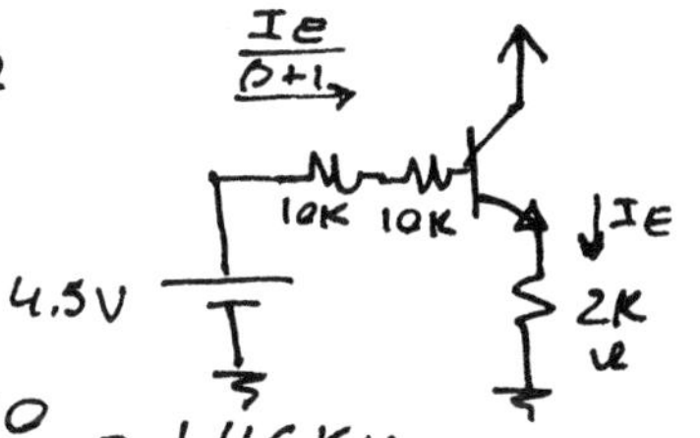

$$r_\pi = \frac{\beta}{g_m} = \frac{100}{68.5} = \underline{\underline{1.46\,K\Omega}}$$

(b) Simplified equivalent circuit:

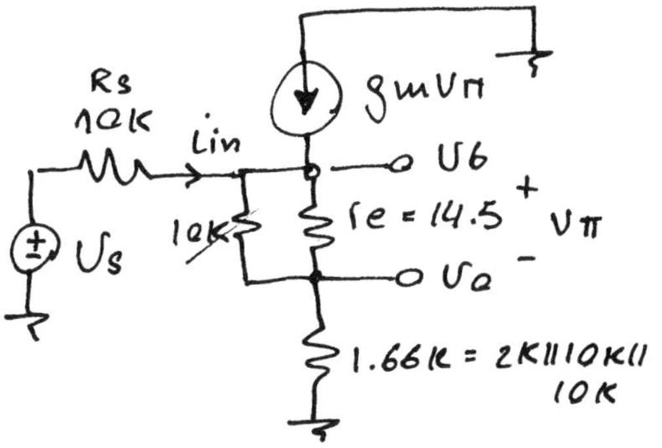

$$V_o \cong 1.66K \times \left(\frac{V_\pi}{14.5}\right) = 114.48\,V_\pi$$

$$\frac{V_\pi}{14.5} = i_{in} + 68\times10^{-3} \times V_\pi$$

$$\rightarrow V_\pi\left(\frac{1}{14.5} - 68\times10^{-3}\right) = i_{in}$$

$$i_{in} = \frac{V_\pi}{1.035K}$$

$$V_b = V_\pi + V_o = 115.48\,V_\pi$$

$$R_i \equiv \frac{V_b}{i_{in}} = \frac{115.48\,V_\pi}{V_\pi/1.035} = \underline{\underline{120\,K\Omega}}$$

$$\frac{V_o}{V_s} = \frac{V_b}{V_s} \cdot \frac{V_o}{V_b} = \frac{R_i}{R_i+R_s} \cdot \frac{114.48\,V_\pi}{115.48\,V_\pi}$$

$$= \frac{120}{120+10} \times \frac{114.48}{115.48} = \underline{\underline{0.92\,V/V}}$$

(c) With C_B open-circuited so that bootstraping is eliminated, we obtain the following equivalent circuit model:

CONT.

C, B, E, $g_m V_\pi$, R_s 10K, v_s, 10K, $r_e = 14.5$, v_o, 20K‖20K = 10K, 2K, R_i

$R_{ib} = (\beta+1)(r_e + 2K)$
$= 203.46\ K\Omega$

$R_i = (10K + 20K \| 20K) \| 203.46K$
$= \underline{\underline{18.21\ K\Omega}}$ (much lower than the value obtained with bootstrap.)

$v_o = \frac{V_\pi}{14.5} \times 2K = 138 \times V_\pi$

$v_b = v_o + v_\pi = (1 + 138) V_\pi = 139\ v_\pi$

$\frac{v_o}{v_s} = \frac{v_b}{v_s} \cdot \frac{v_o}{v_b} = \frac{R_i}{R_s + R_i} \cdot \frac{v_o}{v_b}$

$= \frac{18.21}{10 + 18.21} \cdot \frac{138}{139} = \underline{\underline{0.64}}\ V/V$

Much lower than the value obtained with bootstrapping. This is due to the lower R_i. Bootstrapping raises the component of input resistance due to the base biasing network.

5.148

Refer to the circuit in Fig P5.148

(a) $I_{E2} = \underline{\underline{5 mA}}$
$\beta_1 = 50, \quad \beta_2 = 100$

$I_{E1} = 50\mu + I_{B2}$
$= 50 + \frac{I_{E2}}{\beta_2 + 1} = 50 + \frac{5000}{101}$
$\simeq \underline{\underline{0.1 mA}}$

+9V, +4.5V, 1M‖1M = 0.5 MΩ, Q1, Q2, I_{B1}

$I_{B1} = \frac{0.1 mA}{(50+1)}$
$= 1.96\ \mu A$

$V_{B1} = 4.5 - 0.5 \times 1.96 = \underline{\underline{3.52 V}}$

$V_{B2} = 3.52 - 0.7 = \underline{\underline{2.82}}\ V$

(b) Refer to Fig. P.5.148

$\frac{v_o}{v_{b2}} = \frac{R_L}{R_L + r_{e2}}$

$R_L = 1 K\Omega \quad r_{e2} = \frac{25}{5} = 5 \Omega$

$\frac{v_o}{v_{b2}} = \frac{1}{1 + 0.005} = \underline{\underline{0.995}}\ V/V$

$R_{ib2} = (\beta_2 + 1)(r_{e2} + R_L)$
$= (101) \times (1.005)$
$= \underline{101.5}\ K\Omega$

(c) $\frac{v_{e1}}{v_{b1}} = \frac{R_{ib2}}{R_{ib2} + r_{e1}}$

$r_{e1} = \frac{V_T}{100 \mu A} = 250 \Omega$

$\rightarrow \frac{v_{e1}}{v_{b1}} = \frac{101.5}{101.5 + 0.25} = \underline{\underline{0.997}}\ \frac{V}{V}$

$R_i = 1 M\Omega \| 1 M\Omega \| (\beta_1 + 1)(r_{e1} + R_{ib2})$
$= 1 \| 1 \| 51 \times (0.25 + 101.5) K\Omega$
$= 1 \| 1 \| 5.2\ M\Omega$
$= 0.499\ M\Omega = \underline{\underline{499 K\Omega}}$

(d) $\frac{v_{b1}}{v_s} = \frac{R_i}{R_s + R_i} = \frac{499}{100 + 499} = 0.833\ V/V$

CONT.

(e) $\frac{v_o}{v_s} = \frac{v_{b1}}{v_s} \cdot \frac{v_{e1}}{v_{b1}} \cdot \frac{v_o}{v_{e1}}$

$= 0.833 \times 0.997 \times 0.995$

$= \underline{\underline{0.826}}$ V/V

5.149

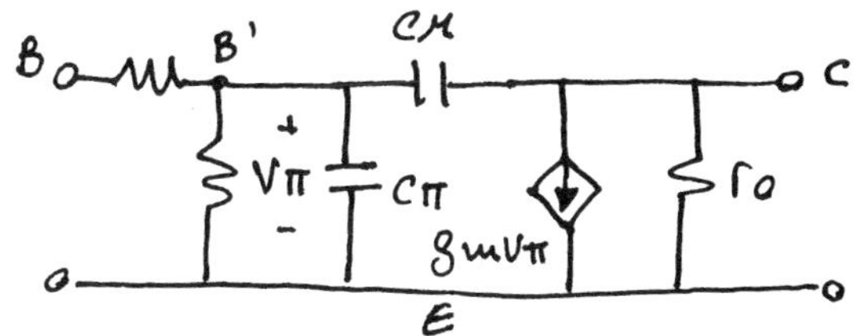

$r_x = \underline{100\,\Omega}$

$g_m = \frac{I_C}{V_T} = \frac{0.5\text{mA}}{25\text{mV}} = \underline{\underline{20\text{ mA/V}}}$

$r_\pi = \frac{\beta_0}{g_m} = \frac{100}{20} = \underline{\underline{5\text{ k}\Omega}}$

$r_o = \frac{V_A}{I_C} = \frac{50\text{V}}{0.5\text{mA}} = \underline{\underline{100\text{ k}\Omega}}$

$C_\mu = \frac{C_{\mu o}}{\left(1 + \frac{V_{CB}}{V_{0c}}\right)^{0.5}} = \frac{30}{\left(1 + \frac{2}{0.75}\right)^{0.5}} = \underline{\underline{15.7}}\text{ fF}$

$C_{je} \simeq 2C_{jeo} = 2 \times 20 = 40\text{ fF}$

$C_{de} = \tau_F g_m = 30 \times 10^{-12} \times 20 \times 10^{-3} = 600\text{ fF}$

$C_\pi = C_{je} + C_{de} = \underline{\underline{0.640}}\text{ pF}$

$f_T = \frac{g_m}{2\pi(C_\pi + C_\mu)} = \frac{20 \times 10^{-3}}{2\pi(0.64 + 0.016) \times 10^{-12}}$

$= \underline{\underline{4.85}}\text{ GHz}$

5.150

$|h_{fe}| \simeq f_T / f$

- At $I_C = 0.2\text{mA}$, $|h_{fe}| = 2.5$ at $f = 500\text{ MHz}$, thus:

$f_T = 2.5 \times 500 = \underline{\underline{1.25}}\text{ GHz}$

- At $I_C = 1.0\text{mA}$, $|h_{fe}| = 11.6$ at $f = 500\text{ MHz}$, thus:

$f_T = 11.6 \times 500 = \underline{\underline{5.8}}\text{ GHz}$

$f_T = \frac{g_m}{2\pi(C_\pi + C_\mu)}$

$C_\pi + C_\mu = \frac{g_m}{2\pi f_T} \rightarrow C_\pi = \frac{g_m}{2\pi f_T} - C_\mu$

$C_\pi\,(I_C = 0.2\text{mA}) = \frac{8 \times 10^{-3}}{2\pi \times 1.25 \times 10^9} - 0.05 \cdot 10^{-12}$

$= 0.9686\text{ pF}$

$C_\pi\,(I_C = 40\text{mA}) = \frac{40 \times 10^{-3}}{2\pi \times 5.8 \times 10^9} - 0.05 \times 10^{-12}$

$= 1.0476\text{ pF}$

Since $C_\pi = C_{je} + \tau_F g_m$,

$C_{je} + 8 \times 10^{-3} \tau_F = 0.9686 \times 10^{-12}$ (1)

$C_{je} + 40 \times 10^{-3} \tau_F = 1.0476 \times 10^{-12}$ (2)

Solving Eqn. (1) and (2) together yields,

$C_{je} = \underline{\underline{0.95\text{pF}}}$, $\tau_F = \underline{\underline{2.47\text{ps}}}$

5.151

$f_T = \frac{g_m}{2\pi(C_\pi + C_\mu)}$

$= \frac{80 \times 10^{-3}}{2\pi(10 + 1) \times 10^{-12}}$

$= \underline{\underline{1.24\text{GHz}}}$

$f_\beta = f_T / \beta_0 = (4.24/150) \times 10^9$

$= \underline{\underline{28.26\text{ MHz}}}$

5.152

At $I_C = 2\text{mA}$ the diffusion part of C_π is: $10 - 2 = 8\text{ pF}$

CONT.

At $I_C = 0.2\,mA$ the diffusion capacitance becomes 0.8 pF and thus,

$$C_\pi = 0.8 + 2 = 2.8\ pF$$

and:

$$f_T = \frac{g_m}{2\pi(C_\pi + C_\mu)}$$

$$= \frac{8\times10^{-3}}{2\pi(2.8+1)\times10^{-12}} = \underline{\underline{335\ MHz}}$$

5.153

$$\omega_T = g_m/(C_\pi + C_\mu)$$

$$2\pi\times5\times10^{9} = \frac{20\times10^{-3}}{(C_\pi+0.1)\times10^{-12}}$$

$$C_\pi + 0.1 = \frac{20}{10\pi} = 0.64\ pF$$

$$C_\pi = \underline{\underline{0.54}}\ pF$$

$$g_m = \underline{\underline{20\,mA/V}}$$

$$r_\pi = \beta/g_m = 150/20 = \underline{\underline{7.5\,k\Omega}}$$

$$f_\beta = f_T/\beta = \frac{5\times10^{9}}{150} = \underline{\underline{33.3\,MHz}}$$

5.154

$|h_{fe}|$ becomes 20 at:

$$f_T/20 = \frac{1\times10^{9}}{20} = \underline{\underline{50}}\ MHz$$

$$f_\beta = f_T/\beta_0 = \frac{1000\,MHz}{200} = \underline{\underline{5\,MHz}}$$

5.155

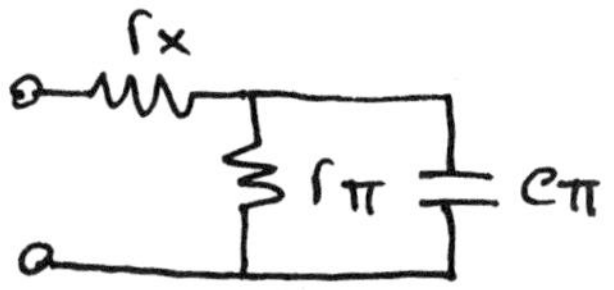

$$Z = r_x + \frac{1}{\frac{1}{r_\pi} + j\omega C_\pi}$$

$$= r_x + \frac{r_\pi}{1 + j\omega C_\pi r_\pi}$$

$$Z = r_x + \frac{r_\pi}{1 + j(\omega/\omega_\beta)}$$

$$= r_x + \frac{r_\pi(1 - j\omega/\omega_\beta)}{1 + \left(\frac{\omega}{\omega_\beta}\right)^2}$$

$$= r_x + \frac{r_\pi}{1 + \left(\frac{\omega}{\omega_\beta}\right)^2} - j\cdot\frac{r_\pi(\omega/\omega_\beta)}{1 + \left(\frac{\omega}{\omega_\beta}\right)^2}$$

$$Re[Z] = r_x + \frac{r_\pi}{1 + \left(\frac{\omega}{\omega_\beta}\right)^2}$$

For $Re[Z]$ to be an estimate of r_x good to within 10% we must keep

$$\frac{r_\pi}{1 + \left(\frac{\omega}{\omega_\beta}\right)^2} \le \frac{r_x}{10}$$

But $r_x \le r_\pi/10$

Thus,

$$\frac{r_\pi}{1 + \left(\frac{\omega}{\omega_\beta}\right)^2} \le \frac{r_\pi}{100}$$

$$1 + \left(\frac{\omega}{\omega_\beta}\right)^2 \ge 100$$

or $\omega \ge 10\,\omega_\beta$ (approx.)

5.156

See completed table below

CONT.

	I_E (mA)	r_e (Ω)	g_m (mA/V)	r_π (KΩ)	β_0
(a)	1	25	40	2.5	100
(b)	1	25	40	3.13	125.3
(c)	0.99	25.3	39.6	2.525	100
(d)	10	2.5	400	0.25	100
(e)	0.1	250	4	25	100
(f)	1.0	25	40	0.25	10
(g)	1.25	20	50	0.20	10

CONT.	f_T (MHz)	C_π (pF)	C_μ (pF)	f_β (MHz)
(a)	400	2	13.9	4
(b)	501.3	2	10.7	4
(c)	400	2	13.8	4
(d)	400	2	157	4
(e)	100	2	4.4	1
(f)	400	2	13.9	40
(g)	800	1	9	80

5.157

From Example 5.18:

$I = 2mA$, $\beta = 100$, $f_T = 800MHz$

$R_B = 50K\Omega$, $R_C = 4K\Omega$, $r_x = 50\Omega$

$V_A = 100$, $C_\mu = 1pF$, $R_{sig} = 5K\Omega$

$R_L = 5K\Omega$

$$g_m = \frac{2mA}{25mV} = 80\frac{mA}{V}$$

$$r_\pi = \frac{\beta_0}{g_m} = \frac{100}{80m} = 1250\Omega$$

$$r_o = \frac{V_A}{I_C} = \frac{100V}{2mA} = 50K\Omega$$

$$C_\pi + C_\mu = \frac{g_m}{\omega_T} = \frac{80\times10^{-3}}{2\pi\times800\times10^6} = 16pF$$

$$C_\mu = 1pF \Rightarrow C_\pi = 15pF$$

$$A_M = \frac{-R_B}{R_B+R_{sig}} \cdot \frac{r_\pi \times g_m R_L'}{(r_\pi + r_x + (R_B // R_{sig})}$$

where $R_L' = r_o // R_C // R_L$

$= (50 // 4 // 5)\ K\Omega$

$= 2.1\ K\Omega$

$$A_M = \frac{-50}{50+5} \cdot \frac{1250 \times 168}{1250 + 50 + (50 // 5) K\Omega}$$

where $168 = g_m \times R_L'$

$= 80\times10^{-3} \times 2.1.10^3$

Then: $A_M = -32.6$

$20 \log |A_M| = 30.3dB$

$$C_{in} = C_\pi + C_\mu (1 + g_m R_L')$$

$= 15 + 1(1 + 168) = 184pF$

$R'_{sig} = r_\pi // [r_x + (R_B // R_{sig})]$

$= 1250 // [50 + (50K // 5K)]$

$= 983\Omega$

$$f_H = \frac{1}{2\pi C_{in} R'_{sig}} = \frac{1}{2\pi\times184\times10^{-12}\times983}$$

$= 880KHz$

Gain-bandwidth product

$GB = |A_M| \times f_H = 32.6 \times 880\times10^3$

$= 29\times10^6$

Previously, in example 5.18

$GB = 39 \times 754\times10^3 = 29\times10^6$

Thus, the designer traded gain for bandwidth by increasing I. However, by doubling I the dissipation increased by a factor of 2, since:

Power $= \underset{2I'}{I} \times V_{supply}$

5.158

Refer to Fig. 5.71(a)

$R_B >> R_{sig}$, $r_x << R_{sig}$

$R_{sig} >> r_\pi$, $g_m R_L' >> 1$

$g_m R_L' C_\mu >> C_\pi$ CONT.

$$A_M = \frac{-R_B}{\underbrace{R_B + R_{sig}}_{④}} \cdot \frac{r_\pi \cdot g_m \cdot R_L'}{r_\pi + r_x + (R_{sig} \| R_B)}$$

① $= R_{sig}$, since $R_B >> R_{sig}$
② $= R_{sig}$, since $R_{sig} >> r_x$
③ $= R_{sig}$, since $R_{sig} >> r_\pi$
④ $= 1$, since $R_B >> R_{sig}$

Thus,

$$A_M = -\frac{r_\pi \cdot g_m \cdot R_L'}{R_{sig}}$$

but $r_\pi = \frac{\beta}{g_m}$

$$\Rightarrow A_M = -\frac{\beta R_L'}{R_{sig}} \quad \text{Q.E.D}$$

(b) $$f_H = \frac{1}{2\pi \cdot C_{in} \cdot R'_{sig}}$$

$$C_{in} = C_\pi + C_\mu (1 + g_m R_L')$$

① $= g_m R_L'$, since $R_L' >> 1$
② $= C_\mu g_m R_L'$, since $C_\mu g_m R_L' >> C_\pi$

$$\Rightarrow C_{in} = C_\mu \cdot g_m \cdot R_L'$$

$$R'_{sig} = r_\pi \,//\, [r_x + (R_B \,//\, R_{sig})]$$

① $= R_{sig}$, since $R_B >> R_{sig}$
② $= R_{sig}$, since $R_{sig} >> r_x$
③ $= r_\pi$, since $R_{sig} >> r_\pi$

$$\Rightarrow R'_{sig} = r_\pi$$

Thus,

$$f_H = \frac{1}{2\pi \cdot C_\mu \cdot g_m \cdot R_L' \cdot r_\pi}$$

Q.E.D.

(c)

$$GB = |A_M| \times f_H = \frac{\beta R_L'}{R_{sig}} \frac{1}{2\pi C_\mu \beta R_L'}$$

$$= \frac{1}{2\pi} \frac{1}{C_\mu \cdot R_{sig}} \quad \text{Q.E.D.}$$

For: $C_\mu = 1pF$ and $R_{sig} = 25K\Omega$

$$GB = \frac{1}{2\pi \times 10^{-12} \times 25 \times 10^3} \cong 6.36 \times 10^6$$

For: $I_C = 1mA$, $\beta = 100$, $R_{sig} = 25K\Omega$

1) If $R_L' = 25K\Omega$

$$A_M = -100 \cdot \frac{25K}{25K} = -100$$

$$20 \log |100| = 40dB$$

$$f_H = \frac{1}{2\pi \times 10^{-12} \times 100 \times 25 \times 10^3} = 63.6KHz$$

2) If $R_L' = 2.5K\Omega$

$$A_M = -100 \cdot \frac{2.5K}{25K} = -10$$

$$20 \log |10| = 20$$

CONT.

$$f_H = \frac{1}{2\pi \times 10^{-12} \times 100 \times 2.5.10^3}$$

$$f_H = \underline{\underline{636}}\ KHz$$

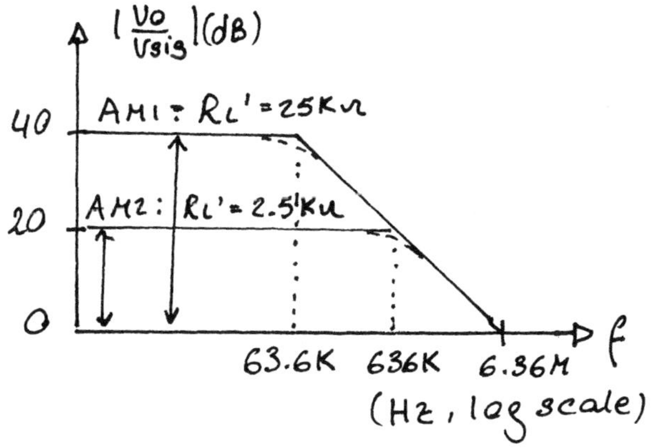

$GP = 6.36 \times 10^6 = A_M \times f_H$

when $A_M = 1 \Rightarrow f_H = \underline{\underline{6.36.10^6}} Hz$

$$R_L' = \frac{1}{2\pi (6.36 \times 10^6)\, C_\mu \times \beta}$$

$C_\mu = 1 \times 10^{-12}$, $\beta = 100$

$R_L' = \underline{\underline{250\Omega}}$

5.159

Refer to Fig. P5.159

$R_{in} = R_1 \,||\, R_2 \,||\, (r_x + r_\pi)$

where $R_1 = 33K\Omega$, $R_2 = 22K\Omega$

Next →

CONT.

$r_x = 50$ and,

$$r_\pi = \frac{\beta_0}{g_m} = \frac{120}{0.3 \times 40} = \frac{120}{12} = 10\,K\Omega$$

$$R_{in} = 33 \parallel 22 \parallel 10.05 = \underline{\underline{5.7\,K\Omega}}$$

$$A_M = -\frac{R_{in}}{R_{in}+R_S} \cdot \frac{r_\pi}{r_\pi + r_x} \cdot g_m (R_C \parallel R_L \parallel r_o)$$

$$= -\frac{5.7}{5.7+5} \cdot \frac{10}{10+0.05} \cdot 12\,(4.7 \parallel 5.6 \parallel 300)$$

$$= \underline{\underline{-16.11}}\ V/V$$

$$R'_{sig} = r_\pi \parallel [r_x + (R_1 \parallel R_2 \parallel R_{sig})]$$

$$= 10K \parallel [50 + (33 \parallel 22 \parallel 5)K]$$

$$= \underline{\underline{2.69\,K\Omega}}$$

$$R'_L = r_o \parallel R_C \parallel R_L = 300 \parallel 4.7 \parallel 5.6 (K\Omega)$$

$$= 2.53\,K\Omega$$

$$C_\pi + C_\mu = \frac{g_m}{2\pi \cdot f_T} = \frac{12 \cdot 10^{-3}}{2\pi \times 700 \cdot 10^6} =$$

$$= 2.73\,pF$$

$$C_\pi = (2.73 - 1)\,pF$$

$$= 1.73\,pF$$

$$C_{in} = C_\pi + C_\mu (1 + g_m R'_L)$$

$$= 1.73p + 1p(1 + 12 \times 2.53)$$

$$= 33\,pF$$

$$f_H = 1/2\pi C_{in} \cdot R'_{sig}$$

$$= 1/2\pi \times 33 \cdot 10^{-12} \times 2.69 \cdot 10^3$$

$$= \underline{\underline{1.79\,MHz}}$$

5.160

$$R_{in} = R_1 \parallel R_2 \parallel r_\pi$$

where $r_\pi = \frac{\beta}{g_m}$ and $g_m = \frac{I_C}{V_T}$

$$g_m = \frac{0.8}{0.025} = 32\,mA/V$$

$$r_\pi = \frac{200}{32} = 6.25\,K\Omega$$

$$R_{in} = 68 \parallel 27 \parallel 6.25 = 4.72\,K\Omega$$

$$R'_L = R_C \parallel R_L = 4.7 \parallel 10 = 3.2\,K\Omega$$

$$A_M = \frac{R_{in}}{R_S + R_{in}} \times -g_m R'_L$$

$$= \frac{-4.72}{10+4.72} \times 32 \times 3.2$$

$$= \underline{\underline{-32.8}}\ V/V$$

$$C_T = C_\pi + C_\mu (1 + g_m R'_L)$$

where $C_\pi + C_\mu = \frac{g_m}{2\pi f_T} = \frac{32 \times 10^{-3}}{2\pi \times 10^9}$

$$= 5.1\,pF$$

$$C_\pi = 5.1 - 0.8 = 4.3\,pF$$

$$C_T = 4.3 + 0.8\,(1 + 32 \times 3.2)$$

$$= 87\,pF$$

The resistance seen by C_T is R_T,

$$R_T = r_\pi \parallel R_1 \parallel R_2 \parallel R_S$$

$$= 6.25 \parallel 68 \parallel 27 \parallel 10 = 3.2\,K\Omega$$

Thus

$$f_H \simeq \frac{1}{2\pi C_T R_T}$$

$$= \frac{1}{2\pi \times 87 \times 10^{-12} \times 3.2 \times 10^3}$$

$$= \underline{\underline{572\,KHz}}$$

5.161

+1.5V

$R_S = 1K$, $R_B = 47K$, $R_C = 1K$, V_b, V_o, V_s, $R_L = 1K$

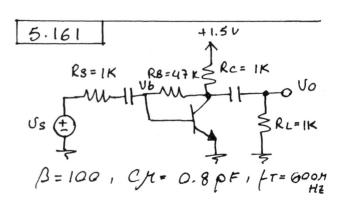

$\beta = 100$, $C_\mu = 0.8\,pF$, $f_T = 600\,MHz$

CONT.

(a) $1.5V = 1K(I_C + I_B) + 47K \cdot I_B + 0.7V$

$= I_C + \frac{I_C}{\beta} + 47\frac{I_C}{\beta} + 0.7$

$I_C = \frac{0.8}{1 + \frac{48}{\beta}} = \frac{0.8}{1 + \frac{48}{100}} = 0.54 mA$

(b) $g_m = \frac{I_C}{V_T} = 40 \times 0.54 = 21.6 \frac{mA}{V}$

$r_\pi = \frac{\beta}{g_m} = \frac{100}{21.6} = 4.63 K\Omega$

(c) $V_o \simeq -g_m (R_C // R_L) \cdot v_b$

$= -21.6 (1 // 1) v_b$

$= -10.8 v_b$

$\Rightarrow \frac{v_o}{v_b} = -10.8 \ V/V$

(d) Using Miller's theorem to find R_i:

$R_i = \frac{47}{1 - \frac{V_o}{v_b}} = \frac{47}{1 + 10.8} = 4K\Omega$

(e) $A_M = \frac{v_b}{v_s} \cdot \frac{v_o}{v_b} = \frac{R_i}{R_s + R_i} \times \frac{v_o}{v_b}$

$= \frac{4K}{1K + 4K} \times -10.8 = -8.64$

(f) $C_\pi + C_\mu = \frac{g_m}{2\pi f_T} = \frac{21.6 \times 10^{-3}}{2\pi \cdot 600 \times 10^6}$

$= 5.73 pF$

$C_\pi = 5.73 - 0.8 = 4.93 pF$

$C_{in} = C_\pi + C_\mu (1 + g_m R_L')$

$= 4.93 + 0.8 (1 + 21.6 \times (1 // 1))$

$= 14.37 pF$

(g) $f_H = \frac{1}{2\pi \cdot C_{in} \cdot R_{in}}$

$R_{in} = R_i // R_s = 4K // 1K$

$= 0.8 K\Omega$

$\Rightarrow f_H = \frac{1}{2\pi \times 14.37 \times 10^{-12} \times 0.8 \times 10^3}$

$f_H = 13.84 MHz.$

5.162

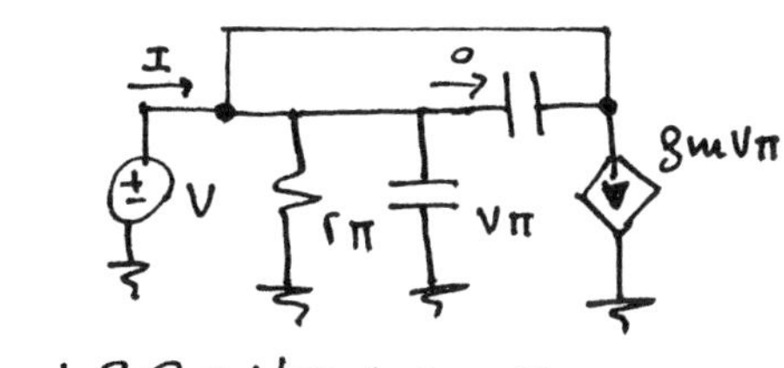

$I = \frac{V}{r_\pi} + sC_\pi V_\pi + g_m V_\pi$

$Y_{in} = (g_m + \frac{1}{r_\pi}) + sC_\pi$

$Z_i = \frac{1}{(g_m + \frac{1}{r_\pi}) + sC_\pi}$

$= \frac{1}{\frac{1}{r_e} + sC_\pi} = \frac{r_e}{1 + sC_\pi r_e}$

$f_T = \frac{g_m}{2\pi (C_\pi + C_\mu)}$

Since C_π contains a component that is proportional to the bias current, it follows that at high currents $C_\pi >> C_\mu$ and

$f_T \simeq \frac{g_m}{2\pi C_\pi} \simeq \frac{1}{2\pi \cdot C_\pi r_e}$

Thus,

$Z_i = \frac{r_e}{1 + s/\omega_T}$ (at high currents)

The phase angle will be $-45°$ at $\omega = \omega_T$, or

$f = f_T = 400 MHz$

CONT.

For a lower bias current so that $C_\pi = C_\mu$,

$$f_T = \frac{1}{4\pi C_\pi r_e}$$

and $$Z_i = \frac{r_e}{1 + \frac{s}{2\omega_T}}$$

$-45°$ angle is obtained at $\omega = 2\omega_T$ or $f = 2f_T$ $= \underline{800\,MHz}$

(Assuming f_T remains constant which is not necessarily true)

5.163

Refer to Fig. P5.159 and Problem 5.159

$\omega_L = \omega_{C1} + \omega_{C2} + \omega_{CE}$

$\omega_{C1} = \frac{1}{C_{C1} \cdot R_{C1}} \Rightarrow f_{C1} = \frac{1}{2\pi C_{C1} \cdot R_{C1}}$

$R_{C1} = R_S + [R_1 \| R_2 \| (r_x + r_\pi)]$

$= 5K + [33K \| 22K \| (50 + 10K)]$

$= 10.7\,K\Omega$

$C_{C1} = 1\,\mu F$

$\Rightarrow f_{C1} = \frac{1}{2\pi \cdot 1\times10^{-6} \times 10.7\cdot10^3}$

$= \underline{14.87\,Hz}$

$f_{C2} = 1/2\pi \cdot C_{C2} \cdot R_{C2}$

$R_{C2} = R_L + (R_C \| r_o)$

$= 5.6K + (4.7K \| 300K)$

$= \underline{10.23\,K\Omega}$

$C_{C2} = 1\,\mu F$

$\Rightarrow f_{C2} = \frac{1}{2\pi \times 1\cdot10^{-6} \times 10.23\cdot10^3}$

$= \underline{15.55\,Hz}$

$f_{CE} = 1/2\pi C_E R_E'$

$R_E' = R_E \| \left[\frac{r_\pi + r_x + (R_1 \| R_2 \| R_S)}{\beta_0 + 1}\right]$

$= 3.9 \| \left(\frac{10K + 50 + (33 \| 22 \| 5)K}{121}\right)$

$= 109.8\,\Omega$

$C_E = 10\,\mu F$

$\Rightarrow f_{CE} = \frac{1}{2\pi \times 10\times10^{-6} \times 109.8}$

$= \underline{144.95\,Hz}$

$f_L = \frac{\omega_L}{2\pi} = 14.87 + 15.55 + 109.8$

$f_L = \underline{140.22\,Hz}$

5.164

Refer to Fig P5.159 and Problem 5.163.

To select C_E so that it contributes 90% of the value of ω_L:

$\frac{1}{2\pi C_E \cdot R_E'} = 0.9 \times 100$

$R_E' = 109.8\,\Omega$ (From problem 5.163)

$\Rightarrow C_E = \frac{1}{2\pi \cdot 109.8 \times 90}$

$= 16.1\,\mu F$ //

To select C_1 so that it contributes 5% of f_L:

$R_{C1} = 10.7\,K\Omega$ (From P 5.163)

$\Rightarrow C_1 = \frac{1}{2\pi \times 10.7\cdot10^3 \times 0.05 \times 100}$

$C_1 = 2.97\,\mu F$ //

CONT.

To select C_2 so that it contributes to 5% of f_L:

$R_{C2} = 10.23\,K\Omega$ (From P5.163)

$$\Rightarrow C_2 = \frac{1}{2\pi \times 10.23 \cdot 10^3 \times 0.05 \times 100}$$

$$= \underline{\underline{3.11\,\mu F}}$$

5.165

Refer to Fig. P5.159.

$$R_{C1} = R_S + [R_B \,//\, (r_x + r_\pi)]$$

$$= 10 + [10 \,//\, (0.1 + 1)]$$

$$= 10.99\,K\Omega$$

$$R_E' = R_E \,//\, \frac{r_\pi + r_x + (R_B // R_S)}{\beta_0 + 1}$$

$$\simeq 1 \,//\, \frac{1 + 0.1 + (10//10)}{100 + 1}$$

$$\simeq 57\,\Omega$$

For C_E and C_{C1} to contribute equally to the determination of f_L,

$$C_E R_E' = C_{C1} \cdot R_{C1}$$

$$\Rightarrow \frac{C_E}{C_{C1}} = \frac{R_{C1}}{R_E'} = \frac{10.99}{0.057} = \underline{\underline{193}}$$

5.166

Refer to Fig P5.166.

(a) At midband frequencies

$$I_B = \frac{V_s}{R_S + r_\pi}$$

$$I_C = \beta \cdot I_b = \frac{\beta V_s}{R_S + r_\pi}$$

$$V_o = -I_C \cdot (R_C // R_L)$$

$$= -\frac{\beta (R_C // R_L)}{R_S + r_\pi} \cdot V_s$$

$$A_M = \frac{V_o}{V_s} = -\beta \frac{(R_C // R_L)}{R_S + r_\pi}$$

(b) Pole due to C_E:

$$\omega_{PE} = \frac{1}{C_E\left(r_e + \frac{R_S}{\beta + 1}\right)} \;//$$

Pole due to C_C:

$$\omega_{PC} = \frac{1}{C_C (R_C + R_L)} \;//$$

Zeros are both at $\underline{\underline{s = 0}}$

(c) $$A(s) = A_M \cdot \frac{s^2}{(s + \omega_{PE})(s + \omega_{PC})}$$

$$A(s) = -\frac{\beta (R_C // R_L)}{R_S + r_\pi} \cdot \frac{s^2}{\left[s + \frac{1}{C_E\left(r_e + \frac{R_S}{\beta+1}\right)}\right] \times \left[s + \frac{1}{C_C (R_C + R_L)}\right]}$$

(d) $$A_M = -\frac{100\,(10//10)}{10 + \frac{100}{40}}$$

$$= -\underline{\underline{40}}\ V/V$$

(e) Since the resistance that forms the pole ω_{PE} is very small, we choose to make ω_{PE} the dominant pole, thus:

$$f_{PE} = f_L = 100 = \frac{1}{2\pi C_E \left(25 + \frac{10K}{101}\right)}$$

$$\Rightarrow C_E = \frac{1}{2\pi \times 100 \times (0.025 + 0.100) \cdot 10^3}$$

$$= \underline{\underline{12.7\,\mu F}}$$

$f_{PC} = 10\ Hz \Rightarrow$

CONT.

$$10\text{ Hz} = \frac{1}{2\pi C_C (R_L + R_C)}$$

$$\Rightarrow C_C = \frac{1}{2\pi \times 10 (10+10)10^3}$$

$$= \underline{0.8\,\mu F}$$

(f)

20dB/Dec

12dB

40dB/Dec

0dB

32dB

10Hz 100Hz f

Unity-gain frequency must be an octave lower than 10 Hz. i.e. at $\underline{5\text{ Hz}}$

(g) $$A(j\omega) = -A_M \frac{\omega^2}{(\omega_{PE}+j\omega)(\omega_{PC}+j\omega)}$$

$$= +40 \frac{\omega^2}{(\omega_{PE}+j\omega)(\omega_{PC}+j\omega)}$$

Thus $\phi = \tan^{-1}\left(\frac{\omega}{\omega_{PE}}\right) - \tan^{-1}\left(\frac{\omega}{\omega_{PC}}\right)$

$$= -\left[\tan^{-1}\frac{f}{f_{PE}} + \tan^{-1}\frac{f}{f_{PC}}\right]$$

$$= -\left[\tan^{-1}\frac{f}{100} + \tan^{-1}\frac{f}{10}\right]$$

Thus at $f = 100\text{ Hz}$,

$$\phi = -[\tan^{-1}1 + \tan^{-1}10]$$

$$\simeq -\underline{129.3^\circ}$$

5.167

Refer to Fig. P5.167

(a) $$I_e = \frac{V_s}{r_e + R_e + \frac{1}{sC_E}}$$

$I_c \simeq I_e$

$$V_o = -R_C I_c = \frac{-R_C}{r_e + R_e + \frac{1}{sC_E}} \cdot V_s$$

$$A(s) \equiv \frac{V_o}{V_s} = \frac{-R_C}{r_e + R_e + \frac{1}{sC_E}}$$

$$= \frac{-R_C}{r_e + R_e} \cdot \frac{s}{s + \frac{1}{C_E(r_e+R_e)}}$$

Thus, $A_M = \frac{-R_C}{r_e + R_e}$

$$\omega_L = \frac{1}{C_E.(r_e+R_e)}$$

(b) A_M is reduced by the factor $\frac{r_e + R_e}{r_e}$

$$= \underline{1 + \frac{R_e}{r_e}}$$

(c) ω_L is reduced by the factor $\left(1 + \frac{R_e}{r_e}\right)$ which is the same as the gain reduction factor. Thus, the value of R_e can be used as the parameter for exercising the gain-bandwidth trade off.

(d) $\underline{R_e = 0}$:

$$|A_M| = \frac{R_C}{r_e} = \frac{10{,}000}{25} = \underline{400\text{ V/V}}$$

$$f_L = \frac{1}{2\pi C_E r_e} = \frac{1}{2\pi \times 100 \times 10^{-6} \times 25}$$

$$= \underline{63.7}\text{ Hz}$$

To lower f_L by a factor of 5 use:

CONT.

$R_e = 4r_e = \underline{\underline{100\,\Omega}}$. The gain is also lowered by a factor of 5 to $\underline{\underline{80}}$ V/V

5.168

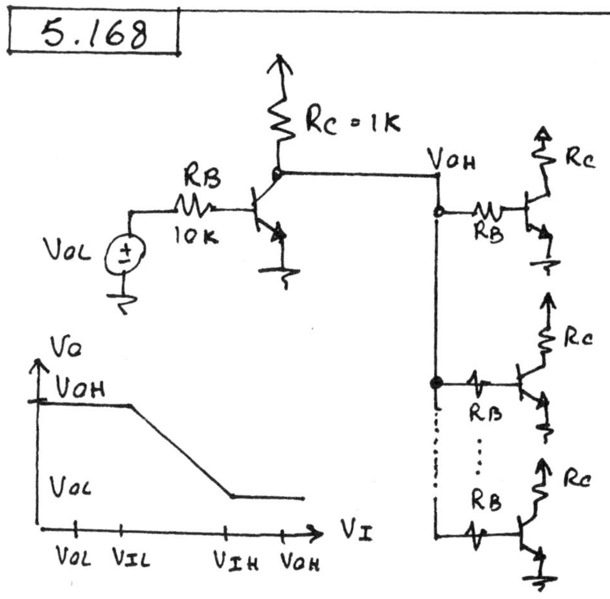

$$V_{IH} = V_{BE} + I_{B(EOS)} R_B$$

$$= 0.7 + \left[\left(\frac{V_{CC} - V_{CEsat}}{R_C}\right)/100\right] R_B$$

$$= 0.7 + \frac{5 - 0.2}{100} \times 10$$

$$= 1.18\text{ V}$$

$\therefore$ For $NM_H = V_{OH} - V_{IH} = 1$

$$V_{OH} = 1 + 1.18 = 2.18\text{ V}$$

$$\therefore V_{OH} = V_{CC} - R_C \cdot \frac{V_{CC} - V_{BE}}{R_C + \frac{R_B}{N}}$$

$$= 5 - 1\left(\frac{5 - 0.7}{1 + \frac{10}{N}}\right) > 2.18$$

$$N < 19.05 \Rightarrow N = \underline{\underline{19}}$$

5.169

(a) Input is low:

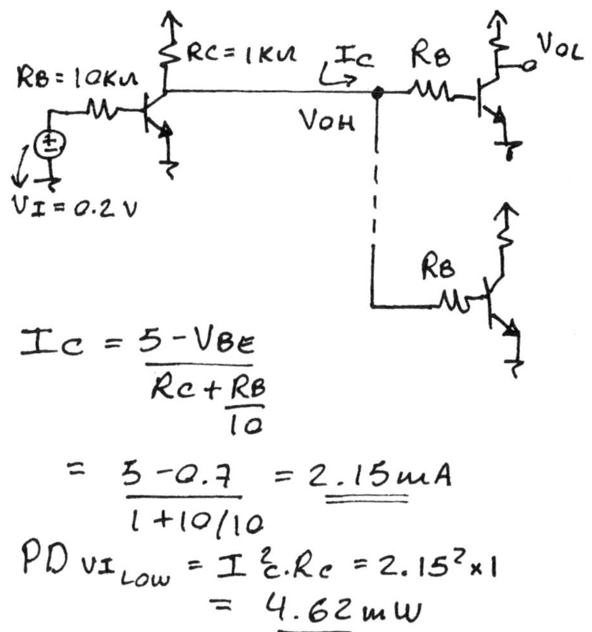

$$I_C = \frac{5 - V_{BE}}{R_C + \frac{R_B}{10}}$$

$$= \frac{5 - 0.7}{1 + 10/10} = \underline{\underline{2.15\text{ mA}}}$$

$$PD_{V_{I_{LOW}}} = I_C^2 \cdot R_C = 2.15^2 \times 1$$

$$= \underline{\underline{4.62\text{ mW}}}$$

(b) Input is high:

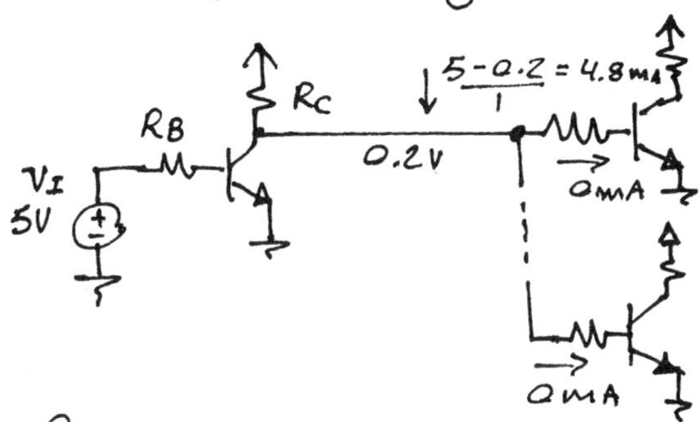

Power dissipated neglecting the base circuit

$$PD_{V_{I\ HIGH}} = V_{CC} I_C$$

$$= 5 \times 4.8 = \underline{\underline{24\text{ mW}}}$$

(c) $PD_{AVG} = \frac{1}{2}(4.62 + 24)$

$$= \underline{\underline{14.31\text{ mW}}}$$

5.170

$\beta_{Forced} = 10 = \dfrac{1.3/R_C}{0.8/R_C+R_B}$ ①

$\dfrac{80}{13} = \dfrac{R_C+R_B}{R_C} = 1 + \dfrac{R_B}{R_C}$

$\dfrac{R_B}{R_C} = \dfrac{67}{13} \Rightarrow R_B = 5.15 R_C$ ②

Total current from supply

$= \dfrac{1.3}{R_C} + \dfrac{0.8}{R_C+R_B}$

From Eqn. ① $\dfrac{0.8}{R_C+R_B} = \dfrac{1.3/R_C}{10}$

Thus $I_{TOTAL} = \dfrac{1.3}{R_C} + \dfrac{1}{10}\dfrac{1.3}{R_C}$

$= \dfrac{1.1 \times 1.3}{R_C}$

$P_{D TOTAL} = I_{TOTAL} \times 1.5V$

$= \dfrac{1.1 \times 1.3}{R_C} \times 1.5 = 1 mW$

$R_C = \dfrac{1.1 \times 1.3 \times 1.5}{1} = 2.145 K\Omega$

Choose $R_C = 2.2 K\Omega$

$R_B = 5.15 \times 2.145 = 11 K\Omega$

5.171

Refer to Fig. P5.171

V_X	V_Y	V_Z
0.2	0.2	5
0.2	5	0.2
5	0.2	0.2
5	5	0.2

These are 4 input combinations. When any input is high, $(V_X \text{ and/or } V_Y)_{high}$, $V_Z = 0.2V$

When both inputs are low, $(V_X \text{ and } V_Y)_{Low} \rightarrow$ V_Z is high

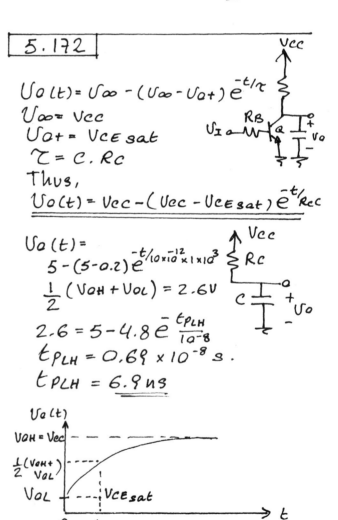

5.172

$V_O(t) = V_\infty - (V_\infty - V_{O+}) e^{-t/\tau}$

$V_\infty = V_{CC}$

$V_{O+} = V_{CE sat}$

$\tau = C . R_C$

Thus,

$V_O(t) = V_{CC} - (V_{CC} - V_{CE sat}) e^{-t/R_C C}$

$V_O(t) =$

$5 - (5-0.2) e^{-t/10 \times 10^{-12} \times 1 \times 10^3}$

$\frac{1}{2}(V_{OH} + V_{OL}) = 2.6V$

$2.6 = 5 - 4.8 e^{-\frac{t_{PLH}}{10^{-8}}}$

$t_{PLH} = 0.69 \times 10^{-8} s.$

$t_{PLH} = 6.9 ns$

5.173

At $t = 0+$

Vcc, I_c, RB, Vcc, Active, I_B, V_{BE}, $\frac{V_{CC}-V_{BE}}{R_B}$, $v_o(t)$

$$I_C = \beta I_B = \beta \cdot \frac{(V_{CC}-V_{BE})}{R_B}$$

$$v_O(0+) = V_{CC}$$

Thus the equivalent circuit for the capacitor discharge will be:

Vcc, Rc, $v_o(t)$, I_c, C $\Longrightarrow$ Rc, $v_o(t)$, $I_cR_c - V_{cc}$, C

$$I_C = \beta \frac{V_{CC}-V_{BE}}{R_B}$$

$v_o(t)$, Vcc, $\frac{1}{2}(V_{OH}+V_{OL})$, 0, t_{PHL}, t, $-(I_CR_C - V_{CC})$

$$= V_{CC} - \beta[(V_{CC}-V_{BE})/R_B] \times R_C$$

$$v_O(t) = v_\infty - (v_\infty - v_{0+})e^{-t/RC}$$

$$= V_{CC} - \beta \frac{(V_{CC}-V_{BE})R_C}{R_B} \cdots$$

$$+ \beta \frac{(V_{CC}-V_{BE})R_C}{R_B} \cdot e^{-t/CR_C}$$

$$= V_{CC} - \beta (V_{CC}-V_{BE}) \frac{R_C}{R_B} (1 - e^{-t/R_C})$$

At $t = t_{PHL}$, $v_O = \frac{1}{2}(V_{OH}+V_{OL})$

$$= \frac{1}{2}(V_{CC} + V_{CEsat})$$

$$= \frac{1}{2}(5 + 0.2) = 2.6\text{V}$$

Thus,

$$2.6 = 5 - 50(5-0.7)\frac{1}{10}\left(1 - e^{-\frac{t_{PHL}}{CR_C}}\right)$$

$$1 - e^{-\frac{t_{PHL}}{CR_C}} = \frac{0.48}{4.3}$$

$$t_{PHL} = 0.118 \times 10 \times 10^{-12} \times 1 \times 10^{3}$$

$$= \underline{\underline{1.18 \text{ ns}}}$$

Comparing this value to that of t_{PLH} found in Problem 5.172 we observe that t_{PHL} is much smaller than t_{PLH}.

$$t_P = \frac{1}{2}(t_{PLH} + t_{PHL})$$

$$= \frac{1}{2}(6.9 + 1.18) = \underline{\underline{4\text{ns}}}$$

Chapter 6 _ Problems

6.1

Assume $0.18\mu m$ CMOS process and refer to Table 6.1:

$\mu_n C_{ox} = 387\ \mu A/V^2$ Also assume $\frac{W}{L} = 10$

Assuming operation in saturation mode:

$V_{ov} = 0.15\ V \Rightarrow I_D = \frac{1}{2} k_n' \frac{W}{L} V_{ov}^2 = \frac{1}{2} \times 387 \times 0.15^2 \times 10$

$I_D = 43.5\ \mu A$

$V_{ov} = 0.4V \Rightarrow I_D = 309.6\ \mu A$

Therefore:

$0.15^V \leq V_{ov} \leq 0.4^V \Rightarrow \underline{43.5\mu A \leq I_D \leq 309.6\mu A}$

Now if we consider the same range of current for I_C of a BJT and we assume an npn transistor in a standard high-voltage process: (refer to Table 6.2)

$I_S = 5 \times 10^{-15} A = 5 \times 10^{-9}\ \mu A$

$I_C = I_S e^{V_{BE}/V_T} \Rightarrow V_{BE} = V_T \ln \frac{I_C}{I_S}$

$I_C = 43.5\ \mu A \Rightarrow V_{BE} = 0.025 \ln \frac{43.5}{5 \times 10^{-9}} = 0.572V$

$I_C = 309.6\ \mu A \Rightarrow V_{BE} = 0.621V$

Therefore:

$43.5\mu A \leq I_C \leq 309.6\mu A \Rightarrow \underline{0.572^V \leq V_{BE} \leq 0.621^V}$

6.2

If the area of the emitter-base junction is changed by a factor of 10, then I_S is changed by the same factor. If V_{BE} is kept constant, then I_C is also changed by the same factor:

$I_C = I_S e^{V_{BE}/V_T}$ $\quad I_S \propto A,\ I_C \propto I_S \Rightarrow I_C \propto A$

$A_2 = 10 A_1 \Rightarrow I_{C2} = 10 I_{C1}$

If I_C is kept constant, then V_{BE} changes:

$I_{S2} = 10 I_{S1} \Rightarrow I_S e^{V_{BE1}/V_T} = 10 I_S e^{V_{BE2}/V_T}$

$e^{\frac{V_{BE1} - V_{BE2}}{V_T}} = 10 \Rightarrow V_{BE1} - V_{BE2} = V_T \ln 10 = 0.058^V$

or $\underline{58mV}$

6.3

$\frac{W}{L} = 10$, $I_D = 100\ \mu A$, $I_D = \frac{1}{2} k' \frac{W}{L} V_{ov}^2$

	0.8µm NMOS	0.8µm PMOS	0.5µm NMOS	0.5µm PMOS	0.25µm NMOS	0.25µm PMOS	0.18µm NMOS	0.18µm PMOS
V_{OV} (V)	0.4	-0.59	0.32	-0.54	0.27	-0.46	0.23	-0.48
V_{GS} (V)	1.1	-1.29	1.02	-1.34	0.7	-1.08	0.71	-0.93

$V_{ov} = \sqrt{\frac{2 I_D}{k_n' \frac{W}{L}}} = \sqrt{\frac{2 \times 100}{k_n' \times 10}} = \sqrt{\frac{20}{k_n'}}$

$V_{GS} = V_t + V_{ov}$

6.4

$|V_{ov}| = 0.25V$, $I_D = 100\ \mu A$

$I_D = \frac{1}{2} k_n' \frac{W}{L} V_{ov}^2 \Rightarrow \frac{W}{L} = \frac{2 I_D}{k_n' V_{ov}^2}$

For NMOS:

$k_n' = 267 \frac{\mu A}{V^2} \Rightarrow \left(\frac{W}{L}\right)_n = \frac{2 \times 100}{267 \times 0.25^2} = 11.98 \simeq \underline{12}$

For PMOS:

$k_n' = 93 \frac{\mu A}{V^2} \Rightarrow \left(\frac{W}{L}\right)_p = \frac{2 \times 100}{93 \times 0.25^2} = 34.4 \simeq \underline{34}$

6.5

$i_{Dn} = i_{Dp} \Rightarrow \frac{1}{2} \mu_n C_{ox} \left(\frac{W}{L}\right)_n V_{ov_n}^2 = \frac{1}{2} \mu_p C_{ox} \left(\frac{W}{L}\right)_p V_{ov_p}^2$ ①

we also have $g_{mn} = g_{mp}$.

$g_m = \frac{2 I_D}{V_{ov}} \Rightarrow V_{ovn} = V_{ovp}$ ②

①, ② $\Rightarrow \frac{(W/L)_p}{(W/L)_n} = \frac{\mu_n}{\mu_p} = \frac{460}{160} = 2.88$

6.6

$g_m = 10\ mA/V$, $V_{ov} = 0.2^V$

$g_m = \frac{2 I_D}{V_{ov}} \Rightarrow I_D = \frac{g_m V_{ov}}{2} = \frac{10 \times 0.2}{2} = \underline{1\ mA}$

$I_D = \frac{1}{2} k_n' \frac{W}{L} V_{ov}^2 \Rightarrow \frac{W}{L} = \frac{2 I_D}{k_n' V_{ov}^2} = \frac{2 \times 1 \times 10^3}{387 \times 0.2^2} = 129.2$

$\underline{\frac{W}{L} \simeq 129}$

for an npn transistor: $g_m = \frac{I_C}{V_T} \Rightarrow I_C = 10 \times 0.025 = \underline{0.25 mA}$

6.7

$\frac{W}{L} = 10 \quad I_D = 100\,\mu A$

	0.8 μm NMOS	0.8 μm PMOS	0.5 μm NMOS	0.5 μm PMOS	0.25 μm NMOS	0.25 μm PMOS	0.18 μm NMOS	0.18 μm PMOS
g_m (mA/V)	0.5	0.34	0.62	0.37	0.73	0.43	0.88	0.41

$g_m = \sqrt{2\mu C_{ox}\frac{W}{L} I_D} = \sqrt{\mu_n C_{ox}}\sqrt{2\times 10\times 100} = 44.7\sqrt{\mu_n C_{ox}}\ \frac{\mu A}{V}$

6.8

$V_{ov} = 0.25\,V$

For an npn transistor: $g_m = \frac{I_C}{V_T} = \frac{0.1}{0.025} = 4\,mA/V$

For an NMOS with the same g_m, i.e. $g_m = 4\,mA/V$

we will have: $g_m = \frac{2I_D}{V_{ov}} \Rightarrow I_D = g_m \times \frac{V_{ov}}{2} = 0.5\,mA$

$\underline{I_D = 0.5\,mA}$

6.9

Assuming large r_o for both transistors, for case (a) we have $r = \frac{1}{g_m} = \frac{1}{\sqrt{2\mu_n C_{ox}\frac{W}{L}I_D}}$

$r = \frac{10^3}{\sqrt{2\times 200\times 10\times 0.1\times 10^3}} = \underline{1.58\,K\Omega}$

for case (b) we have $r = r_\pi \| \frac{1}{g_m} = \frac{\beta}{(\beta+1)g_m}$

$r = \frac{\beta V_T}{(\beta+1)I_C} \simeq \frac{V_T}{I_C} = \frac{0.025}{0.1} = 0.25\,K\Omega$

$\underline{r = 250\,\Omega}$

6.10

$g_m = \frac{2I_D}{V_{ov}} = \frac{2\times 100\times 10^{-3}}{0.5} = 0.4\,mA/V$

$r_o = \frac{V_A}{I_D} = \frac{V_A' L}{I_D} = \frac{25\times 1}{0.1} = \underline{250\,K\Omega}$

$A_0 = g_m r_o = 0.4\times 250 = \underline{100\,V/V}$

$g_m = \mu_n C_{ox}\frac{W}{L}V_{ov} \Rightarrow W = \frac{g_m\times L}{\mu_n C_{ox}\times V_{ov}} = \frac{0.4\times 1}{127\times 10^{-3}\times 0.5}$

$\underline{W = 6.3\,\mu m}$

6.11

$L = 0.3\,\mu m, \quad I_D = 100\,\mu A, \quad V_{ov} = 0.2\,V$

$g_m = \frac{2I_D}{V_{ov}} = \frac{2\times 100\times 10^{-3}}{0.2} = \underline{1\,mA/V}$

$r_o = \frac{V_A}{I_D} = \frac{V_A' \times L}{I_D} = \frac{5\times 0.3}{0.1} = \underline{15\,K\Omega}$

$A_o = g_m r_o = 1\times 15 = \underline{15\,V/V}$

$g_m = \mu_n C_{ox}\frac{W}{L}V_{ov} \Rightarrow W = \frac{g_m\times L}{\mu_n C_{ox} V_{ov}} = \frac{1\times 0.3}{387\times 10^{-3}\times 0.2}$

$\underline{W = 3.88\,\mu m}$

6.12

BJT: $\beta = 100 \quad V_A = 100\,V$

MOSFET: $\mu_n C_{ox} = 200\,\mu A/V^2, \quad \frac{W}{L} = 40, \quad V_A = 10\,V$

Device	BJT		MOSFET	
Bias current (mA)	I_C 0.1	1	I_D 0.1	1
g_m (mA/V)	4	40	1.26	4
r_o (KΩ)	1000	100	100	10
A_o (V/V)	4000	4000	126	40
R_i (KΩ)	25	2.5	∞	∞

$R_i = r_\pi = \beta/g_m$ for BJT

6.13

$L = 0.3\,\mu m, \quad W = 6\,\mu m, \quad V_{ov} = 0.2\,V$

$I_D = \frac{1}{2}\mu_n C_{ox}\frac{W}{L}V_{ov}^2 = \frac{1}{2}\times 387\times\frac{6}{0.3}\times 0.2^2 = 155\,\mu A$

$I_D = 0.155\,mA$

$g_m = \frac{2I_D}{V_{ov}} = 1.55\,mA/V$

$C_{gs} = \frac{2}{3}\frac{W}{L}C_{ox} + C_{ov} = \frac{2}{3}WLC_{ox} + WL_{ov}C_{ox}$

$C_{gs} = \frac{2}{3}\times 6\times 0.3\times 8.6 + 6\times 0.37 = 12.54\,fF$

$C_{gd} = C_{ov}W = 0.37\times 6 = 2.22\,fF$

$f_T = \frac{g_m}{2\pi(C_{gs}+C_{gd})} = \frac{1.55\times 10^{-3}}{2\pi(12.54+2.22)\times 10^{-15}} = \underline{16.7\,GHz}$

If we use the approximation formula:

$f_T \simeq \frac{1.5\mu_n V_{ov}}{2\pi L^2}$ when $C_{gs} \gg C_{gd}, \ C_{gs} \simeq \frac{2}{3}WLC_{ox}$

$f_T \simeq \frac{1.5\times 450\times 10^{-4}\times 0.2}{2\times\pi\times 0.3^2\times 10^{-12}} = \underline{23.9\,GHz}$

The approximation formula overestimates

Cont.

f_T because it ignores $WL_{ov}C_{ox}$ or C_{ov} in C_{gs} and C_{gd} calculation.

6.14

$L = 0.3\,\mu m \quad W = 6\,\mu m \quad V_{ov} = 0.2 \quad C_L = 100\,fF$

$A_o = \frac{2V_A'L}{V_{ov}} = \frac{2 \times 5 \times 0.3}{0.2} = 15\,V/V$

$I_D = \frac{1}{2} k_n' \frac{W}{L} V_{ov}^2 = \frac{1}{2} \times 387 \times \frac{6}{0.3} 0.2^2 = 0.155\,mA$

$r_o = \frac{V_A' L}{I_D} = \frac{5 \times 0.3}{0.155} = 9.7\,k\Omega$

$f_p = \frac{1}{2\pi C_L r_o} = \frac{1}{2\pi \times 9.7 \times 100 \times 10^{-12}} = 164.2\,MHz$

$g_m = \frac{2I_D}{V_{ov}} = 1.55\,mA/V$

$f_T = \frac{g_m}{2\pi C_L} = 2.5\,GHz$

In order to double f_T or equivalently double g_m, $\sqrt{I_D}$ has to be doubled or I_d has to be multiplied by 4: $g_m = \sqrt{2\mu_n C_{ox} \frac{W}{L} I_D}$, $f_T \alpha\, g_m \alpha \sqrt{I_D}$

$I_D = 4 \times 0.155 = 0.62\,mA$

In that case : $A_o \alpha \frac{1}{V_{ov}} \alpha \frac{1}{\sqrt{I_D}} \Rightarrow A_o = 15 \times \frac{1}{2} = 7.5\,V/V$

$f_p \alpha \frac{1}{r_o} \alpha I_D \Rightarrow f_p = 164.2 \times 4 \Rightarrow$

$f_p = 656.8\,MHz$

6.15

$I_c = 10\,\mu A$, High-voltage process:

$g_m = \frac{I_c}{V_T} = \frac{10 \times 10^{-3}}{0.025} = 0.4\,mA/V$

$C_{de} = \tau_F g_m = 0.35 \times 10^{-9} \times 0.4 \times 10^{-3} = 140 \times 10^{-15}\,F = 140\,fF$

$C_{je} = 2C_{jeo} = 2 \times 1 = 2\,pF = 2000\,fF$

$C_\pi = C_{de} + C_{je} = 2140\,fF$

$C_\mu \simeq C_{\mu o} = 0.3\,pF = 300\,fF$

$f_T = \frac{g_m}{2\pi(C_\pi + C_\mu)} = \frac{0.4 \times 10^{-3}}{2\pi(2140+300) \times 10^{-15}} = 26.1\,MHz$

$I_c = 100\,\mu A$, High-Voltage process :

$g_m = 10 \times 0.4 = 4\,mA/V$, $C_{de} = 10 \times 140 = 1400\,fF$

$C_\pi = 3400\,fF \Rightarrow f_T = \frac{4 \times 10^{-3}}{2\pi(3400+300) \times 10^{-15}} = 172.1\,MHz$

$I_c = 10\,\mu A$, Low-Voltage process

$g_m = \frac{10 \times 10^{-3}}{0.025} = 0.4\,mA/V$

$C_{de} = 10 \times 10^{-12} \times 0.4 \times 10^{-3} = 4\,fF$

$C_{je} = 2 \times 5\,fF = 10\,fF$

$C_\pi = C_{de} + C_{je} = 14\,fF$

$C_\mu \simeq C_{\mu o} = 5\,fF$

$f_T = \frac{g_m}{2\pi(C_\mu + C_\pi)} = \frac{0.4 \times 10^{-3}}{2\pi(5+14) \times 10^{-15}} = 3.35\,GHz$

$I_c = 100\,\mu A$, Low-Voltag Process

$g_m = \frac{100 \times 10^{-3}}{0.025} = 4\,mA/V$

$C_{de} = 10 \times 4 = 40\,fF$

$C_\pi = 40 + 10 = 50\,fF$, $C_\mu = 5\,fF$

$f_T = \frac{4 \times 10^{-3}}{2\pi(50+5) \times 10^{-15}} = 11.6\,GHz$

In Summary:

	Standard High-Voltage npn		Standard low-Voltage npn	
	$I_C = 10\,\mu A$	$I_C = 100\,\mu A$	$I_C = 10\,\mu A$	$I_C = 100\,\mu A$
f_T	26.1 MHz	172.1 MHz	3.35 GHz	11.6 GHz

6.16

$L = 1\,\mu m$, $I_D = 100\,\mu A$, $0.8\,\mu m$ - NMOS

a) $V_{ov} = 0.25\,V$

$I_D = \frac{1}{2} k_n' \frac{W}{L} V_{ov}^2 \Rightarrow W = \frac{2L\,I_D}{\mu_n C_{ox} V_{ov}^2} = \frac{2 \times 1 \times 100}{127 \times 0.25^2}$

$\Rightarrow W = 25.2\,\mu m$

$g_m = \frac{2I_D}{V_{ov}} = \frac{2 \times 100 \times 10^{-3}}{0.25} = 0.8\,mA/V$

$r_o = \frac{V_A' L}{I_D} = \frac{25 \times 1}{0.1} = 25\,k\Omega$

$A_o = g_m r_o = 20\,V/V$

$C_{gs} = \frac{2}{3} WLC_{ox} + WL_{ov}C_{ox} = \frac{2}{3} \times 25.2 \times 1 \times 2.3 + 25.2 \times 0.2$

$C_{gs} = 43.68\,fF \simeq 44\,fF$

$C_{gd} = WL_{ov}C_{ox} = 25.2 \times 0.2 = 5.04\,fF$

$f_T = \frac{g_m}{2\pi(C_{gs} + C_{gd})} = \frac{0.8 \times 10^{-3}}{2\pi(44+5.04) \times 10^{-15}} = 2.6\,GHz$

b) $f_T \alpha\, g_m \alpha\, V_{ov}$ therefore in order to double f_T, V_{ov} has to be doubled: $V_{ov} = 0.5\,V$. Consequently,

$W = 25.2/4 = 6.3\,\mu m$ $\qquad r_o, C_{gs}, C_{gd}$ unchanged

$g_m = \frac{0.8}{2} = 0.4\,mA/V$, $\quad A_o = \frac{20}{2} = 10\,V/V$

6.17

$I_C = 1mA \Rightarrow g_m = \frac{I_C}{V_T} = 40\,mA/V$

For pnp:

$C_{de} = \tau_F g_m = 30^n \times 40^m = 1200pF$

$C_{je} = 2C_{jeo} = 2 \times 0.3 = 0.6pF$

$C_\pi = 1200.6pF$

$C_\mu \simeq 1pF$

$\Big\} \Rightarrow f_T = \frac{g_m}{2\pi(C_\pi + C_\mu)} = \frac{40^m}{2\pi(1200.6+1)^{pF}}$

$f_T = 5.3MHz$

For npn:

$C_{de} = \tau_F g_m = 0.35^n \times 40^m = 14\,pF$

$C_{je} = 2 \times 1 = 2pF$

$C_\mu \simeq 0.3pF$

$C_\pi = 14 + 2 = 16pF$

$\Big\} \Rightarrow f_T = \frac{40^m}{2\pi(16+0.3)^{pF}} = 391^{MHz}$

6.18

$A_0 = g_m r_o = \frac{2I_D}{V_{OV}} \times \frac{V_A}{I_D} = \frac{2V_A}{V_{OV}} = \frac{2V'_A L}{V_{OV}}$

Therefore A_0 is only determined by setting values for L and V_{OV}.

$f_T = \frac{g_m}{2\pi(C_{gs}+C_{gd})} = \frac{2I_D/V_{OV}}{2\pi(\frac{2}{3}WLC_{ox}+C_{ov}+C_{ov})}$

If we assume that C_{ov} is very small or equivalently $C_{gs} >> C_{gd}$ and $C_{gs} = \frac{2}{3}WLC_{ox}$:

(replace I_D with $\frac{1}{2}k'_n \frac{W}{L} V_{ov}^2$)

$f_T \simeq \frac{k'_n W/L\, V_{ov}}{2\pi \times \frac{2}{3}WLC_{ox}} = \frac{3\mu_n}{4\pi} V_{ov}/L^2 = \frac{3}{4\pi}\mu_n \frac{V_{ov}}{L^2}$

As we can see f_T can be determined after knowing V_{ov} and L, it is not dependent on either I_D or W.

6.19

$V_{ov} = 0.2V$, $L = 0.2\mu m, 0.3\mu m, 0.4\mu m$

$A_0 = g_m r_o = \frac{2V_A}{V_{ov}} = \frac{2V'_A L}{V_{ov}} = \frac{2\times5\times L}{0.2} = 50L\,V/V$

$f_T \simeq \frac{1.5\mu_n V_{ov}}{2\pi L^2} = \frac{1.5\times450\times10^{-4}\times0.2}{2\times3.14\times L^2\times10^{-12}} = \frac{2.15}{L^2}\,GHz$

L (μm)	0.2	0.3	0.4
A_0 (V/V)	10	15	20
f_T (GHz)	53.75	23.9	13.4

6.20

$L = 0.5\mu m$, $V_{ov} = 0.3V$, $C_L = 1pF$, $f_T = 100MHz$

$f_T = \frac{g_m}{2\pi C_L} \Rightarrow g_m = 2\pi C_L f_T = 2\pi \times 1^p \times 100^M = 628\,\mu A/V$

$g_m = \frac{2I_D}{V_{ov}} \Rightarrow I_D = g_m \times V_{ov}/2 = 628 \times \frac{0.3}{2}$

$I_D = 94.2\mu A$

$I_D = \frac{1}{2}k'_n \frac{W}{L} V_{ov}^2 \Rightarrow W = \frac{2LI_D}{k'_n V_{ov}^2} = \frac{2\times0.5\times94.2}{190\times0.3^2} = 5.51\,\mu m$

$W = 5.51\mu m$

$r_o = \frac{V_A}{I_D} = \frac{V'_A L}{I_D} = \frac{20\times0.5}{94.2\times10^{-3}} = 106.2k\Omega$

$A_0 = g_m r_o = \frac{628}{1000}\times106.2 = 66.7\,V/V$

$f_{3db} = \frac{1}{2\pi C_L r_o} = \frac{1}{2\pi\times1^P\times106.2^K} = 1.5MHz$

6.21

$I_o = I_{REF} = 50\mu A$, $L = 0.5\mu m$, $W = 5\mu m$, $V_t = 0.5^V$

$I_o = I_D = \frac{1}{2}k'_n \frac{W}{L} V_{ov}^2$ $\qquad k'_n = 250\mu A/V^2$

$50 = \frac{1}{2}\times250\times\frac{5}{0.5}(V_{GS}-0.5)^2 \Rightarrow V_{GS} = 0.7V,\ 0.3^V$

$V_{GS} = 0.3V < V_t$ is not acceptable, therefore

$V_{GS} = 0.7V$

$I_D = I_{REF} = \frac{V_{DD}-V_{GS}}{R} \Rightarrow \frac{1.8-0.7}{R} = 0.050 \Rightarrow R = 22k\Omega$

Q_1 and Q_2 have the same V_{GS}. The lowest value of V_o or V_{DS2} is when $V_{DS} = V_{GS} - V_t = 0.7 - 0.5 = 0.2^V$

hence $V_{omin} = 0.2^V$

$r_o = \frac{V_A}{I_D} = \frac{V'_A L}{I_D} = \frac{20\times0.5}{0.05} = 200k\Omega$

$\Delta I_o \simeq \frac{\Delta V_o}{r_o} = \frac{1}{200^K} = 5\mu A \Rightarrow \Delta I_o = 5\mu A$

6.22

$\mu_n C_{ox} = 250\,\mu A/V^2$, $V'_A = 20\,V/\mu m$, $V_t = 0.6V$

$\frac{\Delta I_o}{I_o} = 5\% \Rightarrow \Delta I_o = 5\mu A$ for $\Delta V_o = 1.8 - 0.25 = 1.55^V$

$r_o = \frac{\Delta V_o}{\Delta I_o} = \frac{1.55}{5\mu} = 310\,k\Omega$

$r_o = \frac{V'_A L}{I_D} \Rightarrow L = I_D \times \frac{r_o}{V'_A} = 0.1 \times \frac{310}{20} = 1.55\,\mu m$

$V_{omin} = V_{GS} - V_t = 0.25 \Rightarrow V_{GS} = 0.25 + 0.6 = 0.85^V$

$R = \frac{V_{DD} - V_{GS}}{I_D} = \frac{1.8 - 0.85}{0.1} = 9.5\,k\Omega$

$I_D = \frac{1}{2}\mu_n C_{ox}\frac{W}{L}(V_{GS} - V_t)^2 \Rightarrow W = \frac{2 L I_D}{\mu_n C_{ox}(V_{GS} - V_t)^2}$

$\Rightarrow W = \frac{2 \times 1.55 \times 100}{250\,(0.85 - 0.6)^2} = 19.84\,\mu m$

6.23

$V_{DD} = 1.8V$, $|V_t| = 0.6V$, $\mu_p C_{ox} = 100\,\mu A/V^2$

$I_{REF} = 80\mu A$ $\quad V_{omax} = 1.6^V$

$V_{DS} \le V_{GS} - V_t$

$V_{omax} = V_{DSmax} = V_{GS} - V_t \Rightarrow$

$1.6 - 1.8 = V_{GS} + 0.6 \Rightarrow V_{GS} = -0.8^V$

$\Rightarrow V_G = 1.8 - 0.8 = 1^V$

$R = \frac{V_G}{I_D} = \frac{1}{0.080} = 12.5\,k\Omega$

$I_D = \frac{1}{2}\mu_p C_{ox}(V_{GS} - V_t)^2\frac{W}{L} \Rightarrow W = \frac{2 L I_D}{\mu_p C_{ox}(V_{GS} - V_t)^2}$

$\frac{W}{L} = \frac{2 \times 80}{100(-0.8 + 0.6)^2} = 40$

6.24

$W_2 = 4W_1 \quad L_1 = L_2 \quad V_{ov} = 0.3^V$, $I_{REF} = 20\mu A$

$I_o = I_{REF}\frac{(W/L)_2}{(W/L)_1} = 20 \times 4 = 80\mu A$

$v_{omin} = V_{ov} = 0.3V$

$V_t = 0.5V$. According to Eq. 6.11 $I = \frac{(W/L)_2}{(W/L)_1} I_{REF}\left(1 + \frac{V_o - V_{GS}}{V_{A2}}\right)$

$V_{ov} = V_{GS} - V_t \Rightarrow V_{GS} = 0.3 + 0.5 = 0.8V$

$1 + \frac{V_o - V_{GS}}{25} = 1 \Rightarrow V_o = 0.8V$

Or we could simply say $V_{DS1} = V_{DS2} = v_o$ and

Since $V_{DS1} = V_{GS1} = 0.8^V \Rightarrow V_o = 0.8^V$

$r_{o2} = \frac{V_A}{I_{D2}} = \frac{25}{0.08} = 312.5\,k\Omega$

$r_{o2} = \frac{\Delta V_o}{\Delta I_o} = \frac{1}{\Delta I_o} \Rightarrow \Delta I_o = \frac{1}{312.5^k} = 3.2\mu A$

6.25

Refer to Fig. P6.25

$V_{GS1} = V_{GS2} \Rightarrow \frac{I_{D2}}{I_{D1}} = \frac{(W/L)_2}{(W/L)_1} \Rightarrow I_{D2} = I_{REF}\frac{(W/L)_2}{(W/L)_1}$

$I_{D2} = I_{D3}$

$V_{GS3} = V_{GS4} \Rightarrow I_{D3}/I_{D4} = \frac{(W/L)_3}{(W/L)_4} \Rightarrow \frac{I_{D2}}{I_{D4}} = \frac{(W/L)_3}{(W/L)_4}$

$I_{D4} = I_{REF}\frac{(W/L)_2}{(W/L)_1}\frac{(W/L)_4}{(W/L)_3} = I_o$

6.26

Refer to Fig. P6.26:

$V_{DS2} \le V_{GS2} - V_{tp} \Rightarrow V_{DSmax} = V_{GS2} - V_t$

$(1.3 - 1.5) = V_{GS2} - (-0.6) \Rightarrow V_{GS2} = -0.8V = V_{GS1}$

For Q_1:

$I_{D_1} = 20\mu A = \frac{1}{2} \times 80 \times \frac{W_1}{0.8} \times (-0.8 + 0.6)^2$

$\Rightarrow W_1 = 10\,\mu m$

$I_2 = 100\mu A = 5 I_{REF} \Rightarrow W_2 = 5W_1 = 50\,\mu m$

$I_3 = I_{REF} \Rightarrow W_3 = W_1 = 10\,\mu m$

$I_3 = I_4 \Rightarrow \frac{W_3}{W_4} = \frac{\mu_n}{\mu_p} \Rightarrow W_4 = 10 \times \frac{80}{200} = 4\,\mu m$

For Q_5: $V_{DS5} = V_{GS5} - V_{tn}$ for lowest V_o

$(-1.3 - (-1.5)) = V_{GS5} - 0.6 \Rightarrow V_{GS5} = 0.8^V$

$I_5 = 50\mu A = \frac{1}{2} \times 200 \times \frac{W_5}{0.8} \times (0.8 - 0.6)^2 \Rightarrow W_5 = 10\mu m$

Now we calculate R:

$V_{GS2} = -0.8^V \Rightarrow V_{G2} = 1.5 - 0.8 = 0.7^V$

$R = V_{G2}/I_{REF}$

$R = \frac{0.7}{20\mu A} = 35\,k\Omega$

$R = 35\,k\Omega$

$r_{o2} = \frac{V'_A L}{I_2} = \frac{12 \times 0.8}{0.1} = 96\,k\Omega$, $r_{o5} = \frac{10 \times 0.8}{0.05} = 160\,k\Omega$

6.27

IF the transistor with $W=10$ is diode-connected, then: $I_2 = 100 \times \frac{20}{10} = 200\mu A$

$I_3 = 100 \times \frac{40}{10} = 400\mu A$

IF the transistor with $W=20$ is diode-connected then: $I_2 = 100 \times \frac{10}{20} = 50\mu A$

$I_3 = 100 \times \frac{40}{20} = 200\mu A$

IF the transistor with $W=40$ is diode-connected, then: $I_2 = 100 \times \frac{10}{40} = 25\mu A$

$I_3 = 100 \times \frac{20}{40} = 50\mu A$

So for cases that only one transistor is diode connected, 4 different output currents are possible. (depending on the configuration we choose).

IF 2 transistors are diode-connected; then they act as an equivalent transistor whose width is the sum of the widths of each transistor:

IF $W_{eff} = 10+20$ then $I_o = 100 \times \frac{40}{30} = 133\mu A$

IF $W_{eff} = 20+40$ then $I_o = 100 \times \frac{10}{60} = 16.7\mu A$

IF $W_{eff} = 40+10$ then $I_o = 100 \times \frac{20}{50} = 40\mu A$

So 3 different output currents are possible depending on which two transistors are diode-connected. Now we calculate V_{SG}:

$100 = \frac{1}{2} \times 80 \times \frac{30}{1}(V_{SG}-0.7)^2 \Rightarrow V_{SG} = 1^V$ for $W_{eff} = 30\ \mu m$

all have the same V_{SG} for any given configuration.

For $W_{eff} = 60 \Rightarrow 100 = \frac{1}{2} \times 80 \times \frac{60}{1}(V_{SG}-0.7)^2$

$\Rightarrow V_{SG} = 0.9^V$

for $W_{eff} = 50 \Rightarrow 100 = \frac{1}{2} \times 80 \times \frac{50}{1}(V_{SG}-0.7)^2$

$\Rightarrow V_{SG} = 0.93V$

6.28

$-g_{m2} V_{gs} = g_{m1} V_i$

$\Rightarrow V_{GS} = -\frac{g_{m1}}{g_{m2}} V_i$

$V_o = -R_L g_{m3} V_{gs}$

$V_o = -R_L g_{m3} \frac{-g_{m1}}{g_{m2}} V_i$

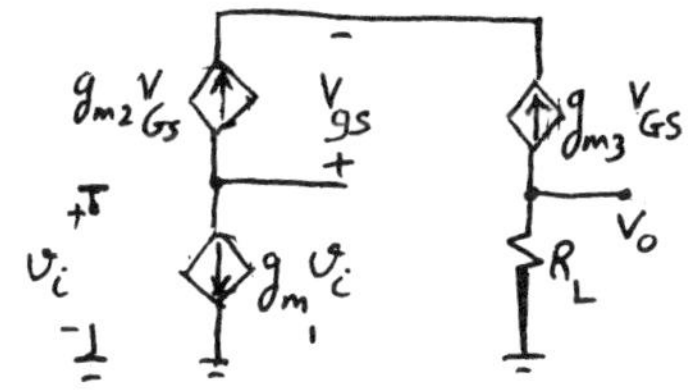

$\frac{V_o}{V_i} = +R_L g_{m1} \frac{g_{m3}}{g_{m2}} = R_L g_{m1} \frac{W_3}{W_2}$

Note that since Q_2, Q_3 have the same V_{GS} and therefore the same V_{OV}: $\frac{g_{m3}}{g_{m2}} = \frac{I_3}{I_2} = \frac{W_3}{W_2}$

6.29

$I_S = 10^{-15} A$

a) $I_{REF} = I_S e^{V_{BE}/V_T} \Rightarrow V_{BE} = V_T \ln \frac{I_{REF}}{I_S}$

$I_{REF} = 10\mu A \Rightarrow V_{BE} = 0.025 \ln \frac{10 \times 10^{-6}}{10^{-15}} = 0.576^V$

$I_{REF} = 10mA \Rightarrow V_{BE} = 0.025 \ln \frac{10 \times 10^{-3}}{10^{-15}} = 0.748^V$

Therefore:

$10\mu A \leq I_{REF} \leq 10mA \Rightarrow 0.576^V \leq V_{BE} \leq 0.748^V$

Since β is very high, I_B is negligible and hence $I_o \approx I_{REF}$: $10\mu A \leq I_o \leq 10\ mA$

b) $I_o = I_{REF} \frac{1}{1 + 2/\beta}$ (Eq. 6.21)

for $0.1^{mA} \leq I_C \leq 5^{mA}$, β remains constant at 100.

$I_{REF} = 10mA \Rightarrow I_o = \frac{10}{1 + \frac{2}{70}} = 9.72mA$

$I_{REF} = 0.1mA \Rightarrow I_o = \frac{0.1}{1 + 2/100} = 0.098mA$

$I_{REF} = 1mA \Rightarrow I_o = \frac{1}{1 + 2/100} = 0.98mA$

$I_{REF} = 10\mu A \Rightarrow$

6.30

$I_{S2} = I_{S1} \times m$, $I_{C1} = I_C$

$I_{REF} = I_C + \frac{I_C}{\beta} + \frac{I_O}{\beta}$ (1)

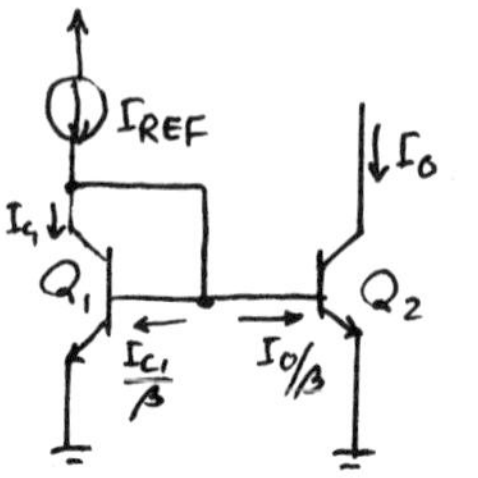

$V_{BE1} = V_{BE2} \Rightarrow$

$V_T \ln \frac{I_C}{I_{S1}} = V_T \ln \frac{I_O}{I_{S2}}$

$\Rightarrow \frac{I_O}{I_C} = \frac{I_{S2}}{I_{S1}} = m \Rightarrow I_C = I_O/m$

by substituting for I_C in (1):

$I_{REF} = \frac{I_O}{m} + \frac{I_O}{m\beta} + \frac{I_O}{\beta} \Rightarrow \frac{I_O}{I_{REF}} = \frac{m}{1+\frac{1}{\beta}+\frac{m}{\beta}}$

$\frac{I_O}{I_{REF}} = \frac{m}{1+\frac{1+m}{\beta}}$

This result is the same as Eq. 6.22.

For large β, $I_O/I_{REF} = m$, with finite β this ratio drops to $I_O/I_{REF} = \frac{m}{1+\frac{1+m}{\beta}}$. To keep the introduced error within 5%: $0.95m = \frac{m}{1+\frac{1+m}{\beta}}$

$\beta_{min} = 80 \Rightarrow 0.95 = \frac{1}{1+\frac{1+m}{80}} \Rightarrow m = 3.21$

6.31

The transfer ratio is the same as Eq. 6.21:

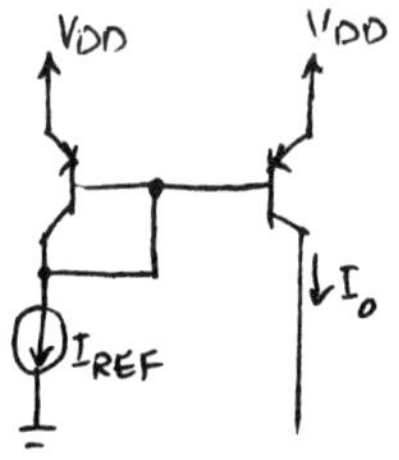

$\frac{I_O}{I_{REF}} = \frac{1}{1+\frac{2}{\beta}}$

$\beta = 20 \Rightarrow \frac{I_O}{I_{REF}} = \frac{1}{1+\frac{2}{20}} = 0.91$

6.32

$I_O = I_{REF} = 2\,mA$

$r_{O2} = \frac{V_{A2}}{I_O} = \frac{90}{2} = 45\,K\Omega$

$r_{O2} = \frac{\Delta V_O}{\Delta I_O} \Rightarrow \frac{10-1}{\Delta I_O} = 45 \Rightarrow \Delta I_O = 0.2\,mA$

$\frac{\Delta I_O}{I_O} = \frac{0.2}{2} = 10\%$ change

6.33

$I_S = 10^{-15} A$ $\beta = 50$

$\frac{I_O}{I_{REF}} = \frac{1}{1+\frac{2}{\beta}} \Rightarrow I_{REF} = I_O(1+\frac{2}{\beta}) = 1\times(1+\frac{2}{50})$

$I_{REF} = 1.02\,mA$

$V_{BE} = V_T \ln \frac{I_O}{I_S(1+V_{CE}/V_A)} = 0.025 \ln \frac{10^{-3}}{10^{-15}(1+\frac{3}{50})} = 0.689^V$

$V_C = V_B = 5 - 0.689 = 4.31V$

$V_C = R \times I_{REF} \Rightarrow R = \frac{4.31}{1.02} = 4.2^{K\Omega}$

$r_O = \frac{V_A}{I_O} = \frac{50}{1} = 50\,K\Omega$

V_{Omax} occurs when Q_2 is on the edge of saturation or $V_{CE} = 0.3^V$. Therefore V_{Omax} is $5 - 0.3 = 4.7^V$

$r_O = \frac{\Delta V_O}{\Delta I_O} \Rightarrow \frac{4.7-(-5)}{\Delta I_O} = 50^K \Rightarrow \Delta I_O = 0.194\,mA$

$\frac{\Delta I_O}{I_O} \times 100 = \frac{0.194}{1} \times 100 = 19.4\%$ change in I_O

6.34

$I_{C1} = I_{C2} = I_{R1}$

$V_{B1} = 10 - 0.7 = 9.3V$, $V_{B2} = -10 + 0.7 = -9.3^V$, $I_{R1} = \frac{9.3+9.3}{10}$

$\Rightarrow I_{R1} = 1.86\,mA = I_{C1} = I_{C2} = I_{C3} = I_{C4} = I_{C5} = I_{C6}$

$V_{C3} = 1.86^m \times 2^K = 3.72V$, $V_{C5} = 0.7V$

$V_{C6} = 5 - 1.86 \times 1 = 3.14V$, $I_{C9} = I_{C8} = I_{C7} = I_{C2} = 1.86^{mA}$

$I_{R4} = 2 \times 1.86 = 3.72\,mA \Rightarrow V_{C7} = -3.72 \times 1 = -3.72V$

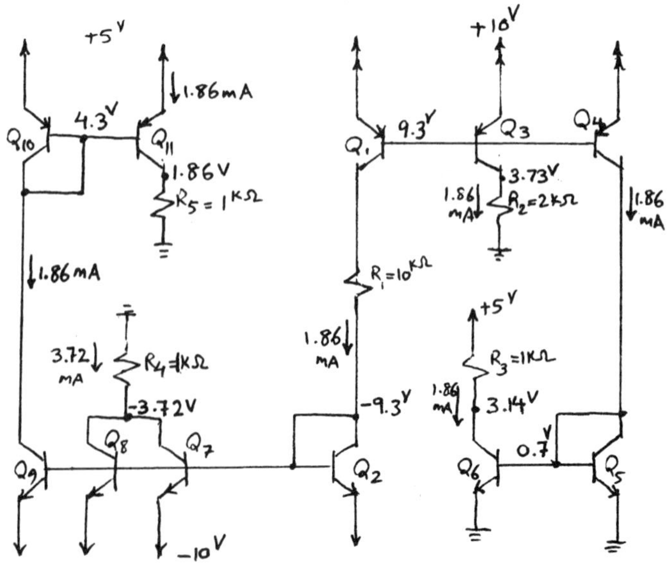

Cont.

$I_{C10} = I_{C9} = 1.86\,mA$

$V_{C9} = V_{C10} = V_{B10} = 5 - 0.7 = 4.3^V$

$I_{C11} = I_{C10} = 1.86\,mA$

$V_{C11} = 1.86 \times 1 = 1.86V$

6.35

a) Refer to Fig. P6.35

$R = 10K\Omega$

$V_1 = -0.7^V \Rightarrow I_{C_1} = \frac{-0.7 - (-10.7)}{10^K} = 1MA$

$I_{C_1} = 1mA$

$V_2 = 5.7 - 0.7 = 5V$

$I = I_{C3} + I_{C4}$, $I_{C3} = I_{C4} = I_{C_1} \Rightarrow I = 2 \times 1 = 2mA$

$V_3 = 0 + 0.7 = 0.7^V$

$V_4 = -10.7 + 1 \times 10^K = -0.7V$

$V_5 = -10.7 + 1 \times \frac{10^K}{2} = -5.7^V$

b) $R = 100K\Omega$

$V_1 = -0.7V \Rightarrow I_{C_1} = \frac{-0.7 + 10.7}{100^K} = 0.1mA$

$I = 2I_{C_1} = 0.2mA$

$V_3 = 0.7V$, $V_2 = 5.7 - 0.7 = 5^V$

$V_4 = -10.7 + \frac{1}{10} \times 100 = -0.7^V$

$V_5 = -10.7 + 0.1 \times \frac{100}{2} = -5.7^V$

6.36

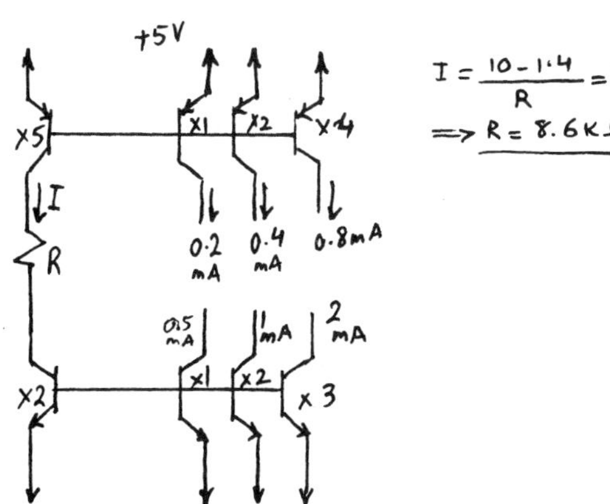

$I = \frac{10 - 1.4}{R} = 1MA$

$\Rightarrow R = 8.6K\Omega$

6.37

a) Refer to Fig. P6.37.

Since Q_1, Q_3 have equal currents, I, and Q_3 and Q_4 have equal V_{BE} which results in equal currents, then Q_2 has the same current as Q_1.

Therefore: $V_{BE1} = V_{BE2} \Rightarrow V_x = V_y = V$

Also, $V_{BE5} = V_{BE3}$

$\Rightarrow I_5 = I_3 = I$

or $I_Z = I$.

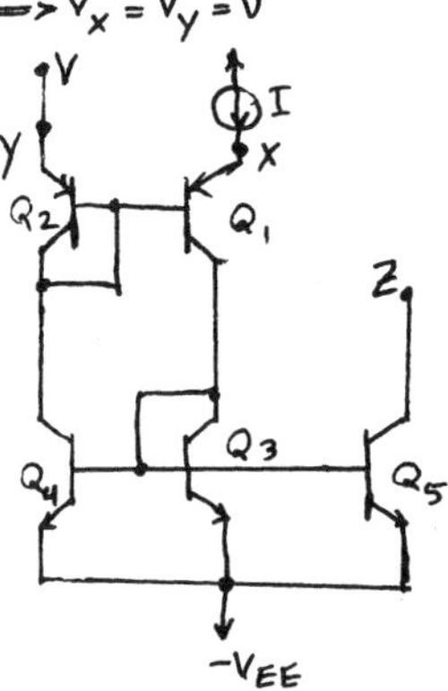

b) $I_2 = I_4$, $I_1 = I_3$

Since $V_{BE4} = V_{BE3}$ then $I_3 = I_4$ and hence $I_1 = I_2$.

From $I_1 = I_2$ we conclude that $V_{BE1} = V_{BE2}$ and

Since $V_{B1} = V_{B2}$ then V_{E1} has to be the same as V_{E2}: $V_{E1} = V_{E2} = 0$

$VX = 0$

$I = \frac{5 - 0}{10^K} = 0.5mA$

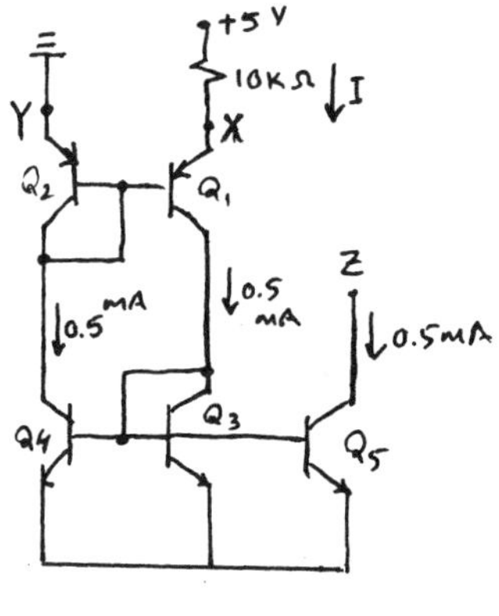

6.38

$40\,db = 20 \log A_o \Rightarrow A_o = 100\,V/V$

$$A(s) = +100 \frac{(1 + S/100 \times 10^6 \times 2\pi)}{(1 + \frac{s}{2\pi \times 10^7})(1 + \frac{s}{2\pi \times 10^6})}$$

$$A(s) = +100 \frac{(1 + \frac{s}{2\pi \times 10^8})}{(1 + \frac{s}{2\pi \times 10^7})(1 + \frac{s}{2\pi \times 10^6})}$$

Eq. 6.36 : $\omega_H = 1 / \sqrt{\left(\left(\frac{1}{2\pi \times 10^7}\right)^2 + \left(\frac{1}{2\pi \times 10^6}\right)^2 - 2\left(\frac{1}{2\pi \times 10^8}\right)^2 \right)}$

$f_H = 0.995\,MHZ$

Cont.

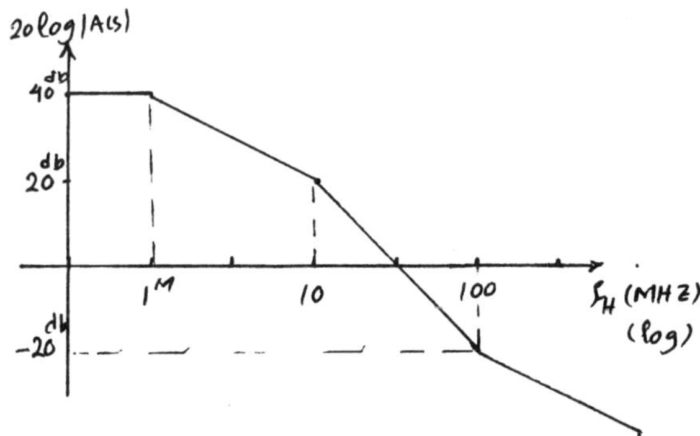

6.39

a) $A(s) = \dfrac{1000}{(1+\frac{s}{2\pi\times10\times10^3})} = \dfrac{1000}{(1+\frac{s}{2\pi\times10^4})}$

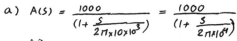

b)

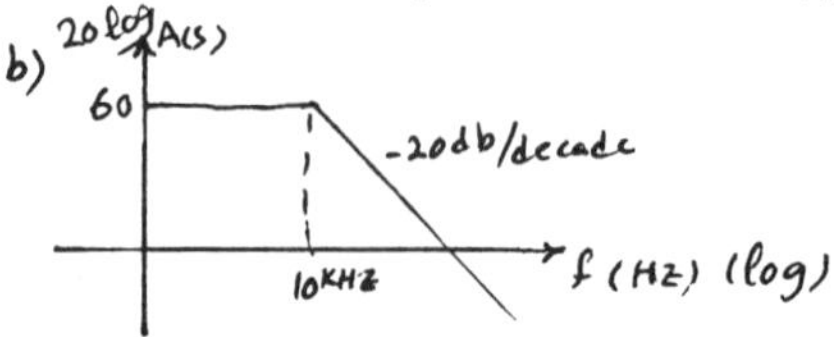

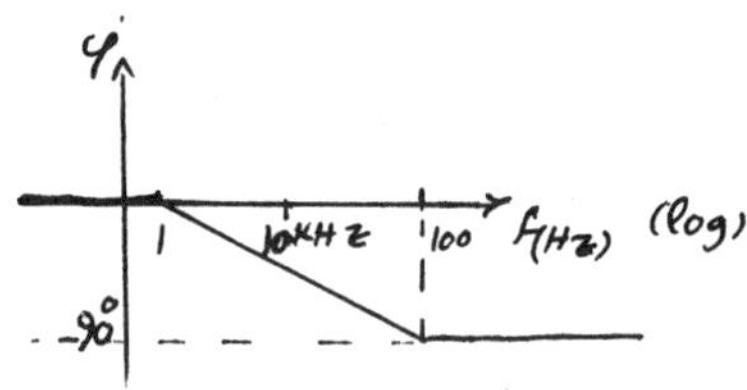

6.40

$f_H(s) = \dfrac{1}{(1+\frac{s}{\omega_{p1}})(1+\frac{s}{\omega_{p2}})}$ $\qquad \omega_{p_1} < \omega_{p2}$

Using dominant-pole approximator: $\omega_H \simeq \omega_{p1}$

Using the root sum of squares formula:

$$\omega_H \simeq \frac{1}{\sqrt{\frac{1}{\omega_{p_1}^2}+\frac{1}{\omega_{p2}^2}}} = \frac{\omega_{p1}}{\sqrt{1+\left(\frac{\omega_{p1}}{\omega_{p2}}\right)^2}}$$

The difference between the two estimates for ω_H is:

$$\Delta\omega_H = \omega_{p_1} - \frac{\omega_{p1}}{\sqrt{1+\left(\frac{\omega_{p1}}{\omega_{p2}}\right)^2}}$$

If $n = \dfrac{\omega_{p2}}{\omega_{p1}}$:

$$\frac{\Delta\omega_H}{\omega_{p1}} = 1 - \frac{1}{\sqrt{1+\frac{1}{n^2}}}$$

for $\dfrac{\Delta\omega_H}{\omega_{p_1}} = 10\% = 0.1 \Rightarrow \underline{n = 2.07}$

for $\dfrac{\Delta\omega_H}{\omega_{p_1}} = 1\% = 0.01 \Rightarrow \underline{n = 7.02}$

6.41

$$A(s) = -100\,\frac{1+s/10^6}{(1+s/10^5)(1+s/10^7)}$$

a) $\omega_H \simeq 10^5$ rad/s

b) $\omega_H \simeq \dfrac{1}{\sqrt{\left(\frac{1}{10^5}\right)^2+\left(\frac{1}{10^7}\right)^2-2\left(\frac{1}{10^6}\right)^2}} = 101$ Krad/s

If the pole at 10^6 rad/s is lowered to 10^5 rad/s, the transfer function becomes:

$A(s) = \dfrac{-100}{1+s/10^7} \Rightarrow f_H = \dfrac{10^7}{2\pi}$ Hz

6.42

$$30^\circ = 3\tan^{-1}\frac{\omega}{\omega_p} = 3\tan^{-1}\frac{10^6}{\omega_p} \Rightarrow \omega_p = 5.67\times10^6 \text{ rad/s}$$

6.43

$$\omega_H \simeq \frac{1}{\tau_{gs}+\tau_{gd}} = \frac{1}{C_{gs}R_{gs}+C_{gd}R_{gd}}$$

$$\omega_H \simeq \frac{1}{C_{gs}R' + C_{gd}(R'+R_L'+g_mR_L'R')} \quad \text{(From Example 6.6)}$$

For $C_{gs} = C_{gd} = 1$ PF, $R_L' = 3.33$ KΩ, $g_m = 4$ mA/V

$$\omega_H = \frac{1}{10^{-12}R' + 10^{-12}(R' + 3.33\times10^3 + 4\times3.33\times R')}$$

To obtain $\omega_H = 2\pi\times150\times10^3$

$$2\pi\times150\times10^3 = \frac{10^{12}}{3.33\times10^3 + 15.32R'} \Rightarrow R' = 69.04\text{K}\Omega$$

$R' = R \,\|\, R_{in} = R \,\|\, 420^K = 69.04 \Rightarrow \underline{R = 82.6\text{K}\Omega}$

6.44

Pole at $\dfrac{1}{2\pi\times10^4\times5\times10^{-12}} = 3.18$ MHz and

Cont.

pole at $\frac{1}{2\pi\times20\times10^3\times2\times10^{-12}} = 3.98\,MHz$

Since both poles are relatively close together, we use the root-sum-of-squares formula:

$$f_H = \frac{1}{\sqrt{\frac{1}{3.18^2}+\frac{1}{3.98^2}}} = 2.5\,MHz$$

Manufactured design:

Poles at $\frac{1}{2\pi\times10^4\times15\times10^{-12}} = 1.06\,MHz$ and at

$$\frac{1}{2\pi\times20\times10^3\times12\times10^{-12}} = 0.66\,MHz$$

$$f_H = \frac{1}{\sqrt{\frac{1}{1.06^2}+\frac{1}{0.66^2}}} = 0.56\,MHz$$

6.45

Refer to solution of Example 6.6:

$$A_M = \frac{V_o}{V_i} = -\frac{R_{in}}{R_{in}+R}\,g_m R_L' = -\frac{1.2}{1.2+0.1}(2\times12)$$

$A_M = -22.2\,V/V$

$R_{gs} = R_{in}\,||\,R = 1.2^M\,||\,0.1^M = 92.3\,k\Omega$

$\tau_{gs} = R_{gs}C_{gs} = 1\times10^{-12}\times92.3\times10^3 = 92.3\,ns$

$R_{gd} = R' + R_L' + g_m R_L' R'$ where $R' = R_{in}||R = 92.3^{k\Omega}$

$R_{gd} = 92.3 + 12 + 2\times12\times92.3 = 2.32\,M\Omega$

$\tau_{gd} = C_{gd}R_{gd} = 1\times10^{-12}\times2.32\times10^6 = 2320\,ns$

$$\omega_H = \frac{1}{\tau_{gs}+\tau_{gd}} = \frac{1}{(92.3+2320)\times10^{-9}} = 414.5\,krad/s$$

$f_H = 66\,KHz$

6.46

a) $V_o = -g_m R_L V_{gs}$ ①

$V_{gs} = V_{sig} - R_s\times g_m V_{gs}$

$V_{gs}(1+g_m R_s) = V_{sig}$

① $\Rightarrow V_o = \frac{-g_m R_L}{1+g_m R_s}V_{sig} \Rightarrow \frac{v_o}{v_{sig}} = \frac{-g_m R_L}{1+g_m R_s}$

b) $V_S = (g_m V_x - i_x)R_S$

$V_G = V_S + V_x = (g_m V_x - i_x)R_S + V_x$

$\Rightarrow i_x = \frac{V_G}{R} = \frac{(1+g_m R_S)}{R}V_x - \frac{i_x R_S}{R}$

$$R_{gs} = \frac{v_x}{i_x} = \frac{1+R_S/R}{\frac{1+g_m R_S}{R}} = \frac{R+R_S}{1+g_m R_S} \quad (R \text{ is } R_{sig})$$

to calculate R_{gd}:

$V_G = -R i_x$

$V_S = -R i_x - V_{gs}$

$V_S = R_S \times g_m V_{gs}$ } $\Rightarrow$

$$R_S g_m V_{gs} = -R i_x - V_{gs} \Rightarrow V_{gs} = \frac{-R i_x}{1+g_m R_S}$$

At D: $i_x = g_m V_{gs} + \frac{V_x - R\times i_x}{R_L}$

substitute V_{gs}: $i_x = -\frac{g_m R i_x}{1+g_m R_S} + \frac{V_x}{R_L} - \frac{R i_x}{R_L}$

$$i_x\left[1+\frac{g_m R}{1+g_m R_S}+\frac{R}{R_L}\right] = \frac{V_x}{R_L}$$

$$R_{gd} = \frac{V_x}{i_x} = R_L + R + \frac{g_m R R_L}{1+g_m R_S} \quad (R \text{ is } R_{sig})$$

c) $R_S = 0$:

$$\frac{v_o}{v_{sig}} = \frac{-4\times5^K}{1+4\times0} = -20\,V/V$$

$R_{gs} = R_{sig} = 100\,k\Omega$

$R_{gd} = 5^K + 100K + 4\times5\times100 = 2105\,K\Omega$

$$\omega_H \simeq \frac{1}{C_{gs}R_{gs}+C_{gd}R_{gd}} = \frac{1}{10^{-12}\times100\times10^3+10^{-12}\times2105\times10^3}$$

$\omega_H \simeq 453.5\,Krad/s$

$|Gain| \times Bandwidth = 20\times453.5 = \underline{9.07\,Mrad/s}$

$R_S = 100^{\Omega}$:

$$\frac{v_o}{v_{sig}} = \frac{-4\times5}{1+4\times0.1} = -14.3\,V/V$$

$$R_{gs} = \frac{100+0.1}{1+4\times0.1} = 71.5\,K\Omega$$

$$R_{gd} = 5+100+\frac{4\times5\times100}{1+4\times0.1} = 1533.6\,K\Omega$$

$$\omega_H = \frac{1}{10^{-12}\times71.5\times10^3+10^{-12}\times1533.6\times10^3} = 623\,krad/s$$

Cont.

$|Gain| \times Bandwidth = 14.3 \times 623^{K} = \underline{8.91 Mrad/s}$

$R_S = 250\Omega$:

$$\frac{v_o}{v_{sig}} = \frac{-4\times5}{1+4\times0.25} = -10 V/V$$

$$R_{gs} = \frac{100+0.25}{1+4\times0.25} = 50.1K\Omega$$

$$R_{gd} = 5+100+\frac{4\times5\times100}{1+4\times0.25} = 1105 \ K\Omega$$

$$\omega_H = \frac{1}{10^{-12}\times50.1\times10^3 + 10^{-12}\times1105\times10^3} = 865.7 Krad/s$$

$|gain| \times Bandwidth = 10 \times 865.7^{K} = \underline{8.66 Mrad/s}$

Summary table:

R_S (Ω)	Gain (V/V)	ω (krad/s)	Gain.BW Product (Mrad/s)
0	-20	453.5	9.07
100	-14.3	623.0	8.91
250	-10	865.7	8.66

The GainxBandwidth is approximately constant.

6.47

$$A_M = \frac{V_o}{V_{sig}} = -\frac{R_{in}}{R_{in}+R_{sig}}(g_m R_L') = -\frac{5}{5+1}(0.3\times100^K)$$

$A_M = -25 V/V$ Now refer to Example 6.6.

$$R_{gs} = R_{in} \| R_{sig} = 5^{M\Omega} \| 1^{M\Omega} = 0.83^{M\Omega}$$

$$\tau_{gs} = R_{gs} C_{gs} = 0.2\times10^{-12}\times0.83\times10^6 = 166.7 ns$$

$$R_{gd} = R' + R_L' + g_m R_L' R'$$

$$R' = R_{in} \| R_{sig} = 0.83 M\Omega$$

$\Rightarrow R_{gd} = 0.83 + 0.1 + 0.83\times0.3\times100$

$$R_{gd} = 25.92 M\Omega$$

$$\tau_{gd} = C_{gd} R_{gd} = 25.92\times10^6\times0.1\times10^{-12} = 2592 ns$$

$$\omega_H \simeq \frac{1}{\tau_{gs}+\tau_{gd}} = \frac{1}{(166.7+2592)\times10^{-9}} = 362.5 \ krad/s$$

$$f_H = 57.7 KHZ$$

6.48

IF we assume that capacitors are perfect open circuits for midband, then:

$$A_M = \frac{V_o}{V_{sig}} = \frac{-R_{in}}{R_{in}+R_{sig}}(g_m R_L') = \frac{-650}{650+150}(5\times10) = 40.6 V/V$$

$$\tau_{gs} = C_{gs} R_{gs} = C_{gs}(R_{in} \| R_{sig}) = 2^P \times (150^K \| 650K)$$

$$\tau_{gs} = 243.75 \ ns$$

$\tau_{gd} = C_{gd} R_{gd}$, Refer to Example 6.6

$$R_{gd} = R' + R_L' + g_m R_L' R'$$

$$R' = 150^K \| 650K = 121.9 K\Omega$$

$\Rightarrow R_{gd} = 121.9 + 10 + 5\times10\times121.9$

$$R_{gd} = 6.2 M\Omega$$

$$\tau_{gd} = C_{gd} R_{gd} = 0.5^P \times 6.2^M = 3100 ns$$

$$\tau_L = R_L' C_L = 10^k \times 3^P = 30 ns$$

$$\omega_H = \frac{1}{\tau_{gs}+\tau_{gd}+\tau_L} = \frac{1}{30^n + 3100^n + 243.75^n} = 296.4 Krad/s$$

$$f_H = 47.2 KHZ$$

6.49

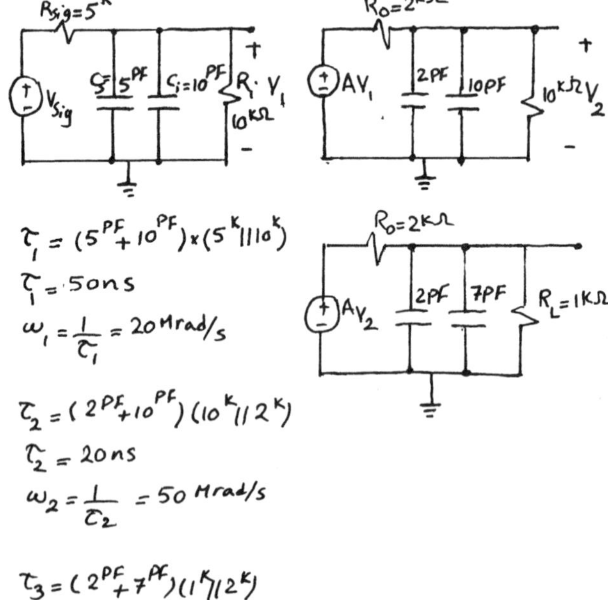

$$\tau_1 = (5^{PF}+10^{PF})\times(5^K \| 10^K)$$

$$\tau_1 = 50 ns$$

$$\omega_1 = \frac{1}{\tau_1} = 20 Mrad/s$$

$$\tau_2 = (2^{PF}+10^{PF})(10^K \| 2^K)$$

$$\tau_2 = 20 ns$$

$$\omega_2 = \frac{1}{\tau_2} = 50 Mrad/s$$

$$\tau_3 = (2^{PF}+7^{PF})(1^K \| 2^K)$$

$$\tau_3 = 6 ns$$

$$\omega_3 = 166 Mrad/s$$

$$\omega_H \simeq \frac{1}{\tau_1+\tau_2+\tau_3} = \frac{1}{50^n+20^n+6^n} = 13.2 MHZ$$

$$\underline{f_H = 2.1 MHZ}$$

6.50

Eq. 6.41 $\omega_H \simeq \frac{1}{\sum RC}$

a) $\omega_H = \frac{1}{20+5+1^{ns}} = 38.46 Mrad/s \Rightarrow f_H = 6.12 MHz$

b) $\omega_H = \frac{1}{\frac{1}{\omega_{P1}} + \frac{1}{\omega_{P2}} + \frac{1}{\omega_{P3}}}$ (Eq. 6.38)

$\omega_H = \frac{2\pi}{\frac{1}{f_1}+\frac{1}{f_2}+\frac{1}{f_3}} = \frac{2\pi}{\frac{1}{50}+\frac{1}{200}+\frac{1}{1000}} = 241.5$ Mrad/s

$f_H = 38.46 MHz$

c) $\omega_H = \frac{1}{\frac{1}{50}+\frac{1}{200}+\frac{1}{1000}} = 38.46 Mrad/s$

$f_H = 6.12 MHz$

d) $\omega_H = \frac{1}{1000+200+200} = 714 Krad/s \Rightarrow f_H = 113.7 KHz$

e) $\omega_H = \frac{1}{0.4+1} = 714 Krad/s \Rightarrow f_H = 113.7 KHz$

f) $\omega_H = \frac{1}{1+0.2+0.15} = 741 Krad/s \Rightarrow f_H = 118 KHz$

g) $\omega_H = \frac{2\pi}{\frac{1}{f_1}+\frac{1}{f_2}+\frac{1}{f_3}+\frac{1}{f_4}} = \frac{2\pi}{\frac{1}{1}+\frac{1}{2}+\frac{1}{5}+\frac{1}{5}}$

$\omega_H = 3.3 Grad/s \Rightarrow f_H = 526 MHz$

6.51

$R_{in} = \frac{R}{1-Gain} = \frac{100}{1-0.95} = 2000 K\Omega = 2 M\Omega$

6.52

Eq. 6.44a $Z_I = Z/_{1-K} \Rightarrow C_I = 0.1\times(1-(-1000))$

$\Rightarrow C_I = 100.1 pF$

$C_o = 0.1\times(\frac{-1}{1000}+1)$

$C_o = 99.9 fF$

(using Miller's Theorem)

$R_{sig} = 1K\Omega$, V_{sig}, C_I, V_i, AV_i, C_o

$V_o = AV_i = A\times V_{sig}\frac{1/C_I s}{R_{sig}+\frac{1}{C_I s}} \Rightarrow \frac{V_o}{V_{sig}} = \frac{A}{1+C_I R_{sig} s}$

$\omega_H = \frac{1}{C_I R_{sig}} = \frac{1}{100.1^p \times 1K} = 9.99 Mrad/s \Rightarrow f_H = 1.59 MHz$

To calculate unity gain frequency:

$|Gain| = 1$

$\frac{V_o}{V_i} = \frac{A}{1+C_I R_{sig} s} = \frac{-1000}{1+100.1\times 10^{-9} s}$ $(s=j\omega)$

$\frac{1000}{\sqrt{1+(100.1\times 10^{-9}\times\omega_T)^2}} = 1 \Rightarrow \omega_T \simeq 10 Grad/s$

$f_T \simeq 1.59 GHz$

As we can see $f_T \simeq f_H \times A$

6.53

Using Miller's Theorem, in each case the capacitance at the input is $C(1-A)$ and the capacitance at the output is $C(1-\frac{1}{A})$. Thus:

a) $A = -1000 V/V$ and $C = 1pF$

$C_i = 1.001 nF$ and $C_o = 1.001 pF$

b) $A = -10 V/V$ and $C = 10pF$

$C_i = 110 pF$ and $C_o = 11 pF$

c) $A = -1 V/V$ and $C = 10pF$

$C_i = 20pF$ and $C_o = 20pF$

d) $A = 1 V/V$ and $C = 10pF$

$C_i = 0pF$ and $C_o = 0pF$

e) $A = 10 V/V$ and $C = 10pF$

$C_i = -90pF$ and $C_o = 9pF$

In (e) the negative capacitance at the input can be used to cancel the effect of the input capacitance of the amplifier.

6.54

a) $R_{in} = \frac{R}{1-A} = \frac{R}{1-2} = -R$ (Miller's theorem)

b) $I_N = \frac{V_{sig}}{R_{sig}}$

R_{sig}, V_{sig}, $R_{in} \equiv I_N$, R_N

$R_N = R_{sig} \| R_{in}$

IF $R_{sig} = R$ then:

$R_N = R \| (-R) = \infty \Rightarrow I_L = I_N = \frac{V_{sig}}{R_{sig}} = \frac{V_{sig}}{R}$

Cont.

KCL at A:

$$\frac{\frac{V_o}{2} - V_{sig}}{R_{sig}} + \frac{v_o}{2} \times CS + \frac{-v_o}{2R} = 0$$

If $R_{sig} = R \Rightarrow \frac{+v_{sig}}{R} = \frac{v_o}{2} CS \Rightarrow \frac{v_o}{v_{sig}} = \frac{2}{RCS}$

6.55

From Table 6.3 we have:

Intrinsic gain $= g_m r_o = \frac{2V_A'L}{V_{ov}}$

$\Rightarrow$ Intrinsic gain $= \frac{2\times10\times1}{0.2} = 100\ V/V$

$g_m r_o = 100\ V/V$

$g_m = \frac{2I_D}{V_{ov}} \Rightarrow I_D = \frac{g_m V_{ov}}{2} = \frac{2\times0.2}{2} = 0.2\ mA$

$g_m = \mu_n C_{ox}\frac{W}{L}V_{ov} \Rightarrow W = \frac{g_m L}{k_n' V_{ov}} = \frac{2\times1\times10^3}{125\times0.2} = 80\ \mu m$

6.56

$A_o = g_m r_o = 100\ V/V = \frac{2V_A'L}{V_{ov}}$

$I_D = \frac{1}{2}k_n'\frac{W}{L}V_{ov}^2 = 100\ \mu A$

$\frac{A_{o2}}{A_{o1}} = \frac{V_{ov1}}{V_{ov2}} = \sqrt{\frac{I_{D1}}{I_{D2}}} \Rightarrow A_{o2} = A_{o1}\sqrt{\frac{I_{D1}}{I_{D2}}}$

Now if I_D is $25\mu A$ then: $A_{o2} = 100\sqrt{\frac{100}{25}} = 200\ V/V$

$I_D = 400\mu A \Rightarrow A_{o2} = 100\sqrt{\frac{100}{400}} = 50\ V/V$

$g_m = \sqrt{2\mu_n C_{ox}\frac{W}{L}I_D} \Rightarrow \frac{g_{m2}}{g_{m1}} = \sqrt{\frac{I_{D2}}{I_{D1}}} \Rightarrow g_{m2} = g_{m1}\sqrt{\frac{I_{D2}}{I_{D1}}}$

For $I_D = 25\mu A$ $g_{m2} = g_{m1}\sqrt{\frac{25}{100}} \Rightarrow g_{m2} = g_{m1}/2$

$I_D = 400\mu A \Rightarrow g_{m2} = g_{m1}\sqrt{\frac{400}{100}} \Rightarrow g_{m2} = 2g_{m1}$

6.57

Refer to Fig. P6.57.

a) if we neglect the current in Feedback resistor: $I_D = 200\mu A = \frac{1}{2}k'\frac{W}{L}V_{ov}^2 = \frac{1}{2}\times2\times V_{ov}^2 \Rightarrow$

$V_{ov}^2 = 200\times10^{-3} \Rightarrow V_{ov} = 0.45V \Rightarrow V_{GS} = V_t + V_{ov}$

$\Rightarrow V_{GS} = 0.5 + 0.45 = 0.95V$

$\Rightarrow V_G = 0.95V \Rightarrow I_F = \frac{0.95}{2M} = 0.48\mu A \ll 200\mu A$

$\Rightarrow V_{DS} = 0.48\times5 = 2.4V \Rightarrow V_{DS} = 2.4V$

So our assumption for neglecting I_F was right.

b) To find the small-signal gain we write a KCL at the output node:

$\frac{v_o}{r_o} + g_m v_{gs} + \frac{v_o - v_i}{R_F} = 0$

$v_{gs} = v_i$

$\frac{v_o}{r_o} + g_m v_i + \frac{v_o}{R_F} - \frac{v_i}{R_F} = 0 \Rightarrow \frac{v_o}{v_i} = \frac{1/R_F - g_m}{\frac{1}{R_F} + \frac{1}{r_o}}$

$g_m = \frac{2I_D}{V_{ov}} = \frac{2\times200\mu}{0.45} = 0.89\ mA/V$

$r_o = \frac{V_A}{I_D} = \frac{20}{0.2} = 100\ k\Omega$

$\frac{v_o}{v_i} = \frac{1/3000 - 0.89}{\frac{1}{3000} + \frac{1}{100}} = -86.1\ V/V$

$V_{DSmin} = V_{GS} - V_t = V_{ov} = 0.45V$

In our case $V_{DS} = 2.4V$, therefore the largest signal at the output is $2.4 - 0.45 = 1.95V$. The corresponding input signal is

$\frac{v_o}{A} = \frac{1.95}{-86.1} = 0.023V$ or 23mV.

c) $R_{in} = \frac{v_x}{i_x}$

$i_x = \frac{v_x}{R_i} + \frac{v_x - v_o}{R_F}$ ①

$\frac{v_x - v_o}{R_F} = \frac{v_o}{r_o} + g_m v_x \Rightarrow v_o = v_x \frac{1/R_F - g_m}{\frac{1}{r_o} + \frac{1}{R_F}}$

① $\Rightarrow i_x = \frac{v_x}{R_i} + \frac{v_x}{R_F} - \frac{v_x}{R_F}\frac{1/R_F - g_m}{1/r_o + \frac{1}{R_F}}$

$\frac{v_x}{i_x} = R_{in} = \frac{1}{\frac{1}{R_i} + \frac{1}{R_F} - \frac{1}{R_F(\frac{R_F}{r_o}+1)} + \frac{g_m}{(\frac{R_F}{r_o}+1)}}$

$R_{in} = R_i \parallel R_F \parallel \frac{1 + R_F/r_o}{g_m - 1/R_F}$

$R_{in} = 2M \parallel 3M \parallel \frac{1 + 3000/100}{0.89 - 1/3000} = 33.9\ k\Omega$

6.58

Refer to Fig. 6.18a and assume Q_2 and Q_3 are matched:

$k'_n = 2.5\,k'_p = 250\,\mu A/V^2$, $V_{An} = |V_{Ap}| = 10^V$

$R_{out} = 100^{k\Omega} = r_{o1} \| r_{o2} = \frac{10}{I_{D1}} \| \frac{10}{I_{D2}}$

Since $I_{D1} = I_{D2} \Rightarrow \frac{10}{I_{D1}} = 200 \Rightarrow I_{D1} = I_{D2} = 0.05\text{ mA}$

$I_{REF} = I_{D2} = 0.05\text{mA}$

Eq. 6.49 : $A_v = -g_{m1}(r_{o1} \| r_{o2}) \Rightarrow -40 = -g_{m1} \times 100^k$

$\Rightarrow g_{m1} = 0.4\text{ mA/V}$

$g_{m1} = \sqrt{2k'_n(\frac{W}{L})_1 I_{D1}} \Rightarrow (\frac{W}{L})_1 = \frac{0.4^2}{2 \times 250 \times 10^{-3} \times 0.05} = 6.4$

$(\frac{W}{L})_1 = 6.4$

$g_{m1} = \frac{2I_{D1}}{V_{ov1}} \Rightarrow V_{ov1} = \frac{2 \times 0.05}{0.4} = 0.25\text{ V}$

If Q_2 and Q_3 have the same V_{ov} as Q_1, then $|V_{ov2}| = 0.25\text{ V}$

$I_{D2} = \frac{1}{2}k'_p(\frac{W}{L})_2 V_{ov}^2 \Rightarrow (\frac{W}{L})_2 = \frac{0.05 \times 2}{100 \times 10^{-3} \times 0.25^2} = 16$

$(\frac{W}{L})_2 = (\frac{W}{L})_3 = 16$ (Q_2 and Q_3 are matched)

6.59

As discussed in Example 6.8, the transfer characteristic of the amplifier over the desired region (segment III) is quite linear. Therefore the DC bias component of the input signal (for maximum output swing) should be chosen at the midpoint between V_{IA} and V_{IB} that is:

input dc bias $= \frac{V_{IA} + V_{IB}}{2} = \frac{0.88 + 0.93}{2} = 0.905\text{V}$

The corresponding amplitude of the resulting output sinusoid is:

output sinusoid amplitude $= \frac{V_{OA} + V_{OB}}{2} = \frac{2.47 + 0.33}{2}$

output sinusoid amplitude $= 1.4\text{V}$

6.60

As discussed in Example 6.8, the transfer characteristic of the amplifier over the region labeled as segment III, is quite linear.

$V_{OA} = V_{DD} - V_{OV3} = 5 - 0.53 = 4.47^V$

Now to find the linear equation for segment III, we can write $i_{D1} = i_{D2}$:

$\frac{1}{2}k'_n(\frac{W}{L})_1(v_I - V_{tn})^2(1 + \frac{v_O}{V_{An}}) = \frac{1}{2}k'_p(\frac{W}{L})_2(V_{SG} - |V_{tp}|)^2(1 + \frac{V_{DD} - v_O}{V_{Ap}})$

$\Rightarrow 200(v_I - 0.6)^2(1 + \frac{v_O}{20}) = 65 \times 0.53^2 \times (1 + \frac{V_{DD} - v_O}{10})$

$\frac{200}{65 \times 0.53^2}(v_I - 0.6)^2 = \frac{1.5 - v_O/10}{1 + \frac{v_O}{20}}$

$7.3(v_I - 0.6)^2 = \frac{1 - v_O/15}{1 + \frac{v_O}{20}} = \frac{1 - 0.067v_O}{1 + 0.05v_O} \simeq 1 - 0.117v_O$

$\Rightarrow v_O = 8.57 - 62.57(v_I - 0.6)^2$ ①

If we substitute for $V_{OA} = 4.47^V$, then $V_{IA} = 0.86^V$

To determine coordinates of B, note that

$V_{IB} - V_{tn} = V_{OB}$ or $V_{IB} - 0.6 = V_{OB}$

substitute in ①:

$V_{OB} = 8.57 - 62.57V_{OB}^2 \Rightarrow V_{OB} = 0.36^V$

$V_{IB} = 0.6 + 0.36 = 0.96\text{V}$

Therefore the linear region is:

$0.86^V \leq V_I \leq 0.96^V$ or $0.36^V \leq V_O \leq 4.47^V$

6.61

Since $V_{An} = |V_{Ap}|$ and the drain current of both Q_1 and Q_2 is equal to I, their output resistances are equal. That is $r_{o1} = r_{o2} = r_o$

The small-signal model of this amplifier is:

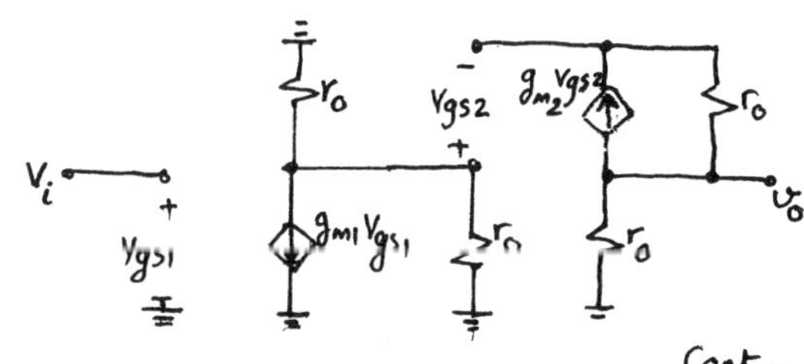

Cont.

$v_{gs2} = -g_{m1} v_{gs1} (r_o \| r_o)$ and $v_{gs1} = v_i$

$v_{gs2} = -g_{m1} \frac{r_o}{2} v_i$

$v_o = -g_{m2} v_{gs2} (r_o \| r_o) = -g_{m2} (-g_{m1} \frac{r_o}{2} v_i) \frac{r_o}{2}$

$A_v = \frac{v_o}{v_i} = g_{m1} g_{m2} \frac{r_o^2}{4}$

6.62

Refer to Fig. 6.18a.

Note that Q_2, Q_3 are not matched:

$I_{D1} = 100\ \mu A$

a) $I_{D2} = I_{D1} = 100\ \mu A$ $\quad \frac{I_{D3}}{I_{D2}} = \frac{(W/L)_3}{(W/L)_2} = \frac{W_3}{W_2}$

(Note that $V_{SG2} = V_{SG3}$)

$\Rightarrow I_{D3} = 100\mu A \frac{10}{40} = 25\mu A \Rightarrow I_{REF} = 25\mu A$

b) By referring to Fig. 6.18d, you notice that in Segment III, both Q_1 and Q_2 are in saturation and the transfer characteristic is quite linear. The output voltage in this segment is limited between V_{OA} and V_{OB}:

Coordinates of point A: $v_{OA} = V_{DD} - V_{OV3}$

$V_{OV3}^2 = \frac{I_{D3}}{\frac{1}{2} k_p' (\frac{W}{L})_3} = \frac{25}{\frac{1}{2} \times 50 \times \frac{10}{1}} = 0.1^V \Rightarrow V_{OV3} = 0.32^V$

$v_{OA} = 3.3 - 0.32 = 2.98^V$

At point B: $V_{OB} = V_{IB} - V_{tn}$

Now we find the transfer equation for the linear section: (Refer to Example 6.8)

$i_{D1} = i_{D2} \Rightarrow$ $\quad$ (Note that $V_{OV2} = V_{OV3}$)

$$k_n' (\frac{W}{L})_1 (v_I - V_{tn})^2 (1 + \frac{v_O}{V_{An}}) = k_p' (\frac{W}{L})_2 V_{OV3}^2 (1 + \frac{V_{DD} - v_O}{|V_{Ap}|})$$

$$100 \times \frac{20}{1} (v_I - 0.8)^2 (1 + \frac{v_O}{100}) = 50 \times \frac{40}{1} \times 0.32^2 (1 + \frac{3.3 - v_O}{50})$$

$$(v_I - 0.8)^2 = 0.32^2 (1.066 - \frac{v_O}{50}) / (1 + \frac{v_O}{100})$$

$$(v_I - 0.8)^2 = 0.11 \left(\frac{1 - 0.019 v_O}{1 + 0.01 v_O} \right) \simeq 0.11 (1 - 0.03 v_O)$$

$$(v_I - 0.8)^2 = 0.11 (1 - 0.03 v_O) \quad (1)$$

Now if we solve for $V_{OB} = V_{IB} - 0.8$

$V_{OB}^2 + 0.0033 v_{OB} - 0.11 = 0 \Rightarrow V_{OB} = 0.33^V$

There fore the extreme values of v_O for which Q_1 and Q_2 are in saturation: $0.33^V \le v_O \le 2.98^V$

c) from (b) we can find V_{IA} and V_{IB}:

$V_{IB} = V_{OB} + V_t = 0.33 + 0.8 = 1.13V$

If we solve (1) for $V_{OA} = 2.98V$ then:

$(V_{IA} - 0.8)^2 = 0.11(1 - 0.03 \times 2.98) \Rightarrow V_{IA} = 1.116^V$

Large-signal voltage gain $= \frac{\Delta v_O}{\Delta V_I} = \frac{2.98 - 0.33}{1.13 - 1.116}$

$\frac{\Delta v_O}{\Delta v_I} = -189.3\ V/V$

d) $v_O = \frac{V_{DD}}{2} = \frac{3.3}{2} = 1.65V$

Differentiating both sides of (1): $(\frac{\partial}{\partial v_i})$

$2(v_i - 0.8) = 0.11 \times (-0.03) \frac{\partial v_O}{\partial v_i}$

$\Rightarrow \frac{\partial v_O}{\partial v_i} = -606.1 (v_i - 0.8)$

For $v_O = 1.65^V$, from (1) we have:

$(v_i - 0.8)^2 = 0.11(1 - 0.03 \times 1.65) \Rightarrow v_i = 1.123^V$

$\left. \frac{\partial v_O}{\partial v_i} \right|_{v_i = 1.123} = -195.8\ V/V$

e) $R_{out} = r_{o1} \| r_{o2}$

$r_{o1} = \frac{V_{An}}{I_{D1}} = \frac{100}{0.1^m} = 1\ M\Omega$

$r_{o2} = \frac{V_{Ap}}{I_{D2}} = \frac{50}{0.1^m} = 500\ k\Omega$

$\Rightarrow R_{out} = 500^K \| 1^M$

$R_{out} = 333\ k\Omega$

$g_{m1} = \sqrt{2 k_n' (\frac{W}{L})_1 I_{D1}} = \sqrt{2 \times 100^M \times \frac{20}{1} \times 100^M} = 0.632\ mA/V$

$A_v = -g_{m1} (r_{o1} \| r_{o2}) = -210.6\ V/V$

6.63

a) When D and G are open, since the gates donot draw any current, therefore no current goes through R and $V_D = V_G$, $I_{D1} = I_{D2}$.

$\frac{1}{2} k_n' (\frac{W}{L})_1 (V_G - (-1.5) - 0.5)^2 = \frac{1}{2} k_p' (\frac{W}{L})_2 (1.5 - V_G - 0.5)^2$

$(V_G + 1)^2 = (1 - V_G)^2 \Rightarrow V_G + 1 = 1 - V_G \Rightarrow V_G = 0$

$I_{D1} = \frac{1}{2} \times 1 \times (0 + 1.5 - 0.5)^2 = 0.5\ mA$

$I_{D1} = I_{D2} = 0.5\ mA$

Cont.

b) $v_o = v_i - R(2g_m v_{gs})$

$v_{gs} = v_i$

$v_o = v_i - 2g_m R v_i$

$A_v = \frac{v_o}{v_i} = 1 - 2g_m R$

$g_m = \frac{2I_D}{V_{GS} - V_t} = \frac{2 \times 0.5}{1.5 - 0.5} = 1\ mA/V$

$A_v = 1 - 2 \times 1 \times 1000 = -1999\ V/V$

c) $r_o = \frac{V_A}{I_D} = \frac{20}{0.5} = 40\,k\Omega$

If we write KCL at D:

$\frac{v_i - v_o}{R} = 2g_m V_{gs} + \frac{v_o}{r_o/2}$, $V_{gs} = v_i$

$\frac{v_i}{R} - 2g_m v_i = \frac{v_o}{R} + \frac{2v_o}{r_o} \Rightarrow \frac{v_o}{v_i} = A_v = \frac{1 - 2g_m R}{1 + \frac{2R}{r_o}}$

or $A_v = \frac{1 - 2 \times 1 \times 1000}{1 + \frac{2 \times 1000}{40}} = -39.2\ V/V$

$R_{in} = \frac{v_x}{i_x}$: $v_x - R i_x = \frac{r_o}{2}(i_x - 2g_m v_{gs})$

$v_{gs} = v_x$

$v_x + 2g_m \frac{r_o}{2} v_x = R i_x + \frac{r_o}{2} i_x \Rightarrow R_{in} = \frac{v_x}{i_x} = \frac{R + r_o/2}{1 + g_m r_o}$

$R_{in} = \frac{1000 + 40/2}{1 + 1 \times 40} = 24.9 \simeq 25\,k\Omega$

d) $\frac{v_o}{v_{sig}} = A_v \frac{R_{in}}{R_{in} + R_{sig}} = -39.2 \times \frac{25^k}{100^k + 25^k}$

$\frac{v_o}{v_{sig}} = -7.84\ V/V$

e) In order for both Q_1 and Q_2 to remain in the saturation region:

$V_{DS} \geq V_{GS} - V_t \Rightarrow V_D + 1.5 \geq V_G + 1.5 - V_t$

$\Rightarrow V_D \geq V_G - 0.5$

For Q_2: $V_{SD} \geq V_{SG} - |V_t| \Rightarrow 1.5 - V_D \geq 1.5 - V_G - 0.5$

$\Rightarrow V_D \leq V_G + 0.5$

In (a) we showed that the gate is at 0 volt, therefore $V_D \geq -0.5$ and $v_D \leq 0.5$

$\Rightarrow -0.5 \leq v_o \leq 0.5$

6.64

$R_i = r_\pi = \frac{\beta}{g_m} = \frac{\beta}{I_C/V_T} = \frac{100}{1} \times 0.025 = 2.5\,k\Omega$

$g_m = \frac{I_C}{V_T} = 40\,mA/V$, $r_o = \frac{V_A}{I_C} = 100\,k\Omega$

$A_{vo} = -g_m r_o = -40 \times 100 = -4000\ V/V$

$R_o = r_o = 100\,k\Omega$

If R_i is multiplied by 4, since $R_i \propto \frac{1}{I_C}$, then I_C has to be divided by 4: $I_C = \frac{1\,mA}{4} = 0.25\,mA$

For $I_C = 0.25\,mA$: $g_m = \frac{0.25}{0.025} = 10\,mA/V$

$r_o = \frac{V_A}{I_C} = \frac{100}{0.25} = 400\,k\Omega$

$A_{vo} = -10 \times 400 = -4000\ V/V$

In general $A_{vo} = -g_m r_o = -\frac{I_C}{V_T} \times \frac{V_A}{I_C} = -\frac{V_A}{V_T}$ and A_{vo} is not dependent on I_C.

$R_o = r_o = 400\,k\Omega$, $R_i = 4 \times 2.5 = 10\,k\Omega$

For $I_C = 1\,mA$ and $R_{sig} = 5\,k\Omega$, $R_L = 500\,k\Omega$

$v_i = v_{sig} \frac{R_{in}}{R_{in} + R_{sig}}$

$G_v = \frac{R_{in}}{R_{in} + R_{sig}} A_{vo} \frac{R_L}{R_L + R_o}$

$G_v = -4000 \times \frac{2.5\,k\Omega}{5 + 2.5} \times \frac{500}{500 + 100} = -1111\ V/V$

$G_v = -1111\ V/V$

For $I_C = 0.25\,mA$: $R_{in} = 10\,k\Omega$, $R_o = 400$

$G_v = \frac{10}{10 + 5} \times (-4000) \times \frac{500}{500 + 400} = -1481.5\ V/V$

$G_v = -1481.5\ V/V$

6.65

a) $I_{REF} = I_{C3} = \frac{3 - V_{BE3}}{23K\Omega}$

$I_{REF} = \frac{3-0.7}{23}$

$I_{REF} = 0.1mA$

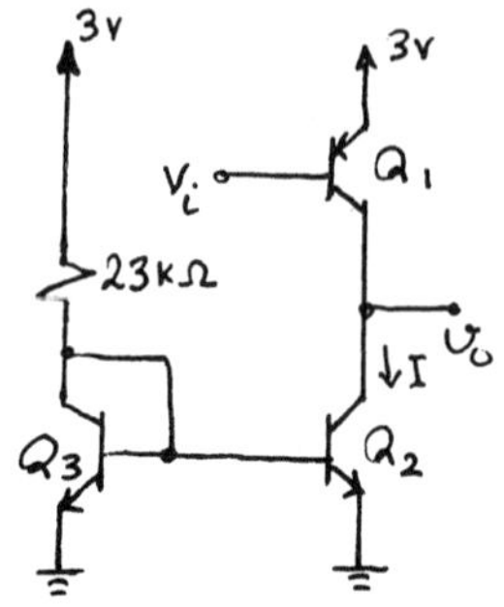

$\frac{I_{C2}}{I_{C3}} = \frac{\text{Area of } Q_2}{\text{Area of } Q_3} = 5$

$\Rightarrow I_{C2} = 5 I_{C3}$

$I_{C2} = I = 0.5mA \Rightarrow I = 0.5mA$

b) $|V_A| = 50V \Rightarrow r_{o1} = \frac{|V_A|}{I} = \frac{50}{0.5} = 100K\Omega$

$r_{o2} = \frac{50}{0.5} = 100K\Omega$

Total resistance at the collector of Q_1 is equal to $r_{o1} \| r_{o2}$, thus: $r_{tot} = 100^K \| 100^K = 50K\Omega$

$r_{tot} = 50K\Omega$

c) $g_{m1} = \frac{I_{C1}}{V_T} = \frac{0.5}{0.025} = 20\,mA/V$

$r_{\pi 1} = \frac{\beta}{g_m} = \frac{50}{20} = 2.5K\Omega$

d) $R_{in} = r_{\pi 1} = 2.5K\Omega$

$R_o = r_{o1} \| r_{o2} = 100^K \| 100^K = 50K\Omega$

$A_v = -g_{m1} R_o = -20 \times 50 = -1000\,V/V$

6.66

$A_M = -g_m R_L' = -5 \times 20 = -100\,V/V$

$C_{in} = C_{gs} + C_{gd}(1 + g_m R_L')$ (Eq. 6.55)

$C_{in} = 2 + 0.1(1 + 5 \times 20) = 12.1PF$

$f_H \simeq \frac{1}{2\pi C_{in} R_{sig}}$ (Eq. 6.54)

$f_H \simeq \frac{1}{2\pi \times 12.1 \times 10^{-12} \times 20k} = 658KHZ$

6.67

Eq. 6.57 : $\tau_H = C_{gs} R_{gs} + C_{gd} R_{gd} + C_L R_{CL}$

$\tau_H = C_{gs} R_{sig} + C_{gd}[R_{sig}(1 + g_m R_L') + R_L'] + C_L R_L'$

$\tau_H = 2^P \times 20^K + 0.1^P[20^K(1 + 5 \times 20) + 20^K] + 1^P \times 20^K$

$\tau_H = 264\,ns$

$f_H \simeq \frac{1}{2\pi \tau_H} = 603\,KHZ$

$A_M = -g_m R_L' = -5 \times 20 = -100\,V/V$

τ_{gs} : 15.1%

τ_{gd} : 77.3%

τ_L : 7.6%

Contribution of each time-constant to the overall τ_H.

If we compare f_H to the one obtained in Problem 6.66, we notice that Problem 6.66 has a larger f_H due to neglecting the time constants of C_L and C_{gs}.

6.68

From Eq. 6.60 we have: $\omega_Z = g_m / C_{gd}$

$\Rightarrow f_Z = \frac{g_m}{2\pi C_{gd}} = \frac{5^m}{2\pi \times 0.1^P} = 7.96GHZ$

f_{P1} and f_{P2} are the poles of the transfer function of equation (6.60), whose denominator is a quadratic polynomial with coefficient of s:

$= [C_{gs} + C_{gd}[1 + g_m R_L']]R_{sig} + (C_L + C_{gd})R_L'$

$= [2 + 0.1(1 + 5 \times 20)]20 + (1 + 0.1) \times 20$

$= 264ns = 264 \times 10^{-9}$ sec

Coefficient of s^2:

$= [(C_L + C_{gd})C_{gs} + C_L C_{gd}] R_{sig} R_L'$

$= [(1 + 0.1)2 + 1 \times 0.1]\, 20^K \times 20^K =$

$= 920 \times 10^{-18} (sec)^2$

Therefore the quadratic equation is:

$1 + 264 \times 10^{-9} s + 920 \times 10^{-18} s^2 = 0$

Denoting the frequencies of the roots of this equation with ω_{P1} and ω_{P2}, we have:

$\omega_{P1} = 3.84 \times 10^6$ rad/s $\Rightarrow f_{P1} = \frac{\omega_{P1}}{2\pi} = 611.15KHZ$

$\omega_{P2} = 283.12 \times 10^6$ rad/s $\Rightarrow f_{P2} = \frac{\omega_{P2}}{2\pi} = 45.06MHZ$

Cont.

Since $f_{P_1} \ll f_{P2}$ and $f_{P_1} \ll f_Z$, a good estimate for f_H is f_{P_1}:

$$f_H \simeq f_{P_1} = 611.15\ \text{KHZ}$$

Approximate value of f_{P_1} obtained using (Eq. 6.66) is:

$$f_{P_1} \cong \frac{1}{2\pi[(C_{gs}+C_{gd}(1+g_m R_L'))R_{sig}+(C_L+C_{gd})R_L']}$$

$f_{P_1} \simeq 603.16$ KHZ

Approximate value of f_{P2} obtained using (Eq. 6.67) is:

$$f_{P_2} = \frac{[C_{gs}+C_{gd}(1+g_m R_L')]R_{sig}+(C_L+C_{gd})R_L'}{2\pi[(C_L+C_{gd})C_{gs}+C_L C_{gd}]R_L' R_{sig}}$$

$f_{P2} = 45.67$ MHz

The estimate of f_{P_1} using Eq. 6.66 is 1.3% lower than the exact value, while the estimate of f_{P2} is about 1.3% higher than its exact value.

6.69

$R_L' = 5\,K\Omega$:

$A_M = -g_m R_L' = -5 \times 5 = -25\ V/V$

using (Eq. 6.66):

$$f_{P1} \cong \frac{1}{2\pi[(C_{gs}+C_{gd}(1+g_m R_L'))R_{sig}+(C_L+C_{gd})R_L']}$$

$$f_{P_1} = \frac{1}{2\pi[(2+0.1\times(1+5\times5))20+(1+0.1)\times5]}$$

$f_{P_1} = 1.63$ MHz

using (Eq. 6.67):

$$f_{P2} = \frac{[C_{gs}+C_{gd}(1+g_m R_L')]R_{sig}+(C_L+C_{gd})R_L'}{[(C_L+C_{gd})C_{gs}+C_L C_{gd}]R_L' R_{sig}\times 2\pi}$$

$$f_{P2} = \frac{(2+0.1(1+5\times5))20+(1+0.1)5}{((1+0.1)\times2+1\times0.1)5\times20\times2\pi}$$

$f_{P2} = 67.5$ MHz

Eq. 6.63: $s_Z = \frac{g_m}{C_{gd}} \Rightarrow f_Z = \frac{g_m}{2\pi C_{gd}} = \frac{5^m}{2\pi 0.1^P} = 7.96$ GHz

$f_{P_1} \ll f_{P2}$ and $f_{P_1} \ll f_Z \Rightarrow f_{P_1}$ is the dominant pole.

$f_H \simeq f_{P_1} = 1.63$ MHz

|Gain| x Bandwidth $= 25 \times 1.63 = 40.75$ MHz

$f_t = |A_M| f_H = 40.75$ MHz

Since $f_{P_1} \ll f_{P_2}$ and $f_{P_1} \ll f_Z$, a dominant pole exists.

$R_L' = 10\,K\Omega$:

$A_M = -5 \times 10 = -50\ V/V$

$$f_{P1} = \frac{1}{2\pi[(2+0.1(1+5\times10))20+(1+0.1)\times10]} = 1.04\ \text{MHz}$$

$$f_{P2} = \frac{(2+0.1(1+5\times10))20+(1+0.1)\times10}{[(1+0.1)2+1\times0.1]10\times20\times2\pi} = 52.96\ \text{MHz}$$

$$f_Z = \frac{5}{2\pi\times0.1} = 7.96\ \text{GHz}$$

$f_{P_1} \ll f_{P2}$, $f_{P_1} \ll f_Z \Rightarrow f_{P_1}$ is the dominant pole and therefore $f_H \simeq f_{P_1} = 1.04$ MHz

$|A_M| \cdot f_H = 50 \times 1.04 = 52$ MHz

Since f_{P_2} is still slightly greater than $|A_M| \cdot f_H$, therfore: $f_T \simeq 52$ MHz

$R_L' = 20\,K\Omega$:

$A_M = -5 \times 20 = -100\ V/V$, from Problem 6.68 we have:

$f_{P_1} = 603.16$ KHZ

$f_{P_2} = 45.67$ MHz

$f_Z = 7.96$ GHz

Again $f_{P_1} \ll f_{P_2}$ and $f_{P_1} \ll f_Z$, therefore f_{P_1} is the dominant pole and f_H can be approximated by f_{P_1}. $f_H \simeq f_{P_1} = 603.16$ KHZ

$|A_M| \cdot f_H = 60.32$ MHz

Since $f_{P2} < |A_M| \cdot f_H$, therefore f_T is smaller than $|A_M| \cdot f_H$

The results are summarized in this table:

R_L'	$5^{K\Omega}$	$10^{K\Omega}$	$20^{K\Omega}$
A_M (V/V)	-25	-50	-100
f_{P1} (MHz)	1.63	1.04	0.60
$\lvert A_M \rvert \cdot f_H$ (MHz)	40.75	52.00	60.32

6.70

Using Eq. 6.70 : $A_M = -\frac{r_\pi}{R_{sig} + r_x + r_\pi}(g_m R_L')$

$r_\pi = \frac{\beta}{g_m} = \frac{100}{20} = 5K\Omega \Rightarrow A_M = -\frac{5}{1 + 0.2 + 5}(20 \times 5)$

$A_M = 80.65\ V/V$

Using Miller's Theorem and Eq. 6.71:

$C_{in} = C_\pi + C_\mu(1 + g_m R_L') = 10 + 0.5(1 + 20 \times 5) = 60.5$ pF

Eq. 6.69: $R_{sig}' = r_\pi \| (R_{sig} + r_x) = 5^K \| (1^K + 0.2)$

$R_{sig}' = 0.97K\Omega$

Eq. 6.72: $f_H \simeq \frac{1}{2\pi C_{in} R_{sig}'} = \frac{1}{2\pi \times 60.5^p \times 0.97^K} \Rightarrow$

$f_H = 2.71MHz$

6.71

From Problem 6.70 we have: $r_\pi = 5K\Omega$ and

$A_M = -80.65\ V/V$, $R_{sig}' = 0.97K\Omega$

Eq. 6.75: $f_Z = \frac{g_m}{2\pi C_\mu} = \frac{20^m}{2\pi \times 0.5^p} = 6.37GHz$

Eq. 6.76: $f_{P_1} \simeq \frac{1}{2\pi\left[(C_\pi + C_\mu(1 + g_m R_L'))R_{sig}' + (C_L + C_\mu)R_L'\right]}$

$f_{P_1} \simeq \frac{1}{2\pi\left[(10 + 0.5(1 + 20\times5))0.97 + 2.5\times5\right]}$

$f_{P_1} = 2.24MHz$

Eq. 6.77: $f_{P2} = \frac{(C_\pi + C_\mu(1 + g_m R_L'))R_{sig}' + (C_L + C_\mu)R_L'}{2\pi\left[C_\pi(C_L + C_\mu) + C_L C_\mu\right]R_{sig}' R_L'}$

$f_{P_2} = \frac{(10 + 0.5(1 + 20\times5))0.97 + 2.5\times5}{2\pi\,(10(2 + 0.5) + 2\times0.5)0.97\times5}$

$f_{P2} = 89.89\ MHz$

Since $f_{P_1} \ll f_{P2}$ and $f_{P_1} \ll f_Z$, we can approximate f_H by f_{P_1} : $f_H \simeq f_{P_1} = 2.24MHz$

If we compare f_H to the results obtained from applying Miller's Theorem in Problem 6.70, then our results are 17% lower.

6.72

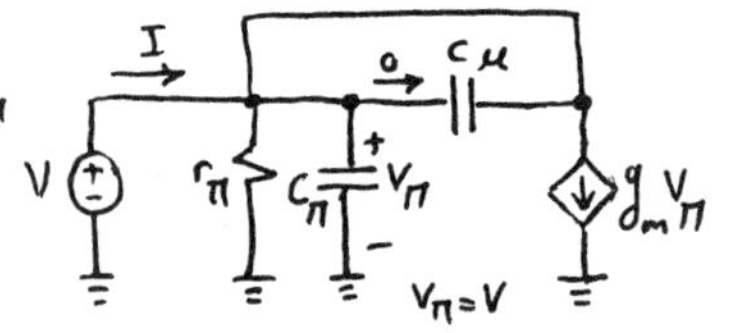

$I = \frac{V}{r_\pi} + sC_\pi V_\pi + g_m V_\pi$

$Y_{in} = (g_m + \frac{1}{r_\pi}) + sC_\pi$

$Z_i = \frac{1}{(g_m + \frac{1}{r_\pi}) + sC_\pi} = \frac{1}{\frac{1}{r_e} + sC_\pi} = \frac{r_e}{1 + sC_\pi r_e}$

$Z_i = \frac{r_e}{1 + sC_\pi r_e}$

$f_T = \frac{g_m}{2\pi(C_\pi + C_\mu)}$ Since $C_\pi = C_{de} + C_{je} = \tau_F g_m + C_{je}$,

Therefore C_π has a component that depends on the bias current and at high currents $C_\pi \gg C_\mu$

and $f_T \simeq \frac{g_m}{2\pi C_\pi} \simeq \frac{1}{2\pi C_\pi r_e}$

Thus: $Z_i \simeq \frac{r_e}{1 + \frac{s}{\omega_T}}$ at high currents

The phase angle will be -45° at $\omega = \omega_T$ or $f = f_T = 400MHz$.

For lower bias currents, so that $C_\pi = C_\mu$

$f_T \simeq \frac{1}{4\pi C_\pi r_e}$ and $Z_i = \frac{r_e}{1 + \frac{s}{2\omega_T}}$ and -45°

phase is obtained at $\omega = 2\omega_T$ or $f = 2f_T = 800$ MHz

(Assuming that f_T remains constant which is not necessarily true!!)

6.73

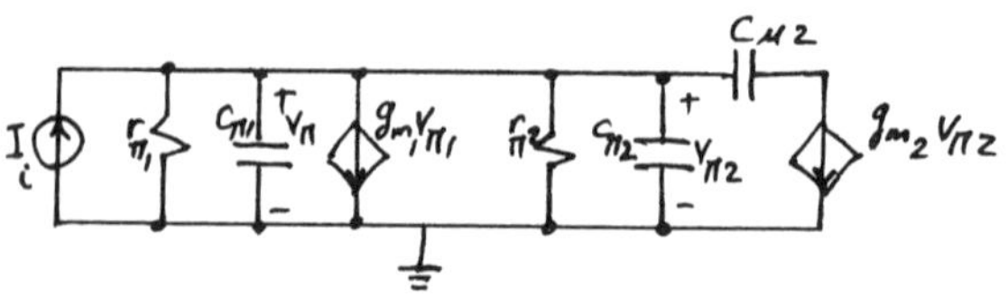

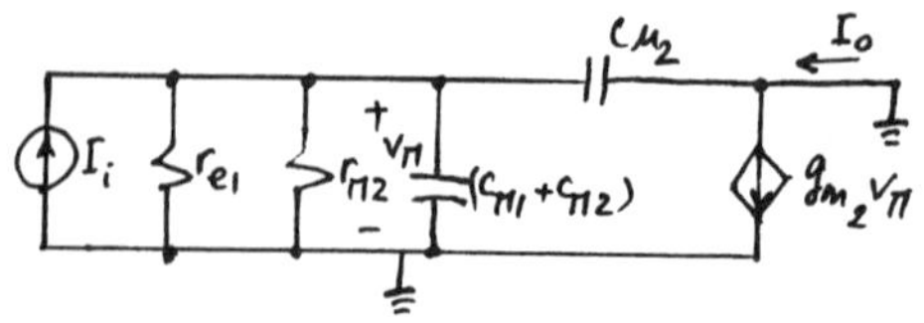

$V_\pi = \frac{I_i}{(\frac{1}{r_{e1}} + \frac{1}{r_{\pi2}}) + s(C_{\pi1} + C_{\pi2} + C_{\mu2})}$

Cont.

$$I_o = g_{m2} V_\pi - C_\mu s V_\pi = \frac{(g_{m2} - C_\mu s) I_i}{\left(\frac{1}{r_{e1}} + \frac{1}{r_{\pi 2}}\right) + s(C_{\pi 1} + C_{\pi 2} + C_{\mu 2})}$$

$$\frac{I_o}{I_i} = \frac{g_{m2} - C_\mu s}{\left(\frac{1}{r_{e1}} + \frac{1}{r_{\pi 2}}\right) + s(C_{\pi 1} + C_{\pi 2} + C_{\mu 2})}$$

$I_{C1} = I_{C2} \Rightarrow r_{\pi 1} = r_{\pi 2}$, $g_{m1} = g_{m2}$, $C_{\pi 1} = C_{\pi 2}$

$$\frac{I_o}{I_i} = \frac{g_m - C_\mu s}{\left(\frac{1}{r_e} + \frac{1}{r_\pi}\right) + (C_\mu + 2C_\pi)s} = \frac{1 - \frac{C_\mu}{g_m} s}{\left(\frac{1}{g_m r_e} + \frac{1}{g_m r_\pi}\right) + s\,\frac{C_\mu + 2C_\pi}{g_m}}$$

$$g_m r_e = \frac{I_C}{V_T} \frac{V_T}{I_E} = \alpha = \frac{\beta}{\beta + 1}$$

$$g_m r_\pi = \beta$$

$$\Rightarrow \frac{I_o}{I_i} = \frac{1 - \frac{C_\mu}{g_m} s}{1 + \frac{1}{\beta} + \frac{1}{\beta} + s(2C_\pi + C_\mu)/g_m}$$

$$\frac{I_o}{I_i} = \frac{1}{1 + 2/\beta} \; \frac{1 - s\, C_\mu/g_m}{1 + s\,\frac{(2C_\pi + C_\mu)}{g_m(1 + \frac{2}{\beta})}}$$

If the circuit is biased at 1mA and $\beta = \infty$, $f_T = 400\,\text{MHz}$ and $C_\mu = 2\,\text{pF}$:

$$g_m = \frac{1}{0.025} = 40\,\text{mA/V}$$

$$f_T = \frac{g_m}{2\pi(C_\pi + C_\mu)} \Rightarrow C_\pi + C_\mu = \frac{40^m}{2\pi \times 400^M} = 15.9\,\text{pF}$$

$$C_\pi = 15.9 - 2 = 13.9\,\text{pF}$$

Pole frequency: $f_p = \dfrac{g_m}{2\pi(2C_\pi + C_\mu)} = \dfrac{40 \times 10^{-3}}{2\pi(2 \times 13.9 + 2)^p}$

$$f_p = 213.74\,\text{MHz}$$

Zero frequency: $f_Z = \dfrac{g_m}{2\pi C_\mu} = \dfrac{40^m}{2\pi \times 2^p} = 3.18\,\text{GHz}$

6.74

$$A_M = -g_m R'_L = -5 \times 20 = -100\,\text{V/V}$$

$$f_H = \frac{1}{2\pi(C_L + C_{gd}) R'_L} = \frac{1}{2\pi(1 + 0.1) \times 20} = 7.23\,\text{MHz}.$$

(Note that in this case there is no R_{sig} and we used Eq. 6.79)

$$f_{3dB} = f_H = 7.23\,\text{MHz}$$

$$f_t = |A_M| \cdot f_H = 723\,\text{MHz}$$

6.75

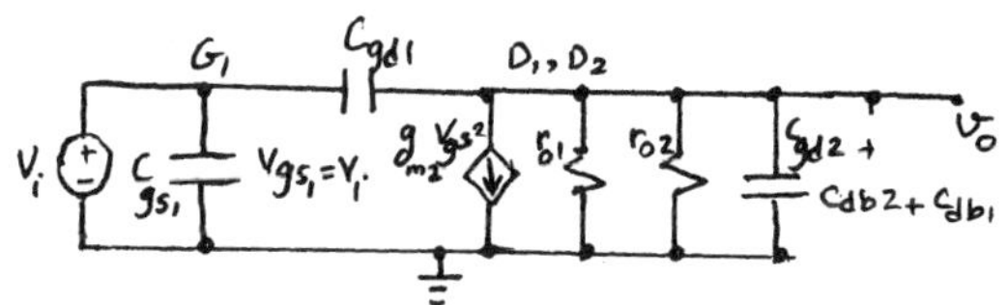

$$g_m = \sqrt{2\mu_n C_{ox} \frac{W}{L} I_D} = \sqrt{2 \times 90 \times \frac{100}{1.6} \times 100} = 1060\,\mu\text{A/V}$$

$$g_m = 1.06\,\text{mA/V}$$

$$r_{o1} = \frac{V_{A1}}{I_{D1}} = \frac{12.8}{0.1} = 128\,\text{k}\Omega$$

$$r_{o2} = \frac{|V_{A2}|}{I_{D2}} = \frac{19.2}{0.1} = 192\,\text{k}\Omega$$

$$\text{DC gain} = -g_{m1}(r_{o1} \| r_{o2}) = -1.06 \times (128 \| 192) = -81.4\,\text{V/V}$$

Total capacitance between output node and ground $= C_{gd2} + C_{db1} + C_{db2} = 0.015 + 0.020 + 0.036$

$$C_L = 0.071\,\text{pF}$$

Write a KCL at output:

$$sC_{gd1}(v_i - v_o) = g_{m1} v_i + \frac{v_o}{r_{o1}} + \frac{v_o}{r_{o2}} + v_o s C_L$$

$$\frac{v_o}{v_i} = -\frac{g_m - sC_{gd1}}{\frac{1}{r_{o1}} + \frac{1}{r_{o2}} + (C_L + C_{gd1})s}$$

Thus: $f_Z = \dfrac{g_m}{2\pi C_{gd1}} = \dfrac{1.06^m}{2\pi \times 0.015^p} = 11.3\,\text{GHz}$

$$f_p = \frac{1}{2\pi} \frac{\frac{1}{r_{o1}} + \frac{1}{r_{o2}}}{C_L + C_{gd1}} = \frac{\frac{1}{128^K} + \frac{1}{192^K}}{2\pi(0.071 + 0.015)^p}$$

$$f_p = 24.1\,\text{MHz}$$

6.76

a) For small C_{gd} and low gain from G to D, we can neglect the miller effect and C_{gd}.

$$\omega_T = \frac{g_m}{C_{gs} + C_{gd}} \simeq \frac{g_m}{C_{gs}} \quad \text{Thus } C_{gs} \simeq \frac{g_m}{\omega_T}$$

b) replace the controlled source $g_{m2} V_{gs2}$ with a resistance $\frac{1}{g_{m2}}$. (Source absorbtion theory)

Cont.

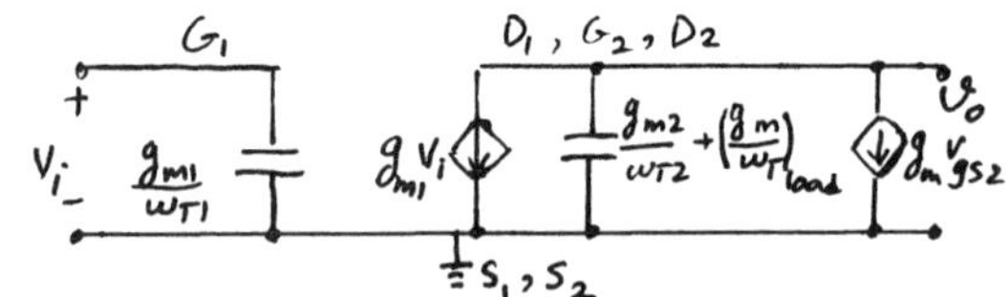

$$V_o = -g_{m1} v_i \frac{1}{g_{m2} + s\left(\frac{g_{m2}}{\omega_{T2}} + \frac{g_{mload}}{\omega_{Tload}}\right)}$$

Since the load device is identical to Q_1,

$g_{mLoad} = g_{m1}$ and $\omega_{Tload} = \omega_{T1} = \omega_T$.

Thus:

$$\frac{v_o}{v_i} = \frac{-g_{m1}/g_{m2}}{1 + \frac{s}{\omega_T}\left(1 + \frac{g_{m1}}{g_{m2}}\right)}$$

$$\frac{g_{m1}}{g_{m2}} = \frac{\mu_n C_{ox}\left(\frac{W}{L}\right)_1 V_{OV}}{\mu_n C_{ox}\left(\frac{W}{L}\right)_2 V_{OV}} = \frac{W_1}{W_2}$$

$$\Rightarrow \frac{v_o}{v_i} = \frac{-A_o}{1 + \frac{s}{\omega_T}(1 + A_o)} \quad \text{where } A_o = \frac{W_1}{W_2} = \frac{g_{m1}}{g_{m2}}$$

c) $A_o = 3\ V/V$, $W_2 = 25\ \mu m$

$A_o = \frac{W_1}{W_2} \Rightarrow W_1 = 3 \times 25 = 75\ \mu m$

$I_{D1} = \frac{1}{2} \mu_n C_{ox} \frac{W_1}{L_1} (V_{GS} - V_t)^2 = \frac{1}{2} \times 200^{\mu} \times \frac{75}{0.5} 0.3^2$

$I_{D1} = 1.35\ mA$

$I_{D2} = \frac{1}{2} \times 200\ \mu A/V^2 \times \frac{25}{0.5} \times 0.3^2 = 0.45\ mA$

Thus: $I = I_{D1} + I_{D2} = 1.35 + 0.45 = 1.8\ mA$

$$f_{3db} = \frac{f_T}{1 + A_o} = \frac{12 \times 10^9}{1 + 3} = 3\ GHz$$

6.77

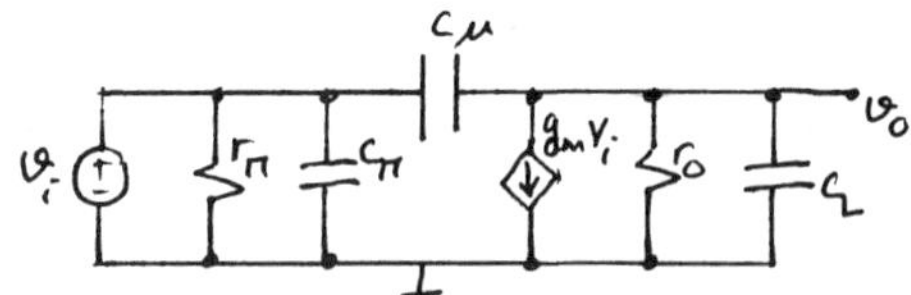

writing a node equation at the output yields:

$$sC_\mu(v_i - v_o) = g_m v_i + \frac{v_o}{r_o} + v_o C_L s$$

$$\frac{v_o}{v_i} = \frac{C_\mu s - g_m}{\frac{1}{r_o} + (C_L + C_\mu)s} = -g_m r_o \frac{1 - sC_\mu/g_m}{1 + s(C_L + C_\mu) r_o}$$

for small C_μ, $\omega C_\mu << g_m$: $\frac{v_o}{v_i} = \frac{-g_m r_o}{1 + s(C_\mu + C_L) r_o}$

For $I_C = 200\ \mu A$, $V_A = 100V$: $g_m = \frac{200^{\mu}}{0.025} = 8\ mA/V$

$r_o = \frac{100}{200} = 0.5\ M\Omega$

Thus DC-Gain $= -g_m r_o = -8 \times 0.5 \times 10^3 = -4000\ V/V$

For $C_L = 1pF$, $C_\mu = 0.2pF$:

$$\omega_{3db} = \frac{1}{(C_L + C_\mu) r_o} = \frac{1}{(1 + 0.2)^p \times 0.5^M} = 1.67\ Mrad/s$$

$f_{3db} = 265.4\ kHz = f_H$

$f_T = |A_o| f_H = 4000 \times 265.4 = 1.06\ GHz$

Bode Plot for $|A|$:

$4000\ V/V = 72\ dB$

Check $f_Z = \frac{g_m}{2\pi C_\mu}$

$$f_Z = \frac{8^m}{2\pi \times 0.2^p} = 6.4\ GHz$$

6.78

$f_T = \frac{g_m}{2\pi(C_L + C_{gd})}$, $g_m = 1\ mA/V$, $f_T = 2\ GHz$

$$\Rightarrow C_L + C_{gd} = \frac{1 \times 10^3}{2\pi \times 2 \times 10^9} = 79.61\ fF$$

To have $f_{T2} = 1\ GHz$, we need:

$$C_L + C_{gd} = \frac{1 \times 10^3}{2\pi \times 1 \times 10^9} = 159.23\ fF$$

Thus we need an additional capacitance of

$159.23 - 79.61 = 79.61\ fF$

6.79

$$g_m = \sqrt{2k_n' \frac{W}{L} I_D} = \sqrt{2 \times \frac{160}{1000} \times \frac{50}{0.5} \times 0.5} = 4\ mA/V$$

$g_{mb} = \chi g_m = 0.2 \times 4 = 0.8\ mA/V$

$r_o = \frac{V_A}{I_D} = \frac{1}{\lambda I_D} = \frac{1}{0.1 \times 0.5} = 20\ K\Omega$

$R_o = r_o = 20\ K\Omega$

$A_{vo} = 1 + (g_m + g_{mb}) r_o = 1 + (4^m + 0.8^m) \times 20^K = 97\ V/V$

$R_{out} = r_o + A_{vo} R_s$ (Eq. 6.102)

$R_{out} = 20^K + 97 \times 20^K = 1.96\ M\Omega$

$R_{in} = \frac{r_o + R_L}{A_{vo}}$ (Eq. 6.86)

$R_{in} = \frac{20 + 20}{97} = 0.41\ K\Omega$

$A_v = A_{vo} \frac{R_L}{R_L + r_o} = 97 \frac{20}{20 + 20} = 48.5\ V/V$

Cont.

$$G_v = \frac{v_o}{v_{sig}} = A_{v_o}\frac{R_L}{R_L + r_o + A_{v_o}R_s} \quad \text{(Eq. 6.95)}$$

$$G_v = \frac{97\times 20^K}{20^K + 20^K + 97\times 20^K} = 0.98\ V/V$$

$$G_{is} = A_{v_o}\frac{R_s}{R_{out}} = 97\times\frac{20}{1960} = 0.99\ A/A$$

$$G_i = G_{is}\frac{R_{out}}{R_{out}+R_L} = 0.99\times\frac{1960}{1960+20} = 0.98\ A/A$$

6.80

$R_L = r_o$

$$A_v = 100\ V/V = \frac{v_o}{v_i} = A_{v_o}\frac{R_L}{R_L+r_o} = A_{v_o}\frac{r_o}{r_o+r_o}$$

$$\Rightarrow A_v = A_{v_o}/2 = 100 \Rightarrow A_{v_o} = 200\ V/V$$

$R_{in} = 2K\Omega$

$$R_{in} = \frac{r_o+R_L}{A_{v_o}} = \frac{r_o+r_o}{A_{v_o}} \Rightarrow 2^K = \frac{2r_o}{200} \Rightarrow r_o = 100\ K\Omega$$

$$r_o = \frac{V_A}{I_D} \Rightarrow I_D = \frac{V_A}{r_o} = \frac{20}{100} = 0.2\ mA$$

$$A_{v_o} = 1 + (g_m+g_{mb})r_o = 1 + (1+\chi)g_m r_o$$

$$200 = 1 + (1+0.2)g_m\times 100 \Rightarrow g_m = 1.66\ mA/V$$

$$g_m = \frac{2I_D}{V_{OV}} \Rightarrow V_{OV} = \frac{2\times 0.2}{1.67} = 0.24^V$$

$$I_D = \frac{1}{2}k_n'\frac{W}{L}V_{OV}^2 \Rightarrow \frac{W}{L} = \frac{2I_D}{k_n' V_{OV}^2} = \frac{2\times 0.2}{0.1\times 0.24^2}$$

$$\frac{W}{L} = 69.4$$

6.81

$$i_{osc} = \frac{R_s i_{sig}}{R_s + (r_o \| \frac{1}{g_m+g_{mb}})} = \frac{R_s}{R_s + \frac{r_o}{1+(g_m+g_{mb})r_o}}\, i_{sig}$$

$$G_{is} = \frac{i_{osc}}{i_{sig}} = \frac{R_s}{R_s + \frac{r_o}{A_{vo}}} = \frac{A_{vo}R_s}{r_o + A_{vo}R_s}$$

Using (6.102) we have: $G_{is} = \frac{A_{vo}R_s}{R_{out}}$

If $A_{vo}R_s \gg r_o$ that is $[1+(g_m+g_{mb})r_o]R_s \gg r_o$

or $(g_m+g_{mb})R_s \gg 1$ then $G_{is} \simeq 1$

6.82

$$R_{in} = \frac{r_o+R_L}{A_{v_o}}, \quad A_{v_o} = 1 + (g_m+g_{mb})r_o$$

$$A_{v_o} \simeq (g_m+g_{mb})r_o \cong A_o$$

$$\Rightarrow R_{in} \simeq \frac{1}{g_m+g_{mb}} + \frac{R_L}{A_o}$$

If $R_L = A_o r_o$ then $R_{in} = \frac{1}{g_m+g_{mb}} + r_o$

6.83

$$g_m = \frac{2I_D}{V_{ov}} = \frac{2\times 1^m}{0.8-0.55} = 8\ mA/V$$

$$g_{mb} = \chi g_m = 0.2\times 8 = 1.6\ mA/V$$

$$r_o = \frac{V_A}{I_D} = \frac{20}{1^m} = 20K\Omega$$

From Eq. 6.101 we have: $R_{out} = r_o + [1+(g_m+g_{mb})r_o]R_s$

$$R_{out} = 20 + (1+(8+1.6)\times 20)R_s = 200K\Omega$$

$R_s = 932.64\Omega$

$$V_{BIAS} = I_D R_s + V_{GS} = 1^m\times 932.64\times 10^{-3} + 0.8 = 1.73^V$$

$V_{BIAS} = 1.73^V$

6.84

$$V_s\times(g_m+g_{mb}) = i_{osc} - \frac{V_s}{r_o}$$

$$\Rightarrow i_{osc} = V_s\left(\frac{1}{r_o} + g_m + g_{mb}\right) \quad (1)$$

$$V_s = V_{sig} - R_s i_{osc}$$

If we substitute V_s in (1):

$$i_{osc} = (V_{sig} - R_s i_{osc})\left(\frac{1}{r_o} + g_m + g_{mb}\right)$$

$$i_{osc}\left(1 + \frac{R_s}{r_o} + R_s(g_m+g_{mb})\right) = V_{sig}\frac{(1+(g_m+g_{mb})r_o)}{r_o}$$

$$i_{osc} = V_{sig}\frac{1+(g_m+g_{mb})r_o}{r_o + R_s + R_s r_o(g_m+g_{mb})}$$

Since $A_{v_o} = G_{vo} = 1 + (g_m+g_{mb})r_o$ (Eq. 6.88)

and $R_{out} = r_o + A_{v_o}R_s$ (Eq. 6.102), then

$$i_{osc} = V_{sig}\frac{A_{v_o}}{R_{out}}$$

6.85

a) $V_{GS2} = V_{GS3} \Rightarrow I_{D2} = I_{D3}$. Also $I_{D1} = I_{D2}$

$\Rightarrow I_{D1} = 100\,\mu A$.

$I_{D1} = \frac{1}{2} k_n' \frac{W}{L} (V_{GS_1} - V_t)^2 \Rightarrow 0.1 = \frac{1}{2} \times 4 \times (V_{GS_1} - 0.8)^2$

$V_{GS_1} - 0.8 = \pm 0.224 \Rightarrow V_{GS_1} = 1.024^V$, $V_{OV_1} = V_{OV_2} = 0.224^V$

$V_{BIAS} = R_S I_{D1} + V_{GS} = 0.05 \times 0.1 + 1.024 = 1.029^V$

b) $g_{m1} = \frac{2 I_{D1}}{V_{OV}} = \frac{0.1 \times 2}{0.224} = 0.89\, mA/V$

$g_{m2} = \sqrt{2 k_n' \frac{W}{L} I_D} = \sqrt{2 \times 4 \times 0.1} = 0.89\, mA/V$

$g_{m3} = g_{m2} = 0.89\, mA/V$

$g_{mb} = \chi g_m \Rightarrow g_{mb1} = g_{mb2} = g_{mb3} = 0.2 \times 0.89 = 0.18\, mA/V$

$r_{o1} = r_{o2} = r_{o3} = \frac{V_A}{I_D} = \frac{20}{0.1} = 200\, k\Omega$

c) $R_{in} = \frac{r_o + R_L}{A_{v_o}}$ where $A_{v_o} = 1 + (g_m + g_{mb}) r_o$

$A_{v_o} = 1 + (0.89 + 0.18)200$

$A_{v_o} = 215\ V/V$

In this case R_L is in fact R_{out} of the Active load in the drain of Q_1, which is $r_{o2} = 200 k\Omega$.

Thus: $R_{in} = \frac{200 + 200}{215} = 1.86\, k\Omega$

d) $R_{out} = r_o + A_{v_o} R_s = 200 + 215 \times 0.05 = 210.75\, k\Omega$

e) $A_v = \frac{v_o}{v_i} = A_{v_o} \frac{R_L}{R_L + r_o} = 215 \frac{200}{200+200} = 107.5\, V/V$

$G_v = A_{v_o} \frac{R_L}{R_L + R_{out}} = 215 \frac{200}{200 + 210.75} = 104.7\, V/V$

f) For Q_2 to stay in saturation region, V_o can go as high as $V_{DD} - V_{OV2}$, Thus $V_o \leq 3.3 - 0.224$

or $V_o \leq 3.076^V$.

For Q_1 to stay in saturation region, we need to have $V_{GD1} \leq V_t \Rightarrow V_{BIAS} - v_o \leq 0.8 \Rightarrow$

$v_o \geq 1.029 - 0.8$ or $v_o \geq 0.229^V$

hence $0.229^V \leq v_o \leq 3.076^V$. This implies that the peak-to-peak output swing is

$3.076 - 0.229 = 2.847^V$.

Since $G_v = \frac{v_o}{v_{sig}} = 104.7$, then the maximum peak-to-peak value of v_{sig} is $\frac{2.847}{104.7}$

that is 27mV.

6.86

From Fig. 6.31b the low-frequency gain $\frac{v_o}{v_{sig}}$ can be written as:

$$\frac{v_o}{v_{sig}} = \frac{\frac{1}{g_m + g_{mb}}}{R_s + \frac{1}{g_m + g_{mb}}} \times (g_m + g_{mb}) \times R_L'$$

$$\frac{v_o}{v_{sig}} = \frac{(g_m + g_{mb}) R_L'}{1 + (g_m + g_{mb}) R_{sig}} = \frac{(5 + 0.2 \times 5) \times 20}{1 + (5 + 0.2 \times 5) \times 1} = 17.14\, V/V$$

$$\frac{v_o}{v_{sig}} = 17.14\, V/V$$

From Eq. 6.105 we have $f_{P1} = \frac{1}{2\pi C_{gs} (R_{sig} \| \frac{1}{g_m + g_{mb}})}$

$$f_{P1} = \frac{1}{2\pi \times 2pF (1K \| \frac{1}{5 + 0.2 \times 5})} = 557\, MHz$$

From Eq. 6.106 we have:

$$f_{P2} = \frac{1}{2\pi (C_{gd} + C_L) R_L'} = \frac{1}{2\pi (0.1^P + 2^P) 20^K} = 3.79\, MHz$$

Since $f_{P2} \ll f_{P1}$, then f_{P2} is the dominant pole and $f_H \simeq f_{P2} = 3.79\, MHz$

6.87

$A_{v_o} = 1 + (g_m + g_{mb}) r_o = 1 + (5 + 0.2 \times 5) 20^K = 121\, V/V$

$R_{out} = r_o + A_{v_o} R_s = 20 + 121 \times 1 = 141\, k\Omega$

$G_v = G_{vo} \frac{R_L}{R_L + R_{out}} = A_{v_o} \frac{R_L}{R_L + R_{out}} = 121 \frac{20^k}{20^k + 141^k}$

$G_v = \frac{v_o}{v_{sig}} = 15\, V/V$

$R_{in} = \frac{r_o + R_L}{A_{v_o}} = \frac{20 + 20}{121} = 0.33\, k\Omega$

$R_{gs} = R_s \| R_{in} = 1^K \| 0.33^K = 0.25\, k\Omega = 250\,\Omega$

$R_{gd} = R_L \| R_{out} = 20^K \| 141^K = 17.5\, k\Omega$

$\tau_H = C_{gs} R_{gs} + (C_L + C_{gd}) R_{gd} = 2^P \times 0.25^K + (2 + 0.1) 17.5^K = 37.25\, nS$

$f_H \simeq \frac{1}{2\pi \tau_H} = 4.27\, MHz$

Comparing with results in Problem 6.86, we notice that gain is reduced by 12.5%, while f_H is increased by 12.7%, when r_o is taken into account.

6.88

$i_o = i_i - v_i/r_\pi$ Eq. 6.110

If we write a KCL at the emitter node, we will have: $\frac{v_i - i_o R_L}{r_o} + \frac{v_i}{r_e} = i_i$

Substitute for i_o from Eq. 6.110, then:

$$\frac{v_i - R_L i_i + R_L/r_\pi \, v_i}{r_o} + \frac{v_i}{r_e} = i_i$$

$$v_i\left(\frac{1}{r_o} + \frac{R_L}{r_\pi r_o} + \frac{1}{r_e}\right) = i_i\left(1 + \frac{R_L}{r_o}\right)$$

$$R_{in} = \frac{v_i}{i_i} = \frac{1 + R_L/r_o}{\frac{1}{r_o} + \frac{1}{r_e} + \frac{R_L}{r_\pi r_o}} = \frac{r_o + R_L}{1 + \frac{r_o}{r_e} + \frac{R_L}{r_\pi}}$$

$r_\pi = (\beta+1) r_e$ therefore:

$$R_{in} = \frac{r_o + R_L}{1 + \frac{r_o}{r_e} + \frac{R_L}{(\beta+1) r_e}}$$

This is the same as equation 6.111.

6.89

$\beta = 100$, $R_{in} \simeq r_e \frac{r_o + R_L}{r_o + R_L/(\beta+1)}$

$$R_{in} \simeq r_e \frac{1 + R_L/r_o}{1 + \frac{R_L}{r_o}\frac{1}{\beta+1}}$$

R_L/r_o	0	1	10	100	1000
R_{in}/r_e	1	1.98	10	50.75	91.83

6.90

$I = 1\text{mA}$, Intrinsic gain $= A_o = \frac{V_A}{V_T} = 2000 \text{V/V}$

$V_A = 2000 V_T = 50\text{V}$

$r_o = \frac{V_A}{I_C} = \frac{50}{1} = 50\text{k}\Omega$

From Eq. 6.112 we have: $R_{in} \simeq r_e \frac{r_o + R_L}{r_o + R_L/(\beta+1)}$

Assuming β is very large so that:

$\frac{R_L}{\beta+1} \ll r_o$, we have: $R_{in} \simeq r_e \frac{r_o + R_L}{r_o}$

For R_{in} to be $2r_e$, we need:

$$2r_e = r_e \frac{r_o + R_L}{r_o} \Rightarrow r_o = R_L \Rightarrow \underline{R_L = 50\text{k}\Omega}$$

6.91

Refer to Fig. 6.34:

write KCL at the emitter:

$$\frac{v}{r_e} + \frac{v}{R_e} = i_x + g_m v \Rightarrow v = \frac{i_x}{\frac{1}{r_e} - g_m + \frac{1}{R_e}}$$

Note that $\frac{1}{r_e} - g_m = \frac{\beta+1}{r_\pi} - g_m = \frac{\beta+1-\beta}{r_\pi} = \frac{1}{r_\pi}$

therefore: $v = \frac{i_x}{\frac{1}{r_\pi} + \frac{1}{R_e}} = (r_\pi \| R_e) i_x$

$v = R_e' i_x$ (1) where $R_e' = r_\pi \| R_e$

Now we write the equation for v:

$v = v_x - (i_x + g_m v) r_o$

if we substitute v from (1):

$R_e' i_x = v_x - i_x r_o - g_m R_e' i_x r_o$

$i_x (R_e' + r_o + g_m R_e' r_o) = v_x$

$R_{in} = \frac{v_x}{i_x} = r_o + (1 + g_m r_o) R_e'$ where $R_e' = R_e \| r_\pi$

(Same as Eq. 6.117)

6.92

Eq. 6.118: $R_{out} \simeq (1 + g_m R_e') r_o$

$\Rightarrow \frac{R_{out}}{r_o} \simeq 1 + g_m R_e'$ where $R_e' = R_e \| r_\pi$

$$R_e' = \frac{r_\pi R_e}{r_\pi + R_e} = \frac{(\beta+1) r_e R_e}{(\beta+1) r_e + R_e} = \frac{(\beta+1) r_e}{\frac{(\beta+1) r_e}{R_e} + 1}$$

$$m = \frac{R_e}{r_e} \Rightarrow R_e' = \frac{(\beta+1) r_e}{\frac{\beta+1}{m} + 1}$$

$$\frac{R_{out}}{r_o} = 1 + g_m \frac{r_e(\beta+1)}{\frac{\beta+1}{m} + 1} = 1 + \frac{g_m \frac{r_\pi}{\beta+1}(\beta+1)}{\frac{\beta+1}{m} + 1}$$

$$\frac{R_{out}}{r_o} = 1 + \frac{\beta}{\frac{\beta+1}{m} + 1} = \frac{(\beta+1)(m+1)}{\beta+1+m} = \frac{101(m+1)}{101+m}$$

$m = \frac{R_e}{r_e}$	1	2	10	$\beta/2$	1000
$\frac{R_{out}}{r_o}$	1.98	2.94	10.01	$\beta/3 + 1$	91.74

6.93

$R_{sig} = 1\text{k}\Omega$, $\beta = 100$, $V_A = 50\text{V}$, $I_C = 0.1\text{mA}$

Cont.

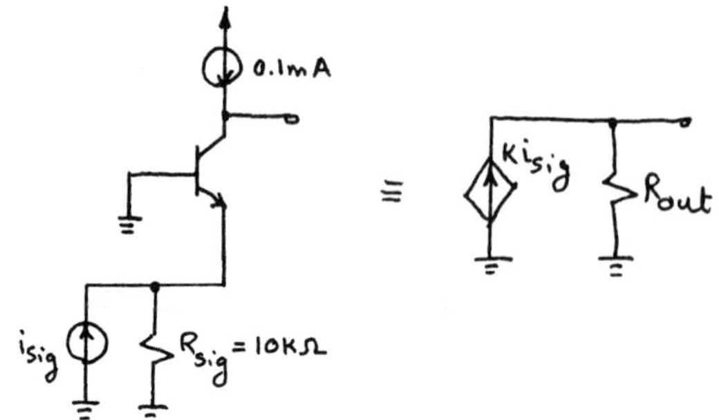

$r_o = \frac{V_A}{I_C} = \frac{50}{0.1^m} = 500K\Omega$

$g_m = \frac{I_C}{V_T} = \frac{0.1}{0.025} = 4\ mA/V$

$r_\pi = \frac{\beta}{g_m} = \frac{100}{4} = 25K\Omega$

The Thevenin equivalent circuit of the source is:

$V_{sig} = R_{sig} i_{sig}$

Thus using equations 6.121 and 6.114 we can write:

$$\frac{V_{out}}{V_{sig}} = \frac{r_\pi}{r_\pi + R_{sig}} A_{v_o} = \frac{r_\pi}{r_\pi + R_{sig}}(1 + g_m r_o)$$

$$\frac{V_{out}}{i_{sig}} = \frac{r_\pi R_{sig}}{r_\pi + R_{sig}}(1 + g_m r_o) = (r_\pi \| R_{sig})(1 + g_m r_o)$$

$$\frac{V_{out}}{i_{sig}} = 14.29 \times 10^6\ V/A$$

From Eq. 6.117a we have $R_{out} = r_o + (1 + g_m r_o) R'_e$ where $R'_e = r_\pi \| R_{sig}$.

$R_{out} = 500^K + (1 + 4^m \times 500^K)(25^K \| 10^K) = 14.79M\Omega$

$R_{out} = 14.79\ M\Omega$

$V_{out} = K i_{sig} R_{out} \Rightarrow K = \frac{V_{out}}{i_{sig}} \times \frac{1}{R_{out}} = \frac{14.29 \times 10^6}{14.79 \times 10^6}$

$\Rightarrow K = 0.97$

6.94

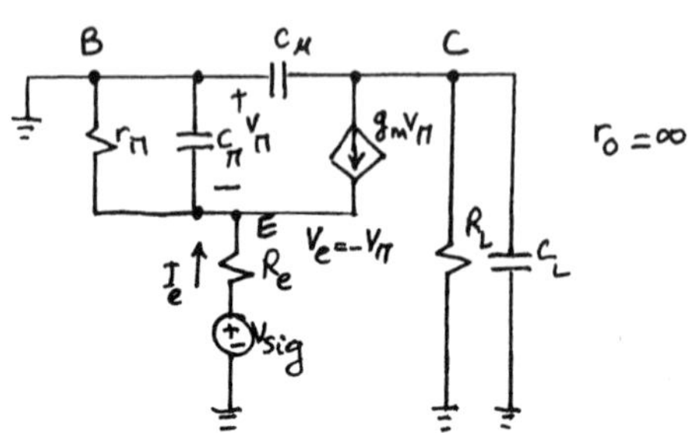

we observe that V_e, the voltage at the emitter is equal to $-V_\pi$. We can write a node equation at the emitter:

$$I_e = -V_\pi\left(\frac{1}{r_\pi} + sC_\pi\right) - g_m V_\pi = V_e\left(\frac{1}{r_\pi} + g_m + sC_\pi\right)$$

Thus, the input admittance looking into the emitter is: $\frac{I_e}{V_e} = \frac{1}{r_\pi} + g_m + sC_\pi = \frac{1}{r_e} + sC_\pi$

Therefore we can replace the transistor at the input of the circuit by this admittance as shown below:

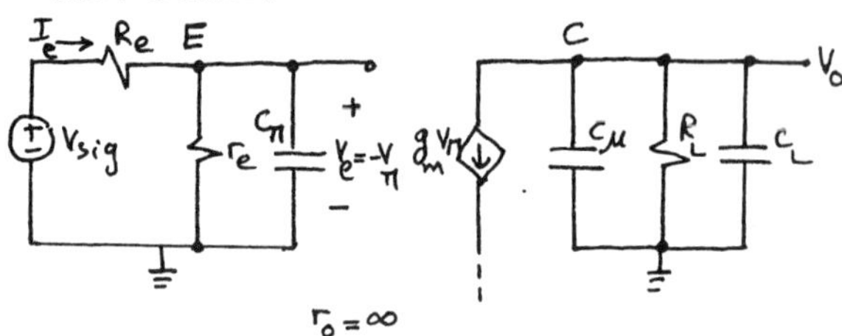

a) As we can see above, the circuit can be seperated into two parts with each part having its own pole.

$$f_{P1} = \frac{1}{2\pi C_\pi (R_e \| r_e)} \quad \text{(input part)}$$

$$f_{P2} = \frac{1}{2\pi (C_\mu + C_L) R_L} \quad \text{(output part)}$$

IF we compare the poles to Eq. 6.105 and 6.106 for MOSFETs, we observe that these equations are the bipolar counterparts of those ones.

$$f_{P1} = \frac{1}{2\pi C_{gs}\left(R_s \| \frac{1}{g_m + g_{mb}}\right)} \quad (6.105)$$

$$f_{P2} = \frac{1}{2\pi (C_{gd} + C_L) R_L} \quad (6.106)$$

b) For $C_\pi = 14pF$, $C_\mu = 2pF$, $C_L = 1pF$, $I_C = 1mA$, $R_{sig} = 1K\Omega$, $R_L = 10K\Omega$. $\Rightarrow g_m = \frac{1}{0.025} = 40\ mA/V$

$$f_{P1} = \frac{1}{2\pi \times 14P\left(1^K \| \frac{100}{40}\right)} \quad \text{(Assuming } \beta = 100\text{)}$$

$f_{P1} = 15.9\ MHz$

$$f_{P2} = \frac{1}{2\pi (2^P + 1^P) 10^K} = 5.3\ MHz$$

$$f_T = \frac{g_m}{2\pi (C_\pi + C_\mu)} = \frac{40^m}{2\pi (14 + 2)^P} = 398.1\ MHz$$

f_T is much greater than the poles. $f_T \gg f_{P1}$

$f_T \gg f_{P2}$

6.95

In Fig. 6.32, if we replace the NMOS with an npn transistor, then we can write:

$R_{be} = R_e \| R_{in}$ where $R_{in} = r_e \dfrac{r_o + R_L}{1 + \frac{r_o}{r_e} + \frac{R_L}{(\beta+1)r_e}}$

$R_{bc} = R_L \| R_{out}$ where $R_{out} = (1 + g_m R'_e) r_o$

Using the open-circuit time constants method to evaluate f_H:

$$f_H = \frac{1}{2\pi \left(C_\pi (R_e \| R_{in}) + (C_L + C_\mu)(R_L \| R_{out}) \right)}$$

6.96

5V, I, 4.3kΩ

$\beta = 100$, $V_A = 100V$

$I_C \simeq I_E = \dfrac{5 - V_{BE}}{4.3^k} = \dfrac{5 - 0.7}{4.3^k} = 1mA$

$I = 1mA$

From Eq. 6.117a we have:

$R_{out} = r_o + (1 + g_m r_o)(r_\pi \| R_e)$

$r_o = \dfrac{V_A}{I_C} = \dfrac{100}{1} = 100k\Omega$

$g_m = \dfrac{I_C}{V_T} = \dfrac{1}{0.025} = 40mA/V$

$r_\pi = \dfrac{\beta}{g_m} = 2.5k\Omega$, $R_e = 4.3k\Omega$

Thus: $R_{out} = 100 + (1 + 40 \times 100)(2.5^k \| 4.3^k)$

$R_{out} = 6.43 M\Omega$

If the collector voltage undergoes a change of 10V while the BJT remains in the active mode, the corresponding change in the collector current is: $\Delta I = \dfrac{\Delta V}{R_{out}} = \dfrac{10}{6.43^M} = 1.56\mu A$

$\Delta I = 1.56\mu A$

6.97

a) $I = \frac{1}{2} k'_n \frac{W}{L} V_{OV}^2 = \frac{1}{2} \times 160^\mu \times 100 \times 0.2^2 = 320\mu A$

b) $g_m = \dfrac{2 I_D}{V_{OV}} = \dfrac{2 \times 0.32}{0.2} = 3.2mA/V = g_{m1} = g_{m2}$

$g_{mb} = \chi g_m = 0.2 \times 3.2 = 0.64mA/V = g_{mb1} = g_{mb2}$

$r_o = \dfrac{V_A}{I_D} = \dfrac{1}{\lambda I_D} = \dfrac{1}{0.05 \times 0.32} = 62.5k\Omega = r_{o1} = r_{o2}$

$A_o = g_m r_o = 3.2^m \times 62.5^k = 200V/V$

$A_{vo2} = 1 + (g_{m2} + g_{mb2}) r_{o2}$ (Eq. 6.123)

$A_{vo2} = 1 + (3.2 + 0.64) \times 62.5 = 241$ V/V

c) $A_{vo} = -A_{o1} A_{vo2}$ (Eq. 6.128)

$A_{vo} = -g_{m1} r_{o1} A_{vo2} = -3.2^m \times 62.5 \times 241 = 48200 V/V$

$A_{vo} = 48200 V/V$

d) $G_m = \dfrac{A_{o1} A_{vo2}}{r_{o2} + A_{vo2} r_{o1}}$ (Eq. 6.130)

$G_m = \dfrac{200 \times 241}{62.5 + 241 \times 62.5} = 3.19 mA/V \simeq g_m$

$R_{out} = r_{o2} + [1 + (g_{m2} + g_{mb2}) r_{o2}] r_{o1}$ (Eq. 6.126)

$R_{out} = 62.5 + [1 + (3.2 + 0.64) 62.5] 62.5$

$R_{out} = 15.125 M\Omega$

e) $A_v = -A_o^2 \dfrac{R_L}{R_L + A_o r_o}$ (Eq. 6.131)

$A_v = -(200)^2 \dfrac{10^M}{10^M + 200 \times \frac{62.5}{1000}^k} = 17778 V/V$

f) $V_{DS} \geq V_{OV}$ for Q_1. Therefore $V_{DSmin1} = V_{OV}$

$V_{DSmin1} = 0.2^V$

$V_{BIAS} = V_{GS2} + V_{DSmin1} = (0.2 + 0.6) + 0.2 = 1^V$

$V_{BIAS} = 1V$

6.98

$v_y = i_x \times r_{o1} = \dfrac{v_x}{R_{out}} \times r_{o1} \Rightarrow \dfrac{v_y}{v_x} = \dfrac{r_{o1}}{R_{out}}$

$\dfrac{v_y}{v_x} = \dfrac{r_{o1}}{r_{o2} + [1 + (g_{m2} + g_{mb2}) r_{o2}] r_{o1}}$

If we use equation 6.127 to approximate R_{out}, then $\dfrac{v_y}{v_x} \simeq \dfrac{r_{o1}}{g_{m2} r_{o2} r_{o1}} = \dfrac{1}{g_{m2} r_{o2}}$

6.99

a) $I = \frac{1}{2} k'_n \frac{W}{L} V_{OV}^2 \Rightarrow$ For same I: $\dfrac{V_{OVb}^2}{V_{OVa}^2} = \dfrac{(\frac{W}{L})_a}{(\frac{W}{L})_b}$

For same I, if $\frac{W}{L}$ is divided by 4, then V_{OV}^2 is multiplied by 4, or equivalently

Cont.

V_{OV} is doubled.

$g_m = \mu_n C_{ox} \frac{W}{L} V_{OV}$. Thus g_m for circuit (b) is half of the one for circuit (a).

$A_o = g_m r_o = \frac{2I_D}{V_{OV}} \times \frac{V_A}{I_D} = \frac{2V_A' L}{V_{OV}}$. Thus if L is multiplied by 4 and V_{OV} is halved, then A_o is doubled for circuit (b).

In summary, for circuit (b), V_{OV} is doubled, g_m is halved, A_o is doubled.

b) Each transistor in circuit (c) has the same overdrive voltage as the one in circuit (a). Referring to Eq. 6.129 and 6.130:

$A_{vo} = -A_o^2 = -(g_m r_o)^2$

$G_m \simeq g_{m1} = g_m$ (Same as circuit (a))

Note that for the transistors in circuit (c) the g_m and r_o are the same as the ones in circuit (a). Thus the intrinsic gain for circuit (c), $A_{vo} = -A_o^2$ where A_o is the intrinsic gain for circuit (a).

In general, circuit (c) has a higher output resistance and for the same V_{OV} of transistors it has lower output swing. The output swing is limited to $2V_{OV}$ on the low side for circuits (b) and (c) while it is only limited to V_{OV} for circuit (a).

6.100

a) $A_M = -g_m R_L' = -5 \times (20^K \| 20^K) = -50\ V/V$

$R_{gs} = R_{sig} = 20\ k\Omega$

$R_{gd} = R_{sig}(1 + g_m R_L') + R_L' = 20^K(1 + 5 \times 20^K \| 20^K) + 20^K \| 20^K$

$R_{gd} = 1030\ k\Omega = 1.03\ M\Omega$

We use Eq. 6.57:

$$\tau_H = C_{gs} R_{gs} + C_{gd} R_{gd} + C_L R_L'$$

$\tau_H = 2^p \times 20^k + 0.2^p \times 1030^k + 1^p \times (20^k \| 20^k) = 256\ ns$

$f_H = \frac{1}{2\pi \tau_H} = 622\ kHz$

$|A_M| \cdot f_H = 31.1\ MHz$

b) For the cascode amplifier:

$A_{01} = g_{m1} r_{01} = 5 \times 20 = 100\ V/V$

$A_{vo2} = 1 + (g_{m2} + g_{mb2}) r_{o2} = 1 + (5 + 0.2 \times 5) \times 20^K$

$A_{vo2} = 121\ V/V$

$R_{out} = r_{o2} + A_{vo2} r_{o1} = 20^K + (121 \times 20^K) = 2.44\ M\Omega$

$A_v = A_{vo} \frac{R_L}{R_L + R_{out}} = -121 \times 100 \times \frac{20}{20 + 2440} = -98.4\ V/V$

Using Eq. 6.137:

$\tau_H = R_{sig}[C_{gs1} + C_{gd1}(1 + g_{m1} R_{d1})] + R_{d1}(C_{gd1} + C_{db1} + C_{gs2}) + (R_L \| R_{out})(C_L + C_{gd2})$

$R_{d1} = r_{o1} \| \left(\frac{1}{g_{m2} + g_{mb2}} + \frac{R_L}{A_{vo2}}\right)$ (Eq. 6.124)

$R_{d1} = 20^K \| \left(\frac{1}{5 + 0.2 \times 5} + \frac{20}{121}\right) = 0.327\ k\Omega$

$\tau_H = 20^K[2 + 0.2(1 + 5 \times 0.327)] + 0.327(0.2^p + 0.2^p + 2) + (20^K \| 2.44^M)(1 + 0.2)$

$\tau_H = 75.1\ ns$

$f_H = \frac{1}{2\pi \tau_H} = 2.12\ MHz$

$|A_v| \cdot f_H = 208.61\ MHz$

6.101

$A_v = 66\,dB = 1995\ V/V$

$A_v = A_{vo} \frac{R_L}{R_L + R_{out}}$ and $R_L = R_{out} \Rightarrow A_v = A_{vo} \times \frac{1}{2}$

$A_{vo} = (1 + g_{m2} r_{o2}) g_{m1} r_{o1} \simeq g_m^2 r_o^2 = \left(\frac{2I_D}{V_{OV}} \times \frac{V_A}{I_D}\right)^2 = \left(\frac{2V_A}{V_{OV}}\right)^2$

$\Rightarrow 1995 = \frac{1}{2} \times \left(\frac{2 \times 10}{V_{OV}}\right)^2 \Rightarrow V_{OV} = 0.317\ V$

$\Rightarrow I_D = \frac{1}{2} k_n' \frac{W}{L} V_{OV}^2 = \frac{1}{2} \times 200 \times 10^{-3} \times 10 \times 0.317^2 = 0.1\ mA$

Since R_{sig} is small:

$\tau_H \simeq (C_L + C_{gd})(R_L \| R_{out})$

$r_o = \frac{V_A}{I_D} = \frac{10}{0.1} = 100\ k\Omega$, $g_m = \frac{2I_D}{V_{OV}} = 0.631\ mA/V$

$R_{out} = A_{vo2} r_{o1} + r_{o2} = (1 + g_m r_o) r_o + r_o$

$R_{out} = 6510\ k\Omega \simeq 6.5\ M\Omega$, $R_L = R_{out}$.

$\tau_H \simeq (1\,pF + 0.1\,pF)\left(\frac{6510^K}{2}\right) = 3580.5\ ns$

$f_H = 44.5\ kHz$

$f_t \simeq |A_v| \cdot f_H = 1995 \times 44.5 = 88.8\ MHz$

If the cascode transistor is removed, then we have a common-source configuration.

Cont.

$A_M = -g_m(r_o \| R_L) = -0.637\,(100^K \| 6510^K)$

$A_M = -62.74$ V/V

$f_H = \dfrac{1}{2\pi(C_L + C_{gd})R_L'} = \dfrac{1}{2\pi(1+0.1)(100^K \| 6510^K)} = 1.47^{MHz}$

$f_H = 1.47$ MHz

$|A_M| \cdot f_H = 92.2$ MHz $\simeq f_T$

Note that the unity-gain stays nearly unchanged. The result is the same as Fig 6.39.

6.102

$R_L = \beta r_o$, $\beta = 100$, $|V_A| = 100^V$, $I = 0.1$ mA

From Fig. 6.40, we can write:

$R_{in} = r_{\pi 1} = \dfrac{\beta}{g_{m1}} = \dfrac{\beta V_T}{I_C} = \dfrac{100 \times 0.025}{0.1} = 25\text{K}\Omega$

$R_{in} = 25\text{ K}\Omega$

$R_{out} \simeq \beta r_{o2}$, $r_{o2} = \dfrac{V_A}{I_C} = \dfrac{100}{0.1} = 1^{M\Omega}$

$R_{out} = 100 \times 1^M = 100\text{ M}\Omega$

$A_{v_o} = -\beta A_{v_{o2}} = -\beta g_{m2} r_{o2} = -100 \times \dfrac{0.1^m}{0.025} \times 1^M \Rightarrow$

$A_{v_o} = -400 \times 10^3$ V/V

$G_m = \dfrac{A_{v_o}}{R_{out}} = \dfrac{-400 \times 10^3}{100 \times 10^6} = 4$ mA/V $\simeq g_m$

$G_m = 4$ mA/V

Since $R_{out} = R_L = \beta r_o$ we have

$\dfrac{v_o}{v_i} = -G_m(R_L \| R_{out}) = -G_m \dfrac{\beta r_o}{2} = -4\,\dfrac{100 \times 1^M}{2}$

$\dfrac{v_o}{v_i} = -200 \times 10^3$ V/V

From Fig. 6.41 the gain of the CE stage is

$A_{CE} = -g_{m1}\left(r_{o1} \| r_{e2}\,\dfrac{r_{o2} + R_L}{r_{o2} + \frac{R_L}{\beta+1}}\right)$

$r_{o1} = r_{o2} = r_o$, $R_L = \beta r_o$

$A_{CE} = -g_{m1}\left(r_o \| r_{e2}\,\dfrac{r_o + \beta r_o}{r_o + \frac{\beta r_o}{r_o + 1}}\right)$

$A_{CE} = -g_{m1}\left(r_o \| r_{e2}\,\dfrac{(\beta+1) r_o}{2 r_o}\right)$

$A_{CE} = -g_{m1}\left(r_o \| \dfrac{r_\pi}{2}\right) = -\dfrac{I}{V_T}\left(1^M \| \dfrac{25^K}{2}\right)$

$A_{CE} = -50$ V/V

6.103

$R_{sig} = 4\text{K}\Omega$, $R_L = 2.4\text{K}\Omega$, $I = 1$ mA , $\beta = 100$, $r_o = 100^{K\Omega}$

Refer to Fig. 6.42:

$A_M = -\dfrac{r_\pi}{r_\pi + r_x + R_{sig}} \times g_m(\beta r_o \| R_L)$

$r_\pi = \dfrac{\beta}{g_m} = \dfrac{100}{1/0.025} = 2.5\text{K}\Omega$ $\quad g_m = \dfrac{I_C}{V_T} = 40$ mA/V

$A_M = -\dfrac{2.5}{2.5 + 0.05 + 4} \times 40 \times (100 \times 100 \| 2.4^K)$

$A_M = -36.6$ V/V

$R_{sig}' = r_\pi \| (r_x + R_{sig}) = 2.5^K \| (0.05 + 4^K)$

$R_{sig}' = 1.55\text{K}\Omega$

$R_{\pi 1} = R_{sig}' = 1.55\text{K}\Omega$

$R_{\mu 1} = R_{sig}'(1 + g_m R_{e1}) + R_{e1}$

$R_{e1} = r_{o1} \| r_{e2}\left(\dfrac{r_o + R_L}{r_o + R_L/\beta+1}\right) = 100^K \| \dfrac{100^K}{101}\left(\dfrac{100 + 2.4}{100 + \frac{2.4}{101}}\right)$

$R_{e1} = 1\text{K}\Omega$

$R_{\mu 1} = 1.55(1 + 40 \times 1) + 1 = 64.55^{K\Omega}$

$R_{out} = \beta r_o = 10\text{M}\Omega$

$\tau_H = C_{\pi 1}R_{\pi 1} + C_{\mu 1}R_{\mu 1} + (C_{cs1} + C_{\pi 2})R_{e1} + (C_L + C_{cs2} + C_{\mu 2})(R_L \| R_{out})$

$\tau_H = 14 \times 1.55 + 2 \times 64.55 + (0 + 14) \times 1 + (0 + 2)(2.4^K \| 10^M)$

$\tau_H = 169.6$ ns

$f_H = 939$ KHz

6.104

a) If we employ Miller's theorem to $C_{\mu 1}$:

$\dfrac{1}{C_{\mu 1}S}\,\dfrac{1}{1-A} = \dfrac{1}{C_\mu S}\,\dfrac{1}{1-(-1)} = \dfrac{1}{2C_\mu S}$

or $2C_\mu$ appears in parallel with C_π. Thus the time constant due to $(C_\pi + 2C_\mu)$ is:

$R_{sig}'(C_\pi + 2C_\mu)$ which results in:

$f_{P1} = \dfrac{1}{2\pi R_{sig}'(C_\pi + 2C_\mu)}$

If we refer to Fig. 6.42, we'll see that the output pole is: $f_{P2} = \dfrac{1}{2\pi(C_L + C_{cs2} + C_{\mu 2})R_L}$

b) $R_{sig} = 1\text{K}\Omega \Rightarrow R_{sig}' = r_\pi \| R_{sig} = \dfrac{100}{1/0.025} \| 1^K$

$\Rightarrow R_{sig}' = 0.714\text{K}\Omega$

$f_{P1} = \dfrac{1}{2\pi \times 0.714^K (5 + 2 \times 1)^P} = 31.85$ MHz

$f_{P2} = \dfrac{1}{2\pi(0 + 0 + 1) \times 10} = 15.9$ MHz

(Assume $R_L = 10$ KΩ)

Cont.

$$f_H = \frac{1}{\sqrt{\left(\frac{1}{f_{P1}}\right)^2 + \left(\frac{1}{f_{P2}}\right)^2}} = 14.2\ \text{MHz}$$

IF $R_{sig} = 10\,k\Omega$:

$R'_{sig} = 2.5^k \parallel 10^k = 2\,k\Omega$

$f_{P1} = \frac{1}{2\pi(5+2)2} = 11.4\ \text{MHz}$

f_{P2} is the same: $f_{P2} = 15.9\ \text{MHz}$

$f_H = 9.26\ \text{MHz}$

6.105

Refer to Fig. 6.43.

$I = \frac{1}{2}\mu_p C_{ox}\frac{W}{L}V_{OV}^2 \Rightarrow 100\mu A = \frac{1}{2}\times 60\,\frac{W}{L}\,0.2^2$

$\Rightarrow \frac{W}{L} = 83.3$

$V_{SG1} = V_{DD} - V_{BIAS1} = 3.3 - V_{BIAS1}$

$V_{OV} = V_{SG1} - |V_{tp}| = 0.2 \Rightarrow 0.2 = 3.3 - V_{BIAS1} - 0.8$

$\Rightarrow V_{BIAS1} = 2.3\,V$

For Maximum swing: $V_{SD1} = V_{OV} \Rightarrow V_{D1} = 3.3 - 0.2 = 3.1^V$

$\Rightarrow V_{D1} = 3.1^V$

then: $V_{SG2} - |V_{tp}| = V_{OV} \Rightarrow 3.1 - V_{BIAS2} - 0.8 = 0.2$

$V_{BIAS2} = 2.1V$

The highest allowable voltage at the output is $V_{DD} - V_{OV} - V_{OV} = 3.3 - 0.2 - 0.2 = 2.9^V$

$R_o \simeq g_{m2}\,r_{o2}\,r_{o1}$ (Eq. 6.141)

$g_m = \frac{I_C}{V_T} = \frac{0.1}{0.025} = 4\,mA/V$ $\qquad r_o = \frac{V_A}{I_C} = \frac{5}{0.1} = 50\,k\Omega$

$R_o = 4 \times 50 \times 50 = 10\,M\Omega$

$R_o = 10\,M\Omega$

6.106

$I_D = 0.2\,mA \Rightarrow g_m = \frac{2I_D}{V_{OV}} = \frac{2\times 0.2}{0.25} = 1.6\,mA/V$

$r_o = \frac{V_A}{I_D} = \frac{5}{0.2} = 25\,k\Omega$

$R_{out} = A_o^2 r_o$, $A_o = g_m r_o = 1.6\times 25 = 40\,V/V$

$R_{out} = 40\times 40\times 25 = 40\,M\Omega$

6.107

$I = 100\mu A$, $V_{BIAS} = 1^V$

a) $I_{C1} = 2I - I_{E2} = 2I - I = I = 100\mu A$

b) $V_X = V_{BIAS} + V_{BE} = 1 + 0.7 = 1.7^V$

c) $g_{m1} = \frac{I}{V_T} = \frac{0.1}{0.025} = 4\,mA/V = g_{m2}$

$r_{o1} = r_{o2} = \frac{V_A}{I_C} = \frac{100}{0.1} = 1\,M\Omega$

d) $V_{Omax} = V_X - V_{CEsat}$

$V_{Omax} = 1.7 - 0.2 = 1.5V$

e) $R_i = r_\pi = \frac{\beta}{g_m} = \frac{100}{4} = 25\,k\Omega$

f) Since the output resistance of the current source is equal to R_o, then the overall output resistance $R_{out} = \frac{R_o}{2}$.

(a)

$R_o = \beta r_o = 100 \times 1^M = 100\,M\Omega \Rightarrow R_{out} = 50\,M\Omega$

g) $A_M \simeq g_m\,(\beta r_o \parallel R_L)$ where $R_L = \beta r_o$

$A_M = g_m\,\frac{\beta r_o}{2} = 4\times 100\times \frac{1^M}{2} = 200\times 10^6\,V/V$

The curren source 2I should ideally have infinite output resistance.

a) $I_1 = 2I - I = I = 100\mu A$

b) $V_X = V_{BIAS} + V_{SG}$

$I = \frac{1}{2}k'_p\frac{W}{L}V_{OV}^2$

$0.1 = \frac{1}{2}\times 2\times V_{OV}^2 \Rightarrow V_{OV} = 0.316^V$

$V_{SG} = |V_t| + V_{OV} = 0.6 + 0.316$

$V_{SG} = 0.916V \Rightarrow V_X = 1^V + 0.916 = 1.916V$

c) $g_{m1} = \frac{0.1}{0.025} = 4\,mA/V$

$g_{m2} = \frac{2I_D}{V_{OV}} = \frac{2\times 0.1}{0.316} = 0.633\,mA/V$

$r_{o1} = \frac{100}{0.1} = 1\,M\Omega$, $r_{o2} = \frac{5}{0.1} = 50\,k\Omega$

d) $V_{Omax} = V_X - V_{SDmin} = V_X - V_{OV} = 1.916 - 0.316 = 1.6^V$

e) $R_i = r_{\pi 1} = \frac{\beta}{g_{m1}} = \frac{100}{4} = 25\,k\Omega$

f) $R_{out} = R_{out1} + [1 + (g_m + g_{mb})R_{out1}]\,r_{o2}$

Cont.

$R_{out1} = r_{o1}$

$R_{out} = r_{o2} + [1+g_{m2}r_{o2}]\, r_{o1} = 50^K + 0.633^m \times 1^M \times 50^K$

$R_{out} = 31.7\ M\Omega$

f) First note that R_{L2} or the output resistance of the current source is equal to R_{out} or $32.6\ M\Omega$

$R_{L2} = 32.6 M\Omega$ or $R_{L2} \simeq g_{m2} r_{o1} r_{o2}$

$\frac{v_x}{v_i} = -g_{m1}\,(r_{o1} \| R_{in2})$

$R_{in2} = \frac{1}{g_{m2}} + \frac{R_{L2}}{A_{o2}} = \frac{1}{g_{m2}} + \frac{g_{m2} r_{o1} r_{o2}}{g_{m2} r_{o2}} = \frac{1}{g_{m2}} + r_{o1}$

$\frac{v_x}{v_i} = -g_{m1}\,(r_{o1} \| (r_{o1} + \frac{1}{g_{m2}})) \simeq -g_{m1} \times \frac{r_{o1}}{2} = -2000 V/V$

$\frac{v_o}{v_x} = (1+g_{m2} r_{o2}) \frac{R_{L2}}{R_{L2} + R_{out}} = (1+0.633\times 50)\frac{1}{2} = 16.3 V/V$

$\frac{v_o}{v_i} = -2000 \times 16.3 = -32600 V/V$

In order for the output resistance of the current-source to reduce the gain by 1%:

$\frac{v_x}{v_i} = -g_{m1}\,(r_{o1} \| (r_{o1} + \frac{1}{g_{m2}}) \| R) \simeq -g_{m1}\,(\frac{r_{o1}}{2} \| R)$

For $\frac{v_x}{v_i} = -\frac{99}{100} \times 2000 = 1980$ we should have:

$\frac{r_{o1}}{2} \| R = \frac{1980}{4} = 495^K \Rightarrow 500^K \| R = 495^K \Rightarrow$

$R = 49.5 M\Omega$ (This is the output resistance of the 2I current-source)

(Note that $\frac{v_o}{v_x}$ did not depend on R.)

a) $I_1 = 2I - I = 100\ \mu A$

b) $V_X = V_{BIAS} + V_{SG}$

$I = \frac{1}{2} k'_p \frac{W}{L} V_{OV}^2 \Rightarrow V_{OV} = 0.316^V$

$V_{SG} = 0.6 + 0.316 = 0.916^V$

$V_X = 1 + 0.916 = 1.916V$

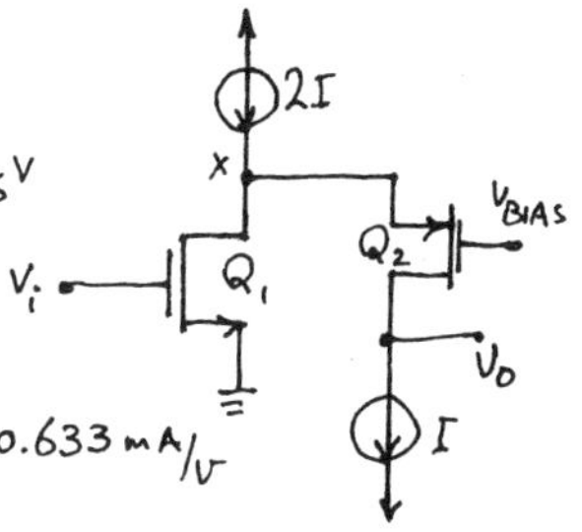

c) $g_{m1} = g_{m2} = \frac{2I_D}{V_{OV}} = \frac{0.1\times 2}{0.316} = 0.633\ mA/V$

$r_{o1} = r_{o2} = \frac{5}{0.1} = 50 K\Omega$

d) $v_{omax} = V_X - V_{DSMin} = V_X - V_{OV} = 1.916 - 0.316 = 1.6^V$

e) $R_{in} = \infty$

f) $R_{out} = r_{o2} + (1+g_{m2} r_{o2}) r_{o1} = 50 + (1+0.633\times 50)\times 50$

$R_{out} = 1682.5K = 1.68 M\Omega = R_{L2} \simeq g_{m2} r_{o2} r_{o1}$

g) $\frac{v_x}{v_i} = -g_{m1}(r_{o1} \| R_{in2})$

$R_{in2} = \frac{1}{g_{m2}} + \frac{R_{L2}}{A_{o2}} \simeq \frac{1}{g_{m2}} + \frac{g_{m2} r_{o2} r_{o1}}{g_{m2} r_{o2}} \simeq \frac{1}{g_{m2}} + r_{o1}$

Note that R_{L2} is the output resistance of the current source I:

$\frac{v_x}{v_i} = -g_{m1}(r_{o1} \| (\frac{1}{g_{m2}} + r_{o1})) = -0.633 \times (50^K \| (\frac{1}{0.633} + 50^K))$

$\frac{v_x}{v_i} = -16.07\ V/V$

$\frac{v_o}{v_x} = (1+g_{m2} r_{o2}) \frac{R_{L2}}{R_{L2} + R_{out}} = (1+0.633\times 50)\times \frac{1}{2}$

$\frac{v_o}{v_x} = 16.32\ V/V$

$\frac{v_o}{v_i} = -262.34\ V/V$

In order to reduce the gain by 1% by introducing non-ideal current-source 2I with output resistance R:

$\frac{v_x}{v_i} = -g_{m1}\,(r_{o1} \| R \| (\frac{1}{g_{m2}} + r_{o1})) = \frac{90}{100} \times 16.07 \simeq 14.46$

$\Rightarrow (50K \| R \| (51.58^K)) = \frac{14.46}{0.633} \Rightarrow R = 227.5^K$

a) $I_1 = 2I - I = I = 0.1 mA$

b) $V_X = V_{BIAS} + 0.7 = 1.7^V$

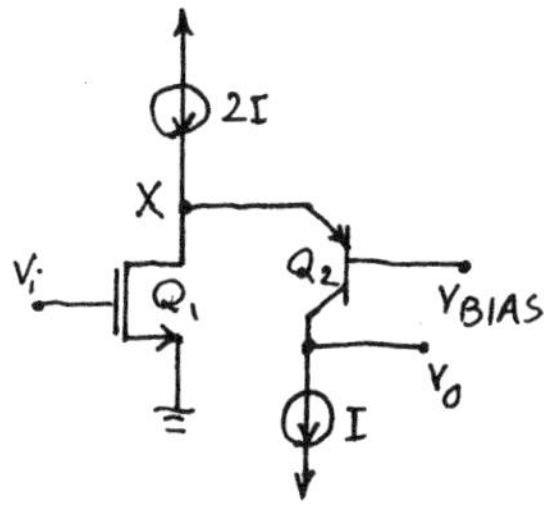

c) $g_{m1} = \frac{2I_D}{V_{OV}}$

$V_{OV}^2 = \frac{0.1}{2\times\frac{1}{2}} \Rightarrow V_{OV} = 0.316 V$

$g_{m1} = 0.633\ mA/V$

$r_{o1} = \frac{5}{0.1} = 50 K\Omega$

$g_{m2} = \frac{0.1}{0.025} = 4\ mA/V$, $r_{o2} = \frac{V_A}{I_C} = \frac{100}{0.1} = 1 M\Omega$

d) $V_{omax} = V_X - V_{CEsat} = V_X - 0.2 = 1.5 V$

e) $R_{in} = \infty$

f) $R_{out} = r_{o2}(1 + g_{m2}(r_{o1} \| r_\pi))$

$R_{out} = 1^M(1 + 4(50^K \| \frac{100^K}{4})) = 67.67\ M\Omega$

$R_{L2} = R_{out}$

g) $\frac{v_x}{v_i} = -g_{m1}(r_{o1} \| R_{in2}) = -g_{m1}(r_{o1} \| \frac{r_{o2} + R_{L2}}{r_{o2} + R_{L2}/\beta+1} r_{e2})$

$r_{e2} = \frac{r_{\pi 2}}{\beta+1} = \frac{\beta}{\beta+1} g_{m2} = 3.96 K\Omega$

$\frac{v_x}{v_i} = -4(50^K \| \frac{1^M + 67.67^M}{1 + \frac{67.67}{101}} \times 3.96^k) = -152.86 V/V$

$\frac{v_o}{v_x} = (1+g_{m2} r_{o2}) \frac{R_{L2}}{R_{L2} + R_{out}} = (1+g_{m2} r_{o2})\frac{1}{2}$

$\frac{v_o}{v_x} = (1 + 4\times 1000)\frac{1}{2} = 2000.5\ V/V$

$\frac{v_o}{v_i} = -152.86 \times 2000.5 = -305.8\times 10^3 V/V$

To redue the gain by 1%, the effect of R in is considered: $0.99 \frac{v_x}{v_i} = 0.99 \times 152.86$

Cont.

$0.99 \times 152.86 = 4 \times (50^K \| \frac{1+67.67}{1+\frac{67.67}{100}} \| R)$

$\Rightarrow R = 3.74 M\Omega$

6.108

$g_m = 2 mA/V \Rightarrow g_{mb} = \chi g_m = 0.2 \times 2 = 0.4 mA/V$

$R_{out} = r_o[1+(g_m+g_{mb})R_s] = 50^K[1+(2+0.4)\times 0.5^K]$

$R_{out} = 110 K\Omega$

$A_{vo} = -g_m r_o = -2\times 50 = -100 V/V$

$A_v = -A_{vo}\frac{R_L}{R_L+R_{out}} = -100\times\frac{50}{50+110} = -31.25 V/V$

Using Eq. 6.144:

$G_m = \frac{g_m}{1+(g_m+g_{mb})R_s} = \frac{2}{1+(2+0.2\times 2)0.5}$

$\Rightarrow G_m = 0.91 mA/V$

$\frac{v_{gs}}{v_i} = \frac{1}{1+(g_m+g_{mb})R_s}\frac{R_L\|R_{out}}{R_L\|r_o}$ (Eq. 6.144)

$\frac{v_{gs}}{v_i} = 0.625V \Rightarrow V_{gs} = 0.625 v_i$

6.109

$\frac{v_{gs}}{v_i} = \frac{1}{3}$, using Eq. 6.144:

$\frac{v_{gs}}{v_i} = \frac{1}{1+(g_m+g_{mb})R_s}\frac{R_L\|R_{out}}{R_L\|r_o}$

$\Rightarrow \frac{1}{3} = \frac{1}{1+(2+0.2\times 2)R_s}\times\frac{50^K\|R_{out}}{50^K\|50K}$ (1)

$R_{out} = r_o[1+(g_m+g_{mb})R_s] = 50(1+2.4R_s)$

$R_{out} = 50(1+2.4\times R_s)$

substitute in (1):

$\frac{1}{3} = \frac{1}{1+2.4R_s}\frac{50K\times 50(1+2.4R_s)}{50+50(1+2.4R_s)}\times\frac{1}{25}$

$\Rightarrow R_s = 1.67 K\Omega$

$\Rightarrow R_{out} = 50(1+2.4\times 1.67) = 250.4 K\Omega$

$A_v = -A_{vo}\frac{R_L}{R_L+R_{out}} = -g_m r_o\frac{R_L}{R_L+R_{out}}$

$A_v = -2\times 50\times\frac{50}{50+250.4} = -16.64 V/V$

6.110

a) $A_M = -A_{vo}\frac{R_L}{R_L+R_{out}} = -g_m r_o\frac{R_L}{R_L+r_o}$

$A_M = -5\times 40\times\frac{40}{40+40} = -100 V/V$

$R_L' = R_L\|R_{out} = R_L\|r_o = 20K\Omega$

$R_{gd} = R_{sig}(1+G_m R_L')$ where $G_m = g_m$ (Eq. 6.148)

$\Rightarrow R_{gd} = 20^K(1+5\times 20) = 2020 K\Omega = 2.02 M\Omega$

$R_{gd} = 2.02 M\Omega$

$R_s = 0 \Rightarrow R_{gs} = R_{sig} = 20K\Omega$

$R_{C_L} = R_L' = 20K\Omega$

$\tau_H = C_{gs}R_{gs} + C_{gd}R_{gd} + C_L R_{C_L}$

$\tau_H = 2^p\times 20^K + 0.1^p\times 2.02^M + 1\times 20^K = 262 ns$

$f_H = \frac{1}{2\pi\tau_H} = 607.8 KHz$

$|A_M|\cdot f_H = +100\times 607.8 = 60.78\times 10^3 KHz = 60.78 MHz$

b) $R_s = 500\Omega$

$R_{out} = r_o[1+(g_m+g_{mb})R_s] = 40[1+(5+1)0.5] = 160K\Omega$

$A_M = -g_m r_o\frac{R_L}{R_L+R_{out}} = -5\times 40\times\frac{40}{40+160} = -40 V/V$

$R_L' = R_L\|R_{out} = 40K\|160K = 32K\Omega$

$R_{gd} = R_{sig}(1+G_m R_L')$

$G_m = \frac{g_m r_o}{r_o[1+(g_m+g_{mb})R_s]}$ (Eq. 6.144)

$G_m = \frac{5\times 40}{40[1+(5+1)0.5]} = 1.25 mA/V$

$R_{gd} = 20K(1+1.25^m\times 32^K) = 820 K\Omega$

$R_{gs} = \frac{R_{sig}+R_s}{1+(g_m+g_{mb})R_s\frac{r_o}{r_o+R_L}}$ (Eq. 6.151)

$R_{gs} = \frac{20K+0.5^K}{1+(5+1)0.5\frac{40}{40+40}} = 8.2K\Omega$

$R_{C_L} = R_L\|R_{out} = R_L' = 32K\Omega$

$\tau_H = C_{gs}R_{gs} + C_{gd}R_{gd} + C_L R_{C_L} = 2\times 8.2 + 0.1\times 820 + 32\times 1$

$\tau_H = 130.4 ns$

$f_H = \frac{1}{2\pi\tau_H} = 1.22 MHz$

$|A_M|\cdot f_H = 48.8 MHz$

6.111

Using Eq. 6.151: $R_{gs} = \frac{R_{sig}+R_s}{1+\underbrace{(g_m+g_{mb})R_s}_{K}\frac{r_o}{r_o+R_L}}$

IF we define:

$K = (g_m+g_{mb})R_s$ and $R_{sig} \gg R_s$

then: $R_{gs} \cong \frac{R_{sig}}{1+K\frac{r_o}{r_o+r_o}} = \frac{R_{sig}}{1+K/2}$

Using Eq. 6.144: $G_m = \frac{g_m}{1+(g_m+g_{mb})R_s} = g_m/(K+1)$

Cont.

$R_L' = R_L \| R_{out}$

$R_{out} = r_o[1+(g_m+g_{mb})R_S] = r_o(1+K)$

$R_L' = r_o \| r_o(1+K) = r_o \dfrac{(1+K)}{2+K}$

Using Eq. 6.148:

$R_{gd} = R_{sig}(1+G_m R_L') + R_L'$

$R_{gd} = R_{sig}\left(1+\dfrac{g_m}{1+K} \times r_o \dfrac{1+K}{2+K}\right) + r_o \dfrac{1+K}{2+K}$

$R_{gd} = R_{sig}\left(1+\dfrac{A_o}{2+K}\right) + r_o \dfrac{1+K}{2+K}$

$R_{C_L} = R_L' = r_o \dfrac{1+K}{2+K}$

$\tau_H = R_{gs}C_{gs} + R_{gd}C_{gd} + R_{C_L}C_L$

$\tau_H = \dfrac{R_{sig}}{1+K/2} C_{gs} + R_{sig}\left(1+\dfrac{A_o}{2+K}\right)C_{gd} + (C_L+C_{gd}) r_o \dfrac{(1+K)}{2+K}$

6.112

$R_{out} = r_o[1+(g_m+g_{mb})R_S] = r_o(1+K)$

$R_{out} = 40(1+K)$

$A_M = -g_m r_o \dfrac{R_L}{R_L+R_{out}} = -5\times 40 \times \dfrac{40}{40+40(1+K)}$

$A_M = -\dfrac{200}{2+K}$

$\tau_H = \dfrac{C_{gs}R_{sig}}{1+K/2} + C_{gd}R_{sig}\left(1+\dfrac{A_o}{2+K}\right) + (C_L+C_{gd})r_o \dfrac{1+K}{2+K}$

(From problem 6.111)

$\tau_H = \dfrac{2^p\, 20^k}{1+K/2} + 0.1^p \times 20^k\left(1+\dfrac{5\times 40}{2+K}\right) + (1^p+0.1^p)40^k \dfrac{1+K}{2+K}$

$\tau_H = \dfrac{80}{2+K} + 2\left(1+\dfrac{200}{2+K}\right) + 44\dfrac{1+K}{2+K}$

$\tau_H = \dfrac{528+46K}{2+K}$ ns

$f_H = \dfrac{1}{2\pi\tau_H} = \dfrac{(2+K)\times 10^3}{2\pi(528+46K)}$ MHz

$f_T \simeq |A_M| \cdot f_H$.

K	A_M (V/V)	f_H (MHz)	$\lvert A_M\rvert \cdot f_H$ (MHz)
0	−100	0.603	60.3
1	−66.67	0.832	55.47
2	−50.00	1.027	51.35
3	−40.00	1.195	47.8
4	−33.33	1.342	44.73
5	−28.57	1.471	42.03
6	−25.00	1.584	39.6
7	−22.22	1.686	37.46
8	−20.00	1.777	35.54
9	−18.18	1.859	33.8
10	−16.67	1.934	32.24
11	−15.38	2.002	30.79
12	−14.28	2.064	29.47
13	−13.33	2.121	28.27
14	−12.5	2.174	26.75
15	−11.76	2.223	26.14

IF $f_H = 2$ MHz, then by looking at the table, $K \simeq 11$. Therefore: $K = 11 = (g_m+g_{mb})R_S \Rightarrow$

$R_S = \dfrac{11}{5+1} = 1.83\,k\Omega$

From the table: $A_M = -15.38$

6.113

a) Eq. 6.156: Gain Bandwidth product $= |A_M| \cdot f_H$

$|A_M| \cdot f_H = \dfrac{1}{2\pi C_{gd} R_{sig}}$

For $C_{gd} = 0.1^{pF}$, $R_{sig} = 10\,K\Omega$:

$|A_M| \cdot f_H = 159.2$ MHz

b) IF $|A_M| = 20$ V/V, then $f_T = 159.2$ MHz,

$f_H = \dfrac{159.2}{20} = 7.96$ MHz

c) $g_m = 5$ mA/V $\quad \chi = 0.2 \Rightarrow g_{mb} = 1$ mA/V

$A_o = 100$ V/V $\quad R_L = 20\,K\Omega$

$r_o = \dfrac{A_o}{g_m} = \dfrac{100}{5} = 20\,K\Omega = R_L$

$A_v = -A_o \dfrac{R_L}{R_L+R_{out}} \Rightarrow 20 = -100 \dfrac{20}{R_{out}+20}$

$\Rightarrow R_{out} = 80\,K\Omega$

$R_{out} = r_o[1+(g_m+g_{mb})R_S] = 20[1+(5+1)R_S] = 80$

$\Rightarrow 1+6R_S = 4 \Rightarrow R_S = 0.5\,K\Omega = 500\,\Omega$

$R_S = 500\,\Omega$

6.114

$R_e = 100\,\Omega$, $I = 0.5$ mA, $\beta = 100$, $V_A = 100$ V, $R_L = r_o$

$R_{in} = (\beta+1)r_e + (\beta+1)R_e \dfrac{1}{1+R_L/r_o}$ when $\dfrac{R_L}{\beta+1} \ll r_o$

(Eq. 6.158)

$r_\pi = \dfrac{\beta}{g_m} = \dfrac{100}{0.5/0.025} = 5\,K\Omega$

$g_m = \dfrac{I}{V_T} = 20$ mA/V

$r_o = \dfrac{V_A}{I_C} = \dfrac{100}{0.5} = 200\,K\Omega$

$r_e = \dfrac{r_\pi}{(\beta+1)} = \dfrac{5^K}{101} = 49.5\,\Omega$

Cont.

$R_{in} = (1+100)0.0495 + (1+100)\times 0.1 \frac{1}{1+1}$

$R_{in} = 10.05\,K\Omega$

$R_o \simeq r_o(1+g_m R_e')$ where $R_e' = R_e \| r_\pi$ (Eq. 6.160)

$R_o \simeq 200(1+20\times(0.1^K \| 5K)) = 592.2K\Omega$

$R_o = 592.2K\Omega$

$A_{v_o} \simeq -g_m r_o = -20\times 200 = -4000\ V/V$

$G_m = -\frac{A_{v_o}}{R_o} = 6.75\ mA/V$

$A_v = A_{v_o}\frac{R_L}{R_L+R_o} = -4000\times\frac{200}{200+592.2} = -1009.8\ V/V$

$A_v = -1009.8\ V/V$

$G_v = \frac{V_o}{V_{sig}} = \frac{R_{in}}{R_{in}+R_{sig}} A_{v_o}\frac{R_L}{R_L+R_o} = \frac{R_{in}}{R_{in}+R_{sig}} A_v$

$G_v = \frac{10.05}{10.05+10}\times(-1009.8) = -506.2\ V/V$

6.115

$g_m = \frac{I}{V_T} = 20\ mA/V$ Assume $R_{sig}=10k\Omega$

$r_\pi = \frac{\beta}{g_m} = 5K\Omega$

$r_e = \frac{5K}{100+1} = 49.5\Omega$

$r_o = \frac{V_A}{I_C} = \frac{100}{0.5} = 200K\Omega$

$R_{in} = (\beta+1)r_e + (\beta+1)R_e\frac{1}{1+R_L/r_o}$ (Eq. 6.158)

$R_{in} = 101\times 49.5 + 101\times 100\times\frac{1}{2} = 10049.5\Omega = 10.05\ K\Omega$

$R_{in}=10.05K\Omega$

$R_i = (\beta+1)(r_e+R_e) = 101\times(49.5+100) = 15.1K\Omega$

$R_i = 15.1K\Omega$

$A_{v_o} = -g_m r_o = -20\times200 = -4000\ V/V$

$G_{vo} = \frac{R_i}{R_i+R_{sig}}A_{v_o} = \frac{-15.1}{15.1+10}\times 4000 = -2406.4\ V/V$

$R_o = r_o(1+g_m R_e')$

$R_e' = R_e\|r_\pi = 0.1^K\|5^K = 0.1K\Omega$

$R_o = 200(1+20\times0.1) = 600k\Omega$

$G_v = \frac{R_{in}}{R_{in}+R_{sig}} A_v = \frac{-10.05}{10.05+10}\times 4000\times\frac{200}{200+600}$

(Note that $A_v = A_{v_o}\frac{R_L}{R_L+R_o}$)

$G_v = -501.25\ V/V$

$G_v = G_{vo}\frac{R_L}{R_L+R_{out}} \Rightarrow 501.25 = 2406.4\frac{200}{200+R_{out}}$

$R_{out} = 760.16K\Omega$

6.116

a) Eq. 6.148: $R_{gd} = R_{sig}(1+G_m R_L') + R_L'$

IF we try to adapt the above formula for BJTs:

Using Miller's Theorem for C_μ, we would have $C_\mu(1+G_m R_L')$ hanging from the base to the ground.

The resistance seen by $C_\mu(1+g_m R_L')$ is: $R_{sig}' = (R_{sig}+r_x)\|R_{in}$

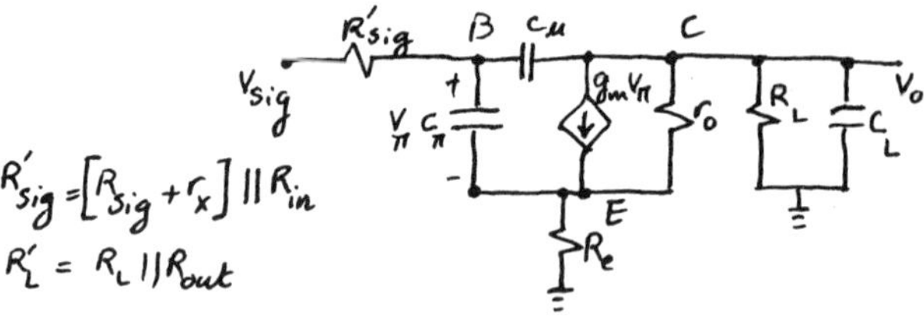

$R_{sig}' = [R_{sig}+r_x]\|R_{in}$

$R_L' = R_L\|R_{out}$

From the collector side, the resistance seen by C_μ is $R_{out}\|R_L$ or R_L'. There fore the total resistance seen by C_μ, R_μ is:

$R_\mu = [(R_{sig}+r_x)\|R_{in}](1+G_m R_L') + R_L'$

The resistance seen by C_L is again $R_L\|R_{out}$ or R_L'. $\Rightarrow R_{C_L} = R_L'$ (Same as Eq.6.150 for MOS)

Now the resistance seen by C_π is r_π parallel to

then we can write: $\tau_H = C_\pi R_\pi + C_\mu R_\mu + C_L R_{C_L}$

b) i) $R_e=0$, $R_{out}\simeq R_o$, $r_\pi = \frac{\beta}{g_m} = 5K\Omega$

$A_M = -\frac{r_\pi}{r_\pi+R_{sig}+r_x}\times g_m R_L'$ Eq. (6.70)

$R_L' = R_L\|R_{out}$, Since $R_{out}\simeq R_o$ the $R_L' = R_L\|R_o$

$R_o = r_o(1+g_m R_e') = r_o(1+g_m r_\pi) = (\beta+1)r_o = 10.1M\Omega$

$R_L' = 5.3^K\|10^M = 5.3^K \Rightarrow A_M = \frac{-5}{5+1+0.2}\times 20\times 5.3 = -85.5\ V/V$

Now using Formulas given in part(a) we can find $f_H = \frac{1}{2\pi\tau_H}$. Since $R_{out}\simeq R_o$, then $R_L' = 5.3K\Omega = R_{C_L}$

Note that For $R_e=0$, we have $R_{in}=r_\pi=5K\Omega$ and $G_m=g_m=20\ mA/V$

Cont.

$R_\mu = ((R_{sig}+r_x)\,||\,r_\pi)(1+g_m R_L') + R_L'$

$R_\mu = ((1+0.2)\,||\,5)(1+20\times 5.3) + 5.3 = 108.85\,K\Omega$

$R_\pi = r_\pi \,||\, (R_{sig}+r_x) = 5^K \,||\, 1.2^K = 0.97\,K\Omega$

$\tau_H = R_\pi C_\pi + R_\mu C_\mu + R_{C_L} C_L = 10\times 0.97 + 0.5\times 108.85 + 2\times 5.3$

$\tau_H = 74.73\,ns \Rightarrow f_H = 2.13\,MHz$, $A_M = -85.5\,V/V$

ii) $R_e = 200\,\Omega$.

$A_{v_o} \cong -g_m r_o$ (Eq. 6.159)

$A_{v_o} = -20\times 100 = -2000\,V/V$

$R_{in} = (\beta+1) r_e + (\beta+1) R_e \dfrac{1}{1+R_L/r_o}$

$R_{in} = 5^K + 101\times 0.2\times \dfrac{1}{1+\frac{5.3}{100}} = 24.18\,K\Omega$

$R_o \cong (1+g_m R_e') r_o$ $\quad R_e' = R_e \,||\, r_\pi$

$R_o \cong (1+20(5^K \,||\, 0.2K))\times 100K = 484.6\,K\Omega$

$G_v = \dfrac{R_{in}}{R_{in}+R_{sig}} A_{vo} \dfrac{R_L}{R_L+R_o} = \dfrac{24.18}{24.18+1}\, 2000\, \dfrac{5.3}{5.3+484.6}$

$G_v = -20.78\,V/V = A_M$

Now to calculate f_H:

$R_L' = R_L \,||\, R_{out}$, $R_{out} \simeq R_o = 484.6\,K\Omega$

$R_L' = 5.3 \,||\, 484.6 = 5.24\,K\Omega = R_{C_L}$

$G_m = \dfrac{g_m}{1+g_m R_e} = \dfrac{20}{1+20\times 0.2} = 4\,mA/V$

$R_\mu = ((R_{sig}+r_x)\,||\,R_{in})(1+G_m R_L') + R_L'$

$R_\mu = (1.2^K \,||\, 24.18^K)(1+4\times 5.24) + 5.24 = 30.35\,K\Omega$

$R_\mu = 30.35\,K\Omega$

$R_\pi = r_\pi \,||\, \dfrac{R_{sig}+r_x+R_e}{1+g_m R_e\left(\frac{r_o}{r_o+R_L}\right)} = 5^K \,||\, \dfrac{1+0.2+0.2}{1+20\times 0.2\,\frac{100}{105.3}}$

$R_\pi = 0.276\,K\Omega$

$\tau_H = 0.276\times 10 + 30.35\times 0.5 + 5.24\times 2 = 28.42\,ns$

$f_H = 5.6\,MHz$, $A_M = -20.78\,V/V$

6.117

a) $I = \frac{1}{2} k_n' \frac{W}{L} V_{ov}^2 = \frac{1}{2}\times 160^\mu \times 100\times (0.5)^2 = 2\,mA$

b) $g_m = \dfrac{2I_D}{V_{ov}} = \dfrac{2\times 2}{0.5} = 8\,mA/V$

$g_{mb} = \chi g_m = 0.2\times 8 = 1.6\,mA/V$

$r_o \simeq \dfrac{1}{\lambda I_D} = \dfrac{1}{0.05\times 2} = 10\,K\Omega$

c) Using Eq. 6.167: $A_{v_o} = \dfrac{g_m r_o}{1+(g_m+g_{mb}) r_o}$

$A_{v_o} = \dfrac{8\times 10}{1+(8+1.6)\times 10} = 0.82\,V/V$

If we use the approximation formula

Eq. 6.168: $A_{v_o} = \dfrac{1}{1+\chi} = 0.83\,V/V$

$R_o = \dfrac{1}{g_m+g_{mb}} \,||\, r_o = \dfrac{1}{8+1.6}^K \,||\, 10^K = 103\,\Omega$

d) with $R_L = 1K\Omega$

$A_v = \dfrac{g_m R_L'}{1+g_m R_L'}$ (Eq. 6.166)

$R_L' = R_L \,||\, r_o \,||\, \dfrac{1}{g_{mb}} = 1^K \,||\, 10^K \,||\, \dfrac{1}{1.6}^K = 370\,\Omega$

$A_v = \dfrac{8^m \times 370\times 10^{-3}}{1+8\times 0.370} = 0.75\,V/V$

6.118

$R_o = r_o \,||\, \dfrac{1}{g_m+g_{mb}} = 20^K + \dfrac{1}{5+1} = 20.17\,K\Omega$

$A_M = \dfrac{v_o}{v_{sig}} = G_v = A_{v_o} \dfrac{R_L}{R_L+R_o} = \dfrac{g_m r_o}{1+(g_m+g_{mb}) r_o} \dfrac{R_L}{R_L+R_o}$

$A_M = \dfrac{5\times 20}{1+(5+1)20} \times \dfrac{20^K}{20^K+20.17^K} = 0.41\,V/V$

$f_Z = \dfrac{g_m}{2\pi C_{gs}}$ (Eq. 6.172) $\Rightarrow f_Z = \dfrac{5}{2\pi\times 2} = 398\,MHz$

$R_{gd} = R_{sig} = 20\,K\Omega$

$R_{gs} = \dfrac{R_{sig}+R_L'}{1+g_m R_L'}$, $R_L' = R_L \,||\, r_o \,||\, \dfrac{1}{g_{mb}} = 20^K \,||\, 20^K \,||\, \dfrac{1}{1}^K$

(Eq. 6.175) $\quad R_L' = 0.91\,K\Omega$

$R_{gs} = \dfrac{20+0.91}{1+5\times 0.91} = 3.77\,K\Omega$

$R_{C_L} = R_L \,||\, R_o = 20K \,||\, 20.17 = 10.04\,K\Omega$

$\tau_H = C_{gd} R_{gd} + C_{gs} R_{gs} + C_L R_{C_L}$

$\tau_H = 0.1\times 20^K + 2\times 3.77 + 1\times 10.04 = 19.58\,ns$

$f_H = \dfrac{1}{2\pi \tau_H} = 8.13\,MHz$

Contribution of each capacitor time constant to the overall τ_H is:

10.2% for C_{gd} , 38.5% for C_{gs} , 51.3% for C_L.

6.119

$R_{gd} = R_{sig}$, $R_{gs} = \dfrac{R_{sig}+R_L'}{1+g_m R_L'} \cong \dfrac{R_{sig}}{1+g_m R_L'}$ if $R_{sig} \gg R_L'$

$\tau_H = R_{gd} C_{gd} + R_{gs} C_{gs} + C_{CL} R_{CL}$

$\tau_H = R_{sig}\left(C_{gd} + \dfrac{C_{gs}}{1+g_m R_L'}\right) \Rightarrow f_H = \dfrac{1}{2\pi R_{sig}\left(C_{gd}+\frac{C_{gs}}{1+g_m R_L'}\right)}$

$R_{C_L} C_{CL}$ is ignored and $R_L' = R_L \,||\, r_o \,||\, \dfrac{1}{g_{mb}}$

$f_H \propto (1+g_m R_L')$ Therefore f_H can be increased

Cont.

by increasing $g_m R'_L$. We also know that $R'_L = R_L \| r_o \| \frac{1}{g_{mb}}$ and hence $R'_L \ll \frac{1}{g_{mb}}$ or $g_m R'_L \ll \frac{g_m}{g_{mb}} = \frac{1}{\chi} \Rightarrow g_m R'_L \ll \frac{1}{\chi}$. Therefore maximum f_H can be achieved when we have $g_m R'_L = \frac{1}{\chi}$:

$$f_H \cong \frac{1}{2\pi R_{sig}\left(C_{gd} + \frac{C_{gs}}{1+\frac{1}{\chi}}\right)} = \frac{1}{2\pi R_{sig}\left(C_{gd} + \frac{\chi C_{gs}}{\chi+1}\right)}$$

For the source follower specified in Problem 6.118: $f_{H_{max}} = \frac{1}{2\pi \times 20^K\left(0.1 + \frac{0.2\times 2}{1+0.2}\right)} = 18.37\text{ MHZ}$

$\chi = \frac{g_{mb}}{g_m} = 0.2$

6.120

Using Eq. 6.178: $f_Z = \frac{1}{2\pi C_\pi r_e}$

$g_m = \frac{I_c}{V_T} = \frac{5}{0.025} = 200\text{mA/V}$

$r_\pi = \frac{\beta}{g_m} = \frac{100}{200} = 0.5\text{K}\Omega$

$r_e = \frac{r_\pi}{\beta+1} = \frac{0.5\text{K}}{100+1} = 4.95\Omega$

$f_T = \frac{g_m}{2\pi(C_\pi+C_\mu)} \Rightarrow (C_\pi+C_\mu) = \frac{200^m}{2\pi\times 800^M} = 39.8\text{ pF}$

$C_\pi = 39.8 - 2 = 37.8\text{pF}$

$f_Z = \frac{1}{2\pi\times 37.8\times 4.95\times 10^{-12}} = 851\text{ MHZ}$

From Tabe 5.6: $A_M = \frac{r_o \| R_L}{\frac{R_{sig}+r_x+r_\pi}{\beta+1} + r_o\|R_L}$

$A_M = \frac{20^K \| 1^K}{\frac{10+0.2+0.5}{101} + (20^K\|1^K)}$

$A_M = \frac{0.95}{0.11+0.95} = 0.9\text{ V/V}$

$R_\mu = R'_{sig} \| (r_\pi + (\beta+1)R'_L)$ (Eq. 6.179)

$R'_{sig} = R_{sig} + r_x = 10 + 0.2 = 10.2\text{K}\Omega$

$R'_L = R_L \| r_o = 1^K \| 20^K = 0.95\text{K}\Omega$

$R_\mu = 10.2\text{K} \| (0.5 + 101\times 0.95) = 9.22\text{K}\Omega$

$R_\pi = \frac{R'_{sig}+R'_L}{1+\frac{R'_{sig}}{r_\pi}+\frac{R'_L}{r_e}} = \frac{10.2\text{K}+0.95^K}{1+\frac{10.2}{0.5}+\frac{0.95}{0.5}\times 101} = 52.3\Omega$

$f_H = \frac{1}{2\pi(C_\mu R_\mu + C_\pi R_\pi)} = \frac{1}{2\pi(2\times 9.22 + 37.8\times 0.0523)}$

$f_H = 7.8\text{MHZ}$

6.121

$g_m = \frac{I_c}{V_T} = \frac{1}{0.025} = 40\text{mA/V}$, $r_e = \frac{\beta}{(\beta+1)g_m} = 25\Omega$

$f_T = \frac{g_m}{2\pi(C_\pi+C_\mu)} = 2\text{GHZ} \Rightarrow C_\pi + C_\mu = 3.18\text{pF}$

$C_\mu = 0.1\text{pF} \Rightarrow C_\pi = 3.08\text{pF}$

$r_\pi = \frac{\beta}{g_m} = 2.5\text{K}\Omega$

$r_o = \frac{V_A}{I_c} = \frac{20}{1} = 20\text{K}\Omega$

r_o is in effect parallel to R_L, so $R'_L = R_L \| r_o$

$R'_L = 1^K \| 20^K = 0.95\text{K}\Omega$. From Table 5.6:

$A_M = \frac{R'_L}{\frac{R_{sig}+r_\pi+r_x}{\beta+1} + R'_L} = \frac{0.95}{\frac{1+2.5+0.1}{101}+0.95} = 0.964\text{ V/V}$

$R_\mu = R'_{sig} \| (r_\pi + (\beta+1)R'_L)$ (Eq. 6.179)

$R'_{sig} = R_{sig} + r_x = 1 + 0.1 = 1.1\text{K}\Omega$

$R_\mu = 1.1^K \| (2.5^K + 101\times 0.95) = 1.08\text{ K}\Omega$

$R_\pi = \frac{R'_{sig}+R'_L}{1+\frac{R'_{sig}}{r_\pi}+\frac{R'_L}{r_e}} = \frac{1.1+0.95}{1+\frac{1.1}{2.5}+\frac{0.95}{0.025}} = 0.052^K = 52\Omega$

$f_H = \frac{1}{2\pi\tau_H} = \frac{1}{2\pi(R_\pi C_\pi + R_\mu C_\mu)} = \frac{1}{2\pi(0.052\times 3.08 + 1.08\times 0.1)}$

$f_H = 593.8\text{ MHZ}$

6.122

$I = 2\text{mA} \Rightarrow g_m = \frac{2}{0.025} = 80\text{ mA/V}$, $r_\pi = \frac{\beta}{g_m} = 1.25^{K\Omega}$

$r_e = \frac{r_\pi}{\beta+1} = 12.4\Omega$

$f_T = \frac{g_m}{2\pi(C_\pi+C_\mu)} \Rightarrow C_\pi + C_\mu = \frac{80^m}{2\pi\times 400\times 10^6} = 31.85\text{ PF}$

$\Rightarrow C_\pi = 31.85 - 2 = 29.85\text{pF}$

$A_M = \frac{R_L}{\frac{R_{sig}+r_\pi}{\beta+1} + R_L} = \frac{1}{\frac{R_{sig}}{101} + 0.0124 + 1} = \frac{1}{1.0124 + \frac{R_{sig}}{101}}$

$R'_L = R_L = 1\text{K}\Omega$, $R'_{sig} = R_{sig} + r_x = R_{sig}$

$R_\mu = R'_{sig} \| [r_\pi + (\beta+1)R'_L]$ (Eq. 6.179)

$R_\mu = R_{sig} \| (1.25 + 101\times 1) = R_{sig} \| 102.25^K$

$R_\pi = \frac{R'_{sig}+R'_L}{1+\frac{R'_{sig}}{r_\pi}+\frac{R'_L}{r_e}}$ (Eq. 6.180)

$R_\pi = \frac{R_{sig}+1\text{K}}{1+0.8R_{sig}+80} = \frac{R_{sig}+1}{0.8R_{sig}+81}$

$f_H = \frac{1}{2\pi(R_\pi C_\pi + R_\mu C_\mu)} = \frac{1}{2\pi(29.85R_\pi + 2R_\mu)}$

a) $R_{sig} = 1^{K\Omega}$: $A_M = 0.978\text{ V/V}$

$R_\mu = 0.99^{K\Omega}$, $R_\pi = 24.4\Omega \Rightarrow f_H = 58.8\text{MHZ}$

b) $R_{sig} = 10\text{K}\Omega$: $A_M = 0.9\text{ V/V}$

$R_\mu = 9.11\text{K}\Omega$, $R_\pi = 124\Omega \Rightarrow f_H = 7.27\text{MHz}$

c) $R_{sig} = 100\text{K}\Omega$: $A_M = 0.499\text{ V/V}$

$R_\mu = 50.6\text{K}\Omega$, $R_\pi = 627\Omega \Rightarrow f_H = 1.34\text{MHZ}$

6.123

Refer to Fig P6.123:

Each of the transistors is operating at a bias current of approximately $100\mu A$. Thus:

$g_m = \frac{0.1}{0.025} = 4mA/V \quad , r_\pi = \frac{100}{4} = 25K\Omega$

$r_e \simeq 250\Omega \quad , r_o = \frac{100}{0.1} = 1M\Omega$

$C_\pi + C_\mu = \frac{g_m}{2\pi f_T} = \frac{4^m}{2\pi \times 400^M} = 1.59pF \Rightarrow C_\pi = 1.39pF$

a) $R_{in} = (\beta+1)\left[r_{e1} + (r_{\pi 2} \| r_{o1})\right]$

$R_{in} = 101\left[250\times10^{-3} + 25^K \| 1M\Omega\right] \simeq 2.5M\Omega$

$A_M = -\frac{R_{in}}{R_{in}+R_{sig}} \times \frac{r_{\pi 2}\|r_{o1}}{r_{e1}+(r_{\pi 2}\|r_{o1})} \times g_{m2} r_{o2}$

$A_M = -\frac{2.5^M}{2.5^M + 0.01} \times \frac{25^K \| 1M}{0.25 + (25^K \| 1M)} \times 4\times 1^M$

$A_M = -3943.6\, V/V$

b) To calculate f_H, refer to Example 6.13:

$R_{\mu 1} = R_{sig} \| R_{in} = 10^K \| 2.5^M \simeq 10K\Omega$

$R_{in2} = r_{\pi 2} \| r_{o1}$

$R_{in2} = 25^K \| 1^M$

$R_{in2} = 24.4K\Omega$

$R_{\pi 1} = \frac{R_{sig} + R_{in2}}{1 + \frac{R_{sig}}{r_{\pi 1}} + \frac{R_{in2}}{r_{e1}}}$

$R_{\pi 1} = \frac{10 + 24.4}{1 + \frac{10}{25} + \frac{24.4}{0.25}} = 0.35K\Omega$

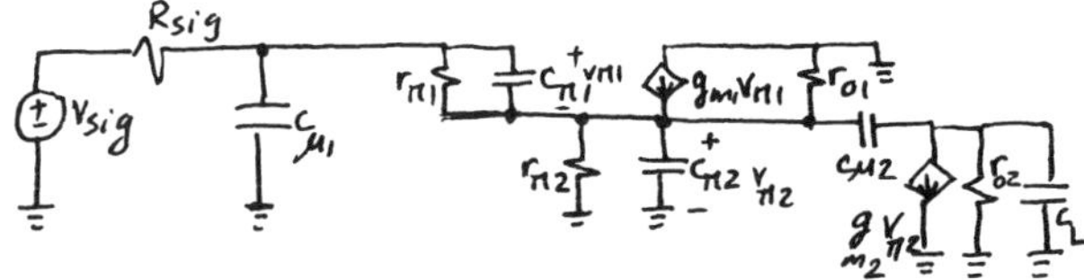

Using Miller's Theorem for $C_{\mu 2}$:

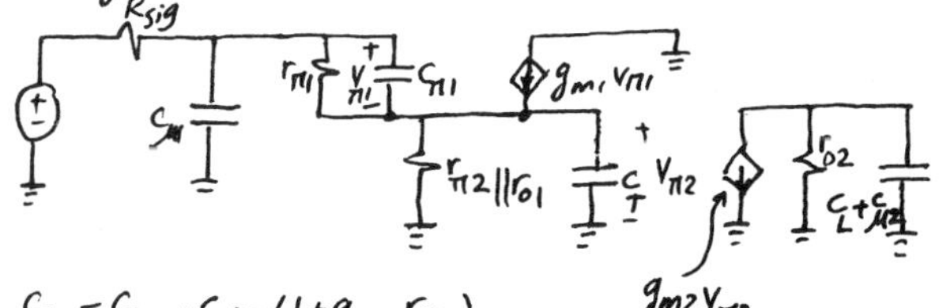

$C_T = C_{\pi 2} + C_{\mu 2}(1 + g_{m2} r_{o2})$

$C_T = 1.39 + 0.2(1 + 4\times 1000) = 801.6pF$

$R_T = r_{\pi 2} \| r_{o1} \| \frac{r_{\pi 1} + R_{sig}}{\beta + 1} = 25^K \| 1000^K \| \frac{25+10}{101}$

$R_T = 342\Omega$

$R_{\mu 2} = r_{o2} = 1000K$

$\tau_H = C_{\mu 1} R_{\mu 1} + C_{\pi 1} R_{\pi 1} + C_T R_T + (C_{\mu 2} + C_L) R_{\mu 2}$

$\tau_H = 0.2\times 10 + 1.39\times 0.35 + 801.6\times 0.342 + (0.2+1)\times 1000$

$\tau_H = 2 + 0.49 + 274.15 + 1200$ ns

Thus $(C_L + C_{\mu 2})R_{\mu 2}$ is the dominating term, The second most significant term is $C_T R_T$. So $(C_L + C_{\mu 2})$ dominates and then C_T or equivalently $C_{\mu 2}$.

$f_H = \frac{1}{2\pi \tau_H} = \frac{1}{2\pi \times 1476.6^n} = 107.8MHz$

c) Increasing the bias currents by a factor of 10:

$g_m = 40mA/V \quad , r_\pi = 2.5K\Omega$

$r_e \simeq 25\Omega \quad , r_o = 100K\Omega$

$C_\pi = C_{je} + C_{de}\times 10 = 0.8 + 0.59\times 10 = 6.7pF$

$C_\mu = 0.2pF$

$R_{in} = 101\left[0.025 + (2.5^K \| 100^K)\right] = 249K\Omega$

R_{in} is almost decreased by a factor of 10.

$A_M = -\frac{249}{249+10} \times \frac{2.5^K \| 100^K}{0.025 + (2.5^K \| 100^K)} \times 4000$

$A_M = -3807\, V/V$

A_M remains almost constant.

$C_T = 6.7 + 0.2(1 + 40\times 100) = 806.9$ (almost Constant)

$R_{\mu 1} = R_{sig} \| R_{in} = 10^K \| 249^K = 9.61K\Omega$

$R_{\mu 1}$ stays almost the same.

$R_T = 2.5^K \| 10^K \| \frac{2.5+10}{101} = 117.8\Omega$

R_T is almost reduced by a factor of 3.

$R_{in2} = r_{\pi 2} \| r_{o1} = 2.44K\Omega$

$R_{\pi 1} = \frac{10^K + 2.44}{1 + \frac{10}{2.5} + \frac{2.44}{0.025}} = 120\Omega$

$R_{\pi 1}$ is almost decreased by a factor of 3.

$R_{\mu 2} = r_{o2} = 100K\Omega$ (decreased by a factor of 10)

$\tau_H = 0.2\times 9.61 + 6.7\times 0.120 + 806.9\times 0.118 + 1.2\times 100$

$\tau_H = 1.92 + 0.8 + 95.2 + 120 = 217.92ns$

Thus the dominant effect, that of the output pole, is reduced by a factor of 10. This occurs because $(C_L + C_{\mu 2})$ remains constant while r_{o2} decreases by a factor of 10. The second most significant factor (that due to C_T or $C_{\mu 2}$ with Miller Effect) also decreases, but only by a factor of 3. The overall result is an increase in f_H.

Cont.

$f_H = \frac{1}{2\pi\tau_H} = \frac{1}{2\pi \times 217.9 \times 10^{-9}} = 730.3\,KHz$

f_H has increased by a factor of nearly 7. Significant increase!

6.124

Note: Although rather long, this is an excellent problem with considerable educational value.

a) DC-Bias

For Q_1: $I_{D1} = \frac{1}{2} k_n' \frac{W}{L} (V_{GS} - V_t)^2$

$0.1 = \frac{1}{2} \times 2 (V_{GS} - 1)^2$

$\Rightarrow V_{GS} = 1.316V$

$I_{D1} \simeq 0.1\,mA$

$I_{C2} \simeq 1\,mA$

See analysis ↗

$1mA = \frac{5-2}{3}$, $R_G = 10M\Omega$, $3K\Omega$, $+5V$, $\simeq 2V$, $0.7 + 1.316 \simeq 2V$, $0.7V$, $\frac{0.7}{6.8K} \simeq 100\mu A$, $6.8K\Omega$

b) For Q_1: $g_{m1} = \sqrt{2\mu_n k_n' \frac{W}{L} I_D} = \sqrt{2 \times 2 \times 0.1} = 0.63\,mA/V$

For Q_2: $g_{m2} = 40\,mA/V$ $\quad r_{\pi 2} = \frac{200}{40} = 5K\Omega$

$C_\pi + C_\mu = \frac{g_m}{2\pi f_T} = \frac{40 \times 10^{-3}}{2\pi \times 600 \times 10^6} = 10.6\,PF$

Since $C_\mu = 0.8\,pF \Rightarrow C_\pi = 9.8\,pF$

c) at midband:

$\frac{V_\pi}{V_i} = \frac{6.8 // 5}{(6.8//5) + \frac{1}{g_{m1}}}$

$\frac{V_\pi}{V_i} = \frac{2.88}{2.88 + \frac{1}{0.63}}$

$\frac{V_o}{V_i} = 0.64\,V/V$

$R_G = 10M\Omega$, $100^K = R_{sig}$, V_i, V_{sig}, Q_1, Q_2, V_π, $1K$, $3K$, $6.8K$, R_{in}, $r_\pi = 5K\Omega$

$\frac{V_o}{V_\pi} = -g_{m2} V_\pi (1^K // 3^K)$ where we have neglected the effect of R_G.

$\frac{V_o}{V_\pi} = -40 \times \frac{3}{4} V_\pi = -30 V_\pi \Rightarrow \frac{V_o}{V_i} = 0.64 \times (-30) = -19.2\,V/V$

$\frac{v_o}{v_i} = -19.2\,V/V$

$R_{in} = \frac{R_G}{1 - \frac{v_o}{v_i}} = \frac{10^M}{1-(-19.2)} = 495K\Omega$

$\frac{V_o}{v_{sig}} = \frac{R_{in}}{R_{sig} + R_{in}} \times \frac{v_o}{v_i} = \frac{495}{495 + 100} \times -19.2 = -16\,V/V$

d) At low frequencies:

$C_1 \rightarrow f_{P1} = \frac{1}{2\pi \times 0.1 \times 10^{-6} \times (100 + 495) \times 10^3} = 2.7\,Hz$

$C_2 \rightarrow f_{P2} = \frac{1}{2\pi \times 1 \times 10^{-6} (3+1) 10^{-3}} = 40\,Hz$

Thus $f_L \simeq 40\,Hz$

e) At high frequencies:

The high frequency equivalent circuit is as follows:

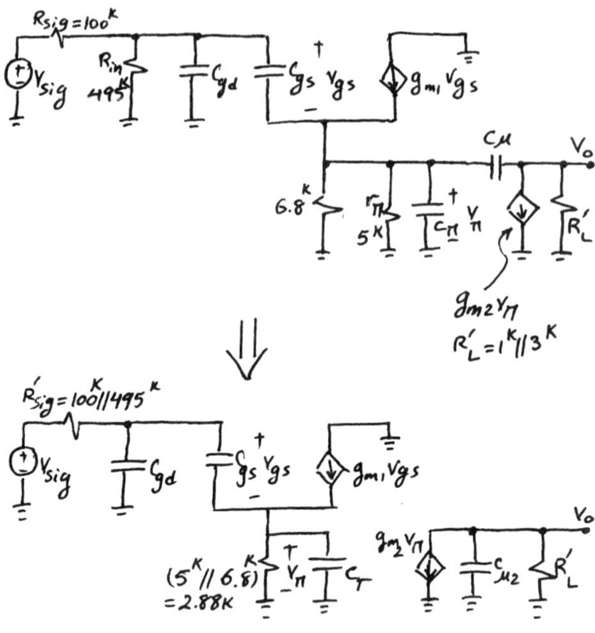

$C_T = C_\pi + C_\mu (1 + g_{m2} R_L') = 9.8 + 0.8(1 + 40 \times \frac{3}{4}) = 34.6\,pF$

$R_{gd} = R_{sig}' = 100K // 495K = 83.2K\Omega$

$R_{gs} = \frac{R_{sig}' + (6.8^K // 5^K)}{1 + g_{m1}(6.8^K // 5^K)} = \frac{83.2 + 2.88}{1 + 0.63 \times 2.88} = 30.6K\Omega$

$R_T = 6.8 // 5 // \frac{1}{g_m} = 1K\Omega$

$R_L' = 0.75\,K\Omega$

$\tau_H = C_{gd} R_{gd} + C_{gs} R_{gs} + C_T R_T + C_\mu R_L'$

$\tau_H = 1 \times 83.2 + 1 \times 30.6 + 34.6 \times 1 + 0.8 \times 0.75 = 149\,ns$

$f_H \simeq \frac{1}{2\pi \tau_H} = 1.07\,MHz$

f) There will no longer be a signal feedback. The left hand side $10M\Omega$ resistor will in effect appear between the input terminal and ground,

Thus: $R_{in} = 10M\Omega$

(a factor of 20 increase)

10^M $\quad 10^M$ $\quad V_o$

Cont.

and correspondingly A_M becomes:

$A_M = \frac{10}{10.1} \times (-19.2) = -19\,V/V$ (an increase from $-16\,V/V$)

Now R'_{sig} becomes approximately $100k\Omega$, as compared to $83.2k\Omega$, and correspondingly R_{gd} becomes $100k\Omega$, and R_{gs} becomes $36.6k\Omega$ while R_T and R'_L remain practically unchanged. Thus τ_H becomes $172.5ns$ and f_H decreases from $1.07MHz$ to $0.92MHz$.

6.125

Refer to Fig. P6.125:

$I_{E2} = 10mA \Rightarrow r_{e2} = 2.5\Omega$, $r_{\pi 2} = 253\Omega$

$I_{E1} = \frac{10}{101} \simeq 0.1mA \Rightarrow r_{e1} = 250\Omega$, $r_{\pi 1} = 25.3K\Omega$

$R_{in} = 101 \times [0.25 + 101(0.0025+1)] = 10.3M\Omega$

$R_{in} = 10.3M\Omega$

$R_{out} = r_{e2} + \frac{1}{\beta_2 + 1}\left[r_{e1} + \frac{R_{sig}}{\beta_1 + 1}\right]$

$R_{out} = 2.5 + \frac{1}{101}\left[250 + \frac{100000}{101}\right] = 14.8\Omega$

Neglecting r_o:

$A_{vo} = 1000\,V/V$, $A_v = \frac{1 \times 1000}{14.8 + 1000} = 0.985\,V/V$

6.126

$I_1 = I_2 = I = 1mA \Rightarrow g_m = 40\,mA/V$, $r_\pi = \frac{120}{40} = 3K\Omega$

$r_e = \frac{3}{121} \simeq 25\Omega$, $C_\pi + C_\mu = \frac{g_m}{2\pi f_T} = \frac{40^m}{2\pi \times 700^M} = 9.1pF$

$\Downarrow$

$C_\pi = 8.6pF$

Using Eq. 6.185:

$A_M = \frac{v_o}{v_{sig}} = \frac{1}{2}\left(\frac{R_{in}}{R_{in} + R_{sig}}\right) g_m R_L$

$R_{in} = 2r_\pi = 2 \times 3K\Omega = 6K\Omega$

$A_M = \frac{1}{2} \times \frac{6}{6+20} \times 40 \times 10^K = 46.15\,V/V$

$f_{P1} = \frac{1}{2\pi(\frac{C_\pi}{2} + C_\mu)(R_{sig} \| 2r_\pi)} = \frac{1}{2\pi(\frac{8.6}{2} + 0.5)(20^K \| 6^K)}$

$f_{P1} = 7.19MHz$

$f_{P2} = \frac{1}{2\pi C_\mu R_L} = \frac{1}{2\pi \times 0.5 \times 10^K} = 31.8\,MHz$

$f_H = \frac{1}{\sqrt{(\frac{1}{f_{P1}})^2 + (\frac{1}{f_{P2}})^2}} = 7.01MHz$

(For f_{P1} and f_{P2} Formulas, refer to Eq. 6.186 and 6.187)

6.127

$v_i = 2v_{gs}$

$v_o = R_L \times g_m v_{gs}$

$\frac{v_o}{v_i} = \frac{v_o}{v_{sig}} = g_m R_L/2$

$A_M = \frac{5 \times 20}{2} = 50\,V/V$

$f_{P1} = \frac{1}{2\pi(\frac{C_{gs}}{2} + C_{gd}) R_{sig}}$

$f_{P1} = \frac{1}{2\pi(1^P + 0.1^P)20^K} = 7.24MHz$

$f_{P2} = \frac{1}{2\pi(C_{gd} + C_L) R_L} = \frac{1}{2\pi(0.1 + 1) \times 20^K} = 7.24MHz$

$f_H = \frac{1}{\sqrt{(\frac{1}{f_{P1}})^2 + (\frac{1}{f_{P2}})^2}} = 5.12MHz$

6.128

All the transistors in this problem are operating at a bias current of 0.5mA and thus have:

$r_e = 50\Omega$, $g_m = 20\,mA/V$, $r_\pi = 5K\Omega$

$C_\pi + C_\mu = \frac{20^m}{2\pi \times 400^M} = 8\,pF$

Since $C_\mu = 2pF \Rightarrow C_\pi = 6pF$, $r_o = \infty$, $r_x = 0$

a) Common-Emitter amplifier:

$R_{sig} = 10K\Omega$, $R_C = 10K\Omega$

$A_M = -\frac{r_\pi}{R_{sig} + r_\pi} g_m R_C = -\frac{5}{10+5} 20 \times 10 = -66.7\,V/V$

$f_H = \frac{1}{2\pi(R_{sig} \| r_\pi)[C_\pi + (1 + g_m R_C) C_\mu]} \Rightarrow$

$f_H = \frac{1}{2\pi(10^K \| 5^K)[6^P + (1 + 20 \times 10)2]} = 117KHz$

b) Cascode:

$A_M = -\frac{\beta_1 \alpha_2 R_C}{R_{sig} + r_\pi} = -\frac{100 \times 0.99 \times 10}{10+5} = -66\,V/V$

Input pole: $f_{P1} = \frac{1}{2\pi(R_{sig} \| r_{\pi 1})(C_{\pi 1} + 2C_{\mu 1})}$

$f_{P1} = \frac{1}{2\pi(10^K \| 5^K)(6+4)^P} = 4.77MHz$

output pole: $f_{P3} = \frac{1}{2\pi C_{\mu 2} R_C} = \frac{1}{2\pi \times 2^P \times 10^K} = 7.96\,MHz$

pole at midband node:

$f_{P2} = \frac{1}{2\pi C_{\pi 2} r_{e2}} = \frac{1}{2\pi \times 6^P \times 50} = 530.5\,MHz$

Very high

Cont.

$$f_H = \frac{1}{\sqrt{\left(\frac{1}{f_{P1}}\right)^2+\left(\frac{1}{f_{P2}}\right)^2}} = 4.1\,MHz$$

c) CC-CB Cascade (Modified diff. amplifier)

$$A_M = \frac{\beta R_c}{R_{sig}+2r_\pi} = \frac{100\times10}{10+10} = 50\,V/V$$

Input pole : $f_{P1} = \frac{1}{2\pi(R_{sig}\|2r_\pi)(C_\pi/2+C_\mu)}$

$$f_{P1} = \frac{1}{2\pi\,(10^k\|10^k)(3+2)^p} = 6.4\,MHz$$

Output pole : $f_{P2} = \frac{1}{2\pi C_{\mu2}R_c} = \frac{1}{2\pi\times2^p\times10^K} = 7.96\,MHz$

Thus: $f_H = \frac{1}{\sqrt{\left(\frac{1}{f_{P1}}\right)^2+\left(\frac{1}{f_{P2}}\right)^2}} = 5\,MHz$

d) CC-CE Cascade :

$$A_M = -\frac{(\beta_1+1)\beta_2 R_c}{R_{sig}+r_{\pi1}+(\beta_1+1)r_{\pi2}} = -\frac{101\times100\times10}{10+5+101\times5} = -194\,V/V$$

Refer to Example 6.13 in :

$R_{\mu1} = (R_{sig}\|R_{in}) = 10^K\|(\beta+1)[r_{e1}+r_{\pi2}]$

$R_{\mu1} = 10^k\|101\times[0.05+5] = 9.81\,K\Omega$

$R_{\pi1} = r_{\pi1}\,\|\,\frac{R_s+r_{\pi2}}{1+g_{m1}r_{\pi2}} = 5\|\frac{10+5}{1+20\times5} = 144\,\Omega$

$R_T = r_{\pi2}\,\|\,\frac{r_{\pi1}+R_{sig}}{\beta+1} = 5^K\|\frac{5+10}{101} = 144\,\Omega$

where $C_T = C_{\pi2}+C_{\mu2}(1+g_{m2}R_c) = 6+2(1+200)$

$C_T = 408\,pF$

$R_{\mu2} = R_c = 10\,K\Omega$

$\tau_H = C_{\mu1}R_{\mu1}+C_{\pi1}R_{\pi1}+C_T R_T+C_{\mu2}R_{\mu2}$

$\tau_H = 2\times9.81+6\times0.144+408\times0.144+2\times10$

$\tau_H = 19.62+0.86+58.75+20 = 99.2\,ns$

$f_H = \frac{1}{2\pi\tau_H} = \frac{1}{2\pi\times99.2^n} = 1.6\,MHz$

e) Folded Cascode :

$$A_M = -\frac{\beta_1\alpha_2 R_c}{R_{sig}+r_{\pi1}} = -\frac{100\times0.99\times10}{10+5} = -66\,V/V$$

Input pole :

$$f_{P1} = \frac{1}{2\pi(R_{sig}\|r_{\pi1})(C_{\pi1}+2C_{\mu1})} = \frac{1}{2\pi(10\|5)(6+4)}$$

$f_{P1} = 4.77\,MHz$

At middle: $f_{P2} = \frac{1}{2\pi C_{\pi2}r_{e2}} = \frac{1}{2\pi\times6^p\times0.05} = 530\,MHz$ very high!

At output : $f_{P3} = \frac{1}{2\pi C_{\mu2}R_c} = \frac{1}{2\pi\times2\times10} \Rightarrow$

$f_{P3} = 7.96\,MHz$

Thus: $f_H \simeq \frac{1}{\sqrt{\frac{1}{4.77^2}+\frac{1}{7.96^2}}} = 4.1\,MHz$

f) CC-CB Cascade:

$$A_M = \frac{(\beta_1+1)\alpha_2 R_c}{R_{sig}+(\beta_1+1)2r_e} = \frac{101\times0.99\times10}{10+101\times0.1} \simeq 50\,V/V$$

Input pole : $f_{P1} = \frac{1}{2\pi(R_{sig}\|2r_\pi)(C_\pi/2+C_\mu)}$

$$f_{P1} = \frac{1}{2\pi(10^K\|10^K)(3^P+2^P)} = 6.4\,MHz$$

Output pole : $f_{P2} = \frac{1}{2\pi R_c C_\mu} = \frac{1}{2\pi\times10^K\times2^P} = 7.96\,MHz$

$$f_H \simeq \frac{1}{\sqrt{\frac{1}{6.4^2}+\frac{1}{7.96^2}}} = 5\,MHz$$

Summary of results:

Configuration	A_M (V/V)	f_H (MHz)	G.B. (MHz)
a) CE	−66.7	0.117	7.8
b) Cascode	−66	4.1	271
c) CC-CB cascade	+50	5.0	250
d) CC-CE cascade	−194	1.6	310
e) Folded cascode	−66	4.1	271
f) CC-CB cascade	+50	5.0	250

6.129

Refer to Fig. 6.58

$I_{REF} = 80\,\mu A = I_4 = I_1 = I_2 = I_3$

All transistors have the same g_m, r_o, V_{ov} values.

$I = \frac{1}{2}k_n'\frac{W}{L}V_{ov}^2 \Rightarrow 0.08 = \frac{1}{2}\times4\times V_{ov}^2 \Rightarrow V_{ov} = 0.2V$

$V_{GS} = V_{ov}+V_t = 0.2+0.5 = 0.7V$

$V_{G1} = V_{GS} = 0.7V = V_{S4} \Rightarrow V_{G4} = 0.7+V_{GS4} = 1.4^V$

$\Rightarrow V_{G3} = 1.4^V \Rightarrow V_{S3} = 1.4^V - V_{GS} = 0.7^V$

$\Rightarrow V_{O\,min} = V_{D3} = V_{S3}+V_{ov} = 0.9V$

As explained, the voltage at the gate of Q_3 is $2V_{GS}$ which implies voltage of $V_{GS} = V_{ov}+V_t$ at the source of Q_3. For minimum allowable voltage at the output, $V_{DS} = V_{ov}$ or equivalently $V_{Omin} = V_{ov}+V_{GS}$

$V_{Omin} = V_{ov}+V_{ov}+V_t = 2V_{ov}+V_t$.

$g_m = \frac{2I_D}{V_{ov}} = \frac{2\times0.08}{0.2} = 0.8\,mA/V$ $r_o = \frac{V_A}{I_D} = \frac{8}{0.08} = 100\,K\Omega$

Using Eq. 6.189: $R_o = r_{o3}+[1+(g_{m3}+g_{mb3})r_{o3}]r_{o2}$

$R_o = 100K+[1+0.8\times100]\times100 = 8.2\,M\Omega$

6.130

$I_{REF} = 25\mu A$, Refer to Fig. 6.58

$I_4 = 25\mu A = I_1$ $\quad W_1 = W_4 = 2\mu m \quad W_2 = W_3 = 40\mu m$

$I_1 = \frac{1}{2} k_n' \frac{W_1}{L_1} V_{OV_1}^2 \Rightarrow 25 = \frac{1}{2} \times 200 \times \frac{2}{1} V_{OV_1}^2 \Rightarrow V_{OV_1} = 0.354\,V$

$V_{OV_1} = V_{OV_2} \Rightarrow \frac{I_2}{I_1} = \frac{(W/L)_2}{(W/L)_1} \Rightarrow I_2 = 25 \times \frac{40}{2} = 500\mu A$

$I_2 = 0.5\,mA = I_3$

$I_O = 0.5\,mA$

$V_{GS_1} = V_{OV_1} + V_t = 0.354 + 0.6 = 0.954\,V$

$V_{G_1} = 0.954\,V$

$V_{G4} = V_{GS1} + V_{GS4}$, Since $I_1 = I_4$ and $W_1 = W_4$ then

$V_{GS1} = V_{GS4} \Rightarrow V_{G4} = 2V_{GS1} = 1.91\,V = V_{G3}$

The lowest possible voltage for the output is when Q_3 has $V_{DS3} = V_{OV3}$ or $V_{omin} = V_{G3} - V_{GS3} + V_{OV3}$

Since $V_{GS1} = V_{GS2}$ and $I_2 = I_3$ then $V_{GS3} = V_{GS1}$

$\Rightarrow V_{omin} = 1.91 - 0.954 + 0.354 = 1.31\,V$

$g_{m2} = g_{m3} = \frac{2I_D}{V_{OV}} = \frac{2 \times 0.5}{0.354} = 2.82\,mA/V$

$r_{o2} = r_{o3} = \frac{V_A}{I_D} = \frac{20}{0.5} = 40\,K\Omega$

Eq. 6.189: $R_o = r_{o3} + [1 + (g_{m3} + g_{mb3}) r_{o3}]\, r_{o2}$

$R_o = 40 + [1 + 2.82^m \times 40^K] \times 40K = 4.6\,M\Omega$

6.131

In the gate of Q_3 we see a small incremental resistance of approximately $\frac{1}{g_m}$ and since the incremental voltages across Q_4, Q_5, Q_6 will be small, we can assume that the gates are grounded. Thus the output resistane will be that of the CG transistor Q_3 which has a resisto $R_{02} = r_{o2} + (1 + (g_{m2} + g_{mb2}) r_{o2}) r_{o1}$ in its source. We use Eq. 6.101:

$R_o = r_{o3} + [1 + (g_{m3} + g_{mb3}) r_{o3}]\, R_{o2}$

$R_o = r_{o3} + [1 + (g_{m3} + g_{mb3}) r_{o3}]\,[r_{o2} + (1 + (g_{m2} + g_{mb2}) r_{o2}) r_{o1}]$

$R_o \simeq r_{o3} + [(g_{m3} r_{o3})]\,[g_{m2} r_{o2} r_{o1}]$

$R_o \simeq r_{o3} + g_{m2} g_{m3} r_{o1} r_{o2} r_{o3} \simeq g_{m2} g_{m3} r_{o1} r_{o2} r_{o3}$

6.132

$V_x = V_{BE3} + V_{BE1} = 1.4\,V$

If I_{REF} in in increased to 1mA or equivalently multiplied by 10, then:

$\frac{I_{C2}}{I_{C1}} = \frac{I_S e^{V_{BE2}/V_T}}{I_S e^{V_{BE1}/V_T}} \Rightarrow 10 = e^{(V_{BE2} - V_{BE1})/V_T}$

$V_{BE2} - V_{BE1} = \Delta V_{BE} = V_T \ln 10 = 0.058\,V$

$\Delta V_{BE} = 0.058\,V \Rightarrow \Delta V_x = 2\Delta V_{BE} = 0.116\,V$

Now we calculate I_o for $V_o = V_x$:

$I_{REF} \simeq I_C = 100\mu A \Rightarrow V_{BE_1} =$

The actual value of $I_o = \frac{I_{REF}}{1 + 2/(\beta^2 + \beta)}$

$\Rightarrow I_o = \frac{100\mu A}{1 + \frac{2}{200^2 + 200}} = 99.995\mu A$

$\frac{\Delta I_o}{I_o} = \frac{0.005}{100} = 5 \times 10^{-5} = 0.005\%$

6.133

Since $Q_{21}, Q_{22}, \cdots Q_{2n}$ are all matched to Q_1:

$I_{o1} = I_{o2} = \cdots = I_{on} = I_o$

The emitter of Q_3 supplies the basecurrent for all trasistors, so $I_{C3} = \frac{(n+1) I_o}{\beta}$

A node equation at the base of Q_3 yields:

$I_{REF} = I_o + \frac{(n+1) I_o}{\beta(\beta+1)}$, Thus : $\frac{I_o}{I_{REF}} = \frac{1}{1 + \frac{n+1}{\beta^2}}$

For a deviation from unity of less than 0.1%: $\frac{99.9}{100} = \frac{1}{1 + \frac{n+1}{\beta^2}} \Rightarrow \frac{n+1}{\beta^2} = \frac{1}{999}$

$\Rightarrow n = \frac{\beta^2}{999} - 1 \Rightarrow n \simeq 9$

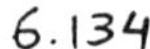

6.134

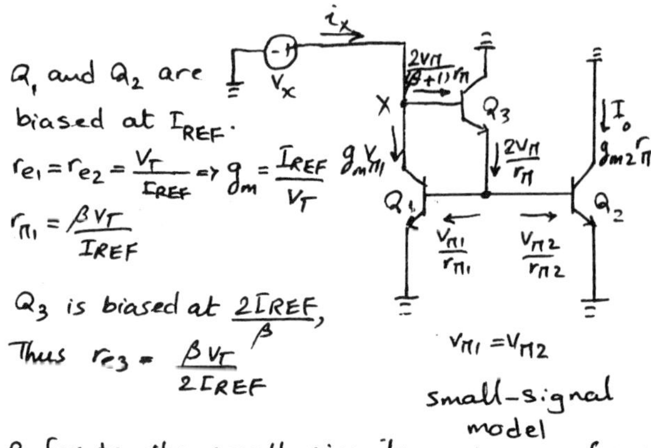

Q_1 and Q_2 are biased at I_{REF}:

$r_{e1} = r_{e2} = \frac{V_T}{I_{REF}} \Rightarrow g_m = \frac{I_{REF}}{V_T}$

$r_{\pi 1} = \frac{\beta V_T}{I_{REF}}$

Q_3 is biased at $\frac{2I_{REF}}{\beta}$,

Thus $r_{e3} = \frac{\beta V_T}{2I_{REF}}$

Refer to the small-circuit analysis performed directly on the circuit. Since the current in the emitter of Q_3 is $\frac{2V_{\pi 1}}{r_{\pi 1}}$, the voltage $V_{\pi 3}$ will be: $V_{\pi 3} = \frac{2V_{\pi 1}}{r_{\pi 1}} \times r_{e3}$.

$V_x = V_{\pi 2} + V_{\pi 1} = \frac{2V_{\pi 1} \cdot r_{e3}}{r_{\pi 1}} + V_{\pi 1} = V_{\pi 1}\left(1 + 2\frac{r_{e3}}{r_{\pi 1}}\right)$

$V_x = V_{\pi 1}\left(1 + 2\frac{\beta V_T}{2I_{REF}} \times \frac{I_{REF}}{\beta V_T}\right) = 2V_{\pi 1}$

and $i_x \simeq g_{m1} V_{\pi 1}$. Thus: $R_{in} = \frac{V_x}{i_x} = \frac{2}{g_{m1}} = \frac{2V_T}{I_{REF}}$

For $I_{REF} = 100\,\mu A \Rightarrow R_{in} = \frac{2 \times 0.025}{0.1} = 0.5\,K\Omega$

6.135

All the output currents are equal to I_o, then we have: $I_{REF} = 2I_o + \frac{11 I_o}{\beta} \Rightarrow \frac{I_o}{I_{REF}} = \frac{1}{2 + 11/\beta}$

I_o is idealy $I_{REF}/2$, For 5% lower I_o:

$0.95 \times \frac{I_{REF}/2}{I_{REF}} = \frac{1}{2 + \frac{11}{\beta}} \Rightarrow \beta = 104.5 \simeq 105$

$\beta = 105$

$I_x = 11 \times \frac{I_o}{\beta}$

IREF, outputs, I_x, I_o, 1, 2, 9

6.136

a) See the analysis on the circuit.

$I_{REF} = I + \frac{\beta+2}{\beta(\beta+1)} I = I\,\frac{\beta^2+2\beta+2}{\beta(\beta+1)}$

$I_{o1} = I_{o2} = \frac{1}{2}\,\frac{\beta+2}{\beta+1}\,I$

$\frac{I_{o1}}{I_{REF}} = \frac{I_{o2}}{I_{REF}} = \frac{1}{2}\,\frac{\beta(\beta+2)}{\beta^2+2\beta+2} = \frac{1}{2} \times \frac{1}{1 + 2/(\beta^2+2\beta)}$

$\frac{I_{o1}}{I_{REF}} \simeq \frac{1}{2}\,\frac{1}{1 + 2/\beta^2}$

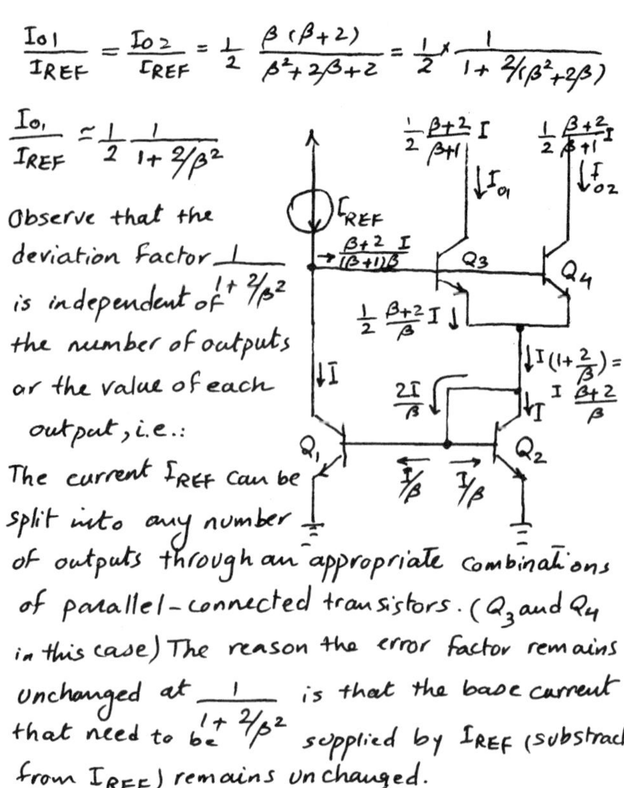

Observe that the deviation factor $\frac{1}{1+2/\beta^2}$ is independent of the number of outputs or the value of each output, i.e.: The current I_{REF} can be split into any number of outputs through an appropriate combinations of parallel-connected transistors. (Q_3 and Q_4 in this case) The reason the error factor remains unchanged at $\frac{1}{1+2/\beta^2}$ is that the base current that need to be supplied by I_{REF} (substract from I_{REF}) remains unchanged.

b) The 7mA reference current can be used to generate three output currents of 1, 2, 4mA by using 3 transistors in parallel having relative area ratios of 1, 2, 4 as shown:

$\frac{I_{o1}}{I_{REF}} = \frac{1}{7}\,\frac{1}{1+2/\beta^2} \Rightarrow I_{o1} = 0.998\,mA$ (1mA ideally)

$\frac{I_{o2}}{I_{REF}} = \frac{2}{7}\,\frac{1}{1+2/\beta^2} \Rightarrow I_{o2} = 1.996\,mA$ (2mA ideally)

$\frac{I_{o3}}{I_{REF}} = \frac{4}{7}\,\frac{1}{1+2/\beta^2} \Rightarrow I_{o3} = 3.992\,mA$ (4mA ideally)

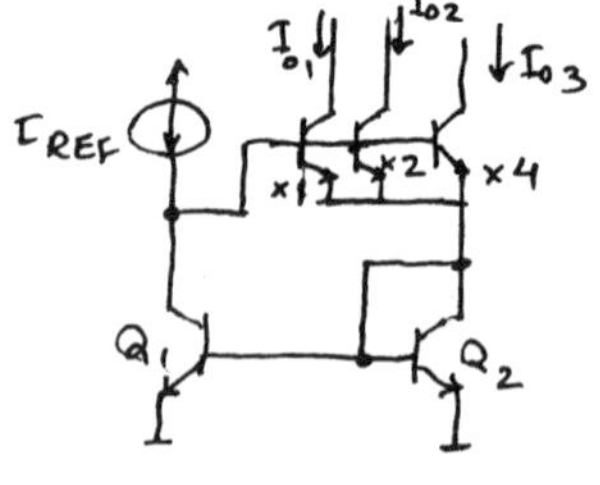

6.137

$I_{REF} = 0.1\,mA = \frac{5 - 0.7 - 0.7 - (-5)}{R}$

$\Rightarrow R = 86\,K\Omega$

V_{omax} is obtained when Q_3 is saturated: $V_{omax} = 5 - 0.7 - 0.2 = 4.1^V$

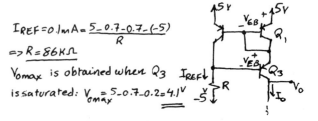

5V, 5V, V_{EB}, Q_1, V_{EB}, Q_3, I_{REF}, R, -5V, V_o, I_o

6.138

$V_{\pi 3} = r_{e3}\left[g_{m1}V_{\pi 1} + \frac{V_{\pi 1}}{r_{\pi 1}} + \frac{V_{\pi 2}}{r_{\pi 2}}\right]$

$V_x = V_{\pi 3} + V_{\pi 1}$

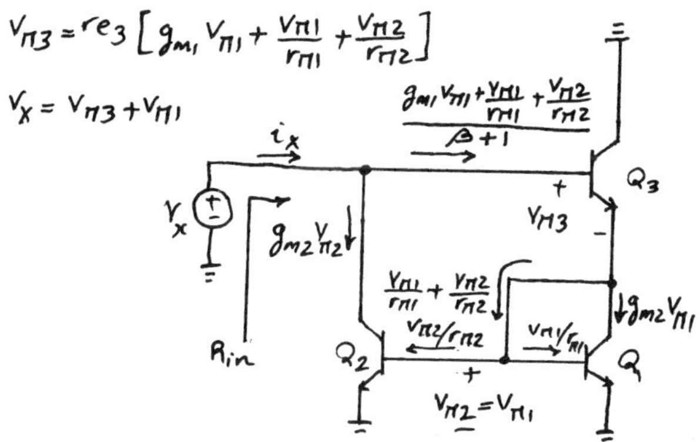

Small-signal analysis

Note that all 3 transistors are biased at I_{REF}

$V_x = V_{\pi 3} + V_{\pi 1} = r_{e3}\left[g_{m1}V_{\pi 1} + \frac{V_{\pi 1}}{r_{\pi 1}} + \frac{V_{\pi 2}}{r_{\pi 2}}\right] + V_{\pi 1}$

$V_x = (r_e g_m + \frac{2r_e}{r_\pi} + 1)V_{\pi 1}$

$V_x = (\frac{\beta}{\beta+1} + \frac{2}{\beta+1} + 1)V_{\pi 1} \simeq 2V_{\pi 1}(1 + \frac{1}{\beta+1}) \simeq 2V_{\pi 1}$

$i_x = g_{m2}V_{\pi 2} + \frac{1}{\beta+1}(g_{m1}V_{\pi 1} + \frac{V_{\pi 1}}{r_{\pi 1}} + \frac{V_{\pi 2}}{r_{\pi 2}})$

$i_x = V_{\pi 1}\left[g_m + \frac{g_m}{\beta+1} + \frac{g_m}{\beta(\beta+1)} + \frac{g_m}{\beta(\beta+1)}\right]$

$i_x \simeq g_m V_{\pi 1}$

Thus: $R_{in} = \frac{V_x}{i_x} = \frac{2V_{\pi 1}}{g_m V_{\pi 1}} = \frac{2}{g_m} = \frac{2V_T}{I_{REF}}$

For $I_{REF} = 100\,\mu A$:

$R_{in} = \frac{2 \times 0.025}{0.1} = 0.5\,k\Omega$

6.139

$R_o = \frac{\beta r_o}{2} = \frac{\beta}{2}\frac{V_A}{I_C} = \frac{100}{2} \times \frac{100}{1} = 5\,M\Omega$

$\Delta I_o = \frac{\Delta V_o}{R_o} = \frac{10\,V}{5\,M} = 2\,\mu A$

$\frac{\Delta I_o}{I_o} = \frac{2\mu}{1m} = 0.2\%$

6.140

$I_{REF} = 25\,\mu A$, $I_2 = I_{REF} = 25\,\mu A$

$V_{OV1} = V_{OV2} \Rightarrow \frac{I_1}{I_2} = \frac{W_1}{W_2} \Rightarrow I_1 = 25\mu \times \frac{2}{40} = 1.25\,\mu A$

$I_o = I_3 = I_1 = 1.25\,\mu A$

Refer to Fig. 6.61a

$V_{OV_2}^2 = \frac{I_2}{\frac{1}{2}k_n'\frac{W}{L}} = \frac{25}{\frac{1}{2} \times 200 \times \frac{40}{1}} = \frac{1}{160} \Rightarrow V_{OV_2} = 0.08\,V$

$\Rightarrow V_{GS2} = V_{GS1} = 0.08 + 0.6 = 0.86\,V$. therefore

$V_{G1} = V_{G2} = 0.86\,V$

$V_{OV3}^2 = \frac{I_3}{\frac{1}{2}k_n'\frac{W}{L}} = \frac{1.25}{\frac{1}{2} \times 200 \times \frac{40}{1}} = \frac{1}{3200} \Rightarrow V_{OV3} = 0.018\,V$

Therefore $V_{GS3} = 0.018 + 0.6 = 0.618\,V$

$\Rightarrow V_{G3} = V_{S3} + V_{GS3} = 0.86 + 0.618 = 1.48\,V$

for V_{Omin}, we have to have $V_{DS3} = V_{OV3} = 0.018\,V$

then: $V_{Omin} = V_{GS1} + V_{DS3min} = 0.86 + 0.018 = 0.88\,V$

For Q_2: $g_{m2} = \frac{2I}{V_{OV}} = \frac{2 \times 25\mu}{0.08} = 625\,\mu A/V = 0.625\,mA/V$

$r_{o2} = \frac{V_A}{I} = \frac{20}{0.025} = 800\,k\Omega$

For Q_3: $g_{m3} = \frac{2 \times 1.25}{0.018} = 139\,\mu A/V = 0.139\,mA/V$

$r_{o3} = \frac{20}{0.00125} = 16\,M\Omega$

Refer to section 6.12.4 in textbook:

$R_o = r_{o3}(g_{m3}r_{o2} + 2) = 16^M(0.139 \times 800^K + 2) = 1811.2\,M\Omega$

6.141

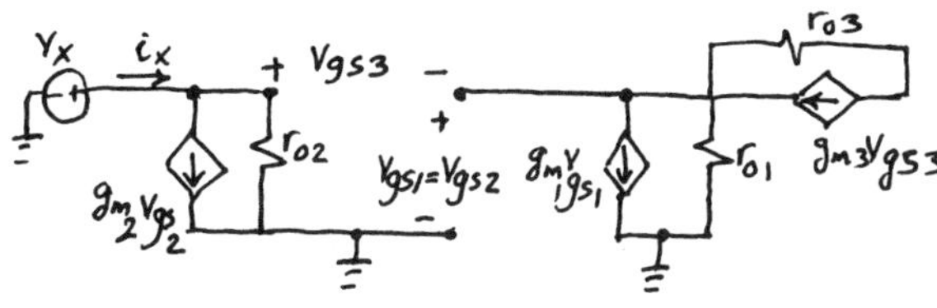

$i_x = g_{m2}V_{gs2}$ ①

$V_{gs2} + V_{gs3} = V_x$.

Since Q_2 and Q_3 have the same parameters and same current, therefore $V_{gs2} = V_{gs3}$

$V_x = 2V_{gs2} \Rightarrow V_{gs2} = \frac{V_x}{2}$

Substitute for V_{gs2} in ①:

$i_x = g_{m2} \times \frac{V_x}{2}$

$R_{in} = \frac{V_x}{i_x} = \frac{2}{g_{m2}}$

6.142

Refer to Fig. 6.61a:

$I_{REF} = 100\mu A$

$V_{DS1} = 2V_{GS}$

$V_{DS2} = V_{GS}$

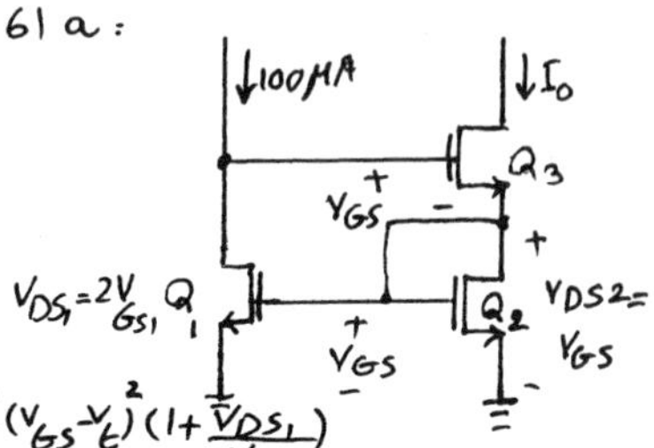

$I_{D1} = I_{REF} = \frac{1}{2}\mu_n C_{ox} \frac{W}{L}(V_{GS}-V_t)^2 (1+\frac{V_{DS_1}}{V_A})$

$100 = \frac{1}{2} \times 2000 (V_{GS}-0.6)^2 (1+\frac{2V_{GS}}{20})$

$1 = (V_{GS}-0.6)^2 (10+V_{GS})$

$V_{GS} \simeq 0.91V$ (by iteration)

$I_o = I_{D2} = \frac{1}{2} \times 2000 (V_{GS}-0.6)^2 (1+\frac{V_{GS}}{20})$

$I_o = 100.47\mu A$

Thus there is $\frac{0.47}{100}$ or 0.5% error. Modifying the circuit as Fig. 6.61C ensures that Q_1 and Q_2 have the same V_{DS} and thus eliminate the above error.

6.143

Refer to Fig. 6.62: $I_{REF} = 100\mu A$ $I_o = 10\mu A$

a) $V_{BE1} = 0.7 + V_t \ln \frac{100}{1000} = 0.642V$

$V_{BE2} = 0.7 + V_t \ln \frac{10}{1000} = 0.585V$

$I_o = \frac{V_{BE1} - V_{BE2}}{R_E} = 10\mu A \Rightarrow R_E = 5.7\ k\Omega$

b) $r_{\pi 2} = (\beta+1)\frac{V_T}{10\mu A} = 503k\Omega >> R_E$

$r_{o2} = \frac{V_A}{I_o} = 10M\Omega \Rightarrow R_o = (1+g_{m2}R_E)r_{o2} = 33M\Omega$

$R_o = 33M\Omega$

$\Delta I_o = \frac{\Delta V_o}{R_o} = \frac{5}{R_o} = 0.15\mu A$

6.144

a) $\frac{I_o}{I_{REF}} = 0.9 \Rightarrow I_o = 90\mu A$

$V_{RE} = V_T \ln \frac{1}{0.9} = 2.63mV$

$R_E = \frac{2.63mV}{90\mu A} = 29.3\Omega$

$r_o = \frac{V_A}{I_o} = 1.11M\Omega$

$g_m = 3.6\ mA/V$

$R_o = (1+g_m R_E)r_o = 1.23M\Omega$ Compare to $r_o = 1.11\ M\Omega$

b) $\frac{I_o}{I_{REF}} = 0.1 \Rightarrow I_o = 10\mu A$

$V_{RE} = V_T \ln 10 = 57.56\ mV$

$R_E = \frac{57.56m}{10\mu} = 5.76k\Omega$

$r_o = \frac{100}{10\mu} = 10M\Omega$

$g_m = 0.4m\ A/V$

$R_o = (1+g_m R_E)r_o = 33M\Omega$ Compare to $r_o = 10M\Omega$

c) $\frac{I_o}{I_{REF}} = 0.01 \Rightarrow I_o = 1\mu A$

$V_{RE} = V_T \ln 100 = 115mV$

$R_E = \frac{115}{1} = 115k\Omega$

$r_o = \frac{100}{1} = 100M\Omega$

$g_m = 0.04\ mA/V$

$R_o = (1+g_m R_E)r_o = 560M\Omega$ Compare to $r_o = 100\ M\Omega$

6.145

$R_o = [1+g_m(R_E \| r_\pi)]r_o$

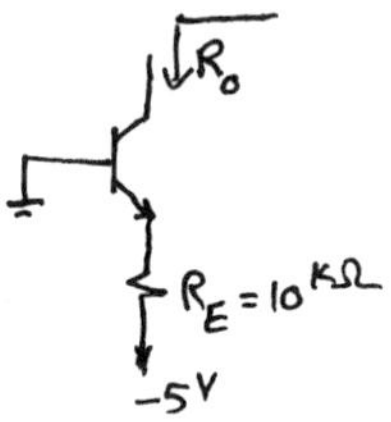

$I_E = \frac{-0.7-(-5)}{R_E} = 0.43mA$

$g_m = \frac{I_C}{V_T} = \frac{0.43}{0.025} = 17.2mA/V$

$r_o = \frac{V_A}{I_C} = \frac{100}{0.43} = 232.6k\Omega$

$r_\pi = \frac{\beta}{g_m} = \frac{100}{17.2} = 5.8k\Omega$

$R_E = 10k\Omega$

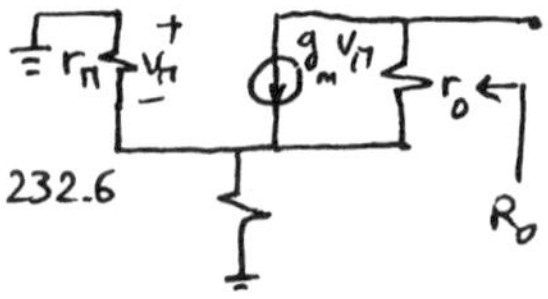

$R_o = [1+(10^k \| 5.8^k) \times 17.2] \times 232.6$

$R_o = 14.92M\Omega$

6.146

a) $I_o = \frac{V_{BE1}+V_{BE2}-V_{BE3}}{R}$

For $I_o = I_{REF}$: $V_{BE1} = V_{BE2} = V_{BE3} \Rightarrow I_o = \frac{V_{BE}}{R}$

For $I_o = 10\mu A$: $V_{BE} = 0.585$ therefore $R = 58.5\ k\Omega$

b) $g_m = \frac{I_C}{V_T} = \frac{10\mu}{0.025} = 0.4mA/V$, $r_\pi = \frac{\beta}{g_m} = 250k\Omega$

$r_o = \frac{V_A}{I_D} = 10M\Omega$

$R_o = [1+g_m(R \| r_\pi)]r_o$

$R_o = [1+0.4(58.5^k \| 250^k)]10^M = 199.6M\Omega$

$R_o \simeq 200M\Omega$

6.147

Q_1 and Q_2 are matched and $I_{D1} = I_{D2} = I$ therefore $V_{GS1} = V_{GS2}$ and this implies that the source of Q_1 and the source of Q_2 have the same voltage. Hence : $IR = V_{BE6} = V_T \ln \frac{I}{I_S}$

$IR = V_T \ln \frac{I}{I_S}$ ①

To find the value of R, first we have to calculate I_S.

For $I_E = 1mA$, $V_{BE} = 0.7^V \Rightarrow 1 = I_S e^{0.7/0.025}$

$\Rightarrow I_S = 6.9 \times 10^{-13} mA$

From ① : $R = \frac{V_T}{I} \ln \frac{I}{I_S} = \frac{0.025}{0.010m} \ln \frac{0.010}{6.9 \times 10^{-13}}$

$R = 58.49 K\Omega \simeq 58.5 K\Omega$

Chapter 7 – Problems

7.1

$V_{DD} = V_{SS} = 2.5V$

$K_n' \frac{W}{L} = 3\frac{mA}{V^2}$; $V_{tn} = 0.7V$

$I = 0.2mA$; $R_D = 5K\Omega$

(a) $V_{OV} = \sqrt{I / K_n' W/L}$

$= \sqrt{0.2/3} = \underline{\underline{0.26V}}$

$V_{GS} = V_{OV} + V_t = 0.26 + 0.7$

$= \underline{\underline{0.96V}}$

(b)

+2.5V
2.5KΩ
2.5KΩ
0.1m
+ 0.96 −
− 0.96 +
0.2mA
−2.5V
① ② ③

① $V_{S1} = V_{S2} = V_{CM} - V_{GS}$

$= 0 - 0.96 = \underline{\underline{-0.96V}}$

② $I_{D1} = I_{D2} = \frac{I}{2} = 0.1mA$

③ $V_{D1} = V_{D2} = V_{DD} - \frac{I}{2} \times R_D$

$= +2.5 - 0.1 \times 2.5 = \underline{\underline{2.25V}}$

(c) If $V_{CM} = +1V$

$V_{S1} = V_{S2} = +1 - 0.96 = \underline{\underline{0.04V}}$

$I_{D1} = I_{D2} = \underline{\underline{0.1mA}}$

$V_{D1} = V_{D2} = \underline{\underline{2.25V}}$

(d) If $V_{CM} = -1V$

$V_{S1} = V_{S2} = -1 - 0.96 = \underline{\underline{-1.96V}}$

$I_{D1} = I_{D2} = \underline{\underline{0.1mA}}$

$V_{D1} = V_{D2} = \underline{\underline{2.25V}}$

(e) $V_{CMAX} = V_t + V_{DD} - \frac{I}{2} R_D$

$= 0.7 + 2.5 - 0.1 \times 2.5 = \underline{\underline{+2.95V}}$

(f) $V_{CMIN} = -V_{SS} + V_{CS} + V_t + V_{OV}$

$= -2.5 + 0.3 + 0.7 + 0.26$

$= \underline{\underline{-1.24V}}$

$V_{S\,min} = V_{CMmin} - V_{GS}$

$= -1.24 - 0.96 = \underline{\underline{-2.2V}}$

7.2

(a) $V_{OV} = -\sqrt{I / K_p'(W/L)}$

$= -\sqrt{0.7/3.5} = \underline{\underline{-0.45V}}$

$V_{GS} = V_{OV} + V_t = -0.45 - 0.8$

$= \underline{\underline{-1.25V}}$

$V_{S1} = V_{S2} = V_G - V_{GS}$

$= 0 + 1.25 = \underline{\underline{+1.25V}}$

$V_{D1} = V_{D2} = \frac{I}{2} \times R_D - V_{DD}$

$= \frac{0.7}{2} \times 2 - 2.5 = \underline{\underline{-1.8V}}$

(b) For Q_1 and Q_2 to remain in saturation:

$V_{DS} \leq V_{GS} - V_t$

$\rightarrow V_{CM} \geq \left(\frac{I}{2} R_D - V_{DD}\right) + V_t$

$V_{CM\,min} = \frac{0.7}{2} \times 2 - 2.5 - 0.8$

$= \underline{\underline{-2.6V}}$

To allow sufficient voltage for the current source to operate properly:

$V_{CM} \leq V_{SS} - V_{CS} + (V_t + V_{OV})$

$\rightarrow V_{CM\,max} = 2.5 - 0.5 - 1.25$

$= 0.75V$ //

7.3

(a) $V_{G2} = 0$, $V_{G1} = v_{id}$
if $I_{D1} = I_{D2} = I/2 = 0.1\,mA$
then $V_{G1} = V_{G2}$
thus $v_{id} = 0V$.

(b) $i_{D1} = \frac{I}{2} + \frac{I}{V_{ov}} \cdot \frac{v_{id}}{2}$

$\rightarrow v_{id} = \left(\frac{2 i_{D1}}{I} - 1\right) \cdot V_{ov}$

For $i_{D1} = 0.15\,mA$

$v_{id} = \left(\frac{2 \times 0.15m}{0.2m} - 1\right) \times 0.26$

$= \underline{0.13V}$

(c) $i_{D1} = 0.2\,mA$ and
$i_{D2} = 0$ (when Q_2 just cuts off)
this occurs at
$v_{id} = +\sqrt{2} \times V_{ov}$
$= +\underline{0.367V}$

(d) $i_{D1} = 0.05\,mA$, $i_{D2} = 1.50\,mA$ } Opposite case to (b)
$\rightarrow v_{id} = -\underline{0.13}V$

(e) $i_{D1} = 0$, $i_{D2} = 0.2\,mA$ when Q_1 just cuts off.
This occurs at $v_{id} = -\sqrt{2} \times V_{ov}$
$= -\underline{0.367}V$

For each case find v_S, v_{D1}, v_{D2}, $v_{D2} - v_{D1}$

$V_{GS} = \sqrt{\frac{2 \times I_D}{K_n W/L}} + V_t$

then $v_S = v_G - v_{GS}$
Solving for each case we obtain the following results

	v_{id} (V)	i_{D1} (mA)	i_{D2} (mA)	v_S (V)	v_{D1} (V)	v_{D2} (V)	$v_{D2}-v_{D1}$ (V)
(c)	+.37	0.2	0	-0.7	1.5	2.5	-1.0
(b)	+.13	0.15	0.05	-0.88	1.75	2.25	-0.5
(a)	0	0.1	0.1	-0.96	2	2	0
(d)	-.13	0.05	0.15	-1.01	2.25	1.75	+0.5
(e)	-.37	0	0.2	-1.07	2.5	1.5	+1.0

7.4

$V_{G2} = 0$
$V_{G1} = v_{id}$

When all the current is on Q_1:
$I = \frac{1}{2}(K_p' \frac{W}{L})(V_{GS1} - V_t)^2$

$\Rightarrow V_{GS1} = V_t + \sqrt{\frac{2I}{K_p' W/L}}$

$= V_t + \sqrt{2}\, V_{ov}$

and V_{GS2} is reduced to V_t
thus $V_S = -V_t$.
Then $v_{id} = v_{GS1} + v_S$
$= V_t + \sqrt{2}\, V_{ov} - V_t = \sqrt{2}\, V_{ov}$
In a similar manner as for the NMOS Differential Amplifier, as v_{id} reaches $-\sqrt{2}\, V_{ov}$ Q_1 turns off and Q_2 on.
Thus the steering range is
$\sqrt{2}\, V_{ov} \leq V_i \leq -(\sqrt{2}\, V_{ov})$

For this particular case
$\sqrt{2} \times -0.45 \leq V_{id} \leq \sqrt{2} \times 0.45$
$-0.63 \leq v_{id} \leq 0.63$
When $V_{id} = -0.63V$,
$i_{D1} = 0.7mA$, $i_{D2} = 0$
$V_S = -V_{t2} = +0.8V$
$V_{D1} = 2K \times 0.7m - 2.5 = -1.1V$

CONT.

$V_{D2} = 0 - 2.5V = \underline{-2.5V}$

When $v_{id} = +0.63$

$i_{D1} = 0$; $i_{D2} = 0.7mA$

$V_S = v_{id} - V_{GS1} = v_{id} - V_{t1}$

$= 0.63 + 0.8 = \underline{1.43}$

$V_{D1} = -2.5V$ //

$V_{D2} = -1.1V$ //

7.5

$V_{G1} = v_{id}$ $\quad i_{D1} = 0.11mA$

$V_{G2} = 0$ $\quad i_{D2} = 0.09mA$

$$I_D = \frac{1}{2} K_n' \frac{W}{L} (V_{GS} - V_t)^2$$

For Q_1:

$$0.11m = \frac{1}{2} 3m (V_{GS1} - 0.7)^2$$

$\rightarrow V_{GS1} = \underline{0.97V}$

For Q_2:

$$0.09m = \frac{1}{2} 3m (V_{GS2} - 0.7)^2$$

$\rightarrow V_{GS2} = \underline{0.94V}$

$V_S = -V_{GS2} = \underline{-0.94V}$

$v_{id} = V_S + V_{GS1} = -0.94 + 0.97$

$= \underline{0.03V}$

$V_{D2} - V_{D1} = 5K(i_{D1} - i_{D2})$

$= 5K(0.11 - 0.09)m$

$= \underline{0.1V}$

thus

$$\frac{V_{D2} - V_{D1}}{v_{id}} = \frac{0.1}{0.03} = \underline{3.33}$$

When $i_{D1} = 0.09mA$ and $i_{D2} = 0.11mA$ is the reverse condition from the case we just studied, thus $v_{id} = \underline{-0.03V}$

7.6

We know that there is a linear relationship between V_{ov} & v_{id} since:

$$V_{ov} = \frac{v_{id}/2}{\sqrt{0.1}}$$

Then from the data in table 7.3 we can tell that for $v_{imax} = 150mV$

$$V_{ov} = 0.2 \times \frac{150}{126} = \underline{0.238V}$$

For W/L: $\frac{W}{L} = \frac{1}{(V_{ov})^2} \cdot \frac{I}{K}$

where I and K are constant thus, for W/L:

$$\left(\frac{W}{L}\right)_2 = \frac{50}{\left(\frac{150}{126}\right)^2} = 35.3$$

For g_m: $g_m = \frac{I}{V_{ov}}$ where I is constant

$$\rightarrow g_{m2} = \frac{g_{m1}}{\left(\frac{150}{126}\right)} = \frac{2}{\frac{150}{126}} = \underline{1.68} \ \frac{mA}{V}$$

7.7

$$\left(\frac{v_{idmax}/2}{V_{ov}}\right)^2 = K$$

$\Rightarrow \underline{2V_{ov}\sqrt{K} = v_{idmax}}$

Q.E.D

$$i_{D1} = \frac{I}{2} + \left(\frac{I}{V_{ov}}\right)\frac{v_{id}}{2}\sqrt{1-K}$$

CONT.

$$i_{D1} = \frac{I}{2} \pm \frac{I}{V_{ov}} \cdot \frac{2V_{ov}\sqrt{K}}{2} \cdot \sqrt{1-K}$$

$$\rightarrow i_{D1} = \frac{I}{2} \pm I\sqrt{K(1-K)}$$

thus $\Delta I = 2I\sqrt{K(1-K)}$

Q.E.D.

For $K = 0.01$

$$\Delta I = 2I\sqrt{0.01(1-0.01)} = 0.198 \times I$$

$$V_{idmax} = 2V_{ov}\sqrt{0.01} = 0.2\,V_{ov}$$

For $K = 0.1$

$$\Delta I = 2I\sqrt{0.1(1-0.1)} = 0.8I$$

$$V_{idmax} = 2V_{ov}\sqrt{0.2} = 0.894\,V_{ov}$$

7.8

$$I_D = \frac{1}{2}\mu_n C_{ox}\frac{W}{L}(V_{GS}-V_t)^2$$

$$\frac{200}{2} = \frac{1}{2} \times 90 \times \frac{100}{1.6}(V_{GS}-0.8)^2$$

$$\Rightarrow V_{GS} = 1.19V$$

$$g_m = \frac{2I_D}{V_{GS}-V_t} = \frac{2\times 100}{(1.19-1)} = 1.06\,\frac{mA}{V}$$

$$V_{id}\Big|_{\substack{\text{full current}\\\text{switching}}} = \sqrt{2}\,(V_{GS}-V_t) = 0.27V$$

To double this value, $V_{GS}-V_t$ must be doubled which means that I_D should be quadrupled. i.e. I changed to: $800\mu A$

7.9

$$g_m = \frac{2I_D}{V_{ov}} \rightarrow 1m = \frac{I}{0.2}$$

$$\rightarrow I = 0.2mA$$

$$I_D = \frac{1}{2}\mu_n C_{ox}\frac{W}{L}V_{ov}^2$$

$$100 = \frac{1}{2} \times 90 \times \frac{W}{L} \times (0.2)^2$$

$$\Rightarrow \frac{W}{L} = 55.6$$

7.10

$$i_D = \frac{1}{2}k_n\frac{W}{L}(V_{GS}-V_t)^2$$

$$50 = \frac{1}{2} \times 400\,(V_{GS}-1)^2$$

$$\Rightarrow V_{GS} = 1.5V$$

For $V_{G1} = V_{G2} = 0$, $V_S = -1.5V$

For $V_{G1} = V_{G2} = 2V$, $V_S = +0.5V$

The drain currents are equal in both cases.

For $V_{G2} = 0$:

To reduce i_{D2} by 10%,

$$i_{D2} = 0.9 \times 50 = 45\mu A$$

$$i_{D1} = 55\mu A$$

$$V_{GS2} = \sqrt{\frac{2\,i_{D2}}{400}} + 1 = 1.47V$$

$$V_{GS1} = \sqrt{\frac{2\times 55}{400}} + 1 = 1.52V$$

Thus, $V_{G1} = V_{GS1} - V_{GS2} = 0.05V$

To increase i_{D2} by 10%

$$i_{D2} = 55\mu A$$

$$i_{D1} = 45\mu A$$

CONT.

$V_{GS2} = 1.52V$

$V_{GS1} = 1.47V$

$\Rightarrow V_{G1} = \underline{\underline{-0.05V}}$

i_{D2}/i_{D1}	i_{D2} (μA)	i_{D1} (μA)	V_{GS2} (V)	V_{GS1} (V)	$V_{G2}-V_{G1}$ (V)
1	50	50	1.5	1.5	0
0.5	33.3	66.7	1.408	1.577	-0.17
0.9	47.4	52.6	1.487	1.513	-0.026
0.99	47.75	50.25	1.4886	1.5012	-0.013

For $i_{D1}/i_{D2} = 20 \Rightarrow i_{D2} = 4.76\mu A$

$i_{D1} = 95.24\mu A$

$V_{GS2} = 1.154V$, $V_{GS1} = 1.690$

Thus $V_{G1} - V_{G2} = \underline{\underline{0.536V}}$

7.11

$$V_{OV} = \sqrt{I/K_n'\frac{W}{L}} = \sqrt{\frac{0.5}{0.25 \times 50}} = \underline{\underline{0.2V}}$$

$$g_m = \frac{I}{V_{OV}} = \frac{0.5mA}{0.2V} = \underline{\underline{2.5\frac{mA}{V}}}$$

$$r_o = \frac{V_A}{I_D} = \frac{10}{(0.5m/2)} = \underline{\underline{40K\Omega}}$$

$$A_d = g_m \times (R_D \| r_o)$$

$$= 2.5\frac{mA}{V}(4K\Omega \| 40K\Omega)$$

$$= \underline{\underline{9.09}}\ V/V$$

7.12

From Eqn. (7.23)

$$\left(\frac{V_{id}/2}{V_{OV}}\right)^2 = 0.1 \rightarrow \left(\frac{0.2/2}{V_{OV}}\right)^2 = 0.1$$

$$\rightarrow V_{OV} = \sqrt{0.1} = \underline{\underline{0.316}}\ V$$

$$g_m = \frac{I}{V_{OV}} \rightarrow 3\frac{mA}{V} = \frac{I}{0.316}$$

$$\rightarrow I = \underline{\underline{0.95mA}}$$

also: $V_{OV} = \sqrt{\dfrac{I}{K_n' W/L}}$

$$\Rightarrow (0.316)^2 = \frac{0.95mA}{0.1\frac{mA}{V^2} \times \left(\frac{W}{L}\right)}$$

$$\rightarrow \frac{W}{L} = \underline{\underline{95}}$$

If $R_D = 5K\Omega \Rightarrow$

$$A_d = g_m R_D = 3\frac{mA}{V} \times 5K\Omega = \underline{\underline{15\frac{V}{V}}}$$

if $V_{id} = 0.2 \Rightarrow V_{od} = V_{id} \times A_d$

$$= 0.2 \times 15 = \underline{\underline{3V}}$$

7.13

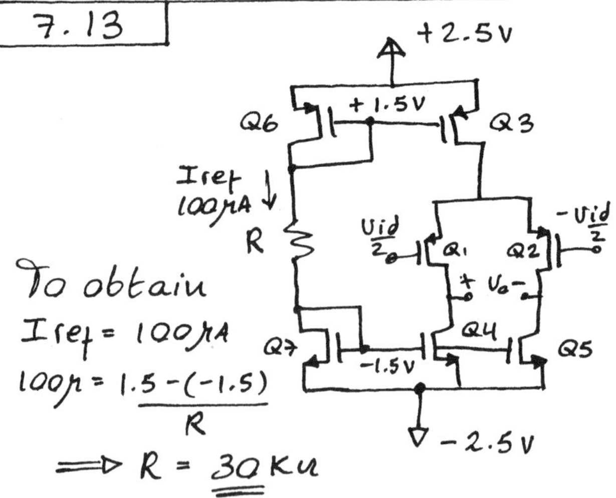

To obtain

$I_{ref} = 100\mu A$

$$100\mu = \frac{1.5 - (-1.5)}{R}$$

$$\Rightarrow R = \underline{\underline{30K\Omega}}$$

$V_{GS\,7,4,5} = -1.5 + 2.5 = \underline{\underline{1V}}$

$V_{OV\,7,4,5} = V_{GS} - V_{tn} = 1 - 0.7 = \underline{\underline{0.3V}}$

$V_{GS\,6,3} = 1.5 - 2.5 = \underline{\underline{-1V}}$

$V_{OV\,6,3} = -1 - (-0.7) = \underline{\underline{-0.3V}}$

The differential half circuit is an active-loaded common-source amplifier.

CONT.

thus, for Q_1, Q_4:

$$v_{o+} = \frac{v_{id}}{2} \times g_{m_1} (r_{o1} \| r_{o4})$$

For Q_2, Q_5:

$$v_{o-} = -\frac{v_{id}}{2} \times g_{m_2} (r_{o2} \| r_{o5})$$

Since $r_{o1} = r_{o2} = r_{o4} = r_{o5} \equiv r_o$

$$v_{o+} - v_{o-} = v_{id} \times g_{m_{1,2}} \times \frac{r_o}{2}$$

$$\rightarrow \frac{v_{o+} - v_{o-}}{v_{id}} = A_d = g_{m_{1,2}} \times \frac{r_o}{2}$$

$$= \frac{g_{m_{1,2}}}{2} \times \frac{V_{An}}{I_{D_{1,2}}}$$

$$= \frac{1}{2} \times \frac{2 I_{D_{1,2}}}{V_{OV}} \times \frac{V_{An}}{I_{D_{1,2}}} = \frac{V_{An}}{V_{OV_{1,2}}}$$

thus:

$$80 = \frac{20}{|V_{OV_{1,2}}|} \rightarrow V_{OV_{1,2}} = -0.25$$

PMOS.

Then $V_{GS_{1,2}} = -0.25 - 0.7$
$= \underline{\underline{-0.95\,V}}$

We have: $I_{D7} = I_{D6} = \underline{\underline{100\,\mu A}}$

If we choose:

$I_{D3} = I_{D6} = 100\,\mu A$ //

then $I_{D1} = I_{D2} = I_{D4} = I_{D5} = \underline{\underline{50\,\mu A}}$

To obtain W/L ratios:

$$I_D = \frac{1}{2} \mu C_{ox} (W/L) V_{OV}^2$$

$$\Longrightarrow \frac{W}{L} = \frac{2 I_D}{\mu C_{ox} V_{OV}^2}$$

where:

$\mu_n C_{ox} = 90\,\mu A/V^2$
$\mu_p C_{ox} = 30\,\mu A/V^2$

For Q_7:

$$\left(\frac{W}{L}\right)_7 = \frac{2 \times 100\mu}{90\mu \times (0.3)^2} = \underline{\underline{24.7}}$$

For Q_4 and Q_5:

$$\left(\frac{W}{L}\right)_{4,5} = \frac{2 \times (100\mu/2)}{90\mu \times (0.3)^2} = \underline{\underline{12.3}}$$

For Q_1 and Q_2:

$$\left(\frac{W}{L}\right)_{1,2} = \frac{2 \times (100\mu/2)}{30\mu \times (0.25)^2} = \underline{\underline{53.3}}$$

For Q_6 and Q_3:

$$\left(\frac{W}{L}\right)_{6,3} = \frac{2 \times 100\mu}{30\mu \times (0.3)^2} = \underline{\underline{74.1}}$$

In summary, the results are:

	Q_1	Q_2	Q_3	Q_4	Q_5	Q_6	Q_7	
$\mu_n C_{ox}$	30	30	30	90	90	30	90	$\mu A/V^2$
I_D	50	50	100	50	50	100	100	μA
V_{OV}	-0.25	-0.25	-0.3	0.3	0.3	-0.3	0.3	V
W/L	53.3	53.3	74.1	12.3	12.3	74.1	24.7	
V_{GS}	-0.95	-0.95	-1	1	1	-1	1	

7.14

(a) $I_{D1} = \frac{1}{2} k_n' \frac{W}{L} (V_{GS_1} - V_t)^2$

$I_{D2} = \frac{1}{2} k_n' \left(2 \times \frac{W}{L}\right) (V_{GS_2} - V_t)^2$

Since $V_{GS} - V_t$ is equal for both transistors:

$$\Rightarrow \frac{I_{D1}}{I_{D2}} = \frac{1}{2}$$

but $I = I_{D1} + I_{D2}$

$\Longrightarrow 2 I_{D1} = I - I_{D2}$

$\underline{\underline{I_{D1} = I/3}}$

$I_{D2} = \underline{\underline{2I/3}}$

(b) $V_{OV} = V_{GS} - V_t$

CONT.

$V_{OV_1} = V_{OV_2} = V_{OV}$

For Q_1: $\frac{I}{3} = \frac{1}{2} k_n' \left(\frac{W}{L}\right) V_{OV}^2$

$$\Rightarrow V_{OV} = \sqrt{\frac{2}{3}\frac{I}{k_n' W/L}}$$

(c) $g_m = \frac{2I_D}{V_{OV}} \rightarrow g_{m1} = \frac{2I}{3V_{OV}}$

$$g_{m2} = \frac{4}{3}\frac{I}{V_{OV}}$$

$$v_{o1} = -g_{m1} \times \frac{v_{id}}{2} \cdot R_D$$
$$= -\frac{2}{3}\frac{I}{V_{OV}} \cdot R_D \cdot v_{id}$$

$$v_{o2} = +g_{m2} \times \frac{v_{id}}{2} \cdot R_D$$
$$= \frac{4}{3}\frac{I}{V_{OV}} \cdot R_D \cdot v_{id}$$

$$\Rightarrow \frac{v_{o2} - v_{o1}}{v_{id}} = \left(\frac{4}{3} + \frac{2}{3}\right)\frac{I}{V_{OV}} \cdot R_D$$
$$= 2 \times \frac{I}{V_{OV}} \cdot R_D$$

7.15

$$V_{OV} = \sqrt{\frac{I}{k_n' W/L}} = \sqrt{\frac{0.2}{3}} = 0.26\text{V}$$

$$g_m = \frac{I}{V_{OV}} = \frac{0.2\text{mA}}{0.26\text{V}} = 0.77\frac{\text{mA}}{\text{V}}$$

(a) Single-ended output:

From Eqn. (7.42)

$$|A_d| = \frac{1}{2} g_m \times R_D = \frac{0.77 \times 10}{2}$$
$$= 3.85 \text{ V/V}$$

From Eqn. (7.41)

$$|A_{cm}| = \frac{R_D}{2R_{SS}} = \frac{10}{2 \times 100} = 0.05 \text{ V/V}$$

$$\text{CMRR} = \left|\frac{A_d}{A_{cm}}\right| = \frac{3.85}{0.05} = 77$$

i.e 37.7 dB

(b) Differential output, and 1% mismatch in R_D's:

Eqn (7.52) $|A_d| = g_m R_D$
$= 0.77 \times 10 = 7.7$ V/V

Eqn. (7.51) $|A_{cm}| = \frac{R_D}{2R_{SS}} \times \left(\frac{\Delta R_D}{R_D}\right)$

$$= \frac{10}{2 \times 100} \times 0.01 = 0.5\text{m V/V}$$

$$\text{CMRR} = \left|\frac{A_d}{A_{cm}}\right| = \frac{7.7}{0.5 \times 10^{-3}} = 15{,}400$$

i.e 83.7 dB

7.16

$$V_{OV} = -\sqrt{\frac{I}{k_p' W/L}} = -\sqrt{\frac{0.7\text{mA}}{3.5\frac{\text{mA}}{\text{V}^2}}}$$
$$= -0.45\text{V}$$

$$g_m = \frac{I}{|V_{OV}|} = \frac{0.7\text{mA}}{0.45\text{V}} = 1.56\frac{\text{mA}}{\text{V}}$$

$$|A_d| = g_m R_D = 1.56 \times 2 = 3.12 \text{ V/V}$$

$$|A_{cm}| = \frac{R_D}{2R_{SS}} \cdot \left(\frac{\Delta R_D}{R_D}\right) = \frac{2}{2 \times 30} \times 0.02$$
$$= 6.7 \times 10^{-4}$$

$$\text{CMRR} = \frac{3.12}{6.7 \times 10^{-4}} = 4680 \rightarrow 73.4 \text{ dB}$$

7.17

(a) $I_{D1} = I_{D2} = \frac{1mA}{2} = 0.5mA$

$I_D = \frac{1}{2} K_n' \frac{W}{L} \cdot V_{ov}^2$

$\Rightarrow 0.5m = \frac{1}{2} \times 2.5m \times V_{ov}^2$

$\rightarrow V_{ov} = 0.632V$

$V_{ov} = V_{GS} - V_t = V_{GS} - 0.7$

$\rightarrow V_{GS} = 0.632 + 0.7 = 1.332V$

To obtain 1mA over $R_{SS} = 1K\Omega$

$V_S = 1m \times 1K = 1V.$

$\rightarrow V_{CM} = V_S + V_{GS} = 1 + 1.332 = \underline{\underline{2.332V}}$

(b) $g_m = \frac{I}{V_{ov}} = \frac{1mA}{0.632V} = 1.6 \frac{mA}{V}$

Eqn. (7.45): $A_d = g_m \cdot R_D$
for $A_d = 8V/V$ $R_D = \frac{8}{1.6m} = \underline{\underline{5K\Omega}}$

(c) At the drains:
$V_{D1} = V_{D2} = 5V - \frac{1mA}{2} \times 5K\Omega = \underline{\underline{+2.5V}}$

(d) Eqn. (7.39):
$\frac{V_{o1}}{V_{icm}} = \frac{-R_D}{\frac{1}{g_m} + 2R_{SS}}$

$\Rightarrow |A_{cm}| = \left|\frac{\Delta V_{o1}}{\Delta V_{cm}}\right| = \frac{5K}{\frac{1}{1.6m} + 2 \times 1K} = \underline{\underline{1.9}}\ V/V$

(e) On the edge of the triode region:
$V_G - V_D = V_t$

gain×ΔVcm, $V_D = 2.5V$, $V_D - \Delta V_{CM} \times gain$, $V_{CM} + \Delta V_{CM}$, ΔVcm, $V_{CM} = 2.332V$

If: $V_G - V_D = V_t$
$\Rightarrow V_{CM} + \Delta V_{CM} - V_D + \Delta V_{CM}|A_{cm}| = V_t.$

$\rightarrow 2.332 + \Delta V_{CM} - 2.5 + \Delta V_{CM} \cdot 1.9 = 0.7$

$2.9\Delta V_{CM} = 0.868$
$\Delta V_{CM} = \underline{\underline{0.3V}}$

7.18

(a) $R_{D1} = R_D + \Delta R/2$
$R_{D2} = R_D - \Delta R/2$
$g_{m1} = g_m + \Delta g_m/2$
$g_{m2} = g_m - \Delta g_m/2$

The following Eqns are still valid:

Eqn. (7.62) $i_{d1} = \frac{g_{m1} v_{icm}}{2 g_m R_{SS}}$

Eqn. (7.63) $i_{d2} = \frac{g_{m2} \cdot v_{icm}}{2 g_m \cdot R_{SS}}$

From which:

$i_{d1} - i_{d2} = (g_{m1} - g_{m2}) \frac{v_{icm}}{2 g_m R_{SS}}$

$= \Delta g_m \frac{v_{icm}}{2 g_m R_{SS}}$ (1)

$i_{d1} + i_{d2} = (g_{m1} + g_{m2}) \frac{v_{icm}}{2 g_m R_{SS}}$

Assuming that Δg_m is small, then
$g_{m1} + g_{m2} \simeq 2 g_m.$

$\rightarrow i_{d1} + i_{d2} = \frac{2 g_m}{2 g_m R_{SS}} \cdot v_{icm} \rightarrow$

CONT.

$$i_{d1} + i_{d2} = \frac{v_{icm}}{R_{SS}} \quad (2)$$

The differential output is:

$$v_{o2} - v_{o1} = -i_{d2} \cdot R_{D2} + i_{d1} \cdot R_{D1}$$

$$= -i_{d2}\left(R_D - \frac{\Delta R_D}{2}\right) + i_{d1}\left(R_D + \frac{\Delta R_D}{2}\right)$$

$$= R_D \cdot (i_{d1} - i_{d2}) + \frac{\Delta R_D}{2}(i_{d1} + i_{d2})$$

Substituting (1) & (2):

$$= R_D \cdot \frac{\Delta g_m}{2 g_m R_{SS}} \cdot v_{icm} + \frac{\Delta R_D}{2} \frac{v_{icm}}{R_{SS}}$$

From which the common-mode gain is:

$$A_{cm} = \frac{v_{o2} - v_{o1}}{v_{icm}} = \frac{R_D}{2R_{SS}} \cdot \frac{\Delta g_m}{g_m} + \frac{\Delta R_D}{2R_{SS}}$$

$$\simeq \frac{R_D}{2R_{SS}} \left[\frac{\Delta g_m}{g_m} + \frac{\Delta R_D}{R_D} \right]$$

Q.E.D.

(b) For $R_D = 5\,k\Omega$, $R_{SS} = 25\,k\Omega$

$\rightarrow A_{cm} = 0.002\,V/V$

For identical R_D's, this common-mode gain is caused by a mismatch in g_m.

Thus, from Eqn. (7.64)

$$A_{cm} = \left(\frac{R_D}{2R_{SS}}\right) \cdot \left(\frac{\Delta g_m}{g_m}\right)$$

$$0.002 = \frac{5}{2 \times 25} \cdot \frac{\Delta g_m}{g_m}.$$

$$\rightarrow \frac{\Delta g_m}{g_m} = 0.02$$

To reduce A_{cm} to zero, create a mismatch in R_D such as:

$$0 = \frac{R_D}{2R_{SS}} \left(\frac{\Delta g_m}{g_m} + \frac{\Delta R_D}{R_D} \right)$$

$$\Rightarrow \frac{\Delta R_D}{R_D} = -\frac{\Delta g_m}{g_m} = -0.02$$

i.e. -2% of R_D

7.19

$$g_m = k_n' \left(\frac{W}{L}\right) \cdot (V_{GS} - V_t)$$

Recalling from calculus that for a function $u = f(x, y)$ the total derivative is:

$$du = \frac{\partial u}{\partial x} \cdot dx + \frac{\partial u}{\partial y} \cdot dy$$

Thus,

$$\Delta g_m = \frac{\partial g_m}{\partial (W/L)} \Delta(W/L) + \frac{\partial g_m}{\partial V_t} \cdot \Delta V_t$$

$$= k_n' (V_{GS} - V_t)(\Delta W/L) - k_n' (W/L) \cdot \Delta V_t$$

thus,

$$\frac{\Delta g_m}{g_m} = \frac{k_n' (V_{GS} - V_t) \Delta (W/L)}{k_n' (W/L)(V_{GS} - V_t)} - \frac{k_n' (W/L) \cdot \Delta V_t}{k_n' (W/L)(V_{GS} - V_t)}$$

$$\frac{\Delta g_m}{g_m} = \frac{\Delta (W/L)}{(W/L)} + \frac{-\Delta V_t}{(V_{GS} - V_t)}$$

but: $V_{GS} - V_t = V_{OV}$

$$\Rightarrow \frac{\Delta g_m}{g_m} = \frac{\Delta (W/L)}{(W/L)} + \frac{-\Delta V_t}{V_{OV}}.$$

Q.E.D

CONT.

If $\frac{\Delta(W/L)}{W/L} = \pm 0.01$; $\Delta V_t = \pm 5\,mV$

$V_{ov} = 0.25\,V$.

then, the worst-case fractional mismatch in g_m is:

$$\frac{\Delta g_m}{g_m} = \left|\frac{\Delta W/L}{W/L}\right| + \left|\frac{\Delta V_t}{V_{ov}}\right|$$

$$= 0.01 + \frac{0.005}{0.25} = \underline{\underline{0.03}}$$

If: $R_D = 5\,K\Omega$, $R_{SS} = 25\,K\Omega$

$I = 1\,mA$

From Eqn. (7.64)

$$A_{cm} = \frac{R_D}{2R_{SS}} \cdot \frac{\Delta g_m}{g_m}$$

$$A_{cm} = \frac{5}{2 \times 25} \times 0.03 = \underline{\underline{0.003}}\ V/V$$

From Eqn. (7.66)

$$CMRR = \frac{2 g_m R_{SS}}{\left(\frac{\Delta g_m}{g_m}\right)}$$

where: $g_m = I / V_{ov}$

$= 1\,mA / 0.25\,V = 4\,mA/V$

$$CMRR = \frac{2 \times 4 \times 25}{0.03} = 6,666.7$$

→ i.e $\underline{\underline{76.5\,dB}}$

7.20

$V_{BE} = 0.7$ @ $i_c = 1\,mA$

→ at $i_c = 0.5\,mA$

$V_{BE} = 0.7 + 25\,m \ln\left(\frac{0.5}{1}\right)$

$= 0.683\,V$

Thus,

$V_E = V_{CM} - V_{BE}$

$= -2 - 0.683 = \underline{\underline{-2.683\,V}}$

$i_{c1} = i_{c2} = \alpha \times 0.5 = \frac{100}{101} \times 0.5$

$= 0.495\,mA$

$V_{c1} = V_{c2} = V_{cc} - i_c R_c$

$= 5 - 0.495 \times 3$

$= +\underline{\underline{3.515\,V}}$

7.21

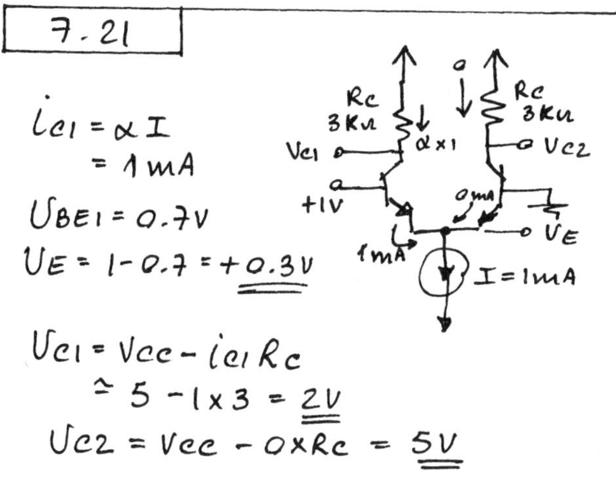

$i_{c1} = \alpha I$

$= 1\,mA$

$V_{BE1} = 0.7\,V$

$V_E = 1 - 0.7 = \underline{\underline{+0.3\,V}}$

$V_{c1} = V_{cc} - i_{c1} R_c$

$\simeq 5 - 1 \times 3 = \underline{\underline{2V}}$

$V_{c2} = V_{cc} - 0 \times R_c = \underline{\underline{5V}}$

7.22

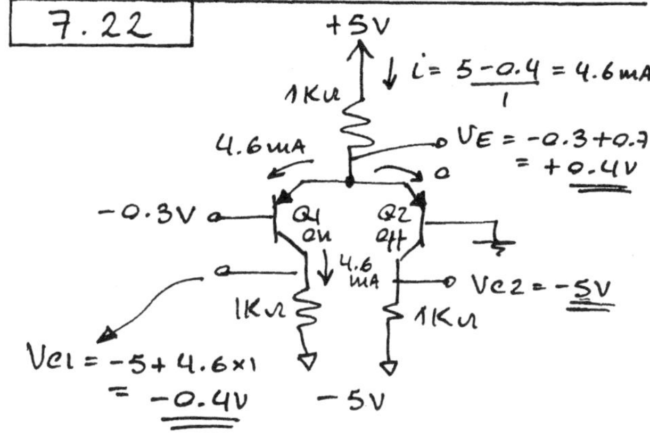

7.23

$$i_{E1} = \frac{I}{1 + e^{(V_{B2} - V_{B1})/V_T}} = 0.8\,I$$

→ $V_{B2} - V_{B1} = V_T \ln 0.25$

$= -34.6\,mV$

Thus, $V_{B1} - V_{B2} = \underline{\underline{34.6\,mV}}$

7.24

(a) $v_{CMmax} = v_{C1,2} = V_{CC} - \frac{I}{2} \cdot R_C$

(b) If the current is steered to Q_1, then

$v_{C1} = V_{CC} - IR_C$, a change of: $-\frac{I}{2} R_C$

$v_{C2} = V_{CC}$, a change of $+\frac{I}{2} R_C$

(c) $v_{CMmax} = 3 = 5 - \frac{I}{2} R_C$

$\Rightarrow IR_C = 4V$

(d) $\frac{I/2}{\beta+1} \leq 2\mu A$

$\Rightarrow I \leq 4(\beta+1)\mu A$

Thus, $I = 4 \times 101 \mu A = 0.404 \mu A$

Select $I = 0.4 mA$

$R_C = \frac{4V}{I} = \frac{4V}{0.4mA} = 10K\Omega$

7.25

$i_{E1} = \frac{I}{1+e^{-v_d/V_T}}$, $v_d = v_{B1} - v_{B2}$

$\frac{\Delta i_{E1}}{I} \equiv \frac{i_{E1} - I/2}{I} = \frac{i_{E1}}{I} - 0.5$

Define normalized Gain

$G_n \equiv \frac{\Delta i_{E1}/I}{v_d}$

v_d (mV)	5	10	20	30	40
G_n	9.97	9.87	9.50	8.95	8.30

Observe that the gain stays relatively constant up to v_d neerly 20 mV. Then it decreases significantly with the increase in signal level. Whenever gain depends on signal level, nonlinear distortion occurs.

7.26

With:

$V_{B1} - V_{B2} = 10mV$

$i_{E1} = \frac{I}{1+e^{-10/25}} = 0.598 I$

Since $i_{E1} + i_{E2} = I$

$i_{E2} = 0.402 I$

For a collector resistance R_C

$v_O = v_{C1} - v_{C2} = (V_{CC} - i_{C1} R_C) - (V_{CC} - i_{C2} R_C)$

$= -(i_{C2} - i_{C1}) R_C$

$= -\alpha (i_{E2} - i_{E1}) R_C$

$\simeq -0.196 I \cdot R_C$

Thus, for

$v_O = 1V$; $0.196 I R_C = 1$

$I R_C = 5.102$

Now $I = 2mA$, thus

$R_C = 2.5K\Omega$

DC (bias) voltage at each collector

$= V_{CC} - \frac{I}{2} R_C = 10 - 1 \times 2.5$

$= 7.5V$

For a $-1V$ output swing, the minimum voltage at each collector is:

CONT.

$7.5 - 0.5 = 7.0V$

Thus, $V_{ICM}|_{max} = \underline{\underline{7V}}$

7.27

(a) $i_{E1} = \dfrac{I}{1 + e^{(V_{B2}-V_{B1})/V_T}}$

$i_{E2} = \dfrac{I}{1 + e^{(V_{B1}-V_{B2})/V_T}}$

For $v_{B1} - v_{B2} = 5mV$, $V_T = 25mV$

$i_{E1} = 0.550I$
$i_{E2} = 0.450I$

Note that $i_{E1} + i_{E2} = I$ as should be the case.

$\alpha \simeq 1$, thus $i_{C1} = 0.55I$
$i_{C2} = 0.45I$

$v_O = v_{C2} - v_{C1} \Rightarrow$

$= (V_{CC} - i_{C2}R_C) - (V_{CC} - i_{C1}R_C)$

$= (i_{C1} - i_{C2})R_C$

$= 0.10\, IR_C$

Voltage gain $A_v = \dfrac{v_O}{v_{B1}-v_{B2}} = \dfrac{0.1IR_C}{0.005}$

$= \underline{\underline{20\, IR_C}}$ V/V

(b) Bias voltage at each collector is: $V_{CC} - \dfrac{I}{2}R_C$.

For an output differential voltage of $0.1IR_C$, each collector should be allowed to fall by $(0.05IR_C)$ below its bias value, thus

$v_{Cmin} = V_{CC} - 0.5IR_C - 0.05IR_C$

$= V_{CC} - 0.55\, IR_C$

Thus for operation in saturation with

$v_{CB} = 0$ (minimum)

$v_{ICM}|_{max} = v_{Cmin} = V_{CC} - 0.55IR_C$

$= V_{CC} - 0.55\dfrac{A_v}{20}$

$= \underline{\underline{V_{CC} - 0.0275\, A_v}}$

Obviously for a given V_{CC}, increasing A_v reduces the maximum allowable v_{ICM}

A_v (V/V)	100	200	300	400
$V_{ICM\,max}$ (V)	$V_{CC}-2.75$	$V_{CC}-5.5$	$V_{CC}-8.25$	$V_{CC}-11$
IR_C (V)	5	10	15	20
R_C (kΩ)	5	10	15	20

Example: For $V_{CC} = 10V$, a gain of 200 can be achieved using $R_C = 10k\Omega$. The corresponding maximum input common-mode voltage is 4.5V. If a gain of 300 is required, it can be achieved by increasing R_C to 15 kΩ, however the maximum permitted input common-mode voltage then becomes 1.75V only.

7.28

$i_{C1} = \alpha i_{E1} \simeq i_{E1} = I/(1 + e^{-v_d/V_T})$

$i_{C2} = \alpha i_{E2} \simeq i_{E2} = I/(1 + e^{v_d/V_T})$

$v_O = v_{C2} - v_{C1} = (V_{CC} - i_{C2}R_C) \ldots$
$- (V_{CC} - i_{C1}R_C)$

$= IR_C\left(\dfrac{1}{1 + e^{-v_d/V_T}} - \dfrac{1}{1 + e^{v_d/V_T}}\right)$

For $IR_C = 5V$:

v_d (mV)	5	10	15	20	25	30	35	40
v_O (mV)	498	987	1457	1900	2311	2685	3022	3320
$\frac{v_O}{v_d}\left(\frac{V}{V}\right)$	99.7	98.7	97.1	95.0	92.4	89.5	86.3	83.0

CONT.

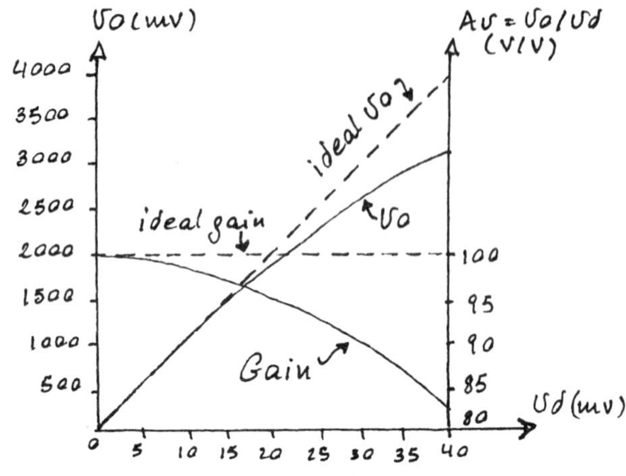

7.29

$I = 6\,mA$

The current will divide between the two transistors in proportion to their emitter areas. Thus with no input,

$I_{E1} = 1.5 I_{E2}$

$I_{E1} + I_{E2} = 2.5 I_{E2} = 6\,mA$

$I_{E2} = 2.4\,mA$

$I_{E1} = 3.6\,mA$

For $\alpha \simeq 1$

$I_{C1} = \underline{3.6\,mA}$

$I_{C2} = \underline{2.4\,mA}$

To equalize the collector currents we apply a difference signal $v_d = v_{B2} - v_{B1}$ whose value can be determined as follows:

$i_{E1} = I_{SE1}\, e^{((v_{B1} - v_E)/V_T)}$

$i_{E2} = I_{SE2}\, e^{((v_{B2} - v_E)/V_T)}$

where $I_{SE1}/I_{SE2} = 1.5$

Now, $i_{E1} = i_{E2}$ when

$1 = 1.5\, e^{(v_{B1} - v_{B2})/V_T}$

$v_d = v_{B2} - v_{B1} = V_T \ln 1.5 = \underline{10.1\,mV}$

7.30

Refer to Fig. P7.30

(a) For v_I sufficiently low so that Q_1 is cut off:

$V_{BE}|_{Q3} = 0.7 + V_T \ln \frac{10}{1}$

$= 0.76\,V$

$V_{OH} = v_O = 5 - \frac{10}{101} \times 1 - V_{BE}$

$= 5 - \frac{10}{101} - 0.76$

$= \underline{4.14\,V}$

+5V, 1KΩ, Q_3, v_O, 10mA

(b) For v_I sufficiently high so that Q_1 is conducting all the current I:

$V_{BE}|_{Q3} = 0.76\,V$

$V_{OL} = v_O = 5 - \left(\frac{10}{101} + 0.99\right) \times 1 - 0.76$

$= \underline{3.15\,V}$

+5V, 1KΩ, $\left(\frac{10}{101} + 0.99\right)$ mA, Q_3, $\frac{10}{101}$, 1mA × α, Q_1, v_O, 10mA

(c) $i_{E1} = I / \left(1 + e^{(v_{B2} - v_{B1})/V_T}\right)$

$= I/100$

$e^{(3.64 - v_I)/V_T} = 99$

$v_I = 3.64 - V_T \ln 99 = \underline{3.525\,V} = V_{IL}$

(d) $i_{E1} = 0.99 I = \dfrac{I}{1 + e^{(3.64 - V_{IH})/V_T}}$

$1 + e^{(3.64 - V_{IH})/V_T} = 1.01$

$V_{IH} = 3.64 + V_T \ln 100 = \underline{3.755\,V}$

(e)

v_O; $V_{OH} = 4.14$; $\frac{1}{2}(V_{OL} + V_{OH}) = 3.64$; $V_{OL} = 3.15$; slope = +1; $NM_H = \underline{0.383\,V}$; $NM_L = \underline{0.375\,V}$; 3.15; $V_{IL} = 3.525\,V$; $V_{IH} = 3.755\,V$; 4.14; v_I

Midpoint: $(V_{IL} + V_{IH})/2 = 3.64\,V$

CONT.

Since the mid-point of the output voltage swing is equal to 3.64V, the output voltage swing is centred on the mid-point of the input range; an ideal choice for it equalizes the noise margins.

7.31

Each device is operating at a current of 150μA = 0.15mA. Thus,

$$g_m = \frac{0.15\,mA}{25\,mV} = \underline{\underline{6\,\frac{mA}{V}}}$$

$$R_{id} = 2(\beta+1)\,r_e = 2\,r_\pi$$

$$= 2 \times \frac{150}{4} = \underline{\underline{75}}\,K\Omega$$

7.32

$R_{id} \geq 10K\Omega$; $A_d = 200\,V/V$;
$\beta \geq 100$; $V_{cc} = 10V$

$$R_{id} = 10^4 = 2r_\pi = 2 \times \frac{100}{g_m}$$

$$\Rightarrow g_m = 20\,mA/V$$

Thus each device is operating at 0.5mA and $I = \underline{\underline{1mA}}$

$$\text{Voltage gain} = g_m . R_c$$

$$200 = 20\,R_c$$

$$\Rightarrow R_c = \underline{\underline{10K\Omega}}$$

7.33

$$\underline{\underline{5mV}}$$

$$r_e = \frac{V_T}{I/2} = \frac{25mV}{50\mu A} = \underline{\underline{500\Omega}}$$

$$\text{Half-circuit gain} = \frac{\alpha R_c}{r_e} \simeq \frac{R_c}{r_e}$$

$$= \frac{10K}{500} = \underline{\underline{20}}\,V/V$$

At one collector we expect a signal of (+100mV) and at the other a signal of (−100mV)

7.34

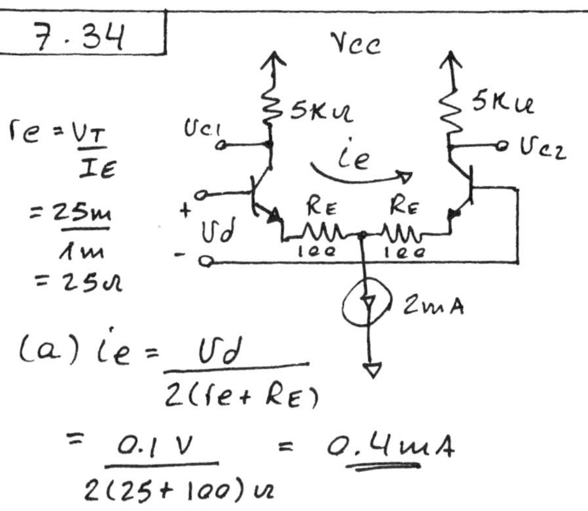

$$r_e = \frac{V_T}{I_E} = \frac{25m}{1m} = 25\Omega$$

(a) $i_e = \dfrac{v_d}{2(r_e + R_E)}$

$$= \frac{0.1\,V}{2(25+100)\,\Omega} = \underline{\underline{0.4mA}}$$

(b) $i_{E1} = 1 + 0.4 = \underline{\underline{1.4}}\,mA$
$i_{E2} = 1 - 0.4 = \underline{\underline{0.6mA}}$

(c) $v_{c1} = -i_e R_c \simeq -0.4 \times 5 = \underline{\underline{-2V}}$
$v_{c2} = \underline{\underline{+2V}}$

(d) $v_{od} = 4V$
$A_d = v_{od}/v_{id} = \dfrac{4}{0.1} = \underline{\underline{40}}\,V/V$

7.35

Rc Rc
− 4V +
+ 95mV − + 95mV −
+ 5mV − RE RE + 5mV
+ 0.2 −
I

$$\frac{2R_c}{2(r_e + R_E)} \simeq \frac{4}{0.2}$$

$$R_c = 20\,(r_e + R_E)$$

$$\frac{R_E}{r_e} = \frac{95}{5} = 19$$

$\beta \geq 200$
$R_{in} \geq 80K$

CONT.

$$R_{in} = 2(\beta+1)(r_e + R_E)$$
$$= 2 \times 201 \times 20 r_e = 80 K\Omega$$
$$\Rightarrow r_e \simeq \frac{80000}{8000} = 10\Omega$$

Thus each device is operating at a current of $\frac{25mV}{10\Omega} = 2.5mA$

$$\Rightarrow I = \underline{5mA}$$

$$R_E = 19 \times 10 = \underline{190\Omega}$$
$$R_C = 20 \times 200 = \underline{4K\Omega}$$

7.36

$\underline{2M\Omega}$

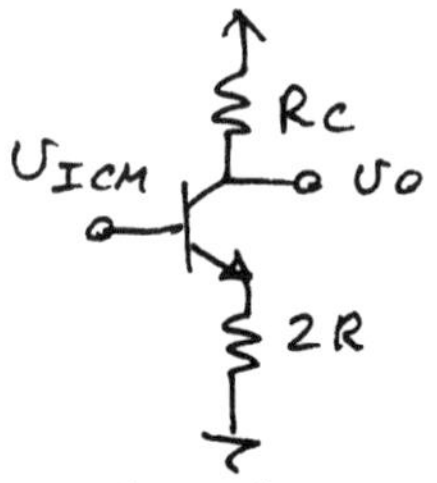

$$\frac{V_o}{V_{ICM}} \simeq \frac{R_C}{2R} = \frac{20K\Omega}{2M\Omega}$$
$$= \underline{0.01} \; V/V \quad \text{or } \underline{-40dB}$$

7.37

Refer to Fig. P7.37

$$\frac{V_o}{V_i} = \frac{\alpha \times 20K\Omega}{(2r_e + 2 \times 200)\,\Omega}$$

Where $r_e = \frac{V_T}{0.5/2} = \frac{0.05V}{0.5mA} = 100\Omega$

$$\frac{V_o}{V_i} \simeq \frac{20\,000}{600} = \underline{33.3V/V}$$

$$R_i = (\beta+1)(2r_e + 2 \times 200)$$
$$= 101 \times 2 \times 300 \simeq \underline{60K\Omega}$$

7.38

Refer to Fig P.7.38

Each transistor is operating at $I_E = 1mA$, thus

$r_e = 25\Omega$ and $r_\pi = 101 \times 25 = 2525\Omega$

$$\frac{V_o}{V_i} = \frac{\alpha \times 7.5K\Omega}{(2r_e + 200)\Omega} \simeq \frac{7500}{250} = \underline{30\,V/V}$$

$$R_i = (\beta+1)(r_e + 200 + r_e) \simeq \underline{25K\Omega}$$

7.39

Refer to Fig P.7.39

(a) As a differential amplifier, the gain is given by

$$\frac{V_o}{V_i} = \frac{\alpha R_C}{2r_e} \quad (1)$$

(b) Transistor Q_1 can be considered as a common-collector stage. It is biased at $I/2$ and has a resistance r_{e2} in its emitter, thus.

$$i_e = \frac{V_i}{r_{e1} + r_{e2}}$$
$$= \frac{V_i}{2r_e} \quad (2)$$

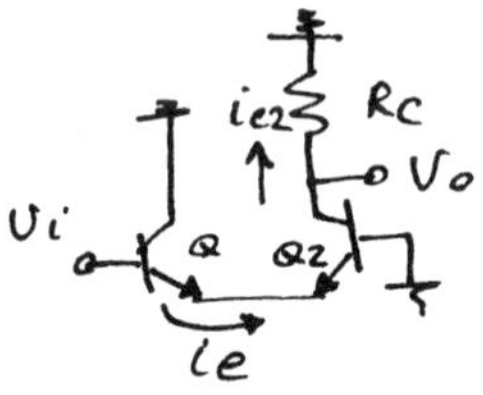

Now, Q_2 is connected in the common-base configuration. It has an input signal current i_e. Thus its collector signal current (in the direction indicated) will be $i_{e2} = \alpha i_e$ (3)

The output voltage will be

$$V_o = i_{e2} \, R_C \quad (4)$$

Combining equations (2) (3) and (4) provides:

CONT.

$\frac{v_o}{v_i} = \frac{\alpha R_C}{2r_e}$, which is identical to the result found above in part (a)

7.40

(a) $V_E = -0.7V$

Thus, $I = \frac{-0.7-(-5)}{4.3} = 1mA$ $r_e = 50\Omega$

$A_d = \frac{v_o}{v_d} = \frac{\alpha \times 2}{2r_e} \simeq \frac{2000}{2\times 50} = \underline{\underline{20}}$ V/V

(b) $A_{cm} \simeq \frac{\alpha R_C}{8.6 + r_e}$

$= \frac{2}{8.65} = \underline{\underline{0.23}}$ V/V

(c) CMRR

CM half circuit

$= 20\log\left|\frac{A_d}{A_{cm}}\right| = 20\log\frac{20}{0.23}$

$= \underline{\underline{38.8 \text{ dB}}}$

(d) $v_d = 2\times 0.005 \sin(2\pi\times 1000t)$

$v_{Icm} = 0.1 \sin(2\pi\times 60t)$

Thus, $v_o = A_d \cdot v_d + A_{cm}\cdot v_{Icm}$

$= 20\times 0.01 \sin 2\pi\times 1000t$

$+ 0.23\times 0.1 \sin(2\pi\times 60t)$

$= \underline{\underline{0.2 \sin 2\pi\times 1000t + 0.023 \sin 2\pi 60t}}$

7.41

Each transistor is biased at 1mA. Thus,

$r_e = 25\Omega$, $g_m = 40mA/V$

$r_o = 100/1 = 100K\Omega$

The differential half-circuit is

$A_d = \frac{v_o}{v_i} = \alpha\frac{[R_C \| (R_L/2)]}{r_e + R_E/2}$

$\simeq \frac{10\|5}{0.025+0.100} = \underline{\underline{26.7}}$ V/V

$R_{id} = 2[R_B \| (\beta+1)(r_e + R_E/2)]$

$= 2[30 \| 101(0.025+0.100)]$

$= \underline{\underline{17.8 \text{ K}\Omega}}$

The common-mode half circuit

$A_{cm} = \frac{v_o}{v_{Icm}} \simeq \frac{10}{300}$

$= \frac{1}{30} = \underline{\underline{0.033}}$ V/V

$2R_{icm} = 30K \| 7.5M\Omega$

$= 30K\Omega$

$R_{icm} = \underline{\underline{15K\Omega}}$

Without the R_B resistors $R_{icm} = \underline{\underline{3.75 \text{ M}\Omega}}$

$\simeq 30 \| 10 = 7.5M\Omega$

7.42

(a) $A_d\big|_{\text{single-ended output}} = \frac{\alpha(R_C \| r_o)}{2r_e}$

where $r_e = \frac{0.025V}{0.25mA} = 100\Omega$

$r_o = \frac{200V}{0.25mA} = 800K\Omega$

$A_d\big|_{\text{single ended}} \simeq \frac{20}{2\times 0.1} = \underline{\underline{100}}$ V/V

(b) $A_d\big|_{\text{diff output}} = 2\times A_d\big|_{\text{single ended}}$

$= \underline{\underline{200}}$ V/V

(c) $R_{id} = 2r_\pi = 2\times 201\times 100$

$= \underline{\underline{40.2K\Omega}}$

CONT.

(d) $A_{cm}\Big|_{\text{single-ended output}} = \dfrac{R_C}{2R}$

$= \dfrac{20}{2000} = \underline{\underline{0.1\ V/V}}$

(e) $A_{cm}\Big|_{\text{diff out}} = 0$

7.43

The output resistance of the current source $(R) = r_o = \dfrac{V_A}{0.1\,mA}$

$= \dfrac{200}{0.1} = 2\,M\Omega$

r_o for each transistor is the pair $= 200/0.05 = 4\,M\Omega$

Thus,

$R_{icm} = (\beta+1)\dfrac{r_o}{2} \,\|\, (\beta+1)R$

$= \dfrac{51 \times 4}{2} \,\|\, 51 \times 2 = \underline{\underline{51\ M\Omega}}$

7.44

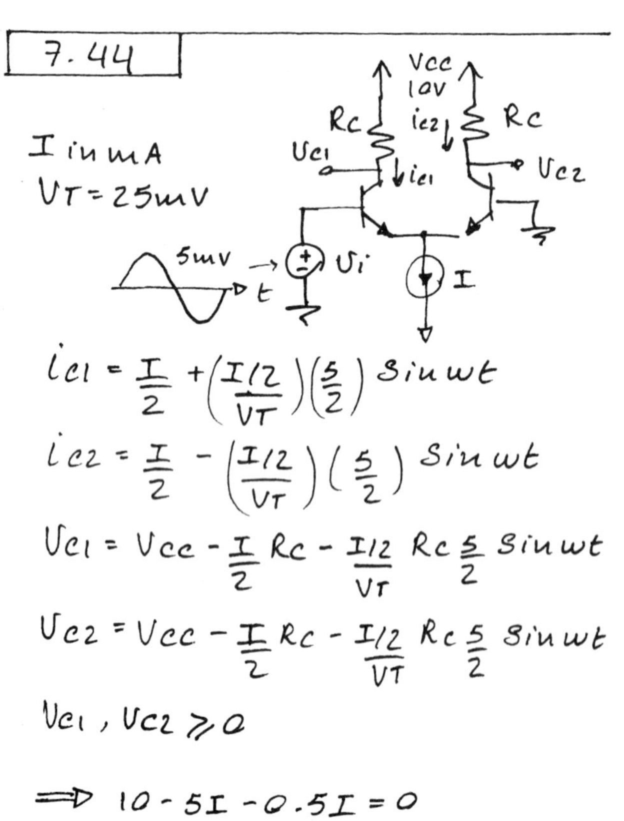

I in mA

$V_T = 25\,mV$

$i_{c1} = \dfrac{I}{2} + \left(\dfrac{I/2}{V_T}\right)\left(\dfrac{5}{2}\right)\sin\omega t$

$i_{c2} = \dfrac{I}{2} - \left(\dfrac{I/2}{V_T}\right)\left(\dfrac{5}{2}\right)\sin\omega t$

$V_{c1} = V_{cc} - \dfrac{I}{2}R_C - \dfrac{I/2}{V_T}R_C\dfrac{5}{2}\sin\omega t$

$V_{c2} = V_{cc} - \dfrac{I}{2}R_C - \dfrac{I/2}{V_T}R_C\dfrac{5}{2}\sin\omega t$

$V_{c1}, V_{c2} \geq 0$

$\Rightarrow 10 - 5I - 0.5I = 0$

$I = \underline{\underline{1.8\,mA}}$

$V_{c1} = V_{c2} = 1V$

$A_d = \dfrac{20K\Omega}{2r_e}$, where

$r_e = \dfrac{25}{0.9} = 27.8\,\Omega$

Thus, $A_d = \underline{\underline{360\,V/V}}$

Vc1: +1V, 0.9V, t
Vc2: 0.9, +1V, t
Vc2 − Vc1: 1.8V, t

$v_{c2} - v_{c1} = \underline{\underline{1.8 \sin\omega t}}$, V

7.45

At $I_C = 1\,mA$, $r_e = 25\,\Omega$

$R_{id} = (\beta+1)\,2r_e = 5.05\,K\Omega < 10K$

$\Rightarrow$ need emitter resistors

In this case:

$R_{id} = (\beta+1)(2r_e + 2R_E) = 10\,K\Omega$

Thus, $R_E = \underline{\underline{25\,\Omega}}$

Rc, Rc, RE, RE, 2mA, Ro

$A_d = 100 = \dfrac{R_C}{2(R_E + r_e)}$

$\Rightarrow R_C = \underline{10K}$

$A_{cm} = 0.1 \geq \dfrac{R_C}{2R_o + R_E + r_e}$

$\Rightarrow R_o \geq \underline{\underline{50\,K\Omega}}$

For ± 2 swing $V_{c1} = V_{c2}$

$= V_{cc} - \dfrac{I}{2}R_C = 2$

$\Rightarrow V_{cc} = 2 + 10^{-3} \times 10^{4} = 12V$

Choose $\underline{\underline{V_{cc} = \pm 15V}}$ although 12V is OK.

$2R_{icm} = (\beta+1)(2R_o + R_E + r_e)$

$\Rightarrow R_{icm} = \underline{\underline{5\,M\Omega}}$

7.46

Taken single-endedly

$$A_{cm_s} = \frac{\alpha R_C}{2R_o}$$

Let collector resistors be R_C & $R_C + \Delta R_C$, then

$$A_{cm} = \frac{\alpha}{2R_o}(R_C + \Delta R_C - R_C)$$

$$= \frac{\alpha \Delta R_C}{2R_o}$$

Which can be written as

$$A_{cm_d} = \frac{\alpha R_C}{2R_o} \cdot \frac{\Delta R_C}{R_C} = A_{cm_s}\frac{\Delta R_C}{R_C}$$

$$CMRR = \frac{A_d}{A_{cmd}} = \frac{2.A_s}{A_{cm_s}\frac{\Delta R_C}{R_C}}$$

$$= \frac{A_s}{A_{cm_s}} \cdot \frac{2}{\frac{\Delta R_C}{R_C}}$$

Thus, $20\log\frac{2}{\frac{\Delta R_C}{R_C}} = 40\,dB$

$\rightarrow \Delta R_C / R_C = \underline{\underline{2\%}}$

7.47

The bias current will split between the two transistors according to their area ratio. Thus the large-area device will carry twice the current of the other device.

That is, the bias currents will be $2I/3$ and $I/3$.

Now with v_{Icm} applied, the CM signal current will ($\rightarrow v_{Icm}/R$) split between Q_1 and Q_2 in the same ratio. This is because their r_e values will be related in the same way. Thus, if Q_1 is the large device r_{e1} will be half the value of r_{e2}.

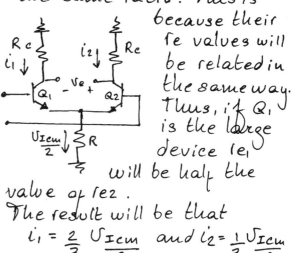

The result will be that

$$i_1 = \frac{2}{3}\frac{v_{Icm}}{R} \text{ and } i_2 = \frac{1}{3}\frac{v_{Icm}}{R}$$

Thus the differential output voltage v_o will be

$$v_o = 0 - i_2R_C - (-i_1R_C) = (i_1 - i_2)R_C$$

$$= \frac{1}{3}\frac{v_{Icm}}{R} \cdot R_C$$

$$A_{cm} = \frac{1}{3}\frac{R_C}{R} = \frac{1}{3} \times \frac{12}{1000} = \underline{\underline{0.004\,V/V}}$$

7.48

For $I = 200\,\mu A$:

$$g_m = \sqrt{2k_n'W/L\,I_D} = \sqrt{2 \times 4 \times 0.1}$$

$$= 0.89\,mA/V$$

$R_D = 10\,K\Omega$

Thus, $A_d = g_m R_D = 10 \times 0.89 = \underline{\underline{8.9\,V/V}}$

$$V_{os} = \frac{(V_{GS} - V_t)}{2} \cdot \frac{\Delta R_D}{R_D}$$

where $\frac{\Delta R_D}{R_D} = 0.02$ (worst case)

and $V_{GS} - V_t = \sqrt{\frac{2I_D}{k_n'W/L}} = \sqrt{\frac{2 \times 0.1}{4}}$

$= \underline{\underline{0.223\,V}}$

CONT.

Thus, $V_{OS} = \frac{1}{2} \times 0.223 \times 0.02$

$= \underline{2.23 mV}$

For $I = 400 \mu A$:

$g_m = \sqrt{2 \times 4 \times 0.2} = 1.265 mA/V$

$A_d = \underline{12.65} V/V$

$V_{OV} = V_{GS} - V_t = 0.316 V$

$V_{OS} = \frac{1}{2} \times 0.316 \times 0.02 = \underline{3.16 mV}$

Thus both A_d and V_{OS} increase by the same ratio since both are proportional to $\sqrt{I}$

7.49

Worst cases: $\Delta V_t = 10 mV$

$\frac{\Delta R_D}{R_D} = 0.04$; $\frac{\Delta(W/L)}{(W/L)} = 0.04$

V_{OS1} (due to ΔR_D) $= \frac{V_{OV}}{2} \frac{\Delta R_D}{R_D} = \frac{0.3 \times 0.04}{2}$

$= 6 mV$

V_{OS2} (due to $\Delta W/L$) $= \frac{V_{OV}}{2} \frac{\Delta W/L}{W/L} = \frac{0.3 \times 0.04}{2}$

$= 6 mV$

V_{OS3} (due to ΔV_t) $= \Delta V_t = \underline{10 mV}$

Since these offsets are not correlated

$V_{OS} = \sqrt{V_{OS1}^2 + V_{OS2}^2 + V_{OS3}^2}$

$V_{OS} = \sqrt{6^2 + 6^2 + 10^2} = \underline{13.11 mV}$

The major contribution is due to the threshold mismatch ΔV_t.

To find the required mismatch ΔR_D that can correct for V_{OS}

$13.11 mV = \frac{V_{OV}}{2} \cdot \frac{\Delta R_D}{R_D}$

$\Rightarrow \frac{\Delta R_D}{R_D} = \frac{2 \times 13.11 mV}{0.3 V}$

$= 0.087$ or $\underline{8.7\%}$

If ΔV_t is reduced by a factor of 10 to 1 mV, V_{OS} reduces to:

$\sqrt{6^2 + 6^2 + 1^2} = 8.54 mV$

and $\frac{\Delta R_D}{R_D} = \frac{2 \times 8.54 mV}{0.3 V} = \underline{5.69\%}$

7.50

$V_{OV} = \sqrt{\frac{I}{k_n' W/L}} = \sqrt{\frac{100}{100 \times 20}} = 0.223 V$

Using Eqn. (7.112) we obtain V_{OS} due to $\Delta R_D / R_D$ as:

$V_{OS} = \frac{V_{OV}}{2} \frac{\Delta R_D}{R_D} = \frac{0.223 \times 0.05}{2}$

$= \underline{5.57 mV}$

From Eqn. (7.117), V_{OS} due to $\Delta(W/L)/(W/L)$ is:

$V_{OS} = \left(\frac{V_{OV}}{2}\right) \frac{\Delta W/L}{W/L} = \frac{0.223 \times 0.05}{2}$

$= \underline{5.57 mV}$

The offset arising from ΔV_t is (Eqn. (7.120)):

$V_{OS} = \Delta V_t = \underline{5 mV}$

Worst case offset is:

$5.57 + 5.57 + 5 = 16.15 mV$

Applying the root-sum-of-squares (Eq. (7.121))

$V_{OS} = \sqrt{2(5.57m)^2 + 5m^2} = \underline{9.33 mV}$

7.51

$$\Delta V_C = \Delta R_C \cdot \frac{I}{2}$$

$$A_d = \frac{R_C}{r_e} = \frac{R_C}{V_T / \frac{I}{2}} = \frac{I R_C}{2 V_T}$$

$$\Rightarrow V_{OS} = \frac{\Delta V_C}{A_d} = \frac{\Delta R_C}{R_C} \cdot V_T$$

$$= 0.1 \times 25 = \underline{\underline{2.5\,mV}}$$

7.52

Eq (7.130) $V_{OS} = V_T \cdot \frac{\Delta I_S}{I_S}$

$$= 25 \times 0.1 = \underline{\underline{2.5\,mV}}$$

7.53

$$\Delta V_C = \Delta R_C \frac{I}{2}$$

$$A_d = \frac{R_C}{r_e + R_E} = \frac{R_C}{\frac{2V_T}{I} + R_E} = \frac{I R_C}{2V_T + I R_E}$$

$$V_{OS} = \frac{\Delta V_C}{A_d} = \underline{\underline{\frac{\Delta R_C}{R_C}\left(V_T + \frac{I R_E}{2}\right)}}$$

7.54

$$\Delta V_C = \alpha_1 \frac{I}{2} R_C - \alpha_2 \frac{I}{2} R_C$$

$$= \frac{I}{2} R_C (\alpha_1 - \alpha_2)$$

$$= \frac{I}{2} R_C \left(\frac{\beta_1}{\beta_1 + 1} - \frac{\beta_2}{\beta_2 + 1}\right)$$

For $\beta_1, \beta_2 \gg 1$

$$\Delta V_C = \frac{I}{2} R_C \cdot \frac{\beta_1 - \beta_2}{\beta_1 \cdot \beta_2}$$

$$= \frac{I}{2} R_C \left(\frac{1}{\beta_2} - \frac{1}{\beta_1}\right)$$

$$A_d = \frac{R_C}{r_e} = \frac{I R_C}{2 V_T}$$

$$V_{OS} = \frac{\Delta V_C}{A_d} = V_T \left(\frac{1}{\beta_2} - \frac{1}{\beta_1}\right) \quad \text{Q.E.D.}$$

For $\beta_1 = 100$ and $\beta_2 = 200$

$$V_{OS} = 25 \left(\frac{1}{200} - \frac{1}{100}\right)$$

$$= \underline{\underline{-125\,\mu V}}$$

7.55

Since the two transistors are matched except for their V_A value, we can express the collector currents when the input terminals are grounded as,

$$I_{C1} = I_C \left(1 + \frac{V_{CE}}{V_{A1}}\right)$$

$$I_{C2} = I_C \left(1 + \frac{V_{CE}}{V_{A2}}\right)$$

where I_C can be determined from

$$I_{C1} + I_{C2} = I$$

$$\Rightarrow I_C = \frac{I}{2 + \frac{V_{CE}}{V_{A1}} + \frac{V_{CE}}{V_{A2}}}$$

Note that for $V_{CE} \ll V_{A1}, V_{A2}$, $I_C \simeq \frac{I}{2}$. Thus, the differential gain A_d can still be written as

$$A_d \simeq \frac{R_C}{r_e} = \frac{I R_C}{2 V_T}$$

The offset voltage at the output can be found from

$$\Delta V_C = V_{C2} - V_{C1} = (I_{C1} - I_{C2}) R_C$$

CONT.

$$= I_C R_C \left(\frac{V_{CE}}{V_{A1}} - \frac{V_{CE}}{V_{A2}} \right)$$

$$\simeq \frac{I}{2} R_C \left(\frac{V_{CE}}{V_{A1}} - \frac{V_{CE}}{V_{A2}} \right)$$

Thus, $V_{OS} = \frac{\Delta V_C}{A_d}$

$$V_{OS} = V_T \left(\frac{V_{CE}}{V_{A1}} - \frac{V_{CE}}{V_{A2}} \right)$$

For $V_{CE} = 10V$, $V_{A1} = 100V$ and $V_{A2} = 300V$

$$V_{OS} = 25 \left(\frac{10}{100} - \frac{10}{300} \right)$$

$$= \underline{\underline{1.7 mV}}$$

7.56

Equating the incremental changes in voltage from ground to emitter on both sides of the pair (and neglecting second-order terms i.e. Δ×Δ terms):

$$\frac{I}{2(\beta+1)} \cdot \frac{\Delta R_S}{2} - \frac{\Delta I}{2(\beta+1)} \cdot R_S - \frac{\Delta I}{2} \cdot r_e$$

$$\simeq \frac{-I}{2(\beta+1)} \frac{\Delta R_S}{2} + \frac{\Delta I}{2(\beta+1)} R_S + \frac{\Delta I}{2} r_e$$

$$\Delta I \left[r_e + \frac{R_S}{\beta+1} \right] = \frac{I}{2(\beta+1)} \cdot \Delta R_S$$

$$\Delta I = \frac{I \, \Delta R_S}{2(\beta+1)} \cdot \frac{1}{r_e + \frac{R_S}{\beta+1}}$$

$$\Delta V_C = -\Delta I \cdot R_C$$

$$= -\frac{I R_C \Delta R_S}{2(\beta+1)} \cdot \frac{1}{r_e + R_S/(\beta+1)}$$

$A_d = R_C / r_e$

Thus, $V_{OS} = \Delta V_C / A_d$

$$= -\frac{I \Delta R_S}{2(\beta+1)} \cdot \frac{r_e}{r_e + \frac{R_S}{\beta+1}}$$

For $\frac{R_S}{\beta+1} \ll r_e$ and $\beta \gg 1$,

$$|V_{OS}| \simeq \frac{I}{2\beta} (\Delta R_S) \quad \text{Q.E.D.}$$

7.57

Refer to Fig. P7.57.

(a) $R_{C1} = 5 \times 1.05 = 5.25 K\Omega$

$R_{C2} = 5 \times 0.95 = 4.75 K\Omega$

Perfect offset nulling will be achieved when x is such that

$R_{C1} + (x \times 1K\Omega) = R_{C2} + (1-x) \times 1K\Omega$

$\Longrightarrow 5.25 + x = 4.75 + 1 - x$

$\Longrightarrow x = \underline{\underline{0.25}}$

(b) $I_{C1} = 1.05 mA$

$I_{C2} = 0.95 mA$

Offset nulling is achieved when x is such that

$1.05(x+5) = 0.95((1-x)+5)$

$x = \underline{\underline{0.225}}$

7.58

$$I_{B\,max} = \frac{I/2}{\beta_{min}+1} = \frac{300}{80+1} = \underline{\underline{3.7 \mu A}}$$

$$I_{B\,min} = \frac{I/2}{\beta_{max}+1} = \frac{300}{200+1} = \underline{\underline{1.5 \mu A}}$$

$$I_{OS} = I_{B\,max} - I_{B\,min} = \underline{\underline{2.2 \mu A}}$$

7.59

+15V

27KΩ 27KΩ

v_{C1} v_{C2}

Q_1 Q_2

$v_{CM}+\frac{v_d}{2}$ $v_{CM}-\frac{v_d}{2}$

0.5mA

−5V

$$v_{C1} \geq v_{CM} + \frac{v_d}{2}$$

$$15 - 0.25\times 27 - g_m \frac{v_d}{2}\times 27 \geq v_{CM} + \frac{v_d}{2}$$

Thus

$$v_{CM}\Big|_{max} = 15 - 0.25\times 27 + -100\times 0.01\times 27 - 0.01$$

$$= \underline{\underline{+5.54\text{V}}}$$

To find v_{CM} min we observe that to keep the current-source transistor in the active region, the voltage at its collector should exceed the collector-emitter saturation voltage (0.2-0.3V). Thus, $v_{E\,min} = -4.7$V and correspondingly $v_{C\,min} = \underline{-4\text{V}}$

7.60

$$I_{E1} = \frac{2}{3}I \text{ and } I_{E2} = \frac{1}{3}I$$

(Q_1 twice the area of Q_2)

$$\Delta V_C = V_{C2} - V_{C1} \simeq \underline{\underline{\frac{1}{3} I R_C}}$$

Nominally,

$$A_d = \frac{R_C}{r_e} = \frac{I R_C}{2V_T}$$

$$V_{OS} = \frac{\Delta V_C}{A_d} = \frac{2}{3}V_T = \underline{\underline{16.7\text{mV}}}$$

Thus, small-signal analysis predicts that a 16.7mV DC voltage applied as $v_{B2} - v_{B1} = 16.7$ would restore the current balance in the pair and reduce ΔV_C to zero.

Using large-signal analysis:

$$i_{E1} = I_{S1} \cdot e^{\frac{v_{B1}-v_E}{V_T}}$$

$$i_{E2} = I_{S2} \cdot e^{\frac{v_{B2}-v_E}{V_T}}$$

Thus,

$$\frac{i_{E1}}{i_{E2}} = \frac{I_{S1}}{I_{S2}} \cdot e^{\frac{v_{B1}-v_{B2}}{V_T}}$$

To restore balance, $i_{E1} = i_{E2}$, thus

$$1 = 2e^{\frac{V_{B1}-V_{B2}}{V_T}}$$

$$\Rightarrow V_{B1} - V_{B2} = -V_T \ln 2$$

$$V_{B2} - V_{B1} = \underline{\underline{17.3\text{mV}}}$$

Nominally

$$I_B = \frac{I/2}{\beta+1} \simeq \frac{100}{2\times 100} = \underline{\underline{0.5\mu A}}$$

But with the imbalance,

$$I_{B1} \simeq \frac{2I/3}{\beta} = \frac{2\times 100}{300} = 0.67\mu A$$

$$I_{B2} = \frac{I/3}{\beta} = \frac{100}{300} = 0.33\mu A$$

$$I_B = \frac{I_{B1}+I_{B2}}{2} = \underline{\underline{0.5\mu A}}$$

$$I_{OS} = |I_{B1} - I_{B2}| = \underline{\underline{0.34\mu A}}$$

7.61

$R_C = 20$KΩ ; $A_d = 90$ V/V

$V_{OS} = \pm 3$mV

Worst case $|V_{OS}|$ is 3mV

$$|V_{OS}| = V_T\left(\frac{\Delta R_C}{R_C}\right) \quad |V_{OS}| = 3\text{mV}$$

$$\Rightarrow \frac{3m\times 20K}{25m} = 2.4K, \underline{\underline{\Delta R_C = 2.4\text{K}\Omega}}$$

CONT.

This is the maximum mismatch that occurs in R_c.
Thus, if the lowest collector resistor is adjusted from $R_{c\,min} + \Delta R$ with ΔR varying between zero and $2.4\,k\Omega$, then the offset would be eliminated!!
This can be achieved with the following circuit:

When R_{c1} & R_{c2} are equal the potentiometer is tuned to the middle point. In the worst case, when either R_c is higher by $2.4\,k\Omega$, the potentiometer is adjusted to one extreme such as to increase the lowest R_c by $2.4\,k\Omega$. In all other cases when ΔR_c is distributed between R_{c1} and R_{c2} the potentiometer is adjusted as in Problem 7.57 above.

7.62

For each transistor $I_D = I/2$.
From Eqn. (7.147): $r_{o2} = r_{o4} = r_o$

$$A_d = \frac{1}{2} g_m r_o$$

but $g_m = \frac{2I_D}{V_{OV}}$ and $r_o = \frac{V_A}{I_D}$

$$\Rightarrow A_d = \frac{1}{2}\left(\frac{2I_D}{V_{OV}}\right)\frac{V_A}{I_D} = \frac{V_A}{V_{OV}}$$

$\rightarrow 80\,V/V = 20\,V/V_{OV}$

$\rightarrow V_{OV} = 20/80 = 0.25\,V$

Finally,

$$I = 2I_D = k\frac{W}{L}V_{OV}^2$$

$$= 3.2\,\frac{mA}{V^2}(0.25\,V)^2 = \underline{\underline{0.2\,mA}}$$

7.63

For all transistors $I_D = I/2$ and all r_o's are equal.

$$V_{OV_{100}} = \sqrt{\frac{2I_D}{k'W/L}} = \sqrt{\frac{100\,\mu A}{0.2\,mA/V^2}} = 0.707\,V$$

$$V_{OV_{200}} = \sqrt{\frac{400\,\mu A}{0.2\,mA/V^2}} = 1.414\,V$$

(a) For $I = 100\,\mu A$:
Range of the differential mode is $-\sqrt{2}\,V_{OV} \le v_{id} \le \sqrt{2}\,V_{OV}$ (as in Eqn. (7.10))
But the range of v_O is limited by the requirement of keeping the transistors in saturation mode.

$V_{O\,min} = V_{G2} - V_t$
$= 0 - V_t$
$= \underline{-V_t}$

$V_{O\,max} = V_{G4} + |V_t|$
$= V_{DD} - |V_{GS}| + |V_t|$
$= 5 - 0.707 = \underline{4.29\,V}$

$$g_m = \frac{2I_D}{V_{OV}} \rightarrow g_{m1} = g_{m2} = \frac{100\,\mu}{0.707}$$

$$= \underline{0.1414\,mA/V}$$

$$r_o = \frac{V_A}{I_D} \Rightarrow r_{o2} = r_{o4} = \frac{20}{50\,\mu} = \underline{\underline{400\,k\Omega}}$$

$$R_o = r_{o2} \| r_{o4} = \frac{1}{2} \times 400\,K = \underline{\underline{200\,k\Omega}}$$

$$A_d = \frac{1}{2} g_m r_o = \frac{1}{2} \times (0.1414\,m)(400\,K)$$

CONT.

$A_d = \underline{\underline{28.28\ V/V}}$

(b) For $I = 400\mu A$:

Linear range of v_O

$V_{O\,min} = -V_t$ //

$V_{O\,max} = (5V - |V_{GS}|) + |V_t|$

$= \underline{\underline{5 - 1 = 4V}}$

$g_m = \dfrac{2 \times 200\mu}{1.414} = \underline{\underline{0.28\ mA/V}}$

$r_o = 20/(400/2) = \underline{\underline{100\,K\Omega}}$

$R_o = 1/2\ 100K\Omega = \underline{\underline{50\,K\Omega}}$

$A_d = \dfrac{1}{2}(0.28m \times 100K) = \underline{\underline{14\ V/V}}$

7.64

Recall from Equ. (7.155)

$CMRR = (g_m r_o)(g_m R_{SS})$

(a) For a simple current mirror

$R_{SS} = r_{o_S} \Rightarrow$ (for $I_D = I/2$)

$CMRR = (g_m r_o)(g_m r_{oS})$

$= \left(\dfrac{2I_D}{V_{OV}} \cdot \dfrac{V_A}{I_D}\right) \cdot \left(\dfrac{2I_D}{V_{OV}} \cdot \dfrac{V_A}{2I_D}\right)$

$= 2 \cdot \dfrac{V_A}{V_{OV}} \cdot \dfrac{V_A}{V_{OV}}$

$= 2\left(\dfrac{V_A}{V_{OV}}\right)^2$ // Q.E.D.

(b) for the modified Wilson current source of Fig. P7.64

$R_{SS} \simeq g_{m7} \cdot r_{o7} \cdot r_{o5}$

$\Rightarrow CMRR = (g_m r_o)(g_m \cdot g_{m7}\, r_{o7}\, r_{o5})$

For $Q_{5,6,7,8}$:

$V_{OV_S} = \sqrt{\dfrac{2I}{K'W/L}}$

while for $Q_{1,2,3,4}$:

$V_{OV} = \sqrt{\dfrac{I}{K'W/L}}$

$\Rightarrow V_{OV_S} = \sqrt{2}\, V_{OV}$.

Thus, (for $I = 2I_D$)

$CMRR = \dfrac{I}{V_{OV}} \cdot \dfrac{V_A}{(I/2)} \cdot \dfrac{I}{V_{OV}} \cdot \dfrac{2I}{\sqrt{2}V_{OV}} \cdot \dfrac{V_A}{I} \cdot \dfrac{V_A}{I}$

$= \dfrac{4}{\sqrt{2}} \dfrac{V_A^3}{V_{OV}^3} = \underline{\underline{2 \cdot \sqrt{2}\, \dfrac{V_A^3}{V_{OV}^3}}}$

For $K'W/L = 10mA/V^2$

$I = 1mA$

$|V_A| = 10V$

$V_{OV} = \sqrt{\dfrac{1mA}{10mA/V^2}} = 0.316\ V$

$\Rightarrow$ For the simple current mirror case:

$CMRR = 2\left(\dfrac{10}{0.316}\right)^2 = 2000 \rightarrow \underline{\underline{66\,dB}}$

For the Wilson source:

$CMRR = 2\sqrt{2}\dfrac{(10)^3}{(0.316)^3} = 89442 \rightarrow \underline{\underline{99\,dB}}$

7.65

$V_{GS_{1-4}} = \sqrt{\dfrac{50\mu \cdot 2}{800\mu}} + 1$

$= 1.35V$

$V_{GS\,5-8} = \sqrt{\dfrac{2 \times 100}{800}} + 1$

$= 1.5V$

CONT.

For $V_{DS} = V_{GS}$

$$-V_{SS} + 2V_{GS_{5-8}} + 2V_{GS_{1-4}} = V_{DD}$$

Thus,

$$V_{DD} + V_{SS} = 2(1.5) + 2(1.35) = \underline{\underline{5.7\,V}}$$

7.66

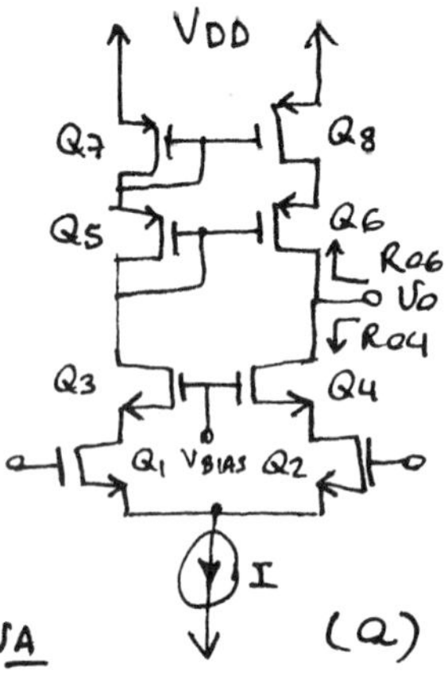

(b)

$$R_{o4} = (g_{m4} r_{o4})\, r_{o2} = g_m r_o^2$$

$$R_{o6} = (g_{m6} r_{o6})\, r_{o8} = g_m r_o^2$$

$$A_d = g_m (R_{o4} \parallel R_{o6}) = g_m \cdot \frac{1}{2} g_m^2 r_o^2$$

$$g_m = \frac{2 I_D}{V_{ov}} \qquad r_o = \frac{V_A}{I_D}$$

thus, $g_m r_o = 2V_A / V_{ov}$

$$\Rightarrow \underline{\underline{A_d = 2(V_A / V_{ov})^2}}$$

Q.E.D.

For $V_{ov} = 0.25\,V$ & $V_A = 20\,V$

$$A_d = 2(20/0.25)^2 = \underline{\underline{12800\,\frac{V}{V}}}$$

7.67

$$R_{id} = (\beta+1)\, 2r_e \; ; \; r_e = \frac{25\,mV}{50\,\mu A} = 500\,\Omega$$

$$\rightarrow R_{id} = 101 \times 1000 = \underline{101\,K\Omega}$$

$$R_o = r_{o4} \parallel r_{o2} = \frac{r_o}{2} \; ; \; r_o = \frac{V_A}{I_C}$$

$$\rightarrow r_o = \frac{160\,V}{50\,\mu A} = 3.2\,M\Omega$$

Thus, $R_o = \underline{\underline{1.6\,M\Omega}}$

$$G_m = g_{m1} = g_{m2} = \frac{50\,\mu A}{25\,mV} = \underline{\underline{2\,mA/V}}$$

$$A_d = G_m R_o = 2 \times 1600 = \underline{\underline{3200\,V/V}}$$

With a subsequent stage having a $100\,K\Omega$ input resistance,

$$A_d = G_m (R_o \parallel 100\,K\Omega) = \underline{\underline{188.2\,V/V}}$$

7.68

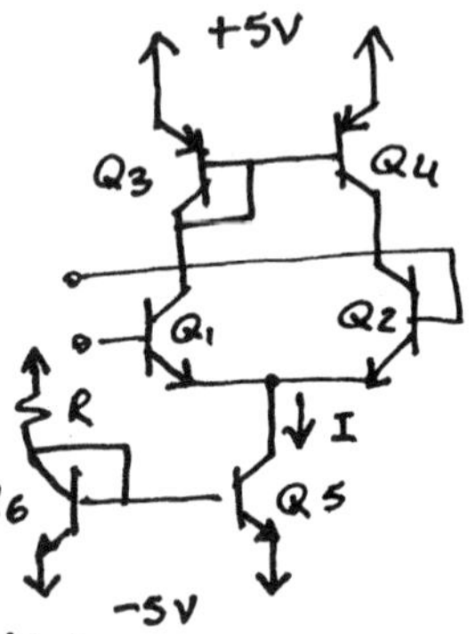

$$G_m = \frac{I/2}{V_T} = 5\,\frac{mA}{V}$$

$$I = \underline{\underline{250\,\mu A}}$$

$$R = \frac{5 - (-5) - V_{BE}}{I} = \frac{9.3}{0.25} = \underline{\underline{37.2\,K\Omega}}$$

$R_{id} = (\beta+1)\, 2r_e$ where,

$$r_e = \frac{V_T}{I/2} = \frac{25\,mV}{0.125\,mA} = 200\,\Omega$$

$$\Rightarrow R_{id} = 151 \times 2 \times 0.2 = \underline{\underline{60.4\,K\Omega}}$$

$$r_o = \frac{V_A}{I_C} = \frac{100}{0.125} = 800\,K\Omega$$

$$R_o = \frac{r_o}{2} = \underline{\underline{400\,K\Omega}}$$

$$A_d = g_m R_o = 5 \times 400 = \underline{\underline{2000\,V/V}}$$

$$I_B = \frac{I/2}{\beta+1} = \frac{125}{151} = \underline{\underline{0.83\,\mu A}}$$

$$v_{ICM}|_{max} \simeq V_{C1} + 0.4\,V = 5 - 0.7 + 0.4 = \underline{\underline{4.7\,V}}$$

$$v_{ICM}|_{min} = V_{BS} - 0.4 + 0.7 = -5 - 0.4 + 0.7 = -4\,V$$

Thus, the input common-mode range is $-4\,V$ to $+4.7\,V$ (where we have assumed that a transistor remains active

CONT.

even through its base-collector junction is forward biased by 0.4V)

$R_{icm} \simeq (\beta+1)\,[r_{o5} \| r_{o1} \| r_{o2}]$

$= 151 \times (400 \| 800 \| 800) = \underline{30 M\Omega}$

7.69

Use $R = 2K\Omega$

Thus,

$I_{REF} = \frac{5-(-5)-0.7}{2}$

$= 4.65 mA$

$V_{BE6} - V_{BE5} = V_T \ln(4.65/0.25)$

$= 73.1 mV$

$R_E = 73.1 mV / 0.25 mA$

$= \underline{292 \Omega}$

The only amplifier parameter that changes is R_{icm}. This is because now R_{o5} is:

$R_{o5} \simeq (1 + g_{m5} R_E)\, r_{o5}$

$= (1 + 10 \times 0.292) \times 400$

$= \underline{1.57 M\Omega}$

Thus, $R_{icm} = (\beta+1)\,[R_{o5} \| r_{o1} \| r_{o2}]$

$= 151\,[1.57 \| 0.80 \| 0.8]$

$= \underline{48 M\Omega}$

7.70

$R_{id} = (\beta+1)\, 2\,(r_e + R_E)$

G_m is still equal to: $g_m = 5 mA/V$

$\Rightarrow I = \underline{250 \mu A}$

(From Problem 7.68 above)

and $r_e = \frac{V_T}{I/2} = 200\Omega$, $r_o = 800 K\Omega$

If $R_{id} = 100 K\Omega \Rightarrow$

$100K = 151 \times 2 \times (200 + R_E)$

$\rightarrow R_E = 131 \Omega$

To obtain A_d:

$A_d = G_m \cdot R_o$ (Eqn. 7.165)

As in the derivation of R_{o2} in Eqn. (7.162), R_{o2} can be found using Eqn. (6.159), but this time noting that R_e at the emitter of Q_2 is:

$r_{e1} + 2R_E$

Thus,

$R_{o2} = r_{o2}\left[1 + g_m\left((r_{e1} + 2R_E) \| r_{\pi 2}\right)\right]$

$R_{o2} = 800K\left[1 + 5m\left((200 + 2 \times 131) \| 30.2K\right)\right]$ ($30.2K = (\beta+1) r_e$)

$R_{o2} = 2620 K\Omega$

$R_o = R_{o2} \| r_{o4}$

$= (2620 \| 800) K = 613 K\Omega$

$\Rightarrow A_d = 5m \times 613K = \underline{3065 V/V}$

7.71

$G_m = g_m = \frac{I/2}{V_T} = \frac{0.5m/2}{25m}$

$g_m = \underline{10 mA/V}$

$R_o = r_{o2} \| r_{o4} = \frac{V_A}{I_{C2}} \| \frac{V_A}{I_{C4}} = \frac{1}{2}\frac{V_A}{I/2}$

$= \frac{120}{0.5m} = \underline{240 K\Omega}$

$A_d = G_m R_o = 10 \times 240 = \underline{2400 \frac{V}{V}}$

$R_{id} = 2 r_\pi \simeq 2 \frac{V_T}{I/2} \beta = \frac{25m \times 150}{0.5m}$

$R_{id} = \underline{7.5 K\Omega}$

For a simple current mirror

CONT.

the output resistance (thus R_{EE}) is r_o

$$\Rightarrow R_{EE} = \frac{V_A}{I} = \frac{120}{0.5m} = 240 K\Omega$$

$$A_{cm} = \frac{-r_{o4}}{\beta_3 R_{EE}} = -\frac{(2\times 240K)}{150\times 240K} = -13.3\ mV/V$$

and, $CMRR = \left|\frac{2400}{-13.3m}\right| = 180{,}451$

i.e. 105 dB

$$\frac{v_i}{v_s} = \frac{R_{id}}{R_{id}+R_s} = \frac{7.5K}{7.5K+10K} = 0.43\ V/V$$

$\Rightarrow$ Overall gain A:

$$A = \frac{v_i}{v_s}\cdot\frac{v_o}{v_i} = 0.43\times 2400 = 1032\ V/V$$

7.72

$$i_o = \frac{v_{icm}}{2R_{EE}} - \frac{v_{icm}}{2R_{EE}}\cdot\frac{1}{1+\frac{2}{\beta_P}}$$

$$= \frac{v_{icm}}{2R_{EE}}\cdot\frac{2/\beta_P}{1+2/\beta_P}$$

$$\simeq \frac{v_{icm}}{\beta_P R_{EE}} \quad (\beta_P \gg 2)$$

Q.E.D.

Thus, $G_{mcm} = \frac{i_o}{v_{icm}} = \frac{1}{\beta_P R_{EE}}$

Q.E.D.

$$CMRR = \frac{G_{md}}{G_{mcm}} = \frac{g_m}{1/\beta_P R_{EE}} = g_m \beta_P R_{EE} = \frac{\beta_P \cdot I R_{EE}}{2V_T}$$

For $I = 0.2mA$, $R_{EE} = 1M\Omega$ and $\beta_P = 25$:

$$CMRR = \frac{\beta_P \cdot I R_{EE}}{2V_T} = \frac{25\times 0.2\times 1000}{2\times 0.025} = 10^5 \text{ i.e. } 100dB$$

7.73

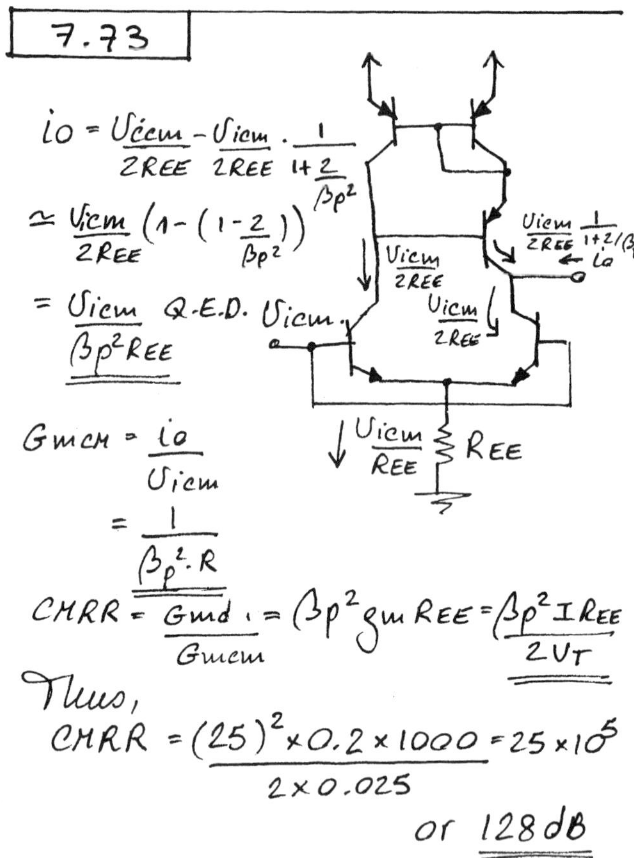

$$i_o = \frac{v_{icm}}{2R_{EE}} - \frac{v_{icm}}{2R_{EE}}\cdot\frac{1}{1+\frac{2}{\beta_P^2}}$$

$$\simeq \frac{v_{icm}}{2R_{EE}}\left(1-\left(1-\frac{2}{\beta_P^2}\right)\right)$$

$$= \frac{v_{icm}}{\beta_P^2 R_{EE}} \quad \text{Q.E.D.}$$

$$G_{mcm} = \frac{i_o}{v_{icm}} = \frac{1}{\beta_P^2\cdot R}$$

$$CMRR = \frac{G_{md}}{G_{mcm}} = \beta_P^2 g_m R_{EE} = \frac{\beta_P^2 I R_{EE}}{2V_T}$$

Thus,

$$CMRR = \frac{(25)^2\times 0.2\times 1000}{2\times 0.025} = 25\times 10^5$$

or 128 dB

7.74

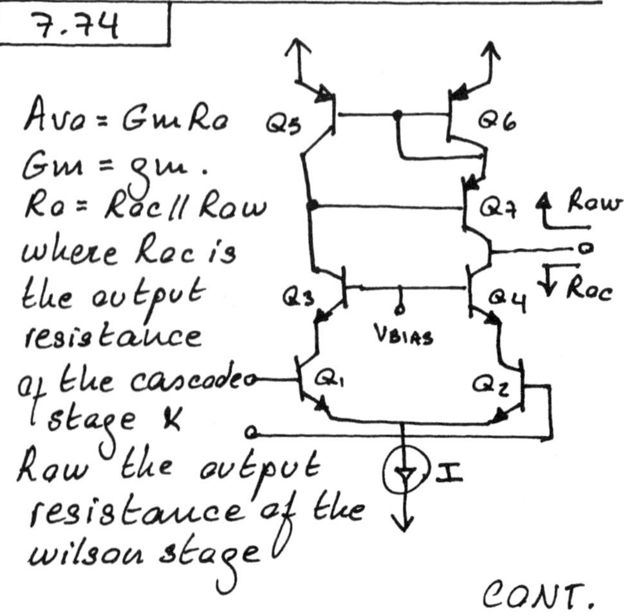

$A_{vo} = G_m R_o$

$G_m = g_m$.

$R_o = R_{oc} \| R_{ow}$ where R_{oc} is the output resistance of the cascode stage & R_{ow} the output resistance of the wilson stage

CONT.

$R_{oc} = \beta r_o$ & $R_{ow} = \frac{\beta r_o}{2}$

$$\Rightarrow R_o = \beta r_o \,\|\, \frac{\beta r_o}{2} = \frac{\beta r_o \cdot \frac{\beta r_o}{2}}{\beta r_o\left(1+\frac{1}{2}\right)} = \frac{\beta r_o}{2\times\frac{3}{2}} = \frac{\beta r_o}{3}$$

$$\Rightarrow A_{vo} = G_m R_o = g_m \frac{\beta r_o}{3} \quad \text{Q.E.D}$$

For: $I = 0.4mA$, $\beta = 100$, $V_A = 120V$

$$A_{vo} = \frac{I/2}{V_T}\cdot\frac{\beta}{3}\cdot\frac{V_A}{I/2} = \frac{\beta}{3}\,\frac{V_A}{V_T} = \frac{100}{3}\times\frac{120V}{25mV} = 160000$$

i.e. 104 dB

7.75

(a) $V_{O\,max} - V_{B7} = 0.4V$

but: $V_{B7} = 5 - 2\times 0.7V = 3.6V$

$\Rightarrow V_{O\,max} = 3.6 + 0.4 = 4.0V$

(b) For a 1.5V max. positive swing the DC bias at the output should be:

$V_{O\,DC} = 4 - 1.5 = 2.5V$

(c) In the edge of saturating Q_4

$V_{BIAS} - V_{O\,min} = 0.4V$

$\rightarrow V_{O\,min} = V_{BIAS} - 0.4V$

If $V_{O\,min}$ is: $V_{O\,DC} - 1.5$

$2.5 - 1.5 = 1V$

$\Rightarrow V_{BIAS} = 1.4V$.

(d) If $V_{BIAS} = 1.4V$

$\Rightarrow V_{C2} = 1.4 - 0.7 = 0.7V$

The upper limit of V_{icm} is such that brings Q_2 to the edge of saturating.

$V_{icm\,min} - V_{C2} = 0.4V$

$\Rightarrow V_{icm\,min} = 0.4 + 0.7 = 1.1V$

7.76

(a)

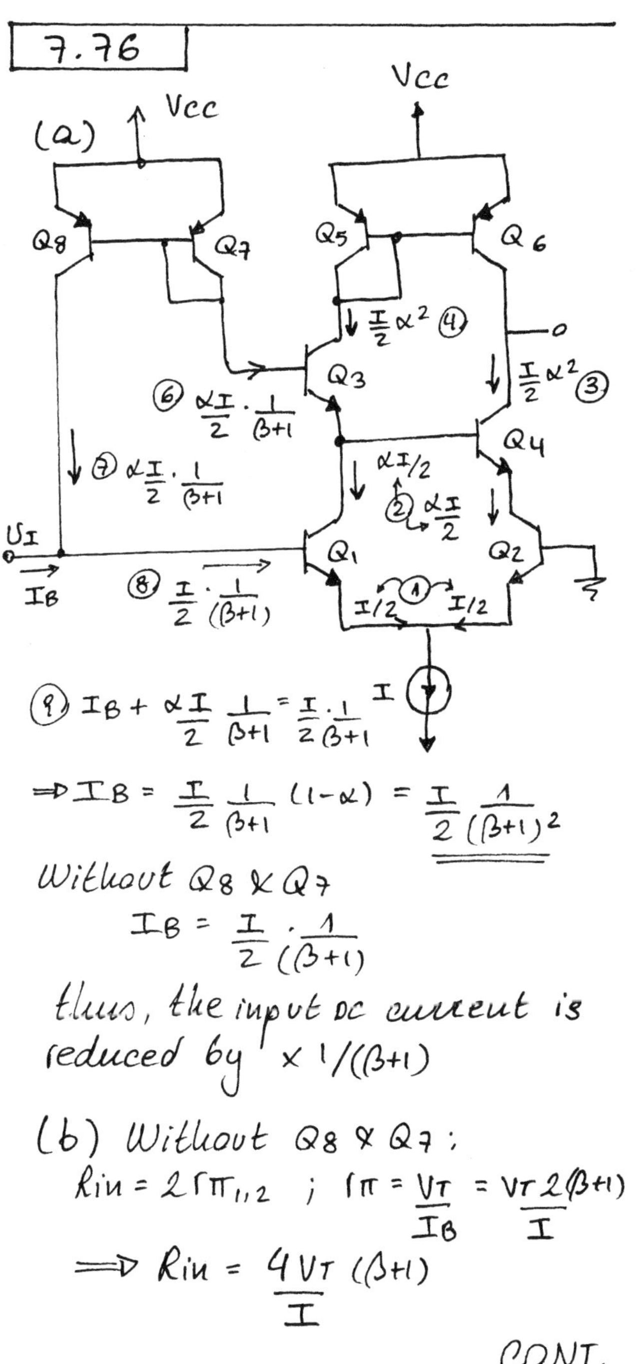

(9) $I_B + \frac{\alpha I}{2}\,\frac{1}{\beta+1} = \frac{I}{2}\cdot\frac{1}{\beta+1}$

$$\Rightarrow I_B = \frac{I}{2}\,\frac{1}{\beta+1}(1-\alpha) = \frac{I}{2}\,\frac{1}{(\beta+1)^2}$$

Without Q_8 & Q_7

$$I_B = \frac{I}{2}\cdot\frac{1}{(\beta+1)}$$

thus, the input DC current is reduced by $\times 1/(\beta+1)$

(b) Without Q_8 & Q_7:

$R_{in} = 2r_{\pi 1,2}$; $r_\pi = \frac{V_T}{I_B} = \frac{V_T\, 2(\beta+1)}{I}$

$$\Rightarrow R_{in} = \frac{4V_T(\beta+1)}{I}$$

CONT.

With Q8 & Q7: $R_{in} = 2r_\pi \| r_{o8}$
where $r_{o8} = \frac{V_A}{\alpha I} \cdot 2(\beta+1)$

$\rightarrow R_{in} = \frac{4V_T}{I}(\beta+1) \| \frac{V_A}{\alpha I} 2(\beta+1)$

Solving, yields:

$$R_{in} = \frac{4V_T}{I}(\beta+1) \cdot \frac{V_A}{2\alpha V_T + V_A}$$

but $\frac{V_A}{2\alpha V_T + V_A} \simeq 1 \Rightarrow R_{in} = \frac{4V_T(\beta+1)}{I}$

Just like when Q8 & Q7 are not present.

7.77

For the Wilson current source

$$\frac{I_4}{I_3} = \frac{1}{1 + 2/\beta_P^2}$$

Since $I_3 = \alpha I/2$

$\rightarrow I_4 = \frac{\alpha I/2}{1 + 2/\beta_P^2}$

$$\Delta i = \frac{\alpha I}{2} - \frac{\alpha I/2}{1 + 2/\beta_P^2}$$

$$= \frac{\alpha I}{2} \cdot \frac{2/\beta_P^2}{1 + 2/\beta_P^2} = \alpha I \cdot \frac{1}{\beta_P^2 + 2}$$

$$\simeq \frac{\alpha I}{\beta_P^2}$$

thus. $V_{OS} = \frac{-\Delta i}{G_m}$ with $G_m = g_m = \frac{\alpha I/2}{V_T}$

$$\Rightarrow V_{OS} = -\frac{\alpha I/\beta_P^2}{\alpha I/2V_T} = \frac{-2V_T}{\beta_P^2}$$

For $\beta = 50$:

$$V_{OS} = -\frac{2 \times 25\,mV}{(50)^2} = -20\,\mu V$$

7.78

To obtain maximum positive swing V_{bias} must be as low as possible. To keep the top current sources out of saturation:

$$V_{CC} - 0.2 - 0.7 = V_{bias\,max}$$
$$V_{bias\,max} = 4.1\,V$$

And: $V_O - V_{bias\,min} = +0.4\,V$

Since $V_O \sim 0 \Rightarrow V_{bias\,min} = -0.4\,V$

$\Rightarrow$ Range of V_{bias} is:

$$(-0.4 \ll V_{bias} \leq 4.1)\,V$$

For: $I = 0.4\,mA$, $\beta_P = 50$, $\beta_N = 150$ & $V_A = 120$

$$G_m = g_{m1} = \frac{0.2\,mA}{25\,mV} = 8\,\frac{mA}{V}$$

For the folded cascode: $R_{o4} = \beta_4 r_{o4}$

For the Wilson mirror: $R_{o5} = \beta_5 \frac{r_{o5}}{2}$

$$\Rightarrow R_o = \left[\beta_4 \cdot r_{o4} \| \beta_5 \cdot \frac{r_{o5}}{2}\right]$$

$$r_{o4} = r_{o5} = 120/0.2\,mA = 600\,K\Omega$$

$$\rightarrow R_o = \left[50 \times 600K \| 150 \times \frac{600K}{2}\right]$$
$$= [30M \| 45M]$$
$$= 18\,M\Omega$$

$$A_d = G_m R_o = 8\,\frac{mA}{V} \times 18\,M\Omega = 144\,000$$

7.79

$K_p' W/L = 6.4\,mA/V^2$

$|V_{Ap}| = 10\,V$

$V_{A\,npn} = 120$.

$$R_o = r_{o2} \| r_{o4} = \frac{V_{Ap}}{I/2} \| \frac{120}{I/2}$$

$$R_o = (10/0.2m) \| (120/0.2m) = 46\,K\Omega$$

$$G_m = g_{m1} = \sqrt{I \times K_p' W/L}$$
$$= \sqrt{0.4\,mA \times 6.4\,mA/V^2}$$

$$\Rightarrow G_m = 1.6\,\frac{mA}{V}$$

$$A_d = G_m \times R_o = 1.6\,\frac{mA}{V} \times 46\,K\Omega$$

$$\longrightarrow A_d = 73.6\;V/V$$

7.80

$K_n'(W/L) = 128\,\mu A/V^2 \times 25 = 3.2\,mA/V^2$

(a) $V_{OV} = \sqrt{\dfrac{I}{K_n' W/L}} = \sqrt{\dfrac{0.2}{3.2}} = \underline{\underline{0.25V}}$

$g_m = \dfrac{I}{V_{OV}} = \dfrac{0.2mA}{0.25V} = \underline{\underline{0.8\dfrac{mA}{V}}}$

(b) $A_d = g_m(R_D \| r_o)$

where $r_o = \dfrac{V_A}{I/2} = \dfrac{20}{0.2/2} = 200K\Omega$

$\Rightarrow A_d = 0.8m \times (20K \| 20K)$

$= \underline{\underline{14.54\ V/V}}$

(c) For a CS amplifier when R_{sig} is low:

$f_H = \dfrac{1}{2\pi (C_L + C_{gd}) R_L'}$

where $R_L' = R_D \| r_o$

$= 20K \| 200K$

$= 18.18K\Omega$

and $C_L' = C_L + C_{db}$

Since for a grounded source terminal C_{db} is in parallel with the load.

$\rightarrow C_L' = 90 + 5 = 95fF$

thus,

$f_H = \dfrac{1}{2\pi(95+5)10^{-15} \times 18.18K}$

$= \underline{\underline{87.54MHz}}$

(d) Using the open-circuit time-constants method for $R_s = 20K\Omega$

$f_H = \dfrac{1}{2\pi \tau_H}$

where $\tau_H = C_{gs} \cdot R_s$
$+ C_{gd}[R_s(1 + g_m R_L') + R_L']$
$+ C_L \cdot R_L'$

thus,

$\tau_H = 30f \times 20K$
$+ 5f\,[20K(1 + 0.8 \times 18.18) + 18.18K]$
$+ (90f + 5f) \times 18.18K$

$\tau_H = 0.6ns$
$+ 1.64ns$
$+ 1.72ns = 3.96ns$

$\Rightarrow f_H = \dfrac{1}{2\pi\, 3.96ns} = \underline{\underline{40.2MHz}}$

7.81

$f_Z = \dfrac{1}{2\pi C_{SS} R_{SS}} = \dfrac{1}{2\pi(0.2p)(100K)} = \underline{\underline{7.95MHz}}$

7.82

(a) Differential half-circuit

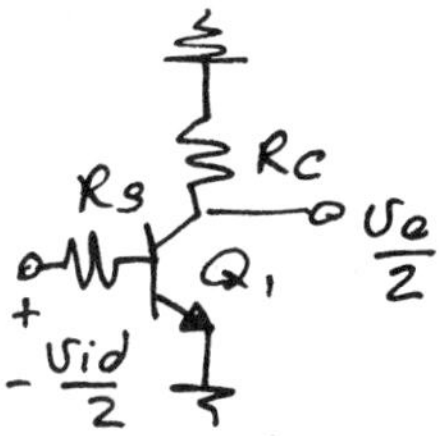

High-frequency equivalent circuit

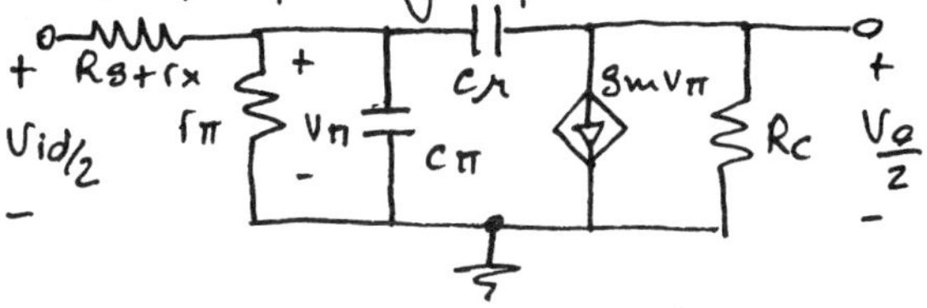

(b) $I_E = 0.5mA \rightarrow g_m = \dfrac{0.5}{25}$

$g_m = 20mA/V$

$r_\pi = \dfrac{\beta}{g_m} = \dfrac{100}{20m} = \underline{\underline{5K\Omega}}$

$C_\pi + C_\mu = \dfrac{20 \times 10^{-3}}{2\pi \times 600 \times 10^6} = 5.3pF$

$C_\mu = 0.5p \Rightarrow C_\pi = 4.8pF$

Now, $R_s = 10K\Omega$, $R_C = 10K\Omega$, $r_x = 100$

For this common-emitter amplifier (Refer to Section...

CONT.

5.9 Eqn. (5.164))

$$\frac{V_o}{V_i} = \frac{-r_\pi}{r_\pi + (R_s + r_x)} \cdot g_m R_c$$

$$= \frac{-5K}{5K + (10K + 100)} \cdot 20m \times 10K$$

$$= -66.22 \text{ V/V}$$

(c) Refer to Section 5.9, Eqn's (5.176), (5.173), (5.168)

From Eqn. (5.176):

$$f_H = \frac{1}{2\pi C_{in} R'_{sig}}$$

$R_s + r_x$, r_π, C_π, C_μ, $g_m V_\pi$, R_c, R'_{sig}, $C_{eq.}$

where: $C_{in} = C_\pi + C_{eq}$

Using Miller's theorem

$C_{eq} = C_\mu (1 + g_m R_c)$

and

$R'_{sig} = (R_s + r_x) \parallel r_\pi$

Thus,

$$f_H = \frac{1}{2\pi \left[(R_s + r_x) \parallel r_\pi\right]\left[C_\pi + C_\mu(1 + g_m R_c)\right]}$$

$$= \frac{1}{2\pi((10K + 100) \parallel 5K)(4.8p + 0.5p(1 + 20 \times 10))}$$

$$= 452 \text{ KHz}$$

$GBW = 66.22 \times 452K = 30 MHz$

7.83

From problem 7.82 above:

$g_m = 20 mA/V$ $r_\pi = 5K\Omega$ $C_\pi = 4.8pF$ $C_\mu = 0.5pF$ $R_s = 5K\Omega$ $R_c = 10K\Omega$ $r_x = 100\Omega$

If we add an emitter resistor, then the equivalent half-circuit is:

R_c, R_s, $+$, $V_s/2$, $-$, R_E, $+$, $V_o/2$, $-$

And the high-frequency equivalent circuit is:

C_μ, $R_s + r_x$, $+$, $V_s/2$, $-$, r_π, $+$ V_π $-$, C_π, $g_m V_\pi$, R_c, $+$ $V_o/2$ $-$, R_E

For this circuit the low-frequency gain is:

$$A_o = \frac{V_o}{V_s} = \frac{-(\beta+1) \cdot (r_e + R_E)}{(R_s + r_x) + (\beta+1)(r_e + R_E)} \times \frac{\alpha R_c}{R_E + r_e}$$

(first factor) obtained by reflecting emitter resistance to the base; (second factor) gain V_o/V_i as in Eq (5.131)

where $r_e = \frac{V_T}{I_E} = \frac{25}{0.5} = 50\Omega$

$$A_o = \frac{-101 \times (50 + 100)}{(10K + 100) + 101 \times (50 + 100)} \times \frac{100}{101} \cdot \frac{10K}{(100 + 50)}$$

$$= -39.6 \text{ V/V}$$

Using the method of open-circuit time constants:

$f_H = 1/(2\pi(C_\pi R_\pi + C_\mu R_\mu))$

CONT.

Refering to the high-frequency equivalent circuit:

$$R_\pi = r_\pi \,\|\, \frac{R_s + r_x + R_E}{1 + g_m R_E}$$

$$= 5K \,\|\, \frac{10K + 100 + 100}{1 + 20m \times 100}$$

$$= 5K \,\|\, 3.4K = 2.02\,K\Omega$$

To obtain R_μ:

$$R_{in} = (\beta + 1)(R_e + r_e)$$

$$= 101 \times (100 + 50) = 15.15\,K\Omega$$

$$G_m = \frac{g_m}{1 + g_m R_e} = \frac{20m}{1 + 20m \times 100}$$

$$= 6.67m.$$

Thus,

$$R_\mu = [(R_s + r_x) \| R_{in}](1 + G_m R_c) + R_c$$

$$= [(10K + 100) \| 15.15K] \cdot (1 + 6.67 \times 10) + 10K$$

$$= 420.26\,K\Omega$$

Thus, $C_\pi R_\pi + C_\mu R_\mu =$

$$= 4.8pF \times 2.02K\Omega + 0.5pF \times 420.26K$$

$$= 219.8ns \simeq 220ns.$$

$$\Rightarrow f_H = \frac{1}{2\pi \times 220ns} = 723.4\,KHz$$

$$GBW = 39.6 \times 723.4K = 28.6\,MHz$$

Approx. the same as in problem 7.82 above.

7.84

From the solution of problem 7.81:

$g_m = 20mA/V$ $\quad r_e = 50\Omega$ $\quad r_\pi = 5K\Omega$

$C_\pi = 4.8pF$ $\quad C_\mu = 0.5pF$ $\quad R_s = 10K\Omega$

$R_c = 10K\Omega$ $\quad r_x = 100$

$$f_H = 1/(2\pi \tau_H) = 1 \times 10^6\,Hz$$

$$\Rightarrow \tau_H = 1/(2\pi \times 10^6)$$

$$= 159.15ns$$

But: $\tau = C_\pi R_\pi + C_\mu R_\mu$

where

$$R_\pi = r_\pi \,\|\, \left(\frac{R_s + r_x + R_e}{1 + g_m R_e}\right)$$

$$= \frac{1}{\frac{1}{r_\pi} + \frac{1 + g_m R_e}{R_s + r_x + R_e}}$$

$$= \frac{1}{\frac{1}{5K} + \frac{1 + g_m R_e}{10.1K + R_e}}$$

$$R_\mu = [(R_s + r_x) \| R_{in}](1 + G_m R_c) + R_c$$

with: $R_{in} = (\beta + 1)(R_e + r_e)$

$G_m = g_m / (1 + g_m R_e)$

thus,

$$R_\mu = \frac{1}{\frac{1}{R_s + r_x} + \frac{1}{(\beta+1)(R_e + r_e)}} \cdot \left(\frac{1 + g_m R_c}{1 + g_m R_e}\right) + R_c$$

$$= \frac{1 + g_m R_c + g_m R_e}{\frac{1 + g_m R_e}{R_s + r_x} + \frac{1 + g_m R_e}{(\beta+1)(R_e + r_e)}} + R_c$$

$$\frac{1 + g_m R_e}{(\beta+1)(R_e + r_e)} = \frac{1 + g_m R_e}{(\beta+1)(R_e + \frac{1}{g_m})}$$

$$= \frac{g_m}{\beta + 1} \cdot \frac{(1 + g_m R_e)}{(1 + g_m R_e)}$$

and, since $r_\pi = \beta / g_m$

$$\Rightarrow \frac{g_m}{\beta + 1} \simeq \frac{1}{r_\pi}$$

thus,

$$R_\mu = \frac{1 + g_m R_c + g_m R_e}{\frac{1 + g_m R_e}{R_s + r_x} + \frac{1}{r_\pi}} + R_c$$

CONT.

$$159.15ns = \frac{4.8pF}{\frac{1}{5K} + \frac{1+g_m R_e}{10.1K + R_e}} + \frac{0.5pF(1+200+g_m R_e)}{\frac{1+g_m R_e}{10.1K} + \frac{1}{5K}} + \underbrace{0.5pF \times 10K\Omega}_{5ns}$$

Assuming that $R_e \ll 10.1K\Omega$

$$154.15ns = \frac{4.8pF + 0.5pF(1+200+g_m R_e)}{\frac{1}{5K} + \frac{1+g_m R_e}{10.1K}}$$

$$(154.15n)\left(\frac{1}{5K} + \frac{1+g_m R_e}{10.1K}\right) = 4.8p + 0.5p(1+g_m R_e) + 100p$$

Solving for $1+g_m R_e$ yields:

$1 + g_m R_e = 5 \Rightarrow R_e = \underline{\underline{200\Omega}}$

The DC gain is:

$$A_o = \frac{-101 \times (50+200)}{(10K+100) + 101 \times (50+200)} \times \frac{100}{101} \times \frac{10K}{(200+50)} = \underline{\underline{-28.28}}\ V/V$$

$R_{in} = 101 \times (200+50) = 25.25K\Omega$

$G_m = 20m/(1+20m \times 200) = 4m$

$R_\mu = ((10K+100)\|25.25K)(1+4\times10)+10K = 305{,}785\Omega$

$R_\pi = 5K \| \frac{10K+100+200}{1+20m\times200} = 1459\Omega$

$\tau_H = 1459 \times 4.8p + 305{,}785 \times 0.5p = 159.8ns$

$f_H = \frac{1}{2\pi\tau_H} = 0.99MHz \sim \underline{\underline{1MHz}}$

$GBW = 28.28 \times 1 = \underline{\underline{28.28\ MHz}}$

7.85

$I = 0.6mA$

$V_{OVn} = 0.3V$

$V_{OVp} = 0.5V$

$V_{An} = |V_{Ap}| = 9V$

$C_m = 0.1pF$

$C_L = 0.2pF$

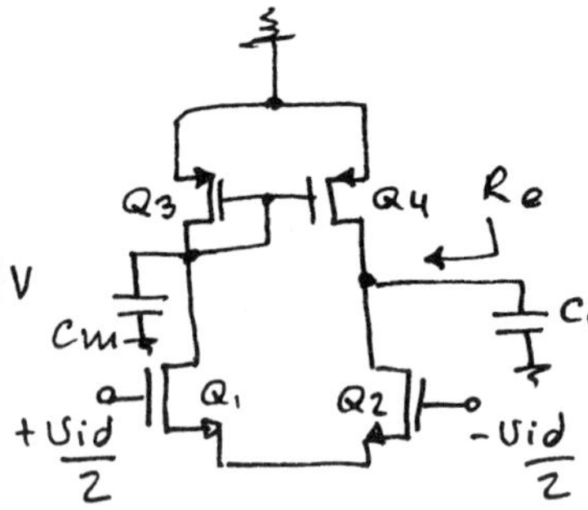

All r_o's are identical:

$$r_o = \frac{V_A}{I_D} = \frac{9}{0.3m} = 30K\Omega.$$

$$g_m = \frac{2I_D}{V_{OV}} \Rightarrow g_{m1,2} = \frac{0.6m}{0.3} = 2\frac{mA}{V}$$

$$g_{m3,4} = \frac{0.6m}{0.5} = 1.2\frac{mA}{V}$$

The low frequency differential gain is:

$$A_d = g_{m1,2}(r_{o2} \| r_{o4}) = 2\frac{mA}{V}(30K \| 30K) = \underline{\underline{30\frac{V}{V}}}$$

From Eqn. (7.192)

$f_{P1} = 1/(2\pi C_L R_o)$

where $R_o = r_{o2} \| r_{o4} = 15K\Omega$

$$f_{P1} = \frac{1}{2\pi \times 0.2p \times 15K} = \underline{\underline{53MHz}}$$

(7.193) $f_{P2} = \frac{g_{m3}}{2\pi C_m} = \frac{1.2mA/V}{2\pi \times 0.1pF} = \underline{\underline{1.9GHz}}$

(7.194) $f_Z = \frac{2g_{m3}}{2\pi C_m} = \frac{2 \times 1.2m}{2\pi \times 0.1p} = \underline{\underline{3.8GHz}}$

7.86

The CMRR will have.. CONT.

poles at $\underline{500\,KHz}$ and at

$$\frac{1}{2\pi \times 10^6 \times 10 \times 10^{-12}} = \underline{15.9\,KHz}$$

7.87

From the solution to Problem 7.82 above:

$g_m = 20\,mA/V$ $r_e = 50\,\Omega$ $r_\pi = 5\,K\Omega$

$C_\pi = 4.8\,pF$ $C_\mu = 0.5\,pF$ $R_s = 10\,K\Omega$

$R_C = 10\,K\Omega$

Here we neglect r_x, i.e $r_x = 0$.

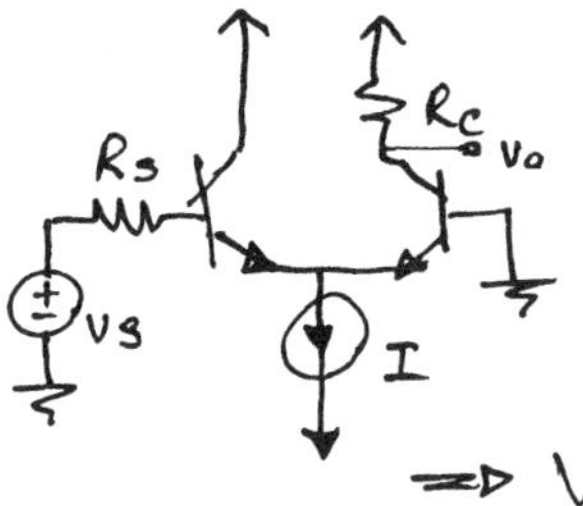

$$V_{b1} = V_s \cdot \frac{2r_\pi}{R_s + 2r_\pi}$$

$$V_o = \frac{V_{b1}}{2r_e}\,\alpha \cdot R_C$$

$$\Rightarrow \frac{V_o}{V_s} = \frac{\alpha R_C}{2r_e} \cdot \frac{2r_\pi}{R_s + 2r_\pi}$$

$$= \frac{\alpha \cdot R_C \cdot r_e(\beta+1)}{r_e(R_s + 2r_\pi)} = \frac{\alpha R_C}{\frac{R_s}{\beta+1} + 2r_e}$$

$$= \frac{\beta R_C}{R_s + 2r_\pi}$$

$$A_o = \frac{V_o}{V_s} = \frac{100 \times 10K}{10K + 2 \times 5K} = \underline{50\,V/V}$$

Using the equivalent circuit of Fig 6.57, i.e.

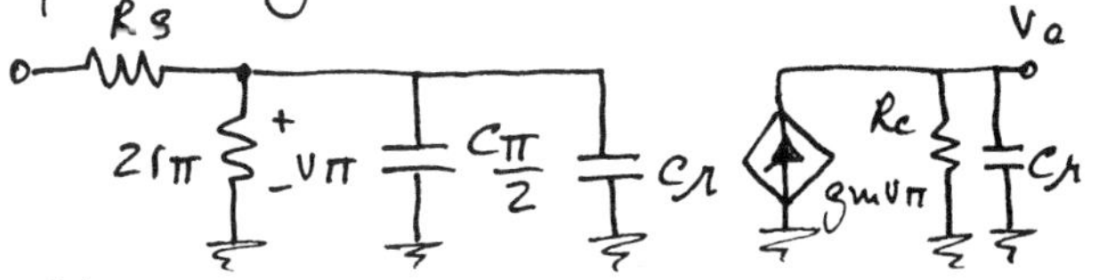

thus,

$$f_{P1} = \frac{1}{2\pi(R_s \| 2r_\pi)\left(\frac{C_\pi}{2} + C_\mu\right)}$$

$$= \frac{1}{2\pi(10K \| 10K)\left(\frac{4.8p}{2} + 0.5p\right)}$$

$$= \underline{11\,MHz}$$

and, $f_{P2} = \dfrac{1}{2\pi R_C C_\mu}$

$$= \frac{1}{2\pi \times 10K \times 0.5p} = 31.8 \sim \underline{32\,MHz}$$

thus,

$$f_H = \frac{1}{\sqrt{\left(\frac{1}{11}\right)^2 + \left(\frac{1}{32}\right)^2}} = \underline{10.4\,MHz}$$

7.88

$$V_{G1} = V_s \cdot \frac{2/g_m}{2/g_m + R_s} \qquad I = \frac{V_{G1}}{2/g_m}$$

$$V_o = I R_D = \frac{V_{G1}}{2/g_m} \times R_D$$

$$= V_s \times \frac{2/g_m}{2/g_m + R_s} \cdot \frac{R_D}{2/g_m}$$

$$= \frac{V_s \cdot R_D}{2/g_m + R_s}$$

$$\Rightarrow A_o = \frac{V_o}{V_s} = \frac{g_m R_D}{2 + g_m R_s}$$

$$g_m = \frac{200\,\mu A}{0.25\,V} = 0.8\,\frac{mA}{V}$$

$$\Rightarrow A_o = \frac{0.8 \times 50}{2 + 0.8 \times 200} = 0.24\,V/V$$

The high-frequency equivalent circuit is:

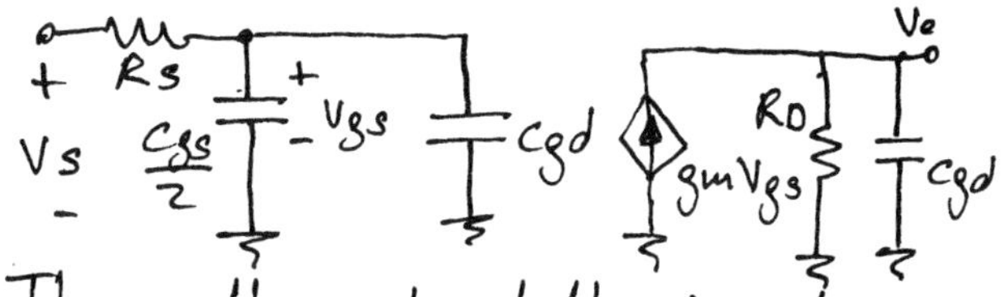

Thus, the pole at the input has a frequency f_{P1}:

CONT.

$$f_{P1} = \frac{1}{2\pi R_s \times \left(\frac{C_{gs}}{2} + C_{gd}\right)}$$

$$= \frac{1}{2\pi \times 200K \times (1/2 + 1)p}$$

$$= \underline{\underline{530\,KHz}}$$

and the pole at the output has a frequency f_{P2}:

$$f_{P2} = \frac{1}{2\pi R_o C_{gd}} = \frac{1}{2\pi \times 50K \times 1p}$$

$$= \underline{\underline{3.18\,MHz}}$$

Thus $f_H \simeq \dfrac{1}{\sqrt{\left(\frac{1}{530K}\right)^2 + \left(\frac{1}{3.18M}\right)^2}}$

$$= \underline{\underline{523\,KHz}}$$

Notice that this low value of f_H is due to the large value of R_s.

$$2\pi f_T = \frac{g_m}{C_\pi + C_\mu} \Rightarrow C_\pi + C_\mu = \frac{4m}{2\pi\, 600M}$$

$C_\pi + C_\mu = 1.06\,pF$

Since $C_\mu = 0.2\,pF \rightarrow C_\pi = 0.86\,pF$

$$f_{P1} = \frac{1}{2\pi (R_s \| 2r_\pi)\left(\frac{C_\pi}{2} + C_\mu\right)}$$

$$= \frac{1}{2\pi (50K \| 2\times 50K)\left(\frac{0.86}{2} + 0.2\right)p}$$

$$= \underline{\underline{7.58\,MHz}}$$

$$f_{P2} = \frac{1}{2\pi R_C C_\mu} = \frac{1}{2\pi\, 50K \times 0.2p}$$

$$= \underline{\underline{15.9\,MHz}}$$

Thus the estimate of f_H is:

$$f_H = \frac{1}{\sqrt{\left(\frac{1}{7.58}\right)^2 + \left(\frac{1}{15.9}\right)^2}}$$

$$= \underline{\underline{6.84\,MHz}}$$

7.89

Refer to Fig. P7.89.
It can be shown that this circuit is equivalent to that of Problem 7.87 above.

$$g_m = \frac{0.1mA}{25mV} = 4\frac{mA}{V} \qquad r_e \simeq 250\Omega$$

$$r_\pi = \frac{\beta}{g_m} = \frac{200}{4mA/V} = 50K\Omega$$

$$A_o = \frac{\alpha R_C}{\frac{R_s}{\beta+1} + 2r_e} = \frac{\frac{200}{201} \times 50K}{\frac{50K}{201} + 2\times 250}$$

$$= \underline{\underline{66.44\ V/V}}$$

7.90

To ensure that the op amp will not have a systematic offset voltage, use Eqn. (7.201)

$$\frac{(W/L)_6}{(W/L)_4} = 2\frac{(W/L)_7}{(W/L)_5}$$

$$\Rightarrow (W/L)_6 = 2 \times \frac{(60/0.5)(10/0.5)}{(60/0.5)}$$

$$\Rightarrow (W/L)_6 = \underline{\underline{(20/0.5)}}$$

Since Q_8 and Q_5 are matched $I = I_{REF} = 225\mu A$
Thus Q_1, Q_2, Q_3 and Q_4

CONT.

each conducts a current equal to $I/2 = 112.5\mu A$.
Since Q_7 is matched to Q_5 and Q_8, the current in Q_7 is equal to $I_{REF} = 225\mu A$.
Finally Q_6 conducts an equal current of $225\mu A$.
Thus,

$I_{D1} = 112.5\mu A$ $\quad I_{D5} = 225\mu A$
$I_{D2} = 112.5\mu A$ $\quad I_{D6} = 225\mu A$
$I_{D3} = 112.5\mu A$ $\quad I_{D7} = 225\mu A$
$I_{D4} = 112.5\mu A$ $\quad I_{D8} = 225\mu A$

To find $|V_{ov}|$ we use:

$$I_D = \frac{1}{2}\mu C_{ox}(W/L)V_{ov}^2$$

then we find $|V_{GS}|$ from:

$$|V_{GS}| = |V_t| + |V_{ov}|$$

this results in:

$|V_{ov1}| = 0.25V$ $\quad |V_{GS1}| = 1V$
$|V_{ov2}| = 0.25V$ $\quad |V_{GS2}| = 1V$
$|V_{ov3}| = 0.25V$ $\quad |V_{GS3}| = 1V$
$|V_{ov4}| = 0.25V$ $\quad |V_{GS4}| = 1V$
$|V_{ov5}| = 0.25V$ $\quad |V_{GS5}| = 1V$
$|V_{ov6}| = 0.25V$ $\quad |V_{GS6}| = 1V$
$|V_{ov7}| = 0.25V$ $\quad |V_{GS7}| = 1V$
$|V_{ov8}| = 0.25V$ $\quad |V_{GS8}| = 1V$

To find g_m: $g_m = 2I_D/|V_{ov}|$
and r_o: $r_o = |V_A|/I_D$
Thus:

$g_{m1} = 0.9\,mA/V$ $\quad r_{o1} = 80K\Omega$
$g_{m2} = 0.9\,mA/V$ $\quad r_{o2} = 80K\Omega$
$g_{m3} = 0.9\,mA/V$ $\quad r_{o3} = 80K\Omega$
$g_{m4} = 0.9\,mA/V$ $\quad r_{o4} = 80K\Omega$
$g_{m5} = 1.8\,mA/V$ $\quad r_{o5} = 40K\Omega$
$g_{m6} = 1.8\,mA/V$ $\quad r_{o6} = 40K\Omega$
$g_{m7} = 1.8\,mA/V$ $\quad r_{o7} = 40K\Omega$
$g_{m8} = 1.8\,mA/V$ $\quad r_{o8} = 40K\Omega$

$$A_1 = -g_{m1}(r_{o2}\|r_{o4})$$
$$= -0.9\frac{mA}{V}(80\|80)K\Omega$$
$$= -\underline{\underline{36}}\,V/V$$
$$A_2 = -g_{m6}(r_{o6}\|r_{o7})$$
$$= -1.8\frac{mA}{V}(40\|40)K\Omega$$
$$= -\underline{\underline{36}}\,V/V$$

Thus, the DC open-loop gain is:

$$A_0 = A_1A_2 = (-36)\times(-36)$$
$$= \underline{\underline{1296}}\ V/V$$
$$\rightarrow \underline{\underline{62.25dB}}$$

Input common-mode range:
lower limit is when the input is such that Q_1 & Q_2 leave the saturation region.

$$V_{D1} = -V_{SS} + V_{GS3}$$
$$= -1.5 + 1 = -0.5V$$
$$V_{in_{CM}min} = -0.5 - |V_{tp}|$$
$$= -0.5 - 0.75 = -\underline{\underline{1.25V}}$$

The upper limit is the value of $V_{in_{CM}}$ at which Q_5 leaves the saturation region.

$$V_{D5_{CM}}max = +1.5 - |V_{ov5}|$$
$$= 1.25V$$
$$\rightarrow V_{in_{CM}}max = 1.25 - 1 = +\underline{\underline{0.25}}V$$

Input range is: -1.25 to $0.25V$

Output voltage range:
V_{omax} is the value at which Q_7 leaves saturation.

$$V_{DD} - |V_{ov7}| = +1.5 - 0.25$$
$$= \underline{\underline{1.25}}V$$

The lowest V_o is the value at which Q_6 leaves saturation
i.e. $-V_{SS} + V_{ov6} = -1.5 + 0.25$
$= -\underline{\underline{1.25}}V$

The output range is
-1.25 to $+1.25V$

7.91

$$I = \frac{1}{2} K V_{ov}^2$$

(a) $V_{ov} = \sqrt{\frac{2I}{K}}$

If K increases by 4 → V_{ov} decreases by $\underline{\underline{1/2}}$

$g_m = 2I/V_{ov} = K.V_{ov}$

→ If K increases by 4 g_m increases by $\underline{\underline{\times 2}}$

(b) $A_1 = g_m R_{o1}$

⇒ A_1 increases ×2 an so A_o

(c) Offsets due to V_t mismatches are unaffected. Others reduced $\times \frac{1}{2}$ since A_o increases ×2

(d) $A_o f_p = f_T \Rightarrow f_p = \frac{f_T}{A_o}$

and is reduced $\times \frac{1}{2}$

⇒ C_c must be doubled.

7.92.

$$I_{D7} = \frac{W_7}{W_8} I_{REF} = \frac{50}{40} \times 90\mu A$$

$$= 112.5\mu A //$$

Output offset current $= I_{D7} - I_{D6}$

$= 112.5 - 90 = 22.5\mu A$

$\Rightarrow V_o = 22.5\mu\,(r_{o6} \parallel r_{o7})$

$r_{o7} = \frac{10}{112.5\mu} = 88.9K\Omega$

$\Rightarrow V_o = 22.5\mu\,(111K \parallel 88.9K)$

$= \underline{\underline{1.11V}}$

$V_{os} = \frac{V_o}{A_o} = \frac{1.11V}{1109} = \underline{\underline{1mV}}$

7.93

Offset current $= I_{D2} - I_{D4}$

$= I_{D3} - I_{D4}$

$I_{D3} = \frac{K}{2}(V_{GS} - V_t)^2$

$I_{D4} = \frac{K}{2}(V_{GS} - (V_t + \Delta V_t))^2$

$I_o = I_{D3} - I_{D4}$

$= \frac{K}{2}\left[(V_{GS} - V_t - V_{GS} + V_t + \Delta V_t) \times (V_{GS} - V_t + V_{GS} - V_t - \Delta V_t)\right]$

$= \Delta V_t . \frac{K}{2}(2V_{GS} - 2V_t - \Delta V_t)$

$\simeq K(V_{GS} - V_t).\Delta V_t$

$I_o = \underline{\underline{g_{m3} \Delta V_t}}$

Recall $I_o = G_{m1}.V_{os}$

and $G_{m1} = g_{m1}$

$\Rightarrow V_{os} = \frac{g_{m3}}{g_{m1}} . \Delta V_t$

For $\Delta V_t = 2mV$

$V_{os} = \frac{0.3m}{0.3m} \times 2m = \underline{\underline{2mV}}$

7.94

From Equ. (7.211) $\omega_t = \frac{G_{m1}}{C_c}$

$G_{m1} = g_{m1} = 1mA/V$

For $f_T = 50MHz$

$C_c = \frac{1mA/V}{2\pi \times 50MHz} = \underline{\underline{3.18pF}}$

From Equ. (7.206)

$f_Z = \frac{G_{m2}}{2\pi C_c}$

$G_{m2} = g_{m6} = 3mA/V$

CONT.

$$f_Z = \frac{3\,mA/V}{2\pi \times 3.18\,pF} = 150.14\,MHz$$

From Eqn. (7.210)

$$f_{P2} = \frac{G_{m2}}{2\pi C_2} = \frac{3\,mA/V}{2\pi \times 3\,pF} = 160\,MHz$$

7.95

(a) $I_{E1} = I_{E2} = 0.1\,mA \simeq I_{E3}, I_{E4}$

$I_{E5} \simeq 1\,mA$ and since the output is held at 0V $I_{E6} = 2\,mA$

(b) $r_{e1} = r_{e2} = \frac{25\,mV}{0.1\,mA} = 250\,\Omega$

$$r_{e5} = \frac{25\,mV}{1\,mA} = 25\,\Omega$$

$$r_{e6} = \frac{25\,mV}{2\,mA} = 12.5\,\Omega$$

For the active loaded differential pair; Recall from Eqn. (7.161) $G_{m1} = g_{m1}$

$$\simeq \frac{1}{r_{e1}} = \frac{1}{250} = 4\,\frac{mA}{V}$$

$R_{o1} = (\beta+1) r_{e5}$ Since all r_o's $= \infty$

$R_{o1} = 101 \times 25 = 2525\,\Omega$

$\Rightarrow A_1 = G_{m1} R_{o1} = 4\,\frac{mA}{V} \times 2525\,\Omega$

$= 10.1\,V/V$

For the common-emitter:

$$A_5 = -g_{m5} \cdot R_{c5} \simeq -\frac{\beta R_L}{r_{e5}} = \frac{-100 \times 10K}{25}$$

$= -40{,}000\,V/V$

For the emitter follower:

$A_6 \simeq 1$

$A_{2nd\text{-}stage} = A_5 \cdot A_6 = -40{,}000\,V/V$

$A = A_1 \cdot A_{2nd\text{-}stage} = 10.1 \times -40{,}000$

$= -404{,}000\,V/V$

(c) Since the dominant low-frequency pole is set by C_c & $r_{\pi 5}$

$$f_P = \frac{1}{2\pi \cdot R_{o1} \underbrace{(A_5+1) C_c}_{\text{by Miller effect}}} = 100\,Hz$$

$\Rightarrow C \cong 1/(2\pi \times 2525 \times 40K \times 100)$

$= 15.76\,pF$

7.96

$I_B = 225\,\mu A$

$\mu_n C_{ox} = 180\,\mu A/V^2$

$\mu_p C_{ox} = 60\,\mu A/V^2$

For Q_8 & Q_9: $W/L = 60/0.5$

$$\Rightarrow |V_{ov}| = \sqrt{\frac{2 I_D}{k_p'(W/L)}}$$

$$|V_{ov}|_{8,9} = \sqrt{\frac{2 \times 225\mu}{60\mu \times 120}} = 0.25\,V$$

then $g_{m8,9} = \frac{2 I_D}{|V_{ov}|} = \frac{2 \times 225\mu}{0.25\,V}$

$= 1.8\,mA/V$

Since g_m of Q_{10}, Q_{11} & Q_{13} are identical to g_m of Q_8 & Q_9 then $V_{ov13} = 0.25\,V$

Thus for Q_{13}

$$(0.25)^2 = \frac{2 \times 225\mu}{180\mu \times (W/L)_{13}}$$

$\rightarrow (W/L)_{13} = 40$ i.e $(20/0.5)$

Since Q_{12} is 4 times as wide as Q_{13}, then

CONT.

$(W/L)_{12} = \frac{4 \times 20}{0.5} = 80/0.5$

$R_B = \frac{2}{\sqrt{2k_n'(W/L)_{12} I_B}} \cdot \left(\sqrt{\frac{(W/L)_{12}}{(W/L)_{13}}} - 1\right)$

$= \frac{2}{\sqrt{2 \times 180\mu \times \frac{80}{0.5} \times 225\mu}} \cdot \left(\sqrt{\frac{80/0.5}{20/0.5}} - 1\right)$ ($\sqrt{4} - 1$)

$\rightarrow R_B = 555.6\,\Omega$

The voltage drop on R_B is:

$555.6 \times 225\mu = 0.125\,V.$

To obtain the gate voltages:
(assume $|V_{tn}| = |V_{tp}| = 0.7V$)

$V_{OV12} = \sqrt{\frac{2 \times 225\mu}{180\mu \times \frac{80}{0.5}}} = 0.125\,V$

$V_{OV12} = V_{GS12} - V_{tn}$

$\rightarrow V_{GS12} = 0.125 + 0.7 = 0.825V$

thus,

$V_{G12,13} = V_{GS12} + I_B R_B - V_{SS}$
$= 0.825 + 0.125 - 1.5$
$= -0.55\,V$

$V_{OV11} = |V_{OV8}| = 0.25V$

$\Rightarrow V_{GS11} = 0.25 + 0.7 = 0.95V$

$V_{G11} = -0.55 + 0.95$

$V_{G11} = V_{G10} = 0.4V$

$V_{G8} = V_{DD} - V_{SG8} = 1.5 + (-0.25 - 0.7)$
$= +0.55\,V$

Finally, from the results above:

$(W/L)_{10} = 20/0.5$
$(W/L)_{11} = 20/0.5$
$(W/L)_{12} = 80/0.5$
$(W/L)_{13} = 20/0.5$

7.97

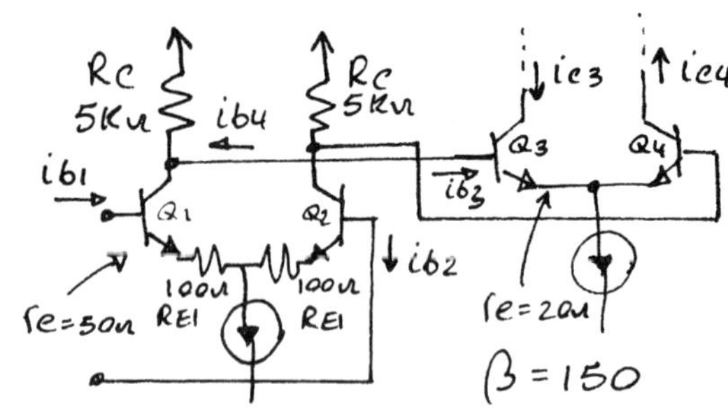

$A_1 = \frac{2R_C \,\|\, R_{id2}}{2(R_{E1} + r_{e1})}$

$R_{id2} = (\beta+1)(2r_{e2}) = 6.04K\Omega$

$\Rightarrow A_1 = 12.5\,V/V$

$A_i = \frac{i_{e4}}{i_{b1}} = \beta_1 \cdot \frac{2R_C}{R_{id2} + 2R_C}\,\beta_4$

$= 1.4 \times 10^4\,A/A$

7.98

$R_{id1} = (\beta+1)(R_{E1} + r_{e1})2$
$= 101 \times (100 + 100)2 = 40.4K\Omega$

increase!

$R_{id2} = (\beta+1)(2)(R_{E2} + r_{e2})$
$= 101 \times (25 + 25)2 = 10.1K\Omega$

$\therefore A_1 = \frac{2(20K) \,\|\, R_{id2}}{2(R_{E1} + r_{e1})} = 20.2\,V/V$ decrease

$A_2 = -\frac{R_3 \,\|\, R_{i3}}{2(R_{E2} + r_{e2})} = -29.6\,V/V$

observe that A_3 and A_4 are unchanged.

$A = A_1 A_2 A_3 A_4$
$= (20.2)(29.6)(6.42)(0.998)$
$= 3823\,V/V$ decrease.

7.99

$R_o \simeq \frac{R_5}{\beta+1} + r_{e8} = R_C'$

CONT.

Thus R_5 affects R_o

We want $R_o' \| 3K = 76$

$\Rightarrow R_o' = 78\,\Omega$

$\Rightarrow R_5 = (78 - r_{e8})(\beta+1)$

$= 7.37\,K\Omega$

$A_3 = -\dfrac{R_5 \| R_{i4}}{r_{e4} + R_4}$; $R_{i4} \simeq 304\,K\Omega$

and $A_3 = -3.09$ V/V

and $A = 8513 \cdot \dfrac{3.09}{6.42} = 4104$ V/V

The gain has been reduced by a factor of 2.07 and can be restored by reducing R_4 by this same factor to increase A_3. Thus $R_4 = 1.11\,K\Omega$

(Note that this is a first order approximation).

7.100

(a) $A_3 = \dfrac{-R_{i4}}{2.325\,K\Omega} = -\dfrac{303.5}{2.325}$

$= -130.5$ V/V

i.e A_3 is increased by $\dfrac{130.5}{6.42}$

$= 20.33$

$\Rightarrow A = 8513 \times 20.33$

$= 173.1 \times 10^3$ V/V

(b) Let the output resistance of the current source be $R \to \infty$

$R_o = 3K \| \left(\dfrac{R}{\beta+1} + r_e\right)$

$= 3K$

The amplifier can be modelled as shown:

v_i — R_{id} — AV_i — R_o — V_o — R_L

Thus,

$A_{LOAD} = \dfrac{A \cdot R_L}{R_L + R_o}$

$= 173.1 \times 10^3 \dfrac{100}{100 + 3000}$

$= 5583$ V/V

For the original amplifier:

$A_{LOAD} = 8513 \times \dfrac{100}{100 + 152} = 3378$ V/V

7.101

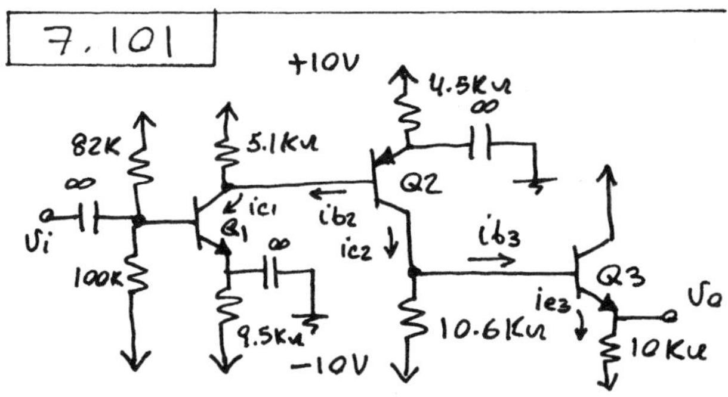

(a) $I_{E1} = \dfrac{\dfrac{20V \times 100K}{82K + 100K} - 0.7}{9.5K + \dfrac{(82K \| 100K)}{\beta+1}}$

$\beta = 100 \Rightarrow I_{E1} = 1.03$ mA

$\alpha = \dfrac{100}{101} \Rightarrow I_{C1} = 1.02$ mA

$V_{C1} \simeq 10V - 1.02mA \times 5.1K\Omega = 4.8V$

$I_{E2} = \dfrac{(10 - 0.7 - 4.8)V}{4.5K\Omega} = 1mA$

$\rightarrow I_{C2} = 0.99$ mA

$V_{C2} \simeq 0.99m \times 10.6K - 10 = 0.5V$

$\Rightarrow V_{O\,DC} = 0.5 - 0.7 = -0.2V$

$I_{E3} = \dfrac{-0.2 - (-10)}{10K} = 0.98$ mA

$\rightarrow I_{C3} = 0.97$ mA

Thus all transistors are operating at $I_C \simeq 1$ mA

CONT.

(b) $R_{in} = 82K \,\|\, 100K \,\|\, r_{\pi 1}$

where $r_{\pi 1} = \dfrac{\beta}{g_{m1}} = \dfrac{100}{40m} = 2.5K\Omega$

$\Rightarrow R_{in} = (82 \,\|\, 100 \,\|\, 2.5)K = \underline{\underline{2.37K\Omega}}$

$R_{out} = 10K \,\|\, \left[r_{e3} + \dfrac{10.6K}{\beta+1}\right]$

$= 10K \,\|\, \left[25 + \dfrac{10.6K}{101}\right]$

$= \underline{\underline{128\Omega}}$

(c) $\dfrac{i_{c1}}{v_i} = g_{m1} = 40mA/V$

$\dfrac{i_{b2}}{i_{c1}} = \dfrac{5.1K}{5.1K + r_{\pi 2}} = \dfrac{5.1}{5.1+2.5} = 0.671\ \dfrac{A}{A}$

$\dfrac{i_{e2}}{i_{b2}} = \beta_2 = 100\ A/A$

$\dfrac{i_{b3}}{i_{c2}} = \dfrac{10.6K}{10.6K + (\beta+1)(r_{e3}+10K)}$

$= 0.01036\ A/A$

$\dfrac{i_{e3}}{i_{b3}} = \beta_3 + 1 = 101$

$v_o = i_{e3} \times 10K$

Thus,

$\dfrac{v_o}{v_i} = 10 \times 101 \times 0.01036 \times 100 \times 0.671 \times 40$

$= \underline{\underline{2.81 \times 10^4\ V/V}}$

(d) $f_{P2} = 1/(2\pi C_2 \cdot R_2)$

where: $R_2 = 5.1K \,\|\, r_{\pi 2}$

$= 5.1K \,\|\, 2.5K = 1.68K\Omega$

$C_2 = C_{\pi 2} + C_{\mu 2}(1 + g_{m2} R_{L2})$

with:

$R_{L2} = 10.6K \,\|\, ((\beta+1)(r_{e3} + 10K)$

$= 10.6K \,\|\, 101 \times (25 + 10K)$

$= 10.5K\Omega.$

$\Rightarrow C_2 = 10p + 2p(1 + 40m \times 10.5K)$

$= 852pF$

$\Rightarrow f_{P2} = \dfrac{1}{2\pi \times 852p \times 10.5K}$

$= \underline{\underline{17.8KHz}}$

7.102

(a)

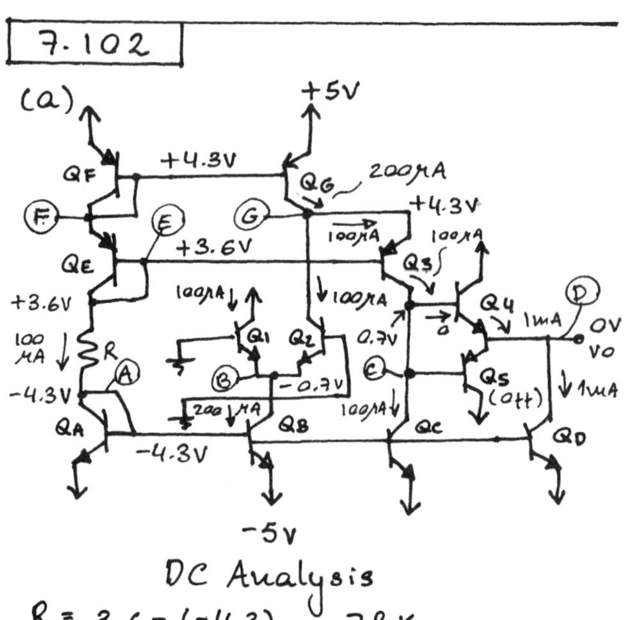

DC Analysis

$R = \dfrac{3.6 - (-4.3)}{100\mu A} = \underline{\underline{79K\Omega}}$

Node voltages:

$V_A = \underline{\underline{-4.3V}}$, $V_B = \underline{\underline{-0.7V}}$

$V_C = \underline{\underline{+0.7V}}$ $V_D = \underline{\underline{0V}}$

$V_E = \underline{\underline{+3.6V}}$ $V_F = \underline{\underline{+4.3V}}$

$V_G = \underline{\underline{+4.3V}}$

(b)

Transistor	I_C (mA)	g_m (mA/V)	r_o (MΩ)
Q_1	0.1	4	2
Q_2	0.1	4	2
Q_3	0.1	4	2
Q_4	1.0	40	0.2
Q_5	0	0	∞
Q_A	0.1		
Q_B	0.2		
Q_C	0.1	....	2
Q_D	1.0	....	0.2
Q_E	0.1		
Q_F	0.1		
Q_G	0.2	....	1

CONT.

(c) Total resistance at collector Q_3 is

$$\simeq \beta_3 r_{o3} \,\|\, r_{oC} \,\|\, (\beta_4+1)(r_{o4} \,\|\, r_{oD})$$

$$= 100 \times 2 \,\|\, 2 \,\|\, 101\,(0.2 \,\|\, 0.2)$$

$$= 1.65\,M\Omega$$

$$\frac{v_{c3}}{v_i} = +g_{m1} \times \frac{1}{2} \times 1.65 \times 10^3 = 3300\ \frac{V}{V}$$

$$\frac{v_o}{v_{c3}} \simeq 1$$

Thus, $\dfrac{v_o}{v_i} \simeq 3300$ V/V (polarity correct)

(d) $R_{in} = 2r_\pi$

$$= 2 \times \frac{100}{4} = 50\,k\Omega$$

$$R_{out} = r_{oD} \,\|\, r_{o4} \,\|\, \left[r_{e4} + \frac{r_{o2C} \,\|\, \beta_3 r_{o3}}{\beta+1}\right]$$

$$= 0.2 \,\|\, 0.2 \,\|\, \left[25.10^{-6} + \frac{2 \,\|\, 100 \times 2}{101}\right]$$

$$\simeq 16.4\,k\Omega$$

(e) $V_{ICM}|_{min} = -4.3 - 0.4 + 0.7 = -4V$

$V_{ICM}|_{max} = V_G + 0.4 = +4.7V$

(f) The voltage at the base of Q_4 can rise to V_{B3}:

$(V_E) + 0.4 = +4V$

before Q_3 saturating. Thus v_o can go up to $+3.3V$

The voltage at the output can go down to $V_{\text{base of } Q_D} + 0.4$

$= V_A - 0.4 = -4.3 - 0.4 = -4.7V$

Thus the linear range at the output is $-4.7V$ to $+3.3V$

(g) At the positive limit of v_o, i.e. $v_o = +3.3V$ and Q_2 just cut off

$$R_L = \frac{3.3V}{9.1mA} = 363\,\Omega$$

(this is the minimum allowed R_L for $+3.3V$ output)

At the negative limit of v_o, i.e. $v_o = -3.3V$ and Q_1 has cut-off. Q_3 will also be cut-off, and Q_4 will cut-off.

Thus,

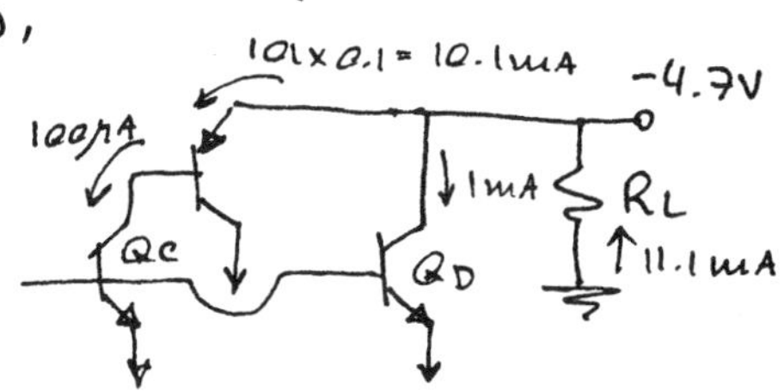

$$R_L = \frac{4.7}{11.1\,mA} = 423\,\Omega$$

This is the minimum allowed R_L for a $-4.7V$ output.

7.103

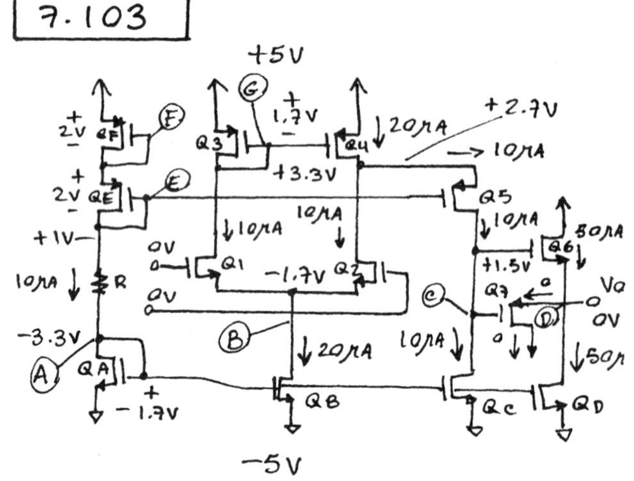

DC analysis.

(a) $I_{REF} = 10\mu A = \frac{1}{2} \times 40 \times \frac{5}{5}(V_{GS_A} - V_t)^2$

$\Rightarrow V_{GS_A} = 1.71V \simeq 1.7V$

$10 = \frac{1}{2} \times 20 \times \frac{5}{5}(V_{GS_{E,F}} - 1)^2$

$\Rightarrow V_{GS_{E,F}} = 2V$

$R = \frac{3 - (-3.3)}{10\mu A} = \underline{\underline{660\,K\Omega}}$

(b) See figure above

$V_{GS1} = V_{GS2} = V_{GSA} \simeq 1.7V$

$V_{GS3} = \sqrt{\frac{2 \times 10}{20 \times \frac{10}{5}}} + 1 = 1.71V \simeq 1.7V$

$V_{GS5} = V_{GS3} = 1.7V$

For Q_6: $50 = \frac{1}{2} \times 40 \times \frac{50}{5}(V_{GS_6} - V_t)^2$

$\Rightarrow V_{GS6} = 1.50V$

$V_A = \underline{\underline{-3.3V}}$	$V_B = \underline{\underline{-1.7V}}$
$V_C = \underline{\underline{+1.5V}}$	$V_D = \underline{\underline{0V}}$
$V_E = \underline{\underline{+1V}}$	$V_F = \underline{\underline{+3V}}$
$V_G = \underline{\underline{+3.3V}}$	$V_H = \underline{\underline{+2.7V}}$

(c)

Transistor	I_D (μA)	V_{GS} (V)	g_m (μA/V)	r_o (MΩ)
Q_1	10	1.7	28.3	5
Q_2	10	1.7	28.3	5
Q_3	10	1.7	28.3	5
Q_4	20	1.7	56.6	2.5
Q_5	10	1.7	28.3	5
Q_6	50	1.5	200	1
Q_7	0	-1.5⊛	0	∞
Q_A	10	1.7	28.3	5
Q_B	20	1.7	56.6	2.5
Q_C	10	1.7	28.3	5
Q_D	50	1.7	141.4	1
Q_E	10	2	20	5
Q_F	10	2	20	5

⊛ cut-off.

(d)

Total resistance at the drain of Q_5, R, is:

$R = (g_{m5} r_{o5})(r_{o4} \| r_{o2}) \| r_{oC}$

$= [(28.3 \times 5)(2.5 \| 2)] \| 5$

$= 4.9\,M\Omega$

Thus, $\frac{v_{ds}}{v_i} = g_{m1} R$

$= 28.3 \times 4.9 = 138.7\ V/V$

and $\frac{v_o}{v_{ds}} = \frac{(r_{oD} \| r_{o6})}{(r_{oD} \| r_{o6}) + \frac{1}{g_{m6}}}$

$= \frac{(1 \| 1)}{(1 \| 1) + \frac{1}{200}} \simeq 1$

$\frac{v_o}{v_i} = \underline{\underline{138.7}}\ V/V$

$R_{in} = \infty$

$R_{out} = r_{oD} \| r_{o6} \| 1/g_{m6}$

$= 1 \| 1 \| 1/200\ M\Omega$

$\simeq \underline{\underline{5K\Omega}}$

CONT.

(e) $v_{ICM}|_{max} = V_G + V_t$
$= +\underline{\underline{4.3V}}$

$v_{ICM}|_{min} = V_{GS1} + V_{Bmin}$
$= V_{GS1} + V_A - V_t$
$= 1.7 - 3.3 - 1 = \underline{\underline{-2.6V}}$

(f) $v_{O max} = v_{C max} - V_{GS6}$
$= V_E + |V_t| - V_{GS6}$
$= +1 + 1 - 1.5 = \underline{\underline{+0.5V}}$

$v_{O min} = V_A - V_t = -3.3 - 1 = \underline{\underline{-4.3V}}$

(g) Q6 cuts off thus:

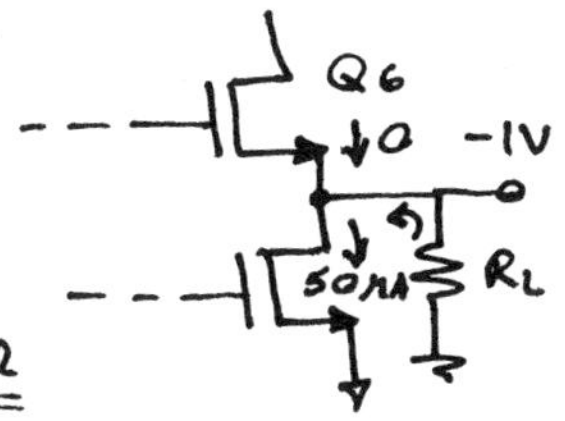

$\frac{1V}{R_L} = 50\mu A$

$R_L = \frac{1V}{50\mu A} = \underline{\underline{20\,K\Omega}}$

Q6, 0, −1V, 50μA, R_L

(h) Maximum possible voltage at drain of Q5 is +2V.
At this value we have:

$I_D = 50\mu A + \frac{\hat{V}_O}{2}\,mA$

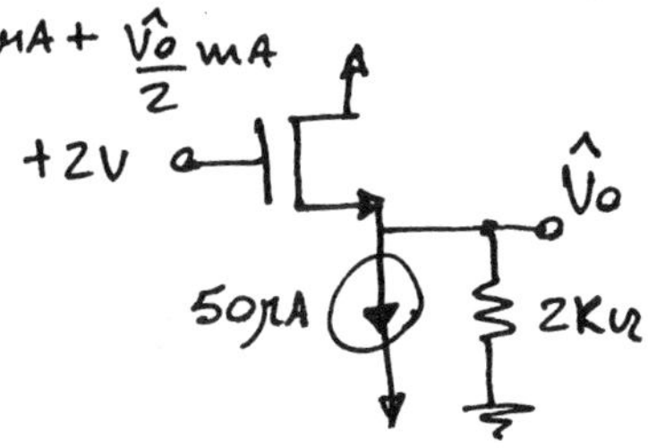

+2V, $\hat{V}_O$, 50μA, 2KΩ

$I_D = \frac{1}{2}\mu_n C_{ox}\frac{W}{L}(2 - \hat{V}_O - V_t)^2$

$\Longrightarrow \hat{V}_O \simeq 0.17V$

For the lowest possible output, the circuit becomes

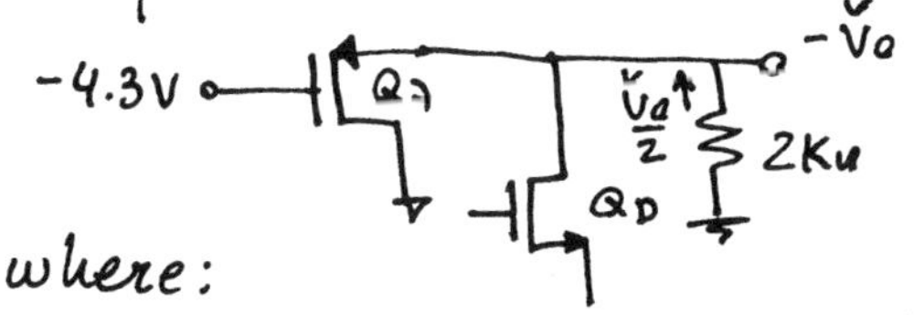

−4.3V, Q7, $-\check{V}_O$, $\frac{\check{V}_O}{2}$, 2KΩ, QD

where:
Q6 cuts off and Q7 conducts

$I_D = \frac{\check{V}_O}{2} - 0.05mA$

$= \frac{1}{2}\mu_p C_{ox}\left(\frac{100}{5}\right)(-\hat{V}_O + 4.3 - 1)^2$ ← $(W/L)_7$

$\Longrightarrow \hat{V}_O = 1.45V$

That is, the range of v_O is $\underline{\underline{-1.45V}}$ to $+\underline{\underline{0.17V}}$

Chapter 8 – Problems

8.1

$$A_F = \frac{A}{1+A\beta} = 100$$

$$A\beta = \frac{10^5}{100} - 1 = 999$$

$$\Rightarrow \beta = \frac{999}{10^5} = 9.99\times10^{-3}$$

$$A = 10^3, \quad A_F = \frac{10^3}{1+10^3(9.99\times10^{-3})}$$

$$= 90.99$$

$$\frac{\Delta A_f}{A_f} = \frac{90.99-100}{100} \Rightarrow -9\%$$

8.2

(b) $A_f = 10 = \dfrac{100}{1+100\beta} \Rightarrow \beta = 90\times10^{-3}$

$$\beta = \frac{R_1}{R_1+R_2} \Rightarrow \frac{R_2}{R_1} = \frac{1}{\beta} - 1 = 10.11$$

(c) amount of feedback $= 1 + A\beta$

$$= 1 + 100(90\times10^{-3}) = 10 \equiv 20\text{dB}$$

(d) $V_o = 10V_s = 10V$

$$V_f = \beta V_o = 90\times10^{-3}\times10 = 0.9\text{ V}$$

$$V_i = V_s - V_f = 1-0.9 = 0.1V$$

(e) $A_f = \dfrac{80}{1+80(90\times10^{-3})} = 9.756$

$$\frac{\Delta A_f}{A_f} = \frac{9.756-10}{10} \Rightarrow -2.44\%$$

8.3

(b) $A_f = 10^3 = 10^3/(1+10^3\beta)$

$$\Rightarrow \beta = 900\times10^{-6}$$

$$\frac{R_2}{R_1} = \frac{1}{\beta} - 1 = 1110$$

(c) $1 + A\beta = 1 + 10^3(900\times10^{-6}) = 10 \equiv 20\text{dB}$

(d) $V_s = 0.01V$

$$V_o = 10^3 V_s = 10V$$

$$V_f = \beta V_o = 900\times10^{-6}\times10 \equiv 9\text{mV}$$

$$V_i = V_s - V_f = 1\text{mV}$$

(e) $A = 8000$

$$A_f = \frac{8000}{1+8000(900\times10^{-6})} = 975.6$$

$$\frac{\Delta A_f}{A_f} = \frac{975.6-1000}{1000} \equiv -2.44\%$$

8.4

All output voltage is fed back $\therefore \beta = 1$

$$A_f = \frac{100}{1+100\times1} = 0.99$$

$$1 + A\beta = 1 + 100\times1 = 101 \equiv 40.1\text{dB}$$

$$V_o = 0.99V_s = 0.99V$$

$$V_i = V_s - V_o\beta = 1-0.99 = 10\text{mV}$$

$$A = 90 \Rightarrow A_f = \frac{90}{1+90\times1} \approx 0.989$$

$$\frac{\Delta A_f}{A_f} = \frac{0.989-0.99}{0.99} \equiv -0.1\%$$

8.5

$$V_s = 1V, \quad V_i = 10\text{mV}, \quad V_o = 10V$$

$$A_o = (V_o/V_i) = 10V/10\text{mA} = 1000$$

$$V_i = V_s - V_f = V_s - \beta V_o = 1 - \beta 10$$

$$\Rightarrow \beta = (1-10^{-2})10 = 0.099$$

8.6

$$A_F = \frac{A_o}{1+A_o\beta} = \frac{1}{1/A_o + \beta} = \frac{1}{\beta(1+1/A_o\beta)}$$

so A_f + $1/\beta$ will be within $x\%$ when

$$1/(A_o\beta) = 0.01\times x$$

(a) For 1%: $A_o\beta = 1/0.01 = 100$

Many possible solutions.

Let $A_o = 10^5$ + $A_o\beta = 100 \Rightarrow \beta = 10^{-3}$

(b) For 5%: $A_o\beta = 1/0.05 = 20$

Let $A_o = 10^5$ + $A_o\beta = 20 \Rightarrow \beta = 2\times10^{-4}$

(c) For 10%: $A_o\beta = 1/0.1 = 10$

Let $A_o = 10^5$ + $A_o\beta = 10 \Rightarrow \beta = 10^{-4}$

(d) For 50%: $A_o\beta = 1/0.5 = 2$

Let $A_o = 10^5$: $A_o\beta = 2 \Rightarrow \beta = 2\times10^{-5}$

% error	A_o	$A_o\beta$	$1+A_o\beta$
1	10^5	100	101
5	10^5	20	21
10	10^5	10	11
50	10^5	2	3

8.7.

$0 \le \beta \le 1$ Linear

(a) For $A_o = 1$: $A_{f_1} = \frac{A_o}{1+A_o\beta} = \frac{1}{1+0} = 1\,V/V$

$A_{f_2} = \frac{1}{1+1\times0.5} = 0.667\,V/V$

$A_{f_3} = \frac{1}{1+1\times1} = 0.5\,V/V$

(b) For $A_o = 10$: $A_{f_1} = \frac{10}{1+0} = 10\,V/V$

$A_{f_2} = \frac{10}{1+10/2} = 1.6\,V/V$

$A_{f_3} = \frac{10}{1+10\times1} = 0.909\,V/V$

(c) For $A_o = 100$: $A_{f_1} = \frac{100}{1+0} = 100\,V/V$

$A_{f_2} = \frac{100}{1+100/2} = 1.96\,V/V$

$A_{f_3} = \frac{100}{1+100} = 0.99\,V/V$

(d) For $A_o = 10^4$: $A_{f_1} = \frac{10^4}{1+0} = 10^4\,V/V$

$A_{f_2} = \frac{10^4}{1+10^4/2} = 1.99\,V/V$

$A_{f_3} = \frac{10^4}{1+10^4} = 0.9999\,V/V$

8.8

A_o : 2mV → 10V

$A_o = 10V/(2\times10^{-3})V = 5000 \doteq 74\,dB$

A_F : 200mV → 10V

$A_F = (10/(200) = 500 \doteq 54\,dB$

$A_F = \frac{A_o}{1+\beta A_o} = \frac{5000}{1+5000\beta} = 500$

$\Rightarrow 1+5000\beta = 10$

$\Rightarrow \beta = 9/5000 = 0.0018 \equiv -54\,dB$

$(1+A_o\beta) = 10 \equiv 20\,dB$

$A_o\beta = 5000\,(9/5000) = 9 \equiv 19.08\,dB.$

8.9

$\frac{dA_f}{A_f} = \frac{1}{1+A\beta}\cdot\frac{dA}{A}$

$\frac{dA_f/A_f}{dA/A} = \frac{1}{1+A\beta} \equiv -20\,dB$

$\Rightarrow 1+A\beta = +20\,dB \equiv 10$

$\therefore A\beta = 9$

Require $\frac{1}{1+A\beta} = \frac{1}{2} \Rightarrow A\beta = 1$

8.10

$A_o \equiv 1000 \pm 30\%$ want $A_f = 100 \pm 1\%$

To reduce % change in A_o we need

$\frac{1}{1+A\beta} = \frac{1}{30} \Rightarrow A_f = \frac{1000}{30} < 100$

For single stage $A_F = \frac{A}{1+A\beta_1} \Rightarrow \frac{1}{1+A\beta_1} = \frac{100}{1000} = \frac{1}{10}$

For two stages (identical) $A_2 = \frac{A}{1+A\beta_2} = 10$

$\Rightarrow (1+A\beta_2) = 1000/10 = 100$

Thus each stage has $\pm 30/100\,\% = \pm 0.3\%$

But two such stages may give $\pm 0.6\%$ OK

[3 stages $A_3 = A_f^{1/3} = 100^{1/3}$

$(1+A_o\beta_3) = 1000/100^{1/3} \doteq 215$

Now each stage has $\pm 0.14\%$]

8.11

Gain desensitivity factor $= [1+A_o\beta]$

$A_f = \frac{A_o}{1+A_o\beta} = \frac{10^5}{1+10^5\beta} = 10^3$

$\Rightarrow \beta = 0.99 \times 10^{-3}$

From $\frac{dA_F}{A_f} = \frac{1}{1+A\beta} \cdot \frac{dA}{A}$

(a) When A drops 10%

then A_F drops by $\frac{10\%}{1+A_o\beta} = \frac{10\%}{10^5/10^3} = 0.1\%$

But using $A_F = \frac{A}{1+A\beta} = \frac{0.9\times10^5}{1+(0.9\times10^5)(99/10^5)}$

$= 998.89$

Corresponding % $\frac{1000-998.89}{1000} \equiv 0.11\%$ (close)

(b) When A drops 30%

then A_F drops by $\frac{30\%}{100} = 0.3\%$

while $A_F = \frac{0.7\times10^5}{1+(0.7\times10^5)(99/10^5)} = 995.7$

and $\frac{1000-995.7}{1000} \equiv 0.43\%$ (not so close)

8.12

$A(s) = A_m \frac{s}{s+\omega_L}$

$A_f(s) = \frac{A_m \frac{s}{s+\omega_L}}{1+\frac{A_m s}{s+\omega_L}\beta} = \frac{A_m s}{s+\omega_L+A_m\beta s}$

$= \frac{A_m}{1+A_m\beta} \cdot \frac{s}{s+\frac{\omega_L}{1+A_m\beta}}$

Thus $A_{mf} = \frac{A_m}{1+A_m\beta}$

$\omega_{Lf} = \frac{\omega_L}{1+A_m\beta}$

Both decreased by same amount

8.13

Worst case: $A_{F1} = \frac{A_o}{1+A_o\beta} = 9.8$ (down 2%)

full battery: $A_{F2} = \frac{2A_o}{1+2A_o\beta} = 10$

from A_{F1}: $1+A_o\beta = A_o/9.8$

$\therefore \beta = \frac{1}{9.8} - \frac{1}{A_o}$

Then $A_{F2} = \frac{2A_o}{1+2A_o\left[\frac{1}{9.8}-\frac{1}{A_o}\right]} = 10$

$\Rightarrow 1+2A_o\left[\frac{1}{9.8}-\frac{1}{A_o}\right] = \frac{2A_o}{10}$

$\Rightarrow 2A_o\left[\frac{1}{9.8}-\frac{1}{A_o}\right] = 2-1$

$\therefore 2A_o = 490$

[Check $\frac{2A_o}{1+2A_o\left[\frac{1}{9.8}-\frac{2}{2A_o}\right]} = 10$ ← β const

$\frac{A_o}{1+A_o\left[\frac{1}{9.8}-\frac{1}{A_o}\right]} = 9.8$]

8.14

$A_f = \frac{A_o}{1+A_o\beta} = 10 = \frac{100}{1+100\beta}$

$\therefore (1+A_o\beta) = 100/10 = 10$

$f_L' = f_L/(1+A_o\beta) = 100/10 = 10\text{ Hz}$

$f_h' = f_H(1+A_o\beta) = 10K\times10 = 100\text{ KHz}$

8.15

Need pole at 0.5 MHz. any others should be well above for stability.

$f_H = 10\text{ KHz}$ and $f_{Hf} = 500\text{ KHz}$

$\therefore$ need $(1+A\beta) = 500/10 = 50$

Thus $A_{f1} = \frac{A}{1+A\beta} = \frac{1000}{50} = 20$

and $\beta = (50-1)/1000 = 4.9\times10^{-3}$

Subsequent stages must provide gain

$A_{Total}/A_{f1} = 1000/20 = 50$

Let $f_{H2} = 3\times f_{H1} = 3\times500 = 1500\text{ KHz}$

needs $(1+A\beta_2) = \frac{1500}{10} = 150$

Then $A_{f2} = \frac{1000}{150} = 6.67$

and $\beta_2 = \frac{150-1}{1000} = 0.149$

This leaves stage 3 with $A_{f3} = \frac{50}{6.67} = 7.5$

This gain needs $1 + A\beta = \frac{1000}{7.5} = 133.3$

and $f_{Hf3} = 10K(1+A\beta) = 1.33\,MHz$

this is safely well above 500 KHz

and $\beta_3 = 0.132$

The actual f_{Hf} is given by

$$f_{Hf} = \frac{1}{\left[\left(\frac{1}{0.5}\right)^2 + \left(\frac{1}{1.5}\right)^2 + \left(\frac{1}{1.33}\right)^2\right]}$$

$\equiv 0.447\,MHz$ close!

Using a 4 stage amplifier would allow lower stage gains and higher f_H's

8.16

$$V_o = V_S \frac{A_1 A_2}{1 + A_1 A_2 \beta} + V_N \frac{A_1}{1 + A_1 A_2 \beta}$$

and $\frac{S}{N} = \frac{V_S}{V_N} A_2$

if we raise V_S by factor A_2 then S/N is improved by A_2 also

To reduce V_N from $\pm 1V$ to $\pm 1mV$ we must have $A_2 = 1000$, then

$$V_{os} = \frac{A_1 A_2}{1 + A_1 A_2 \beta} = \frac{1000 \times 0.9}{1 + 1000 \times 0.9\beta} = 10$$

$\Rightarrow \beta = 0.099$

if $A_2 = 100$; $V_n = \pm 10mV$

$$V_{os} = \frac{100 \times 0.9}{1 + 100 \times 0.9\beta} = 10$$

$\Rightarrow \beta = 0.089$

if $A_2 = 10$: $V_n = \pm 100mV$

$A_1 A_2 = 0.9 \times 10$ & $V_o = 9V < 10V$!!

8.17

$$A_F = \frac{A_o}{1 + A_o\beta}$$

$$\frac{\Delta A_F / A_F}{\Delta A_o / A_o} = \frac{1}{1 + A_o\beta}$$

Then for 1% change in A_F for 90% change in A_o

$1 + A_o\beta = 90 \Rightarrow A_o\beta = 89$

Then $A_F = 100 = \frac{A_o}{90} \Rightarrow A_o = 9000$

Let $A_o^* \to 10 \times A_o$ then for $\beta = 89/A_o$

$$A_F = \frac{9000 \times 10}{1 + (9000 \times 10)(89/9000)} = 101.12$$

Therefore, select $A_o \geq 9 \times 10^4$

and $\beta = 89/A_o$

if $A_o^* \to 100 A_o$ & same β.

$$A_F = \frac{9000 \times 100}{1 + (9000 \times 100)(89/9000)}$$

$\approx 9000 \times 100 / 8900$

≈ 101.12

if $A_o^* \to \infty$

$A_F \Rightarrow A_o/(A_o\beta) \Rightarrow 1/\beta$.

$\approx 9000/89 = 101.12$

Ideally select A_o as high as possible

8.18

A_1 has f_{H1} high, A_2 has $A_m = 10V/V$ with $f_L = 80Hz$, $f_H = 8KHz$.

$$A_F = \frac{A_1 A_2}{1 + A_1 A_2 \beta} = 100$$

Require $f_{Hf} = 40KHz = 8(1 + A_1 A_2 \beta)$

$\therefore 1 + A_1 A_2 \beta = 40/8 = 5$

and $A_F = \frac{A_1 A_2}{5} = 100 \Rightarrow A_1 A_2 = 500$

$\Rightarrow A_1 = 500/A_2 = 500/10 = 50$

$1 + A_1 A_2 \beta = 5 \Rightarrow \beta = 4/A_1 A_2 = 4/500$

$\therefore \beta = 0.08$

$f_{Lf} = f_L/(1 + A_1 A_2 \beta) = 80/5 = 16Hz$

8.19

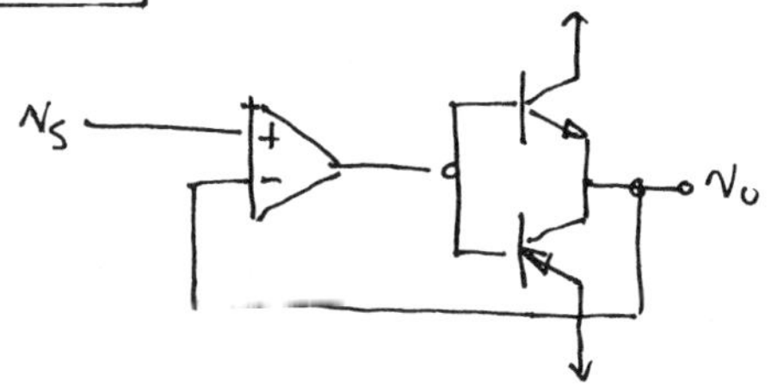

Dead band with be narrowed by the factor $1 + A\beta = 1 + A$ since $\beta = 1$ and since $A \gg 1$, $1 + A \to A$

$\therefore$ new limits are $\pm \frac{0.7}{A} = \pm \frac{0.7}{100}$

$= \pm 7\text{mV}$

New slope $\equiv$ gain $= A_f = \frac{A}{1+A}$

$\Rightarrow \frac{100}{1+100} = 0.99$

8.20

For $A = v_o/v_i = 10^2$ (select lowest A_o) to reduce % change in gain by factor of 10

$1 + A\beta = 10 \Rightarrow \beta = 9/10^2$

For $A_2 = 10^2$: $A_{2F} = 10^2/10 = 10$

For $A_1 = 10^3$: $A_{1F} = \frac{A}{1 + A\beta}$

$\therefore A_{2F} = \frac{10^3}{1 + 10^3(9/10^2)} = \frac{10^3}{91} = 10.98$

For $A_3 = 0$: stays saturated

[For 10mV in and $A_1 = 10^3$, $v_o = 10\text{ V}$?
For 10mV in and $A_2 = 10^2$, $v_o = 1\text{ V}$]

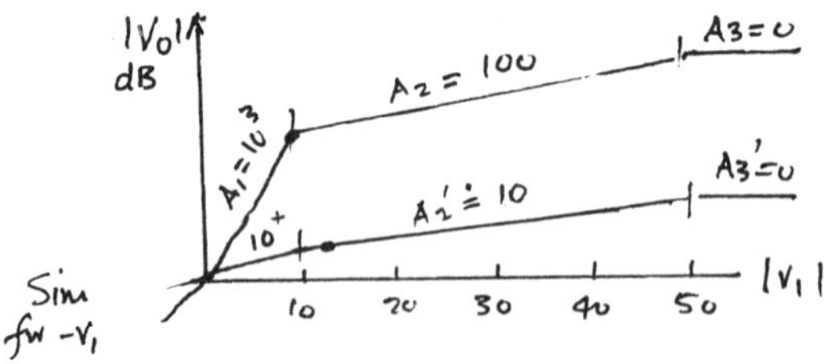

8.21

$V_s = 100\text{mV}$, $V_f = 95\text{mV}$, $V_o = 10\text{V}$

$V_1 = V_s - V_f = 100 - 95 = 5\text{mV}$

$A_V = \frac{V_o}{V_1} = \frac{10\text{V}}{5\text{ mV}} = 2 \times 10^3 \text{ V/V}$

$\beta = \frac{V_f}{V_o} = \frac{95 \times 10^{-3}}{10} = 9.5 \times 10^{-3} \text{ V/V} = 0.0095$

8.22

$I_s = 100\mu\text{A}$, $I_F = 90\mu\text{A}$, $I_o = 10\text{mA}$

$A \equiv \frac{I_o}{I_s - I_F} = \frac{10 \times 10^{-3}}{(100 - 90) \times 10^{-6}} = 10^3 \text{ A/A} = 1\text{ mA}/\mu\text{A}$

$\beta \equiv \frac{I_F}{I_o} = \frac{90\mu\text{A}}{10\text{mA}} = 9\mu\text{A}/\text{mA}$

8.23

For r_{o1}, r_{o2}, R_s very large in Fig 8.5

At G1: $I_s = I_F$ since $I_{G1} = 0$

$V_{G1} = [R_1 + R_2] I_s + I_o R$

$V_{D1} = -g_{m1} V_{g1} R_{L1}$

$= -g_{m1} R_L [I_s(R_1 + R_2) + I_o R_1]$

$I_o = g_{m2}(V_{D1} - V_{S2})$

$= g_{m2}[(-g_{m1} R_{L1})[I_s(R_1 + R_2) - I_s R_1 - I_o R_1]$

$\therefore \frac{I_o}{I_s} = -\frac{g_{m2}[g_{m1} R_{L1}(R_1 + R_2)]}{1 + g_{m2} g_{m1} R_{L1} R_1 + g_{m2} R_1}$

$A_f = -\frac{R_1 + g_{m1} R_{L1}(R_1 + R_2)}{R_1 + 1/g_{m2} + g_{m1} R_{L1} R_1}$

QED

$R_{IN} \equiv \frac{V_s}{I_s} = I_s[R_1 + R_2] + I_o R_1$

$= I_s[R_1 + R_2 + A_f R_1]$

$\therefore R_{IN} = R_1 + R_2 + A_f R_1$

QED

(b) When $g_{m1} R_{L1} = 100$, $R_L = 10K$, $R_2 = 90K$ and $g_{m2} = 5\text{mA/V}$

$A_f = -\frac{R_1 + g_{m1} R_{L1}(R_1 + R_2)}{R_1 + 1/g_{m2} + g_{m1} R_{L1} R_1}$

$= -\frac{10^4 + 100(100 \times 10^3)}{10^4 + 1000/5 + 100 \times 10^4}$

$= -9.91 \text{ A/A} \approx -10 \text{ A/A}$

$R_{IN} = 10K + 90K - 10K \times 10K \approx 0$

or $10K + 90K - 9.91 \times 10K$

$= 100 - 99.1K = 0.9K$

(c) $\beta \equiv \frac{I_F}{I_o} = \frac{R_1}{R_1+R_2} \Rightarrow \beta = \frac{10}{10+90} = \frac{1}{10}$

from above $A_F \approx -10 \equiv -1/\beta$

8.24

series - series

$V_S = 100mV$, $V_f = 95mV$, $I_o = 10mA$

$V_I = V_S - V_f = 100 - 95 = 5mV$

$V_f = \beta I_o \Rightarrow \beta = \frac{95mV}{10mA} = 9.5V/A$

$A \equiv \frac{I_o}{V_I} = \frac{10mA}{5mV} = 2\,mA/V$

$A_F \equiv \frac{I_o}{V_S} = \frac{A}{1+AB} = \frac{2}{1+2(9.5)} = 0.1mA/V$

8.25

shunt- shunt

$I_S = 100\mu A$, $I_F = 95mA$, $V_o = 10V$

$A \equiv \frac{V_o}{I_I} = \frac{V_o}{I_S - I_F} = \frac{10}{(100-95)\times 10^6} = 2\times 10^6\, V/A$

$\beta \equiv \frac{I_F}{V_o} = \frac{95}{1V}\frac{\mu A}{V} = 9.5\mu A/V$

8.26

(a)

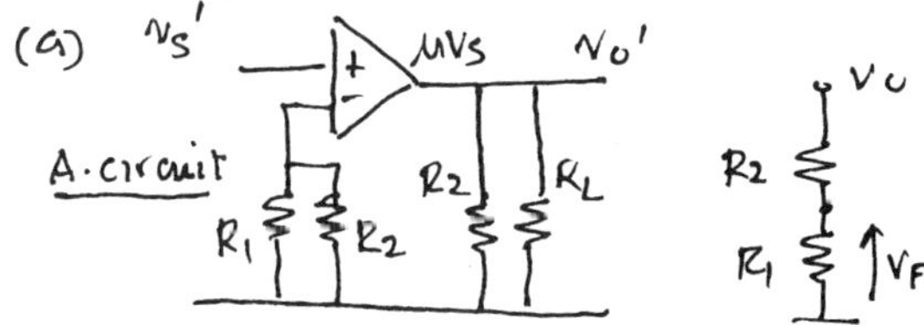

series - shunt; sample V, return V_F

$\beta \equiv \frac{V_F}{V_o} = \frac{R_1}{R_1+R_2}$

$A \equiv \frac{V_o'}{V_s'} = \mu$

$A_F \equiv \frac{V_o}{V_s} = \frac{A}{1+A\beta} = \frac{\mu}{1+\mu \frac{R_1}{R_1+R_2}}$

$\Rightarrow \frac{R_1+R_2}{R_1} = 1/\beta$

(b)

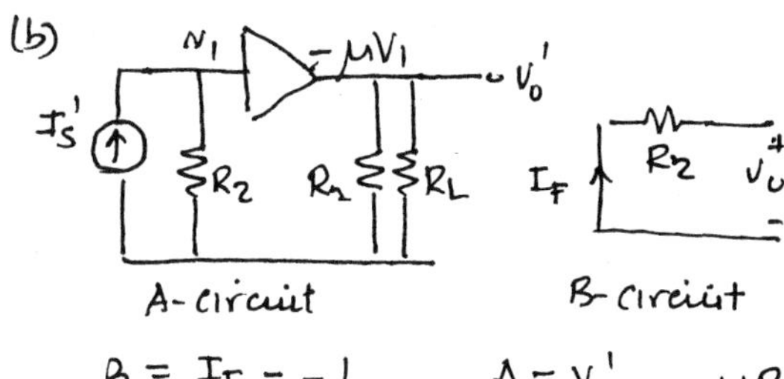

$\beta \equiv \frac{I_F}{V_o} = -\frac{1}{R_2}$ $\qquad A \equiv \frac{V_o'}{I_s'} = -\mu R_2$

$A_F \equiv \frac{V_o}{I_s} = \frac{A}{1+A\beta} = \frac{-\mu R_2}{1+\mu R_2/R_2} = -\frac{\mu R_2}{1+\mu}$

$\Rightarrow -R_2 = -1/\beta$

(c)

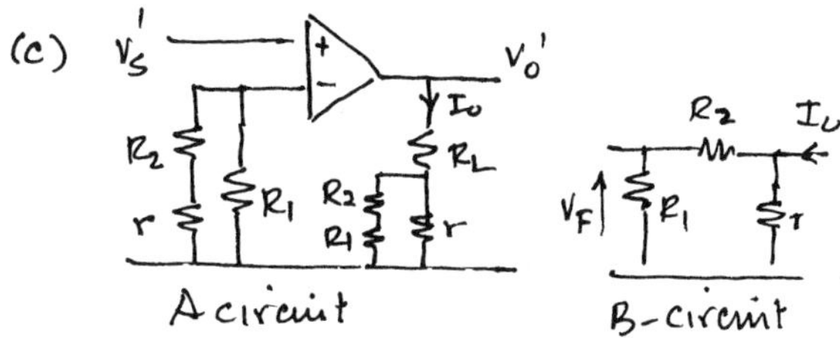

series - series.

$B \equiv \frac{V_F}{I_o} = \frac{R_1 r}{R_1+R_2+r}$ $\qquad$ let $r^* = r\|(R_1+R_2)$

$A \equiv \frac{I_o}{I_s} = \frac{\mu}{R_L + r\|(R_1+R_2)} = \frac{\mu}{R_L+r^*}$

$A_F \equiv \frac{A}{1+A\beta} = \frac{\frac{\mu}{R_L+r^*}}{1+\frac{\mu}{R+r^*}\cdot\frac{R_1 r}{R_1+R_2+r}}$

(d)

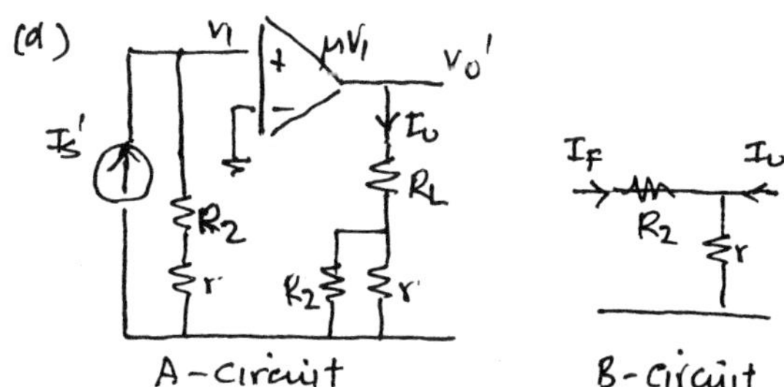

shunt-series: return I_F sample I_o

$\beta \equiv \frac{I_F}{I_o} = \frac{r}{R_2+r}$, $\quad A \equiv \frac{V_o'}{I_s'} = \frac{\mu(R_2+r)}{R_L+r\|R_2}$

$$\frac{I_o}{I_s} = \frac{\dfrac{\mu(R_2+r)}{R_L + r \| R_2}}{1 + \dfrac{\mu(R_2+r)}{R_L + r\|R_2} \cdot \dfrac{r}{r+R_2}}$$

$$\underset{\mu \gg 1}{\Rightarrow} \quad \frac{1}{r/(r+R_2)} \equiv \frac{1}{\beta}$$

Section 8.4 The Series-Shunt Feedback Amplifier

8.27

$$A_F = \frac{A}{1+A\beta} = \frac{10^3 \times 2}{1 + 2.10^3 \times 0.1} = 9.95 \text{ V}$$

$$R_{if} = R_i(1+A\beta) = 1(201) = 201 \text{ K}\Omega$$

$$R_{of} = R_o/(1+A\beta) = 1/201 = 4.975 \text{ K}\Omega$$

8.28

Here R_o is lowered by amount of feedback

i.e. $(1+A\beta) = 80$

$\Rightarrow A\beta = 79$

$R_o = R_{of}(1+A\beta) = 100 \times 80 = 8 \text{ K}\Omega$

8.29

$A_o = 10^4 \text{ V/V} \qquad f_t = 1 \text{ MHz}$

20 dB/dec roll off $\therefore f_p = f_t/A_o = 100 \text{ Hz}$

$$A(f) = \frac{A_o}{1 + jf/f_p} = \frac{10^4}{1 + jf/100}$$

$$\text{loop gain} = A(f)\beta = \frac{10^4(0.1)}{1+jf/100} = \frac{10^5}{jf+100}$$

$$Z_{if} = R_i(1+A\beta) = 10^4\left(1 + \frac{10^3}{1+jf/100}\right)$$

$$= 10^4 + \frac{10^7}{1+jf/100} \equiv R_A + \frac{R_B}{1+sR_BC}$$

Thus

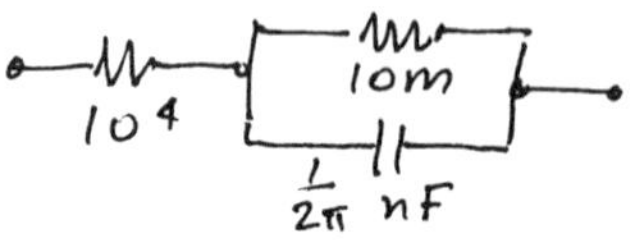

$$Z_{of} = \frac{R_o}{(1+A\beta)} = \frac{10^3}{1 + 10^3/(1+jf/100)}$$

$$\therefore Z_{of} = \frac{1}{\dfrac{1}{10^3} + \dfrac{1}{1+jf/100}}$$

$$= \frac{1}{\dfrac{1}{R_A} + \dfrac{1}{R_B + j\omega L}}$$

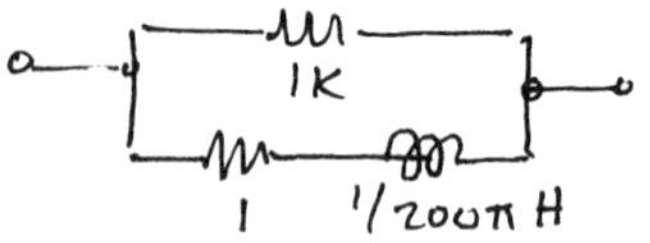

At $f = 10^3$

$$|Z_{if}(10^3)| = \left|10^4 + \frac{10^7}{1 + j\,10^3/100}\right| = 1 \text{ M}\Omega$$

$$|Z_{of}(10^3)| = \left|\frac{1}{\dfrac{1}{10^3} + \dfrac{1}{1+j\,10^2}}\right| = 10\,\Omega$$

At $f = 10^5$

$$|Z_{if}(10^5)| = \left|10^4 + \frac{10^7}{1+j\,10^5/100}\right| = 14.1 \text{ K}\Omega$$

$$|Z_{of}(10^5)| = \left|\frac{1}{\dfrac{1}{10^3} + \dfrac{1}{1+j10^3}}\right| = 700\,\Omega$$

8.30

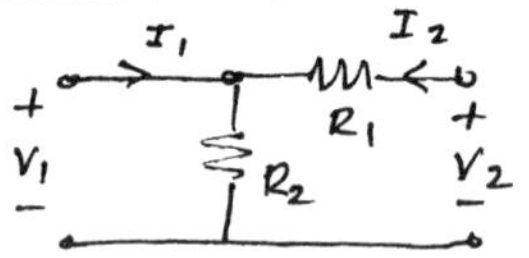

$V_1 = h_{11} I_1 + h_{12} V_2$

$I_2 = h_{21} I_1 + h_{22} V_2$

(a) $h_{11} = \left.\dfrac{V_1}{I_1}\right|_{V_2=0} = R_1 \| R_2 = \dfrac{R_1 R_2}{R_1+R_2}\ \Omega$

$h_{12} = \left.\dfrac{V_1}{V_2}\right|_{I_1=0} = \dfrac{R_2}{R_1+R_2}\ \text{V/V}$

$h_{21} = \left.\dfrac{I_2}{I_1}\right|_{V_2=0} = -\dfrac{R_2}{R_1+R_2}\ \text{A/A}$

$h_{22} = \left.\dfrac{I_2}{V_2}\right|_{I_1=0} = \dfrac{1}{R_1+R_2}\ ℧$

(b) $\beta = \left.\dfrac{V_1}{V_2}\right|_{I_1=0} = h_{12} = \dfrac{R_2}{R_1+R_2}$

$\Rightarrow \dfrac{R_1}{R_2} = \dfrac{1}{\beta} - 1 = 99$

$\Rightarrow R_2 = R_1/99 = 10.1\,\Omega$

Thus $h_{11} = 10\,\Omega$; $h_{12} = 0.01$ V/V

$h_{21} = -0.01$ A/A; $h_{22} = 0.99 \times 10^{-3}\ ℧$

(c)

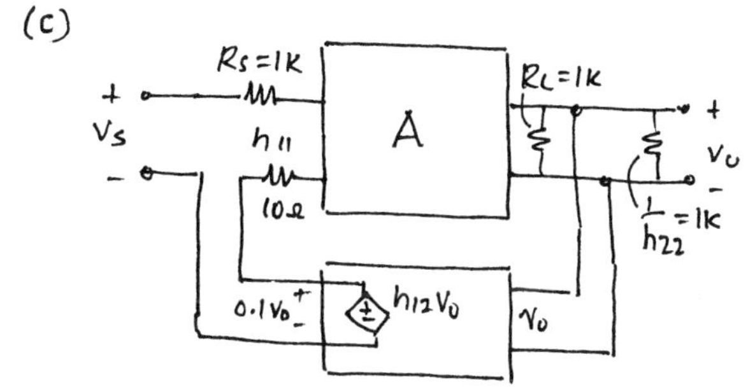

$R_i = R_s + r_{e_1} + \frac{(R_1 \| R_2)}{1+\beta}$

$= 100 + 250 + 909/101 = 700\Omega$

$R_o = R_L \| (R_1 + R_2) \| (r_e + r_{o1}/\beta+1)$

$= 2K \| 10K \| \infty = 1.67\,K\Omega$

$A = \frac{r_o \| (\beta+1)(r_{e3} + 10K\|2K)}{R_s + r_{e1}} \times \frac{2K\|10K}{2K\|10K + r_{e3}}$

$\beta = \frac{R_1}{R_1 + R_2} = \frac{1}{1+10} = 0.0909\ V/V$

$A_F = \frac{A}{1+A\beta} \doteq \frac{\frac{16.4}{0.7} \cdot \frac{1.67}{1.69}}{1 + 23.15 \times 0.0909}$

$= 7.45\ V/V$

$R_{if} = R_{IN}(1 + A\beta) = 700(3.1) = 2170\,\Omega$

Neglecting base currents since $\beta >> 1$

DC voltage at input $= I_{B1}R_S = 0.01V$

DC voltage at output $= 0V$

8.31

$R_{of} = \frac{R_o}{1+A\beta} = \frac{1000}{1+A\beta} = 100\Omega$

$\Rightarrow 1 + A\beta = 1000/100 = 10$

$\Rightarrow A_F = \frac{A}{1+AB} = \frac{100}{10} = 10\,V/V$

if $\beta = 1$: $R_{of} \Rightarrow \frac{R_o}{1+A} = \frac{1000}{1+10^4} = 9.9\Omega$

8.32

Redraw

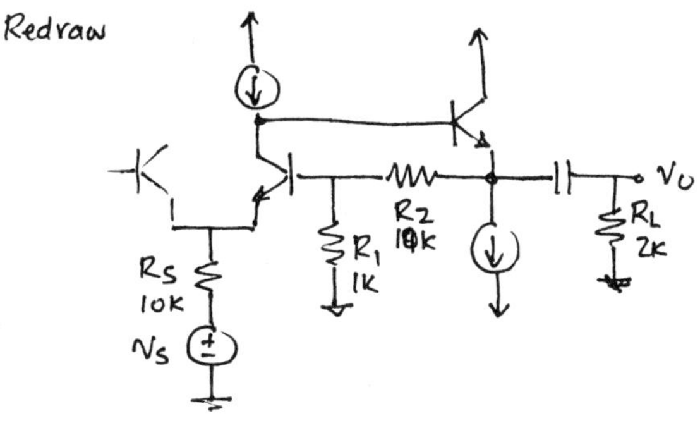

$r_{e1} = V_T/I_E = 25mV/0.1mA \approx 250\Omega$

$r_{e2} = 25/1 = 25\Omega$

A-circuit

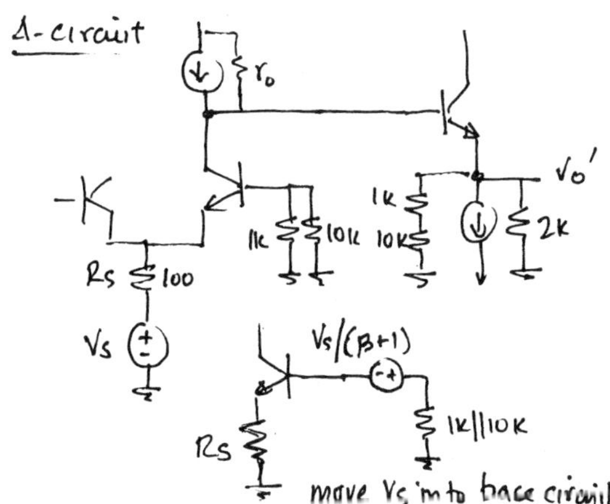

8.33

$Q_3 + Q_4$ form current multiplier $\times 120/40 = \times 3$

$g_{m1} = 2\sqrt{\frac{1}{2}120(20/1)\,100} \doteq 693\,\mu A/V$

$g_{m5} = 2\sqrt{\frac{1}{2}\,60\,(20/1)\,1000} \doteq 1550\,\mu A/V$

$g_{m3} = 2\sqrt{\frac{1}{2}\,60\,(40/1)\,100} \doteq 693\,\mu A/V$

$g_{m4} = 2\sqrt{\frac{1}{2}\,60(120/1)\,300} \doteq 2078\,\mu A/V$

$g_{m2} = g_{m1} = 693\,\mu A/V$

$r_{o1} = 24/100 \Rightarrow 240\,K\Omega$

$r_{o2} = 24/100 \Rightarrow 240\,K\Omega$

$r_{o3} = 24/100 \Rightarrow 240\,K\Omega$

$r_{o4} = 24/300 \Rightarrow 80\,K\Omega$

$r_{o5} = 24/1000 \Rightarrow 24\,K$

(c) open loop gain $A\beta \approx g_{m1}(r_{o1}\|r_{o3}) \times 1$

$\left[(3 \times g_{m1})\left(\frac{r_{o4}}{3}\right) \equiv g_{m1} r_{o1} \right]$

$\beta = 1$: i.e. $A \approx g_{m1}(r_{o2}\|r_{o3})$

$\Rightarrow A \approx 693 \times 120 \times 10^{-3} = 84$

(d) $A_F = \frac{A}{1+A\beta} = \frac{84}{1+84} = 0.988\ V/V$

$R_o = r_{o5} \| r_{o5} = 12\ K$

$R_{of} = R_o/(1+A\beta) = 12/85 = 140\ \Omega$

(e) To obtain $V_o/V_s = 5$ we could change direct connection from Q5S to Q2G by voltage divider $R_1/(R_1+R_2)$ to change β from 1 to 1/5.3 then

$A_F = \frac{84}{1+84 \times 1/5.3} = \frac{84}{16.8} = 5$

Now $1+A\beta = 16.8$

$R_{of}'' = R_o/(1+A\beta') = 12/16.8 = 714\ \Omega$

8.34

Since $V_{G1} = 0 = V_{G2}$

$\Rightarrow V_{E3} = V_o = 0$ and $V_{B3} = +0.7V$

$g_{m1} = g_{m2} = 2\sqrt{\frac{1}{2}k'(W/L)I}$

$= 2\sqrt{\frac{1}{2} 1 (0.5)} = 1\ mA/V$

$r_{e3} = V_T/5mA = 5\ \Omega$

$r_o = V_A/I = 100/0.5 = 200K$

A-circuit

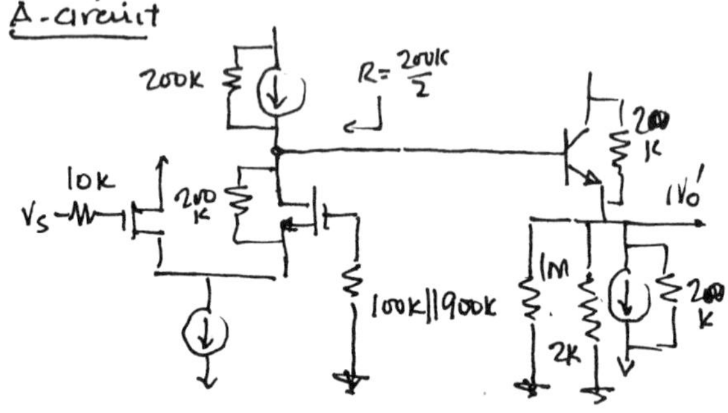

$R_i = \infty$

$R_o = 1M \| 2K \| \frac{200K}{2} \| (r_{e3} + \frac{200K}{\beta+1}) = 622.2\ \Omega$

$A = \frac{100 \| (\beta+1)(r_{e3} + 1M \| 10K \| 2K)}{2/g_m}$

$\times \frac{1.66K}{r_{e3}+1.66K} = 31.3\ V/V$

$\beta = \frac{100}{100+900} = 0.1\ V/V$

$A_F = \frac{A}{1+A\beta} = \frac{31.3}{1+31.3 \times 0.1} = 7.58\ V/V$

$R_{if} = \infty \quad \therefore R_{IN} = \infty$

$R_{of} = \frac{R_o}{1+A\beta} = \frac{622.2}{1+3.13} = 150.6\ \Omega$

$R_{of} = R_{OUT} \| R_L \Rightarrow R_{out} = 163\ \Omega$

8.35

(a) A-circuit   B-circuit

RC1, RC2, V_s', V_o', $R_E \| R_F$, $R_E + R_F$ R_F, V_f, R_E, V_o'

(b) For $A\beta \gg 1$: $A_F = \frac{A}{1+A\beta} \Rightarrow \frac{1}{\beta}$

$\beta = \frac{R_E}{R_E+R_F} \Rightarrow A_f \approx \frac{R_E+R_F}{R_E} = 1 + \frac{R_F}{R_E}$

(c) $R_E = 50\ \Omega$

$\Rightarrow A_F = 1 + R_F/R_E = 25\ V/V$

$\Rightarrow R_F/R_E = 24$ and $R_F = 24 R_E = 1.2K\Omega$

(d) $I_{Q1} = 1mA$, $I_{Q2} = 2mA$, $I_3 = 5mA$

$\beta = 100$

$r_{e1} = \frac{25mV}{1mA} = 25\ \Omega$, $r_{e2} = 12.5\ \Omega$, $r_{e3} = 5\ \Omega$

$A_1 = \frac{-R_{C1} \| r_{\pi 2}}{r_{e1} + R_E \| R_F} = -10$

$\Rightarrow R_{C1} \| r_{\pi 2} = 10(25 + 50 \| 1.2K) = 730\ \Omega$

$\Rightarrow R_{C1} = 1.75K\Omega$

$A_2 = \frac{-R_{C2} \| [\beta(r_{e1} + R_E + R_F)]}{r_{e2}} = -5$

$\Rightarrow R_{C2} \| 125.5K = 5 \times 12.5 = 62.5\ \Omega$

$\Rightarrow R_{C2} = 62.8.1\ \Omega$

$A_3 = \frac{R_E + R_F}{r_{e3} + R_E + R_F} = \frac{1.25}{1.255} = 0.996\ V/V$

(e) $\therefore A_1 A_2 A_3 = 10 \times 50 \times 0.996 = 498\ V/V$

$A\beta = 498(50/1250) = 19.92$

$A_F = \frac{A}{1+A\beta} = 23.8\ V/V$

(f) $R_i = (\beta+1)(r_{e1} + R_E \| R_F)$

$\therefore R_i = 101(25+4.8) = 7.37K\Omega$

$R_{if} = R_1(1+19.92) = 154K\Omega$

$R_o = 1.25K \| (r_{e3} + R_{c2}\|101) = 11.12\Omega$

$R_{OF} = \frac{R_o}{1+19.92} = 0.53\Omega$

$R_i = R_S + R_{id} + R_1 = 30K\Omega$

$R_{if} = R_1(1+A\beta) = 120.8K\Omega$

$R_{in} = R_{if} - R_S = 110.8K\Omega$

$R_o = R_L + R_{op} + R_2 = 110.1K\Omega$

$R_{of} = R_o(1+A\beta) = 443.4K\Omega$

$R_{out} = R_{of} - R_L = 433.4K\Omega$

Section 8.5
Series-Series Feedback

8.36

For $A\beta \gg 1$: $A_F = \frac{A}{1+A\beta} \Rightarrow \frac{1}{\beta} \equiv A_F^*$

Here, $\beta = \frac{V_f}{I_o} = \frac{R_{E2} \times R_{E1}}{R_{E2}+R_F+R_{E1}}$

$= \frac{100\times100}{100+640+100} = 11.9\Omega$

$\frac{I_o}{V_s} \approx A_F^* \equiv \frac{1}{\beta} = 84\,\mu\frac{A}{V}$ (cf. 83.7)

$\frac{V_o}{V_s} = \frac{-I_oR_{C3}}{V_s} = -84\times0.6$

$= -50.4\,V/V$ (cf. 50.21)

8.37

$A = G_m = 100mA/V$ $\quad\beta = 0.1V/mA$

$r_{in} = 10K$, $r_{out} = 100K$

A-circuit

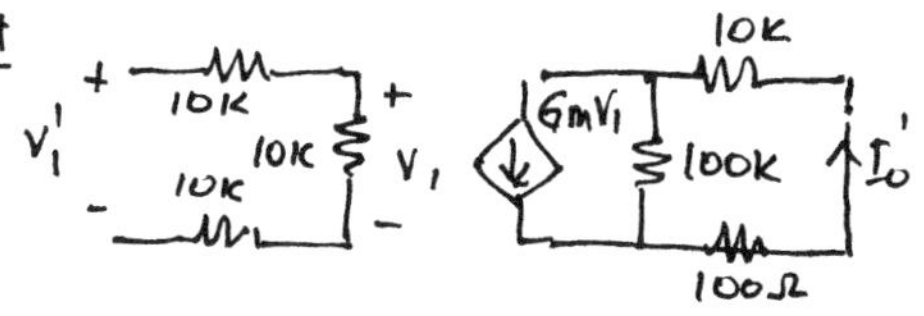

[Circuit: V_1' source through 10K and 10K resistors to 10K resistor across V_1; G_mV_1 current source in parallel with 100k and 10K, 100Ω; output current I_o']

$V_i = V_1' \frac{10}{10+10+10} = V_1'/3$

$I_o' = \frac{G_mV_1\,100}{100+10+0.1} = 30.28V_1'\,mA$

$A = \frac{I_o'}{V_i'} = 30.28\,mA/V$

$A_F = \frac{A}{1+A\beta} = \frac{30.28}{1+30.28(0.1)} = 7.52\frac{mA}{V}$

8.38

For $A\beta \gg 1$: $I_o/V_S \approx 1/\beta$

R_E samples I_o/α and $V_E = I_oR_E/\alpha$

$\beta = \frac{V_F}{I_v} = R_E$ $\;\; A_F \rightarrow \frac{1}{\beta} = \frac{1}{R_E}$

QED

A-circuit

[Circuit: $R_S = 10K$, V_1', R_{id} 100K across V_1, $R_E = 1K$, R_{in}; $100V_1$ source, I_b, r_π, βI_b, r_o, $R_o \rightarrow$, I_o', 1K]

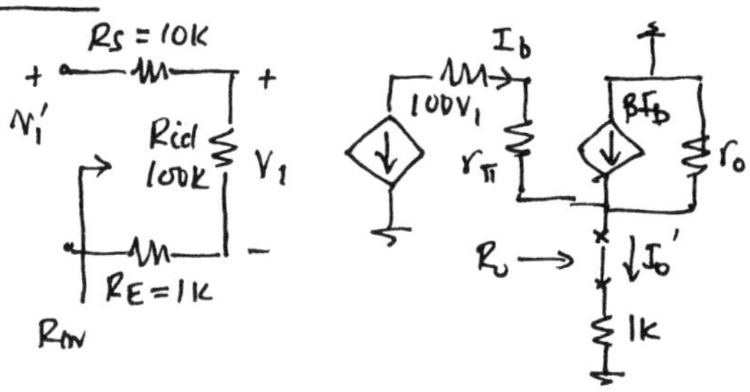

For $\frac{I_o}{V_S} = 1\frac{mA}{V} \approx \frac{1}{\beta} \approx \frac{1}{R_E} \Rightarrow R_E = 1K\Omega$

$V_1 = V_1' \frac{R_{id}}{R_S+R_{id}+R_E} = \frac{100V_1'}{100+10+1} = 0.9V_1'$

$I_b = \frac{G_{m1}V_1}{1+r_\pi+(1+\beta)(r_o\|R_E)}$

$I_o' = \frac{(\beta+1)I_b(R_E\|r_o)}{R_E}$

$\frac{I_o'}{V_1'} = \frac{1}{R_E}\frac{(\beta+1)(R_E\|r_o)\,100\times0.9}{1+r_\pi+(1+\beta)(R_E\|r_o)} \longrightarrow \frac{1}{R_E}$

$A = \frac{I_o'}{V_1'} = \frac{1}{1}\cdot\frac{101(1\times100)\times90}{1+2.5+101(1\|100)} = 87\,mA/V$

$A_F = \frac{I_E}{V_S} = \frac{87}{1+87\times1} = 0.989\,mA/V$

Thus $\frac{I_o}{V_S} = \frac{\beta}{\beta+1}\cdot\frac{I_E}{V_S} = 0.989\times0.99$

$= 0.98\,mA/V$

$R_i = 10 + 100 + 1 = 111\,K\Omega$

$R_{if} = 111 \times 88 = 9768\,K\Omega$

$R_{IN} = R_{if} - R_s = 9758\,K \approx 9.8\,M\Omega$

R_o (looking into xx in A circuit)

$$= R_E \| r_o + \frac{1K + r_\pi}{\beta + 1}$$

$$= (1\|100) + \frac{1 + 2.5}{101} = 1.025\,K\Omega$$

R_{if} (in amplifier circuit) $= 1.025 \times 88$

$= 90.2\,K\Omega$

$R_{OUT} = r_o\left[1 + g_m\left(90.2 \| r_\pi\right)\right] = 9.83\,K\Omega$

8.39

$I_1 = 0.2\,mA/2 = 100\,\mu A$

$I_5 = 0.8\,mA = 800\,\mu A$

$g_{m1} = 2\sqrt{\tfrac{1}{2}\, 20\, (36/10)\, 100} = 120\,\mu A/V$

$g_{m5} = 2\sqrt{\tfrac{1}{2}\, 20\, (200/10)\, 800} = 800\,\mu A/V$

$r_{o1} = V_A / I_1 = 100\,V/100\,\mu A = 1\,M\Omega$

$r_{o5} = V_A / I_5 = 100\,V/800\,\mu A = 125\,K\Omega$

$r_{e1} = V_T / I_1 = 25\,mV/100\,\mu A = 250\,\Omega$

$r_{e5} = V_T / I_5 = 25\,mV/800\,\mu A = 31.25\,\Omega$

A-circuit

$$\beta = \frac{V_F}{I_o} = \frac{I_o R_E}{I_o} = R_E \ V/A$$

$$A \equiv \frac{I_o'}{V_s'} \simeq \frac{g_{m1} r_{o2}}{2} \cdot \frac{1}{R_E \| r_{o5}}$$

$$= \frac{120 \times 1000}{2} \times \frac{1}{9.25} = 6.48\,mA/V$$

$$\Rightarrow \frac{I_o}{V_s} = \frac{6.48 \times 10^{-3}}{1 + 6.48\,(10)} = 98.5\,\mu A/V$$

$$\frac{V_o}{V_{os}} = \frac{I_o (R_E \| r_{o5})}{V_{os}} = 98.5\,\frac{\mu A}{V}\ 9.25\,\frac{V}{mA}$$

$$= 0.911\,V/V \approx 1\,V/V$$

8.40

$$A_f \equiv \frac{I_o}{V_s} \qquad \text{(Assume } (R_1 + R_2) \gg r\text{)}$$

$$\beta \equiv \frac{V_F}{I_o} = R_1 \frac{r}{r + R_1 + R_2} \quad V/A$$

A-circuit

$$\frac{I_o'}{V_s'} = \frac{A}{r + R_L}$$

$$A_F = \frac{I_o}{V_s} = \frac{A}{1 + A\beta} \doteq \frac{A/(r + R_L)}{1 + \dfrac{A}{r + R_L} \cdot \dfrac{r R_1}{r + R_1 + R_2}}$$

$$= \frac{A}{(R_L + r) + \dfrac{A r R_1}{r + R_1 + R_2}}$$

$$\xrightarrow[A \gg 1]{} \frac{r + R_1 + R_2}{r R_1} \equiv \frac{1}{\beta}$$

(a) when $\mu = 10^5\,V/V$ and $R_1 = 100\,\Omega$

$$A_F = \frac{\mu = 10^5}{(1K + 100) + \dfrac{10^5 \cdot 100 \cdot 100}{100 + 100 + 1K}}$$

$$= \frac{10^5}{1100 + \dfrac{10^5 \cdot 10^4}{1200}} = 0.12\,\frac{mA}{V}$$

$$\frac{1}{\beta} = \frac{100 + 100 + 1000}{100 \times 100} = \frac{1200}{10^4} = 0.12\,\frac{mA}{V}$$

$R_i' = R_{id} + R_1 = 10K + 100 = 10.1\,K\Omega$

$R_{if} = R_i'(1 + A\beta) = R_i'(1 + \tfrac{1}{\beta} \cdot \beta) = 20.2K$

$R_o = r_o \| (R_L + (r \| R_2))$

$= 100 \| (1050) = 91.3\,\Omega$

$R_{of} = R_o'(1 + A\beta) = 2 \times 91.3 = 182.6\,\Omega$

(b) $\beta = V_F / I_o = I_o r / I_o = r\ A/V = 100\,\Omega$

$A_F = 10^4 / [(1100) + 10^4 \cdot 100] \approx 1/100$

$A_F\beta = 1$ and $(1 + A\beta) = 2$

$R_i = R_i'(1 + A\beta) \to \infty$

$R_{of} = 2(100 \| 1025) = 182.2\,\Omega$

Section 8.6 The Shunt-shunt and Shunt Series Feedback Amplifiers

8.41

A-circuit

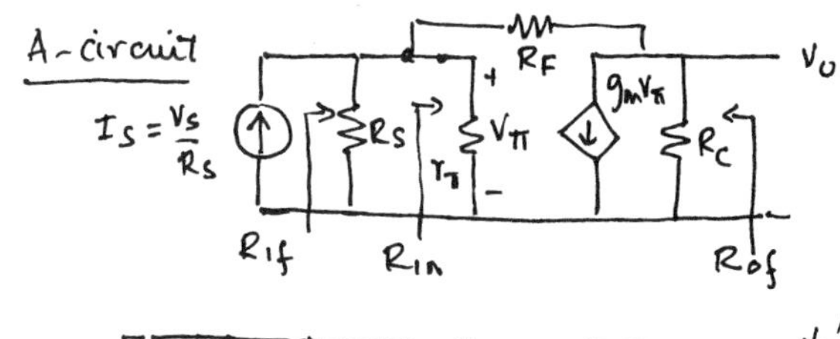

I_s', R_s, R_F, r_π, V_π, $g_m V_\pi$, R_F, R_C, V_o'

$$V_\pi = I_s (R_s \| R_F \| r_\pi)$$

$$A \equiv \frac{V_o'}{I_s'} = -g_m (R_s \| R_F \| r_\pi)(R_F \| R_C)$$

B-Circuit

I_F, R_F, V_o

$$\beta \equiv \frac{I_F}{V_o} = -\frac{1}{R_F}$$

$$A_F \equiv \frac{V_o}{I_s} = \frac{A}{1+A\beta} = \frac{-g_m R^*}{1 + g_m R^*/R_F}$$

$$\rightarrow \frac{-g_m R^*}{g_m R^*} \cdot R_F = -R_F$$

$$A_V = \frac{V_o}{I_o R_s} \simeq \frac{-R_F}{R_s} \qquad \text{QED}$$

For $R_S = 10K$, $R_F = 47K$, $R_C = 4.7K$

$$A_V \approx -\frac{47K}{10} = -4.7 \text{ V/V}$$

For $A_V = -7.5 \text{ V/V}$

$$R_F = A_V R_S = 7.5 \times 10 = 75K$$

8.42

A-Circuit

$I_s = V_s/R_x$, R_x, R_F, V_i, r_o, μV_i, R_F, R_L, V_o', R_{in}, R_o

B-circuit

I_F, R_2, V_o

$$\beta \equiv \frac{I_F}{V_D} = -\frac{1}{R_F}$$

Here let $R_x = R_s + R_1$ and $\mu = g_m r_o$

$$A \equiv \frac{V_o'}{I_s'} = -[R_x \| R_2][R_2 \| r_o] g_m$$

$$A_F = \frac{V_o}{I_s} = \frac{A}{1+A\beta} = \frac{-g_m [R_x \| R_2][R_2 \| r_o]}{1 + (A)(-1/R_2)}$$

$$= \frac{-g_m \frac{R_x R_2}{R_x + R_2} \cdot \frac{R_2 r_o}{R_2 + r_o}}{1 + g_m \frac{R_x R_2}{R_x + R_2} \cdot \frac{R_2 r_o}{R_2 + r_o}\left[\frac{-1}{R_2}\right]}$$

$$\xrightarrow{A \text{ large}} -R_2 = -\frac{1}{\beta}$$

Thus $\frac{V_o}{I_s R_x} = \frac{V_o}{V_1} = \frac{1}{\beta R_x} = \frac{-R_2}{R_1 + R_s}$

$$R_{IN}' = (R_s + R_1) \| R_2 = (10K + 1M) \| 4.7M$$
$$= 1.01M \| 4.7M = 0.83 M\Omega$$

Now $g_m = \frac{2 I_D}{V_{ov}} = \frac{2 \times 1}{0.8 - 0.6} = \frac{2}{0.2} = 10 \text{ mA/V}$

$$r_o = \frac{V_A}{I} = \frac{30V}{1mA} = 30K\Omega$$

Hence
$$A = -[(R_s + R_1) \| R_2][R_2 \| r_o] g_m$$
$$= -10 [1.01M \| 4.7M][4.7M \| 0.03M]$$
$$= -10 [831K][29K] \text{ V/mA}$$
$$= -24100$$

$$1 + A\beta = 1 + \left(24100 \times \frac{1}{4700}\right) = 6.13$$

$$A_F = \frac{-24100}{6.13} = -3931 K\Omega$$

Hence $\frac{V_o}{V_1} = \frac{V_o}{I_s(R_s + R_1)} = \frac{-3931K}{1010} \approx -3.89 \text{ V/V}$

$$R_o' = R_2 \| r_o = 4.7M \| 30K = 29.8K\Omega$$
$$R_{of} = R_o'/(1 + A\beta) = 29.8/6.13 = 4.86K$$
$$R_1' = (R_s + R_1) \| R_2 = 1010 \| 4.7 = 0.83M\Omega$$
$$R_{if} = R_i/(1 + A\beta) = 0.83/6.13 = 136K\Omega$$
$$R_{IN} = R_{if} - R_s = 136K - 10K = 126K$$

8.43

$G_m = 100 \text{ mA/V}$, $R_{input} = 1K$, $R_{output} = 1K$

$R_1 = 10K$, $R_2 = 100$, $R_S = 10K$, $R_L = 1K$

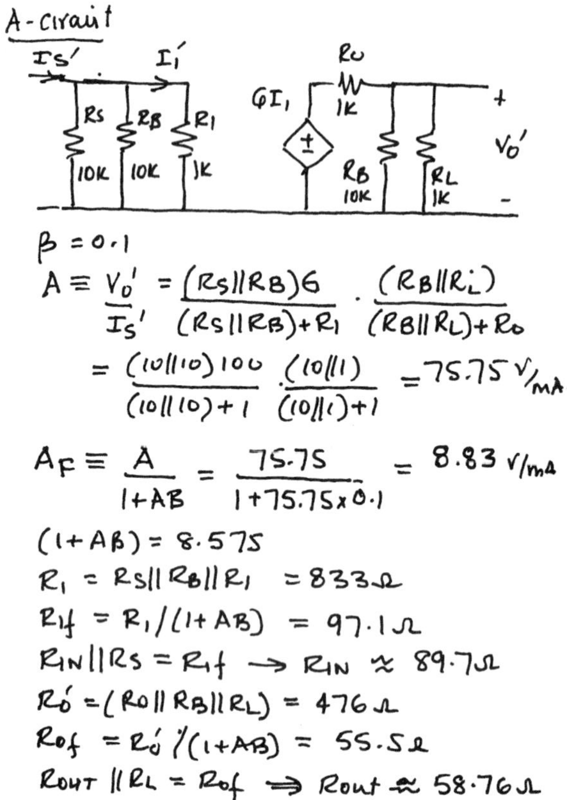

$\beta = 0.1$

$$A \equiv \frac{V_o'}{I_s'} = \frac{(R_S \| R_B) G}{(R_S \| R_B) + R_1} \cdot \frac{(R_B \| R_L)}{(R_B \| R_L) + R_o}$$

$$= \frac{(10 \| 10)\, 100}{(10 \| 10) + 1} \cdot \frac{(10 \| 1)}{(10 \| 1) + 1} = 75.75\ \text{V/mA}.$$

$$A_F \equiv \frac{A}{1 + A\beta} = \frac{75.75}{1 + 75.75 \times 0.1} = 8.83\ \text{V/mA}$$

$(1 + A\beta) = 8.575$

$R_1 = R_S \| R_B \| R_1 = 833\Omega$

$R_{if} = R_1 / (1 + A\beta) = 97.1\Omega$

$R_{IN} \| R_S = R_{if} \rightarrow R_{IN} \approx 89.7\Omega$

$R_o' = (R_o \| R_B \| R_L) = 476\Omega$

$R_{of} = R_o' / (1 + A\beta) = 55.5\Omega$

$R_{OUT} \| R_L = R_{of} \Rightarrow R_{out} \approx 58.76\Omega$

8.44

Here shunt-series: sample I_o, return I_F

Because Q_2 is voltage-driven device, unlike BJT we find $I_o = V_o / R$ as voltage follower.

RD
Io'
Q2
Q1
Vo'
Is'
RF
Rs
IF
RF
Rs
RF
Rs

A-circuit   B-network

$$A \equiv \frac{I_o'}{I_s'} = \frac{-g_{m1} R_D (R_S + R_F)}{(R_S \| R_F)} = -555.5$$

$$\beta \equiv \frac{I_F}{I_o} = \frac{R_S}{R_F + R_S} = \frac{10}{100} = 0.1$$

$$A_F \equiv \frac{I_o}{I_s} = \frac{\dfrac{-g_{m1} R_D (R_F + R_S)}{(R_S \| R_F)}}{1 + \dfrac{g_{m1} R_D (R_F + R_S)}{(R_S \| R_F)} \cdot \dfrac{R_S}{(R_F + R_S)}}$$

$$\longrightarrow \frac{R_F + R_S}{R_S} = 1 + R_F / R_S$$

$A\beta = 56 \Rightarrow 1 + A\beta = 57$

$$A_F = \frac{-555.5}{1 + 555.5 \times 0.1} = -9.8\ \text{A/A}$$

$R_1' = R_F + R_S = 100\text{K}$

$R_{if} = R_1' / (1 + A\beta) = 1.58\ \text{K}\Omega$

$R_{of} = (R_F \| R_S) + r_{o2}$

$= 9\text{K} + 20\text{K} = 29\text{K}\Omega$

$R_o \approx r_o (1 + g_m R_{of}) =$

$= 20\text{K}(1 + 5 \times 29)$

$\approx 2.92\ \text{M}\Omega$

8.45

Here shunt-shunt: sample V_o return I_f

RD
Q2
Q1
Vo'
Is'
RF
RF
Rs
IF
RF
Vo
+
-

A-circuit   B-Network

$$A \equiv \frac{V_o'}{I_s'} = -g_{m1} R_D \cdot R_F \quad (\text{neglecting } r_o\text{'s})$$

$$\beta \equiv \frac{I_F}{V_o} = -1/R_F, \qquad A\beta = g_m R_D$$

$$A_F \equiv \frac{V_o}{I_s} = \frac{A}{1 + A\beta} = \frac{-g_{m1} R_D R_F}{1 + g_{m1} R_D R_F / R_F}$$

$$= \frac{-5 \times 10 \times 90}{1 + 5 \times 10}$$

$$= -4500/51 = -88.2\ \text{V/A}$$

$R_i' = R_F = 90\text{K} \longrightarrow R_{if} = R'/(1 + A\beta) = 1.76\text{K}\Omega$

$R_o' = r_o \| R_F \| R_S = 20 \| 90 \| 20 = 6.2\text{K}\Omega$

$\Rightarrow R_{of} = R_o' / (1 + A\beta) = 121.6\Omega$

8.46

A-circuit

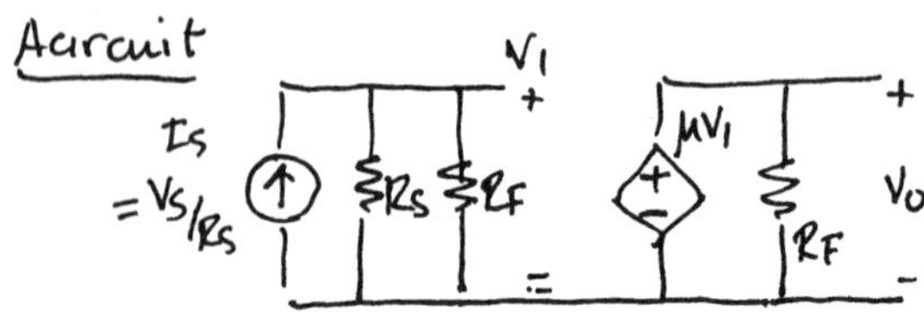

$\beta = -1/R_F$

$V_o = -\mu V_1 R_F$

$V_1 = I_s(R_s \| R_F)$

$\therefore V_o = -\mu I_s (R_s \| R_F) R_F$

$$A \equiv \frac{V_o}{I_s} = -\mu \frac{R_s R_F}{(R_s+R_F)} . R_F$$

$$A_F = \frac{A}{1+A\beta} = \frac{-\mu (R_s \| R_F) R_F}{1+\mu (R_s\|R_F) R_F/R_F} = -\frac{\mu (R_s\|R_F) R_F}{1+\mu(R_s\|R_F)}$$

$$\frac{V_o}{V_s} = \frac{V_o}{I_s R_s} = \frac{-\mu(R_s . R_F)R_F}{[(R_s+R_F)+\mu(R_s R_F)]R_s}$$

$$\underset{\mu>>1}{\Longrightarrow} -\frac{\mu(R_s R_F)}{\mu(R_s R_F)} . \frac{R_F}{R_s} = -\frac{R_F}{R_s}$$

QED

(b) For circuit of Fig P8.46(b)

$$I_B = \frac{V_{cc}\frac{R}{10+15} - V_{BE}}{(15\|10) + (4.7\times101)}$$

$$= \frac{15(10/25) - 0.7}{6+474.7} = \frac{5.3\text{V}}{480.7\text{K}} \approx 0.011 \text{ mA}$$

$I_c = 100 I_B = 1.1\text{mA}$

$r_e = V_T/I = 22.6\Omega$

$r_\pi = (\beta+1)r_e = 2.286\text{K}$

$g_m = \beta/r_\pi = 100/2.286\text{K} \rightarrow 43.7\text{mA/V}$

$R_{IN} = (15\|10\|2.286)\text{K} \doteq 1.5\text{K}$

$R_C \| R_B \| r_\pi \doteq (7.5\|6\|2.286) \doteq 1.35\text{K}$

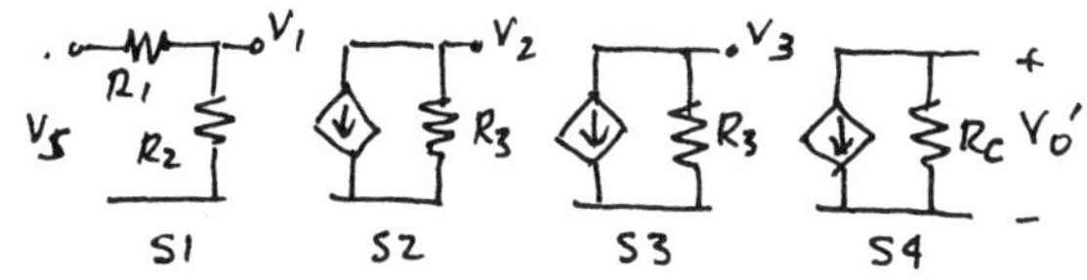

For S1: $R_1 = R_s$, $R_2 = R_{IN}$

$\therefore V_1 = 1.5/11.5\, V_s = 0.13 V_s$

For S2: $R_3 = R_C\|R_B\|r_\pi$

$\therefore \frac{V_2}{V_1} = -g_m R_3 \doteq -43.7\times 1.35 \doteq -59$ V/V

For S3: Same as S2 $\therefore \frac{V_3}{V_2} = -59$ V/V

For S4: $\frac{V_o}{V_3} = -g_m R_c = -43.7\times 7.5 = -327.75$ V/V

$\frac{V_o}{V_s} = -0.13\times 59\times 59\times 327.75$

$\rightarrow \frac{V_o}{V_s} = -1.488\times 10^5$

Because we have ignored r_o etc let us estimate $V_o/V_s = -1\times 10^5$ which is quite large.

Then $A_F = \frac{A}{1+A\beta} \approx 100$ needed

$$= \frac{10^5}{1+10^5\beta} \approx \frac{1}{\beta} = 100$$

Select R_F so that $R_F/R_s = 100$

$\rightarrow R_F = 100\times 10\text{K} = 1\text{M}\Omega$

We can ignore loading effect of R_F in A-circuit. R_L will cause loading of R_C & $V_L = (R_L/(R_C+R_L))V_o$

$= (1/8.5) = 0.11 V_o$

Now $A_o \approx 10^4$.

$$A_F' = \frac{10^4}{1+10^4/100} \doteq 99.00.$$

8.47

(a) To lower R_{IN} and raise R_{out}

SHUNT - SERIES

(b) To raise R_{IN} and R_{out}

SERIES - SERIES

(c) To lower R_{IN} and R_{out}

SHUNT - SHUNT

8.48

A-circuit

Vs', Rs 1K, RF 1K, Rid, μV_1, 100K, RE, RL, Vo'

β-circuit

Vo, IF, 100K, RF 1K, RE 1K

$$\beta = \frac{I_F}{V_o} = -\frac{100\text{K}\|1\text{K}}{100\text{K}\|1\text{K}+1\text{K}} \cdot \frac{1}{100\text{K}}$$

$= -4.98\,\mu\text{A/V}$

$V_1 = (1 \| 100 \| 1.99) \times 10^3 I_s' = 661.2 I_s'$

$$\frac{V_o'}{I_s'} = -661.2\,(10)^4 \,\frac{2K \| 100.5K}{2K \| 100.5K + 1K}$$

$$= -4.38 \times 10^6 = -4.38\,M\Omega$$

$$\Rightarrow \frac{V_o}{V_s} = \frac{A}{1+A\beta} = \frac{-4.38 \times 10^6}{1 + 4.38 \times 4.98}$$

$$= -191.86 \times 10^3 \; V/A$$

$$\frac{V_o}{V_s} = \frac{V_o}{I_s R_s} = -191.9 \; V/V$$

$R_1 = (1 \| 100 \| 1.99)K = 661.2K$

$R_{if} = R_i / 22.8 = 29 = R_{in} \| R_s$

$\Rightarrow R_{in} = 29.9\,\Omega$

$R_o = 1K \| 100.5K \| 2K = 662.2\,\Omega$

$R_{of} = R_{out} \| R_L = R_o / 22.8 = 29\,\Omega$

$\Rightarrow R_{out} = 29.5\,\Omega$

8.49

$\beta = -1/R_F = -10^{-4}$

$I_{E1} = I_{E2} = 1mA \rightarrow r_{e1} = r_{e2} = 25\,\Omega$

$r_{\pi 1} = r_{\pi 2} = 2.5K\Omega$, $r_o = 100K$

A - circuit

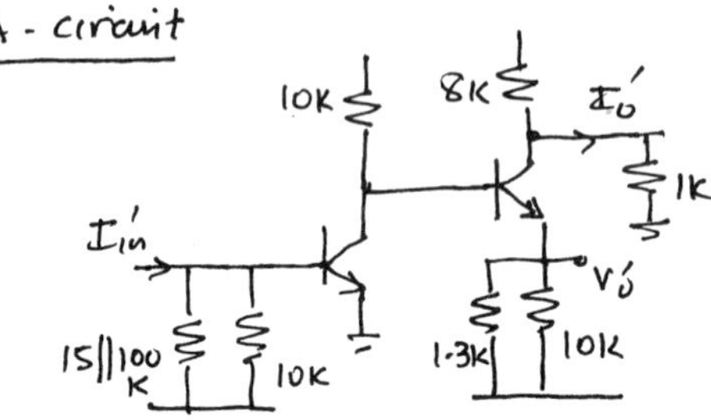

$$\mu = \frac{V_{E2}}{V_{B1}} = \frac{-10K \| r_o \| \beta(1.3K \| 100K)}{25} \times \frac{1.3K \| 10K}{25 + 1.3K \| 10K}$$

$$= -337\,(0.979) = -330 \; V/V$$

$$\frac{V_o'}{I_{in}'} = (15K \| 100K \| 10K \| 2.5K)(-330)$$

$$= -572 \; K\Omega$$

$$\Rightarrow \frac{V_o}{I_{IN}} = \frac{-572}{1 + 572(0.1)} = -9.83 \; K\Omega$$

$R_{if} = 15K \| 100K \| 10K \| 2.5K = 1.73K$

$R_{in} = R_{if} = 1.73K / 58.2 = 29.7\,\Omega$

$$I_{out} = \frac{8K}{1K + 8K} \cdot \frac{V_o}{1.3K \| 10K} = 0.773 \times 10^{-3} V_o$$

$$\Rightarrow \frac{I_o}{I_{IN}} = -9.83 \times 0.773 = -7.6 \; A/A$$

cf. Ex 8.4 $\quad \frac{I_o}{I_{in}} = -7.57 A/A$, $R_{in} = 30\,\Omega$

8.50

$I_o / I_s = 100\,A/A$, $R_{in} = 1K$, $R_{out} = 10K$

$\beta = 0.1 \qquad$ shunt - series topology

$$A_o = \frac{I_o'}{I_s'} = 100$$

$$A_F = \frac{A_o}{1 + A_o\beta} = \frac{100}{1 + 100(0.1)} = 9.09 \; A/A$$

$R_i = 1K\Omega$

$\Rightarrow R_{if} = R_i/(1 + A\beta) = 90.9\,\Omega$

$R_o = 10K\Omega$

$\Rightarrow R_{of} = R_o(1 + A\beta) = 110K\Omega$

8.51

Neglect I_{B2}; $I_{B1} \approx \frac{200}{100} = 2\mu A$

$V_{BE} = 0.7V \quad \therefore V_{B1} = +0.7V$

But no d.c. component in V_s

$\therefore I_{RS}$ (into V_s) $= 0.7/10K = 0.07\,\mu A$

Thus $I_F = I_{RS} + I_{B1} = 0.07 + 0.002$
$= 0.072\,mA$

$V_{E2} = 0.7 + 10 \times 0.072 = 0.7 + 0.72 = 1.42V$

$I_{C2} = 1.42/140 + 0.072 = 10.2\,mA$

$I_{B2} = I_{E2}/(\beta+1) = 0.1\,\mu A \approx \frac{1}{2}\, 200\mu A$

Iterate:

$$I_{B1} = \frac{200\mu A - 100\mu A}{100} = 0.001\,\mu A$$

$V_{E2} = 0.7 + 10 \times 0.073 = 1.41V$

$I_{E2} = 1.41/0.140 + 0.071 = 10.1\,mA$

$I_{B2} = 10.1/101 = 100\,\mu A \quad \therefore I_{C2} = 10mA$

$V_{B2} = V_{E2} + V_{BE} = 1.41 + 0.7 = 2.11V$

$V_o = 10 - 10 \times 500\,\Omega = +5V$

$r_{e1} = \frac{V_T}{I_E} = \frac{25mV}{0.1mA} = 250\Omega$, $r_{e2} = 25\Omega$

$\beta = \frac{R_E}{R_f+R_E} = \frac{140}{10K+140} \approx 0.0138$

A-circuit

$V_{B1} = 10.14k \| 10k \| \beta(250) I_s'$

$= 4.2\times10^3\ I_s'$

$\Rightarrow \frac{I_o'}{I_s'} = \frac{4.2\times10^3(\beta+1)(r_{e2}+10K\|140)}{250} \times \frac{1}{(r_{e2}+10K\|140)}$

$= 1.69\times10^3$ A/A

$A_F \equiv \frac{I_o}{I_s} = \frac{A}{1+A\beta} = \frac{1.69\times10^3}{1+1.69\times10^3\times(0.0138)}$

$= 69.6$

$\Rightarrow \frac{V_o}{V_s} = \frac{I_o R_L}{I_s R_s} = \frac{500}{10^4}\cdot 69.6 = 3.5\ V/V$

$R_i = 10K \| 10.14K \| 25K = 4.2\ K\Omega$

$R_{if} = R_i/(1+A\beta) = \frac{4.2}{1+23.3} = 172.8\Omega$

$\Rightarrow = R_{IN}\|R_s \Rightarrow R_{IN} = 175.8\Omega$

8.52

A-CIRCUIT

$A \equiv \frac{V_o'}{I_s'} = [R_D \| X_C] = R_D$ kΩ

Neglecting loading of C's

$A_F = \frac{A}{1+A\beta}$

B-CIRCUIT

$\beta = \frac{C_1}{C_1+C_2}\, g_{mf}$

$= \frac{1}{10}\times 1\ mA/V$

Here, $g_{m1} = 5mA/V$ $g_{mf} = 1mA/V$

$R_D = 10K$

Thus $A = 10K\Omega$

$A_F = \frac{10K}{1+10K(0.1)} = 5K\Omega$

$R_{IN} = (R_D \| r_o) \rightarrow R_D$

$R_{if} = R_D/(1+A\beta)$ shunt $= \frac{R_D}{2} = 5K\Omega$

$R_{out} = R_D/(1+A\beta) = \frac{R_D}{2} = 5K\Omega$

Section 8.7
Determining the Loop Gain

8.53

A-CIRCUIT

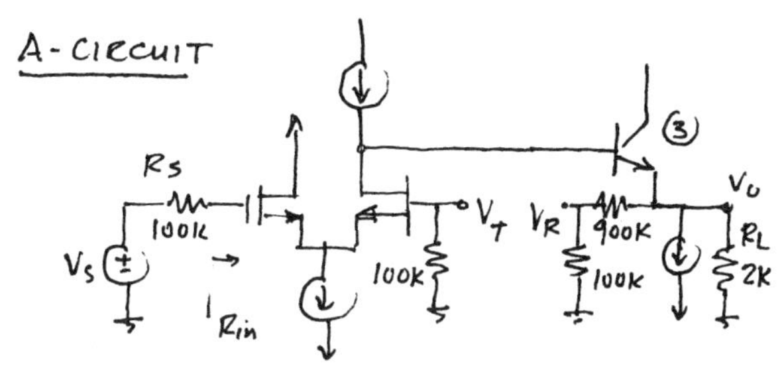

$V_R = -\frac{[100K\|(\beta+1)[r_{e2}+(1M\|\frac{20K}{2}\|R_L)]]}{2/g_m} \times \frac{r_{e2}}{[r_{e2}+(1M\|\frac{20K}{2}\|R_L)]}$

$= -\frac{[100K\|168K]}{2} \times \frac{5}{1665} = -31.25$

$R_o = [1M\|\frac{20K}{2}\|R_L\|[\frac{200K}{2}/\beta + r_{e3}]]$

$= 623\Omega$

which agree with previous solution

8.54

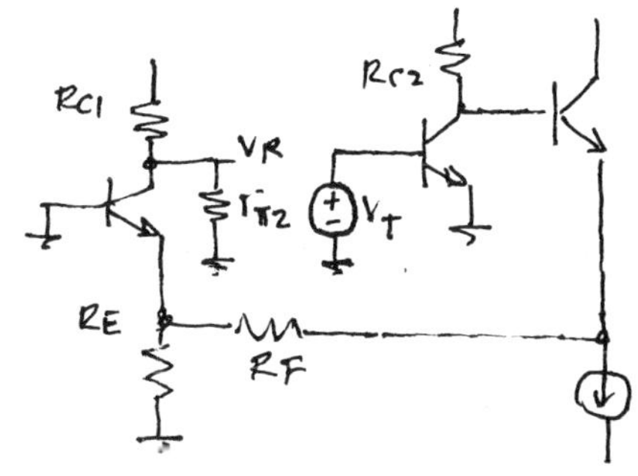

$V_R = V_{c1}$

$V_{c1} = -I_{c1}(R_C \| r_\pi)$

$I_{c1} = (\beta/(\beta+1)) I_{E1} = I_F R_E/(R_E + r_{e1})$

$I_F = I_2 = (\beta+1) I_{B1}$

$$I_{B1} = \frac{R_C I_{C2}}{R_C + (\beta+1)[r_{e3} + R_F + (R_E \| r_{e1})]}$$

$I_{C2} = g_m V_T$

Combining

$$\frac{V_R}{V_T} = \frac{g_{m2} R_C (\beta+1)}{R_C + (\beta+1)[r_{e3} + R_F + (R_E \| r_{e1})]} \times \frac{\alpha R_E (R_C \| r_{\pi 2})}{R_E + r_{e1}}$$

QED

8.55

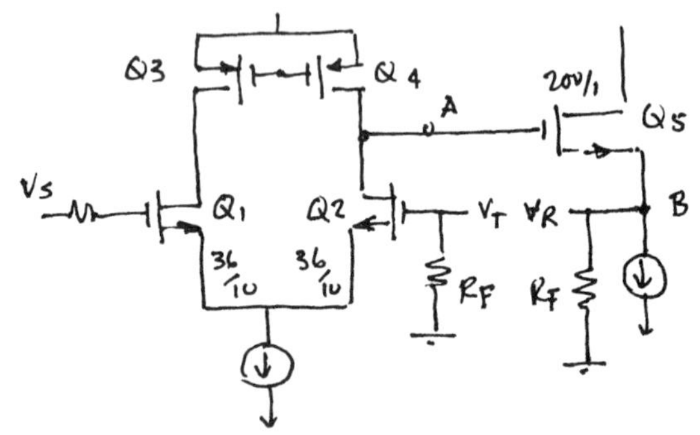

$V_T = V_{G2}$ and $V_S \to 0$

$V_A = -g_{m2}(r_{o2} \| r_{o4})$

$$V_B = V_A \frac{(R_F \| r_{o5})}{(R_F \| r_{o5}) + 1/g_{m5}}$$

$$A\beta = -\frac{V_T}{V_R} = +\frac{g_{m2}(r_{o2} \| r_{o4})(R_F \| r_{o5})}{(R_F \| r_{o5}) + 1/g_{m5}}$$

QED

8.56

Break

R_1, R_2, V_T, V_R, R_o, R_{id}, R_L, $R_2 + R_1 \| R_{id}$ same load, $V_S = 0$

$$\frac{V_R}{V_T} = \frac{-\mu\, R_L \| (R_2 + R_1 \| R_{id})}{R_o + R_L \| (R_2 + R_1 \| R_{id})} \cdot \frac{R_1 \| R_{id}}{R_2 + R_1 \| R_{id}}$$

For $(R_1 + R_2) \gg R_L$ and $R_{id} \gg R_1$

and $R_o \ll R_L$

$$\frac{V_R}{V_T} \approx -\mu \cdot 1 \cdot \frac{R_1}{R_1 + R_2} = \frac{-\mu R_1}{R_1 + R_2}$$

$= -A\beta.$

(b) $I_S = 0$

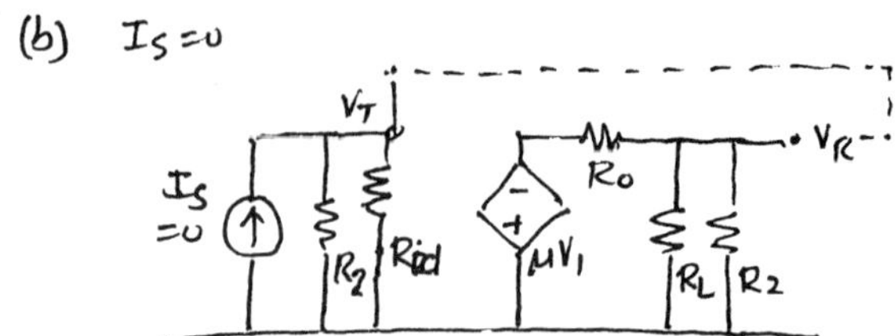

$$\frac{V_R}{V_T} = -\mu \frac{R_L \| R_2}{R_o + R_L \| R_2}$$

For $R_o \ll R_L$

$$\frac{V_R}{V_T} \approx -\mu = -A\beta.$$

(c)

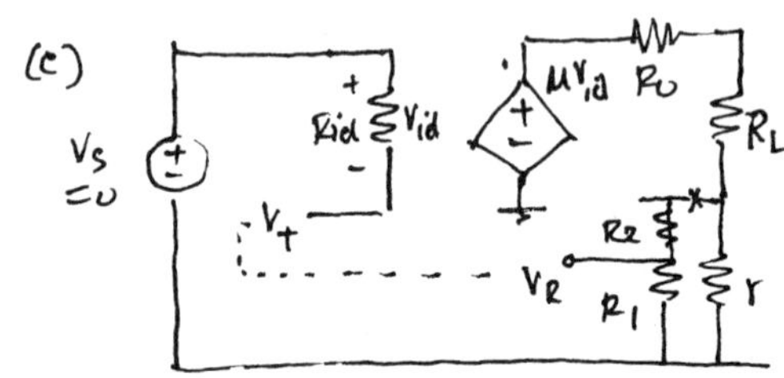

$$\frac{V_R}{V_T} = \frac{-\mu \cdot [r \| (R_1 + R_2)]}{R_o + R_L + [r \| (R_1 + R_2)]} \cdot \frac{R_1}{R_1 + R_2}$$

For $r \ll (R_1 + R_2)$ and $R_o \ll R_L$

$$\frac{V_R}{V_T} \approx \frac{-\mu r}{R_L + r} \cdot \frac{R_1}{R_1 + R_2} = -A\beta$$

[Note: breaking at R_2-r is simpler

$$\frac{V_R}{V_T} = \frac{-\mu R_1}{R_{id} \| R_1 + R_2} \cdot \frac{r}{R_o + R_L + r}$$]

(d)

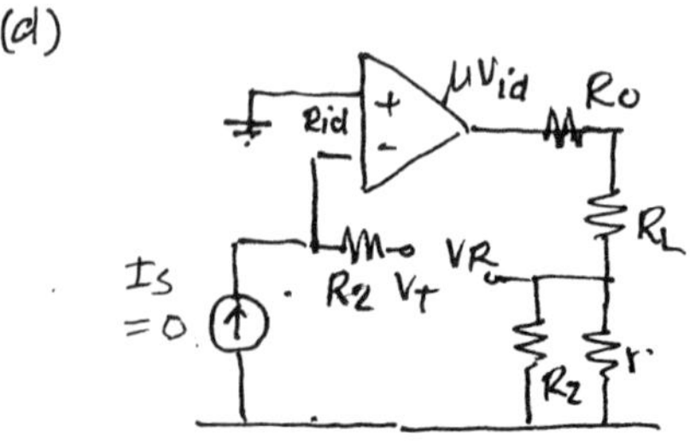

$$V_{id} = V_T \frac{R_{id}}{R_2 + R_{id}}$$

$$\frac{V_R}{V_T} = - \frac{R_{id}}{R_{id}+R_2} \mu \frac{(r \| R_2)}{(r\|R_2)+R_L+R_o}$$

$$A\beta = \mu \frac{r}{r+R_L} \quad \text{for } R_{id} \gg R_1, R_2,\ R_o \ll R_L$$

8.57

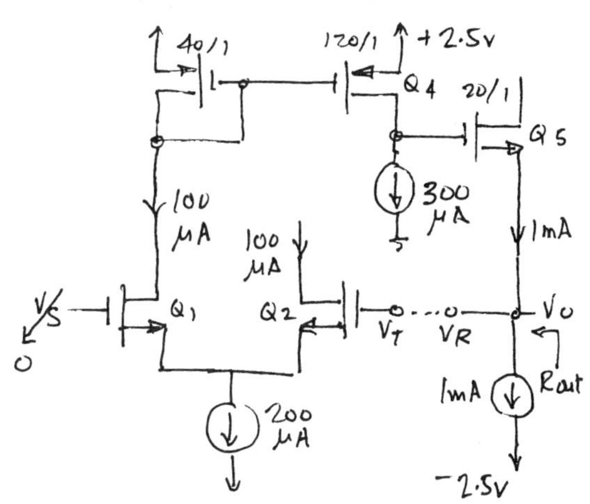

From P8.33 $g_{m1} = g_{m2} = 693\,\mu A/V$

$g_{m3} = 693\,\mu A/V$, $g_{m4} = 2078\,\mu A/V$

$g_{m5} = 1550\,\mu A/V$. $r_{o5} = 24\,K\Omega$

$r_{o1} = r_{o2} = r_{o3} = 240\,K\Omega$, $r_{o4} = 80\,K\Omega$

$R_{out} = (r_{o5} \| r_{o5}) = 12\,K\Omega$

$i_{c1} = -g_{m1} V_T$

$i_{c4} = -3 g_{m1} V_T$

$v_{c4} = -3 g_{m1} V_T (r_{o4} \| r_{o4})$

$v_{s5} = v_{c4} = V_R$

$$\therefore \frac{V_R}{V_T} = -3 g_{m1} \frac{(r_{o1} \| r_{o1})}{3} = -\frac{g_{m1} r_{o1}}{2}$$

$$A\beta = -\frac{V_R}{V_T} = \frac{g_{m1} r_{o1}}{2} = 83.16$$

$R_o = (r_{o5} \| r_{o5}) = 12\,K\Omega$

8.58

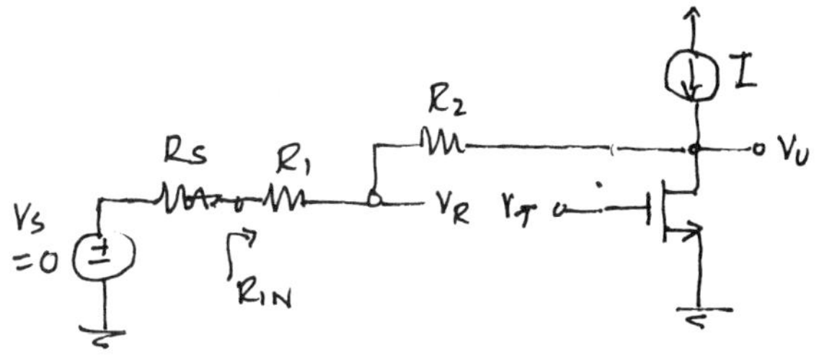

Apply V_T with $V_s = 0$

$$\frac{V_O}{V_T} = -g_m [R_2 + R_1 + R_s] \| r_o$$

$$\frac{V_R}{V_T} = -g_m [[R_2 + R_1 + R_s] \| r_o] \frac{[R_1 + R_s]}{[R_2 + R_1 + R_s]}$$

$$= -\frac{g_m (R_1 + R_s)\, r_o}{R_1 + R_2 + R_s + r_o}$$

$$= -A\beta$$

For $I = 1mA$, $V_{GS} = 0.8V$, $V_t = 0.6V$

$V_A = 30V$, $R_s = 10K$, $R_1 = 1M$, $R_2 = 4.7M$.

$$g_m = \frac{2I}{V_{ov}} = \frac{2}{0.8-0.6} = \frac{2}{0.2} = 10\,mA/V$$

$r_o = V_A / I = 30V/1mA = 30K\Omega$

Then
$$A\beta = \frac{10\times10^{-3}(10K + 10^3 K)\,30K}{1M + 4.7M + 10K + 30K}$$

$$= \frac{10 \times 1010 \times 30 \times 10^{-3}}{5740}$$

$$= 52.8 \times 10^{-3}$$

8.59

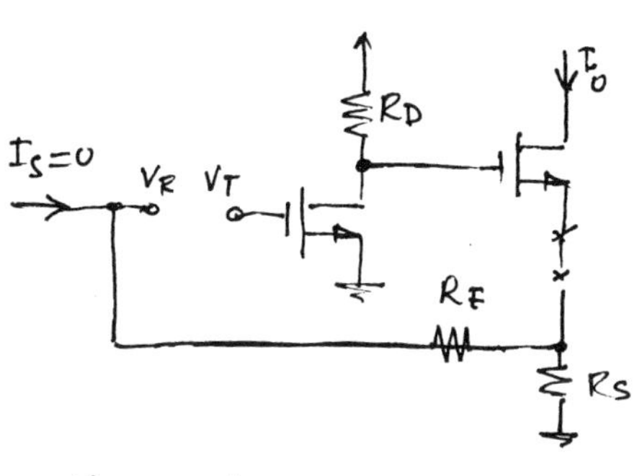

$$\frac{V_R}{V_T} = (g_{m2} R_S)(-g_{m1} R_D)$$

$$= -g_{m1} g_{m2} R_D R_S = -A\beta.$$

8.60

RD
C_1
I_1
I_2
V_R
V_T
$I_S = 0$
C_2

$$\frac{V_R}{V_T} = \frac{\frac{1}{sC_2}}{\frac{1}{sC_2} + \frac{1}{sC_1}} \cdot g_{m2} R_D$$

$$-A\beta = \frac{C_1}{C_1 + C_2} \cdot g_{m2} R_D$$

since $I_1 = I_2$

Section 8.8
The Stability Problem

8.61

$$A(s) = \frac{10^5}{1+s/100}$$

$$Ang(A) = -\tan^{-1}\frac{w}{100} - 2\tan^{-1}\frac{w}{10^4}$$

at w_{180}: $Ang(A) = -180°$ for $w_{180} \gg 100$

$$\Rightarrow 180° = 90° + 2\tan^{-1}\left[\frac{w_{180}}{10^4}\right]$$

hence $\tan^{-1}\frac{w_{180}}{10^4} = \frac{90°}{2}$

i.e. $\frac{w_{180}}{10^4} = \tan(45°) = 1$

$\therefore w_{180} = 10^4$ rad/s

$$|AB| = \frac{10^5 B}{\sqrt{1+(10^4/10^2)^2}} \cdot \frac{1}{(\sqrt{1+1})^2} = 1$$

$$\Rightarrow B = 0.002$$

$$A_f(0) = \frac{10^5}{1+10^5(0.002)} \approx 500 V/V$$

8.62

ω	Ang(A)	$\lvert A \rvert B1$	$\lvert A \rvert B2$
0	0	10^5	10^2
10^2	45	7.07×10^4	70.7
10^3	95.7	9.85×10^3	9.85
10^4	180	500	0.5
∞	0	0	0

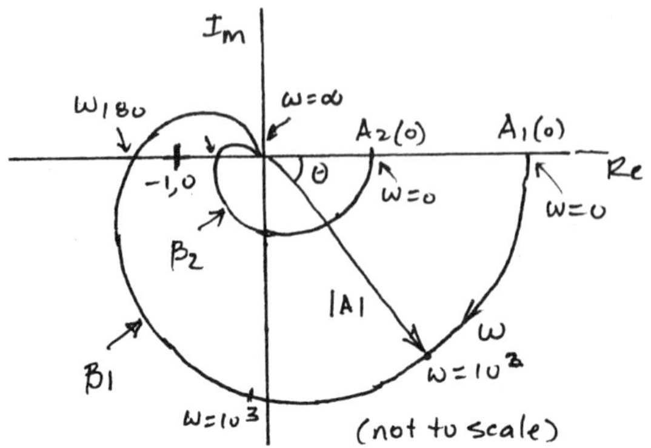

8.63

$$A(s) = \frac{10^3}{1+s/10^4}$$

$$\beta(s) = \frac{K}{(1+s/10^4)^2}$$

$$Ang(A\beta) = -\tan^{-1}\frac{w}{10^4} - 2\tan^{-1}\frac{w}{10^4}$$

$$= -3\tan^{-1}\frac{w}{10^4}$$

For 180° $w_{180} = \sqrt{3}\times10^4$ rad/s

For $|A\beta(w_{180})| < 1$

$$\frac{10^3}{\sqrt{1+(\sqrt{3})^2}} \cdot \frac{K}{1+(\sqrt{3})^2} < 1$$

$$\Rightarrow K < 0.008$$

8.64

$$A(s) = \frac{1000}{(1+s/10^4)(1+s/10^5)^2}$$

and β is independent of frequency

$$Ang(A) = -\tan^{-1}\frac{w}{10^4} - 2\tan^{-1}\frac{w}{10^5}$$

try $w = 10^4$: $\theta = 45° + 2\times5.7 = 56.4°$

try $w = 10^5$: $\theta = 84.2° + 2\times45 = 174.2°$

Iteration yields $w \approx 1.1\times10^5$ rad/s

For oscillations: $|A\beta(w_{180})| \geq 1$

$$\frac{\beta\, 10^3}{(\sqrt{1+11^2})(\sqrt{1+1.1^2})^2} \geq 1$$

$$\Rightarrow \beta \geq 0.0244$$

Section 8.9
Effect of Feedback on Amplifier Poles

8.65

$$A(jf) = \frac{(10\times10^6)/10^4}{1+jf/10^4}$$

$\therefore \; A(jf) = \dfrac{10^3}{1 + jf/10^4}$

$\beta = 0.1$ independent of frequency

$A_f(jf) = \dfrac{10^3}{1+10^3(0.1)} \cdot \dfrac{1}{1 + \dfrac{jf}{10^4(1+10^3(0.1))}}$

$= \dfrac{9.9}{1 + jf/(101 \times 10^4)}$

$A_f(0) = 9.9$ V/V

$f_{pf} = 10^4(101) = 1.01$ MHz

for $\dfrac{f}{f_{pf}} \gg 1$: $A_f \approx 9.9 \dfrac{10^4(101)}{f}$

for $A_f = 1$: $f = f_t = 10$ MHz

Pole is shifted by $(1 + A(0)\beta) = 101$

8.66

$A(jf) = \dfrac{10^3}{[1 + jf/10^4)(1 + jf/10^5)}$

(a) closed-loop poles given by

$1 + A(f)B = 0$

using $p = jf$

$p^2 + p(10^4 + 10^5) + (1 + 10^3 B)10^4 . 10^5 = 0$

i.e. $p^2 + (1.1 \times 10^5)p + 10^9(1 + 10^3\beta) = 0$

compare terms with

$(p + f_{pf})^2 = p^2 + 2f_{pf}.p + f_{pf}^2$

$2f_{pf} = (1.1 \times 10^5)$

$(1 + 10^3\beta) \times 10^9 = f_{pf}^2$

$\Rightarrow f_{pf} = 5.5 \times 10^5$

and $(1 + 10^3\beta) = 3.025 \Rightarrow \beta = 2.025 \times 10^{-3}$

(b) $A_f(0) = \dfrac{10^3}{1 + 10^3(2.025 \times 10^3)} = 330.6$ V/V

@ 55 KHz : $|A(f)\beta| = 1$

$\Rightarrow A_f(55\text{KHz}) = \dfrac{A_f(0)}{1 + |A\beta|} = 165.3$ V/V

(c) from $s^2 + (\omega_0/Q)s + \omega_0^2$ cf. above

$Q = \dfrac{f_{pf}}{2f_{pf}} = \dfrac{1}{2}$

(d) $p^2 + 1.1 \times 10^5 p + (1 + 10^3\beta) = 0$

$\Rightarrow p^2 + 1.1 \times 10^5 p + 21.25 \times 10^9 = 0$

$p = \dfrac{-1.1 \times 10^5 \pm \sqrt{(1.1 \times 10^5)^2 - 4(1)(21.25 \times 10^9)}}{2}$

$= \dfrac{-1.1 \times 10^5 \pm j\,2.7 \times 10^5}{2}$

$= -5.5 \times 10^4 \pm j1.35 \times 10^5$ Hz

$Q = \dfrac{|P|}{2(5.5 \times 10^4)} = \dfrac{\sqrt{(5.5 \times 10^4)^2 + (1.35 \times 10^5)^2}}{1.1 \times 10^5}$

$= 1.33$

8.67

$A(jf) = \dfrac{10^3}{(1 + jf/10)(1 + jf/f_p)}$

$A_f(0) = \dfrac{10^3}{1 + 10^3 B} = 100$

$\Rightarrow \beta = 9 \times 10^{-3}$ V/V

Maximally flat when $Q = 0.707 = 1/\sqrt{2}$

from

$p^2 + p(f_1 + f_2) + (1 + A_0\beta)(f_1 \times f_2) = 0$

$Q = \dfrac{\sqrt{(1 + A_0\beta) f_1 f_2}}{f_1 + f_2}$

$\Rightarrow \dfrac{\sqrt{(1 + 10^3\beta)\, 10^3 f_{pf}}}{10^3 + f_{pf}} = \dfrac{1}{\sqrt{2}}$

$\Rightarrow f_{pf}^2 + (2 \times 10^3) f_{pf} + 10^6 = 2(1 + 10^3\beta)^2 10^6$

$\Rightarrow f_{pf} = \dfrac{18 \times 10^3 \pm \sqrt{(18 \times 10^3)^2 - 4(10^6)}}{2}$

$= 17.94$ KHz.

8.68

$i = \dfrac{sCV_r}{10} \Rightarrow V = V_r + sCRV_r$

$= V_r(1 + sCR)$

By KCL: $\Sigma i = 0$ at V

$$\Rightarrow \frac{(KV_t - V)}{1/sC} = \frac{V}{R} + i = \frac{V}{R} + \frac{sCV_r}{10}$$

$$KsCV_t - sCV_r(1+sCR) =$$
$$= V_r\left(\frac{1+sCR}{R} + \frac{sC}{10}\right)$$

Collecting terms:

$$KsCV_tR = V_r[(sCR)^2 + 2.1\,sCR + 1]$$

Thus $L(s) \triangleq -V_r/V_t$

$$= \frac{-K/CR\ s}{s^2 + \frac{2.1s}{CR} + \frac{1}{CR^2}}$$

$$\equiv \frac{-K/CR\ \ s}{s^2 + \left(\frac{\omega_0}{Q}\right)s + \omega_0^2}$$

from which $\omega_0 = 1/CR$

$$Q = \frac{1}{2.1 - K}$$

poles coincide when $Q = 1/2$

$$\Rightarrow K = 2.1 - 2 = 0.1$$

maximally flat when $Q = 1/\sqrt{2}$

$$\Rightarrow K = 2.1 - 1.414 = 0.686$$

Oscillates when $Q \to \infty$

$$\Rightarrow K = 2.1$$

8.69

$$A(jf) = \frac{K}{1 + jf/10^7}$$

for $\beta = 1$: $A\beta = \frac{K^3\beta}{(1+jf/10^7)^3} = \frac{K^3}{(1+jf/10^7)^3}$

For oscillations:

$|A\beta| \geq 1$ at $Ang(A\beta) = 180°$

i.e. $+3\tan^{-1}\left(\frac{f_{180}}{10^7}\right) = 60°$

$$\Rightarrow f_{180} = \sqrt{3} \times 10^7\ \text{HZ}$$

$$\therefore \left[\frac{K}{\sqrt{1+(\sqrt{3})^2}}\right]^3 \geq 1 \Rightarrow K \geq 2$$

$$f_{osc} = f_{180} = 17.3\ \text{MHZ}.$$

Section 8.10
Stability Study using Bode Plots

8.70

$$A(f) = \frac{10^5}{1 + jf/10}$$

for $\beta = 1$: $A(f)\beta = \frac{10^5}{1+jf/10}$

for $f \gg 10$: $|A\beta| \approx 10^5 \cdot \frac{10}{f_1}$

$$\Rightarrow f_1 = 1\text{MHZ}$$

at f_1: phase margin $= 180° - \tan^{-1}\frac{10^6}{10}$
$$= 90°$$

8.71

$$A(f) = \frac{10^5}{(1+jf/10)(1+jf/10^4)}$$

$$A\beta(0) = 10^5\beta$$

$$A_F(0) = 100 = \frac{10^5}{1+10^5\beta} \Rightarrow \beta \approx 0.01$$

$$|A\beta| = 1 \Rightarrow |1 + jf/10| \cdot |1 + jf/10^4|$$
$$= 10^5\beta = 10^3$$

$$(1 + f^2/10^2)(1+f^2/10^8) = 10^6$$

$$f^4 + f^2(10^8 + 10^2) - (10^8)(10^2)10^6 = 0$$

$$f^2 \approx \frac{-10^8 + \sqrt{10^{16} + 4\times 10^{16}}}{2}$$

$$\Rightarrow 61.8 \times 10^6 \Rightarrow f = 7.86\ \text{KHZ}$$

Phase margin

$$= 180 - \left(\tan^{-1}\frac{7.86\times10^3}{10} + \tan^{-1}\frac{7.86}{10}\right)$$
$$\approx 180 - 90° - 38.16° = 51.8°$$

For PM $\geq 45°$: $\tan^{-1} f_1/10^4 \leq 45°$

$$\Rightarrow f_1 \leq 10^4$$

thus

$$|A\beta| = 1 = \frac{10^5\beta}{\sqrt{(1+(10^3)^2}\cdot\sqrt{2}}$$

$$\Rightarrow \beta = \sqrt{2}/100 = 0.0141$$

8.72

$$|1+e^{-j\theta}| = |1+\cos\theta - j\sin\theta|$$
$$= [(1+\cos\theta)^2 + (\sin\theta)^2]$$
$$= [1+2\cos\theta+\cos^2\theta+1-\cos^2\theta]^{1/2}$$
$$= \sqrt{2}\,(1+\cos\theta)^{1/2}$$

for 5%: $1+\cos\theta = \dfrac{1}{1.05^2(2)} = 0.4535$

$\theta = 123.13^\circ$ and $PM = 180-\theta = 56.87^\circ$

for 10%: $1+\cos\theta = \dfrac{1}{1.1^2(2)} = -0.586$

$\theta = 125.93^\circ$ and $PM = 54.07^\circ$

for 0.1dB $\equiv 10^{0.1/20} = 1.0116$

$\cos\theta = \dfrac{1}{2(1.0116)^2} - 1 = -0.5114$

$\theta = 120.76^\circ$ and $PM = 59.24^\circ$

for 1dB $\equiv 10^{1/20} = 1.122$

$\cos\theta = \dfrac{1}{2(1.122)^2} - 1 = -0.6028$

$\theta = 127.07^\circ$ and $PM = 52.93^\circ$

8.73

$$A(jf) = \frac{10^5}{\left(1+\frac{jf}{10^5}\right)\left(1+\frac{jf}{3.16\times10^5}\right)\left(1+\frac{jf}{10^6}\right)}$$

Assume β independent of frequency

For 45° PM: $\theta = 180-45$

$$\tan^{-1}\frac{f_1}{10^5} + \tan^{-1}\frac{f_1}{3.16\times10^5} + \tan^{-1}\frac{f_1}{10^6} = 135^\circ$$

Solve $\Rightarrow f_1 = 3.16\times10^5$ Hz

$$|A\beta(f_1)| = 1 = \frac{10^5\beta}{\sqrt{1+(3.16)^2}\cdot\sqrt{2}\cdot\sqrt{1+(0.316)^2}}$$

$\Rightarrow \beta = 49\times10^{-6}$

$$A_f(0) = \frac{10^5}{1+10^5(4.9\times10^{-6})} = 16.9\times10^3$$

8.74

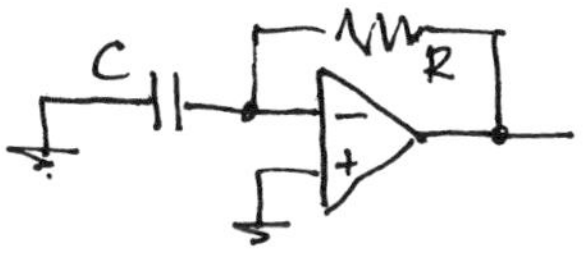

$$\beta = \frac{1/sC}{R+1/sC} = \frac{1}{1+sCR}$$

$$\beta(f) = \frac{1}{1+j2\pi fCR}$$

$$A(jf) = \frac{10^3}{\left(1+\frac{jf}{10^6}\right)\left(1+\frac{jf}{10^7}\right)}$$

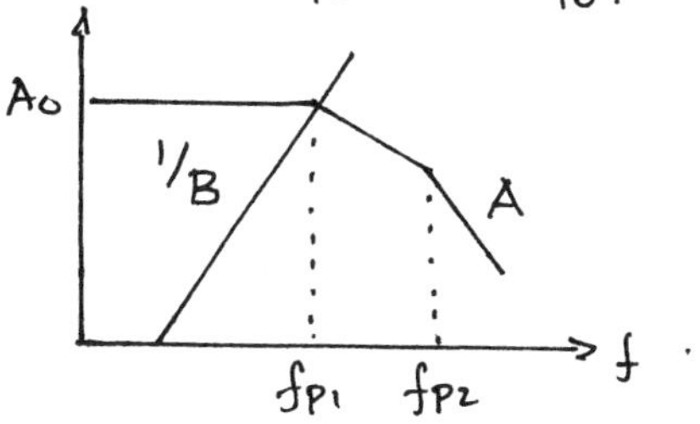

From sketch, we need

$$A_0\,\frac{1}{2\pi f_{p1}CR} = 1 = A\beta$$

$$\Rightarrow RC = \frac{A_0}{2\pi f_{p1}} = \frac{10^3}{2\pi\times10^6} = 159.2\,\mu s$$

At 1MHz $\;Ang(\beta) = -90^\circ$

$Ang(A) = -\tan^{-1}1 - \tan^{-1}0.1$

$= -45 - 5.7 = -50.7^\circ$

$\therefore PM = 180 - (90+50.7) = 39.3^\circ$

Gain margin exists at ω_{180}

Then $\tan^{-1}\frac{f}{10^6} + \tan^{-1}\frac{f}{10^7} = 90^\circ$

$\therefore f_{180} = \sqrt{10^6\cdot10^7}$ = geometric mean

$= 3.16$ MHz

$|A\beta(f_{180})| = 20\log|A| - 20\log|1/\beta|$

$|A|$ has fallen 10db, $|\beta|$ has risen 10dB

thus $GM = +10 - (-10) = 20$ dB

8.75

For 90° PM:

$$\tan^{-1}\frac{f_1}{10^5} + \tan^{-1}\frac{f_1}{10^6} + \tan^{-1}\frac{f_1}{10^7} = 90^\circ$$

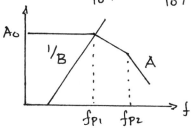

From graph $f_1 = 3\times10^5$ Hz

thus $71.6 + 16.7 + 1.72 = 89.9°$ (close)

$$|A(f_1)| = \frac{10^5}{\sqrt{1+3^2}\cdot\sqrt{1.0.3^2}\cdot\sqrt{1+0.03^2}} = 30.28\times10^3$$

$|AB| = 1 \Rightarrow \beta = 33.0\times10^{-6}$

$$\therefore A_f(0) = \frac{10^5}{1+10^5\beta} = 2.32\times10^4$$

For PM = 45° $f_1 \approx 10^6$ Hz from graph

thus $84.3 + 45 + 5.7 = 135°$ (ok)

$$|A(f_2)| = \frac{10^5}{\sqrt{1+10^2}\cdot\sqrt{2}\sqrt{1+0.1^2}} = 7\times10^3$$

$|A\beta| = 1 \Rightarrow \beta = 1.43\times10^{-4}$

$$\therefore A_f(0) = \frac{10^5}{1+10^5\beta} = 6.54\times10^3$$

Section 8.11
Frequency Compensation

8.76

$f_1 = 2$ MHz

$A_0 = 80$ dB $\equiv 10^4$

$\Rightarrow f_p = f_1/A = (2\times10^6)/10^4 = 200$ Hz

8.77

$f_{p1} = 2$ MHz, $f_{p2} = 10$ MHz

$A_0 = 80$ dB $\equiv 10^4$

$$f_D = \frac{f_p}{A_0} = \frac{10\times10^6}{10^4} = 10^3 \text{ Hz}$$

$f'_D = 1/(C_x + C_c)2\pi R_x \rightarrow C \times \frac{2\times10^6}{10^3} = 2000C$

8.78

$$\omega_{p1} = \frac{1}{10CR}, \quad \omega_{p2} = \frac{1}{CR}$$

$$\omega'_{p1} \approx \frac{1}{g_m R_2 C_f R_1} = \frac{1}{\frac{100}{R}\cdot R \cdot C_f R} = \frac{1}{100 C_f R}$$

$$\omega'_{p2} \approx \frac{g_m C_f}{C_1C_2 + C_f(C_1+C_2)} = \frac{g_m C_f}{C_1(C_2+C_f)+C_fC_2}$$

for $C_f \gg C_2 = C$

$$\omega'_{p2} \approx \frac{g_m}{C_1+C_2} = \frac{100}{R(C_1+C_2)} = \frac{9.1}{CR}$$

8.79

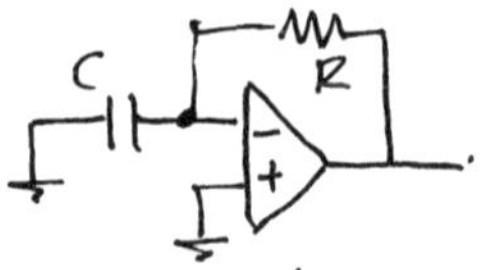

$A_0 = 10^4$

poles at $10^5, 10^6, 10^7$ Hz

For $\beta = 1$, f_p must be kept $\times10^4$ lower than lowest amplifier pole at 10^5 Hz

$$\Rightarrow f_p = \frac{10^5}{10^4} = 10 \text{ Hz}$$

$$f_p = \frac{1}{2\pi CR} \quad \text{and } R = 1\text{M}\Omega$$

$$\Rightarrow C = \frac{1}{2\pi\,10^6\,(10)} = 15.9 \text{nF}$$

8.80

$A_0 = 80$ dB $\equiv 10^4$

$$f_{p1} = 10^5 = \frac{1}{2\pi C_1R_1} \Rightarrow R_1 = \frac{1}{2\pi f_{p1}C_1}$$

$$\Rightarrow R_1 = \frac{1}{2\pi\cdot10^5\cdot(150\times10^{-12})} = 10.62\text{K}\Omega$$

$$f_{p2} = 10^6 = \frac{1}{2\pi C_2R_2}$$

$$\Rightarrow R_2 = \frac{1}{2\pi\,10^6(5\times10^{-12})} = 31.85\text{K}\Omega$$

Assuming $f'_{p2} \gg f_{p3}$

$$f'_{p1} = \frac{f_{p3}}{10^4} = \frac{2\times10^6}{10^4} = 200 \text{Hz}$$

$$\text{and } f'_{p1} = \frac{1}{2\pi g_m R_1R_2C_f} \Rightarrow C_f = \frac{1}{2\pi g_m R_1R_2 f'_{p1}}$$

$$\therefore C_f = \frac{1}{2\pi(40\times10^{-3})(10.62\times10^3)(31.85\times10^3)200}$$

$= 58.8$ pF

$$f'_{p2} = \frac{1}{2\pi}\frac{g_m C_f}{C_1C_2 + C_f(C_1+C_2)}$$

$$= \frac{1}{2\pi}\frac{40\times10^{-3}(58.8\times10^{-12})}{(150\times5)10^{-24} + 58.8(155)10^{-24}} = 38.8\text{MHz}$$

8.81

$$\beta(s) = \frac{1}{1+sCR} = \frac{1}{1+10^{-3}s}$$

(a) $$A\beta(s) = \frac{10^5}{1+s/10}\cdot\frac{1}{1+s/10^3}$$

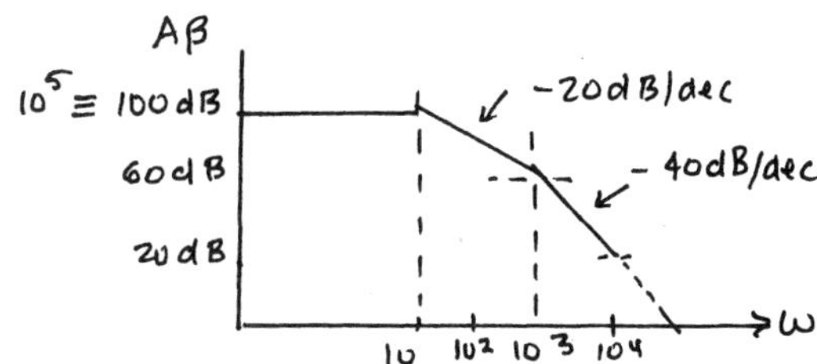

(b) From plot $|A\beta| = 20\,\text{dB}$ at $10^4 = \omega$

Hence $|A\beta| = 1$ at 31.6 krad/s ($\tfrac{1}{2}$dec)

(c) $$A_f(s) = \frac{\dfrac{10^5}{1+s/10}}{1+\dfrac{10^5}{1+s/10}\cdot\dfrac{1}{1+s/10^3}}$$

$$= \frac{10^5(1+s/10^3)}{(1+s/10)(1+s/10^3)+10^5}$$

$$\therefore A_f(s) = \frac{1+s/10^3}{1+s/10^6+s^2/10^9} = \frac{10^6 s+10^9}{s^2+10^2 s+10^9}$$

Zera at $s = -10^3$ rad/s

Poles at $$\frac{-10^3 \pm \sqrt{10^6 - 4\times 10^9}}{2}$$

$$= \frac{-10^3 \pm j\,63.2\times 10^3}{2}$$

$$= -500 \pm j\,31.6\times 10^3 \text{ rad/s}$$

$\omega_0 = 31.6$ krad/s

$Q = 31\text{-}6$

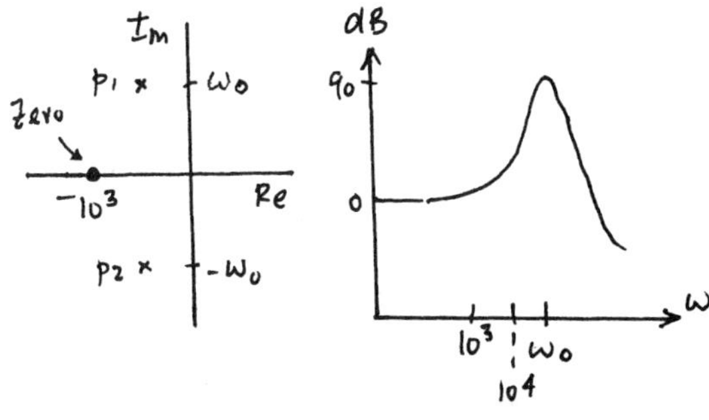

Chapter 9 – Problems

Section 9.1: The two-stage CMOS op-Amp

9.1

$V_{ICM}(max) \leq V_{DD} - |V_{tp}| - |V_{ov1}| - |V_{ov5}|$
$\leq +2.5 - 0.7 - 0.3 - 0.3$
$\leq +1.2V$

$V_{ICM}(min) \geq -V_{SS} + V_{ov3} + V_{tn} - |V_{tp}|$
$\geq -2.5 + 0.3 (+0.7 - 0.7)$
$\geq -2.2V$

$-V_{SS} + V_{ov6} \leq V_O \leq V_{DD} - |V_{ov7}|$
$-2.5 + 0.3 \leq V_O \leq +2.5 - 0.3$
$-2.2V \leq V_O \leq +2.2V$

9.2

$V_A' = 25V/\mu m$, $|V_P'| = 20V/\mu m$, $L = 0.8\mu m$
Hence $V_A = 20V$ and $|V_P| = 16V$
For all devices $V_{ov} = 0.25V$

$A = A_1 A_2 = G_{m1}(r_{o2} \| r_{o4}) G_{m2}(r_{o6} \| r_{o7})$

$[r_{op} \| r_{on}] = \left[\frac{V_A}{I} \times \frac{V_P}{I}\right] \times \frac{I}{V_A + V_P} = \left[\frac{V_A \| V_P}{I}\right]$

For A_2: $R_o = \frac{8.89}{I} \rightarrow \frac{8.89V}{0.4mA} = 22.2K\Omega$

To avoid systematic output d.c. offset

$$\frac{(W/L)_6}{(W/L)_4} = \frac{2(W/L)_7}{(W/L)_5}$$

Since Q_5, Q_6, Q_7 carry I and Q_4 only $I/2$ satisfy requirement by making Q_4 have $(W/L)/2$

Since $g_m = \sqrt{2(\mu C_{ox})(W/L) I} = 2K(V_{ov})$

$g_{m1} = 2I_1/V_{ov} = 0.4mA/0.25V = 1.6mA/V$

$g_{m6} = 2I_6/V_{ov} = 3.2mA/V$

$\therefore A = (1.6)(44.4)(3.2)(22.2) = 5047 V/V$

For unity gain amplifier

$$A_F = \frac{A}{1+A\beta} = \frac{5047}{1+5047\beta} = 1$$

Thus $(1 + A\beta) = 5047$
Then $R_{of} = R_o / (1 + A\beta)$
$= 22.2K / 5047 \approx 4.4\Omega$

9.3

$A = A_1 A_2 = G_{m1}(r_{o2} \| r_{o4}) G_{m2}(r_{o6} \| r_{o7})$

$$= \frac{2I_1}{V_{ov}} \cdot \frac{1}{2}\frac{V_A}{I_1} \cdot \frac{2I_2}{V_{ov}} \cdot \frac{1}{2}\frac{V_A}{I_2}$$

$$= \left[\frac{V_A}{V_{ov}}\right]^2 = 2500$$

where $V_A = 10V/\mu m \times 1\mu m = 10V$

Hence $V_{ov} = V_A/50 = 10/50 = 0.2V$

9.4

From $\frac{(W/L)_6}{(W/L)_4} = \frac{2(W/L)_7}{(W/L)_5}$

$(W/L)_6 = \frac{2(60/0.5)(10/0.5)}{(60/0.5)} = (20/0.5)$

$I_{Dp} = \frac{1}{2}[60](W/L)(V_{GS} - V_T)^2$

Given $I_{ref} = 225\mu A = \frac{1}{2} 60 (60/0.5)[V_{ov}]^2$
$\Rightarrow V_{ov} = 15/60 = 0.25V$

Given $V_T = 0.75V$
$\Rightarrow V_{GS} = V_T + V_{ov} = 0.25 + 0.75 = 1V$

Since Q_5, Q_7, Q_8 have same (W/L)
$\Rightarrow I_1 = I_2 = I_3 = I_4 = \frac{1}{2} I_5 = \frac{1}{2} I_8$

For Q_1Q_2: $I_1 = \frac{1}{2} 60 (30/0.5)(V_{ov})^2 = 112.5mA$
$\Rightarrow V_{ov1} = 0.25V$
$\Rightarrow V_{GS1} = V_T + V_{ov1} = 1.0V$

For Q_3Q_4: $I_3 = \frac{1}{2} 180 (10/0.5)(V_{ov})^2 = 112.5 mA$
$\Rightarrow V_{ov3} = 0.25V$
$\Rightarrow V_{GS3} = 0.25 + 0.75 = 1V$

For Q_6: $I_6 = \frac{1}{2} 180 (20/0.5)(V_{ov})^2 = 225mA$
$\Rightarrow V_{ov6} = 0.35V$
$\Rightarrow V_{GS6} = 0.35 + 0.75 = 1.1V$

$r_o = \frac{V_A\ V}{I\ mA} = r_o\ K\Omega$ and $|V_A| = 9V$

$g_m = \frac{2I}{V_{ov}}$ and $|V_{ov}|$ vary

Q	K ($\mu A/V^2$)	I (μA)	V_{OV} (V)	V_{GS} (V)	g_m (mA/V)	r_o (kΩ)
Q1	60	112.5	0.25	0.81	0.9	80
Q2	60	112.5	0.25	0.81	0.9	80
Q3	180	180	0.25	1.0	0.9	80
Q4	180	180	0.25	1.0	0.9	80
Q5	60	225	0.25	1.0	1.8	40
Q6	180	225	0.35	1.1	1.285	40
Q7	60	60	0.25	1.0	1.8	40
Q8	60	60	0.25	1.0	1.8	40

$A_1 = G_{m1}(r_{o2} \| r_{o4}) = 0.9 \times 40 = 36$ V/V

$A_2 = G_{m6}(r_{o6} \| r_{o7}) = 1.285 \times 20 = 25.7$ V/V

$A = A_1 A_2 = 36 \times 25.7 = 925$ V/V

$V_{ICM(max)} \leq V_{DD} - |V_{tp}| - |V_{OV1}| - |V_{OV5}|$

$\leq 1.5 - 0.75 - 0.25 - 0.25$

$\leq +0.25$ V

$V_{ICM(min)} \geq -V_{SS} + V_{OV3} + V_{tn} - |V_{tp}|$

$\geq -1.5 + 0.25 = -1.25$ V

$-V_{SS} + V_{OV6} \leq V_O \leq +V_{DD} - |V_{OV7}|$

$-1.5 + 0.35 \leq V_O \leq +1.5 - 0.25$

$-1.15\text{ V} \leq V_O \leq +1.25\text{ V}$

9.5

$G_{m1} = 0.3$ mA/V $\quad G_{m2} = 0.6$ mA/V $\quad C_2 = 1$p

$r_{o2} = r_{o4} = 222$ kΩ $\quad r_{o6} = r_{o7} = 111$ kΩ

(a) $f_{P2} = \dfrac{G_{m2}}{2\pi C_2} = \dfrac{0.6\times10^{-3}}{2\pi\, 10^{-12}} = 95.5$ MHz

(b) $R = \dfrac{1}{G_{m2}} = \dfrac{1}{0.6} = 1.66$ kΩ

(c) For PM = 80°: $\tan^{-1}\dfrac{f_t}{f_{P2}} = 10°$

$f_t = f_{P2}\tan 10°$

$= 95.5 \times 0.176 = 16.84$ MHz

$C_C = \dfrac{G_{m1}}{2\pi f_t} = \dfrac{0.3\times10^{-3}}{2\pi\, 16.84\times10^6} \Rightarrow 2.83$ pF

$A = A_1 A_2 = G_{m1}(r_{o2} \| r_{o4})\, G_{m2}(r_{o6} \| r_{o7})$

$= 0.3 \times 111 \times 0.6 \times 55.5$

$= 1109 \equiv 60.8$ dB

Dominant pole $f_{P1} = f_t/|A|$

Thus f_{P1} is approx 3 decades below f_t i.e. at 16.84 kHz providing uniform 20 dB/dec slope down to f_t.

(d) $f_t = \dfrac{G_{m1}}{2\pi C_C}$ $\quad\therefore$ to double f_t, halve C_C

$C_{C(new)} = 1.4$ pF

$\tan^{-1}\dfrac{f_t}{f_P} = \tan^{-1}\dfrac{33.7}{95.5} = 19.4°$

The zero must be moved to reduce the $19.1 - 10 = 9.4°$

$\tan^{-1}\dfrac{f_t}{f_Z} = 9.4° \rightarrow \dfrac{f_t}{f_Z} = 0.16$

$\Rightarrow f_Z = 0.16 f_t = 0.16 \times 33.7$

$= 5.6$ MHz

$f_Z = \dfrac{1}{2\pi C_C\left[R - \frac{1}{G_m}\right]} \rightarrow \left[R - \dfrac{1}{G_m}\right] = \dfrac{1}{2\pi f_Z C_C}$

Hence $[R - 1/G_m] = \dfrac{10^{12}\cdot10^{-6}}{2\pi\; 5.6\times1.4}$

$R = 1.67 + 20.3 = 21.97$ kΩ

9.6

Two-stage amp with $C_2 = 1$ pF

$f_t = 100$ MHz, PM = 75°

For PM = 75°: $\tan^{-1}\dfrac{f_t}{f_{P2}} = 15°$

$\therefore f_{P2} = f_t \tan 15° = 3.73 f_t = 373$ MHz

$f_{P2} = \dfrac{G_{m6}}{2\pi C_2} = \dfrac{1}{2\pi R_2\, 10^{-12}} = 373$ MHz

$\Rightarrow R_2 = \dfrac{10^{12}}{2\pi(373\times10^6)} = 426\,\Omega$

$\Rightarrow G_{m6} = \dfrac{1}{R_2} = 2.35\times10^{-3}$ mA/V

To move zero to infinity $R = \dfrac{1}{G_{m6}} = 426\,\Omega$

$SR = \dfrac{I}{C_C} = \dfrac{200\,\mu A}{C_C}$

$SR = 2\pi f_t V_{OV1} = 2\pi\, 10^8 \times 0.2 = 1.26\times10^8$ V/s

$\rightarrow C_C = \dfrac{200\times10^{-6}}{1.26\times10^8} \Rightarrow 1.6$ pF

9.7

$G_{m1} = 1mA/V$, $G_{m2} = 2mA/V$, $R = 500\Omega$

$$f_t \approx \frac{G_{m1}}{2\pi C_c} \Rightarrow C_c = \frac{G_{m1}}{2\pi f_t} = \frac{1\times10^{-3}}{2\pi\, 100\times10^6} \Rightarrow 1.59pF$$

$$\text{For } \frac{1}{G_{m2}} - R = \frac{10^3}{2} - 500 = 0$$

Zero has been moved to ∞

For PM = 60: $f_t = f_{P2} \tan(90-30)°$

$\Rightarrow f_{P2} = f_t / \tan 30° = 173\ MHz$

$$C_2 \approx \frac{G_{m2}}{2\pi f_{P2}} = \frac{2\times10^{-3}}{2\pi\ 173\times10^6} = 1.84pF$$

9.8

$SR = 60V/\mu s$ $\qquad f_t = 50\ MHz$

(a) $SR = 2\pi f_t V_{OV1}$

$\Rightarrow V_{OV1} = SR/2\pi f_t$

$$= \frac{60\times10^6}{2\pi(50\times10^6)} \approx 0.2V$$

(b) $SR = \frac{I}{C_c} \Rightarrow C_c = \frac{100\mu A}{60\times10^6} = 1.67pF$

(c) $I = \frac{1}{2}\mu C_{ox}(W/L)[V_{ov}]^2$

$$\Rightarrow \left(\frac{W}{L}\right) = \frac{2I}{50[0.2]^2} = \frac{100}{1}$$

9.9

Invert circuit leaving V_{DD} + V_{SS} and reverse all arrows on FETS.

Section 9.2 The Folded-Cascode OP AMP

9.10

From circuit: Current drawn from V_{DD} rail $= 2I_B$ = Current returned to V_{SS} rail

$\therefore$ Power $= (V_{DD}+V_{SS}) \times 2I_B \Rightarrow$

$1mW = (1.65+1.65)\times 2I_B$

$$\therefore I_B = \frac{1\ mW}{4\times1.65\ V} = 151.5\mu A$$

$\Rightarrow I = I_B/1.2 = 126.3\mu A$

9.11

V_{BIAS1}: V_S can rise to $V_{DD} + V_T - V_{OV}$
V_{D3} can rise to $V_{S3} - V_{OV}$

$\therefore V_{BIAS1} = V_{DD} - V_{OV10} - V_{OV4} + V_T$
$= 1.65 - 0.2 - 0.2 + 0.5$
$= 1.75V$

$V_{BIAS2} = V_{DD} - V_{OV10}$
$= +1.65 - 0.2 = +1.45V$

$V_{BIAS3} = -V_{SS} + V_{OV11}$
$= -1.65 + 0.2 = -1.45V$

$V_{ICM(max)} = V_{DD} - |V_{OV9}| + V_{tn}$
$= +1.65 - 0.2 + 0.5$
$= +1.75V$

$V_{ICM(min)} = -V_{SS} + V_{OV11} + V_{OV1} + V_T$
$= -1.65 + 0.2 + 0.2 + 0.5$
$= -0.75V$

$-V_{SS} + 2V_{OV} + V_T \le V_O \le +V_{DD} - 2V_{OV}$
$-1.65 + 0.4 + 0.5 \le V_O \le +1.65 - 0.4$
$-0.75V \le V_O \le +1.25V$

9.12

$I = 125\mu A$, $I_B = 150\mu A$. $V_t = 0.2V$

For Q_9, Q_{10}: $I_B = 150\mu A$

$I = \frac{1}{2}(\mu C_{ox})(W/L)(V_{ov})^2$

$150 = \frac{1}{2}\ 90\ (W/L)(0.2)^2$

$\Rightarrow (W/L)_{9,10} = (83.33/1)$

For Q_1, Q_2: $I = 125\mu A/2$

$\frac{1}{2}.125 = \frac{1}{2}\ 250(W/L)(0.2)^2$

$\Rightarrow (W/L)_{1,2} = (12.5/1)$

For Q_{11}: $I = 125\mu A$

$125 = \frac{1}{2} 250 (W/L)(0.2)^2$

$\Rightarrow (W/L)_{11} = (25/1)$

For $Q_3 Q_4$: $I = 125\mu A/2$

$\frac{1}{2} 125 = \frac{1}{2} 60 (W/L)(0.2)^2$

$\Rightarrow (W/L)_{3,4} = (52/1)$

For Q_5, Q_6, Q_7, Q_8: $I = 125\mu A/2$

$\frac{1}{2} 125 = \frac{1}{2} 250 (W/L)(0.2)^2$

$\Rightarrow (W/L)_{5,6,7,8} = (12.5/1)$

9.13

(1) $SR = \frac{I}{C_L} \Rightarrow I = SR \times C_L$
$= 10\times10^6 \times 10\times10^{-12}$
$= 100\mu A$

(2) $I_B = 1.2\,I = 120\mu A$

(3) $f_p = \frac{1}{2\pi C_L R_o}$, $f_t = \frac{G_m}{2\pi C_L}$

$G_m = \frac{2I/2}{V_{ov}} = \frac{100\mu A}{0.2V} = 0.5 mA/V$

$f_t = \frac{0.5\times10^{-3}}{2\pi\ 10\times10^{-12}} = 7.96\ MHz$

(4) $R_{of} \approx \frac{1}{G_m} = \frac{1000}{0.5} \equiv 2K\Omega$

$A_v = f_t/f_p = G_m R_o$

But $R_{of} = \frac{R_o}{1+G_m R_o} \Rightarrow R_o = 2M\Omega$

$\therefore A_v = 0.5\times10^{-3} \times 2\times10^6 = 1000$

$f_p = f_t/1000 = 7.96\ KHz$

(5) $\theta = -\tan^{-1}\frac{f_t}{f_p} - 2\left(\tan^{-1}\frac{f_t}{f_{p2}}\right)$

$PM @ f_{p2} = 90^\circ - \tan^{-1}\frac{f_t}{f_{p2}}$

$= 90^\circ - \tan^{-1}\left[\frac{7.96 MHz}{25 MHz}\right]$

$= 90^\circ - 17.7^\circ = 72.3^\circ$

(6) For $PM = 75^\circ$: $\tan^{-1}\frac{f_t}{f_{p2}} = 15^\circ$

Thus $f_t^* = f_{p2} \tan 15^\circ$
$= 25 MHz \times 0.27 = 6.7 MHz$

$\frac{f_t^*}{f_t} = \frac{C_L}{C_L^*} \Rightarrow \frac{6.7}{7.96} = \frac{10 pF}{C_L^*}$

$\Rightarrow C_L^* = C_L \frac{7.96}{6.7} = C_L \times 1.19$

$\therefore$ Increase C_L by 19%.

(7) $SR^* = \frac{I}{C_L^*} \Rightarrow \frac{SR}{1.19} = 8.4 V/\mu s$

9.14

$G_m = \frac{I}{V_{ov}}$ $r_o = \frac{V_A}{I}$

$R_o = R_{o4} \| R_{o6}$

$R_{o4} = g_{m4} r_{o4} (r_{o2} \| r_{o10})$

$R_{o6} = g_{m6} (r_{o6} \| r_{o8})$

$A = G_{m1} R_o$

Q_{10}: $I = I_B \Rightarrow g_{m10} = 0.75 mA/V$
$r_{o10} = 66.6 K\Omega$

Q_x: $I = 125\mu A \Rightarrow g_x = 0.31 mA/V$
$r_{ox} = 160 K\Omega$

$\therefore R_{o4} = 0.31 \times 160 \times 47K = 2359K$

$R_{o6} = 0.31 \times 160 \times 160K = 8000K$

$R_o = 2.35 \| 8.0 = 1.8 M\Omega$

$A = g_{m1} R_o = 0.31 \times 1800 = 558$

$\frac{V_o}{V_i} = 1 + \frac{(1/sC)}{(1/9sC)} = 1 + \frac{9sC}{sC} = 10$

$\therefore \beta = 1/10 = 0.1$

$A_F = \frac{A}{1+A\beta} = \frac{558}{1+558\times0.1} = 9.8$

$R_{of} = \frac{R_o}{1+A\beta} = \frac{1.8}{56.8} = 31.69 K\Omega$

9.15

Rail-to-Rail OP Amp

All $V_{ov} = 0.2V$ All $V_T = 0.5V$

$V_{DD} = V_{SS} = 1.65V$

$V_{icm}(max) = V_{DD} - |V_{ov9}| + V_T$

$= 1.65 - 0.2 + 0.5 = +1.95V$

$V_{icm}(min) = -V_{SS} + V_{ov11} + V_{ov1} + V_T$

$= -1.65 + 0.2 + 0.2 + 0.5$

$= -0.75V$

(a) For NMOS stage

$-0.75V \le V_{icm} \le +1.95V$

(b) For PMOS stage (symetrical)

$-1.95V \le V_{icm} \le +0.75V$

(c) For both active

$-0.75V \le V_{icm} \le +0.75V$

(d) Overall

$-1.95V \le V_{icm} \le +1.95V$

9.16

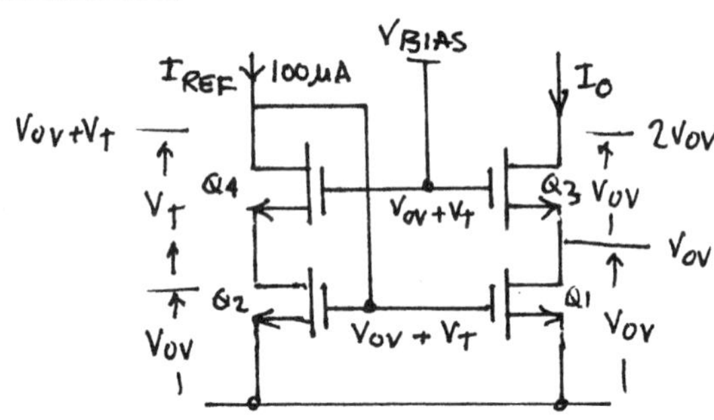

All same $k(W/L)$ $\therefore I_o \approx I_{REF}$

All same $r_o = V_A/I = 10V/100\mu A = 100K\Omega$

$I = \frac{1}{2}k[W/L][V_{ov}]^2$

$V_{D3} = V_o$: $V_o(min) = 2V_{ov}$

R_o (looking into Q3D and assuming I_o current source is ideal, $r_o' = \infty$)

$= r_{o3}(1 + g_{m3}r_{o1}) \approx g_m r_o^2$

$g_m = \frac{2I}{V_{ov}} = \frac{2 \times 100}{0.2} \Rightarrow 1mA/V$

Then $R_o \approx g_m r_o^2 = 1 \times 10^5 \times 10^5$

$\approx 10^4 M\Omega$

9.17

$A = 80dB$ $f_t = 10MHz$ $C_L = 10pF$

$I_B = I$ All same $|V_{ov}|$, $L = 1\mu m$

$|V_A| = 20V$

$g_m = \frac{2I}{V_{ov}}$ and $f_t = \frac{g_m}{2\pi C_L}$

$A = g_{m1}[g_{m4}r_{o4}(r_{o2}||r_{o10})] || [g_{m6}r_{o6}r_{o8}]$

Consider Q_1: $I_1 = \frac{1}{2}k_n[W/L]_1[V_{ov}]^2$

$= \frac{1}{2}200[W/L]_1[V_{ov}]^2$

$Q_1 Q_2 Q_5 Q_6 Q_7 Q_8$ are same

$g_{m1} = \frac{2I_1}{V_{ov}}$ and $r_{o1} = \frac{V_A}{I_1}$

Consider Q_3, Q_4: $I_3 = \frac{1}{2}k_p[W/L]_3[V_{ov}]^2$

$I_1 = \frac{1}{2}\frac{200}{2.5}[W/L]_3[V_{ov}]^2$

$\Rightarrow [W/L]_3 = 2.5[W/L]_1$

$g_{m3,4} = g_{m1}$ and $r_{o3,4} = r_{o1}$

Consider $Q_9 Q_{10}$: $I_{10} = \frac{1}{2}k_p[W/L]_{10}[V_{ov}]^2$

$2I_1 = \frac{1}{2}\frac{200}{2.5}[W/L]_{10}[V_{ov}]^2$

$\Rightarrow [W/L]_{10} = 5[W/L]_1$

$g_{m9,10} = 2g_{m1}$ and $r_{o9,10} = r_{o1}/2$

Consider Q_{11}: $I_{11} = \frac{1}{2}k_n[W/L]_{11}[V_{ov}]^2$

$2I_1 = \frac{1}{2}200[W/L]_{11}[V_{ov}]^2$

$\Rightarrow [W/L]_{11} = 2[W/L]_1$

$g_{m11} = 2g_{m1}$ and $r_{o11} = r_{o1}/2$

Thus $A = g_{m1}[g_{m1}r_{o1}(r_{o1}||\frac{r_{o1}}{2})] || [g_{m1}r_{o1}r_{o1}]$

$= g_{m1}[g_{m1}r_{o1}](r_{o1}||\frac{r_{o1}}{2}||r_{o1}]$

$10^4 = \frac{1}{4}g_{m1}g_{m1}r_{o1}r_{o1}$

$\Rightarrow g_{m1}r_{o1} = 200$

Now $g_{m1}r_{o1} = \frac{2I}{V_{ov}} \cdot \frac{V_A}{I}$

$\Rightarrow V_{ov} = 2V_A/200 = 2(20)/200 = 0.2V$

Hence $g_{m1} = \frac{1}{2\pi f_t C_L} = 0.628 mA/V$

$\rightarrow r_{o1} = 200/g_{m1} = 318K\Omega$

$$g_m = \frac{2I}{V_{ov}} \Rightarrow I = \frac{g_m V_{ov}}{2}$$

$$\Rightarrow I_1 = \frac{g_{m1} V_{ov}}{2} = \frac{0.628\ mA/V \times 0.2V}{2}$$

$$= 62.8\ \mu A$$

$$SR = 2\pi f_t V_{ov} = 2\pi\, 10\times10^6 \times 0.2 = 12.5\ V/\mu s$$

$Q_1 Q_2 Q_5 Q_6 Q_7 Q_8$:

$$I = \tfrac{1}{2} k_n [W/L][V_{ov}]^2$$

$$62.8 = \tfrac{1}{2}\, 200\, [W/L][V_{ov}]^2$$

$$\Rightarrow [W/L]_1 = 15.7$$

For Q_3, Q_4: $I = \frac{1}{2}\frac{200}{2.5}[W/L][V_{ov}]^2$

$$62.8 = \tfrac{1}{2}\tfrac{200}{2.5}[W/L][0.2]^2$$

$$\Rightarrow [W/L]_3 = 2.5[W/L]_1 = 39.25$$

For Q_9, Q_{10}:

$$[W/L]_9 = 5[W/L]_1 = 78.5$$

For Q_{11}:

$$[W/L]_{11} = 2[W/L]_1 = 31.4$$

For $L = 1\mu m$: $W_x = [W/L]_x\ \mu m$

$\therefore$ Width for $Q_1, Q_2, Q_5, Q_6, Q_7, Q_8$ $= 15.7\ \mu m$

for Q_3, Q_4 $= 39.25\ \mu m$

for Q_9, Q_{10} $= 78.5\ \mu m$

for Q_{11} $= 31.4\ \mu m$

9.18

Simply invert circuit relative to V_{DD}, V_{SS} and reverse all arrows on FETs

9.19

Model for CG stage:

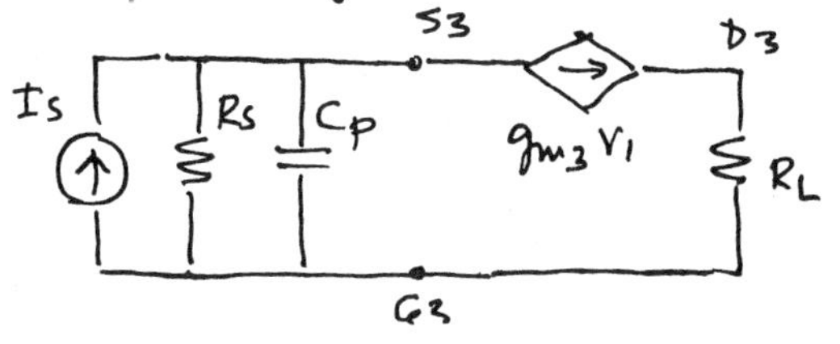

at S_3: $I_s - g_{m3}V_1 - V_1\frac{(1+sC_pR_s)}{R_s} = 0$

$$\Rightarrow \frac{V_1}{I_s} = \frac{1}{g_{m3} + (1+sC_pR_s)/R_s}$$

$$= \frac{1}{(g_{m3} + 1/R_s) + sC_pR_s/R_s}$$

3dB down when

$$sC_p = (g_{m3} + 1/R_s)$$

$$\Rightarrow f_{P3} = \frac{(g_{m3} + 1/R_s)}{2\pi C_p} \approx \frac{g_{m3}}{2\pi C_p}$$

QED.

Also $f_t \approx \frac{g_{m1}}{2\pi C_c}$

$$PM = 180 - \phi_{total} = 90° - \tan^{-1}\frac{f_t}{f_2}$$

For $PM = 75°$: $\frac{f_t}{f_2} = \tan 15° = 0.27$

$$\Rightarrow \frac{f_t}{f_2} = \frac{C_p}{C_L} = 0.27$$

$$\Rightarrow C_p = 0.27\,C_L$$

20 9

$V_{BE} = V_T \ln \frac{I_C}{I_S}$

$g_m = \frac{I_C}{V_T}$

$r_e = \frac{1}{g_m}$

$r_\pi = \beta r_e$

$r_o = \frac{V_A}{I_C}$

$r_\mu = \beta \frac{10 V_A}{I_C}$

DEVICE	V_{BE} (mV)	g_m (mA/V)	r_e (Ω)	r_π (kΩ)	r_o (MΩ)	r_μ (GΩ)
$Q_1 - Q_2$, $Q_5 - Q_6$	517	0.38	2632	526	13.2	26.3
Q_{16}	530	0.648	1543	309	7.72	15.4
Q_{17}	618	22	45.5	9.09	0.227	0.45

21 9

$I_3 = I_1 \sqrt{\frac{I_{S3} I_{S4}}{I_{S1} I_{S2}}}$

$= 154 \sqrt{\frac{10^{-14} \cdot 10^{-14}}{3\times10^{-14} \cdot 6\times10^{-14}}}$

$= \underline{\underline{36.3\ \mu A}}$

22 9

$I_{E\,TOT} = 0.73\ mA$

$I_{EA} = 0.25(0.73) = 0.1825\ mA$

$I_{EB} = 0.75(0.73) = 0.5475\ mA$

$V_{EB_A} = V_T \ln \frac{0.1825\times10^{-3}}{0.25\times10^{-14}} = \underline{\underline{0.625\ V}}$

$g_{mA} = \frac{I_C}{V_T} \simeq \frac{I_E}{V_T} = \underline{\underline{7.3\ mA/V}}$

$r_{eA} = \frac{\alpha}{g_{mA}} = \underline{\underline{134.3\ \Omega}}$

$r_{\pi A} = (\beta+1) r_{eA} = \underline{\underline{6.85\ k\Omega}}$

$r_{oA} = \frac{V_A}{I_{CA}} = \underline{\underline{274\ k\Omega}}$

$V_{EB_B} = V_{EB_A} = \underline{\underline{0.625\ V}}$

$g_{mB} = \frac{0.5475}{25} = \underline{\underline{21.9\ mA/V}}$

$r_{eB} = \frac{\alpha}{g_{mB}} = \underline{\underline{44.7\ \Omega}}$

$r_{\pi B} = (\beta+1) r_{eB} = \underline{\underline{2.28\ k\Omega}}$

$r_{oB} = \underline{\underline{91.3\ k\Omega}}$

23 9.

Let $V_{B2} = 0$

For breakdown $v_{ID} = v_{B1} - v_{B2}$
$> V_{BE1} + V_{BE2} + 7 + 50$

or $v_{ID} \geq \underline{\underline{58.4\ V}}$

24 9

$V_{SG1} + V_{GS2} = V_{SG4} + V_{GS3}$

Since V_t's are equal

$\sqrt{\frac{I_1}{K_1}} + \sqrt{\frac{I_1}{K_2}} = \sqrt{\frac{I_3}{K_4}} + \sqrt{\frac{I_3}{K_3}}$

$\sqrt{I_1}\left[\frac{1}{\sqrt{K_1}} + \frac{1}{\sqrt{K_2}}\right] = \sqrt{I_3}\left[\frac{1}{\sqrt{K_4}} + \frac{1}{\sqrt{K_3}}\right]$

or $\sqrt{\frac{I_1}{I_3}} = \underline{\underline{\frac{\sqrt{\frac{1}{K_3}} + \sqrt{\frac{1}{K_4}}}{\sqrt{\frac{1}{K_1}} + \sqrt{\frac{1}{K_2}}}}}$

$K_1 = K_2$, $K_3 = K_4 = 16 K_1$

$\sqrt{I_1} = \sqrt{I_3} \sqrt{\frac{K_2}{K_4}}$

or $I_1 = I_3 \frac{K_2}{K_4}$

$= \frac{I_3}{16} = \underline{\underline{100\ \mu A}}$

25 9

As $V_{BE} = 0.7$

$I_{ref} = \frac{5 - 1.4 - (-5)}{R_S}$

$= \underline{\underline{220.5\ \mu A}}$

At this current level
$V_{BE} = V_T \ln \frac{220.5\times10^{-3}}{10^{-14}} = 595\ mV$

$\Rightarrow I_{ref} = \frac{10 - 2(0.595)}{39K} = \underline{\underline{226\ \mu A}}$

For $I_{ref} = 0.73\ mA$, $V_{BE} = 0.625\ V$

$R_S = \frac{10 - 2(0.625)}{0.73\times10^{-3}} = \underline{\underline{12\ k\Omega}}$

26 9

Assume $\beta_p >> 1$

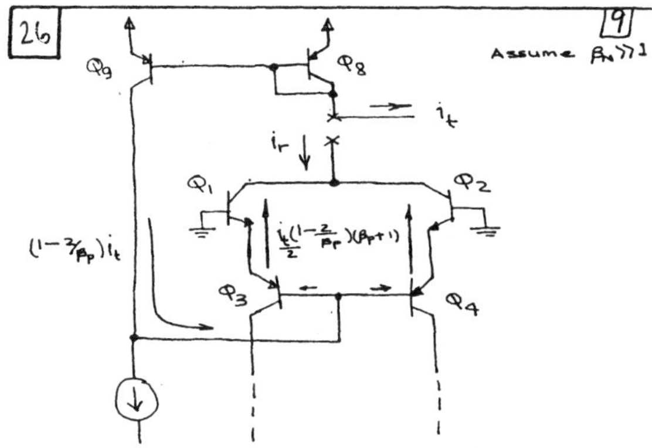

27 19

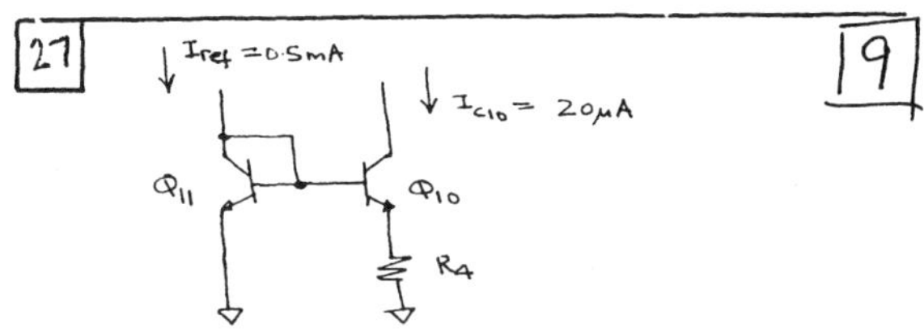

$$I_{C10} R_4 = V_{BE11} - V_{BE10}$$

$$= V_T \ln \frac{I_{ref}}{I_{C10}}$$

$$\therefore R_4 = \frac{25 \times 10^{-3}}{20 \times 10^{-6}} \ln \frac{0.5 \times 10^{-3}}{20 \times 10^{-6}} = \underline{\underline{4.02\ k\Omega}}$$

$$V_{BE11} = V_T \ln \frac{0.5 \times 10^{-3}}{10^{-14}} = \underline{\underline{616\ mV}}$$

$$V_{BE10} = V_T \ln \frac{20 \times 10^{-6}}{10^{-14}} = \underline{\underline{535\ mV}}$$

28 9

Assume $\beta_p >> 1$

$$I_{C10} \doteq \frac{2I}{1 + 2/\beta_p} + \frac{2I}{\beta_p}$$

$$\simeq 2I\left(1 - \frac{2}{\beta_p} + \frac{1}{\beta_p}\right)$$

$$= 2I(1 - 1/\beta_p)$$

$$\Rightarrow I \simeq \frac{I_{C10}}{2}\left(1 + \frac{1}{\beta_p}\right)$$

Thus $\frac{1}{\beta_p} = 0.1 \Rightarrow \underline{\underline{\beta_p = 10}}$

Without the above assumption and using the exact relationship $\beta_p = 11.14$.

29 19

In this case

$$\frac{4I}{1 + 2/\beta_p} + \frac{2I}{\beta_p} = I_{C10}$$

For $\beta_p >> 1$

$$I_{C10} \simeq 4I \quad \text{or} \quad I = \underline{\underline{4.75\mu A}}$$

To correct we need $I_{C10} = 38\mu A$

$$\Rightarrow R_4 = \frac{V_T}{I_{C10}} \ln \frac{0.73 mA}{I_{C10}} = \underline{\underline{1.94\ k\Omega}}$$

30 9

At $I = 9.5\mu A$

$$V_{BE5} = V_{BE6} = 517\ mV$$

and $V_{B6} = V_{BE6} + IR_2$

$$= 526.5 mV$$

If R_2 is shorted $V_{BE6} = V_{B6} = 526.5 mV$

and $I_{C6} = I_S e^{V_{BE6}/V_T}$

$$= \underline{\underline{14\mu A}}$$

31 19

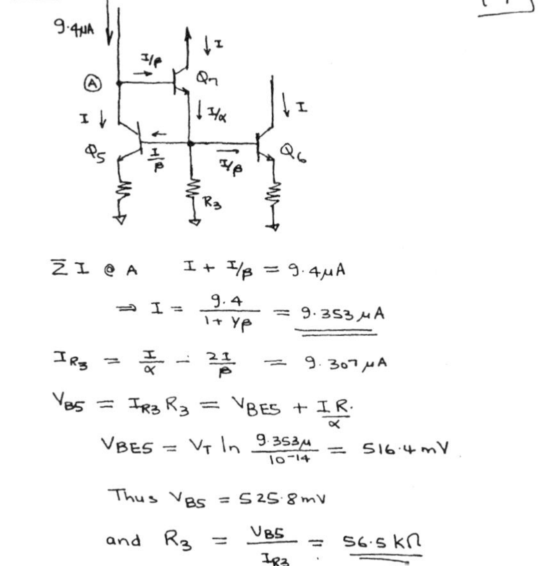

ΣI @ A $\quad I + I/\beta = 9.4\mu A$

$$\Rightarrow I = \frac{9.4}{1 + 1/\beta} = \underline{\underline{9.353\mu A}}$$

$$I_{R_3} = \frac{I}{\alpha} - \frac{2I}{\beta} = 9.307\mu A$$

$$V_{B5} = I_{R3} R_3 = V_{BE5} + \frac{IR}{\alpha}$$

$$V_{BE5} = V_T \ln \frac{9.353\mu}{10^{-14}} = 516.4 mV$$

Thus $V_{B5} = 525.8 mV$

and $R_3 = \frac{V_{B5}}{I_{R3}} = \underline{\underline{56.5\ k\Omega}}$

32 9

Assume equal collector current

$$I_{C1} = I_{C2} = 9.5\mu A$$

$$I_{os} = I_{B1} - I_{B2} = \frac{I_{C1}}{\beta_1} - \frac{I_{C2}}{\beta_2}$$

$$= \underline{\underline{15.7 nA}}$$

$$I_B = \frac{1}{2}(I_{B1} + I_{B2})$$

$$= \underline{\underline{55.4\mu A}}$$

33 $I_B = 40\,nA$, $I_{os} = 4nA$

Thus, base currents are

$I_{B1} = (I_B \pm \frac{I_{os}}{2})$

$= \frac{9.5}{\beta_N}\ \mu A$

$\hat{\beta}_N = \frac{9.5\,\mu A}{38\,nA} = \underline{\underline{250}}$

$\check{\beta}_N = \frac{9.5\,\mu A}{42\,nA} = \underline{\underline{226}}$

$\Rightarrow \Delta\beta_N = 24$

$\bar{\beta}_N = \frac{\hat{\beta}_N + \check{\beta}_N}{2} = 238$

34 $I_{C1} + I_{C2} = 19\,\mu A$

Mirror forces $I_{C2} = 0.9\, I_{C1}$

Thus $I_{C1} = \frac{19}{1.9}\,\mu A = 10\,\mu A$

and $I_{C2} = 9\,\mu A$

$V_{os} = \Delta V_{BE}$

$= V_{BE1} - V_{BE2}$

$= V_T \ln \frac{10}{9} = \underline{\underline{2.63\ mV}}$

9

35 At $I_{C17} = 550\,\mu A$, $V_{BE17} = 618\,mV$

$I_{B17} \simeq \frac{550}{200} = 2.75\,\mu A$

$\Rightarrow I_{C16} = 9.5\,\mu A = I_{B17} + \frac{I_{E17} R_8 + V_{BE17}}{R_9}$

or $R_9 = 99.7\,k\Omega$

36 Neglecting base currents

$I_{C18} = I_{C19} = \frac{180}{2} = 90\,\mu A$

$V_{BE18} = V_T \ln \frac{90 \times 10^{-6}}{10^{-14}} = 573\,mV$

Thus $R_{10} = \frac{V_{BE18}}{I_{C19}} = \underline{\underline{6.37\ k\Omega}}$

$I_{C14} = 3 \times 10^{-14}\, e^{573/26} = \underline{\underline{270\,\mu A}} = I_{C20}$

37

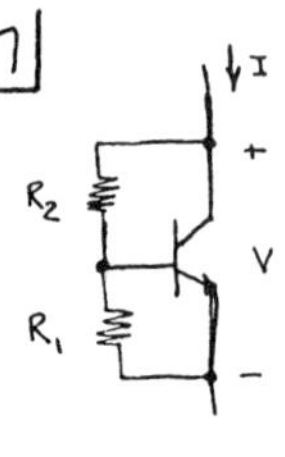

$V_{R_1} = V_{BE}$

If we ignore base currents

$V_{R_1} = \frac{R_1}{R_1 + R_2} V$

and $V = V_{BE}\left(\frac{R_1 + R_2}{R_1}\right)$

$V = V_{BE}\left(1 + \frac{R_2}{R_1}\right)$

38 $I_{V_{CC}} = I_{C12} + I_{C13A} + I_{C13B} + I_{C14} + I_{C9} + I_{C8} + I_{C7} + I_{C16}$

$= (730 + 180 + 550 + 154 + 19 + 19 + 10.5 + 16.2)\ \mu A$

$= \underline{\underline{1.68\ mA}}$

$P_{DISS} = P_Q = I_{V_{CC}}(V_{CC} + V_{EE})$

$= 1.68(15 + 15)\ mW$

$= \underline{\underline{50.4\ mW}}$

9

39 Series connection of devices assures the same bias currents

$R_{id} = (\beta + 1)(6 r_e)$

$r_e = \frac{V_T}{9.5\,\mu A} = 2.63\,k\Omega$

$R_{id} = \underline{\underline{3.17\ M\Omega}}$

$i_e = \frac{v_{id}}{6 r_e}$; $i_o = 2 i_e$

$\Rightarrow G_{m1} = \frac{i_o}{v_{id}} = \frac{2}{6 r_e} = \frac{1}{3 r_e}$

$= \underline{\underline{127\ \mu A/V}}$

$R_{o4} = r_o(1 + g_m(R_E \| r_\pi))$

$g_m \simeq 1/r_e$

$R_E = 2 r_e = 5.26\,k\Omega$

$r_\pi = (\beta_p + 1)\, r_e = 134\,k\Omega$

Thus $R_{o4} = \underline{\underline{15.4\ M\Omega}}$

$R_{o6} = 18.2\,M\Omega$ (from text)

$R_{o1} = R_{o4} \| R_{o6} = \underline{\underline{8.34\,M\Omega}}$

$G_{m1} R_{o1} = 127 \times 8.34 = 1059\ V/V$

See gain decreases due to negative feedback.

40 $R_o = r_{o6}(1 + g_{m6}(R_2 \| r_{\pi 6}))$

– need to double the second factor

Since $r_{\pi 6} \gg R_2$

$R_{oc} \simeq r_{o6}(1 + g_{m6} R_2)$

Thus

$1 + g_{m6} R_2' = 2(1 + g_{m6} R_2)$

$g_{m6} = \frac{1}{2.63\,k\Omega}$, $R_2 = 1K$

$\Rightarrow R_2' = \underline{\underline{4.63\ k\Omega}}$

41 $I_{C5} = I_{C6} = I_{C7}$

$\Rightarrow r_{e5} = r_{e6} = r_{e7} = 2.63\,k\Omega$

(a) $v_{b6} = (r_{e6} + R_2)\,i_e = \underline{\underline{4.63\,k\Omega \times i_e}}$

(b) $R_B = 50k \,\|\, r_{\pi 5} \,\|\, r_{\pi 6}$

$= 45.1\,k\Omega$

$\Rightarrow i_{e7} = \dfrac{v_{b6}}{R_B} = \underline{\underline{0.103\, i_e}}$

(c) $i_{b7} = \dfrac{i_{e7}}{\beta+1} = \dfrac{0.103}{201} i_e = \underline{\underline{510\,\mu A \times i_e}}$

(d) $v_{b7} = v_{b6} + r_{e7}\, i_{e7}$

$= (4.63\,k\Omega + 2.63\,k\Omega \times 0.103)\, i_e$

$= \underline{\underline{4.9\,k\Omega \times i_e}}$

(e) $R_{in} = \dfrac{v_{b7}}{i_e} = \underline{\underline{4.9\,k\Omega}}$

---

42 From Eqn 9.85

$$\frac{\Delta I}{I} = \frac{\Delta R}{R + \Delta R + r_e}$$

$$\Delta I \simeq G_{m1} V_{os} = \frac{V_{os}}{2 r_e}$$

Thus

$$\frac{V_{os}}{2 r_e I} = \frac{\Delta R}{R + \Delta R + r_e} \quad ; \quad r_e I = V_T$$

$$\frac{V_{os}}{2V_T} = \frac{\Delta R}{R}\left[\frac{1}{1 + \frac{r_e}{R} + \frac{\Delta R}{R}}\right] \quad (*)$$

$$\frac{V_{os}}{2V_T}\left(1 + \frac{r_e}{R}\right) = \frac{\Delta R}{R}\left(1 - \frac{V_{os}}{2V_T}\right)$$

$$\underline{\underline{\frac{\Delta R}{R} = \frac{V_{os}}{2V_T}\, \frac{1 + r_e/R}{1 - \frac{V_{os}}{2V_T}}}}$$

(b) $V_{os} = 5\,mV$, $r_e = 2.63\,k\Omega$, $R = 1\,k\Omega$

$$\frac{\Delta R}{R} = \frac{5}{2(25)}\, \frac{1 + 2.63}{1 - \frac{5}{2(25)}} = \underline{\underline{0.40}}$$

(c) R_2 completely shorted

$\Rightarrow \dfrac{\Delta R}{R} = -1$

From $(*)$ $\quad \dfrac{V_{os}}{2V_T} = -1 \dfrac{1}{r_e/R}$

$\Rightarrow V_{os} = \underline{\underline{-19\,mV}}$ $\quad$ (or $19\,mV$)

---

43 Current in the collector of Q_3 remains unchanged at $9.5\,\mu A$.

Thus $\quad I_{E3} = I_{E4} = \frac{51}{50}\, 9.5\,\mu A = 9.69\,\mu A$

$I_{C4} = \frac{25}{26}\, I_{E4} = 9.317\,\mu A$

$\Rightarrow \Delta I = 9.5 - 9.317 = 0.183\,\mu A$

$V_{os} = \dfrac{\Delta I}{G_{m1}} = 2 r_e \Delta I = 2(2.63\,k\Omega)(0.183\,\mu A)$

$= \underline{\underline{0.96\,mV}}$

---

44 $\qquad$ 9

---

45 Working with Common-mode half circuits

$$i_{c3} = \frac{\beta_P}{\beta_P + 1}\; \frac{v_{icm}}{r_{e1} + r_{e2} + \frac{2R_o}{\beta_P + 1}}$$

$$= \frac{\beta_P\, v_{icm}}{(\beta_P + 1)(r_{e1} + r_{e2}) + 2R_o}$$

Similarly

$$i_{c4} = \frac{K\beta_P\, v_{icm}}{(K\beta_P + 1)(r_{e1} + r_{e2}) + 2R_o}$$

---

46 $\qquad$ 9

Recall that for a resistor degenerated mirror the current gain is given, approximately, by

$$\frac{i_o}{i_I} \approx \frac{R_{E1}}{R_{E2}} \qquad \text{where } R_E \text{ is the total emitter resistance}$$

(a) R_1 shorted

$$\frac{i_o}{i_I} = \frac{r_{e1}}{r_{e2} + R_2} = \frac{2.63}{2.63 + 1} = 0.72$$

Thus $i_o = (1.72)\, i_e$ $\quad$ (cf. $2 i_e$)

Since the output impedance is unaffected, the gain is thus reduced by $\underline{\underline{14\%}}$ $\left(\frac{1.72 - 2}{2}\right)$

(b) R_2 shorted

$$\frac{i_o}{i_I} = \frac{r_{e1} + R_1}{r_{e2}} = \frac{3.63}{2.63} = 1.38$$

$\Rightarrow i_o = (1 + 1.38)\, i_e = 2.38\, i_e$

Note that the output of the mirror decreases because of the lack of degeneration from R_2. Neglecting this, since R_{o1} is largely determined by R_{i2},

$\Rightarrow$ gain increases by $\frac{2.38 - 2}{2} = \underline{19\%}$

(c) Current gain remains at unity. Thus $i_o = 2 i_e$ and gain is unaffected.

---

47 $\qquad$ 9

Since r_μ is typically very large we will ignore its effect

$r_{o1} = \dfrac{125}{9.5\,\mu A} = 13.16\,M\Omega$

$r_{o3} = \dfrac{50}{9.5\,\mu A} = 5.26\,M\Omega$

$$R_i = (\beta_n + 1)\left[r_{e1} + \left(r_{e2} + \frac{2R_o}{\beta_P + 1}\right) \| \, r_{o1} \,\|\, r_{o3}\right]$$

$$= 201\left[2.63K + \left(2.63K + \frac{4.8M}{51}\right) \| \, 13.16M \,\|\, 5.26M\right]$$

$= \underline{\underline{19.5\,M\Omega}}$

Since $R_i = 2 R_{icm}$

$R_{icmf} = \dfrac{R_i}{2}(1 + \beta_P)$ $\qquad$ See problem #8

$= \underline{\underline{497\,M\Omega}}$

48

$R_{i2} = (\beta+1)\left[r_{e16} + R_{i17} \| R_9\right]$

$r_{e16} = 1.54\,k\Omega$

$r_{e17} = 45.5\,\Omega$

$R_{i17} = 201\,(45.5 + 50) = 19.2\,k\Omega$

$\Rightarrow R_{i2} = 201\,(1.54 + 19.2 \| 50)\,k\Omega = \underline{\underline{3.1\,M\Omega}}$

$v_{b17} = \dfrac{R_{i17} \| R_9}{r_{e16} + R_{i17} \| R_9}\, v_{i2}$

$= 0.9\, v_{i2}$

$i_{c17} = \dfrac{\alpha}{r_{e17} + R_8}\, 0.9\, v_{i2}$

$\Rightarrow G_{m2} = \dfrac{\alpha\,(0.9)}{45.5 + 50} = \underline{\underline{9.38\,mA/V}}$

49

$R_{o17} = 787\,k\Omega$

$I_{13B} = 550\,\mu A$

$g_{m13B} = 22\,mA/V \;;\; r_{\pi 13B} = (\beta+1)/g_m = 2.32\,k\Omega$

$r_o = \dfrac{50}{550\,\mu A} = 90.9\,k\Omega$

$R_{o13B} = r_o\left(1 + g_m (R_E \| r_\pi)\right)$

$= 90.9\left[1 + 22(R_E \| 2.32)\right]$

$= 787$

$\Rightarrow R_E \| 2.32 = 0.348$

and $\dfrac{1}{R_E} = 2.44$ or $R_E = 0.410\,k\Omega$

$\underline{R_E = 410\,\Omega}$

Current $\dfrac{R_{E12}}{R_E} = \dfrac{550\,\mu A}{730\,\mu A} \Rightarrow \underline{\underline{R_{E12} = 309\,\Omega}}$

$\dfrac{R_{E13A}}{R_E} = \dfrac{550}{180} = \underline{\underline{1.25\,k\Omega}}$

50

9

$\hat{v}_o = V_{CC} - V_{CEsat\,13A} - V_{BE14}$

$= \underline{\underline{4.2\,V}}$

$\check{v}_o \simeq -V_{EE} + V_{CEsat\,17} + V_{BE23} + V_{BE20}$

$= -5 + 0.2 + 0.6 + 0.6$

$= \underline{\underline{-3.6\,V}}$

51

With Q_{23} removed, current in Q_{17} increases to $730\,\mu A$. This changes G_{m2}

$r_{e17} = \dfrac{V_T}{730\,\mu A} = 34.2\,\Omega$

$\Rightarrow G_{m2} \simeq 0.923\, \dfrac{\alpha}{100 + 34.2} = 6.8\,mA/V$

Because $r_{o17} \gg r_{o13B}$, R_{o2} remains virtually unchanged at $81\,k\Omega$

$R_{i3} = (\beta_P + 1) R_L \| r_{o13A} = 74\,k\Omega$

$\Rightarrow A_2 = -6.8\,(81)\, \dfrac{74}{74 + 81} = \underline{\underline{-263\,V/V}}$

52

Ignore base currents of Q_{15}

$180\,\mu A = I_{C15} + \dfrac{I}{\beta+1}$, where $I = I_{R6}$

$I_{C15} = I_S\, e^{V_{BE}/V_T}$ where $V_{BE} = I R_6$

Thus $I = \dfrac{V_T}{27} \ln\left[\dfrac{180\,\mu A - \frac{I}{201}}{I_S}\right]$

9

$= 191{,}422\,V/V$

$\equiv \underline{\underline{105.6\,dB}}$

output current is limited to $\pm 20\,mA$ (see problems 35 and 36)

$\Rightarrow |V_o| < 20\,mA\,(200)$

$\underline{\underline{|V_o| < 4\,V}}$

To obtain a seed solution, let $I = 0$ on RHS

$\Rightarrow I = \dfrac{V_T}{27} \ln \dfrac{180\,\mu A}{10^{-14}} = 21.9\,mA$

iterating $\quad I = \underline{\underline{21.0\,mA}}$

53

Maximum output current of the 1st stage $= 19\,\mu A$

$\Rightarrow I_{C22} = 19\,\mu A \Rightarrow V_{BE22} = V_{BE24} = 534\,mV$

$\Rightarrow I_{R11} = \dfrac{534}{50} = 10.7\,\mu A$

$\therefore I_{C21} = (19 + 10.7)\,\mu A = 29.7\,\mu A$

and $V_{BE21} = 545.3\,mV$

$V_{BE21} = I R_7 \Rightarrow I = \underline{\underline{20.2\,mA}}$

A simple doubling of R_7

54 $\frac{v_o}{v_i} = \frac{243,147}{0.97} = 250,667\ V/V \cong \underline{108\ dB}$

$\frac{R_L}{R_o + R_L} = 0.9 \implies R_o = R_L\left(\frac{1}{0.97} - 1\right)$

or $\underline{R_o = 61.9\Omega}$

$\left.\frac{v_o}{v_i}\right|_{R_L=200} = 250,667 \cdot \frac{200}{200 + 61.9}$

55 80° PM says that 2nd pole introduces 10° of phase shift at 1MHz

i.e. $\tan^{-1}\frac{f_t}{f_{p2}} = 10°$

or $\underline{f_{p2} = 5.67\ MHz}$

56 Each pole adds 5° of phase shift

$\tan^{-1}\frac{10^6}{f_{2,3}} = 5°$

$\Rightarrow \underline{f_{2,3} = 11.4\ MHz}$

57 Consider Bode plot

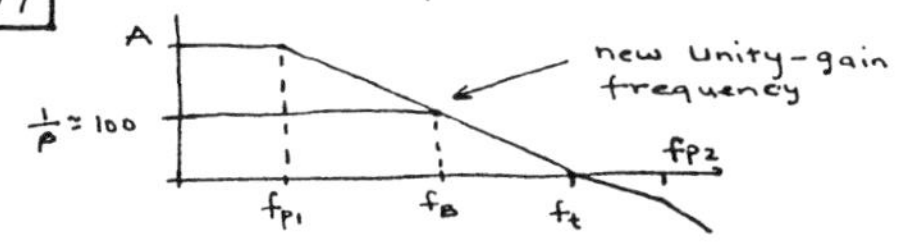

85° of closed-loop phase margin

$\Rightarrow \tan^{-1}\frac{f_B}{f_{p2}} = 5°$

or $\underline{f_B = 437\ kHz}$

Recalling the "broadbanding" effect of negative feedback, we get

$f_B = f_{p1}(1 + A\beta) \simeq f_{p1}A\beta$

Loop gain $A\beta = 2.43 \times 10^5 \cdot \frac{1}{100} = 2.43 \times 10^3$

$\Rightarrow \underline{f_{p1} = 180\ Hz}$

$f_B = \frac{G_{m1}}{2\pi C_c} - 437\ kHz$

$\Rightarrow C_c = \frac{1}{5.26 \times 10^{-8}\,(2\pi)\,437 \times 10^3}$

$= \underline{0.69\ pF}$

9.58 dominant pole $f_p = \frac{1}{2\pi R(AC_c)}$; $A = 1000$

with single pole response

$A_o f_p = f_t \implies f_p = \frac{5 \times 10^6}{10^6} = 5\ Hz$

$\Rightarrow R = \frac{1}{2\pi(5)\,1000\,(50pF)} = \underline{637\ k\Omega}$

9.59

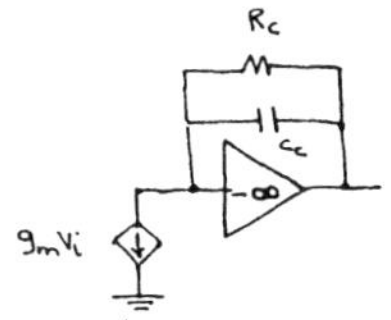

DC, gain is

$A_o = G_{m1}R_c$
$= 10 \times 10^{-3} \times 10^8$
$= \underline{10^6\ V/V}$

$f_p = \frac{1}{2\pi R_c C_c} = \frac{1}{2\pi \times 10^8\ 50 \times 10^{-12}}$

$= 31.8\ Hz$

$f_t = A_o f_p$

$= 31.8\ MHz$

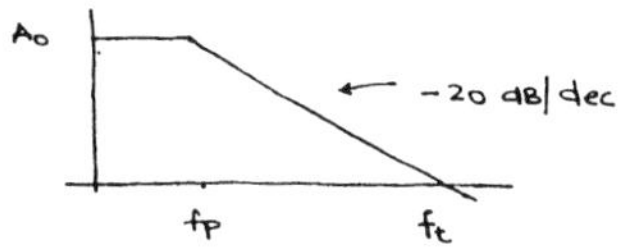

$SR = \frac{2I}{C_c}$

$G_{m1} = \frac{I}{2V_T} \implies 2I = 4G_{m1}V_T$

$SR = \frac{4G_{m1}V_T}{C_c} = \frac{4\,(10 \times 10^{-3})(25 \times 10^{-3})}{50 \times 10^{-12}}$

$= \underline{20\ V/\mu s}$

9.60 $SR = 10\ V/\mu s$

$|V_{omax}| = 10\ V$

For sine wave of amplitude, V_{omax}, maximum rate of change

$\left.\frac{dv_o}{dt}\right|_{max} = \omega_m \hat{V}_o = 2\pi f_m \hat{V}_o$

Thus

$2\pi f_m \hat{V}_o = SR$

$\Rightarrow f_m = \frac{10 \times 10^6}{2\pi\,(10)} = \underline{159.2\ kHz}$

$f_t = \frac{SR}{2\pi(4V_T)} = \frac{10 \times 10^6}{2\pi\,(4)\,25 \times 10^{-3}}$

$= \underline{15.9\ MHz}$

9.61 $I_{E1} = I_{E2} = 50\mu A \simeq I_{E3} = I_{E4}$

$I_{E5} = 1mA$; $V_{BE5} = V_{BE6}$

$\therefore$ $I_{E6} = 1mA = I_{E7}$

$r_{e1} = r_{e2} = 500\Omega$

$r_{e5} = r_{e6} = r_{e7} = 25\Omega$

$G_{m1} = 2\left(\frac{1}{2r_{e1}}\right) = 2mA/V$

$R_{o1} = (\beta+1)(r_{e5} \| r_{e6})$

$= 1.25 k\Omega$

and $A_1 = G_{m1}R_{o1} = 2.5 V/V$

9.62 $\frac{1}{2}$ LSB must be less than 1%

i.e $\frac{1}{2}\frac{1}{2^N} \le \frac{1}{100} \Rightarrow N \ge 5.6$

$\therefore$ N = 6 bits

Resolution $\frac{10V}{2^6} = 0.156V$

For same resolution need 7 bits

Still 7 bits

resolution $= \frac{15}{2^7} = 0.117V$

$Q = \frac{1}{2}LSB = \frac{1}{2}\frac{15}{2^7} = 0.059V$

9.63 9

9.64 Require error in MSB $\le \frac{1}{2}$ LSB

$$\frac{V}{R} - \frac{V}{R(1+\frac{x}{100})} \le \frac{1}{2}\frac{V}{2^{N-1}R}$$

$$\frac{1+\frac{x}{100}-1}{1+\frac{x}{100}} \le \frac{1}{2^N} \quad \text{or } \frac{x}{100}(2^N-1) \le 1$$

$$\Rightarrow x = \frac{1}{2^N-1} \times 100$$

N = 2 $x = 33.3\%$

N = 4 $x = 6.67\%$

N = 8 $x = 0.39\%$

9.65 Since V_{BE}'s are equal, collector currents are scaled with respect to emitter areas

$I_1 + I_1 + I_2 + I_3 + I_4 = I$

$I_1(1+1+2+4+8) = I \Rightarrow I_1 = I/16$

$I_2 = I/8$

$I_3 = I/4$ $\quad I_4 = \frac{I}{2}$

9.66 Circuit is sketched below 9

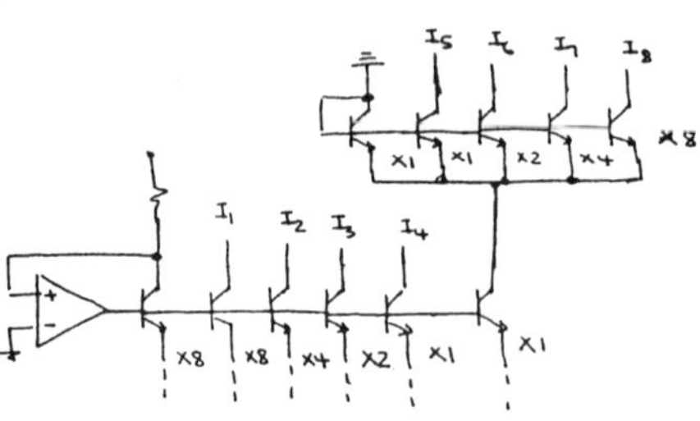

$A_{tot} = 8+8+4+2+1+1+1+1+2+4+8$

$= 40$

For 8-bits binary weighted

$A_{tot} = (1 + 1+2+4+\dots+2^{N-1} + 2^{N-1})$

Q_t $\quad Q_{ref}$

$= 2^{N-1}+1 \quad \{1+2+4+\dots+2^{N-1}\}$

$= 2^{N-1}+1 + 2^N - 1 = 2^{N-1}+2^N$

N = 8

$A_{tot} = 2^7 + 2^8 = 384$

9.67 9

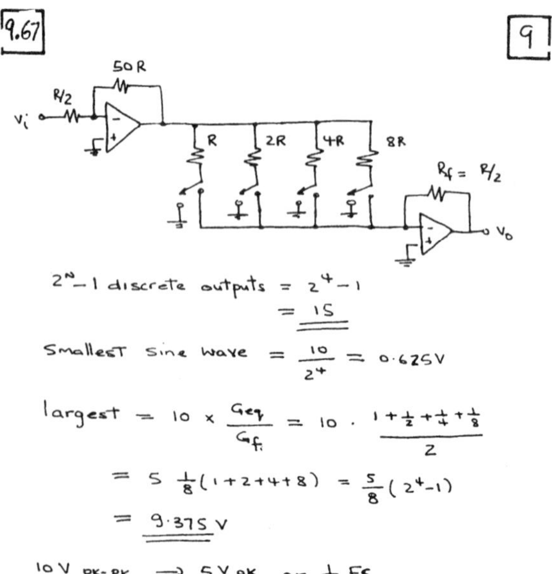

2^N-1 discrete outputs $= 2^4-1$

$= 15$

Smallest sine wave $= \frac{10}{2^4} = 0.625V$

largest $= 10 \times \frac{G_{eq}}{G_f} = 10 \cdot \frac{1+\frac{1}{2}+\frac{1}{4}+\frac{1}{8}}{2}$

$= 5 \cdot \frac{1}{8}(1+2+4+8) = \frac{5}{8}(2^4-1)$

$= 9.375V$

10V pk-pk $\Rightarrow$ 5V pk or $\frac{1}{2}$ FS

$\therefore$ D = 1000

9.68

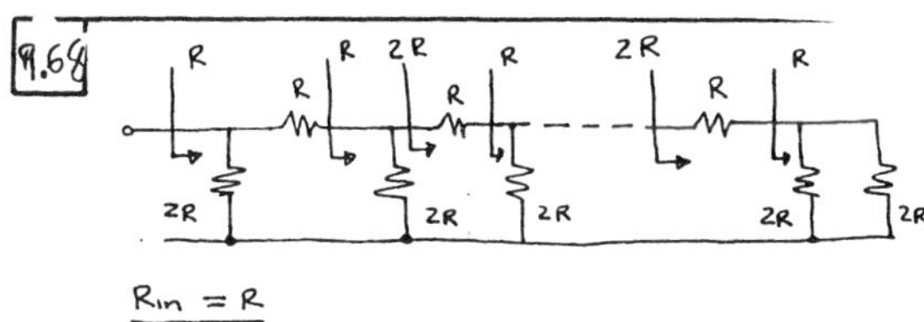

$R_{in} = R$

9.69 $T_c = \frac{1}{f_{clk}} = 1\mu s$

$T_1 = 2^{12} T_c = 4.096\,ms$

$T = T_1 + T_2 = T_1(1 + \frac{V_A}{V_{ref}})$

$= 2T_1 = 8.19\,ms$

$V_{peak} = 10 = \frac{V_A}{\tau} T_1$

$\Rightarrow \tau = \frac{V_A}{V_{peak}} T_1 = 4.096\,ms$

$\Delta\tau = -1\%$ and causes a -1% change in V_{peak}

$\Rightarrow V_{peak} = 9.9V$

No. Final count does not depend on τ

9.70 $2^N - 1$ Comparators $= 15$

comparators are biased 1LSB apart starting from $\frac{1}{2}$ LSB.

Thus, $V_{ref\,n} = (\frac{2n-1}{2})$ LSB

$1\,LSB = \frac{10}{2^4} = 0.625V$

and references are 0.3125, 0.9375, 1.5625, 2.1875, 2.8125 ..... 9.0625

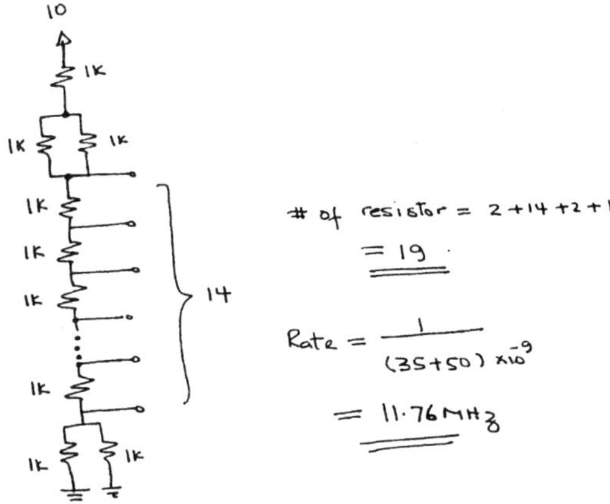

# of resistor $= 2+14+2+1$

$= 19$.

$Rate = \frac{1}{(35+50)\times10^{-9}}$

$= 11.76\,MHz$

(a) (0000..... 000) (15 in all) and $(0000)_2$

(b) all comparators with references less than 5.1V will produces ones
Let's find 'em

$(\frac{2n-1}{2})\,0.625 \geq 5.1 \Rightarrow n \geq 8.6$

Thus the lowest 8 outputs will be ones.
(000000011111111) and $(0100)_2$

(c) Full scale input (1111.... 111) and $(1111)_2$

10

Chapter 10 - Problems

1 Ideal 3V logic implies:

$V_{OH} = V_{DD} = \underline{3.0V}$; $V_{OL} = \underline{0.0V}$;

$V_{th} = V_{DD}/2 = 3.0/2 = \underline{1.5V}$;

$V_{IL} = V_{DD}/2 = \underline{1.5V}$; $V_{IH} = V_{DD}/2 = \underline{1.5V}$

$NM_H = V_{OH} - V_{IH} = 3.0 - 1.5 = \underline{1.5V}$

$NM_L = V_{IL} - V_{OL} = 1.5 - 0.0 = \underline{1.5V}$

The gain in the transition region is:

$(V_{OH} - V_{OL})/(V_{IH} - V_{IL}) =$

$(3.0 - 0.0)/(1.5 - 1.5) = 3/0 = \underline{\infty V/V}$

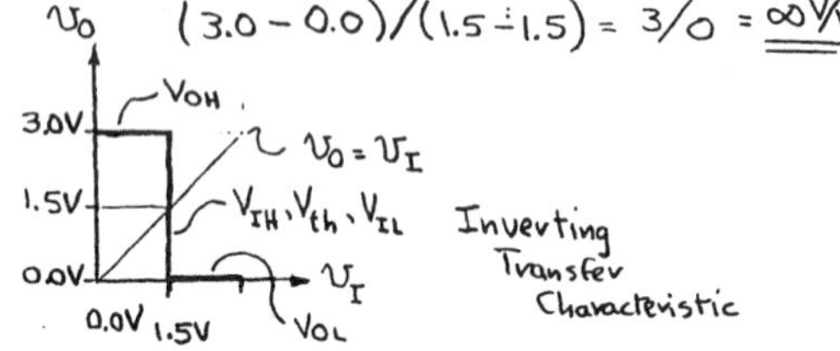

2 10

Nearly ideal 3.3V logic, assumed ideal:

→ $V_{OH} = \underline{3.3V}$, $V_{OL} = \underline{0.0V}$, $V_{th} = 0.4(3.3) = \underline{1.32V}$

Now, at V_{th}, $v_O = v_I$, so to reach $v_O = 1.32V$

the required input is $1.32/(-50) = -26.4mV$

Thus, $V_{IL} = 1.32 - 26.4 \times 10^{-3} = \underline{1.294V}$

Likewise, $V_{IH} = 1.32 + (3.3 - 1.32)/50 = \underline{1.360V}$

Best possible noise margins are:

$NM_H = V_{OH} - V_{IH} = 3.30 - 1.360 = \underline{1.940V}$

$NM_L = V_{IL} - V_{OL} = 1.294 - 0.0 = \underline{1.294V}$

For noise margins only 7/10 of these, and V_{OH}, V_{OL} still ideal:

$V_{IH} = 3.3 - 0.7(1.940) = \underline{1.942V}$, and

$V_{IL} = 0.0 + 0.7(1.294) = \underline{0.906V}$

Correspondingly, the large-signal voltage gain is:

$G = (3.3 - 0.0)/(0.906 - 1.942) = \underline{-3.18V/V}$

3 10

Here, $V_{OH} = 1.2V$, and $V_{OL} = 0.0V$.

Also, $V_{IH} - V_{IL} \leq 1.2/3 = 0.4V$ ---(1)

Now, the noise margins are "within 30% of one other". Thus $NM_H = (1 \pm 0.3)NM_L$ or $NM_L = (1 \pm 0.3)NM_H$. Thus to remain "within" either $NM_H = 1.3NM_L$, or $NM_L = 1.3NM_H$, in which case, either $NM_L = 0.769\,NM_H$, or $NM_H = 0.769\,NM_L$

For the former case: $0.769(V_{OH} - V_{IH}) = (V_{IL} - V_{OL})$

or $0.769(1.2 - V_{IH}) = V_{IL} - 0$, whence

$V_{IL} = 0.923 - 0.769\,V_{IH}$

Now, from (1), $V_{IH} = V_{IL} + 0.4$.

Thus, $V_{IL} = 0.923 - 0.769(V_{IL} + 0.4) = 0.615 - 0.769V_{IL}$

and $V_{IL} = 0.615/1.769 = \underline{0.349V}$,

whence $V_{IH} = 0.4 + 0.349 = \underline{0.749V}$. 10

Alternatively, $NM_H = 0.769\,NM_L$, and

$(V_{OH} - V_{IH}) = 0.769(V_{IL} - V_{OL})$, or

$1.2 - V_{IH} = 0.769\,V_{IL} - 0$, and $V_{IH} = 1.2 - 0.769V_{IL}$

With (1), $V_{IL} + 0.4 = 1.2 - 0.769V_{IL}$, and

$1.769V_{IL} = 0.8$, whence $V_{IL} = \underline{0.452V}$

and $V_{IH} = 0.4 + 0.452 = \underline{0.852V}$

Thus, overall, $V_{OH} = \underline{1.2V}$, $V_{OL} = \underline{0.0V}$,

V_{IH} ranges from $\underline{0.749V}$ to $\underline{0.852V}$, and

V_{IL} ranges from $\underline{0.349V}$ to $\underline{0.452V}$, in which case, margins can be as low as:

$NM_L = V_{IL} - V_{OL} = \underline{0.349V}$, and

$NM_H = V_{OH} - V_{IH} = 1.2 - 0.852 = \underline{0.348V}$,

and as high as $\underline{0.452V}$, and $\underline{0.451V}$.

4 **10**

a) Generally, $t_p = (t_{PHL} + t_{PLH})/2$,

but due to current ratio, $t_{PHL} = 0.5\, t_{PLH}$.

Thus $1.5\, t_{PLH} = 2(1.2\text{ ns})$, whence

$t_{PLH} = 2.4/1.5 = \underline{1.6\text{ ns}}$, and $t_{PHL} = \underline{0.8\text{ ns}}$

Check: $t_p = (1.6 + 0.8)/2 = 1.2\text{ ns}$ ✓

b) Generally, $t_p = CV/I = kC$

Originally, $1.2 = kC$ --- (1)

Then, $1.7(1.2) = k(C+1)$ --- (2)

Dividing (2/1): $1.7 = (C+1)/C$

Thus, $1.7C = C+1$, $0.7C = 1$, $C = \underline{1.43\text{ pF}}$

(the combined load and output capacitances)

c) With the load inverter removed:

$0.6(1.2) = k(1.43 - C_{in})$ --- (3)

Dividing (3/1): $0.6 = (1.43 - C_{in})/1.43$

Thus, $C_{in} = 1.43(1 - 0.6) = \underline{0.57\text{ pF}}$; $C_{out} = 1.43 - 0.57 = \underline{0.86\text{ pF}}$

5 **10**

Average static current (at 50% duty cycle) is $(0+40)/2 = 20\,\mu A$

When switching at 100 MHz, current is $150\,\mu A$

Dynamic current is $(150 - 20) = 130\,\mu A$, and

dynamic power is $P_D = (3.3V)(130\,\mu A) = 429\,\mu W$.

But, $P_D = fCV_{DD}^2$, and $C = 429 \times 10^{-6}/(100 \times 10^6 \times 3.3^2)$

Thus $C_{eq} = C = \underline{0.394\text{ pF}}$

6 **10**

Now, $P_D = fCV_{DD}^2$. Thus for V_{DD} reduced from 5V to 3.3V, power reduces by a factor $(3.3/5)^2 =$ $\underline{0.436}$. For frequency reduced by a factor $(3.3/5)$, additional power saved is $((5-3.3)/5) \times 10^{-3} \times 0.436 =$ $\underline{1.48\text{ mW}}$.

Check: First reduction is to $10(0.436) = 4.36\text{ mW}$, and the second to $4.36(3.3/5) = \underline{2.88\text{ mW}}$, by $4.36 - 2.88 = 1.48\text{ mW}$ ✓

7 **10**

a) For current proportional to V_{DD}, reduction in current (and frequency) is to $3.3/5 = \underline{0.66}$ or 66% of previous (or by 34%)

b) For current proportional to V_{DD}^2, reduction is to $(3.3/5)^2 = \underline{0.436}$ or 44% of previous (or by 56%).

For case a), delay increases by $1/0.66 = 1.515$ times, and power reduces (from Eq 13.4) to $0.66 \times (3.3/5)^2 = 0.287$ of previous, for a net change in DP to $1.515 \times 0.287 =$ $\underline{0.435}$ of previous, a net decrease by 56.5%

For case b), delay increases by $1/0.436 = \underline{2.29}$ times, and power reduces to $0.436(3.3/5)^2 = 0.190$ of previous, for a net change in DP to $2.29 \times 0.190 = \underline{0.435}$ of previous (by 56.5%)

8 **10**

a) For current proportional to $(V_{DD} - V_t)$

i) For $V_t = 1V$: Change in current to $(3.3-1)/(5.0-1) = 0.575$, or $\underline{57.5\%}$ of previous, a decrease of 42.5%

Change in delay to $1/0.575 = 1.74$ or an increase of 74%.

Change in frequency to 0.575 or 57.5% of previous.

Change in dynamic power due to change in both voltage and frequency is to $0.575 \times (3.3/5)^2 = 0.250$ or $\underline{25\%}$ of previous.

Change in DP is to $1.74 \times 0.25 = 0.435$, or $\underline{43.5\%}$ of previous, a decrease of 56.5%

ii) For $V_t = 0.5V$:

Current: to $(3.3-0.5)/(5.0-0.5) = 0.622$, to $\underline{62.2\%}$

(next)

18 10

Delay : to $1/0.622 = 1.61$, to 161%

Frequency : to 62.2%

Dynamic Power : to $0.622(3.3/5.0)^2 = 0.271$, to 27.1%

Delay-Power (DP) : to $1.61 \times 0.271 = 0.436$, to 43.6%

(iii) For $V_t = 0V$ (as in P10.7)

Current: to $3.3/5.0 = 0.66$, to 66%

Delay: to $1/0.66 = 1.515$, to 151.5%

Frequency : to $1/1.515 = 0.66$, to 66%

Dynamic Power : to $0.66(3.3/5.0)^2 = 0.287$, to 28.7%

Delay-Power (DP): to $1.515 \times 0.287 = 0.435$, to 43.5%

b) For current proportional to $(V_{DD} - V_t)^2$

i) For $V_t = 1V$:

Current : to $[(3.3-1)/(5.0-1)]^2 = 0.331$, to 33.1%

Delay : to $1/0.331 = 3.02$, to 302%

(next)

18 10

Frequency : to $1/3.02 = 0.331$, to 33.1%

Dynamic Power: to $0.331(3.3/5.0)^2 = 0.144$, to 14.4%

Delay-Power (DP): to $3.02 \times 0.144 = 0.435$, to 43.5%

ii) For $V_t = 0.5V$:

Current: to $[(3.3-0.5)/(5.0-0.5)]^2 = 0.387$, to 38.7%

Delay : to $1/0.387 = 2.58$, to 258%

Frequency: to $1/2.58 = 0.387$, to 38.7%

Dynamic Power: to $0.387(3.3/5.0)^2 = 0.168$, to 16.9%

Delay-Power (DP) : to $2.58 \times 0.168 = 0.435$, to 43.5%

(iii) For $V_t = 0V$ (as in P10.7):

Current : to $[3.3/5.0]^2 = 0.436$, to 43.6%

Delay : to $1/0.436 = 2.30$, to 230%

Frequency: to $1/2.30 = 0.436$, to 43.6%

Dynamic Power: to $0.436(3.3/5.0)^2 = 0.190$, to 19.0%

Delay-Power (DP): to $2.30 \times 0.190 = 0.437$, to 43.7%

---

9 10

Chip area changes to $0.9 \times 0.9 = 0.81$ of previous

Current changes only due to oxide change, since W/L change cancels, to $1/0.9 = 1.11$ of previous

Effective Capacitance : half changes by 0.9^2 to 0.81
half changes by $0.9^2/0.9$ to 0.90

Net Capacitance changes by $(0.81 + 0.90)/2 = 0.855$

Propagation Delay changes by $(1/1.111)0.855 = 0.770$

Max. Operating Frequency changes by $1/0.77 = 1.30$

Dyn. Power Diss. (at max. freq.) changes by $1.30 \times 0.855 = 1.11$

Delay-Power Product changes by $0.770 \times 1.11 = 0.86$

Performance (f/A) changes by $1.30/0.81 = 1.60$

Now, if the supply voltage is reduced by 10%, (but V_t is not, remaining at $0.2V_{DD}$),

$(V_{DD} - V_t)$ changes to $(0.9V_{DD} - 0.2V_{DD}) = 0.7V_{DD}$
or to $(0.7V_{DD}/0.8V_{DD}) = 0.875$ of previous.

9 10

Thus, current changes by $(1/0.9)(0.875)^2 = 0.858$

Assume capacitances are voltage-independant.

Thus, propagation delay changes by $(1/0.851)0.855 = 1.005$

Maximum frequency changes by $1/1.005 = 0.995$

Dyn. Power changes by $(0.995)(0.855)(0.9)^2 = 0.689$

Delay-Power product changes by $1.005(0.689) = 0.693$

Performance (f/A) changes by $0.995/0.81 = 1.23$

10 | 10

The results depend on whether the gates are inverting or non-inverting.

For inverting gates, the timing diagram is:

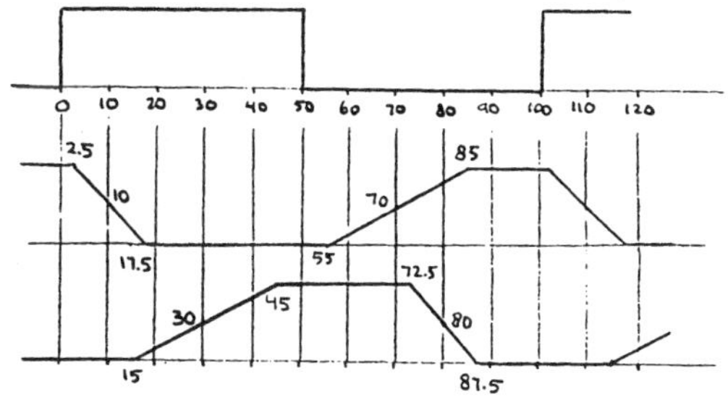

Note: For simplicity, 0% to 100% (rather than 10% to 90%) both in the diagram above and calculation to follow:

For inverting gates (as shown, above):

a) For a rising input, time to 90% change of output of second gate is $10 + 20 + 30/2 = \underline{45\,ns}$

b) For a falling input, time to 90% change

10 | 10

of output of 2nd gate is $20 + 10 + 15/2 = \underline{37.5\,ns}$

For non-inverting gates:

a) Time to 90% rise is $10 + 10 + 15/2 = \underline{27.5\,ns}$

b) Time to 90% fall is $20 + 20 + 30/2 = \underline{55\,ns}$

The propagation delay for these gates is

$t_P = (t_{PHL} + t_{PLH})/2 = (10 + 20)/2 = \underline{15\,ns}$

11 | 10

Note that this question ignores the possibility of dynamic power dissipation:

Average propagation delay is $t_P = (50+70)/2 = 60\,ns$

Average power loss at 50% duty cycle $= (1+0.5)/2 = 0.75\,mW$

Delay-Power product is $DP = 60\times10^{-9} \times 0.75\times10^{-3}$

or $DP = 45\times10^{-12}\,J = \underline{45\,pJ}$

10.12

10.13

14 | 10

For $v_I = 1.5\,V$, the NMOS operates in triode mode, while the PMOS is cutoff.

$r_{DSn} = [k_n(v_I - V_t)]^{-1} = [100\times10^{-6}(1.5-0.5)]^{-1} = \underline{10\,k\Omega}$

Thus $v_O = 100\times10^{-3}\times10^4/(10^4+10^5) = \underline{9.09\,mV}$

For $v_I = -1.5\,V$, the PMOS operates with

$r_{DSp} = [k_p(|v_I| - V_t)] = [(100\times10^{-6}/10)(1.5-0.5)]^{-1} = \underline{10^5\,\Omega}$

Thus $v_O = 100\times10^{-3}\times10^5/(10^5+10^5) = \underline{50\,mV}$

15 | 10

At V_{th}, both transistors operate in saturation with $V_{th} = v_I = v_O = v$,

at which $i_D = \frac{k_n}{2}(v - V_{tn})^2 = \frac{k_p}{2}(V_{DD} - v - |V_{tp}|)^2$

Thus, $V_{DD} - v - |V_{tp}| = \sqrt{k_n/k_p}\,(v - V_{tn})$.

$v(1 + (k_n/k_p)^{1/2}) = V_{DD} - |V_{tp}| + V_{tn}(k_n/k_p)^{1/2}$

Thus $v = V_{th} = \dfrac{V_{DD} - |V_{tp}| + V_{tn}(k_n/k_p)^{1/2}}{1 + (k_n/k_p)^{1/2}}$

as presented in Equation 10.8

10.16

10.17

10.18

19 | 10

NMOS width is $8\,\mu m$; PMOS width is $8(2.5) = 20\,\mu m$

Total output capacitance $= (20+8)2 + 50 = \underline{106\,fF}$.

Now, $t_P = 1.6C/[k'(W/L)V_{DD}]$

Thus $t_P = 1.6(106)10^{-15}/[75\times10^{-6}(8/0.8)3.3] = \underline{68.5\,ps}$

20 | 10

From Eq. 10.14, 10.15, 10.16:

$$i_{DnlAv} = \tfrac{1}{2}\left[\tfrac{1}{2} k_n'(W/L)_n (V_{DD}(1-\alpha))^2 + k_n'(W/L)_n (V_{DD}(1-\alpha)V_{DD}/2 - \tfrac{1}{2}(V_{DD}/2)^2\right]$$
$$= \tfrac{1}{2} k_n'(W/L)_n \left[V_{DD}^2(1-\alpha)^2/2 + V_{DD}^2(1-\alpha)/2 - V_{DD}^2/8\right]$$
$$= \tfrac{1}{4} k_n'(W/L)_n V_{DD}^2 \left[(1-\alpha)^2 + (1-\alpha) - \tfrac{1}{4}\right]$$

Now, see the term [ ] becomes:

$$[1 - 2\alpha + \alpha^2 + 1 - \alpha - \tfrac{1}{4}] = \alpha^2 - 3\alpha + 1.75$$

Thus, $t_{PHL} = (CV_{DD}/2)/i_{DnlAv} = \dfrac{4C/2}{k_n'(W/L)_n V_{DD}(\alpha^2 - 3\alpha + 1.75)}$

or $t_{PHL} = \dfrac{2C}{k_n'(W/L)_n V_{DD}(\alpha^2 - 3\alpha + 1.75)}$

Check for $\alpha = 0.2$: $(\alpha^2 - 3\alpha + 1.75) = 0.2^2 - 3(0.2) + 1.75 = 1.19$

and the multiplier in the numerator is $\frac{2}{1.19} = 1.68 \simeq 1.7$ ✓

20 | 10

For $\alpha = 0.1$, $\alpha^2 - 3\alpha + 1.75 = 0.1^2 - 3(0.1) + 1.75 = 1.46$,

and the multiplier becomes $2/1.46 = \underline{1.37}$

For $\alpha = 0.5$, $\alpha^2 - 3\alpha + 1.75 = 0.5^2 - 3(0.5) + 1.75 = 0.50$

and the multiplier becomes $2/0.5 = \underline{4.00}$

10.21

22 | 10

Dynamic Power is $P_D = fCV_{DD}^2$; Static Power is P_S.

Now, $9.0 = P_S + 120 \times 10^6 C\, 5^2$

and $4.7 = P_S + 50 \times 10^6 C\, 5^2$

Subtracting, $4.3 = 70 \times 10^6 C\,(25)$

Whence $C = 4.3/(25 \times 70 \times 10^6) = \underline{2457\text{ pF}}$

and $P_S = 9.0 - 120 \times 10^6 (25)\, 2457 \times 10^{-12} = 9.0 - 7.37 = \underline{1.63W}$

For 70% of the gates active, total gates $= 0.7 \times 10^6$

Capacitance per gate is $2457 \times 10^{-12}/(0.7 \times 10^6) = \underline{3.5fF}$

10.23

10.24

25 | 10

For $Y = \overline{A + B(C+D)}$, the PDN can be drawn directly, and then the PUN as direct dual:

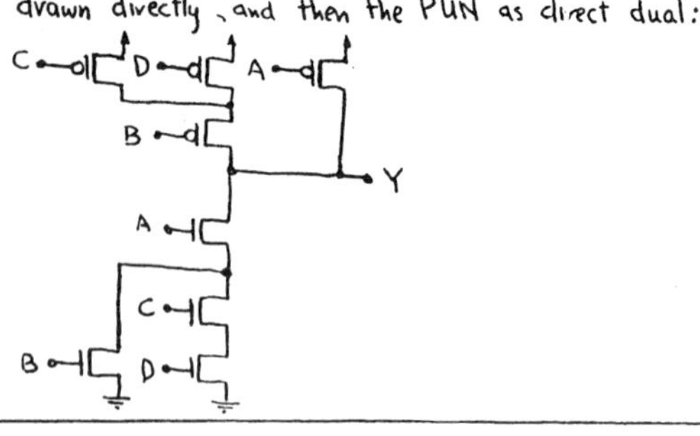

26 | 10

$Y = \bar{A}BC + A\bar{B}C + AB\bar{C}$

A very direct implimentation would need:

$2(3 \times 3) = 18$ MOS for the gate itself, plus $3 \times 2 = 6$ for the required inverters, for a total of $\underline{24}$ transistors.

For the PUN directly: (Inverting variables)

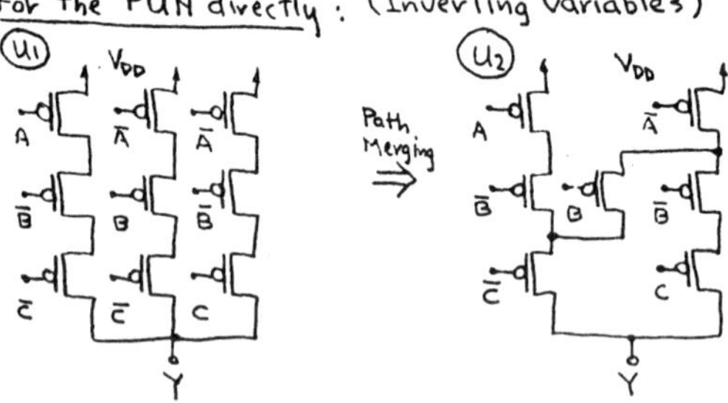

For the PDN, $Y = \bar{A}BC + A\bar{B}C + AB\bar{C}$.

Correspondingly: $\bar{Y} = \overline{\bar{A}BC + A\bar{B}C + AB\bar{C}}$
$= \overline{\bar{A}BC} \cdot \overline{A\bar{B}C} \cdot \overline{AB\bar{C}} = (A + \bar{B} + \bar{C})(\bar{A} + B + \bar{C})(\bar{A} + \bar{B} + C)$ ①
$= ABC + A\bar{B}\bar{C} + \bar{A}B\bar{C} + \bar{A}\bar{B}\bar{C} + \bar{A}\bar{B}C$, and replicating $\bar{A}\bar{B}\bar{C}$,

$\bar{Y} = ABC + \bar{A}\bar{B} + \bar{A}\bar{C} + \bar{B}\bar{C}$ --- ②
(next)

26 | 10

For the PDN directly:

From ①, $\overline{Y} = (A+\overline{B}+\overline{C})(\overline{A}+B+\overline{C})(\overline{A}+\overline{B}+C)$

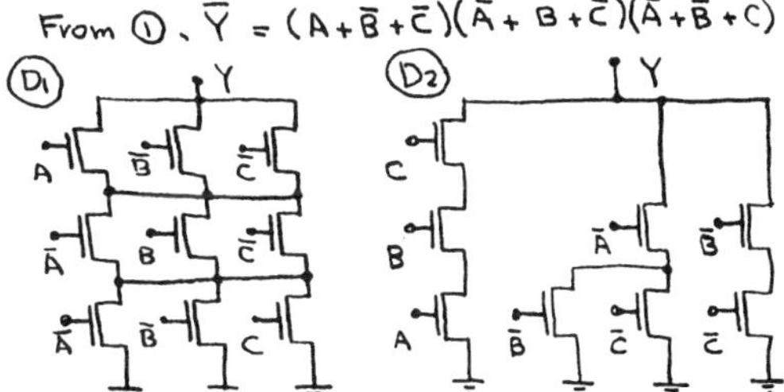

From ②, $\overline{Y} = ABC + \overline{A}\overline{B} + \overline{A}\overline{C} + \overline{B}\overline{C}$, see (D2) where path merging is includeded

For the PDN from the PUN (U1)

Simply, replacing series connections with parallel connections:

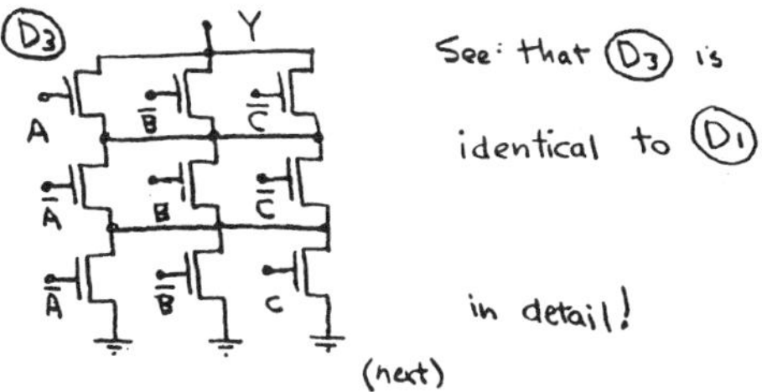

See: that (D3) is identical to (D1) in detail!

(next)

10

(D4) (reordered (D1))

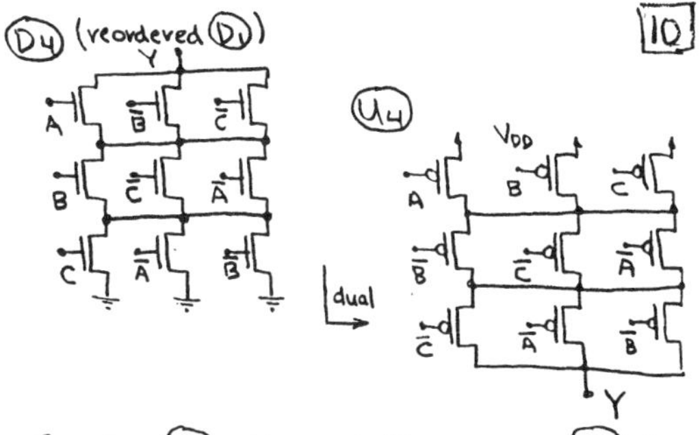

See that (U4), while not the same as (U1), is highly related, having some variable exchange in the middle and right columns.

Clearly, there are lots of variations of the completely-connected array.

27 | 10

$Y = A\overline{B} + \overline{A}B \rightarrow \overline{Y} = \overline{A\overline{B} + \overline{A}B} = \overline{A\overline{B}} \cdot \overline{\overline{A}B}$

or $\overline{Y} = (\overline{A}+B)(A+\overline{B}) = AB + \overline{A}\overline{B}$

PUN for $Y = A\overline{B} + \overline{A}B$: | PDN dual to (U1):

(next)

27 | 10

For the PUN from the PDN (D1)

See that both parallel and series connections can be identified for transformation, but that the series (vertical) ones (as drawn)* contain variables and their complements, and are therefore always open.

Thus converting parallel paths to serial ones

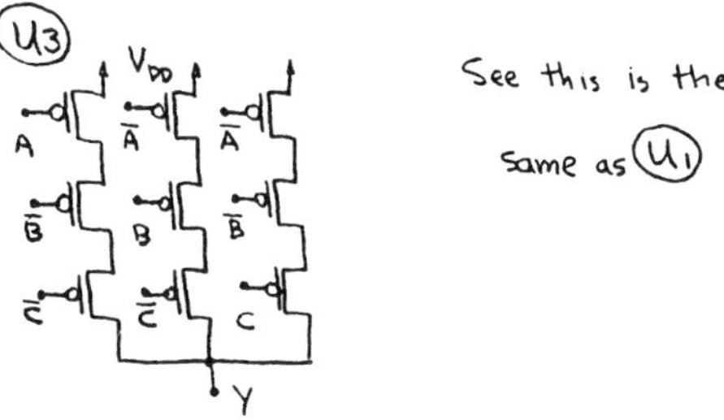

See this is the same as (U1)

*Note, however that (D1) can be redrawn as shown, then its colums (series links) converted to rows (parallel links of a PUN:

10

PDN for $\overline{Y} = AB + \overline{A}\overline{B}$: | PUN dual to (D2)

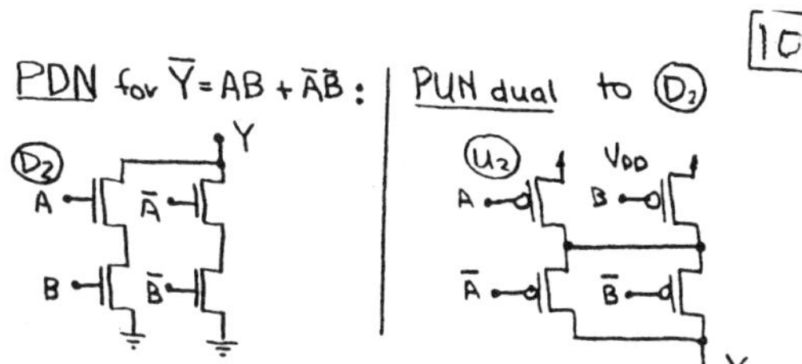

The two circuits required are (U1) with (D1) and (U2) with (D2)

28 | 10

$Y = AB + \overline{A}\overline{B}$. Directly, the PUN is as follows:

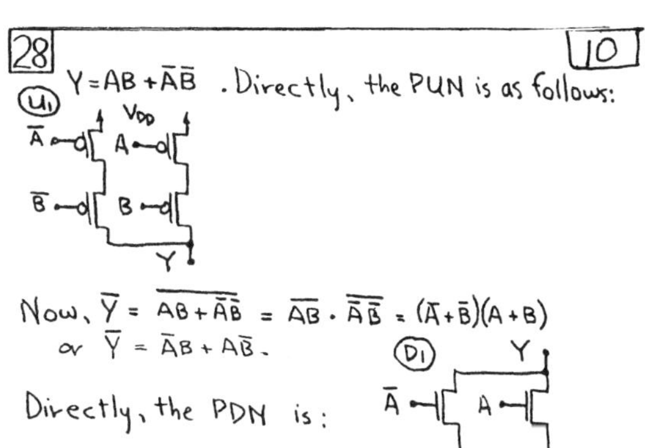

Now, $\overline{Y} = \overline{AB + \overline{A}\overline{B}} = \overline{AB} \cdot \overline{\overline{A}\overline{B}} = (\overline{A}+\overline{B})(A+B)$

or $\overline{Y} = \overline{A}B + A\overline{B}$.

Directly, the PDN is:

29 **10**

$Y = ABC + \bar{A}\bar{B}\bar{C}$. Directly, the PUN is as shown below:

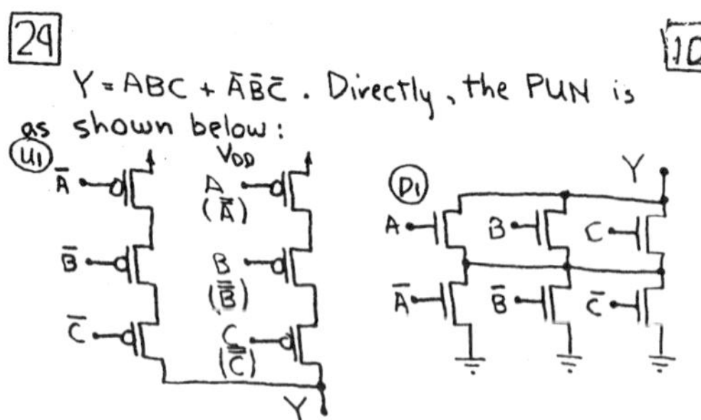

The corresponding dual PDN is shown above right.

30 **10**

Even-parity circuit: $\bar{Y} = \bar{A}BC + A\bar{B}C + AB\bar{C} + \bar{A}\bar{B}\bar{C}$

PDN, directly, is:

It uses 12 transistors

31 **10**

For output high with odd parity:

$Y = A\bar{B}\bar{C} + \bar{A}B\bar{C} + \bar{A}\bar{B}C + ABC = (A\bar{B} + \bar{A}B)\bar{C} + (\bar{A}\bar{B} + AB)C$

Directly the PUN is:

[Note that two transistors could be eliminated by combining the A and $\bar{A}$ inputs in a non-planar topology]

Using duality, the PDN becomes:

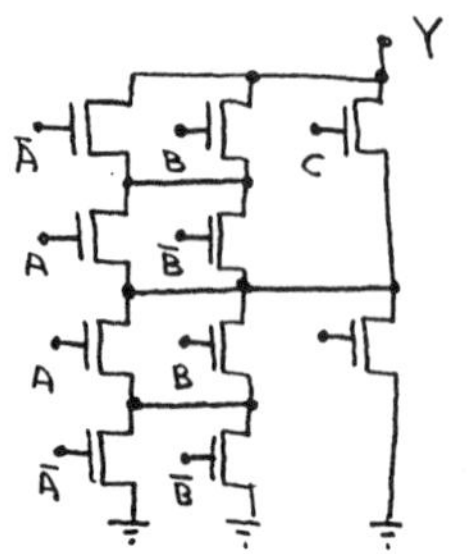

31 **10**

PDN reduced to 10 transistors:

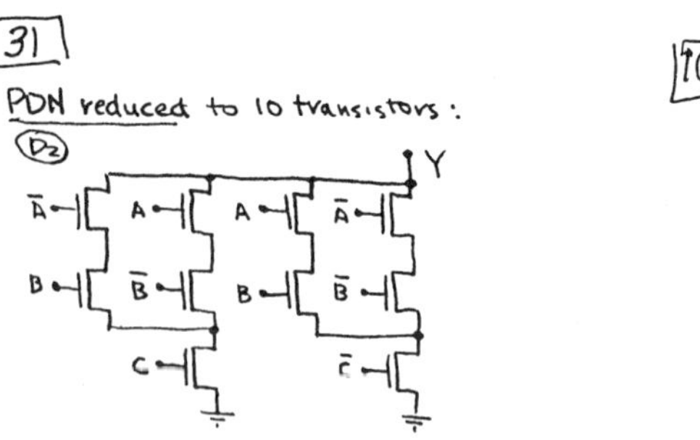

PDN reduced to 8 transistors: (X and X are joined)

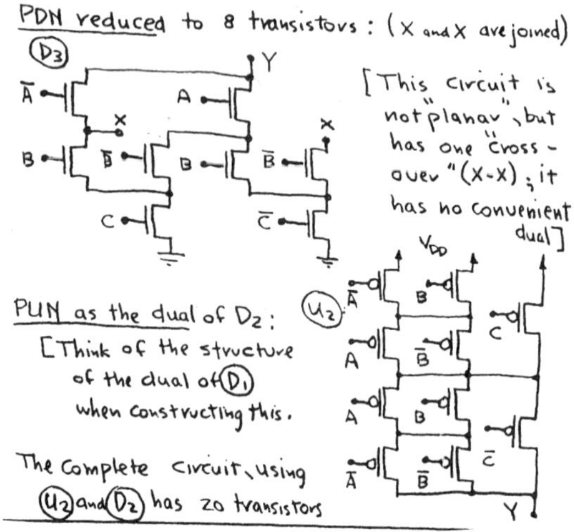

[This circuit is "not planar", but has one "cross-over" (X-X); it has no convenient dual]

PUN as the dual of D_2:

[Think of the structure of the dual of (D_1) when constructing this.

The complete circuit, using (U_2) and (D_2) has 20 transistors

32 **10**

Sum, $S = A\bar{B}\bar{C} + \bar{A}B\bar{C} + \bar{A}\bar{B}C + ABC$

Carry $C_o = AB\bar{C} + A\bar{B}C + \bar{A}BC + ABC$

$= AB + AC + BC = A(B + C) + BC$

Create the PUNs directly, simplifying that for S as in P10.30 above, as

$S = \bar{A}(B\bar{C} + \bar{B}C) + A(\bar{B}\bar{C} + BC)$

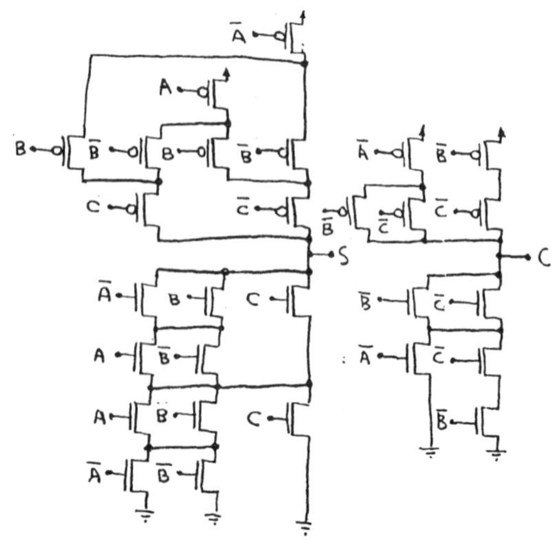

33 [10]

For matched-inverter equivalence of the circuit in Fig 13.14 : $p_A = p$; $p_B = p_C = p_D = 2p$; and $n_A = n_B = 2n$; $n_C = n_D = 2(2n) = 4n$.

10.34

35 [10]

Ignore the capacitances of the transistors themselves: For the matched NAND, $t_{PLH} = t_{PHL} = t_P$.

For the "uncompensated NAND, $t_{PLH} = t_P$, $t_{PHL} = t_P/4$.

Thus t_{PLH} are the same, but t_{PHL} is 4 times greater with no matching.

36

<u>For design a)</u>, there are $2(6) + 2 = \underline{14}$ transistors:

All 7 NMOS use $(W/L)_n = n$
1 PMOS uses $(W/L)_p = p$
6 PMOS use $(W/L)_p = 6p$

Total Area $= 7(1.2)0.8 + 1(3.6)0.8 + 6(6)(3.6)0.8 = \underline{113.3\,\mu m^2}$

<u>For design b)</u>, there are $2(3)2 + 1(2)2 = \underline{16}$ transistors:

6 NMOS use $(W/L)_n = n$
6 PMOS use $(W/L)_p = 3p$
2 PMOS use $(W/L)_p = p$
2 NMOS use $(W/L)_n = 2n$

Total equivalent devices is $6n + 18p + 2p + 2n = 10n + 20p$

Total equivalent area is $[10 + 3(20)]n = 70n$, and

Total Area $= 70(1.2)0.8 = \underline{67.2\,\mu m^2}$, or $\underline{59\%}$ of a)

---

37 [10]

Corresponding to a matched inverter characterized by n and p where $k_p = k_n = k$, the two-input NOR uses transistors n and $2p$ where $k_p = 2k_n$

a) <u>For A grounded</u>, V_{thB} occurs near $V_{DD}/2$, with Q_{PB} and Q_{NB} in saturation and Q_{PA} in triode. Let $V_{th} = v$, and the voltage <u>across</u> Q_{PA} be x.

Thus $i_D = k_p[(5-1)x - x^2/2]$
and $i_D = \frac{1}{2}k_p(5 - x - v - 1)^2$
and $i_D = \frac{1}{2}k_n(v-1)^2$

For $k_p = 2k_n$, $i_D = 2k_n(4x - x^2/2) = k_n(8x - x^2)$ --- (1)

and $i_D = k_n(4 - x - v)$ --- (2)

and $i_D = \frac{1}{2}k_n(v-1)^2$ --- (3)

From 2) 3) : $\pm(v-1)(0.707) = 4 - x - v$

Thus, $1.707v = 4.707 - x$ or $x = 4.707 - 1.707v$: (or) $0.293v = 3.293 - x$ or $x = 3.293 - 0.293v$

Now $x \approx 0$, in which case, $v = 4.707/1.707 \approx 2.38$ (ok)
or $v = 3.293/0.293 = 11.2$ (Clearly too large)

Thus $x = 4.707 - 1.707v$ --- (4)

[10]

Now, from 1), 3): $(v-1)^2 = 2(8x - x^2)$

With 4), $v^2 - 2v + 1 = 16(4.707 - 1.707v) - 2(4.707 - 1.707v)^2$

or $v^2 - 2v + 1 = 75.32 - 27.32v - 44.31 + 32.13v - 5.83v^2$

or $6.83v^2 + v(-2 + 27.32 - 32.13) + (1 - 75.32 + 44.31) = 0$

or $6.83v^2 - 6.81v - 30.01 = 0$

whence $v = (--6.81 \pm (6.81^2 - 4(6.83)30.01)^{1/2})/2(6.83)$

$= (6.81 \pm 29.43)/13.66 = 2.65V$

<u>Check</u>: [$>2.5V$ probably OK since one PMOS is full on]

Thus $V_{th} = \underline{2.65V}$

b) For A and B joined, the PMOS can be approximated as a single device with twice the length, for which the width is twice that in a matched inverter. Thus, for the equivalent PMOS device, $(W/L)_{peq} = p$ and $k_p = k$. For each of the two NMOS, $(W/L)_n = n$ and $k_n = k$.

Thus at $V_{th} = v$ with all devices in saturation:

(next)

10

$i_D = 2k/2(v-1)^2 = k/2(5-v-1)^2$,

$2(v-1)^2 = (4-v)^2$, and $\pm\sqrt{2}(v-1) = (4-v)$

Thus, $1.414v - 1.414 = 4 - v$, $2.414v = 5.414$,

whence $V_{th} = v = \underline{2.24V}$

See this is reduced from the single-input value (of 2.65V)!

Note that this fact can be used to control the relative threshold of multiple gates connected to a single fanout node, in order to guarantee operation sequence for slowly changing signals.

38

For the resistor load, the output voltage rising is $v_O = V_{DD}(1-e^{-t/R_DC})$

This reaches $V_{DD}/2$ when $(1-e^{-t/R_DC}) = 0.5$,

or $e^{-t/R_DC} = 0.5$, and $t = -R_DC \ln 0.5 = 0.693R_DC$

Thus $t_{PLH} = \underline{0.693R_DC}$

For the current load, $I = V_{DD}/R_D$, the output reaches $V_{DD}/2$ when $t = (CV_{DD}/2)/I$,

or $t = (CV_{DD}/2)/(V_{DD}/R_D) = 0.5R_DC$

(next)

10

Now, $k_n = k_n'(W/L)_n = 75\times10^{-6}(1.2/0.8) = 112.5\mu A/V^2$

and $k_p = 0.256 \times 112.5 = 28.8\mu A/V^2$

Now, $k_p'(W/L)_p = 28.8$; Thus $(W/L)_p = 28.8/(75/3) = \underline{1.152}$

For $L_p = 0.8\mu m$, $W_p = 1.152(0.8) = \underline{0.922\mu m}$

In summary, for this inverter: $k_n = \underline{112.5\mu A/V^2}$

$k_p = \underline{28.8\mu A/V^2}$

and $r = k_n/k_p = \underline{3.91}$

Now, $V_{OH} = \underline{+5.0V}$

For $V_{OL} = v \approx 0$, $i_D = k_p/2(v_{GS}-V_t)^2$

or $i_D = (28.8/2)(5.0-0.8)^2 = 254\mu A$

For the NMOS, $254 = 112.5[(5-0.8)v - v^2/2]$

or $v^2 - 8.4v + 4.52 = 0$,

whence $v = (--8.4 \pm (8.4^2 - 4(1)4.52)^{1/2})/2$

$= (8.40 \pm 7.24)/2 = 0.58V$

Thus $V_{OL} = \underline{0.58V}$

10

Thus $t_{PLH} = \underline{0.5R_DC}$, a reduction to $(0.50/0.69)\times100 = \underline{72.5\%}$ or by $\frac{0.69-0.50}{0.69}\times100 = \underline{27.5\%}$

39

Here $V_{DD}/4 = 5/4 = 1.25V$

Now, for v_O rising, the NMOS is cutoff, and the PMOS is in triode mode with:

$i_{D_P} = k_p[(v_{SG} - V_t)v_{SD} - v_{SD}^2/2]$, and here

$i_D = k_p[(5-0.8)(5-1.25) - (5-1.25)^2/2]$

$= k_p(15.75 - 7.03) = \underline{8.72k_p}$

Now, for v_O falling, the net current extracted from the load is $i_{D_n} - i_{D_p}$ which should be i_{D_P}

Thus $i_{D_n} = 2i_{D_P} = 2(8.72)k_p$, for triode operation where $i_{D_n} = k_n[(5-0.8)1.25 - 1.25^2/2]$

Overall, $i_{D_n} = 2(8.72)k_p = k_n(5.25-0.78) = 4.47k_n$

Thus $k_p = (4.47/(2(8.72)))k_n = \underline{0.256k_n}$

Check using Eq 10.39, where $r = k_n/k_p = 3.91$:

$V_{OL} = (V_{DD}-V_t)[1-(1-1/r)^{1/2}]$

$= (5-0.8)[1-(1-1/3.91)^{1/2}] = \underline{0.577V}$ ✓

From Eq 13.35, $V_{IL} = V_t + (V_{DD}-V_t)/(r(r+1))^{1/2}$

$= 0.8 + 4.2/(3.91\times4.91)^{1/2} = \underline{1.76V}$

From Eq 13.38, $V_{IH} = V_t + (2/\sqrt{3r})(V_{DD}-V_t)$

$= 0.8 + (2/\sqrt{3(3.91)})4.2 = \underline{3.25V}$

From Eq 13.36, $V_M = V_t + (V_{DD}-V_t)/(r+1)^{1/2}$

$= 0.8 + 4.2/(4.91)^{1/2} = \underline{2.70V}$

Now, $NM_H = V_{OH} - V_{IH} = 5.00 - 3.25 = \underline{1.75V}$

and $NM_L = V_{IL} - V_{OL} = 1.76 - 0.58 = \underline{1.18V}$

---

40 **10**

Note that a design with $r = 2$ sacrifices V_{OL} for improved propagation-delay symmetry.

From Eq. 10.39, $V_{OL} = (V_{DD} - V_t)[1 - (1 - 1/r)^{1/2}]$
$= (5 - 0.8)[1 - (1 - 1/2)^{1/2}] = \underline{1.23V}$

Now, $k_n'(W/L)_n = 75(1.2/0.8) = r k_p = 2(75/3)(W/L)_p$
Thus, $(W/L)_p = (70/25) \times 1/2 \times (1.2/0.8) = \underline{(1.8/0.8)}$

Now $C_{gs_n} = 1.2(1.5) = 1.8fF$; $C_{gd_n} = 1.2(0.5) = \underline{0.6fF}$
Thus $C_{in} = 1.8 + 0.6 = \underline{2.4fF}$.

Now, the output capacitance includes C_{db} and C_{gd} of the load plus C_{db} and twice C_{gd} of the switch.

Thus, the total capacitance of the output and the loading gate is:

$$C_{out} = 1.8(0.5 + 2.0) + 1.2(2.0 + 2(0.5)) + 2.4 = \underline{10.5fF}$$

Now, from Eq. 10.43, $t_{PLH} = 1.7C/(k_p V_{DD})$,
and $t_{PLH} = 1.7 \times 10.5 \times 10^{-15}/[(75/3)10^{-6}(1.8/0.8)5] = \underline{63.5ps}$

Now, from Eq. 10.44, $t_{PHL} = 1.7C/[k_n(1 - 0.46/r)V_{DD}]$
and $t_{PHL} = 1.7 \times 10.5 \times 10^{-15}/[75 \times 10^{-6}(1.2/0.8)(1 - 0.46/2)5] = \underline{41.2ps}$

Thus $t_p = (t_{PLH} + t_{PHL})/2 = (63.5 + 41.2)/2 = \underline{52.4ps}$

Thus $(2r+1)^2(r-1) = (r+1)^3$
and $(4r^2 + 4r + 1)(r - 1) = (r^2 + 2r + 1)(r + 1)$
$4r^3 + 4r^2 + r - 4r^2 - 4r - 1 = r^3 + 2r^2 + r + r^2 + 2r + 1$
$4r^3 - 3r - 1 = r^3 + 3r^2 + 3r + 1$
$3r^3 - 3r^2 - 6r - 2 = 0$
$r^3 - r^2 - 2r - 2/3 = 0$

Test $r^3 - r^2 - 2r - 2/3$
$r = 2 \rightarrow 8 - 4 - 4 - 2/3 = -2/3$
$r = 3 \rightarrow 27 - 9 - 6 - 2/3 = 11\,1/3$
$r = 2.1 \rightarrow 2.1^3 - 2.1^2 - 4.2 - .67 = 9.26 - 4.4 - 4.2 - .67 = -0.01$

Thus $r = \underline{2.1}$, for which:
$NM_L = 0.8 - (5.0 - 0.8)[1 - (1 - 1/2.1)^{1/2} - (2.1(2.1+1))^{-1/2}]$
$= 0.8 - 4.2[1 - 0.724 - 0.392] = \underline{1.29V}$

41

From Eq. 10.41, $NM_L = V_t - (V_{DD} - V_t)[1 - (1 - \frac{1}{r})^{1/2} - (r(r+1))^{-1/2}]$

Now, $\partial NM_L/\partial r = -(V_{DD} - V_t)[-\frac{1}{2}(1 - \frac{1}{r})^{-1/2}(-- \frac{1}{r^2}) - (-\frac{1}{2}(r(r+1))^{-3/2}(2r+1))]$

Maximum occurs where:
$-\frac{1}{2}(1 - 1/r)^{-1/2}(1/r^2) = -\frac{1}{2}(r(r+1))^{-3/2}(2r+1)$

Square both sides: $(1 - 1/r)^{-1}/r^4 = r^{-3}(r+1)^{-3}(2r+1)^2$
or $\frac{1}{1 - 1/r} = \frac{r}{(r+1)^3}(2r+1)^2 = \frac{r}{r-1}$

10.42

43 **10**

From Eq. 10.41, $NM_H = (V_{DD} - V_t)(1 - 2/\sqrt{3r})$

This is zero, when $1 - 2/\sqrt{3r} = 0$,
or $\sqrt{3r} = 2$, or $3r = 4$, or $r = \underline{1.33}$

For $r = 1$, $NM_H = 4.2(1 - 2/\sqrt{3}) = \underline{-0.65V}$

For $r = 2$, $NM_H = 4.2(1 - 2/\sqrt{6}) = \underline{0.77V}$

For $r = 4$, $NM_H = 4.2(1 - 2/\sqrt{12}) = \underline{1.78V}$

For $r = 8$, $NM_H = 4.2(1 - 2/\sqrt{24}) = \underline{2.48V}$

For $r = 16$, $NM_H = 4.2(1 - 2/\sqrt{48}) = \underline{2.99V}$

But, what about NM_L? (For $r = 16$, it is 0.92V)

44

From Eq. 10.41 and 10.42, noise margins are equal when
$V_t - (V_{DD} - V_t)[1 - (1 - \frac{1}{r})^{1/2} - 1/(r(r+1))^{1/2}] = (V_{DD} - V_t)[1 - 2/(3r)^{1/2}]$

or $V_t/(V_{DD} - V_t) = 2 - 2/(3r)^{1/2} - (1 - 1/r)^{1/2} - 1/(r(r+1))^{1/2}$ --- (1)

Here $V_t/(V_{DD} - V_t) = 0.8/(5.0 - 0.8) = \underline{0.1904}$
Try various values of r to solve (1):

10

For $r = 2$, $f(r) = 2 - 2/6^{1/2} - (1 - 1/2)^{1/2} - 1/(2(3))^{1/2}$
$= 2 - 0.816 - 0.707 - 0.408 = \underline{0.069}$

For $r = 3$, $f(r) = 2 - 2/9^{1/2} - (1 - 1/3)^{1/2} - 1/(3(4))^{1/2}$
$= 2 - 0.667 - 0.816 - 0.289 = \underline{0.228}$

Try $r = 2.8$, $f(r) = 2 - 2/(3(2.8))^{1/2} - (1 - 1/2.8)^{1/2} - 1/(2.8(3.8))^{1/2}$
$= 2 - 0.690 - 0.802 - 0.307 = \underline{0.201}$

Try $r = 2.7$, $f(r) = 2 - 2/(3(2.7))^{1/2} - (1 - 1/2.7)^{1/2} - (1/(2.7(3.7)))^{1/2}$
$= 2 - 0.703 - 0.793 - 0.316 = \underline{0.188}$

Conclude $r \approx \underline{2.72}$, for which the margins are:
$NM = NM_H = NM_L = (V_{DD} - V_t)(1 - 2/(3r)^{1/2})$
$= 4.2(1 - 2/(3(2.72))^{1/2}) = \underline{1.26V}$

10.45

46

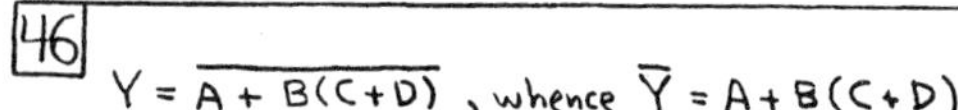

$Y = \overline{A + B(C+D)}$, whence $\overline{Y} = A + B(C+D)$

Thus the PDN can be formed directly as shown:

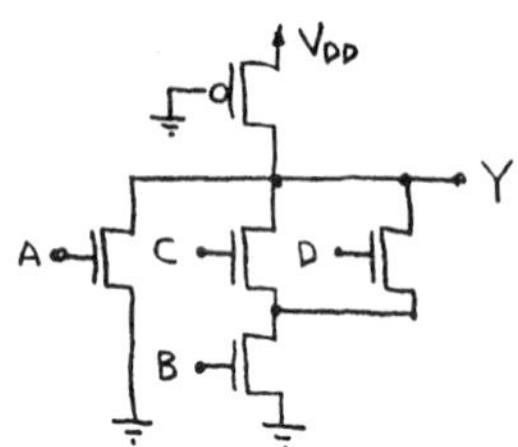

Now, $t_{PLH}/t_{PHL} = (k_n/k_p)(1-0.46/r)$
$= r(1-0.46/r) = 2.72 - 0.46 = \underline{\underline{2.26}}$

Now, $t_{PLH} = 1.7(1\times10^{-12})/(25\times10^{-6}(1.3^2/0.8)5) = \underline{\underline{8.24ns}}$

and $t_{PHL} = 8.24/2.26 = \underline{\underline{3.65ns}}$

and $t_P = (8.24 + 3.65)/2 = \underline{\underline{5.95ns}}$

Now, dynamic power is approximately fCV_{DD}^2, since the output swing is not quite V_{DD}.

For equal static and dynamic power
$f \times 1 \times 10^{-12} \times 5^2 = 1.82\times10^{-3}$
whence $f = 1.82\times10^{-3}/(25\times10^{-12}) = \underline{\underline{72.8MHz}}$,
for which the period is $1/(72.8\times10^6) = \underline{\underline{13.7ns}}$

Now, for transition times in the same proportion as propagation delays $t_{TLH}/t_{THL} = 8.24/3.65 \approx 2.26$

Now, for full output swing, there must be time for 2 full transitions in each cycle:
Thus $t_{THL} \approx 13.7/(1+2.26) = \underline{\underline{4.19ns}}$, and
$t_{TLH} \approx 4.19(2.26) = \underline{\underline{9.47ns}}$

Since these values are of the same order as the propagation delays, full swing operation is likely <u>not</u> possible at 72.8MHz.

47 **10**

For an Exclusive OR, $Y = A\overline{B} + \overline{A}B$, and

$\overline{Y} = \overline{A\overline{B} + \overline{A}B} = \overline{A\overline{B}} \cdot \overline{\overline{A}B} = (\overline{A}+B)(A+\overline{B})$

or $\overline{Y} = \overline{A}\overline{B} + AB$

The PDN results directly:

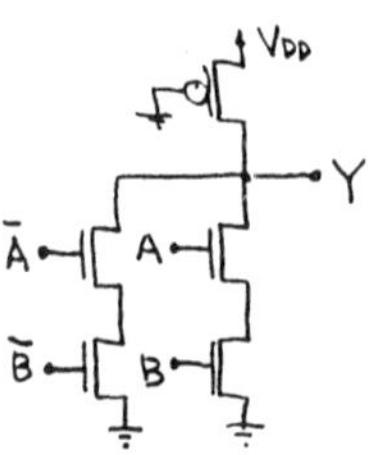

48

For a pseudo-NMOS NOR gate, independant of the number of inputs, the worst-case value of V_{OL} occurs for one input high (and a single NMOS conducting

From Eq. 10.39, $V_{OL} = (V_{DD}-V_t)[1-(1-1/r)^{1/2}]$,
for which $0.2 = (5.0-0.8)[1-(1-1/r)^{1/2}]$,
and $(1-1/r)^{1/2} = 1 - 0.2/4.2 = 0.952$
Thus $1 - 1/r = 0.907$
$1/r = 0.093$, and $r = \underline{10.76}$
Thus $k_n/k_p = 10.76 = 75(1.8/1.2)/(25(W/L)_p)$
Thus $(W/L)_p = (75/25)(1.8/1.2)/10.76 = \underline{0.418}$
Thus for $W_p = 1.8\mu m$, $L_p = 1.8/0.418 = \underline{4.31\mu m}$
and $(W/L)_p = \underline{(1.8/4.31)}$

49 **10**

For a), see directly that $X = 1 \cdot \overline{A} = \overline{A}$

and $Y = X \cdot \overline{B} = \underline{\underline{\overline{A} \cdot \overline{B}}}$

For b), see directly that $Y = \underline{\underline{\overline{A} \cdot \overline{B}}}$

For each circuit, node Y nominally satisfies both conditions. However in a), with A high and B low, Y is not pulled down completely to ground, but remains at V_{tp}, due to the PMOS threshold. Circuit b) does not have this problem, but node X is floating for A, B both high. However, X is not an output node. The body effect makes this worse! Notice that b) is exactly a complementary CMOS NOR gate for which $Y = \overline{A} \cdot \overline{B} = \underline{\underline{\overline{A+B}}}$

For V_{DD} replaced by an inverter driven by C,
$Y = \overline{C}(\overline{A} \cdot \overline{B}) = \overline{A} \cdot \overline{B} \cdot \overline{C} = \underline{\underline{\overline{A+B+C}}}$,
a 3-input NOR (for both a) and b)).

Practically speaking, however, there is a problem because, as noted above, the series PMOS do not operate well with a low input. In fact Y is pulled down only to one threshold drop below ground, when C is high.

50 **10**

With these reversals, the output is high when B is low or when A is low.

Thus, $Y = \overline{A} + \overline{B}$, or $\underline{Y = \overline{A \cdot B}}$, a <u>NAND</u> function

Circuit b), being a fully complementary CMOS gate, functions ideally. However circuit b) provides a relatively low high output for A low with B high. In this case, the output becomes $V_{DD} - V_{tn}$, where V_{tn} is raised from V_{t0n} by the body effect.

51

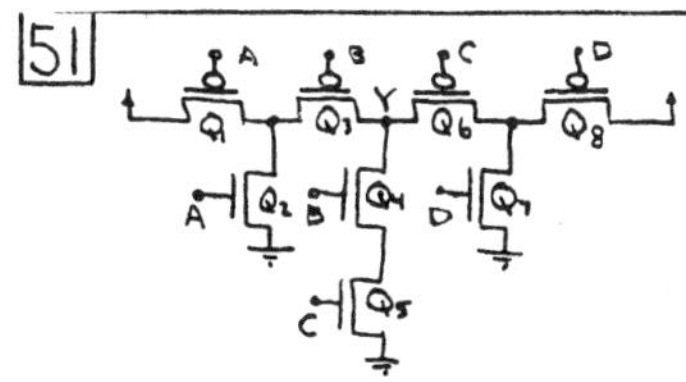

There are several deficiencies :

First, there are two possibilities for short-circuits when B and C are low and A and D are complementary, in which case current flows (for example) through Q_1, Q_3, Q_6, Q_7. This problem occurs for $\bar{A}\bar{B}\bar{C}D$ and $A\bar{B}\bar{C}\bar{D}$.

10

Second, node Y is pulled down to V_t (increased by the body effect) when B or C are low with A or D high, respectively. In this respect, this circuit is like that in Fig P13.49a.

With respect to Y as a function of A,B,C,D, the result will depend on device sizing. For Q_2, Q_7, relatively weaker than the PMOS, $Y = \bar{A}\bar{B} + \bar{C}\bar{D}$. A better circuit is shown below:

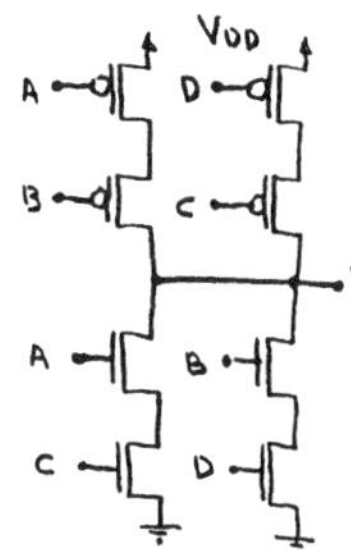

For Q_3, Q_6 relatively weaker, $\bar{Y} = A + D + BC$ and $Y = \overline{A+D} \cdot \overline{BC}$
$= \bar{A}\bar{D}(\bar{B}+\bar{C})$
$= \bar{A}\bar{B}\bar{D} + \bar{A}\bar{C}\bar{D}$

With input E driving an inverter to replace both V_{DD} connections, $Y = \bar{E}(\bar{A}\bar{B} + \bar{C}\bar{D})$ is possible, barring all the other problems this circuit brings.

52

10

For the switch gate and the input both at 3.3V, the switch output is

$$V_{OH} = 3.3 - V_t \text{, where } V_t = V_{to} + \gamma\left(\sqrt{V_{OH}+2\phi_f} - \sqrt{2\phi_f}\right)$$

Thus $V_{OH} = 3.3 - 0.8 - 0.5(V_{OH}+0.6)^{1/2} + 0.5(0.6)^{1/2}$
$= 2.89 - 0.5(V_{OH}+0.6)^{1/2}$
or $2V_{OH} - 5.77 = (V_{OH}+0.6)^{1/2}$

Squaring, $V_{OH} + 0.6 = 4V_{OH}^2 - 23.1V_{OH} + 32.29$
$4V_{OH}^2 - 24.1V_{OH} + 31.69 = 0$, and

$V_{OH} = (-\,-24.1 \pm \sqrt{24.1^2 - 4(4)31.69})/8$
or $V_{OH} = (24.1 \pm 8.59)/8 = \underline{1.94V}$ (and a larger one)

Now, for input low and switch gate high, $V_{OL} = \underline{0V}$

For the inverter current: For $V_{OH} = 1.94V$, the PMOS is in saturation with

$$i_D = \tfrac{1}{2}\mu_p C_{ox}(W/L)_p(3.3-1.94-0.8)^2 = \tfrac{1}{2}(25(10^{-6})3(\tfrac{1.2}{0.8})(0.56))^2$$

Thus, the static inverter current is $\underline{17.64\mu A}$

For t_{PLH}: Here, $v_O = 0$, $V_t = V_{to} = 0.8V$, $V_{GS} = V_{DD} = 3.3V$.
Thus, initially, $i_D(0) = \tfrac{1}{2}(75)(1.2/0.8)(3.3-0.8)^2 = \underline{351.6\mu A}$
At $v_O = 3.3/2 = 1.65V$,
$V_t = 0.8 + 0.5(1.65+0.6)^{1/2} - 0.5(0.6)^{1/2} = \underline{1.16V}$

and $i_D(t_p) = \tfrac{1}{2}(75)(1.2/0.8)(3.3-1.65-1.16)^2 = \underline{13.5\mu A}$

Thus $i_{D_{av}} = (351.6 + 13.5)/2 = \underline{183\mu A}$

and $t_{PLH} = (CV_{DD}/2)/i_{D_{av}} = 100(10^{-15})1.65/(183\times10^{-6}) = \underline{0.90ns}$

For t_{PHL}: Here, $V_t = V_{to}$, and the initial current is

$$i_D(0) = \tfrac{1}{2}(75)(1.2/0.8)(3.3-0.8)^2 = \underline{352\mu A}$$

At half swing, operation is in triode mode, and

$$i_D(t_p) = 75(1.2/0.8)[(3.3-0.8)1.65 - 1.65^2/2] = \underline{311\mu A}$$

Thus, the average current is $(352 + 311)/2 = \underline{332\mu A}$

and $t_{PHL} = 100(10^{-15})1.65/311\times10^{-6} = \underline{\underline{0.53ns}}$

53 10

a) For the inverter, with $v_{O2} = 3.3 - 0.8 = 2.5V$, Q_H is in saturation and the current is

$i_{DH} = ½(75)(1.2/0.8)(v_{O1} - 0.8)^2 = 56.25(v_{O1} - 0.8)^2$,

and Q_P is in triode mode with current

$i_{DP} = (75/3)(3.6/0.8)[(3.3 - v_{O1} - 0.8)0.8 - 0.8^2/2]$

or $i_{DP} = 100[(2.5 - v_{O1})0.8 - 0.8^2/2] = 80(2.5 - v_{O1}) - 32$

Now $i_{DH} = i_{DP}$, or $56.25(v_{O1} - 0.8)^2 = 80(2.5 - v_{O1}) - 32$

Thus, $v_{O1}^2 - 1.6v_{O1} + 0.64 = 3.56 - 1.422v_{O1} - 0.569$

or $v_{O1}^2 - 0.178v_{O1} - 2.35 = 0$,

whence $v_{O1} = [- -0.178 \pm (0.178^2 - 4(-2.35))^{½}]/2$
$= [0.178 \pm 3.07]/2 = \underline{1.62V}$

For Q_1, $V_t = 0.8 + 0.5[(1.62 + 0.6)^{1/2} - 0.6^{1/2}] = \underline{1.158V}$

Capacitor charging current:

At V_{O1}: $i_D = ½(75)(1.2/0.8)(3.3 - 1.62 - 1.158)^2 = \underline{15.3\mu A}$

At 0V: $i_D = ½(75)(1.2/0.8)(3.3 - 0.8)^2 = \underline{351.6\mu A}$

Average: $i_D = (15.3 + 351.6)/2 = \underline{183\mu A}$

Now, $t_{PLH} = 20 \times 10^{-15} \times 1.62 / 183 \times 10^{-6} = \underline{177 ps}$

b) For the inverter, $V_{IH} = 1/8(5(3.3) - 2(0.8)) = \underline{1.86V}$

For this value, $i_{D1} = 75(1.2/0.8)[(3.3 - 0.8)1.86 - 1.86^2/2 = \underline{328\mu A}$

Current in $Q_R = (75/3)(W/L)_R(3.3 - 0.8)^2 = 328/2$

for which $(W/L)_R = 1.05$, and for $L_R = 0.8\mu m$, $(W/L)_R = (0.84/0.8)$

For t_{PHL}: For Q_1 with source grounded and gate at V_{DD}, while Q_R has a source-to-gate voltage V_{DD}:

Initially, at $v_{O1} = V_{DD}$, $i_{DR} = 0$, since $v_{SDR} = 0V$,
and $i_{D1} = ½(75)(1.2/0.8)(3.3 - 0.8)^2 = \underline{352\mu A}$

At $v_{O1} = V_{IH}$, $i_{DR} = 25(0.84/0.80)[(3.3 - 0.8)(3.3 - 1.86) - (3.3 - 1.86)^2/2]$
$= \underline{67.3\mu A}$

and $i_{D1} = 75(1.2/0.8)[(3.3 - 0.8)1.86 - 1.86^2/2 = \underline{328\mu A}$

whence average current for capacitor charging is
$i_{Dav} = (352 + 328 - 67.3)/2 = \underline{306\mu A}$

Thus, $t_{PHL} = 20 \times 10^{-15}(3.3 - 1.86)/(306 \times 10^{-6}) = \underline{94ps}$

54

ⓐ Need $\overline{Y} = AB + \overline{A}\overline{B}$. In direct analogy to Fig 10.31:

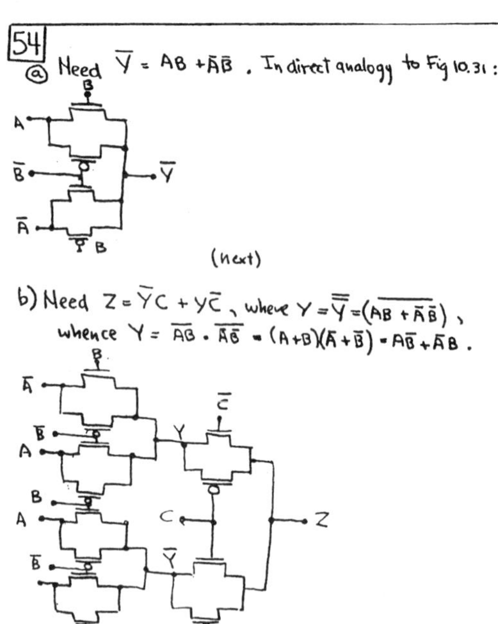

b) Need $Z = \overline{Y}C + Y\overline{C}$, where $Y = \overline{\overline{Y}} = (\overline{AB + \overline{A}\overline{B}})$,
whence $Y = \overline{AB} \cdot \overline{\overline{A}\overline{B}} = (\overline{A} + \overline{B})(A + B) = A\overline{B} + \overline{A}B$.

55

Need a CPL circuit for $Y = A\overline{B} + \overline{A}B$ and

$\overline{Y} = AB + \overline{A}\overline{B}$. {See Exercise 13.9b}

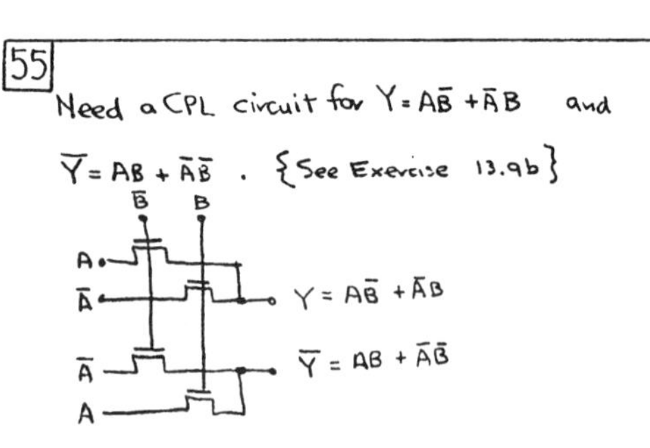

56 10

Require a CPL for $Z = ABC$ and $\overline{Z} = \overline{ABC} = \overline{A} + \overline{B} + \overline{C}$

Extend Fig 10.32 to 3 variables by dealing in pairs, creating $Y = AB$, then $Z = YC$ with $\overline{Y} = \overline{A} + \overline{B}$, then $\overline{Z} = \overline{Y} + \overline{C}$

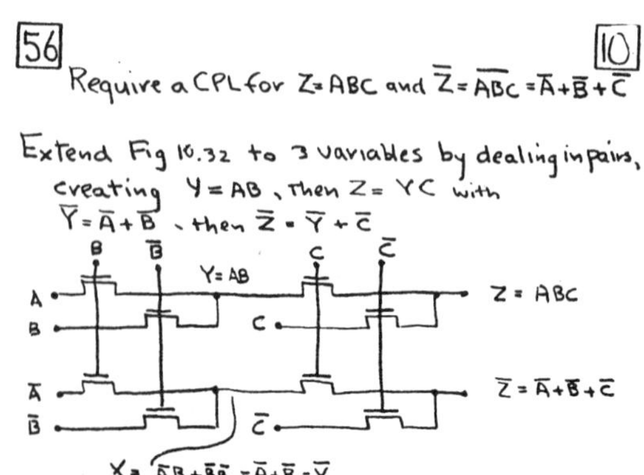

57

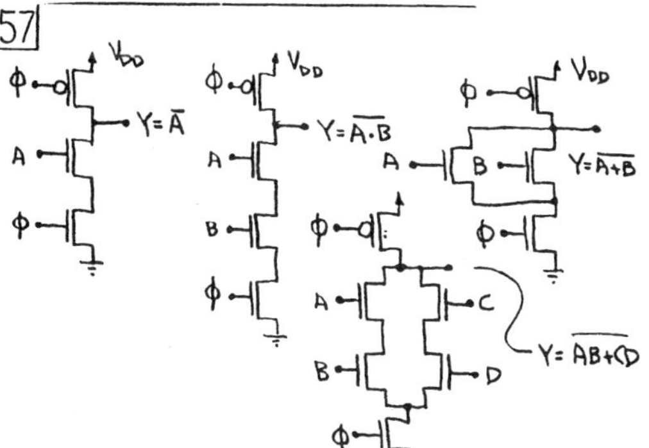

58 10

At $v_Y = 0.3V$, $i_{DP} = \frac{1}{2}(75/3)(2.4/0.8)(3.0-0.8)^2 = \underline{181.5 \mu A}$

At $v_Y = 2.7V$, $i_{DP} = (75/3)(2.4/0.8)[(3.0-0.8)0.3 - 0.3^2/2]$
$= \underline{46.1 \mu A}$

Thus $i_{Dav} = (181.5 + 46.1)/2 = \underline{114 \mu A}$

and $t_{TLH} = t_r = 15 \times 10^{-15}(2.7-0.3)/(114 \times 10^{-6}) = \underline{316\ ps}$

59

For 3 NMOS in series, the equivalent length is $3L_n = 3(0.8) = 2.4 \mu m$

Thus for $v_Y = 3.0V$, $i_D = \frac{1}{2}(75)(1.2/2.4)(3.0-0.8)^2 = \underline{90.8 \mu A}$

and for $v_Y = 1.5V$, $i_D = 75(1.2/2.4)[(3.0-0.8)1.5 - 1.5^2/2] = \underline{81.6 \mu A}$

Thus $i_{Dav} = (90.8 + 81.6)/2 = \underline{86.2 \mu A}$

and $t_{PHL} = 15 \times 10^{-15}(3.0/2)/(86.2 \times 10^{-6}) = \underline{261\ ps}$

60 10

a) $C_1 = 5fF$:

Now, for v_{C1} rising to $V_{DD} - V_t = 5 - 1 = 4V$, and assuming Q_1 continues to conduct, v_Y will fall by an amount $(C_1/C_L)(\Delta v_{C1}) = 5/30(4) = \underline{0.67V}$ to $5.0 - 0.67 = 4.33V$. Since this exceeds 4.0, the assumption that Q_1 continues to conduct is verified. Thus v_Y drops by $\underline{0.67V}$

Note that if the body effect is included, it will likely to be impossible to raise v_{C1} to 4V. Thus 0.67V is the largest possible change.

b) $C_1 = 10fF$:

In view of the previous analysis, assume that ultimately $v_Y = v_{C1} = v$. Now, the charge change in each capacitor is the same:

$Q = CV \rightarrow 10(v - 0) = 30(5 - v)$

and $10v = 150 - 30v$, $40v = 150$, and $\underline{v = 3.75V}$

Thus v_Y drops by $5 - 3.75 = \underline{1.25V}$ to 3.75V

61 10

For a 0.5V change, $t = C\Delta V/I_\ell = 30(10^{-15})0.5/10^{-12} = \underline{15\ ms}$

Since the precharge interval is much shorter than the evaluate, the period of the minimum clocking frequency can be as great as 15 ms, for which $f_{min} = 1/(15 \times 10^{-3}) = \underline{67\ Hz}$

62

In each cycle, the output must charge completely, then discharge to at least the threshold of the succeeding gate.

From Exercise 10.10, the precharge time is about $t_r = 0.4\ ns$. From Exercise 10.11, the discharge time is about $t_{PHL} = 0.5\ ns$.

Thus, the maximum allowed clocking frequency can be as high as $1/(0.4+0.5) \times 10^{-9} = \underline{1.1\ GHz}$

Chapter 11 Problems

1 **11**

For the inverters, using Eq 13.8, :

$$V_{th} = \left[V_{DD} - |V_{tp}| + (k_n/k_p)^{1/2} V_{tn}\right] / \left(1 + (k_n/k_p)^{1/2}\right)$$
$$= (5 - 1 + \sqrt{2.5} \times 1)/(1 + \sqrt{2.5}) = 5.58/2.58 = \underline{2.16V}$$

Current from the PMOS, for which $v_{SG} = 5V$ with $v_D = 2.16V$, is $i_D = (100/2.5)(2/1)[(5-1)(5-2.16) - (5-2.16)^2/2]$, or $i_D = \underline{586 \mu A}$

To sustain this from the NMOS,

$$586 = 100(W/L)_{eq}[(5-1)2.16 - 2.16^2/2],$$

whence $(W/L)_{eq} = 5.86/6.31 = \underline{0.929}$

Thus, for Q_5, Q_6 in series, $(W/L)_{5,6} = 2(0.929) = \underline{1.86}$ and $W_5 = W_6 = \underline{1.86 \mu m}$ for $L_{5,6} = 1 \mu m$.

2 **11**

Initially, the current from the PMOS (say Q_4) is essentially zero, since v_{SP} is 0.
Thus, for $v_Q = 5V$, $i_{D_{7,8}} = \frac{1}{2}(100)(4/(2 \times 1))(5-1)^2 = \underline{1600 \mu A}$

Now, for $v_Q = V_{DD}/2 = 2.5V$:

$$i_{D_{7,8}} = 100(4/2)[(5-1)2.5 - 2.5^2/2] = \underline{1375 \mu A}$$
$$i_{D_4} = (100/2.5)(2/1)[(5-1)2.5 - 2.5^2/2] = \underline{550 \mu A}$$

Thus, the average discharge current is $(1600 - 0 + 1375 - 550)/2 = \underline{1212 \mu A}$

and $t_{PHL} = 30 \times 10^{-15} \times 2.5/1212 \times 10^{-6} = \underline{61.9 ps}$

Now, $t_{PLH} = 1.7C/(k_p'(W/L)_p V_{DD}) = \dfrac{1.7 \times 30 \times 10^{-15}}{(100/2.5)(2/1)5} = \underline{127.5 ps}$

Thus the minimum set/reset pulse length is

$127.5 + 61.9 = \underline{189 ps}$

3 **11**

Assume that Q_2 is conducting, Q_1 is cut off, and Q_5 is conducting. Now to lower $\bar{Q}$ to $V_{DD}/2 = 2.5V$, Q_2 and Q_5 must be matched, since equivalent control voltages are applied to each.

Thus $k_2 = k_5$, and $25(W/L)_2 = 75/2$, whence $(W/L)_2 = (W/L)_p = \underline{6}$

4 **11**

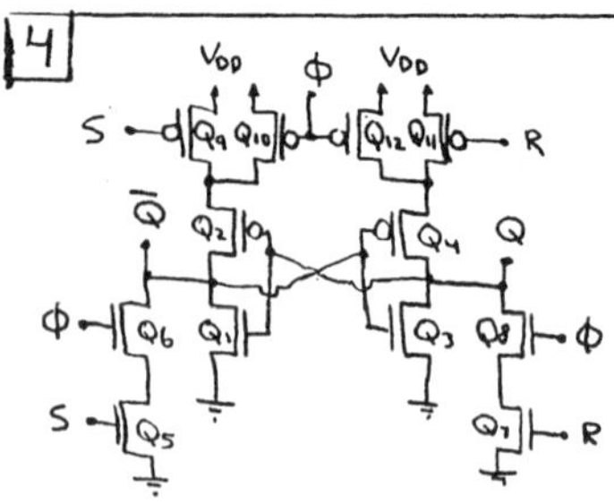

As noted, this fully-complementary circuit uses 12 transistors. However, a 10-T version exists in which Q_{10} and Q_{12} are omitted.

Note further that an effective 9-T version exists, in which Q_6 and Q_8 are moved below Q_5 and Q_7, then merged into a single grounded-source device. Note that all of the designs can employ the latter idea to reduce the transistor count by 1. See the sketch following:

(next)

4 **11**

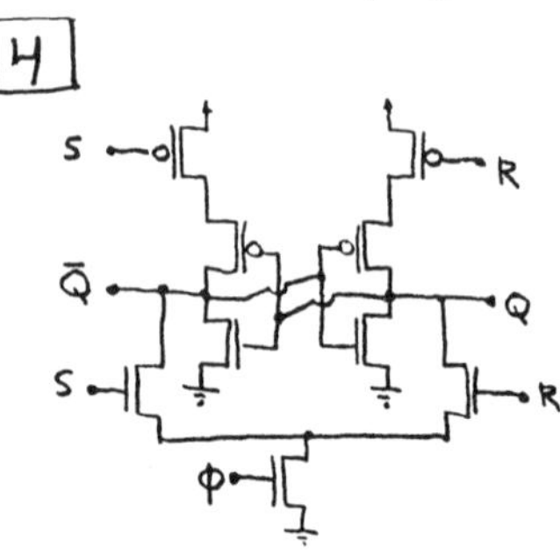

This circuit suffers only from the fact that unclocked changes in S and R have a secondary impact on $Q\bar{Q}$ since raising S or R disconnects $Q\bar{Q}$ from V_{DD}. In some applications this may lead to system noise sensitivity in which case one or the other or both of Q_{10}, Q_{12} (in the previous sketch) may be added.

5 **11**

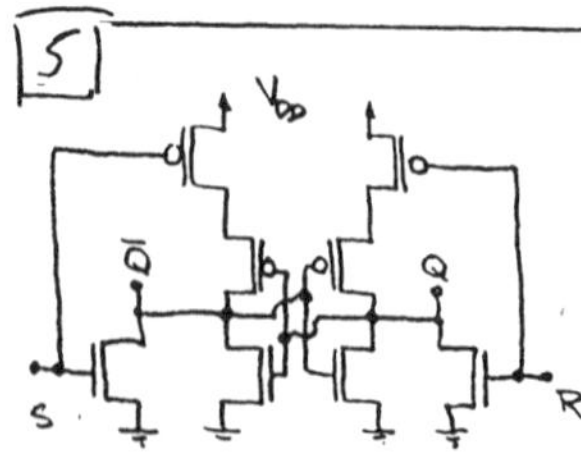

See that this is a complementary CMOS NOR-gate implimentation.

6 11

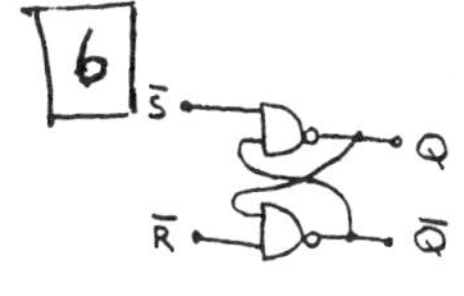

R	S	$\bar{R}$	$\bar{S}$	Q_{n+1}
0	0	1	1	Q_n
0	1	1	0	1
1	0	0	1	0
1	1	0	0	Not Used

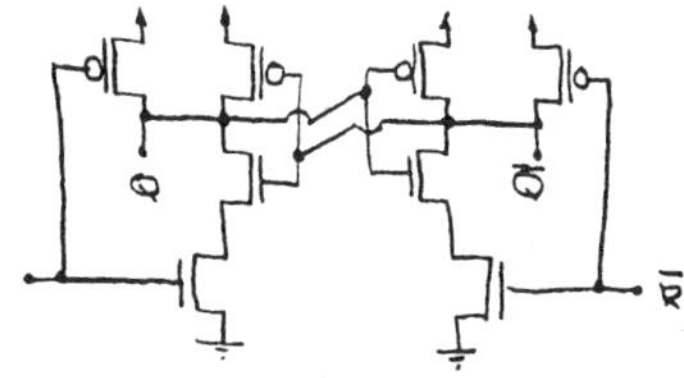

In this circuit, $\bar{R}$ and $\bar{S}$ rest high. Either going low forces the corresponding output to go high. [Thus, $\bar{S}=0 \rightarrow S=1 \rightarrow Q=1$]

If both $\bar{S}$ and $\bar{R}$ are low, the outputs Q and $\bar{Q}$ are no longer complementary, but are both high. Following this, the output retains the state defined by the last of $\bar{R}, \bar{S}$ to go high. If both $\bar{R}$ and $\bar{S}$ go high simultaneously, the outputs Q and $\bar{Q}$ oscillate, initially in phase, with a phase difference, initiated by noise, which

6 11

grows increasingly rapidly until the outputs "snap" into a complementary state of 0/1 or 1/0. During the oscillation, the period is somewhat less than twice the propagation delay of the basic inverter, and the amplitude is a fraction of the full logic swing. This fraction depends on the ratio of propagation time to transition time of the inverters, being larger for faster logic transitions and longer pure propagation delays. Generally speaking, the outputs are usually nearly sinusoidal.

---

7 11

Note that the devices are matched, with

a. $k_n = k_p = 20(12/6) = 40\mu A/V^2$, and $|V_t| = 1V$.

For $v_I = 0V, 5V$: one device is on, one off ; $v_O = \underline{5V}, \underline{0V}$

For $v_I = 1V, 4V$: one on, one off ; $v_O = \underline{5V}, \underline{0V}$

For $v_I = 1.5V, 3.5V$: one in saturation, one in triode mode.

$i_D = \frac{1}{2}(40)(1.5-1)^2 = 40[(5-1.5)v_O - v_O^2/2]$

Thus $0.125 = 2.5v_O - v_O^2/2$

or $v_O^2 - 5v_O + 0.25 = 0$

and $v_O = [--5 \pm \sqrt{5^2 - 4(0.25)}]/2$

$= (5 \pm 4.8989)/2 = 0.05V$

Thus $v_O = \underline{0.05V}$ or $\underline{4.95V}$

For $v_I = 2.0V, 3.0V$:

$\frac{1}{2}(2-1)^2 = (5-2-1)v_O - v_O^2/2$

or $(2-1)^2 = 2 \times 2 v_O - v_O^2$

and $v_O^2 - 4v_O + 1 = 0$

whence $v_O = (--4 \pm \sqrt{4^2 - 4(1)})/2$

$= (4 \pm 3.464)/2 = 0.27V$

Thus $v_O = \underline{0.27V}$ or $\underline{4.73V}$

For $v_I = 2.25V$ or $2.75V$;

$(2.25-1)^2 = 2(5-2.25-1)v_O - v_O^2$

$1.5625 = 3.5 - v_O - v_O^2$

$v_O^2 - 3.5v_O + 1.5625 = 0$

7 11

whence $v_O = (--3.5 \pm \sqrt{3.5^2 - 4(1.5625)})/2$

$= (3.5 \pm 2.45)/2 = 0.525$

Thus $v_O = \underline{0.525V}$ or $\underline{4.475V}$

b. For $v_I = 2.5V$, $v_O = \underline{2.5V}$, by symmetry.

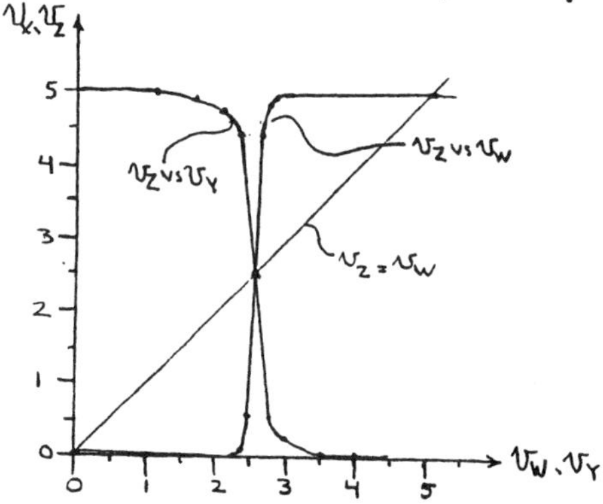

Now, having plotted v_Z versus v_Y (or v_X versus v_W), use the graph to find v_Z versus v_W:
Work backwards: first v_Z, then $v_Y = v_X$, then v_W.

For $v_Z = \underline{2.5V}$, $v_Y = v_X = v_W = \underline{2.5V}$

For $v_Z = \underline{4.4V}$, $v_Y = 2.25V$; for $v_X = 2.25V$, $v_W = \underline{2.55V}$

For $v_Z = \underline{4.9V}$, $v_Y = 1.50V$; for $v_X = 1.50V$, $v_W = \underline{2.65V}$

(next)

7 **11**

Plot these on the sketch above. See that the v_Z versus v_W characteristic crosses the

C. $v_Z = v_W$ line at:

point A : (0, 0)

point B : (2.5, 2.5)

point C : (5, 5)

At point B, the current flow in each inverter is:

$i_D = \frac{1}{2}(40)(2.5-1)^2 = \underline{45\,\mu A}$

where for each transistor, $r_o = 100/(45 \times 10^{-6}) = 2.22\,M\Omega$

and $g_m = 2(\frac{1}{2})40(2.5-1) = \underline{60\,\mu A/V}$

Thus for each inverter operating at (2.5, 2.5), the voltage gain is $-(g_m + g_m)(r_o \| r_o) = -g_m r_o = -60\times10^{-6} \times 2.22 \times 10^{6} = \underline{133\,V/V}$

Thus an estimate of the slope of the v_Z versus v_W curve at B is $(133)^2 = \underline{17.7 \times 10^3\,V/V}$

Correspondingly a lower bound on the width of the transition region is $(5-0)/(17.7\times10^3)$, or $\underline{0.28\,mV}$, that is $\pm 0.14\,mV$ around 2.5V.

Alternatively, a more detailed analysis can be done in a number of ways. For example, consider a sample-point analysis:

From earlier work above, for $v_Z = 4.95V$, $v_Y = 1.5V$

Now to produce $v_X = v_Y = 1.5V$, v_W must be around 2.5V with a value determined by

7 **11**

the PMOS in saturation and the NMOS in triode mode sharing the current i_D:

Thus $i_D = \frac{1}{2}(40)(5 - v - 1)^2 = 40[(v-1)1.5 - 1.5^2/2]$

or $(4-v)^2 = 2(v-1)1.5 - 1.5^2$

or $16 - 8v + v^2 = 3v - 3 - 2.25$

or $v^2 - 11v + 21.25 = 0$

whence $v = (--11 \pm (11^2 - 4(21.25))^{1/2})/2$

$= (11 \pm 6)/2 = 2.5V$

While this seems strange at first, it is simply an indicator that for $v_X = 1.5V$, the NMOS is actually just at the edge of saturation, such that the saturation-current expression still applies, for which the saturation currents of both devices are equal when $v_W = 2.5V$. Thus the output resistance and finite current gain must be considered (as we did earlier)

For the gain of $-133\,V/V$ found there, to produce a 1V change in v_X (from 2.5 to 1.5V) would require $1/133$ V, or 7.5 mV (or 15 mV total) on v_W. Recall that this 15 mV change on v_W moves v_Z from 0.05 to 4.95V!

We conclude that the transition region is more like 15 mV than a fraction of a mV.

Note that this approach uses a combination of large- and small-signal modelling to better approximate the large-signal nonlinearities of logic.

8 **11**

The approximate transfer characteristic of each inverter passes through points: (0, 5), (2.0, 4.6), (2.42, 0.4), (5, 0).

For the linear centre segment between (2.0, 2.46), (2.42, 0.4) an equation is $v_O = a - b v_I$

Here: $4.60 = a - 2.00b$, and

$0.40 = a - 2.42b$

Subtract: $4.20 = 0.42b \rightarrow b = \underline{10}$

Now, $4.60 = a - 2(10) \rightarrow a = 4.6 + 20 = \underline{24.6}$

Check: $0.4 = 24.6 - 2.42(10)$ ✓

Thus the middle part of the characteristic is

$v_O = 24.6 - 10 v_I$

For each device, $v_O = v_I = v$, when

$v = 24.6 - 10v$, or $11v = 24.6$, or $v = \underline{2.236V}$

where the gain is $\Delta v_O / \Delta v_I = -b = \underline{-10\,V/V}$

Thus point B on the open-loop characteristic is $v_W = v_Z = \underline{2.236V}$, where the loop gain can be approximated to be at least $(-10)^2 = \underline{100\,V/V}$

The open-loop characteristic reaches $v_O = 5V$, where $v_I = 2.236 + (5 - 2.236)/100 = \underline{2.264V}$

and it reaches 0V where

$v_I = 2.236 - 2.236/100 = \underline{2.214V}$

8 **11**

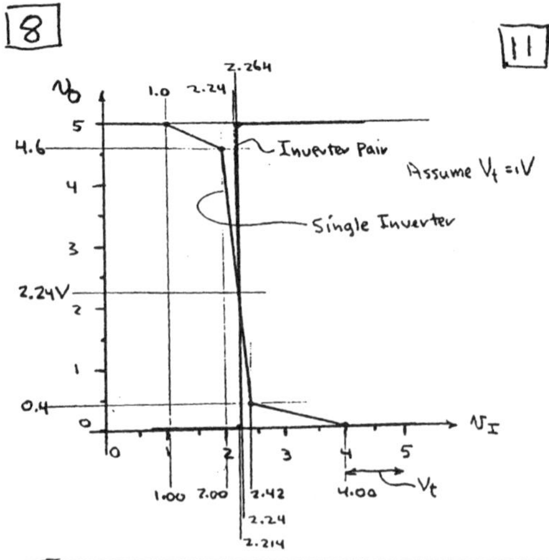

9 **11**

From Exercise , T = 0.69 RC. From the equation just preceding the exercise, a 2% error introduced by R_{on} implies that:

$R + R_{on} = 1.02R$

and $R = R_{on}/0.02 = 1\,k\Omega/0.02 = \underline{50\,k\Omega}$,

for which $C = T/0.69R = 1\times10^{-3}/(0.69 \times 50\times10^3) = \underline{30\,nF}$

10 — 11

From page 1025, for the monostable:

$$T = C(R + R_{on}) \ln\left[R/(R+R_{on})(V_{DD}/(V_{DD}-V_{th}))\right]$$

Neglecting R_{on}, $T = \underline{CR \ln(V_{DD}/(V_{DD}-V_{th}))}$

Now, $V_{th} = (0.5 \pm 0.1)V_{DD}$, and
$\ln(1/(1-0.5)) = 0.693$, $\ln(1/(1-0.4)) = 0.511$, and
$\ln(1/(1-0.6)) = 0.916$

Thus T can vary from $(0.511/0.693)100 = \underline{73.7\%}$
to $(0.916/0.693)100 = \underline{132.2\%}$ of nominal

or by $\underline{-26.3\%}$ to $\underline{+32.2\%}$ of nominal.

11 — 11

$T = C(R + R_{on}) \ln\left[R/(R+R_{on}) \times V_{DD}/(V_{DD}-V_{th})\right]$
$= 0.001\times10^{-6}(10^4 + 200)\ln\left[10^4/(10^4+200) \times 10/(10-5)\right]$
$= 1.02\times10^{-5} \ln 1.96 = 1.04\times10^{-5}\text{s} = \underline{10.4\mu s}$

Now, $\Delta V_1 = V_{DD}R/(R+R_{on}) = 10\times10^4/(1.02\times10^4) = \underline{9.80V}$
and $\Delta V_2 = V_{DD} + V_{D1} - V_{th} = 10 + 0.5 - 5 = \underline{5.70V}$

Change in V_{O1} is due to the variable current in R flowing in R_{on}.

Initially, the voltage across R_{on} is $V_{DD}R_{on}/(R+R_{on})$
Finally, it is $(V_{DD} - V_{DD}/2)R_{on}/(R+R_{on})$

11 — 11

Thus, the total change is $(V_{DD}/2)R_{on}/(R+R_{on})$
or $(10/2)200/(10^4+200) = \underline{0.098V}$

The peak sinking current through G1 occurs as v_{O1} rises and current flows through R_{on} and the diode. Its peak value is
$(V_{DD} - V_{th} - V_D)/R_{on} = (10-5-0.7)/200 = \underline{21.5mA}$

12 — 11

$T = C(R + R_{on})\ln\left[R/(R+R_{on}) \times V_{DD}/(V_{DD}-V_{th})\right]$
Thus $1 = 10^6(R+100)\ln\left[R/(R+100) \times 1/(1-0.4)\right]$

Now for R_{on} ignored:
$R = 10^6/\ln[1/0.6] = 1.958 M\Omega$.
Clearly, for this R, R_{on} can be neglected, and $R = \underline{1.96M\Omega}$

13 — 11

a) From Ex 13.17, $T = CR\ln\left[V_{DD}/(V_{DD}-V_{th}) \times V_{DD}/V_{th}\right]$
Now, for $V_{th} = V_{DD}/2$, $T = CR\ln(2\times2) = \underline{1.39\,CR}$

b) For $f_o = 100kHz$, $T = 1/10^5 = 1.39CR$
whence $CR = 0.721\times10^{-5}$

For a choice of $R = \underline{10k\Omega}$, $C = 0.721\times10^{-5}\times10^{-4} = \underline{721pF}$

14 — 11

From Exercise 13.17, $T = CR\ln\left(\frac{V_{DD}}{V_{DD}-V_{th}} \times \frac{V_{DD}}{V_{th}}\right)$

For $V_{th} = 0.5V_{DD}$, $f_o = 1/T = [CR\ln 4]^{-1} = 0.721/CR$

For $V_{th} = 0.6V_{DD}$, $f_o = [CR\ln(1/0.4 \times 1/0.6)]^{-1} = 0.701/CR$

For $V_{th} = 0.4V_{DD}$, $f_o = [CR\ln(1/0.6 \times 1/0.4)]^{-1} = 0.701/CR$

Thus, for V_{th}/V_{DD} over the range 0.4 to 0.6, f_o varies from $(0.701/0.721)100 = \underline{97.2\%}$ of nominal, through nominal to 97.2% of nominal again.

Note that the frequency is relatively constant, independant of V_{th} variation. This is due to compensation of one half-cycle by the next, each varying quite strongly but in opposite directions.

15 — 11

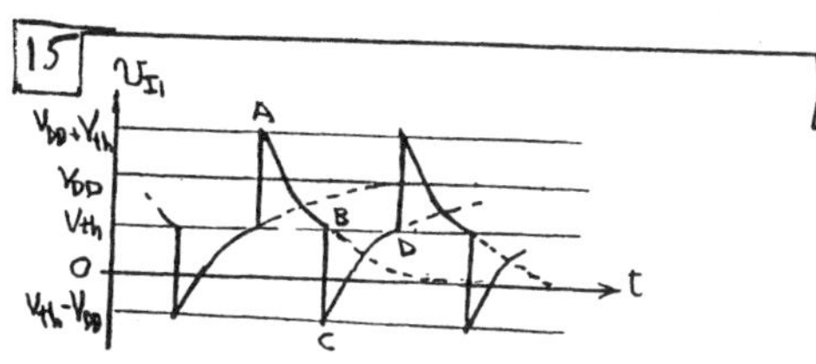

At Point A: $v_I = V_{DD} + V_{th}$, and thereafter
$v_I = (V_{DD} + V_{th} - 0)e^{-t/RC} + 0$

15 — 11

At Point B: $v_I = V_{th}$ and $t = t_1$, where
$V_{th} = (V_{DD} - V_{th})e^{-t_1/RC}$
or $t_1 = -RC\ln[V_{th}/(V_{DD}+V_{th})]$
or $t_1 = RC[(V_{DD}+V_{th})/V_{th}]$

At Point C: $v_I = V_{th} - V_{DD}$, and thereafter
$v_I = (V_{th} - V_{DD} - V_{DD})e^{-t/RC} + V_{DD}$

At Point D: $v_I = V_{th}$ at $t = t_2$, where
$V_{th} = (V_{th} - 2V_{DD})e^{-t_2/RC} + V_{DD}$
or $t_2 = -RC\ln[(V_{DD}-V_{th})/(2V_{DD}-V_{th})]$
or $t_2 = RC\ln[(2V_{DD}-V_{th})/(V_D - V_{th})]$

Thus, the period $T = t_1 + t_2$
whence $T = RC\left[\ln\frac{V_{DD}+V_{th}}{V_{th}} + \ln\frac{2V_{DD}-V_{th}}{V_{DD}-V_{th}}\right]$

or $T = RC\ln\left[\frac{V_{DD}+V_{th}}{V_{th}} \times \frac{2V_{DD}-V_{th}}{V_{DD}-V_{th}}\right]$, as suggested.

16 **11**

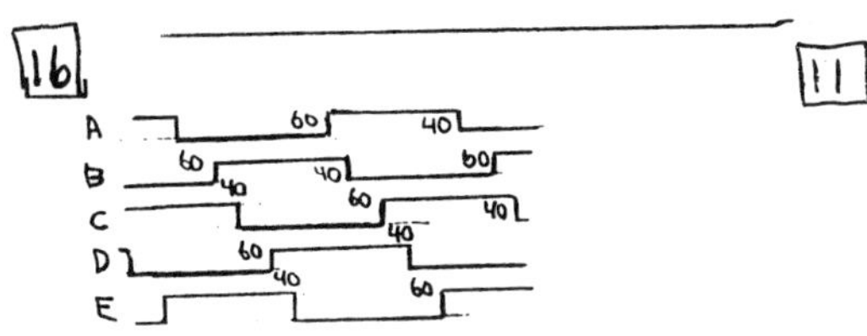

16 **11**

Waveform A is low for 60+40+60+40+60 = <u>260ns</u>
and high for 40+60+40+60+40 = <u>240ns</u>

Period is (260 + 240) = <u>500ns</u>
Frequency is 1/500 = <u>2MHz</u>

Percentage of cycle for which output is high is <u>48%</u>
<u>Check</u>: $t_p = (60+40)/2 = 50ns$ ✓

17 **11**

For 11 inverters, there are 2(11) = 22 transitions whose average length is $t_p = (1/20\times10^6)/22 = \underline{2.27ns}$

18 **11**

A 1Mb array requires n address bits where
$2^n = 10^6$, or $n\log_{10}2 = 6$, $n = 6/\log_{10}2 = 19.93$
Thus 20 bits are needed to address every cell.

For 16-bit words, $2^4 = 16$ and 4 bits are not needed.
Thus 20−4 = <u>16 bits</u> of address are sufficient.

<u>Check</u>: $m = \log_{10}(10^6/16)/\log_{10}(2) = 4.796/0.301 = 15.93$
Use 16 ✓

<u>Note</u>: A "1Mb array" actually holds $2^{20} = 1024^2 = 1{,}048{,}576$ cells.

19 **11**

Since the array is square, there are about $(10^6)^{1/2}$ or 1000 word lines. More precisely, there are $(2^{20})^{1/2} = 2^{10} = \underline{1024}$ word lines.

For a 1-Mb square array, there are also 1024 bit lines.

Thus a straightforward design would use <u>1024</u> sense amplifiers/drivers.

Now for dynamic storage on C operating at f with voltage V, the dynamic power is $P_D = fCV^2$.
Thus $C = P_D/(fV^2) = 500\times10^{-3}/(1/200\times10^9\times5^2)$
or C = <u>4000pF</u>

If 90% of the dynamic loss is in the array, the array capacitance is 4000(0.9) = <u>3600pF</u>
For 16 bit lines selected and active at a time, the capacitance per bit line could be 3600/16 = <u>225pF</u>
which is 225/1024 = <u>220fF/bit</u>.
For voltage reduction to 3V, power is reduced by a factor $3^2/5^2 = \underline{0.36}$. For the same power level the array could be larger by a factor 1/0.36 = <u>2.8 times</u>

20 **11**

Cell area = $\frac{1}{2}(21\times10^{-3}\times31\times10^{-3})/10^9 = \underline{0.326\mu m^2}$.
For 2 cells, the area is $0.65\mu m^2$, for which (if square) the side length is $\sqrt{0.65} = \underline{0.807\mu m}$
and the cell dimension is about <u>0.4μm by 0.8μm</u>

21 **11**

The cell area is $10^9\times0.38\times10^{-6}\times0.76\times10^{-6} = 0.289\times10^{-3}m^2$

The chip area is $19\times10^{-3}\times38\times10^{-3} = \underline{0.722\times10^{-3}m^2}$

Thus the peripheral circuits and interconnect occupy $(0.722-0.289)10^{-3} = 0.433mm^2$,
or (433/722)×100 = <u>60%</u> of the chip area.

22 **11**

For 16 blocks, <u>4 bits</u> of block address are needed. The arrays are each of 256M/16 = 16Mbit size. If each block array is square, it has 4096 rows and 4096 columns. It needs $\log_2 4096$ = <u>12 bits</u> each of row and column address bits, for a total of 2(12) + 4 = <u>28 bits</u> in all.

<u>Check</u>: $2^{28} = 268\times10^6$, or 256M bits. ✓

23 [14]

a) For W high and $\bar{B}$ high, with $\bar{Q}$ originally low and the gate of Q_1 high:

At $v_{\bar{Q}} = 2.5V$, from Eq. 5.30, $V_t = V_{t0} + \gamma[\sqrt{V_{SB}+2\phi_f} - \sqrt{2\phi_f}]$
or $V_t = 0.8 + 0.5[\sqrt{2.5+0.6} - \sqrt{0.6}] = \underline{1.293V}$

Now, for $I_5 = I_1$, $\frac{1}{2}k_{n5}(v_{GS5} - V_t)^2 = \frac{1}{2}k_{n1}(v_{GS1} - V_t)^2$
or $(W/L)_5(2.5-1.293)^2 = n(5-0.8)^2$
or $(W/L)_5 = n(4.2)^2/1.207^2 = \underline{12.1n}$.

b) For W high and B low, with Q originally high and the gate of Q_6 high:

At $v_Q = 2.5V$, with Q_4 in saturation and Q_6 in triode:
$\frac{1}{2}k_{p4}(5-0.8)^2 = k_{n6}[(5-0.8)2.5 - 2.5^2/2]$
or $\frac{1}{2}n(4.2)^2 = (W/L)_6[7.375]$
or $(W/L)_6 = n(4.2^2)/(2(7.375)) = \underline{1.20n}$

c) Clearly, for a very conservative design, one in which both write mechanisms work, the larger of the choices in a) b) would be used, namely $(W/L)_5 = (W/L)_6 = 12.1\ n$, very large!

In practice, this design needs a lot of area, and, instead, a conservative version of b) would likely be used, with $(W/L)_5 = (W/L)_6$ larger than 1.2n, say by a factor of 2, perhaps $\underline{2.4n}$

d. [11]

For $(W/L)_6 = 12.1n$, and matched inverters,
$k_4' = 50\times10^{-6}n = 50\times10^{-6}\times2 = 100\mu A/V^2$,
and $k_6' = 1210\mu A/V^2$

For $v_Q = 5V$: $v_{SD4} = 0$ and $i_{D4} = 0$, but
$i_{D6} = \frac{1}{2}(1210\times10^{-6})(5-0.8)^2 = \underline{10.7mA}$

For $v_Q = 2.5V$:
$i_{D6} = 1210\times10^{-6}[(5-0.8)2.5 - 2.5^2/2] = \underline{8.92mA}$
and $i_{D4} = 100\times10^{-6}[(5-0.8)2.5 - 2.5^2/2] = \underline{0.74mA}$

Thus, the average discharge current is $(10.7 - 0 + 8.92 - 0.74)/2 = \underline{9.44\ mA}$

Thus, the discharge time is $t = 50\times10^{-15}\times2.5/9.44\times10^{-3} = \underline{13.2ps}$

For a minimum-size design with $(W/L)_6 = 2.4n$, the average current would be:
$[(2.4/1.2)[10.7-0+8.92]-0.74]/2 = \underline{1.58mA}$
and the time to $V_{DD}/2$ would be
$t = 50\times10^{-15}\times2.5/1.58\times10^{-3} = \underline{79ps}$

24 [11]

a) Initially, $I_5 = \frac{1}{2}k'_{n5}(v_{GS} - V_t)^2 = \frac{1}{2}(50)10^{-6}(10/2)(5-0.8)^2$
or $I_5 = 2.205mA$, and $I_1 = 0$, since $v_{DS1} = 0$.
Thus, $I_{CQ} = 2.205 - 0 = \underline{2.205mA}$

b) At the end of the interval, $v_{\bar{Q}} = 2.5V$, and
$V_{t5} = V_t + \gamma[\sqrt{v_{SB}+2\phi_f} - \sqrt{2\phi_f}]$
$= 1.0 + 0.5[\sqrt{2.5+0.6} - \sqrt{0.6}] = \underline{1.49V}$

Correspondingly, $I_5 = \frac{1}{2}(50)10^{-6}(10/2)(2.5-1.49)^2 = \underline{127.5\ \mu A}$
Now, Q_1 is in triode mode, with
$I_1 = 50\times10^{-6}(4/2)[(5-1)2.5 - 2.5^2/2] = \underline{687.5\mu A}$
Thus $I_{C\bar{Q}} = 127.5 - 687.5 = \underline{-560\mu A}$, negative!
Thus we see that Q_5 is incapable of raising $\bar{Q}$ to 2.5V. This is the case even if the body effect is ignored (check this for interest!).
(You may also be interested in calculating the highest voltage to which $\bar{Q}$ can be raised.)

c) The average value is not relevant.

d) Δt is ∞ as defined here!
Note that in Exercise 11.9, $\Delta t = 69.4\ ps$, during which time the voltage on the gate of Q_1 reduces, the current in Q_1 reduces and the current in Q_2 increases, allowing Q_5 to become more effective.

25 [11]

For $v_W = 2.5V$, $v_B = 2.5V$, $v_{\bar{Q}} = 0V$:
$i_5 = \frac{1}{2}k_n'(v_{GS} - V_t)^2 = \frac{1}{2}(50)10^{-6}(10/2)(2.5-1)^2 = \underline{281\mu A}$

However, the voltage at Q will rise due to this current where $281 = 50(4/2)[(5-1)v_{Q_1} - v_{Q_1}^2/2]$.
Thus $v_{Q_1}^2 - 8v_{Q_1} + 5.62 = 0$,
whence $v_{Q_1} = (--8 \pm \sqrt{8^2 - 4(5.62)})/2$
$= (8 \pm 6.44)/2 = \underline{0.78V}$

This reduces i_5 to $i_5 = \frac{1}{2}(50)(10/2)(2.5-0.78-1)^2 = \underline{64.8\mu A}$
for which $v_{Q_1}^2 - 8v_{Q_1} + (0.648\times2) = 0$
whence $v_{Q_1} = (--8 \pm \sqrt{8^2 - 4(0.648\times2)})/2$
$= \underline{0.165V}$

Try $v_{Q_1} = (0.165 + 0.78)/2 = \underline{0.47V}$
for which $i_5 = 125(1.5 - 0.47)^2 = \underline{133\mu A}$
and $v_{Q_1} = (8 \pm \sqrt{64 - 4(1.33\times2)})/2 = \underline{0.347V}$

Try $v_{Q_1} = (0.347 + 0.47)/2 = 0.41 \approx 0.40V$
for which $i_5 = 125(1.5 - 0.40)^2 = \underline{151\mu A}$
and $v_{Q_1} = (8 \pm \sqrt{64 - 4(151\times2)})/2 = 0.397 \approx \underline{0.40V}$
with $i_5 = \underline{150\mu A}$

Now, for the bit-line voltage to reduce by 0.2V, it will take

$t = (1\times10^{-12}\times0.2)/150\times10^{-6} = \underline{1.33ns}$

26 | **11**

The bit-line capacitance is $[n(2)+20] = 2n + 20\,fF$

For a 5V supply and precharge to $V_{DD}/2$, there is a 2.5V change on the 50fF cell capacitor to produce a 0.1V bit-line signal.

Thus $0.1 = 2.5(50/(2n+20+50))$

or $2n + 70 = 2.5(50)/0.1 = 1250$

or $2n = 1180$ and $n = 590 > 512$

Thus we can use 512 rows for which $\log_2 512 = 9$ or 9 bits addressing is needed.

For a sense amplifier of 5× greater gain

$0.1/5 = 125/(2n+70)$

$2n + 70 = 5(125)/0.1 = 6250$

$2n = 6180$, and $n = 3090 > 2048$

To address 2048 rows requires 11 bits

---

27 | **11**

If the memory array has n columns, it has 2n rows and $2n^2$ cells

Refresh time is $2n(20)10^{-9}s = (1.00 - 0.98)8 \times 10^{-3}s$

whence $n = 0.02 \times 8\times10^{-3}/40\times10^{-9} = 4000$

The corresponding memory capacity is $2n^2 = 2(4000)^2$

or 32 Mbits

---

28 | **11**

For a 1Mb square array there are 1024 rows and 1024 columns.

Thus the bit-line capacitance is $10^{-15}(1024\times1 + 12)$

or 1.036 pF

When storing a '1', the voltage on C_S is $(V_{DD} - V_t)$ or $(5 - 1.5) = 3.5V$. With precharge to $V_{DD}/2 = 2.5V$, the change in voltage on $C_S = 3.5 - 2.5 = 1.0V$,

For $C_S = 25fF$, the bit-line voltage resulting is $25/(25+1036)^{\times 1} = 23.6\,mV$

When storing a '0', the voltage on C_S is 0V and the change is $2.5 - 0 = 2.5V$ with a resulting bit-line signal of $25/(25+1036)^{\times 2.5} = 58.9\,mV$

---

29 | **11**

For leakage current I, the voltage change on C in time T is $V = IT/C$

Correspondingly, $1 = I \times 10\times10^{-3}/20\times10^{-15}$, and the maximum leakage is $I = 20\times10^{-15}/10\times10^{-3} = 2pA$

---

30 | **11**

Pattern the solution after the approach used in the solution of Example 11.3:

For the bit-line output to reach $0.9\,V_{DD} = 2.7V$ from $V_{DD}/2 = 1.5V$ in 2ns for an initial bit-line signal of $0.1/2 = 0.05V$:

$2.7 = 1.5 + 0.05e^{2/\tau}$

whence $2/\tau = \ln[(2.7-1.5)/0.05] = 3.178$

and $\tau = 2/3.178 = 0.629\,ns$

Thus $C/G_m = 0.629\times10^{-9}s$, and $G_m = 1\times10^{-12}/(0.629\times10^{-9})$
$= 1.589\,mA/V$

For matched inverters, $g_{mn} = g_{mp} = G_m/2 = 1.589/2 = 0.795\,mA/V$

Now, $g_m = k'(W/L)(V_{GS} - V_t)$

and $0.795\times10^{-3} = 100\times10^{-6}(W/L)_n[3.0/2 - 0.8)$

Thus $(W/L)_n = 0.795\times10^{-3}/(100\times10^{-6})/0.7 = 11.36$

Now, for devices assumed to have length $L = 1\mu m$ (or, alternatively, for each micron of device length)
$W_n = 11.36\mu m$ and $W_p = 3(11.36) = 34.1\mu m$

Now, for a differential input signal of 0.2V (and 0.1V on each bit-line), the response time is t, where $2.7 = 1.5 + 0.1e^{t/0.629}$,

whence $t = 0.629\ln(2.7-1.5)/0.1 = 1.56\,ns$

31 | 11

Note that for the inverters,

$k_n = k_n'(W/L)_n = 100(6/1.5) = 400\mu A/V^2$

$k_p = k_p'(W/L)_p = (100/2.5)(15/1.5) = 400\mu A/V^2$

Thus we see that the inverters are matched.

Generally, $i_D = \frac{1}{2} k_n (v_{GS} - V_t)^2$, and

$g_m = \partial i_D / \partial v_{GS} = k_n (v_{GS} - V_t)$

Now, at $v_{GS} = v_B = V_{DD}/2 = 3.3/2 = 1.65V$,

$g_{mn} = 400(1.65 - 0.8) = 340\mu A/V$

Thus $G_m = g_{mn} + g_{mp} = 2(340) = \underline{680\mu A/V}$

For a bit-line capacitance of 0.8pF, $\tau = C/G_m$

or $\tau = 0.8 \times 10^{-12} / 680 \times 10^{-6} = \underline{1.176ns}$

Now, for $0.9V_{DD}$ reached in 2ns, for a signal Δv,

$0.9(3.3) = 1.65 + \Delta v\, e^{2/1.176}$

or $\Delta v = (2.97 - 1.65)/5.478 = 0.241V$

Thus the initial voltage between B lines must be $2(0.241) = \underline{0.482V}$

If an additional 1ns is allowed: $t = 2 + 1 = 3ns$,

and $\Delta v = (2.97 - 1.65)/e^{3/1.176} = 0.103V$

allowing a signal to be used of $2(0.103) = \underline{0.206V}$

Now, with the original bit-line signal of 0.241V, and a delay of 3ns:

31 | 11

$2.97 = 1.65 + 0.241 e^{3/\tau}$

and $e^{3/\tau} = (2.97 - 1.65)/0.241 = 5.477$

$3/\tau = \ln(5.477) = 1.7006$

whence $\tau = 3/1.7006 = 1.764ns$

Thus $C = G_m \tau = 680 \times 10^{-6} \times 1.764 \times 10^{-9} = \underline{1.20pF}$

This is an increase (from 0.8pF) of $\left(\frac{1.2 - 0.8}{0.8}\right)100 = \underline{50\%}$

For the longer line, the initial delay to establish a suitable signal becomes 150% of 5ns = $\underline{7.5ns}$

32 | 11

a) For an initial difference between bit lines of ΔV, each bit-line signal is $\Delta V/2$.

For the rising line: $v_B = V_{DD}/2 + (\Delta V/2) e^{t/(C_B/G_m)}$

whence $e^{t/(C_B/G_m)} = (2/\Delta V)(0.9 - 0.5)V_{DD}$

$= 0.8V_{DD}/\Delta V$

Taking base-e logarithms and cross-multiplying,

$t = t_d = (C_B/G_m)\ln(0.8V_{DD}/\Delta V)$ as stated.

b) For a reduction of t_d to ½ the original value, G_m must be doubled. To do this, double the width of all transistors (or increase by a factor of $\underline{2.0\times}$)

(next)

32 | 11

c) Now, for $V_{DD} = 5V$, $\Delta V = 0.2V$, with the original design, $t_d = C_B/G_m \ln(0.8(5)/0.2)$

or $t_d = 3.00\, C_B/G_m$

For the modified design, $\Delta V = 0.2/4 = 0.05V$

and $t_d = C_B/G_m' \ln(0.8(5)/0.05) = 4.38\, C_B/G_m'$

For these to be equal, $3.00/G_m = 4.38/G_m'$

and $G_m' = 1.46\, G_m$.

Thus the transistors must be made 46% wider (or be increased by a factor of $\underline{1.46\times}$)

33 | 11

For the DRAM arrangement, the signal is applied to only one side; Thus in comparison to the SRAM treatment, the applied signal is only half as large.

Now, the specification must be met for either a '0' or a '1' stored. The worst case is a differential signal of 40mV (corresponding to a single-side signal of 20mV)

Thus $2.0 = 20 \times 10^{-3} e^{5/\tau}$, and $5 = \tau \ln(2/20\times10^{-3})$,

or $5 = \tau \ln(100) = 4.605\tau$, whence $\tau = \underline{1.086ns}$

(next)

33 | 11

For a 1pF bit-line capacitance, $G_m = C/\tau$

or $G_m = 1\times10^{-12}/1.086\times10^{-9} = 0.921mA/V$,

with $0.921 = 0.46mA/V$ from each transistor.

Now, for the n-channel device, $g_m = k_n'(W/L)_n(V_{GS} - V_t)$

or $0.46\times10^{-3} = 100\times10^{-6}(W/L)_n(2.5 - 1)$

Thus $(W/L)_n = (0.46/0.1)/1.5 = \underline{3.07}$

For matched inverters, $(W/L)_p = 2.5(3.07) = \underline{7.68}$

When a '1' is read, the response time will be

$t = \tau \ln(2/20\times10^{-3}) = 1.086 \ln 100 = \underline{5ns}$

(Note: this is as designed!)

When a '0' is read,

$t = 1.086\times10^{-9}\ln(2/((100/2)\times10^{-3})) = \underline{4.01ns}$

34 | 11

Here $2^n = 512$, $n\log_{10}2 = \log_{10}512$, $n = 2.709/0.301 = \underline{9.00}$

Thus the number of bits is $\underline{9}$

The decoder has $\underline{512}$ output lines, one of which is active (high). The NOR array requires true and complement input lines for each bit: $2\times9 = \underline{18}$

Each row uses 9 NMOS for a total of $9\times512 = \underline{4608}$ NMOS and $\underline{512}$ PMOS, for a total of $\underline{5120}$ transistors.

35 **11**

For a 256Kbit square array, there are $(256 \times 1024)^{1/2} = 512$ rows and columns

Number of column-address bits is $\log_2 512 = \underline{\underline{9}}$

Two multiplexors are needed, since both true and complement bit lines are required. For each multiplexor, there are 512 output lines.

For each (half) multiplexor, 512 NMOS are needed for a total of $\underline{\underline{1024}}$ NMOS pass gates.
For the 512 output NOR decoder itself, $512 \times 9 = \underline{\underline{4608}}$ NMOS and $\underline{\underline{512}}$ PMOS are needed.
The address-bit inverters need $\underline{\underline{9}}$ NMOS and $\underline{\underline{9}}$ PMOS
Overall, the need is for $1024 + 4608 + 9 = \underline{\underline{5641}}$ NMOS and $512 + 9 = \underline{\underline{521}}$ PMOS, for a total of $\underline{\underline{6162}}$ transistors.

---

36 **11**

From the solution to P11.35 above, a square 256K-bit array has 512 rows and 512 columns for which 9 row and 9 column address bits are needed

Check: $2^{9+9} = 2^{18} = 262144$ ✓

For the tree of Fig 11.28, 9 levels of pass gates are needed.

36 **11**

The total number of pass gates is
$N = 2+4+8+16+32+64+128+256+512$
See that $N = 2 + 2(N-512)$, or $N = 2 + 2N - 1024$, whence $N = 1022$
Thus a tree column decoder for 9 bits needs $\underline{\underline{1022}}$ pass transistors

For true and complement bit lines, a total of $2(1022) = \underline{\underline{2044}}$ pass transistors are needed. Compare this with the number required beyond the input inverters in P11.35 namely $6162 - 18 = \underline{6144}$

---

37 **11**

Note from Fig 11.29 that the output is high if no word is selected. Thus, logically, high must correspond to logic 0 (and no transistor, as noted in the text).

Correspondingly, the words stored in Fig 11.29 are 0100, 0000, 1000, 1001, 0101, 0001, 0110, and 0010.

---

38 **11**

Need $Z = X \times Y$

X	Y	Z	
00	00	0000	W_0
00	01	0000	W_1
00	10	0000	W_2
00	11	0000	W_3
01	00	0000	W_4
01	01	0001	W_5
01	10	0010	W_6
01	11	0011	W_7
10	00	0000	W_8
10	01	0010	W_9
10	10	0100	W_{10}
10	11	0110	W_{11}
11	00	0000	W_{12}
11	01	0011	W_{13}
11	10	0110	W_{14}
11	11	1001	W_{15}

V_{DD}

D_4 D_3 D_2 D_1

Note that a total of 14 NMOS and 4 PMOS are used.

---

39 **11**

a) For the PMOS, with $V_B = 2.5V$
$$i_D = (90/3)10^{-6}(12/1.2)[(5-1)2.5 - 2.5^2/2]$$
$$= 30 \times 10^{-6}(10)[4(2.5) - 2.5^2/2] = 2.0625 mA$$

Thus the average charging current is $\underline{\underline{2.06 mA}}$

Time for precharge, $t = CV/I$
whence $t = 1 \times 10^{-12}(5-0)/(2.06 \times 10^{-3}) = \underline{\underline{2.42 ns}}$

b) For the word-line rise, $\tau = RC = 5 \times 10^3 \times 2 \times 10^{-12} = \underline{10 ns}$

Here, $v_W = 5(1 - e^{-t/10})$
Thus the rise time (10% to 90%) is essentially the time t to 90%, where
$$0.9(5) = 5(1 - e^{-t/10})$$
$$e^{-t/10} = 0.1$$
and $t = -10 \ln(0.1) = \underline{\underline{23 ns}}$

At the end of one time constant, $t = \tau = 10 ns$, and $v_W = 5(1 - e^{-10/10}) = \underline{\underline{3.16V}}$

For discharge, $i_{Dav} = \frac{1}{2} k_n'(W/L)_n (v_{GS} - V_t)^2$
$= \frac{1}{2}(90)(3/1.2)(3.16-1)^2 = \underline{\underline{525 \mu A}}$
Thus, the bit-line voltage will lower by 1V in about $\Delta t = C\Delta V / i_{Dav} = 1 \times 10^{-12} \times 1 / (525 \times 10^{-6})$
$= \underline{\underline{1.90 ns}}$

40 11

(a) $V_{OH} = 0 - 0.75 = -0.75V$

$V_{OL} = 0 - 0.75 - IR = -(0.75 + IR)$

(b) $V_{th} = -(IR/2 + 0.75) = -(0.75 + IR/2)$

(c) For $i = 0.99I$, $v_{BE} = 750 + 25\ln 0.99 = 750$ mV

$i = 0.01I$, $v_{BE} = 750 + 25\ln 0.01 = 635$ mV

For $0.99I$ in Q_R, $v_I = -(0.75 + \frac{IR}{2}) - (0.750 - 0.635)$

$= -(0.875 + IR/2)$

(d) For $0.01I$ in Q_R, $v_I = -(0.75 + IR/2) + 0.115$

$= -(0.635 + IR/2)$

(e) $V_{IH} = -(0.635 + IR/2)$

$V_{IL} = -(0.875 + IR/2)$

(f) $NM_H = -0.75 - -(0.635 + IR/2)$

$= IR/2 - 0.115$

$NM_L = -(0.875 + IR/2) - -(0.75 + IR)$

$= IR/2 - 0.115$

(g) $V_{IH} - V_{IL} = -(0.635 + IR/2) - -(0.875 + IR/2)$

That is ; $IR/2 - 0.115 = 0.230$

and $IR = 2(0.345) = 0.690V$

(h) $V_{OH} = \underline{-0.75V}$; $V_{OL} = -0.75 - 0.69 = \underline{-1.44V}$;

$V_{IL} = -(0.875 + 0.345) = \underline{-1.22V}$;

$V_{IH} = -(0.635 + 0.345) = \underline{-0.98V}$;

$V_R = -(0.750 + 0.345) = \underline{-1.095V}$.

41 11

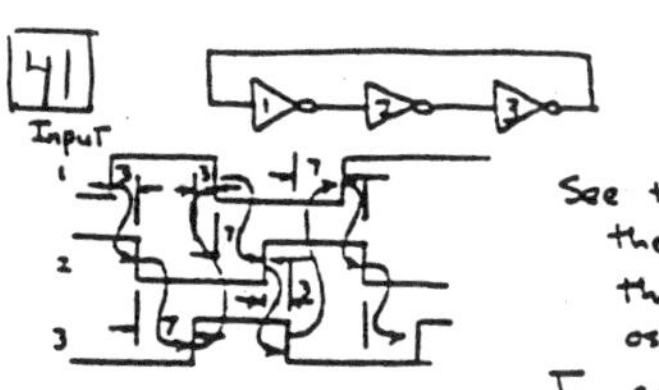

See that once started, the process continues; that is we have an oscillation.

In each cycle, each gate output rises and falls.

Thus the period is $3(3+7) = 30$ ns

Frequency is $1/30 = 33.3$ MHz.

Any output is high for $3+7+3 = 13$ ns

and low for $7+3+7 = \underline{17}$ ns

Check : 30 ns ✓

42 11

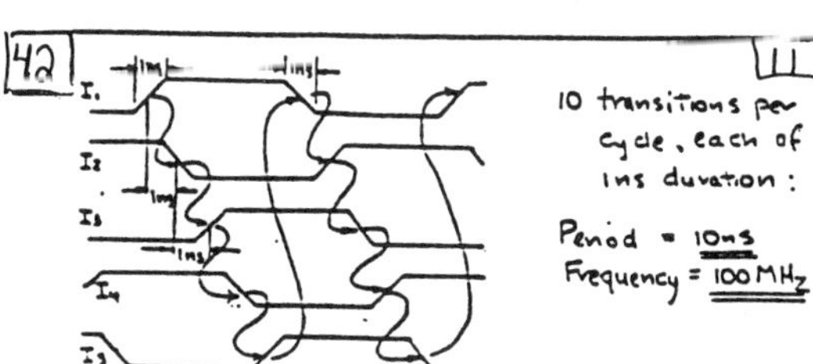

10 transitions per cycle, each of 1ns duration :

Period = $\underline{10ns}$

Frequency = $\underline{100 MHz}$

43 11

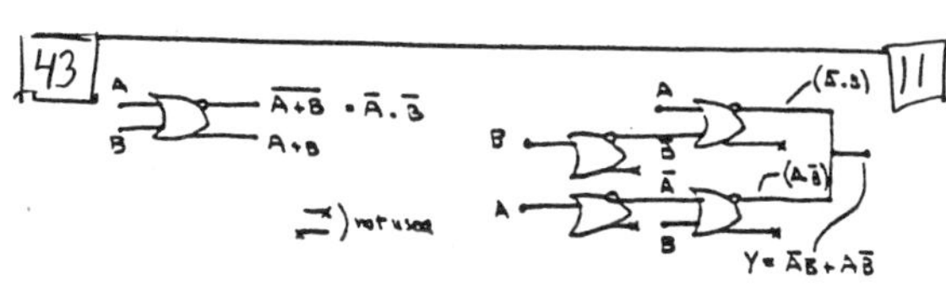

44 11

For $v_I = V_{IL} = -1.435V$, $I_E = 3.97$, $v_O = -1.77V$

For $v_I = V_R = -1.32V$, $I_E = 4.00mA$, $v_O = -1.31V$

For $v_I = V_{IH} = -1.205V$, $I_E = 4.12mA$, $v_O = -0.88V$

At x : $I_A = 0.01(3.97) = 39.7\mu A \rightarrow r_e = \frac{25\times10^{-3}}{39.7\times10^{-6}} = 630\Omega$

$I_B = 0.99(3.97) = 3.93mA \rightarrow r_e = \frac{25}{3.93} = 6.4\Omega$

$I_{E2} = \frac{2-1.77}{50} = 4.6mA \rightarrow r_e = \frac{25}{4.6} = 5.4\Omega$

and $R_{in2} = 101(50+5.4) = 5.6k\Omega$

$\therefore$ gain $= \frac{50}{50+5.4} \times (5.6k\Omega \| .245) \times \frac{100}{101} \frac{1}{630+6.4} = \underline{0.329 V/V}$

At m: $I_A = I_B = 2mA \rightarrow r_e = 25/2 = 12.5\Omega$

$I_{E2} = \frac{2-1.31}{50} = 13.8mA \rightarrow r_e = 1.81\Omega$

and $R_{in2} = 101(50+1.81) = 5.23k\Omega$

$\therefore$ gain $= \frac{50}{50+1.81} \times (5.2k\Omega \| .245) \times \frac{100}{101} \times \frac{1}{12.5+12.5} = \underline{8.94 V/V}$

At y : $I_A = 0.99(4.12) = 4.08mA \rightarrow r_e = \frac{25}{4.08} = 6.13\Omega$

$I_B = 0.01(4.12) = .041mA \rightarrow r_e = \frac{25}{.041} = 610\Omega$

$I_{E2} = \frac{2-0.88}{50} = 22.4mA \rightarrow r_e = 1.1\Omega$

and $R_{in2} = 101(50+1.1) = 5.2k\Omega$

$\therefore$ gain $= \frac{50}{50+1.1} \times (5.2 \| 0.245) \times \frac{100}{101} \times \frac{1}{610+6} = \underline{0.368 V/V}$

45 11

Assume I_E is constant at 4mA.

(a) Currents are : 3.6mA and 0.4mA

$\therefore$ Emitter-Base voltage difference $= V_T \ln \frac{3.6}{0.4}$

or $= 25\ln 9 = 54.9$ mV.

Thus $V_{IL} = -1.32 - .055 = \underline{-1.375V}$

$V_{IH} = -1.32 + .055 = \underline{-1.265V}$

(b) Currents are : $4(0.999) = 3.996mA$ and $.001 \times 4 = 4\mu A$

$\therefore$ Emitter-Base voltage difference $= V_T \ln \frac{3.996}{.004}$

or $= 25\ln 999 = 173$ mV.

Thus $V_{IL} = -1.32 - 0.173 = \underline{-1.493V}$

$V_{IH} = -1.32 + 0.173 = \underline{-1.147V}$

46 [11]

$I_{OL} = \frac{2-1.77}{50} = 4.6\text{ mA}$

$P_D = 2(4.6) = 9.2\text{ mW}$

$I_{OH} = \frac{2-0.88}{50} = 22.4\text{ mA}$

$P_D = 2(22.4) = 44.8\text{ mW}$

For the logic-gate core $P_D \simeq 4\text{mA} \times 5.2\text{V} = 20.8\text{mW}$.

Thus total $P_D = 9.2 + 44.8 + 20.8 = \underline{74.8\text{mW}}$ (ignoring R_3)

---

47 [11]

$NM_H = 0.325\text{V}$, of which 50% is 162mV, for $\beta = 100$ and $V_{BE2} = 0.83\text{V}$, $I_{E2} = 22.4\text{mA}$.

Approximately:

$-2 + \frac{50}{50 + \frac{245}{\beta+1}} \cdot (2 - 0.83) = -0.88 - 0.162$

or $\frac{50(1.17)}{50 + \frac{245}{\beta+1}} = 0.958$

$50 + \frac{245}{\beta+1} = 61.06$

whence $\beta = \frac{245}{11.06} - 1 = \underline{21.2}$

Check: For $v_O = -.88 - .162 = -1.042\text{ V}$

$I_{E2} = \frac{2-1.042}{50} = 19.2\text{ mA}$

and $V_T \ln \frac{22.4}{19.2} = 3.85\text{ mV}$ — OK, since small, can ignore.

---

48 [11]

v, -0.88 V, C, -1.77V, -2V, $R = 50$, -2V

$v = -0.88 + (.88 - 2)(1 - e^{-t/RC})$

or $v = -2 + 1.12\, e^{-t/50C}$

After 1ns, $v = -1.77\text{V}$

ie $-1.77 = -2 + 1.12\, e^{-1/50C}$

or $e^{-1/50C} = \frac{2-1.77}{1.12}$ and $-1/50C = -1.583$

Thus $C = \frac{10^{-9}}{50(1.583)} = 12.6 \times 10^{-12}\text{ F} = \underline{12.6\text{ pF}}$

---

49 [11]

$v = 2/3 \times 30\text{ cm/ns} = 20\text{ cm/ns}$

Ratio, $\frac{\text{Rise Time}}{\text{Return Time}} = 5/1 = \frac{3.5}{2L/20}$

$L = \frac{3.5\text{ns} \times 20\text{ cm/ns}}{5 \times 2} = \underline{7\text{ cm}}$

---

50 [11]

$V_{OL} \approx 0.7\text{V}$; $V_{OH} = +5.0\text{V}$

More precisely, for $V_{OL} = v$.

+5V, 2.5kΩ, v, 0.7V, $18k \| 51 = 353\Omega$

$v = \frac{0.353}{2.5 + .353} \cdot (5 - 0.7) + 0.7 = 1.23\text{V}$

ie $V_{OL} = \underline{1.23\text{V}}$

Logically: A is high if one of A or B and one of C or D are high.

That is $\underline{A = (A+B)\cdot(C+D)}$

---

51 [11]

For $v_I = v_O = V_{DD}/2 = 5/2 = 2.5\text{V}$,

$i_{DN} = \frac{1}{2} k_n' (W/L)_n (v_{GS} - V_t)^2$

$= \frac{1}{2}(100)10^{-6}(2/1)(2.5 - 0.7 - 1)^2 = \underline{64\,\mu\text{A}}$

Now, the collector current of $Q_2 = \beta i_B = \beta i_{DN}$

$= 100(64 \times 10^{-6}) = \underline{6.4\text{ mA}}$

Correspondingly, the totem-pole current is

$i_{EQ2} = (6400 + 64)10^{-6} = \underline{6.46\text{mA}}$

Now, for $i_{EQ1} = i_{EQ2}$, $i_{DP} = i_{DN} = 64\,\mu\text{A}$

Thus $64 = \frac{1}{2}(100/2.5)(W/L)_p(5 - 2.5 - 0.7 - 1)^2$

whence $(W/L)_p = 2.5(2/1) = (\underline{5\mu\text{m}/1\mu\text{m}})$

Note that the latter could be seen directly!

52

At the threshold V_{th}, $v_O = v_I = V_{th} = v$, and the two MOS operate in saturation with equal currents. Thus $\frac{1}{2}(100/2.5)(2/1)(5 - v - 0.7 - 1)^2 = \frac{1}{2}(100)(2/1)(v - 0.7 - 1)^2$.

Thus, $(3.3 - v)^2 = 2.5(v - 1.7)^2$

and $(3.3 - v) = \pm\sqrt{2.5}(v - 1.7)$.

Usefully, $(3.3 - v) = (1.58v - 2.69)$,

whence $2.58v = 5.99$, and $v = V_{th} = \underline{2.32\text{V}}$

For this value, $i_{DN} = \frac{1}{2}(100)(2/1)(2.32 - 0.7 - 1)^2 = 38.4\,\mu\text{A}$

and the totem-pole current is $(\beta+1) i_{DN}$

or $101(38.4)10^{-6} = \underline{3.88\text{mA}}$

53 **11**

The problem as stated is very general, and, correspondingly, its solution can be long and complex. Assume for simplicity that the specifications of Problem 11.57 apply, with matched MOS having $(W/L)_p = 2.5(W/L)_n$.

For R_2: With $v_{DS} = V_t/3 = 1/3 = 0.333$ V

$i_{D_N} = 100(10^{-6})(2/1)[(5-0.7-1)0.3\dot{3} - 0.3\dot{3}^2/2] = \underline{209\,\mu A}$

Now, if 50% of this is lost in R_2,

$R_2 = 0.7/(0.50 \times 209) = \underline{6.70\ k\Omega}$

Now if 20% is lost in R_2,

$R_2 = 0.7/(0.20 \times 209) = \underline{16.7\ k\Omega}$

For R_1:

$i_{D_P} = (100/2.5)10^{-6}/2.5(2/1))[(5-0-1)0.3\dot{3} - 0.3\dot{3}^2/2] = \underline{256\,\mu A}$

Now, if 50% of this is lost in R_1,

$R_1 = (5-0.333)/(0.5 \times 256) = \underline{36.5\ k\Omega}$

Now, if 20% is lost in R_1

$R_1 = 2.5(36.5) = \underline{91.1\ k\Omega}$

In comparison:

For the 50% case, $R_1/R_2 = 36.5/6.70 = \underline{5.45}$

For the 20% case, $R_1/R_2 = 91.1/16.7 = \underline{5.45}$

(Why should their equality be obvious?)

Thus, in general $R_1/R_2 = \underline{5.45}$

---

54 **11**

For t_{PLH}:

At $v_O = 0$ V, $i_{DP} = 1/2(100/2.5)(2/1)(5-0-1)^2 = \underline{640\,\mu A}$

At $v_O = 2.5$ V, $i_{DP} = (100/2.5)(2/1)[(5-1)2.5 - 2.5^2/2] = \underline{550\,\mu A}$

Thus $i_{DPav} = (640 + 550)/2 = \underline{595\,\mu A}$

and $i_{Oav} = (100+1)595 = \underline{60.1\ mA}$

Thus $t_{PLH} = CV/I = 2\times10^{-12} \times 2.5/(60.1\times10^{-3}) = \underline{83.2\ ps}$

For t_{PHL}:

At $v_O = 5.0$ V, $i_{DN} = 1/2(100)(2/1)(5-0.7-1)^2 = \underline{1.09\ mA}$

At $v_O = 2.5$ V, $i_{DN} = 100(2/1)[(5-0.7-1)(2.5-0.7) - (2.5-0.7)^2/2] = \underline{864\,\mu A}$

Thus $i_{DNav} = (1089 + 864)/2 = \underline{977\,\mu A}$

and $i_{Oav} = 101(977\times10^{-6}) = \underline{98.6\ mA}$

Thus $t_{PHL} = CV/I = 2\times10^{-12}(2.5)/(98.6\times10^{-3}) = \underline{50.7\ ps}$

Thus $t_P = (83.2 + 50.7)/2 = \underline{67.0\ ps}$

Note that this solution embodies two assumptions:

1) Internal capacitances can be neglected.
2) Transitions are from ideal 0V and 5V output-signal levels.

If outputs of $(5-0.7) = 4.3$ V and $(0+0.7) = 0.7$ V apply, t_P becomes about

$67\times(2.5-0.7)/2.5 = \underline{48\ ps}$

55

For the circuit in Fig 11.45e, $R_1 = R_2 = 5\,k\Omega$ robs the base of some of its drive current, namely $0.7/5\times10^3 = \underline{140\,\mu A}$.

Using results from the solution of P11.52 above:

For t_{PLH}: $i_{Bav} = 595 - 140 = \underline{455\,\mu A}$

and $i_{Oav} = 101(455\times10^{-6}) = \underline{46.0\ mA}$

Thus $t_{PLH} = 2\times10^{-12} \times 2.5/4.6\times10^{-3} = \underline{108.7\ ps}$

For t_{PLH}: $i_{Bav} = 977 - 140 = \underline{837\,\mu A}$

and $i_{Oav} = 101(837\times10^{-6}) = \underline{84.5\ mA}$

Thus $t_{PHL} = 2\times10^{-12} \times 2.5/84.5\times10^{-3} = \underline{59.2\ ps}$

Thus $t_P = (59.2 + 108.7)/2 = \underline{84\ ps}$

---

56 **11**

For the BiCMOS NAND of Fig 11.47 to have a dynamic response somewhat like that of the inverter of Fig 11.45 e):

$$(W/L)_{PA} = (W/L)_{PB} = (W/L)_P$$

and $(W/L)_{NA} = (W/L)_{NB} = 2(W/L)_N$

---

57 **11**

A BiCMOS 2-input NOR is as shown:

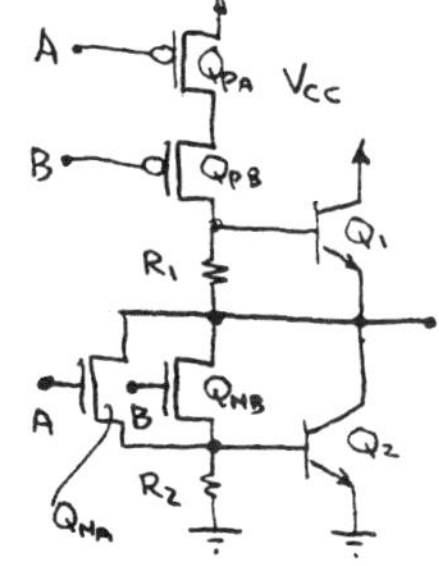

In terms of the basic matched inverter:

$(W/L)_{PA} = (W/L)_{PB} = 2(W/L)_P$

$(W/L)_{NA} = (W/L)_{NB} = (W/L)_N$

where $(W/L)_P$ and $(W/L)_N$ characterize the inverter.

CHAPTER 12 - PROBLEMS

12.1

$T(s) = \frac{\omega_o}{s+\omega_o}$ $T(j\omega) = \frac{\omega_o}{j\omega+\omega_o}$

$$|T(j\omega)| = \frac{\omega_o}{\sqrt{\omega_o^2+\omega^2}}$$

$$\phi(\omega) \triangleq \text{Tan}^{-1}\left[\frac{\text{Im}(T(j\omega))}{\text{Re}(T(j\omega))}\right]$$

$$= -\tan^{-1} \omega/\omega_o$$

$$G = 20\log_{10}|T(j\omega)|$$

$$A = -20\log_{10}|T(j\omega)|$$

ω	$\lvert T(j\omega)\rvert$ [V/V]	G [dB]	A [dB]	ϕ °
0	1	0	0	0
$0.5\omega_o$	0.8944	-0.97	0.97	-26.57
ω_o	0.7071	-3.01	3.01	-45.0
$2\omega_o$	0.4472	-6.99	6.99	-63.43
$5\omega_o$	0.1961	-14.1	14.1	-78.69
$10\omega_o$	0.0995	-20.0	20.0	-84.29
$100\omega_o$	0.010	-40.0	40.0	-89.43

12.2

$$T(s) = \frac{1}{(s+1)(s^2+s+1)}$$

$$= \frac{1}{s^3+2s^2+2s+1}$$

$$T(j\omega) = \left[j(2\omega-\omega^3)+(1-2\omega^2)\right]^{-1}$$

$$|T(j\omega)| = \left[(2\omega-\omega^3)^2+(1-2\omega^2)^2\right]^{-1/2}$$

$$= \left[4\omega^2-4\omega^4+\omega^6+1-4\omega^2+4\omega^4\right]^{-1/2}$$

$$= \left[1+\omega^6\right]^{-1/2}$$

$$= \frac{1}{\sqrt{1+\omega^6}}$$

For Phase Angle :

$$\phi(\omega) = \tan^{-1}\left[\frac{\text{Im}[T(j\omega)]}{\text{Re}[T(j\omega)]}\right]$$

$$= -\tan^{-1}\left[\frac{2\omega-\omega^3}{1-2\omega^2}\right]$$

For $\omega = 0.1$:

$$|T(j\omega)| = (1+0.1^6)^{-1/2} \cong 1$$

$$\phi(\omega) = -11.5° = -0.20 \text{ rad}$$

For $\omega = 1$ rad/s :

$$|T(j\omega)| = (1+1^6)^{-1/2} = 1/\sqrt{2} = 0.707$$

$$\phi = -\tan^{-1}(1/-1) = -135° = 2.356 \text{ rad}$$

Note : $G = -3$dB

Also : $\tan^{-1}(-1) = -45°$ or $-135°$
$\tan^{-1}(-1/1) = -45°$
$\tan^{-1}(1/-1) = -135°$

For $\omega = 10$ rad/s :

$$|T(j\omega)| = (1+10^6)^{-1/2} = 0.001$$

$$\phi = -\tan^{-1}\left[\frac{2(10)-10^3}{1-2(10^2)}\right]$$

CONT.

$$= -\tan^{-1}\left[\frac{-980}{-199}\right]$$

$$= -\left[180° + \tan^{-1}\left(\frac{980}{199}\right)\right]$$

$$= -258.5°$$

$$= \underline{\underline{4.512 \text{ rad}}}$$

Now consider an input of $A\sin\omega t$ to $T(s)$. The output is then given by:

$$A\,|T(j\omega)|\,\sin(\omega t + \phi(\omega))$$

Using this result, the output to each of the following inputs will be:

INPUT	OUTPUT
$2\sin(0.1t)$	$2\sin(0.1t - 0.2)$ ie $2 \times 1 = 2$
$2\sin(1t)$	$\sqrt{2}\sin(t - 2.356)$ i.e. $2 \times 1/\sqrt{2} = \sqrt{2}$
$2\sin(10t)$	$2\times10^{-3}\sin(10t - 4.512)$

12.3

12.4

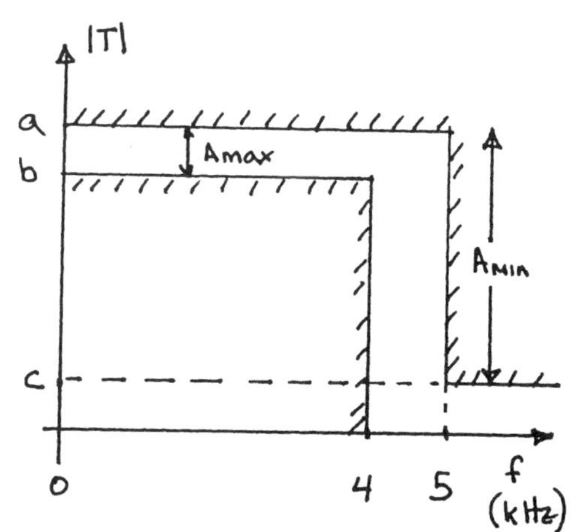

Note $|T|$ is shown in a linear scale but Amax and Amin are in dB

From the problem

$\frac{a}{b} = 1.1$, $c = 0.1\%\,a$ or $\frac{c}{a} = 0.001$

$$A_{max} = 20\log_{10} a - 20\log_{10} b$$

$$= 20\log_{10} a/b$$

$$= 20\log_{10}(1.1)$$

$$= \underline{\underline{0.83\text{dB}}}$$

$$A_{min} = 20\log_{10} a - 20\log_{10} c$$

$$= 20\log_{10}(a/c)$$

$$= 20\log_{10}(0.001)$$

$$= \underline{\underline{60\text{dB}}}$$

$$\text{Selectivity} = \frac{\omega_s}{\omega_p} = \frac{2\pi 5}{2\pi 4} = \underline{\underline{1.25}}$$

12.5

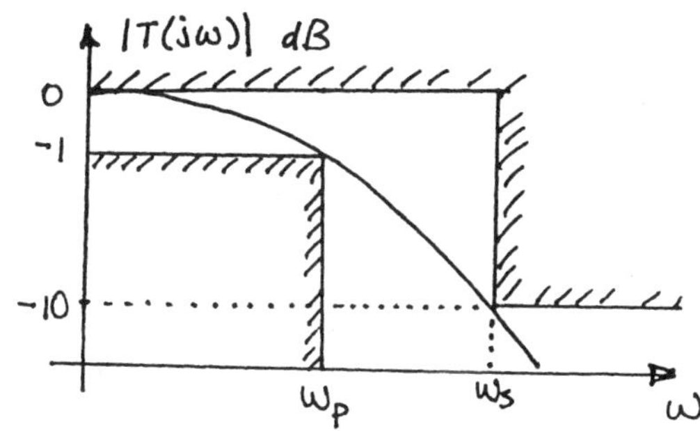

$$T(s) = \frac{k}{1+s\tau}$$

If $\tau = 1$s & the DC gain $= 1$ then $\underline{k = 1}$

$$= \frac{1}{1+1s}$$

$$|T(j\omega)| = \frac{1}{\sqrt{1+\omega^2}}$$

At the passband edge:

CONT.

$$|T(j\omega_p)| = \frac{1}{\sqrt{1+\omega_p^2}} = 10^{-1/20}$$

$$\therefore \omega_p = \underline{\underline{0.5088 \text{ rad/s}}}$$

At the stopband edge:

$$|T(j\omega_s)| = \frac{1}{\sqrt{1+\omega_s^2}} = 10^{-10/20}$$

$$\therefore \omega_s = \underline{\underline{3 \text{ rad/s}}}$$

$$\text{Selectivity} = \frac{\omega_s}{\omega_p} = \frac{3}{0.5088} = \underline{\underline{5.9}}$$

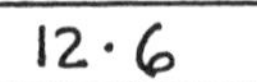

Passband is defined by : $f \geq 2\text{kHz}$

$\Rightarrow \omega_p = 2\pi(2000)$ rad/s

Stopband is defined by: $f \leq 1\text{kHz}$

$\Rightarrow \omega_s = 2\pi(1000)$ rad/s

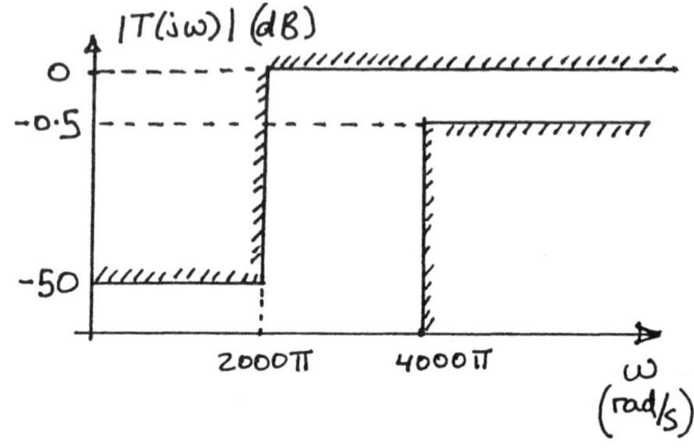

Note we assumed a maximum transmission of 0 dB.

12.7

Passband: $f \in \{[0, 10\text{kHz}] \cup [20\text{kHz}, \infty]\}$

Stopband: $f \in [12\text{kHz}, 16\text{kHz}]$

$A_{max} = 1\text{dB}$, $A_{min} = 40\text{dB}$

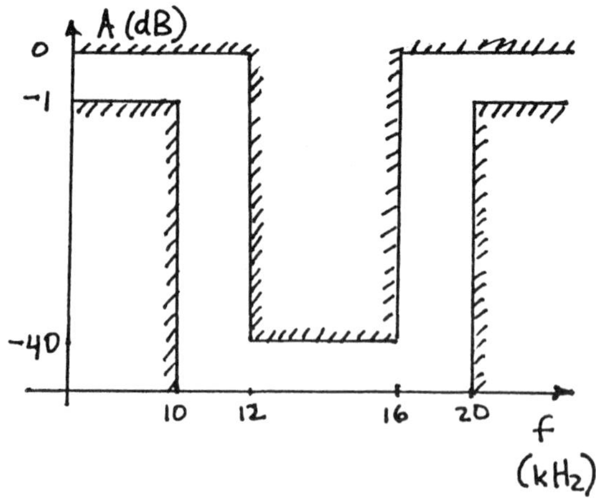

12.8

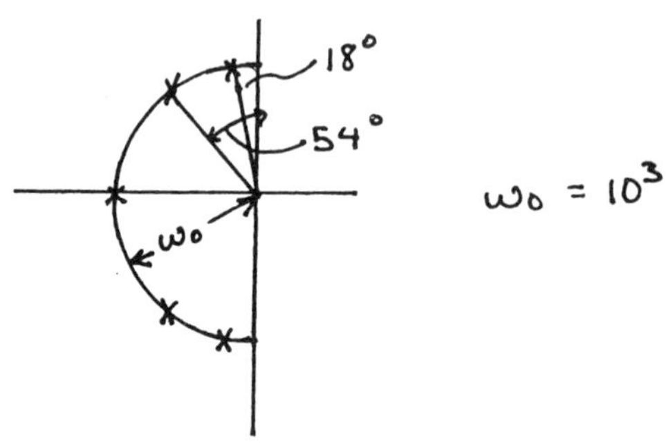

$\omega_0 = 10^3$

Poles at 18°:

$$P_1 = \omega_0\left(-\cos(90° - 18°) \pm j\sin(90° - 18°)\right)$$

$$= \omega_0\left(-\cos 72° \pm j\sin 72°\right)$$

$$= \omega_0(-0.309 \pm j\,0.951)$$

Poles at 54°

$$P_2 = \omega_0(-\cos 36° \pm j\sin 36°)$$

$$= \omega_0(-0.809 \pm j\,0.588)$$

Poles on Real Axis

$$P_3 = -\omega_0$$

CONT.

Note: Given a pair of poles

$P_i = \omega_0(-\cos\alpha \pm j\sin\alpha)$,

introduces a second order term as follows:

$(s+\omega_0\cos\alpha - j\omega_0\sin\alpha)(s+\omega_0\cos\alpha + j\omega_0\sin\alpha)$

$= s^2 + s(\omega_0\cos\alpha - j\omega_0\sin\alpha + \omega_0\cos\alpha + j\omega_0\sin\alpha)$

$+ \omega_0^2[\cos^2\alpha + j\cos\alpha\sin\alpha - j\cos\alpha\sin\alpha + \sin^2\alpha]$

$= s^2 + s(2\omega_0\cos\alpha) + \omega_0^2$

So for P_1 we get a term:

$s^2 + s(2\omega_0 0.309) + \omega_0^2$

$= s^2 + 0.618\omega_0 s + \omega_0^2$

For P_2 we get:

$s^2 + 1.618\omega_0 s + \omega_0^2$

For P_3: $(s+\omega_0)$

$\therefore$ The denominator of $T(s)$ is given by

$$D(s) = (s+\omega_0)(s^2 + 0.618\omega_0 s + \omega_0^2) \times (s^2 + 1.618\omega_0 s + \omega_0^2)$$

Case (a) - If all the zeros are @ ∞, the numerator is a constant

$|T(0)| = \frac{k}{\omega_0^5} = 1$ for unity gain at DC

$\therefore k = \omega_0^5$

$$T(s) = \frac{k}{D(s)} = \frac{\omega_0^5}{D(s)}$$

where $D(s)$ is given above.

Case (b) - For all zeros at 0, the numerator is given by ks^5

A $s = j\infty$ $|T(s \to j\infty)| = \frac{k}{1} = 1$

$$\therefore T(s) = \frac{s^5}{D(s)}$$

12.9

Poles at -1 and $-0.5 \pm j0.8$ gives a denominator:

$D(s) = (s+1)(s+0.5-j0.8)(s+0.5+j0.8)$

$= (s+1)(s^2 + 2(0.5)s + 0.5^2 + 0.8^2)$

$= (s+1)(s^2 + s + 0.89)$

Zeros at ∞ and $\pm j2$ give a numerator:

$N(s) = k(s+j2)(s-j2) = k(s^2+4)$

Note there is one zero at ∞ because Degree $(D(s))$ - Degree $(N(s)) = 1$

$T(s) = \frac{k(s^2+4)}{(s+1)(s^2+s+0.89)}$

$|T(j0)| = \frac{k(4)}{0.89} = 1$ $\because$ DC gain $= 1$

$\Rightarrow k = 0.2225$

$$\therefore T(s) = \frac{0.2225(s^2+4)}{(s+1)(s^2+s+0.89)}$$

12.10

Numerator is given by

$$a_7 (s-0)(s^2+(10^3)^2)(s^2+(3\times10^3)^2)\times(s^2+(6\times10^3)^2)$$

$$= a_7 s(s^2+10^6)(s^2+9\times10^6)(s^2+36\times10^6)$$

Degree of Numerator $\triangleq M = 7$

Degree of Denominator $\triangleq N$

Given that there is one zero at ∞:

$$N - M = 1 \Rightarrow \underline{\underline{N = 8}}$$

$$\therefore T(s) = \frac{a_7 s(s^2+10^6)(s^2+9\times10^6)(s^2+36\times10^6)}{s^8 + b_7 s^7 + b_6 s^6 + \cdots + b_0}$$

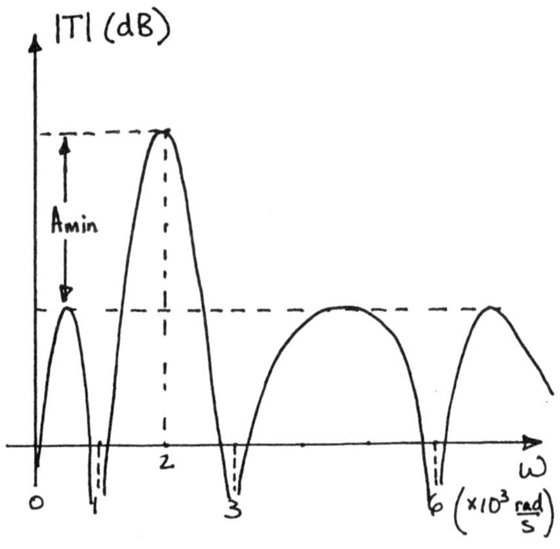

12.11

1 Ω, 2H, 1F, 1F, 1Ω; V_i, V_1, V_o; nodes ①, ②, ③

The easiest way to solve the circuit is to use nodal analysis at nodes ①, ②, ③

At node ③ $\Sigma I = 0$

$$\frac{V_o}{1} + \frac{V_o}{1/s} + \frac{V_o - V_1}{2s} = 0$$

$$\therefore V_1 = V_o(2s^2 + 2s + 1) \qquad Eq.(a)$$

At node ② $\Sigma I = 0$

$$\frac{V_1 - V_i}{1} + \frac{V_1}{1/s} + \frac{V_1 - V_o}{2s} = 0$$

$$\therefore V_1(2s^2+2s+1) = V_o + 2sV_i \qquad Eq.(b)$$

(a) → (b)

$$V_o(2s^2+2s+1)^2 = V_o + 2sV_i$$

$$V_o(4s^4 + s^3(4+4) + s^2(2+4+2) + s(2+2) + 1) = V_o + 2sV_i$$

$$\frac{V_o(s)}{V_i(s)} \triangleq T(s) = \frac{2s}{4s^4+8s^3+8s^2+4s}$$

$$\underline{\underline{T(s) = \frac{0.5}{s^3+2s^2+2s+1}}}$$

Poles are given by:

$$s^3 + 2s^2 + 2s + 1 = 0$$

$$(s+1)(s^2+s+1) = 0$$

$\therefore$ Poles are $s = \underline{-1}$ and

$$\underline{\underline{s = -\frac{1}{2} \pm j\sqrt{3}/2}}$$

12.12

$A_{max} = 1\text{dB}$, $A_{min} = 20\text{dB}$, $\omega_s/\omega_p = 1.3$

Using:

$$A(\omega_s) = 10\log\left[1+\epsilon^2\left(\frac{\omega_s}{\omega_p}\right)^{2N}\right] \qquad Eq(12.15)$$

$$= A_{min}$$

CONT

$\epsilon = \left[10^{1/10} - 1\right]^{1/2} = 0.5088$

$A_{min} = 10 \log\left[1 + \epsilon^2 \left(\frac{\omega_s}{\omega_p}\right)^{2N}\right]$

$10^{A_{MIN}/10} - 1 = \epsilon^2 \left(\frac{\omega_s}{\omega_p}\right)^{2N}$

$\log\left(10^{A_{MIN}/10} - 1\right) = \log\left(\epsilon^2 \left(\frac{\omega_s}{\omega_p}\right)^{2N}\right)$

$N = \frac{\log\left[\left(10^{A_{MIN}/10} - 1\right)/\epsilon^2\right]}{2 \log(\omega_s/\omega_p)}$

$= 11.3 \Rightarrow$ choose $N = 12$

The actual value of stopband attenuation can be calculated using the integer value of N:

$A(\omega_s) = 10 \log\left[1 + \epsilon^2 \left(\frac{\omega_s}{\omega_p}\right)^{2N}\right]$; $N = 12$

$= 27.35\,\text{dB}$ actual attenuation

If the stopband specs are to be met exactly we need to find Amax.

Eq. 12.15 can be rearranged to give

$\epsilon^2 = \frac{10^{A_{min}/10} - 1}{(\omega_s/\omega_p)^{2N}}$ $\quad A_{min} = 20$, $N = 12$

$= 0.1824$

$\therefore A_{max} = 10 \log(1 + \epsilon^2)$

$= 0.73\,\text{dB}$

12.13

$N = 7$, $A_{max} = 3\,\text{dB}$

We want attenuation at $\omega = 1.6\,\omega_p$ or $\frac{\omega}{\omega_p} = 1.6$

$\epsilon = \sqrt{10^{A_{max}/10} - 1} = 0.998$

$A(\omega) = 10 \log\left[1 + \epsilon^2 \left(\frac{\omega}{\omega_p}\right)^{2N}\right]$

$= 10 \log\left[1 + 0.998^2 (1.6)^{14}\right]$

$= 28.56\,\text{dB}$

12.14

$\omega_p = 10^3$ rad/s, $N = 5$

$A_{max} = 1\,\text{dB} \Rightarrow \epsilon = 0.5088$

Find solution graphically

$\frac{\pi}{2N}$, $\frac{\pi}{N}$, $\frac{\pi}{N}$, $\omega_p\left(\frac{1}{\epsilon}\right)^{\frac{1}{N}}$

$P_1 = \omega_p \left(\frac{1}{\epsilon}\right)^{1/N} \measuredangle \left(\frac{\pi}{2} + \frac{\pi}{2N}\right)$

$= 873.59 \measuredangle (6\pi/10)$

$= 873.59\left[\cos(6\pi/10) \pm j \sin\left(\frac{6\pi}{10}\right)\right]$

$= -269.96 \pm j\,830.84$

$P_2 = 873.59 \measuredangle \left[\frac{\pi}{2} + \frac{\pi}{2N} + \frac{\pi}{N}\right]$

$= -706.75 \pm j\,513.49$

$P_3 = 873.59 \measuredangle \pi = -873.59$

12.15

$f_p = 10\,\text{kHz}$, $f_s = 15\,\text{kHz}$ $\quad \frac{\omega_s}{\omega_p} = 1.5$ $\quad A_{min} = 15\,\text{dB}$, $A_{max} = 2\,\text{dB}$

$\epsilon^2 = 10^{A_{max}/10} - 1 \Rightarrow \epsilon = 0.76478$

CONT.

Manipulation Eq (12.15) we get:

$$N = \frac{\log\left[\left(10^{A_{min}/10}-1\right)/\epsilon^2\right]}{2\log(\omega_s/\omega_p)} = 4.88$$

∴ Use $\underline{N = 5}$

Finding naturalmodes graphically :-

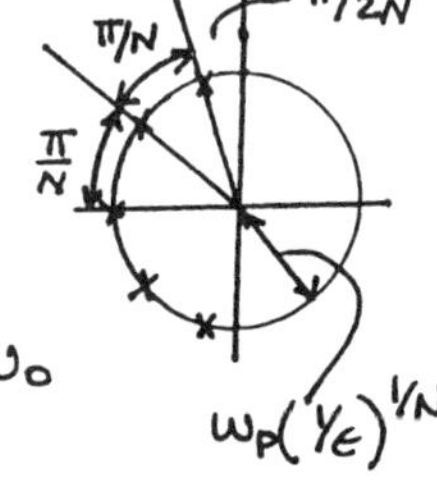

$$\text{radius} = \omega_p\left(\frac{1}{\epsilon}\right)^{1/N} \triangleq \omega_0$$

$$\omega_0 = 6.629\times10^4$$

$$P_1 = \omega_0 \angle (\pi/2 + \pi/2N) = \omega_0 \angle (6\pi/10)$$

$$= \omega_0\left(\cos\left(\frac{6\pi}{10}\right) \pm j\sin\left(\frac{6\pi}{10}\right)\right)$$

$$= \underline{\omega_0(-0.309 \pm j0.951)}$$

$$P_2 = \omega_0\left(\cos\frac{8\pi}{10} \pm j\sin\frac{8\pi}{10}\right)$$

$$= \underline{\omega_0(-0.809 \pm j0.588)}$$

$$P_3 = \omega_0(\cos\pi \pm j\sin\pi) = \underline{-\omega_0}$$

Given a natural mode $-\alpha \pm j\beta$, the following term results

$$(s+\alpha+j\beta)(s+\alpha-j\beta)$$

$$= s^2 + 2\alpha s + \alpha^2 + \beta^2$$

$$= \underline{s^2 + 2\mathrm{Re}[P]s + |P|^2}$$

Also, note that for a Butterworth, all natural modes have a magnitude of ω_0.

P_1 yields: $s^2 + 0.618\omega_0 s + \omega_0^2$

P_2 yields: $s^2 + 1.618\omega_0 s + \omega_0^2$

P_3 yields: $s + \omega_0$

$$\therefore T(s) = \frac{k}{(s+\omega_0)(s^2+.618\omega_0 s+\omega_0^2)} \times \frac{1}{s^2+1.618\omega_0 s+\omega_0^2}$$

For unity dc gain

$$|T(j0)| = \frac{k}{\omega_0^5} = 1 \Rightarrow k = \omega_0^5$$

$$\therefore T(s) = \frac{\omega_0^5}{(s+\omega_0)(s^2+.618\omega_0 s+\omega_0^2)} \times \frac{1}{(s^2+1.618\omega_0 s+\omega_0^2)}$$

For attenuation at 20 kHz use Eq (12.15) with $\frac{\omega_s}{\omega_p} = \frac{20}{10} = 2$

$$A(\omega_s) = 10\log\left[1+\epsilon^2\left(\frac{\omega_s}{\omega_p}\right)^{2N}\right]$$

$$= \underline{27.8\,\text{dB}}$$

12.16

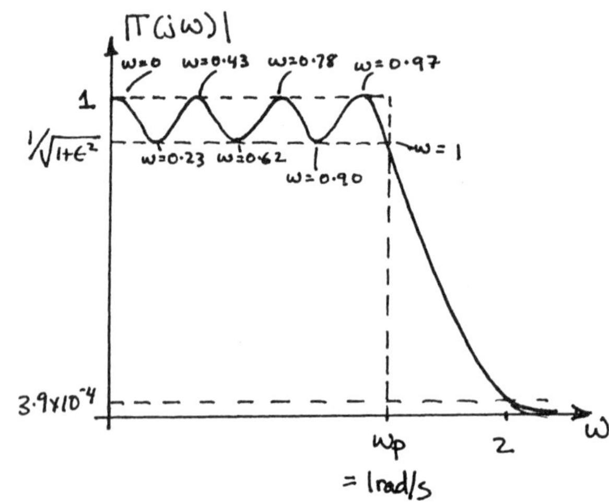

Given $A_{max} = 1\,\text{dB} \Rightarrow \epsilon = 0.5088$

CONT.

Using Eq 12·18

$$|T(j\omega)| = \left[1+\epsilon^2 \cos^2\left(N \cos^{-1}\left(\frac{\omega}{\omega_p}\right)\right)\right]^{-1/2}$$

for $\omega \le \omega_p$

If $|T(j\omega)| = 1$

$$1 = 1+\epsilon^2 \cos^2\left(N\cos^{-1}(\omega/\omega_p)\right) \quad \omega_p = 1$$

$$N\cos^{-1}(\omega/1) = \cos^{-1}(0)$$

$$\cos^{-1}(\omega) = \frac{2i+1}{2N}\pi$$

ω's repeat after this value

$$\therefore \omega_i = \cos\left[\frac{2i+1}{2N}\pi\right] \quad i = 0, 1, \ldots, \frac{N-1}{2}$$

$\omega_0 = 0.9749$
$\omega_1 = 0.7818$
$\omega_2 = 0.4339$
$\omega_3 = 0$

ω values at which $|T| = 1$

note $\omega_4 = -0.4339 = -\omega_2$!

If $|T| = 1/\sqrt{1+\epsilon^2}$, then

$$1/\sqrt{1+\epsilon^2} = \left[1+\epsilon^2\cos^2\left(N\cos^{-1}(\omega/\omega_p)\right)\right]^{-\frac{1}{2}}$$

$$1 = \cos\left(N\cos^{-1}\left(\frac{\omega}{\omega_p}\right)\right)$$

$$N\cos^{-1}(\omega) = \cos^{-1}(0)$$

$$= i\pi \quad i = 0, 1, 2 \ldots$$

$$\omega_i = \cos\left[\frac{i\pi}{N}\right] \quad i = 0, 1, 2, \ldots \frac{N}{2}$$

$\omega_0 = 1.0$
$\omega_1 = 0.9010$
$\omega_2 = 0.6235$
$\omega_3 = 0.2252$

ω values at which $|T| = (1+\epsilon^2)^{-1/2}$

Note $\omega_4 = -0.2252 = -\omega_3$!

To find $|T(j2)|$ use Eq(12·19) since $\omega > \omega_p$

$$|T(j\omega)| = \left[1+\epsilon^2\cosh^2\left(N\cosh^{-1}\left(\frac{\omega}{\omega_p}\right)\right)\right]^{-\frac{1}{2}}$$

$$= \left[1+0.5088^2\cosh^2\left(7\cosh^{-1} 2\right)\right]^{-\frac{1}{2}}$$

$$= \underline{\underline{3.898 \times 10^{-4} \text{ V/V}}}$$

$$|T|_{dB} = \underline{\underline{-68.2 \text{ dB}}}$$

For roll-off consider

$$T(s) = \frac{k}{s^7 + b_6 s^6 + \ldots + b_0}$$

for $\omega \gg \omega_p$ $\quad T(j\omega) \simeq \frac{k}{\omega^7}$

$\therefore$ Rolloff is $\frac{1}{2^7}$ or $20\log(1/2^7)$ per octave $= \underline{\underline{-42 \text{ dB/octave}}}$.

12.17

$\omega_s/\omega_p = 2 \quad A_{max} = 1\text{dB} \Rightarrow \epsilon = \sqrt{10^{\frac{A_{max}}{10}} - 1} = 0.5088$

$$|T_B| = \left[1+\epsilon^2(\omega_s/\omega_p)^{2N}\right]^{-1/2}$$

$$|T_c| = \left[1+\epsilon^2\cosh^2\left(N\cosh^{-1}\left[\frac{\omega_s}{\omega_p}\right]\right)\right]^{-1/2}$$

$$|T_B| = 6.13 \times 10^{-2} \Rightarrow \underline{\underline{-24.3\text{dB}}}$$

$$|T_c| = 5.43 \times 10^{-3} \Rightarrow \underline{\underline{-45.3\text{dB}}}$$

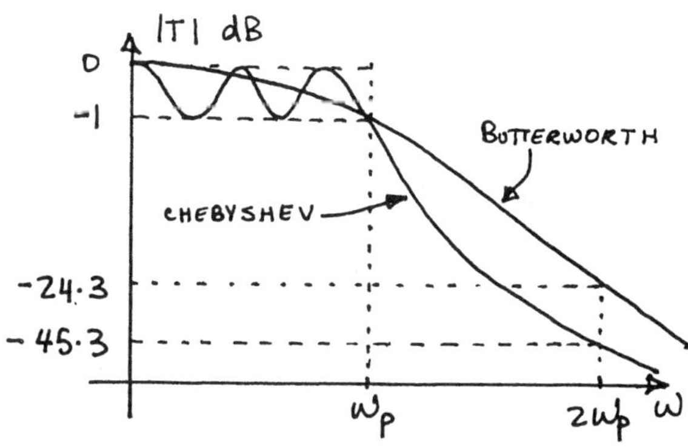

12.18

$f_p = 3.4\,kHz$ $\quad A_{max} = 1dB \Rightarrow \epsilon = 0.5088$

$f_s = 4\,Khz$ $\quad A_{min} = 35\,dB$

$\omega_s/\omega_p = 1.176$

Using Eq (12.22) :

$$A(\omega_s) = 10\log\left[1 + \epsilon^2 \cosh^2\left(N\cosh^{-1}\left(\frac{\omega_s}{\omega_p}\right)\right)\right]$$

& trying different values for N

N	$A(\omega_s)$
8	28.8 dB
9	33.9 dB
10	38.98 dB

$\therefore$ Use $N = 10$

Excess attenuation = 39 − 35 = 4 dB

Poles are given by:

$$P_k = -\omega_p \sin\left(\frac{2k-1}{N}\cdot\frac{\pi}{2}\right)\sinh\left(\frac{1}{N}\sinh^{-1}\left(\frac{1}{\epsilon}\right)\right)$$

$$+j\,\omega_p \cos\left(\frac{2k-1}{N}\cdot\frac{\pi}{2}\right)\cosh\left(\frac{1}{N}\sinh^{-1}\left(\frac{1}{\epsilon}\right)\right)$$

for $k = 1, 2, \ldots, N$.

Since $\epsilon = 0.5088$ and $N = 10$

$$\sinh\left(1/N \sinh^{-1}(1/\epsilon)\right) = 0.1433$$

$$\cosh\left(1/N \sinh^{-1}(1/\epsilon)\right) = 1.010$$

$$\therefore P_1 = \omega_p\left[-0.1433\sin\left(\frac{\pi}{20}\right) + j1.010\cos\left(\frac{\pi}{20}\right)\right]$$

$$= \omega_p(-0.0224 + j0.9978)$$

$$P_2 = \omega_p(-0.0650 + j0.900)$$

$$P_3 = \omega_p(-0.1013 + j0.7143)$$

$$P_4 = \omega_p(-0.1277 + j0.4586)$$

$$P_5 = \omega_p(-0.1415 + j0.1580)$$

Now it should be realized that the remaining poles are complex conjugates of the above.

Pole-pair P_1 & P_1^* give a factor:

$$s^2 + 2(0.0224)\omega_p s + \omega_p^2(0.0224^2 + 0.9978^2)$$

$$= s^2 + 0.0448\omega_p s + 1.023\omega_p^2$$

i.e. this factor is from $(s-p_1)(s-p_1^*)$

P_2 yields: $s^2 + 0.130\omega_p s + 0.902\omega_p^2$

P_3 yields: $s^2 + 0.203\omega_p s + 0.721\omega_p^2$

P_4 yields: $s^2 + 0.255\omega_p s + 0.476\omega_p^2$

P_5 yields: $s^2 + 0.283\omega_p s + 0.212\omega_p^2$

Now T(s) is given by

$$T(s) = \frac{k\omega_p^{10}}{\epsilon\, 2^9 (s-P_1)(s-P_1^*)\cdots(s-P_5)(s-P_5^*)}$$

where the second order terms of the denominator are given above.

k is the dc gain

$\therefore$ we want the dc gain to be

$$k = \frac{1}{1+\epsilon^2} = 0.8913$$

$$\omega_p = 2\pi \times 3400$$

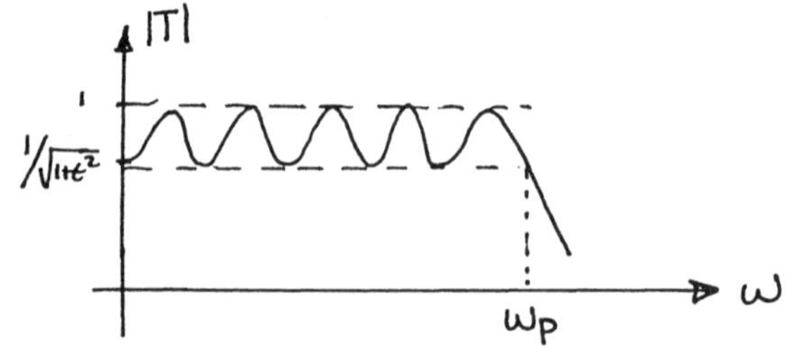

12.19

$f_0 = 10\text{kHz}$ DC gain = 10 $R_{in} = 10\text{k}\Omega$

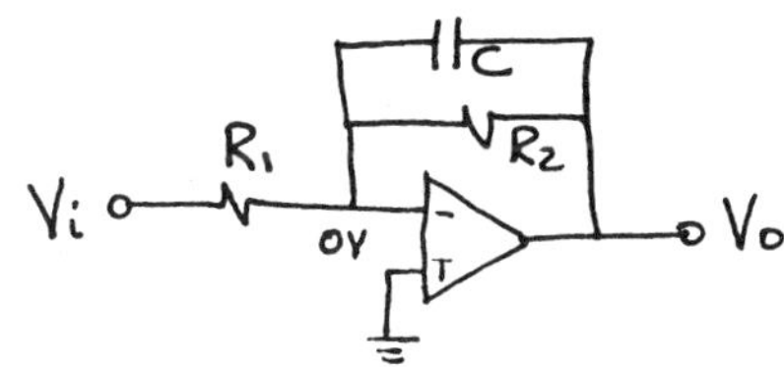

$R_{in} = R_1 = \underline{10\text{k}\Omega}$

$\text{DC gain} = -R_2/R_1 = -10$

$R_2 = 10R_1 = \underline{100\text{k}\Omega}$

$R_2 C = 1/\omega_0$

$C = \frac{1}{\omega_0 R_2} = \frac{1}{2\pi\, 10^4 \times 100\times10^3}$

$= \underline{0.159\text{nF}}$

12.20

$f_0 = 100\text{kHz}$ $R_i(\infty) = 100\text{k}\Omega$ $|T(\infty)| = 1$

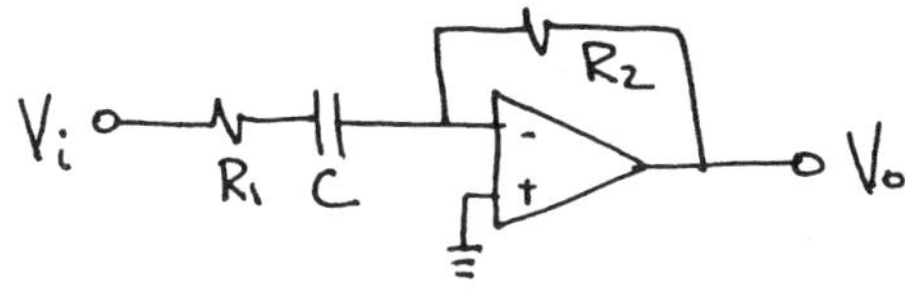

$R_i(\infty) = R_1 = \underline{100\text{k}\Omega}$

$|T(\infty)| = R_2/R_1 = 1$

$R_2 = R_1 = \underline{100\text{k}\Omega}$

$CR_1 = 1/\omega_0$

$C = \frac{1}{\omega_0 R_1} = \frac{1}{2\pi\, 100\times10^3 \times 100\times10^3}$

$= \underline{15.9\text{nF}}$

12.21

Use general first-order circuit:

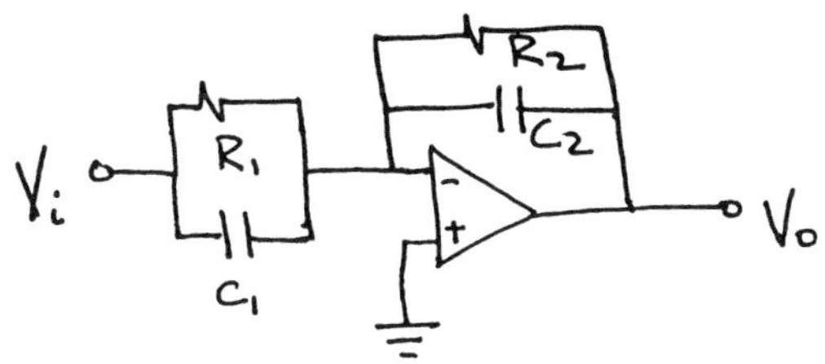

- Zero at 1kHz ; Pole at 100kHz
- $|T(0)| = 1$; $R_i(0) = 1\text{k}\Omega$

Thus: $R_i(\text{DC}) = R_1 = \underline{1\text{k}\Omega}$

$T(\text{DC}) = -R_2/R_1 = -1$

$R_2 = R_1 = \underline{1\text{k}\Omega}$

For a pole at 100kHz

$C_2 R_2 = \frac{1}{\omega_0} \Rightarrow C_2 = \frac{1}{2\pi f_0 R_2}$

$= \underline{1.59\text{nF}}$

For the circuit $T(s) = \frac{a_1 s + a_0}{s + \omega_0}$

Thus the zero at $-a_0/a_1 = -2\pi\, 10^3$

$C_1 R_1 = a_1/a_0$

$C_1 = \frac{1}{2\pi\, 10^3 R_1}$

$= \underline{159\text{nF}}$

High freq gain $= \frac{-C_1}{C_2} = \underline{-100}$

$= \underline{40\text{dB}}$

12.22

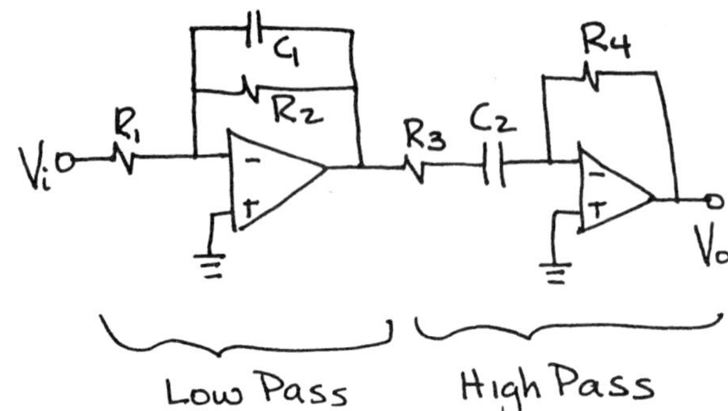

Low Pass   High Pass

gain $= 10^{12/20} = 3.98 \approx \underline{4}$

want $R_i = R_1$ large

$\therefore R_1 = \underline{\underline{100k\Omega}}$

Total gain $= A_{LP} A_{HP} = 4$

$A_{LP} = -R_2/R_1 \Rightarrow R_2 = -A_{LP} R_1$ and

$R_2 \leq 100k\Omega$

$\therefore$ make $A_{LP} = -1 \quad A_{HP} = -4$

$R_2 = \underline{\underline{100k\Omega}}$

$$R_2 C_1 = \frac{1}{\omega_{0,LP}}$$

$$C_1 = \frac{1}{2\pi f_{0,LP} R_2} = \frac{1}{2\pi(10\times10^3)\,100\times10^3}$$

$$= \underline{\underline{0.159nF}}$$

$A_{HP} = -R_4/R_3 = -4$

$R_4 = 4R_3$

make $R_4 = \underline{\underline{100k\Omega}}$

$R_3 = \underline{\underline{25k\Omega}}$

Now $R_3 C_2 = 1/\omega_{o,HP}$

$$C_2 = \frac{1}{2\pi f_{o,HP} R_3}$$

$$= \frac{1}{2\pi(100\times10^3)\,25\times10^3}$$

$$= \underline{63.7\,nF}$$

12.23

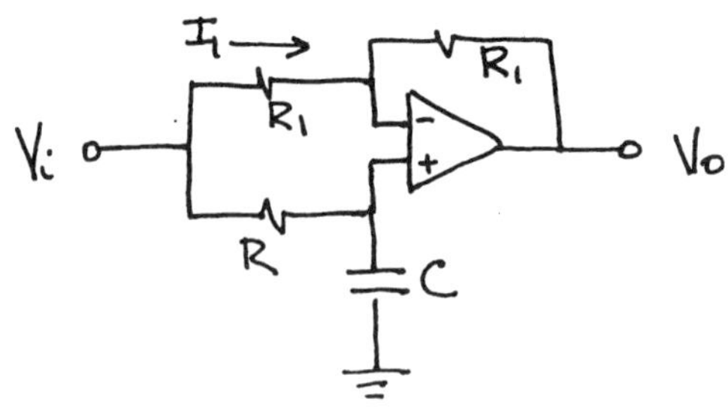

At +ve terminal

$$V_+ = \frac{1/sC}{1/sC + R_1} V_i$$

$$= \frac{1}{1+s\tau} V_i \qquad \underline{\tau = RC}$$

$V_- = V_+$ due to virtual short between terminals.

$$\therefore I_1 = \left(V_i - \frac{1}{1+s\tau}V_i\right)\frac{1}{R_1}$$

$$V_o = V_- - I_1 R_1$$

$$= \frac{V_i}{1+s\tau} - \left(V_i - \frac{V_i}{1+s\tau}\right)\frac{R_1}{R_1}$$

$$\frac{V_o}{V_i} = \frac{1-(1+s\tau)+1}{1+s\tau} = \frac{1-s\tau}{1+s\tau}$$

$$= \frac{\omega_0 - s}{\omega_0 + s} \qquad \omega_0 = \frac{1}{\tau}$$

$$= -\frac{s-\omega_0}{s+\omega_0} = T(s)$$

$$T(j\omega) = -\frac{j\omega - \omega_0}{j\omega + \omega_0}$$

$$\phi(\omega) = 180° + \tan^{-1}\left(\frac{\omega}{-\omega_0}\right) - \tan^{-1}\left(\frac{\omega}{\omega_0}\right)$$

$$= 360° - 2\tan^{-1}\left(\frac{\omega}{\omega_0}\right) \qquad \therefore \tan^{-1}\left(\frac{\omega}{-\omega_0}\right) = 180 - \tan^{-1}\left(\frac{\omega}{\omega_0}\right)$$

$$= -2\tan^{-1}(\omega/\omega_0)$$

Now this equation can be rearranged:

$$\frac{\omega}{\omega_0} = \tan(-\phi/2) \Leftarrow \omega_0 = \frac{1}{\tau} = \frac{1}{RC}$$

$$RC\omega = \tan(-\phi/2)$$

CONT.

$\therefore R = \dfrac{\tan(-\phi/2)}{C\omega} = 10^4 \tan(-\phi/2)$

$\phi = -30°, -60°, -90°, -120°, -150°$

$R = 2.68\,k\Omega,\ 5.77\,k\Omega,\ 10\,k\Omega,\ 17.32\,k\Omega,\ 37.32\,k\Omega$

12.24

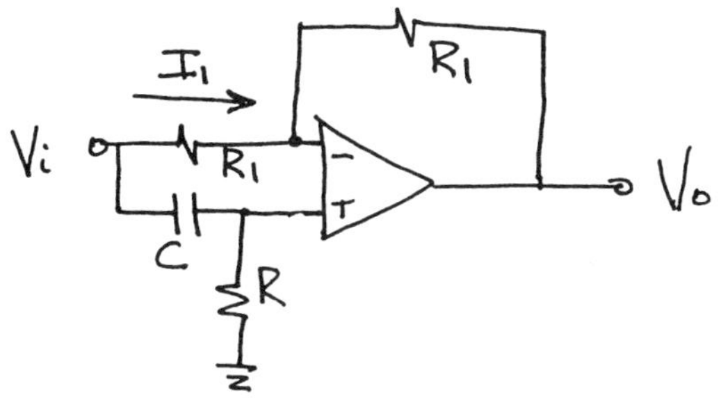

$V_+ = \dfrac{R}{R + 1/sC} V_i = \dfrac{s}{s+\omega_0} V_i$

where $\omega_0 = \dfrac{1}{RC}$

$I_1 = \dfrac{V_i - (s/s+\omega_0)V_i}{R_1}$

$V_o = \dfrac{s}{s+\omega_0} V_i - I_1 R_1$

$= \dfrac{s}{s+\omega_0} V_i - V_i\left(1 - \dfrac{s}{s+\omega_0}\right)$

$\dfrac{V_o}{V_i} = \dfrac{2s - s - \omega_0}{s+\omega_0} = \dfrac{s-\omega_0}{s+\omega_0}$

Now:

NOTE $\tan^{-1}\dfrac{\omega}{-\omega_0} = 180° - \tan^{-1}\dfrac{\omega}{\omega_0}$

$\phi(\omega) = \tan^{-1}\left(\dfrac{\omega}{-\omega_0}\right) - \tan^{-1}\left(\dfrac{\omega}{\omega_0}\right)$

$= 180 - \tan^{-1}\left(\dfrac{\omega}{\omega_0}\right) - \tan^{-1}\left(\dfrac{\omega}{\omega_0}\right)$

$= 180 - 2\tan^{-1}\left(\dfrac{\omega}{\omega_0}\right)$

Clearly $\phi(0) = 180°$ & $\phi(\omega \to \infty) = 0°$

12.25

Low Pass $\omega_0 = 10^3$ rad/s

$Q = 1$

DC gain $= 1$

$T(s) = \dfrac{a_0}{s^2 + s\dfrac{\omega_0}{Q} + \omega_0^2}$

$T(0) = a_0/\omega_0^2 = 1$

$a_0 = \omega_0^2 = 10^6$

$\therefore T(s) = \dfrac{10^6}{s^2 + 10^3 s + 10^6}$

$\omega_{max} = \omega_0 \sqrt{1 - 1/2Q^2}$

$= \dfrac{\omega_0}{\sqrt{2}}$

$= 0.707$ rad/s

$|T_{max}| = \dfrac{|a_0|\,Q}{\omega_0^2\sqrt{1 - \frac{1}{4Q^2}}} \leftarrow a_0 = \omega_0^2$

$= \dfrac{|\omega_0^2|}{\omega_0^2\sqrt{3/4}}$

$= 2/\sqrt{3}$

$= 1.15$ V/V

$= 1.21$ dB

12.26

$\omega_p = 1$ rad/s

$A_{max} = 3$ dB

$10^{-3/20} = 0.708 \cong \dfrac{1}{\sqrt{2}}$

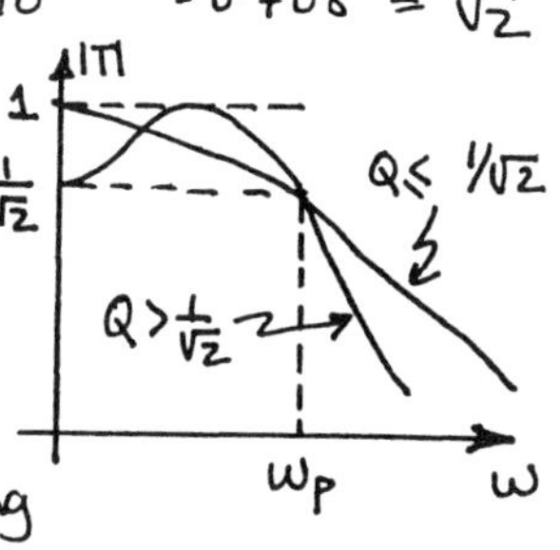

There are many Q-values which may be used

$Q \le 1/\sqrt{2}$ - no peaking

$Q > 1/\sqrt{2}$ - peaking

CONT.

Solution 1 $\quad Q \leq 1/\sqrt{2}$

For $Q = 1/\sqrt{2}$ the response is maximally flat. Because this is desirable, use: $Q = \frac{1}{\sqrt{2}}$

$$T(s) = \frac{a_0}{s^2 + s\omega_0\sqrt{2} + \omega_0^2}$$

$$|T(0)| = \frac{a_0}{\omega_0^2} = 1$$

$$a_0 = \omega_0^2$$

$$|T(j1)|^2 = \frac{\omega_0^2}{(\omega_0^2 - 1) + 2\omega_0^2} = \left(\frac{1}{\sqrt{2}}\right)^2$$

$$\omega_0 = 1 \text{ rad/s}$$

$$\therefore T_1(s) = \frac{1}{s^2 + \sqrt{2}s + 1}$$

Solution 2 $\quad Q > 1/\sqrt{2}$

From the figure: $|T(0)| = 1/\sqrt{2} = \frac{a_0}{\omega_0^2}$

$$\therefore a_0 = \omega_0^2/\sqrt{2}$$

Now $|T|_{max} = \frac{|a_0|Q}{\omega_0^2\sqrt{1 - 1/4Q^2}} = 1$

$$\frac{Q}{\sqrt{2}\sqrt{1 - 1/4Q^2}} = 1$$

$$Q = \sqrt{2}\sqrt{1 - 1/4Q^2}$$

$$\therefore Q^2 = 2\left(1 - \frac{1}{4Q^2}\right)$$

$$= 2 - \frac{1}{2Q^2}$$

$$Q^4 - 2Q^2 + \frac{1}{2} = 0$$

Solving for Q^2 gives:-

$$Q^2 = 1 \pm \sqrt{2}$$

ASIDE:

$\because Q > 1/\sqrt{2}$

$Q^2 > 1/2$

$4Q^2 > 2$

$\frac{1}{4Q^2} < \frac{1}{2}$

$\therefore 1 - \frac{1}{4Q^2} > 1/2$

$\therefore \left|1 - \frac{1}{4Q^2}\right| = 1 - \frac{1}{4Q^2}$

$\Rightarrow Q = 0.5412$ or 1.3066

$\therefore Q > \frac{1}{\sqrt{2}}$ $\quad$ use $Q = 1.3066$

Now at the passband edge

$$|T(j1)| = 1/\sqrt{2}$$

$$|T(j1)|^2 = \frac{(\omega_0^2/\sqrt{2})^2}{(\omega_0^2 - 1)^2 + \frac{\omega_0^2}{Q^2}} = \frac{1}{2}$$

$$\frac{\omega_0^4}{2} = \frac{1}{2}\left[\omega_0^4 - 2\omega_0^2 + 1 + \frac{\omega_0^2}{Q^2}\right]$$

$$\omega_0^2(2 - 1/Q^2) = 1$$

$$\omega_0 = 0.841$$

$$\therefore T_2(s) = \frac{\omega_0^2/\sqrt{2}}{s^2 + \omega_0/Q\, s + \omega_0^2}$$

$$= \frac{0.5}{s^2 + 0.644s + 0.707}$$

If $\omega_s = 2$

$|T_1(j2)| = 0.242 \quad |T_2(j2)| = 0.1414$

$\therefore A_{MIN,1} = -12.3\text{dB} \quad A_{MIN,2} = -17\text{dB}$

12.27

V_2 lags V_1 by $120°$

V_3 lags V_2 by $120°$

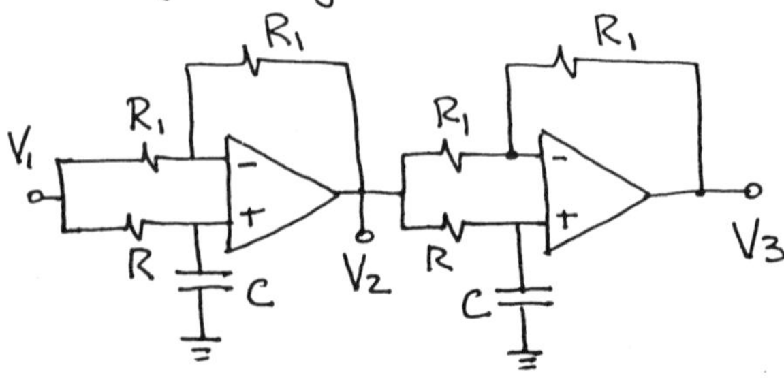

$\omega = 2\pi 60$ & $C = 1\,\mu F$

$T(s) = \frac{s - \omega_0}{s + \omega_0} \quad \omega_0 = \frac{1}{RC}$

CONT.

$$\phi(\omega) = 180^\circ + \tan^{-1}\left(\frac{\omega}{-\omega_o}\right) - \tan^{-1}\left(\frac{\omega}{\omega_o}\right)$$

Sub: $\tan\left(\frac{\omega}{-\omega_o}\right) = 180 - \tan^{-1}\left(\frac{\omega}{\omega_o}\right)$

$$\Rightarrow \phi(\omega) = -2\tan(\omega/\omega_o)$$

Now $\phi = -120^\circ$ at $\omega = 2\pi 60$

$$-120 = -2\tan(\omega RC)$$

$$-60 = -\tan^{-1}(2\pi 60 \times R \times 10^{-6})$$

$$\underline{\underline{R = 4.59\ k\Omega}}$$

R_1 can be arbitrarily chosen

use $\underline{\underline{R_1 = 10k\Omega}}$

12.28

Natural Modes:

$$-\frac{1}{2} \pm j\frac{\sqrt{3}}{2}$$

$$\omega_o = \sqrt{\left(\tfrac{1}{2}\right)^2 + \left(\tfrac{\sqrt{3}}{2}\right)^2}$$

$$= \underline{1.0}$$

$$\frac{\omega_o}{2Q} = \frac{1}{2} \Rightarrow \frac{\omega_o}{Q} = 1$$

$$T(s) = \frac{a_2 s^2}{s^2 + s\frac{\omega_o}{Q} + \omega_o^2} = \frac{a_2 s^2}{s^2 + s + 1}$$

$$|T(j\infty)| = a_2 = 1$$

$$\therefore\ T(s) = \frac{s^2}{s^2 + s + 1}$$

12.29

For a 2nd-order bandpass

$$T(s) = \frac{a_1 s}{s^2 + s\frac{\omega_o}{Q} + \omega_o^2}$$

$$T(j\omega) = \frac{j\omega a_1}{(\omega_o^2 - \omega^2) + \frac{j\omega\omega_o}{Q}}$$

$$|T(j\omega)| = \frac{a_1\omega}{\left[(\omega_o^2 - \omega^2)^2 + \frac{\omega^2\omega_o^2}{Q^2}\right]^{1/2}}$$

Part (a):

$$|T(j\omega_1)| = |T(j\omega_2)|$$

$$\frac{a_1\omega_1}{\sqrt{(\omega_o^2 - \omega_1^2)^2 + \left(\frac{\omega_1\omega_o}{Q}\right)^2}} = \frac{a_1\omega_2}{\sqrt{(\omega_o^2 - \omega_2^2)^2 + \left(\frac{\omega_2\omega_o}{Q}\right)^2}}$$

$$\omega_1^2\left[(\omega_o^2 - \omega_2^2)^2 + \left(\frac{\omega_2\omega_o}{Q}\right)^2\right] = \omega_2^2\left[(\omega_o^2 - \omega_1^2)^2 + \left(\frac{\omega_1\omega_o}{Q}\right)^2\right]$$

$$\omega_1^2(\omega_o^4 - 2\omega_o^2\omega_2^2 + \omega_2^4) = \omega_2^2(\omega_o^4 - 2\omega_o^2\omega_1^2 + \omega_1^4)$$

$$\omega_1^2\omega_o^4 + \omega_1^2\omega_2^4 = \omega_2^2\omega_o^4 + \omega_2^2\omega_1^4$$

$$\omega_o^4(\omega_1^2 - \omega_2^2) = \omega_2^2\omega_1^4 - \omega_1^2\omega_2^4$$

$$\omega_o^4(\omega_1^2 - \omega_2^2) = \omega_2^2\omega_1^2(\omega_1^2 - \omega_2^2)$$

$$\omega_o^4 = \omega_1^2\omega_2^2$$

$$\underline{\underline{\omega_o^2 = \omega_1\omega_2}}\quad \text{Q.E.D.}$$

CONT.

For Fig. 12.4: $\omega_{p1} = 8100$ rad/s
$\omega_{p2} = 10.000$ rad/s
$A_{max} = 1$ dB

$$\omega_0^2 = (8100)(10\,000)$$

$$\omega_0 = \underline{\underline{9000 \text{ rad/s}}}$$

$$|T(j\omega_{p1})| = |T(j\omega_{p2})| = 10^{-1/20} = \underline{0.8913}$$

$$|T(j\omega_0)| = \frac{\omega_0 a_1}{\sqrt{(\omega_0^2 - \omega_0^2)^2 + \left(\frac{\omega_0^2}{Q}\right)^2}} = 1$$

$$\Rightarrow \frac{\omega_0 a_1}{\omega_0^2/Q} = 1$$

$$\therefore \frac{Q a_1}{\omega_0} = 1 \Rightarrow a_1 = \frac{\omega_0}{Q}$$

$$|T(j\omega_{p1})|^2 = |T(j0.9\omega_0)|^2 = 0.8913^2$$

$$\frac{(\omega_0/Q)^2(0.9\omega_0)^2}{(\omega_0^2 - (0.9\omega_0)^2)^2 + \left(\frac{0.9\omega_0}{Q}\right)^2} = 0.8913^2$$

$$\left(\frac{\omega_0}{Q}(.9\omega_0)\right)^2 = 0.8913^2\left[(\omega_0^2 - (0.9\omega_0)^2)^2 + \left(\frac{0.9\omega_0}{Q}\right)^2\right]$$

$$\frac{0.81\omega_0^4}{Q^2} = 0.8913^2\left[\omega_0^4(1-0.81)^2 + \frac{0.81\omega_0^4}{Q^2}\right]$$

$$\frac{0.81\omega_0^4}{Q^2}(1-0.8913^2) = 0.8913^2\omega_0^4 \times (1-0.81)^2$$

SUB $\omega_0 = 9000$ gives

$$\underline{\underline{Q = 2.41}}$$

Now $a_1 = \frac{\omega_0}{Q} = 0.415\omega_0$

$$\therefore T(s) = \underline{\underline{\frac{0.415\omega_0 s}{s^2 + 0.415\omega_0 s + \omega_0^2}}}$$

If $\omega_{s1} = 3000$ rad/s

$$|T(j3000)| = \frac{0.415\omega_0(3000)}{\sqrt{(\omega_0^2 - 3000^2)^2 + (\omega_0 3000 \times .415)^2}}$$

$$= 0.1537$$

$$\therefore A_{MIN} = -20\log(0.1537)$$

$$= \underline{\underline{16.3\text{dB}}}$$

Now ω_{s1} and ω_{s2} are geometrically symmetrical about ω_0:

$$\omega_{s1}\omega_{s2} = \omega_0^2$$

$$\omega_{s2} = \frac{9000^2}{3000}$$

$$= \underline{\underline{27\,000 \text{ rad/s}}}$$

12.30

From exercise 12.15

$$Q = \frac{\omega_0}{BW\sqrt{10^{A/10} - 1}} \quad \begin{cases} \omega_0 = 2\pi(60) \\ BW = 2\pi 6 \\ A = 20\text{dB} \end{cases}$$

$$= \underline{\underline{1.005}}$$

$$T(s) = a_2 \frac{s^2 + \omega_0^2}{s^2 + s\frac{\omega_0}{Q} + \omega_0^2}$$

$$|T(0)| = \frac{a_2\omega_0^2}{\omega_0^2} = 1 \leftarrow \text{DC Gain}$$

CONT.

$\underline{a_2 = 1}$

$$T(s) = \frac{s^2 + (2\pi 60)^2}{s^2 + s\dfrac{2\pi 60}{1.005} + (2\pi 60)^2}$$

$$T(s) = \frac{s + 1.421 \times 10^5}{s^2 + 375.1s + 1.421 \times 10^5}$$

12.31

FOR ALL PASS:

$$T(s) = a_2 \frac{s^2 - s\,\omega_0/Q + \omega_0^2}{s^2 + s\,\omega_0/Q + \omega_0^2}$$

If zero frequency < pole frequency

$$T(s) = a_2 \frac{s^2 - s\,\omega_n/Q + \omega_n^2}{s^2 + s\,\omega_0/Q + \omega_0^2} \quad \omega_n < \omega_0$$

At DC: $|T| = a_2 \dfrac{\omega_n^2}{\omega_0^2}$ where $\dfrac{\omega_n^2}{\omega_0^2} < 1$

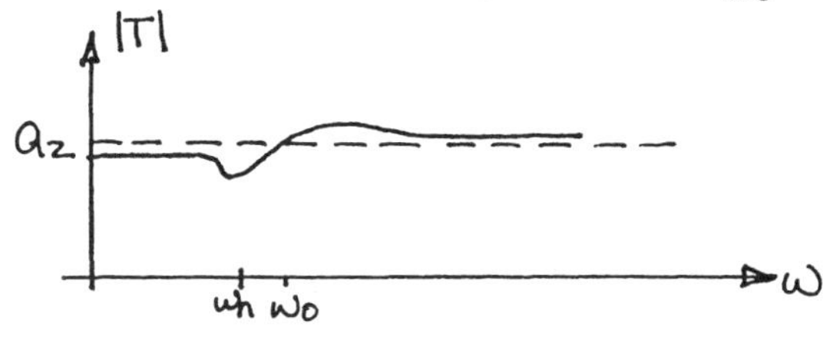

If zero frequency > pole frequency then $\omega_n > \omega_0$

At DC: $|T| = a_2\, \omega_n^2/\omega_0^2$ where $\dfrac{\omega_n^2}{\omega_0^2} > 1$

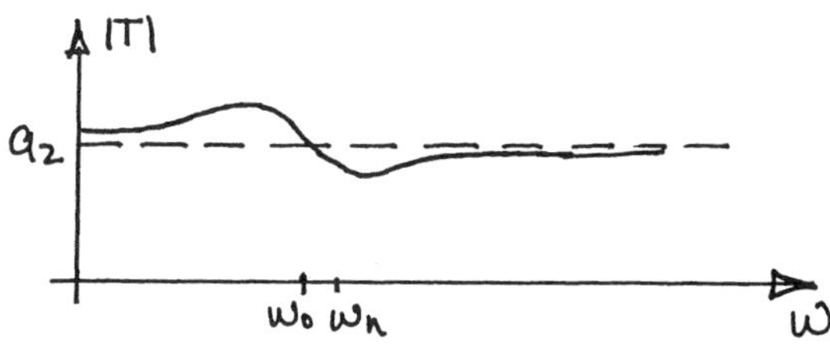

12.32

$$T(s) = \frac{s^2 - s\,\omega_0/Q' + \omega_0^2}{s^2 + s\,\omega_0/Q_0 + \omega_0^2}\, a_2$$

Zero Q < Pole Q $\Rightarrow Q' < Q_0$

At $\omega = \omega_0$:

$$|T| = \frac{a_2\, \omega_0^2/Q'}{\omega_0^2/Q_0} = \frac{a_2 Q_0}{Q'} > a_2$$

If $Q' > Q_0$

$$|T(j\omega_0)| = \frac{a_2 Q_0}{Q'} < a_2$$

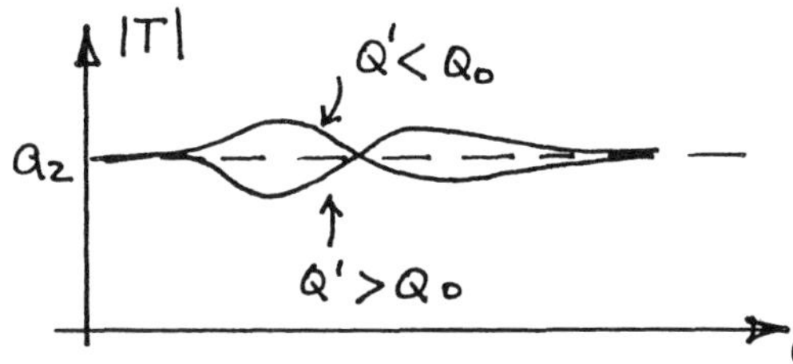

12.33

$\omega_0 = 10^4$ rad/s, $Q = 2$, $R = 10\,k\Omega$

$$\omega_0 = \frac{1}{\sqrt{LC}} \Rightarrow \omega_0^2 = \frac{1}{LC} \quad \Leftarrow L = \frac{R^2 C}{Q^2}$$

$$= \frac{Q^2}{R^2 C^2}$$

$$C = \frac{Q}{R\omega_0} = \underline{20\,nF}$$

$$L = \frac{1}{C\omega_0^2} = \underline{500\,mH}$$

12.34

$\omega_0 = 1/\sqrt{LC}$

If $L' = 1.01L$

$\omega_0' = (1.01LC)^{-1/2}$

$= 0.9950 \frac{1}{\sqrt{LC}}$

$= 0.9950\,\omega_0$

$\therefore \Delta\omega_0 = -0.5\%$

If $C' = 1.01C$

$\omega_0' = 0.9950\,\omega_0$

$\Delta\omega_0' = -0.5\%$

Changing R has no effect on ω_0

12.35

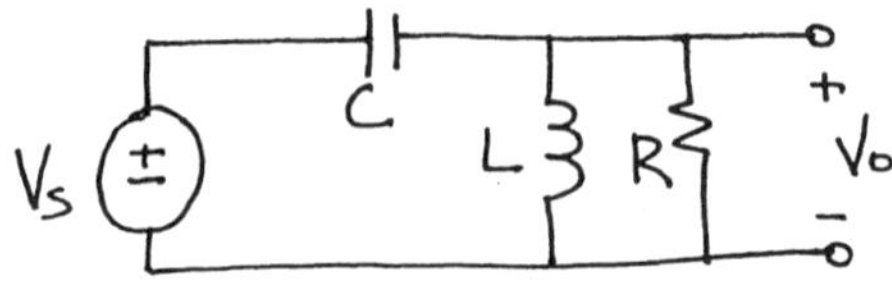

Use voltage divider rule:

$$V_o = \frac{Z_{R\parallel L}}{Z_{R\parallel L} + Z_C} V_s$$

$$\frac{V_o}{V_s} = \frac{\left(\frac{1}{R} + \frac{1}{sL}\right)^{-1}}{\left(\frac{1}{R} + \frac{1}{sL}\right)^{-1} + \frac{1}{sC}}$$

$$= \frac{sC}{\left(\frac{1}{sL} + \frac{1}{R}\right) + sC}$$

$$\therefore T(s) = \frac{V_o(s)}{V_s(s)} = \frac{s^2}{s^2 + s/RC + \frac{1}{LC}}$$

12.36

Low Pass: $\omega_0 = 10^5$, $C = 0.1\,\mu F$

$Q = 1/\sqrt{2}$

$Q = \omega_0 CR$

$R = \frac{Q}{\omega_0 C}$

$= \frac{1}{\sqrt{2} \times 10^5 \times 0.1 \times 10^{-6}}$

$= 70.7\,\Omega$

$\omega_0 = 1/\sqrt{LC}$

$L = \frac{1}{\omega_0^2 C}$

$= 1\,mH$

12.37

Vi —R— C, L — Vo $\Rightarrow$ Vi —2R— C, L, 2R — Vo

$\omega_0 = \frac{1}{\sqrt{LC}}$

$Q = \omega_0 CR$

$A_{MID} = 1$

$\omega_0 = 1/\sqrt{LC}$

$Q = \omega_0 C(2R \parallel 2R)$

$= \omega_0 CR$

$A_{MID} = \frac{2R}{2R+2R} = 1/2$

12.38

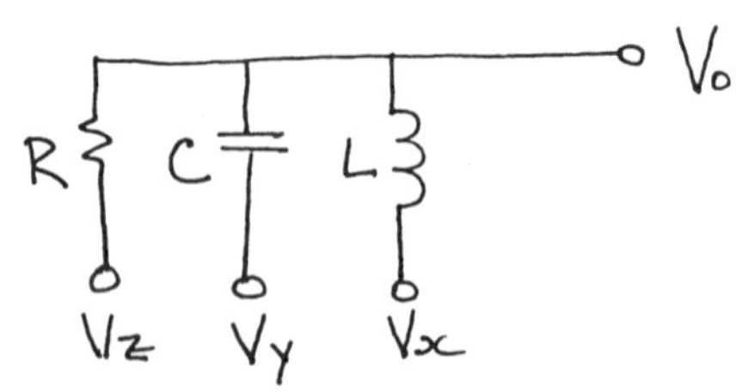

$V_o/V_z \big|_{V_y = V_x = 0} = T_{BP}(s)$

$V_o/V_y \big|_{V_x = V_z = 0} = T_{HP}(s)$

CONT.

$$\left.\frac{V_o}{V_x}\right|_{V_y = V_z = 0} = T_{LP}(S)$$

Using superposition

$$V_o = \frac{V_o}{V_x}V_x + \frac{V_o}{V_y}V_y + \frac{V_o}{V_z}V_z$$

$$= T_{LP}V_x + T_{HP}V_y + T_{BP}V_z$$

$$= \frac{\frac{1}{LC}V_x + s^2 V_y + \frac{s}{RC}V_z}{s^2 + s/RC + 1/LC}$$

$$\therefore V_o = V_x \frac{1/LC}{s^2 + s/RC + 1/LC} + V_y \frac{s^2}{s^2 + s/RC + 1/LC} + V_z \frac{s/RC}{s^2 + s/RC + 1/LC}$$

12.39

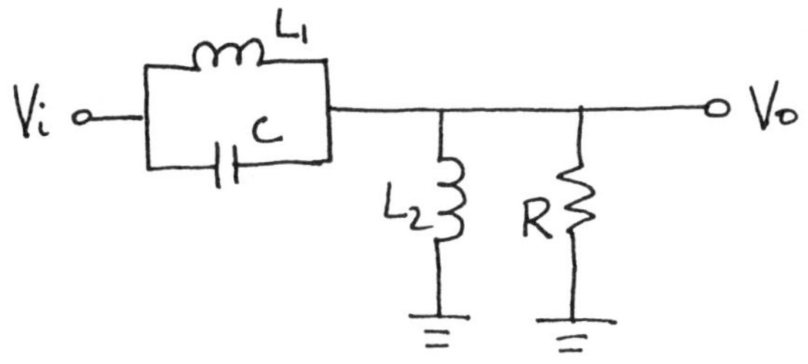

From Eq 12.46

$$T(s) = \frac{s^2 + 1/L_1C}{s^2 + s\,(1/CR) + \frac{1}{(L_1 \| L_2)C}}$$

Required notch $\omega_n^2 = \frac{1}{L_1 C} = (0.9\omega_0)^2$

but:

$$\omega_0^2 = \frac{1}{(L_1 \| L_2)C} \qquad \text{where } L_1 \| L_2 = \frac{1}{L_1^{-1} + L_2^{-1}} = \frac{L_1 L_2}{L_1 + L_2}$$

$$= \frac{L_1 + L_2}{L_1 L_2 C}$$

$$= \frac{L_1 + L_2}{L_2}(.9\omega_0)^2$$

$$1 = \left(\frac{L_1}{L_2} + 1\right)0.9^2$$

$$\therefore L_1/L_2 = \frac{1}{0.9^2} - 1 = \underline{\underline{0.2346}}$$

For $\omega \ll \omega_0$:-

$$|T| \cong \frac{1/L_1C}{1/(L_1 \| L_2)C} = \frac{L_2}{L_1 + L_2}$$

i.e. inductors dominate!

For $\omega \gg \omega_0$ L_1 & L_2 are "open" C is shorted

$$\underline{\underline{|T| \cong 1}}$$

12.40

$L = C_4 R_1 R_3 R_5 / R_2$

Choose $\underline{\underline{R_1 = R_2 = R_3 = R_5 = 10\,k\Omega}}$

$\therefore L = C_4 \times 10^8$

For:

$L = 10\,H = C_4 \times 10^8 \Rightarrow \underline{\underline{C_4 = 100\,nF}}$

$L = 1\,H \Rightarrow \underline{\underline{C_4 = 10\,nF}}$

$L = 0.1\,H \Rightarrow \underline{\underline{C_4 = 1\,nF}}$

12.41

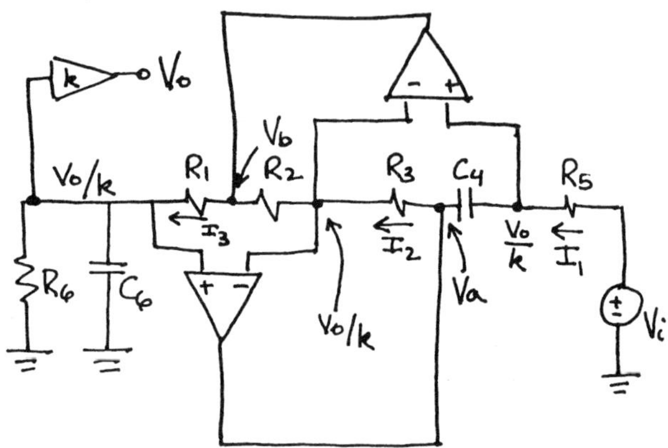

Because of the virtual short at opamp input terminals, the voltages are : V_o/k

$$I_1 = \frac{V_i - V_o/k}{R_5}$$

$$V_a = V_o/k - I_1/sC_4$$

$$= V_o/k - \frac{V_i - V_o/k}{sC_4 R_5}$$

$$= \frac{V_o(sC_4R_5 + 1) - kV_i}{sC_4R_5k}$$

$$I_2 = \frac{V_a - V_o/k}{R_3}$$

$$= -\frac{V_i}{sC_4R_3R_5} + \frac{V_o/k}{sC_4R_3R_5}$$

$$V_b = \frac{V_o}{k} - I_2R_2$$

$$= \frac{V_o}{k} - \frac{V_o/k - V_i}{sC_4R_3R_5/R_2}$$

$$I_3 = \frac{V_b - V_o/k}{R_1} = \frac{V_i - V_o/k}{sC_4R_1R_3R_5/R_2}$$

Now, I_3 flows only into R_6 & C_6 since for ideal opamps, $R_{IN} = \infty$!

$$\therefore \frac{V_o}{k} = I_3\left(R_6 \parallel \frac{1}{sC_6}\right)$$

Let $L \triangleq \dfrac{R_1R_3R_5C_4}{R_2}$

So: $I_3 = \dfrac{V_i - V_o/k}{sL}$ &

$$V_o/k = I_3\left(R_6 \parallel \frac{1}{sC_6}\right)$$

$$\frac{V_o}{k} = \frac{V_i - V_o/k}{sL} \cdot \frac{1}{1/R_6 + sC_6}$$

$$= \frac{V_i - V_o/k}{sL/R_6 + s^2LC_6}$$

$$V_o/k\left(1 + sL/R_6 + s^2LC_6\right) = V_i$$

$$\frac{V_o}{V_i} = \frac{k/LC_6}{s^2 + \frac{s}{R_6C_6} + \frac{1}{LC_6}}$$

Recall $L = R_1R_3R_5C_4/R_2$

$$\therefore \frac{V_o}{V_i} = \frac{\frac{kR_2}{C_4R_1R_3R_5}}{s^2 + \frac{s}{R_6C_6} + \frac{R_2}{C_6C_4R_1R_3R_5}}$$

12.42

$$A_{max} = 10\log(1+\epsilon^2) = 3\text{dB}$$

$$\therefore \epsilon = 0.998 \cong 1$$

$$\omega_0 = \omega_p = 10^4$$

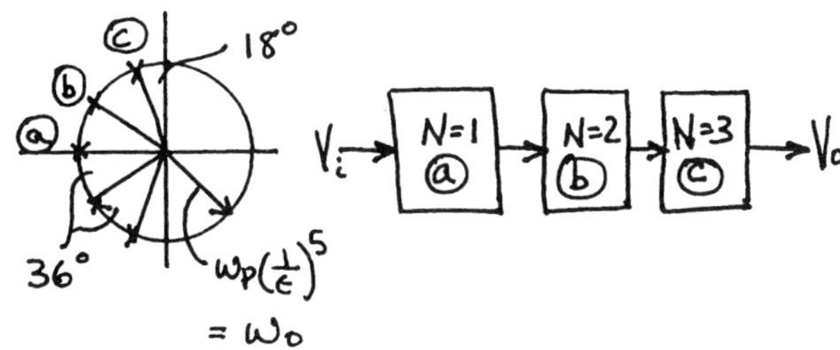

For circuit ⓐ use fig 12.13(a)

DC Gain = 1 = R_2/R_1 => $R_1 = R_2 = 10\,k\Omega$

$CR_2 = 1/\omega_0$ => $C = 1/R_2\omega_0 = \frac{1}{10^4 10^4}$ = 10 nF

For circuit ⓑ use Fig. 12.22(a)

$\omega_0 = 10^4$ rad/s

$\frac{\omega_0}{2Q} = \omega_0 \cos 36°$

$Q = \frac{1}{2\cos 36} = 0.618$

From Table 12.1

$$T(s) = \frac{\dfrac{kR_2}{C_4C_6R_1R_3R_5}}{s^2 + s/C_6R_6 + \dfrac{R_2}{C_4C_6R_1R_3R_5}}$$

$\omega_0^2 = \frac{R_2}{C_4C_6R_1R_3R_5}$ Let $R_1 = R_3 = R_5 = R_2 = R$, $C_4 = C_6 = C$

$\omega_0^2 = \frac{1}{R^2C^2}$ USE $C_4 = C_6 = 100\,nF$

$\therefore R = \frac{1}{\omega_0 C}$ => $R_1 = R_2 = R_3 = R_5 = 1\,k\Omega$

Now using:

$\frac{\omega_0}{Q} = \frac{1}{C_6R_6}$ & $Q = 0.618$

$R_6 = \frac{Q}{C_6\omega_0} = 618\,\Omega$

For circuit ⓒ use Fig 12.22(a)

$\omega_0 = 10^4$ which is the same as for circuit ⓑ.

$\therefore C_4 = C_6 = 100\,nF$
$R_1 = R_2 = R_3 = R_5 = 1\,k\Omega$

Now: $Q = \frac{1}{2\cos 72°} = 1.618$

$R_6 = Q/\omega_0 C_6 = 1.618\,k\Omega$

12.43

$f_0 = 4\,kHz$ $f_N = 5\,kHz$ $Q = 10$

now $C_4 = 10\,nF$ and $k = 1 \equiv$ dc gain

$\omega_0 = \left[C_4(C_{61}+C_{62})R_1R_3R_5/R_2\right]^{-1/2}$

$C_{61} + C_{62} = C_6$

Choose $C_6 = C_4 = 10\,nF$ & $R_1 = R_3 = R_5 = R_2 = R$

$\therefore \omega_0 = (C_4C_6R^2)^{-1/2}$

$R = \frac{1}{\omega_0 C_4}$

=> $R_1 = R_3 = R_5 = R_2 = 3.979\,k\Omega$

$\omega_n = (C_4C_{61}R^2)^{-1/2}$

$C_{61} = \frac{1}{\omega_n^2R^2C_4}$ => $C_{61} = 6.4\,nF$

& $C_{62} = 3.6\,nF$

$Q = R_6\sqrt{\frac{C_{61}+C_{62}}{C_4} \cdot \frac{R_2}{R_1R_3R_5}}$

$= R_6\sqrt{\frac{1}{R_1^2}} = R_6/R_1$ => $R_6 = 39.79\,k\Omega$

12.44

From Fig. 12.16 (g) $\phi = 180°$ at f_o !

∴ Use $f_o = 1\,kHz$ $Q = 1$

$$\omega_o^2 = \frac{R_2}{C_4 C_6 R_1 R_3 R_5}$$

LET $C = C_4 = C_6 = 1nF$

$R_1 = R_3\ R_5 = R_2 = R$

$$= 1/C^2 R^2$$

$$R = 1/\omega_o C = 159.16\,k\Omega = R_1 = R_3 = R_5 = R_2$$

$$\frac{\omega_o}{Q} = \frac{1}{R_6 C_6} \Rightarrow R_6 = \frac{Q}{C_6 \omega_o}$$

$$= \frac{1}{10^{-9}\,2\pi 10^3}$$

$$\therefore\ R_6 = 159.16\,k\Omega$$

12.45

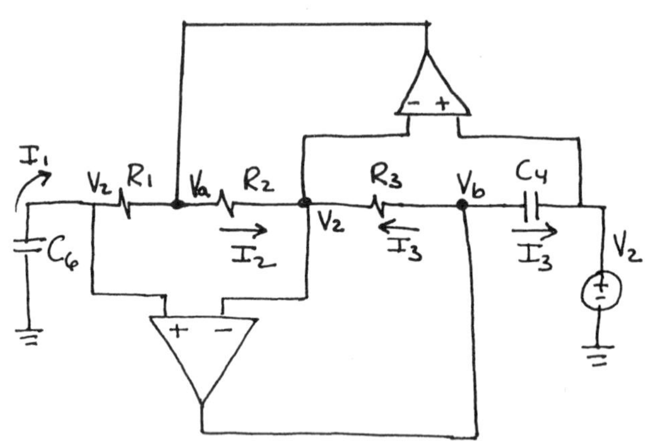

Because of virtual short at opamp input terminals all nodes are at V_2 !

$$I_1 = -sC_6 V_2$$

Since no current goes into the opamp input terminals we have:

$$V_a = V_2 - I_1 R_1$$
$$= V_2 (1 + sC_6 R_1)$$

$$I_2 = (V_a - V_2)\frac{1}{R_2}$$
$$= \frac{sC_6 R_1}{R_2}$$

$$V_b = V_2 - I_2 R_3$$
$$= V_2 - \frac{sC_6 R_1 R_3}{R_2}$$

$$I_3 = (V_b - V_2)\,sC_4$$
$$= -\frac{s^2 C_4 C_6 R_1 R_3}{R_2} V_2$$

Now the voltage source sees an input impedance given by:

$$Z_{in} = -V_2/I_3 = \frac{R_2}{s^2 C_4 C_6 R_1 R_3}$$

As required.

for $s = j\omega \Rightarrow s^2 = -\omega^2$

$$Z_{in}(j\omega) = \frac{-R_2}{C_4 C_6 R_1 R_3} \cdot \frac{1}{\omega^2}$$

$$= -R(\omega)$$ i.e. A PURE NEGATIVE RESISTANCE!

12.46

For LPN: See Table 12.1 & Fig. 12.22 (e)

$$T(s) = k\frac{C_{61}}{C_{61}+C_{62}} \cdot \frac{s^2 + \frac{R_2}{C_4 C_{61} R_1 R_3 R_5}}{s^2 + s/C_6 R_6 + \frac{R_2}{C_4 (C_{61}+C_{62}) R_1 R_3 R_5}}$$

At DC → s = 0

$$T(0) = k\frac{C_{61}}{C_{61}+C_{62}} \quad \frac{R_2/C_4 C_6 R_1 R_3 R_5}{R_2/C_4 (C_{61}+C_{62}) R_1 R_3 R_5}$$

CONT.

$\Rightarrow T(0) = \underline{k \triangleq \text{DC Gain}}!$

Note that $C_{61} + C_{62}$ is the total capacitance across R_6

$\therefore \; C_6 = C_{61} + C_{62}$

$$\frac{\omega_n^2}{\omega_o^2} = \frac{R_2/C_4 C_{61} R_1 R_3 R_5}{R_2/C_4 (C_{61}+C_{62}) R_1 R_3 R_5}$$

$$= \frac{C_{61}}{C_{61}+C_{62}}$$

$$\frac{\omega_n^2}{\omega_o^2} = \frac{C_{61}}{C_6}$$

$$\therefore \; \underline{C_{61} = C_6 \left(\frac{\omega_n}{\omega_o}\right)^2 = C\left(\frac{\omega_n}{\omega_o}\right)^2}$$

Clearly from $T(s)$ above:

$$\omega_n^2 = R_2/C_4 C_{61} R_1 R_3 R_5$$

$$\Rightarrow \omega_n = \sqrt{\frac{R_2}{C_4 C_{61} R_1 R_3 R_5}}$$

$$\omega_o^2 = R_2/C_4 C_6 R_1 R_3 R_5$$

$$\Rightarrow \omega_o = \sqrt{\frac{R_2}{C_4 (C_{61}+C_{62}) R_1 R_3 R_5}}$$

12.47

For HPN: See Table 12.1 & Fig 11.22(f)

$$T(s) = k\,\frac{s^2 + \left(R_2/C_4 C_6 R_1 R_3 R_{61}\right)}{s^2 + s/C_6 R_6 + \left(\frac{R_2}{C_4 C_6 R_1 R_3}\right)\left(\frac{1}{R_{51}} + \frac{1}{R_{52}}\right)}$$

clearly: $\omega_n = \sqrt{\frac{R_2}{C_4 C_6 R_1 R_3 R_{51}}}$

$$\omega_o = \sqrt{\frac{R_2}{C_4 C_6 R_1 R_3}\left(\frac{1}{R_{51}} + \frac{1}{R_{52}}\right)}$$

At high frequencies $s \to \infty$

$T(\infty) = k \triangleq \underline{\text{high freq gain}}.$

Observe that the equivalent resistance at the -ve terminal of A_1 is:

$\frac{1}{R_5} = \frac{1}{R_{51}} + \frac{1}{R_{52}}$ AND

for the resonator (table 12.1)

$$R_5 = 1/\omega_o C \Rightarrow \frac{1}{R_5} = \omega_o C$$

$$\frac{\omega_o^2}{\omega_n^2} = \frac{R_2/C_4 C_6 R_1 R_3 R_5}{R_2/C_4 C_6 R_1 R_3 R_{51}} \Rightarrow \underline{R_{51} = R_5 \frac{\omega_o^2}{\omega_n^2}}$$

Now $\frac{1}{R_5} = \frac{1}{R_5 \frac{\omega_o^2}{\omega_n^2}} + \frac{1}{R_{52}}$

$$\frac{1}{R_{52}} = \frac{1}{R_5}\left[1 - \frac{\omega_n^2}{\omega_o^2}\right]$$

$$R_{52} = \frac{R_5}{1 - \omega_n^2/\omega_o^2}$$

12.48

$$T(s) = \frac{0.4508\,(s^2 + 1.6996)}{(s+0.7294)(s^2 + 0.2786s + 1.0504)}$$

PART (a) Replace s with s/ω_p

$$T(s) = \frac{0.4508\left(s^2/\omega_p^2 + 1.6996\right)}{\left(\frac{s}{\omega_p} + 0.7294\right)\left(\frac{s^2}{\omega_p^2} + \frac{0.2786s}{\omega_p} + 1.0504\right)}$$

CONT.

$$T(S) = \frac{0.4508\,\omega_p\,(s^2 + 1.6996\,\omega_p^2)}{(s + 0.7294\omega_p)(s^2 + 0.2786\omega_p s + 1.0504\,\omega_p^2)}$$

SUB $\omega_p = 10^4$ rad/s

$$T(S) = \frac{4508\,(s^2 + 1.6996\times10^8)}{(s+7294)(s^2 + 2786s + 1.0504\times10^8)}$$

Part (b)

First decompose T(s) into 1st-and 2nd-order sections with unity DC gain!

$$T_1(S) = \frac{k_1}{S+7294} \qquad T_1(0) = \frac{k_1}{7294} = 1$$

$$\Rightarrow k_1 = 7294$$

Now $k_1 k_2 = 4508 \Rightarrow k_2 = 0.6180$

$$\therefore T_2(S) = \frac{0.6180\,(s^2 + 1.6996\times10^8)}{s^2 + 2786s + 1.0504\times10^8}$$

As a check:

$$T_2(0) = \frac{0.6180(1.6996\times10^8)}{1.0504\times10^8} = 1.000$$

As EXPECTED!

$\therefore T(S) = T_1(S) \cdot T_2(S)$

For first-order section use FIG 12.13(a)

$\omega_0 = 7294$ rad/s DC Gain = 1

Let $C = 10nF$

$R_1 = R_2 = \frac{1}{\omega_0 C} \Rightarrow R_1 = R_2 = 13.71k\Omega$

For second-order section

$\omega_n^2 = 1.6996\times10^8 \Rightarrow \omega_n = 13.037\times10^3$

$\omega_0^2 = 1.0504\times10^8 \Rightarrow \omega_0 = 10.249\times10^3$

$\frac{\omega_0}{Q} = 2786 \Rightarrow Q = 3.6787$

For LPN use table 12.1 and Fig 12.22(e)

Make $R_1 = R_2 = R_3 = R_5 = R$ and $C = C_4 = C_6 = 10nF$

$R = \frac{1}{\omega_0 C} = 9.757k\Omega = R_1 = R_2 = R_3 = R_5$

$\omega_n^2 = \frac{1}{CR^2C_{61}} \Rightarrow C_{61} = 6.18nF$

$C_{62} = C - C_{61} = 3.82nF$

$$Q = R_6\sqrt{\frac{C_6}{C_4}\frac{R_2}{R_1R_3R_5}} = \frac{R_6}{\sqrt{R^2}} = \frac{R_6}{R}$$

$\therefore R_6 = RQ = (9.757\times10^3)(3.6787) = 35.89k\Omega$

12.49

$f_0 = 1kHz$

The 3dB bandwidth for a 2nd order filter is given by:

$B = \omega_0/Q \Rightarrow Q = \frac{2\pi10^3}{2\pi50} = 20$

Choose $C = 10nF$

$R = \frac{1}{\omega_0 C} = 15.92k\Omega$

Use $R_1 = R_f = 10k\Omega$

CONT.

$$\frac{R_3}{R_2} = 2Q - 1 = 39$$

choose $R_2 = 10\,k\Omega$ $\quad R_3 = 390\,k\Omega$

Now $T(s) = \dfrac{-k\omega_0 s}{s^2 + s\frac{\omega_0}{Q} + \omega_0^2}$

$$\Rightarrow |T(j\omega_0)| = \frac{k\omega_0^2}{\omega_0^2/Q} = kQ$$

but $k = 2 - 1/Q = 1.95$

$\therefore$ Centre-freq gain $= kQ = \underline{\underline{39}}$

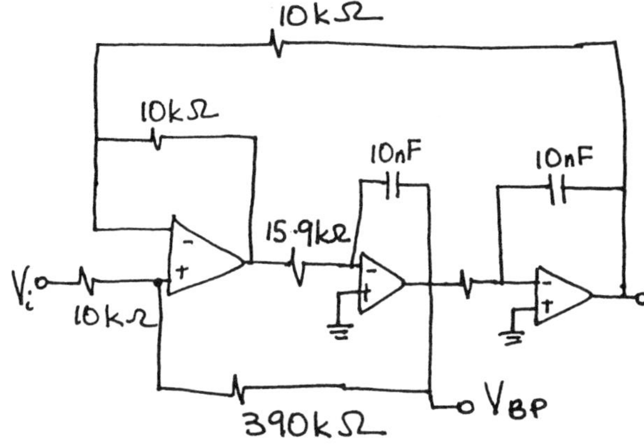

12.50

$R_L = R_H = R_B/Q \Rightarrow R_B = QR_H$

$R_L = R_H$

Using Eq 12.66:

$$\frac{V_o}{V_i} = -k\,\frac{\frac{R_F}{R_H}s^2 - s\left(\frac{R_F}{R_B}\right)\omega_0 + \left(\frac{R_F}{R_L}\right)\omega_0^2}{s^2 + s\frac{\omega_0}{Q} + \omega_0^2}$$

$$= -k\frac{R_F}{R_H}\,\frac{s^2 - \frac{\omega_0}{Q}s + \omega_0^2}{s^2 + \frac{\omega_0}{Q}s + \omega_0^2}$$

Flat Gain $= -k\,R_F/R_H$

Part (b) – $\omega_0 = 10^4$ rad/s $\quad Q = 2 \quad$ Flat Gain $= 10$

choose $C = 10\,nF \Rightarrow R = \dfrac{1}{\omega_0 C} = \underline{\underline{10\,k\Omega}}$

choose $R_F = R_1 = 10\,k\Omega$

$\dfrac{R_3}{R_2} = 2Q - 1 = 3 \Rightarrow R_2 = 10\,k\Omega$

$R_3 = 30\,k\Omega$

Now $k = 2 - 1/Q = 1.5$

$\therefore$ Flat Gain $= 10 = (1.5)\dfrac{R_F}{R_H}$

$$\therefore \frac{R_H}{R_F} = 0.15$$

choose $R_F = 100\,k\Omega$

$R_H = R_L = 15\,k\Omega$

$R_B = QR_H = \underline{\underline{30\,k\Omega}}$

12.51

Note ω_n does not depend on R or C

From eq. 12.67:

$$\frac{R_H}{R_L} = \left(\frac{\omega_n}{\omega_0}\right)^2$$

$\therefore \omega_n = \omega_0\sqrt{\dfrac{R_H}{R_L}}$ $\quad$ Nominally $R_H = R_L \pm 1\%$

Thus:

$\omega_n' = \omega_0\sqrt{\dfrac{1.01}{0.99}} = 1.01\,\omega_0$ $\qquad \omega_n'' = \omega_0\sqrt{\dfrac{0.99}{1.01}} = 0.99\,\omega_0$

$\therefore \omega_n$ can deviate from ω_0 by $\underline{\underline{\pm 1\%}}$

12.52

Use Tow Thomas to realize a LPN (Fig 12.26)

$\omega_0 = 10^4 \quad \omega_n = 1.2\omega_0 \quad Q = 10 \quad \text{DC Gain} = 1$

$C = 10nF \quad r = 20k\Omega$

$R = \frac{1}{\omega_0 C} = \underline{\underline{10k\Omega}}$

from P. 16(e):

$\text{DC Gain} = a_2 \frac{\omega_n^2}{\omega_0^2} = 1$

$a_2 \frac{1.2^2 \omega_0^2}{\omega_0^2} = 1$

$a_2 = \frac{1}{1.2^2} = \text{HF. Gain}$

$C_1 = C a_2 = \frac{10 \times 10^{-9}}{1.2^2} = \underline{\underline{6.94nF}}$

$R_2 = \frac{R(\omega_0/\omega_n)^2}{\text{HF Gain}} = R\left(\frac{1}{1.2}\right)^2 \times (1.2)^2$

$= R = \underline{\underline{10k\Omega}}$

$\underline{\underline{R_1 = R_3 = \infty}}$

12.53

For all pass:

$$T(s) = \frac{-s^2\left(\frac{C_1}{C}\right) + s\frac{1}{C}\left(\frac{1}{R_1} - \frac{r}{RR_3}\right) + \frac{1}{C^2RR_2}}{s^2 + s\frac{\omega_0}{Q} + \omega_0^2}$$

$$\omega_z^2 = \frac{1}{C^2RR_2} \cdot \frac{C}{C_1} \Rightarrow \omega_z = \frac{1}{C\sqrt{RR_2}} \cdot \sqrt{\frac{C}{C_1}}$$

$$Q_z = \frac{\omega_z}{\frac{1}{C}\left(\frac{1}{R_1} - \frac{r}{RR_3}\right)\frac{C}{C_1}}$$

$$Q_z = \frac{\sqrt{\frac{1}{C^2RR_2}\frac{C}{C_1}}}{\frac{1}{C}\left(\frac{1}{R_1} - \frac{r}{RR_3}\right)\left(\frac{C}{C_1}\right)}$$

$$= \frac{1}{\sqrt{RR_2}\left(\frac{1}{R_1} - \frac{r}{RR_3}\right)\sqrt{\frac{C}{C_1}}}$$

For All Pass $R_1 \to \infty$

To adjust Q_z, trim r or R_2 (independent of ω_z!)

Now $\omega_0 = \frac{1}{CR}$ so do not trim R or C!

Note if we trim R_2 or C_1 to adjust ω_z, this will also affect Q_z. So the options are:

For ω_z: (a) trim R_2 AND (r or R_3) to maintain the value of Q_z

OR

(b) trim C_1, and r or R_3

Prefer not to trim a capacitor so use (a)!

12.54

$$T(s) = \frac{0.4508(s^2 + 1.6996)}{(s + 0.7294)(s^2 + 0.2786s + 1.0504)}$$

Part (a) Replace s with s/ω_p
$\omega_p = 10^4$ rad/s.

$$T(s) = \frac{0.4508\left(s^2/\omega_p^2 + 1.6996\right)}{\left(\frac{s}{\omega_p} + 0.7294\right)\left(\frac{s^2}{\omega_p^2} + \frac{0.2786s}{\omega_p} + 1.0504\right)}$$

CONT.

$$T(s) = \frac{0.4508\,\omega_p(s^2 + 1.6996\,\omega_p^2)}{(s+0.7294\,\omega_p)(s^2+0.2786\,\omega_p s + 1.0504\,\omega_p^2)}$$

$$= \frac{4508(s^2 + 1.6996\times10^8)}{(s+7294)(s^2+2786s+1.0504\times10^8)}$$

For First Order Section use Fig 12.13(a)

$\omega_0 = 7294$ Dc gain $=1$

choose $C = 10nF$

$R_1 = R_2 = \frac{1}{\omega_0 C} \Rightarrow R_1 = R_2 = 13.71k\Omega$

For Second Order Section - use Fig 12.26

$\omega_n^2 = 1.6996\times10^8 \Rightarrow \omega_n = 13.037\times10^3$

$\omega_0^2 = 1.0504\times10^8 \Rightarrow \omega_0 = 10.249\times10^3$

$\frac{\omega_0}{Q} = 2786 \Rightarrow Q = 3.6787$

Dc gain $=1$

For Tow Thomas LPN use table 12.2

choose $C = 10nF$

$R = \frac{1}{\omega_0 C} = \frac{1}{10.249\times10^3\ 10\times10^{-9}}$

$= 9.757k\Omega$

choose $r = 20k\Omega$

Now from Fig 12.16 (e):

$T(0) = a_2\frac{\omega_n^2}{\omega_0^2} = 1 \Rightarrow a_2 = \frac{\omega_0^2}{\omega_n^2} = 0.618$

$\therefore$ HF gain $= a_2 = 0.618$

$C_1 = C \times HF gain \Rightarrow C_1 = 6.18nF$

$R_2 = R(\omega_0/\omega_n)^2$

$R_2 = 0.618R \Rightarrow R_2 = 6.03k\Omega$

$R_1 = R_3 = \infty$ $QR = 35.89k\Omega$

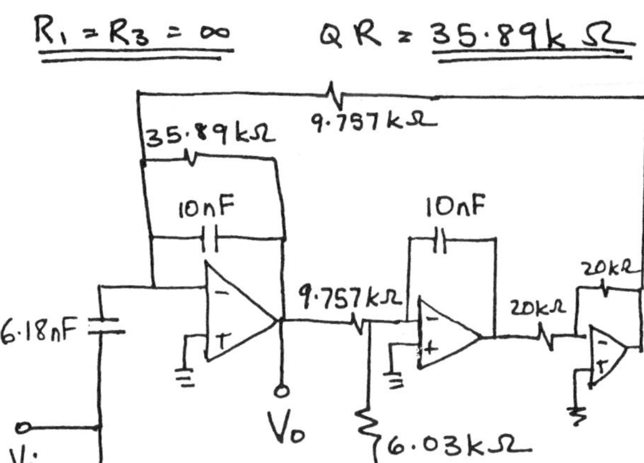

12.55

Make $C_1 = C_2 = 1nF = C$

$\omega_0 = \frac{1}{\sqrt{C_1C_2R_3R_4}}$ $Q = \left[\frac{\sqrt{C_1C_2R_3R_4}}{R_3}\left(\frac{1}{C_1}+\frac{1}{C_2}\right)\right]^{-1}$

Let $R_3 = R$ $m = 4Q^2$

$R_4 = \frac{R}{m}$ $= 4/2 = 2$

$\therefore \omega_0 = \frac{1}{\sqrt{C^2R^2/2}} = \frac{\sqrt{2}}{RC} = 10^4$

$\Rightarrow R = \frac{\sqrt{2}}{10^4 10^{-9}} = 141.42k\Omega$

$R_3 = 141.4k\Omega$

$R_4 = \frac{R_3}{2} \Rightarrow R_4 = 70.7k\Omega$

12.56

For Fig 12.28 (a)

$$t(s) = \frac{s^2 + s\left(\frac{1}{C_1} + \frac{1}{C_2}\right)\frac{1}{R_3} + \frac{1}{C_1C_2R_3R_4}}{s^2 + s\left(\frac{1}{C_1R_3} + \frac{1}{C_1R_4} + \frac{1}{C_2R_3}\right) + \frac{1}{C_1C_2R_3R_4}}$$

But $C_1 = C_2 = C$ & $R_3 = R_4 = R$, $RC = \tau$

$$\therefore t(s) = \frac{s^2 + s\,{}^2/RC + {}^1/R^2C^2}{s^2 + s\,{}^3/RC + \frac{1}{R^2C^2}}$$

$$= \frac{s^2 + s\,{}^2/\tau + {}^1/\tau^2}{s^2 + s\frac{3}{\tau} + {}^1/\tau^2}$$

Zeros defined by $\omega_z = 1/\tau$
$Q_z = \frac{1}{2}$

$\Rightarrow$ Double Root at $s = -1/\tau$

Poles of $t(s)$ are given by the quadratic formula:

$$s = \frac{-3}{2\tau} \pm \frac{\sqrt{5}}{2\tau} = \frac{-3 \pm \sqrt{5}}{2\tau}$$

i.e. two roots on the negative real axis

If the network is placed in the negative feedback path of an ideal amplifier ($A = \infty$) then the poles are given by the zeros of $t(s)$:

Closed loop poles:

$$s = -1/\tau \quad (\text{multiplicity} = 2)$$

12.57

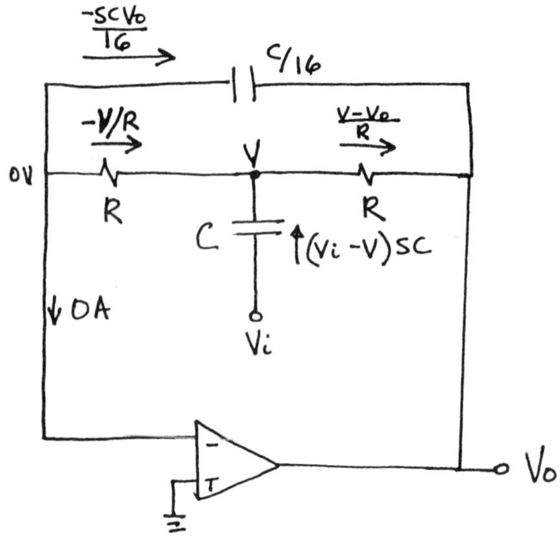

Note first $-\frac{sCV_o}{16} = -\frac{V}{R}$

$$V = -\frac{sCRV_o}{16}$$

ΣI at V

$$-\frac{V}{R} + sC(V_i - V) - \frac{V - V_o}{R} = 0$$

$$\frac{sCV_oR}{R\,16} + sCV_i + \frac{s^2C^2RV_o}{16} + \frac{sCV_o}{16} + \frac{V_o}{R} = 0$$

mult by: $16R$ and let $RC = \tau$

$$s\tau V_o + 16\tau V_i s + s^2\tau^2 V_o + s\tau V_o + 16V_o = 0$$

$$V_o\left[s^2\tau^2 + s \times 2\tau + 16\right] = -16s\tau V_i$$

$$\therefore \frac{V_o}{V_i} = -\frac{16s\tau}{s^2\tau^2 + 2\tau s + 16}$$

$$\therefore T(s) = \frac{s\,{}^{16}/RC}{s^2 + s\,{}^2/RC + {}^{16}/R^2C^2}$$

CONT.

Let $\omega_0^2 = \frac{16}{(RC)^2} \Rightarrow \omega_0 = \frac{4}{RC}$

$\frac{\omega_0}{Q} = \frac{2}{RC} \Rightarrow Q = \frac{RC\omega_0}{2} = 2$

$$\therefore \frac{V_o}{V_i} = \frac{-4\omega_0 s}{s^2 + s\frac{\omega_0}{Q} + \omega_0^2}$$

$\left.|T|\right|_{s=0} = 0$, $\left.|T|\right|_{s=\infty} = 0$ } Bandpass

$|T(j\omega_0)| = 4/\tfrac{1}{2} = 8\,\frac{V}{V}$ = CENTRE FREQ GAIN

12.58

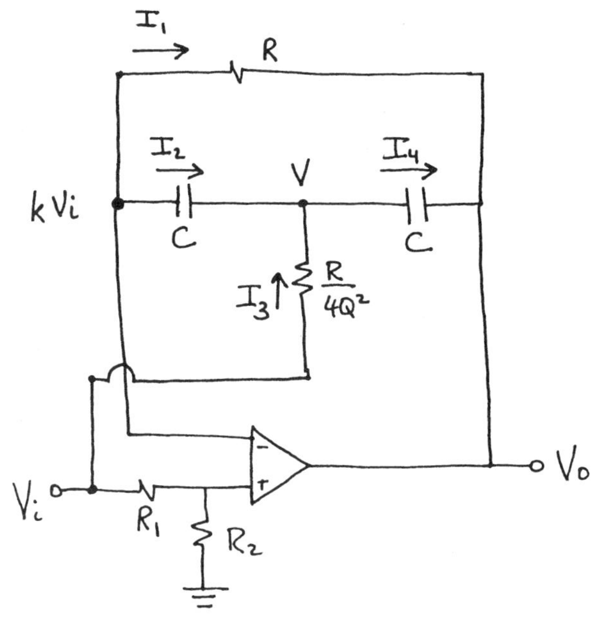

$RC = 2Q/\omega_0$

$k = \frac{R_2}{R_1+R_2}$

$V_{+ve} = V_{-ve} = kV_i$ due to virtual short

$I_1 = -I_2$

$$\frac{kV_i - V_o}{R} = \frac{V - kV_i}{1}(sC)$$

$$V = \frac{1}{sCR}\left(kV_i - V_o + sCRkV_i\right)$$

ΣI at $V = 0$

$I_2 + I_3 - I_4 = 0$

$$sC(kV_i - V) + \frac{4Q^2}{R}(V_i - V) - sC(V - V_o) = 0$$

$$sC\left(kV_i - \frac{kV_i}{sCR} + \frac{V_o}{sCR} - kV_i\right)$$
$$+ \frac{4Q^2}{R}\left(V_i - \frac{kV_i}{sCR} + \frac{V_o}{sCR} - kV_i\right)$$
$$- sC\left(\frac{kV_i}{sCR} - \frac{V_o}{sCR} + kV_i - V_o\right) = 0$$

$\Rightarrow$

$$-\frac{kV_i}{R} + \frac{V_o}{R}$$
$$+ \frac{4Q^2}{R}\left(V_i - \frac{kV_i}{sCR} + \frac{V_o}{sCR} - kV_i\right)$$
$$- \frac{sC}{R}\left(\frac{kV_i}{sC} - \frac{V_o}{sC} + kRV_i - V_oR\right) = 0$$

$\Rightarrow$ SUB $CR = \frac{2Q}{\omega_0}$ & $R = \frac{2Q}{C\omega_0}$

$$-kV_i + V_o$$
$$+ 4Q^2V_i - \frac{2Q\,kV_i\,\omega_0}{s} + \frac{V_o\,\omega_0\,2Q}{s} - 4Q^2kV_i$$
$$-kV_i + V_o \quad sk\,2Q/\omega_0\,V_i + sV_o\frac{2Q}{\omega_0} - 0$$

$$V_o\left[1+\frac{2Q\omega_0}{s}+1+\frac{2Qs}{\omega_0}\right]$$

$$=V_i\left[k-4Q^2+\frac{2kQ\omega_0}{s}+4Q^2k+k+\frac{2kQs}{\omega_0}\right]$$

$$\Rightarrow V_o\left[s^2\frac{2Q}{\omega_0}+2s+2Q\omega_0\right]=V_i\left[s^2\frac{2kQ}{\omega_0}+s(4Q^2k-4Q^2+2k)+2kQ\omega_0\right]$$

$$\Rightarrow \frac{V_o}{V_i}=\frac{s^2\frac{2kQ}{\omega_0}+s(4Q^2k-4Q^2+2k)+2kQ\omega_0}{s^2\frac{2Q}{\omega_0}+2s+2Q\omega_0}$$

$$=k\,\frac{s^2+s\frac{\omega_0}{Q}\left(2Q^2-\frac{2Q^2}{k}+1\right)+\omega_0^2}{s^2+s\frac{\omega_0}{Q}+\omega_0^2}$$

Recall $k=\frac{R_2}{R_1+R_2}$ and $\frac{1}{k}=1+\frac{R_1}{R_2}$

$$\Rightarrow \frac{V_o}{V_i}=\left(\frac{R_2}{R_1+R_2}\right)\frac{s^2+s\frac{\omega_0}{Q}\left(1-\frac{R_1}{R_2}\cdot 2Q^2\right)+\omega_0^2}{s^2+s\frac{\omega_0}{Q}+\omega_0^2}$$

$$\therefore T(s)=\frac{R_2}{R_1+R_2}\;\frac{s^2+s\frac{\omega_0}{Q}\left(1-\frac{2Q^2R_1}{R_2}\right)+\omega_0^2}{s^2+s\frac{\omega_0}{Q}+\omega_0^2}$$

For All Pass we want

$$T(s)\propto\frac{s^2+\frac{\omega_0}{Q}(-1)s+\omega_0^2}{s^2+s\frac{\omega_0}{Q}+\omega_0^2}$$

$$\Rightarrow 1-\frac{2Q^2R_1}{R_2}=-1$$

$$2Q^2\frac{R_1}{R_2}=2$$

$$\frac{R_1}{R_2}=\frac{1}{Q^2}$$

$$\therefore \frac{R_2}{R_1}=Q^2 \quad \& \quad \frac{R_2}{R_1+R_2}=\frac{R_2/R_1}{1+R_2/R_1}=\frac{Q^2}{1+Q^2}$$

For Notch:

$$1-2Q^2\frac{R_1}{R_2}=0$$

$$\frac{R_1}{R_2}=\frac{1}{2Q^2}$$

$$\frac{R_2}{R_1}=2Q^2 \quad \& \quad \frac{R_2}{R_1+R_2}=\frac{2Q^2}{1+2Q^2}$$

12.59

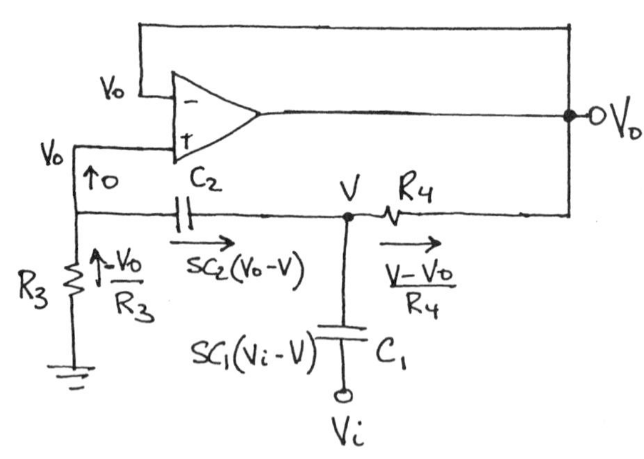

∵ No current can't flow into the terminal

$$-\frac{V_o}{R_3}=sC_2(V_o-V)$$

$$V=V_o\left(1+\frac{1}{sC_2R_3}\right)$$

CONT.

ΣI @ $V = 0$

$$-\frac{V_o}{R_3} + \frac{V_i - V}{1} sC_1 = \frac{V - V_o}{R_4}$$

$$V_o\left[-\frac{1}{R_3}+\frac{1}{R_4}\right] + V\left[-sC_1 - \frac{1}{R_4}\right] = -sC_1V_i$$

$$V_o[R_4 - R_3] + V[sC_1R_3R_4 + R_3] = V_i sC_1R_3R_4$$

$$V_o(R_4-R_3) + V_o\left(1+\frac{1}{sC_2R_3}\right)(sC_1R_3R_4 + R_3) = sC_1R_3R_4V_i$$

$$V_o\left(R_4 - \cancel{R_3} + sC_1R_3R_4 + \cancel{R_3} + \frac{C_1}{C_2}R_4 + \frac{1}{sC_2}\right) = sC_1R_3R_4V_i$$

$$V_o\left(s^2C_1C_2R_3R_4 + sC_1R_4 + sC_2R_4 + 1\right) = s^2C_1R_3R_4C_2V_i$$

$$\therefore \frac{V_o}{V_i} = \frac{s^2C_1C_2R_3R_4}{s^2C_1C_2R_3R_4 + sC_1R_4 + sC_2R_4 + 1}$$

$$= \frac{s^2}{s^2 + s\left(\frac{1}{C_2R_3} + \frac{1}{C_1R_3}\right) + \frac{1}{C_1C_2R_3R_4}}$$

Note $|T(0)| = 0$, $|T(\infty)| = 1$ $\therefore$ HighPass, High Freq Gain $= 1\,\frac{V}{V}$

3dB freq $= 10^3$ rad/s, $Q = \frac{1}{\sqrt{2}}$ for max flat.

$\therefore \omega_0 = 10^3 \frac{rad}{s}$ $C_1 = C_2 = 10nF$

clearly $\omega_0^2 = \frac{1}{C_1C_2R_3R_4}$ and

$$\frac{\omega_0}{Q} = \frac{1}{C_2R_3} + \frac{1}{C_1R_3} = \frac{C_1 + C_2}{C_1C_2R_3}$$

$$= \frac{2C}{C^2R_3} = \frac{2}{CR_3} = \sqrt{2}\times 10^3$$

$$R_3 = \frac{2}{10\times10^{-9}\times10^3\times\sqrt{2}}$$

$$R_3 = 141.4\ k\Omega$$

$$R_4 = \frac{1}{\omega_0^2 C_1C_2R_3} \Rightarrow R_4 = 70.7k\Omega$$

12.60

$A_{max} = 3dB$

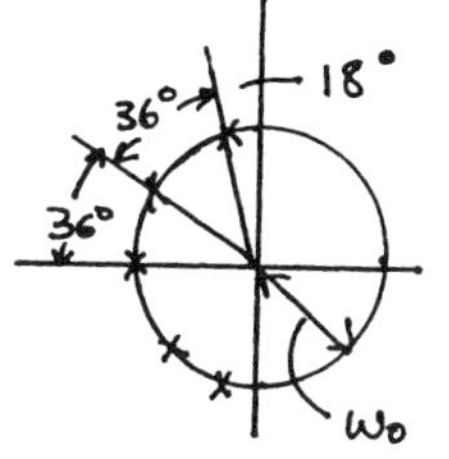

$$\epsilon = \left(10^{3/10} - 1\right)^{-\frac{1}{2}} \cong 1$$

$$\omega_0 = \omega_p\left(\frac{1}{\epsilon}\right)^{1/N} = \omega_p = 2\pi\,5000 = 10^4\pi$$

$$Q_1 = \frac{1}{2\cos 36} = 0.618$$

$$Q_2 = \frac{1}{2\cos 72} =$$

For first order section:

$\omega_0 = 10^4\pi$ dc gain $= 1$

From 12.13 (a)

$R_1 = R_2 = 10k\Omega$

$$C = \frac{1}{\omega_0 R_2} = \frac{1}{10^4\pi\,10^4} = 3.18\ nF$$

Second-Order Section $Q = 0.618$:

from 12.34 (c): $m = 4Q^2 = 1.528$

CONT.

$$RC = \frac{2Q}{\omega_0} \qquad \text{let } R_1 = R_2 = 10k\Omega$$

$$C = \frac{2Q}{\omega_0 R} \Rightarrow C_4 = C = 3.93nF$$

$$C_3 = \frac{C}{m} = 2.57nF$$

Second Order Section $Q = 1.618$:

$$C = \frac{2Q}{\omega_0 R} = 10.3nF \qquad m = 4Q^2 = 10.472$$

$$\Rightarrow R_1 = R_2 = 10k\Omega$$

$$C_4 = C = 10.3nF$$

$$C_3 = \frac{C}{m} = 0.984nF$$

12.61

For a bandpass filter

$$T(s) = \frac{\omega_0/Q\, s}{s^2 + s\omega_0/Q + \omega_0^2}$$

centre freq. gain = 1

complementary transfer function:

$$T' = 1 - T = \frac{s^2 + \omega_0^2}{s^2 + s\omega_0/Q + \omega_0^2} \equiv \text{NOTCH!}$$

From Fig 12.18(d)

R
V_i V_o
C L

$\Rightarrow$ INTERCHANGE V_i & gnd to get:

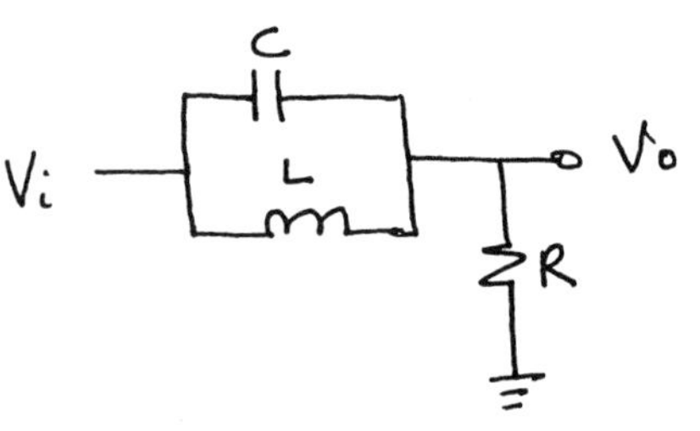

this is the same as Fig 12.18(e)

12.62

for Fig 12.18(d):

$$T(s) = \frac{s/RC}{s^2 + s/RC + 1/LC}$$

$$\omega_0 = \frac{1}{\sqrt{LC}} \qquad Q = R\sqrt{\frac{C}{L}}$$

FOR ω_0

$$\frac{\partial \omega_0}{\partial L} = \frac{\partial (LC)^{-1/2}}{\partial L} = -\frac{1}{2}L^{-3/2}C^{-1/2} = -\frac{\omega_0}{2L}$$

$$\frac{\partial \omega_0}{\partial C} = -\frac{\omega_0}{2C}$$

$$\frac{\partial \omega_0}{\partial R} = 0$$

$$\therefore S_L^{\omega_0} = \frac{\partial \omega_0}{\partial L}\frac{L}{\omega_0} = -1/2$$

$$S_C^{\omega_0} = \frac{\partial \omega_0}{\partial C} \times \frac{C}{\omega_0} = \frac{-1}{2}$$

$$S_R^{\omega_0} = \frac{\partial \omega_0}{\partial R}\frac{R}{\omega_0} = 0$$

FOR Q

$$\frac{\partial Q}{\partial L} = \frac{R\sqrt{C}}{L\sqrt{L}}\left(\frac{-1}{2}\right)$$

$$\frac{\partial Q}{\partial C} = \frac{1}{2}\frac{R}{\sqrt{LC}} = \frac{1}{2}\frac{R\sqrt{C}}{C\sqrt{L}} = \frac{Q}{2C}$$

$$\frac{\partial Q}{\partial R} = \sqrt{C/L} = \frac{R}{R}\sqrt{C/L} = Q/R$$

CONT.

$$S_L^Q = \frac{-Q}{2L} \times \frac{L}{Q} = -\frac{1}{2}$$

$$S_C^Q = \frac{Q}{2C} \times \frac{C}{Q} = \frac{1}{2}$$

$$S_R^Q = \frac{Q}{R} \cdot \frac{R}{Q} = 1$$

12.63

$y = uv$

$$S_x^y = \frac{\partial(uv)}{\partial x} \frac{x}{uv}$$

$$= v\frac{\partial u}{\partial x}\frac{x}{uv} + u\frac{\partial v}{\partial x}\frac{x}{uv}$$

$$= \frac{\partial u}{\partial x}\frac{x}{u} + \frac{\partial v}{\partial x}\frac{x}{v}$$

$$= S_x^u + S_x^v$$

Part (b) $y = u/v$

$$S_x^y = \frac{\partial y}{\partial x} \cdot \frac{x}{y} = \frac{\partial(u/v)}{\partial x} \frac{xv}{u}$$

$$= \frac{1}{v}\frac{\partial u}{\partial x}\frac{vx}{u} + \frac{-u}{v^2}\frac{\partial v}{\partial x} \cdot \frac{xv}{u}$$

$$= \frac{\partial u}{\partial x}\frac{x}{u} - \frac{\partial v}{\partial x}\frac{x}{v}$$

$$= S_x^u - S_x^v$$

Part (c) $y = ku$

$$S_x^y = \frac{\partial y}{\partial x} \cdot \frac{x}{y} = \frac{\partial(ku)}{\partial x}\frac{x}{ku}$$

$$= k\frac{\partial u}{\partial x}\frac{x}{ku}$$

$$= \frac{\partial u}{\partial x}\frac{x}{u}$$

$$= S_x^u$$

Part (d) $y = u^n$

$$S_x^y = \frac{\partial y}{\partial x} \cdot \frac{x}{y} = \frac{\partial(u^n)}{\partial x}\frac{x}{u^n}$$

$$= nu^{n-1}\frac{\partial u}{\partial x} \cdot \frac{x}{u^n}$$

$$= n\frac{\partial u}{\partial x}\frac{x}{u} = nS_x^u$$

Part (e) $y = f_1(u)$ $u = f_2(x)$

$$S_x^y = \frac{\partial y}{\partial x}\frac{x}{y} = \frac{\partial f_1(u)}{\partial x}\frac{x}{f_1(u)}$$

$$= \frac{\partial f_1(u)}{\partial x}\frac{\partial u}{\partial x} \cdot \frac{x}{f_1(u)}$$

$$= \frac{\partial f_1}{\partial u} \cdot \frac{\partial f_2}{\partial x} \cdot \frac{x}{f_1} \cdot \frac{u}{u}$$

But $u = f_2$

$$\therefore S_x^y = \frac{\partial f_1}{\partial u} \cdot \frac{\partial f_2}{\partial x} \cdot \frac{x}{f_1} \cdot \frac{u}{f_2}$$

$$= \frac{\partial f_1}{\partial u} \cdot \frac{u}{f_1} \cdot \frac{\partial f_2}{\partial x}\frac{x}{f_2}$$

$$= S_u^{f_1} \cdot S_x^{f_2}$$

$$= S_u^y \, S_x^u$$

12.64

Since the characteristic equation of FIG. 12.33(b) is the same as for FIG. 12.29 the poles are given by EQ 12.86:

$$s^2 + s\frac{\omega_0}{Q}\left[1 + \frac{2Q^2}{A+1}\right] + \omega_0^2 = 0$$

This is because 12.33(b) is based on the complementary transform of 12.29 and hence the pole

locations are preserved!
Now the actual ω_o and Q are given by:

$$\omega_{o,a} = \omega_o \quad \text{and} \quad Q_a = \frac{Q}{1 + \frac{2Q^2}{(A+1)}}$$

Thus from Ex. 12.3

$$S_A^{\omega_{o,a}} = 0$$

$$S_A^{Q_a} = \frac{A}{A+1} \quad \frac{2Q^2/(A+1)}{1 + 2Q^2/(A+1)}$$

$$\therefore S_A^{Q_a} \cong \frac{2Q^2}{A}$$

$$\frac{\partial Q}{\partial C_4} = \frac{Q}{2C_4} \Rightarrow S_{C_4}^{Q} = +1/2$$

$$\frac{\partial Q}{\partial R_1} = \frac{1/\sqrt{R_1} - \sqrt{R_1}/R_2}{R_1\left(\frac{1}{\sqrt{R_1}} + \sqrt{R_1}/R_2\right)} \cdot \frac{Q}{2}$$

$$= \frac{\sqrt{R_2/R_1} - \sqrt{R_1/R_2}}{R_1\left(\sqrt{\frac{R_2}{R_1}} + \sqrt{\frac{R_1}{R_2}}\right)} \cdot \frac{Q}{2}$$

$$\therefore S_{R_1}^{Q} = \frac{\sqrt{R_2/R_1} - \sqrt{R_1/R_2}}{\sqrt{R_2/R_1} + \sqrt{R_1/R_2}}$$

if $R_1 = R_2 \Rightarrow S_{R_1}^{Q} = 0$ &

$$S_{R_2}^{Q} = 0$$

12.65

If $R_1 = R_2$, then from (12.77) & (12.78)

$$\omega_o = \frac{1}{\sqrt{C_3 C_4 R_1 R_2}}$$

$$Q = \frac{1}{\sqrt{C_3 C_4 R_1 R_2}\left(\frac{1}{C_4}\right)\left(\frac{1}{R_1} + \frac{1}{R_2}\right)}$$

$$\frac{\partial \omega_o}{\partial C_3} = \frac{-1}{2C_3\sqrt{C_3 C_4 R_1 R_2}}$$

$$\frac{\partial \omega_o}{\partial C_3} = \frac{-1}{2C_3\sqrt{C_3 C_4 R_1 R_2}} = -\frac{\omega_o}{2C_3}$$

$$S_{C_3}^{\omega_o} = \frac{\partial \omega_o}{\partial C_3}\frac{C_3}{\omega_o} = -\frac{1}{2}$$

clearly $S_{C_3}^{\omega_o} = S_{C_4}^{\omega_o} = S_{R_1}^{\omega_o} = S_{R_2}^{\omega_o} = -1/2$

$$\frac{\partial Q}{\partial C_3} = \frac{-1}{2C_3\sqrt{C_3 C_4 R_1 R_2}\left(\frac{1}{C_4}\right)\left(\frac{1}{R_1} + \frac{1}{R_2}\right)} = \frac{-Q}{2C_3}$$

$$\therefore S_{C_3}^{Q} = -1/2$$

12.66

From table 12.1

$$\omega_o = \frac{1}{\sqrt{C_4 C_6 R_1 R_3 R_5 / R_2}}$$

$$Q = R_6\sqrt{\frac{C_6}{C_4}\frac{R_2}{R_1 R_3 R_5}}$$

$$\frac{\partial \omega_o}{\partial C_4} = \frac{-\omega_o}{2C_4}$$

$$\therefore S_{C_4}^{\omega_o} = \frac{-\omega_o}{2C_4} \times \frac{C_4}{\omega_o} = -1/2$$

Similarly $S_{C_6}^{\omega_o} = S_{R_1}^{\omega_o} = S_{R_3}^{\omega_o} = S_{R_5}^{\omega_o} = -\frac{1}{2}$

$$\frac{\partial \omega}{\partial R_2} = \frac{\omega_o}{2R_2} \Rightarrow S_{R_2}^{\omega_o} = \frac{1}{2}$$

Now for Q:

$$\frac{\partial Q}{\partial R_6} = \frac{Q}{R_6} \Rightarrow S_{R_6}^{Q} = \frac{\partial Q}{\partial R_6}\frac{R_6}{Q} = +1$$

CONT.

$$\frac{\partial Q}{\partial C_6} = \frac{Q}{2C_6} \Rightarrow S_{C_6}^{Q} = S_{R_2}^{Q} = \underline{\underline{+\frac{1}{2}}}$$

$$\frac{\partial Q}{\partial C_4} = \frac{-Q}{2C_4} \Rightarrow S_{C_4}^{Q} = S_{R_1,R_3,R_5}^{Q} = \underline{\underline{-\frac{1}{2}}}$$

12.67

$$R_{eq} = \frac{T_c}{C_1} = \frac{1/100\times10^3}{C_1}$$

for 1pF $\rightarrow R_{eq} = 10^{-5}/10^{-12} = \underline{\underline{10\,M\Omega}}$

for 10pF $\rightarrow R_{eq} = 10^{-5}/10\times10^{-12} = \underline{\underline{1M\Omega}}$

12.68

charge transferred $\Rightarrow Q = CV$

$= 10^{-12}(1)$

$= \underline{\underline{1pC}}$

For $f_o = 100$kHz, average current is given by:

$$I_{AVE} = \frac{Q}{T} = 1pC \times \frac{1}{100\times10^3}$$

$$= \underline{\underline{0.1\,\mu A}}$$

For each clock cycle, the output will change by the same amount as the change in voltage across C_2!

$$\therefore \Delta V = Q/C_2 = \frac{1pC}{10pF} = \underline{\underline{0.1V}}$$

For $\Delta V = 0.1V$ for each clock cycle, the amplifier will saturate in

$$\#\text{cycles} = \frac{10V}{0.1V} = \underline{\underline{100\text{ cycles}}}$$

$$\text{slope} = \frac{\Delta V}{\Delta t} = \frac{10V}{(100\text{ cycles})(1/100\times10^3)}$$

$$= \underline{\underline{10^4\ \frac{V}{S}}}$$

12.69

$f_c = 400$kHz $f_o = 10$kHz $Q = 20$

$C_1 = C_2 = 20pF = C$

$$C_3 = C_4 = \omega_o T_c C = 2\pi(10^4)\frac{1}{400\times10^3}20\times10^{-12}$$

$$= \underline{\underline{3.14pF}}$$

$$C_5 = \frac{\omega_o T_c C}{Q}$$

$$= \frac{C_3}{Q} = \underline{\underline{0.157\,pF}}$$

$$C_6 = \frac{\omega_o T_c C}{Q} \times \text{centre frequency gain}$$

$$= \underline{\underline{0.157\ pF}}$$

Note that the clock frequency has doubled. Hence the period, T_c, is halved. Therefore, for the same integrating capacitors, the resistors (switched capacitors) will change by the factor of 2. So compensate for this by changing the switched caps by a factor of 1/2.

12.70

Ex 12-31 for $Q = 40$ $f_c = 200$kHz
$f_o = 10$kHz

$C_1 = C_2 = 20pF = C$

$C_3 = C_4 = \omega_o T_c C$

$$= 2\pi(10^4)\left(\frac{1}{200\times10^3}\right)20\times10^{-12}$$

$$= \underline{\underline{6.28\,pF}}$$

CONT.

$$C_5 = \frac{\omega_0 T_c C}{Q} = \frac{C_3}{Q} = \underline{\underline{0.157\text{pF}}}$$

$$C_6 = \frac{\omega_0 T_c C}{Q} = C_5 = \underline{\underline{0.157\text{ pF}}}$$

12.71

$\omega_0 = 10^4$, $Q = 1/\sqrt{2}$, $f_c = 100\text{kHz}$

DC gain $\Rightarrow \frac{R_4}{R_6} \Rightarrow \frac{C_6}{C_4} = 1$

$C_1 = C_2 = 10\text{pF}$

$$C_3 = C_4 = C_6 = \omega_0 T_c C$$
$$= 10^4\left(\frac{1}{100 \times 10^3}\right) 10 \times 10^{-12}$$
$$= \underline{\underline{1\text{pF}}}$$

$$C_8 = C_4/Q = \underline{\underline{1.41\text{pF}}}$$

12.72

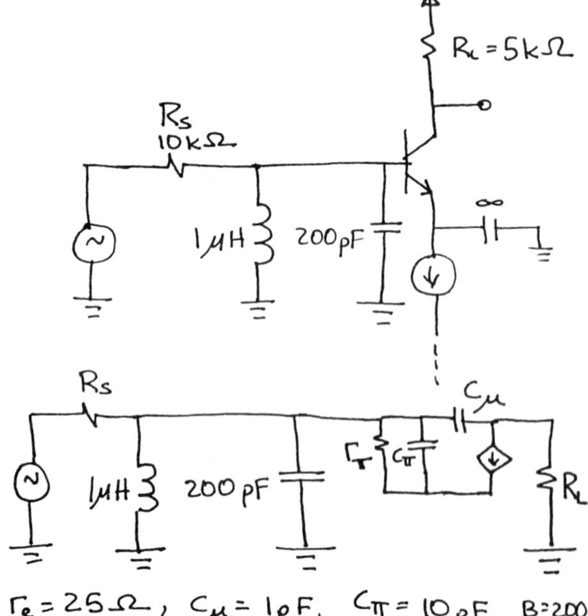

$r_e = 25\Omega$, $C_\mu = 1\text{pF}$, $C_\pi = 10\text{pF}$, $\beta = 200$

$r_\pi = (\beta+1) r_e = 5.025\text{ k}\Omega$

From base to collector

$$\frac{V_c}{V_b} = -\frac{\beta}{\beta+1} \cdot \frac{R_L}{r_e} = -199 = k$$

Total capacitance at base

$$C_T = C_\pi + 200\text{p} + C_\mu(1-k) \quad \leftarrow \text{Miller Effect}$$
$$= 10 + 200 + 1(1+199)$$
$$= 410\text{pF}$$

$$\therefore \omega_0 = \frac{1}{\sqrt{LC}}$$
$$= \frac{1}{\sqrt{10^{-6} \times 410 \times 10^{-12}}}$$
$$= \underline{\underline{49.4 \times 10^6 \text{ rad/s}}}$$

Centre frequency gain =

$$\frac{r_\pi}{R_s + r_\pi} \cdot k$$
$$= \frac{5.025}{10 + 5.025} \times -199$$
$$= \underline{\underline{-66.6 \text{ V/V}}}$$

$$BW = \frac{1}{RC}$$
$$= \frac{1}{(R_s \| r_\pi)\, 410\text{pF}}$$
$$= \underline{\underline{729 \times 10^3 \frac{\text{rad}}{\text{s}}}}$$

$$Q = \frac{\omega_0}{BW}$$
$$= 49.4/0.7293$$
$$= \underline{\underline{67.7}}$$

12.73

$$Q_0 = \frac{R_p}{\omega_0 L} \Rightarrow R_p = Q_0 \omega_0 L$$

$$= 200(2\pi 10^6)(10\times10^{-6})$$

$$= \underline{\underline{12.57\,k\Omega}}$$

$$\omega_0 = \frac{1}{\sqrt{LC}} \Rightarrow C = \frac{1}{\omega_0^2 L}$$

$$= \frac{1}{(2\pi 10^6)^2\, 10\times10^{-6}}$$

$$= \underline{\underline{2.533\,nF}}$$

$$B = \frac{1}{RC} \qquad R_r = \frac{1}{(2\pi\times10\times10^3)(2.533\times10^{-9})}$$

$$= 6.283\,k\Omega$$

$$\therefore \frac{1}{R_1} + \frac{1}{R_p} = \frac{1}{R_r}$$

$$\Rightarrow R_1 = \underline{\underline{12.57\,k\Omega}} \quad \text{i.e. } R_1 // R_p = R_r$$

12.74

$$f_0 = \frac{1}{2\pi\sqrt{LC}}$$

$$= \left(2\pi\sqrt{(36\times10^{-6})\,10^{-9}}\right)^{-1}$$

$$= \underline{\underline{838.8\,kHz}}$$

36μH, 2, 1, 1kΩ, 1nF

$$R_p = n^2 R$$

$$= 9\,(1k\Omega)$$

$$= 9k\Omega$$

$$Q = R_p/\omega_0 L$$

$$= \frac{9\times10^3}{2\pi\; 838.8\times10^3 \times 36\times10^{-6}}$$

$$= \underline{\underline{47.4}}$$

12.75

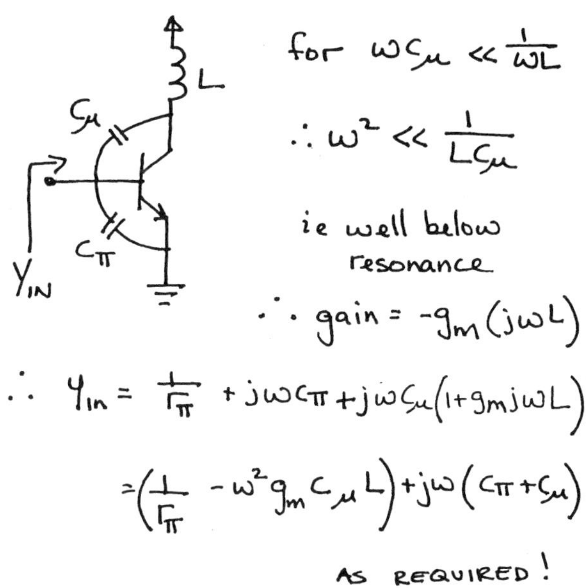

for $\omega C_\mu \ll \frac{1}{\omega L}$

$\therefore \omega^2 \ll \frac{1}{LC_\mu}$

ie well below resonance

$\therefore$ gain $= -g_m(j\omega L)$

$$\therefore Y_{in} = \frac{1}{r_\pi} + j\omega C_\pi + j\omega C_\mu(1 + g_m j\omega L)$$

$$= \left(\frac{1}{r_\pi} - \omega^2 g_m C_\mu L\right) + j\omega(C_\pi + C_\mu)$$

AS REQUIRED!

12.76

From Fig 12.16 (c):

$$T(s) = \frac{a_1 s}{s^2 + s\frac{\omega_0}{Q} + \omega_0^2}$$

$$T(j\omega) = \frac{ja_1\omega}{\omega_0^2 - \omega^2 + j\frac{\omega\omega_0}{Q}}$$

$$T(j\omega_0) = \frac{ja_1\omega_0}{j\omega_0^2/Q} = \frac{a_1 Q}{\omega_0}$$

$$|T(j\omega)| = a_1\omega\left[(\omega_0^2-\omega^2)^2 + \left(\frac{\omega\omega_0}{Q}\right)^2\right]^{-\frac{1}{2}}$$

$$= \frac{a_1\omega\; Q/\omega\omega_0}{\sqrt{1 + Q^2\left(\frac{\omega_0^2-\omega^2}{\omega_0\omega}\right)^2}}$$

Now $\omega = \omega_0 + \delta\omega$, $\frac{\delta\omega}{\omega_0} \ll 1$

and $\omega^2 \simeq \omega_0^2\left(1 + \frac{2\delta\omega}{\omega_0}\right)$

CONT.

so $\omega_0^2 - \omega^2 = -2\delta\omega\omega_0$

$$\therefore |T(j\omega)| \cong \frac{a_1 Q/\omega_0}{\sqrt{1 + Q^2\left(\frac{2\delta\omega}{\omega}\right)^2}}$$

for $Q >> 1$: $Q^2\left(\frac{2\delta\omega}{\omega}\right)^2 \cong Q^2\left(\frac{2\delta\omega}{\omega_0}\right)$

$\therefore \omega \approx \omega_0$!

$$\Rightarrow |T(j\omega)| \cong \frac{|T(j\omega_0)|}{\sqrt{1 + Q^2\left(\frac{2\delta\omega}{\omega_0}\right)^2}}$$

$$= \frac{|T(j\omega_0)|}{\sqrt{1 + 4Q^2\frac{\delta\omega}{\omega_0}}}$$

For N bandpass sections, synchronously tuned in cascade, half power is given by:

$$\left(\frac{1}{\sqrt{1 + 4Q^2\left(\frac{\delta\omega}{\omega_0}\right)^2}}\right)^N = \frac{1}{\sqrt{2}}$$

$$\left(1 + 4Q^2\left(\frac{\delta\omega}{\omega_0}\right)^2\right)^N = 2$$

$$4Q^2\left(\frac{\delta\omega}{\omega_0}\right)^2 = 2^{1/N} - 1$$

$$\delta\omega = \frac{\omega_0}{2Q}\sqrt{2^{1/N} - 1}$$

$\therefore$ Bandwidth:

$$B = 2\delta\omega = \frac{\omega_0}{Q}\sqrt{2^{1/N} - 1}$$

12.77

For first order low pass:

$$T(s) = \frac{\omega_0'}{s + \omega_0'} \qquad |T(j\omega)| = \frac{\omega_0'}{\sqrt{\omega^2 + \omega_0'^2}}$$

for a bandpass response around ω_0 with $\omega_0' = \frac{\omega_0}{2Q}$:

$$|T(j\omega)| \cong \frac{\omega_0/2Q}{(\delta\omega)^2 + \left(\frac{\omega_0}{2Q}\right)^2}$$

$$= \frac{\omega_0/2Q}{\frac{\omega_0}{2Q}\sqrt{\left(\frac{2Q}{\omega_0}\right)^2(\delta\omega)^2 + 1}}$$

$$= \frac{1}{\sqrt{1 + 4Q^2(\delta\omega/\omega_0)^2}}$$

Now at $\omega = \omega_0$ or $\delta\omega = 0$ $|r(j\omega_0)| = 1$, then

$$T(j\omega) \approx \frac{|T(j\omega_0)|}{\sqrt{1 + 4Q^2\left(\frac{\delta\omega}{\omega_0}\right)^2}}$$

Part (b)

For N synchronously tuned sections in cascade; 3dB bandwidth is given by:

$$\left(|T|/|T_0|\right)^N = \frac{1}{\sqrt{2}}$$

$$\left(|T|/|T_0|\right)^2 = \frac{1}{2^{1/N}} \quad \text{OR}$$

$$1 + 4Q^2\left(\frac{\delta\omega}{\omega_0}\right)^2 = 2^{1/N} \quad \text{OR}$$

$$2\delta\omega = \frac{\omega_0}{Q}\sqrt{2^{1/N} - 1} \qquad (12.110)$$

CONT.

Thus: $|T(j\omega)|_{overall} = |T(j\omega)|^N$

$$= \frac{|T(j\omega_0)|_{overall}}{\left[1+4Q^2\left(\frac{\delta\omega}{\omega_0}\right)^2\right]^{N/2}}$$

NOTE $Q = \frac{\omega_0}{B}\sqrt{2^{1/N}-1}$

$$= \frac{|T(j\omega_0)|_{overall}}{\left(1+4\frac{\omega_0^2}{B^2}(2^{1/N}-1)\left(\frac{\delta\omega}{\omega_0}\right)^2\right)^{\frac{N}{2}}}$$

$$= \frac{|T(j\omega_0)|_{overall}}{\left[1+4(2^{1/N}-1)\left(\frac{\delta\omega}{B}\right)^2\right]^{\frac{N}{2}}}$$

Part (c) (i)

for bandwidth = 2B, ie $\delta\omega = \pm B$

$$Att = -20\log\left(1+4(2^{1/N}-1)(1)\right)^{-N/2}$$

$$= -10N\log\left(1+2^{2+1/N}-4\right)$$

$$= 10N\log\left(2^{2+1/N}-3\right)$$

N =	1	2	3	4	5
Att (dB)	6.70	8.49	9.28	9.79	10.13

Part (c) (ii)

3 dB bandwidth $\delta\omega = \pm B/2$

30dB bandwidth $\frac{\delta\omega}{B} = x$

$$-30 = -20\frac{N}{2}\log\left(1+4(2^{1/N}-1)x^2\right)$$

$$3 = N\log\left(1+4(2^{1/N}-1)x^2\right)$$

$$x = \left[\frac{10^{3/N}-1}{4(2^{1/N}-1)}\right]^{1/2}$$

Ratio of 30 dB to 3dB

$$BW = \frac{2Bx}{B} = 2x$$

N =	1	2	3	4	5
Ratio =	31.6	8.6	5.7		4.5

12.78

See fig 12.48 and Eq 12.115 and 12.116

(a) For the narrowband approximation variation of Ω around 0 is equivalent to $\delta\omega$ around ω_0. Thus, a low-pass maximally flat filter of bandwidth B/2 and order N for which $|T| = \left[\left(1+\Omega/_{B/2}\right)^{2N}\right]^{-\frac{1}{2}}$

is transformed to a band-pass maximally flat filter of bandwidth B/2 and order 2N, and centre frequency ω_0, for which:

$$|T| = \left(1+\left(\frac{\delta\omega}{B/2}\right)^{2N}\right)^{-1/2}$$

(b) For bandwidth 2B, $\delta\omega = B$ &

$$|T| = \left(1+\left(\frac{B}{B/2}\right)^{2N}\right)^{-1/2}$$

$$= \left(1+2^{2N}\right)^{-1/2} \quad \text{thus:}$$

N	1	2	3	4	5
|T|	0.447	0.242	0.124	0.062	0.031
|T|dB	-6.99	-12.3	-18.1	-24.1	-30.1

For 30 dB bandwidth,

$$-30 = 20\log x \Rightarrow x = 10^{-3/2} = \frac{1}{31.6}$$

$$\therefore \quad 1 + \left(\frac{\delta' \omega}{B/2}\right)^{2N} = (31.6)^2$$

$$\left(\frac{\delta' \omega}{B/2}\right)^{2N} = 999 - 1 = 998$$

Now the ratio of 30dB to 3dB bandwidths is

$$\text{ratio} = \frac{2\delta'\omega}{B} = \frac{\delta'\omega}{B/2} = 998^{\frac{1}{2N}}$$

N	1	2	3	4	5
ratio	31.6	5.62	3.16	2.37	1.99

12.79

$A_{max} = 3\text{dB} \Rightarrow \epsilon = \sqrt{10^{A_{max}/10} - 1} \approx 1$

Poles of low pass prototype are given by FIG 12.10(c)

Poles: $-\omega_p$, $\omega_p\left(-\tfrac{1}{2} \pm j\sqrt{3}/2\right)$

Make $\omega_p = B/2$

$\Rightarrow$ poles: $\left\{-\frac{B}{2},\ +\frac{B}{2}\left(-\frac{1}{2} \pm j\frac{\sqrt{3}}{2}\right)\right\}$

Using the low-pass to bandpass transformation:

Poles of the bandpass filter:

$-\frac{B}{2} \pm j\omega_0$,

$-\frac{B}{4} \pm j\left(\frac{\sqrt{3}}{4}B + \omega_0\right)$ and

$-\frac{B}{4} \pm j\left(\frac{\sqrt{3}}{4}B - \omega_0\right)$

For the three circuits:

① $\omega_{01} = \omega_0$ $\quad B_1 = B$ $\quad Q_1 = \omega_0/B$

② $\omega_{02} \cong \frac{\sqrt{3}}{4}B + \omega_0$ $\quad B_2 = \frac{B}{2}$ $\quad Q_2 \cong \frac{2\omega_0}{B}$

③ $\omega_{03} \cong \frac{\sqrt{3}}{4}B - \omega_0$ $\quad B_3 = \frac{B}{2}$ $\quad Q_3 \cong \frac{2\omega_0}{B}$

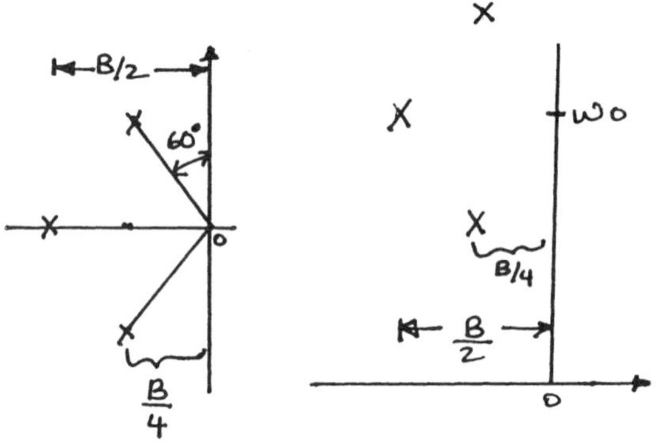

CHAPTER 13 - PROBLEMS

13.1

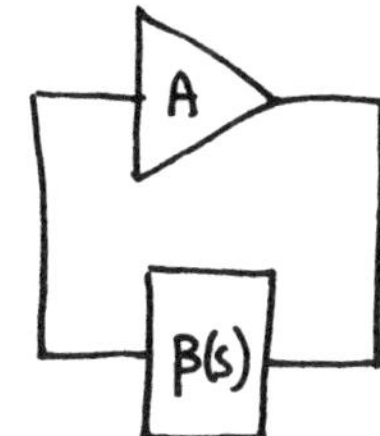

$$A = A_0 > 0$$

$$\beta(s) = \frac{k\frac{\omega_0}{Q}s}{s^2 + s\frac{\omega_0}{Q} + \omega_0^2}$$

(a) for oscillations $1 - A\beta(s) = 0$

$$\therefore A_0 k \frac{\omega_0}{Q} s = s^2 + s\frac{\omega_0}{Q} + \omega_0^2$$

$$\omega_0^2 - \omega^2 = j\omega\left(\frac{\omega_0}{Q}\right)(A_0k - 1)$$

at the freq of oscillation, both Real & Imaginary parts are 0.

$\therefore$ $\underline{\omega = \omega_0}$ & $\underline{A_0K = 1}$

(b)

$$L(j\omega) \triangleq A\beta(j\omega) = \frac{AK\frac{\omega_0}{Q}j\omega}{(\omega_0^2 - \omega^2) + j\omega\left(\frac{\omega_0}{Q}\right)}$$

$$\therefore \phi(\omega) = 90° - \tan^{-1}\left(\frac{\omega\omega_0/Q}{\omega_0^2 - \omega^2}\right)$$

Now $\frac{\partial}{\partial x}\tan^{-1}v = \frac{1}{1+v^2}\cdot\frac{\partial v}{\partial x}$

$$\therefore \frac{\partial\phi}{\partial\omega} = \frac{1}{1+\left(\frac{\omega\omega_0/Q}{\omega_0^2-\omega^2}\right)^2}\cdot\frac{\partial}{\partial\omega}\left(\frac{\omega\omega_0/Q}{\omega_0^2-\omega^2}\right)$$

$$= \frac{-(\omega_0^2-\omega^2)^2}{(\omega_0^2-\omega^2)+\left(\frac{\omega\omega_0}{Q}\right)^2}\cdot\left[\frac{\frac{\omega_0}{Q}(\omega_0^2-\omega^2) - 2\omega\frac{\omega\omega_0}{Q}}{(\omega_0^2-\omega^2)^2}\right]$$

$$\left.\frac{d\phi}{\partial\omega}\right|_{\omega=\omega_0} = \frac{-1}{\omega_0^4/Q^2}\cdot\frac{2\omega_0^3}{Q}$$

$$= -\frac{2Q}{\omega_0}$$

(c) $$\Delta\omega_0 = \frac{\Delta\phi}{\partial\phi/\partial\omega} = \frac{\Delta\phi}{-2Q/\omega_0}$$

$$= \frac{-\Delta\phi\,\omega_0}{2Q}$$

$\therefore$ Per unit change in ω_0 is given by

$$\frac{\Delta\omega_0}{\omega_0} = -\frac{\Delta\phi}{2Q}$$

13.2

For the circuit of problem 13.1, the poles, which are the zeros of the characteristic equation, are given by:

$$1 - L(s) = 0$$

$$L(s) = 1$$

$$\frac{AK\left(\frac{\omega_0}{Q}\right)s}{s^2 + s\frac{\omega_0}{Q} + \omega_0^2} = 1$$

$$s^2 + s\frac{\omega_0}{Q}(1 - AK) + \omega_0^2 = 0$$

$\therefore$ Poles are at:

$$s = \frac{-\frac{\omega_0}{Q}(1-AK) \pm \sqrt{\left(\frac{\omega_0}{Q}\right)^2(1-AK)^2 - 4\omega_0^2}}{2}$$

$$= -\omega_0\left[\frac{1-AK}{2Q} \pm \sqrt{\left(\frac{1-AK}{2Q}\right)^2 - 1}\right]$$

$$= -\omega_0\left(\frac{1-AK}{2Q}\right)\left[1 \pm j\sqrt{\left(\frac{2Q}{1-AK}\right)^2 - 1}\right]$$

CONT.

Radial distance of ω_0 =>

$$|s^2| = \omega_0^2 \left(\frac{1-AK}{2Q}\right)^2 \left[1 + \left(\frac{2Q}{1-AK}\right)^2 - 1\right]$$

$$= \omega_0^2$$

$\therefore |s| = \omega_0$ ~ independent of A or K!

(a) For poles on $j\omega$-axis => real part = 0

$\therefore -(1-AK) = 0 \Rightarrow AK = 1$

(b) For poles in RHS => Real Part > 0

$$-(1-AK) > 0$$
$$AK > 1$$

13.3

Assume ideal opamp.

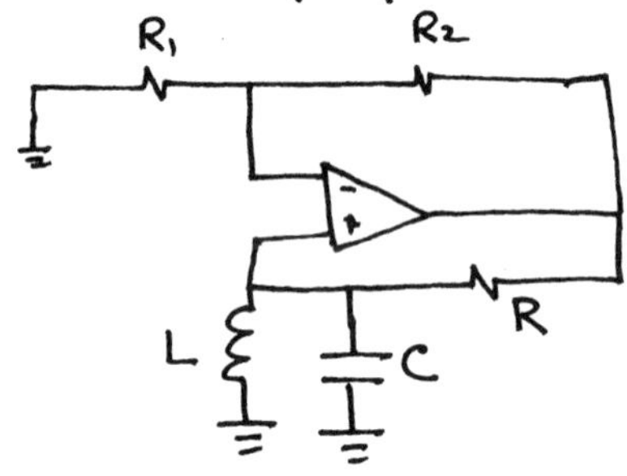

At resonance $\omega_0 = \frac{1}{\sqrt{LC}}$

$K = 1$

Thus A must also be 1 (or slightly higher).

For $R_1 = 10k\Omega$, $R_2 = 100\Omega$

$$A = 1 + \frac{R_2}{R_1} = 1.01$$

(a) If $L' = 1.01L$

$$\omega_0' = \frac{1}{\sqrt{1.01}}\omega_0 = 0.995\omega_0$$

or $-\frac{1}{2}\%$

(b) If C changes by +1%

=> ω_0 changes by $-\frac{1}{2}\%$

(c) If R changes by +1%

=> ω_0 ~ unchanged

13.4

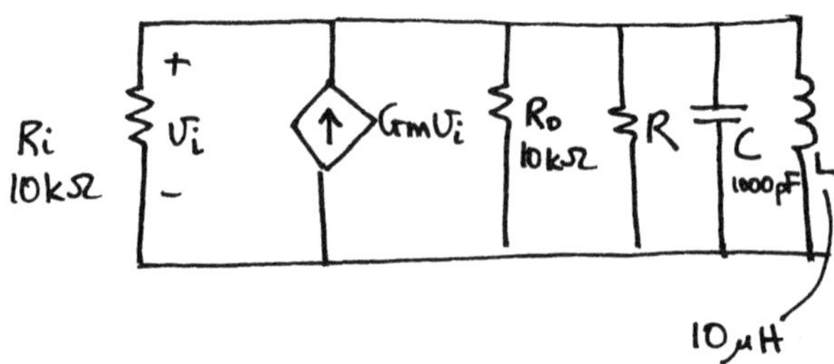

For resonator: $\omega_0 = \frac{1}{\sqrt{LC}} = 10^7 \frac{rad}{s}$

$$\frac{\omega_0}{Q} = \frac{1}{RC}$$

$$R = \frac{Q}{\omega_0 C} = \frac{100}{10^7 \times 1000 \times 10^{-12}} = 10k\Omega$$

Oscillations will occur at $\omega_0 = 10^7 \frac{rad}{s}$ when $G_m(R_i \| R_o \| R) = 1$ i.e. gain = 1

$$\therefore G_m = \frac{1}{10k \| 10k \| 10k} = \frac{3}{10^4} = 300 \frac{\mu A}{V}$$

13.5

At ω_0 $A\beta = 1$

If $\beta(\omega_0)$ is −20 dB with a phase

CONT.

shift of 180° then clearly A should have a gain of 20dB (i.e. $A(\omega_0) = 10$) with a phase shift of ± 180°

i.e. $\underline{A = -10}$

13.6

From Eq(13.8)

$$L_- = -V\frac{R_3}{R_2} - V_D\left(1+\frac{R_3}{R_2}\right)$$

$$6 = 10\frac{R_3}{R_2} + 0.7\left(1+\frac{R_3}{R_2}\right)$$

$$= 10.7\frac{R_3}{R_2} + 0.7$$

$\frac{R_3}{R_2} = 0.495$ By symmetry $\frac{R_4}{R_5} = 0.495$

Use $\underline{R_2 = R_5 = 10k\Omega}$

$\therefore \underline{R_3 = R_4 \cong 5k\Omega}$

Slope of limiting characteristic

$$= \frac{R_4}{R_5} = 0.1$$

$$\therefore R_1 = \frac{1}{0.1}R_4 = \underline{50k\Omega}$$

13.7

For V_B connected via R_B to the virtual ground, a current $= \frac{V_B}{R_B}$ flows into the node. To compensate, v_I must be moved by Δv_I, in a direction opposite to V_B to produce a current ⟹

$$\frac{\Delta v_I}{R_1} = \frac{-V_B}{R_B}$$

$$\therefore \Delta v_I = -\frac{R_1}{R_B}V_B$$

$V_D = 0$ ~ assumed

$$L_- = -5 = -15\,R_3/R_2$$

$$\frac{R_3}{R_2} = \frac{1}{3} = R_4/R_5$$

Given $R_{in} = 100k\Omega \Rightarrow R_1 = \underline{100k\Omega}$

$$\text{Slope} = R_4/R_1 \leq 0.05$$

$$R_4 \leq R_1 \times 0.05$$

$R_4 \leq 5k\Omega \Rightarrow$ Let $\underline{R_4 = 4.3k\Omega}$

$\therefore R_3 = R_4 \Rightarrow$ $\underline{R_3 = 4.3k\Omega}$

$$R_2 = R_5 = 3R_4 = 12.9k\Omega$$

For standard resistance values:

$R_2 = R_5 = \underline{13k\Omega}$

$$\therefore L_- = -15\frac{R_3}{R_2} = -15 \times \frac{4.3}{12.9}$$

$$= -4.96V \cong -5V$$

Offset is $+5V \Rightarrow$ Use $V_B = -15V$

and $5 = R_1/R_B\; 15$

$\therefore R_B = 3R_1 = \underline{300k\Omega}$

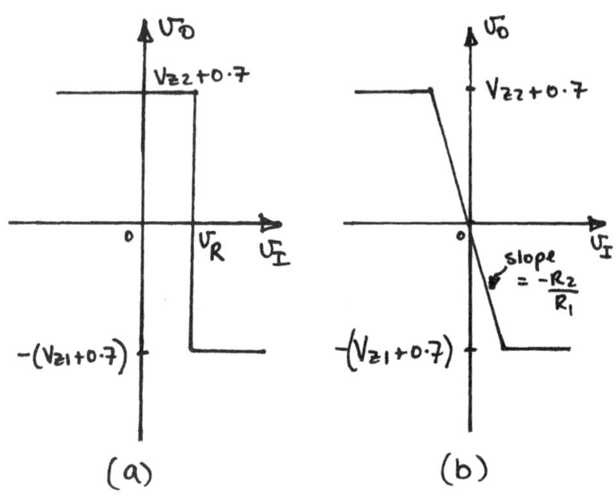

$$\omega_0^2 = \frac{1}{R^2C^2} \Rightarrow \omega_0 = \frac{1}{RC}$$

$$\frac{\omega_0}{Q} = \frac{3}{RC} \Rightarrow Q = \frac{1}{3}$$

For centre frequency gain:

$$s = j\omega_0 = j/RC$$

$$\therefore \left.\frac{V_a}{V_o}\right|_{s=j/RC} = \frac{\frac{1}{RC}\, j/RC}{-\frac{1}{R^2C^2} + \frac{3}{RC}\left(\frac{j}{RC}\right) + \frac{1}{R^2C^2}}$$

$$= \frac{1}{3} = \text{centre freq. gain}$$

13.9

$$\frac{V_a}{V_o} = \frac{1/sC \,\|\, R}{1/sC \,\|\, R + 1/sC + R}$$

$$= \frac{\left(R/sC\right)/\left(\frac{1}{sC}+R\right)}{\frac{(R/sC)}{\left(\frac{1}{sC}+R\right)} + \frac{1}{sC} + R}$$

$$= \frac{\frac{R}{sC}}{\frac{R}{sC} + \left(\frac{1}{sC}+R\right)^2} \times \frac{s^2C^2}{s^2C^2}$$

$$= \frac{sCR}{sCR + (1+sCR)^2}$$

$$= \frac{sCR}{sCR + 1 + 2sCR + s^2C^2R^2}$$

$$= \frac{\frac{1}{RC}s}{s^2 + s\frac{3}{RC} + \frac{1}{R^2C^2}}$$

Note $\frac{V_a}{V_o}$ has zeros at 0 and ∞

i.e. A Bandpass!

13.10

$$L(j\omega) = \frac{1 + R_2/R_1}{3 + j\left(\omega CR - \frac{1}{\omega CR}\right)} \quad \text{Eq}(13.11)$$

$$\phi(\omega) = -\tan^{-1}\left(\frac{\omega CR - \frac{1}{\omega CR}}{3}\right)$$

using $\frac{\partial \tan^{-1} v}{\partial x} = \frac{1}{1+v^2}\frac{\partial v}{\partial x}$

$$\frac{\partial\phi}{\partial\omega} = \frac{-1}{1+\left(\frac{\omega CR - \frac{1}{\omega CR}}{3}\right)^2} \cdot \frac{1}{3}\left(CR + \frac{1}{\omega^2 CR}\right)$$

$$\left.\frac{\partial\phi}{\partial\omega}\right|_{\omega=\frac{1}{RC}} = -\frac{1}{3}(CR + CR) = -\frac{2}{3}CR$$

for $\Delta\phi = -0.1$ rad

$$\Delta\omega_0 = \frac{\Delta\phi}{\partial\phi/\partial\omega} = \frac{-0.1}{-2/3\,\frac{1}{\omega_0}}$$

$$= 0.15\,\omega_0$$

∴ New frequency of oscillation

$$= 1.15\,\omega_0 = \frac{1.15}{RC}$$

13.11

Using Eq(13.10)

$$L(s) = \frac{1 + R_2/R_1}{3 + sCR + 1/sCR}$$

Poles of closed loop given by: $L(s) = 1$

$$1 + R_2/R_1 = 3 + sCR + \frac{1}{sCR}$$

$$0 = s^2 + \frac{s}{RC}\left(2 - \frac{R_2}{R_1}\right) + \frac{1}{R^2C^2}$$

$$Q = \frac{1}{\left(2 - R_2/R_1\right)}$$

for $Q = \infty$ ~ poles on $j\omega$ axis
~ $R_2/R_1 = 2$

For poles in R.H.P. $R_2/R_1 > 2$

13.12

From Fig(13.5) assuming resistance of limiting network is very low

At positive peak

$$v_O = \left(1 + \frac{20.3k}{10k}\right) v_I = 3.03\, v_I \quad (1)$$

$$v_O - \left[\frac{R_5}{R_5 + R_6} \cdot \left(v_O - (-15)\right)\right] - 0.7 = v_I \quad (2)$$

Now for $10V_{p\text{-}p}$ out

$\hat{v}_O = 5V$

$$\hat{v}_I = \frac{5}{3.03} = 1.65\,V$$

Using (2) $R_5 = 1k\Omega$

$$5 - \left(\frac{1}{1+R_6} \cdot (V_O + 15)\right) - 0.7 = 1.65$$

$$\frac{20}{1+R_6} = 2.65$$

$$R_6 = \frac{20}{2.65} - 1$$

$$R_6 = 6.5\,k\Omega = R_3$$

If $R_3 = R_6 = \infty$ from (2)

$$v_O - \left(\frac{1}{1+\infty}(V_O + 15)\right) - 0.7 = \frac{v_O}{3.03}$$

$$v_O - 0.7 = \frac{v_O}{3.3}$$

$v_O = 1.04\,V$

∴ opp output is $2v_O = 2.08\,V_{p\text{-}p}$.

13.13

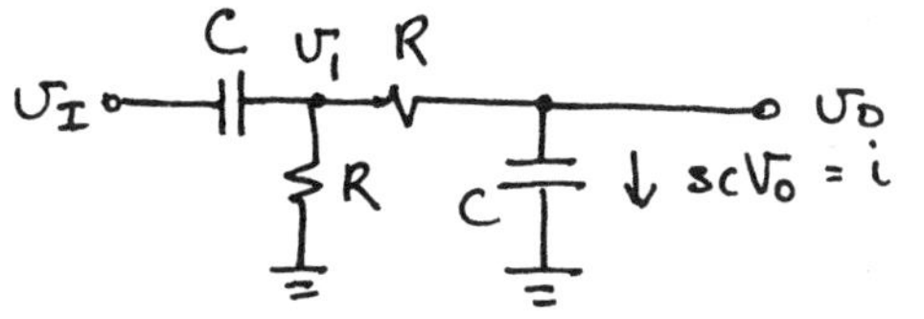

$$\frac{v_1 - v_O}{R} = sCv_O \Rightarrow v_1 = v_O(1 + sCR)$$

ΣI at v_1

$$\frac{v_1}{R} + sC(v_1 - v_I) + sC\,v_O = 0$$

$$v_O(1 + sCR) + sCR(v_O + v_O sCR) - sCRv_I + sCRv_O = 0$$

$$v_O(1 + sCR + sCR + s^2C^2R^2 + sCR) = sCR\,v_I$$

CONT.

$$\beta(s) \triangleq \frac{V_o}{V_I} = \frac{sCR}{s^2C^2R^2 + 3sCR + 1}$$

$$= \frac{1}{3 + sCR + 1/sCR}$$

From FigP(13.B) $\quad A = 1 + R_2/R_1$

$$\beta(j\omega) = \frac{1}{3 + j\left(\omega CR - \frac{1}{\omega CR}\right)}$$

Zero phase when $\omega CR = \frac{1}{\omega CR}$

$$\underline{\underline{\omega = \frac{1}{CR}}}$$

$$\left|\beta(\omega = 1/RC)\right| = \frac{1}{3}$$

for oscillations $\quad 1 + R_2/R_1 \geq 3$

$$\Rightarrow \underline{\underline{\frac{R_2}{R_1} \geq 2}}$$

$$\underline{\underline{L(s) = A\beta = \frac{1 + R_2/R_1}{3 + sCR + \frac{1}{sCR}}}}$$

$$\underline{\underline{L(j\omega) = \frac{1 + R_2/R_1}{3 + j\left(\omega CR - \frac{1}{\omega CR}\right)}}}$$

13.14

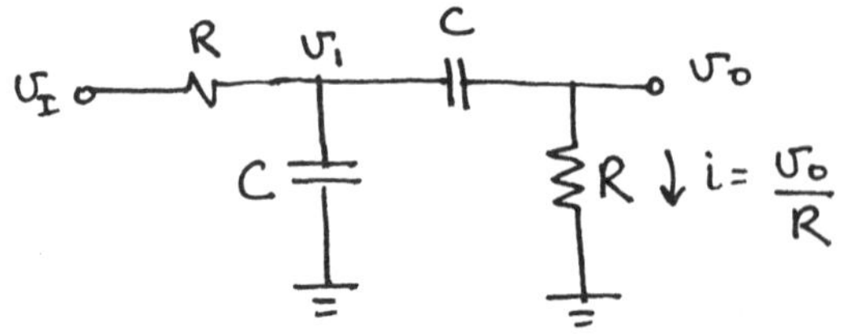

$$sC(V_1 - V_O) = \frac{V_O}{R}$$

$$V_1 = V_O\left(1 + \frac{1}{sRC}\right)$$

ΣI at V_1:

$$\frac{-V_I + V_1}{R} + sCV_1 + \frac{V_O}{R} = 0 \quad \leftarrow \text{SUB FOR } V_1 \text{ and mult by } \frac{1}{sC}$$

$$\frac{1}{sCR}\left[-V_I + V_O\left(1 + sCR\right)\right] + V_O\left(1 + \frac{1}{sCR}\right) + \frac{V_O}{sCR} = 0$$

$$V_O\left[\frac{1}{sCR} + 1 + \frac{1}{sCR} + \frac{1}{s^2C^2R^2} + \frac{1}{sCR}\right] = \frac{V_I}{sCR}$$

$$\frac{V_O}{V_I} \triangleq \beta(s) = \frac{1}{3 + sCR + \frac{1}{sCR}}$$

$$\beta(j\omega) = \frac{1}{3 + j\left(\omega CR - \frac{1}{\omega CR}\right)}$$

phase is zero when $\omega_0 CR = \frac{1}{\omega_0 CR}$

$$\underline{\underline{\omega_0 = \frac{1}{RC}}}$$

$$\beta(j\omega_0) = \frac{1}{3}$$

∴ For oscillations

$$A = 1 + \frac{R_2}{R_1} \geq 3$$

$$\underline{\underline{\frac{R_2}{R_1} \geq 2}}$$

$$\underline{\underline{L(s) = \frac{1 + R_2/R_1}{3 + sCR + 1/sCR}}}$$

$$\underline{\underline{L(j\omega) = \frac{1 + R_2/R_1}{3 + j\left(\omega CR - \frac{1}{\omega CR}\right)}}}$$

13.15

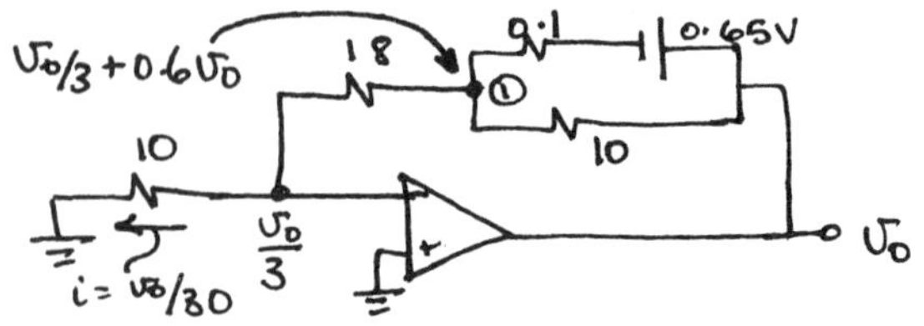

CONT.

ΣI at node ①

$$\frac{v_o}{30} = \frac{v_o - \frac{v_o}{3} - 0.6v_o}{10}$$

$$+ \frac{v_o - 0.65 - \frac{v_o}{3} - 0.6v_o}{0.1}$$

$$= 0.00666\,v_o + 0.666\,v_o - 0.65$$

$$v_o = 10.156\ V$$

$\therefore$ max. output $= \underline{\underline{20.3V\ p\text{-}p}}$

13.16

$$\omega_0 = \frac{1}{RC} = 2\pi 10^4 \qquad R = 10k\Omega$$

$$C = \frac{1}{10^4 \times 2\pi \times 10^4} \Rightarrow \underline{\underline{C \cong 1.6nF}}$$

Now from Eq (13.11)

$$\beta(j\omega) = \left[3 + j\left(\omega CR - \frac{1}{\omega CR}\right)\right]^{-1}$$

$$\therefore \phi(\omega) = -\tan^{-1}\left(\frac{\omega CR - \frac{1}{\omega CR}}{3}\right)$$

Using $\frac{\partial \tan^{-1} v}{\partial x} = \frac{1}{1+v^2}\frac{\partial v}{\partial x}$ we get

$$\frac{\partial \phi(\omega)}{\partial \omega} = \frac{-1}{1 + \left(\frac{\omega CR - \frac{1}{\omega CR}}{3}\right)^2}\left[\frac{RC + 1/\omega^2 RC}{3}\right]$$

At $\omega = \omega_0 = \frac{1}{RC}$ $\quad \frac{\partial\phi(\omega)}{\partial\omega} = -\frac{2}{3}RC$

Now $5.7^\circ \approx 0.1$ rad $\quad$ (lag $= -0.1$ rad)

$$\therefore \Delta\omega_0 = \frac{-0.1}{-2/3\,RC} \approx 0.15\,\omega_0 = 1.5kHz$$

$\therefore$ New frequency of oscillation $= 8.5kHz$

To restore operation:

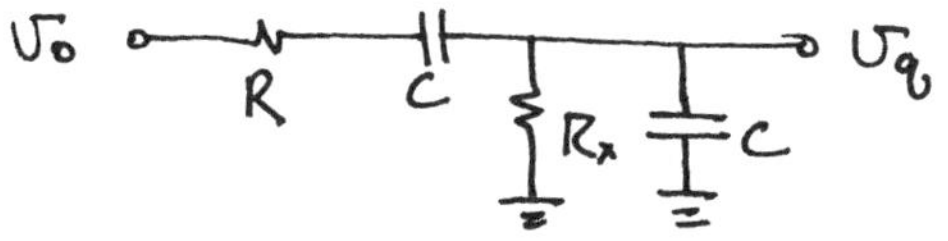

$$\beta(s) = \frac{R_x \| \frac{1}{sC}}{R_x \| \frac{1}{sC} + R + \frac{1}{sC}}$$

$$= \frac{\frac{R_x/sC}{R_x + 1/sC}}{\frac{R_x/sC}{R_x + \frac{1}{sC}} + R + \frac{1}{sC}}$$

$$= \frac{R_x/sC}{R_x/sC + RR_x + \frac{R}{sC} + \frac{R_x}{sC} + \frac{1}{s^2C^2}}$$

$$\therefore \beta(s) = \frac{1}{2 + \frac{R}{R_x} + sCR + \frac{1}{sCR_x}}$$

$$\phi = -\tan^{-1}\left(\frac{\omega CR - \frac{1}{\omega R_x C}}{2 + R/R_x}\right)$$

Now it is required that $\phi = 5.7^\circ$ at $\omega = \omega_0$! where $\omega_0 = 1/RC$

$$\therefore \omega_0 RC - \frac{1}{\omega_0 R_x C} = \left(2 + \frac{R}{R_x}\right)\tan(-5.7)$$

$$1 - \frac{1}{\omega_0 R_x C} = (2 + R/R_x)(-0.1)$$

$$1 + 0.2 = \frac{1}{\omega_0 R_x C} - 0.1\frac{R}{R_x}$$

CONT.

$$R_x = \frac{1/\omega_0 C - 0.1R}{1.2}$$

given:
$\omega_0 = 2\pi 10^4$
$C = 1.6\times10^{-9}$
$R = 10^4$

$R_x = \underline{\underline{7.5\,k\Omega}}$

Now:

$$\beta(j\omega_0) = \frac{1}{2 + 10/7.5 + j\left(1 - 1/\omega_0 C R_x\right)}$$

$$= \left(3.333 - j\,0.326\right)^{-1}$$

$$|\beta(j\omega_0)| = \frac{1}{3.35}$$

$\therefore\ 1 + R_2/R_1 = 3.35$ for oscillations

$\frac{R_2}{R_1} = \underline{\underline{2.35}}$ (not 2 as before)

13.17

(3) $v + \frac{1}{sC}\frac{v}{R} = v\left(1 + \frac{1}{sCR}\right)$

(4) $\frac{v\left(1 + \frac{1}{sCR}\right)}{R} = \frac{v}{R} + \frac{v}{sCR^2}$

(5) $i_s = \frac{2v}{R} + \frac{v}{sCR^2}$

(6) (3) + (5)/sC

$$= v + \frac{v}{sCR} + \frac{1}{sC}\left(2\frac{v}{R} + \frac{v}{sCR^2}\right)$$

$$= v + \frac{3v}{sCR} + \frac{v}{s^2C^2R^2}$$

(7) $\frac{v}{R} + \frac{3v}{sCR^2} + \frac{v}{s^2C^2R^3}$

(8) = (5) + (7)

$$= \frac{3v}{R} + \frac{4v}{sCR^2} + \frac{v}{s^2C^2R^3}$$

(9) $v_0 =$ (6) $+ \frac{1}{sC}$ (8)

$$= v + \frac{3v}{sCR} + \frac{v}{s^2C^2R^2} + \frac{3v}{sCR} + \frac{4v}{s^2C^2R^2} + \frac{v}{s^3C^3R^3}$$

$$= v + \frac{6v}{sCR} + \frac{5v}{s^2C^2R^2} + \frac{v}{s^3C^3R^3}$$

Now loop gain $\equiv$

$$L(s) = -\frac{v_0'}{v_0}$$

$$\therefore L(s) = \frac{R_f/R\ v}{v\left(1 + \frac{6}{sCR} + \frac{5}{s^2C^2R^2} + \frac{1}{s^3C^3R^3}\right)}$$

$$= \frac{s^3 R_f/R}{s^3 + \frac{6s^2}{RC} + \frac{5s}{C^2R^2} + \frac{1}{C^3R^3}}$$

$$L(j\omega) = \frac{-j\omega^3 R_f/R}{\frac{1}{C^3R^3} - \frac{6\omega^2}{RC} + j\left(\frac{5\omega}{C^2R^2} - \omega^3\right)}$$

CONT.

$L(j\omega)$ is real if

$$\frac{6\omega_0^2}{RC} = \frac{1}{R^3C^3}$$

$$\omega_0 = \frac{1}{\sqrt{6}\,RC}$$

$$L(j\omega_0) = \frac{\omega_0^2\, R_f/R}{-\omega_0^2 + 5/R^2C^2}$$

$$= \frac{R_f/R\ \omega_0^2}{-\omega_0^2 + 30\omega_0^2}$$

$$= \frac{R_f/R}{29}$$

Now Loop Gain = 1 if $R_f = 29R$

∴ Minimum value for $R_f = 29R$

Given $C = 16\text{nF}$, $R = 10\text{k}\Omega$

$$f_0 = \frac{1}{2\pi\sqrt{6}\ 16\times10^{-9}\ 10\times10^{3}}$$

$$= 406.1\text{Hz}$$

or $f_0 = \dfrac{0.065}{RC}$

13.18

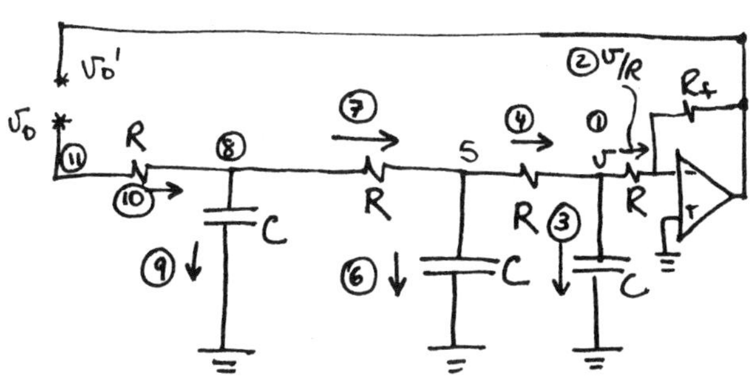

(3) $i = scv$ (4) $scv + v/R$

(5) $v + \left(scv + \frac{v}{R}\right)R = 2v + scRv$

(6) $2scv + s^2c^2Rv$

(7) = (6)+(4) $= 3scv + s^2c^2Rv + v/R$

(8) $2v + scRv + v + 3scRv + s^2c^2R^2v$
$= 3v + 4scRv + s^2c^2R^2v$

(9) $3scv + 4s^2c^2Rv + s^3c^3R^2v$

(10) = (7) + (9)
$= 6scv + 5s^2c^2Rv + \frac{v}{R} + s^3c^3R^2v$

(11) = (8) + (10)×R
$v_0 = 4v + 10scRv + 6s^2c^2R^2v + s^3c^3R^3v$

Now:

$$L(s) = -\frac{v_0'}{v_0} = \frac{v\ R_f/R}{v\left(s^3c^3R^3 + 6s^2c^2R^2 + 10scR + 4\right)}$$

$$= \frac{R_f/R}{s^3c^3R^3 + 6s^2c^2R^2 + 10scR + 4}$$

$$L(j\omega) = \frac{R_f/R}{(4 - 6\omega^2c^2R^2) + j(10\omega cR - \omega^3R^3c^3)}$$

$L(j\omega)$ is purely real if

$$10\omega_0 cR = \omega_0^3R^3c^3$$

$$\omega_0 = \frac{1}{\sqrt{10}}\ \frac{1}{RC}$$

Given $R = 10\text{k}\Omega$, $f_0 = 10\text{kHz}$

$$C = \frac{1}{\sqrt{10}\times10^4\times2\pi10^4}$$

$$= 0.503\text{nF}$$

CONT.

Now,

$$|L(j\omega_0)| = \frac{R_f/R}{4 - 6\omega_0^2 R^2 C^2} \quad \text{Sub for } \omega_0$$

$$= \frac{R_f/R}{4 - 6\frac{1}{10R^2C^2}R^2C^2}$$

$$= \frac{R_f/R}{4 - 6/10} \geq 1$$

$$\therefore \; R_f/R \geq 3.4$$

$$R_f \geq 34\,k\Omega$$

13.19

for 2nd integrator

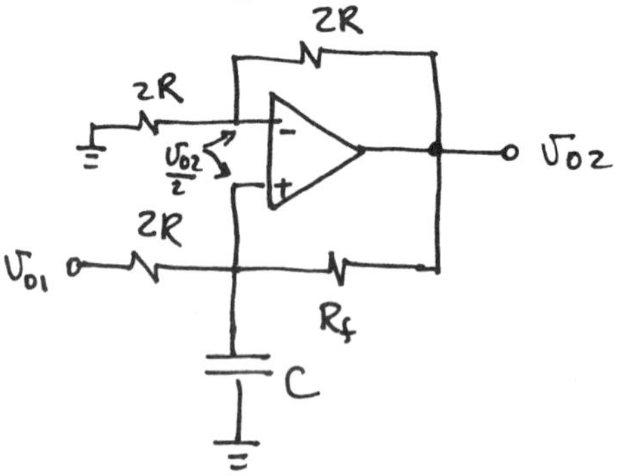

From the voltage divider around the upper branch: $V_+ = V_- = \frac{1}{2}V_{o2}$

$\Sigma I = 0$ at +ve input

$$\frac{\frac{1}{2}V_{o2} - V_{o1}}{2R} + sC\frac{V_{o2}}{2} + \frac{\frac{V_{o2}}{2} - V_{o2}}{R_f} = 0$$

$$\frac{V_{o2} - 2V_{o1}}{2R} + sCV_{o2} - \frac{V_{o2}}{R_f} = 0 \qquad R_f = \frac{2R}{1+\Delta}$$

$$V_{o2}\left(\frac{1}{2R} + sC - \frac{1+\Delta}{2R}\right) = \frac{V_{o1}}{R}$$

$$V_{o2}\left(sCR - \frac{\Delta}{2}\right) = V_{o1}$$

$$\therefore \frac{V_{o2}}{V_{o1}} = \frac{1}{sCR - \Delta/2}$$

Now: $\frac{V_{o1}}{V_x} = \frac{-1}{sCR} \quad \therefore L(s) = \frac{-1/sCR}{sCR - \Delta/2}$

Characteristic equation $L(s) = 1$

$$\therefore s^2C^2R^2 - \frac{sCR\Delta}{2} + 1 = 0$$

$\therefore$ Poles are

$$S_p = \frac{\frac{RC\Delta}{2} \pm \sqrt{\frac{R^2C^2\Delta^2}{4} - 4C^2R^2}}{2R^2C^2}$$

$$= \frac{\Delta/2 \pm 2j\sqrt{1 - (\Delta/4)^2}}{2RC}$$

for $\Delta \ll 1$ $\quad \left(1 - \left(\frac{\Delta}{4}\right)^2\right)^{1/2} \approx \left(1 - \frac{1}{2}\left(\frac{\Delta}{4}\right)^2\right)$

$$\therefore S_p \cong \left[\Delta/2 \pm j2\left(1 - \frac{1}{2}\left(\frac{\Delta}{4}\right)^2\right)\right]\frac{1}{2RC}$$

$$= \frac{\frac{\Delta}{2} \pm j\left(2 - \left(\frac{\Delta}{4}\right)^2\right)}{2RC}$$

Now:

$Re[S_p] > 0 \Rightarrow$ Poles in R.H.P.!

for $\Delta \ll 1$

$$S_p \cong \frac{\Delta/2 \pm j2}{2RC} = \frac{1}{RC}\left(\frac{\Delta}{4} \pm j\right)$$

Q.E.D.

13.20

The transmission of the filter normalized to the centre frequency, ω_0 is:

$$|T(j\omega)| = \frac{\omega\omega_0/Q}{(\omega_0^2-\omega^2)^2 + \frac{\omega^2\omega_0^2}{Q^2}}$$

$$= \frac{1/Q\left(\frac{\omega_0}{\omega}\right)}{\left(\left(\frac{\omega_0}{\omega}\right)^2-1\right)^2 + \frac{1}{Q^2}\left(\frac{\omega_0}{\omega}\right)^2}$$

Relative to the amplitude of the fondamental

(a) The second harmonic $=0$

(b) The third harmonic

$$= \frac{1}{3}\ \frac{1/20 \times 1/3}{\left(\frac{1}{9}-1\right)^2 + \left(\frac{1}{20}\right)^2\left(\frac{1}{9}\right)} = 6.25\times10^{-3}$$

(c) The fifth harmonic

$$= \frac{1}{5}\ \frac{\frac{1}{20}\times\frac{1}{5}}{\left(\frac{1}{25}-1\right)^2 + \left(\frac{1}{20}\right)^2\left(\frac{1}{25}\right)} = 2.08\times10^{-3}$$

(d) The 4th harmonic $= 6^{th} = 10^{th} = 0$

7th harmonic $= 1.04\times10^{-3}$

9th " $= 0.625\times10^{-3}$

$\therefore$ $\frac{\text{RMS of } 2^{nd} \text{ to } 10^{th} \text{ harmoie}}{\text{RMS of fundamental}}$ is

$$\left[6.25^2 + 2.08^2 + 1.04^2 + 0.625^2\right]^{1/2}\times10^{-3}$$

$= \underline{6.7\times10^{-3}}$ OR $\underline{0.7\%}$

13.21

Consider the small signal models for each circuit. Assume r_π very large:

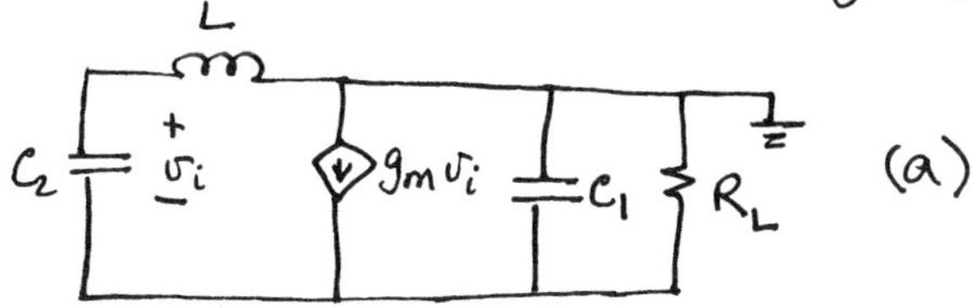

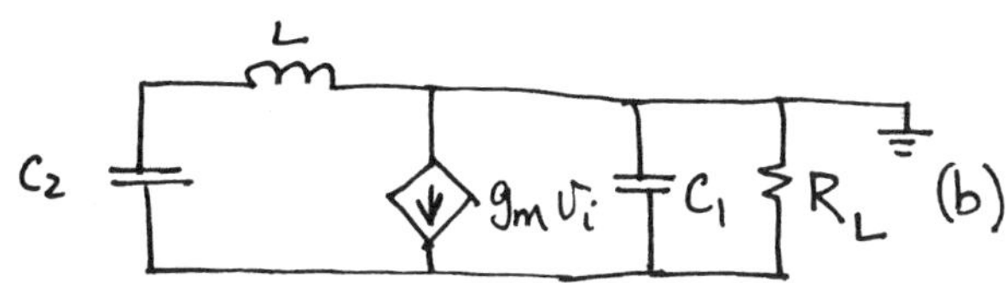

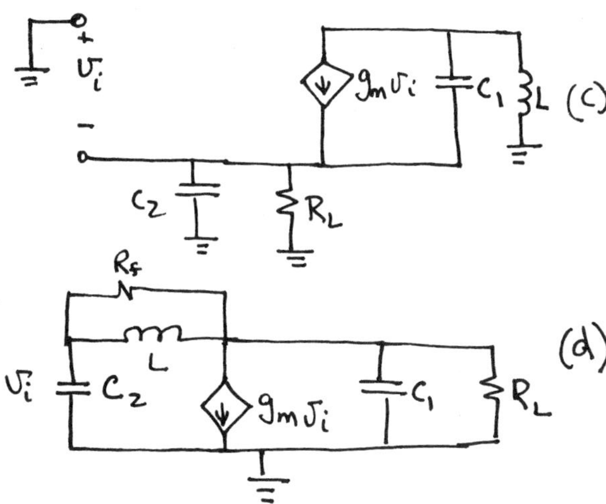

Given $R_f >> \omega_0 L$, circuits (a), (b) and (d) are the same except for the reference (ground) node.

For circuits (a), (b) & (d)

- Break the loop at v_i and assume unit return.

CONT.

$$V_a = 1 + sC_2 sL$$
$$= 1 + s^2 C_2 L$$

$\Sigma I = 0$ at V_a

$$g_m + sC_2 + sC_1(1 + s^2 C_2 L) + \frac{(1 + s^2 C_2 L)}{R} = 0$$

$$\therefore g_m + \frac{1}{R} + s(C_1 + C_2) + \frac{s^2 C_2 L}{R} + s^3 C_1 C_2 L = 0$$

This is the characteristic equation.
For $s = j\omega$:

$$g_m + \frac{1}{R} - \frac{\omega^2 C_2 L}{R} + j\left((C_1 + C_2)\omega - \omega^3 C_1 C_2 L\right) = 0$$

IMAGINARY PART = 0:

$$C_1 + C_2 = \omega^2 C_1 C_2 L$$
$$\omega = \sqrt{\frac{C_1 + C_2}{C_1 C_2 L}} \equiv \text{Frequency of Oscillation}$$

REAL PART = 0

$$g_m + \frac{1}{R} = \frac{\omega^2 C_2 L}{R}$$
$$g_m R = \left(\frac{C_1 + C_2}{C_1 C_2 L}\right) C_2 L - 1$$
$$g_m R = \frac{C_2}{C_1} \equiv \text{LIMIT ON GAIN}$$

FOR CIRCUIT (c)

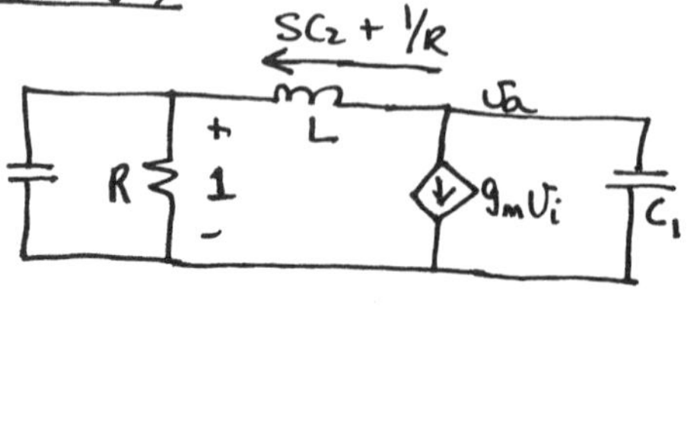

[Circuit: C_2, R, $+\ 1\ -$, inductor L with current $sC_2 + 1/R$, node V_a, dependent source $g_m V_i$, C_1]

$$V_a = \left(sC_2 + \frac{1}{R}\right) sL + 1$$

$\Sigma I = 0$ at V_a, $V_i = 1$

$$g_m + sC_2 + \frac{1}{R} + sC_1\left[sL\left(sC_2 + \frac{1}{R}\right) + 1\right] = 0$$

$$g_m + \frac{1}{R} + sC_2 + s^3 C_1 C_2 L + \frac{s^2 C_1 L}{R} + sC_1 = 0$$

THE CHARACTERISTIC EQUATION ≡

$$g_m + \frac{1}{R} + s(C_1 + C_2) + \frac{s^2 C_1 L}{R} + s^3 C_1 C_2 L = 0$$

Note this is the same as above, with $C_1 \leftrightarrow C_2$

$$\therefore \omega_0 = \sqrt{\frac{C_1 + C_2}{C_1 C_2 L}} \quad \text{and} \quad g_m R = \frac{C_1}{C_2}$$

13.22

(a) frequency of oscillations $\omega_0 = \frac{1}{\sqrt{LC}}$

gain ≥ 1 gain $= \frac{R_C}{2r_e} = \frac{R_C}{2V_T/I/2}$

$$= \frac{IR_C}{4V_T}$$

for $V_T = 0.025\,V$ then

$$IR_C \geq 4V_T$$

$$R_C \geq 0.1/I \quad \text{for oscillations to start.}$$

(b) for $R_C = \frac{1}{I}$ (kΩ) we have

$$\text{gain} = \frac{1/I}{2\left(\frac{2V_T}{I}\right)} = \frac{1}{4 \times 0.025} = 10$$

CONT.

Oscillations will start (10>1) and grow until Q_1, Q_2, go into cutoff. Output will go from V_{CC} to $V_{CC} - IR_C = V_{CC} - 1$. Therefore, output will be 1V p-p. Fundamental has a p-p amplitude of $\frac{4}{\pi} = 1.27$ Vp-p.

13.23

From Exercise 13.10,

$L = 0.52H$

$C_s = 0.012\,pF$

$C_p = 4pF$

From (13.27)

$$C_{eq} = \frac{C_s\left(C_p + \frac{C_1C_2}{C_1+C_2}\right)}{C_s + C_p + \frac{C_1C_2}{C_1+C_2}}$$

$C_2 = 10pF \qquad C_1 = 1$ to $10pF$

$$C_L = \frac{0.012\left(4 + \frac{10\times1}{10+1}\right)}{\left(0.012 + 4 + \frac{10}{11}\right)} = 0.01197pF$$

$$C_H = \frac{0.012\left(4 + \frac{10\times10}{10+10}\right)}{\left(0.012 + 4 + \frac{100}{20}\right)} = 0.01198pF$$

$$\therefore f_{OH} = \frac{1}{2\pi\left[0.52\times0.01197\times10^{-12}\right]^{1/2}}$$

$= \underline{2.0172\text{ MHz}}$

$f_{OL} = \left[2\pi\left(0.52\times0.01198\times10^{-12}\right)^{1/2}\right]^{-1}$

$= 2.01612$ MHz

Difference = 1kHz!

13.24

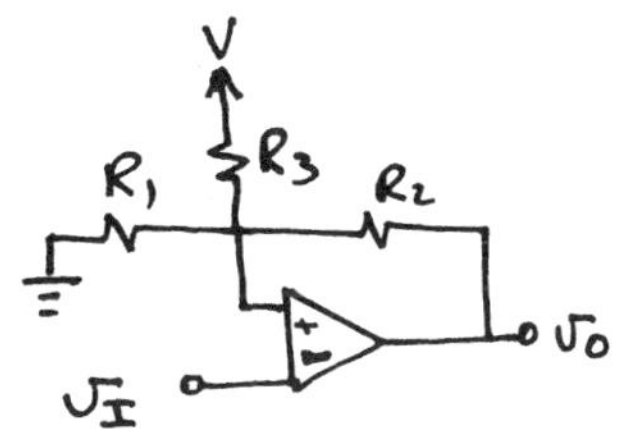

ΣI at +ve node:

$$\frac{V_{TH}}{R_1} = \frac{V - V_{TH}}{R_3} + \frac{L_+ - V_{TH}}{R_2}$$

$$V_{TH}\left(\frac{1}{R_1} + \frac{1}{R_2} + \frac{1}{R_3}\right) = \frac{V}{R_3} + \frac{L_+}{R_2}$$

$$V_{TH} = \left(V/R_3 + L_+/R_2\right)\left(\frac{1}{R_1} + \frac{1}{R_2} + \frac{1}{R_3}\right)^{-1}$$

$$= \left(\frac{V}{R_3} + \frac{L_+}{R_2}\right) R_1 \| R_2 \| R_3$$

Similarly

$$V_{TL} = \left(\frac{V}{R_3} + \frac{L_-}{R_2}\right)\left(R_1 \| R_2 \| R_3\right)$$

(b) Now

$$V_{TH} = 5.1 = \left(\frac{15}{R_3} + \frac{13}{R_2}\right)\left(\frac{1}{10} + \frac{1}{R_2} + \frac{1}{R_3}\right)^{-1}$$

$$\frac{5.1}{10} + \frac{5.1}{R_2} + \frac{5.1}{R_3} = \frac{15}{R_3} + \frac{13}{R_L}$$

$$0.51 = \frac{7.9}{R_2} + \frac{9.9}{R_3} \quad (1)$$

AND

$$V_{TL} = 4.9 = \left(\frac{15}{R_3} - \frac{13}{R_2}\right)\left(\frac{1}{10} + \frac{1}{R_2} + \frac{1}{R_3}\right)^{-1}$$

$$0.49 = \frac{-17.9}{R_2} + \frac{10.1}{R_3} \quad (2)$$

$(1)\times\frac{10.1}{9.9} \Rightarrow 0.52 = \frac{8.06}{R_2} + \frac{10.1}{R_3}$

$(2) \Rightarrow 0.49 = -\frac{17.9}{R_2} + \frac{10.1}{R_3}$

SUBTRACT TO GET ⇓

CONT.

$$0.52 - 0.49 = \frac{8.06 + 17.9}{R_2}$$

$$R_2 = \frac{25.96}{0.0303} = 856.8\,k\Omega$$

$$\frac{10.1}{R_3} = 0.49 + \frac{17.9}{856.8}$$

$$R_3 \cong 19.8\,k\Omega$$

13.25

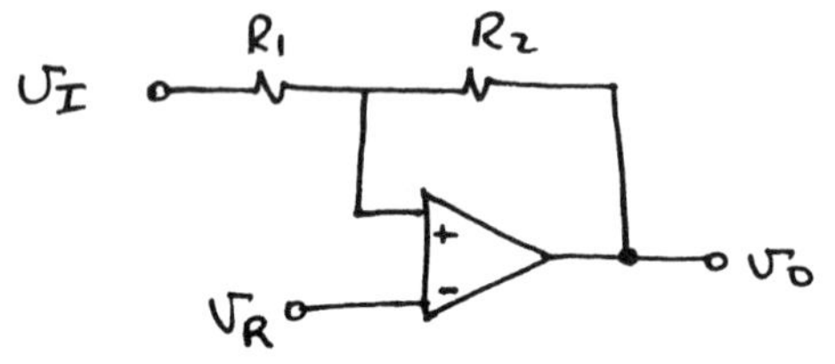

for $v_I = v_{TL}$ and $v_O = L_+$ initially

$$\frac{L_+ - v_R}{R_2} = \frac{v_R - v_{TL}}{R_1}$$

$$v_{TL} = v_R - \frac{R_1}{R_2} v_R - \frac{R_1}{R_2} L_+$$

$$\therefore\ v_{TL} = v_R\left(1 - \frac{R_1}{R_2}\right) - \frac{R_1}{R_2} L_+$$

Similarly

$$\frac{L_- - v_R}{R_2} = \frac{v_R - v_{TH}}{R_1}$$

$$v_{TH} = v_R\left(1 + R_2/R_1\right) - \frac{R_1}{R_2} L_-$$

(b) Given $L_+ = -L_- = V$

$R_1 = 10\,k\Omega$

$V_{TL} = 0$

$V_{TH} = V/10$

Substituting these values we get:

$$0 = V_R(1 + 10/R_2) - 10/R_2\,V \quad (1)$$

$$\frac{V}{10} = V_R(1 + 10/R_2) + 10/R_2\,V \quad (2)$$

(1) − (2) $\quad -\frac{V}{10} = -\frac{20}{R_2} V$

$$R_2 = 200\,k\Omega$$

$$0 = V_R(1 + 10/200) - \frac{10}{200} V$$

$$V_R = \frac{10/200\,V}{1 + 10/200} = 47.62\,mV$$

13.26

Output levels = $\pm 0.7V$

Threshold levels = $\pm \frac{10}{10+60} \times 0.7 = 0.1V$

$$i_{D,max} = \frac{12 - 0.7}{10} - \frac{0.7}{10+60} = 1.12\,mA$$

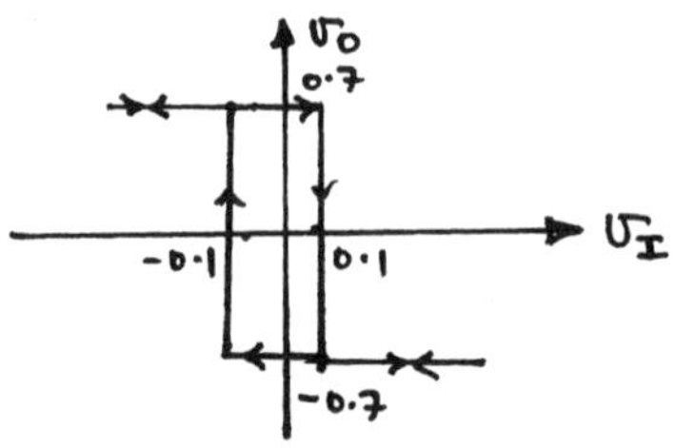

13.27

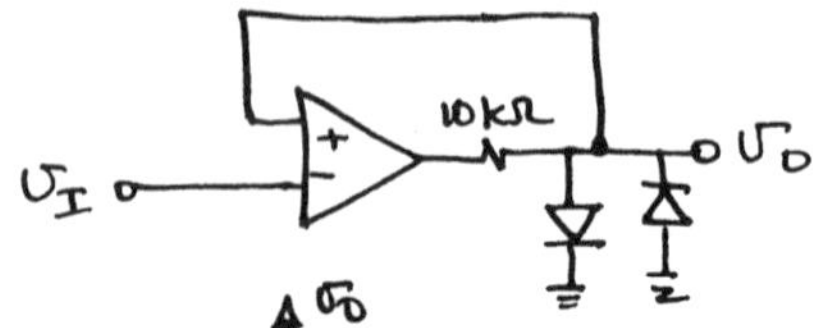

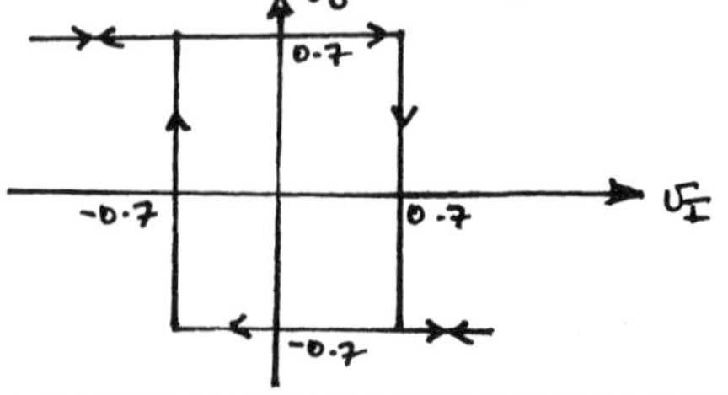

13.28

(a) A 0.5V peak sine wave, is not large enough to change the state of the circuit. Hence, the output will be either +12V or −12V at DC.

(b) The 1.1 V peak will change the state when

$$1.1 \sin\theta = 1$$

$$\theta = 65.4^\circ$$

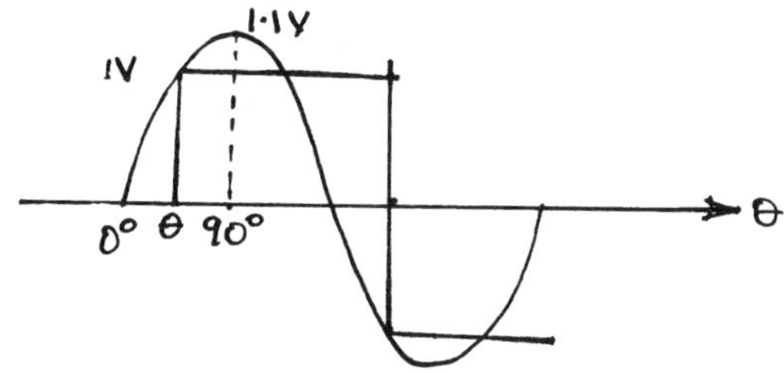

∴ The output is a symmetric square wave at frequency f, and lags the sine wave by an angle of 65.4°. The square wave has a swing of ±12V.

Since $V_{TH} = -V_{TL} = 1V$, if the average shifts by an amount so either the +ve or −ve swing is <1V, then no change of state will occur. Clearly, if the shift is 0.1V, the output will be a DC voltage.

13.29

For $L+ = -L- = 7.5V$

$\underline{V_Z = 6.8V}$ with $V_D = 0.7V$.

For $V_{TH} = -V_{TL} = 7.5V \Rightarrow R_1 = R_2$

for $v_I = 0$ $I_{R_2} = 0.1mA = \frac{7.5}{R_1 + R_2}$

$\Rightarrow R_1 = R_2 = \underline{37.5\ k\Omega}$

$$I_D = 1mA = \frac{12-7.5}{R} - \frac{7.5}{2R_1}$$

$$1 = \frac{4.5}{R} + 0.1$$

$$R = \underline{4.1k\Omega}$$

13.30

$$T = 2\tau \ln\frac{1+\beta}{1-\beta} \qquad \beta = \frac{R_1}{R_1+R_2} = \frac{10}{26}$$

$$T = 2(10\times10^{-9})(62\times10^{3})\ln\left(\frac{1+10/26}{1-10/26}\right)$$

$$T = 1.006ms \Rightarrow f = \underline{994.5Hz}$$

13.31

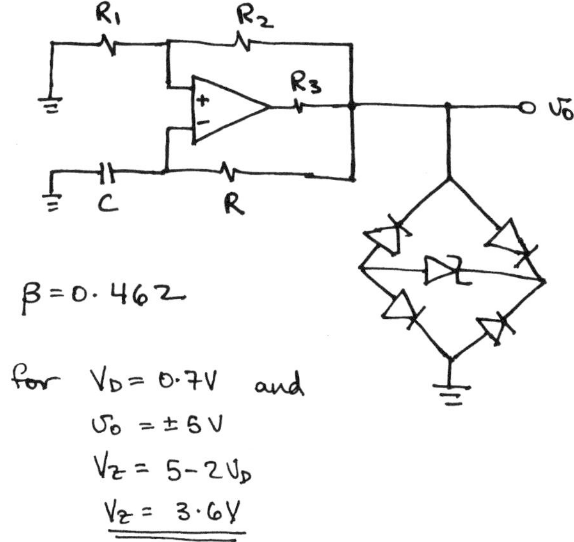

$\beta = 0.462$

for $V_D = 0.7V$ and

$v_O = \pm 5V$

$V_Z = 5 - 2V_D$

$\underline{V_Z = 3.6V}$

CONT.

$$T = 2\tau \ln\left(\frac{1+\beta}{1-\beta}\right)$$

$$10^{-3} = 2\tau \ln\left(\frac{1.462}{1-0.462}\right) \Rightarrow \tau = 0.5\,\text{msec}$$

$$\tau = RC \qquad \Rightarrow R = \tau/C = \underline{50\,k\Omega}$$

Thresholds $= \pm\, 0.462 \times 5 = \pm\, 2.31V$

Average current in R in ½ cycle:

$$I \cong \frac{1}{R}\left(\frac{5-2.31+2.31+5}{2}\right)$$

$$= \frac{5}{R} = \frac{5}{50k} = \underline{0.1\,mA}$$

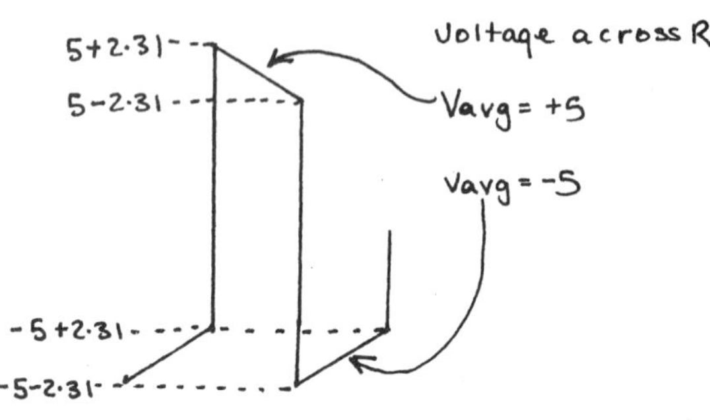

$$R_1 + R_2 = \frac{5V}{0.1mA} = 50\,k\Omega$$

$$\frac{R_1}{R_1+R_2} = 0.462 \rightarrow R_1 = 50(0.462)$$
$$= 23.1\,k\Omega$$

$$\therefore R_2 = 26.9\,k\Omega$$

$$1 = \frac{13-5}{R_3} - 0.1 - 0.1$$

$= I_{DIODE}$

USE $\underline{R_1 = 23\,k\Omega}$

$\underline{R_2 = 27\,k\Omega}$

$$R_3 = \frac{8}{1.2}$$
$$= \underline{6.67\,k\Omega}$$

13.32

From Fig 13.23(b), for ±5V outputs

$V_Z = 5 - 2V_{DIODE} = 5 - 1.4 = \underline{3.6V}$

For ±5V out:

$R_1 = R_2 \qquad L_+ = -L_- = 5V$

$V_{TH} = -V_{TL} = 5V$

Max current in feedback network $= 0.2mA$

$$\therefore\ 0.2 = \frac{5}{R_1+R_2} \Rightarrow \underline{R_1 = R_2 = 25\,k\Omega}$$

Max diode current $= 1mA$

$$\therefore\ \frac{13-5}{R_Z} = (0.2+1)\,mA$$

$$R_Z = \frac{8}{1.2} = \underline{6.67\,k\Omega}$$

Now from Fig 13.25 (c)

$$\text{slope} = \frac{-L_-}{RC} = \frac{V_{TH} - V_{TL}}{T/2}$$

for $f = 1kHz$
$T = 10^{-3}s$
$\underline{C = 0.01\,\mu F}$

$$\frac{5}{RC} = \frac{10}{10^{-3}/2} \Rightarrow \underline{R = 25\,k\Omega}$$

13.33

For 15 Vpp output $v_Z = 15/2 - 0.7$
$= \underline{6.8V}$

For the integrator:

INPUT ... +7.5 ... −7.5

v_C ... 7.5/2 ... −7.5/2

i.e. V_C should ramp between V_{TH} & V_{TL}!

CONT.

$$v_C(t_1) = \frac{1}{RC}\int_{t_0}^{t_1} v\,dt + v_C(t_0)$$ — v is a square wave

$$\frac{7.5}{2} = \frac{1}{RC}(t_1-t_0)(7.5-(-7.5)) - \frac{7.5}{2}$$

$(t_1 - t_0) = \frac{T}{2}$

$$7.5 = \frac{1}{RC}\frac{T}{2}(15)$$

$$1 = \frac{T}{RC} \Rightarrow R = \frac{T}{C} = \frac{1}{fC}$$

$$= \frac{1}{10^4(0.5\times10^{-9})}$$

$\therefore$ $R = R_{1-6} = 200k\Omega$

Minimum zener current = 1mA

$$\frac{13-7.5}{R_7} = 1 + \frac{7.5}{R_1+R_2} + \frac{7.5-V_C}{R_5}$$

Maximum current into the integrator when $V_C = -\frac{7.5}{2}$

$$\therefore \quad \frac{5.5}{7.5} = 1 + \frac{7.5}{400} + \frac{11.25}{200}$$

$\therefore$ $R_7 = 5.12k\Omega \overset{use}{\Rightarrow} R_7 = 5.1k\Omega$

Integrator output is triangular, with period = 100μs and ±7.5V peaks. (ie. 2x voltage at capacitor)

13.34

See sketches that follow:

$$V_A(t=T) = -V_{ref} = -(L_+-L_-)e^{-T/RC}$$

$$\frac{V_{ref}}{L_+-L_-} = e^{-T/RC}$$

$$T = -RC\ln\left(\frac{V_{ref}}{L_+-L_-}\right) = RC\ln\left(\frac{L_+-L_-}{V_{ref}}\right)$$

Q.E.D.

Trigger: v

v_B: 0, $-V_{ref}$

v_A: 0, $-V_{ref}$, $-(L_+-L_-)$; $V_A = -(L_+-L_-)e^{-t/RC}$

v_O:

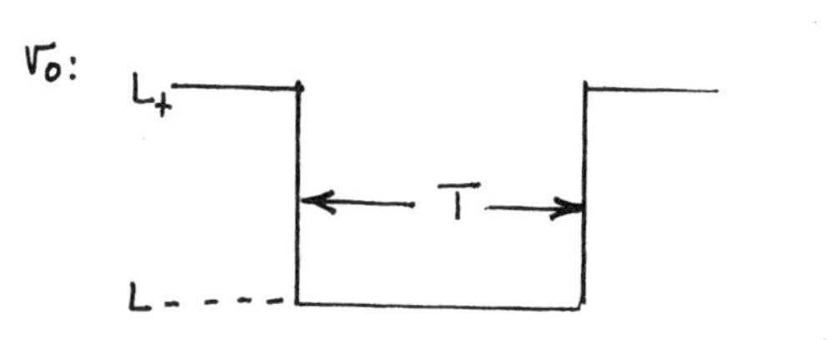

13.35

For recovery, v_B goes from βL_- to L_+ until D_1 conducts at $V_{D1} = 0.7V$

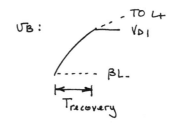

v_B: TO L_+, V_{D1}, βL_-, $T_{recovery}$

For recovery

$$v_B = -0.1(12) + (12+1.2)(1-e^{-t/\tau})$$

$$= 12 - 13.2e^{-t/\tau}$$

At $T_{recovery}$:

$$V_{D1} = 12 - 13.2e^{-T_r/\tau} \qquad \tau = R_3C_1$$

$$T_r = -R_3C_1\ln\left(\frac{V_{D1}-12}{13.2}\right)$$

$$= -(6171)(0.1\times10^{-6})\ln\left(\frac{11.3}{13.2}\right)$$

$= 96\mu s$

13.36

choose $C_1 = 1nF$ $\quad$ $C_2 = 0.1nF$

$R_1 = R_2 = 100k\Omega \Rightarrow \beta \doteq \frac{1}{2}$

$$T \cong C_1 R_3 \ln\left(\frac{0.7+13}{-13(0.5-1)}\right)$$

$$10^{-4} = 10^{-9} R_3 \ln\left(\frac{13.7}{13(0.5)}\right)$$

$R_3 = 134.1k\Omega$

Need $R_4 \gg R_1 \Rightarrow$ choose $R_4 = 470k\Omega$

Min trigger voltage $= \beta L_+ - V_{D2} + V_{D1}$
$= 6.5V$

For recovery

$$v_B = 13 - (13 - \beta L_-)e^{-t/\tau}$$

$$= 13 - 19.5e^{-t/\tau} = 0.7$$

$$\therefore t_{recovery} = -\tau \ln\left(\frac{12.3}{19.5}\right)$$

$$= (134.1\times10^3)(10^{-9})(-0.4608)$$

$$= 61.8\mu s$$

13.37

For $v_I > \frac{2}{3}V_{cc}$ comp-1 = "1" and comp-2 = "0" and flip flop is reset. I.E. $v_O = 0V$. Now v_O will not change until $v_I = \frac{1}{3}V_{cc}$, when comp-2 = "1" and comp-1 = "0" and FF is set: I.E. $V_O = V_{cc}$

For $\frac{1}{3}V_{cc} < v_I < \frac{2}{3}V_{cc}$, comp-1 = comp-2 = "0" and no change of state will occur.

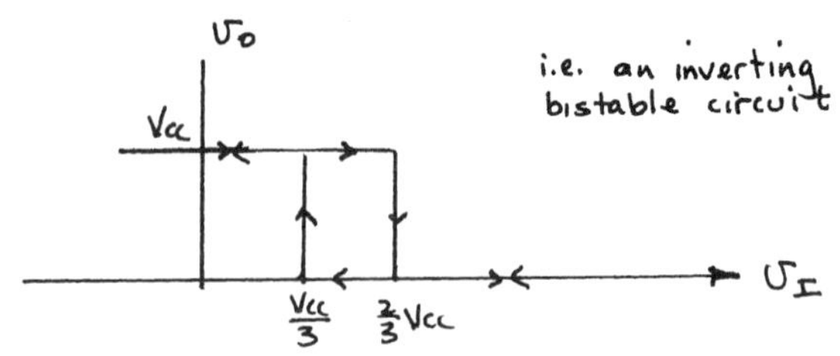

13.38

(a) $\quad C = 1nF$

$v_C = V_{cc}(1 - e^{-t/\tau})$ $\quad$ where $\tau = RC$

Pulse width of $10\mu s$ when $v_C = V_{TH} = \frac{2}{3}V_{cc}$

$$\therefore \frac{2}{3} = 1 - e^{-t/RC} \qquad t = T = 10\mu s$$

$$-\frac{T}{RC} = \ln\left(\frac{1}{3}\right) \Rightarrow R = \frac{-T}{C\ln(1/3)}$$

$$= 9.1k\Omega$$

(b) for $T = 20\mu s$ $\quad$ $R = 9.1k\Omega$, $C = 1nF$

$$\therefore V_{TH} = 15(1 - e^{-T/RC})$$

$$= 15\left(1 - e^{\frac{-20\times10^{-6}}{9.1\times10^3\times10^{-9}}}\right)$$

$$= 13.3V$$

13.39

$C = 680pF$ $\quad$ $f = 50kHz$

$T = 20\mu s = T_H + T_L$

For 75% Duty $\quad T_H = 15\mu s$
$\qquad\qquad\qquad\quad T_L = 5\mu s$

From Eq (13.43) we have:

CONT.

$T_L = CR_B \ln 2$

$\therefore R_B = \dfrac{5\times10^{-6}}{680\times10^{-12}\ln(2)} = \underline{\underline{10.6\ k\Omega}}$

From Eq (13.41)

$T_H = C(R_A+R_B)\ln(2)$

$R_A = \dfrac{15\times10^{-6}}{680\times10^{-12}\ln(2)} - 10.6\times10^3$

$= \underline{\underline{21.2\ k\Omega}}$

13.40

V_{TH} - - - - -

V_{TL} - - -

For the rise:

$V_C = V_{CC} - (V_{CC} - V_{TL})\, e^{-t/C(R_A+R_B)}$

$V_{TH} = V_{CC} - (V_{CC} - V_{TL})\, e^{-T_H/C(R_A+R_B)}$

$\dfrac{V_{CC} - V_{TH}}{V_{CC} - V_{TL}} = e^{-T_H/C(R_A+R_B)}$

$\underline{\underline{T_H = C(R_A+R_B)\ln\left(\dfrac{V_{CC}-V_{TL}}{V_{CC}-V_{TH}}\right)}}$

For exponential fall:

$v_C = V_{TH}\, e^{-t/CR_B}$

$\therefore V_{TL} = V_{TH}\, e^{-T_L/CR_B}$

$T_L = CR_B \ln\left(\dfrac{V_{TH}}{V_{TL}}\right)$

for $V_{TH} = 2V_{TL} \Rightarrow T_L = CR_B\ln(2)$

(b) $C = 1\text{nF}$, $R_A = 7.2\text{k}\Omega$, $R_B = 3.6\text{k}\Omega$
$V_{CC} = 5V$ $\quad V_{TH,\,EXT} = 0$

$\therefore T_H + T_L = T = \ln 2\,(R_A + 2R_B)C$

$T = 9.98\,\mu s \rightarrow \underline{\underline{f = 100\text{kHz}}}$

Duty cycle $= \dfrac{T_H}{T_H+T_L} = \dfrac{R_A+R_B}{R_A+2R_B} = 0.75$

$\Rightarrow \underline{\underline{75\%}}$

(c) $V_{CC} = 5V$, $\quad V_{TH} = \frac{2}{3}\times5 = \frac{10}{3} = 3.33\,V$

for 1V input $\quad V_{TH}' = 4.33V$

$V_{TL}' = \frac{1}{2}V_{TH}' = 2.17V$

$T_H' = 10^{-9}(3.6+7.2)\times10^3 \ln\left(\dfrac{5-2.17}{5-4.33}\right)$

$= \underline{15.6\,\mu s}$

$T_L' = 10^{-9}\times3.6\times10^3 \ln 2 = \underline{2.5\,\mu s}$

$\therefore f = \dfrac{1}{(15.6+2.5)10^{-6}} = \underline{\underline{55.2\text{kHz}}}$

duty cycle $= \dfrac{15.6}{2.5+15.6} = \underline{\underline{86.2\%}}$

for 1V input $\quad V_{TH}'' = 2.33$

$V_{TL}'' = 1.17$

$\therefore T_H'' = 10^{-9}(3.6\ \ 7.2)10^3 \ln\left(\dfrac{5-1.17}{5-2.33}\right)$

$= 3.92\,\mu s$

$T_L'' + T_L' = 2.5\,\mu s$

$\therefore f = \dfrac{10^6}{(3.92+2.5)} = \underline{\underline{156\text{ kHz}}}$

duty cycle $= \dfrac{3.92}{2.5+3.92} = \underline{\underline{61\%}}$

13.41

for sine wave: $v_O = 0.7 \sin \omega t$

slope at zero crossing

$0.7\omega \cos \omega t \big|_{t=1/f}$

$= 0.7\omega = 0.7(2\pi f)$

Slope of triangular wave with peak of V volts and period 1/f is:

$$\frac{V - -V}{T/2} = \frac{2V}{\frac{1}{2f}} = 4Vf$$

Equating the slopes:

$4Vf = 0.7(2\pi f)$

$V = 0.7 \times \frac{2\pi}{4} = \underline{\underline{1.0996}}$

Now $R = \frac{1.0996 - 0.7}{1\text{mA}} = 399.6\,\Omega$

$\cong \underline{\underline{400\,\Omega}}$

v_D changes by 0.1 V per decade change in current.

$\therefore v_O = 0.7 + 0.1 \log\left(\frac{1\text{mA}}{I_x}\right)$

$\Rightarrow i_x = 10^{\frac{v_O - 0.7}{0.1}}$ mA

For output of 0.7 V $i_x = 1$mA and

$\theta = 90°$

Error = 0%

output = 0.65 $i_x = 0.316$mA

$\therefore v_I = 0.65 + 0.316 \times 0.4$

$= 0.7765$V

$\theta = \frac{0.7765}{1.0996} \times 90° = 63.6°$

output = 0.6V $i_x = 0.1$mA

$v_I = 0.6 + 0.1 \times 0.4 = 0.640$V

$\theta = \frac{0.640}{1.0996} \times 90 = 52.4°$

out = 0.55V $i_x = 0.0316$mA

$v_I = 0.55 + 0.0316 \times 0.4$

$= 0.563$V

$\theta = \frac{0.563}{1.0996} \times 90 = 46.1°$

OUT = 0.5V $i_x = 0.01$mA

$v_I = 0.5 + 0.01 \times 0.4 = 0.504$V

$\theta = \frac{0.504}{1.0996}\ 90° = 41.3°$

OUT (V)	
0.4	32.8°
0.3	24.6°
0.2	16.4°
0.1	8.2°
0	0°

Summarizing in Table Form:

v_O (V)	θ	$0.7\sin\theta$	% Error
0.7	90	0.7	0
0.65	63.6	0.627	3.7
0.6	52.4	0.554	8.2
0.55	46.1	0.504	9.1
0.50	41.3	0.462	8.2
0.40	32.8	0.379	5.5
0.30	24.6	0.291	3.0
0.20	16.4	0.198	1.2
0.10	8.2	0.0998	0.1
0.00	0	0.0	0.0

13.42

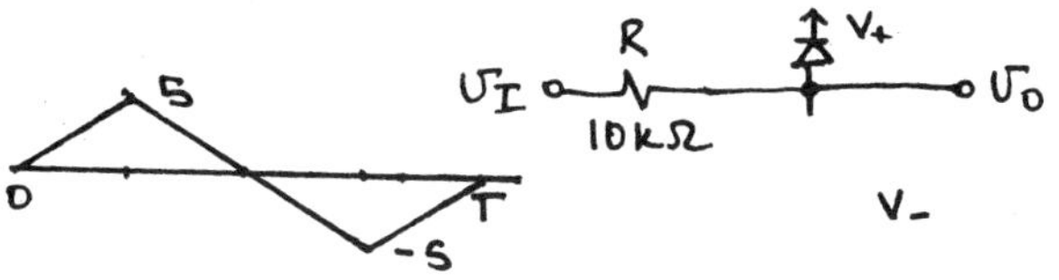

$v_O = A \sin \frac{2\pi}{T} t$

Slope of v_O at $t = 0$:

$$\frac{\partial v_O}{\partial t} = A\frac{2\pi}{T} \cos\left(\frac{2\pi}{T}\right)t \Big|_{t=0}$$

$$= \frac{A\,2\pi}{T} \equiv \text{SLOPE AT ZERO CROSSING}$$

Slope of Δ-wave $= \frac{5}{T/4} = \frac{20}{T}$

$$\therefore \frac{20}{T} = \frac{A\,2\pi}{T}$$

$$\underline{A = 3.18\text{ V}}$$

∴ Clamp voltage:

$$V_+ = -V_- = 3.18 - 0.7$$

$$= 2.48 = \underline{\underline{2.5\text{V}}}$$

13.43

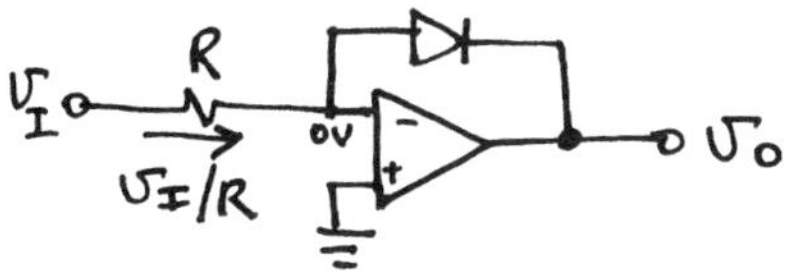

→ $v_I > 0$

→ voltage across diode is $-v_O$

$$i_D = \frac{v_I}{R} = I_S e^{-v_O/nV_T}$$

$$-\frac{v_O}{nV_T} = \ln\left(\frac{v_I}{RI_S}\right)$$

$$v_O = -nV_T \ln\left(\frac{V_I}{RI_S}\right), \quad v_I > 0$$

Q.E.D.

13.44

from P13.43

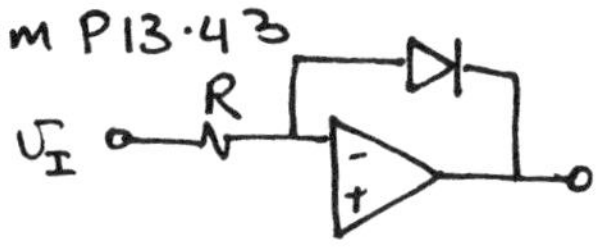

$v_O = -nV_T \ln\left(\frac{v_I}{RI_S}\right)$

Now.

$$V_A = -nV_T \ln \frac{V_1}{RI_S} \qquad R = 1\text{k}\Omega$$

$$V_B = -nV_T \ln \frac{V_2}{RI_S} \qquad V_1, V_2 > 0$$

$$V_C = +nV_T \ln \frac{1}{RI_S}$$

$$V_D = -(V_A + V_B + V_C)$$

$$= nV_T\left(\ln\left[\frac{V_1}{RI_S} \times \frac{V_2}{RI_S} \times \frac{RI_S}{1}\right]\right)$$

$$= nV_T \ln\left(\frac{V_1 V_2}{RI_S}\right)$$

$$i_{D4} = I_S e^{V_D/nV_T}$$

$$= I_S \times \frac{V_1 V_2}{RI_S} = \frac{V_1 V_2}{R}$$

$$v_O = -i_{D4} R = -\frac{V_1 V_2}{R} \times R$$

$$\therefore \underline{\underline{v_O = -V_1 V_2}} \qquad \text{ANALOG MULTIPLIER}$$

To check $\underline{v_1 = 0.5, \quad v_2 = 2}$

$$I_{D1} = 0.5\text{mA} \rightarrow V_A = -0.7 + nV_T \ln\left(\frac{0.5}{1}\right)$$

$$= 0.7 + 2(0.025)\ln\left(\tfrac{1}{2}\right)$$

$$= -0.6653\text{V}$$

$$I_{D2} = 2\text{mA} \rightarrow V_B = (0.7 + 0.05\ln(2))(-1)$$

$$= -0.7347\text{V}$$

$$I_{D3} = 1\text{mA} \rightarrow V_C = 0.700\text{V}$$

$$V_D = -(-0.6653 - 0.7347 + 0.7) = 0.7\text{V}$$

CONT.

$V_D = V_{D4} = 0.7V \Rightarrow I_{D4} = 1mA$

$\therefore v_O = -1V$ i.e. $2 \times 0.5 = 1$

For $v_1 = 3$, $v_2 = 2$:

$I_{D1} = 3mA \rightarrow V_A = -(0.7 + 0.05\ln 3) = -0.7549V$
$I_{D2} = 2mA \rightarrow V_A = -(0.7 + 0.05\ln 2) = -0.7347V$
$I_{D3} = 1mA \rightarrow V_C = 0.7V$

$\therefore V_D = V_{D4} = -(V_A + V_B + V_C) = +0.7896V$

$$\therefore \frac{I_{D4}}{1mA} = \frac{I_S e^{V_D/0.05}}{I_S e^{0.7/0.05}}$$

$$I_{D4} = e^{\frac{0.7986-0.7}{0.05}} = 6mA$$

$\therefore v_O = -6V$ i.e. $2 \times 3 = 6$.

For squarer: $v_1 = 2$ through $\frac{1}{2}k\Omega$ resistor

$I_{D1} = 4mA \rightarrow V_A = -(0.7 + 0.05\ln 4) = -0.7693$

$V_D = -(-0.7693) = 0.7693V$

$$I_{D4} = e^{\frac{0.7693-0.7}{0.05}} = 3.999mA$$

$\therefore v_O = -3.999V$ ie. $2^2 = 4$.

13.45

Say $V_{BE} = \tilde{V}_D$ @ I $\quad n = 1$

for $v_O = 0.25\,V_T$:

$$I_R = \frac{0.25V_T}{R} = \frac{0.25V_T}{\frac{2.5V_T}{I}} = \frac{I}{10}$$

$$V_{BE1} = \tilde{V}_D + nV_T \ln\left(\frac{I + I/10}{I}\right) \cong \tilde{V}_D + V_T\ln(1.1)$$

$$V_{BE2} = \tilde{V}_D + nV_T \ln\left(\frac{I - I/10}{I}\right) \cong \tilde{V}_D + V_T\ln(0.9)$$

$$V_I = -V_{BE2} + V_O + V_{BE1}$$
$$= V_T[\ln(1.1) + 0.25 - \ln(0.9)]$$
$$= \underline{0.451V_T}$$

For $v_O = 0.5V_T$

$$I_R = \frac{0.5\,I}{2.5} = 0.2I$$

$$V_I = V_T[\ln(1.2) + 0.5 - \ln(0.8)]$$
$$= \underline{0.905V_T}$$

$V_O = V_T$ $\quad I_R = 0.4I$

$$V_I = V_T[\ln 1.4 + 1 - \ln 0.6]$$
$$= \underline{1.847V_T}$$

$V_O = 1.5V_T$ $\quad I_R = 0.6I$

$$V_I = V_T(\ln 1.6 + 1.5 - \ln 0.4)$$
$$= \underline{2.886V_T}$$

$V_O = 2V_T$ $\quad I_R = 0.8I$

$$V_I = V_T(\ln 1.8 + 2 - \ln 0.2) = \underline{4.197V_T}$$

$V_O = 2.4V_T$ $\quad I_R = 0.96I$

$$V_I = V_T(\ln 1.96 + 2.4 - \ln 0.04) = \underline{6.292V_T}$$

$V_O = 2.42V_T$ $\quad I_R = 0.968I$

$$V_I = V_T(\ln 1.968 + 2.42 - \ln 0.032)$$
$$= \underline{6.519V_T}$$

CONT.

Ideal curve given by

$$v_O = 2.42\,V_T \sin\left(\frac{v_I}{6.6\,V_T} \times 90°\right)$$

$$\frac{v_I}{V_T} = \frac{6.6}{90}\sin^{-1}\left(\frac{v_O}{2.42\,V_T}\right)$$

v_O/V_T	0.25	0.50	1.00	1.50	2.00	2.40	2.42
v_I/V_T	0.451	0.905	1.85	2.89	4.20	6.29	6.52
v_I/V_T (ideal)	0.435	0.874	1.79	2.81	4.09	6.06	6.60

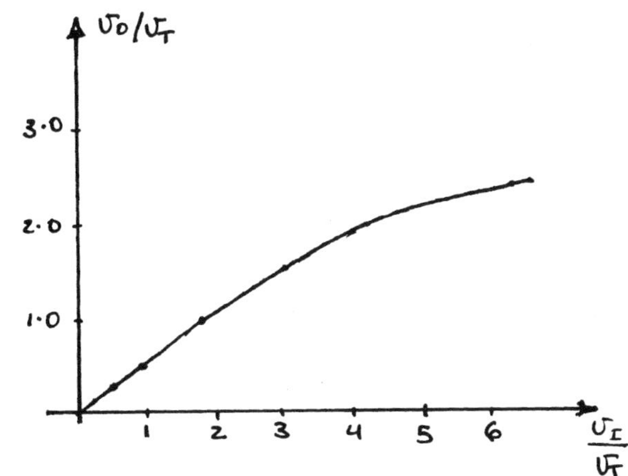

13.46

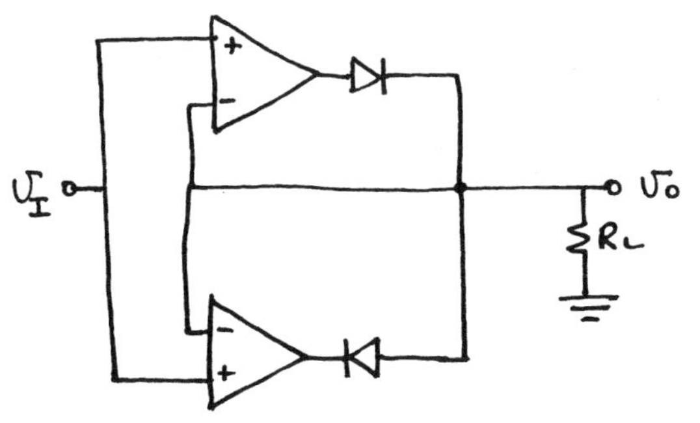

for high opamp gain $v_O = v_I$

$\therefore$ v_O is also a 10Vpp sine wave.

for $v_I > 0$

$$v_O = v_I - \frac{0.7}{A}$$

$$\cong v_I$$

with D_1 "on"
D_2 "off"

for $v_I < 0$

$$v_O = v_I + \frac{0.7}{A}$$

$$\cong v_I$$

with D_1 "off"
D_2 "on"

13.47

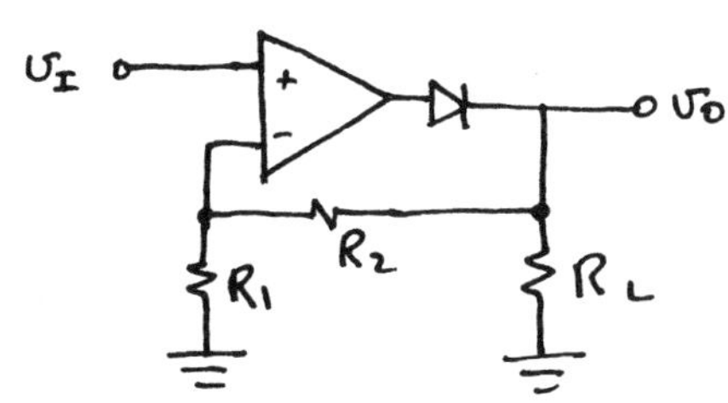

for $v_I \geq 0$

$$v_O = \left(1 + \frac{R_2}{R_1}\right) v_I$$

for a gain of 2 $\quad$ $R_1 = R_2 = 10\,k\Omega$

for v_I = 10Vpp sine wave $v_O \Rightarrow$

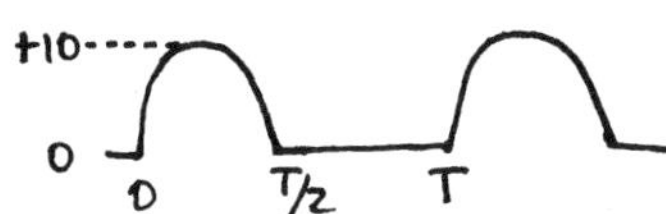

$$Avg = \frac{1}{T}\int_0^{T/2} 10 \sin \frac{2\pi}{T} t \, dt$$

$$= \frac{1}{T}\frac{T}{2\pi} \cos \frac{2\pi}{T} t \times (-10) \Big|_{t=0}^{T/2}$$

$$= \frac{-10}{2\pi}\left(\cos \pi - \cos 0\right)$$

$$= 10/\pi = \underline{3.18\,V}$$

13.48

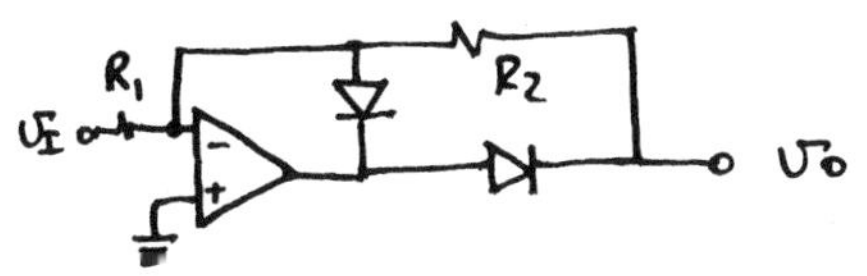

CONT.

for $v_I < 0 \Rightarrow v_O = -R_2/R_1$

$R_{in} = R_1 = 100k\Omega$ $\therefore$ $R_2 = 200k\Omega$

13.49

for high R_{in}, use $R_1 = 1M\Omega$

AC gain is given by R_2/R_1

$\Rightarrow R_2 = 1M\Omega$

Now for 1Vrms sine, peak is 1.414V.

The value of V_1 is then $\frac{1.414}{\pi} = 0.450V$

For 10V out at second stage

gain (dc) $= \frac{10}{0.450} = 22.2$

$\therefore R_4/R_3 = 22.2$

& choose $\frac{1}{2\pi R_4 C} = 10Hz$ (i.e. corner frequency)

To make C small, make $R_4 = 1M\Omega$

$\therefore C = 15.9nF$

$R_3 = \frac{1M\Omega}{22.2} = 45k\Omega$

13.50

At the +ve terminal $V_+ = -5V$

for $v_I > -5$ D_1 is "ON" and faces virtual short. $\therefore V_- = -5$, and no current will flow in feedback R.

$\therefore v_O = -5V$

for $v_I < -5$ - D_1 "off" and

$$\frac{v_O}{v_I} = \frac{-5-v_I}{R} = \frac{v_O+5}{R}$$

$\Rightarrow v_O = -v_I -$

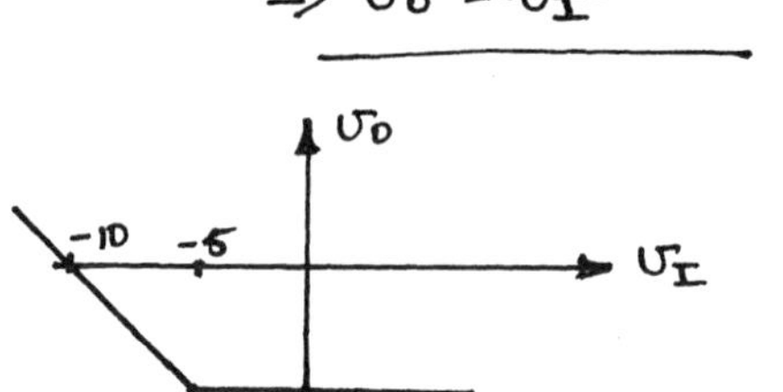

13.51

for $v_I < 0$

D_2 "off"

$\frac{v_{O1}}{v_I} = -1$

$\frac{v_{O2}}{v_I} = 0$

for $v_I > 0$

D_1 ~ "off"

$\frac{v_{O1}}{v_I} = 0$

$\frac{v_{O2}}{v_I} = -1$

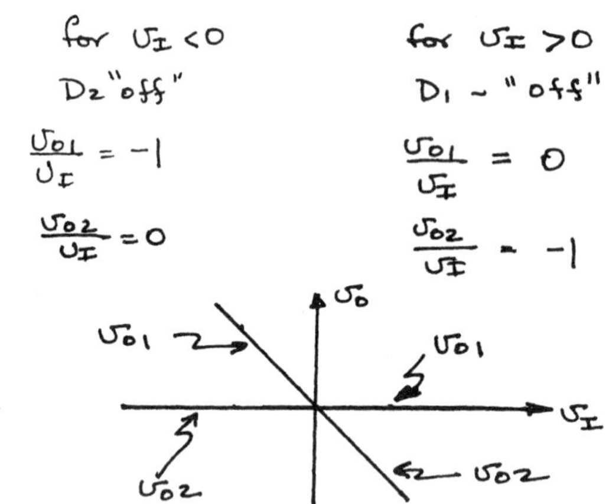

13.52

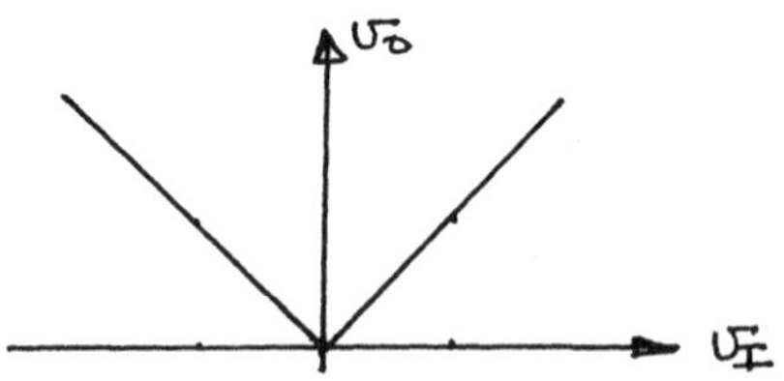

For $v_I < 0$ ~ Diode is on, and cathode is forced to ≈ 0V.

$\therefore v_O/v_I = -1$

For $v_I > 0$ ~ Diode is off, and the cathode now follows v_I since no current flows in Resistor. So v_O must follow

CONT.

v_I so that no current flows in feedback resistor.

$\therefore \frac{v_O}{v_I} = +1$

13.53

Simply place the LED in the feedback path.

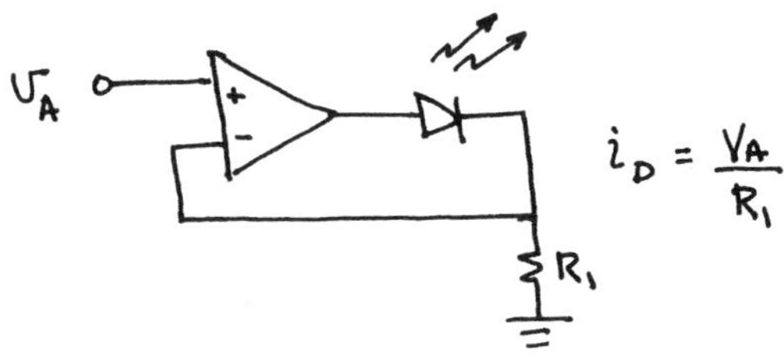

$i_D = \frac{V_A}{R_1}$

13.54

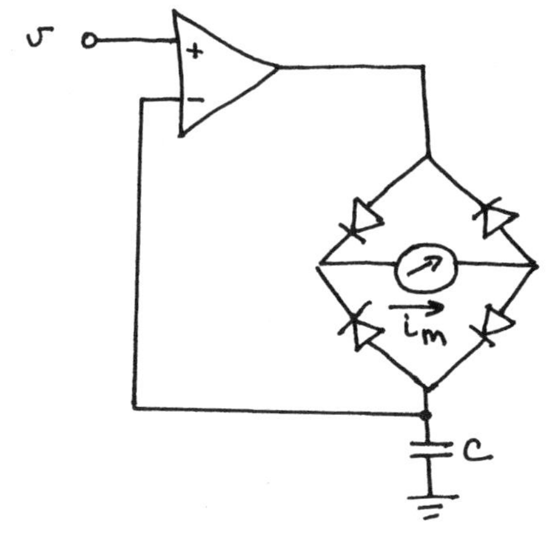

$i_m = |i_C| = C\frac{dv}{dt}$ using $R = 1k\Omega$

$i_m = |i_R| = \frac{|v|}{R} = \frac{|v|}{1k\Omega} \Rightarrow i_m = |V|$ mA

Now $v = V \sin 2\pi 60t$

$\Rightarrow i_m = C \times 2\pi 60 |\cos(2\pi 60t)|$

for equivalence:

$\frac{V}{10^3}|\sin 2\pi 60t| = 2\pi 60 VC\, |\cos 2\pi 60t|$

$\therefore C = \frac{1}{2\pi 60 10^3} = \underline{\underline{2.65\mu F}}$

At 120Hz: $i_m = 2\pi\ 120 VC\, |\cos 2\pi 60t|$

$i_{m_{120}} = 2\, i_{m_{60}}$

At 180Hz: $i_{m_{180}} = 3 i_{m_{60}}$

For Δ-wave

with R_1

$i_m = 1mA$, $R = 1k\Omega$

peak

0 T/2 T

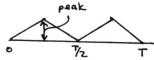

∴ Full wave rectified wave has average voltage = 1V. ∴ $V_{peak} = 2V$

with C: $\text{slope} = \frac{V_{peak}}{T/4} = 4V_{peak}f$

$= 4 \times 2 \times 60 = 480$

Now: current through the capacitor will be a square wave (50% duty cycle)

Peak current $= 2.65 \times 10^{-6} \times 480$

$= 1.27$ mA.

$\therefore i_m = i_{avg} = \underline{\underline{1.27\,mA}}$

13.55

10V pulses of 10μs, and large C_{LOAD}, will cause the op amp to current limit.

Charge transfered in one pulse:

$Q = (10mA)(10\mu s)$

$= 10^{-7}$ C

CONT

Voltage change per pulse:

$$\Delta V = Q/C = \frac{10^{-7}}{10\times10^{-6}} = 10\text{mV}$$

After:		V_c
	1 pulse	$V_c = 10\text{mV}$
	2 pulses	20mV
	10 pulses	100mV

to reach		require
	0.5V	50 pulses
	1.0V	100 "
	2.0V	200 "

13.56

For 1Vpp, peak detector output =
$V_o = 0.5\text{V}$.

Ripple Voltage $= (1\%)\, 0.5 = 5\text{mV}$

Total leakage $= 10 + 1 = 11\text{nA}$

∴ total charge lost:

$$\Delta Q = 11\text{nA} \times 1\text{ms} = 11\text{pC}$$

∴ Required capacitance:

$$C = \frac{Q}{\Delta V} = \frac{11\times10^{-12}}{5\times10^{-3}} = \underline{\underline{2.2\text{nF}}}$$

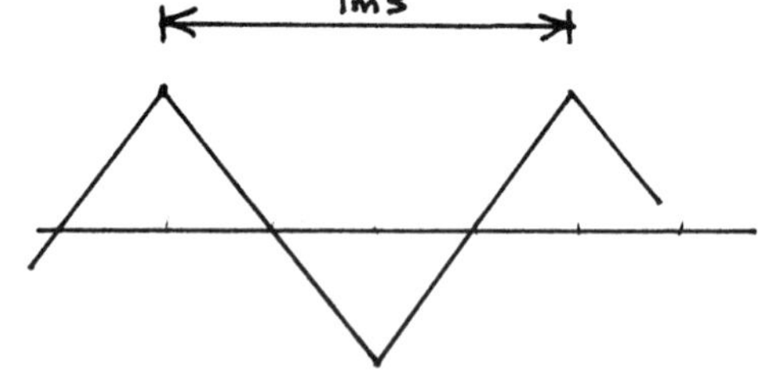

Chapter 14 _ Problems

14

14.1 Refer to Fig. 14.2 (page 1232). The upper limit of the output voltage is determined by the saturation of Q_1 as

$$v_{O\,max} = V_{CC} - V_{CE\,sat}$$
$$= 5 - 0.3 = \underline{\underline{4.7\ V}}$$

The corresponding input is

$$v_I = 4.7 + 0.7 = \underline{\underline{5.4\ V}}$$

The bias current I is

$$I = \frac{0 - (-V_{CC} + V_{BE2})}{R}$$
$$= \frac{5 - 0.7}{1} = 4.3\ \text{mA}$$

The lower limit of v_O is determined by either Q_1 cutting off,

$$\frac{-v_O}{R_L} = I \Rightarrow v_O = -4.3\ V$$

or by Q_2 saturating,

$$v_O = -V_{CC} + V_{CE\,sat} = -4.7\ V$$

Obviously, $v_{O\,min} = \underline{\underline{-4.3\ V}}$

and the corresponding input is

$$v_I = -4.3 + 0.7 = \underline{\underline{-3.6\ V}}$$

14

If the emitter-base junction area of Q_3 is made twice as large as that of Q_2, I becomes one half its previous value,

$$I = \frac{4.3}{2} = 2.15\ \text{mA}$$

and thus the lower limit of v_O changes to

$$v_{O\,min} = -IR_L = \underline{\underline{-2.15\ V}}.$$

The corresponding value of v_I is

$$v_I = -2.15 + 0.7 = \underline{\underline{-1.45\ V}}$$

The upper limit does not change.

14.2 First we determine the bias current I as follows:

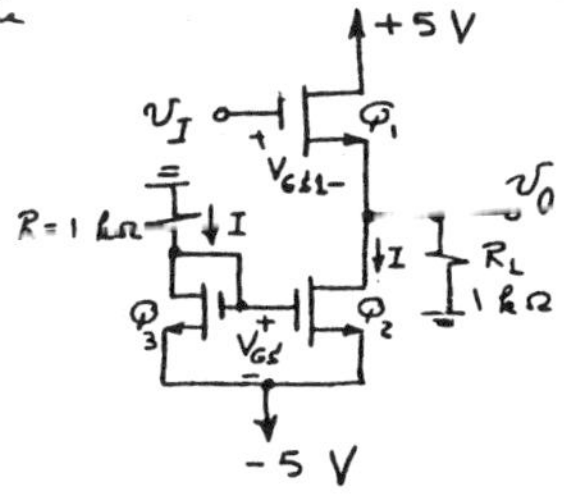

$$I = \tfrac{1}{2}\mu_n C_{ox}\tfrac{W}{L}(V_{GS} - V_t)^2$$

But $V_{GS} = 5 - IR$
$= 5 - I$

Thus

$$I = \tfrac{1}{2}\mu_n C_{ox}\tfrac{W}{L}(5 - I - V_t)^2$$
$$I = 10\,(5 - I - 1)^2$$
$$\Rightarrow I^2 - 8.1 I + 16 = 0$$
$$I = 3.416\ \text{mA} \quad \text{and } V_{GS} = 1.584\ V$$

14

The upper limit on v_O is determined by Q_1 leaving the saturation region (and entering the triode region). This occurs when v_I exceeds V_{D1} by V_t volts,

$$v_{I\,max} = 5 + 1 = \underline{+6\ V}$$

To obtain the corresponding value of v_O we must find the corresponding value of V_{GS1}, as follows:

$$v_O = v_I - V_{GS1}$$
$$i_L = \frac{v_O}{R_L} = \frac{v_I - V_{GS1}}{1}$$
$$= v_I - V_{GS1} = 6 - V_{GS1}$$
$$i_1 = I + i_L$$
$$= 3.416 + 6 - V_{GS1}$$
$$= 9.416 - V_{GS1}$$

But $i_1 = \tfrac{1}{2}\mu_n C_{ox}\tfrac{W}{L}(V_{GS1} - V_t)^2$

Thus, $9.416 - V_{GS1} = 10\,(V_{GS1} - 1)^2$

$$\Rightarrow V_{GS1}^2 - 1.9\,V_{GS1} + 0.0584 = 0$$
$$V_{GS1} = 1.869\ V$$
$$v_{O\,max} = 6 - 1.869$$
$$= \underline{\underline{+4.131\ V}}$$

14

The lower limit of v_O is determined either by Q_1 cutting off,

$$v_O = -IR_L = -3.416 \times 1 = -3.416\ V$$

or by Q_2 leaving saturation,

$$v_O = V_{G2} - V_t$$
$$= -5 + 1.584 - 1 = -4.416\ V$$

Thus, $v_{O\,min} = \underline{\underline{-3.416\ V}}$

The corresponding value of v_I is determined by noting that since Q_1 is on the verge of cut-off, $V_{GS1} = V_t = 1\ V$ and

$$v_I = -3.416 + 1 = \underline{\underline{-2.416\ V}}$$

14.3 Refer to Fig. 14.2. With $V_{CC} = +9\ V$, the upper limit on v_O is $+8.7\ V$, which is greater than the required value of $+7\ V$. To obtain a lower limit of $-7\ V$ we select I so that

$IR_L = 7$ **14**

$\Rightarrow \; I = 7 \text{ mA}$

Since we are provided with four devices, we can minimize the total supply current by paralleling two devices to form Q_2 as shown below. The resulting supply current will be $3 \times \frac{I}{2}$ rather than $2I$ which is the value obtained in the circuit of Fig. 14.2. Thus the supply current is 10.5 mA. The value of R is found from

$$R = \frac{8.3 \text{ V}}{3.5 \text{ mA}} = \underline{2.37 \text{ k}\Omega}$$

In a practical design we would select a standard value for R that results in I somewhat larger than 7 mA. Say, $R = 2.2 \text{ k}\Omega$

14.4 Refer to Fig. 14.2. For a load resistance of 100 Ω and v_O ranging between −5 V and +5 V, the maximum current through Q_1 is **14**

$$I + \frac{5}{0.1} = I + 50, \text{ mA}$$

and the minimum current is

$$I - \frac{5}{0.1} = I - 50, \text{ mA}.$$

For a current ratio of 10,

$$\frac{I+50}{I-50} = 10$$

$$\Rightarrow \; I = 61.1 \text{ mA}$$

$$R = \frac{9.3 \text{ V}}{61.1 \text{ mA}} = \underline{152 \; \Omega}$$

The incremental voltage gain is

$$A_v = \frac{R_L}{R_L + r_{e1}}$$

For $R_L = 100 \; \Omega$;

At $v_O = +5$ V, $I_{E1} = 61.1 + 50 = 111.1$ mA

$$r_{e1} = \frac{25}{111.1} = 0.225 \; \Omega$$

$$A_v = \frac{100}{100 + 0.225} = \underline{0.998 \text{ V/V}}$$

At $v_O = 0$ V, $I_{E1} = 61.1$ mA **14**

$$r_{e1} = \frac{25}{61.1} = 0.409 \; \Omega$$

$$A_v = \frac{100}{100.409} = \underline{0.996 \text{ V/V}}$$

At $v_O = -5$ V, $I_{E1} = 61.1 - 50 = 11.1$ mA

$$r_{e1} = \frac{25}{11.1} = 2.25 \; \Omega$$

$$A_v = \frac{100}{102.25} = \underline{0.978 \text{ V/V}}$$

Thus the incremental gain changes by $0.998 - 0.978 = 0.02$ or about $\underline{\underline{2\%}}$ over the range of v_O.

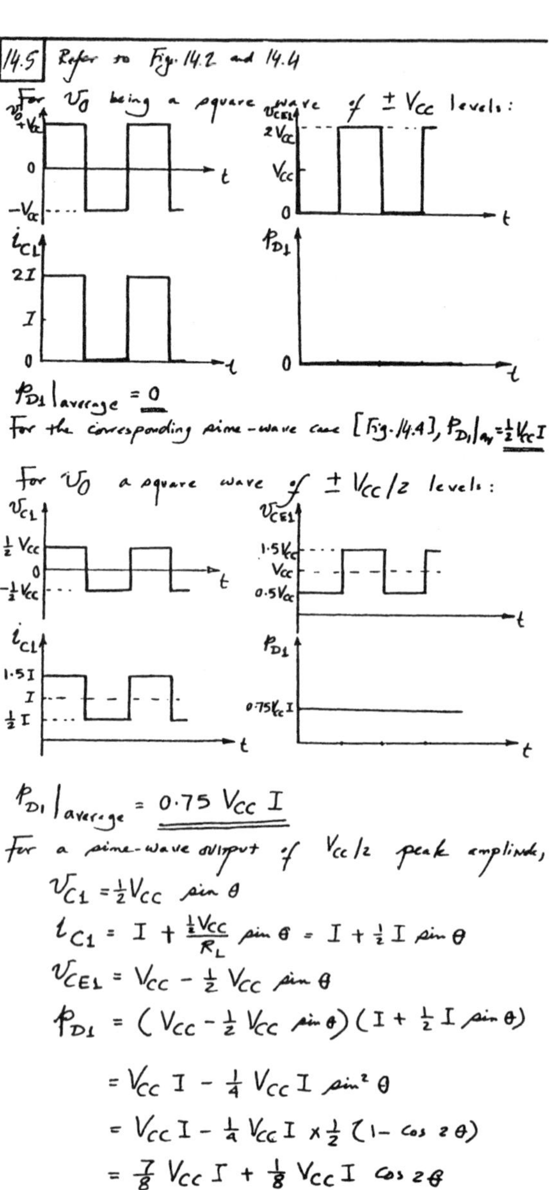

14.5 Refer to Fig. 14.2 and 14.4

For v_O being a square wave of $\pm V_{CC}$ levels:

$$P_{D1}\big|_{average} = \underline{0}$$

For the corresponding sine-wave case [Fig. 14.4], $P_{D1}\big|_{av} = \frac{1}{2} V_{CC} I$

For v_O a square wave of $\pm V_{CC}/2$ levels:

$$P_{D1}\big|_{average} = \underline{0.75 \, V_{CC} I}$$

For a sine-wave output of $V_{CC}/2$ peak amplitude,

$$v_{C1} = \tfrac{1}{2} V_{CC} \sin\theta$$

$$i_{C1} = I + \frac{\frac{1}{2}V_{CC}}{R_L} \sin\theta = I + \tfrac{1}{2} I \sin\theta$$

$$v_{CE1} = V_{CC} - \tfrac{1}{2} V_{CC} \sin\theta$$

$$P_{D1} = (V_{CC} - \tfrac{1}{2} V_{CC} \sin\theta)(I + \tfrac{1}{2} I \sin\theta)$$

$$= V_{CC} I - \tfrac{1}{4} V_{CC} I \sin^2\theta$$

$$= V_{CC} I - \tfrac{1}{4} V_{CC} I \times \tfrac{1}{2}(1 - \cos 2\theta)$$

$$= \tfrac{7}{8} V_{CC} I + \tfrac{1}{8} V_{CC} I \cos 2\theta$$

$$P_{D1}\big|_{average} = \underline{\underline{\tfrac{7}{8} V_{CC} I}}$$

14.6 In all cases, the average voltage across Q_2 is equal to V_{CC}. Thus, since Q_2 conducts a constant current I, its average power dissipation is $\underline{\underline{V_{CC} I}}$. **14**

14.7 $V_{CC} = 16, 12, 10$ and 8 V

$I = 100$ mA $R_L = 100\ \Omega$

$\hat{V}_o = 8$ V

$$\eta = \frac{1}{4}\left(\frac{\hat{V}_o}{IR_L}\right)\left(\frac{\hat{V}_o}{V_{CC}}\right)$$

$$= \frac{1}{4}\left(\frac{8}{10}\right)\left(\frac{8}{V_{CC}}\right) = \frac{1.6}{V_{CC}}$$

V_{CC}	16	12	10	8
η	10%	13.3%	16%	20%

14.8 Refer to Fig. P14.8. The bias current I is

$$I = \frac{1}{2}\mu_n C_{ox}\frac{W}{L}(0-(-2))^2 = 10 \times 4 = 40 \text{ mA}$$

For $R_L = \infty$, the maximum v_O is limited by the saturation of Q_1 to $V_{CC} - V_{CE\,sat} = \underline{\underline{+4.7\text{ V}}}$, and the minimum v_O is limited by Q_2 leaving saturation to $-V_{CC} + |V_t| = \underline{\underline{-3\text{ V}}}$.

For $R_L = 100\ \Omega$, the maximum v_O is still $\underline{\underline{+4.7\text{V}}}$, and the minimum v_O is still $\underline{\underline{-3\text{ V}}}$ because at this level the current in R_L is -30 mA and thus Q_1 is still conducting 10 mA. (That is, the cut off limit of $-40\text{ mA} \times 100\ \Omega = -4$ V is not applicable since Q_2 would have entered the triode region.) **14**

To obtain a 1-V peak sine-wave output, R_L must be limited to $(\leq)$

$$R_L = \frac{1\text{ V}}{40\text{ mA}} = \underline{\underline{25\ \Omega}}$$

For this case,

$$\eta = \frac{1}{4}\left(\frac{\hat{V}_o}{IR_L}\right)\left(\frac{\hat{V}_o}{V_{CC}}\right)$$

$$= \frac{1}{4}\left(\frac{1}{0.04 \times 25}\right)\left(\frac{1}{5}\right)$$

$$= \underline{\underline{5\ \%}}$$

14.9 Refer to Figs. 14.6 and 14.7. A 10% loss in peak amplitude is obtained when the amplitude of the input signal is $\underline{\underline{5\text{ V}}}$. **14**

14.10 With v_I sufficiently positive so that Q_N is conducting, the situation shown obtains. Thus,

$$(v_I - v_O) \times 100 = v_O + 0.7$$

$$\Rightarrow v_O = \frac{1}{1.01}(v_I - 0.007)$$

This relationship applies for $v_I \geq 0.007$. Similarly, for v_I sufficiently negative so that Q_P conducts, the voltage at the output of the amplifier becomes $v_O - 0.7$, thus

$$(v_I - v_O) \times 100 = v_O - 0.7$$

$$\Rightarrow v_O = \frac{1}{1.01}(v_I + 0.007)$$

This relationship applies for $v_I \leq -0.007$ **14**

The result is the transfer characteristic

Without the feedback arrangement, the deadband becomes ± 700 mV and the slope change a little (to nearly +1 V/V).

14.11 With $R_L = \infty$ and $v_I = +5$ V, v_O will be $v_I - V_{GS1} = v_I - V_t = 4$ V (since the current is virtually zero and thus $V_{GS} \simeq V_t$). Thus the resulting peak output voltage will be $\underline{4\ V}$.

$\sin\theta = \frac{1}{5}$

$\Rightarrow \theta = 11.54°$

Cross-over interval $= 4\theta$

Fraction of Cycle $= \frac{4\theta}{360°} = \underline{12.8\%}$

+5, +1, 0, −1, θ, −5, ωt

For $v_I = +5$ V and $v_O = +2.5$ V, $V_{GS} = 2.5$ V, thus

+10 V, +5, i_D, v_O = 2.5 V, R_L — 14

$$i_D = \tfrac{1}{2}\mu C_{ox}\tfrac{W}{L}(V_{GS} - V_t)^2$$

$$= 0.1\,(2.5-1)^2 = 0.225 \text{ mA}$$

Thus, $R_L = \frac{2.5}{0.225} = \underline{11.1\ k\Omega}$

14.12 For $V_{CC} = 10$ V and $R_L = 100\,\Omega$, the maximum sine-wave output power occurs when $\hat{V}_o = V_{CC}$ and is

$$P_{L\max} = \frac{1}{2}\frac{V_{CC}^2}{R_L}$$

$$= \frac{1}{2} \times \frac{100}{100} = \underline{0.5\ W}$$

Correspondingly,

$$P_{S+} = P_{S-} = \frac{1}{\pi}\frac{\hat{V}_o}{R_L}V_{CC}$$

$$= \frac{1}{\pi} \times \frac{10}{100} \times 10 = 0.318 \text{ W}$$

for a total supply power of

$$P_S = 2 \times 0.318 = \underline{0.637\ W}$$

The power conversion efficiency η is

$$\eta = \frac{P_L}{P_S} \times 100 = \frac{0.5}{0.637} \times 100 = \underline{78.5\%}$$

For $\hat{V}_o = 5$ V, — 14

$$P_L = \frac{1}{2}\frac{\hat{V}_o^2}{R_L} = \frac{1}{2} \times \frac{25}{100} = \underline{\frac{1}{8}\ W}$$

$$P_{S+} = P_{S-} = \frac{1}{\pi}\frac{\hat{V}_o}{R_L}V_{CC}$$

$$= \frac{1}{\pi} \times \frac{5}{100} \times 10 = \frac{1}{2\pi}$$

$$P_S = \frac{1}{\pi}\text{ W} = \underline{0.318\ W}$$

$$\eta = \frac{1/8}{1/\pi}\times 100 = \frac{\pi}{8} \times 100 = \underline{39.3\%}$$

14.13 $V_{CC} = 5$ V

For maximum η,

$$\hat{V}_o = V_{CC} = \underline{5\ V}$$

The output voltage that results in maximum device dissipation is given by Eq. (9.20),

$$\hat{V}_o = \frac{2}{\pi}V_{CC}$$

$$= \frac{2}{\pi} \times 5 = \underline{3.18\ V}$$

If operation is always at full output voltage, $\eta = 78.5\%$ and thus

$$P_{dissipation} = (1-\eta)P_S$$

$$= (1-\eta)\frac{P_L}{\eta} = \frac{1-0.785}{0.785}P_L = 0.274\,P_L$$

$$P_{dissipation}/\text{device} = \frac{1}{2} \times 0.274\,P_L = 0.137\,P_L$$

For a rated device dissipation of 1 W, and using a factor of 2 safety margin, — 14

$$P_{dissipation}/\text{device} = 0.5 \text{ W}$$

$$= 0.137\,P_L$$

$$\Rightarrow P_L = 3.65 \text{ W}$$

$$3.65 = \frac{1}{2} \times \frac{25}{R_L}$$

$\Rightarrow R_L = \underline{3.425\ \Omega}$ (i.e. $R_L \geq 3.425\,\Omega$)

The corresponding output power (i.e. greatest possible output power) is $\underline{3.65\ W}$.

If operation is allowed at $\hat{V}_o = \frac{1}{2}V_{CC} = 2.5$ V,

$$\eta = \frac{\pi}{4}\frac{\hat{V}_o}{V_{CC}} \quad \text{(Eq. 9.15)}$$

$$= \frac{\pi}{4} \times \frac{1}{2} = 0.393$$

$$P_{dissipation}/\text{device} = \frac{1}{2}\frac{1-\eta}{\eta}P_L = 0.772\,P_L$$

$$0.5 = 0.772\,P_L$$

$$\Rightarrow P_L = \underline{0.647\ W}$$

$$= \frac{1}{2}\frac{2.5^2}{R_L}$$

$\Rightarrow R_L = \underline{4.83\ \Omega}$ (i.e. $\geq 4.83\,\Omega$)

14.14 14

$P_L = \frac{1}{2}\frac{\hat{V}_o^2}{R_L}$

$100 = \frac{1}{2}\frac{\hat{V}_o^2}{16}$

$\hat{V}_o = 56.6 \text{ V}$

$V_{CC} = 56.6 + 6 = 60.6 \longrightarrow \underline{\underline{61 \text{ V}}}$

Peak current from each supply $= \frac{\hat{V}_o}{R_L} = \frac{56.6}{16} = \underline{\underline{3.54 \text{ A}}}$

$P_{S+} = P_{S-} = \frac{1}{\pi} \times 3.54 \times 61$

Thus, $P_S = \frac{2}{\pi} \times 3.54 \times 61$

$= \underline{\underline{137.4 \text{ W}}}$

$\eta \equiv \frac{P_L}{P_S} = \frac{100}{137.4} = \underline{\underline{73\%}}$

Using Eq. (9.22),

$P_{DNmax} = P_{DPmax} = \frac{V_{CC}^2}{\pi^2 R_L} = \frac{61^2}{\pi^2 \times 16}$

$= \underline{\underline{23.6 \text{ W}}}$

14.15 14

$P_L = \frac{\hat{V}_o^2}{R_L}$

$P_{S+} = P_{S-} = \frac{1}{2}\left(\frac{\hat{V}_o}{R_L}\right) V_{SS}$

$P_S = \frac{\hat{V}_o}{R_L} V_{SS}$

$\eta \equiv \frac{P_L}{P_S} = \frac{\hat{V}_o^2/R_L}{\hat{V}_o V_{SS}/R_L} = \frac{\hat{V}_o}{V_{SS}}$

$\eta_{max} = 1$ (100%), obtained for $\hat{V}_o = V_{SS}$

$P_{Lmax} = \frac{V_{SS}^2}{R_L}$

$P_{dissipation} = P_S - P_L$

$= \frac{\hat{V}_o}{R_L} V_{SS} - \frac{\hat{V}_o^2}{R_L}$

$\frac{\partial P_{dissipation}}{\partial \hat{V}_o} = \frac{V_{SS}}{R_L} - \frac{2\hat{V}_o}{R_L}$

$= 0$ for $\underline{\underline{\hat{V}_o = \frac{V_{SS}}{2}}}$

Correspondingly, $\eta = \frac{V_{SS}/2}{V_{SS}} = \frac{1}{2}$ or $\underline{\underline{50\%}}$

14.16

$A_v = \frac{R_L}{R_L + R_{out}}$

where, $R_{out} = \frac{r_e}{2} = \frac{V_T}{2 I_Q}$

For $A_v \geq 0.99$ with $R_L \geq 100\ \Omega$,

$0.99 = \frac{100}{100 + R_{out}} \Rightarrow R_{out} = 1\ \Omega$

$\frac{V_T}{2 I_Q} = 1 \Rightarrow I_Q = \underline{\underline{12.5 \text{ mA}}}$

$V_{BB} = 2 V_{BE}$

$= 2\left[0.7 + V_T \ln \frac{12.5}{100}\right]$

$= \underline{\underline{1.296 \text{ V}}}$

14.17 14

$R_{out} = \frac{1}{g_m} \,//\, \frac{1}{g_m} = \frac{1}{2 g_m} = 10\ \Omega$

$\Rightarrow g_m = \frac{1}{20} = 50 \text{ mA/V}$

But, $g_m = \mu C_{ox}\frac{W}{L}(V_{GS} - V_t)$

$50 = 2 \times 100\,(V_{GS} - 1)$

$\Rightarrow V_{GS} = 1.25 \text{ V}$

$V_{GG} = \underline{\underline{2.5 \text{ V}}}$

14.18

$I_Q = \underline{1 \text{ mA}}$

For output of -1 V, $i_L = -\frac{1}{100} = -10$ mA

Using Eq. (14.27),

$i_N^2 - i_L i_N - I_Q^2 = 0$

$i_N^2 + 10\, i_N - 1 = 0$

$i_N = 0.1$ mA

$i_P = \underline{10.1 \text{ mA}}$

Thus v_{EBp} increases by $V_T \ln \frac{10.1}{1} = 0.06$ V

and the input step must be $\underline{\underline{-1.06 \text{ V}}}$

Largest possible positive output from 6 to 10, i.e. $\underline{4 \text{ V}}$.

Largest negative output from 6 to 0, i.e. $\underline{\underline{6 \text{ V}}}$

14.19 14

$R_{out} = r_e/2 = 10\ \Omega$

$\Rightarrow r_e = 20\ \Omega$

$I_Q = \frac{V_T}{r_e} = \frac{25}{20} = 1.25 \text{ mA}$

Thus, $n = \frac{1.25}{0.1} = \underline{\underline{12.5}}$

14.20 $I_Q \simeq I_{bias} = 0.5$ mA, neglecting the base current of Q_N. More precisely,

$$I_Q = I_{bias} - \frac{I_Q}{\beta+1}$$

$$\Rightarrow I_Q = \frac{I_{bias}}{1+\frac{1}{\beta+1}} \simeq 0.98 \times 0.5 = \underline{\underline{0.49 \text{ mA}}}$$

The largest positive output is obtained when all of I_{bias} flows into the base of Q_N, resulting in

$$v_O = (\beta_N + 1) I_{bias} R_L$$
$$= 51 \times 0.5 \times 100\Omega = \underline{\underline{2.55 \text{ V}}}$$

The largest possible negative output voltage is limited by the saturation of Q_P to $-10 + V_{ECsat} = \underline{\underline{-10 \text{ V}}}$.

To achieve a maximum positive output of 10 V without changing I_{bias}, β_N must be

$$10 = (\beta_N + 1) \times 0.5 \times 100\Omega$$

14

$$\Rightarrow \beta_N = \underline{\underline{199}}$$

Alternatively, if β_N is held at 50, I_{bias} must be increased so that

$$10 = 51 \times I_{bias} \times 100\Omega$$

$$\Rightarrow I_{bias} = \underline{\underline{1.96 \text{ mA}}}$$

for which,

$$I_Q = \frac{I_{bias}}{1+\frac{1}{\beta+1}} = \underline{\underline{1.92 \text{ mA}}}$$

14.21 At 20°C, $I_Q = 1 \text{ mA} = I_S e^{(0.6/0.025)}$

$$\Rightarrow I_S \text{ (at 20°C)} = 3.78 \times 10^{-11} \text{ mA}$$

At 70°C, $I_S = 3.78 \times 10^{-11} \times (1.14)^{50}$

$$= 2.64 \times 10^{-8} \text{ mA}$$

At 70°C, $V_T = 25 \; \frac{273+70}{273+20} = 29.3$ mV

Thus, $I_Q \text{ (at 70°C)} = 2.64 \times 10^{-8} \; e^{0.6/0.0293}$

$$= \underline{\underline{20.7 \text{ mA}}}$$

Additional current $= 20.7 - 1 = 19.7$ mA

Additional power $= 2 \times 20 \times 19.7 = \underline{\underline{788 \text{ mW}}}$

Additional temperature rise $= 10 \times 0.788 = \underline{\underline{7.9°\text{C}}}$

At 77.9°: $V_T = \frac{25}{293}(273 + 77.9) = 29.9$ mV

$$I_Q = 3.78 \times 10^{-11} \times (1.14)^{57.9} \; e^{(0.6/0.0299)}$$

14

$$= \underline{\underline{37.6 \text{ mA}}}$$

etc., etc.

14.22 Refer to Fig. P14.22.

$$0.100 = 1 \; (V_{GS3} - 1)^2$$

$$V_{GS3} = 1.316 \text{ V} = |V_{GS4}|$$

$$A_v = \frac{R_L}{R_L + R_{out}} = 0.99$$

$$\frac{1}{1 + R_{out}} = 0.99$$

$$\Rightarrow R_{out} = 0.01 \text{ k}\Omega = 10\,\Omega$$

But $R_{out} = (1/g_{m1}) \,//\, (1/g_{m2})$

$$= \frac{1}{2g_m} \Rightarrow g_m = 50 \text{ mA/V}$$

$$50 = k_1 (V_{GS1} - V_t)$$

$$= k_1 (1.316 - 1)$$

$$\Rightarrow k_1 = \frac{50}{0.316} = 158.2 \text{ mA/V}^2$$

$$n = \frac{158.2}{2} = \underline{\underline{79.1}}$$

14.23 Since the peak positive output current is 200 mA, the base current of Q_N can be as high as $\frac{200}{\beta_N+1} = \frac{200}{51} \simeq 4$ mA. We select $I_{bias} = 5$ mA, thus providing the multiplier with a minimum current of 1 mA.

14

Under quiescent conditions ($v_O = 0$ and $i_L = 0$) the base current of Q_N can be neglected. Selecting $I_R = 0.5$ mA leaves $I_{C1} = 4.5$ mA.

To obtain a quiescent current of 2 mA in the output transistors, V_{BB} should be

$$V_{BB} = 2 V_T \ln \frac{2 \times 10^{-3}}{10^{-13}} = 1.19 \text{ V}$$

Thus

$$R_1 + R_2 = \frac{V_{BB}}{I_R} = \frac{1.19}{0.5} = 2.38 \text{ k}\Omega$$

At a collector current of 4.5 mA, Q_1 has

$$V_{BE1} = V_T \ln \frac{4.5 \times 10^{-3}}{10^{-14}} = 0.671 \text{ V}$$

The value of R_1 can now be determined as

$$R_1 = \frac{0.671}{0.5} = \underline{\underline{1.34 \text{ k}\Omega}}$$

and

$$R_2 = 2.38 - 1.34 = \underline{\underline{1.04 \text{ k}\Omega}}$$

14.24 (a) $V_{BE} = 0.7$ V at 1 mA

14

At 0.5 mA, $V_{BE} = 0.7 + 0.025 \ln \frac{0.5}{1} = 0.683$ V

Thus $R_1 = \frac{0.683}{0.5} = \underline{1.365\ k\Omega}$

and $R_2 = \underline{1.365\ k\Omega}$

(b) For $I_{bias} = 2$ mA, I_C increases to nearly 1.5 mA for which

$V_{BE} = 0.7 + 0.025 \ln \frac{1.5}{1} = 0.710$ V

Note that $I_R = \frac{0.710}{1.365} = 0.52$ mA is very nearly equal to the assumed value of 0.50 mA, thus no further iterations are required.

$V_{BB} = 2 V_{BE} = \underline{1.420\ V}$

(c) For $I_{bias} = 10$ mA, assume that I_R remains constant at 0.5 mA, thus $I_{C1} = 9.5$ mA

and

$V_{BE} = 0.7 + 0.025 \ln \frac{9.5}{1} = 0.756$ V

at which

$I_R = \frac{0.756}{1.365} = 0.554$ mA

Thus,

$I_{C1} = 10 - 0.554 = 9.45$ mA

and $V_{BE} = 0.7 + 0.025 \ln \frac{9.45}{1} = 0.756$ V

Thus $V_{BB} = 2 \times 0.756 = \underline{1.512\ V}$

(d) Now for $\beta = 100$,

14

$I_{R1} = \frac{0.756}{1.365} = 0.554$ mA

$I_{R2} = 0.554 + \frac{9.45}{101} = 0.648$ mA

$I_C = 10 - 0.648 = 9.352$ mA

Thus, $V_{BE} = 0.7 + 0.025 \ln \frac{9.352}{1} = 0.756$ V

$V_{BB} = 0.756 + I_{R2} R_2$

$= 0.756 + 0.648 \times 1.365$

$= \underline{1.641\ V}$

14.25 Power rating $= \frac{130 - 30}{2} = \underline{50\ W}$

$I_{Cav} \leq \frac{50}{20} = \underline{2.5\ A}$

14.26 $\theta_{JA} = \frac{150 - 25}{0.2} = 625\ °C/W = \underline{0.625\ °C/mW}$

At 70°C, Power rating $= \frac{150 - 70}{0.625} = \underline{128\ mW}$

$T_J = 50 + 0.625 \times 100 = \underline{112.5°C}$

14.27 $T_J \leq 50 + 3 \times 30 = \underline{140°C}$

$V_{BE} = 800 - 2 \times (140 - 25) = 570$ mV

$= \underline{0.57\ V}$

14.28 (a) $\theta_{JA} = \frac{T_{Jmax} - T_{A0}}{P_{D0}}$

14

$= \frac{100 - 25}{2} = \underline{37.5°C/W}$

(b) At $T_A = 50°C$,

$P_{Dmax} = \frac{T_{Jmax} - T_A}{\theta_{JA}}$

$= \frac{100 - 50}{37.5} = \underline{1.33\ W}$

(c) $T_J = 25° + 37.5 \times 1 = \underline{62.5°C}$

14.29 $T_C - T_A = \theta_{CA} P_D$

$= (\theta_{CS} + \theta_{SA}) P_D$

$\Rightarrow P_D = \frac{T_C - T_A}{\theta_{CS} + \theta_{SA}} = \frac{90 - 30}{0.5 + 0.1} = \underline{100\ W}$

$T_J - T_C = \theta_{JC} P_D$

$130 - 90 = \theta_{JC} \times 100$

$\Rightarrow \theta_{JC} = \underline{0.4°C/W}$

14.30 $\theta_{JC} = \frac{T_J - T_C}{P_D} = \frac{180° - 50°}{50} = \underline{2.6\ °C/W}$

$T_J - T_S = \theta_{JS} P_D$

$180° - T_S = (\theta_{JC} + \theta_{CS}) P_D$

$\Rightarrow T_S = 180 - (2.6 + 0.6) \times 30 = \underline{84°}$

$T_S - T_A = \theta_{SA} P_D$

14

$84 - 39 = \theta_{SA} \times 30$

$\Rightarrow \theta_{SA} = \underline{1.5\ °C/W}$

Required heat-sink length $= \frac{4.5\ °C/W/cm}{1.5\ °C/W}$

$= \underline{3\ cm}$

14.31 $r_\pi = \frac{n V_T}{I_B} = \frac{2 \times 25 \times 10^{-3}}{0.5} = 0.1\ \Omega$

$r_x \simeq r_i - r_\pi = 0.98 - 0.1 = \underline{0.85\ \Omega}$

14.32 $V_{B'E} = V_{BE} - I_B r_x$

$= 1.05 - 0.19 \times 0.8$

$= 0.898$ V (at $I_C = 5$ A)

For $I_C = 2$ A,

$V_{B'E} = 0.898 + m V_T \ln \frac{2}{5}$

$= 0.898 + 2 \times 0.025 \ln\left(\frac{2}{5}\right)$

$= 0.852$ V

$V_{BE} = 0.852 + I_B r_x$

$= 0.852 + \left(0.19 \times \frac{2}{5}\right) \times 0.8$

$= \underline{0.913\ V}$

14.33 (a) For $R_L = \infty$: 14

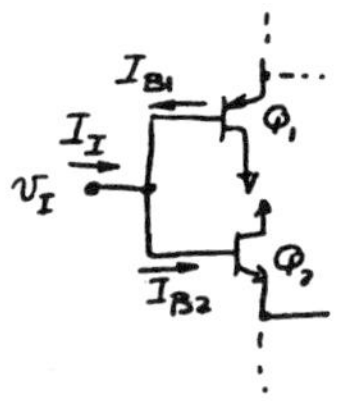

At $v_I = 0$ V,

$$I_{B1} = I_{B2} = \frac{2.87}{200}$$

$$I_I \equiv I_{B2} - I_{B1} = \underline{\underline{0}}$$

At $v_I = +10$ V,

$$I_{B1} = \frac{0.88}{200}\ \text{mA} = 4.4\ \mu A$$

$$I_{B2} = \frac{4.87}{200}\ \text{mA} = 24.4\ \mu A$$

$$I_I \equiv I_{B2} - I_{B1} = \underline{\underline{20\ \mu A}}$$

At $v_I = -10$ V,

$$I_{B1} = \frac{4.87}{200}\ \text{mA} = 24.4\ \mu A$$

$$I_{B2} = \frac{0.88}{200}\ \text{mA} = 4.4\ \mu A$$

$$I_I \equiv I_{B2} - I_{B1} = \underline{\underline{-20\ \mu A}}$$

(b) For $R_L = 100\ \Omega$:

At $v_I = 0$, $I_I = \underline{\underline{0}}$

At $v_I = +10$ V,

$$I_{B1} = \frac{0.38}{200} = 1.9\ \mu A$$

$$I_{B2} = \frac{4.87}{200} = 24.4\ \mu A$$

$$I_I \equiv I_{B2} - I_{B1} = \underline{\underline{22.5\ \mu A}}$$

At $v_I = -10$ V, $I_I = \underline{\underline{-22.5\ \mu A}}$

14.34 For the design process: 14

At $v_I = +5$ V, the voltage across R_1, v_{R1}, is

$$v_{R1} = 10 - 5 - \overset{V_{EB1}}{0.7} = 4.3\ \text{V}$$

and, $i_{R1} = 2 \times 10$ mA (to allow for i_{B3} of as much as 10 mA and only a 2 to 1 variation in i_{E1})

Thus, $R_1 = \frac{4.3}{20} = \underline{\underline{215\ \Omega}}$

Correspondingly,

$$R_2 = \underline{\underline{215\ \Omega}}$$

Now, for $v_I = 0$ and $V_{EB1} \simeq 0.7$ V

$$i_{R1} = \frac{10 - 0.7 - 0}{215} = 43.3\ \text{mA}$$

for which $v_{EB1} = 700 + 25 \ln \frac{43.3}{10} = 736.6$ mV

Since $I_Q = 40$ mA and $I_{S3} = 3 I_{S1}$,

$$v_{BE3} = 700 + 25 \ln \frac{40}{3 \times 10} = 707.2\ \text{mV}$$

Thus, $R_3 = \frac{736.6 - 707.2}{40\ \text{mA}} = \underline{\underline{0.74\ \Omega}}$

Correspondingly,

$$R_4 = \underline{\underline{0.74\ \Omega}}$$

Output Resistance at $v_I = 0$: 14

As a result of 2 paths to output,

$$R_{out} = \frac{1}{2}\left(R_3 + r_{e3} + \frac{r_{e1}}{\beta_3 + 1}\right)$$

where

$$r_{e1} \simeq r_{e3} = \frac{25\ \text{mV}}{40\ \text{mA}} = 0.625\ \Omega$$

Thus, $R_{out} = \frac{1}{2}\left(0.74 + 0.625 + \frac{0.625}{51}\right)$

$$= \underline{\underline{0.69\ \Omega}}$$

Output Voltage for $v_I = +1$ V and $R_L = 2\ \Omega$

To start the solution, let $v_O \simeq 1$ V,

thus $i_L \simeq \frac{1}{2} = 0.5$ A $= 500$ mA

$$i_{B3} \simeq \frac{500}{50} = 10\ \text{mA}$$

$$i_{E1} \simeq \frac{10 - 1 - 0.7}{0.215} - 10 = 28.6\ \text{mA}$$

for which,

$$v_{EB1} = 0.7 + 0.025 \ln \frac{28.6}{10} = 0.726\ \text{V}$$

$$v_{B3} = 1 + 0.726 = 1.726\ \text{V}$$

Assuming $i_{E4} \simeq 0$,

$$v_{BE3} \simeq 700 + 25 \ln \frac{500}{3 \times 10} = 770\ \text{mV}$$

Thus, $i_L = \frac{1.726 - 0.770}{0.74 + 2} = 0.349$ A

for which 14

$$v_O = 2 \times 0.349 = 0.698\ \text{V}$$

Thus, the voltage drop across the series combination of R_4 and the EB junction of Q_4 can be found as follows,

$$V_{B4} = V_{E4} \simeq 1 - 0.74 = 0.26\ \text{V}$$

leaving a drop across the series combination of R_4 and EBJ4 of $0.698 - 0.26 = 0.438$. It follows that $i_{E4} \simeq 0$, as assumed.

Iterate again:

$$i_L \simeq 0.35\ \text{A}$$

$$i_{B3} \simeq \frac{0.35}{51} \simeq 7\ \text{mA}$$

$$i_{E1} \simeq \frac{10 - 1 - 0.73}{0.215} - 7 = 31.5\ \text{mA}$$

$$v_{EB3} = 0.7 + 0.025 \ln \frac{31.5}{10} = 0.729\ \text{V}$$

$$v_{B3} = 1 + 0.729 = 1.729\ \text{V}$$

$$v_{BE3} = 0.7 + 25 \ln \frac{350}{3 \times 10} = 0.761\ \text{V}$$

$$i_L = \frac{1.729 - 0.761}{0.74 + 2} = 0.353\ \text{A}$$

$$v_O = 2 \times 0.353$$ 14

$$= \underline{\underline{0.706\ \text{V}}}$$

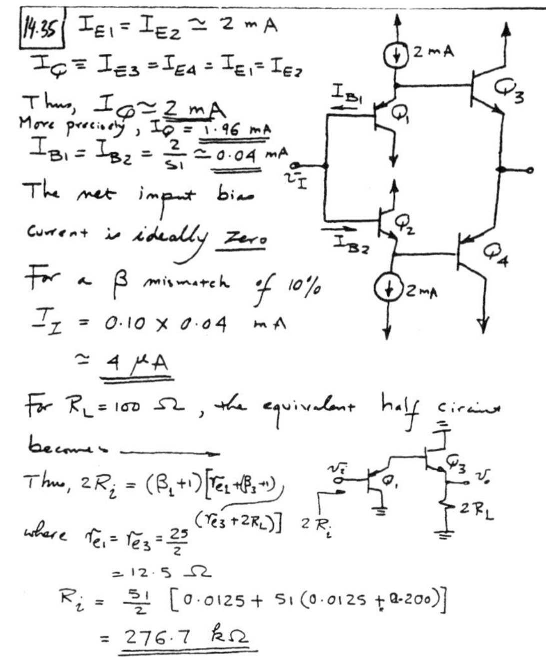

14.35 $I_{E1} = I_{E2} \simeq 2\ mA$

$I_Q = I_{E3} = I_{E4} = I_{E1} = I_{E2}$

Thus, $I_Q \simeq \underline{2\ mA}$

More precisely, $I_Q = \underline{1.96\ mA}$

$I_{B1} = I_{B2} = \frac{2}{51} \simeq \underline{0.04\ mA}$

The net input bias current is ideally $\underline{zero}$

For a β mismatch of 10%

$I_I = 0.10 \times 0.04\ mA$

$\simeq \underline{4\ \mu A}$

For $R_L = 100\ \Omega$, the equivalent half circuit becomes ⟶

Thus, $2R_i = (\beta_1+1)[r_{e1} + (\beta_3+1)(r_{e3}+2R_L)]$

where $r_{e1} = r_{e3} = \frac{25}{2} = 12.5\ \Omega$

$R_i = \frac{51}{2}[0.0125 + 51(0.0125 + 0.200)]$

$= \underline{276.7\ k\Omega}$

14

The gain can be determined using the equivalent half-circuit, as follows:

$$A_v = \frac{v_o}{v_i} = \frac{2R_L}{2R_L + r_{e3} + \frac{r_{e1}}{\beta_3+1}}$$

$$= \frac{2 \times 100}{2 \times 100 + 12.5 + \frac{12.5}{51}} = \underline{0.94\ V/V}$$

14.36

$i_C = \beta_1 i_B + \beta_2(\beta_1+1)\,i_B$

$$\beta_{eq} \equiv \frac{i_C}{i_B} = \underline{\beta_1 + \beta_2(\beta_1+1)}$$

$$\simeq \underline{\beta_1\beta_2} \quad (\text{for } \beta_1 \gg 1 \text{ and } \beta_2 \gg 1)$$

If the composite transistor is operated at a collector current I_C, then

$$I_B = \frac{I_C}{\beta_1 + \beta_2(\beta_1+1)}$$

$$I_{C1} = \frac{\beta_1}{\beta_1 + \beta_2(\beta_1+1)} I_C \simeq I_C/\beta_2$$

14

$$I_{C2} = \frac{\beta_2(\beta_1+1)I_C}{\beta_1 + \beta_2(\beta_1+1)} \simeq I_C$$

Thus,

$$V_{BE2} = V_{BE}(1\ mA) + V_T \ln\left(\frac{I_C}{1\ mA}\right)$$

$$V_{BE1} = V_{BE}(1\ mA) + V_T \ln\left(\frac{I_C}{\beta_2 \times 1\ mA}\right)$$

$$V_{BEeq} = V_{BE1} + V_{BE2}$$

$$= \underline{2V_{BE}(1\ mA) + V_T \ln\left(\frac{I_C^2}{\beta_2}\right)}$$

$$r_{e2} = \frac{V_T}{I_{E2}} \simeq \frac{V_T}{I_{C2}} \simeq \frac{V_T}{I_C}$$

$$r_{e1} = \frac{V_T}{I_{E1}} \simeq \frac{V_T}{I_{C1}} \simeq \frac{V_T}{I_C/\beta_2} = \beta_2 r_{e2}$$

$$r_{\pi eq} = (\beta_1+1)[r_{e1} + (\beta_2+1)r_{e2}]$$

$$\simeq \beta_1[\beta_2 r_{e2} + \beta_2 r_{e2}]$$

$$= \underline{2\beta_1\beta_2 r_{e2}} \simeq \underline{2\beta_1\beta_2\left(\frac{V_T}{I_C}\right)}$$

To determine g_{meq}, apply a signal v_{be} and find the corresponding current i_c,

$$i_c = i_{c1} + i_{c2} = g_{m1} v_{be1} + g_{m2} v_{be2}$$

$$= g_{m1} v_{be} \frac{r_{e1}}{r_{e1} + (\beta_2+1)r_{e2}} + g_{m2} \frac{(\beta_2+1)r_{e2}}{r_{e1} + (\beta_2+1)r_{e2}}$$

14

$$i_c \simeq v_{be}\frac{1}{2\beta_2 r_{e2}} + \frac{\beta_2}{2\beta_2 r_{e2}} v_{be}$$

$$g_{meq} \equiv \frac{i_c}{v_{be}} \simeq \frac{1}{2r_{e2}} = \underline{\frac{1}{2}\frac{I_C}{V_T}}$$

For operation at $I_C = 10\ mA$:

$\beta_{eq} \simeq \beta_1\beta_2 = 50^2 = \underline{2500}$

$V_{BEeq} = 1.4 + 0.025 \ln \frac{10^2}{50}$

$= \underline{1.42\ V}$

$r_{\pi eq} \simeq 2 \times 50 \times 50 \times \frac{25}{10} = \underline{12.5\ k\Omega}$

$g_{meq} \simeq \frac{1}{2} \times \frac{10}{25} = \underline{200\ mA/V}$

14.37 DC analysis:

$5 = I_{C2} \times 1\ k\Omega + \frac{I_{C2}}{\beta^2} \times 1\ M\Omega + 1.4$

$$I_{C2} = \frac{3.6}{1 + \frac{1000}{100 \times 100}}$$

$= 3.3\ mA$

$I_{C1} \simeq \frac{3.3}{100} = 0.033\ mA$

$V_C \simeq \underline{1.7\ V}$

14.37 CONT'D 14

$$i_c = g_{m2} v_{be2}$$

$$= g_{m2} \frac{(\beta_2+1) r_{e2}}{r_{e1} + (\beta_2+1) r_{e2}} v_i$$

But $r_{e1} = \frac{V_T}{I_{E1}} = \frac{V_T}{I_{E2}/(\beta_2+1)} = r_{e2}(\beta_2+1)$

Thus, $i_c \simeq \frac{v_i}{2 r_{e2}}$

$$g_{meq} \simeq \frac{i_c}{v_i} = \frac{1}{2 r_{e2}}$$

$$= \frac{1}{2 \times (25/3.3)} = \underline{\underline{66\ \text{mA/V}}}$$

$$v_o \simeq -i_c \times 1\,k\Omega$$

$$= -g_{meq} v_i \times 1\,k\Omega$$

$$\frac{v_o}{v_i} = -66 \times 1 = \underline{\underline{-66\ \text{V/V}}}$$

To find R_{in},

$$i_i = i_{b1} + i_{1\text{-}M\Omega\ \text{resistor}}$$

$$\simeq \frac{i_c}{\beta^2} + \frac{v_i - v_o}{1\,M\Omega}$$

$$= \frac{v_i}{2\beta^2 r_{e2}} + \frac{67 v_i}{1\,M\Omega}$$

$$= v_i \left[\frac{1}{2 \times 100^2 \times \frac{25}{3.3}} + \frac{67}{1\,M\Omega}\right] = 73.6\, v_i\ (\mu A)$$

$$R_{in} \equiv \frac{v_i}{i_i} = \frac{1}{73.6}\,M\Omega = \underline{\underline{13.6\ k\Omega}}$$

14

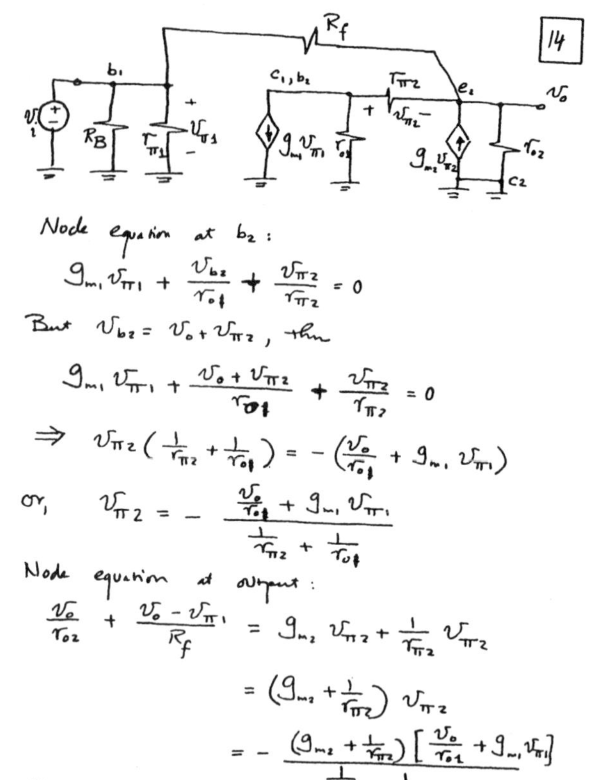

Node equation at b_2:

$$g_{m1} v_{\pi 1} + \frac{v_{b2}}{r_{o1}} + \frac{v_{\pi 2}}{r_{\pi 2}} = 0$$

But $v_{b2} = v_o + v_{\pi 2}$, then

$$g_{m1} v_{\pi 1} + \frac{v_o + v_{\pi 2}}{r_{o1}} + \frac{v_{\pi 2}}{r_{\pi 2}} = 0$$

$$\Rightarrow v_{\pi 2}\left(\frac{1}{r_{\pi 2}} + \frac{1}{r_{o1}}\right) = -\left(\frac{v_o}{r_{o1}} + g_{m1} v_{\pi 1}\right)$$

or, $$v_{\pi 2} = -\frac{\frac{v_o}{r_{o1}} + g_{m1} v_{\pi 1}}{\frac{1}{r_{\pi 2}} + \frac{1}{r_{o1}}}$$

Node equation at output:

$$\frac{v_o}{r_{o2}} + \frac{v_o - v_{\pi 1}}{R_f} = g_{m2} v_{\pi 2} + \frac{1}{r_{\pi 2}} v_{\pi 2}$$

$$= \left(g_{m2} + \frac{1}{r_{\pi 2}}\right) v_{\pi 2}$$

$$= -\frac{\left(g_{m2} + \frac{1}{r_{\pi 2}}\right)\left[\frac{v_o}{r_{o1}} + g_{m1} v_{\pi 1}\right]}{\frac{1}{r_{\pi 2}} + \frac{1}{r_{o1}}}$$

Substituting $v_{\pi 1} = v_i$ and collecting terms, results in

14

$$v_o\left[\frac{1}{r_{o2}} + \frac{1}{R_f} + \frac{\left(g_{m2} + \frac{1}{r_{\pi 2}}\right)}{r_{o1}\left(\frac{1}{r_{\pi 2}} + \frac{1}{r_{o1}}\right)}\right]$$

$$= -v_i\left[\frac{g_{m1}\left(g_{m2} + \frac{1}{r_{\pi 2}}\right)}{\frac{1}{r_{\pi 2}} + \frac{1}{r_{o1}}} - \frac{1}{R_f}\right]$$

$$\frac{v_o}{v_i} = -\frac{\frac{g_{m1}\left(g_{m2} + \frac{1}{r_{\pi 2}}\right)}{\frac{1}{r_{\pi 2}} + \frac{1}{r_{o1}}} - \frac{1}{R_f}}{\frac{1}{r_{o2}} + \frac{1}{R_f} + \frac{g_{m2} + \frac{1}{r_{\pi 2}}}{r_{o1}\left(\frac{1}{r_{\pi 2}} + \frac{1}{r_{o1}}\right)}}$$

Since $r_{\pi 2} \ll r_{o1}$,

$$\frac{v_o}{v_i} \simeq -\frac{g_{m1}(g_{m2} r_{\pi 2} + 1) - \frac{1}{R_f}}{\frac{1}{r_{o2}} + \frac{1}{R_f} + \frac{1}{r_{o1}}(g_{m2} r_{\pi 2} + 1)}$$

$$= -\frac{g_{m1}(\beta_2+1) - \frac{1}{R_f}}{\frac{1}{r_{o2}} + \frac{1}{R_f} + \frac{1}{r_{o1}}(\beta_2+1)}$$

Since $\frac{1}{R_f} \ll g_{m1}(\beta_2+1)$

$$\frac{v_o}{v_i} \simeq -\frac{g_{m1}(\beta_2+1)}{\left(\frac{1}{r_{o2}} + \frac{1}{R_f}\right) + \frac{1}{r_{o1}}(\beta_2+1)}$$

Substituting $\beta_2 = \beta_N$ and noting that $\beta_N \gg 1$,

$$\frac{v_o}{v_i} \simeq -g_{m1} \frac{1}{\frac{1}{\beta_N}\left(\frac{1}{r_{o2}} + \frac{1}{R_f}\right) + \frac{1}{r_{o1}}}$$

$$= -g_{m1}\left[r_{o1} \,//\, \beta_N (r_{o2} // R_f)\right]$$

Q.E.D.

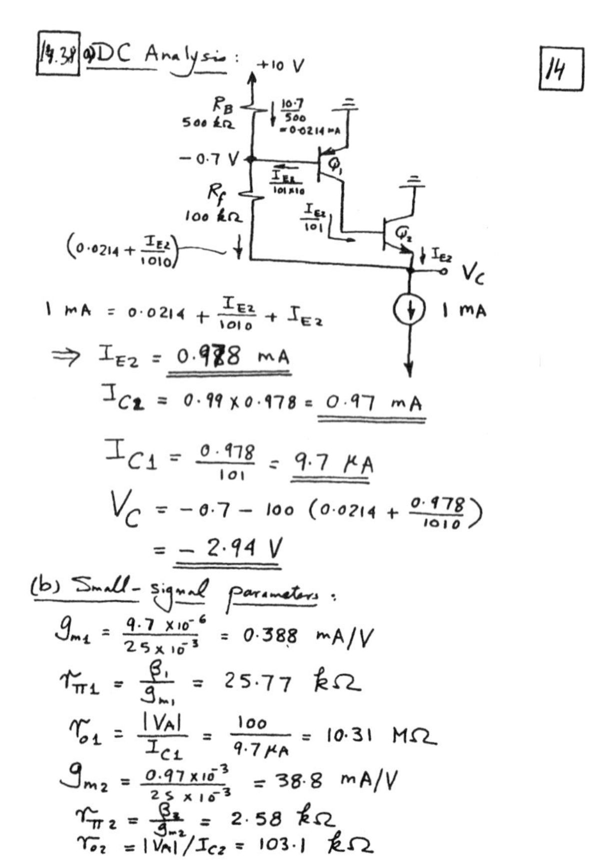

14.38 (a) DC Analysis: 14

$$1\ \text{mA} = 0.0214 + \frac{I_{E2}}{1010} + I_{E2}$$

$$\Rightarrow I_{E2} = \underline{\underline{0.978\ \text{mA}}}$$

$$I_{C2} = 0.99 \times 0.978 = \underline{\underline{0.97\ \text{mA}}}$$

$$I_{C1} = \frac{0.978}{101} = \underline{\underline{9.7\ \mu A}}$$

$$V_C = -0.7 - 100\left(0.0214 + \frac{0.978}{1010}\right)$$

$$= \underline{\underline{-2.94\ V}}$$

(b) Small-signal parameters:

$$g_{m1} = \frac{9.7 \times 10^{-6}}{25 \times 10^{-3}} = 0.388\ \text{mA/V}$$

$$r_{\pi 1} = \frac{\beta_1}{g_{m1}} = 25.77\ k\Omega$$

$$r_{o1} = \frac{|V_A|}{I_{C1}} = \frac{100}{9.7\,\mu A} = 10.31\ M\Omega$$

$$g_{m2} = \frac{0.97 \times 10^{-3}}{25 \times 10^{-3}} = 38.8\ \text{mA/V}$$

$$r_{\pi 2} = \frac{\beta_2}{g_{m2}} = 2.58\ k\Omega$$

$$r_{o2} = |V_A|/I_{C2} = 103.1\ k\Omega$$

14.38 CONT'D 14

(c) $\frac{v_o}{v_i} = -0.388\,[10.31 \times 10^3 // 100\,(103.1 // 100)]$

$= -1320$ V/V

$R_{in} = R_B // r_{\pi 1} // \left[v_i / \frac{v_i - v_o}{R_f} \right]$

$= 500 // 25.77 // \left[\frac{R_f}{1 - \frac{v_o}{v_i}} \right]$

$= 500 // 25.77 // \frac{100}{1 + 1320}$

$= 500 // 25.77 // 0.0757$

$= 75.5\ \Omega$

14.39 Refer to Fig. 14.27 (p. 1258).

The quiescent current through Q_2 and Q_4 is to be 2 mA. Thus

$V_{BE2} = V_{BE4} = 0.7 + 0.025 \ln\left(\frac{2}{10}\right) = 0.660$ V

For Q_1 and Q_3, $I_C \simeq \frac{2}{100} = 0.02$ mA, thus

$V_{BE1} = V_{EB3} = 0.7 + 0.025 \ln \frac{0.02}{1} = 0.602$ V

$I_{B1} = \frac{20\ \mu A}{100} = 0.2\ \mu A$

$I_{bias} = 100 \times 0.2 = 20\ \mu A$

$I_{R_1, R_2} = \frac{1}{10} \times 20\ \mu A = 2\ \mu A$

$I_{C5} = 20 - 2 = 18\ \mu A$

$V_{BE5} = 0.7 + 0.025 \ln \frac{0.018}{1} = 0.600$ V

$V_{BB} = V_{BE1} + V_{BE2} + V_{EB3} = 1.864$ V

14

$R_1 + R_2 = \frac{1.864}{2\ \mu A} = 932$ kΩ

$R_1 = \frac{0.600}{2\ \mu A} = 300$ kΩ

$R_2 = 932 - 300 = 632$ kΩ

For $v_O = -10$ V and $R_L = 1$ kΩ:

$i_L = \frac{-10}{1} = -10$ mA

Assume that the current through Q_2 becomes almost zero, thus

$I_{C4} = 10$ mA

i.e. the current through Q_4 increases by a factor of 5. It follows that the current through Q_3 must increase by the same factor, thus V_{EB3} becomes

$V_{EB3} = 0.602 + 0.025 \ln 5$

$= 0.642$ V (an increase of 0.04 V)

Let us check the current through Q_2. Since we assumed Q_1 and Q_2 to be almost cut off, all of I_{bias} now flows through the V_{BE} multiplier, an increase of 0.2 μA. Assuming that most of this increase occurs in I_{C5}, V_{BE5} becomes

$V_{BE5} = 0.7 + 0.025 \ln \frac{0.0182}{1} = 0.600$ V

Thus the voltage across the V_{BE}-multiplier remains approximately constant and the voltage $(V_{BE1} + V_{BE2})$ decreases by the same value that V_{EB3} increases by. That is

14

$V_{BE1} + V_{BE2} = 0.660 + 0.602 - 0.04$

Since the current through each of Q_1 and Q_2 decreases by the same factor (call it m),

$0.025 \ln m + 0.025 \ln m = -0.04$ V

$\Rightarrow m = 0.45$

Thus $I_{C2} = 0.45 \times 2 = 0.9$ mA

New iteration: $I_{C4} = 10.9$ mA (an increase by a factor $\simeq 5.5$).

$V_{EB3} = 0.602 + 0.025 \ln 5.5$

$= 0.645$ V

$v_I \simeq -10.645$ V

For $v_O = +10$ V and $R_L = 1$ kΩ:

Assume that Q_4 is now conducting a negligible current. Thus, $I_{C2} \simeq I_L = 10$ mA. i.e. the current through each of Q_1 and Q_2 increase by a factor of 5. Thus

14

$V_{BE2} = 0.66 + 0.025 \ln 5$

$= 0.700$ V

$V_{BE1} = 0.602 + 0.025 \ln 5 = 0.642$ V

$I_{B1} = 5 \times 0.2 = 1\ \mu A$

Thus the current through the multiplier becomes 19 μA, and assuming that most of the decrease occurs in I_{C5},

$V_{BE5} = 0.7 + 0.025 \ln \frac{0.017}{1}$

$= 0.598$ V

Thus the voltage across the multiplier becomes

$V_{BB} = 0.598 \times \frac{932}{300} = 1.858$ V

It follows that V_{EB3} becomes

$V_{EB3} = 1.858 - 0.700 - 0.642 = 0.516$ V

i.e. V_{EB3} decreases by $0.600 - 0.516 = 0.084$ V and correspondingly I_{C3} decreases by a factor of $e^{\frac{-0.084}{0.025}} = 0.035$. Thus I_{C4} becomes $0.035 \times 2 = 0.07$ mA, close to the zero value assumed. Thus no further iteration are required and

$v_I \simeq 10 + 0.7 + 0.642 - 1.858$

$= +9.484$ V

14.40 Refer to Exercise 14.13 and Fig. 14.28.

Now Q_5 has $I_S = 10^{-13}$ A. Thus,

$$2 \times 10^{-3} = 10^{-13}\, e^{V_{BE}/V_T}$$

$$V_{BE} = 0.025 \ln \frac{2\times10^{-3}}{10^{-13}} = 0.593\ \text{V}$$

$$R_{E1} = \frac{0.593}{150\ \text{mA}} \simeq \underline{\underline{4\ \Omega}}$$

For a normal peak current of 100 mA, the voltage drop across R_{E1} is $\underline{400\ \text{mV}}$ and its collector current is $10^{-13}\, e^{400/25} = \underline{\underline{0.89\ \mu A}}$

14.41 Refer to Exercise 14.13 and Fig. 14.28.

$$2 \times 10^{-3} = 10^{-14}\, e^{V_{BE}/V_T}$$

$$\Rightarrow V_{BE} = 0.650\ \text{V}$$

$$R_{E1} = \frac{0.650\ \text{V}}{50\ \text{mA}} = \underline{13\ \Omega}$$

For a peak output current of 33.3 mA,

$$V_{BE} = 13 \times 33.3 = \underline{433\ \text{mV}}$$

$$I_{C5} = 10^{-14}\, e^{433/25} = \underline{\underline{0.33\ \mu A}}$$

14.42 Refer to Fig. P.14.42. **14**

$$2 \times 10^{-3} = 10^{-14}\, e^{V_{EB5}/V_T}$$

$$V_{EB5} = 0.025 \ln(2\times10^{11}) = 0.650\ \text{V}$$

$$R = \frac{0.650\ \text{V}}{150\ \text{mA}} = \underline{\underline{4.3\ \Omega}}$$

For a peak output current of 100 mA,

$$V_{EB5} = \underline{430\ \text{mV}}$$

$$I_{C5} = 10^{-14}\, e^{430/25} = \underline{\underline{0.3\ \mu A}}$$

14.43 Refer to Fig. 14.29 (p.1260)

At 125°C, $V_Z = 6.8 + (125-25)\times 2 = 7.0$ V

$$V_{E1} = 7.0 - (0.7 - 100\times0.002) = 6.5\ \text{V}$$

$$V_{BE2} = 0.5\ \text{V}$$

$$R_2 = \frac{0.5\ \text{V}}{100\ \mu A} = \underline{\underline{5\ k\Omega}}$$

$$R_1 = \frac{6.5-0.5}{100\ \mu A} = \underline{\underline{60\ k\Omega}}$$

At 25°C, $V_Z = 6.8$ V, $V_{E1} = 6.8 - 0.7 = 6.1$ V

$$V_{B2} = 6.1 \times \frac{5}{60+5} = 0.469\ \text{V}$$

$$I_{C2} = 100\, e^{(469-700)/25} = \underline{\underline{0.01\ \mu A}}$$

14.44 Refer to the circuit of Fig. 14.30 (p 1262). **14**

Resistors R_2 and R_3 control the gain,

$$A_v = -\frac{2R_2}{R_3} \qquad \text{(See analysis on pages 681-682)}$$

Resistor R_3 controls the gain alone. Resistor R_2 affects both the gain and the dc output level. To see the latter point, Equate I_3 and I_4 from equations (9.43) and (9.44) to obtain

$$\frac{V_S - 3V_{EB}}{R_1} = \frac{V_O - 2V_{EB}}{R_2}$$

$$\Rightarrow V_O = 2V_{EB} + \frac{R_2}{R_1}V_S - \frac{3R_2}{R_1}V_{EB} = \frac{R_2}{R_1}V_S + \left(2 - \frac{3R_2}{R_1}\right)V_{EB}$$

For $V_O \simeq \frac{1}{3}V_S$, select $\frac{R_2}{R_1} = \frac{1}{3}$

$$R_2 = \frac{R_1}{3} = \frac{50}{3} = \underline{\underline{16.7\ k\Omega}}$$

To keep the gain unchanged, we must change R_3 so that

$$\frac{2R_2}{R_3} = 50$$

$$R_3 = \frac{2\times(50/3)}{50} = \frac{2}{3} = \underline{\underline{0.67\ k\Omega}}$$

14.45 Refer to the circuit in Fig. 14.30 **14**

$V_{B1} \simeq 0$

$V_{E1} \simeq +0.7$ V

$V_{E3} \simeq +1.4$ V

$V_{C10} = 20 - 0.7 = 19.3$ V

$$I_{E3} = \frac{19.3 - 1.4}{50} = 0.358\ \text{mA}$$

$$I_{B3} = I_{E1} = \frac{0.358}{21} = 17.05\ \mu A$$

$$I_{B1} = \frac{17.05}{21} = 0.81\ \mu A$$

$V_{B1} = 0.81\ \mu A \times 150\ k\Omega = 0.122\ \text{V} \simeq 0$ ✓

i.e. $I_{E1} = I_{E2} \simeq \underline{17\ \mu A}$

$I_{E3} = I_{E4} \simeq \underline{358\ \mu A}$

$I_{E5} \simeq I_{E6} = \frac{20}{21} \times 358 = \underline{\underline{341\ \mu A}}$

$I_{R_2} = I_{R_1} = 358\ \mu A$

$V_O = 0.12 + 1.4 + 25\ k\Omega \times 0.358\ \text{mA}$

$= \underline{\underline{10.5\ \text{V}}}$

14.46 The equivalent common-mode half circuit is approximately

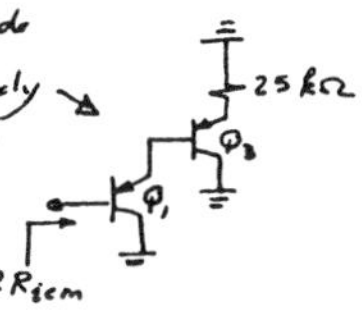

From the results of Problem 14.45 above: $I_{E1} = 17\ \mu A$

$I_{E3} = 0.358\ mA$

$2R_{icm}$ 25 kΩ Q_1 Q_3

$$r_{o1} = \frac{|V_A|}{I_{C1}}$$

14

Assume $|V_A|$ for all transistors = 100 V.

$$r_{o1} \simeq \frac{100}{17\mu A} = 5.9\ M\Omega$$

$$r_{o3} = \frac{100}{0.358} = 279.3\ k\Omega$$

Assume $r_\mu = 10\beta r_o$

$$r_{\mu 1} = 10 \times 20 \times 5.9 = 1180\ M\Omega$$

$$r_{\mu 3} = 10 \times 20 \times 279.3 = 55.9\ M\Omega$$

Now,

$$2R_{icm} = r_{\mu 1} \,//\, (\beta+1)\left[r_{o1} \,//\, r_{\mu 3} \,//\, (\beta+1)(25 \,//\, r_{o3})\right]$$

$$= 1180\ M\Omega \,//\, 21\left[5.9 \,//\, 55.9 \,//\, 21(25//279.3)\times 10^{-3}\right] M\Omega$$

$$\simeq 21 \times 21 \times (25//279.3)\ k\Omega$$

$$= 10.1\ M\Omega$$

If we neglect all r_o's and r_μ's

$$2R_{icm} \simeq (\beta+1)^2 \times 25\ k\Omega$$

$$= 11\ M\Omega \quad \text{(close enough!)}$$

$$R_{icm} = 5.5\ M\Omega$$

To find R_{id} we use the equivalent differential half-circuit →

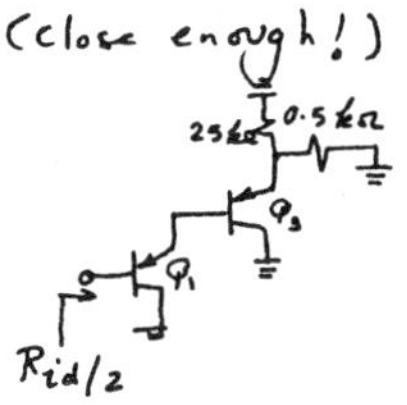

25 kΩ 0.5 kΩ Q_1 Q_3 $R_{id}/2$

$$R_{id}/2 = (\beta+1)\left[r_{e1} + (\beta+1)(r_{e3} + 0.5//25)\right]$$

14

$$r_{e1} = \frac{25\ mV}{17\ \mu A} = 1.47\ k\Omega$$

$$r_{e3} = \frac{25\ mV}{0.358\ mA} = 69.8\ \Omega$$

$$R_{id} = 2 \times 21\left[1.47 + 21(0.0698 + 0.49)\right]$$

$$= 556\ k\Omega$$

To find $G_{m1} \equiv \frac{2i}{v_{id}}$

25 kΩ 0.5 kΩ Q_1 Q_3 Q_5 Q_6 $-v_{id}/2$ i $2i$

$$i_{c3} = \alpha_3 i_{e3}$$

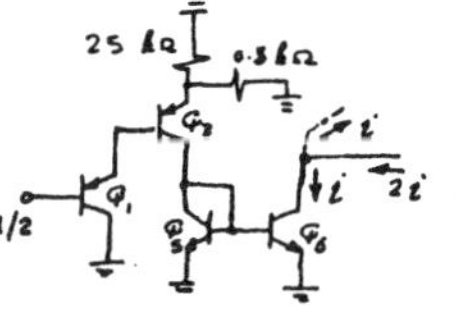

$$= \alpha_3 \frac{v_{id}/2}{(0.5//25) + r_{e3} + \frac{r_{e1}}{\beta+1}}$$

$$= \frac{20}{21} \times \frac{1}{2} \frac{v_{id}}{(0.5//25) + 0.069 + \frac{1.47}{21}}$$

$$= 0.757\ v_{id}$$

$$i = \frac{i_{c3}}{1 + \frac{2}{\beta_{npn}}} = 0.98\ i_{c3} = 0.98 \times 0.757\ v_{id}$$

$$G_{m1} \equiv \frac{2i}{v_{id}} = 2 \times 0.98 \times 0.757 = 1.5\ mA/V$$

14.47 Using Fig. 14.32 (p 1264) for 8 Ω load, we see that $V_S = 16$ V allows more than 1.5 W power dissipation for some input signals. Thus we use

$$V_S = 14\ V$$

For THD = 3%, $P_{L\max} = 1.9\ W$

$$1.9 = V_o^2/R_L = V_o^2/8$$

14

$$V_o = \sqrt{8 \times 1.9}$$

Peak-to-Peak output sinusoid $= 2\sqrt{2}\sqrt{8\times 1.9} = 11\ V$

14.48 $f_t \simeq \dfrac{G_{m1}}{2\pi C}$

$$= \frac{1.6 \times 10^{-3}}{2\pi \times 10 \times 10^{-12}} = 25.5\ MHz$$

With feedback that results in a closed-loop gain of 50,

$$f_b = \frac{25.5}{50} = 509\ kHz$$

14.49 Refer to Fig. 14.33

For $i_L = 1$ A, $i_{C5} \simeq 1$ A and $i_{B5} = \frac{1A}{50} = 20$ mA

For $i_L = 20$ mA, $i_{E3} = 0.9 \times 20 = 18$ mA, (90%)

$i_{C5} = 2$ mA $\quad i_{B5} = \frac{2}{50} = 0.04$ mA

$i_{C3} = \frac{50}{51} \times 18 = 17.65$ mA

$i_{R3} = i_{C3} - i_{B5} = 17.61$ mA

Thus, $R_3 = \frac{0.7}{17.61} = 39.8 \simeq 40\ \Omega$

Similarly, $R_4 = 40\ \Omega$

Since $i_{B5} \leq 20$ mA, $i_{C3} \leq \frac{0.7V}{40\ \Omega} + 20$ mA

i.e. $i_{C3} \leq 37.5$ mA

$$i_{B3} \leq \frac{37.5}{50} = 0.75\ mA$$

14

Allowing for a factor of safety of 2, we select R_1 so that the current through it is 1.5 mA. Now, for $V_o = 11$ V, $V_{E1} = 11.7$ V,

$$R_1 = \frac{15 - 11.7}{1.5} = 2.2\ k\Omega$$

Similarly,

$$R_2 = 2.2\ k\Omega$$

14.50 For $v_I \geq 0$:

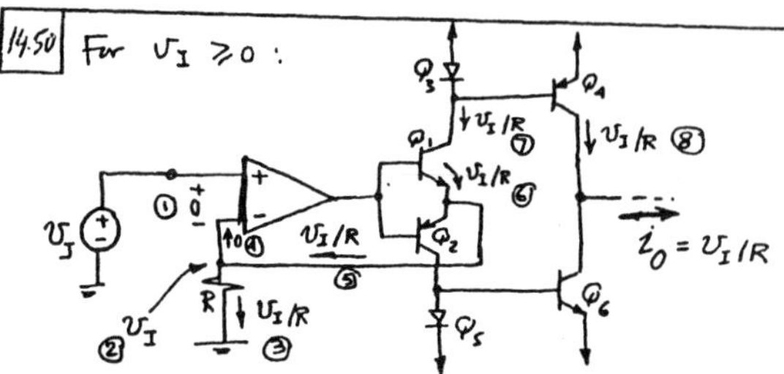

For $v_I \leq 0$, Q_2, Q_5 and Q_6 take the roles of Q_1, Q_3 and Q_4, respectively.

For $\beta = 100$, $i_{C1} = 0.99\, v_I / R$ and

$$i_{C4} = i_{C1} \frac{1}{1+\frac{2}{\beta}} = 0.99 \frac{v_I}{R} \frac{1}{1+\frac{2}{100}} = 0.97 \frac{v_I}{3}$$

That is, the current is lower than the ideal value by approximately **3 %**.

14.51 Refer to Fig. 14.34

$$\frac{v_O}{v_I} = 2K = 2\left(1+\frac{R_2}{R_1}\right) = 10$$

$$\Rightarrow \frac{R_2}{R_1} = 4$$

$$R_2 = \underline{40\ \text{k}\Omega}$$

Also, $K = \frac{R_4}{R_3} = 5$

$$\Rightarrow R_4 = \underline{50\ \text{k}\Omega}$$

14.52

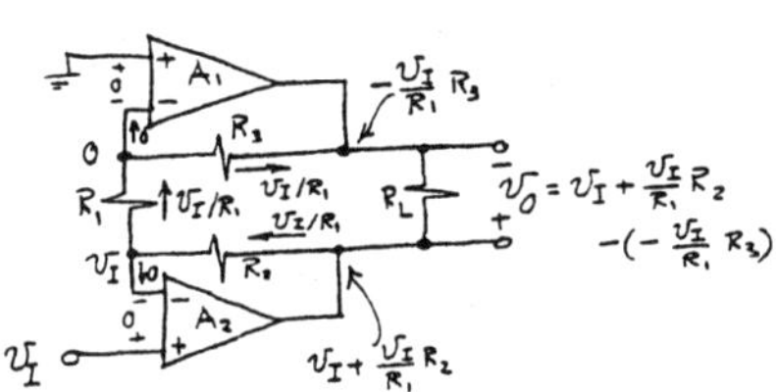

$$\frac{v_O}{v_I} = 1 + \frac{R_2}{R_1} + \frac{R_3}{R_1}$$

The largest sine-wave output is obtained when the voltage at the output of one op amp (say A_2) is +13 V and that at the output of the other op amp (A_1) is −13 V, resulting in a **26-V** peak output (**52 V** peak-to-peak).

For $\frac{v_O}{v_I} = 10 = 1 + \frac{R_2 + R_3}{R_1}$

and selecting $R_1 = 1\ \text{k}\Omega$

$R_2 + R_3 = 9\ \text{k}\Omega$

Now to keep the outputs complementary

$$\frac{R_3}{R_1} = 1 + \frac{R_2}{R_1}$$

$\Rightarrow R_3 = 1 + R_2$

Thus $R_2 = \underline{4\ \text{k}\Omega}$ & $R_3 = \underline{5\ \text{k}\Omega}$

14

14.53 Normally, (Eq. 14.46)

$$i_D = \frac{1}{2} \mu_n C_{ox} \frac{W}{L} (v_{GS} - V_t)^2$$

For velocity saturation, (Eq. 14.47)

$$i_D = \frac{1}{2} C_{ox} W U_{sat} (v_{GS} - V_t)$$

Equating at the two expressions (at the boundary) gives $\mu_n \frac{1}{L} (v_{GS} - V_t) = U_{sat}$

$$\Rightarrow L = \frac{\mu_n (v_{GS} - V_t)}{U_{sat}}$$

For $\mu_n = 500\ \text{cm}^2/\text{V.s.}$, $v_{GS} = 5$ V, $V_t = 2$ V, $U_{sat} = 5 \times 10^6$ cm/s,

$$L = \frac{500\ (5-2)}{5 \times 10^6} = 3 \times 10^{-4}\ \text{cm} = \underline{3\ \mu\text{m}}$$

Velocity saturation begins at

14

$$i_D = \frac{1}{2} C_{ox} W U_{sat} (v_{GS} - V_t)$$

$$= \frac{1}{2} \times 400 \times 10^{-6} \times 10^5 \times 10^{-6} \times 5 \times 10^6 \times 10^{-2} (5-2)$$

$$= \underline{3\ \text{A}}$$

At high currents,

$$g_m \equiv \frac{\partial i_D}{\partial v_{GS}} = \frac{1}{2} C_{ox} W U_{sat}$$

$$= \frac{1}{2} \times 400 \times 10^{-6} \times 10^5 \times 10^{-6} \times 5 \times 10^6 \times 10^{-2}$$

$$= \underline{1\ \text{A/V}}$$

14

14.54 Refer to the circuit in Fig. 14.38

For $I_{Q_N} = I_{Q_P} = 10\,mA = \mu C_{ox}\frac{W}{L}(V_{GS} - V_t)^2$

$$10 = 100\,(V_{GS} - 2)^2$$

$$\Rightarrow V_{GS} = 2.32\,V$$

$$V_R = 2|V_{GS}| = 4.63\,V$$

For $I_R = 10\,mA$,

$$R = \frac{4.63\,V}{10\,mA} = \underline{\underline{463\,\Omega}}$$

$$V_{BB} = 4.63 + 4 \times 0.7 = 7.43\,V$$

$$I_{R_2} = I_{R_4} = \frac{100}{2} = 50\,\mu A$$

$$R_2 = R_4 = \frac{700\,mV}{50\,\mu A} = \underline{\underline{14\,k\Omega}}$$

Now, since V_{G6} changes by $2 \times -3\,mV/°C = -6\,mV/°C$ while V_{BE1}, V_{BE2}, V_{BE3} and V_{BE4} remain constant, V_{BB} changes by $-6\,mV/°C$. But the voltage across the Q_5 multiplier remains constant. Thus the voltage across the Q_6 multiplier should by made to change by $-6\,mV/°C$ which can be achieved by making 14

$$1 + \frac{R_3}{R_4} = 3$$

$$\Rightarrow R_3 = 2R_4 = \underline{\underline{28\,k\Omega}}$$

The voltage across the Q_5 multiplier is

$$V_{BB} - 3V_{BE6} = 7.43 - 2.1 = 5.33\,V$$

Thus,

$$5.33 = \left(1 + \frac{R_1}{R_2}\right) \times 0.7$$

$$\Rightarrow \frac{R_1}{R_2} = 6.61$$

But $R_2 = 14\,k\Omega$, thus

$$R_1 = \underline{\underline{92.6\,k\Omega}}$$

List of Transparency Masters

Chapter 1
1.13, 1.14, Table 1.1, 1.23, 1.24

Chapter 2
2.3, 2.6, 2.8, 2.10, 2.11, 2.12, 2.13, 2.15, 2.16, 2.17, 2.18, 2.19, 2.20, 2.22, 2.26, 2.27, 2.39, 2.44

Chapter 3
3.1, 3.3, 3.10, 3.11, 3.12, 3.13, 3.15, 3.16, 3.17, Table 3.1, 3.20, 3.21, 3.22, 3.25, 3.27, 3.29, 3.35, 3.45, 3.50

Chapter 4
4.1, 4.2, 4.3, 4.4, 4.5, 4.6, 4.8, 4.9, 4.11, 4.12, 4.13, 4.14, 4.15, 4.16, 4.26, 4.28, 4.30, 4.36, Table 4.2, 4.43, 4.44, 4.46, Table 4.4, 4.47, 4.49, 4.50, 4.51, 4.52, 4.55, 4.56, 4.57

Chapter 5
5.1, 5.2, Table 5.1, 5.3, 5.4, 5.18, 5.19, 5.21, 5.22, 5.23, Table 5.3, 5.26, 5.27, 5.28, 5.30, 5.48, 5.49, 5.53, 5.54, Table 5.4, 5.60, 5.61, 5.62, 5.63, 5.64, 5.65, Table 5.6, 5.71, 5.72, 5.73

Chapter 6
Table 6.1, Table 6.2, Table 6.3, 6.5, 6.6, 6.18, 6.26, 6.27, 6.28, 6.29, 6.30, 6.31, 6.33, 6.34, 6.35, 6.36, 6.37, 6.39, 6.40, 6.43, 6.44, 6.47, 6.50, 6.52, 6.60, 6.61

Chapter 7
7.6, 7.7, 7.8, 7.9, 7.10, 7.13, 7.14, 7.15, 7.16, 7.17, 7.18, 7.28, 7.29, 7.30, 7.31, 7.32, 7.35, 7.36, 7.37, 7.39, 7.40, 7.43, 7.44, 7.45, 7.46, 7.47, 7.48, 7.49, 7.50

Chapter 8
8.4, 8.8, 8.9, 8.10, 8.11, 8.12, 8.13, 8.14, 8.15, 8.16, 8.17, 8.18, 8.19, 8.20, 8.22, 8.23, 8.24, 8.25, 8.36, 8.37, 8.38

Chapter 9
9.1, 9.8, 9.9, 9.11, 9.12, 9.13, 9.14, 9.16, 9.17, 9.19, 9.20, 9.24, 9.28, 9.29, 9.30, 9.36, 9.38, 9.39, 9.40, 9.44, 9.46

Chapter 10
10.2, 10.3, 10.4, 10.5, 10.12, 10.13, 10.16, 10.17, 10.21, 10.33, 10.34, 10.35, 10.36, 10.37

Chapter 11
11.1, 11.3, 11.5, 11.7. 11.8, 11.10, 11.13, 11.15, 11.16, 11.17, 11.18, 11.23, 11.24, 11.25, 11.26, 11.27, 11.28, 11.29, 11.30, 11.31, 11.32, 11.34, 11.36, 11.37, 11.39, 11.45

Chapter 12
12.1, 12.2, 12.5, 12.6, 12.8, 12.9, 12.10, 12.12, 12.13, 12.14, 12.16, 12.18, 12.20, 12.22, 12.26, 12.37, 12.47, 12.48

Chapter 13
13.3, 13.5, 13.6, 13.8, 13.9, 13.10, 13.11, 13.19, 13.22, 13.24, 13.25, 13.26, 13.28, 13.29, 13.30

Chapter 14
14.1, 14.2, 14.3, 14.5, 14.7, 14.15, 14.16, 14.30, 14.31, 14.33, 14.34, 14.35, 14.36, 14.38

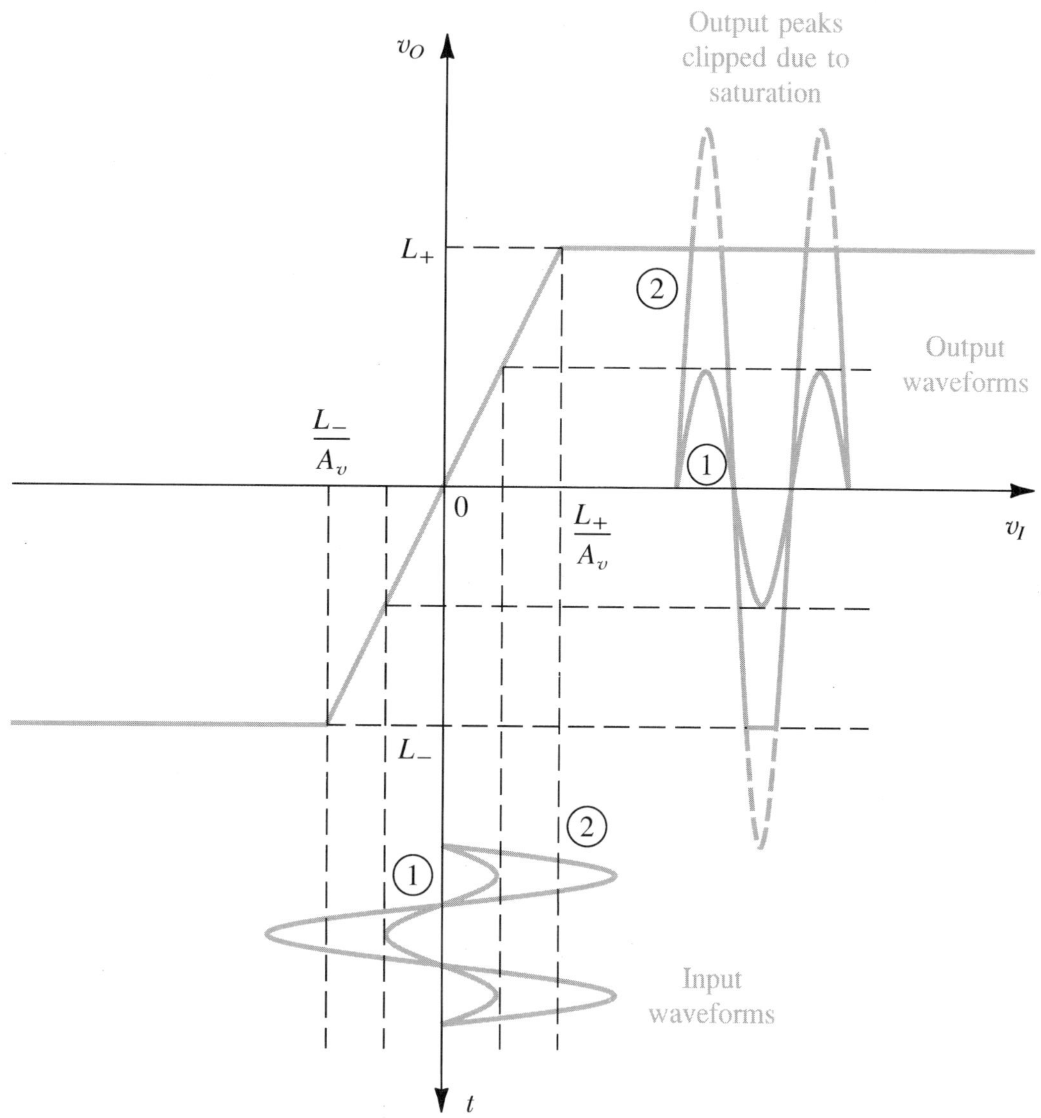

FIGURE 1.13 An amplifier transfer characteristic that is linear except for output saturation.

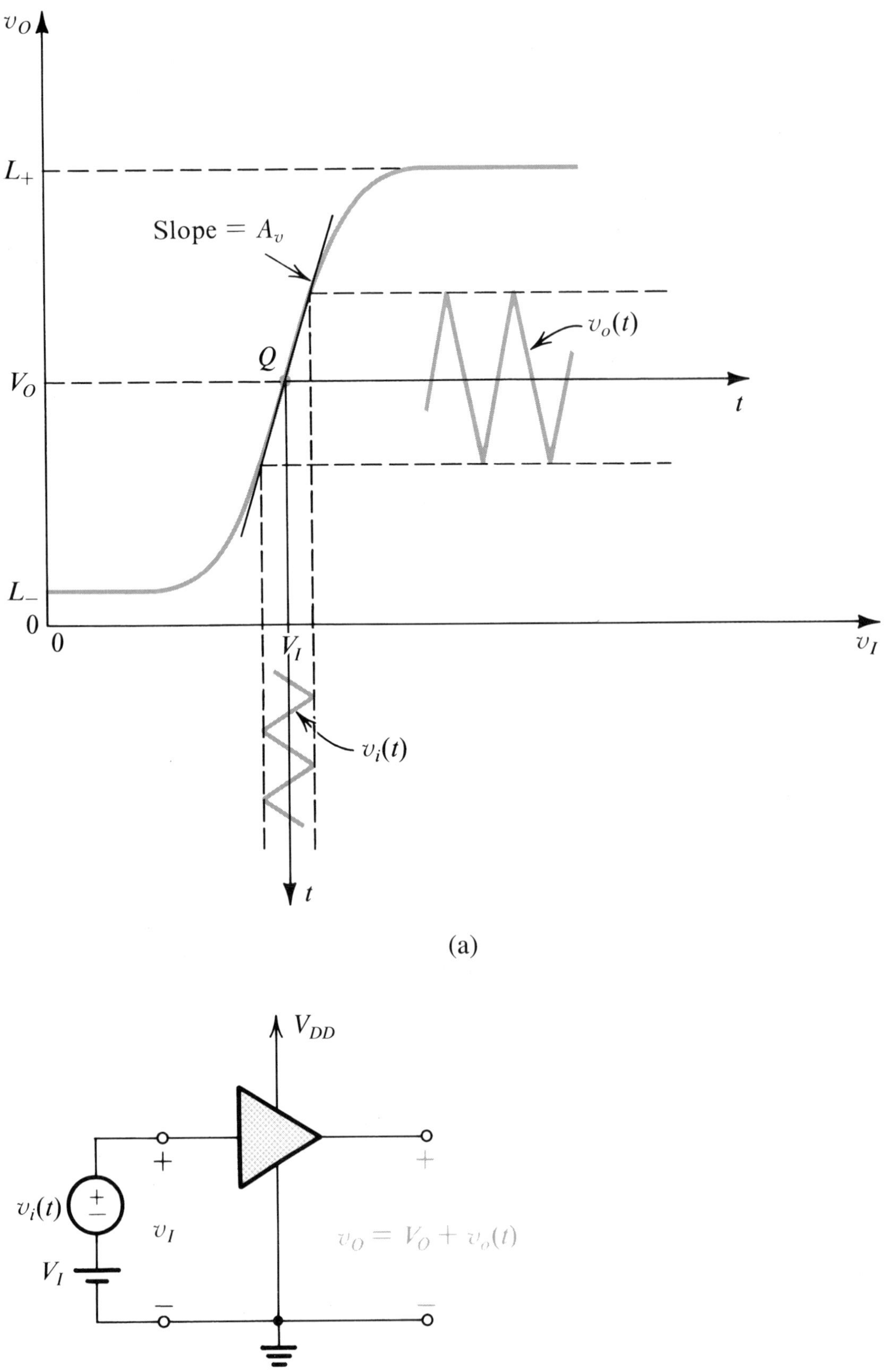

FIGURE 1.14 **(a)** An amplifier transfer characteristic that shows considerable nonlinearity. **(b)** To obtain linear operation the amplifier is biased as shown, and the signal amplitude is kept small. Observe that this amplifier is operated from a single power supply, V_{DD}.

TABLE 1.1 The Four Amplifier Types

Type	Circuit Model	Gain Parameter	Ideal Characteristics
Voltage Amplifier	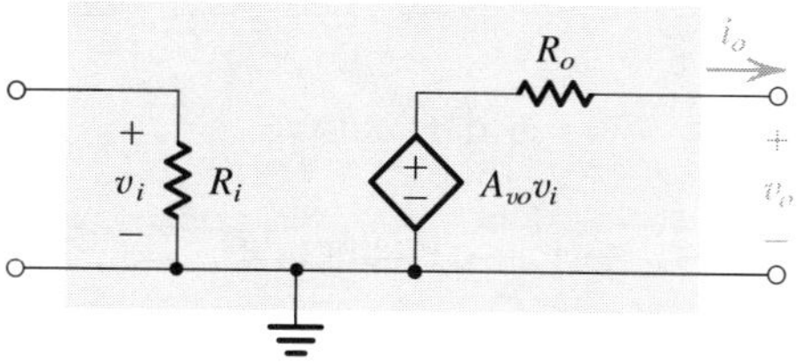	Open-Circuit Voltage Gain $A_{vo} \equiv \left.\frac{v_o}{v_i}\right\|_{i_o=0}$ (V/V)	$R_i = \infty$ $R_o = 0$
Current Amplifier	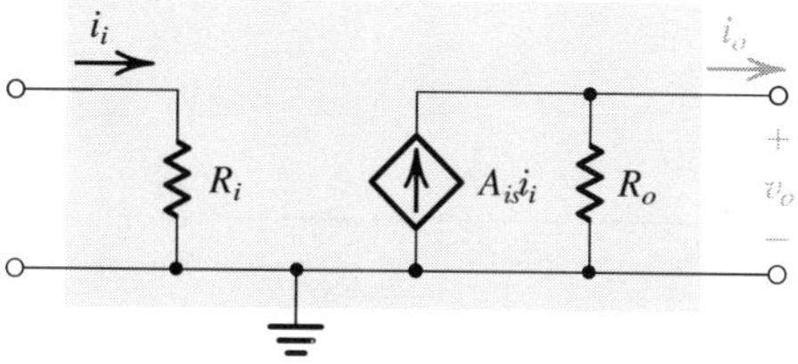	Short-Circuit Current Gain $A_{is} \equiv \left.\frac{i_o}{i_i}\right\|_{v_o=0}$ (A/A)	$R_i = 0$ $R_o = \infty$
Transconductance Amplifier	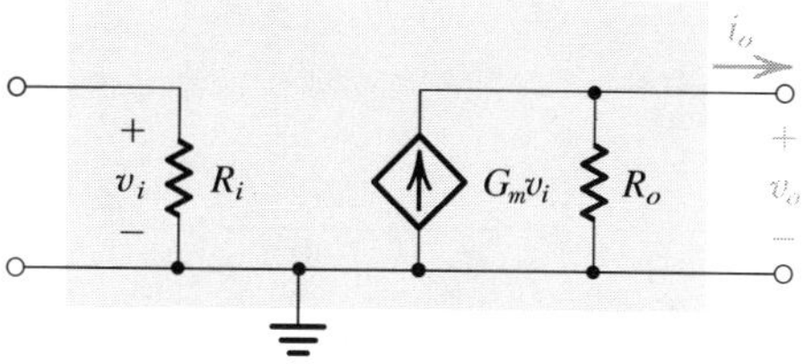	Short-Circuit Transconductance $G_m \equiv \left.\frac{i_o}{v_i}\right\|_{v_o=0}$ (A/V)	$R_i = \infty$ $R_o = \infty$
Transresistance Amplifier	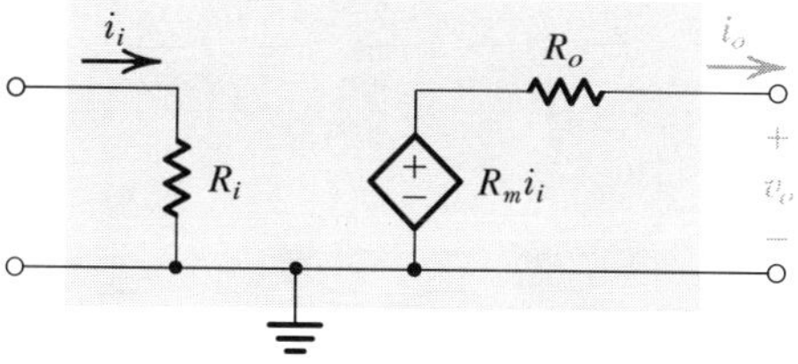	Open-Circuit Transresistance $R_m \equiv \left.\frac{v_o}{i_i}\right\|_{i_o=0}$ (V/A)	$R_i = 0$ $R_o = 0$

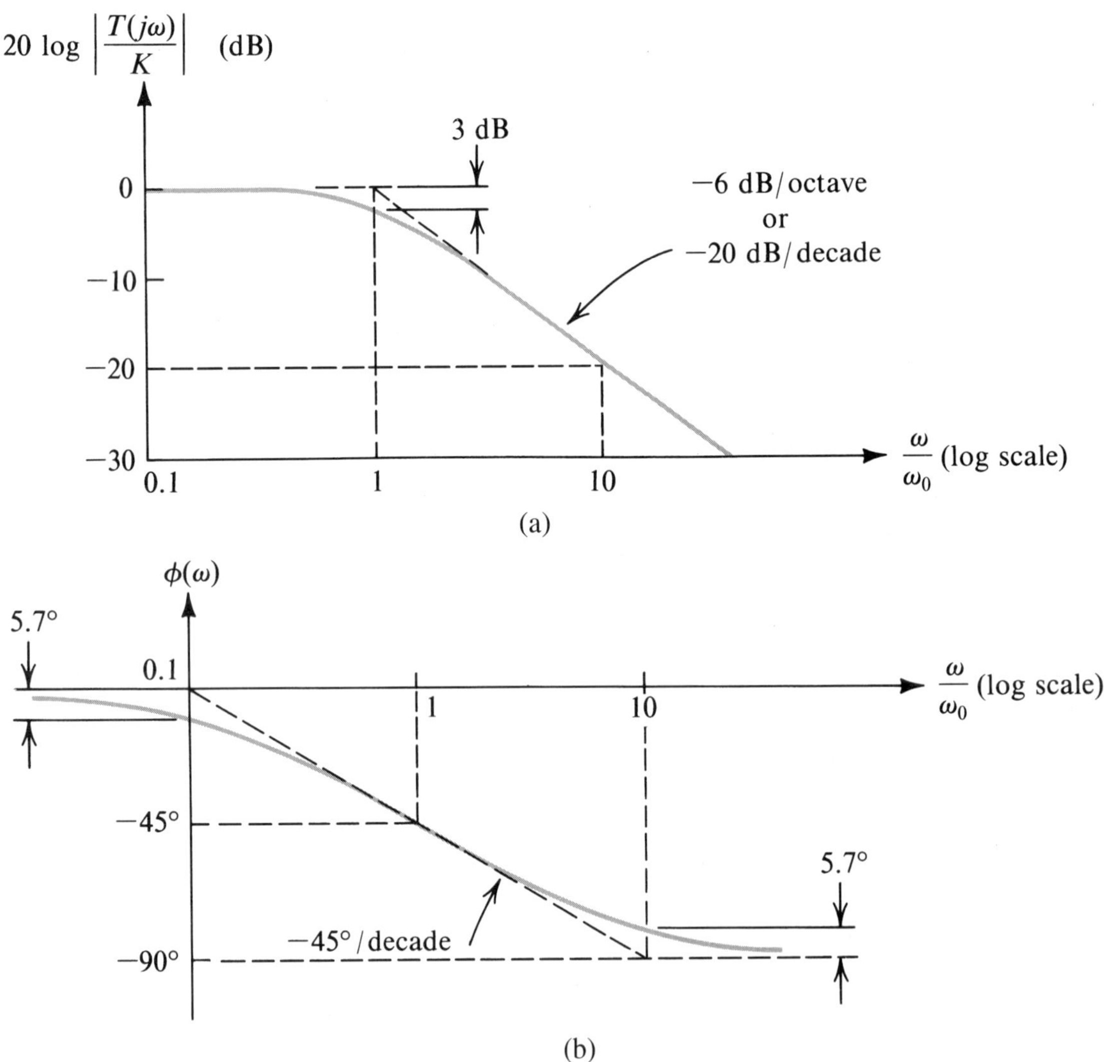

FIGURE 1.23 **(a)** Magnitude and **(b)** phase response of STC networks of the low-pass type.

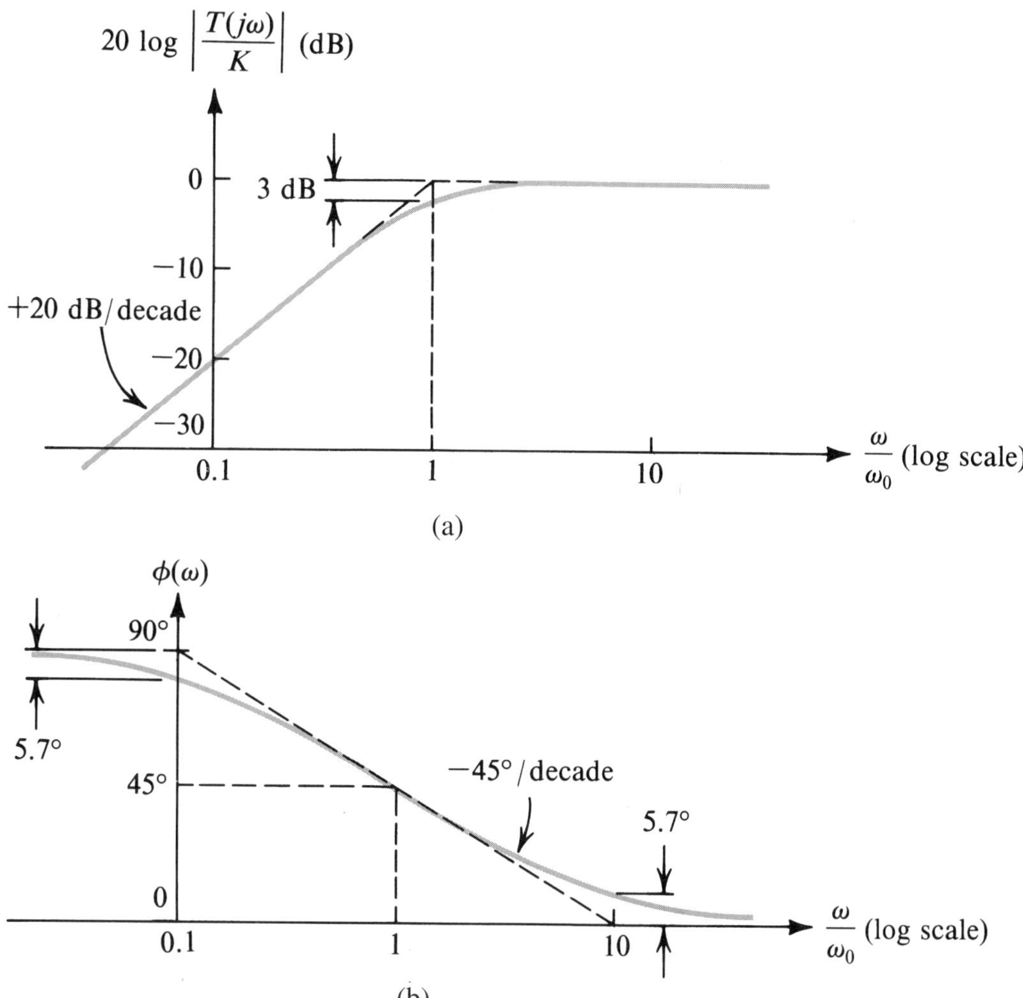

FIGURE 1.24 **(a)** Magnitude and **(b)** phase response of STC networks of the high-pass type.

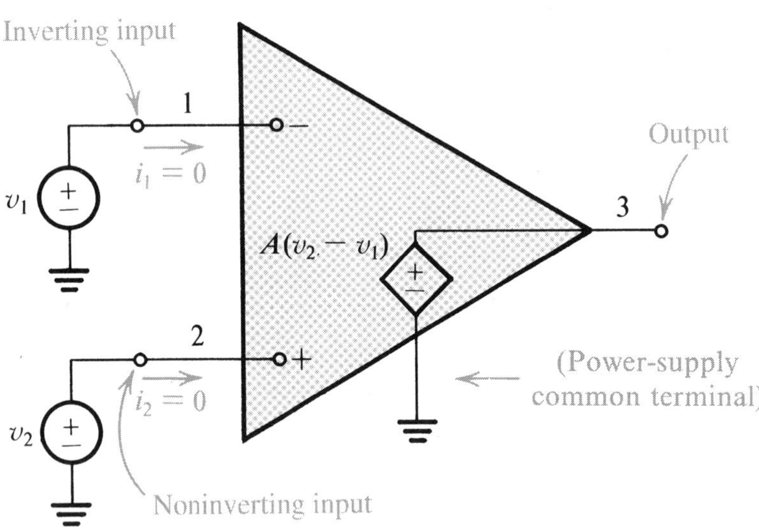

FIGURE 2.3 Equivalent circuit of the ideal op amp.

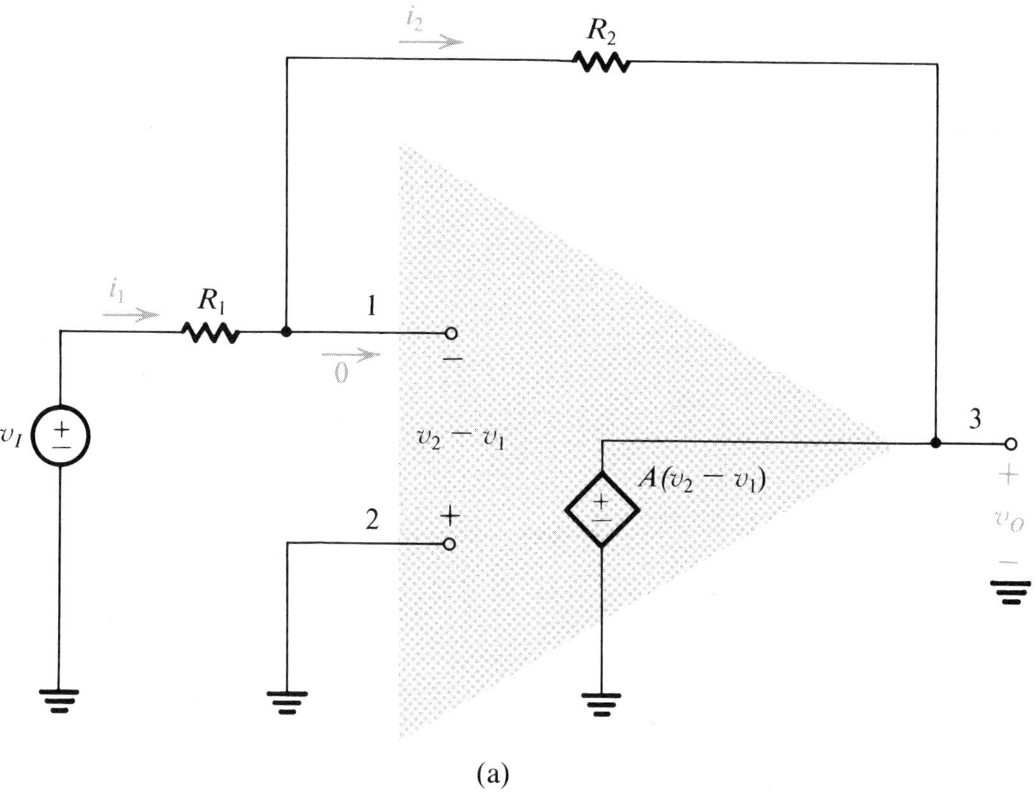

(a)

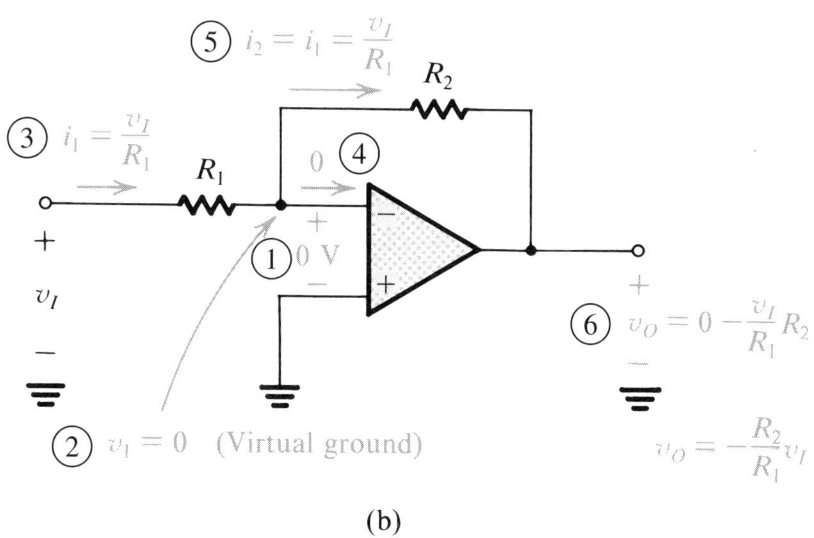

(b)

FIGURE 2.6 Analysis of the inverting configuration. The circled numbers indicate the order of the analysis steps.

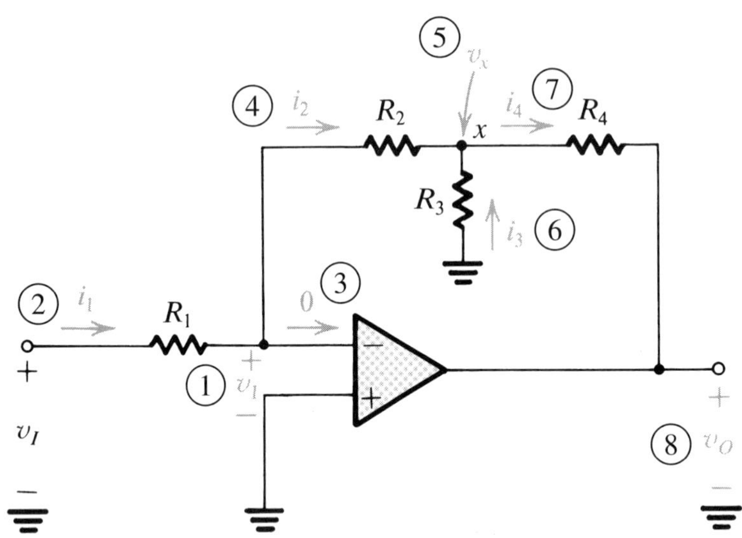

FIGURE 2.8 Circuit for Example 2.2. The circled numbers indicate the sequence of the steps in the analysis.

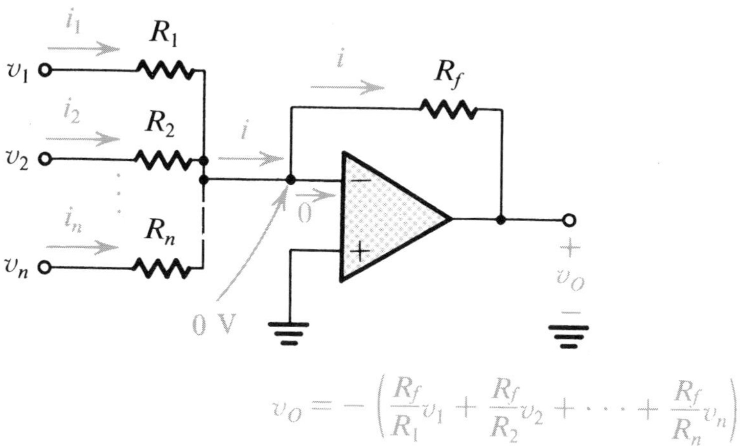

FIGURE 2.10 A weighted summer.

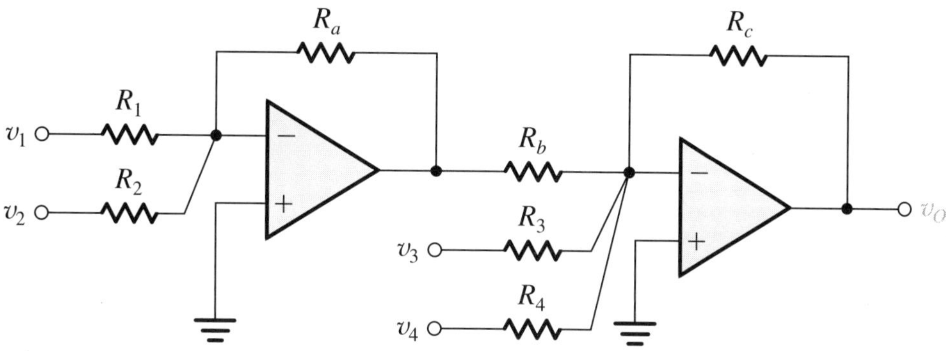

FIGURE 2.11 A weighted summer capable of implementing summing coefficients of both signs.

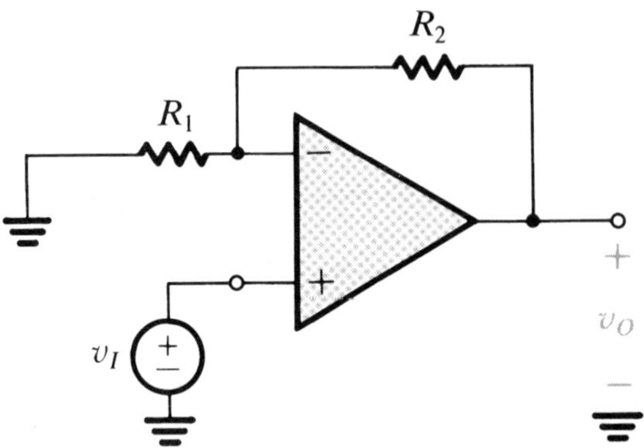

FIGURE 2.12 The noninverting configuration.

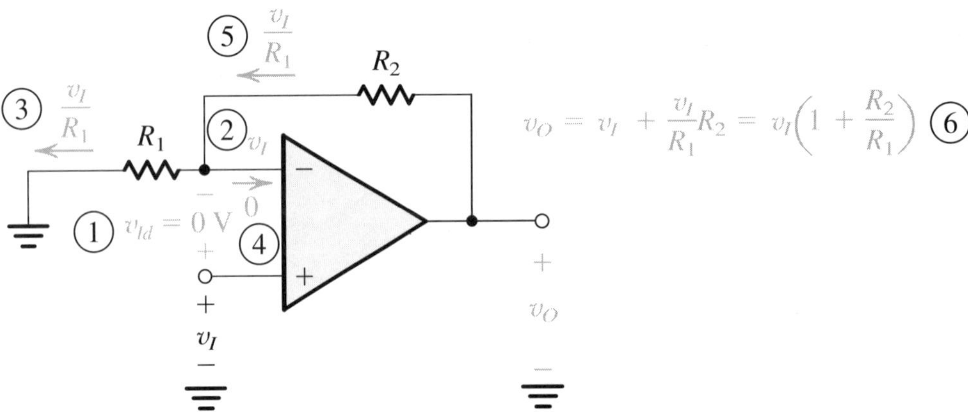

FIGURE 2.13 Analysis of the noninverting circuit. The sequence of the steps in the analysis is indicated by the circled numbers.

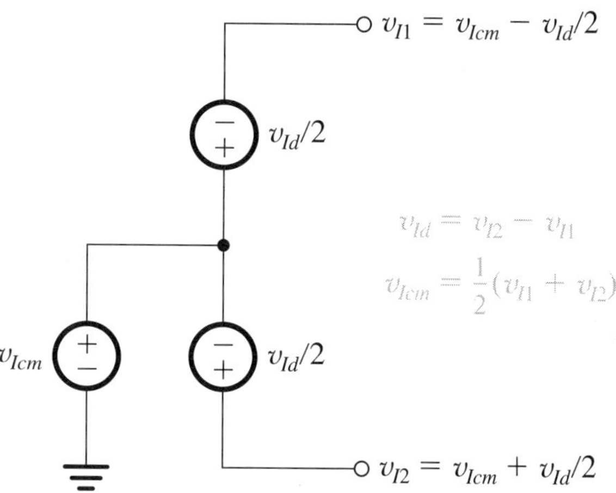

FIGURE 2.15 Representing the input signals to a differential amplifier in terms of their differential and common-mode components.

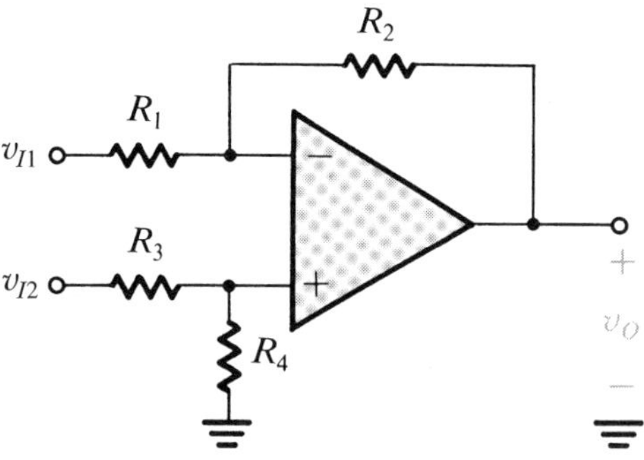

FIGURE 2.16 A difference amplifier.

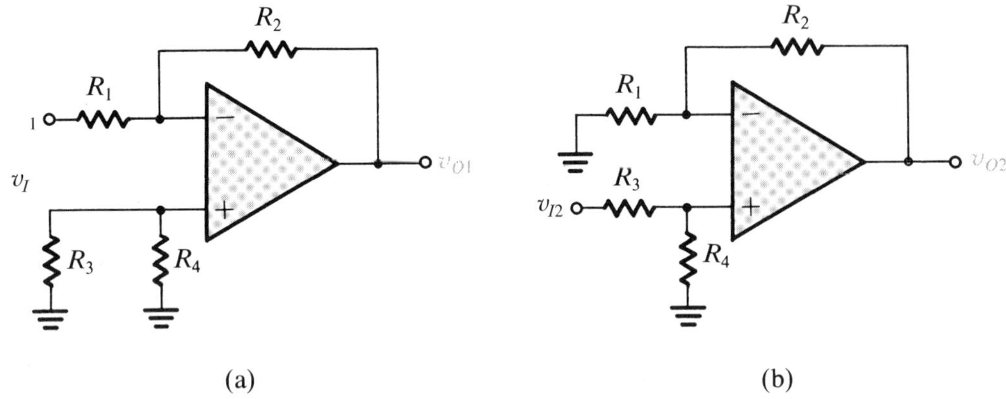

FIGURE 2.17 Application of superposition to the analysis of the circuit of Fig. 2.16.

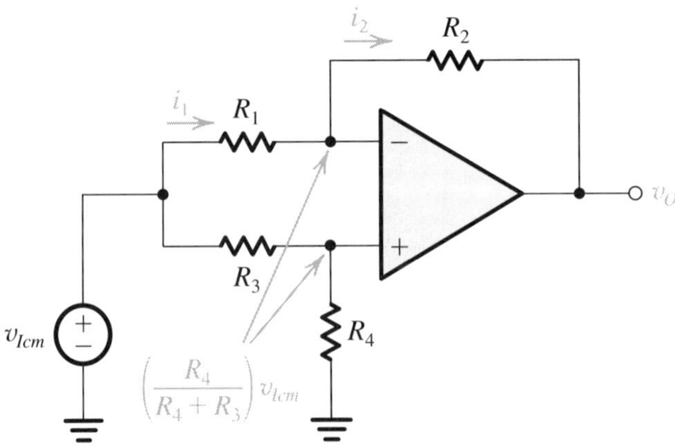

FIGURE 2.18 Analysis of the difference amplifier to determine its common-mode gain $A_{cm} \equiv v_O / v_{Icm}$.

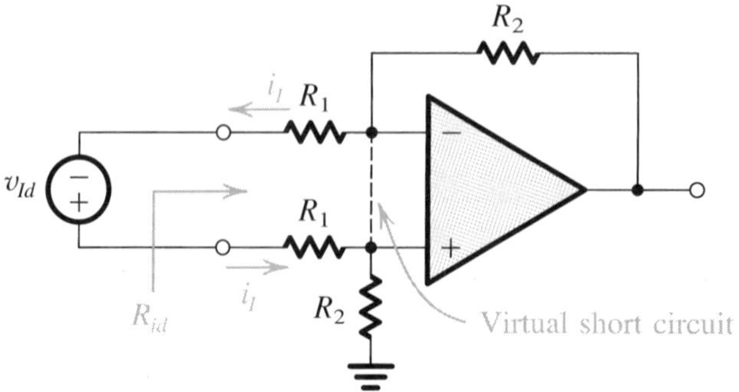

FIGURE 2.19 Finding the input resistance of the difference amplifier for the case $R_3 = R_1$ and $R_4 = R_2$.

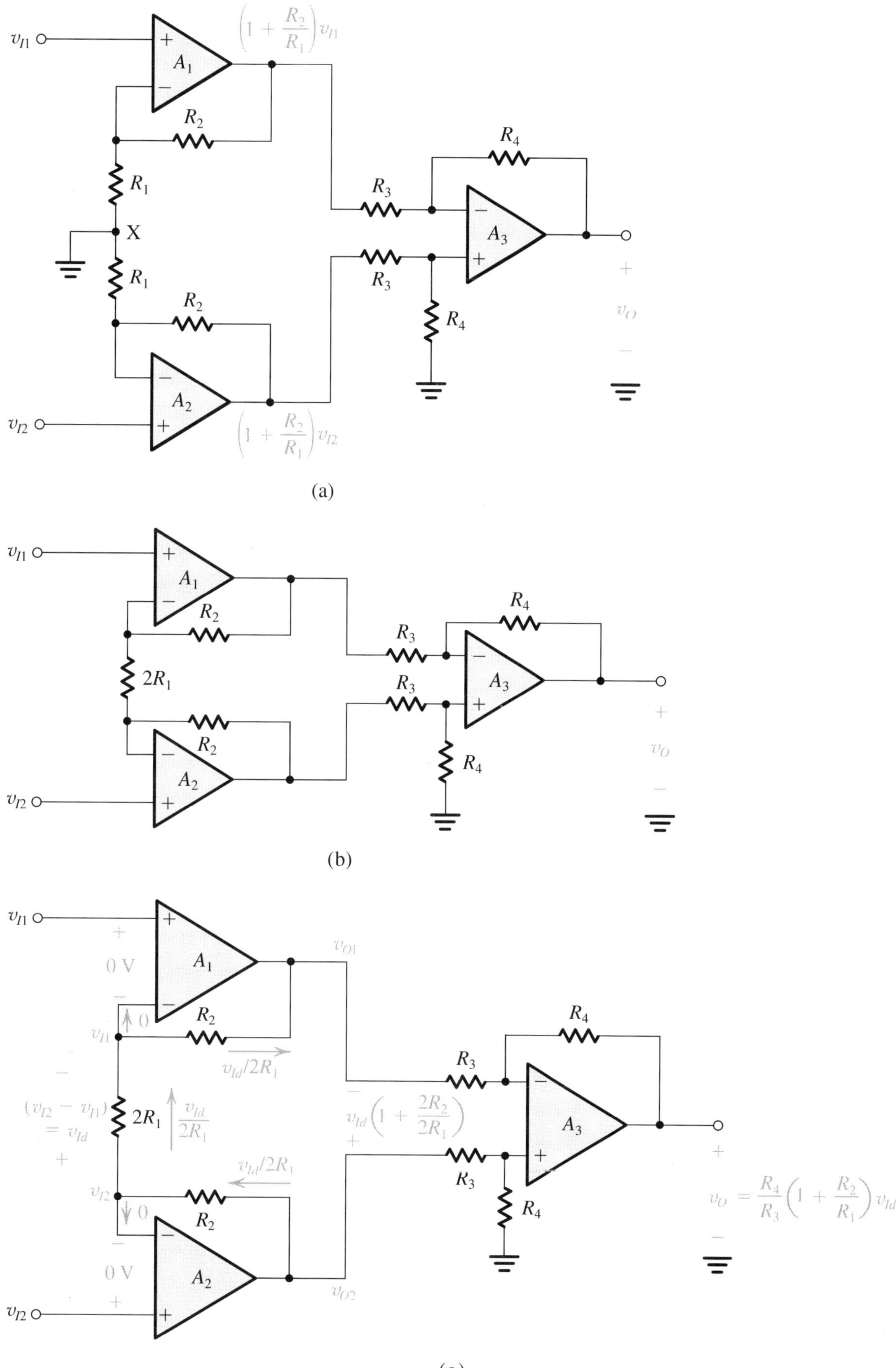

FIGURE 2.20 A popular circuit for an instrumentation amplifier; **(a)** Initial approach to the circuit; **(b)** The circuit in **(a)** with the connection between node X and ground removed and the two resistors R_1 and R_1 lumped together. This simple wiring change dramatically improves performance; **(c)** Analysis of the circuit in' **(b)** assuming ideal op amps.

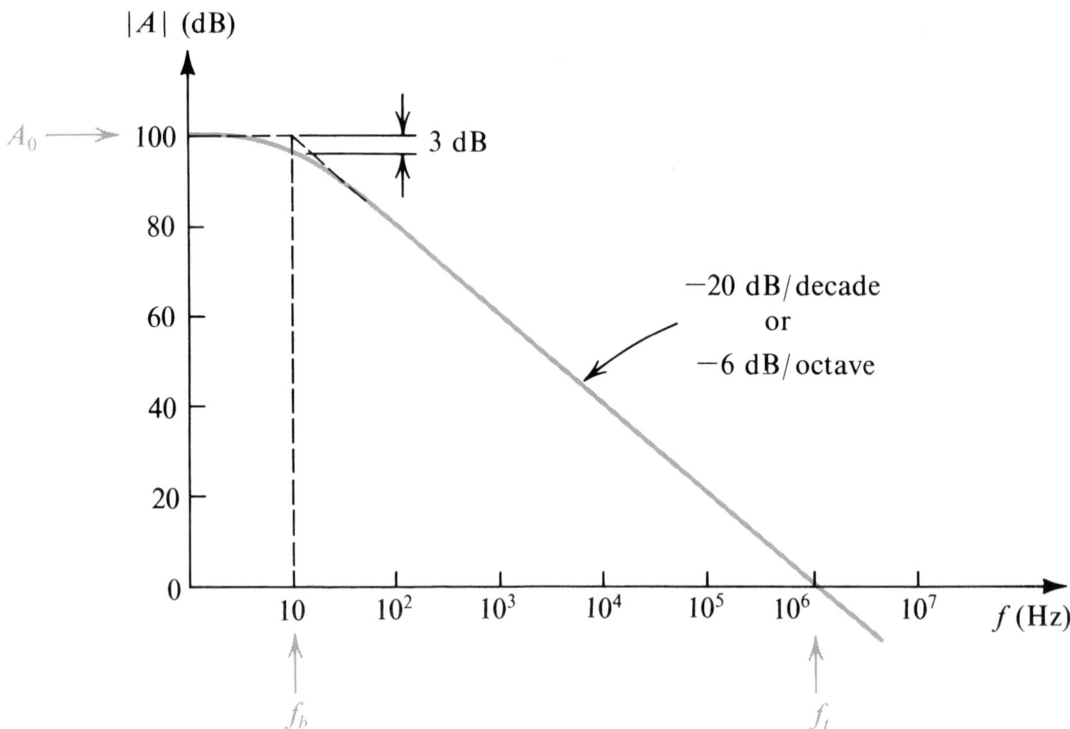

FIGURE 2.22 Open-loop gain of a typical general-purpose internally compensated op amp.

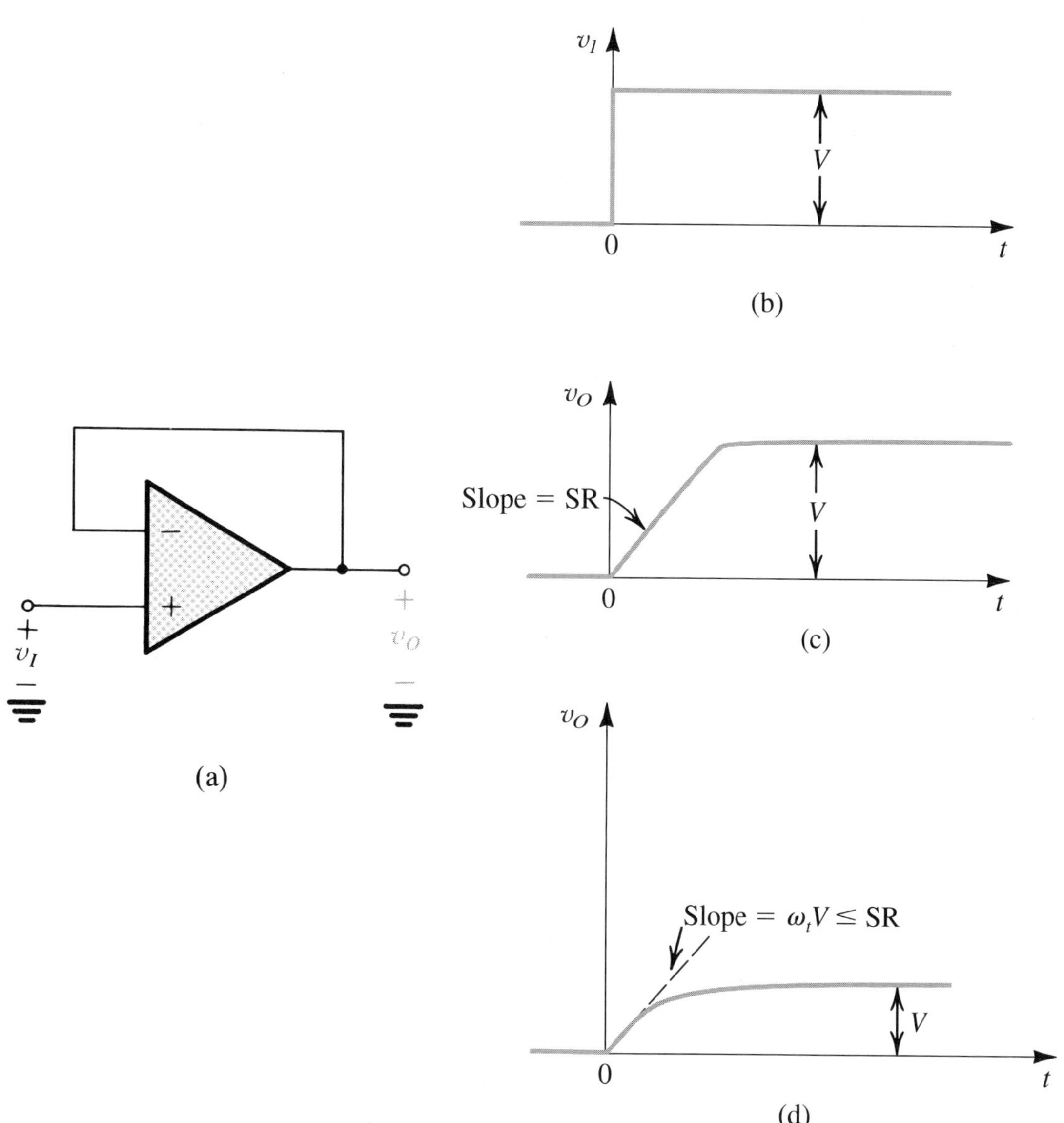

FIGURE 2.26 **(a)** Unity-gain follower. **(b)** Input step waveform. **(c)** Linearly rising output waveform obtained when the amplifier is slew rate limited. **(d)** Exponentially rising output waveform obtained when V is sufficiently small so that the initial slope ($\omega_t V$) is smaller than or equal to SR.

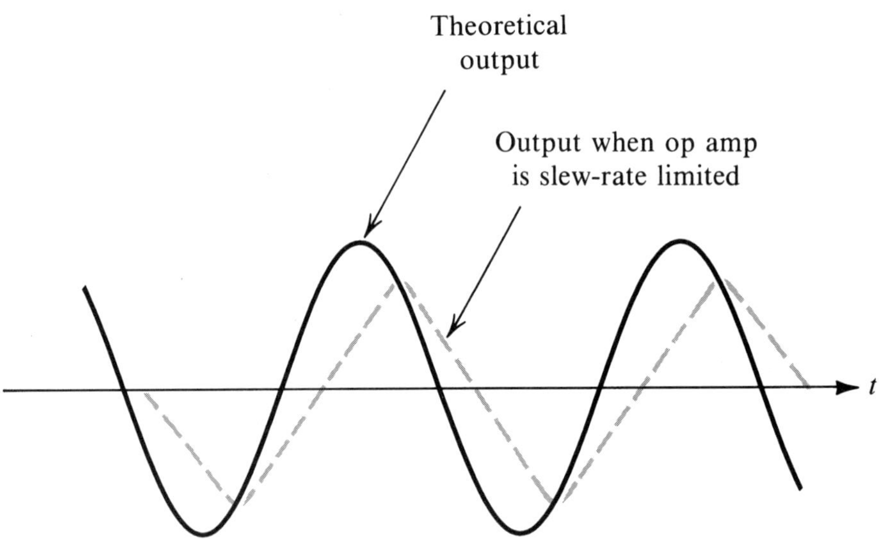

FIGURE 2.27 Effect of slew-rate limiting on output sinusoidal waveforms.

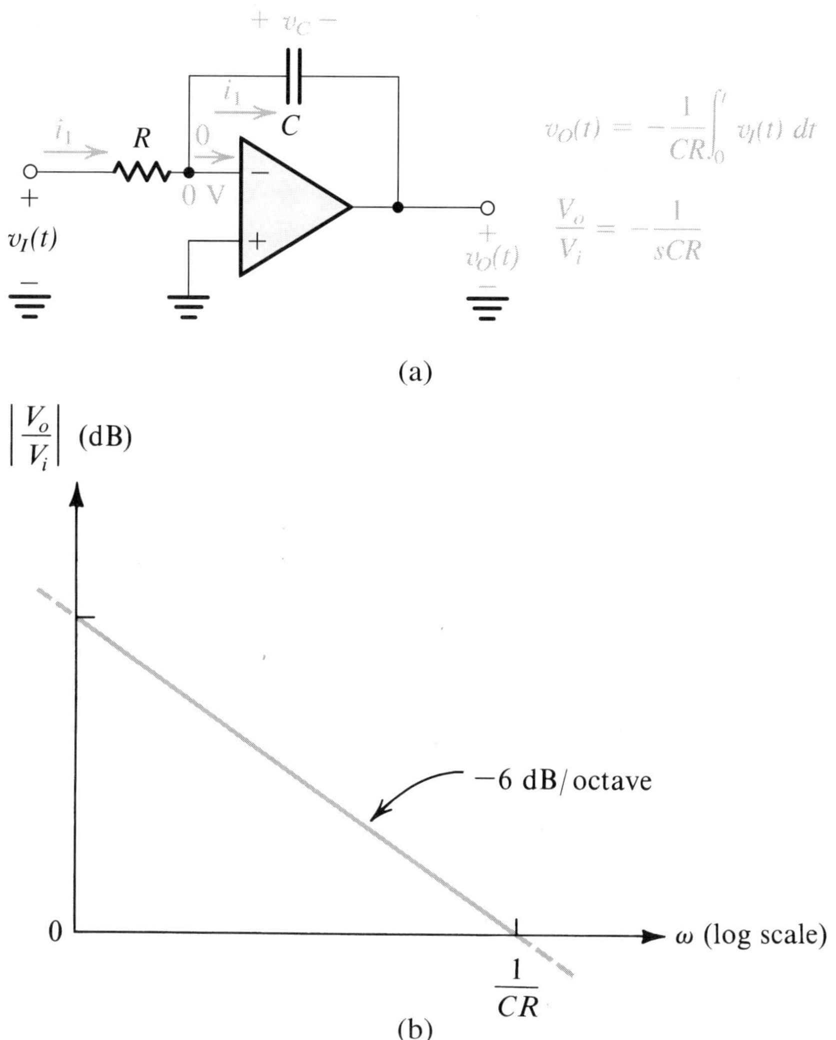

FIGURE 2.39 **(a)** The Miller or inverting integrator. **(b)** Frequency response of the integrator.

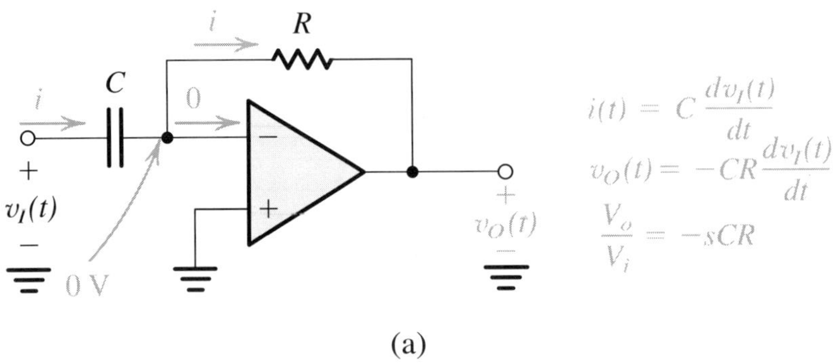

(a)

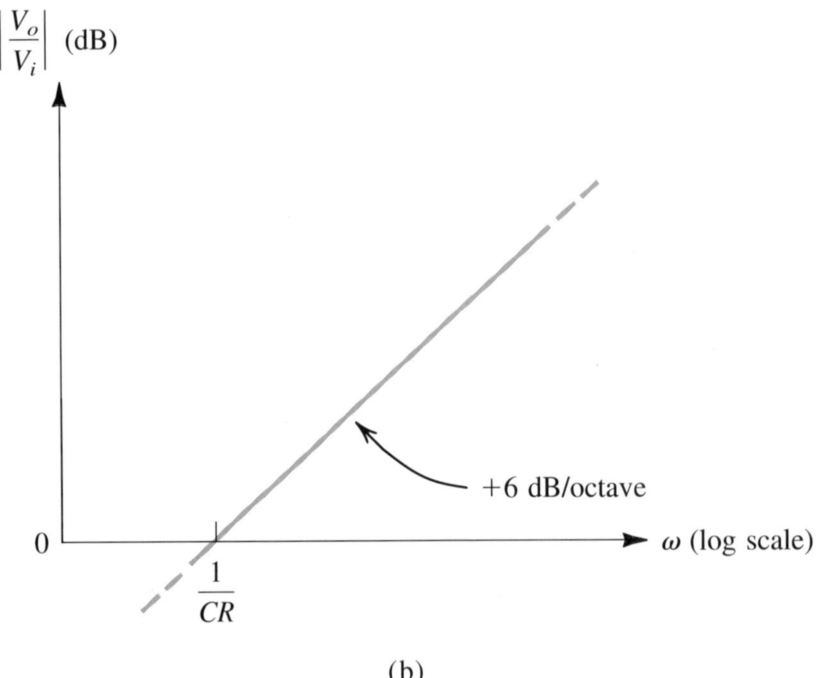

(b)

FIGURE 2.44 **(a)** A differentiator. **(b)** Frequency response of a differentiator with a time-constant *CR*.

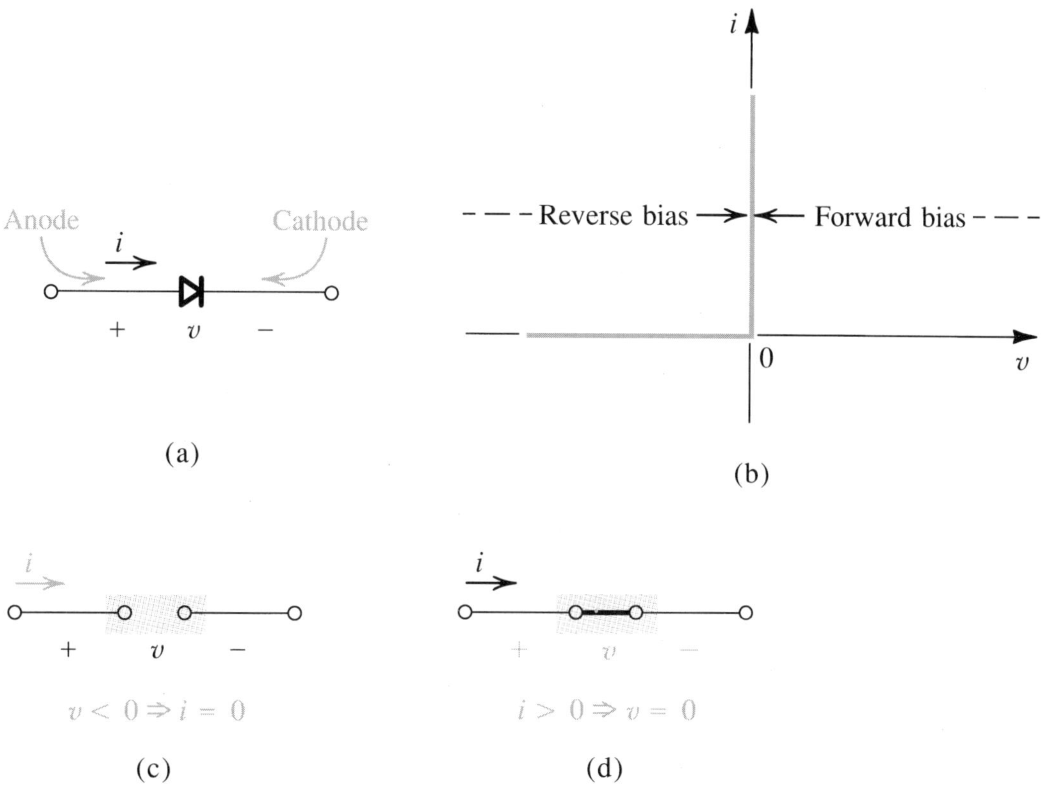

FIGURE 3.1 The ideal diode: **(a)** diode circuit symbol; **(b)** i–v characteristic; **(c)** equivalent circuit in the reverse direction; **(d)** equivalent circuit in the forward direction.

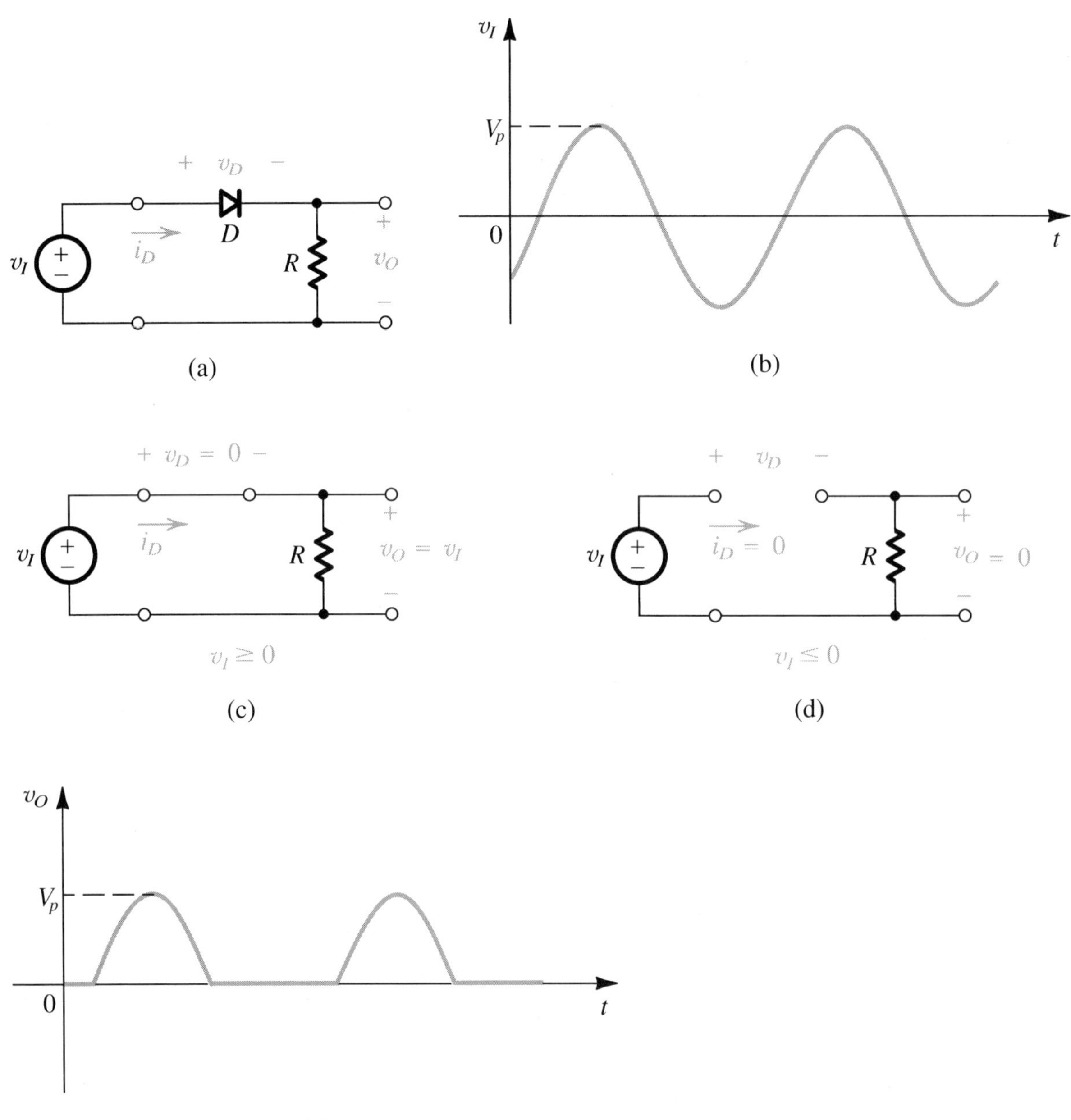

FIGURE 3.3 **(a)** Rectifier circuit. **(b)** Input waveform. **(c)** Equivalent circuit when $v_I \geq 0$. **(d)** Equivalent circuit when $v_I \leq 0$. **(e)** Output waveform.

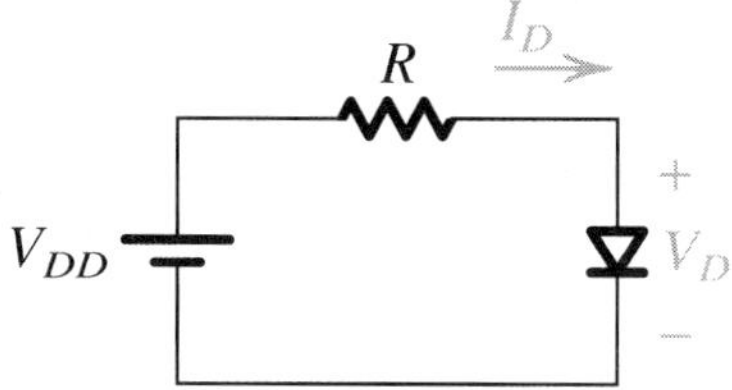

FIGURE 3.10 A simple circuit used to illustrate the analysis of circuits in which the diode is forward conducting.

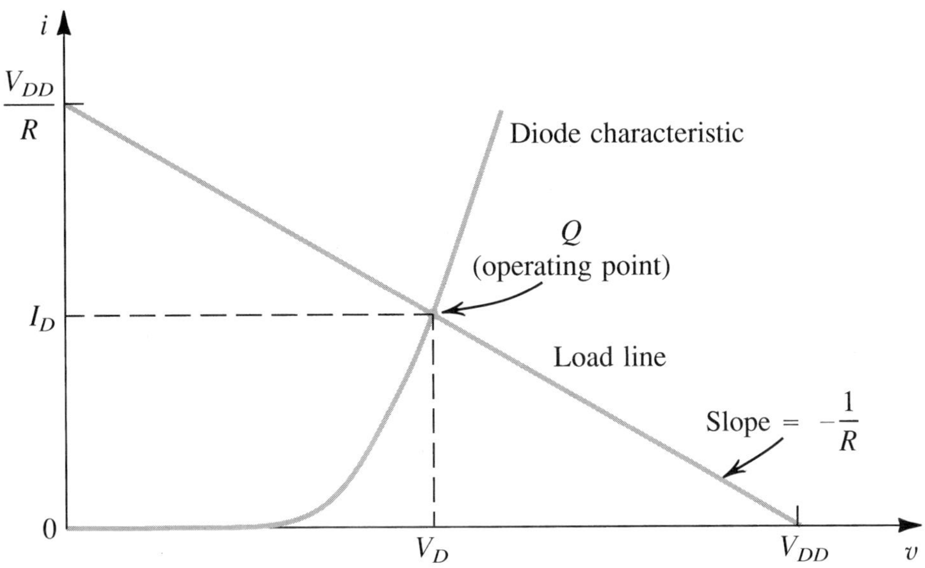

FIGURE 3.11 Graphical analysis of the circuit in Fig. 3.10 using the exponential diode model.

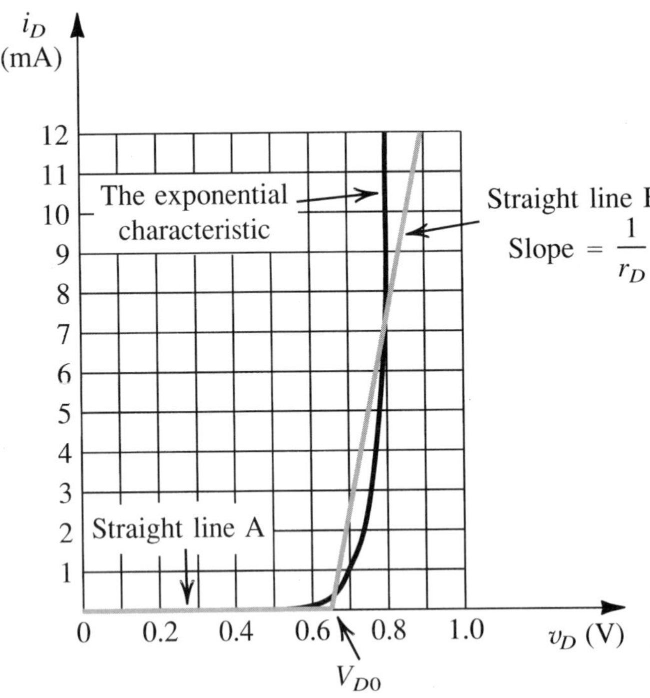

FIGURE 3.12 Approximating the diode forward characteristic with two straight lines: the piecewise-linear model.

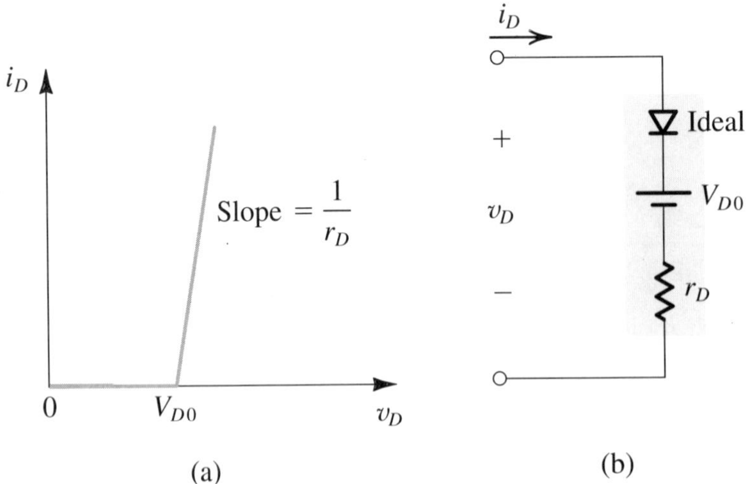

FIGURE 3.13 Piecewise-linear model of the diode forward characteristic and its equivalent circuit representation.

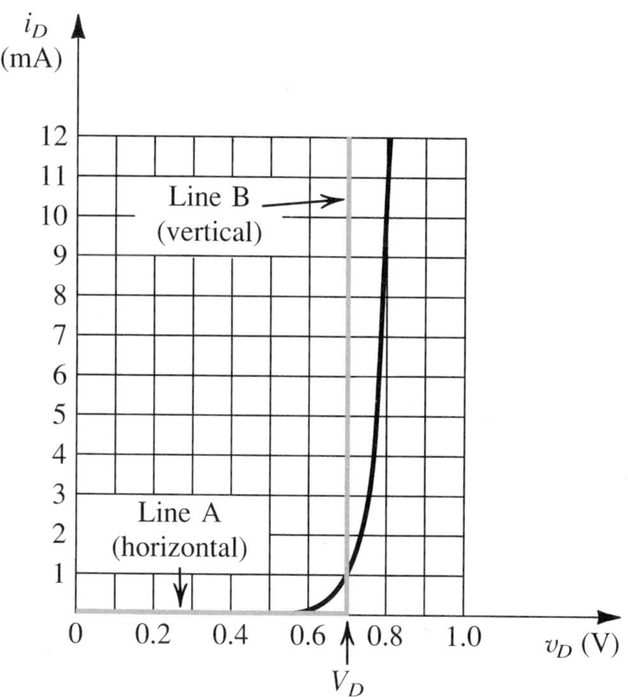

FIGURE 3.15 Development of the constant-voltage-drop model of the diode forward characteristics. A vertical straight line (B) is used to approximate the fast-rising exponential. Observe that this simple model predicts V_D to within ±0.1 V over the current range of 0.1 mA to 10 mA.

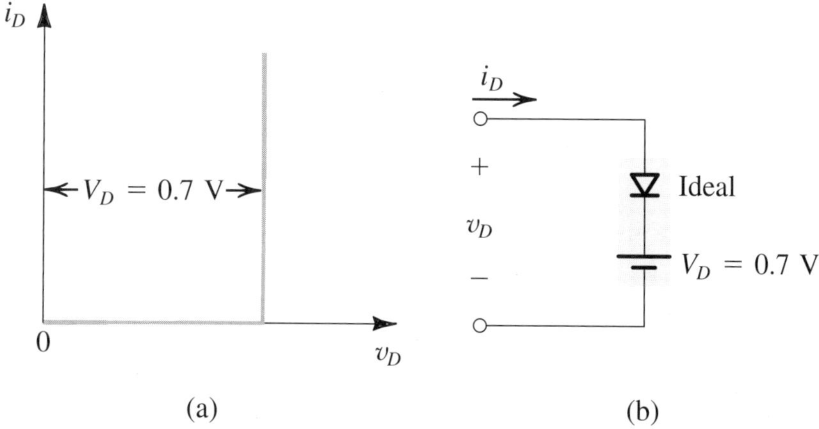

FIGURE 3.16 The constant-voltage-drop model of the diode forward characteristics and its equivalent-circuit representation.

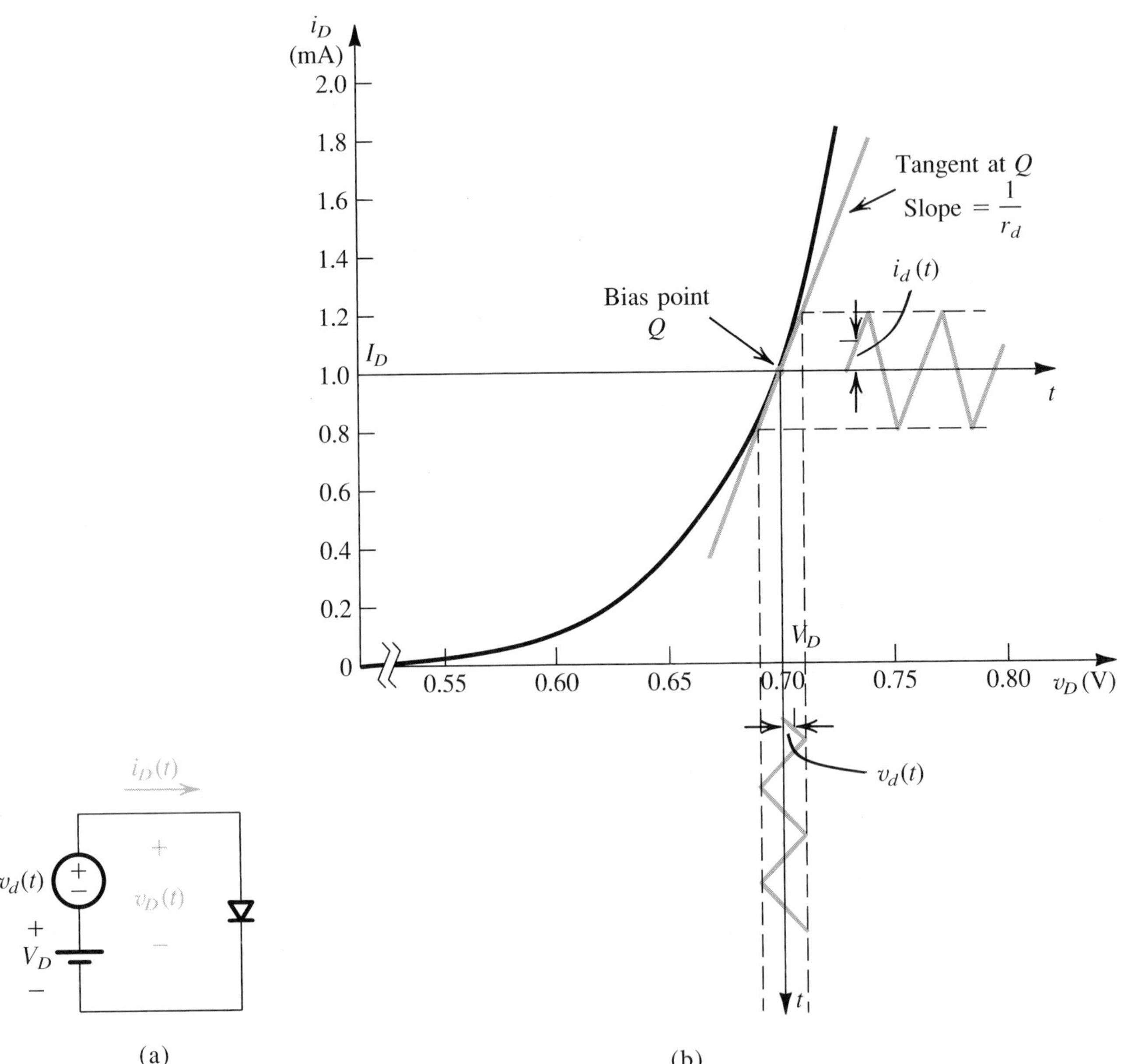

FIGURE 3.17 Development of the diode small-signal model. Note that the numerical values shown are for a diode with $n = 2$.

TABLE 3.1 Modeling the Diode Forward Characteristic

Model	Graph	Equations	Circuit	Comments
Exponential	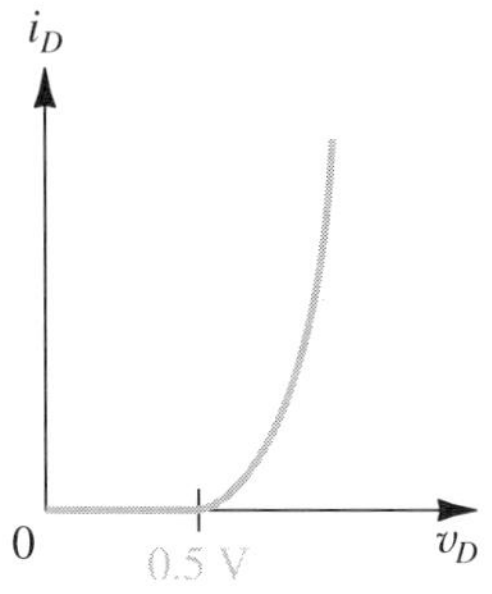 i_D vs. v_D; 0; 0.5 V	$i_D = I_S e^{v_D/nV_T}$ $v_D = 2.3nV_T \log\left(\frac{i_D}{I_S}\right)$ $V_{D2} - V_{D1} = 2.3nV_T \log\left(\frac{I_{D2}}{I_{D1}}\right)$ $2.3nV_T = 60$ mV for $n = 1$ $2.3nV_T = 120$ mV for $n = 2$	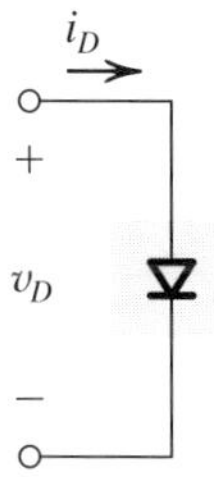 i_D; +; v_D; −	$I_S = 10^{-12}$ A to 10^{-15} A, depending on junction area $V_T \cong 25$ mV $n = 1$ to 2 Physically based and remarkably accurate model Useful when accurate analysis is needed
Piecewise-linear (battery-plus-resistance)	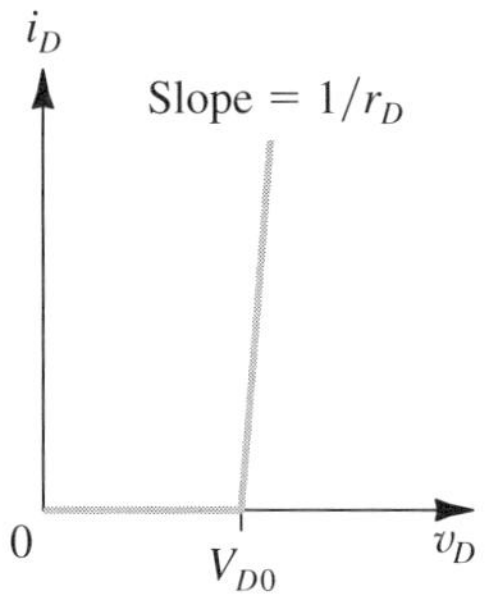 i_D; Slope $= 1/r_D$; 0; V_{D0}; v_D	For $v_D \le V_{D0}$: $i_D = 0$ For $v_D \ge V_{D0}$: $i_D = \frac{1}{r_D}(v_D - V_{D0})$	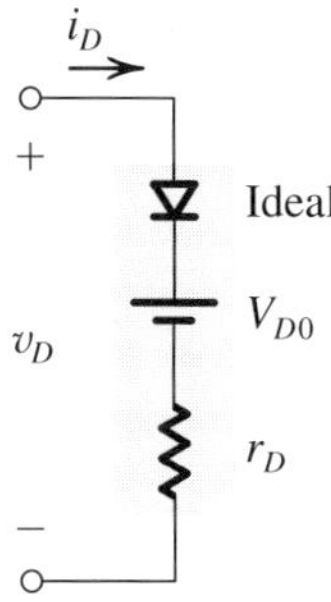 i_D; +; v_D; −; Ideal; V_{D0}; r_D	Choice of V_{D0} and r_D is determined by the current range over which the model is required. For the amount of work involved, not as useful as the constant-voltage-drop model. Used only infrequently.
Constant-voltage-drop (or the "0.7-V model")	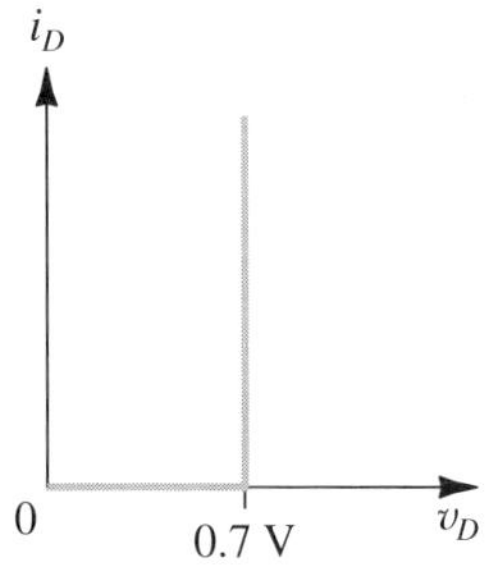 i_D; 0; 0.7 V; v_D	For $i_D > 0$: $v_D = 0.7$ V	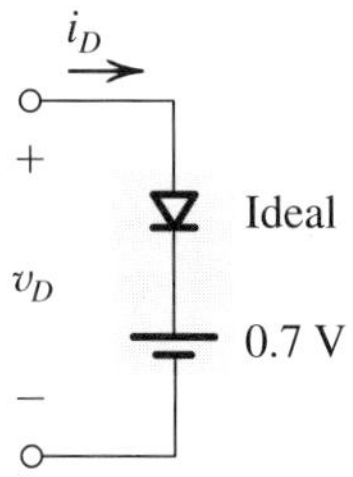 i_D; +; v_D; −; Ideal; 0.7 V	Easy to use and very popular for the quick, hand analysis that is essential in circuit design.
Ideal-diode	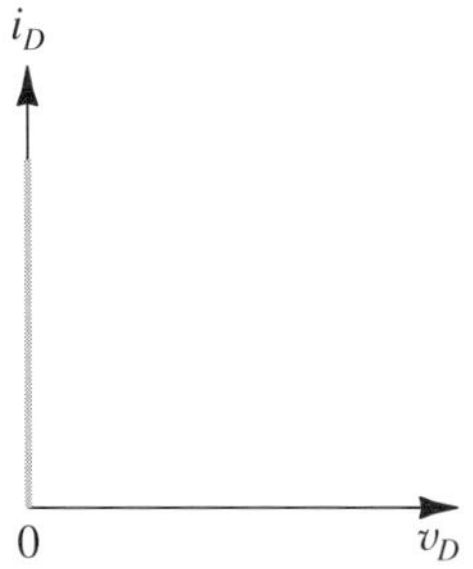 i_D; 0; v_D	For $i_D > 0$: $v_D = 0$	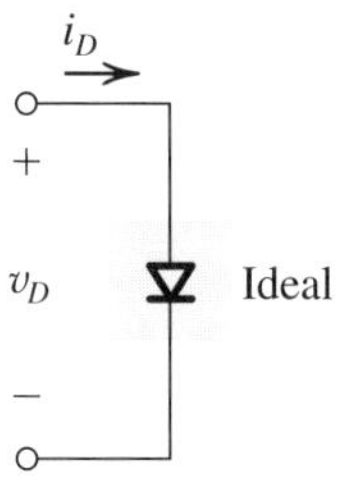 i_D; +; v_D; −; Ideal	Good for determining which diodes are conducting and which are cutoff in a multiple-diode circuit. Good for obtaining very approximate values for diode currents, especially when the circuit voltages are much greater than V_D.
Small-signal	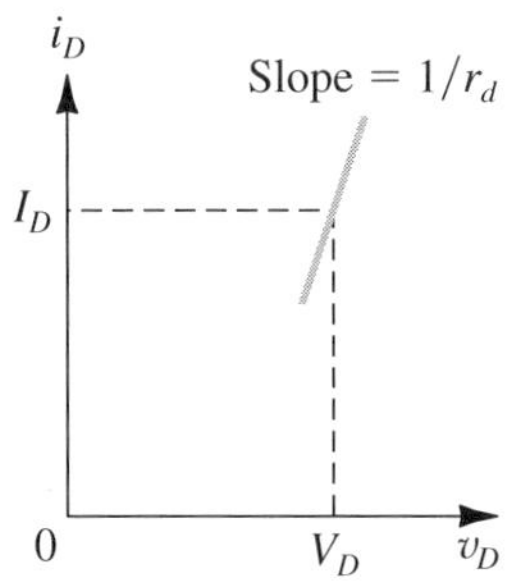 i_D; Slope $= 1/r_d$; I_D; 0; V_D; v_D	For small signals superimposed on V_D and I_D: $i_d = v_d/r_d$ $r_d = nV_T/I_D$ (For $n = 1$, v_d is limited to 5 mV; for $n = 2$, 10 mV)	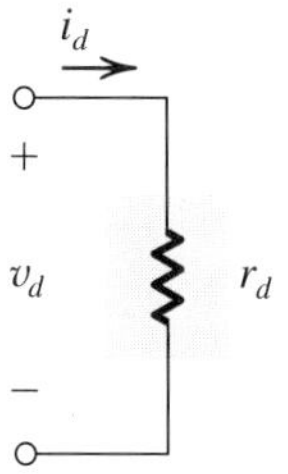 i_d; +; v_d; −; r_d	Useful for finding the signal component of the diode voltage (e.g., in the voltage-regulator application). Serves as the basis for small-signal modeling of transistors (Chapters 4 and 5).

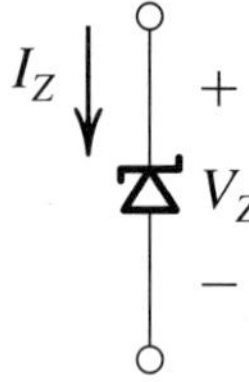

FIGURE 3.20 Circuit symbol for a zener diode.

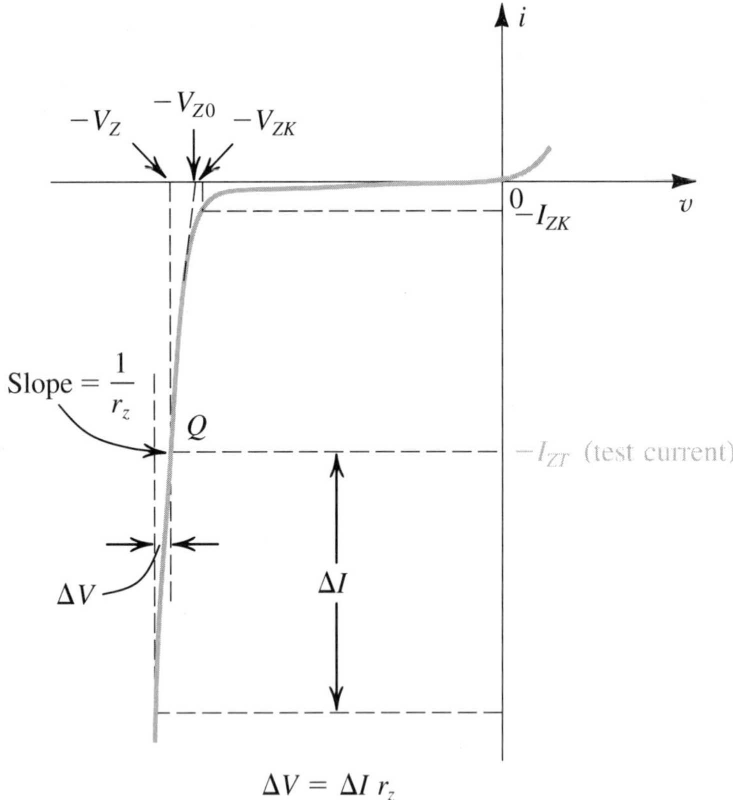

FIGURE 3.21 The diode i–v characteristic with the breakdown region shown in some detail.

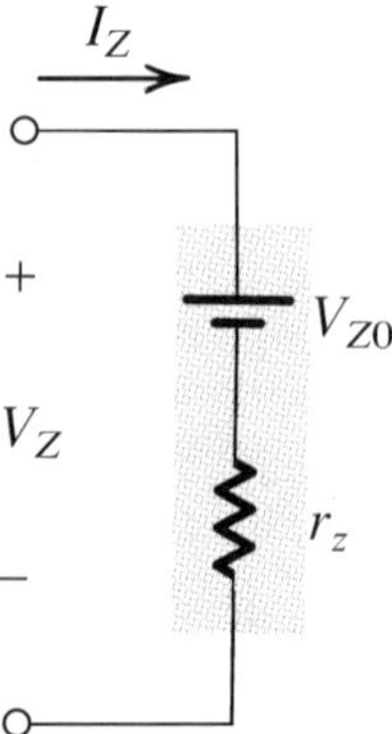

FIGURE 3.22 Model for the zener diode.

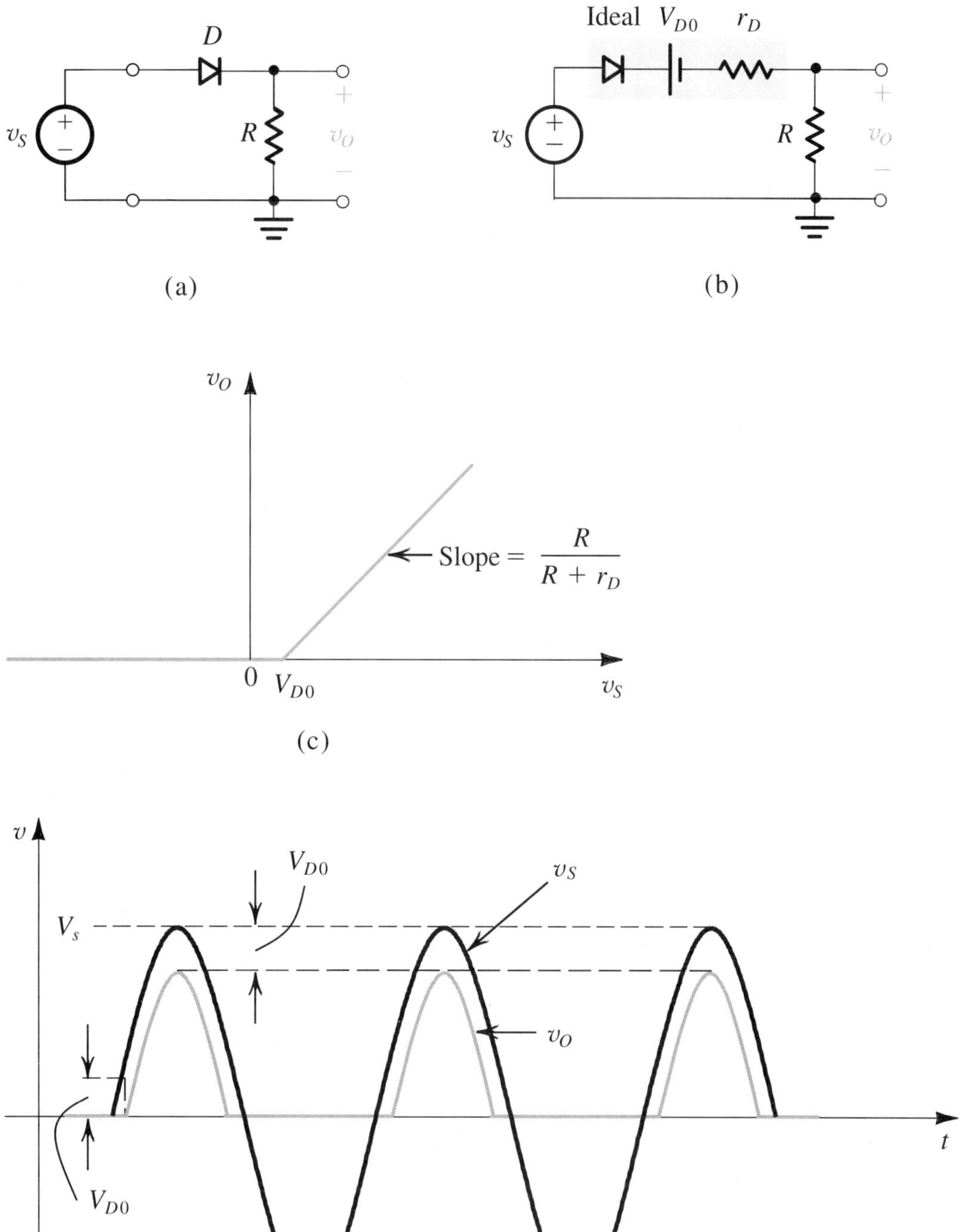

FIGURE 3.25 **(a)** Half-wave rectifier. **(b)** Equivalent circuit of the half-wave rectifier with the diode replaced with its battery-plus-resistance model. **(c)** Transfer characteristic of the rectifier circuit. **(d)** Input and output waveforms, assuming that $r_D \ll R$.

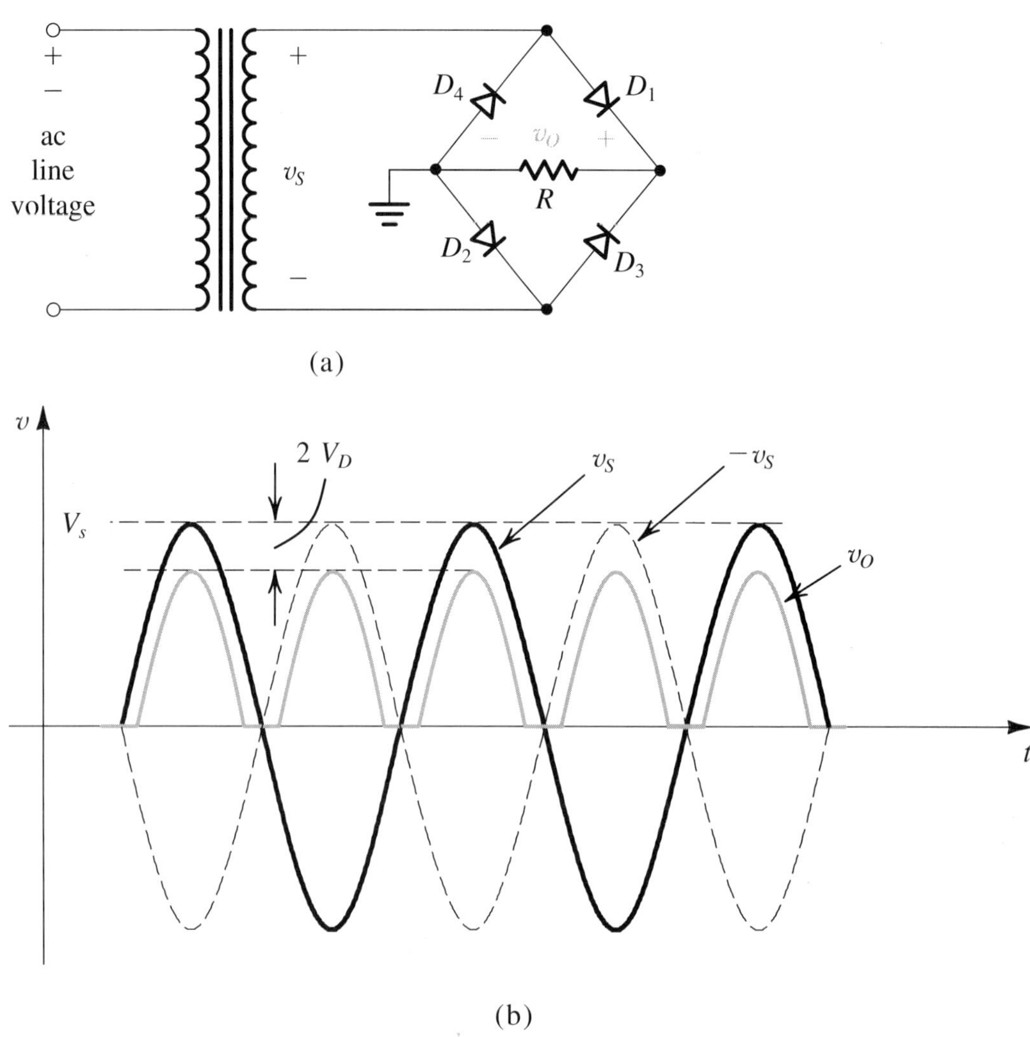

FIGURE 3.27 The bridge rectifier: **(a)** circuit; **(b)** input and output waveforms.

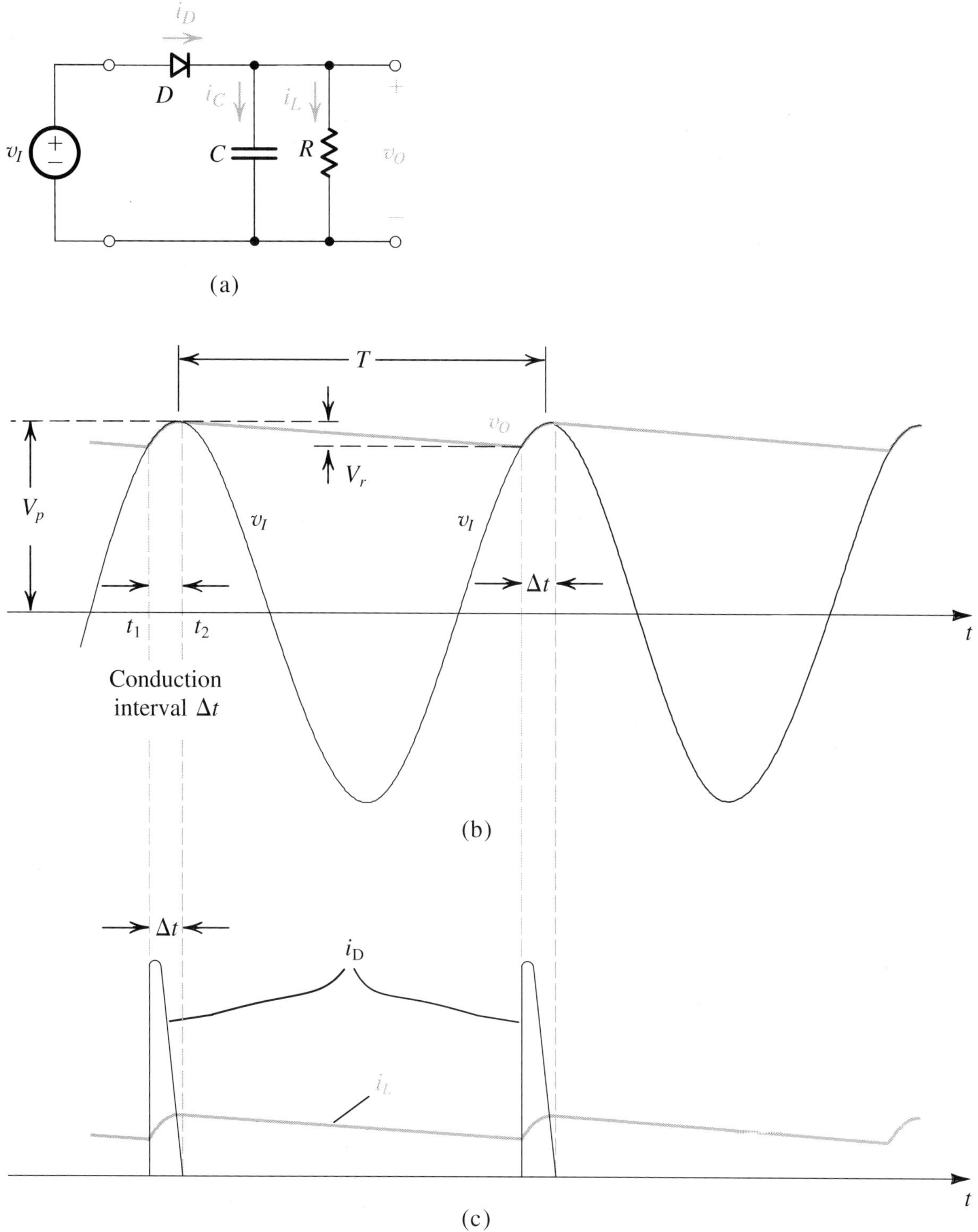

FIGURE 3.29 Voltage and current waveforms in the peak rectifier circuit with $CR \gg T$. The diode is assumed ideal.

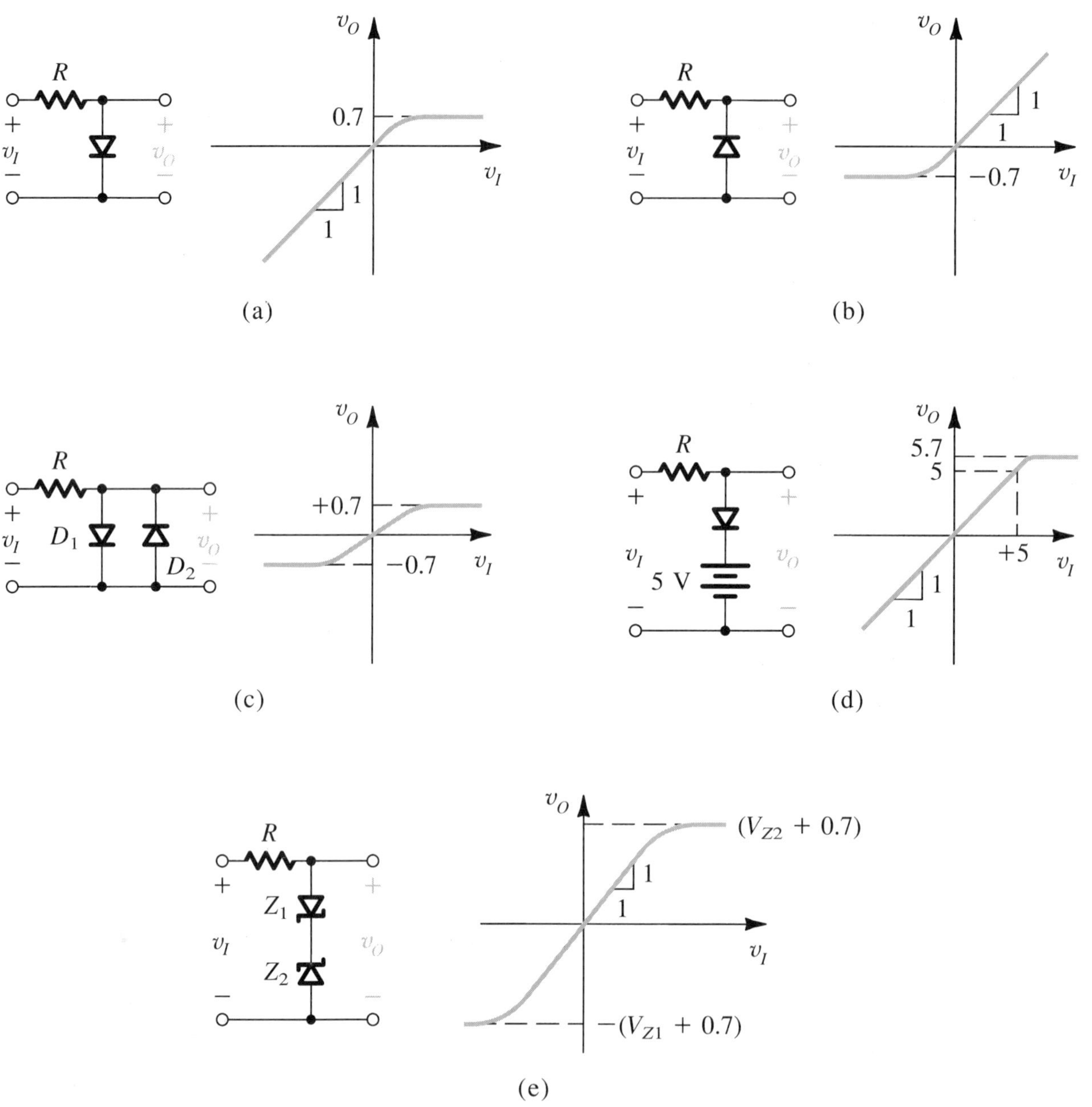

FIGURE 3.35 A variety of basic limiting circuits.

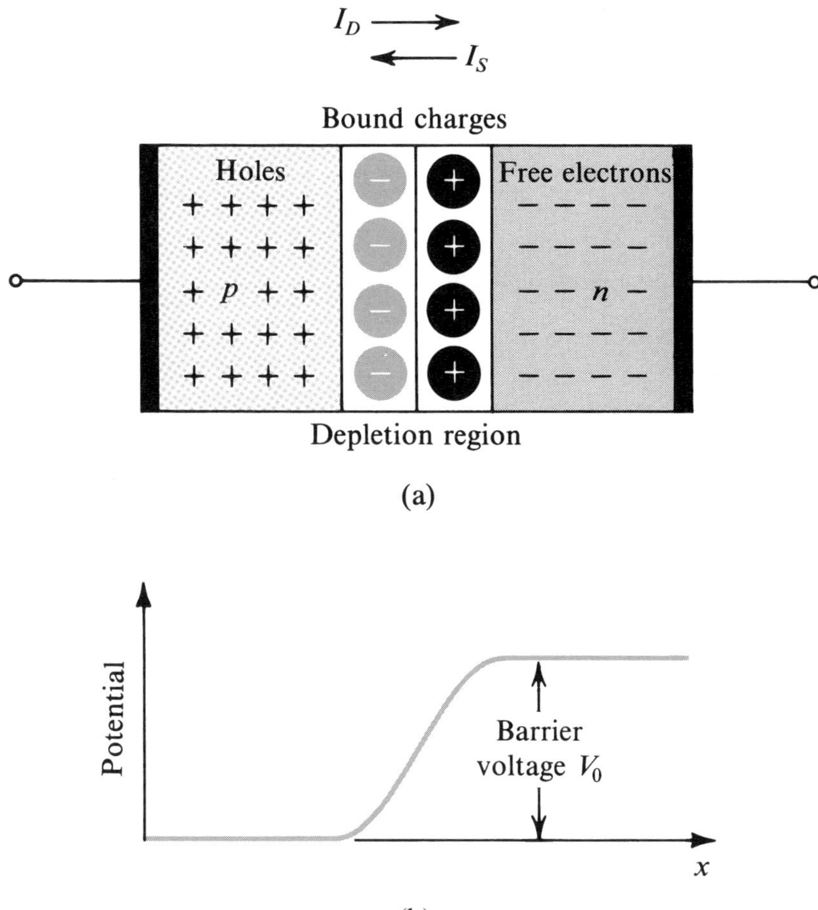

FIGURE 3.45 **(a)** The *pn* junction with no applied voltage (open-circuited terminals). **(b)** The potential distribution along an axis perpendicular to the junction.

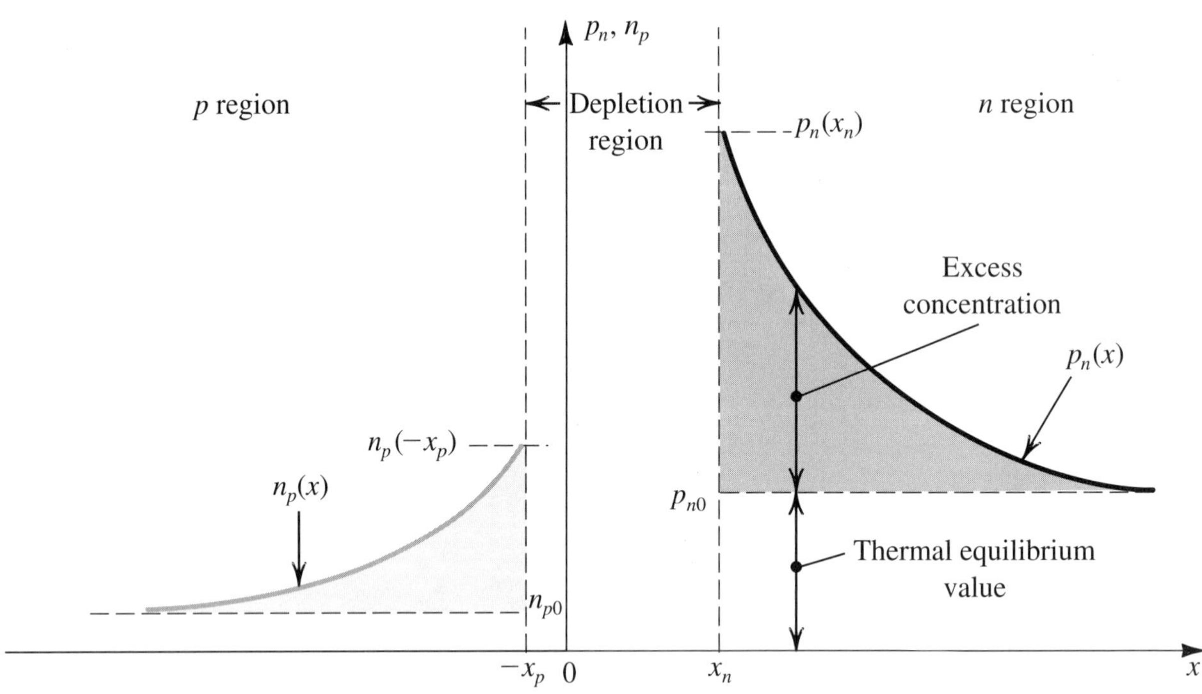

FIGURE 3.50 Minority-carrier distribution in a forward-biased *pn* junction. It is assumed that the *p* region is more heavily doped than the *n* region; $N_A \gg N_D$.

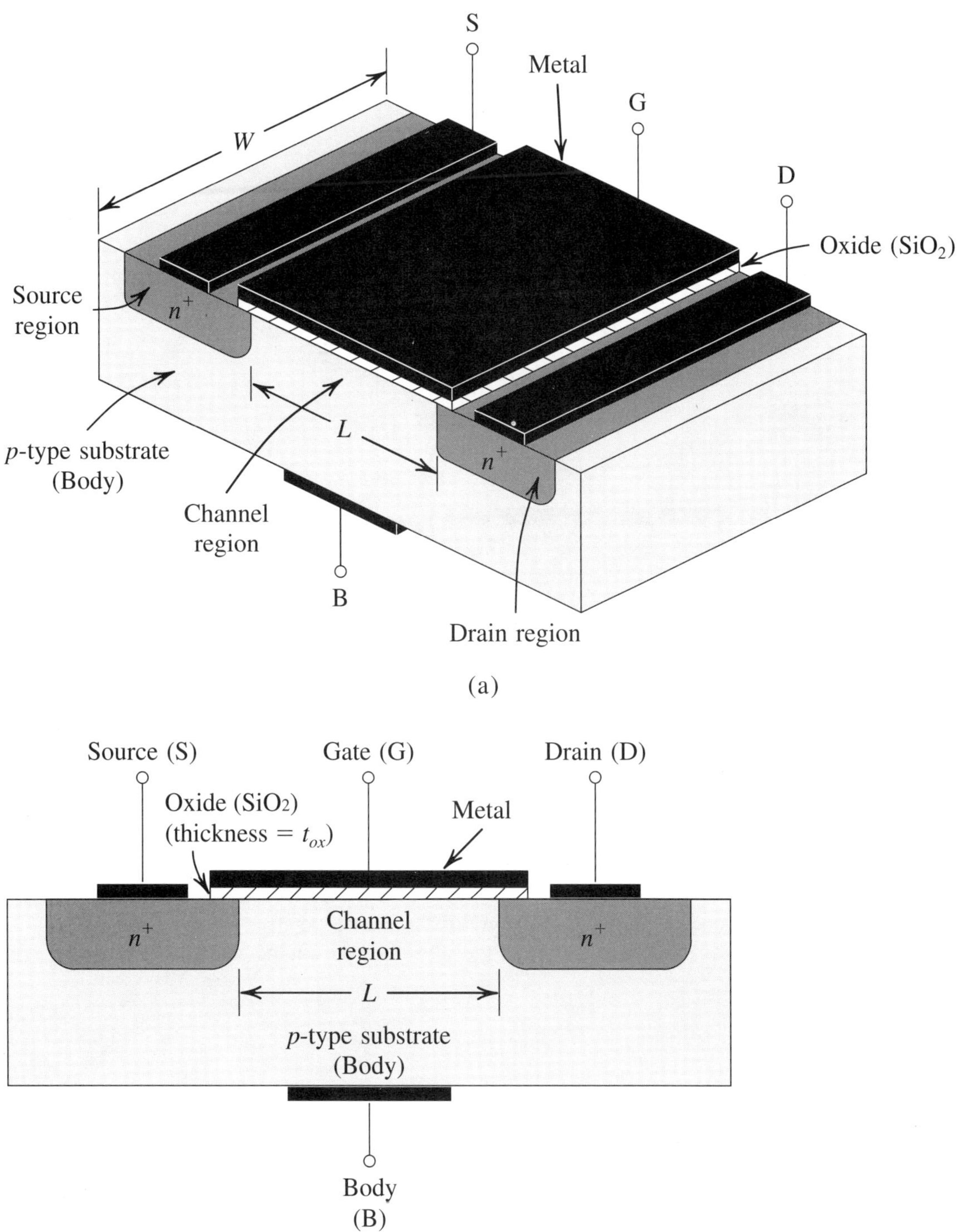

FIGURE 4.1 Physical structure of the enhancement-type NMOS transistor: **(a)** perspective view; **(b)** cross-section. Typically $L = 0.1$ to $3\ \mu$m, $W = 0.2$ to $100\ \mu$m, and the thickness of the oxide layer (t_{ox}) is in the range of 2 to 50 nm.

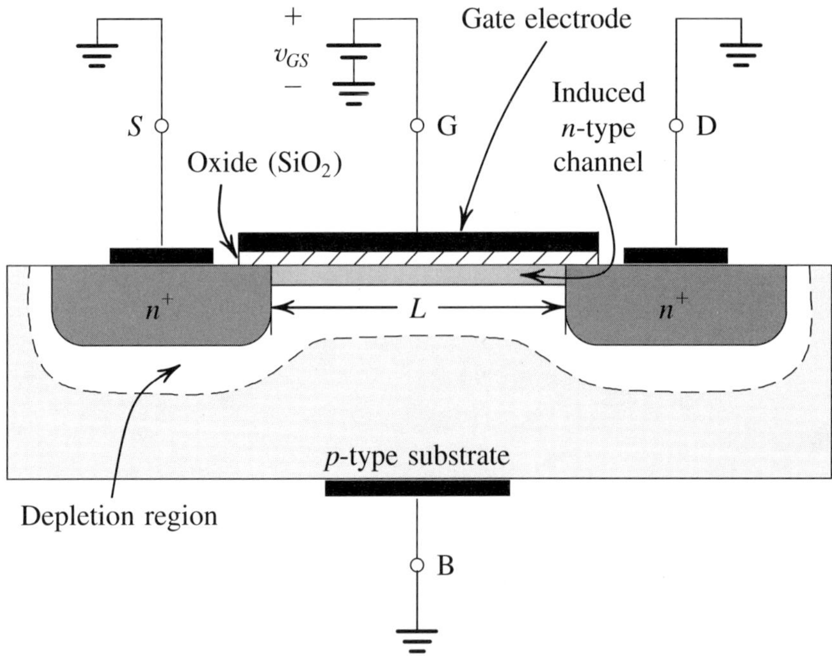

FIGURE 4.2 The enhancement-type NMOS transistor with a positive voltage applied to the gate. An n channel is induced at the top of the substrate beneath the gate.

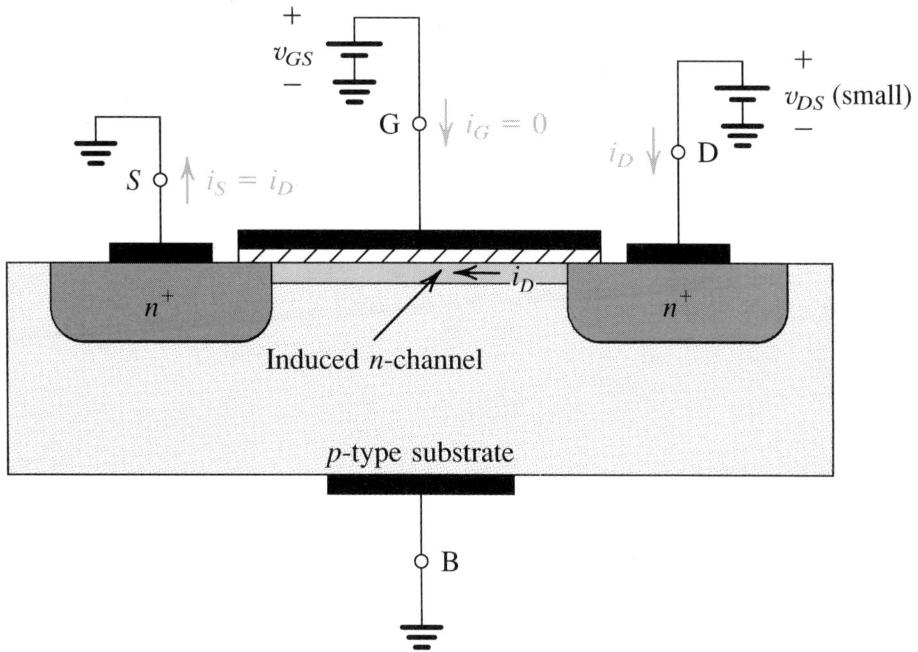

FIGURE 4.3 An NMOS transistor with $v_{GS} > V_t$ and with a small v_{DS} applied. The device acts as a resistance whose value is determined by v_{GS}. Specifically, the channel conductance is proportional to $v_{GS} - V_t$, and thus i_D is proportional to $(v_{GS} - V_t)v_{DS}$. Note that the depletion region is not shown (for simplicity).

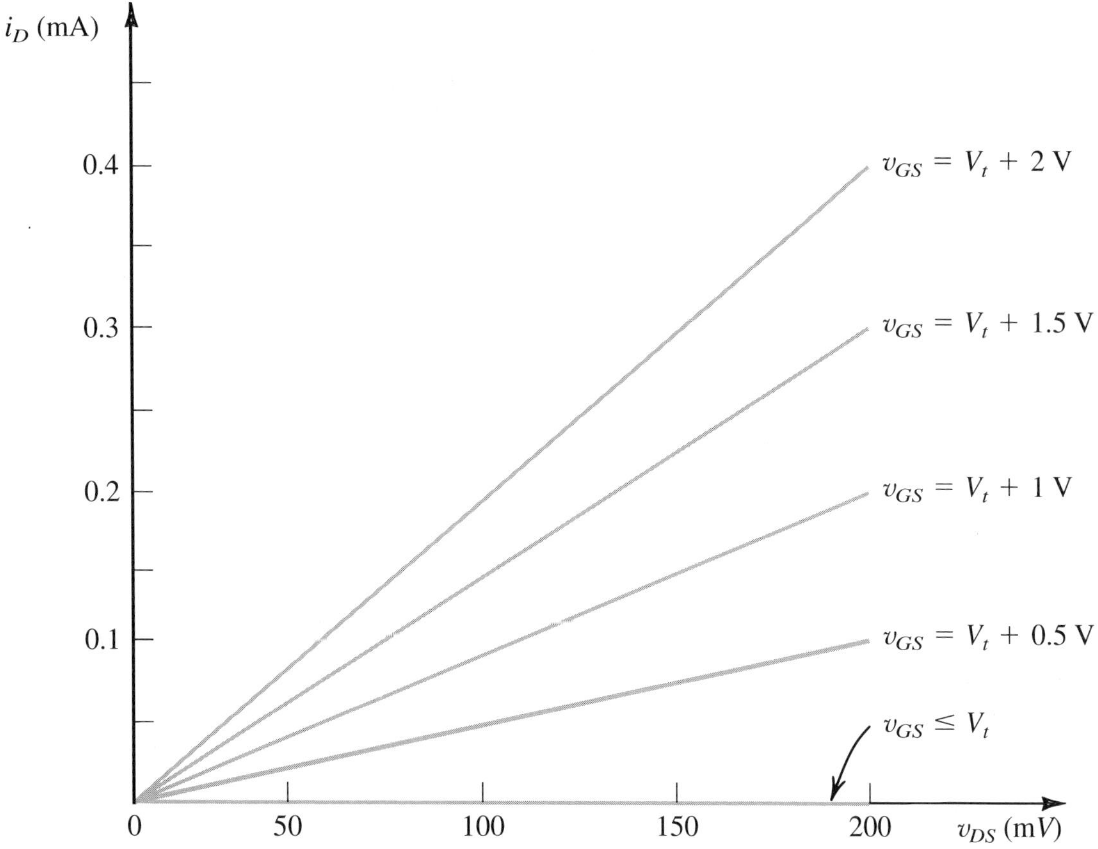

FIGURE 4.4 The i_D–v_{DS} characteristics of the MOSFET in Fig. 4.3 when the voltage applied between drain and source, v_{DS}, is kept small. The device operates as a linear resistor whose value is controlled by v_{GS}.

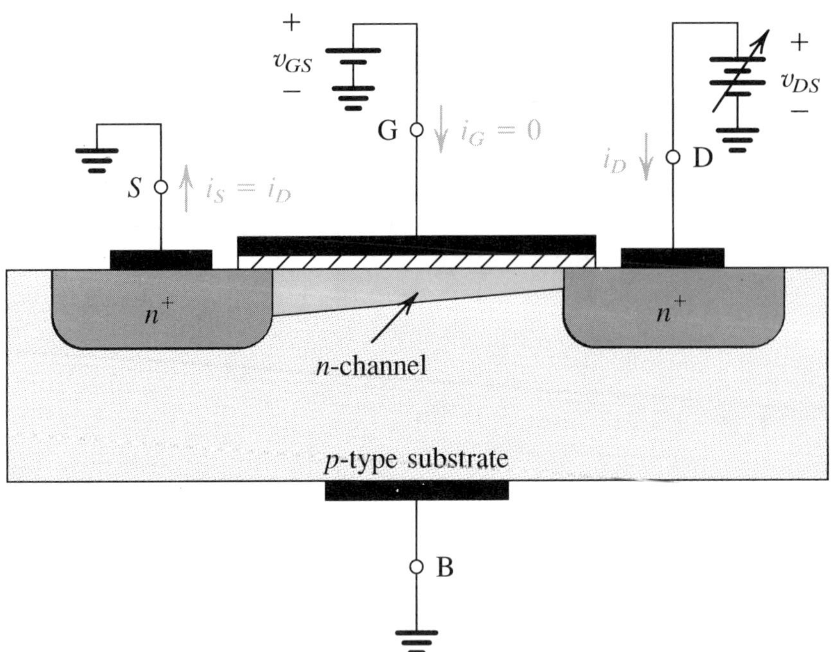

FIGURE 4.5 Operation of the enhancement NMOS transistor as v_{DS} is increased. The induced channel acquires a tapered shape, and its resistance increases as v_{DS} is increased. Here, v_{GS} is kept constant at a value > V_t.

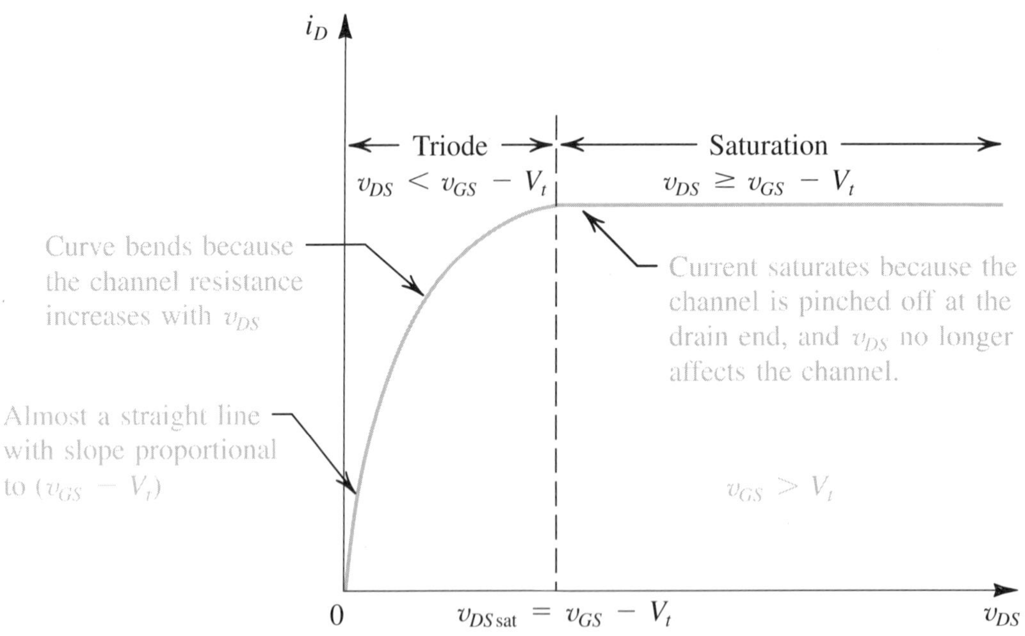

FIGURE 4.6 The drain current i_D versus the drain-to-source voltage v_{DS} for an enhancement-type NMOS transistor operated with $v_{GS} > V_t$.

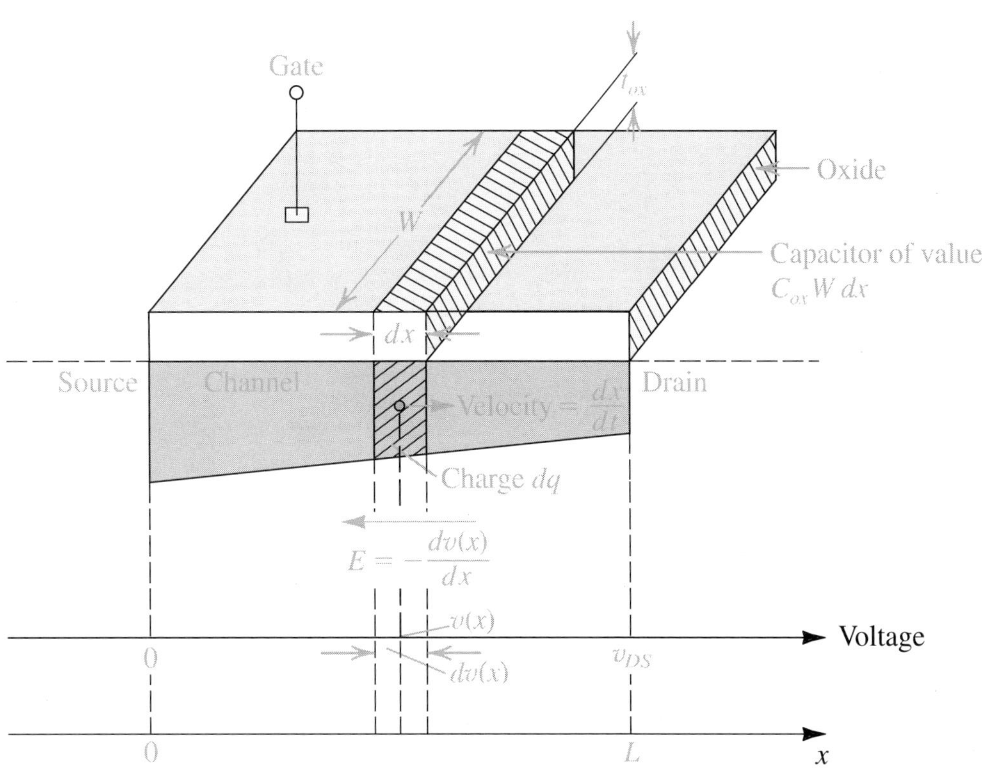

FIGURE 4.8 Derivation of the i_D–v_{DS} characteristic of the NMOS transistor.

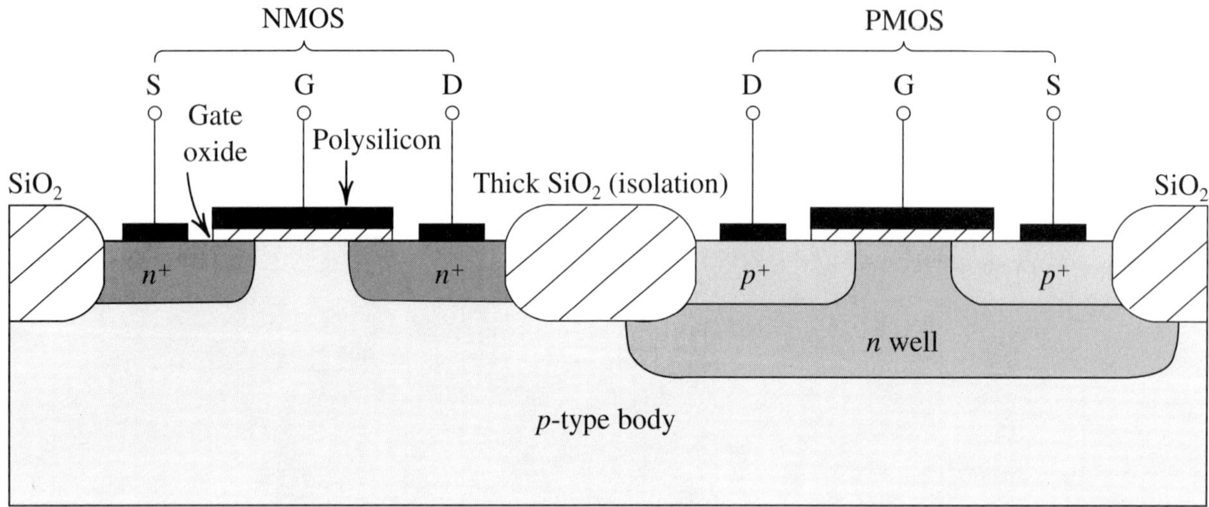

FIGURE 4.9 Cross-section of a CMOS integrated circuit. Note that the PMOS transistor is formed in a separate *n*-type region, known as an *n* well. Another arrangement is also possible in which an *n*-type body is used and the *n* device is formed in a *p* well. Not shown are the connections made to the *p*-type body and to the *n* well; the latter functions as the body terminal for the *p*-channel device.

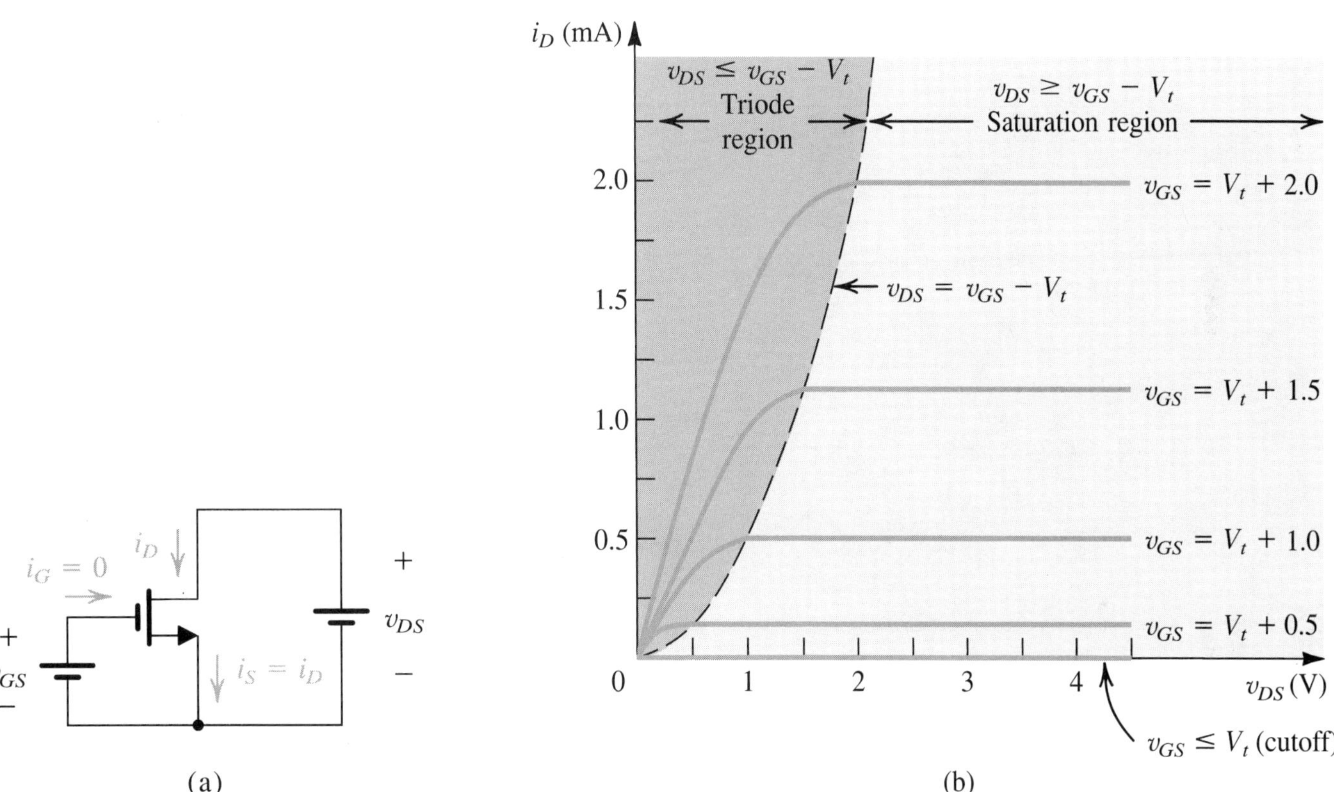

FIGURE 4.11 **(a)** An n-channel enhancement-type MOSFET with v_{GS} and v_{DS} applied and with the normal directions of current flow indicated. **(b)** The i_D–v_{DS} characteristics for a device with $k_n'(W/L) = 1.0$ mA/V^2.

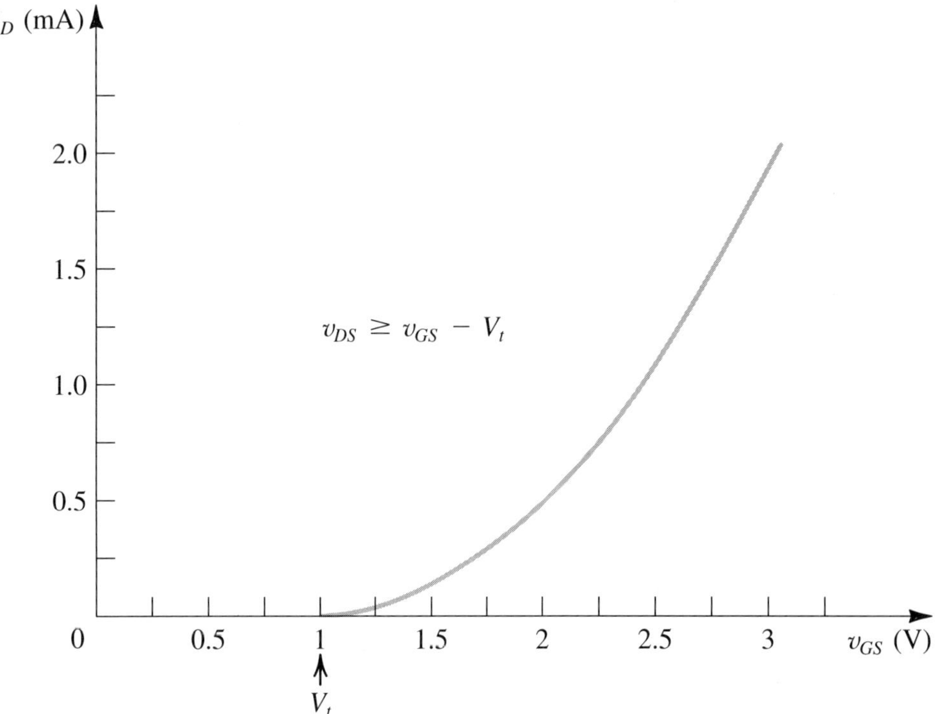

FIGURE 4.12 The i_D–v_{GS} characteristic for an enhancement-type NMOS transistor in saturation ($V_t = 1$ V, $k_n' W/L = 1.0$ mA/V^2).

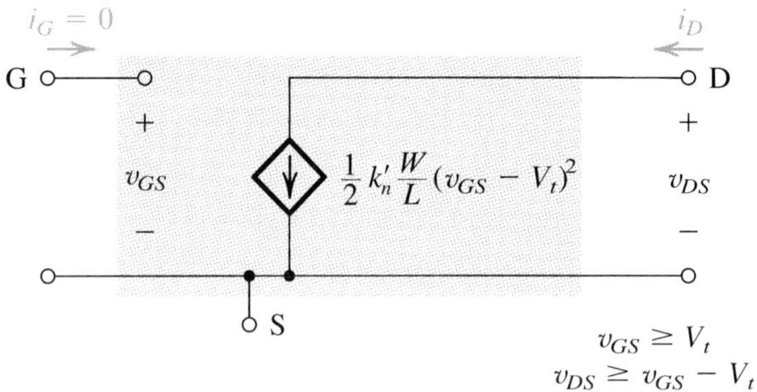

FIGURE 4.13 Large-signal equivalent-circuit model of an n-channel MOSFET operating in the saturation region.

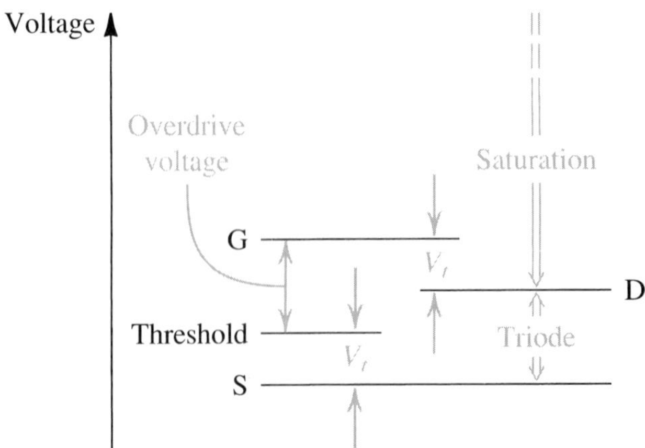

FIGURE 4.14 The relative levels of the terminal voltages of the enhancement NMOS transistor for operation in the triode region and in the saturation region.

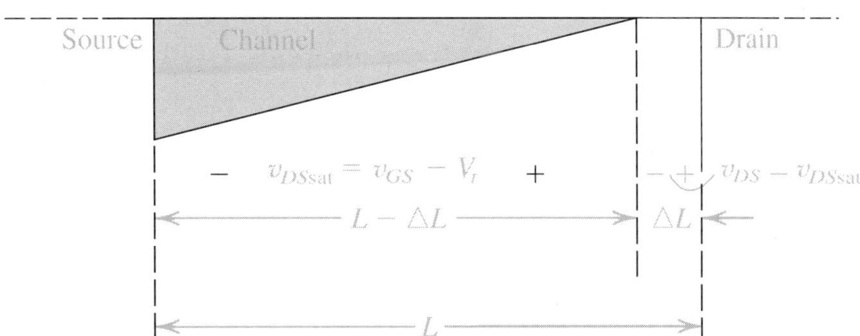

FIGURE 4.15 Increasing v_{DS} beyond v_{DSsat} causes the channel pinch-off point to move slightly away from the drain, thus reducing the effective channel length (by ΔL).

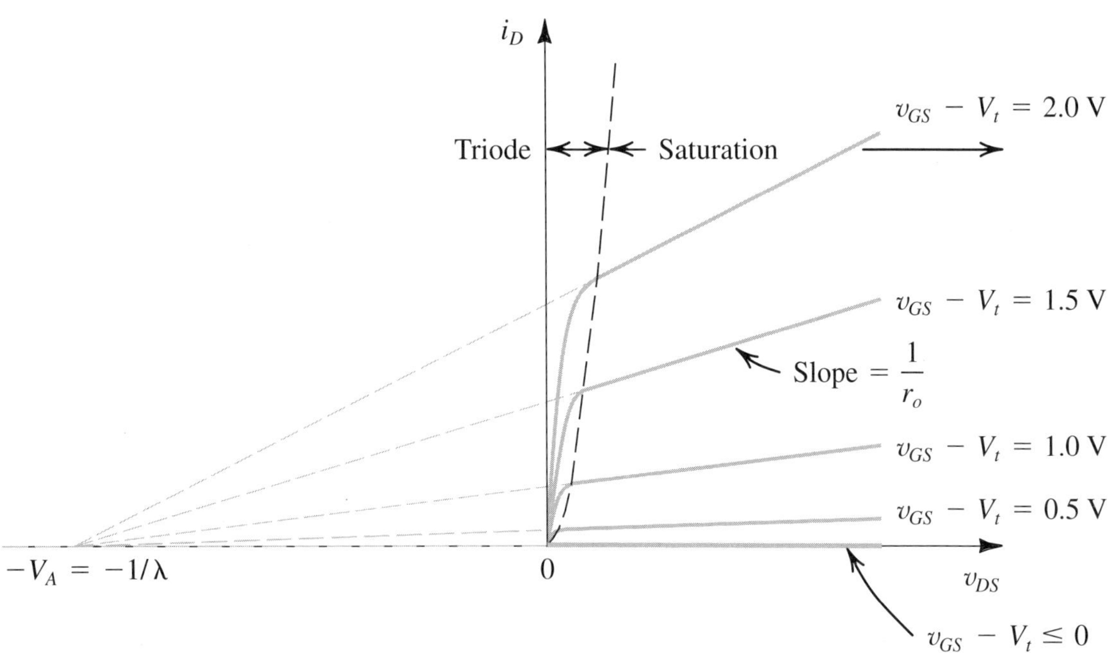

FIGURE 4.16 Effect of v_{DS} on i_D in the saturation region. The MOSFET parameter V_A depends on the process technology and, for a given process, is proportional to the channel length L.

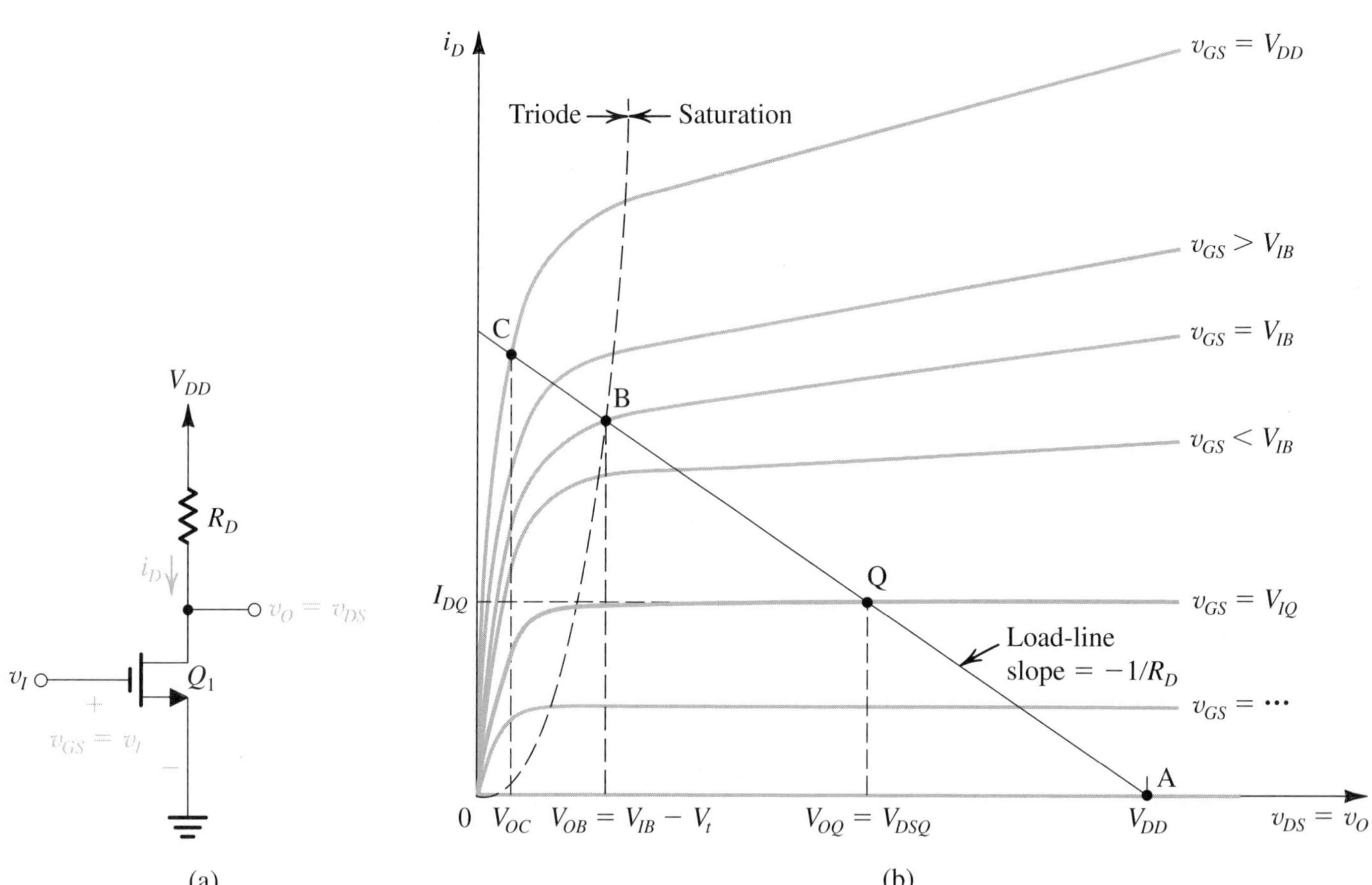

FIGURE 4.26 **(a)** Basic structure of the common-source amplifier. **(b)** Graphical construction to determine the transfer characteristic of the amplifier in (a).

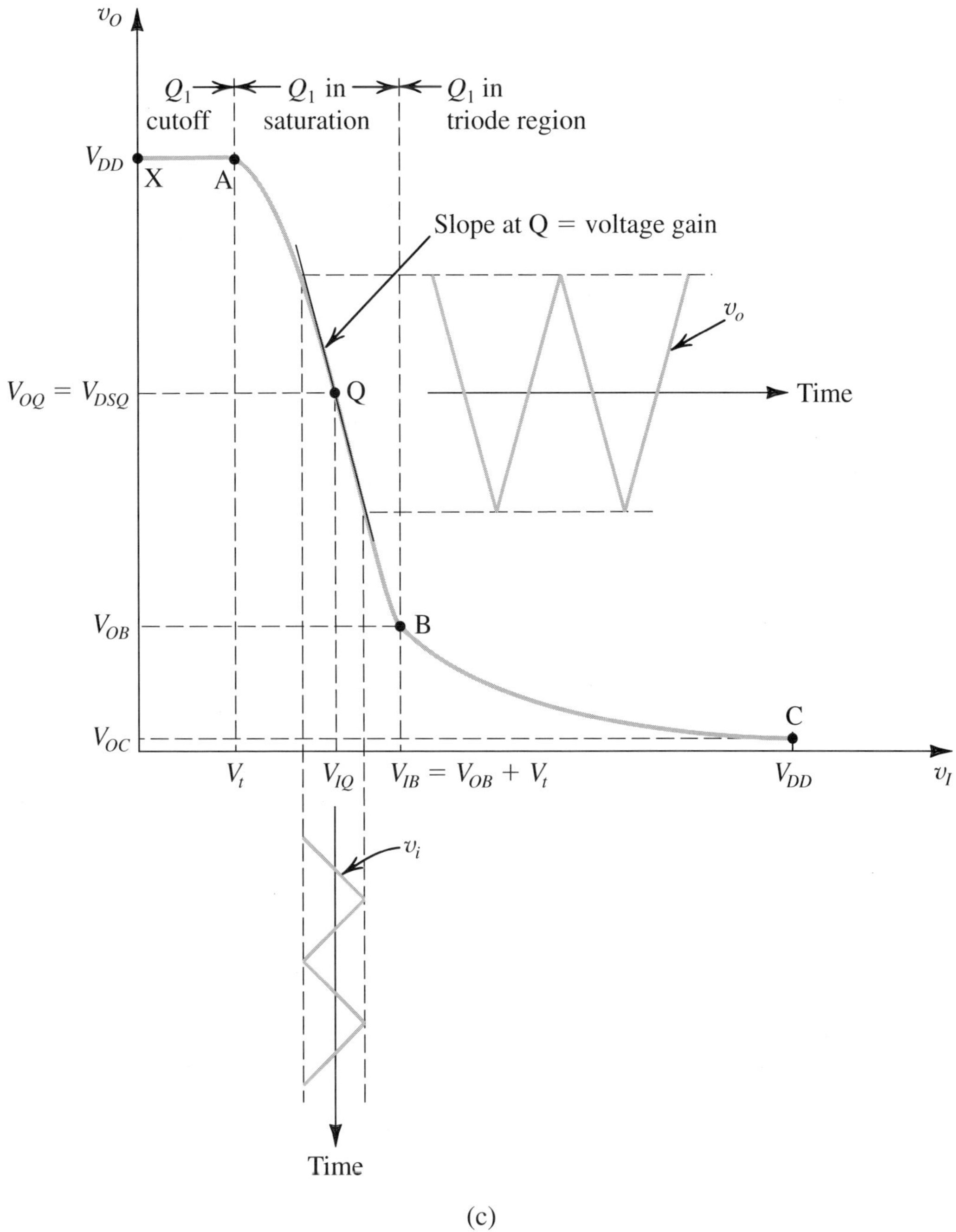

FIGURE 4.26 *(Continued)* **(c)** Transfer characteristic showing operation as an amplifier biased at point Q.

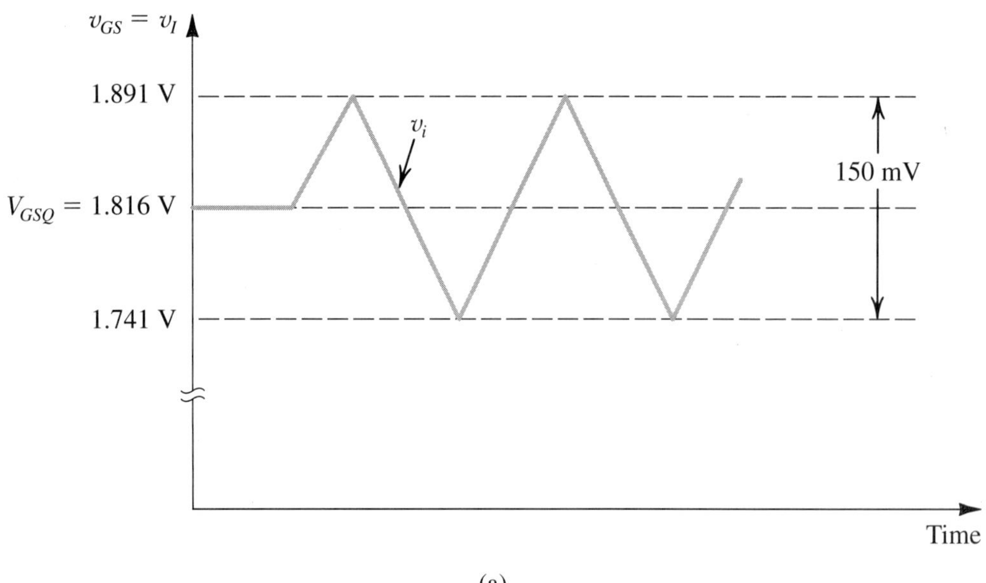

(a)

(b)

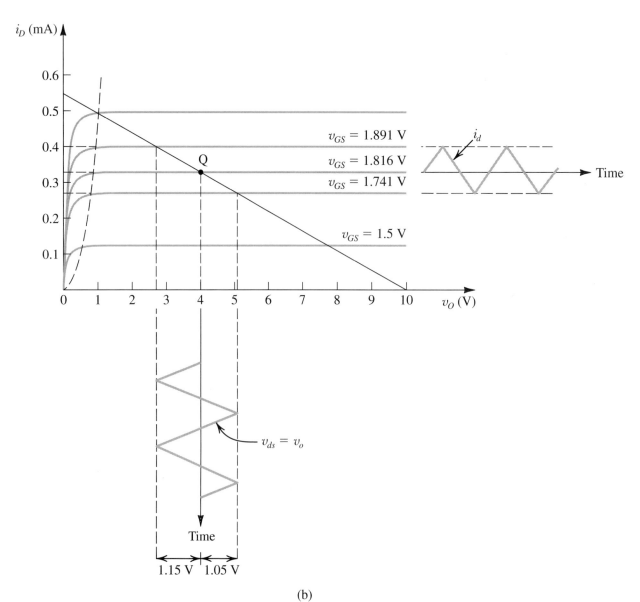

FIGURE 4.28 Example 4.8.

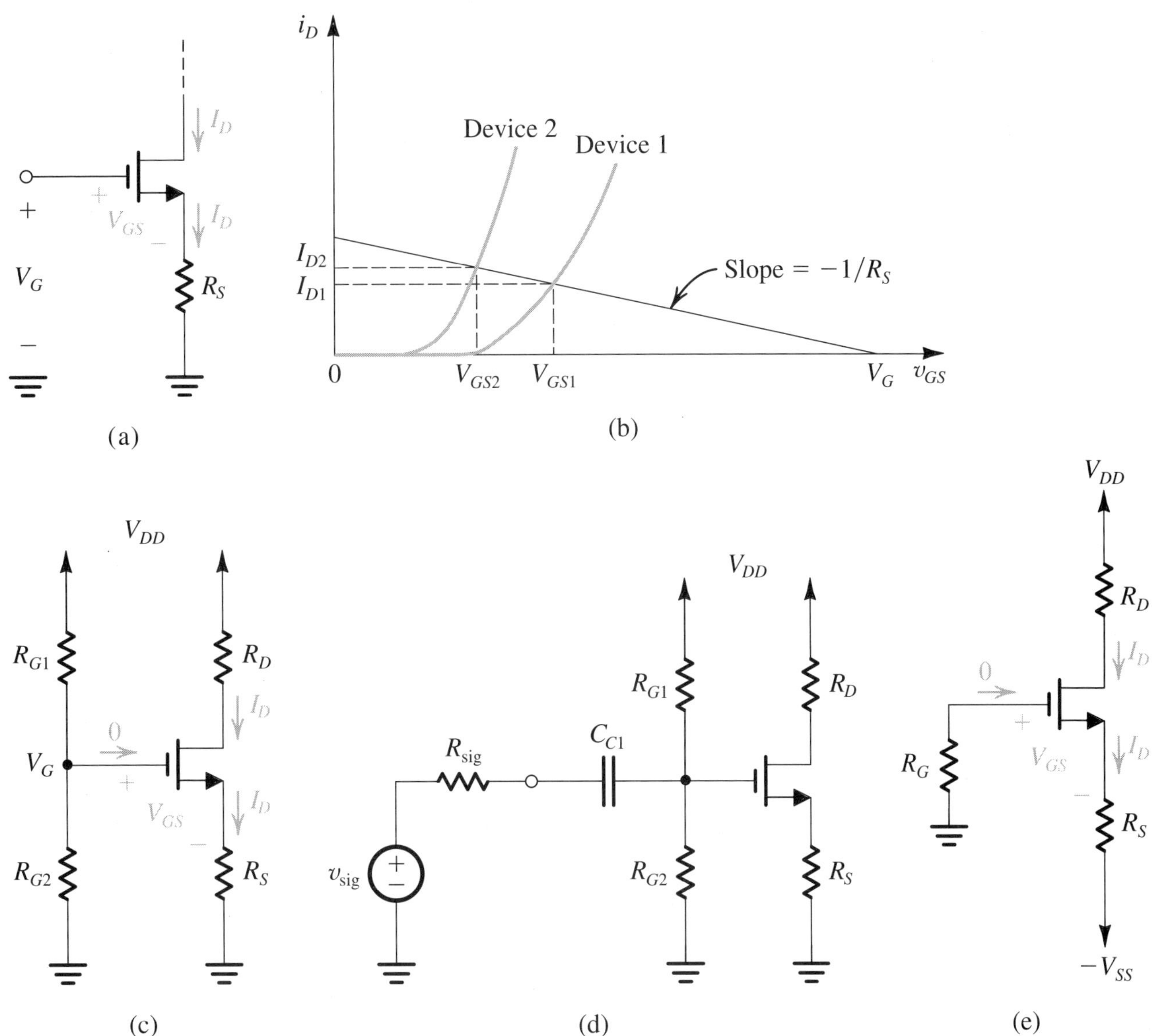

FIGURE 4.30 Biasing using a fixed voltage at the gate, V_G, and a resistance in the source lead, R_S: **(a)** basic arrangement; **(b)** reduced variability in I_D; **(c)** practical implementation using a single supply; **(d)** coupling of a signal source to the gate using a capacitor C_{C1}; **(e)** practical implementation using two supplies.

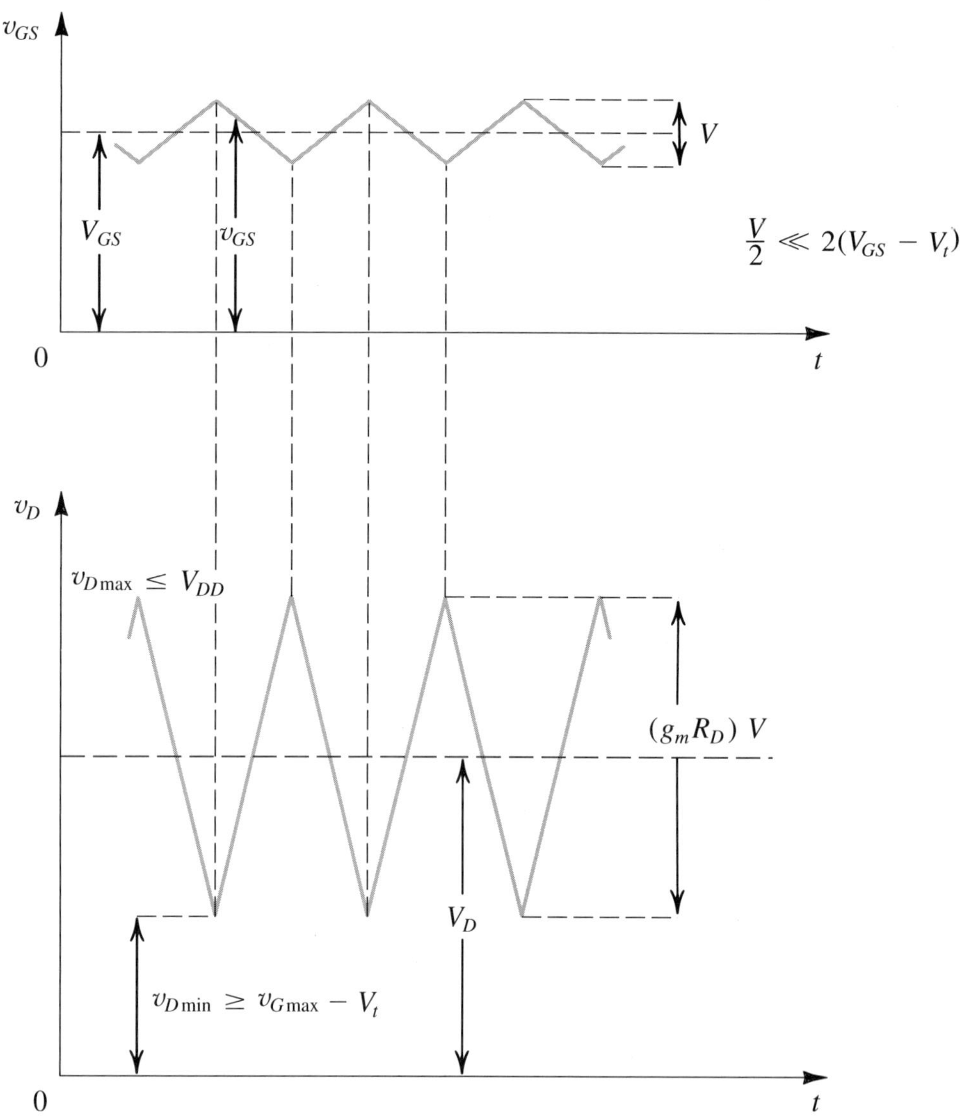

FIGURE 4.36 Total instantaneous voltages v_{GS} and v_D for the circuit in Fig. 4.34.

TABLE 4.2 Small-Signal Equivalent-Circuit Models for the MOSFET

Small-Signal Parameters

NMOS transistors:

- Transconductance:

$$g_m = \mu_n C_{ox} \frac{W}{L} V_{OV} = \sqrt{2\mu_n C_{ox} \frac{W}{L} I_D} = \frac{2I_D}{V_{OV}}$$

- Output resistance:

$$r_o = V_A / I_D = 1/\lambda I_D$$

- Body transconductance:

$$g_{mb} = \chi g_m = \frac{\gamma}{2\sqrt{2\phi_f + V_{SB}}} g_m$$

PMOS transistors:

Same formulas as for NMOS *except* using $|V_{OV}|$, $|V_A|$, $|\lambda|$, $|\gamma|$, $|V_{SB}|$, and $|\chi|$ and replacing μ_n with μ_p.

Small-Signal Equivalent Circuit Models when $|V_{SB}| = 0$ (i.e., No Body Effect)

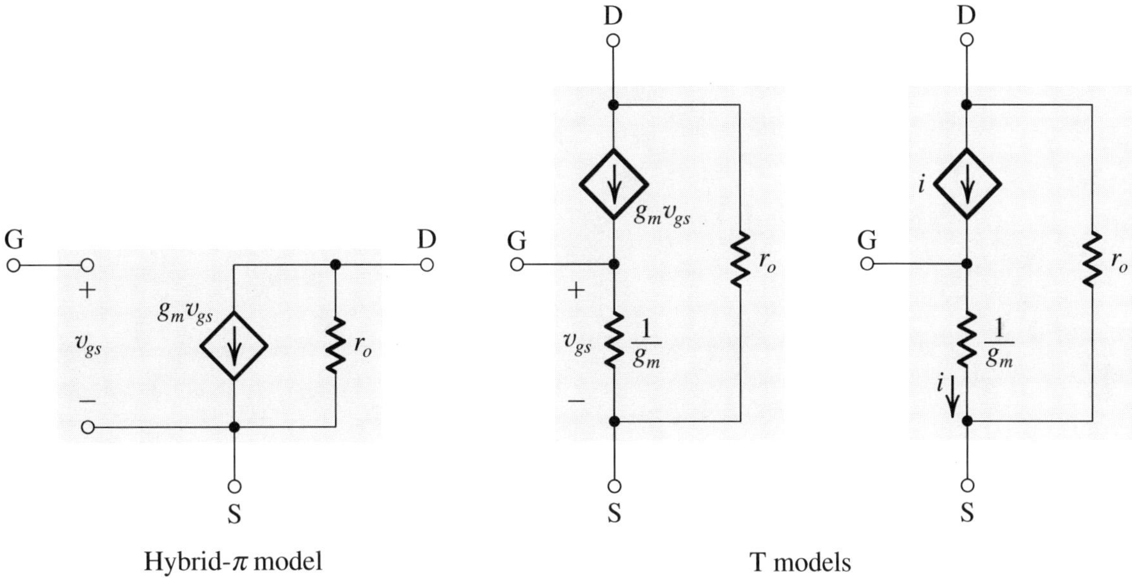

Hybrid-π model

T models

Small-Signal Circuit Model when $|V_{SB}| \neq 0$ (i.e., Including the Body Effect)

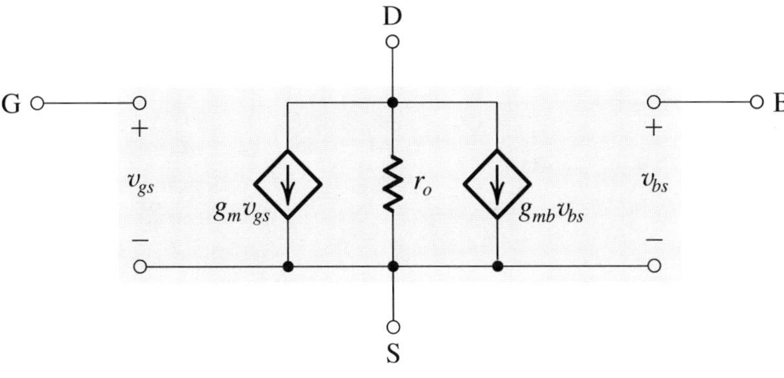

Hybrid π model

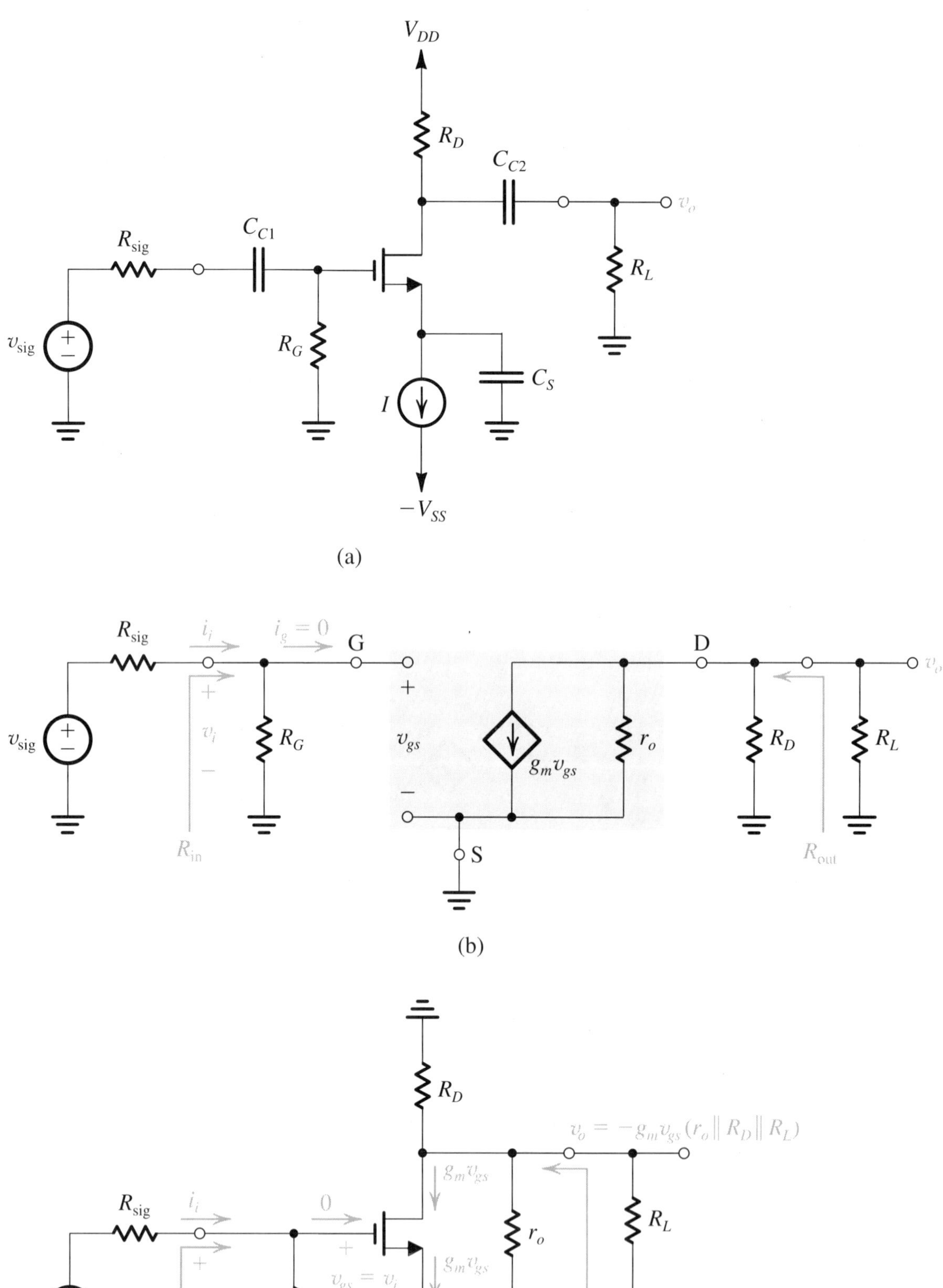

FIGURE 4.43 **(a)** Common-source amplifier based on the circuit of Fig. 4.42. **(b)** Equivalent circuit of the amplifier for small-signal analysis. **(c)** Small-signal analysis performed directly on the amplifier circuit with the MOSFET model implicitly utilized.

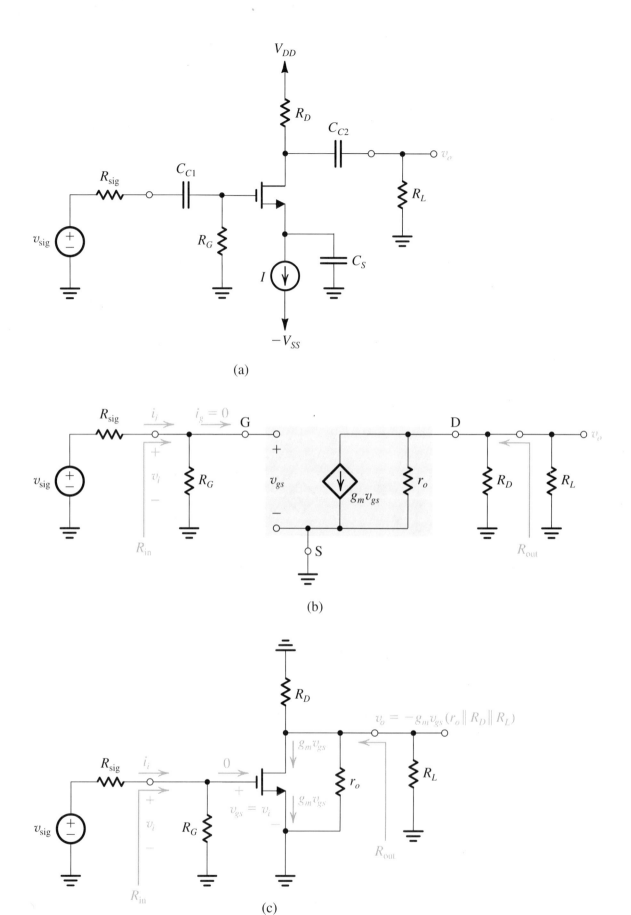

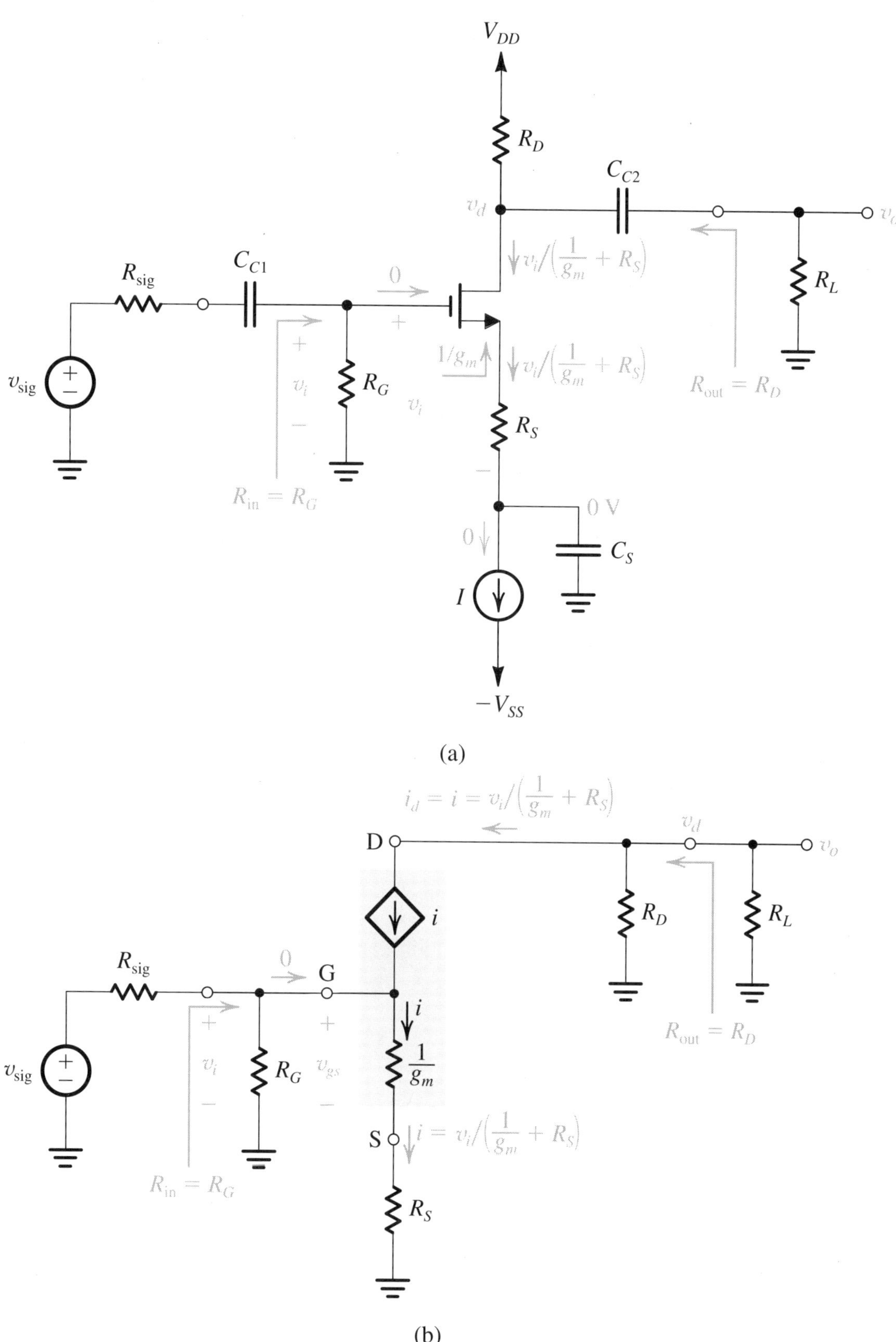

FIGURE 4.44 **(a)** Common-source amplifier with a resistance R_S in the source lead. **(b)** Small-signal equivalent circuit with r_o neglected.

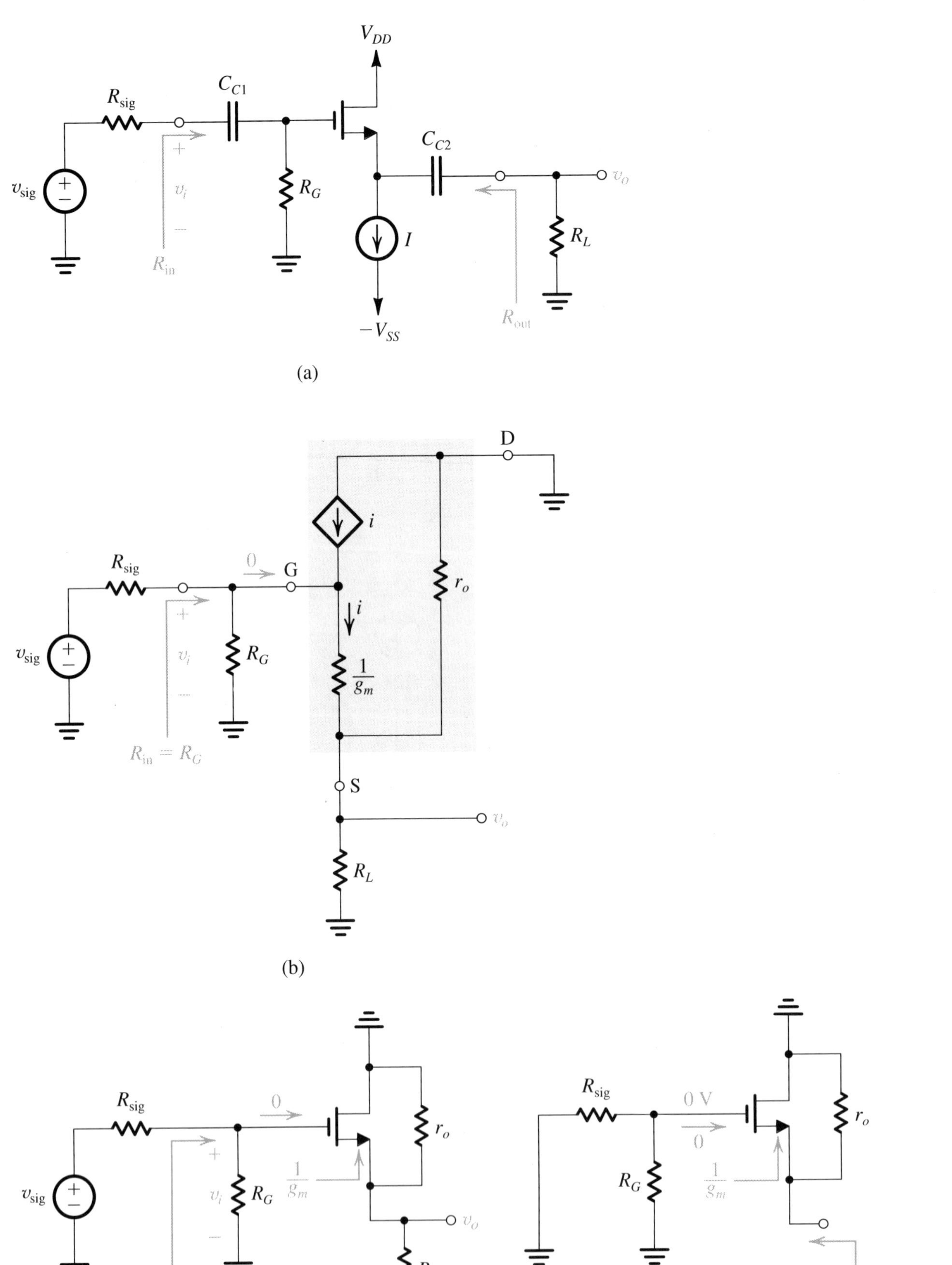

FIGURE 4.46 **(a)** A common-drain or source-follower amplifier. **(b)** Small-signal equivalent-circuit model. **(c)** Small-signal analysis performed directly on the circuit. **(d)** Circuit for determining the output resistance R_{out} of the source follower.

TABLE 4.4 Characteristics of Single-Stage Discrete MOS Amplifiers

Common-Source

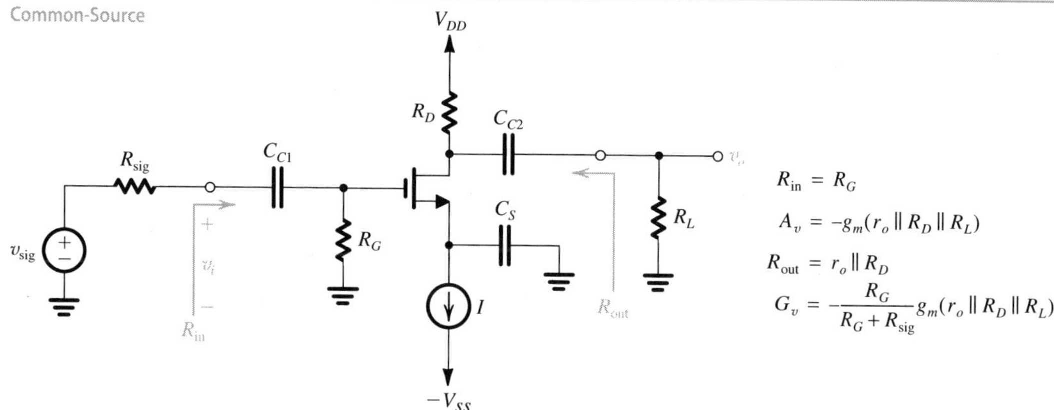

$$R_{in} = R_G$$
$$A_v = -g_m(r_o \parallel R_D \parallel R_L)$$
$$R_{out} = r_o \parallel R_D$$
$$G_v = -\frac{R_G}{R_G + R_{sig}} g_m(r_o \parallel R_D \parallel R_L)$$

Common-Source with Source Resistance

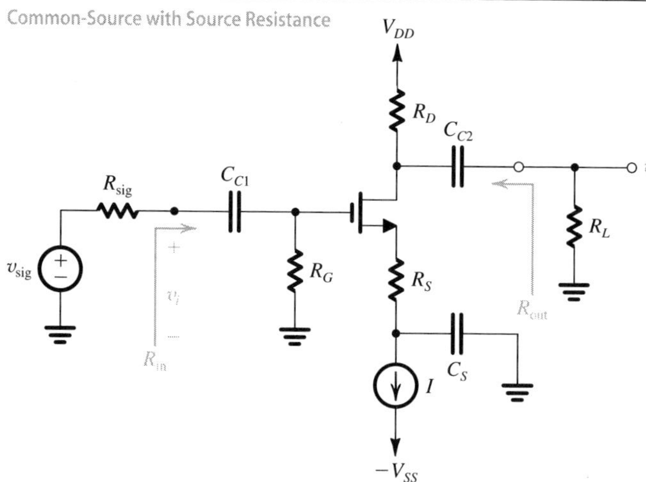

Neglecting r_o:

$$R_{in} = R_G$$
$$A_v = -\frac{R_D \parallel R_L}{\frac{1}{g_m} + R_S} = -\frac{g_m(R_D \parallel R_L)}{1 + g_m R_S}$$
$$R_{out} = R_D$$
$$G_v = -\frac{R_G}{R_G + R_{sig}} \frac{g_m(R_D \parallel R_L)}{1 + g_m R_S}$$
$$\frac{v_{gs}}{v_i} = \frac{1}{1 + g_m R_S}$$

Common Gate

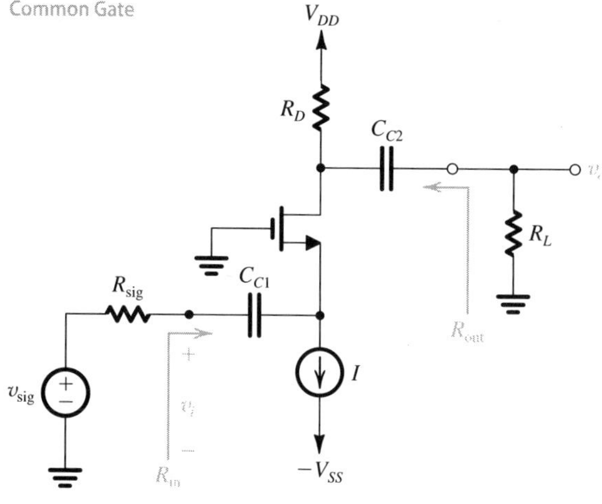

Neglecting r_o:

$$R_{in} = \frac{1}{g_m}$$
$$A_v = g_m(R_D \parallel R_L)$$
$$R_{out} = R_D$$
$$G_v = \frac{1}{1 + g_m R_{sig}} g_m(R_D \parallel R_L)$$

Common-Drain or Source Follower

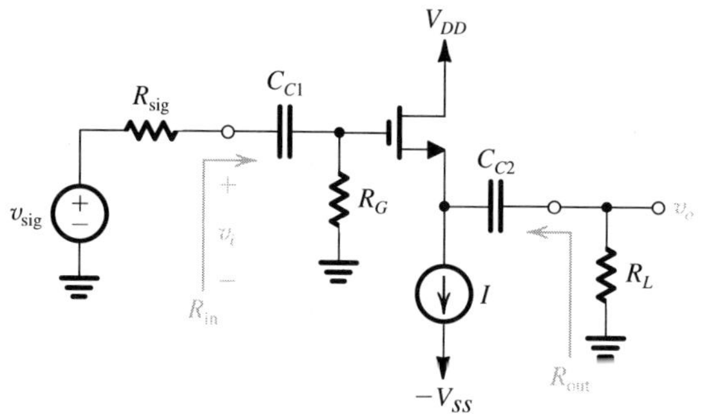

$$R_{in} = R_G$$
$$A_v = \frac{r_o \parallel R_L}{(r_o \parallel R_L) + \frac{1}{g_m}}$$
$$R_{out} = r_o \parallel \frac{1}{g_m} \cong \frac{1}{g_m}$$
$$G_v = \frac{R_G}{R_G + R_{sig}} \frac{r_o \parallel R_L}{(r_o \parallel R_L) + \frac{1}{g_m}}$$

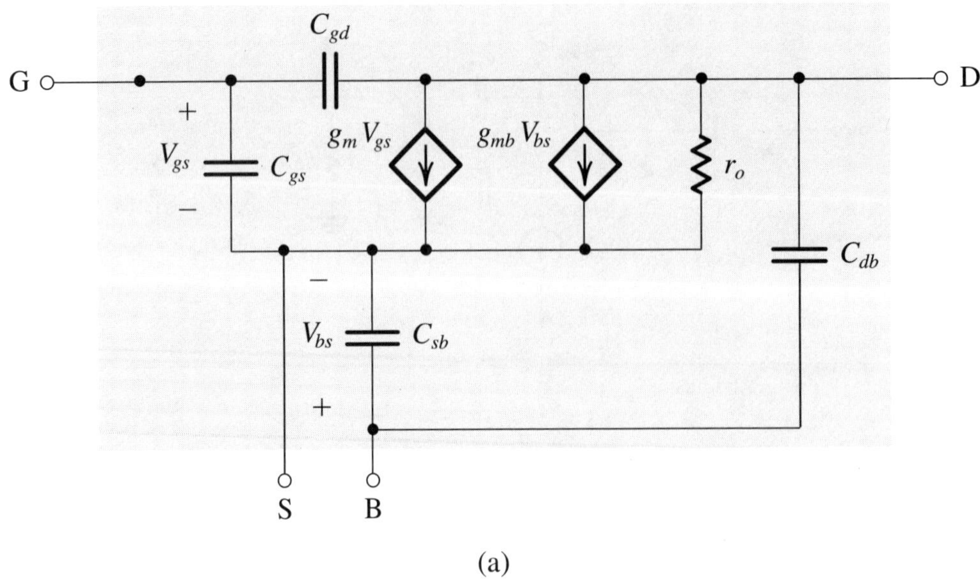

(a)

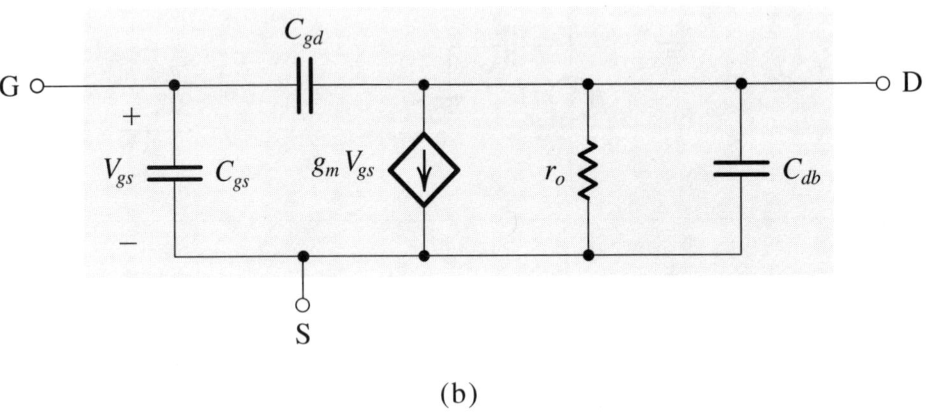

(b)

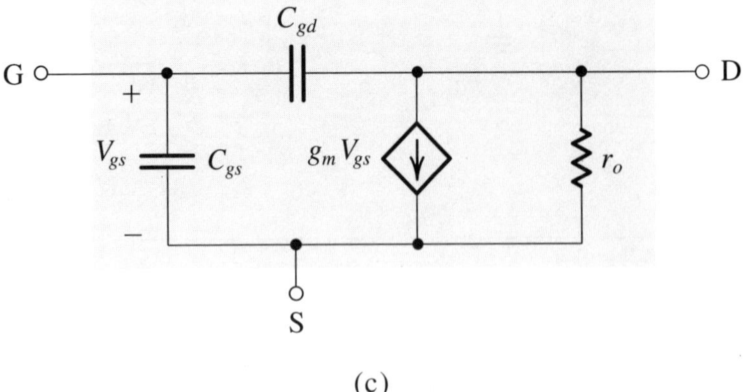

(c)

FIGURE 4.47 **(a)** High-frequency equivalent circuit model for the MOSFET. **(b)** The equivalent circuit for the case in which the source is connected to the substrate (body). **(c)** The equivalent circuit model of (b) with C_{db} neglected (to simplify analysis).

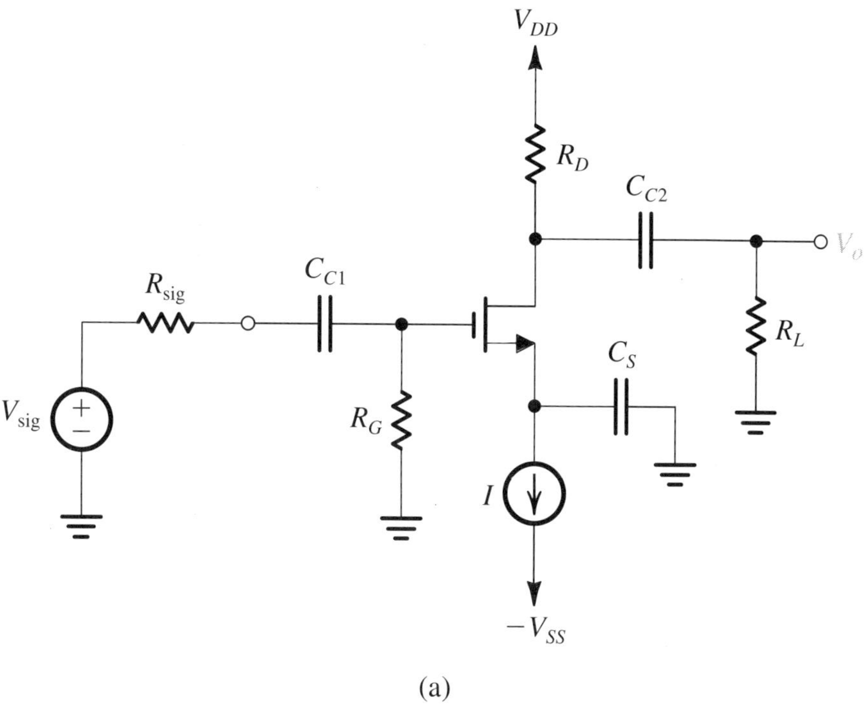

(a)

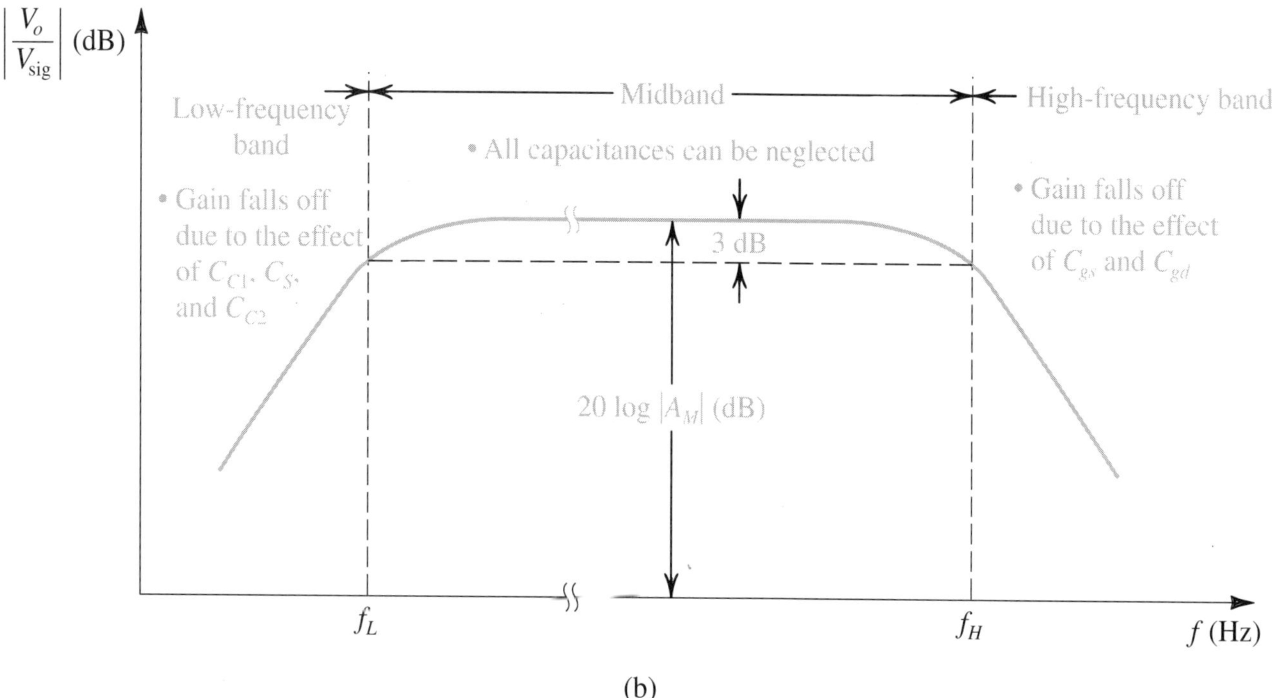

(b)

FIGURE 4.49 **(a)** Capacitively coupled common-source amplifier. **(b)** A sketch of the frequency response of the amplifier in (a) delineating the three frequency bands of interest.

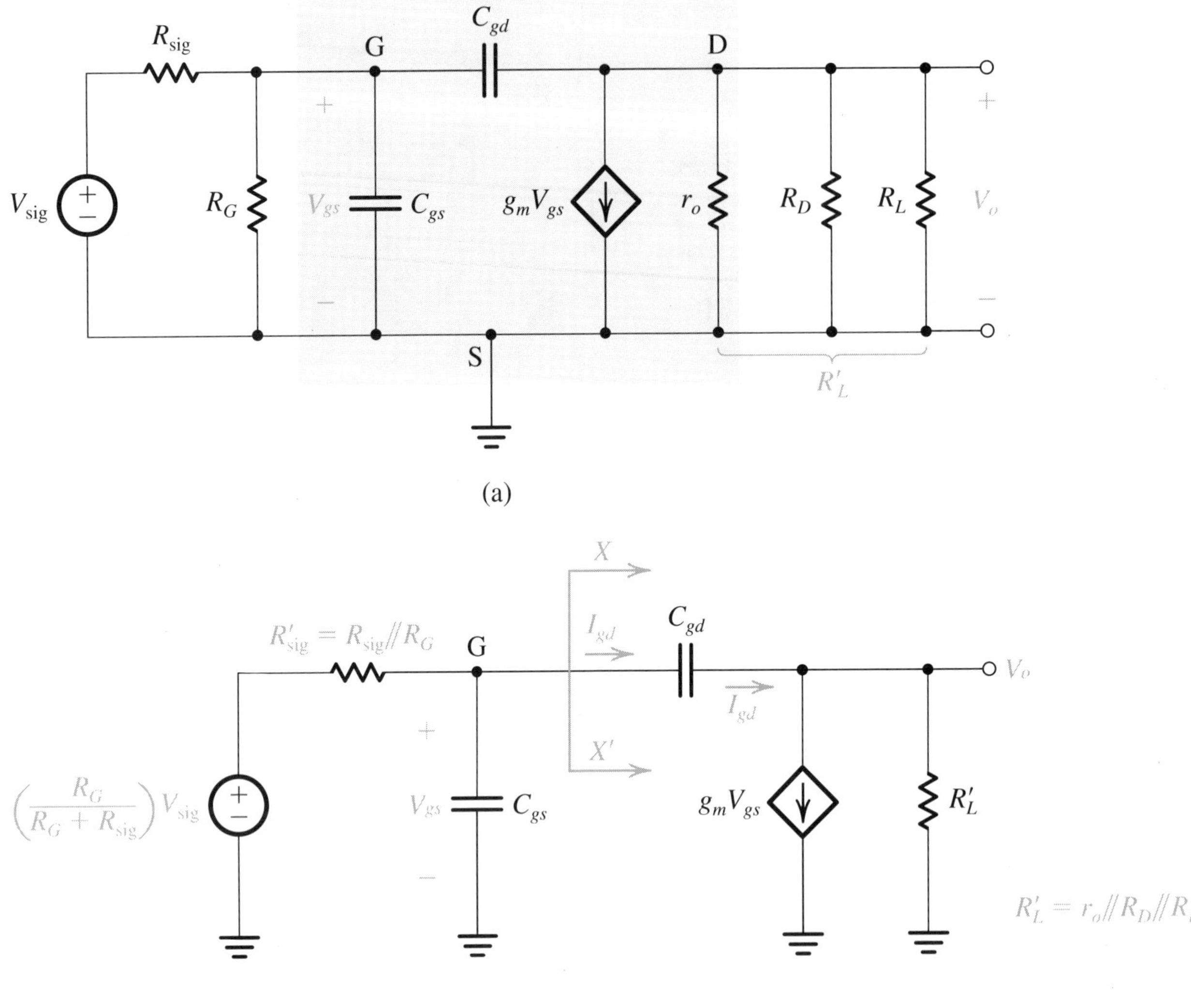

FIGURE 4.50 Determining the high-frequency response of the CS amplifier: **(a)** equivalent circuit; **(b)** the circuit of (a) simplified at the input and the output;

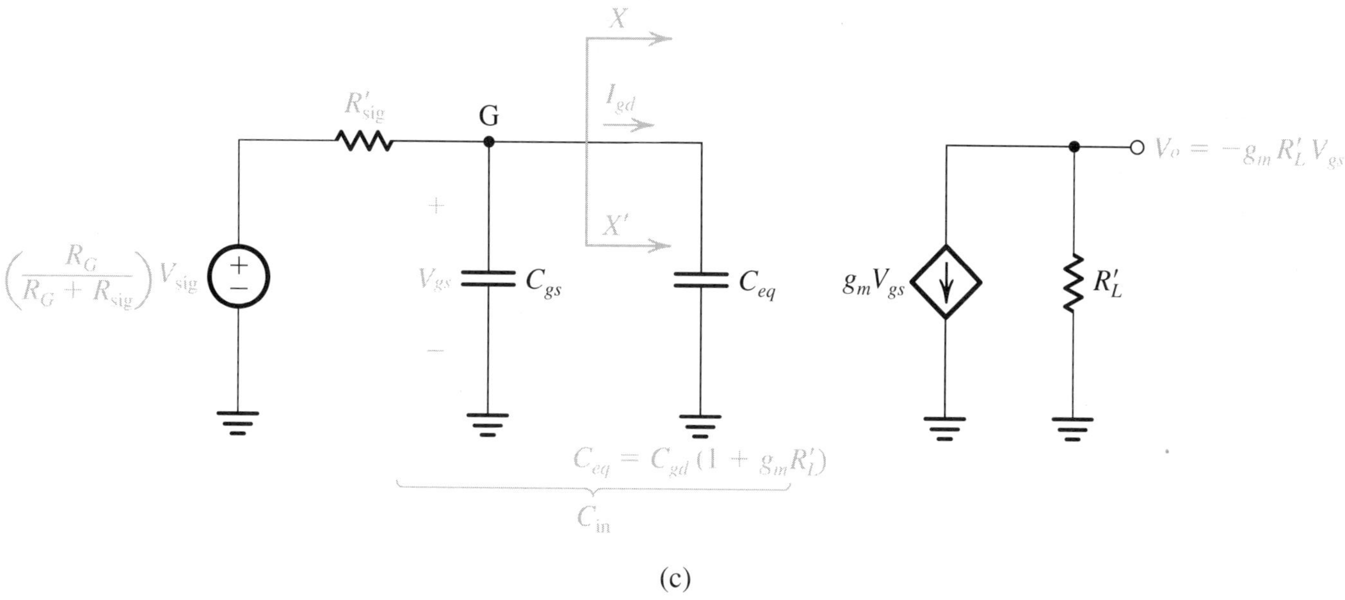

(c)

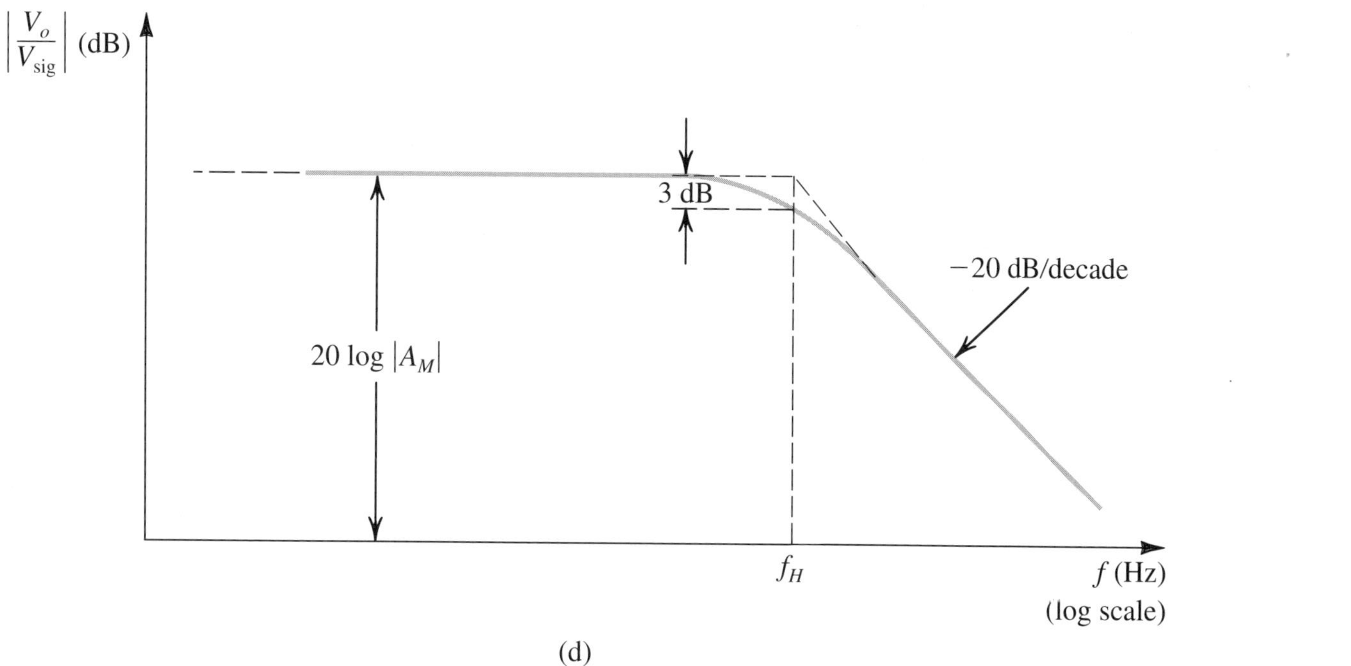

(d)

FIGURE 4.50 *(Continued)* **(c)** the equivalent circuit with C_{gd} replaced at the input side with the equivalent capacitance C_{eq}; **(d)** the frequency response plot, which is that of a low-pass single-time-constant circuit.

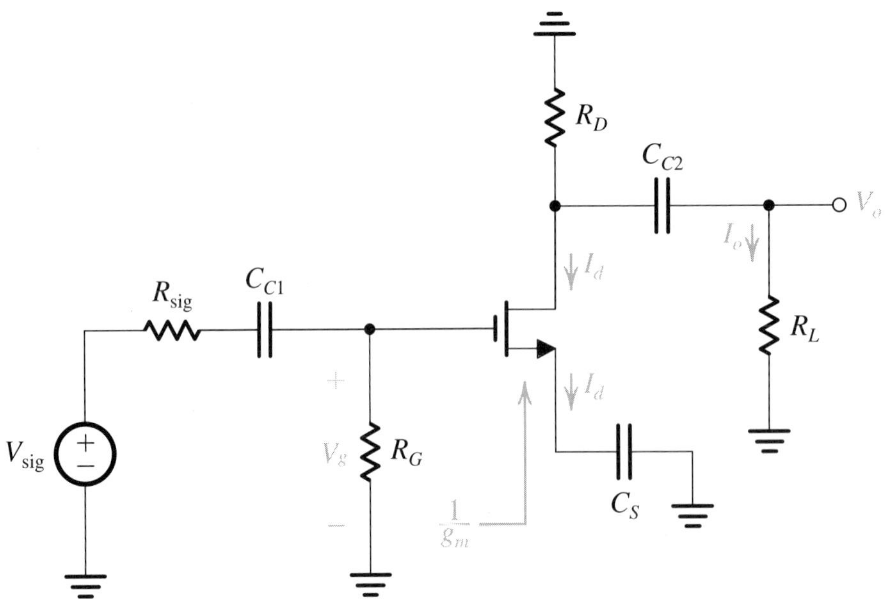

FIGURE 4.51 Analysis of the CS amplifier to determine its low-frequency transfer function. For simplicity, r_o is neglected.

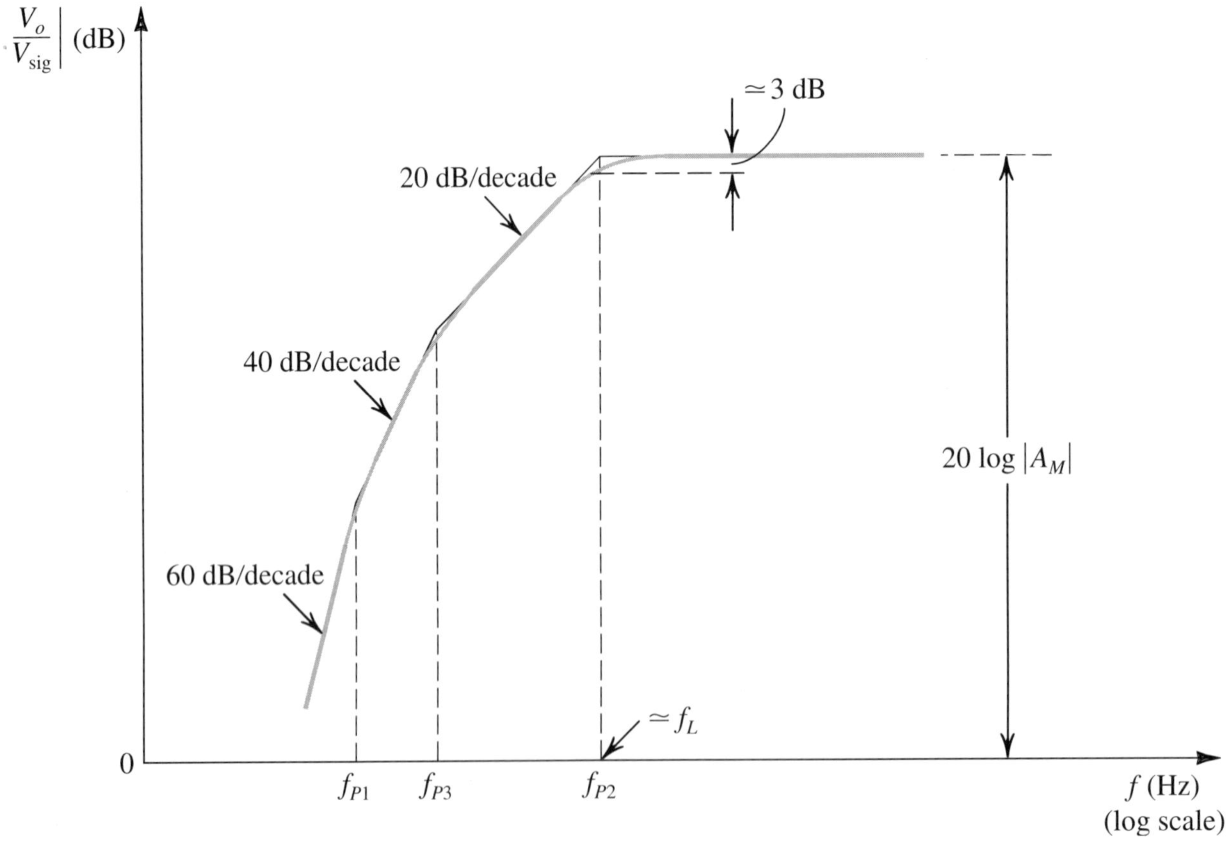

FIGURE 4.52 Sketch of the low-frequency magnitude response of a CS amplifier for which the three break frequencies are sufficiently separated for their effects to appear distinct.

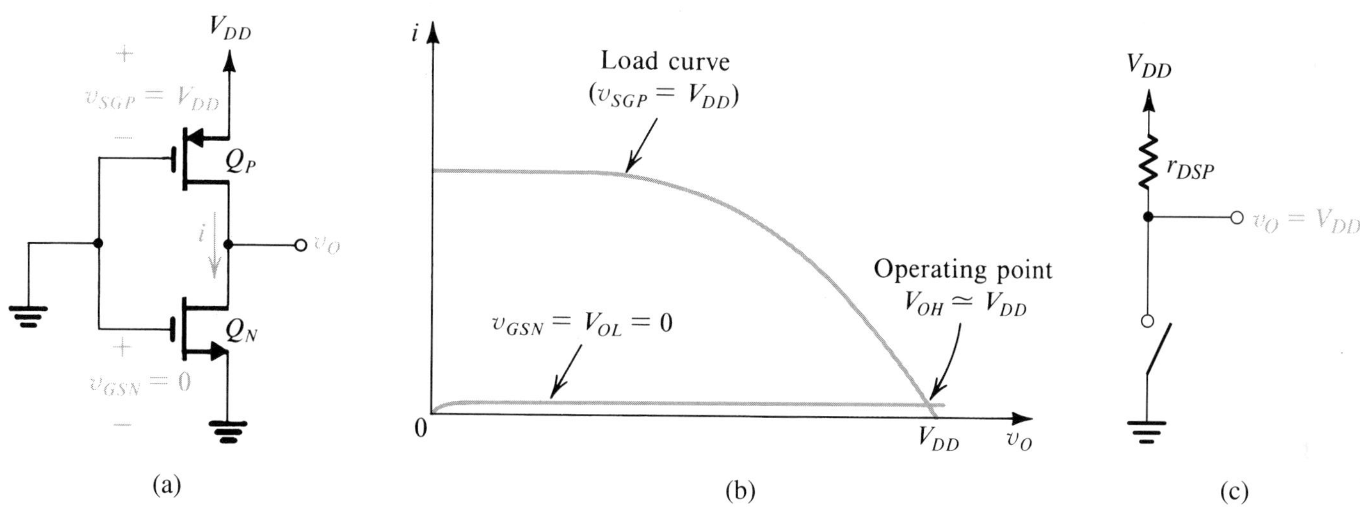

FIGURE 4.55 Operation of the CMOS inverter when v_I is low: **(a)** circuit with $v_I = 0$ V (logic-0 level, or V_{OL}); **(b)** graphical construction to determine the operating point; **(c)** equivalent circuit.

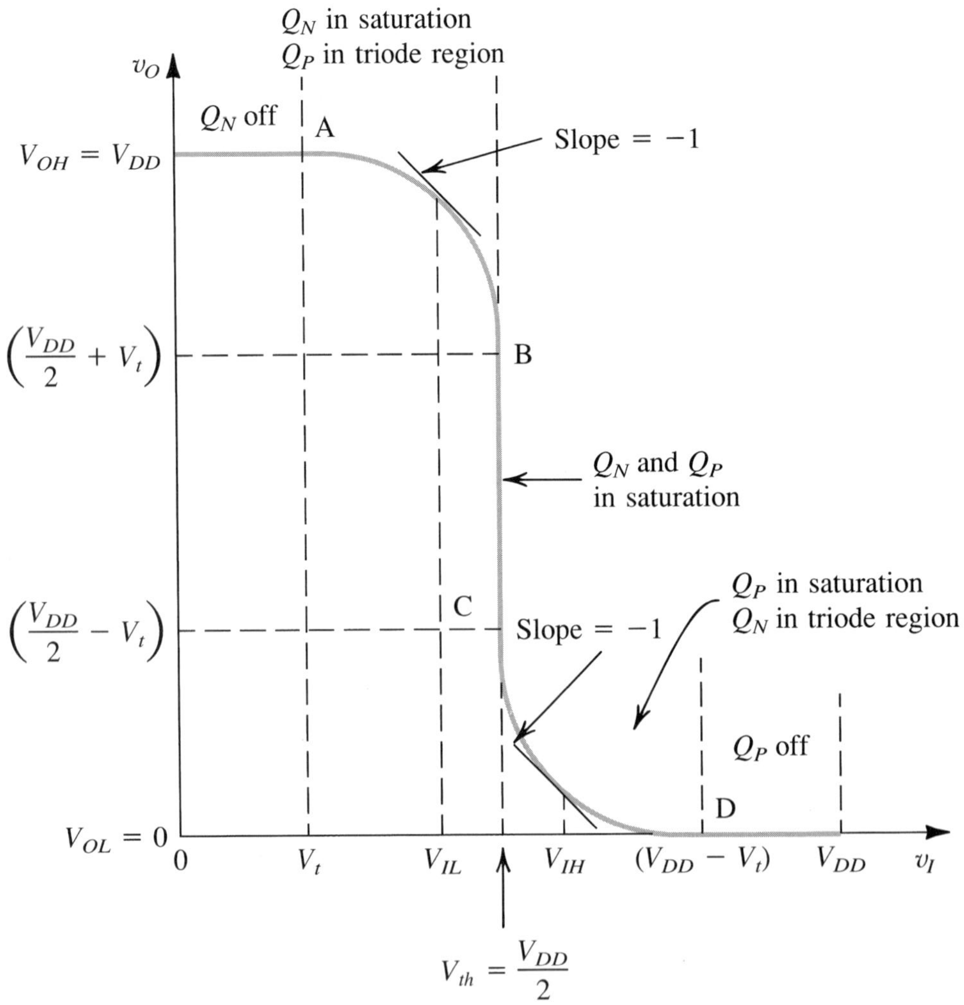

FIGURE 4.56 The voltage transfer characteristic of the CMOS inverter.

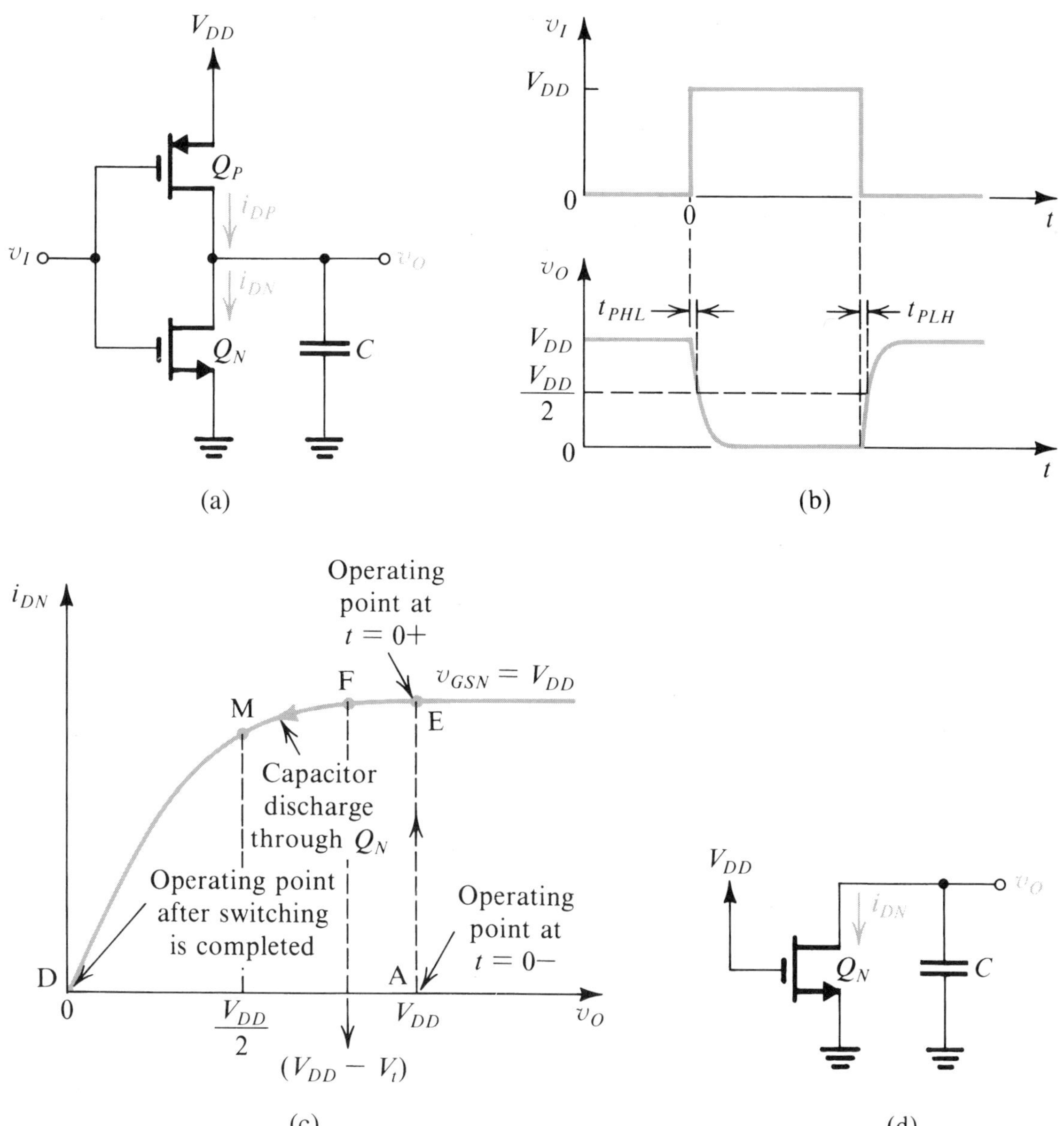

FIGURE 4.57 Dynamic operation of a capacitively loaded CMOS inverter: **(a)** circuit; **(b)** input and output waveforms; **(c)** trajectory of the operating point as the input goes high and C discharges through Q_N; **(d)** equivalent circuit during the capacitor discharge.

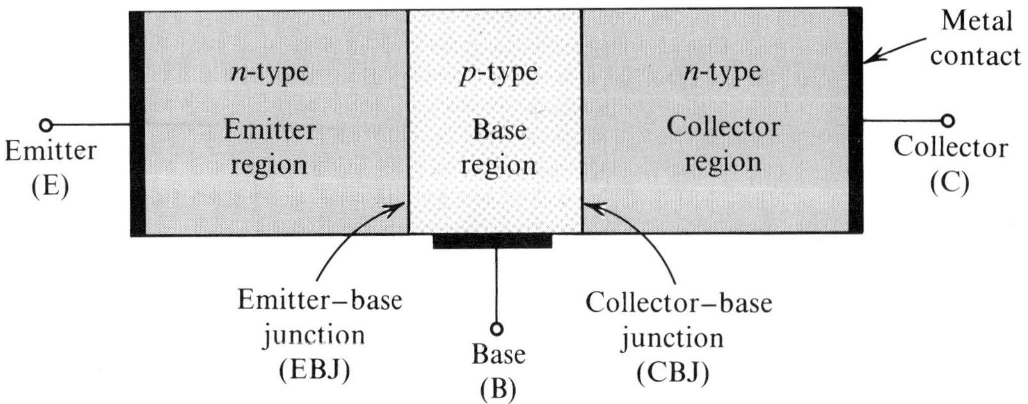

FIGURE 5.1 A simplified structure of the *npn* transistor.

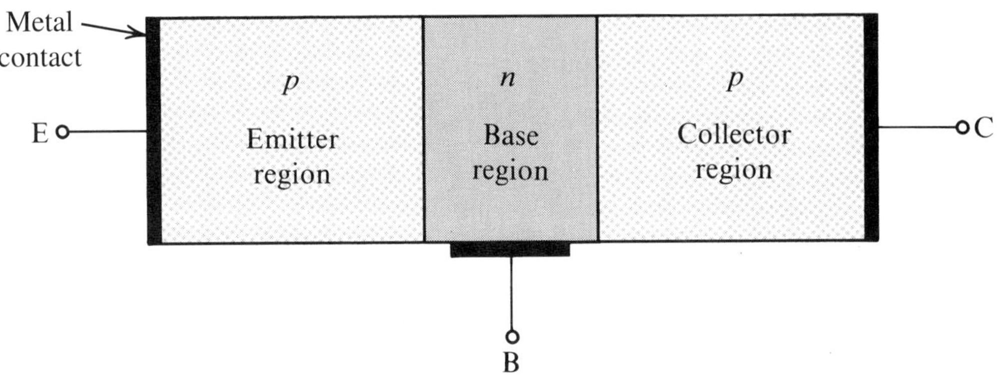

FIGURE 5.2 A simplified structure of the *pnp* transistor.

TABLE 5.1 **BJT Modes of Operation**

Mode	EBJ	CBJ
Cutoff	Reverse	Reverse
Active	Forward	Reverse
Reverse active	Reverse	Forward
Saturation	Forward	Forward

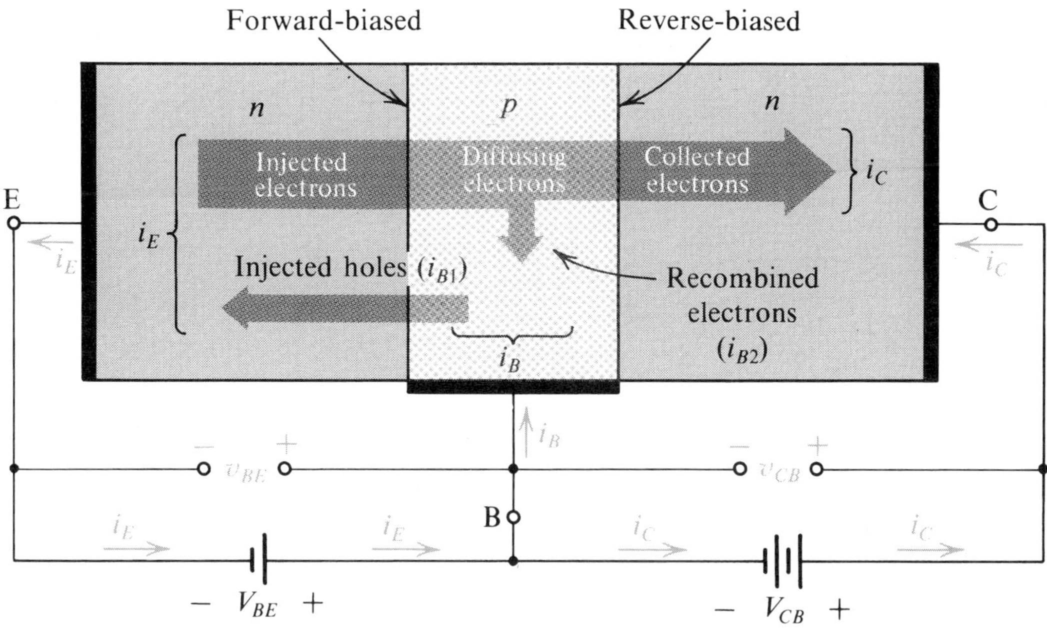

FIGURE 5.3 Current flow in an *npn* transistor biased to operate in the active mode. (Reverse current components due to drift of thermally generated minority carriers are not shown.)

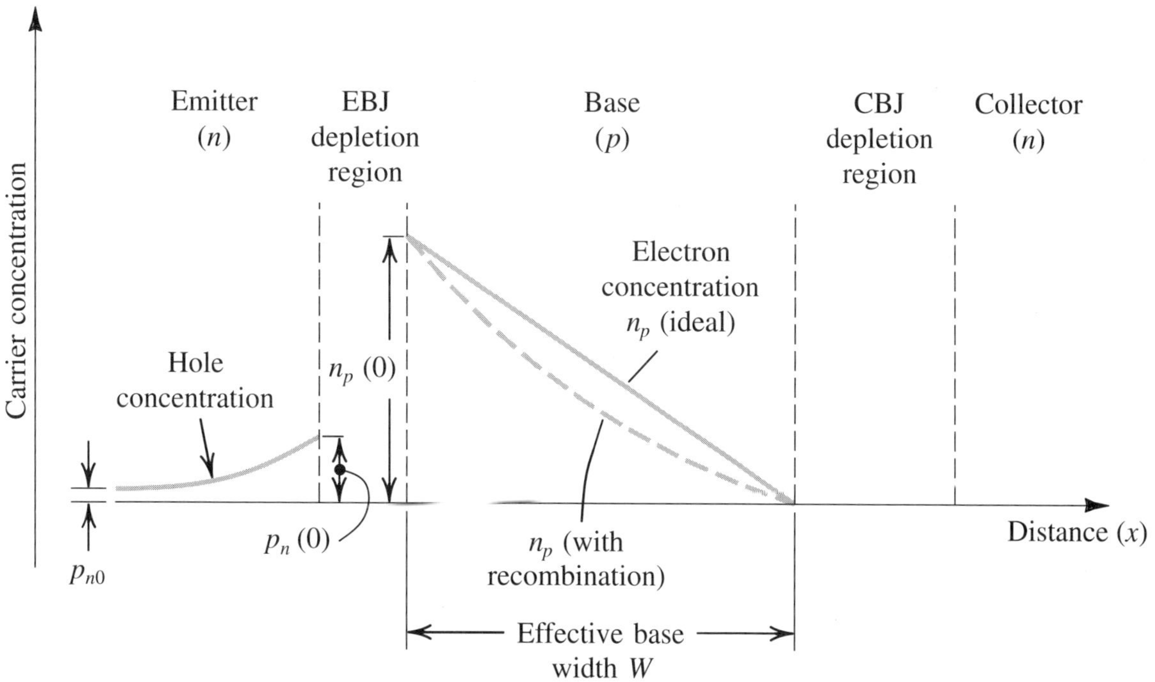

FIGURE 5.4 Profiles of minority-carrier concentrations in the base and in the emitter of an *npn* transistor operating in the active mode: $v_{BE} > 0$ and $v_{CB} \geq 0$.

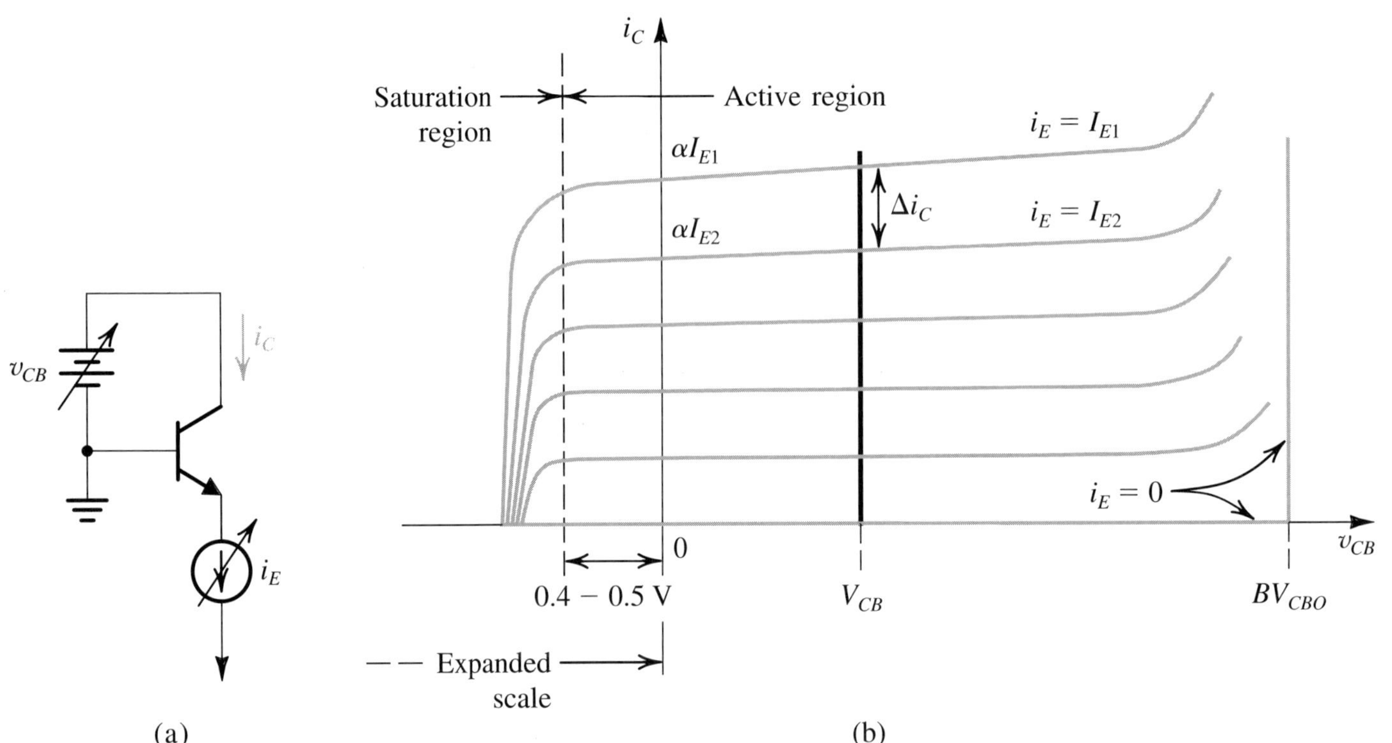

FIGURE 5.18 The i_C–v_{CB} characteristics of an *npn* transistor.

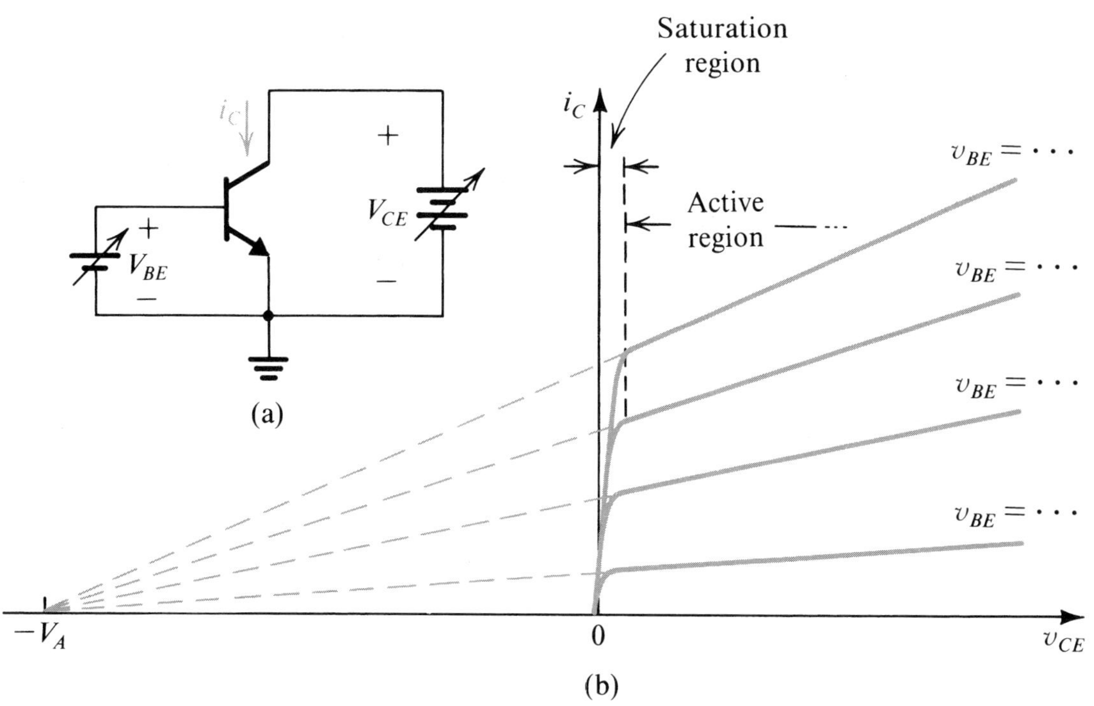

FIGURE 5.19 **(a)** Conceptual circuit for measuring the i_C–v_{CE} characteristics of the BJT. **(b)** The i_C–v_{CE} characteristics of a practical BJT.

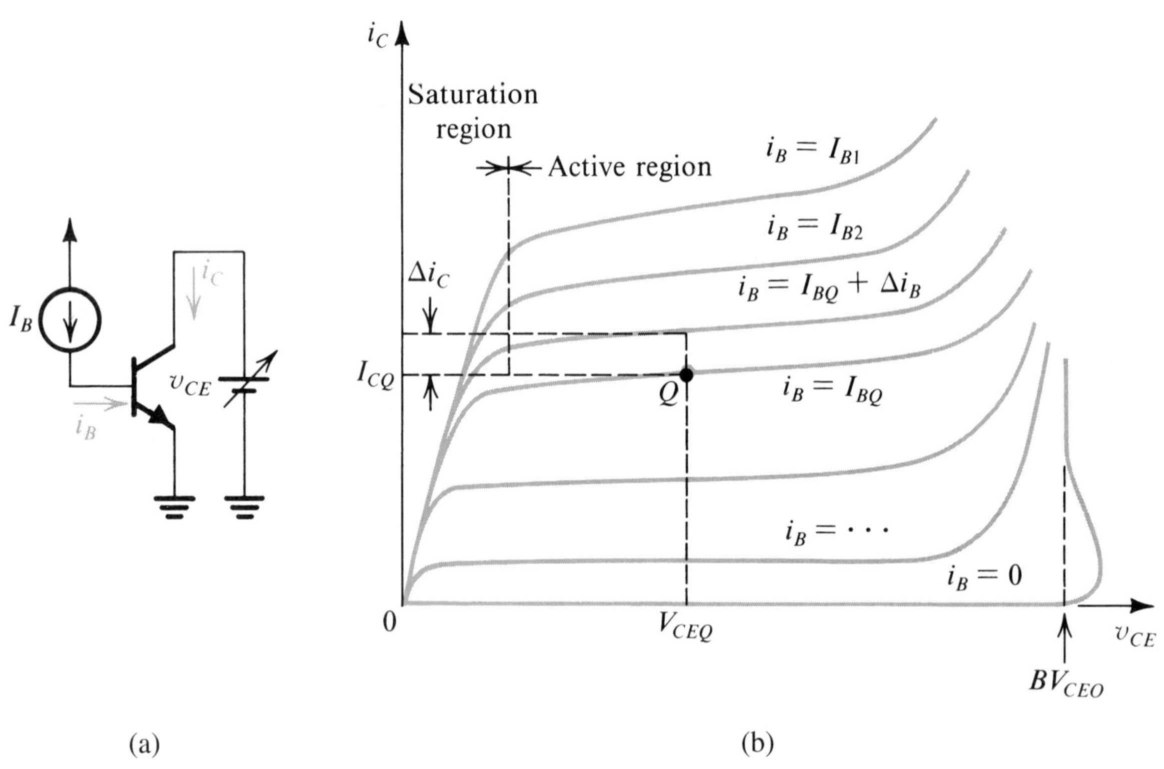

FIGURE 5.21 Common-emitter characteristics. Note that the horizontal scale is expanded around the origin to show the saturation region in some detail.

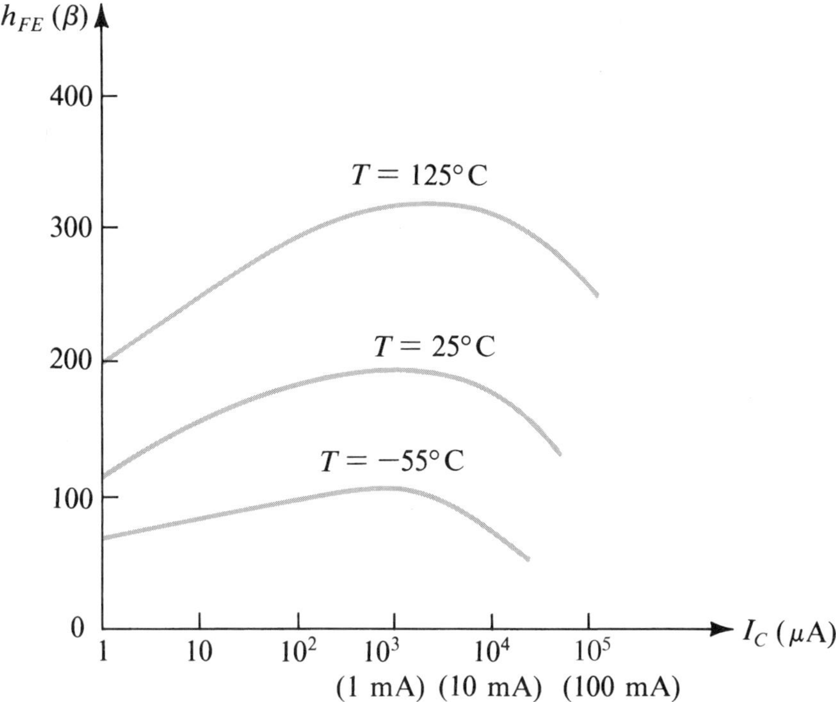

FIGURE 5.22 Typical dependence of β on I_C and on temperature in a modern integrated-circuit *npn* silicon transistor intended for operation around 1 mA.

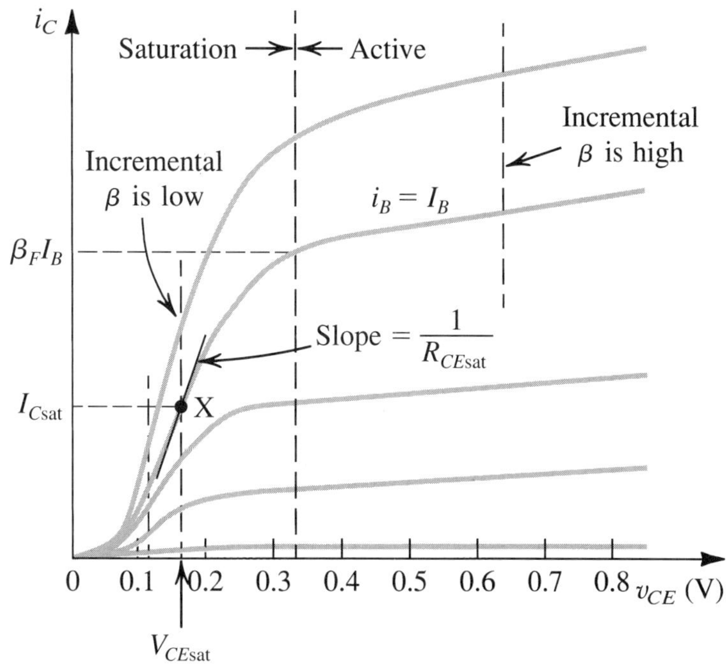

FIGURE 5.23 An expanded view of the common-emitter characteristics in the saturation region.

TABLE 5.3 Summary of the BJT Current-Voltage Characteristics

Circuit Symbol and Directions of Current Flow

	npn Transistor	*pnp* Transistor
	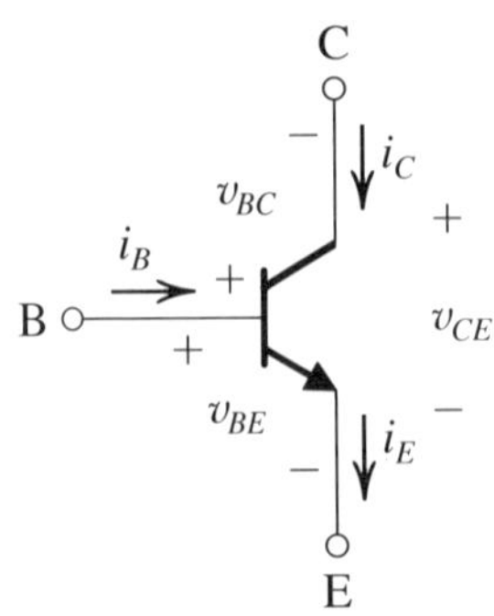	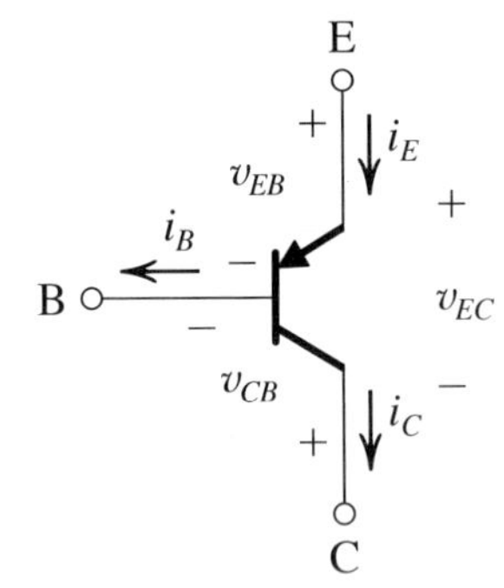

Operation in the Active Mode (for Amplifier Application)

	npn	*pnp*
Conditions:		
1. EBJ Forward Biased	$v_{BE} > V_{BE\text{on}}$; $V_{BE\text{on}} \cong 0.5$ V Typically, $v_{BE} = 0.7$ V	$v_{EB} > V_{EB\text{on}}$; $V_{EB\text{on}} \cong 0.5$ V Typically, $v_{EB} = 0.7$ V
2. CBJ Reversed Biased	$v_{BC} \leq V_{BC\text{on}}$; $V_{BC\text{on}} \cong 0.4$ V $\Rightarrow v_{CE} \geq 0.3$ V	$v_{CB} \leq V_{CB\text{on}}$; $V_{CB\text{on}} \cong 0.4$ V $\Rightarrow v_{EC} \geq 0.3$ V
Current-Voltage Relationships	$i_C = I_S e^{v_{BE}/V_T}$	$i_C = I_S e^{v_{EB}/V_T}$

$$i_B = i_C/\beta \quad \Leftrightarrow \quad i_C = \beta i_B$$

$$i_E = i_C/\alpha \quad \Leftrightarrow \quad i_C = \alpha i_E$$

$$\beta = \frac{\alpha}{1-\alpha} \quad \Leftrightarrow \quad \alpha = \frac{\beta}{\beta+1}$$

Large-Signal Equivalent-Circuit Model (Including the Early Effect)

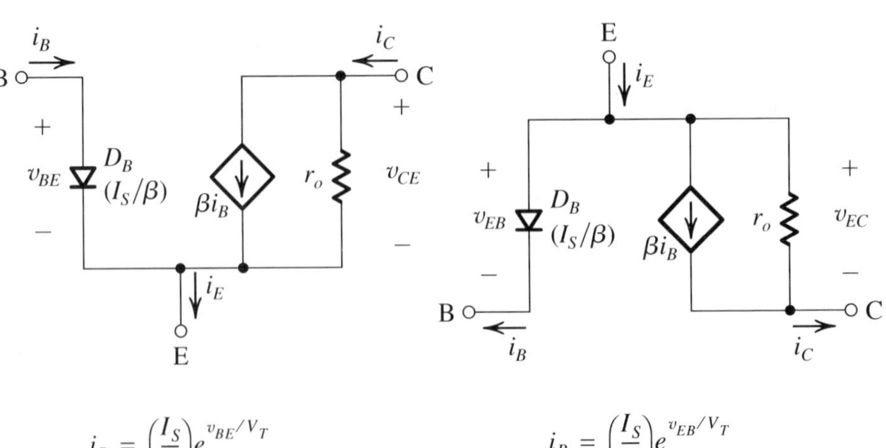

npn	*pnp*
$i_B = \left(\frac{I_S}{\beta}\right)e^{v_{BE}/V_T}$	$i_B = \left(\frac{I_S}{\beta}\right)e^{v_{EB}/V_T}$
$i_C = I_S e^{v_{BE}/V_T}\left(1 + \frac{v_{CE}}{V_A}\right)$	$i_C = I_S e^{v_{EB}/V_T}\left(1 + \frac{v_{EC}}{\lvert V_A \rvert}\right)$
$r_o = V_A/(I_S e^{V_{BE}/V_T})$	$r_o = \lvert V_A \rvert/(I_S e^{V_{EB}/V_T})$

Ebers-Moll Model

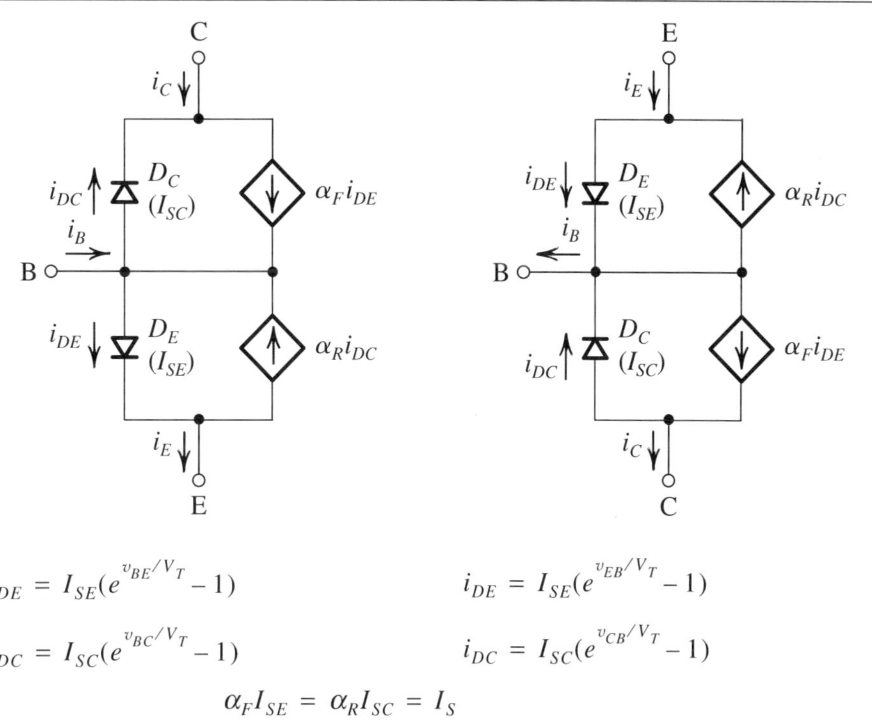

$$i_{DE} = I_{SE}(e^{v_{BE}/V_T} - 1) \qquad i_{DE} = I_{SE}(e^{v_{EB}/V_T} - 1)$$

$$i_{DC} = I_{SC}(e^{v_{BC}/V_T} - 1) \qquad i_{DC} = I_{SC}(e^{v_{CB}/V_T} - 1)$$

$$\alpha_F I_{SE} = \alpha_R I_{SC} = I_S$$

$$\frac{I_{SC}}{I_{SE}} = \frac{\alpha_F}{\alpha_R} = \frac{\text{CBJ Area}}{\text{EBJ Area}}$$

Operation in the Saturation Mode

Conditions:

1. EBJ Forward-Biased: $v_{BE} > V_{BEon}$; $V_{BEon} \cong 0.5$ V; Typically, $v_{BE} = 0.7$–0.8 V | $v_{EB} > V_{EBon}$; $V_{EBon} \cong 0.5$ V; Typically, $v_{EB} = 0.7$–0.8 V

2. CBJ Forward-Biased: $v_{BC} \geq V_{BCon}$; $V_{BCon} \cong 0.4$ V; Typically, $v_{BC} = 0.5$–0.6 V; $\Rightarrow v_{CE} = V_{CEsat} = 0.1$–$0.2$ V | $v_{CB} \geq V_{CBon}$; $V_{CBon} \cong 0.4$ V; Typically, $v_{CB} = 0.5$–0.6 V; $\Rightarrow v_{EC} = V_{ECsat} = 0.1$–$0.2$ V

Currents

$$I_{Csat} = \beta_{forced} I_B$$

$$\beta_{forced} \leq \beta_F, \qquad \frac{\beta_F}{\beta_{forced}} = \text{Overdrive factor}$$

Equivalent Circuits

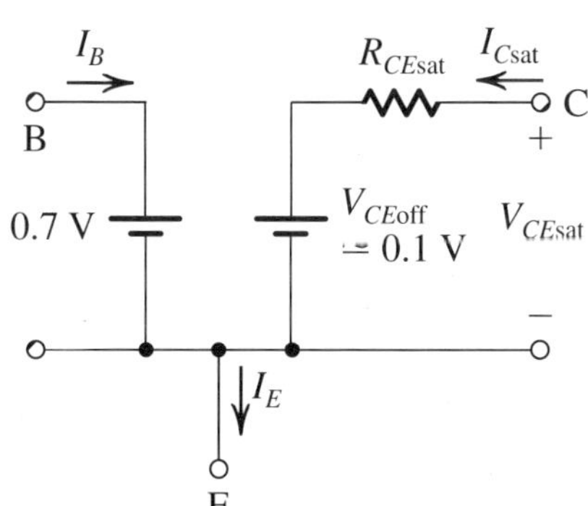

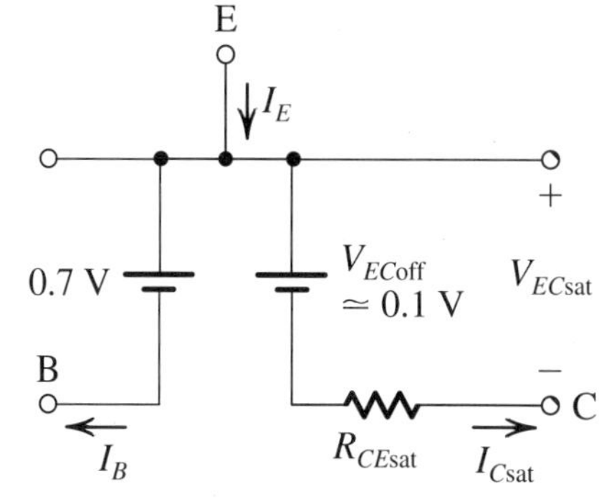

$$|V_{CEsat}| = V_T \ln\left[\frac{1 + (\beta_{forced} + 1)/\beta_F}{1 - \beta_{forced}/\beta_F}\right]$$

For $\beta_{forced} = \beta_F/2$: $\qquad R_{CEsat} = 1/10\beta_F I_B$

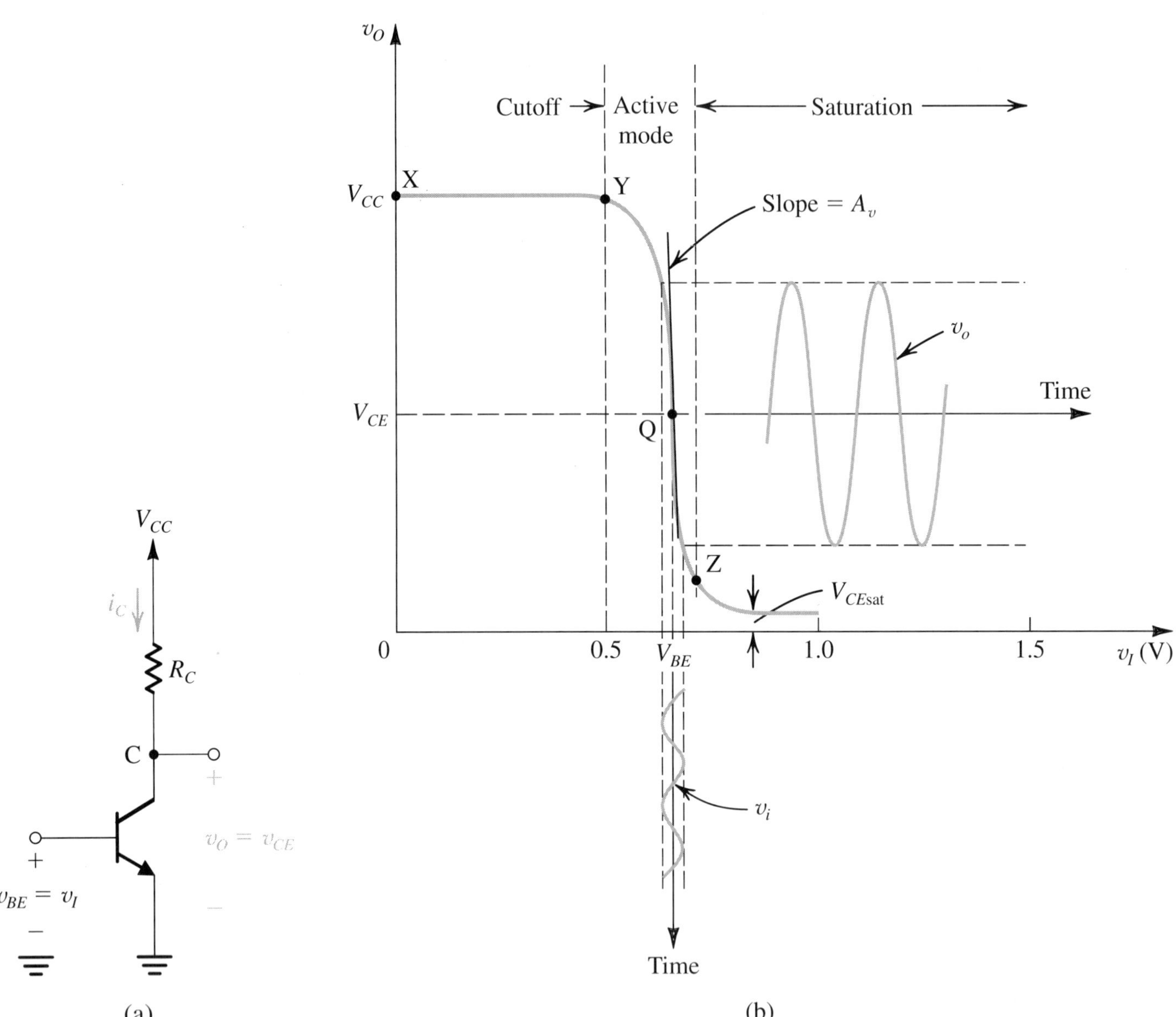

FIGURE 5.26 **(a)** Basic common-emitter amplifier circuit. **(b)** Transfer characteristic of the circuit in (a). The amplifier is biased at a point Q, and a small voltage signal v_i is superimposed on the dc bias voltage V_{BE}. The resulting output signal v_o appears superimposed on the dc collector voltage V_{CE}. The amplitude of v_o is larger than that of v_i by the voltage gain A_v.

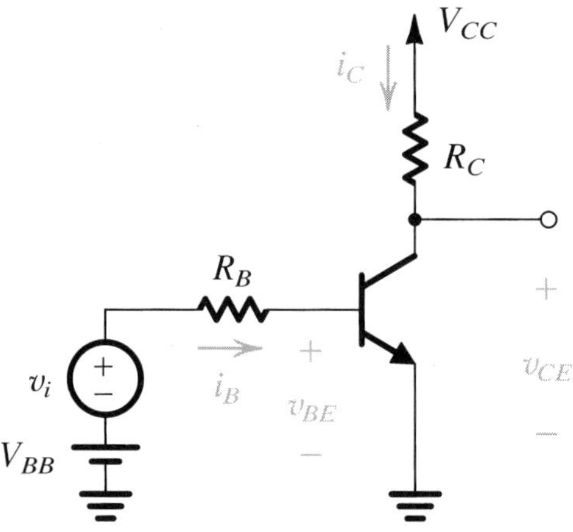

FIGURE 5.27 Circuit whose operation is to be analyzed graphically.

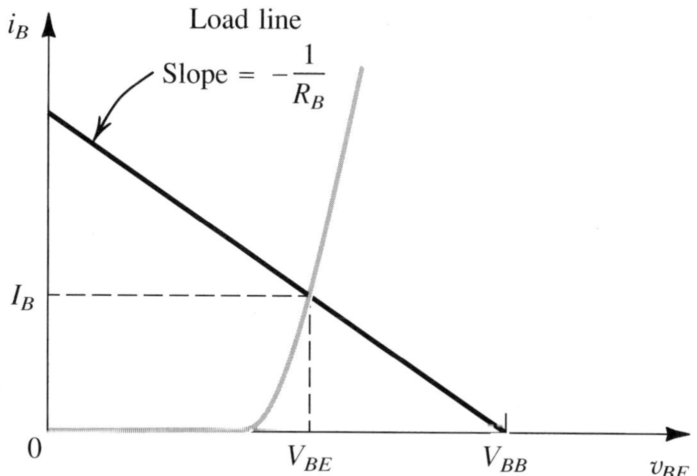

FIGURE 5.28 Graphical construction for the determination of the dc base current in the circuit of Fig. 5.27.

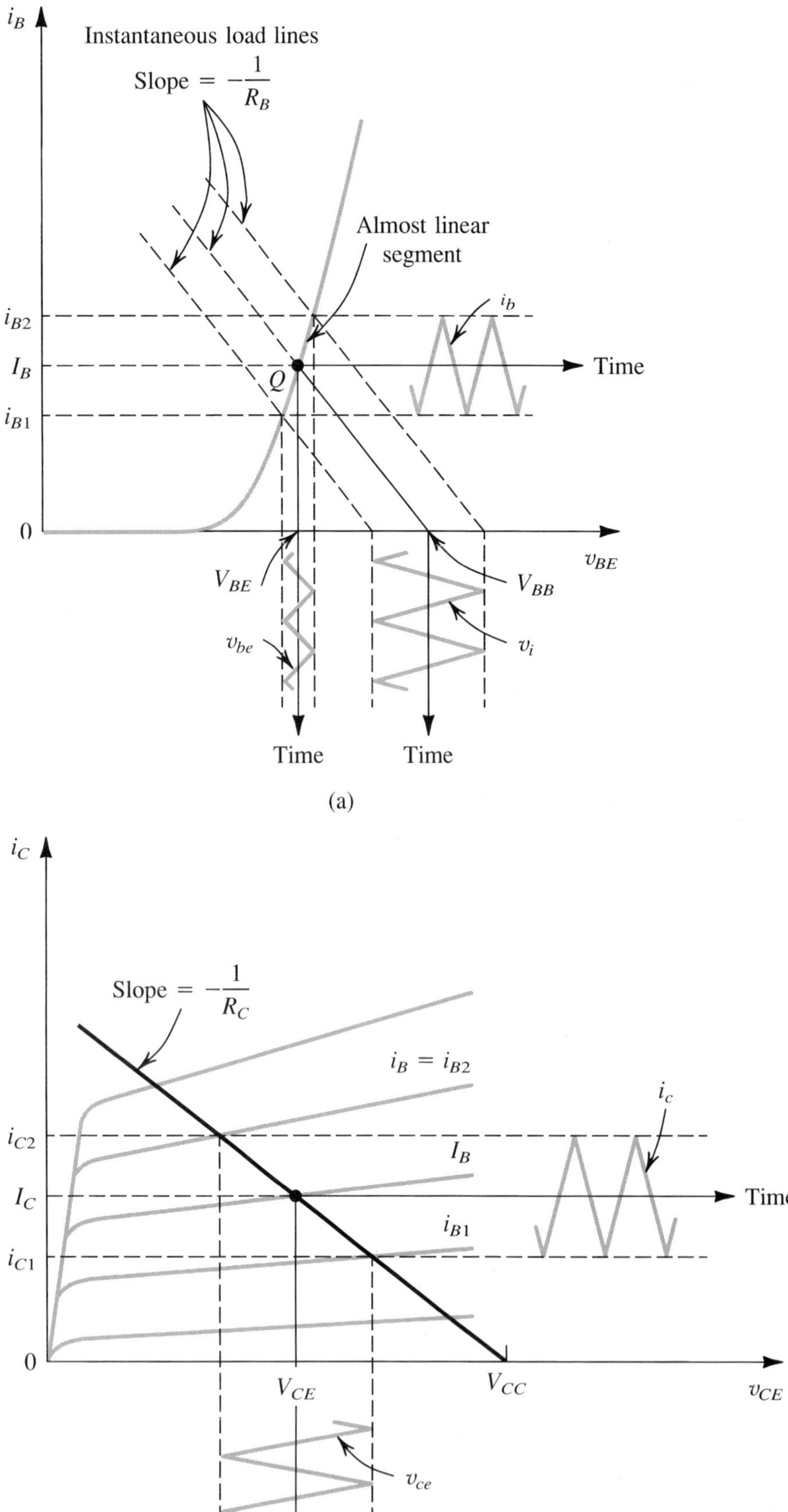

FIGURE 5.30 Graphical determination of the signal components v_{be}, i_b, i_c, and v_{ce} when a signal component v_i is superimposed on the dc voltage V_{BB} (see Fig. 5.27).

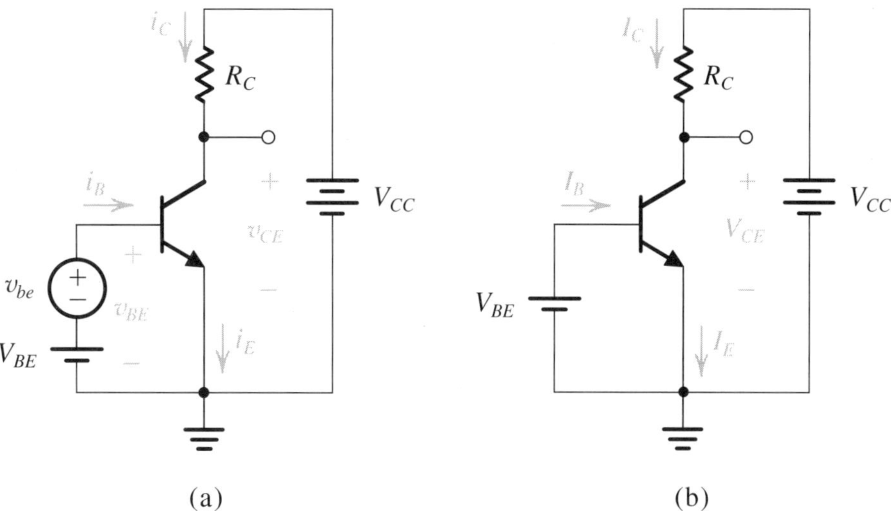

FIGURE 5.48 **(a)** Conceptual circuit to illustrate the operation of the transistor as an amplifier. **(b)** The circuit of (a) with the signal source v_{be} eliminated for dc (bias) analysis.

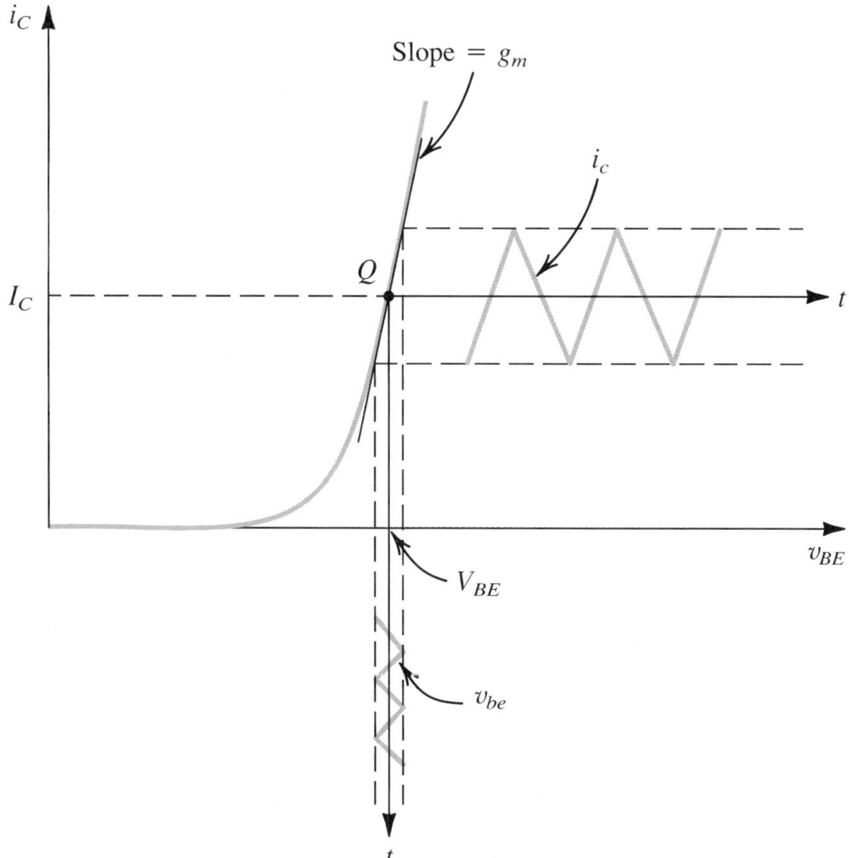

FIGURE 5.49 Linear operation of the transistor under the small-signal condition: A small signal v_{be} with a triangular waveform is superimposed on the dc voltage V_{BE}. It gives rise to a collector signal current i_c, also of triangular waveform, superimposed on the dc current I_C. Here, $i_c = g_m v_{be}$, where g_m is the slope of the i_C–v_{BE} curve at the bias point Q.

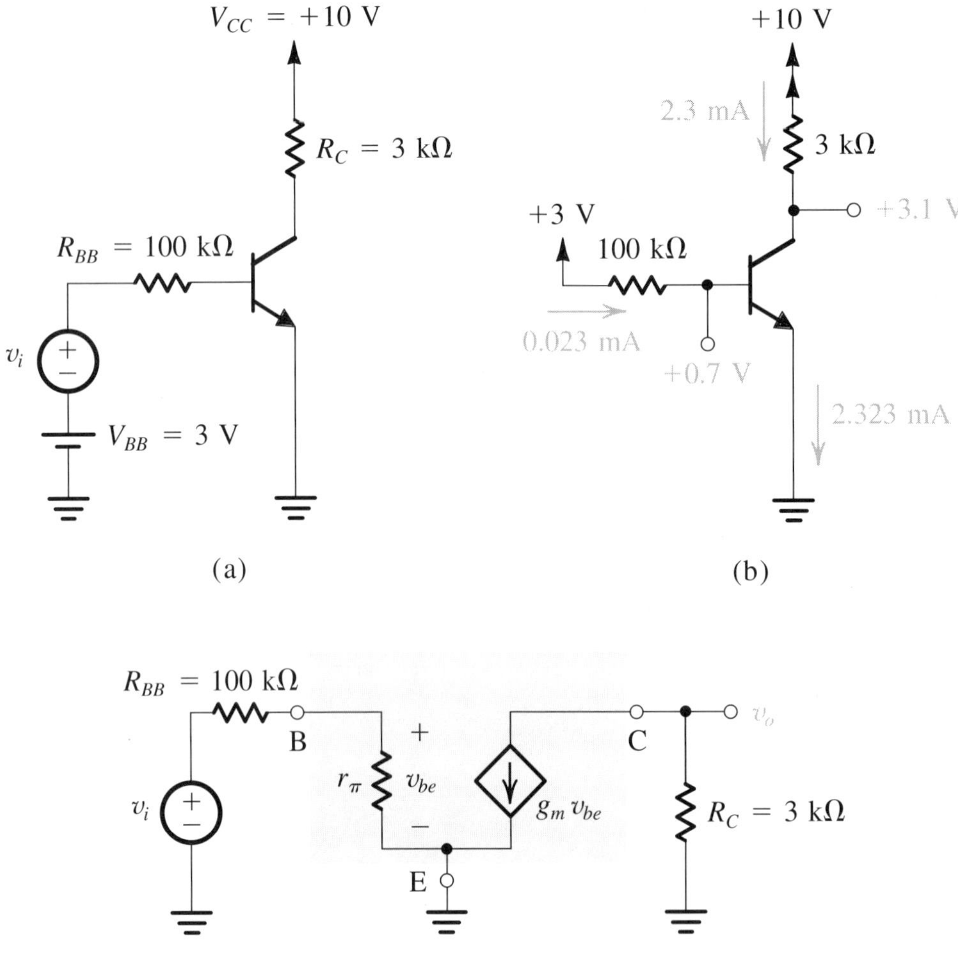

FIGURE 5.53 Example 5.14: **(a)** circuit; **(b)** dc analysis; **(c)** small-signal model.

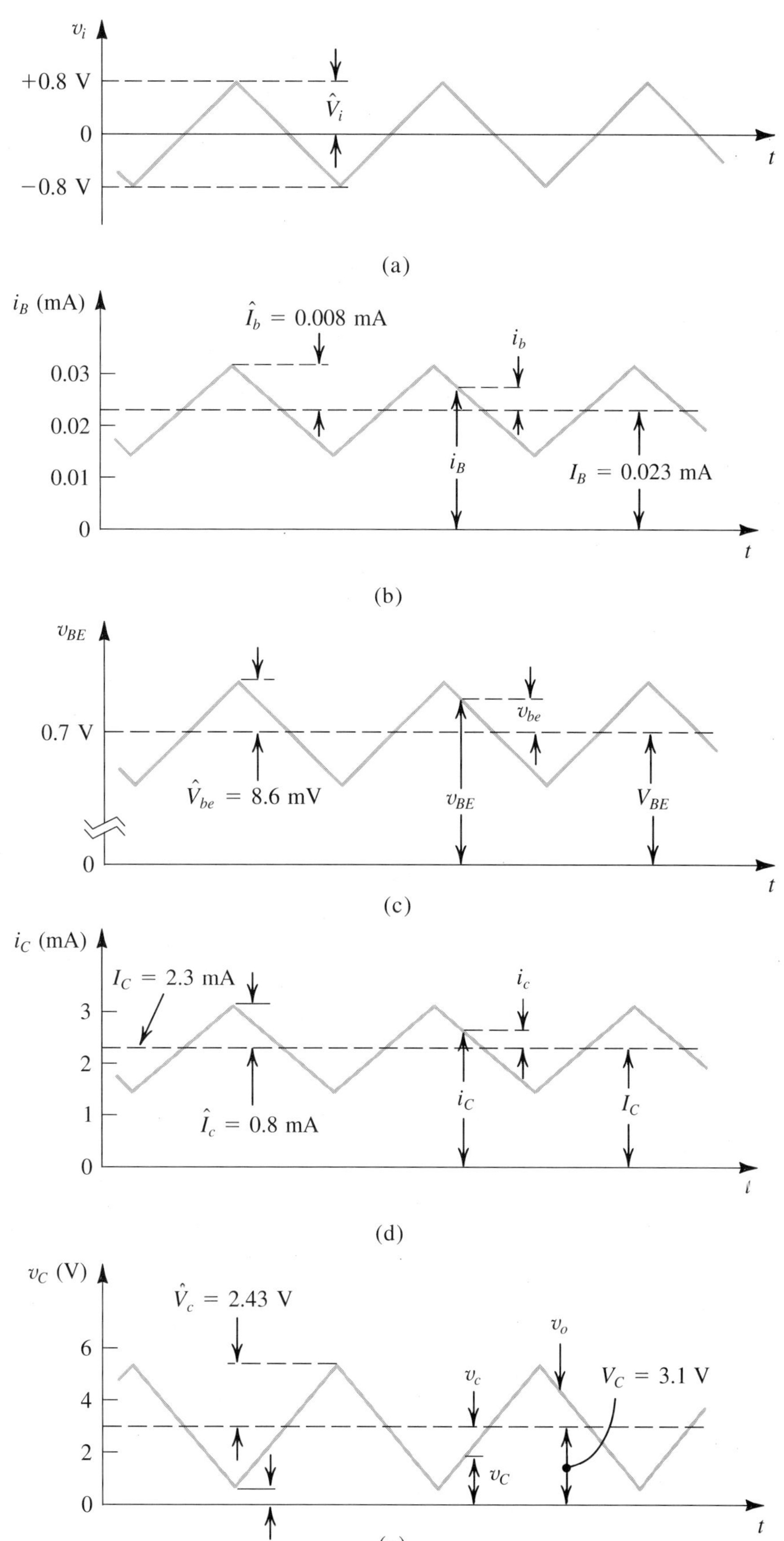

FIGURE 5.54 Signal waveforms in the circuit of Fig. 5.53.

TABLE 5.4 Small-Signal Models of the BJT

Hybrid-π Model

- ($g_m v_\pi$) Version
- (βi_b) Version

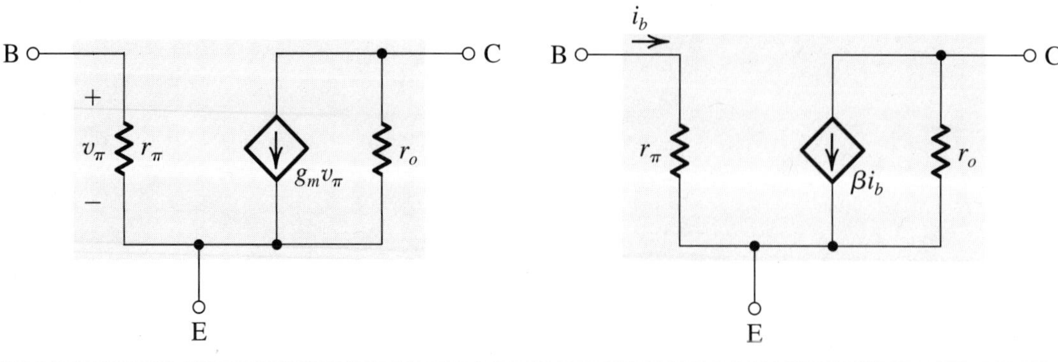

T Model

- ($g_m v_\pi$) Version
- (αi) Version

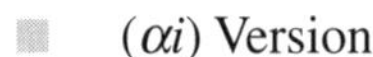

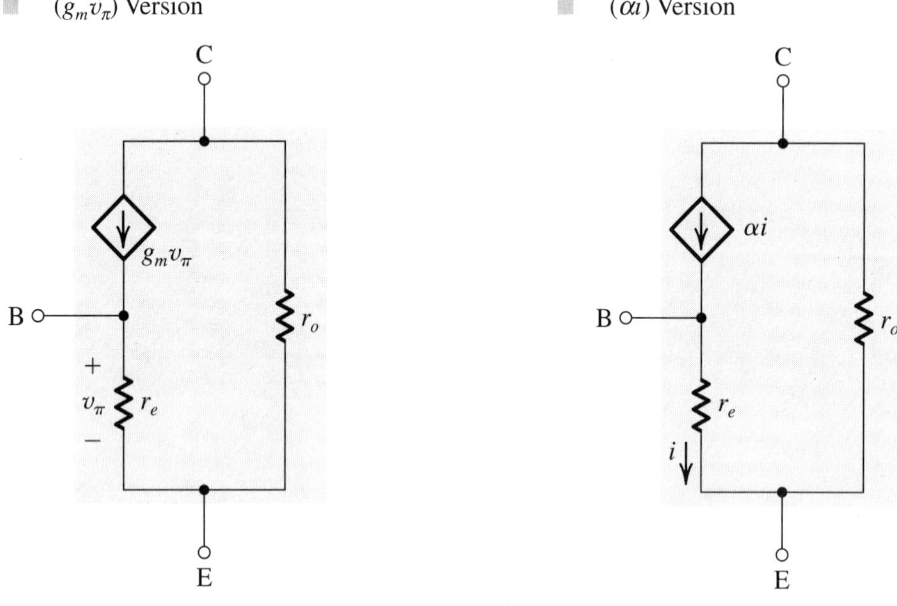

Model Parameters in Terms of DC Bias Currents

$$g_m = \frac{I_C}{V_T} \qquad r_e = \frac{V_T}{I_E} = \alpha\left(\frac{V_T}{I_C}\right) \qquad r_\pi = \frac{V_T}{I_B} = \beta\left(\frac{V_T}{I_C}\right) \qquad r_o = \frac{|V_A|}{I_C}$$

In Terms of g_m

$$r_e = \frac{\alpha}{g_m} \qquad r_\pi = \frac{\beta}{g_m}$$

In Terms of r_e

$$g_m = \frac{\alpha}{r_e} \qquad r_\pi = (\beta + 1)r_e \qquad g_m + \frac{1}{r_\pi} = \frac{1}{r_e}$$

Relationships Between α and β

$$\beta = \frac{\alpha}{1-\alpha} \qquad \alpha = \frac{\beta}{\beta+1} \qquad \beta + 1 = \frac{1}{1-\alpha}$$

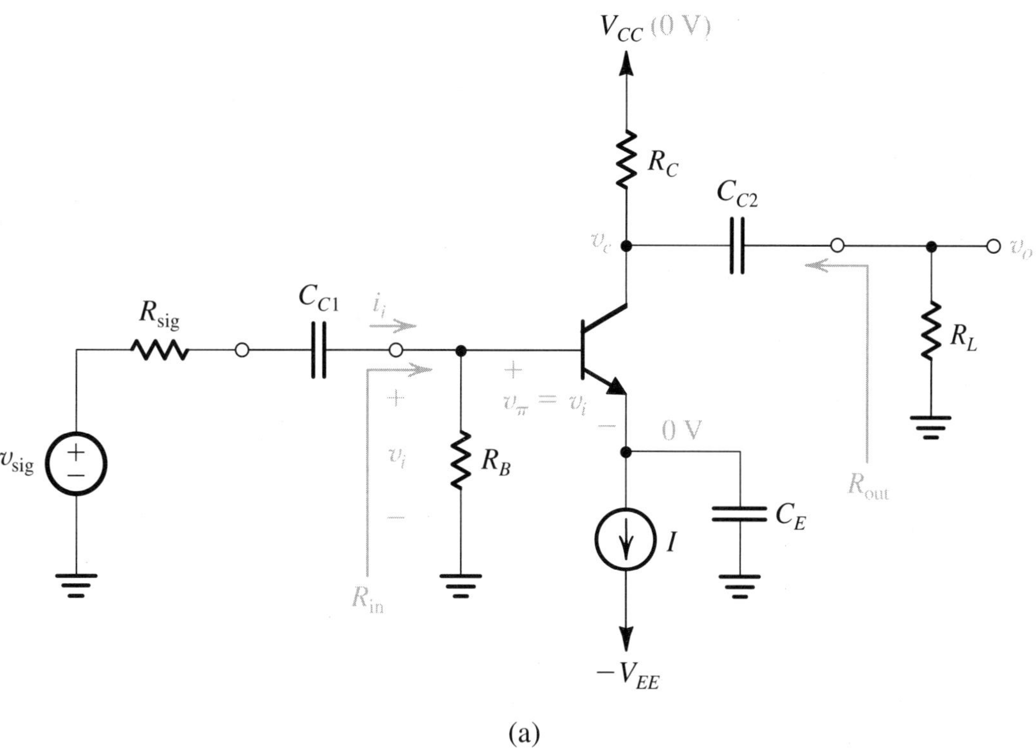

(a)

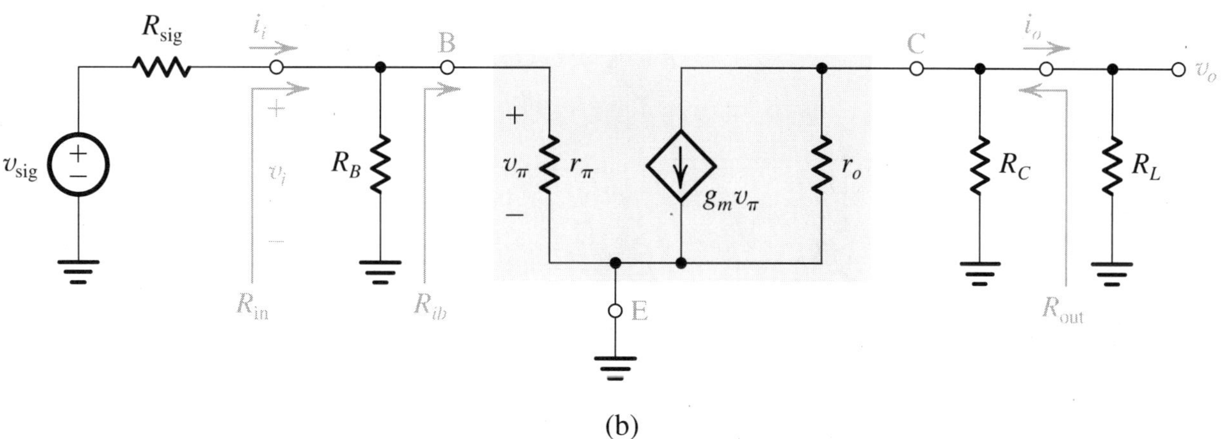

(b)

FIGURE 5.60 **(a)** A common-emitter amplifier using the structure of Fig. 5.59. **(b)** Equivalent circuit obtained by replacing the transistor with its hybrid-π model.

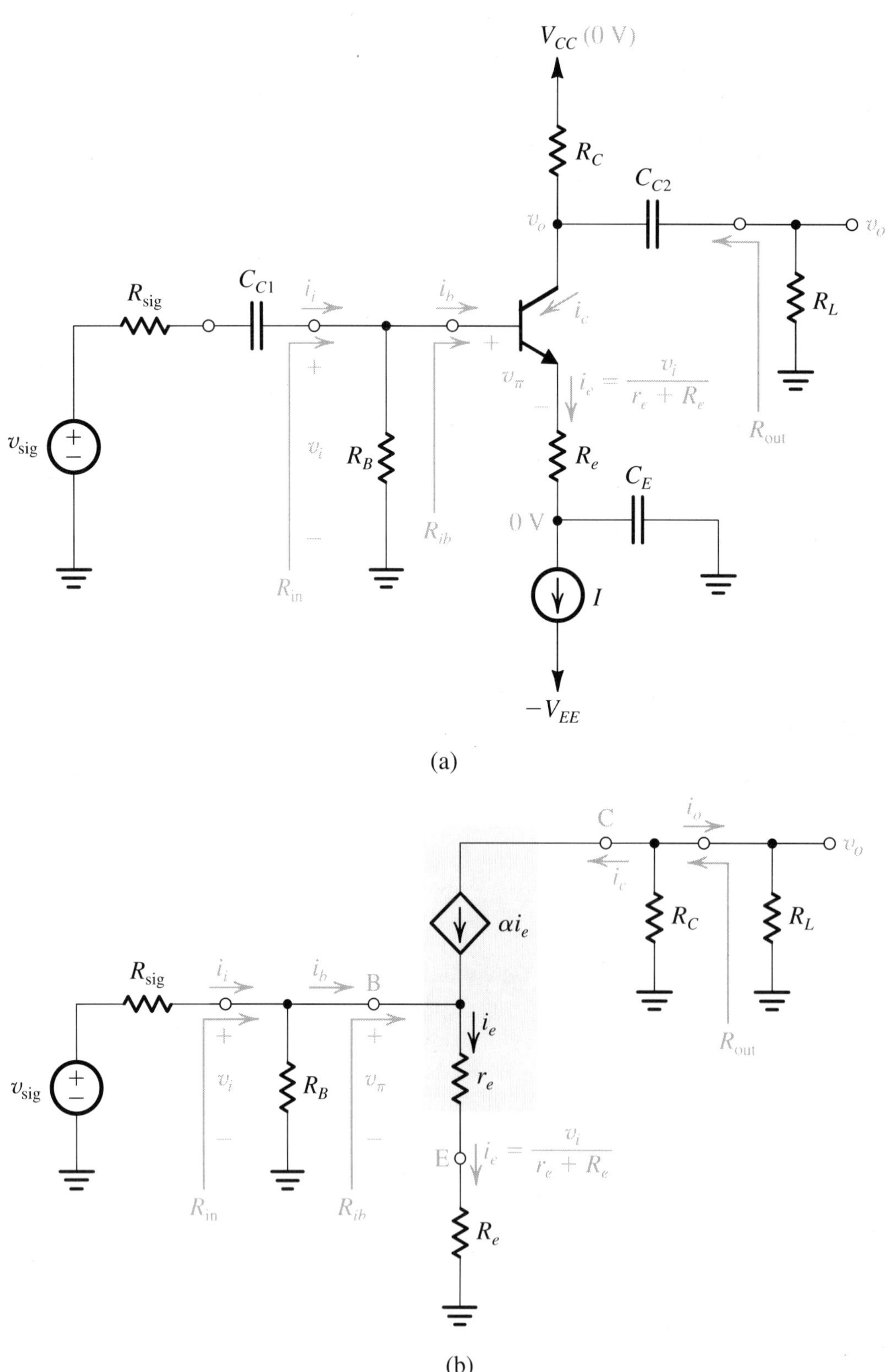

FIGURE 5.61 **(a)** A common-emitter amplifier with an emitter resistance R_e. **(b)** Equivalent circuit obtained by replacing the transistor with its T model.

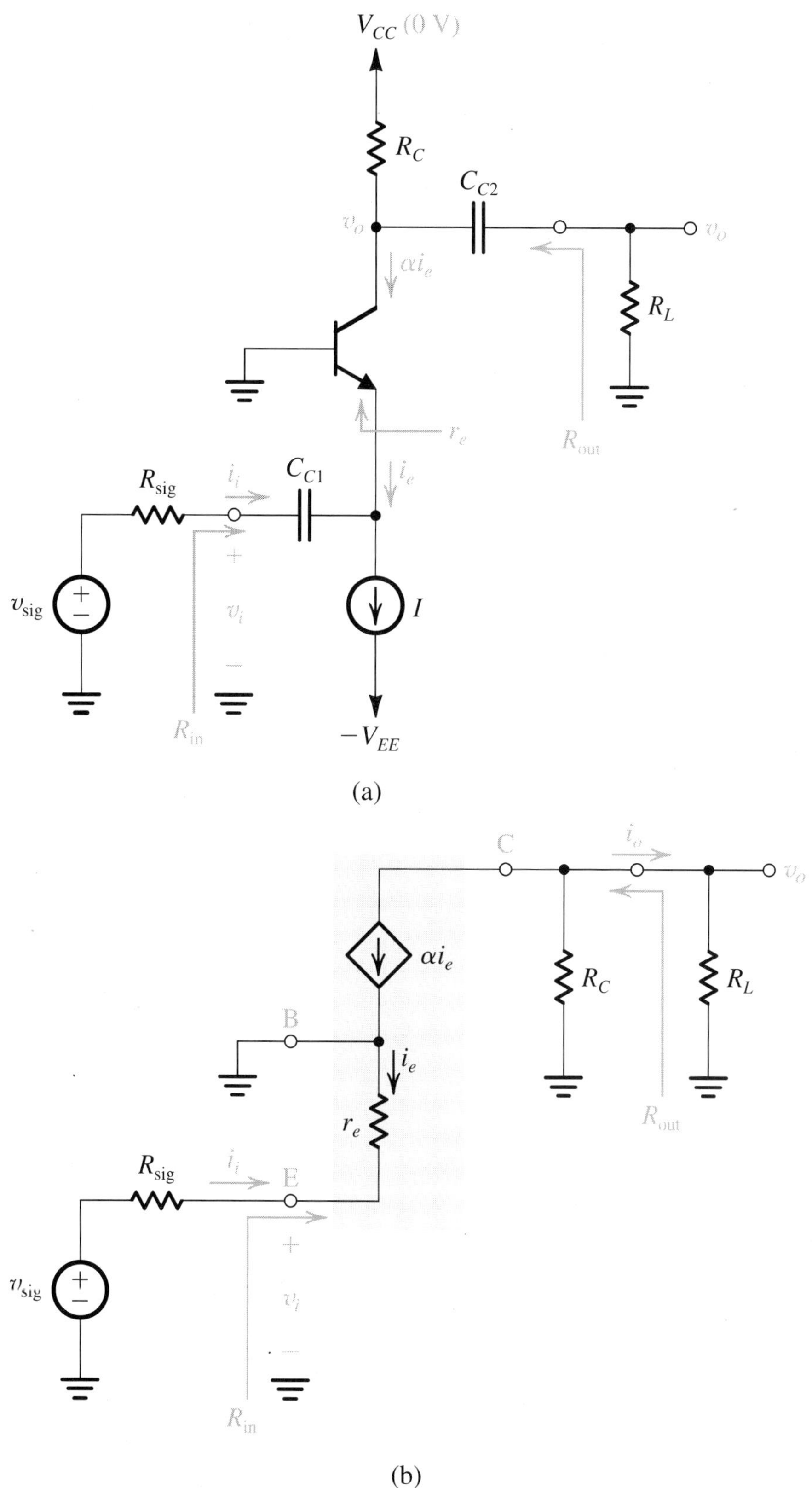

FIGURE 5.62 **(a)** A common-base amplifier using the structure of Fig. 5.59. **(b)** Equivalent circuit obtained by replacing the transistor with its T model.

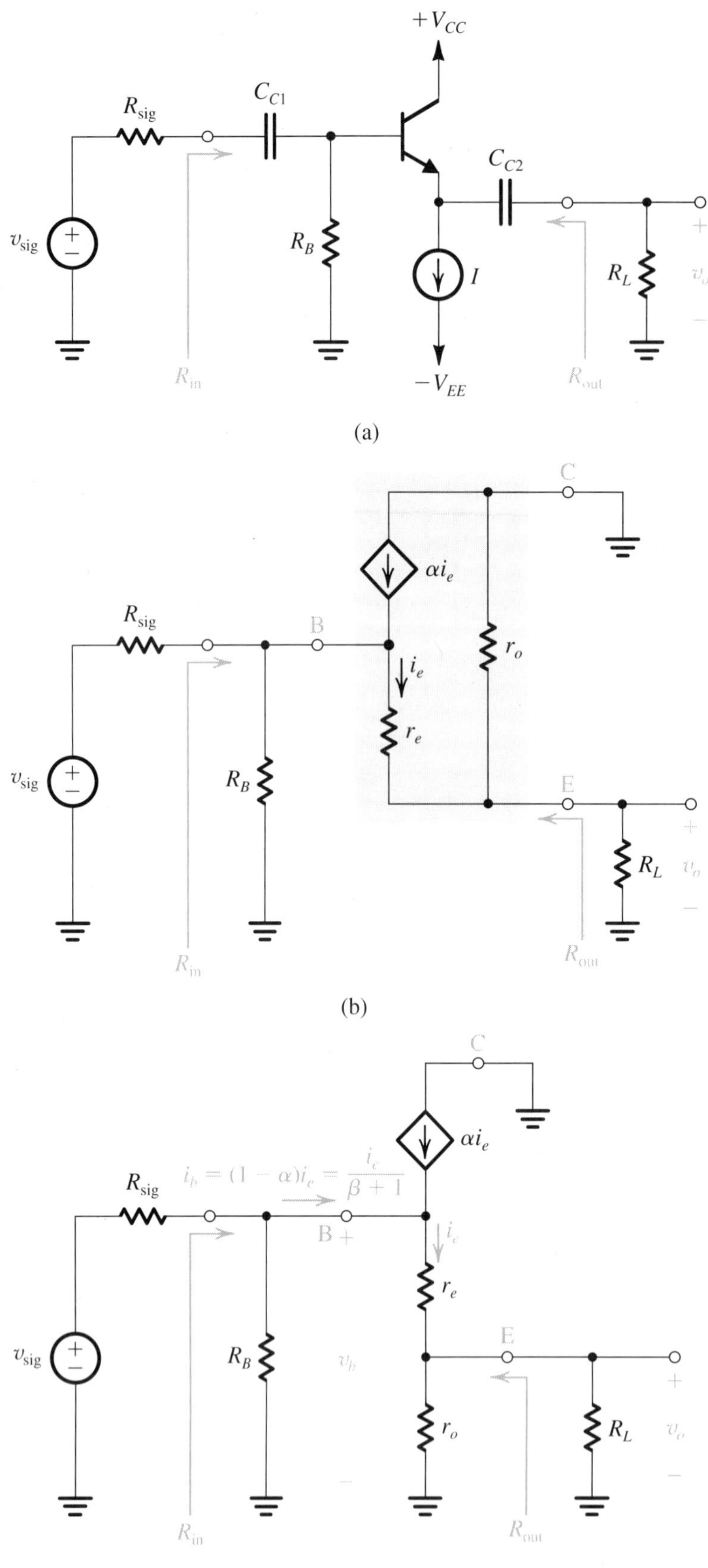

FIGURE 5.63 **(a)** An emitter-follower circuit based on the structure of Fig. 5.59. **(b)** Small-signal equivalent circuit of the emitter follower with the transistor replaced by its T model augmented with r_o. **(c)** The circuit in (b) redrawn to emphasize that r_o is in parallel with R_L. This simplifies the analysis considerably.

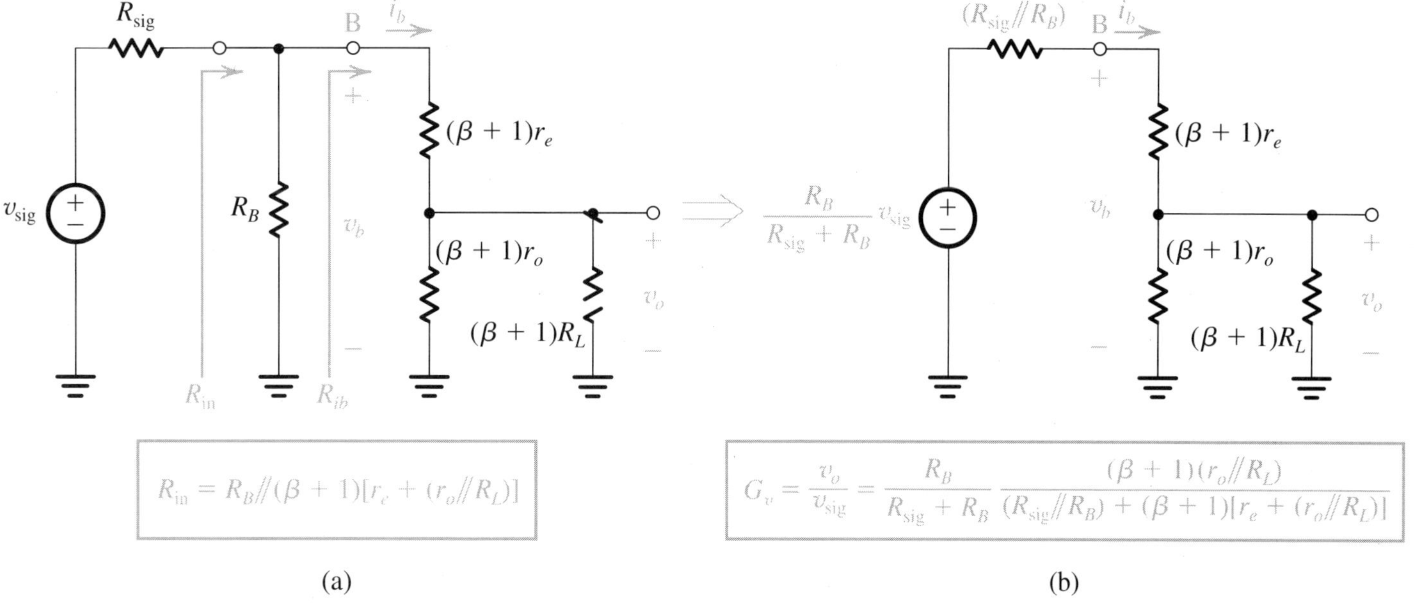

FIGURE 5.64 **(a)** An equivalent circuit of the emitter follower obtained from the circuit in Fig. 5.63(c) by reflecting all resistances in the emitter to the base side. **(b)** The circuit in (a) after application of Thévenin theorem to the input circuit composed of v_{sig}, R_{sig}, and R_B.

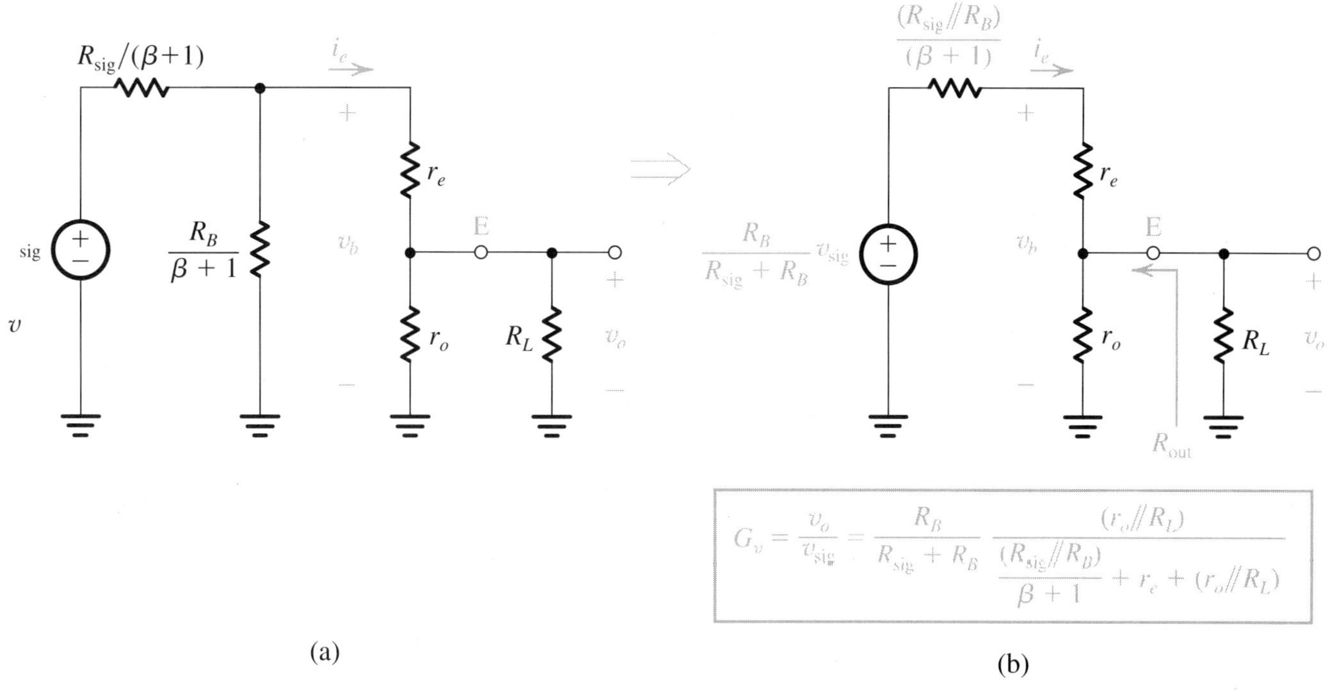

FIGURE 5.65 **(a)** An alternate equivalent circuit of the emitter follower obtained by reflecting all base-circuit resistances to the emitter side. **(b)** The circuit in (a) after application of Thévenin theorem to the input circuit composed of v_{sig}, $R_{sig}/(\beta + 1)$, and $R_B/(\beta + 1)$.

TABLE 5.6 Characteristics of Single-Stage Discrete BJT Amplifiers

Common Emitter

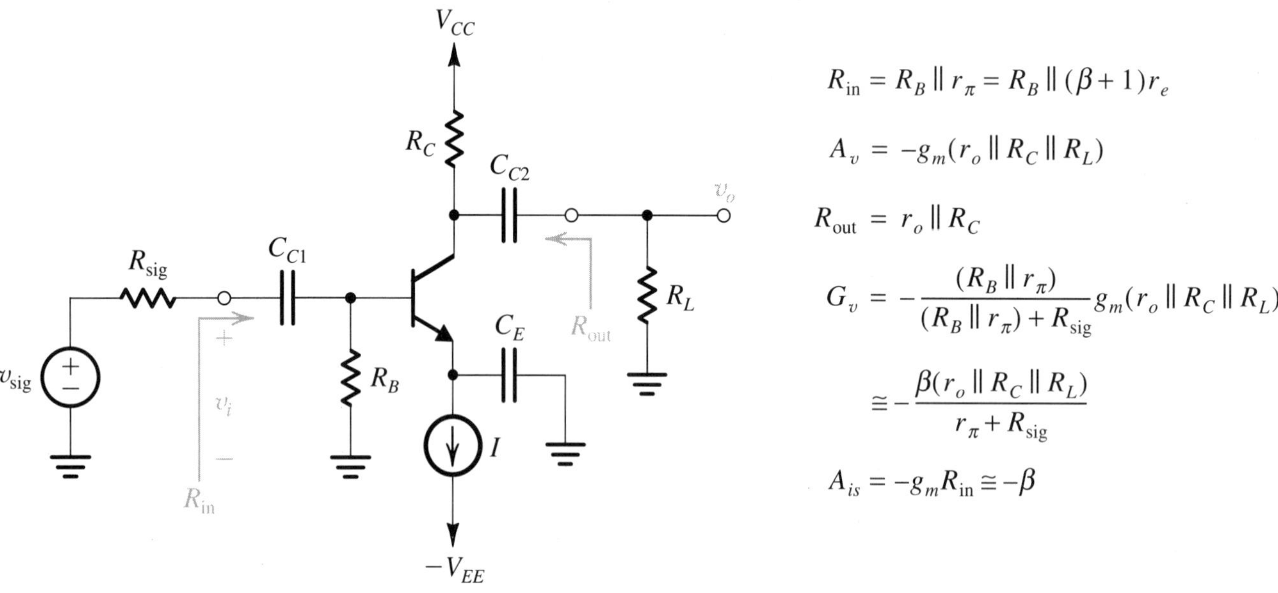

$$R_{in} = R_B \parallel r_\pi = R_B \parallel (\beta + 1) r_e$$

$$A_v = -g_m (r_o \parallel R_C \parallel R_L)$$

$$R_{out} = r_o \parallel R_C$$

$$G_v = -\frac{(R_B \parallel r_\pi)}{(R_B \parallel r_\pi) + R_{sig}} g_m (r_o \parallel R_C \parallel R_L)$$

$$\cong -\frac{\beta (r_o \parallel R_C \parallel R_L)}{r_\pi + R_{sig}}$$

$$A_{is} = -g_m R_{in} \cong -\beta$$

Common Emitter with Emitter Resistance

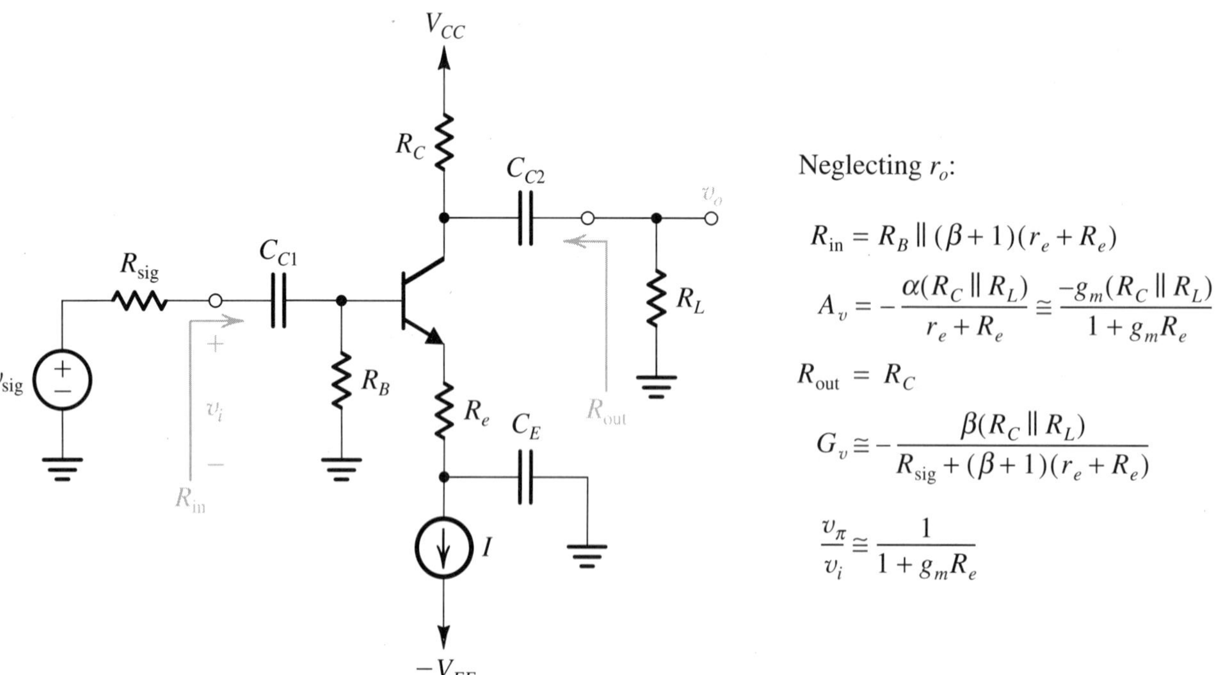

Neglecting r_o:

$$R_{in} = R_B \parallel (\beta + 1)(r_e + R_e)$$

$$A_v = -\frac{\alpha (R_C \parallel R_L)}{r_e + R_e} \cong \frac{-g_m (R_C \parallel R_L)}{1 + g_m R_e}$$

$$R_{out} = R_C$$

$$G_v \cong -\frac{\beta (R_C \parallel R_L)}{R_{sig} + (\beta + 1)(r_e + R_e)}$$

$$\frac{v_\pi}{v_i} \cong \frac{1}{1 + g_m R_e}$$

Common Base

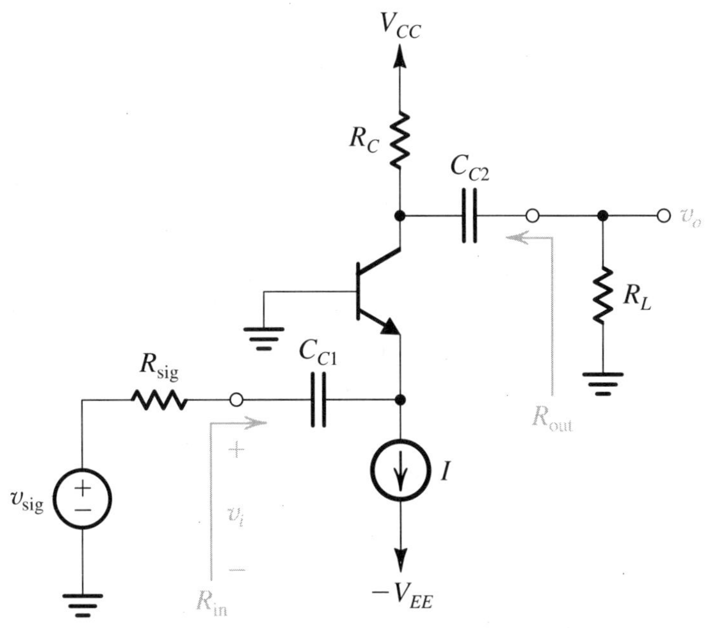

Neglecting r_o:

$$R_{in} = r_e$$

$$A_v = g_m(R_C \parallel R_L)$$

$$R_{out} = R_C$$

$$G_v = \frac{\alpha(R_C \parallel R_L)}{R_{sig} + r_e}$$

$$A_{is} \cong \alpha$$

Common Collector or Emitter Follower

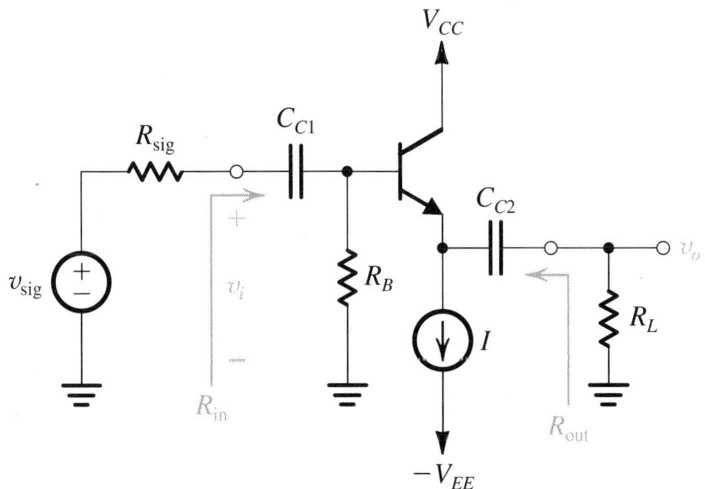

$$R_{in} = R_B \parallel (\beta+1)[r_e + (r_o \parallel R_L)]$$

$$A_v = \frac{(r_o \parallel R_L)}{(r_o \parallel R_L) + r_e}$$

$$R_{out} = r_o \parallel \left[r_e + \frac{R_{sig} \parallel R_B}{\beta+1}\right]$$

$$G_v = \frac{R_B}{R_B + R_{sig}} \frac{(r_o \parallel R_L)}{\frac{R_{sig} \parallel R_B}{\beta+1} + r_e + (r_o \parallel R_L)}$$

$$A_{is} \cong \beta+1$$

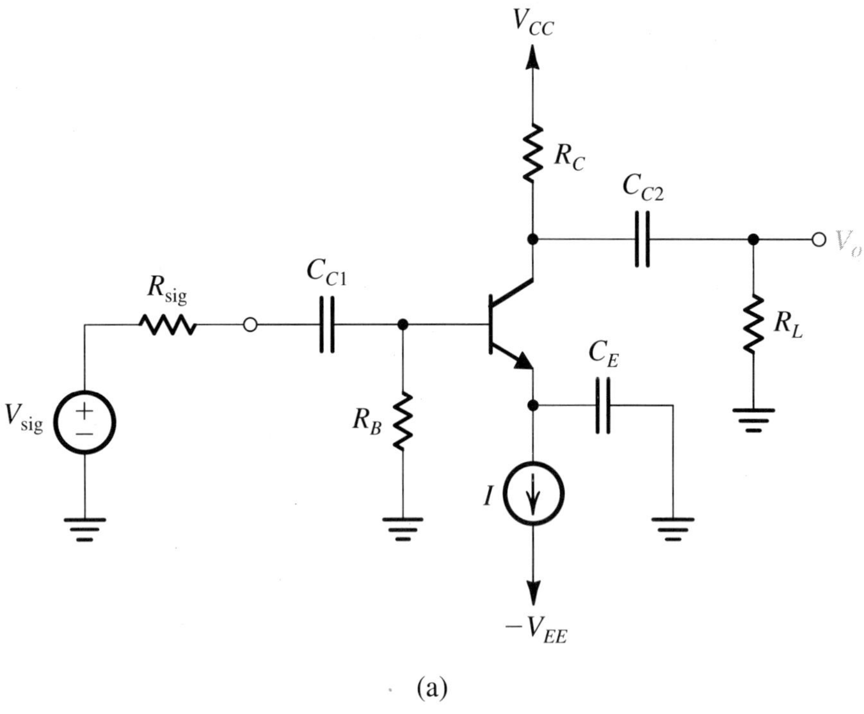

(a)

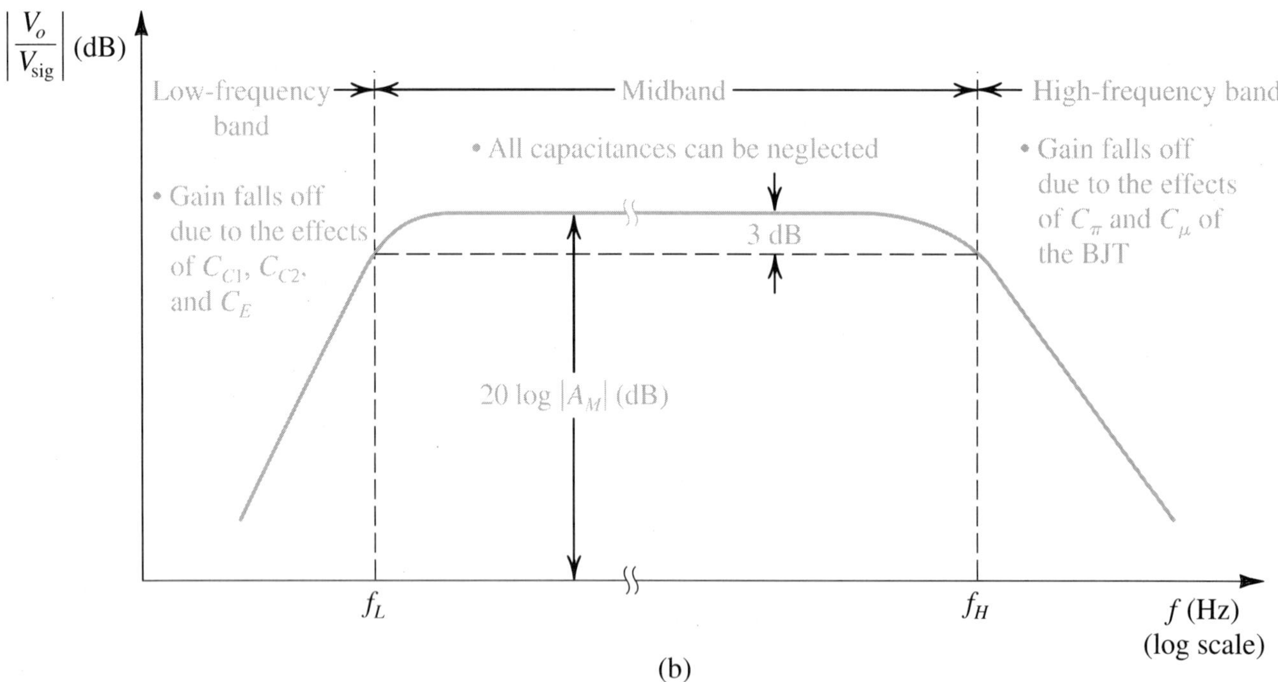

(b)

FIGURE 5.71 **(a)** Capacitively coupled common-emitter amplifier. **(b)** Sketch of the magnitude of the gain of the CE amplifier versus frequency. The graph delineates the three frequency bands relevant to frequency-response determination.

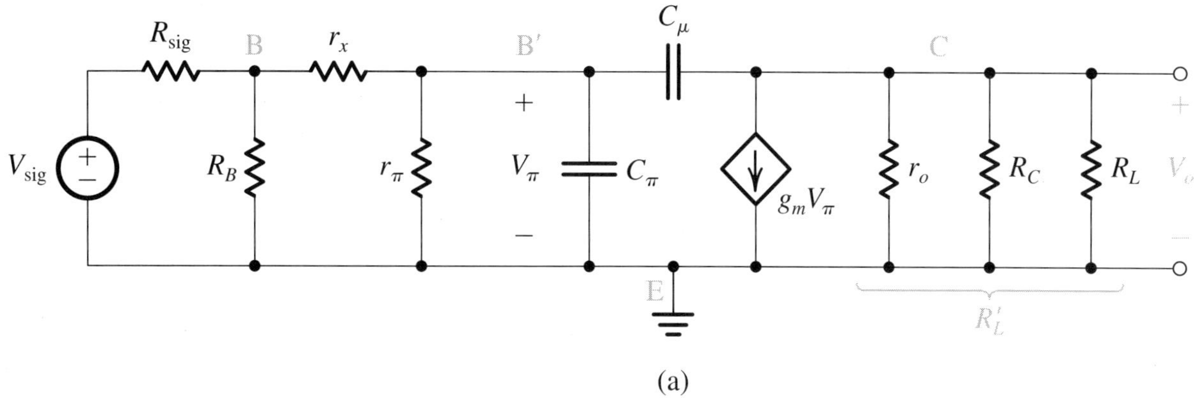

(a)

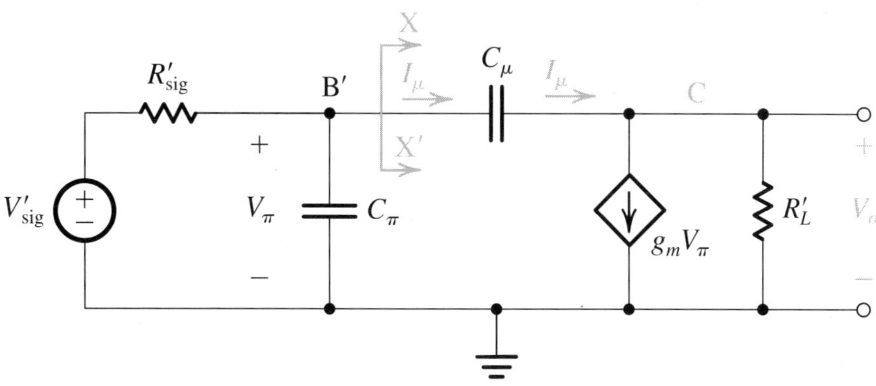

$$V'_{sig} = V_{sig} \frac{R_B}{R_B + R_{sig}} \frac{r_\pi}{r_\pi + r_x + (R_{sig} /\!/ R_B)}$$

$$R'_L = r_o /\!/ R_C /\!/ R_L$$

$$R'_{sig} = r_\pi /\!/ [r_x + (R_B /\!/ R_{sig})]$$

(b)

X
R'_{sig}
B′
I_μ
C
X′
V'_{sig}
V_π
C_π
C_{eq}
$g_m V_\pi$
R'_L
V_o
C_{in}

$$C_{in} = C_\pi + C_{eq} = C_\pi + C_\mu(1 + g_m R'_L)$$

$$V_o = -g_m R'_L V_\pi$$

(c)

FIGURE 5.72 Determining the high-frequency response of the CE amplifier: **(a)** equivalent circuit; **(b)** the circuit of (a) simplified at both the input side and the output side; **(c)** equivalent circuit with C_μ replaced at the input side with the equivalent capacitance C_{eq};

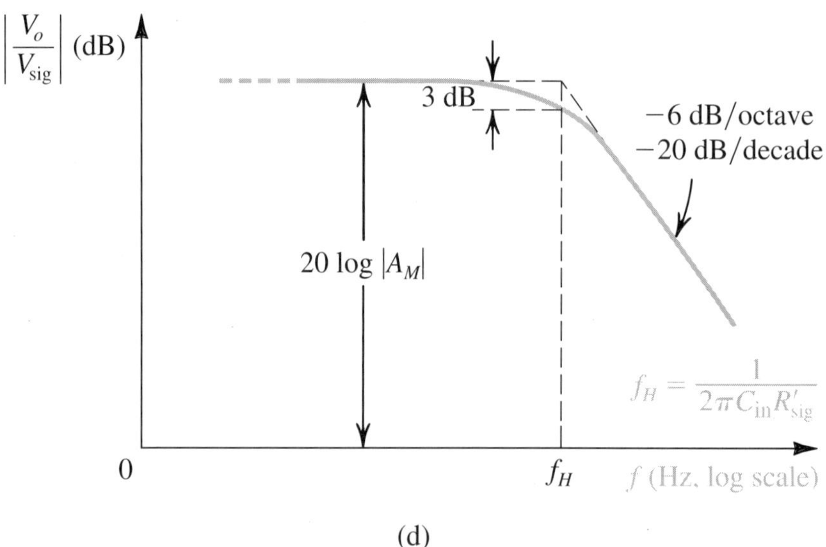

FIGURE 5.72 *(Continued)* **(d)** sketch of the frequency-response plot, which is that of a low-pass STC circuit.

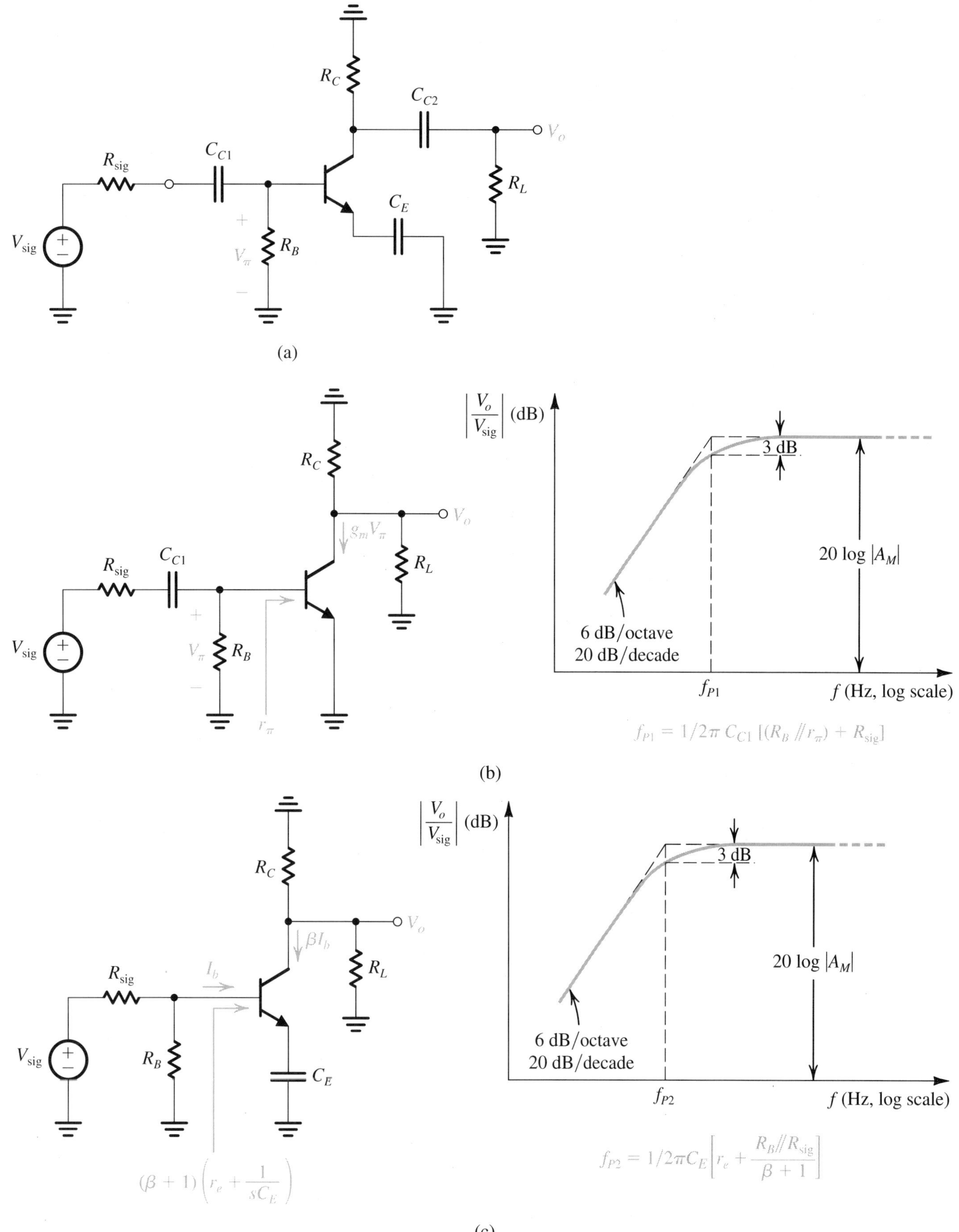

FIGURE 5.73 Analysis of the low-frequency response of the CE amplifier: **(a)** amplifier circuit with dc sources removed; **(b)** the effect of C_{C1} is determined with C_E and C_{C2} assumed to be acting as perfect short circuits; **(c)** the effect of C_E is determined with C_{C1} and C_{C2} assumed to be acting as perfect short circuits;

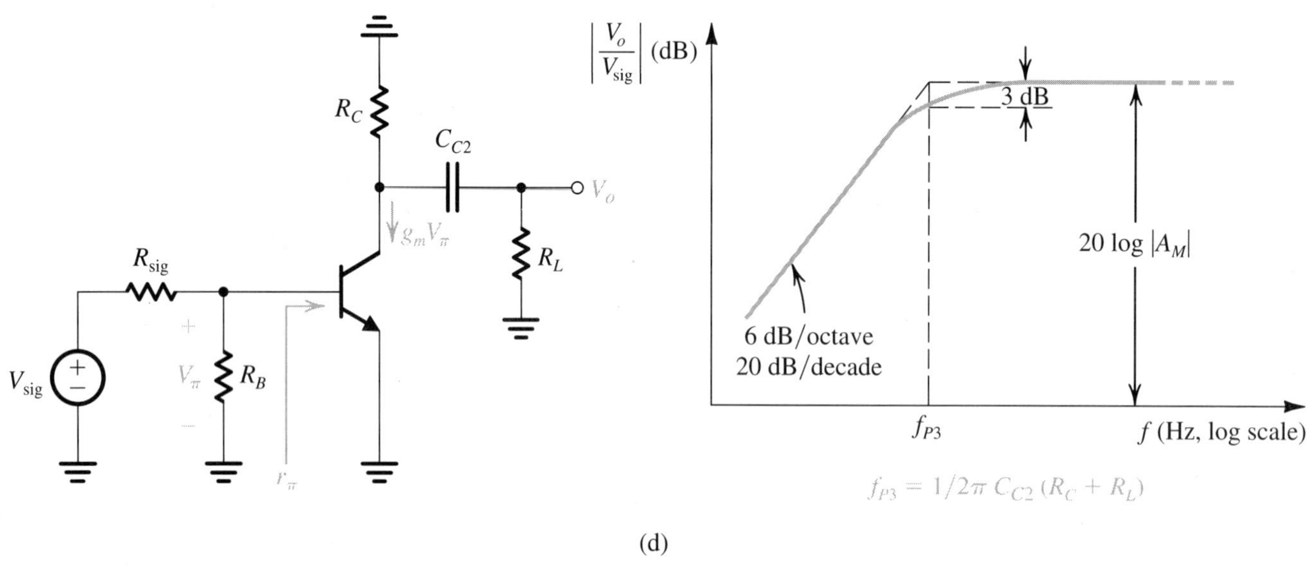

(d)

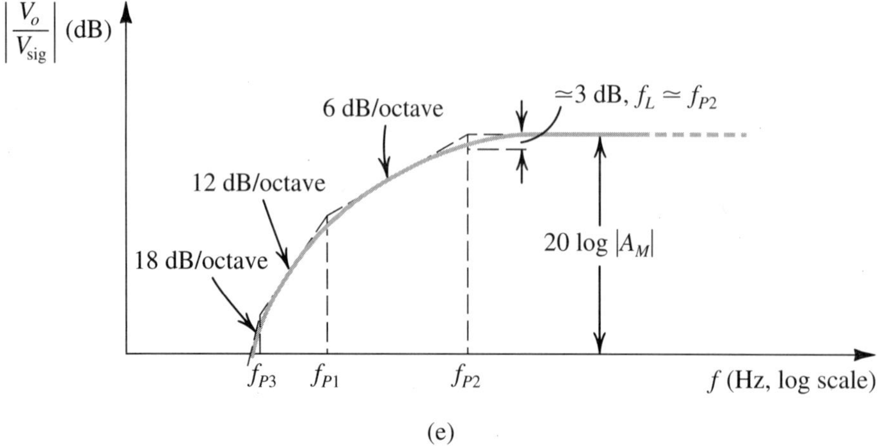

(e)

FIGURE 5.73 *(Continued)* **(d)** the effect of C_{C2} is determined with C_{C1} and C_E assumed to be acting as perfect short circuits; **(e)** sketch of the low-frequency gain under the assumptions that C_{C1}, C_E, and C_{C2} do not interact and that their break (or pole) frequencies are widely separated.

TABLE 6.1 Typical Values of CMOS Device Parameters

	0.8 μm		0.5 μm		0.25 μm		0.18 μm	
Parameter	**NMOS**	**PMOS**	**NMOS**	**PMOS**	**NMOS**	**PMOS**	**NMOS**	**PMOS**
t_{ox} (nm)	15	15	9	9	6	6	4	4
C_{ox} (fF/μm^2)	2.3	2.3	3.8	3.8	5.8	5.8	8.6	8.6
μ (cm^2/V·s)	550	250	500	180	460	160	450	100
μC_{ox} (μA/V^2)	127	58	190	68	267	93	387	86
V_{t0} (V)	0.7	–0.7	0.7	–0.8	0.43	–0.62	0.48	–0.45
V_{DD} (V)	5	5	3.3	3.3	2.5	2.5	1.8	1.8
$\lvert V_A' \rvert$ (V/μm)	25	20	20	10	5	6	5	6
C_{ov} (fF/μm)	0.2	0.2	0.4	0.4	0.3	0.3	0.37	0.33

TABLE 6.2 Typical Parameter Values for BJTs[1]

	Standard High-Voltage Process		Advanced Low-Voltage Process	
Parameter	***npn***	**Lateral *pnp***	***npn***	**Lateral *pnp***
A_E (μm^2)	500	900	2	2
I_S (A)	5×10^{-15}	2×10^{-15}	6×10^{-18}	6×10^{-18}
β_0 (A/A)	200	50	100	50
V_A (V)	130	50	35	30
V_{CE0} (V)	50	60	8	18
τ_F	0.35 ns	30 ns	10 ps	650 ps
C_{je0}	1 pF	0.3 pF	5 fF	14 fF
$C_{\mu 0}$	0.3 pF	1 pF	5 fF	15 fF
r_x (Ω)	200	300	400	200

[1]Adapted from Gray et al. (2000); see Bibliography.

TABLE 6.3 Comparison of the MOSFET and the BJT

	NMOS	npn
Circuit Symbol	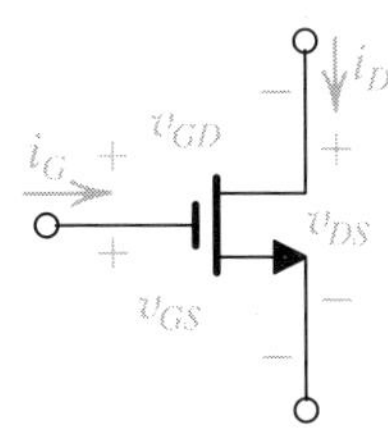	i_C, v_{BC}, i_B, v_{CE}, v_{BE}
To Operate in the Active Mode, Two Conditions Have To Be Satisfied	(1) Induce a channel: $v_{GS} \geq V_t, \quad V_t = 0.5 - 0.7 \text{ V}$ Let $v_{GS} = V_t + v_{OV}$ (2) Pinch-off channel at drain: $v_{GD} < V_t$ or equivalently, $v_{DS} \geq V_{OV}, \quad V_{OV} = 0.2 - 0.3 \text{ V}$	(1) Forward-bias EBJ: $v_{BE} \geq V_{BE\text{on}}, \quad V_{BE\text{on}} \cong 0.5 \text{ V}$ (2) Reverse-bias CBJ: $v_{BC} < V_{BC\text{on}}, \quad V_{BC\text{on}} \cong 0.4 \text{ V}$ or equivalently, $v_{CE} \geq 0.3 \text{ V}$
Current-Voltage Characteristics in the Active Region	$i_D = \frac{1}{2}\mu_n C_{ox}\frac{W}{L}(v_{GS} - V_t)^2\left(1 + \frac{v_{DS}}{V_A}\right)$ $= \frac{1}{2}\mu_n C_{ox}\frac{W}{L}v_{OV}^2\left(1 + \frac{v_{DS}}{V_A}\right)$ $i_G = 0$	$i_C = I_S e^{v_{BE}/V_T}\left(1 + \frac{v_{CE}}{V_A}\right)$ $i_B = i_C/\beta$
Low-Frequency Hybrid-π Model	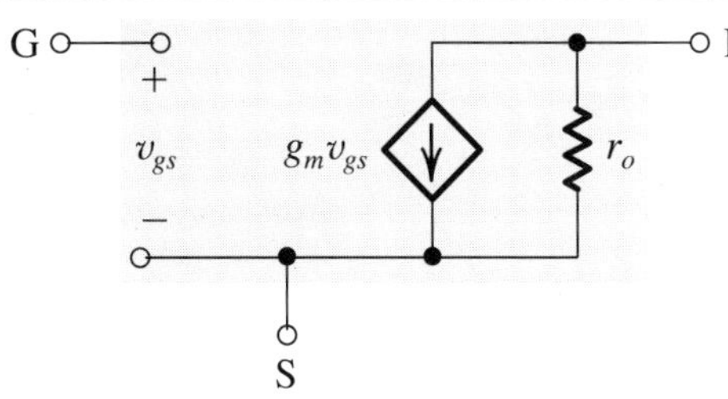	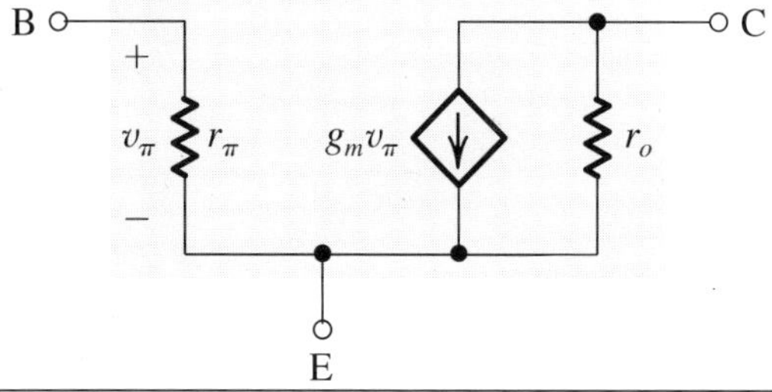
Low-Frequency T Model	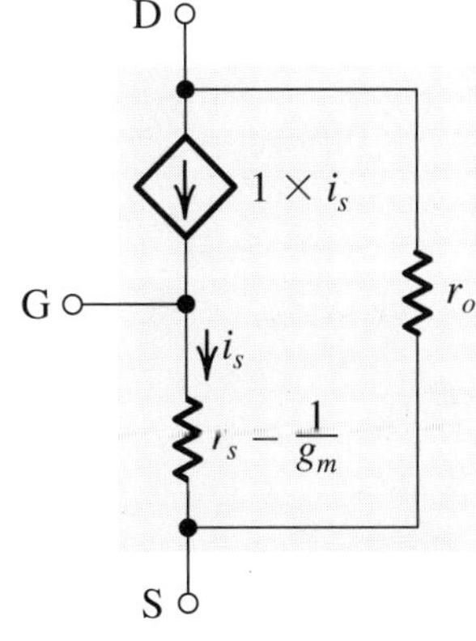	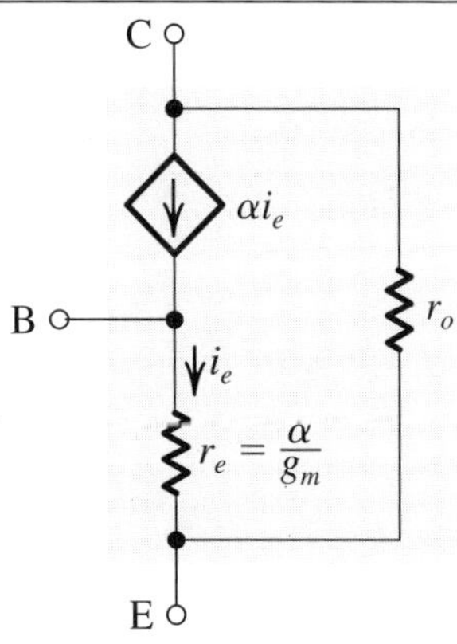
Transconductance g_m	$g_m = I_D/(V_{OV}/2)$ $g_m = (\mu_n C_{ox})\left(\frac{W}{L}\right)V_{OV}$ $g_m = \sqrt{2(\mu_n C_{ox})\left(\frac{W}{L}\right)I_D}$	$g_m = I_C/V_T$

TABLE 6.3 Comparison of the MOSFET and the BJT *(Continued)*

	NMOS	*npn*
Output Resistance r_o	$r_o = V_A/I_D = \frac{V_A' L}{I_D}$	$r_o = V_A/I_C$
Intrinsic Gain $A_0 \equiv g_m r_o$	$A_0 = V_A/(V_{OV}/2)$ $A_0 = \frac{2V_A' L}{V_{OV}}$ $A_0 = \frac{V_A'\sqrt{2\mu_n C_{ox} WL}}{\sqrt{I_D}}$	$A_0 = V_A/V_T$
Input Resistance with Source (Emitter) Grounded	∞	$r_\pi = \beta/g_m$
High-Frequency Model	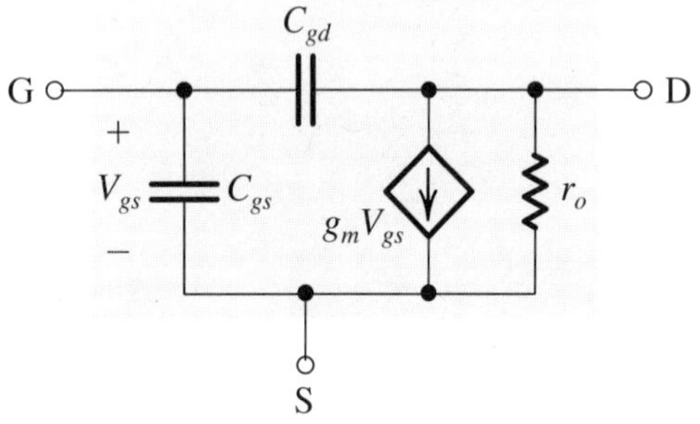 G, D, S, C_{gd}, C_{gs}, $+\ V_{gs}\ -$, $g_m V_{gs}$, r_o	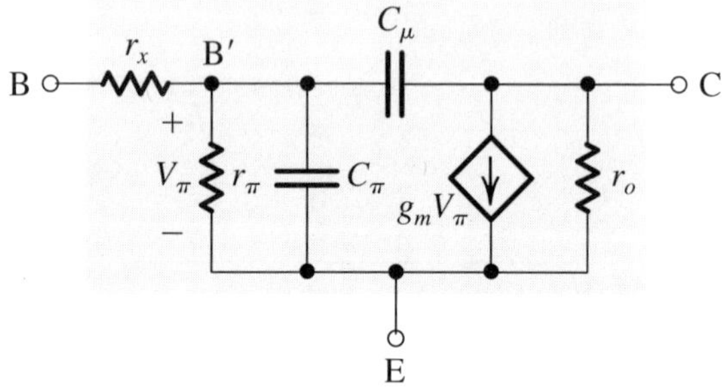 B, B′, C, E, r_x, C_μ, $+\ V_\pi\ -$, r_π, C_π, $g_m V_\pi$, r_o
Capacitances	$C_{gs} = \frac{2}{3}WLC_{ox} + WL_{ov}C_{ox}$ $C_{gd} = WL_{ov}C_{ox}$	$C_\pi = C_{de} + C_{je}$ $C_{de} = \tau_F g_m$ $C_{je} \cong 2C_{je0}$ $C_\mu = C_{\mu 0}\Big/\left[1 + \frac{V_{CB}}{V_{C0}}\right]^m$
Transition Frequency f_T	$f_T = \frac{g_m}{2\pi(C_{gs} + C_{gd})}$ For $C_{gs} \gg C_{gd}$ and $C_{gs} \cong \frac{2}{3}WLC_{ox}$, $f_T \cong \frac{1.5\mu_n V_{OV}}{2\pi L^2}$	$f_T = \frac{g_m}{2\pi(C_\pi + C_\mu)}$ For $C_\pi \gg C_\mu$ and $C_\pi \cong C_{de}$, $f_T \cong \frac{2\mu_n V_T}{2\pi W_B^2}$
Design Parameters	$I_D, V_{OV}, L, \frac{W}{L}$	I_C, V_{BE}, A_E (or I_S)
Good Analog Switch?	Yes, because the device is symmetrical and thus the i_D–v_{DS} characteristics pass directly through the origin.	No, because the device is asymmetrical with an offset voltage V_{CEoff}.

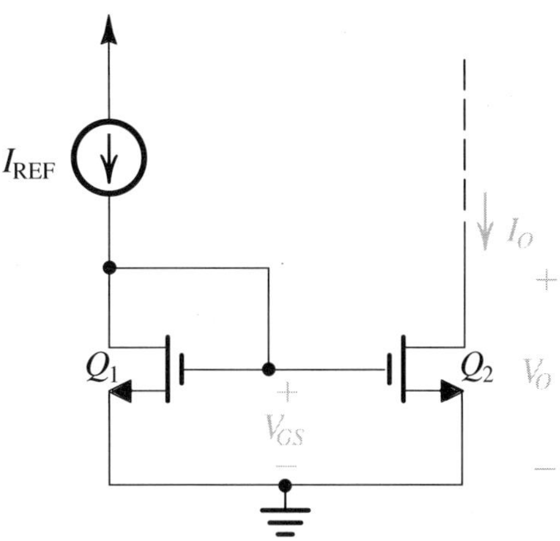

FIGURE 6.5 Basic MOSFET current mirror.

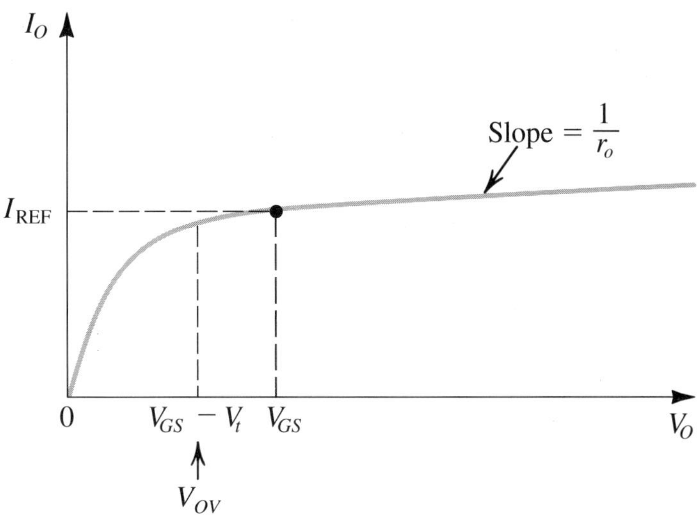

FIGURE 6.6 Output characteristic of the current source in Fig. 6.4 and the current mirror of Fig. 6.5 for the case Q_2 is matched to Q_1.

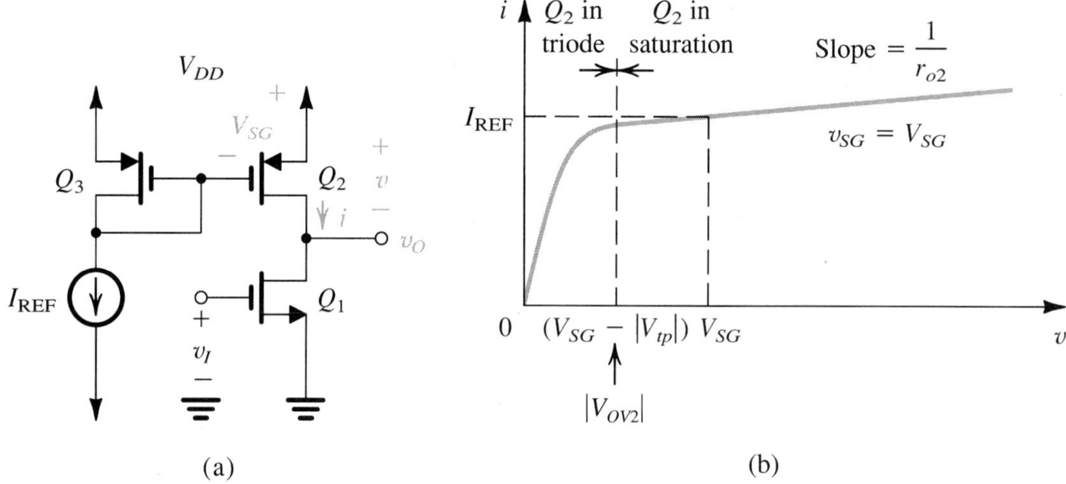

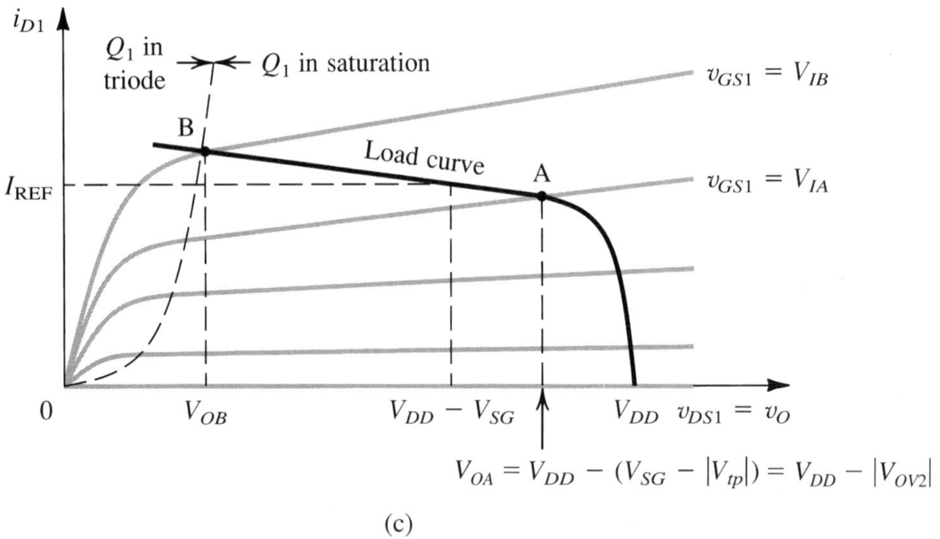

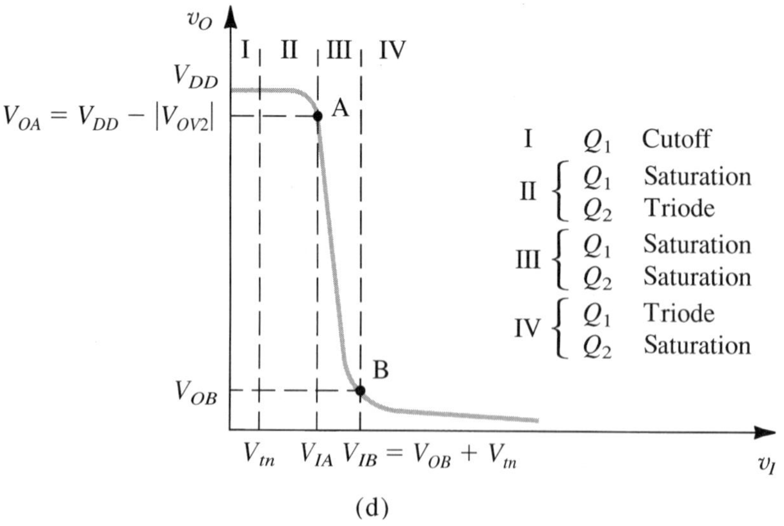

FIGURE 6.18 The CMOS common-source amplifier: **(a)** circuit; **(b)** i–v characteristic of the active-load Q_2; **(c)** graphical construction to determine the transfer characteristic; and **(d)** transfer characteristic.

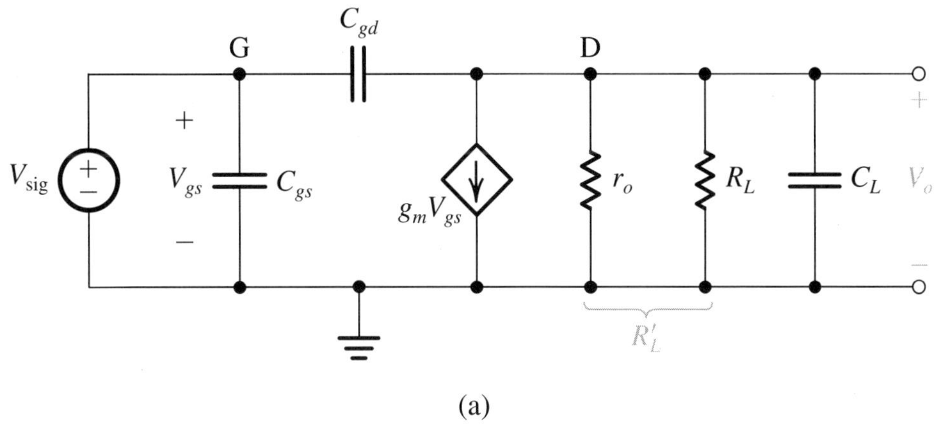

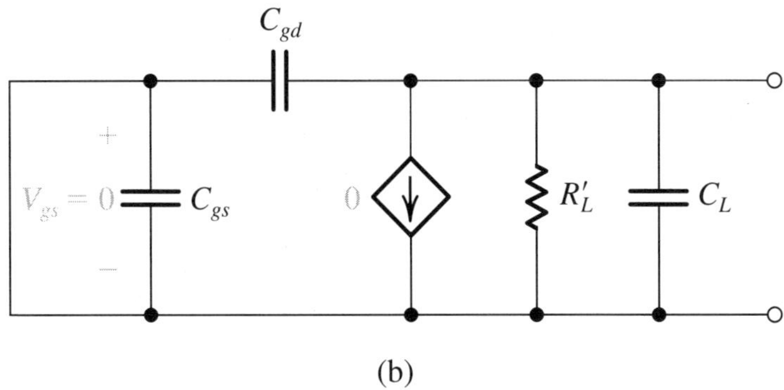

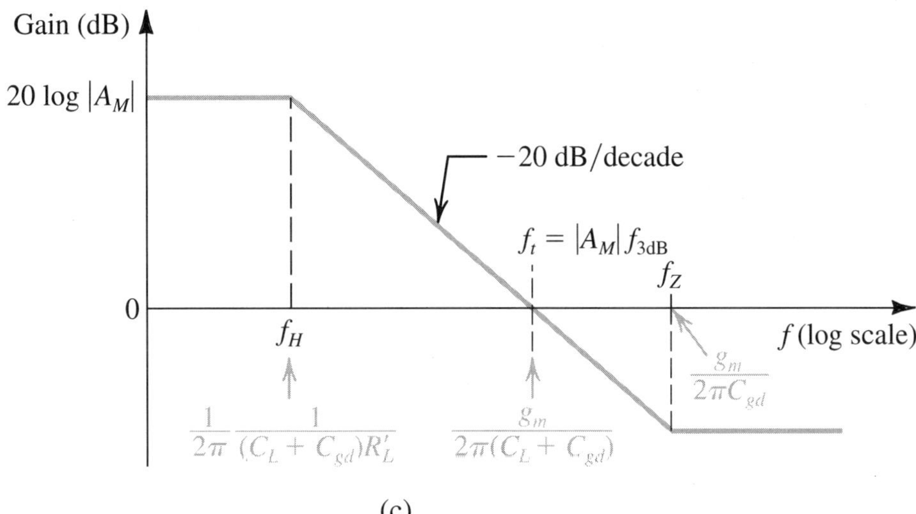

FIGURE 6.26 **(a)** High-frequency equivalent circuit of a CS amplifier fed with a signal source having a very low (effectively zero) resistance. **(b)** The circuit with V_{sig} reduced to zero. **(c)** Bode plot for the gain of the circuit in (a).

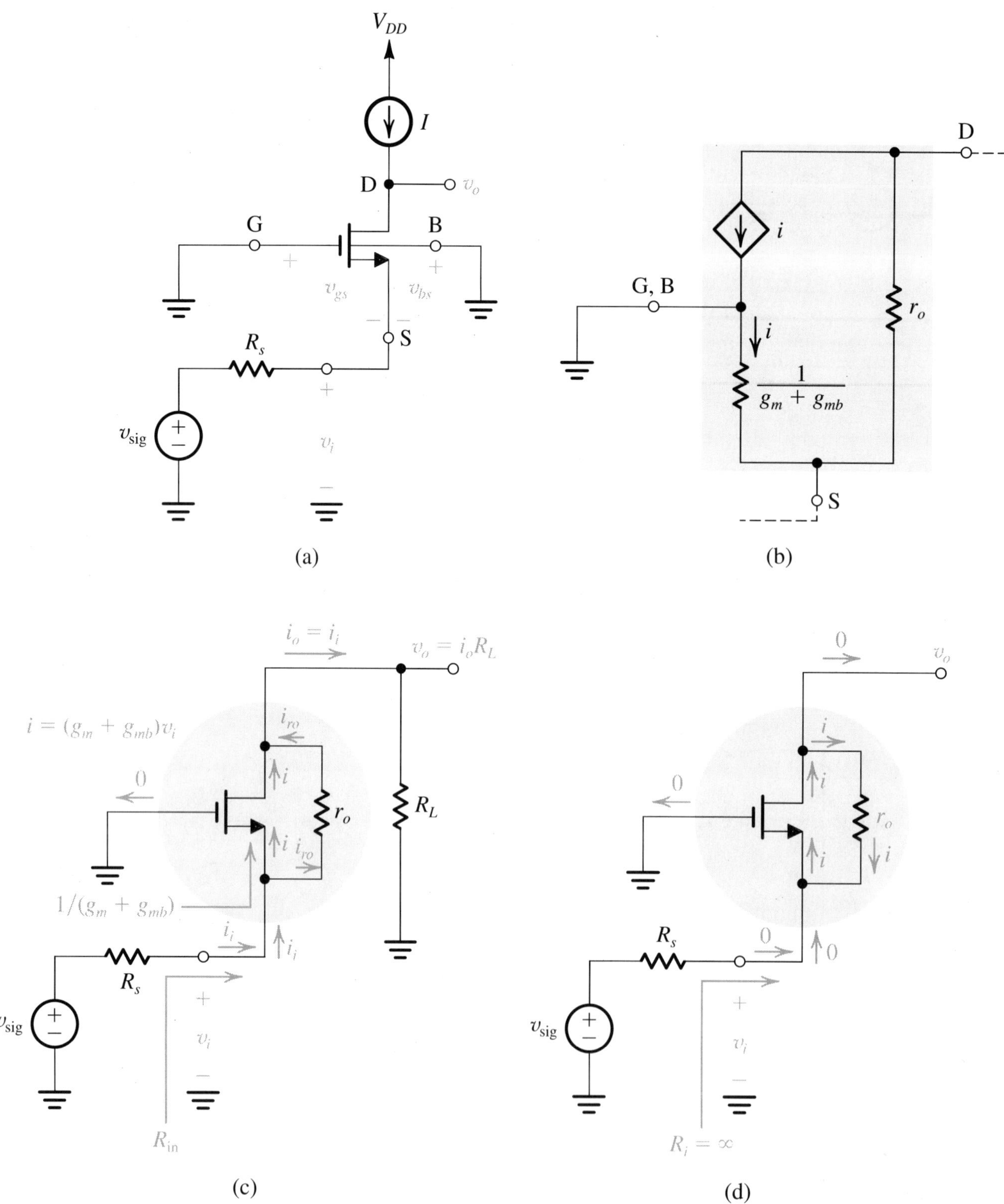

FIGURE 6.27 **(a)** Active-loaded common-gate amplifier. **(b)** MOSFET equivalent circuit for the CG case in which the body and gate terminals are connected to ground. **(c)** Small-signal analysis performed directly on the circuit diagram with the T model of (b) used implicitly. **(d)** Operation with the output open-circuited.

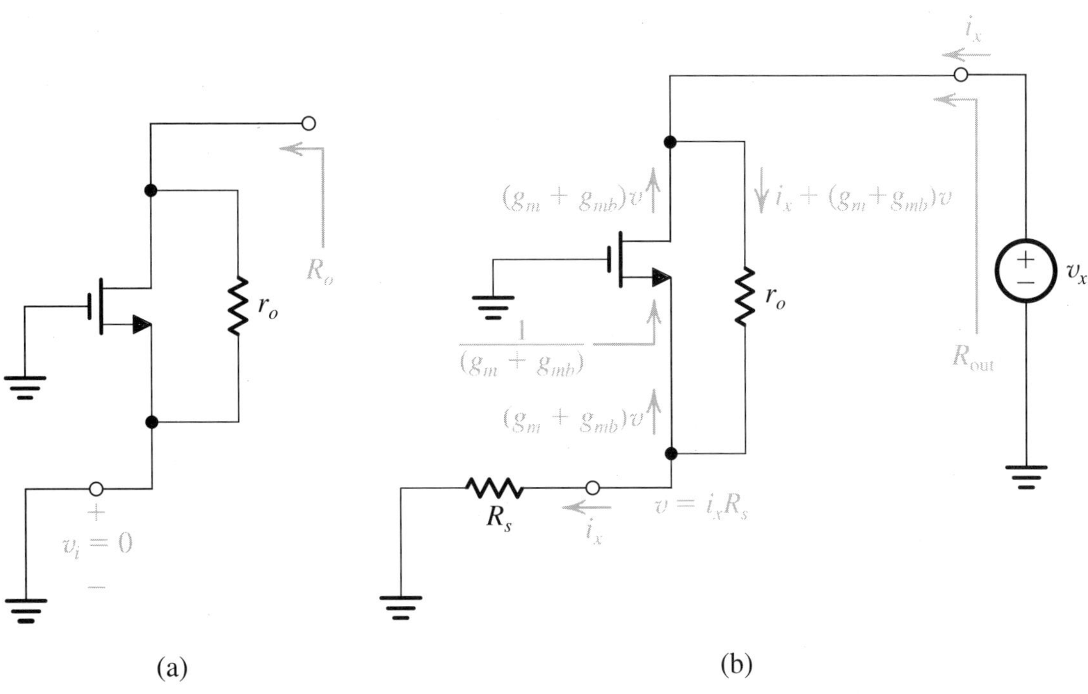

FIGURE 6.28 **(a)** The output resistance R_o is found by setting $v_i = 0$. **(b)** The output resistance R_{out} is obtained by setting $v_{sig} = 0$.

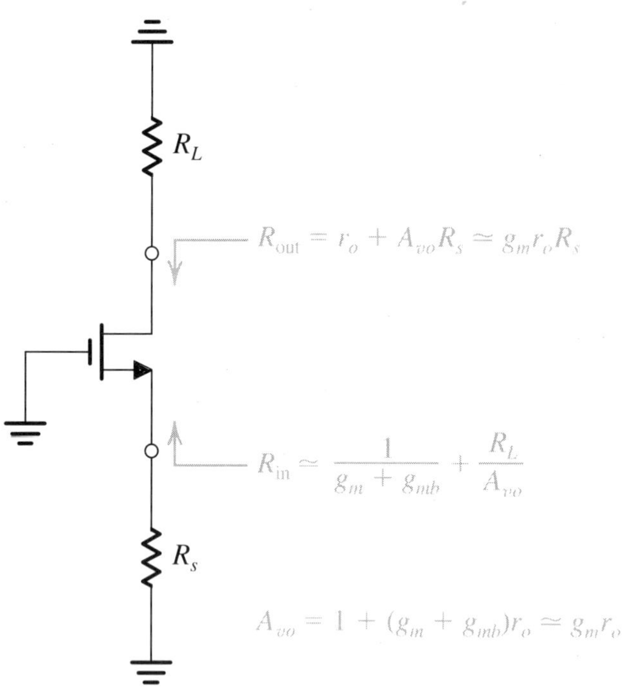

R_L
$R_{out} = r_o + A_{vo}R_s \simeq g_m r_o R_s$
$R_{in} \simeq \frac{1}{g_m + g_{mb}} + \frac{R_L}{A_{vo}}$
R_s
$A_{vo} = 1 + (g_m + g_{mb})r_o \simeq g_m r_o$

FIGURE 6.29 The impedance transformation property of the CG configuration.

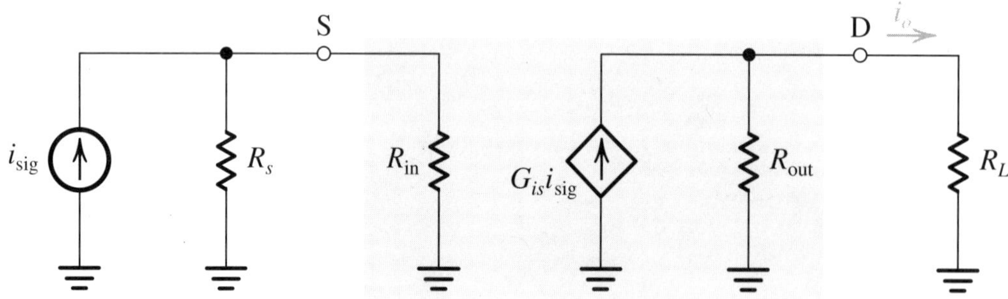

FIGURE 6.30 Equivalent circuit of the CG amplifier illustrating its application as a current buffer. R_{in} and R_{out} are given in Fig. 6.29, and $G_{is} = A_{vo}(R_s/R_{out}) \simeq 1$.

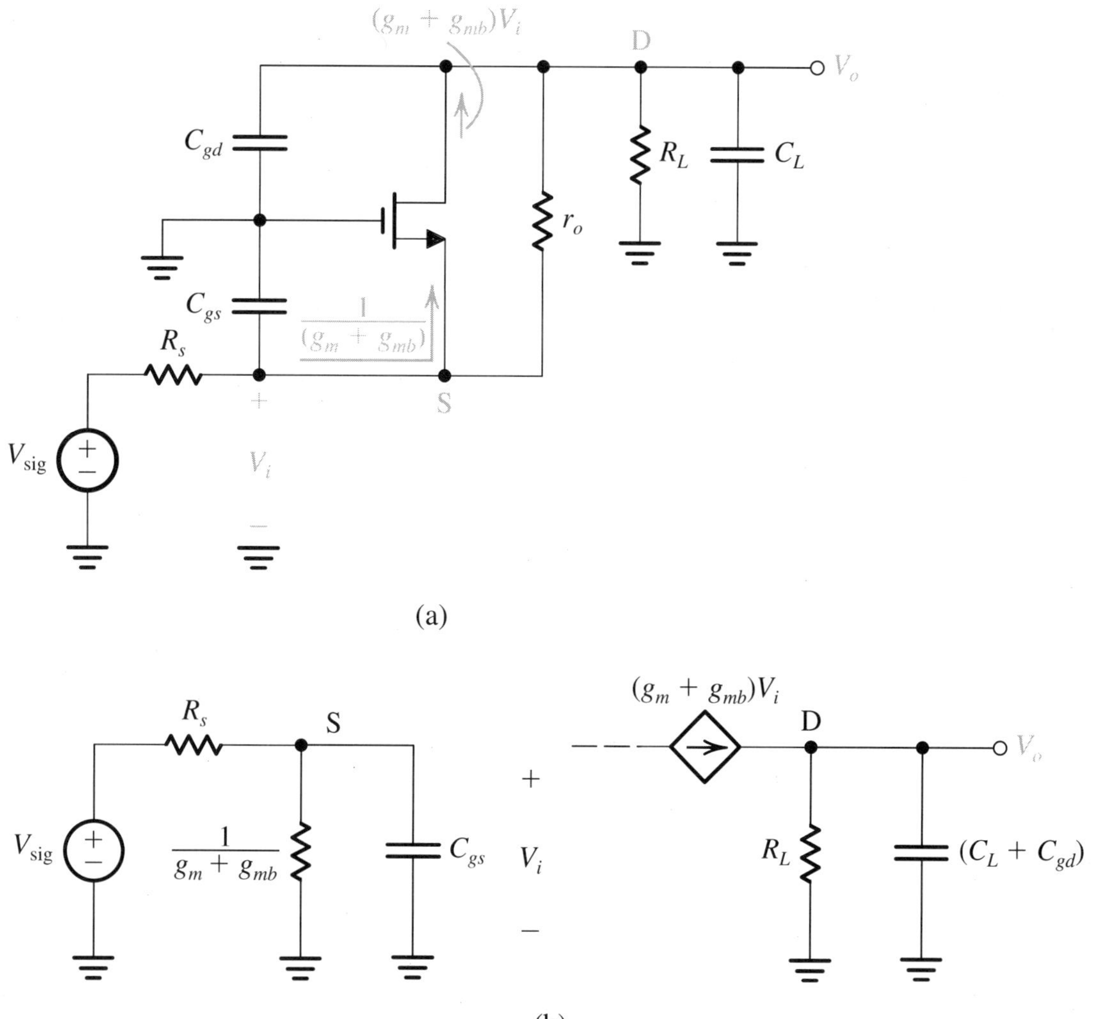

FIGURE 6.31 **(a)** The common-gate amplifier with the transistor internal capacitances shown. A load capacitance C_L is also included. **(b)** Equivalent circuit for the case in which r_o is neglected.

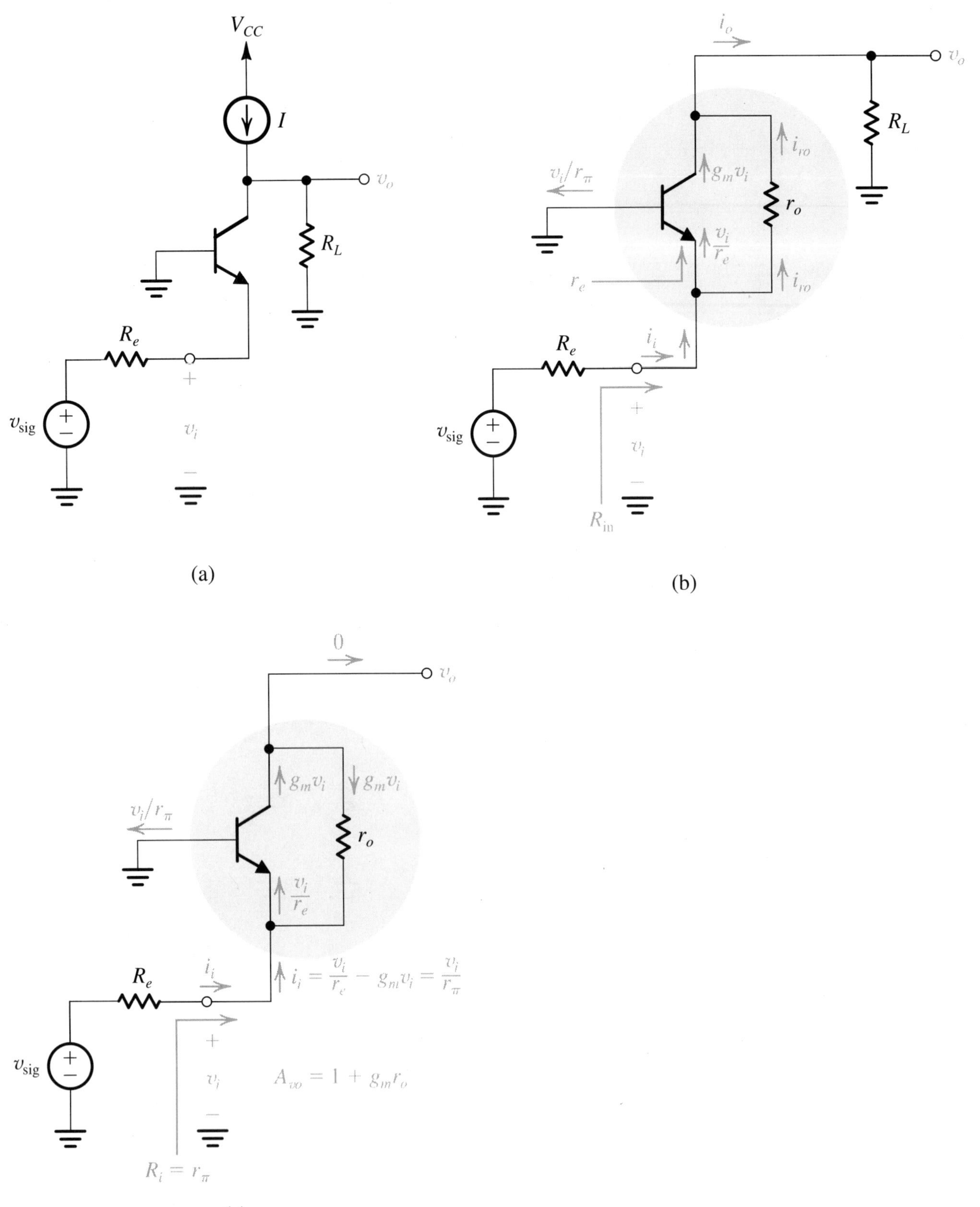

FIGURE 6.33 **(a)** Active-loaded common-base amplifier. **(b)** Small-signal analysis performed directly on the circuit diagram with the BJT T model used implicitly. **(c)** Small-signal analysis with the output open-circuited.

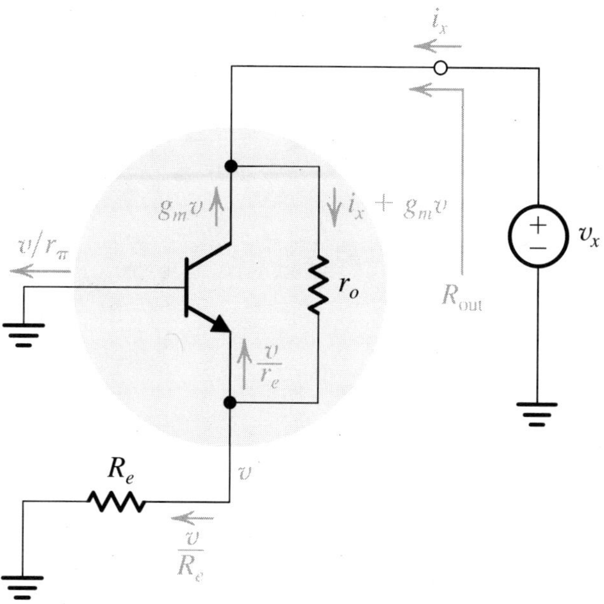

FIGURE 6.34 Analysis of the CB circuit to determine R_{out}. Observe that the current i_x that enters the transistor must equal the sum of the two currents v/r_π and v/R_e that leave the transistor; that is, $i_x = v/r_\pi + v/R_e$.

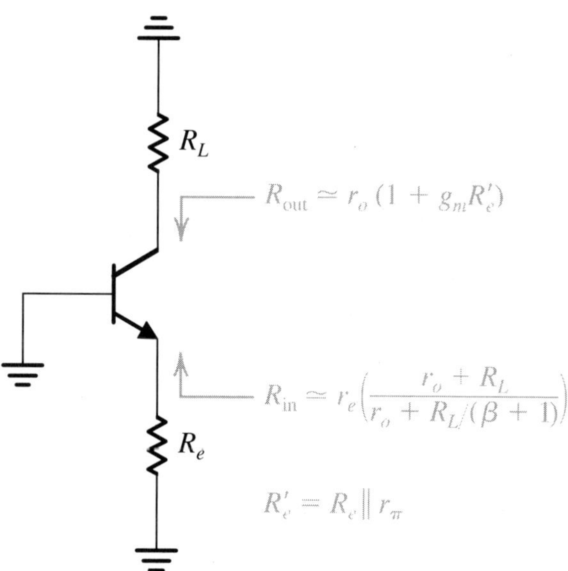

R_L
$R_{out} \simeq r_o(1 + g_m R'_e)$
$R_{in} \simeq r_e\left(\frac{r_o + R_L}{r_o + R_L/(\beta + 1)}\right)$
R_e
$R'_e = R_e \| r_\pi$

FIGURE 6.35 Input and output resistances of the CB amplifier.

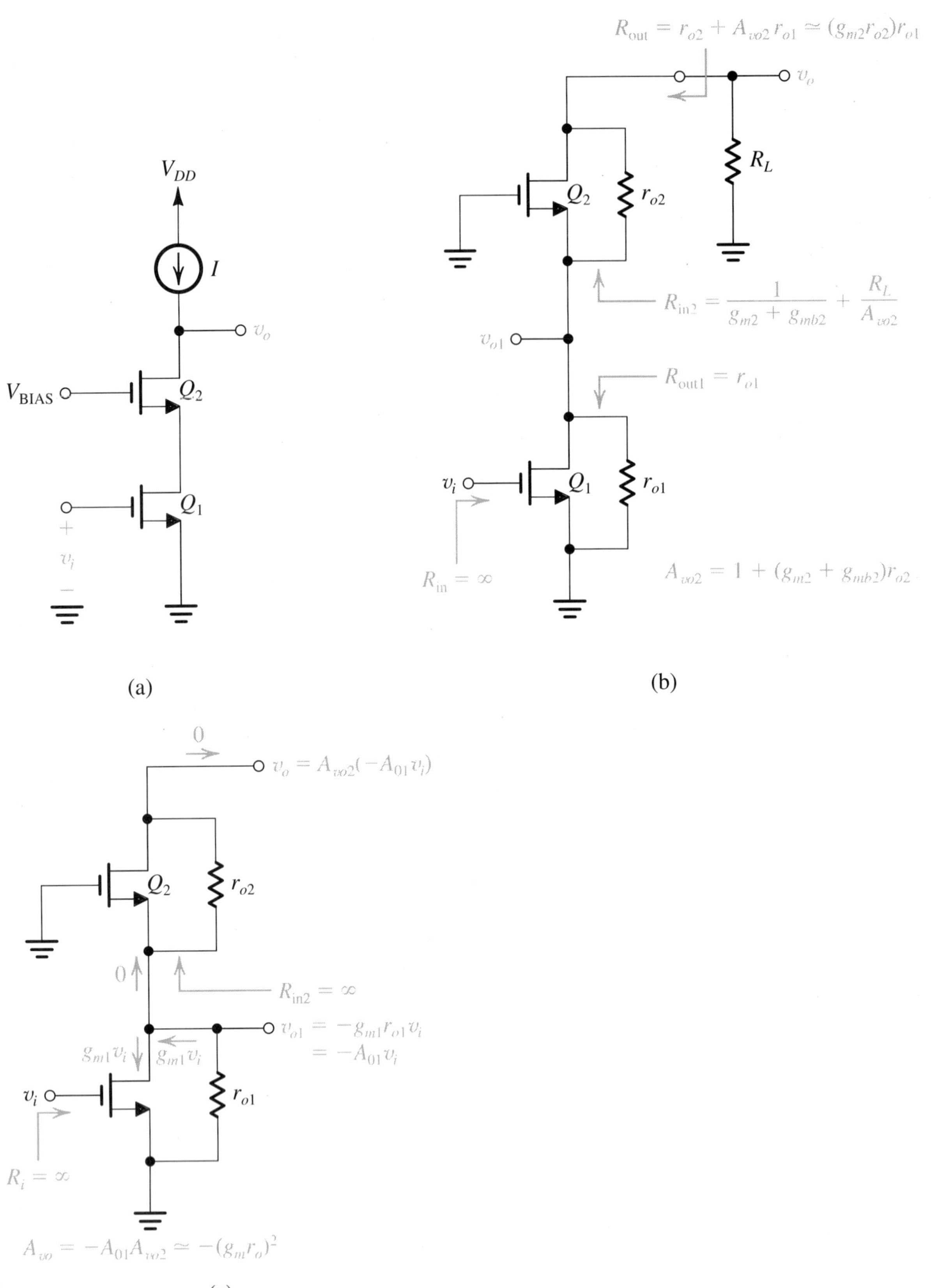

FIGURE 6.36 **(a)** The MOS cascode amplifier. **(b)** The circuit prepared for small-signal analysis with various input and output resistances indicated. **(c)** The cascode with the output open-circuited.

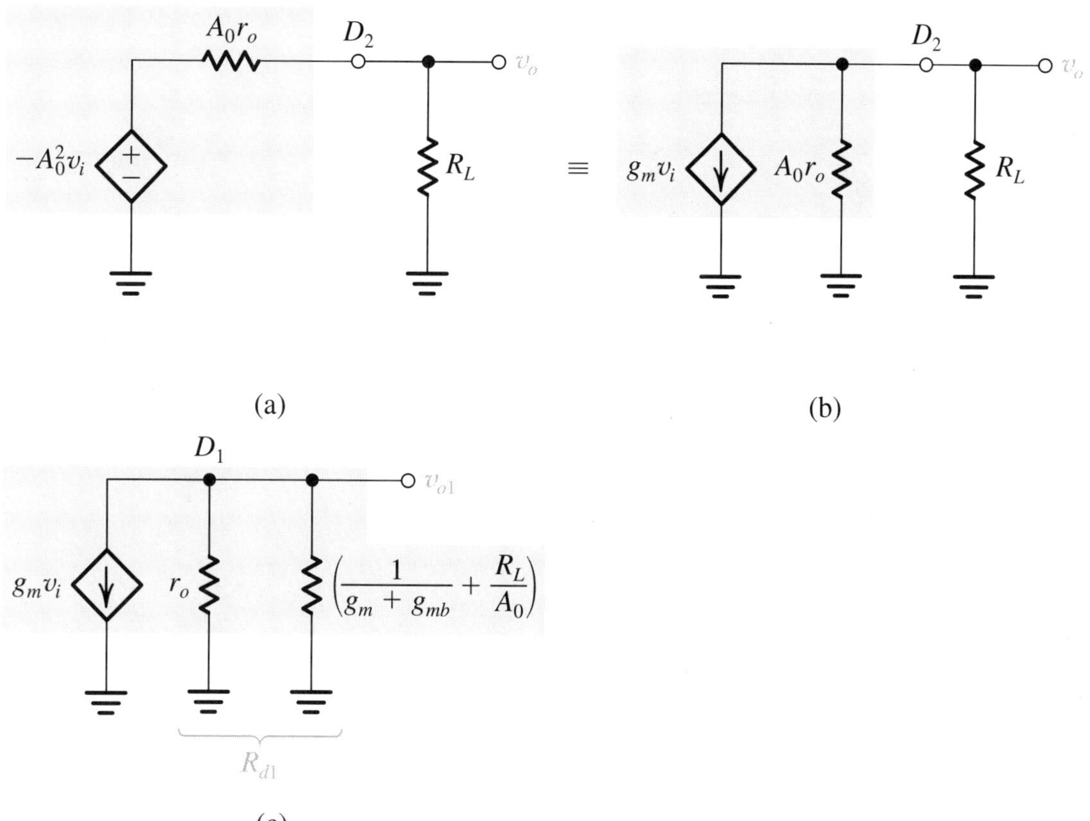

FIGURE 6.37 (**a** and **b**) Two equivalent circuits for the output of the cascode amplifier. Either circuit can be used to determine the gain $A_v = v_o/v_i$, which is equal to G_v because $R_{in} = \infty$ and thus $v_i = v_{sig}$. (**c**) Equivalent circuit for determining the voltage gain of the CS stage, Q_1.

	Common Source	Cascode
Circuit	V_i, r_o, R_L, C_L, V_o, $R'_L = R_L \| r_o$	V_i, $A_0 r_o$, $A_0 R_L$, C_L, V_o
DC Gain	$-g_m R'_L$	$-A_0 g_m R'_L$
f_{3dB}	$\dfrac{1}{2\pi(C_L + C_{gd})R'_L}$	$\dfrac{1}{2\pi(C_L + C_{gd})A_0 R'_L}$
f_t	$\dfrac{g_m}{2\pi(C_L + C_{gd})}$	$\dfrac{g_m}{2\pi(C_L + C_{gd})}$

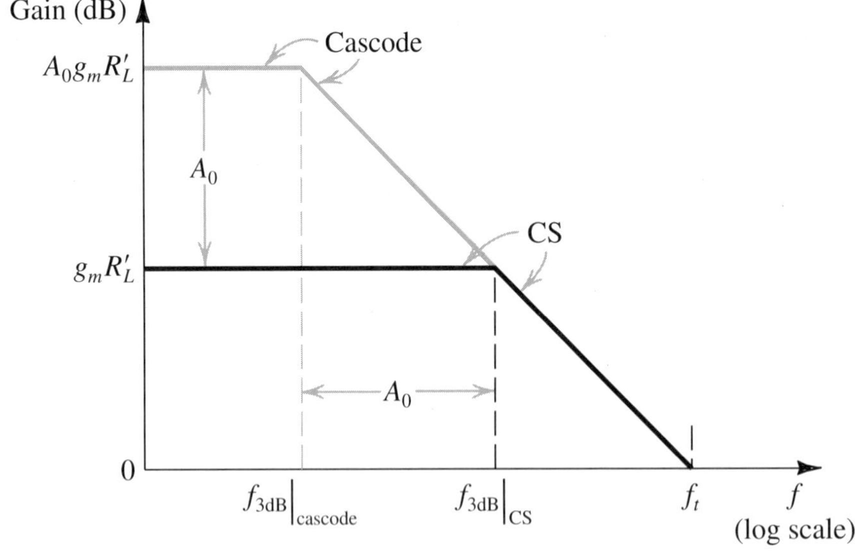

FIGURE 6.39 Effect of cascoding on gain and bandwidth in the case $R_{sig} = 0$. Cascoding can increase the dc gain by the factor A_0 while keeping the unity-gain frequency constant. Note that to achieve the high gain, the load resistance must be increased by the factor A_0.

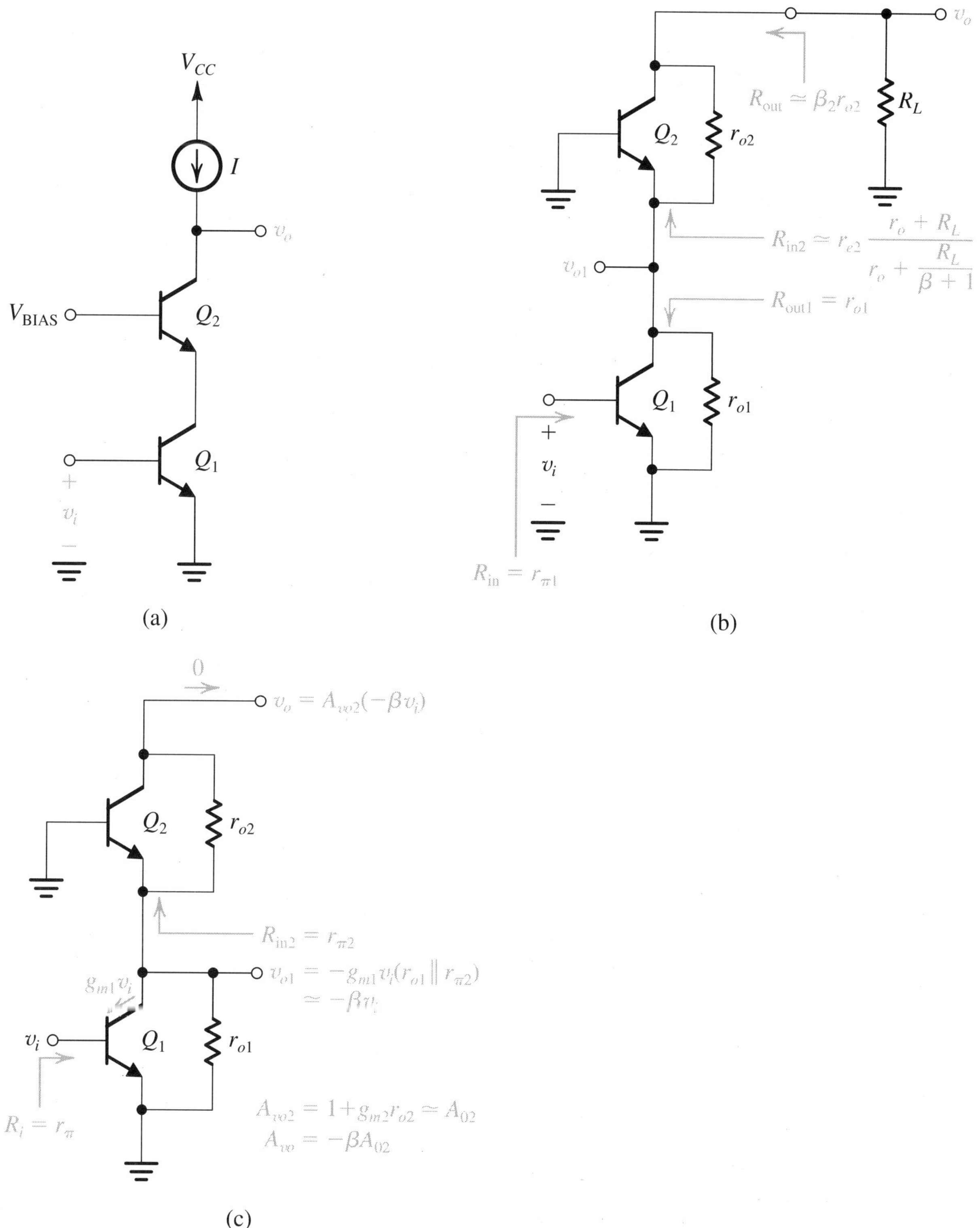

FIGURE 6.40 **(a)** The BJT cascode amplifier. **(b)** The circuit prepared for small-signal analysis with various input and output resistances indicated. Note that r_x is neglected. **(c)** The cascode with the output open-circuited.

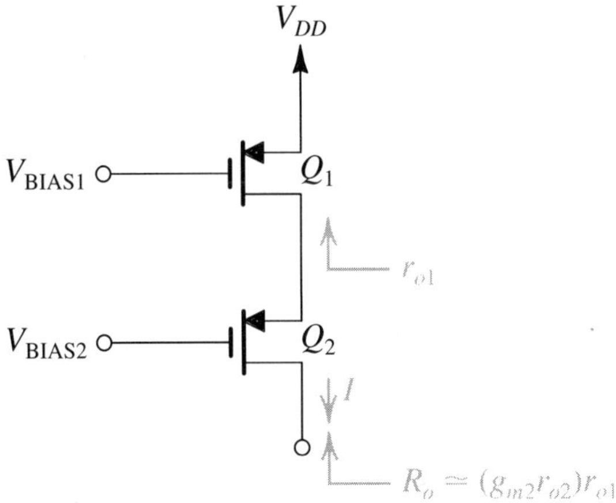

FIGURE 6.43 A cascode current-source.

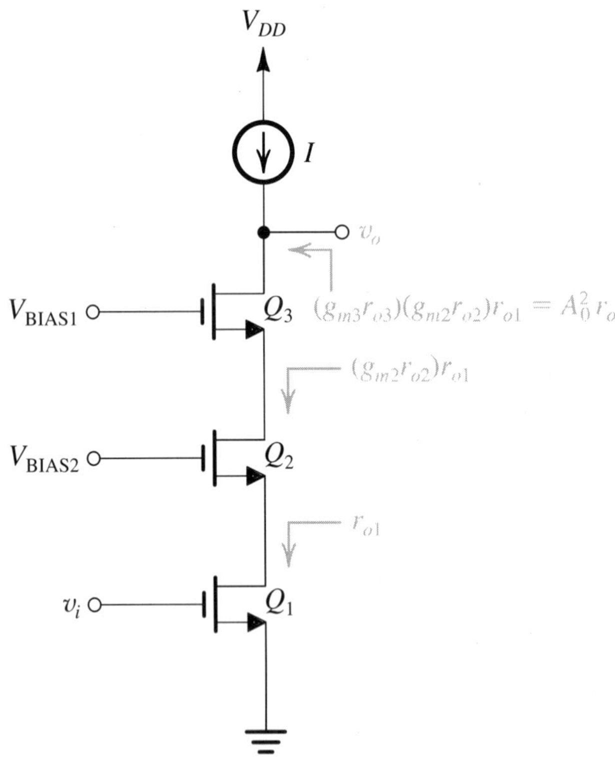

FIGURE 6.44 Double cascoding.

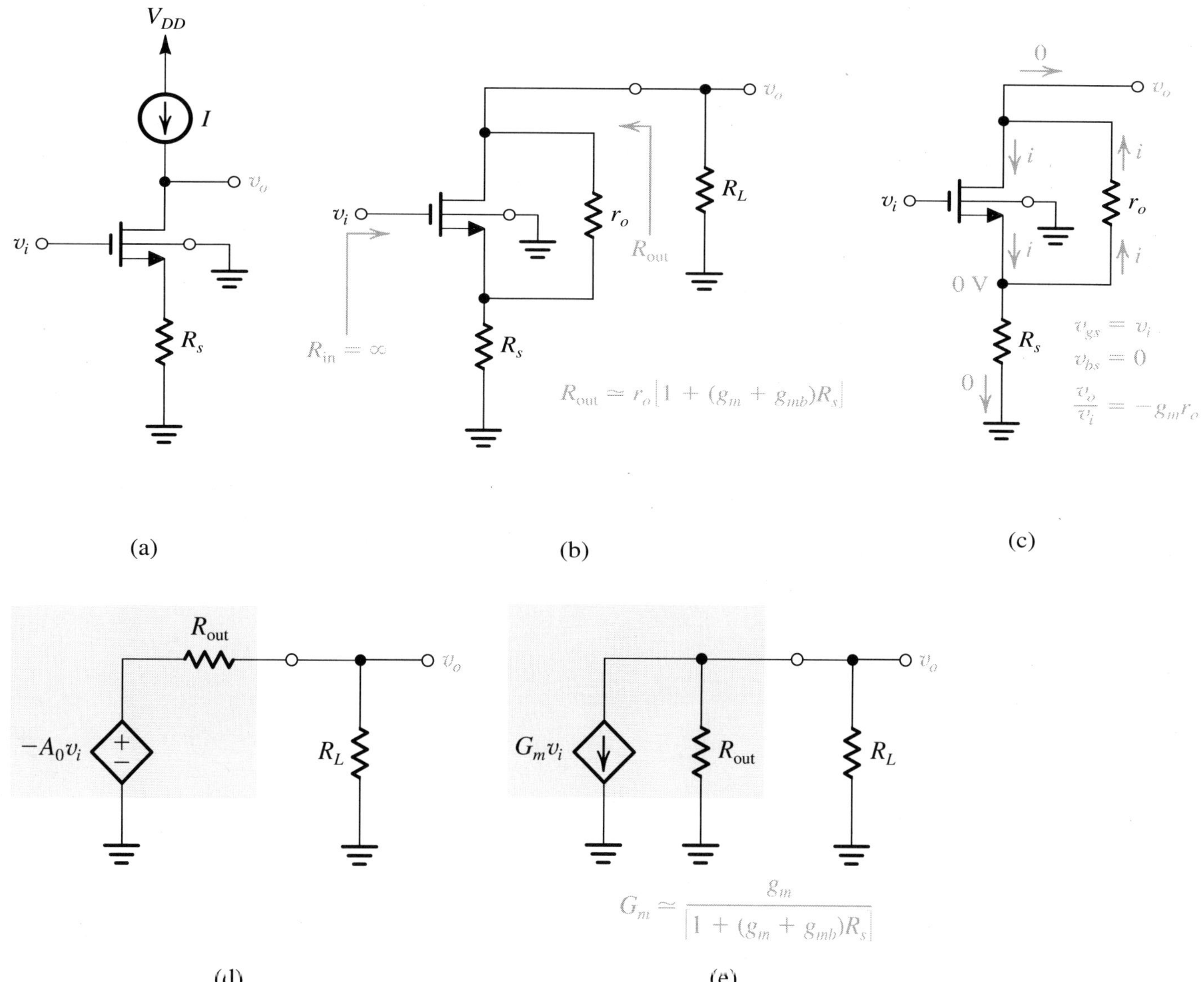

FIGURE 6.47 **(a)** A CS amplifier with a source-degeneration resistance R_s. **(b)** Circuit for small-signal analysis. **(c)** Circuit with the output open to determine A_{vo}. **(d)** Output equivalent circuit. **(e)** Another output equivalent circuit in terms of G_m.

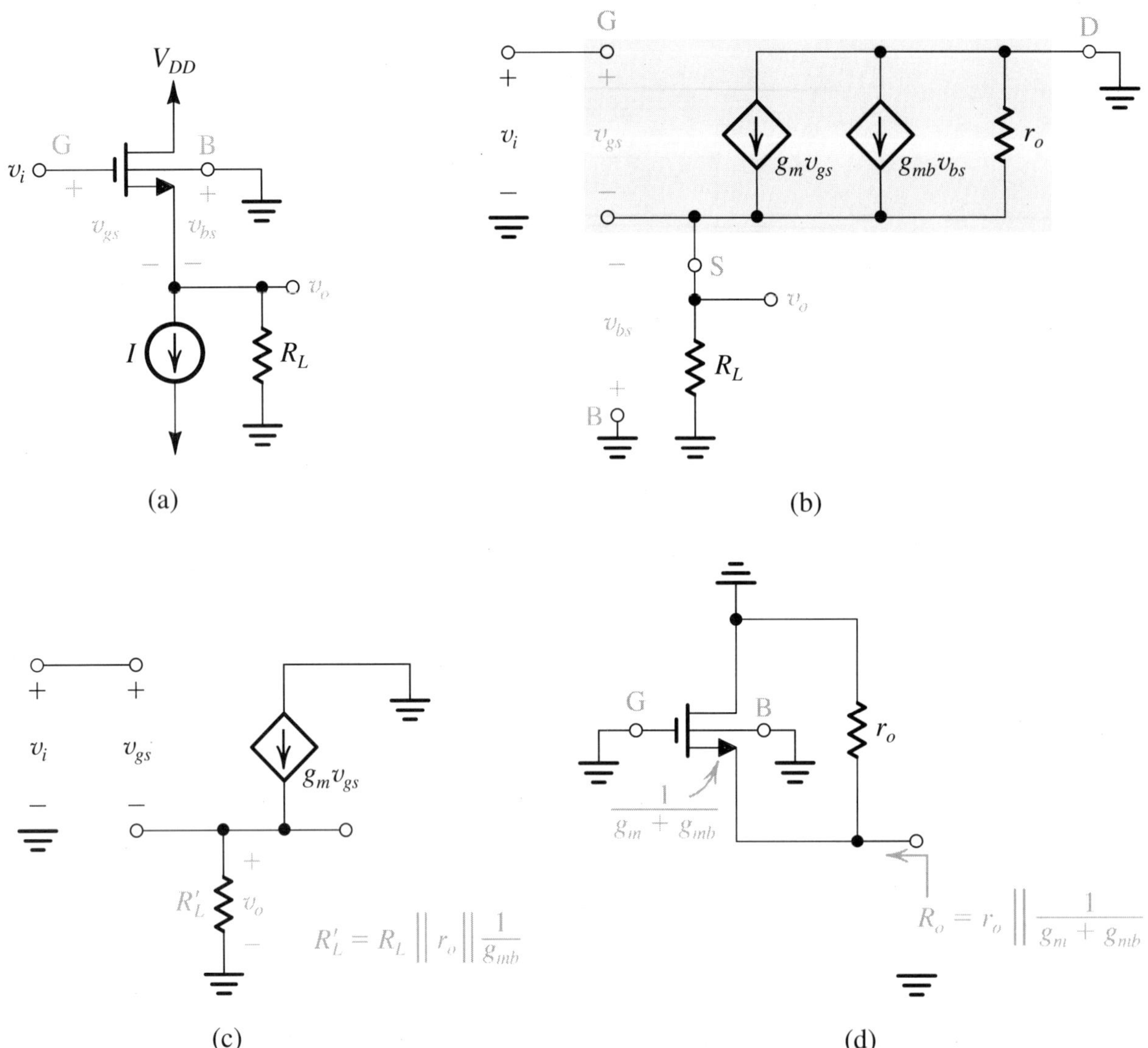

FIGURE 6.50 **(a)** An IC source follower. **(b)** Small-signal equivalent-circuit model of the source follower. **(c)** A simplified version of the equivalent circuit. **(d)** Determining the output resistance of the source follower.

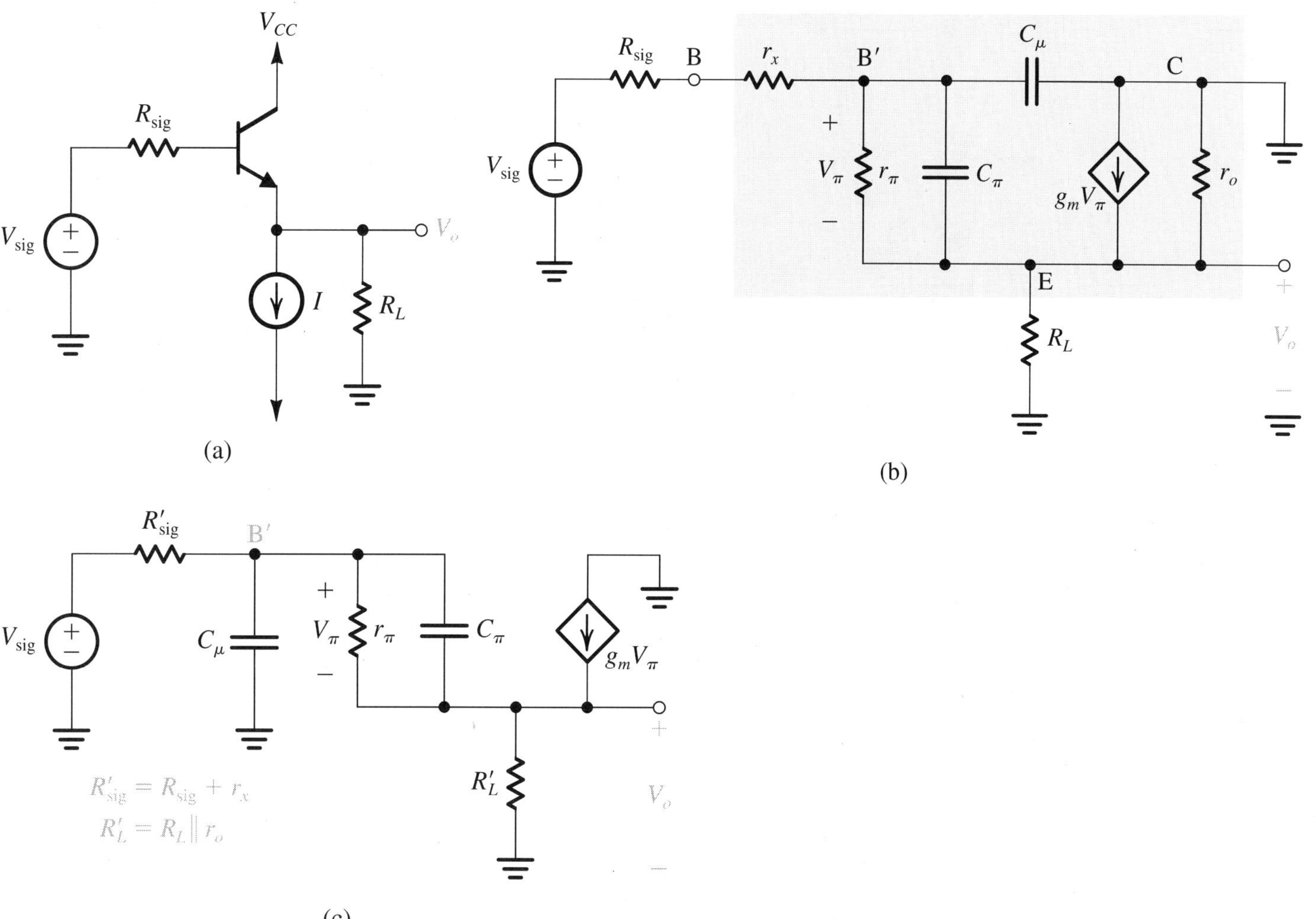

FIGURE 6.52 **(a)** Emitter follower. **(b)** High-frequency equivalent circuit. **(c)** Simplified equivalent circuit.

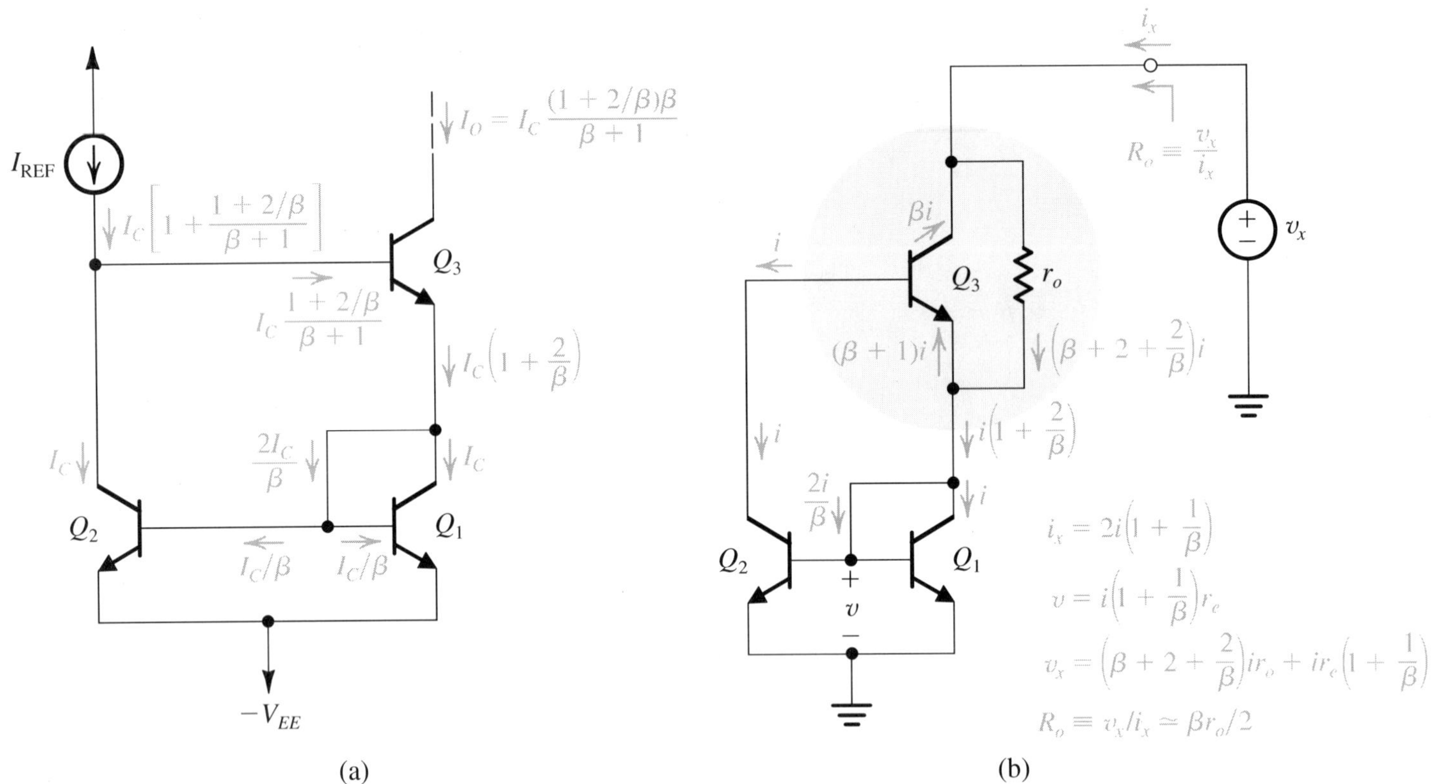

FIGURE 6.60 The Wilson bipolar current mirror: **(a)** circuit showing analysis to determine the current transfer ratio; and **(b)** determining the output resistance. Note that the current i_x that enters Q_3 must equal the sum of the currents that leave it, $2i$.

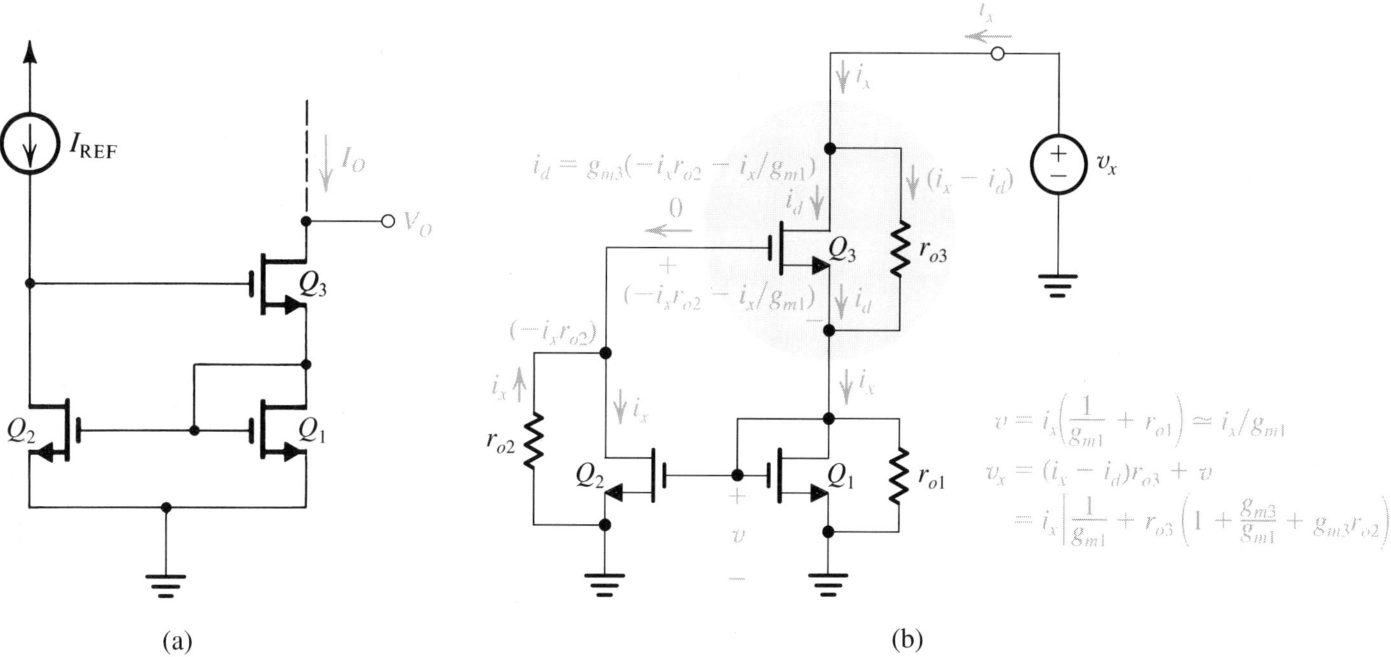

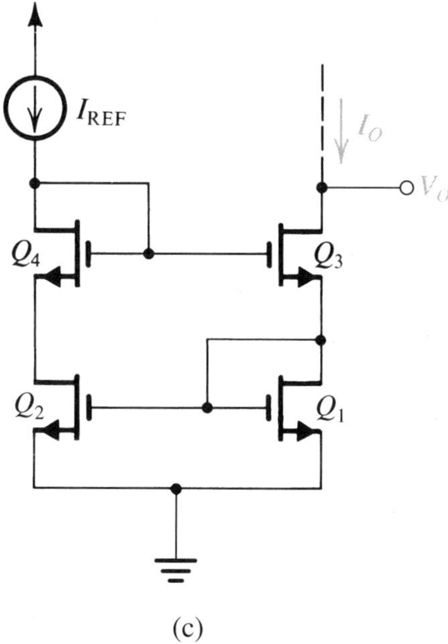

FIGURE 6.61 The Wilson MOS mirror: **(a)** circuit; **(b)** analysis to determine output resistance; and **(c)** modified circuit.

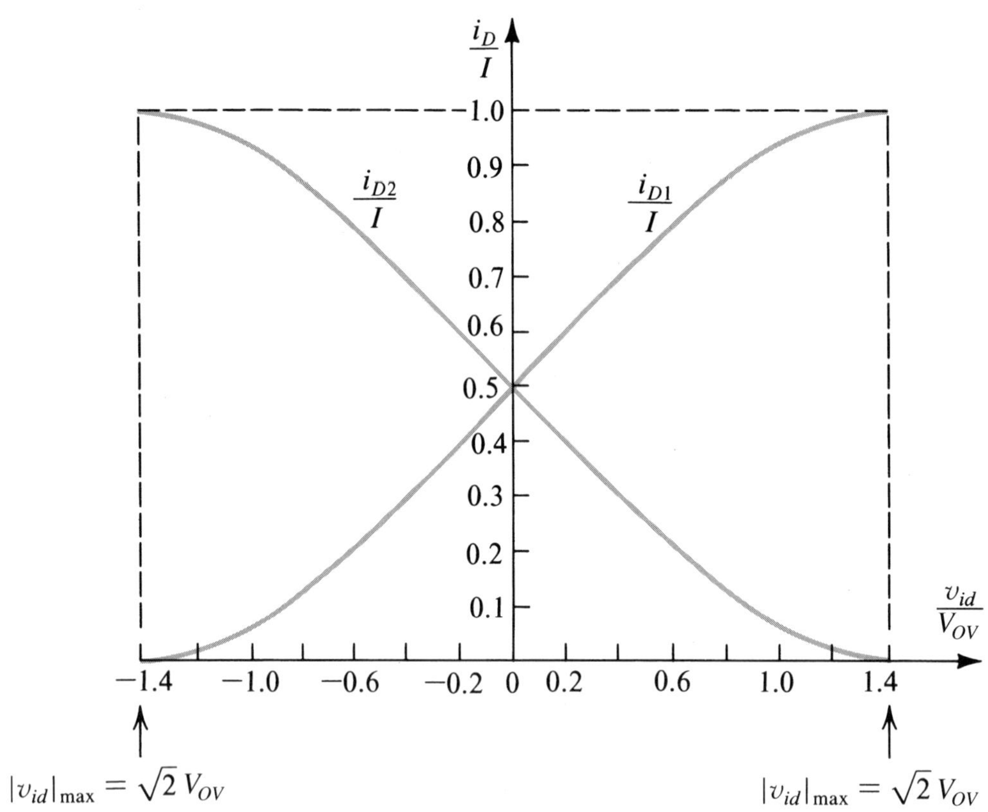

FIGURE 7.6 Normalized plots of the currents in a MOSFET differential pair. Note that V_{OV} is the overdrive voltage at which Q_1 and Q_2 operate when conducting drain currents equal to $I/2$.

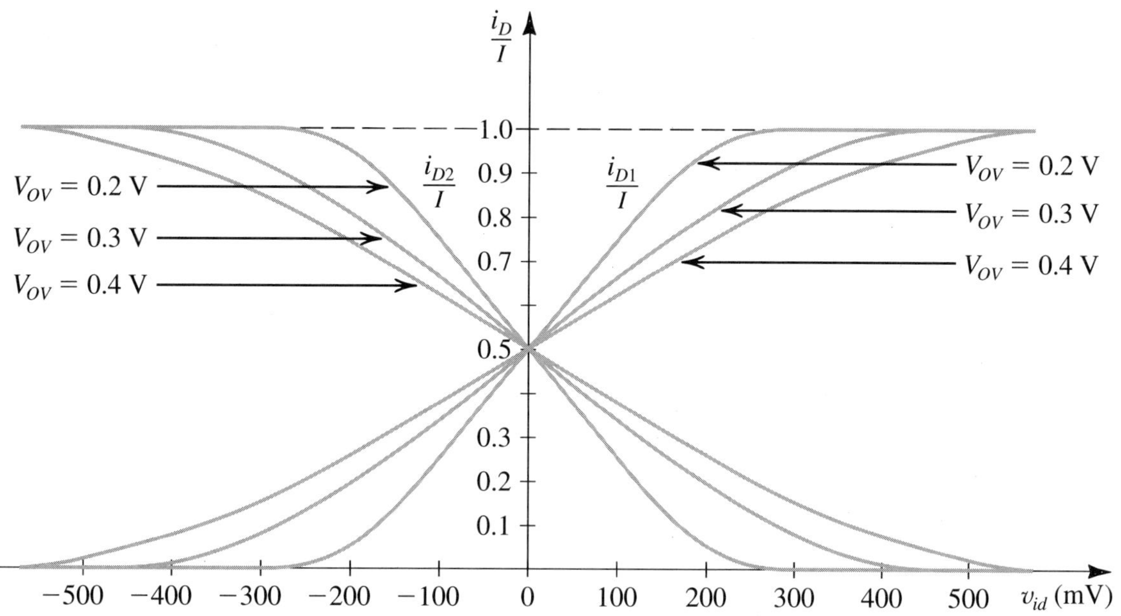

FIGURE 7.7 The linear range of operation of the MOS differential pair can be extended by operating the transistor at a higher value of V_{OV}.

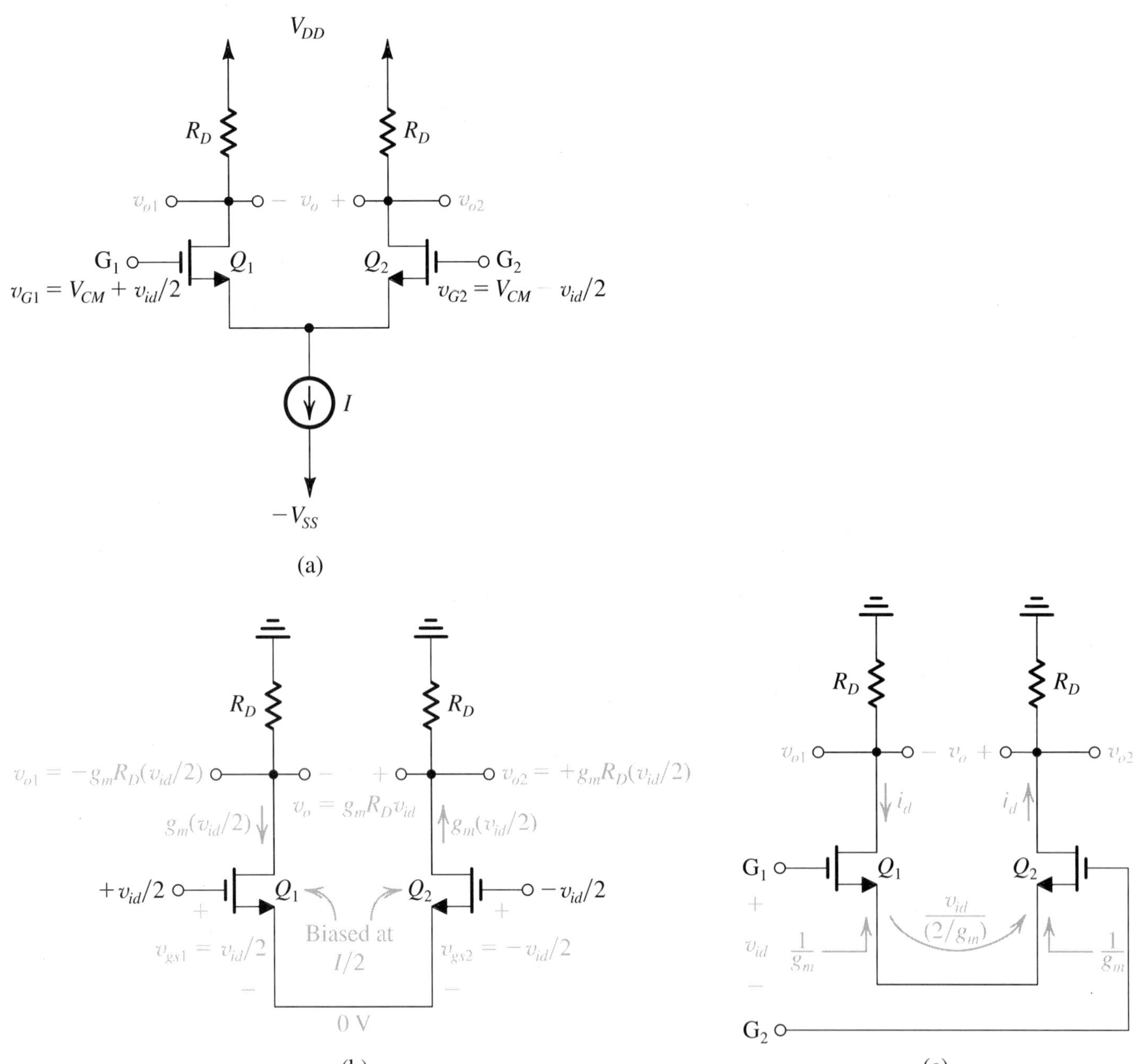

FIGURE 7.8 Small-signal analysis of the MOS differential amplifier: **(a)** The circuit with a common-mode voltage applied to set the dc bias voltage at the gates and with v_{id} applied in a complementary (or balanced) manner. **(b)** The circuit prepared for small-signal analysis. **(c)** An alternative way of looking at the small-signal operation of the circuit.

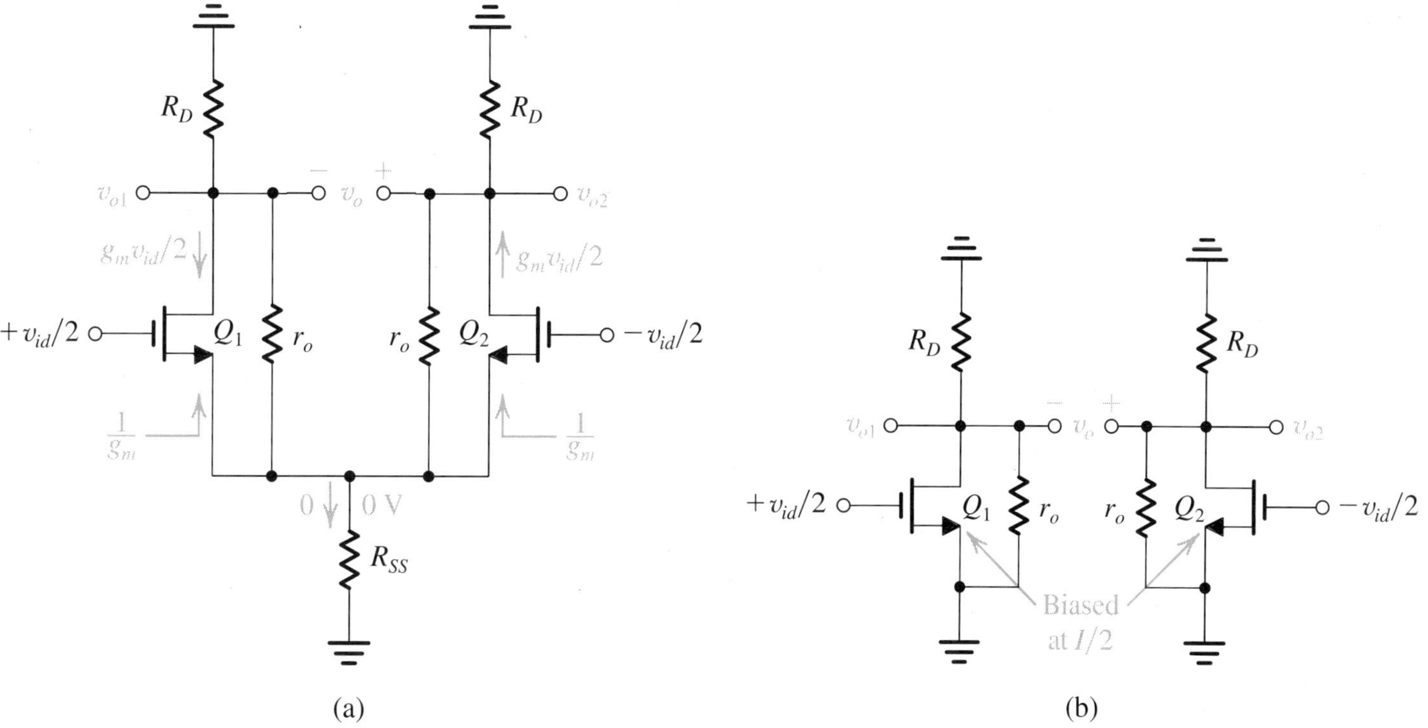

FIGURE 7.9 **(a)** MOS differential amplifier with r_o and R_{SS} taken into account. **(b)** Equivalent circuit for determining the differential gain. Each of the two halves of the differential amplifier circuit is a common-source amplifier, known as its differential "half-circuit."

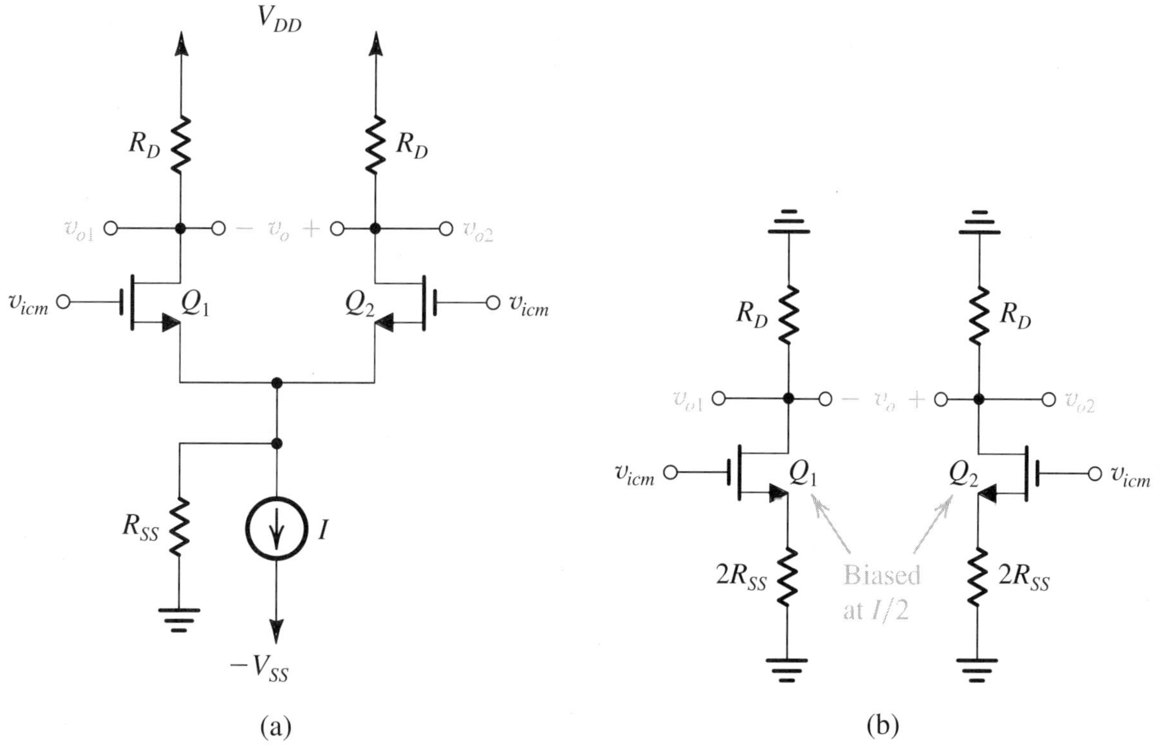

FIGURE 7.10 **(a)** The MOS differential amplifier with a common-mode input signal v_{icm}. **(b)** Equivalent circuit for determining the common-mode gain (with r_o ignored). Each half of the circuit is known as the "common-mode half-circuit."

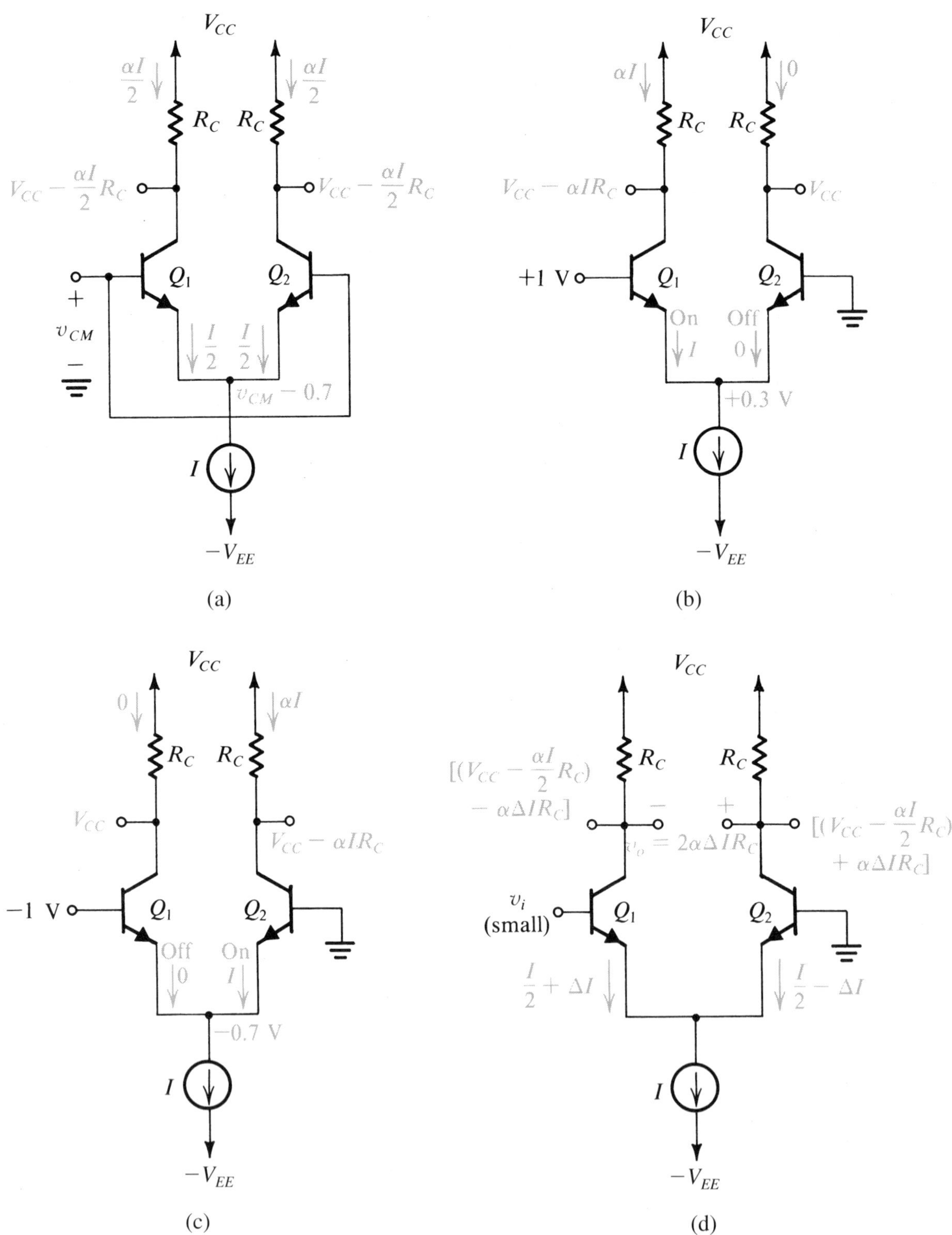

FIGURE 7.13 Different modes of operation of the BJT differential pair: **(a)** The differential pair with a common-mode input signal v_{CM}. **(b)** The differential pair with a "large" differential input signal. **(c)** The differential pair with a large differential input signal of polarity opposite to that in (b). **(d)** The differential pair with a small differential input signal v_i. Note that we have assumed the bias current source I to be ideal (i.e., it has an infinite output resistance) and thus I remains constant with the change in v_{CM}.

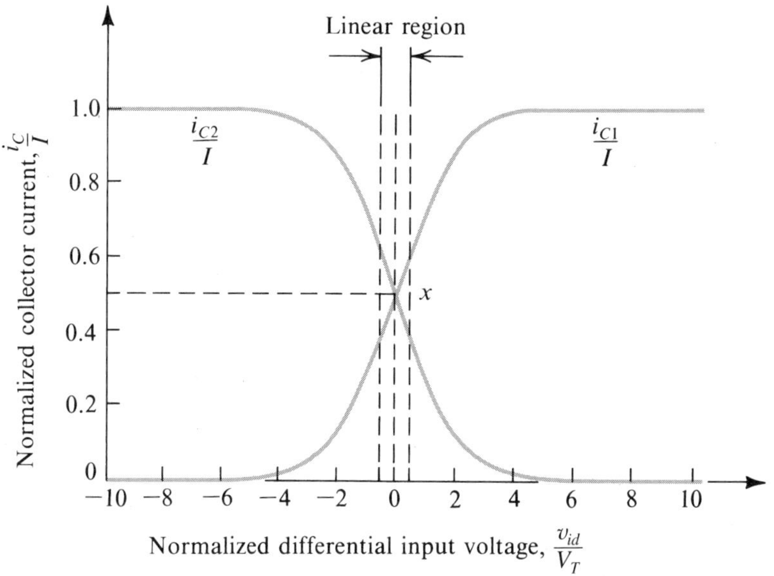

FIGURE 7.14 Transfer characteristics of the BJT differential pair of Fig. 7.12 assuming $\alpha \simeq 1$.

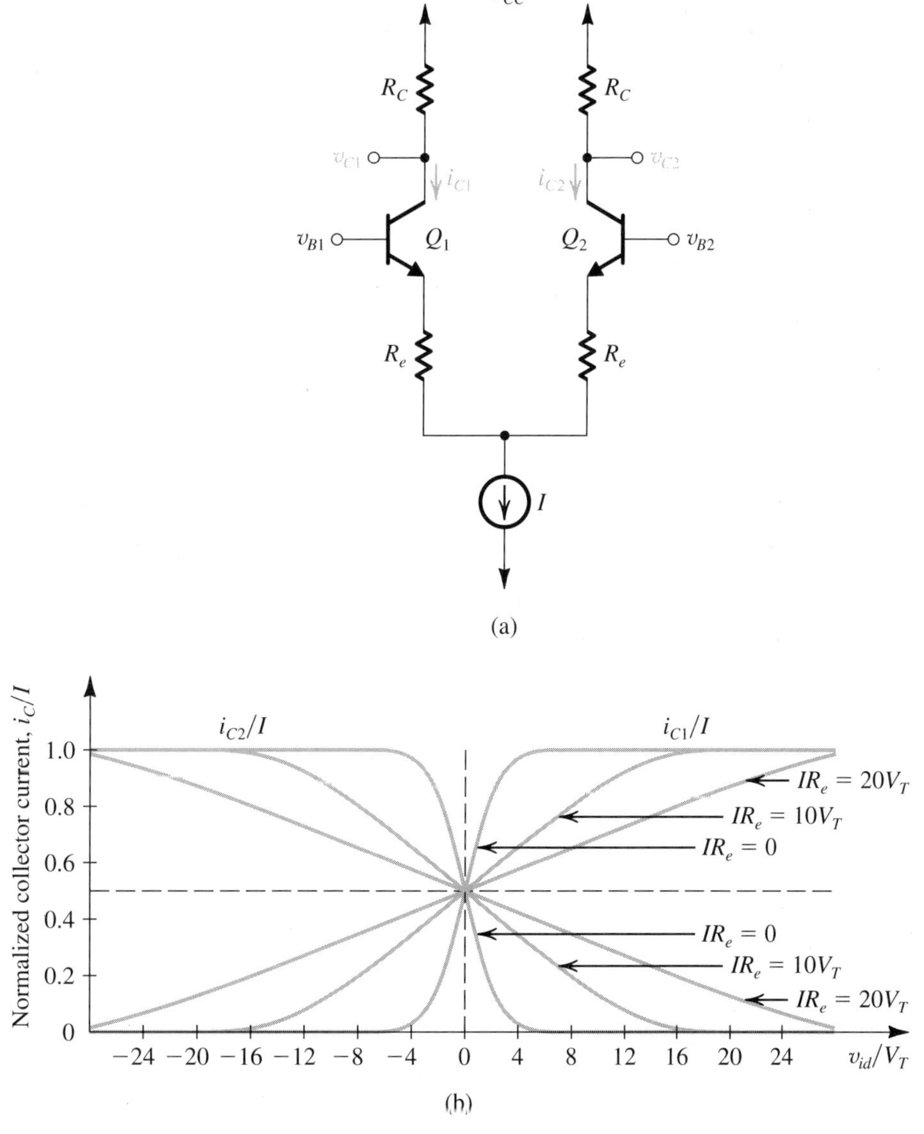

FIGURE 7.15 The transfer characteristics of the BJT differential pair **(a)** can be linearized **(b)** (i.e., the linear range of operation can be extended) by including resistances in the emitters.

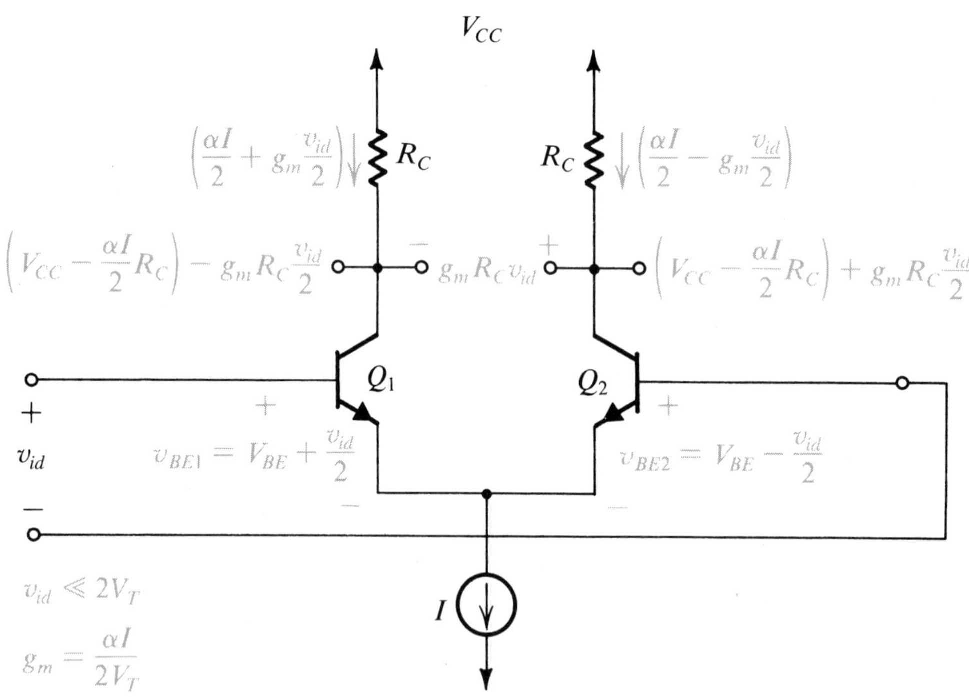

FIGURE 7.16 The currents and voltages in the differential amplifier when a small differential input signal v_{id} is applied.

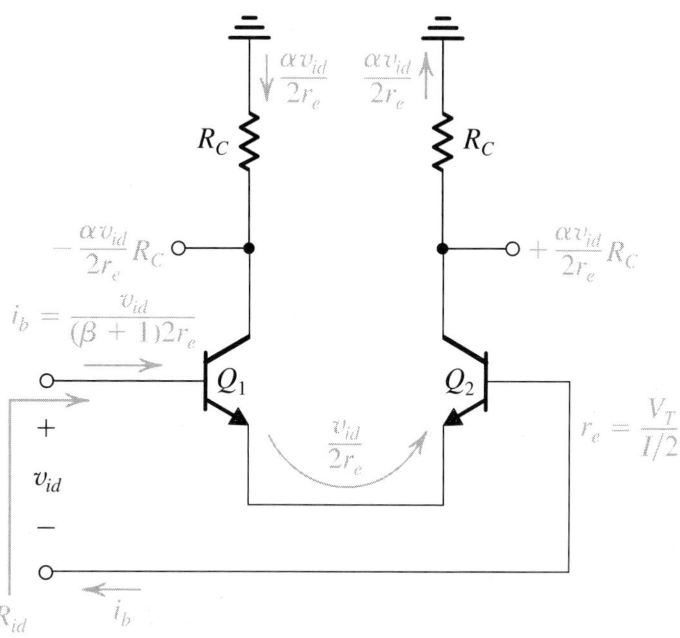

FIGURE 7.17 A simple technique for determining the signal currents in a differential amplifier excited by a differential voltage signal v_{id}; dc quantities are not shown.

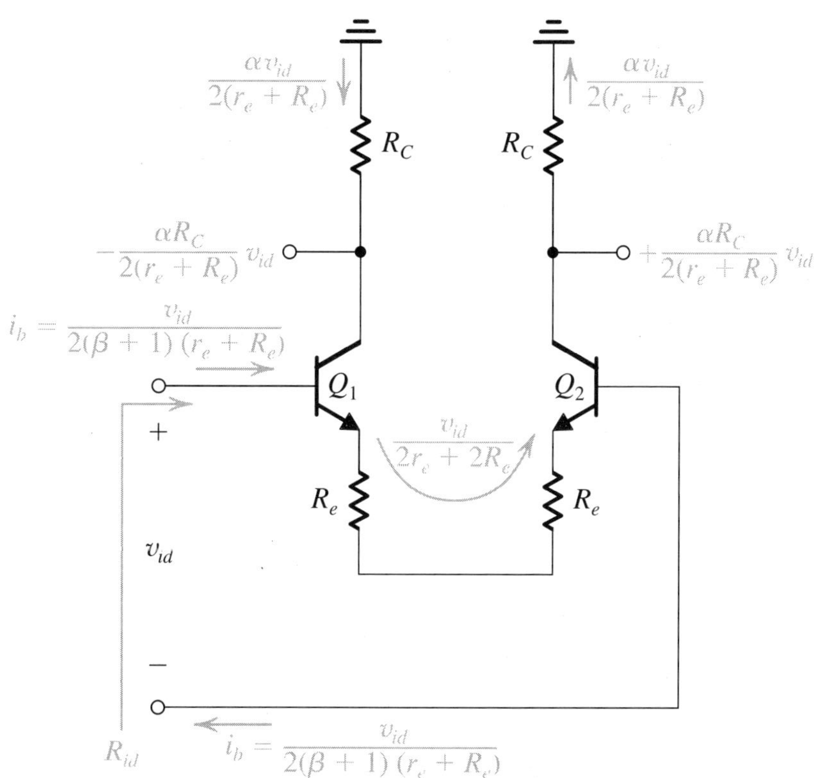

FIGURE 7.18 A differential amplifier with emitter resistances. Only signal quantities are shown (in color).

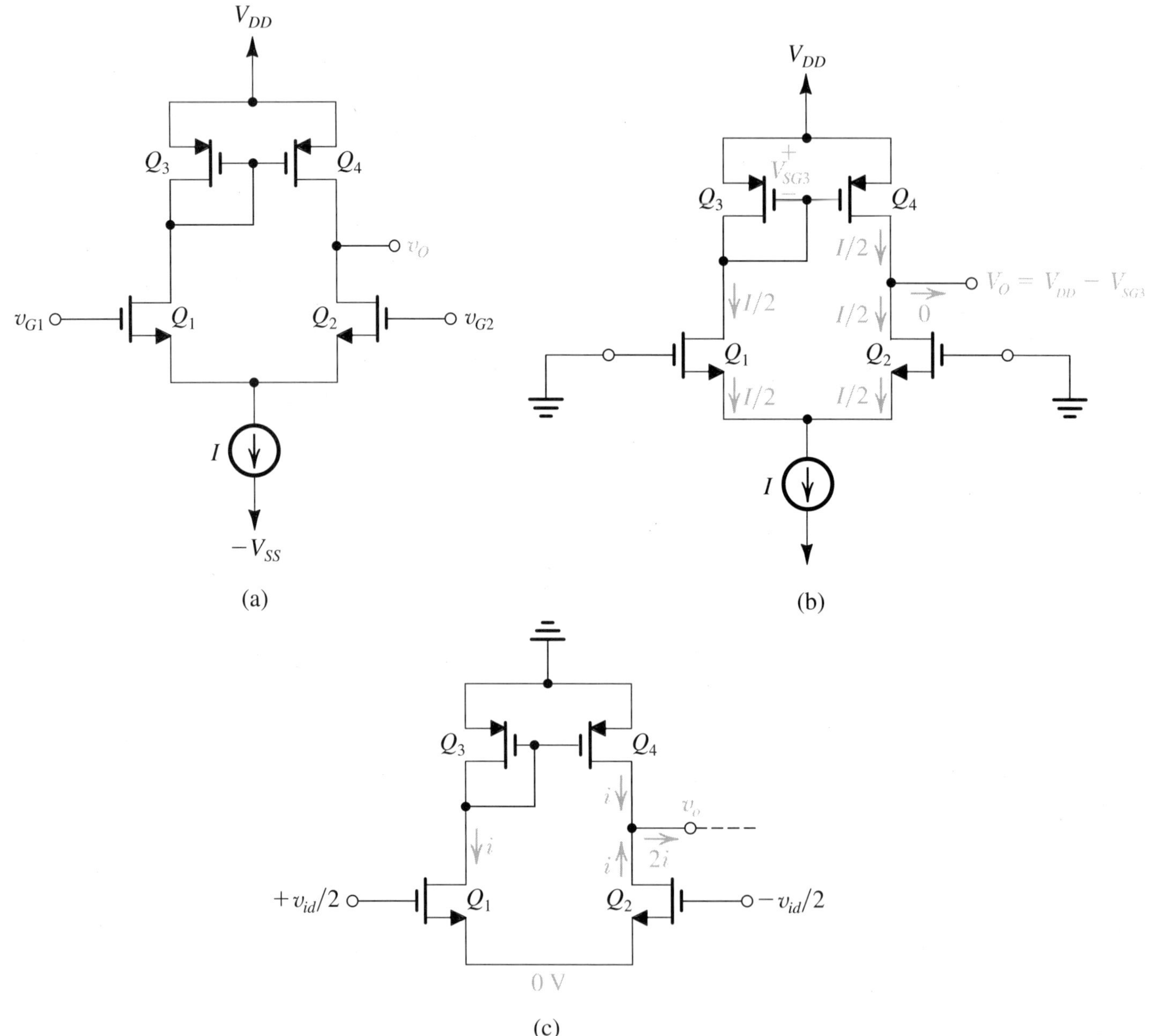

FIGURE 7.28 **(a)** The active-loaded MOS differential pair. **(b)** The circuit at equilibrium assuming perfect matching. **(c)** The circuit with a differential input signal applied, neglecting the r_o of all transistors.

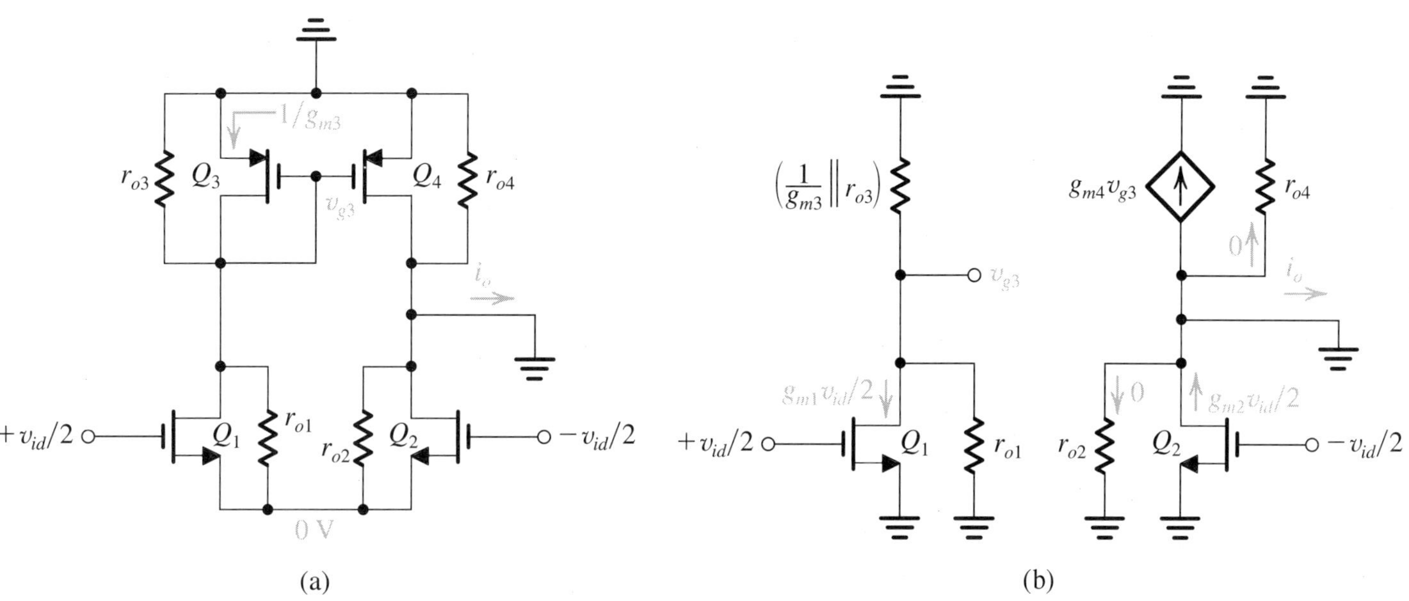

FIGURE 7.29 Determining the short-circuit transconductance $G_m \equiv i_o/v_{id}$ of the active-loaded MOS differential pair.

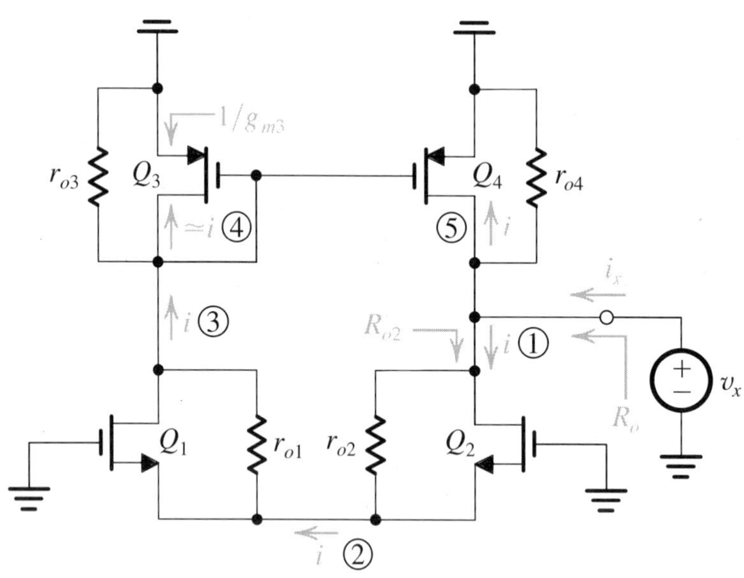

FIGURE 7.30 Circuit for determining R_o. The circled numbers indicate the order of the analysis steps.

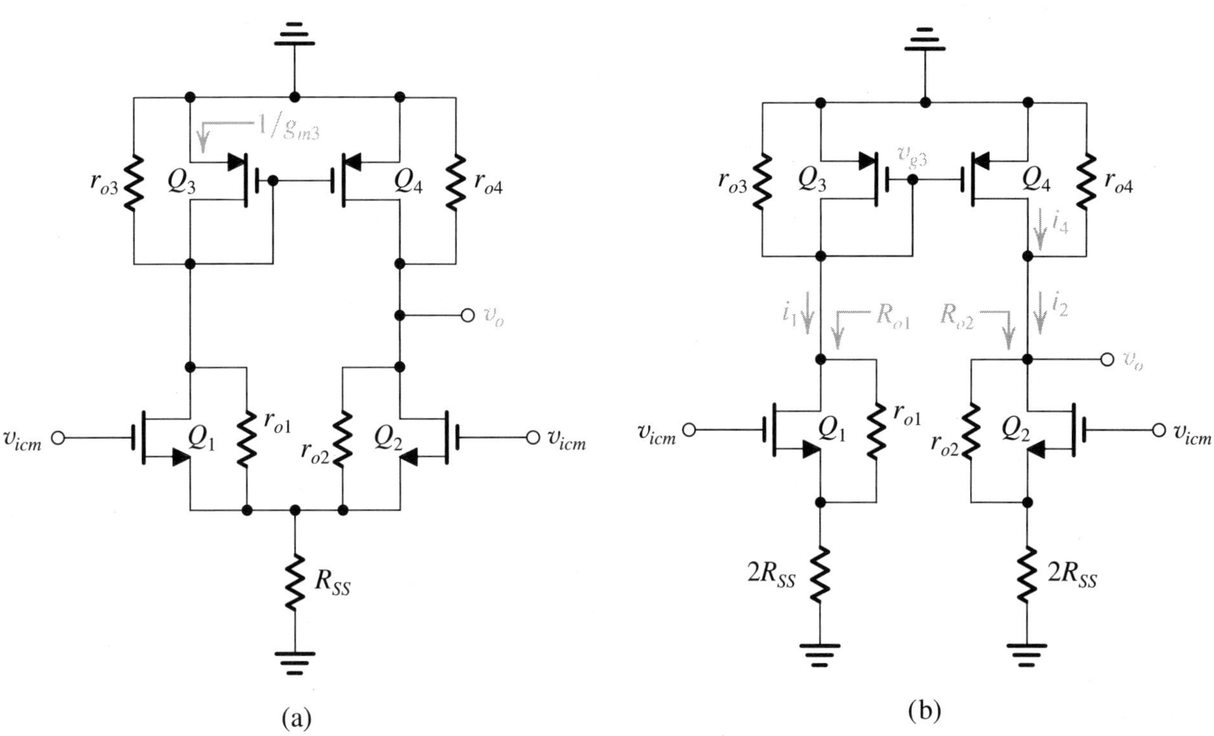

FIGURE 7.31 Analysis of the active-loaded MOS differential amplifier to determine its common-mode gain.

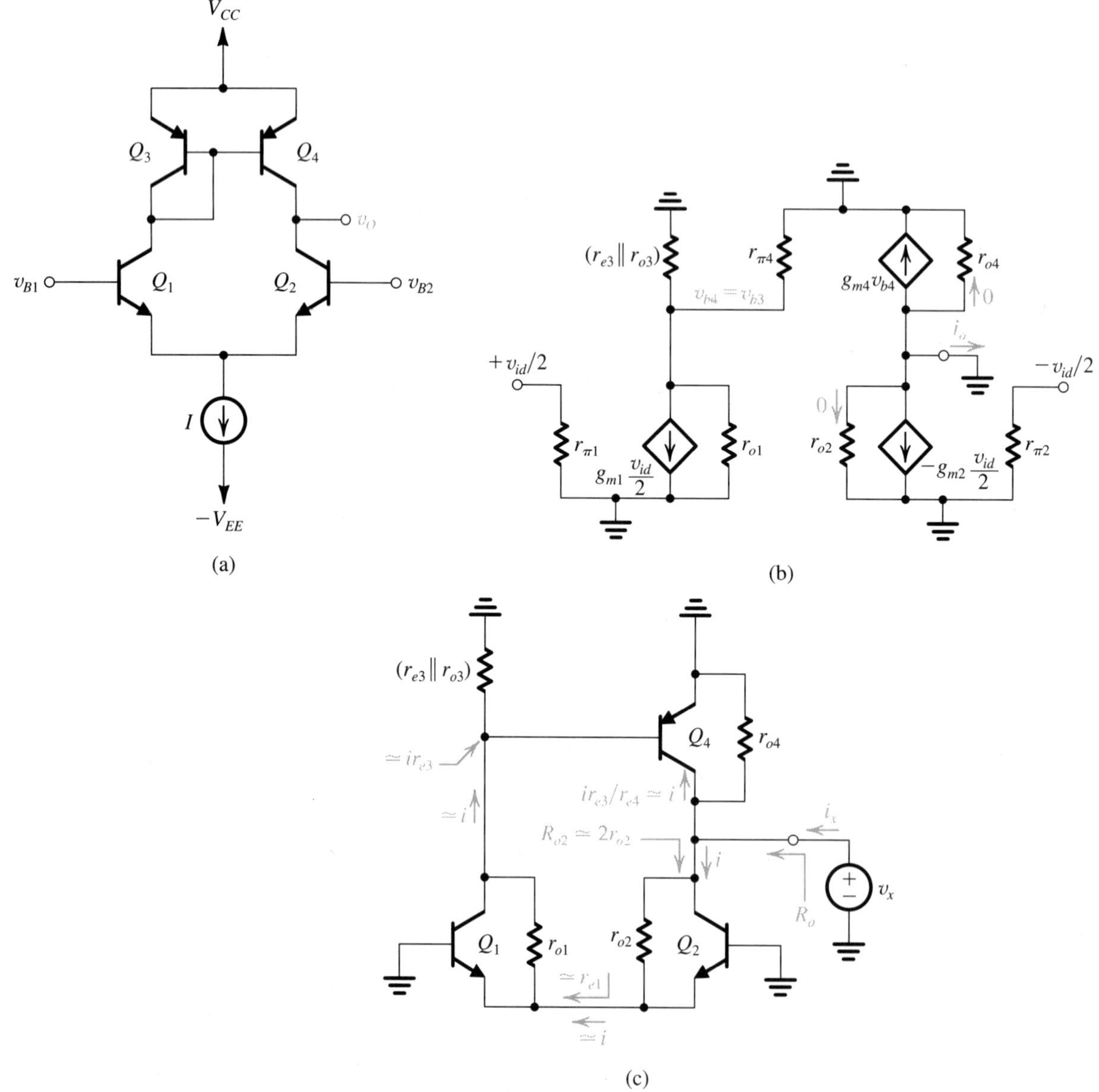

FIGURE 7.32 **(a)** Active-loaded bipolar differential pair. **(b)** Small-signal equivalent circuit for determining the transconductance $G_m \equiv i_o/v_{id}$. **(c)** Equivalent circuit for determining the output resistance $R_o \equiv v_x/i_x$.

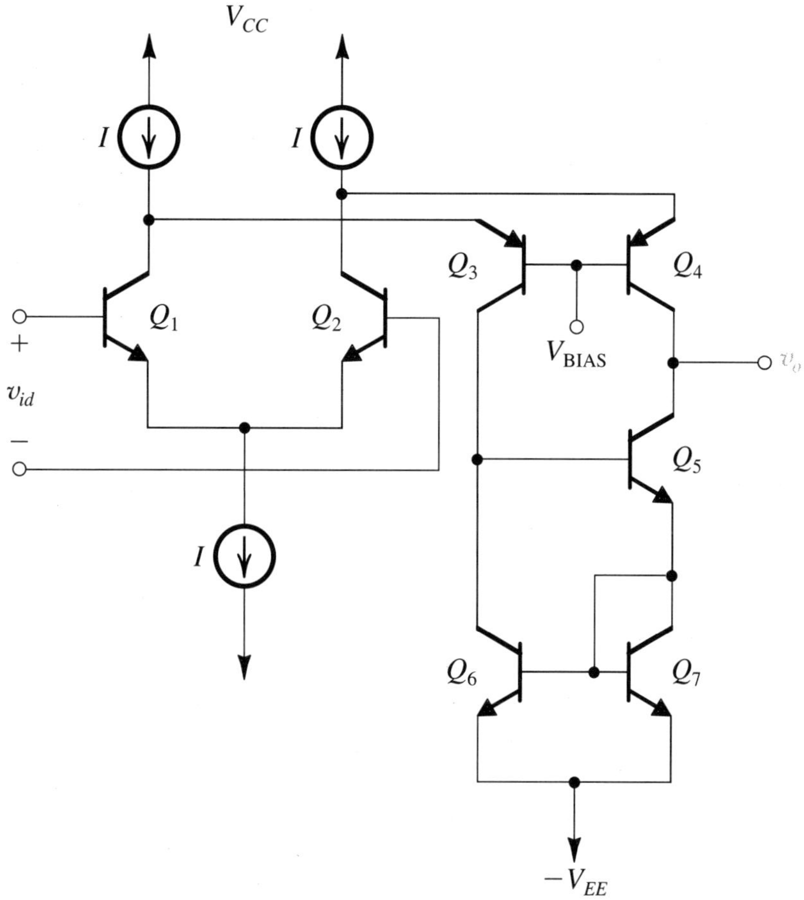

FIGURE 7.35 An active-loaded bipolar differential amplifier employing a folded cascode stage (Q_3 and Q_4) and a Wilson current mirror load (Q_5, Q_6, and Q_7).

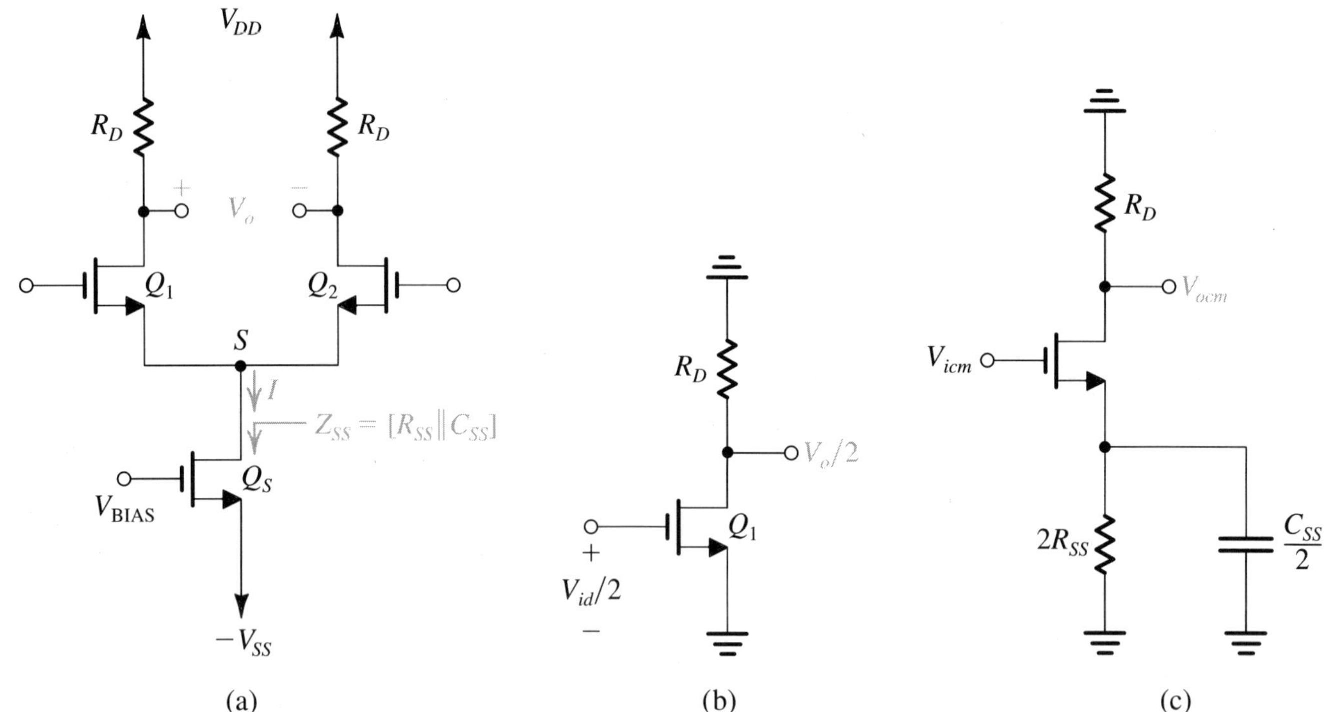

FIGURE 7.36 **(a)** A resistively loaded MOS differential pair with the transistor supplying the bias current explicitly shown. It is assumed that the total impedance between node S and ground, Z_{SS}, consists of a resistance R_{SS} in parallel with a capacitance C_{SS}. **(b)** Differential half-circuit. **(c)** Common-mode half-circuit.

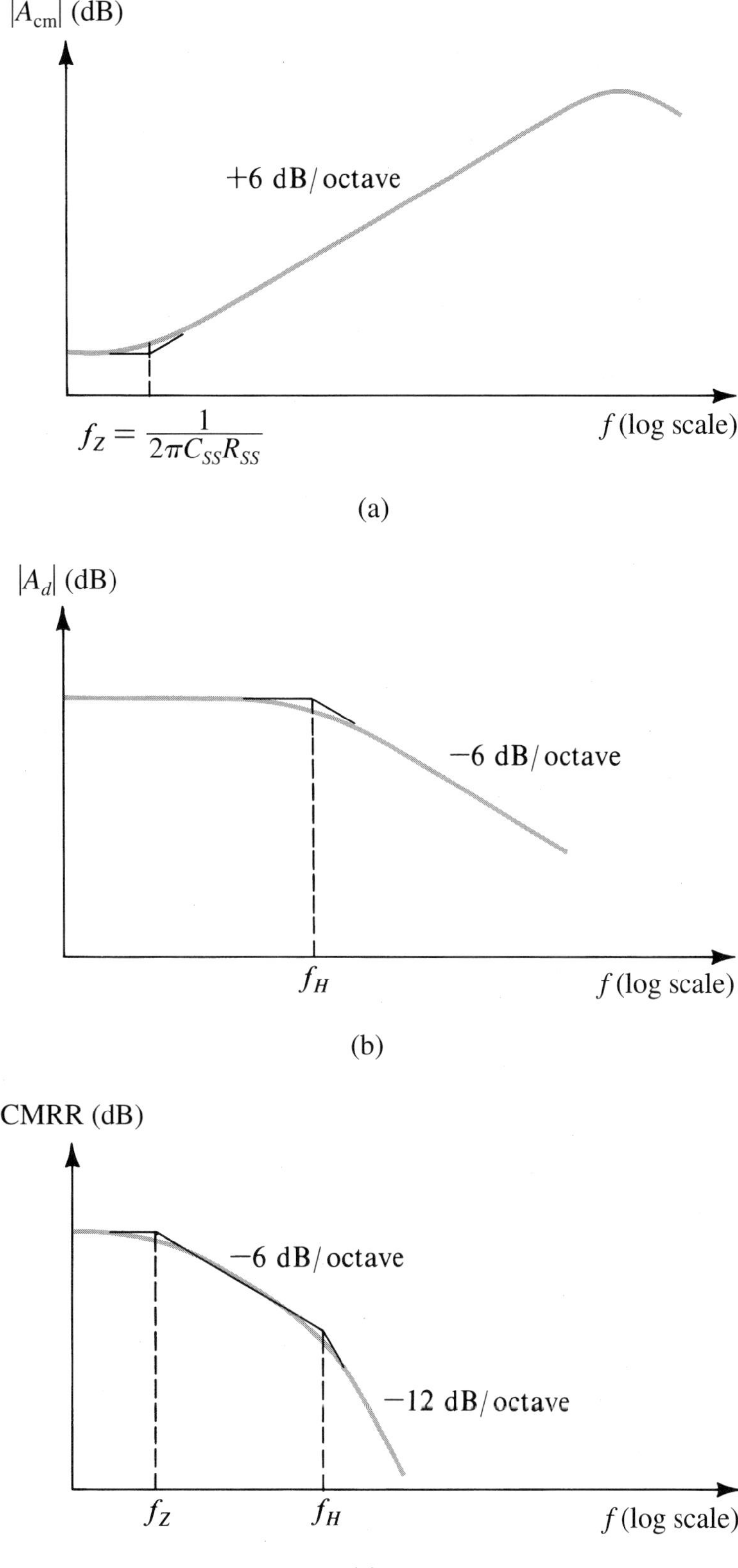

FIGURE 7.37 Variation of **(a)** common-mode gain, **(b)** differential gain, and **(c)** common-mode rejection ratio with frequency.

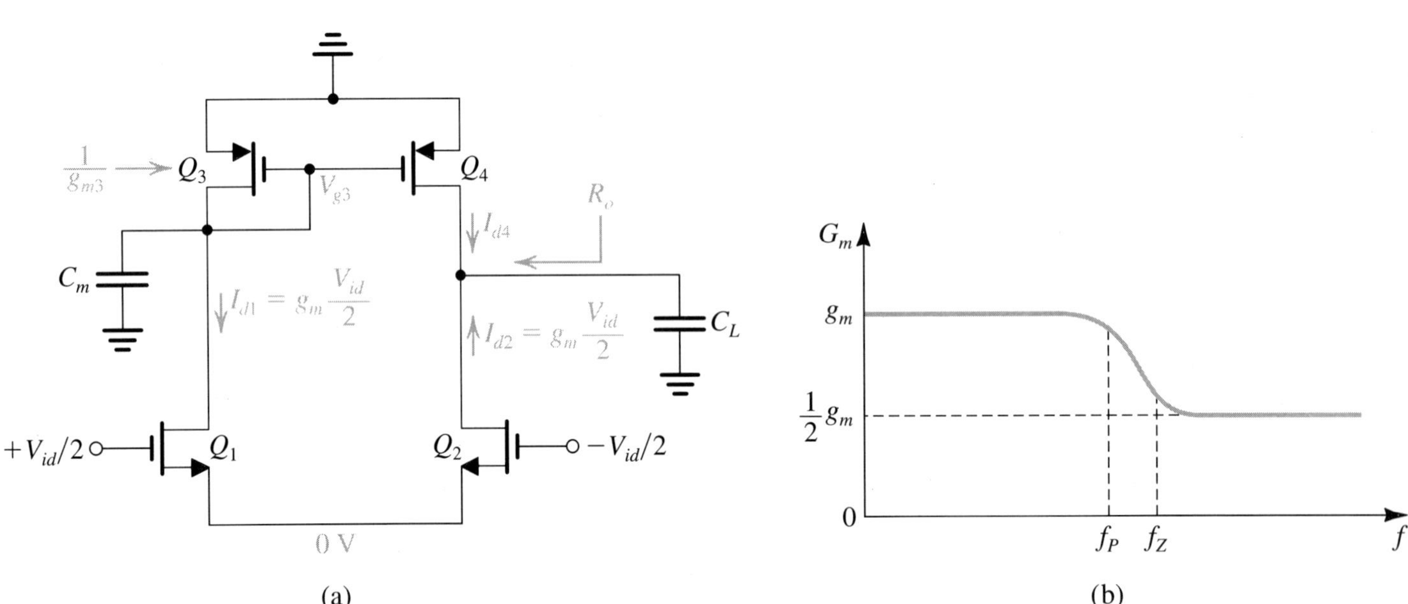

FIGURE 7.39 **(a)** Frequency–response analysis of the active-loaded MOS differential amplifier. **(b)** The overall transconductance G_m as a function of frequency.

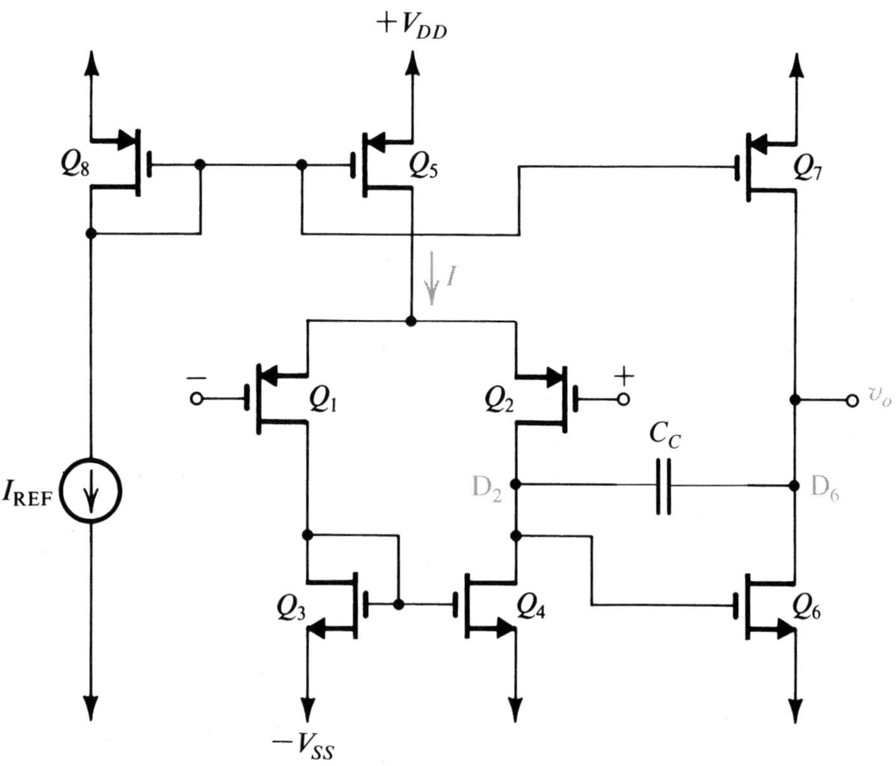

FIGURE 7.40 Two-stage CMOS op-amp configuration.

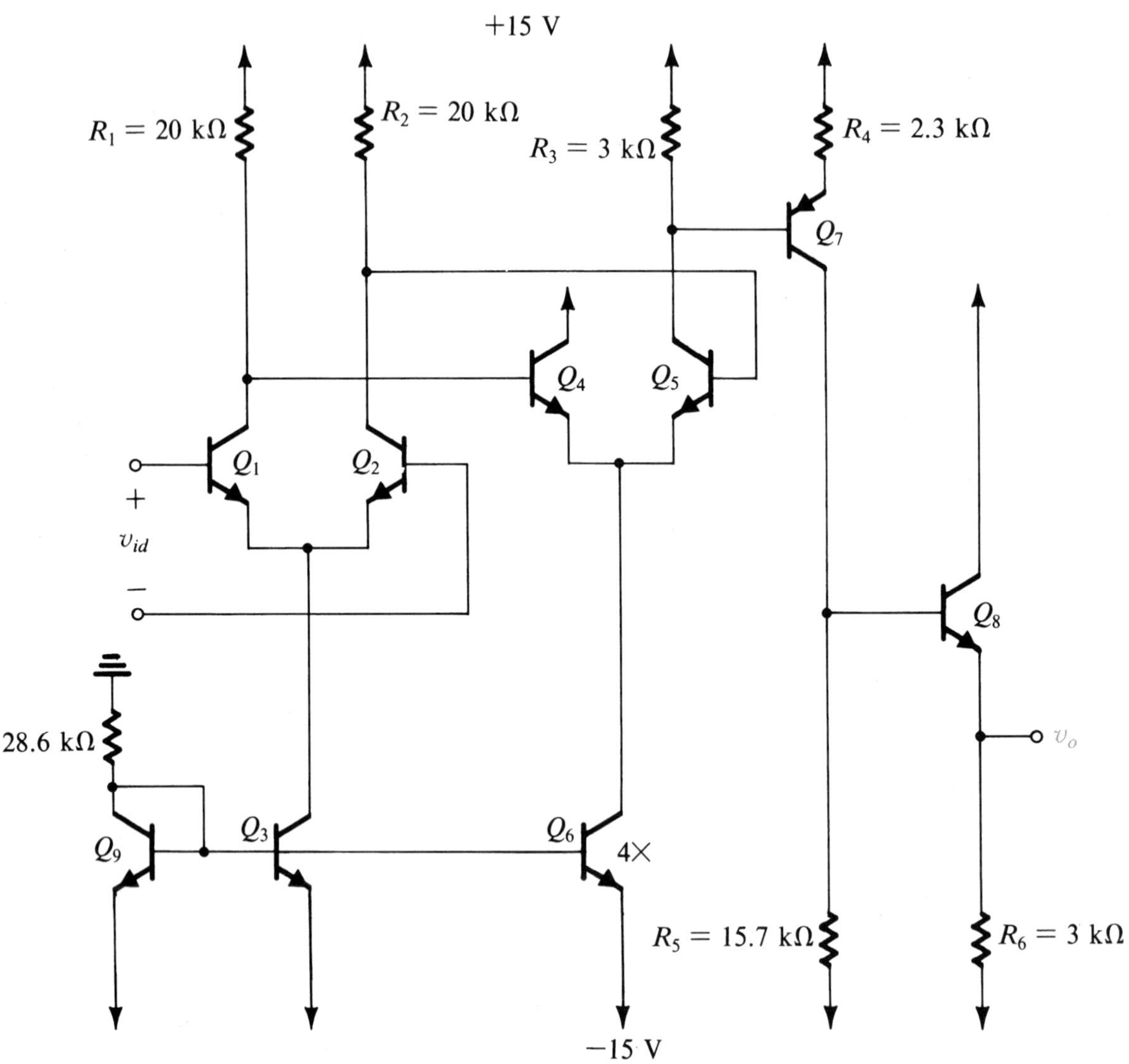

FIGURE 7.43 A four-stage bipolar op amp.

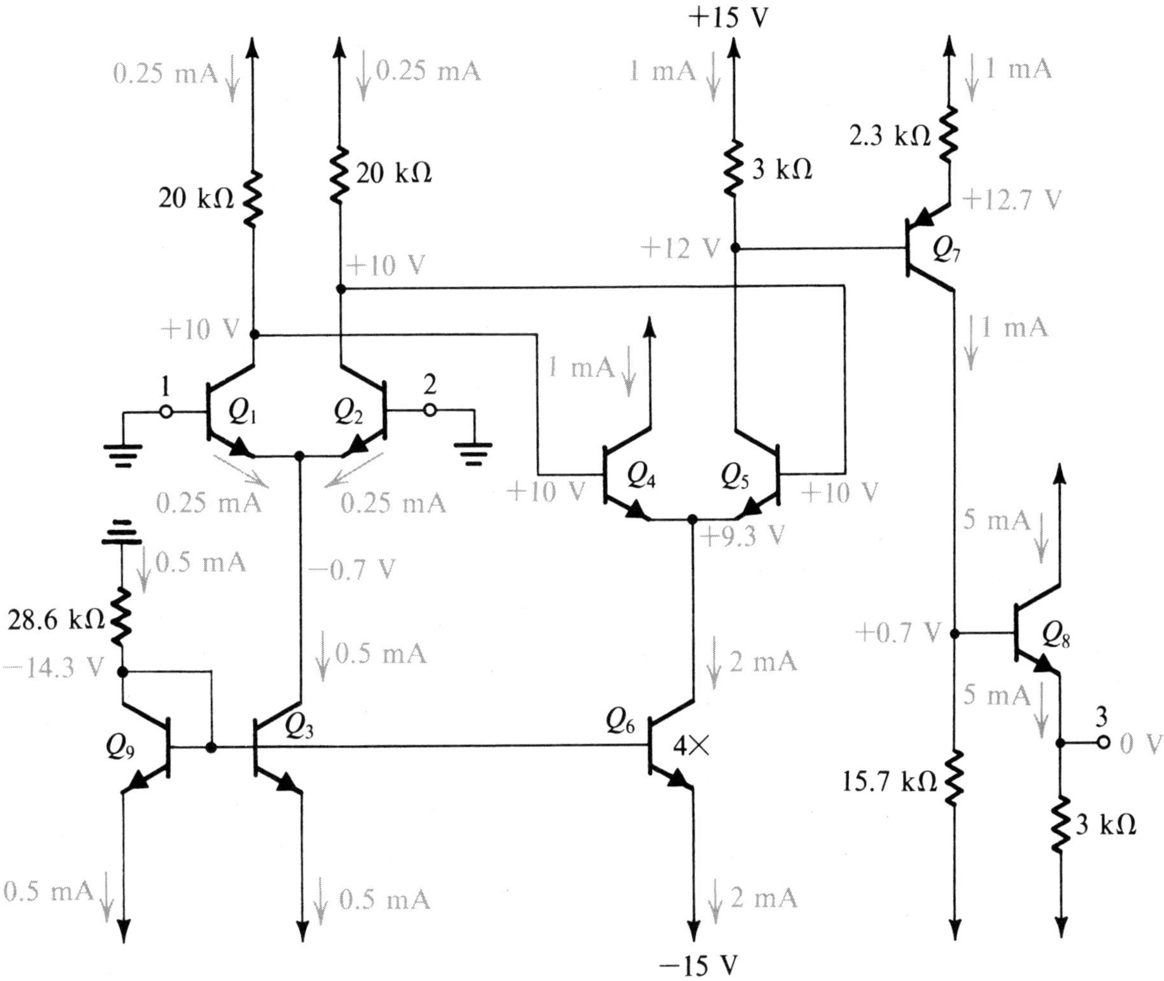

FIGURE 7.44 Circuit for Example 7.4.

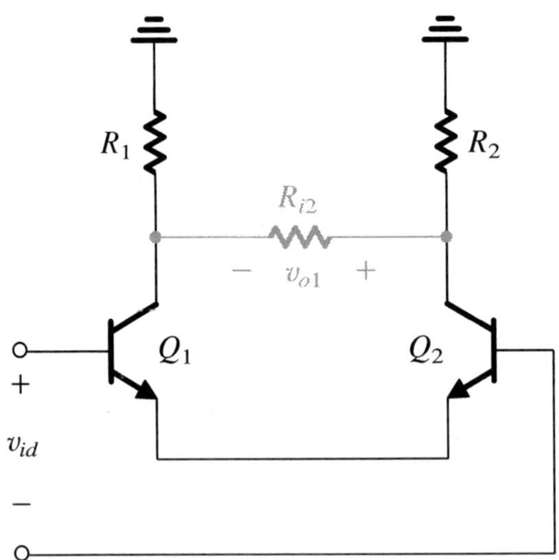

FIGURE 7.45 Equivalent circuit for calculating the gain of the input stage of the amplifier in Fig. 7.43.

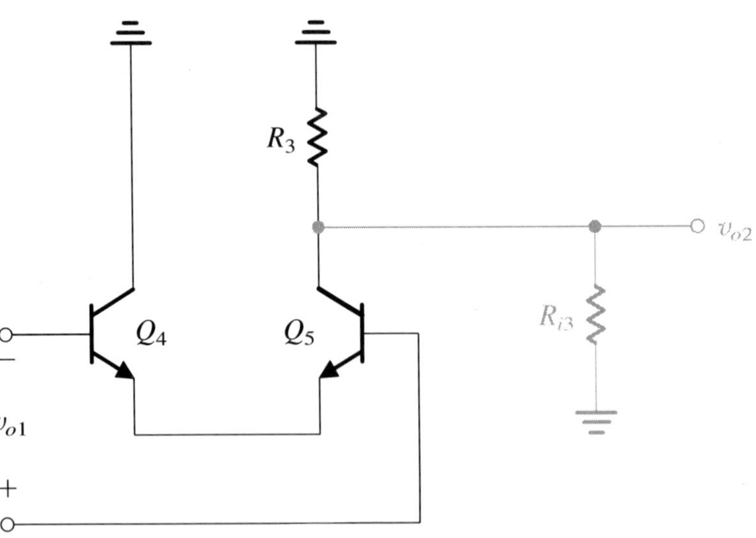

FIGURE 7.46 Equivalent circuit for calculating the gain of the second stage of the amplifier in Fig. 7.43.

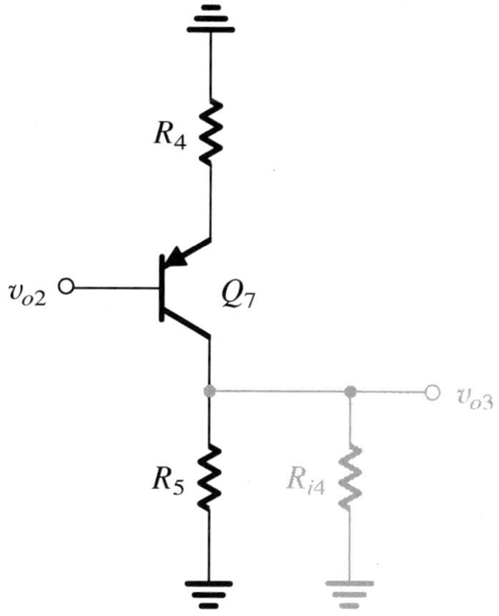

FIGURE 7.47 Equivalent circuit for evaluating the gain of the third stage in the amplifier circuit of Fig. 7.43.

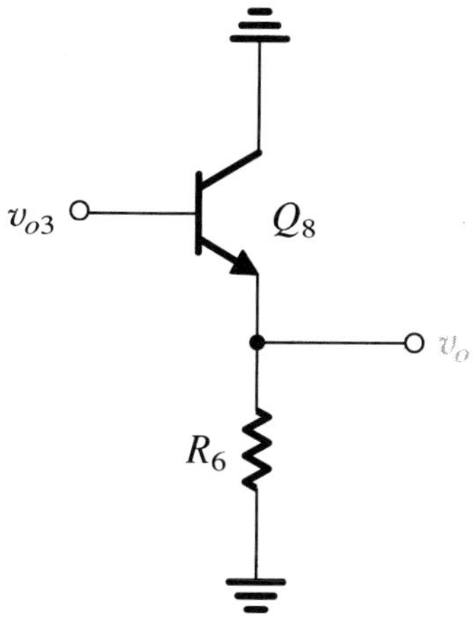

FIGURE 7.48 Equivalent circuit of the output stage of the amplifie circuit of Fig. 7.43.

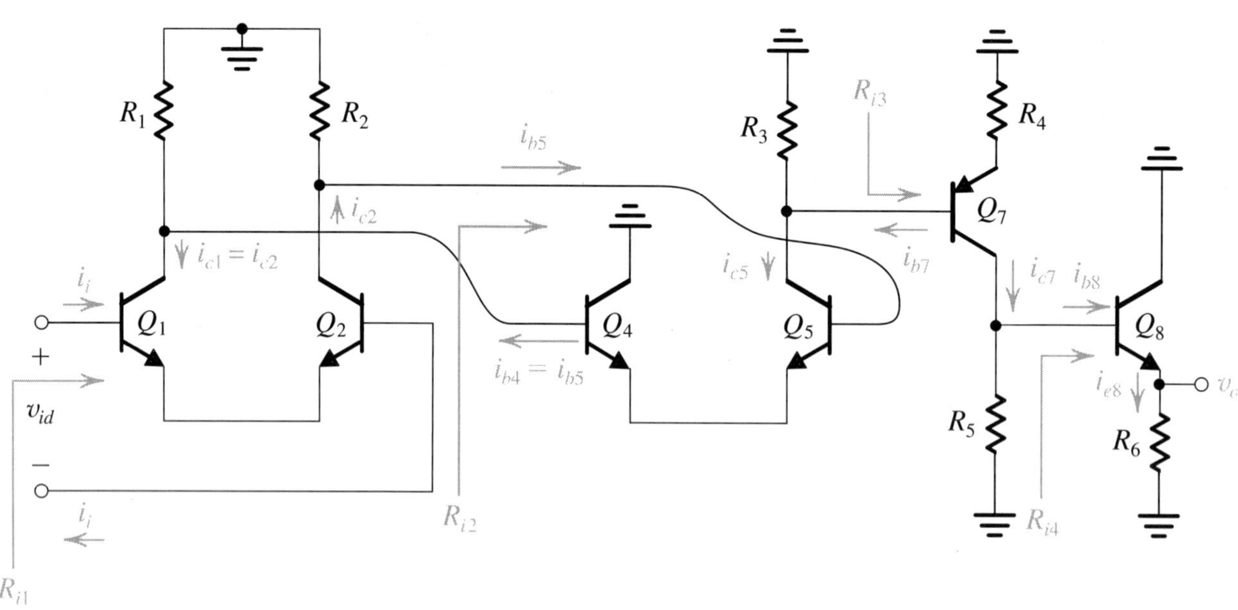

FIGURE 7.49 The circuit of the multistage amplifier of Fig. 7.43 prepared for small-signal analysis. Indicated are the signal currents throughout the amplifier and the input resistances of the four stages.

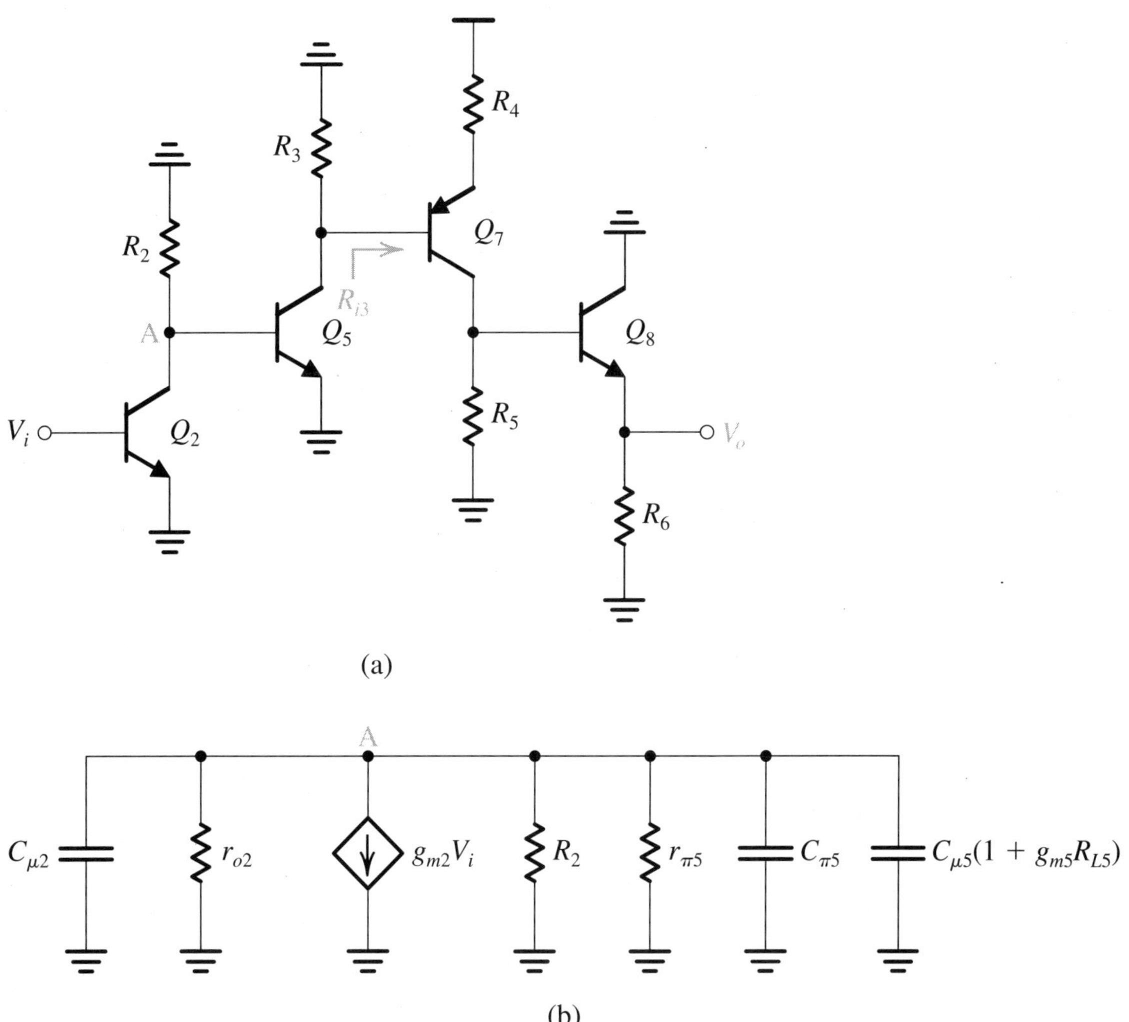

FIGURE 7.50 **(a)** Approximate equivalent circuit for determining the high-frequency response of the op amp of Fig. 7.43. **(b)** Equivalent circuit of the interface between the output of Q_2 and the input of Q_5.

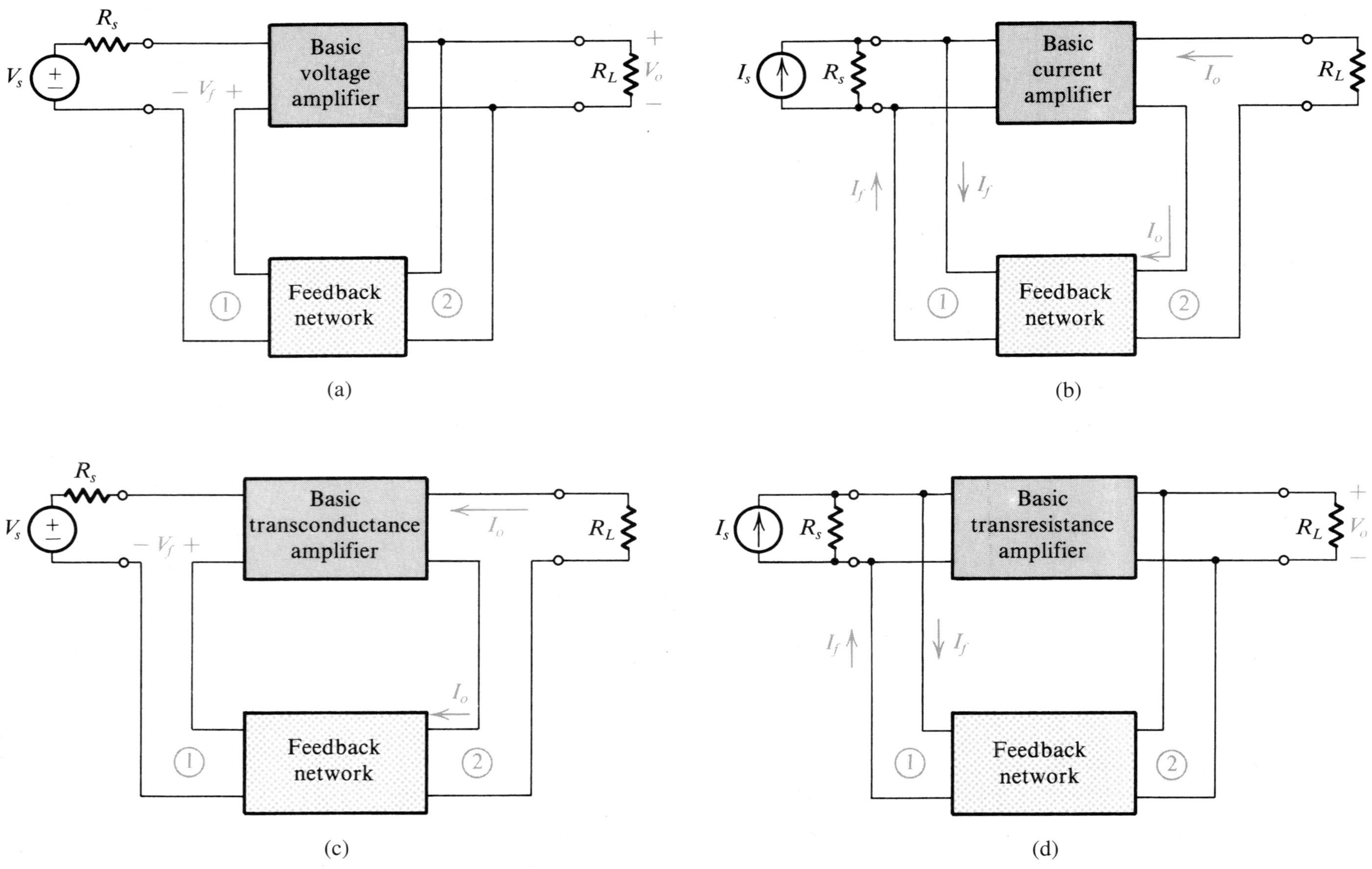

FIGURE 8.4 The four basic feedback topologies: **(a)** voltage-mixing voltage-sampling (series–shunt) topology; **(b)** current-mixing current-sampling (shunt–series) topology; **(c)** voltage-mixing current-sampling (series–series) topology; **(d)** current-mixing voltage-sampling (shunt–shunt) topology.

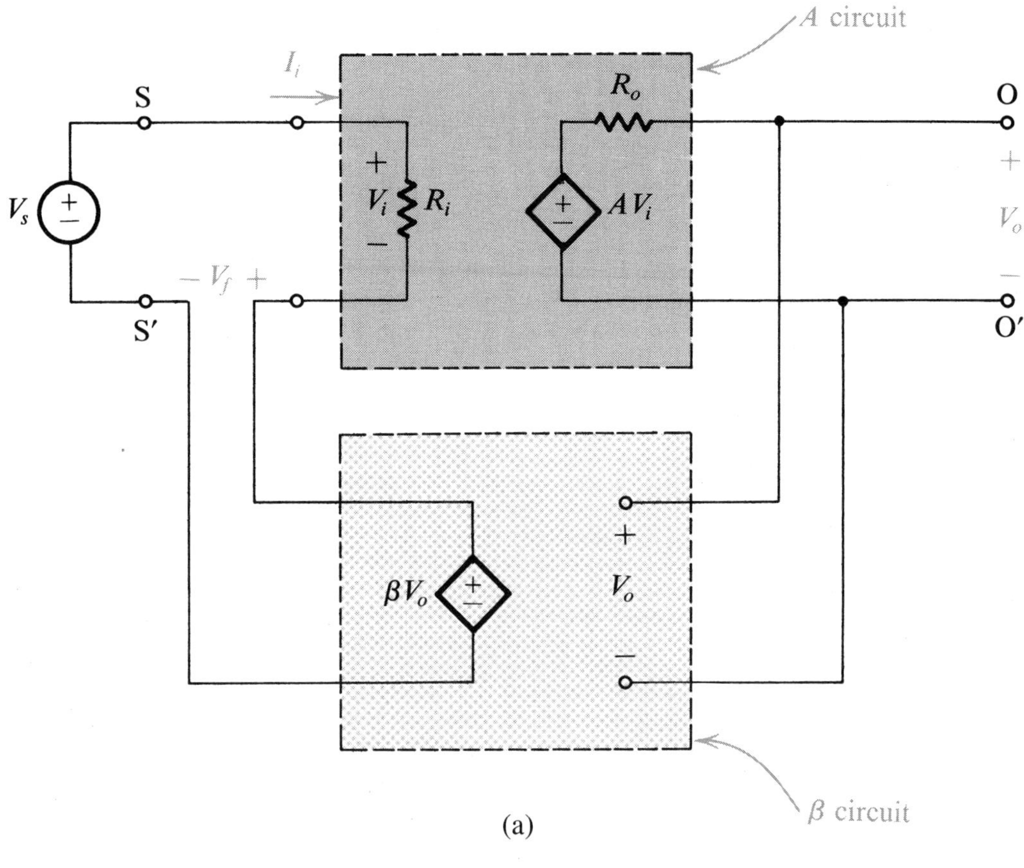

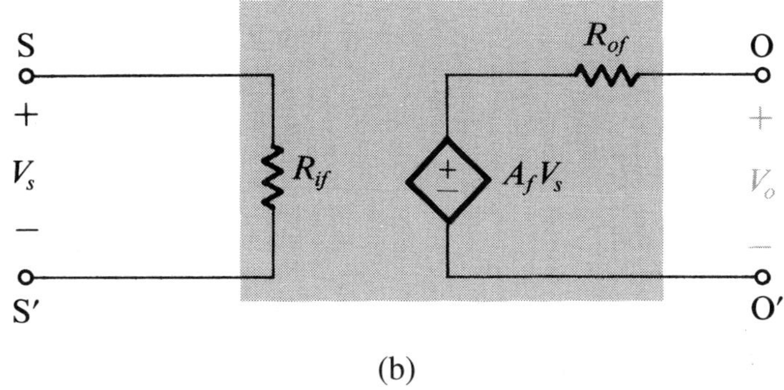

FIGURE 8.8 The series–shunt feedback amplifier: **(a)** ideal structure and **(b)** equivalent circuit.

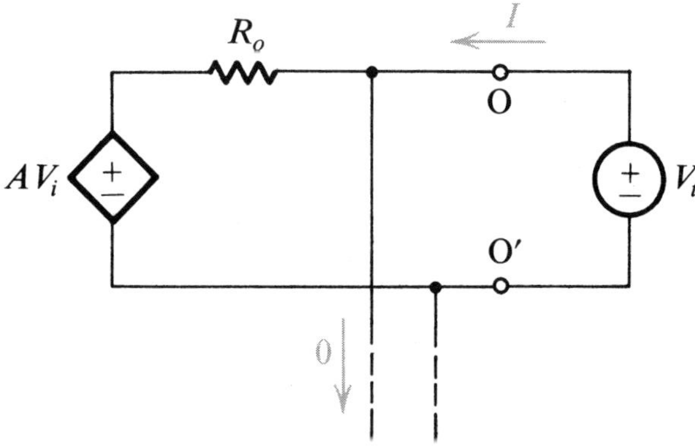

FIGURE 8.9 Measuring the output resistance of the feedback amplifier of Fig. 8.8(a): $R_{of} \equiv V_t/I$.

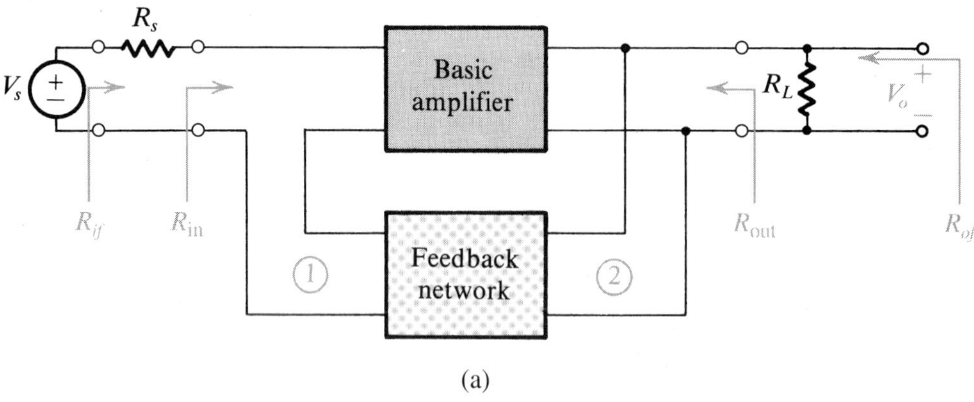

(a)

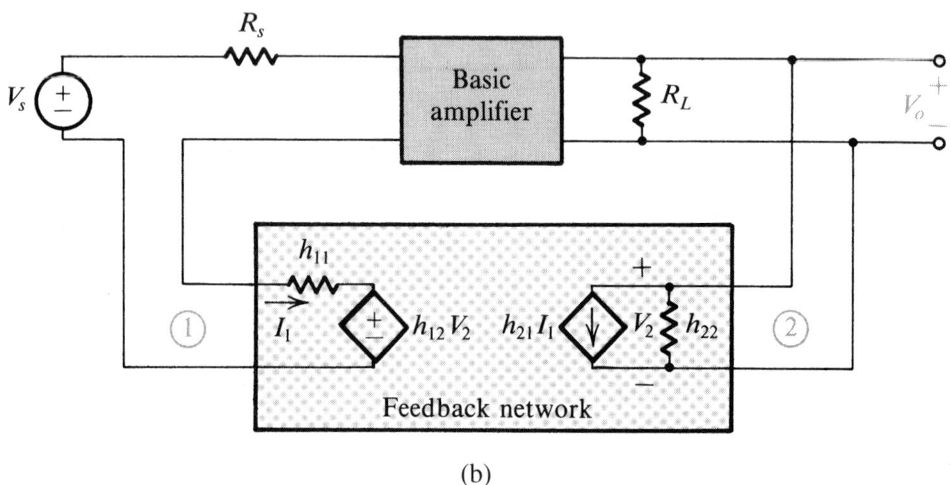

(b)

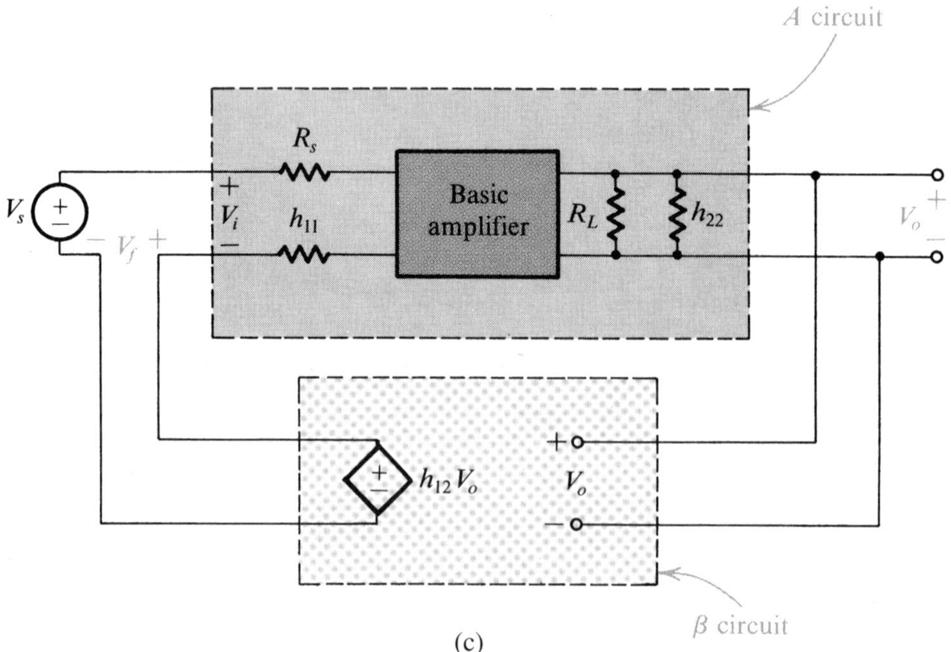

(c)

FIGURE 8.10 Derivation of the A circuit and β circuit for the series–shunt feedback amplifier. **(a)** Block diagram of a practical series–shunt feedback amplifier. **(b)** The circuit in (a) with the feedback network represented by its h parameters. **(c)** The circuit in (b) with h_{21} neglected.

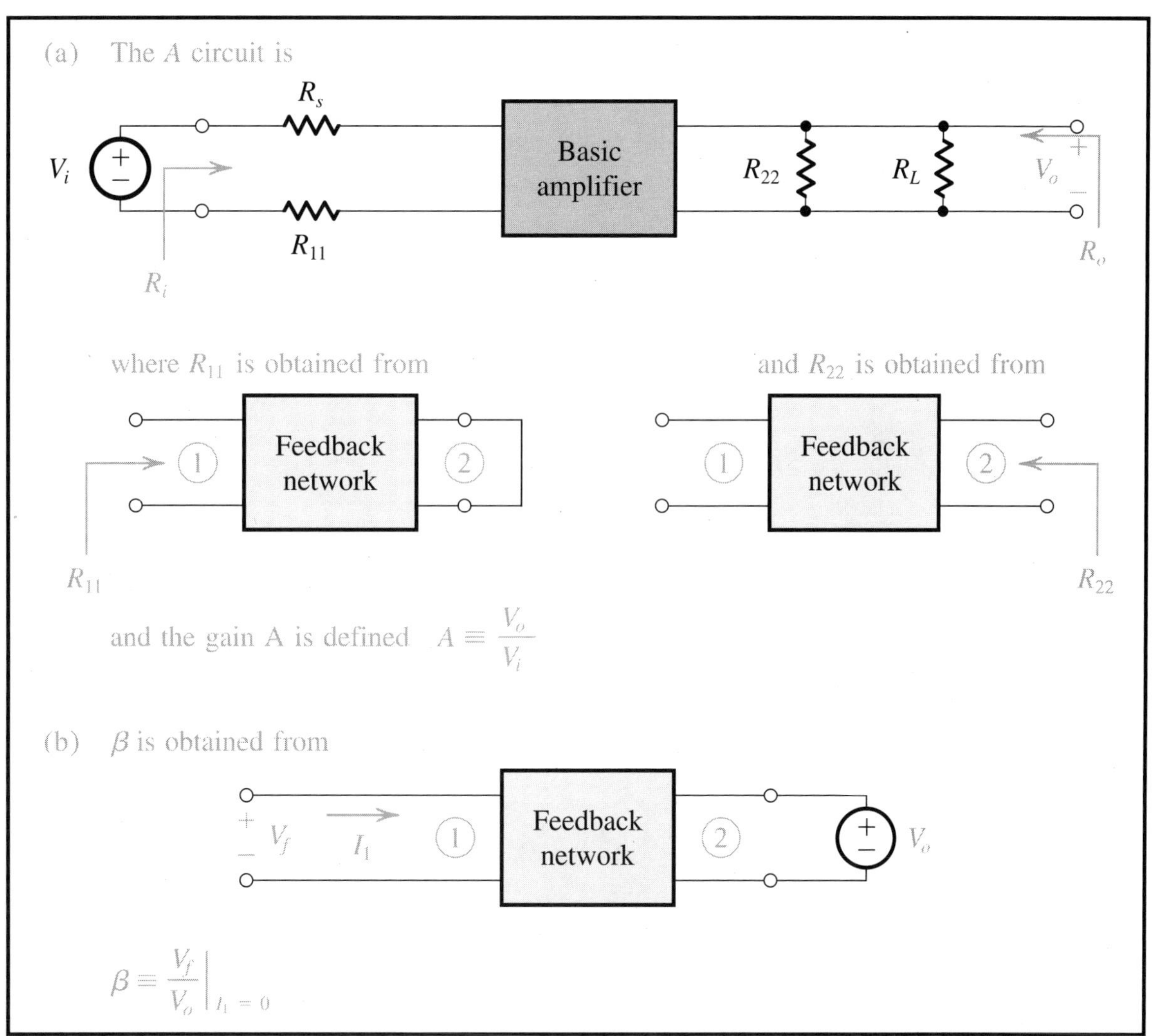

FIGURE 8.11 Summary of the rules for finding the A circuit and β for the voltage-mixing voltage-sampling case of Fig. 8.10(a).

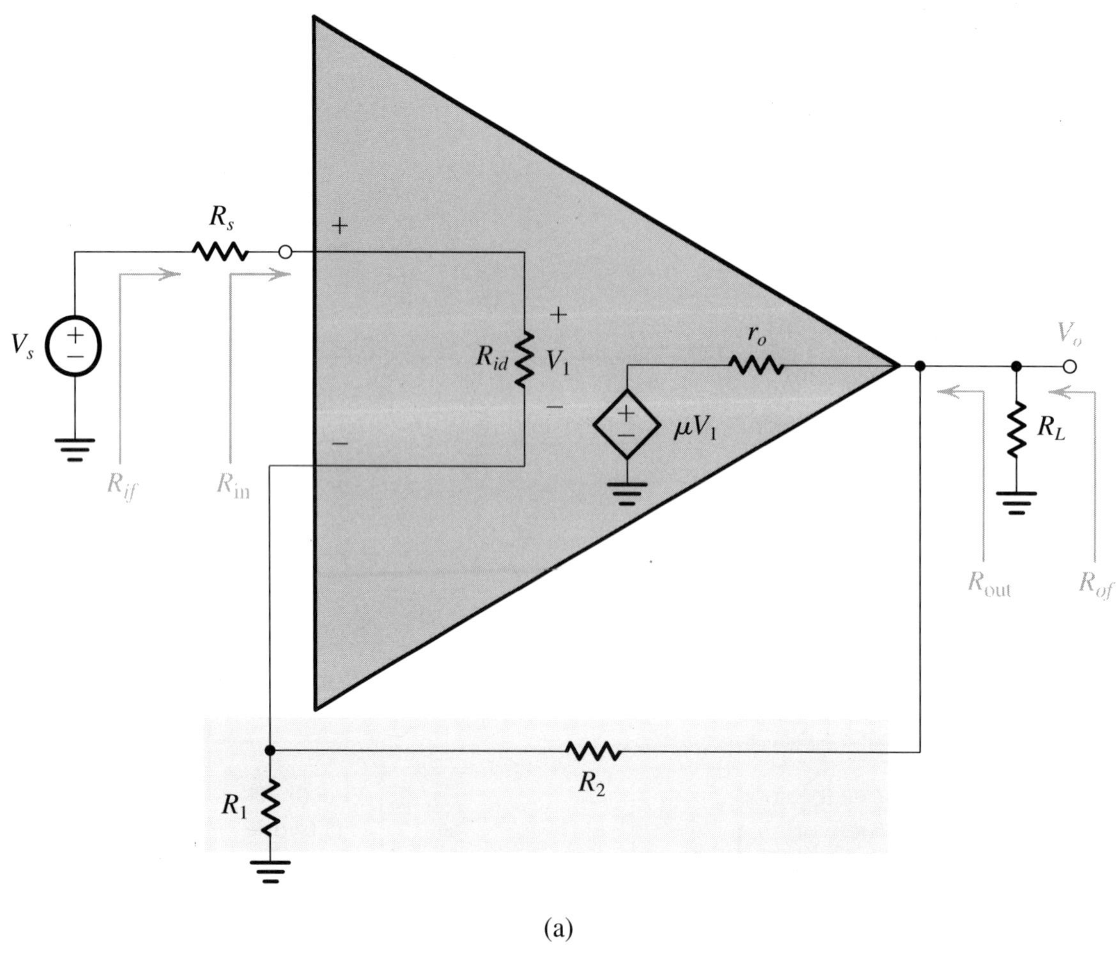

(a)

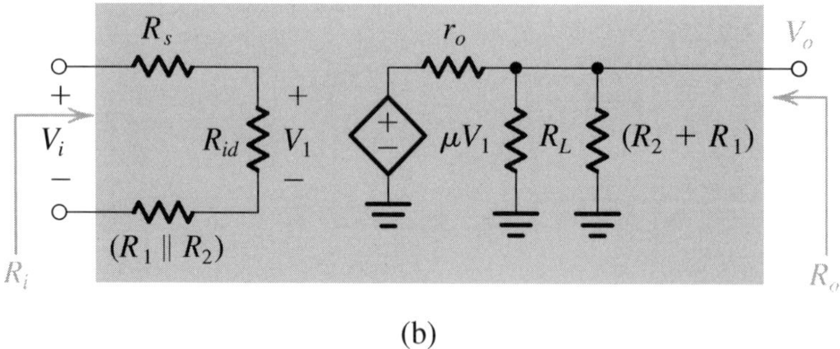

(b)

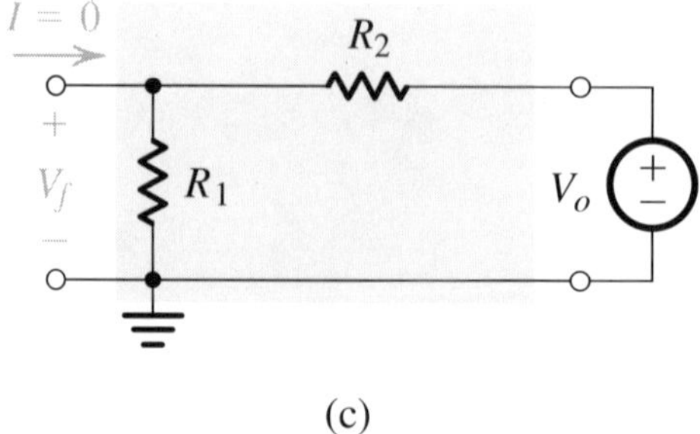

(c)

FIGURE 8.12 Circuits for Example 8.1.

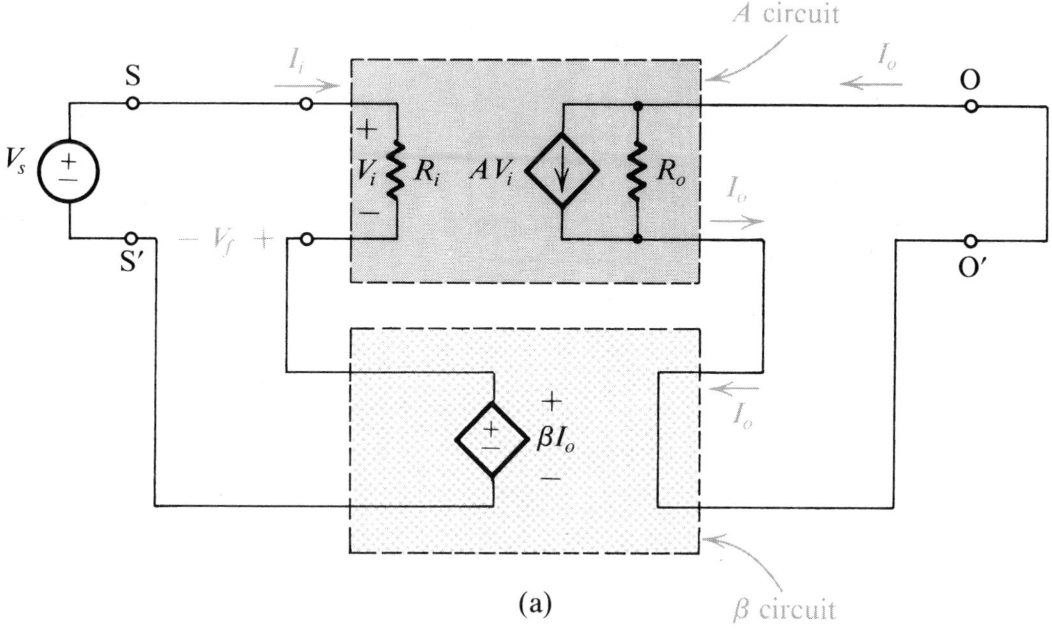

(a)

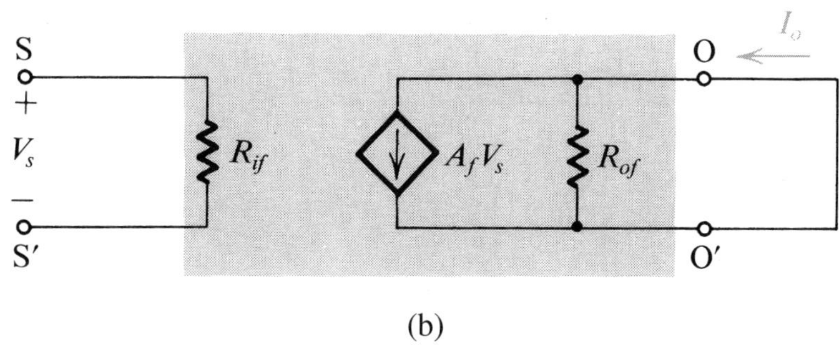

(b)

FIGURE 8.13 The series–series feedback amplifier: **(a)** ideal structure and **(b)** equivalent circuit.

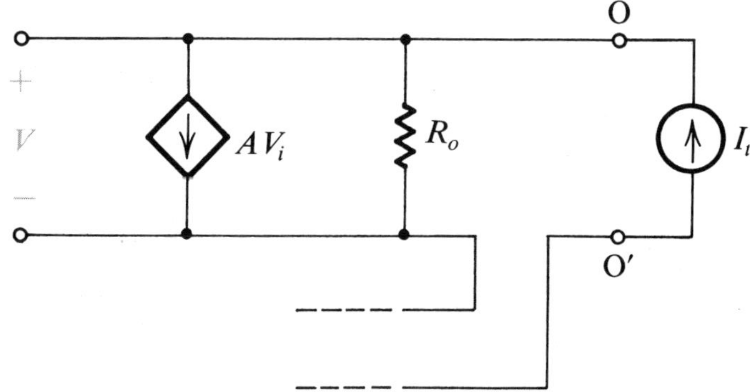

FIGURE 8.14 Measuring the output resistance R_{of} of the series–series feedback amplifier.

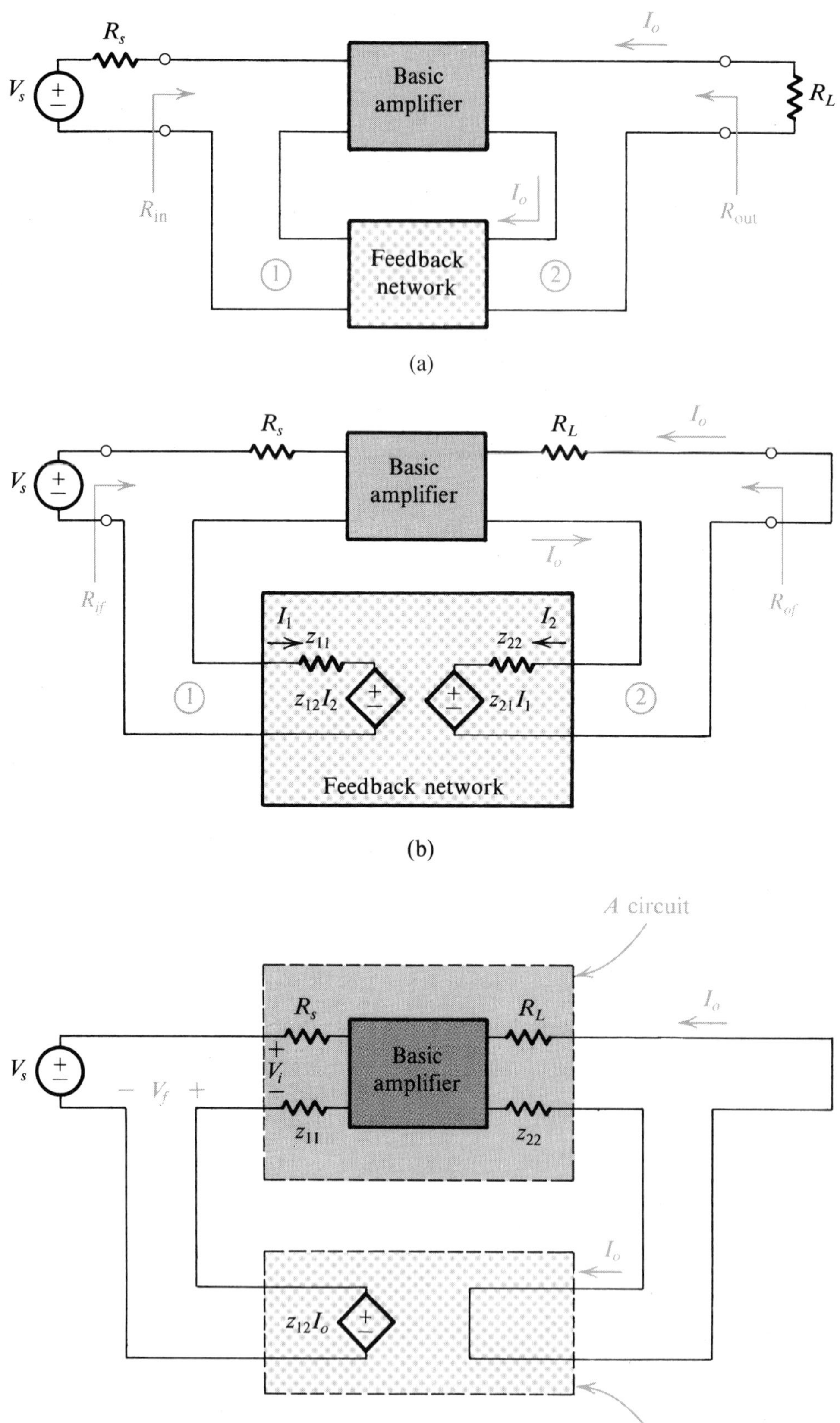

FIGURE 8.15 Derivation of the A circuit and the β circuit for series–series feedback amplifiers. **(a)** A series–series feedback amplifier. **(b)** The circuit of (a) with the feedback network represented by its z parameters. **(c)** A redrawing of the circuit in (b) with z_{21} neglected.

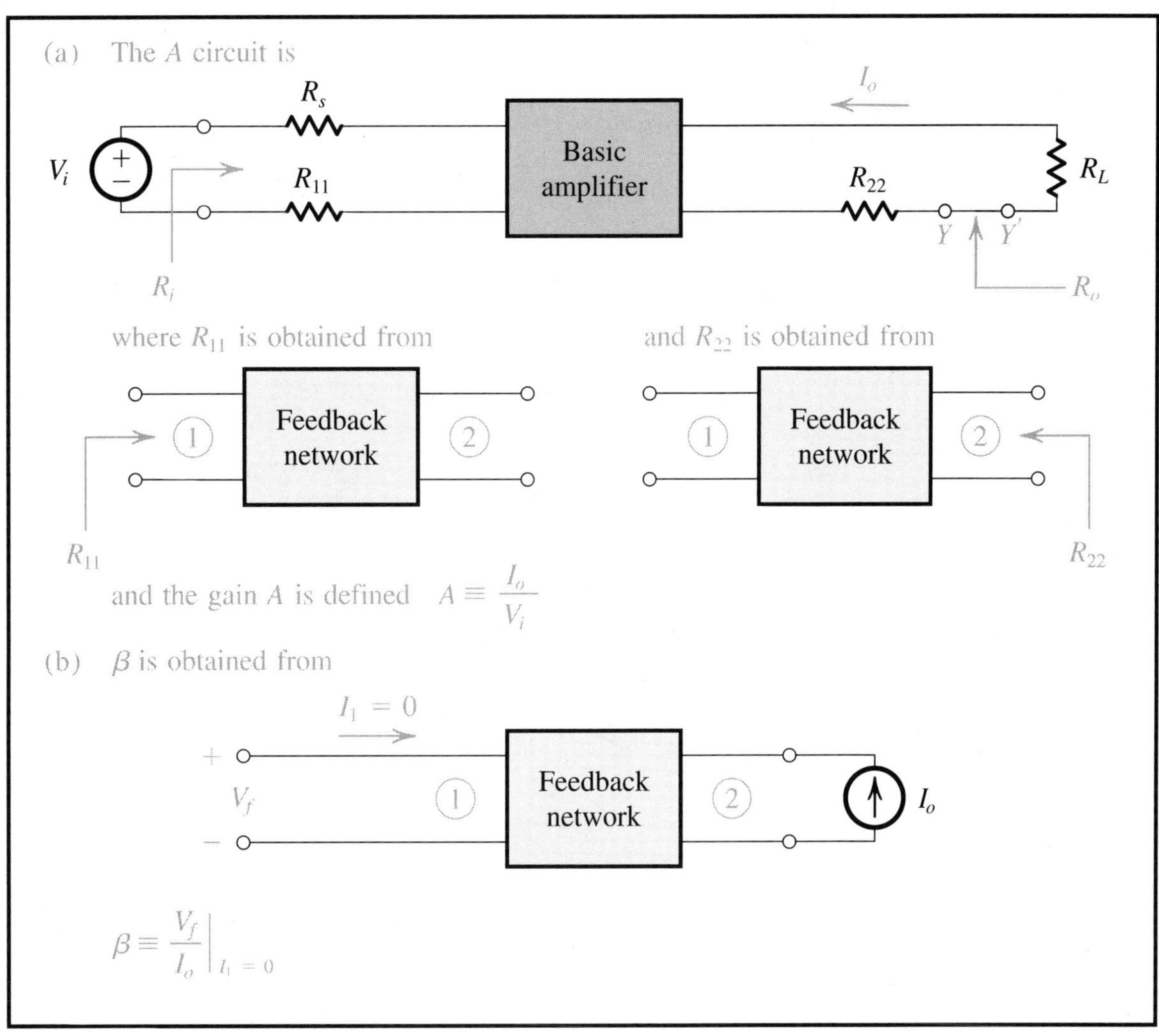

FIGURE 8.16 Finding the A circuit and β for the voltage-mixing current-sampling (series–series) case.

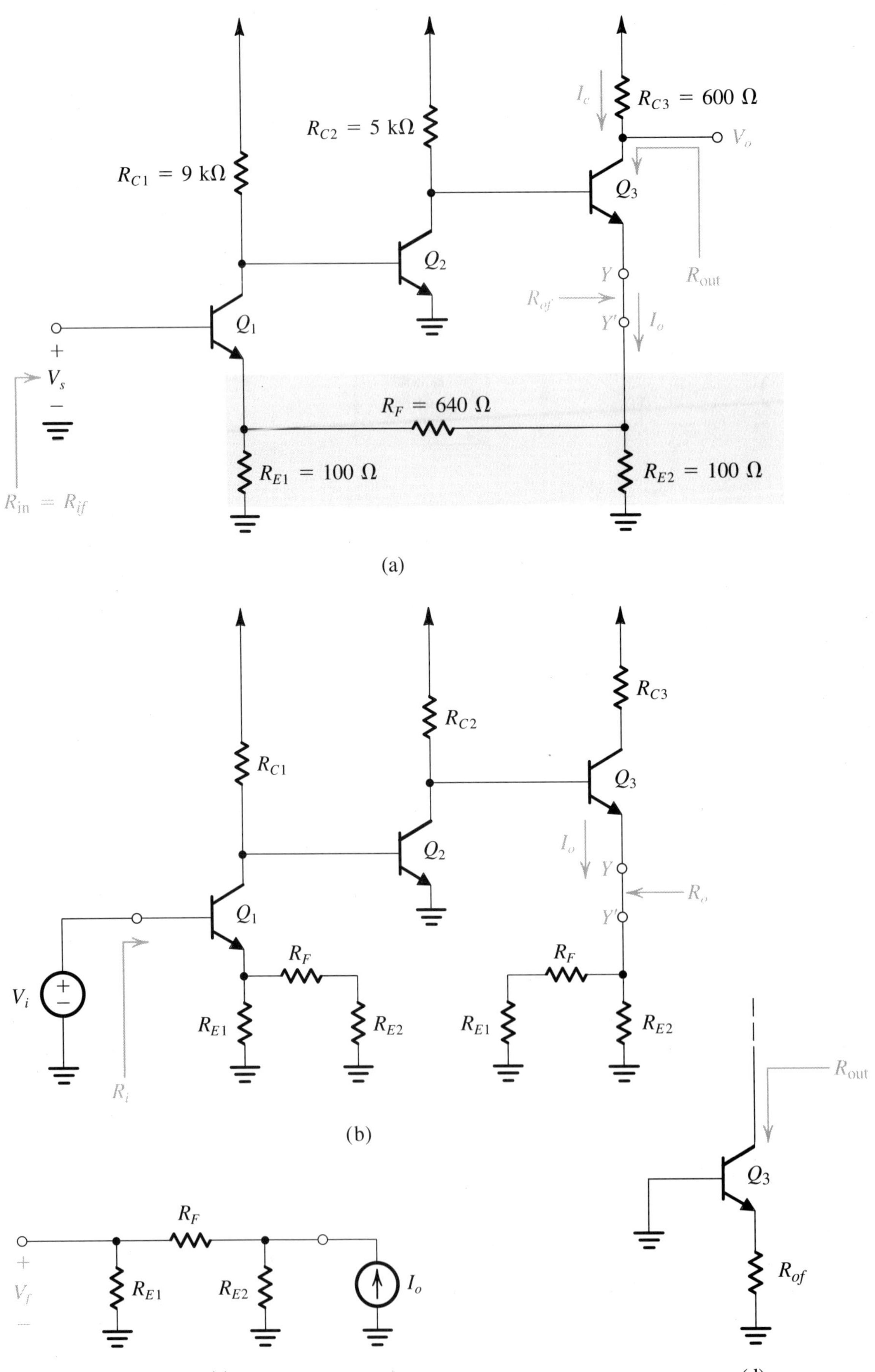

FIGURE 8.17 Circuits for Example 8.2.

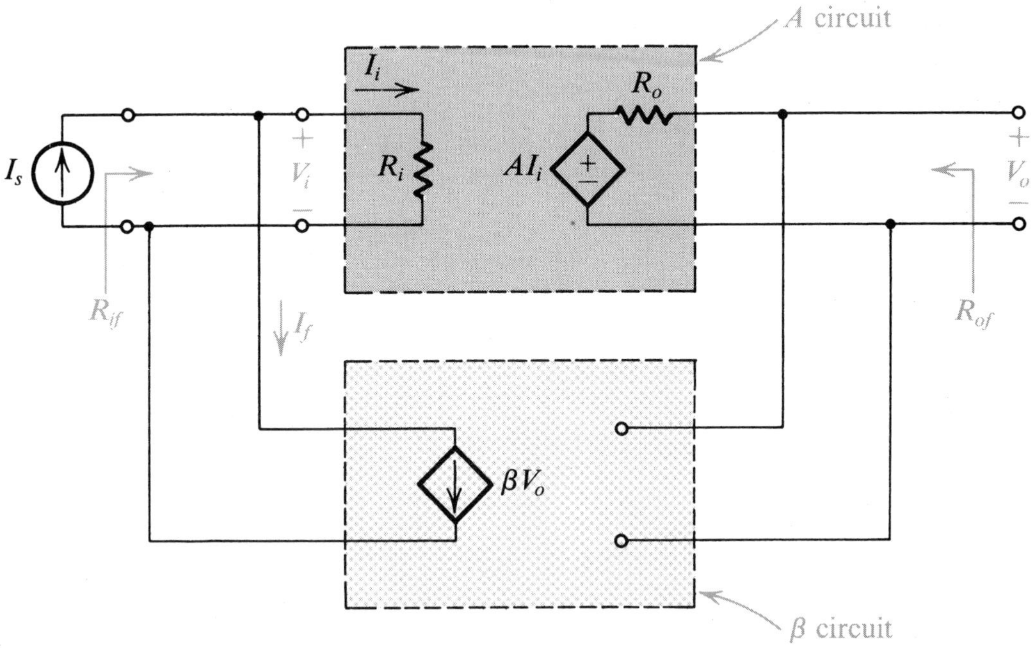

FIGURE 8.18 Ideal structure for the shunt–shunt feedback amplifier.

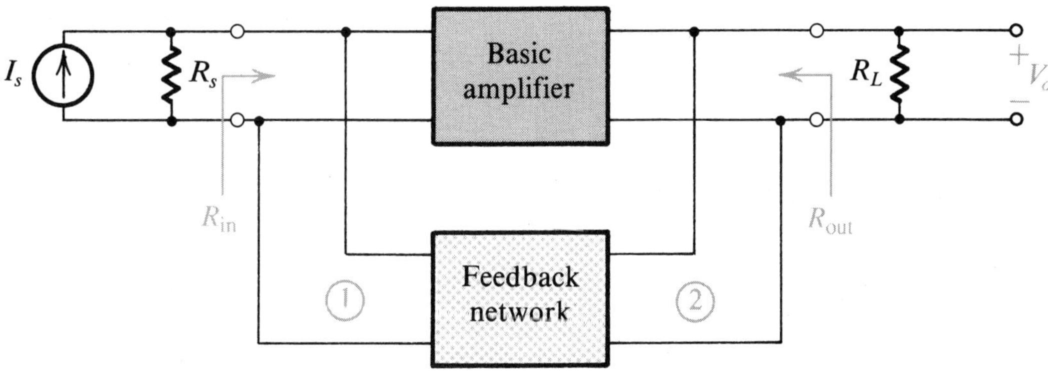

FIGURE 8.19 Block diagram for a practical shunt–shunt feedback amplifier.

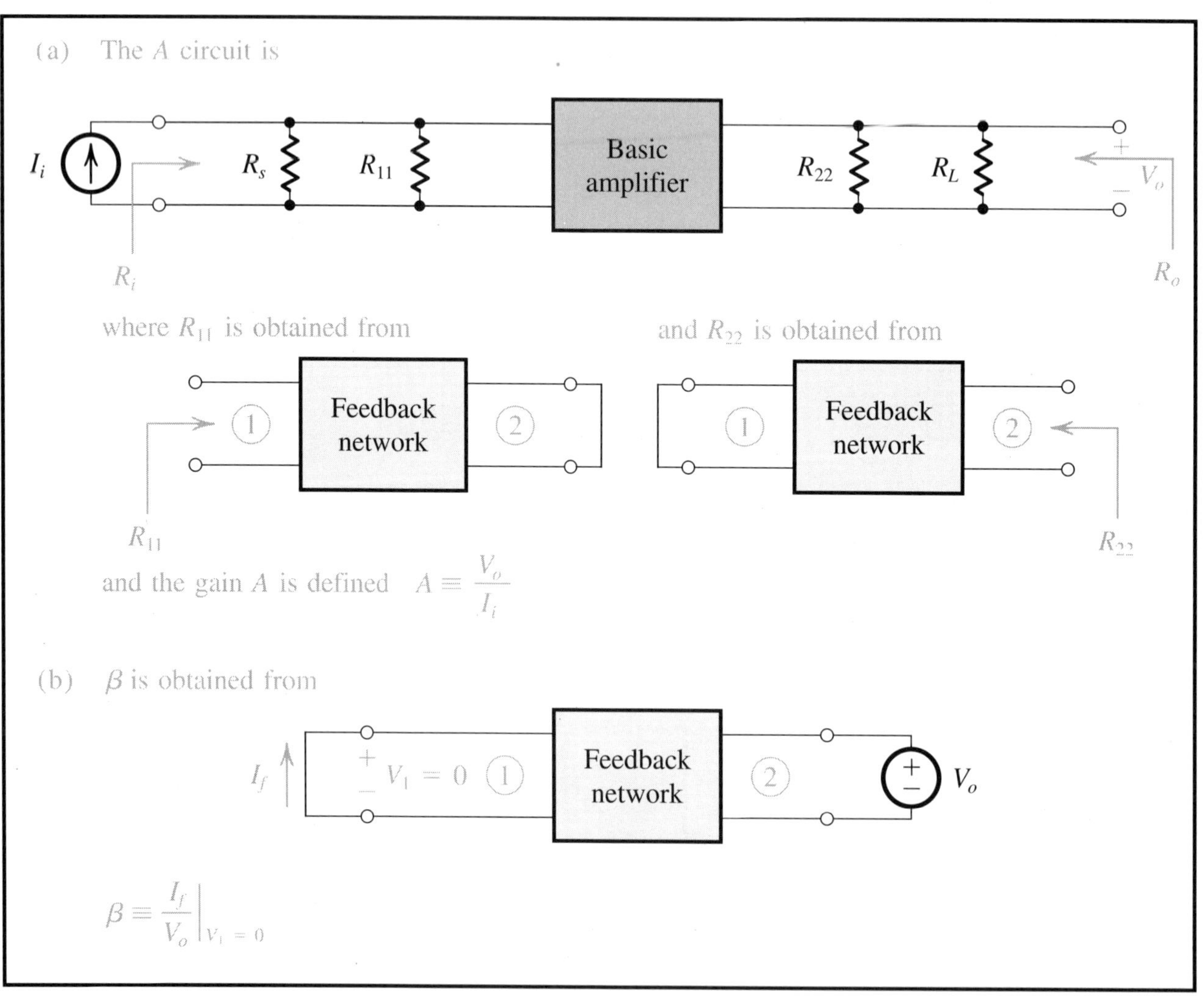

FIGURE 8.20 Finding the A circuit and β for the current-mixing voltage-sampling (shunt–shunt) feedback amplifier in Fig. 8.19.

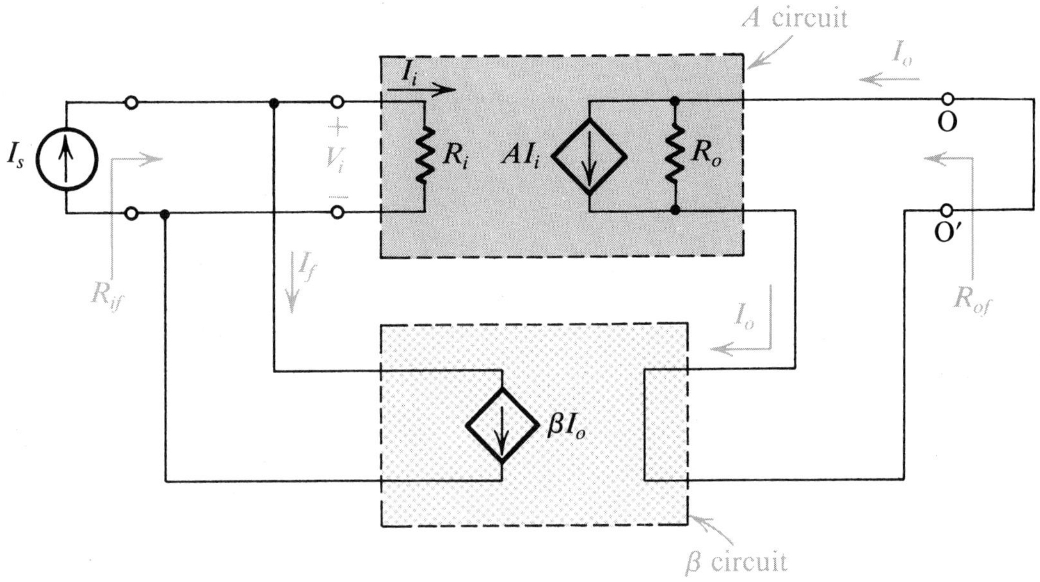

FIGURE 8.22 Ideal structure for the shunt–series feedback amplifier.

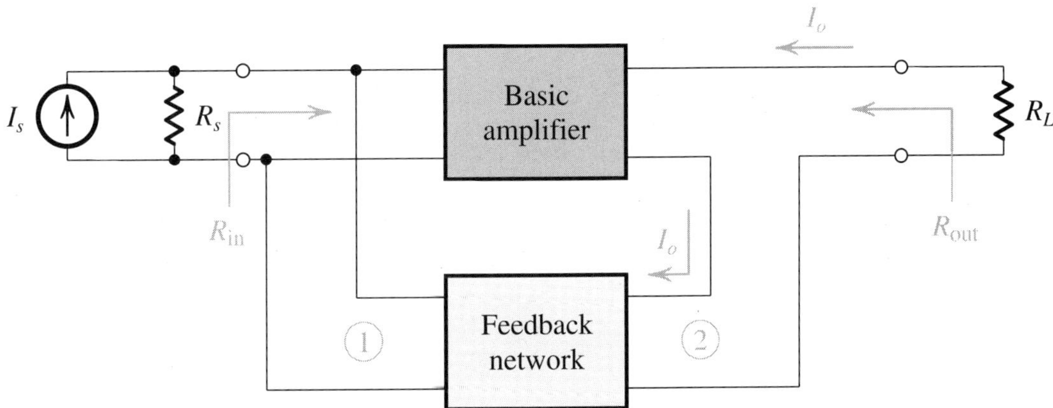

FIGURE 8.23 Block diagram for a practical shunt–series feedback amplifier.

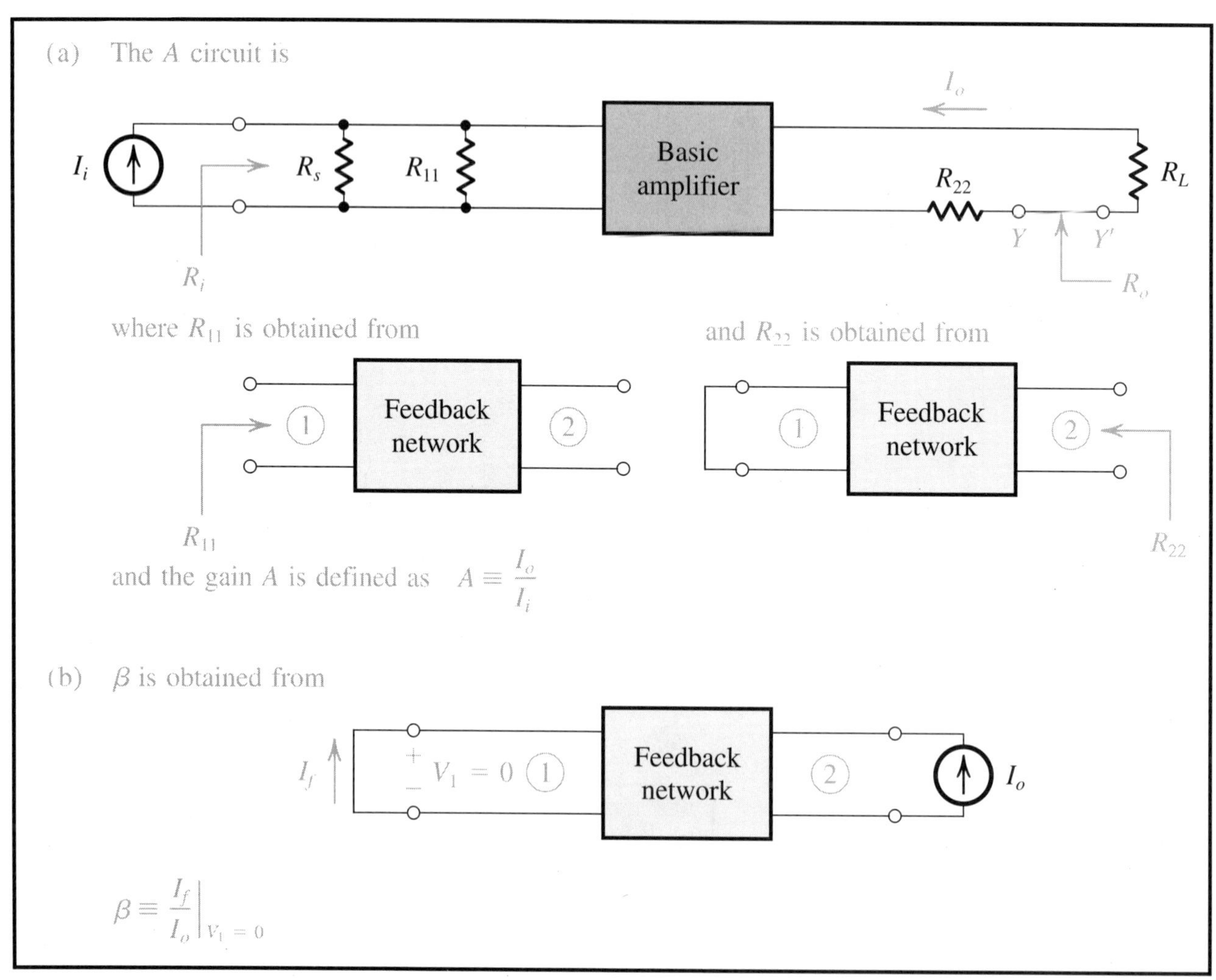

FIGURE 8.24 Finding the A circuit and β for the current-mixing current-sampling (shunt–series) feedback amplifier of Fig. 8.23.

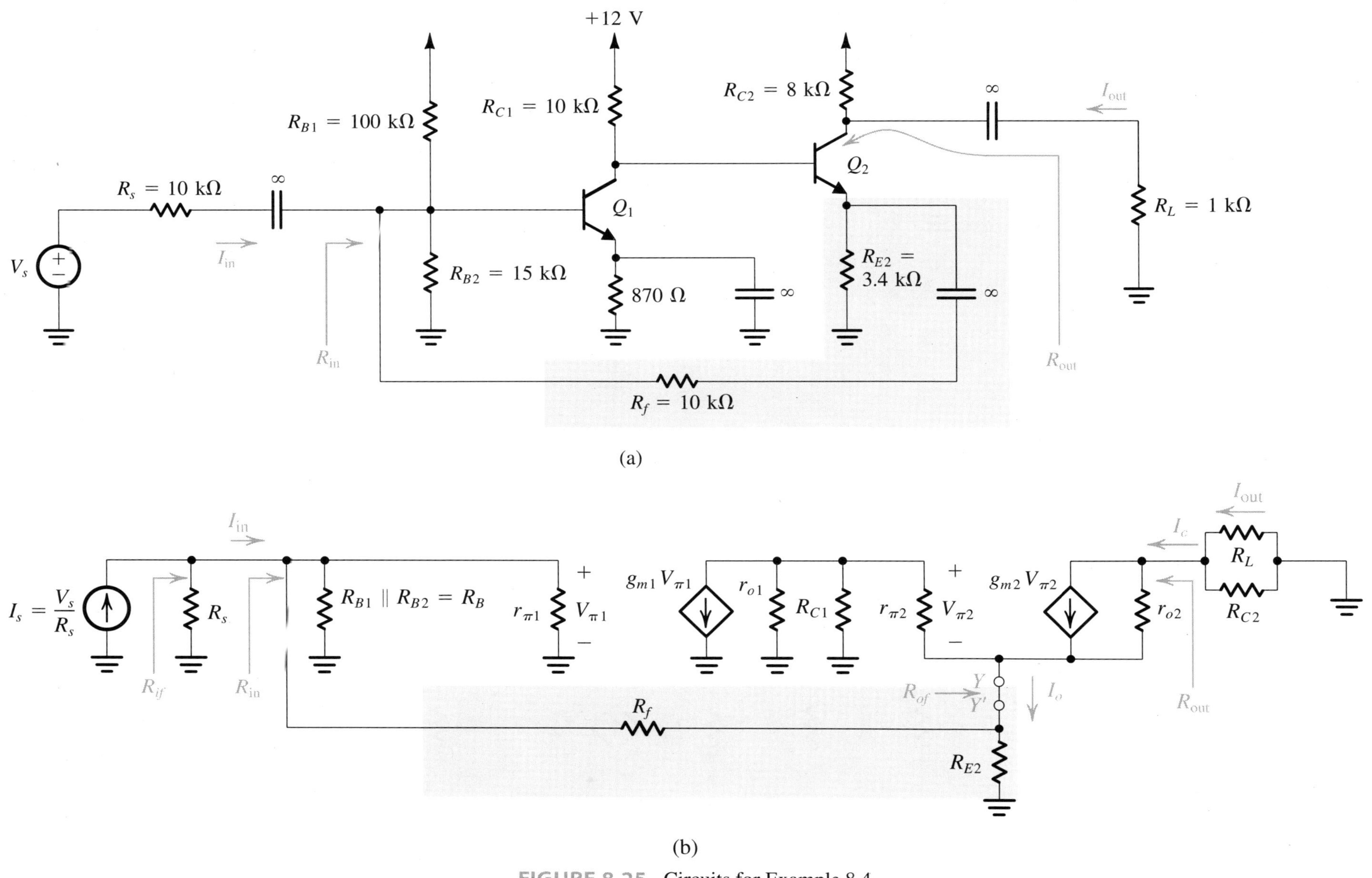

FIGURE 8.25 Circuits for Example 8.4.

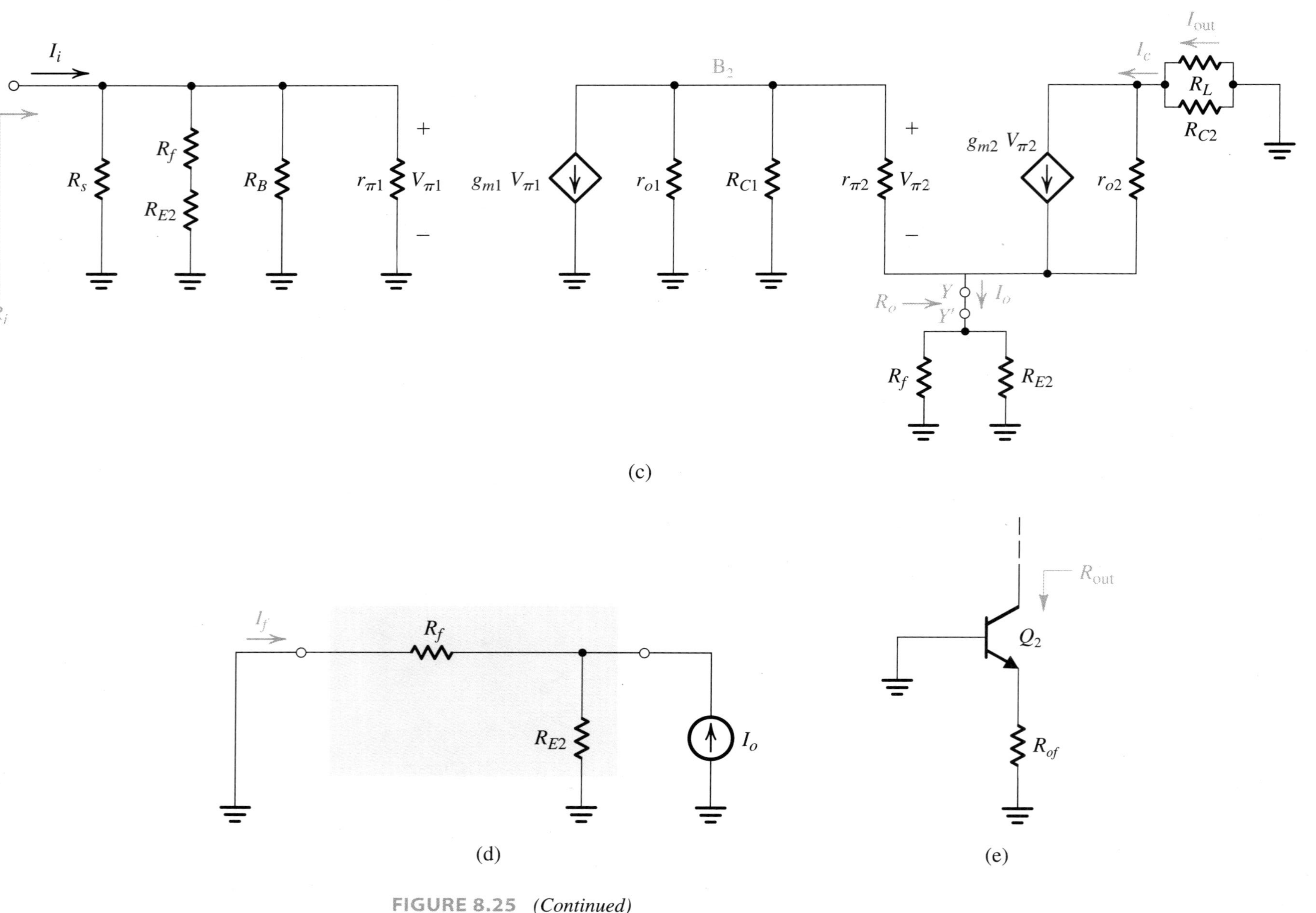

FIGURE 8.25 *(Continued)*

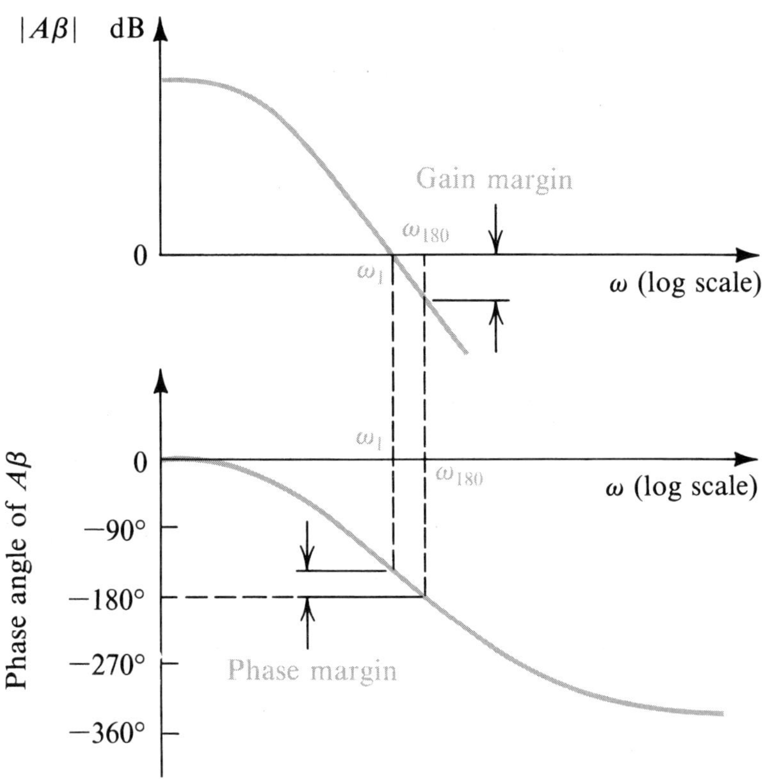

FIGURE 8.36 Bode plot for the loop gain $A\beta$ illustrating the definitions of the gain and phase margins.

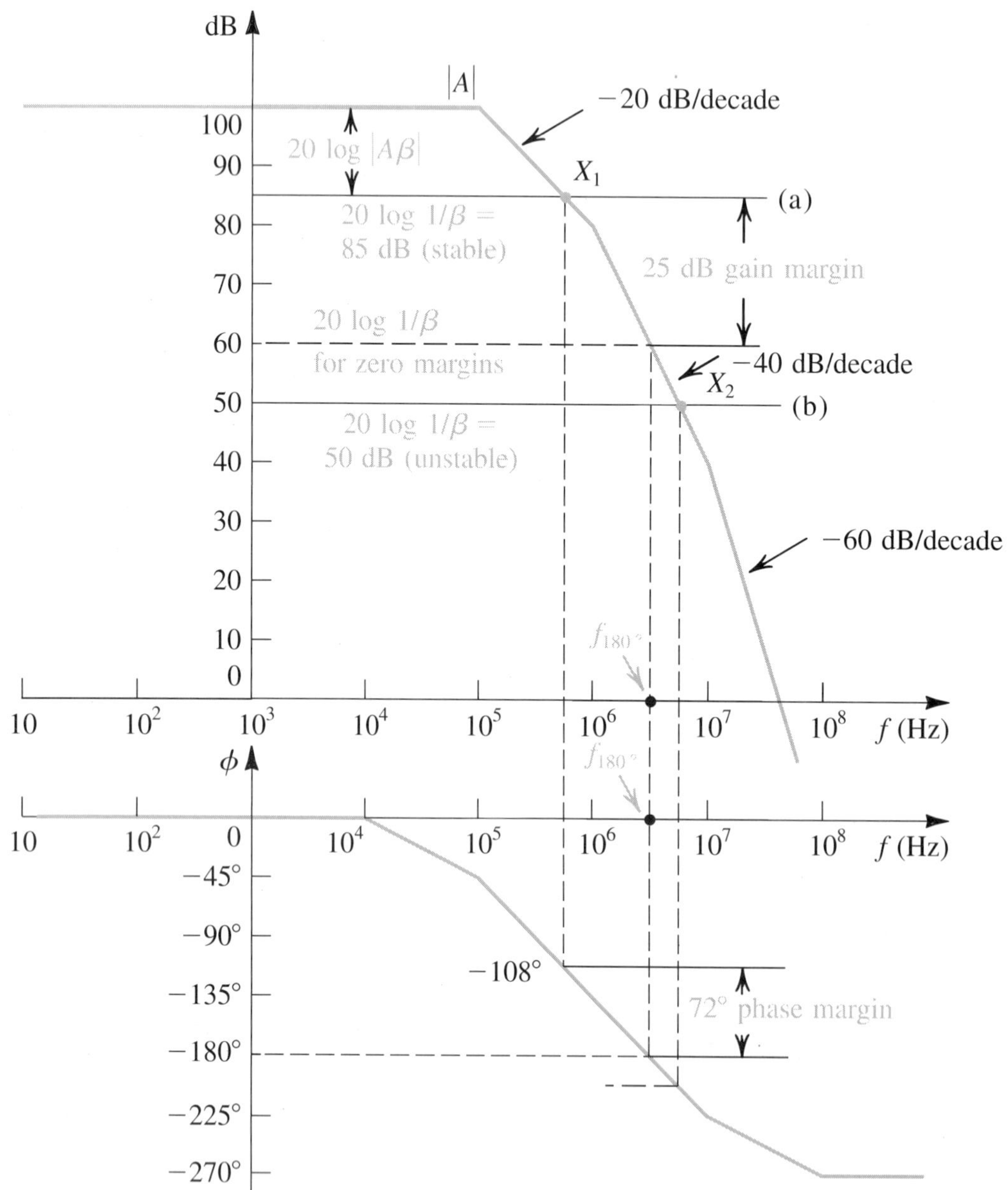

FIGURE 8.37 Stability analysis using Bode plot of $|A|$.

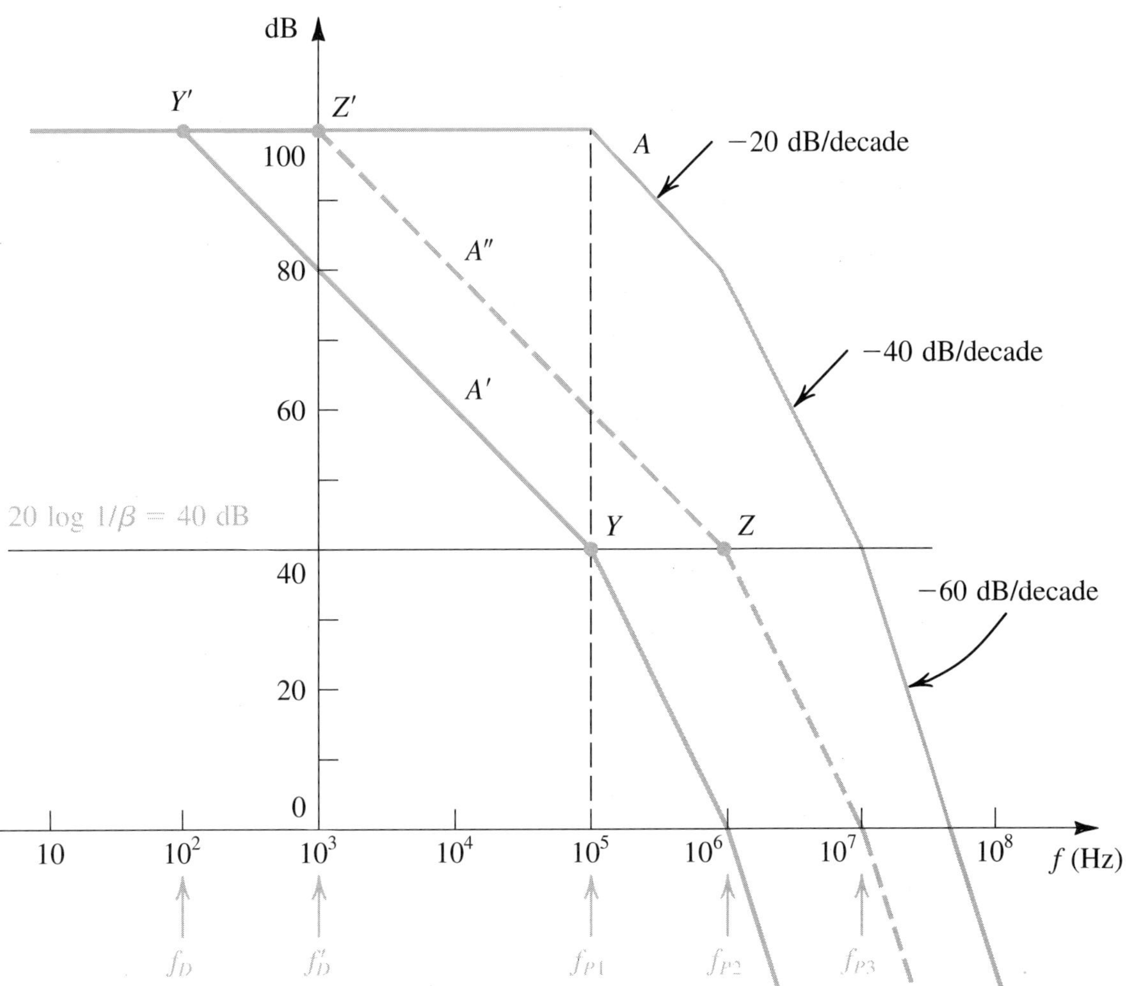

FIGURE 8.38 Frequency compensation for $\beta = 10^{-2}$. The response labeled A' is obtained by introducing an additional pole at f_D. The A'' response is obtained by moving the original low-frequency pole to f'_D.

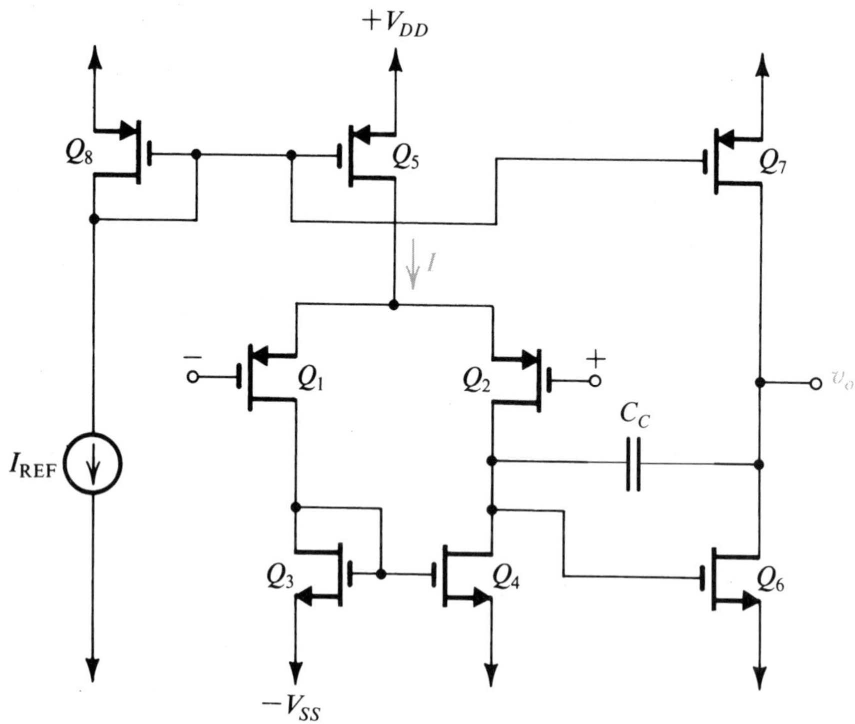

FIGURE 9.1 The basic two-stage CMOS op-amp configuration.

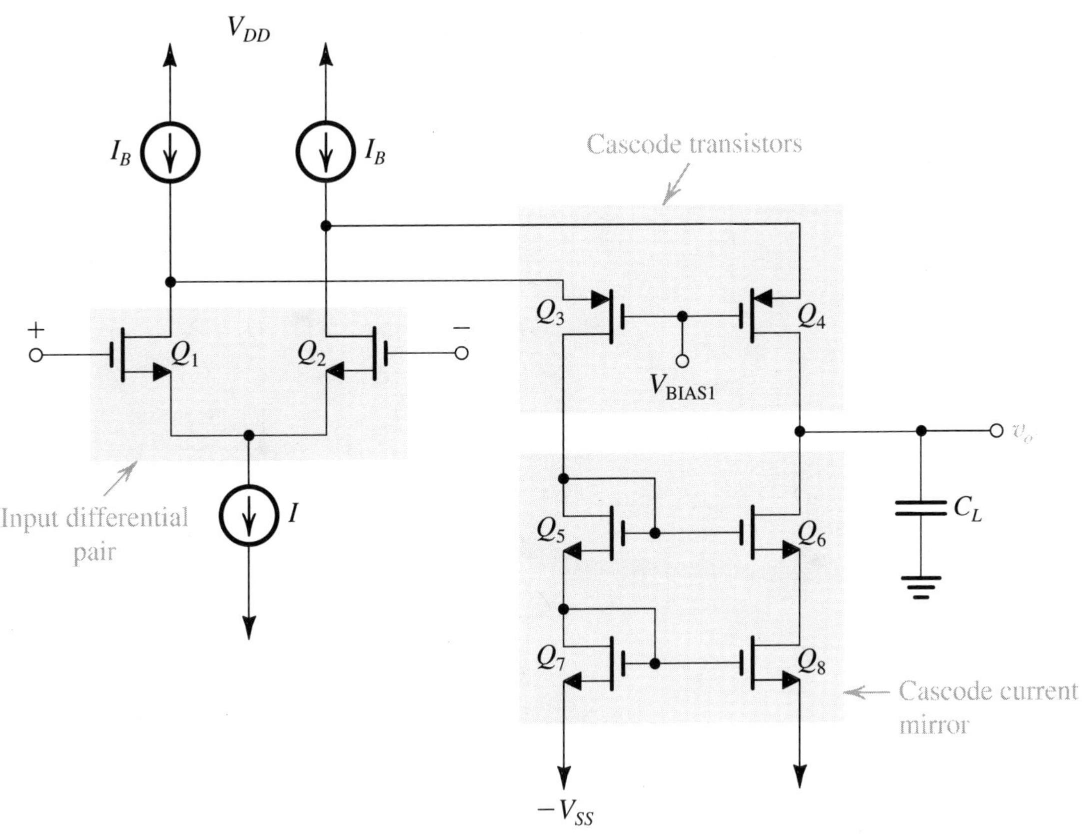

FIGURE 9.8 Structure of the folded-cascode CMOS op amp.

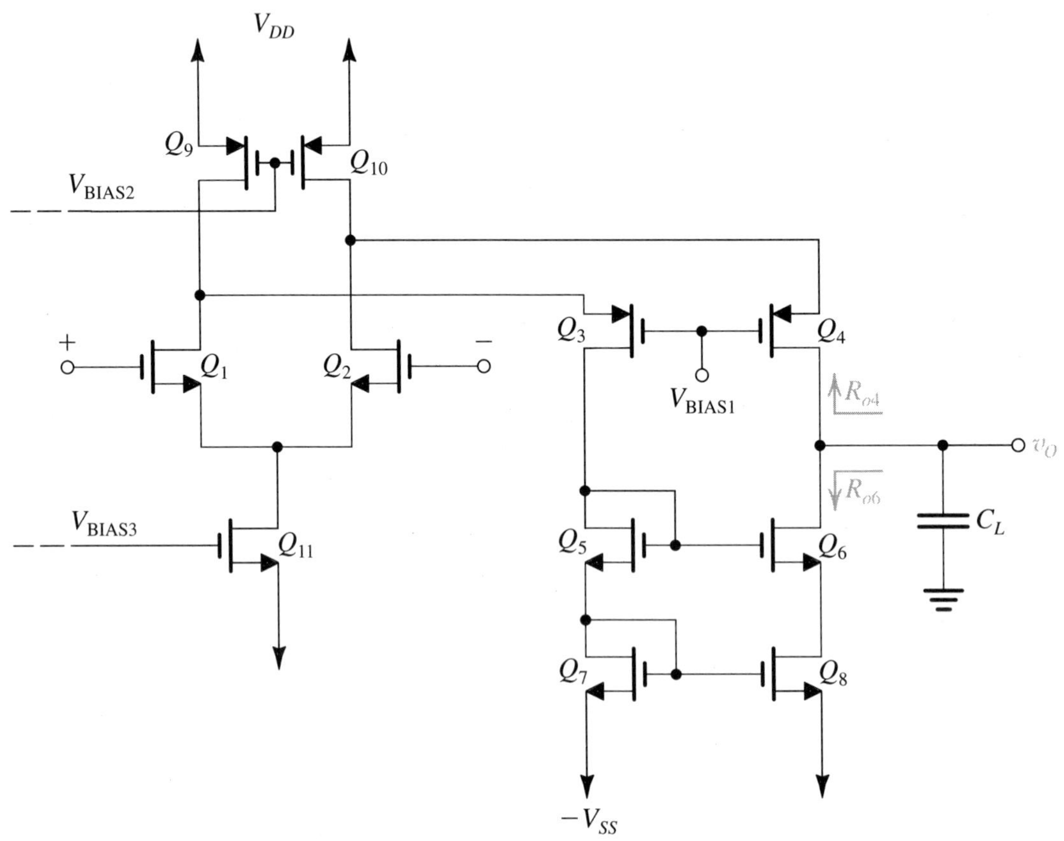

FIGURE 9.9 A more complete circuit for the folded-cascode CMOS amplifier of Fig. 9.8.

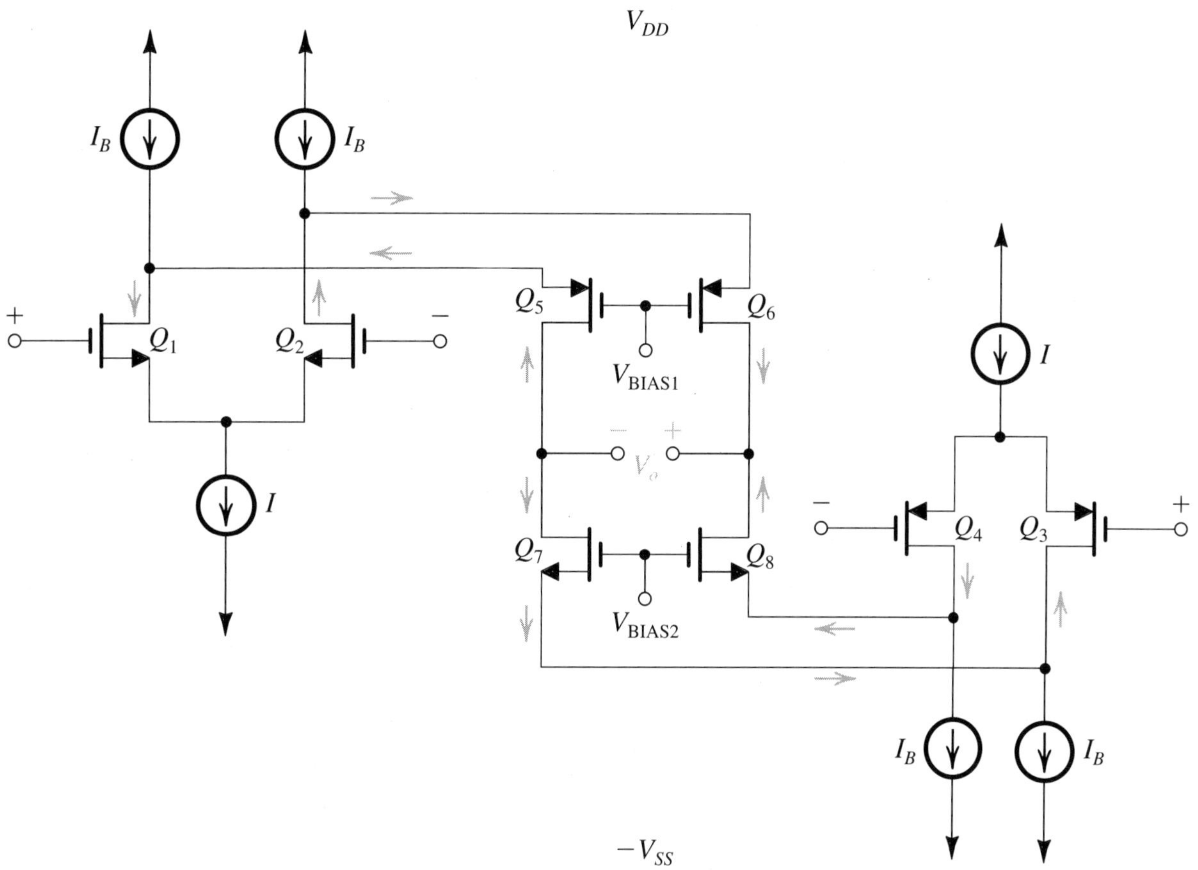

FIGURE 9.11 A folded-cascode op amp that employs two parallel complementary input stages to achieve rail-to-rail input common-mode operation. Note that the two "+" terminals are connected together and the two "–" terminals are connected together.

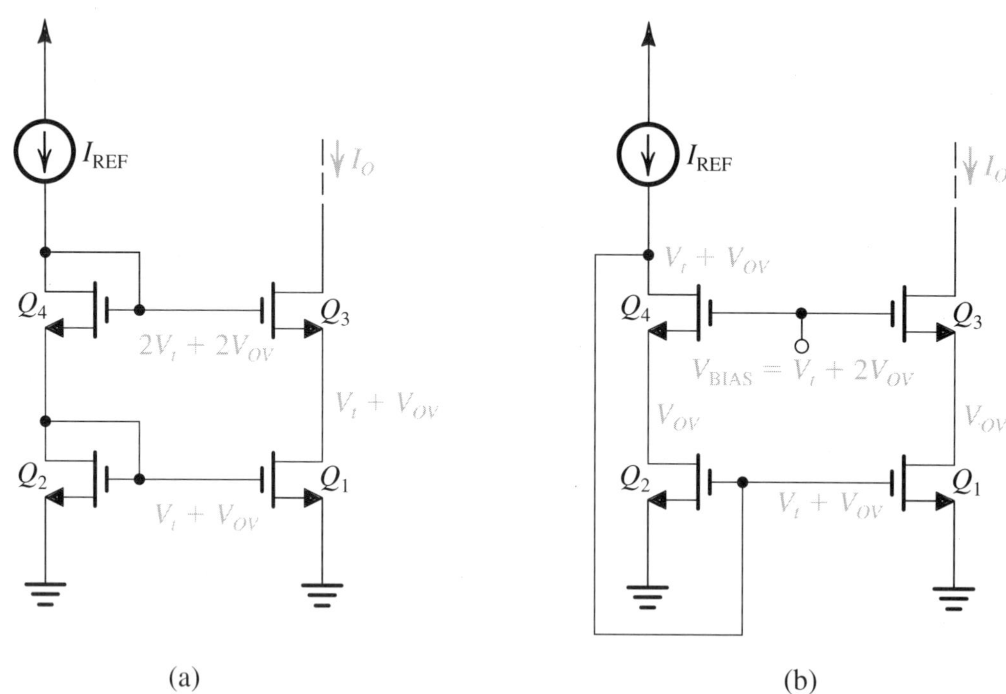

FIGURE 9.12 **(a)** Cascode current mirror with the voltages at all nodes indicated. Note that the minimum voltage allowed at the output is $V_t + V_{OV}$. **(b)** A modification of the cascode mirror that results in the reduction of the minimum output voltage to V_{OV}. This is the wide-swing current mirror.

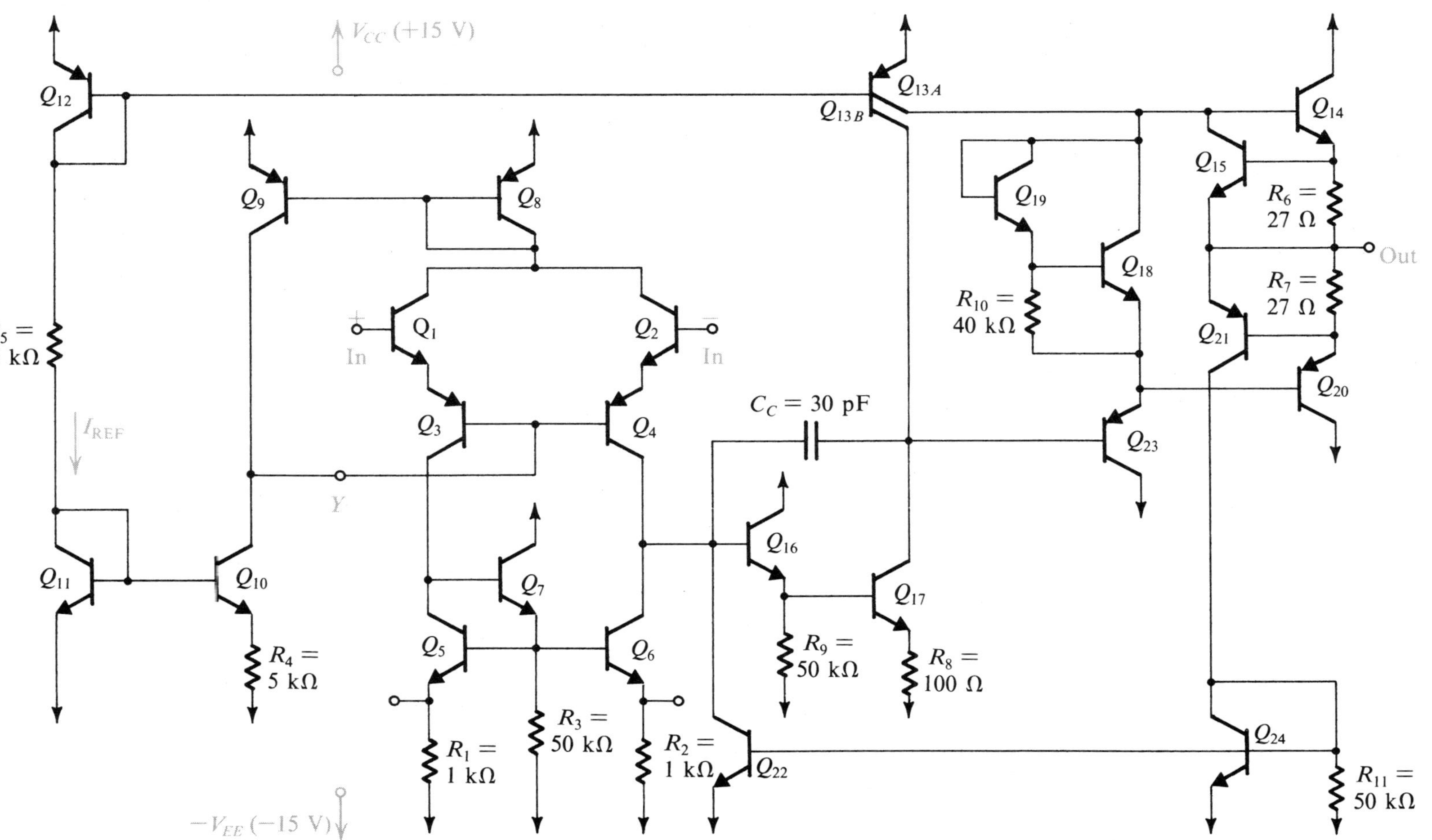

FIGURE 9.13 The 741 op-amp circuit. Q_{11}, Q_{12}, and R_5 generate a reference bias current, I_{REF}. Q_{10}, Q_9, and Q_8 bias the input stage, which is composed of Q_1 to Q_7. The second gain stage is composed of Q_{16} and Q_{17} with Q_{13B} acting as active load. The class AB output stage is formed by Q_{14} and Q_{20} with biasing devices Q_{13A}, Q_{18}, and Q_{19}, and an input buffer Q_{23}. Transistors Q_{15}, Q_{21}, Q_{24}, and Q_{22} serve to protect the amplifier against output short circuits and are normally cut off.

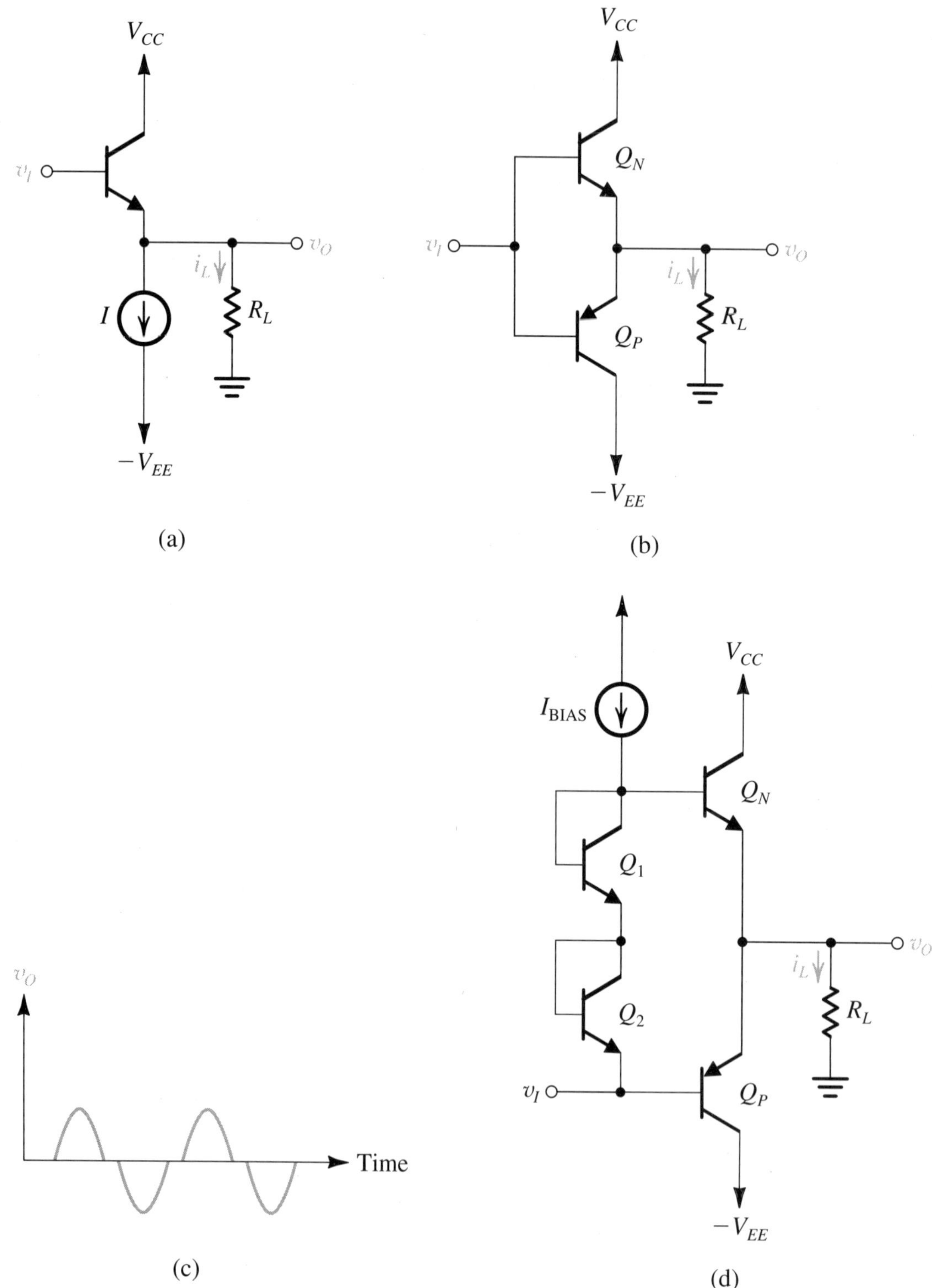

FIGURE 9.14 **(a)** The emitter follower is a class A output stage. **(b)** Class B output stage. **(c)** The output of a class B output stage fed with an input sinusoid. Observe the crossover distortion. **(d)** Class AB output stage.

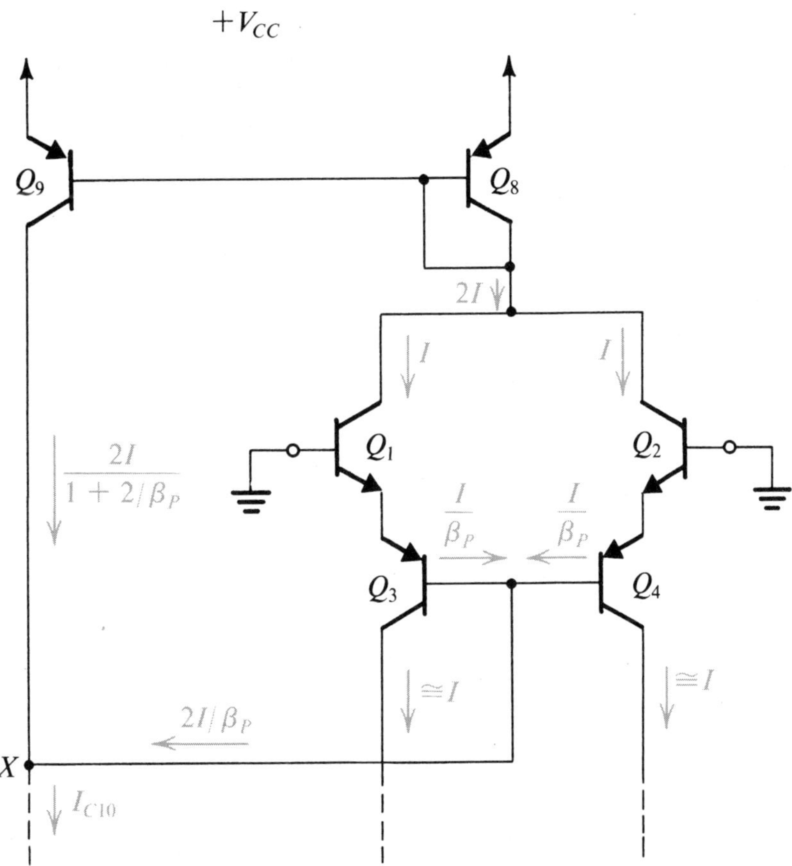

FIGURE 9.16 The dc analysis of the 741 input stage.

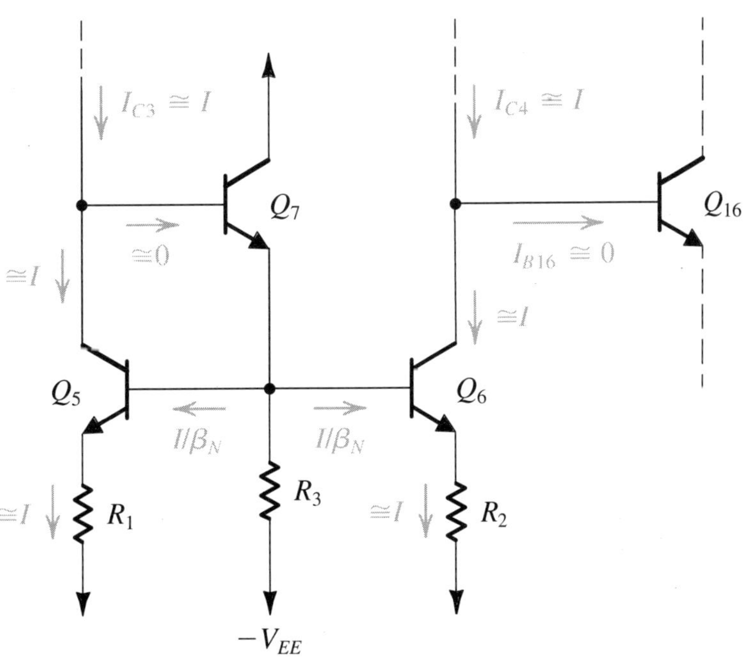

FIGURE 9.17 The dc analysis of the 741 input stage, continued.

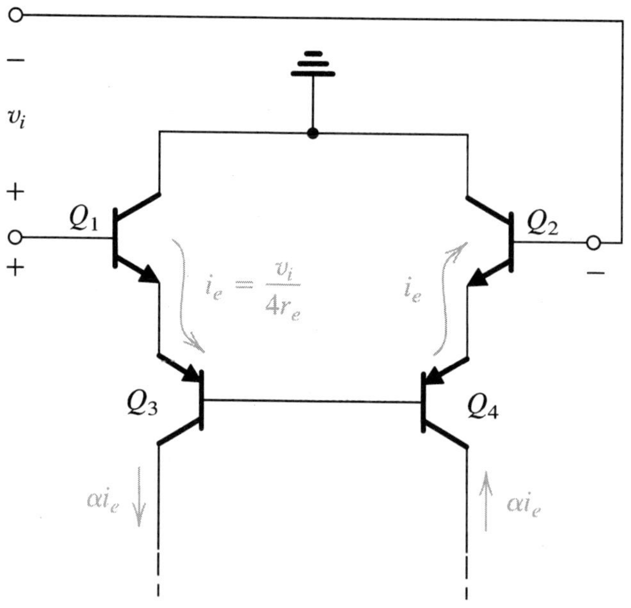

FIGURE 9.19 Small-signal analysis of the 741 input stage.

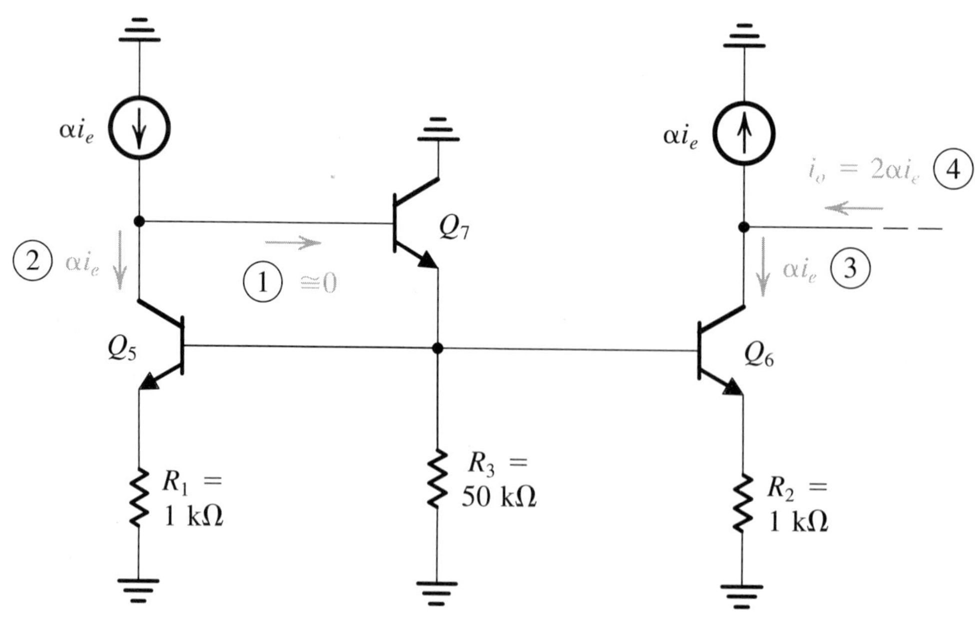

FIGURE 9.20 The load circuit of the input stage fed by the two complementary current signals generated by Q_1 through Q_4 in Fig. 9.19. Circled numbers indicate the order of the analysis steps.

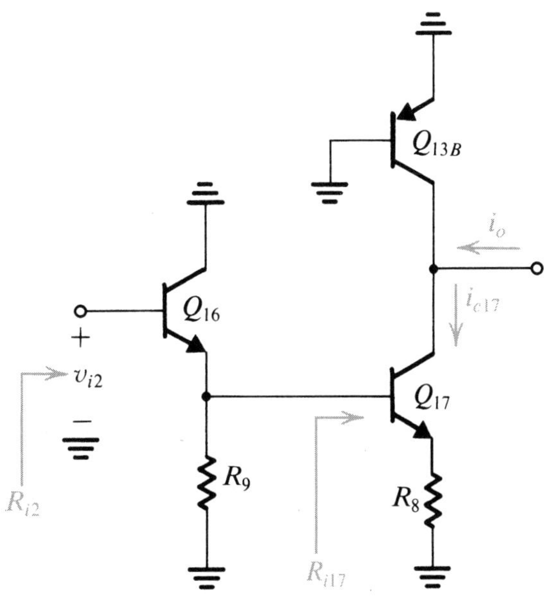

FIGURE 9.24 The 741 second stage prepared for small-signal analysis.

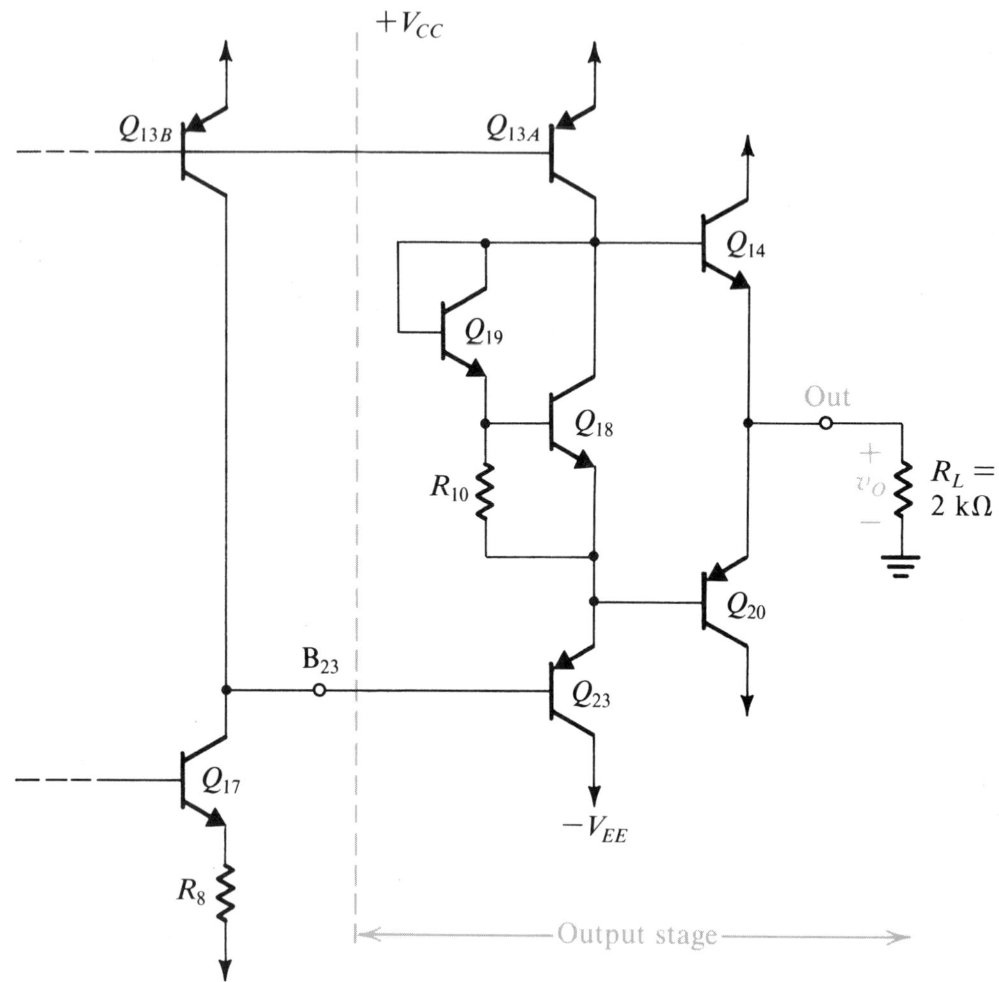

FIGURE 9.28 The 741 output stage.

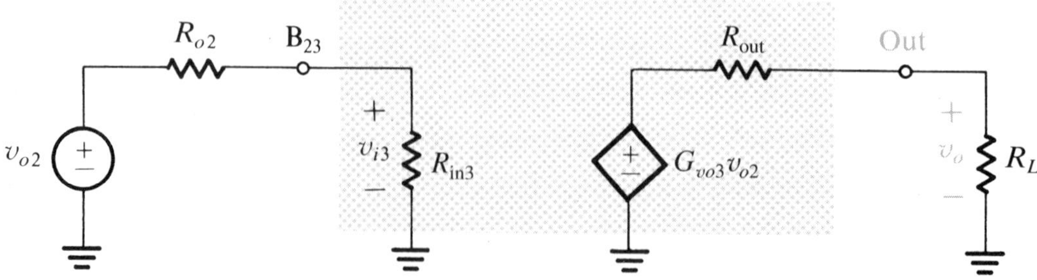

FIGURE 9.29 Model for the 741 output stage. This model is based on the amplifier equivalent circuit presented in Table 5.5 as "Equivalent Circuit C."

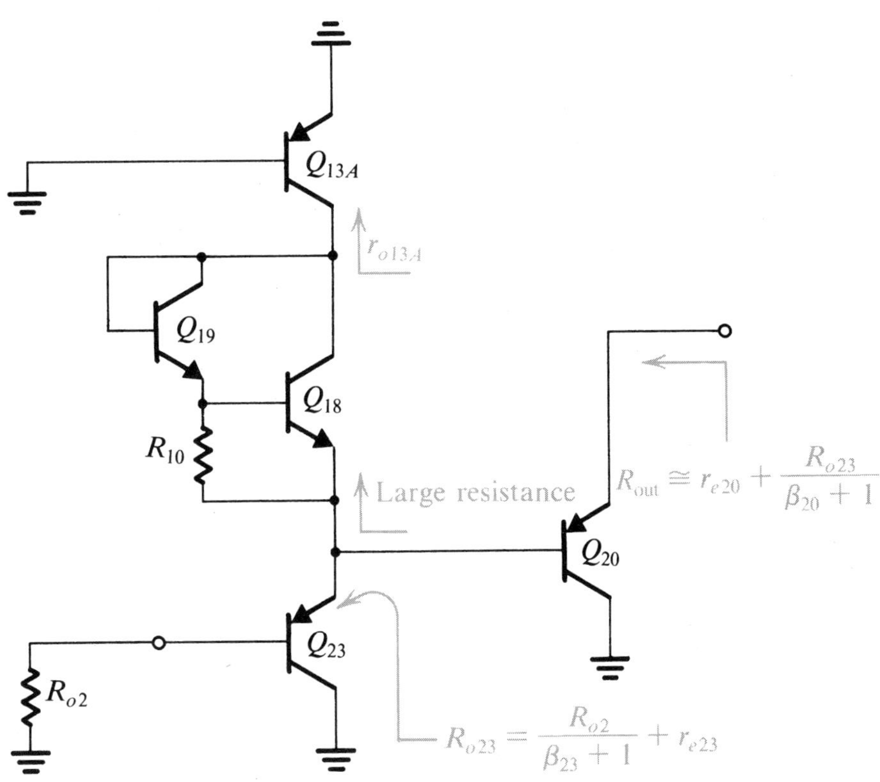

FIGURE 9.30 Circuit for finding the output resistance R_{out}.

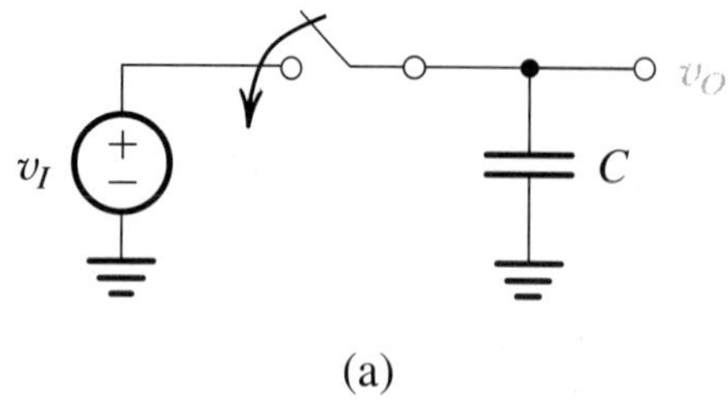

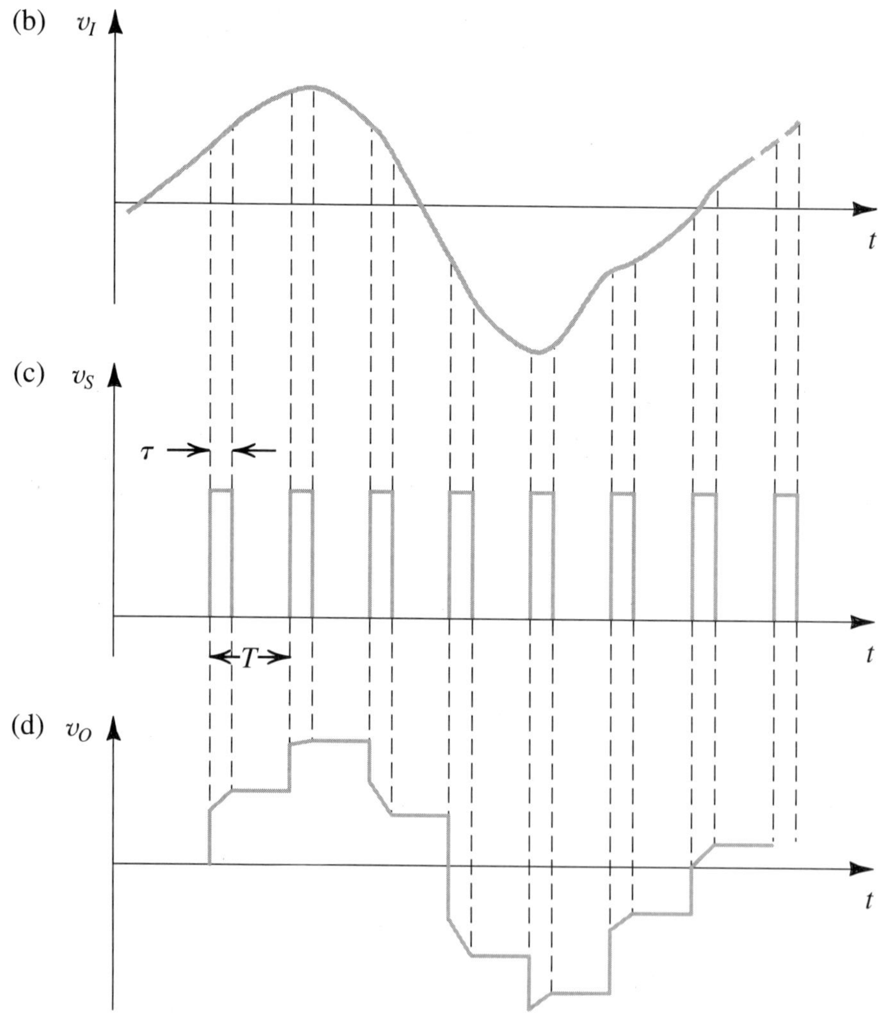

FIGURE 9.36 The process of periodically sampling an analog signal. **(a)** Sample-and-hold (S/H) circuit. The switch closes for a small part (τ seconds) of every clock period (T). **(b)** Input signal waveform. **(c)** Sampling signal (control signal for the switch). **(d)** Output signal (to be fed to A/D converter).

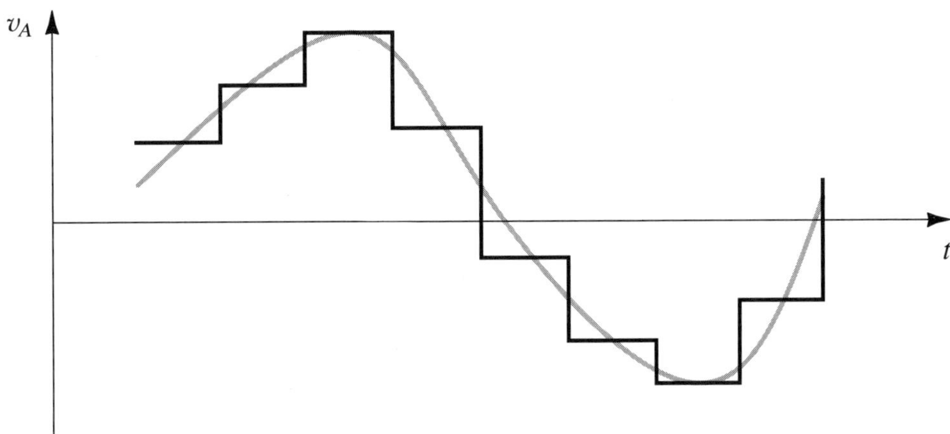

FIGURE 9.38 The analog samples at the output of a D/A converter are usually fed to a sample-and-hold circuit to obtain the staircase waveform shown. This waveform can then be filtered to obtain the smooth waveform, shown in color. The time delay usually introduced by the filter is not shown.

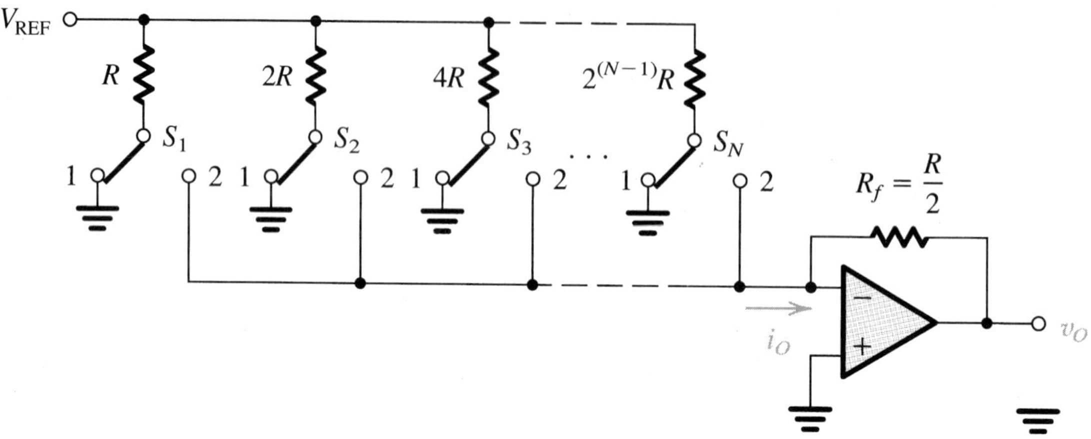

FIGURE 9.39 An N-bit D/A converter using a binary-weighted resistive ladder network.

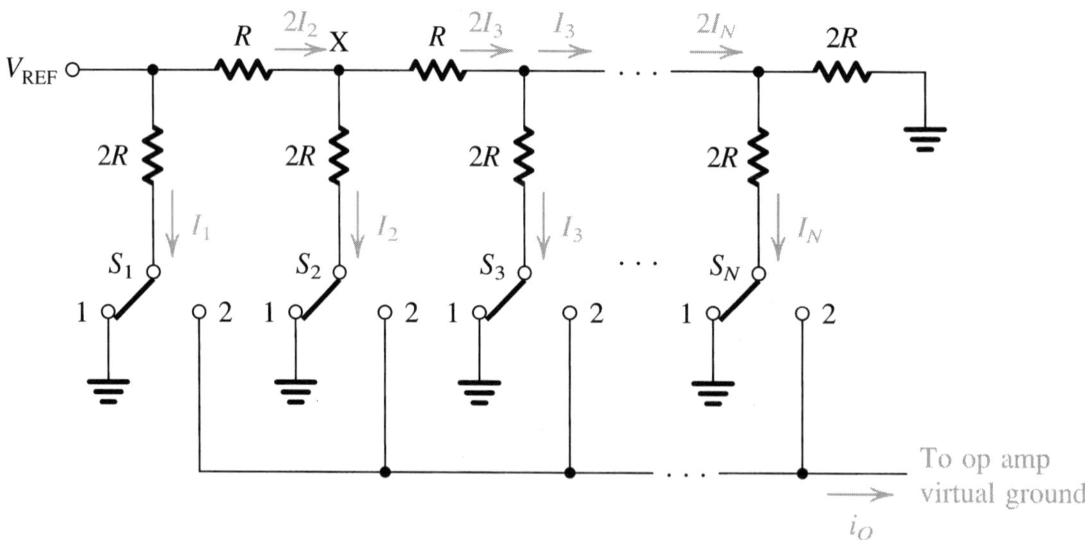

FIGURE 9.40 The basic circuit configuration of a DAC utilizing an R-2R ladder network.

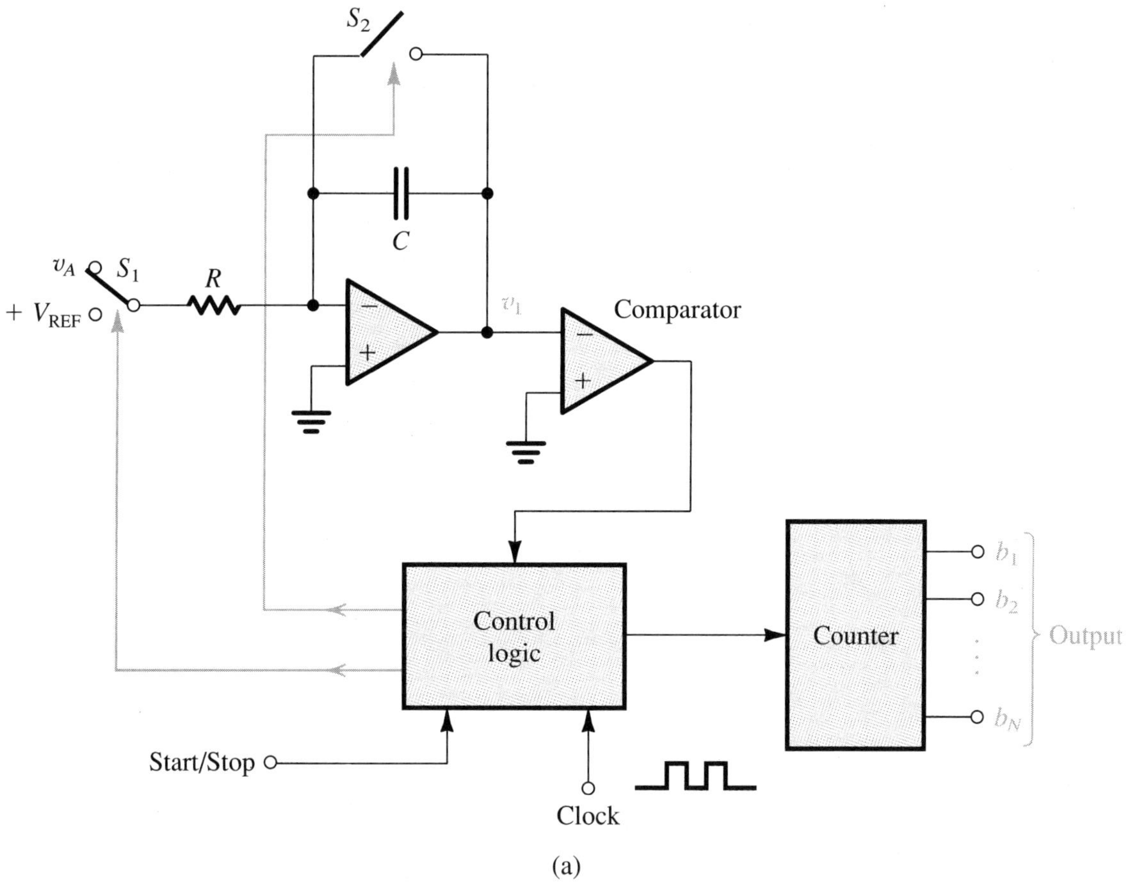

(a)

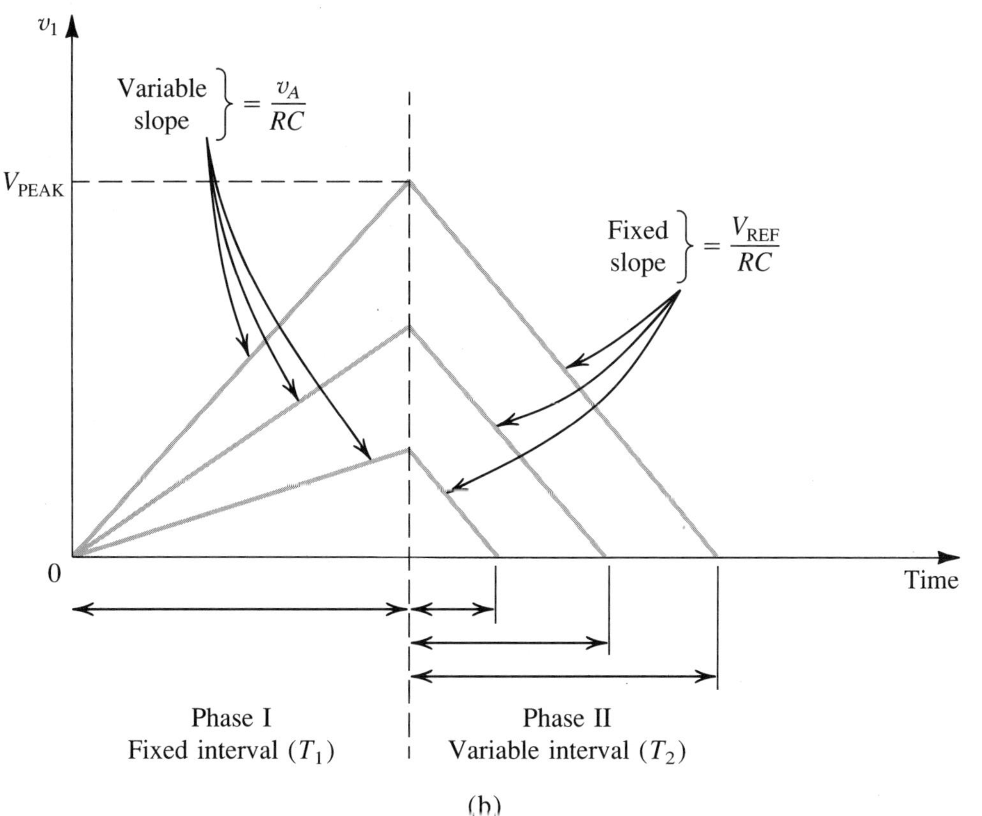

(b)

FIGURE 9.44 The dual-slope A/D conversion method. Note that v_A is assumed to be negative.

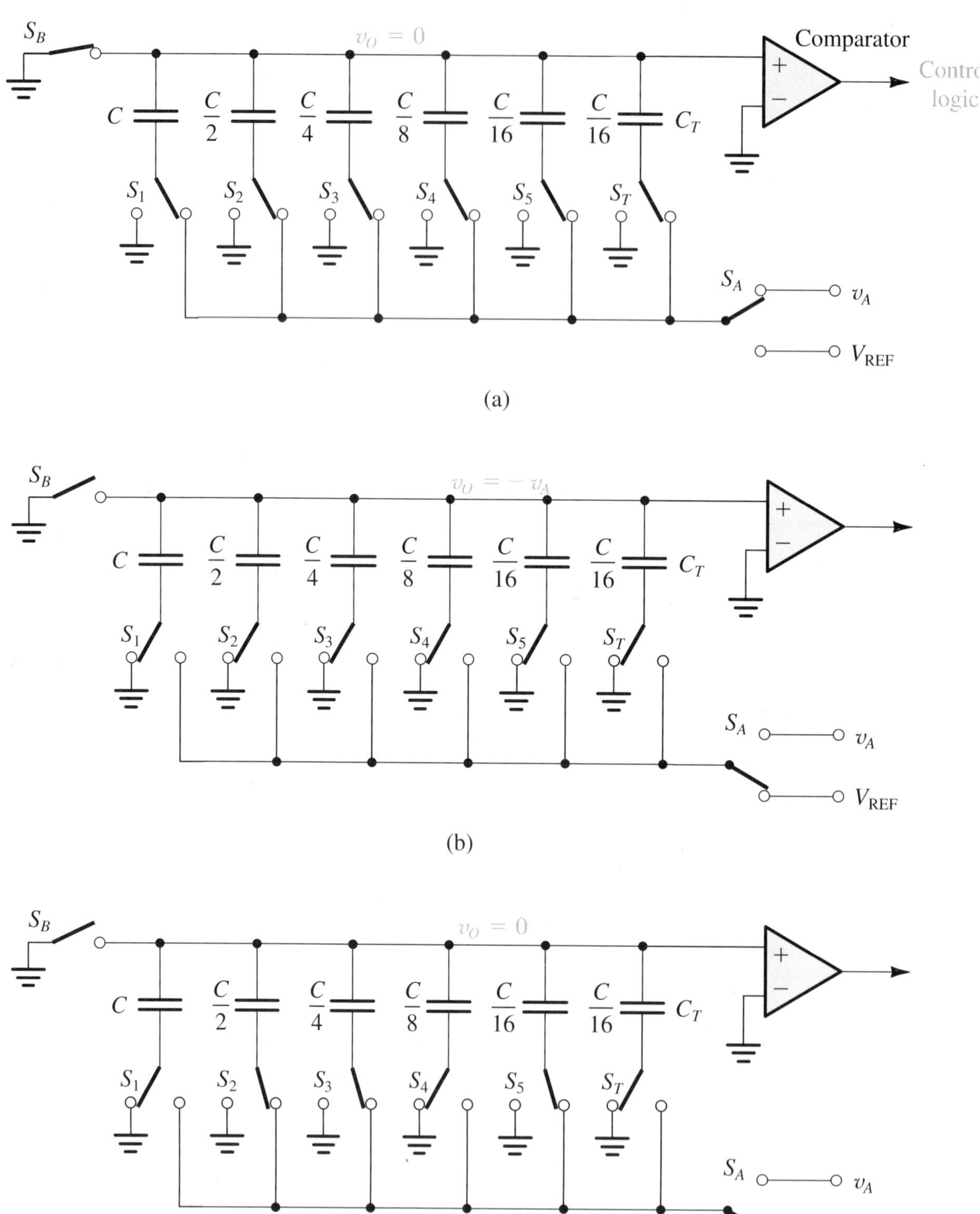

FIGURE 9.46 Charge-redistribution A/D converter suitable for CMOS implementation: **(a)** sample phase, **(b)** hold phase, and **(c)** charge-redistribution phase.

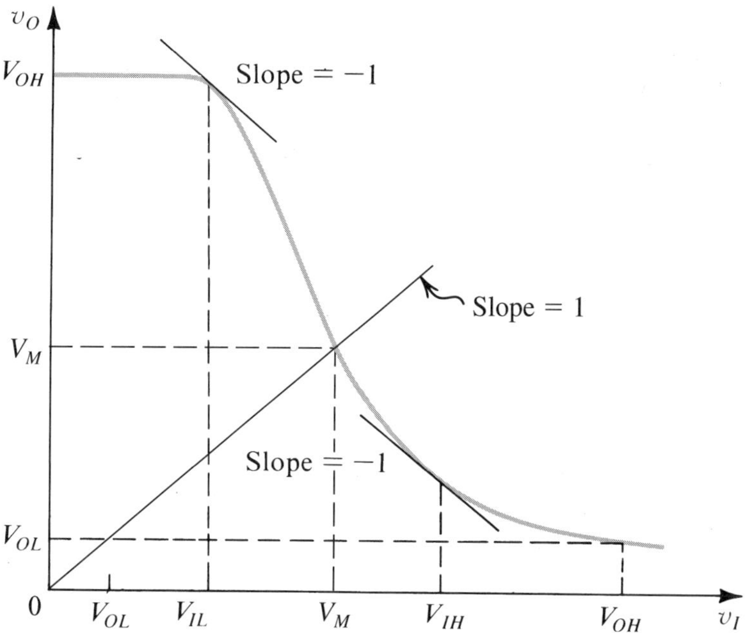

FIGURE 10.2 Typical voltage transfer characteristic (VTC) of a logic inverter, illustrating the definition of the critical points.

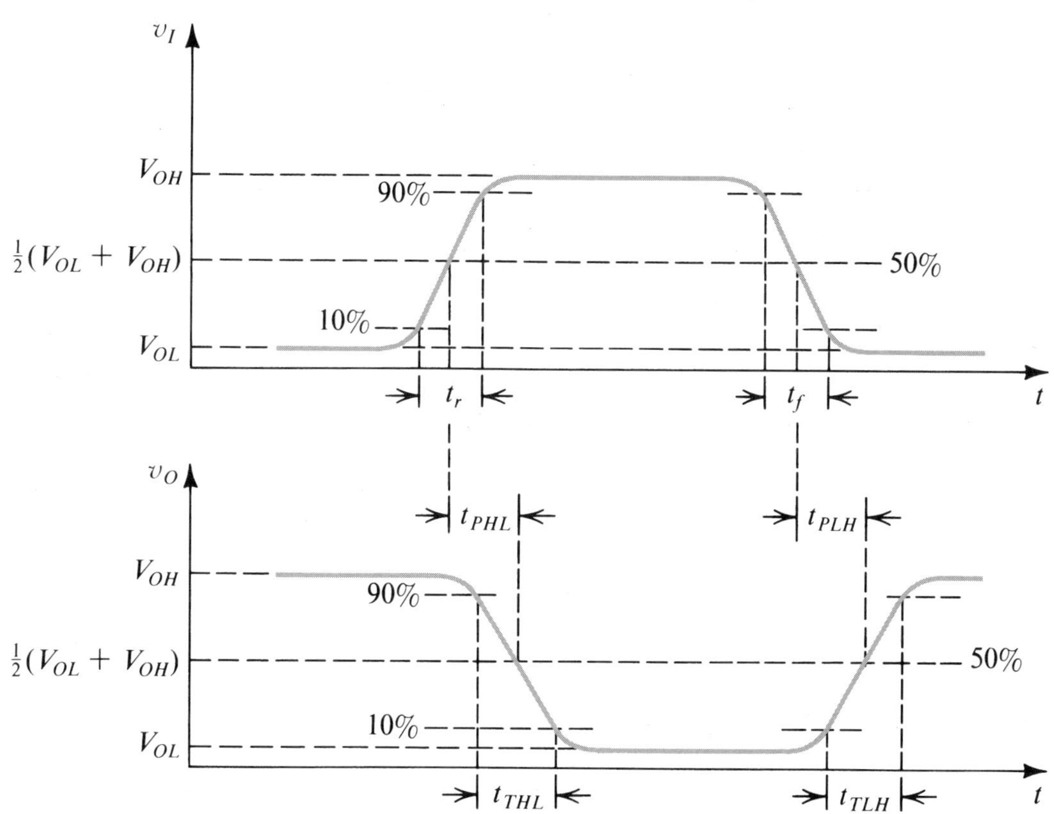

FIGURE 10.3 Definitions of propagation delays and switching times of the logic inverter.

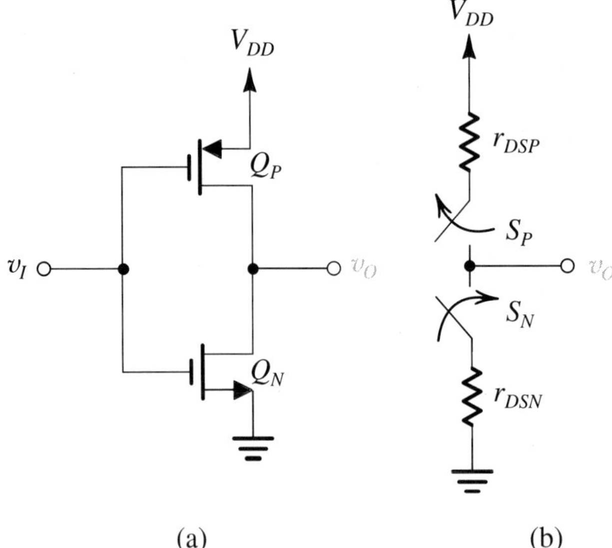

FIGURE 10.4 **(a)** The CMOS inverter and **(b)** its representation as a pair of switches operated in a complementary fashion.

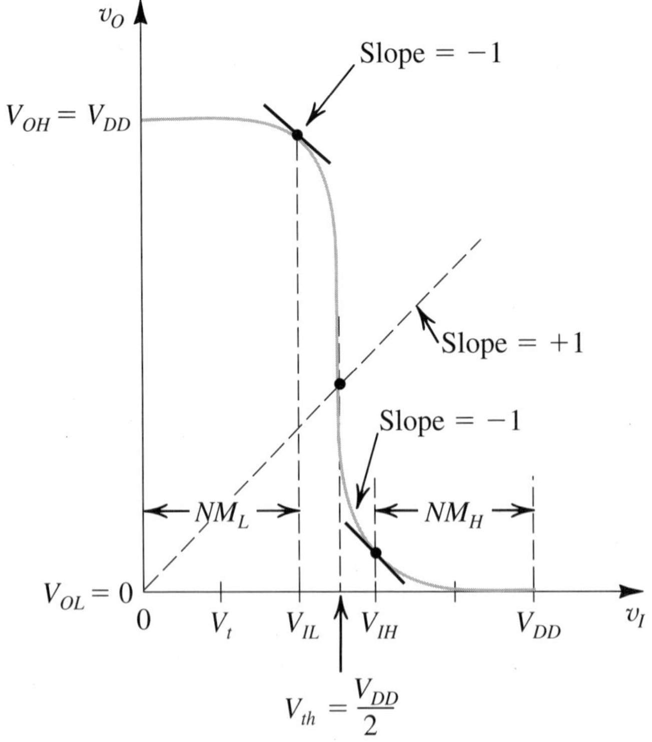

FIGURE 10.5 The voltage transfer characteristic (VTC) of the CMOS inverter when Q_N and Q_P are matched.

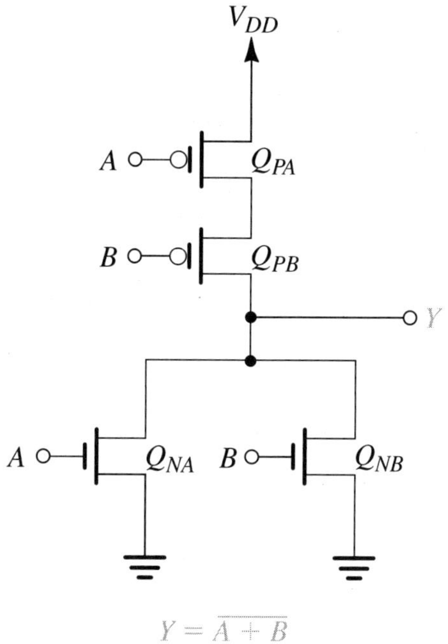

FIGURE 10.12 A two-input CMOS NOR gate.

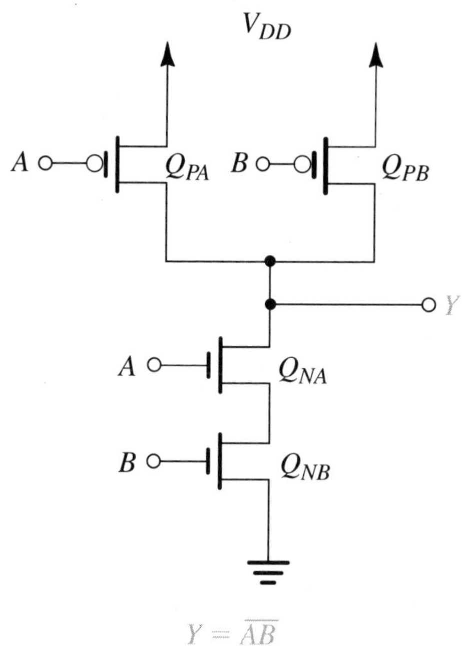

FIGURE 10.13 A two-input CMOS NAND gate.

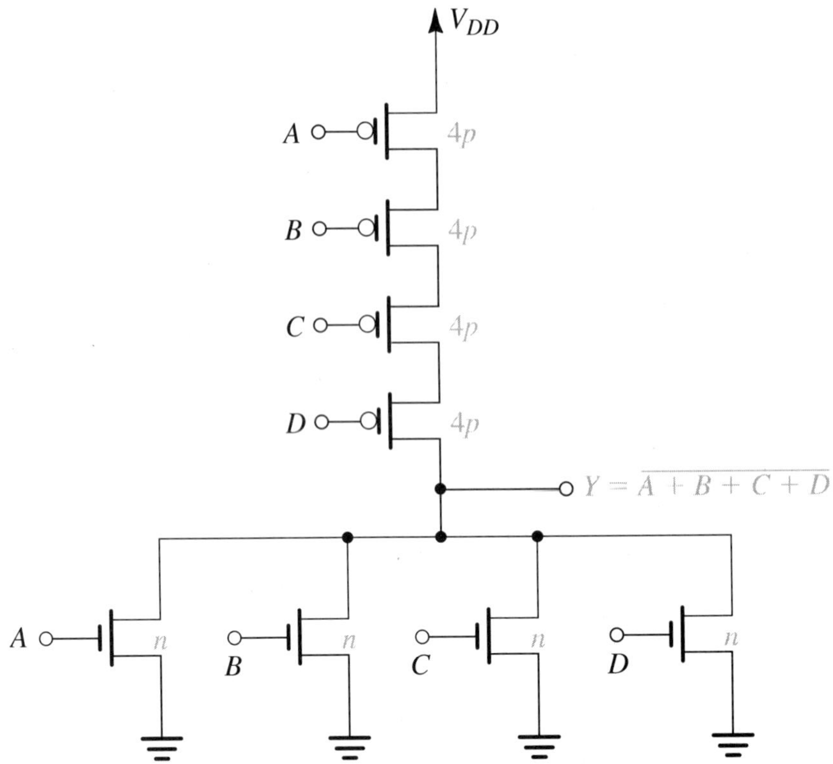

FIGURE 10.16 Proper transistor sizing for a four-input NOR gate. Note that n and p denote the (W/L) ratios of Q_N and Q_P, respectively, of the basic inverter.

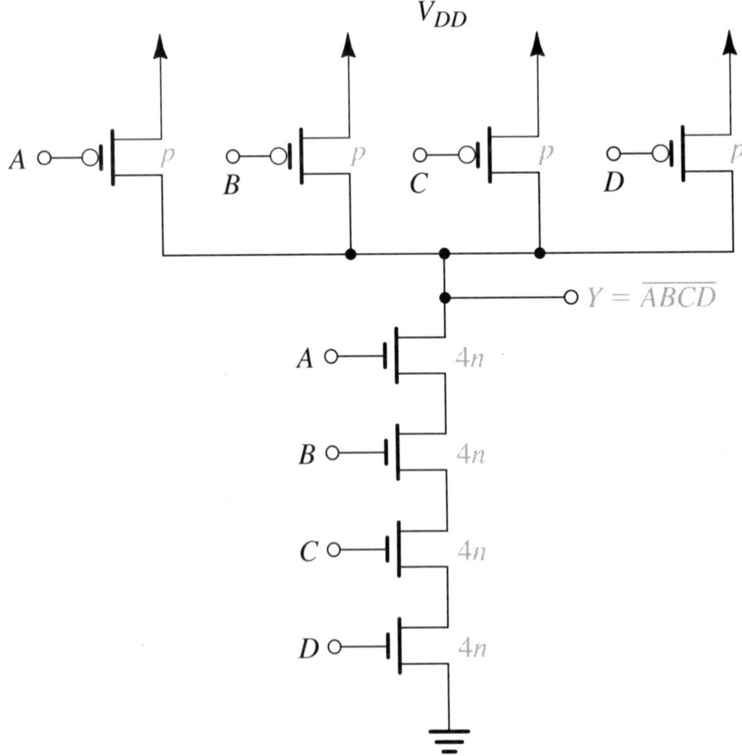

FIGURE 10.17 Proper transistor sizing for a four-input NAND gate. Note that n and p denote the (W/L) ratios of Q_N and Q_P, respectively, of the basic inverter.

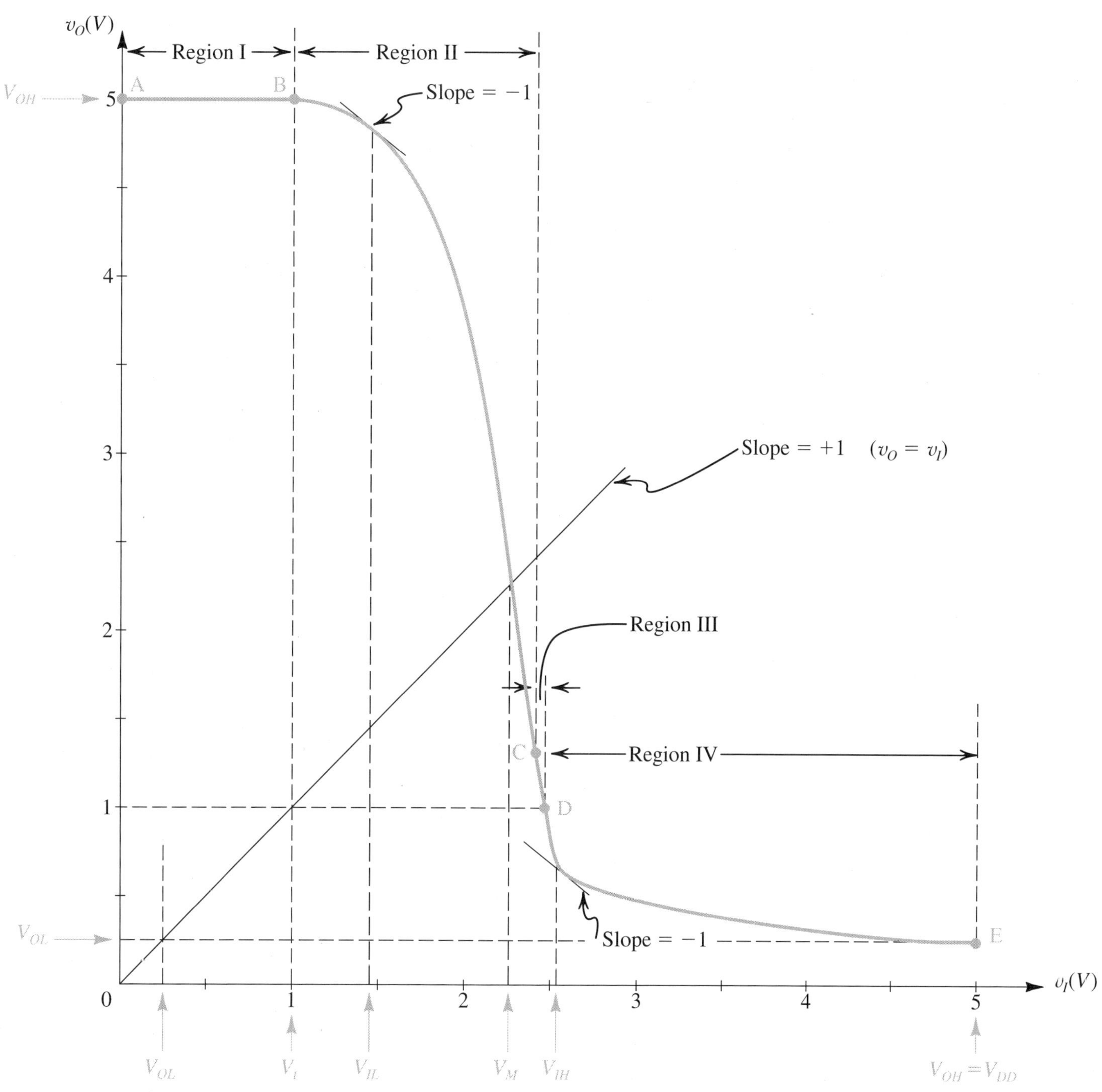

FIGURE 10.21 VTC for the pseudo-NMOS inverter. This curve is plotted for $V_{DD} = 5$ V, $V_{tn} = -V_{tp} = 1$ V, and $r = 9$.

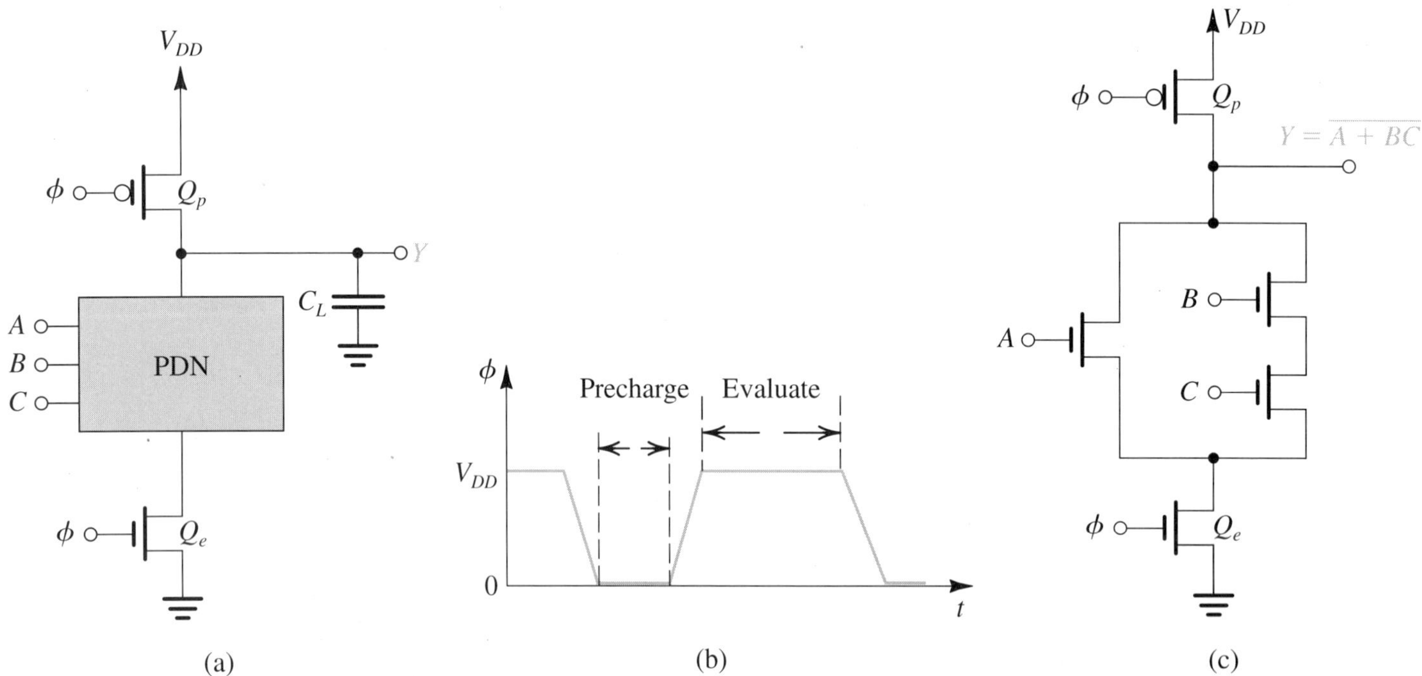

FIGURE 10.33 (a) Basic structure of dynamic-MOS logic circuits. (b) Waveform of the clock needed to operate the dynamic logic circuit. (c) An example circuit.

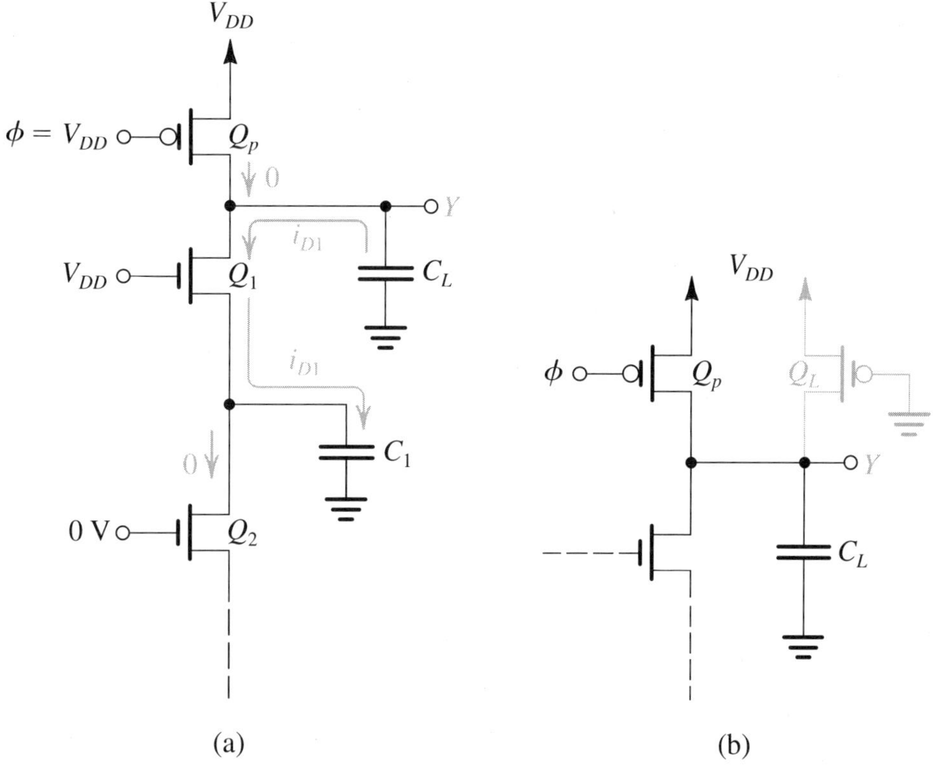

FIGURE 10.34 **(a)** Charge sharing. **(b)** Adding a permanently turned-on transistor Q_L solves the charge-sharing problem at the expense of static power dissipation.

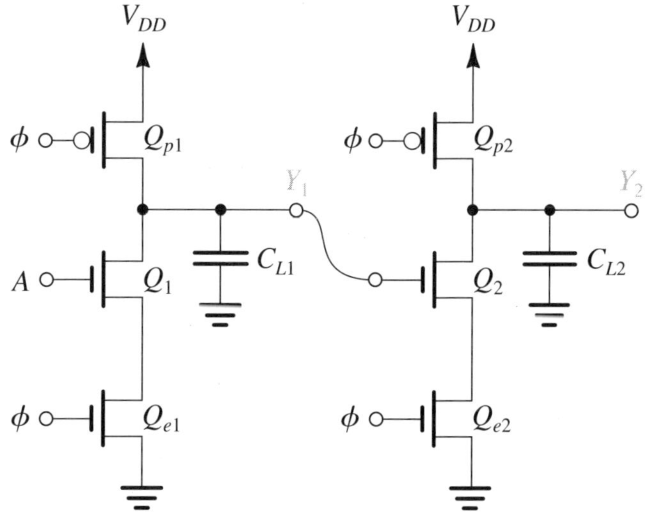

FIGURE 10.35 Two single-input dynamic logic gates connected in cascade. With the input A high, during the evaluation phase C_{L2} will partially discharge and the output at Y_2 will fall lower than V_{DD}, which can cause logic malfunction.

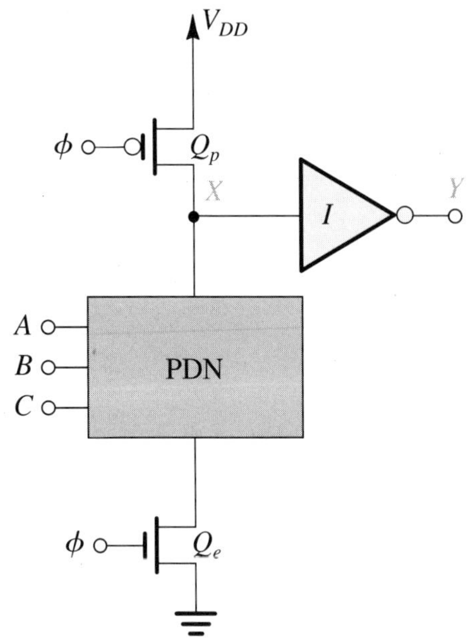

FIGURE 10.36 The Domino CMOS logic gate. The circuit consists of a dynamic-MOS logic gate with a static-CMOS inverter connected to the output. During evaluation, Y either will remain low (at 0 V) or will make one 0-to-1 transition (to V_{DD}).

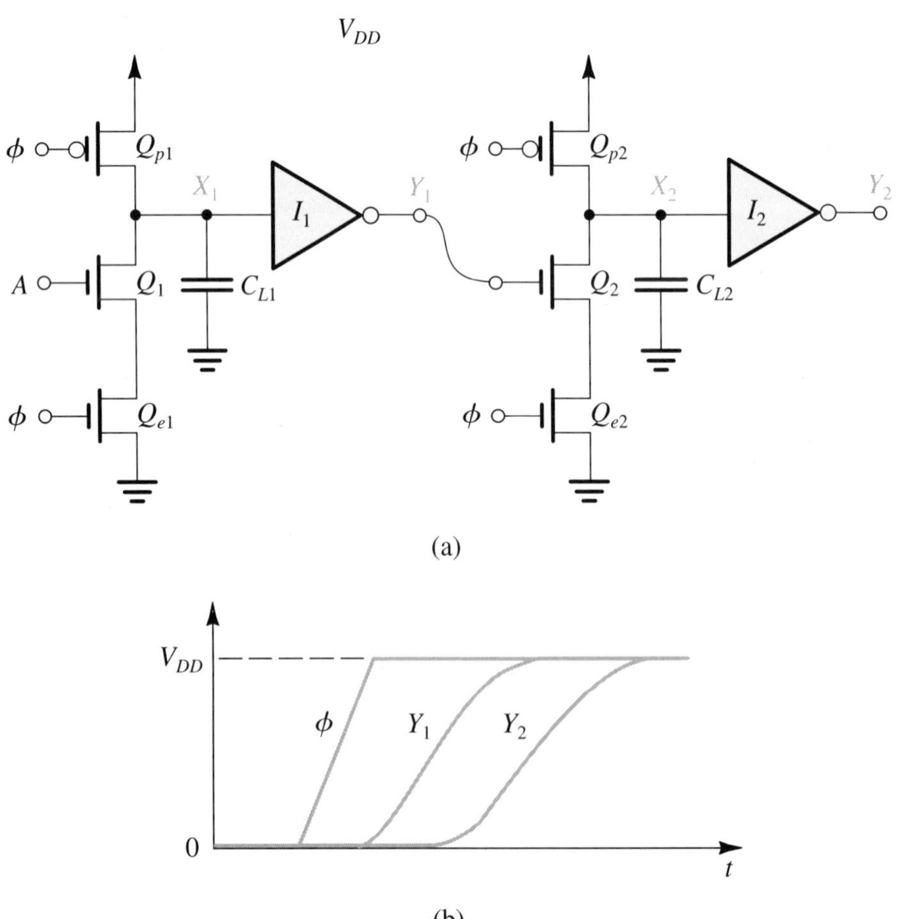

FIGURE 10.37 **(a)** Two single-input Domino CMOS logic gates connected in cascade. **(b)** Waveforms during the evaluation phase.

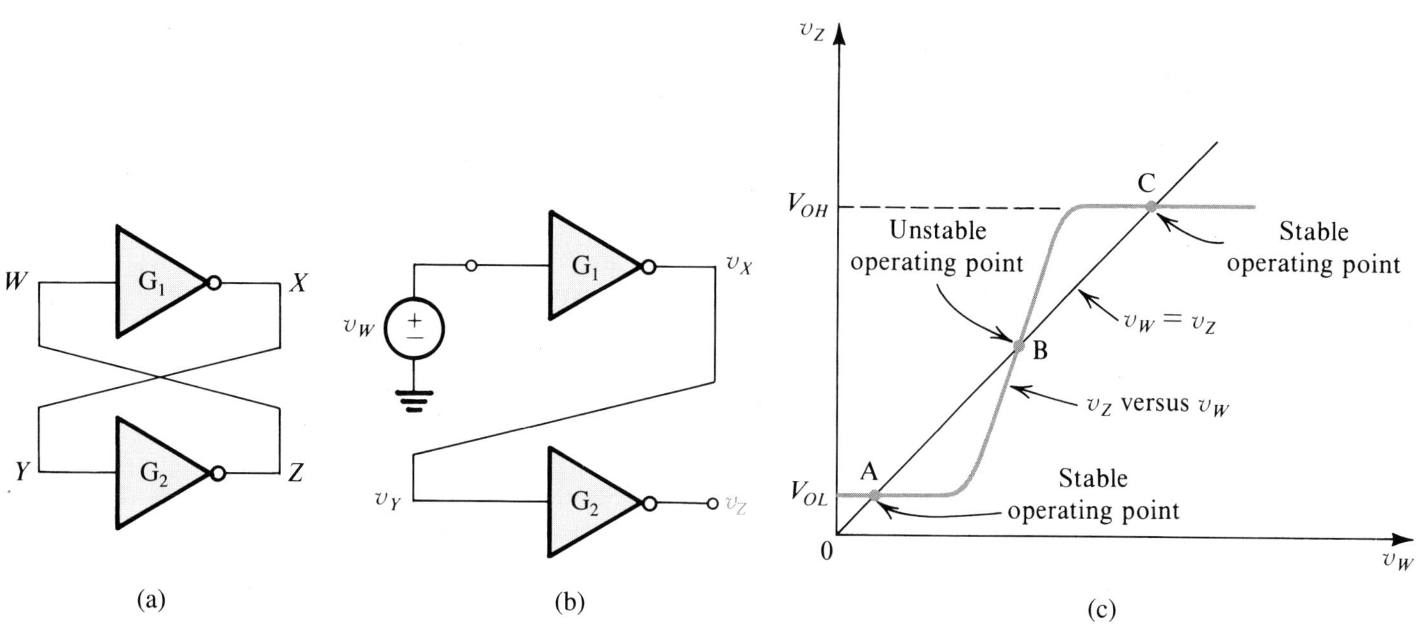

FIGURE 11.1 **(a)** Basic latch. **(b)** The latch with the feedback loop opened. **(c)** Determining the operating point(s) of the latch.

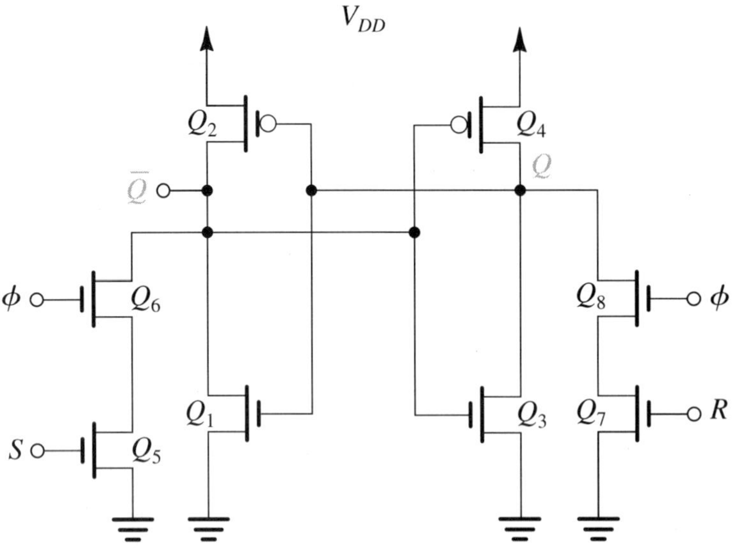

FIGURE 11.3 CMOS implementation of a clocked SR flip-flop. The clock signal is denoted by ϕ.

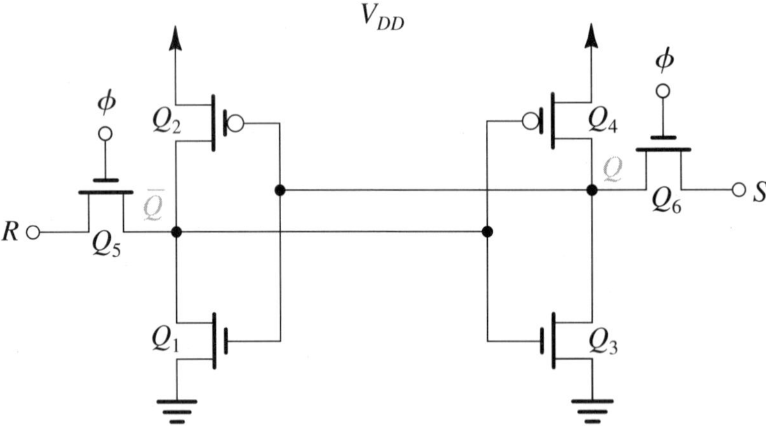

FIGURE 11.5 A simpler CMOS implementation of the clocked SR flip-flop. This circuit is popular as the basic cell in the design of static random-access memory (SRAM) chips.

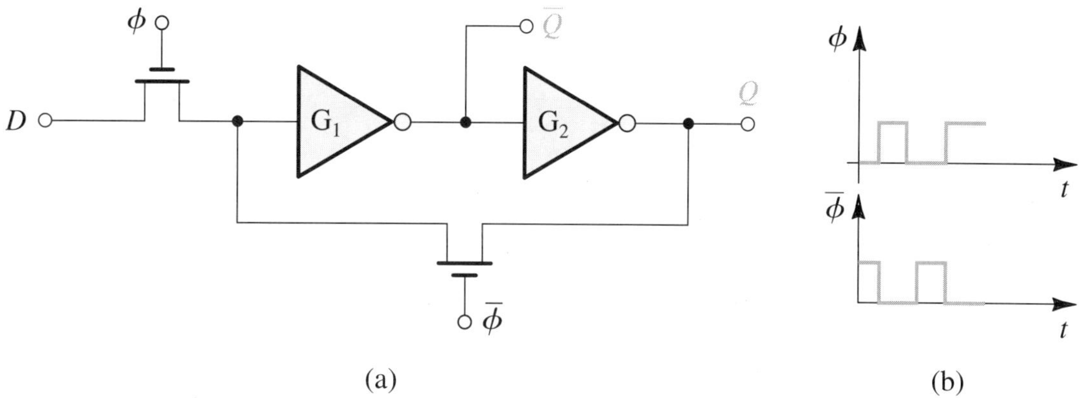

FIGURE 11.7 A simple implementation of the D flip-flop. The circuit in **(a)** utilizes the two-phase nonoverlapping clock whose waveforms are shown in **(b).**

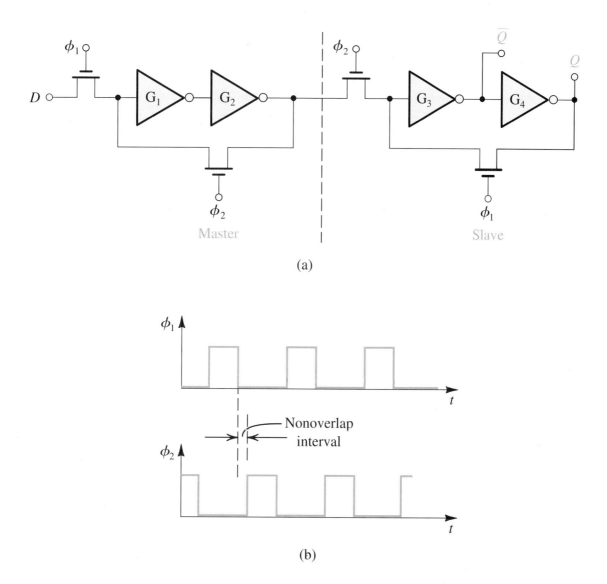

FIGURE 11.8 **(a)** A master–slave D flip flop. The switches can be, and usually are, implemented with CMOS transmission gates. **(b)** Waveforms of the two-phase nonoverlapping clock required.

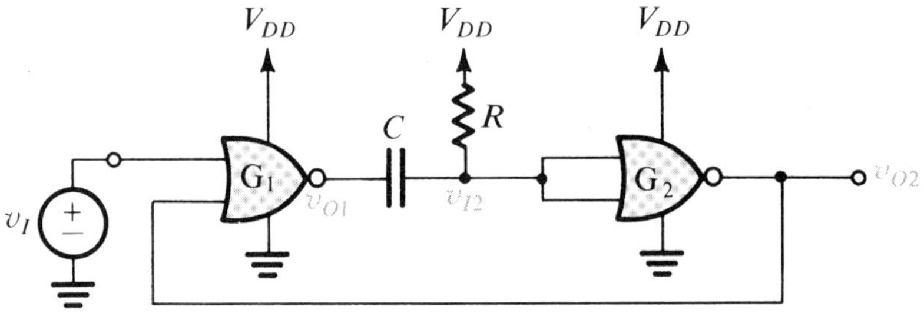

FIGURE 11.10 A monostable circuit using CMOS NOR gates. Signal source v_I supplies positive trigger pulses.

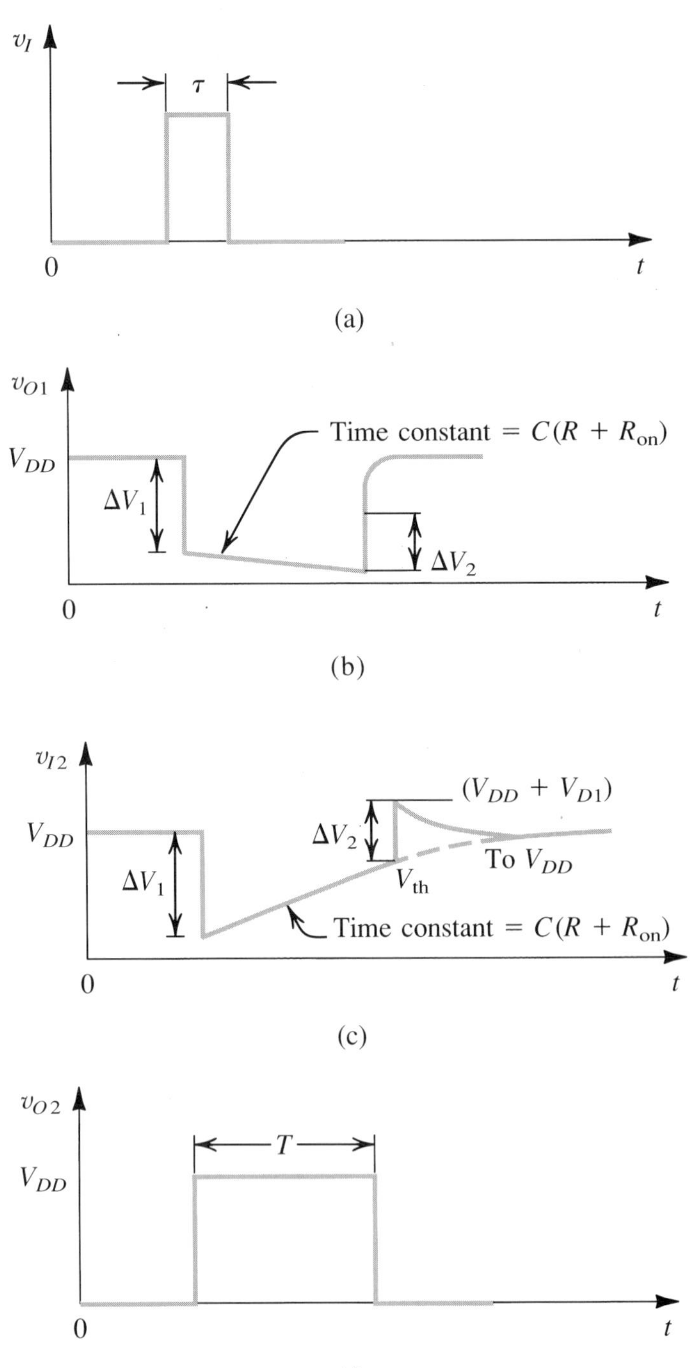

FIGURE 11.13 Timing diagram for the monostable circuit in Fig. 11.10.

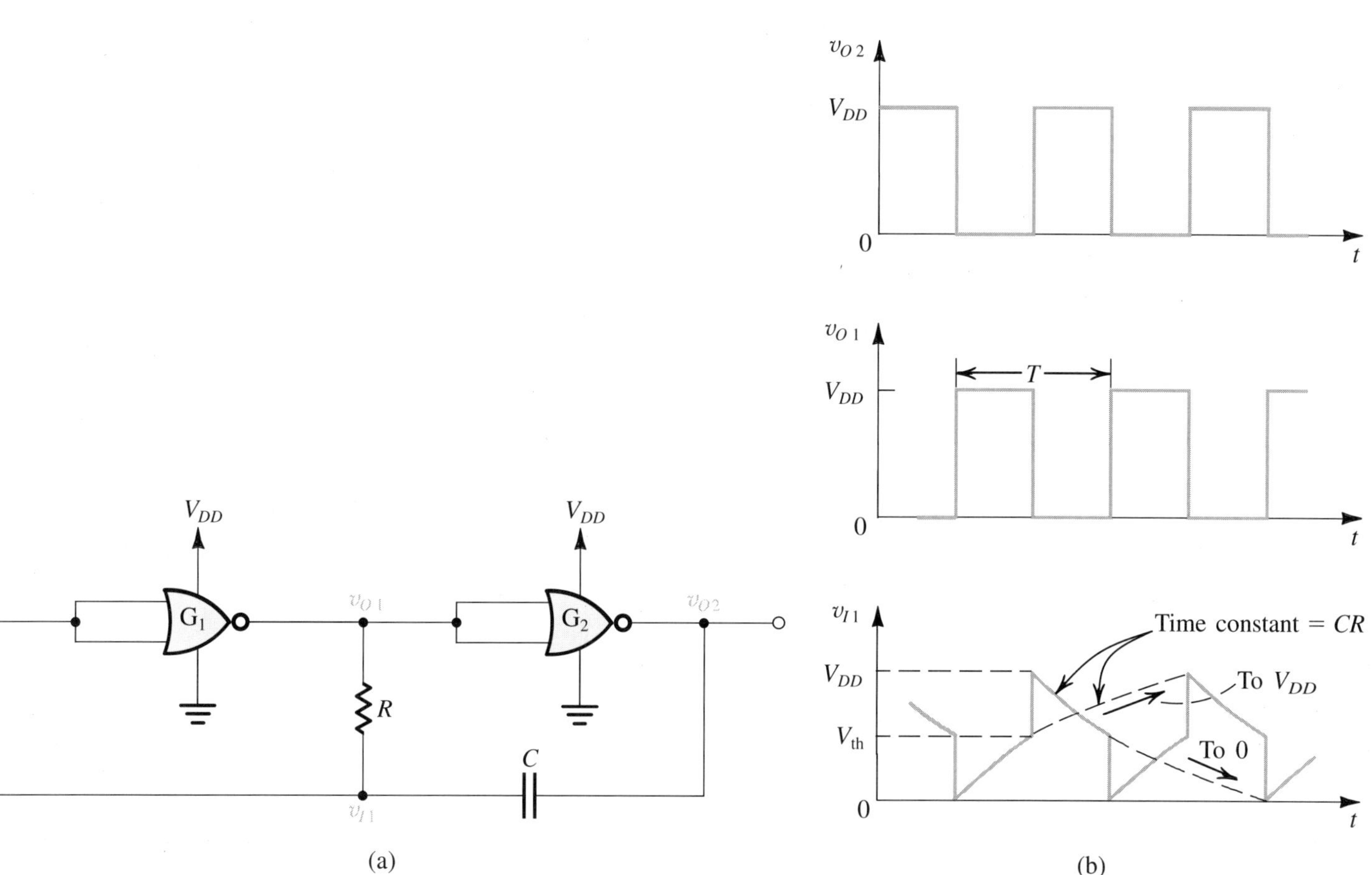

FIGURE 11.15 **(a)** A simple astable multivibrator circuit using CMOS gates. **(b)** Waveforms for the astable circuit in (a). The diodes at the gate input are assumed to be ideal and thus to limit the voltage v_{I1} to 0 and V_{DD}.

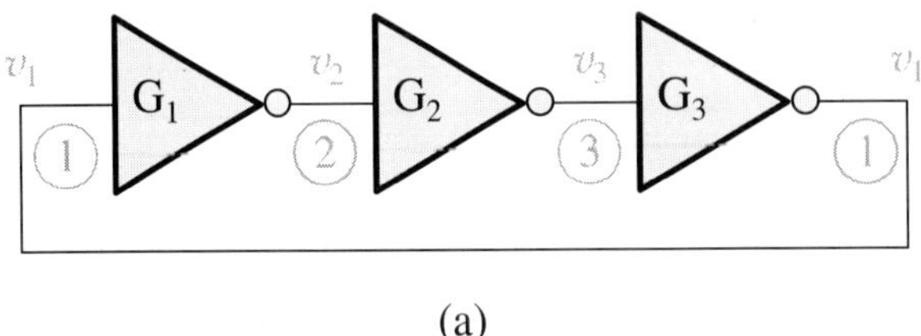

(a)

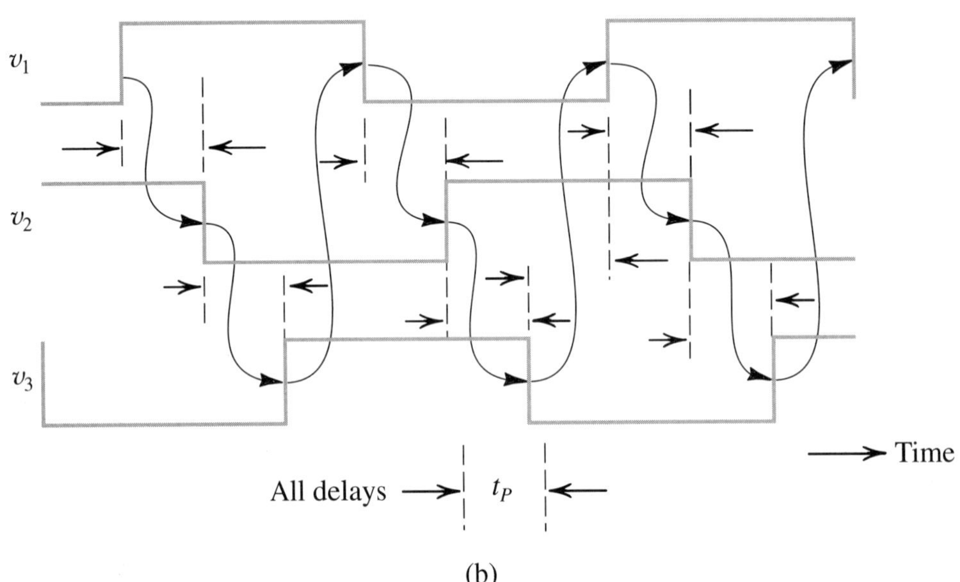

(b)

FIGURE 11.16 **(a)** A ring oscillator formed by connecting three inverters in cascade. (Normally at least five inverters are used.) **(b)** The resulting waveform. Observe that the circuit oscillates with frequency $1/6t_P$.

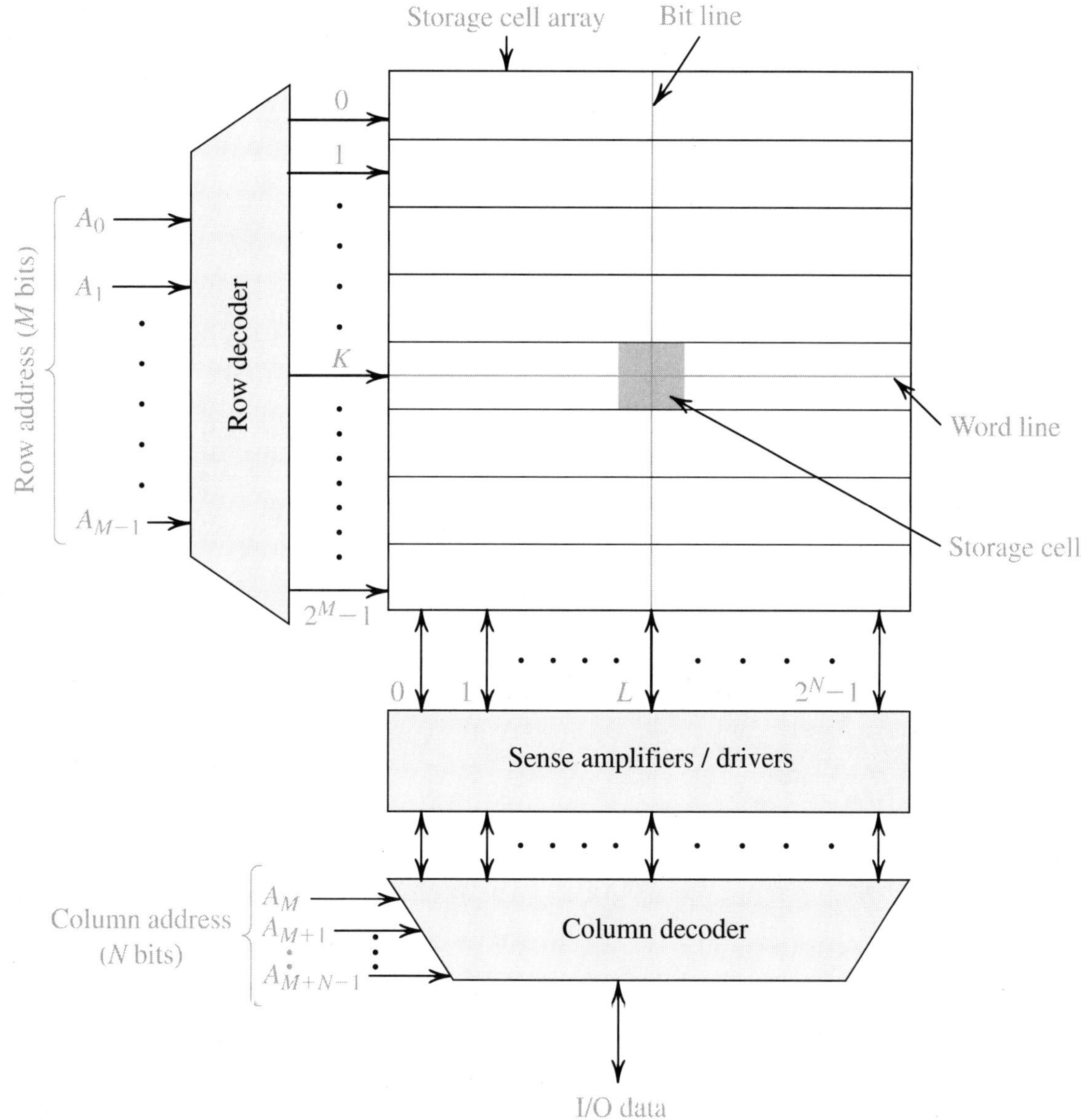

FIGURE 11.17 A 2^{M+N}-bit memory chip organized as an array of 2^M rows × 2^N columns.

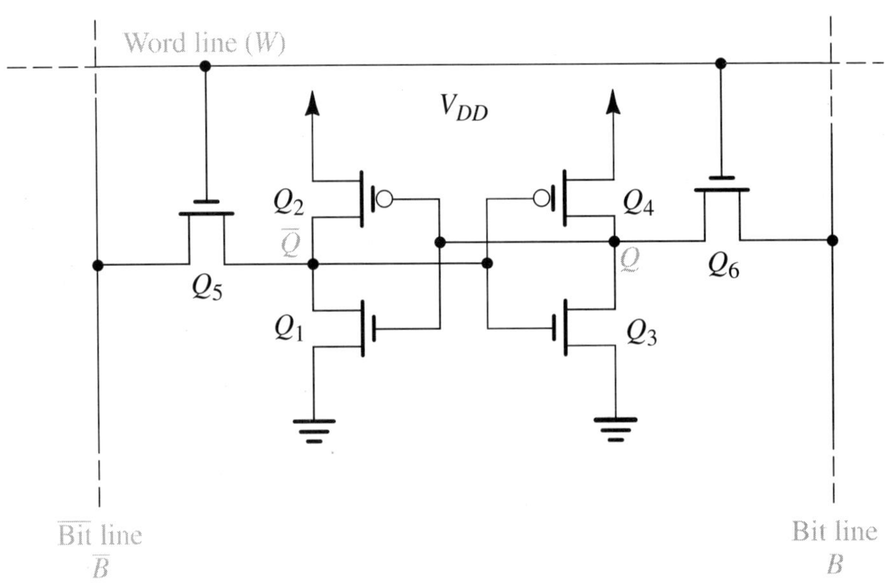

FIGURE 11.18 A CMOS SRAM memory cell.

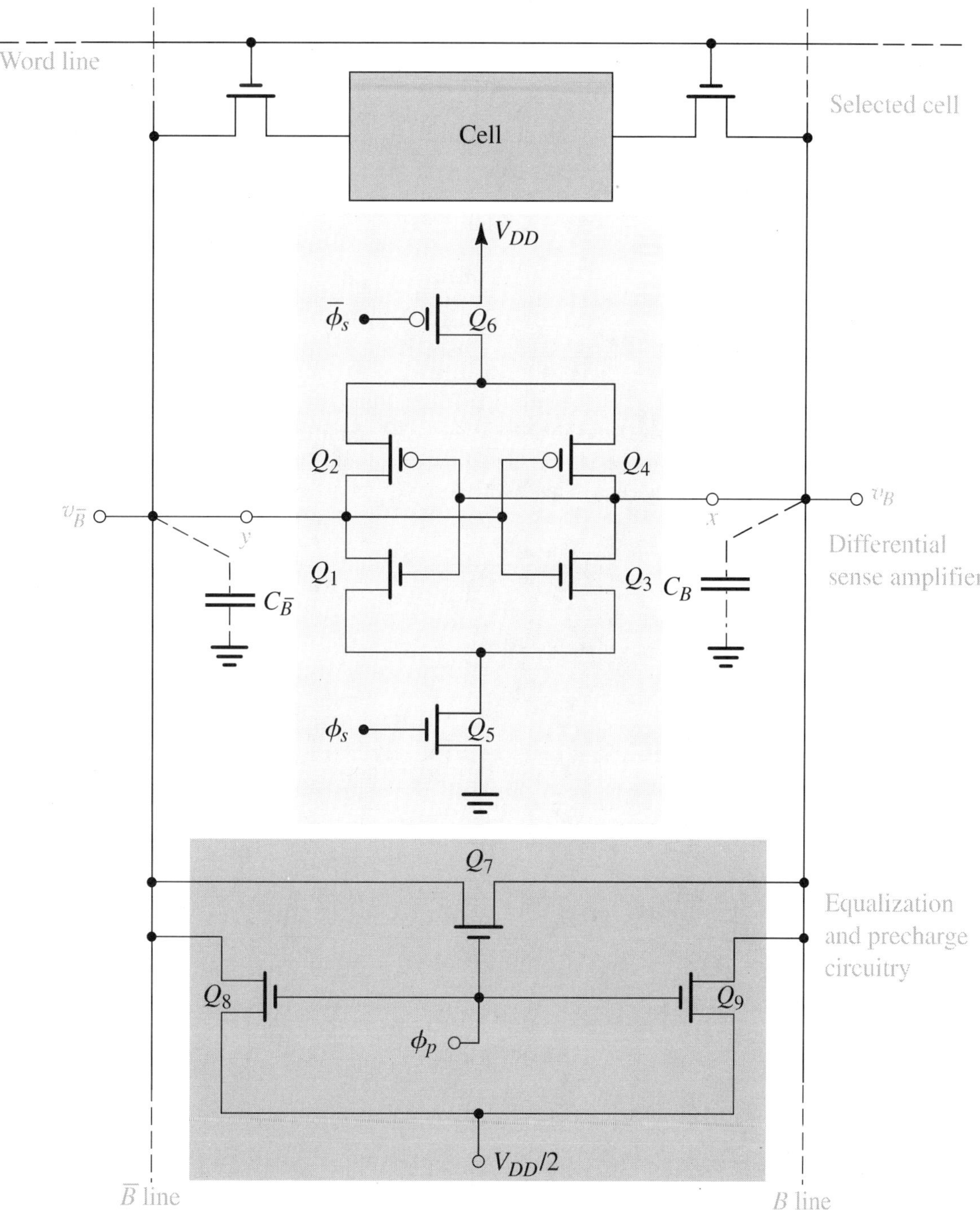

FIGURE 11.23 A differential sense amplifier connected to the bit lines of a particular column. This arrangement can be used directly for SRAMs (which utilize both the B and $\overline{B}$ lines). DRAMs can be turned into differential circuits by using the "dummy cell" arrangement shown in Fig. 11.25.

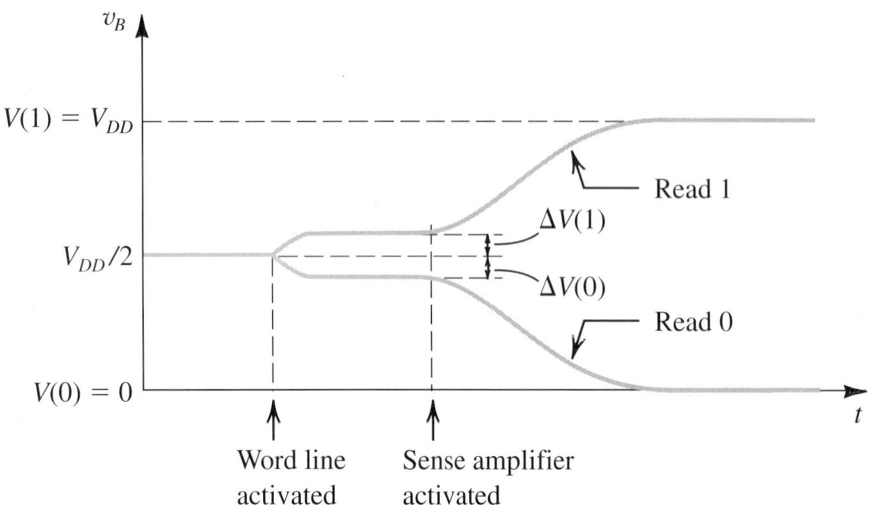

FIGURE 11.24 Waveforms of v_B before and after the activation of the sense amplifier. In a read-1 operation, the sense amplifier causes the initial small increment $\Delta V(1)$ to grow exponentially to V_{DD}. In a read-0 operation, the negative $\Delta V(0)$ grows to 0. Complementary signal waveforms develop on the $\overline{B}$ line.

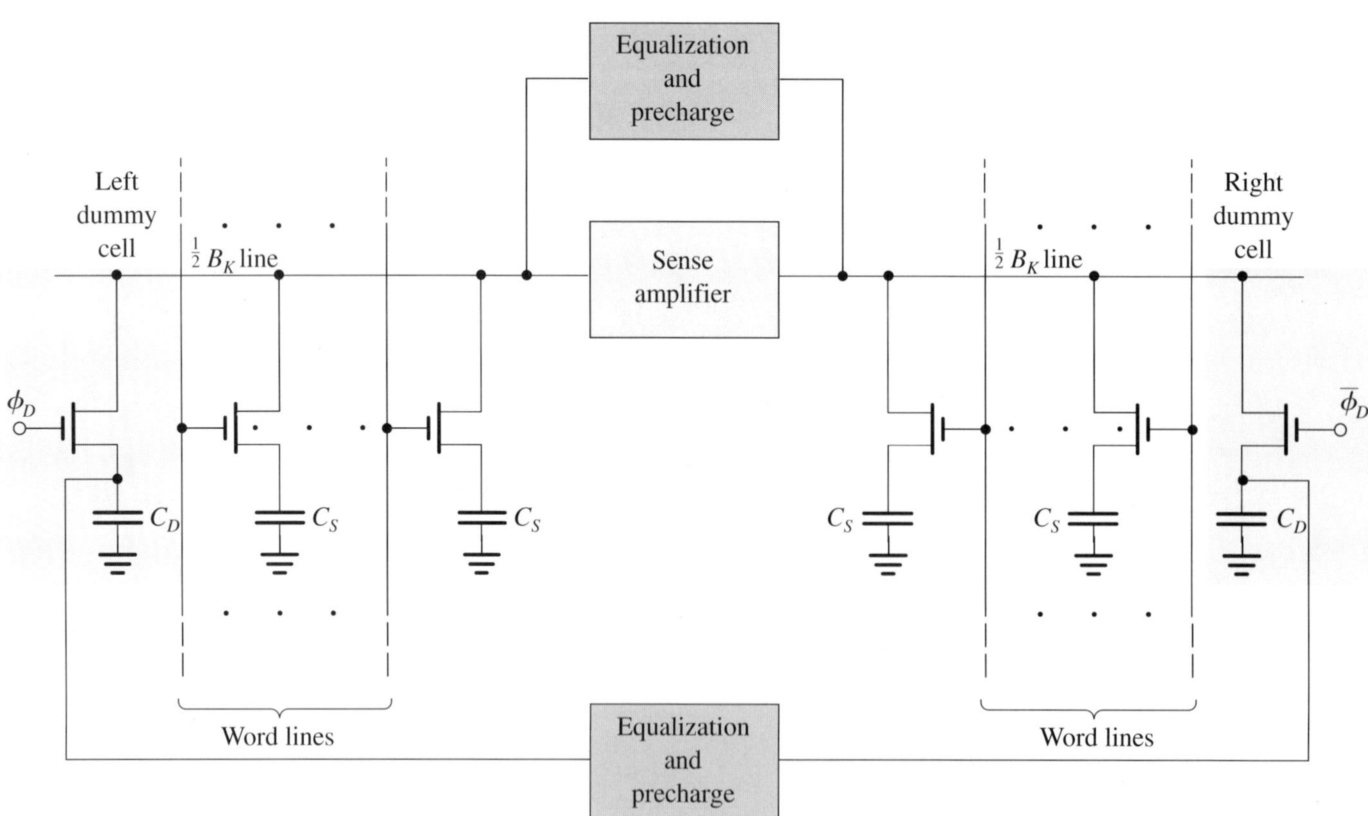

FIGURE 11.25 An arrangement for obtaining differential operation from the single-ended DRAM cell. Note the dummy cells at the far right and far left.

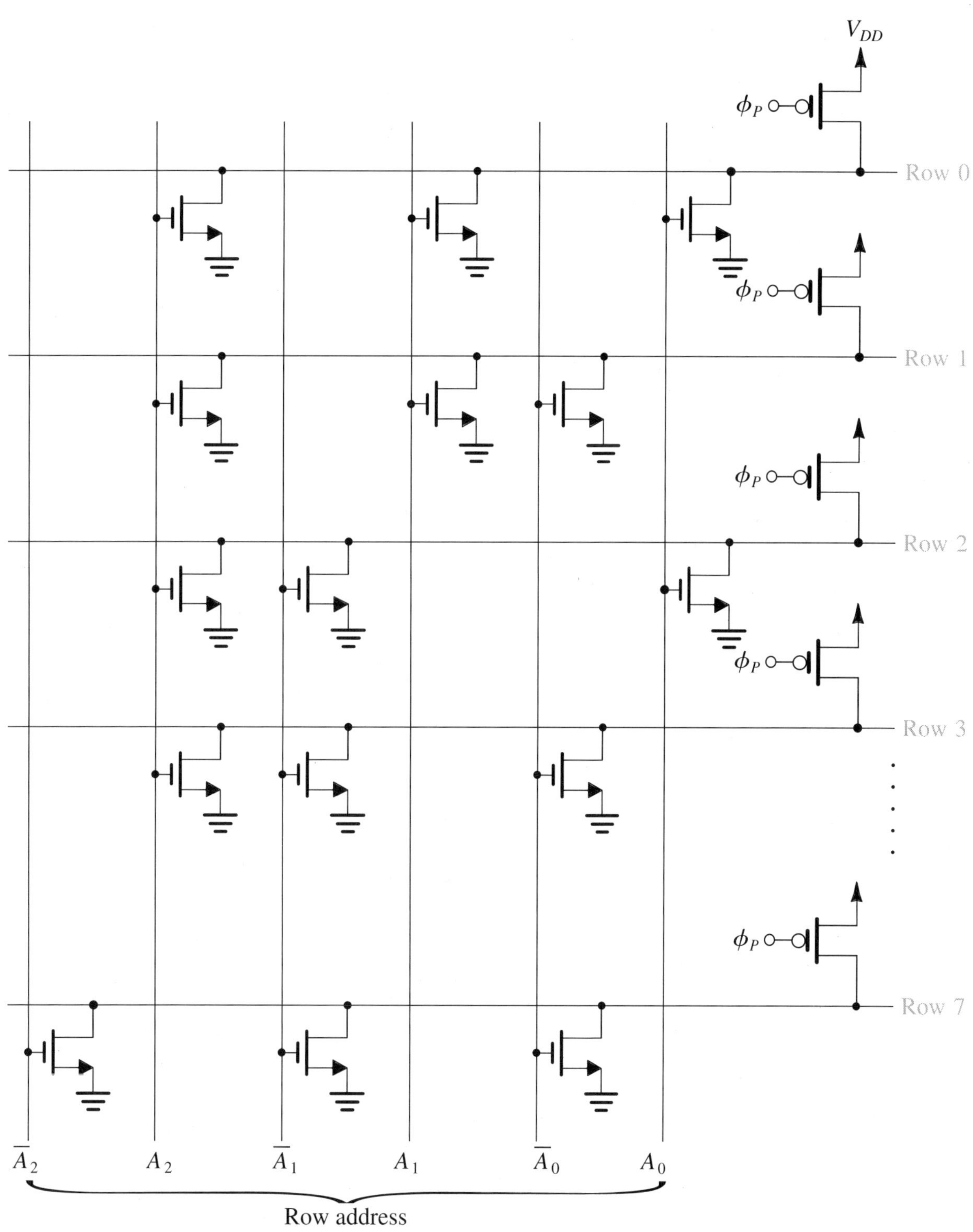

FIGURE 11.26 A NOR address decoder in array form. One out of eight lines (row lines) is selected using a 3-bit address.

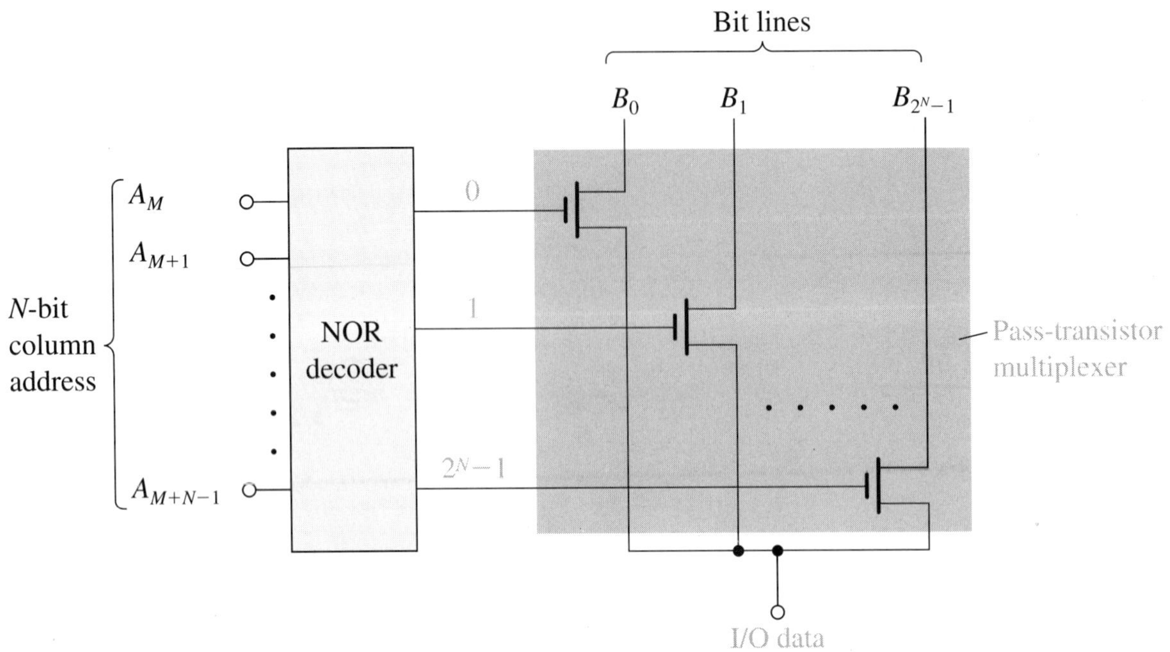

FIGURE 11.27 A column decoder realized by a combination of a NOR decoder and a pass-transistor multiplexer.

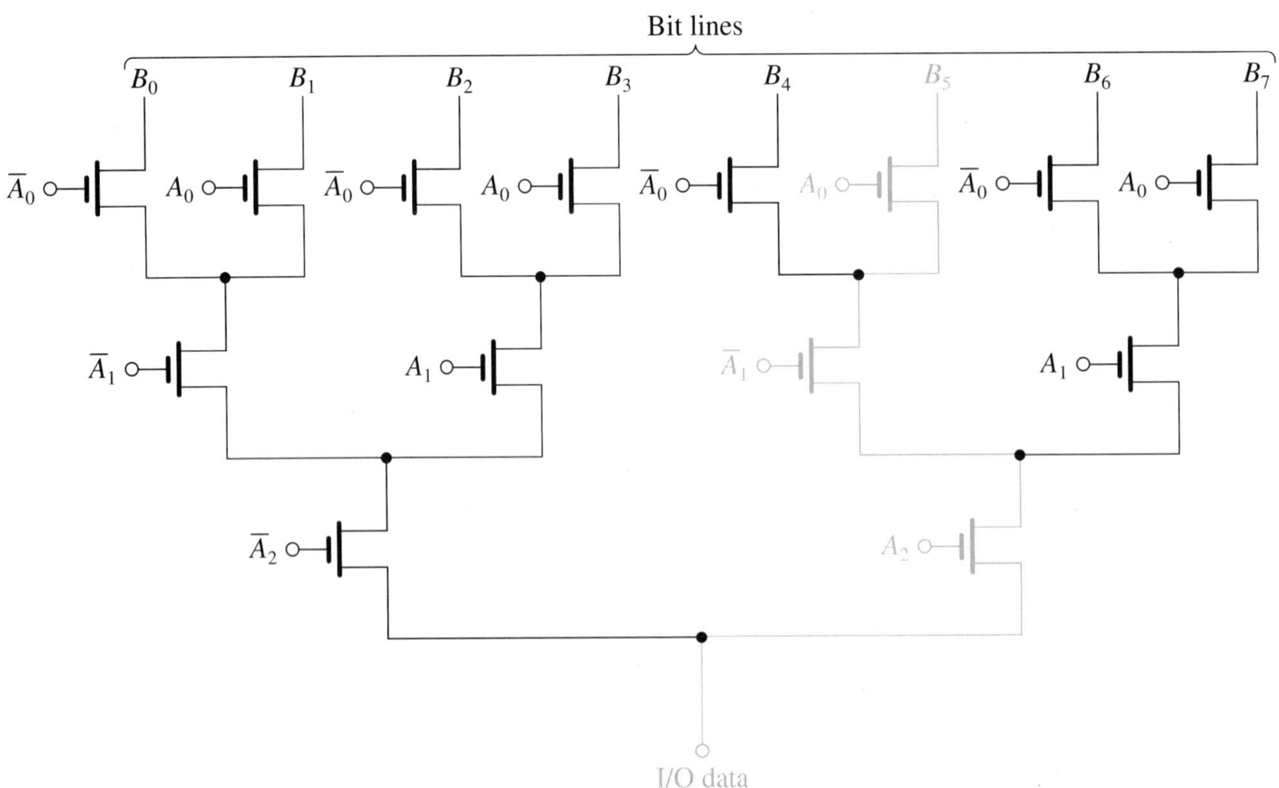

FIGURE 11.28 A tree column decoder. Note that the colored path shows the transistors that are conducting when $A_0 = 1$, $A_1 = 0$, and $A_2 = 1$, the address that results in connecting B_5 to the data line.

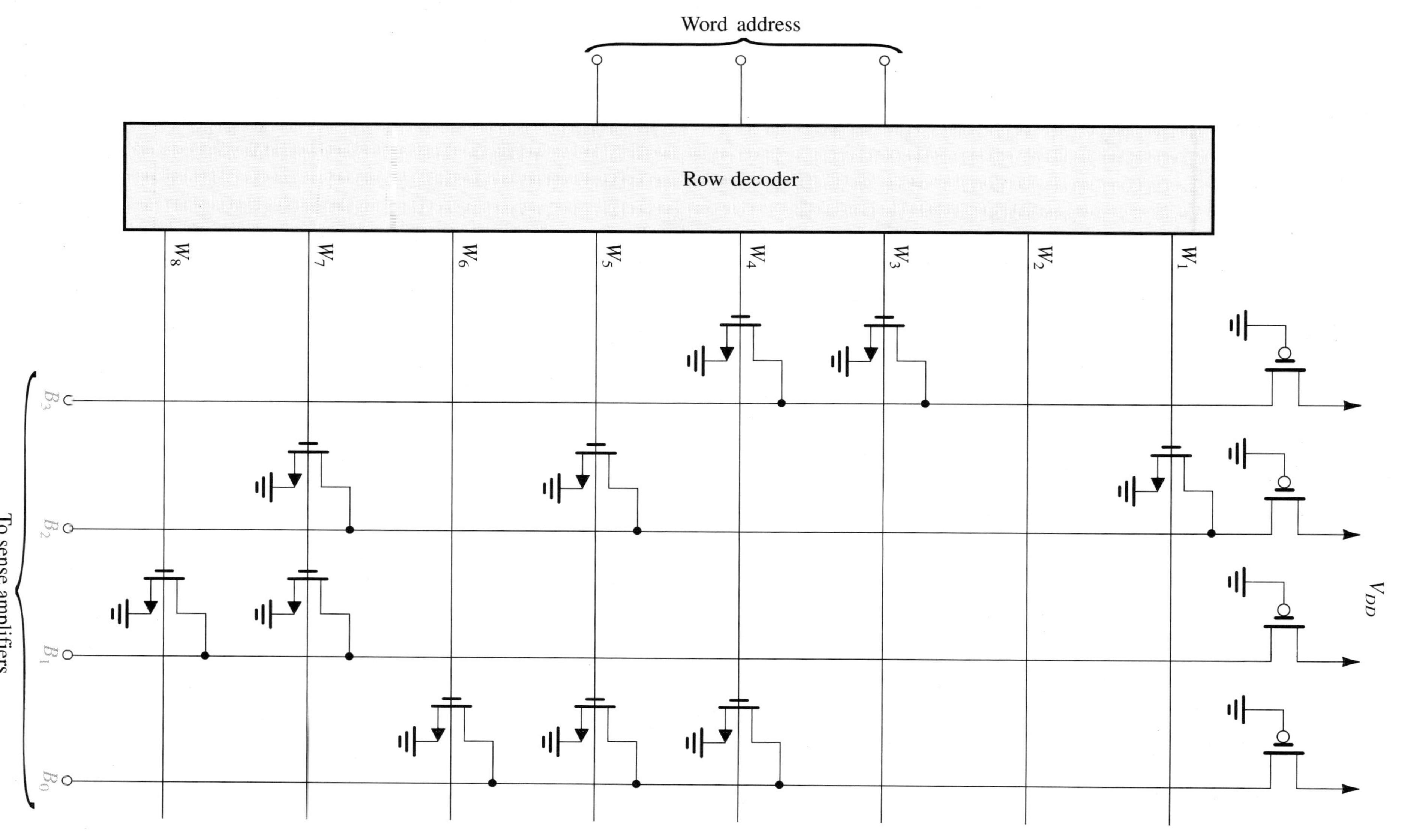

FIGURE 11.29 A simple MOS ROM organized as 8 words × 4 bits.

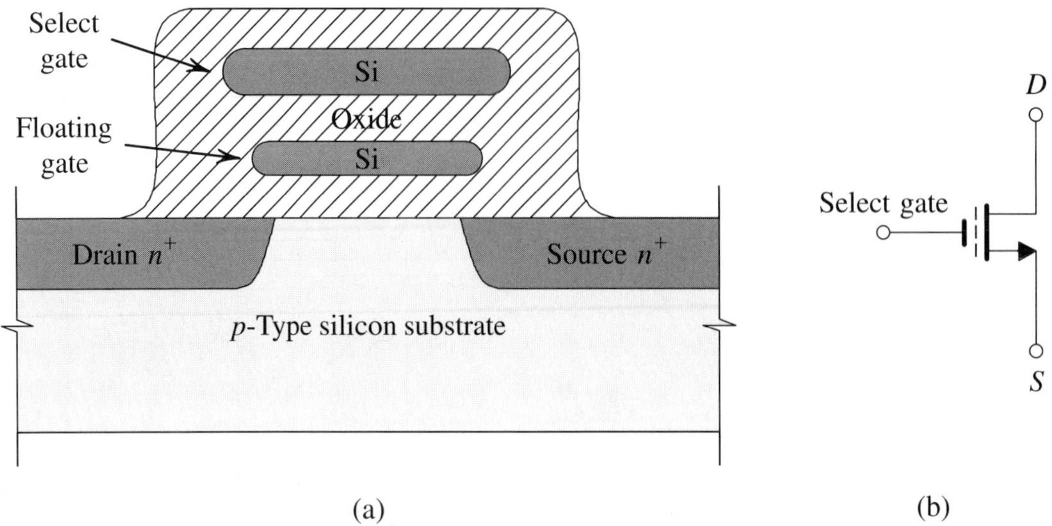

FIGURE 11.30 **(a)** Cross section and **(b)** circuit symbol of the floating-gate transistor used as an EPROM cell.

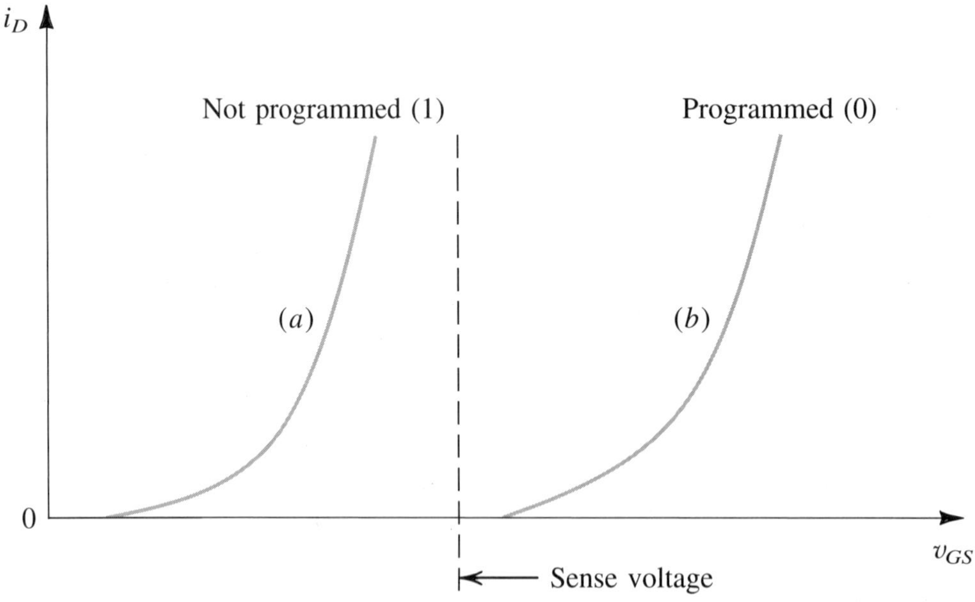

FIGURE 11.31 Illustrating the shift in the i_D–v_{GS} characteristic of a floating-gate transistor as a result of programming.

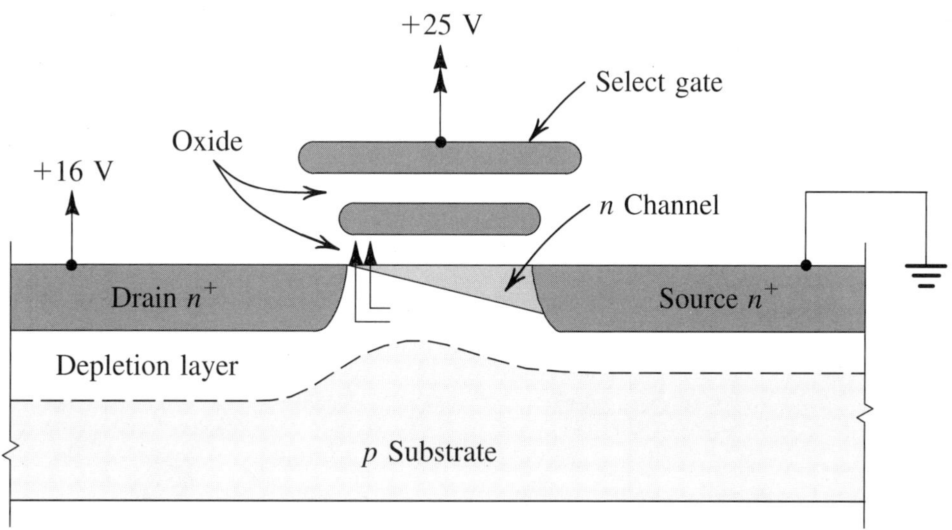

FIGURE 11.32 The floating-gate transistor during programming.

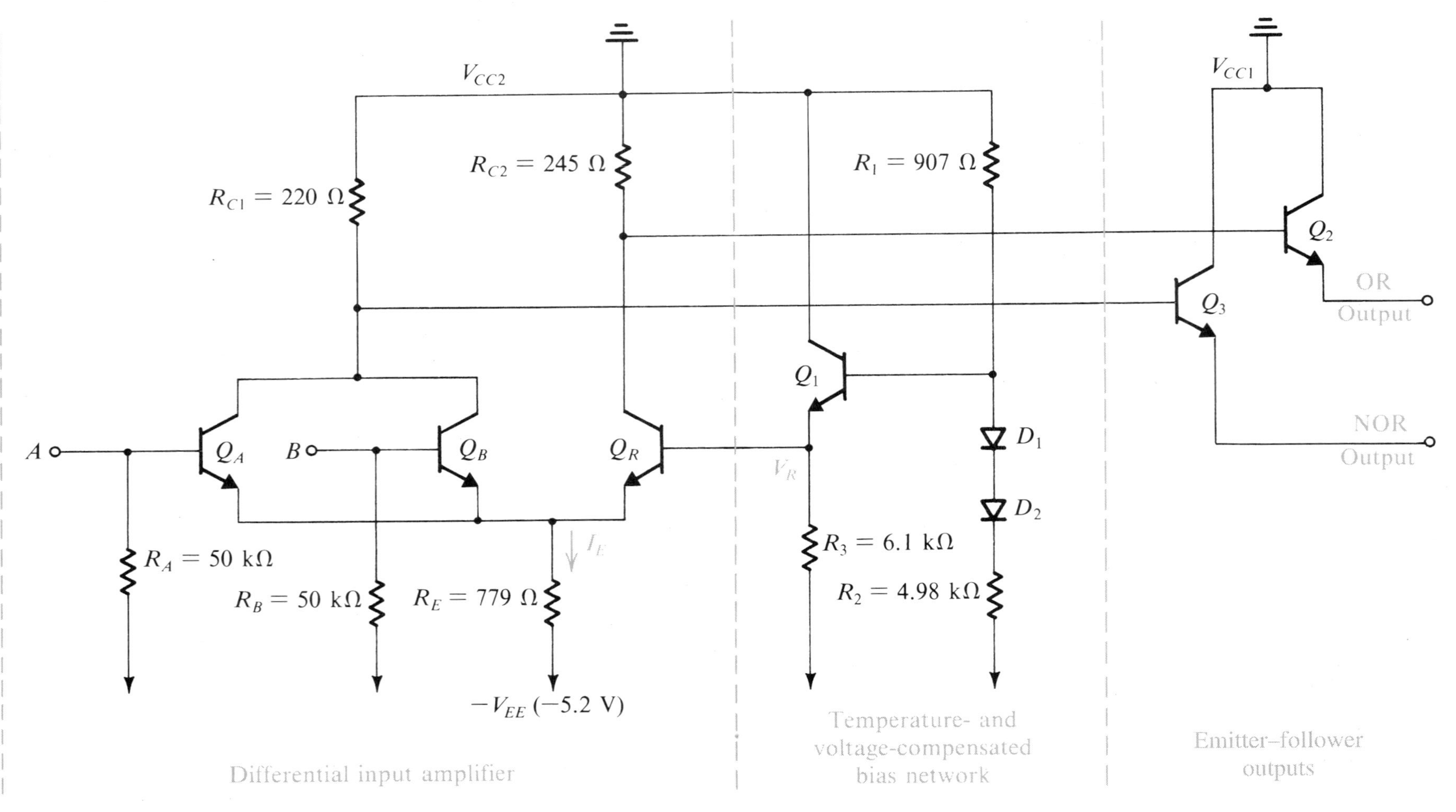

FIGURE 11.34 Basic circuit of the ECL 10K logic-gate family.

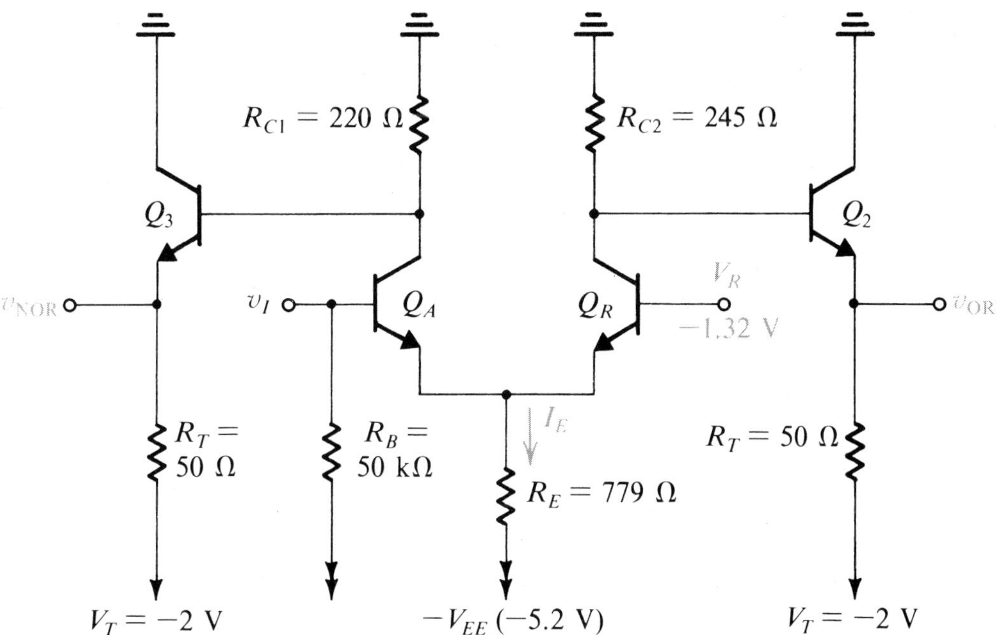

FIGURE 11.36 Simplified version of the ECL gate for the purpose of finding transfer characteristics.

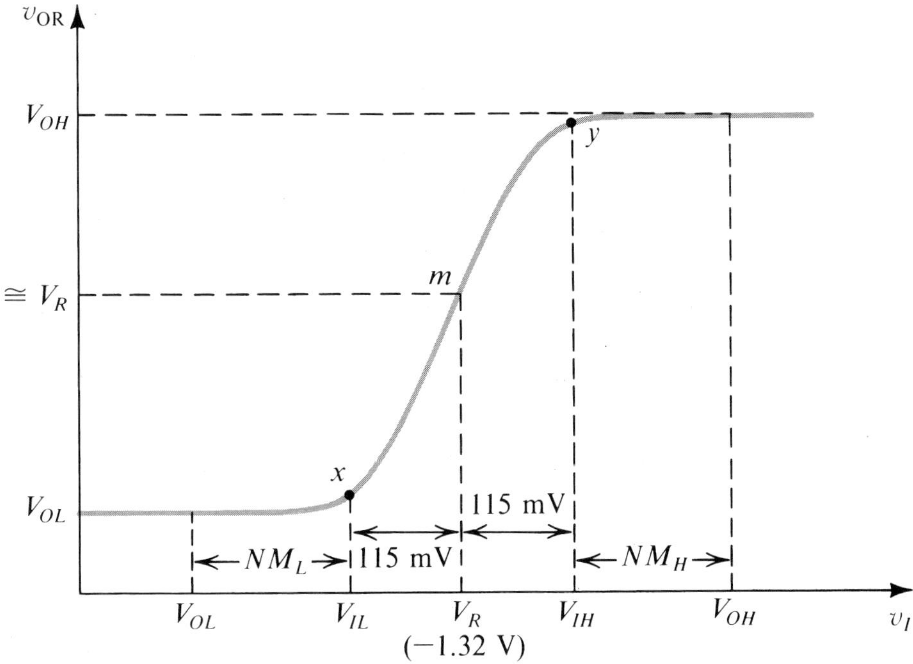

FIGURE 11.37 The OR transfer characteristic v_{OR} versus v_I, for the circuit in Fig. 11.36.

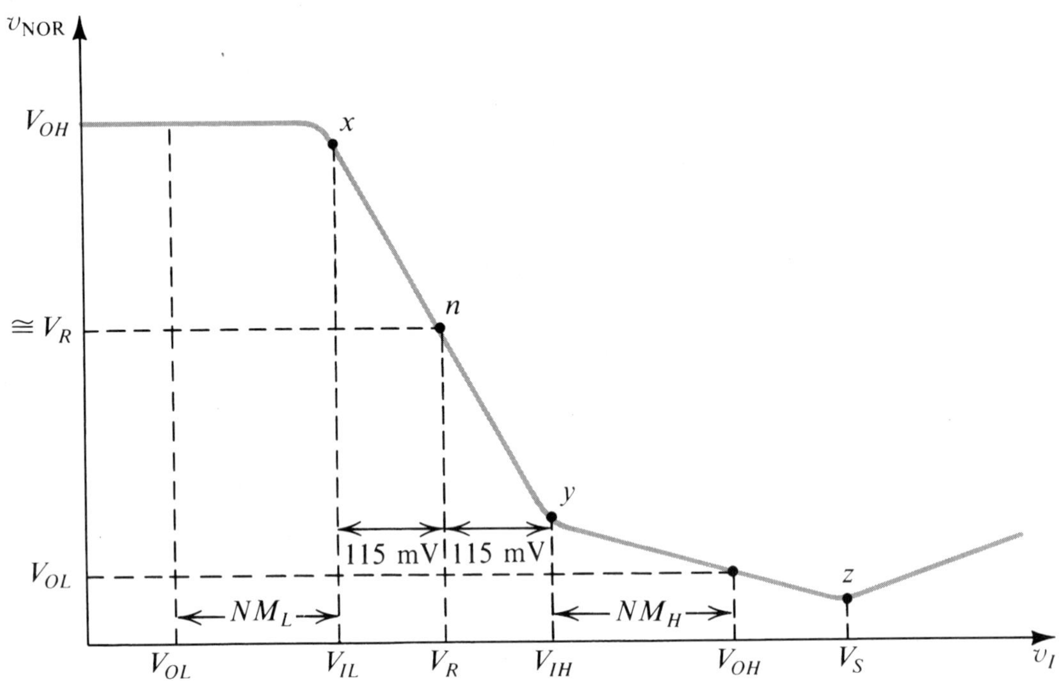

FIGURE 11.39 The NOR transfer characteristic, v_{NOR} versus v_I, for the circuit in Fig. 11.36.

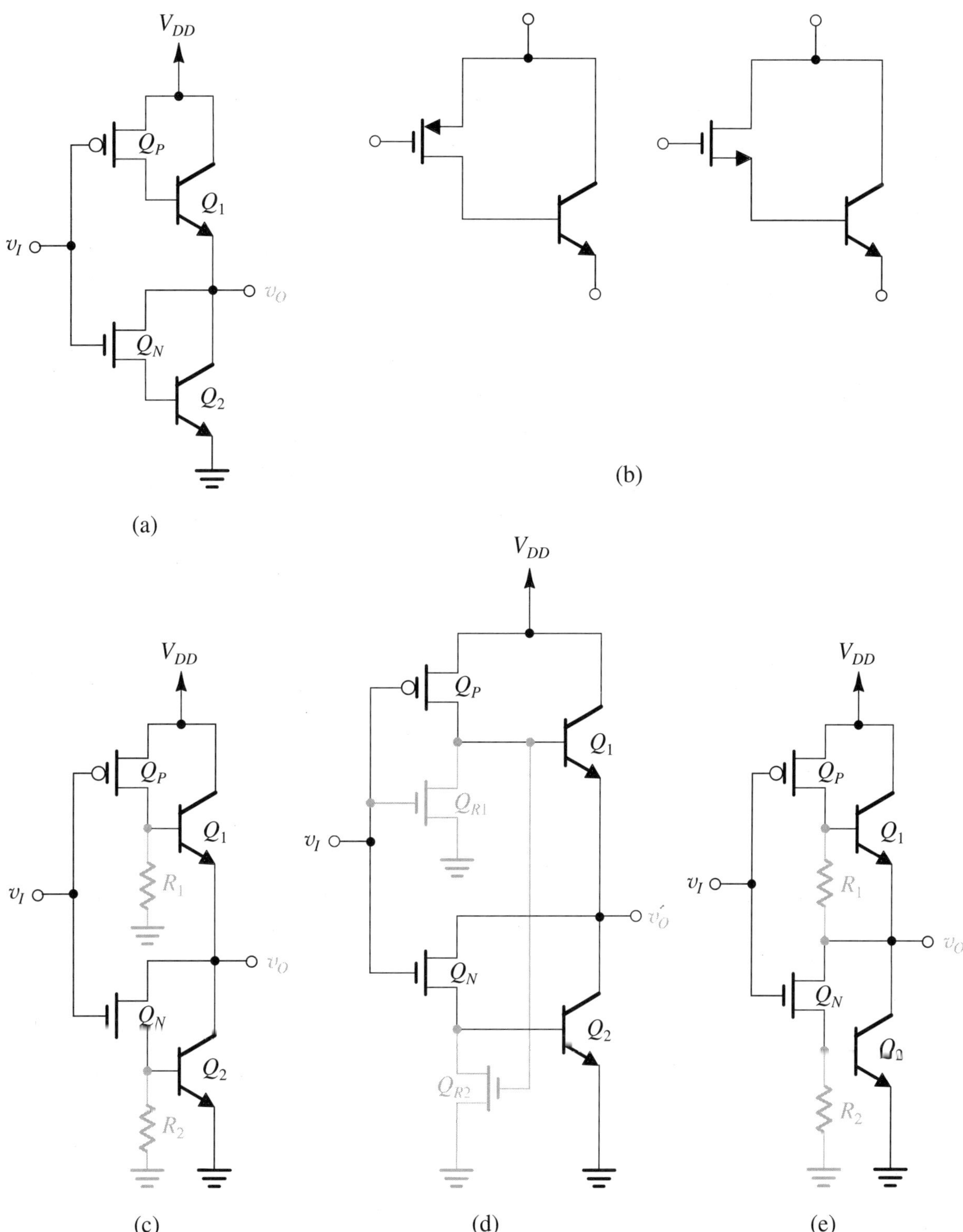

FIGURE 11.45 Development of the BiCMOS inverter circuit. **(a)** The basic concept is to use an additional bipolar transistor to increase the output current drive of each of Q_N and Q_P of the CMOS inverter. **(b)** The circuit in (a) can be thought of as utilizing these composite devices. **(c)** To reduce the turn-off times of Q_1 and Q_2, "bleeder resistors" R_1 and R_2 are added. **(d)** Implementation of the circuit in (c) using NMOS transistors to realize the resistors. **(e)** An improved version of the circuit in (c) obtained by connecting the lower end of R_1 to the output node.

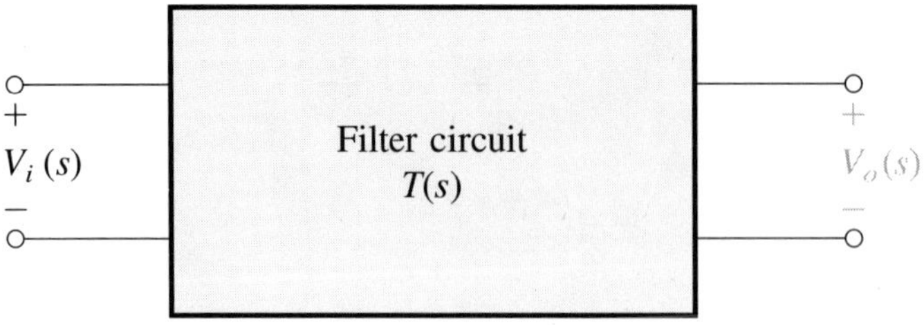

FIGURE 12.1 The filters studied in this chapter are linear circuits represented by the general two-port network shown. The filter transfer function $T(s) \equiv V_o(s)/V_i(s)$.

$|T|$ 1 0 ← Passband →← Stopband — — ω_p ω

(a) Low-pass (LP)

$|T|$ 1 0 ← Stopband →← Passband — — ω_p ω

(b) High-pass (HP)

$|T|$ 1 0 ← Lower stopband →← Passband →← Upper stopband — — ω_{p1} ω_{p2} ω

(c) Bandpass (BP)

$|T|$ 1 0 ← Lower passband →← Stopband →← Upper passband — — ω_{s1} ω_{s2} ω

(d) Bandstop (BS)

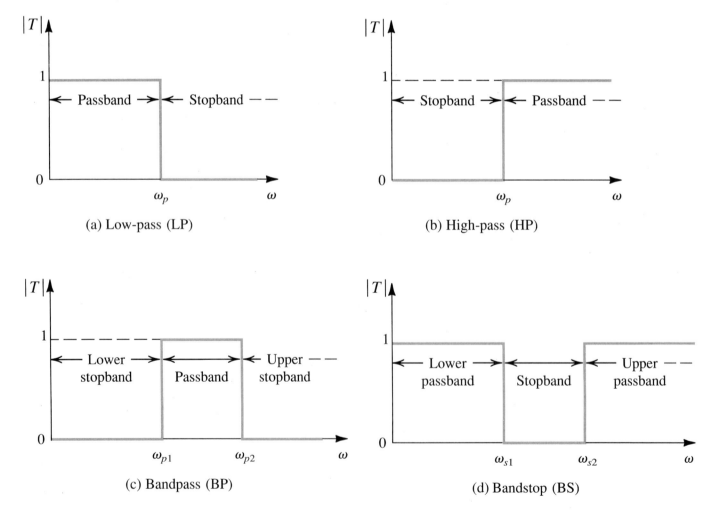

FIGURE 12.2 Ideal transmission characteristics of the four major filter types: **(a)** low-pass (LP), **(b)** high-pass (HP), **(c)** bandpass (BP), and **(d)** bandstop (BS).

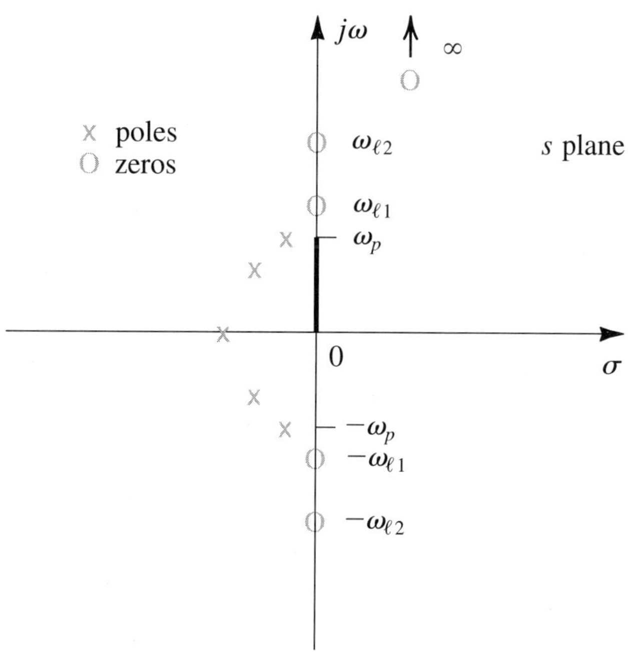

FIGURE 12.5 Pole–zero pattern for the low-pass filter whose transmission is sketched in Fig. 12.3. This is a fifth-order filter ($N = 5$).

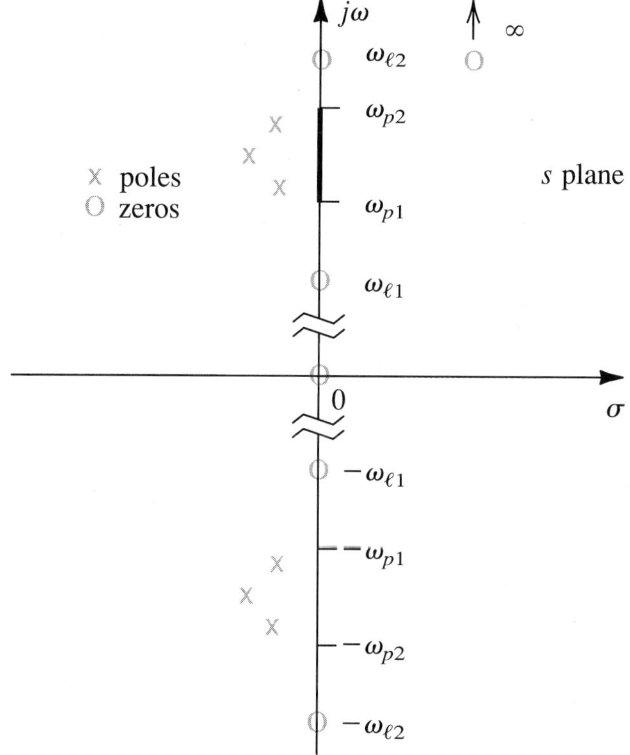

FIGURE 12.6 Pole–zero pattern for the bandpass filter whose transmission function is shown in Fig. 12.4. This is a sixth-order filter ($N = 6$).

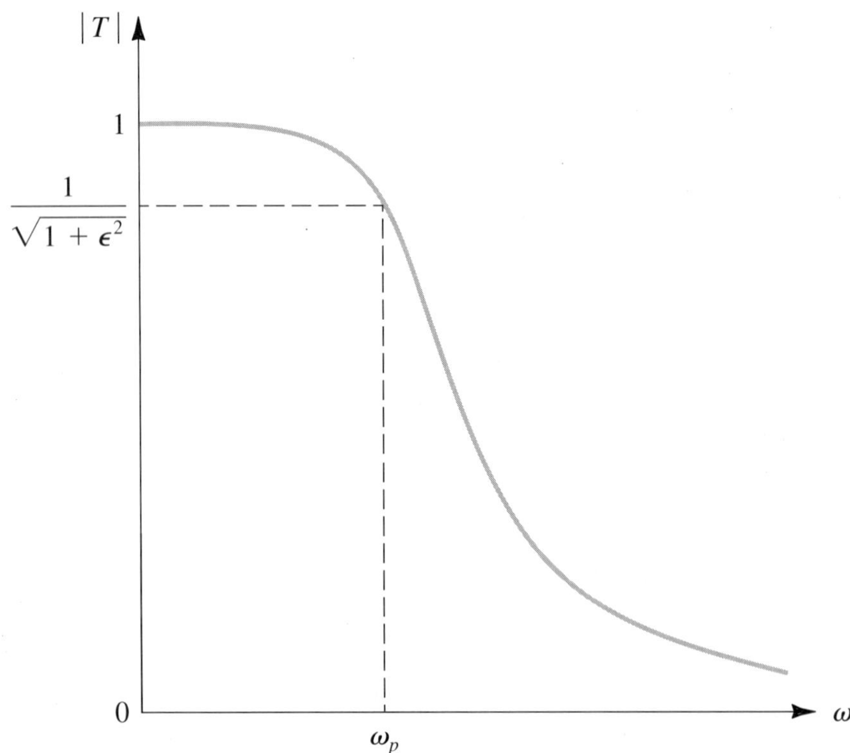

FIGURE 12.8 The magnitude response of a Butterworth filter.

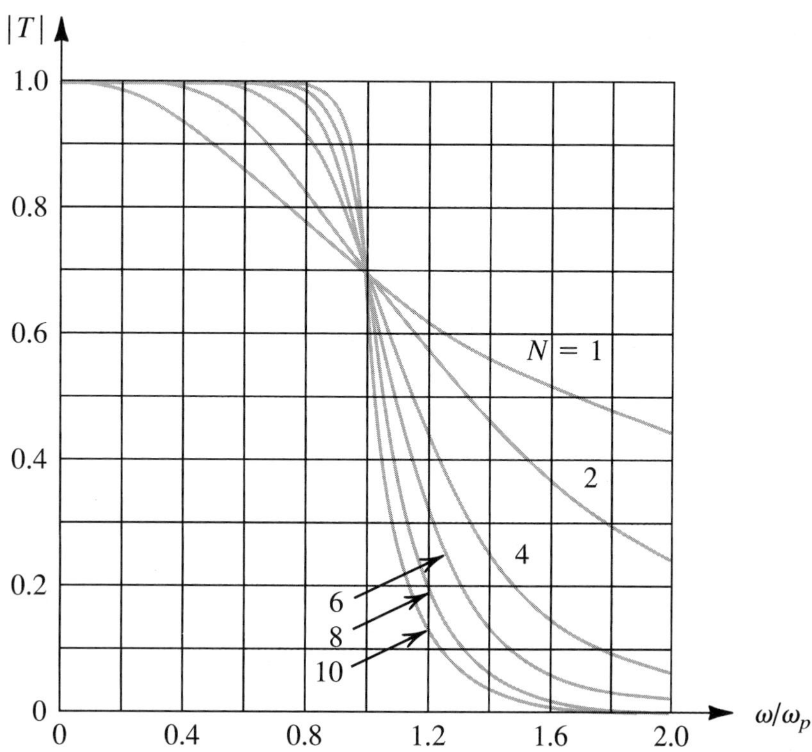

FIGURE 12.9 Magnitude response for Butterworth filters of various order with $\epsilon = 1$. Note that as the order increases, the response approaches the ideal brick-wall type of transmission.

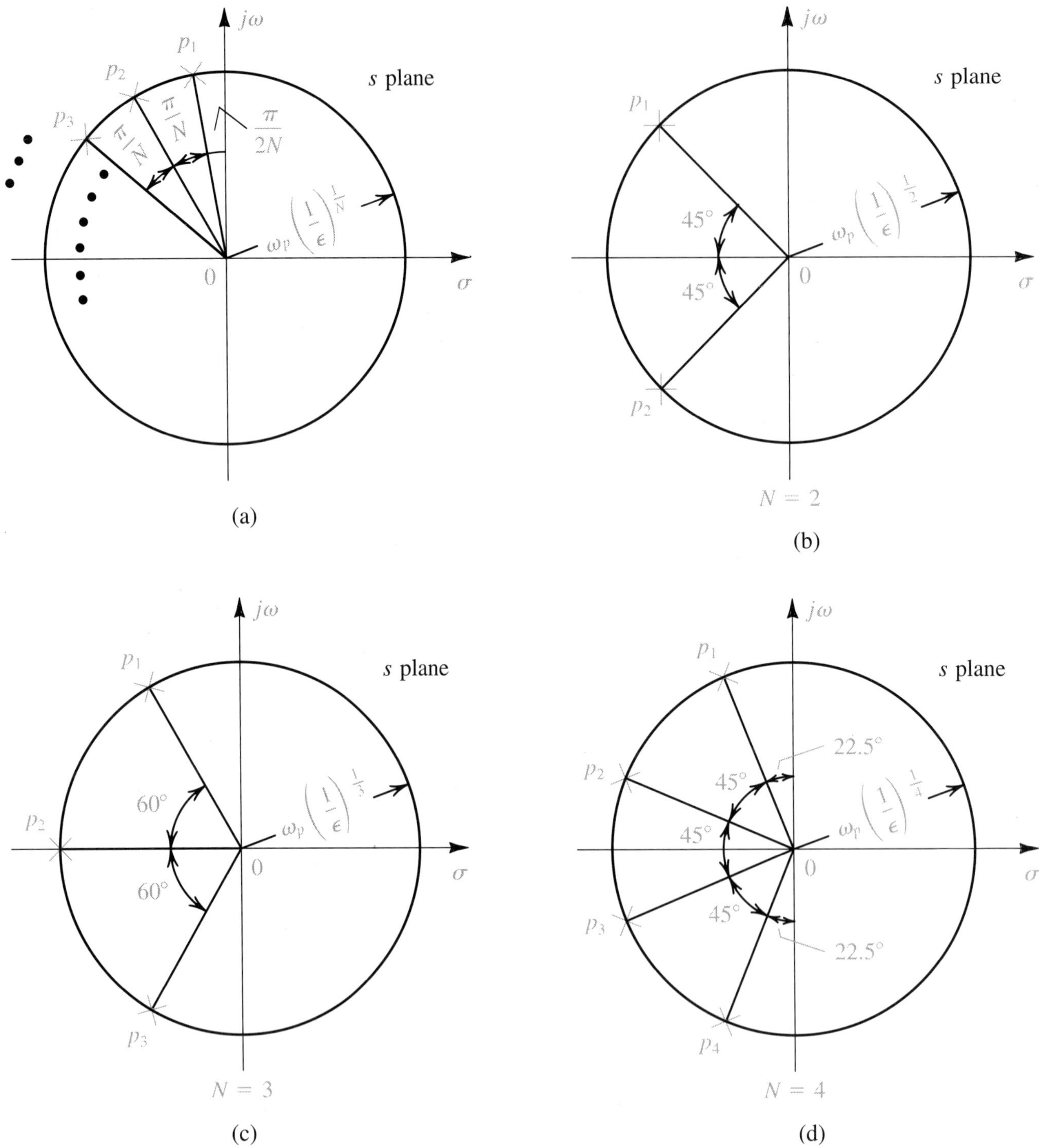

FIGURE 12.10 Graphical construction for determining the poles of a Butterworth filter of order N. All the poles lie in the left half of the s plane on a circle of radius $\omega_0 = \omega_p(1/\epsilon)^{1/N}$, where ϵ is the passband deviation parameter ($\epsilon = \sqrt{10^{A_{\max}/10} - 1}$): **(a)** the general case, **(b)** $N = 2$, **(c)** $N = 3$, and **(d)** $N = 4$.

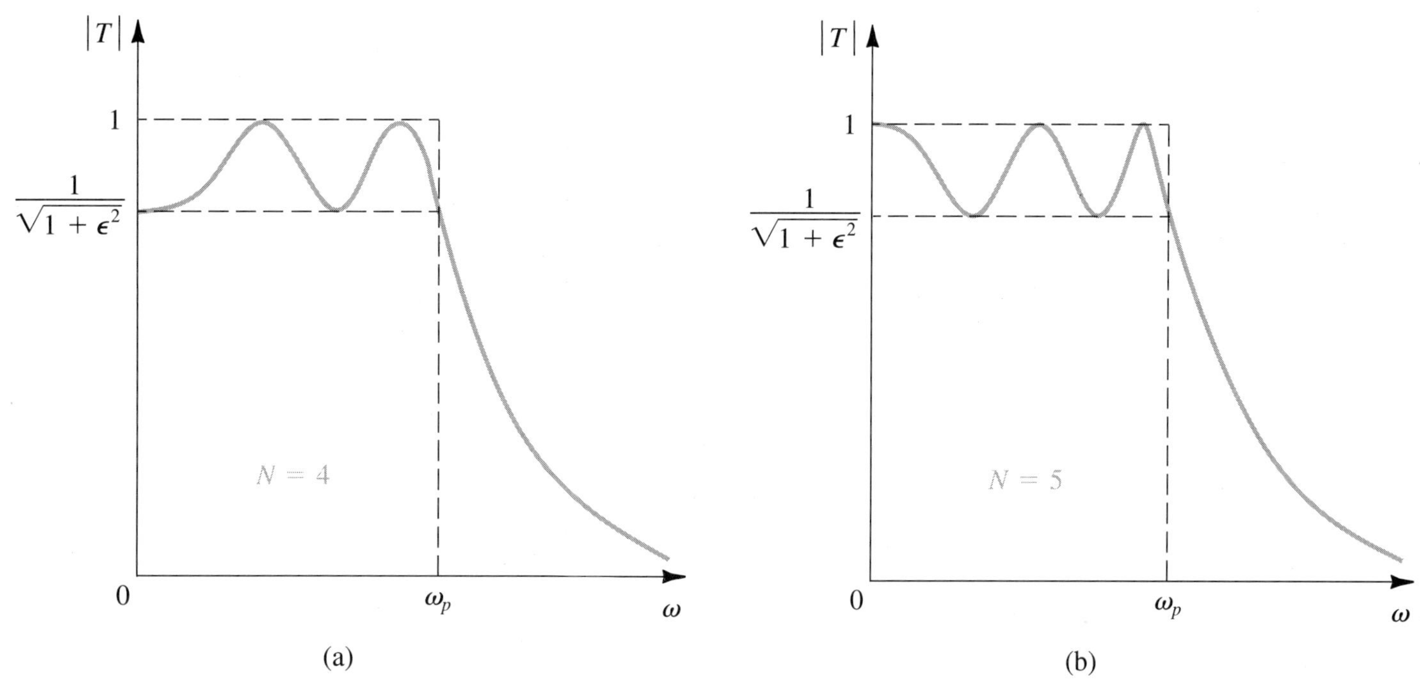

FIGURE 12.12 Sketches of the transmission characteristics of representative **(a)** even-order and **(b)** odd-order Chebyshev filters.

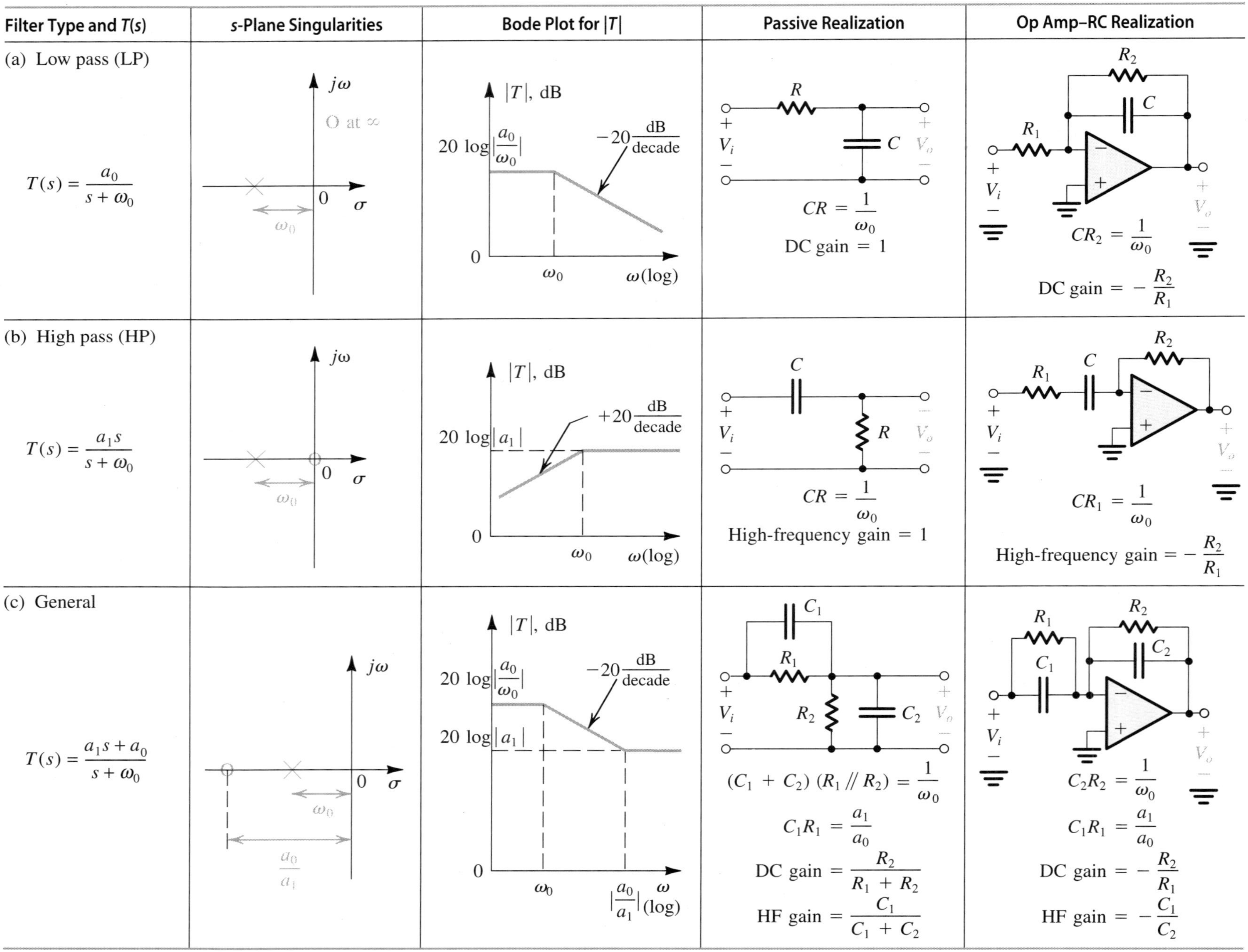

Filter Type and $T(s)$	s-Plane Singularities	Bode Plot for $\|T\|$	Passive Realization	Op Amp–RC Realization
(a) Low pass (LP) $T(s) = \frac{a_0}{s + \omega_0}$	$j\omega$; O at ∞; ω_0; 0; σ	$\|T\|$, dB; $20 \log\left\|\frac{a_0}{\omega_0}\right\|$; $-20\frac{\text{dB}}{\text{decade}}$; 0; ω_0; $\omega(\log)$	R; C; V_i; V_o; $CR = \frac{1}{\omega_0}$; DC gain = 1	R_1; R_2; C; V_i; V_o; $CR_2 = \frac{1}{\omega_0}$; DC gain $= -\frac{R_2}{R_1}$
(b) High pass (HP) $T(s) = \frac{a_1 s}{s + \omega_0}$	$j\omega$; ω_0; 0; σ	$\|T\|$, dB; $20 \log\|a_1\|$; $+20\frac{\text{dB}}{\text{decade}}$; 0; ω_0; $\omega(\log)$	C; R; V_i; V_o; $CR = \frac{1}{\omega_0}$; High-frequency gain = 1	R_1; C; R_2; V_i; V_o; $CR_1 = \frac{1}{\omega_0}$; High-frequency gain $= -\frac{R_2}{R_1}$
(c) General $T(s) = \frac{a_1 s + a_0}{s + \omega_0}$	$j\omega$; ω_0; $\frac{a_0}{a_1}$; 0; σ	$\|T\|$, dB; $20 \log\left\|\frac{a_0}{\omega_0}\right\|$; $20 \log\|a_1\|$; $-20\frac{\text{dB}}{\text{decade}}$; 0; ω_0; $\left\|\frac{a_0}{a_1}\right\|$; ω (log)	C_1; R_1; R_2; C_2; V_i; V_o; $(C_1 + C_2)(R_1 // R_2) = \frac{1}{\omega_0}$; $C_1R_1 = \frac{a_1}{a_0}$; DC gain $= \frac{R_2}{R_1 + R_2}$; HF gain $= \frac{C_1}{C_1 + C_2}$	R_1; C_1; R_2; C_2; V_i; V_o; $C_2R_2 = \frac{1}{\omega_0}$; $C_1R_1 = \frac{a_1}{a_0}$; DC gain $= -\frac{R_2}{R_1}$; HF gain $= -\frac{C_1}{C_2}$

FIGURE 12.13 First-order filters.

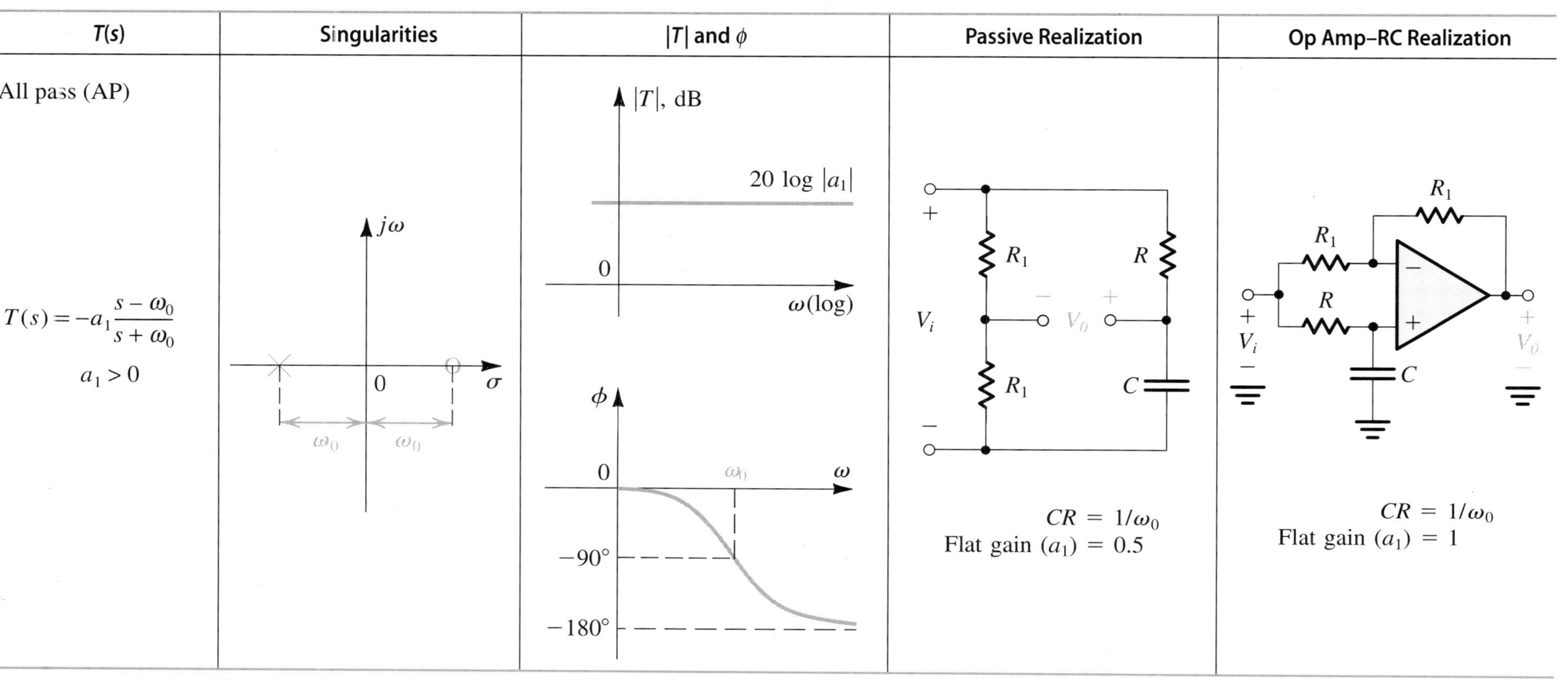

FIGURE 12.14 First-order all-pass filter.

Filter Type and $T(s)$	s-Plane Singularities	$\lvert T \rvert$
(a) Low pass (LP) $T(s) = \dfrac{a_0}{s^2 + s\dfrac{\omega_0}{Q} + \omega_0^2}$ DC gain $= \dfrac{a_0}{\omega_0^2}$	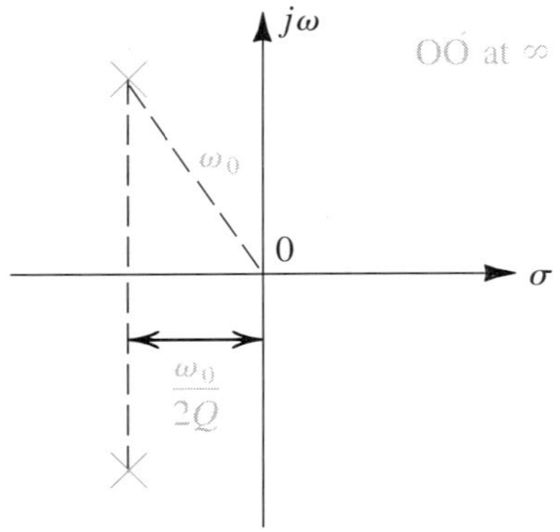 $j\omega$, OO at ∞, ω_0, 0, σ, $\dfrac{\omega_0}{2Q}$	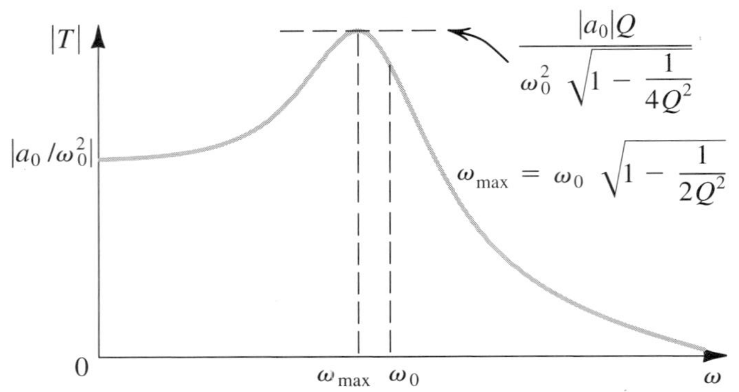 $\lvert T \rvert$, $\lvert a_0/\omega_0^2 \rvert$, $\dfrac{\lvert a_0 \rvert Q}{\omega_0^2 \sqrt{1 - \dfrac{1}{4Q^2}}}$, $\omega_{max} = \omega_0 \sqrt{1 - \dfrac{1}{2Q^2}}$, 0, ω_{max}, ω_0, ω
(b) High pass (HP) $T(s) = \dfrac{a_2 s^2}{s^2 + s\dfrac{\omega_0}{Q} + \omega_0^2}$ High-frequency gain $= a_2$	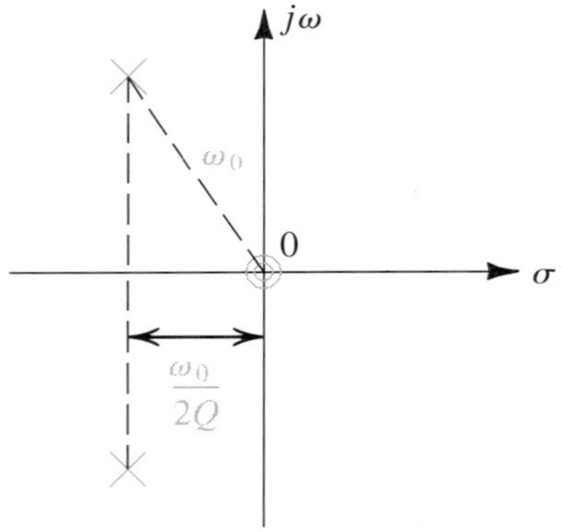 $j\omega$, ω_0, 0, σ, $\dfrac{\omega_0}{2Q}$	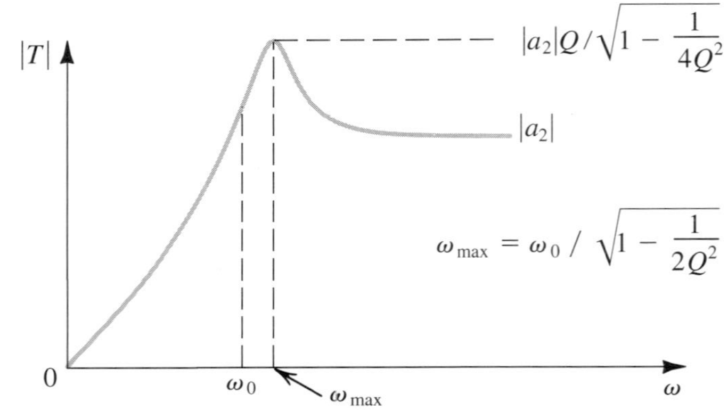 $\lvert T \rvert$, $\lvert a_2 \rvert Q / \sqrt{1 - \dfrac{1}{4Q^2}}$, $\lvert a_2 \rvert$, $\omega_{max} = \omega_0 / \sqrt{1 - \dfrac{1}{2Q^2}}$, 0, ω_0, ω_{max}, ω
(c) Bandpass (BP) $T(s) = \dfrac{a_1 s}{s^2 + s\dfrac{\omega_0}{Q} + \omega_0^2}$ Center-frequency gain $= \dfrac{a_1 Q}{\omega_0}$	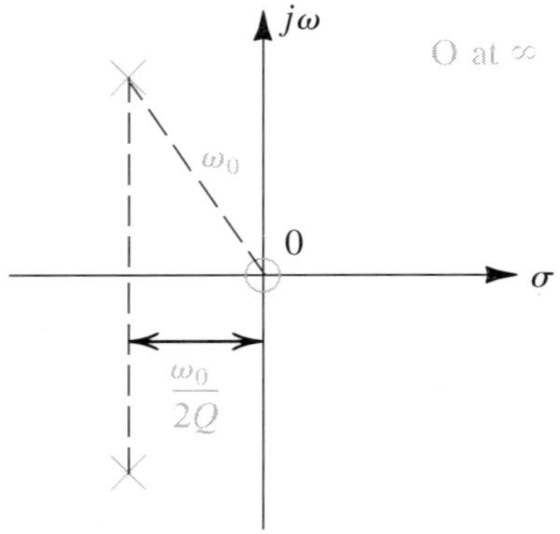 $j\omega$, O at ∞, ω_0, 0, σ, $\dfrac{\omega_0}{2Q}$	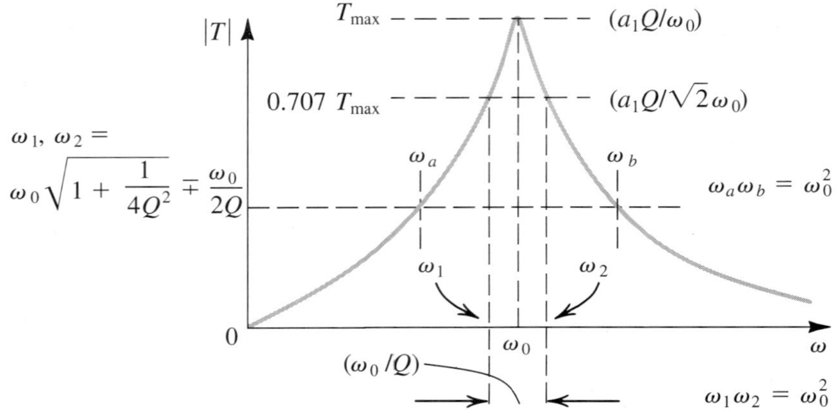 $\lvert T \rvert$, T_{max} — $(a_1 Q/\omega_0)$, $0.707\ T_{max}$ — $(a_1 Q/\sqrt{2}\omega_0)$, $\omega_1, \omega_2 = \omega_0 \sqrt{1 + \dfrac{1}{4Q^2}} \mp \dfrac{\omega_0}{2Q}$, ω_a, ω_b, $\omega_a \omega_b = \omega_0^2$, ω_1, ω_2, 0, ω_0, ω, (ω_0/Q), $\omega_1 \omega_2 = \omega_0^2$

FIGURE 12.16 Second-order filtering functions.

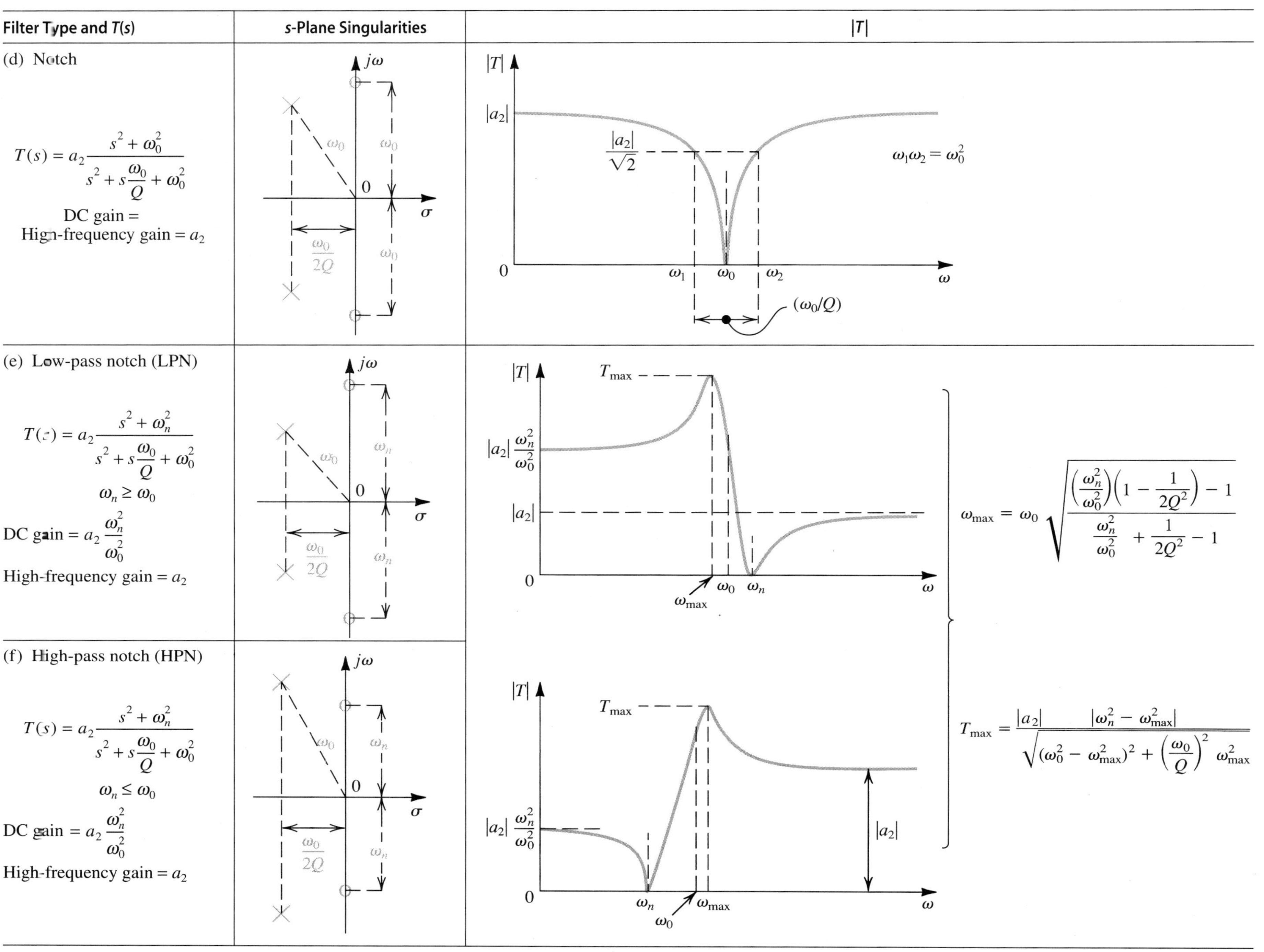

Filter Type and $T(s)$
s-Plane Singularities
$|T|$
(d) Notch
$T(s) = a_2 \frac{s^2 + \omega_0^2}{s^2 + s\frac{\omega_0}{Q} + \omega_0^2}$
DC gain =
High-frequency gain = a_2
$j\omega$
σ
0
ω_0
$\frac{\omega_0}{2Q}$
$|T|$
$|a_2|$
$\frac{|a_2|}{\sqrt{2}}$
$\omega_1\omega_2 = \omega_0^2$
ω_1
ω_0
ω_2
ω
(ω_0/Q)
(e) Low-pass notch (LPN)
$T(s) = a_2 \frac{s^2 + \omega_n^2}{s^2 + s\frac{\omega_0}{Q} + \omega_0^2}$
$\omega_n \geq \omega_0$
DC gain = $a_2 \frac{\omega_n^2}{\omega_0^2}$
High-frequency gain = a_2
ω_n
T_{max}
$|a_2| \frac{\omega_n^2}{\omega_0^2}$
ω_{max}
$\omega_{max} = \omega_0 \sqrt{\frac{\left(\frac{\omega_n^2}{\omega_0^2}\right)\left(1 - \frac{1}{2Q^2}\right) - 1}{\frac{\omega_n^2}{\omega_0^2} + \frac{1}{2Q^2} - 1}}$
(f) High-pass notch (HPN)
$T(s) = a_2 \frac{s^2 + \omega_n^2}{s^2 + s\frac{\omega_0}{Q} + \omega_0^2}$
$\omega_n \leq \omega_0$
DC gain = $a_2 \frac{\omega_n^2}{\omega_0^2}$
High-frequency gain = a_2
$T_{max} = \frac{|a_2| \; |\omega_n^2 - \omega_{max}^2|}{\sqrt{(\omega_0^2 - \omega_{max}^2)^2 + \left(\frac{\omega_0}{Q}\right)^2 \omega_{max}^2}}$

FIGURE 12.16 (Continued)

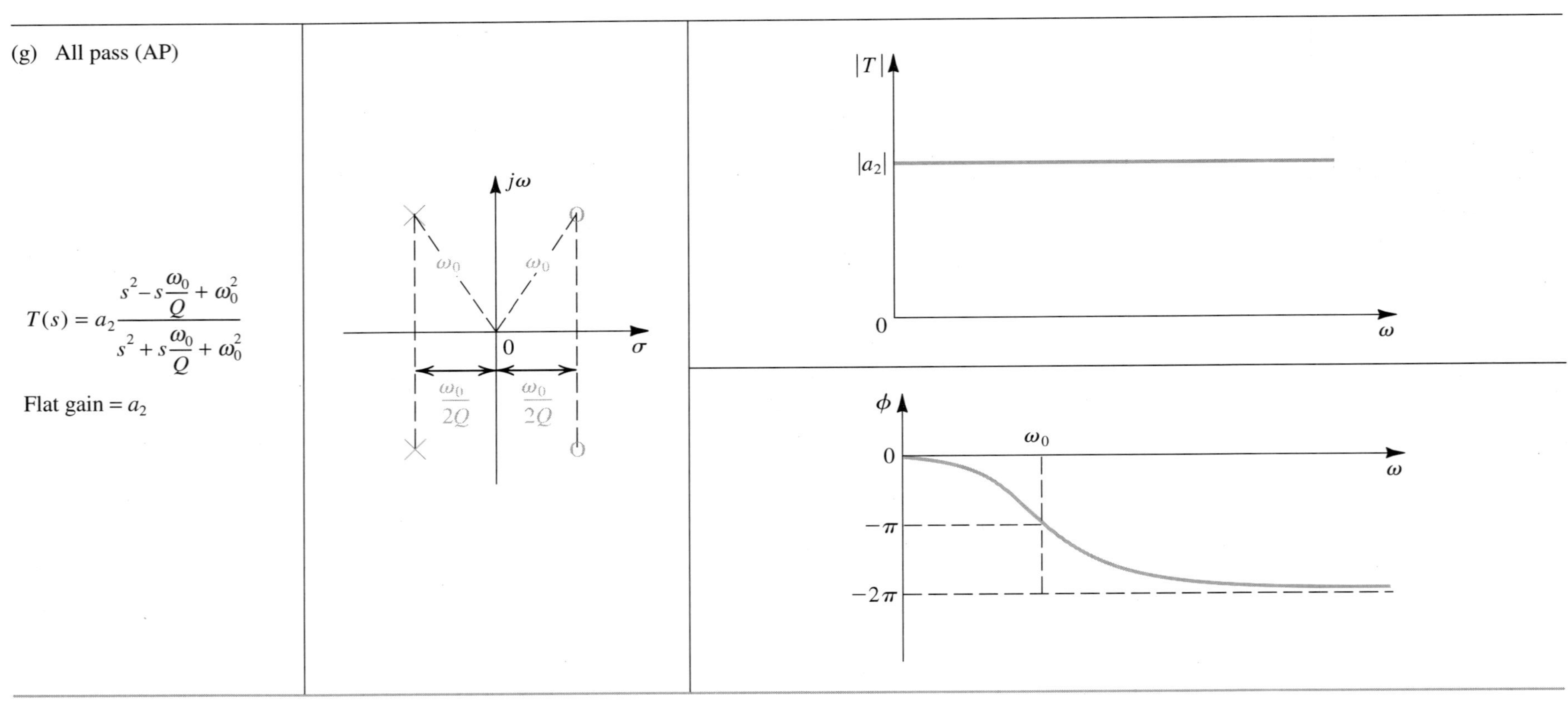

FIGURE 12.16 *(Continued)*

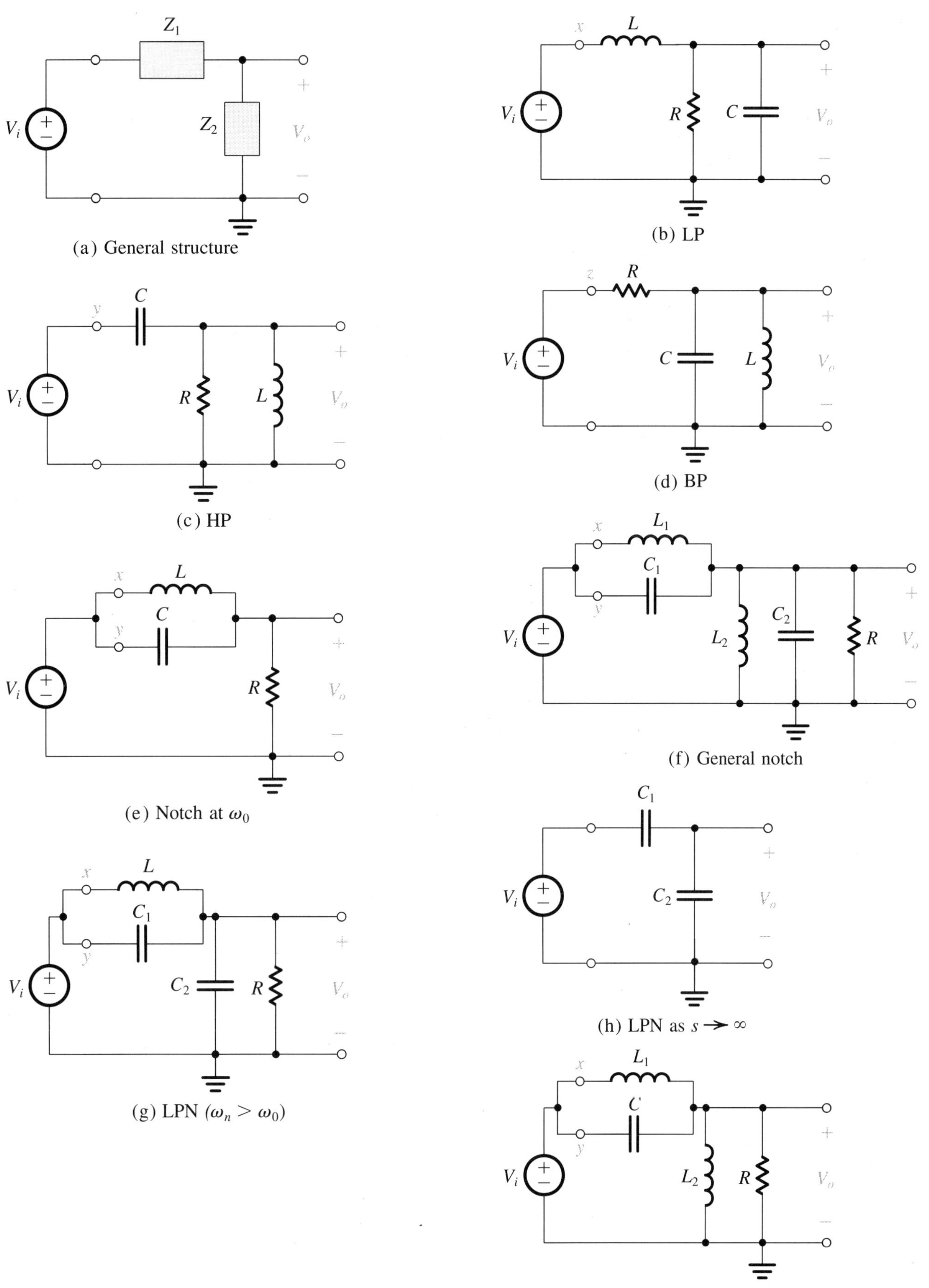

FIGURE 12.18 Realization of various second-order filter functions using the LCR resonator of Fig. 12.17(b): **(a)** general structure, **(b)** LP, **(c)** HP, **(d)** BP, **(e)** notch at ω_0, **(f)** general notch, **(g)** LPN $(\omega_n \geq \omega_0)$, **(h)** LPN as $s \to \infty$, **(i)** HPN $(\omega_n < \omega_0)$.

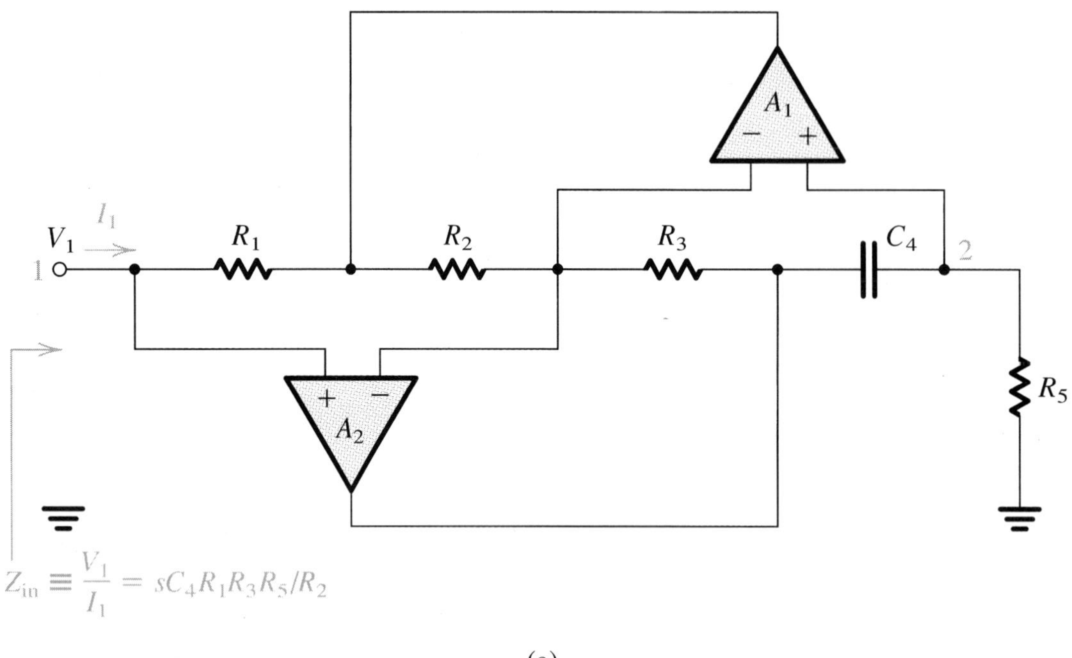

(a)

(14) $\left[V_1 - \frac{V_1R_2}{sC_4R_5R_3}\right]$

(15) $\frac{V_1R_2}{sC_4R_5R_3R_1}$

(13) $V_1/sC_4R_5R_3$

(12) 0

(7) 0

(1) V_1

(2)

(3) V_1 R_3

(4) C_4

(5) V_1

(8) V_1/R_5

(6) $\frac{V_1}{R_5}$

(10) $\frac{V_1}{sC_4R_5R_3}$

(11) 0

(16) 0

(17) $I_1 = \frac{V_1R_2}{sC_4R_5R_3R_1}$

(9) $V_1 + \frac{1}{sC_4}\frac{V_1}{R_5}$

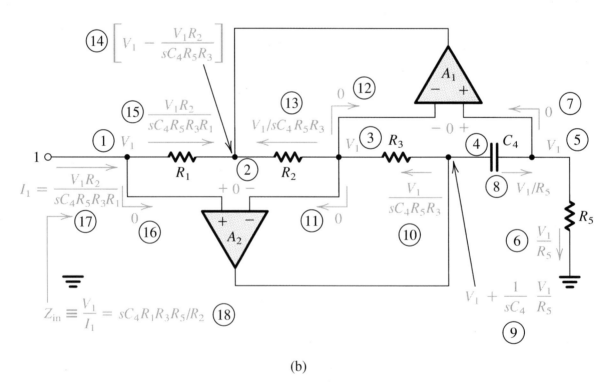

(18) $Z_{in} \equiv \frac{V_1}{I_1} = sC_4R_1R_3R_5/R_2$

(b)

FIGURE 12.20 **(a)** The Antoniou inductance-simulation circuit. **(b)** Analysis of the circuit assuming ideal op amps. The order of the analysis steps is indicated by the circled numbers.

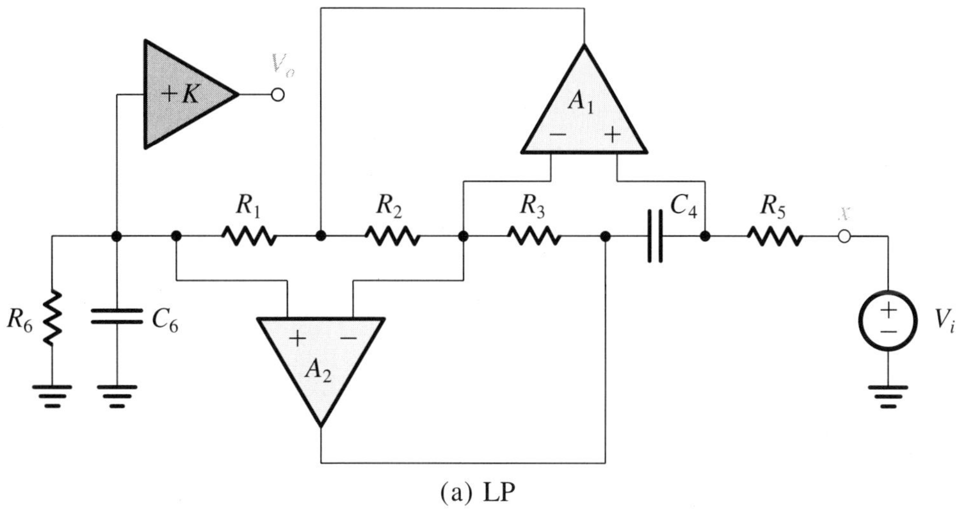

(a) LP

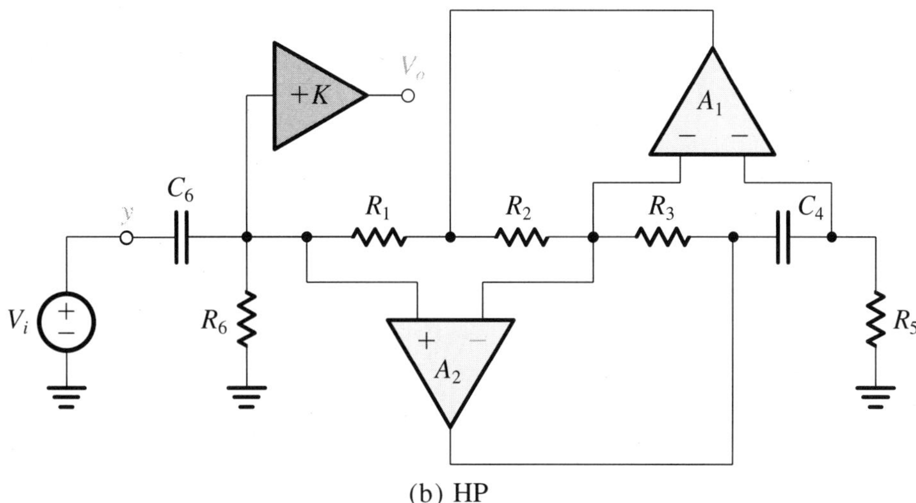

(b) HP

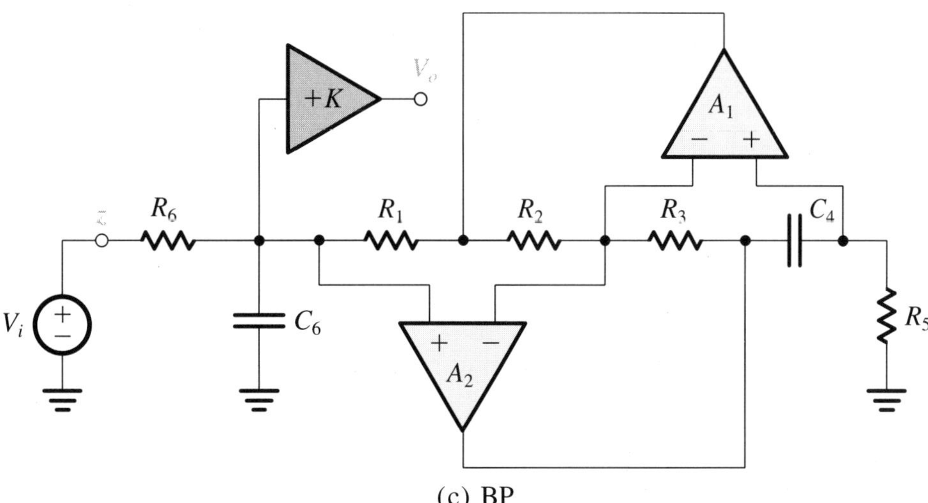

(c) BP

FIGURE 12.22 Realizations for the various second-order filter functions using the op amp–RC resonator of Fig. 12.21(b): **(a)** LP, **(b)** HP, **(c)** BP,

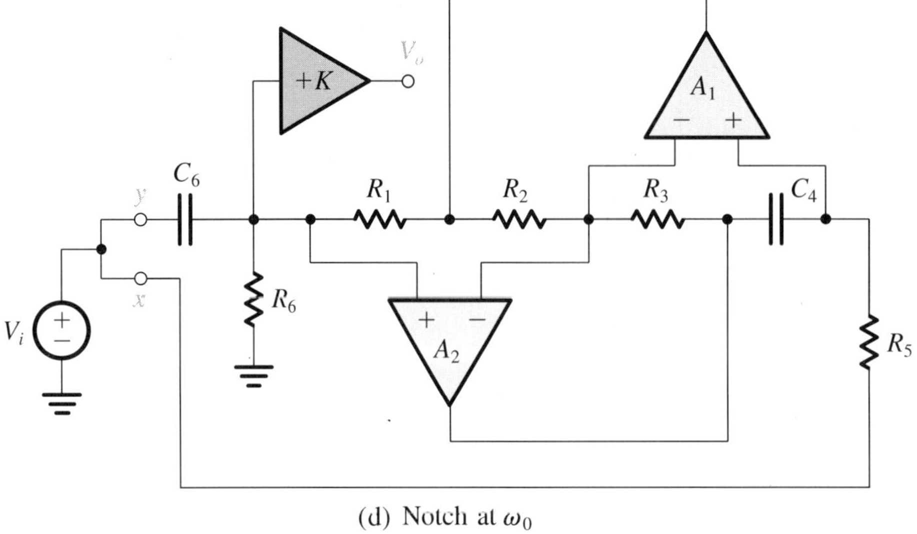

(d) Notch at ω_0

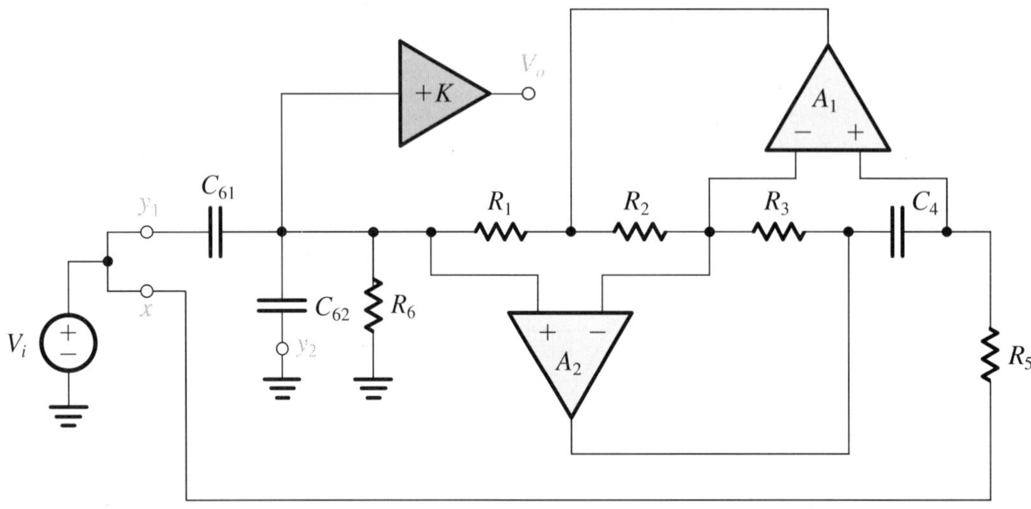

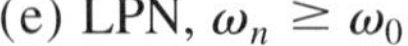

(e) LPN, $\omega_n \geq \omega_0$

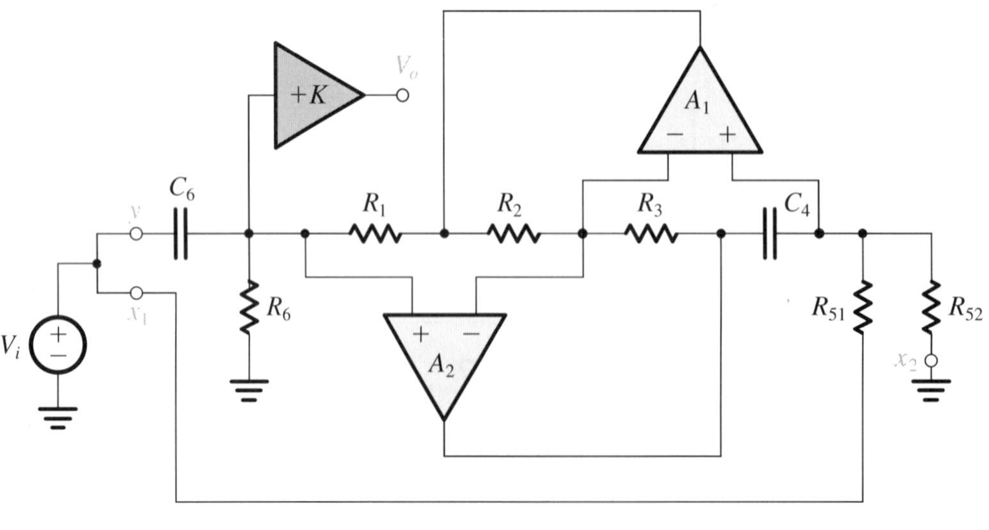

(f) HPN, $\omega_n \leq \omega_0$

FIGURE 12.22 *(Continued)* **(d)** notch at ω_0, **(e)** LPN, $\omega_n \geq \omega_0$, **(f)** HPN, $\omega_n \leq \omega_0$, and

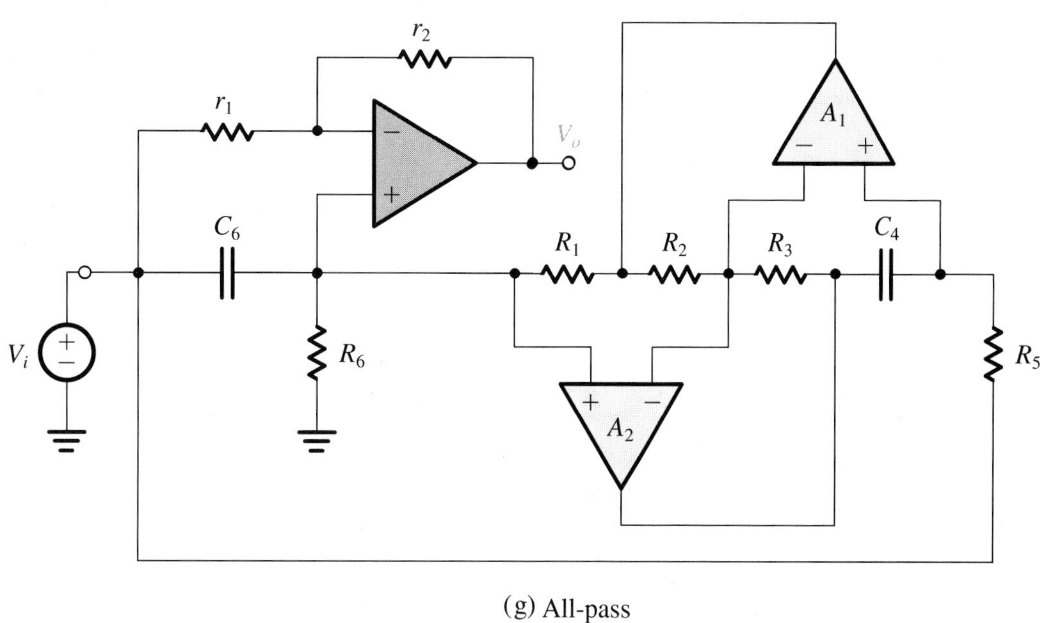

(g) All-pass

FIGURE 12.22 *(Continued)* **(g)** all pass. The circuits are based on the LCR circuits in Fig. 12.18. Design equations are given in Table 12.1.

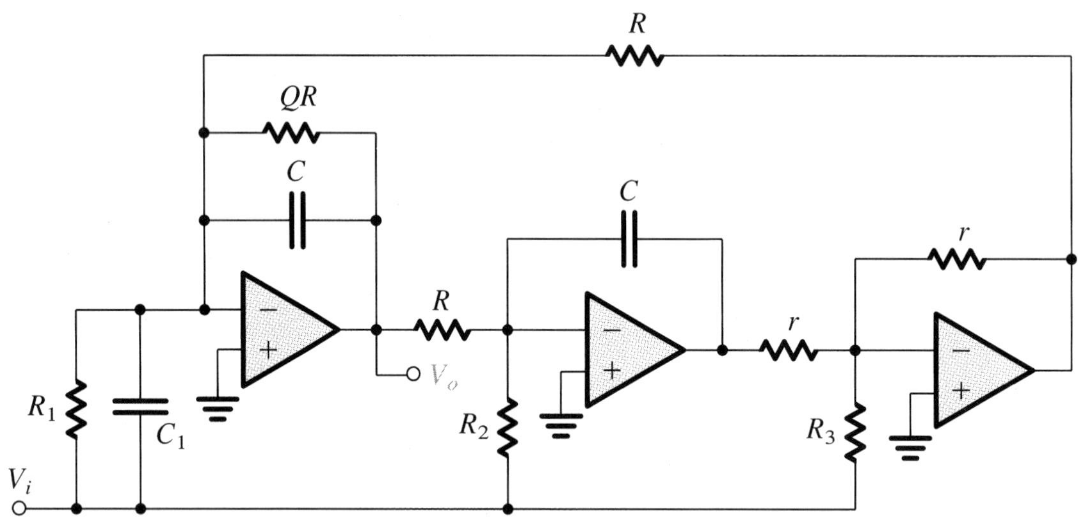

FIGURE 12.26 The Tow–Thomas biquad with feedforward. The transfer function of Eq. (12.68) is realized by feeding the input signal through appropriate components to the inputs of the three op amps. This circuit can realize all special second-order functions. The design equations are given in Table 12.2.

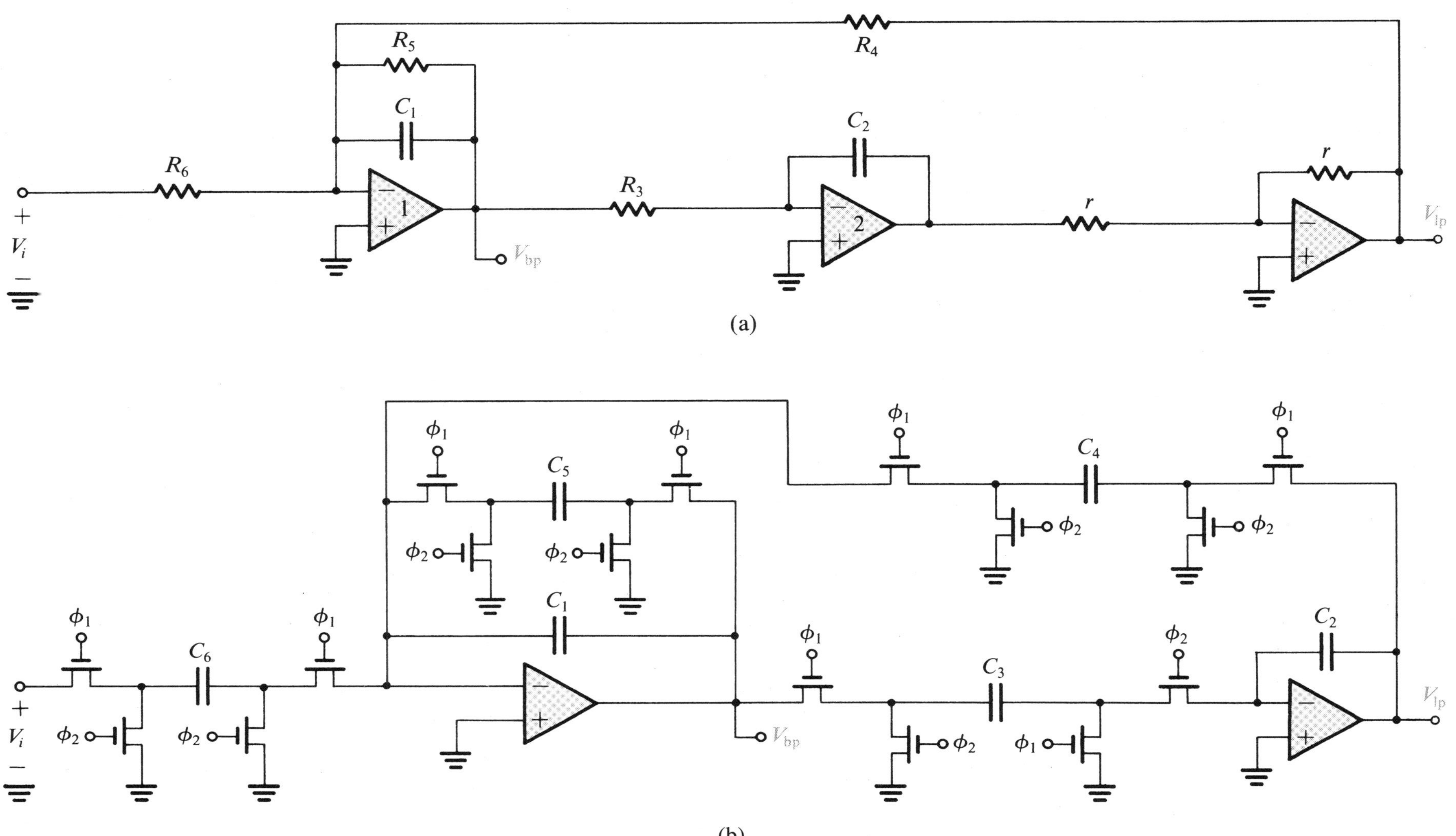

FIGURE 12.37 **(a)** A two-integrator-loop active-RC biquad and **(b)** its switched-capacitor counterpart.

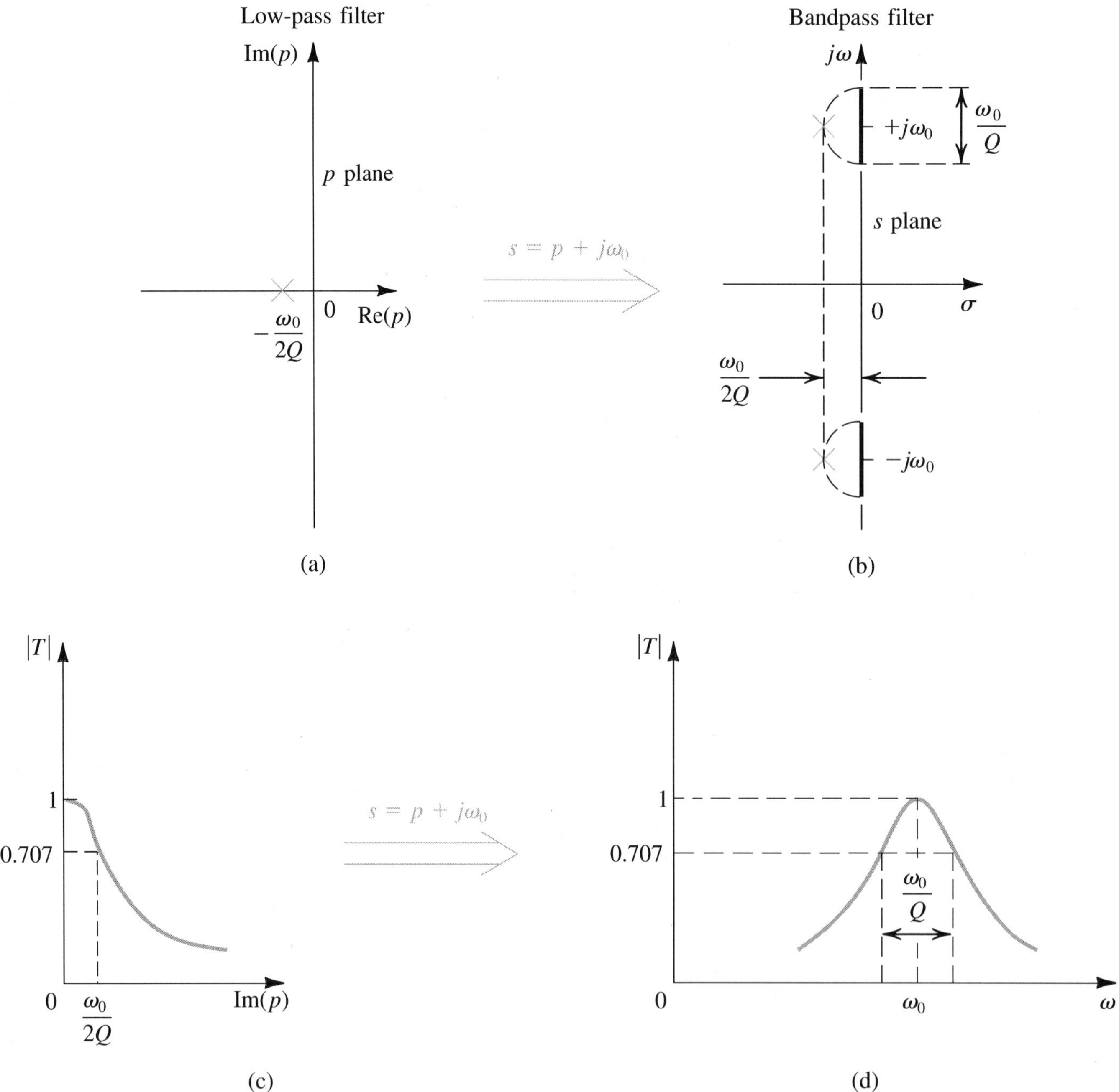

FIGURE 12.47 Obtaining a second-order narrow-band bandpass filter by transforming a first-order low-pass filter. **(a)** Pole of the first-order filter in the p plane. **(b)** Applying the transformation $s = p + j\omega_0$ and adding a complex-conjugate pole results in the poles of the second-order bandpass filter. **(c)** Magnitude response of the first-order low-pass filter. **(d)** Magnitude response of the second-order bandpass filter.

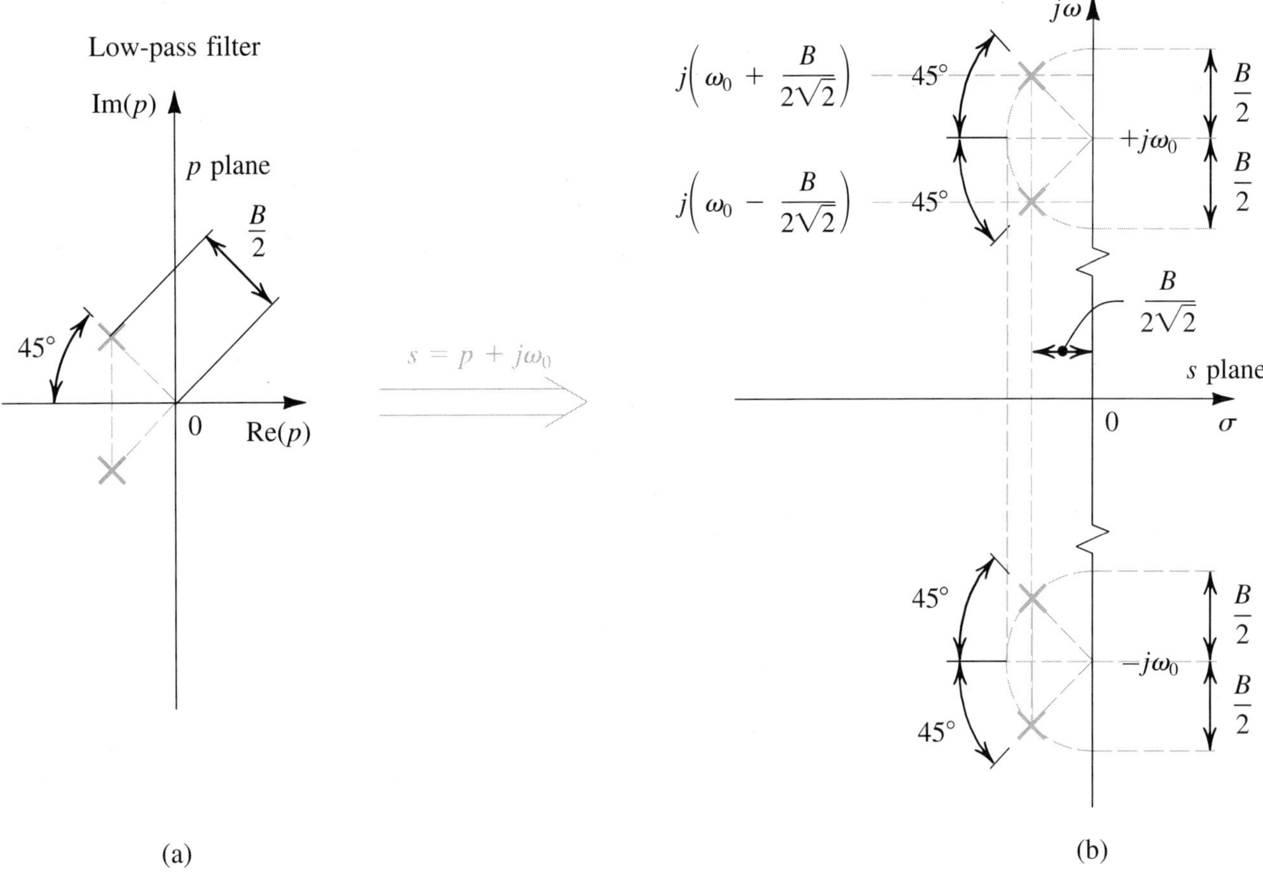

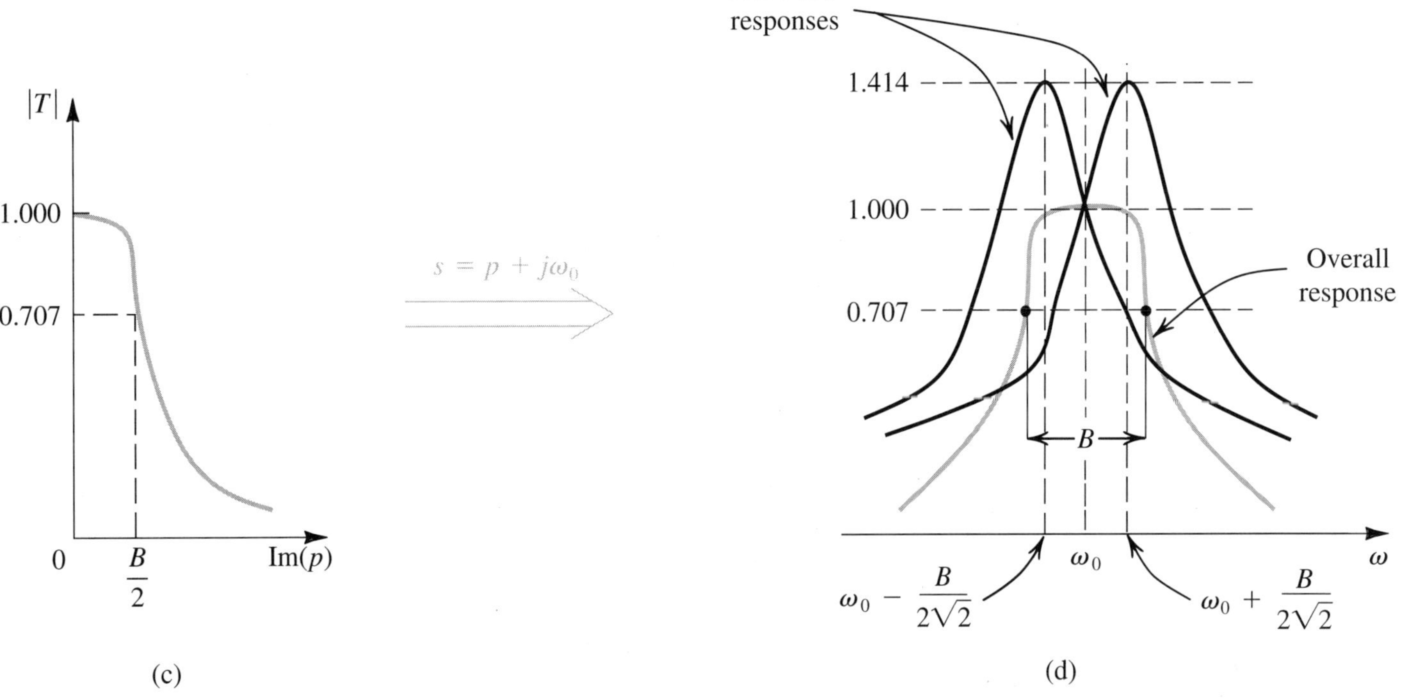

FIGURE 12.48 Obtaining the poles and the frequency response of a fourth-order stagger-tuned narrow-band bandpass amplifier by transforming a second-order low-pass maximally flat response.

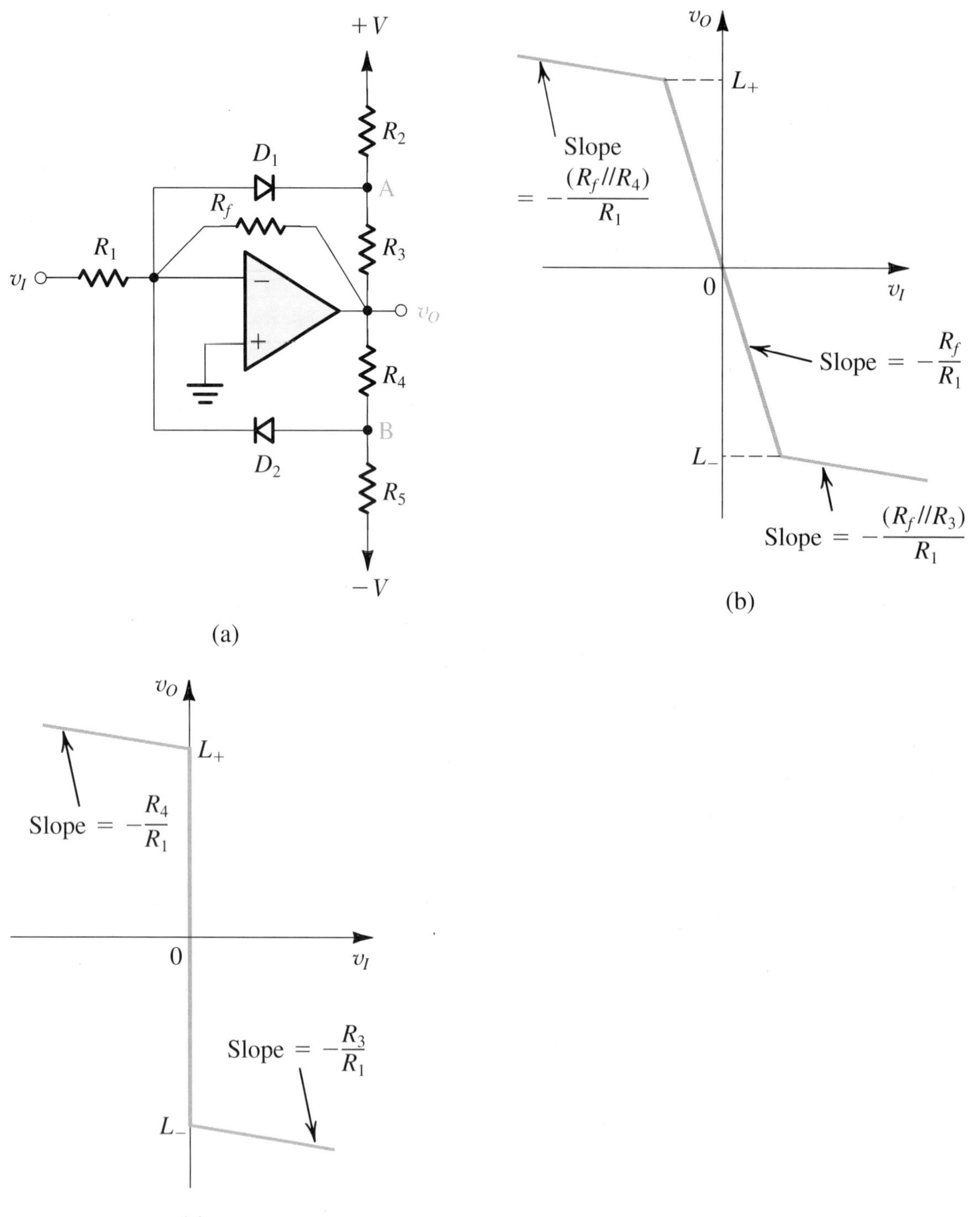

FIGURE 13.3 **(a)** A popular limiter circuit. **(b)** Transfer characteristic of the limiter circuit; L_- and L_+ are given by Eqs. (13.8) and (13.9), respectively. **(c)** When R_f is removed, the limiter turns into a comparator with the characteristic shown.

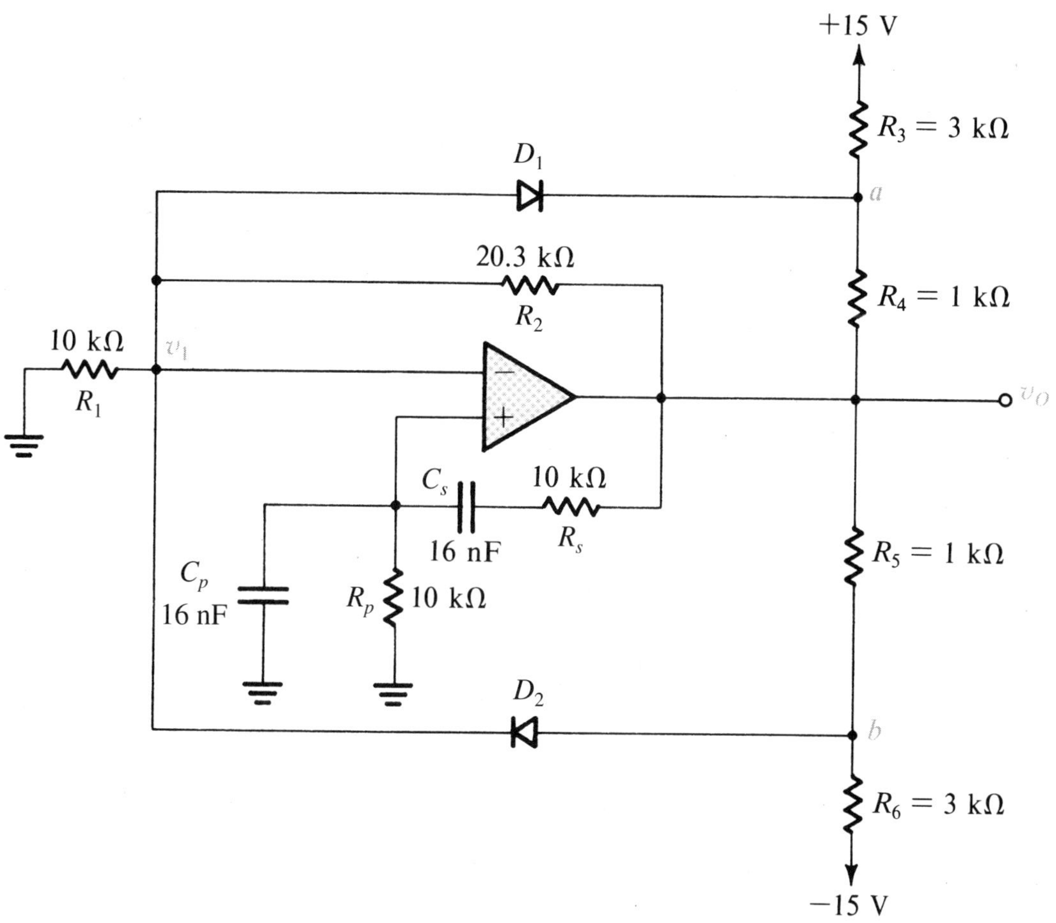

FIGURE 13.5 A Wien-bridge oscillator with a limiter used for amplitude control.

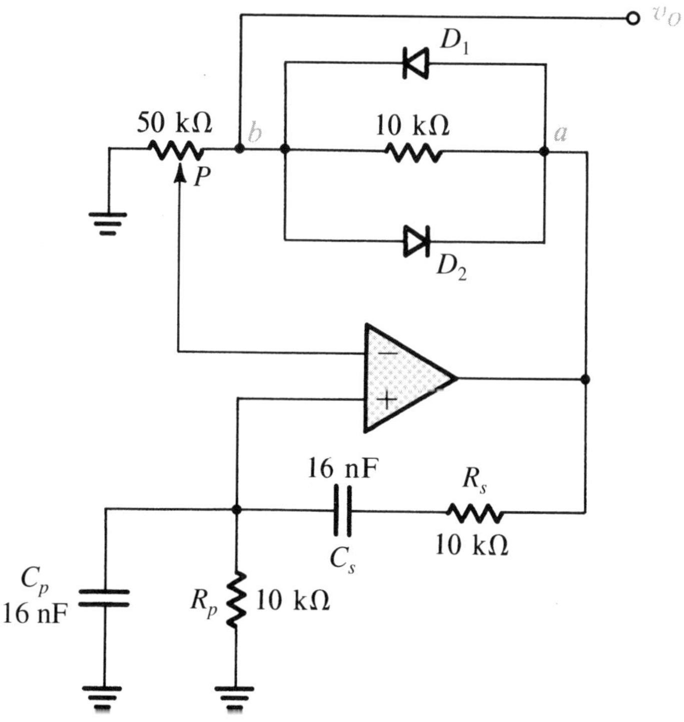

FIGURE 13.6 A Wien-bridge oscillator with an alternative method for amplitude stabilization.

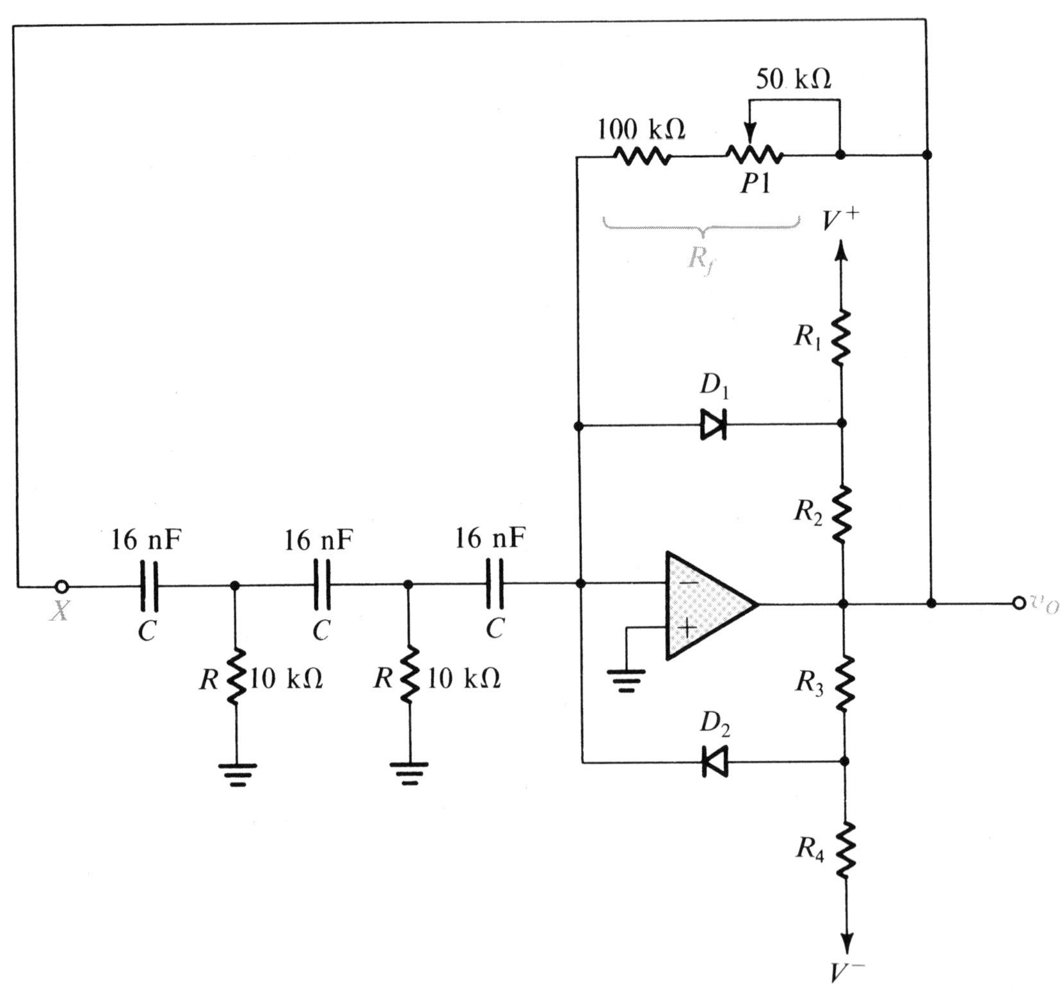

FIGURE 13.8 A practical phase-shift oscillator with a limiter for amplitude stabilization.

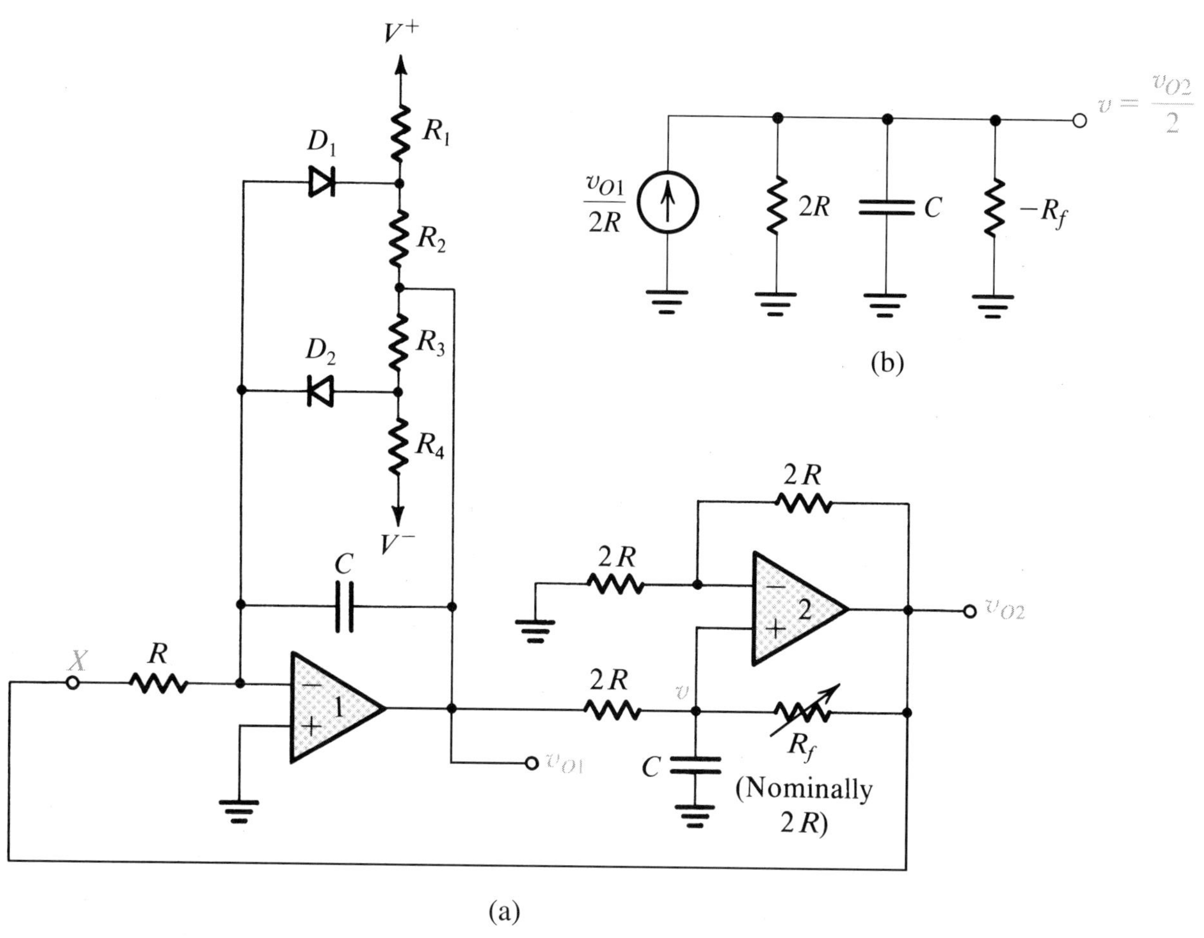

FIGURE 13.9 **(a)** A quadrature-oscillator circuit. **(b)** Equivalent circuit at the input of op amp 2.

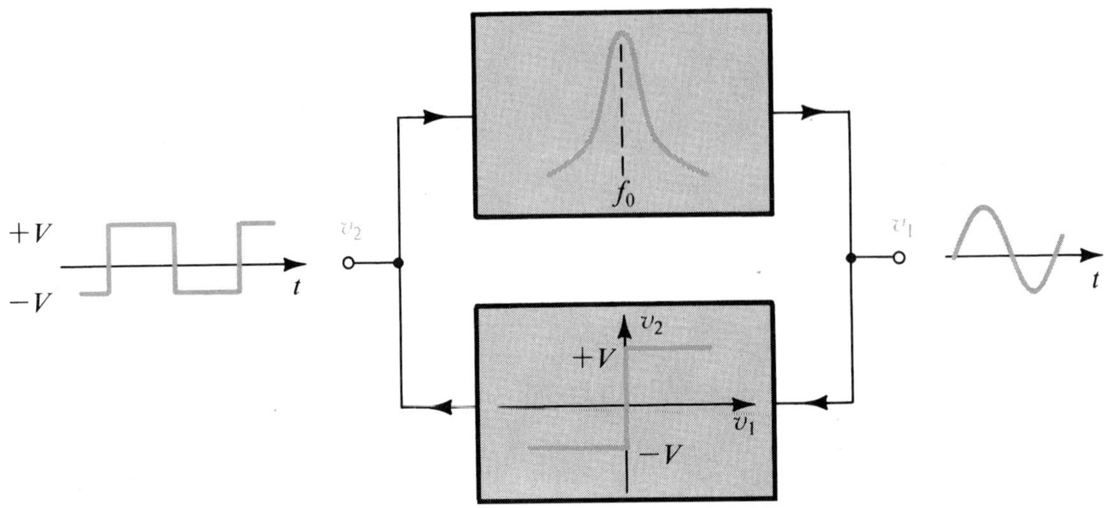

FIGURE 13.10 Block diagram of the active-filter-tuned oscillator.

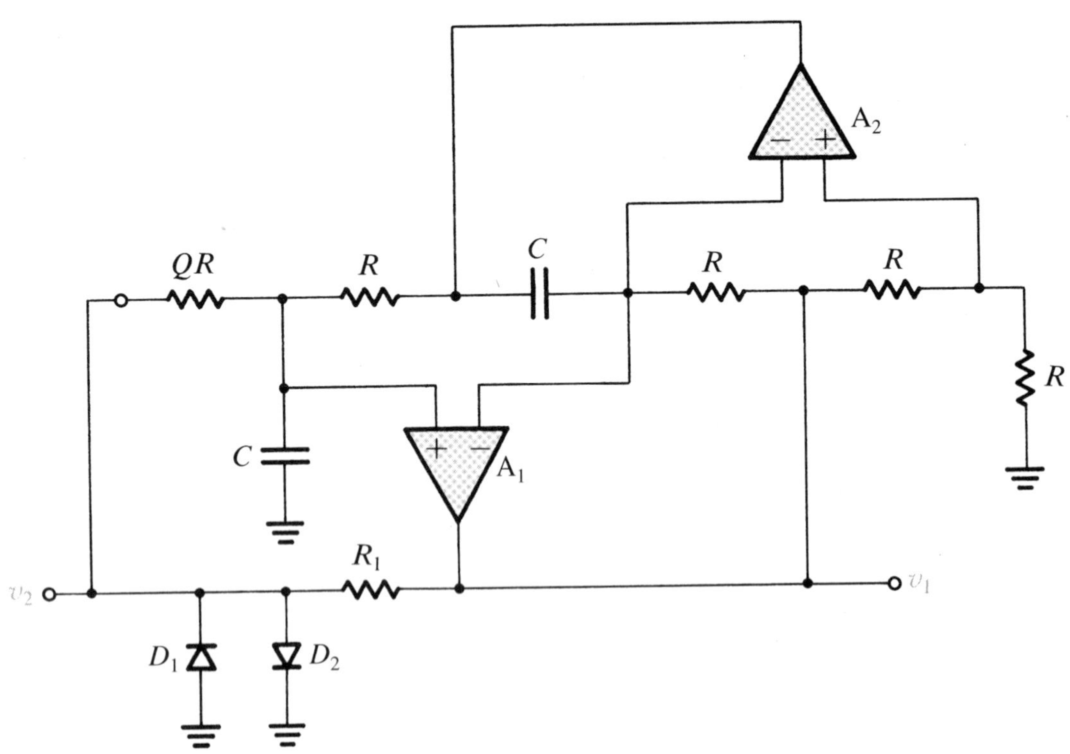

FIGURE 13.11 A practical implementation of the active-filter-tuned oscillator.

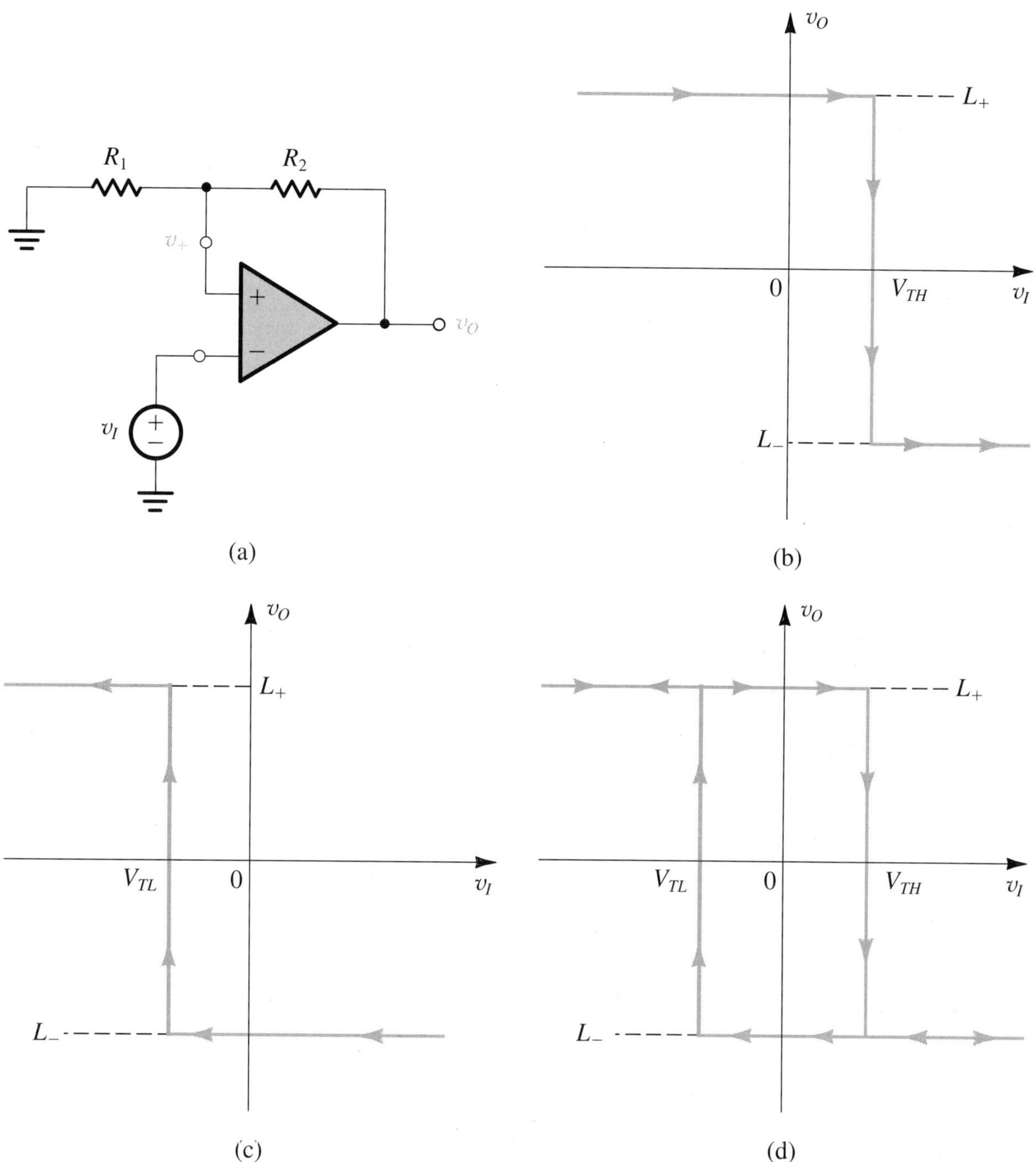

FIGURE 13.19 **(a)** The bistable circuit of Fig. 13.17 with the negative input terminal of the op amp disconnected from ground and connected to an input signal v_I. **(b)** The transfer characteristic of the circuit in (a) for increasing v_I. **(c)** The transfer characteristic for decreasing v_I. **(d)** The complete transfer characteristics.

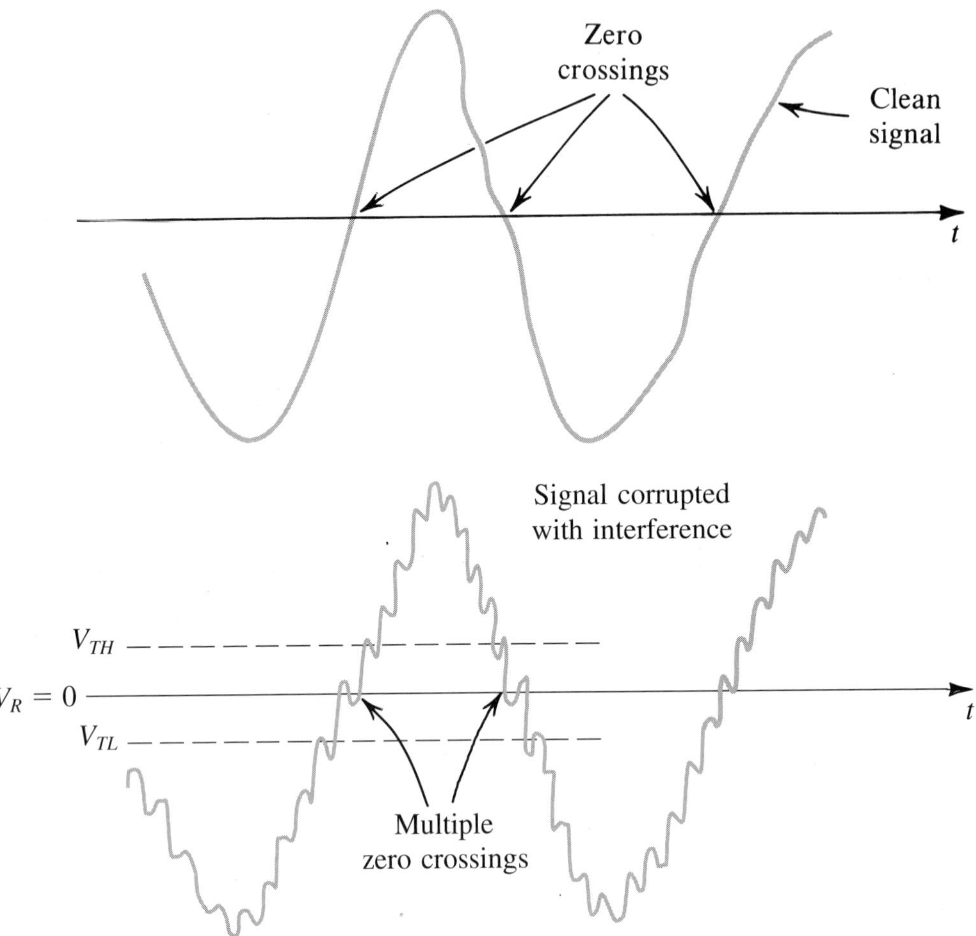

FIGURE 13.22 Illustrating the use of hysteresis in the comparator characteristics as a means of rejecting interference.

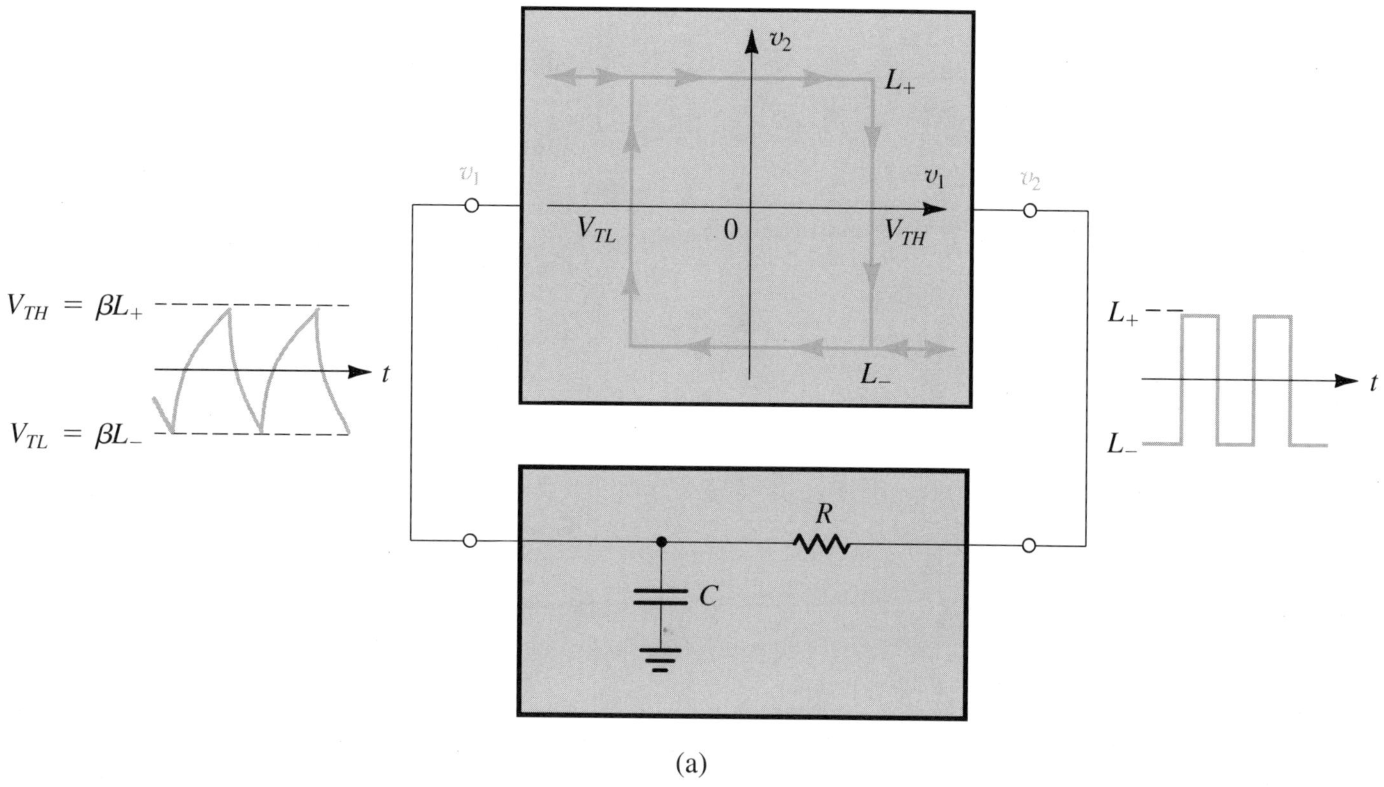

(a)

FIGURE 13.24 **(a)** Connecting a bistable multivibrator with inverting transfer characteristics in a feedback loop with an RC circuit results in a square-wave generator.

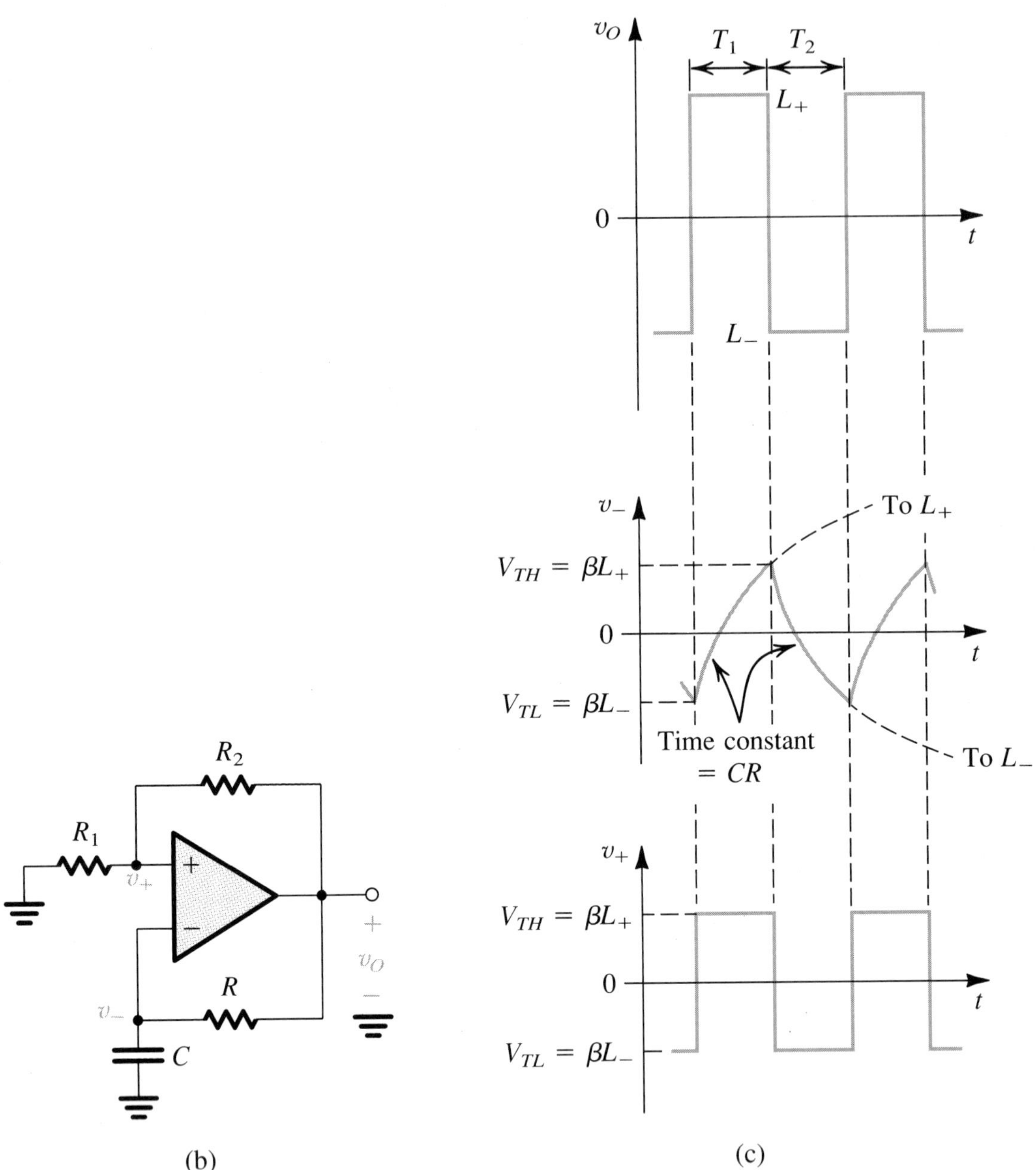

FIGURE 13.24 *(Continued)* **(b)** The circuit obtained when the bistable multivibrator is implemented with the circuit of Fig. 13.19(a). **(c)** Waveforms at various nodes of the circuit in (b). This circuit is called an astable multivibrator.

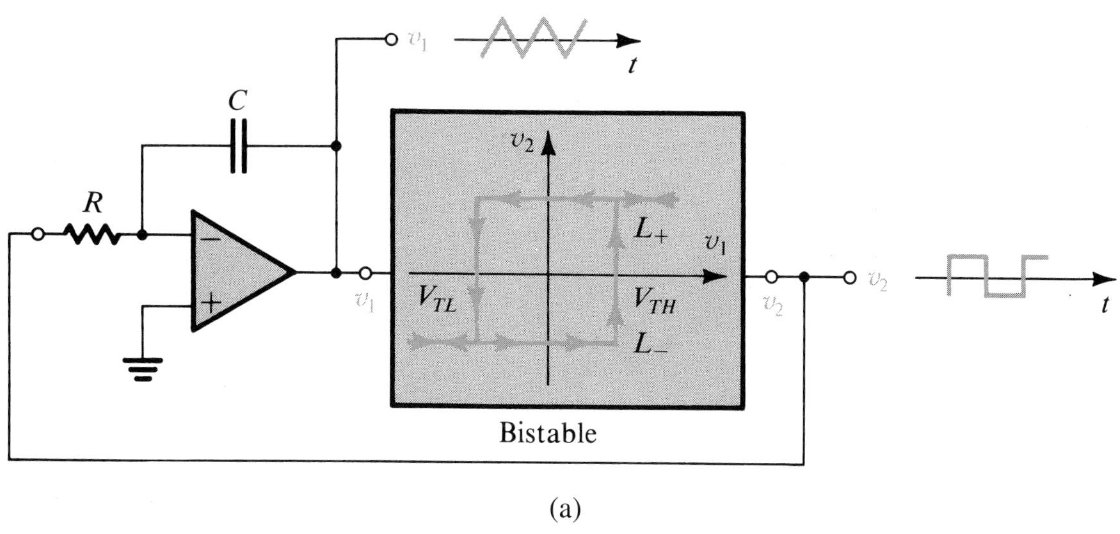

(a)

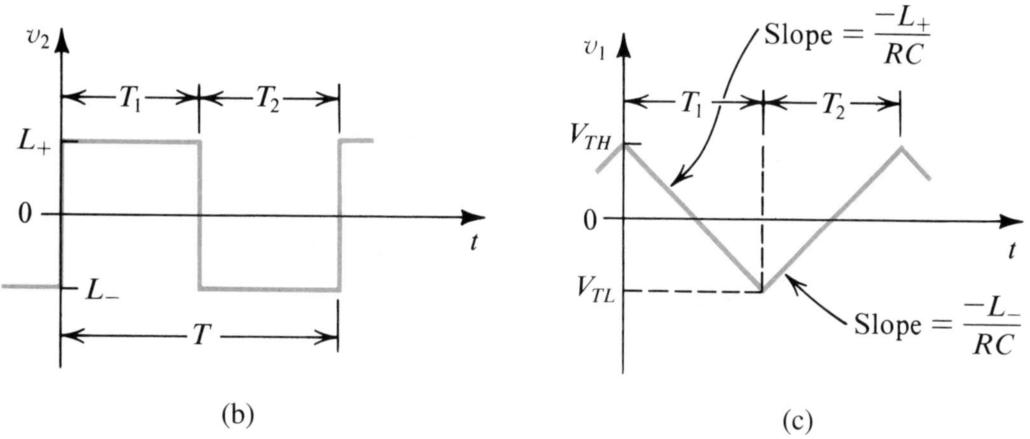

(b) (c)

FIGURE 13.25 A general scheme for generating triangular and square waveforms.

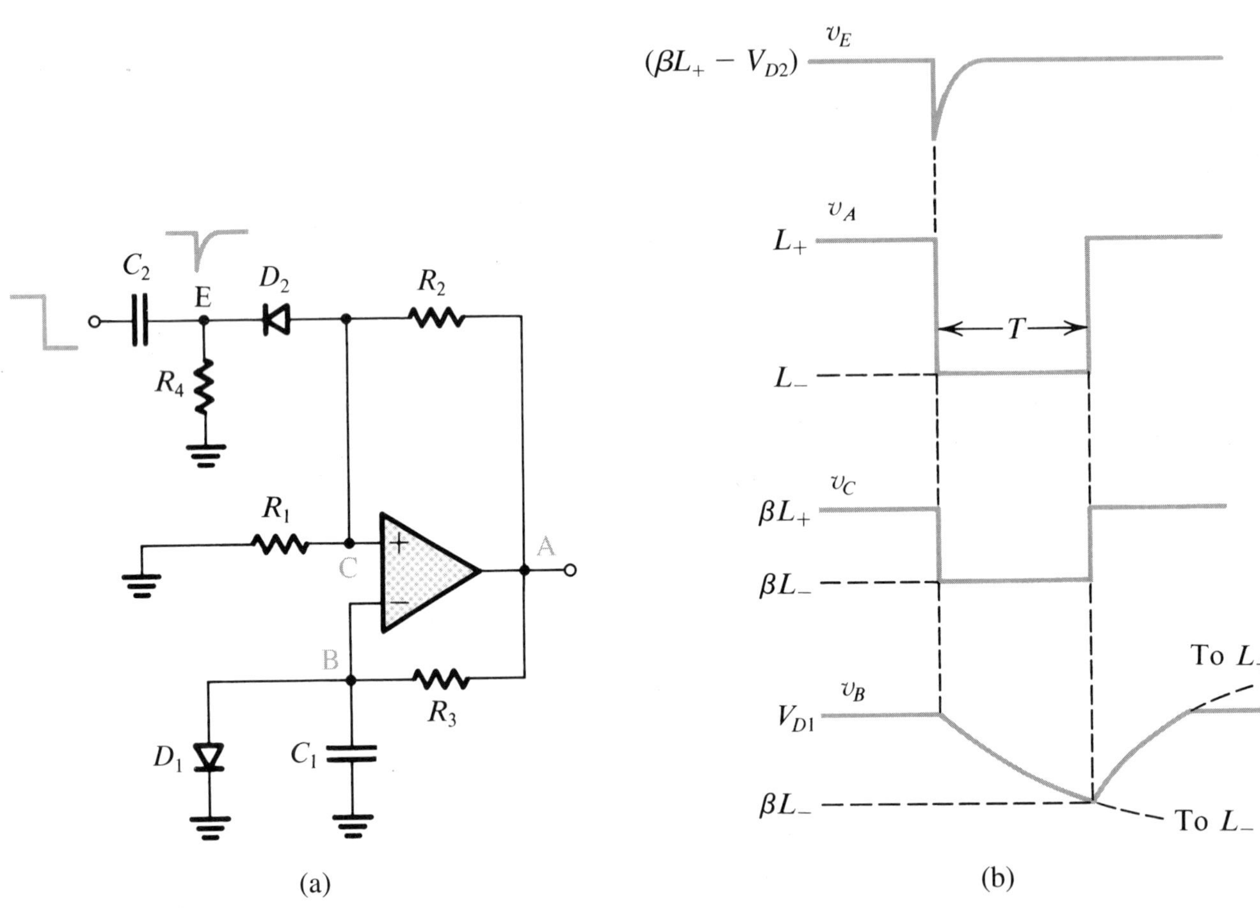

FIGURE 13.26 **(a)** An op-amp monostable circuit. **(b)** Signal waveforms in the circuit of (a).

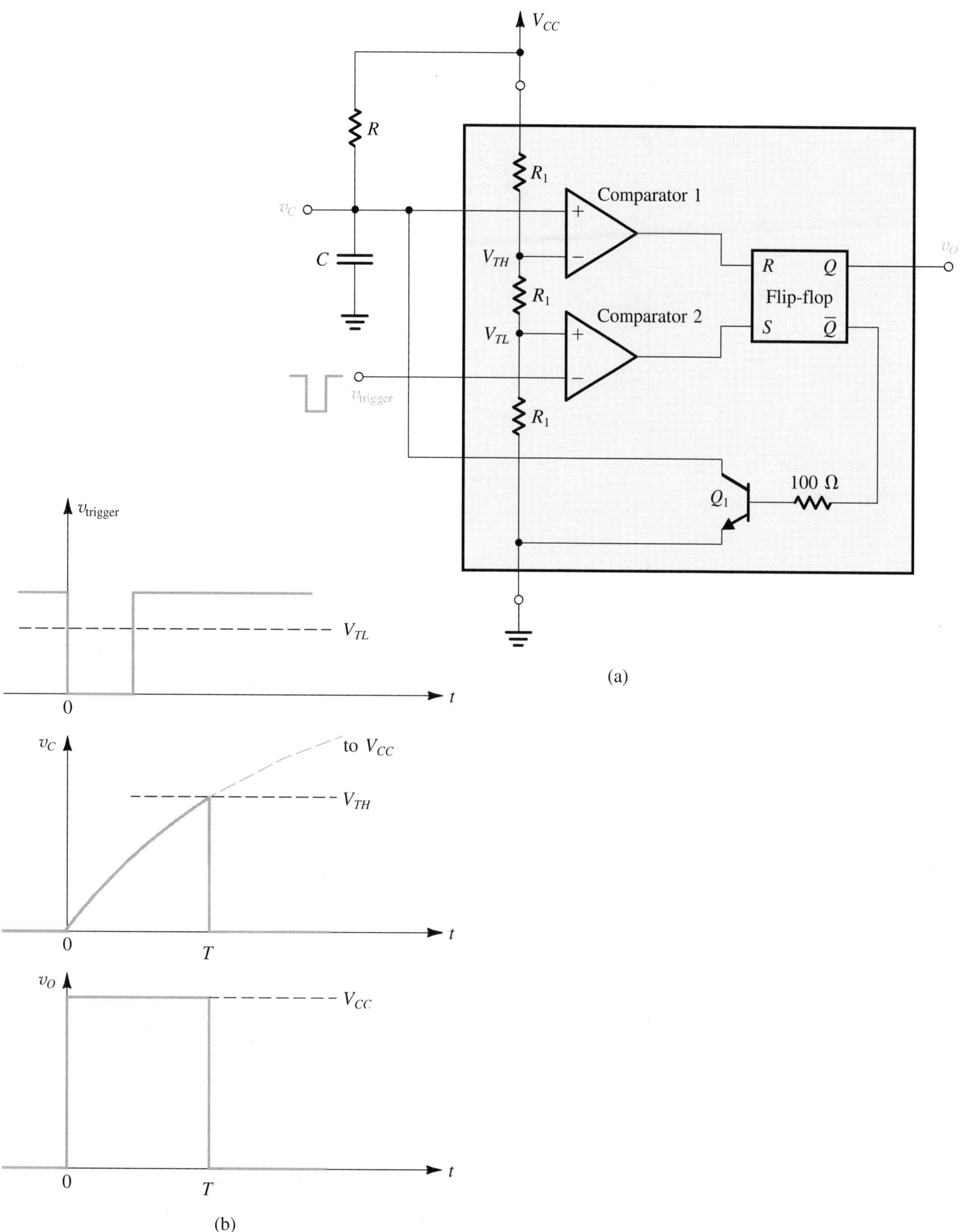

FIGURE 13.28 **(a)** The 555 timer connected to implement a monostable multivibrator. **(b)** Waveforms of the circuit in (a).

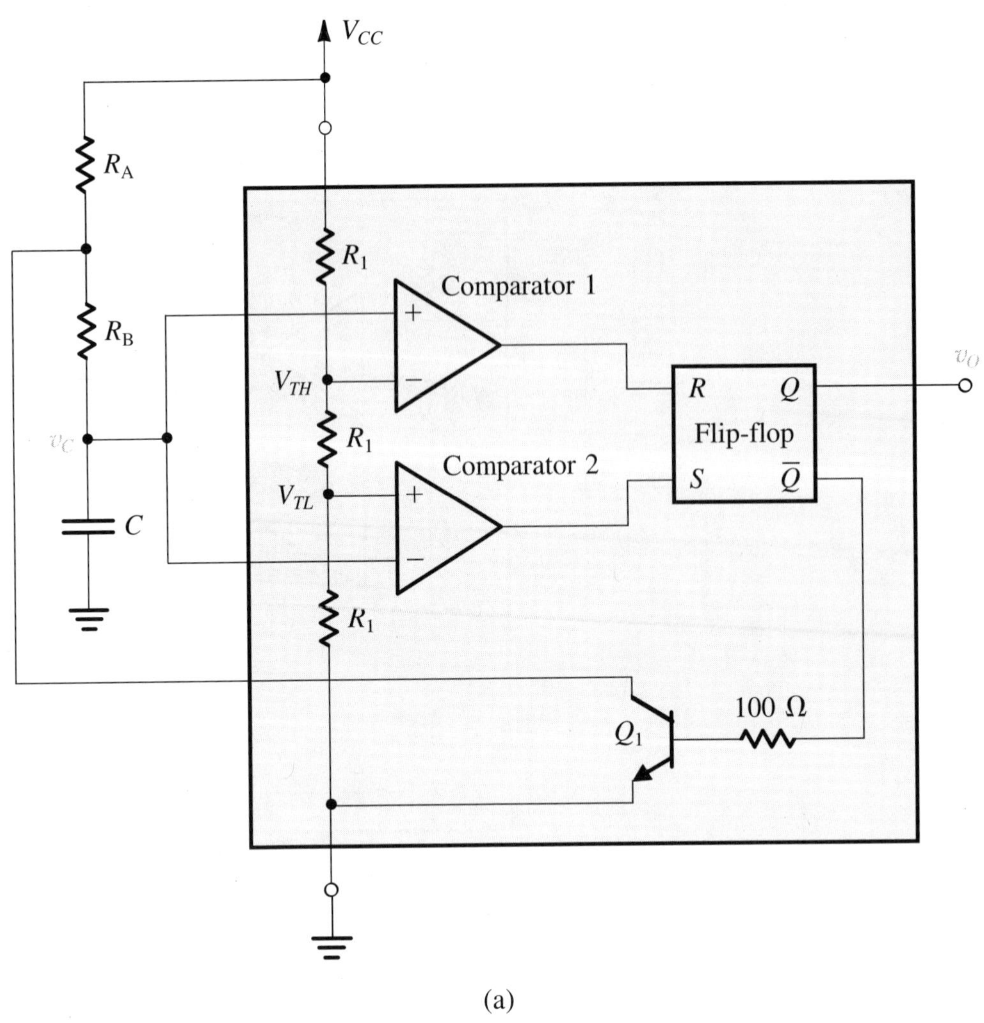

(a)

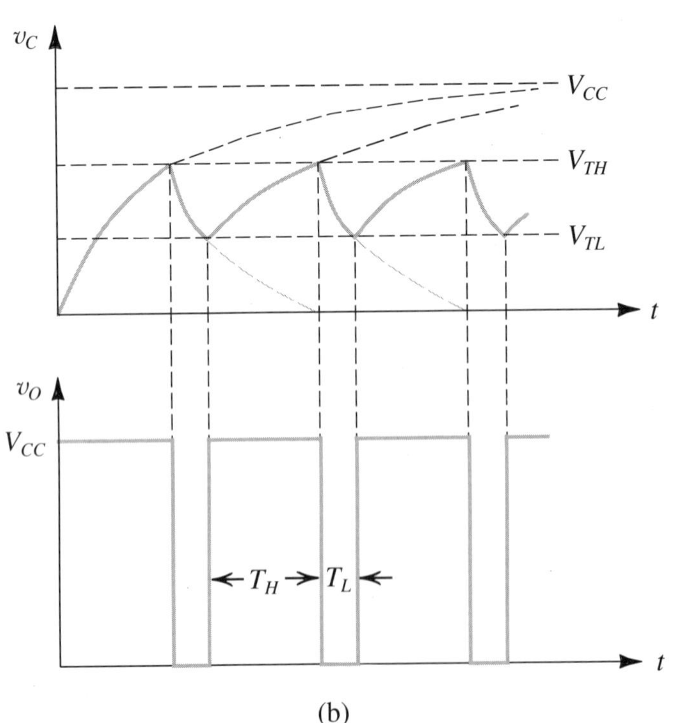

(b)

FIGURE 13.29 **(a)** The 555 timer connected to implement an astable multivibrator. **(b)** Waveforms of the circuit in (a).

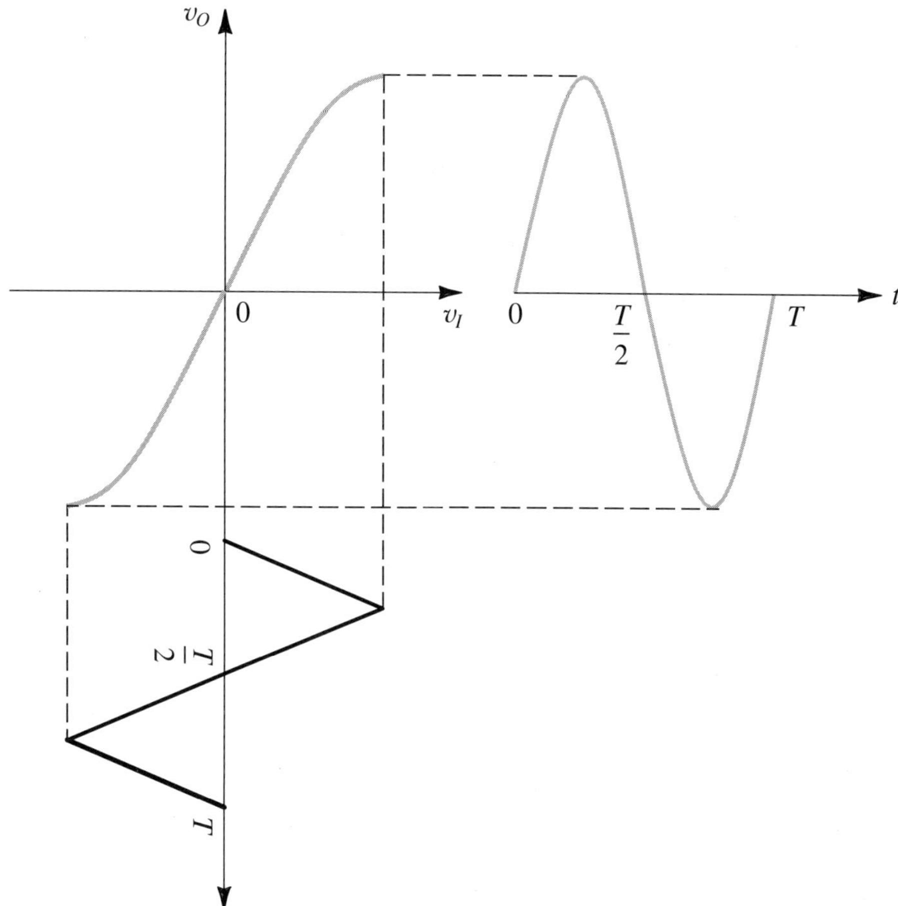

FIGURE 13.30 Using a nonlinear (sinusoidal) transfer characteristic to shape a triangular waveform into a sinusoid.

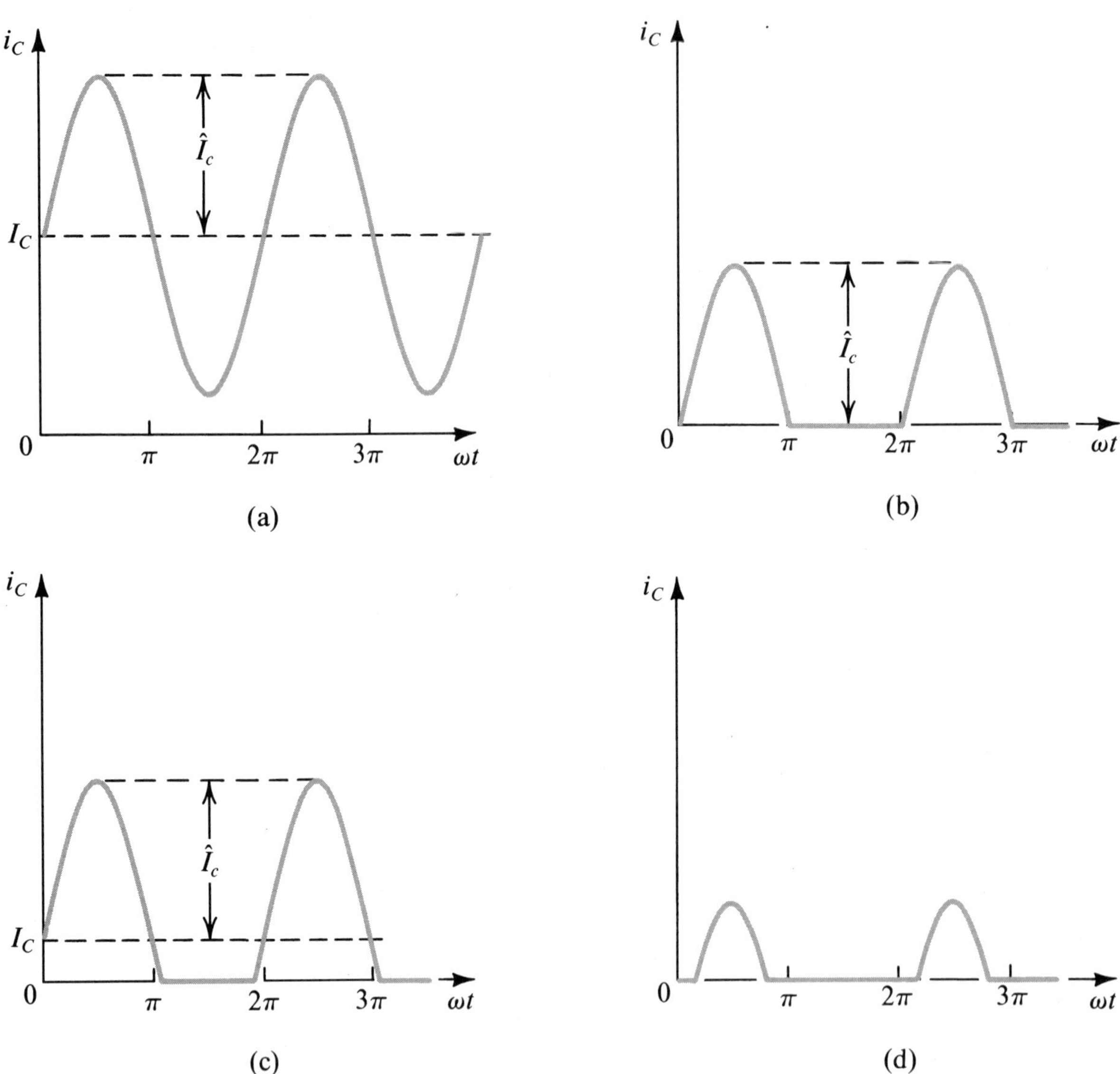

FIGURE 14.1 Collector current waveforms for transistors operating in **(a)** class A, **(b)** class B, **(c)** class AB, and **(d)** class C amplifier stages.

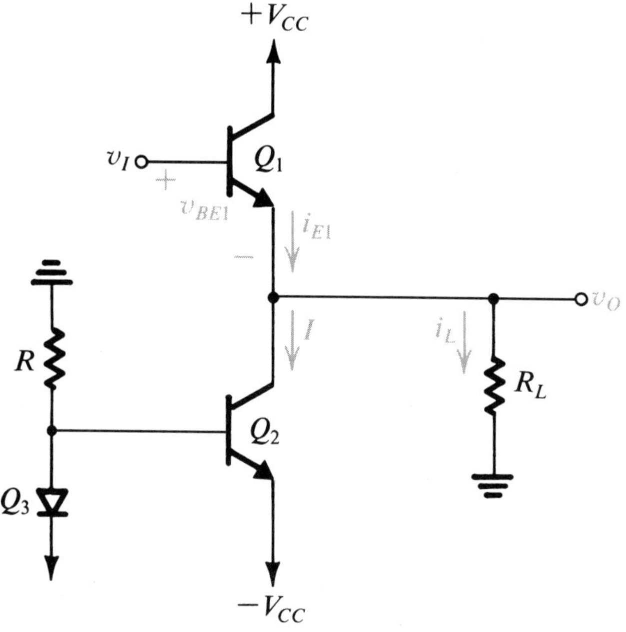

FIGURE 14.2 An emitter follower (Q_1) biased with a constant current I supplied by transistor Q_2.

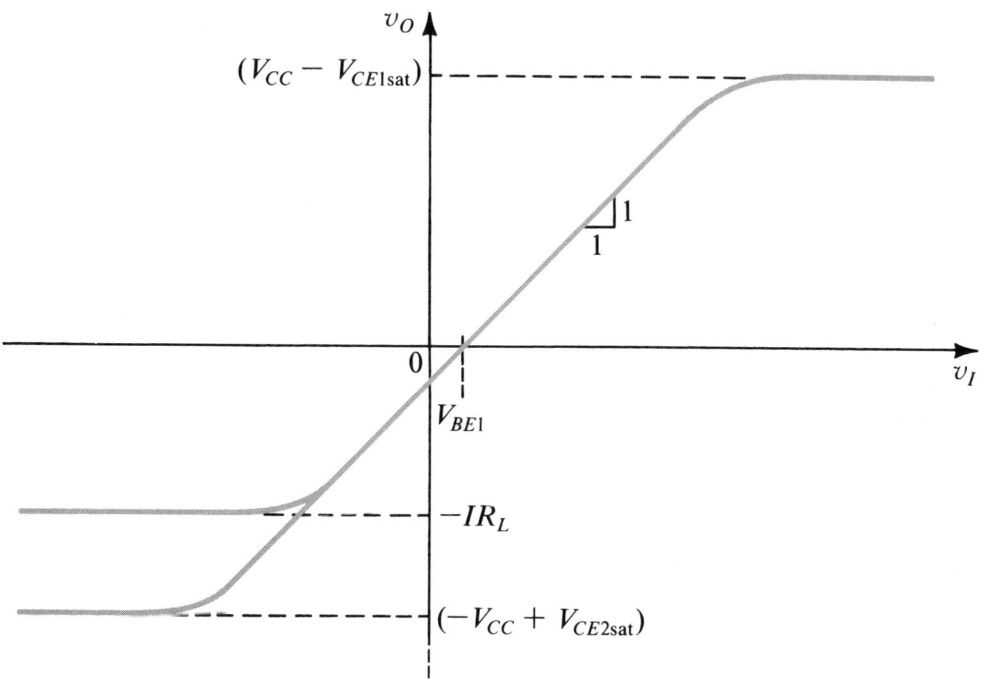

FIGURE 14.3 Transfer characteristic of the emitter follower in Fig. 14.2. This linear characteristic is obtained by neglecting the change in v_{BE1} with i_L. The maximum positive output is determined by the saturation of Q_1. In the negative direction, the limit of the linear region is determined either by Q_1 turning off or by Q_2 saturating, depending on the values of I and R_L.

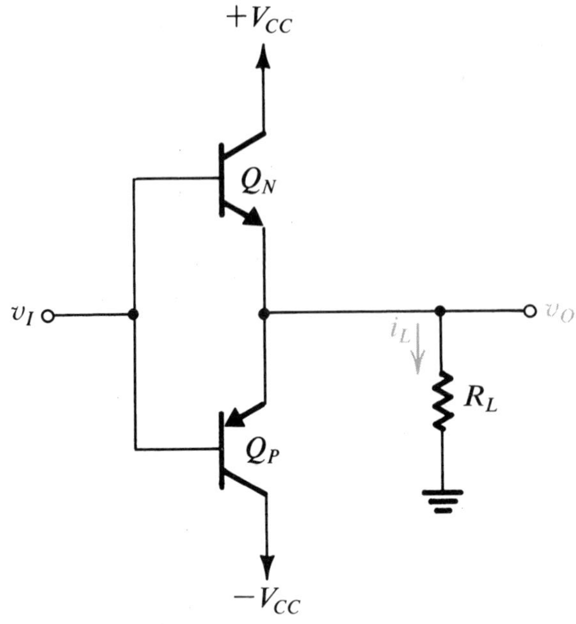

FIGURE 14.5 A class B output stage.

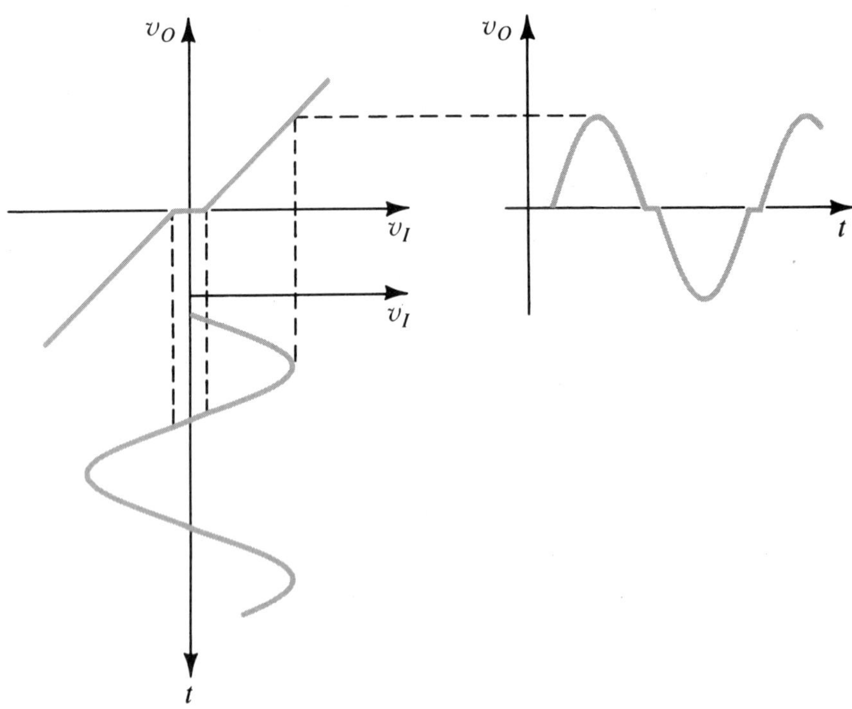

FIGURE 14.7 Illustrating how the dead band in the class B transfer characteristic results in crossover distortion.

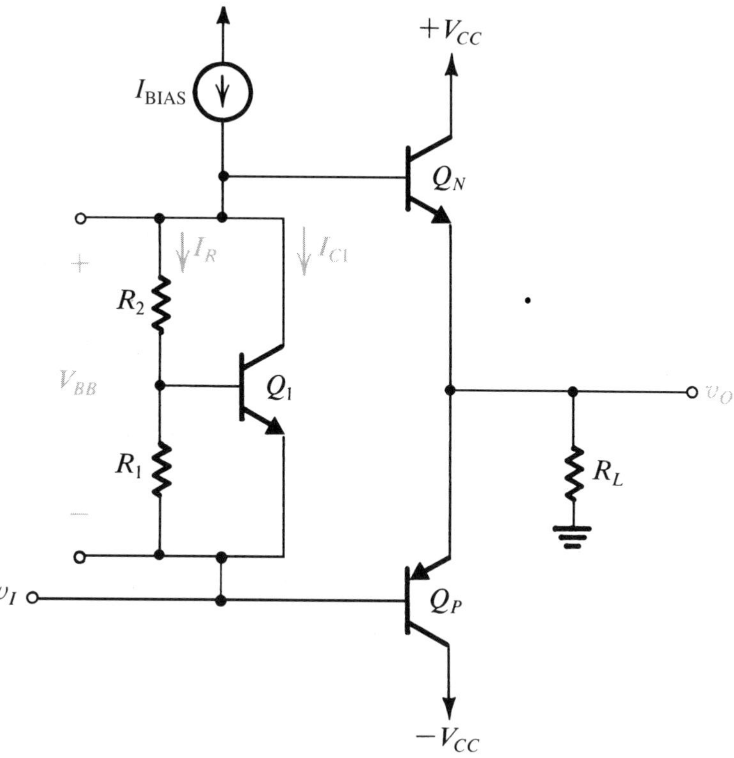

FIGURE 14.15 A class AB output stage utilizing a V_{BE} multiplier for biasing.

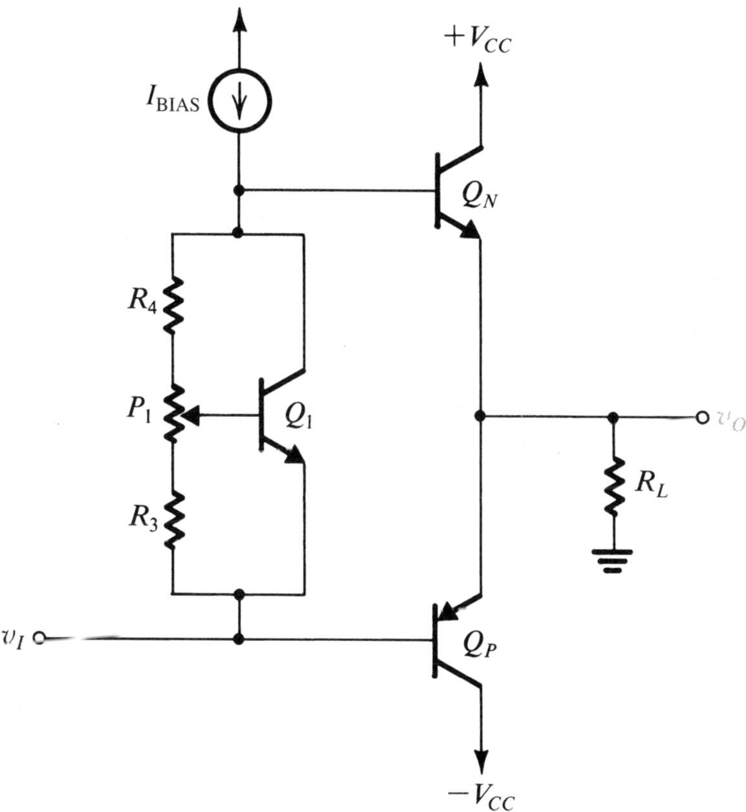

FIGURE 14.16 A discrete-circuit class AB output stage with a potentiometer used in the V_{BE} multiplier. The potentiometer is adjusted to yield the desired value of quiescent current in Q_N and Q_P.

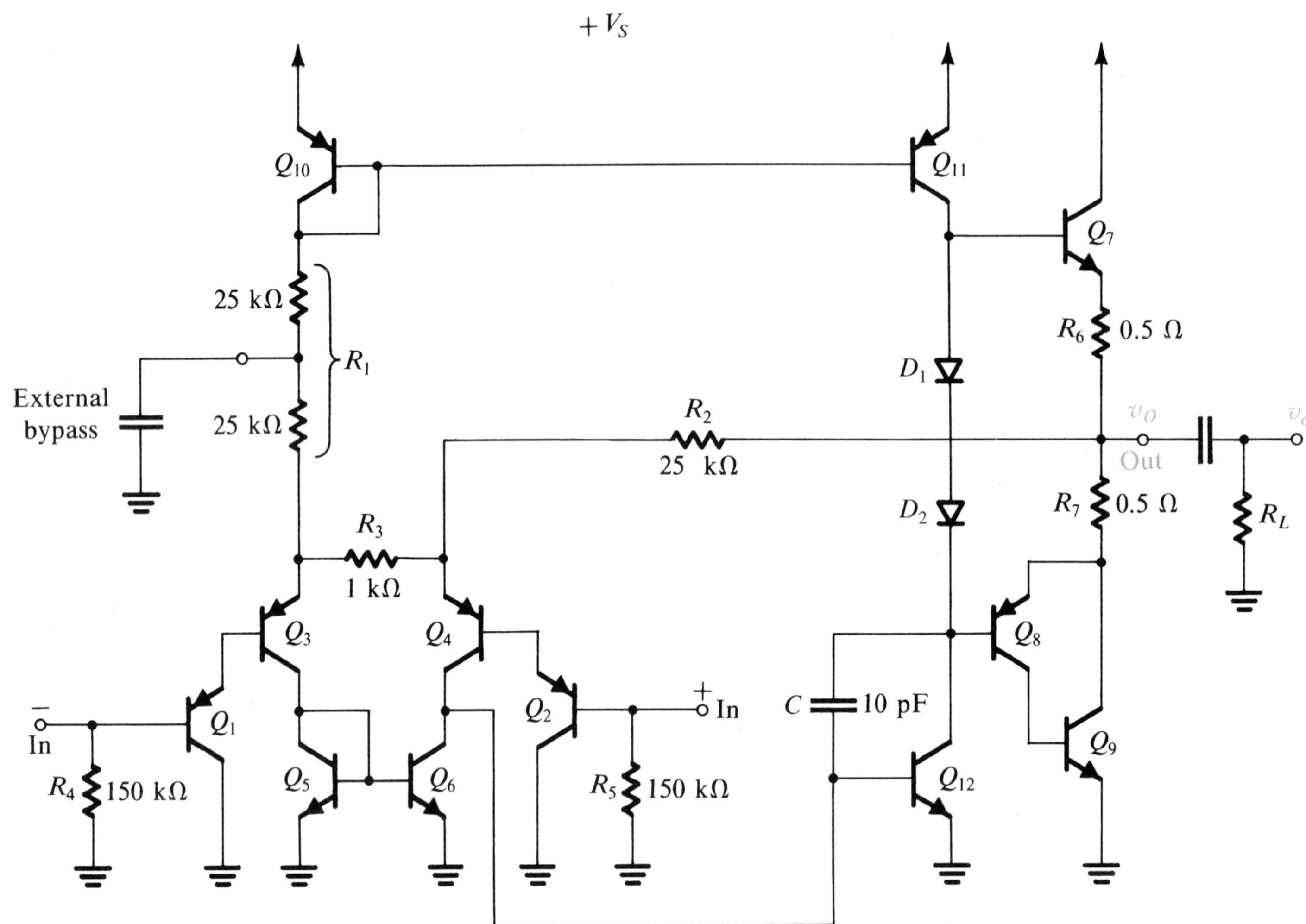

FIGURE 14.30 The simplified internal circuit of the LM380 IC power amplifier. (Courtesy National Semiconductor Corporation.)

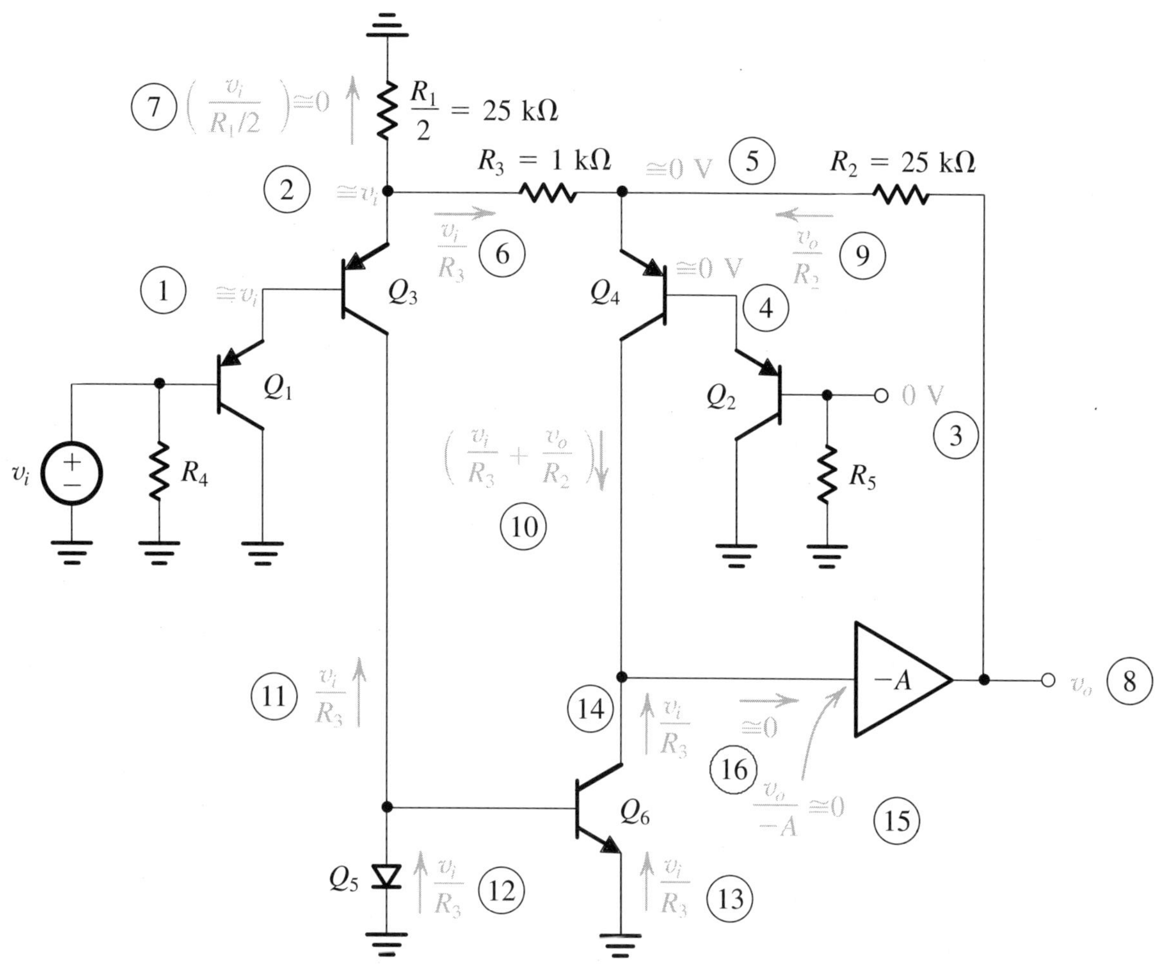

FIGURE 14.31 Small-signal analysis of the circuit in Fig. 14.30. The circled numbers indicate the order of the analysis steps.

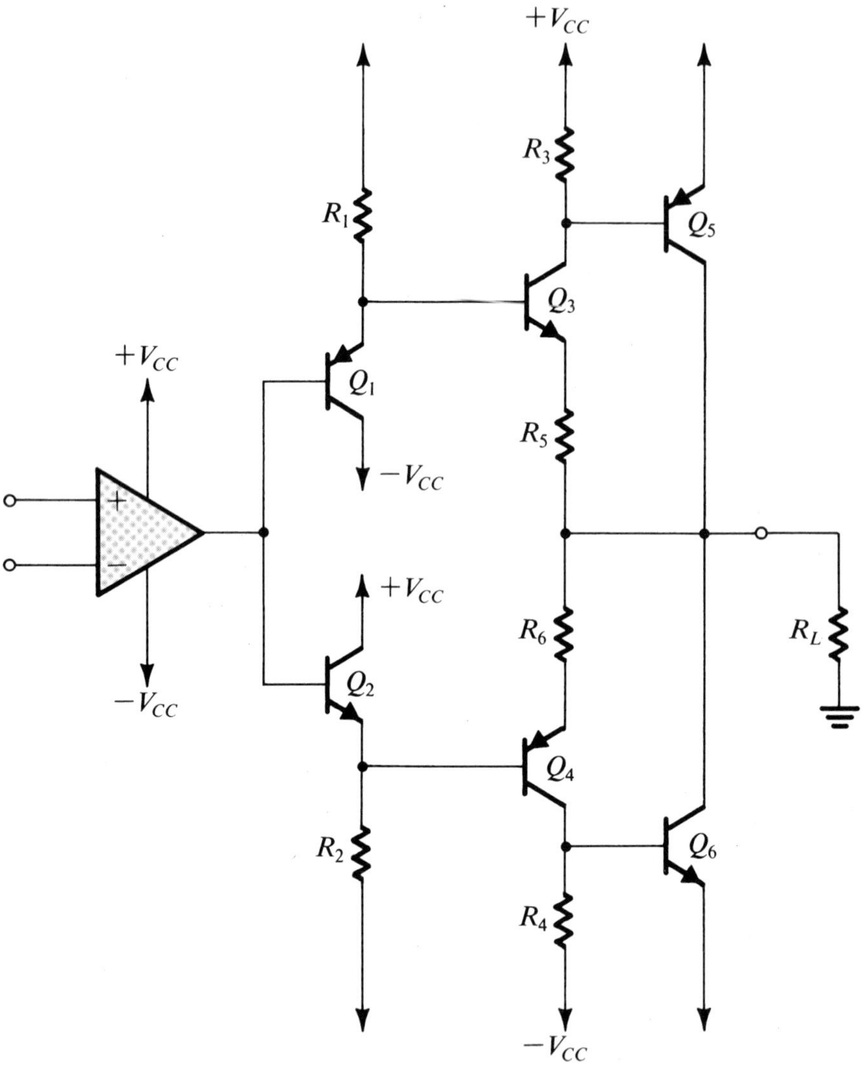

FIGURE 14.33 Structure of a power op amp. The circuit consists of an op amp followed by a class AB buffer similar to that discussed in Section 14.7.1. The output current capability of the buffer, consisting of Q_1, Q_2, Q_3, and Q_4, is further boosted by Q_5 and Q_6.

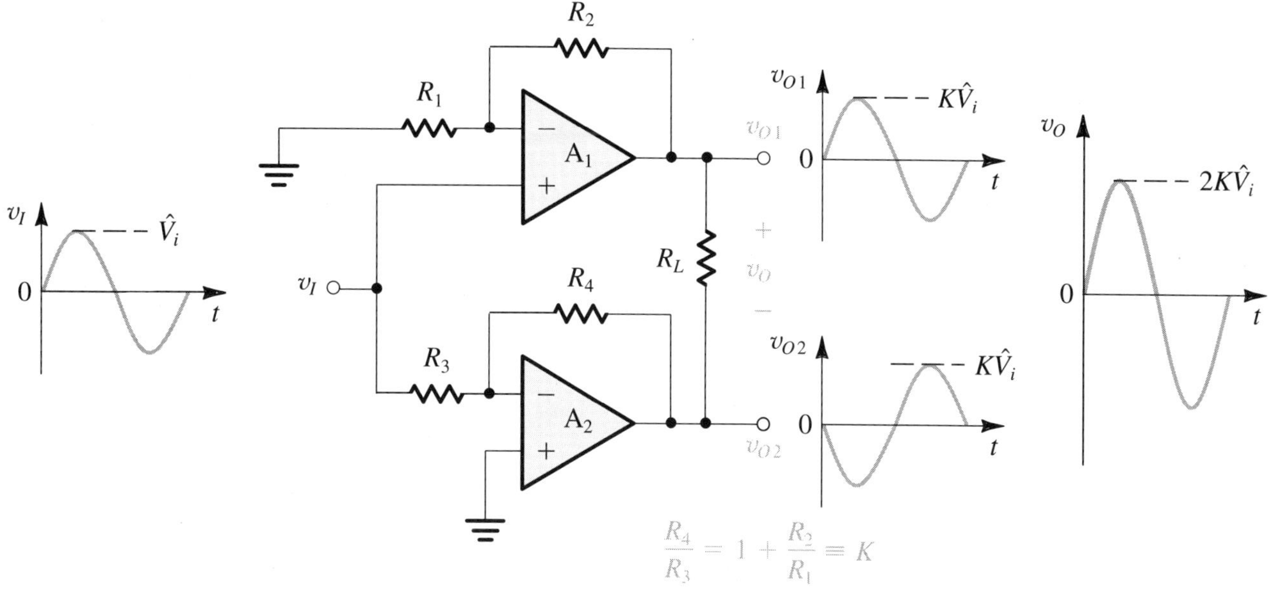

FIGURE 14.34 The bridge amplifier configuration.

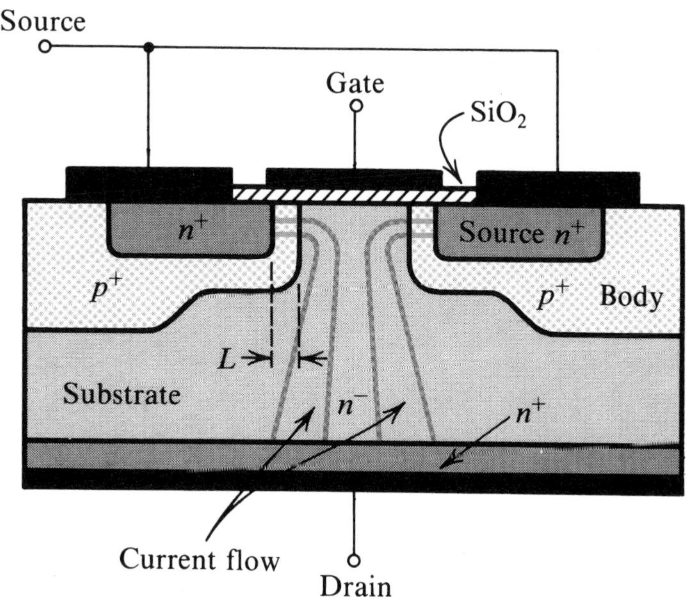

FIGURE 14.35 Double-diffused vertical MOS transistor (DMOS).

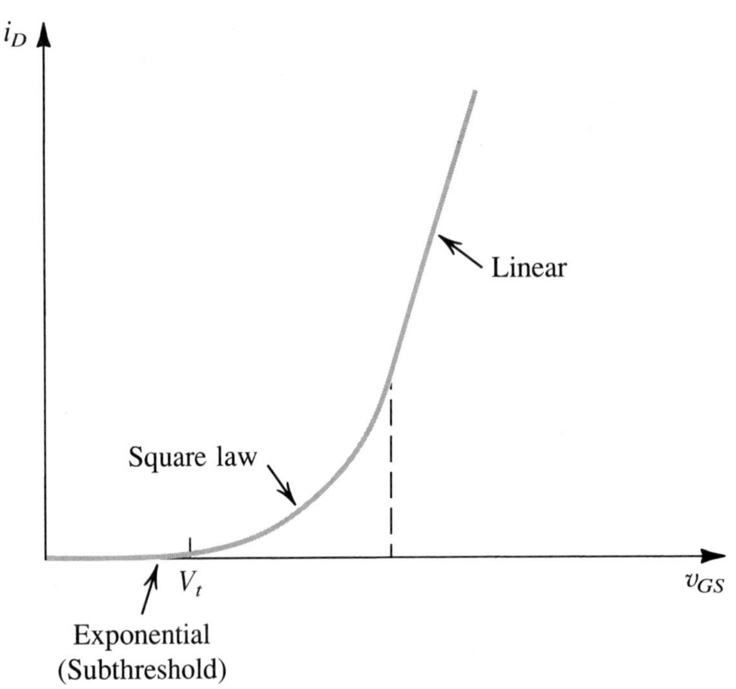

FIGURE 14.36 Typical i_D–v_{GS} characteristic for a power MOSFET.

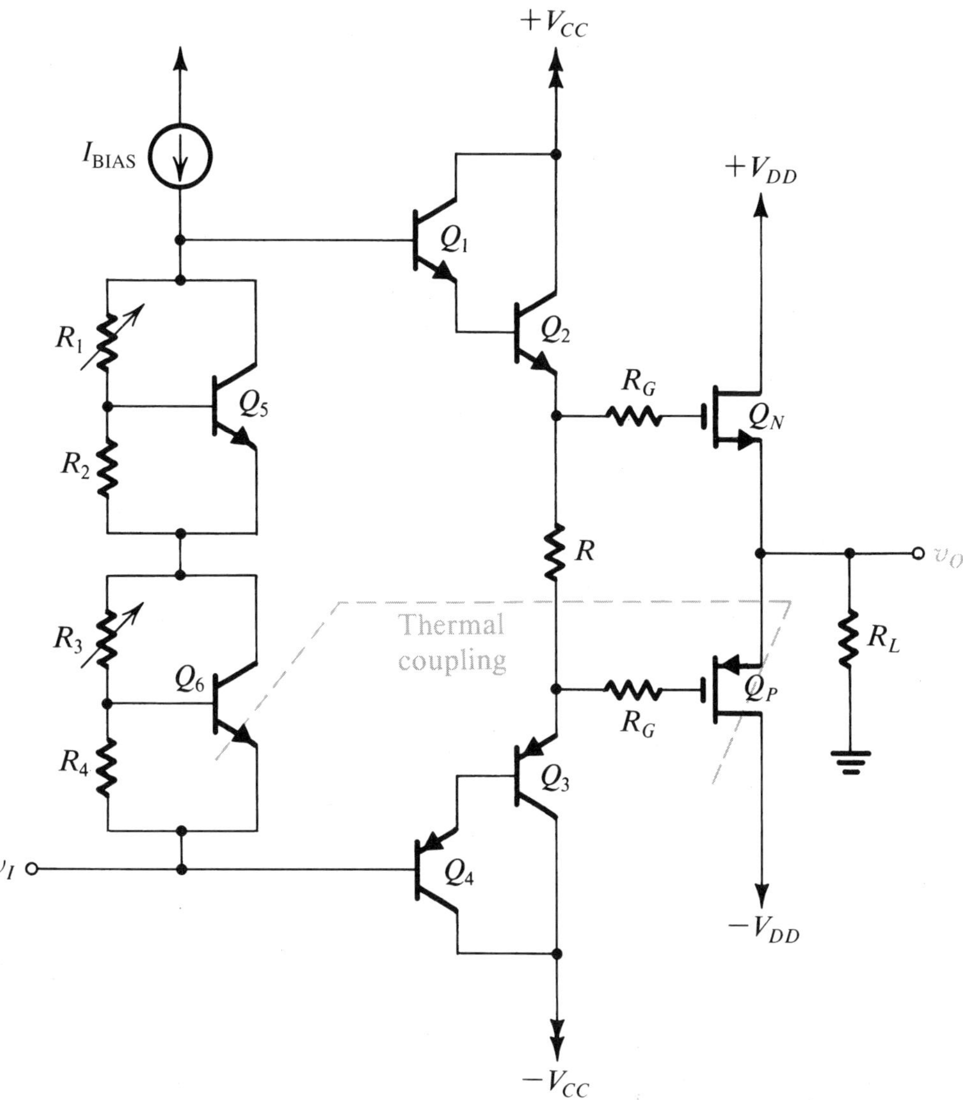

FIGURE 14.38 A class AB amplifier with MOS output transistors and BJT drivers. Resistor R_3 is adjusted to provide temperature compensation while R_1 is adjusted to yield the desired value of quiescent current in the output transistors. Resistors R_G are used to suppress parasitic oscillations at high frequencies. Typically, $R_G = 100\ \Omega$.